One-Stop Inter

Log on to

www.in.algebra1.com

Indiana Online Study Tools

- Extra Examples
- Self-Check Quizzes
- Vocabulary Review
- Chapter Test Practice
- ISTEP+/GQE Pratice

Online Research

- WebQuest Projects
- USA TODAY Activities
- Career Links
- Data Updates

Parent & Student Study Guide Workbook

- Printable Worksheets

Graphing Calculator Keystrokes

- Calculator Keystrokes for other calculators

For more information on these resources, see page 1.

Indiana Edition

Algebra 1

Contents

New York, New York Columbus, Ohio Chicago, Illinois Peoria, Illinois Woodland Hills, California

ISBN: 0-07-860391-9 (*Indiana Student Edition*)

Indiana Teacher Advisory Board

Bill Anderson
Former Mathematics Teacher
Zionsville Community Schools
Zionsville, Indiana

Eric D. Erdman
Mathematics Department Chairperson
Northrop High School
Fort Wayne, Indiana

David Ferris
Mathematics Teacher
Noblesville High School
Noblesville, Indiana

Linda Gibson
Mathematics Department Chairperson
Anderson High School
Anderson, Indiana

Jill Henady
Mathematics Teacher and Department Chairperson
Rensselaer Central High School
Rensselaer, Indiana

Robert Lovell
Mathematics Department Chairperson
North Side High School
Fort Wayne, Indiana

Jeannine McBride
Mathematics Teacher
Bloomfield High School
Bloomfield, Indiana

Jenny Miller
Mathematics Teacher and Department Chairperson
Tuttle Middle School
Crawfordsville, Indiana

Ronald C. Myer
Mathematics Teacher
Indian Springs Middle School
Columbia City, Indiana

Linda Rohleder
Mathematics Teacher
Jasper Middle School
Jasper, Indiana

Steve Sachs
Mathematics Department Chairperson
Lawrence North High School
Lawrence, Indiana

Richard R. Soendlin
Mathematics Department Chairperson
Arsenal Technical High School
Indianapolis, Indiana

Mike Sturm
Mathematics Teacher
Terre Haute North Vigo
Terre Haute, Indiana

Reading Consultant

Lynn T. Havens
Director, Project CRISS
Kalispell, Montana

Image Credits: **i** Courtesy of the Office of Indiana Tourism; **ii** PhotoDisc

Algebra 1 Academic Standards

Standard 1 OPERATIONS WITH REAL NUMBERS

Students simplify and compare expressions. They use rational exponents, and simplify square roots.

A1.1.1	Compare real number expressions.
A1.1.2	Simplify square roots using factors.
A1.1.3	Understand and use the distributive, associative, and commutative properties.
A1.1.4	Use the laws of exponents for rational exponents.
A1.1.5	Use dimensional (unit) analysis to organize conversions and computations.

Standard 2 LINEAR EQUATIONS AND INEQUALITIES

Students solve linear equations and inequalities in one variable. They solve word problems that involve linear equations, inequalities, or formulas.

A1.2.1	Solve linear equations.
A1.2.2	Solve equations and formulas for a specified variable.
A1.2.3	Find solution sets of linear inequalities when possible numbers are given for the variable.
A1.2.4	Solve linear inequalities using properties of order.
A1.2.5	Solve combined linear inequalities.
A1.2.6	Solve word problems that involve linear equations, formulas, and inequalities.

Standard 3 RELATIONS AND FUNCTIONS

Students sketch and interpret graphs representing given situations. They understand the concept of a function and analyze the graphs of functions.

A1.3.1	Sketch a reasonable graph for a given relationship.
A1.3.2	Interpret a graph representing a given situation.
A1.3.3	Understand the concept of a function, decide if a given relation is a function, and link equations to functions.
A1.3.4	Find the domain and range of a relation.

Standard 4 GRAPHING LINEAR EQUATIONS AND INEQUALITIES

Students graph linear equations and inequalities in two variables. They write equations of lines and find and use the slope and y-intercept of lines. They use linear equations to model real data.

A1.4.1	Graph a linear equation.

A1.4.2	Find the slope, x-intercept and y-intercept of a line given its graph, its equation, or two points on the line.
A1.4.3	Write the equation of a line in slope-intercept form. Understand how the slope and y-intercept of the graph are related to the equation.
A1.4.4	Write the equation of a line given appropriate information.
A1.4.5	Write the equation of a line that models a data set and use the equation (or the graph of the equation) to make predictions. Describe the slope of the line in terms of the data, recognizing that the slope is the rate of change.
A1.4.6	Graph a linear inequality in two variables.

Standard 5 PAIRS OF LINEAR EQUATIONS AND INEQUALITIES

Students solve pairs of linear equations using graphs and using algebra. They solve pairs of linear inequalities using graphs. They solve word problems involving pairs of linear equations.

A1.5.1	Use a graph to estimate the solution of a pair of linear equations in two variables.
A1.5.2	Use a graph to find the solution set of a pair of linear inequalities in two variables.
A1.5.3	Understand and use the substitution method to solve a pair of linear equations in two variables.
A1.5.4	Understand and use the addition or subtraction method to solve a pair of linear equations in two variables.
A1.5.5	Understand and use multiplication with the addition or subtraction method to solve a pair of linear equations in two variables.
A1.5.6	Use pairs of linear equations to solve word problems.

Standard 6 POLYNOMIALS

Students add, subtract, multiply, and divide polynomials. They factor quadratics.

A1.6.1	Add and subtract polynomials.
A1.6.2	Multiply and divide monomials.
A1.6.3	Find powers and roots of monomials (only when the answer has an integer exponent).
A1.6.4	Multiply polynomials.
A1.6.5	Divide polynomials by monomials.
A1.6.6	Find a common monomial factor in a polynomial.
A1.6.7	Factor the difference of two squares and other quadratics.
A1.6.8	Understand and describe the relationships among the solutions of an equation, the zeros of a function, the x-intercepts of a graph, and the factors of a polynomial expression.

Standard 7 ALGEBRAIC FRACTIONS

Students simplify algebraic ratios and solve algebraic proportions.

A1.7.1	Simplify algebraic ratios.

A1.7.2	Solve algebraic proportions.
Standard 8	**QUADRATIC, CUBIC, AND RADICAL EQUATIONS**
Students graph and solve quadratic and radical equations. They graph cubic equations.	
A1.8.1	Graph quadratic, cubic, and radical equations.
A1.8.2	Solve quadratic equations by factoring.
A1.8.3	Solve quadratic equations in which a perfect square equals a constant.
A1.8.4	Complete the square to solve quadratic equations.
A1.8.5	Derive the quadratic formula by completing the square.
A1.8.6	Solve quadratic equations by using the quadratic formula.
A1.8.7	Use quadratic equations to solve word problems.
A1.8.8	Solve equations that contain radical expressions.
A1.8.9	Use graphing technology to find approximate solutions of quadratic and cubic equations.
Standard 9	**MATHEMATICAL REASONING AND PROBLEM SOLVING**
Students use a variety of strategies to solve problems.	
A1.9.1	Use a variety of problem solving strategies, such as drawing a diagram, making a chart, guess-and-check, solving a simpler problem, writing an equation, and working backwards.
A1.9.2	Decide whether a solution is reasonable in the context of the original situation.
Students develop and evaluate mathematical arguments and proofs.	
A1.9.3	Use the properties of the real number system and the order of operations to justify the steps of simplifying functions and solving equations.
A1.9.4	Understand that the logic of equation solving begins with the assumption that the variable is a number that satisfies the equation, and that the steps taken when solving equations create new equations that have, in most cases, the same solution set as the original. Understand that similar logic applies to solving systems of equations simultaneously.
A1.9.5	Decide whether a given algebraic statement is true always, sometimes, or never (statements involving linear or quadratic expressions, equations, or inequalities).
A1.9.6	Distinguish between inductive and deductive reasoning, identifying and providing examples of each.
A1.9.7	Identify the hypothesis and conclusion in a logical deduction.
A1.9.8	Use counterexamples to show that statements are false, recognizing that a single counterexample is sufficient to prove a general statement false.

Prepare for the GQE

Countdown To GQE

Pages IN2–IN26 of this text include a section called **Countdown to GQE**. Each page contains 7 problems that are just like those on the ISTEP+/GQE. You should plan to complete one page each week to help you prepare for the test.

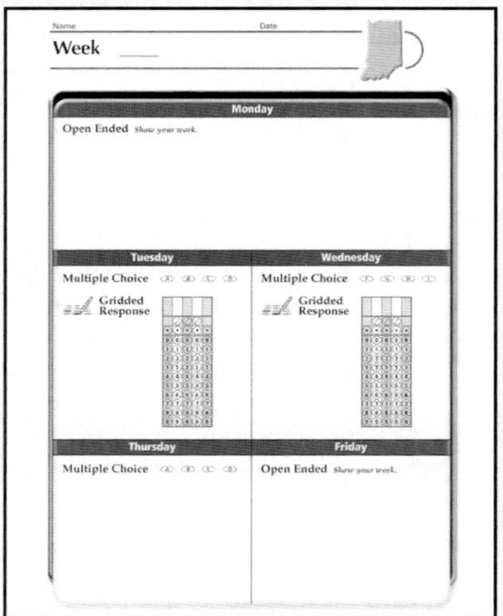

Plan to spend a few minutes each day working on the GQE problem(s) for that day unless your teacher asks you to do otherwise. For each week, the individual days have the same type(s) of problem. If you have difficulty with any problem, you can refer to the lesson that is referenced in parentheses after the problem.

Monday	Open Ended
Tuesday	Multiple Choice, Gridded Response
Wednesday	Multiple Choice, Gridded Response
Thursday	Multiple Choice
Friday	Open Ended

Your teacher can provide you with an answer sheet to record your work and your answers for each week. A printable worksheet is also available at in.algebra1.com. At the end of the week, your teacher may want you to turn in the answer sheet.

ISTEP+/GQE Practice and Sample Test Workbook

The **ISTEP+/GQE Practice and Sample Test Workbook** contains a diagnostic test, practice for each of the GQE Academic Standards, and two sample tests.

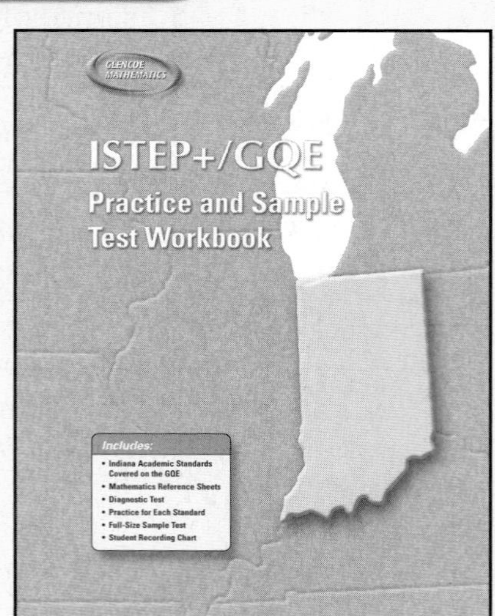

As you practice and master each objective, you can record your progress in the Student Recording Chart in your workbook.

Your teacher may also ask you to take a sample test at various points throughout the year to see if you're ready to take the real GQE.

Your Textbook

Your textbook contains many opportunities for you to get ready for the GQE every day. Take advantage of these so you don't need to cram before the test

- **Each lesson** contains at least two Standardized Test Practice problems. You can use these problems every day to keep your GQE skills sharp. The **Chapter Practice Test** also includes a Standardized Test Practice problem.

- **Worked-out examples** in each chapter show you step-by-step solutions of Standardized Test Practice problems. Just like the practice problems, these include multiple-choice, gridded-response, and open-ended formats. **Test-Taking Tips** are also included.

- Two pages of **Standardized Test Practice** are included at the end of each chapter. These problems may cover any of the content up to and including the chapter they follow.

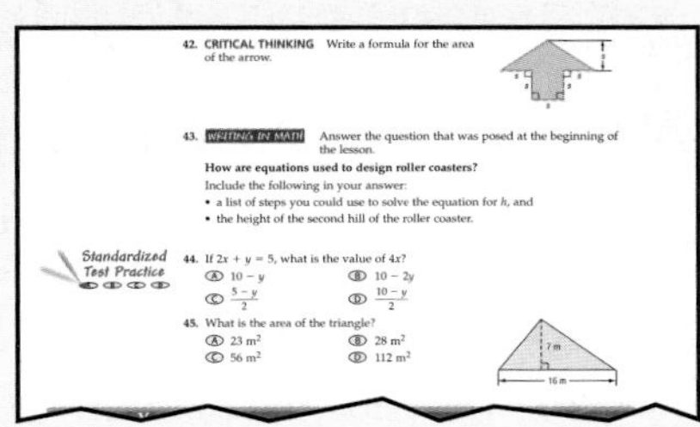

42. CRITICAL THINKING Write a formula for the area of the arrow.

43. WRITING IN MATH Answer the question that was posed at the beginning of the lesson.

How are equations used to design roller coasters?

Include the following in your answer:
- a list of steps you could use to solve the equation for h, and
- the height of the second hill of the roller coaster.

Standardized Test Practice

44. If $2x + y = 5$, what is the value of $4x$?
(A) $10 - y$ (B) $10 - 2y$
(C) $\frac{5-y}{2}$ (D) $\frac{10-y}{2}$

45. What is the area of the triangle?
(A) $23\ m^2$ (B) $28\ m^2$
(C) $56\ m^2$ (D) $112\ m^2$

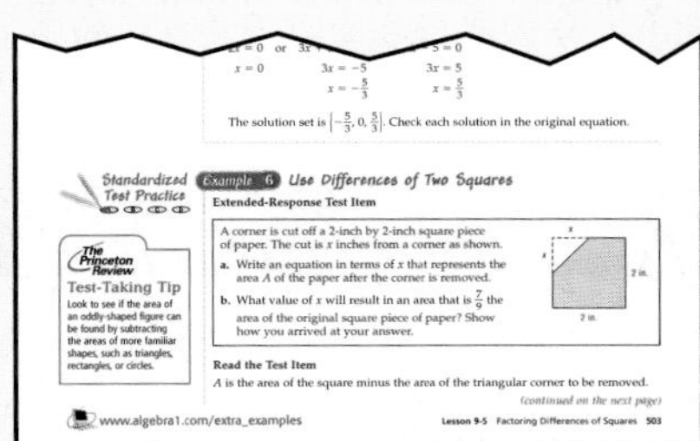

$x = 0$ $3x = -5$ $3x = 5$

$x = -\frac{5}{3}$ $x = \frac{5}{3}$

The solution set is $\left\{-\frac{5}{3}, 0, \frac{5}{3}\right\}$. Check each solution in the original equation.

Standardized Test Practice

Example 6 *Use Differences of Two Squares*

Extended-Response Test Item

A corner is cut off a 2-inch by 2-inch square piece of paper. The cut is x inches from a corner as shown.

a. Write an equation in terms of x that represents the area A of the paper after the corner is removed.

b. What value of x will result in an area that is $\frac{7}{9}$ the area of the original square piece of paper? Show how you arrived at your answer.

Read the Test Item

A is the area of the square minus the area of the triangular corner to be removed.

(continued on the next page)

The Princeton Review

Test-Taking Tip

Look to see if the area of an oddly-shaped figure can be found by subtracting the areas of more familiar shapes, such as triangles, rectangles, or circles.

www.algebra1.com/extra_examples

Lesson 9-5 Factoring Differences of Squares 503

Test-Taking Tips

- Go to bed early the night before the test. You will think more clearly after a good night's rest.
- Read each problem carefully and think about ways to solve the problem before you try to answer the question.
- Relax. Most people get nervous when taking a test. It's natural. Just do your best.
- Answer questions you are sure about first. If you do not know the answer to a question, skip it and go back to that question later.
- Become familiar with common formulas and when they should be used.
- Think positively. Some problems may seem hard to you, but you may be able to figure out what to do if you read each question carefully.
- If no figure is provided, draw one. If one is furnished, mark it up to help you solve the problem.
- When you have finished each problem, reread it to make sure your answer is reasonable.
- Make sure that the number of the question on the answer sheet matches the number of the question on which you are working in your test booklet.

Countdown to GQE

Week 1

Monday

Open Ended A certain type of bacteria doubles every second. For your biology experiment at Notre Dame you start with 4 bacteria cells. You record the results for eight seconds. *(Lesson 1-1)*

Part A Make a table showing your results.

Part B Describe the pattern.

Part C How many bacteria cells are there after 8 seconds? (Assume that no cells die off in this time period).

Tuesday

Multiple Choice Evaluate $3 + 8 \times 7 + 2^3$. *(Lesson 1-2)*

Ⓐ 65 Ⓑ 67
Ⓒ 83 Ⓓ 85

Gridded Response In the figure, what is the area of the region outside the inner circle and inside the outer circle? *(Geometry)*

Area of Circle = πr^2
Use 3.14 for π.

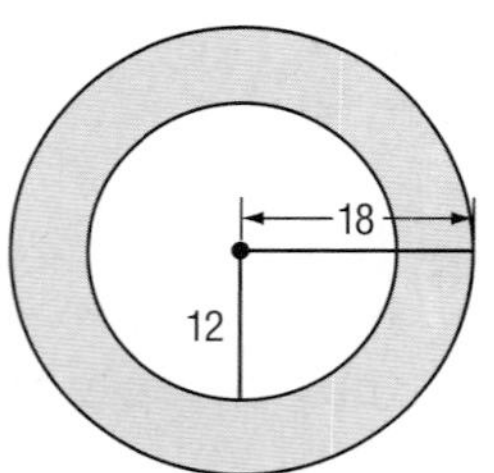

Wednesday

Multiple Choice Which expression could be used to calculate the expenses for John for 8 weeks? *(Lesson 1-5)*

John's Weekly Expenses			
Bus Money	Lunch	Snacks	Video Games
\$6.50	\$15.25	\$5.00	\$8.00

Ⓐ $6.50 + 15.25 + 5 + 8$
Ⓑ $8(6.50) + 15.25 + 5 + 8$
Ⓒ $8(6.50 + 15.25 + 5 + 8)$
Ⓓ $6.50 + 15.25 + 5 + 8(8)$

Gridded Response Evaluate $\frac{3}{4}[4 \div (7 - 4)]$. *(Lesson 1-4)*

Thursday

Multiple Choice Which is the hypothesis of the statement *I will go to the movies with you on Sunday? (Lesson 1-7)*

Ⓐ I will go to the movies with you.
Ⓑ It is Sunday.
Ⓒ I will not go to the movies with you.
Ⓓ It is not Sunday.

Friday

Open Ended The graph below corresponds to Mrs. Garcia's shopping trip to the mall by car. What was most likely happening between 3:00 and 5:00 P.M.? *(Lesson 1-8)*

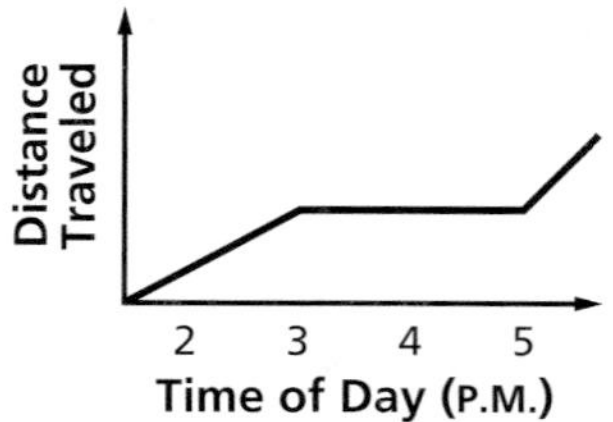

Week 2

Monday

Open Ended An Indiana high school student conducted a survey on what subject students liked most in high school. The results of this survey are given in the frequency table. Use the data to make a bar graph to show the results of the survey. Be sure you use appropriate titles and labels, use a proper scale, and carefully graph the data in the frequency table. *(Lesson 1-9)*

What High School Subject Do You Like Best?

Subject	Number of Students
Math	卌 II
English	卌 IIII
Science	卌
History	III
Physical Education	卌 卌 II

Tuesday

Multiple Choice Simplify $|4 - 8|$. *(Lesson 2–1)*

Ⓐ -12 Ⓑ -4
Ⓒ 4 Ⓓ 12

Gridded Response Simplify $-2 - (-4)$. *(Lesson 2-2)*

Wednesday

Multiple Choice Simplify $\frac{-36c + (-18b)}{9}$. *(Lesson 2-4)*

Ⓐ $-4c - 18b$ Ⓑ $-4c + 2b$
Ⓒ $-4c - 2b$ Ⓓ $-36c - 2b$

Gridded Response Evaluate $4x - 2xy$ if $x = \frac{3}{4}$ and $y = -2\frac{1}{2}$. *(Lesson 2-3)*

Thursday

Multiple Choice What is the approximate circumference of a circle with a diameter of 14.2 centimeters? *(Geometry)*

Circumference $= 2\pi r$ or πd
Use 3.14 for π.

Ⓐ 22.3 centimeters Ⓑ 38.4 centimeters
Ⓒ 44.6 centimeters Ⓓ 158.3 centimeters

Friday

Open Ended Box lunches were packed, but not marked, for a student field trip. Each box lunch contained a sandwich. Three bags contained a cheese sandwich, 2 bags contained roast beef, and 7 bags contained ham and cheese. *(Lesson 2-6)*

Part A What is the probability of choosing a box lunch with a cheese sandwich?

Part B What is the probability of choosing a box lunch with a roast beef sandwich?

Part C What is the probability of choosing a box lunch with a ham and cheese sandwich?

Countdown to GQE

Week 3

Monday

Open Ended Salvador and his friends drove to the Indiana State Fair in Indianapolis. It cost \$20 for gas and parking and \$35 for each all-day pass. Together they spent \$230. *(Lesson 3-1)*

Part A Write an equation to find the number of people x who went to the Indiana State Fair.

Part B How many people, including Salvador, went to the State Fair?

Tuesday

Multiple Choice Which equation represents *two times a number decreased by 10 equals 40?* *(Lesson 3-1)*

Ⓐ $2n - 10 = 40$

Ⓑ $10 - 2n = 40$

Ⓒ $2(n - 10) = 40$

Ⓓ $5 - n = 20$

Gridded Response Thirty-four subtracted from a number is -5. Find the number. *(Lesson 3-2)*

Wednesday

Multiple Choice Solve $x + \frac{3}{8} = \frac{1}{6}$. *(Lesson 3-2)*

Ⓐ $-\frac{5}{24}$

Ⓑ -1

Ⓒ $\frac{5}{24}$

Ⓓ 1

Gridded Response Paul is saving money for a trip to Angola, Indiana. Each week he doubles the amount that he saves. If it takes him 40 weeks to save all the money he needs, after how many weeks will he have 50% of the amount needed? *(Lesson 3-3)*

Thursday

Multiple Choice Solve $3(x - 18) = 5(x - 12)$. *(Lesson 3-5)*

Ⓐ 2

Ⓑ 3

Ⓒ 4

Ⓓ 5

Friday

Open Ended A ball was dropped from a height of 30 feet. Each time the ball bounced, it reached a maximum height that was $\frac{3}{5}$ of its previous height. *(Lesson 2-3)*

Part A How high did the ball bounce after the second bounce?

Part B After which bounce did the ball reach a height less than 1 foot?

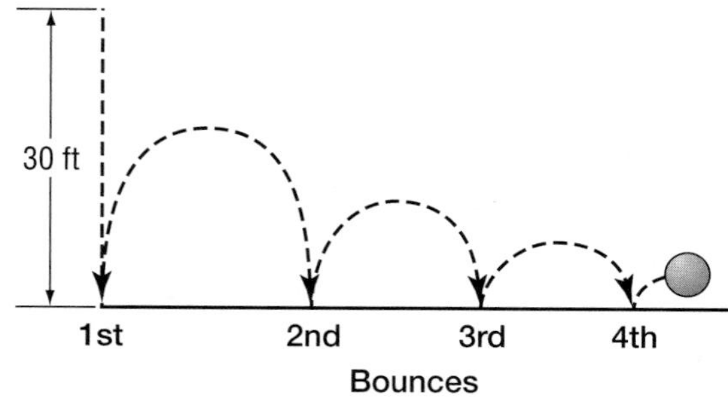

Week 4

Monday

Open Ended The distance a spring stretches is directly proportional to the force applied. *(Lesson 3-6)*

Part A If a force of 30 pounds stretches a spring 14 inches, how far will a force of 45 pounds stretch the spring?

Part B How far will a force of 75 pounds stretch the spring?

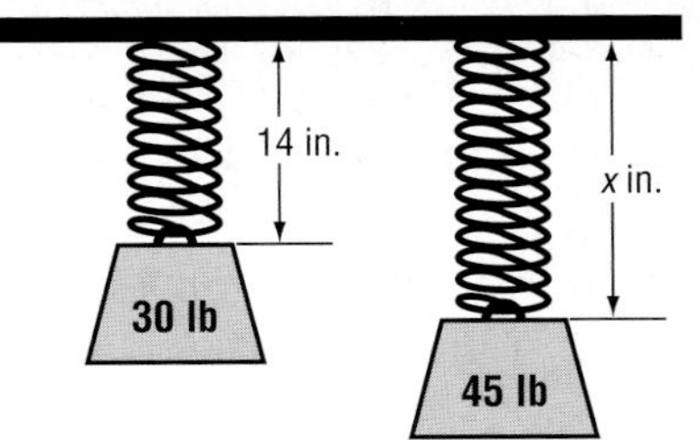

Tuesday

Multiple Choice Enzi entered the DINO Mountain Bike Race near Richmond, Indiana. For each gear setting on his bike there is a ratio of the number of turns on the pedals to the number of turns on the wheels. If 4th gear has a pedal-to-wheel turns ratio of 4 to 9, how many times do the pedals turn when the wheels turn 81 times? *(Lesson 3-6)*

(A) 27 (B) 36
(C) 48 (D) 64

Gridded Response The interest on an investment varies directly as the rate of simple interest. If the interest is \$48 when the interest rate is 5%, find the interest when the rate is 3%. *(Lesson 3-6)*

Wednesday

Multiple Choice Solve $2bx - b = -7$ for x. *(Lesson 3-8)*

(A) $\frac{-7 + b}{2}$ (B) $\frac{-7 + b}{2b}$
(C) -3 (D) $\frac{b + 7}{2b}$

Gridded Response A store offers two successive discounts of 20% and 10%. If the original cost of a coat was \$80, what is the sale price? *(Lesson 3-7)*

Thursday

Multiple Choice The first long-distance automobile race in the U.S. took place on May 30, 1911, at the Indianapolis Motor Speedway. Roy Harroun won the race averaging about 75 miles per hour. About how many feet per second is 75 mph? *(Lesson 3-7)*

(A) 88 ft/sec (B) 92 ft/sec
(C) 100 ft/sec (D) 110 ft/sec

Friday

Open Ended Marissa is studying chemistry at Taylor University. She needs to mix 20 liters of 40% acid solution with a 70% acid solution to get a mixture that is 50% acid. How many liters of the 70% acid solution should she use? *(Lesson 3-9)*

Week 5

Monday

Open Ended On a gridded map of Indiana, Indianapolis, was located at (6, 3) and Columbus was located at (10, −5). *(Lesson 4-1)*

Part A Graph the locations of Indianapolis and Columbus on a coordinate plane. Let A represent Indianapolis and B, Columbus, and draw $\overline{AB}$.

Part B In which quadrants are points A and B located?

Part C If the coordinates of the midpoint of a line segment whose endpoints are (a, b) and (c, d) are $\left(\frac{a+c}{2}, \frac{b+d}{2}\right)$, then what are the coordinates of the midpoint of $\overline{AB}$?

Tuesday

Multiple Choice Write the ordered pair for a point that is 8 units to the left of the origin and lies on the x-axis. *(Lesson 4-1)*

Ⓐ (0, −8)　Ⓑ (−8, 0)

Ⓒ (0, 8)　Ⓓ (8, 0)

Gridded Response Suppose $\triangle ABC$ has coordinates $A(3, 6)$, $B(6, 2)$, and $C(9, 3)$. What is the y-coordinate of the image of C when $\triangle ABC$ is reflected over the y-axis? *(Lesson 4-2)*

Wednesday

Multiple Choice Which of the following is demonstrated by the figures? *(Lesson 4-2)*

Ⓐ reflection (flip)

Ⓑ translation (slide)

Ⓒ rotation (turn)

Ⓓ none of these

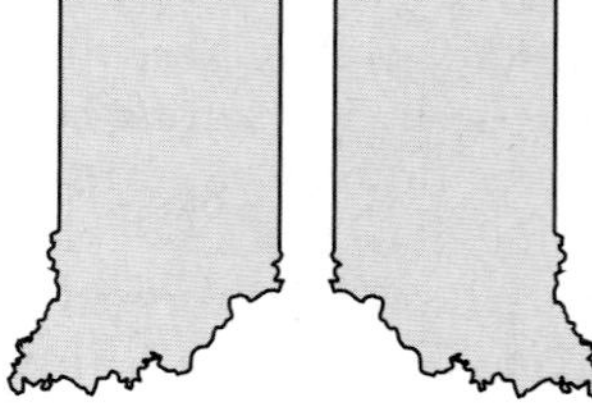

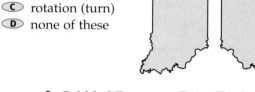

Gridded Response Point Z is located in quadrant II. The absolute value of both its x- and y-coordinates is 15. What is its y-coordinate? *(Lesson 4-1)*

Thursday

Multiple Choice Which figure represents a 90° clockwise rotation of $\triangle ACB$? *(Lesson 4-2)*

Ⓐ

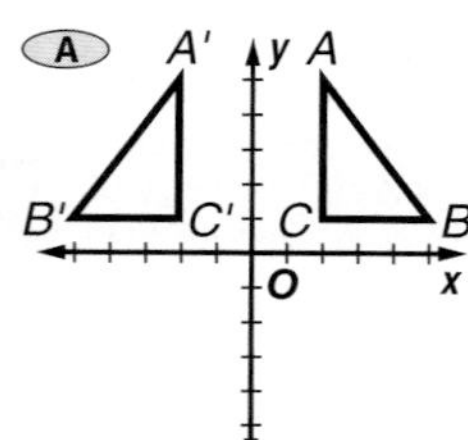

Ⓑ

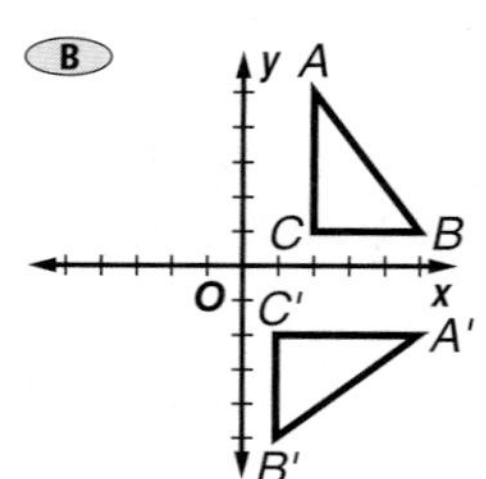

Ⓒ

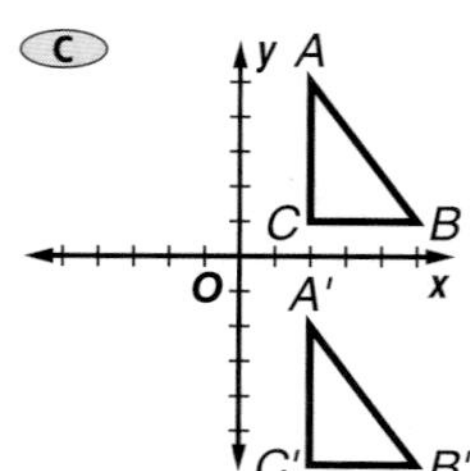

Ⓓ

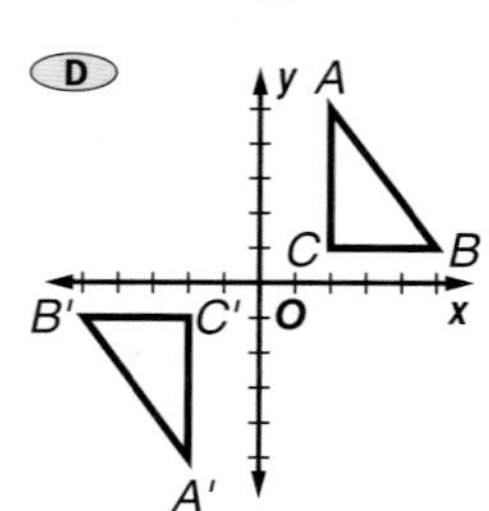

Friday

Open Ended Triangle MNL has vertices $M(4, 5)$, $N(6, 3)$, and $L(-2, 4)$. *(Lesson 4-2)*

Part A Find the coordinates of the image of $\triangle MNL$ after it is rotated 90° counterclockwise about the origin.

Part B Graph $\triangle MNL$ and its rotated image. Label the corresponding vertices M', N', and L'.

Week 6

Monday

Open Ended Pedro was conducting an experiment in biology class in which he measured the volume of the air in a balloon at various temperatures. He kept his results in the table as shown. *(Lesson 4-3)*

Temperature and Volume Experiment												
Temperature (°C)	−2	−1	0	1	2	3	4	5	6	7	8	9
Volume (cubic inches)	3	3.5	4	4.5	5	5.5	6	6.5	7	7.5	8	8.5

Part A Determine the domain and range of the data.

Part B Graph the data.

Part C Pedro determined that the equation for his data is $y = \frac{1}{2}x + 4$ where x is the temperature and y is the volume. Find the values of y when $x = 10, 12, 15$, and 30.

Tuesday

Multiple Choice Which of these equations is linear? *(Lesson 4-5)*

I. $3x - 7y = 2x + 8$

II. $2x - 8 = y^2$

(A) I only
(B) II only
(C) Both are linear.
(D) Neither is linear.

Gridded Response If $f(x) = 3x - 1$ and $g(x) = 2x^2 + 1$, find $f(3) + g(0)$. *(Lesson 4-6)*

Wednesday

Multiple Choice Find the next three terms of the arithmetic sequence. *(Lesson 4-7)*

18, 24, 30, 36, 42, …

(A) 48, 50, 52
(B) 48, 51, 54
(C) 48, 53, 58
(D) 48, 54, 60

Gridded Response What is the 15th term in the arithmetic sequence? *(Lesson 4-7)*

12, 23, 34, 45, …

Thursday

Multiple Choice What is the area of rhombus *ABCD*? *(Geometry)*

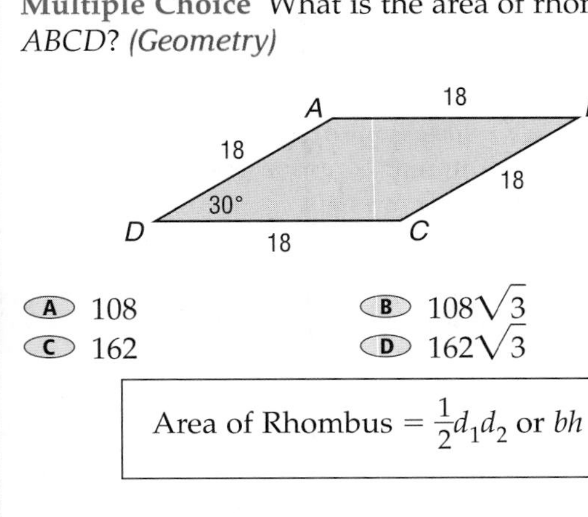

(A) 108
(B) $108\sqrt{3}$
(C) 162
(D) $162\sqrt{3}$

Area of Rhombus $= \frac{1}{2}d_1d_2$ or bh

Friday

Open Ended Certain sets of numbers can be represented by geometric patterns. These are examples of triangular numbers. *(Lesson 4-8)*

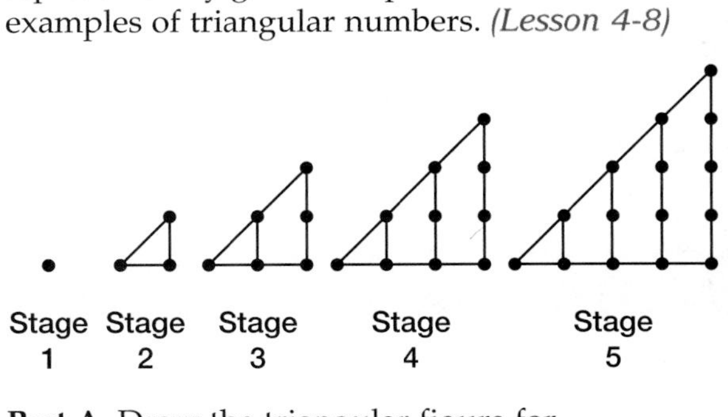

Part A Draw the triangular figure for Stage 6.

Part B What numbers are represented by Stages 1 through 6?

Part C Use the pattern to determining the triangular number for Stage 10.

Week 7

Monday

Open Ended The graph shows the amount Shakeel saved from 1980–2000. *(Lesson 5-1)*

Part A Find the rate of change for 1980–2000.

Part B Explain the meaning of the slope for the years 1980–2000.

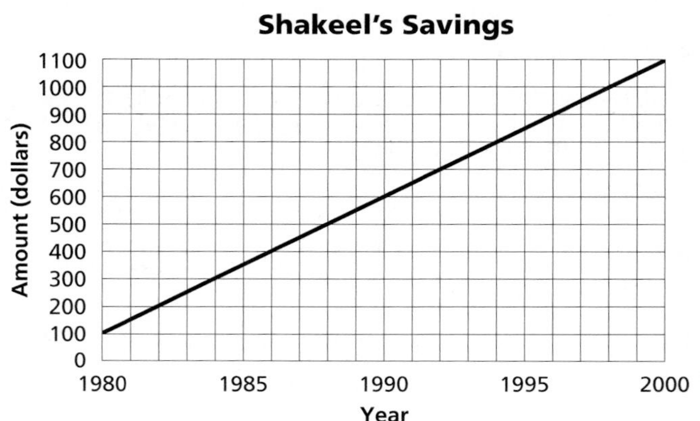

Tuesday

Multiple Choice What is the slope of the line that contains points at (a, b) and (m, n)? *(Lesson 5-1)*

(A) $\dfrac{n-b}{m-a}$ (B) $\dfrac{n-m}{b-a}$

(C) $\dfrac{n-b}{a-m}$ (D) $\dfrac{a-m}{b-n}$

Gridded Response Find the value of r for the line that contains (6, 2) and (8, r) and has slope of 2. *(Lesson 5-1)*

Wednesday

Multiple Choice Which graph is the graph of $y = -\frac{3}{5}x$? *(Lesson 5-2)*

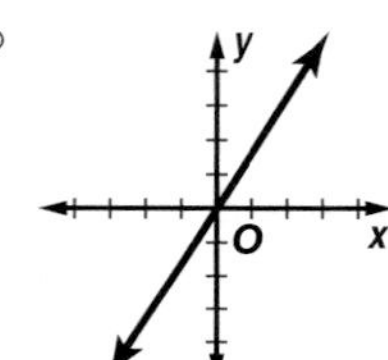

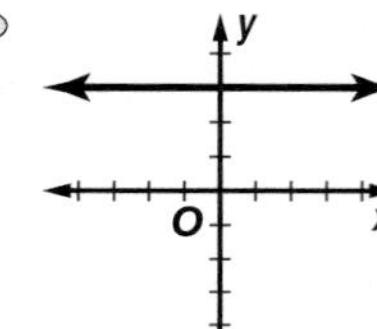

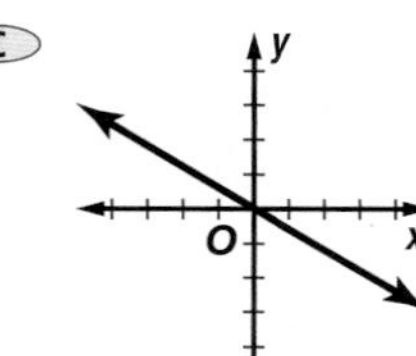

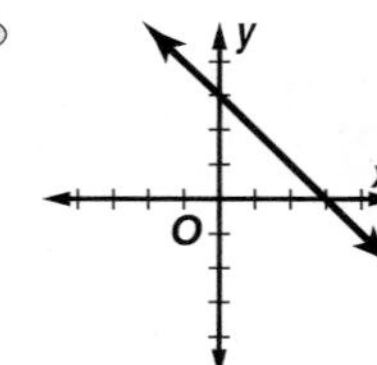

Gridded Response Find the slope of the line that contains points at (1, 1) and (4, 3). *(Lesson 5-1)*

Thursday

Multiple Choice Write the equation of the line with a slope of $-\frac{3}{5}$ and a y-intercept of 3. *(Lesson 5-3)*

(A) $y = -\frac{3}{5}x + 3$ (B) $y = \frac{5}{3}x + 3$

(C) $y = \frac{3}{5}x + 3$ (D) $y = \frac{3}{5}x$

Friday

Open Ended Allyson and Sancia were preparing for the Northwestern Indiana Regional Science and Engineering Fair. On one of their charts they showed data that forms a line. The line has an x-intercept of 4 and a y-intercept of 6. What is the equation of the line? *(Lesson 5-4)*

Week 8

Monday

Open Ended Line ℓ contains points $A(7, 4)$ and $B(11, 9)$. *(Lesson 5-4)*

Part A Find the slope of line ℓ.

Part B Find the equation of line ℓ.

Tuesday

Multiple Choice Write an equation of the line that contains points at (8, 5) and (9, 6). *(Lesson 5-4)*

Ⓐ $y = x - 3$ Ⓑ $y = x + 3$
Ⓒ $y = -x - 3$ Ⓓ $y = -x + 3$

Gridded Response The equation of line ℓ is $2y + 7 = 6(x - 4)$. Find its slope. *(Lesson 5-5)*

Wednesday

Multiple Choice Write the slope-intercept form of an equation of the line that contains (2, 5) and is parallel to the graph of $y = 6x - 2$. *(Lesson 5-6)*

Ⓐ $y = 6x - 28$ Ⓑ $y = -\frac{1}{6}x - 28$
Ⓒ $y = -\frac{1}{6}x - 7$ Ⓓ $y = 6x - 7$

Gridded Response $\ell_1 \perp \ell_2$. If the equation of ℓ_1 is $y = -\frac{2}{3}x + 8$, what is the slope of ℓ_2? *(Lesson 5-6)*

Thursday

Multiple Choice Write the equation of the line that contains (4, 1) and is parallel to the y-axis. *(Lesson 5-6)*

Ⓐ $y = 1$ Ⓑ $y = 4$
Ⓒ $x = 1$ Ⓓ $x = 4$

Friday

Open Ended The chart shows the weights of various lengths of pipes. *(Lesson 5-7)*

Length (ft)	20	22	25	26	29	30	31
Weight (lb)	18	29	17	25	22	24	26

Part A Draw a scatter plot with length on the x-axis and weight on the y-axis.

Part B Determine the type of correlation. Explain your answer.

Week 9

Monday

Open Ended Marco is trying to qualify for the Hoosier Invitational in gymnastics. He is competing in five events vying for best all-around. At this meet, Jacob is the present leader in the competition. He has completed all 5 events with a total score of 46 points. Marco's scores for his first 4 events are 9.1, 9.2, 9.0, and 9.3. *(Lesson 6-1)*

Part A Write an inequality to determine the minimum number of points Marco needs on his final event to surpass Jacob.

Part B What are all the scores that Marco can score to surpass Jacob?

Tuesday

Multiple Choice Which inequality best describes the graph below? *(Lesson 6-1)*

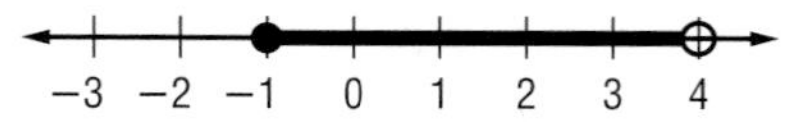

Ⓐ $-1 \geq x > 4$

Ⓑ $-1 \leq x < 4$

Ⓒ $-1 > x > 4$

Ⓓ $-1 < x \leq 4$

Gridded Response Mr. Sparrow set his car's trip odometer to 0 before driving to Holiday World. As he passed a sign saying "Holiday World 32 miles," he checked his odometer. It displayed 95.4. What is the minimum total distance Mr. Sparrow will have traveled when he arrives at his destination? *(Lesson 6-1)*

Wednesday

Multiple Choice What solid will the pattern form when folded along the dotted lines? (Note: All edges are equal.) *(Geometry)*

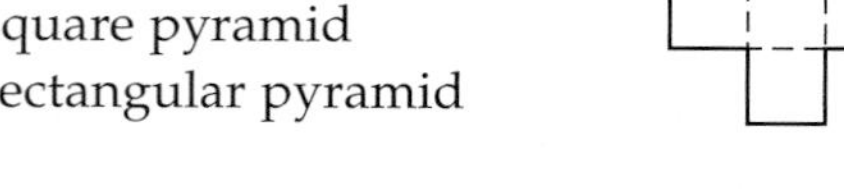

Ⓐ triangular prism

Ⓑ cube

Ⓒ square pyramid

Ⓓ rectangular pyramid

Gridded Response Benita's mother is a real estate broker in Gary. She earns a monthly salary of $800 plus a 3% commission on every home sold. What is the minimum amount of home sales needed for her to earn at least $2,900 per month? *(Lesson 6-1)*

Thursday

Multiple Choice Stephanie has test scores of 84, 80, 68, 72, and 98. If x represents the score on her next test, which inequality would guarantee that she will have at least an average score of 81? *(Lesson 6-1)*

Ⓐ $x < 84$

Ⓑ $x \leq 84$

Ⓒ $x > 84$

Ⓓ $x \geq 84$

Friday

Open Ended *(Lesson 6-3)*

Part A Solve $8(n - 1) > 5n + 2$.

Part B Graph the solution set.

Week 10

Monday

Open Ended Peter is designing a game in which a player can win if after a series of moves that player scores at least 30 points or no more than 18 points. *(Lesson 6-4)*

Part A Write the inequalities to express how a player can win this game.

Part B Graph the solution set.

Tuesday

Multiple Choice Solve the inequality $4 < 2y - 2 < 10$. *(Lesson 6-4)*

(A) $3 < y$ (B) $y < 6$
(C) $3 > y$ or $y < 6$ (D) $3 < y < 6$

Gridded Response Solve $|a + 20| = 4$. What is the absolute value of the difference of the solutions to $|a + 20| = 4$? *(Lesson 6-5)*

Wednesday

Multiple Choice Which is the graph of the inequality $|6x + 3| < 15$? *(Lesson 6-5)*

(A) −3 −2 −1 0 1 2

(B) −3 −2 −1 0 1 2

(C) −3 −2 −1 0 1 2

(D) −3 −2 −1 0 1 2

Gridded Response Which number in the following set of numbers is not a solution of $|3x - 1| < 8$? *(Lesson 6-5)*

$$\{-1, 0, 1, 1\tfrac{1}{2}, 2, 3\}$$

Thursday

Multiple Choice Graph the inequality $x + 2 > y$. *(Lesson 6-6)*

(A)

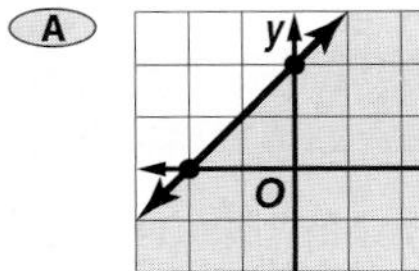
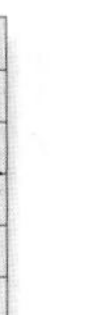

(B)

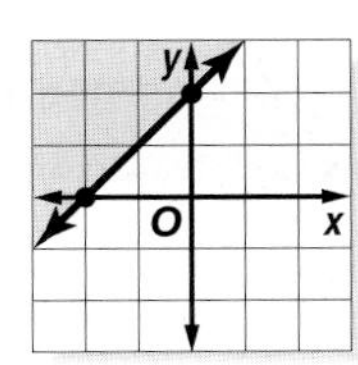

(C)

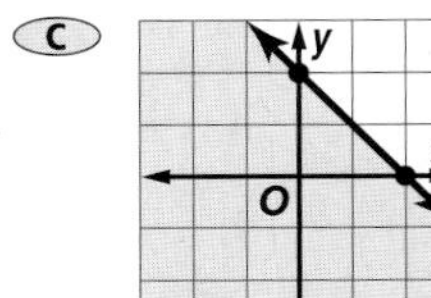

(D)

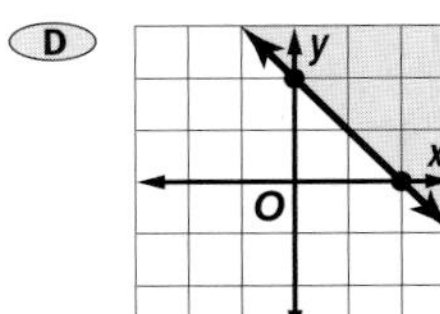

Friday

Open Ended Beth attends South Bend High School and is practicing to qualify for the Academic Team. In the preparation materials given to her by her math teacher was the inequality $|2m + 1| < 8$. *(Lesson 6-5)*

Part A Explain how Beth should solve this inequality.

Part B Solve the inequality and graph the solution set.

Countdown to GQE

Week 11

Monday

Open Ended Refer to the system of equations at the right.

$$y = x - 6$$
$$y = \frac{2}{3}x - 4$$

Part A Graph the system of equations.

Part B Determine whether the system has no solution, one solution, or infinitely many solutions.

Part C If the system can be solved, name the solution.

Tuesday

Multiple Choice Which graph represents a system of equations with no solution? *(Lesson 7-1)*

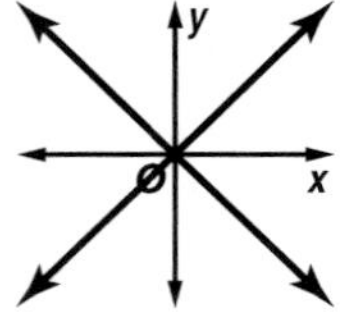

Ⓑ
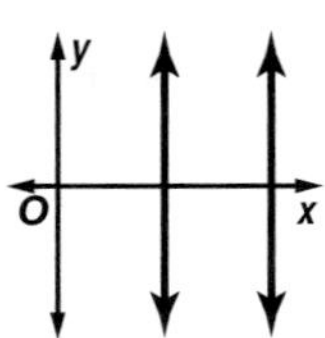

Ⓒ
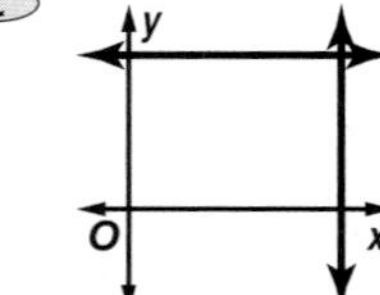

Ⓓ
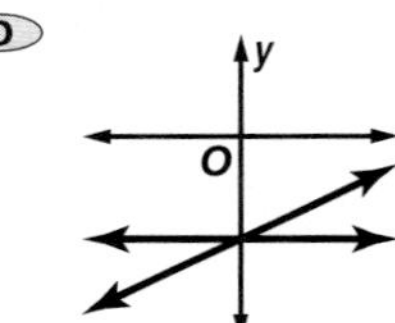

Gridded Response Rob and Carlo own a towing service in Evansville. Last week, they answered a total of 174 emergency road calls. If Rob answered 94 more calls than Carlo did, how many calls did Rob answer? *(Lesson 7-2)*

Wednesday

Multiple Choice Use substitution to solve the system of equations. *(Lesson 7-2)*

$$y = 4x$$
$$2x + 3y = 14$$

Ⓐ (1, 4) Ⓑ (4, 1)
Ⓒ (−1, 4) Ⓓ (4, −1)

Gridded Response The sum of two numbers is 170, and their difference is 110. What is the larger of the two numbers? *(Lesson 7-3)*

Thursday

Multiple Choice A car tire has a radius of 9 inches. How many inches does the car travel in one revolution of the tire? *(Geometry)*

Circumference of Circle = $2\pi r$ or πd Use 3.14 for π.

Ⓐ 2.87 Ⓑ 5.73
Ⓒ 28.26 Ⓓ 56.52

Friday

Open Ended Seven times a number minus two times a second number is −12. The sum of the numbers is 24. Find both numbers. *(Lesson 7-3)*

Week 12

Monday

Open Ended Traveling against the wind, a jet flies 648 miles in 2 hours from New York City to Indianapolis non-stop. The return trip, traveling with the wind, takes $1\frac{3}{4}$ hours. *(Lesson 7-4)*

Part A Find the speed of the jet itself with no wind to the nearest mile per hour.

Part B Find the wind speed.

Tuesday

Multiple Choice Which could be the first step in solving this system of equations? *(Lesson 7-2)*

$$y = 7x$$
$$y - x = 15$$

A $7x = y - x$ B $7x - x = 15$

C $y - x - 7x = 15$ D $y = 7(15 + x)$

Gridded Response Use elimination to solve the system of equations.

$$5x + 4y = 12$$
$$3x + 5y = 15$$

What is the y-coordinate of the solution? *(Lesson 7-4)*

Wednesday

Multiple Choice Solve $\frac{x}{3} - \frac{3y}{4} = -\frac{1}{2}$ and $\frac{x}{6} + \frac{y}{8} = \frac{3}{4}$. *(Lesson 7-4)*

A (4, 1) B (3, 2)

C (−1, 5) D (6, 10)

Gridded Response The solution of the system $4x + 5y = -4$ and $x - y = b$ is $(4, a)$. Find the value of b. *(Lesson 7-4)*

Thursday

Multiple Choice Solve the system of equations. *(Lesson 7-4)*

$$3x - 7y = 1$$
$$-5x + 4y = 4$$

A $\left(-\frac{32}{23}, -\frac{17}{23}\right)$ B $\left(-6, \frac{17}{7}\right)$

C $\left(2, \frac{7}{2}\right)$ D $(0, 1)$

Friday

Open Ended Solve the system of inequalities $y \geq -2x + 4$ and $x > 2$ by graphing. *(Lesson 7-5)*

Countdown to GQE

Week 13

Monday

Open Ended On the Friday before classes started at Purdue University, the bookstore sold \$11,004 in calculus books. A total of 112 calculus books were sold. A new calculus book cost \$105 and a used calculus book cost \$78. *(Lesson 7-2)*

Part A Write an equation to represent the number or calculus books sold and the amount of money collected.

Part B How many of each kind of calculus book was sold?

Tuesday

Multiple Choice Which expression is a binomial? *(Lesson 8-1)*

(A) $6x$

(B) $\frac{6x + 3}{2}$

(C) $x + 5$

(D) $3x^2 + 2x + 1$

Gridded Response Simplify $\left(\frac{2a^3b^{-2}}{2^{-1}a^{-16}b^3}\right)^0$, where $a \neq 0$ and $b \neq 0$. *(Lesson 8-2)*

Wednesday

Multiple Choice Simplify $\left(\frac{2a^2b^4}{4a^3b}\right)^2$, where $a \neq 0$ and $b \neq 0$. *(Lesson 8-2)*

(A) $\frac{b^6}{2a^2}$

(B) $\frac{b^6}{4a^2}$

(C) $\frac{b^{14}}{2a^2}$

(D) $\frac{b^{14}}{4a^2}$

Gridded Response What is the degree of $7 + d^6 - b^3c^2d^3 + b^4$? *(Lesson 8-4)*

Thursday

Multiple Choice The mass of one atom of hydrogen, the lightest element, is 0.00000000000000000000000675 gram. Write this number in scientific notation. *(Lesson 8-3)*

(A) 6.75×10^{-24}

(B) 6.75×10^{-23}

(C) 67.5×10^{-27}

(D) 67.5×10^{-22}

Friday

Open Ended A sculptor made the piece of art shown below, which is to be displayed in the Richmond Art Museum. Write a polynomial to represent the area of the shaded region of the figure. *(Lesson 8-4)*

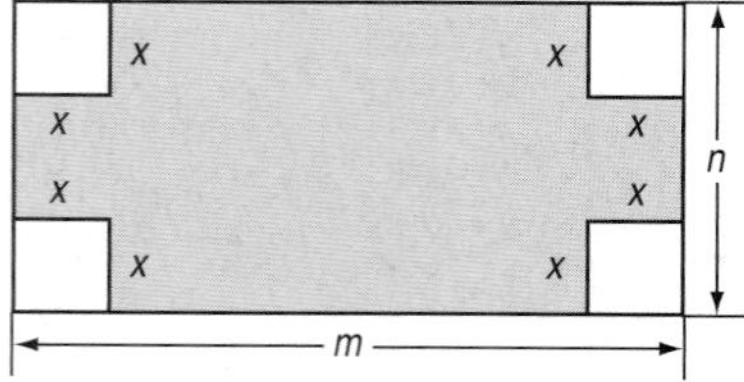

Week 14

Monday

Open Ended The perimeter of the triangle shown is $10x^2 + 5x$. *(Lesson 8-5)*

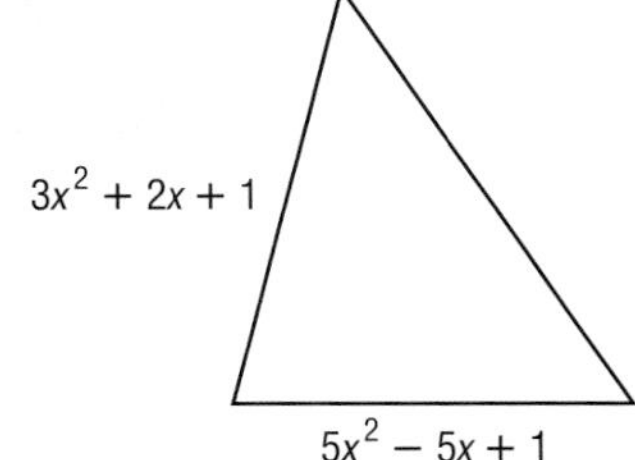

Part A Find the algebraic expression for the length of the third side.

Part B If $x = 3$, what is the length of each side and the perimeter of the triangle?

Tuesday

Multiple Choice Subtract $(3a^2 + 2a - 1)$ from $(5a^2 - 6a + 2)$. *(Lesson 8-5)*

A) $2a^2 - 8a + 3$
B) $-2a^2 + 8a - 3$
C) $2a^2 + 8a + 1$
D) $-2a^2 - 8a - 1$

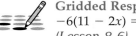

Gridded Response Solve the equation $-6(11 - 2x) = 8(2x - 5) - 38$ *(Lesson 8-6)*

Wednesday

Multiple Choice Find the product $-2mn(9m^2 + 3mn + 6n^2)$. *(Lesson 8-6)*

A) $-18m^3n + 3mn + 6n^2$
B) $-18m^3n - 6m^2n^2 - 12mn^3$
C) $-18m^3n + 6m^2n^2 + 12mn^3$
D) $-30m^3n - 6m^2n^2$

Gridded Response What is the coefficient of the x term in the product $(7x + 3)(x + 4)$? *(Lesson 8-7)*

Thursday

Multiple Choice Which expression is equivalent to $\pi r^2 + 2\pi rh$? *(Lesson 8-7)*

A) $\pi r(r + 2h)$
B) $(\pi + h)(r - h)$
C) $(\pi + 2)(r^2 + h)$
D) $\pi r^2(1 + 2h)$

Friday

Open Ended Mrs. Awandi bought a house in Indiana. She wants to plant grass on the ground around the house as shown on the drawing. Write a polynomial that represents the area of her lawn. *(Lesson 8-8)*

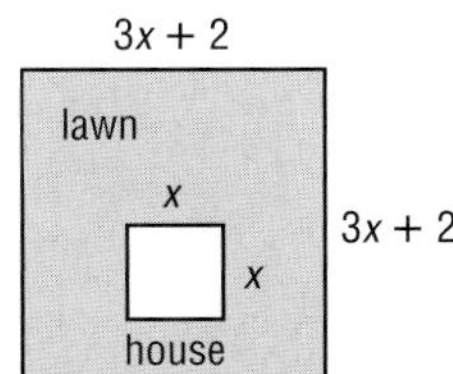

Countdown to GQE

Week 15

Monday

Open Ended The area of a rectangle is 32 square inches and the length and width are both whole numbers. *(Lesson 9-1)*

Part A What are the dimensions of the rectangle that has a maximum perimeter?

Part B What is the maximum perimeter of the rectangle?

Tuesday

Multiple Choice Which number is neither prime nor composite? *(Lesson 9-1)*

(A) 1 (B) 2
(C) 3 (D) 39

Gridded Response Find the greatest common factor of 45 and 60. *(Lesson 9-1)*

Wednesday

Multiple Choice What is the least common multiple of 18 and 27? *(Lesson 9-1)*

(A) 3 (B) 6
(C) 9 (D) 54

Gridded Response Mrs. Whitefeather and Mr. Hasaki work as inspectors on an assembly line at a factory in Gary. Mrs. Whitefeather checks every 8th item while Mr. Hasaki checks every 12th item. If 1,200 items passed by each of them, how many items were checked by both of the inspectors? *(Lesson 9-1)*

Thursday

Multiple Choice Factor $27a^2b + 9b^5$. *(Lesson 9-2)*

(A) $27(a^2b + 9b^5)$ (B) $9(3a^2b + b^5)$
(C) $9b(3a^2 + b^4)$ (D) $9b^2(3a^2 + b^3)$

Friday

Open Ended Find two consecutive odd integers that have a product of 323. *(Lesson 9-3)*

Week 16

Monday

Open Ended A ball is thrown in the air with an initial velocity of 20 feet per second from a height of 4 feet as shown. When the ball reached a height of 10 feet, it hit a branch on a tree. The height of the ball can be determined at any time by the equation $h = -16t^2 + vt + s$, where h is the height in feet, v is the initial upward velocity in feet per second, t is the time in seconds, and s is the starting height. *(Lesson 9-4)*

Part A How long was the ball in the air before it hit the tree branch?

Part B When you solve a quadratic equation, you sometimes get two solutions. Explain why you chose the solution that you did.

Tuesday

Multiple Choice Factor $6x^2 - 11x - 10$. *(Lesson 9-4)*

Ⓐ $(3x - 2)(2x + 5)$
Ⓑ $(3x + 2)(2x - 5)$
Ⓒ $(3x - 5)(2x - 2)$
Ⓓ $(3x + 5)(2x + 2)$

Gridded Response What is the greater value of t if $2t^2 + 10t - 48 = 0$? *(Lesson 9-4)*

Wednesday

Multiple Choice Solve $12d^3 - 108d = 0$. *(Lesson 9-5)*

Ⓐ $d = 0$
Ⓑ $d = 0$ or 3
Ⓒ $d = 0$ or -3
Ⓓ $d = 0, -3,$ or 3

Gridded Response If a car skids on dry concrete, the Indiana State Highway Patrol can use the formula $\frac{1}{24}s^2 = d$ to approximate the speed s of a vehicle in miles per hour given the length d of the skid marks in feet. If the skid marks on dry concrete are 96 feet long, how fast was the car traveling when the brakes were applied? *(Lesson 9-5)*

Thursday

Multiple Choice Solve $3x^2 + 8x + 4 = 0$. *(Lesson 9-6)*

Ⓐ $x = -2$ or $-\frac{2}{3}$
Ⓑ $x = -2$ or $\frac{2}{3}$
Ⓒ $x = 2$ or $-\frac{2}{3}$
Ⓓ $x = 2$ or $\frac{2}{3}$

Friday

Open Ended If an object is propelled from ground level at an initial velocity of 96 feet per second, after t seconds its height h in feet is given by the formula $h = -16t^2 + 96t$. *(Lesson 9-6)*

Part A After how many seconds will its height be 128 feet?

Part B After how many seconds will it hit the ground?

Countdown to GQE

Week 17

Monday

Open Ended Sarah is building a rectangular dog pen for her dog Mopsy. She has 60 feet of fencing to use as the perimeter of the pen. *(Lesson 10-1)*

Part A Make a list of possible whole number dimensions for the pen for the 60-foot perimeter. Draw and label a diagram for the pen that has a width of x.

Part B Write an equation for the area of the pen using x as the width.

Part C What is the greatest area of the pen? Explain your answer.

Tuesday

Multiple Choice What is the axis of symmetry of the graph of $y = 3x^2 - 6x - 14$? *(Lesson 10-1)*

Ⓐ $x = -6$ Ⓑ $x = -1$
Ⓒ $x = 1$ Ⓓ $x = 6$

Gridded Response If $x = a$ is the equation of the axis of symmetry for the graph of $y = x^2 - x - 6$, what is the value of a? *(Lesson 10-2)*

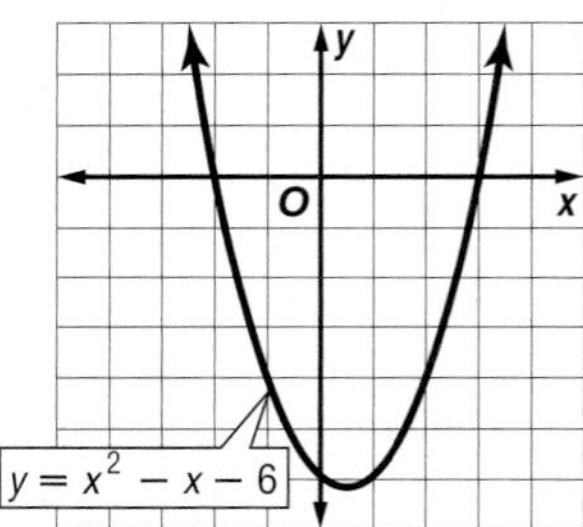

Wednesday

Multiple Choice Which equation will not have a maximum value? *(Lesson 10-1)*

Ⓐ $y = -6x^2 + 3x - 4$
Ⓑ $y = -18 + 3x + x^2$
Ⓒ $y = -x^2 + 4$
Ⓓ $y = 9 - 3x - x^2$

Gridded Response What is the minimum value for y in $y = x^2 - 2x + 7$? *(Lesson 10-1)*

Thursday

Multiple Choice The sum of two numbers is 9, and their product is 14. What is three times the larger number? *(Lesson 10-2)*

Ⓐ 9 Ⓑ 12
Ⓒ 18 Ⓓ 21

Friday

Open Ended The 49th Indiana Volunteer Infantry reenacts Civil War battles. In a mock battle, a cannon was fired. The cannonball was fired with an initial speed of 32 feet per second, and the muzzle was set at 45°. The equation relating the height above the cannon y to the horizontal distance traveled x is $y = x - \frac{1}{32}x^2$. What was the cannonball's maximum height above the cannon if all measurements are in feet? *(Lesson 10-2)*

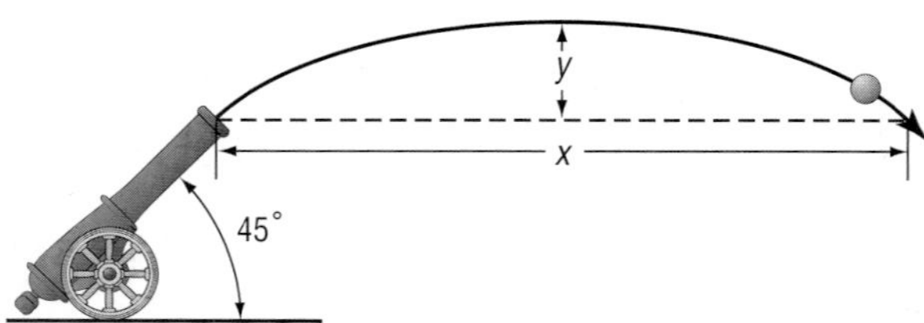

Week 18

Monday

Open Ended The path of debris from a firework display on a calm day can be modeled by the equation $h = -0.05x^2 + x + 6$, where h is the height and x is the horizontal distance in feet. How far away from the launch site will the debris land? *(Lesson 10-3)*

Tuesday

Multiple Choice Which one of the following steps is an appropriate way to begin solving the quadratic equation $2x^2 - 4x = 9$ by completing the square? *(Lesson 10-3)*

- (A) Add 4 to both sides of the equation.
- (B) Factor the left side as $2x(x - 2)$.
- (C) Factor the left side as $x(2x - 4)$.
- (D) Divide both sides by 2.

Gridded Response The discriminant for an equation in the form $ax^2 + bx + c = 0$ is defined as $b^2 - 4ac$. What is the value of the discriminant of $x^2 + 5x - 7 = 0$? *(Lesson 10-4)*

Wednesday

Multiple Choice Solve $x^2 - 4x - 15 = 0$ by completing the square. Round to the nearest tenth, if necessary. *(Lesson 10-3)*

- (A) -2.4 or 6.4
- (B) -3.1 or 7.2
- (C) 4.1 or -1.7
- (D) 2.7 or 1.4

Gridded Response Determine the number of solutions of $x^2 - 6x + 5 = 0$. *(Lesson 10-4)*

Thursday

Multiple Choice Jermaine invested \$8,000 at an Indiana bank at 3% interest compounded daily. About how much would his investment be worth after 5 years? Use the formula $A = P\left(1 + \frac{r}{n}\right)^{nt}$ where A = amount, P = principal, r = rate, t = time in years, and n = number of times compounded. *(Lesson 10-6)*

- (A) \$9,295
- (B) \$9,387
- (C) \$9,510
- (D) \$10,120

Friday

Open Ended Imagine you have a huge sheet of paper $\frac{1}{1000}$ inch thick. You cut the sheet in half and put 1 piece on top of the other. You cut these 2 pieces in half and put the 4 pieces together in a pile. Then you cut that pile in half and repeat the pattern. You do this until you have cut the pile 30 times. *(Lesson 10-7)*

Part A Complete a table comparing the number of cuts and the number of resulting pieces.

Part B What is the pattern? How many pieces are there after the 30th cut?

Part C About how high will the pile be in miles?

Countdown to GQE

Week 19

Monday

Open Ended Jordan says that $\sqrt{317}$ is closer to 18. Gabriella says that it is closer to 17. *(Lesson 11-1)*

Part A Who is correct?

Part B Explain your answer.

Tuesday

Multiple Choice Use the information in the figure to find AB, if $AC = 24$ centimeters. *(Lesson 11-1)*

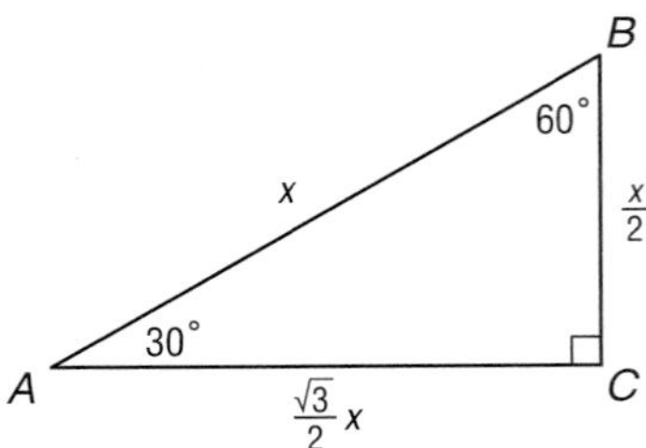

(A) 12 centimeters
(B) $16\sqrt{3}$ centimeters
(C) $16\sqrt{6}$ centimeters
(D) $24\sqrt{3}$ centimeters

Gridded Response The leg of a 45°-45°-90° triangle can be determined by dividing the hypotenuse of the triangle by $\sqrt{2}$. If the length of the hypotenuse of a 45°-45°-90° triangle is 16 cm, what is the length of each leg? Round to the nearest tenth of a centimeter, if necessary. *(Lesson 11-1)*

Wednesday

Multiple Choice Simplify $\frac{\sqrt{7}}{\sqrt{6}}$. *(Lesson 11-1)*

(A) $\frac{\sqrt{42}}{6}$
(B) $\frac{6\sqrt{7}}{6}$
(C) $\frac{3\sqrt{14}}{6}$
(D) $\frac{\sqrt{42}}{12}$

Gridded Response Amber is a member of the Northwest Indiana Woodworkers Association. She is making a small triangular wall shelf. She uses Heron's formula to find the area of the piece of wood shown below. Heron's formula is $S = \frac{1}{2}(a + b + c)$ where a, b, and c are the length of the sides of the triangle and $A = \sqrt{S(S-a)(S-b)(S-c)}$. What is the area of the shelf? Round to the nearest tenth, if necessary. *(Lesson 11-2)*

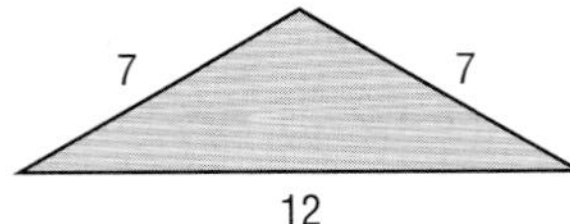

Thursday

Multiple Choice Find the difference of $18\sqrt{7}$ and $4\sqrt{28}$. *(Lesson 11-2)*

(A) $2\sqrt{7}$
(B) $14\sqrt{21}$
(C) $3\sqrt{14}$
(D) $10\sqrt{7}$

Friday

Open Ended Find the sum $\sqrt{18} + 3\sqrt{12} - \sqrt{8}$. *(Lesson 11-2)*

Countdown to GQE

Week 20

Monday

Open Ended Solve $\sqrt{x + 3} = 2x - 9$. *(Lesson 11-3)*

Tuesday

Multiple Choice Which of the lengths of a triangle below would not form a right triangle? *(Lesson 11-4)*

(A) 3, 4, 5

(B) $\frac{3}{16}, \frac{4}{16}, \frac{5}{16}$

(C) 5, 12, 13

(D) 7, 24, 26

Gridded Response Solve $\sqrt{x} = 7$. *(Lesson 11-3)*

Wednesday

Multiple Choice Find the distance between $A(3, 5)$ and $B(-6, 7)$. Round to the nearest tenth, if necessary. *(Lesson 11-5)*

(A) $-\frac{2}{9}$

(B) 3.6

(C) 9.2

(D) 12.1

Gridded Response In the figure, $\triangle ABE \sim \triangle ACD$. Find CD in centimeters. *(Lesson 11-6)*

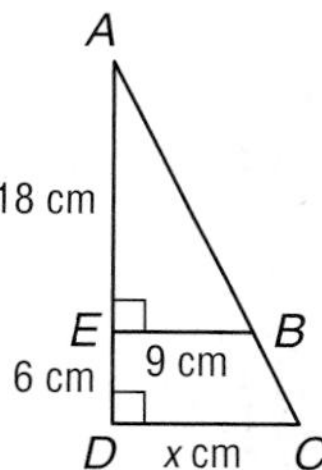

Thursday

Multiple Choice When the sun's angle of elevation is 70°, a building casts a shadow 25 meters long. Approximately how tall is the building? *(Lesson 11-7)*

(A) 24 meters

(B) 42 meters

(C) 69 meters

(D) 82 meters

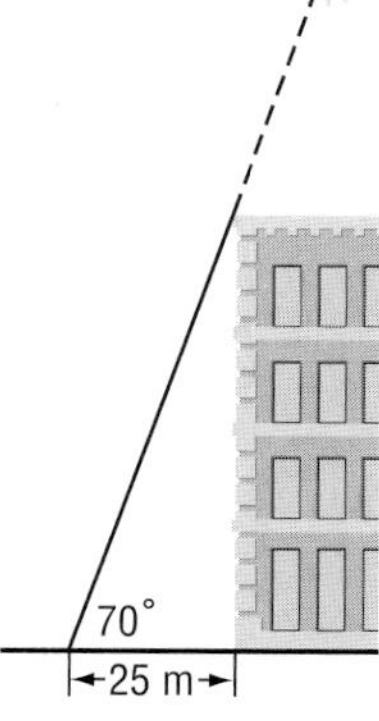

Friday

Open Ended Carlotta Cortez is piloting a small cargo plane bound for Grissom Air Force Base. She determined that she was flying 3,500 feet parallel to the ground and that the ground distance to the landing strip was 9,000 feet. Approximately what is her plane's angle of depression to the landing strip, represented by $\angle A$? Round to the nearest degree. *(Lesson 11-7)*

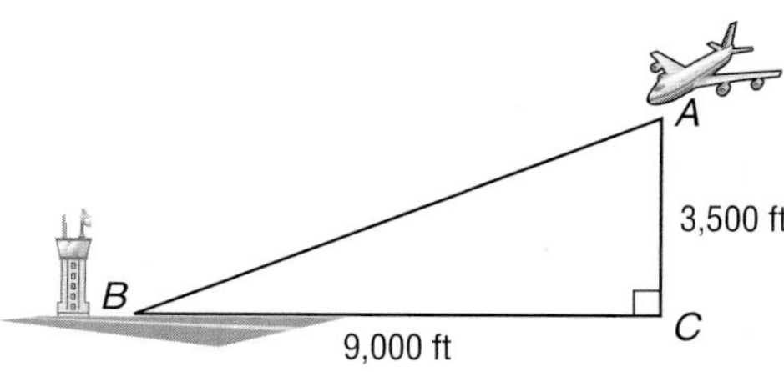

Countdown to GQE

Week 21

Monday

Open Ended Use the figure to calculate the approximate height of French Park Falls. Round to the nearest tenth if necessary. Explain how you found this height. *(Lesson 11-7)*

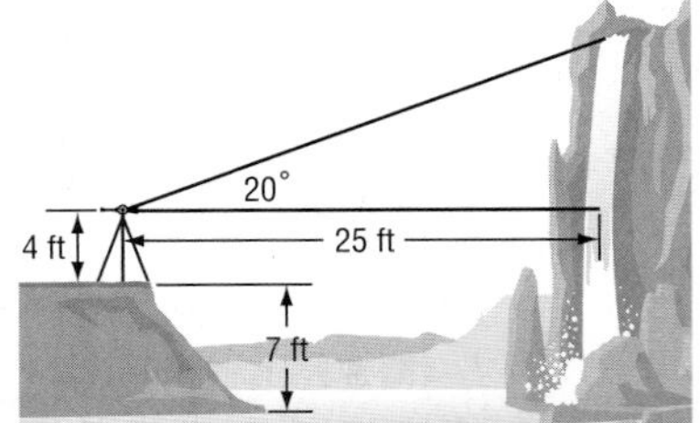

Tuesday

Multiple Choice If y varies inversely as x, and $y = 4$ when $x = 12$, find x when $y = 6$. *(Lesson 12-1)*

(A) 6
(B) 7
(C) 8
(D) 9

Gridded Response Indiana University Press determined that the cost of printing a certain book varies inversely as the number produced. If 10,000 books are printed, the cost is $2 per unit. Find the cost per unit to print 25,000 books. *(Lesson 12-1)*

Wednesday

Multiple Choice State the excluded values for which $\frac{x^2 - 36}{x^2 + 3x + 2}$ is undefined. *(Lesson 12-2)*

(A) 1
(B) -2
(C) -1 and -2
(D) 1 and 2

Gridded Response Determine the constant of variation if y varies inversely as x and $y = 6.5$ when $x = 4.2$. *(Lesson 12-1)*

Thursday

Multiple Choice The length of a violin string varies inversely as the frequency of its vibrations. A violin string 10 inches long vibrates at a frequency of 512 cycles per second. Find the frequency of a 16-inch string. *(Lesson 12-1)*

(A) 240 cycles per second
(B) 320 cycles per second
(C) 360 cycles per second
(D) 480 cycles per second

Friday

Open Ended Find the product $\frac{p^2 + 4p - 5}{p^2 + 7p + 10} \cdot \frac{p + 4}{p - 1}$. *(Lesson 12-3)*

Week 22

Monday

Open Ended Katie is studying engineering at Indiana University. She knows that by using a lever and fulcrum heavy objects can be lifted more easily. She determined that the expression $\frac{W(L-x)}{x}$ represents the weight of an object that can be lifted if W pounds of force are applied to a lever L inches long with the fulcrum placed x inches from the object. Her father weighs 180 pounds. What is the maximum weight he can lift using an 8-foot lever with its fulcrum placed 18 inches from the heavy object? *(Lesson 12-4)*

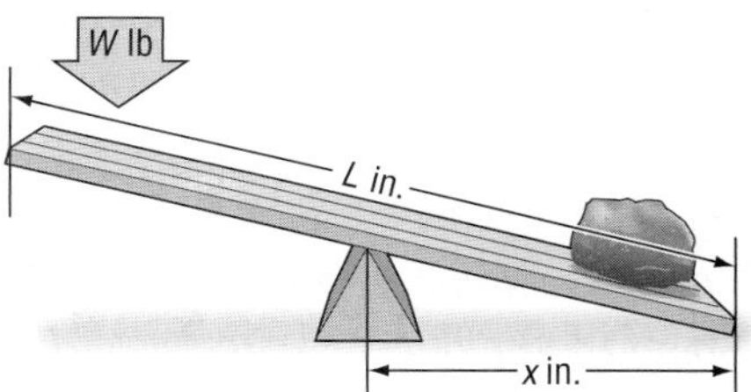

Tuesday

Multiple Choice Which expression is NOT equivalent to the reciprocal of $\frac{x^2-9y^2}{x+3y}$? *(Lesson 12-4)*

Ⓐ $\frac{1}{x-3y}$

Ⓑ $\frac{1}{\frac{x^2-9y^2}{x+3y}}$

Ⓒ $\frac{-1}{3y-x}$

Ⓓ $\frac{1}{x}-\frac{1}{3y}$

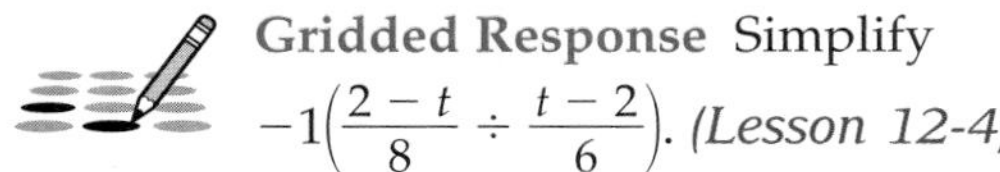

Gridded Response Simplify $-1\left(\frac{2-t}{8} \div \frac{t-2}{6}\right)$. *(Lesson 12-4)*

Wednesday

Multiple Choice Find the quotient of $\frac{2m^2+7m+3}{m^2-9} \div \frac{2m^2+11m+5}{m^2-3m}$. *(Lesson 12-5)*

Ⓐ $\frac{-m}{m+5}$

Ⓑ $\frac{m}{m+5}$

Ⓒ $-m^2+5m$

Ⓓ $\frac{2m}{m-9}$

Gridded Response An Indiana farmer has planned on using 1730 liters of diluted fertilizer per square kilometer. How many liters would that be per square meter? Round to the nearest thousandth liter. *(Lesson 12-4)*

Thursday

Multiple Choice Find $\frac{4x+1}{x-2}+\frac{3x+1}{x-2}$. *(Lesson 12-6)*

Ⓐ $\frac{7x+2}{x-4}$

Ⓑ $\frac{7x+2}{x-2}$

Ⓒ $\frac{7x+1}{x-2}$

Ⓓ $\frac{7x+2}{2x}$

Friday

Open Ended Explain how to solve $\frac{-4}{m^2-8m+12}=\frac{m}{m-6}+\frac{1}{m-2}$. Then solve. *(Lesson 12-9)*

Countdown to GQE

Week 23

Monday

Open Ended Jamaal and Christian were deciding on how to do a survey to predict who will be elected president of the student government association. They came up with 4 ideas. *(Lesson 13-1)*

1. Ask all freshmen only.
2. Ask all the teachers.
3. Publish a survey in the school newspaper and ask students to reply.
4. Stand by the doorway in the cafeteria and ask every 10th student their choice.

Part A Which choice is best?

Part B Explain your answer.

Tuesday

Multiple Choice State the dimension of $[3 \ 1 \ -2 \ 4]$. *(Lesson 13-2)*

(A) 1 by 4 (B) 4 by 1
(C) 2 by 2 (D) 3 by 1

Gridded Response In a 4-by-5 matrix, how many columns are there? *(Lesson 13-2)*

Wednesday

Multiple Choice Matrices *A* and *B* give the number of pants by size and color at two men's stores. Determine which matrix represents the sum of *A* and *B*. *(Lesson 13-2)*

Matrix A

	32″	34″	36″	38″
blue	30	26	10	6
black	30	24	12	4

Matrix B

32″	34″	36″	38″
25	4	17	30
26	18	20	32

(A) $\begin{bmatrix} 55 & 30 & 27 & 36 \\ 56 & 42 & 32 & 36 \end{bmatrix}$ (B) $\begin{bmatrix} 5 & 22 & -7 & -24 \\ 4 & 6 & -8 & -28 \end{bmatrix}$

(C) $\begin{bmatrix} 55 & 30 & 27 & 36 \\ 30 & 24 & 12 & 4 \end{bmatrix}$ (D) $\begin{bmatrix} 30 & 26 & 10 & 6 \\ 56 & 42 & 32 & 36 \end{bmatrix}$

Gridded Response Solve $\sqrt{y+8} - 1 = 2$. *(Lesson 13-3)*

Thursday

Multiple Choice Suppose *M* and *N* are 3-by-3 matrices. If $M + N = M$, which of the following is true? *(Lesson 13-3)*

(A) $N = \begin{bmatrix} 1 & 1 & 1 \\ 1 & 1 & 1 \\ 1 & 1 & 1 \end{bmatrix}$ (B) $N = \begin{bmatrix} 0 & 0 & 0 \\ 0 & 0 & 0 \\ 0 & 0 & 0 \end{bmatrix}$

(C) $N = \begin{bmatrix} 1 & 0 & 0 \\ 0 & 1 & 0 \\ 0 & 0 & 1 \end{bmatrix}$ (D) $N = \begin{bmatrix} 0 & 0 & 1 \\ 0 & 1 & 0 \\ 1 & 0 & 0 \end{bmatrix}$

Friday

Open Ended The diagram is a network representing cities between which there are direct air routes. Write a matrix to describe all routes. Use a 0 to represent no direct route and a 1 to represent a direct route. *(Lesson 13-2)*

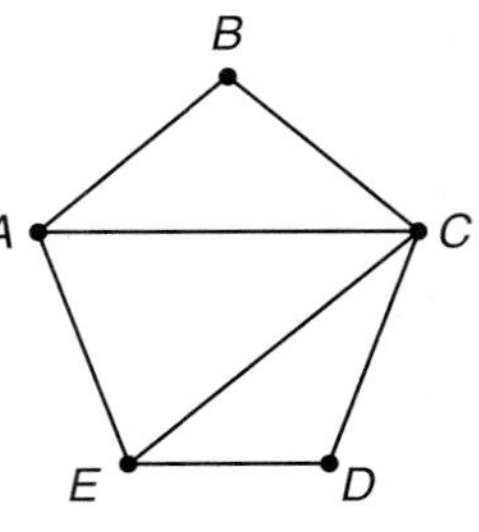

Week 24

Monday

Open Ended The box-and-whisker plot below represents the ages of the members of a senior citizens club in Kokomo. Use the graph to answer these questions. Explain your reasoning. *(Lesson 13-5)*

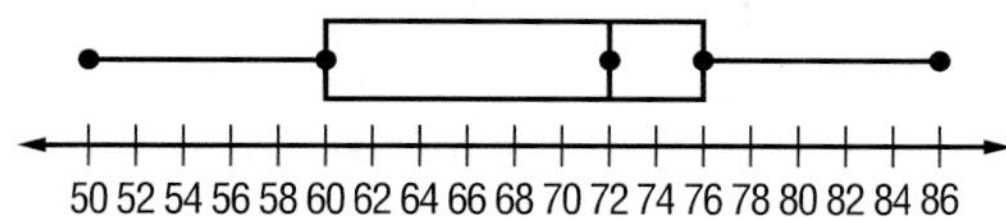

Part A What is the age of the youngest member of the club?

Part B What values are the lower and upper quartiles?

Part C What is the median?

Tuesday

Multiple Choice Which is the median for the set of data? *(Lesson 13-4)*

{84, 86, 48, 72, 36, 98}

(A) 72 (B) 78
(C) 84 (D) 86

Gridded Response Identify the outlier, if any, in the set of data shown. *(Lesson 13-4)*

{15, 22, 22, 26, 27, 29, 30, 31, 32, 36, 50}

Wednesday

Multiple Choice If $A = \begin{bmatrix} 2 & -4 \\ -3 & 0 \end{bmatrix}$ and $C = \begin{bmatrix} 1 & 0 \\ 0 & 1 \end{bmatrix}$, find $A - 5C$. *(Lesson 13-2)*

(A) $\begin{bmatrix} 1 & -4 \\ -3 & 1 \end{bmatrix}$ (B) $\begin{bmatrix} -3 & -4 \\ -3 & -5 \end{bmatrix}$

(C) $\begin{bmatrix} -3 & -5 \\ -3 & -4 \end{bmatrix}$ (D) $\begin{bmatrix} 3 & 4 \\ 3 & 5 \end{bmatrix}$

Gridded Response Find the interquartile range for the set of data. *(Lesson 13-4)*

{27, 37, 21, 54, 47, 35}

Thursday

Multiple Choice What is the range of the data? *(Lesson 13-4)*

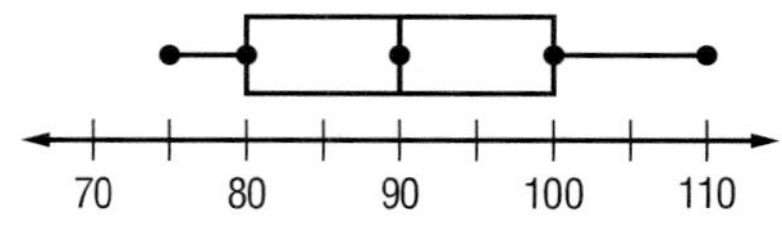

(A) 30 (B) 35
(C) 35.5 (D) 40

Friday

Open Ended The data below represents the ages of the first 8 people to ride the Tilt-a-Whirl on opening day at the Indiana State Fair. *(Lesson 13-5)*

18, 20, 17, 32, 15, 17, 28, 16

Part A Draw a box-and-whisker plot for the data. Explain your reasoning.

Part B What is the range and the interquartile range?

Part C What is the median?

Week 25

Monday

Open Ended Mr. Samuelson owns a deli in Indiana where he offers a choice of 8 meats, 5 types of cheese, and 3 types of bread. *(Lesson 14-1)*

Part A How many different types of sandwiches does he offer?

Part B Mr. Samuelson wants to add one more choice either to meats, cheese, or bread. Where should he add the choice in order to get the maximum number of choices? Explain.

Tuesday

Multiple Choice What is the value of 6!? *(Lesson 14-1)*

Ⓐ 21 Ⓑ 120
Ⓒ 720 Ⓓ 5040

Gridded Response New Indiana license plates are composed of 2 numbers, followed by one letter, and then 3 more numbers. How many thousands of license plates can be made if the first digit cannot be 0? *(Lesson 14-1)*

Wednesday

Multiple Choice Evaluate ${}_{10}C_2$. *(Lesson 14-2)*

Ⓐ 45 Ⓑ 50
Ⓒ 55 Ⓓ 60

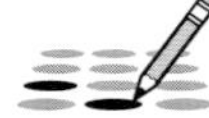

Gridded Response How many different teams of 3 players can be chosen from 8 players? *(Lesson 14-2)*

Thursday

Multiple Choice The chart shows the possible sums when two dice are rolled. What is the probability that the sum is 8 or less? *(Lesson 14-5)*

	1	2	3	4	5	6
1	2	3	4	5	6	7
2	3	4	5	6	7	8
3	4	5	6	7	8	9
4	5	6	7	8	9	10
5	6	7	8	9	10	11
6	7	8	9	10	11	12

Ⓐ $\frac{5}{36}$ Ⓑ $\frac{2}{9}$
Ⓒ $\frac{13}{18}$ Ⓓ $\frac{5}{12}$

Friday

Open Ended Scott flipped three different coins 200 times and recorded his results in the table below. *(Lesson 14-5)*

Results of Tossing 3 Coins 200 Times			
HHH	26	TTT	21
HHT	18	TTH	41
HTH	15	THT	20
HTT	29	THH	30

Part A What was Scott's experimental probability of getting a THT on a flip of 3 coins?

Part B What is the *theoretical* probability of getting a THT?

Part C Why aren't both probabilities the same?

GLENCOE MATHEMATICS

Indiana Edition

Algebra 1

Holliday

Cuevas

Moore-Harris

Carter

Marks

Casey

Day

Hayek

New York, New York
Columbus, Ohio
Chicago, Illinois
Peoria, Illinois
Woodland Hills, California

The McGraw·Hill Companies

The Standardized Test Practice features in this book were aligned and verified by The Princeton Review, the nation's leader in test preparation. Through its association with McGraw-Hill, The Princeton Review offers the best way to help students excel on standardized assessments.

The Princeton Review is not affiliated with Princeton University or Educational Testing Service.

Send all inquiries to:
Glencoe/McGraw-Hill
8787 Orion Place
Columbus, OH 43240-4027

ISBN: 0-07-860391-9

Printed in the United States of America.

7 8 9 10 055/071 12 11 10 09 08 07

Contents in Brief

Authors

Berchie Holliday, Ed.D.
Former Mathematics Teacher
Northwest Local School District
Cincinnati, OH

Gilbert J. Cuevas, Ph.D.
Professor of Mathematics Education
University of Miami
Miami, FL

Beatrice Moore-Harris
Educational Specialist
Bureau of Education and Research
League City, TX

John A. Carter, Ph.D.
Director of Mathematics
Adlai E. Stevenson High School
Lincolnshire, IL

Authors

Daniel Marks, Ed.D.
Associate Professor of Mathematics
Auburn University at Montgomery
Montgomery, AL

Ruth M. Casey
Mathematics Teacher Department Chair
Anderson County High School
Lawrenceburg, KY

Roger Day, Ph.D.
Associate Professor of Mathematics
Illinois State University
Normal, IL

Linda M. Hayek
Mathematics Teacher
Ralston Public Schools
Omaha, NE

Contributing Authors

USA TODAY
The USA TODAY Snapshots®, created by USA TODAY®, help students make the connection between real life and mathematics.

Dinah Zike
Educational Consultant
Dinah-Might Activities, Inc.
San Antonio, TX

Content Consultants

Each of the Content Consultants reviewed every chapter and gave suggestions for improving the effectiveness of the mathematics instruction.

Mathematics Consultants

Gunnar E. Carlsson, Ph.D.
Consulting Author
Professor of Mathematics
Stanford University
Stanford, CA

Ralph L. Cohen, Ph.D.
Consulting Author
Professor of Mathematics
Stanford University
Stanford, CA

Alan G. Foster
Former Mathematics Teacher & Department Chairperson
Addison Trail High School
Addison, IL

Les Winters
Instructor
California State University, Northridge
Northridge, CA

William Collins
Director, The Sisyphus Math Learning Center
East Side Union High School District
San Jose, CA

Dora Swart
Mathematics Teacher
W.F. West High School
Chehalis, WA

David S. Daniels
Former Mathematics Chair
Longmeadow High School
Longmeadow, MA

Mary C. Enderson, Ph.D.
Associate Professor of Mathematics
Middle Tennessee State University
Murfreesboro, TN

Gerald A. Haber
Consultant, Mathematics Standards and Professional Development
New York, NY

Angiline Powell Mikle
Assistant Professor Mathematics Education
Texas Christian University
Fort Worth, TX

C. Vincent Pané, Ed.D.
Associate Professor of Education/ Coordinator of Secondary & Special Subjects Education
Molloy College
Rockville Centre, NY

Reading Consultant

Lynn T. Havens
Director of Project CRISS
Kalispell School District
Kalispell, MT

Teacher Reviewers

Each Teacher Reviewer reviewed at least two chapters of the Student Edition, giving feedback and suggestions for improving the effectiveness of the mathematics instruction.

Susan J. Barr
Department Chair/Teacher
Dublin Coffman High School
Dublin, OH

Diana L. Boyle
Mathematics Teacher, 6–8
Judson Middle School
Salem, OR

Judy Buchholtz
Math Department Chair/Teacher
Dublin Scioto High School
Dublin, OH

Holly A. Budzinski
Mathematics Department Chairperson
Green Hope High School
Morrisville, NC

Rusty Campbell
Mathematics Instructor/Chairperson
North Marion High School
Farmington, WV

Nancy M. Chilton
Mathematics Teacher
Louis Pizitz Middle School
Birmingham, AL

Teacher Reviewers

Lisa Cook
Mathematics Teacher
Kaysville Junior High School
Kaysville, UT

Bonnie Daigh
Mathematics Teacher
Eudora High School
Eudora, KS

Carol Seay Ferguson
Mathematics Teacher
Forestview High School
Gastonia, NC

Carrie Ferguson
Teacher
West Monroe High School
West Monroe, LA

Melissa R. Fetzer
Teacher/Math Chairperson
Hollidaysburg Area Junior High School
Hollidaysburg, PA

Diana Flick
Mathematics Teacher
Harrisonburg High School
Harrisonburg, VA

Kathryn Foland
Teacher/Subject Area Leader
Ben Hill Middle School
Tampa, FL

Celia Foster
Assistant Principal Mathematics
Grover Cleveland High School
Ridgewood, NY

Patricia R. Franzer
Secondary Math Instructor
Celina City Schools
Celina, OH

Candace Frewin
Teacher on Special Assignment
Pinellas County Schools
Largo, FL

Larry T. Gathers
Mathematics Teacher
Springfield South High School
Springfield, OH

Maureen M. Grant
Mathematics Teacher/Department Chair
North Central High School
Indianapolis, IN

Marie Green
Mathematics Teacher
Anthony Middle School
Manhattan, KS

Vicky S. Hamen
High School Math Teacher
Celina High School
Celina, OH

Kimberly A. Hepler
Mathematics Teacher
S. Gordon Stewart Middle School
Fort Defiance, VA

Deborah L. Hewitt
Mathematics Teacher
Chester High School
Chester, NY

Marilyn S. Hughes
Mathematics Department Chairperson
Belleville West High School
Belleville, IL

Larry Hummel
Mathematics Department Chairperson
Central City High School
Central City, NE

William Leschensky
Former Mathematics Teacher
Glenbard South High School
College of DuPage
Glen Ellyn, IL

Sharon Linamen
Mathematics Teacher
Lake Brantley High School
Altamonte Springs, FL

Patricia Lund
Mathematics Teacher
Divide County High School
Crosby, ND

Marilyn Martau
Mathematics Teacher (Retired)
Lakewood High School
Lakewood, OH

Kathy Massengill
Mathematics Teacher
Midlothian High School
Midlothian, VA

Marie Mastandrea
District Mathematics Coordinator
Amity Regional School District #5
Woodbridge, CT

Laurie Newton
Teacher
Crossler Middle School
Salem, OR

James Leo Oliver
Teacher of the Emotionally Impaired
Lakeview Junior High School
Battle Creek, MI

Shannon Collins Pan
Department of Mathematics
Waverly High School
Waverly, NY

Cindy Plunkett
Math Educator
E.M. Pease Middle School
San Antonio, TX

Ann C. Raymond
Teacher
Oak Ave. Intermediate School
Temple City, CA

Sandy Schoff
Math Curriculum Coordinator K–12
Anchorage School District
Anchorage, AK

Susan E. Sladowski
Assistant Principal–Mathematics
Bayside High School
Bayside, NY

Paul E. Smith
Teacher/Consultant
Plaza Park Middle School
Evansville, IN

Dr. James Henry Snider
Teacher–Math Dept. Chair/Curriculum & Technology Coordinator
Nashville School of the Arts
Nashville, TN

Diane Stilwell
Mathematics Teacher/Technology Coordinator
South Middle School
Morgantown, WV

Richard P. Strausz
Math and Technology Coordinator
Farmington Schools
Farmington, MI

Patricia Taepke
Mathematics Teacher and BTSA Trainer
South Hills High School
West Covina, CA

C. Arthur Torell
Mathematics Teacher and Supervisor
Summit High School
Summit, NJ

Lou Jane Tynan
Mathematics Department Chair
Sacred Heart Model School
Louisville, KY

Julia Dobbins Warren
Mathematics Teacher
Mountain Brook Junior High School
Birmingham, AL

Jo Amy Wynn
Mathematics Teacher
Captain Shreve High School
Shreveport, LA

Rosalyn Zeid
Mathematics Supervisor
Union Township School District
Union, NJ

Teacher Advisory Board and Field Test Schools

Teacher Advisory Board

Glencoe/McGraw-Hill wishes to thank the following teachers for their feedback on Glencoe *Algebra*. They were instrumental in providing valuable input toward the development of this program.

Mary Jo Ahler
Mathematics Teacher
Davis Drive Middle School
Apex, NC

David Armstrong
Mathematics Facilitator
Huntington Beach Union High School District
Huntington Beach, CA

Berta Guillen
Mathematics Department Chairperson
Barbara Goleman Senior High School
Miami, FL

Bonnie Johnston
Academically Gifted Program Coordinator
Valley Springs Middle School
Arden, NC

JoAnn Lopykinski
Mathematics Teacher
Lincoln Way East High School
Frankfort, IL

David Lorkiewicz
Mathematics Teacher
Lockport High School
Lockport, IL

Norma Molina
Ninth Grade Success Initiative Campus Coordinator
Holmes High School
San Antonio, TX

Sarah Morrison
Mathematics Department Chairperson
Northwest Cabarrus High School
Concord, NC

Raylene Paustian
Mathematics Curriculum Coordinator
Clovis Unified School District
Clovis, CA

Tom Reardon
Mathematics Department Chairperson
Austintown Fitch High School
Youngstown, OH

Guy Roy
Mathematics Coordinator
Plymouth Public Schools
Plymouth, MA

Jenny Weir
Mathematics Department Chairperson
Felix Verela Sr. High School
Miami, FL

Field Test Schools

Glencoe/McGraw-Hill wishes to thank the following schools that field-tested pre-publication manuscript during the 2001–2002 school year. They were instrumental in providing feedback and verifying the effectiveness of this program.

Northwest Cabarrus High School
Concord, NC

Davis Drive Middle School
Apex, NC

Barbara Goleman Sr. High School
Miami, FL

Lincoln Way East High School
Frankfort, IL

Scotia-Glenville High School
Scotia, NY

Wharton High School
Tampa, FL

UNIT 1

Table of Contents

Expressions and Equations 2

Chapter 1 The Language of Algebra 4

Lesson 1-7, p. 41

Prerequisite Skills

FOLDABLES™ **Study Organizer** 5

Reading and Writing Mathematics

Standardized Test Practice

Snapshots 3, 27, 50, 53

Chapter 2 Real Numbers 66

Lesson 2-4, p. 87

Chapter 3 Solving Linear Equations 118

Prerequisite Skills

FOLDABLES™

Reading and Writing Mathematics

Standardized Test Practice

Lesson 3-4, p. 142

UNIT 2

Linear Functions 188

Chapter 4 Graphing Relations and Functions 190

Lesson 4-5, p. 222

Chapter 5 Analyzing Linear Equations 254

Prerequisite Skills

FOLDABLES™

Reading and Writing Mathematics

Standardized Test Practice

Lesson 5-2, p. 266

Unit 2

Chapter 6 Solving Linear Inequalities 316

Prerequisite Skills

FOLDABLES™

Reading and Writing Mathematics

Standardized Test Practice

Lesson 6-1, p. 322

Chapter 7 Solving Systems of Linear Equations and Inequalities 366

Lesson 7-2, p. 380

UNIT 3

Polynomials and Nonlinear Functions

Chapter 8 Polynomials

WebQuest Internet Project

Lesson 8-2, p. 422

Prerequisite Skills

Study Organizer 409

Reading and Writing Mathematics

Standardized Test Practice

Snapshots 407, 427

Chapter 9 Factoring 472

Lesson 9-5, p. 505

Chapter 10 Quadratic and Exponential Functions 522

Lesson 10-4, p. 551

UNIT 4

Radical and Rational Functions

Chapter 11 Radical Expressions and Triangles

WebQuest Internet Project

Prerequisite Skills

FOLDABLES™

Reading and Writing Mathematics

Standardized Test Practice

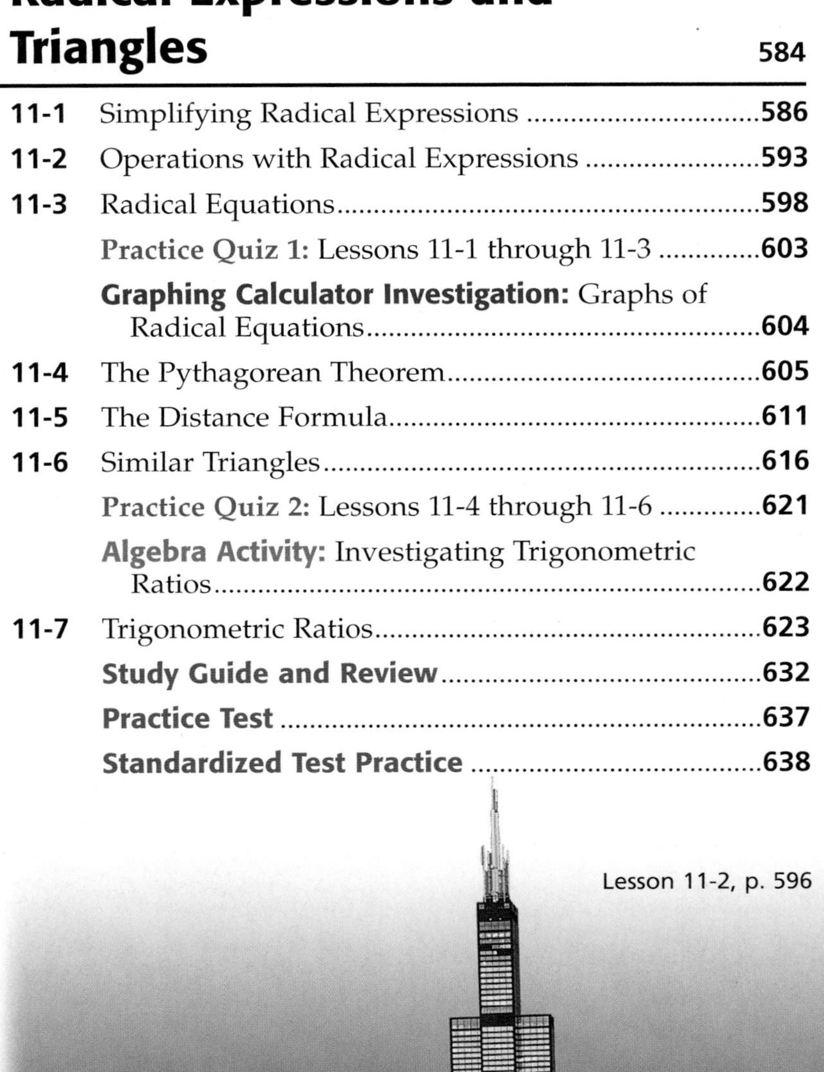

Lesson 11-2, p. 596

Chapter 12 Rational Expressions and Equations 640

Lesson 12-5, p. 670

UNIT 5

Data Analysis 704

Chapter 13 Statistics 706

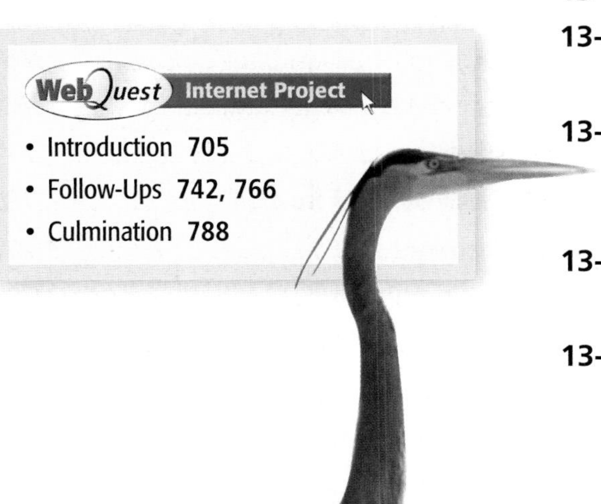

Lesson 13-5, p. 738

Prerequisite Skills

Reading and Writing Mathematics

Standardized Test Practice

Chapter 14 Probability 752

Prerequisite Skills

FOLDABLES™

Reading and Writing Mathematics

Standardized Test Practice

USA TODAY

Student Handbook

Skills

Reference

Lesson 14-1, p. 756

One-Stop Internet Resources

Need extra help or information? Log on to math.glencoe.com or any of the Web addresses below to learn more.

Online Study Tools

- www.algebra1.com/extra_examples shows you additional worked-out examples that mimic the ones in your book.
- www.algebra1.com/self_check_quiz provides you with a practice quiz for each lesson that grades itself.
- www.algebra1.com/vocabulary_review lets you check your understanding of the terms and definitions used in each chapter.
- www.algebra1.com/chapter_test allows you to take a self-checking test before the actual test.
- www.algebra1.com/standardized_test is another way to brush up on your standardized test-taking skills.

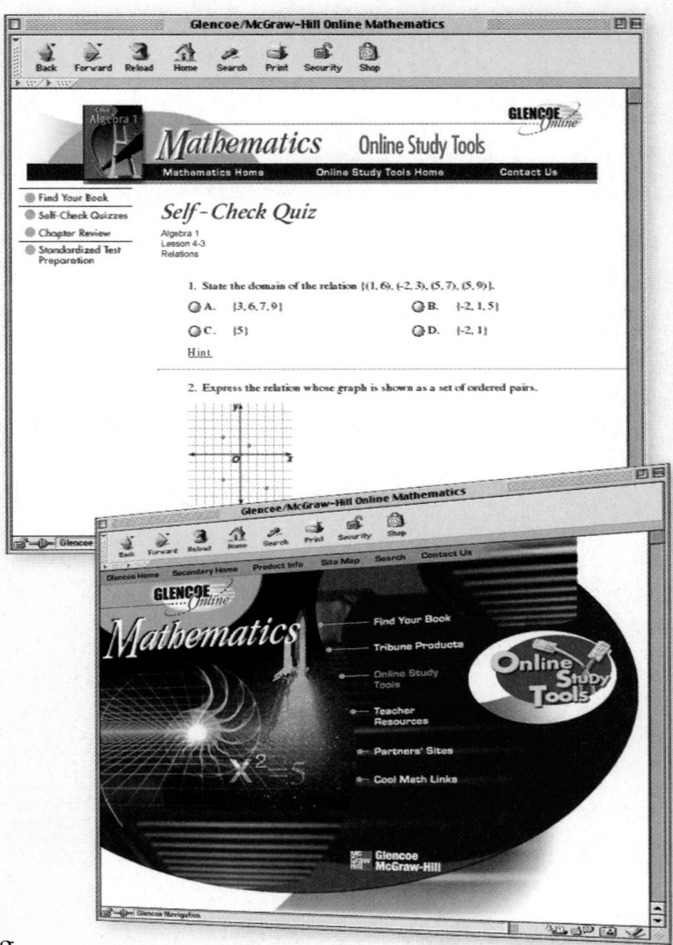

Research Options

- www.algebra1.com/webquest walks you step-by-step through a long-term project using the Web. One WebQuest for each unit is explored using the mathematics from that unit.
- www.algebra1.com/usa_today provides activities related to the concept of the lesson as well as up-to-date Snapshot data.
- www.algebra1.com/careers links you to additional information about interesting careers.
- www.algebra1.com/data_update links you to the most current data available for subjects such as basketball and family.

Calculator Help

- www.algebra1.com/other_calculator_keystrokes provides you with keystrokes other than the TI-83 Plus used in your textbook.

UNIT 1

Expressions and Equations

You can use algebraic expressions and equations to model and analyze real-world situations. In this unit, you will learn about expressions, equations, and graphs.

WebQuest Internet Project

Can You Fit 100 Candles on a Cake?

Source: *USA TODAY,* January, 2001

"The mystique of living to be 100 will be lost by the year 2020 as 100th birthdays become commonplace, predicts Mike Parker, assistant professor of social work, University of Alabama, Tuscaloosa, and a gerontologist specializing in successful aging. He says that, in the 21st century, the fastest growing age group in the country will be centenarians—those who live 100 years or longer." In this project, you will explore how equations, functions, and graphs can help represent aging and population growth.

Log on to www.algebra1.com/webquest. Begin your WebQuest by reading the Task.

Then continue working on your WebQuest as you study Unit 1.

Lesson	1-9	2-6	3-6
Page	55	100	159

USA TODAY Snapshots®

Longer lives ahead

Projected life expectancy for American men and women born in these years:

Men Women

Year	1999	1999	2025	2025	2050	2050
	Men	Women	Men	Women	Men	Women
Life expectancy	74 years	80 years	78 years	84 years	81 years	87 years

Source: U.S. Census Bureau

By James Abundis and Quin Tian, USA TODAY

Chapter 1

The Language of Algebra

What You'll Learn

- **Lesson 1-1** Write algebraic expressions.
- **Lessons 1-2 and 1-3** Evaluate expressions and solve open sentences.
- **Lessons 1-4 through 1-6** Use algebraic properties of identity and equality.
- **Lesson 1-7** Use conditional statements and counterexamples.
- **Lessons 1-8 and 1-9** Interpret graphs of functions and analyze data in statistical graphs.

Key Vocabulary

- variable (p. 6)
- order of operations (p. 11)
- identity (p. 21)
- like terms (p. 28)
- counterexample (p. 38)

Why It's Important

In every state and in every country, you find unique and inspiring architecture. Architects can use algebraic expressions to describe the volume of the structures they design. A few of the shapes these buildings can resemble are a rectangle, a pentagon, or even a pyramid. *You will find the amount of space occupied by a pyramid in Lesson 1-2.*

Getting Started

Prerequisite Skills To be successful in this chapter, you'll need to master these skills and be able to apply them in problem-solving situations. Review these skills before beginning Chapter 1.

For Lessons 1-1, 1-2, and 1-3 — **Multiply and Divide Whole Numbers**

Find each product or quotient.

1. $8 \cdot 8$ **2.** $4 \cdot 16$ **3.** $18 \cdot 9$ **4.** $23 \cdot 6$

5. $57 \div 3$ **6.** $68 \div 4$ **7.** $\frac{72}{3}$ **8.** $\frac{90}{6}$

For Lessons 1-1, 1-2, 1-5, and 1-6 — **Find Perimeter**

Find the perimeter of each figure. *(For review, see pages 813 and 814.)*

9.

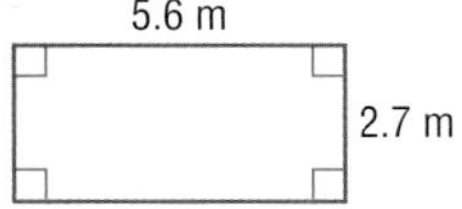

10.

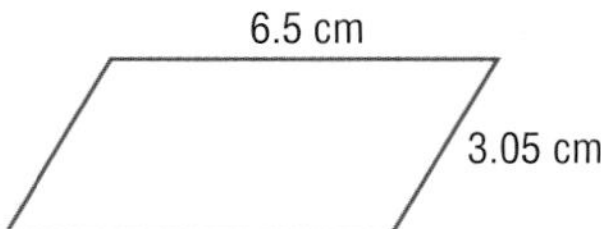

11.

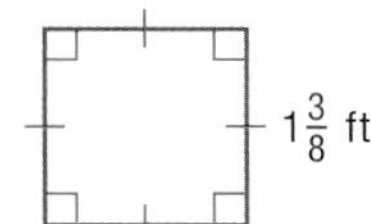

12.

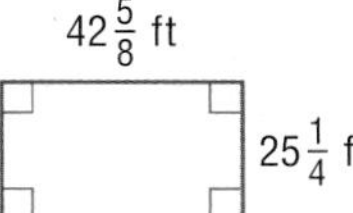

For Lessons 1-5 and 1-6 — **Multiply and Divide Decimals and Fractions**

Find each product or quotient. *(For review, see pages 800 and 801.)*

13. $6 \cdot 1.2$ **14.** $0.5 \cdot 3.9$ **15.** $3.24 \div 1.8$ **16.** $10.64 \div 1.4$

17. $\frac{3}{4} \cdot 12$ **18.** $1\frac{2}{3} \cdot \frac{3}{4}$ **19.** $\frac{5}{16} \div \frac{9}{12}$ **20.** $\frac{5}{6} \div \frac{2}{3}$

Make this Foldable to help you organize information about algebraic properties. Begin with a sheet of notebook paper.

Step 1 Fold

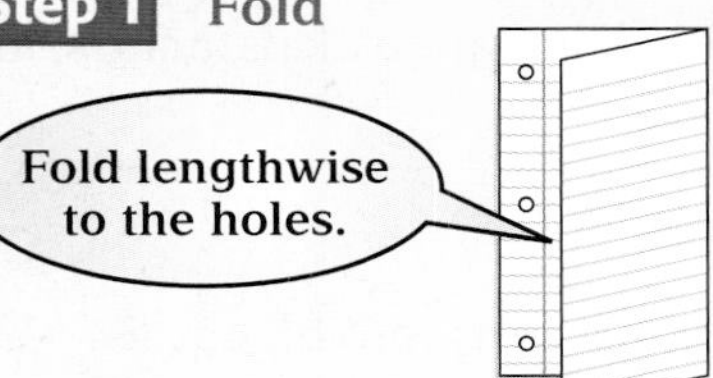

Step 2 Cut

Cut along the top line and then cut 9 tabs.

Step 3 Label

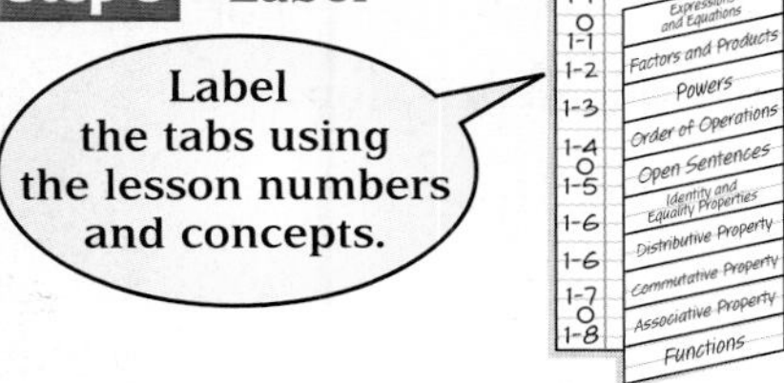

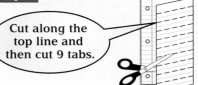

Reading and Writing Store the Foldable in a 3-ring binder. As you read and study the chapter, write notes and examples under the tabs.

1-1 Variables and Expressions

What You'll Learn

- Write mathematical expressions for verbal expressions.
- Write verbal expressions for mathematical expressions.

What expression can be used to find the perimeter of a baseball diamond?

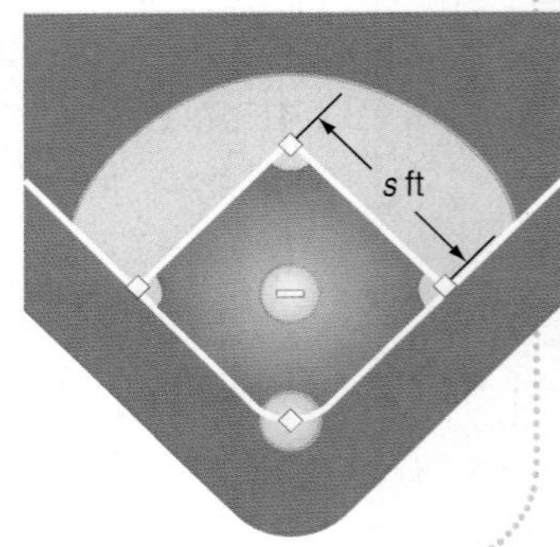

A baseball infield is a square with a base at each corner. Each base lies the same distance from the next one. Suppose s represents the length of each side of the square. Since the infield is a square, you can use the expression 4 times s, or $4s$ to find the perimeter of the square.

Vocabulary

- variables
- algebraic expression
- factors
- product
- power
- base
- exponent
- evaluate

WRITE MATHEMATICAL EXPRESSIONS In the algebraic expression $4s$, the letter s is called a variable. In algebra, **variables** are symbols used to represent unspecified numbers or values. Any letter may be used as a variable. *The letter s was used above because it is the first letter of the word side.*

An **algebraic expression** consists of one or more numbers and variables along with one or more arithmetic operations. Here are some examples of algebraic expressions.

$$5x \qquad 3x - 7 \qquad 4 + \frac{p}{q} \qquad m \times 5n \qquad 3ab \div 5cd$$

In algebraic expressions, a raised dot or parentheses are often used to indicate multiplication as the symbol $\times$ can be easily mistaken for the letter x. Here are several ways to represent the product of x and y.

$$xy \qquad x \cdot y \qquad x(y) \qquad (x)y \qquad (x)(y)$$

In each expression, the quantities being multiplied are called **factors**, and the result is called the **product**.

It is often necessary to translate verbal expressions into algebraic expressions.

Example 1 Write Algebraic Expressions

Write an algebraic expression for each verbal expression.

a. eight more than a number n

The words *more than* suggest addition.

eight	more than	a number n
8	+	n

Thus, the algebraic expression is $8 + n$.

b. 7 less the product of 4 and a number x

Less implies subtract, and *product* implies multiply. So the expression can be written as $7 - 4x$.

c. one third of the size of the original area a

The word *of* implies multiply, so the expression can be written as $\frac{1}{3}a$ or $\frac{a}{3}$.

An expression like x^n is called a **power** and is read "x to the nth power." The variable x is called the **base**, and n is called the **exponent**. The exponent indicates the number of times the base is used as a factor.

Study Tip

Reading Math
When no exponent is shown, it is understood to be 1. For example, $a = a^1$.

Symbols	Words	Meaning
3^1	3 to the first power	3
3^2	3 to the second power or 3 squared	$3 \cdot 3$
3^3	3 to the third power or 3 cubed	$3 \cdot 3 \cdot 3$
3^4	3 to the fourth power	$3 \cdot 3 \cdot 3 \cdot 3$
$2b^6$	2 times b to the sixth power	$2 \cdot b \cdot b \cdot b \cdot b \cdot b \cdot b$
x^n	x to the nth power	$\underbrace{x \cdot x \cdot x \cdot \ldots \cdot x}_{n \text{ factors}}$

By definition, for any nonzero number x, $x^0 = 1$.

Example 2 Write Algebraic Expressions with Powers

Write each expression algebraically.

a. the product of 7 and m to the fifth power

$7m^5$

b. the difference of 4 and x squared

$4 - x^2$

To **evaluate** an expression means to find its value.

Example 3 Evaluate Powers

Evaluate each expression.

a. 2^6

$2^6 = 2 \cdot 2 \cdot 2 \cdot 2 \cdot 2 \cdot 2$ Use 2 as a factor 6 times.

$= 64$ Multiply.

b. 4^3

$4^3 = 4 \cdot 4 \cdot 4$ Use 4 as a factor 3 times.

$= 64$ Multiply.

WRITE VERBAL EXPRESSIONS Another important skill is translating algebraic expressions into verbal expressions.

Example 4 Write Verbal Expressions

Write a verbal expression for each algebraic expression.

a. $4m^3$

the product of 4 and m to the third power

b. $c^2 + 21d$

the sum of c squared and 21 times d

c. 5^3

five to the third power or five cubed

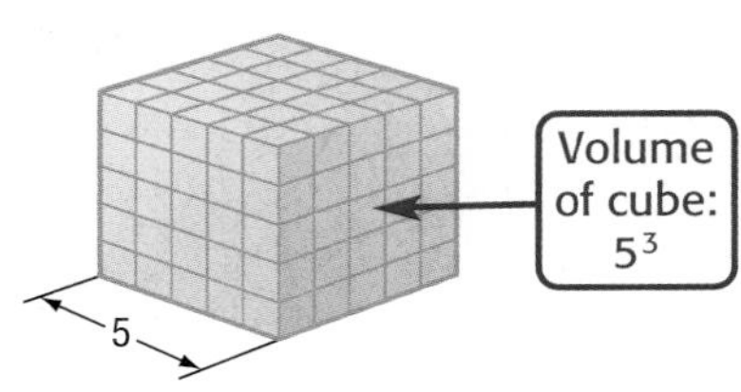

Check for Understanding

Concept Check

1. **Explain** the difference between an algebraic expression and a verbal expression.
2. **Write** an expression that represents the perimeter of the rectangle.
3. **OPEN ENDED** Give an example of a variable to the fifth power.

ℓ

w

Guided Practice

Write an algebraic expression for each verbal expression.

4. the sum of j and 13
5. 24 less than three times a number

Evaluate each expression.

6. 9^2
7. 4^4

Write a verbal expression for each algebraic expression.

8. $4m^4$
9. $\frac{1}{2}n^3$

Application

10. **MONEY** Lorenzo bought several pounds of chocolate-covered peanuts and gave the cashier a \$20 bill. Write an expression for the amount of change he will receive if p represents the cost of the peanuts.

Practice and Apply

Homework Help

For Exercises	See Examples
11–18	1, 2
21–28	3
31–42	4

Extra Practice

See page 820.

Write an algebraic expression for each verbal expression.

11. the sum of 35 and z
12. the sum of a number and 7
13. the product of 16 and p
14. the product of 5 and a number
15. 49 increased by twice a number
16. 18 and three times d
17. two-thirds the square of a number
18. one-half the cube of n

19. **SAVINGS** Kendra is saving to buy a new computer. Write an expression to represent the amount of money she will have if she has s dollars saved and she adds d dollars per week for the next 12 weeks.

20. **GEOMETRY** The area of a circle can be found by multiplying the number π by the square of the radius. If the radius of a circle is r, write an expression that represents the area of the circle.

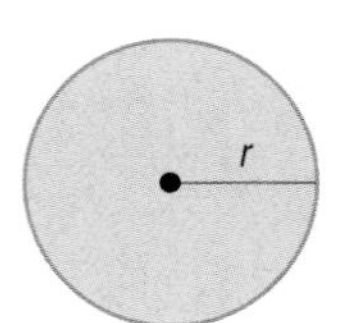

Evaluate each expression.

21. 6^2
22. 8^2
23. 3^4
24. 6^3
25. 3^5
26. 15^3
27. 10^6
28. 100^3

29. **FOOD** A bakery sells a dozen bagels for \$8.50 and a dozen donuts for \$3.99. Write an expression for the cost of buying b dozen bagels and d dozen donuts.

30. **TRAVEL** Before starting her vacation, Sari's car had 23,500 miles on the odometer. She drives an average of m miles each day for two weeks. Write an expression that represents the mileage on Sari's odometer after her trip.

Write a verbal expression for each algebraic expression.

31. $7p$	**32.** $15r$	**33.** 3^3	**34.** 5^4
35. $3x^2 + 4$	**36.** $2n^3 + 12$	**37.** $a^4 \cdot b^2$	**38.** $n^3 \cdot p^5$
39. $\frac{12z^2}{5}$	**40.** $\frac{8g^3}{4}$	**41.** $3x^2 - 2x$	**42.** $4f^5 - 9k^3$

43. **PHYSICAL SCIENCE** When water freezes, its volume is increased by one-eleventh. In other words, the volume of ice equals the sum of the volume of the water and the product of one-eleventh and the volume of the water. If x cubic centimeters of water is frozen, write an expression for the volume of the ice that is formed.

44. **GEOMETRY** The surface area of a rectangular prism is the sum of:

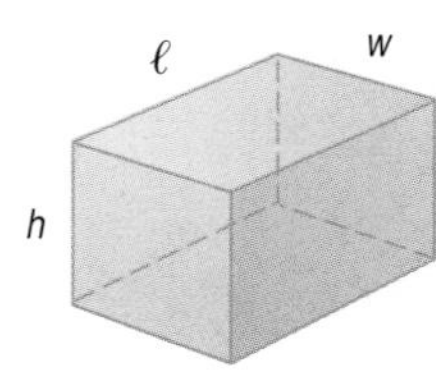

- the product of twice the length ℓ and the width w,
- the product of twice the length and the height h, and
- the product of twice the width and the height.

Write an expression that represents the surface area of a prism.

More About. . .

Recycling

In 2000, about 30% of all waste was recycled.

Source: U.S. Environmental Protection Agency

45. **RECYCLING** Each person in the United States produces approximately 3.5 pounds of trash each day. Write an expression representing the pounds of trash produced in a day by a family that has m members. **Source:** *Vitality*

46. **CRITICAL THINKING** In the square, the variable a represents a positive whole number. Find the value of a such that the area and the perimeter of the square are the same.

47. WRITING IN MATH Answer the question that was posed at the beginning of the lesson.

What expression can be used to find the perimeter of a baseball diamond?

Include the following in your answer:

- two different verbal expressions that you can use to describe the perimeter of a square, and
- an algebraic expression other than $4s$ that you can use to represent the perimeter of a square.

Standardized Test Practice

48. What is 6 more than 2 times a certain number x?

Ⓐ $2x - 6$ Ⓑ $2x$ Ⓒ $6x - 2$ Ⓓ $2x + 6$

49. Write $4 \cdot 4 \cdot 4 \cdot c \cdot c \cdot c \cdot c$ using exponents.

Ⓐ $3^4 4^c$ Ⓑ $4^3 c^4$ Ⓒ $(4c)^7$ Ⓓ $4c$

Maintain Your Skills

Getting Ready for the Next Lesson

PREREQUISITE SKILL **Evaluate each expression.**

*(To review **operations with fractions**, see pages 798–801.)*

50. $14.3 + 1.8$	**51.** $10 - 3.24$	**52.** 1.04×4.3	**53.** $15.36 \div 4.8$
54. $\frac{1}{3} + \frac{2}{5}$	**55.** $\frac{3}{4} - \frac{1}{6}$	**56.** $\frac{3}{8} \times \frac{4}{9}$	**57.** $\frac{7}{10} \div \frac{3}{5}$

www.algebra1.com/self_check_quiz

Reading Mathematics

Translating from English to Algebra

You learned in Lesson 1-1 that it is often necessary to translate words into algebraic expressions. Generally, there are "clue" words such as *more than*, *times*, *less than*, and so on, which indicate the operation to use. These words also help to connect numerical data. The table shows a few examples.

Words	Algebraic Expression
four times x plus y	$4x + y$
four times the sum of x and y	$4(x + y)$
four times the quantity x plus y	$4(x + y)$

Notice that all three expressions are worded differently, but the first expression is the only one that is different algebraically. In the second expression, parentheses indicate that the *sum*, $x + y$, is multiplied by four. In algebraic expressions, terms grouped by parentheses are treated as one quantity. So, $4(x + y)$ can also be read as *four times the quantity x plus y*.

Words that may indicate parentheses are *sum*, *difference*, *product*, and *quantity*.

Reading to Learn

Read each verbal expression aloud. Then match it with the correct algebraic expression.

1. nine divided by 2 plus n

2. four divided by the difference of n and six

3. n plus five squared

4. three times the quantity eight plus n

5. nine divided by the quantity 2 plus n

6. three times eight plus n

7. the quantity n plus five squared

8. four divided by n minus six

a. $(n + 5)^2$
b. $4 \div (n - 6)$
c. $9 \div 2 + n$
d. $3(8) + n$
e. $4 \div n - 6$
f. $n + 5^2$
g. $9 \div (2 + n)$
h. $3(8 + n)$

Write each algebraic expression in words.

9. $5x + 1$

10. $5(x + 1)$

11. $3 + 7x$

12. $(3 + x) \cdot 7$

13. $(6 + b) \div y$

14. $6 + (b \div y)$

1-2 Order of Operations

What You'll Learn

- Evaluate numerical expressions by using the order of operations.
- Evaluate algebraic expressions by using the order of operations.

Vocabulary

- order of operations

How is the monthly cost of internet service determined?

Nicole is signing up with a new internet service provider. The service costs \$4.95 a month, which includes 100 hours of access. If she is online for more than 100 hours, she must pay an additional \$0.99 per hour. Suppose Nicole is online for 117 hours the first month. The expression $4.95 + 0.99(117 - 100)$ represents what Nicole must pay for the month.

@home.net
$4.95 per month*
- includes 100 free hours
- accessible anywhere**
*0.99 per hour after 100 hours
**Requires v.95 net modem

EVALUATE RATIONAL EXPRESSIONS Numerical expressions often contain more than one operation. A rule is needed to let you know which operation to perform first. This rule is called the **order of operations**.

Key Concept — *Order of Operations*

Step 1 Evaluate expressions inside grouping symbols.

Step 2 Evaluate all powers.

Step 3 Do all multiplications and/or divisions from left to right.

Step 4 Do all additions and/or subtractions from left to right.

Example 1 *Evaluate Expressions*

Evaluate each expression.

a. $3 + 2 \cdot 3 + 5$

$3 + 2 \cdot 3 + 5 = 3 + 6 + 5$	Multiply 2 and 3.
$= 9 + 5$	Add 3 and 6.
$= 14$	Add 9 and 5.

b. $15 \div 3 \cdot 5 - 4^2$

$15 \div 3 \cdot 5 - 4^2 = 15 \div 3 \cdot 5 - 16$	Evaluate powers.
$= 5 \cdot 5 - 16$	Divide 15 by 3.
$= 25 - 16$	Multiply 5 by 5.
$= 9$	Subtract 16 from 25.

Grouping symbols such as parentheses ( ), brackets [ ], and braces { } are used to clarify or change the order of operations. They indicate that the expression within the grouping symbol is to be evaluated first.

Study Tip

Grouping Symbols
When more than one grouping symbol is used, start evaluating within the innermost grouping symbols.

Example 2 Grouping Symbols

Evaluate each expression.

a. $\mathbf{2(5) + 3(4 + 3)}$

$2(5) + 3(4 + 3) = 2(5) + 3(7)$ Evaluate inside grouping symbols.

$= 10 + 21$ Multiply expressions left to right.

$= 31$ Add 10 and 21.

b. $\mathbf{2[5 + (30 \div 6)^2]}$

$2[5 + (30 \div 6)^2] = 2[5 + (5)^2]$ Evaluate innermost expression first.

$= 2[5 + 25]$ Evaluate power inside grouping symbol.

$= 2[30]$ Evaluate expression in grouping symbol.

$= 60$ Multiply.

A fraction bar is another type of grouping symbol. It indicates that the numerator and denominator should each be treated as a single value.

Example 3 Fraction Bar

Evaluate $\dfrac{6 + 4^2}{3^2 \cdot 4}$.

$\dfrac{6 + 4^2}{3^2 \cdot 4}$ means $(6 + 4^2) \div (3^2 \cdot 4)$.

$\dfrac{6 + 4^2}{3^2 \cdot 4} = \dfrac{6 + 16}{3^2 \cdot 4}$ Evaluate the power in the numerator.

$= \dfrac{22}{3^2 \cdot 4}$ Add 6 and 16 in the numerator.

$= \dfrac{22}{9 \cdot 4}$ Evaluate the power in the denominator.

$= \dfrac{22}{36}$ or $\dfrac{11}{18}$ Multiply 9 and 4 in the denominator. Then simplify.

EVALUATE ALGEBRAIC EXPRESSIONS Like numerical expressions, algebraic expressions often contain more than one operation. Algebraic expressions can be evaluated when the values of the variables are known. First, replace the variables with their values. Then, find the value of the numerical expression using the order of operations.

Example 4 Evaluate an Algebraic Expression

Evaluate $\mathbf{a^2 - (b^3 - 4c)}$ **if** $\mathbf{a = 7, b = 3,}$ **and** $\mathbf{c = 5}$**.**

$a^2 - (b^3 - 4c) = 7^2 - (3^3 - 4 \cdot 5)$ Replace *a* with 7, *b* with 3, and *c* with 5.

$= 7^2 - (27 - 4 \cdot 5)$ Evaluate 3^3.

$= 7^2 - (27 - 20)$ Multiply 4 and 5.

$= 7^2 - 7$ Subtract 20 from 27.

$= 49 - 7$ Evaluate 7^2.

$= 42$ Subtract.

Career Choices

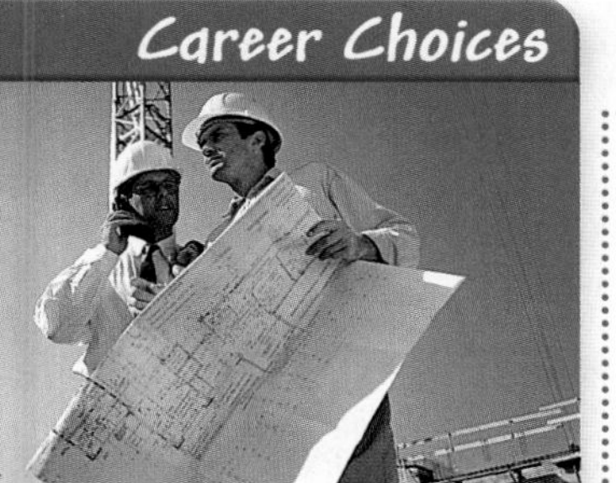

Architect

Architects must consider the function, safety, and needs of people, as well as appearance when they design buildings.

Online Research For more information about a career as an architect, visit: www.algebra1.com/careers

Example 5 Use Algebraic Expressions

ARCHITECTURE The Pyramid Arena in Memphis, Tennessee, is the third largest pyramid in the world. The area of its base is 360,000 square feet, and it is 321 feet high. The volume of any pyramid is one third of the product of the area of the base B and its height h.

a. Write an expression that represents the volume of a pyramid.

one third	of	the product of area of base and height	
$\frac{1}{3}$	$\times$	$(B \cdot h)$	or $\frac{1}{3}Bh$

b. Find the volume of the Pyramid Arena.

Evaluate $\frac{1}{3}(Bh)$ for $B = 360{,}000$ and $h = 321$.

$\frac{1}{3}(Bh) = \frac{1}{3}(360{,}000 \cdot 321)$ $\quad B = 360{,}000$ and $h = 321$

$= \frac{1}{3}(115{,}560{,}000)$ $\quad$ Multiply 360,000 by 321.

$= \frac{115{,}560{,}000}{3}$ $\quad$ Multiply $\frac{1}{3}$ by 115,560,000.

$= 38{,}520{,}000$ $\quad$ Divide 115,560,000 by 3.

The volume of the Pyramid Arena is 38,520,000 cubic feet.

Check for Understanding

Concept Check

1. **Describe** how to evaluate $8[6^2 - 3(2 + 5)] \div 8 + 3$.
2. **OPEN ENDED** Write an expression involving division in which the first step in evaluating the expression is addition.
3. **FIND THE ERROR** Laurie and Chase are evaluating $3[4 + (27 \div 3)]^2$.

Laurie	Chase
$3[4 + (27 \div 3)]^2 = 3(4 + 9^2)$	$3[4 + (27 \div 3)]^2 = 3(4 + 9)^2$
$= 3(4 + 81)$	$= 3(13)^2$
$= 3(85)$	$= 3(169)$
$= 255$	$= 507$

Who is correct? Explain your reasoning.

Guided Practice

Evaluate each expression.

4. $(4 + 6)7$
5. $50 - (15 + 9)$
6. $29 - 3(9 - 4)$
7. $[7(2) - 4] + [9 + 8(4)]$
8. $\frac{(4 \cdot 3)^2 \cdot 5}{9 + 3}$
9. $\frac{3 + 2^3}{5^2(4)}$

Evaluate each expression if $g = 4$, $h = 6$, $j = 8$, and $k = 12$.

10. $hk - gj$
11. $2k + gh^2 - j$
12. $\frac{2g(h - g)}{gh - j}$

Application

SHOPPING For Exercises 13 and 14, use the following information.
A computer store has certain software on sale at 3 for \$20.00, with a limit of 3 at the sale price. Additional software is available at the regular price of \$9.95 each.

13. Write an expression you could use to find the cost of 5 software packages.
14. How much would 5 software packages cost?

www.algebra1.com/extra_examples

Practice and Apply

Homework Help

For Exercises	See Examples
15–28	1–3
29–31	5
32–39	4, 5

Extra Practice

See page 820.

Evaluate each expression.

15. $(12 - 6) \cdot 2$ **16.** $(16 - 3) \cdot 4$ **17.** $15 + 3 \cdot 2$

18. $22 + 3 \cdot 7$ **19.** $4(11 + 7) - 9 \cdot 8$ **20.** $12(9 + 5) - 6 \cdot 3$

21. $12 \div 3 \cdot 5 - 4^2$ **22.** $15 \div 3 \cdot 5 - 4^2$ **23.** $288 \div [3(9 + 3)]$

Evaluate each expression.

24. $390 \div [5(7 + 6)]$ **25.** $\dfrac{2 \cdot 8^2 - 2^2 \cdot 8}{2 \cdot 8}$ **26.** $\dfrac{4 \cdot 6^2 - 4^2 \cdot 6}{4 \cdot 6}$

27. $\dfrac{[(8 + 5)(6 - 2)^2] - (4 \cdot 17 \div 2)}{[(24 \div 2) \div 3]}$ **28.** $6 - \left[\dfrac{2 + 7}{3} - (2 \cdot 3 - 5)\right]$

29. **GEOMETRY** Find the area of the rectangle when $n = 4$ centimeters.

ENTERTAINMENT **For Exercises 30 and 31, use the following information.**
Derrick and Samantha are selling tickets for their school musical. Floor seats cost \$7.50 and balcony seats cost \$5.00. Samantha sells 60 floor seats and 70 balcony seats, Derrick sells 50 floor seats and 90 balcony seats.

30. Write an expression to show how much money Samantha and Derrick have collected for tickets.

31. Evaluate the expression to determine how much they collected.

Evaluate each expression if $x = 12$, $y = 8$, and $z = 3$.

32. $x + y^2 + z^2$ **33.** $x^3 + y + z^3$

34. $3xy - z$ **35.** $4x - yz$

36. $\dfrac{2xy - z^3}{z}$ **37.** $\dfrac{xy^2 - 3z}{3}$

38. $\left(\dfrac{x}{y}\right)^2 - \dfrac{3y - z}{(x - y)^2}$ **39.** $\dfrac{x - z^2}{y \div x} + \dfrac{2y - x}{y^2 \div 2}$

40. **BIOLOGY** Most bacteria reproduce by dividing into identical cells. This process is called *binary fission*. A certain type of bacteria can double its numbers every 20 minutes. Suppose 100 of these cells are in one culture dish and 250 of the cells are in another culture dish. Write and evaluate an expression that shows the total number of bacteria cells in both dishes after 20 minutes.

BUSINESS **For Exercises 41–43, use the following information.**
Mr. Martinez is a sales representative for an agricultural supply company. He receives a salary and monthly commission. He also receives a bonus each time he reaches a sales goal.

41. Write a verbal expression that describes how much Mr. Martinez earns in a year if he receives four equal bonuses.

42. Let e represent earnings, s represent his salary, c represent his commission, and b represent his bonus. Write an algebraic expression to represent his earnings if he receives four equal bonuses.

43. Suppose Mr. Martinez's annual salary is \$42,000 and his average commission is \$825 each month. If he receives four bonuses of \$750 each, how much does he earn in a year?

44. **CRITICAL THINKING** Choose three numbers from 1 to 6. Write as many expressions as possible that have different results when they are evaluated. You must use all three numbers in each expression, and each can only be used once.

45. **WRITING IN MATH** Answer the question that was posed at the beginning of the lesson.

How is the monthly cost of internet service determined?

Include the following in your answer:

- an expression for the cost of service if Nicole has a coupon for $25 off her base rate for her first six months, and
- an explanation of the advantage of using an algebraic expression over making a table of possible monthly charges.

46. Find the perimeter of the triangle using the formula $P = a + b + c$ if $a = 10$, $b = 12$, and $c = 17$.

Ⓐ 39 mm Ⓑ 19.5 mm

Ⓒ 60 mm Ⓓ 78 mm

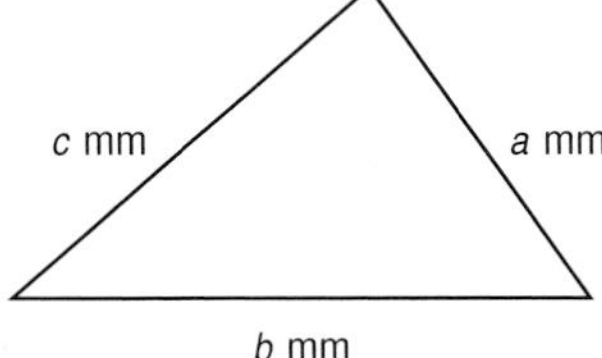

47. Evaluate $(5 - 1)^3 + (11 - 2)^2 + (7 - 4)^3$.

Ⓐ 586 Ⓑ 172 Ⓒ 106 Ⓓ 39

EVALUATING EXPRESSIONS Use a calculator to evaluate each expression.

48. $\frac{0.25x^2}{7x^3}$ if $x = 0.75$ 49. $\frac{2x^2}{x^2 - x}$ if $x = 27.89$ 50. $\frac{x^3 + x^2}{x^3 - x^2}$ if $x = 12.75$

Maintain Your Skills

Mixed Review

Write an algebraic expression for each verbal expression. *(Lesson 1-1)*

51. the product of the third power of a and the fourth power of b
52. six less than three times the square of y
53. the sum of a and b increased by the quotient of b and a
54. four times the sum of r and s increased by twice the difference of r and s
55. triple the difference of 55 and the cube of w

Evaluate each expression. *(Lesson 1-1)*

56. 2^4 57. 12^1 58. 8^2 59. 4^4

Write a verbal expression for each algebraic expression. *(Lesson 1-1)*

60. $5n + \frac{n}{2}$ 61. $q^2 - 12$ 62. $\frac{(x + 3)}{(x - 2)^2}$ 63. $\frac{x^3}{9}$

Getting Ready for the Next Lesson

PREREQUISITE SKILL **Find the value of each expression.**
*(To review **operations with decimals and fractions**, see pages 798–801.)*

64. $0.5 - 0.0075$ 65. $5.6 + 1.612$ 66. $14.9968 \div 5.2$ 67. $2.3(6.425)$

68. $4\frac{1}{8} - 1\frac{1}{2}$ 69. $\frac{3}{5} + 2\frac{5}{7}$ 70. $\frac{5}{6} \cdot \frac{4}{5}$ 71. $8 \div \frac{2}{9}$

1-3 Open Sentences

What You'll Learn

- Solve open sentence equations.
- Solve open sentence inequalities.

How can you use open sentences to stay within a budget?

The Daily News sells garage sale kits. The Spring Creek Homeowners Association is planning a community garage sale, and their budget for advertising is \$135. The expression $15.50 + 5n$ can be used to represent the cost of purchasing $n + 1$ kits. The open sentence $15.50 + 5n \leq 135$ can be used to ensure that the budget is met.

HAVING A
GARAGE
SALE?
Garage sale kit includes:
• Weekend ad
• Signs
• Announcements
• Balloons
• Price stickers
• Sales sheet
COMPLETE PACKAGE
\$15.50
Additional kits available for \$5.00 each

Vocabulary

- open sentence
- solving an open sentence
- solution
- equation
- replacement set
- set
- element
- solution set
- inequality

SOLVE EQUATIONS A mathematical statement with one or more variables is called an **open sentence**. An open sentence is neither true nor false until the variables have been replaced by specific values. The process of finding a value for a variable that results in a true sentence is called **solving the open sentence**. This replacement value is called a **solution** of the open sentence. A sentence that contains an equals sign, =, is called an **equation**.

A set of numbers from which replacements for a variable may be chosen is called a **replacement set**. A **set** is a collection of objects or numbers. It is often shown using braces, { }, and is usually named by a capital letter. Each object or number in the set is called an **element**, or member. The **solution set** of an open sentence is the set of elements from the replacement set that make an open sentence true.

Example 1 Use a Replacement Set to Solve an Equation

Find the solution set for each equation if the replacement set is {3, 4, 5, 6, 7}.

a. $6n + 7 = 37$

Replace n in $6n + 7 = 37$ with each value in the replacement set.

n	$6n + 7 = 37$	True or False?
3	$6(3) + 7 \stackrel{?}{=} 37 \rightarrow 25 \neq 37$	false
4	$6(4) + 7 \stackrel{?}{=} 37 \rightarrow 31 \neq 37$	false
5	$6(5) + 7 \stackrel{?}{=} 37 \rightarrow 37 = 37$	true ✓
6	$6(6) + 7 \stackrel{?}{=} 37 \rightarrow 43 \neq 37$	false
7	$6(7) + 7 \stackrel{?}{=} 37 \rightarrow 49 \neq 37$	false

Since $n = 5$ makes the equation true, the solution of $6n + 7 = 37$ is 5.

The solution set is {5}.

b. $5(x + 2) = 40$

Replace x in $5(x + 2) = 40$ with each value in the replacement set.

x	$5(x + 2) = 40$	True or False?
3	$5(3 + 2) \stackrel{?}{=} 40 \rightarrow 25 \neq 40$	false
4	$5(4 + 2) \stackrel{?}{=} 40 \rightarrow 30 \neq 40$	false
5	$5(5 + 2) \stackrel{?}{=} 40 \rightarrow 35 \neq 40$	false
6	$5(6 + 2) \stackrel{?}{=} 40 \rightarrow 40 = 40$	true ✓
7	$5(7 + 2) \stackrel{?}{=} 40 \rightarrow 45 \neq 40$	false

The solution of $5(x + 2) = 40$ is 6. The solution set is {6}.

You can often solve an equation by applying the order of operations.

Example 2 Use Order of Operations to Solve an Equation

Solve $\frac{13 + 2(4)}{3(5 - 4)} = q$.

$\frac{13 + 2(4)}{3(5 - 4)} = q$ Original equation

$\frac{13 + 8}{3(1)} = q$ Multiply 2 and 4 in the numerator. Subtract 4 from 5 in the denominator.

$\frac{21}{3} = q$ Simplify.

$7 = q$ Divide. The solution is 7.

Study Tip

Reading Math

Inequality symbols are read as follows.

$<$ *is less than*

$\leq$ *is less than or equal to*

$>$ *is greater than*

$\geq$ *is greater than or equal to*

SOLVE INEQUALITIES An open sentence that contains the symbol $<$, $\leq$, $>$, or $\geq$ is called an **inequality**. Inequalities can be solved in the same way as equations.

Example 3 Find the Solution Set of an Inequality

Find the solution set for $18 - y < 10$ if the replacement set is {7, 8, 9, 10, 11, 12}.

Replace y in $18 - y < 10$ with each value in the replacement set.

y	$18 - y < 10$	True or False?
7	$18 - 7 \stackrel{?}{<} 10 \rightarrow 11 \nless 10$	false
8	$18 - 8 \stackrel{?}{<} 10 \rightarrow 10 \nless 10$	false
9	$18 - 9 \stackrel{?}{<} 10 \rightarrow 9 < 10$	true ✓
10	$18 - 10 \stackrel{?}{<} 10 \rightarrow 8 < 10$	true ✓
11	$18 - 11 \stackrel{?}{<} 10 \rightarrow 7 < 10$	true ✓
12	$18 - 12 \stackrel{?}{<} 10 \rightarrow 6 < 10$	true ✓

The solution set for $18 - y < 10$ is {9, 10, 11, 12}.

Example 4 Solve an Inequality

FUND-RAISING Refer to the application at the beginning of the lesson. How many garage sale kits can the association buy and stay within their budget?

Explore The association can spend no more than \$135. So the situation can be represented by the inequality $15.50 + 5n \leq 135$.

(continued on the next page)

Plan Since no replacement set is given, estimate to find reasonable values for the replacement set.

Solve Start by letting $n = 10$ and then adjust values up or down as needed.

$15.50 + 5n \leq 135$ Original inequality

$15.50 + 5(10) \leq 135$ $n = 10$

$15.50 + 50 \leq 135$ Multiply 5 and 10.

$65.50 \leq 135$ Add 15.50 and 50.

The estimate is too low. Increase the value of n.

n	$15.50 + 5n \leq 135$	Reasonable?
20	$15.50 + 5(20) \stackrel{?}{\leq} 135 \rightarrow 115.50 \leq 135$	too low
25	$15.50 + 5(25) \stackrel{?}{\leq} 135 \rightarrow 140.50 \not\leq 135$	too high
23	$15.50 + 5(23) \stackrel{?}{\leq} 135 \rightarrow 130.50 \leq 135$	almost
24	$15.50 + 5(24) \stackrel{?}{\leq} 135 \rightarrow 135.50 \leq 135$	too high

Examine The solution set is {0, 1, 2, 3, …, 21, 22, 23}. In addition to the first kit, the association can buy as many as 23 additional kits. So, the association can buy as many as 1 + 23 or 24 garage sale kits and stay within their budget.

Study Tip

Reading Math
In {1, 2, 3, 4, …}, the three dots are an *ellipsis*. In math, an ellipsis is used to indicate that numbers continue in the same pattern.

Check for Understanding

Concept Check

1. **Describe** the difference between an expression and an open sentence.
2. **OPEN ENDED** Write an inequality that has a solution set of {8, 9, 10, 11, …}.
3. **Explain** why an open sentence always has at least one variable.

Guided Practice

Find the solution of each equation if the replacement set is {10, 11, 12, 13, 14, 15}.

4. $3x - 7 = 29$
5. $12(x - 8) = 84$

Find the solution of each equation using the given replacement set.

6. $x + \frac{2}{5} = 1\frac{3}{20}; \left\{\frac{1}{4}, \frac{1}{2}, \frac{3}{4}, 1, 1\frac{1}{4}\right\}$
7. $7.2(x + 2) = 25.92$; {1.2, 1.4, 1.6, 1.8}

Solve each equation.

8. $4(6) + 3 = x$
9. $w = \frac{14 - 8}{2}$

Find the solution set for each inequality using the given replacement set.

10. $24 - 2x \geq 13$; {0, 1, 2, 3, 4, 5, 6}
11. $3(12 - x) - 2 \leq 28$; {1.5, 2, 2.5, 3}

Application

NUTRITION **For Exercises 12 and 13, use the following information.**
A person must burn 3500 Calories to lose one pound of weight.

12. Write an equation that represents the number of Calories a person would have to burn a day to lose four pounds in two weeks.
13. How many Calories would the person have to burn each day?

Practice and Apply

Homework Help

For Exercises	See Examples
14–25	1
26–28	4
29–36	2
37–44	3

Extra Practice

See page 820.

Find the solution of each equation if the replacement sets are a = {0, 3, 5, 8, 10} and b = {12, 17, 18, 21, 25}.

14. $b - 12 = 9$
15. $34 - b = 22$
16. $3a + 7 = 31$
17. $4a + 5 = 17$
18. $\frac{40}{a} - 4 = 0$
19. $\frac{b}{3} - 2 = 4$

Find the solution of each equation using the given replacement set.

20. $x + \frac{7}{4} = \frac{17}{8}; \left\{\frac{1}{8}, \frac{3}{8}, \frac{5}{8}, \frac{7}{8}\right\}$
21. $x + \frac{7}{12} = \frac{25}{12}; \left\{\frac{1}{2}, 1, 1\frac{1}{2}, 2\right\}$
22. $\frac{2}{5}(x + 1) = \frac{8}{15}; \left\{\frac{1}{6}, \frac{1}{3}, \frac{1}{2}, \frac{2}{3}\right\}$
23. $2.7(x + 5) = 17.28$; {1.2, 1.3, 1.4, 1.5}
24. $16(x + 2) = 70.4$; {2.2, 2.4, 2.6, 2.8}
25. $21(x + 5) = 216.3$; {3.1, 4.2, 5.3, 6.4}

MOVIES For Exercises 26–28, use the table and the following information.
The Conkle family is planning to see a movie. There are two adults, a daughter in high school, and two sons in middle school. They do not want to spend more than $30.

Admission Prices		
	Evening	Matinee
Adult	$7.50	All Seats $4.50
Student	$4.50	
Child	$4.50	
Senior	$3.50	

26. The movie theater charges the same price for high school and middle school students. Write an inequality to show the cost for the family to go to the movies.
27. How much will it cost for the family to see a matinee?
28. How much will it cost to see an evening show?

Solve each equation.

29. $14.8 - 3.75 = t$
30. $a = 32.4 - 18.95$
31. $y = \frac{12 \cdot 5}{15 - 3}$
32. $g = \frac{15 \cdot 6}{16 - 7}$
33. $d = \frac{7(3) + 3}{4(3 - 1)} + 6$
34. $a = \frac{4(14 - 1)}{3(6) - 5} + 7$
35. $p = \frac{1}{4}[7(2^3) + 4(5^2) - 6(2)]$
36. $n = \frac{1}{8}[6(3^2) + 2(4^3) - 2(7)]$

Find the solution set for each inequality using the given replacement set.

37. $a - 2 < 6$; {6, 7, 8, 9, 10, 11}
38. $a + 7 < 22$; {13, 14, 15, 16, 17}
39. $\frac{a}{5} \geq 2$; {5, 10, 15, 20, 25}
40. $\frac{2a}{4} \leq 8$; {12, 14, 16, 18, 20, 22}
41. $4a - 3 \geq 10.6$; {3.2, 3.4, 3.6, 3.8, 4}
42. $6a - 5 \geq 23.8$; {4.2, 4.5, 4.8, 5.1, 5.4}
43. $3a \leq 4; \left\{0, \frac{1}{3}, \frac{2}{3}, 1, 1\frac{1}{3}\right\}$
44. $2b < 5; \left\{1, 1\frac{1}{2}, 2, 2\frac{1}{2}, 3\right\}$

More About. . .

Food

During a lifetime, the average American drinks 15,579 glasses of milk, 6220 glasses of juice, and 18,995 glasses of soda.

Source: *USA TODAY*

FOOD For Exercises 45 and 46, use the information about food at the left.

45. Write an equation to find the total number of glasses of milk, juice, and soda the average American drinks in a lifetime.
46. How much milk, juice, and soda does the average American drink in a lifetime?

MAIL ORDER For Exercises 47 and 48, use the following information.
Suppose you want to order several sweaters that cost $39.00 each from an online catalog. There is a $10.95 charge for shipping. You have $102.50 to spend.

47. Write an inequality you could use to determine the maximum number of sweaters you can purchase.
48. What is the maximum number of sweaters you can buy?

49. **CRITICAL THINKING** Describe the solution set for x if $3x \leq 1$.

50. **WRITING IN MATH** Answer the question that was posed at the beginning of the lesson.

How can you use open sentences to stay within a budget?

Include the following in your answer:
- an explanation of how to use open sentences to stay within a budget, and
- examples of real-world situations in which you would use an inequality and examples where you would use an equation.

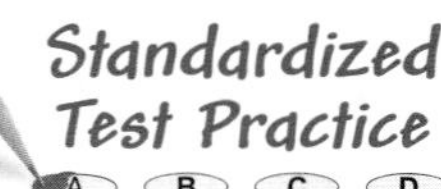

51. Find the solution set for $\frac{(5 \cdot n)^2 + 5}{(9 \cdot 3^2) - n} < 28$ if the replacement set is {5, 7, 9, 11, 13}.

Ⓐ {5} Ⓑ {5, 7} Ⓒ {7} Ⓓ {7, 9}

52. Which expression has a value of 17?

Ⓐ $(9 \times 3) - 63 \div 7$ Ⓑ $6(3 + 2) \div (9 - 7)$

Ⓒ $27 \div 3 + (12 - 4)$ Ⓓ $2[2(6 - 3)] - 5$

Maintain Your Skills

Mixed Review

Write an algebraic expression for each verbal expression. Then evaluate each expression if $r = 2$, $s = 5$, and $t = \frac{1}{2}$. *(Lesson 1-2)*

53. r squared increased by 3 times s

54. t times the sum of four times s and r

55. the sum of r and s times the square of t

56. r to the fifth power decreased by t

Evaluate each expression. *(Lesson 1-2)*

57. $5^3 + 3(4^2)$ 58. $\frac{38 - 12}{2 \cdot 13}$ 59. $[5(2 + 1)]^4 + 3$

Getting Ready for the Next Lesson

PREREQUISITE SKILL Find each product. Express in simplest form.
*(To review **multiplying fractions**, see pages 800 and 801.)*

60. $\frac{1}{6} \cdot \frac{2}{5}$ 61. $\frac{4}{9} \cdot \frac{3}{7}$ 62. $\frac{5}{6} \cdot \frac{15}{16}$ 63. $\frac{6}{14} \cdot \frac{12}{18}$

64. $\frac{8}{13} \cdot \frac{2}{11}$ 65. $\frac{4}{7} \cdot \frac{4}{9}$ 66. $\frac{3}{11} \cdot \frac{7}{16}$ 67. $\frac{2}{9} \cdot \frac{24}{25}$

Practice Quiz 1 *Lessons 1-1 through 1-3*

Write a verbal expression for each algebraic expression. *(Lesson 1-1)*

1. $x - 20$ 2. $5n + 2$ 3. a^3 4. $n^4 - 1$

Evaluate each expression. *(Lesson 1-2)*

5. $6(9) - 2(8 + 5)$ 6. $4[2 + (18 \div 9)^3]$ 7. $9(3) - 4^2 + 6^2 \div 2$ 8. $\frac{(5 - 2)^2}{3(4 \cdot 2 - 7)}$

9. Evaluate $\frac{5a^2 + c - 2}{6 + b}$ if $a = 4$, $b = 5$, and $c = 10$. *(Lesson 1-2)*

10. Find the solution set for $2n^2 + 3 \leq 75$ if the replacement set is {4, 5, 6, 7, 8, 9}. *(Lesson 1-3)*

1-4 Identity and Equality Properties

What You'll Learn

- Recognize the properties of identity and equality.
- Use the properties of identity and equality.

Vocabulary

- additive identity
- multiplicative identity
- multiplicative inverses
- reciprocal

How are identity and equality properties used to compare data?

During the college football season, teams are ranked weekly. The table shows the last three rankings of the top five teams for the 2000 football season. The open sentence below represents the change in rank of Oregon State from December 11 to the final rank.

	Dec. 4	Dec. 11	Final Rank
University of Oklahoma	1	1	1
University of Miami	2	2	2
University of Washington	4	3	3
Oregon State University	5	4	4
Florida State University	3	5	5

Rank on December 11, 2000	plus	increase in rank	equals	final rank for 2000 season.
4	$+$	r	$=$	4

The solution of this equation is 0. Oregon State's rank changed by 0 from December 11 to the final rank. In other words, $4 + 0 = 4$.

IDENTITY AND EQUALITY PROPERTIES The sum of any number and 0 is equal to the number. Thus, 0 is called the **additive identity**.

Key Concept — *Additive Identity*

- **Words** For any number a, the sum of a and 0 is a.
- **Symbols** $a + 0 = 0 + a = a$
- **Examples** $5 + 0 = 5,\ 0 + 5 = 5$

There are also special properties associated with multiplication. Consider the following equations.

$$7 \cdot n = 7$$

The solution of the equation is 1. Since the product of any number and 1 is equal to the number, 1 is called the **multiplicative identity**.

$$9 \cdot m = 0$$

The solution of the equation is 0. The product of any number and 0 is equal to 0. This is called the **Multiplicative Property of Zero**.

$$\frac{1}{3} \cdot 3 = 1$$

Two numbers whose product is 1 are called **multiplicative inverses** or **reciprocals**. Zero has no reciprocal because any number times 0 is 0.

Key Concept — Multiplication Properties

Property	Words	Symbols	Examples
Multiplicative Identity	For any number a, the product of a and 1 is a.	$a \cdot 1 = 1 \cdot a = a$	$12 \cdot 1 = 12$, $1 \cdot 12 = 12$
Multiplicative Property of Zero	For any number a, the product of a and 0 is 0.	$a \cdot 0 = 0 \cdot a = 0$	$8 \cdot 0 = 0$, $0 \cdot 8 = 0$
Multiplicative Inverse	For every number $\frac{a}{b}$, where $a, b \neq 0$, there is exactly one number $\frac{b}{a}$ such that the product of $\frac{a}{b}$ and $\frac{b}{a}$ is 1.	$\frac{a}{b} \cdot \frac{b}{a} = \frac{b}{a} \cdot \frac{a}{b} = 1$	$\frac{2}{3} \cdot \frac{3}{2} = \frac{6}{6} = 1$, $\frac{3}{2} \cdot \frac{2}{3} = \frac{6}{6} = 1$

Example 1 Identify Properties

Name the property used in each equation. Then find the value of n.

a. $42 \cdot n = 42$

Multiplicative Identity Property

$n = 1$, since $42 \cdot 1 = 42$.

b. $n + 0 = 15$

Additive Identity Property

$n = 15$, since $15 + 0 = 15$.

c. $n \cdot 9 = 1$

Multiplicative Inverse Property

$n = \frac{1}{9}$, since $\frac{1}{9} \cdot 9 = 1$.

There are several properties of equality that apply to addition and multiplication. These are summarized below.

Key Concept — Properties of Equality

Property	Words	Symbols	Examples
Reflexive	Any quantity is equal to itself.	For any number a, $a = a$.	$7 = 7$, $2 + 3 = 2 + 3$
Symmetric	If one quantity equals a second quantity, then the second quantity equals the first.	For any numbers a and b, if $a = b$, then $b = a$.	If $9 = 6 + 3$, then $6 + 3 = 9$.
Transitive	If one quantity equals a second quantity and the second quantity equals a third quantity, then the first quantity equals the third quantity.	For any numbers a, b, and c, if $a = b$ and $b = c$, then $a = c$.	If $5 + 7 = 8 + 4$ and $8 + 4 = 12$, then $5 + 7 = 12$.
Substitution	A quantity may be substituted for its equal in any expression.	If $a = b$, then a may be replaced by b in any expression.	If $n = 15$, then $3n = 3 \cdot 15$.

USE IDENTITY AND EQUALITY PROPERTIES The properties of identity and equality can be used to justify each step when evaluating an expression.

Example 2 Evaluate Using Properties

Evaluate $2(3 \cdot 2 - 5) + 3 \cdot \frac{1}{3}$. Name the property used in each step.

$2(3 \cdot 2 - 5) + 3 \cdot \frac{1}{3} = 2(6 - 5) + 3 \cdot \frac{1}{3}$	Substitution; $3 \cdot 2 = 6$
$= 2(1) + 3 \cdot \frac{1}{3}$	Substitution; $6 - 5 = 1$
$= 2 + 3 \cdot \frac{1}{3}$	Multiplicative Identity; $2 \cdot 1 = 2$
$= 2 + 1$	Multiplicative Inverse; $3 \cdot \frac{1}{3} = 1$
$= 3$	Substitution; $2 + 1 = 3$

Check for Understanding

Concept Check

1. **Explain** whether 1 can be an additive identity.
2. **OPEN ENDED** Write two equations demonstrating the Transitive Property of Equality.
3. **Explain** why 0 has no multiplicative inverse.

Guided Practice

Name the property used in each equation. Then find the value of *n*.

4. $13n = 0$
5. $17 + 0 = n$
6. $\frac{1}{6}n = 1$
7. Evaluate $6(12 - 48 \div 4)$. Name the property used in each step.
8. Evaluate $\left(15 \cdot \frac{1}{15} + 8 \cdot 0\right) \cdot 12$. Name the property used in each step.

Application

HISTORY For Exercises 9–11, use the following information.
On November 19, 1863, Abraham Lincoln delivered the famous Gettysburg Address. The speech began "Four score and seven years ago, . . ."

9. Write an expression to represent four score and seven. (*Hint*: A score is 20.)
10. Evaluate the expression. Name the property used in each step.
11. How many years is four score and seven?

Practice and Apply

Homework Help

For Exercises	See Examples
12–19	1
20–23	1, 2
24–29	2
30–35	1, 2

Extra Practice

See page 821.

Name the property used in each equation. Then find the value of *n*.

12. $12n = 12$
13. $n \cdot 1 = 5$
14. $8 \cdot n = 8 \cdot 5$
15. $0.25 + 1.5 = n + 1.5$
16. $8 = n + 8$
17. $n + 0 = \frac{1}{3}$
18. $1 = 2n$
19. $4 \cdot \frac{1}{4} = n$
20. $(9 - 7)(5) = 2(n)$
21. $3 + (2 + 8) = n + 10$
22. $n\left(5^2 \cdot \frac{1}{25}\right) = 3$
23. $6\left(\frac{1}{2} \cdot n\right) = 6$

Evaluate each expression. Name the property used in each step.

24. $\frac{3}{4}[4 \div (7 - 4)]$
25. $\frac{2}{3}[3 \div (2 \cdot 1)]$
26. $2(3 \cdot 2 - 5) + 3 \cdot \frac{1}{3}$
27. $6 \cdot \frac{1}{6} + 5(12 \div 4 - 3)$
28. $3 + 5(4 - 2^2) - 1$
29. $7 - 8(9 - 3^2)$

FUND-RAISING **For Exercises 30 and 31, use the following information.**
The spirit club at Central High School is selling items to raise money. The profit the club earns on each item is the difference between what an item sells for and what it costs the club to buy.

School Spirit Items		
Item	**Cost**	**Selling Price**
Pennant	$3.00	$5.00
Button	$1.00	$2.50
Cap	$6.00	$10.00

30. Write an expression that represents the profit for 25 pennants, 80 buttons, and 40 caps.
31. Evaluate the expression, indicating the property used in each step.

MILITARY PAY **For Exercises 32 and 33, use the table that shows the monthly base pay rates for the first five ranks of enlisted personnel.**

Monthly Basic Pay Rates by Grade, Effective July 1, 2001								
	Years of Service							
Grade	<2	>2	>3	>4	>6	>8	>10	>12
E-5	1381.80	1549.20	1623.90	1701.00	1779.30	1888.50	1962.90	2040.30
E-4	1288.80	1423.80	1500.60	1576.20	1653.00	1653.00	1653.00	1653.00
E-3	1214.70	1307.10	1383.60	1385.40	1385.40	1385.40	1385.40	1385.40
E-2	1169.10	1169.10	1169.10	1169.10	1169.10	1169.10	1169.10	1169.10
E-1	1042.80	1042.80	1042.80	1042.80	1042.80	1042.80	1042.80	1042.80

Source: U.S. Department of Defense

32. Write an equation using addition that shows the change in pay for an enlisted member at grade E-2 from 3 years of service to 12 years.
33. Write an equation using multiplication that shows the change in pay for someone at grade E-4 from 6 years of service to 10 years.

FOOTBALL **For Exercises 34–36, use the table that shows the base salary and various bonus plans for the NFL from 2002–2005.**

34. Suppose a player rushed for 12 touchdowns in 2002 and another player scored 76 points that same year. Write an equation that compares the two salaries and bonuses.
35. Write an expression that could be used to determine what a team owner would pay in base salaries and bonuses in 2004 for the following:
    - eight players who keep their weight under 240 pounds and are involved in at least 35% of the offensive plays,
    - three players who score 12 rushing touchdowns and score 76 points, and
    - four players who gain 1601 yards of total offense and average 4.5 yards per carry.
36. Evaluate the expression you wrote in Exercise 35. Name the property used in each step.

NFL Salaries and Bonuses	
Year	**Base Salary**
2002	$350,000
2003	375,000
2004	400,000
2005	400,000
Goal	**Bonus**
Involved in 35% of offensive plays	$50,000
Average 4.5 yards per carry	50,000
12 rushing touchdowns	50,000
12 receiving touchdowns	50,000
76 points scored	50,000
1601 yards of total offense	50,000
Keep weight below 240 lb	100,000
Goal—Rushing Yards	**Bonus**
1600 yards	$1 million
1800 yards	1.5 million
2000 yards	2 million
2100 yards	2.5 million

Source: *ESPN Sports Almanac*

More About. . .

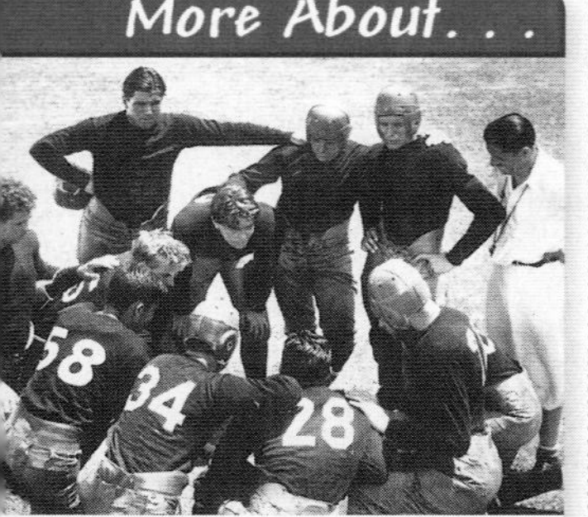

Football

Nationally organized football began in 1920 and originally included five teams. In 2002, there were 32 teams.

Source: www.infoplease.com

Online Research **Data Update** Find the most recent statistics for a professional football player. What were his base salary and bonuses? Visit www.algebra1.com/data_update to learn more.

37. **CRITICAL THINKING** The Transitive Property of Inequality states that if $a < b$ and $b < c$, then $a < c$. Use this property to determine whether the following statement is *sometimes*, *always*, or *never* true.

If $x > y$ and $z > w$, then $xz > yw$.

Give examples to support your answer.

38. **WRITING IN MATH** Answer the question that was posed at the beginning of the lesson.

How are identity and equality properties used to compare data?

Include the following in your answer:

- a description of how you could use the Reflexive or Symmetric Property to compare a team's rank for any two time periods, and
- a demonstration of the Transitive Property using one of the team's three rankings as an example.

39. Which equation illustrates the Symmetric Property of Equality?

(A) If $a = b$, then $b = a$.
(B) If $a = b$, $b = c$, then $a = c$.
(C) If $a = b$, then $b = c$.
(D) If $a = a$, then $a + 0 = a$.

40. The equation $(10 - 8)(5) = (2)(5)$ is an example of which property of equality?

(A) Reflexive
(B) Substitution
(C) Symmetric
(D) Transitive

Extending the Lesson

The sum of any two whole numbers is always a whole number. So, the set of whole numbers {0, 1, 2, 3, ...} is said to be closed under addition. This is an example of the **Closure Property**. State whether each of the following statements is *true* or *false*. If false, justify your reasoning.

41. The set of whole numbers is closed under subtraction.
42. The set of whole numbers is closed under multiplication.
43. The set of whole numbers is closed under division.

Maintain Your Skills

Mixed Review

Find the solution set for each inequality using the given replacement set. *(Lesson 1-3)*

44. $10 - x > 6$; {3, 5, 6, 8}
45. $4x + 2 < 58$; {11, 12, 13, 14, 15}
46. $\frac{x}{2} \geq 3$; {5.8, 5.9, 6, 6.1, 6.2, 6.3}
47. $8x \leq 32$; {3, 3.25, 3.5, 3.75, 4}
48. $\frac{7}{10} - 2x < \frac{3}{10}$; $\left\{\frac{1}{2}, \frac{1}{3}, \frac{1}{4}, \frac{1}{5}, \frac{1}{6}\right\}$
49. $2x - 1 \leq 2$; $\left\{1\frac{1}{4}, 2, 3, 3\frac{1}{2}\right\}$

Evaluate each expression. *(Lesson 1-2)*

50. $(3 + 6) \div 3^2$
51. $6(12 - 7.5) - 7$
52. $20 \div 4 \cdot 8 \div 10$
53. $\frac{(6 + 2)^2}{16} + 3(9)$
54. $[6^2 - (2 + 4)2]3$
55. $9(3) - 4^2 + 6^2 \div 2$

56. Write an algebraic expression for the sum of twice a number squared and 7. *(Lesson 1-1)*

Getting Ready for the Next Lesson

PREREQUISITE SKILL Evaluate each expression.
*(To review **order of operations**, see Lesson 1-2.)*

57. $10(6) + 10(2)$
58. $(15 - 6) \cdot 8$
59. $12(4) - 5(4)$
60. $3(4 + 2)$
61. $5(6 - 4)$
62. $8(14 + 2)$

1-5 The Distributive Property

What You'll Learn

- Use the Distributive Property to evaluate expressions.
- Use the Distributive Property to simplify expressions.

How can the Distributive Property be used to calculate quickly?

Vocabulary

- term
- like terms
- equivalent expressions
- simplest form
- coefficient

Instant Replay Video Games sells new and used games. During a Saturday morning sale, the first 8 customers each bought a bargain game and a new release. To calculate the total sales for these customers, you can use the Distributive Property.

Sale Prices	
Used Games	\$9.95
Bargain Games	\$14.95
Regular Games	\$24.95
New Releases	\$34.95

EVALUATE EXPRESSIONS There are two methods you could use to calculate the video game sales.

Method 1

$$\underbrace{8(14.95)}_{\text{sales of bargain games}} \underbrace{+}_{\text{plus}} \underbrace{8(34.95)}_{\text{sales of new releases}}$$

$= 119.60 + 279.60$

$= 399.20$

Method 2

$$\underbrace{8}_{\text{number of customers}} \underbrace{\times}_{\text{times}} \underbrace{(14.95 + 34.95)}_{\text{each customer's purchase price}}$$

$= 8(49.90)$

$= 399.20$

Either method gives total sales of \$399.20 because the following is true.

$$8(14.95) + 8(34.95) = 8(14.95 + 34.95)$$

This is an example of the **Distributive Property**.

Key Concept — Distributive Property

- **Symbols** For any numbers a, b, and c,
  $a(b + c) = ab + ac$ and $(b + c)a = ba + ca$ and
  $a(b - c) = ab - ac$ and $(b - c)a = ba - ca$.
- **Examples**
  $3(2 + 5) = 3 \cdot 2 + 3 \cdot 5$ $\quad$ $4(9 - 7) = 4 \cdot 9 - 4 \cdot 7$
  $3(7) = 6 + 15$ $\quad$ $4(2) = 36 - 28$
  $21 = 21$ ✓ $\quad$ $8 = 8$ ✓

Notice that it does not matter whether a is placed on the right or the left of the expression in the parentheses.

The Symmetric Property of Equality allows the Distributive Property to be written as follows.

$$\text{If } a(b + c) = ab + ac, \text{ then } ab + ac = a(b + c).$$

Example 1 Distribute Over Addition

Rewrite 8(10 + 4) using the Distributive Property. Then evaluate.

$8(10 + 4) = 8(10) + 8(4)$ Distributive Property

$= 80 + 32$ Multiply.

$= 112$ Add.

Example 2 Distribute Over Subtraction

Rewrite (12 − 3)6 using the Distributive Property. Then evaluate.

$(12 - 3)6 = 12 \cdot 6 - 3 \cdot 6$ Distributive Property

$= 72 - 18$ Multiply.

$= 54$ Subtract.

Log on for:

- Updated data
- More activities on the Distributive Property

www.algebra1.com/usa_today

Example 3 Use the Distributive Property

CARS The Morris family owns two cars. In 1998, they drove the first car 18,000 miles and the second car 16,000 miles. Use the graph to find the total cost of operating both cars.

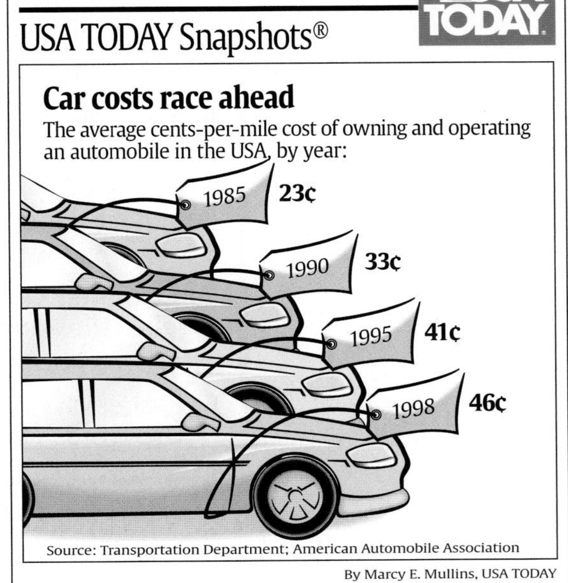

Use the Distributive Property to write and evaluate an expression.

$0.46(18{,}000 + 16{,}000)$ Distributive Prop.

$= 8280 + 7360$ Multiply.

$= 15{,}640$ Add.

It cost the Morris family $15,640 to operate their cars.

The Distributive Property can be used to simplify mental calculations.

Example 4 Use the Distributive Property

Use the Distributive Property to find each product.

a. $15 \cdot 99$

$15 \cdot 99 = 15(100 - 1)$ Think: $99 = 100 - 1$

$= 15(100) - 15(1)$ Distributive Property

$= 1500 - 15$ Multiply.

$= 1485$ Subtract.

b. $35\left(2\frac{1}{5}\right)$

$35\left(2\frac{1}{5}\right) = 35\left(2 + \frac{1}{5}\right)$ Think: $2\frac{1}{5} = 2 + \frac{1}{5}$

$= 35(2) + 35\left(\frac{1}{5}\right)$ Distributive Property

$= 70 + 7$ Multiply.

$= 77$ Add.

SIMPLIFY EXPRESSIONS You can use algebra tiles to investigate how the Distributive Property relates to algebraic expressions.

Algebra Activity

The Distributive Property

Consider the product $3(x + 2)$. Use a product mat and algebra tiles to model $3(x + 2)$ as the area of a rectangle whose dimensions are 3 and $(x + 2)$.

Step 1 Use algebra tiles to mark the dimensions of the rectangle on a product mat.

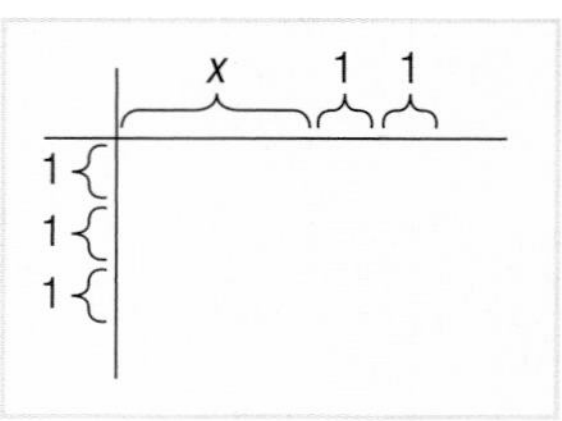

Step 2 Using the marks as a guide, make the rectangle with the algebra tiles. The rectangle has 3 x-tiles and 6 1-tiles. The area of the rectangle is $x + 1 + 1 + x + 1 + 1 + x + 1 + 1$ or $3x + 6$. Therefore, $3(x + 2) = 3x + 6$.

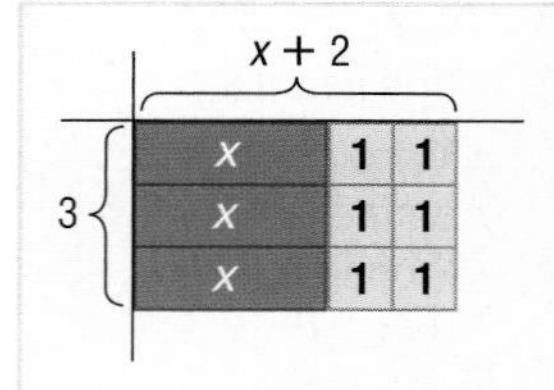

Model and Analyze

Find each product by using algebra tiles.

1. $2(x + 1)$ **2.** $5(x + 2)$ **3.** $2(2x + 1)$

Tell whether each statement is *true* or *false*. Justify your answer with algebra tiles and a drawing.

4. $3(x + 3) = 3x + 3$ **5.** $x(3 + 2) = 3x + 2x$

6. Rachel says that $3(x + 4) = 3x + 12$, but José says that $3(x + 4) = 3x + 4$. Use words and models to explain who is correct and why.

You can apply the Distributive Property to algebraic expressions.

Study Tip

Reading Math
The expression $5(g - 9)$ is read *5 times the quantity g minus 9* or *5 times the difference of g and 9.*

Example 5 Algebraic Expressions

Rewrite each product using the Distributive Property. Then simplify.

a. $5(g - 9)$

$5(g - 9) = 5 \cdot g - 5 \cdot 9$ Distributive Property

$= 5g - 45$ Multiply.

b. $3(2x^2 + 4x - 1)$

$3(2x^2 + 4x - 1) = (3)(2x^2) + (3)(4x) - 3(1)$ Distributive Property

$= 6x^2 + 12x - 3$ Simplify.

A **term** is a number, a variable, or a product or quotient of numbers and variables. For example, y, p^3, $4a$, and $5g^2h$ are all terms. **Like terms** are terms that contain the same variables, with corresponding variables having the same power.

$2x^2 + 6x + 5$

three terms

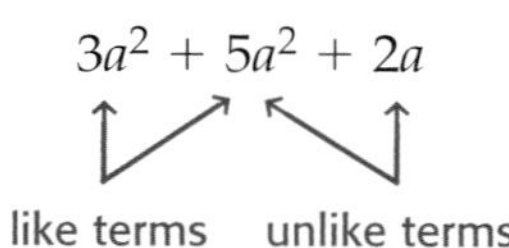

The Distributive Property and the properties of equality can be used to show that $5n + 7n = 12n$. In this expression, $5n$ and $7n$ are like terms.

$$5n + 7n = (5 + 7)n \quad \text{Distributive Property}$$
$$= 12n \quad \text{Substitution}$$

The expressions $5n + 7n$ and $12n$ are called **equivalent expressions** because they denote the same number. An expression is in **simplest form** when it is replaced by an equivalent expression having no like terms or parentheses.

Example 6 Combine Like Terms

Simplify each expression.

a. $15x + 18x$

$15x + 18x = (15 + 18)x$ Distributive Property

$= 33x$ Substitution

b. $10n + 3n^2 + 9n^2$

$10n + 3n^2 + 9n^2 = 10n + (3 + 9)n^2$ Distributive Property

$= 10n + 12n^2$ Substitution

Study Tip

Like Terms
Like terms may be defined as terms that are the same or vary only by the coefficient.

The **coefficient** of a term is the numerical factor. For example, in $17xy$, the coefficient is 17, and in $\frac{3y^2}{4}$, the coefficient is $\frac{3}{4}$. In the term m, the coefficient is 1 since $1 \cdot m = m$ by the Multiplicative Identity Property.

Check for Understanding

Concept Check

1. **Explain** why the Distributive Property is sometimes called The Distributive Property of Multiplication Over Addition.
2. **OPEN ENDED** Write an expression that has five terms, three of which are like terms and one term with a coefficient of 1.
3. **FIND THE ERROR** Courtney and Ben are simplifying $4w^4 + w^4 + 3w^2 - 2w^2$.

Courtney	Ben
$4w^4 + w^4 + 3w^2 - 2w^2$	$4w^4 + w^4 + 3w^2 - 2w^2$
$= (4 + 1)w^4 + (3 - 2)w^2$	$= (4)w^4 + (3 - 2)w^2$
$= 5w^4 + 1w^2$	$= 4w^4 + 1w^2$
$= 5w^4 + w^2$	$= 4w^4 + w^2$

Who is correct? Explain your reasoning.

Guided Practice

Rewrite each expression using the Distributive Property. Then simplify.

4. $6(12 - 2)$
5. $2(4 + t)$
6. $(g - 9)5$

Use the Distributive Property to find each product.

7. $16(102)$
8. $\left(3\frac{1}{17}\right)(17)$

Simplify each expression. If not possible, write *simplified*.

9. $13m + m$
10. $3(x + 2x)$
11. $14a^2 + 13b^2 + 27$
12. $4(3g + 2)$

Application **COSMETOLOGY** **For Exercises 13 and 14, use the following information.**
Ms. Curry owns a hair salon. One day, she gave 12 haircuts. She earned \$19.95 for each and received an average tip of \$2 for each haircut.

13. Write an expression to determine the total amount she earned.

14. How much did Ms. Curry earn?

Practice and Apply

Homework Help

For Exercises	See Examples
15–18	1, 2
19–28	5
29, 30, 37–41	3
31–36	4
42–53	6

Extra Practice

See page 821.

Rewrite each expression using the Distributive Property. Then simplify.

15. $8(5 + 7)$ **16.** $7(13 + 12)$ **17.** $12(9 - 5)$

18. $13(10 - 7)$ **19.** $3(2x + 6)$ **20.** $8(3m + 4)$

21. $(4 + x)2$ **22.** $(5 + n)3$ **23.** $28\left(y - \frac{1}{7}\right)$

24. $27\left(2b - \frac{1}{3}\right)$ **25.** $a(b - 6)$ **26.** $x(z + 3)$

27. $2(a - 3b + 2c)$ **28.** $4(8p + 4q - 7r)$

OLYMPICS **For Exercises 29 and 30, use the following information.**
At the 2000 Summer Olympics in Australia, about 110,000 people attended events at Olympic Stadium each day while another 17,500 fans were at the aquatics center.

29. Write an expression you could use to determine the total number of people at Olympic Stadium and the Aquatic Center over 4 days.

30. What was the attendance for the 4-day period?

Use the Distributive Property to find each product.

31. $5 \cdot 97$ **32.** $8 \cdot 990$ **33.** $17 \cdot 6$

34. $24 \cdot 7$ **35.** $18\left(2\frac{1}{9}\right)$ **36.** $48\left(3\frac{1}{6}\right)$

More About. . .

Olympics

The first modern Olympics were held in Athens, Greece, in 1896. The games featured 43 events and included 14 nations. The 2000 Olympics featured 300 events and included 199 nations.

Source: www.olympic.org

COMMUNICATIONS **For Exercises 37 and 38, use the following information.**
A public relations consultant keeps a log of all contacts made by e-mail, telephone, and in person. In a typical week, she averages 5 hours using e-mail, 12 hours of meeting in person, and 18 hours on the telephone.

37. Write an expression that could be used to predict how many hours she will spend on these activities over the next 12 weeks.

38. How many hours should she plan for contacting people for the next 12 weeks?

INSURANCE **For Exercises 39–41, use the table that shows the monthly cost of a company health plan.**

Available Insurance Plans—Monthly Charge			
Coverage	Medical	Dental	Vision
Employee	\$78	\$20	\$12
Family (additional coverage)	\$50	\$15	\$7

39. Write an expression that could be used to calculate the cost of medical, dental, and vision insurance for an employee for 6 months.

40. How much does it cost an employee to get all three types of insurance for 6 months?

41. How much would an employee expect to pay for individual and family medical and dental coverage per year?

Simplify each expression. If not possible, write *simplified*.

42. $2x + 9x$ **43.** $4b + 5b$ **44.** $5n^2 + 7n$

45. $3a^2 + 14a^2$ **46.** $12(3c + 4)$ **47.** $15(3x - 5)$

48. $6x^2 + 14x - 9x$ **49.** $4y^3 + 3y^3 + y^4$ **50.** $6(5a + 3b - 2b)$

51. $5(6m + 4n - 3n)$ **52.** $x^2 + \frac{7}{8}x - \frac{x}{8}$ **53.** $a + \frac{a}{5} + \frac{2}{5}a$

54. **CRITICAL THINKING** The expression $2(\ell + w)$ may be used to find the perimeter of a rectangle. What are the length and width of a rectangle if the area is $13\frac{1}{2}$ square units and the length of one side is $\frac{1}{5}$ the measure of the perimeter?

55. **WRITING IN MATH** Answer the question that was posed at the beginning of the lesson.

How can the Distributive Property be used to calculate quickly?

Include the following in your answer:

- a comparison of the two methods of finding the total video game sales.

56. Simplify $3(x + y) + 2(x + y) - 4x$.

Ⓐ $5x + y$ Ⓑ $9x + 5y$ Ⓒ $5x + 9y$ Ⓓ $x + 5y$

57. If $a = 2.8$ and $b = 4.2$, find the value of c in the equation $c = 7(2a + 3b)$.

Ⓐ 18.2 Ⓑ 238.0 Ⓒ 127.4 Ⓓ 51.8

Maintain Your Skills

Mixed Review

Name the property illustrated by each statement or equation. *(Lesson 1-4)*

58. If $7 \cdot 2 = 14$, then $14 = 7 \cdot 2$. **59.** $8 + (3 + 9) = 8 + 12$

60. $mnp = 1mnp$ **61.** $3\left(5^2 \cdot \frac{1}{25}\right) = 3 \cdot 1$

62. $\left(\frac{3}{4}\right)\left(\frac{4}{3}\right) = 1$ **63.** $32 + 21 = 32 + 21$

PHYSICAL SCIENCE For Exercises 64 and 65, use the following information. Sound travels 1129 feet per second through air. *(Lesson 1-3)*

64. Write an equation that represents how many feet sound can travel in 2 seconds when it is traveling through air.

65. How far can sound travel in 2 seconds when traveling through air?

Evaluate each expression if $a = 4$, $b = 6$, and $c = 3$. *(Lesson 1–2)*

66. $3ab - c^2$ **67.** $8(a - c)^2 + 3$ **68.** $\frac{6ab}{c(a + 2)}$ **69.** $(a + c)\left(\frac{a + b}{2}\right)$

Getting Ready for the Next Lesson

PREREQUISITE SKILL Find the area of each figure.
*(To review **finding area**, see pages 813 and 814.)*

70.

71.

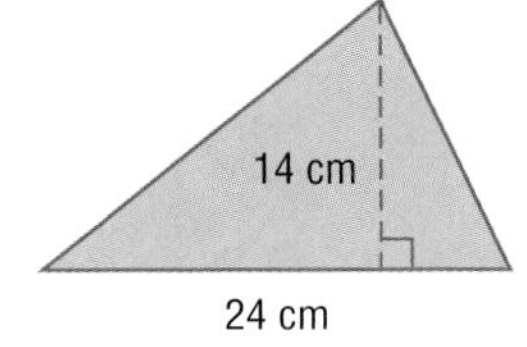

72.

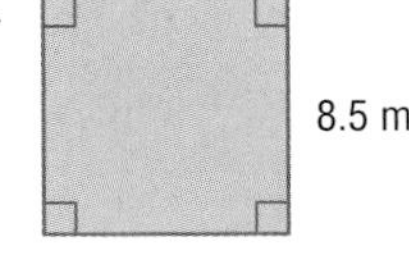

1-6 Commutative and Associative Properties

What You'll Learn

- Recognize the Commutative and Associative Properties.
- Use the Commutative and Associative Properties to simplify expressions.

How can properties help you determine distances?

The South Line of the Atlanta subway leaves Five Points and heads for Garnett, 0.4 mile away. From Garnett, West End is 1.5 miles. The distance from Five Points to West End can be found by evaluating the expression $0.4 + 1.5$. Likewise, the distance from West End to Five Points can be found by evaluating the expression $1.5 + 0.4$.

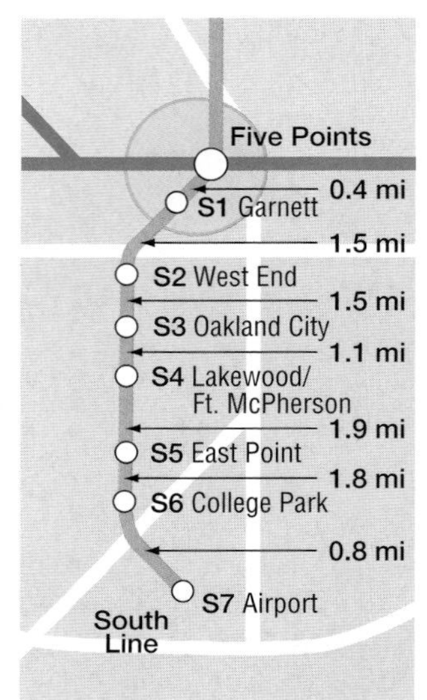

COMMUTATIVE AND ASSOCIATIVE PROPERTIES In the situation above, the distance from Five Points to West End is the same as the distance from West End to Five Points. This distance can be represented by the following equation.

The distance from Five Points to West End	equals	the distance from West End to Five Points.
$0.4 + 1.5$	$=$	$1.5 + 0.4$

This is an example of the **Commutative Property**.

Key Concept — *Commutative Property*

- **Words** The order in which you add or multiply numbers does not change their sum or product.
- **Symbols** For any numbers a and b, $a + b = b + a$ and $a \cdot b = b \cdot a$.
- **Examples** $5 + 6 = 6 + 5$, $3 \cdot 2 = 2 \cdot 3$

An easy way to find the sum or product of numbers is to group, or associate, the numbers using the **Associative Property**.

Key Concept — *Associative Property*

- **Words** The way you group three or more numbers when adding or multiplying does not change their sum or product.
- **Symbols** For any numbers a, b, and c, $(a + b) + c = a + (b + c)$ and $(ab)c = a(bc)$.
- **Examples** $(2 + 4) + 6 = 2 + (4 + 6)$, $(3 \cdot 5) \cdot 4 = 3 \cdot (5 \cdot 4)$

Example 1 Multiplication Properties

Evaluate $8 \cdot 2 \cdot 3 \cdot 5$.

You can rearrange and group the factors to make mental calculations easier.

$$\begin{aligned} 8 \cdot 2 \cdot 3 \cdot 5 &= 8 \cdot 3 \cdot 2 \cdot 5 && \text{Commutative } (\times) \\ &= (8 \cdot 3) \cdot (2 \cdot 5) && \text{Associative } (\times) \\ &= 24 \cdot 10 && \text{Multiply.} \\ &= 240 && \text{Multiply.} \end{aligned}$$

Example 2 Use Addition Properties

TRANSPORTATION **Refer to the application at the beginning of the lesson. Find the distance between Five Points and Lakewood/Ft. McPherson.**

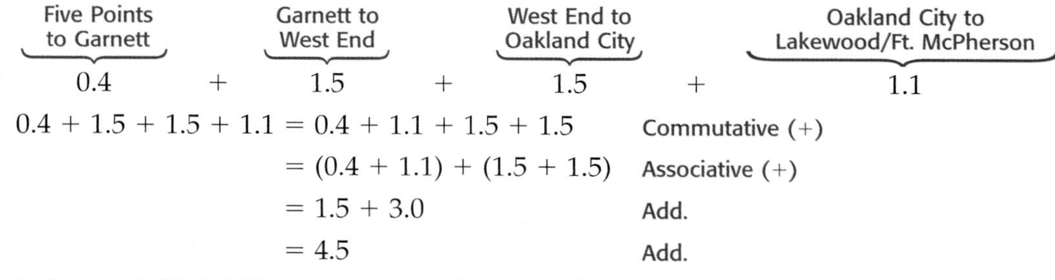

$$\underbrace{0.4}_{\text{Five Points to Garnett}} + \underbrace{1.5}_{\text{Garnett to West End}} + \underbrace{1.5}_{\text{West End to Oakland City}} + \underbrace{1.1}_{\text{Oakland City to Lakewood/Ft. McPherson}}$$

$$\begin{aligned} 0.4 + 1.5 + 1.5 + 1.1 &= 0.4 + 1.1 + 1.5 + 1.5 && \text{Commutative } (+) \\ &= (0.4 + 1.1) + (1.5 + 1.5) && \text{Associative } (+) \\ &= 1.5 + 3.0 && \text{Add.} \\ &= 4.5 && \text{Add.} \end{aligned}$$

Lakewood/Ft. McPherson is 4.5 miles from Five Points.

More About. . .

Transportation

New York City has the most extensive subway system, covering 842 miles of track and serving about 4.3 million passengers per day.

Source: *The Guinness Book of Records*

SIMPLIFY EXPRESSIONS The Commutative and Associative Properties can be used with other properties when evaluating and simplifying expressions.

Concept Summary — *Properties of Numbers*

The following properties are true for any numbers a, b, and c.

Properties	Addition	Multiplication
Commutative	$a + b = b + a$	$ab = ba$
Associative	$(a + b) + c = a + (b + c)$	$(ab)c = a(bc)$
Identity	0 is the identity. $a + 0 = 0 + a = a$	1 is the identity. $a \cdot 1 = 1 \cdot a = a$
Zero	—	$a \cdot 0 = 0 \cdot a = 0$
Distributive	$a(b + c) = ab + ac$ and $(b + c)a = ba + ca$	
Substitution	If $a = b$, then a may be substituted for b.	

Example 3 Simplify an Expression

Simplify $3c + 5(2 + c)$.

$$\begin{aligned} 3c + 5(2 + c) &= 3c + 5(2) + 5(c) && \text{Distributive Property} \\ &= 3c + 10 + 5c && \text{Multiply.} \\ &= 3c + 5c + 10 && \text{Commutative } (+) \\ &= (3c + 5c) + 10 && \text{Associative } (+) \\ &= (3 + 5)c + 10 && \text{Distributive Property} \\ &= 8c + 10 && \text{Substitution} \end{aligned}$$

Example 4 Write and Simplify an Expression

Use the expression *four times the sum of a and b increased by twice the sum of a and 2b.*

a. Write an algebraic expression for the verbal expression.

four times the sum of a and b	increased by	twice the sum of a and $2b$
$4(a + b)$	$+$	$2(a + 2b)$

b. Simplify the expression and indicate the properties used.

$4(a + b) + 2(a + 2b) = 4(a) + 4(b) + 2(a) + 2(2b)$	Distributive Property
$= 4a + 4b + 2a + 4b$	Multiply.
$= 4a + 2a + 4b + 4b$	Commutative (+)
$= (4a + 2a) + (4b + 4b)$	Associative (+)
$= (4 + 2)a + (4 + 4)b$	Distributive Property
$= 6a + 8b$	Substitution

Check for Understanding

Concept Check

1. **Define** the Associative Property in your own words.
2. **Write** a short explanation as to whether there is a Commutative Property of Division.
3. **OPEN ENDED** Write examples of the Commutative Property of Addition and the Associative Property of Multiplication using 1, 5, and 8 in each.

Guided Practice

Evaluate each expression.

4. $14 + 18 + 26$
5. $3\frac{1}{2} + 4 + 2\frac{1}{2}$
6. $5 \cdot 3 \cdot 6 \cdot 4$
7. $\frac{5}{6} \cdot 16 \cdot 9\frac{3}{4}$

Simplify each expression.

8. $4x + 5y + 6x$
9. $5a + 3b + 2a + 7b$
10. $\frac{1}{4}q + 2q + 2\frac{3}{4}q$
11. $3(4x + 2) + 2x$
12. $7(ac + 2b) + 2ac$
13. $3(x + 2y) + 4(3x + y)$
14. Write an algebraic expression for *half the sum of p and 2q increased by three-fourths q*. Then simplify, indicating the properties used.

Application

15. **GEOMETRY** Find the area of the large triangle if each smaller triangle has a base measuring 5.2 centimeters and a height of 4.5 centimeters.

Practice and Apply

Evaluate each expression.

16. $17 + 6 + 13 + 24$
17. $8 + 14 + 22 + 9$
18. $4.25 + 3.50 + 8.25$
19. $6.2 + 4.2 + 4.3 + 5.8$
20. $6\frac{1}{2} + 3 + \frac{1}{2} + 2$
21. $2\frac{3}{8} + 4 + 3\frac{3}{8}$
22. $5 \cdot 11 \cdot 4 \cdot 2$
23. $3 \cdot 10 \cdot 6 \cdot 3$
24. $0.5 \cdot 2.4 \cdot 4$
25. $8 \cdot 1.6 \cdot 2.5$
26. $3\frac{3}{7} \cdot 14 \cdot 1\frac{1}{4}$
27. $2\frac{5}{8} \cdot 24 \cdot 6\frac{2}{3}$

Homework Help

For Exercises	See Examples
16–29	1, 2
30, 31	2
32–43	3
44–47	4

Extra Practice
See page 821.

TRAVEL **For Exercises 28 and 29, use the following information.**
Hotels often have different rates for weeknights and weekends. The rates of one hotel are listed in the table.

28. If a traveler checks into the hotel on Friday and checks out the following Tuesday morning, what is the total cost of the room?

29. Suppose there is a sales tax of $5.40 for weeknights and $5.10 for weekends. What is the total cost of the room including tax?

Hotel Rates	
Weeknights (M–F)	$72
Weekends	$63
Weekly (5 weeknights)	$325

ENTERTAINMENT **For Exercises 30 and 31, use the following information.**
A video store rents new release videos for $4.49, older videos for $2.99, and DVDs for $3.99. The store also sells its used videos for $9.99.

30. Write two expressions to represent the total sales of a clerk after renting 2 DVDs, 3 new releases, 2 older videos, and selling 5 used videos.

31. What are the total sales of the clerk?

Simplify each expression.

32. $4a + 2b + a$
33. $2y + 2x + 8y$
34. $x^2 + 3x + 2x + 5x^2$
35. $4a^3 + 6a + 3a^3 + 8a$
36. $6x + 2(2x + 7)$
37. $5n + 4(3n + 9)$
38. $3(x + 2y) + 4(3x + y)$
39. $3.2(x + y) + 2.3(x + y) + 4x$
40. $3(4m + n) + 2m$
41. $6(0.4f + 0.2g) + 0.5f$
42. $\frac{3}{4} + \frac{2}{3}(s + 2t) + s$
43. $2p + \frac{3}{5}\left(\frac{1}{2}p + 2q\right) + \frac{2}{3}$

Write an algebraic expression for each verbal expression. Then simplify, indicating the properties used.

44. twice the sum of s and t decreased by s

45. five times the product of x and y increased by $3xy$

46. the product of six and the square of z, increased by the sum of seven, z^2, and 6

47. six times the sum of x and y squared decreased by three times the sum of x and half of y squared

48. **CRITICAL THINKING** Tell whether the Commutative Property *always*, *sometimes*, or *never* holds for subtraction. Explain your reasoning.

49. **WRITING IN MATH** Answer the question that was posed at the beginning of the lesson.

How can properties help you determine distances?

Include the following in your answer:

- an expression using the Commutative and Associative Properties that you could use to easily determine the distance from the airport to Five Points, and
- an explanation of how the Commutative and Associative Properties are useful in performing calculations.

Stop	Distance from Previous Stop
Five Points	0
Garnett	0.4
West End	1.5
Oakland City	1.5
Lakewood/ Ft. McPherson	1.1
East Point	1.9
College Park	1.8
Airport	0.8

www.algebra1.com/self_check_quiz

Standardized Test Practice

50. Simplify $6(ac + 2b) + 2ac$.

(A) $10ab + 2ac$ (B) $12ac + 20b$ (C) $8ac + 12b$ (D) $12abc + 2ac$

51. Which property can be used to show that the areas of the two rectangles are equal?

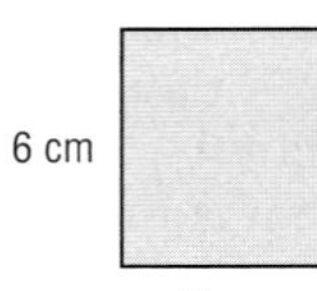

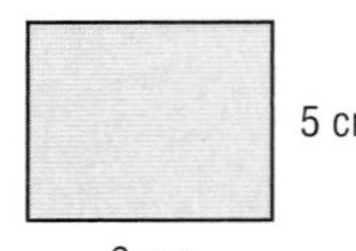

(A) Associative
(B) Commutative
(C) Distributive
(D) Reflexive

Maintain Your Skills

Mixed Review

Simplify each expression. *(Lesson 1-5)*

52. $5(2 + x) + 7x$
53. $3(5 + 2p)$
54. $3(a + 2b) - 3a$
55. $7m + 6(n + m)$
56. $(d + 5)f + 2f$
57. $t^2 + 2t^2 + 4t$

58. Name the property used in each step. *(Lesson 1-4)*

$$\begin{aligned} 3(10 - 5 \cdot 2) + 21 \div 7 &= 3(10 - 10) + 21 \div 7 \\ &= 3(0) + 21 \div 7 \\ &= 0 + 21 \div 7 \\ &= 0 + 3 \\ &= 3 \end{aligned}$$

Evaluate each expression. *(Lesson 1-2)*

59. $12(5) - 6(4)$
60. $7(0.2 + 0.5) - 0.6$
61. $8[6^2 - 3(2 + 5)] \div 8 + 3$

Getting Ready for the Next Lesson

PREREQUISITE SKILL Evaluate each expression for the given value of the variable.
*(To review **evaluating expressions**, see Lesson 1-2.)*

62. If $x = 4$, then $2x + 7 =$ _?_.
63. If $x = 8$, then $6x + 12 =$ _?_.
64. If $n = 6$, then $5n - 14 =$ _?_.
65. If $n = 7$, then $3n - 8 =$ _?_.
66. If $a = 2$, and $b = 5$, then $4a + 3b =$ _?_.

Practice Quiz 2 — Lessons 1-4 through 1-6

Write the letters of the properties given in the right-hand column that match the examples in the left-hand column.

1. $28 + 0 = 28$
2. $(18 - 7)6 = 11(6)$
3. $24 + 15 = 15 + 24$
4. $8 \cdot 5 = 8 \cdot 5$
5. $(9 + 3) + 8 = 9 + (3 + 8)$
6. $1(57) = 57$
7. $14 \cdot 0 = 0$
8. $3(13 + 10) = 3(13) + 3(10)$
9. If $12 + 4 = 16$, then $16 = 12 + 4$.
10. $\frac{2}{5} \cdot \frac{5}{2} = 1$

a. Distributive Property
b. Multiplicative Property of 0
c. Substitution Property of Equality
d. Multiplicative Identity Property
e. Multiplicative Inverse Property
f. Reflexive Property of Equality
g. Associative Property
h. Symmetric Property of Equality
i. Commutative Property
j. Additive Identity Property

1-7 Logical Reasoning

What You'll Learn

- Identify the hypothesis and conclusion in a conditional statement.
- Use a counterexample to show that an assertion is false.

How is logical reasoning helpful in cooking?

Popcorn is a popular snack with 16 billion quarts consumed in the United States each year. The directions at the right can help you make perfect popcorn. If the popcorn burns, then the heat was too high or the kernels heated unevenly.

Stovetop Popping

To pop popcorn on a stovetop, you need:
- A 3- to 4-quart pan with a loose lid that allows steam to escape
- Enough popcorn to cover the bottom of the pan, one kernel deep
- 1/4 cup of oil for every cup of kernels (Don't use butter!)

Heat the oil to 400–460 degrees Fahrenheit (if the oil smokes, it is too hot). Test the oil on a couple of kernels. When they pop, add the rest of the popcorn, cover the pan, and shake to spread the oil. When the popping begins to slow, remove the pan from the stovetop. The heated oil will pop the remaining kernels.

Source: Popcorn Board

Vocabulary

- conditional statement
- if-then statement
- hypothesis
- conclusion
- deductive reasoning
- counterexample

CONDITIONAL STATEMENTS The statement *If the popcorn burns, then the heat was too high or the kernels heated unevenly* is called a conditional statement. **Conditional statements** can be written in the form *If A, then B*. Statements in this form are called **if-then statements**.

If *A*, then *B*.

If the popcorn burns, then the heat was too high or the kernels heated unevenly.

The part of the statement immediately following the word *if* is called the **hypothesis**.

The part of the statement immediately following the word *then* is called the **conclusion**.

Study Tip

Reading Math
Note that "if" is not part of the hypothesis and "then" is not part of the conclusion.

Example 1 Identify Hypothesis and Conclusion

Identify the hypothesis and conclusion of each statement.

a. If it is Friday, then Madison and Miguel are going to the movies.

Recall that the hypothesis is the part of the conditional following the word *if* and the conclusion is the part of the conditional following the word *then*.

Hypothesis: it is Friday

Conclusion: Madison and Miguel are going to the movies

b. If $4x + 3 > 27$, then $x > 6$.

Hypothesis: $4x + 3 > 27$

Conclusion: $x > 6$

Sometimes a conditional statement is written without using the words *if* and *then*. But a conditional statement can always be rewritten as an if-then statement. For example, the statement *When it is not raining, I ride my bike* can be written as *If it is not raining, then I ride my bike.*

Example 2 Write a Conditional in If-Then Form

Identify the hypothesis and conclusion of each statement. Then write each statement in if-then form.

a. I will go to the ball game with you on Saturday.

Hypothesis: it is Saturday

Conclusion: I will go to the ball game with you

If it is Saturday, then I will go to the ball game with you.

b. For a number x such that $6x - 8 = 16$, $x = 4$.

Hypothesis: $6x - 8 = 16$

Conclusion: $x = 4$

If $6x - 8 = 16$, then $x = 4$.

DEDUCTIVE REASONING AND COUNTEREXAMPLES

Deductive reasoning is the process of using facts, rules, definitions, or properties to reach a valid conclusion. Suppose you have a true conditional and you know that the hypothesis is true for a given case. Deductive reasoning allows you to say that the conclusion is true for that case.

Example 3 Deductive Reasoning

Study Tip

Common Misconception

Suppose the conclusion of a conditional is true. This does not mean that the hypothesis is true. Consider the conditional "If it rains, Annie will stay home." If Annie stays home, it does not necessarily mean that it is raining.

Determine a valid conclusion that follows from the statement "If two numbers are odd, then their sum is even" for the given conditions. If a valid conclusion does not follow, write *no valid conclusion* and explain why.

a. The two numbers are 7 and 3.

7 and 3 are odd, so the hypothesis is true.

Conclusion: The sum of 7 and 3 is even.

CHECK $7 + 3 = 10$ ✓ The sum, 10, is even.

b. The sum of two numbers is 14.

The conclusion is true. If the numbers are 11 and 3, the hypothesis is true also. However, if the numbers are 8 and 6, the hypothesis is false. There is no way to determine the two numbers. Therefore, there is no valid conclusion.

Not all if-then statements are always true or always false. Consider the statement "If Luke is listening to CDs, then he is using his portable CD player." Luke may be using his portable CD player. However, he could also be using a computer, a car CD player, or a home CD player.

To show that a conditional is false, we can use a counterexample. A **counterexample** is a specific case in which a statement is false. It takes only one counterexample to show that a statement is false.

Example 4 Find Counterexamples

Find a counterexample for each conditional statement.

a. If you are using the Internet, then you own a computer.

You could be using the Internet on a computer at a library.

b. If the Commutative Property holds for multiplication, then it holds for division.

$2 \div 1 \stackrel{?}{=} 1 \div 2$

$2 \neq 0.5$

Standardized Test Practice

Example 5 Find a Counterexample

Multiple-Choice Test Item

> Which numbers are counterexamples for the statement below?
>
> *If $x \div y = 1$, then x and y are whole numbers.*
>
> Ⓐ $x = 2, y = 2$ Ⓑ $x = 0.25, y = 0.25$
>
> Ⓒ $x = 1.2, y = 0.6$ Ⓓ $x = 6, y = 3$

Read the Test Item

Find the values of x and y that make the statement false.

Solve the Test Item

Replace x and y in the equation $x \div y = 1$ with the given values.

Ⓐ $x = 2, y = 2$

$2 \div 2 \stackrel{?}{=} 1$

$1 = 1$ ✓

The hypothesis is true and both values are whole numbers. The statement is true.

Ⓑ $x = 0.25, y = 0.25$

$0.25 \div 0.25 \stackrel{?}{=} 1$

$1 = 1$ ✓

The hypothesis is true, but 0.25 is not a whole number. Thus, the statement is false.

Ⓒ $x = 1.2, y = 0.6$

$1.2 \div 0.6 \stackrel{?}{=} 1$

$2 \neq 1$

The hypothesis is false, and the conclusion is false. However, this is not a counterexample. A counterexample is a case where the hypothesis is true and the conclusion is false.

Ⓓ $x = 6, y = 3$

$6 \div 3 \stackrel{?}{=} 1$

$2 \neq 1$

The hypothesis is false. Therefore, this is not a counterexample.

The only values that prove the statement false are $x = 0.25$ and $y = 0.25$. So these numbers are counterexamples. The answer is B.

> **The Princeton Review**
>
> **Test-Taking Tip**
>
> Since choice B is the correct answer, you can check your result by testing the other values.

Check for Understanding

Concept Check

1. **OPEN ENDED** Write a conditional statement and label its hypothesis and conclusion.
2. **Explain** why counterexamples are used.
3. **Explain** how deductive reasoning is used to show that a conditional is true or false.

Guided Practice

Identify the hypothesis and conclusion of each statement.

4. If it is January, then it might snow.
5. If you play tennis, then you run fast.
6. If $34 - 3x = 16$, then $x = 6$.

Identify the hypothesis and conclusion of each statement. Then write the statement in if-then form.

7. Lance watches television when he does not have homework.
8. A number that is divisible by 10 is also divisible by 5.
9. A rectangle is a quadrilateral with four right angles.

Determine a valid conclusion that follows from the statement *If the last digit of a number is 2, then the number is divisible by 2* for the given conditions. If a valid conclusion does not follow, write *no valid conclusion* and explain why.

10. The number is 10,452.
11. The number is divisible by 2.
12. The number is 946.

Find a counterexample for each statement.

13. If Anna is in school, then she has a science class.
14. If you can read 8 pages in 30 minutes, then you can read any book in a day.
15. If a number x is squared, then $x^2 > x$.
16. If $3x + 7 \geq 52$, then $x > 15$.

17. Which number is a counterexample for the statement $x^2 > x$?

(A) 1 (B) 4 (C) 5 (D) 8

Practice and Apply

Homework Help

For Exercises	See Examples
18–23	1
24–29	2
30–35	3
36–43	4

Extra Practice

See page 822.

Identify the hypothesis and conclusion of each statement.

18. If both parents have red hair, then their children have red hair.
19. If you are in Hawaii, then you are in the tropics.
20. If $2n - 7 > 25$, then $n > 16$.
21. If $4(b + 9) \leq 68$, then $b \leq 8$.
22. If $a = b$, then $b = a$.
23. If $a = b$ and $b = c$, then $a = c$.

Identify the hypothesis and conclusion of each statement. Then write the statement in if-then form.

24. The trash is picked up on Monday.
25. Greg will call after school.
26. A triangle with all sides congruent is an equilateral triangle.
27. The sum of the digits of a number is a multiple of 9 when the number is divisible by 9.
28. For $x = 8$, $x^2 - 3x = 40$.
29. $4s + 6 > 42$ when $s > 9$.

Determine whether a valid conclusion follows from the statement ***If a VCR costs less than $150, then Ian will buy one*** **for the given condition. If a valid conclusion does not follow, write** ***no valid conclusion*** **and explain why.**

30. A VCR costs $139.

31. A VCR costs $99.

32. Ian will not buy a VCR.

33. The price of a VCR is $199.

34. A DVD player costs $229.

35. Ian bought 2 VCRs.

Find a counterexample for each statement.

36. If you were born in Texas, then you live in Texas.

37. If you are a professional basketball player, then you play in the United States.

38. If a baby is wearing blue clothes, then the baby is a boy.

39. If a person is left-handed, then each member of that person's family is left-handed.

40. If the product of two numbers is even, then both numbers must be even.

41. If a whole number is greater than 7, then two times the number is greater than 16.

42. If $4n - 8 \geq 52$, then $n > 15$.

43. If $x \cdot y = 1$, then x or y must equal 1.

More About. . .

Groundhog Day

Groundhog Day has been celebrated in the United States since 1897. The most famous groundhog, Punxsutawney Phil, has seen his shadow about 85% of the time.

Source: www.infoplease.com

GEOMETRY For Exercises 44 and 45, use the following information.
If points P, Q, and R lie on the same line, then Q is between P and R.

44. Copy the diagram. Label the points so that the conditional is true.

45. Copy the diagram. Provide a counterexample for the conditional.

46. RESEARCH On Groundhog Day (February 2) of each year, some people say that if a groundhog comes out of its hole and sees its shadow, then there will be six more weeks of winter weather. However, if it does not see its shadow, then there will be an early spring. Use the Internet or another resource to research the weather on Groundhog Day for your city for the past 10 years. Summarize your data as examples or counterexamples for this belief.

NUMBER THEORY For Exercises 47–49, use the following information.
Copy the Venn diagram and place the numbers 1 to 25 in the appropriate places on the diagram.

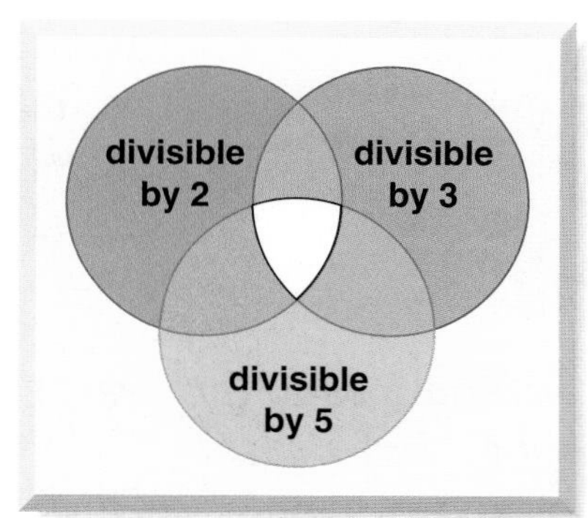

47. What conclusions can you make about the numbers and where they appear on the diagram?

48. What conclusions can you form about numbers that are divisible by 2 and 3?

49. Find a counterexample for your conclusions, if possible.

www.algebra1.com/self_check_quiz

50. **CRITICAL THINKING** Determine whether the following statement is always true. If it is not, provide a counterexample.

*If the mathematical operation * is defined for all numbers a and b as* $a * b = a + 2b$, *then the operation * is commutative.*

51. **WRITING IN MATH** Answer the question that was posed at the beginning of the lesson.

How is logical reasoning helpful in cooking?

Include the following in your answer:

- the hypothesis and conclusion of the statement *If you have small, underpopped kernels, then you have not used enough oil in your pan,* and
- examples of conditional statements used in cooking food other than popcorn.

Standardized Test Practice

52. **GRID IN** What value of n makes the following statement true?
If $14n - 12 \geq 100$, *then* $n \geq$ __?__.

53. If # is defined as $\#x = \frac{x^3}{2}$, what is the value of #4?

(A) 8 (B) 16 (C) 32 (D) 64

Maintain Your Skills

Mixed Review

Simplify each expression. *(Lesson 1-6)*

54. $2x + 5y + 9x$
55. $a + 9b + 6b$
56. $\frac{3}{4}g + \frac{2}{5}f + \frac{5}{8}g$
57. $4(5mn + 6) + 3mn$
58. $2(3a + b) + 3b + 4$
59. $6x^2 + 5x + 3(2x^2) + 7x$

60. **ENVIRONMENT** According to the U.S. Environmental Protection Agency, a typical family of four uses 100 gallons of water flushing the toilet each day, 80 gallons of water showering and bathing, and 8 gallons of water using the bathroom sink. Write two expressions that represent the amount of water a typical family of four uses for these purposes in d days. *(Lesson 1-5)*

Name the property used in each expression. Then find the value of n. *(Lesson 1-4)*

61. $1(n) = 64$
62. $12 + 7 = 12 + n$
63. $(9 - 7)5 = 2n$
64. $\frac{1}{4}n = 1$
65. $n + 18 = 18$
66. $36n = 0$

Solve each equation. *(Lesson 1-3)*

67. $5(7) + 6 = x$
68. $7(4^2) - 6^2 = m$
69. $p = \frac{22 - (13 - 5)}{28 \div 2^2}$

Write an algebraic expression for each verbal expression. *(Lesson 1-1)*

70. the product of 8 and a number x raised to the fourth power
71. three times a number n decreased by 10
72. twelve more than the quotient of a number a and 5

Getting Ready for the Next Lesson

PREREQUISITE SKILL **Evaluate each expression. Round to the nearest tenth.**
*(To review **percents**, see pages 802 and 803.)*

73. 40% of 90
74. 23% of 2500
75. 18% of 950
76. 38% of 345
77. 42.7% of 528
78. 67.4% of 388

1-8 Graphs and Functions

What You'll Learn

- Interpret graphs of functions.
- Draw graphs of functions.

Vocabulary

- function
- coordinate system
- *x*-axis
- *y*-axis
- origin
- ordered pair
- *x*-coordinate
- *y*-coordinate
- independent variable
- dependent variable
- relation
- domain
- range

How can real-world situations be modeled using graphs and functions?

Many athletes suffer concussions as a result of sports injuries. The graph shows the relationship between blood flow to the brain and the number of days after the concussion. The graph shows that as the number of days increases, the percent of blood flow increases.

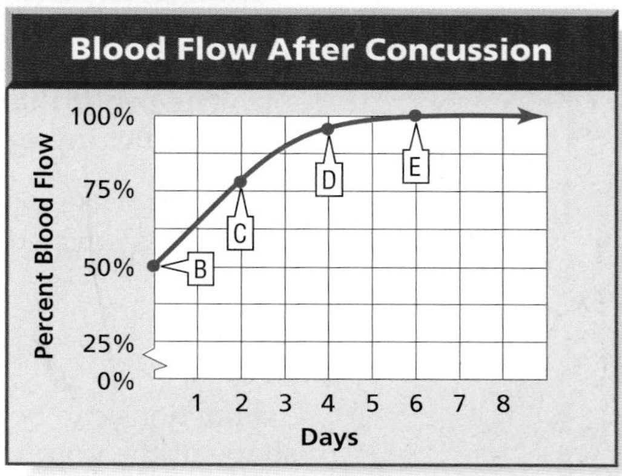

Blood Flow After Concussion
Percent Blood Flow
100%
75%
50%
25%
0%
B
C
D
E
1 2 3 4 5 6 7 8
Days

Source: *Scientific American*

INTERPRET GRAPHS The return of normal blood flow to the brain is said to be a function of the number of days since the concussion. A **function** is a relationship between input and output. In a function, the output depends on the input. There is exactly one output for each input.

A function is graphed using a **coordinate system**. It is formed by the intersection of two number lines, the *horizontal axis* and the *vertical axis*.

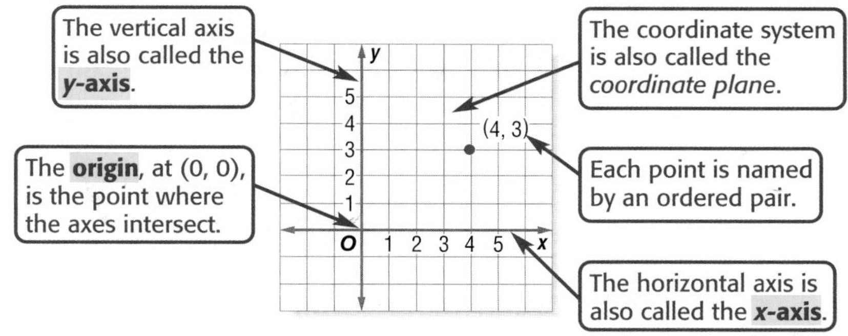

The vertical axis is also called the **y-axis**.
The coordinate system is also called the *coordinate plane*.
The **origin**, at (0, 0), is the point where the axes intersect.
Each point is named by an ordered pair.
The horizontal axis is also called the **x-axis**.
y
x
O
(4, 3)
1 2 3 4 5
1 2 3 4 5

Each input x and its corresponding output y can be represented on a graph using ordered pairs. An **ordered pair** is a set of numbers, or *coordinates*, written in the form (x, y). The x value, called the **x-coordinate**, corresponds to the x-axis and the y value, or **y-coordinate**, corresponds to the y-axis.

Example 1 Identify Coordinates

SPORTS MEDICINE Refer to the application above. Name the ordered pair at point *C* and explain what it represents.

Point C is at 2 along the x-axis and about 80 along the y-axis. So, its ordered pair is (2, 80). This represents 80% normal blood flow 2 days after the injury.

In Example 1, the percent of normal blood flow depends on the number of days from the injury. Therefore, the number of days from the injury is called the **independent variable** or *quantity*, and the percent of normal blood flow is called the **dependent variable** or *quantity*. Usually the independent variable is graphed on the horizontal axis and the dependent variable is graphed on the vertical axis.

Example 2 Independent and Dependent Variables

Identify the independent and dependent variables for each function.

a. In general, the average price of gasoline slowly and steadily increases throughout the year.

Time is the independent variable as it is unaffected by the price of gasoline, and the price is the dependent quantity as it is affected by time.

b. The profit that a business makes generally increases as the price of their product increases.

In this case, price is the independent quantity. Profit is the dependent quantity as it is affected by the price.

Functions can be graphed without using a scale on either axis to show the general shape of the graph that represents a function.

Example 3 Analyze Graphs

a. The graph at the right represents the speed of a school bus traveling along its morning route. Describe what is happening in the graph.

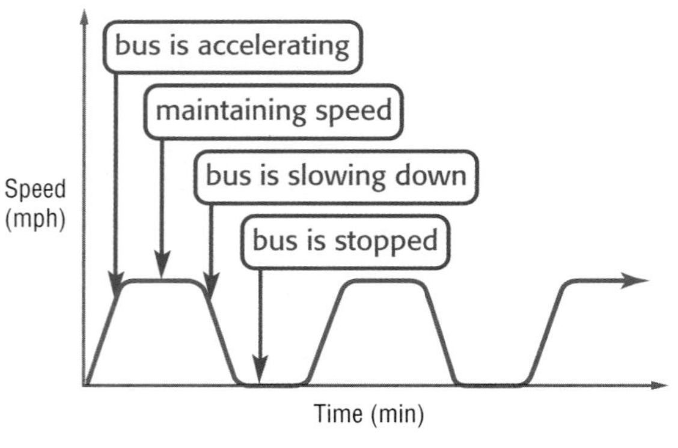

At the origin, the bus is stopped. It accelerates and maintains a constant speed. Then it begins to slow down, eventually stopping. After being stopped for a short time, the bus accelerates again. The starting and stopping process repeats continually.

b. Identify the graph that represents the altitude of a space shuttle above Earth, from the moment it is launched until the moment it lands.

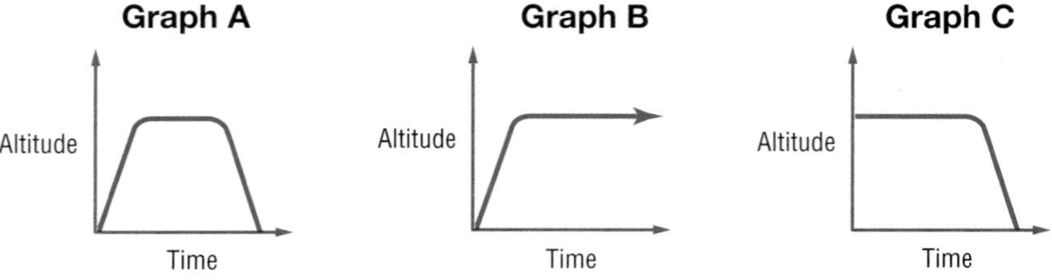

Before it takes off, the space shuttle is on the ground. It blasts off, gaining altitude until it reaches space where it orbits Earth at a constant height until it comes back to Earth. Graph A shows this situation.

DRAW GRAPHS Graphs can be used to represent many real-world situations.

Example 4 Draw Graphs

An electronics store is having a special sale. For every two DVDs you buy at the regular price of $29 each, you get a third DVD free.

a. Make a table showing the cost of buying 1 to 5 DVDs.

Number of DVDs	1	2	3	4	5
Total Cost ($)	29	58	58	87	116

b. Write the data as a set of ordered pairs.

The ordered pairs can be determined from the table. The number of DVDs is the independent variable, and the total cost is the dependent variable. So, the ordered pairs are (1, 29), (2, 58), (3, 58), (4, 87), and (5, 116).

c. Draw a graph that shows the relationship between the number of DVDs and the total cost.

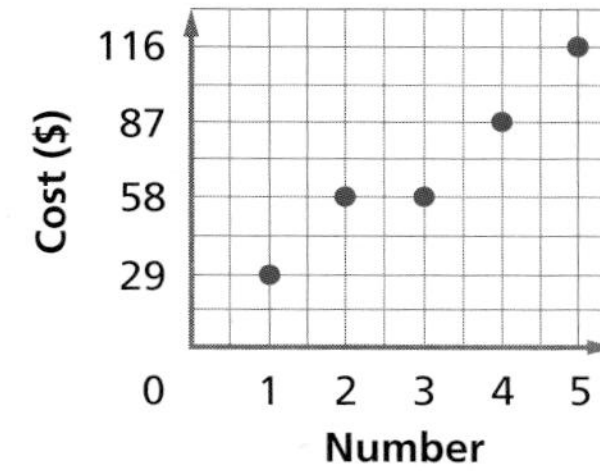

Study Tip

Different Representations
Example 4 illustrates several of the ways data can be represented—tables, ordered pairs, and graphs.

A set of ordered pairs, like those in Example 4, is called a **relation**. The set of the first numbers of the ordered pairs is the **domain**. The domain contains all values of the independent variable. The set of second numbers of the ordered pairs is the **range** of the relation. The range contains all values of the dependent variable.

Example 5 Domain and Range

JOBS **Rasha earns $6.75 per hour working up to 4 hours each day after school. Her weekly earnings are a function of the number of hours she works.**

a. Identify a reasonable domain and range for this situation.

The domain contains the number of hours Rasha works each week. Since she works up to 4 hours each weekday, she works up to 5×4 or 20 hours a week. Therefore, a reasonable domain would be values from 0 to 20 hours. The range contains her weekly earnings from $0 to $20 \times \$6.75$ or $135. Thus, a reasonable range is $0 to $135.

b. Draw a graph that shows the relationship between the number of hours Rasha works and the amount she earns each week.

Graph the ordered pairs (0, 0) and (20, 135). Since she can work any amount of time up to 20 hours, connect the two points with a line to include those points.

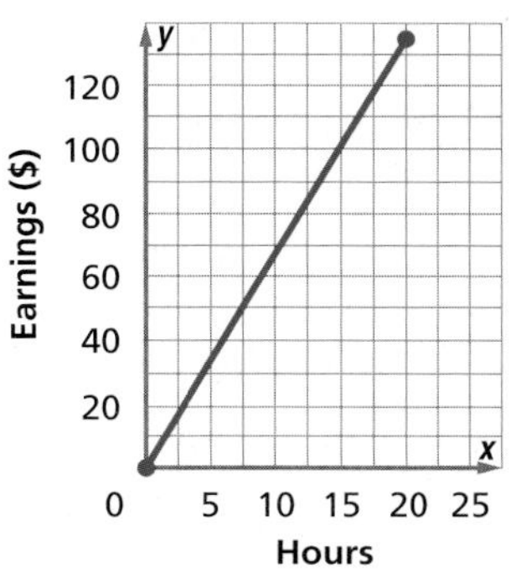

Check for Understanding

Concept Check

1. **Explain** why the order of the numbers in an ordered pair is important.
2. **Describe** the difference between dependent and independent variables.
3. **OPEN ENDED** Give an example of a relation. Identify the domain and range.

Guided Practice

4. The graph at the right represents Alexi's speed as he rides his bike. Give a description of what is happening in the graph.

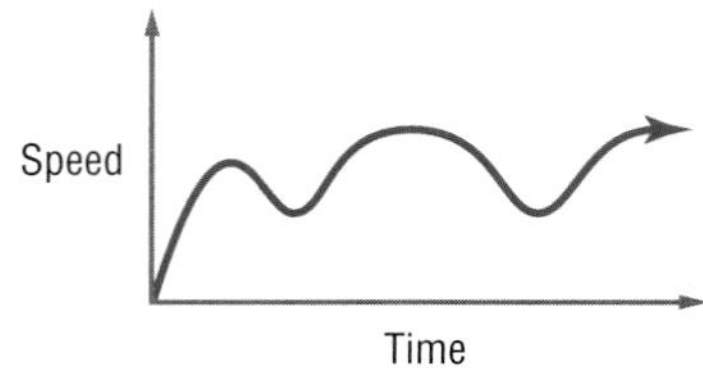

5. Identify the graph that represents the height of a skydiver just before she jumps from a plane until she lands.

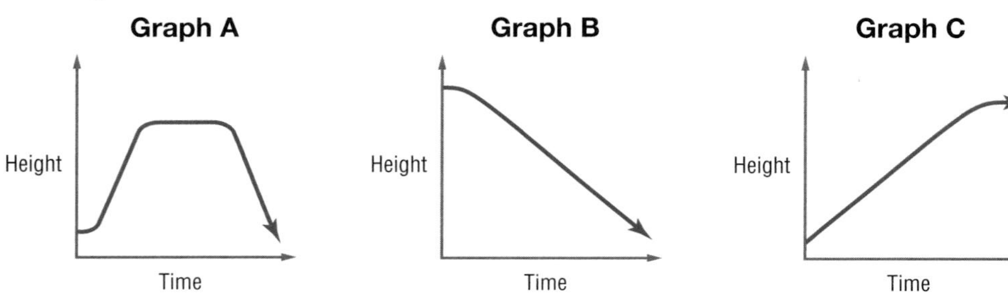

Applications

PHYSICAL SCIENCE **For Exercises 6–8, use the table and the information.** During an experiment, the students of Ms. Roswell's class recorded the height of an object above the ground at several intervals after it was dropped from a height of 5 meters. Their results are in the table below.

Time (s)	0	0.2	0.4	0.6	0.8	1
Height (cm)	500	480	422	324	186	10

6. Identify the independent and dependent variables.
7. Write a set of ordered pairs representing the data in the table.
8. Draw a graph showing the relationship between the height of the falling object and time.

9. **BASEBALL** Paul is a pitcher for his school baseball team. Draw a reasonable graph that shows the height of the baseball from the ground from the time he releases the ball until the time the catcher catches the ball. Let the horizontal axis show the time and the vertical axis show the height of the ball.

Practice and Apply

Homework Help

For Exercises	See Examples
10, 11	2
12, 13	3
14–21	4, 5

Extra Practice

See page 822.

10. The graph below represents Michelle's temperature when she was sick. Describe what is happening in the graph.

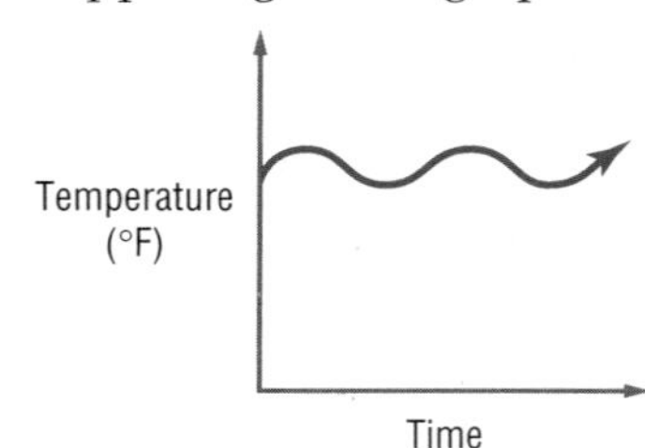

11. The graph below represents the balance in Rashaad's checking account. Describe what is happening in the graph.

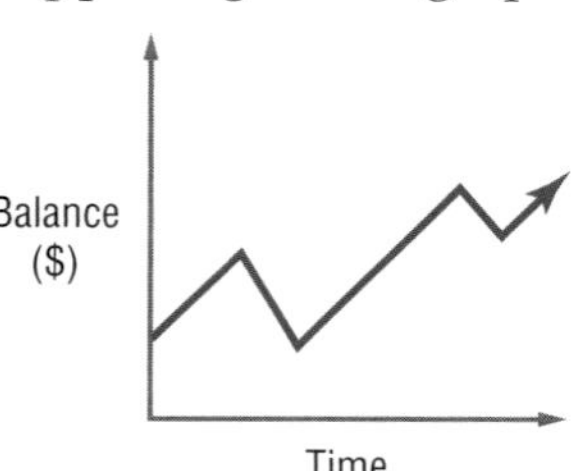

12. **TOYS** Identify the graph that displays the speed of a radio-controlled car as it moves along and then hits a wall.

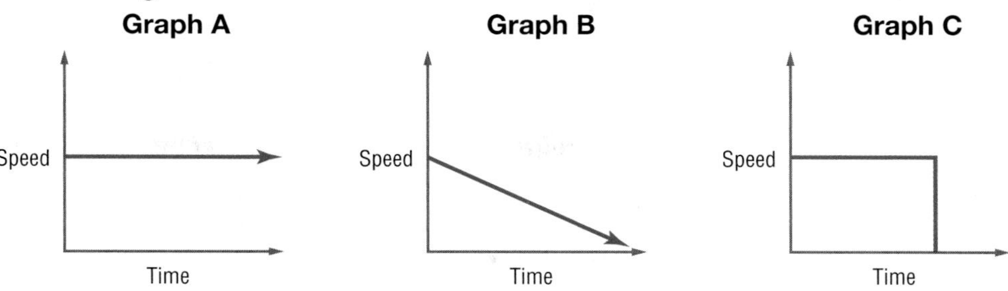

13. **INCOME** In general, as a person gets older, their income increases until they retire. Which of the graphs below represents this?

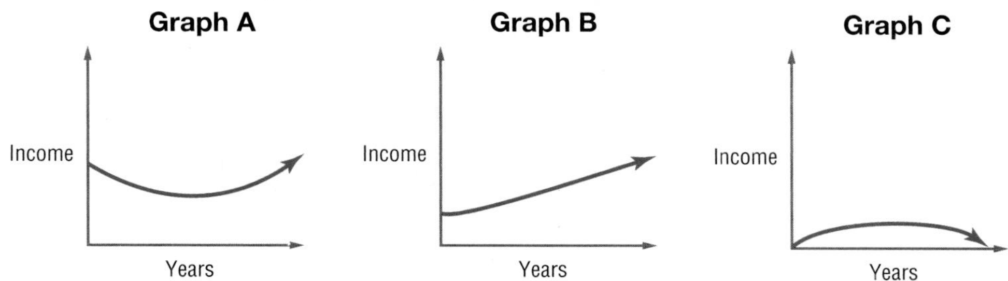

TRAVEL For Exercises 14–16, use the table that shows the charges for short-term parking at an airport.

Time (h)	0–1:59	2–3:59	4–5:59	6–11:59	12–24
Cost ($)	1	2	4	5	30
	After 24 hours: an additional $15 per day or portion thereof				

14. Write the ordered pairs with whole-number coordinates that represent the cost of parking for up to 36 hours.
15. Draw a graph to show the cost of parking for up to 36 hours.
16. What is the cost of parking if you arrive on Monday at 7:00 A.M. and depart on Tuesday at 9:00 P.M.?

GEOMETRY For Exercises 17–19, use the table that shows the relationships between the sum of the measures of the interior angles of convex polygons and the number of sides of the polygons.

Polygon	triangle	quadrilateral	pentagon	hexagon	heptagon
Sides	3	4	5	6	7
Interior Angle Sum	180	360	540	720	900

17. Identify the independent and dependent variables.
18. Draw a graph of the data.
19. Use the data to predict the sum of the measures of the interior angles for an octagon, nonagon, and decagon.
20. **CARS** A car was purchased new in 1970. The owner has taken excellent care of the car, and it has relatively low mileage. Draw a reasonable graph to show the value of the car from the time it was purchased to the present.
21. **CHEMISTRY** When ice is exposed to temperatures above 32°F, it begins to melt. Draw a reasonable graph showing the relationship between the temperature of a block of ice as it is removed from a freezer and placed on a counter at room temperature. (*Hint*: The temperature of the water will not exceed the temperature of its surroundings.)

More About. . .

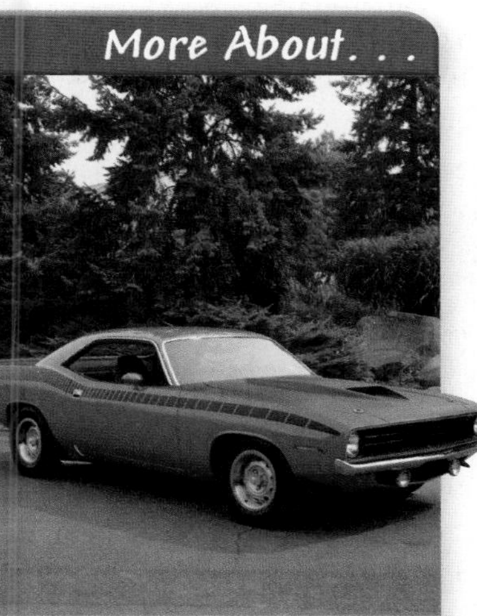

Cars

Most new cars lose 15 to 30 percent of their value in the first year. After about 12 years, more popular cars tend to increase in value.

Source: *Consumer Guide*

22. **CRITICAL THINKING** Mallory is 23 years older than Lisa.

  a. Draw a graph showing Mallory's age as a function of Lisa's age for the first 40 years of Lisa's life.

  b. Find the point on the graph when Mallory is twice as old as Lisa.

23. **WRITING IN MATH** Answer the question that was posed at the beginning of the lesson.

How can real-world situations be modeled using graphs and functions?

Include the following in your answer:

- an explanation of how the graph helps you analyze the situation,
- a summary of what happens during the first 24 hours from the time of a concussion, and
- an explanation of the time in which significant improvement occurs.

24. The graph shows the height of a model rocket shot straight up. How many seconds did it take for the rocket to reach its maximum height?

  (A) 3 (B) 4 (C) 5 (D) 6

25. Andre owns a computer backup service. He charges his customers $2.50 for each backup CD. His expenses include $875 for the CD recording equipment and $0.35 for each blank CD. Which equation could Andre use to calculate his profit p for the recording of n CDs?

  (A) $p = 2.15n - 875$ (B) $p = 2.85 + 875$

  (C) $p = 2.50 - 875.65$ (D) $p = 875 - 2.15n$

Maintain Your Skills

Mixed Review

Identify the hypothesis and conclusion of each statement. *(Lesson 1-7)*

26. You can send e-mail with a computer.

27. The express lane is for shoppers who have 9 or fewer items.

28. Name the property used in each step. *(Lesson 1-6)*

$$\begin{aligned} ab(a + b) &= (ab)a + (ab)b \\ &= a(ab) + (ab)b \\ &= (a \cdot a)b + a(b \cdot b) \\ &= a^2b + ab^2 \end{aligned}$$

Name the property used in each statement. Then find the value of n. *(Lesson 1-4)*

29. $(12 - 9)(4) = n(4)$

30. $7(n) = 0$

31. $n(87) = 87$

Getting Ready for the Next Lesson

32. **PREREQUISITE SKILL** Use the information in the table to construct a bar graph. *(To review **making bar graphs**, see pages 806 and 807.)*

U.S. Commercial Radio Stations by Format, 2000

Format	country	adult contemporary	news/talk	oldies	rock
Number	2249	1557	1426	1135	827

Source: *The World Almanac*

Algebra Activity

A Follow-Up of Lesson 1-8

Investigating Real-World Functions

The table shows the number of students enrolled in elementary and secondary schools in the United States for the given years.

Year	Enrollment (thousands)	Year	Enrollment (thousands)
1900	15,503	1970	45,550
1920	21,578	1980	41,651
1940	25,434	1990	40,543
1960	36,807	1998	46,327

Source: *The World Almanac*

Step 1 On grid paper, draw a vertical and horizontal axis as shown. Make your graph large enough to fill most of the sheet. Label the horizontal axis 0 to 120 and the vertical axis 0 to 60,000.

Step 2 To make graphing easier, let x represent the number of years since 1900. Write the eight ordered pairs using this method. The first will be (0, 15,503).

Step 3 Graph the ordered pairs on your grid paper.

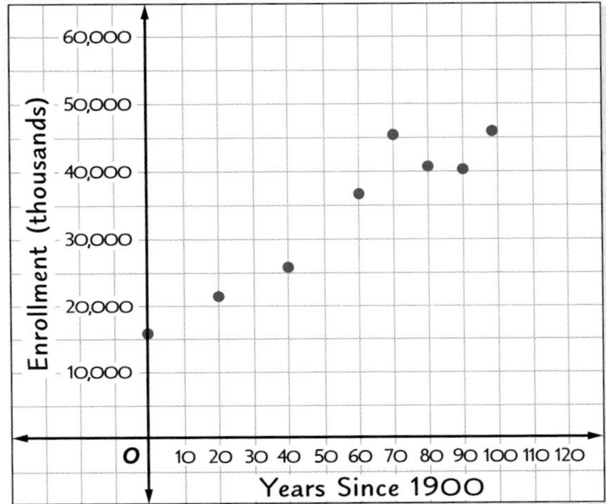

Analyze

1. Use your graph to estimate the number of students in elementary and secondary school in 1910 and in 1975.
2. Use your graph to estimate the number of students in elementary and secondary school in 2020.

Make a Conjecture

3. Describe the methods you used to make your estimates for Exercises 1 and 2.
4. Do you think your prediction for 2020 will be accurate? Explain your reasoning.
5. Graph this set of data, which shows the number of students per computer in U.S. schools. Predict the number of students per computer in 2010. Explain how you made your prediction.

Year	Students per Computer	Year	Students per Computer	Year	Students per Computer	Year	Students per Computer
1984	125	1988	32	1992	18	1996	10
1985	75	1989	25	1993	16	1997	7.8
1986	50	1990	22	1994	14	1998	6.1
1987	37	1991	20	1995	10.5	1999	5.7

Source: *The World Almanac*

1-9 Statistics: Analyzing Data by Using Tables and Graphs

What You'll Learn

- Analyze data given in tables and graphs (bar, line, and circle).
- Determine whether graphs are misleading.

Vocabulary

- bar graph
- data
- circle graph
- line graph

Why are graphs and tables used to display data?

For several weeks after Election Day in 2000, data regarding the presidential vote counts changed on a daily basis. The bar graph at the right illustrates just how close the election was at one point and the importance of each vote in the election. The graph allows you to compare the data visually.

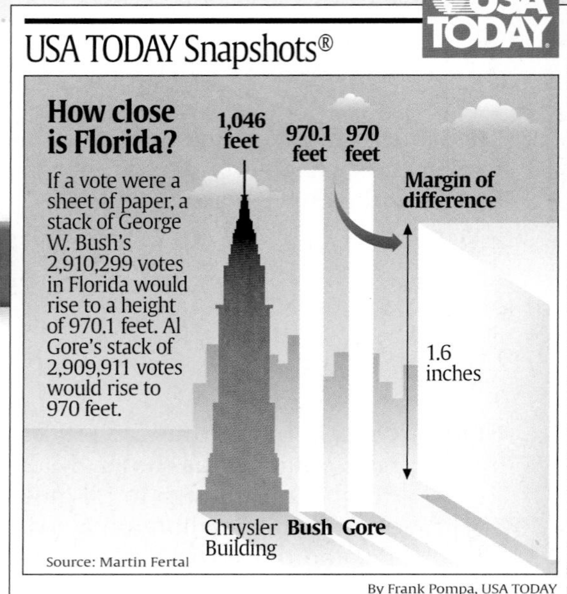

ANALYZE DATA A **bar graph** compares different categories of numerical information, or **data**, by showing each category as a bar whose length is related to the frequency. Bar graphs can also be used to display multiple sets of data in different categories at the same time. Graphs with multiple sets of data always have a key to denote which bars represent each set of data.

Example 1 Analyze a Bar Graph

The table shows the number of men and women participating in NCAA championship sports programs from 1995 to 1999.

NCAA Championship Sports Participation 1995–1999				
Year	'95–'96	'96–'97	'97–'98	'98–'99
Men	206,366	199,375	200,031	207,592
Women	125,268	129,295	133,376	145,832

Source: NCAA

This same data is displayed in a bar graph.

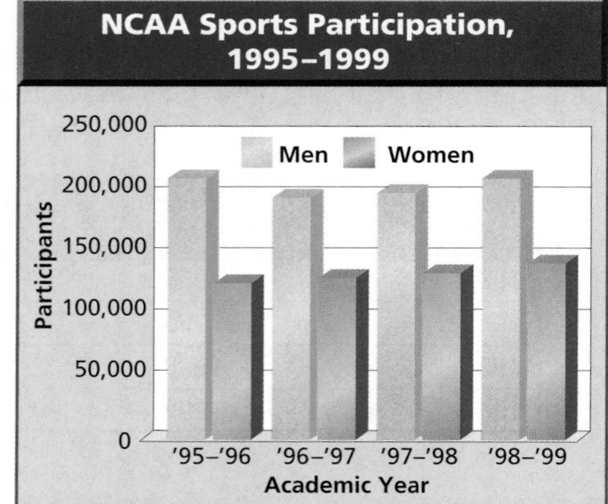

a. Describe the general trend shown in the graph.

The graph shows that the number of men has remained fairly constant while the number of women has been increasing.

> **Study Tip**
>
> *Graphs and Tables*
> Graphs are useful for visualizing data and for estimations. Tables are used when you need precise data for computation.

b. Approximately how many more men than women participated in sports during the 1997–1998 school year?

The bar for the number of men shows about 200,000 and the bar for the women shows about 130,000. So, there were approximately 200,000–130,000 or 70,000 more men than women participating in the 1997–1998 school year.

c. What was the total participation among men and women in the 1998–1999 academic year?

Since the table shows the exact numbers, use the data in it.

Number of men	plus	number of women	equals	total participation.
207,592	+	145,832	=	353,424

There was a total of 353,424 men and women participating in sports in the 1998–1999 academic year.

Study Tip

Reading Math
In everyday life, circle graphs are sometimes called *pie graphs* or *pie charts*.

Another type of graph used to display data is a circle graph. A **circle graph** compares parts of a set of data as a percent of the whole set. The percents in a circle graph should always have a sum of 100%.

Example 2 Analyze a Circle Graph

A recent survey asked drivers in several cities across the United States if traffic in their area had gotten better, worse, or had not changed in the past five years. The results of the survey are displayed in the circle graph.

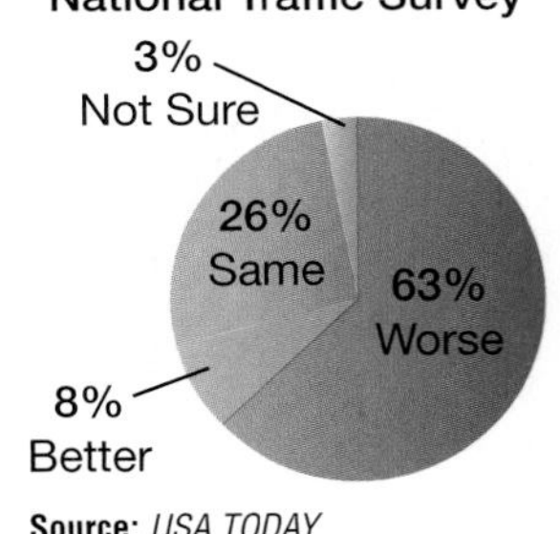

Source: *USA TODAY*

a. If 4500 people were surveyed, how many felt that traffic had improved in their area?

The section of the graph representing people who said traffic is better is 8% of the circle, so find 8% of 4500.

8%	of	4500	equals	360.
0.08	×	4500	=	360

360 people said that traffic was better.

b. If a city with a population of 647,000 is representative of those surveyed, how many people could be expected to think that traffic conditions are worse?

63% of those surveyed said that traffic is worse, so find 63% of 647,000.

$0.63 \times 647{,}000 = 407{,}610$

Thus, 407,610 people in the city could be expected to say that traffic conditions are worse.

A third type of graph used to display data is a line graph. **Line graphs** are useful when showing how a set of data changes over time. They can also be helpful when making predictions.

Example 3 Analyze a Line Graph

EDUCATION **Refer to the line graph below.**

a. Estimate the change in enrollment between 1995 and 1999.

The enrollment for 1995 is about 14.25 million, and the enrollment for 1999 is about 14.9 million. So, the change in enrollment is 14.9 − 14.25 or 0.65 million.

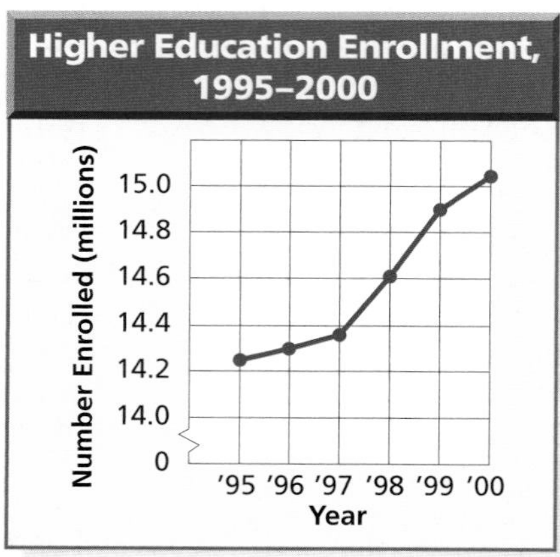

Source: U.S. National Center for Educational Statistics

b. If the rate of growth between 1998 and 1999 continues, predict the number of people who will be enrolled in higher education in the year 2005.

Based on the graph, the increase in enrollment from 1998 to 1999 is 0.3 million. So, the enrollment should increase by 0.3 million per year.

$14.9 + 0.3(6) = 14.9 + 1.8$ Multiply the annual increase, 0.3, by the number of years, 6.

$= 16.7$ Enrollment in 2005 should be about 16.7 million.

Career Choices

Professor

A college professor may teach by lecturing to several hundred students at a time or by supervising students in small groups in a laboratory. Often they also do their own research to expand knowledge in their field.

Online Research For information about a career as a professor, visit: www.algebra1.com/careers

Concept Summary — Statistical Graphs

Type of Graph	bar graph	circle graph	line graph
When to Use	to compare different categories of data	to show data as parts of a whole set of data	to show the change in data over time

MISLEADING GRAPHS Graphs are very useful for displaying data. However, graphs that have been constructed incorrectly can be confusing and can lead to false assumptions. Many times these types of graphs are mislabeled, incorrect data is compared, or the graphs are constructed to make one set of data appear greater than another set. Here are some common ways that a graph may be misleading.

- Numbers are omitted on an axis, but no break is shown.
- The tick marks on an axis are not the same distance apart or do not have the same-sized intervals.
- The percents on a circle graph do not have a sum of 100.

Example 4 Misleading Graphs

AUTOMOBILES **The graph shows the number of sport-utility vehicle (SUV) sales in the United States from 1990 to 1999. Explain how the graph misrepresents the data.**

The vertical axis scale begins at 1 million. This causes the appearance of no vehicles sold in 1990 and 1991, and very few vehicles sold through 1994.

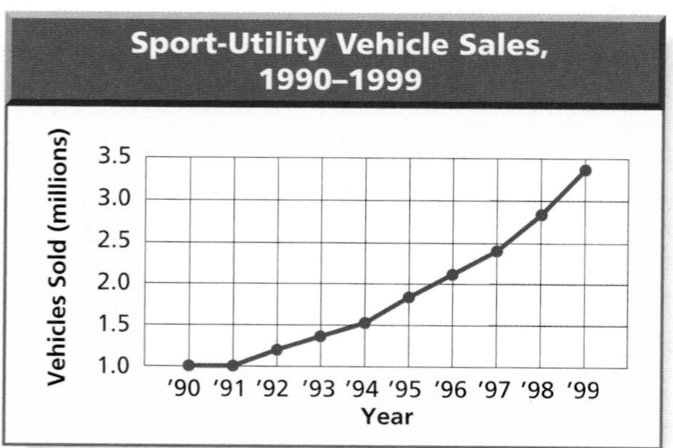

Source: *The World Almanac*

Check for Understanding

Concept Check

1. **Explain** the appropriate use of each type of graph.
   - circle graph
   - bar graph
   - line graph
2. **OPEN ENDED** Find a real-world example of a graph in a newspaper or magazine. Write a description of what the graph displays.
3. **Describe** ways in which a circle graph could be drawn so that it is misleading.

Guided Practice

SPORTS For Exercises 4 and 5, use the following information. There are 321 NCAA Division I schools. The graph at the right shows the sports that are offered at the most Division I schools.

4. How many more schools participate in basketball than in golf?
5. What sport is offered at the fewest schools?

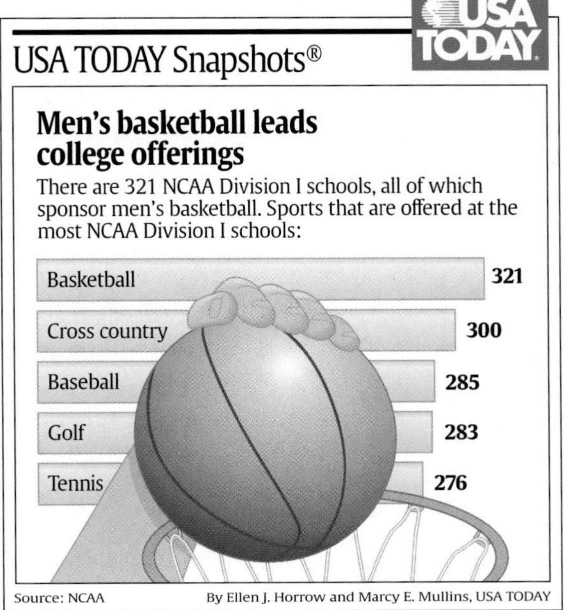
USA TODAY
USA TODAY Snapshots®
Men's basketball leads college offerings
There are 321 NCAA Division I schools, all of which sponsor men's basketball. Sports that are offered at the most NCAA Division I schools:
Basketball
321
Cross country
300
Baseball
285
Golf
283
Tennis
276
Source: NCAA
By Ellen J. Horrow and Marcy E. Mullins, USA TODAY

EDUCATION For Exercises 6–9, use the table that shows the number of foreign students as a percent of the total college enrollment in the United States.

Country of Origin	Total Student Enrollment (%)
Australia	0.02
Canada	0.15
France	0.04
Germany	0.06
Italy	0.22
Spain	0.03
United Kingdom	0.05

Source: *Statistical Abstract of the United States*

6. There were about 14.9 million students enrolled in colleges in 1999. How many of these students were from Germany?
7. How many more students were from Canada than from the United Kingdom in 1999?
8. Would it be appropriate to display this data in a circle graph? Explain.
9. Would a bar or a line graph be more appropriate to display these data? Explain.

HOME ENTERTAINMENT For Exercises 10 and 11, refer to the graph.

10. Describe why the graph is misleading.
11. What should be done so that the graph displays the data more accurately?

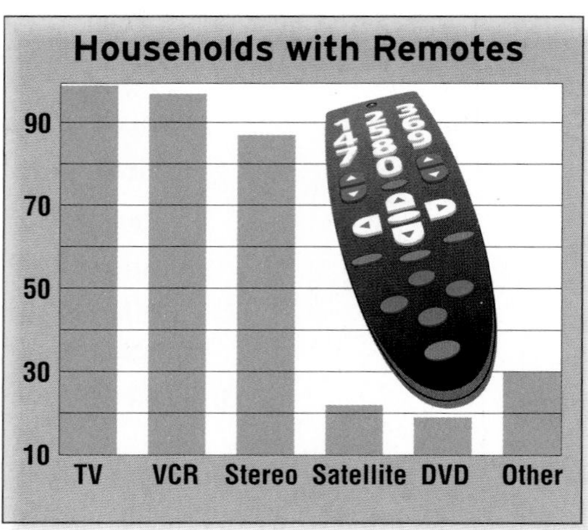

Practice and Apply

Homework Help

For Exercises	See Examples
12, 13	1
14, 15	2
16	3, 4
17	2–4

Extra Practice

See page 822.

VIDEOGRAPHY For Exercises 12 and 13, use the table that shows the average cost of preparing one hour of 35-millimeter film versus one hour of digital video.

12. What is the total cost of using 35-millimeter film?
13. Estimate how many times as great the cost of using 35-millimeter film is as using digital video.

35 mm, editing video	
Film stock	$3110.40
Processing	621.00
Prep for telecine	60.00
Telecine	1000.00
Tape stock	73.20
Digital, editing on video	
Tape stock (original)	$10.00
Tape stock (back up)	10.00

When People Buy Books

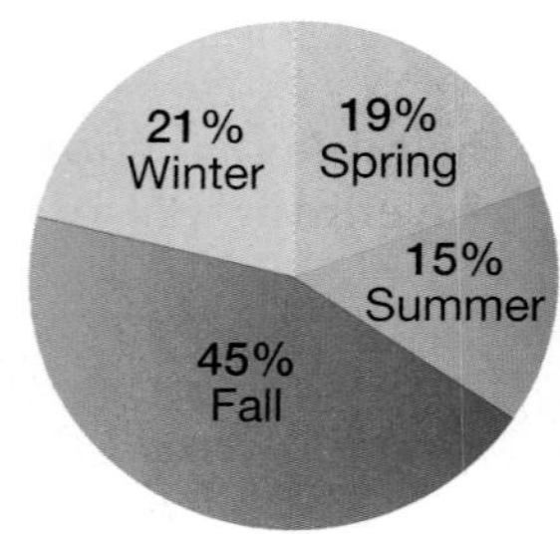

Source: *USA TODAY*

BOOKS For Exercises 14 and 15, use the graph that shows the time of year people prefer to buy books.

14. Suppose the total number of books purchased for the year was 25 million. Estimate the number of books purchased in the spring.
15. Suppose the manager of a bookstore has determined that she sells about 15,000 books a year. Approximately how many books should she expect to sell during the summer?

16. **ENTERTAINMENT** The line graph shows the number of cable television systems in the United States from 1995 to 2000. Explain how the graph misrepresents the data.

Cable Television Systems, 1995–2000

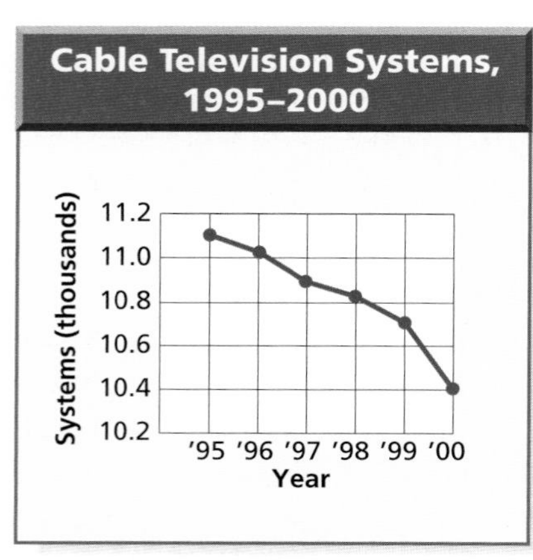

Data Source: *The World Almanac*

17. **FOOD** Oatmeal can be found in 80% of the homes in the United States. The circle graph shows favorite oatmeal toppings. Is the graph misleading? If so, explain why and tell how the graph can be fixed so that it is not misleading.

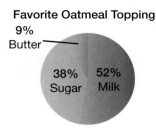

Favorite Oatmeal Topping

9% Butter
38% Sugar
52% Milk

Data Source: NPD Group for Quaker Oats

A graph of the number of people over 65 in the U.S. for the years since 1900 will help you predict trends. Visit www.algebra1.com/webquest to continue work on your WebQuest project.

18. **CRITICAL THINKING** The table shows the percent of United States households owning a color television for the years 1980 to 2000.

a. Display the data in a line graph that shows little increase in ownership.

b. Draw a line graph that shows a rapid increase in the number of households owning a color television.

c. Are either of your graphs misleading? Explain.

Households with Color Televisions

Year	Percent
1980	83
1985	91
1990	98
1995	99
2000	99

Source: *The World Almanac*

19. WRITING IN MATH Answer the question that was posed at the beginning of the lesson.

Why are graphs and tables used to display data?

Include the following in your answer:

- a description of how to use graphs to make predictions, and
- an explanation of how to analyze a graph to determine whether the graph is misleading.

20. According to the graph, the greatest increase in temperature occurred between which two days?

(A) 1 and 2 (B) 6 and 7
(C) 2 and 3 (D) 5 and 6

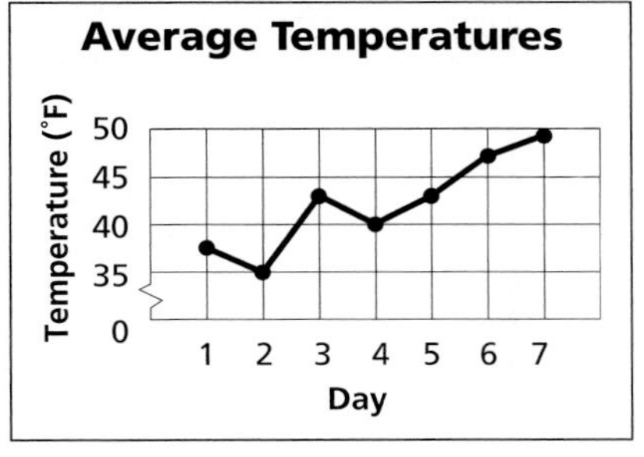

21. A graph that is primarily used to show the change in data over time is called a

(A) circle graph. (B) bar graph.
(C) line graph. (D) data graph.

Maintain Your Skills

Mixed Review

22. **PHYSICAL FITNESS** Pedro likes to exercise regularly. On Mondays, he walks two miles, runs three miles, sprints one-half of a mile, and then walks for another mile. Sketch a graph that represents Mitchell's heart rate during his Monday workouts. *(Lesson 1-8)*

Find a counterexample for each statement. *(Lesson 1-7)*

23. If $x \leq 12$, then $4x - 5 \leq 42$.

24. If $x > 1$, then $x < \frac{1}{x}$.

25. If the perimeter of a rectangle is 16 inches, then each side is 4 inches long.

Simplify each expression. *(Lesson 1-6)*

26. $7a + 5b + 3b + 3a$

27. $4x^2 + 9x + 2x^2 + x$

28. $\frac{1}{2}n + \frac{2}{3}m + \frac{1}{2}m + \frac{1}{3}n$

Spreadsheet Investigation

A Follow-Up of Lesson 1-9

Statistical Graphs

You can use a computer spreadsheet program to display data in different ways. The data is entered into a table and then displayed in your chosen type of graph.

Example

Use a spreadsheet to make a line graph of the data on sports equipment sales.

In-line Skating and Wheel Sports Equipment Sales								
Year	1990	1992	1993	1994	1995	1996	1997	1998
Sales (million $)	150	268	377	545	646	590	562	515

Source: National Sporting Goods Association

Step 1 Enter the data in a spreadsheet. Use Column A for the years and Column B for the sales.

Step 2 Select the data to be included in your graph. Then use the graph tool to create the graph.

The spreadsheet will allow you to change the appearance of the graph by adding titles and axis labels, adjusting the scales on the axes, changing colors, and so on.

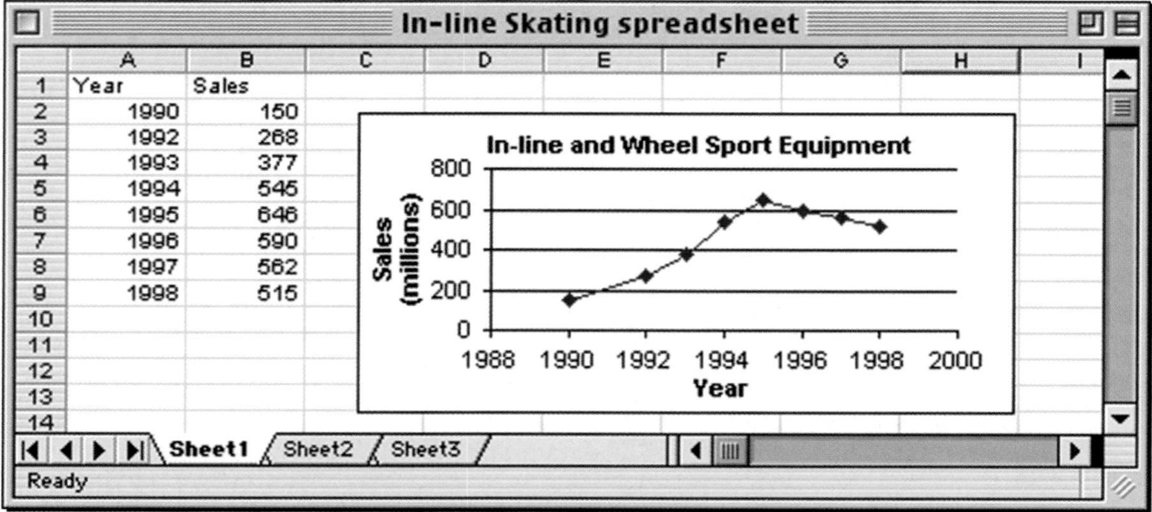

Exercises

For Exercises 1–3, use the data on snowmobile sales in the table below.

Snowmobile Sales								
Year	1990	1992	1993	1994	1995	1996	1997	1998
Sales (million $)	322	391	515	715	910	974	975	957

Source: National Sporting Goods Association

1. Use a spreadsheet program to create a line graph of the data.
2. Use a spreadsheet program to create a bar graph of the data.
3. Adjust the scales on each of the graphs that you created. Is it possible to create a misleading graph using a spreadsheet program? Explain.

Chapter 1

Study Guide and Review

Vocabulary and Concept Check

additive identity (p. 21)
algebraic expression (p. 6)
Associative Property (p. 32)
bar graph (p. 50)
base (p. 7)
circle graph (p. 51)
Closure Property (p. 25)
coefficient (p. 29)
Commutative Property (p. 32)
conclusion (p. 37)
conditional statement (p. 37)
coordinate system (p. 43)
coordinates (p. 43)
counterexample (p. 38)
data (p. 50)
deductive reasoning (p. 38)
dependent quantity (p. 44)
dependent variable (p. 44)
Distributive Property (p. 26)
domain (p. 45)
element (p. 16)
equation (p. 16)
equivalent expressions (p. 29)
exponent (p. 7)
factors (p. 6)
function (p. 43)
horizontal axis (p. 43)
hypothesis (p. 37)
if-then statement (p. 37)
independent quantity (p. 44)
independent variable (p. 44)
inequality (p. 17)
like terms (p. 28)
line graph (p. 51)
multiplicative identity (p. 21)
Multiplicative Inverse Property (p. 22)
multiplicative inverses (p. 21)
Multiplicative Property of Zero (p. 21)
open sentence (p. 16)
order of operations (p. 11)
ordered pair (p. 43)
origin (p. 43)
power (p. 7)
product (p. 6)
range (p. 45)
reciprocal (p. 21)
Reflexive Property of Equality (p. 22)
relation (p. 45)
replacement set (p. 16)
set (p. 16)
simplest form (p. 29)
solution (p. 16)
solution set (p. 16)
solving an open sentence (p. 16)
Substitution Property of Equality (p. 22)
Symmetric Property of Equality (p. 22)
term (p. 28)
Transitive Property of Equality (p. 22)
variables (p. 6)
vertical axis (p. 43)
x-axis (p. 43)
x-coordinate (p. 43)
y-axis (p. 43)
y-coordinate (p. 43)

Choose the letter of the property that best matches each statement.

1. For any number a, $a + 0 = 0 + a = a$.
2. For any number a, $a \cdot 1 = 1 \cdot a = a$.
3. For any number, a, $a \cdot 0 = 0 \cdot a = 0$.
4. For any nonzero number a, there is exactly one number $\frac{1}{a}$ such that $\frac{1}{a} \cdot a = a \cdot \frac{1}{a} = 1$.
5. For any number a, $a = a$.
6. For any numbers a and b, if $a = b$, then $b = a$.
7. For any numbers a and b, if $a = b$, then a may be replaced by b in any expression.
8. For any numbers a, b, and c, if $a = b$ and $b = c$, then $a = c$.
9. For any numbers a, b, and c, $a(b + c) = ab + ac$.
10. For any numbers a, b, and c, $a + (b + c) = (a + b) + c$.

a. Additive Identity Property
b. Distributive Property
c. Commutative Property
d. Associative Property
e. Multiplicative Identity Property
f. Multiplicative Inverse Property
g. Multiplicative Property of Zero
h. Reflexive Property
i. Substitution Property
j. Symmetric Property
k. Transitive Property

Lesson-by-Lesson Review

1-1 Variables and Expressions

See pages 6–9.

Concept Summary

- Variables are used to represent unspecified numbers or values.
- An algebraic expression contains letters and variables with an arithmetic operation.

www.algebra1.com/vocabulary_review

Examples

1 **Write an algebraic expression for *the sum of twice a number x and fifteen.***

twice a number x	sum of	fifteen	
$2x$	$+$	15	The algebraic expression is $2x + 15$.

2 **Write a verbal expression for $4x^2 - 13$.**

Four times a number x squared minus thirteen.

Exercises **Write an algebraic expression for each verbal expression.**
See Examples 1 and 2 on pages 6 and 7.

11. a number x to the fifth power

12. five times a number x squared

13. the sum of a number x and twenty-one

14. the difference of twice a number x and 8

Evaluate each expression. *See Example 3 on page 7.*

15. 3^3

16. 2^5

17. 5^4

Write a verbal expression for each algebraic expression. *See Example 4 on page 7.*

18. $2p^2$

19. $3m^5$

20. $\frac{1}{2} + 2$

1-2 Order of Operations

See pages 11–15.

Concept Summary

- Expressions must be simplified using the order of operations.

  Step 1 Evaluate expressions inside grouping symbols.

  Step 2 Evaluate all powers.

  Step 3 Do all multiplications and/or divisions from left to right.

  Step 4 Do all additions and/or subtractions from left to right.

Example

Evaluate $x^2 - (y + 2)$ if $x = 4$ and $y = 3$.

$x^2 - (y + 2) = 4^2 - (3 + 2)$ Replace x with 4 and y with 3.

$= 4^2 - 5$ Add 3 and 2.

$= 16 - 5$ Evaluate power.

$= 11$ Subtract 5 from 16.

Exercises **Evaluate each expression.** *See Examples 1–3 on pages 11 and 12.*

21. $3 + 2 \cdot 4$

22. $\frac{(10 - 6)}{8}$

23. $18 - 4^2 + 7$

24. $8(2 + 5) - 6$

25. $4(11 + 7) - 9 \cdot 8$

26. $288 \div [3(9 + 3)]$

27. $16 \div 2 \cdot 5 \cdot 3 \div 6$

28. $6(4^3 + 2^2)$

29. $(3 \cdot 1)^3 - \frac{(4 + 6)}{(5 \cdot 2)}$

Evaluate each expression if $x = 3$, $t = 4$, and $y = 2$. *See Example 4 on page 12.*

30. $t^2 + 3y$

31. xty^3

32. $\frac{ty}{x}$

33. $x + t^2 + y^2$

34. $3ty - x^2$

35. $8(x - y)^2 + 2t$

1-3 Open Sentences

See pages 16–20.

Concept Summary

- Open sentences are solved by replacing the variables in an equation with numerical values.
- Inequalities like $x + 2 \geq 7$ are solved the same way that equations are solved.

Example **Solve $5^2 - 3 = y$.**

$5^2 - 3 = y$ Original equation

$25 - 3 = y$ Evaluate the power.

$22 = y$ Subtract 3 from 25.

The solution is 22.

Exercises **Solve each equation.** *See Example 2 on page 17.*

36. $x = 22 - 13$ **37.** $y = 4 + 3^2$ **38.** $m = \frac{64 + 4}{17}$

39. $x = \frac{21 - 3}{12 - 3}$ **40.** $a = \frac{14 + 28}{4 + 3}$ **41.** $n = \frac{96 \div 6}{8 \div 2}$

42. $b = \frac{7(4 \cdot 3)}{18 \div 3}$ **43.** $\frac{6(7) - 2(3)}{4^2 - 6(2)}$ **44.** $y = 5[2(4) - 1^3]$

Find the solution set for each inequality if the replacement set is {4, 5, 6, 7, 8}. *See Example 3 on page 17.*

45. $x + 2 > 7$ **46.** $10 - x < 7$ **47.** $2x + 5 \geq 15$

1-4 Identity and Equality Properties

See pages 21–25.

Concept Summary

- Adding zero to a quantity or multiplying a quantity by one does not change the quantity.
- Using the Reflexive, Symmetric, Transitive, and Substitution Properties along with the order of operations helps in simplifying expressions.

Example **Evaluate $36 + 7 \cdot 1 + 5(2 - 2)$. Name the property used in each step.**

$36 + 7 \cdot 1 + 5(2 - 2) = 36 + 7 \cdot 1 + 5(0)$ Substitution

$= 36 + 7 + 5(0)$ Multiplicative Identity

$= 36 + 7$ Multiplicative Prop. of Zero

$= 43$ Substitution

Exercises **Evaluate each expression. Name the property used in each step.** *See Example 2 on page 23.*

48. $2[3 \div (19 - 4^2)]$ **49.** $\frac{1}{2} \cdot 2 + 2[2 \cdot 3 - 1]$ **50.** $4^2 - 2^2 - (4 - 2)$

51. $1.2 - 0.05 + 2^3$ **52.** $(7 - 2)(5) - 5^2$ **53.** $3(4 \div 4)^2 - \frac{1}{4}(8)$

1-5 The Distributive Property

See pages 26–31.

Concept Summary

- For any numbers a, b, and c, $a(b + c) = ab + ac$ and $(b + c)a = ba + ca$.
- For any numbers a, b, and c, $a(b - c) = ab - ac$ and $(b - c)a = ba - ca$.

Examples

1 Rewrite $5(t + 3)$ using the Distributive Property. Then simplify.

$5(t + 3) = 5(t) + 5(3)$ Distributive Property

$= 5t + 15$ Multiply.

2 Simplify $2x^2 + 4x^2 + 7x$.

$2x^2 + 4x^2 + 7x = (2 + 4)x^2 + 7x$ Distributive Property

$= 6x^2 + 7x$ Substitution

Exercises **Rewrite each product using the Distributive Property. Then simplify.** *See Examples 1 and 2 on page 27.*

54. $2(4 + 7)$ **55.** $8(15 - 6)$ **56.** $4(x + 1)$

57. $3\left(\frac{1}{3} - p\right)$ **58.** $6(a + b)$ **59.** $8(3x - 7y)$

Simplify each expression. If not possible, write *simplified*. *See Example 6 on page 29.*

60. $4a + 9a$ **61.** $4np + 7mp$ **62.** $3w - w + 4v - 3v$

63. $3m + 5m + 12n - 4n$ **64.** $2p(1 + 16r)$ **65.** $9y + 3y - 5x$

1-6 Commutative and Associative Properties

See pages 32–36.

Concept Summary

- For any numbers a and b, $a + b = b + a$ and $a \cdot b = b \cdot a$.
- For any numbers a, b and c, $(a + b) + c = a + (b + c)$ and $(ab)c = a(bc)$.

Example

Simplify $3x + 7xy + 9x$.

$3x + 7xy + 9x = 3x + 9x + 7xy$ Commutative (+)

$= (3 + 9)x + 7xy$ Distributive Property

$= 12x + 7xy$ Substitution

Exercises **Simplify each expression.** *See Example 3 on page 33.*

66. $3x + 4y + 2x$ **67.** $7w^2 + w + 2w^2$ **68.** $3\frac{1}{2}m + \frac{1}{2}m + n$

69. $6a + 5b + 2c + 8b$ **70.** $3(2 + 3x) + 21x$ **71.** $6(2n - 4) + 5n$

Write an algebraic expression for each verbal expression. Then simplify, indicating the properties used. *See Example 4 on page 34.*

72. five times the sum of x and y decreased by $2x$

73. twice the product of p and q increased by the product of p and q

74. six times a plus the sum of eight times b and twice a

75. three times the square of x plus the sum of x squared and seven times x

1-7 Logical Reasoning

See pages 37–42.

Concept Summary

- Conditional statements can be written in the form *If A, then B*, where A is the hypothesis and B is the conclusion.
- One counterexample can be used to show that a statement is false.

Example **Identify the hypothesis and conclusion of the statement *The trumpet player must audition to be in the band.* Then write the statement in if-then form.**

Hypothesis: a person is a trumpet player

Conclusion: the person must audition to be in the band

If a person is a trumpet player, then the person must audition to be in the band.

Exercises **Identify the hypothesis and conclusion of each statement. Then, write each statement in if-then form.** *See Example 2 on page 38.*

76. School begins at 7:30 A.M.

77. Triangles have three sides.

Find a counterexample for each statement. *See Example 4 on page 39.*

78. If $x > y$, then $2x > 3y$.

79. If $a > b$ and $a > c$, then $b > c$.

1-8 Graphs and Functions

See pages 43–48.

Concept Summary

- Graphs can be used to represent a function and to visualize data.

Example **A computer printer can print 12 pages of text per minute.**

a. Make a table showing the number of pages printed in 1 to 5 minutes.

Time (min)	1	2	3	4	5
Pages	12	24	36	48	60

b. Sketch a graph that shows the relationship between time and the number of pages printed.

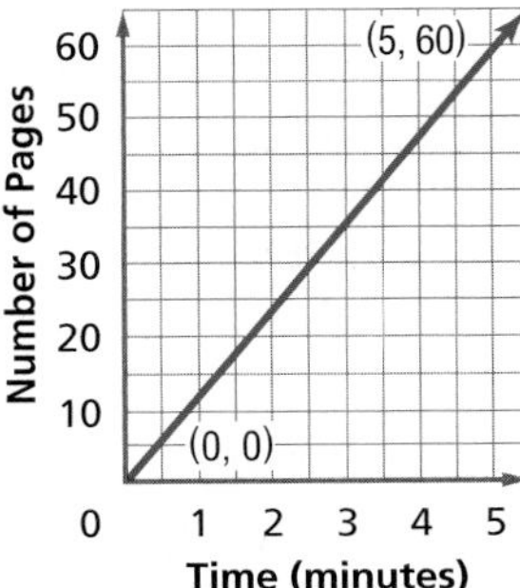

Exercises

80. Identify the graph that represents the altitude of an airplane taking off, flying for a while, then landing. *See Example 3 on page 44.*

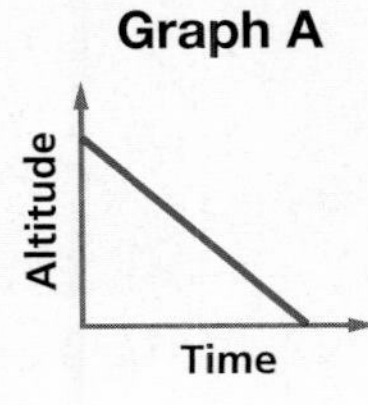

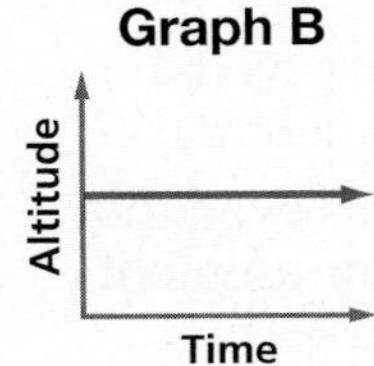

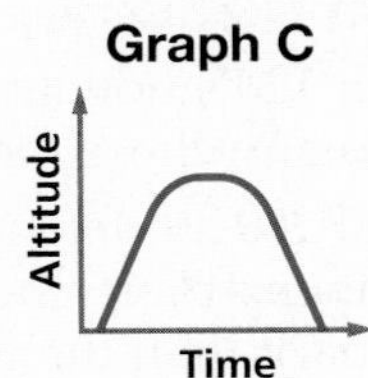

• Extra Practice, see pages 820–822.
• Mixed Problem Solving, see page 853.

81. Sketch a reasonable graph that represents the amount of helium in a balloon if it is filled until it bursts. *See Examples 3–5 on pages 44 and 45.*

For Exercises 82 and 83, use the following information.
The planet Mars takes longer to orbit the sun than does Earth. One year on Earth is about 0.54 year on Mars. *See Examples 4 and 5 on page 45.*

82. Construct a table showing the relationship between years on Earth and years on Mars.

83. Draw a graph showing the relationship between Earth years and Mars years.

1-9 Statistics: Analyzing Data by Using Tables and Graphs

See pages 50–55.

Concept Summary

- Bar graphs are used to compare different categories of data.
- Circle graphs are used to show data as parts of a whole set of data.
- Line graphs are used to show the change in data over time.

Example

The bar graph shows ways people communicate with their friends.

a. About what percent of those surveyed chose e-mail as their favorite way to talk to friends?

The bar for e-mail is about halfway between 30% and 40%. Thus, about 35% favor e-mail.

b. What is the difference in the percent of people favoring letters and those favoring the telephone?

The bar for those favoring the telephone is at 60%, and the bar for letters is about 20%. So, the difference is $60 - 20$ or 40%.

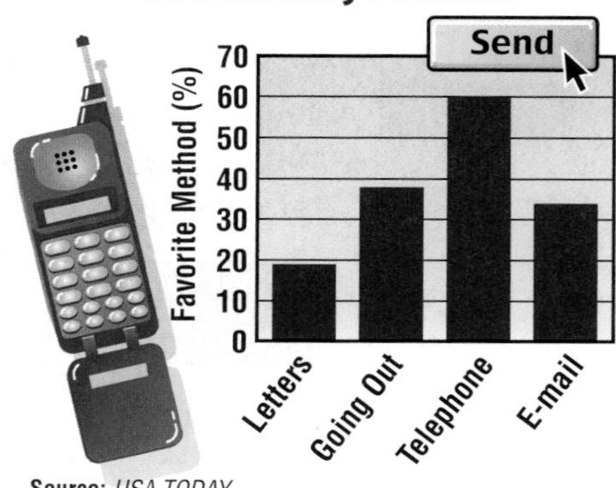

Exercises

CLASS TRIP For Exercises 84 and 85, use the circle graph and the following information.
A survey of the ninth grade class asked members to indicate their choice of locations for their class trip. The results of the survey are displayed in the circle graph. *See Example 2 on page 51.*

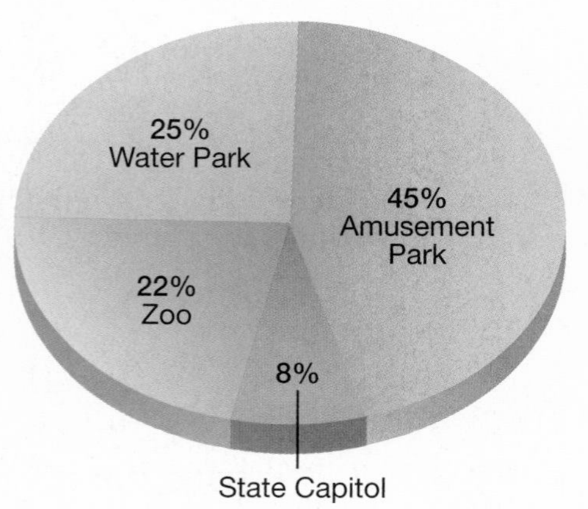

84. If 120 students were surveyed, how many chose the amusement park?

85. If 180 students were surveyed, how many more chose the amusement park than the water park?

Chapter 1 Practice Test

Vocabulary and Concepts

Choose the letter of the property that best matches each statement.

1. For any number a, $a = a$.
2. For any numbers a and b, if $a = b$, then b may be replaced by a in any expression or equation.
3. For any numbers a, b, and c, if $a = b$ and $b = c$, then $a = c$.

a. Substitution Property of Equality
b. Symmetric Property of Equality
c. Transitive Property of Equality
d. Reflexive Property of Equality

Skills and Applications

Write an algebraic expression for each verbal expression.

4. the sum of a number x and 13
5. the difference of 7 and a number x squared

Simplify each expression.

6. $5(9 + 3) - 3 \cdot 4$
7. $12 \cdot 6 \div 3 \cdot 2 \div 8$

Evaluate each expression if $a = 2$, $b = 5$, $c = 3$, and $d = 1$.

8. $a^2b + c$
9. $(cd)^3$
10. $(a + d)c$

Solve each equation.

11. $y = (4.5 + 0.8) - 3.2$
12. $4^2 - 3(4 - 2) = x$
13. $\frac{2^3 - 1^3}{2 + 1} = n$

Evaluate each expression. Name the property used in each step.

14. $3^2 - 2 + (2 - 2)$
15. $(2 \cdot 2 - 3) + 2^2 + 3^2$

Rewrite each expression in simplest form.

16. $2m + 3m$
17. $4x + 2y - 2x + y$
18. $3(2a + b) - 5a + 4b$

Find a counterexample for each conditional statement.

19. If you run fifteen minutes today, then you will be able to run a marathon tomorrow.
20. If $x \leq 6$, then $2x - 3 < 9$.

Sketch a reasonable graph for each situation.

21. A basketball is shot from the free throw line and falls through the net.
22. A nickel is dropped on a stack of pennies and bounces off.

ICE CREAM **For Exercises 23 and 24, use the following information.** A school survey at West High School determined the favorite flavors of ice cream are chocolate, vanilla, butter pecan, and bubble gum. The results of the survey are displayed in the circle graph.

23. If 200 students were surveyed, how many more chose chocolate than vanilla?
24. What was the total percent of students who chose either chocolate or vanilla?
25. **STANDARDIZED TEST PRACTICE** Which number is a counterexample for the statement below?

If a is a prime number, then a is odd.

Ⓐ 5 Ⓑ 4 Ⓒ 3 Ⓓ 2

Chapter 1 Standardized Test Practice

Part 1 Multiple Choice

Record your answers on the answer sheet provided by your teacher or on a sheet of paper.

1. The Maple Grove Warehouse measures 800 feet by 200 feet. If $\frac{3}{4}$ of the floor space is covered, how many square feet are *not* covered? (Prerequisite Skill)

(A) 4000 (B) 40,000
(C) 120,000 (D) 160,000

2. The radius of a circular flower garden is 4 meters. How many meters of edging will be needed to surround the garden? (Prerequisite Skill)

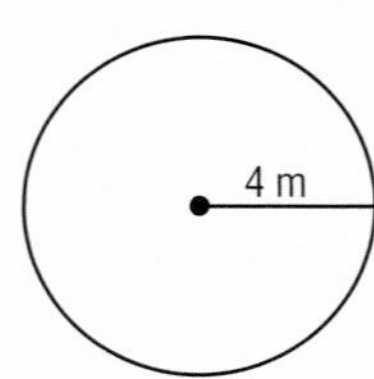

(A) 7.14 m (B) 12.56 m
(C) 25.12 m (D) 20.24 m

3. The Johnson family spends about \$80 per week on groceries. Approximately how much do they spend on groceries per year? (Prerequisite Skill)

(A) \$400 (B) \$4000
(C) \$8000 (D) \$40,000

4. Daria is making 12 party favors for her sister's birthday party. She has 50 stickers, and she wants to use as many of them as possible. If she puts the same number of stickers in each bag, how many stickers will she have left over? (Prerequisite Skill)

(A) 2 (B) 4 (C) 6 (D) 8

The Princeton Review Test-Taking Tip

Questions 1, 3, and 8 Read each question carefully. Be sure you understand what the question asks. Look for words like *not*, *estimate*, and *approximately*.

5. An auto repair shop charges \$36 per hour, plus the cost of replaced parts. Which of the following expressions can be used to calculate the total cost of repairing a car, where h represents the number of hours of work and the cost of replaced parts is \$85? (Lesson 1-1)

(A) $36 + h + 85$ (B) $(85 \times h) + 36$
(C) $36 + 85 \times h$ (D) $(36 \times h) + 85$

6. Which expression is equivalent to $3(2x + 3) + 2(x + 1)$? (Lessons 1-5 and 1-6)

(A) $7x + 8$ (B) $8x + 4$
(C) $8x + 9$ (D) $8x + 11$

7. Find a counterexample for the following statement. (Lesson 1-7)
If x is a positive integer, then x^2 is divisible by 2.

(A) 2 (B) 3 (C) 4 (D) 6

8. The circle graph shows the regions of birth of foreign-born persons in the United States in 2000. According to the graph, which statement is *not* true? (Lesson 1-9)

Regions of Birth

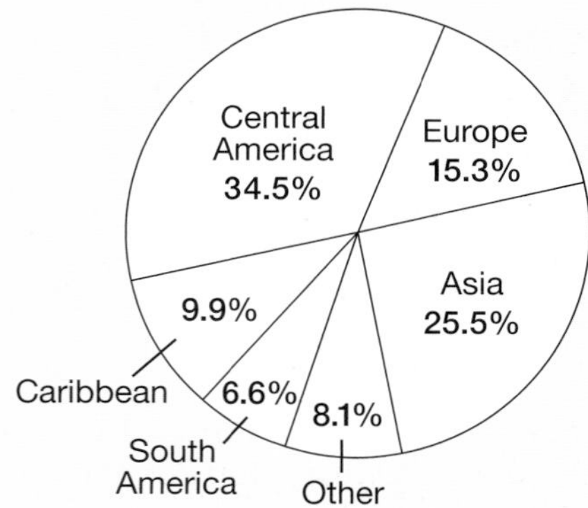

(A) More than $\frac{1}{3}$ of the foreign-born population is from Central America.

(B) More foreign-born people are from Asia than Central America.

(C) About half of the foreign-born population comes from Central America or Europe.

(D) About half of the foreign-born population comes from Central America, South America, or the Caribbean.

Part 2 Short Response/Grid In

Record your answers on the answer sheet provided by your teacher or on a sheet of paper.

9. There are 32 students in the class. Five eighths of the students are girls. How many boys are in the class? (Prerequisite Skill)

10. Tonya bought two paperback books. One book cost \$8.99 and the other \$13.99. Sales tax on her purchase was 6%. How much change should she receive if she gives the clerk \$25? (Prerequisite Skill)

11. According to the bar graph of the home runs hit by two baseball players, in which year was the difference between the numbers of home runs hit by the two players the least? (Prerequisite Skill)

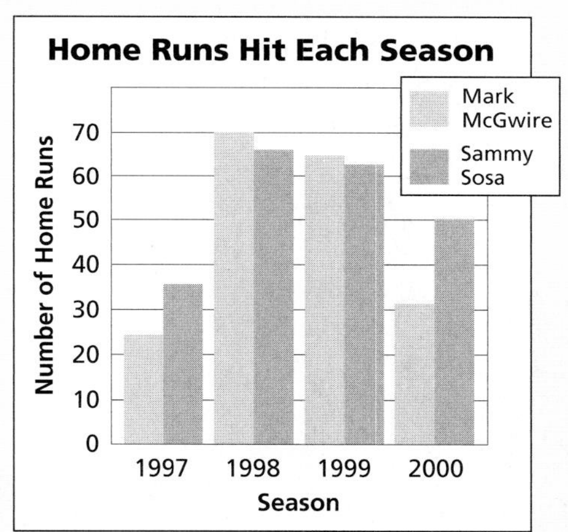

Part 3 Quantitative Comparison

Compare the quantity in Column A and the quantity in Column B. Then determine whether:

Ⓐ **the quantity in Column A is greater,**

Ⓑ **the quantity in Column B is greater,**

Ⓒ **the two quantities are equal, or**

Ⓓ **the relationship cannot be determined from the information given.**

	Column A	Column B	
12.	15% of 80	25% of 50	(Prerequisite Skill)
13. 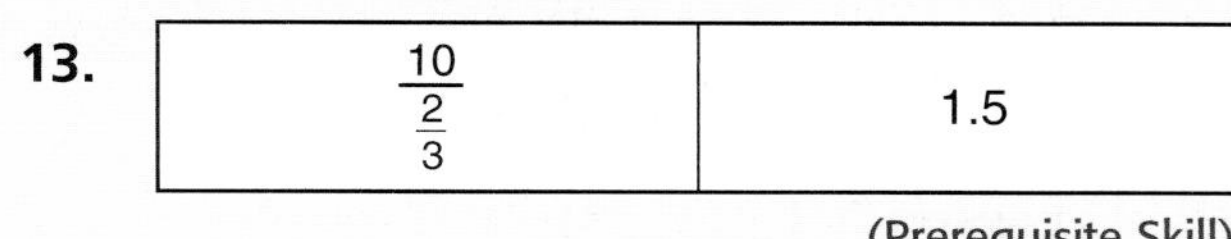	$\frac{10}{\frac{2}{3}}$	1.5	(Prerequisite Skill)
14. 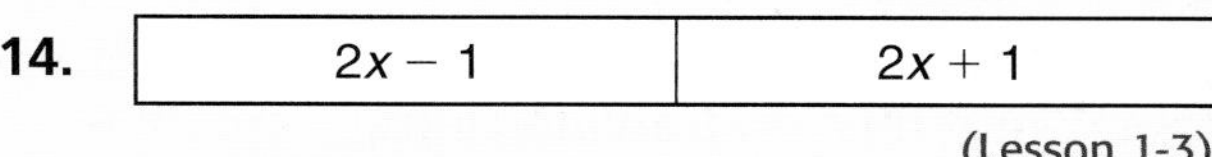	$2x - 1$	$2x + 1$	(Lesson 1-3)
15.	$\frac{1}{4}(a + b)c$	$\frac{ac + bc}{4}$	(Lesson 1-5)
16.	$(26 \times 39) + (39 \times 13)$	$(39)^2$	(Lesson 1-5)

Part 4 Open Ended

Record your answers on a sheet of paper. Show your work.

17. Workers are draining water from a pond. They have an old pump and a new pump. The graphs below show how each pump drains water. (Lesson 1-8)

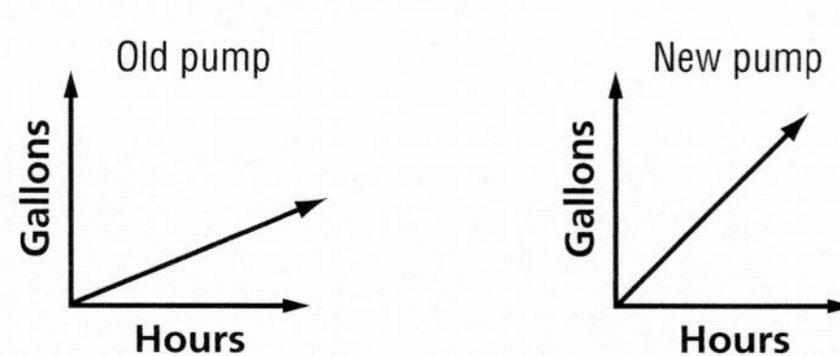

a. Describe how the old and new pumps are different in the amount of water they pump per hour.

b. Draw a graph that shows the gallons pumped per hour by both pumps at the same time.

c. Explain what the graph below tells about how the water is pumped out.

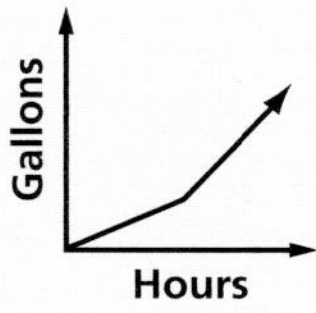

Chapter

2 Real Numbers

What You'll Learn

- **Lesson 2-1** Classify and graph rational numbers.
- **Lessons 2-2 through 2-4** Add, subtract, multiply, and divide rational numbers.
- **Lesson 2-5** Display and interpret statistical data on line graphs and stem-and-leaf plots.
- **Lesson 2-6** Determine simple probability and odds.
- **Lesson 2-7** Find square roots and compare real numbers.

Key Vocabulary

- **rational number** (p. 68)
- **absolute value** (p. 69)
- **probability** (p. 96)
- **square root** (p.103)
- **real number** (p. 104)

Why It's Important

The ability to work with real numbers lays the foundation for further study in mathematics and allows you to solve a variety of real-world problems. For example, temperatures in the United States vary greatly from cold arctic regions to warm tropical regions. You can use real numbers and absolute value to compare these temperature extremes. *You will use absolute value and real numbers to compare temperatures in Lessons 2-1 and 2-2.*

Getting Started

Prerequisite Skills To be successful in this chapter, you'll need to master these skills and be able to apply them in problem-solving situations. Review these skills before beginning Chapter 2.

For Lessons 2-1 through 2-5 — Operations with Decimals and Fractions

Perform the indicated operation. *(For review, see pages 798 and 799.)*

1. $2.2 + 0.16$ **2.** $13.4 - 4.5$ **3.** $6.4 \cdot 8.8$ **4.** $76.5 \div 4.25$

5. $\frac{1}{4} + \frac{2}{3}$ **6.** $\frac{1}{2} - \frac{1}{3}$ **7.** $\frac{5}{4} \cdot \frac{3}{10}$ **8.** $\frac{4}{9} \div \frac{1}{3}$

For Lessons 2-1 through 2-5 — Evaluate Expressions

Evaluate each expression if $a = 2$, $b = \frac{1}{4}$, $x = 7$, and $y = 0.3$. *(For review, see Lesson 1-2.)*

9. $3a - 2$ **10.** $2x + 5$ **11.** $8(y + 2.4)$ **12.** $4(b + 2)$

13. $a - \frac{1}{2}$ **14.** $b + 3$ **15.** xy **16.** $y(a \div b)$

For Lesson 2-5 — Find Mean, Median, and Mode

Find the mean, median, and mode for each set of data. *(For review, see pages 818 and 819.)*

17. 2, 4, 7, 9, 12, 15 **18.** 23, 23, 23, 12, 12, 14 **19.** 7, 19, 2, 7, 4, 9

For Lesson 2-7 — Square Numbers

Simplify. *(For review, see Lesson 1-1.)*

20. 11^2 **21.** 0.9^2 **22.** $\left(\frac{2}{3}\right)^2$ **23.** $\left(\frac{4}{5}\right)^2$

FOLDABLES™
Study Organizer

Make this Foldable to collect examples and notes about operations with real numbers. Begin with a sheet of grid paper.

Step 1 Fold

Fold the short sides to meet in the middle.

Step 2 Fold Again

Fold the top to the bottom.

Step 3 Cut

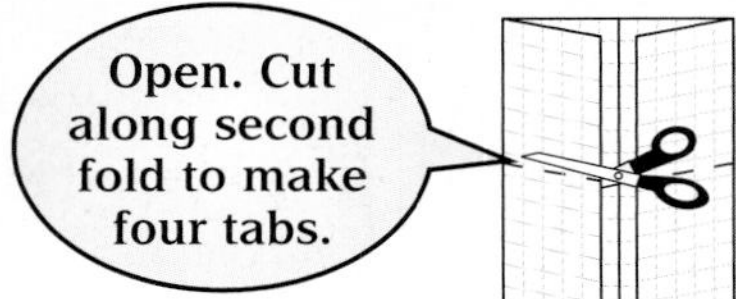
Open. Cut along second fold to make four tabs.

Step 4 Label

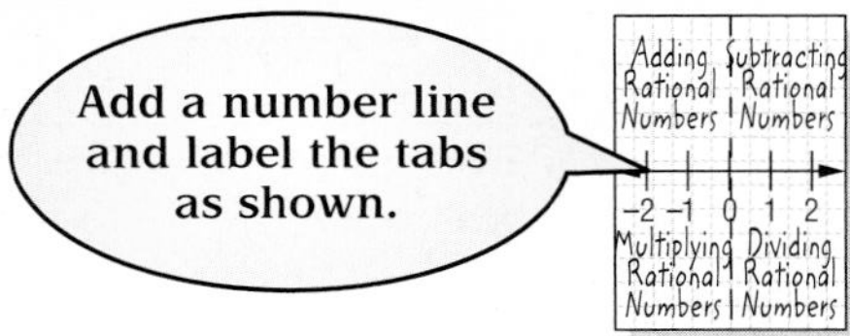
Add a number line and label the tabs as shown.
Adding Rational Numbers
Subtracting Rational Numbers
–2 –1 0 1 2
Multiplying Rational Numbers
Dividing Rational Numbers

Reading and Writing As you read and study the chapter, use the number line to help you solve problems. Write examples and notes under each tab.

2-1 Rational Numbers on the Number Line

What You'll Learn

- Graph rational numbers on a number line.
- Find absolute values of rational numbers.

How can you use a number line to show data?

A river's level rises and falls depending on rainfall and other conditions. The table shows the percent of change in river depths for various rivers in Texas over a 24-hour period. You can use a number line to graph these values and compare the changes in each river.

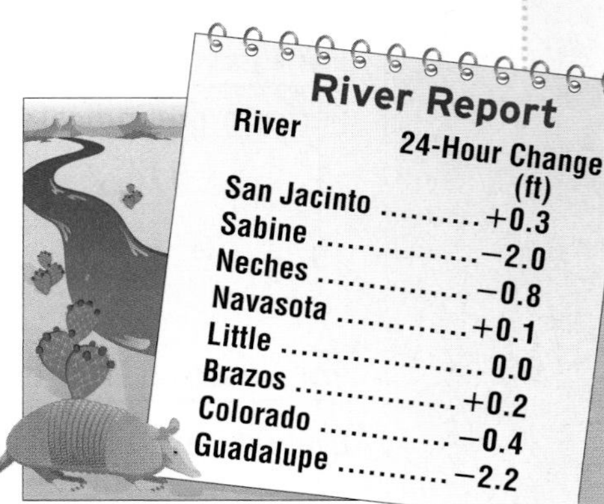

River Report
River
24-Hour Change (ft)
San Jacinto ....+0.3
Sabine ....−2.0
Neches ....−0.8
Navasota ....+0.1
Little 0.0
Brazos ....+0.2
Colorado ....−0.4
Guadalupe ....−2.2

Vocabulary

- natural number
- whole number
- integers
- positive number
- negative number
- rational number
- infinity
- graph
- coordinate
- absolute value

GRAPH RATIONAL NUMBERS A number line can be used to show the sets of **natural numbers**, **whole numbers**, and **integers**. Values greater than 0, or **positive numbers**, are listed to the right of 0, and values less than 0, or **negative numbers**, are listed to the left of 0.

natural numbers: 1, 2, 3, ...

whole numbers: 0, 1, 2, 3, ...

integers: ..., −3, −2, −1, 0, 1, 2, 3, ...

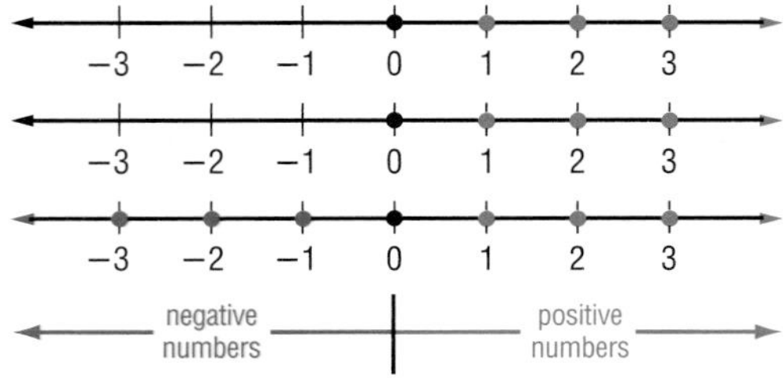

Study Tip

Number Line
Although only a portion of the number line is shown, the arrowheads indicate that the line and the set continue to **infinity**, which means that they never end.

Another set of numbers you can display on a number line is the set of rational numbers. A **rational number** is any number that can be written in the form $\frac{a}{b}$, where a and b are integers and $b \neq 0$. Some examples of rational numbers are shown below.

$$\frac{1}{2} \quad \frac{-2}{3} \quad \frac{17}{5} \quad \frac{15}{-3} \quad \frac{-14}{-11} \quad \frac{3}{1}$$

A rational number can also be expressed as a decimal that terminates, or as a decimal that repeats indefinitely.

$$0.5 \quad -0.\overline{3} \quad 3.4 \quad 2.6767\ldots \quad -5 \quad 1.\overline{27} \quad -1.23568994141\ldots$$

Concept Summary — Rational Numbers

Natural Numbers	{1, 2, 3, ...}
Whole Numbers	{0, 1, 2, 3, ...}
Integers	{..., −2, −1, 0, 1, 2, ...}
Rational Numbers	numbers that can be expressed in the form $\frac{a}{b}$, where a and b are integers and $b \neq 0$

Rational Numbers
Integers
Whole Numbers
Natural Numbers

Later in this chapter, you will be introduced to numbers that are not rational.

To **graph** a set of numbers means to draw, or plot, the points named by those numbers on a number line. The number that corresponds to a point on a number line is called the **coordinate** of that point.

Example 1 Identify Coordinates on a Number Line

Name the coordinates of the points graphed on each number line.

a.

−5 −4 −3 −2 −1 0 1 2 3 4 5

The dots indicate each point on the graph.
The coordinates are $\{-4, -3, -2, 1, 2\}$.

b.

−1 −0.5 0 0.5 1 1.5 2 2.5 3

The bold arrow on the right means that the graph continues indefinitely in that direction. The coordinates are $\{1, 1.5, 2, 2.5, 3, \ldots\}$.

Example 2 Graph Numbers on a Number Line

Graph each set of numbers.

a. $\{\ldots, -4, -2, 0, 2, 4, 6\}$

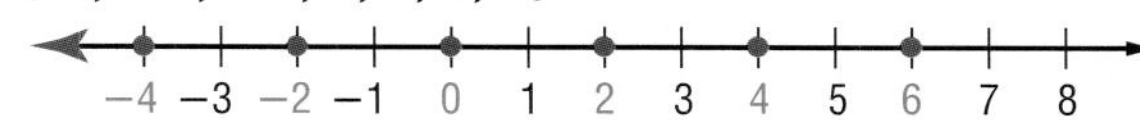

b. $\left\{-\frac{4}{3}, -\frac{1}{3}, \frac{2}{3}, \frac{5}{3}\right\}$

$-\frac{5}{3}$ $-\frac{4}{3}$ -1 $-\frac{2}{3}$ $-\frac{1}{3}$ 0 $\frac{1}{3}$ $\frac{2}{3}$ 1 $\frac{4}{3}$ $\frac{5}{3}$ 2 $\frac{7}{3}$

c. {integers less than −3 or greater than or equal to 5}

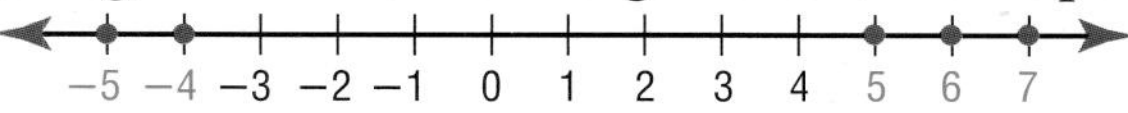

ABSOLUTE VALUE On a number line, 4 is four units from zero in the positive direction, and −4 is four units from zero in the negative direction. This number line illustrates the meaning of **absolute value**.

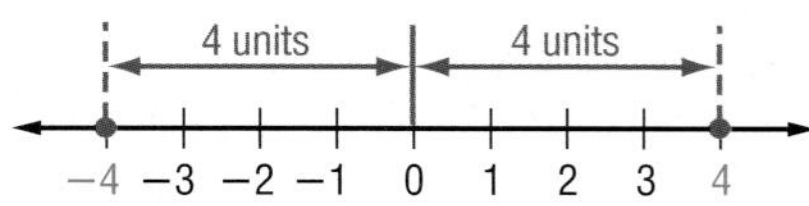

Key Concept — Absolute Value

- **Words** The absolute value of any number n is its distance from zero on a number line and is written as $|n|$.
- **Examples** $|-4| = 4$ $|4| = 4$

Since distance cannot be less than zero, absolute values are always greater than or equal to zero.

Study Tip

Reading Math
$|-7| = 7$ is read *the absolute value of negative 7 equals 7.*

Example 3 Absolute Value of Rational Numbers

Find each absolute value.

a. $|-7|$

−7 is seven units from zero in the negative direction.

$|-7| = 7$

b. $\left|\frac{7}{9}\right|$

$\frac{7}{9}$ is seven-ninths unit from zero in the positive direction.

$\left|\frac{7}{9}\right| = \frac{7}{9}$

You can also evaluate expressions involving absolute value. The absolute value bars serve as grouping symbols.

Example 4 Expressions with Absolute Value

Evaluate $15 - |x + 4|$ if $x = 8$.

$15 - \|x + 4\|$	$= 15 - \|8 + 4\|$	Replace x with 8.
	$= 15 - \|12\|$	$8 + 4 = 12$
	$= 15 - 12$	$\|12\| = 12$
	$= 3$	Simplify.

Check for Understanding

Concept Check

1. **Tell** whether the statement is *sometimes*, *always*, or *never* true. *An integer is a rational number.*
2. **Explain** the meaning of absolute value.
3. **OPEN ENDED** Give an example where absolute values are used in a real-life situation.

Guided Practice

Name the coordinates of the points graphed on each number line.

4. (number line from −4 to 6 with points at −2, 1, 2, 5)

5. (number line from $-\frac{11}{2}$ to $-\frac{1}{2}$: $-\frac{11}{2}$, $-\frac{10}{2}$, $-\frac{9}{2}$, $-\frac{8}{2}$, $-\frac{7}{2}$, $-\frac{6}{2}$, $-\frac{5}{2}$, $-\frac{4}{2}$, $-\frac{3}{2}$, $-\frac{2}{2}$, $-\frac{1}{2}$)

Graph each set of numbers.

6. $\{-4, -2, 1, 5, 7\}$
7. $\{-2.8, -1.5, 0.2, 3.4\}$
8. $\left\{-\frac{1}{2}, 0, \frac{1}{4}, \frac{2}{5}, \frac{5}{3}\right\}$
9. {integers less than or equal to −4}

Find each absolute value.

10. $|-2|$
11. $|18|$
12. $|2.5|$
13. $\left|-\frac{5}{6}\right|$

Evaluate each expression if $x = 18$, $y = 4$, and $z = -0.76$.

14. $57 - |x + 34|$
15. $19 + |21 - y|$
16. $|z| - 0.26$

Application

17. **NUMBER THEORY** Copy the Venn diagram at the right. Label the remaining sets of numbers. Then place the numbers $-3, -13, 0, 53, \frac{2}{3}, -\frac{1}{5}$, 0.33, 40, 2.98, −49.98, and $-\frac{5}{2}$ in the most specific categories.

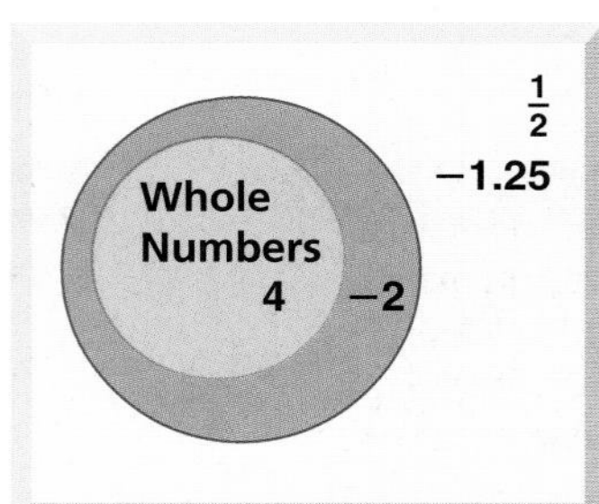

Practice and Apply

Homework Help

For Exercises	See Examples
18–23	1
24–33	2
34–41	3
42–44, 58, 59	2, 3
45–56	4

Extra Practice

See page 823.

Name the coordinates of the points graphed on each number line.

18. Number line: −5 −4 −3 −2 −1 0 1 2 3 4 5

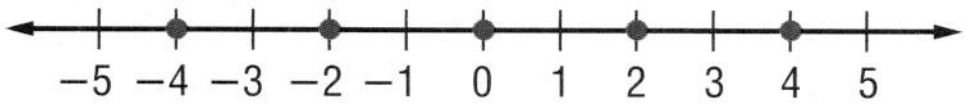

19. Number line: −8 −7 −6 −5 −4 −3 −2 −1 0 1 2

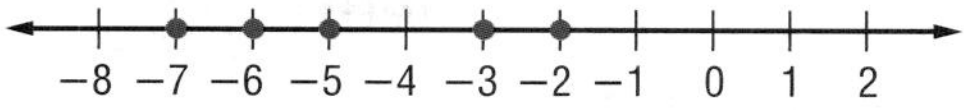

20. Number line: −4 −3 −2 −1 0 1 2 3 4 5 6

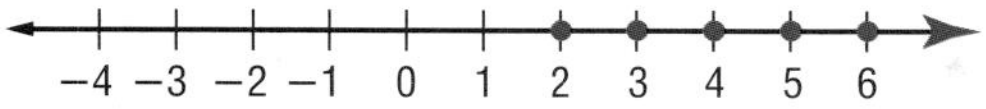

21. Number line: 0 0.2 0.4 0.6 0.8 1.0 1.2 1.4 1.6 1.8 2.0

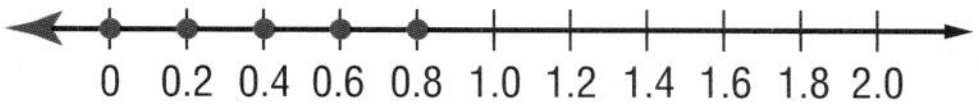

22. Number line: -2 $-\frac{5}{3}$ $-\frac{4}{3}$ -1 $-\frac{2}{3}$ $-\frac{1}{3}$ 0 $\frac{1}{3}$ $\frac{2}{3}$ 1 $\frac{4}{3}$

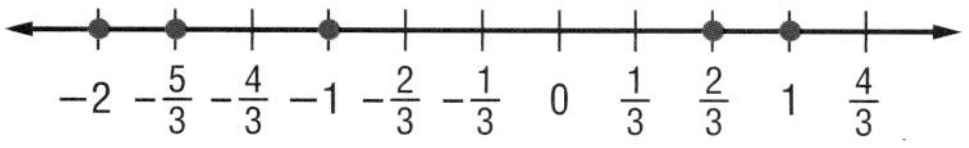

23. Number line: 0 $\frac{1}{5}$ $\frac{2}{5}$ $\frac{3}{5}$ $\frac{4}{5}$ 1 $\frac{6}{5}$ $\frac{7}{5}$ $\frac{8}{5}$ $\frac{9}{5}$ 2

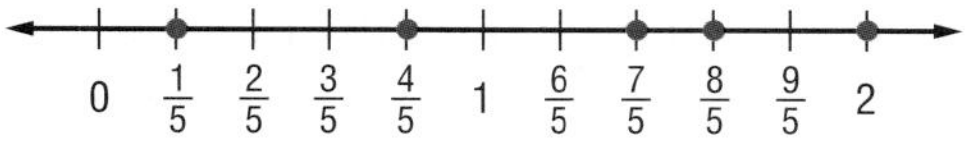

Career Choices

Demographer

A demographer analyzes the size, nature, and movement of human populations. Many demographers specialize in one area such as health, housing, education, agriculture, or economics.

Online Research

For information about a career as a demographer, visit www.algebra1.com/careers

Graph each set of numbers.

24. $\{-4, -2, -1, 1, 3\}$

25. $\{0, 2, 5, 6, 9\}$

26. $\{-5, -4, -3, -2, \ldots\}$

27. $\{\ldots, -2, 0, 2, 4, 6\}$

28. $\{-8.4, -7.2, -6.0, -4.8\}$

29. $\{-2.4, -1.6, -0.8, 0, \ldots\}$

30. $\left\{\ldots, -\frac{2}{3}, -\frac{1}{3}, 0, \frac{1}{3}, \frac{2}{3}, \ldots\right\}$

31. $\left\{-3\frac{2}{5}, -2\frac{1}{5}, -1\frac{4}{5}, -\frac{4}{5}, 1\right\}$

32. {integers less than -7 or greater than -1}

33. {integers greater than -5 and less than 9}

Find each absolute value.

34. $|-38|$

35. $|10|$

36. $|97|$

37. $|-61|$

38. $|3.9|$

39. $|-6.8|$

40. $\left|-\frac{23}{56}\right|$

41. $\left|\frac{35}{80}\right|$

POPULATION **For Exercises 42–44, refer to the table below.**

Population of Various Counties, 1990–1999			
County	**Percent Change**	**County**	**Percent Change**
Kings, NY	−1.4	Wayne, MI	−0.2
Los Angeles, CA	5.3	Philadelphia, PA	−10.6
Cuyahoga, OH	−2.9	Suffolk, NY	4.7
Santa Clara, CA	10.0	Alameda, CA	8.5
Cook, IL	1.7	New York, NY	4.3

Source: *The World Almanac*

42. Use a number line to order the percents of change from least to greatest.

43. Which population had the greatest percent increase or decrease? Explain.

44. Which population had the least percent increase or decrease? Explain.

Evaluate each expression if $a = 6$, $b = \frac{2}{3}$, $c = \frac{5}{4}$, $x = 12$, $y = 3.2$, and $z = -5$.

45. $48 + |x - 5|$

46. $25 + |17 + x|$

47. $|17 - a| + 23$

48. $|43 - 4a| + 51$

49. $|z| + 13 - 4$

50. $28 - 13 + |z|$

51. $6.5 - |8.4 - y|$

52. $7.4 + |y - 2.6|$

53. $\frac{1}{6} + \left|b - \frac{7}{12}\right|$

54. $\left(b + \frac{1}{2}\right) - \left|-\frac{5}{6}\right|$

55. $|c - 1| + \frac{2}{5}$

56. $|-c| + \left(2 + \frac{1}{2}\right)$

57. CRITICAL THINKING Find all values for x if $|x| = -|x|$.

More About. . .

Weather

The lowest temperature ever recorded in the world was $-129°F$ at the Soviet Antarctica station of Vostok.

Source: *The World Almanac*

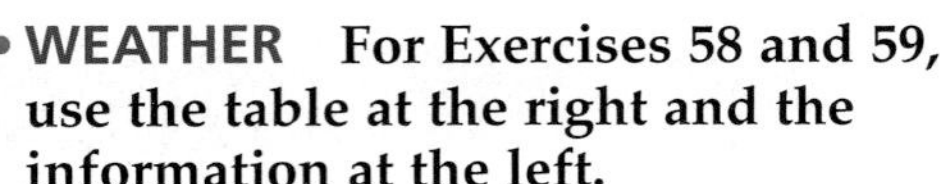

WEATHER **For Exercises 58 and 59, use the table at the right and the information at the left.**

Same Day Low Temperatures for Certain U.S. Cities

City	Low Temperature (°F)
Bismarck, ND	−11
Caribou, ME	−5
Chicago, IL	−4
Fairbanks, AK	−9
International Falls, MN	−13
Kansas City, MO	7
Sacramento, CA	34
Shreveport, LA	33

Source: *The World Almanac*

58. Draw a number line and graph the set of numbers that represents the low temperatures for these cities.

59. Write the absolute value of the low temperature for each city.

60. WRITING IN MATH Answer the question that was posed at the beginning of the lesson.

How can you use a number line to show data?

Include the following in your answer:

- an explanation of how to choose the range for a number line, and
- an explanation of how to tell which river had the greatest increase or decrease.

Standardized Test Practice

61. Which number is a natural number?

Ⓐ -2.5 Ⓑ $5 - |5|$ Ⓒ $-|3 + 5|$ Ⓓ $|-8| - 2$

62. Which sentence is *not* true?

Ⓐ All natural numbers are whole numbers.

Ⓑ Natural numbers are positive numbers.

Ⓒ Every whole number is a natural number.

Ⓓ Zero is neither positive nor negative.

Maintain Your Skills

Mixed Review

SALES **For Exercises 63–65, refer to the graph.** *(Lesson 1-9)*

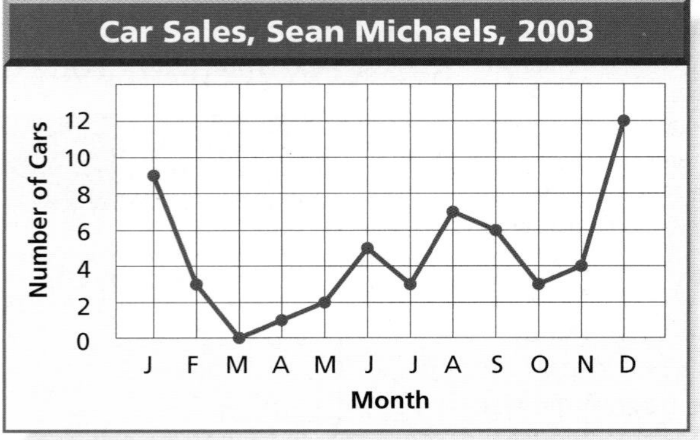

63. In which month did Mr. Michaels have the greatest sales?

64. Between which two consecutive months did the greatest change in sales occur?

65. In which months were sales equal?

66. **ENTERTAINMENT** Juanita has the volume on her stereo turned up. When her telephone rings, she turns the volume down. After she gets off the phone, she returns the volume to its previous level. Sketch a reasonable graph to show the volume of Juanita's stereo during this time. *(Lesson 1-8)*

Simplify each expression. *(Lesson 1-6)*

67. $8x + 2y + x$ **68.** $7(5a + 3b) - 4a$ **69.** $4[1 + 4(5x + 2y)]$

Getting Ready for the Next Lesson

PREREQUISITE SKILL **Find each sum or difference.**
(To review ***addition and subtraction of fractions****, see pages 798 and 799.)*

70. $\frac{3}{8} + \frac{1}{8}$ **71.** $\frac{7}{12} - \frac{3}{12}$ **72.** $\frac{7}{10} + \frac{1}{5}$ **73.** $\frac{3}{8} + \frac{2}{3}$

74. $\frac{5}{6} + \frac{1}{2}$ **75.** $\frac{3}{4} - \frac{1}{3}$ **76.** $\frac{9}{15} - \frac{1}{2}$ **77.** $\frac{7}{9} - \frac{7}{18}$

2-2 Adding and Subtracting Rational Numbers

What You'll Learn

- Add integers and rational numbers.
- Subtract integers and rational numbers.

Vocabulary

- opposites
- additive inverses

How can a number line be used to show a football team's progress?

In one series of plays during Super Bowl XXXV, the New York Giants received a five-yard penalty before completing a 13-yard pass.

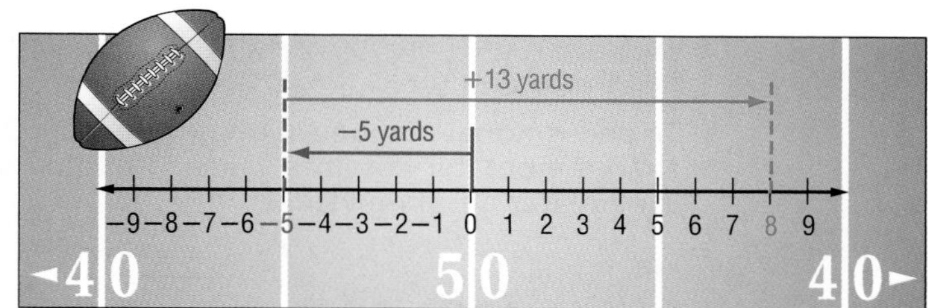

The number line shows the yards gained during this series of plays. The total yards gained was 8 yards.

ADD RATIONAL NUMBERS The number line above illustrates how to add integers on a number line. You can use a number line to add any rational numbers.

Example 1 Use a Number Line to Add Rational Numbers

Use a number line to find each sum.

a. $-3 + (-4)$

Step 1 Draw an arrow from 0 to −3.

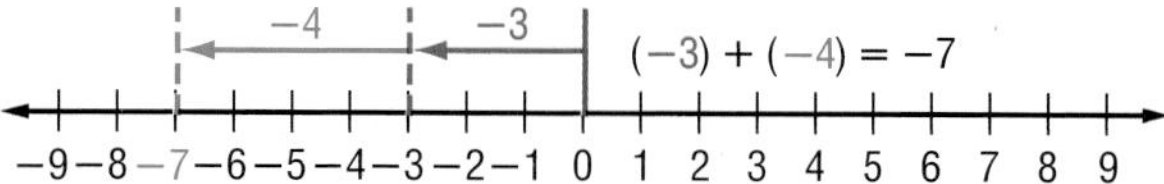

Step 2 Then draw a second arrow 4 units to the left to represent adding −4.

Step 3 The second arrow ends at the sum −7. So, $-3 + (-4) = -7$.

b. $2.5 + (-3.5)$

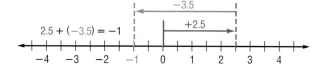

Step 1 Draw an arrow from 0 to 2.5.

−3.5
+2.5
2.5 + (−3.5) = −1
−4 −3 −2 −1 0 1 2 3 4

Step 2 Then draw a second arrow 3.5 units to the left.

Step 3 The second arrow ends at the sum −1. So, $2.5 + (-3.5) = -1$.

You can use absolute value to add rational numbers.

Same Signs

+ +	− −
$3 + 5 = 8$	$-3 + (-5) = -8$
3 and 5 are positive, so the sum is positive.	−3 and −5 are negative, so the sum is negative.

Different Signs

+ −	− +
$3 + (-5) = -2$	$-3 + 5 = 2$
Since −5 has the greater absolute value, the sum is negative.	Since 5 has the greater absolute value, the sum is positive.

The examples above suggest the following rules for adding rational numbers.

Key Concept — *Addition of Rational Numbers*

- To add rational numbers with the *same sign*, add their absolute values. The sum has the same sign as the addends.
- To add rational numbers with *different signs*, subtract the lesser absolute value from the greater absolute value. The sum has the same sign as the number with the greater absolute value.

Example 2 Add Rational Numbers

Find each sum.

a. $\mathbf{-11 + (-7)}$

$$\begin{aligned} -11 + (-7) &= -(|-11| + |-7|) && \text{Both numbers are negative, so the sum is negative.} \\ &= -(11 + 7) \\ &= -18 \end{aligned}$$

b. $\mathbf{\frac{7}{16} + \left(-\frac{3}{8}\right)}$

$$\begin{aligned} \frac{7}{16} + \left(-\frac{3}{8}\right) &= \frac{7}{16} + \left(-\frac{6}{16}\right) && \text{The LCD is 16. Replace } -\tfrac{3}{8} \text{ with } -\tfrac{6}{16}. \\ &= +\left(\left|\frac{7}{16}\right| - \left|-\frac{6}{16}\right|\right) && \text{Subtract the absolute values.} \\ &= +\left(\frac{7}{16} - \frac{6}{16}\right) && \text{Since the number with the greater absolute value is } \tfrac{7}{16}\text{, the sum is positive.} \\ &= \frac{1}{16} \end{aligned}$$

SUBTRACT RATIONAL NUMBERS Every positive rational number can be paired with a negative rational number. These pairs are called **opposites**.

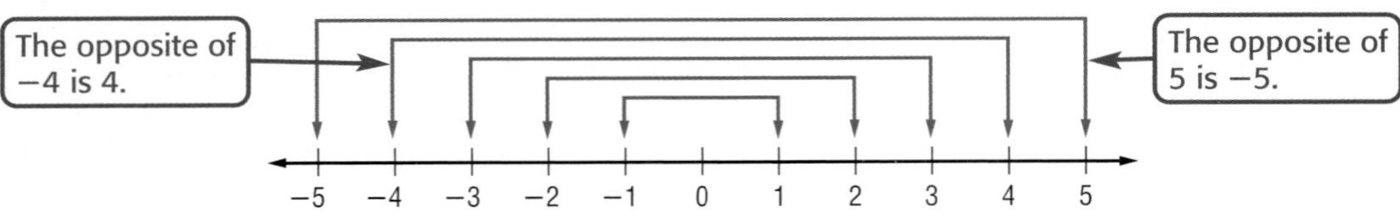
The opposite of −4 is 4.
The opposite of 5 is −5.
−5 −4 −3 −2 −1 0 1 2 3 4 5

A number and its opposite are **additive inverses** of each other. When you add two opposites, the sum is always 0.

Study Tip

Additive Inverse

Since $0 + 0 = 0$, zero is its own additive inverse.

Key Concept — *Additive Inverse Property*

- **Words** The sum of a number and its additive inverse is 0.
- **Symbols** For every number a, $a + (-a) = 0$.
- **Examples** $2 + (-2) = 0$ $-4.25 + 4.25 = 0$ $\frac{1}{3} + \left(-\frac{1}{3}\right) = 0$

Additive inverses can be used when you subtract rational numbers.

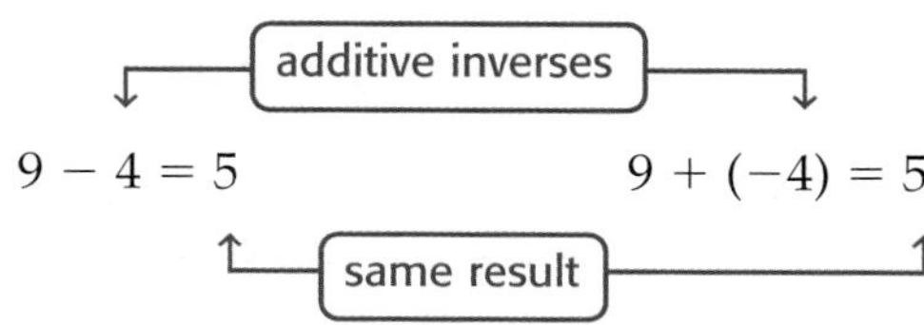

This example suggests that subtracting a number is equivalent to adding its inverse.

Key Concept — *Subtraction of Rational Numbers*

- **Words** To subtract a rational number, add its additive inverse.
- **Symbols** For any numbers a and b, $a - b = a + (-b)$.
- **Examples** $8 - 15 = 8 + (-15)$ or -7
  $-7.6 - 12.3 = -7.6 + (-12.3)$ or -19.9

Career Choices

Stockbroker

Stockbrokers perform various duties, including buying or selling stocks, bonds, mutual funds, or other financial products for an investor.

Online Research For information about a career as a stockbroker, visit: www.algebra1.com/careers

Example 3 *Subtract Rational Numbers to Solve a Problem*

STOCKS During a five-day period, a telecommunications company's stock price went from \$17.82 to \$15.36 per share. Find the change in the price of the stock.

Explore The stock price began at \$17.82 and ended at \$15.36. You need to determine the change in price for the week.

Plan Subtract to find the change in price.

ending price	minus	beginning price
15.36	−	17.82

Solve

$$15.36 - 17.82 = 15.36 + (-17.82)$$ To subtract 17.82, add its inverse.

$$= -(|-17.82| - |15.36|)$$ Subtract the absolute values.

$$= -(17.82 - 15.36)$$ The absolute value of −17.82 is greater, so the result is negative.

$$= -2.46$$

The price of the stock changed by −\$2.46.

Examine The problem asks for the change in a stock's price from the beginning of a week to the end. Since the change was negative, the price dropped. This makes sense since the ending price is less than the beginning price.

Check for Understanding

Concept Check

1. **OPEN ENDED** Write a subtraction expression using rational numbers that has a difference of $-\frac{2}{5}$.
2. **Describe** how to subtract real numbers.
3. **FIND THE ERROR** Gabriella and Nick are subtracting fractions.

Gabriella	Nick
$\left(-\frac{4}{9}\right) - \left(-\frac{2}{3}\right) = \left(-\frac{4}{9}\right) - \left(-\frac{6}{9}\right)$	$\left(-\frac{4}{9}\right) - \left(-\frac{2}{3}\right) = \left(-\frac{4}{9}\right) - \left(-\frac{6}{9}\right)$
$= \left(-\frac{4}{9}\right) + \left(\frac{6}{9}\right)$	$= \left(-\frac{4}{9}\right) + \left(-\frac{6}{9}\right)$
$= \left(\frac{6}{9} - \frac{4}{9}\right)$	$= -\left(\frac{6}{9} + \frac{4}{9}\right)$
$= \frac{2}{9}$	$= -\frac{10}{9}$

Who is correct? Explain your reasoning.

Guided Practice

Find each sum.

4. $-15 + (-12)$
5. $-24 + (-45)$
6. $38.7 + (-52.6)$
7. $-4.62 + (-12.81)$
8. $\frac{4}{7} + \left(-\frac{1}{2}\right)$
9. $-\frac{5}{12} + \frac{8}{15}$

Find each difference.

10. $18 - 23$
11. $12.7 - (-18.4)$
12. $(-3.86) - 1.75$
13. $-32.25 - (-42.5)$
14. $-\frac{2}{9} - \frac{3}{10}$
15. $\left(-\frac{7}{10}\right) - \left(-\frac{11}{12}\right)$

Application

16. **WEATHER** The highest recorded temperature in the United States was in Death Valley, California, while the lowest temperature was recorded at Prospect Creek, Alaska. What is the difference between these two temperatures?

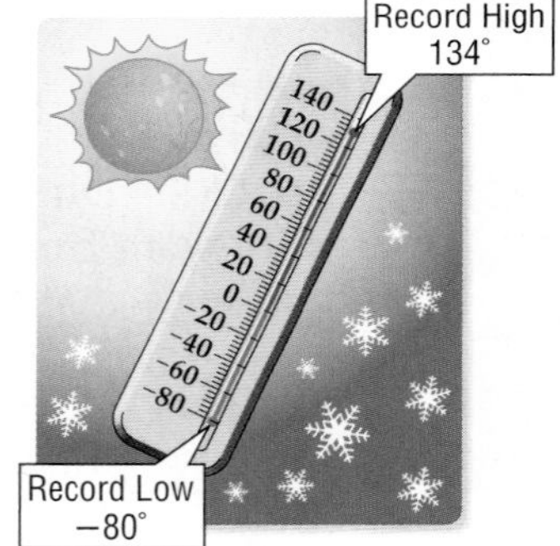

Practice and Apply

Homework Help

For Exercises	See Examples
17–38	1, 2
39–62	3

Extra Practice

See page 823.

Find each sum.

17. $-8 + 13$
18. $-11 + 19$
19. $41 + (-63)$
20. $80 + (-102)$
21. $-77 + (-46)$
22. $-92 + (-64)$
23. $-1.6 + (-3.8)$
24. $-32.4 + (-4.5)$
25. $-38.9 + 24.2$
26. $-7.007 + 4.8$
27. $43.2 + (-57.9)$
28. $38.7 + (-61.1)$
29. $\frac{6}{7} + \frac{2}{3}$
30. $\frac{3}{18} + \frac{6}{17}$
31. $-\frac{4}{11} + \frac{3}{5}$
32. $-\frac{2}{5} + \frac{17}{20}$
33. $-\frac{4}{15} + \left(-\frac{9}{16}\right)$
34. $-\frac{16}{40} + \left(-\frac{13}{20}\right)$
35. Find the sum of $4\frac{1}{8}$ and $-1\frac{1}{2}$.
36. Find the sum of $1\frac{17}{50}$ and $-3\frac{17}{25}$.

37. **GAMES** Sarah was playing a computer trivia game. Her scores for round one were +100, +200, +500, −300, +400, and −500. What was her total score at the end of round one?

38. **FOOTBALL** The Northland Vikings' offense began a drive from their 20-yard line. They gained 6 yards on the first down, lost 8 yards on the second down, then gained 3 yards on third down. What yard line were they on at fourth down?

Find each difference.

39. $-19 - 8$
40. $16 - (-23)$
41. $9 - (-24)$
42. $12 - 34$
43. $22 - 41$
44. $-9 - (-33)$
45. $-58 - (-42)$
46. $79.3 - (-14.1)$
47. $1.34 - (-0.458)$
48. $-9.16 - 10.17$
49. $67.1 - (-38.2)$
50. $72.5 - (-81.3)$
51. $-\frac{1}{6} - \frac{2}{3}$
52. $\frac{1}{2} - \frac{4}{5}$
53. $-\frac{7}{8} - \left(-\frac{3}{16}\right)$
54. 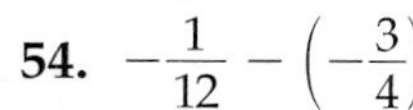 $-\frac{1}{12} - \left(-\frac{3}{4}\right)$
55. $2\frac{1}{4} - 6\frac{1}{3}$
56. $5\frac{3}{10} - 1\frac{31}{50}$

More About. . .

Golf

In the United States, there are more than 16,000 golf courses played by 26 million people each year.

Source: Encarta Online

GOLF For Exercises 57–59, use the following information.
In golf, scores are based on *par*. Par 72 means that a golfer should hit the ball 72 times to complete 18 holes of golf. A score of 67, or 5 under par, is written as −5. A score of 3 over par is written as +3. At the Masters Tournament (par 72) in April, 2001, Tiger Woods shot 70, 66, 68, and 68 during four rounds of golf.

57. Use integers to write his score for each round as over or under par.
58. Add the integers to find his overall score.
59. Was his score under or over par? Would you want to have his score? Explain.

Online Research Data Update Find the most recent winner of the Masters Tournament. What integer represents the winner's score for each round as over or under par? What integer represents the winner's overall score? Visit www.algebra1.com/data_update to learn more.

STOCKS For Exercises 60–62, refer to the table that shows the weekly closing values of the stock market for an eight-week period.

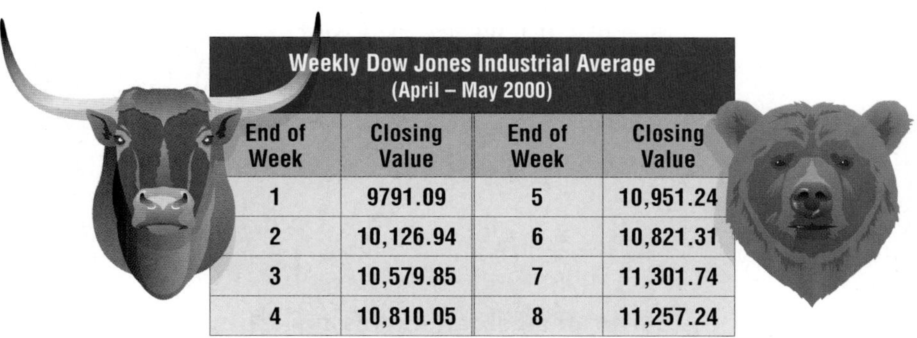

Weekly Dow Jones Industrial Average (April – May 2000)

End of Week	Closing Value	End of Week	Closing Value
1	9791.09	5	10,951.24
2	10,126.94	6	10,821.31
3	10,579.85	7	11,301.74
4	10,810.05	8	11,257.24

Source: *The Wall Street Journal*

60. Find the change in value from week 1 to week 8.
61. Which week had the greatest change from the previous week?
62. Which week had the least change from the previous week?

63. **CRITICAL THINKING** Tell whether the equation $x + |x| = 0$ is *always*, *sometimes*, or *never* true. Explain.

64. **WRITING IN MATH** Answer the question that was posed at the beginning of the lesson.

How can a number line be used to show a football team's progress?

Include the following in your answer:

- an explanation of how you could use a number line to determine the yards gained or lost by the Giants on their next three plays, and
- a description of how to determine the total yards gained or lost without using a number line.

65. What is the value of n in $-57 - n = -144$?

(A) -201 (B) -87 (C) 87 (D) 201

66. Which expression is equivalent to $5 - (-8)$?

(A) $(-5) + 8$ (B) $8 + 5$ (C) $8 - 5$ (D) $5 - 8$

Maintain Your Skills

Mixed Review

Evaluate each expression if $x = 4.8$, $y = -7.4$, and $z = 10$. *(Lesson 2-1)*

67. $12.2 + |8 - x|$ 68. $|y| + 9.4 - 3$ 69. $24.2 - |18.3 - z|$

For Exercises 70 and 71, refer to the graph. *(Lesson 1-9)*

70. If you wanted to make a circle graph of the data, what additional category would you have to include so that the circle graph would not be misleading?

71. Construct a circle graph that displays the data accurately.

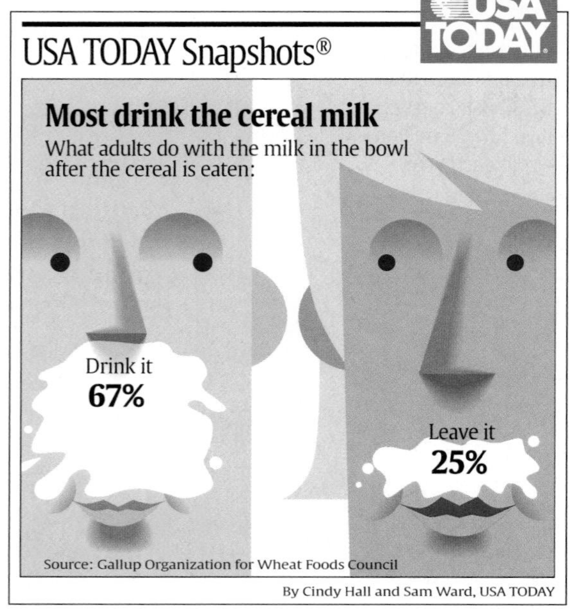

Find the solution sets for each inequality if the replacement sets are $a = \{2, 3, 4, 5, 6\}$, $b = \{0.3, 0.4, 0.5, 0.6, 0.7\}$, and $c = \left\{\frac{1}{4}, \frac{1}{2}, \frac{3}{4}, 1, 1\frac{1}{4}\right\}$. *(Lesson 1-3)*

72. $b + 1.3 \geq 1.8$

73. $3a - 5 > 7$

74. $c + \frac{1}{2} < 2\frac{1}{4}$

Write an algebraic expression for each verbal phrase. *(Lesson 1-1)*

75. eight less than the square of q

76. 37 less than 2 times a number k

Getting Ready for the Next Lesson

PREREQUISITE SKILL Find each product.

*(To review **multiplication of fractions**, see pages 800 and 801.)*

77. $\frac{1}{2} \cdot \frac{2}{3}$ 78. $\frac{1}{4} \cdot \frac{2}{5}$ 79. $\frac{3}{4} \cdot \frac{5}{6}$

80. $4 \cdot \frac{3}{5}$ 81. $8 \cdot \frac{5}{8}$ 82. $\frac{7}{9} \cdot 12$

2-3 Multiplying Rational Numbers

What You'll Learn

- Multiply integers.
- Multiply rational numbers.

How do consumers use multiplication of rational numbers?

Stores often offer coupons to encourage people to shop in their stores. The receipt shows a purchase of four CDs along with four coupons for $1.00 off each CD. How could you determine the amount saved by using the coupons?

CD SHOP	
CD	13.99
CD	12.99
CD	14.99
CD	14.99
COUPON	−1.00
COUPON	−1.00
COUPON	−1.00
COUPON	−1.00
TAX	0.31
TOTAL DUE	53.27
CASH	55.00
CHANGE	1.73

MULTIPLY INTEGERS One way to find the savings from the coupons is to use repeated addition.

$$-\$1.00 + (-\$1.00) + (-\$1.00) + (-\$1.00) = -\$4.00$$

An easier way to find the savings would be to multiply $-\$1.00$ by 4.

$$4(-\$1.00) = -\$4.00$$

Suppose the coupons were expired and had to be removed from the total. You can represent this by multiplying $-\$1.00$ by -4.

$$(-4)(-\$1.00) = \$4.00$$

In other words, $4.00 would be added back to the total.

These examples suggest the following rules for multiplying integers.

Study Tip

Multiplying Integers
When multiplying integers, if there are an even number of negative integers, the product is positive. If there are an odd number of negative integers, the product is negative.

Key Concept — *Multiplication of Integers*

- **Words** The product of two numbers having the *same sign* is positive. The product of two numbers having *different signs* is negative.
- **Examples** $(-12)(-7) = 84$ same signs → positive product
  $15(-8) = -120$ different signs → negative product

Example 1 Multiply Integers

Find each product.

a. $\mathbf{4(-5)}$

$4(-5) = -20$ different signs → negative product

b. $\mathbf{(-12)(-14)}$

$(-12)(-14) = 168$ same signs → positive product

You can simplify expressions by applying the rules of multiplication.

Example 2 Simplify Expressions

Simplify the expression $4(-3y) - 15y$.

$4(-3y) - 15y$	$= 4(-3)y - 15y$	Associative Property (×)
	$= -12y - 15y$	Substitution
	$= (-12 - 15)y$	Distributive Property
	$= -27y$	Simplify.

MULTIPLY RATIONAL NUMBERS Multiplying rational numbers is similar to multiplying integers.

Example 3 Multiply Rational Numbers

Find $\left(-\frac{3}{4}\right)\left(\frac{3}{8}\right)$.

$\left(-\frac{3}{4}\right)\left(\frac{3}{8}\right) = -\frac{9}{32}$ different signs → negative product

Example 4 Multiply Rational Numbers to Solve a Problem

BASEBALL Fenway Park, home of the Boston Red Sox, is the oldest ball park in professional baseball. It has a seating capacity of about 34,000. Determine the approximate total ticket sales for a sold-out game.

- Updated data
- More activities on writing equations www.algebra1.com/usa_today

To find the approximate total ticket sales, multiply the number of tickets sold by the average price.

$34{,}000 \cdot 24.05 = 817{,}770$

same signs → positive product

The total ticket sales for a sold-out game are about \$817,770.

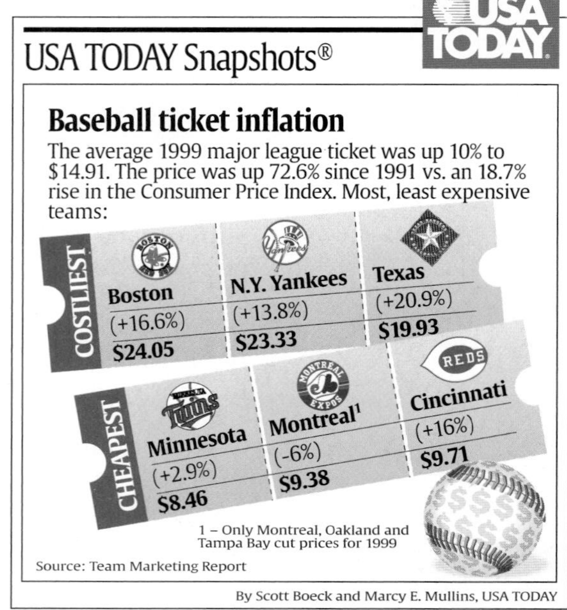

USA TODAY
USA TODAY Snapshots®
Baseball ticket inflation
The average 1999 major league ticket was up 10% to $14.91. The price was up 72.6% since 1991 vs. an 18.7% rise in the Consumer Price Index. Most, least expensive teams:
COSTLIEST
Boston (+16.6%) $24.05
N.Y. Yankees (+13.8%) $23.33
Texas (+20.9%) $19.93
CHEAPEST
Minnesota (+2.9%) $8.46
Montreal[1] (-6%) $9.38
Cincinnati (+16%) $9.71
1 – Only Montreal, Oakland and Tampa Bay cut prices for 1999
Source: Team Marketing Report
By Scott Boeck and Marcy E. Mullins, USA TODAY

You can evaluate expressions that contain rational numbers.

Example 5 Evaluate Expressions

Evaluate $n^2\left(-\frac{5}{8}\right)$ if $n = -\frac{2}{5}$.

$n^2\left(-\frac{5}{8}\right)$	$= \left(-\frac{2}{5}\right)^2\left(-\frac{5}{8}\right)$	Substitution
	$= \left(\frac{4}{25}\right)\left(-\frac{5}{8}\right)$	$\left(-\frac{2}{5}\right)^2 = \left(-\frac{2}{5}\right)\left(-\frac{2}{5}\right)$ or $\frac{4}{25}$
	$= -\frac{20}{200}$ or $-\frac{1}{10}$	different signs → negative product

In Lesson 1-4, you learned about the Multiplicative Identity Property, which states that any number multiplied by 1 is equal to the number. Another important property is the Multiplicative Property of -1.

Key Concept — *Multiplicative Property of -1*

- **Words** The product of any number and -1 is its additive inverse.
- **Symbols** For any number a, $-1(a) = a(-1) = -a$.
- **Examples** $(-1)(4) = (4)(-1) = -4$ $(-1)(-2.3) = (-2.3)(-1) = 2.3$

Check for Understanding

Concept Check

1. **List** the conditions under which the product ab is negative. Give examples to support your answer.
2. **OPEN ENDED** Describe a real-life situation in which you would multiply a positive rational number by a negative rational number. Write a corresponding multiplication expression.
3. **Explain** why the product of two negative numbers is positive.

Guided Practice

Find each product.

4. $(-6)(3)$
5. $(5)(-8)$
6. $(4.5)(2.3)$
7. $(-8.7)(-10.4)$
8. $\left(\frac{5}{3}\right)\left(-\frac{2}{7}\right)$
9. $\left(-\frac{4}{9}\right)\left(\frac{7}{15}\right)$

Simplify each expression.

10. $5s(-6t)$
11. $6x(-7y) + (-15xy)$

Evaluate each expression if $m = -\frac{2}{3}$, $n = \frac{1}{2}$, and $p = -3\frac{3}{4}$.

12. $6m$
13. np
14. $n^2(m + 2)$

Application

15. **NATURE** The average worker honeybee makes about $\frac{1}{12}$ teaspoon of honey in its lifetime. How much honey do 675 honeybees make?

Practice and Apply

Homework Help

For Exercises	See Examples
16–33	1, 3
34–39	2
40, 41	4
42–49	5
50–54	4

Extra Practice

See page 823.

Find each product.

16. $5(18)$
17. $8(22)$
18. $-12(15)$
19. $-24(8)$
20. $-47(-29)$
21. $-81(-48)$
22. $\left(\frac{4}{5}\right)\left(\frac{3}{8}\right)$
23. $\left(\frac{5}{12}\right)\left(\frac{4}{9}\right)$
24. $\left(-\frac{3}{5}\right)\left(\frac{5}{6}\right)$
25. $\left(-\frac{2}{5}\right)\left(\frac{6}{7}\right)$
26. $\left(-3\frac{1}{5}\right)\left(-7\frac{1}{2}\right)$
27. $\left(-1\frac{4}{5}\right)\left(-2\frac{1}{2}\right)$
28. $7.2(0.2)$
29. $6.5(0.13)$
30. $(-5.8)(2.3)$
31. $(-0.075)(6.4)$
32. $\frac{3}{5}(-5)(-2)$
33. $\frac{2}{11}(-11)(-4)$

Simplify each expression.

34. $6(-2x) - 14x$
35. $5(-4n) - 25n$
36. $5(2x - x)$
37. $-7(3d + d)$
38. $-2a(-3c) + (-6y)(6r)$
39. $7m(-3n) + 3s(-4t)$

STOCK PRICES **For Exercises 40 and 41, use the table that lists the closing prices of a company's stock over a one-week period.**

40. What was the change in price of 35 shares of this stock from day 2 to day 3?

41. If you bought 100 shares of this stock on day 1 and sold half of them on day 4, how much money did you gain or lose on those shares?

Closing Stock Price ($)	
Day	**Price**
1	64.38
2	63.66
3	61.66
4	61.69
5	62.34

Evaluate each expression if $a = -2.7$, $b = 3.9$, $c = 4.5$, and $d = -0.2$.

42. $-5c^2$
43. $-2b^2$
44. $-4ab$
45. $-5cd$
46. $ad - 8$
47. $ab - 3$
48. $d^2(b - 2a)$
49. $b^2(d - 3c)$

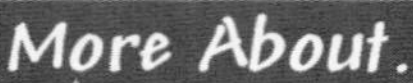

Civics

The Marine Corps War Memorial in Washington, D.C., is dedicated to all Marines who have defended the United States since 1775. It is the most famous memorial that is centered around the flag.

Source: The United States National Park Service

50. **CIVICS** In a United States flag, the length of the union is $\frac{2}{5}$ of the fly, and the width is $\frac{7}{13}$ of the hoist. If the fly is 6 feet, how long is the union?

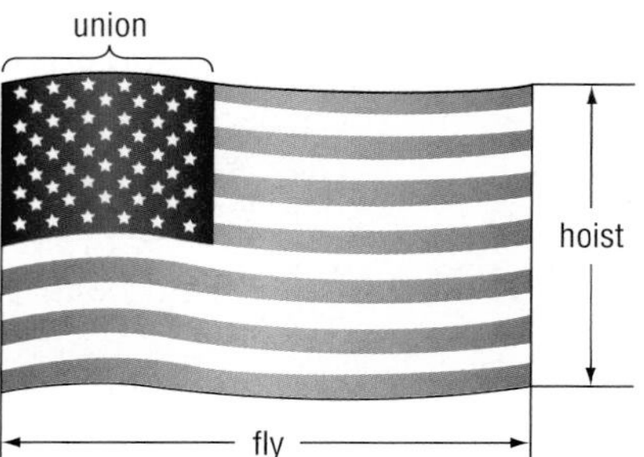

51. **COMPUTERS** The price of a computer dropped $34.95 each month for 7 months. If the starting price was $1450, what was the price after 7 months?

52. **BALLOONING** The temperature drops about 2°F for every rise of 530 feet in altitude. Per Lindstrand achieved the altitude record of 64,997 feet in a hot-air balloon over Laredo, Texas, on June 6, 1988. About how many degrees difference was there between the ground temperature and the air temperature at that altitude? **Source:** *The Guinness Book of Records*

ECOLOGY **For Exercises 53 and 54, use the following information.**
Americans use about 2.5 million plastic bottles every hour.
Source: www.savethewater.com

53. About how many plastic bottles are used in one day?

54. About how many bottles are used in one week?

55. **CRITICAL THINKING** An even number of negative numbers is multiplied. What is the sign of the product? Explain your reasoning.

56. **WRITING IN MATH** Answer the question that was posed at the beginning of the lesson.

How do consumers use multiplication of rational numbers?

Include the following in your answer:

- an explanation of why the amount of a coupon is expressed as a negative value, and
- an explanation of how you could use multiplication to find your total discount if you bought 3 CDs for $13.99 each and there was a discount of $1.50 on each CD.

Standardized Test Practice

57. Which expression can be simplified as $-8xy$?

(A) $2y - 4x$ (B) $-2x(4y)$ (C) $(-4)^2xy$ (D) $-4x(-2y)$

58. Find the value of m if $m = -2ab$, $a = -4$, and $b = 6$.

(A) 8 (B) 48 (C) 12 (D) -48

Maintain Your Skills

Mixed Review **Find each sum or difference.** *(Lesson 2-2)*

59. $-6.5 + (-5.6)$ 60. $\frac{4}{5} + \left(-\frac{3}{4}\right)$ 61. $42 - (-14)$ 62. $-14.2 - 6.7$

Graph each set of numbers on a number line. *(Lesson 2-1)*

63. $\{\ldots, -3, -1, 1, 3, 5\}$ 64. $\{-2.5, -1.5, 0.5, 4.5\}$ 65. $\{-1, -\frac{1}{3}, \frac{2}{3}, 2\}$

66. Identify the graph below that best represents the following situation. Brandon has a deflated balloon. He slowly fills the balloon up with air. Without tying the balloon, he lets it go. *(Lesson 1-8)*

a.

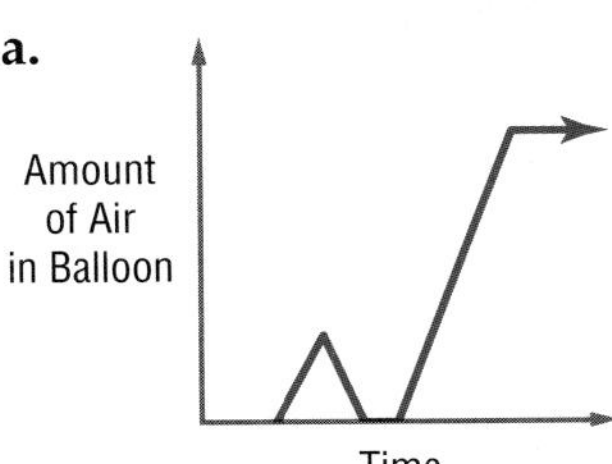

b.

Amount of Air in Balloon

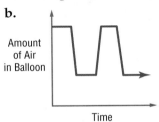

Time

c.

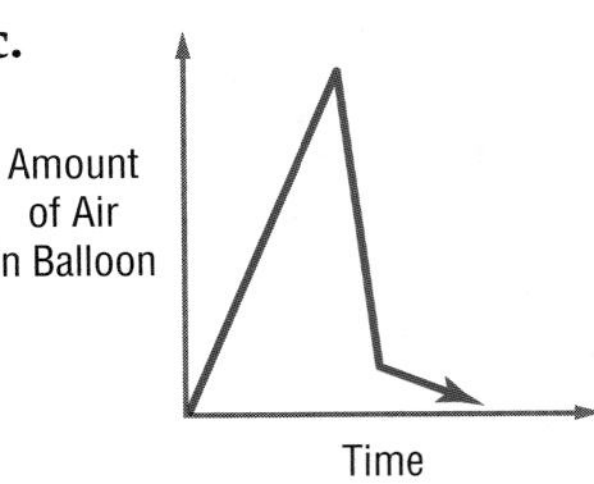

Write a counterexample for each statement. *(Lesson 1-7)*

67. If $2x - 4 \geq 6$, then $x > 5$. 68. If $|a| > 3$, then $a > 3$.

Getting Ready for the Next Lesson **PREREQUISITE SKILL** **Find each quotient.**

*(To review **division of fractions**, see pages 800 and 801.)*

69. $\frac{5}{8} \div 2$ 70. $\frac{2}{3} \div 4$ 71. $5 \div \frac{3}{4}$ 72. $1 \div \frac{2}{5}$

73. $\frac{1}{2} \div \frac{3}{8}$ 74. $\frac{7}{9} \div \frac{5}{6}$ 75. $\frac{4}{5} \div \frac{6}{5}$ 76. $\frac{7}{8} \div \frac{2}{3}$

Practice Quiz 1

Lessons 2-1 through 2-3

1. Name the set of points graphed on the number line. *(Lesson 2-1)*

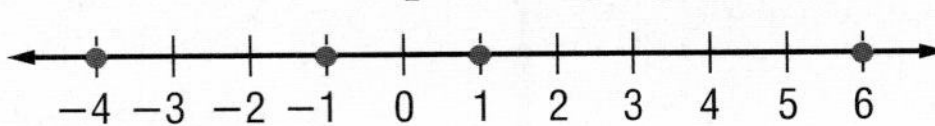

2. Evaluate $32 - |x + 8|$ if $x = 15$. *(Lesson 2-1)*

Find each sum or difference. *(Lesson 2-2)*

3. $-15 + 7$ 4. $27 - (-12)$ 5. $-6.05 + (-2.1)$ 6. $-\frac{3}{4} - \left(-\frac{2}{5}\right)$

Find each product. *(Lesson 2-3)*

7. $-9(-12)$ 8. $(3.8)(-4.1)$

9. Simplify $(-8x)(-2y) + (-3y)(z)$. *(Lesson 2-3)*

10. Evaluate $mn + 5$ if $m = 2.5$ and $n = -3.2$. *(Lesson 2-3)*

2-4 Dividing Rational Numbers

What You'll Learn

- Divide integers.
- Divide rational numbers.

How can you use division of rational numbers to describe data?

Each year, many sea turtles are stranded on the Texas Gulf Coast. The number of sea turtles stranded from 1997 to 2000 and the changes in number from the previous years are shown in the table. The following expression can be used to find the *mean* change per year of the number of stranded turtles.

$$\text{mean} = \frac{(-127) + 54 + (-65)}{3}$$

Stranded Sea Turtles Texas Gulf Coast

Year	Number of Turtles	Change
1997	523	——
1998	396	−127
1999	450	+54
2000	385	−65

Source: www.ridleyturtles.org

DIVIDE INTEGERS Since multiplication and division are inverse operations, the rule for finding the sign of the quotient of two numbers is similar to the rule for finding the sign of a product of two numbers.

Key Concept — *Division of Integers*

- **Words** The quotient of two numbers having the *same sign* is positive. The quotient of two numbers having *different signs* is negative.
- **Examples** $(-60) \div (-5) = 12$ same signs → positive quotient
  $32 \div (-8) = -4$ different signs → negative quotient

Example 1 *Divide Integers*

Find each quotient.

a. $\mathbf{-77 \div 11}$

$-77 \div 11 = -7$ negative quotient

b. $\mathbf{\frac{-51}{-3}}$

$\frac{-51}{-3} = -51 \div (-3)$ Divide.

$= 17$ positive quotient

When simplifying fractions, recall that the fraction bar is a grouping symbol.

Example 2 *Simplify Before Dividing*

Simplify $\mathbf{\frac{-3(-12 + 8)}{7 + (-5)}}$.

$\frac{-3(-12 + 8)}{7 + (-5)} = \frac{-3(-4)}{7 + (-5)}$ Simplify the numerator first.

$= \frac{12}{7 + (-5)}$ Multiply.

$= \frac{12}{2}$ or 6 same signs → positive quotient

DIVIDE RATIONAL NUMBERS The rules for dividing positive and negative integers also apply to division with rational numbers. Remember that to divide by any nonzero number, multiply by the reciprocal of that number.

Example 3 Divide Rational Numbers

Find each quotient.

a. 245.66 ÷ (−14.2)

$245.66 \div (-14.2) = -17.3$ Use a calculator. different signs → negative quotient

b. $-\frac{2}{5} \div \frac{1}{4}$

$-\frac{2}{5} \div \frac{1}{4} = -\frac{2}{5} \cdot \frac{4}{1}$ Multiply by $\frac{4}{1}$, the reciprocal of $\frac{1}{4}$.

$= -\frac{8}{5}$ or $-1\frac{3}{5}$ different signs → negative quotient

Example 4 Divide Rational Numbers to Solve a Problem

ARCHITECTURE The Pentagon in Washington, D.C., has an outside perimeter of 4608 feet. Find the length of each outside wall.

To find the length of each wall, divide the perimeter by the number of sides.

$4608 \div 5 = 921.6$ same signs → positive quotient

The length of each outside wall is 921.6 feet.

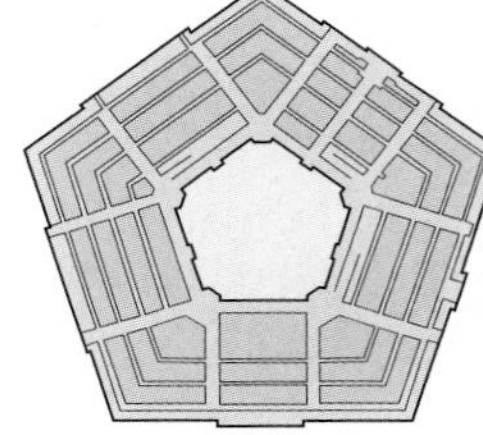

The Pentagon

More About. . .

Architecture

The Pentagon is one of the world's largest office buildings. It contains 131 stairways, 19 escalators, 13 elevators, 284 restrooms, and 691 drinking fountains.

Source: www.infoplease.com

You can use the Distributive Property to simplify fractional expressions.

Example 5 Simplify Algebraic Expressions

Simplify $\frac{24 - 6a}{3}$.

$\frac{24 - 6a}{3} = (24 - 6a) \div 3$ The fraction bar indicates division.

$= (24 - 6a)\left(\frac{1}{3}\right)$ Multiply by $\frac{1}{3}$, the reciprocal of 3.

$= 24\left(\frac{1}{3}\right) - 6a\left(\frac{1}{3}\right)$ Distributive Property

$= 8 - 2a$ Simplify.

Example 6 Evaluate Algebraic Expressions

Evaluate $\frac{ab}{c^2}$ **if** $a = -7.8$, $b = 5.2$, **and** $c = -3$. **Round to the nearest hundredth.**

$\frac{ab}{c^2} = \frac{(-7.8)(5.2)}{(-3)^2}$ Replace a with −7.8, b with 5.2, and c with −3.

$= \frac{-40.56}{9}$ Find the numerator and denominator separately.

≈ -4.51 Use a calculator. different signs → negative quotient

Check for Understanding

Concept Check

1. **Compare and contrast** multiplying and dividing rational numbers.
2. **OPEN ENDED** Find a value for x if $\frac{1}{x} > x$.
3. **Explain** how to divide any rational number by another rational number.

Guided Practice

Find each quotient.

4. $96 \div (-6)$
5. $-36 \div 4$
6. $-64 \div 5$
7. $64.4 \div 2.5$
8. $-\frac{2}{3} \div 12$
9. $-\frac{2}{3} \div \frac{4}{5}$

Simplify each expression.

10. $\frac{25 + 3}{-4}$
11. $\frac{-650a}{10}$
12. $\frac{6b + 18}{-2}$

Evaluate each expression if $a = 3$, $b = -4.5$, and $c = 7.5$. Round to the nearest hundredth.

13. $\frac{2ab}{-ac}$
14. $\frac{cb}{4a}$
15. $-\frac{a}{b} \div \frac{a}{c}$

Application

16. **ONLINE SHOPPING** During the 2000 holiday season, the sixth most visited online shopping site recorded 419,000 visitors. This is eight times as many visitors as in 1999. About how many visitors did the site have in 1999?

Practice and Apply

Homework Help

For Exercises	See Examples
17–36	1, 3
37–44	5
45, 46, 55–57	4
47–54	6

Extra Practice

See page 824.

Find each quotient.

17. $-64 \div (-8)$
18. $-78 \div (-4)$
19. $-78 \div (-1.3)$
20. $108 \div (-0.9)$
21. $42.3 \div (-6)$
22. $68.4 \div (-12)$
23. $-23.94 \div 10.5$
24. $-60.97 \div 13.4$
25. $-32.25 \div (-2.5)$
26. $-98.44 \div (-4.6)$
27. $-\frac{1}{3} \div 4$
28. $-\frac{3}{4} \div 12$
29. $-7 \div \frac{3}{5}$
30. $-5 \div \frac{2}{7}$
31. $\frac{16}{36} \div \frac{24}{60}$
32. $-\frac{24}{56} \div \frac{31}{63}$
33. $\frac{14}{32} \div \left(-\frac{12}{25}\right)$
34. $\frac{80}{25} \div \left(-\frac{2}{3}\right)$
35. Find the quotient of -74 and $-\frac{5}{3}$.
36. Find the quotient of -156 and $-\frac{3}{8}$.

Simplify each expression.

37. $\frac{81c}{9}$
38. $\frac{105g}{5}$
39. $\frac{8r + 24}{-8}$
40. $\frac{7h + 35}{-7}$
41. $\frac{40a - 50b}{2}$
42. $\frac{42c - 18d}{3}$
43. $\frac{-8f + (-16g)}{8}$
44. $\frac{-5x + (-10y)}{5}$

45. **CRAFTS** Hannah is making pillows. The pattern states that she needs $1\frac{3}{4}$ yards of fabric for each pillow. If she has $4\frac{1}{2}$ yards of fabric, how many pillows can she make?

46. **BOWLING** Bowling centers in the United States made \$2,800,000,000 in 1990. Their receipts in 1998 were \$2,764,000,000. What was the average change in revenue for each of these 8 years? **Source:** U.S. Census Bureau

Evaluate each expression if $m = -8$, $n = 6.5$, $p = 3.2$, and $q = -5.4$. Round to the nearest hundredth.

47. $\frac{mn}{p}$ **48.** $\frac{np}{m}$ **49.** $mq \div np$ **50.** $pq \div mn$

51. $\frac{n+p}{m}$ **52.** $\frac{m+p}{q}$ **53.** $\frac{m-2n}{-n+q}$ **54.** $\frac{p-3q}{-q-m}$

55. BUSINESS The president of a small business is looking at her profit/loss statement for the past year. The loss in income for the last year was $23,985. On average, what was the loss per month last year?

More About. . .

Jewelry

The discovery of gold at Sutter's Mill early in 1848 brought more than 40,000 prospectors to California within two years.

Source: www.infoplease.com

JEWELRY For Exercises 56 and 57, use the following information.
The gold content of jewelry is given in karats. For example, 24-karat gold is pure gold, and 18-karat gold is $\frac{18}{24}$ or 0.75 gold.

56. What fraction of 10-karat gold is pure gold? What fraction is not gold?

57. If a piece of jewelry is $\frac{2}{3}$ gold, how would you describe it using karats?

58. CRITICAL THINKING What is the least positive integer that is divisible by all whole numbers from 1 to 9?

59. WRITING IN MATH Answer the question that was posed at the beginning of the lesson.

How can you use division of rational numbers to describe data?

Include the following in your answer:

- an explanation of how you could use the mean of a set of data to describe changes in the data over time, and
- reasons why you think the change from year to year is not consistent.

Standardized Test Practice

60. If the rod is cut as shown, how many inches long will each piece be?

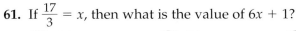

6.25 ft

(A) 0.625 in. (B) 1.875 in.
(C) 5.2 in. (D) 7.5 in.

61. If $\frac{17}{3} = x$, then what is the value of $6x + 1$?

(A) 32 (B) 33 (C) 35 (D) 44

Maintain Your Skills

Mixed Review

Find each product. *(Lesson 2-3)*

62. $-4(11)$ **63.** $-2.5(-1.2)$ **64.** $\frac{1}{4}(-5)$ **65.** $1.6(0.3)$

Find each difference. *(Lesson 2-2)*

66. $8 - (-6)$ **67.** $15 - 21$ **68.** $-7.5 - 4.8$ **69.** $-\frac{5}{8} - \left(-\frac{1}{6}\right)$

70. Name the property illustrated by $2(1.2 + 3.8) = 2 \cdot 5$.

Simplify each expression. If not possible, write *simplified*. *(Lesson 1-5)*

71. $8b + 12(b + 2)$ **72.** $6(5a + 3b - 2b)$ **73.** $3(x + 2y) - 2y$

Getting Ready for the Next Lesson

PREREQUISITE SKILL Find the mean, median, and mode for each set of data.
*(To review **mean**, **median**, and **mode**, see pages 818 and 819.)*

74. 40, 34, 40, 28, 38 **75.** 3, 9, 0, 2, 11, 8, 14, 3

76. 1.2, 1.7, 1.9, 1.8, 1.2, 1.0, 1.5 **77.** 79, 84, 81, 84, 75, 73, 80, 78

2-5 Statistics: Displaying and Analyzing Data

What You'll Learn

- Interpret and create line plots and stem-and-leaf plots.
- Analyze data using mean, median, and mode.

How are line plots and averages used to make decisions?

Vocabulary

- line plot
- frequency
- stem-and-leaf plot
- back-to-back stem-and-leaf plot
- measures of central tendency

How many people do you know with the same first name? Some names are more popular than others. The table below lists the top five most popular names for boys and girls born in each decade from 1950 to 1999.

Top Five First Names of America

Decade					
1950-59 (Boys)	Michael	James	Robert	John	David
1950-59 (Girls)	Deborah	Mary	Linda	Patricia	Susan
1960-69 (Boys)	Michael	John	David	James	Robert
1960-69 (Girls)	Lisa	Deborah	Mary	Karen	Michelle
1970-79 (Boys)	Michael	Christopher	Jason	David	James
1970-79 (Girls)	Jennifer	Michelle	Amy	Melissa	Kimberly
1980-89 (Boys)	Michael	Christopher	Matthew	Joshua	David
1980-89 (Girls)	Jessica	Jennifer	Ashley	Sarah	Amanda
1990-99 (Boys)	Michael	Christopher	Matthew	Joshua	Nicholas
1990-99 (Girls)	Ashley	Jessica	Sarah	Brittany	Emily

Source: *The World Almanac*

To help determine which names appear most frequently, these data could be displayed graphically.

CREATE LINE PLOTS AND STEM-AND-LEAF PLOTS In some cases, data can be presented using a **line plot**. Most line plots have a number line labeled with a scale to include all the data. Then an × is placed above a data point each time it occurs to represent the **frequency** of the data.

Example 1 Create a Line Plot

Draw a line plot for the data.

-2 4 3 2 6 10 7 4 -2 0 10 8 7 10 7 4 -1 9 -1 3

Step 1 The value of the data ranges from -2 to 10, so construct a number line containing those points.

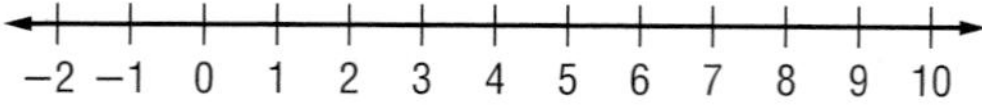

Step 2 Then place an × above a number each time it occurs.

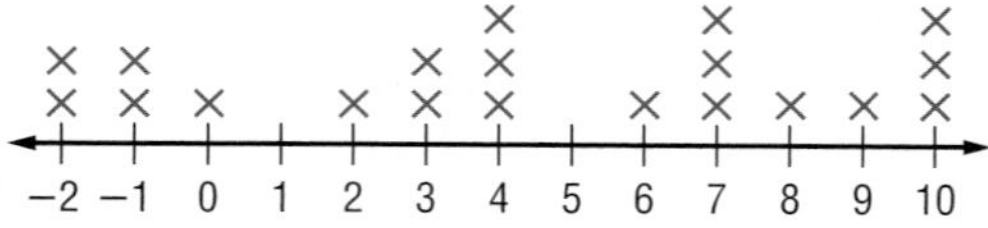

More About. . .

Animals

Whereas the cheetah is the fastest land animal, the fastest marine animal is the sailfish. It is capable of swimming 68 miles per hour.

Source: *The Top 10 of Everything*

Line plots are a convenient way to organize data for comparison.

Example 2 Use a Line Plot to Solve a Problem

ANIMALS The speeds (mph) of 20 of the fastest land animals are listed below.

45	70	43	45	32	42	40	40	35	50
40	35	61	48	35	32	50	36	50	40

Source: *The World Almanac*

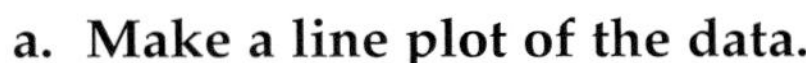

a. Make a line plot of the data.
The lowest value is 30, and the highest value is 70, so use a scale that includes those values. Place an × above each value for each occurrence.

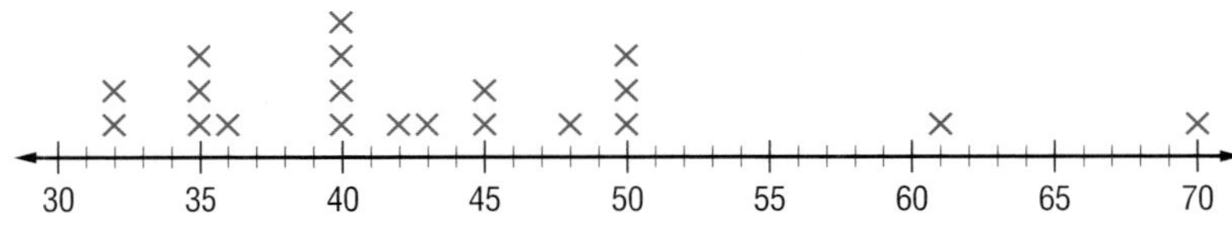

b. Which speed occurs most frequently?
Looking at the line plot, we can easily see that 40 miles per hour occurs most frequently.

Another way to organize and display data is by using a **stem-and-leaf plot**. In a stem-and-leaf plot, the greatest common place value is used for the *stems*. The numbers in the next greatest place value are used to form the *leaves*. In Example 2, the greatest place value is tens. Thus, 32 miles per hour would have a stem of 3 and a leaf of 2. A complete stem-and-leaf plot for the data in Example 2 is shown below.

Stem	Leaf
3	2 2 5 5 5 6
4	0 0 0 0 2 3 5 5 8
5	0 0 0
6	1
7	0 *3|2 = 32* ← key

Study Tip

Stem-and-Leaf Plots

A key is included on stem-and-leaf plots to indicate what the stems and leaves represent when read.

Example 3 Create a Stem-and-Leaf Plot

Use the data below to make a stem-and-leaf plot.

108	104	86	82	80	72	70	62	64	68	84	64	98	96	98
103	87	65	83	79	97	96	112	62	80	62	83	76	66	97

The greatest common place value is tens, so the digits in the tens place are the stems.

Stem	Leaf
6	2 2 2 4 4 5 6 8
7	0 2 6 9
8	0 0 2 3 3 4 6 7
9	6 6 7 7 8 8
10	3 4 8
11	2 *10|3 = 103*

A **back-to-back stem-and-leaf plot** can be used to compare two related sets of data.

Example 4 Back-to-Back Stem-and-Leaf Plot

Mrs. Evans wants to compare recent test scores from her two algebra classes. The table shows the scores for both classes.

a. Make a stem-and-leaf plot to compare the data.
To compare the data, we can use a back-to-back stem-and-leaf plot. Since the data represent similar measurements, the plot will share a common stem.

Class 1	Stem	Class 2
8 8 7	6	2
9 9 9 8 7 6 4 2	7	5 5 6 6 8 8
9 9 7 6 6 5 3 2	8	0 0 3 3 5 5 6 6 7 8 8 9
8 4 3 3 3 0	9	2 2 5 5 6
7|6 = 67	10	0 *6|2 = 62*

b. What is the difference between the highest score in each class?
100 − 98 or 2 points

c. Which class scored higher overall on the test?
Looking at the scores of 80 and above, we see that class 2 has a greater number of scores at or above 80 than class 1.

ANALYZE DATA When analyzing data, it is helpful to have one number that describes the set of data. Numbers known as **measures of central tendency** are often used to describe sets of data because they represent a centralized, or middle, value. Three of the most commonly used measures of central tendency are the mean, median, and mode.

When you use a measure of central tendency to describe a set of data, it is important that the measure you use best represents all of the data.

- Extremely high or low values can affect the mean, while not affecting the median or mode.
- A value with a high frequency can cause the mode to be misleading.
- Data that is clustered with a few values separate from the cluster can cause the median to be too low or too high.

Example 5 Analyze Data

Study Tip

Look Back
To review **finding mean**, **median**, and **mode**, see pages 818 and 819.

Which measure of central tendency best represents the data?

Stem	Leaf
7	7 8 9
8	2 2 2 2 2 3 4 4 6
9	
10	8
11	6 8 *7|9 = 0.79*

Determine the mean, median, and mode.

The mean is about 0.88. Add the data and divide by 15.

The median is 0.82. The middle value is 0.82.

The mode is 0.82. The most frequent value is 0.82.

Either the median or the mode best represent the set of data since both measures are located in the center of the majority of the data. In this instance, the mean is too high.

Example 6 Determine the Best Measure of Central Tendency

PRESIDENTS The numbers below show the ages of the U.S. Presidents since 1900 at the time they were inaugurated. Which measure of central tendency best represents the data?

42	**51**	**56**	**55**	**51**	**54**	**51**	**60**	**62**
43	**55**	**56**	**61**	**52**	**69**	**64**	**46**	**54**

The mean is about 54.6. Add the data and divide by 18.

The median is 54.5. The middle value is 54.5.

The mode is 51. The most frequent value is 51.

The mean or the median can be used to best represent the data. The mode for the data is too low.

Check for Understanding

Concept Check

1. **Explain** why it is useful to find the mean, median, and mode of a set of data.
2. Mitchell says that a line plot and a line graph are the same thing. Show that he is incorrect.
3. **OPEN ENDED** Write a set of data for which the median is a better representation than the mean.

Guided Practice

4. Use the data to make a line plot.

22 19 14 15 14 21 19 16 22 19 10 15 19 14 19

For Exercises 5–7, use the list that shows the number of hours students in Mr. Ricardo's class spent online last week.

7	**4**	**7**	**11**	**3**	**1**	**5**	**10**	**10**	**0**	**9**	**4**	**0**	**14**	**13**	**4**
11	**3**	**1**	**12**	**0**	**9**	**13**	**14**	**7**	**6**	**10**	**5**	**12**	**0**	**6**	**5**

5. Make a line plot of the data.
6. Which value occurs most frequently?
7. Does the mean, median, or mode best represent the data? Explain.
8. Use the data to make a stem-and-leaf plot.

68 66 68 88 76 71 88 93 86 64 73 80 81 72 68

For Exercises 9 and 10, use the data in the stem-and-leaf plot.

9. What is the difference between the least and greatest values?
10. Which measure of central tendency best describes the data? Explain.

Stem	Leaf
9	3 5 5
10	2 2 5 8
11	5 8 8 9 9 9
12	0 1 7 8 9

$9|3 = 9.3$

Application

BUILDINGS For Exercises 11–13, use the data below that represents the number of stories in the 25 tallest buildings in the world.

88	**88**	**110**	**88**	**80**	**69**	**102**	**78**	**70**	**54**	**80**	**85**	
83	**100**	**60**	**90**	**77**	**55**	**73**	**55**	**56**	**61**	**75**	**64**	**105**

11. Make a stem-and-leaf plot of the data.
12. Which value occurs most frequently?
13. Does the mode best describe the set of data? Explain.

Practice and Apply

Homework Help

For Exercises	See Examples
14–18	1, 2
20–22, 28, 29, 32, 33, 35, 36	3, 4
19, 23–27, 30, 31, 34, 37	5, 6

Extra Practice

See page 824.

Use each set of data to make a line plot.

14. 43 36 48 52 41 54 45 48 49 52 35 44 53 46 38 41 53

15. 1.0 −1.5 1.5 2.0 −1.5 2.1 −2.0 2.4 1.5 −1.4
2.5 1.4 −1.2 1.3 1.0 2.2 2.3 −1.2 −1.5 2.1

BASKETBALL **For Exercises 16–19, use the table that shows the seeds, or rank, of the NCAA men's basketball Final Four from 1991 to 2001.**

16. Make a line plot of the data.

17. How many of the teams in the Final Four were *not* number 1 seeds?

18. How many teams were seeded higher than third? (*Hint*: Higher seeds have lesser numerical value.)

19. Which measure of central tendency best describes the data? Explain.

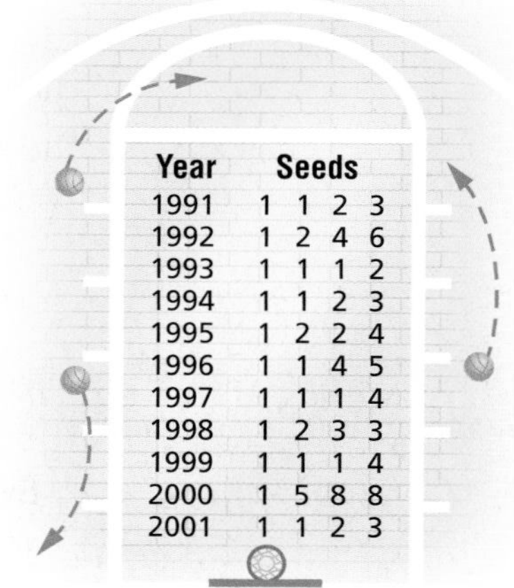

Year	Seeds
1991	1 1 2 3
1992	1 2 4 6
1993	1 1 1 2
1994	1 1 2 3
1995	1 2 2 4
1996	1 1 4 5
1997	1 1 1 4
1998	1 2 3 3
1999	1 1 1 4
2000	1 5 8 8
2001	1 1 2 3

Source: www.espn.com

Use each set of data to make a stem-and-leaf plot.

20. 6.5 6.3 6.9 7.1 7.3 5.9 6.0 7.0 7.2 6.6 7.1 5.8

21. 31 30 28 26 22 34 26 31 47 32 18 33 26 23 18 29

WEATHER **For Exercises 22–24, use the list of the highest recorded temperatures in each of the 50 states.**

112	100	128	120	134	118	106	110	109	112
100	118	117	116	118	121	114	114	105	109
107	112	114	115	118	117	118	125	106	110
122	108	110	121	113	120	119	111	104	111
120	113	120	117	105	110	118	112	114	114

Source: *The World Almanac*

22. Make a stem-and-leaf plot of the data.

23. Which temperature occurs most frequently?

24. Does the mode best represent the data? Explain.

25. RESEARCH Use the Internet or another source to find the total number of each CD sold over the past six months to reach number one. Which measure of central tendency best describes the average number of top selling CDs sold? Explain.

GEOLOGY **For Exercises 26 and 27, refer to the stem-and-leaf plot that shows the magnitudes of earthquakes occurring in 2000 that measured at least 5.0 on the Richter scale.**

Stem	Leaf
5	1 2 2 3 4 8 8 9 9
6	1 1 2 3 4 5 6 7 7 8
7	0 1 1 2 2 3 5 5 5 6 8 8
8	0 0 2 $5\|1 = 5.1$

Source: National Geophysical Data Center

26. What was the most frequent magnitude of these earthquakes?

27. Which measure of central tendency best describes this set of data? Explain.

OLYMPICS **For Exercises 28–31, use the information in the table that shows the number of medals won by the top ten countries during the 2000 Summer Olympics in Sydney, Australia.**

Sydney Olympics Total Medals by Country

Country	Gold	Silver	Bronze	Total
United States	40	24	33	97
Russia	32	28	28	88
China	28	16	15	59
Australia	16	25	17	58
Germany	13	17	26	56
France	13	14	11	38
Italy	14	8	13	35
Cuba	11	11	7	29
Britain	11	10	7	28
Korea	8	10	10	28

Source: www.espn.com

28. Make a line plot showing the number of gold medals won by the countries.
29. How many countries won fewer than 25 gold medals?
30. What was the median number of gold medals won by a country?
31. Is the median the best measure to describe this set of data? Explain.

CARS **For Exercises 32–34, use the list of the fuel economy of various vehicles in miles per gallon.**

25	28	29	30	24	28	29	31	34	30
33	47	34	43	33	36	37	29	30	30
29	26	29	22	23	19	18	20	23	21
20	20	19	16	18	21	20	19	28	20

Source: United States Environmental Protection Agency

32. Make a stem-and-leaf plot of the data.
33. How many of the vehicles get more than 25 miles per gallon?
34. Which measure of central tendency would you use to describe the fuel economy of the vehicles? Explain your reasoning.

More About. . .

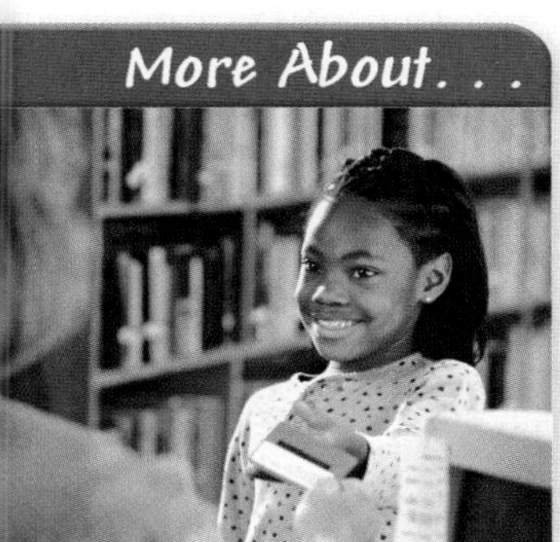

Education

In 1848, the Boston Public Library became the first public library to allow users to borrow books and materials.

Source: The Boston Public Library

EDUCATION **For Exercises 35–37, use the table that shows the top ten public libraries in the United States by population served.**

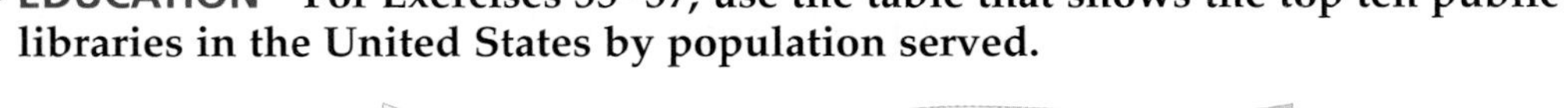

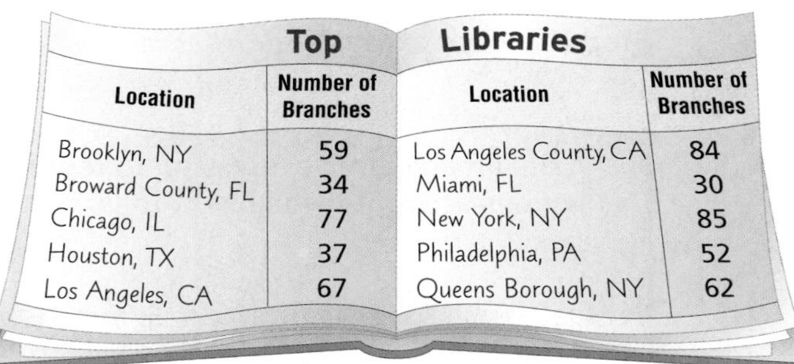

Top Libraries

Location	Number of Branches	Location	Number of Branches
Brooklyn, NY	59	Los Angeles County, CA	84
Broward County, FL	34	Miami, FL	30
Chicago, IL	77	New York, NY	85
Houston, TX	37	Philadelphia, PA	52
Los Angeles, CA	67	Queens Borough, NY	62

35. Make a stem-and-leaf plot to show the number of library branches.
36. Which interval has the most values?
37. What is the mode of the data?

38. **CRITICAL THINKING** Construct a set of twelve numbers with a mean of 7, a median of 6, and a mode of 8.

SALARIES For Exercises 39–41, refer to the bar graph that shows the median income of males and females based on education levels.

39. What are the differences between men's and women's salaries at each level of education?

40. What do these graphs say about the difference between salaries and education levels?

41. Why do you think that salaries are usually represented by the median rather than the mean?

Education (25 or older) — Male / Female

	Male	Female
High school graduate	\$33,184	\$23,061
Some college	\$39,221	\$27,757
Bachelor's degree	\$60,201	\$41,747
Doctoral degree	\$81,687	\$60,079

Source: *USA TODAY*

42. WRITING IN MATH Answer the question that was posed at the beginning of the lesson.

How are line plots and averages used to make decisions?

Include the following in your answer:

- a line plot to show how many male students in your class have the most popular names for the decade in which they were born, and
- a convincing argument that explains how you would use this information to sell personalized T-shirts.

Standardized Test Practice

For Exercises 43 and 44, refer to the line plot.

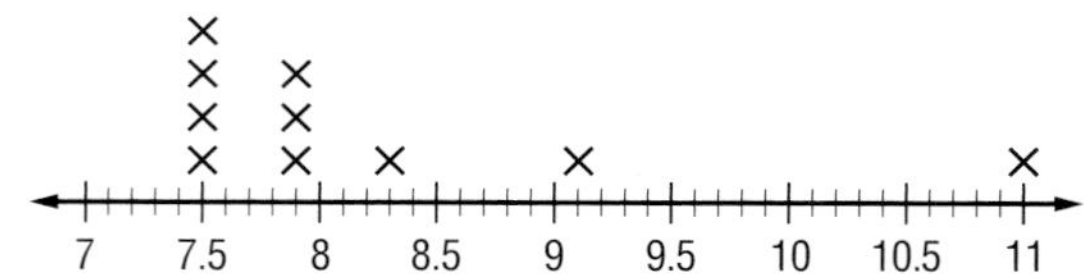

43. What is the average wingspan for these types of butterflies?

Ⓐ 7.6 in. Ⓑ 7.9 in.

Ⓒ 8.2 in. Ⓓ 9.1 in.

44. Which sentence is *not* true?

Ⓐ The difference between the greatest and least wingspan is 3.5 inches.

Ⓑ Most of the wingspans are in the 7.5 inch to 8.5 inch interval.

Ⓒ Most of the wingspans are greater than 8 inches.

Ⓓ The mode of the data is 7.5 inches.

Maintain Your Skills

Mixed Review

Find each quotient. *(Lesson 2-4)*

45. $56 \div (-14)$ **46.** $-72 \div (-12)$ **47.** $-40.5 \div 3$ **48.** $102 \div 6.8$

Simplify each expression. *(Lesson 2-3)*

49. $-2(6x) - 5x$ **50.** $3x(-7y) - 4x(5y)$ **51.** $5(3t - 2t) - 2(4t)$

52. Write an algebraic expression to represent the amount of money in Kara's savings account if she has d dollars and adds x dollars per week for 12 weeks. *(Lesson 1-1)*

Evaluate each expression if $x = 5$, $y = 16$, and $z = 9$. *(Lesson 1-2)*

53. $y - 3x$ **54.** $xz \div 3$ **55.** $2x - x + (y \div 4)$ **56.** $\frac{x^2 - z}{2y}$

Getting Ready for the Next Lesson

PREREQUISITE SKILL **Write each fraction in simplest form.**

(To review ***simplifying fractions****, see pages 798 and 799.)*

57. $\frac{12}{18}$ **58.** $\frac{54}{60}$ **59.** $\frac{21}{30}$ **60.** $\frac{42}{48}$

61. $\frac{32}{64}$ **62.** $\frac{28}{52}$ **63.** $\frac{16}{36}$ **64.** $\frac{84}{90}$

Reading Mathematics

Interpreting Statistics

The word *statistics* is associated with the collection, analysis, interpretation, and presentation of numerical data. Sometimes, when presenting data, *notes* and *unit indicators* are included to help you interpret the data.

Headnotes give information about the table as a whole.

If the numerical data are too large, *unit indicators* are used to save space.

Public Elementary and Secondary School Enrollment, 1994–1998
[(in thousands) 44,111 represents 44,111,000.]
As of fall year, Kindergarten includes nursery schools.

Grade	1994	1995	1996	1997	1998, prel.
Pupils enrolled	**44,111**	**44,840**	**45,611**	**46,127**	**46,535**
Kindergarten and grades 1 to 8	31,898	32,341	32,764	33,073	33,344
Kindergarten	4047	4173	4202	4198	4171
First	3593	3671	3770	3755	3727
Second	3440	3507	3600	3689	3682
Third	3439	3445	3524	3597	3696
Fourth	3426	3431	3454	3507	3592
Fifth	3372	3438	3453	3458	3520
Sixth	3381	3395	3494	3492	3497
Seventh	3404	3422	3464	3520	3530
Eighth	3302	3356	3403	3415	3480
Unclassified[1]	494	502	401	442	460
Grades 9 to 12	12,213	12,500	12,847	13,054	13,191
Ninth	3604	3704	3801	3819	3856
Tenth	3131	3237	3323	3376	3382
Eleventh	2748	2826	2930	2972	3018
Twelfth	2488	2487	2586	2673	2724
Unclassified[1]	242	245	206	214	211

[1] Includes ungraded and special education.
Source: U.S. Census Bureau

Footnotes give information about specific items within the table.

Suppose you need to find the number of students enrolled in the 9th grade in 1997. The following steps can be used to determine this information.

Step 1 Locate the number in the table. The number that corresponds to 1997 and 9th grade is 3819.

Step 2 Determine the unit indicator. The *unit indicator* is thousands.

Step 3 If the unit indicator is not 1 unit, multiply to find the data value. In this case, multiply 3819 by 1000.

Step 4 State the data value. The number of students enrolled in the 9th grade in 1997 was 3,819,000.

Reading to Learn

Use the information in the table to answer each question.

1. Describe the data.
2. What information is given by the footnote?
3. How current is the data?
4. What is the unit indicator?
5. How many acres of state parks and recreation areas does New York have?
6. Which of the states shown had the greatest number of visitors? How many people visited that state's parks and recreation areas in 1999?

State Parks and Recreation Areas for Selected States, 1999

State	Acreage (1000)	Visitors (1000)[1]
United States	**12,916**	**766,842**
Alaska	3291	3855
California	1376	76,736
Florida	513	14,645
Indiana	178	18,652
New York	1016	61,960
North Carolina	158	13,269
Oregon	94	38,752
South Carolina	82	9563
Texas	628	21,446

Source: U.S. Census Bureau

[1] Includes overnight visitors.

2-6 Probability: Simple Probability and Odds

What You'll Learn

- Find the probability of a simple event.
- Find the odds of a simple event.

Vocabulary

- probability
- simple event
- sample space
- equally likely
- odds

Why is probability important in sports?

A basketball player is at the free throw line. Her team is down by one point. If she makes an average of 75% of her free throws, what is the probability that she will tie the game with her first shot?

PROBABILITY One way to describe the likelihood of an event occurring is with probability. The **probability** of a **simple event**, like a coin landing heads up when it is tossed, is a ratio of the number of favorable outcomes for the event to the total number of possible outcomes of the event. The probability of an event can be expressed as a fraction, a decimal, or a percent.

Suppose you wanted to find the probability of rolling a 4 on a die. When you roll a die, there are six possible outcomes, 1, 2, 3, 4, 5, or 6. This list of all possible outcomes is called the **sample space**. Of these outcomes, only one, a 4, is favorable. So, the probability of rolling a 4 is $\frac{1}{6}$, $0.1\overline{6}$, or about 16.7%.

Study Tip

Reading Math
$P(a)$ *is read the probability of a.*

Key Concept — Probability

The probability of an event a can be expressed as

$$P(a) = \frac{\text{number of favorable outcomes}}{\text{total number of possible outcomes}}.$$

Example 1 Find Probabilities of Simple Events

a. Find the probability of rolling an even number on a die.

There are six possible outcomes. Three of the outcomes are favorable. That is, three of the six outcomes are even numbers.

Sample space: 1, 2, 3, 4, 5, 6

3 even numbers → $\frac{3}{6}$ ← 6 total possible outcomes

So, $P(\text{even number}) = \frac{3}{6}$ or $\frac{1}{2}$.

b. A bowl contains 5 red chips, 7 blue chips, 6 yellow chips, and 10 green chips. One chip is randomly drawn. Find P(blue).

There are 7 blue chips and 28 total chips.

$P(\text{blue chip}) = \frac{7}{28}$ ← number of favorable outcomes / ← number of possible outcomes

$= \frac{1}{4}$ or 0.25 Simplify.

The probability of selecting a blue chip is $\frac{1}{4}$ or 25%.

c. A bowl contains 5 red chips, 7 blue chips, 6 yellow chips, and 10 green chips. One chip is randomly drawn. Find P(red or yellow).

There are 5 ways to pick a red chip and 6 ways to pick a yellow chip. So there are 5 + 6 or 11 ways to pick a red or a yellow chip.

$P(\text{red or yellow}) = \frac{11}{28}$ ← number of favorable outcomes / ← number of possible outcomes

≈ 0.39 Divide.

The probability of selecting a red chip or a yellow chip is $\frac{11}{28}$ or about 39%.

d. A bowl contains 5 red chips, 7 blue chips, 6 yellow chips, and 10 green chips. One chip is randomly drawn. Find P(not green).

There are 5 + 7 + 6 or 18 chips that are not green.

$P(\text{not green}) = \frac{18}{28}$ ← number of favorable outcomes / ← number of possible outcomes

≈ 0.64 Divide.

The probability of selecting a chip that is not green is $\frac{9}{14}$ or about 64%.

Study Tip

Reading Math
Inclusive means that the end values are included.

Notice that the probability that an event will occur is somewhere between 0 and 1 inclusive. If the probability of an event is 0, that means that it is impossible for the event to occur. A probability equal to 1 means that the event is certain to occur.

When there are n outcomes and the probability of each one is $\frac{1}{n}$, we say that the outcomes are **equally likely**. For example, when you roll a die, the 6 possible outcomes are equally likely because each outcome has a probability of $\frac{1}{6}$.

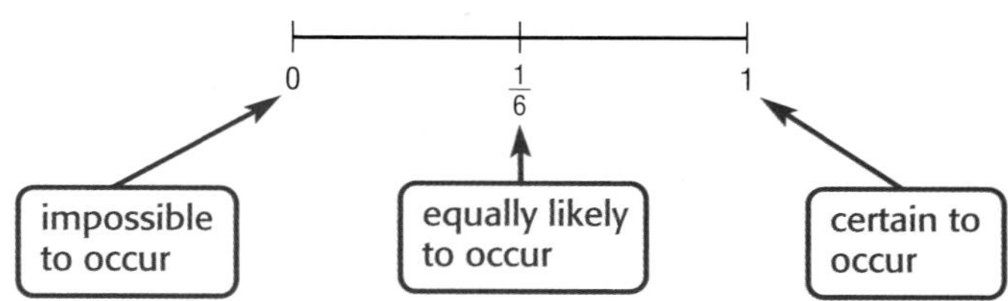

ODDS Another way to express the chance of an event occurring is with **odds**.

Key Concept — Odds

The odds of an event occurring is the ratio that compares the number of ways an event can occur (successes) to the number of ways it cannot occur (failures).

Study Tip

Odds
Odds are usually written in the form *number of successes : number of failures.*

Example 2 Odds of an Event

Find the odds of rolling a number less than 3.

There are 6 possible outcomes, 2 are successes and 4 are failures.

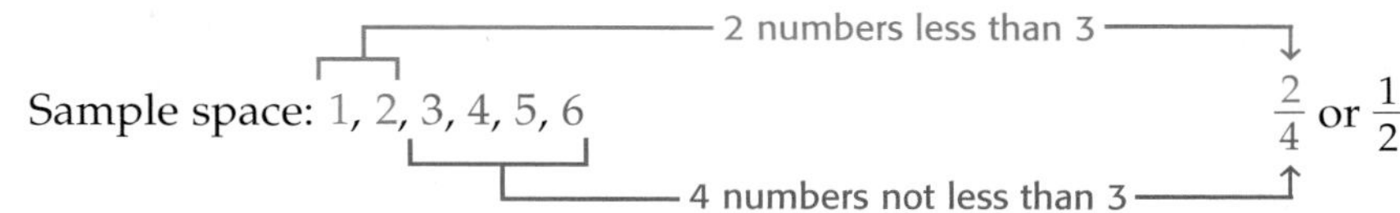

So, the odds of rolling a number less than three are $\frac{1}{2}$ or 1:2.

The odds *against* an event occurring are the odds that the event will *not* occur.

Study Tip

In this text, a *standard deck of cards* always indicates 52 cards in 4 suits of 13 cards each.

Example 3 Odds Against an Event

A card is selected at random from a standard deck of 52 cards. What are the odds against selecting a 3?

There are four 3s in a deck of cards, and there are $52 - 4$ or 48 cards that are not a 3.

odds against a 3 $= \frac{48}{4}$ ← number of ways to *not* pick a 3

The odds against selecting a 3 from a deck of cards are 12:1.

Example 4 Probability and Odds

WEATHER **A weather forecast states that the probability of rain the next day is 40%. What are the odds that it will rain?**

The probability that it will rain is 40%, so the probability that it will not rain is 60%.

odds of rain = 40:60 or 2:3

The odds that it will rain tomorrow are 2:3.

Check for Understanding

Concept Check

1. OPEN ENDED Give an example of an impossible event, a certain event, and an equally likely event when a die is rolled.
2. **Describe** how to find the odds of an event occurring if the probability that the event will occur is $\frac{3}{5}$.
3. FIND THE ERROR Mark and Doug are finding the probability of picking a red card from a standard deck of cards.

Mark	Doug
P(red card) $= \frac{26}{26}$ or $\frac{1}{1}$	P(red card) $= \frac{26}{52}$ or $\frac{1}{2}$

Who is correct? Explain your reasoning.

Guided Practice **A card is selected at random from a standard deck of cards. Determine each probability.**

4. $P(5)$
5. P(red 10)
6. P(odd number)
7. P(queen of hearts or jack of diamonds)

Find the odds of each outcome if the spinner is spun once.

8. multiple of 3
9. even number less than 8
10. odd number or blue
11. red or yellow

Application NUMBER THEORY **One of the factors of 48 is chosen at random.**

12. What is the probability that the chosen factor is not a multiple of 4?
13. What is the probability that the number chosen has 4 and 6 as two of its factors?

Practice and Apply

Homework Help

For Exercises	See Examples
14–35, 51, 54, 56	1
36–47, 52, 53, 55	2, 3
48, 49	4

Extra Practice

See page 824.

One coin is randomly selected from a jar containing 70 nickels, 100 dimes, 80 quarters, and 50 1-dollar coins. Find each probability.

14. *P*(quarter)
15. *P*(dime)
16. *P*(nickel or dollar)
17. *P*(quarter or nickel)
18. *P*(value less than \$1.00)
19. *P*(value greater than \$0.10)
20. *P*(value at least \$0.25)
21. *P*(value at most \$1.00)

Two dice are rolled, and their sum is recorded. Find each probability.

22. *P*(sum less than 7)
23. *P*(sum less than 8)
24. *P*(sum is greater than 12)
25. *P*(sum is greater than 1)
26. *P*(sum is between 5 and 10)
27. *P*(sum is between 2 and 9)

One of the polygons is chosen at random. Find each probability.

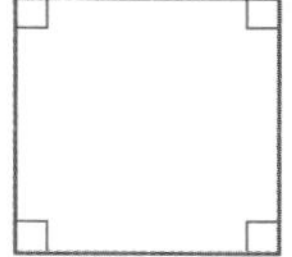 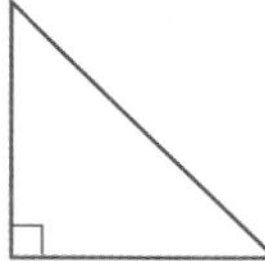 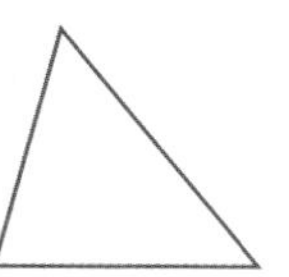 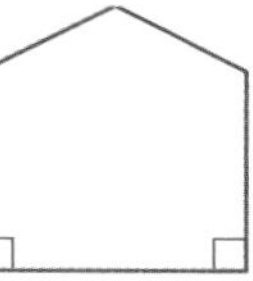  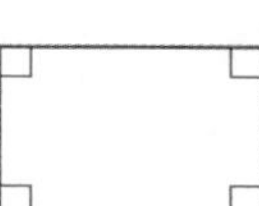

28. *P*(triangle)
29. *P*(pentagon)
30. *P*(not a triangle)
31. *P*(not a quadrilateral)
32. *P*(more than three sides)
33. *P*(more than one right angle)

34. If a person's birthday is in April, what is the probability that it is the 29th?

35. If a person's birthday is in July, what is the probability that it is after the 16th?

Find the odds of each outcome if a computer randomly picks a letter in the name *The United States of America.*

36. the letter *a*
37. the letter *t*
38. a vowel
39. a consonant
40. an uppercase letter
41. a lowercase vowel

More About...

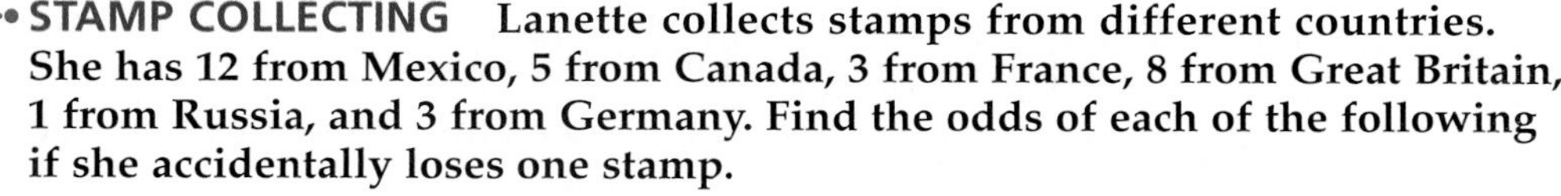

Stamp Collecting

Stamp collecting can be a very inexpensive hobby. Most stamp collectors start by saving stamps from letters, packages, and postcards.

Source: United States Postal Service

STAMP COLLECTING **Lanette collects stamps from different countries. She has 12 from Mexico, 5 from Canada, 3 from France, 8 from Great Britain, 1 from Russia, and 3 from Germany. Find the odds of each of the following if she accidentally loses one stamp.**

42. the stamp is from Canada
43. the stamp is from Mexico
44. the stamp is not from France
45. the stamp is not from a North American country
46. the stamp is from Germany or Russia
47. the stamp is from Canada or Great Britain

48. If the probability that an event will occur is $\frac{3}{7}$, what are the odds that it will occur?

49. If the probability that an event will occur is $\frac{2}{3}$, what are the odds against it occurring?

50. **CONTESTS** Every Tuesday, Mike's Submarine Shop has a business card drawing for a free lunch. Four coworkers from InvoAccounting put their business cards in the bowl for the drawing. If there are 80 cards in the bowl, what are the odds that one of the coworkers will win a free lunch?

GAMES For Exercises 51–53, use the following information.
A game piece is randomly placed on the board shown at the right by blindfolded players.

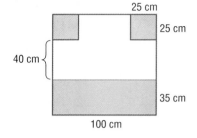

25 cm
25 cm
40 cm
35 cm
100 cm

51. What is the probability that a game piece is placed on a shaded region?

52. What are the odds against placing a game piece on a shaded region?

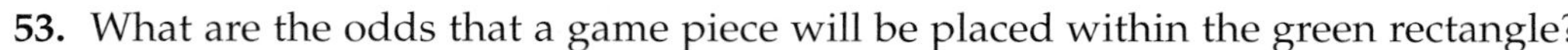

53. What are the odds that a game piece will be placed within the green rectangle?

More About. . .

Baseball

The record for the most home runs in a single season is 84. It was set by Joshua Gibson of the Homestead Grays in 1934.

Source: National Baseball Hall of Fame

BASEBALL For Exercises 54–56, use the following information.
The stem-and-leaf plot shows the number of home runs hit by the top major league baseball players in the 2000 season. **Source:** www.espn.com

Stem	Leaf
3	0 0 0 0 1 1 1 1 1 1 1 2 2 2 3
	3 4 4 4 5 5 5 6 6 6 7 7 8 8 9
4	0 1 1 1 1 2 2 3 3 3 4 4 7 7 9
5	0 $3 \mid 0 = 30$

54. What is the probability that one of these players picked at random hit more than 35 home runs?

55. What are the odds that a randomly selected player hit fewer than 45 home runs?

56. If a player batted 439 times and hit 38 home runs, what is the probability that the next time the player bats he will hit a home run?

CONTESTS For Exercises 57 and 58, use the following information.
A fast-food restaurant is holding a contest in which the grand prize is a new sports car. Each customer is given a game card with their order. The contest rules state that the odds of winning the grand prize are 1:1,000,000.

57. For any randomly-selected game card, what is the probability that it is the winning game card for the grand prize?

58. Do your odds of winning the grand prize increase significantly if you have several game cards? Explain.

You can use real-world data to find the probability that a person will live to be 100. Visit www.algebra1.com/webquest to continue work on your WebQuest project.

59. **CRITICAL THINKING** Three coins are tossed, and a tail appears on at least one of them. What is the probability that at least one head appears?

60. **WRITING IN MATH** Answer the question that was posed at the beginning of the lesson.

Why is probability important in sports?

Include the following in your answer:

- examples of two sports in which probability is used and an explanation of each sport's importance, and
- examples of methods other than probability used to show chance.

61. If the probability that an event will occur is $\frac{12}{25}$, what are the odds that the event will *not* occur?

(A) 12:13 (B) 13:12 (C) 13:25 (D) 25:12

62. What is the probability that a number chosen at random from the domain $\{-6, -5, -4, -3, -2, -1, 0, 1, 2, 3, 4, 5, 6, 7, 8\}$ will satisfy the inequality $3x + 2 \leq 17$?

(A) 20% (B) 27% (C) 73% (D) 80%

Maintain Your Skills

Mixed Review

63. **WEATHER** The following data represents the average daily temperature in Fahrenheit for Sacramento, California, for two weeks during the month of May. Organize the data using a stem-and-leaf plot. *(Lesson 2-5)*

58.3	64.3	66.7	65.1	68.7	67.0	69.3
70.0	72.8	77.4	77.4	73.2	75.8	65.5

Evaluate each expression if $a = -\frac{1}{3}$, $b = \frac{2}{5}$, and $c = \frac{1}{2}$. *(Lesson 2-4)*

64. $b \div c$ 65. $2a \div b$ 66. $\frac{ab}{c}$

Find each sum. *(Lesson 2-2)*

67. $4.3 + (-8.2)$ 68. $-12.2 + 7.8$ 69. $-\frac{1}{4} + \left(-\frac{3}{8}\right)$ 70. $\frac{7}{12} + \left(-\frac{5}{6}\right)$

Find each absolute value. *(Lesson 2-1)*

71. $|4.25|$ 72. $|-8.4|$ 73. $\left|-\frac{2}{3}\right|$ 74. $\left|\frac{1}{6}\right|$

Getting Ready for the Next Lesson

PREREQUISITE SKILL **Evaluate each expression.**
*(To review **evaluating expressions**, see Lesson 1-2.)*

75. 6^2 76. 17^2 77. $(-8)^2$ 78. $(-11.5)^2$

79. 1.6^2 80. $\left(\frac{5}{12}\right)^2$ 81. $\left(-\frac{4}{9}\right)^2$ 82. $\left(-\frac{16}{15}\right)^2$

Practice Quiz 2

Lessons 2-4 through 2-6

Find each quotient. *(Lesson 2-4)*

1. $-136 \div (-8)$ 2. $15 \div \left(-\frac{3}{8}\right)$ 3. $(-46.8) \div 4$

Simplify each expression. *(Lesson 2-4)*

4. $\frac{3a + 9}{3}$ 5. $\frac{4x + 32}{4}$ 6. $\frac{15n - 20}{-5}$

7. State the scale you would use to make a line plot for the following data. Then draw the line plot. *(Lesson 2-5)*

1.9	1.1	3.2	5.0	4.3	2.7	2.5	1.1	1.4	1.8	1.8	1.6
4.3	2.9	1.4	1.7	3.6	2.9	1.9	0.4	1.3	0.9	0.7	1.9

Determine each probability if two dice are rolled. *(Lesson 2-6)*

8. $P(\text{sum of } 10)$ 9. $P(\text{sum} \geq 6)$ 10. $P(\text{sum} < 10)$

Algebra Activity

A Follow-Up of Lesson 2-6

Investigating Probability and Pascal's Triangle

Collect the Data

- If a family has one child, you know that the child is either a boy or a girl. You can make a simple table to show this type of family.

1 boy	1 girl
B	G

You can see that there are 2 possibilities for a one-child family.

- If a family has two children, the table below shows the possibilities for two children, including the order of birth. For example, BG means that a boy is born first and a girl second.

2 boys, 0 girls	1 boy, 1 girl	0 boys, 2 girls
BB	BG	GG
	GB	

There are 4 possibilities for the two-child family: BB, BG, GB, or GG.

Analyze the Data

3 boys	2 boys, 1 girl	1 boy, 2 girls	3 girls
BBB	BBG	BGG	GGG

1. Copy and complete the table that shows the possibilities for a three-child family.
2. Make your own table to show the possibilities for a four-child family.
3. List the total number of possibilities for a one-child, two-child, three-child, and four-child family. How many possibilities do you think there are for a five-child family? a six-child family? Describe the pattern of the numbers you listed.
4. Find the probability that a three-child family has 2 boys and 1 girl.
5. Find the probability that a four-child family has 2 boys and 2 girls.

Make a Conjecture

6. Blaise Pascal was a French mathematician who lived in the 1600s. He is known for this triangle of numbers, called Pascal's triangle, although the pattern was known by other mathematicians before Pascal's time.

				1					Row 0
			1		1				Row 1
		1		2		1			Row 2
	1		3		3		1		Row 3
1		4		6		4		1	Row 4

   Explain how Pascal's triangle relates to the possibilities for the make-up of families. (*Hint*: The first row indicates that there is 1 way to have 0 children.)

7. Use Pascal's triangle to find the probability that a four-child family has 1 boy.

2-7 Square Roots and Real Numbers

What You'll Learn

- Find square roots.
- Classify and order real numbers.

Vocabulary

- square root
- perfect square
- radical sign
- principal square root
- irrational numbers
- real numbers
- rational approximations

How can using square roots determine the surface area of the human body?

In the 2000 Summer Olympics, Australian sprinter Cathy Freeman wore a special running suit that covered most of her body. The surface area of the human body may be found using the formula below, where height is measured in centimeters and weight is in kilograms.

$$\text{Surface Area} = \sqrt{\frac{\text{height} \times \text{weight}}{3600}} \text{ square meters}$$

The symbol $\sqrt{}$ designates a square root.

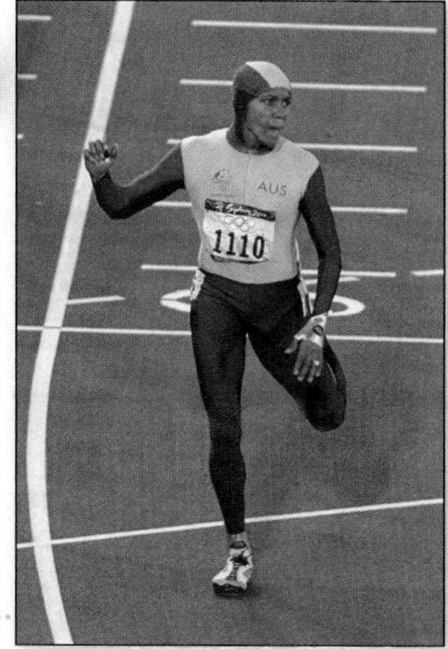

SQUARE ROOTS A **square root** is one of two equal factors of a number. For example, one square root of 64 is 8 since $8 \cdot 8$ or 8^2 is 64. Another square root of 64 is -8 since $(-8) \cdot (-8)$ or $(-8)^2$ is also 64. A number like 64, whose square root is a rational number is called a **perfect square**.

The symbol $\sqrt{}$, called a **radical sign**, is used to indicate a nonnegative or **principal square root** of the expression under the radical sign.

Study Tip

Reading Math

$\pm\sqrt{64}$ is read *plus or minus the square root of 64.*

$\sqrt{64} = 8$ ← $\sqrt{64}$ indicates the *principal* square root of 64.

$-\sqrt{64} = -8$ ← $-\sqrt{64}$ indicates the *negative* square root of 64.

$\pm\sqrt{64} = \pm 8$ ← $\pm\sqrt{64}$ indicates *both* square roots of 64.

Note that $-\sqrt{64}$ is not the same as $\sqrt{-64}$. The notation $-\sqrt{64}$ represents the negative square root of 64. The notation $\sqrt{-64}$ represents the square root of -64, which is not a real number since no real number multiplied by itself is negative.

Example 1 Find Square Roots

Find each square root.

a. $-\sqrt{\dfrac{49}{256}}$

$-\sqrt{\dfrac{49}{256}}$ represents the negative square root of $\dfrac{49}{256}$.

$\dfrac{49}{256} = \left(\dfrac{7}{16}\right)^2 \rightarrow -\sqrt{\dfrac{49}{256}} = -\dfrac{7}{16}$

b. $\pm\sqrt{0.81}$

$\pm\sqrt{0.81}$ represents the positive and negative square roots of 0.81.

$0.81 = 0.9^2$ and $0.81 = (-0.9)^2$

$\pm\sqrt{0.81} = \pm 0.9$

CLASSIFY AND ORDER NUMBERS Recall that rational numbers are numbers that can be expressed as terminating or repeating decimals, or in the form $\frac{a}{b}$, where a and b are integers and $b \neq 0$.

As you have seen, the square roots of perfect squares are rational numbers. However, numbers such as $\sqrt{3}$ and $\sqrt{24}$ are the square roots of numbers that are not perfect squares. Numbers like these cannot be expressed as a terminating or repeating decimal.

$$\sqrt{3} = 1.73205080\ldots$$

$$\sqrt{24} = 4.89897948\ldots$$

Numbers that cannot be expressed as terminating or repeating decimals, or in the form $\frac{a}{b}$, where a and b are integers and $b \neq 0$, are called **irrational numbers**. Irrational numbers and rational numbers together form the set of **real numbers**.

Concept Summary — *Real Numbers*

The set of real numbers consists of the set of rational numbers and the set of irrational numbers.

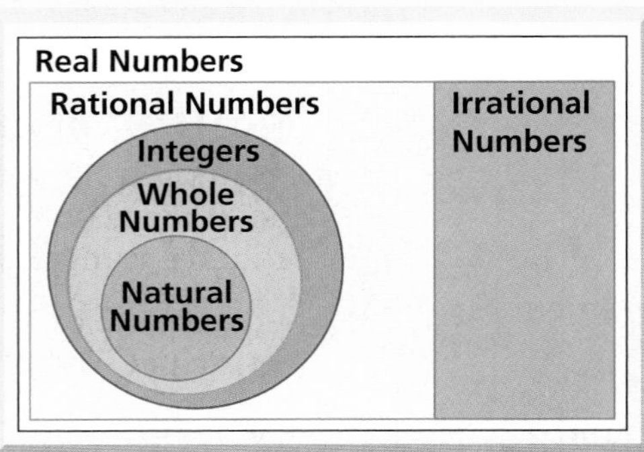

Example 2 *Classify Real Numbers*

Name the set or sets of numbers to which each real number belongs.

a. $\frac{5}{22}$

Because 5 and 22 are integers and $5 \div 22 = 0.2272727\ldots$, which is a repeating decimal, this number is a rational number.

b. $\sqrt{121}$

Because $\sqrt{121} = 11$, this number is a natural number, a whole number, an integer, and a rational number.

c. $\sqrt{56}$

Because $\sqrt{56} = 7.48331477\ldots$, which is not a repeating or terminating decimal, this number is irrational.

d. $-\frac{36}{4}$

Because $-\frac{36}{4} = -9$, this number is an integer and a rational number.

Study Tip

Common Misconception

Pay close attention to the placement of a negative sign when working with square roots. $\sqrt{-121}$ is undefined for real numbers since no real number multiplied by itself can result in a negative product.

In Lesson 2-1 you graphed rational numbers on a number line. However, the rational numbers alone do not complete the number line. By including irrational numbers, the number line is complete. This is illustrated by the **Completeness Property** which states that each point on the number line corresponds to exactly one real number.

Recall that inequalities like $x < 7$ are open sentences. To solve the inequality, determine what replacement values for x make the sentence true. This can be shown by the solution set {all real numbers less than 7}. Not only does this set include integers like 5 and -2, but it also includes rational numbers like $\frac{3}{8}$ and $-\frac{12}{13}$ and irrational numbers like $\sqrt{40}$ and π.

Example 3 Graph Real Numbers

Graph each solution set.

a. $x > -2$

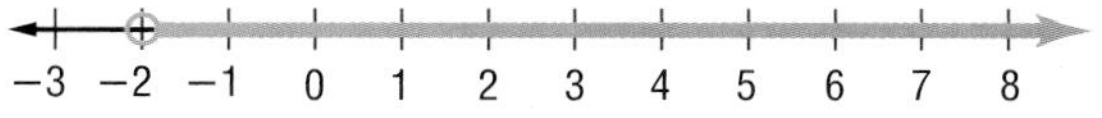

−3 −2 −1 0 1 2 3 4 5 6 7 8

The heavy arrow indicates that all numbers to the right of -2 are included in the graph. The circle at -2 indicates -2 is *not* included in the graph.

b. $a \leq 4.5$

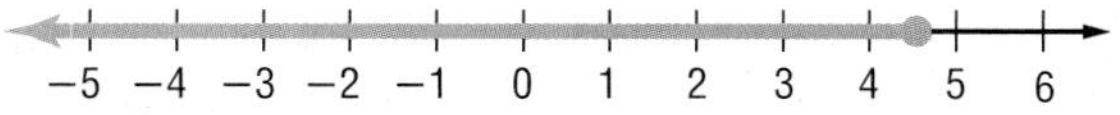

−5 −4 −3 −2 −1 0 1 2 3 4 5 6

The heavy arrow indicates that all points to the left of 4.5 are included in the graph. The dot at 4.5 indicates that 4.5 *is* included in the graph.

To express irrational numbers as decimals, you need to use a rational approximation. A **rational approximation** of an irrational number is a rational number that is close to, but not equal to, the value of the irrational number. For example, a rational approximation of $\sqrt{2}$ is 1.41 when rounded to the nearest hundredth.

Example 4 Compare Real Numbers

Replace each ● with <, >, or = to make each sentence true.

a. $\sqrt{19}$ ● $3.\overline{8}$

Find two perfect squares closest to $\sqrt{19}$ and write an inequality.

$16 < 19 < 25$ — 19 is between 16 and 25.

$\sqrt{16} < \sqrt{19} < \sqrt{25}$ — Find the square root of each number.

$4 < \sqrt{19} < 5$ — $\sqrt{19}$ is between 4 and 5.

Since $\sqrt{19}$ is between 4 and 5, it must be greater than $3.\overline{8}$.
So, $\sqrt{19} > 3.\overline{8}$.

b. $7.\overline{2}$ ● $\sqrt{52}$

You can use a calculator to find an approximation for $\sqrt{52}$.

$\sqrt{52} = 7.211102551\ldots$

$7.\overline{2} = 7.222\ldots$

Therefore, $7.\overline{2} > \sqrt{52}$.

You can write a set of real numbers in order from greatest to least or from least to greatest. To do so, find a decimal approximation for each number in the set and compare.

Example 5 Order Real Numbers

Write $2.\overline{63}$, $-\sqrt{7}$, $\frac{8}{3}$, $\frac{53}{-20}$ in order from least to greatest.

Write each number as a decimal.

$2.\overline{63} = 2.6363636\ldots$ or about 2.636.

$-\sqrt{7} = -2.64575131\ldots$ or about -2.646.

$\frac{8}{3} = 2.66666666\ldots$ or about 2.667.

$\frac{53}{-20} = -2.65$

$-2.65 < -2.646 < 2.636 < 2.667$

The numbers arranged in order from least to greatest are $\frac{53}{-20}$, $-\sqrt{7}$, $2.\overline{63}$, $\frac{8}{3}$.

You can use rational approximations to test the validity of some algebraic statements involving real numbers.

Standardized Test Practice

A B C D

Example 6 Rational Approximation

Multiple-Choice Test Item

For what value of x is $\frac{1}{\sqrt{x}} > \sqrt{x} > x$ true?

(A) $\frac{1}{2}$ (B) 0 (C) -2 (D) 3

Read the Test Item

The expression $\frac{1}{\sqrt{x}} > \sqrt{x} > x$ is an open sentence, and the set of choices $\left\{\frac{1}{2}, 0, -2, 3\right\}$ is the replacement set.

Solve the Test Item

Replace x in $\frac{1}{\sqrt{x}} > \sqrt{x} > x$ with each given value.

(A) $x = \frac{1}{2}$

$\frac{1}{\sqrt{\frac{1}{2}}} \stackrel{?}{>} \sqrt{\frac{1}{2}} \stackrel{?}{>} \frac{1}{2}$ Use a calculator.

$1.41 > 0.71 > 0.5$ ✓ True

(B) $x = 0$

$\frac{1}{\sqrt{0}} \stackrel{?}{>} \sqrt{0} \stackrel{?}{>} 0$

False; $\frac{1}{\sqrt{0}}$ is not a real number.

(C) $x = -2$

$\frac{1}{\sqrt{-2}} \stackrel{?}{>} \sqrt{-2} \stackrel{?}{>} -2$

False; $\frac{1}{\sqrt{-2}}$ and $\sqrt{-2}$ are not real numbers.

(D) $x = 3$

$\frac{1}{\sqrt{3}} \stackrel{?}{>} \sqrt{3} \stackrel{?}{>} 3$ Use a calculator.

~~$0.58 > 1.73 > 3$~~ False

The inequality is true for $x = \frac{1}{2}$, so the correct answer is A.

The Princeton Review

Test-Taking Tip

You could stop when you find that A is a solution. But testing the other values is a good check.

Check for Understanding

Concept Check

1. **Tell** whether the square root of any real number is *always*, *sometimes* or *never* positive. Explain your answer.
2. **OPEN ENDED** Describe the difference between rational numbers and irrational numbers. Give examples of both.
3. **Explain** why you cannot evaluate $\sqrt{-25}$ using real numbers.

Guided Practice

Find each square root. If necessary, round to the nearest hundredth.

4. $-\sqrt{25}$
5. $\sqrt{1.44}$
6. $\pm\sqrt{\frac{16}{49}}$
7. $\sqrt{32}$

Name the set or sets of numbers to which each real number belongs.

8. $-\sqrt{64}$
9. $\frac{8}{3}$
10. $\sqrt{28}$
11. $\frac{56}{7}$

Graph each solution set.

12. $x < -3.5$
13. $x \geq -7$

Replace each ● with <, >, or = to make each sentence true.

14. 0.3 ● $\frac{1}{3}$
15. $\frac{2}{9}$ ● $0.\overline{2}$
16. $\frac{1}{6}$ ● $\sqrt{6}$

Write each set of numbers in order from least to greatest.

17. $\frac{1}{8}, \sqrt{\frac{1}{8}}, 0.\overline{15}, -15$
18. $\sqrt{30}, 5\frac{4}{9}, 13, \frac{1}{\sqrt{30}}$

Standardized Test Practice

19. For what value of a is $-\sqrt{a} < -\frac{1}{\sqrt{a}}$ true?

Ⓐ $\frac{1}{3}$ Ⓑ -4 Ⓒ 2 Ⓓ 1

Practice and Apply

Homework Help

For Exercises	See Examples
20–31, 50, 51	1
32–49	2
52–57	3
58–63	4
64–69	5

Extra Practice

See page 825.

Find each square root. If necessary, round to the nearest hundredth.

20. $\sqrt{49}$
21. $\sqrt{81}$
22. $\sqrt{5.29}$
23. $\sqrt{6.25}$
24. $-\sqrt{78}$
25. $-\sqrt{94}$
26. $\pm\sqrt{\frac{36}{81}}$
27. $\pm\sqrt{\frac{100}{196}}$
28. $\sqrt{\frac{9}{14}}$
29. $\sqrt{\frac{25}{42}}$
30. $\pm\sqrt{820}$
31. $\pm\sqrt{513}$

Name the set or sets of numbers to which each real number belongs.

32. $-\sqrt{22}$
33. $\frac{36}{6}$
34. $\frac{1}{3}$
35. $-\frac{5}{12}$
36. $\sqrt{\frac{82}{20}}$
37. $-\sqrt{46}$
38. $\sqrt{10.24}$
39. $\frac{-54}{19}$
40. $-\frac{3}{4}$
41. $\sqrt{20.25}$
42. $\frac{18}{3}$
43. $\sqrt{2.4025}$
44. $\frac{-68}{35}$
45. $\frac{6}{11}$
46. $\sqrt{5.5696}$
47. $\sqrt{\frac{78}{42}}$
48. $-\sqrt{9.16}$
49. π

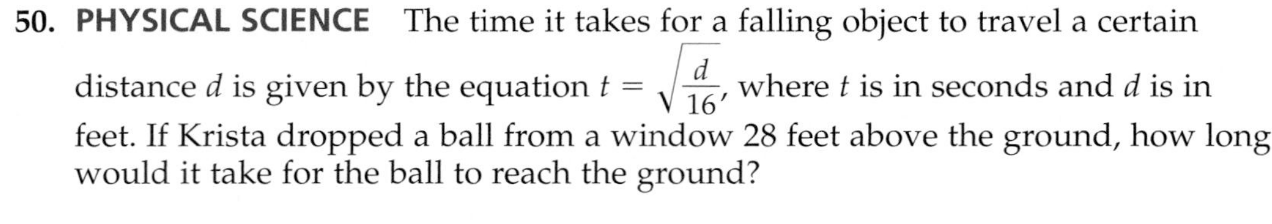

50. **PHYSICAL SCIENCE** The time it takes for a falling object to travel a certain distance d is given by the equation $t = \sqrt{\frac{d}{16}}$, where t is in seconds and d is in feet. If Krista dropped a ball from a window 28 feet above the ground, how long would it take for the ball to reach the ground?

More About. . .

Tourism

Built in 1758, the Sambro Island Lighthouse at Halifax Harbor is the oldest operational lighthouse in North America.

Source: Canadian Coast Guard

51. **LAW ENFORCEMENT** Police can use the formula $s = \sqrt{24d}$ to estimate the speed s of a car in miles per hour by measuring the distance d in feet a car skids on a dry road. On his way to work, Jerome skidded trying to stop for a red light and was involved in a minor accident. He told the police officer that he was driving within the speed limit of 35 miles per hour. The police officer measured his skid marks and found them to be $43\frac{3}{4}$ feet long. Should the officer give Jerome a ticket for speeding? Explain.

Graph each solution set.

52. $x > -12$
53. $x \leq 8$
54. $x \geq -10.2$
55. $x < -0.25$
56. $x \neq -2$
57. $x \neq \pm\sqrt{36}$

Replace each ● with <, >, or = to make each sentence true.

58. $5.\overline{72}$ ● $\sqrt{5}$
59. $2.\overline{63}$ ● $\sqrt{8}$
60. $\frac{1}{7}$ ● $\frac{1}{\sqrt{7}}$
61. $\frac{2}{3}$ ● $\frac{2}{\sqrt{3}}$
62. $\frac{1}{\sqrt{31}}$ ● $\frac{\sqrt{31}}{31}$
63. $\frac{\sqrt{2}}{2}$ ● $\frac{1}{2}$

Write each set of numbers in order from least to greatest.

64. $\sqrt{0.42}, 0.\overline{63}, \frac{\sqrt{4}}{3}$
65. $\sqrt{0.06}, 0.\overline{24}, \frac{\sqrt{9}}{12}$
66. $-1.\overline{46}, 0.2, \sqrt{2}, -\frac{1}{6}$
67. $-4.\overline{83}, 0.4, \sqrt{8}, -\frac{3}{8}$
68. $-\sqrt{65}, -6\frac{2}{5}, -\sqrt{27}$
69. $\sqrt{122}, 7\frac{4}{9}, \sqrt{200}$

TOURISM For Exercises 70–72, use the following information.

The formula to determine the distance d in miles that an object can be seen on a clear day on the surface of a body of water is $d = 1.4\sqrt{h}$, where h is the height in feet of the viewer's eyes above the surface of the water.

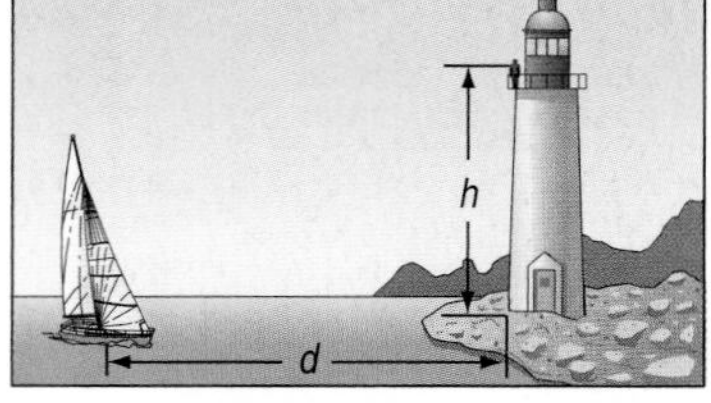

70. A charter plane is used to fly tourists on a sightseeing trip along the coast of North Carolina. If the plane flies at an altitude of 1500 feet, how far can the tourists see?

71. Dillan and Marissa are parasailing while on vacation. Marissa is 135 feet above the ocean while Dillan is 85 feet above the ocean. How much farther can Marissa see than Dillan?

72. The observation deck of a lighthouse stands 120 feet above the ocean surface. Can the lighthouse keeper see a boat that is 17 miles from the lighthouse? Explain.

73. **CRITICAL THINKING** Determine when the following statements are all true for real numbers q and r.

a. $q^2 > r^2$
b. $\frac{1}{q} < \frac{1}{r}$
c. $\sqrt{q} > \sqrt{r}$
d. $\frac{1}{\sqrt{q}} < \frac{1}{\sqrt{r}}$

GEOMETRY For Exercises 74–76, use the table.

Squares		
Area (units²)	Side Length	Perimeter
1		
4		
9		
16		
25		

74. Copy and complete the table. Determine the length of each side of each square described. Then determine the perimeter of each square.

75. Describe the relationship between the lengths of the sides and the area.

76. Write an expression you can use to find the perimeter of a square whose area is a units^2.

77. WRITING IN MATH Answer the question that was posed at the beginning of the lesson.

How can using square roots determine the surface area of the human body?

Include the following in your answer:

- an explanation of the order of operations that must be followed to calculate the surface area of the human body,
- a description of other situations in which you might need to calculate the surface area of the human body, and
- examples of real-world situations involving square roots.

78. Which point on the number line is closest to $-\sqrt{7}$?

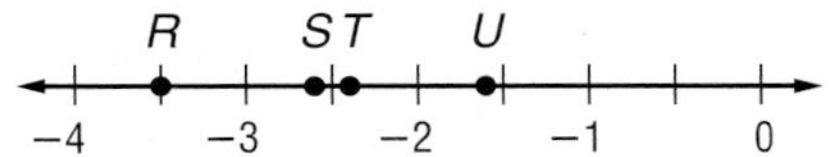

(A) R (B) S
(C) T (D) U

79. Which of the following is a true statement?

(A) $-\frac{6}{3} > \frac{3}{6}$ (B) $-\frac{3}{6} > -\frac{6}{3}$ (C) $-\frac{3}{6} < -\frac{6}{3}$ (D) $\frac{6}{3} < \frac{3}{6}$

Maintain Your Skills

Mixed Review **Find the odds of each outcome if a card is randomly selected from a standard deck of cards.** *(Lesson 2-6)*

80. red 4

81. even number

82. against a face card

83. against an ace

84. AUTO RACING Jeff Gordon's finishing places in the 2000 season races are listed below. Which measure of central tendency best represents the data? Explain. *(Lesson 2-5)*

34 10 28 9 8 8 25 4 1 11 14 10 32 14 8 1 4
10 5 3 33 23 36 23 4 1 6 9 5 39 4 2 7 7

Simplify each expression. *(Lesson 2-3)*

85. $4(-7) - 3(11)$

86. $3(-4) + 2(-7)$

87. $1.2(4x - 5y) - 0.2(-1.5x + 8y)$

88. $-4x(y - 2z) + x(6z - 3y)$

Chapter 2 Study Guide and Review

Vocabulary and Concept Check

absolute value (p. 69)
additive inverses (p. 74)
back-to-back stem-and-leaf plot (p. 89)
Completeness Property (p. 105)
coordinate (p. 69)
equally likely (p. 97)
frequency (p. 88)
graph (p. 69)
infinity (p. 68)
integers (p. 68)
irrational number (p. 104)
line plot (p. 88)
measures of central tendency (p. 90)
natural number (p. 68)
negative number (p.68)
odds (p. 97)
opposites (p. 74)
perfect square (p. 103)
positive number (p. 68)
principal square root (p. 103)
probability (p. 96)
radical sign (p. 103)
rational approximation (p. 105)
rational number (p. 68)
real number (p. 104)
sample space (p. 96)
simple event (p. 96)
square root (p. 103)
stem-and-leaf plot (p. 89)
whole number (p. 68)

State whether each sentence is *true* or *false*. If false, replace the underlined term or number to make a true sentence.

1. The absolute value of -26 is <u>26</u>.
2. Terminating decimals are <u>rational</u> numbers.
3. The principal square root of 144 is <u>12</u>.
4. $-\sqrt{576}$ is an <u>irrational number</u>.
5. 225 is a <u>perfect square</u>.
6. <u>-3.1</u> is an integer.
7. <u>0.666</u> is a repeating decimal.
8. The product of two numbers with different signs is <u>negative</u>.

Lesson-by-Lesson Review

2-1 Rational Numbers on the Number Line

See pages 68–72.

Concept Summary

- A set of numbers can be graphed on a number line by drawing points.
- To evaluate expressions with absolute value, treat the absolute value symbols as grouping symbols.

Example Graph $\{\ldots, -5, -4, -3\}$.

The bold arrow means that the graph continues indefinitely in that direction.

Exercises **Graph each set of numbers.** *See Example 2 on page 69.*

9. $\{5, 3, -1, -3\}$
10. $\left\{-1\frac{1}{2}, -\frac{1}{2}, \frac{1}{2}, 1\frac{1}{2}, \ldots\right\}$
11. {integers less than -4 and greater than or equal to 2}

Evaluate each expression if $x = -4$, $y = 8$, and $z = -9$. *See Example 4 on page 70.*

12. $32 - |y - 3|$
13. $3|x| - 7$
14. $4 + |z|$
15. $46 - y|x|$

www.algebra1.com/vocabulary_review

2-2 Adding and Subtracting Rational Numbers

See pages 73–78.

Concept Summary

- To add rational numbers with the *same* sign, add their absolute values. The sum has the same sign as the addends.
- To add rational numbers with *different* signs, subtract the lesser absolute value from the greater absolute value. The sum has the same sign as the number with the greater absolute value.
- To subtract a rational number, add its additive inverse.

Examples

1 Find $-4 + (-3)$.

$-4 + (-3)$

$= -(|-4| + |-3|)$ Both numbers are negative, so the sum is negative.

$= -(4 + 3)$

$= -7$

2 Find $12 - 18$.

$12 - 18$

$= 12 + (-18)$ To subtract 18, add its inverse.

$= -(|-18| - |12|)$ The absolute value of 18 is greater, so the result is negative.

$= -(18 - 12)$

$= -6$

Exercises **Find each sum or difference.** *See Examples 1–3 on pages 73–75.*

16. $4 + (-4)$ **17.** $2 + (-7)$ **18.** $-0.8 + (-1.2)$

19. $-3.9 + 2.5$ **20.** $-\frac{1}{4} + \left(-\frac{1}{8}\right)$ **21.** $\frac{5}{6} + \left(-\frac{1}{3}\right)$

22. $-2 - 10$ **23.** $9 - (-7)$ **24.** $1.25 - 0.18$

25. $-7.7 - (-5.2)$ **26.** $\frac{9}{2} - \left(-\frac{1}{2}\right)$ **27.** $-\frac{1}{8} - \left(-\frac{2}{3}\right)$

2-3 Multiplying Rational Numbers

See pages 79–83.

Concept Summary

- The product of two numbers having the same sign is positive.
- The product of two numbers having different signs is negative.

Example

Multiply $\left(-2\frac{1}{7}\right)\left(3\frac{2}{3}\right)$.

$\left(-2\frac{1}{7}\right)\left(3\frac{2}{3}\right) = \frac{-15}{7} \cdot \frac{11}{3}$ Write as improper fractions.

$= \frac{-55}{7}$ or $-7\frac{6}{7}$ Simplify.

Exercises **Find each product.** *See Examples 1 and 3 on pages 79 and 80.*

28. $(-11)(9)$ **29.** $12(-3)$ **30.** $-8.2(4.5)$

31. $-2.4(-3.6)$ **32.** $\frac{3}{4} \cdot \frac{7}{12}$ **33.** $\left(-\frac{1}{3}\right)\left(-\frac{9}{10}\right)$

Simplify each expression. *See Example 2 on page 80.*

34. $8(-3x) + 12x$ **35.** $-5(-2n) - 9n$ **36.** $-4(6a) - (-3)(-7a)$

2-4 Dividing Rational Numbers

See pages 84–87.

Concept Summary

- The quotient of two positive numbers is positive.
- The quotient of two negative numbers is positive.
- The quotient of a positive number and a negative number is negative.

Example **Simplify $\frac{-3(4)}{-2-3}$.**

$\frac{-3(4)}{-2-3} = \frac{-12}{-2-3}$ Simplify the numerator.

$= \frac{-12}{-5}$ Simplify the denominator.

$= 2\frac{2}{5}$ same signs → positive quotient

Exercises **Find each quotient.** *See Examples 1–3 on pages 84 and 85.*

37. $\frac{-54}{6}$
38. $-\frac{74}{8}$
39. $21.8 \div (-2)$
40. $-7.8 \div (-6)$
41. $-15 \div \left(\frac{3}{4}\right)$
42. $\frac{21}{24} \div \frac{1}{3}$

Simplify each expression. *See Example 5 on page 85.*

43. $\frac{14 - 28x}{-7}$
44. $\frac{-5 + 25x}{5}$
45. $\frac{-4x + 24y}{4}$

Evaluate each expression if $x = -4$, $y = 2.4$, and $z = 3$. *See Example 6 on page 85.*

46. $xz - 2y$
47. $-2\left(\frac{2y}{z}\right)$
48. $\frac{2x - z}{4} + 3y$

2-5 Statistics: Displaying and Analyzing Data

See pages 88–94.

Concept Summary

- A set of numerical data can be displayed in a line plot or stem-and-leaf plot.
- A measure of central tendency represents a centralized value of a set of data. Examine each measure of central tendency to choose the one most representative of the data.

Examples 1 **Draw a line plot for the data.**

2 8 6 4 5 9 13 12 5 2 5 5 2

The value of the data ranges from 2 to 13. Construct a number line containing those points. Then place an × above a number each time it occurs.

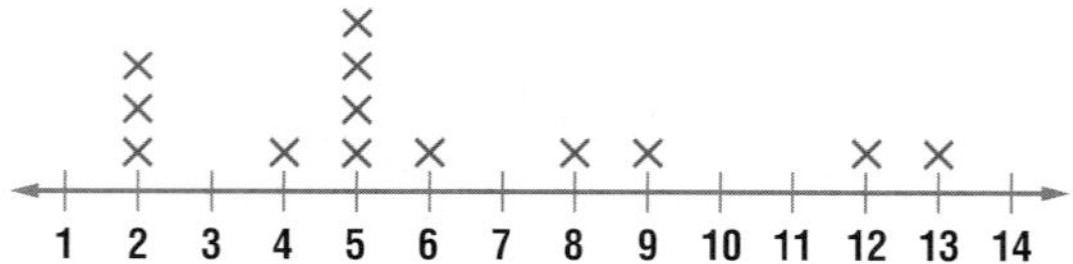

2 **SCHOOL Melinda's scores on the 25-point quizzes in her English class are 20, 21, 12, 21, 22, 22, 22, 21, 20, 20, and 21. Which measure of central tendency best represents her grade?**

mean: 20.2 Add the data and divide by 11.

median: 21 The middle value is 21.

mode: 21 The most frequent value is 21.

The median and mode are both representative of the data. The mean is less than most of the data.

Exercises

49. Draw a line plot for the data. Then make a stem-and-leaf plot.
See Examples 1–3 on pages 88 and 89.

28	17	16	18	19	21	26	15
19	19	16	14	21	12	26	17
30	17	13	18	14	22	20	12
19	19	15	12	15	21	15	17

50. BUSINESS Of the 42 employees at Pirate Printing, four make $6.50 an hour, sixteen make $6.75 an hour, six make $6.85 an hour, thirteen make $7.25 an hour, and three make $8.85 an hour. Which measure best describes the average wage? Explain. *See Examples 5 and 6 on pages 90 and 91.*

51. HOCKEY Professional hockey uses a point system based on wins, losses and ties, to determine teams' rank. The stem-and-leaf plot shows the number of points earned by each of the 30 teams in the National Hockey League during the 2000–2001 season. Which measure of central tendency best describes the average number of points earned? Explain.
See Example 5 on page 90.

Stem	Leaf
11	1 1 8
10	0 3 6 9
9	0 0 0 2 3 5 6 6 8
8	0 8 8
7	0 1 1 2 3
6	0 6 6 8
5	2 9 *11|1 = 111*

2-6 Probability: Simple Probability and Odds

See pages 96–101.

Concept Summary

- The probability of an event a can be expressed as
  $P(a) = \frac{\text{number of favorable outcomes}}{\text{total number of possible outcomes}}$.
- The odds of an event can be expressed as the ratio of the number of successful outcomes to the number of unsuccessful outcomes.

Examples 1 **Find the probability of randomly choosing the letter *I* in the word *MISSISSIPPI*.**

$P(\text{letter I}) = \frac{4}{11}$ ← number of favorable outcomes / ← number of possible outcomes

≈ 0.36

The probability of choosing an I is $\frac{4}{11}$ or about 36%.

• Extra Practice, see pages 823–825.
• Mixed Problem Solving, see page 854.

2 **Find the odds that you will randomly select a letter that is *not* S in the word *MISSISSIPPI*.**

number of successes : number of failures = 7:4

The odds of not selecting an S are 7:4.

Exercises **Find the probability of each outcome if a computer randomly chooses a letter in the word *REPRESENTING*.** *See Example 1 on pages 96 and 97.*

52. $P(\text{S})$ **53.** $P(\text{E})$ **54.** $P(\text{not N})$ **55.** $P(\text{R or P})$

Find the odds of each outcome if you randomly select a coin from a jar containing 90 pennies, 75 nickels, 50 dimes, and 30 quarters.
See Examples 2 and 3 on pages 97 and 98.

56. a dime **57.** a penny **58.** *not* a nickel **59.** a nickel or a dime

2-7 Square Roots and Real Numbers

See pages 103–109.

Concept Summary

- A square root is one of two equal factors of a number.
- The symbol $\sqrt{}$ is used to indicate the nonnegative square root of a number.

Example **Find $\sqrt{169}$.**

$\sqrt{169}$ represents the square root of 169.

$169 = 13^2 \rightarrow \sqrt{169} = 13$

Exercises **Find each square root. If necessary, round to the nearest hundredth.**
See Example 1 on page 103.

60. $\sqrt{196}$ **61.** $\pm\sqrt{1.21}$ **62.** $-\sqrt{160}$ **63.** $\pm\sqrt{\frac{4}{225}}$

Name the set or sets of numbers to which each real number belongs.
See Example 2 on page 104.

64. $\frac{16}{25}$ **65.** $\frac{\sqrt{64}}{2}$ **66.** $-\sqrt{48.5}$

Replace each ● with <, >, or = to make each sentence true. *See Example 4 on page 105.*

67. $\frac{1}{8}$ ● $\frac{1}{\sqrt{49}}$ **68.** $\sqrt{\frac{2}{3}}$ ● $\frac{4}{9}$ **69.** $\sqrt{\frac{3}{4}}$ ● $\sqrt{\frac{1}{3}}$

70. WEATHER Meteorologists can use the formula $t = \sqrt{\frac{d^3}{216}}$ to estimate the amount of time t in hours a storm of diameter d will last. Suppose the eye of a hurricane, which causes the greatest amount of destruction, is 9 miles in diameter. To the nearest tenth of an hour, how long will the worst part of the hurricane last?
See Example 1 on pages 103 and 104.

Chapter 2

Practice Test

Vocabulary and Concepts

Choose the correct term to complete each sentence.

1. The *(absolute value, square)* of a number is its distance from zero on a number line.
2. A number that can be written as a fraction where the numerator and denominator are integers and the denominator does not equal zero is a *(repeating, rational)* number.
3. The list of all possible outcomes is called the *(simple event, sample space)*.

Skills and Applications

Evaluate each expression.

4. $-|x| - 38$ if $x = -2$
5. $34 - |x + 21|$ if $x = -7$
6. $-12 + |x - 8|$ if $x = 1.5$

Find each sum or difference.

7. $-19 + 12$
8. $-21 - (-34)$
9. $16.4 + (-23.7)$
10. $6.32 - (-7.41)$
11. $-\frac{7}{16} + \frac{3}{8}$
12. $-\frac{7}{12} - \left(-\frac{5}{9}\right)$

Find each quotient or product.

13. $-5(19)$
14. $-56 \div (-7)$
15. $96 \div (-0.8)$
16. $(-7.8)(5.6)$
17. $-\frac{1}{8} \div -5$
18. $-\frac{15}{32} \div \frac{3}{4}$

Simplify each expression.

19. $5(-3x) - 12x$
20. $7(6h - h)$
21. $-4m(-7n) + (3d)(-4c)$
22. $\frac{36k}{4}$
23. $\frac{9a + 27}{-3}$
24. $\frac{70x - 30y}{-5}$

Find each square root. If necessary, round to the nearest hundredth.

25. $-\sqrt{64}$
26. $\sqrt{3.61}$
27. $\pm\sqrt{\frac{16}{81}}$

Replace each ● with <, >, or = to make each sentence true.

28. $\frac{1}{\sqrt{3}}$ ● $\frac{1}{3}$
29. $\sqrt{\frac{1}{2}}$ ● $\frac{8}{11}$
30. $\sqrt{0.56}$ ● $\frac{\sqrt{3}}{2}$

STATISTICS For Exercises 31 and 32, use the following information.
The height, in inches, of the students in a health class are 65, 63, 68, 66, 72, 61, 62, 63, 59, 58, 61, 74, 65, 63, 71, 60, 62, 63, 71, 70, 59, 66, 61, 62, 68, 69, 64, 63, 70, 61, 68, and 67.

31. Make a line plot of the data.
32. Which measure of central tendency best describes the data? Explain.

33. **STANDARDIZED TEST PRACTICE** During a 20-song sequence on a radio station, 8 soft-rock, 7 hard-rock, and 5 rap songs are played at random. Assume that all of the songs are the same length. What is the probability that when you turn on the radio, a hard-rock song will be playing?

   (A) $\frac{1}{4}$ (B) $\frac{7}{20}$ (C) $\frac{2}{5}$ (D) $\frac{13}{20}$ (E) $\frac{7}{10}$

Chapter 2 Standardized Test Practice

Part 1 Multiple Choice

Record your answers on the answer sheet provided by your teacher or on a sheet of paper.

1. Darryl works 9 days at the State Fair and earns \$518.40. If he works 8 hours each day, what is his hourly pay? (Prerequisite Skill)

Ⓐ \$6.48 Ⓑ \$7.20
Ⓒ \$30.50 Ⓓ \$57.60

2. The graph below shows how many toy trains are assembled at a factory at the end of 10-minute intervals. What is the best prediction for the number of products assembled per hour? (Prerequisite Skill)

Ⓐ 80
Ⓑ 100
Ⓒ 120
Ⓓ 130

Toy Train Production

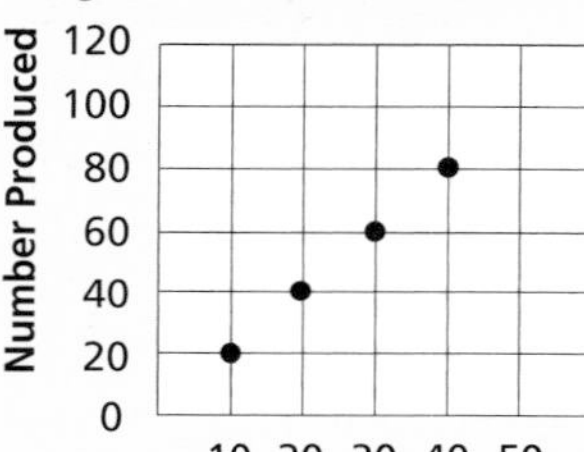

3. Which graph shows the integers greater than -2 and less than or equal to 3? (Lesson 2-1)

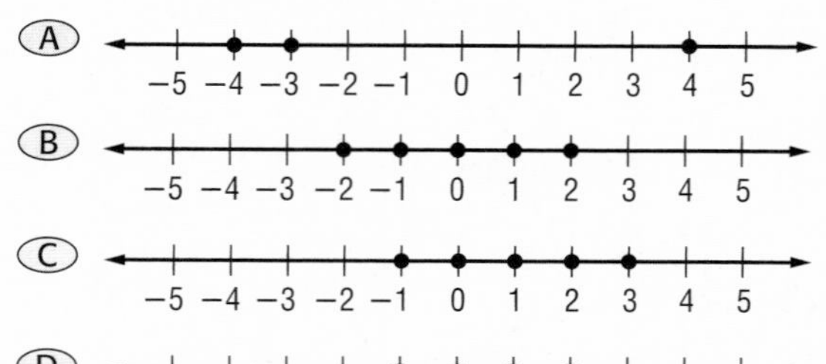

The Princeton Review Test-Taking Tip

Question 1 If you don't know how to solve a problem, eliminate the answer choices you know are incorrect and then guess from the remaining choices. Even eliminating only one answer choice greatly increases your chance of guessing the correct answer.

4. Which number is the greatest? (Lesson 2-1)

Ⓐ $|-4|$ Ⓑ $|4|$
Ⓒ $|7|$ Ⓓ $|-9|$

5. What is $-3.8 + 4.7$? (Lesson 2-2)

Ⓐ 0.9 Ⓑ -0.9
Ⓒ 8.5 Ⓓ -8.5

6. Simplify $3(-2m) - 7m$. (Lesson 2-3)

Ⓐ $-12m$ Ⓑ $-m$
Ⓒ $-2m$ Ⓓ $-13m$

7. Which statement about the stem-and-leaf plot is *not* true? (Lesson 2-5)

Stem	Leaf
3	1 1 5 6 8 8
4	2 2 2 4
5	0 0
6	0 3 7 8 9 9
7	4 $7\|4 = 74$

Ⓐ The greatest value is 74.
Ⓑ The mode is 42.
Ⓒ Seven of the values are greater than 50.
Ⓓ The least value is 38.

8. There are 4 boxes. If you choose a box at random, what are the odds that you will choose the one box with a prize? (Lesson 2-6)

Ⓐ 1:3 Ⓑ 1:4
Ⓒ 3:1 Ⓓ 3:4

9. Which point on the number line is closest to $\sqrt{10}$? (Lesson 2-7)

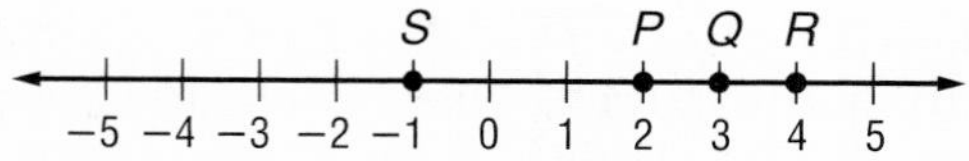

Ⓐ point *P* Ⓑ point *Q*
Ⓒ point *R* Ⓓ point *S*

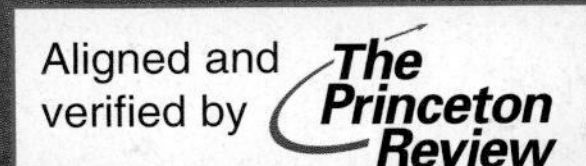

Part 2 Short Response/Grid In

Record your answers on the answer sheet provided by your teacher or on a sheet of paper.

10. Ethan needs to wrap a label around a jar of homemade jelly so that there is no overlap. Find the length of the label. (Prerequisite Skill)

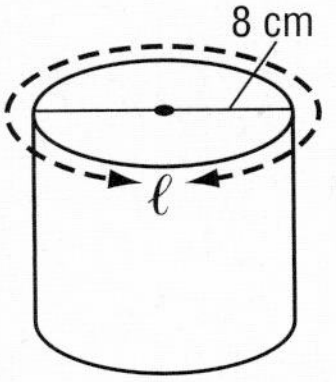

11. Evaluate $\frac{5 - 1}{4 + 12 \div 3 \times 2}$. (Lesson 1-2)

12. Find the solution of $4m - 3 = 9$ if the replacement set is $\{0, 2, 3, 5\}$. (Lesson 1-3)

13. Write an algebraic expression for *2p plus three times the difference of m and n.* (Lesson 1-6)

14. State the hypothesis in the statement *If $3x + 3 > 24$, then $x > 7$.* (Lesson 1-7)

Part 3 Quantitative Comparison

Compare the quantity in Column A to the quantity in Column B. Then determine whether:

- Ⓐ **the quantity in Column A is greater,**
- Ⓑ **the quantity in Column B is greater,**
- Ⓒ **the quantities are equal, or**
- Ⓓ **the relationship cannot be determined from the information given.**

	Column A	Column B
15.	$x > 0$	
	$\|x\|$	$\|-x\|$

(Lesson 2-1)

16. $x > y > 0$

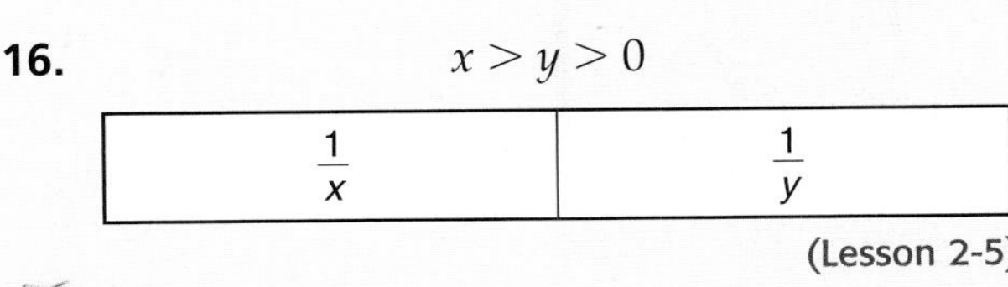

Column A	Column B
$\frac{1}{x}$	$\frac{1}{y}$

(Lesson 2-5)

	Column A	Column B
17.	$\frac{1}{n} > 1$	
	1	n

(Lesson 2-6)

18. $a^2 = 49$

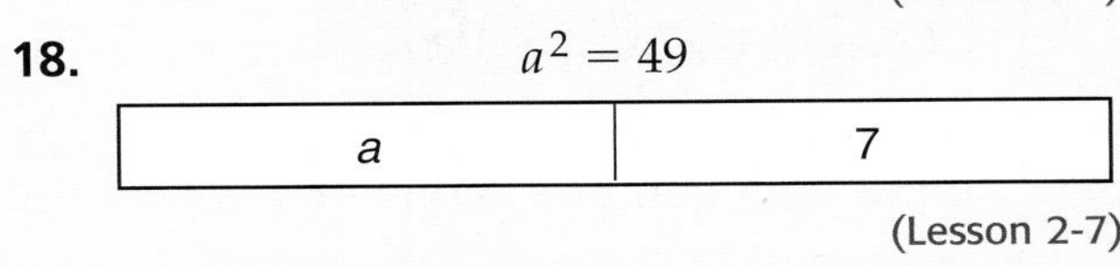

Column A	Column B
a	7

(Lesson 2-7)

Part 4 Open Ended

Record your answers on a sheet of paper. Show your work.

19. Mia has created the chart below to compare the three cellular phone plans she is considering. (Lessons 2-2 and 2-3)

Plan	Monthly Fee	Cost/Minute
A	\$5.95	\$0.30
B	\$12.95	\$0.10
C	\$19.99	\$0.08

a. Write an algebraic expression that Mia can use to figure the monthly cost of each plan. Use C for the total monthly cost, m for the cost per minute, x for the monthly fee, and y for the minutes used per month.

b. If Mia uses 150 minutes of calls each month, which plan will be least expensive? Explain.

20. The stem-and-leaf plot lists the annual profit for seven small businesses. (Lesson 2-5)

Stem	Leaf
3	2 9
4	1 1 3 5
5	0

a. Explain how the absence of a key could lead to misinterpreting the data.

b. How do the keys below affect how the data should be interpreted?

$3|2 = 3.2$ $3|2 = 0.32$

www.algebra1.com/standardized_test

Chapter 3

Solving Linear Equations

What You'll Learn

- **Lesson 3-1** Translate verbal sentences into equations and equations into verbal sentences.
- **Lessons 3-2 through 3-6** Solve equations and proportions.
- **Lesson 3-7** Find percents of change.
- **Lesson 3-8** Solve equations for given variables.
- **Lesson 3-9** Solve mixture and uniform motion problems.

Key Vocabulary

- **equivalent equations** (p. 129)
- **identity** (p. 150)
- **proportion** (p. 155)
- **percent of change** (p. 160)
- **mixture problem** (p. 171)

Why It's Important

Linear equations can be used to solve problems in every facet of life from planning a garden, to investigating trends in data, to making wise career choices. One of the most frequent uses of linear equations is solving problems involving mixtures or motion. For example, in the National Football League, a quarterback's passing performance is rated using an equation based on a mixture, or weighted average, of five factors, including passing attempts and completions. *You will learn how this rating system works in Lesson 3-9.*

Getting Started

Prerequisite Skills To be successful in this chapter, you'll need to master these skills and be able to apply them in problem-solving situations. Review these skills before beginning Chapter 3.

For Lesson 3-1 — Write Mathematical Expressions

Write an algebraic expression for each verbal expression. *(For review, see Lesson 1-1.)*

1. five greater than half of a number t
2. the product of seven and s divided by the product of eight and y
3. the sum of three times a and the square of b
4. w to the fifth power decreased by 37
5. nine times y subtracted from 95
6. the quantity of r plus six divided by twelve

For Lesson 3-4 — Use the Order of Operations

Evaluate each expression. *(For review, see Lesson 1-2.)*

7. $3 \cdot 6 - \frac{12}{4}$ **8.** $5(13 - 7) - 22$ **9.** $5(7 - 2) - 3^2$ **10.** $\frac{2 \cdot 6 - 4}{2}$

11. $(25 - 4) \div (2^2 - 1)$ **12.** $36 \div 4 - 2 + 3$ **13.** $\frac{19 - 5}{7} + 3$ **14.** $\frac{1}{4}(24) - \frac{1}{2}(12)$

For Lesson 3-7 — Find the Percent

Find each percent. *(For review, see pages 802 and 803.)*

15. Five is what percent of 20? **16.** What percent of 300 is 21?

17. What percent of 5 is 15? **18.** Twelve is what percent of 60?

19. Sixteen is what percent of 10? **20.** What percent of 50 is 37.5?

Make this Foldable to help you organize information about solving linear equations. Begin with 4 sheets of plain $8\frac{1}{2}$" by 11" paper.

Step 1 Fold

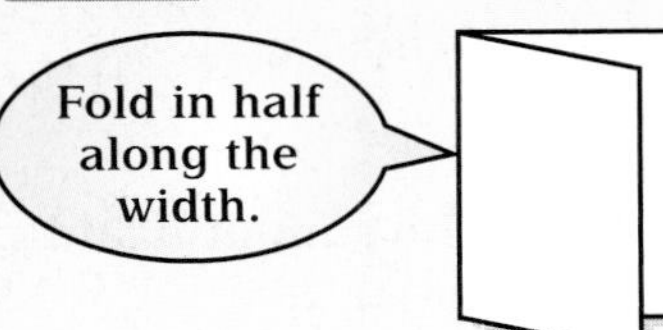

Fold in half along the width.

Step 2 Open and Fold Again

Fold the bottom to form a pocket. Glue edges.

Step 3 Repeat Steps 1 and 2

Repeat three times and glue all four pieces together.

Step 4 Label

Label each pocket. Place an index card in each pocket.
Solving Linear Equations

Reading and Writing As you read and study the chapter, you can write notes and examples on each index card.

3-1 Writing Equations

What You'll Learn

- Translate verbal sentences into equations.
- Translate equations into verbal sentences.

Vocabulary

- four-step problem-solving plan
- defining a variable
- formula

How are equations used to describe heights?

The Statue of Liberty sits on a pedestal that is 154 feet high. The height of the pedestal and the statue is 305 feet. If s represents the height of the statue, then the following equation represents the situation.

$$154 + s = 305$$

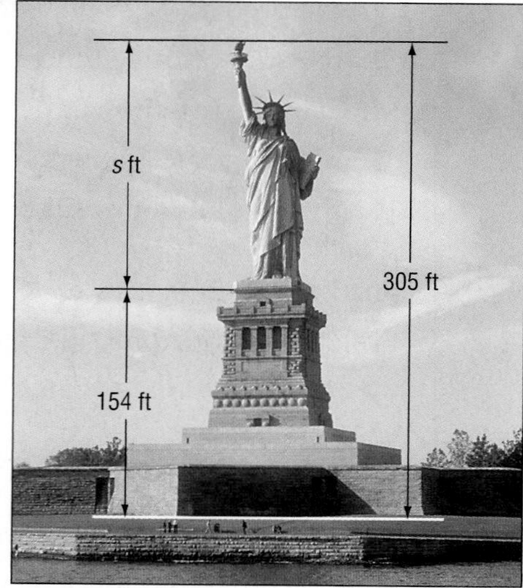

Source: *World Book Encyclopedia*

WRITE EQUATIONS When writing equations, use variables to represent the unspecified numbers or measures referred to in the sentence or problem. Then write the verbal expressions as algebraic expressions. Some verbal expressions that suggest the *equals sign* are listed below.

- is
- equals
- is equal to
- is the same as
- is as much as
- is identical to

Study Tip

Look Back
To review **translating verbal expressions to algebraic expressions**, see Lesson 1-1.

Example 1 Translate Sentences into Equations

Translate each sentence into an equation.

a. Five times the number a is equal to three times the sum of b and c.

Five	times	a	is equal to	three	times	the sum of b and c.
5	$\times$	a	$=$	3	$\times$	$(b + c)$

The equation is $5a = 3(b + c)$.

b. Nine times y subtracted from 95 equals 37.

Rewrite the sentence so it is easier to translate.

95	less	nine times y	equals	37.
95	$-$	$9y$	$=$	37

The equation is $95 - 9y = 37$.

Using the **four-step problem-solving plan** can help you solve any word problem.

Key Concept — *Four-Step Problem-Solving Plan*

Step 1 Explore the problem.

Step 2 Plan the solution.

Step 3 Solve the problem.

Step 4 Examine the solution.

Each step of the plan is important.

Step 1 Explore the Problem
To solve a verbal problem, first read the problem carefully and explore what the problem is about.

- Identify what information is given.
- Identify what you are asked to find.

Step 2 Plan the Solution
One strategy you can use to solve a problem is to write an equation. Choose a variable to represent one of the unspecific numbers in the problem. This is called **defining a variable**. Then use the variable to write expressions for the other unspecified numbers in the problem. *You will learn to use other strategies throughout this book.*

Step 3 Solve the Problem
Use the strategy you chose in Step 2 to solve the problem.

Step 4 Examine the Solution
Check your answer in the context of the original problem.

- Does your answer make sense?
- Does it fit the information in the problem?

Study Tip

Reading Math
In a verbal problem, the sentence that tells what you are asked to find usually contains *find*, *what*, *when*, or *how*.

More About. . .

Ice Cream
The first ice cream plant was established in 1851 by Jacob Fussell. Today, 2,000,000 gallons of ice cream are produced in the United States each day.
Source: *World Book Encyclopedia*

Example 2 *Use the Four-Step Plan*

ICE CREAM **Use the information at the left. In how many days can 40,000,000 gallons of ice cream be produced in the United States?**

Explore You know that 2,000,000 gallons of ice cream are produced in the United States each day. You want to know how many days it will take to produce 40,000,000 gallons of ice cream.

Plan Write an equation to represent the situation. Let d represent the number of days needed to produce the ice cream.

2,000,000	times	the number of days	equals	40,000,000.
2,000,000	$\times$	d	$=$	40,000,000

Solve $2{,}000{,}000d = 40{,}000{,}000$ Find d mentally by asking, "What number times 2,000,000 equals 40,000,000?"

$d = 20$

It will take 20 days to produce 40,000,000 gallons of ice cream.

Examine If 2,000,000 gallons of ice cream are produced in one day, $2{,}000{,}000 \times 20$ or 40,000,000 gallons are produced in 20 days. The answer makes sense.

A **formula** is an equation that states a rule for the relationship between certain quantities. Sometimes you can develop a formula by making a model.

Algebra Activity

Surface Area

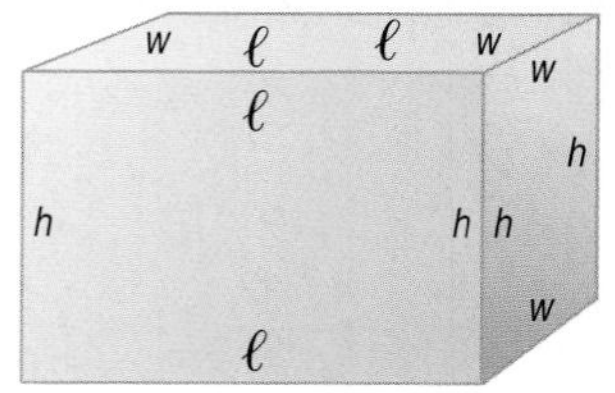

- Mark each side of a rectangular box as the length ℓ, the width w, or the height h.
- Use scissors to cut the box so that each surface or face of the box is a separate piece.

Analyze

1. Write an expression for the area of the front of the box.
2. Write an expression for the area of the back of the box.
3. Write an expression for the area of one side of the box.
4. Write an expression for the area of the other side of the box.
5. Write an expression for the area of the top of the box.
6. Write an expression for the area of the bottom of the box.
7. The surface area of a rectangular box is the sum of all the areas of the faces of the box. If S represents surface area, write a formula for the surface area of a rectangular box.

Make a Conjecture

8. If s represents the length of the side of a cube, write a formula for the surface area of a cube.

Example 3 Write a Formula

Translate the sentence into a formula.

The perimeter of a rectangle equals two times the length plus two times the width.

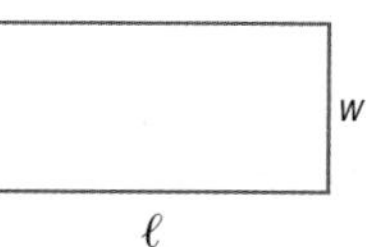

Words Perimeter equals two times the length plus two times the width.

Variables Let P = perimeter, ℓ = length, and w = width.

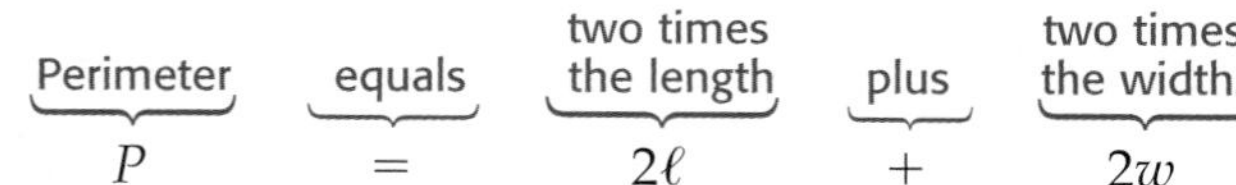

Formula $P \quad = \quad 2\ell \quad + \quad 2w$

The formula for the perimeter of a rectangle is $P = 2\ell + 2w$.

WRITE VERBAL SENTENCES You can also translate equations into verbal sentences or make up your own verbal problem if you are given an equation.

Study Tip

Look Back
To review **translating algebraic expressions to verbal expressions**, see Lesson 1-1.

Example 4 Translate Equations into Sentences

Translate each equation into a verbal sentence.

a. $3m + 5 = 14$

$3m$	$+$	5	$=$	14
Three times m	plus	five	equals	fourteen.

b. $w + v = y^2$

$\underbrace{w + v}$ $\underbrace{=}$ $\underbrace{y^2}$

The sum of w and v equals the square of y.

Example 5 Write a Problem

Write a problem based on the given information.

a = Rafael's age **$a + 5$ = Tierra's age** **$a + 2(a + 5) = 46$**

You know that a represents Rafael's age and $a + 5$ represents Tierra's age. The equation adds a plus twice $(a + 5)$ to get 46. A sample problem is given below.

Tierra is 5 years older than Rafael. The sum of Rafael's age and twice Tierra's age equals 46. How old is Rafael?

Check for Understanding

Concept Check

1. **List** the four steps used in solving problems.
2. **Analyze** the following problem.

   Misae has \$1900 in the bank. She wishes to increase her account to a total of \$3500 by depositing \$30 per week from her paycheck. Will she reach her savings goal in one year?

   a. How much money did Misae have in her account at the beginning?

   b. How much money will Misae add to her account in 10 weeks? in 20 weeks?

   c. Write an expression representing the amount added to the account after w weeks have passed.

   d. What is the answer to the question? Explain.
3. **OPEN ENDED** Write a problem that can be answered by solving $x + 16 = 30$.

Guided Practice

Translate each sentence into an equation.

4. Two times a number t decreased by eight equals seventy.
5. Five times the sum of m and n is the same as seven times n.

Translate each sentence into a formula.

6. The area A of a triangle equals one half times the base b times the height h.
7. The circumference C of a circle equals the product of two, pi, and the radius r.

Translate each equation into a verbal sentence.

8. $14 + d = 6d$
9. $\frac{1}{3}b - \frac{3}{4} = 2a$

10. Write a problem based on the given information.

    c = cost of a suit $c - 25 = 150$

Application

WRESTLING For Exercises 11 and 12, use the following information.
Darius is training to prepare for wrestling season. He weighs 155 pounds now. He wants to gain weight so that he starts the season weighing 160 pounds.

11. If g represents the number of pounds he wants to gain, write an equation to represent the situation.
12. How many pounds does Darius need to gain to reach his goal?

Practice and Apply

Homework Help

For Exercises	See Examples
13–22	1
23–28	3
29–38	4
39, 40	5
41–51	2

Extra Practice

See page 825.

Translate each sentence into an equation.

13. Two hundred minus three times x is equal to nine.
14. The sum of twice r and three times s is identical to thirteen.
15. The sum of one-third q and 25 is as much as twice q.
16. The square of m minus the cube of n is sixteen.
17. Two times the sum of v and w is equal to two times z.
18. Half of the sum of nine and p is the same as p minus three.
19. The number g divided by the number h is the same as seven more than twice the sum of g and h.
20. Five-ninths the square of the sum of a, b, and c equals the sum of the square of a and the square of c.

21. **GEOGRAPHY** The Pacific Ocean covers about 46% of Earth. If P represents the area of the Pacific Ocean and E represents the area of Earth, write an equation for this situation.

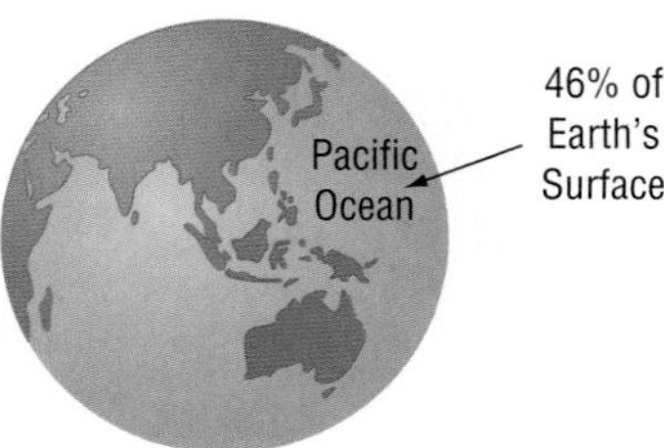

Source: *World Book Encyclopedia*

22. **GARDENING** Mrs. Patton is planning to place a fence around her vegetable garden. The fencing costs \$1.75 per yard. She buys f yards of fencing and pays \$3.50 in tax. If the total cost of the fencing is \$73.50, write an equation to represent the situation.

Translate each sentence into a formula.

23. The area A of a parallelogram is the base b times the height h.

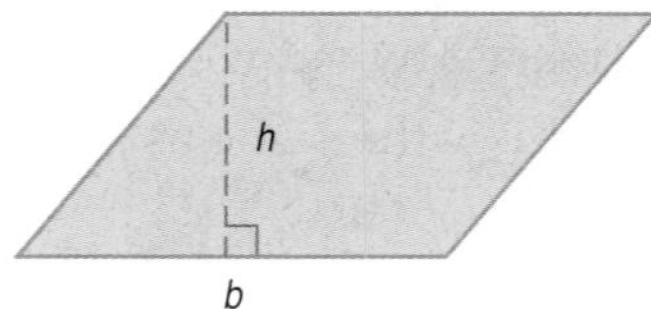

24. The volume V of a pyramid is one-third times the product of the area of the base B and its height h.

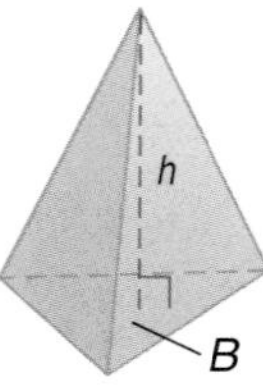

25. The perimeter P of a parallelogram is twice the sum of the lengths of the two adjacent sides, a and b.

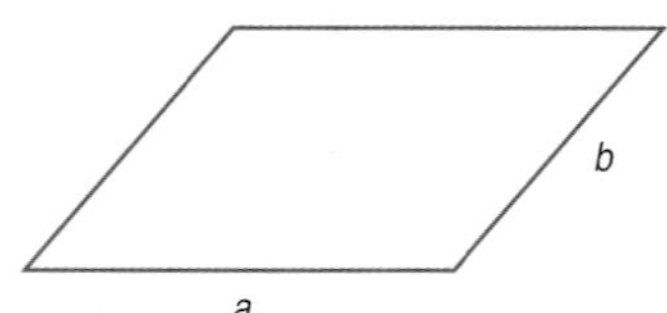

26. The volume V of a cylinder equals the product of π, the square of the radius r of the base, and the height.

27. In a right triangle, the square of the measure of the hypotenuse c is equal to the sum of the squares of the measures of the legs, a and b.
28. The temperature in degrees Fahrenheit F is the same as nine-fifths of the degrees Celsius C plus thirty-two.

Translate each equation into a verbal sentence.

29. $d - 14 = 5$
30. $2f + 6 = 19$
31. $k^2 + 17 = 53 - j$
32. $2a = 7a - b$
33. $\frac{3}{4}p + \frac{1}{2} = p$
34. $\frac{2}{5}w = \frac{1}{2}w + 3$
35. $7(m + n) = 10n + 17$
36. $4(t - s) = 5s + 12$

37. **GEOMETRY** If a and b represent the lengths of the bases of a trapezoid and h represents its height, then the formula for the area A of the trapezoid is $A = \frac{1}{2}h(a + b)$. Write the formula in words.

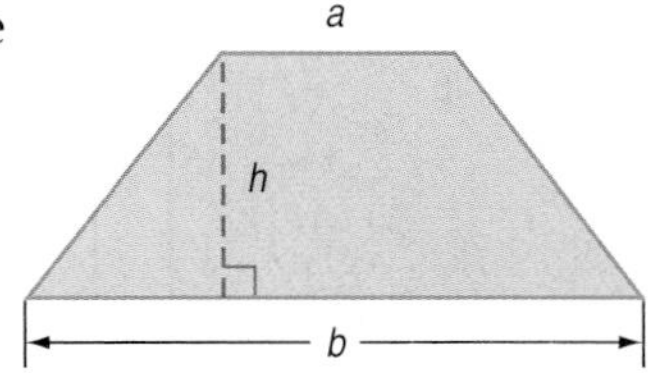

38. **SCIENCE** If r represents rate, t represents time, and d represents distance, then $rt = d$. Write the formula in words.

Write a problem based on the given information.

39. y = Yolanda's height in inches
$y + 7$ = Lindsey's height in inches
$2y + (y + 7) = 193$

40. p = price of a new backpack
$0.055p$ = tax
$p + 0.055p = 31.65$

GEOMETRY For Exercises 41 and 42, use the following information.
The volume V of a cone equals one-third times the product of π, the square of the radius r of the base, and the height h.

41. Write the formula for the volume of a cone.
42. Find the volume of a cone if r is 10 centimeters and h is 30 centimeters.

GEOMETRY For Exercises 43 and 44, use the following information.
The volume V of a sphere is four-thirds times π times the radius r of the sphere cubed.

43. Write a formula for the volume of a sphere.
44. Find the volume of a sphere if r is 4 inches.

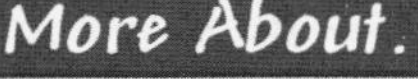

Literature
More than 50 movies featuring Tarzan have been made. The first, *Tarzan of the Apes*, in 1918, was among the first movies to gross over $1 million.

Source: www.tarzan.org

LITERATURE For Exercises 45–47, use the following information.
Edgar Rice Burroughs is the author of the *Tarzan of the Apes* stories. He published his first Tarzan story in 1912. Some years later, the town in southern California where he lived was named Tarzana.

45. Let y represent the number of years after 1912 that the town was named Tarzana. Write an expression for the year the town was named.
46. The town was named in 1928. Write an equation to represent the situation.
47. Use what you know about numbers to determine the number of years between the first Tarzan story and the naming of the town.

TELEVISION For Exercises 48–51, use the following information.
During a highly rated one-hour television program, the entertainment portion lasted 15 minutes longer than 4 times the advertising portion.

48. If a represents the time spent on advertising, write an expression for the entertainment portion.
49. Write an equation to represent the situation.
50. Use your equation and the guess-and-check strategy to determine the number of minutes spent on advertising. Choose different values of a and evaluate to find the solution.
51. Time the entertainment and advertising portions of a one-hour television program you like to watch. Describe what you found. Are the results of this problem similar to your findings?

52. **CRITICAL THINKING** The surface area of a prism is the sum of the areas of the faces of the prism. Write a formula for the surface area of the triangular prism at the right.

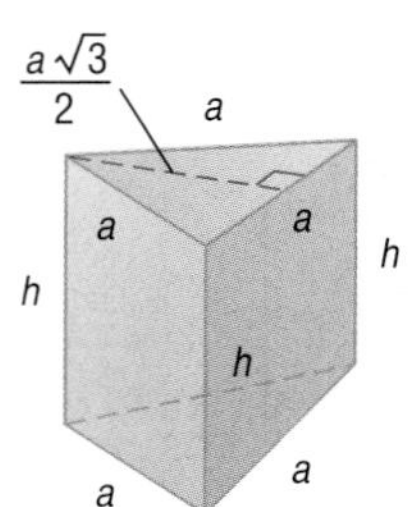

53. **WRITING IN MATH** Answer the question that was posed at the beginning of the lesson.

How are equations used to describe heights?

Include the following in your answer:

- an equation relating the Sears Tower, which is 1454 feet tall; the twin antenna towers on top of the building, which are a feet tall; and a total height, which is 1707 feet, and
- an equation representing the height of a building of your choice.

54. Which equation represents the following sentence?
One fourth of a number plus five equals the number minus seven.

(A) $\frac{1}{4}n + 7 = n - 5$ (B) $\frac{1}{4}n + 5 = n - 7$

(C) $4n + 7 = n - 5$ (D) $4n + 5 = n - 7$

55. Which sentence can be represented by $7(x + y) = 35$?

(A) Seven times x plus y equals 35.
(B) One seventh of the sum of x and y equals 35.
(C) Seven plus x and y equals 35.
(D) Seven times the sum of x and y equals 35.

Maintain Your Skills

Mixed Review

Find each square root. Use a calculator if necessary. Round to the nearest hundredth if the result is not a whole number or a simple fraction. *(Lesson 2-7)*

56. $\sqrt{8100}$ 57. $-\sqrt{\frac{25}{36}}$ 58. $\sqrt{90}$ 59. $-\sqrt{55}$

Find the probability of each outcome if a die is rolled. *(Lesson 2-6)*

60. a 6 61. an even number 62. a number greater than 2

Simplify each expression. *(Lesson 1-5)*

63. $12d + 3 - 4d$ 64. $7t^2 + t + 8t$ 65. $3(a + 2b) + 5a$

Evaluate each expression. *(Lesson 1-2)*

66. $5(8 - 3) + 7 \cdot 2$ 67. $6(4^3 + 2^2)$ 68. $7(0.2 + 0.5) - 0.6$

Getting Ready for the Next Lesson

PREREQUISITE SKILL **Find each sum or difference.**
*(To review **operations with fractions**, see pages 798 and 799.)*

69. $5.67 + 3.7$ 70. $0.57 + 2.8$ 71. $5.28 - 3.4$ 72. $9 - 7.35$

73. $\frac{2}{3} + \frac{1}{5}$ 74. $\frac{1}{6} + \frac{2}{3}$ 75. $\frac{7}{9} - \frac{2}{3}$ 76. $\frac{3}{4} - \frac{1}{6}$

Algebra Activity

A Preview of Lesson 3-2

Solving Addition and Subtraction Equations

You can use algebra tiles to solve equations. To solve an equation means to find the value of the variable that makes the equation true. After you model the equation, the goal is to get the x tile by itself on one side of the mat using the rules stated below.

Rules for Equation Models
You can remove or add the same number of identical algebra tiles to each side of the mat without changing the equation.
One positive tile and one negative tile of the same unit are a **zero pair**. Since $1 + (-1) = 0$, you can remove or add zero pairs to the equation mat without changing the equation.

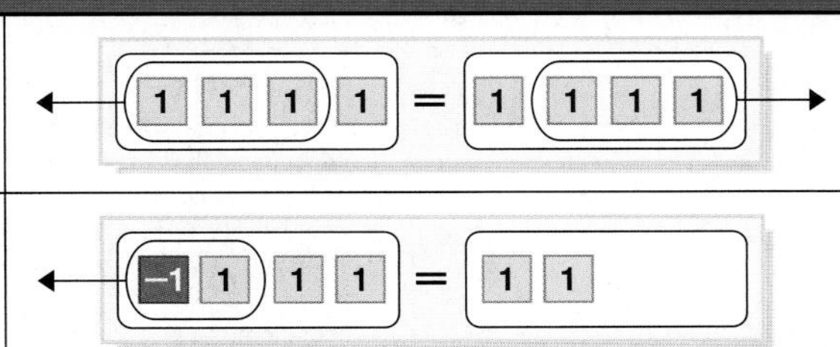

Use an equation model to solve $x - 3 = 2$.

Step 1 *Model the equation.*

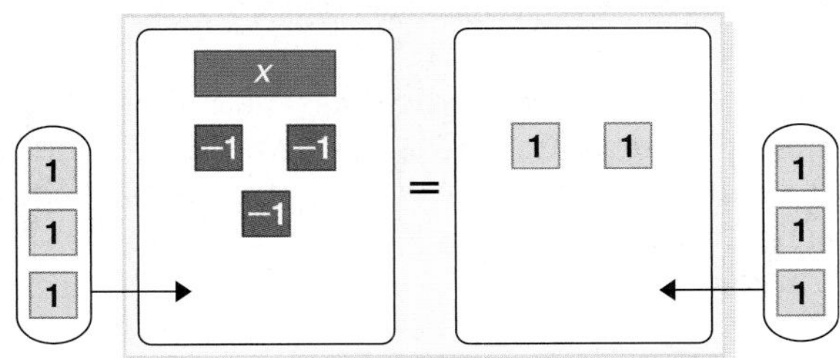

$$x - 3 = 2$$

$$x - 3 + 3 = 2 + 3$$

Place 1 x tile and 3 negative 1 tiles on one side of the mat. Place 2 positive 1 tiles on the other side of the mat. Then add 3 positive 1 tiles to each side.

Step 2 *Isolate the x term.*

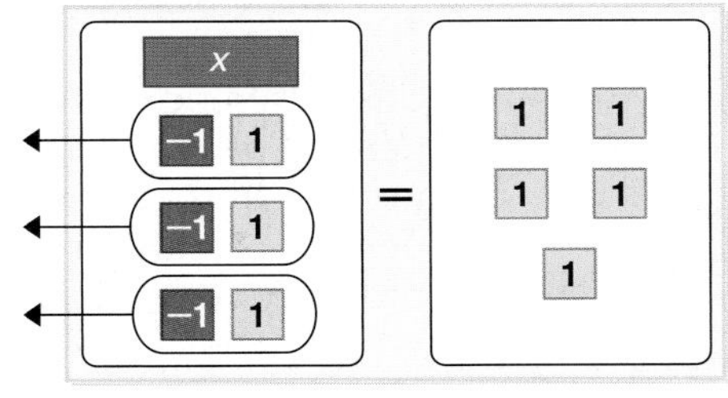

$$x = 5$$

Group the tiles to form zero pairs. Then remove all the zero pairs. The resulting equation is $x = 5$.

Model and Analyze

Use algebra tiles to solve each equation.

1. $x + 5 = 7$

2. $x + (-2) = 28$

3. $x + 4 = 27$

4. $x + (-3) = 4$

5. $x + 3 = -4$

6. $x + 7 = 2$

Make a Conjecture

7. If $a = b$, what can you say about $a + c$ and $b + c$?

8. If $a = b$, what can you say about $a - c$ and $b - c$?

3-2 Solving Equations by Using Addition and Subtraction

What You'll Learn

- Solve equations by using addition.
- Solve equations by using subtraction.

Vocabulary
- equivalent equation
- solve an equation

How can equations be used to compare data?

The graph shows some of the fastest-growing occupations from 1992 to 2005.

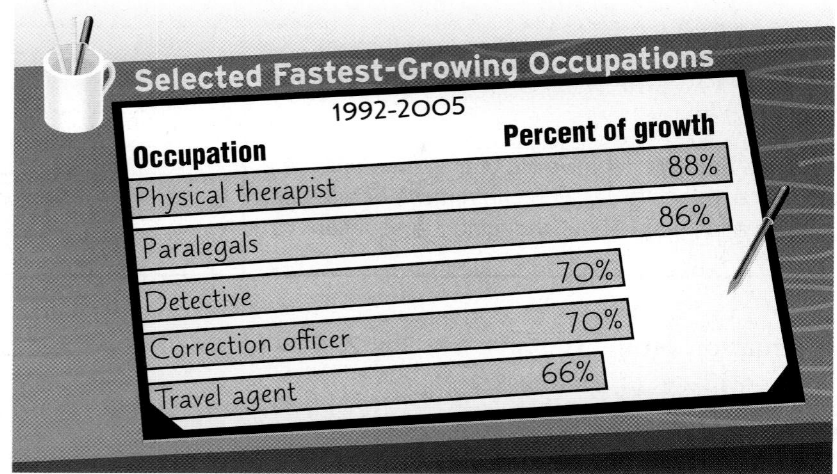

Source: Bureau of Labor Statistics

The difference between the percent of growth for medical assistants and the percent of growth for travel agents in these years is 5%. An equation can be used to find the percent of growth expected for medical assistants. If m is the percent of growth for medical assistants, then $m - 66 = 5$. You can use a property of equality to find the value of m.

SOLVE USING ADDITION Suppose your school's boys' soccer team has 15 members and the girls' soccer team has 15 members. If each team adds 3 new players, the number of members on the boys' and girls' teams would still be equal.

$15 = 15$	Each team has 15 members before adding the new players.
$15 + 3 = 15 + 3$	Each team adds 3 new members.
$18 = 18$	Each team has 18 members after adding the new members.

This example illustrates the **Addition Property of Equality**.

Key Concept — Addition Property of Equality

- **Words** If an equation is true and the same number is added to each side, the resulting equation is true.
- **Symbols** For any numbers a, b, and c, if $a = b$, then $a + c = b + c$.
- **Examples**

$$7 = 7 \qquad\qquad 14 = 14$$
$$7 + 3 = 7 + 3 \qquad\qquad 14 + (-6) = 14 + (-6)$$
$$10 = 10 \qquad\qquad 8 = 8$$

If the same number is added to each side of an equation, then the result is an equivalent equation. **Equivalent equations** have the same solution.

$t + 3 = 5$	The solution of this equation is 2.
$t + 3 + 2 = 5 + 2$	Using the Addition Property of Equality, add 2 to each side.
$t + 5 = 7$	The solution of this equation is also 2.

Study Tip

Reading Math
Remember that x means $1 \cdot x$. The coefficient of x is 1.

To **solve an equation** means to find all values of the variable that make the equation a true statement. One way to do this is to isolate the variable having a coefficient of 1 on one side of the equation. You can sometimes do this by using the Addition Property of Equality.

Example 1 Solve by Adding a Positive Number

Solve $m - 48 = 29$. Then check your solution.

$m - 48 = 29$	Original equation
$m - 48 + 48 = 29 + 48$	Add 48 to each side.
$m = 77$	$-48 + 48 = 0$ and $29 + 48 = 77$

To check that 77 is the solution, substitute 77 for m in the original equation.

CHECK	$m - 48 = 29$	Original equation
	$77 - 48 \stackrel{?}{=} 29$	Substitute 77 for m.
	$29 = 29$ ✓	Subtract.

The solution is 77.

Example 2 Solve by Adding a Negative Number

Solve $21 + q = -18$. Then check your solution.

$21 + q = -18$	Original equation
$21 + q + (-21) = -18 + (-21)$	Add -21 to each side.
$q = -39$	$21 + (-21) = 0$ and $-18 + (-21) = -39$

CHECK	$21 + q = -18$	Original equation
	$21 + (-39) \stackrel{?}{=} -18$	Substitute -39 for q.
	$-18 = -18$ ✓	Add.

The solution is -39.

SOLVE USING SUBTRACTION Similar to the Addition Property of Equality, there is a **Subtraction Property of Equality** that may be used to solve equations.

Key Concept — *Subtraction Property of Equality*

- **Words** If an equation is true and the same number is subtracted from each side, the resulting equation is true.
- **Symbols** For any numbers a, b, and c, if $a = b$, then $a - c = b - c$.
- **Examples**

$17 = 17$	$3 = 3$
$17 - 9 = 17 - 9$	$3 - 8 = 3 - 8$
$8 = 8$	$-5 = -5$

Example 3 Solve by Subtracting

Solve 142 + d = 97. Then check your solution.

$142 + d = 97$ Original equation

$142 + d - 142 = 97 - 142$ Subtract 142 from each side.

$d = -45$ $142 - 142 = 0$ and $97 - 142 = -45$

CHECK $142 + d = 97$ Original equation

$142 + (-45) \stackrel{?}{=} 97$ Substitute -45 for d.

$97 = 97$ ✓ Add.

The solution is -45.

Remember that subtracting a number is the same as adding its inverse.

Example 4 Solve by Adding or Subtracting

Solve $g + \frac{3}{4} = -\frac{1}{8}$ in two ways.

Method 1 Use the Subtraction Property of Equality.

$g + \frac{3}{4} = -\frac{1}{8}$ Original equation

$g + \frac{3}{4} - \frac{3}{4} = -\frac{1}{8} - \frac{3}{4}$ Subtract $\frac{3}{4}$ from each side.

$g = -\frac{7}{8}$ $\frac{3}{4} - \frac{3}{4} = 0$ and $-\frac{1}{8} - \frac{3}{4} = -\frac{1}{8} - \frac{6}{8}$ or $-\frac{7}{8}$

The solution is $-\frac{7}{8}$.

Method 2 Use the Addition Property of Equality.

$g + \frac{3}{4} = -\frac{1}{8}$ Original equation

$g + \frac{3}{4} + \left(-\frac{3}{4}\right) = -\frac{1}{8} + \left(-\frac{3}{4}\right)$ Add $-\frac{3}{4}$ to each side.

$g = -\frac{7}{8}$ $\frac{3}{4} + \left(-\frac{3}{4}\right) = 0$ and $-\frac{1}{8} + \left(-\frac{3}{4}\right) = -\frac{1}{8} + \left(-\frac{6}{8}\right)$ or $-\frac{7}{8}$

The solution is $-\frac{7}{8}$.

Example 5 Write and Solve an Equation

Write an equation for the problem. Then solve the equation and check your solution.

A number increased by 5 is equal to 42. Find the number.

A number	increased by	5	is equal to	42.
n	+	5	=	42

$n + 5 = 42$ Original equation

$n + 5 - 5 = 42 - 5$ Subtract 5 from each side.

$n = 37$ $5 - 5 = 0$ and $42 - 5 = 37$

CHECK $n + 5 = 42$ Original equation

$37 + 5 \stackrel{?}{=} 42$ Substitute 37 for n.

$42 = 42$ ✓

The solution is 37.

Study Tip

Checking Solutions
You should always check your solution in the context of the original problem. For instance, in Example 5, is 37 increased by 5 equal to 42? The solution checks.

More About. . .

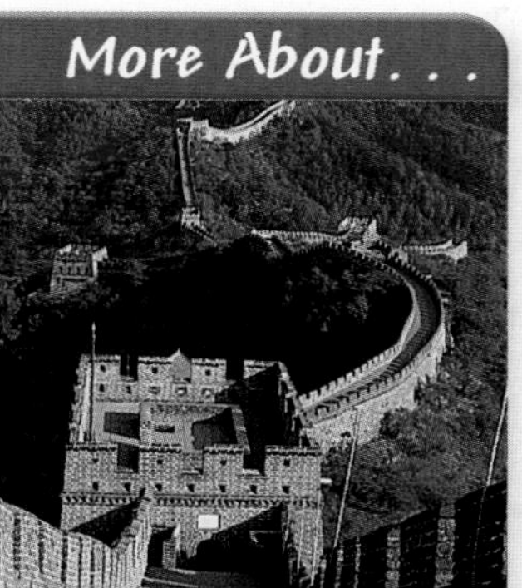

History

The first emperor of China, Qui Shi Huangdi, ordered the building of the Great Wall of China to protect his people from nomadic tribes that attacked and looted villages. By 204 B.C., this wall guarded 1000 miles of China's border.

Source: *National Geographic World*

Example 6 Write an Equation to Solve a Problem

HISTORY Refer to the information at the right.

In the fourteenth century, the part of the Great Wall of China that was built during Qui Shi Huangdi's time was repaired, and the wall was extended. When the wall was completed, it was 2500 miles long. How much of the wall was added during the 1300s?

Amount added

1000 mi

2500 mi

Source: *National Geographic World*

Words The original length plus the additional length equals 2500.

Variable Let a = the additional length.

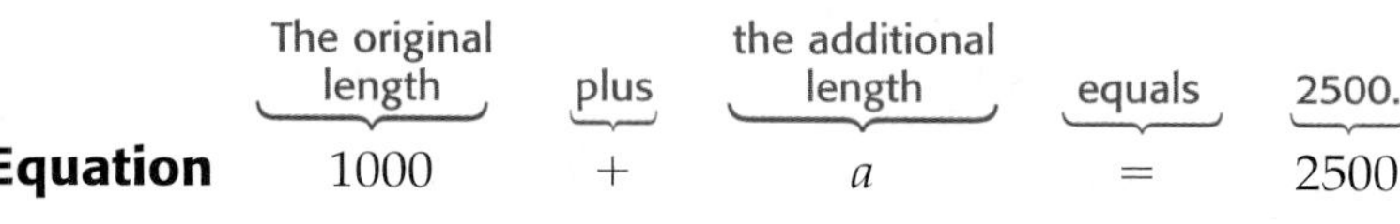
The original length
plus
the additional length
equals
2500.

Equation $1000 + a = 2500$

$$1000 + a = 2500$$ Original equation

$$1000 + a - 1000 = 2500 - 1000$$ Subtract 1000 from each side.

$$a = 1500$$ $1000 - 1000 = 0$ and $2500 - 1000 = 1500$.

The Great Wall of China was extended 1500 miles in the 1300s.

Check for Understanding

Concept Check

1. **OPEN ENDED** Write three equations that are equivalent to $n + 14 = 27$.
2. **Compare and contrast** the Addition Property of Equality and the Subtraction Property of Equality.
3. **Show** two ways to solve $g + 94 = 75$.

Guided Practice

Solve each equation. Then check your solution.

4. $t - 4 = -7$
5. $p + 19 = 6$
6. $15 + r = 71$
7. $104 = y - 67$
8. $h - 0.78 = 2.65$
9. $\frac{2}{3} + w = 1\frac{1}{2}$

Write an equation for each problem. Then solve the equation and check your solution.

10. Twenty-one subtracted from a number is -8. Find the number.
11. A number increased by -37 is -91. Find the number.

Application

CARS For Exercises 12–14, use the following information.

The average time it takes to manufacture a car in the United States is equal to the average time it takes to manufacture a car in Japan plus 8.1 hours. The average time it takes to manufacture a car in the United States is 24.9 hours.

12. Write an addition equation to represent the situation.
13. What is the average time to manufacture a car in Japan?
14. The average time it takes to manufacture a car in Europe is 35.5 hours. What is the difference between the average time it takes to manufacture a car in Europe and the average time it takes to manufacture a car in Japan?

Practice and Apply

Homework Help

For Exercises	See Examples
15–40	1–4
41–48	5
51–64	6

Extra Practice
See page 825.

Solve each equation. Then check your solution.

15. $v - 9 = 14$
16. $s - 19 = -34$
17. $g + 5 = 33$
18. $18 + z = 44$
19. $a - 55 = -17$
20. $t - 72 = -44$
21. $-18 = -61 + d$
22. $-25 = -150 + q$
23. $r - (-19) = -77$
24. $b - (-65) = 15$
25. $18 - (-f) = 91$
26. $125 - (-p) = 88$
27. $-2.56 + c = 0.89$
28. $k + 0.6 = -3.84$
29. $-6 = m + (-3.42)$
30. $6.2 = -4.83 + y$
31. $t - 8.5 = 7.15$
32. $q - 2.78 = 4.2$
33. $x - \frac{3}{4} = \frac{5}{6}$
34. $a - \frac{3}{5} = -\frac{7}{10}$
35. $-\frac{1}{2} + p = \frac{5}{8}$
36. $\frac{2}{3} + r = -\frac{4}{9}$
37. $\frac{2}{3} = v + \frac{4}{5}$
38. $\frac{2}{5} = w + \frac{3}{4}$

39. If $x - 7 = 14$, what is the value of $x - 2$?

40. If $t + 8 = -12$, what is the value of $t + 1$?

GEOMETRY **For Exercises 41 and 42, use the rectangle at the right.**

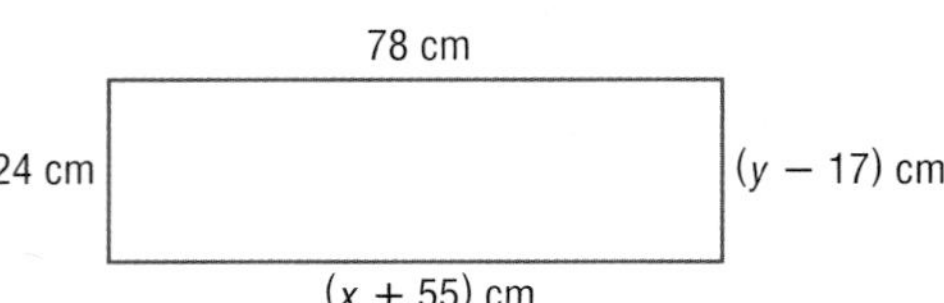

41. Write an equation you could use to solve for x and then solve for x.

42. Write an equation you could use to solve for y and then solve for y.

Write an equation for each problem. Then solve the equation and check your solution.

43. Eighteen subtracted from a number equals 31. Find the number.

44. What number decreased by 77 equals -18?

45. A number increased by -16 is -21. Find the number.

46. The sum of a number and -43 is 102. What is the number?

47. What number minus one-half is equal to negative three-fourths?

48. The sum of 19 and 42 and a number is equal to 87. What is the number?

49. Determine whether $x + x = x$ is *sometimes*, *always*, or *never* true. Explain your reasoning.

50. Determine whether $x + 0 = x$ is *sometimes*, *always*, or *never* true. Explain your reasoning.

GAS MILEAGE **For Exercises 51–55, use the following information.**
A midsize car with a 4-cylinder engine goes 10 miles more on a gallon of gasoline than a luxury car with an 8-cylinder engine. A midsize car consumes one gallon of gas for every 34 miles driven.

51. Write an addition equation to represent the situation.

52. How many miles does a luxury car travel on a gallon of gasoline?

53. A subcompact car with a 3-cylinder engine goes 13 miles more than a luxury car on one gallon of gas. How far does a subcompact car travel on a gallon of gasoline?

54. How many more miles does a subcompact travel on a gallon of gasoline than a midsize car?

55. Estimate how many miles a full-size car with a 6-cylinder engine goes on one gallon of gasoline. Explain your reasoning.

HISTORY For Exercises 56 and 57, use the following information.
Over the years, the height of the Great Pyramid at Giza, Egypt, has decreased.

56. Write an addition equation to represent the situation.

57. What was the decrease in the height of the pyramid?

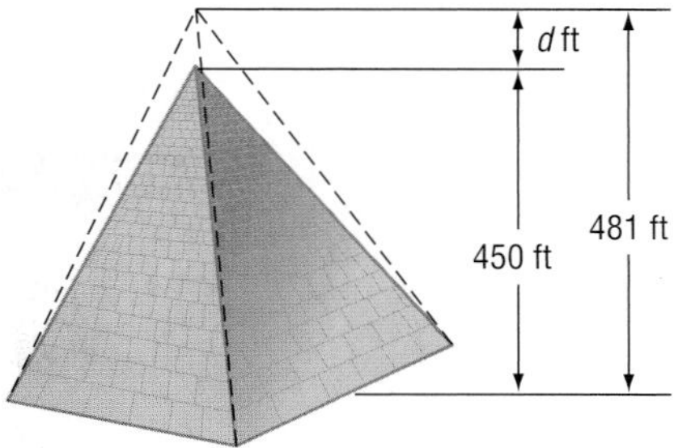

Source: *World Book Encyclopedia*

LIBRARIES For Exercises 58–61, use the graph at the right to write an equation for each situation. Then solve the equation.

58. How many more volumes does the Library of Congress have than the Harvard University Library?

59. How many more volumes does the Harvard University Library have than the New York Public Library?

60. How many more volumes does the Library of Congress have than the New York Public Library?

61. What is the total number of volumes in the three largest U.S. libraries?

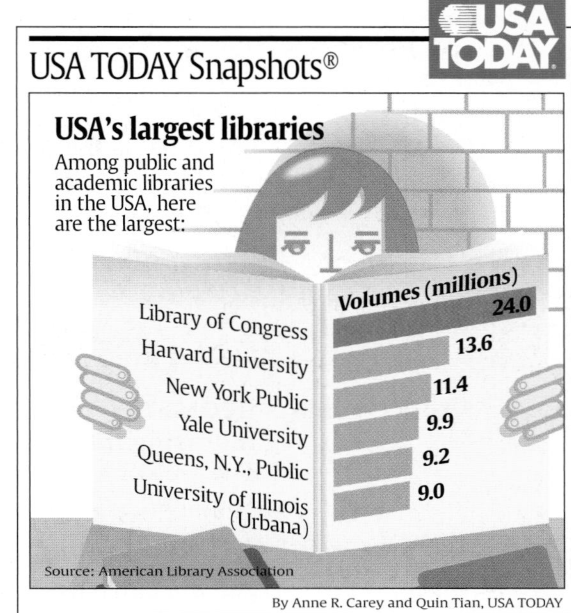

ANIMALS For Exercises 62–64, use the information below to write an equation for each situation. Then solve the equation.
Wildlife authorities monitor the population of animals in various regions. One year's deer population in Dauphin County, Pennsylvania, is shown in the graph below.

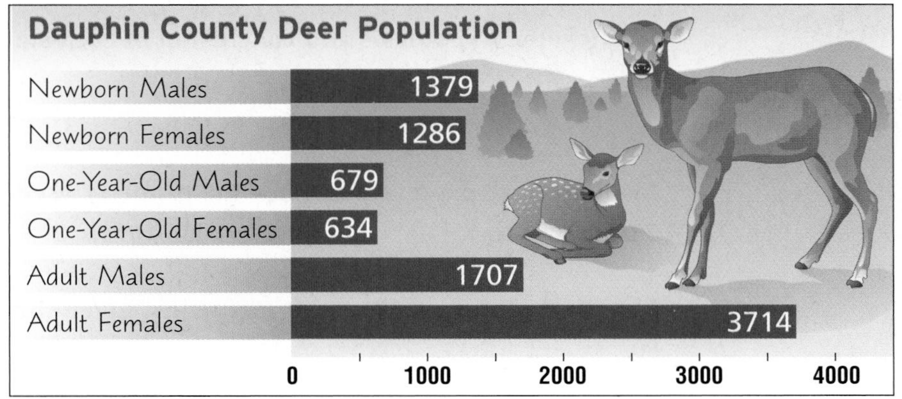

Source: www.visi.com

62. How many more newborns are there than one-year-olds?

63. How many more females are there than males?

64. What is the total deer population?

65. **CRITICAL THINKING** If $a - b = x$, what values of a, b, and x would make the equation $a + x = b + x$ true?

66. **WRITING IN MATH** Answer the question that was posed at the beginning of the lesson.

How can equations be used to compare data?

Include the following in your answer:

- an explanation of how to solve the equation to find the growth rate for medical assistants, and
- a sample problem and related equation using the information in the graph.

67. Which equation is *not* equivalent to $b - 15 = 32$?

Ⓐ $b + 5 = 52$ Ⓑ $b - 20 = 27$
Ⓒ $b - 13 = 30$ Ⓓ $b = 47$

68. What is the solution of $x - 167 = -52$?

Ⓐ 115 Ⓑ -115
Ⓒ 219 Ⓓ -219

Maintain Your Skills

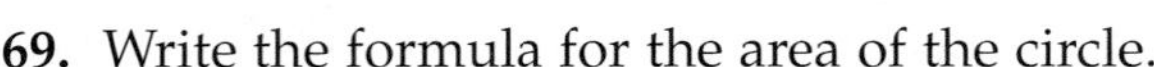

Mixed Review

GEOMETRY For Exercises 69 and 70, use the following information.
The area of a circle is the product of π times the radius r squared. *(Lesson 3-1)*

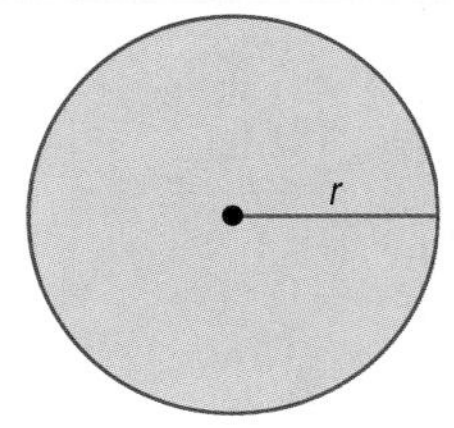

69. Write the formula for the area of the circle.

70. If a circle has a radius of 16 inches, find its area.

Replace each ● with >, <, or = to make the sentence true. *(Lesson 2-7)*

71. $\frac{1}{2}$ ● $\sqrt{2}$ 72. $\frac{3}{4}$ ● $\frac{2}{3}$ 73. 0.375 ● $\frac{3}{8}$

Use each set of data to make a stem-and-leaf plot. *(Lesson 2-5)*

74. 54, 52, 43, 41, 40, 36, 35, 31, 32, 34, 42, 56

75. 2.3, 1.4, 1.7, 1.2, 2.6, 0.8, 0.5, 2.8, 4.1, 2.9, 4.5, 1.1

Identify the hypothesis and conclusion of each statement. *(Lesson 1-7)*

76. For $y = 2$, $4y - 6 = 2$.

77. There is a science quiz every Friday.

Evaluate each expression. Name the property used in each step. *(Lesson 1-4)*

78. $4(16 \div 4^2)$ 79. $(2^5 - 5^2) + (4^2 - 2^4)$

Find the solution set for each inequality, given the replacement set. *(Lesson 1-3)*

80. $3x + 2 > 2$; {0, 1, 2} 81. $2y^2 - 1 > 0$; {1, 3, 5}

Getting Ready for the Next Lesson

PREREQUISITE SKILL Find each product or quotient.
(To review ***operations with fractions****, see pages 800 and 801.)*

82. 6.5×2.8 83. 70.3×0.15 84. $17.8 \div 2.5$ 85. $0.33 \div 1.5$

86. $\frac{2}{3} \times \frac{5}{8}$ 87. $\frac{5}{9} \times \frac{3}{10}$ 88. $\frac{1}{2} \div \frac{2}{5}$ 89. $\frac{8}{9} \div \frac{4}{15}$

3-3 Solving Equations by Using Multiplication and Division

What You'll Learn

- Solve equations by using multiplication.
- Solve equations by using division.

How can equations be used to find how long it takes light to reach Earth?

It may look like all seven stars in the Big Dipper are the same distance from Earth, but in fact, they are not. The diagram shows the distance between each star and Earth.

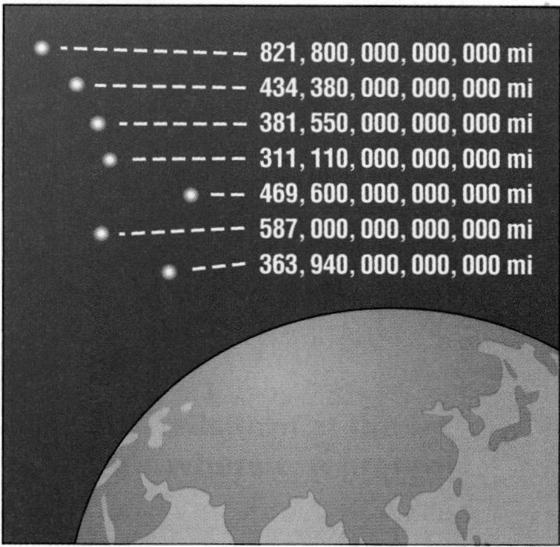

Source: *National Geographic World*

Light travels at a rate of about 5,870,000,000,000 miles per year. In general, the rate at which something travels times the time equals the distance ($rt = d$). The following equation can be used to find the time it takes light to reach Earth from the closest star in the Big Dipper.

$$rt = d$$

$$5{,}870{,}000{,}000{,}000t = 311{,}110{,}000{,}000{,}000$$

SOLVE USING MULTIPLICATION To solve equations such as the one above, you can use the **Multiplication Property of Equality**.

Key Concept — *Multiplication Property of Equality*

- **Words** If an equation is true and each side is multiplied by the same number, the resulting equation is true.
- **Symbols** For any numbers a, b, and c, if $a = b$, then $ac = bc$.
- **Examples**

$6 = 6$	$9 = 9$	$10 = 10$
$6 \times 2 = 6 \times 2$	$9 \times (-3) = 9 \times (-3)$	$10 \times \frac{1}{2} = 10 \times \frac{1}{2}$
$12 = 12$	$-27 = -27$	$5 = 5$

Example 1 *Solve Using Multiplication by a Positive Number*

Solve $\frac{t}{30} = \frac{7}{10}$. Then check your solution.

$\frac{t}{30} = \frac{7}{10}$ Original equation

$30\left(\frac{t}{30}\right) = 30\left(\frac{7}{10}\right)$ Multiply each side by 30.

$t = 21$ $\frac{t}{30}(30) = t$ and $\frac{7}{10}(30) = 21$

(continued on the next page)

CHECK $\frac{t}{30} = \frac{7}{10}$ Original equation

$\frac{21}{30} \stackrel{?}{=} \frac{7}{10}$ Substitute 21 for t.

$\frac{7}{10} = \frac{7}{10}$ ✓ The solution is 21.

Example 2 Solve Using Multiplication by a Fraction

Solve $\left(2\frac{1}{4}\right)g = 1\frac{1}{2}$.

$\left(2\frac{1}{4}\right)g = 1\frac{1}{2}$ Original equation

$\left(\frac{9}{4}\right)g = \frac{3}{2}$ Rewrite each mixed number as an improper fraction.

$\frac{4}{9}\left(\frac{9}{4}\right)g = \frac{4}{9}\left(\frac{3}{2}\right)$ Multiply each side by $\frac{4}{9}$, the reciprocal of $\frac{9}{4}$.

$g = \frac{12}{18}$ or $\frac{2}{3}$ Check this result.

The solution is $\frac{2}{3}$.

Example 3 Solve Using Multiplication by a Negative Number

Solve $42 = -6m$.

$42 = -6m$ Original equation

$-\frac{1}{6}(42) = -\frac{1}{6}(-6m)$ Multiply each side by $-\frac{1}{6}$, the reciprocal of -6.

$-7 = m$ Check this result.

The solution is -7.

You can write an equation to represent a real-world problem. Then use the equation to solve the problem.

More About. . .

Space Travel

On July 20, 1969, Neil Armstrong stepped on the surface of the moon. On the moon, his suit and life-support backpacks weighed about 33 pounds.

Source: NASA

Example 4 Write and Solve an Equation Using Multiplication

SPACE TRAVEL Refer to the information about space travel at the left. The weight of anything on the moon is about one-sixth its weight on Earth. What was the weight of Neil Armstrong's suit and life-support backpacks on Earth?

Words One sixth times the weight on Earth equals the weight on the moon.

Variable Let w = the weight on Earth.

	One sixth	times	the weight on Earth	equals	the weight on the moon.
Equation	$\frac{1}{6}$	$\cdot$	w	$=$	33

$\frac{1}{6}w = 33$ Original equation

$6\left(\frac{1}{6}w\right) = 6(33)$ Multiply each side by 6.

$w = 198$ $\frac{1}{6}(6) = 1$ and $33(6) = 198$

The weight of Neil Armstrong's suit and life-support backpacks on Earth was about 198 pounds.

SOLVE USING DIVISION The equation in Example 3, $42 = -6m$, was solved by multiplying each side by $-\frac{1}{6}$. The same result could have been obtained by dividing each side by -6. This method uses the **Division Property of Equality**.

Key Concept — Division Property of Equality

- **Words** If an equation is true and each side is divided by the same nonzero number, the resulting equation is true.
- **Symbols** For any numbers a, b, and c, with $c \neq 0$, if $a = b$, then $\frac{a}{c} = \frac{b}{c}$.
- **Examples**

$$15 = 15 \qquad 28 = 28$$
$$\frac{15}{3} = \frac{15}{3} \qquad \frac{28}{-7} = \frac{28}{-7}$$
$$5 = 5 \qquad -4 = -4$$

Example 5 Solve Using Division by a Positive Number

Solve $13s = 195$. Then check your solution.

$13s = 195$	Original equation
$\frac{13s}{13} = \frac{195}{13}$	Divide each side by 13.
$s = 15$	$\frac{13s}{13} = s$ and $\frac{195}{13} = 15$

CHECK

$13s = 195$	Original equation
$13(15) \stackrel{?}{=} 195$	Substitute 15 for s.
$195 = 195$ ✓	

The solution is 15.

Study Tip

Alternative Method

You can also solve equations like those in Examples 5, 6, and 7 by using the Multiplication Property of Equality. For instance, in Example 6, you could multiply each side by $-\frac{1}{3}$.

Example 6 Solve Using Division by a Negative Number

Solve $-3x = 12$.

$-3x = 12$	Original equation
$\frac{-3x}{-3} = \frac{12}{-3}$	Divide each side by -3.
$x = -4$	$\frac{-3x}{-3} = x$ and $\frac{12}{-3} = -4$

The solution is -4.

Example 7 Write and Solve an Equation Using Division

Write an equation for the problem below. Then solve the equation.
Negative eighteen times a number equals −198.

Negative eighteen	times	a number	equals	−198.
-18	$\times$	n	$=$	-198

$-18n = -198$	Original equation
$\frac{-18n}{-18} = \frac{-198}{-18}$	Divide each side by −18.
$n = 11$	Check this result.

The solution is 11.

Check for Understanding

Concept Check

1. **OPEN ENDED** Write a multiplication equation that has a solution of -3.

2. **Explain** why the Multiplication Property of Equality and the Division Property of Equality can be considered the same property.

3. **FIND THE ERROR** Casey and Juanita are solving $8n = -72$.

Casey	Juanita
$8n = -72$	$8n = -72$
$8n(8) = -72(8)$	$\frac{8n}{8} = \frac{-72}{8}$
$n = -576$	$n = -9$

Who is correct? Explain your reasoning.

Guided Practice

Solve each equation. Then check your solution.

4. $-2g = -84$
5. $\frac{t}{7} = -5$
6. $\frac{a}{36} = \frac{4}{9}$
7. $\frac{4}{5}k = \frac{8}{9}$
8. $3.15 = 1.5y$
9. $\left(3\frac{1}{4}\right)p = 2\frac{1}{2}$

Write an equation for each problem. Then solve the equation.

10. Five times a number is 120. What is the number?
11. Two fifths of a number equals -24. Find the number.

Application

12. **GEOGRAPHY** The discharge of a river is defined as the width of the river times the average depth of the river times the speed of the river. At one location in St. Louis, the Mississippi River is 533 meters wide, its speed is 0.6 meter per second, and its discharge is 3198 cubic meters per second. How deep is the Mississippi River at this location?

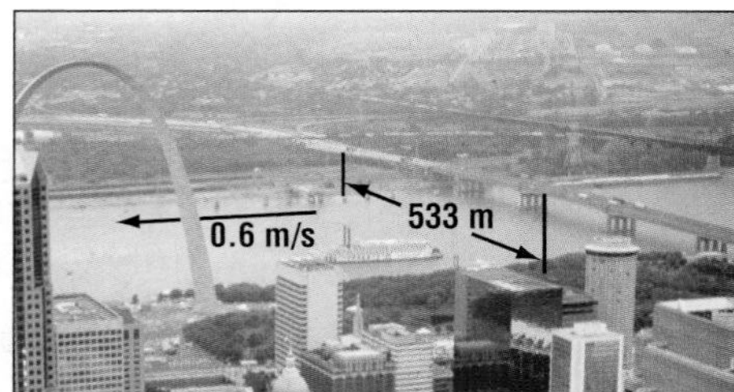

Practice and Apply

Homework Help

For Exercises	See Examples
13–32	1–3, 5, 6
33–38	7
39–49	4

Extra Practice
See page 826.

Solve each equation. Then check your solution.

13. $-5r = 55$
14. $8d = 48$
15. $-910 = -26a$
16. $-1634 = 86s$
17. $\frac{b}{7} = -11$
18. $-\frac{v}{5} = -45$
19. $\frac{2}{3}n = 14$
20. $\frac{2}{5}g = -14$
21. $\frac{g}{24} = \frac{5}{12}$
22. $\frac{z}{45} = \frac{2}{5}$
23. $1.9f = -11.78$
24. $0.49k = 6.272$
25. $-2.8m = 9.8$
26. $-5.73q = 97.41$
27. $\left(-2\frac{3}{5}\right)t = -22$
28. $\left(3\frac{2}{3}\right)x = -5\frac{1}{2}$
29. $-5h = -3\frac{2}{3}$
30. $3p = 4\frac{1}{5}$
31. If $4m = 10$, what is the value of $12m$?
32. If $15b = 55$, what is the value of $3b$?

Write an equation for each problem. Then solve the equation.

33. Seven times a number equals −84. What is the number?

34. Negative nine times a number is −117. Find the number.

35. One fifth of a number is 12. Find the number.

36. Negative three eighths times a number equals 12. What is the number?

37. Two and one half times a number equals one and one fifth. Find the number.

38. One and one third times a number is −4.82. What is the number?

GENETICS For Exercises 39–41, use the following information.
Research conducted by a daily U.S. newspaper has shown that about one seventh of people in the world are left-handed.

39. Write a multiplication equation relating the number of left-handed people ℓ and the total number of people p.

40. About how many left-handed people are there in a group of 350 people?

41. If there are 65 left-handed people in a group, about how many people are in that group?

42. **WORLD RECORDS** In 1993, a group of people in Utica, New York, made a very large round jelly doughnut which broke the world record for doughnut size. It weighed 1.5 tons and had a circumference of 50 feet. What was the diameter of the doughnut? (*Hint*: $C = \pi d$)

BASEBALL For Exercises 43–45, use the following information.
In baseball, if all other factors are the same, the speed of a four-seam fastball is faster than a two-seam fastball. The distance from the pitcher's mound to home plate is 60.5 feet.

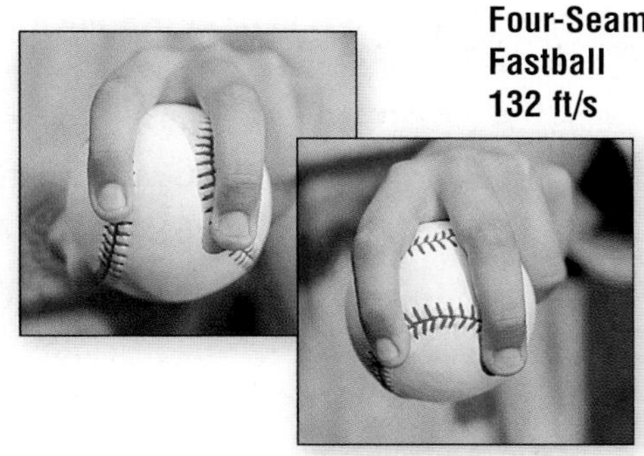

Source: *Baseball and Mathematics*

43. How long does it take a two-seam fastball to go from the pitcher's mound to home plate? Round to the nearest hundredth. (*Hint*: $rt = d$)

44. How long does it take a four-seam fastball to go from the pitcher's mound to home plate? Round to the nearest hundredth.

45. How much longer does it take for a two-seam fastball to reach home plate than a four-seam fastball?

PHYSICAL SCIENCE For Exercises 46–49, use the following information.
In science lab, Devin and his classmates are asked to determine how many grams of hydrogen and how many grams of oxygen are in 477 grams of water. Devin used what he learned in class to determine that for every 8 grams of oxygen in water, there is 1 gram of hydrogen.

46. If x represents the number of grams of hydrogen, write an expression to represent the number of grams of oxygen.

47. Write an equation to represent the situation.

48. How many grams of hydrogen are in 477 grams of water?

49. How many grams of oxygen are in 477 grams of water?

50. **CRITICAL THINKING** If $6y - 7 = 4$, what is the value of $18y - 21$?

51. **WRITING IN MATH** Answer the question that was posed at the beginning of the lesson.

How can equations be used to find how long it takes light to reach Earth?

Include the following in your answer:

- an explanation of how to find the length of time it takes light to reach Earth from the closest star in the Big Dipper, and
- an equation describing the situation for the farthest star in the Big Dipper.

Standardized Test Practice

52. The rectangle at the right is divided into 5 identical squares. If the perimeter of the rectangle is 48 inches, what is the area of each square?

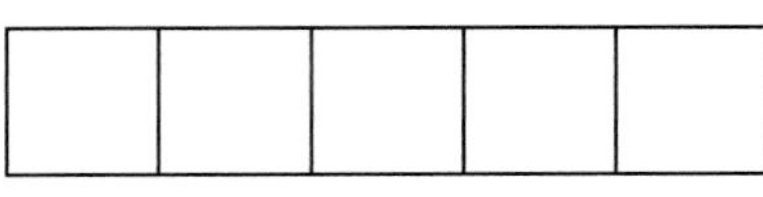

(A) 4 in^2 (B) 9.8 in^2 (C) 16 in^2 (D) 23.04 in^2

53. Which equation is equivalent to $4t = 20$?

(A) $-2t = -10$ (B) $t = 80$ (C) $2t = 5$ (D) $-8t = 40$

Maintain Your Skills

Mixed Review

Solve each equation. Then check your solution. *(Lesson 3-2)*

54. $m + 14 = 81$ 55. $d - 27 = -14$ 56. $17 - (-w) = -55$

57. Translate the following sentence into an equation. *(Lesson 3-1)*
Ten times a number a is equal to 5 times the sum of b and c.

Find each product. *(Lesson 2-3)*

58. $(-5)(12)$ 59. $(-2.93)(-0.003)$ 60. $(-4)(0)(-2)(-3)$

Graph each set of numbers on a number line. *(Lesson 2-1)*

61. $\{-4, -3, -1, 3\}$ 62. {integers between -6 and 10}
63. {integers less than -4} 64. {integers less than 0 and greater than -6}

Name the property illustrated by each statement. *(Lesson 1-6)*

65. $67 + 3 = 3 + 67$ 66. $(5 \cdot m) \cdot n = 5 \cdot (m \cdot n)$

Getting Ready for the Next Lesson

PREREQUISITE SKILL Use the order of operations to find each value.
*(To review the **order of operations**, see Lesson 1-2.)*

67. $2 \times 8 + 9$ 68. $24 \div 3 - 8$ 69. $\frac{3}{8}(17 + 7)$ 70. $\frac{15 - 9}{26 + 12}$

Practice Quiz 1 — *Lessons 3-1 through 3-3*

GEOMETRY For Exercises 1 and 2, use the following information.
The surface area S of a sphere equals four times π times the square of the radius r. *(Lesson 3-1)*

1. Write the formula for the surface area of a sphere.
2. What is the surface area of a sphere if the radius is 7 centimeters?

Solve each equation. Then check your solution. *(Lessons 3-2 and 3-3)*

3. $d + 18 = -27$ 4. $m - 77 = -61$ 5. $-12 + a = -36$ 6. $t - (-16) = 9$
7. $\frac{2}{3}p = 18$ 8. $-17y = 391$ 9. $5x = -45$ 10. $-\frac{2}{5}d = -10$

Algebra Activity

A Preview of Lesson 3-4

Solving Multi-Step Equations

You can use an equation model to solve multi-step equations.

Solve $3x + 5 = -7$.

Step 1 *Model the equation.*

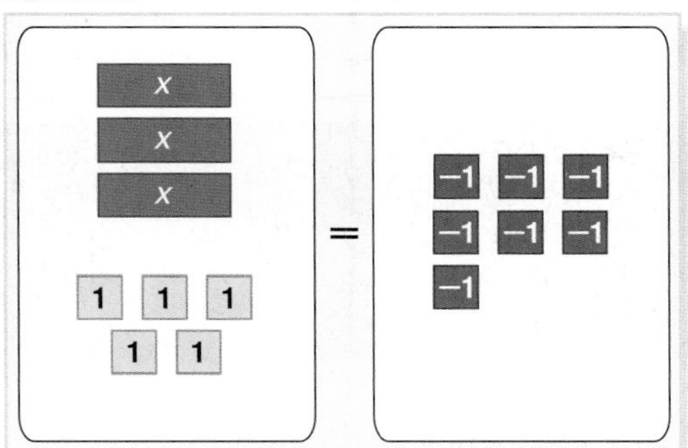

$3x + 5 = -7$

Place 3 x tiles and 5 positive 1 tiles on one side of the mat. Place 7 negative 1 tiles on the other side of the mat.

Step 2 *Isolate the x term.*

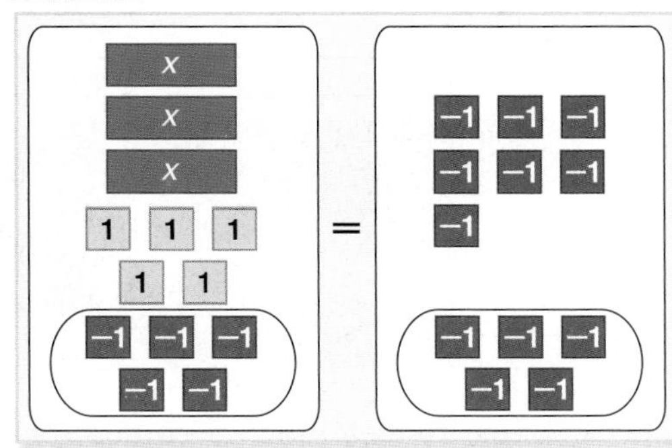

$3x + 5 - 5 = -7 - 5$

Since there are 5 positive 1 tiles with the x tiles, add 5 negative 1 tiles to each side to form zero pairs.

Step 3 *Remove zero pairs.*

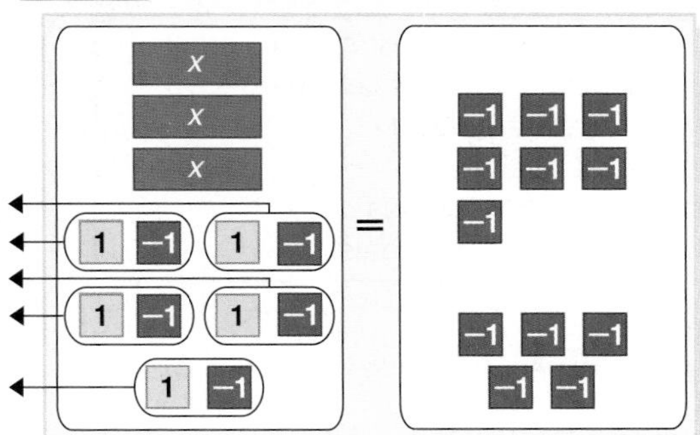

$3x = -12$

Group the tiles to form zero pairs and remove the zero pairs.

Step 4 *Group the tiles.*

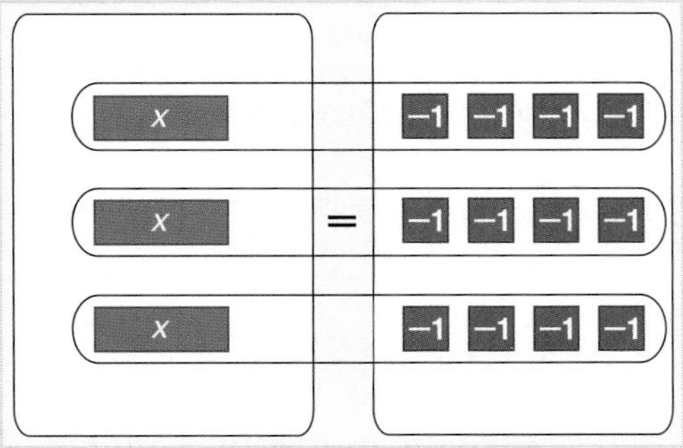

$\frac{3x}{3} = \frac{-12}{3}$

Separate the tiles into 3 equal groups to match the 3 x tiles. Each x tile is paired with 4 negative 1 tiles. Thus, $x = -4$.

Model **Use algebra tiles to solve each equation.**

1. $2x - 3 = -9$ **2.** $3x + 5 = 14$ **3.** $3x - 2 = 10$ **4.** $-8 = 2x + 4$

5. $3 + 4x = 11$ **6.** $2x + 7 = 1$ **7.** $9 = 4x - 7$ **8.** $7 + 3x = -8$

9. MAKE A CONJECTURE What steps would you use to solve $7x - 12 = -61$?

3-4 Solving Multi-Step Equations

What You'll Learn

- Solve problems by working backward.
- Solve equations involving more than one operation.

Vocabulary

- work backward
- multi-step equations
- consecutive integers
- number theory

How can equations be used to estimate the age of an animal?

An American alligator hatchling is about 8 inches long. These alligators grow about 12 inches per year. Therefore, the expression $8 + 12a$ represents the length in inches of an alligator that is a years old.

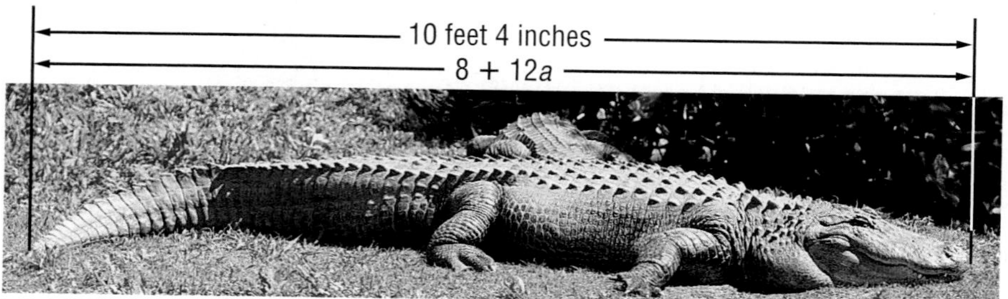

Since 10 feet 4 inches equals $10(12) + 4$ or 124 inches, the equation $8 + 12a = 124$ can be used to estimate the age of the alligator in the photograph. Notice that this equation involves more than one operation.

WORK BACKWARD **Work backward** is one of many *problem-solving strategies* that you can use. Here are some other problem-solving strategies.

Problem-Solving Strategies	
draw a diagram	solve a simpler (or similar) problem
make a table or chart	eliminate the possibilities
make a model	look for a pattern
guess and check	act it out
check for hidden assumptions	list the possibilities
use a graph	identify the subgoals

Example 1 Work Backward to Solve a Problem

Solve the following problem by working backward.

After cashing her paycheck, Tara paid her father the $20 she had borrowed. She then spent half of the remaining money on a concert ticket. She bought lunch for $4.35 and had $10.55 left. What was the amount of the paycheck?

Start at the end of the problem and undo each step.

Statement	Undo the Statement
She had $10.55 left.	$10.55
She bought lunch for $4.35.	$10.55 + $4.35 = $14.90
She spent half of the money on a concert ticket.	$14.90 × 2 = $29.80
She paid her father $20.	$29.80 + $20.00 = $49.80

The paycheck was for $49.80. *Check this answer in the context of the problem.*

SOLVE MULTI-STEP EQUATIONS To solve equations with more than one operation, often called **multi-step equations**, undo operations by working backward.

Study Tip

Solving Multi-Step Equations
When solving a multi-step equation, "undo" the operations in reverse of the order of operations.

Example 2 Solve Using Addition and Division

Solve $7m - 17 = 60$. Then check your solution.

$7m - 17 = 60$	Original equation
$7m - 17 + 17 = 60 + 17$	Add 17 to each side.
$7m = 77$	Simplify.
$\frac{7m}{7} = \frac{77}{7}$	Divide each side by 7.
$m = 11$	Simplify.

CHECK

$7m - 17 = 60$	Original equation
$7(11) - 17 \stackrel{?}{=} 60$	Substitute 11 for m.
$77 - 17 \stackrel{?}{=} 60$	Multiply.
$60 = 60$ ✓	The solution is 11.

You have seen a multi-step equation in which the first, or *leading*, coefficient is an integer. You can use the same steps if the leading coefficient is a fraction.

Example 3 Solve Using Subtraction and Multiplication

Solve $\frac{t}{8} + 21 = 14$. Then check your solution.

$\frac{t}{8} + 21 = 14$	Original equation
$\frac{t}{8} + 21 - 21 = 14 - 21$	Subtract 21 from each side.
$\frac{t}{8} = -7$	Simplify.
$8\left(\frac{t}{8}\right) = 8(-7)$	Multiply each side by 8.
$t = -56$	Simplify.

CHECK

$\frac{t}{8} + 21 = 14$	Original equation
$\frac{-56}{8} + 21 \stackrel{?}{=} 14$	Substitute -56 for t.
$-7 + 21 \stackrel{?}{=} 14$	Divide.
$14 = 14$ ✓	The solution is -56.

Example 4 Solve Using Multiplication and Addition

Solve $\frac{p - 15}{9} = -6$.

$\frac{p - 15}{9} = -6$	Original equation
$9\left(\frac{p - 15}{9}\right) = 9(-6)$	Multiply each side by 9.
$p - 15 = -54$	Simplify.
$p - 15 + 15 = -54 + 15$	Add 15 to each side.
$p = -39$	The solution is -39.

Example 5 Write and Solve a Multi-Step Equation

Write an equation for the problem below. Then solve the equation.
Two-thirds of a number minus six is −10.

Two-thirds	of	a number	minus	six	is	−10.
$\frac{2}{3}$	$\cdot$	n	$-$	6	$=$	-10

$\frac{2}{3}n - 6 = -10$ Original equation

$\frac{2}{3}n - 6 + 6 = -10 + 6$ Add 6 to each side.

$\frac{2}{3}n = -4$ Simplify.

$\frac{3}{2}\left(\frac{2}{3}n\right) = \frac{3}{2}(-4)$ Multiply each side by $\frac{3}{2}$.

$n = -6$ Simplify.

The solution is −6.

Consecutive integers are integers in counting order, such as 7, 8, and 9. Beginning with an even integer and counting by two will result in *consecutive even integers*. For example, −4, −2, 0, and 2 are consecutive even integers. Beginning with an odd integer and counting by two will result in *consecutive odd integers*. For example, −3, −1, 1, 3 and 5 are consecutive odd integers. The study of numbers and the relationships between them is called **number theory**.

Example 6 Solve a Consecutive Integer Problem

Study Tip

Representing Consecutive Integers

You can use the same expressions to represent either consecutive even integers or consecutive odd integers. It is the value of *n*—odd or even—that differs between the two expressions.

NUMBER THEORY **Write an equation for the problem below. Then solve the equation and answer the problem.**

Find three consecutive even integers whose sum is −42.

Let n = the least even integer.

Then $n + 2$ = the next greater even integer, and

$n + 4$ = the greatest of the three even integers.

The sum of three consecutive even integers	is	−42.
$n + (n + 2) + (n + 4)$	$=$	-42

$n + (n + 2) + (n + 4) = -42$ Original equation

$3n + 6 = -42$ Simplify.

$3n + 6 - 6 = -42 - 6$ Subtract 6 from each side.

$3n = -48$ Simplify.

$\frac{3n}{3} = \frac{-48}{3}$ Divide each side by 3

$n = -16$ Simplify.

$n + 2 = -16 + 2$ or -14 $n + 4 = -16 + 4$ or -12

The consecutive even integers are −16, −14, and −12.

CHECK −16, −14, and −12 are consecutive even integers.
$-16 + (-14) + (-12) = -42$ ✓

Check for Understanding

Concept Check

1. **OPEN ENDED** Give two examples of multi-step equations that have a solution of -2.
2. **List** the steps used to solve $\frac{w + 3}{5} - 4 = 6$.
3. **Write** an expression for the odd integer before odd integer n.
4. Justify each step.

$$\frac{4 - 2d}{5} + 3 = 9$$

Step	
$\frac{4 - 2d}{5} + 3 - 3 = 9 - 3$	a. ?
$\frac{4 - 2d}{5} = 6$	b. ?
$\frac{4 - 2d}{5}(5) = 6(5)$	c. ?
$4 - 2d = 30$	d. ?
$4 - 2d - 4 = 30 - 4$	e. ?
$-2d = 26$	f. ?
$\frac{-2d}{-2} = \frac{26}{-2}$	g. ?
$d = -13$	h. ?

Guided Practice

Solve each problem by working backward.

5. A number is multiplied by seven, and then the product is added to 13. The result is 55. What is the number?
6. **LIFE SCIENCE** A bacteria population triples in number each day. If there are 2,187,000 bacteria on the seventh day, how many bacteria were there on the first day?

Solve each equation. Then check your solution.

7. $4g - 2 = -6$
8. $18 = 5p + 3$
9. $\frac{3}{2}a - 8 = 11$
10. $\frac{b + 4}{-2} = -17$
11. $0.2n + 3 = 8.6$
12. $3.1y - 1.5 = 5.32$

Write an equation and solve each problem.

13. Twelve decreased by twice a number equals –34. Find the number.
14. Find three consecutive integers whose sum is 42.

Application

15. **WORLD CULTURES** The English alphabet contains 2 more than twice as many letters as the Hawaiian alphabet. How many letters are there in the Hawaiian alphabet?

Practice and Apply

Solve each problem by working backward.

16. A number is divided by 4, and then the quotient is added to 17. The result is 25. Find the number.
17. Nine is subtracted from a number, and then the difference is multiplied by 5. The result is 75. What is the number?

Homework Help

For Exercises	See Examples
16–21	1
22–41	2–4
42–54	5, 6

Extra Practice

See page 826.

Solve each problem by working backward.

18. **GAMES** In the Trivia Bowl, each finalist must answer four questions correctly. Each question is worth twice as much as the question before it. The fourth question is worth \$6000. How much is the first question worth?

19. **ICE SCULPTING** Due to melting, an ice sculpture loses one-half its weight every hour. After 8 hours, it weighs $\frac{5}{16}$ of a pound. How much did it weigh in the beginning?

20. **FIREFIGHTING** A firefighter spraying water on a fire stood on the middle rung of a ladder. The smoke lessened, so she moved up 3 rungs. It got too hot, so she backed down 5 rungs. Later, she went up 7 rungs and stayed until the fire was out. Then, she climbed the remaining 4 rungs and went into the building. How many rungs does the ladder have?

21. **MONEY** Hugo withdrew some money from his bank account. He spent one third of the money for gasoline. Then he spent half of what was left for a haircut. He bought lunch for \$6.55. When he got home, he had \$13.45 left. How much did he withdraw from the bank?

Solve each equation. Then check your solution.

22. $5n + 6 = -4$
23. $7 + 3c = -11$
24. $15 = 4a - 5$
25. $-63 = 7g - 14$
26. $\frac{c}{-3} + 5 = 7$
27. $\frac{y}{5} + 9 = 6$
28. $3 - \frac{a}{7} = -2$
29. $-9 - \frac{p}{4} = 5$
30. $\frac{t}{8} - 6 = -12$
31. $\frac{m}{-5} + 6 = 31$
32. $\frac{17 - s}{4} = -10$
33. $\frac{-3j - (-4)}{-6} = 12$
34. $-3d - 1.2 = 0.9$
35. $-2.5r - 32.7 = 74.1$
36. $-0.6 + (-4a) = -1.4$
37. $\frac{p}{-7} - 0.5 = 1.3$
38. $3.5x + 5 - 1.5x = 8$
39. $\frac{9z + 4}{5} - 8 = 5.4$

40. If $3a - 9 = 6$, what is the value of $5a + 2$?
41. If $2x + 1 = 5$, what is the value of $3x - 4$?

Write an equation and solve each problem.

42. Six less than two thirds of a number is negative ten. Find the number.
43. Twenty-nine is thirteen added to four times a number. What is the number?
44. Find three consecutive odd integers whose sum is 51.
45. Find three consecutive even integers whose sum is −30.
46. Find four consecutive integers whose sum is 94.
47. Find four consecutive odd integers whose sum is 8.

48. **BUSINESS** Adele Jones is on a business trip and plans to rent a subcompact car from Speedy Rent-A-Car. Her company has given her a budget of \$60 per day for car rental. What is the maximum distance Ms. Jones can drive in one day and still stay within her budget?

Speedy Rent-A-Car Price List

Subcompact
\$14.95 per day plus \$0.10 per mile

Compact
\$19.95 per day plus \$0.12 per mile

Full Size
\$22.95 per day plus \$0.15 per mile

49. **GEOMETRY** The measures of the three sides of a triangle are consecutive even integers. The perimeter of the triangle is 54 centimeters. What are the lengths of the sides of the triangle?

More About. . .

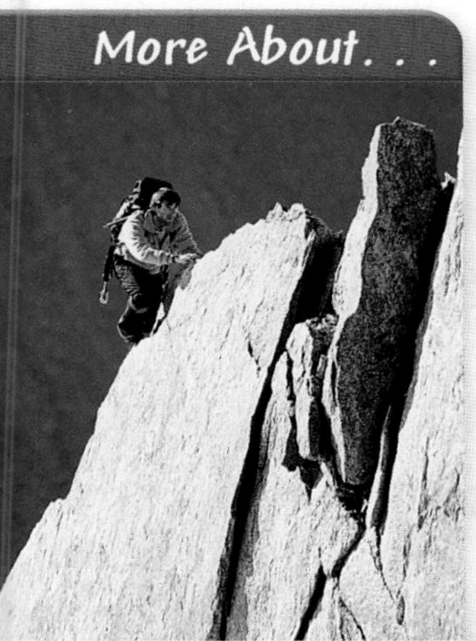

Mountain Climbing

Many mountain climbers experience altitude sickness caused by a decrease in oxygen. Climbers can acclimate themselves to these higher altitudes by camping for one or two weeks at various altitudes as they ascend the mountain.

Source: *Shape*

50. **MOUNTAIN CLIMBING** A general rule for those climbing more than 7000 feet above sea level is to allow a total of $\left(\frac{a - 7000}{2000} + 2\right)$ weeks of camping during the ascension. In this expression, a represents the altitude in feet. If a group of mountain climbers have allowed for 9 weeks of camping in their schedule, how high can they climb without worrying about altitude sickness?

SHOE SIZE For Exercises 51 and 52, use the following information.
If ℓ represents the length of a person's foot in inches, the expression $2\ell - 12$ can be used to estimate his or her shoe size.

51. What is the approximate length of the foot of a person who wears size 8?

52. Measure your foot and use the expression to determine your shoe size. How does this number compare to the size of shoe you are wearing?

53. **SALES** Trever Goetz is a salesperson who is paid a monthly salary of $500 plus a 2% commission on sales. How much must Mr. Goetz sell to earn $2000 this month?

54. **GEOMETRY** A rectangle is cut from the corner of a 10-inch by 10-inch of paper. The area of the remaining piece of paper is $\frac{4}{5}$ of the area of the original piece of paper. If the width of the rectangle removed from the paper is 4 inches, what is the length of the rectangle?

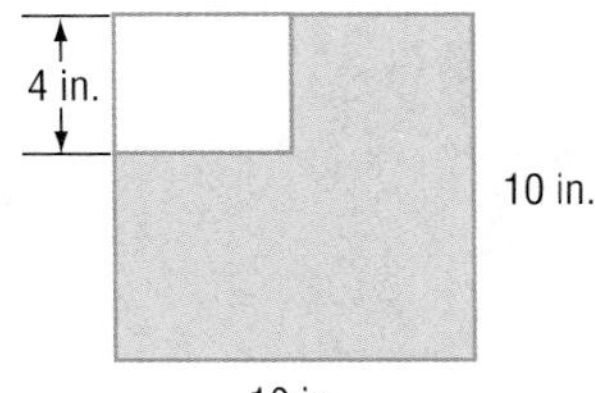

55. **CRITICAL THINKING** Determine whether the following statement is *sometimes, always,* or *never* true.

The sum of two consecutive even numbers equals the sum of two consecutive odd numbers.

56. **WRITING IN MATH** Answer the question that was posed at the beginning of the lesson.

How can equations be used to estimate the age of an animal?

Include the following in your answer:

- an explanation of how to solve the equation representing the age of the alligator, and
- an estimate of the age of the alligator.

Standardized Test Practice

57. Which equation represents the following problem?

Fifteen minus three times a number equals negative twenty-two. Find the number.

(A) $3n - 15 = -22$ (B) $15 - 3n = -22$

(C) $3(15 - n) = -22$ (D) $3(n - 15) = -22$

58. Which equation has a solution of -5?

(A) $2a - 6 = 4$ (B) $3a + 7 = 8$

(C) $\frac{3a - 7}{4} = 2$ (D) $\frac{3}{5}a + 19 = 16$

EQUATION SOLVER **You can use a graphing calculator to solve equations that are rewritten as expressions that equal zero.**

Step 1 Write the equation so that one side is equal to 0.

Step 2 On a TI-83 Plus, press MATH and choose 0, for solve.

Step 3 Enter the equation after 0=. Use ALPHA to enter the variables. Press ENTER.

Step 4 Press ALPHA [SOLVE] to reveal the solution. Use the ▲ key to begin entering a new equation.

Use a graphing calculator to solve each equation.

59. $0 = 11y + 33$ **60.** $\frac{w + 2}{5} - 4 = 0$ **61.** $6 = -12 + \frac{h}{-7}$

62. $\frac{p - (-5)}{-2} = 6$ **63.** $0.7 = \frac{r - 0.8}{6}$ **64.** $4.91 + 7.2t = 38.75$

Maintain Your Skills

Mixed Review

Solve each equation. Then check your solution. *(Lesson 3-3)*

65. $-7t = 91$ **66.** $\frac{r}{15} = -8$ **67.** $-\frac{2}{3}b = -1\frac{1}{2}$

TRANSPORTATION **For Exercises 68 and 69, use the following information.** In the year 2000, there were 18 more models of sport utility vehicles than there were in the year 1990. There were 47 models of sport utility vehicles in 2000. *(Lesson 3-2)*

68. Write an addition equation to represent the situation.

69. How many models of sport utility vehicles were there in 1990?

Find the odds of each outcome if you spin the spinner at the right. *(Lesson 2-6)*

70. spinning a number divisible by 3

71. spinning a number equal to or greater than 5

72. spinning a number less than 7

Find each quotient. *(Lesson 2-4)*

73. $-\frac{6}{7} \div 3$ **74.** $\frac{\frac{2}{3}}{8}$ **75.** $\frac{-3a + 16}{4}$ **76.** $\frac{15t - 25}{-5}$

Use the Distributive Property to find each product. *(Lesson 1-5)*

77. $17 \cdot 9$ **78.** $13(101)$ **79.** $16\left(1\frac{1}{4}\right)$ **80.** $18\left(2\frac{1}{9}\right)$

Write an algebraic expression for each verbal expression. *(Lesson 1-1)*

81. the product of 5 and m plus half of n

82. the quantity 3 plus b divided by y

83. the sum of 3 times a and the square of b

Getting Ready for the Next Lesson

PREREQUISITE SKILL **Simplify each expression.**
*(To review **simplifying expressions**, see Lesson 1-5.)*

84. $5d - 2d$ **85.** $11m - 5m$

86. $8t + 6t$ **87.** $7g - 15g$

88. $-9f + 6f$ **89.** $-3m + (-7m)$

3-5 Solving Equations with the Variable on Each Side

What You'll Learn

- Solve equations with the variable on each side.
- Solve equations involving grouping symbols.

Vocabulary

- identity

How can an equation be used to determine when two populations are equal?

In 1995, there were 18 million Internet users in North America. Of this total, 12 million were male, and 6 million were female. During the next five years, the number of male Internet users on average increased 7.6 million per year, and the number of female Internet users increased 8 million per year. If this trend continues, the following expressions represent the number of male and female Internet users x years after 1995.

Male Internet Users: $12 + 7.6x$

Female Internet Users: $6 + 8x$

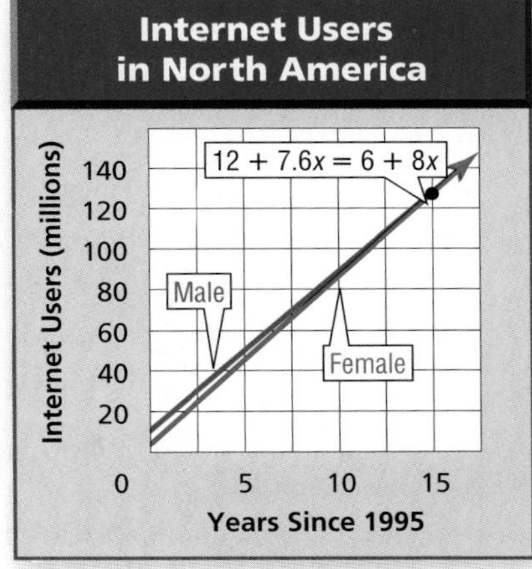

The equation $12 + 7.6x = 6 + 8x$ represents the time at which the number of male and female Internet users are equal. Notice that this equation has the variable x on each side.

VARIABLES ON EACH SIDE Many equations contain variables on each side. To solve these types of equations, first use the Addition or Subtraction Property of Equality to write an equivalent equation that has all of the variables on one side.

Example 1 Solve an Equation with Variables on Each Side

Solve $-2 + 10p = 8p - 1$. Then check your solution.

$-2 + 10p = 8p - 1$	Original equation
$-2 + 10p - 8p = 8p - 1 - 8p$	Subtract $8p$ from each side.
$-2 + 2p = -1$	Simplify.
$-2 + 2p + 2 = -1 + 2$	Add 2 to each side.
$2p = 1$	Simplify.
$\frac{2p}{2} = \frac{1}{2}$	Divide each side by 2.
$p = \frac{1}{2}$ or 0.5	Simplify.

CHECK

$-2 + 10p = 8p - 1$	Original equation
$-2 + 10(0.5) \stackrel{?}{=} 8(0.5) - 1$	Substitute 0.5 for p.
$-2 + 5 \stackrel{?}{=} 4 - 1$	Multiply.
$3 = 3$ ✓	The solution is $\frac{1}{2}$ or 0.5.

GROUPING SYMBOLS When solving equations that contain grouping symbols, first use the Distributive Property to remove the grouping symbols.

> **Study Tip**
>
> *Look Back*
> To review the **Distributive Property**, see Lesson 1-5.

Example 2 Solve an Equation with Grouping Symbols

Solve $4(2r - 8) = \frac{1}{7}(49r + 70)$. Then check your solution.

$4(2r - 8) = \frac{1}{7}(49r + 70)$	Original equation
$8r - 32 = 7r + 10$	Distributive Property
$8r - 32 - 7r = 7r + 10 - 7r$	Subtract $7r$ from each side.
$r - 32 = 10$	Simplify.
$r - 32 + 32 = 10 + 32$	Add 32 to each side.
$r = 42$	Simplify.

CHECK

$4(2r - 8) = \frac{1}{7}(49r + 70)$	Original equation
$4[2(42) - 8] \stackrel{?}{=} \frac{1}{7}[49(42) + 70]$	Substitute 42 for r.
$4(84 - 8) \stackrel{?}{=} \frac{1}{7}(2058 + 70)$	Multiply.
$4(76) \stackrel{?}{=} \frac{1}{7}(2128)$	Add and subtract.
$304 = 304 \checkmark$	

The solution is 42.

Some equations with the variable on each side may have no solution. That is, there is no value of the variable that will result in a true equation.

Example 3 No Solutions

Solve $2m + 5 = 5(m - 7) - 3m$.

$2m + 5 = 5(m - 7) - 3m$	Original equation
$2m + 5 = 5m - 35 - 3m$	Distributive Property
$2m + 5 = 2m - 35$	Simplify.
$2m + 5 - 2m = 2m - 35 - 2m$	Subtract $2m$ from each side.
$5 = -35$	This statement is false.

Since $5 = -35$ is a false statement, this equation has no solution.

An equation that is true for every value of the variable is called an **identity**.

Example 4 An Identity

Solve $3(r + 1) - 5 = 3r - 2$.

$3(r + 1) - 5 = 3r - 2$	Original equation
$3r + 3 - 5 = 3r - 2$	Distributive Property
$3r - 2 = 3r - 2$	Reflexive Property of Equality

Since the expressions on each side of the equation are the same, this equation is an identity. The statement $3(r + 1) - 5 = 3r - 2$ is true for all values of r.

Concept Summary — Steps for Solving Equations

Step 1 Use the Distributive Property to remove the grouping symbols.

Step 2 Simplify the expressions on each side of the equals sign.

Step 3 Use the Addition and/or Subtraction Properties of Equality to get the variables on one side of the equals sign and the numbers without variables on the other side of the equals sign.

Step 4 Simplify the expressions on each side of the equals sign.

Step 5 Use the Multiplication or Division Property of Equality to solve.
- If the solution results in a false statement, there is no solution of the equation.
- If the solution results in an identity, the solution is all numbers.

Standardized Test Practice

Example 5 Use Substitution to Solve an Equation

Multiple-Choice Test Item

Solve $2(b - 3) + 5 = 3(b - 1)$.

(A) -2 (B) 2 (C) -3 (D) 3

Read the Test Item

You are asked to solve an equation.

Solve the Test Item

You can solve the equation or substitute each value into the equation and see if it makes the equation true. We will solve by substitution.

A
$$2(b - 3) + 5 = 3(b - 1)$$
$$2(-2 - 3) + 5 \stackrel{?}{=} 3(-2 - 1)$$
$$2(-5) + 5 \stackrel{?}{=} 3(-3)$$
$$-10 + 5 \stackrel{?}{=} -9$$
$$-5 \neq -9$$

B
$$2(b - 3) + 5 = 3(b - 1)$$
$$2(2 - 3) + 5 \stackrel{?}{=} 3(2 - 1)$$
$$2(-1) + 5 \stackrel{?}{=} 3(1)$$
$$-2 + 5 \stackrel{?}{=} 3$$
$$3 = 3 \quad \checkmark$$

Since the value 2 results in a true statement, you do not need to check -3 and 3. The answer is B.

The Princeton Review — Test-Taking Tip

If you are asked to solve a complicated equation, it sometimes takes less time to check each possible answer rather than to actually solve the equation.

Check for Understanding

Concept Check

1. **Determine** whether each solution is correct. If the solution is not correct, find the error and give the correct solution.

a.
$$2(g + 5) = 22$$
$$2g + 5 = 22$$
$$2g + 5 - 5 = 22 - 5$$
$$2g = 17$$
$$\frac{2g}{2} = \frac{17}{2}$$
$$g = 8.5$$

b.
$$5d = 2d - 18$$
$$5d - 2d = 2d - 18 - 2d$$
$$3d = -18$$
$$\frac{3d}{3} = \frac{-18}{3}$$
$$d = -6$$

c.
$$-6z + 13 = 7z$$
$$-6z + 13 - 6z = 7z - 6z$$
$$13 = z$$

2. **Explain** how to determine whether an equation is an identity.

3. **OPEN ENDED** Find a counterexample to the statement *all equations have a solution.*

Guided Practice

4. Justify each step.

$$6n + 7 = 8n - 13$$
$$6n + 7 - 6n = 8n - 13 - 6n \quad \textbf{a.}\ \underline{\ ?\ }$$
$$7 = 2n - 13 \quad \textbf{b.}\ \underline{\ ?\ }$$
$$7 + 13 = 2n - 13 + 13 \quad \textbf{c.}\ \underline{\ ?\ }$$
$$20 = 2n \quad \textbf{d.}\ \underline{\ ?\ }$$
$$\frac{20}{2} = \frac{2n}{2} \quad \textbf{e.}\ \underline{\ ?\ }$$
$$10 = n \quad \textbf{f.}\ \underline{\ ?\ }$$

Solve each equation. Then check your solution.

5. $20c + 5 = 5c + 65$
6. $\frac{3}{8} - \frac{1}{4}t = \frac{1}{2}t - \frac{3}{4}$
7. $3(a - 5) = -6$
8. $7 - 3r = r - 4(2 + r)$
9. $6 = 3 + 5(d - 2)$
10. $\frac{c + 1}{8} = \frac{c}{4}$
11. $5h - 7 = 5(h - 2) + 3$
12. $5.4w + 8.2 = 9.8w - 2.8$

Standardized Test Practice

13. Solve $75 - 9t = 5(-4 + 2t)$.

(A) -5 (B) -4 (C) 4 (D) 5

Practice and Apply

Homework Help

For Exercises	See Examples
14–48	1–4
51, 52	5

Extra Practice

See page 826.

Justify each step.

14.
$$\frac{3m - 2}{5} = \frac{7}{10}$$
$$\frac{3m - 2}{5}(10) = \frac{7}{10}(10) \quad \textbf{a.}\ \underline{\ ?\ }$$
$$(3m - 2)2 = 7 \quad \textbf{b.}\ \underline{\ ?\ }$$
$$6m - 4 = 7 \quad \textbf{c.}\ \underline{\ ?\ }$$
$$6m - 4 + 4 = 7 + 4 \quad \textbf{d.}\ \underline{\ ?\ }$$
$$6m = 11 \quad \textbf{e.}\ \underline{\ ?\ }$$
$$\frac{6m}{6} = \frac{11}{6} \quad \textbf{f.}\ \underline{\ ?\ }$$
$$m = 1\frac{5}{6} \quad \textbf{g.}\ \underline{\ ?\ }$$

15.
$$v + 9 = 7v + 9$$
$$v + 9 - v = 7v + 9 - v \quad \textbf{a.}\ \underline{\ ?\ }$$
$$9 = 6v + 9 \quad \textbf{b.}\ \underline{\ ?\ }$$
$$9 - 9 = 6v + 9 - 9 \quad \textbf{c.}\ \underline{\ ?\ }$$
$$0 = 6v \quad \textbf{d.}\ \underline{\ ?\ }$$
$$\frac{0}{6} = \frac{6v}{6} \quad \textbf{e.}\ \underline{\ ?\ }$$
$$0 = v \quad \textbf{f.}\ \underline{\ ?\ }$$

Solve each equation. Then check your solution.

16. $3 - 4q = 10q + 10$
17. $3k - 5 = 7k - 21$
18. $5t - 9 = -3t + 7$
19. $8s + 9 = 7s + 6$
20. $\frac{3}{4}n + 16 = 2 - \frac{1}{8}n$
21. $\frac{1}{4} - \frac{2}{3}y = \frac{3}{4} - \frac{1}{3}y$
22. $8 = 4(3c + 5)$
23. $7(m - 3) = 7$
24. $6(r + 2) - 4 = -10$
25. $5 - \frac{1}{2}(x - 6) = 4$
26. $4(2a - 1) = -10(a - 5)$
27. $4(f - 2) = 4f$
28. $3(1 + d) - 5 = 3d - 2$
29. $2(w - 3) + 5 = 3(w - 1)$

30. $\frac{3}{2}y - y = 4 + \frac{1}{2}y$

31. $3 + \frac{2}{5}b = 11 - \frac{2}{5}b$

32. $\frac{1}{4}(7 + 3g) = -\frac{g}{8}$

33. $\frac{1}{6}(a - 4) = \frac{1}{3}(2a + 4)$

34. $28 - 2.2x = 11.6x + 262.6$

35. $1.03p - 4 = -2.15p + 8.72$

36. $18 - 3.8t = 7.36 - 1.9t$

37. $13.7v - 6.5 = -2.3v + 8.3$

38. $2[s + 3(s - 1)] = 18$

39. $-3(2n - 5) = 0.5(-12n + 30)$

40. One half of a number increased by 16 is four less than two thirds of the number. Find the number.

41. The sum of one half of a number and 6 equals one third of the number. What is the number?

42. **NUMBER THEORY** Twice the greater of two consecutive odd integers is 13 less than three times the lesser number. Find the integers.

43. **NUMBER THEORY** Three times the greatest of three consecutive even integers exceeds twice the least by 38. What are the integers?

44. **HEALTH** When exercising, a person's pulse rate should not exceed a certain limit, which depends on his or her age. This maximum rate is represented by the expression $0.8(220 - a)$, where a is age in years. Find the age of a person whose maximum pulse is 152.

More About. . .

Energy

One British Thermal Unit (BTU) is the amount of energy needed to raise the temperature of one pound of water 1°F. If a heating system is 100% efficient, one cubic foot of natural gas provides 1000 BTU.

Source: *World Book Encyclopedia*

45. **HARDWARE** Traditionally, nails are given names such as 2-penny, 3-penny, and so on. These names describe the lengths of the nails. What is the name of a nail that is $2\frac{1}{2}$ inches long?

x-penny nail

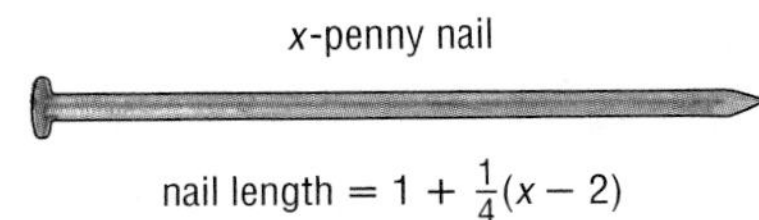

nail length $= 1 + \frac{1}{4}(x - 2)$

Source: *World Book Encyclopedia*

46. **TECHNOLOGY** About 4.9 million households had one brand of personal computers in 2001. The use of these computers grew at an average rate of 0.275 million households a year. In 2001, about 2.5 million households used another type of computer. The use of these computers grew at an average rate of 0.7 million households a year. How long will it take for the two types of computers to be in the same number of households?

47. **GEOMETRY** The rectangle and square shown below have the same perimeter. Find the dimensions of each figure.

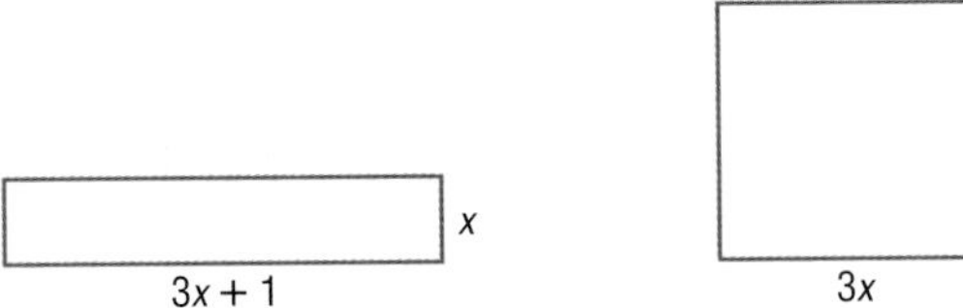

48. **ENERGY** Use the information on energy at the left. The amount of energy E in BTUs needed to raise the temperature of water is represented by the equation $E = w(t_f - t_O)$. In this equation, w represents the weight of the water in pounds, t_f represents the final temperature in degrees Fahrenheit, and t_O represents the original temperature in degrees Fahrenheit. A 50-gallon water heater is 60% efficient. If 10 cubic feet of natural gas are used to raise the temperature of water with the original temperature of 50°F, what is the final temperature of the water? (One gallon of water weighs about 8 pounds.)

49. **CRITICAL THINKING** Write an equation that has one or more grouping symbols, the variable on each side of the equals sign, and a solution of -2.

50. **WRITING IN MATH** Answer the question that was posed at the beginning of the lesson.

How can an equation be used to determine when two populations are equal?

Include the following in your answer:

- a list of the steps needed to solve the equation,
- the year when the number of female Internet users will equal the number of male Internet users according to the model, and
- an explanation of why this method can be used to predict future events.

Standardized Test Practice

51. Solve $8x - 3 = 5(2x + 1)$.

Ⓐ 4 Ⓑ 2 Ⓒ −2 Ⓓ −4

52. Solve $5n + 4 = 7(n + 1) - 2n$.

Ⓐ 0 Ⓑ −1 Ⓒ no solution Ⓓ all numbers

Maintain Your Skills

Mixed Review

Solve each equation. Then check your solution. *(Lesson 3-4)*

53. $\frac{2}{9}v - 6 = 14$

54. $\frac{x - 3}{7} = -2$

55. $5 - 9w = 23$

HEALTH For Exercises 56 and 57, use the following information.

Ebony burns 4.5 Calories per minute pushing a lawn mower. *(Lesson 3-3)*

56. Write a multiplication equation representing the number of Calories C burned if Ebony pushes the lawn mower for m minutes.

57. How long will it take Ebony to burn 150 Calories mowing the lawn?

Use each set of data to make a line plot. *(Lesson 2-5)*

58. 13, 15, 11, 15, 16, 17, 12, 12, 13, 15, 16, 15

59. 22, 25, 19, 21, 22, 24, 22, 25, 28, 21, 24, 22

Find each sum or difference. *(Lesson 2-2)*

60. $-10 + (-17)$

61. $-12 - (-8)$

62. $6 - 14$

Write a counterexample for each statement. *(Lesson 1-7)*

63. If the sum of two numbers is even, then both addends are even.

64. If you are baking cookies, you will need chocolate chips.

Evaluate each expression when $a = 5$, $b = 8$, $c = 7$, $x = 2$, and $y = 1$. *(Lesson 1-2)*

65. $\frac{3a^2}{b + c}$

66. $x(a + 2b) - y$

67. $5(x + 2y) - 4a$

Getting Ready for the Next Lesson

PREREQUISITE SKILL Simplify each fraction.

(To review ***simplifying fractions****, see pages 798 and 799.)*

68. $\frac{12}{15}$

69. $\frac{28}{49}$

70. $\frac{36}{60}$

71. $\frac{8}{120}$

72. $\frac{108}{9}$

73. $\frac{28}{42}$

74. $\frac{16}{40}$

75. $\frac{19}{57}$

3-6 Ratios and Proportions

What You'll Learn

- Determine whether two ratios form a proportion.
- Solve proportions.

Vocabulary

- ratio
- proportion
- extremes
- means
- rate
- scale

How are ratios used in recipes?

The ingredients in the recipe will make 4 servings of honey frozen yogurt. Keri can use ratios and equations to find the amount of each ingredient needed to make enough yogurt for her club meeting.

Honey Frozen Yogurt	
2 cups 2% milk	2 eggs, beaten
$\frac{3}{4}$ cup honey	2 cups plain low-fat
1 dash salt	yogurt
	1 tablespoon vanilla

RATIOS AND PROPORTIONS A **ratio** is a comparison of two numbers by division. The ratio of x to y can be expressed in the following ways.

$$x \text{ to } y \qquad x{:}y \qquad \frac{x}{y}$$

Ratios are often expressed in simplest form. For example, the recipe above states that for 4 servings you need 2 cups of milk. The ratio of servings to milk may be written as 4 to 2, 4:2, or $\frac{4}{2}$. Written in simplest form, the ratio of servings to milk can be written as 2 to 1, 2:1, or $\frac{2}{1}$.

Study Tip

Reading Math
A ratio that is equivalent to a whole number is written with a denominator of 1.

Suppose you wanted to double the recipe to have 8 servings. The amount of milk required would be 4 cups. The ratio of servings to milk is $\frac{8}{4}$. When this ratio is simplified, the ratio is $\frac{2}{1}$. Notice that this ratio is equal to the original ratio.

$$\frac{4}{2} \overset{\div 2}{=} \frac{2}{1} \qquad \frac{8}{4} \overset{\div 4}{=} \frac{2}{1}$$

An equation stating that two ratios are equal is called a **proportion**. So, we can state that $\frac{4}{2} = \frac{8}{4}$ is a proportion.

Example 1 Determine Whether Ratios Form a Proportion

Determine whether the ratios $\frac{4}{5}$ and $\frac{24}{30}$ form a proportion.

$$\frac{4}{5} \overset{\div 1}{=} \frac{4}{5} \qquad \frac{24}{30} \overset{\div 6}{=} \frac{4}{5}$$

The ratios are equal. Therefore, they form a proportion.

Another way to determine whether two ratios form a proportion is to use cross products. If the cross products are equal, then the ratios form a proportion.

Example 2 Use Cross Products

Use cross products to determine whether each pair of ratios form a proportion.

a. $\frac{0.4}{0.8}, \frac{0.7}{1.4}$

$\frac{0.4}{0.8} \stackrel{?}{=} \frac{0.7}{1.4}$ Write the equation.

$0.4(1.4) \stackrel{?}{=} 0.8(0.7)$ Find the cross products.

$0.56 = 0.56$ Simplify.

The cross products are equal, so $\frac{0.4}{0.8} = \frac{0.7}{1.4}$. Since the ratios are equal, they form a proportion.

b. $\frac{6}{8}, \frac{24}{28}$

$\frac{6}{8} \stackrel{?}{=} \frac{24}{28}$ Write the equation.

$6(28) \stackrel{?}{=} 8(24)$ Find the cross products.

$168 \neq 192$ Simplify.

The cross products are not equal, so $\frac{6}{8} \neq \frac{24}{28}$. The ratios do not form a proportion.

Study Tip

Cross Products
When you find cross products, you are said to be *cross multiplying*.

In the proportion $\frac{0.4}{0.8} = \frac{0.7}{1.4}$ above, 0.4 and 1.4 are called the **extremes**, and 0.8 and 0.7 are called the **means**.

Key Concept — Means-Extremes Property of Proportion

- **Words** In a proportion, the product of the extremes is equal to the product of the means.
- **Symbols** If $\frac{a}{b} = \frac{c}{d}$, then $ad = bc$.
- **Examples** Since $\frac{2}{4} = \frac{1}{2}$, $2(2) = 4(1)$ or $4 = 4$.

SOLVE PROPORTIONS You can write proportions that involve a variable. To solve the proportion, use cross products and the techniques used to solve other equations.

Example 3 Solve a Proportion

Solve the proportion $\frac{n}{15} = \frac{24}{16}$.

$\frac{n}{15} = \frac{24}{16}$ Original equation

$16(n) = 15(24)$ Find the cross products.

$16n = 360$ Simplify.

$\frac{16n}{16} = \frac{360}{16}$ Divide each side by 16.

$n = 22.5$ Simplify.

The ratio of two measurements having different units of measure is called a **rate**. For example, a price of \$1.99 per dozen eggs, a speed of 55 miles per hour, and a salary of \$30,000 per year are all rates. Proportions are often used to solve problems involving rates.

Example 4 Use Rates

BICYCLING Trent goes on a 30-mile bike ride every Saturday. He rides the distance in 4 hours. At this rate, how far can he ride in 6 hours?

Explore Let m represent the number of miles Trent can ride in 6 hours.

Plan Write a proportion for the problem.

$$\begin{matrix} \text{miles} \rightarrow \\ \text{hours} \rightarrow \end{matrix} \frac{30}{4} = \frac{m}{6} \begin{matrix} \leftarrow \text{miles} \\ \leftarrow \text{hours} \end{matrix}$$

Solve

$\frac{30}{4} = \frac{m}{6}$ Original proportion

$30(6) = 4(m)$ Find the cross products.

$180 = 4m$ Simplify.

$\frac{180}{4} = \frac{4m}{4}$ Divide each side by 4.

$45 = m$ Simplify.

Examine If Trent rides 30 miles in 4 hours, he rides 7.5 miles in 1 hour. So, in 6 hours, Trent can ride 6×7.5 or 45 miles. The answer is correct.

Since the rates are equal, they form a proportion. So, Trent can ride 45 miles in 6 hours.

A ratio or rate called a **scale** is used when making a model or drawing of something that is too large or too small to be conveniently drawn at actual size. The scale compares the model to the actual size of the object using a proportion. Maps and blueprints are two commonly used scale drawings.

Example 5 Use a Scale Drawing

CRATER LAKE The scale of a map for Crater Lake National Park is 2 inches = 9 miles. The distance between Discovery Point and Phantom Ship Overlook on the map is about $1\frac{3}{4}$ inches. What is the distance between these two places?

Let d represent the actual distance.

$$\begin{matrix} \text{scale} \rightarrow \\ \text{actual} \rightarrow \end{matrix} \frac{2}{9} = \frac{1\frac{3}{4}}{d} \begin{matrix} \leftarrow \text{scale} \\ \leftarrow \text{actual} \end{matrix}$$

$2(d) = 9\left(1\frac{3}{4}\right)$ Find the cross products.

$2d = \frac{63}{4}$ Simplify.

$2d \div 2 = \frac{63}{4} \div 2$ Divide each side by 2.

$d = \frac{63}{8}$ or $7\frac{7}{8}$ Simplify.

The actual distance is about $7\frac{7}{8}$ miles.

More About. . .

Crater Lake

Crater Lake is a volcanic crater in Oregon that was formed by an explosion 42 times the blast of Mount St. Helens.

Source: travel.excite.com

Check for Understanding

Concept Check

1. **OPEN ENDED** Find an example of ratios used in advertisements.
2. **Explain** the difference between a ratio and a proportion.
3. **Describe** how to solve a proportion if one of the ratios contains a variable.

Guided Practice

Use cross products to determine whether each pair of ratios form a proportion. Write *yes* or *no*.

4. $\frac{4}{11}, \frac{12}{33}$
5. $\frac{16}{17}, \frac{8}{9}$
6. $\frac{2.1}{3.5}, \frac{0.5}{0.7}$

Solve each proportion. If necessary, round to the nearest hundredth.

7. $\frac{3}{4} = \frac{6}{x}$
8. $\frac{a}{45} = \frac{5}{15}$
9. $\frac{0.6}{1.1} = \frac{n}{8.47}$

Application

10. **TRAVEL** The Lehmans' minivan requires 5 gallons of gasoline to travel 120 miles. How much gasoline will they need for a 350-mile trip?

Practice and Apply

Homework Help

For Exercises	See Examples
11–18	1, 2
19–30	3
31, 32	4
33, 34	5

Extra Practice

See page 827.

Use cross products to determine whether each pair of ratios form a proportion. Write *yes* or *no*.

11. $\frac{3}{2}, \frac{21}{14}$
12. $\frac{8}{9}, \frac{12}{18}$
13. $\frac{2.3}{3.4}, \frac{3.0}{3.6}$
14. $\frac{4.2}{5.6}, \frac{1.68}{2.24}$
15. $\frac{21.1}{14.4}, \frac{1.1}{1.2}$
16. $\frac{5}{2}, \frac{4}{1.6}$

SPORTS For Exercises 17 and 18, use the graph at the right.

17. Write a ratio of the number of gold medals won to the total number of medals won for each country.
18. Do any two of the ratios you wrote for Exercise 17 form a proportion? If so, explain the real-world meaning of the proportion.

USA TODAY Snapshots®

USA stands atop all-time medals table

The USA, which led the 2000 Summer Olympics with 97 medals, has dominated the medal standings over the years. The all-time Summer Olympics medal standings:

	Gold	Silver	Bronze	Total
USA	**871**	**659**	**586**	**2,116**
USSR/Russia[1]	498	409	371	1,278
Germany[2]	374	392	416	1,182
Great Britain	180	233	225	638
France	188	193	217	598
Italy	179	143	157	479
Sweden	136	156	177	469

1 – Competed as the Unified Team in 1992 after the breakup of the Soviet Union

2 – Totals include medals won by both East and West Germany.

Source: *The Ultimate Book of Sports Lists*

By Ellen J. Horrow and Marcy E. Mullins, USA TODAY

Solve each proportion. If necessary, round to the nearest hundredth.

19. $\frac{4}{x} = \frac{2}{10}$
20. $\frac{1}{y} = \frac{3}{15}$
21. $\frac{6}{5} = \frac{x}{15}$
22. $\frac{20}{28} = \frac{n}{21}$
23. $\frac{6}{8} = \frac{7}{a}$
24. $\frac{16}{7} = \frac{9}{b}$
25. $\frac{1}{0.19} = \frac{12}{n}$
26. $\frac{2}{0.21} = \frac{8}{n}$
27. $\frac{2.405}{3.67} = \frac{s}{1.88}$
28. $\frac{7}{1.066} = \frac{z}{9.65}$
29. $\frac{6}{14} = \frac{7}{x - 3}$
30. $\frac{5}{3} = \frac{6}{x + 2}$

A percent of increase or decrease can be used to describe trends in populations. Visit www.algebra1.com/webquest to continue work on your WebQuest project.

31. **WORK** Seth earns \$152 in 4 days. At that rate, how many days will it take him to earn \$532?

32. **DRIVING** Lanette drove 248 miles in 4 hours. At that rate, how long will it take her to drive an additional 93 miles?

33. **BLUEPRINTS** A blueprint for a house states that 2.5 inches equals 10 feet. If the length of a wall is 12 feet, how long is the wall in the blueprint?

34. **MODELS** A collector's model racecar is scaled so that 1 inch on the model equals $6\frac{1}{4}$ feet on the actual car. If the model is $\frac{2}{3}$ inch high, how high is the actual car?

35. **PETS** A research study shows that three out of every twenty pet owners got their pet from a breeder. Of the 122 animals cared for by a veterinarian, how many would you expect to have been bought from a breeder?

36. **CRITICAL THINKING** Consider the proportion $a:b:c = 3:1:5$. What is the value of $\frac{2a + 3b}{4b + 3c}$? (*Hint*: Choose different values of a, b, and c for which the proportion is true and evaluate the expression.)

37. **WRITING IN MATH** Answer the question that was posed at the beginning of the lesson.

How are ratios used in recipes?

Include the following in your answer:
- an explanation of how to use a proportion to determine how much honey is needed if you use 3 eggs, and
- a description of how to alter the recipe to get 5 servings.

38. Which ratio is *not* equal to $\frac{9}{12}$?

Ⓐ $\frac{18}{24}$ Ⓑ $\frac{3}{4}$ Ⓒ $\frac{15}{20}$ Ⓓ $\frac{18}{27}$

39. In the figure at the right, $x:y = 2:3$ and $y:z = 3:5$. If $x = 10$, find the value of z.

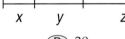

Ⓐ 15 Ⓑ 20 Ⓒ 25 Ⓓ 30

Maintain Your Skills

Mixed Review

Solve each equation. Then check your solution. *(Lessons 3-4 and 3-5)*

40. $8y - 10 = -3y + 2$
41. $17 + 2n = 21 + 2n$
42. $-7(d - 3) = -4$
43. $5 - 9w = 23$
44. $\frac{m}{-5} + 6 = 31$
45. $\frac{z - 7}{5} = -3$

Find each product. *(Lesson 2-3)*

46. $(-7)(-6)$
47. $\left(-\frac{8}{9}\right)\left(\frac{9}{8}\right)$
48. $\left(\frac{3}{7}\right)\left(\frac{3}{7}\right)$
49. $(-0.075)(-5.5)$

Find each absolute value. *(Lesson 2-1)*

50. $|-33|$
51. $|77|$
52. $|2.5|$
53. $|-0.85|$

54. Sketch a reasonable graph for the temperature in the following statement. *In August, you enter a hot house and turn on the air conditioner.* *(Lesson 1-8)*

Getting Ready for the Next Lesson

PREREQUISITE SKILL **Find each percent.** *(To review **percents**, see pages 802 and 803.)*

55. Eighteen is what percent of 60?
56. What percent of 14 is 4.34?
57. Six is what percent of 15?
58. What percent of 2 is 8?

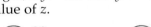

3-7 Percent of Change

What You'll Learn

- Find percents of increase and decrease.
- Solve problems involving percents of change.

Vocabulary

- percent of change
- percent of increase
- percent of decrease

How can percents describe growth over time?

Phone companies began using area codes in 1947. The graph shows the number of area codes in use in different years. The growth in the number of area codes can be described by using a percent of change.

Source: Associated Press

PERCENT OF CHANGE When an increase or decrease is expressed as a percent, the percent is called the **percent of change**. If the new number is greater than the original number, the percent of change is a **percent of increase**. If the new number is less than the original, the percent of change is a **percent of decrease**.

Example 1 Find Percent of Change

State whether each percent of change is a percent of increase or a percent of decrease. Then find each percent of change.

a. original: 25
new: 28

Find the *amount* of change. Since the new amount is greater than the original, the percent of change is a percent of increase.

$28 - 25 = 3$

Find the percent using the original number, 25, as the base.

$$\begin{array}{r} \text{change} \rightarrow \\ \text{original amount} \rightarrow \end{array} \frac{3}{25} = \frac{r}{100}$$

$$3(100) = 25(r)$$

$$300 = 25r$$

$$\frac{300}{25} = \frac{25r}{25}$$

$$12 = r$$

The percent of increase is 12%.

b. original: 30
new: 12

The percent of change is a percent of decrease because the new amount is less than the original. Find the change.

$30 - 12 = 18$

Find the percent using the original number, 30, as the base.

$$\begin{array}{r} \text{change} \rightarrow \\ \text{original amount} \rightarrow \end{array} \frac{18}{30} = \frac{r}{100}$$

$$18(100) = 30(r)$$

$$1800 = 30r$$

$$\frac{1800}{30} = \frac{30r}{30}$$

$$60 = r$$

The percent of decrease is 60%.

Study Tip

Look Back
To review the **percent proportion**, see page 834.

More About. . .

Football

On November 12, 1892, the Allegheny Athletic Association paid William "Pudge" Heffelfinger $500 to play football. This game is considered the start of professional football.

Source: *World Book Encyclopedia*

Example 2 Find the Missing Value

FOOTBALL The field used by the National Football League (NFL) is 120 yards long. The length of the field used by the Canadian Football League (CFL) is 25% longer than the one used by the NFL. What is the length of the field used by the CFL?

Let ℓ = the length of the CFL field. Since 25% is a percent of increase, the length of the NFL field is less than the length of the CFL field. Therefore, $\ell - 120$ represents the amount of change.

change → / original amount → $\frac{\ell - 120}{120} = \frac{25}{100}$	Percent proportion
$(\ell - 120)(100) = 120(25)$	Find the cross products.
$100\ell - 12{,}000 = 3000$	Distributive Property
$100\ell - 12{,}000 + 12{,}000 = 3000 + 12{,}000$	Add 12,000 to each side.
$100\ell = 15{,}000$	Simplify.
$\frac{100\ell}{100} = \frac{15{,}000}{100}$	Divide each side by 100.
$\ell = 150$	Simplify.

The length of the field used by the CFL is 150 yards.

SOLVE PROBLEMS Two applications of percent of change are sales tax and discounts. Sales tax is a tax that is added to the cost of the item. It is an example of a percent of increase. Discount is the amount by which the regular price of an item is reduced. It is an example of a percent of decrease.

Example 3 Find Amount After Sales Tax

SALES TAX A concert ticket costs $45. If the sales tax is 6.25%, what is the total price of the ticket?

The tax is 6.25% of the price of the ticket.

6.25% of \$45 = 0.0625 × 45   6.25% = 0.0625

= 2.8125   Use a calculator.

Round \$2.8125 to \$2.81.
Add this amount to the original price.

\$45.00 + \$2.81 = \$47.81

The total price of the ticket is \$47.81.

Example 4 Find Amount After Discount

DISCOUNT A sweater is on sale for 35% off the original price. If the original price of the sweater is $38, what is the discounted price?

The discount is 35% of the original price.

35% of \$38 = 0.35 × 38   35% = 0.35

= 13.30   Use a calculator.

Subtract \$13.30 from the original price.

\$38.00 − \$13.30 = \$24.70

The discounted price of the sweater is \$24.70.

Check for Understanding

Concept Check

1. **Compare and contrast** percent of increase and percent of decrease.
2. **OPEN ENDED** Give a counterexample to the statement *The percent of change must always be less than 100%.*
3. **FIND THE ERROR** Laura and Cory are writing proportions to find the percent of change if the original number is 20 and the new number is 30.

Laura	Cory
Amount of change: 30 − 20 = 10	Amount of change: 30 − 20 = 10
$\frac{10}{20} = \frac{r}{100}$	$\frac{10}{30} = \frac{r}{100}$

Who is correct? Explain your reasoning.

Guided Practice

State whether each percent of change is a percent of increase or a percent of decrease. Then find each percent of change. Round to the nearest whole percent.

4. original: 72
   new: 36
5. original: 45
   new: 50
6. original: 14
   new: 16
7. original: 150
   new: 120

Find the total price of each item.

8. software: $39.50
   sales tax: 6.5%
9. compact disc: $15.99
   sales tax: 5.75%

Find the discounted price of each item.

10. jeans: $45.00
    discount: 25%
11. book: $19.95
    discount: 33%

Application

EDUCATION **For Exercises 12 and 13, use the following information.**
According to the Census Bureau, the average income of a person with a bachelor's degree is $40,478, and the average income of a person with a high school diploma is $22,895.

12. Write an equation that could be used to find the percent of increase in average income for a person with a high school diploma to average income for a person with a bachelor's degree.
13. What is the percent of increase?

Practice and Apply

Homework Help

For Exercises	See Examples
14–27	1
28–30, 46, 47	2
31–36	3
37–42	4
43–45	3, 4

Extra Practice

See page 827.

State whether each percent of change is a percent of increase or a percent of decrease. Then find each percent of change. Round to the nearest whole percent.

14. original: 50
    new: 70
15. original: 25
    new: 18
16. original: 66
    new: 30
17. original: 58
    new: 152
18. original: 13.7
    new: 40.2
19. original: 15.6
    new: 11.4
20. original: 132
    new: 150
21. original: 85
    new: 90
22. original: 32.5
    new: 30
23. original: 9.8
    new: 12.1
24. original: 40
    new: 32.5
25. original: 25
    new: 21.5

26. **THEME PARKS** In 1990, 253 million people visited theme parks in the United States. In 2000, the number of visitors increased to 317 million people. What was the percent of increase?

27. **MILITARY** In 1987, the United States had 2 million active-duty military personnel. By 2000, there were only 1.4 million active-duty military personnel. What was the percent of decrease?

28. The percent of increase is 16%. If the new number is 522, find the original number.

29. **FOOD** In order for a food to be marked "reduced fat," it must have at least 25% less fat than the same full-fat food. If one ounce of reduced fat chips has 6 grams of fat, what is the least amount of fat in one ounce of regular chips?

30. **TECHNOLOGY** From January, 1996, to January, 2001, the number of internet hosts increased by 1054%. There were 109.6 million internet hosts in January, 2001. Find the number of internet hosts in January, 1996.

Career Choices

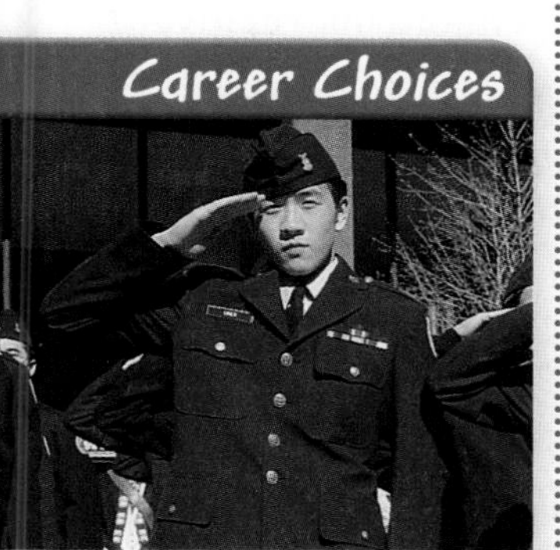

Military

A military career can involve many different duties like working in a hospital, programming computers, or repairing helicopters. The military provides training and work in these fields and others for the Army, Navy, Marine Corps, Air Force, Coast Guard, and the Air and Army National Guard.

Online Research
For information about a career in the military, visit: www.algebra1.com/careers

Find the total price of each item.

31. umbrella: \$14.00
tax: 5.5%

32. backpack: \$35.00
tax: 7%

33. candle: \$7.50
tax: 5.75%

34. hat: \$18.50
tax: 6.25%

35. clock radio: \$39.99
tax: 6.75%

36. sandals: \$29.99
tax: 5.75%

Find the discounted price of each item.

37. shirt: \$45.00
discount: 40%

38. socks: \$6.00
discount: 20%

39. watch: \$37.55
discount: 35%

40. gloves: \$24.25
discount: 33%

41. suit: \$175.95
discount: 45%

42. coat: \$79.99
discount: 30%

Find the final price of each item.

43. lamp: \$120.00
discount: 20%
tax: 6%

44. dress: \$70.00
discount: 30%
tax: 7%

45. camera: \$58.00
discount: 25%
tax: 6.5%

POPULATION **For Exercises 46 and 47, use the following table.**

Country	1997 Population (billions)	Projected Percent of Increase for 2050
China	1.24	22.6%
India	0.97	57.8%
United States	0.27	44.4%

Source: *USA TODAY*

46. What are the projected 2050 populations for each country in the table?

47. Which of these three countries is projected to be the most populous in 2050?

48. **RESEARCH** Use the Internet or other reference to find the tuition for the last several years at a college of your choice. Find the percent of change for the tuition during these years. Predict the tuition for the year you plan to graduate from high school.

49. **CRITICAL THINKING** Are the following expressions *sometimes*, *always*, or *never* equal? Explain your reasoning.

$x\%$ of y $\qquad\qquad$ $y\%$ of x

50. **WRITING IN MATH** Answer the question that was posed at the beginning of the lesson.

How can percents describe growth over time?

Include the following in your answer:

- the percent of increase in the number of area codes from 1996 to 1999, and
- an explanation of why knowing a percent of change can be more informative than knowing how much the quantity changed.

51. The number of students at Franklin High School increased from 840 to 910 over a 5-year period. Which proportion represents the percent of change?

(A) $\frac{70}{910} = \frac{r}{100}$ (B) $\frac{70}{840} = \frac{r}{100}$ (C) $\frac{r}{910} = \frac{70}{100}$ (D) $\frac{r}{840} = \frac{70}{100}$

52. The list price of a television is $249.00. If it is on sale for 30% off the list price, what is the sale price of the television?

(A) $74.70 (B) $149.40 (C) $174.30 (D) $219.00

Maintain Your Skills

Mixed Review

Solve each proportion. *(Lesson 3-6)*

53. $\frac{a}{45} = \frac{3}{15}$ 54. $\frac{2}{3} = \frac{8}{d}$ 55. $\frac{5.22}{13.92} = \frac{t}{48}$

Solve each equation. Then check your solution. *(Lesson 3-5)*

56. $6n + 3 = -3$ 57. $7 + 5c = -23$ 58. $18 = 4a - 2$

Find each quotient. *(Lesson 2-4)*

59. $\frac{2}{5} \div 4$ 60. $-\frac{4}{5} \div \frac{2}{3}$ 61. $-\frac{1}{9} \div \left(-\frac{3}{4}\right)$

State whether each equation is *true* or *false* for the value of the variable given. *(Lesson 1-3)*

62. $a^2 + 5 = 17 - a, a = 3$ 63. $2v^2 + v = 65, v = 5$

64. $8y - y^2 = y + 10, y = 4$ 65. $16p - p = 15p, p = 2.5$

Getting Ready for the Next Lesson

PREREQUISITE SKILL **Solve each equation. Then check your solution.** *(To review **solving equations**, see Lesson 3-5.)*

66. $-43 - 3t = 2 - 6t$ 67. $7y + 7 = 3y - 5$ 68. $7(d - 3) - 2 = 5$

69. $6(p + 3) = 4(p - 1)$ 70. $-5 = 4 - 2(a - 5)$ 71. $8x - 4 = -10x + 50$

Practice Quiz 2

Lessons 3-4 through 3-7

Solve each equation. Then check your solution. *(Lessons 3-4 and 3-5)*

1. $-3x - 7 = 18$ 2. $5 = \frac{m - 5}{4}$ 3. $4h + 5 = 11$

4. $5d - 6 = 3d + 9$ 5. $7 + 2(w + 1) = 2w + 9$ 6. $-8(4 + 9r) = 7(-2 - 11r)$

Solve each proportion. *(Lesson 3-6)*

7. $\frac{2}{10} = \frac{1}{a}$ 8. $\frac{3}{5} = \frac{24}{x}$ 9. $\frac{y}{4} = \frac{y + 5}{8}$

10. **POSTAGE** In 1975, the cost of a first-class stamp was 10¢. In 2001, the cost of a first-class stamp became 34¢. What is the percent of increase in the price of a stamp? *(Lesson 3-7)*

Reading Mathematics

Sentence Method and Proportion Method

Recall that you can solve percent problems using two different methods. With either method, it is helpful to use "clue" words such as *is* and *of*. In the sentence method, *is* means equals and *of* means multiply. With the proportion method, the "clue" words indicate where to place the numbers in the proportion.

Sentence Method

15% of 40 is what number?

$0.15 \cdot 40 = ?$

Proportion Method

15% of 40 is what number?

$\frac{(\text{is})\ P}{(\text{of})\ B} = \frac{R(\text{percent})}{100} \rightarrow \frac{P}{40} = \frac{15}{100}$

You can use the proportion method to solve percent of change problems. In this case, use the proportion $\frac{\text{difference}}{\text{original}} = \frac{\%}{100}$. When reading a percent of change problem, or any other word problem, look for the important numerical information.

Example **In chemistry class, Kishi heated 20 milliliters of water. She let the water boil for 10 minutes. Afterward, only 17 milliliters of water remained, due to evaporation. What is the percent of decrease in the amount of water?**

$\frac{\text{difference}}{\text{original}} = \frac{\%}{100} \rightarrow \frac{20 - 17}{20} = \frac{r}{100}$ Percent proportion

$\frac{3}{20} = \frac{r}{100}$ Simplify.

$3(100) = 20(r)$ Find the cross products.

$300 = 20r$ Simplify.

$\frac{300}{20} = \frac{20r}{20}$ Divide each side by 20.

$15 = r$ Simplify.

There was a 15% decrease in the amount of water.

Reading to Learn

Give the original number and the amount of change. Then write and solve a percent proportion.

1. Monsa needed to lose weight for wrestling. At the start of the season, he weighed 166 pounds. By the end of the season, he weighed 158 pounds. What is the percent of decrease in Monsa's weight?
2. On Carla's last Algebra test, she scored 94 points out of 100. On her first Algebra test, she scored 75 points out of 100. What is the percent of increase in her score?
3. In a catalog distribution center, workers processed an average of 12 orders per hour. After a reward incentive was offered, workers averaged 18 orders per hour. What is the percent of increase in production?

3-8 Solving Equations and Formulas

What You'll Learn

- Solve equations for given variables.
- Use formulas to solve real-world problems.

Vocabulary

- dimensional analysis

How are equations used to design roller coasters?

Ron Toomer designs roller coasters, including the Magnum XL-200. This roller coaster starts with a vertical drop of 195 feet and then ascends a second shorter hill. Suppose when designing this coaster, Mr. Toomer decided he wanted to adjust the height of the second hill so that the coaster would have a speed of 49 feet per second when it reached its top.

If we ignore friction, the equation $g(195 - h) = \frac{1}{2}v^2$ can be used to find the height of the second hill. In this equation, g represents the acceleration due to gravity (32 feet per second squared), h is the height of the second hill, and v is the velocity of the coaster when it reaches the top of the second hill.

SOLVE FOR VARIABLES Some equations such as the one above contain more than one variable. At times, you will need to solve these equations for one of the variables.

Example 1 Solve an Equation for a Specific Variable

Solve $3x - 4y = 7$ for y.

$3x - 4y = 7$	Original equation
$3x - 4y - 3x = 7 - 3x$	Subtract $3x$ from each side.
$-4y = 7 - 3x$	Simplify.
$\frac{-4y}{-4} = \frac{7 - 3x}{-4}$	Divide each side by -4.
$y = \frac{7 - 3x}{-4}$ or $\frac{3x - 7}{4}$	Simplify.

The value of y is $\frac{3x - 7}{4}$.

It is sometimes helpful to use the Distributive Property to isolate the variable for which you are solving an equation or formula.

Example 2 Solve an Equation for a Specific Variable

Solve $2m - t = sm + 5$ for m.

$2m - t = sm + 5$	Original equation
$2m - t - sm = sm + 5 - sm$	Subtract *sm* from each side.
$2m - t - sm = 5$	Simplify.
$2m - t - sm + t = 5 + t$	Add *t* to each side.
$2m - sm = 5 + t$	Simplify.
$m(2 - s) = 5 + t$	Use the Distributive Property.
$\frac{m(2 - s)}{2 - s} = \frac{5 + t}{2 - s}$	Divide each side by $2 - s$.
$m = \frac{5 + t}{2 - s}$	Simplify.

The value of m is $\frac{5 + t}{2 - s}$. Since division by 0 is undefined, $2 - s \neq 0$ or $s \neq 2$.

USE FORMULAS Many real-world problems require the use of formulas. Sometimes solving a formula for a specific variable will help you solve the problem.

Example 3 Use a Formula to Solve Problems

WEATHER **Use the information about the Kansas City hailstorm at the left. The formula for the circumference of a circle is $C = 2\pi r$, where C represents circumference and r represents radius.**

a. Solve the formula for r.

$C = 2\pi r$	Formula for circumference
$\frac{C}{2\pi} = \frac{2\pi r}{2\pi}$	Divide each side by 2π.
$\frac{C}{2\pi} = r$	Simplify.

b. Find the radius of one of the largest hailstones that fell on Kansas City in 1898.

$\frac{C}{2\pi} = r$	Formula for radius
$\frac{9.5}{2\pi} = r$	$C = 9.5$
$1.5 \approx r$	The largest hailstones had a radius of about 1.5 inches.

More About. . .

Weather

On May 14, 1898, a severe hailstorm hit Kansas City. The largest hailstones were 9.5 inches in circumference. Windows were broken in nearly every house in the area.

Source: National Weather Service

When using formulas, you may want to use dimensional analysis. **Dimensional analysis** is the process of carrying units throughout a computation.

Example 4 Use Dimensional Analysis

PHYSICAL SCIENCE **The formula $s = \frac{1}{2}at^2$ represents the distance s that a free-falling object will fall near a planet or the moon in a given time t. In the formula, a represents the acceleration due to gravity.**

a. Solve the formula for a.

$s = \frac{1}{2}at^2$	Original formula
$\frac{2}{t^2}(s) = \frac{2}{t^2}\left(\frac{1}{2}at^2\right)$	Multiply each side by $\frac{2}{t^2}$.
$\frac{2s}{t^2} = a$	Simplify.

b. A free-falling object near the moon drops 20.5 meters in 5 seconds. What is the value of *a* for the moon?

$a = \frac{2s}{t^2}$ Formula for *a*

$a = \frac{2(20.5\text{ m})}{(5\text{ s})^2}$ $s = 20.5$ m and $t = 5$ s.

$a = \frac{1.64\text{ m}}{\text{s}^2}$ or 1.64 m/s^2 Use a calculator.

The acceleration due to gravity on the moon is 1.64 meters per second squared.

Check for Understanding

Concept Check

1. **List** the steps you would use to solve $ax - y = az + w$ for a.
2. **Describe** the possible values of t if $s = \frac{r}{t - 2}$.
3. **OPEN ENDED** Write a formula for A, the area of a geometric figure such as a triangle or rectangle. Then solve the formula for a variable other than A.

Guided Practice

Solve each equation or formula for the variable specified.

4. $-3x + b = 6x$, for x
5. $-5a + y = -54$, for a
6. $4z + b = 2z + c$, for z
7. $\frac{y + a}{3} = c$, for y
8. $p = a(b + c)$, for a
9. $mw - t = 2w + 5$, for w

Application

GEOMETRY For Exercises 10–12, use the formula for the area of a triangle.

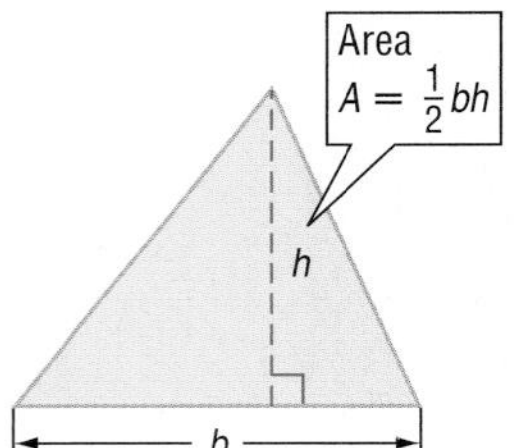

10. Find the area of a triangle with a base of 18 feet and a height of 7 feet.
11. Solve the formula for h.
12. What is the height of a triangle with area of 28 square feet and base of 8 feet?

Practice and Apply

Homework Help

For Exercises	See Examples
13–30	1, 2
31–41	3, 4

Extra Practice

See page 827.

Solve each equation or formula for the variable specified.

13. $5g + h = g$, for g
14. $8t - r = 12t$, for t
15. $y = mx + b$, for m
16. $v = r + at$, for a
17. $3y + z = am - 4y$, for y
18. $9a - 2b = c + 4a$, for a
19. $km + 5x = 6y$, for m
20. $4b - 5 = -t$, for b
21. $\frac{3ax - n}{5} = -4$, for x
22. $\frac{5x + y}{a} = 2$, for a
23. $\frac{by + 2}{3} = c$, for y
24. $\frac{6c - t}{7} = b$, for c
25. $c = \frac{3}{4}y + b$, for y
26. $\frac{3}{5}m + a = b$, for m
27. $S = \frac{n}{2}(A + t)$, for A
28. $p(t + 1) = -2$, for t
29. $at + b = ar - c$, for a
30. $2g - m = 5 - gh$, for g

Write an equation and solve for the variable specified.

31. Five less than a number t equals another number r plus six. Solve for t.

32. Five minus twice a number p equals six times another number q plus one. Solve for p.

33. Five eighths of a number x is three more than one half of another number y. Solve for y.

GEOMETRY **For Exercises 34 and 35, use the formula for the area of a trapezoid.**

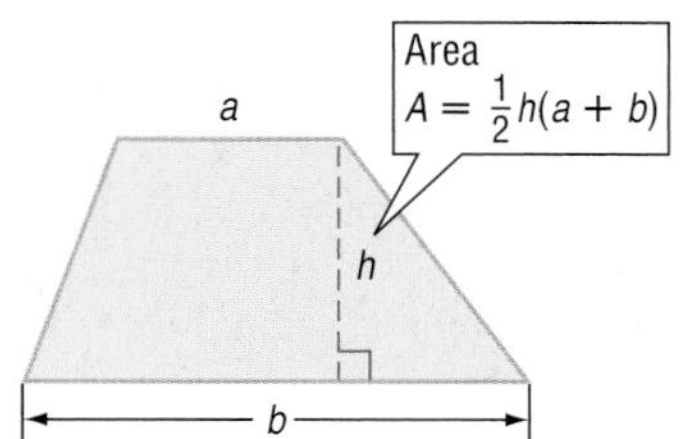

34. Solve the formula for h.

35. What is the height of a trapezoid with an area of 60 square meters and bases of 8 meters and 12 meters?

WORK **For Exercises 36 and 37, use the following information.**

The formula $s = \frac{w - 10e}{m}$ is often used by placement services to find keyboarding speeds. In the formula, s represents the speed in words per minute, w represents the number of words typed, e represents the number of errors, and m represents the number of minutes typed.

36. Solve the formula for e.

37. If Miguel typed 410 words in 5 minutes and received a keyboard speed of 76 words per minute, how many errors did he make?

FLOORING **For Exercises 38 and 39, use the following information.**

The formula $P = \frac{1.2W}{H^2}$ represents the amount of pressure exerted on the floor by the heel of a shoe. In this formula, P represents the pressure in pounds per square inch, W represents the weight of a person wearing the shoe in pounds, and H is the width of the heel of the shoe in inches.

38. Solve the formula for W.

39. Find the weight of the person if the heel is 3 inches wide and the pressure exerted is 30 pounds per square inch.

40. **ROCKETRY** In the book *October Sky*, high school students were experimenting with different rocket designs. One formula they used was $R = \frac{S + F + P}{S + P}$, which relates the mass ratio R of a rocket to the mass of the structure S, the mass of the fuel F, and the mass of the payload P. The students needed to determine how much fuel to load in the rocket. How much fuel should be loaded in a rocket whose basic structure and payload each have a mass of 900 grams, if the mass ratio is to be 6?

41. **PACKAGING** The Yummy Ice Cream Company wants to package ice cream in cylindrical containers that have a volume of 5453 cubic centimeters. The marketing department decides the diameter of the base of the containers should be 20 centimeters. How tall should the containers be? (*Hint*: $V = \pi r^2 h$)

42. CRITICAL THINKING Write a formula for the area of the arrow.

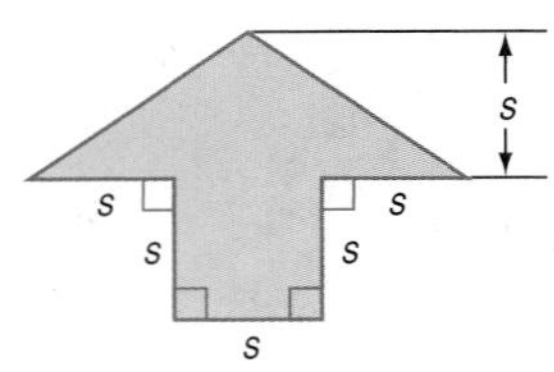

43. WRITING IN MATH Answer the question that was posed at the beginning of the lesson.

How are equations used to design roller coasters?

Include the following in your answer:

- a list of steps you could use to solve the equation for h, and
- the height of the second hill of the roller coaster.

Standardized Test Practice

44. If $2x + y = 5$, what is the value of $4x$?

(A) $10 - y$ (B) $10 - 2y$

(C) $\frac{5 - y}{2}$ (D) $\frac{10 - y}{2}$

45. What is the area of the triangle?

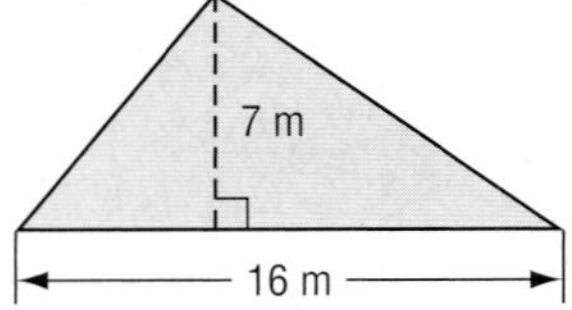

(A) 23 m^2 (B) 28 m^2

(C) 56 m^2 (D) 112 m^2

Maintain Your Skills

Mixed Review

Find the discounted price of each item. *(Lesson 3-7)*

46. camera: \$85.00 discount: 20%

47. scarf: \$15.00 discount: 35%

48. television: \$299.00 discount: 15%

Solve each proportion. *(Lesson 3-6)*

49. $\frac{2}{9} = \frac{5}{a}$

50. $\frac{15}{32} = \frac{t}{8}$

51. $\frac{x + 1}{8} = \frac{3}{4}$

Write the numbers in each set in order from least to greatest. *(Lesson 2-7)*

52. $\frac{1}{4}, \sqrt{\frac{1}{4}}, 0.\overline{5}, 0.2$

53. $\sqrt{5}, 3, \frac{2}{3}, 1.1$

Find each sum or difference. *(Lesson 2-2)*

54. $2.18 + (-5.62)$

55. $-\frac{1}{2} - \left(-\frac{3}{4}\right)$

56. $-\frac{2}{3} - \frac{2}{5}$

Name the property illustrated by each statement. *(Lesson 1-4)*

57. $mnp = 1mnp$

58. If $6 = 9 - 3$, then $9 - 3 = 6$.

59. $32 + 21 = 32 + 21$

60. $8 + (3 + 9) = 8 + 12$

Getting Ready for the Next Lesson

PREREQUISITE SKILL **Use the Distributive Property to rewrite each expression without parentheses.** *(To review the **Distributive Property**, see Lesson 1-5.)*

61. $6(2 - t)$

62. $(5 + 2m)3$

63. $-7(3a + b)$

64. $\frac{2}{3}(6h - 9)$

65. $-\frac{3}{5}(15 - 5t)$

66. $0.25(6p + 12)$

3-9 Weighted Averages

What You'll Learn

- Solve mixture problems.
- Solve uniform motion problems.

Vocabulary

- weighted average
- mixture problem
- uniform motion problem

How are scores calculated in a figure skating competition?

In individual figure skating competitions, the score for the long program is worth twice the score for the short program. Suppose Olympic gold medal winner Ilia Kulik scores 5.5 in the short program and 5.8 in the long program at a competition. His final score is determined using a weighted average.

$$\frac{5.5(1) + 5.8(2)}{1 + 2} = \frac{5.5 + 11.6}{3}$$

$$= \frac{17.1}{3} \text{ or } 5.7 \quad \text{His final score would be 5.7.}$$

MIXTURE PROBLEMS Ilia Kulik's average score is an example of a weighted average. The **weighted average** M of a set of data is the sum of the product of the number of units and the value per unit divided by the sum of the number of units.

Mixture problems are problems in which two or more parts are combined into a whole. They are solved using weighted averages.

Example 1 Solve a Mixture Problem with Prices

TRAIL MIX **Assorted dried fruit sells for \$5.50 per pound. How many pounds of mixed nuts selling for \$4.75 per pound should be mixed with 10 pounds of dried fruit to obtain a trail mix that sells for \$4.95 per pound?**

Let w = the number of pounds of mixed nuts in the mixture. Make a table.

	Units (lb)	Price per Unit (lb)	Total Price
Dried Fruit	10	\$5.50	5.50(10)
Mixed Nuts	w	\$4.75	$4.75w$
Trail Mix	$10 + w$	\$4.95	$4.95(10 + w)$

Price of dried fruit	plus	price of nuts	equals	price of trail mix.
5.50(10)	+	$4.75w$	=	$4.95(10 + w)$

$5.50(10) + 4.75w = 4.95(10 + w)$ — Original equation

$55.00 + 4.75w = 49.50 + 4.95w$ — Distributive Property

$55.00 + 4.75w - 4.75w = 49.50 + 4.95w - 4.75w$ — Subtract $4.75w$ from each side.

$55.00 = 49.50 + 0.20w$ — Simplify.

$55.00 - 49.50 = 49.50 + 0.20w - 49.50$ — Subtract 49.50 from each side.

$5.50 = 0.20w$ — Simplify.

$\frac{5.50}{0.20} = \frac{0.20w}{0.20}$ — Divide each side by 0.20.

$27.5 = w$ — Simplify.

27.5 pounds of nuts should be mixed with 10 pounds of dried fruit.

Sometimes mixture problems are expressed in terms of percents.

Example 2 Solve a Mixture Problem with Percents

SCIENCE A chemistry experiment calls for a 30% solution of copper sulfate. Kendra has 40 milliliters of 25% solution. How many milliliters of 60% solution should she add to obtain the required 30% solution?

Let x = the amount of 60% solution to be added. Make a table.

	Amount of Solution (mL)	Amount of Copper Sulfate
25% Solution	40	0.25(40)
60% Solution	x	$0.60x$
30% Solution	$40 + x$	$0.30(40 + x)$

Study Tip

Mixture Problems
When you organize the information in mixture problems, remember that the final mixture must contain the sum of the parts in the correct quantities and at the correct percents.

Write and solve an equation using the information in the table.

Amount of copper sulfate in 25% solution	plus	amount of copper sulfate in 60% solution	equals	amount of copper sulfate in 30% solution.
$0.25(40)$	$+$	$0.60x$	$=$	$0.30(40 + x)$

$0.25(40) + 0.60x = 0.30(40 + x)$ Original equation

$10 + 0.60x = 12 + 0.30x$ Distributive Property

$10 + 0.60x - 0.30x = 12 + 0.30x - 0.30x$ Subtract $0.30x$ from each side.

$10 + 0.30x = 12$ Simplify.

$10 + 0.30x - 10 = 12 - 10$ Subtract 10 from each side.

$0.30x = 2$ Simplify.

$\frac{0.30x}{0.30} = \frac{2}{0.30}$ Divide each side by 0.30.

$x \approx 6.67$ Simplify.

Kendra should add 6.67 milliliters of the 60% solution to the 40 milliliters of the 25% solution.

UNIFORM MOTION PROBLEMS Motion problems are another application of weighted averages. **Uniform motion problems** are problems where an object moves at a certain speed, or rate. The formula $d = rt$ is used to solve these problems. In the formula, d represents distance, r represents rate, and t represents time.

Example 3 Solve for Average Speed

TRAVEL On Alberto's drive to his aunt's house, the traffic was light, and he drove the 45-mile trip in one hour. However, the return trip took him two hours. What was his average speed for the round trip?

To find the average speed for each leg of the trip, rewrite $d = rt$ as $r = \frac{d}{t}$.

Going

$r = \frac{d}{t}$

$= \frac{45 \text{ miles}}{1 \text{ hour}}$ or 45 miles per hour

Returning

$r = \frac{d}{t}$

$= \frac{45 \text{ miles}}{2 \text{ hours}}$ or 22.5 miles per hour

You may think that the average speed of the trip would be $\frac{45 + 22.5}{2}$ or 33.75 miles per hour. However, Alberto did not drive at these speeds for equal amounts of time. You must find the weighted average for the trip.

Round Trip

$M = \frac{45(1) + 22.5(2)}{1 + 2}$ Definition of weighted average

$= \frac{90}{3}$ or 30 Simplify.

Alberto's average speed was 30 miles per hour.

Sometimes a table is useful in solving uniform motion problems.

Example 4 Solve a Problem Involving Speeds of Two Vehicles

SAFETY Use the information about sirens at the left. A car and an emergency vehicle are heading toward each other. The car is traveling at a speed of 30 miles per hour or about 44 feet per second. The emergency vehicle is traveling at a speed of 50 miles per hour or about 74 feet per second. If the vehicles are 1000 feet apart and the conditions are ideal, in how many seconds will the driver of the car first hear the siren?

More About. . .

Safety

Under ideal conditions, a siren can be heard from up to 440 feet. However, under normal conditions, a siren can be heard from only 125 feet.

Source: U.S. Department of Transportation

Draw a diagram. The driver can hear the siren when the total distance traveled by the two vehicles equals 1000 − 440 or 560 feet.

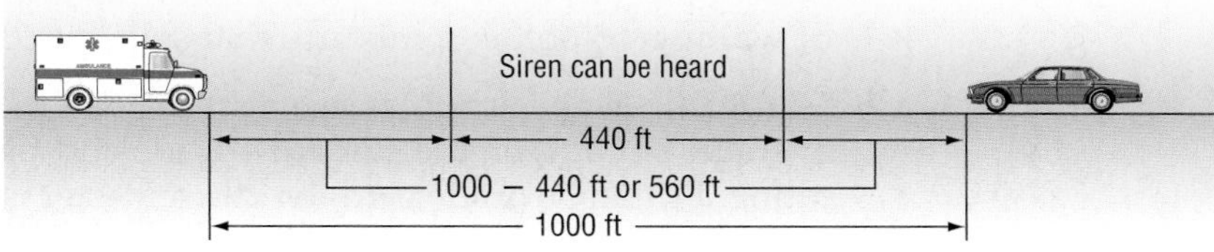

Let t = the number of seconds until the driver can hear the siren. Make a table of the information.

	r	t	$d = rt$
Car	44	t	$44t$
Emergency Squad	74	t	$74t$

Write an equation.

Distance traveled by car	plus	distance traveled by emergency vehicle	equals	560 feet.
$44t$	$+$	$74t$	$=$	560

Solve the equation.

$44t + 74t = 560$ Original equation

$118t = 560$ Simplify.

$\frac{118t}{118} = \frac{560}{118}$ Divide each side by 118.

$t \approx 4.75$ Round to the nearest hundredth.

The driver of the car will hear the siren in about 4.75 seconds.

Check for Understanding

Concept Check

1. **OPEN ENDED** Give a real-world example of a weighted average.
2. **Write** the formula used to solve uniform motion problems and tell what each letter represents.
3. **Make a table** that can be used to solve the following problem.
Lakeisha has $2.55 in dimes and quarters. She has 8 more dimes than quarters. How many quarters does she have?

Guided Practice

FOOD For Exercises 4–7, use the following information.
How many quarts of pure orange juice should Michael add to a 10% orange drink to create 6 quarts of a 40% orange juice mixture? Let p represent the number of quarts of pure orange juice he should add to the orange drink.

4. Copy and complete the table representing the problem.

	Quarts	Amount of Orange Juice
10% Juice	$6 - p$	
100% Juice	p	
40% Juice		

5. Write an equation to represent the problem.
6. How much pure orange juice should Michael use?
7. How much 10% juice should Michael use?
8. **BUSINESS** The Nut Shoppe sells walnuts for $4.00 a pound and cashews for $7.00 a pound. How many pounds of cashews should be mixed with 10 pounds of walnuts to obtain a mixture that sells for $5.50 a pound?
9. **GRADES** Many schools base a student's grade point average, or GPA, on the student's grade and the class credit rating. Brittany's grade card for this semester is shown. Find Brittany's GPA if a grade of A equals 4 and a B equals 3.
10. **CYCLING** Two cyclists begin traveling in the same direction on the same bike path. One travels at 20 miles per hour, and the other travels at 14 miles per hour. When will the cyclists be 15 miles apart?

Grade Card

Class	Credit Rating	Grade
Algebra 1	1	A
Science	1	B
English	1	A
Spanish	1	B
Phys. Ed.	$\frac{1}{2}$	A

Practice and Apply

Homework Help

For Exercises	See Examples
11–18, 22–25, 27–29, 33	1, 2
19–21, 26, 30–32, 34	3, 4

Extra Practice
See page 828.

BUSINESS For Exercises 11–14, use the following information.
Cookies Inc. sells peanut butter cookies for $6.50 per dozen and chocolate chip cookies for $9.00 per dozen. Yesterday, they sold 85 dozen more peanut butter cookies than chocolate chip cookies. The total sales for both types of cookies were $4055.50. Let p represent the number of dozens of peanut butter cookies sold.

11. Copy and complete the table representing the problem.

	Number of Dozens	Price per Dozen	Total Price
Peanut Butter Cookies	p		
Chocolate Chip Cookies	$p - 85$		

12. Write an equation to represent the problem.
13. How many dozen peanut butter cookies were sold?
14. How many dozen chocolate chip cookies were sold?

METALS For Exercises 15–18, use the following information.
In 2000, the international price of gold was \$270 per ounce, and the international price of silver was \$5 per ounce. Suppose gold and silver were mixed to obtain 15 ounces of an alloy worth \$164 per ounce. Let g represent the amount of gold used in the alloy.

15. Copy and complete the table representing the problem.

	Number of Ounces	Price per Ounce	Value
Gold	g		
Silver	$15 - g$		
Alloy			

16. Write an equation to represent the problem.
17. How much gold was used in the alloy?
18. How much silver was used in the alloy?

TRAVEL For Exercises 19–21, use the following information.
Two trains leave Pittsburgh at the same time, one traveling east and the other traveling west. The eastbound train travels at 40 miles per hour, and the westbound train travels at 30 miles per hour. Let t represent the amount of time since their departure.

19. Copy and complete the table representing the situation.

	r	t	$d = rt$
Eastbound Train			
Westbound Train			

20. Write an equation that could be used to determine when the trains will be 245 miles apart.
21. In how many hours will the trains be 245 miles apart?

22. **FUND-RAISING** The Madison High School marching band sold gift wrap. The gift wrap in solid colors sold for \$4.00 per roll, and the print gift wrap sold for \$6.00 per roll. The total number of rolls sold was 480, and the total amount of money collected was \$2340. How many rolls of each kind of gift wrap were sold?

23. **COFFEE** Charley Baroni owns a specialty coffee store. He wants to create a special mix using two coffees, one priced at \$6.40 per pound and the other priced at \$7.28 per pound. How many pounds of the \$7.28 coffee should he mix with 9 pounds of the \$6.40 coffee to sell the mixture for \$6.95 per pound?

24. **FOOD** Refer to the graphic at the right. How much whipping cream and 2% milk should be mixed to obtain 35 gallons of milk with 4% butterfat?

25. **METALS** An alloy of metals is 25% copper. Another alloy is 50% copper. How much of each alloy should be used to make 1000 grams of an alloy that is 45% copper?

26. **TRAVEL** An airplane flies 1000 miles due east in 2 hours and 1000 miles due south in 3 hours. What is the average speed of the airplane?

27. **SCIENCE** Hector is performing a chemistry experiment that requires 140 milliliters of a 30% copper sulfate solution. He has a 25% copper sulfate solution and a 60% copper sulfate solution. How many milliliters of each solution should he mix to obtain the needed solution?

28. **CAR MAINTENANCE** One type of antifreeze is 40% glycol, and another type of antifreeze is 60% glycol. How much of each kind should be used to make 100 gallons of antifreeze that is 48% glycol?

29. **GRADES** In Ms. Martinez's science class, a test is worth three times as much as a quiz. If a student has test grades of 85 and 92 and quiz grades of 82, 75, and 95, what is the student's average grade?

30. **RESCUE** A fishing trawler has radioed the Coast Guard for a helicopter to pick up an injured crew member. At the time of the emergency message, the trawler is 660 kilometers from the helicopter and heading toward it. The average speed of the trawler is 30 kilometers per hour, and the average speed of the helicopter is 300 kilometers per hour. How long will it take the helicopter to reach the trawler?

31. **ANIMALS** A cheetah is 300 feet from its prey. It starts to sprint toward its prey at 90 feet per second. At the same time, the prey starts to sprint at 70 feet per second. When will the cheetah catch its prey?

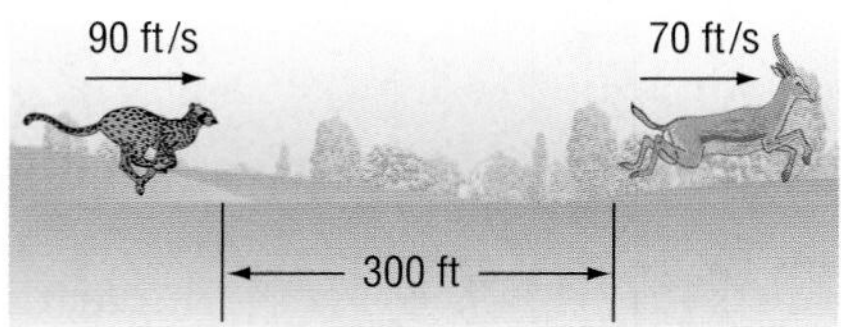

32. **TRACK AND FIELD** A sprinter has a bad start, and his opponent is able to start 1 second before him. If the sprinter averages 8.2 meters per second and his opponent averages 8 meters per second, will he be able to catch his opponent before the end of the 200-meter race? Explain.

33. **CAR MAINTENANCE** A car radiator has a capacity of 16 quarts and is filled with a 25% antifreeze solution. How much must be drained off and replaced with pure antifreeze to obtain a 40% antifreeze solution?

34. **TRAVEL** An express train travels 80 kilometers per hour from Ironton to Wildwood. A local train, traveling at 48 kilometers per hour, takes 2 hours longer for the same trip. How far apart are Ironton and Wildwood?

35. **FOOTBALL** NFL quarterbacks are rated for their passing performance by a type of weighted average as described in the formula below.
$R = [50 + 2000(C \div A) + 8000(T \div A) - 10{,}000(I \div A) + 100(Y \div A)] \div 24$
In this formula,
   - R represents the rating,
   - C represents number of completions,
   - A represents the number of passing attempts,
   - T represents the number of touchdown passes,
   - I represents the number of interceptions, and
   - Y represents the number of yards gained by passing.

   In the 2000 season, Daunte Culpepper had 297 completions, 474 passing attempts, 33 touchdown passes, 16 interceptions, and 3937 passing yards. What was his rating for that year?

Online Research **Data Update** What is the current passing rating for your favorite quarterback? Visit www.algebra1.com/data_update to get statistics on quarterbacks.

36. **CRITICAL THINKING** Write a mixture problem for the equation $1.00x + 0.28(40) = 0.40(x + 40)$.

37. **WRITING IN MATH** Answer the question that was posed at the beginning of the lesson.

How are scores calculated in a figure skating competition?

Include the following in your answer:

- an explanation of how a weighted average can be used to find a skating score, and
- a demonstration of how to find the weighted average of a skater who received a 4.9 in the short program and a 5.2 in the long program.

38. Eula Jones is investing $6000 in two accounts, part at 4.5% and the remainder at 6%. If d represents the number of dollars invested at 4.5%, which expression represents the amount of interest earned in one year by the account paying 6%?

(A) $0.06d$ (B) $0.06(d - 6000)$
(C) $0.06(d + 6000)$ (D) $0.06(6000 - d)$

39. Todd drove from Boston to Cleveland, a distance of 616 miles. His breaks, gasoline, and food stops took 2 hours. If his trip took 16 hours altogether, what was his average speed?

(A) 38.5 mph (B) 40 mph (C) 44 mph (D) 47.5 mph

Maintain Your Skills

Mixed Review

Solve each equation for the variable specified. *(Lesson 3-8)*

40. $3t - 4 = 6t - s$, for t

41. $a + 6 = \frac{b-1}{4}$, for b

State whether each percent of change is a percent of increase or a percent of decrease. Then find the percent of change. Round to the nearest whole percent. *(Lesson 3-7)*

42. original: 25
new: 14

43. original: 35
new: 42

44. original: 244
new: 300

45. If the probability that an event will occur is $\frac{2}{3}$, what are the odds that the event will occur? *(Lesson 2-6)*

Simplify each expression. *(Lesson 2-3)*

46. $(2b)(-3a)$

47. $3x(-3y) + (-6x)(-2y)$

48. $5s(-6t) + 2s(-8t)$

Name the set of numbers graphed. *(Lesson 2-1)*

49. [number line: −2 −1 0 1 2 3 4 5 6 7]

50. [number line: −1 0 1 2 3 4 5 6 7 8]

WebQuest Internet Project

Can You Fit 100 Candles on a Cake?

It's time to complete your project. Use the information and data you have gathered about living to be 100 to prepare a portfolio or Web page. Be sure to include graphs and/or tables in the presentation.

www.algebra1.com/webquest

Spreadsheet Investigation

A Follow-Up of Lesson 3-9

Finding a Weighted Average

You can use a computer spreadsheet program to calculate weighted averages. A spreadsheet allows you to make calculations and print almost anything that can be organized in a table.

The basic unit in a spreadsheet is called a **cell**. A cell may contain numbers, words, or a formula. Each cell is named by the column and row that describe its location. For example, cell B4 is in column B, row 4.

Example

Greta Norris manages the Java Roaster Coffee Shop. She has entered the price per pound and the number of pounds sold in October for each type of coffee in a spreadsheet. What was the average price per pound of coffee sold?

October Sales

	A	B	C	D
1	Product	Price per Pound	Pounds Sold	Income
2	Hawaiian Cafe	16.95	59	=B2*C2
3	Mocha Java	12.59	85	=B3*C3
4	House Blend	10.75	114	=B4*C4
5	Decaf Espresso	10.15	75	=B5*C5
6	Breakfast Blend	11.25	93	=B6*C6
7	Italian Roast	9.95	55	=B7*C7
8	Total		=SUM(C2:C7)	=SUM(D2:D7)
9	Weighted Average	=D8 / C8		
10				

Sheet1 / Sheet2 / Sheet3

The spreadsheet shows the formula that will calculate the weighted average. The formula multiplies the price of each product by its volume and calculates its sum for all the products. Then it divides that value by the sum of the volume for all products together. To the nearest cent, the weighted average of a pound of coffee is $11.75.

Exercises

For Exercises 1–4, use the spreadsheet of coffee prices.

November Sales

Product	Pounds Sold
Hawaiian Cafe	56
Mocha Java	97
House Blend	124
Decaf Espresso	71
Breakfast Blend	69
Italian Roast	45

1. What is the average price of a pound of coffee for the November sales shown in the table at the right?
2. How does the November weighted average change if all of the coffee prices are increased by $1.00?
3. How does the November weighted average change if all of the coffee prices are increased by 10%?
4. Find the weighted average of a pound of coffee if the shop sold 50 pounds of each type of coffee. How does the weighted average compare to the average of the per-pound coffee prices? Explain.

Chapter 3

Study Guide and Review

Vocabulary and Concept Check

Addition Property of Equality (p. 128)
consecutive integers (p. 144)
defining a variable (p. 121)
dimensional analysis (p. 167)
Division Property of Equality (p. 137)
equivalent equation (p. 129)
extremes (p. 156)
formula (p. 122)
four-step problem-solving plan (p. 121)
identity (p. 150)
means (p. 156)
mixture problem (p. 171)
Multiplication Property of Equality (p. 135)
multi-step equations (p. 143)
number theory (p. 144)
percent of change (p. 160)
percent of decrease (p. 160)
percent of increase (p. 160)
proportion (p. 155)
rate (p. 157)
ratio (p. 155)
scale (p. 157)
solve an equation (p. 129)
Subtraction Property of Equality (p. 129)
uniform motion problem (p. 172)
weighted average (p. 171)
work backward (p. 142)

Choose the correct term to complete each sentence.

1. According to the *(Addition, Multiplication)* Property of Equality, if $a = b$, then $a + c = b + c$.
2. A *(means, ratio)* is a comparison of two numbers by division.
3. A rate is the ratio of two measurements with *(the same, different)* units of measure.
4. The first step in the four-step problem-solving plan is to *(explore, solve)* the problem.
5. $2x + 1 = 2x + 1$ is an example of a(n) *(identity, formula)*.
6. An equivalent equation for $3x + 5 = 7$ is $(3x = 2, 3x = 12)$.
7. If the original amount was 80 and the new amount is 90, then the percent of *(decrease, increase)* is 12.5%.
8. *(Defining the variable, Dimensional analysis)* is the process of carrying units throughout a computation.
9. The *(weighted average, rate)* of a set of data is the sum of the product of each number in the set and its weight divided by the sum of all the weights.
10. An example of consecutive integers is *(8 and 9, 8 and 10)*.

Lesson-by-Lesson Review

3-1 Writing Equations

See pages 120–126.

Concept Summary

- Variables are used to represent unknowns when writing equations.
- Formulas given in sentence form can be written as algebraic equations.

Example **Translate the following sentence into an equation.**
The sum of x and y equals 2 plus two times the product of x and y.

The sum of x and y	equals	2	plus	two times the product of x and y.
$x + y$	$=$	2	$+$	$2xy$

The equation is $x + y = 2 + 2xy$.

Exercises **Translate each sentence into an equation.** *See Example 1 on page 120.*

11. Three times a number n decreased by 21 is 57.
12. Four minus three times z is equal to z decreased by 2.
13. The sum of the square of a and the cube of b is 16.
14. Translate the equation $16 - 9r = r$ into a verbal sentence. *See Example 4 on pages 122 and 123.*

3-2 Solving Equations by Using Addition and Subtraction

See pages 128–134.

Concept Summary

- **Addition Property of Equality** For any numbers a, b, and c, if $a = b$, then $a + c = b + c$.
- **Subtraction Property of Equality** For any numbers a, b, and c, if $a = b$, then $a - c = b - c$.

Example **Solve $x - 13 = 45$. Then check your solution.**

$x - 13 = 45$ Original equation

$x - 13 + 13 = 45 + 13$ Add 13 to each side.

$x = 58$ Simplify.

CHECK $x - 13 = 45$ Original equation

$58 - 13 \stackrel{?}{=} 45$ Substitute 58 for x.

$45 = 45$ ✓ Simplify. The solution is 58.

Exercises **Solve each equation. Then check your solution.**
See Examples 1–4 on pages 129 and 130.

15. $r - 21 = -37$
16. $14 + c = -5$
17. $27 = 6 + p$
18. $b + (-14) = 6$
19. $d - (-1.2) = -7.3$
20. $r + \left(-\frac{1}{2}\right) = -\frac{3}{4}$

3-3 Solving Equations by Using Multiplication and Division

See pages 135–140.

Concept Summary

- **Multiplication Property of Equality** For any numbers a, b, and c, if $a = b$, then $ac = bc$.
- **Division Property of Equality** For any numbers a, b, and c, with $c \neq 0$, if $a = b$, then $\frac{a}{c} = \frac{b}{c}$.

Example **Solve $\frac{4}{9}t = -72$.**

$\frac{4}{9}t = -72$ Original equation

$\frac{9}{4}\left(\frac{4}{9}t\right) = \frac{9}{4}(-72)$ Multiply each side by $\frac{9}{4}$.

$t = -162$ Simplify.

CHECK $\frac{4}{9}t = -72$ Original equation

$\frac{4}{9}(-162) \stackrel{?}{=} -72$ Substitute -162 for t.

$-72 = -72$ ✓ Simplify.

The solution is -162.

Exercises **Solve each equation. Then check your solution.**
See Examples 1–3 on pages 135 and 136.

21. $6x = -42$ **22.** $-7w = -49$ **23.** $\frac{3}{4}n = 30$

24. $-\frac{3}{5}y = -50$ **25.** $\frac{5}{2}a = -25$ **26.** $5 = \frac{r}{2}$

3-4 Solving Multi-Step Equations

See pages 142–148.

Concept Summary

- Multi-step equations can be solved by undoing the operations in reverse of the order of operations.

Example **Solve $34 = 8 - 2t$. Then check your solution.**

$34 = 8 - 2t$	Original equation
$34 - 8 = 8 - 2t - 8$	Subtract 8 from each side.
$26 = -2t$	Simplify.
$\frac{26}{-2} = \frac{-2t}{-2}$	Divide each side by −2.
$-13 = t$	Simplify.

CHECK

$34 = 8 - 2t$	Original equation
$34 \stackrel{?}{=} 8 - 2(-13)$	Substitute −13 for t.
$34 = 34$ ✓	The solution is −13.

Exercises **Solve each equation. Then check your solution.**
See Examples 2–4 on page 143.

27. $4p - 7 = 5$ **28.** $6 = 4v + 2$ **29.** $\frac{y}{3} + 6 = -45$

30. $\frac{c}{-4} - 8 = -42$ **31.** $\frac{4d + 5}{7} = 7$ **32.** $\frac{7n + (-1)}{8} = 8$

3-5 Solving Equations with the Variable on Each Side

See pages 149–154.

Concept Summary

Steps for Solving Equations

Step 1 Use the Distributive Property to remove the grouping symbols.

Step 2 Simplify the expressions on each side of the equals sign.

Step 3 Use the Addition and/or Subtraction Properties of Equality to get the variables on one side of the equals sign and the numbers without variables on the other side of the equals sign.

Step 4 Simplify the expressions on each side of the equals sign.

Step 5 Use the Multiplication and/or Division Properties of Equalities to solve.

Example **Solve $\frac{3}{4}q - 8 = \frac{1}{4}q + 9$.**

$\frac{3}{4}q - 8 = \frac{1}{4}q + 9$	Original equation
$\frac{3}{4}q - 8 - \frac{1}{4}q = \frac{1}{4}q + 9 - \frac{1}{4}q$	Subtract $\frac{1}{4}q$ from each side.
$\frac{1}{2}q - 8 = 9$	Simplify.
$\frac{1}{2}q - 8 + 8 = 9 + 8$	Add 8 to each side.
$\frac{1}{2}q = 17$	Simplify.
$2\left(\frac{1}{2}q\right) = 2(17)$	Multiply each side by 2.
$q = 34$	Simplify.

The solution is 34.

Exercises **Solve each equation. Then check your solution.**
See Examples 1–4 on pages 149 and 150.

33. $n - 2 = 4 - 2n$
34. $3t - 2(t + 3) = t$
35. $3 - \frac{5}{6}y = 2 + \frac{1}{6}y$
36. $\frac{x - 2}{6} = \frac{x}{2}$
37. $2(b - 3) = 3(b - 1)$
38. $8.3h - 2.2 = 6.1h - 8.8$

3-6 Ratios and Proportions

See pages 155–159.

Concept Summary

- A ratio is a comparison of two numbers by division.
- A proportion is an equation stating that two ratios are equal.
- A proportion can be solved by finding the cross products.

If $\frac{a}{b} = \frac{c}{d}$, then $ad = bc$.

Example **Solve the proportion $\frac{8}{7} = \frac{a}{1.75}$.**

$\frac{8}{7} = \frac{a}{1.75}$	Original equation
$8(1.75) = 7(a)$	Find the cross products.
$14 = 7a$	Simplify.
$\frac{14}{7} = \frac{7a}{7}$	Divide each side by 7.
$2 = a$	Simplify.

Exercises **Solve each proportion.** *See Example 3 on page 156.*

39. $\frac{6}{15} = \frac{n}{45}$
40. $\frac{x}{11} = \frac{35}{55}$
41. $\frac{12}{d} = \frac{20}{15}$
42. $\frac{14}{20} = \frac{21}{m}$
43. $\frac{2}{3} = \frac{b + 5}{9}$
44. $\frac{6}{8} = \frac{9}{s - 4}$

3-7 Percent of Change

See pages 160–164.

Concept Summary

- The proportion $\frac{\text{amount of change}}{\text{original amount}} = \frac{r}{100}$ is used to find percents of change.

Example **Find the percent of change. original: $120**
new: $114

First, subtract to find the amount of change.

$120 − $114 = $6 Note that since the new amount is less than the original, the percent of change will be a percent of decrease.

Then find the percent using the original number, 120, as the base.

change → original amount → $\frac{6}{120} = \frac{r}{100}$	Percent proportion
$6(100) = 120(r)$	Find the cross products.
$600 = 120r$	Simplify.
$\frac{600}{120} = \frac{120r}{120}$	Divide each side by 120.
$5 = r$	Simplify.

The percent of decrease is 5%.

Exercises **State whether each percent of change is a percent of increase or a percent of decrease. Then find the percent of change. Round to the nearest whole percent.** *See Example 1 on page 160.*

45. original: 40
new: 32

46. original: 50
new: 88

47. original: 35
new: 37.1

48. Find the total price of a book that costs $14.95 plus 6.25% sales tax.
See Example 3 on page 161.

49. A T-shirt priced at $12.99 is on sale for 20% off. What is the discounted price?
See Example 4 on page 161.

3-8 Solving Equations and Formulas

See pages 166–170.

Concept Summary

- For equations with more than one variable, you can solve for one of the variables by using the same steps as solving equations with one variable.

Example **Solve $\frac{x+y}{b} = c$ for x.**

$\frac{x+y}{b} = c$	Original equation
$b\left(\frac{x+y}{b}\right) = b(c)$	Multiply each side by b.
$x + y = bc$	Simplify.
$x + y - y = bc - y$	Subtract y from each side.
$x = bc - y$	Simplify.

• Extra Practice, see pages 825–828.
• Mixed Problem Solving, see page 855.

Exercises **Solve each equation or formula for the variable specified.**
See Examples 1 and 2 on pages 166 and 167.

50. $5x = y$, for x

51. $ay - b = c$, for y

52. $yx - a = cx$, for x

53. $\frac{2y - a}{3} = \frac{a + 3b}{4}$, for y

3-9 Weighted Averages

See pages 171–177.

Concept Summary

- The weighted average of a set of data is the sum of the product of each number in the set and its weight divided by the sum of all the weights.
- The formula $d = rt$ is used to solve uniform motion problems.

Example

SCIENCE **Mai Lin has a 35 milliliters of 30% solution of copper sulfate. How much of a 20% solution of copper sulfate should she add to obtain a 22% solution?**

Let x = amount of 20% solution to be added. Make a table.

	Amount of Solution (mL)	Amount of Copper Sulfate
30% Solution	35	0.30(35)
20% Solution	x	$0.20x$
22% Solution	$35 + x$	$0.22(35 + x)$

$0.30(35) + 0.20x = 0.22(35 + x)$ Write and solve an equation.

$10.5 + 0.20x = 7.7 + 0.22x$ Distributive Property

$10.5 + 0.20x - 0.20x = 7.7 + 0.22x - 0.20x$ Subtract $0.20x$ fom each side.

$10.5 = 7.7 + 0.02x$ Simplify.

$10.5 - 7.7 = 7.7 + 0.02x - 7.7$ Subtract 7.7 from each side.

$2.8 = 0.02x$ Simplify.

$\frac{2.8}{0.02} = \frac{0.02x}{0.02}$ Divide each side by 0.02.

$140 = x$ Simplify.

Mai Lin should add 140 milliliters of the 20% solution.

Exercises

54. **COFFEE** Ms. Anthony wants to create a special blend using two coffees, one priced at $8.40 per pound and the other at $7.28 per pound. How many pounds of the $7.28 coffee should she mix with 9 pounds of the $8.40 coffee to sell the mixture for $7.95 per pound? *See Example 1 on page 171.*

55. **TRAVEL** Two airplanes leave Dallas at the same time and fly in opposite directions. One airplane travels 80 miles per hour faster than the other. After three hours, they are 2940 miles apart. What is the speed of each airplane? *See Example 3 on pages 172 and 173.*

Chapter 3 Practice Test

Vocabulary and Concepts

Choose the correct term to complete each sentence.

1. The study of numbers and the relationships between them is called *(consecutive, number)* theory.
2. An equation that is true for *(every, only one)* value of the variable is called an identity.
3. When a new number is *(greater than, less than)* the original number, the percent of change is called a percent of increase.

Skills and Applications

Translate each sentence into an equation.

4. The sum of twice x and three times y is equal to thirteen.
5. Two thirds of a number is negative eight fifths.

Solve each equation. Then check your solution.

6. $-15 + k = 8$
7. $-1.2x = 7.2$
8. $k - 16 = -21$
9. $\frac{t - 7}{4} = 11$
10. $\frac{3}{4}y = -27$
11. $-12 = 7 - \frac{y}{3}$
12. $t - (-3.4) = -5.3$
13. $-3(x + 5) = 8x + 18$
14. $5a = 125$
15. $\frac{r}{5} - 3 = \frac{2r}{5} + 16$
16. $0.1r = 19$
17. $-\frac{2}{3}z = -\frac{4}{9}$
18. $-w + 11 = 4.6$
19. $2p + 1 = 5p - 11$
20. $25 - 7w = 46$

Solve each proportion.

21. $\frac{36}{t} = \frac{9}{11}$
22. $\frac{n}{4} = \frac{3.25}{52}$
23. $\frac{5}{12} = \frac{10}{x - 1}$

State whether each percent of change is a percent of increase or a percent of decrease. Then find the percent of change. Round to the nearest whole percent.

24. original: 45
new: 9
25. original: 12
new: 20

Solve each equation or formula for the variable specified.

26. $h = at - 0.25vt^2$, for a
27. $a(y + 1) = b$, for y

28. **SALES** Suppose the Central Perk coffee shop sells a cup of espresso for \$2.00 and a cup of cappuccino for \$2.50. On Friday, Destiny sold 30 more cups of cappuccino than espresso for a total of \$178.50 worth of espresso and cappuccino. How many cups of each were sold?

29. **BOATING** *The Yankee Clipper* leaves the pier at 9:00 A.M. at 8 knots (nautical miles per hour). A half hour later, *The River Rover* leaves the same pier in the same direction traveling at 10 knots. At what time will *The River Rover* overtake *The Yankee Clipper*?

30. **STANDARDIZED TEST PRACTICE** If $\frac{4}{5}$ of $\frac{3}{4} = \frac{2}{5}$ of $\frac{x}{4}$, find the value of x.

(A) 12 (B) 6 (C) 3 (D) $\frac{3}{2}$

www.algebra1.com/chapter_test

Chapter 3 Standardized Test Practice

Part 1 Multiple Choice

Record your answers on the answer sheet provided by your teacher or on a sheet of paper.

1. Bailey planted a rectangular garden that is 6 feet wide by 15 feet long. What is the perimeter of the garden? (Prerequisite Skill)

(A) 21 ft (B) 27 ft

(C) 42 ft (D) 90 ft

2. Which of the following is true about 65 percent of 20? (Prerequisite Skill)

(A) It is greater than 20.

(B) It is less than 10.

(C) It is less than 20.

(D) Can't tell from the information given

3. For a science project, Kelsey measured the height of a plant grown from seed. She made the bar graph below to show the height of the plant at the end of each week. Which is the most reasonable estimate of the plant's height at the end of the sixth week? (Lesson 1-8)

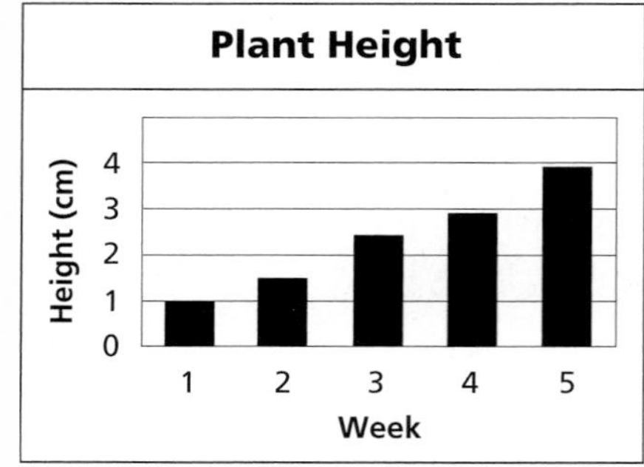

(A) 2 to 3.5 cm (B) 4 to 5.5 cm

(C) 6 to 7 cm (D) 8 to 8.5 cm

4. WEAT predicted a 25% chance of snow. WFOR said the chance was 1 in 4. Myweather.com showed the chance of snow as $\frac{1}{5}$, and Allweather.com listed the chance as 0.3. Which forecast predicted the greatest chance of snow? (Lesson 2-7)

(A) WEAT (B) WFOR

(C) Myweather.com (D) Allweather.com

5. Amber owns a business that transfers photos to CD-ROMs. She charges her customers \$24.95 for each CD-ROM. Her expenses include \$575 for equipment and \$0.80 for each blank CD-ROM. Which of these equations could be used to calculate her profit p for creating n CD-ROMs? (Lesson 3-1)

(A) $p = (24.95 - 0.8)n - 575$

(B) $p = (24.95 + 0.8)n + 575$

(C) $p = 24.95n - 574.2$

(D) $p = 24.95n + 575$

6. Which of the following equations has the same solution as $8(x + 2) = 12$? (Lesson 3-4)

(A) $8x + 2 = 12$

(B) $x + 2 = 4$

(C) $8x = 10$

(D) $2x + 4 = 3$

7. Eduardo is buying pizza toppings for a birthday party. His recipe uses 8 ounces of shredded cheese for 6 servings. How many ounces of cheese are needed for 27 servings? (Lesson 3-6)

(A) 27 (B) 32

(C) 36 (D) 162

8. The sum of x and $\frac{1}{y}$ is 0, and y does not equal 0. Which of the following is true? (Lesson 3-8)

(A) $x = -y$ (B) $\frac{x}{y} = 0$

(C) $x = 1 - y$ (D) $x = -\frac{1}{y}$

Test-Taking Tip

Questions 2, 6, 8 Always read every answer choice, particularly in questions that ask, "Which of the following is true?"

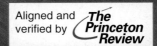

Part 2 Short Response/Grid In

Record your answers on the answer sheet provided by your teacher or on a sheet of paper.

9. Let $x = 2$ and $y = -3$. Find the value of $\frac{x(xy + 5)}{4}$. (Lesson 1-2)

10. Use the formula $F = \frac{9}{5}C + 32$ to convert temperatures from Celsius (C) to Fahrenheit (F). If it is $-5°$ Celsius, what is the temperature in degrees Fahrenheit? (Lesson 2-3)

11. Darnell keeps his cotton socks folded in pairs in his drawer. Five pairs are black, 2 pairs are navy, and 1 pair is brown. In the dark, he pulls out one pair at random. What are the odds that it is black? (Lesson 2-6)

12. The sum of the ages of the Kruger sisters is 39. Their ages can be represented as three consecutive integers. What is the age of the middle sister? (Lesson 3-4)

13. On a car trip, Tyson drove 65 miles more than half the number of miles Pete drove. Together they drove 500 miles. How many miles did Tyson drive? (Lesson 3-4)

14. Solve $7(x + 2) + 4(2x - 3) = 47$ for x. (Lesson 3-5)

15. A bookshop sells used hardcover books with a 45% discount. The price of a book was \$22.95 when it was new. What is the discounted price for that book? (Lesson 3-7)

Part 3 Quantitative Comparison

Compare the quantity in Column A and the quantity in Column B. Then determine whether:

- Ⓐ **the quantity in Column A is greater,**
- Ⓑ **the quantity in Column B is greater,**
- Ⓒ **the two quantities are equal, or**
- Ⓓ **the relationship cannot be determined from the information given.**

	Column A	Column B
16.	$\lvert a \rvert$	$\lvert -a \rvert$
17.	solution of $3x + 7 = 10$	solution of $4y - 2 = 6$
18.	the percent of increase from \$75 to \$100	the percent of increase from \$150 to \$200

(16: Lesson 2-1) (17: Lesson 3-4) (18: Lesson 3-7)

Part 4 Open Ended

Record your answers on a sheet of paper.

19. Kirby's pickup truck travels at a rate of 6 miles every 10 minutes. Nola's SUV travels at a rate of 15 miles every 25 minutes. The speed limit on this street is 40 mph. Is either vehicle or are both vehicles exceeding the speed limit? Explain. (Lesson 3-6)

20. A chemist has one solution of citric acid that is 20% acid and another solution of citric acid that is 80% acid. She plans to mix these solutions together to make 200 liters of a solution that is 50% acid. (Lesson 3-9)

a. Complete the table to show the liters of 20% and 80% solutions that will be used to make the 50% solution. Use x to represent the number of liters of the 80% solution that will be used to make the 50% solution.

	Liters of Solution	Liters of Acid
20% Solution		
80% Solution	x	
50% Solution	200	0.50(200)

b. Write an equation that represents the number of liters of acid in the solution.

c. How many liters of the 20% solution and how many of the 80% solution will the chemist need to mix together to make 200 liters of a 50% solution?

www.algebra1.com/standardized_test

UNIT 2

Linear Functions

Many real-world situations such as Olympic race times can be represented using functions. In this unit, you will learn about linear functions and equations.

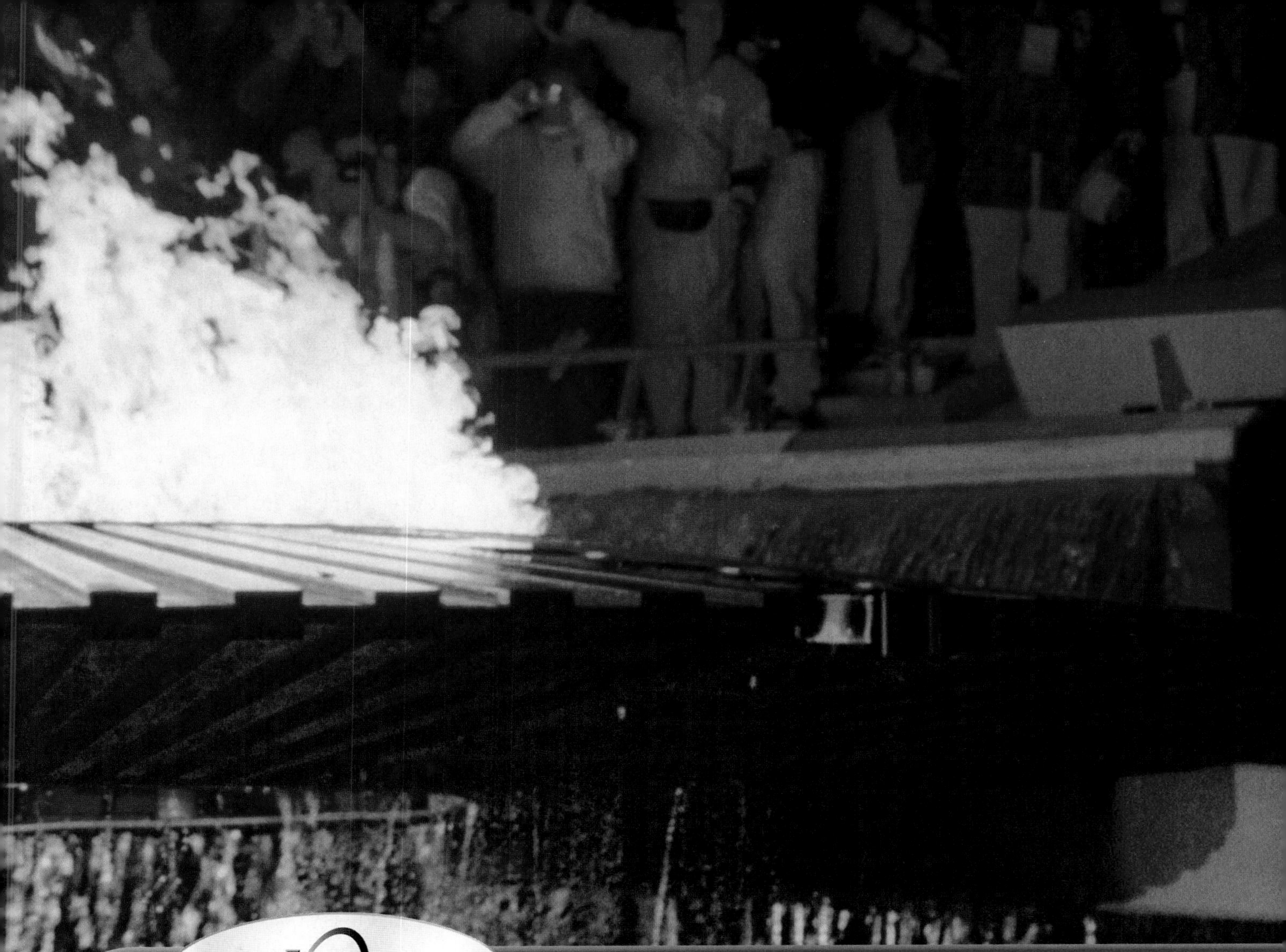

WebQuest Internet Project

The Spirit of the Games

The first Olympic Games featured only one event— a foot race. The 2004 Games will include thousands of competitors in about 300 events. In this project, you will explore how linear functions can be illustrated by the Olympics.

Log on to www.algebra1.com/webquest. Begin your WebQuest by reading the Task.

Then continue working on your WebQuest as you study Unit 2.

Lesson	4-6	5-7	6-6	7-1
Page	230	304	357	373

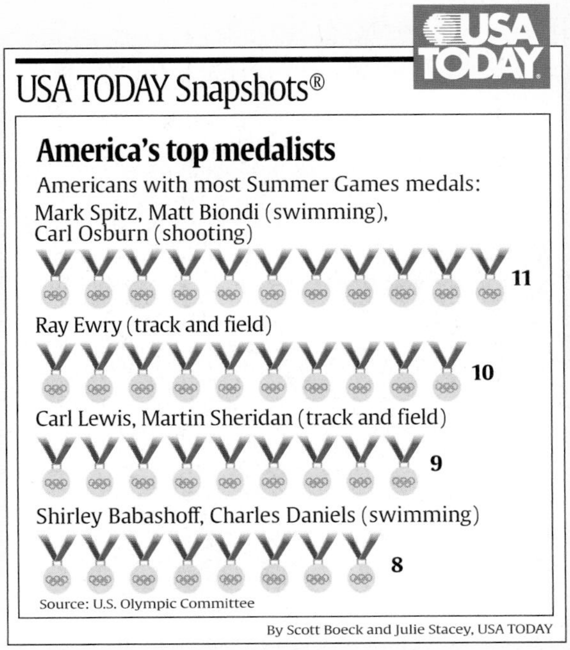

America's top medalists

Americans with most Summer Games medals:

Mark Spitz, Matt Biondi (swimming), Carl Osburn (shooting) 11

Ray Ewry (track and field) 10

Carl Lewis, Martin Sheridan (track and field) 9

Shirley Babashoff, Charles Daniels (swimming) 8

Source: U.S. Olympic Committee

By Scott Boeck and Julie Stacey, USA TODAY

Chapter 4 Graphing Relations and Functions

What You'll Learn

- **Lessons 4-1, 4-4, and 4-5** Graph ordered pairs, relations, and equations.
- **Lesson 4-2** Transform figures on a coordinate plane.
- **Lesson 4-3** Find the inverse of a relation.
- **Lesson 4-6** Determine whether a relation is a function.
- **Lessons 4-7 and 4-8** Look for patterns and write formulas for sequences.

Key Vocabulary

- coordinate plane (p. 192)
- transformation (p. 197)
- inverse (p. 206)
- function (p. 226)
- arithmetic sequence (p. 233)

Why It's Important

The concept of a function is used throughout higher mathematics, from algebra to calculus. A function is a rule or a formula. You can use a function to describe real-world situations like converting between currencies. For example, if you are in Mexico, you can calculate that an item that costs 100 pesos is equivalent to about 11 U.S. dollars. *You will learn how to convert different currencies in Lesson 4-4.*

Getting Started

Prerequisite Skills To be successful in this chapter, you'll need to master these skills and be able to apply them in problem-solving situations. Review these skills before beginning Chapter 4.

For Lesson 4-1 — Graph Real Numbers

Graph each set of numbers. *(For review, see Lesson 2-1.)*

1. $\{1, 3, 5, 7\}$ **2.** $\{-3, 0, 1, 4\}$ **3.** $\{-8, -5, -2, 1\}$ **4.** $\left\{\frac{1}{2}, 1, 1\frac{1}{2}, 2\right\}$

For Lesson 4-2 — Distributive Property

Rewrite each expression using the Distributive Property. *(For review, see Lesson 1-5.)*

5. $3(7 - t)$ **6.** $-4(w + 2)$ **7.** $-5(3b - 2)$ **8.** $\frac{1}{2}(2z + 4)$

For Lessons 4-4 and 4-5 — Solve Equations for a Specific Variable

Solve each equation for y. *(For review, see Lesson 3-8.)*

9. $2x + y = 1$ **10.** $x = 8 - y$ **11.** $6x - 3y = 12$

12. $2x + 3y = 9$ **13.** $9 - \frac{1}{2}y = 4x$ **14.** $\frac{y + 5}{3} = x + 2$

For Lesson 4-6 — Evaluate Expressions

Evaluate each expression if $a = -1$, $b = 4$, and $c = -3$. *(For review, see Lesson 2-3.)*

15. $a + b - c$ **16.** $2c - b$ **17.** $c - 3a$

18. $3a - 6b - 2c$ **19.** $8a + \frac{1}{2}b - 3c$ **20.** $6a + 8b + \frac{2}{3}c$

Make this Foldable to help you organize your notes about graphing relations and functions. Begin with four sheets of grid paper.

Step 1 Fold

Fold each sheet of grid paper in half from top to bottom.

Step 2 Cut and Staple

Cut along fold. Staple the eight half-sheets together to form a booklet.

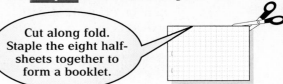

Step 3 Cut Tabs into Margin

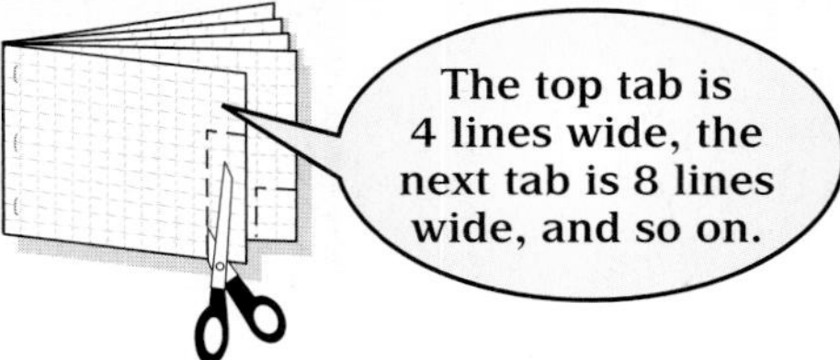

The top tab is 4 lines wide, the next tab is 8 lines wide, and so on.

Step 4 Label

The Coordinate Plane
4-1
4-2
Label each of the tabs with a lesson number.

Reading and Writing As you read and study the chapter, use each page to write notes and to graph examples.

4-1 The Coordinate Plane

What You'll Learn

- Locate points on the coordinate plane.
- Graph points on a coordinate plane.

How do archaeologists use coordinate systems?

Underwater archaeologists use a grid system to map excavation sites of sunken ships. The grid is used as a point of reference on the ocean floor. The coordinate system is also used to record the location of objects they find. Knowing the position of each object helps archaeologists reconstruct how the ship sank and where to find other artifacts.

Vocabulary

- axes
- origin
- coordinate plane
- *y*-axis
- *x*-axis
- *x*-coordinate
- *y*-coordinate
- quadrant
- graph

IDENTIFY POINTS In mathematics, points are located in reference to two perpendicular number lines called **axes**.

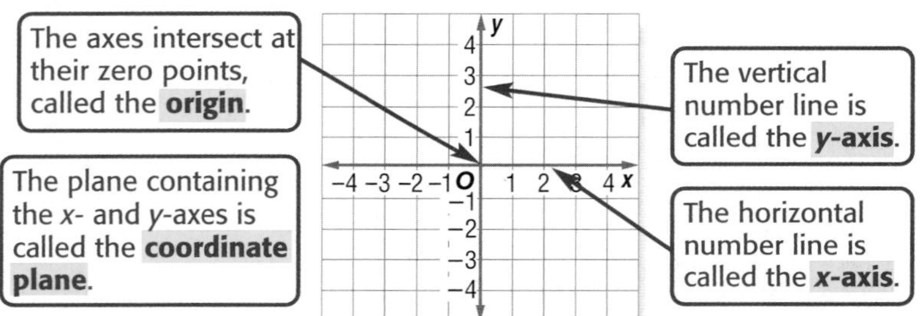
The axes intersect at their zero points, called the **origin**.
The plane containing the *x*- and *y*-axes is called the **coordinate plane**.
The vertical number line is called the **y-axis**.
The horizontal number line is called the **x-axis**.
y
4
3
2
1
−4 −3 −2 −1 O 1 2 3 4 x
−1
−2
−3
−4

Study Tip

Reading Math
The *x*-coordinate is called the *abscissa*. The *y*-coordinate is called the *ordinate*.

Points in the coordinate plane are named by ordered pairs of the form (x, y). The first number, or **x-coordinate**, corresponds to the numbers on the x-axis. The second number, or **y-coordinate**, corresponds to the numbers on the y-axis. The origin, labeled O, has coordinates $(0, 0)$.

Example 1 Name an Ordered Pair

Write the ordered pair for point G.

- Follow along a vertical line through the point to find the x-coordinate on the x-axis. The x-coordinate is -4.
- Follow along a horizontal line through the point to find the y-coordinate on the y-axis. The y-coordinate is 3.
- So, the ordered pair for point G is $(-4, 3)$. This can also be written as $G(-4, 3)$.

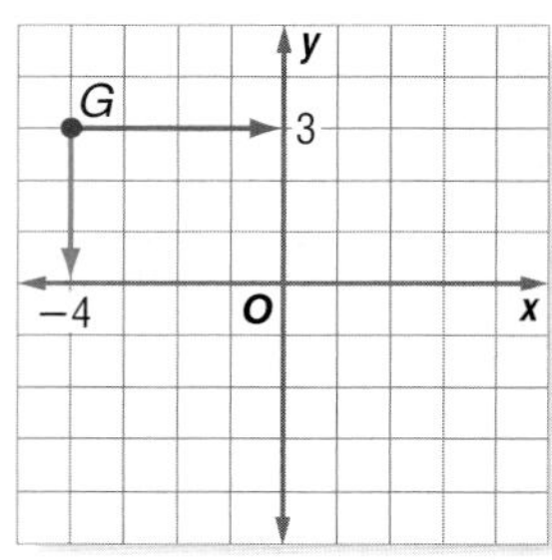
y
G
3
−4
O
x

Unless marked otherwise, you can assume that each division on the axes represents 1 unit.

The x-axis and y-axis separate the coordinate plane into four regions, called **quadrants**. Notice which quadrants contain positive and negative x-coordinates and which quadrants contain positive and negative y-coordinates. The axes are not located in any of the quadrants.

Quadrant II (−, +) | Quadrant I (+, +)
Quadrant III (−, −) | Quadrant IV (+, −)

Example 2 Identify Quadrants

Write ordered pairs for points *A*, *B*, *C*, and *D*. Name the quadrant in which each point is located.

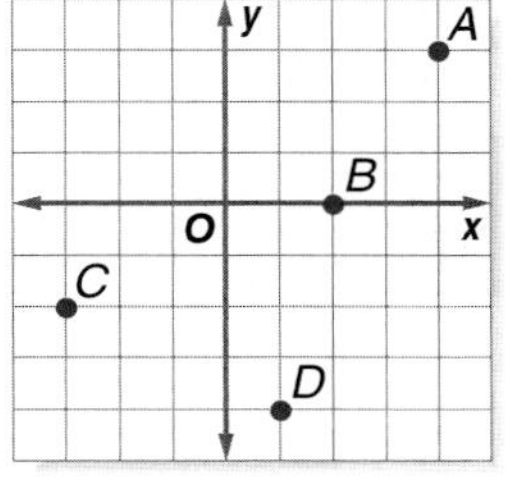

Use a table to help find the coordinates of each point.

Point	x-Coordinate	y-Coordinate	Ordered Pair	Quadrant
A	4	3	(4, 3)	I
B	2	0	(2, 0)	none
C	−3	−2	(−3, −2)	III
D	1	−4	(1, −4)	IV

GRAPH POINTS To **graph** an ordered pair means to draw a dot at the point on the coordinate plane that corresponds to the ordered pair. This is sometimes called *plotting a point*. When graphing an ordered pair, start at the origin. The x-coordinate indicates how many units to move right (positive) or left (negative). The y-coordinate indicates how many units to move up (positive) or down (negative).

Example 3 Graph Points

Plot each point on a coordinate plane.

a. $R(-4, 1)$

- Start at the origin.
- Move left 4 units since the x-coordinate is -4.
- Move up 1 unit since the y-coordinate is 1.
- Draw a dot and label it R.

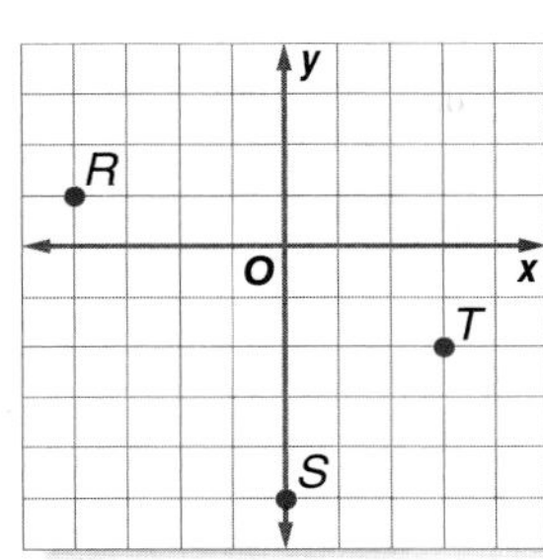

b. $S(0, -5)$

- Start at the origin.
- Since the x-coordinate is 0, the point will be located on the y-axis.
- Move down 5 units.
- Draw a dot and label it S.

c. $T(3, -2)$

- Start at the origin.
- Move right 3 units and down 2 units.
- Draw a dot and label it T.

More About...

Geography

The prime meridian, 0° longitude, passes through London's Greenwich Observatory. A metal marker indicates its exact location.

Source: www.encarta.msn.com

Example 4 Use a Coordinate System

GEOGRAPHY Latitude and longitude lines form a system of coordinates to designate locations on Earth. Latitude lines run east and west and are the first coordinate of the ordered pairs. Longitude lines run north and south and are the second coordinate of the ordered pairs.

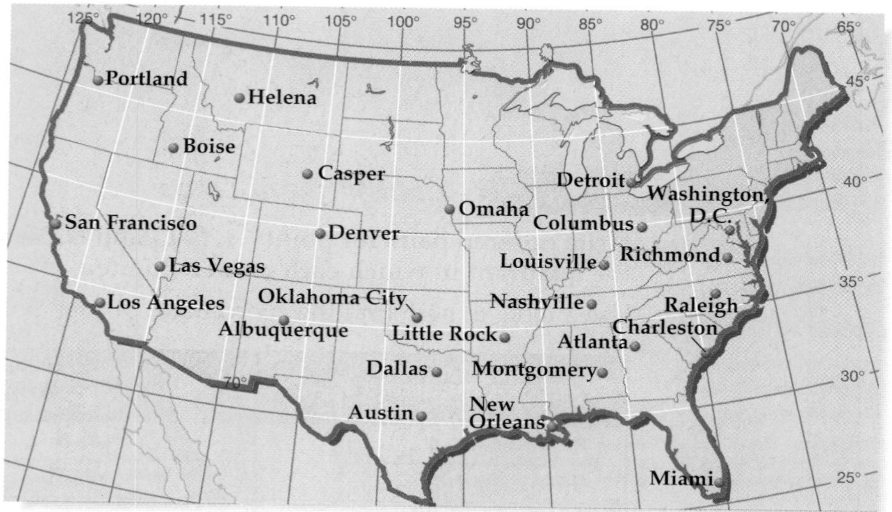

a. Name the city at (40°, 105°).
Locate the latitude line at 40°. Follow the line until it intersects with the longitude line at 105°. The city is Denver.

b. Estimate the latitude and longitude of Washington, D.C.
Locate Washington, D.C., on the map. It is close to 40° latitude and 75° longitude. There are 5° between each line, so a good estimate is 39° for the latitude and 77° for the longitude.

Check for Understanding

Concept Check

1. **Draw** a coordinate plane. Label the origin, x-axis, y-axis, and the quadrants.
2. **Explain** why $(-1, 4)$ does not name the same point as $(4, -1)$.
3. **OPEN ENDED** Give the coordinates of a point for each quadrant in the coordinate plane.

Guided Practice

Write the ordered pair for each point shown at the right. Name the quadrant in which the point is located.

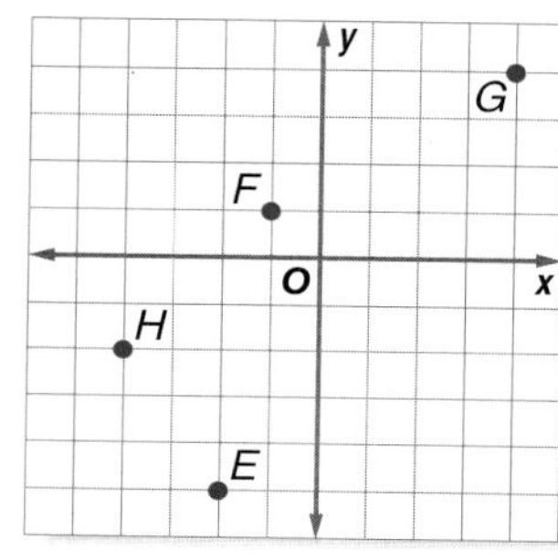

4. E
5. F
6. G
7. H

Plot each point on a coordinate plane.

8. $J(2, 5)$
9. $K(-1, 4)$
10. $L(0, -3)$
11. $M(-2, -2)$

Application

12. **ARCHITECTURE** Chun Wei has sketched the southern view of a building. If A is located on a coordinate system at $(-40, 10)$, locate the coordinates of the other vertices.

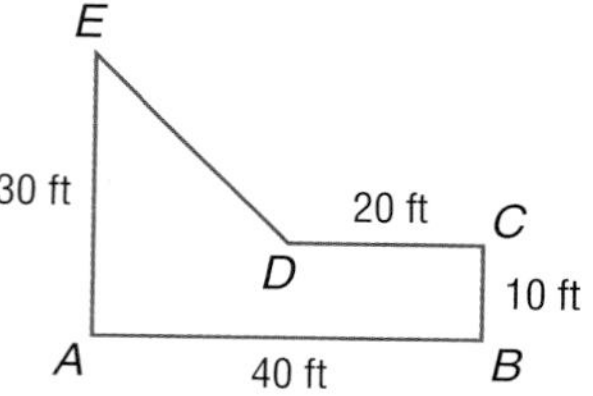

Practice and Apply

Homework Help

For Exercises	See Examples
13–24, 39	1, 2
25–36	3
37, 38, 40–43	4

Extra Practice

See page 828.

Write the ordered pair for each point shown at the right. Name the quadrant in which the point is located.

13. N 14. P
15. Q 16. R
17. S 18. T
19. U 20. V
21. W 22. Z

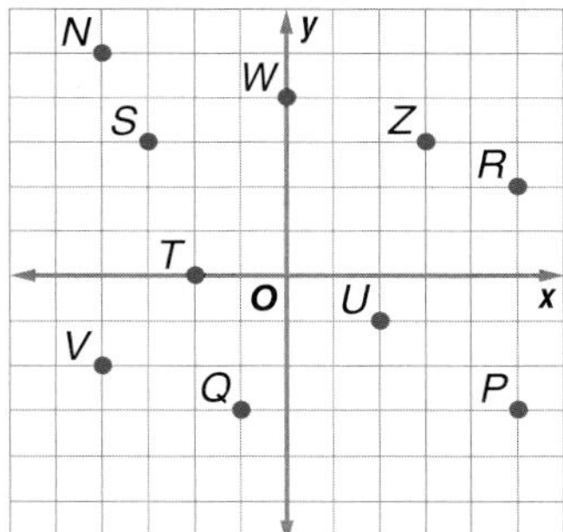

23. Write the ordered pair that describes a point 12 units down from and 7 units to the right of the origin.

24. Write the ordered pair for a point that is 9 units to the left of the origin and lies on the x-axis.

Plot each point on a coordinate plane.

25. $A(3, 5)$ 26. $B(-2, 2)$ 27. $C(4, -2)$ 28. $D(0, -1)$
29. $E(-2, 5)$ 30. $F(-3, -4)$ 31. $G(4, 4)$ 32. $H(-4, 4)$
33. $I(3, 1)$ 34. $J(-1, -3)$ 35. $K(-4, 0)$ 36. $L(2, -4)$

GEOGRAPHY For Exercises 37 and 38, use the map on page 194.

37. Name two cities that have approximately the same latitude.
38. Name two cities that have approximately the same longitude.

39. **ARCHAEOLOGY** The diagram at the right shows the positions of artifacts found on the ocean floor. Write the coordinates of the location for each object: coins, plate, goblet, and vase.

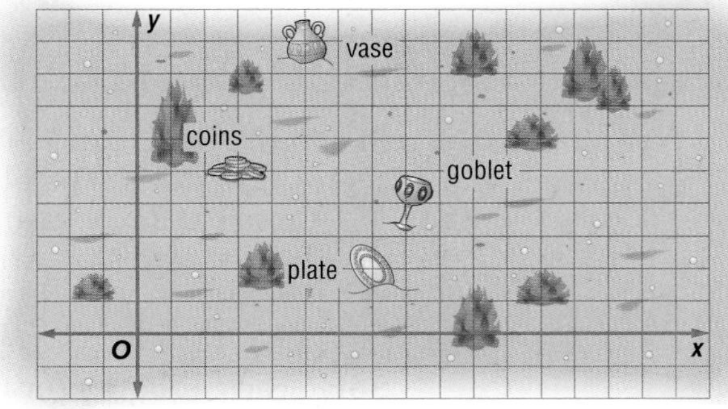

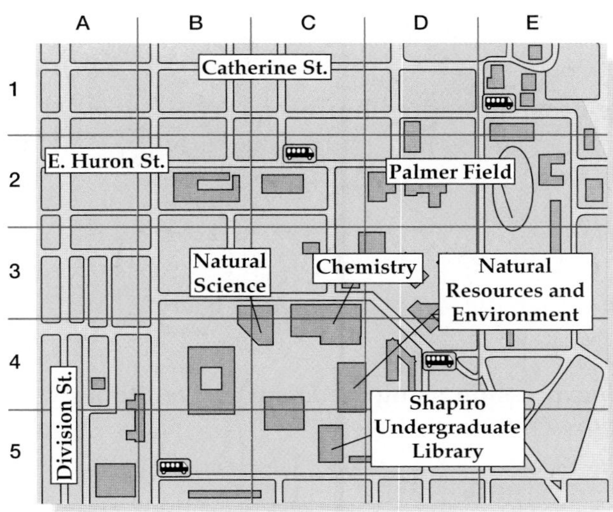

MAPS For Exercises 40–43, use the map at the left. On many maps, letters and numbers are used to define a region or sector. For example, Palmer Field is located in sector E2. Rogelio is a guide for new students at the University of Michigan. He has selected campus landmarks to show the students.

40. In what sector is the Undergraduate Library?
41. In what sector are most of the science buildings?
42. Which street goes from sector (A, 2) to (D, 2)?
43. Name the sectors that have bus stops.

44. **CRITICAL THINKING** Describe the possible locations, in terms of quadrants or axes, for the graph of (x, y) given each condition.
    a. $xy > 0$ b. $xy < 0$ c. $xy = 0$

45. **WRITING IN MATH** Answer the question that was posed at the beginning of the lesson.

How do archaeologists use coordinate systems?

Include the following in your answer:

- an explanation of how dividing an excavation site into sectors can be helpful in excavating a site, and
- a reason why recording the exact location of an artifact is important.

For Exercises 46 and 47, refer to the figure at the right.

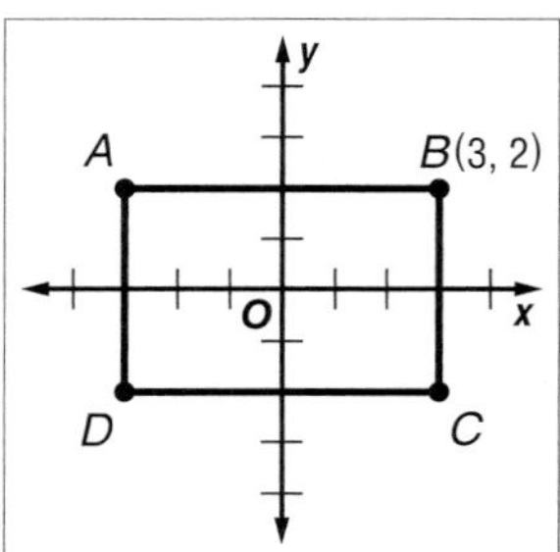

46. *ABCD* is a rectangle with its center at the origin. If the coordinates of vertex *B* are (3, 2), what are the coordinates of vertex *A*?

Ⓐ (−3, −2) Ⓑ (3, −2)
Ⓒ (−3, 2) Ⓓ (3, 2)

47. What is the length of $\overline{AD}$?

Ⓐ 6 units Ⓑ 4 units
Ⓒ 5 units Ⓓ 3 units

Extending the Lesson

The midpoint of a line segment is the point that lies exactly halfway between the two endpoints. The midpoint of a line segment whose endpoints are at (*a*, *b*) and (*c*, *d*) is at $\left(\frac{a+c}{2}, \frac{b+d}{2}\right)$. Find the midpoint of each line segment whose endpoints are given.

48. (7, 1) and (−3, 1)
49. (5, −2) and (9, −8)
50. (−4, 4) and (4, −4)

Maintain Your Skills

Mixed Review

51. **AIRPLANES** At 1:30 P.M., an airplane leaves Tucson for Baltimore, a distance of 2240 miles. The plane flies at 280 miles per hour. A second airplane leaves Tucson at 2:15 P.M. and is scheduled to land in Baltimore 15 minutes before the first airplane. At what rate must the second airplane travel to arrive on schedule? *(Lesson 3-9)*

Solve each equation or formula for the variable specified. *(Lesson 3-8)*

52. $3x + b = 2x + 5$ for x
53. $10c = 2(2d + 3c)$ for d
54. $6w - 3h = b$ for h
55. $\frac{3(a-t)}{4} = 2t$ for t

Find each square root. Round to the nearest hundredth if necessary. *(Lesson 2-7)*

56. $-\sqrt{81}$
57. $\sqrt{63}$
58. $\sqrt{180}$
59. $-\sqrt{256}$

Evaluate each expression. *(Lesson 2-1)*

60. $52 + |18 - 7|$
61. $|81 - 47| + 17$
62. $42 - |60 - 74|$
63. $36 - |15 - 21|$
64. $|10 - 16 + 27|$
65. $|38 - 65 - 21|$

Getting Ready for the Next Lesson

PREREQUISITE SKILL **Rewrite each expression using the Distributive Property. Then simplify.** *(To review the **Distributive Property**, see Lesson 1-5.)*

66. $4(x + y)$
67. $-1(x + 3)$
68. $3(1 - 6y)$
69. $-3(2x - 5)$
70. $\frac{1}{3}(2x + 6y)$
71. $\frac{1}{4}(5x - 2y)$

4-2 Transformations on the Coordinate Plane

What You'll Learn

- Transform figures by using reflections, translations, dilations, and rotations.
- Transform figures on a coordinate plane by using reflections, translations, dilations, and rotations.

Vocabulary

- transformation
- preimage
- image
- reflection
- translation
- dilation
- rotation

How are transformations used in computer graphics?

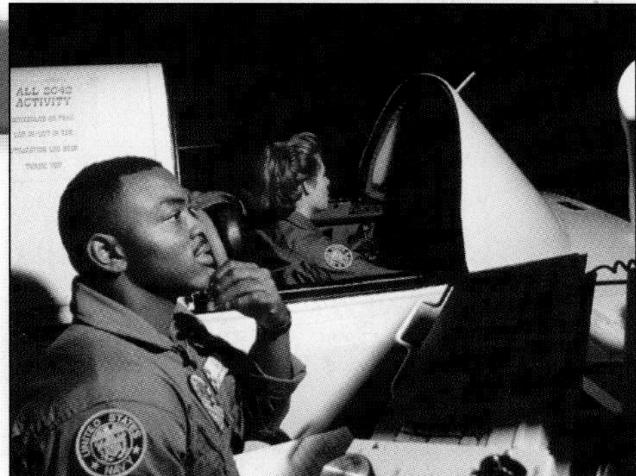

Computer programs can create movements that mimic real-life situations. A new CD-ROM-based flight simulator replicates an actual flight experience so closely that the U.S. Navy is using it for all of their student aviators. The movements of the on-screen graphics are accomplished by using mathematical transformations.

TRANSFORM FIGURES **Transformations** are movements of geometric figures. The **preimage** is the position of the figure before the transformation, and the **image** is the position of the figure after the transformation.

reflection
a figure is flipped over a line

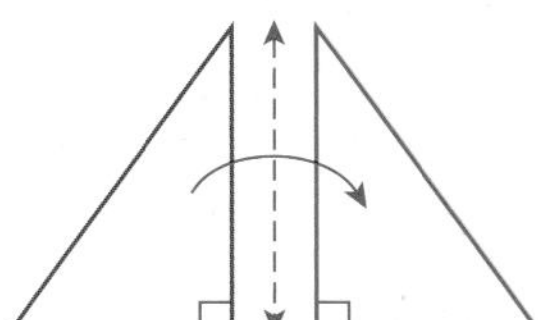
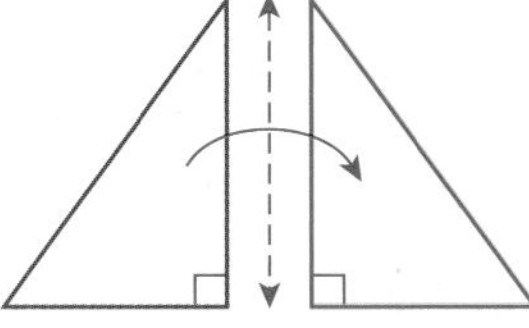

translation
a figure is slid in any direction

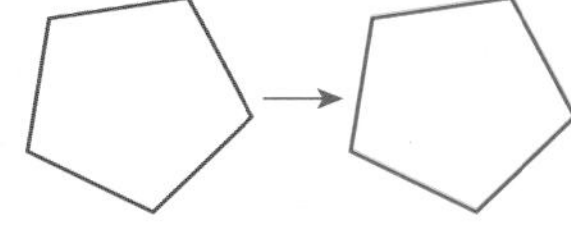

dilation
a figure is enlarged or reduced

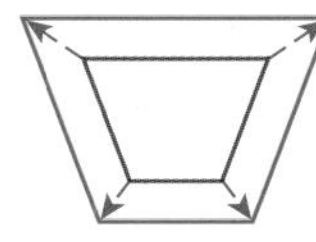

rotation
a figure is turned around a point

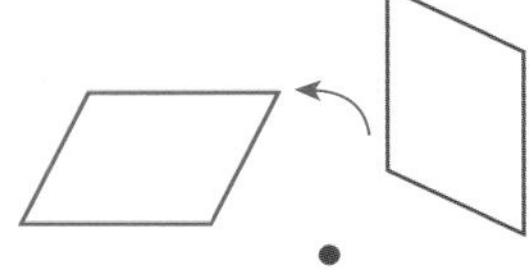

Example 1 Identify Transformations

Identify each transformation as a *reflection, translation, dilation,* or *rotation.*

a.

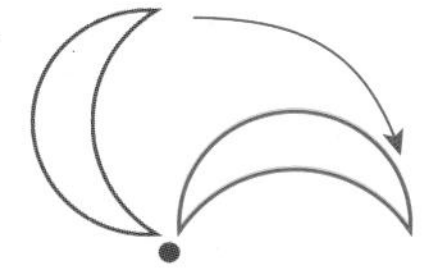

b.

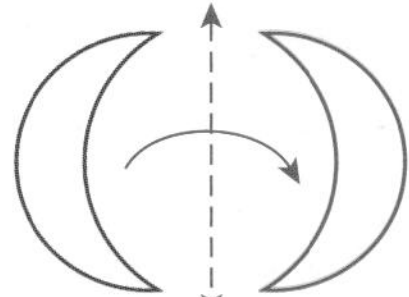

c.

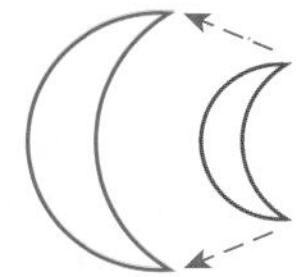

d.

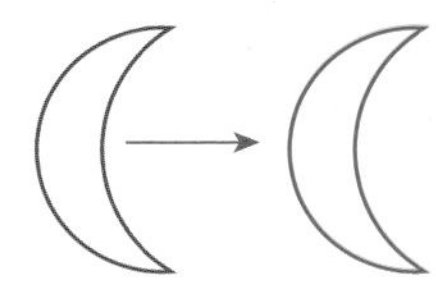

a. The figure has been turned around a point. This is a rotation.

b. The figure has been flipped over a line. This is a reflection.

c. The figure has been increased in size. This is a dilation.

d. The figure has been shifted horizontally to the right. This is a translation.

TRANSFORM FIGURES ON THE COORDINATE PLANE You can perform transformations on a coordinate plane by changing the coordinates of the points on a figure. The points on the translated figure are indicated by the prime symbol ′ to distinguish them from the original points.

Key Concept *Transformations on the Coordinate Plane*

Name	Words	Symbols	Model
Reflection	To reflect a point over the x-axis, multiply the y-coordinate by -1.	reflection over x-axis: $(x, y) \rightarrow (x, -y)$	$(-x, y)$, (x, y), O, x, y, $(x, -y)$
	To reflect a point over the y-axis, multiply the x-coordinate by -1.	reflection over y-axis: $(x, y) \rightarrow (-x, y)$	
Translation	To translate a point by an ordered pair (a, b), add a to the x-coordinate and b to the y-coordinate.	$(x, y) \rightarrow (x + a, y + b)$	y, $(x + a, y + b)$, (x, y), O
Dilation	To dilate a figure by a scale factor k, multiply both coordinates by k. If $k > 1$, the figure is enlarged. If $0 < k < 1$, the figure is reduced.	$(x, y) \rightarrow (kx, ky)$	y, (kx, ky), (x, y), O, x
Rotation	To rotate a figure 90° *counterclockwise* about the origin, switch the coordinates of each point and then multiply the new first coordinate by -1.	90° rotation: $(x, y) \rightarrow (-y, x)$	y, $(-y, x)$, (x, y), O, x, $(-x, -y)$
	To rotate a figure 180° about the origin, multiply both coordinates of each point by -1.	180° rotation: $(x, y) \rightarrow (-x, -y)$	

Study Tip

Reading Math
The *vertices* of a polygon are the endpoints of the angles.

Example 2 Reflection

A parallelogram has vertices $A(-4, 3)$, $B(1, 3)$, $C(0, 1)$, and $D(-5, 1)$.

a. Parallelogram $ABCD$ is reflected over the x-axis. Find the coordinates of the vertices of the image.

To reflect the figure over the x-axis, multiply each y-coordinate by -1.

$(x, y) \rightarrow (x, -y)$	$(x, y) \rightarrow (x, -y)$
$A(-4, 3) \rightarrow A'(-4, -3)$	$C(0, 1) \rightarrow C'(0, -1)$
$B(1, 3) \rightarrow B'(1, -3)$	$D(-5, 1) \rightarrow D'(-5, -1)$

The coordinates of the vertices of the image are $A'(-4, -3)$, $B'(1, -3)$, $C'(0, -1)$, and $D'(-5, -1)$.

Study Tip

Reading Math
Parallelogram $ABCD$ and its image $A'B'C'D'$ are said to be **symmetric** about the x-axis. The x-axis is called the **line of symmetry**.

b. Graph parallelogram $ABCD$ and its image $A'B'C'D'$.

Graph each vertex of the parallelogram $ABCD$. Connect the points.

Graph each vertex of the reflected image $A'B'C'D'$. Connect the points.

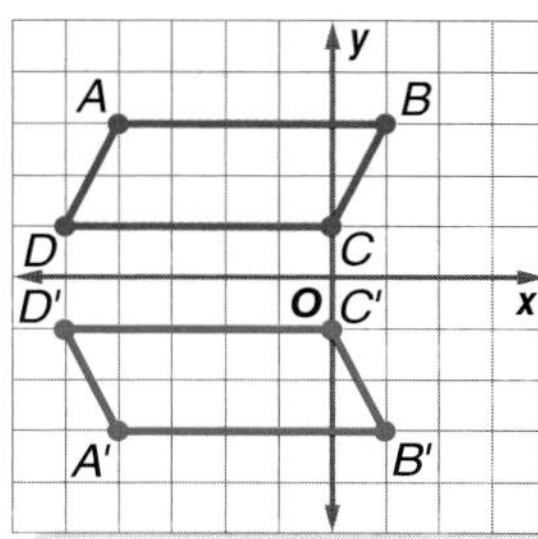

Example 3 Translation

Triangle ABC has vertices $A(-2, 3)$, $B(4, 0)$, and $C(2, -5)$.

a. Find the coordinates of the vertices of the image if it is translated 3 units to the left and 2 units down.

To translate the triangle 3 units to the left, add -3 to the x-coordinate of each vertex. To translate the triangle 2 units down, add -2 to the y-coordinate of each vertex.

$$(x, y) \rightarrow (x - 3, y - 2)$$

$$A(-2, 3) \rightarrow A'(-2 - 3, 3 - 2) \rightarrow A'(-5, 1)$$

$$B(4, 0) \rightarrow B'(4 - 3, 0 - 2) \rightarrow B'(1, -2)$$

$$C(2, -5) \rightarrow C'(2 - 3, -5 - 2) \rightarrow C'(-1, -7)$$

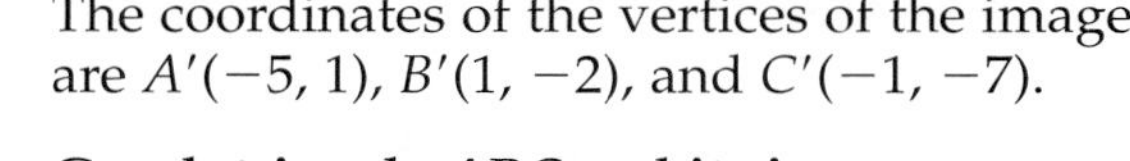
The coordinates of the vertices of the image are $A'(-5, 1)$, $B'(1, -2)$, and $C'(-1, -7)$.

b. Graph triangle ABC and its image.

The preimage is $\triangle ABC$.

The translated image is $\triangle A'B'C'$.

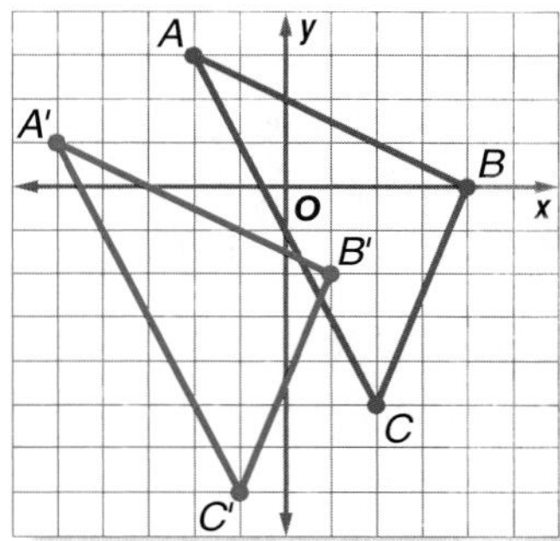

Example 4 Dilation

A trapezoid has vertices $L(-4, 1)$, $M(1, 4)$, $N(7, 0)$, and $P(-3, -6)$.

a. Find the coordinates of the dilated trapezoid $L'M'N'P'$ if the scale factor is $\frac{3}{4}$.

To dilate the figure multiply the coordinates of each vertex by $\frac{3}{4}$.

$$(x, y) \rightarrow \left(\frac{3}{4}x, \frac{3}{4}y\right)$$

$$L(-4, 1) \rightarrow L'\left(\frac{3}{4} \cdot (-4), \frac{3}{4} \cdot 1\right) \rightarrow L'\left(-3, \frac{3}{4}\right)$$

$$M(1, 4) \rightarrow M'\left(\frac{3}{4} \cdot 1, \frac{3}{4} \cdot 4\right) \rightarrow M'\left(\frac{3}{4}, 3\right)$$

$$N(7, 0) \rightarrow N'\left(\frac{3}{4} \cdot 7, \frac{3}{4} \cdot 0\right) \rightarrow N'\left(5\frac{1}{4}, 0\right)$$

$$P(-3, -6) \rightarrow P'\left(\frac{3}{4} \cdot (-3), \frac{3}{4} \cdot (-6)\right) \rightarrow P'\left(-2\frac{1}{4}, -4\frac{1}{2}\right)$$

The coordinates of the vertices of the image are $L'\left(-3, \frac{3}{4}\right)$, $M'\left(\frac{3}{4}, 3\right)$, $N'\left(5\frac{1}{4}, 0\right)$, and $P'\left(-2\frac{1}{4}, -4\frac{1}{2}\right)$.

(continued on the next page)

b. Graph the preimage and its image.

The preimage is trapezoid $LMNP$.

The image is trapezoid $L'M'N'P'$.

Notice that the image has sides that are three-fourths the length of the sides of the original figure.

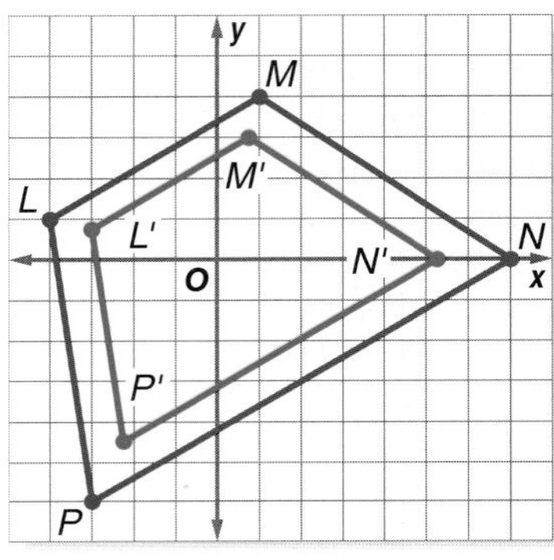

Example 5 Rotation

Triangle XYZ has vertices $X(1, 5)$, $Y(5, 2)$, and $Z(-1, 2)$.

a. Find the coordinates of the image of $\triangle XYZ$ after it is rotated 90° counterclockwise about the origin.

To find the coordinates of the vertices after a 90° rotation, switch the coordinates of each point and then multiply the new first coordinate by -1.

$(x, y) \rightarrow (-y, x)$

$X(1, 5) \rightarrow X'(-5, 1)$

$Y(5, 2) \rightarrow Y'(-2, 5)$

$Z(-1, 2) \rightarrow Z'(-2, -1)$

b. Graph the preimage and its image.

The image is $\triangle XYZ$.

The rotated image is $\triangle X'Y'Z'$.

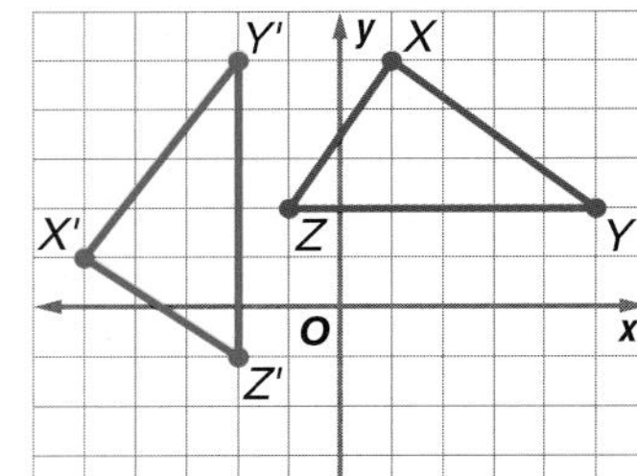

Check for Understanding

Concept Check

1. **Compare and contrast** the size, shape, and orientation of a preimage and an image for each type of transformation.
2. **OPEN ENDED** Draw a figure on the coordinate plane. Then show a dilation of the object that is an enlargement and a dilation of the object that is a reduction.

Guided Practice

Identify each transformation as a *reflection, translation, dilation,* or *rotation.*

3. 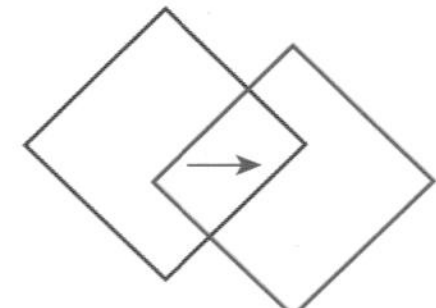

4. 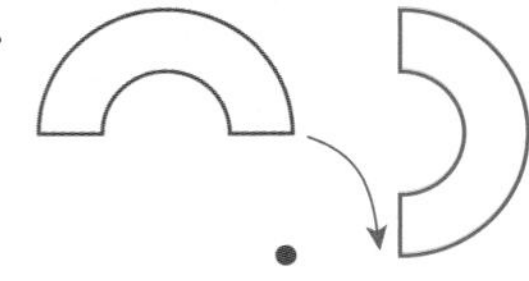

Find the coordinates of the vertices of each figure after the given transformation is performed. Then graph the preimage and its image.

5. triangle PQR with $P(1, 2)$, $Q(4, 4)$, and $R(2, -3)$ reflected over the x-axis
6. quadrilateral $ABCD$ with $A(4, 2)$, $B(4, -2)$, $C(-1, -3)$, and $D(-3, 2)$ translated 3 units up
7. parallelogram $EFGH$ with $E(-1, 4)$, $F(5, -1)$, $G(2, -4)$, and $H(-4, 1)$ dilated by a scale factor of 2
8. triangle JKL with $J(0, 0)$, $K(-2, -5)$, and $L(-4, 5)$ rotated 90° counterclockwise about the origin

Application **NAVIGATION** **For Exercises 9 and 10, use the following information.**
A ship was heading on a chartered route when it was blown off course by a storm. The ship is now ten miles west and seven miles south of its original destination.

9. Using a coordinate grid, make a drawing to show the original destination A and the current position B of the ship.

10. Using coordinates (x, y) to represent the original destination of the ship, write an ordered pair to show its current location.

Practice and Apply

Homework Help

For Exercises	See Examples
11–16, 37, 38	1
17–36	2–5

Extra Practice

See page 828.

Identify each transformation as a *reflection, translation, dilation,* or *rotation.*

11. 

12. 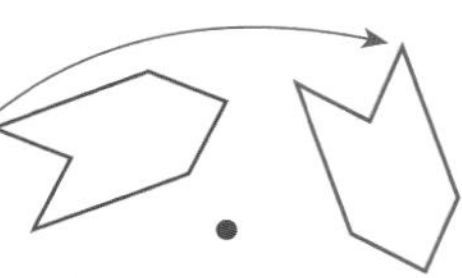

13. 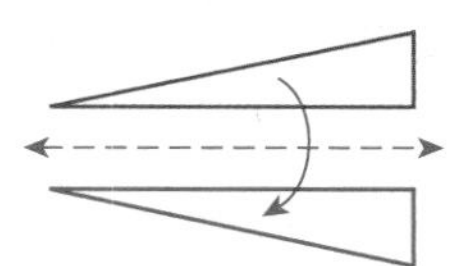

14. 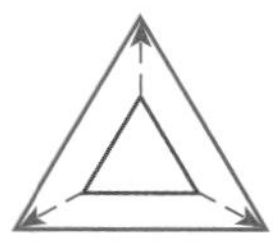

15. 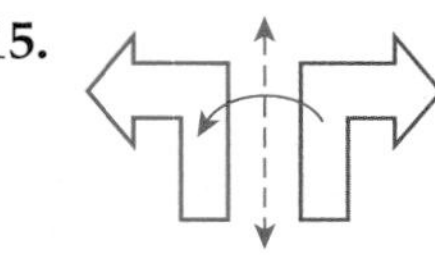

16. 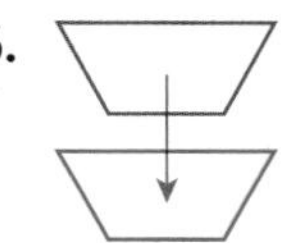

For Exercises 17–26, complete parts a and b.

a. Find the coordinates of the vertices of each figure after the given transformation is performed.

b. Graph the preimage and its image.

17. triangle RST with $R(2, 0)$, $S(-2, -3)$, and $T(-2, 3)$ reflected over the y-axis

18. trapezoid $ABCD$ with $A(2, 3)$, $B(5, 3)$, $C(6, 1)$, and $D(-2, 1)$ reflected over the x-axis

19. quadrilateral $RSTU$ with $R(-6, 3)$, $S(-4, 2)$, $T(-1, 5)$, and $U(-3, 7)$ translated 8 units right

20. parallelogram $MNOP$ with $M(-6, 0)$, $N(-4, 3)$, $O(-1, 3)$, and $P(-3, 0)$ translated 3 units right and 2 units down

21. trapezoid $JKLM$ with $J(-4, 2)$, $K(-2, 4)$, $L(4, 4)$, and $M(-4, -4)$ dilated by a scale factor of $\frac{1}{2}$

22. square $ABCD$ with $A(-2, 1)$, $B(2, 2)$, $C(3, -2)$, and $D(-1, -3)$ dilated by a scale factor of 3

23. triangle FGH with $F(-3, 2)$, $G(2, 5)$, and $H(6, 3)$ rotated 180° about the origin

24. quadrilateral $TUVW$ with $T(-4, 2)$, $U(-2, 4)$, $V(0, 2)$, and $W(-2, -4)$ rotated 90° counterclockwise about the origin

25. parallelogram $WXYZ$ with $W(-1, 2)$, $X(3, 2)$, $Y(0, -4)$, and $Z(-4, -4)$ reflected over the y-axis, then rotated 180° about the origin

26. pentagon $PQRST$ with $P(0, 5)$, $Q(3, 4)$, $R(2, 1)$, $S(-2, 1)$, and $T(-3, 4)$ reflected over the x-axis, then translated 2 units left and 1 unit up

ANIMATION **For Exercises 27–29, use the diagram at the right.**
An animator places an arrow representing an airplane on a coordinate grid. She wants to move the arrow 2 units right and then reflect it across the x-axis.

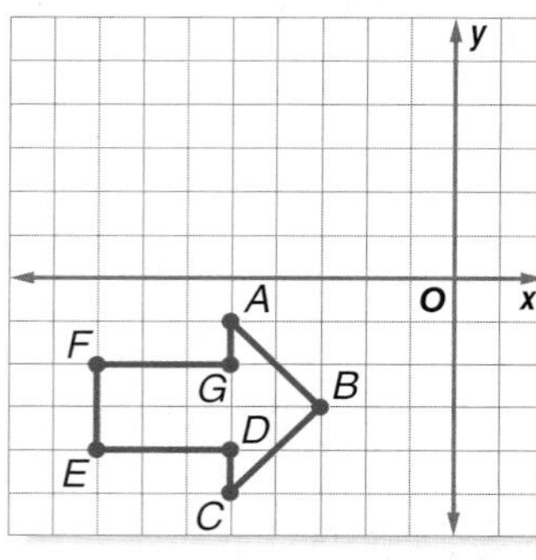

27. Write the coordinates for the vertices of the arrow.

28. Find the coordinates of the final position of the arrow.

29. Graph the image.

30. Trapezoid $JKLM$ with $J(-6, 0)$, $K(-1, 5)$, $L(-1, 1)$, and $M(-3, -1)$ is translated to $J'K'L'M'$ with $J'(-3, -2)$, $K'(2, 3)$, $L'(2, -1)$, $M'(0, -3)$. Describe this translation.

31. Triangle QRS with vertices $Q(-2, 6)$, $R(8, 0)$, and $S(6, 4)$ is dilated. If the image $Q'R'S'$ has vertices $Q'(-1, 3)$, $R'(4, 0)$, and $S'(3, 2)$, what is the scale factor?

32. Describe the transformation of parallelogram $WXYZ$ with $W(-5, 3)$, $X(-2, 5)$, $Y(0, 3)$, and $Z(-3, 1)$ if the coordinates of its image are $W'(5, 3)$, $X'(2, 5)$, $Y'(0, 3)$, and $Z'(3, 1)$.

33. Describe the transformation of triangle XYZ with $X(2, -1)$, $Y(-5, 3)$, and $Z(4, 0)$ if the coordinates of its image are $X'(1, 2)$, $Y'(-3, -5)$, and $Z'(0, 4)$.

DIGITAL PHOTOGRAPHY **For Exercises 34–36, use the following information.**
Soto wants to enlarge a digital photograph that is 1800 pixels wide and 1600 pixels high (1800×1600) by a scale factor of $2\frac{1}{2}$.

34. What will be the dimensions of the new digital photograph?

35. Use a coordinate grid to draw a picture representing the 1800×1600 digital photograph. Place one corner of the photograph at the origin and write the coordinates of the other three vertices.

36. Draw the enlarged photograph and write its coordinates.

More About. . .

Digital Photography
Digital photographs contain hundreds of thousands or millions of tiny squares called pixels.
Source: www.shortcourses.com

ART **For Exercises 37 and 38, use the following information.**
On grid paper, draw an octagon like the one shown.

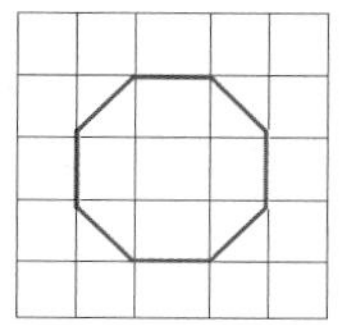

37. Reflect the octagon over each of its sides. Describe the pattern that results.

38. Could this same pattern be drawn using any of the other transformations? If so, which kind?

39. **CRITICAL THINKING** Make a conjecture about the coordinates of a point (x, y) that has been rotated 90° *clockwise* about the origin.

40. **CRITICAL THINKING** Determine whether the following statement is *sometimes*, *always*, or *never* true.

A reflection over the x-axis followed by a reflection over the y-axis gives the same result as a rotation of 180°.

41. WRITING IN MATH Answer the question that was posed at the beginning of the lesson.

How are transformations used in computer graphics?

Include the following in your answer:

- examples of movements that could be simulated by transformations, and
- types of other industries that might use transformations in computer graphics to simulate movement.

42. The coordinates of the vertices of quadrilateral $QRST$ are $Q(-2, 4)$, $R(3, 7)$, $S(4, -2)$, and $T(-5, -3)$. If the quadrilateral is moved up 3 units and right 1 unit, which point below has the correct coordinates?

(A) $Q'(1, 5)$ (B) $R'(4, 4)$ (C) $S'(5, 1)$ (D) $T'(-6, 0)$

43. x is $\frac{2}{3}$ of y and y is $\frac{1}{4}$ of z. If $x = 14$, then $z =$

(A) 48. (B) 72. (C) 84. (D) 96.

Extending the Lesson

Graph the image of each figure after a reflection over the graph of the given equation. Find the coordinates of the vertices.

44. $x = 0$

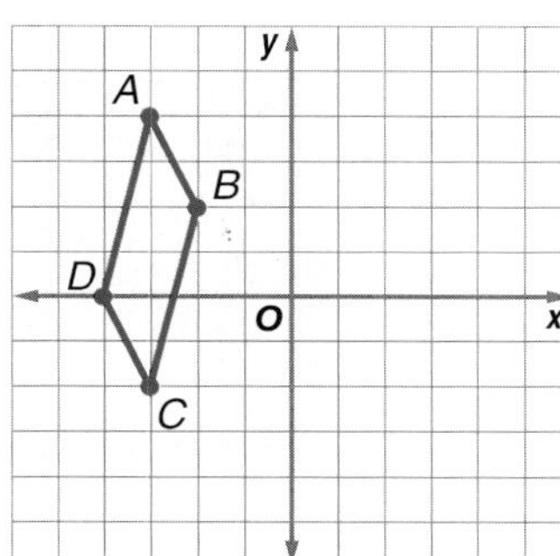

45. $y = -3$

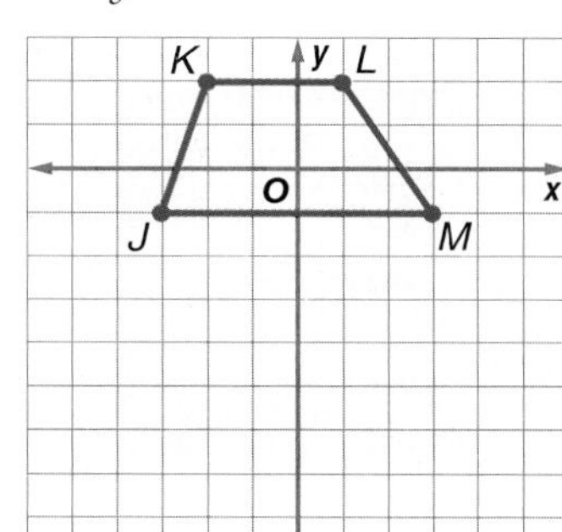

46. $y = x$

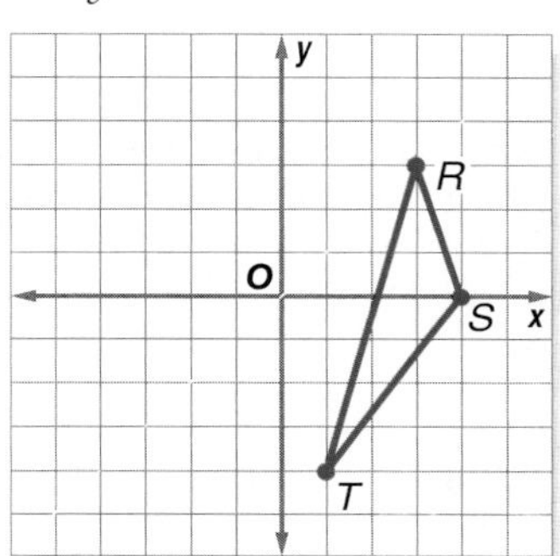

Maintain Your Skills

Mixed Review

Plot each point on a coordinate plane. *(Lesson 4-1)*

47. $A(2, -1)$

48. $B(-4, 0)$

49. $C(1, 5)$

50. $D(-1, -1)$

51. $E(-2, 3)$

52. $F(4, -3)$

53. **CHEMISTRY** Jamaal needs a 25% solution of nitric acid. He has 20 milliliters of a 30% solution. How many milliliters of a 15% solution should he add to obtain the required 25% solution? *(Lesson 3-9)*

Two dice are rolled and their sum is recorded. Find each probability. *(Lesson 2-6)*

54. P(sum is less than 9)

55. P(sum is greater than 10)

56. P(sum is less than 7)

57. P(sum is greater than 4)

Getting Ready for the Next Lesson

PREREQUISITE SKILL Write a set of ordered pairs that represents the data in the table. *(To review **ordered pairs**, see Lesson 1-8.)*

58.

Number of toppings	1	2	3	4	5	6
Cost of large pizza ($)	9.95	11.45	12.95	14.45	15.95	17.45

59.

Time (minutes)	0	5	10	15	20	25	30
Temperature of boiled water as it cools (°C)	100	90	81	73	66	60	55

Graphing Calculator Investigation

A Preview of Lesson 4-3

Graphs of Relations

You can represent a relation as a graph using a TI-83 Plus graphing calculator.

Graph the relation {(3, 7), (−8, 12), (−5, 7), (11, −1)}.

Step 1 *Enter the data.*

- Enter the x-coordinates in L1 and the y-coordinates in L2.

KEYSTROKES: [STAT] [ENTER] 3 [ENTER] −8 [ENTER] −5 [ENTER] 11 [ENTER] [▶] 7 [ENTER] 12 [ENTER] 7 [ENTER] −1 [ENTER]

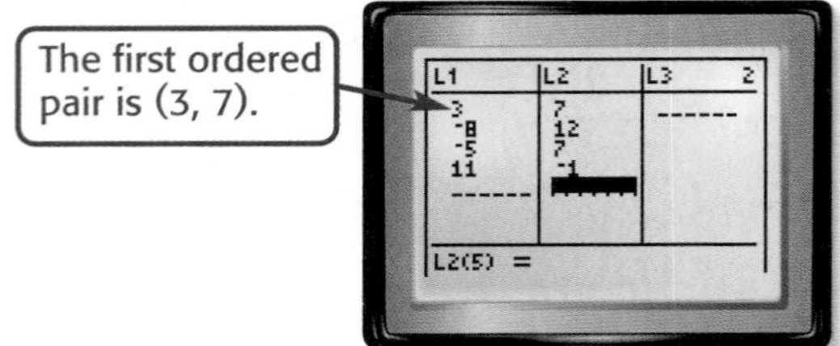

Step 2 *Format the graph.*

- Turn on the statistical plot.

KEYSTROKES: [2nd] [STAT PLOT] [ENTER] [ENTER]

- Select the scatter plot, L1 as the Xlist and L2 as the Ylist.

KEYSTROKES: [▼] [ENTER] [▼] [2nd] [L1] [ENTER] [2nd] [L2] [ENTER]

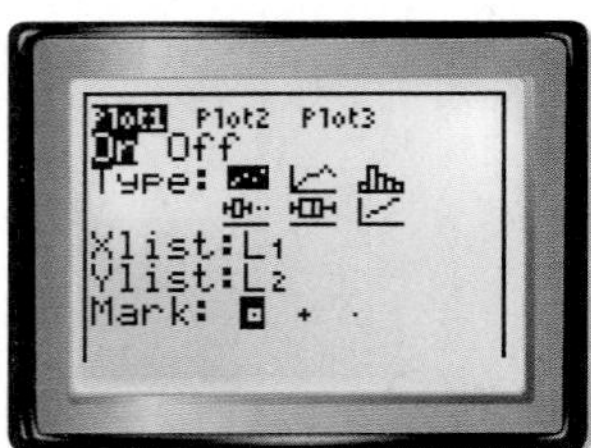

Step 3 *Choose the viewing window.*

- Be sure you can see all of the points.
[−10, 15] scl: 1 by [−5, 15] scl: 1

KEYSTROKES: [WINDOW] −10 [ENTER] 15 [ENTER] 1 [ENTER] −5 [ENTER] 15 [ENTER] 1

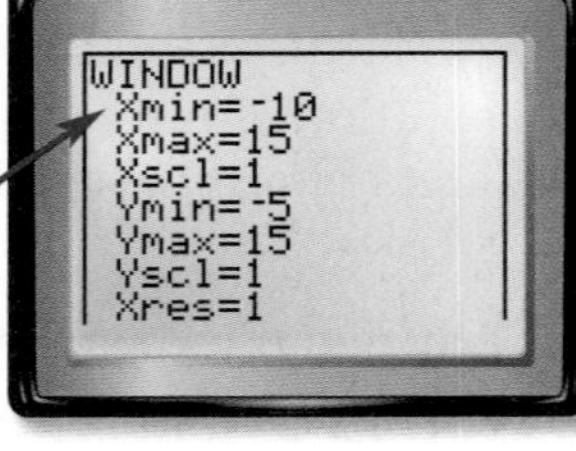

Step 4 *Graph the relation.*

- Display the graph.

KEYSTROKES: [GRAPH]

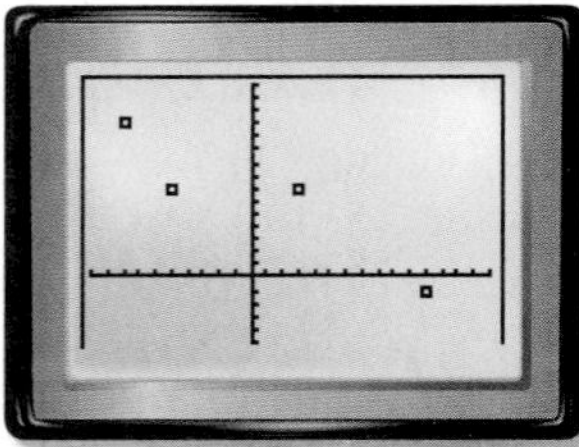

[−10, 15] scl: 1 by [−5, 15] scl: 1

Exercises

Graph each relation. Sketch the result.

1. {(10, 10), (0, −6), (4, 7), (5, −2)}

2. {(−4, 1), (3, −5), (4, 5), (−5, 1)}

3. {(12, 15), (10, −16), (11, 7), (−14, −19)}

4. {(45, 10), (23, 18), (22, 26), (35, 26)}

5. **MAKE A CONJECTURE** How are the values of the domain and range used to determine the scale of the viewing window?

 www.algebra1.com/other_calculator_keystrokes

4-3 Relations

What You'll Learn

- Represent relations as sets of ordered pairs, tables, mappings, and graphs.
- Find the inverse of a relation.

Vocabulary

- mapping
- inverse

How can relations be used to represent baseball statistics?

Ken Griffey, Jr.'s, batting statistics for home runs and strikeouts can be represented as a set of ordered pairs. These statistics are shown in the table at the right, where the first coordinates represent the number of home runs and the second coordinates represent the number of strikeouts. You can plot the ordered pairs on a graph to look for patterns in the distribution of the points.

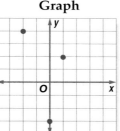

Ken Griffey, Jr.		
Year	Home Runs	Strikeouts
1994	40	73
1995	17	53
1996	49	104
1997	56	121
1998	56	121
1999	48	108
2000	40	117
2001	22	72

REPRESENT RELATIONS Recall that a *relation* is a set of ordered pairs. A relation can be represented by a set of ordered pairs, a table, a graph, or a **mapping**. A mapping illustrates how each element of the domain is paired with an element in the range. Study the different representations of the same relation below.

Study Tip

Look Back
To review **relations**, see Lesson 1-8.

Ordered Pairs

(1, 2)
(−2, 4)
(0, −3)

Table

x	y
1	2
−2	4
0	−3

Graph

Mapping

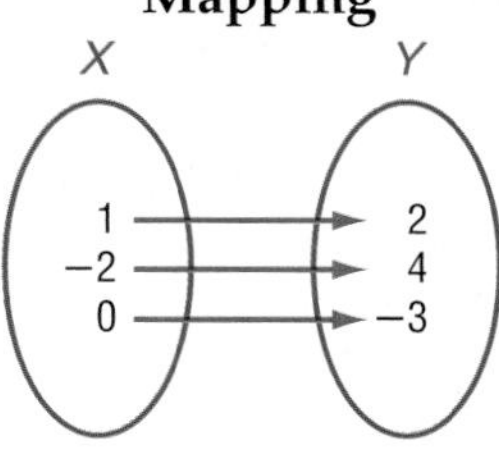

Example 1 Represent a Relation

a. Express the relation {(3, 2), (−1, 4), (0, −3), (−3, 4), (−2, −2)} as a table, a graph, and a mapping.

Table

List the set of x-coordinates in the first column and the corresponding y-coordinates in the second column.

x	y
3	2
−1	4
0	−3
−3	4
−2	−2

Graph

Graph each ordered pair on a coordinate plane.

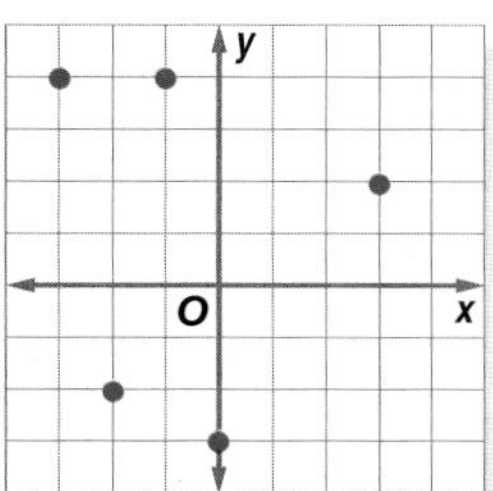

(continued on the next page)

Study Tip

Domain and Range

When writing the elements of the domain and range, if a value is repeated, you need to list it only once.

Mapping

List the x values in set X and the y values in set Y. Draw an arrow from each x value in X to the corresponding y value in Y.

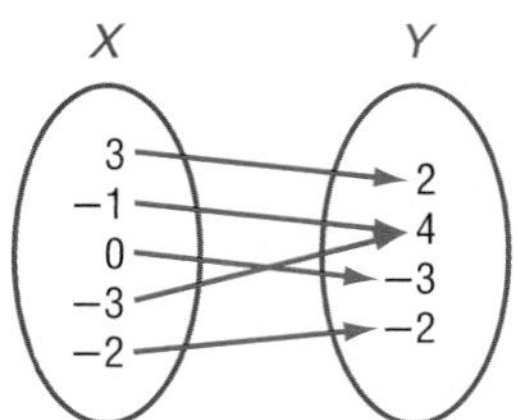

b. Determine the domain and range.
The domain for this relation is $\{-3, -2, -1, 0, 3\}$.
The range is $\{-3, -2, 2, 4\}$.

When graphing relations that represent real-life situations, you may need to select values for the x- or y-axis that do not begin with 0 and do not have units of 1.

Example 2 Use a Relation

BALD EAGLES In 1990, New York purchased 12,000 acres for the protection of bald eagles. The table shows the number of eagles observed in New York during the annual mid-winter bald eagle survey from 1993 to 2000.

Bald Eagle Survey								
Year	1993	1994	1995	1996	1997	1998	1999	2000
Number of Eagles	102	116	144	174	175	177	244	350

Source: New York Department of Environmental Conservation

More About. . .

Bald Eagles

The bald eagle is not really bald. Its name comes from the Old English meaning of bald, "having white feathers on the head."

Source: *Webster's Dictionary*

a. Determine the domain and range of the relation.
The domain is {1993, 1994, 1995, 1996, 1997, 1998, 1999, 2000}.
The range is {102, 116, 144, 174, 175, 177, 244, 350}.

b. Graph the data.

- The values of the x-axis need to go from 1993 to 2000. It is not practical to begin the scale at 0. Begin at 1992 and extend to 2001 to include all of the data. The units can be 1 unit per grid square.
- The values on the y-axis need to go from 102 to 350. In this case, it is possible to begin the scale at 0. Begin at 0 and extend to 400. You can use units of 50.

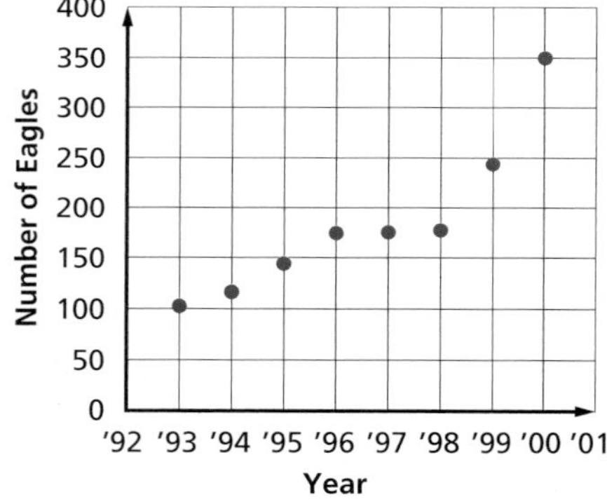

c. What conclusions might you make from the graph of the data?
The number of eagles has increased each year. This may be due to the efforts of those who are protecting the eagles in New York.

INVERSE RELATIONS The **inverse** of any relation is obtained by switching the coordinates in each ordered pair.

Key Concept *Inverse of a Relation*

Relation Q is the inverse of relation S if and only if for every ordered pair (a, b) in S, there is an ordered pair (b, a) in Q.

Relation	Inverse
(2, 5)	(5, 2)
(−3, 2)	(2, −3)
(6, 7)	(7, 6)
(5, −1)	(−1, 5)

Notice that the domain of a relation becomes the range of the inverse and the range of a relation becomes the domain of the inverse.

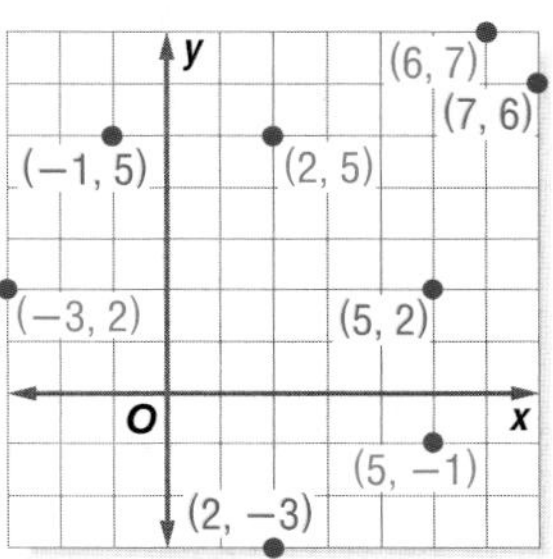

Example 3 Inverse Relation

Express the relation shown in the mapping as a set of ordered pairs. Then write the inverse of the relation.

Relation Notice that both 2 and 3 in the domain are paired with −4 in the range.
{(2, −4), (3, −4), (5, −7), (6, −8)}

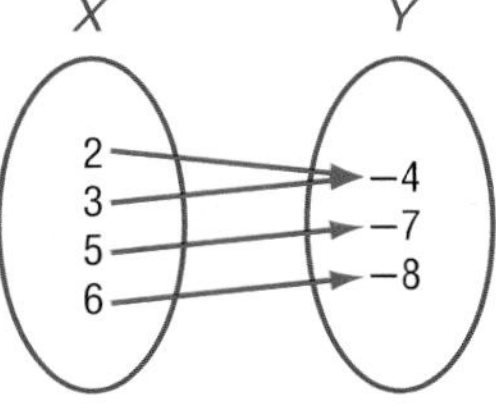

Inverse Exchange x and y in each ordered pair to write the inverse relation.
{(−4, 2), (−4, 3), (−7, 5), (−8, 6)}

The mapping of the inverse is shown at the right. Compare this to the mapping of the relation.

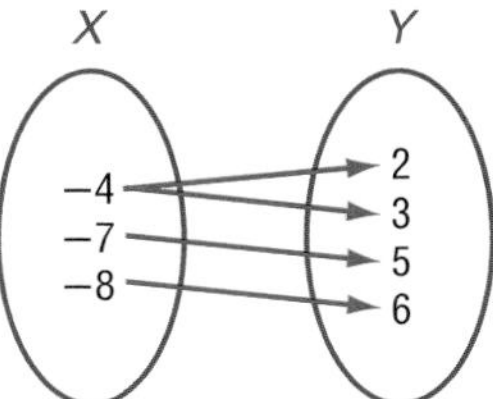

Algebra Activity

Relations and Inverses

- Graph the relation {(3, 4), (−2, 5), (−4, −3), (5, −6), (−1, 0), (0, 2)} on grid paper using a colored pencil. Connect the points in order using the same colored pencil.
- Use a different colored pencil to graph the inverse of the relation, connecting the points in order.
- Fold the grid paper through the origin so that the positive y-axis lies on top of the positive x-axis. Hold the paper up to a light so that you can see all of the points you graphed.

Analyze

1. What do you notice about the location of the points you graphed when you looked at the folded paper?
2. Unfold the paper. Describe the transformation of each point and its inverse.
3. What do you think are the ordered pairs that represent the points on the fold line? Describe these in terms of x and y.

Make a Conjecture

4. How could you graph the inverse of a function without writing ordered pairs first?

Check for Understanding

Concept Check

1. **Describe** the different ways a relation can be represented.
2. **OPEN ENDED** Give an example of a set of ordered pairs that has five elements in its domain and four elements in its range.
3. **State** the relationship between the domain and range of a relation and the domain and range of its inverse.

Guided Practice

Express each relation as a table, a graph, and a mapping. Then determine the domain and range.

4. $\{(5, -2), (8, 3), (-7, 1)\}$
5. $\{(6, 4), (3, -3), (-1, 9), (5, -3)\}$
6. $\{(7, 1), (3, 0), (-2, 5)\}$
7. $\{(-4, 8), (-1, 9), (-4, 7), (6, 9)\}$

Express the relation shown in each table, mapping, or graph as a set of ordered pairs. Then write the inverse of the relation.

8.

x	y
3	−2
−6	7
4	3
−6	5

9.

x	y
−4	9
2	5
−2	−2
11	12

10.
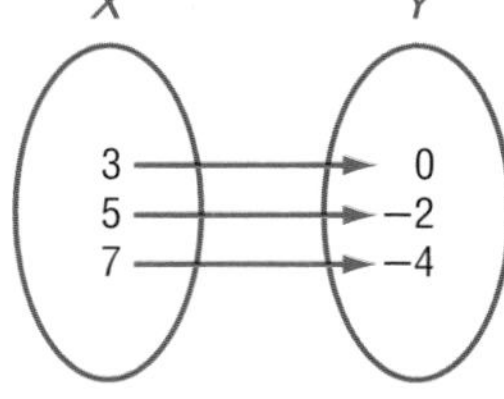

11.
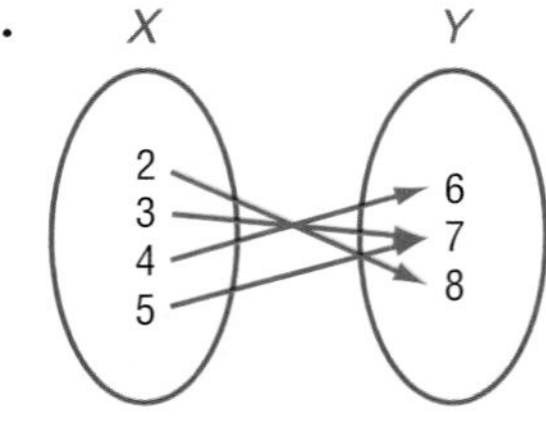

12.
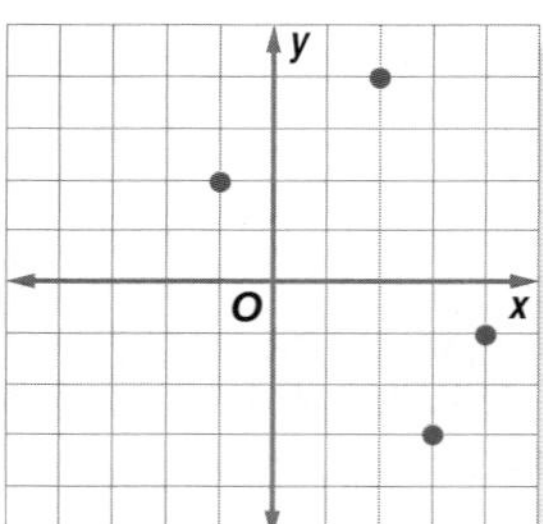

13.
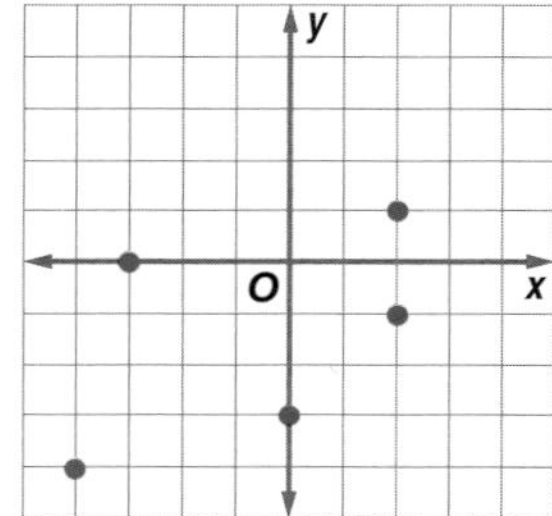

Application

TECHNOLOGY **For Exercises 14–17, use the graph of the average number of students per computer in U.S. public schools.**

14. Name three ordered pairs from the graph.
15. Determine the domain of the relation.
16. What are the least value and the greatest value in the range?
17. What conclusions can you make from the graph of the data?

Online Research **Data Update** What is the average number of students per computer in your state? Visit www.algebra1.com/data_update to learn more.

Average Number of Students per Computer

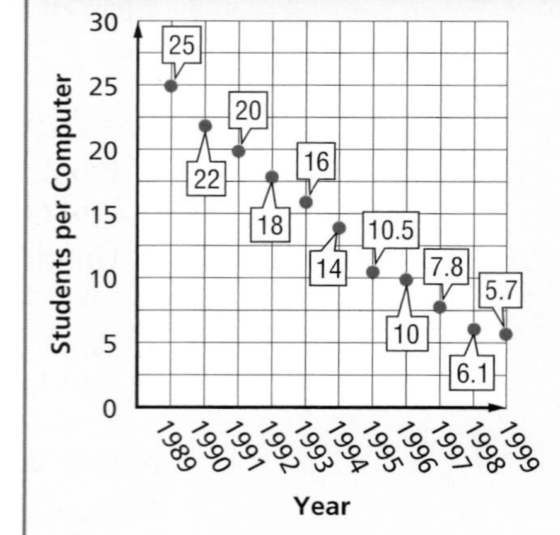

Source: Quality Education Data

Practice and Apply

Homework Help

For Exercises	See Examples
18–25	1
26–37	3
38–48	2

Extra Practice

See page 829.

Express each relation as a table, a graph, and a mapping. Then determine the domain and range.

18. $\{(4, 3), (1, -7), (1, 3), (2, 9)\}$
19. $\{(5, 2), (-5, 0), (6, 4), (2, 7)\}$
20. $\{(0, 0), (6, -1), (5, 6), (4, 2)\}$
21. $\{(3, 8), (3, 7), (2, -9), (1, -9)\}$
22. $\{(4, -2), (3, 4), (1, -2), (6, 4)\}$
23. $\{(0, 2), (-5, 1), (0, 6), (-1, 9)\}$
24. $\{(3, 4), (4, 3), (2, 2), (5, -4), (-4, 5)\}$
25. $\{(7, 6), (3, 4), (4, 5), (-2, 6), (-3, 2)\}$

Express the relation shown in each table, mapping, or graph as a set of ordered pairs. Then write the inverse of the relation.

26.

x	y
1	2
3	4
5	6
7	8

27.

x	y
0	3
−5	2
4	7
−3	2

28.

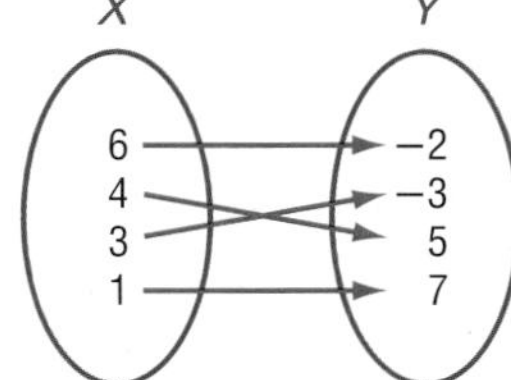

29.

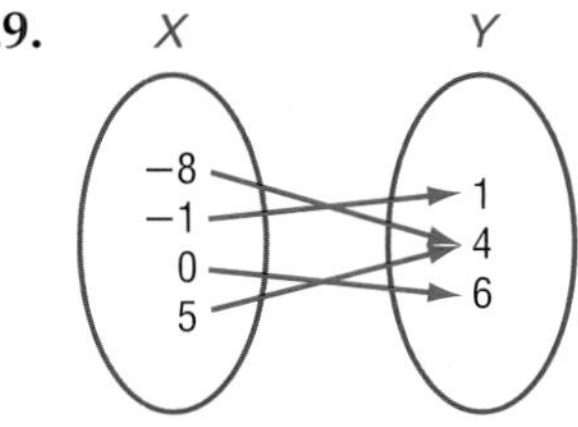

30.

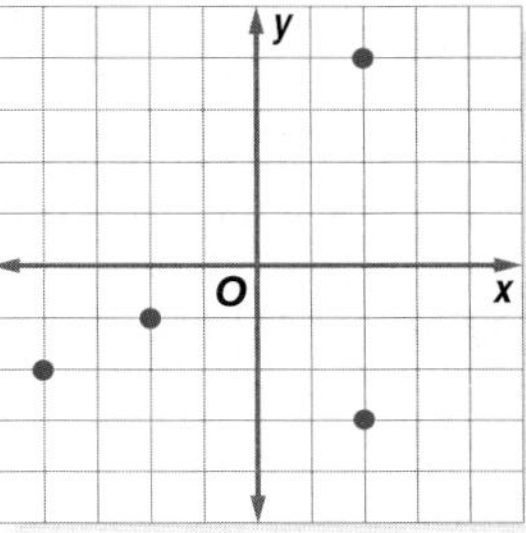

31.

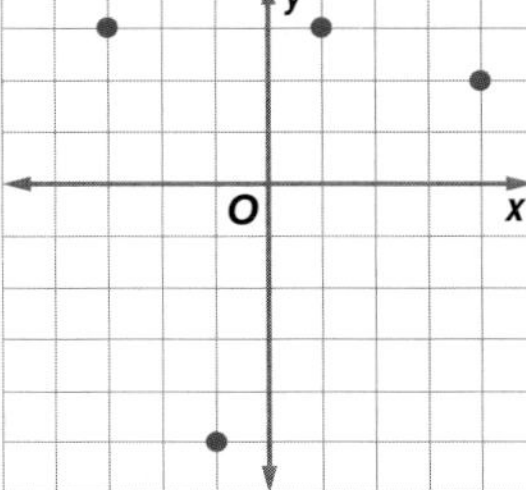

32.

x	y
0	0
4	7
8	10.5
12	13
16	14.5

33.

x	y
1	16.50
1.75	28.30
2.5	49.10
3.25	87.60
4	103.40

34.

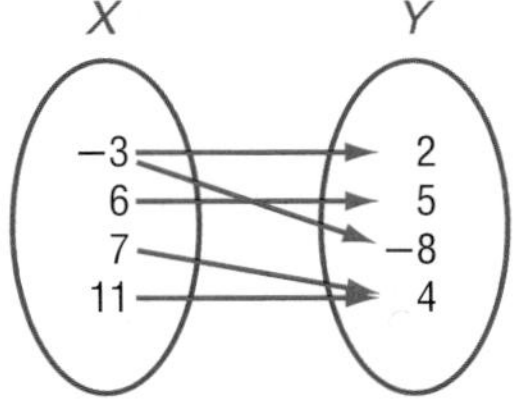

35.

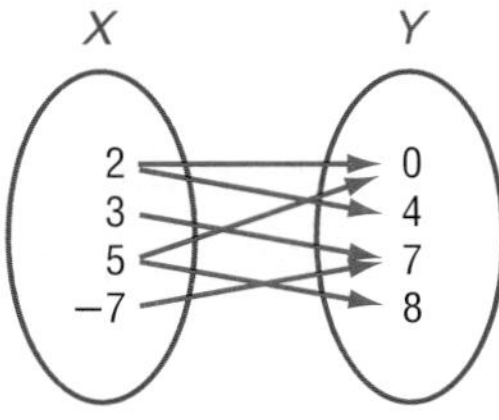

36.

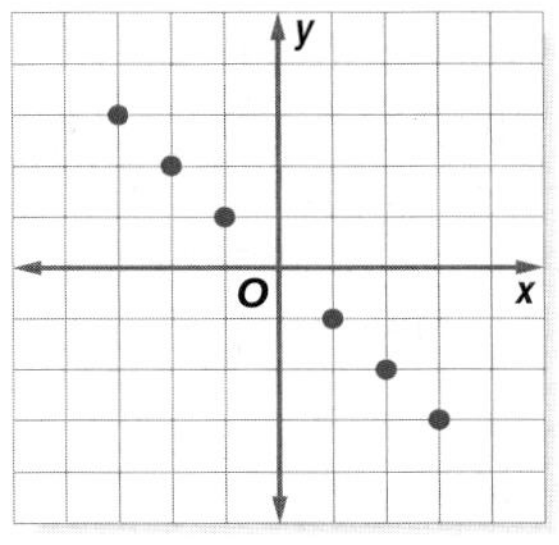

37.

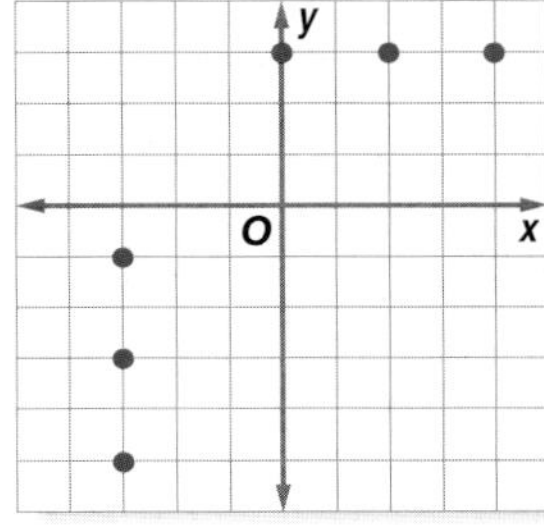

COOKING **For Exercises 38–40, use the table that shows the boiling point of water at various altitudes.** Many recipes have different cooking times for high altitudes. This is due to the fact that water boils at a lower temperature in higher altitudes.

Altitude (feet)	Boiling Point of Water (°F)
0	212.0
1000	210.2
2000	208.4
3000	206.5
5000	201.9
10,000	193.7

Source: Stevens Institute of Technology

38. Graph the relation.
39. Write the inverse as a set of ordered pairs.
40. How could you estimate your altitude by finding the boiling point of water at your location?

FOOD **For Exercises 41–43, use the graph that shows the annual production of corn from 1991–2000.**

41. Estimate the domain and range of the relation.

42. Which year had the lowest production? the highest?

43. Describe any pattern you see.

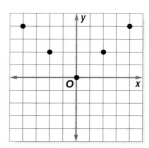

USA TODAY Snapshots®

Farmers growing bumper crop

U.S. farmers are predicted to produce a billion more bushels of corn than last year. Annual production:

Billions of bushels

10.4

7.5

10
9
8
7
6
0

'91 '93 '95 '97 '99 2000

Source: National Agricultural Statistics Service

By Suzy Parker, USA TODAY

HEALTH **For Exercises 44–48, use the following information.**
A person's muscle weight is about 2 pounds of muscle for each 5 pounds of body weight.

44. Make a table to show the relation between body and muscle weight for people weighing 100, 105, 110, 115, 120, 125, and 130 pounds.

45. What are the domain and range?

46. Graph the relation.

47. What are the domain and range of the inverse?

48. Graph the inverse relation.

49. **CRITICAL THINKING** Find a counterexample to disprove the following.

The domain of relation F contains the same elements as the range of relation G. The range of relation F contains the same elements as the domain of relation G. Therefore, relation G must be the inverse of relation F.

50. WRITING IN MATH Answer the question that was posed at the beginning of the lesson.

How can relations be used to represent baseball statistics?

Include the following in your answer:

- a graph of the relation of the number of Ken Griffey, Jr.'s, home runs and his strikeouts, and
- an explanation of any relationship between the number of home runs hit and the number of strikeouts.

For Exercises 51 and 52, use the graph at the right.

51. State the domain and range of the relation.

Ⓐ D = {0, 2, 4}; R = {−4, −2, 0, 2, 4}

Ⓑ D = {−4, −2, 0, 2, 4}; R = {0, 2, 4}

Ⓒ D = {0, 2, 4}; R = {−4, −2, 0}

Ⓓ D = {−4, −2, 0, 2, 4}; R = {−4, −2, 0, 2, 4}

y

O

x

52. **SHORT RESPONSE** Graph the inverse of the relation.

For Exercises 53–56, use a graphing calculator.
a. Graph each relation.
b. State the WINDOW settings that you used.
c. Write the coordinates of the inverse. Then graph the inverse.
d. Name the quadrant in which each point of the relation and its inverse lies.

53. {(0, 10), (2, −8), (6, 6), (9, −4)}

54. {(−1, 18), (−2, 23), (−3, 28), (−4, 33)}

55. {(35, 12), (48, 25), (60, 52)}

56. {(−92, −77), (−93, 200), (19, −50)}

Maintain Your Skills

Mixed Review

Identify each transformation as a *reflection, translation, dilation,* or *rotation.* *(Lesson 4-2)*

57. 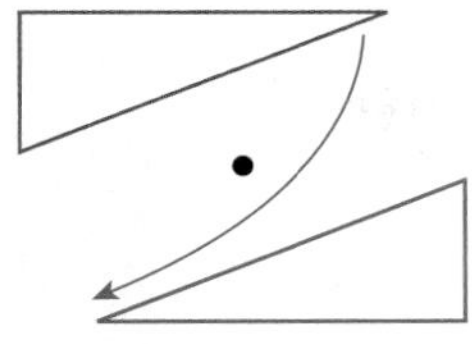

58. 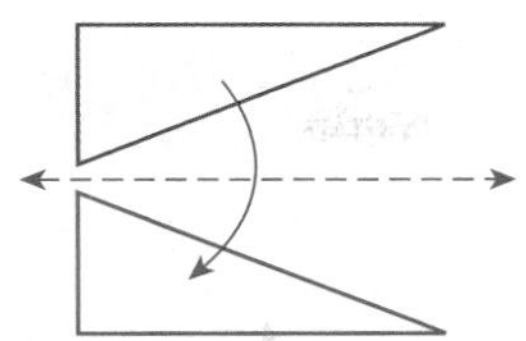

59. 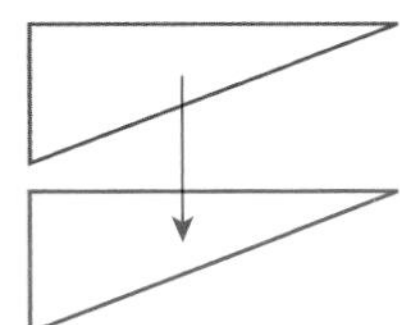

Write the ordered pair for each point shown at the right. Name the quadrant in which the point is located. *(Lesson 4-1)*

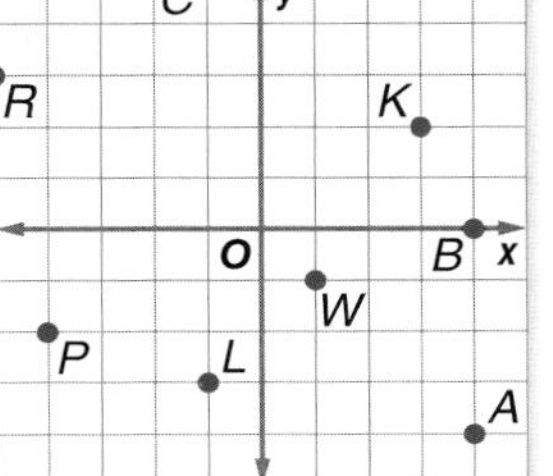

60. A **61.** K

62. L **63.** W

64. B **65.** P

66. R **67.** C

68. HOURLY PAY Dominique earns \$9.75 per hour. Her employer is increasing her hourly rate to \$10.15 per hour. What is the percent of increase in her salary? *(Lesson 3-7)*

Simplify each expression. *(Lesson 2-4)*

69. $72 \div 9$ **70.** $105 \div 15$

71. $3 \div \frac{1}{3}$ **72.** $16 \div \frac{1}{4}$

73. $\frac{54n + 78}{6}$ **74.** $\frac{98x - 35y}{7}$

Getting Ready for the Next Lesson

PREREQUISITE SKILL Find the solution set for each equation if the replacement set is {3, 4, 5, 6, 7, 8}. *(To review **solution sets**, see Lesson 1-3.)*

75. $a + 15 = 20$ **76.** $r - 6 = 2$

77. $9 = 5n - 6$ **78.** $3 + 8w = 35$

79. $\frac{g}{3} + 15 = 17$ **80.** $\frac{m}{5} + \frac{3}{5} = 2$

Practice Quiz 1 — Lessons 4-1 through 4-3

Plot each point on a coordinate plane. *(Lesson 4-1)*

1. $Q(2, 3)$ **2.** $R(-4, -4)$ **3.** $S(5, -1)$ **4.** $T(-1, 3)$

Find the coordinates of the vertices of each figure after the given transformation is performed. Then graph the preimage and its image. *(Lesson 4-2)*

5. triangle ABC with $A(4, 8)$, $B(7, 5)$, and $C(2, -1)$ reflected over the x-axis

6. quadrilateral $WXYZ$ with $W(1, 0)$, $X(2, 3)$, $Y(4, 1)$, and $Z(3, -3)$ translated 5 units to the left and 4 units down

State the domain, range, and inverse of each relation. *(Lesson 4-3)*

7. $\{(1, 3), (4, 6), (2, 3), (1, 5)\}$ **8.** $\{(-2, 6), (0, 3), (4, 2), (8, -5)\}$

9. $\{(11, 5), (15, 3), (-8, 22), (11, 31)\}$ **10.** $\{(-5, 8), (-1, 0), (-1, 4), (2, 7), (6, 3)\}$

4-4 Equations as Relations

What You'll Learn

- Use an equation to determine the range for a given domain.
- Graph the solution set for a given domain.

Vocabulary

- equation in two variables
- solution

Why are equations of relations important in traveling?

During the summer, Eric will be taking a trip to England. He has saved \$500 for his trip, and he wants to find how much that will be worth in British pounds sterling. The exchange rate today is 1 dollar = 0.69 pound. Eric can use the equation $p = 0.69d$ to convert dollars d to pounds p.

SOLVE EQUATIONS The equation $p = 0.69d$ is an example of an **equation in two variables**. A **solution** of an equation in two variables is an ordered pair that results in a true statement when substituted into the equation.

Example 1 Solve Using a Replacement Set

Find the solution set for $y = 2x + 3$, given the replacement set $\{(-2, -1), (-1, 3), (0, 4), (3, 9)\}$.

Make a table. Substitute each ordered pair into the equation.

x	y	$y = 2x + 3$	True or False?
−2	−1	$-1 = 2(-2) + 3$ $-1 = -1$	true ✓
−1	3	$3 = 2(-1) + 3$ $3 = 1$	false
0	4	$4 = 2(0) + 3$ $4 = 3$	false
3	9	$9 = 2(3) + 3$ $9 = 9$	true ✓

The ordered pairs $(-2, -1)$ and $(3, 9)$ result in true statements. The solution set is $\{(-2, -1), (3, 9)\}$.

Since the solutions of an equation in two variables are ordered pairs, the equation describes a relation. So, in an equation involving x and y, the set of x values is the domain, and the corresponding set of y values is the range.

Study Tip

Variables
Unless the variables are chosen to represent real quantities, when variables other than x and y are used in an equation, assume that the letter that comes first in the alphabet is the domain.

Example 2 Solve Using a Given Domain

Solve $b = a + 5$ if the domain is $\{-3, -1, 0, 2, 4\}$.

Make a table. The values of a come from the domain. Substitute each value of a into the equation to determine the values of b in the range.

a	$a + 5$	b	(a, b)
−3	−3 + 5	2	(−3, 2)
−1	−1 + 5	4	(−1, 4)
0	0 + 5	5	(0, 5)
2	2 + 5	7	(2, 7)
4	4 + 5	9	(4, 9)

The solution set is $\{(-3, 2), (-1, 4), (0, 5), (2, 7), (4, 9)\}$.

GRAPH SOLUTION SETS You can graph the ordered pairs in the solution set for an equation in two variables. The domain contains values represented by the *independent variable*. The range contains the corresponding value represented by the *dependent variable*.

Study Tip

Look Back
To review **independent and dependent variables**, see Lesson 1-8.

Example 3 Solve and Graph the Solution Set

Solve $4x + 2y = 10$ if the domain is $\{-1, 0, 2, 4\}$. Graph the solution set.

First solve the equation for y in terms of x. This makes creating a table of values easier.

$4x + 2y = 10$ Original equation

$4x + 2y - 4x = 10 - 4x$ Subtract $4x$ from each side.

$2y = 10 - 4x$ Simplify.

$\frac{2y}{2} = \frac{10 - 4x}{2}$ Divide each side by 2.

$y = 5 - 2x$ Simplify.

Substitute each value of x from the domain to determine the corresponding values of y in the range.

x	$5 - 2x$	y	(x, y)
−1	5 − 2(−1)	7	(−1, 7)
0	5 − 2(0)	5	(0, 5)
2	5 − 2(2)	1	(2, 1)
4	5 − 2(4)	−3	(4, −3)

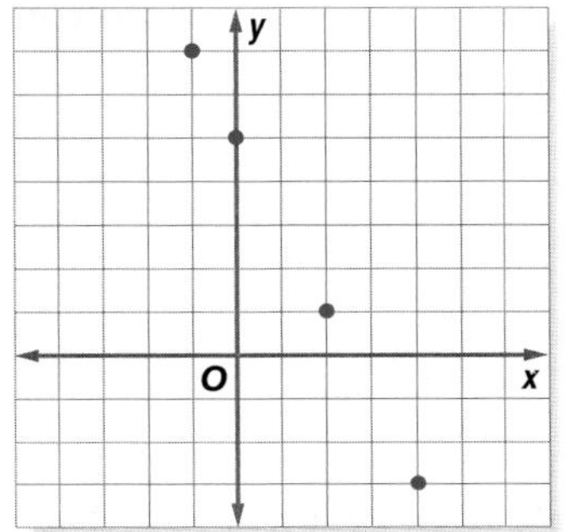

Graph the solution set $\{(-1, 7), (0, 5), (2, 1), (4, -3)\}$.

When you solve an equation for a given variable, that variable becomes the dependent variable. That is, its value depends upon the domain values chosen for the other variable.

Example 4 Solve for a Dependent Variable

Refer to the application at the beginning of the lesson. Eric has made a list of the expenses he plans to incur while in England. Use the conversion rate to find the equivalent U.S. dollars for these amounts given in pounds (£) and graph the ordered pairs.

Explore In the equation $p = 0.69d$, d represents U.S. dollars and p represents British pounds. However, we are given values in pounds and want to find values in dollars. Solve the equation for d since the values of d depend on the given values of p.

$p = 0.69d$ Original equation

$\frac{p}{0.69} = \frac{0.69d}{0.69}$ Divide each side by 0.69.

$1.45p = d$ Simplify and round to the nearest hundredth.

(continued on the next page)

Plan The values of p, {40, 30, 15, 6}, are the domain. Use the equation $d = 1.45p$ to find the values for the range.

Solve Make a table of values. Substitute each value of p from the domain to determine the corresponding values of d. Round to the nearest dollar.

p	$1.45p$	d	(p, d)
40	1.45(40)	58.00	(40, 58)
30	1.45(30)	43.50	(30, 44)
15	1.45(15)	21.75	(15, 22)
6	1.45(6)	8.70	(6, 9)

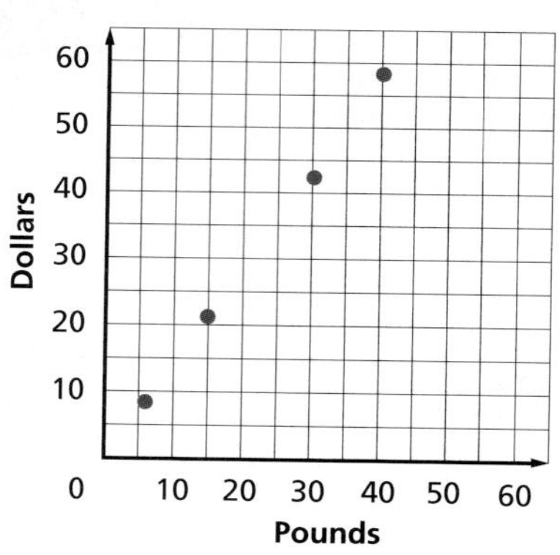

Graph the ordered pairs. Notice that the values for the independent variable p are graphed along the horizontal axis, and the values for dependent variable d are graphed along the vertical axis.

Examine Look at the values in the range. The cost in dollars is higher than the cost in pounds. Do the results make sense?

Expense	Pounds	Dollars
Hotel	40	58
Meals	30	43
Entertainment	15	22
Transportation	6	9

Check for Understanding

Concept Check

1. **Describe** how to find the domain of an equation if you are given the range.
2. **OPEN ENDED** Give an example of an equation in two variables and state two solutions for your equation.
3. **FIND THE ERROR** Malena says that (5, 1) is a solution of $y = 2x + 3$. Bryan says it is not a solution.

Malena	Bryan
$y = 2x + 3$	$y = 2x + 3$
$5 = 2(1) + 3$	$1 = 2(5) + 3$
$5 = 5$	$1 \neq 13$

Who is correct? Explain your reasoning.

Guided Practice

Find the solution set for each equation, given the replacement set.

4. $y = 3x + 4$; {(−1, 1), (2, 10), (3, 12), (7, 1)}
5. $2x - 5y = 1$; {(−7, −3), (7, 3), (2, 1), (−2, −1)}

Solve each equation if the domain is {−3, −1, 0, 2}.

6. $y = 2x - 1$
7. $y = 4 - x$
8. $2y + 2x = 12$
9. $3x + 2y = 13$

Solve each equation for the given domain. Graph the solution set.

10. $y = 3x$ for $x = \{-3, -2, -1, 0, 1, 2, 3\}$
11. $2y = x + 2$ for $x = \{-4, -2, 0, 2, 4\}$

Application **JEWELRY For Exercises 12 and 13, use the following information.**
Since pure gold is very soft, other metals are often added to it to make an alloy that is stronger and more durable. The relative amount of gold in a piece of jewelry is measured in karats. The formula for the relationship is $g = \frac{25k}{6}$, where k is the number of karats and g is the percent of gold in the jewelry.

12. Find the percent of gold if the domain is $\{10, 14, 18, 24\}$. Make a table of values and graph the function.

13. How many karats are in a ring that is 50% gold?

Practice and Apply

Homework Help

For Exercises	See Examples
14–19	1
20–31	2
32–39	3
40–45	4

Extra Practice
See page 829.

Find the solution set for each equation, given the replacement set.

14. $y = 4x + 1$; $\{(2, -1), (1, 5), (9, 2), (0, 1)\}$

15. $y = 8 - 3x$; $\{(4, -4), (8, 0), (2, 2), (3, 3)\}$

16. $x - 3y = -7$; $\{(-1, 2), (2, -1), (2, 4), (2, 3)\}$

17. $2x + 2y = 6$; $\{(3, 0), (2, 1), (-2, -1), (4, -1)\}$

18. $3x - 8y = -4$; $\{(0, 0.5), (4, 1), (2, 0.75), (2, 4)\}$

19. $2y + 4x = 8$; $\{(0, 2), (-3, 0.5), (0.25, 3.5), (1, 2)\}$

Solve each equation if the domain is $\{-2, -1, 1, 3, 4\}$.

20. $y = 4 - 5x$ **21.** $y = 2x + 3$ **22.** $x = y + 4$

23. $x = 7 - y$ **24.** $6x - 3y = 18$ **25.** $6x - y = -3$

26. $8x + 4y = 12$ **27.** $2x - 2y = 0$ **28.** $5x - 10y = 20$

29. $3x + 2y = 14$ **30.** $x + \frac{1}{2}y = 8$ **31.** $2x - \frac{1}{3}y = 4$

Solve each equation for the given domain. Graph the solution set.

32. $y = 2x + 3$ for $x = \{-3, -2, -1, 1, 2, 3\}$

33. $y = 3x - 1$ for $x = \{-5, -2, 1, 3, 4\}$

34. $3x - 2y = 5$ for $x = \{-3, -1, 2, 4, 5\}$

35. $5x + 4y = 8$ for $x = \{-4, -1, 0, 2, 4, 6\}$

36. $\frac{1}{2}x + y = 2$ for $x = \{-4, -1, 1, 4, 7, 8\}$

37. $y = \frac{1}{4}x - 3$ for $x = \{-4, -2, 0, 2, 4, 6\}$

38. The domain for $3x + y = 8$ is $\{-1, 2, 5, 8\}$. Find the range.

39. The range for $2y - x = 6$ is $\{-4, -3, 1, 6, 7\}$. Find the domain.

TRAVEL For Exercises 40 and 41, use the following information.
Heinrich and his brother live in Germany. They are taking a trip to the United States and have been checking the average temperatures in different U.S. cities for the month they will be traveling. They are unfamiliar with the Fahrenheit scale, so they would like to convert the temperatures to Celsius. The equation $F = 1.8C + 32$ relates the temperature in degrees Celsius C to degrees Fahrenheit F.

City	Temperature (°F)
New York	34
Chicago	23
San Francisco	55
Miami	72
Washington, D.C.	40

40. Solve the equation for C.

41. Find the temperatures in degrees Celsius for each city.

GEOMETRY For Exercises 42–44, use the following information.
The equation for the perimeter of a rectangle is $P = 2\ell + 2w$. Suppose the perimeter of rectangle $ABCD$ is 24 centimeters.

42. Solve the equation for ℓ.

43. State the independent and dependent variables.

44. Choose five values for w and find the corresponding values of ℓ.

45. ANTHROPOLOGY When the remains of ancient people are discovered, usually only a few bones are found. Anthropologists can determine a person's height by using a formula that relates the length of the tibia T (shin bone) to the person's height H, both measured in centimeters. The formula for males is $H = 81.7 + 2.4T$ and for females is $H = 72.6 + 2.5T$. Copy and complete the tables below. Then graph each set of ordered pairs.

Male		
Length of Tibia (cm)	Height (cm)	(T, H)
30.5		
34.8		
36.3		
37.9		

Female		
Length of Tibia (cm)	Height (cm)	(T, H)
30.5		
34.8		
36.3		
37.9		

Career Choices

Forensic Anthropologist

Forensic anthropologists assist police investigations. They can determine the age and stature of a victim by examining dental wear on the teeth and measuring certain bones.

Online Research For information about a career as a forensic anthropologist, visit: www.algebra1.com/careers

46. RESEARCH Choose a country that you would like to visit. Use the Internet or other reference to find the cost of various services such as hotels, meals, and transportation. Use the currency exchange rate to determine how much money in U.S. dollars you will need on your trip.

47. CRITICAL THINKING Find the domain values of each relation if the range is $\{0, 16, 36\}$.

a. $y = x^2$ **b.** $y = |4x| - 16$ **c.** $y = |4x - 16|$

48. CRITICAL THINKING Select five values for the domain and find the range of $y = x + 4$. Then look at the range and domain of the inverse relation. Make a conjecture about the equation that represents the inverse relation.

49. WRITING IN MATH Answer the question that was posed at the beginning of the lesson.

Why are equations of relations important in traveling?

Include the following in your answer:

- an example of how you would keep track of how much you were spending in pounds and the equivalent amount in dollars, and
- an explanation of your spending power if the currency exchange rate is 0.90 pound compared to one U.S. dollar or 1.04 pounds compared to one dollar.

Standardized Test Practice

50. If $3x - y = 18$ and $y = 3$, then $x =$

Ⓐ 4. Ⓑ 5. Ⓒ 6. Ⓓ 7.

51. If the perimeter of a rectangle is 14 units and the area is 12 square units, what are the dimensions of the rectangle?

Ⓐ 2×6 Ⓑ 3×3

Ⓒ 3×4 Ⓓ 1×12

Graphing Calculator

TABLE FEATURE You can enter selected x values in the TABLE feature of a graphing calculator, and it will calculate the corresponding y values for a given equation. To do this, enter an equation into the Y= list. Go to TBLSET and highlight Ask under the Independent variable. Now you can use the TABLE function to enter any domain value and the corresponding range value will appear in the second column.

Use a graphing calculator to find the solution set for the given equation and domain.

52. $y = 3x - 4; x = \{-11, 15, 23, 44\}$

53. $y = -6.5x + 42; x = \{-8, -5, 0, 3, 7, 12\}$

54. $y = 3x + 12$ for $x = \{0.4, 0.6, 1.8, 2.2, 3.1\}$

55. $y = 1.4x - 0.76$ for $x = \{-2.5, -1.75, 0, 1.25, 3.33\}$

Maintain Your Skills

Mixed Review

Express the relation shown in each table, mapping, or graph as a set of ordered pairs. Then write the inverse of the relation. *(Lesson 4-3)*

56.

x	y
4	9
3	−2
1	5
−4	2

57. 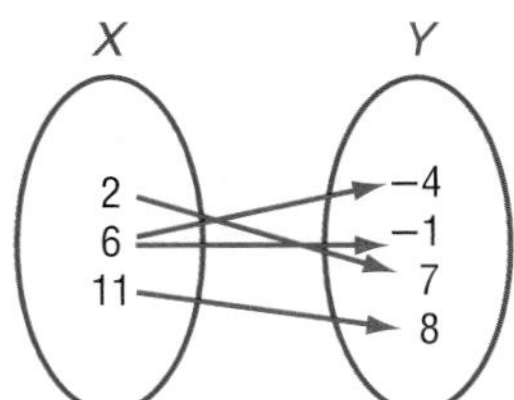

58. 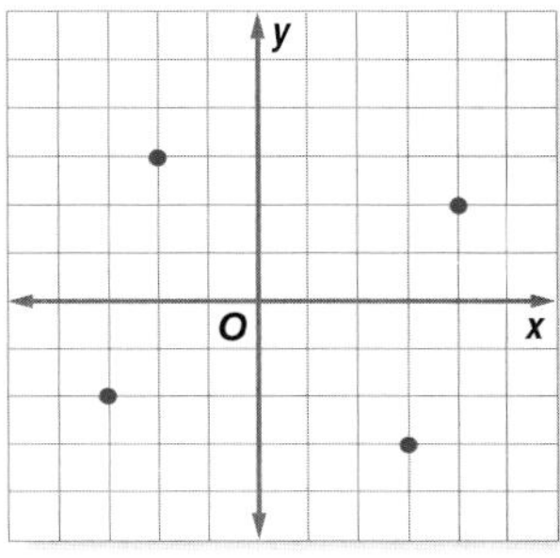

Find the coordinates of the vertices of each figure after the given transformation is performed. Then graph the preimage and its image. *(Lesson 4-2)*

59. triangle XYZ with $X(-6, 4)$, $Y(-5, 0)$, and $Z(3, 3)$ reflected over the y-axis

60. quadrilateral $QRST$ with $Q(2, 2)$, $R(3, -3)$, $S(-1, -4)$ and $T(-4, -3)$ rotated 90° counterclockwise about the origin

Use cross products to determine whether each pair of ratios forms a proportion. Write *yes* or *no*. *(Lesson 3-6)*

61. $\frac{6}{15}, \frac{18}{45}$

62. $\frac{11}{12}, \frac{33}{34}$

63. $\frac{8}{22}, \frac{20}{55}$

64. $\frac{6}{8}, \frac{3}{4}$

65. $\frac{3}{5}, \frac{9}{25}$

66. $\frac{26}{35}, \frac{12}{15}$

Identify the hypothesis and conclusion of each statement. *(Lesson 1-7)*

67. If it is hot, then we will go swimming.

68. If you do your chores, then you get an allowance.

69. If $3n - 7 = 17$, then $n = 8$.

70. If $a > b$ and $b > c$, then $a > c$.

Getting Ready for the Next Lesson

PREREQUISITE SKILL **Solve each equation.** *(To review **solving equations**, see Lesson 3-4.)*

71. $a + 15 = 20$

72. $r - 9 = 12$

73. $-4 = 5n + 6$

74. $3 - 8w = 35$

75. $\frac{g}{4} + 2 = 5$

76. $\frac{m}{5} + \frac{3}{5} = 2$

4-5 Graphing Linear Equations

What You'll Learn

- Determine whether an equation is linear.
- Graph linear equations.

How can linear equations be used in nutrition?

Vocabulary

- linear equation
- standard form
- *x*-intercept
- *y*-intercept

Nutritionists recommend that no more than 30% of a person's daily caloric intake come from fat. Each gram of fat contains nine Calories. To determine the most grams of fat *f* you should have, find the total number of Calories *C* you consume each day and use the equation $f = 0.3\left(\frac{C}{9}\right)$ or $f = \left(\frac{C}{30}\right)$. The graph of this equation shows the maximum number of grams of fat you can consume based on the total number of Calories consumed.

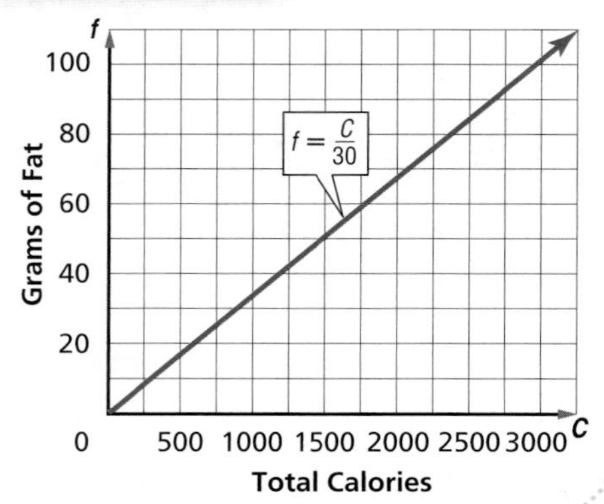

IDENTIFY LINEAR EQUATIONS A **linear equation** is the equation of a line. Linear equations can often be written in the form $Ax + By = C$. This is called the **standard form** of a linear equation.

Key Concepts *Standard Form of a Linear Equation*

The standard form of a linear equation is

$$Ax + By = C,$$

where $A \geq 0$, A and B are not both zero, and A, B, and C are real numbers.

Example 1 Identify Linear Equations

Determine whether each equation is a linear equation. If so, write the equation in standard form.

a. $y = 5 - 2x$

First rewrite the equation so that both variables are on the same side of the equation.

$y = 5 - 2x$	Original equation
$y + 2x = 5 - 2x + 2x$	Add $2x$ to each side.
$2x + y = 5$	Simplify.

The equation is now in standard form where $A = 2$, $B = 1$, and $C = 5$. This is a linear equation.

b. $2xy - 5y = 6$

Since the term $2xy$ has two variables, the equation cannot be written in the form $Ax + By = C$. Therefore, this is not a linear equation.

c. $3x + 9y = 15$

Since the GCF of 3, 9, and 15 is not 1, the equation is not written in standard form. Divide each side by the GCF.

$3x + 9y = 15$	Original equation
$3(x + 3y) = 15$	Factor the GCF.
$\frac{3(x + 3y)}{3} = \frac{15}{3}$	Divide each side by 3.
$x + 3y = 5$	Simplify.

The equation is now in standard form where $A = 1$, $B = 3$, and $C = 5$.

d. $\frac{1}{3}y = -1$

To write the equation with integer coefficients, multiply each term by 3.

$\frac{1}{3}y = -1$	Original equation
$3\left(\frac{1}{3}\right)y = 3(-1)$	Multiply each side of the equation by 3.
$y = -3$	Simplify.

The equation $y = -3$ can be written as $0x + y = -3$. Therefore, it is a linear equation in standard form where $A = 0$, $B = 1$, and $C = -3$.

GRAPH LINEAR EQUATIONS The graph of a linear equation is a line. The line represents all the solutions of the linear equation. Also, every ordered pair on this line satisfies the equation.

Example 2 Graph by Making a Table

Graph $x + 2y = 6$.

In order to find values for y more easily, solve the equation for y.

$x + 2y = 6$	Original equation
$x + 2y - x = 6 - x$	Subtract x from each side.
$2y = 6 - x$	Simplify.
$\frac{2y}{2} = \frac{6 - x}{2}$	Divide each side by 2.
$y = 3 - \frac{1}{2}x$	Simplify.

Select five values for the domain and make a table. Then graph the ordered pairs.

x	$3 - \frac{1}{2}x$	y	(x, y)
−2	$3 - \frac{1}{2}(-2)$	4	(−2, 4)
0	$3 - \frac{1}{2}(0)$	3	(0, 3)
2	$3 - \frac{1}{2}(2)$	2	(2, 2)
4	$3 - \frac{1}{2}(4)$	1	(4, 1)
6	$3 - \frac{1}{2}(6)$	0	(6, 0)

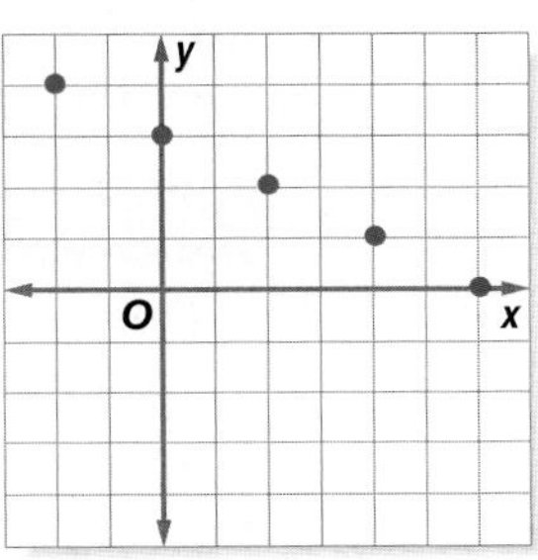

(continued on the next page)

Study Tip

Graphing Equations

When you graph an equation, use arrows at both ends to show that the graph continues. You should also label the graph with the equation.

When you graph the ordered pairs, a pattern begins to form. The domain of $y = 3 - \frac{1}{2}x$ is the set of all real numbers, so there are an infinite number of solutions of the equation. Draw a line through the points. This line represents all of the solutions of $y = 3 - \frac{1}{2}x$.

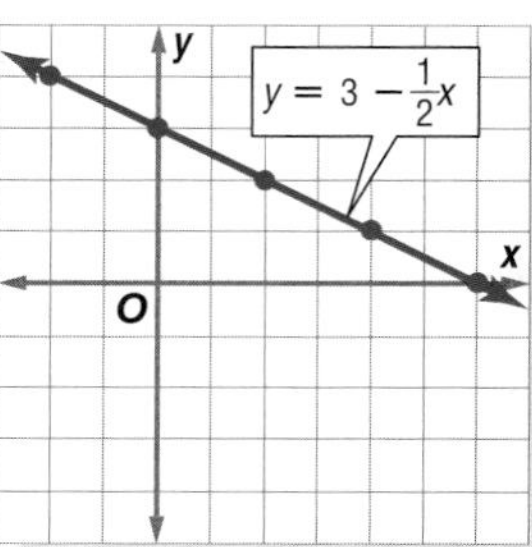

Example 3 Use the Graph of a Linear Equation

PHYSICAL FITNESS Carlos swims every day. He burns approximately 10.6 Calories per minute when swimming laps.

a. Graph the equation $C = 10.6t$, where C represents the number of Calories burned and t represents the time in minutes spent swimming.

Select five values for t and make a table. Graph the ordered pairs and connect them to draw a line.

t	$10.6t$	C	(t, C)
10	10.6(10)	106	(10, 106)
15	10.6(15)	159	(15, 159)
20	10.6(20)	212	(20, 212)
30	10.6(30)	318	(30, 318)

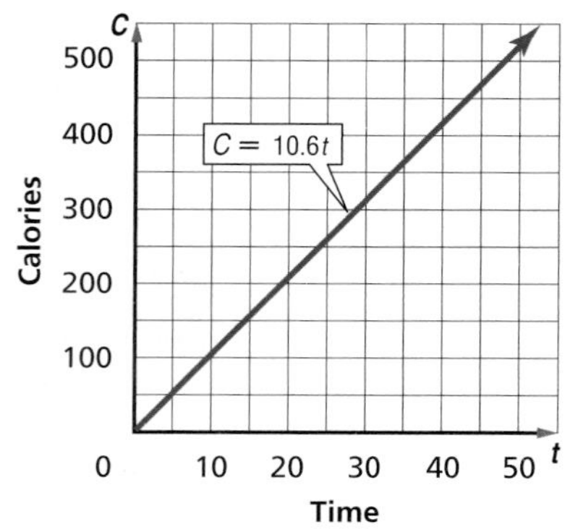

More About...

Physical Fitness

In a triathlon competition, athletes swim 1.5 kilometers, bicycle 40 kilometers, and run 10 kilometers.

Source: www.usatriathlon.org

b. Suppose Carlos wanted to burn 350 Calories. Approximately how long should he swim?

Since any point on the line is a solution of the equation, use the graph to estimate the value of the x-coordinate in the ordered pair that contains 350 as the y-coordinate. The ordered pair (33, 350) appears to be on the line so Carlos should swim for 33 minutes to burn 350 Calories. *Check this solution algebraically by substituting (33, 350) into the original equation.*

Since two points determine a line, a simple method of graphing a linear equation is to find the points where the graph crosses the x-axis and the y-axis. The x-coordinate of the point at which it crosses the x-axis is the **x-intercept**, and the y-coordinate of the point at which the graph crosses the y-axis is called the **y-intercept**.

Example 4 Graph Using Intercepts

Determine the x-intercept and y-intercept of $3x + 2y = 9$. Then graph the equation.

To find the x-intercept, let $y = 0$.

$3x + 2y = 9$	Original equation
$3x + 2(0) = 9$	Replace y with 0.
$3x = 9$	Divide each side by 3.
$x = 3$	

To find the y-intercept, let $x = 0$.

$3x + 2y = 9$	Original equation
$3(0) + 2y = 9$	Replace x with 0.
$2y = 9$	Divide each side by 2.
$y = 4.5$	

The x-intercept is 3, so the graph intersects the x-axis at (3, 0). The y-intercept is 4.5, so the graph intersects the y-axis at (0, 4.5). Plot these points. Then draw a line that connects them.

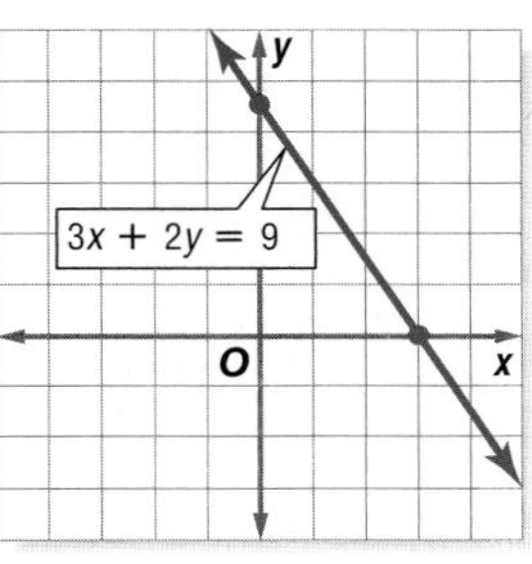

Check for Understanding

Concept Check

1. **Explain** how the graph of $y = 2x + 1$ for the domain {1, 2, 3, 4} differs from the graph of $y = 2x + 1$ for the domain of all real numbers.
2. **OPEN ENDED** Give an example of a linear equation in the form $Ax + By = C$ for each of the following conditions.
   a. $A = 0$ **b.** $B = 0$ **c.** $C = 0$
3. **Explain** how to graph an equation using the x- and y-intercepts.

Guided Practice

Determine whether each equation is a linear equation. If so, write the equation in standard form.

4. $x + y^2 = 25$ **5.** $3y + 2 = 0$

6. $\frac{3}{5}x - \frac{2}{5}y = 5$ **7.** $x + \frac{1}{y} = 7$

Graph each equation.

8. $x = 3$ **9.** $x - y = 0$ **10.** $y = 2x + 8$

11. $y = -3 - x$ **12.** $x + 4y = 10$ **13.** $4x + 3y = 12$

Application

TAXI FARE For Exercises 14 and 15, use the following information.
A taxi company charges a fare of \$2.25 plus \$0.75 per mile traveled. The cost of the fare c can be described by the equation $c = 0.75m + 2.25$, where m is the number of miles traveled.

14. Graph the equation.

15. If you need to travel 18 miles, how much will the taxi fare cost?

Practice and Apply

Homework Help

For Exercises	See Examples
16–25	1
26–45	2, 4
46–56	3

Extra Practice
See page 829.

Determine whether each equation is a linear equation. If so, write the equation in standard form.

16. $3x = 5y$ **17.** $6 - y = 2x$

18. $6xy + 3x = 4$ **19.** $y + 5 = 0$

20. $7y = 2x + 5x$ **21.** $y = 4x^2 - 1$

22. $\frac{3}{x} + \frac{4}{y} = 2$ **23.** $\frac{x}{2} = 10 + \frac{2y}{3}$

24. $7n - 8m = 4 - 2m$ **25.** $3a + b - 2 = b$

Graph each equation.

26. $y = -1$ **27.** $y = 2x$ **28.** $y = 5 - x$

29. $y = 2x - 8$ **30.** $y = 4 - 3x$ **31.** $y = x - 6$

32. $x = 3y$ **33.** $x = 4y - 6$ **34.** $x - y = -3$

35. $x + 3y = 9$ **36.** $4x + 6y = 8$ **37.** $3x - 2y = 15$

Graph each equation.

38. $1.5x + y = 4$
39. $2.5x + 5y = 75$
40. $\frac{1}{2}x + y = 4$
41. $x - \frac{2}{3}y = 1$
42. $\frac{4x}{3} = \frac{3y}{4} + 1$
43. $y + \frac{1}{3} = \frac{1}{4}x - 3$
44. Find the x- and y-intercept of the graph of $4x - 7y = 14$.
45. Write an equation in standard form of the line with an x-intercept of 3 and a y-intercept of 5.

GEOMETRY **For Exercises 46–48, refer to the figure.** The perimeter P of a rectangle is given by $2\ell + 2w = P$, where ℓ is the length of the rectangle and w is the width.

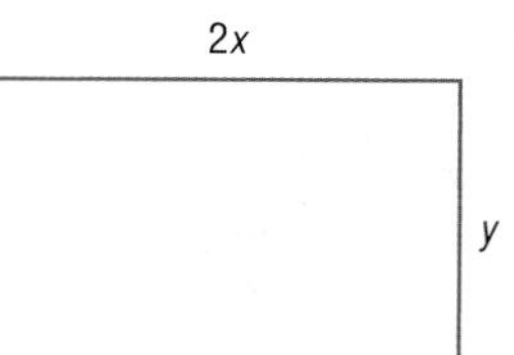

46. If the perimeter of the rectangle is 30 inches, write an equation for the perimeter in standard form.
47. What are the x- and y-intercepts of the graph of the equation?
48. Graph the equation.

METEOROLOGY **For Exercises 49–51, use the following information.** As a thunderstorm approaches, you see lightning as it occurs, but you hear the accompanying sound of thunder a short time afterward. The distance d in miles that sound travels in t seconds is given by the equation $d = 0.21t$.

49. Make a table of values.
50. Graph the equation.
51. Estimate how long it will take to hear the thunder from a storm 3 miles away.

BIOLOGY **For Exercises 52 and 53, use the following information.** The amount of blood in the body can be predicted by the equation $y = 0.07w$, where y is the number of pints of blood and w is the weight of a person in pounds.

52. Graph the equation.
53. Predict the weight of a person whose body holds 12 pints of blood.

OCEANOGRAPHY **For Exercises 54–56, use the information at left and below.** Under water, pressure increases 4.3 pounds per square inch (psi) for every 10 feet you descend. This can be expressed by the equation $p = 0.43d + 14.7$, where p is the pressure in pounds per square inch and d is the depth in feet.

54. Graph the equation.
55. Divers cannot work at depths below about 400 feet. What is the pressure at this depth?
56. How many times as great is the pressure at 400 feet as the pressure at sea level?
57. **CRITICAL THINKING** Explain how you can determine whether a point at (x, y) is *above*, *below*, or *on* the line given by $2x - y = 8$ without graphing it. Give an example of each.
58. **WRITING IN MATH** Answer the question that was posed at the beginning of the lesson.

How can linear equations be used in nutrition?

Include the following in your answer:

- an explanation of how you could use the Nutrition Information labels on packages to limit your fat intake, and
- an equation you could use to find how many grams of protein you should have each day if you wanted 10% of your diet to consist of protein. (*Hint*: Protein contains 4 Calories per gram.)

More About. . .

Oceanography

How heavy is air? The atmospheric pressure is a measure of the weight of air. At sea level, air pressure is 14.7 pounds per square inch.

Source: www.brittanica.com

Standardized Test Practice

59. Which point lies on the line given by $y = 3x - 5$?

(A) (1, −2) (B) (0, 5) (C) (1, 2) (D) (4, 3)

60. In the graph at the right, (0, 1) and (4, 3) lie on the line. Which ordered pair also lies on the line?

(A) (1, 1) (B) (2, 2)

(C) (3, 3) (D) (4, 4)

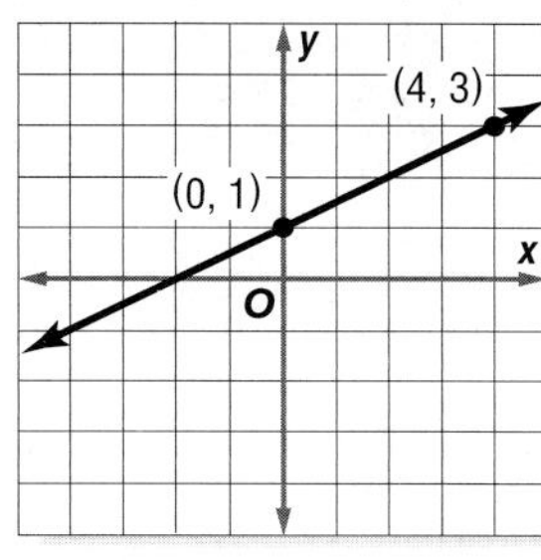

Maintain Your Skills

Mixed Review

Solve each equation if the domain is {−3, −1, 2, 5, 8}. *(Lesson 4-4)*

61. $y = x - 5$
62. $y = 2x + 1$
63. $3x + y = 12$
64. $2x - y = -3$
65. $3x - \frac{1}{2}y = 6$
66. $-2x + \frac{1}{3}y = 4$

Express each relation as a table, a graph, and a mapping. Then determine the domain and range. *(Lesson 4-3)*

67. {(3, 5), (−4, −1), (−3, 2), (3, 1)}
68. {(4, 0), (2, −3), (−1, −3), (4, 4)}
69. {(1, 4), (3, 0), (−1, −1), (3, 5)}
70. {(4, 5), (2, 5), (4, −1), (3, 2)}

Solve each equation. Then check your solution. *(Lesson 3-5)*

71. $2(x - 2) = 3x - (4x - 5)$
72. $3a + 8 = 2a - 4$
73. $3n - 12 = 5n - 20$
74. $6(x + 3) = 3x$

ANIMALS **For Exercises 75–78, use the table below that shows the average life spans of 20 different animals.** *(Lesson 2-5)*

Animal	Life Span (years)	Animal	Life Span (years)	Animal	Life Span (years)
Baboon	20	Lion	15	Squirrel	10
Camel	12	Monkey	15	Tiger	16
Cow	15	Mouse	3	Wolf	5
Elephant	40	Opossum	1	Zebra	15
Fox	7	Pig	10		
Gorilla	20	Rabbit	5		
Hippopotamus	25	Sea Lion	12		
Kangaroo	7	Sheep	12		

75. Make a line plot of the average life spans of the animals in the table.
76. How many animals live between 7 and 16 years?
77. Which number occurred most frequently?
78. How many animals live at least 20 years?

Getting Ready for the Next Lesson

PREREQUISITE SKILL **Evaluate each expression.**
(To review ***evaluating expressions****, see Lesson 1-2.)*

79. $19 + 5 \cdot 4$
80. $(25 - 4) \div (2^2 - 1^3)$
81. $12 \div 4 + 15 \cdot 3$
82. $12(19 - 15) - 3 \cdot 8$
83. $6(4^3 + 2^2)$
84. $7[4^3 - 2(4 + 3)] \div 7 + 2$

Graphing Calculator

A Follow-Up of Lesson 4-5

Graphing Linear Equations

The power of a graphing calculator is the ability to graph different types of equations accurately and quickly. Often linear equations are graphed in the standard viewing window. The **standard viewing window** is $[-10, 10]$ by $[-10, 10]$ with a scale of 1 on both axes. To quickly choose the standard viewing window on a TI-83 Plus, press ZOOM 6.

Example 1

Graph $2x - y = 3$ on a TI-83 Plus graphing calculator.

Step 1 *Enter the equation in the Y= list.*

- The Y= list shows the equation or equations that you will graph.
- Equations must be entered with the y isolated on one side of the equation. Solve the equation for y, then enter it into the calculator.

$2x - y = 3$	Original equation
$2x - y - 2x = 3 - 2x$	Subtract $2x$ from each side.
$-y = -2x + 3$	Simplify.
$y = 2x - 3$	Multiply each side by -1.

KEYSTROKES: Y= 2 X,T,θ,*n* − 3

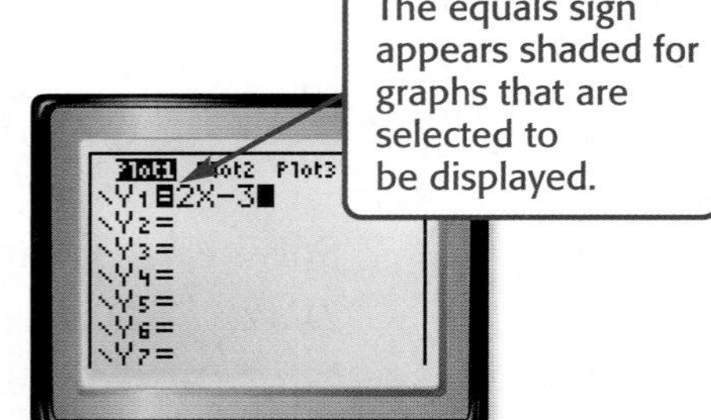
The equals sign appears shaded for graphs that are selected to be displayed.
Plot1 Plot2 Plot3
\Y1=2X-3
\Y2=
\Y3=
\Y4=
\Y5=
\Y6=
\Y7=

Step 2 *Graph the equation in the standard viewing window.*

Graph the selected equations.

KEYSTROKES: ZOOM 6

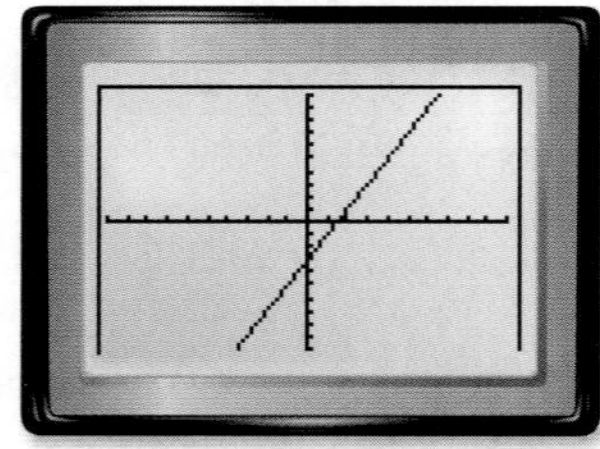

$[-10, 10]$ scl: 1 by $[-10, 10]$ scl: 1

Notice that the graph of $2x - y = 3$ above is a complete graph because all of these points are visible.

Sometimes a complete graph is not displayed using the standard viewing window. A **complete graph** includes all of the important characteristics of the graph on the screen. These include the origin, and the x- and y-intercepts.

When a complete graph is not displayed using the standard viewing window, you will need to change the viewing window to accommodate these important features. You can use what you have learned about intercepts to help you choose an appropriate viewing window.

www.algebra1.com/other_calculator_keystrokes

Investigation

Example 2

Graph $y = 3x - 15$ on a graphing calculator.

Step 1 *Enter the equation in the Y= list and graph in the standard viewing window.*

Clear the previous equation from the Y= list. Then enter the new equation and graph.

KEYSTROKES: [Y=] [CLEAR] 3 [X,T,θ,n] [−] 15 [ZOOM] 6

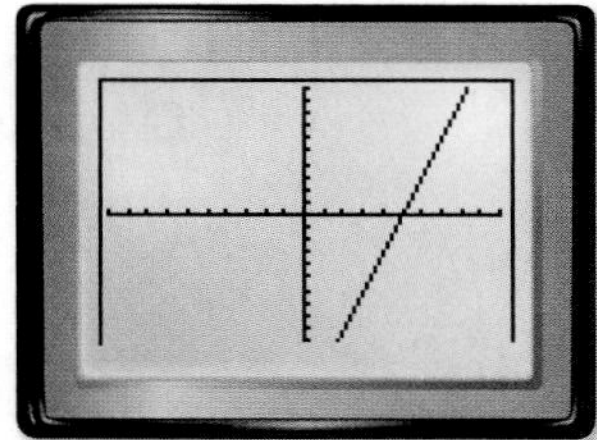

[−10, 10] scl: 1 by [−10, 10] scl: 1

Step 2 *Modify the viewing window and graph again.*

The origin and the x-intercept are displayed in the standard viewing window. But notice that the y-intercept is outside of the viewing window. Find the y-intercept.

$y = 3x - 15$ Original equation

$y = 3(0) - 15$ Replace x with 0.

$y = -15$ Simplify.

Since the y-intercept is -15, choose a viewing window that includes a number less than -15. The window $[-10, 10]$ by $[-20, 5]$ with a scale of 1 on each axis is a good choice.

KEYSTROKES: [WINDOW] −10 [ENTER] 10 [ENTER] 1 [ENTER]
−20 [ENTER] 5 [ENTER] 1 [GRAPH]

This window allows the complete graph, including the y-intercept, to be displayed.

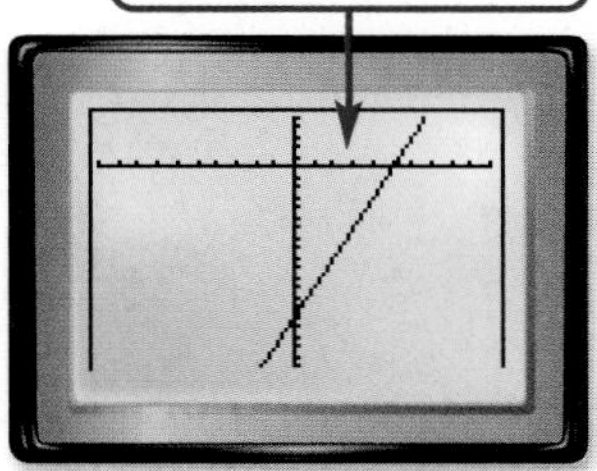

[−10, 10] scl: 1 by [−20, 5] scl: 1

Exercises

Use a graphing calculator to graph each equation in the standard viewing window. Sketch the result.

1. $y = x + 2$ **2.** $y = 4x + 5$ **3.** $y = 6 - 5x$

4. $2x + y = 6$ **5.** $x + y = -2$ **6.** $x - 4y = 8$

Graph each linear equation in the standard viewing window. Determine whether the graph is complete. If the graph is not complete, choose a viewing window that will show a complete graph and graph the equation again.

7. $y = 5x + 9$ **8.** $y = 10x - 6$ **9.** $y = 3x - 18$

10. $3x - y = 12$ **11.** $4x + 2y = 21$ **12.** $3x + 5y = -45$

For Exercises 13–15, consider the linear equation $y = 2x + b$.

13. Choose several different positive and negative values for b. Graph each equation in the standard viewing window.

14. For which values of b is the complete graph in the standard viewing window?

15. How is the value of b related to the y-intercept of the graph of $y = 2x + b$?

4-6 Functions

What You'll Learn

- Determine whether a relation is a function.
- Find function values.

Vocabulary

- function
- vertical line test
- function notation

How are functions used in meteorology?

The table shows barometric pressures and temperatures recorded by the National Climatic Data Center over a three-day period.

Pressure (millibars)	1013	1006	997	995	995	1000	1006	1011	1016	1019
Temperature (°C)	3	4	10	13	8	4	1	−2	−6	−9

Notice that when the pressure is 995 and 1006 millibars, there is more than one value for the temperature.

IDENTIFY FUNCTIONS Recall that relations in which each element of the domain is paired with exactly one element of the range are called **functions**.

Study Tip

Functions
In a function, knowing the value of x tells you the value of y.

Key Concept — *Function*

A function is a relation in which each element of the domain is paired with *exactly* one element of the range.

Example 1 *Identify Functions*

Determine whether each relation is a function. Explain.

a.

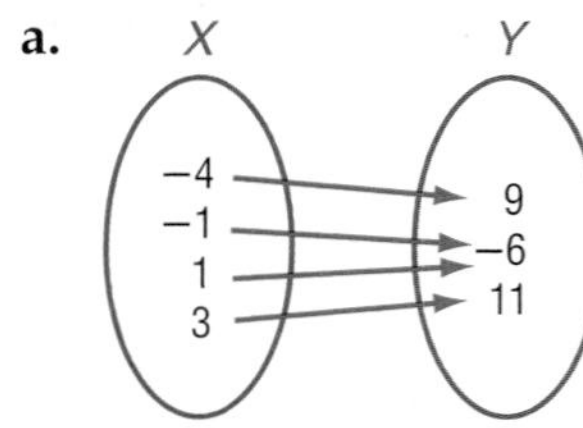

This mapping represents a function since, for each element of the domain, there is only one corresponding element in the range. It does not matter if two elements of the domain are paired with the same element in the range.

b.

x	y
−3	6
2	5
3	1
2	4

This table represents a relation that is not a function. The element 2 in the domain is paired with both 5 and 4 in the range. If you are given that x is 2, you cannot determine the value of y.

c. **{(−2, 4), (1, 5), (3, 6), (5, 8), (7, 10)}**

Since each element of the domain is paired with exactly one element of the range, this relation is a function. If you are given that x is -3, you can determine that the value of y is 6 since 6 is the only value of y that is paired with $x = 3$.

You can use the **vertical line test** to see if a graph represents a function. If no vertical line can be drawn so that it intersects the graph more than once, then the graph is a function. If a vertical line can be drawn so that it intersects the graph at two or more points, the relation is not a function.

Function

Not a Function

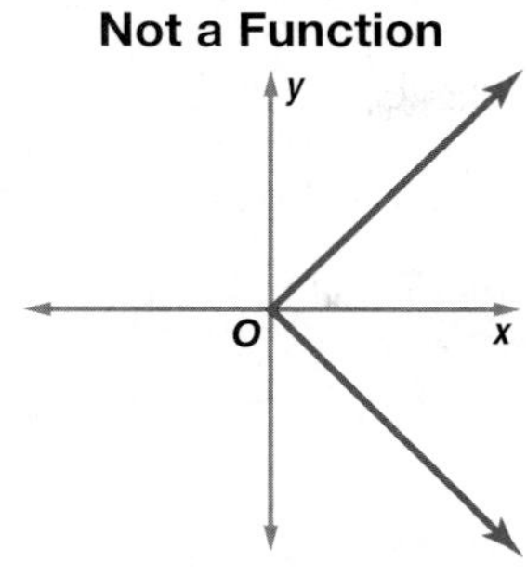

Function

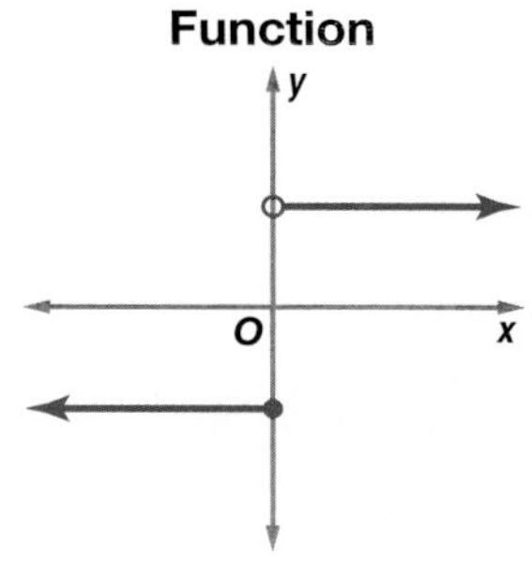

One way to perform the vertical line test is to use a pencil.

Example 2 Equations as Functions

Determine whether $2x - y = 6$ is a function.

Graph the equation using the x- and y-intercepts.

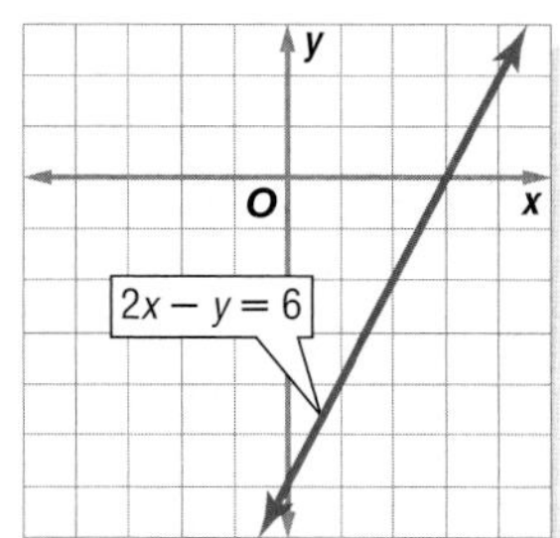

Since the equation is in the form $Ax + By = C$, the graph of the equation will be a line. Place your pencil at the left of the graph to represent a vertical line. Slowly move the pencil to the right across the graph.

For each value of x, this vertical line passes through no more than one point on the graph. Thus, the line represents a function.

FUNCTION VALUES Equations that are functions can be written in a form called **function notation**. For example, consider $y = 3x - 8$.

equation	function notation
$y = 3x - 8$	$f(x) = 3x - 8$

Study Tip

Reading Math
The symbol $f(x)$ is read *f of x*.

In a function, x represents the elements of the domain, and $f(x)$ represents the elements of the range. Suppose you want to find the value in the range that corresponds to the element 5 in the domain. This is written $f(5)$ and is read "f of 5." The value $f(5)$ is found by substituting 5 for x in the equation.

Example 3 Function Values

If $f(x) = 2x + 5$, find each value.

a. $f(-2)$

$f(-2) = 2(-2) + 5$ Replace x with -2.

$= -4 + 5$ Multiply.

$= 1$ Add.

b. $f(1) + 4$

$f(1) + 4 = [2(1) + 5] + 4$ Replace x with 1.

$= 7 + 4$ Simplify.

$= 11$ Add.

c. $f(x + 3)$

$f(x + 3) = 2(x + 3) + 5$ Replace x with $x + 3$.

$= 2x + 6 + 5$ Distributive Property

$= 2x + 11$ Simplify.

The functions we have studied thus far have been linear functions. However, many functions are not linear. You can find the value of these functions in the same way.

Example 4 Nonlinear Function Values

Study Tip

Reading Math
Other letters such as g and h can be used to represent functions, for example, $g(x)$ or $h(z)$.

If $h(z) = z^2 + 3z - 4$, find each value.

a. $h(-4)$

$h(-4) = (-4)^2 + 3(-4) - 4$ Replace z with -4.

$= 16 - 12 - 4$ Multiply.

$= 0$ Simplify.

b. $h(5a)$

$h(5a) = (5a)^2 + 3(5a) - 4$ Replace z with $5a$.

$= 25a^2 + 15a - 4$ Simplify.

c. $2[h(g)]$

$2[h(g)] = 2[(g)^2 + 3(g) - 4]$ Evaluate $h(g)$ by replacing z with g.

$= 2(g^2 + 3g - 4)$ Multiply the value of $h(g)$ by 2.

$= 2g^2 + 6g - 8$ Simplify.

On some standardized tests, an arbitrary symbol may be used to represent a function.

Standardized Test Practice

Example 5 Nonstandard Function Notation

Multiple-Choice Test Item

If $\ll x \gg = x^2 - 4x + 2$, then $\ll 3 \gg =$

(A) -2. (B) -1. (C) 1. (D) 2.

Read the Test Item

The symbol $\ll x \gg$ is just a different notation for $f(x)$.

Solve the Test Item

Replace x with 3.

$\ll x \gg = x^2 - 4x + 2$ Think: $\ll x \gg = f(x)$

$\ll 3 \gg = (3)^2 - 4(3) + 2$ Replace x with 3.

$= 9 - 12 + 2$ or -1 The answer is B.

Test-Taking Tip
If the nonstandard function notation confuses you, replace the arbitrary symbol with $f(x)$.

Check for Understanding

Concept Check

1. **Study** the following set of ordered pairs that describe a relation between x and y: $\{(1, -1), (-1, 2), (4, -3), (3, 2), (-2, 4), (3, -3)\}$. Is y a function of x? Is x a function of y? Explain your answer.
2. **OPEN ENDED** Define a function using nonstandard function notation.
3. **Find a counterexample** to disprove the following statement.
All linear equations are functions.

Guided Practice

Determine whether each relation is a function.

4.

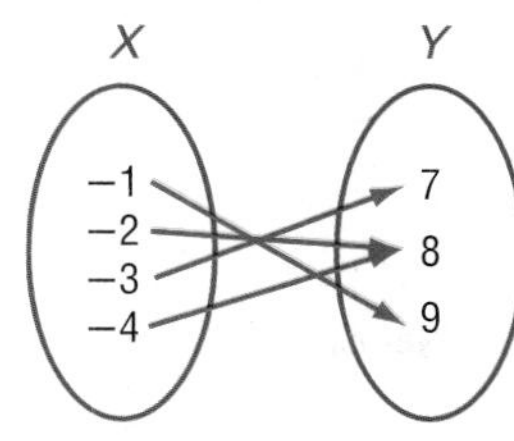

5.

x	y
−3	0
2	1
2	4
6	5

6. $\{(24, 1), (21, 4), (3, 22), (24, 5)\}$

7. $y = x + 3$

8.

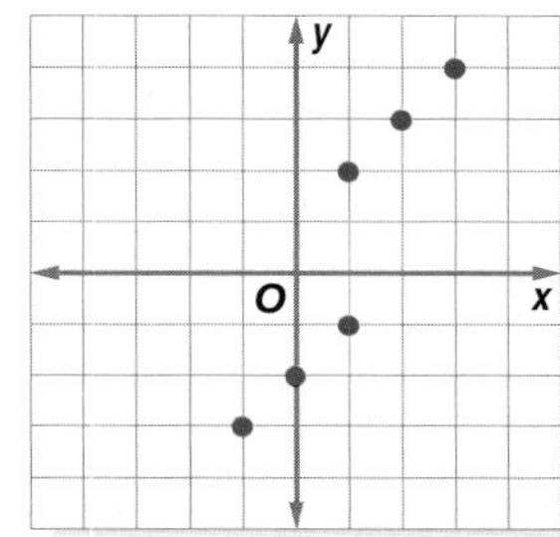

9.

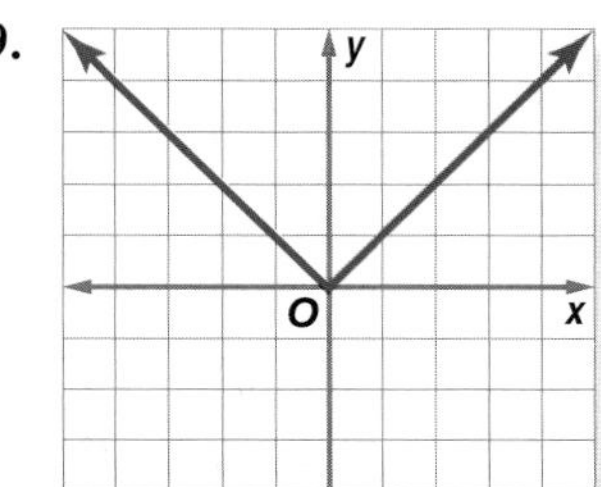

If $f(x) = 4x - 5$ and $g(x) = x^2 + 1$, find each value.

10. $f(2)$

11. $g(-1)$

12. $f(c)$

13. $g(t) - 4$

14. $f(3a^2)$

15. $f(x + 5)$

16. If $x^{**} = 2x - 1$, then $5^{**} - 2^{**} =$

Ⓐ 3. Ⓑ 4. Ⓒ 5. Ⓓ 6.

Practice and Apply

Homework Help

For Exercises	See Examples
17–31, 44	1, 2
32–43, 45–51	3–5

Extra Practice

See page 830.

Determine whether each relation is a function.

17.

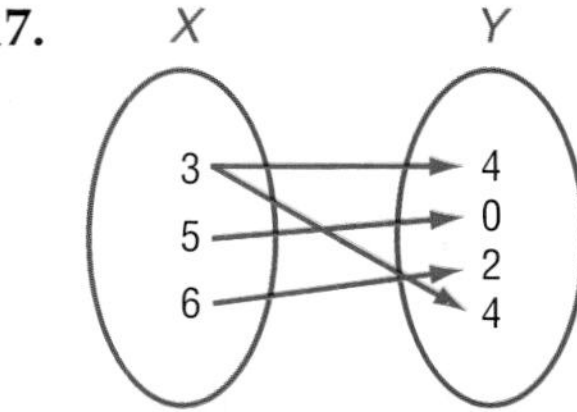

18.

X Y

1 → 4
−2 → 7
3 → 9
−4 → 6

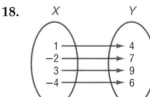

19.

x	y
2	7
4	9
5	5
8	−1

20.

x	y
−9	−5
−4	0
3	6
7	1
6	−5
3	2

21.

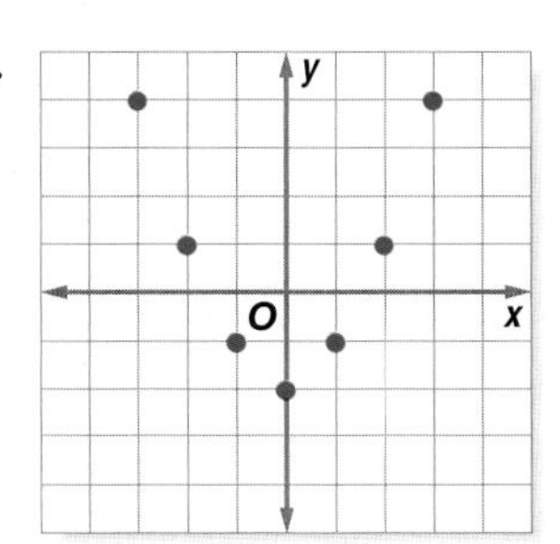

22.

23. $\{(5, -7), (6, -7), (-8, -1), (0, -1)\}$

24. $\{(4, 5), (3, -2), (-2, 5), (4, 7)\}$

25. $y = -8$

26. $x = 15$

27. $y = 3x - 2$

28. $y = 3x + 2y$

Determine whether each relation is a function.

29. 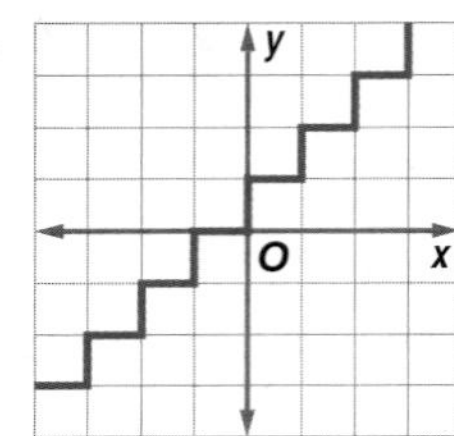

30. 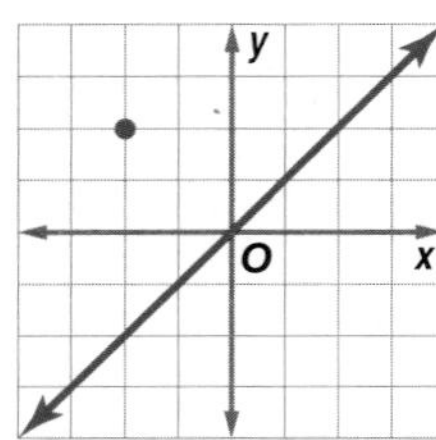

31. 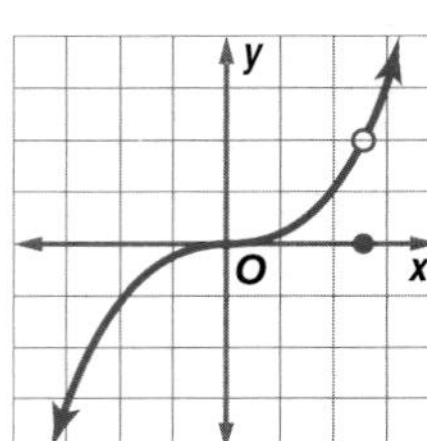

A graph of the winning Olympic swimming times will help you determine whether the winning time is a function of the year. Visit www.algebra1.com/webquest to continue work on your WebQuest project.

If $f(x) = 3x + 7$ and $g(x) = x^2 - 2x$, find each value.

32. $f(3)$
33. $f(-2)$
34. $g(5)$
35. $g(0)$
36. $g(-3) + 1$
37. $f(8) - 5$
38. $g(2c)$
39. $f(a^2)$
40. $f(k + 2)$
41. $f(2m - 5)$
42. $3[g(x) + 4]$
43. $2[f(x^2) - 5]$

44. PARKING The rates for a parking garage are as follows: \$2.00 for the first hour; \$2.75 for the second hour; \$3.50 for the third hour; \$4.25 for the fourth hour; and \$5.00 for any time over four hours. Choose the graph that best represents the information given and determine whether the graph represents a function. Explain your reasoning.

a. 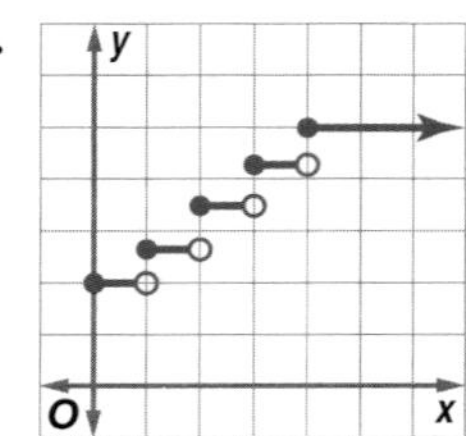

b. 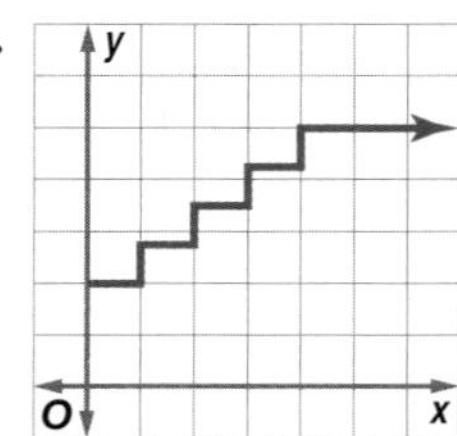

c. 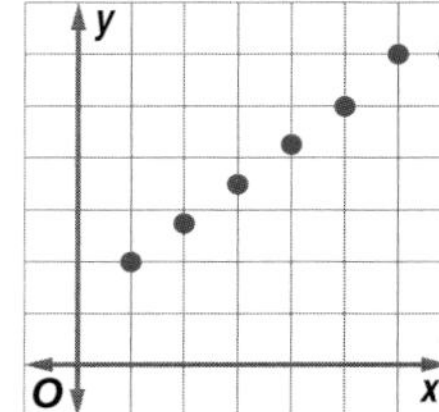

More About. . .

Climate

Earth's average land surface temperature has risen 0.8–1.0°F in the last century. Scientists believe it could rise 1–4.5°F in the next fifty years and 2.2–10°F in the next century.

Source: Environmental Protection Agency

CLIMATE For Exercises 45–48, use the following information.
The temperature of the atmosphere decreases about 5°F for every 1000 feet increase in altitude. Thus, if the temperature at ground level is 77°F, the temperature at a given altitude is found by using the equation $t = 77 - 0.005h$, where h is the height in feet.

45. Write the equation in function notation.

46. Find $f(100)$, $f(200)$, and $f(1000)$.

47. Graph the equation.

48. Use the graph of the function to determine the temperature at 4000 feet.

EDUCATION For Exercises 49–51, use the following information.
The National Assessment of Educational Progress tests 4th, 8th, and 12th graders in the United States. The average math test scores for 17-year-olds can be represented as a function of the science scores by $f(s) = 0.8s + 72$, where $f(s)$ is the math score and s is the science score.

49. Graph this function.

50. What is the science score that corresponds to a math score of 308?

51. Krista scored 260 in science and 320 in math. How does her math score compare to the average score of other students who scored 260 in science? Explain your answer.

52. CRITICAL THINKING State whether the following is *sometimes*, *always*, or *never* true.

The inverse of a function is also a function.

53. **WRITING IN MATH** Answer the question that was posed at the beginning of the lesson.

How are functions used in meteorology?

Include the following in your answer:

- a description of the relationship between pressure and temperature, and
- an explanation of whether the relation is a function.

Standardized Test Practice

54. If $f(x) = 20 - 2x$, find $f(7)$.

Ⓐ 6 Ⓑ 7 Ⓒ 13 Ⓓ 14

55. If $f(x) = 2x$, which of the following statements must be true?

I. $f(3x) = 3[f(x)]$

II. $f(x + 3) = f(x) + 3$

III. $f(x^2) = [f(x)]^2$

Ⓐ I only Ⓑ II only Ⓒ I and II only Ⓓ I, II, and III

Maintain Your Skills

Mixed Review

Graph each equation. *(Lesson 4-5)*

56. $y = x + 3$ 57. $y = 2x - 4$ 58. $2x + 5y = 10$

Find the solution set for each equation, given the replacement set. *(Lesson 4-4)*

59. $y = 5x - 3$; $\{(3, 12), (1, -2), (-2, -7), (-1, -8)\}$

60. $y = 2x + 6$; $\{(3, 0), (-1, 4), (6, 0), (5, -1)\}$

61. **RUNNING** Adam is training for an upcoming 26-mile marathon. He can run a 10K race (about 6.2 miles) in 45 minutes. If he runs the marathon at the same pace, how long will it take him to finish? *(Lesson 3-6)*

Name the property used in each equation. Then find the value of *n*. *(Lesson 1-4)*

62. $16 = n + 16$ 63. $3.5 + 6 = n + 6$ 64. $\frac{3}{5}n = \frac{3}{5}$

Getting Ready for the Next Lesson

PREREQUISITE SKILL **Find each difference.**
*(To review **subtracting integers**, see Lesson 2-2.)*

65. $12 - 16$ 66. $-5 - (-8)$ 67. $16 - (-4)$

68. $-9 - 6$ 69. $\frac{3}{4} - \frac{1}{8}$ 70. $3\frac{1}{2} - \left(-1\frac{2}{3}\right)$

Practice Quiz 2 — Lessons 4-4 through 4-6

Solve each equation if the domain is $\{-3, -1, 0, 2, 4\}$. *(Lesson 4-4)*

1. $y = x + 5$ 2. $y = 3x + 4$ 3. $x + 2y = 8$

Graph each equation. *(Lesson 4-5)*

4. $y = x - 2$ 5. $3x + 2y = 6$

Determine whether each relation is a function. *(Lesson 4-6)*

6. $\{(3, 4), (5, 3), (-1, 4), (6, 2)\}$ 7. $\{(-1, 4), (-2, 5), (7, 2), (3, 9), (-2, 1)\}$

If $f(x) = 3x + 5$, find each value. *(Lesson 4-6)*

8. $f(-4)$ 9. $f(2a)$ 10. $f(x + 2)$

Spreadsheet Investigation

A Preview of Lesson 4-7

Number Sequences

You can use a spreadsheet to generate number sequences and patterns. The simplest type of sequence is one in which the difference between successive terms is constant. This type of sequence is called an **arithmetic sequence**.

Example

Use a spreadsheet to generate a sequence of numbers from an initial value of 10 to 90 with a fixed interval of 8.

Step 1 Enter the initial value 10 in cell A1.

Step 2 Highlight the cells in column A. Under the Edit menu, choose the Fill option and then Series.

Step 3 A command box will appear on the screen asking for the Step value and the Stop value. The Step value is the fixed interval between each number, which in this case is 8. The Stop value is the last number in your sequence, 90. Enter these numbers and click OK. The column is filled with the numbers in the sequence from 10 to 90 at intervals of 8.

	A	B
1	10	
2	18	
3	26	
4	34	
5	42	
6	50	
7	58	
8	66	
9	74	
10	82	
11	90	
12		
13		
14		

Exercises

For Exercises 1–5, use a sequence of numbers from 7 to 63 with a fixed interval of 4.

1. Use a spreadsheet to generate the sequence. Write the numbers in the sequence.
2. How many numbers are in the sequence?

MAKE A CONJECTURE Let a_n represent each number in a sequence if n is the position of the number in the sequence. For example, a_1 = the first number in the sequence, a_2 = the second number, a_3 = the third number, and so on.

3. Write a formula for a_2 in terms of a_1. Write similar formulas for a_3 and a_4 in terms of a_1.
4. Look for a pattern. Write an equation that can be used to find the nth term of a sequence.
5. Use the equation from Exercise 4 to find the 21st term in the sequence.

4-7 Arithmetic Sequences

What You'll Learn

- Recognize arithmetic sequences.
- Extend and write formulas for arithmetic sequences.

How are arithmetic sequences used to solve problems in science?

A probe to measure air quality is attached to a hot-air balloon. The probe has an altitude of 6.3 feet after the first second, 14.5 feet after the next second, 22.7 feet after the third second, and so on. You can make a table and look for a pattern in the data.

Time (s)	1	2	3	4	5	6	7	8
Altitude (ft)	6.3	14.5	22.7	30.9	39.1	47.3	55.5	63.7

+ 8.2 + 8.2 + 8.2 + 8.2 + 8.2 + 8.2 + 8.2

Vocabulary
- sequence
- terms
- arithmetic sequence
- common difference

RECOGNIZE ARITHMETIC SEQUENCES A **sequence** is a set of numbers in a specific order. The numbers in the sequence are called **terms**. If the difference between successive terms is constant, then it is called an **arithmetic sequence**. The difference between the terms is called the **common difference**.

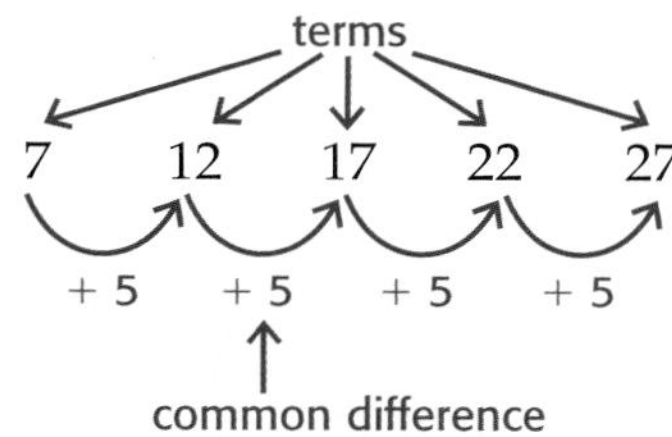

Key Concept — *Arithmetic Sequence*

An arithmetic sequence is a numerical pattern that increases or decreases at a constant rate or value called the common difference.

Example 1 Identify Arithmetic Sequences

Determine whether each sequence is arithmetic. Justify your answer.

a. 1, 2, 4, 8, ...

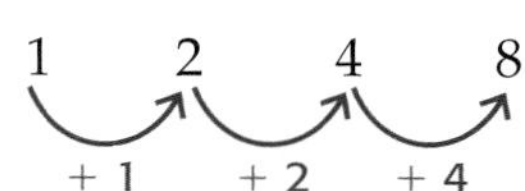

This is not an arithmetic sequence because the difference between terms is not constant.

b. $\frac{1}{2}, \frac{1}{4}, 0, -\frac{1}{4}, \ldots$

$\frac{1}{2} \quad \frac{1}{4} \quad 0 \quad -\frac{1}{4}$

$-\frac{1}{4} \quad -\frac{1}{4} \quad -\frac{1}{4}$

This is an arithmetic sequence because the difference between terms is constant.

Study Tip

Reading Math
The three dots after the last number in a sequence are called an *ellipsis*. The ellipsis indicates that there are more terms in the sequence that are not listed.

WRITE ARITHMETIC SEQUENCES You can use the common difference of an arithmetic sequence to find the next term in the sequence.

Key Concept *Writing Arithmetic Sequences*

- **Words** Each term of an arithmetic sequence after the first term can be found by adding the common difference to the preceding term.
- **Symbols** An arithmetic sequence can be found as follows

$$a_1, a_1 + d, a_2 + d, a_3 + d, \ldots,$$

where d is the common difference, a_1 is the first term, a_2 is the second term, and so on.

Example 2 Extend a Sequence

Find the next three terms of the arithmetic sequence 74, 67, 60, 53, ...

Find the common difference by subtracting successive terms.

74 67 60 53 ? ? ?

−7 −7 −7 −7 −7 −7

The common difference is −7.

Add −7 to the last term of the sequence to get the next term in the sequence. Continue adding −7 until the next three terms are found.

53 46 39 32

−7 −7 −7

The next three terms are 46, 39, 32.

Each term in an arithmetic sequence can be expressed in terms of the first term a_1 and the common difference d.

Term	Symbol	In Terms of a_1 and d	Numbers
first term	a_1	a_1	8
second term	a_2	$a_1 + d$	$8 + 1(3) = 11$
third term	a_3	$a_1 + 2d$	$8 + 2(3) = 14$
fourth term	a_4	$a_1 + 3d$	$8 + 3(3) = 17$
⋮	⋮	⋮	⋮
nth term	a_n	$a_1 + (n - 1)d$	$8 + (n - 1)(3)$

Study Tip

Reading Math

The formula for the nth term of an arithmetic sequence is called a *recursive formula*. This means that each succeeding term is formulated from one or more of the previous terms.

The following formula generalizes this pattern and can be used to find any term in an arithmetic sequence.

Key Concept *nth Term of an Arithmetic Sequence*

The nth term a_n of an arithmetic sequence with first term a_1 and common difference d is given by

$$a_n = a_1 + (n - 1)d,$$

where n is a positive integer.

Example 3 Find a Specific Term

Find the 14th term in the arithmetic sequence 9, 17, 25, 33, …

In this sequence, the first term, a_1, is 9. You want to find the 14th term, so $n = 14$. Find the common difference.

The common difference is 8.

Use the formula for the nth term of an arithmetic sequence.

$a_n = a_1 + (n - 1)d$	Formula for the nth term
$a_{14} = 9 + (14 - 1)8$	$a_1 = 9, n = 14, d = 8$
$a_{14} = 9 + 104$	Simplify.
$a_{14} = 113$	The 14th term in the sequence is 113.

Example 4 Write an Equation for a Sequence

Consider the arithmetic sequence 12, 23, 34, 45, …

a. Write an equation for the nth term of the sequence.

In this sequence, the first term, a_1, is 12. Find the common difference.

12 23 34 45

+ 11 + 11 + 11 The common difference is 11.

Use the formula for the nth term to write an equation.

$a_n = a_1 + (n - 1)d$	Formula for nth term
$a_n = 12 + (n - 1)11$	$a_1 = 12, d = 11$
$a_n = 12 + 11n - 11$	Distributive Property
$a_n = 11n + 1$	Simplify.

CHECK For $n = 1$, $11(1) + 1 = 12$.
For $n = 2$, $11(2) + 1 = 23$.
For $n = 3$, $11(3) + 1 = 34$, and so on.

b. Find the 10th term in the sequence.

Replace n with 10 in the equation written in part **a**.

$a_n = 11n + 1$	Equation for the nth term
$a_{10} = 11(10) + 1$	Replace n with 10.
$a_{10} = 111$	Simplify.

c. Graph the first five terms of the sequence.

n	$11n + 1$	a_n	(n, a_n)
1	11(1) + 1	12	(1, 12)
2	11(2) + 1	23	(2, 23)
3	11(3) + 1	34	(3, 34)
4	11(4) + 1	45	(4, 45)
5	11(5) + 1	56	(5, 56)

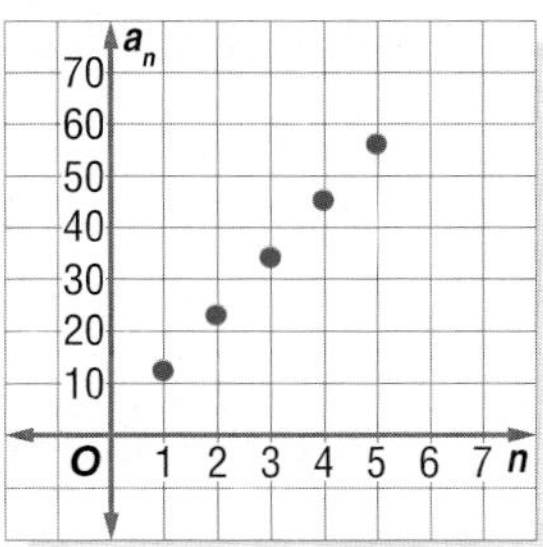

Notice that the points fall on a line. The graph of an arithmetic sequence is linear.

Check for Understanding

Concept Check

1. **OPEN ENDED** Write an arithmetic sequence whose common difference is -10.
2. **Find** the common difference and the first term in the sequence defined by $a_n = 5n + 2$.
3. **FIND THE ERROR** Marisela and Richard are finding the common difference for the arithmetic sequence $-44, -32, -20, -8$.

Marisela	Richard
$-32 - (-44) = 12$	$-44 - (-32) = -12$
$-20 - (-32) = 12$	$-32 - (-20) = -12$
$-8 - (-20) = 12$	$-20 - (-8) = -12$

Who is correct? Explain your reasoning.

Guided Practice

Determine whether each sequence is an arithmetic sequence. If it is, state the common difference.

4. 24, 16, 8, 0, …
5. 3, 6, 12, 24, …

Find the next three terms of each arithmetic sequence.

6. 7, 14, 21, 28, …
7. 34, 29, 24, 19, …

Find the nth term of each arithmetic sequence described.

8. $a_1 = 3, d = 4, n = 8$
9. $a_1 = 10, d = -5, n = 21$
10. 23, 25, 27, 29, … for $n = 12$
11. $-27, -19, -11, -3, \ldots$ for $n = 17$

Write an equation for the nth term of each arithmetic sequence. Then graph the first five terms of the sequence.

12. 6, 12, 18, 24, …
13. 12, 17, 22, 27, …

Application

14. **FITNESS** Latisha is beginning an exercise program that calls for 20 minutes of walking each day for the first week. Each week thereafter, she has to increase her walking by 7 minutes a day. Which week of her exercise program will be the first one in which she will walk over an hour a day?

Practice and Apply

Homework Help

For Exercises	See Examples
15–20, 43, 44	1
21–26	2
27–38, 45–49, 54, 55	3
39–42, 50–53	4

Extra Practice

See page 830.

Determine whether each sequence is an arithmetic sequence. If it is, state the common difference.

15. 7, 6, 5, 4, …
16. 10, 12, 15, 18, …
17. $9, 5, -1, -5, \ldots$
18. $-15, -11, -7, -3, \ldots$
19. $-0.3, 0.2, 0.7, 1.2, \ldots$
20. 2.1, 4.2, 8.4, 17.6, …

Find the next three terms of each arithmetic sequence.

21. 4, 7, 10, 13, …
22. 18, 24, 30, 36, …
23. $-66, -70, -74, -78, \ldots$
24. $-31, -22, -13, -4, \ldots$
25. $2\frac{1}{3}, 2\frac{2}{3}, 3, 3\frac{1}{3}, \ldots$
26. $\frac{7}{12}, 1\frac{1}{3}, 2\frac{1}{12}, 2\frac{5}{6}, \ldots$

Find the nth term of each arithmetic sequence described.

27. $a_1 = 5, d = 5, n = 25$

28. $a_1 = 8, d = 3, n = 16$

29. $a_1 = 52, d = 12, n = 102$

30. $a_1 = 34, d = 15, n = 200$

31. $a_1 = \frac{5}{8}, d = \frac{1}{8}, n = 22$

32. $a_1 = 1\frac{1}{2}, d = 2\frac{1}{4}, n = 39$

33. $-9, -7, -5, -3, \ldots$ for $n = 18$

34. $-7, -3, 1, 5, \ldots$ for $n = 35$

35. $0.5, 1, 1.5, 2, \ldots$ for $n = 50$

36. $5.3, 5.9, 6.5, 7.1, \ldots$ for $n = 12$

37. 200 is the __?__th term of 24, 35, 46, 57, …

38. -34 is the __?__th term of 30, 22, 14, 6, …

Write an equation for the nth term of each arithmetic sequence. Then graph the first five terms in the sequence.

39. $-3, -6, -9, -12, \ldots$

40. $8, 9, 10, 11, \ldots$

41. $2, 8, 14, 20, \ldots$

42. $-18, -16, -14, -12, \ldots$

43. Find the value of y that makes $y + 4, 6, y, \ldots$ an arithmetic sequence.

44. Find the value of y that makes $y + 8, 4y + 6, 3y, \ldots$ an arithmetic sequence.

GEOMETRY For Exercises 45 and 46, use the diagram below that shows the perimeter of the pattern consisting of trapezoids.

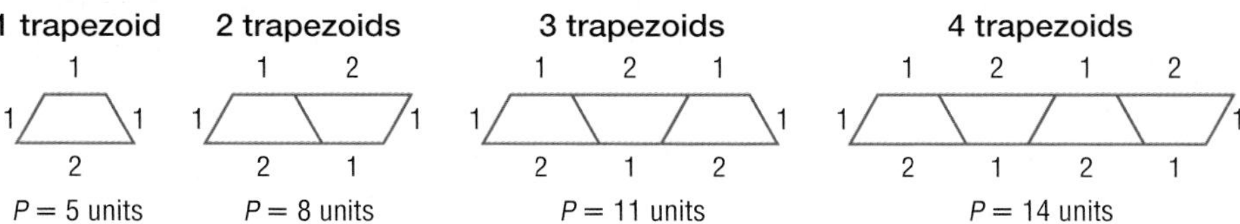

45. Write a formula that can be used to find the perimeter of a pattern containing n trapezoids.

46. What is the perimeter of the pattern containing 12 trapezoids?

More About. . .

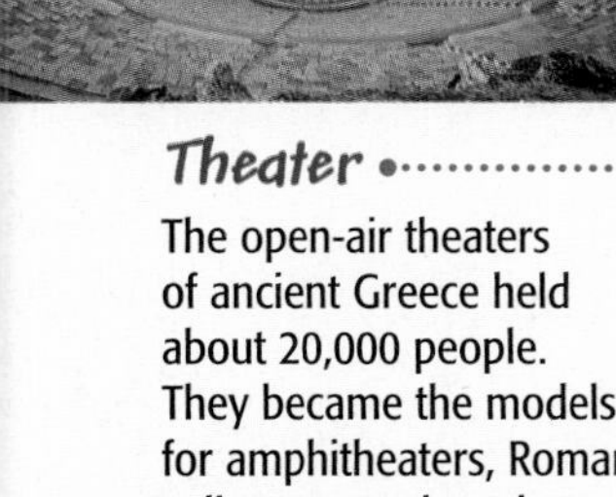

Theater

The open-air theaters of ancient Greece held about 20,000 people. They became the models for amphitheaters, Roman coliseums, and modern sports arenas.

Source: www.encarta.msn.com

THEATER For Exercises 47–49, use the following information.
The Coral Gables Actors' Playhouse has 76 seats in the last row of the orchestra section of the theater, 68 seats in the next row, 60 seats in the next row, and so on. There are 7 rows of seats in the section. On opening night, 368 tickets were sold for the orchestra section.

47. Write a formula to find the number of seats in any given row of the orchestra section of the theater.

48. How many seats are in the first row?

49. Was this section oversold?

PHYSICAL SCIENCE For Exercises 50–53, use the following information.
Taylor and Brooklyn are recording how far a ball rolls down a ramp during each second. The table below shows the data they have collected.

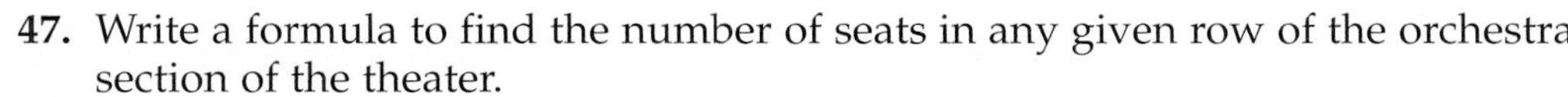

Time (s)	1	2	3	4	5	6
Distance traveled (cm)	9	13	17	21	25	29

50. Do the distances traveled by the ball form an arithmetic sequence? Justify your answer.

51. Write an equation for the sequence.

52. How far will the ball travel during the 35th second?

53. Graph the sequence.

GAMES **For Exercises 54 and 55, use the following information.**
Contestants on a game show win money by answering 10 questions. The value of each question increases by \$1500.

54. If the first question is worth \$2500, find the value of the 10th question.

55. If the contestant answers all ten questions correctly, how much money will he or she win?

56. **CRITICAL THINKING** Is $2x + 5, 4x + 5, 6x + 5, 8x + 5 \ldots$ an arithmetic sequence? Explain your answer.

57. **CRITICAL THINKING** Use an arithmetic sequence to find how many multiples of 7 are between 29 and 344.

58. **WRITING IN MATH** Answer the question that was posed at the beginning of the lesson.

How are arithmetic sequences used to solve problems in science?

Include the following in your answer:

- a formula for the arithmetic sequence that represents the altitude of the probe after each second, and
- an explanation of how you could use this information to predict the altitude of the probe after 15 seconds.

59. Luis puts \$25 a week into a savings account from his part-time job. If he has \$350 in savings now, how much will he have 12 weeks from now?

(A) \$600 (B) \$625 (C) \$650 (D) \$675

60. In an arithmetic sequence a_n, if $a_1 = 2$ and $a_4 = 11$, find a_{20}.

(A) 40 (B) 59 (C) 78 (D) 97

Maintain Your Skills

Mixed Review

If $f(x) = 3x - 2$ and $g(x) = x^2 - 5$, find each value. *(Lesson 4-6)*

61. $f(4)$ **62.** $g(-3)$ **63.** $2[f(6)]$

Determine whether each equation is a linear equation. If so, write the equation in standard form. *(Lesson 4-5)*

64. $x^2 + 3x - y = 8$ **65.** $y - 8 = 10 - x$ **66.** $2y = y + 2x - 3$

Translate each sentence into an algebraic equation. *(Lesson 3-1)*

67. Two hundred minus three times x is equal to nine.

68. The sum of twice r and three times s is identical to thirteen.

Find each product. *(Lesson 2-3)*

69. $7(-3)$ **70.** $-11 \cdot 15$ **71.** $-8(-1.5)$

72. $6\left(\frac{2}{3}\right)$ **73.** $\left(-\frac{5}{8}\right)\left(\frac{4}{7}\right)$ **74.** $5 \cdot 3\frac{1}{2}$

Getting Ready for the Next Lesson

PREREQUISITE SKILL **Write the ordered pair for each point shown at the right.**
*(To review **graphing points**, see Lesson 4-1.)*

75. H **76.** J

77. K **78.** L

79. M **80.** N

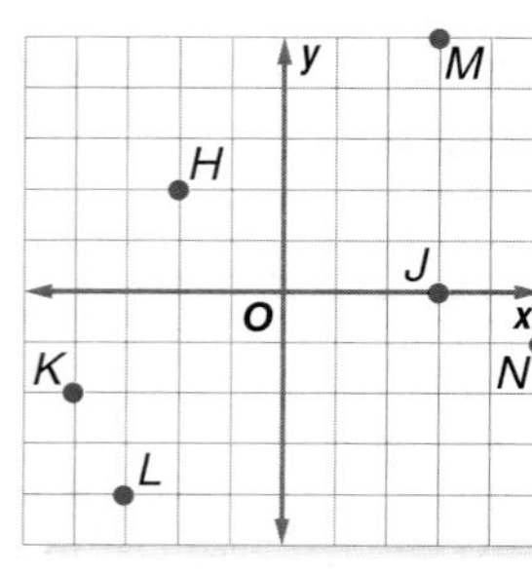

Reading Mathematics

Reasoning Skills

Throughout your life, you have used reasoning skills, possibly without even knowing it. As a child, you used inductive reasoning to conclude that your hand would hurt if you touched the stove while it was hot. Now, you use inductive reasoning when you decide, after many trials, that one of the worst ways to prepare for an exam is by studying only an hour before you take it. **Inductive reasoning** is used to derive a general rule after observing many individual events.

Inductive reasoning involves . . .

- observing many examples
- looking for a pattern
- making a conjecture
- checking the conjecture
- discovering a likely conclusion

With **deductive reasoning**, you use a general rule to help you decide about a specific event. You come to a conclusion by accepting facts. There is no conjecturing involved. Read the two statements below.

1) If a person wants to play varsity sports, he or she must have a C average in academic classes.

2) Jolene is playing on the varsity tennis team.

If these two statements are accepted as facts, then the obvious conclusion is that Jolene has at least a C average in her academic classes. This is an example of deductive reasoning.

Reading to Learn

1. Explain the difference between *inductive* and *deductive* reasoning. Then give an example of each.

2. When Sherlock Holmes reaches a conclusion about a murderer's height because he knows the relationship between a man's height and the distance between his footprints, what kind of reasoning is he using? Explain.

3. When you examine a sequence of numbers and decide that it is an arithmetic sequence, what kind of reasoning are you using? Explain.

4. Once you have found the common difference for an arithmetic sequence, what kind of reasoning do you use to find the 100th term in the sequence?

5. **a.** Copy and complete the following table.

3^1	3^2	3^3	3^4	3^5	3^6	3^7	3^8	3^9
3	9	27						

b. Write the sequence of numbers representing the numbers in the ones place.

c. Find the number in the ones place for the value of 3^{100}. Explain your reasoning. State the type of reasoning that you used.

6. A sequence contains all numbers less than 50 that are divisible by 5. You conclude that 35 is in the sequence. Is this an example of inductive or deductive reasoning? Explain.

4-8 Writing Equations from Patterns

What You'll Learn

- Look for a pattern.
- Write an equation given some of the solutions.

Vocabulary

- look for a pattern
- inductive reasoning

Why is writing equations from patterns important in science?

Water is one of the few substances that expands when it freezes. The table shows different volumes of water and the corresponding volumes of ice.

Volume of Water (ft^3)	11	22	33	44	55
Volume of Ice (ft^3)	12	24	36	48	60

The relation in the table can be represented by a graph. Let w represent the volume of water, and let c represent the volume of ice. When the ordered pairs are graphed, they form a linear pattern. This pattern can be described by an equation.

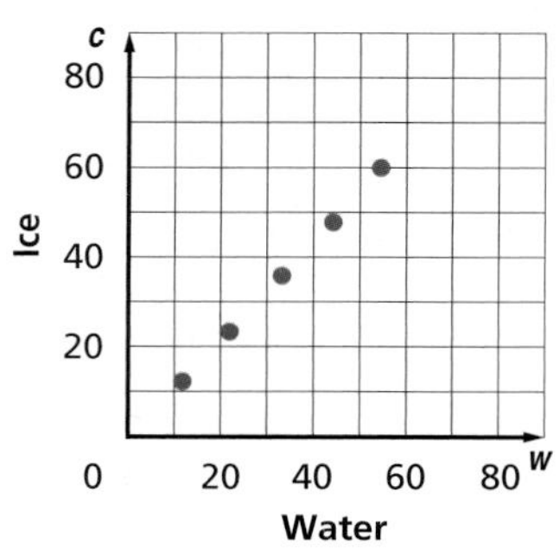

Study Tip

Look Back
To review **deductive reasoning**, see Lesson 1-7.

LOOK FOR PATTERNS A very useful problem-solving strategy is **look for a pattern**. When you make a conclusion based on a pattern of examples, you are using **inductive reasoning**. Recall that *deductive reasoning* uses facts, rules, or definitions to reach a conclusion.

Example 1 Extend a Pattern

Study the pattern below.

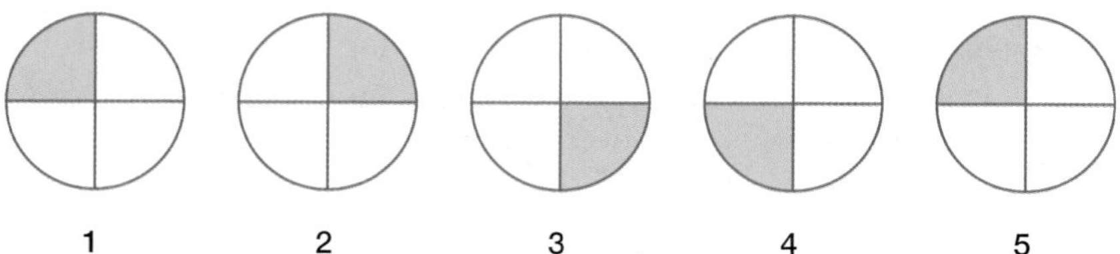

a. Draw the next three figures in the pattern.

The pattern consists of circles with one-fourth shaded. The section that is shaded is rotated in a clockwise direction. The next three figures are shown.

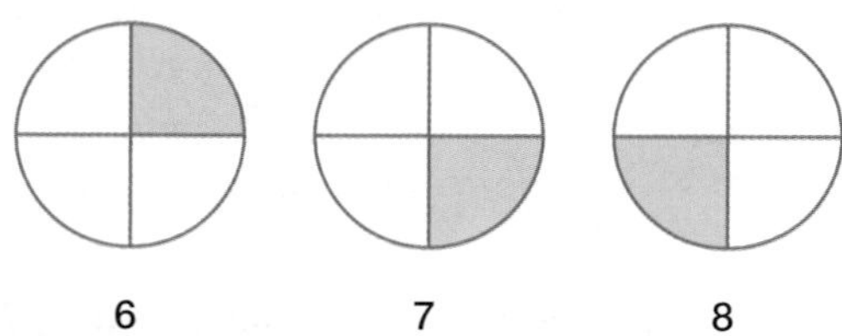

b. Draw the 27th circle in the pattern.

The pattern repeats every fourth design. Therefore designs 4, 8, 12, 16, and so on, will all be the same. Since 24 is the greatest number less than 27 that is a multiple of 4, the 25th circle in the pattern will be the same as the first circle.

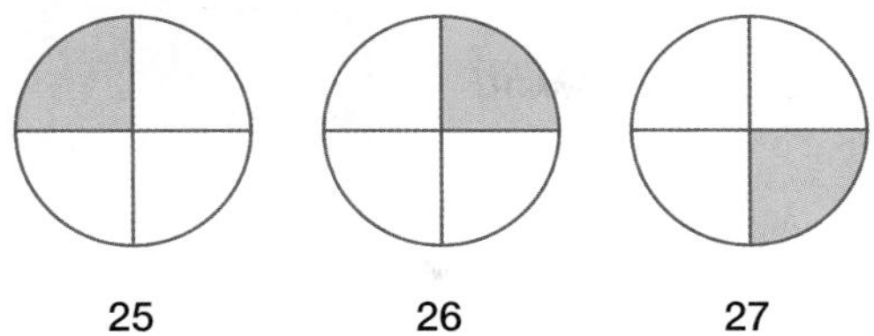

Other sequences besides arithmetic sequences can follow a pattern.

Example 2 Patterns in a Sequence

Find the next three terms in the sequence 3, 6, 12, 24, … .

Study the pattern in the sequence.

3 → 6 → 12 → 24

+ 3, + 6, + 12

You can use inductive reasoning to find the next term in a sequence. Notice the pattern 3, 6, 12, … The difference between each term doubles in each successive term. To find the next three terms in the sequence, continue doubling each successive difference. Add 24, 48, and 96.

3 → 6 → 12 → 24 → 48 → 96 → 192

+ 3, + 6, + 12, + 24, + 48, + 96

The next three terms are 48, 96, and 192.

Algebra Activity

Looking for Patterns

- You will need several pieces of string.
- Loop a piece of string around one of the cutting edges of the scissors and cut. How many pieces of string do you have as a result of this cut? Discard those pieces.
- Use another piece of string to make 2 loops around the scissors and cut. How many pieces of string result?
- Continue making loops and cutting until you see a pattern.

Analyze

1. Describe the pattern and write a sequence that describes the number of loops and the number of pieces of string.
2. Write an expression that you could use to find the number of pieces of string you would have if you made *n* loops.
3. How many pieces of string would you have if you made 20 loops?

WRITE EQUATIONS Sometimes a pattern can lead to a general rule. If the relationship between the domain and range of a relation is linear, the relationship can be described by a linear equation.

Example 3 Write an Equation from Data

FUEL ECONOMY **The table below shows the average amount of gas Rogelio's car uses depending on how many miles he drives.**

Gallons of gasoline	1	2	3	4	5
Miles driven	28	56	84	112	140

a. Graph the data. What conclusion can you make about the relationship between the number of gallons used and the number of miles driven?

The graph shows a linear relationship between the number of gallons used g and the number of miles driven m.

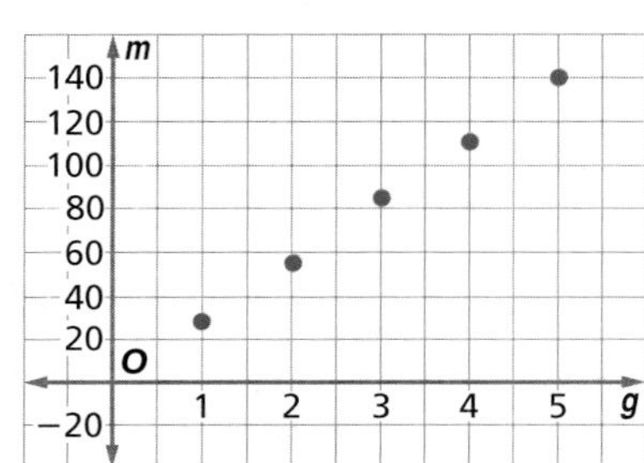

b. Write an equation to describe this relationship.

Look at the relationship between the domain and range to find a pattern that can be described by an equation.

+1 +1 +1 +1

Gallons of gasoline	1	2	3	4	5
Miles driven	28	56	84	112	140

+28 +28 +28 +28

Since this is a linear relationship, the ratio of the range values to the domain values is constant. The difference of the values for g is 1, and the difference of the values for m is 28. This suggests that $m = 28g$. Check to see if this equation is correct by substituting values of g into the equation.

CHECK If $g = 1$, then $m = 28(1)$ or 28. ✓
If $g = 2$, then $m = 28(2)$ or 56. ✓
If $g = 3$, then $m = 28(3)$ or 84. ✓

The equation checks. Since this relation is also a function, we can write the equation as $f(g) = 28g$, where $f(g)$ represents the number of miles driven.

More About. . .

Fuel Economy

Not all cars use gasoline. Many alternative fuels are being used today in place of fossil fuels like oil. In 2001, there were 456,306 alternative fuel vehicles (AFVs) on the road in the U.S.

Source: U.S. Department of Energy

Example 4 Write an Equation with a Constant

Write an equation in function notation for the relation graphed at the right.

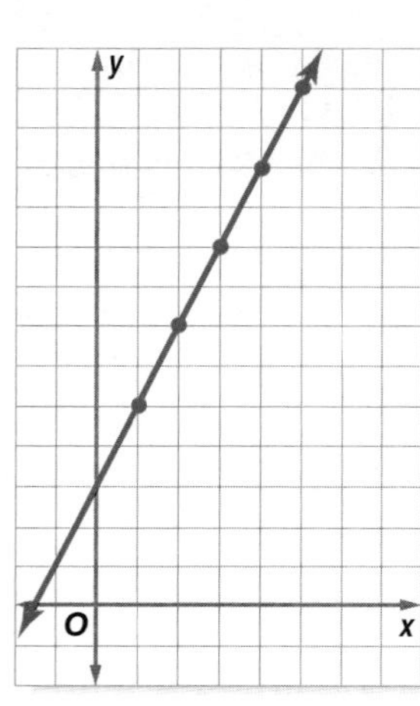

Make a table of ordered pairs for several points on the graph.

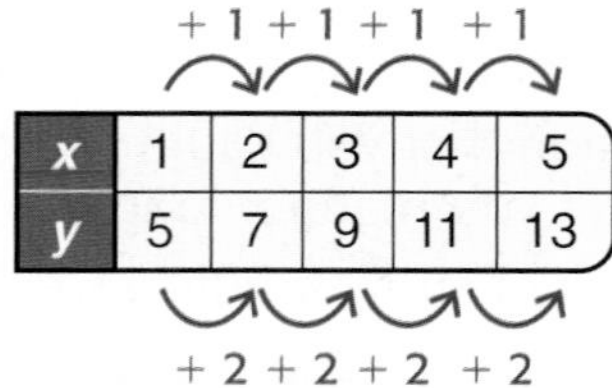

+1 +1 +1 +1

x	1	2	3	4	5
y	5	7	9	11	13

+2 +2 +2 +2

The difference of the x values is 1, and the difference of the y values is 2. The difference in y values is twice the difference of x values. This suggests that $y = 2x$. Check this equation.

CHECK If $x = 1$, then $y = 2(1)$ or 2. But the y value for $x = 1$ is 5. This is a difference of 3. Try some other values in the domain to see if the same difference occurs.

x	1	2	3	4	5
$2x$	2	4	6	8	10
y	5	7	9	11	13

y is always 3 more than 2*x*.

This pattern suggests that 3 should be added to one side of the equation in order to correctly describe the relation. Check $y = 2x + 3$.

If $x = 2$, then $y = 2(2) + 3$ or 7.
If $x = 3$, then $y = 2(3) + 3$, or 9.

Thus, $y = 2x + 3$ correctly describes this relation. Since this relation is also a function, we can write the equation in function notation as $f(x) = 2x + 3$.

Check for Understanding

Concept Check

1. **Explain** how you can use inductive reasoning to write an equation from a pattern.
2. **OPEN ENDED** Write a sequence for which the first term is 4 and the second term is 8. Explain the pattern that you used.
3. **Explain** how you can determine whether an equation correctly represents a relation given in a table.

Guided Practice

4. Find the next two items for the pattern. Then find the 16th figure in the pattern.

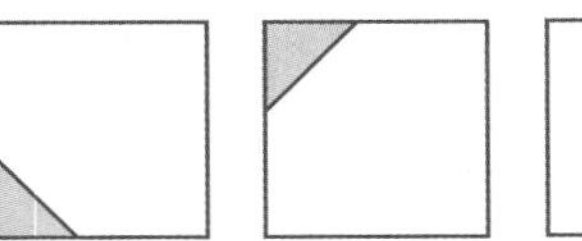

Find the next three terms in each sequence.

5. 1, 2, 4, 7, 11, …
6. 5, 9, 6, 10, 7, 11, …

Write an equation in function notation for each relation.

7. 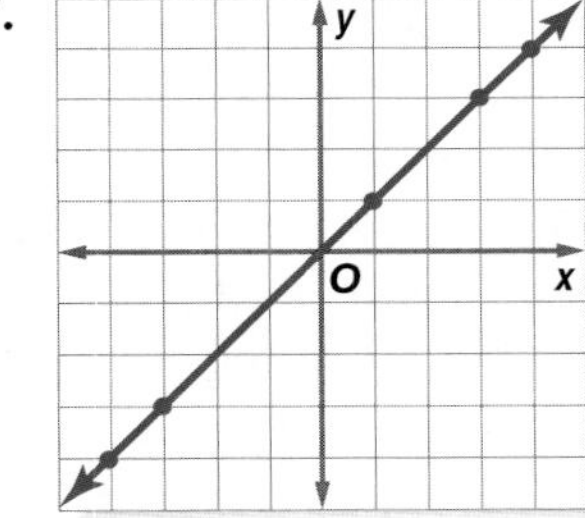

8. 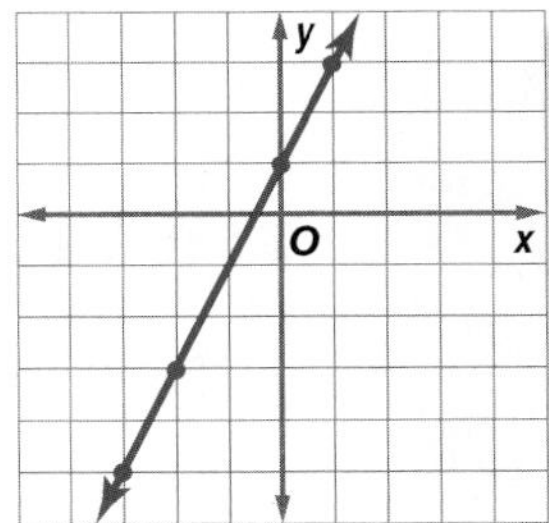

Application

GEOLOGY **For Exercises 9–11, use the table below that shows the underground temperature of rocks at various depths below Earth's surface.**

Depth (km)	1	2	3	4	5	6
Temperature (°C)	55	90	125	160	195	230

9. Graph the data.
10. Write an equation in function notation for the relation.
11. Find the temperature of a rock that is 10 kilometers below the surface.

Practice and Apply

Homework Help

For Exercises	See Examples
12, 13, 26	1
14–19, 27, 28	2
20–25	4
29, 30	5

Extra Practice

See page 830.

Find the next two items for each pattern. Then find the 21st figure in the pattern.

12.

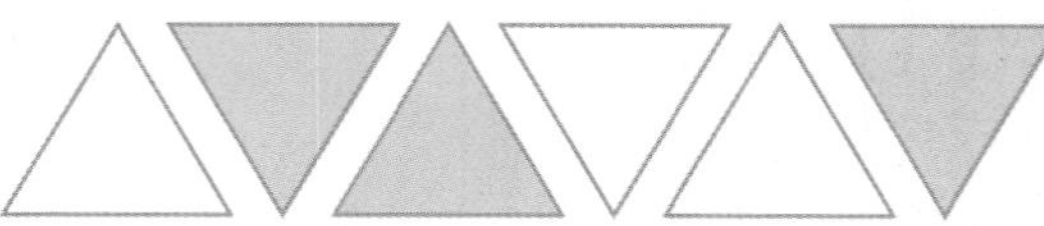

13.

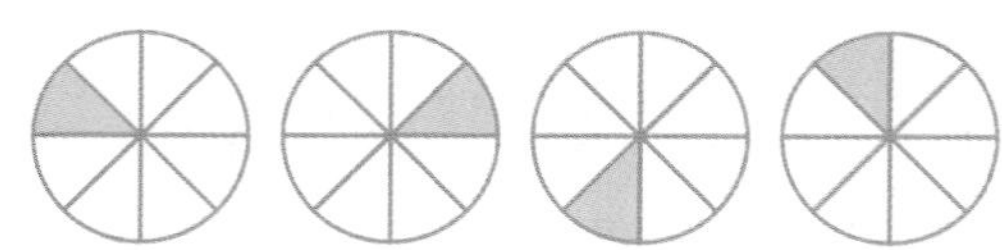

Find the next three terms in each sequence.

14. 0, 2, 6, 12, 20, …

15. 9, 7, 10, 8, 11, 9, 12, …

16. 1, 4, 9, 16, …

17. 0, 2, 5, 9, 14, 20, …

18. $a + 1, a + 2, a + 3, \ldots$

19. $x + 1, 2x + 1, 3x + 1, \ldots$

Write an equation in function notation for each relation.

20.

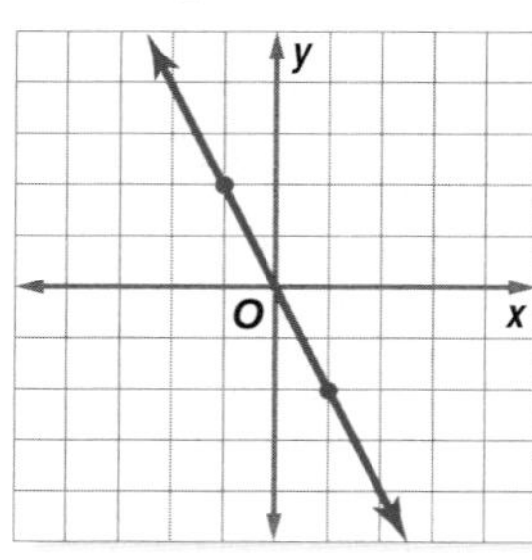

21.

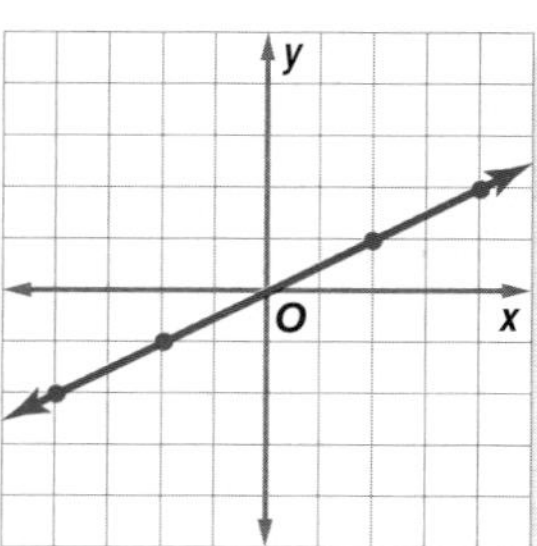

22.

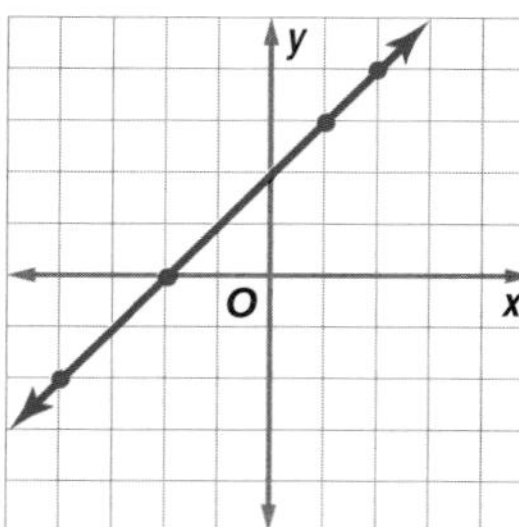

23.

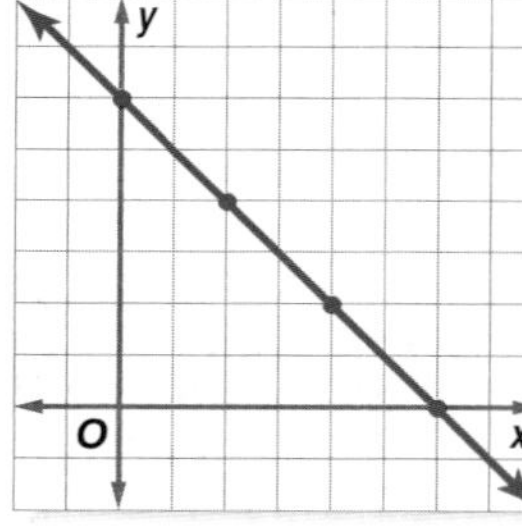

24.

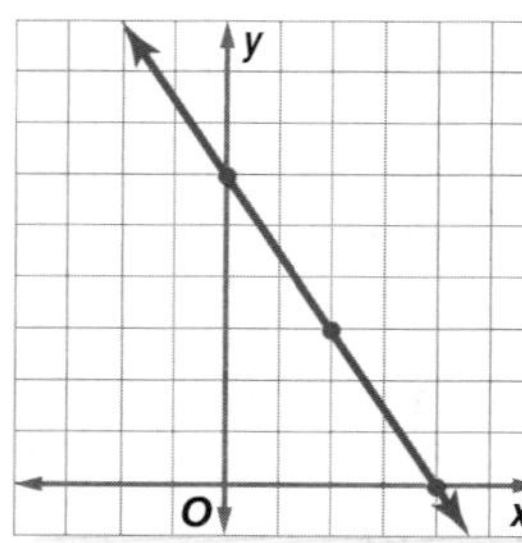

25.

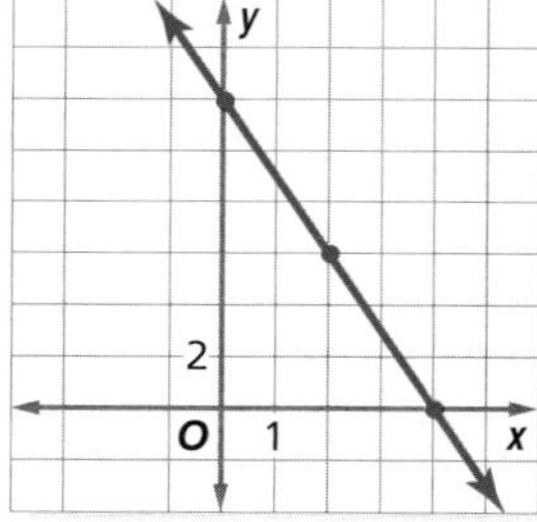

26. **TRAVEL** On an island cruise in Hawaii, each passenger is given a flower chain. A crew member hands out 3 red, 3 blue, and 3 green chains in that order. If this pattern is repeated, what color chain will the 50th person receive?

More About. . .

Number Theory

Fibonacci numbers occur in many areas of nature, including pine cones, shell spirals, flower petals, branching plants, and many fruits and vegetables.

NUMBER THEORY **For Exercises 27 and 28, use the following information.**
In 1201, Leonardo Fibonacci introduced his now famous pattern of numbers called the Fibonacci sequence.

$$1, 1, 2, 3, 5, 8, 13, \ldots$$

Notice the pattern in this sequence. After the second number, each number in the sequence is the sum of the two numbers that precede it. That is $2 = 1 + 1$, $3 = 2 + 1$, $5 = 3 + 2$, and so on.

27. Write the first 12 terms of the Fibonacci sequence.

28. Notice that every third term is divisible by 2. What do you notice about every fourth term? every fifth term?

FITNESS For Exercises 29 and 30, use the table below that shows the maximum heart rate to maintain, for different ages, during aerobic activities such as running, biking, or swimming.

Age (yr)	20	30	40	50	60	70
Pulse rate (beats/min)	175	166	157	148	139	130

Source: Ontario Association of Sport and Exercise Sciences

29. Write an equation in function notation for the relation.

30. What would be the maximum heart rate to maintain in aerobic training for a 10-year old? an 80-year old?

CRITICAL THINKING For Exercises 31–33, use the following information.
Suppose you arrange a number of regular pentagons so that only one side of each pentagon touches. Each side of each pentagon is 1 centimeter.

1 pentagon

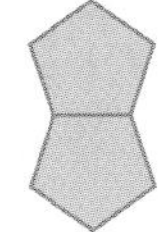
2 pentagons

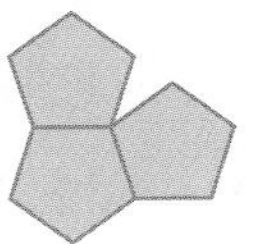
3 pentagons

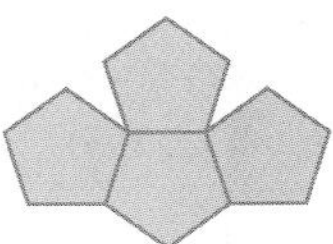
4 pentagons

31. For each arrangement of pentagons, compute the perimeter.

32. Write an equation in function form to represent the perimeter $f(n)$ of n pentagons.

33. What is the perimeter if 24 pentagons are used?

34. WRITING IN MATH Answer the question that was posed at the beginning of the lesson.

Why is writing equations from patterns important in science?

Include the following in your answer:

- an explanation of the relationship between the volume of water and the volume of ice, and
- a reasonable estimate of the size of a container that had 99 cubic feet of water, if it was going to be frozen.

35. Find the next two terms in the sequence 3, 4, 6, 9, … .

(A) 12, 15 (B) 13, 18 (C) 14, 19 (D) 15, 21

36. After P pieces of candy are divided equally among 5 children, 4 pieces remain. How many would remain if $P + 4$ pieces of candy were divided equally among the 5 children?

(A) 0 (B) 1 (C) 2 (D) 3

Maintain Your Skills

Mixed Review **Find the next three terms of each arithmetic sequence.** *(Lesson 4-7)*

37. 1, 4, 7, 10, …

38. 9, 5, 1, −3, …

39. −25, −19, −13, −7, …

40. 22, 34, 46, 58, …

41. Determine whether the relation graphed at the right is a function. *(Lesson 4-6)*

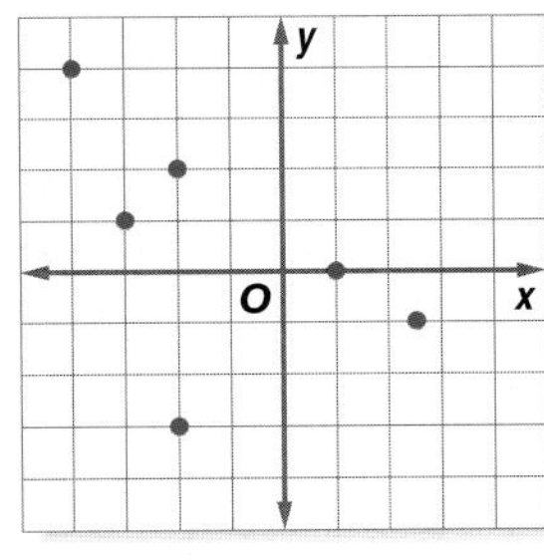

42. GEOGRAPHY The world's tallest waterfall is Angel Falls in Venezuela at 3212 feet. It is 102 feet higher than Tulega Falls in South Africa. How high is Tulega Falls? *(Lesson 3-2)*

Study Guide and Review

Vocabulary and Concept Check

Choose the letter of the term that best matches each statement or phrase.

1. In the coordinate plane, the axes intersect at the __?__.
2. A(n) __?__ is a set of ordered pairs.
3. A(n) __?__ flips a figure over a line.
4. In a coordinate system, the __?__ is a horizontal number line.
5. In the ordered pair, $A(2, 7)$, 7 is the __?__.
6. The coordinate axes separate a plane into four __?__.
7. A(n) __?__ has a graph that is a nonvertical straight line.
8. In the relation $\{(4, -2), (0, 5), (6, 2), (-1, 8)\}$, the __?__ is $\{-1, 0, 4, 6\}$.
9. A(n) __?__ enlarges or reduces a figure.
10. In a coordinate system, the __?__ is a vertical number line.

a. domain
b. dilation
c. linear function
d. reflection
e. origin
f. quadrants
g. relation
h. *x*-axis
i. *y*-axis
j. *x*-coordinate
k. *y*-coordinate

Lesson-by-Lesson Review

4-1 The Coordinate Plane

See pages 192–196.

Concept Summary

- The first number, or *x*-coordinate, of an ordered pair corresponds to the numbers on the *x*-axis.
- The second number, or *y*-coordinate, corresponds to the numbers on the *y*-axis.

Example

Plot $T(3, -2)$ on a coordinate plane. Name the quadrant in which the point is located.

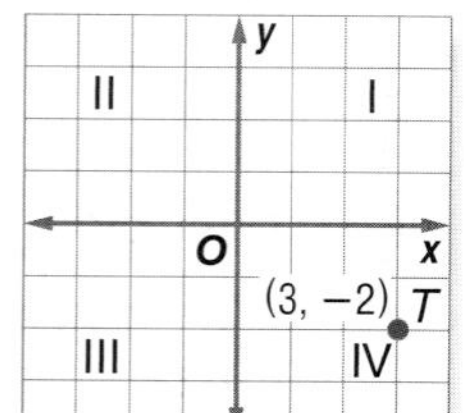

$T(3, -2)$ is located in Quadrant IV.

Exercises **Plot each point on a coordinate plane.** *See Example 3 on page 193.*

11. $A(4, 2)$
12. $B(-1, 3)$
13. $C(0, -5)$
14. $D(-3, -2)$
15. $E(-4, 0)$
16. $F(2, -1)$

www.algebra1.com/vocabulary_review

4-2 Transformations on the Coordinate Plane

See pages 197–203.

Concept Summary

- A reflection is a flip.
- A translation is a slide.
- A dilation is a reduction or enlargement.
- A rotation is a turn.

Example **A quadrilateral with vertices $W(1, 2)$, $X(2, 3)$, $Y(5, 2)$, and $Z(2, 1)$ is reflected over the y-axis. Find the coordinates of the vertices of the image. Then graph quadrilateral $WXYZ$ and its image $W'X'Y'Z'$.**

Multiply each x-coordinate by –1.

$W(1, 2) \rightarrow W'(-1, 2)$ $\quad$ $Y(5, 2) \rightarrow Y'(-5, 2)$

$X(2, 3) \rightarrow X'(-2, 3)$ $\quad$ $Z(2, 1) \rightarrow Z'(-2, 1)$

The coordinates of the image are $W'(-1, 2)$, $X'(-2, 3)$, $Y'(-5, 2)$, and $Z'(-2, 1)$.

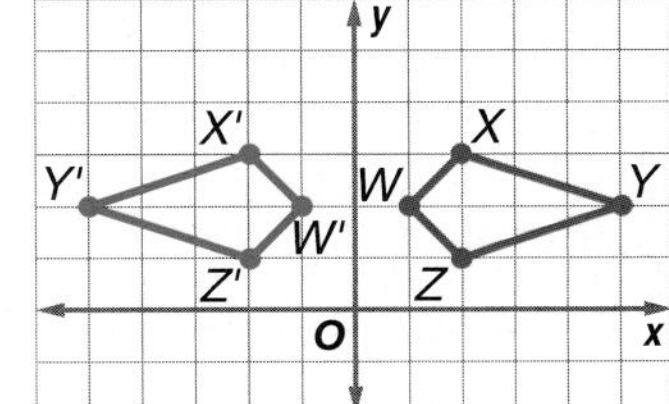

Exercises **Find the coordinates of the vertices of each figure after the given transformation is performed. Then graph the preimage and its image.**
See Examples 2–5 on pages 198–200.

17. triangle ABC with $A(3, 3)$, $B(5, 4)$, and $C(4, -3)$ reflected over the x-axis
18. quadrilateral $PQRS$ with $P(-2, 4)$, $Q(0, 6)$, $R(3, 3)$, and $S(-1, -4)$ translated 3 units down
19. parallelogram $GHIJ$ with $G(2, 2)$, $H(6, 0)$, $I(6, 2)$, and $J(2, 4)$ dilated by a scale factor of $\frac{1}{2}$
20. trapezoid $MNOP$ with $M(2, 0)$, $N(4, 3)$, $O(6, 3)$, and $P(8, 0)$ rotated 90° counterclockwise about the origin

4-3 Relations

See pages 205–211.

Concept Summary

- A relation can be expressed as a set of ordered pairs, a table, a graph, or a mapping.

Example **Express the relation {(3, 2), (5, 3), (4, 3), (5, 2)} as a table, a graph, and a mapping.**

Table

List the set of x-coordinates and corresponding y-coordinates.

x	y
3	2
5	3
4	3
5	2

Graph

Graph each ordered pair on a coordinate plane.

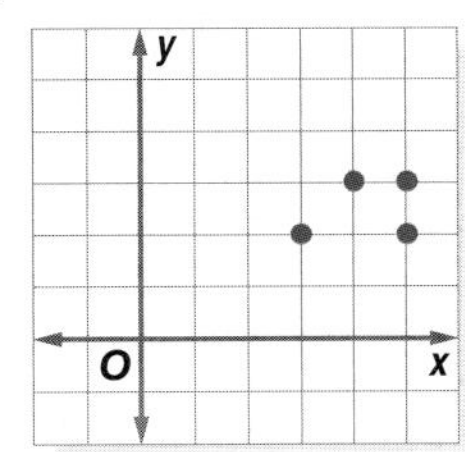

Mapping

List the x and y values. Draw arrows to show the relation.

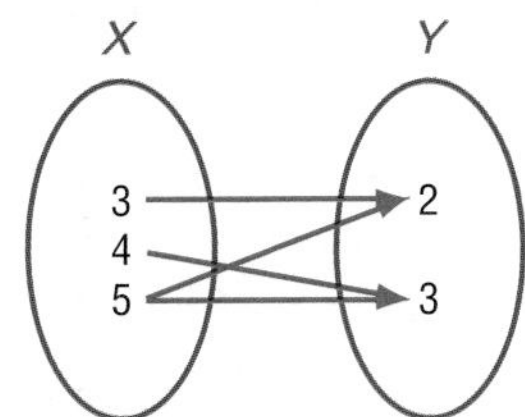

Exercises **Express each relation as a table, a graph, and a mapping. Then determine the domain and range.** *See Example 1 on page 205.*

21. $\{(-2, 6), (3, -2), (3, 0), (4, 6)\}$ **22.** $\{(-1, 0), (3, 0), (6, 2)\}$

23. $\{(3, 8), (9, 3), (-3, 8), (5, 3)\}$ **24.** $\{(2, 5), (-3, 1), (4, -2), (2, 3)\}$

4-4 Equations as Relations

See pages 212–217.

Concept Summary

- In an equation involving x and y, the set of x values is the domain, and the corresponding set of y values is the range.

Solve $2x + y = 8$ if the domain is $\{3, 2, 1\}$. Graph the solution set.

First solve the equation for y in terms of x.

$2x + y = 8$ Original equation

$y = 8 - 2x$ Subtract $2x$ from each side.

Substitute each value of x from the domain to determine the corresponding values of y in the range. Then graph the solution set $\{(3, 2), (2, 4), (1, 6)\}$.

x	$8 - 2x$	y	(x, y)
3	8 − 2(3)	2	(3, 2)
2	8 − 2(2)	4	(2, 4)
1	8 − 2(1)	6	(1, 6)

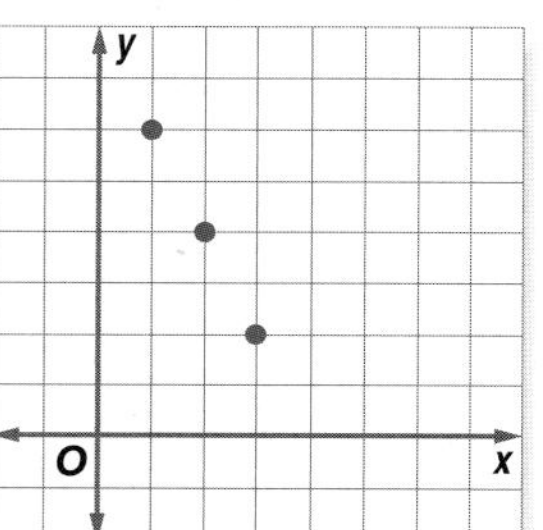

Exercises **Solve each equation if the domain is $\{-4, -2, 0, 2, 4\}$. Graph the solution set.** *See Example 3 on page 213.*

25. $y = x - 9$ **26.** $y = 4 - 2x$ **27.** $4x - y = -5$

28. $2x + y = 8$ **29.** $3x + 2y = 9$ **30.** $4x - 3y = 0$

4-5 Graphing Linear Equations

See pages 218–223.

Concept Summary

- Standard form: $Ax + By = C$, where $A \geq 0$ and A and B are not both zero
- To find the x-intercept, let $y = 0$. To find the y-intercept, let $x = 0$.

Example

Determine the x- and y-intercepts of $3x - y = 4$. Then graph the equation.

To find the x-intercept, let $y = 0$.

$3x - y = 4$ Original equation

$3x - 0 = 4$ Replace y with 0.

$3x = 4$ Simplify.

$x = \frac{4}{3}$ Divide each side by 3.

To find the y-intercept, let $x = 0$.

$3x - y = 4$ Original equation

$3(0) - y = 4$ Replace x with 0.

$-y = 4$ Simplify.

$y = -4$ Divide each side by −1.

The x-intercept is $\frac{4}{3}$, so the graph intersects the x-axis at $\left(\frac{4}{3}, 0\right)$.

The y-intercept is -4, so the graph intersects the y-axis at $(0, -4)$.

Plot these points, then draw a line that connects them.

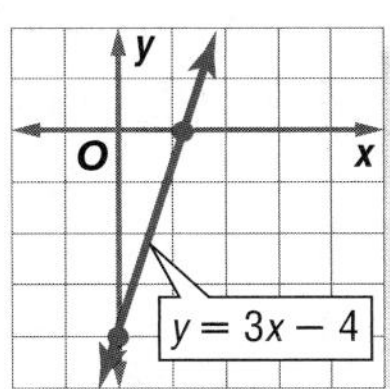

Exercises **Graph each equation.** *See Examples 2 and 4 on pages 219 and 220.*

31. $y = -x + 2$
32. $x + 5y = 4$
33. $2x - 3y = 6$
34. $5x + 2y = 10$
35. $\frac{1}{2}x + \frac{1}{3}y = 3$
36. $y - \frac{1}{3} = \frac{1}{3}x + \frac{2}{3}$

4-6 Functions

See pages 226–231.

Concept Summary

- A relation is a function if each element of the domain is paired with exactly one element of the range.
- Substitute values for x to determine $f(x)$ for a specific value.

Examples

1 **Determine whether the relation {(0, −4), (1, −1), (2, 2), (6, 3)} is a function.**

Since each element of the domain is paired with exactly one element of the range, the relation is a function.

2 **If $g(x) = 2x - 1$, find $g(-6)$.**

$g(-6) = 2(-6) - 1$ Replace x with -6.
$= -12 - 1$ Multiply.
$= -13$ Subtract.

Exercises **Determine whether each relation is a function.** *See Example 1 on page 226.*

37.

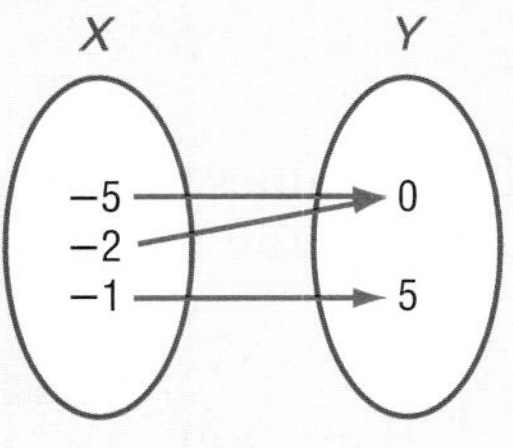

38.

x	y
5	3
1	4
−6	5
1	6
−2	7

39. {(2, 3), (−3, −4), (−1, 3)}

If $g(x) = x^2 - x + 1$, find each value. *See Examples 3 and 4 on pages 227 and 228.*

40. $g(2)$
41. $g(-1)$
42. $g\left(\frac{1}{2}\right)$
43. $g(5) - 3$
44. $g(a)$
45. $g(-2a)$

4-7 Arithmetic Sequences

See pages 233–238.

Concept Summary

- An arithmetic sequence is a numerical pattern that increases or decreases at a constant rate or value called the common difference.
- To find the next term in an arithmetic sequence, add the common difference to the last term.

Chapter 4 For More ...
• Extra Practice, see pages 828–830.
• Mixed Problem Solving, see page 856.

Example **Find the next three terms of the arithmetic sequence 10, 23, 36, 49, ...**

Find the common difference.

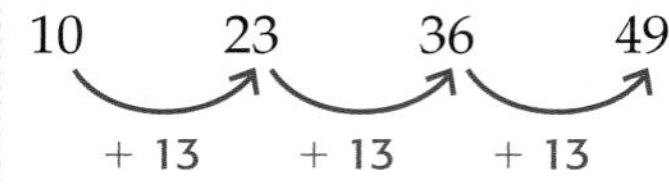

So, $d = 13$.

Add 13 to the last term of the sequence to get the next term. Continue adding 13 until the next three terms are found.

49 62 75 88
+ 13 + 13 + 13

The next three terms are 62, 75, and 88.

Exercises **Find the next three terms of each arithmetic sequence.**
See Example 2 on page 234.

46. 9, 18, 27, 36, ...
47. 6, 11, 16, 21, ...
48. 10, 21, 32, 43, ...
49. 14, 12, 10, 8, ...
50. −3, −11, −19, −27, ...
51. −35, −29, −23, −17, ...

4-8 Writing Equations from Patterns

See pages 240–245.

Concept Summary

- Look for a pattern in data. If the relationship between the domain and range is linear, the relationship can be described by an equation.

Example **Write an equation in function notation for the relation graphed at the right.**

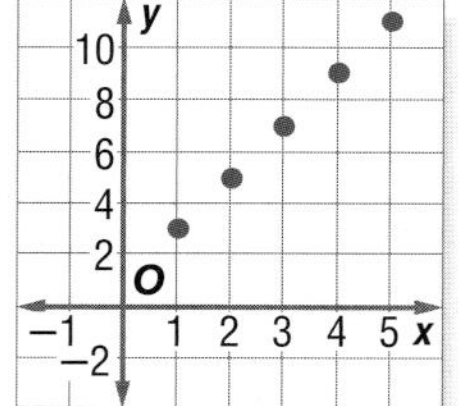

Make a table of ordered pairs for several points on the graph.

x	1	2	3	4	5
y	3	5	7	9	11

The difference in y values is twice the difference of x values. This suggests that $y = 2x$. However, $3 \neq 2(1)$. Compare the values of y to the values of $2x$.

x	1	2	3	4	5
2x	2	4	6	8	10
y	3	5	7	9	11

y is always 3 more than 2*x*.

The difference between y and $2x$ is always 1. So the equation is $y = 2x + 1$. Since this relation is also a function, it can be written as $f(x) = 2x + 1$.

Exercises **Write an equation in function notation for each relation.**
See Example 4 on pages 242 and 243.

52.

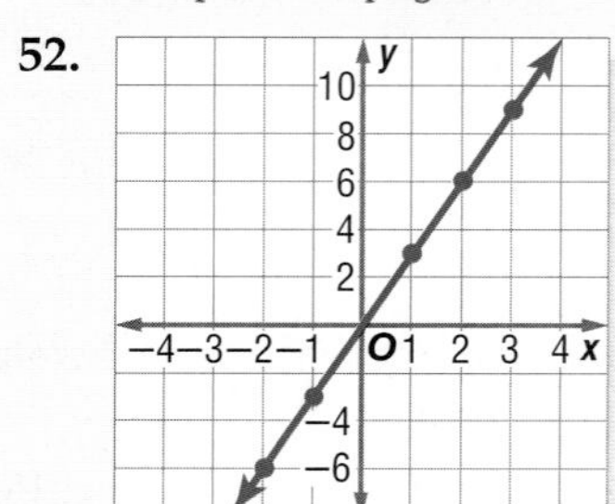

53.

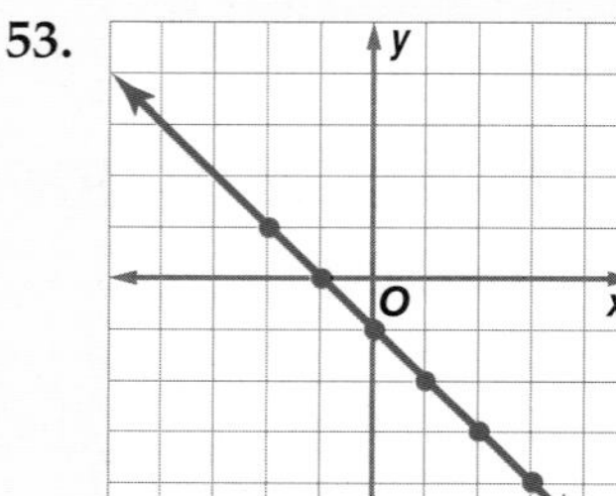

Chapter 4 Practice Test

Vocabulary and Concepts

Choose the letter that best matches each description.

1. a figure turned around a point
2. a figure slid horizontally, vertically, or both
3. a figure flipped over a line

a. reflection
b. rotation
c. translation

Skills and Applications

4. Graph $K(0, -5)$, $M(3, -5)$, and $N(-2, -3)$.
5. Name the quadrant in which $P(25, 1)$ is located.

For Exercises 6 and 7, use the following information.
A parallelogram has vertices $H(-2, -2)$, $I(-4, -6)$, $J(-5, -5)$, and $K(-3, -1)$.

6. Reflect parallelogram $HIJK$ over the y-axis and graph its image.
7. Translate parallelogram $HIJK$ up 2 units and graph its image.

Express the relation shown in each table, mapping, or graph as a set of ordered pairs. Then write the inverse of the relation.

8.

x	$f(x)$
0	−1
2	4
4	5
6	10

9.

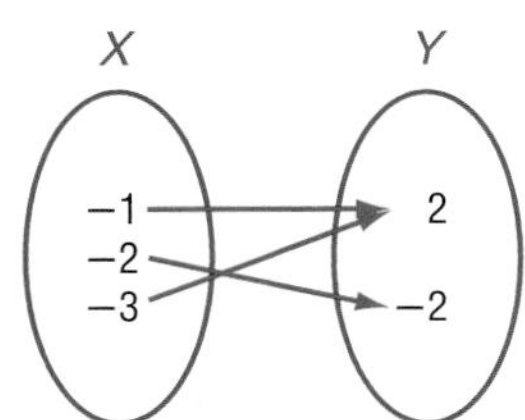

10.

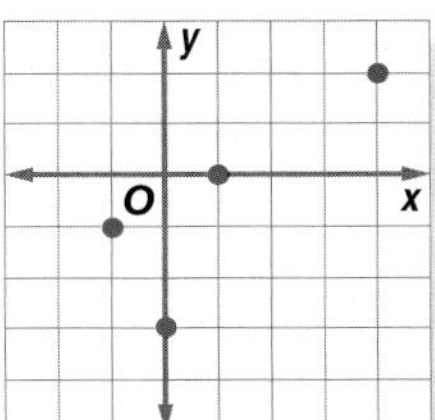

Solve each equation if the domain is {−2, −1, 0, 2, 4}. Graph the solution set.

11. $y = -4x + 10$
12. $3x - y = 10$
13. $\frac{1}{2}x - y = 5$

Graph each equation.

14. $y = x + 2$
15. $x + 2y = -1$
16. $-3x = 5 - y$

Determine whether each relation is a function.

17. {(2, 4), (3, 2), (4, 6), (5, 4)}
18. {(3, 1), (2, 5), (4, 0), (3, −2)}
19. $8y = 7 + 3x$

If $f(x) = -2x + 5$ and $g(x) = x^2 - 4x + 1$, find each value.

20. $g(-2)$
21. $f\left(\frac{1}{2}\right)$
22. $g(3a) + 1$
23. $f(x + 2)$

Determine whether each sequence is an arithmetic sequence. If it is, state the common difference.

24. 16, 24, 32, 40, …
25. 99, 87, 76, 65, …
26. 5, 17, 29, 41, …

Find the next three terms in each sequence.

27. 5, −10, 15, −20, 25, …
28. 5, 5, 6, 8, 11, 15, …

29. **TEMPERATURE** The equation to convert Celsius temperature to Kelvin temperature is $K = C + 273$. Solve the equation for C. State the independent and dependent variables. Choose five values for K and their corresponding values for C.

30. **STANDARDIZED TEST PRACTICE** If $f(x) = 3x - 2$, find $f(8) - f(-5)$.

Ⓐ 7 Ⓑ 9 Ⓒ 37 Ⓓ 39

Chapter 4 Standardized Test Practice

Part 1 Multiple Choice

Record your answers on the answer sheet provided by your teacher or on a sheet of paper.

1. The number of students in Highview School is currently 315. The school population is predicted to increase by 2% next year. According to the prediction, how many students will attend next year? (Prerequisite Skill)

Ⓐ 317 Ⓑ 321
Ⓒ 378 Ⓓ 630

2. In 2001, two women skied 1675 miles in 89 days across the land mass of Antarctica. They still had to ski 508 miles across the Ross Ice Shelf to reach McMurdo Station. About what percent of their total distance remained? (Prerequisite Skill)

Ⓐ 2% Ⓑ 17%
Ⓒ 23% Ⓓ 30%

3. Only 2 out of 5 students surveyed said they eat five servings of fruits or vegetables daily. If there are 470 students in a school, how many would you predict eat five servings of fruits or vegetables daily? (Lesson 2-6)

Ⓐ 94 Ⓑ 188
Ⓒ 235 Ⓓ 282

4. Solve $13x = 2(5x + 3)$ for x. (Lesson 3-4)

Ⓐ 0 Ⓑ 2 Ⓒ 3 Ⓓ 4

The Princeton Review Test-Taking Tip

Questions 4 and 14 Some multiple-choice questions ask you to solve an equation or inequality. You can check your solution by replacing the variable in the equation or inequality with your answer. The answer choice that results in a true statement is the correct answer.

5. The circle shown below passes through points at (1, 4),(−2, 1), (−5, 4), and (−2, 7). Which point represents the center of the circle? (Lesson 4-1)

Ⓐ (−2, −4)
Ⓑ (−2, 4)
Ⓒ (−4, 2)
Ⓓ (4, −2)

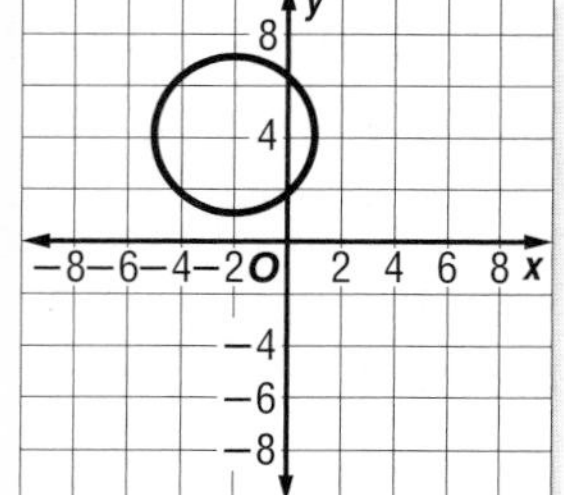

6. Which value of x would cause the relation $\{(2, 5), (x, 8), (7, 10)\}$ *not* to be a function? (Lesson 4-4)

Ⓐ 1 Ⓑ 2 Ⓒ 5 Ⓓ 8

7. Which ordered pair (x, y) is a solution of $3x + 4y = 12$? (Lesson 4-4)

Ⓐ (−2, 4) Ⓑ (0, −3)
Ⓒ (1, 2) Ⓓ (4, 0)

8. Which missing value for y would make this relation a linear relation? (Lesson 4-7)

Ⓐ −2
Ⓑ 0
Ⓒ 1
Ⓓ 2

x	y
1	−3
2	−1
3	?
4	3

9. Which equation describes the data in the table? (Lesson 4-8)

Ⓐ $y = -2x + 1$
Ⓑ $y = x + 1$
Ⓒ $y = -x + 3$
Ⓓ $y = x - 5$

x	y
−2	5
1	2
4	−1
6	−3

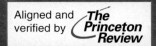

Aligned and verified by The Princeton Review

Part 2 Short Response/Grid In

Record your answers on the answer sheet provided by your teacher or on a sheet of paper.

10. The lengths of the corresponding sides of these two rectangles are proportional. What is the width w? (Lesson 2-6)

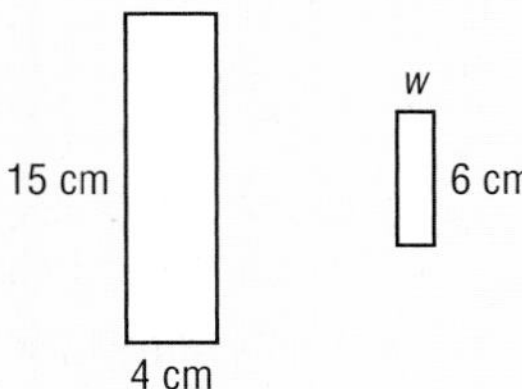

11. The PTA at Fletcher's school sold raffle tickets for a television set. Two thousand raffle tickets were sold. Fletcher's family bought 25 raffle tickets. What is the probability that his family will win the television? Express the answer as a percent. (Lesson 2-7)

12. The sum of three integers is 52. The second integer is 3 more than the first. The third integer is 1 more than twice the first. What are the integers? (Lessons 3-1 and 3-4)

13. Solve $5(x - 2) - 3(x + 4) = 10$ for x. (Lesson 3-4)

14. A CD player originally cost $160. It is now on sale for $120. What is the percent of decrease in its price? (Lesson 3-5)

15. A swimming pool holds 1800 cubic feet of water. It is 6 feet deep and 20 feet long. How many feet wide is the pool? ($V = \ell wh$) (Lesson 3-8)

16. Garth used toothpicks to form a pattern of triangles as shown below. If he continues this pattern, what is the total number of toothpicks that he will use to form a pattern of 7 triangles? (Lessons 4-7 and 4-8)

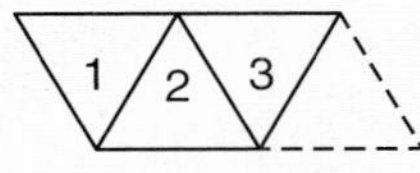

Part 3 Quantitative Comparison

Compare the quantity in Column A and in Column B. Then determine whether:

- Ⓐ **the quantity in Column A is greater,**
- Ⓑ **the quantity in Column B is greater,**
- Ⓒ **the two quantities are equal, or**
- Ⓓ **the relationship cannot be determined from the information given.**

	Column A	Column B
17.	$4^2 \div 16(2 + 5) \cdot 3$	$\frac{60 - 2^3 \cdot 3 + 6}{4^3 - 62}$
18.	$\left(\frac{2}{3}\right)\left(\frac{15}{8}\right)\left(\frac{1}{9}\right)$	$7\left(\frac{3}{4}\right)\left(\frac{1}{14}\right)$
19.	x if $6x - 15 = -3x + 75$	y if $3y - 32 = 7y - 74$
20.	$f(-10)$ if $f(x) = 37 + 10x$	$g(-15)$ if $g(x) = 9x - 7$

(17: Lesson 1-2) (18: Lesson 2-3) (19: Lesson 3-5) (20: Lesson 4-6)

Part 4 Open Ended

Record your answers on a sheet of paper. Show your work.

21. Latoya bought 48 one-foot-long sections of fencing. She plans to use the fencing to enclose a rectangular area for a garden. (Lesson 3-8)

a. Using ℓ for the length and w for the width of the garden, write an equation for its perimeter.

b. If the length ℓ in feet and width w in feet are whole numbers, what is the greatest possible area of this garden?

Chapter 5

Analyzing Linear Equations

What You'll Learn

- **Lesson 5-1** Find the slope of a line.
- **Lesson 5-2** Write direct variation equations.
- **Lessons 5-3 through 5-5** Write linear equations in slope-intercept and point-slope forms.
- **Lesson 5-6** Write equations for parallel and perpendicular lines.
- **Lesson 5-7** Draw a scatter plot and write the equations of a line of fit.

Key Vocabulary

- slope (p. 256)
- rate of change (p. 258)
- direct variation (p. 264)
- slope-intercept form (p. 272)
- point-slope form (p. 286)

Why It's Important

Linear equations are used to model a variety of real-world situations. The concept of slope allows you to analyze how a quantity changes over time.

You can use a linear equation to model the cost of the space program. The United States began its exploration of space in January, 1958, when it launched its first satellite into orbit. In the 1970s, NASA developed the space shuttle to reduce costs by inventing the first reusable spacecraft. *You will use a linear equation to model the cost of the space program in Lesson 5-7.*

Getting Started

Prerequisite Skills To be successful in this chapter, you'll need to master these skills and be able to apply them in problem-solving situations. Review these skills before beginning Chapter 5.

For Lesson 5-1 — Simplify Fractions

Simplify. *(For review, see pages 798 and 799.)*

1. $\frac{2}{10}$ **2.** $\frac{8}{12}$ **3.** $\frac{2}{-8}$ **4.** $\frac{-4}{8}$

5. $\frac{-5}{-15}$ **6.** $\frac{-7}{-28}$ **7.** $\frac{9}{3}$ **8.** $\frac{18}{12}$

For Lesson 5-2 — Evaluate Expressions

Evaluate $\frac{a-b}{c-d}$ for each set of values. *(For review, see Lesson 1-2.)*

9. $a = 6, b = 5, c = 8, d = 4$

10. $a = 5, b = -1, c = 2, d = -1$

11. $a = -2, b = 1, c = 4, d = 0$

12. $a = 8, b = -2, c = -1, d = 1$

13. $a = -3, b = -3, c = 4, d = 7$

14. $a = \frac{1}{2}, b = \frac{3}{2}, c = 7, d = 9$

For Lessons 5-3 through 5-7 — Identify Points on a Coordinate Plane

Write the ordered pair for each point.
(For review, see Lesson 4-1.)

15. J **16.** K

17. L **18.** M

19. N **20.** P

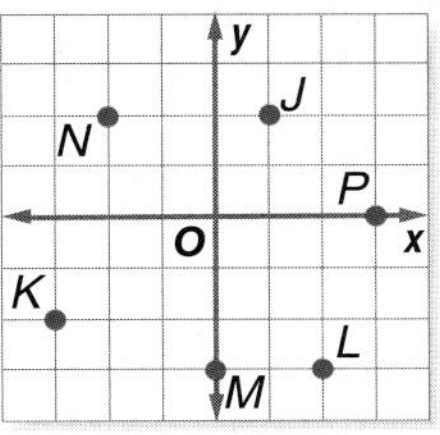

Make this Foldable to help you organize information about writing linear equations. Begin with four sheets of grid paper.

Step 1 Fold and Cut

Fold each sheet of grid paper in half along the width. Then cut along the crease.

Step 2 Staple

Staple the eight half-sheets together to form a booklet.

Step 3 Cut Tabs

Cut seven lines from the bottom of the top sheet, six lines from the second sheet, and so on.

Step 4 Label

Label each of the tabs with a lesson number. The last tab is for the vocabulary.

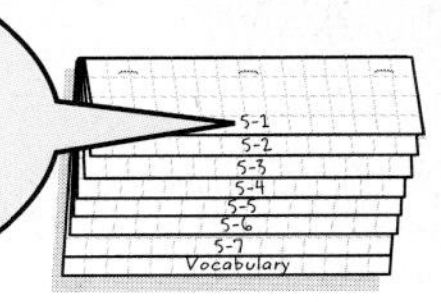

Reading and Writing As you read and study the chapter, use each page to write notes and to graph examples for each lesson.

5-1 Slope

What You'll Learn

- Find the slope of a line.
- Use rate of change to solve problems.

Vocabulary

- slope
- rate of change

Why is slope important in architecture?

The slope of a roof describes how steep it is. It is the number of units the roof rises for each unit of run. In the photo, the roof rises 8 feet for each 12 feet of run.

$$\text{slope} = \frac{\text{rise}}{\text{run}}$$
$$= \frac{8}{12} \text{ or } \frac{2}{3}$$

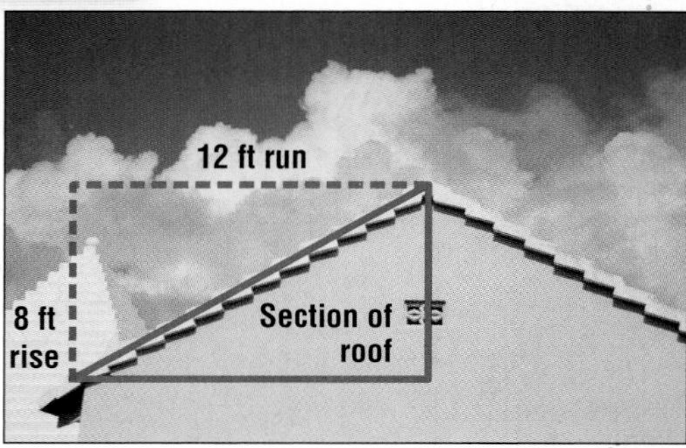

FIND SLOPE The **slope** of a line is a number determined by any two points on the line. This number describes how steep the line is. The greater the absolute value of the slope, the steeper the line. Slope is the ratio of the change in the y-coordinates (rise) to the change in the x-coordinates (run) as you move from one point to the other.

The graph shows a line that passes through (1, 3) and (4, 5).

$$\text{slope} = \frac{\text{rise}}{\text{run}}$$
$$= \frac{\text{change in } y\text{-coordinates}}{\text{change in } x\text{-coordinates}}$$
$$= \frac{5-3}{4-1} \text{ or } \frac{2}{3}$$

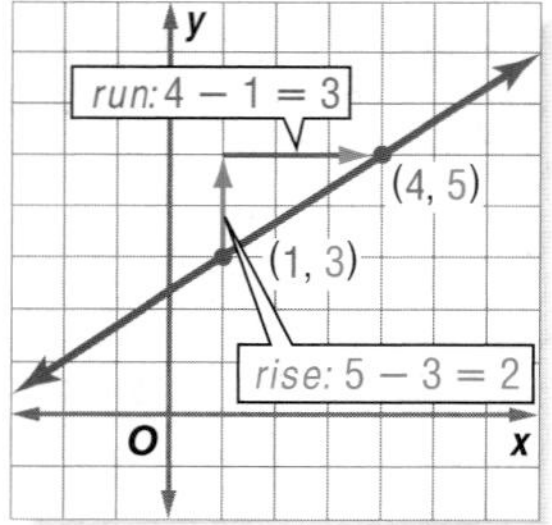

So, the slope of the line is $\frac{2}{3}$.

Study Tip

Reading Math

In x_1, the 1 is called a *subscript*. It is read *x sub 1*.

Key Concept — *Slope of a Line*

- **Words** The slope of a line is the ratio of the rise to the run.
- **Symbols** The slope m of a nonvertical line through any two points, (x_1, y_1) and (x_2, y_2), can be found as follows.

$$m = \frac{y_2 - y_1}{x_2 - x_1} \quad \begin{array}{l}\leftarrow \text{change in } y \\ \leftarrow \text{change in } x\end{array}$$

- **Model**

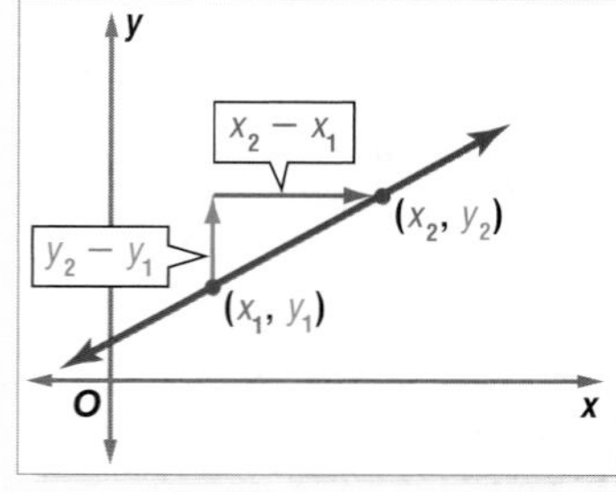

Example 1 Positive Slope

Find the slope of the line that passes through (−1, 2) and (3, 4).

Let $(-1, 2) = (x_1, y_1)$ and $(3, 4) = (x_2, y_2)$.

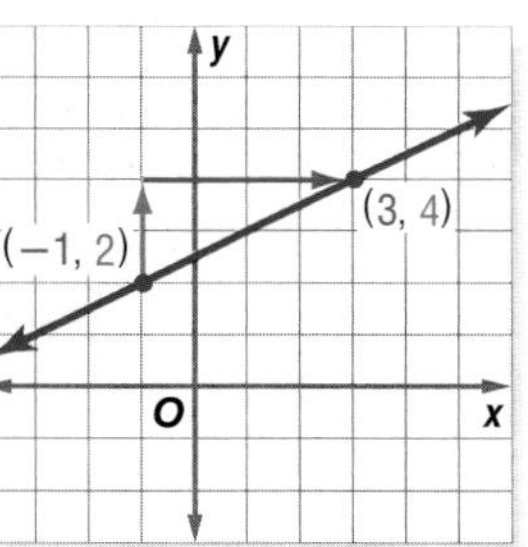

$m = \dfrac{y_2 - y_1}{x_2 - x_1}$ rise/run

$= \dfrac{4 - 2}{3 - (-1)}$ Substitute.

$= \dfrac{2}{4}$ or $\dfrac{1}{2}$ Simplify.

The slope is $\frac{1}{2}$.

Study Tip

Common Misconception

It may make your calculations easier to choose the point on the left as (x_1, y_1). However, *either point* may be chosen as (x_1, y_1).

Example 2 Negative Slope

Find the slope of the line that passes through (−1, −2) and (−4, 1).

Let $(-1, -2) = (x_1, y_1)$ and $(-4, 1) = (x_2, y_2)$.

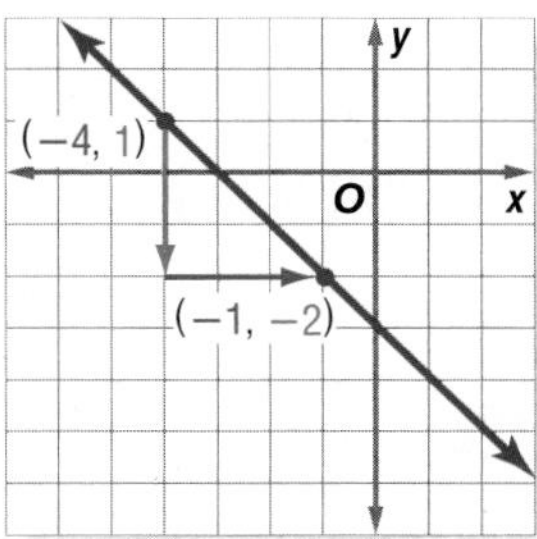

$m = \dfrac{y_2 - y_1}{x_2 - x_1}$ rise/run

$= \dfrac{1 - (-2)}{-4 - (-1)}$ Substitute.

$= \dfrac{3}{-3}$ or -1 Simplify.

The slope is -1.

Example 3 Zero Slope

Find the slope of the line that passes through (1, 2) and (−1, 2).

Let $(1, 2) = (x_1, y_1)$ and $(-1, 2) = (x_2, y_2)$.

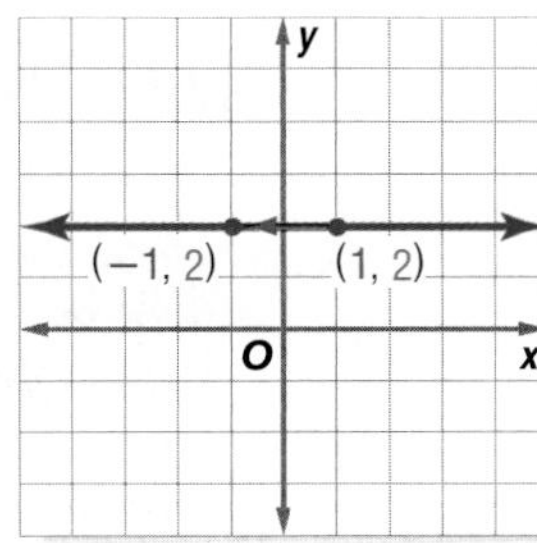

$m = \dfrac{y_2 - y_1}{x_2 - x_1}$ rise/run

$= \dfrac{2 - 2}{-1 - 1}$ Substitute.

$= \dfrac{0}{-2}$ or 0 Simplify.

The slope is zero.

Example 4 Undefined Slope

Find the slope of the line that passes through (1, −2) and (1, 3).

Let $(1, -2) = (x_1, y_1)$ and $(1, 3) = (x_2, y_2)$.

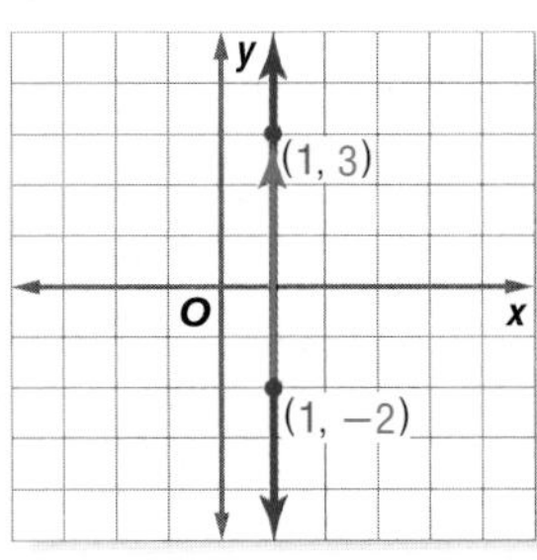

$m = \dfrac{y_2 - y_1}{x_2 - x_1}$ rise/run

$= \dfrac{3 - (-2)}{1 - 1}$ or $\dfrac{5}{0}$ (crossed out)

Since division by zero is undefined, the slope is undefined.

Concept Summary — *Classifying Lines*

Positive Slope

line slopes up from left to right

Negative Slope

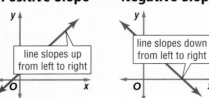

line slopes down from left to right

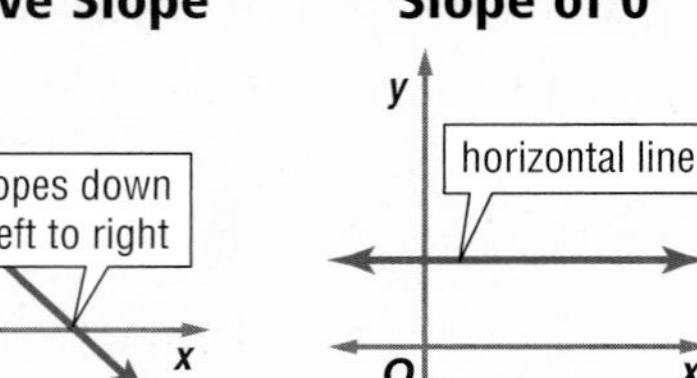

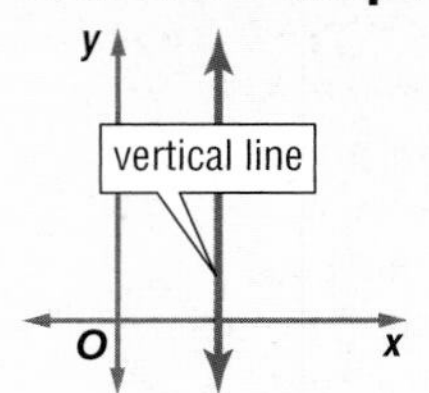

If you know the slope of a line and the coordinates of one of the points on a line, you can find the coordinates of other points on the line.

Example 5 Find Coordinates Given Slope

Find the value of r so that the line through $(r, 6)$ and $(10, -3)$ has a slope of $-\frac{3}{2}$.

Let $(r, 6) = (x_1, y_1)$ and $(10, -3) = (x_2, y_2)$.

$$m = \frac{y_2 - y_1}{x_2 - x_1}$$ Slope formula

$$-\frac{3}{2} = \frac{-3 - 6}{10 - r}$$ Substitute.

$$-\frac{3}{2} = \frac{-9}{10 - r}$$ Subtract.

$$-3(10 - r) = 2(-9)$$ Find the cross products.

$$-30 + 3r = -18$$ Simplify.

$$-30 + 3r + 30 = -18 + 30$$ Add 30 to each side.

$$3r = 12$$ Simplify.

$$\frac{3r}{3} = \frac{12}{3}$$ Divide each side by 3.

$$r = 4$$ Simplify.

Study Tip

Look Back

To review **cross products**, see Lesson 3-6.

RATE OF CHANGE Slope can be used to describe a rate of change. The **rate of change** tells, on average, how a quantity is changing over time.

Example 6 Find a Rate of Change

DINING OUT The graph shows the amount spent on food and drink at U.S. restaurants in recent years.

a. Find the rates of change for 1980–1990 and 1990–2000.

Use the formula for slope.

$$\frac{\text{rise}}{\text{run}} = \frac{\text{change in quantity}}{\text{change in time}} \quad \begin{matrix}\leftarrow \text{billion \$} \\ \leftarrow \text{years}\end{matrix}$$

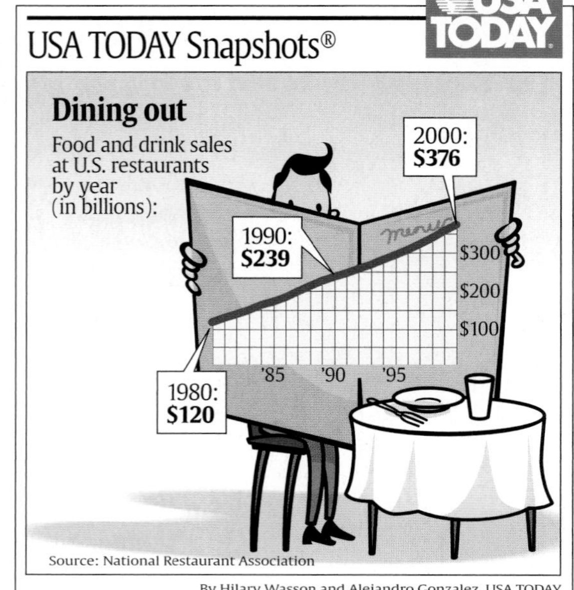

Log on for:
- Updated data
- More activities on rate of change www.algebra1.com/usa_today

1980–1990: $\frac{\text{change in quantity}}{\text{change in time}} = \frac{239 - 120}{1990 - 1980}$ Substitute.

$= \frac{119}{10}$ or 11.9 Simplify.

Spending on food and drink increased by $119 billion in a 10-year period for a rate of change of $11.9 billion per year.

1990–2000: $\frac{\text{change in quantity}}{\text{change in time}} = \frac{376 - 239}{2000 - 1990}$ Substitute.

$= \frac{137}{10}$ or 13.7 Simplify.

Over this 10-year period, spending increased by $137 billion, for a rate of change of $13.7 billion per year.

b. Explain the meaning of the slope in each case.

For 1980–1990, on average, $11.9 billion more was spent each year than the last. For 1990–2000, on average, $13.7 billion more was spent each year than the last.

c. How are the different rates of change shown on the graph?

There is a greater vertical change for 1990–2000 than for 1980–1990. Therefore, the section of the graph for 1990–2000 has a steeper slope.

Check for Understanding

Concept Check

1. **Explain** how you would find the slope of the line at the right.

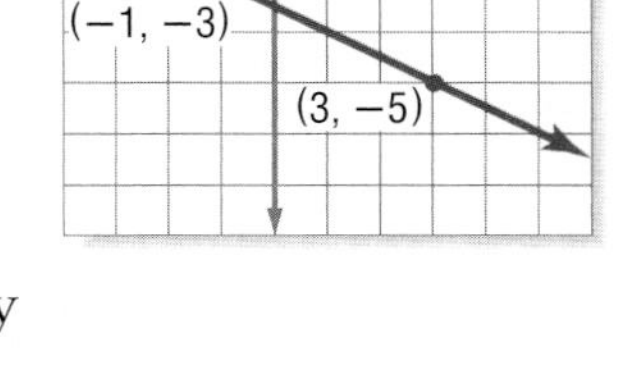

2. **OPEN ENDED** Draw the graph of a line having each slope.
   a. positive slope **b.** negative slope
   c. slope of 0 **d.** undefined slope
3. **Explain** why the formula for determining slope using the coordinates of two points does not apply to vertical lines.
4. **FIND THE ERROR** Carlos and Allison are finding the slope of the line that passes through (2, 6) and (5, 3).

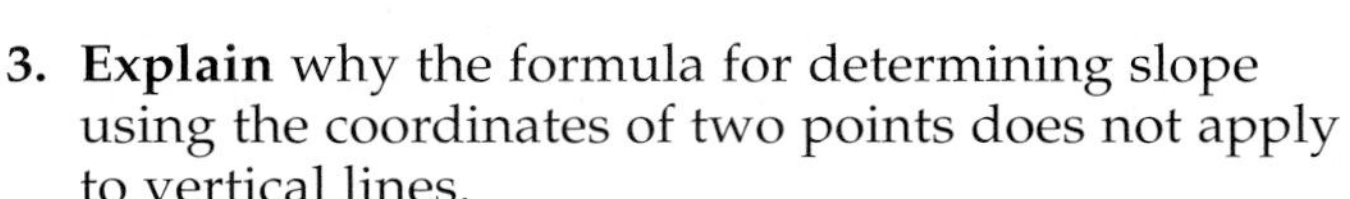

Carlos

$\frac{3-6}{5-2} = \frac{-3}{3}$ or -1

Allison

$\frac{6-3}{5-2} = \frac{3}{3}$ or 1

Who is correct? Explain your reasoning.

Guided Practice

Find the slope of the line that passes through each pair of points.

5. (1, 1), (3, 4)
6. (0, 0), (5, 4)
7. (−2, 2), (−1, −2)
8. (7, −4), (9, −1)
9. (3, 5), (−2, 5)
10. (−1, 3), (−1, 0)

Find the value of r so the line that passes through each pair of points has the given slope.

11. $(6, -2), (r, -6), m = 4$
12. $(9, r), (6, 3), m = -\frac{1}{3}$

Application

CABLE TV **For Exercises 13 and 14, use the graph at the right.**

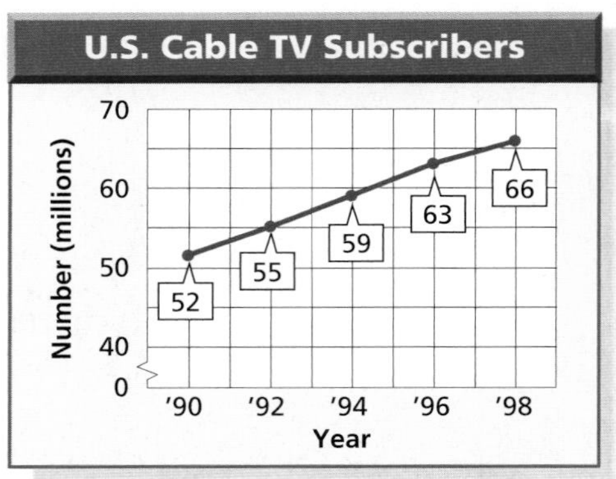

13. Find the rate of change for 1990–1992.
14. Without calculating, find a 2-year period that had a greater rate of change than 1990–1992. Explain your reasoning.

Practice and Apply

Homework Help

For Exercises	See Examples
15–34	1–4
41–48	5
53–57	6

Extra Practice

See page 831.

Find the slope of the line that passes through each pair of points.

15.
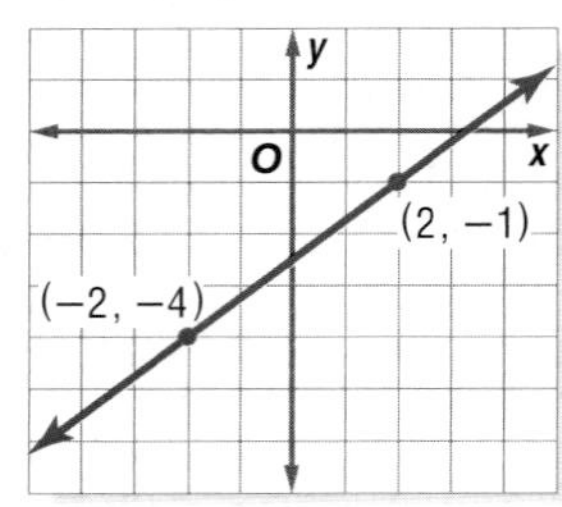

16.
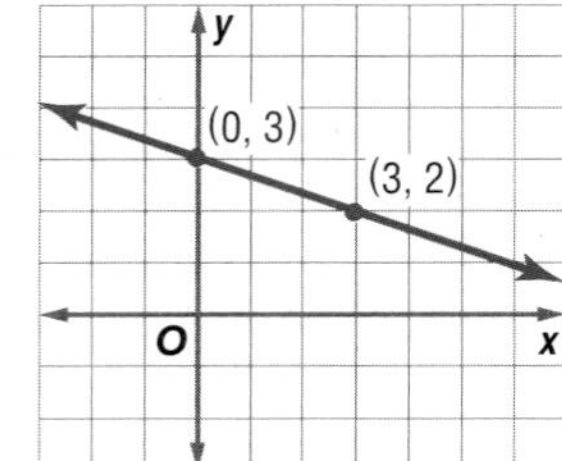

17. (−4, −1), (−3, −3)
18. (−3, 3), (1, 3)
19. (−2, 1), (−2, 3)
20. (2, 3), (9, 7)
21. (5, 7), (−2, −3)
22. (−3, 6), (2, 4)
23. (−3, −4), (5, −1)
24. (2, −1), (5, −3)
25. (−5, 4), (−5, −1)
26. (2, 6), (−1, 3)
27. (−2, 3), (8, 3)
28. (−3, 9), (−7, 6)
29. (−8, 3), (−6, 2)
30. (−2, 0), (1, −1)
31. (4.5, −1), (5.3, 2)
32. (0.75, 1), (0.75, −1)
33. $\left(2\frac{1}{2}, -1\frac{1}{2}\right), \left(-\frac{1}{2}, \frac{1}{2}\right)$
34. $\left(\frac{3}{4}, 1\frac{1}{4}\right), \left(-\frac{1}{2}, -1\right)$

ARCHITECTURE **Use a ruler to estimate the slope of each roof.**

35.

36.

37. Find the slope of the line that passes through the origin and (r, s).
38. What is the slope of the line that passes through (a, b) and $(a, -b)$?
39. **PAINTING** A ladder reaches a height of 16 feet on a wall. If the bottom of the ladder is placed 4 feet away from the wall, what is the slope of the ladder as a positive number?

40. **PART-TIME JOBS** In 1991, the federal minimum wage rate was \$4.25 per hour. In 1997, it was increased to \$5.15. Find the annual rate of change in the federal minimum wage rate from 1991 to 1997.

Find the value of *r* so the line that passes through each pair of points has the given slope.

41. $(6, 2), (9, r), m = -1$
42. $(4, -5), (3, r), m = 8$
43. $(5, r), (2, -3), m = \frac{4}{3}$
44. $(-2, 7), (r, 3), m = \frac{4}{3}$
45. $\left(\frac{1}{2}, -\frac{1}{4}\right), \left(r, -\frac{5}{4}\right), m = 4$
46. $\left(\frac{2}{3}, r\right), \left(1, \frac{1}{2}\right), m = \frac{1}{2}$
47. $(4, r), (r, 2), m = -\frac{5}{3}$
48. $(r, 5), (-2, r), m = -\frac{2}{9}$

49. **CRITICAL THINKING** Explain how you know that the slope of the line through $(-4, -5)$ and $(4, 5)$ is positive without calculating.

HEALTH **For Exercises 50–52, use the table that shows Karen's height from age 12 to age 20.**

Age (years)	12	14	16	18	20
Height (inches)	60	64	66	67	67

50. Make a broken-line graph of the data.
51. Use the graph to determine the two-year period when Karen grew the fastest. Explain your reasoning.
52. Explain the meaning of the horizontal section of the graph.

SCHOOL **For Exercises 53–55, use the graph that shows public school enrollment.**

53. For which 5-year period was the rate of change the greatest? When was the rate of change the least?
54. Find the rate of change from 1985 to 1990.
55. Explain the meaning of the part of the graph with a negative slope.

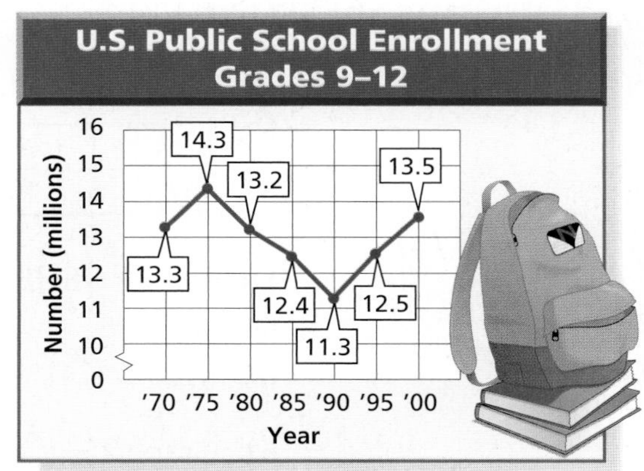

56. **RESEARCH** Use the Internet or other reference to find the population of your city or town in 1930, 1940, . . . , 2000. For which decade was the rate of change the greatest?

57. **CONSTRUCTION** The slope of a stairway determines how easy it is to climb the stairs. Suppose the vertical distance between two floors is 8 feet 9 inches. Find the total run of the ideal stairway in feet and inches.

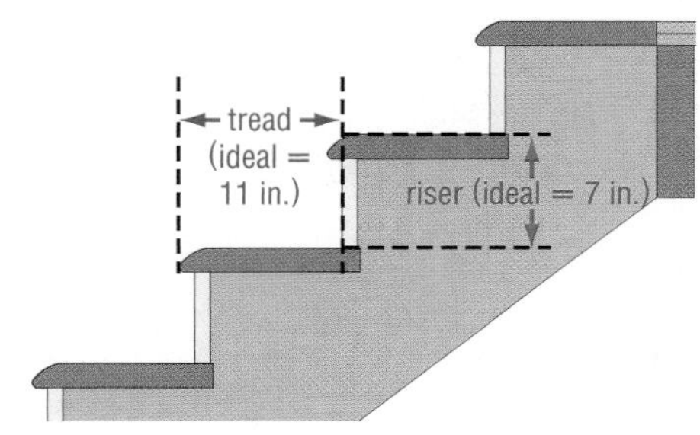

58. **WRITING IN MATH** Answer the question that was posed at the beginning of the lesson.

Why is slope important in architecture?

Include the following in your answer:

- an explanation of how to find the slope of a roof, and
- a comparison of the appearance of roofs with different slopes.

Standardized Test Practice

59. The slope of the line passing through (5, −4) and (5, −10) is

Ⓐ positive. Ⓑ negative. Ⓒ zero. Ⓓ undefined.

60. The slope of the line passing through (a, b) and (c, d) is

Ⓐ $\frac{d-c}{b-a}$. Ⓑ $\frac{b-d}{a-c}$. Ⓒ $\frac{d-b}{a-c}$. Ⓓ $\frac{a-c}{b-d}$.

Extending the Lesson

61. Choose four different pairs of points from those labeled on the graph. Find the slope of the line using the coordinates of each pair of points. Describe your findings.

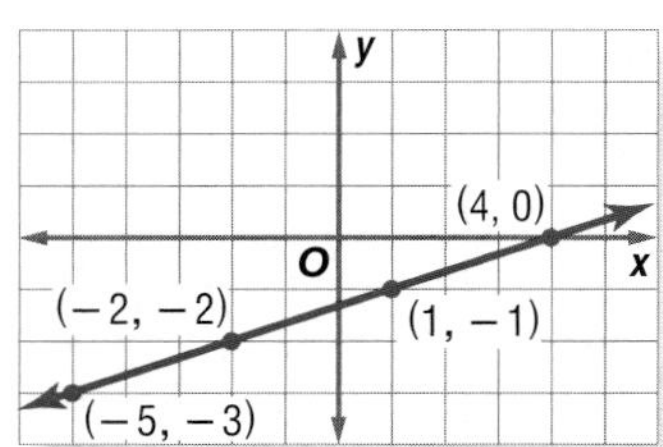

62. **MAKE A CONJECTURE** Determine whether $Q(2, 3)$, $R(-1, -1)$, and $S(-4, -2)$ lie on the same line. Explain your reasoning.

Maintain Your Skills

Mixed Review

Write an equation for each function. *(Lesson 4-8)*

63.

x	1	2	3	4	5
f(x)	5	10	15	20	25

64.

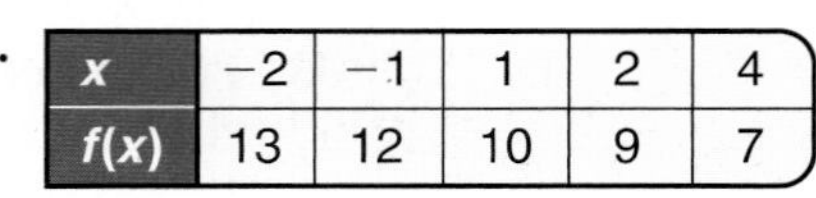

x	−2	−1	1	2	4
f(x)	13	12	10	9	7

Determine whether each relation is a function. *(Lesson 4-6)*

65. $y = -15$

66. $x = 5$

67. {(1, 0), (1, 4), (−1, 1)}

68. {(6, 3), (5, −2), (2, 3)}

69. Graph $x - y = 0$. *(Lesson 4-4)*

70. What number is 40% of 37.5? *(Prerequisite Skill)*

Find each product. *(Lesson 2-4)*

71. 7(−3)

72. (−4)(−2)

73. (9)(−4)

74. (−8)(3.7)

75. $\left(-\frac{7}{8}\right)\left(\frac{1}{3}\right)$

76. $\left(\frac{1}{4}\right)\left(\frac{1}{2}\right)(-14)$

Getting Ready for the Next Lesson

PREREQUISITE SKILL **Find each quotient.**

*(To review **dividing fractions**, see pages 800 and 801.)*

77. $6 \div \frac{2}{3}$

78. $12 \div \frac{1}{4}$

79. $10 \div \frac{3}{8}$

80. $\frac{1}{2} \div \frac{1}{3}$

81. $\frac{3}{4} \div \frac{1}{6}$

82. $\frac{3}{4} \div 6$

83. $18 \div \frac{7}{8}$

84. $\frac{3}{8} \div \frac{2}{5}$

85. $2\frac{2}{3} \div \frac{1}{4}$

Reading Mathematics

Mathematical Words and Everyday Words

You may have noticed that many words used in mathematics are also used in everyday language. You can use the everyday meaning of these words to better understand their mathematical meaning. The table shows two mathematical words along with their everyday and mathematical meanings.

Word	Everyday Meaning	Mathematical Meaning
expression	**1.** something that expresses or communicates in words, art, music, or movement **2.** the manner in which one expresses oneself, especially in speaking, depicting, or performing	one or more numbers or variables along with one or more arithmetic operations
function	**1.** the action for which one is particularly fitted or employed **2.** an official ceremony or a formal social occasion **3.** something closely related to another thing and dependent on it for its existence, value, or significance	a relationship in which the output depends upon the input

Source: *The American Heritage Dictionary of the English Language*

Notice that the mathematical meaning is more specific, but related to the everyday meaning. For example, the mathematical meaning of *expression* is closely related to the first everyday definition. In mathematics, an expression communicates using symbols.

Reading to Learn

1. How does the mathematical meaning of *function* compare to the everyday meaning?
2. **RESEARCH** Use the Internet or other reference to find the everyday meaning of each word below. How might these words apply to mathematics? Make a table like the one above and note the mathematical meanings that you learn as you study Chapter 5.
   a. slope **b.** intercept **c.** parallel

5-2 Slope and Direct Variation

What You'll Learn

- Write and graph direct variation equations.
- Solve problems involving direct variation.

Vocabulary

- direct variation
- constant of variation
- family of graphs
- parent graph

How is slope related to your shower?

A standard showerhead uses about 6 gallons of water per minute. If you graph the ordered pairs from the table, the slope of the line is 6.

x (minutes)	y (gallons)
0	0
1	6
2	12
3	18
4	24

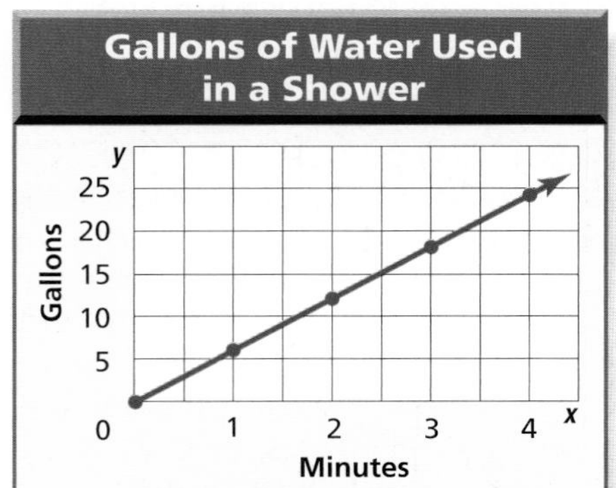

The equation is $y = 6x$. The number of gallons of water y depends *directly* on the amount of time in the shower x.

DIRECT VARIATION A **direct variation** is described by an equation of the form $y = kx$, where $k \neq 0$. We say that *y varies directly with x* or *y varies directly as x*. In the equation $y = kx$, k is the **constant of variation**.

Example 1 Slope and Constant of Variation

Name the constant of variation for each equation. Then find the slope of the line that passes through each pair of points.

a.

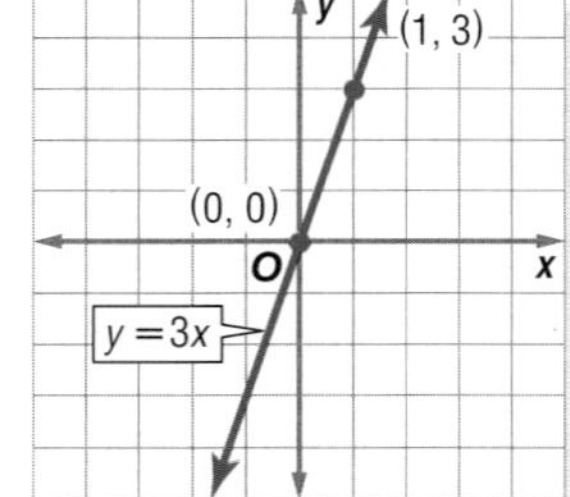

The constant of variation is 3.

$m = \frac{y_2 - y_1}{x_2 - x_1}$ Slope formula

$m = \frac{3 - 0}{1 - 0}$ $(x_1, y_1) = (0, 0)$ $(x_2, y_2) = (1, 3)$

$m = 3$ The slope is 3.

b.

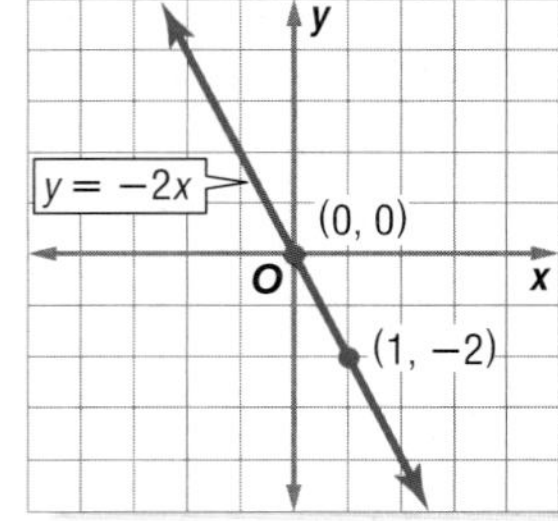

The constant of variation is -2.

$m = \frac{y_2 - y_1}{x_2 - x_1}$ Slope formula

$m = \frac{-2 - 0}{1 - 0}$ $(x_1, y_1) = (0, 0)$ $(x_2, y_2) = (1, -2)$

$m = -2$ The slope is -2.

Compare the constant of variation with the slope of the graph for each example. Notice that the slope of the graph of $y = kx$ is k.

The ordered pair (0, 0) is a solution of $y = kx$. Therefore, the graph of $y = kx$ passes through the origin. You can use this information to graph direct variation equations.

Example 2 Direct Variation with $k > 0$

Graph $y = 4x$.

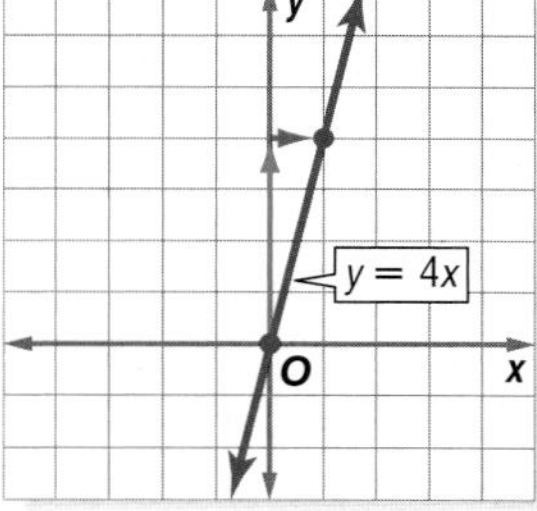

Step 1 Write the slope as a ratio.

$4 = \frac{4}{1}$ $\frac{\text{rise}}{\text{run}}$

Step 2 Graph (0, 0).

Step 3 From the point (0, 0), move up 4 units and right 1 unit. Draw a dot.

Step 4 Draw a line containing the points.

Example 3 Direct Variation with $k < 0$

Graph $y = -\frac{1}{3}x$.

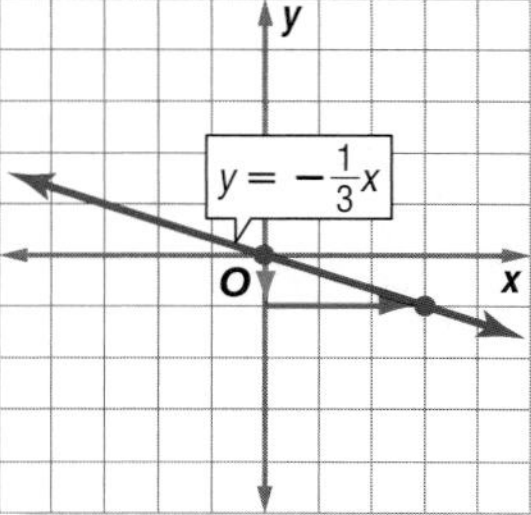

Step 1 Write the slope as a ratio.

$-\frac{1}{3} = \frac{-1}{3}$ $\frac{\text{rise}}{\text{run}}$

Step 2 Graph (0, 0).

Step 3 From the point (0, 0), move down 1 unit and right 3 units. Draw a dot.

Step 4 Draw a line containing the points.

A **family of graphs** includes graphs and equations of graphs that have at least one characteristic in common. The **parent graph** is the simplest graph in a family.

Graphing Calculator Investigation

Family of Graphs

The calculator screen shows the graphs of $y = x$, $y = 2x$, and $y = 4x$.

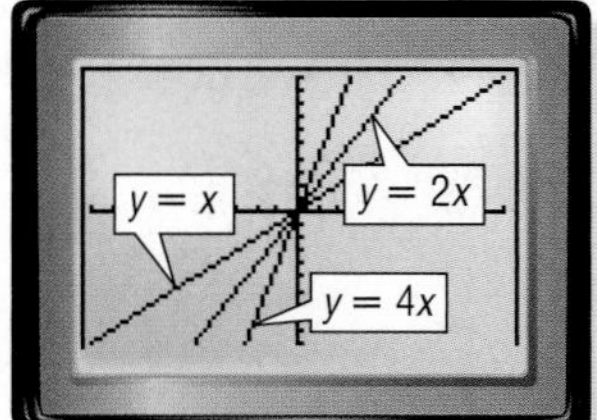

[−10, 10] scl: 1 by [−10, 10] scl: 1

Think and Discuss

1. Describe any similarities among the graphs.
2. Describe any differences among the graphs.
3. Write an equation whose graph has a steeper slope than $y = 4x$. Check your answer by graphing $y = 4x$ and your equation.
4. Write an equation whose graph lies between the graphs of $y = x$ and $y = 2x$. Check your answer by graphing the equations.
5. Write a description of this family of graphs. What characteristics do the graphs have in common? How are they different?
6. The equations whose graphs are in this family are all of the form $y = mx$. How does the graph change as the absolute value of m increases?

Concept Summary — Direct Variation Graphs

- Direct variation equations are of the form $y = kx$, where $k \neq 0$.
- The graph of $y = kx$ always passes through the origin.

• The slope can be positive. $k > 0$	• The slope can be negative. $k < 0$
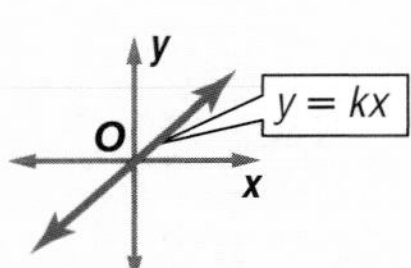	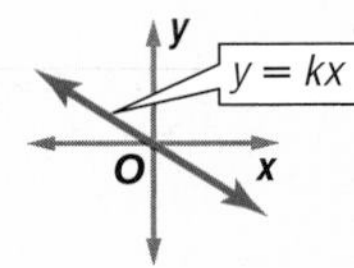

If you know that y varies directly as x, you can write a direct variation equation that relates the two quantities.

Example 4 Write and Solve a Direct Variation Equation

Suppose y varies directly as x, and $y = 28$ when $x = 7$.

a. Write a direct variation equation that relates x and y.

Find the value of k.

$y = kx$ Direct variation formula

$28 = k(7)$ Replace y with 28 and x with 7.

$\frac{28}{7} = \frac{k(7)}{7}$ Divide each side by 7.

$4 = k$ Simplify.

Therefore, $y = 4x$.

b. Use the direct variation equation to find x when $y = 52$.

$y = 4x$ Direct variation equation

$52 = 4x$ Replace y with 52.

$\frac{52}{4} = \frac{4x}{4}$ Divide each side by 4.

$13 = x$ Simplify.

Therefore, $x = 13$ when $y = 52$.

SOLVE PROBLEMS One of the most common uses of direct variation is the formula for distance, $d = rt$. In the formula, distance d varies directly as time t, and the rate r is the constant of variation.

Example 5 Direct Variation Equation

BIOLOGY A flock of snow geese migrated 375 miles in 7.5 hours.

a. Write a direct variation equation for the distance flown in any time.

Words The distance traveled is 375 miles, and the time is 7.5 hours.

Variables Let r = rate.

	Distance	equals	rate	times	time.
Equation	375 mi	=	r	×	7.5 h

Solve for the rate.

$375 = r(7.5)$ Original equation

$\frac{375}{7.5} = \frac{r(7.5)}{7.5}$ Divide each side by 7.5.

$50 = r$ Simplify.

Therefore, the direct variation equation is $d = 50t$.

More About. . .

Biology

Snow geese migrate more than 3000 miles from their winter home in the southwest United States to their summer home in the Canadian arctic.

Source: Audubon Society

b. Graph the equation.

The graph of $d = 50t$ passes through the origin with slope 50.

$m = \frac{50}{1}$ $\frac{\text{rise}}{\text{run}}$

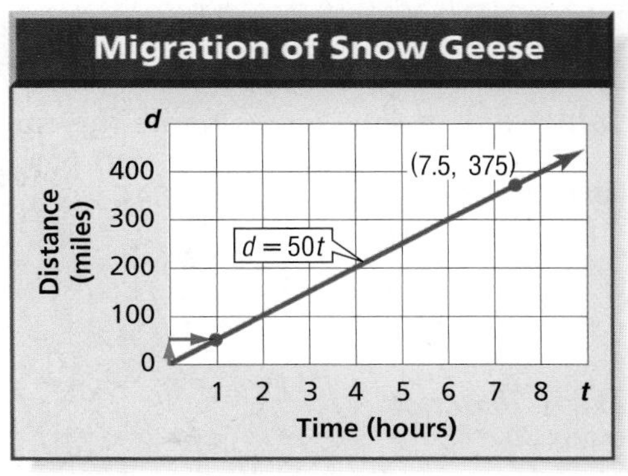

c. Estimate how many hours of flying time it would take the geese to migrate 3000 miles.

$d = 50t$ Original equation

$3000 = 50t$ Replace d with 3000.

$\frac{3000}{50} = \frac{50t}{50}$ Divide each side by 50.

$t = 60$ Simplify.

At this rate, it will take 60 hours of flying time to migrate 3000 miles.

Check for Understanding

Concept Check

1. **OPEN ENDED** Write a general equation for *y varies directly as x*.
2. **Choose** the equations that represent direct variations. Then find the constant of variation for each direct variation.

   a. $15 = rs$ b. $4a = b$ c. $z = \frac{1}{3}x$ d. $s = \frac{9}{t}$
3. **Explain** how the constant of variation and the slope are related in a direct variation equation.

Guided Practice

Name the constant of variation for each equation. Then determine the slope of the line that passes through each pair of points.

4. 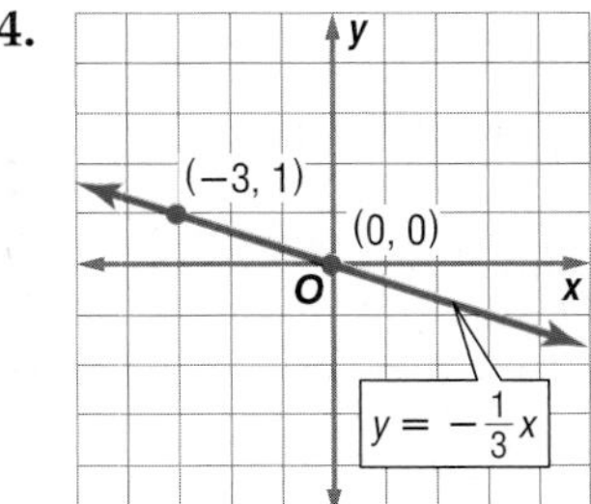

5. 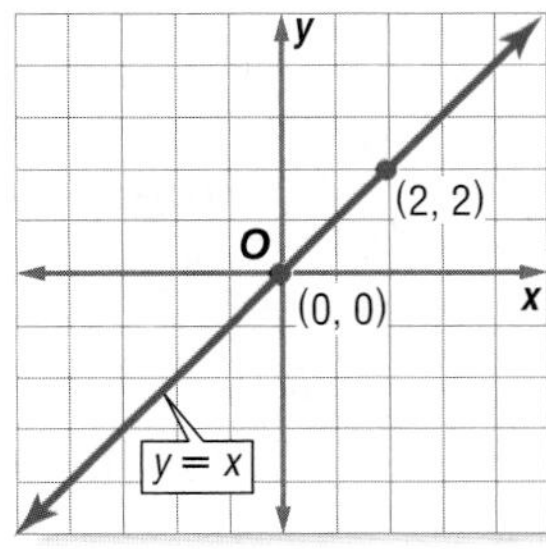

Graph each equation.

6. $y = 2x$
7. $y = -3x$
8. $y = \frac{1}{2}x$

Write a direct variation equation that relates x and y. Assume that y varies directly as x. Then solve.

9. If $y = 27$ when $x = 6$, find x when $y = 45$.
10. If $y = 10$ when $x = 9$, find x when $y = 9$.
11. If $y = -7$ when $x = -14$, find y when $x = 20$.

Application

JOBS **For Exercises 12–14, use the following information.**
Suppose you work at a job where your pay varies directly as the number of hours you work. Your pay for 7.5 hours is $45.

12. Write a direct variation equation relating your pay to the hours worked.
13. Graph the equation.
14. Find your pay if you work 30 hours.

Practice and Apply

Homework Help

For Exercises	See Examples
15–32	1–3
33–42	4
43–46, 52–55	5

Extra Practice

See page 831.

Name the constant of variation for each equation. Then determine the slope of the line that passes through each pair of points.

15. 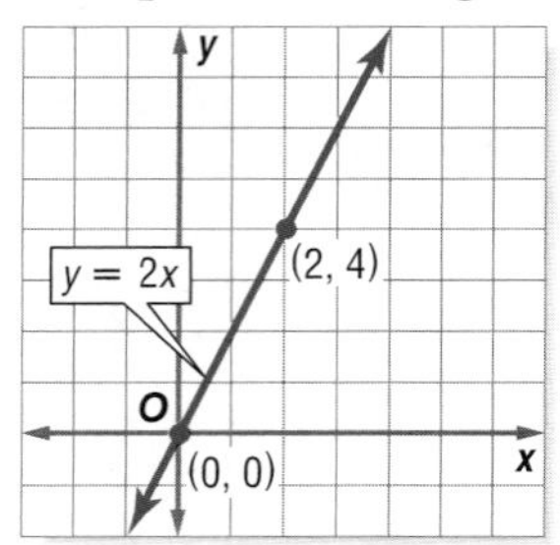

16. 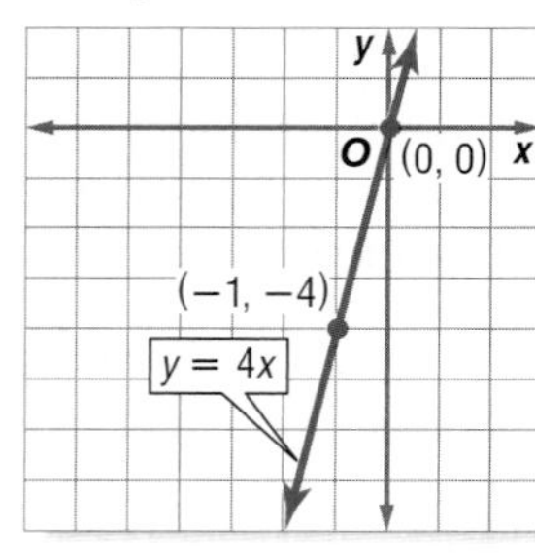

17. 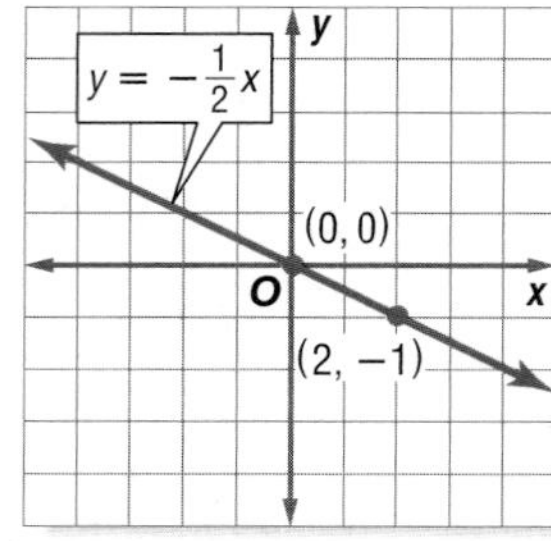

18. 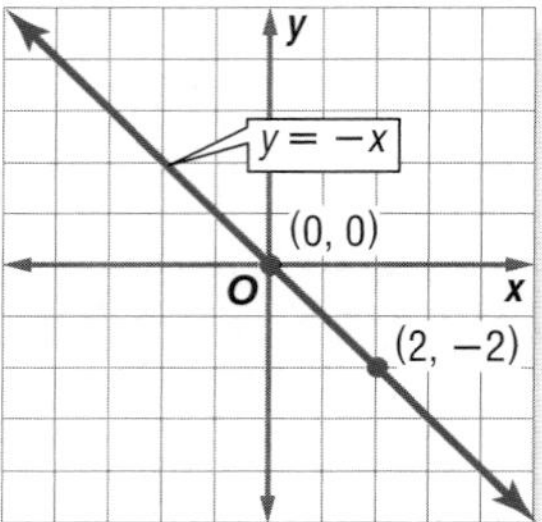

19. 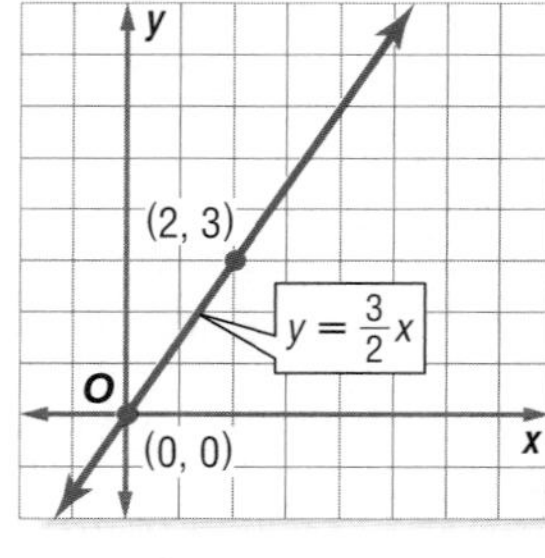

20. 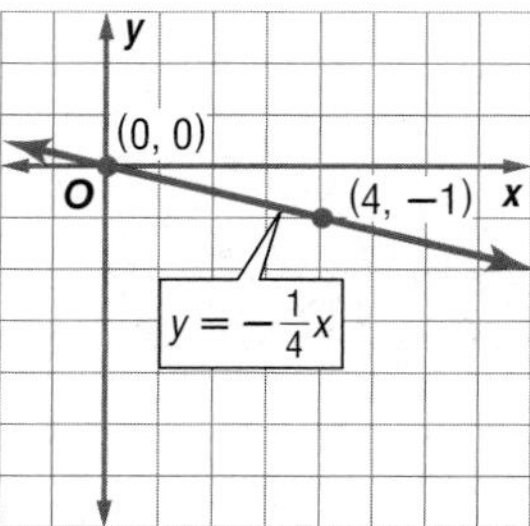

Graph each equation.

21. $y = x$
22. $y = 3x$
23. $y = -x$
24. $y = -4x$
25. $y = \frac{1}{4}x$
26. $y = \frac{3}{5}x$
27. $y = \frac{5}{2}x$
28. $y = \frac{7}{5}x$
29. $y = \frac{1}{5}x$
30. $y = -\frac{2}{3}x$
31. $y = -\frac{4}{3}x$
32. $y = -\frac{9}{2}x$

Write a direct variation equation that relates x and y. Assume that y varies directly as x. Then solve.

33. If $y = 8$ when $x = 4$, find y when $x = 5$.
34. If $y = 36$ when $x = 6$, find x when $y = 42$.
35. If $y = -16$ when $x = 4$, find x when $y = 20$.
36. If $y = -18$ when $x = 6$, find x when $y = 6$.
37. If $y = 4$ when $x = 12$, find y when $x = -24$.
38. If $y = 12$ when $x = 15$, find x when $y = 21$.
39. If $y = 2.5$ when $x = 0.5$, find y when $x = 20$.
40. If $y = -6.6$ when $x = 9.9$, find y when $x = 6.6$.
41. If $y = 2\frac{2}{3}$ when $x = \frac{1}{4}$, find y when $x = 1\frac{1}{8}$.
42. If $y = 6$ when $x = \frac{2}{3}$, find x when $y = 12$.

Write a direct variation equation that relates the variables. Then graph the equation.

43. **GEOMETRY** The circumference C of a circle is about 3.14 times the diameter d.
44. **GEOMETRY** The perimeter P of a square is 4 times the length of a side s.
45. **SEWING** The total cost is C for n yards of ribbon priced at \$0.99 per yard.
46. **RETAIL** Kona coffee beans are \$14.49 per pound. The total cost of p pounds is C.

47. **CRITICAL THINKING** Suppose y varies directly as x. If the value of x is doubled, what happens to the value of y? Explain.

BIOLOGY Which line in the graph represents the sprinting speeds of each animal?

48. elephant, 25 mph
49. reindeer, 32 mph
50. lion, 50 mph
51. grizzly bear, 30 mph

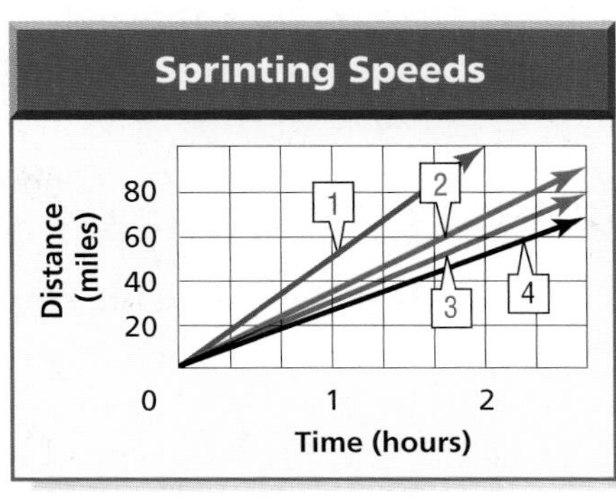

Career Choices

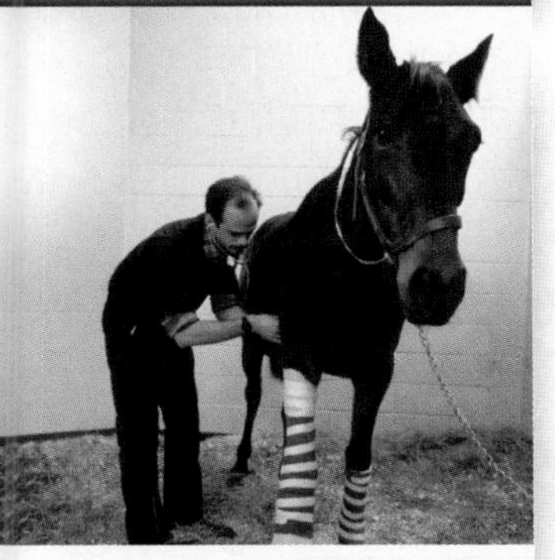

Veterinary Medicine

Veterinarians compare the age of an animal to the age of a human on the basis of bone and tooth growth.

Online Research For information about a career as a veterinarian, visit: www.algebra1.com/careers

SPACE For Exercises 52 and 53, use the following information.
The weight of an object on the moon varies directly with its weight on Earth. With all of his equipment, astronaut Neil Armstrong weighed 360 pounds on Earth, but weighed only 60 pounds on the moon.

52. Write an equation that relates weight on the moon m with weight on Earth e.
53. Suppose you weigh 138 pounds on Earth. What would you weigh on the moon?

ANIMALS For Exercises 54 and 55, use the following information.
Most animals age more rapidly than humans do. The chart shows equivalent ages for horses and humans.

Horse age (x)	0	1	2	3	4	5
Human age (y)	0	3	6	9	12	15

54. Write an equation that relates human age to horse age.
55. Find the equivalent horse age for a human who is 16 years old.

56. **WRITING IN MATH** Answer the question that was posed at the beginning of the lesson.

How is slope related to your shower?

Include the following in your answer:

- an equation that relates the number of gallons y to the time spent in the shower x for a low-flow showerhead that uses only 2.5 gallons of water per minute, and
- a comparison of the steepness of the graph of this equation to the graph at the top of page 264.

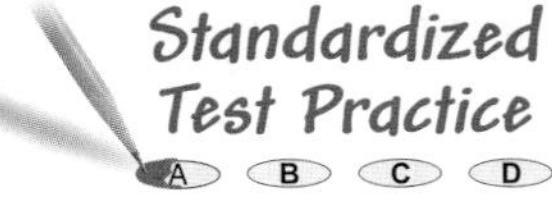

57. Which equation best describes the graph at the right?

(A) $y = 2x$ (B) $y = -2x$
(C) $y = \frac{1}{2}x$ (D) $y = -\frac{1}{2}x$

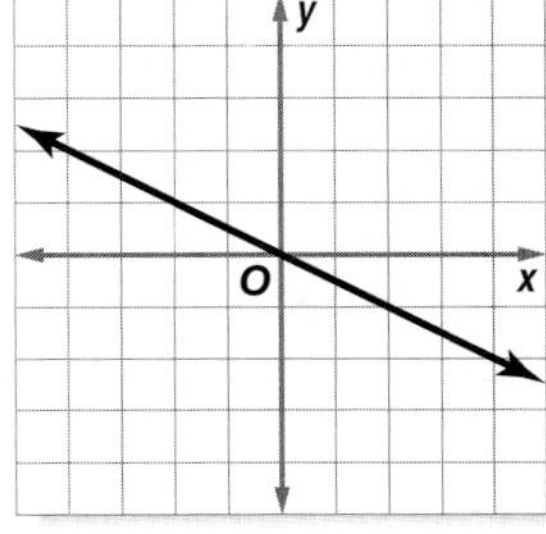

58. Which equation does *not* model a direct variation?

(A) $y = 4x$ (B) $y = 22x$
(C) $y = 3x + 1$ (D) $y = \frac{1}{2}x$

Graphing Calculator

FAMILIES OF GRAPHS For Exercises 59–62, use the graphs of $y = -1x$, $y = -2x$, and $y = -4x$, which form a family of graphs.

59. Graph $y = -1x$, $y = -2x$, and $y = -4x$ on the same screen.
60. How are these graphs similar to the graphs in the Graphing Calculator Investigation on page 265? How are they different?

61. Write an equation whose graph has a steeper slope than $y = -4x$.

62. **MAKE A CONJECTURE** Explain how you can tell without graphing which of two direct variation equations has the graph with a steeper slope.

Maintain Your Skills

Mixed Review **Find the slope of the line that passes through each pair of points.** *(Lesson 5-1)*

63.
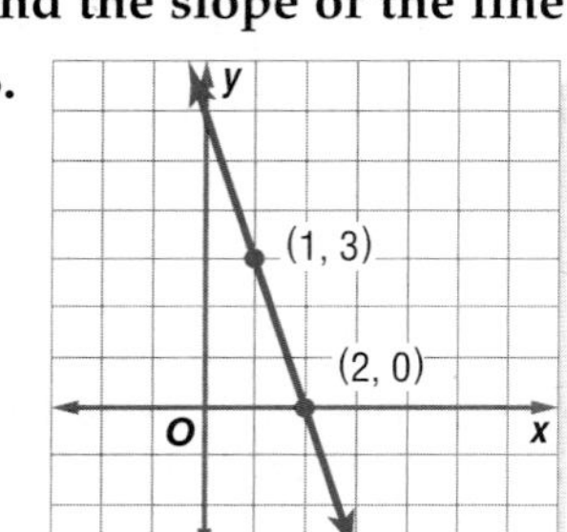

64.
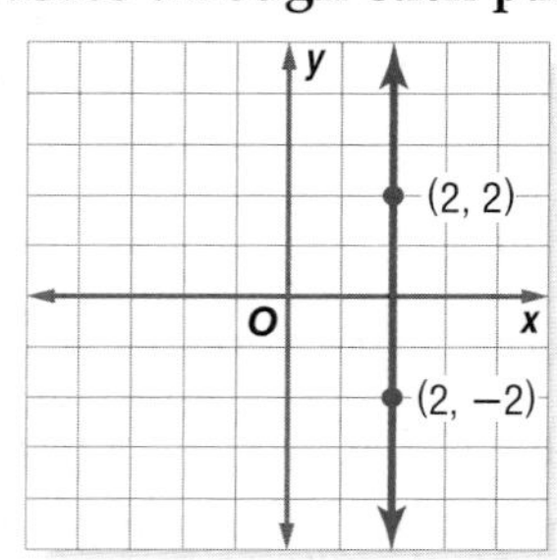

65.
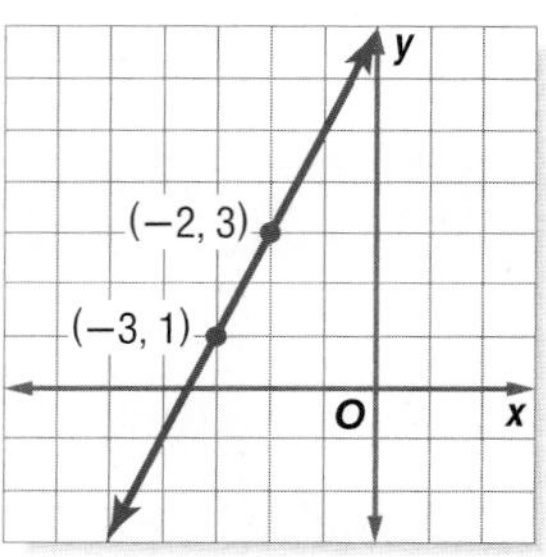

66. Find the value of r so that the line that passes through (1, 7) and (r, 3) has a slope of 2. *(Lesson 5-1)*

Each table below represents points on a linear graph. Copy and complete each table. *(Lesson 4-8)*

67.

x	0	1	2	3	4	5
y	1		9	13		21

68.

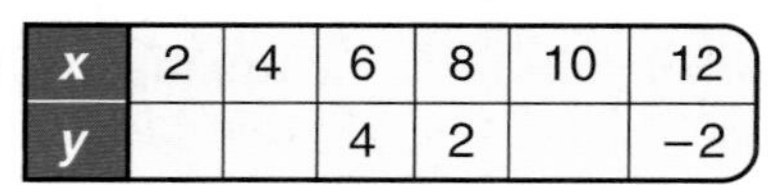

x	2	4	6	8	10	12
y			4	2		−2

Add or subtract. *(Lesson 2-2)*

69. $15 + (-12)$ 70. $8 - (-5)$ 71. $-9 - 6$ 72. $-18 - 12$

Getting Ready for the Next Lesson **PREREQUISITE SKILL** **Solve each equation for y.** *(To review **rewriting equations**, see Lesson 3-8.)*

73. $-3x + y = 8$ 74. $2x + y = 7$ 75. $4x = y + 3$

76. $2y = 4x + 10$ 77. $9x + 3y = 12$ 78. $x - 2y = 5$

Practice Quiz 1 — Lessons 5-1 and 5-2

Find the slope of the line that passes through each pair of points. *(Lesson 5-1)*

1. (−4, −6), (−3, −8) 2. (8, 3), (−11, 3) 3. (−4, 8), (5, 9) 4. (0, 1), (7, 11)

Find the value of r so the line that passes through each pair of points has the given slope. *(Lesson 5-1)*

5. (5, −3), (r, −5), $m = 2$ 6. (6, r), (−4, 9), $m = \frac{3}{2}$

Graph each equation. *(Lesson 5-2)*

7. $y = -7x$ 8. $y = \frac{3}{4}x$

Write a direct variation equation that relates x and y. Assume that y varies directly as x. Then solve. *(Lesson 5-2)*

9. If $y = 24$ when $x = 8$, find y when $x = -3$. 10. If $y = -10$ when $x = 15$, find x when $y = -6$.

Algebra Activity

A Preview of Lesson 5-3

Investigating Slope-Intercept Form

Collect the Data

- Cut a small hole in a top corner of a plastic sandwich bag. Loop a long rubber band through the hole.
- Tape the free end of the rubber band to the desktop.
- Use a centimeter ruler to measure the distance from the desktop to the end of the bag. Record this distance for 0 washers in the bag using a table like the one below.

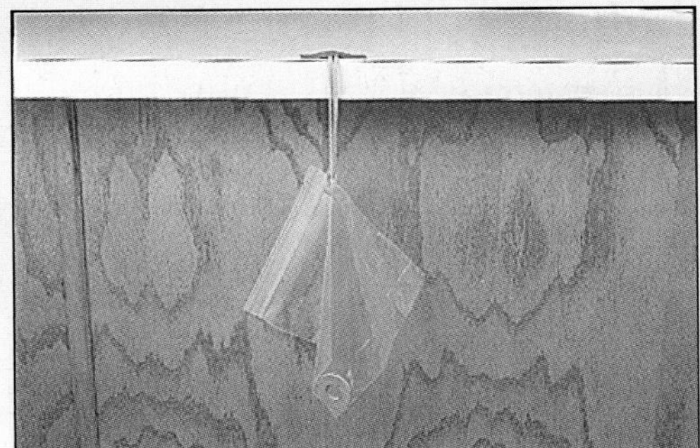

Number of Washers x	Distance y
0	
1	

- Place one washer in the plastic bag. Then measure and record the new distance from the desktop to the end of the bag.
- Repeat the experiment, adding different numbers of washers to the bag. Each time, record the number of washers and the distance from the desktop to the end of the bag.

Analyze the Data

1. The domain contains values represented by the independent variable, washers. The range contains values represented by the dependent variable, distance. On grid paper, graph the ordered pairs (washers, distance).
2. Write a sentence that describes the points on the graph.
3. Describe the point that represents the trial with no washers in the bag.
4. The rate of change can be found by using the formula for slope.

   $$\frac{\text{rise}}{\text{run}} = \frac{\text{change in distance}}{\text{change in number of washers}}$$

   Find the rate of change in the distance from the desktop to the end of the bag as more washers are added.
5. Explain how the rate of change is shown on the graph.

Make a Conjecture

The graph shows sample data from a rubber band experiment. Draw a graph for each situation.

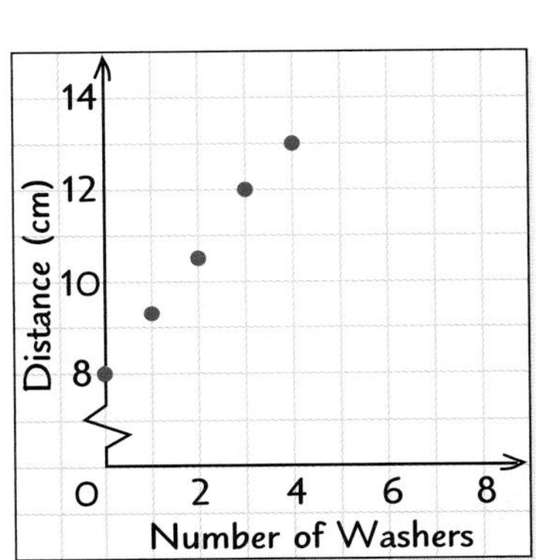

6. A bag that hangs 10.5 centimeters from the desktop when empty and lengthens at the rate of the sample.
7. A bag that has the same length when empty as the sample and lengthens at a faster rate.
8. A bag that has the same length when empty as the sample and lengthens at a slower rate.

5-3 Slope-Intercept Form

What You'll Learn

- Write and graph linear equations in slope-intercept form.
- Model real-world data with an equation in slope-intercept form.

Vocabulary

- slope-intercept form

How is a y-intercept related to a flat fee?

A cellular phone service provider charges $0.10 per minute plus a flat fee of $5.00 each month.

x (minutes)	y (dollars)
0	5.00
1	5.10
2	5.20
3	5.30
4	5.40
5	5.50
6	5.60
7	5.70

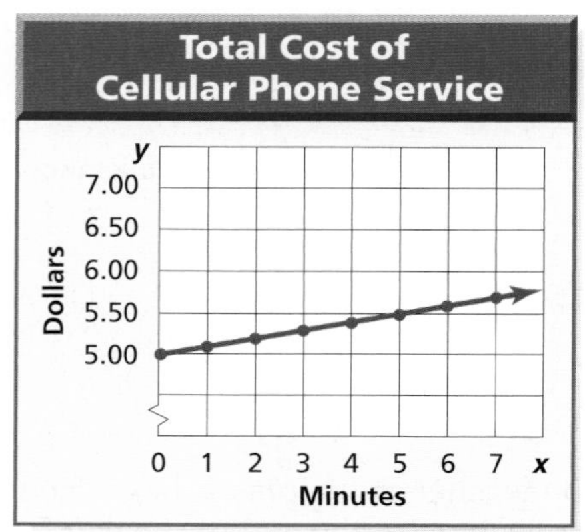

The slope of the line is 0.1. It crosses the y-axis at (0, 5).
The equation of the line is $y = 0.1x + 5$.

charge per minute, $0.10 ↑ ↑ flat fee, $5.00

SLOPE-INTERCEPT FORM An equation of the form $y = mx + b$ is in **slope-intercept form**. When an equation is written in this form, you can identify the slope and y-intercept of its graph.

Key Concept *Slope-Intercept Form*

- **Words** The linear equation $y = mx + b$ is written in slope-intercept form, where m is the slope and b is the y-intercept.
- **Symbols** $y = mx + b$

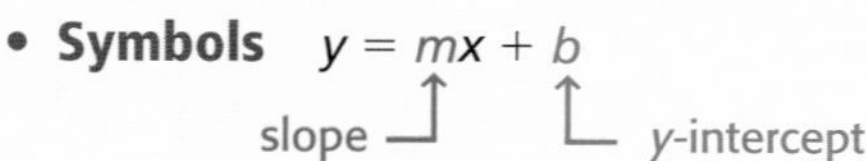

- **Model**

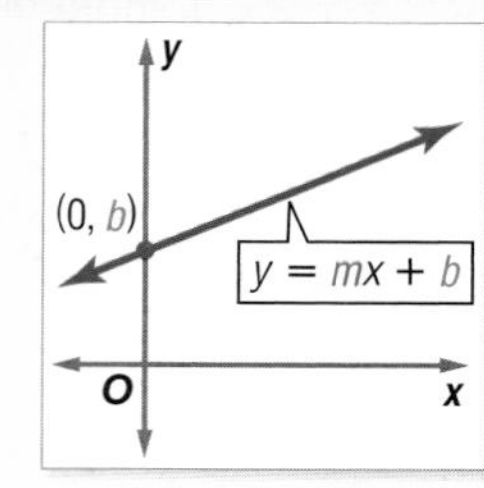

Study Tip

Look Back
To review **intercepts**, see Lesson 4-5.

Example 1 Write an Equation Given Slope and y-Intercept

Write an equation of the line whose slope is 3 and whose y-intercept is 5.

$y = mx + b$ Slope-intercept form

$y = 3x + 5$ Replace m with 3 and b with 5.

Example 2 Write an Equation Given Two Points

Write an equation of the line shown in the graph.

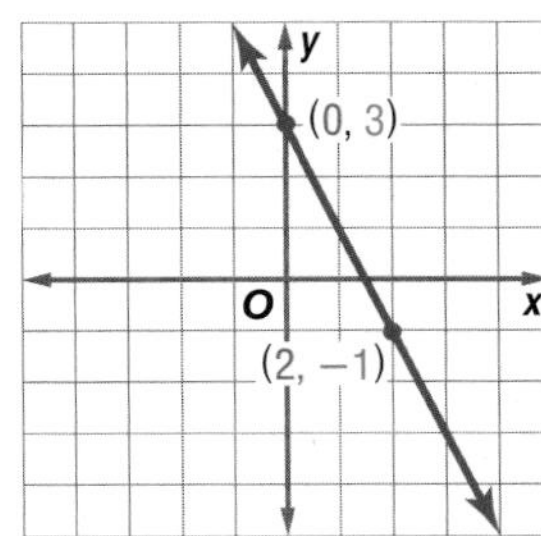

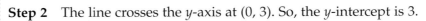

Step 1 You know the coordinates of two points on the line. Find the slope. Let $(x_1, y_1) = (0, 3)$ and $(x_2, y_2) = (2, -1)$.

$m = \dfrac{y_2 - y_1}{x_2 - x_1}$ rise/run

$m = \dfrac{-1 - 3}{2 - 0}$ $x_1 = 0, x_2 = 2$; $y_1 = 3, y_2 = -1$

$m = \dfrac{-4}{2}$ or -2 Simplify.

The slope is -2.

Step 2 The line crosses the y-axis at (0, 3). So, the y-intercept is 3.

Step 3 Finally, write the equation.

$y = mx + b$ Slope-intercept form

$y = -2x + 3$ Replace m with -2 and b with 3.

The equation of the line is $y = -2x + 3$.

Study Tip

Vertical Lines
The equation of a vertical line *cannot* be written in slope-intercept form. *Why?*

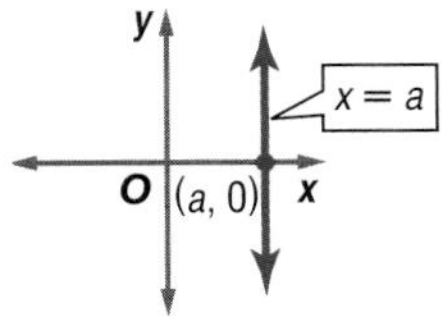

Horizontal Lines
The equation of a horizontal line *can* be written in slope-intercept form as $y = 0x + b$ or $y = b$.

y = b
(0, b)
O
x

One advantage of the slope-intercept form is that it allows you to graph an equation quickly.

Example 3 Graph an Equation in Slope-Intercept Form

Graph $y = -\frac{2}{3}x + 1$.

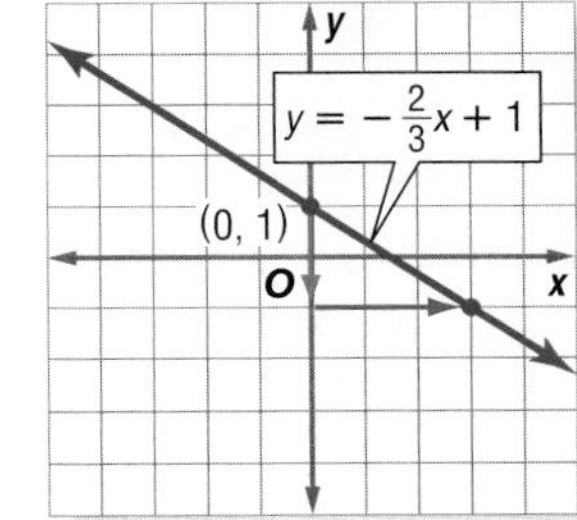

Step 1 The y-intercept is 1. So, graph (0, 1).

Step 2 The slope is $-\frac{2}{3}$ or $\frac{-2}{3}$. rise/run

From (0, 1), move down 2 units and right 3 units. Draw a dot.

Step 3 Draw a line connecting the points.

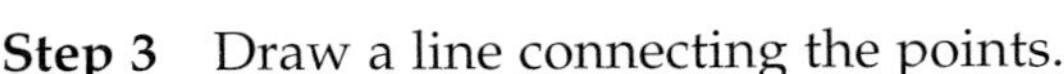

Example 4 Graph an Equation in Standard Form

Graph $5x - 3y = 6$.

Step 1 Solve for y to find the slope-intercept form.

$5x - 3y = 6$ Original equation

$5x - 3y - 5x = 6 - 5x$ Subtract $5x$ from each side.

$-3y = 6 - 5x$ Simplify.

$-3y = -5x + 6$ $6 - 5x = 6 + (-5x)$ or $-5x + 6$

$\dfrac{-3y}{-3} = \dfrac{-5x + 6}{-3}$ Divide each side by -3.

$\dfrac{-3y}{-3} = \dfrac{-5x}{-3} + \dfrac{6}{-3}$ Divide each term in the numerator by -3.

$y = \frac{5}{3}x - 2$ Simplify.

(continued on the next page)

Step 2 The y-intercept of $y = \frac{5}{3}x - 2$ is -2. So, graph $(0, -2)$.

Step 3 The slope is $\frac{5}{3}$. From $(0, -2)$, move up 5 units and right 3 units. Draw a dot.

Step 4 Draw a line containing the points.

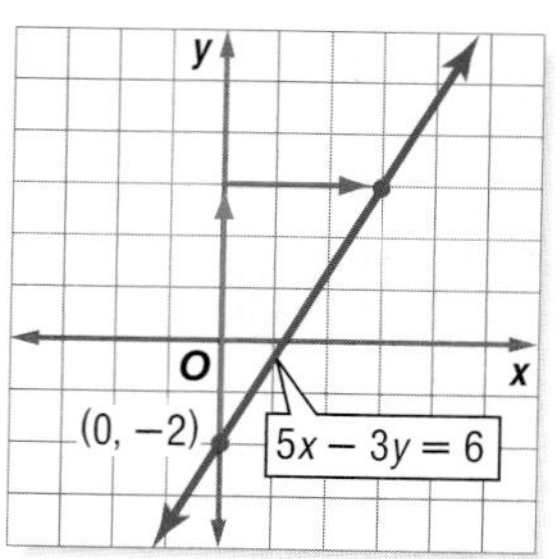

MODEL REAL-WORLD DATA If a quantity changes at a constant rate over time, it can be modeled by a linear equation. The y-intercept represents a starting point, and the slope represents the rate of change.

Example 5 Write an Equation in Slope-Intercept Form

AGRICULTURE The natural sweeteners used in foods include sugar, corn sweeteners, syrup, and honey. Use the information at the left about natural sweeteners.

More About. . .

Agriculture

In 1989, each person in the United States consumed an average of 133 pounds of natural sweeteners.

Source: USDA *Agricultural Outlook*

a. The amount of natural sweeteners consumed has increased by an average of 2.6 pounds per year. Write a linear equation to find the average consumption of natural sweeteners in any year after 1989.

Words The consumption increased 2.6 pounds per year, so the rate of change is 2.6 pounds per year. In the first year, the average consumption was 133 pounds.

Variables Let C = average consumption.
Let n = number of years after 1989.

Equation

Average consumption	equals	rate of change	times	number of years after 1989	plus	amount at start.
C	$=$	2.6	$\cdot$	n	$+$	133

b. Graph the equation.

The graph passes through $(0, 133)$ with slope 2.6.

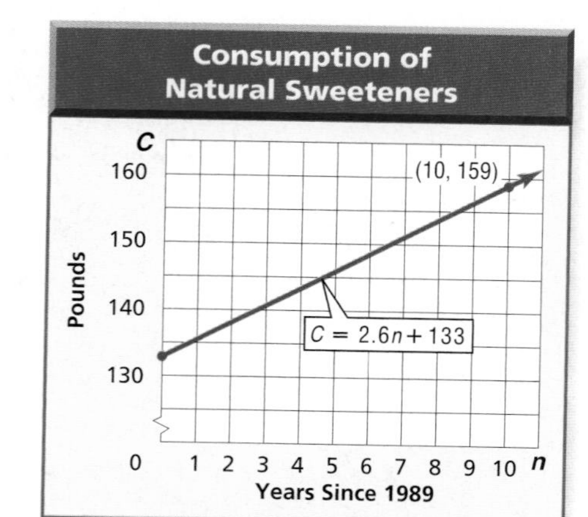

c. Find the number of pounds of natural sweeteners consumed by each person in 1999.

The year 1999 is 10 years after 1989. So, $n = 10$.

$C = 2.6n + 133$ Consumption equation

$C = 2.6(10) + 133$ Replace n with 10.

$C = 159$ Simplify.

So, the average person consumed 159 pounds of natural sweeteners in 1999.

CHECK Notice that $(10, 159)$ lies on the graph.

Check for Understanding

Concept Check

1. **OPEN ENDED** Write an equation for a line with a slope of 7.
2. **Explain** why equations of vertical lines cannot be written in slope-intercept form, but equations of horizontal lines can.
3. **Tell** which part of the slope-intercept form represents the rate of change.

Guided Practice

Write an equation of the line with the given slope and y-intercept.

4. slope: -3, y-intercept: 1
5. slope: 4, y-intercept: -2

Write an equation of the line shown in each graph.

6. 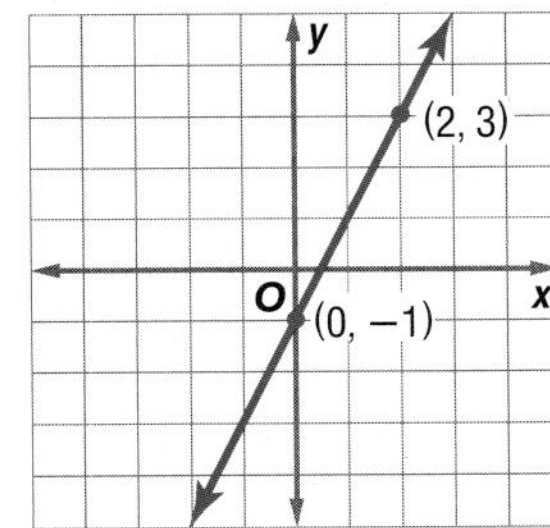

7. 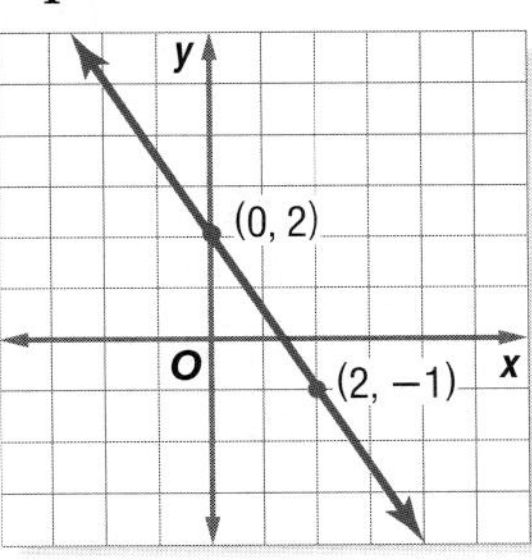

Graph each equation.

8. $y = 2x - 3$
9. $y = -3x + 1$
10. $2x + y = 5$

Application

MONEY For Exercises 11–13, use the following information.
Suppose you have already saved $50 toward the cost of a new television set. You plan to save $5 more each week for the next several weeks.

11. Write an equation for the total amount T you will have w weeks from now.
12. Graph the equation.
13. Find the total amount saved after 7 weeks.

Practice and Apply

Homework Help

For Exercises	See Examples
14–19	1
20–27	2
28–39	3, 4
40–43	5

Extra Practice

See page 831.

Write an equation of the line with the given slope and y-intercept.

14. slope: 2, y-intercept: -6
15. slope: 3, y-intercept: -5
16. slope: $\frac{1}{2}$, y-intercept: 3
17. slope: $-\frac{3}{5}$, y-intercept: 0
18. slope: -1, y-intercept: 10
19. slope: 0.5; y-intercept: 7.5

Write an equation of the line shown in each graph.

20. 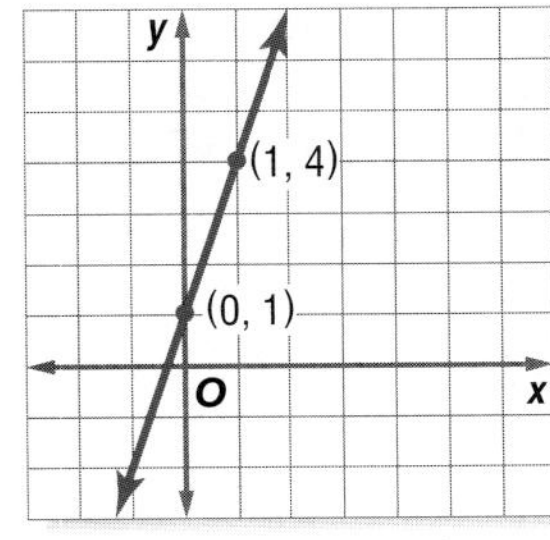

21. 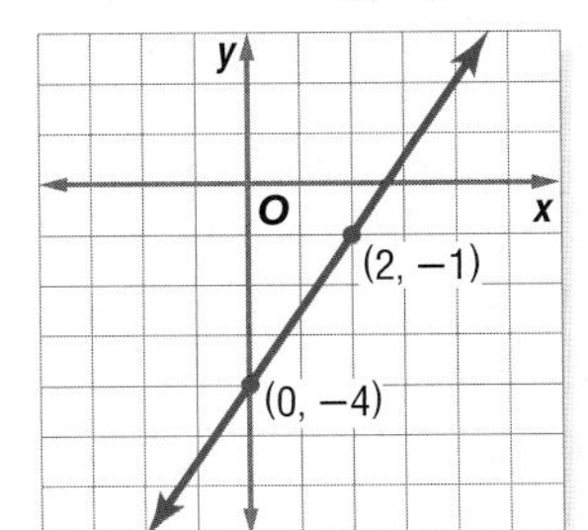

22. 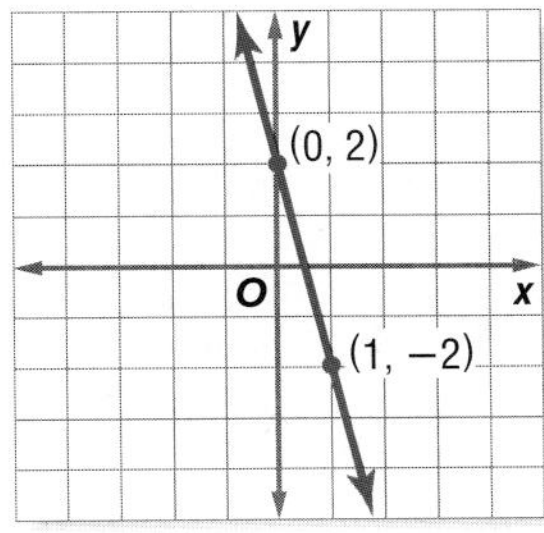

Write an equation of the line shown in each graph.

23. 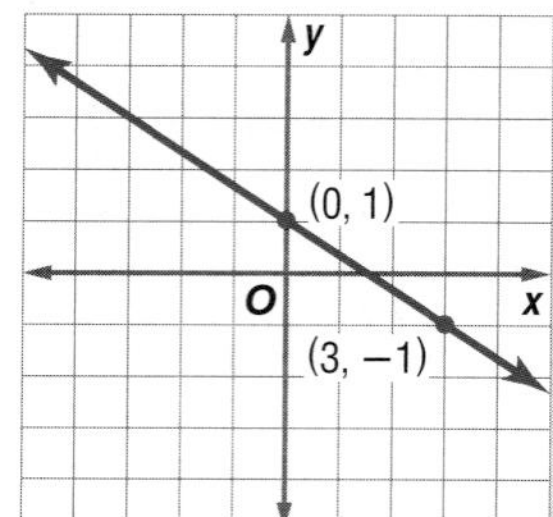

24. 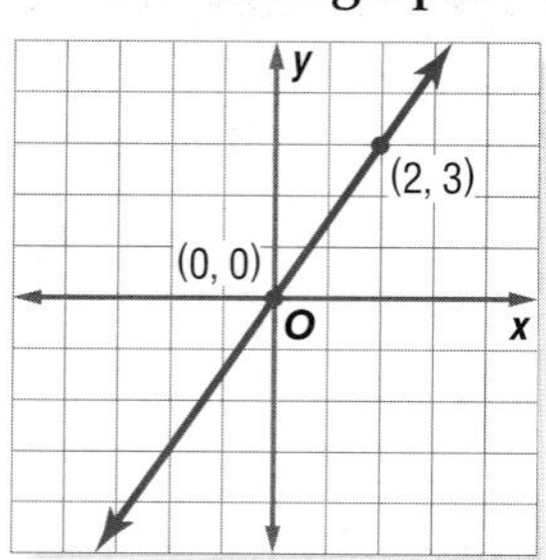

25. 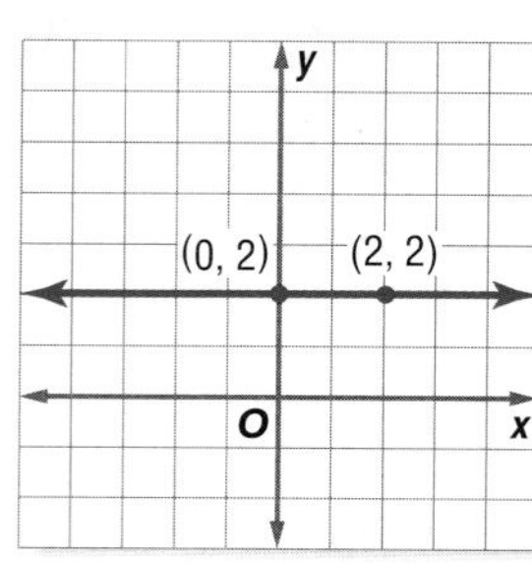

26. Write an equation of a horizontal line that crosses the y-axis at $(0, -5)$.

27. Write an equation of a line that passes through the origin with slope 3.

Graph each equation.

28. $y = 3x + 1$	**29.** $y = x - 2$	**30.** $y = -4x + 1$
31. $y = -x + 2$	**32.** $y = \frac{1}{2}x + 4$	**33.** $y = -\frac{1}{3}x - 3$
34. $3x + y = -2$	**35.** $2x - y = -3$	**36.** $3y = 2x + 3$
37. $-2y = 6x - 4$	**38.** $2x + 3y = 6$	**39.** $4x - 3y = 3$

Write a linear equation in slope-intercept form to model each situation.

40. You rent a bicycle for \$20 plus \$2 per hour.

41. An auto repair shop charges \$50 plus \$25 per hour.

42. A candle is 6 inches tall and burns at a rate of $\frac{1}{2}$ inch per hour.

43. The temperature is 15° and is expected to fall 2° each hour during the night.

44. **CRITICAL THINKING** The equations $y = 2x + 3$, $y = 4x + 3$, $y = -x + 3$, and $y = -10x + 3$ form a family of graphs. What characteristic do their graphs have in common?

SALES For Exercises 45 and 46, use the following information and the graph at the right.
In 1991, book sales in the United States totaled \$16 billion. Sales increased by about \$1 billion each year until 1999.

45. Write an equation to find the total sales S for any number of years t since 1991.

46. If the trend continues, what will sales be in 2005?

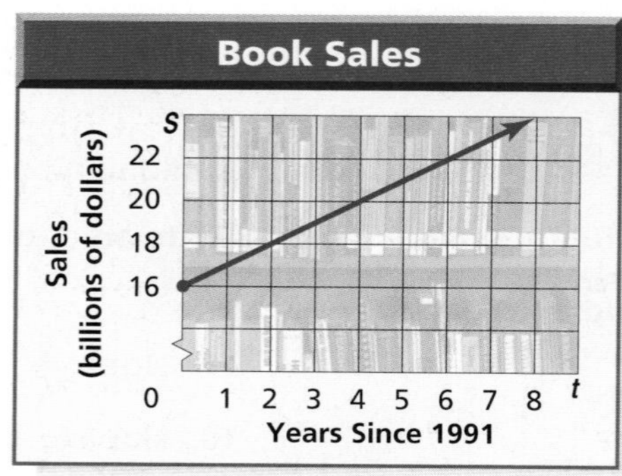

Source: Association of American Publishers

TRAFFIC For Exercises 47–49, use the following information.
In 1966, the traffic fatality rate in the United States was 5.5 fatalities per 100 million vehicle miles traveled. Between 1966 and 1999, the rate decreased by about 0.12 each year.

47. Write an equation to find the fatality rate R for any number of years t since 1966.

48. Graph the equation.

49. Find the fatality rate in 1999.

50. **WRITING IN MATH** Answer the question that was posed at the beginning of the lesson.

How is a y-intercept related to a flat fee?

Include the following in your answer:

- the point at which the graph would cross the y-axis if your cellular phone service provider charges a rate of \$0.07 per minute plus a flat fee of \$5.99,
- and a description of a situation in which the y-intercept of its graph is \$25.

Standardized Test Practice

51. Which equation does *not* have a y-intercept of 5?

Ⓐ $2x = y - 5$ Ⓑ $3x + y = 5$

Ⓒ $y = x + 5$ Ⓓ $2x - y = 5$

52. Which situation below is modeled by the graph?

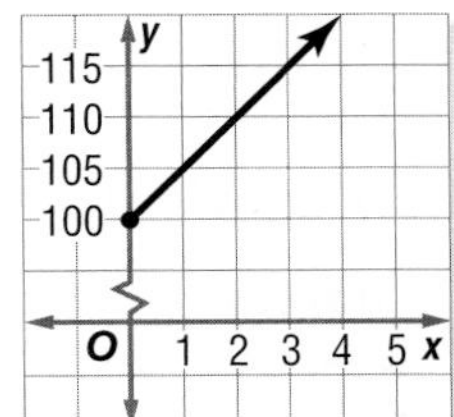

Ⓐ You have \$100 and plan to spend \$5 each week.

Ⓑ You have \$100 and plan to save \$5 each week.

Ⓒ You need \$100 for a new CD player and plan to save \$5 each week.

Ⓓ You need \$100 for a new CD player and plan to spend \$5 each week.

Extending the Lesson

53. The standard form of a linear equation is $Ax + By = C$, where A, B, and C are integers, $A \geq 0$, and A and B are not both zero. Solve $Ax + By = C$ for y. *Your answer is written in slope-intercept form.*

54. Use the slope-intercept equation in Exercise 53 to write expressions for the slope and y-intercept in terms of A, B, and C.

55. Use the expressions in Exercise 54 to find the slope and y-intercept of each equation.

a. $2x + y = -4$ **b.** $3x + 4y = 12$ **c.** $2x - 3y = 9$

Maintain Your Skills

Mixed Review

Write a direct variation equation that relates x and y. Assume that y varies directly as x. Then solve. *(Lesson 5-2)*

56. If $y = 45$ when $x = 60$, find x when $y = 8$.

57. If $y = 15$ when $x = 4$, find y when $x = 10$.

Find the slope of the line that passes through each pair of points. *(Lesson 5-1)*

58. $(-3, 0), (-4, 6)$ 59. $(3, -1), (3, -4)$ 60. $(5, -5), (9, 2)$

61. Write the numbers $2.5, \frac{3}{4}, -0.5, \frac{7}{8}$ in order from least to greatest. *(Lesson 2-7)*

Solve each equation. *(Lesson 1-3)*

62. $x = \frac{15 - 9}{2}$ 63. $3(7) + 2 = b$ 64. $q = 6^2 - 2^2$

Getting Ready for the Next Lesson

PREREQUISITE SKILL Find the slope of the line that passes through each pair of points. *(To review* ***slope****, see Lesson 5-1.)*

65. $(-1, 2), (1, -2)$ 66. $(5, 8), (-2, 8)$ 67. $(1, -1), (10, -13)$

Graphing Calculator

A Follow-Up of Lesson 5-3

Families of Linear Graphs

A family of people is a group of people related by birth, marriage, or adoption. Recall that a *family of graphs* includes graphs and equations of graphs that have at least one characteristic in common.

Families of linear graphs fall into two categories—those with the same slope and those with the same y-intercept. A graphing calculator is a useful tool for studying a group of graphs to determine whether they form a family.

Example 1

Graph $y = x$, $y = x + 4$, and $y = x - 2$ in the standard viewing window. Describe any similarities and differences among the graphs. Write a description of the family.

Enter the equations in the Y= list as Y1, Y2, and Y3. Then graph the equations.

KEYSTROKES: *Review graphing on pages 224 and 225.*

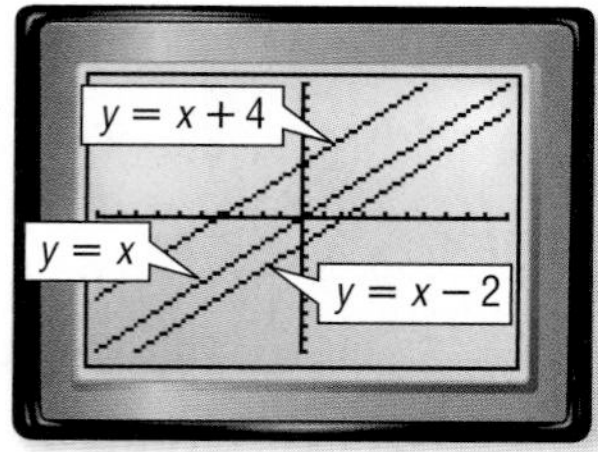

[−10, 10] scl: 1 by [−10, 10] scl: 1

- The graph of $y = x$ has a slope of 1 and a y-intercept of 0.
- The graph of $y = x + 4$ has a slope of 1 and a y-intercept of 4.
- The graph of $y = x - 2$ has a slope of 1 and a y-intercept of -2.

Notice that the graph of $y = x + 4$ is the same as the graph of $y = x$, moved 4 units up. Also, the graph of $y = x - 2$ is the same as the graph of $y = x$, moved 2 units down. All graphs have the same slope and different intercepts.

Because they all have the same slope, this family of graphs can be described as linear graphs with a slope of 1.

Example 2

Graph $y = x + 1$, $y = 2x + 1$, and $y = -\frac{1}{3}x + 1$ in the standard viewing window. Describe any similarities and differences among the graphs. Write a description of the family.

Enter the equations in the Y= list and graph.

y = 2x + 1
y = x + 1
$y = -\frac{1}{3}x + 1$

[−10, 10] scl: 1 by [−10, 10] scl: 1

- The graph of $y = x + 1$ has a slope of 1 and a y-intercept of 1.
- The graph of $y = 2x + 1$ has a slope of 2 and a y-intercept of 1.
- The graph of $y = -\frac{1}{3}x + 1$ has a slope of $-\frac{1}{3}$ and a y-intercept of 1.

These graphs have the same intercept and different slopes. This family of graphs can be described as linear graphs with a y-intercept of 1.

www.algebra1.com/other_calculator_keystrokes

Investigation

Nonlinear functions can also be defined in terms of a family of graphs. Consider the **absolute value function** $y = |x|$, where $y \geq 0$ for all values of x.

Example 3

Graph $y = |x|$, $y = |x| + 2$, and $y = |x + 3|$ in the standard viewing window. Describe any similarities and differences among the graphs.

Enter the equations in the Y= list and graph.

KEYSTROKES: [Y=] [MATH] [▶] 1 [X,T,θ,n] [ENTER]
[MATH] [▶] 1 [X,T,θ,n] [)] [+] 2 [ENTER]
[MATH] [▶] 1 [X,T,θ,n] [+] 3 [)] [ZOOM] 6

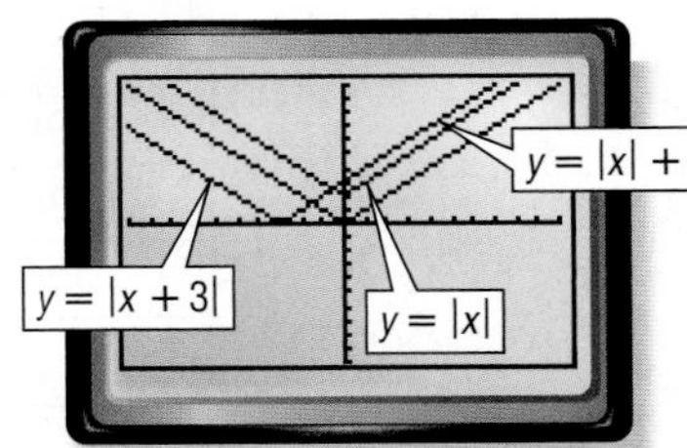

[−10, 10] scl: 1 by [−10, 10] scl: 1

- The graph of $y = |x|$ is v-shaped with its vertex at the origin.
- The graph of $y = |x| + 2$ is v-shaped with its vertex at (0, 2).
- The graph of $y = |x + 3|$ is v-shaped with its vertex at (−3, 0).

These graphs have the same shape, but they are positioned in different places on the coordinate plane.

Exercises

Graph each set of equations on the same screen. Describe any similarities or differences among the graphs. If the graphs are part of the same family, describe the family.

1. $y = -4$
   $y = 0$
   $y = 7$
2. $y = -x + 1$
   $y = 2x + 1$
   $y = \frac{1}{4}x + 1$
3. $y = x + 4$
   $y = 2x + 4$
   $y = 2x - 4$
4. $y = \frac{1}{2}x + 2$
   $y = \frac{1}{3}x + 3$
   $y = \frac{1}{4}x + 4$
5. $y = -2x - 2$
   $y = 2x - 2$
   $y = \frac{1}{2}x - 2$
6. $y = 3x$
   $y = 3x + 6$
   $y = 3x - 7$
7. $y = |x|$
   $y = -3|x|$
   $y = |-3x|$
8. $y = |x|$
   $y = |x| + 3$
   $y = |x| - 2$
9. $y = |x|$
   $y = |2x| + 4$
   $y = 3|x| - 5$
10. **MAKE A CONJECTURE** Write a paragraph explaining how the values of m and b in the slope-intercept form affect the graph of a linear equation.
11. Families of graphs are also called **classes of functions**. Describe the similarities and differences in the class of functions $f(x) = x + c$, where c is any real number.
12. Describe the similarities and differences in the classes of functions $f(x) = |x| + c$, where c is any real number.

5-4

Writing Equations in Slope-Intercept Form

What You'll Learn

- Write an equation of a line given the slope and one point on a line.
- Write an equation of a line given two points on the line.

Vocabulary

- linear extrapolation

How can slope-intercept form be used to make predictions?

In 1995, the population of Orlando, Florida, was about 175,000. At that time, the population was growing at a rate of about 2000 per year.

x (year)	y (population)
⋮	⋮
1994	173,000
1995	175,000
1996	177,000
⋮	⋮

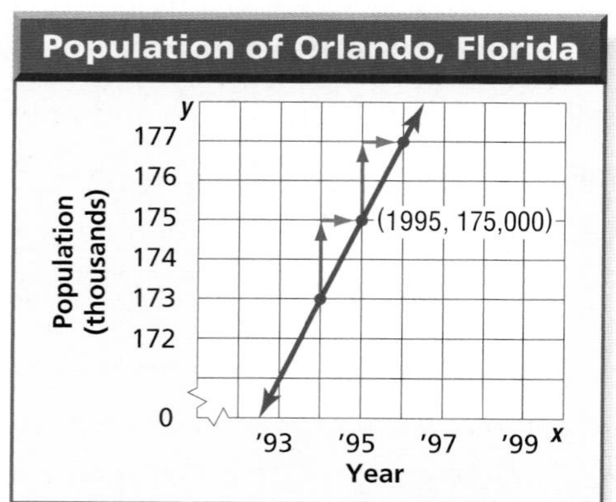

If you could write an equation based on the slope, 2000, and the point (1995, 175,000), you could predict the population for another year.

WRITE AN EQUATION GIVEN THE SLOPE AND ONE POINT You have learned how to write an equation of a line when you know the slope and a specific point, the y-intercept. The following example shows how to write an equation when you know the slope and any point on the line.

Example 1 Write an Equation Given Slope and One Point

Write an equation of a line that passes through (1, 5) with slope 2.

Step 1 The line has slope 2. To find the y-intercept, replace m with 2 and (x, y) with (1, 5) in the slope-intercept form. Then, solve for b.

$y = mx + b$	Slope-intercept form
$5 = 2(1) + b$	Replace m with 2, y with 5, and x with 1.
$5 = 2 + b$	Multiply.
$5 - 2 = 2 + b - 2$	Subtract 2 from each side.
$3 = b$	Simplify.

Step 2 Write the slope-intercept form using $m = 2$ and $b = 3$.

$y = mx + b$	Slope-intercept form
$y = 2x + 3$	Replace m with 2 and b with 3.

Therefore, the equation is $y = 2x + 3$.

CHECK You can check your result by graphing $y = 2x + 3$ on a graphing calculator. Use the CALC menu to verify that it passes through (1, 5).

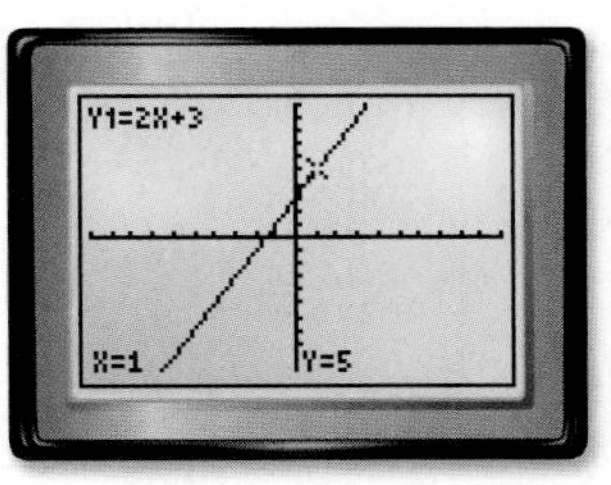

[−10, 10] scl: 1 by [−10, 10] scl: 1

WRITE AN EQUATION GIVEN TWO POINTS Sometimes you do not know the slope of a line, but you know two points on the line. In this case, find the slope of the line. Then follow the steps in Example 1.

Standardized Test Practice

Example 2 Write an Equation Given Two Points

Multiple-Choice Test Item

The table of ordered pairs shows the coordinates of the two points on the graph of a function. Which equation describes the function?

x	y
−3	−1
6	−4

(A) $y = -\frac{1}{3}x - 2$ (B) $y = 3x - 2$

(C) $y = -\frac{1}{3}x + 2$ (D) $y = \frac{1}{3}x - 2$

Read the Test Item

The table represents the ordered pairs (−3, −1) and (6, −4).

Solve the Test Item

Step 1 Find the slope of the line containing the points. Let $(x_1, y_1) = (-3, -1)$ and $(x_2, y_2) = (6, -4)$.

$m = \frac{y_2 - y_1}{x_2 - x_1}$ Slope formula

$m = \frac{-4 - (-1)}{6 - (-3)}$ $x_1 = -3, x_2 = 6, y_1 = -1, y_2 = -4$

$m = \frac{-3}{9}$ or $-\frac{1}{3}$ Simplify.

Step 2 You know the slope and two points. Choose one point and find the y-intercept. In this case, we chose (6, −4).

$y = mx + b$ Slope-intercept form

$-4 = -\frac{1}{3}(6) + b$ Replace m with $-\frac{1}{3}$, x with 6, and y with −4.

$-4 = -2 + b$ Multiply.

$-4 + 2 = -2 + b + 2$ Add 2 to each side.

$-2 = b$ Simplify.

Step 3 Write the slope-intercept form using $m = -\frac{1}{3}$ and $b = -2$.

$y = mx + b$ Slope-intercept form

$y = -\frac{1}{3}x - 2$ Replace m with $-\frac{1}{3}$ and b with −2.

Therefore, the equation is $y = -\frac{1}{3}x - 2$. The answer is A.

The Princeton Review

Test-Taking Tip

You can check your result by graphing. The line should pass through (−3, −1) and (6, −4).

Example 3 Write an Equation to Solve a Problem

BASEBALL **In the middle of the 1998 baseball season, Mark McGwire seemed to be on track to break the record for most runs batted in. After 40 games, McGwire had 45 runs batted in. After 86 games, he had 87 runs batted in. Write a linear equation to estimate the number of runs batted in for any number of games that season.**

More About. . .

Baseball

Mark McGwire is best known for breaking Roger Maris' single-season home run record of 61. In the 1998 season, McGwire hit 70 home runs.

Source: *USA TODAY*

Explore You know the number of runs batted in after 40 and 86 games.

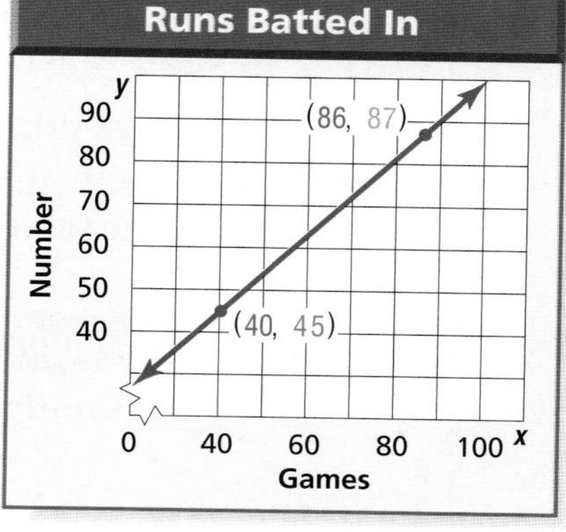

Plan Let x represent the number of games. Let y represent the number of runs batted in. Write an equation of the line that passes through (40, 45) and (86, 87).

Solve Find the slope.

$m = \frac{y_2 - y_1}{x_2 - x_1}$ Slope formula

$m = \frac{87 - 45}{86 - 40}$ Let $(x_1, y_1) = (40, 45)$ and $(x_2, y_2) = (86, 87)$.

$m = \frac{42}{46}$ or about 0.91 Simplify.

Choose (40, 45) and find the y-intercept of the line.

$y = mx + b$ Slope-intercept form

$45 = 0.91(40) + b$ Replace m with 0.91, x with 40, and y with 45.

$45 = 36.4 + b$ Multiply.

$45 - 36.4 = 36.4 + b - 36.4$ Subtract 36.4 from each side.

$8.6 = b$ Simplify.

Write the slope-intercept form using $m = 0.91$, and $b = 8.6$.

$y = mx + b$ Slope-intercept form

$y = 0.91x + 8.6$ Replace m with 0.91 and b with 8.6.

Therefore, the equation is $y = 0.91x + 8.6$.

Examine Check your result by substituting the coordinates of the point not chosen, (86, 87), into the equation.

$y = 0.91x + 8.6$ Original equation

$87 \stackrel{?}{=} 0.91(86) + 8.6$ Replace y with 87 and x with 86.

$87 \stackrel{?}{=} 78.26 + 8.6$ Multiply.

$87 \approx 86.86$ ✓ The slope was rounded, so the answers vary slightly.

Concept Summary — Writing Equations

Given the Slope and One Point

Step 1 Substitute the values of m, x, and y into the slope-intercept form and solve for b.

Step 2 Write the slope-intercept form using the values of m and b.

Given Two Points

Step 1 Find the slope.

Step 2 Choose one of the two points to use.

Step 3 Then, follow the steps for writing an equation given the slope and one point.

When you use a linear equation to predict values that are beyond the range of the data, you are using **linear extrapolation**.

Example 4 Linear Extrapolation

SPORTS The record for most runs batted in during a single season is 190. Use the equation in Example 3 to decide whether a baseball fan following the 1998 season would have expected McGwire to break the record in the 162 games played that year.

$y = 0.91x + 8.6$ Original equation

$y = 0.91(162) + 8.6$ Replace x with 162.

$y \approx 156$ Simplify.

Since the record is 190 runs batted in, a fan would have predicted that Mark McGwire would not break the record.

Be cautious when making a prediction using just two given points. The model may be *approximately* correct, but still give inaccurate predictions. For example, in 1998, Mark McGwire had 147 runs batted in, which was nine less than the prediction.

Check for Understanding

Concept Check

1. **Compare and contrast** the process used to write an equation given the slope and one point with the process used for two points.

2. **OPEN ENDED** Write an equation in slope-intercept form of a line that has a y-intercept of 3.

3. **Tell** whether the statement is *sometimes*, *always*, or *never* true. Explain. *You can write the equation of a line given its x- and y-intercepts.*

Guided Practice

Write an equation of the line that passes through each point with the given slope.

4. $(4, -2)$, $m = 2$
5. $(3, 7)$, $m = -3$
6. $(-3, 5)$, $m = -1$

Write an equation of the line that passes through each pair of points.

7. $(5, 1)$, $(8, -2)$
8. $(6, 0)$, $(0, 4)$
9. $(5, 2)$, $(-7, -4)$

Standardized Test Practice

10. The table of ordered pairs shows the coordinates of the two points on the graph of a line. Which equation describes the line?

x	y
−5	2
0	7

(A) $y = x + 7$
(B) $y = x - 7$
(C) $y = -5x + 2$
(D) $y = 5x + 2$

Practice and Apply

Homework Help

For Exercises	See Examples
11–18	1
19–29	2
34–39	3, 4

Extra Practice

See page 832.

Write an equation of the line that passes through each point with the given slope.

11.

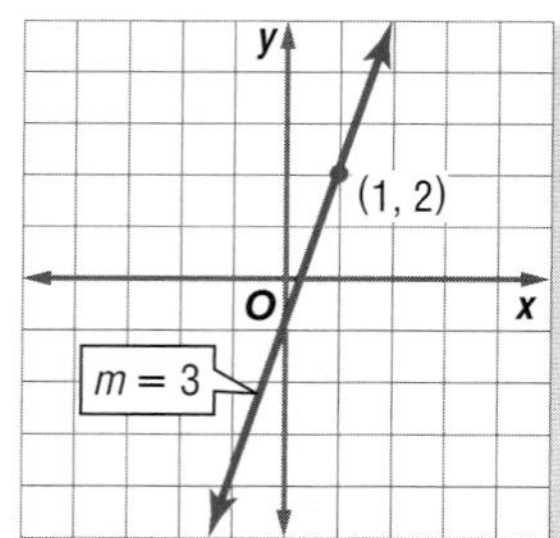

12.

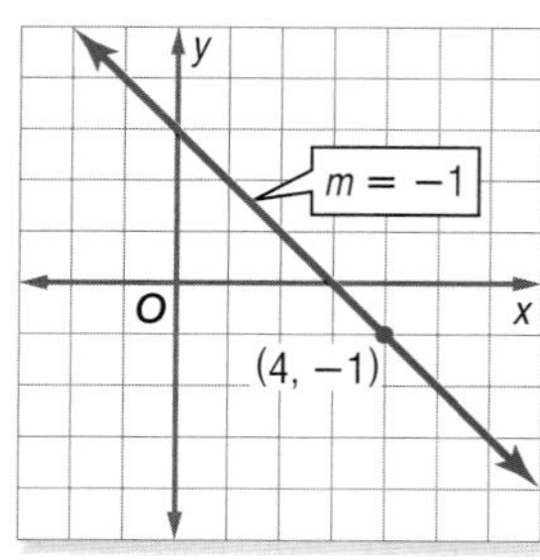

13. $(5, -2)$, $m = 3$

14. $(5, 4)$, $m = -5$

15. $(3, 0)$, $m = -2$

16. $(5, 3)$, $m = \frac{1}{2}$

17. $(-3, -1)$, $m = -\frac{2}{3}$

18. $(-3, -5)$, $m = -\frac{5}{3}$

Write an equation of the line that passes through each pair of points.

19.

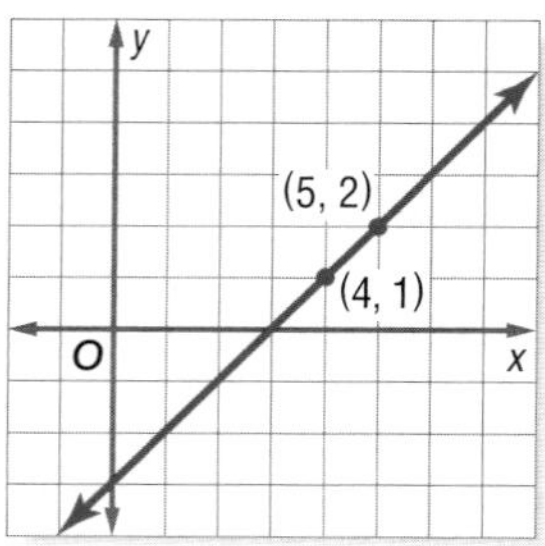

20.

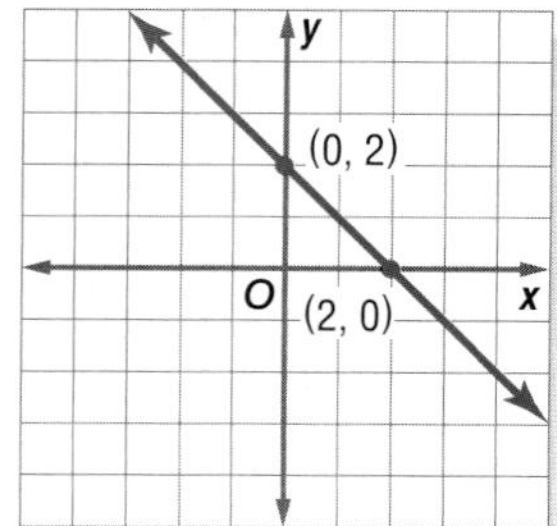

21. $(4, 2)$, $(-2, -4)$

22. $(3, -2)$, $(6, 4)$

23. $(-1, 3)$, $(2, -3)$

24. $(2, -2)$, $(3, 2)$

25. $(7, -2)$, $(-4, -2)$

26. $(0, 5)$, $(-3, 5)$

27. $(1, 1)$, $(7, 4)$

28. $(5, 7)$, $(0, 6)$

29. $\left(-\frac{5}{4}, 1\right)$, $\left(-\frac{1}{4}, \frac{3}{4}\right)$

Write an equation of the line that has each pair of intercepts.

30. x-intercept: -3, y-intercept: 5

31. x-intercept: 3, y-intercept: 4

32. x-intercept: 6, y-intercept: 3

33. x-intercept: 2, y-intercept: -2

MARRIAGE AGE **For Exercises 34–37, use the information in the graphic.**

34. Write a linear equation to predict the median age that men marry M for any year t.

35. Use the equation to predict the median age of men who marry for the first time in 2005.

36. Write a linear equation to predict the median age that women marry W for any year t.

37. Use the equation to predict the median age of women who marry for the first time in 2005.

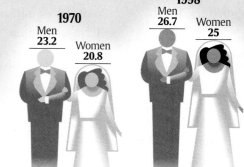

USA TODAY Snapshots®

Waiting on weddings

Couples are marrying later. The median age of men and women who tied the knot for the first time in 1970 and 1998:

1970
Men **23.2**
Women **20.8**

1998
Men **26.7**
Women **25**

Source: Census Bureau, March 2000

By Hilary Wasson and Sam Ward, USA TODAY

POPULATION **For Exercises 38 and 39, use the data at the top of page 280.**

38. Write a linear equation to find Orlando's population for any year.

39. Predict what Orlando's population will be in 2010.

40. **CANOE RENTAL** If you rent a canoe for 3 hours, you will pay \$45. Write a linear equation to find the total cost C of renting the canoe for h hours.

CANOE RENTALS
DAILY RATE PLUS \$10 PER HOUR

For Exercises 41–43, consider line ℓ that passes through (14, 2) and (28, 6).

41. Write an equation for line ℓ.

42. What is the slope of line ℓ?

43. Where does line ℓ intersect the x-axis? the y-axis?

44. **CRITICAL THINKING** The x-intercept of a line is p, and the y-intercept is q. Write an equation of the line.

45. **WRITING IN MATH** Answer the question that was posed at the beginning of the lesson.

How can slope-intercept form be used to make predictions?

Include the following in your answer:

- a definition of linear extrapolation, and
- an explanation of how slope-intercept form is used in linear extrapolation.

46. Which is an equation for the line with slope $\frac{1}{3}$ through $(-2, 1)$?

Ⓐ $y = \frac{1}{3}x + 1$ Ⓑ $y = \frac{1}{3}x + \frac{5}{3}$ Ⓒ $y = \frac{1}{3}x - \frac{5}{3}$ Ⓓ $y = \frac{1}{3}x + \frac{1}{3}$

47. About 20,000 fewer babies were born in California in 1996 than in 1995. In 1995, about 560,000 babies were born. Which equation can be used to predict the number of babies y (in thousands), born x years after 1995?

Ⓐ $y = 20x + 560$ Ⓑ $y = -20x + 560$

Ⓒ $y = -20x - 560$ Ⓓ $y = 20x - 560$

Maintain Your Skills

Mixed Review **Graph each equation.** *(Lesson 5-3)*

48. $y = 3x - 2$ 49. $x + y = 6$ 50. $x + 2y = 8$

51. **HEALTH** Each time your heart beats, it pumps 2.5 ounces of blood through your heart. Write a direct variation equation that relates the total volume of blood V with the number of times your heart beats b. *(Lesson 5-2)*

State the domain of each relation. *(Lesson 4-3)*

52. $\{(0, 8), (9, -2), (4, 2)\}$ 53. $\{(-2, 1), (5, 1), (-2, 7), (0, -3)\}$

Replace each ● with $<$, $>$, or $=$ to make a true sentence. *(Lesson 2-7)*

54. $-3 \bullet -5$ 55. $4 \bullet \frac{16}{3}$ 56. $\frac{3}{4} \bullet \frac{2}{3}$

Getting Ready for the Next Lesson **PREREQUISITE SKILL** **Find each difference.**
*(To review **subtracting integers**, see Lesson 2-2.)*

57. $4 - 7$ 58. $5 - 12$ 59. $2 - (-3)$

60. $-1 - 4$ 61. $-7 - 8$ 62. $-5 - (-2)$

5-5 Writing Equations in Point-Slope Form

What You'll Learn

- Write the equation of a line in point-slope form.
- Write linear equations in different forms.

Vocabulary

- point-slope form

How can you use the slope formula to write an equation of a line?

The graph shows a line with slope 2 that passes through (3, 4). Another point on the line is (x, y).

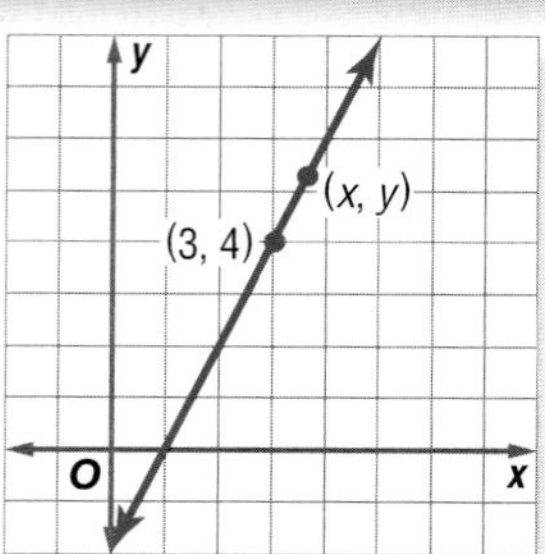

$m = \frac{y_2 - y_1}{x_2 - x_1}$	Slope formula
$2 = \frac{y - 4}{x - 3}$	$(x_2, y_2) = (x, y)$ $(x_1, y_1) = (3, 4)$
$2(x - 3) = \frac{y - 4}{x - 3}(x - 3)$	Multiply each side by $(x - 3)$.
$2(x - 3) = y - 4$	Simplify.
$y - 4 = 2(x - 3)$	Symmetric Property of Equality

In $y - 4 = 2(x - 3)$: 4 is the *y*-coordinate, 2 is the slope, 3 is the *x*-coordinate.

POINT-SLOPE FORM The equation above was generated using the coordinates of a known point and the slope of the line. It is written in **point-slope form**.

Key Concept — *Point-Slope Form*

- **Words** The linear equation $y - y_1 = m(x - x_1)$ is written in point-slope form, where (x_1, y_1) is a given point on a nonvertical line and m is the slope of the line.
- **Symbols** $y - y_1 = m(x - x_1)$

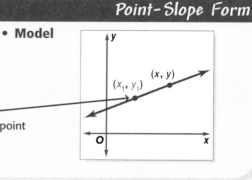

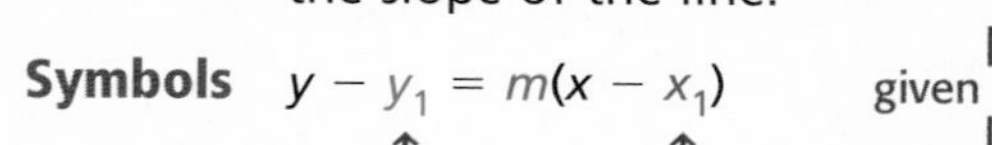

- **Model**

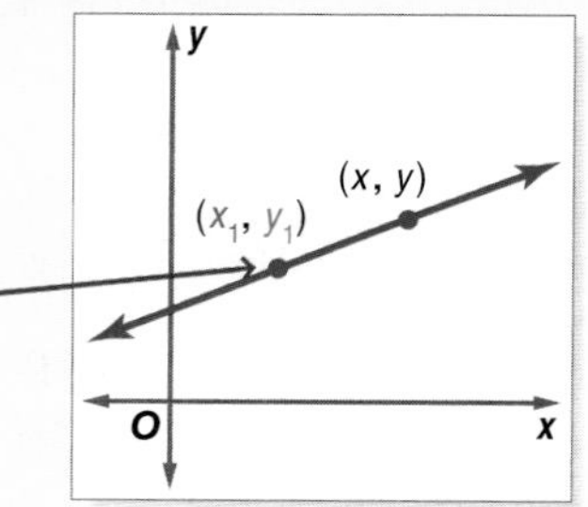

Study Tip

Point-Slope Form
Remember, (x_1, y_1) represents the *given* point, and (x, y) represents *any other* point on the line.

Example 1 Write an Equation Given Slope and a Point

Write the point-slope form of an equation for a line that passes through (−1, 5) with slope −3.

$y - y_1 = m(x - x_1)$	Point-slope form
$y - 5 = -3[x - (-1)]$	$(x_1, y_1) = (-1, 5)$
$y - 5 = -3(x + 1)$	Simplify.

Therefore, the equation is $y - 5 = -3(x + 1)$.

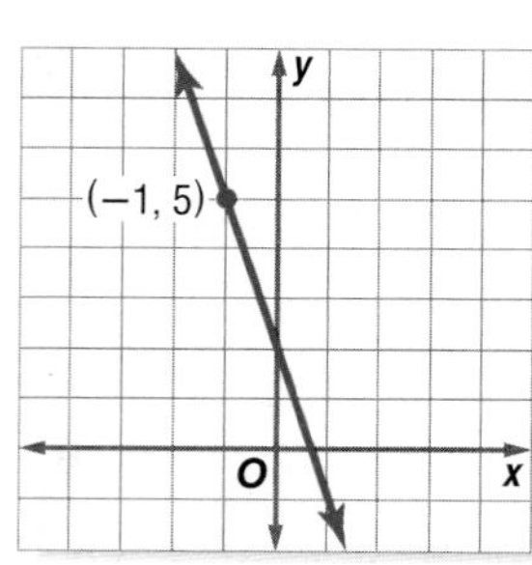

Vertical lines cannot be written in point-slope form because the slope is undefined. However, since the slope of a horizontal line is 0, horizontal lines can be written in point-slope form.

Example 2 Write an Equation of a Horizontal Line

Write the point-slope form of an equation for a horizontal line that passes through (6, −2).

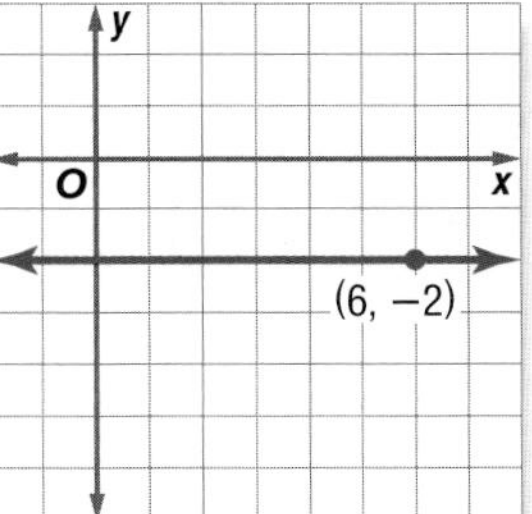

$y - y_1 = m(x - x_1)$ Point-slope form

$y - (-2) = 0(x - 6)$ $(x_1, y_1) = (6, -2)$

$y + 2 = 0$ Simplify.

Therefore, the equation is $y + 2 = 0$.

FORMS OF LINEAR EQUATIONS You have learned about three of the most common forms of linear equations.

Concept Summary — *Forms of Linear Equations*

Form	Equation	Description
Slope-Intercept	$y = mx + b$	m is the slope, and b is the y-intercept.
Point-Slope	$y - y_1 = m(x - x_1)$	m is the slope and (x_1, y_1) is a given point.
Standard	$Ax + By = C$	A and B are not both zero. Usually A is nonnegative and A, B, and C are real numbers.

Study Tip

Look Back
To review **standard form**, see Lesson 4-5.

Linear equations in point-slope form can be written in slope-intercept or standard form.

Example 3 Write an Equation in Standard Form

Write $y + 5 = -\frac{5}{4}(x - 2)$ in standard form.

In standard form, the variables are on the left side of the equation. A, B, and C are all integers.

$y + 5 = -\frac{5}{4}(x - 2)$ Original equation

$4(y + 5) = 4\left(-\frac{5}{4}\right)(x - 2)$ Multiply each side by 4 to eliminate the fraction.

$4y + 20 = -5(x - 2)$ Distributive Property

$4y + 20 = -5x + 10$ Distributive Property

$4y + 20 - 20 = -5x + 10 - 20$ Subtract 20 from each side.

$4y = -5x - 10$ Simplify.

$4y + 5x = -5x - 10 + 5x$ Add $5x$ to each side.

$5x + 4y = -10$ Simplify.

The standard form of the equation is $5x + 4y = -10$.

Example 4 Write an Equation in Slope-Intercept Form

Write $y - 2 = \frac{1}{2}(x + 5)$ in slope-intercept form.

In slope-intercept form, y is on the left side of the equation. The constant and x are on the right side.

$y - 2 = \frac{1}{2}(x + 5)$	Original equation
$y - 2 = \frac{1}{2}x + \frac{5}{2}$	Distributive Property
$y - 2 + 2 = \frac{1}{2}x + \frac{5}{2} + 2$	Add 2 to each side.
$y = \frac{1}{2}x + \frac{9}{2}$	$2 = \frac{4}{2}$ and $\frac{4}{2} + \frac{5}{2} = \frac{9}{2}$

The slope-intercept form of the equation is $y = \frac{1}{2}x + \frac{9}{2}$.

You can draw geometric figures on a coordinate plane and use the point-slope form to write equations of the lines.

Example 5 Write an Equation in Point-Slope Form

Study Tip

Geometry

The *hypotenuse* is the side of a right triangle opposite the right angle.

GEOMETRY **The figure shows right triangle ABC.**

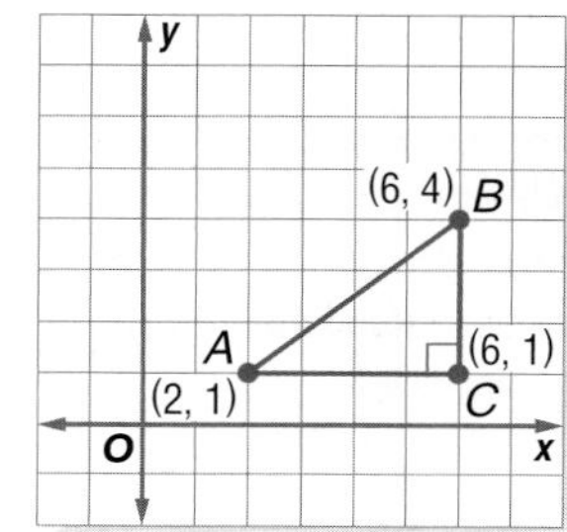

a. Write the point-slope form of the line containing the hypotenuse $\overline{AB}$.

Step 1 First, find the slope of $\overline{AB}$.

$m = \frac{y_2 - y_1}{x_2 - x_1}$	Slope formula
$= \frac{4 - 1}{6 - 2}$ or $\frac{3}{4}$	$(x_1, y_1) = (2, 1)$, $(x_2, y_2) = (6, 4)$

Step 2 You can use either point for (x_1, y_1) in the point-slope form.

Method 1 Use (6, 4).	**Method 2** Use (2, 1).
$y - y_1 = m(x - x_1)$	$y - y_1 = m(x - x_1)$
$y - 4 = \frac{3}{4}(x - 6)$	$y - 1 = \frac{3}{4}(x - 2)$

b. Write each equation in standard form.

$y - 4 = \frac{3}{4}(x - 6)$	Original equation	$y - 1 = \frac{3}{4}(x - 2)$
$4(y - 4) = 4\left(\frac{3}{4}\right)(x - 6)$	Multiply each side by 4.	$4(y - 1) = 4\left(\frac{3}{4}\right)(x - 2)$
$4y - 16 = 3(x - 6)$	Multiply.	$4y - 4 = 3(x - 2)$
$4y - 16 = 3x - 18$	Distributive Property	$4y - 4 = 3x - 6$
$4y = 3x - 2$	Add to each side.	$4y = 3x - 2$
$-3x + 4y = -2$	Subtract $3x$ from each side.	$-3x + 4y = -2$
$3x - 4y = 2$	Multiply each side by -1.	$3x - 4y = 2$

Regardless of which point was used to find the point-slope form, the standard form results in the same equation.

Check for Understanding

Concept Check

1. **Explain** what x_1 and y_1 in the point-slope form of an equation represent.
2. **FIND THE ERROR** Tanya and Akira wrote the point-slope form of an equation for a line that passes through $(-2, -6)$ and $(1, 6)$. Tanya says that Akira's equation is wrong. Akira says they are both correct.

Tanya	Akira
$y + 6 = 4(x + 2)$	$y - 6 = 4(x - 1)$

Who is correct? Explain your reasoning.

3. **OPEN ENDED** Write an equation in point-slope form. Then write an equation for the same line in slope-intercept form.

Guided Practice

Write the point-slope form of an equation for a line that passes through each point with the given slope.

4.

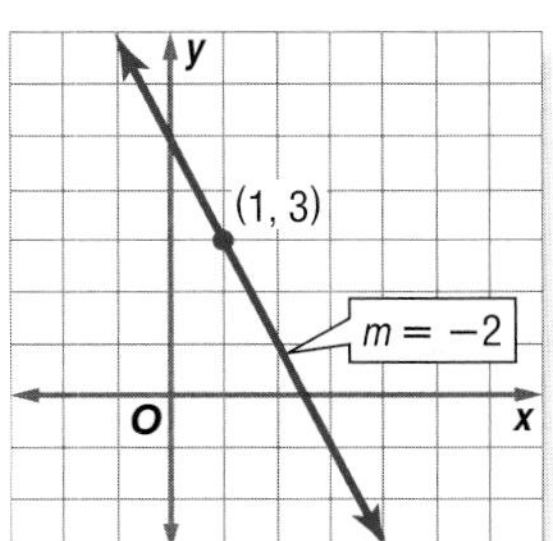

5.

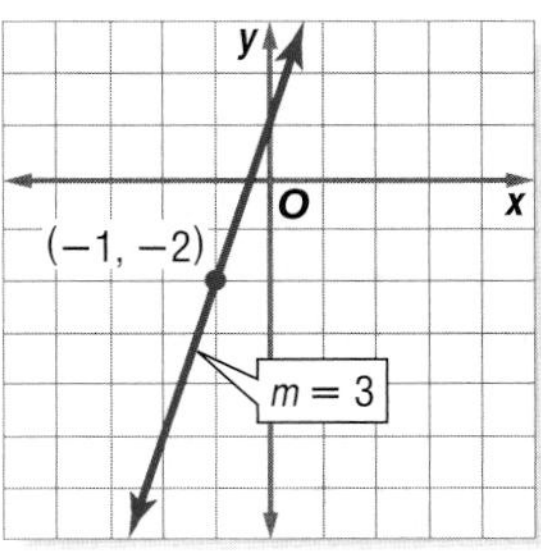

6.

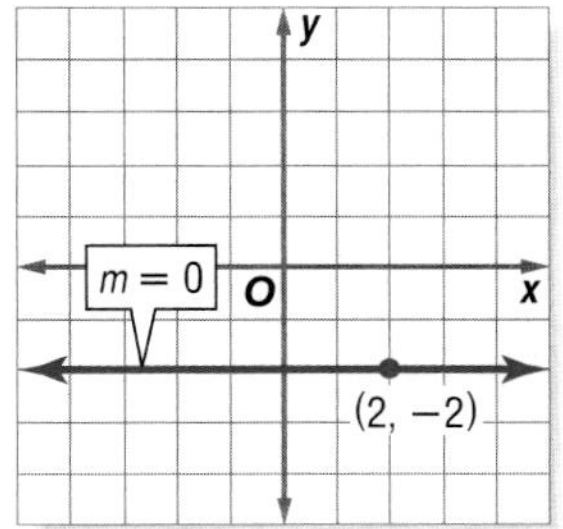

Write each equation in standard form.

7. $y - 5 = 4(x + 2)$
8. $y + 3 = -\frac{3}{4}(x - 1)$
9. $y - 3 = 2.5(x + 1)$

Write each equation in slope-intercept form.

10. $y + 6 = 2(x - 2)$
11. $y + 3 = -\frac{2}{3}(x - 6)$
12. $y - \frac{7}{2} = \frac{1}{2}(x - 4)$

Application

GEOMETRY For Exercises 13 and 14, use parallelogram *ABCD*.
A parallelogram has opposite sides parallel.

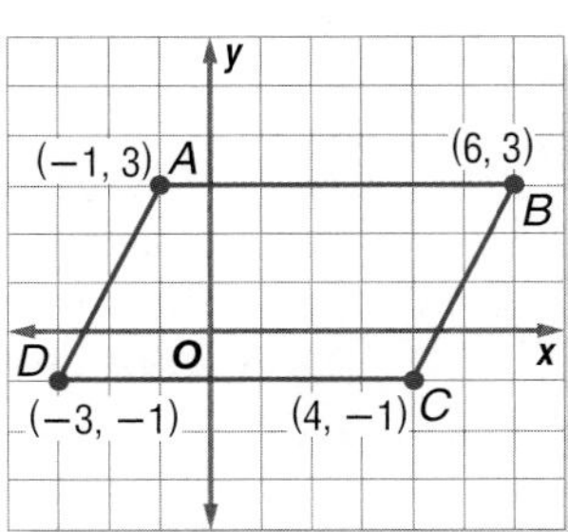

13. Write the point-slope form of the line containing $\overline{AD}$.
14. Write the standard form of the line containing $\overline{AD}$.

Practice and Apply

Homework Help

For Exercises	See Examples
15–26	1
27–28	2
29–40	3
41–52	4

Extra Practice
See page 832.

Write the point-slope form of an equation for a line that passes through each point with the given slope.

15. $(3, 8), m = 2$
16. $(-4, -3), m = 1$
17. $(-2, 4), m = -3$
18. $(-6, 1), m = -4$
19. $(-3, 6), m = 0$
20. $(9, 1), m = \frac{2}{3}$
21. $(8, -3), m = \frac{3}{4}$
22. $(-6, 3), m = -\frac{2}{3}$
23. $(1, -3), m = -\frac{5}{8}$
24. $(9, -5), m = 0$
25. $(-4, 8), m = \frac{7}{2}$
26. $(1, -4), m = -\frac{8}{3}$

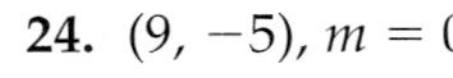

27. Write the point-slope form of an equation for a horizontal line that passes through $(5, -9)$.

28. A horizontal line passes through $(0, 7)$. Write the point-slope form of its equation.

Write each equation in standard form.

29. $y - 13 = 4(x - 2)$
30. $y + 3 = 3(x + 5)$
31. $y - 5 = -2(x + 6)$
32. $y + 3 = -5(x + 1)$
33. $y + 7 = \frac{1}{2}(x + 2)$
34. $y - 1 = \frac{5}{6}(x - 4)$
35. $y - 2 = -\frac{2}{5}(x - 8)$
36. $y + 4 = -\frac{1}{3}(x - 12)$
37. $y + 2 = \frac{5}{3}(x + 6)$
38. $y + 6 = \frac{3}{2}(x - 4)$
39. $y - 6 = 1.3(x + 7)$
40. $y - 2 = -2.5(x - 1)$

Write each equation in slope-intercept form.

41. $y - 2 = 3(x - 1)$
42. $y - 5 = 6(x + 1)$
43. $y + 2 = -2(x - 5)$
44. $y - 1 = -7(x - 3)$
45. $y + 3 = \frac{1}{2}(x + 4)$
46. $y - 1 = \frac{2}{3}(x + 9)$
47. $y + 3 = -\frac{1}{4}(x + 2)$
48. $y - 5 = -\frac{2}{5}(x + 15)$
49. $y + \frac{1}{2} = x - \frac{1}{2}$
50. $y - \frac{1}{3} = -2\left(x + \frac{1}{3}\right)$
51. $y + \frac{1}{4} = -3\left(x + \frac{1}{2}\right)$
52. $y + \frac{3}{5} = -4\left(x - \frac{1}{2}\right)$

53. Write the point-slope form, slope-intercept form, and standard form of an equation for a line that passes through $(5, -3)$ with slope 10.

54. Line ℓ passes through $(1, -6)$ with slope $\frac{3}{2}$. Write the point-slope form, slope-intercept form, and standard form of an equation for line ℓ.

BUSINESS **For Exercises 55–57, use the following information.**
A home security company provides security systems for \$5 per week, plus an installation fee. The total fee for 12 weeks of service is \$210.

55. Write the point-slope form of an equation to find the total fee y for any number of weeks x.

56. Write the equation in slope-intercept form.

57. What is the flat fee for installation?

More About. . .

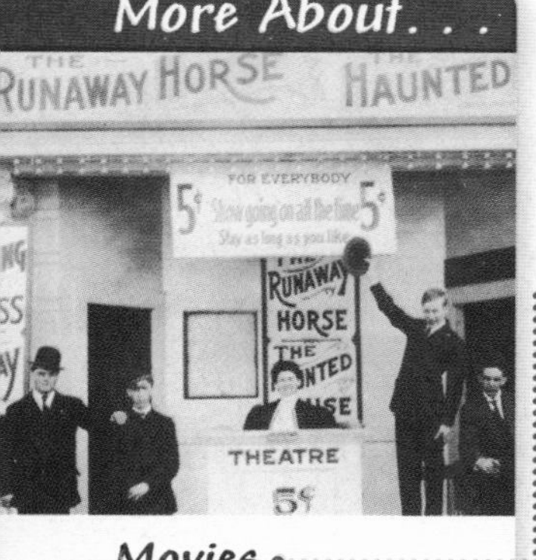

Movies

In 1907, movie theaters were called nickelodeons. There were about 5000 movie screens, and the average movie ticket cost 5 cents.

Source: National Association of Theatre Owners

MOVIES **For Exercises 58–60, use the following information.**
Between 1990 and 1999, the number of movie screens in the United States increased by about 1500 each year. In 1996, there were 29,690 movie screens.

58. Write the point-slope form of an equation to find the total number of screens y for any year x.

59. Write the equation in slope-intercept form.

60. Predict the number of movie screens in the United States in 2005.

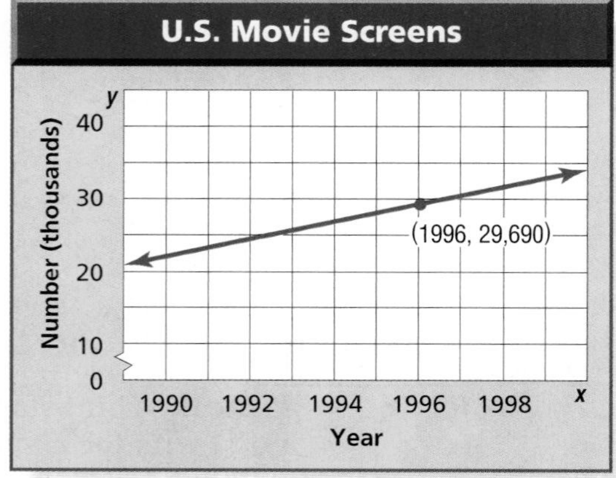

Source: Motion Picture Association of America

Online Research **Data Update** What has happened to the number of movie screens since 1999? Visit www.algebra1.com/data_update to learn more.

GEOMETRY **For Exercises 61–63, use square *PQRS*.**

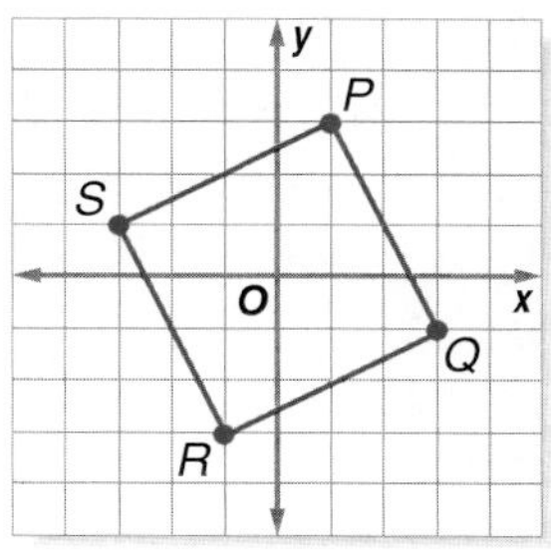

61. Write a point-slope equation of the line containing each side.

62. Write the slope-intercept form of each equation.

63. Write the standard form of each equation.

64. **CRITICAL THINKING** A line contains the points (9, 1) and (5, 5). Write a convincing argument that the same line intersects the x-axis at (10, 0).

65. **WRITING IN MATH** Answer the question that was posed at the beginning of the lesson.

How can you use the slope formula to write an equation of a line?

Include the following in your answer:

- an explanation of how you can use the slope formula to write the point-slope form.

Standardized Test Practice

66. Which equation represents a line that *neither* passes through (0, 1) *nor* has a slope of 3?

Ⓐ $-2x + y = 1$ Ⓑ $y + 1 = 3(x + 6)$

Ⓒ $y - 3 = 3(x - 6)$ Ⓓ $x - 3y = -15$

67. **OPEN ENDED** Write the slope-intercept form of an equation of a line that passes through (2, −5).

Extending the Lesson

For Exercises 68–71, use the graph at the right.

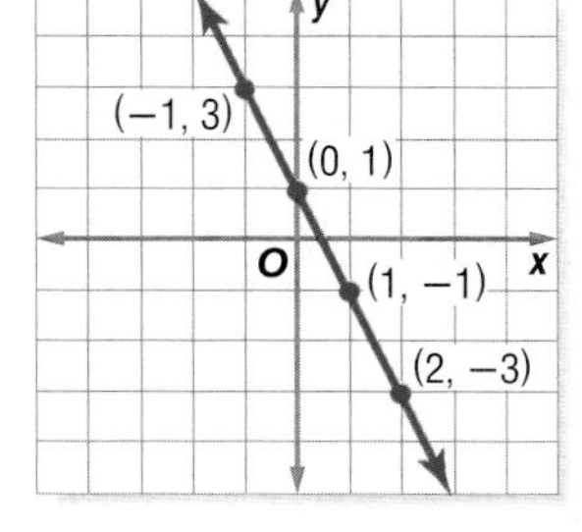

68. Choose three different pairs of points from the graph. Write the slope-intercept form of the line using each pair.

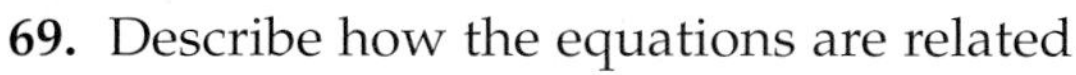

69. Describe how the equations are related.

70. Choose a different pair of points from the graph and predict the equation of the line determined by these points. Check your conjecture by finding the equation.

71. **MAKE A CONJECTURE** What conclusion can you draw from this activity?

Maintain Your Skills

Mixed Review

Write the slope-intercept form of an equation of the line that satisfies each condition. *(Lessons 5-3 and 5-4)*

72. slope −2 and y-intercept −5

73. passes through (−2, 4) with slope 3

74. passes through (2, −4) and (0, 6)

75. a horizontal line through (1, −1)

Solve each equation. *(Lesson 3-4)*

76. $4a - 5 = 15$

77. $7 + 3c = -11$

78. $\frac{2}{9}v - 6 = 14$

79. Evaluate $(25 - 4) \div (2^2 - 1^3)$. *(Lesson 1-3)*

Getting Ready for the Next Lesson

PREREQUISITE SKILL **Write the multiplicative inverse of each number.**

*(For review of **multiplicative inverses**, see pages 800 and 801.)*

80. 2

81. 10

82. 1

83. −1

84. $\frac{2}{3}$

85. $-\frac{1}{9}$

86. $\frac{5}{2}$

87. $-\frac{2}{3}$

5-6 Geometry: Parallel and Perpendicular Lines

What You'll Learn

- Write an equation of the line that passes through a given point, parallel to a given line.
- Write an equation of the line that passes through a given point, perpendicular to a given line.

Vocabulary

- parallel lines
- perpendicular lines

How can you determine whether two lines are parallel?

The graphing calculator screen shows a family of linear graphs whose slope is 1. Notice that the lines do not appear to intersect.

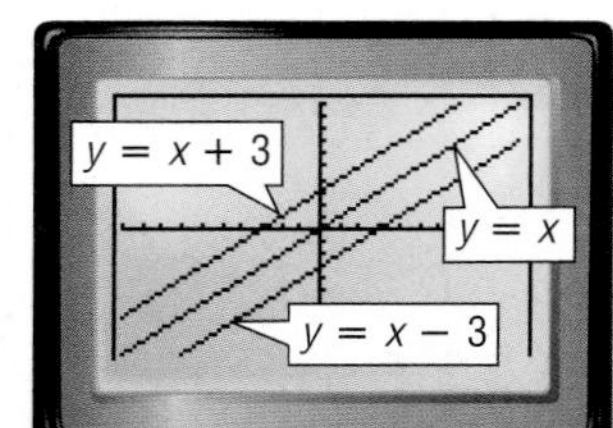

PARALLEL LINES Lines in the same plane that do not intersect are called **parallel lines**. Parallel lines have the same slope.

Key Concept — *Parallel Lines in a Coordinate Plane*

- **Words** Two nonvertical lines are parallel if they have the same slope. All vertical lines are parallel.
- **Model**

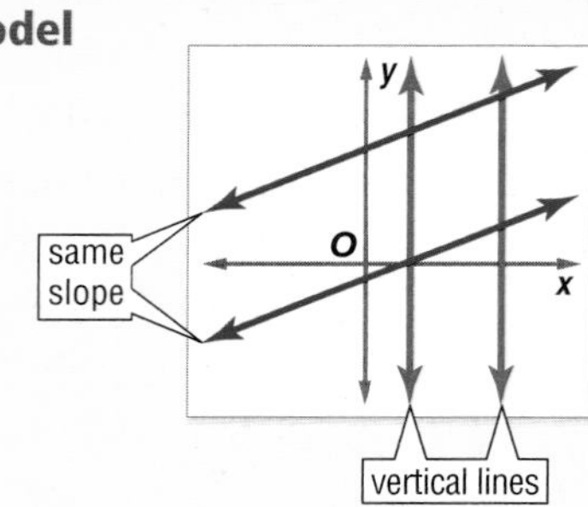

You can write the equation of a line parallel to a given line if you know a point on the line and an equation of the given line.

Example 1 Parallel Line Through a Given Point

Write the slope-intercept form of an equation for the line that passes through (−1, −2) and is parallel to the graph of $y = -3x - 2$.

The line parallel to $y = -3x - 2$ has the same slope, −3. Replace m with −3, and (x_1, y_1) with (−1, −2) in the point-slope form.

$y - y_1 = m(x - x_1)$	Point-slope form
$y - (-2) = -3[x - (-1)]$	Replace m with −3, y with −2, and x with −1.
$y + 2 = -3(x + 1)$	Simplify.
$y + 2 = -3x - 3$	Distributive Property
$y + 2 - 2 = -3x - 3 - 2$	Subtract 2 from each side.
$y = -3x - 5$	Write the equation in slope-intercept form.

Therefore, the equation is $y = -3x - 5$.

CHECK You can check your result by graphing both equations. The lines appear to be parallel. The graph of $y = -3x - 5$ passes through $(-1, -2)$.

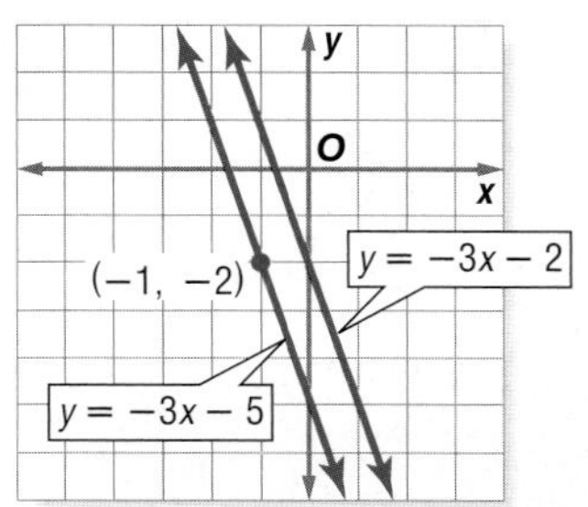

PERPENDICULAR LINES Lines that intersect at right angles are called **perpendicular lines**. There is a relationship between the slopes of perpendicular lines.

Algebra Activity

Perpendicular Lines

Model

- A scalene triangle is one in which no two sides are equal. Cut out a scalene right triangle *ABC* so that $\angle C$ is a right angle. Label the vertices and the sides as shown.
- Draw a coordinate plane on grid paper. Place $\triangle ABC$ on the coordinate plane so that *A* is at the origin and side *b* lies along the positive *x*-axis.

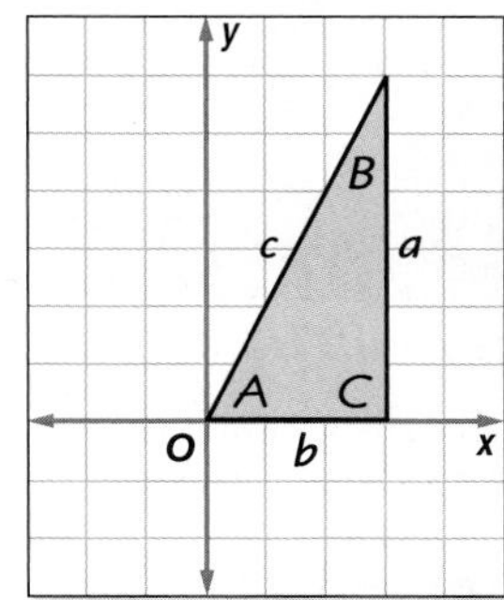

Analyze

1. Name the coordinates of *B*.
2. What is the slope of side *c*?
3. Rotate the triangle 90° counterclockwise so that *A* is still at the origin and side *b* is along the positive *y*-axis. Name the coordinates of *B*.
4. What is the slope of side *c*?
5. Repeat the activity for two other different scalene triangles.
6. For each triangle and its rotation, what is the relationship between the first position of side *c* and the second?
7. For each triangle and its rotation, describe the relationship between the coordinates of *B* in the first and second positions.
8. Describe the relationship between the slopes of *c* in each position.

Make a Conjecture

9. Describe the relationship between the slopes of any two perpendicular lines.

Key Concept — *Perpendicular Lines in a Coordinate Plane*

- **Words** Two nonvertical lines are perpendicular if the product of their slopes is −1. That is, the slopes are *opposite reciprocals* of each other. Vertical lines and horizontal lines are also perpendicular.
- **Model**

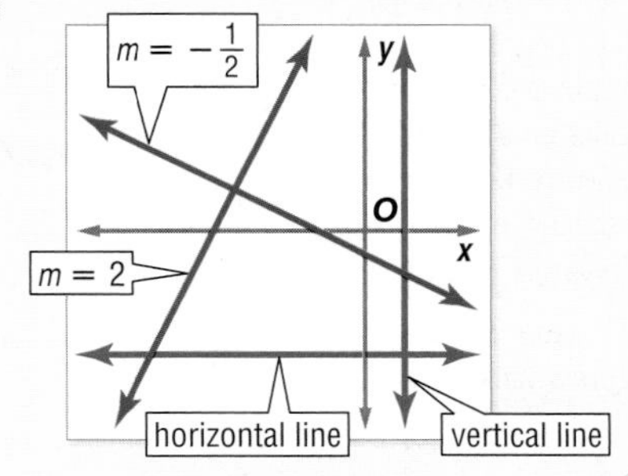

Example 2 Determine Whether Lines are Perpendicular

KITES The outline of a kite is shown on a coordinate plane. Determine whether $\overline{AC}$ is perpendicular to $\overline{BD}$.

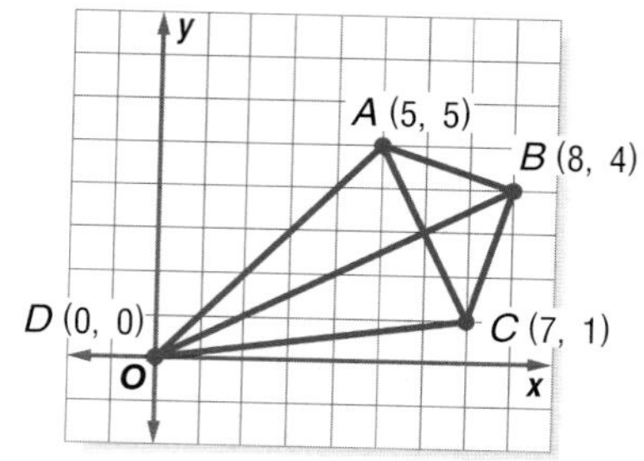

Find the slope of each segment.

Slope of $\overline{AC}$: $m = \frac{5-1}{5-7}$ or -2

Slope of $\overline{BD}$: $m = \frac{4-0}{8-0}$ or $\frac{1}{2}$

The line segments are perpendicular because $\frac{1}{2}(-2) = -1$.

More About. . .

Kites

In India, kite festivals mark *Makar Sankranti*, when the Sun moves into the northern hemisphere.

Source: www.cam-india.com

You can write the equation of a line perpendicular to a given line if you know a point on the line and the equation of the given line.

Example 3 Perpendicular Line Through a Given Point

Write the slope-intercept form for an equation of a line that passes through $(-3, -2)$ and is perpendicular to the graph of $x + 4y = 12$.

Step 1 Find the slope of the given line.

$x + 4y = 12$	Original equation
$x + 4y - x = 12 - x$	Subtract 1x from each side.
$4y = -1x + 12$	Simplify.
$\frac{4y}{4} = \frac{-1x + 12}{4}$	Divide each side by 4.
$y = -\frac{1}{4}x + 3$	Simplify.

Step 2 The slope of the given line is $-\frac{1}{4}$. So, the slope of the line perpendicular to this line is the opposite reciprocal of $-\frac{1}{4}$, or 4.

Step 3 Use the point-slope form to find the equation.

$y - y_1 = m(x - x_1)$	Point-slope form
$y - (-2) = 4[x - (-3)]$	$(x_1, y_1) = (-3, -2)$ and $m = 4$
$y + 2 = 4(x + 3)$	Simplify.
$y + 2 = 4x + 12$	Distributive Property
$y + 2 - 2 = 4x + 12 - 2$	Subtract 2 from each side.
$y = 4x + 10$	Simplify.

Therefore, the equation of the line is $y = 4x + 10$.

CHECK You can check your result by graphing both equations on a graphing calculator. Use the CALC menu to verify that $y = 4x + 10$ passes through $(-3, -2)$.

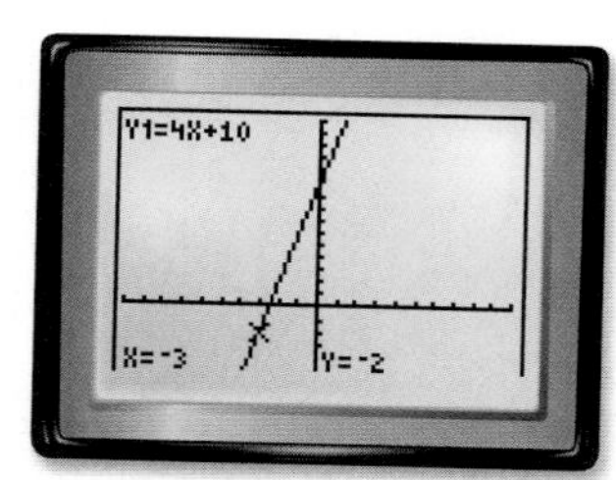

[−15.16..., 15.16...] scl: 1 by [−10, 10] scl: 1

Study Tip

Graphing Calculator

The lines will not appear to be perpendicular on a graphing calculator if the scales on the axes are not set correctly. After graphing, press ZOOM 5 to set the axes for a correct representation.

Example 4 Perpendicular Line Through a Given Point

Write the slope-intercept form for an equation of a line perpendicular to the graph of $y = -\frac{1}{3}x + 2$ and passes through the x-intercept of that line.

Step 1 Find the slope of the perpendicular line. The slope of the given line is $-\frac{1}{3}$, therefore a perpendicular line has slope 3 because $-\frac{1}{3} \cdot 3 = -1$.

Step 2 Find the x-intercept of the given line.

$y = -\frac{1}{3}x + 2$ Original equation

$0 = -\frac{1}{3}x + 2$ Replace y with 0.

$-2 = -\frac{1}{3}x$ Subtract 2 from each side.

$6 = x$ Multiply each side by -3.

The x-intercept is at (6, 0).

Step 3 Substitute the slope and the given point into the point-slope form of a linear equation. Then write the equation in slope-intercept form.

$y - y_1 = m(x - x_1)$ Point-slope form

$y - 0 = 3(x - 6)$ Replace x with 6, y with 0, and m with 3.

$y = 3x - 18$ Distributive Property

Check for Understanding

Concept Check

1. **Explain** how to find the slope of a line that is perpendicular to the line shown in the graph.
2. **OPEN ENDED** **Give an example** of two numbers that are negative reciprocals.
3. **Define** *parallel lines* and *perpendicular lines*.

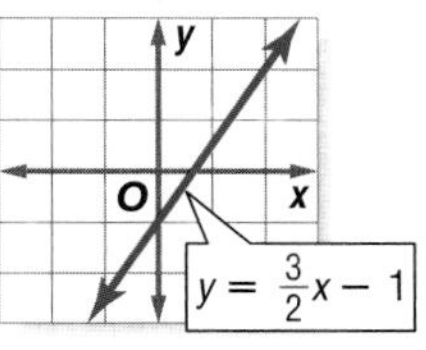

Guided Practice **Write the slope-intercept form of an equation of the line that passes through the given point and is parallel to the graph of each equation.**

4.

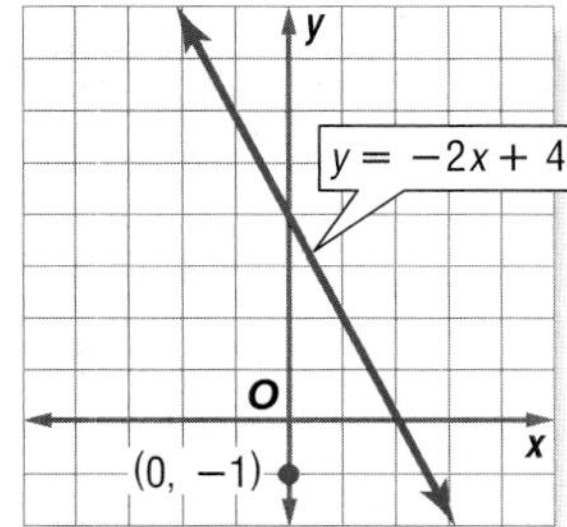

5.

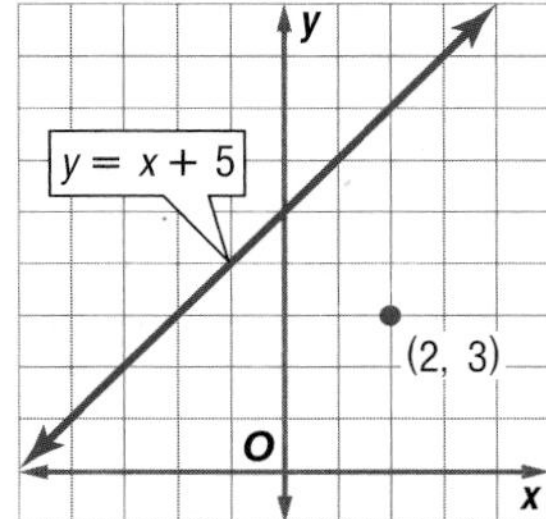

6. $(1, -3)$, $y = 2x - 1$
7. $(-2, 2)$, $-3x + y = 4$
8. **GEOMETRY** Quadrilateral $ABCD$ has vertices $A(-2, 1)$, $B(3, 3)$, $C(5, 7)$, and $D(0, 5)$. Determine whether $\overline{AC}$ is perpendicular to $\overline{BD}$.

Write the slope-intercept form of an equation that passes through the given point and is perpendicular to the graph of each equation.

9. $(-3, 1)$, $y = \frac{1}{3}x + 2$
10. $(6, -2)$, $y = \frac{3}{5}x - 4$
11. $(2, -2)$, $2x + y = 5$

Application

12. **GEOMETRY** The line with equation $y = 3x - 4$ contains side $\overline{AC}$ of right triangle ABC. If the vertex of the right angle C is at (3, 5), what is an equation of the line that contains side $\overline{BC}$?

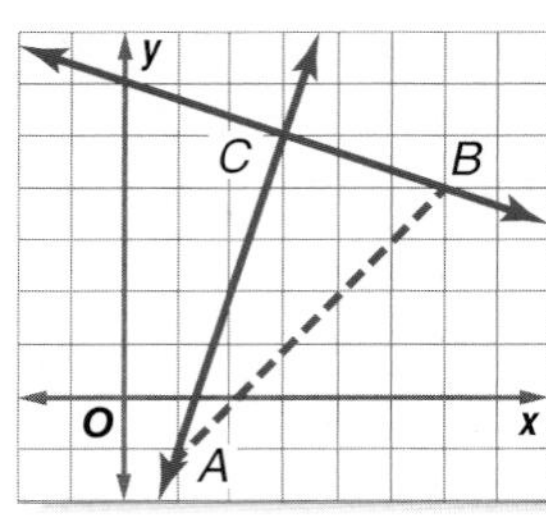

Practice and Apply

Homework Help

For Exercises	See Examples
13–24	1
26	2
28–39	3, 4

Extra Practice

See page 832.

Write the slope-intercept form of an equation of the line that passes through the given point and is parallel to the graph of each equation.

13. $(2, -7), y = x - 2$
14. $(2, -1), y = 2x + 2$
15. $(-3, 2), y = x - 6$
16. $(4, -1), y = 2x + 1$
17. $(-5, -4), y = \frac{1}{2}x + 1$
18. $(3, 3), y = \frac{2}{3}x - 1$
19. $(-4, -3), y = -\frac{1}{3}x + 3$
20. $(-1, 2), y = -\frac{1}{2}x - 4$
21. $(-3, 0), 2y = x - 1$
22. $(2, 2), 3y = -2x + 6$
23. $(-2, 3), 6x + y = 4$
24. $(2, 2), 3x - 4y = -4$

25. **GEOMETRY** A *parallelogram* is a quadrilateral in which opposite sides are parallel. Is $ABCD$ a parallelogram? Explain.

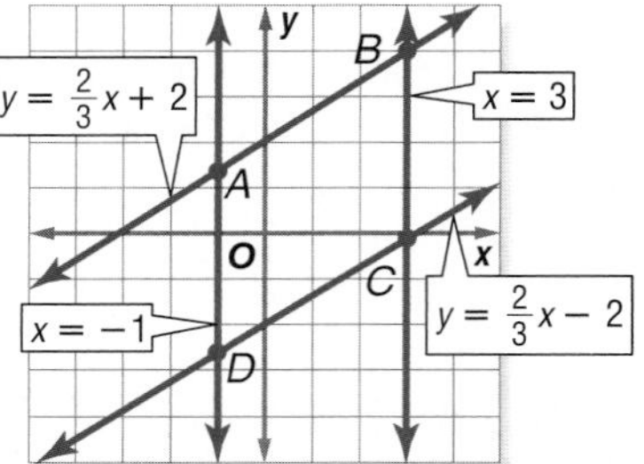

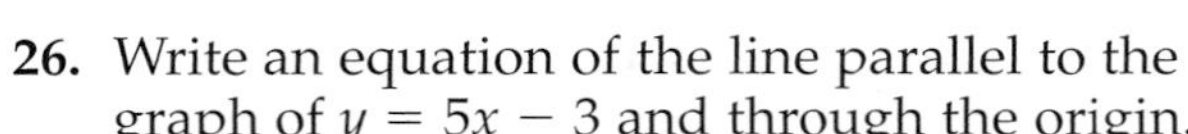

26. Write an equation of the line parallel to the graph of $y = 5x - 3$ and through the origin.

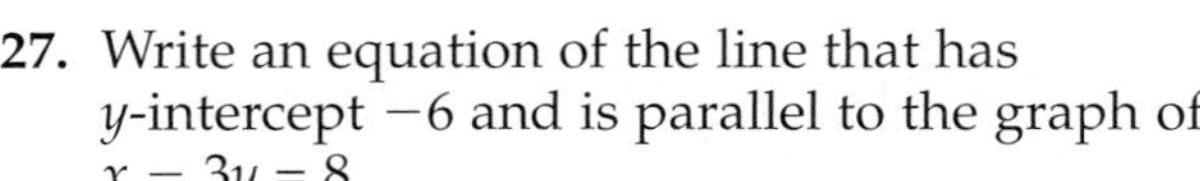

27. Write an equation of the line that has y-intercept -6 and is parallel to the graph of $x - 3y = 8$.

Write the slope-intercept form of an equation that passes through the given point and is perpendicular to the graph of each equation.

28. $(-2, 0), y = x - 6$
29. $(1, 1), y = 4x + 6$
30. $(-3, 1), y = -3x + 7$
31. $(0, 5), y = -8x + 4$
32. $(1, -3), y = \frac{1}{2}x + 4$
33. $(4, 7), y = \frac{2}{3}x - 1$
34. $(0, 4), 3x + 8y = 4$
35. $(-2, 7), 2x - 5y = 3$
36. $(6, -1), 3y + x = 3$
37. $(0, -1), 5x - y = 3$
38. $(8, -2), 5x - 7 = 3y$
39. $(3, -3), 3x + 7 = 2x$

40. Find an equation of the line that has a y-intercept of -2 and is perpendicular to the graph of $3x + 6y = 2$.

41. Write an equation of the line that is perpendicular to the line through (9, 10) and (3, −2) and passes through the x-intercept of that line.

Determine whether the graphs of each pair of equations are *parallel*, *perpendicular*, or *neither*.

42. $y = -2x + 11$; $y + 2x = 23$
43. $3y = 2x + 14$; $2x - 3y = 2$
44. $y = -5x$; $y = 5x - 18$

45. **GEOMETRY** The diagonals of a square are segments that connect the opposite vertices. Determine the relationship between the diagonals $\overline{AC}$ and $\overline{BD}$ of square $ABCD$.

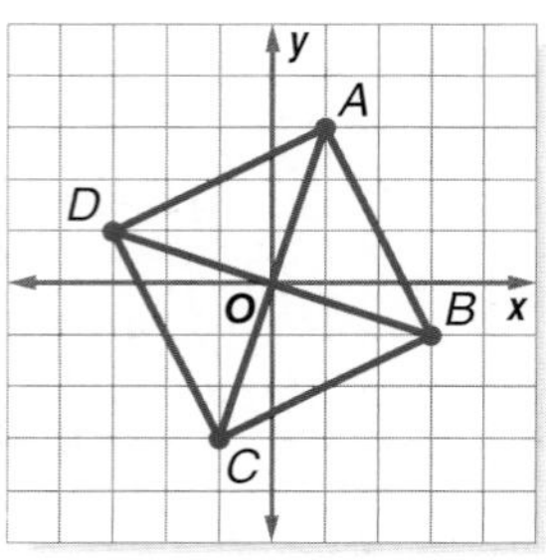

46. **CRITICAL THINKING** What is a if the lines with equations $y = ax + 5$ and $2y = (a + 4)x - 1$ are parallel?

47. **WRITING IN MATH** Answer the question that was posed at the beginning of the lesson.

How can you determine whether two lines are parallel?

Include the following in your answer:

- an equation whose graph is parallel to the graph of $y = -5x$, with an explanation of your reasoning, and
- an equation whose graph is perpendicular to the graph of $y = -5x$, with an explanation of your reasoning.

Standardized Test Practice

48. What is the slope of a line perpendicular to the graph of $3x + 4y = 24$?

(A) $-\frac{4}{3}$ (B) $-\frac{3}{4}$ (C) $\frac{3}{4}$ (D) $\frac{4}{3}$

49. How can the graph of $y = 3x + 4$ be used to graph $y = 3x + 2$?

(A) Move the graph of the line right 2 units.

(B) Change the slope of the graph from 4 to 2.

(C) Change the y-intercept from 4 to 2.

(D) Move the graph of the line left 2 units.

Maintain Your Skills

Mixed Review

Write the point-slope form of an equation for a line that passes through each point with the given slope. *(Lesson 5-5)*

50. $(3, 5)$, $m = -2$ 51. $(-4, 7)$, $m = 5$ 52. $(-1, -3)$, $m = -\frac{1}{2}$

TELEPHONE **For Exercises 53 and 54, use the following information.**
An international calling plan charges a rate per minute plus a flat fee. A 10-minute call to the Czech Republic costs \$3.19. A 15-minute call costs \$4.29. *(Lesson 5-4)*

53. Write a linear equation in slope-intercept form to find the total cost C of an m-minute call.

54. Find the cost of a 12-minute call.

Getting Ready for the Next Lesson

PREREQUISITE SKILL **Write the slope-intercept form of an equation of the line that passes through each pair of points.** *(To review* ***slope-intercept form****, see Lesson 5-4.)*

55. $(5, -1), (-3, 3)$ 56. $(0, 2), (8, 0)$ 57. $(2, 1), (3, -4)$

58. $(5, 5), (8, -1)$ 59. $(6, 9), (4, 9)$ 60. $(-6, 4), (2, -2)$

Practice Quiz 2

Lessons 5-3 through 5-6

Write the slope-intercept form for an equation of the line that satisfies each condition.

1. slope 4 and y-intercept -3 *(Lesson 5-3)*
2. passes through $(1, -3)$ with slope 2 *(Lesson 5-4)*
3. passes through $(-1, -2)$ and $(1, 3)$ *(Lesson 5-4)*
4. parallel to the graph of $y = 2x - 2$ and passes through $(-2, 3)$ *(Lesson 5-6)*
5. Write $y - 4 = \frac{1}{2}(x + 3)$ in standard form and in slope-intercept form. *(Lesson 5-5)*

5-7 Statistics: Scatter Plots and Lines of Fit

What You'll Learn

- Interpret points on a scatter plot.
- Write equations for lines of fit.

Vocabulary

- scatter plot
- positive correlation
- negative correlation
- line of fit
- best-fit line
- linear interpolation

How do scatter plots help identify trends in data?

The points of a set of real-world data do not always lie on one line. But, you may be able to draw a line that seems to be close to all the points.

The line in the graph shows a linear relationship between the year x and the number of bushels of apples y. As the years increase, the number of bushels of apples also increases.

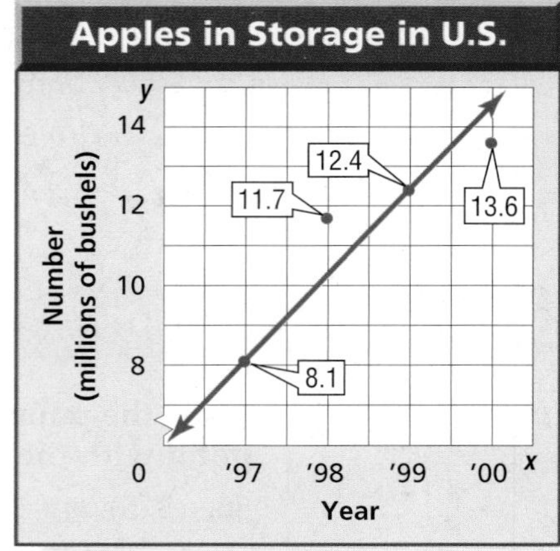

Source: U.S. Apple Association

INTERPRET POINTS ON A SCATTER PLOT A **scatter plot** is a graph in which two sets of data are plotted as ordered pairs in a coordinate plane. Scatter plots are used to investigate a relationship between two quantities.

- In the first graph below, there is a **positive correlation** between x and y. That is, as x increases, y increases.
- In the second graph below, there is a **negative correlation** between x and y. That is, as x increases, y decreases.
- In the third graph below, there is *no correlation* between x and y. That is, x and y are not related.

If the pattern in a scatter plot is linear, you can draw a line to summarize the data. This can help identify trends in the data.

Key Concept — Scatter Plots

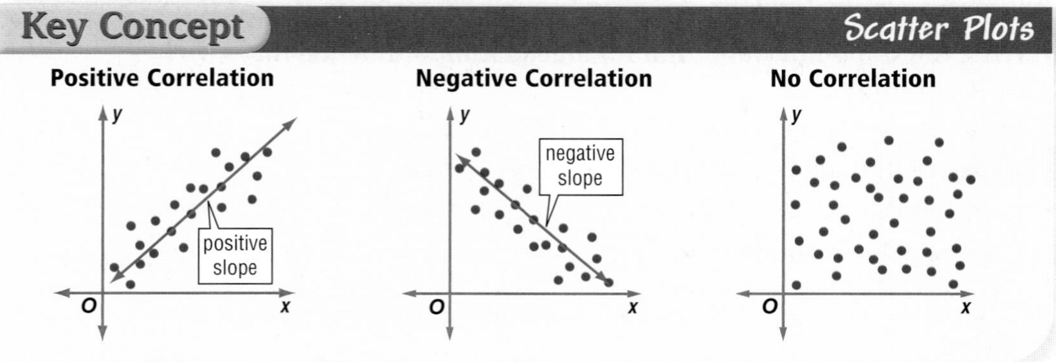

Example 1 Analyze Scatter Plots

Determine whether each graph shows a *positive correlation*, a *negative correlation*, or *no correlation*. If there is a positive or negative correlation, describe its meaning in the situation.

a. NUTRITION **The graph shows fat grams and Calories for selected choices at a fast-food restaurant.**

The graph shows a positive correlation. As the number of fat grams increases, the number of Calories increases.

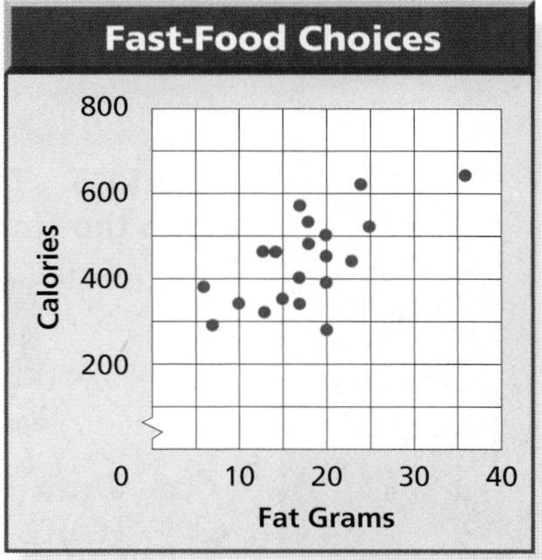

Source: Olen Publishing Co.

b. CARS **The graph shows the weight and the highway gas mileage of selected cars.**

The graph shows a negative correlation. As the weight of the automobile increases, the gas mileage decreases.

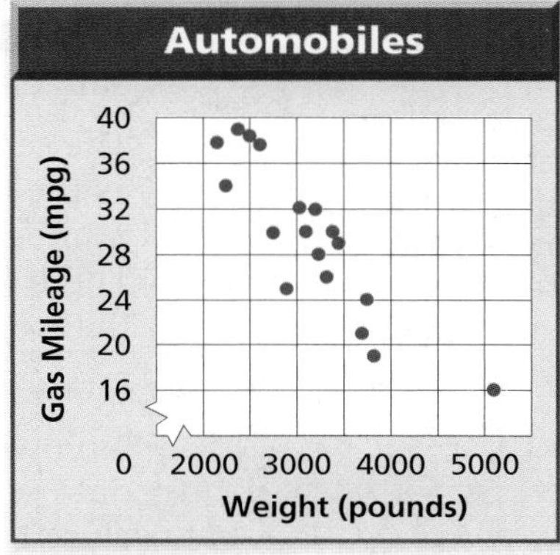

Source: Yahoo!

Is there a relationship between the length of a person's foot and his or her height? Make a scatter plot and then look for a pattern.

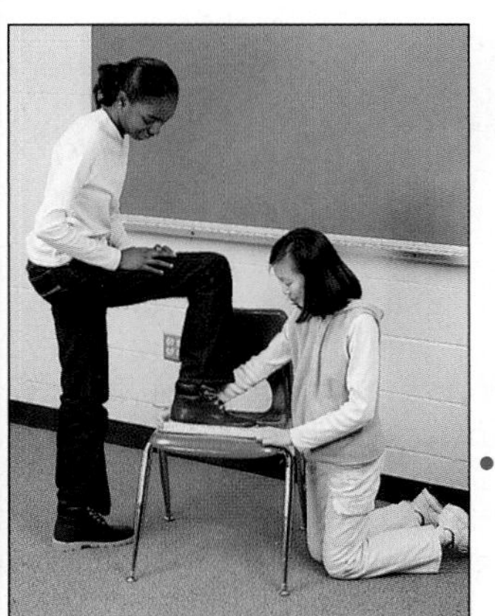

Algebra Activity

Making Predictions

Collect the Data

- Measure your partner's foot and height in centimeters. Then trade places.
- Add the points (foot length, height) to a class scatter plot.

Analyze the Data

1. Is there a correlation between foot length and height for the members of your class? If so, describe it.
2. Draw a line that summarizes the data and shows how the height changes as the foot length changes.

Make a Conjecture

3. Use the line to predict the height of a person whose foot length is 25 centimeters. Explain your method.

LINES OF FIT If the data points do not all lie on a line, but are close to a line, you can draw a **line of fit**. This line describes the trend of the data. Once you have a line of fit, you can find an equation of the line.

In this lesson, you will use a graphical method to find a line of fit. In the follow-up to Lesson 5-7, you will use a graphing calculator to find a line of fit. The calculator uses a statistical method to find the line that most closely approximates the data. This line is called the **best-fit line**.

Example 2 Find a Line of Fit

BIRDS The table shows an estimate for the number of bald eagle pairs in the United States for certain years since 1985.

Years since 1985	3	5	7	9	11	14
Bald Eagle Pairs	2500	3000	3700	4500	5000	5800

Source: U.S. Fish and Wildlife Service

a. Draw a scatter plot and determine what relationship exists, if any, in the data.

Let the independent variable x be the number of years since 1985, and let the dependent variable y be the number of bald eagle pairs.

The scatter plot seems to indicate that as the number of years increases, the number of bald eagle pairs increases. There is a positive correlation between the two variables.

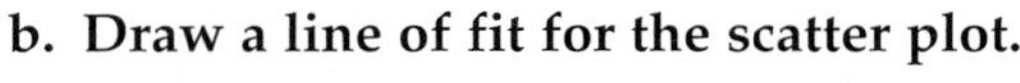

b. Draw a line of fit for the scatter plot.

No one line will pass through all of the data points. Draw a line that passes close to the points. A line of fit is shown in the scatter plot at the right.

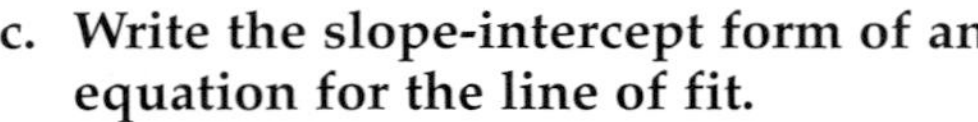

c. Write the slope-intercept form of an equation for the line of fit.

The line of fit shown above passes through the data points (3, 2500) and (11, 5000).

Step 1 Find the slope.

$m = \frac{y_2 - y_1}{x_2 - x_1}$ Slope formula

$m = \frac{5000 - 2500}{11 - 3}$ Let $(x_1, y_1) = (3, 2500)$ and $(x_2, y_2) = (11, 5000)$.

$m = \frac{2500}{8}$ or 312.5 Simplify.

Step 2 Use $m = 312.5$ and either the point-slope form or the slope-intercept form to write the equation. You can use either data point. We chose (3, 2500).

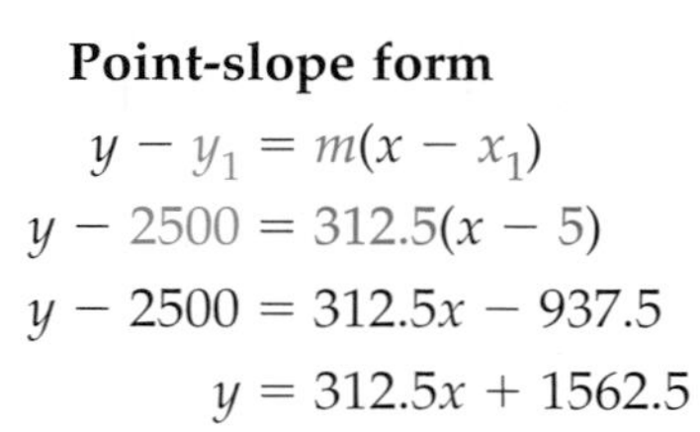

Point-slope form

$y - y_1 = m(x - x_1)$
$y - 2500 = 312.5(x - 5)$
$y - 2500 = 312.5x - 937.5$
$y = 312.5x + 1562.5$

Slope-intercept form

$y = mx + b$
$2500 = 312.5(3) + b$
$2500 = 937.5 + b$
$1562.5 = b$
$y = 312.5x + 1562.5$

Using either method, $y = 312.5x + 1562.5$.

More About. . .

Birds

The bald eagle was listed as an endangered species in 1963, when the number of breeding pairs had dropped below 500.

Source: U.S. Fish and Wildlife Service

Study Tip

Lines of Fit

When you use the graphical method, the line of fit is an approximation. So, you may draw another line of fit using other points that is equally valid. Some valid lines of fit may not contain any of the data points.

CHECK Check your result by substituting (11, 5000) into $y = 312.5x + 1562.5$.

$y = 312.5x + 1562.5$	Line of fit equation
$5000 \stackrel{?}{=} 312.5(11) + 1562.5$	Replace x with 11 and y with 5000.
$5000 \stackrel{?}{=} 3437.5 + 1562.5$	Multiply.
$5000 = 5000$ ✓	Add.

The solution checks.

In Lesson 5-4, you learned about linear extrapolation, which is predicting values that are *outside* the range of the data. You can also use a linear equation to predict values that are *inside* the range of the data. This is called **linear interpolation**.

Example 3 Linear Interpolation

BIRDS Use the equation for the line of fit in Example 2 to estimate the number of bald eagle pairs in 1998.

Use the equation $y = 312.5x + 1562.5$, where x is the number of years since 1985 and y is the number of bald eagle pairs.

$y = 312.5x + 1562.5$	Original equation
$y = 312.5(13) + 1562.5$	Replace x with 1998 – 1985 or 13.
$y = 5625$	Simplify.

There were about 5625 bald eagle pairs in 1998.

Check for Understanding

Concept Check

1. **Explain** how to determine whether a scatter plot has a positive or negative correlation.
2. **OPEN ENDED** Sketch scatter plots that have each type of correlation.
   a. positive **b.** negative **c.** no correlation
3. **Compare and contrast** linear interpolation and linear extrapolation.

Guided Practice

Determine whether each graph shows a *positive correlation*, a *negative correlation*, or *no correlation*. If there is a positive or negative correlation, describe its meaning in the situation.

4. 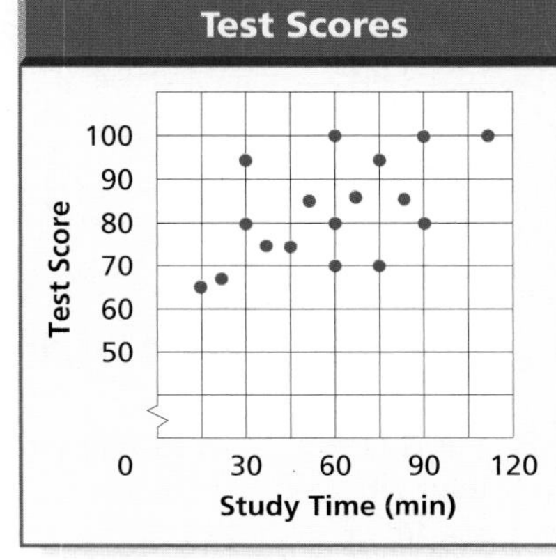

5. 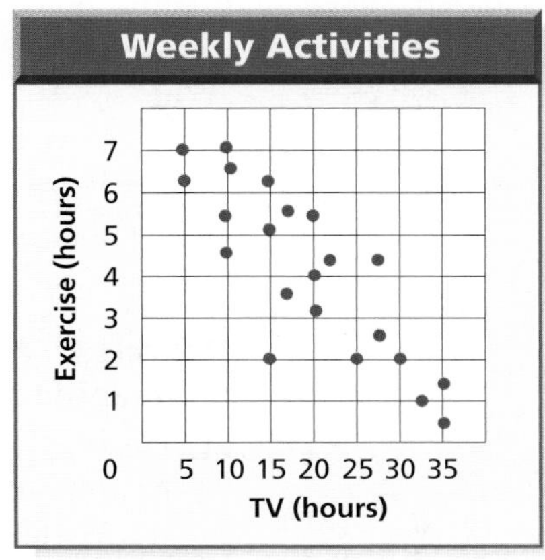

Application **BIOLOGY** **For Exercises 6–9, use the table that shows the average body temperature in degrees Celsius of 9 insects at a given air temperature.**

	Temperature (°C)								
Air	25.7	30.4	28.7	31.2	31.5	26.2	30.1	31.5	18.2
Body	27.0	31.5	28.9	31.0	31.5	25.6	28.4	31.7	18.7

6. Draw a scatter plot and determine what relationship exists, if any, in the data.
7. Draw a line of fit for the scatter plot.
8. Write the slope-intercept form of an equation for the line of fit.
9. Predict the body temperature of an insect if the air temperature is 40.2°F.

Practice and Apply

Homework Help

For Exercises	See Examples
10–13	1
14–33	2, 3

Extra Practice

See page 833.

Determine whether each graph shows a *positive correlation*, a *negative correlation*, or *no correlation*. If there is a positive or negative correlation, describe its meaning in the situation.

10. 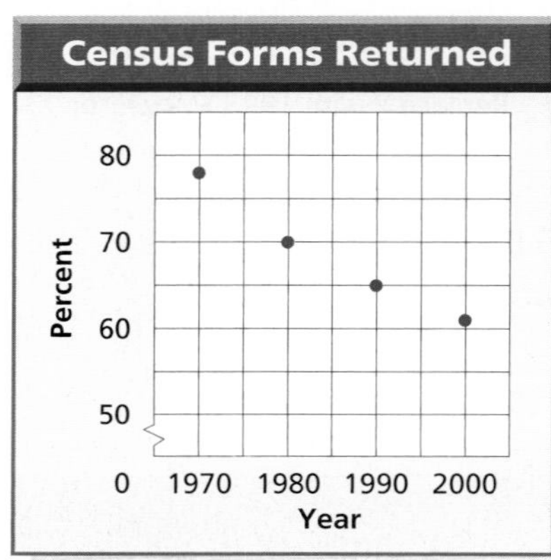

Source: U.S. Census Bureau

11. 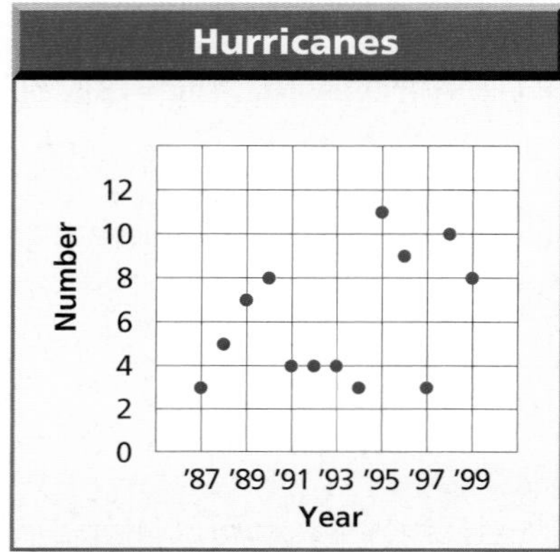

Source: *USA TODAY*

12. 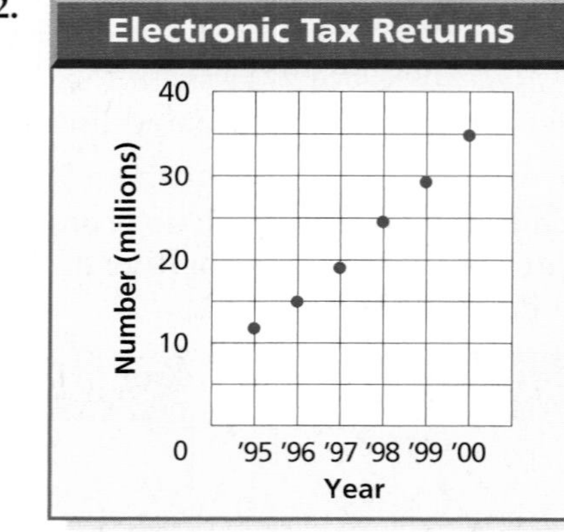

Source: IRS

13. 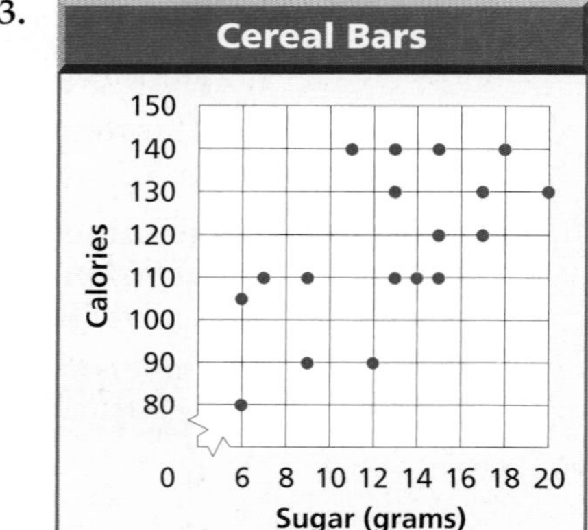

Source: *Vitality*

FARMING **For Exercises 14 and 15, refer to the graph at the top of page 298 about apple storage.**

14. Use the points (1997, 8.1) and (1999, 12.4) to write the slope-intercept form of an equation for the line of fit.
15. Predict the number of bushels of apples in storage in 2002.

USED CARS For Exercises 16 and 17, use the scatter plot that shows the ages and prices of used cars from classified ads.

16. Use the points (2, 9600) and (5, 6000) to write the slope-intercept form of an equation for the line of fit shown in the scatter plot.
17. Predict the price of a car that is 7 years old.

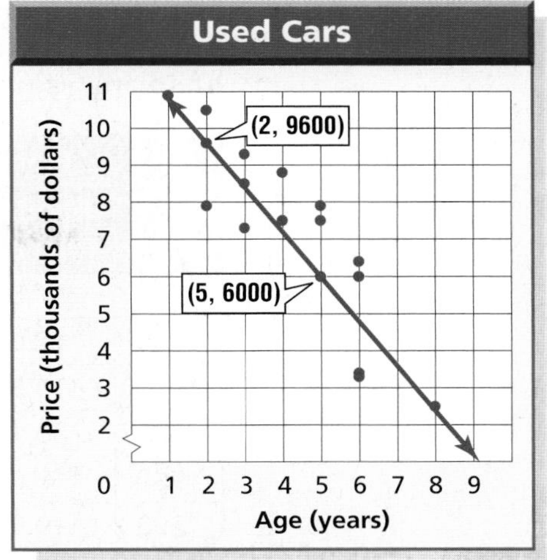

Source: *Columbus Dispatch*

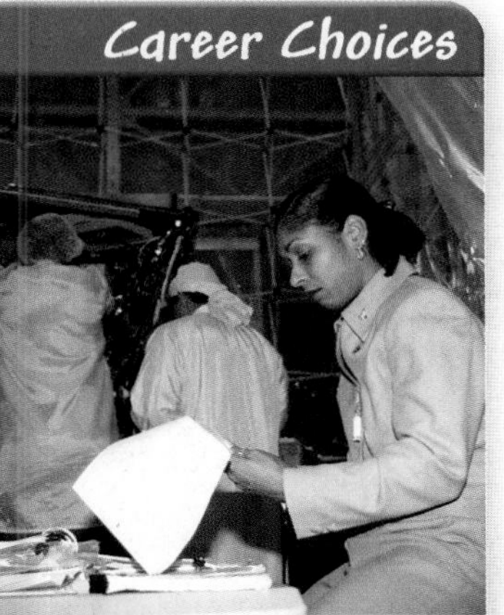

Career Choices

Aerospace Engineer

Aerospace engineers design, develop, and test aircraft and spacecraft. Many specialize in a particular type of aerospace product, such as commercial airplanes, military fighter jets, helicopters, or spacecraft.

Online Research For information about a career as an aerospace engineer, visit: www.algebra1.com/careers

PHYSICAL SCIENCE For Exercises 18–23, use the following information. Hydrocarbons like methane, ethane, propane, and butane are composed of only carbon and hydrogen atoms. The table gives the number of carbon atoms and the boiling points for several hydrocarbons.

Hydrocarbons			
Name	Formula	Number of Carbon Atoms	Boiling Point (°C)
Ethane	C_2H_6	2	−89
Propane	C_3H_8	3	−42
Butane	C_4H_{10}	4	−1
Hexane	C_6H_{12}	6	69
Octane	C_8H_{18}	8	126

18. Draw a scatter plot comparing the numbers of carbon atoms to the boiling points.
19. Draw a line of fit for the data.
20. Write the slope-intercept form of an equation for the line of fit.
21. Predict the boiling point for methane (CH_4), which has 1 carbon atom.
22. Predict the boiling point for pentane (C_5H_{12}), which has 5 carbon atoms.
23. The boiling point of heptane is 98.4°C. Use the equation of the line of fit to predict the number of carbon atoms in heptane.

SPACE For Exercises 24–28, use the table that shows the amount the United States government has spent on space and other technologies in selected years.

Federal Spending on Space and Other Technologies								
Year	1980	1985	1990	1995	1996	1997	1998	1999
Spending (billions of dollars)	4.5	6.6	11.6	12.6	12.7	13.1	12.9	12.4

Source: U.S. Office of Management and Budget

24. Draw a scatter plot and determine what relationship, if any, exists in the data.
25. Draw a line of fit for the scatter plot.
26. Let x represent the number of years since 1980. Let y represent the spending in billions of dollars. Write the slope-intercept form of the equation for the line of fit.
27. Predict the amount that will be spent on space and other technologies in 2005.
28. The government projects spending of $14.3 billion in space and other technologies in 2005. How does this compare to your prediction?

FORESTRY **For Exercises 29–33, use the table that shows the number of acres burned by wildfires in Florida each year and the corresponding number of inches of spring rainfall.**

Florida's Burned Acreage and Spring Rainfall					
Year	Rainfall (inches)	Acres (thousands)	Year	Rainfall (inches)	Acres (thousands)
1988	17.5	194	1994	18.1	180
1989	12.0	645	1995	16.3	46
1990	14.0	250	1996	20.4	94
1991	30.1	87	1997	18.5	146
1992	16.0	83	1998	22.2	507
1993	19.6	80	1999	12.7	340

Source: Florida Division of Forestry

29. Draw a scatter plot with rainfall on the x-axis and acres on the y-axis.
30. Draw a line of fit for the data.
31. Write the slope-intercept form of an equation for the line of fit.
32. In 2000, there was only 8.25 inches of spring rainfall. Estimate the number of acres burned by wildfires in 2000.
33. In 1998, there was 22.2 inches of rainfall, yet 507,000 acres were burned. Where was this data graphed in the scatter plot? How did this affect the line of fit?

Online Research **Data Update** What has happened to the number of acres burned by wildfires in Florida since 1999? Visit www.algebra1.com/data_update to learn more.

WebQuest

You can use a line of fit to describe the trend in winning Olympic times. Visit www.algebra1.com/webquest to continue work on your WebQuest project.

34. **CRITICAL THINKING** A test contains 20 true-false questions. Draw a scatter plot that shows the relationship between the number of correct answers x and the number of incorrect answers y.

RESEARCH **For Exercises 35 and 36, choose a topic to research that you believe may be correlated, such as arm span and height. Find existing data or collect your own.**

35. Draw a line of fit for the data.
36. Use the line to make a prediction about the data.

37. **WRITING IN MATH** Answer the question that was posed at the beginning of the lesson.

How do scatter plots help identify trends in data?

Include the following in your answer:

- a scatter plot that shows a person's height and his or her age, with a description of any trends, and
- an explanation of how you could use the scatter plot to predict a person's age given his or her height.

Standardized Test Practice

38. Which graph is the best example of data that show a negative linear relationship between the variables x and y?

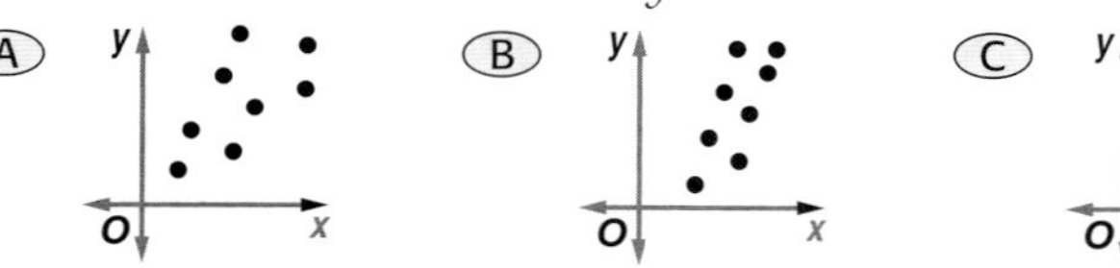

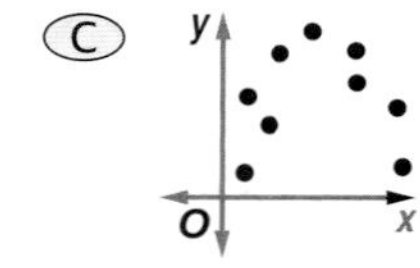

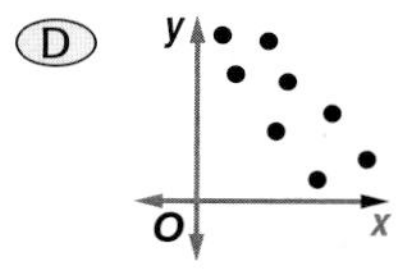

39. Choose the equation for the line that best fits the data in the table at the right.

(A) $y = x + 4$
(B) $y = 2x + 3$
(C) $y = 7$
(D) $y = 4x - 5$

x	y
1	5
2	7
3	7
4	11

Extending the Lesson

GEOGRAPHY For Exercises 40–44, use the following information.
The *latitude* of a place on Earth is the measure of its distance from the equator.

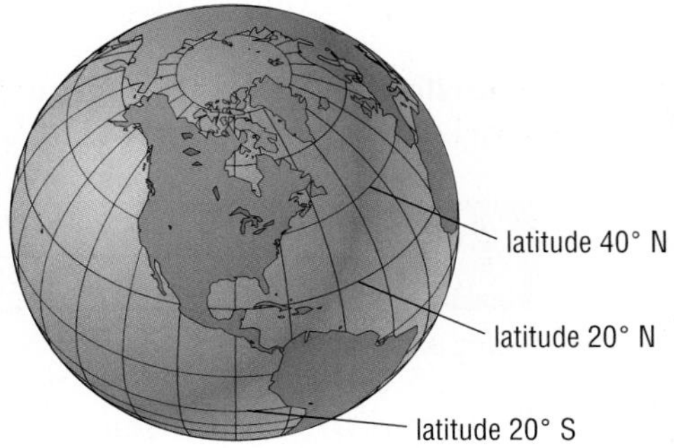

40. **MAKE A CONJECTURE** What do you think is the relationship between a city's latitude and its January temperature?

41. **RESEARCH** Use the Internet or other reference to find the latitude of 15 cities in the northern hemisphere and the corresponding January mean temperatures.

42. Make a scatter plot and draw a line of fit for the data.

43. Write an equation for the line of fit.

44. **MAKE A CONJECTURE** Find the latitude of your city and use the equation to predict its mean January temperature. Check your prediction by using another source such as the newspaper.

Maintain Your Skills

Mixed Review

Write the slope-intercept form of an equation for the line that satisfies each condition. *(Lesson 5-6)*

45. parallel to the graph of $y = -4x + 5$ and passes through $(-2, 5)$

46. perpendicular to the graph of $y = 2x + 3$ and passes through $(0, 0)$

Write the point-slope form of an equation for a line that passes through each point with the given slope. *(Lesson 5-5)*

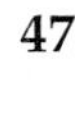

47.
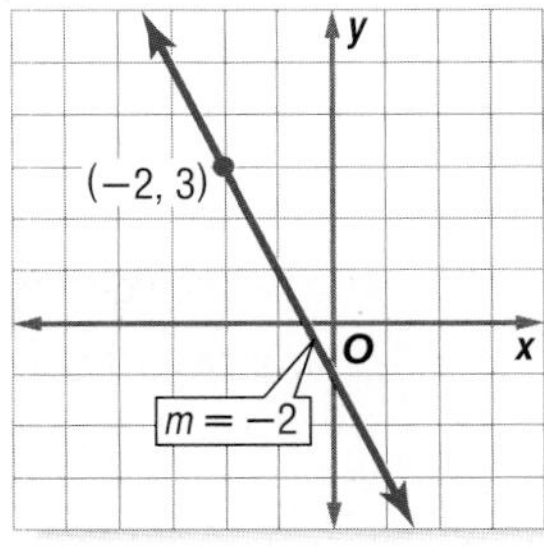

48.
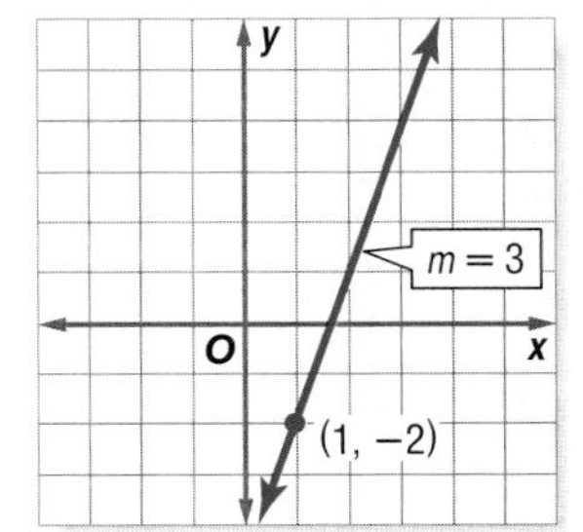

49.
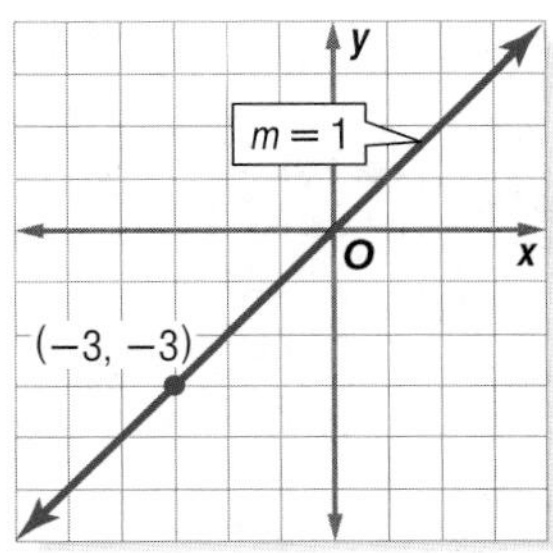

Find the x- and y-intercepts of the graph of each equation. *(Lesson 4-5)*

50. $3x + 4y = 12$
51. $2x - 5y = 8$
52. $y = 3x + 6$

Solve each equation. Then check your solution. *(Lesson 3-4)*

53. $\frac{r + 7}{-4} = \frac{r + 2}{6}$
54. $\frac{n - (-4)}{-3} = 7$
55. $\frac{2x - 1}{5} = \frac{4x - 5}{7}$

Graphing Calculator

A Follow-Up of Lesson 5-7

Regression and Median-Fit Lines

One type of equation of best-fit you can find is a linear **regression equation**. Linear regression is sometimes called the *method of least squares*.

EARNINGS The table shows the average hourly earnings of U.S. production workers for selected years.

Year	1960	1965	1970	1975	1980	1985	1990	1995	1999
Earnings	$2.09	2.46	3.23	4.53	6.66	8.57	10.01	11.43	13.24

Source: Bureau of Labor Statistics

Find and graph a linear regression equation. Then predict the average hourly earnings in 2010.

Step 1 *Find a regression equation.*

- Enter the years in L1 and the earnings in L2.

  KEYSTROKES: *Review entering a list on page 204.*

- Find the regression equation by selecting LinReg(*ax*+*b*) on the STAT CALC menu.

  KEYSTROKES: STAT ▶ 4 ENTER

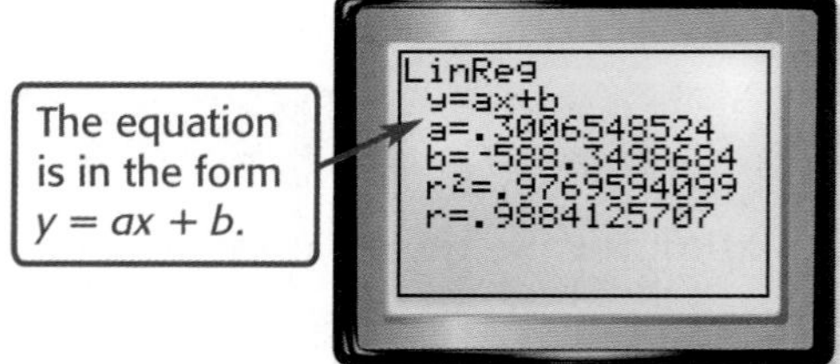

The equation is about $y = 0.30x - 588.35$.

r is the **linear correlation coefficient**. The closer the absolute value of r is to 1, the better the equation models the data. Because the r value is close to 1, the model fits the data well.

Step 2 *Graph the regression equation.*

- Use STAT PLOT to graph the scatter plot.

  KEYSTROKES: *Review statistical plots on page 204.*

- Copy the equation to the Y= list and graph.

  KEYSTROKES: Y= VARS 5 ▶ ▶ 1 GRAPH

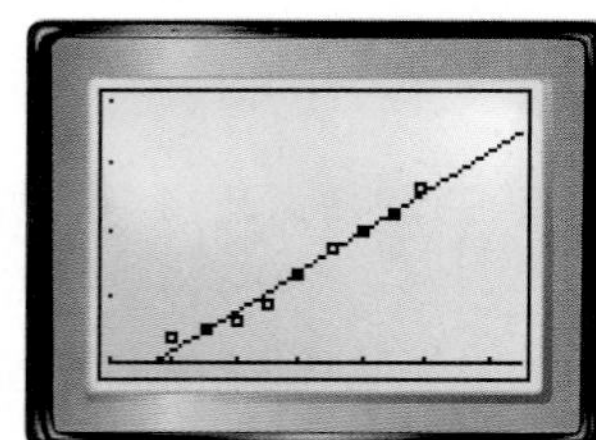

[1950, 2020] scl: 10 by [0, 20] scl: 5

The *residual* is the difference between actual and predicted data. The predicted earnings in 1970 using this model were $3.94. (To calculate, press 2nd [CALC] 1 1970 ENTER.) So the residual for 1970 was $3.94 – $3.23 or $0.71.

Step 3 *Predict using the regression equation.*

- Find y when $x = 2010$ using value on the CALC menu.

  KEYSTROKES: 2nd [CALC] 1 2010 ENTER

According to the regression equation, the average hourly earnings in 2010 will be about $15.97.

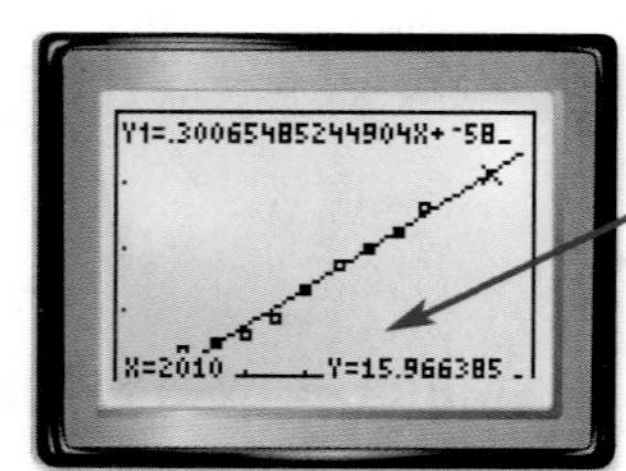

 www.algebra1.com/other_calculator_keystrokes

Investigation

A second type of best-fit line that can be found using a graphing calculator is a **median-fit line**. The equation of a median-fit line is calculated using the medians of the coordinates of the data points.

Find and graph a median-fit equation for the data on hourly earnings. Then predict the average hourly earnings in 2010. Compare this prediction to the one made using the regression equation.

Step 1 *Find a median-fit equation.*

- The data are already in Lists 1 and 2. Find the median-fit equation by using Med-Med on the STAT CALC menu.

KEYSTROKES: STAT ▶ 3 ENTER

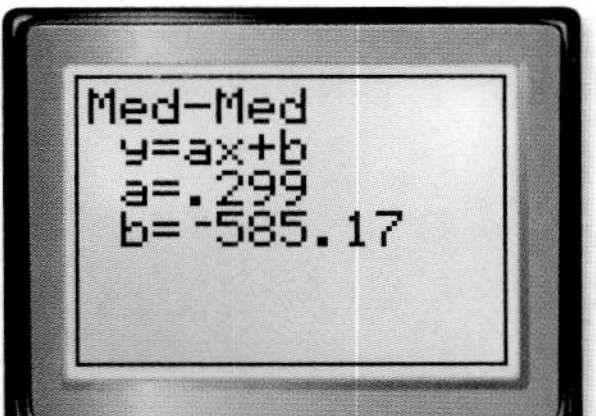

The median-fit equation is $y = 0.299x - 585.17$.

Step 2 *Graph the median-fit equation.*

- Copy the equation to the Y= list and graph.

KEYSTROKES: Y= CLEAR VARS 5 ▶ ▶ 1 GRAPH

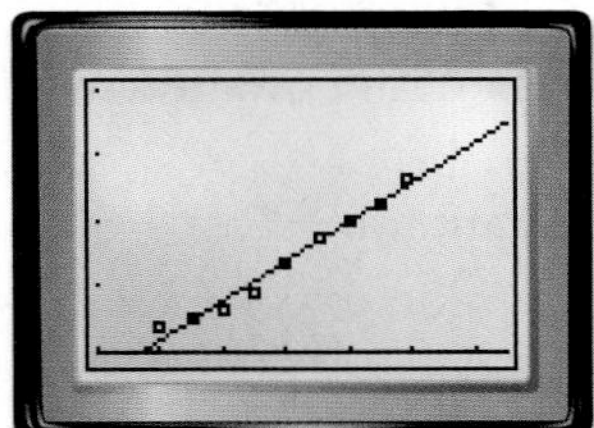

[1950, 2010] scl: 10 by [0, 20] scl: 5

Step 3 *Predict using the median-fit equation.*

KEYSTROKES: 2nd [CALC] 1 2010 ENTER

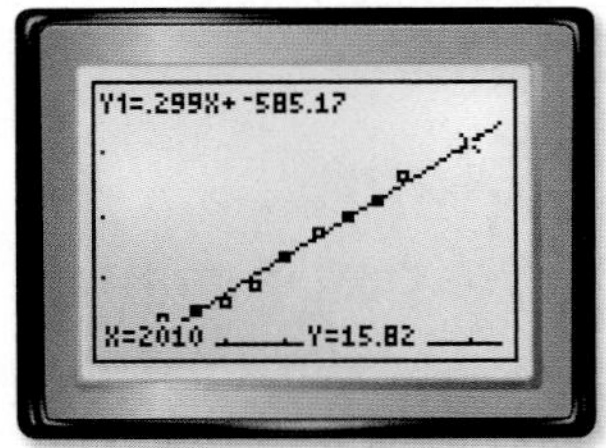

According to the median-fit equation, the average hourly earnings in 2010 will be about $15.82. This is slightly less than the predicted value found using the regression equation.

Exercises

Refer to the data on bald eagles in Example 2 on pages 300 and 301.

1. Find regression and median-fit equations for the data.
2. What is the correlation coefficient of the regression equation? What does it tell you about the data?
3. Use the regression and median-fit equations to predict the number of bald eagle pairs in 1998. Compare these to the number found in Example 3 on page 301.

For Exercises 4 and 5, use the table that shows the number of votes cast for the Democratic presidential candidate in selected North Carolina counties in the 1996 and 2000 elections.

4. Find regression and median-fit equations for the data.
5. In 1996, New Hanover County had 22,839 votes for the Democratic candidate. Use the regression and median-fit equations to estimate the number of votes for the Democratic candidate in that county in 2000. How do the predictions compare to the actual number of 29,292?

1996	2000
14,447	16,284
19,458	19,281
28,674	30,921
31,658	38,545
32,739	38,626
46,543	52,457
49,186	53,907
69,208	80,787
103,429	126,911
103,574	123,466

Source: NC State Board of Elections

Chapter 5

Study Guide and Review

Vocabulary and Concept Check

best-fit line (p. 300)
constant of variation (p. 264)
direct variation (p. 264)
family of graphs (p. 265)
linear extrapolation (p. 283)
linear interpolation (p. 301)
line of fit (p. 300)
negative correlation (p. 298)
parallel lines (p. 292)
parent graph (p. 265)
perpendicular lines (p. 293)
point-slope form (p. 286)
positive correlation (p. 298)
rate of change (p. 258)
scatter plot (p. 298)
slope (p. 256)
slope-intercept form (p. 272)

Exercises **Choose the correct term to complete each sentence.**

1. An equation of the form $y = kx$, where $k \neq 0$, describes a *(direct variation, linear extrapolation)*.
2. The ratio of *(rise, run)*, or vertical change, to the *(rise, run)*, or horizontal change, as you move from one point on a line to another, is the slope of a nonvertical line.
3. The lines with equations $y = -2x + 7$ and $y = -2x - 6$ are *(parallel, perpendicular)* lines.
4. The equation $y - 2 = -3(x - 1)$ is written in *(point-slope, slope-intercept)* form.
5. The equation $y = -\frac{1}{3}x + 6$ is written in *(slope-intercept, standard)* form.
6. The *(x-intercept, y-intercept)* of the equation $-x - 4y = 2$ is $-\frac{1}{2}$.

Lesson-by-Lesson Review

5-1 *Slope*

See pages 256–262.

Concept Summary

- The slope of a nonvertical line is the ratio of the rise to the run.
- $m = \dfrac{y_2 - y_1}{x_2 - x_1}$

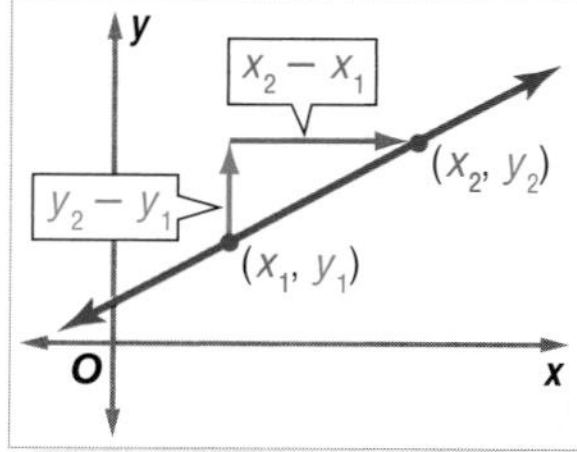

Example **Determine the slope of the line that passes through (0, −4) and (3, 2).**

Let $(0, -4) = (x_1, y_1)$ and $(3, 2) = (x_2, y_2)$.

$m = \dfrac{y_2 - y_1}{x_2 - x_1}$ Slope formula

$m = \dfrac{2 - (-4)}{3 - 0}$ $x_1 = 0, x_2 = 3, y_1 = -4, y_2 = 2$

$m = \dfrac{6}{3}$ or 2 Simplify.

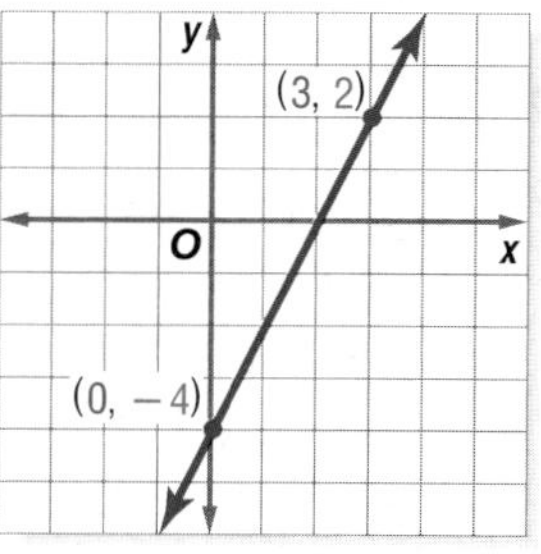

Exercises **Find the slope of the line that passes through each pair of points.**
See Examples 1–4 on page 257.

7. (1, 3), (−2, −6)
8. (0, 5), (6, 2)
9. (−6, 4), (−6, −2)
10. (8, −3), (−2, −3)
11. (2.9, 4.7), (0.5, 1.1)
12. $\left(\frac{1}{2}, 1\right), \left(-1, \frac{2}{3}\right)$

5-2 Slope and Direct Variation

See pages 264–270.

Concept Summary

- A direct variation is described by an equation of the form $y = kx$, where $k \neq 0$.
- In $y = kx$, k is the constant of variation. It is also the slope of the related graph.

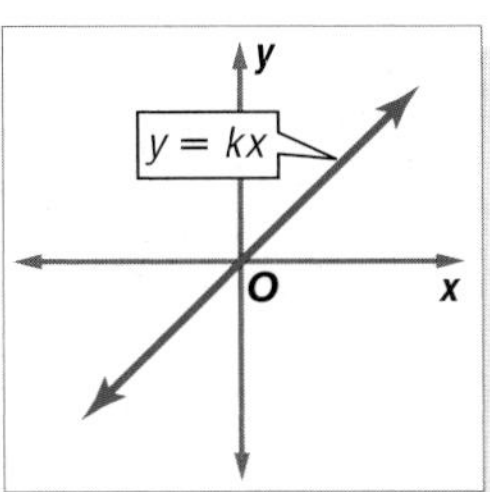

Example **Suppose y varies directly as x, and $y = -24$ when $x = 8$. Write a direct variation equation that relates x and y.**

$y = kx$	Direct variation equation
$-24 = k(8)$	Replace y with -24 and x with 8.
$\frac{-24}{8} = \frac{k(8)}{8}$	Divide each side by 8.
$-3 = k$	Simplify.

Therefore, $y = -3x$.

Exercises **Graph each equation.** *See Examples 2 and 3 on page 265.*

13. $y = 2x$ **14.** $y = -4x$ **15.** $y = \frac{1}{3}x$

16. $y = -\frac{1}{4}x$ **17.** $y = \frac{3}{2}x$ **18.** $y = -\frac{4}{3}x$

Suppose y varies directly as x. Write a direct variation equation that relates x and y. *See Example 4 on page 266.*

19. $y = -6$ when $x = 9$ **20.** $y = 15$ when $x = 2$ **21.** $y = 4$ when $x = -4$

22. $y = -6$ when $x = -18$ **23.** $y = -10$ when $x = 5$ **24.** $y = 7$ when $x = -14$

5-3 Slope-Intercept Form

See pages 272–277.

Concept Summary

- The linear equation $y = mx + b$ is written in slope-intercept form, where m is the slope, and b is the y-intercept.
- Slope-intercept form allows you to graph an equation quickly.

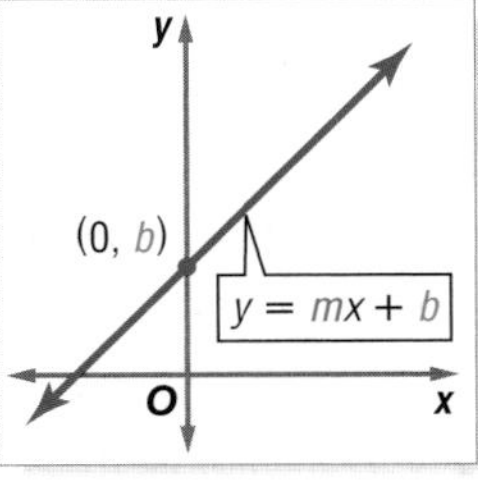

Example **Graph $-3x + y = -1$.**

$-3x + y = -1$	Original equation
$-3x + y + 3x = -1 + 3x$	Add $3x$ to each side.
$y = 3x - 1$	Simplify.

Step 1 The y-intercept is -1. So, graph $(0, -1)$.

Step 2 The slope is 3 or $\frac{3}{1}$. From $(0, -1)$, move up 3 units and right 1 unit. Then draw a line.

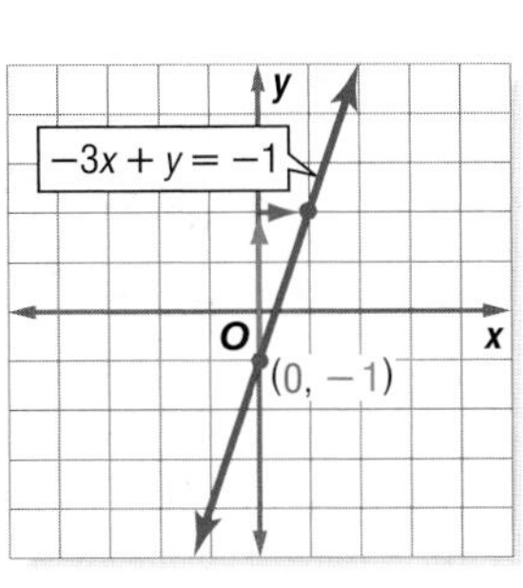

Exercises **Write an equation of the line with the given slope and y-intercept.**
See Examples 1 and 2 on pages 272 and 273.

25. slope: 3, y-intercept: 2
26. slope: 1, y-intercept: -3
27. slope: 0, y-intercept: 4
28. slope: $\frac{1}{3}$, y-intercept: 2
29. slope: 0.5, y-intercept: -0.3
30. slope: -1.3, y-intercept: 0.4

Graph each equation. *See Examples 3 and 4 on pages 273 and 274.*

31. $y = 2x + 1$
32. $y = -x + 5$
33. $y = \frac{1}{2}x + 3$
34. $y = -\frac{4}{3}x - 1$
35. $5x - 3y = -3$
36. $6x + 2y = 9$

5-4 Writing Equations in Slope-Intercept Form

See pages 280–285.

Concept Summary

- To write an equation given the slope and one point, substitute the values of m, x, and y into the slope-intercept form and solve for b. Then, write the slope-intercept form using the values of m and b.
- To write an equation given two points, find the slope. Then follow the steps above.

Example **Write an equation of a line that passes through $(-2, -3)$ with slope $\frac{1}{2}$.**

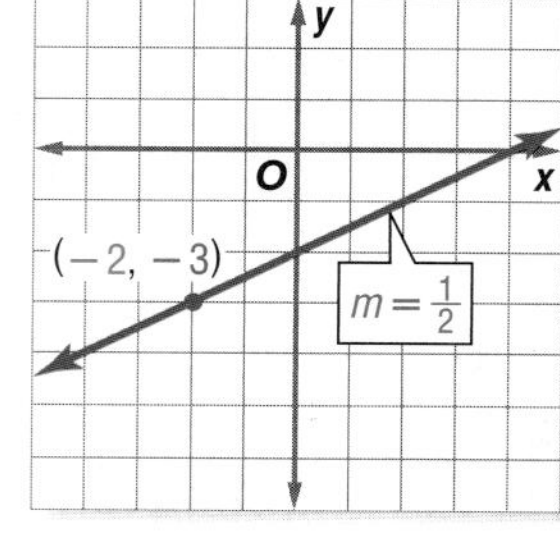

$y = mx + b$	Slope-intercept form
$-3 = \frac{1}{2}(-2) + b$	Replace m with $\frac{1}{2}$, y with -3, and x with -2.
$-3 = -1 + b$	Multiply.
$-3 + 1 = -1 + b + 1$	Add 1 to each side.
$-2 = b$	Simplify.

Therefore, the equation is $y = \frac{1}{2}x - 2$.

Exercises **Write an equation of the line that satisfies each condition.**
See Examples 1 and 2 on pages 280 and 281.

37. passes through $(-3, 3)$ with slope 1
38. passes through $(0, 6)$ with slope -2
39. passes through $(1, 6)$ with slope $\frac{1}{2}$
40. passes through $(4, -3)$ with slope $-\frac{3}{5}$
41. passes through $(-4, 2)$ and $(1, 12)$
42. passes through $(5, 0)$ and $(4, 5)$
43. passes through $(8, -1)$ with slope 0
44. passes through $(4, 6)$ and has slope 0

5-5 Writing Equations in Point-Slope Form

See pages 286–291.

Concept Summary

- The linear equation $y - y_1 = m(x - x_1)$ is written in point-slope form, where (x_1, y_1) is a given point on a nonvertical line and m is the slope.

Example **Write the point-slope form of an equation for a line that passes through (−2, 5) with slope 3.**

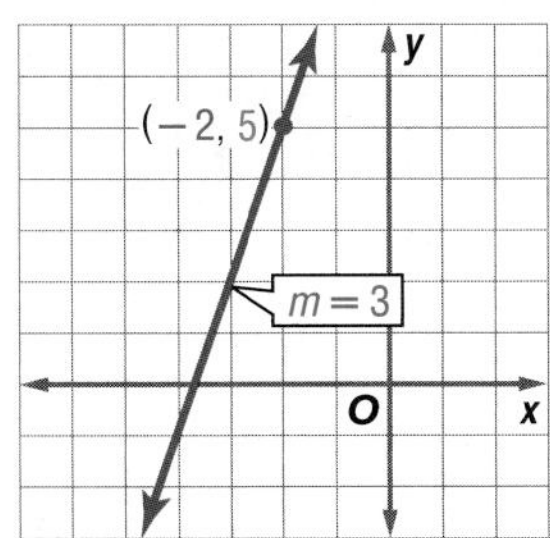

$y - y_1 = m(x - x_1)$ Use the point-slope form.

$y - 5 = 3[x - (-2)]$ $(x_1, y_1) = (-2, 5)$

$y - 5 = 3(x + 2)$ Subtract.

Exercises **Write the point-slope form of an equation for a line that passes through each point with the given slope.** *See Example 2 on page 287.*

45. $(4, 6), m = 5$ **46.** $(-1, 4), m = -2$ **47.** $(5, -3), m = \frac{1}{2}$

48. $(1, -4), m = -\frac{5}{2}$ **49.** $\left(\frac{1}{4}, -2\right), m = 3$ **50.** $(4, -2), m = 0$

Write each equation in standard form. *See Example 3 on page 287.*

51. $y - 1 = 2(x + 1)$ **52.** $y + 6 = \frac{1}{3}(x - 9)$ **53.** $y + 4 = 1.5(x - 4)$

5-6 Geometry: Parallel and Perpendicular Lines

See pages 292–297.

Concept Summary

- Two nonvertical lines are parallel if they have the same slope.

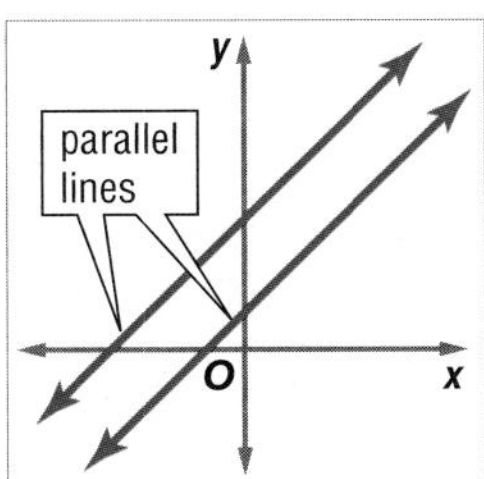

- Two nonvertical lines are perpendicular if the product of their slopes is −1.

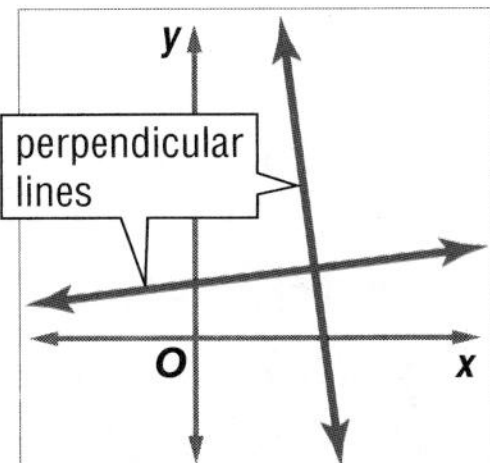

Example **Write the slope-intercept form for an equation of the line that passes through (5, −2) and is parallel to $y = 2x + 7$.**

The line parallel to $y = 2x + 7$ has the same slope, 2.

$y - y_1 = m(x - x_1)$ Point-slope form

$y - (-2) = 2(x - 5)$ Replace m with 2, y with −2, and x with 5.

$y + 2 = 2x - 10$ Simplify.

$y = 2x - 12$ Subtract 2 from each side.

• Extra Practice, see pages 831–833.
• Mixed Problem Solving, see page 857.

Exercises **Write the slope-intercept form for an equation of the line parallel to the graph of the given equation and passing through the given point.** *See Example 1 on page 292.*

54. $y = 3x - 2$, (4, 6) **55.** $y = -2x + 4$, (6, −6) **56.** $y = -6x - 1$, (1, 2)
57. $y = \frac{5}{12}x + 2$, (0, 4) **58.** $4x - y = 7$, (2, −1) **59.** $3x + 9y = 1$, (3, 0)

Write the slope-intercept form for an equation of the line perpendicular to the graph of the given equation and passing through the given point. *See Example 3 on page 294.*

60. $y = 4x + 2$, (1, 3) **61.** $y = -2x - 7$, (0, −3) **62.** $y = 0.4x + 1$, (2, −5)
63. $2x - 7y = 1$, (−4, 0) **64.** $8x - 3y = 7$, (4, 5) **65.** $5y = -x + 1$, (2, −5)

5-7 Statistics: Scatter Plots and Lines of Fit

See pages 298–305.

Concept Summary

- If y increases as x increases, then there is a positive correlation between x and y.
- If y decreases as x increases, then there is a negative correlation between x and y.
- If there is no relationship between x and y, then there is no correlation between x and y.
- A line of fit describes the trend of the data.
- You can use the equation of a line of fit to make predictions about the data.

Positive Correlation

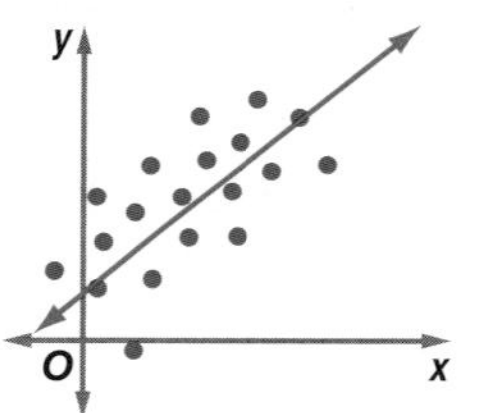

Negative Correlation

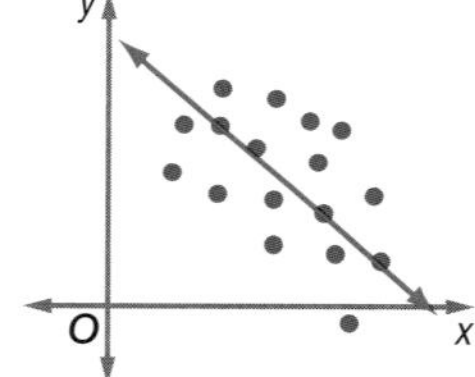

No Correlation

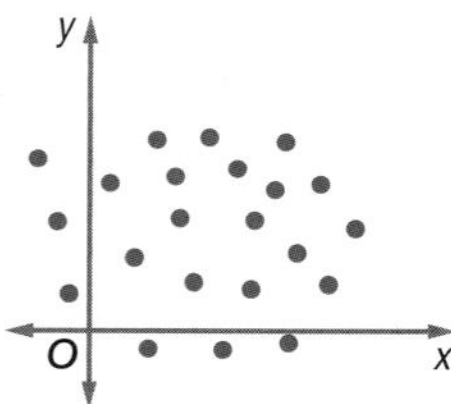

Exercises **For Exercises 66–70, use the table that shows the length and weight of several humpback whales.** *See Examples 2 and 3 on pages 300 and 301.*

Length (ft)	40	42	45	46	50	52	55
Weight (long tons)	25	29	34	35	43	45	51

66. Draw a scatter plot with length on the x-axis and weight on the y-axis.

67. Draw a line of fit for the data.

68. Write the slope-intercept form of an equation for the line of fit.

69. Predict the weight of a 48-foot humpback whale.

70. Most newborn humpback whales are about 12 feet in length. Use the equation of the line of fit to predict the weight of a newborn humpback whale. Do you think your prediction is accurate? Explain.

Chapter 5

Practice Test

Vocabulary and Concepts

1. **Explain** why the equation of a vertical line cannot be in slope-intercept form.
2. **Draw** a scatter plot that shows a positive correlation.
3. **Name** the part of the slope-intercept form that represents the rate of change.

Skills and Applications

Find the slope of the line that passes through each pair of points.

4. $(5, 8), (-3, 7)$
5. $(5, -2), (3, -2)$
6. $(6, -3), (6, 4)$

7. **BUSINESS** A web design company advertises that it will design and maintain a website for your business for \$9.95 per month. Write a direct variation equation to find the total cost C for any number of months m.

Graph each equation.

8. $y = 3x - 1$
9. $y = 2x + 3$
10. $2x + 3y = 9$

11. **WEATHER** The temperature is 16°F at midnight and is expected to fall 2° each hour during the night. Write the slope-intercept form of an equation to find the temperature T for any hour h after midnight.

Suppose y varies directly as x. Write a direct variation equation that relates x and y.

12. $y = 6$ when $x = 9$
13. $y = -12$ when $x = 4$
14. $y = -8$ when $x = 8$

Write the slope-intercept form of an equation of the line that satisfies each condition.

15. has slope -4 and y-intercept 3
16. passes through $(-2, -5)$ and $(8, -3)$
17. parallel to $3x + 7y = 4$ and passes through $(5, -2)$
18. a horizontal line passing through $(5, -8)$
19. perpendicular to the graph of $5x - 3y = 9$ and passes through the origin

20. Write the point-slope form of an equation for a line that passes through $(-4, 3)$ with slope -2.

ANIMALS For Exercises 21–24, use the table that shows the relationship between dog years and human years.

Dog Years	1	2	3	4	5	6	7
Human Years	15	24	28	32	37	42	47

21. Draw a scatter plot and determine what relationship, if any, exists in the data.
22. Draw a line of fit for the scatter plot.
23. Write the slope-intercept form of an equation for the line of fit.
24. Determine how many human years are comparable to 13 dog years.

25. **STANDARDIZED TEST PRACTICE** A line passes through $(0, 4)$ and $(3, 0)$. Which equation does *not* represent the equation of this line?

(A) $y - 4 = -\frac{4}{3}(x - 0)$
(B) $y = -\frac{4}{3}x + 3$
(C) $\frac{x}{3} + \frac{y}{4} = 1$
(D) $y - 0 = -\frac{4}{3}(x - 3)$
(E) $4x + 3y = 12$

Chapter 5

Standardized Test Practice

Part 1 Multiple Choice

Record your answers on the answer sheet provided by your teacher or on a sheet of paper.

1. If a person's weekly salary is $\$x$ and she saves $\$y$, what fraction of her weekly salary does she spend? (Lesson 1-1)

(A) $\frac{x}{y}$ (B) $\frac{x-y}{x}$

(C) $\frac{x-y}{y}$ (D) $\frac{y-x}{x}$

2. Evaluate $-2x + 7y$ if $x = -5$ and $y = 4$. (Lesson 2-6)

(A) 38 (B) 43

(C) 227 (D) 243

3. Find x, if $5x + 6 = 10$. (Lesson 3-3)

(A) $-\frac{5}{4}$ (B) $\frac{1}{10}$

(C) $\frac{5}{16}$ (D) $\frac{4}{5}$

4. According to the data in the table, which of the following statements is true? (Lesson 3-7)

Age	Frequency
8	1
10	3
14	2
16	1
17	2

(A) mean age = median age

(B) mean age > median age

(C) mean age < median age

(D) median age < mode age

5. What relationship exists between the x- and y-coordinates of each of the data points shown in the table? (Lesson 4-8)

x	y
−3	4
−2	3
0	1
1	0
3	−2
5	−4

(A) x and y are opposites.

(B) The sum of x and y is 2.

(C) The y-coordinate is 1 more than the square of the x-coordinate.

(D) The y-coordinate is 1 more than the opposite of the x-coordinate.

6. What is the y-intercept of the line with equation $\frac{x}{3} - \frac{y}{2} = 1$? (Lesson 4-5)

(A) -3 (B) -2

(C) $\frac{2}{3}$ (D) $\frac{3}{2}$

7. Find the slope of a line that passes through (2, 4) and (−4, 7). (Lesson 5-1)

(A) $-\frac{1}{2}$ (B) $\frac{1}{2}$

(C) -2 (D) 2

8. Which equation represents the line that passes through (3, 7) and (10, 21)? (Lesson 5-4)

(A) $x + y = 10$ (B) $y = \frac{1}{2}x + \frac{11}{2}$

(C) $y = 2x + 1$ (D) $y = 3x - 2$

9. Choose the equation of a line parallel to the graph of $y = 3x + 4$. (Lesson 5-6)

(A) $y = -\frac{1}{3}x + 4$ (B) $y = -3x + 4$

(C) $y = -x + 1$ (D) $y = 3x + 5$

Part 2 Short Response/Grid In

Record your answers on the answer sheet provided by your teacher or on a sheet of paper.

10. While playing a game with her friends, Ellen scored 12 points less than twice the lowest score. She scored 98. What was the lowest score in the game? (Lesson 3-4)

11. The graph of $3x + 2y = 3$ is shown at the right. What is the y-intercept? (Lesson 5-3)

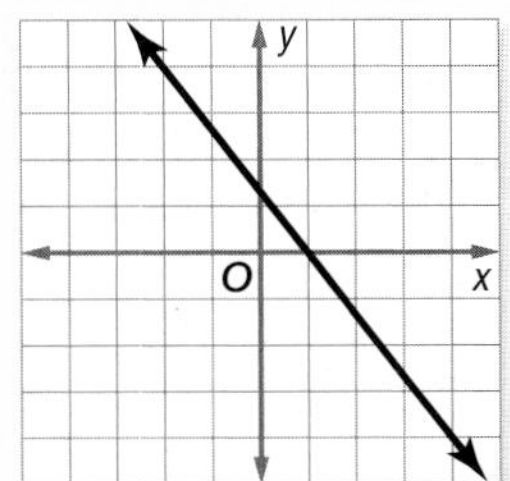

12. The table of ordered pairs shows the coordinates of some of the points on the graph of a function.

x	y
−1	6
0	4
1	2
2	0
3	−2

What is the y-coordinate of a point $(5, y)$ that lies on the graph of the function? (Lesson 5-4)

13. The equation $y - 3 = -2(x + 5)$ is written in point-slope form. What is the slope of the line? (Lesson 5-5)

Part 3 Quantitative Comparison

Compare the quantity in Column A and the quantity in Column B. Then determine whether:

Ⓐ **the quantity in Column A is greater,**

Ⓑ **the quantity in Column B is greater,**

Ⓒ **the two quantities are equal, or**

Ⓓ **the relationship cannot be determined from the information given.**

	Column A	Column B
14.	$2(x + 6)$	$2x + 6$

(Lesson 1-2)

	Column A	Column B
15.	$\lvert x \rvert$	$\lvert x + 1 \rvert$

(Lesson 2-1)

	Column A	Column B
16.	the slope of any nonvertical line	the slope of a line parallel to the line in Column A

(Lesson 5-6)

	Column A	Column B
17.	the slope of $y = -2x$	the slope of the line perpendicular to $y = -2x$

(Lesson 5-6)

Part 4 Open Ended

Record your answers on a sheet of paper. Show your work.

18. A friend wants to enroll for cellular phone service. Three different plans are available. (Lesson 5-3)

Plan 1 charges \$0.59 per minute.

Plan 2 charges a monthly fee of \$10, plus \$0.39 per minute.

Plan 3 charges a monthly fee of \$59.95.

a. For each plan, write an equation that represents the monthly cost C for m number of minutes per month.

b. Graph each of the three equations.

c. Your friend expects to use 100 minutes per month. In which plan do you think that your friend should enroll? Explain.

The Princeton Review Test-Taking Tip

Questions 14–17 Before you choose answer A, B, or C on quantitative comparison questions, ask yourself: "Is this always the case?" If not, mark D.

 www.algebra1.com/standardized_test

Chapter 6

Solving Linear Inequalities

What You'll Learn

- **Lessons 6-1 through 6-3** Solve linear inequalities.
- **Lesson 6-4** Solve compound inequalities and graph their solution sets.
- **Lesson 6-5** Solve absolute value equations and inequalities.
- **Lesson 6-6** Graph inequalities in the coordinate plane.

Key Vocabulary

- **set-builder notation** (p. 319)
- **compound inequality** (p. 339)
- **intersection** (p. 339)
- **union** (p. 340)
- **half-plane** (p. 353)

Why It's Important

Inequalities are used to represent various real-world situations in which a quantity must fall within a range of possible values. For example, figure skaters and gymnasts frequently want to know what they need to score to win a competition. That score can be represented by an inequality. *You will learn how a competitor can determine what score is needed to win in Lesson 6-1.*

Getting Started

Prerequisite Skills To be successful in this chapter, you'll need to master these skills and be able to apply them in problem-solving situations. Review these skills before beginning Chapter 6.

For Lessons 6-1 and 6-3 — Solve Equations

Solve each equation. *(For review, see Lessons 3-2, 3-4, and 3-5.)*

1. $t + 31 = 84$ **2.** $b - 17 = 23$ **3.** $18 = 27 + f$ **4.** $d - \frac{2}{3} = \frac{1}{2}$

5. $3r - 45 = 4r$ **6.** $5m + 7 = 4m - 12$ **7.** $3y + 4 = 16$ **8.** $2a + 5 - 3a = 4$

9. $\frac{1}{2}k - 4 = 7$ **10.** $4.3b + 1.8 = 8.25$ **11.** $6s - 12 = 2(s + 2)$ **12.** $n - 3 = \frac{n+1}{2}$

For Lesson 6-5 — Evaluate Absolute Values

Find each value. *(For review, see Lesson 2-1.)*

13. $|-8|$ **14.** $|20|$ **15.** $|-30|$ **16.** $|-1.5|$

17. $|14 - 7|$ **18.** $|1 - 16|$ **19.** $|2 - 3|$ **20.** $|7 - 10|$

For Lesson 6-6 — Graph Equations with Two Variables

Graph each equation. *(For review, see Lesson 4-5.)*

21. $2x + 2y = 6$ **22.** $x - 3y = -3$ **23.** $y = 2x - 3$ **24.** $y = -4$

25. $x = -\frac{1}{2}y$ **26.** $3x - 6 = 2y$ **27.** $15 = 3(x + y)$ **28.** $2 - x = 2y$

Make this Foldable to record information about solving linear inequalities. Begin with two sheets of notebook paper.

Step 1 Fold and Cut

Fold
in half along the
width. Cut along fold
from edges to
margin.

Step 2 Fold a New Paper and Cut

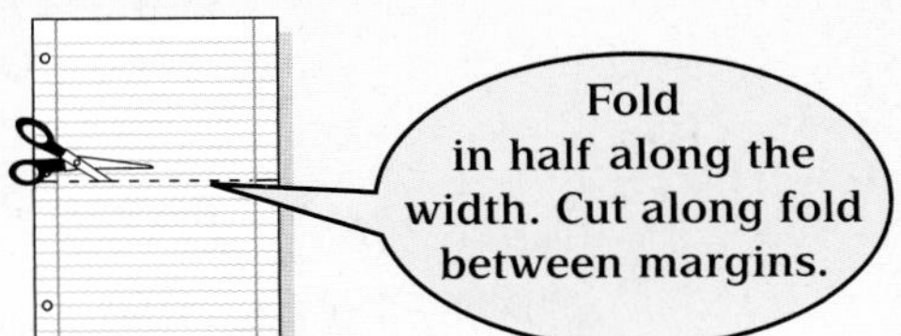

Fold
in half along the
width. Cut along fold
between margins.

Step 3 Fold

Insert first sheet
through second sheet
and align folds.

Step 4 Label

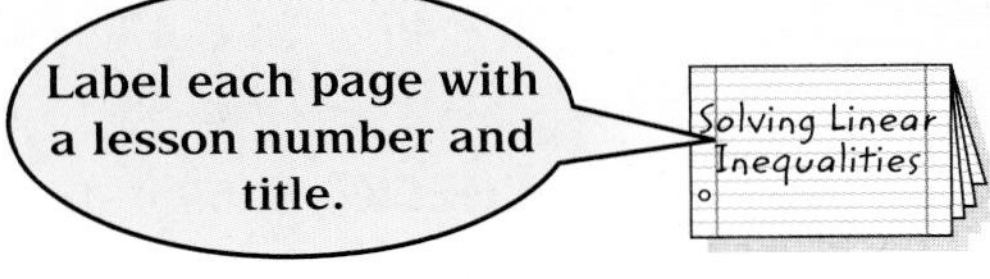

Label each page with
a lesson number and
title.
Solving Linear
Inequalities

Reading and Writing As you read and study the chapter, fill the journal with notes, diagrams, and examples of linear inequalities.

6-1 Solving Inequalities by Addition and Subtraction

What You'll Learn

- Solve linear inequalities by using addition.
- Solve linear inequalities by using subtraction.

Vocabulary

- set-builder notation

How are inequalities used to describe school sports?

In the 1999–2000 school year, more high schools offered girls' track and field than girls' volleyball.

$$14{,}587 > 13{,}426$$

If 20 schools added girls' track and field and 20 schools added girls' volleyball the next school year, there would still be more schools offering girls' track and field than schools offering girls' volleyball.

$$14{,}587 + 20 \underline{\quad ? \quad} 13{,}426 + 20$$

$$14{,}607 > 13{,}446$$

USA TODAY Snapshots®

USA TODAY
Girls gear up for high school sports
High school girls are playing sports in record numbers, almost 2.7 million in the 1999-2000 school year. Most popular girls sports by number of schools offering each program:

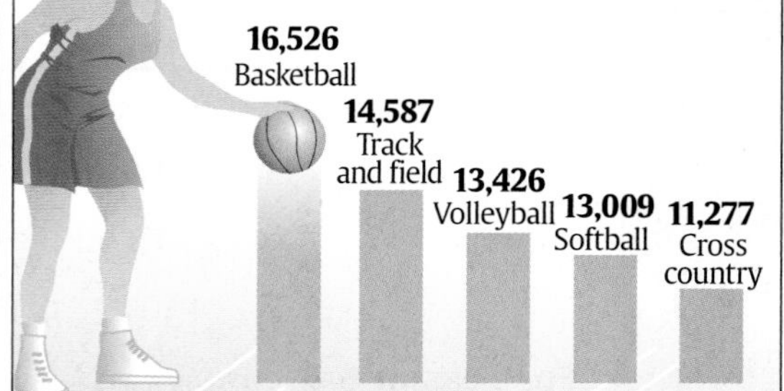

Source: National Federation of State High School Associations

By Ellen J. Horrow and Alejandro Gonzalez, USA TODAY

Study Tip

Look Back
To review **inequalities**, see Lesson 1-3.

SOLVE INEQUALITIES BY ADDITION Recall that statements with greater than ($>$), less than ($<$), greater than or equal to ($\geq$), or less than or equal to ($\leq$) are *inequalities*. The sports application illustrates the **Addition Property of Inequalities**.

Key Concept — *Addition Property of Inequalities*

- **Words** If any number is added to each side of a true inequality, the resulting inequality is also true.
- **Symbols** For all numbers a, b, and c, the following are true.
  1. If $a > b$, then $a + c > b + c$.
  2. If $a < b$, then $a + c < b + c$.
- **Example**
  $2 < 7$
  $2 + 6 < 7 + 6$
  $8 < 13$

This property is also true when $>$ and $<$ are replaced with $\geq$ and $\leq$.

Example 1 Solve by Adding

Solve $t - 45 \leq 13$. Then check your solution.

$t - 45 \leq 13$	Original inequality
$t - 45 + 45 \leq 13 + 45$	Add 45 to each side.
$t \leq 58$	This means all numbers less than or equal to 58.

CHECK Substitute 58, a number less than 58, and a number greater than 58.

Let $t = 58$.	Let $t = 50$.	Let $t = 60$.
$58 - 45 \overset{?}{\leq} 13$	$50 - 45 \overset{?}{\leq} 13$	$60 - 45 \overset{?}{\leq} 13$
$13 \leq 13$ ✓	$5 \leq 13$ ✓	$15 \not\leq 13$

The solution is the set {all numbers less than or equal to 58}.

Study Tip

Reading Math

$\{t|t \leq 58\}$ is read *the set of all numbers t such that t is less than or equal to 58.*

The solution of the inequality in Example 1 was expressed as a set. A more concise way of writing a solution set is to use **set-builder notation**. The solution in set-builder notation is $\{t \mid t \leq 58\}$.

The solution to Example 1 can also be represented on a number line.

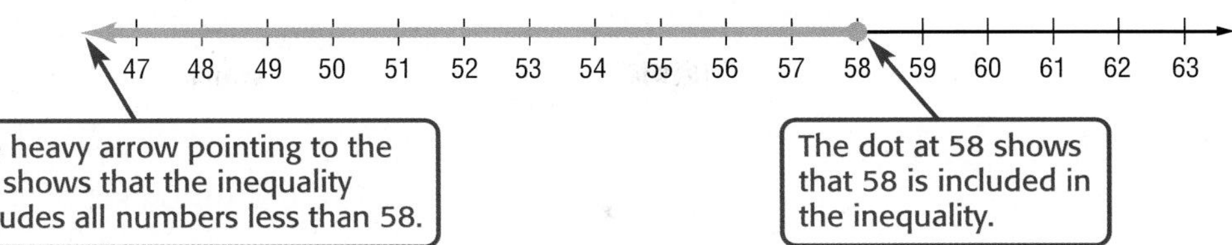

Example 2 Graph the Solution

Solve $7 < x - 4$. Then graph it on a number line.

$7 < x - 4$	Original inequality
$7 + 4 < x - 4 + 4$	Add 4 to each side.
$11 < x$	Simplify.

Since $11 < x$ is the same as $x > 11$, the solution set is $\{x \mid x > 11\}$.

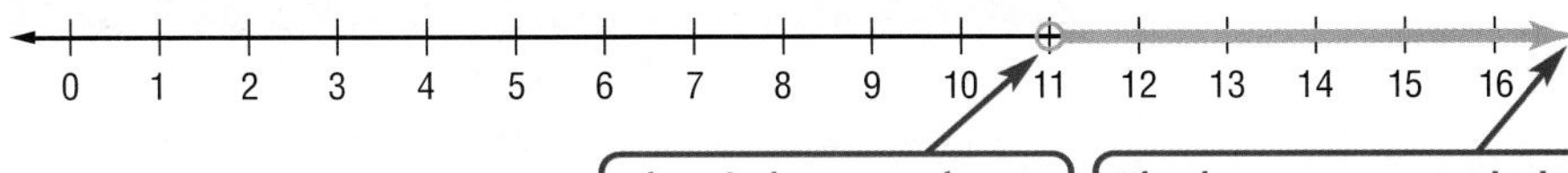

The circle at 11 shows that 11 is *not* included in the inequality.

The heavy arrow pointing to the right shows that the inequality includes all numbers greater than 11.

SOLVE INEQUALITIES BY SUBTRACTION Subtraction can also be used to solve inequalities.

Key Concept — Subtraction Property of Inequalities

- **Words** If any number is subtracted from each side of a true inequality, the resulting inequality is also true.
- **Symbols** For all numbers *a*, *b*, and *c*, the following are true.
  1. If $a > b$, then $a - c > b - c$.
  2. If $a < b$, then $a - c < b - c$.
- **Example** $17 > 8$, $17 - 5 > 8 - 5$, $12 > 3$

This property is also true when $>$ and $<$ are replaced with $\geq$ and $\leq$.

Example 3 Solve by Subtracting

Solve $19 + r \geq 16$. Then graph the solution.

$19 + r \geq 16$	Original inequality
$19 + r - 19 \geq 16 - 19$	Subtract 19 from each side.
$r \geq -3$	Simplify.

The solution set is $\{r \mid r \geq -3\}$.

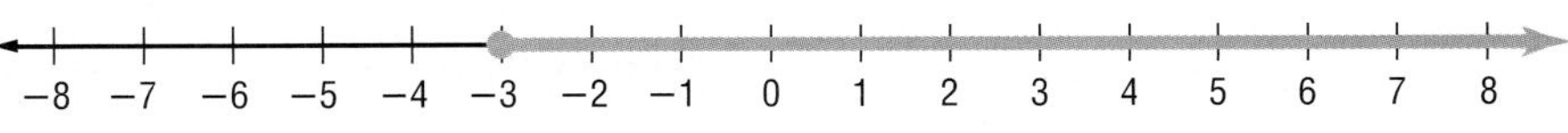

Terms with variables can also be subtracted from each side to solve inequalities.

Example 4 Variables on Both Sides

Solve $5p + 7 > 6p$. Then graph the solution.

$5p + 7 > 6p$	Original inequality
$5p + 7 - 5p > 6p - 5p$	Subtract $5p$ from each side.
$7 > p$	Simplify.

Since $7 > p$ is the same as $p < 7$, the solution set is $\{p \mid p < 7\}$.

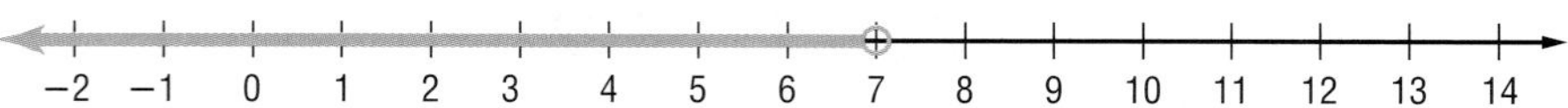

Verbal problems containing phrases like *greater than* or *less than* can often be solved by using inequalities. The following chart shows some other phrases that indicate inequalities.

Inequalities			
$<$	$>$	$\leq$	$\geq$
• less than • fewer than	• greater than • more than	• at most • no more than • less than or equal to	• at least • no less than • greater than or equal to

Example 5 Write and Solve an Inequality

Write an inequality for the sentence below. Then solve the inequality.

Four times a number is no more than three times that number plus eight.

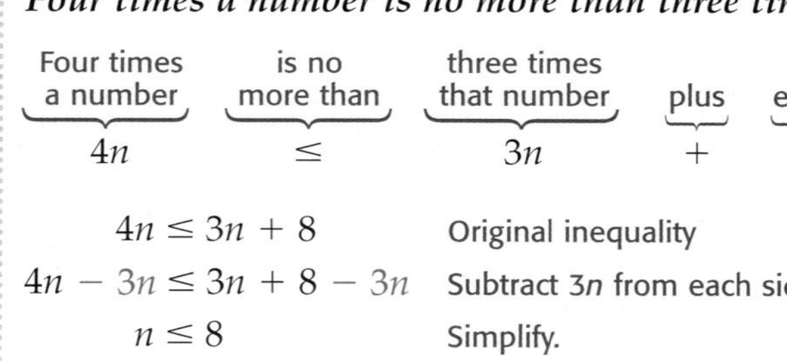

$4n \leq 3n + 8$	Original inequality
$4n - 3n \leq 3n + 8 - 3n$	Subtract $3n$ from each side.
$n \leq 8$	Simplify.

The solution set is $\{n \mid n \leq 8\}$.

More About. . .

Olympics

Yulia Barsukova of the Russian Federation won the gold medal in rhythmic gymnastics at the 2000 Summer Olympics in Sydney, and Yulia Raskina of Belarus won the silver medal.

Source: www.olympic.org

Example 6 Write an Inequality to Solve a Problem

OLYMPICS Yulia Raskina scored a total of 39.548 points in the four events of rhythmic gymnastics. Yulia Barsukova scored 9.883 in the rope competition, 9.900 in the hoop competition, and 9.916 in the ball competition. How many points did Barsukova need to score in the ribbon competition to surpass Raskina and win the gold medal?

Words Barsukova's total must be greater than Raskina's total.

Variable Let r = Barsukova's score in the ribbon competition.

	Barsukova's total	is greater than	Raskina's total.
Inequality	$9.883 + 9.900 + 9.916 + r$	$>$	39.548

Solve the inequality.

$9.883 + 9.900 + 9.916 + r > 39.548$	Original inequality
$29.699 + r > 39.548$	Simplify.
$29.699 + r - 29.699 > 39.548 - 29.699$	Subtract 29.699 from each side.
$r > 9.849$	Simplify.

Barsukova needed to score more than 9.849 points to win the gold medal.

Check for Understanding

Concept Check

1. **OPEN ENDED** List three inequalities that are equivalent to $y < -3$.
2. **Compare and contrast** the graphs of $a < 4$ and $a \leq 4$.
3. **Explain** what $\{b \mid b \geq -5\}$ means.

Guided Practice

4. Which graph represents the solution of $m + 3 > 7$?

a. 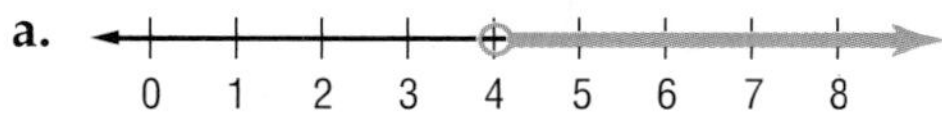

b. 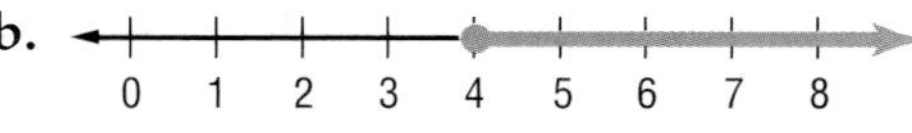

c. 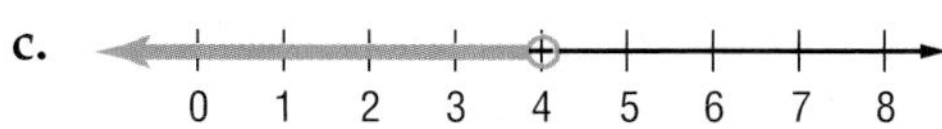

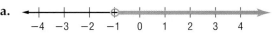

d. 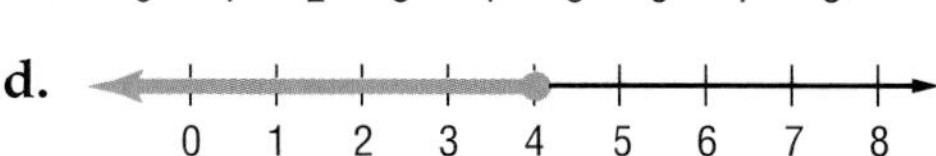

Solve each inequality. Then check your solution, and graph it on a number line.

5. $a + 4 < 2$
6. $9 \leq b + 4$
7. $t - 7 \geq 5$
8. $y - 2.5 > 3.1$
9. $5.2r + 6.7 \geq 6.2r$
10. $7p \leq 6p - 2$

Define a variable, write an inequality, and solve each problem. Then check your solution.

11. A number decreased by 8 is at most 14.
12. A number plus 7 is greater than 2.

Application

13. **HEALTH** Chapa's doctor recommended that she limit her fat intake to no more than 60 grams per day. This morning, she ate two breakfast bars with 3 grams of fat each. For lunch she ate pizza with 21 grams of fat. If she follows her doctor's advice, how many grams of fat can she have during the rest of the day?

Practice and Apply

Homework Help

For Exercises	See Examples
14–39	1–4
40–45	5
46–55	6

Extra Practice

See page 833.

Match each inequality with its corresponding graph.

14. $x - 3 \geq -2$
15. $x + 7 \leq 6$
16. $4x > 3x - 1$
17. $8 + x < 9$
18. $5 \leq x + 6$
19. $x - 1 > 0$

a. −4 −3 −2 −1 0 1 2 3 4

b. −4 −3 −2 −1 0 1 2 3 4

c. 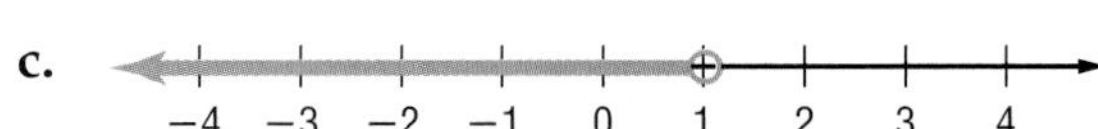

d. −4 −3 −2 −1 0 1 2 3 4

e. −4 −3 −2 −1 0 1 2 3 4

f. −4 −3 −2 −1 0 1 2 3 4

Solve each inequality. Then check your solution, and graph it on a number line.

20. $t + 14 \geq 18$	**21.** $d + 5 \leq 7$	**22.** $n - 7 < -3$
23. $s - 5 > -1$	**24.** $5 < 3 + g$	**25.** $4 > 8 + r$
26. $-3 \geq q - 7$	**27.** $2 \leq m - 1$	**28.** $2y > -8 + y$
29. $3f < -3 + 2f$	**30.** $3b \leq 2b - 5$	**31.** $4w \geq 3w + 1$
32. $v - (-4) > 3$	**33.** $a - (-2) \leq -3$	**34.** $-0.23 < h - (-0.13)$
35. $x + 1.7 \geq 2.3$	**36.** $a + \frac{1}{4} > \frac{1}{8}$	**37.** $p - \frac{2}{3} \leq \frac{4}{9}$

38. If $d + 5 \geq 17$, then complete each inequality.

a. $d \geq$ __?__ **b.** $d +$ __?__ ≥ 20 **c.** $d - 5 \geq$ __?__

39. If $z - 2 \leq 10$, then complete each inequality.

a. $z \leq$ __?__ **b.** $z -$ __?__ ≤ 5 **c.** $z + 4 \leq$ __?__

Define a variable, write an inequality, and solve each problem. Then check your solution.

40. The sum of a number and 13 is at least 27.

41. A number decreased by 5 is less than 33.

42. Thirty is no greater than the sum of a number and −8.

43. Twice a number is more than the sum of that number and 14.

44. The sum of two numbers is at most 18, and one of the numbers is −7.

45. Four times a number is less than or equal to the sum of three times the number and −2.

46. **BIOLOGY** Adult Nile crocodiles weigh up to 2200 pounds. If a young Nile crocodile weighs 157 pounds, how many pounds might it be expected to gain in its lifetime?

47. **ASTRONOMY** There are at least 200 billion stars in the Milky Way. If 1100 of these stars can be seen in a rural area without the aid of a telescope, how many stars in the galaxy cannot be seen in this way?

More About. . .

Biology

One common species of bees is the honeybee. A honeybee colony may have 60,000 to 80,000 bees.

Source: Penn State, Cooperative Extension Service

48. **BIOLOGY** There are 3500 species of bees and more than 600,000 species of insects. How many species of insects are not bees?

49. **BANKING** City Bank requires a minimum balance of \$1500 to maintain free checking services. If Mr. Hayashi knows he must write checks for \$1300 and \$947, how much money should he have in his account before writing the checks?

50. **GEOMETRY** The length of the base of the triangle at the right is less than the height of the triangle. What are the possible values of x?

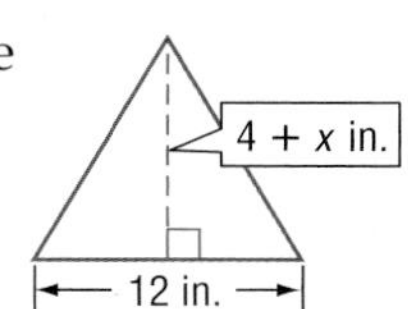

51. **SHOPPING** Terrell has \$65 to spend at the mall. He bought a T-shirt for \$18 and a belt for \$14. If Terrell still wants to buy a pair of jeans, how much can he spend on the jeans?

52. **SOCCER** The Centerville High School soccer team plays 18 games in the season. The team has a goal of winning at least 60% of its games. After the first three weeks of the season, the team has won 4 games. How many more games must the team win to meet their goal?

53. **CRITICAL THINKING** Determine whether each statement is *always*, *sometimes*, or *never* true.

a. If $a < b$ and $c < d$, then $a + c < b + d$.

b. If $a < b$ and $c < d$, then $a + c \geq b + d$.

c. If $a < b$ and $c < d$, then $a - c = b - d$.

HEALTH For Exercises 54 and 55, use the following information.
Hector's doctor told him that his cholesterol level should be below 200. Hector's cholesterol is 225.

54. Let p represent the number of points Hector should lower his cholesterol. Write an inequality with $225 - p$ on one side.

55. Solve the inequality.

56. **WRITING IN MATH** Answer the question that was posed at the beginning of the lesson.

How are inequalities used to describe school sports?

Include the following in your answer:

- an inequality describing the number of schools needed to add girls' track and field so that the number is greater than the number of schools currently participating in girls' basketball.

57. Which inequality is *not* equivalent to $x \leq 12$?

(A) $x - 7 \leq 5$ (B) $x + 4 \leq 16$ (C) $x - 1 \leq 13$ (D) $12 \geq x$

58. Which statement is modeled by $n + 6 \geq 5$?

(A) The sum of a number and six is at least five.
(B) The sum of a number and six is at most five.
(C) The sum of a number and six is greater than five.
(D) The sum of a number and six is no greater than five.

Maintain Your Skills

Mixed Review

59. Would a scatter plot for the relationship of a person's height to the person's grade on the last math test show a *positive*, *negative*, or *no correlation*? *(Lesson 5-7)*

Write an equation in slope-intercept form of the line that passes through the given point and is parallel to the graph of each equation. *(Lesson 5-6)*

60. $(1, -3)$; $y = 3x - 2$
61. $(0, 4)$; $x + y = -3$
62. $(-1, 2)$; $2x - y = 1$

Find the next two terms in each sequence. *(Lesson 4-8)*

63. 7, 13, 19, 25, …
64. 243, 81, 27, 9, …
65. 3, 6, 12, 24, …

Solve each equation if the domain is {−1, 3, 5}. *(Lesson 4-4)*

66. $y = -2x$
67. $y = 7 - x$
68. $2x - y = 6$

Getting Ready for the Next Lesson

PREREQUISITE SKILL Solve each equation.
*(For review of **multiplication and division equations**, see Lesson 3-3.)*

69. $6g = 42$
70. $\frac{t}{9} = 14$
71. $\frac{2}{3}y = 14$
72. $3m = 435$
73. $\frac{4}{7}x = 28$
74. $5.3g = 11.13$
75. $\frac{a}{3.5} = 7$
76. $8p = 35$

Algebra Activity

A Preview of Lesson 6-2

Solving Inequalities

You can use algebra tiles to solve inequalities.

Solve $-2x \geq 6$.

Step 1 *Model the inequality.*

Use a self-adhesive note to cover the equals sign on the equation mat. Then write a ≥ symbol on the note. Model the inequality.

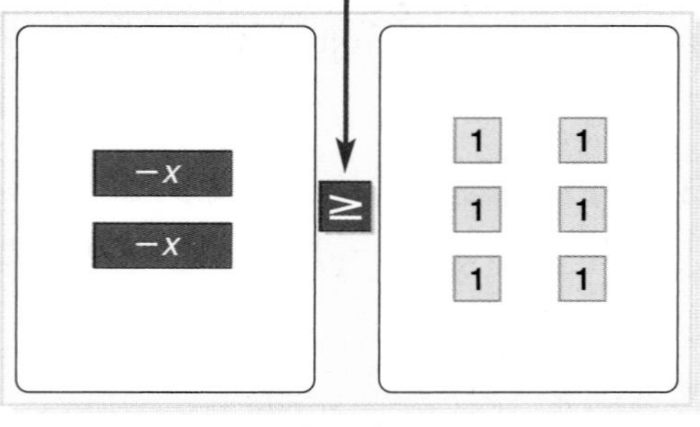

$-2x \geq 6$

Step 2 *Remove zero pairs.*

Since you do not want to solve for a negative x tile, eliminate the negative x tiles by adding 2 positive x tiles to each side. Remove the zero pairs.

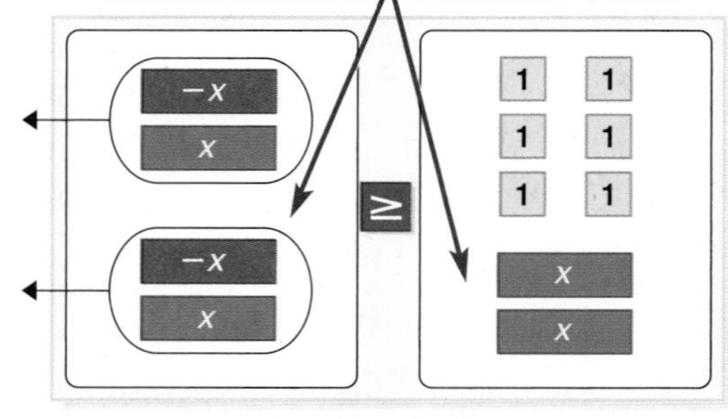

$-2x + 2x \geq 6 + 2x$

Step 3 *Remove zero pairs.*

Add 6 negative 1 tiles to each side to isolate the x tiles. Remove the zero pairs.

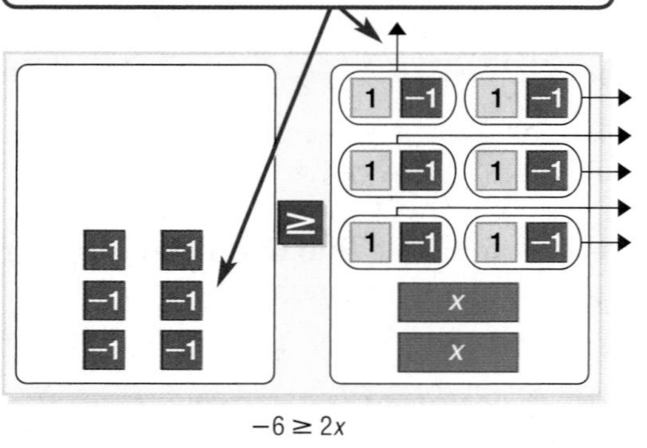

$-6 \geq 2x$

Step 4 *Group the tiles.*

Separate the tiles into 2 groups.

−1 −1 −1 ≥ x

−1 −1 −1 ≥ x

$-3 \geq x$ or $x \leq -3$

Model and Analyze

Use algebra tiles to solve each inequality.

1. $-4x < 12$ **2.** $-2x > 8$ **3.** $-3x \geq -6$ **4.** $-5x \leq -5$

5. In Exercises 1–4, is the coefficient of x in each inequality positive or negative?

6. Compare the inequality symbols and locations of the variable in Exercises 1–4 with those in their solutions. What do you find?

7. Model the solution for $2x \geq 6$. What do you find? How is this different from solving $-2x \geq 6$?

6-2 Solving Inequalities by Multiplication and Division

What You'll Learn

- Solve linear inequalities by using multiplication.
- Solve linear inequalities by using division.

Why are inequalities important in landscaping?

Isabel Franco is a landscape architect. To beautify a garden, she plans to build a decorative wall of either bricks or blocks. Each brick is 3 inches high, and each block is 12 inches high. Notice that $3 < 12$.

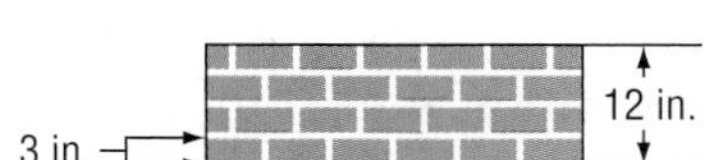

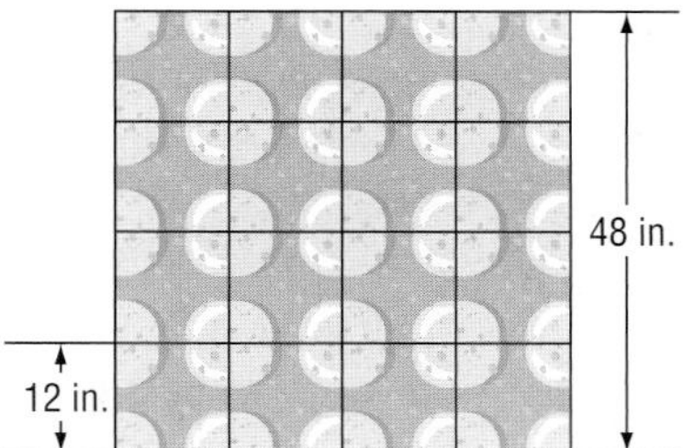

A wall 4 bricks high would be lower than a wall 4 blocks high.

$$3 \times 4 \ \underline{\ ?\ } \ 12 \times 4$$
$$12 < 48$$

SOLVE INEQUALITIES BY MULTIPLICATION If each side of an inequality is multiplied by a positive number, the inequality remains true.

$8 > 5$		$5 < 9$	
$8(2) \ \underline{\ ?\ } \ 5(2)$	Multiply each side by 2.	$5(4) \ \underline{\ ?\ } \ 9(4)$	Multiply each side by 4.
$16 > 10$		$20 < 36$	

This is *not* true when multiplying by negative numbers.

$5 > 3$		$-6 < 8$	
$5(-2) \ \underline{\ ?\ } \ 3(-2)$	Multiply each side by -2.	$-6(-5) \ \underline{\ ?\ } \ 8(-5)$	Multiply each side by -5.
$-10 < -6$		$30 > -40$	

If each side of an inequality is multiplied by a negative number, the direction of the inequality symbol changes. These examples illustrate the **Multiplication Property of Inequalities**.

Key Concept — *Multiplying by a Positive Number*

- **Words** If each side of a true inequality is multiplied by the same positive number, the resulting inequality is also true.
- **Symbols** If a and b are any numbers and c is a positive number, the following are true.
  If $a > b$, then $ac > bc$, and if $a < b$, then $ac < bc$.

Key Concept — Multiplying by a Negative Number

- **Words** If each side of a true inequality is multiplied by the same negative number, the direction of the inequality symbol must be *reversed* so that the resulting inequality is also true.
- **Symbols** If a and b are any numbers and c is a negative number, the following are true.
  If $a > b$, then $ac < bc$, and if $a < b$, then $ac > bc$.

This property also holds for inequalities involving $\geq$ and $\leq$.

You can use this property to solve inequalities.

Example 1 Multiply by a Positive Number

Solve $\frac{b}{7} \geq 25$. Then check your solution.

$\frac{b}{7} \geq 25$	Original inequality
$(7)\frac{b}{7} \geq (7)25$	Multiply each side by 7. Since we multiplied by a positive number, the inequality symbol stays the same.
$b \geq 175$	

CHECK To check this solution, substitute 175, a number less than 175, and a number greater than 175 into the inequality.

Let $b = 175$.	Let $b = 140$.	Let $b = 210$.
$\frac{175}{7} \overset{?}{\geq} 25$	$\frac{140}{7} \overset{?}{\geq} 25$	$\frac{210}{7} \overset{?}{\geq} 25$
$25 \geq 25$ ✓	$20 \not\geq 25$	$30 \geq 25$ ✓

The solution set is $\{b \mid b \geq 175\}$.

Study Tip

Common Misconception

A negative sign in an inequality does not necessarily mean that the direction of the inequality should change. For example, when solving $\frac{x}{6} > -3$, do *not* change the direction of the inequality.

Example 2 Multiply by a Negative Number

Solve $-\frac{2}{5}p < -14$.

$-\frac{2}{5}p < -14$	Original inequality
$\left(-\frac{5}{2}\right)\left(-\frac{2}{5}p\right) > \left(-\frac{5}{2}\right)(-14)$	Multiply each side by $-\frac{5}{2}$ and change $<$ to $>$.
$p > 35$	The solution set is $\{p \mid p > 35\}$.

Example 3 Write and Solve an Inequality

Write an inequality for the sentence below. Then solve the inequality.

One fourth of a number is less than −7.

One fourth	of	a number	is less than	−7.
$\frac{1}{4}$	$\times$	n	$<$	-7

$\frac{1}{4}n < -7$	Original inequality
$(4)\frac{1}{4}n < (4)(-7)$	Multiply each side by 4 and do not change the inequality's direction.
$n < -28$	The solution set is $\{n \mid n < -28\}$.

SOLVE INEQUALITIES BY DIVISION

Dividing each side of an inequality by the same number is similar to multiplying each side of an equality by the same number. Consider the inequality $6 < 15$.

Divide each side by 3.

$$6 < 15$$
$$6 \div 3 \ \underline{\ ?\ }\ 15 \div 3$$
$$2 < 5$$

Since each side is divided by a positive number, the direction of the inequality symbol remains the same.

Divide each side by -3.

$$6 < 15$$
$$6 \div (-3) \ \underline{\ ?\ }\ 15 \div (-3)$$
$$-2 > -5$$

Since each side is divided by a negative number, the direction of the inequality symbol is reversed.

These examples illustrate the **Division Property of Inequalities**.

Key Concept — Dividing by a Positive Number

- **Words** If each side of a true inequality is divided by the same positive number, the resulting inequality is also true.
- **Symbols** If a and b are any numbers and c is a positive number, the following are true.

  If $a > b$, then $\frac{a}{c} > \frac{b}{c}$, and if $a < b$, then $\frac{a}{c} < \frac{b}{c}$.

Dividing by a Negative Number

- **Words** If each side of a true inequality is divided by the same negative number, the direction of the inequality symbol must be *reversed* so that the resulting inequality is also true.
- **Symbols** If a and b are any numbers and c is a negative number, the following are true.

  If $a > b$, then $\frac{a}{c} < \frac{b}{c}$, and if $a < b$, then $\frac{a}{c} > \frac{b}{c}$.

This property also holds for inequalities involving $\geq$ and $\leq$.

Example 4 Divide by a Positive Number

Solve $14h > 91$.

$14h > 91$ Original inequality

$\frac{14h}{14} > \frac{91}{14}$ Divide each side by 14 and do not change the direction of the inequality sign.

$h > 6.5$

CHECK

Let $h = 6.5$.	Let $h = 7$.	Let $h = 6$.
$14h > 91$	$14h > 91$	$14h > 91$
$14(6.5) \overset{?}{>} 91$	$14(7) \overset{?}{>} 91$	$14(6) \overset{?}{>} 91$
$91 \not> 91$	$98 > 91$ ✓	$84 \not> 91$

The solution set is $\{h \mid h > 6.5\}$.

Since dividing is the same as multiplying by the reciprocal, there are two methods to solve an inequality that involve multiplication.

Example 5 Divide by a Negative Number

Solve $-5t \geq 275$ using two methods.

Method 1 Divide.

$-5t \geq 275$ Original inequality

$\frac{-5t}{-5} \leq \frac{275}{-5}$ Divide each side by -5 and change $\geq$ to $\leq$.

$t \leq -55$ Simplify.

Method 2 Multiply by the multiplicative inverse.

$-5t \geq 275$ Original inequality

$\left(-\frac{1}{5}\right)(-5t) \leq \left(-\frac{1}{5}\right)275$ Multiply each side by $-\frac{1}{5}$ and change $\geq$ to $\leq$.

$t \leq -55$ Simplify.

The solution set is $\{t \mid t \leq -55\}$.

You can use the Multiplication Property and the Division Property for Inequalities to solve standardized test questions.

Standardized Test Practice

A B C D

Example 6 The Word "not"

Multiple-Choice Test Item

Which inequality does *not* have the solution $\{y \mid y \leq -5\}$?

(A) $-7y \geq 35$ (B) $2y \leq -10$ (C) $\frac{7}{5}y \geq -7$ (D) $-\frac{y}{4} \geq \frac{5}{4}$

Read the Test Item

You want to find the inequality that does *not* have the solution set $\{y \mid y \leq -5\}$.

Solve the Test Item

Consider each possible choice.

(A) $-7y \geq 35$

$\frac{-7y}{-7} \leq \frac{35}{-7}$

$y \leq -5$ ✓

(B) $2y \leq -10$

$\frac{2y}{2} \leq \frac{-10}{2}$

$y \leq -5$ ✓

(C) $\frac{7}{5}y \geq -7$

$\left(\frac{5}{7}\right)\frac{7}{5}y \geq \left(\frac{5}{7}\right)(-7)$

$y \geq -5$

(D) $-\frac{y}{4} \geq \frac{5}{4}$

$(-4)\left(-\frac{y}{4}\right) \leq (-4)\frac{5}{4}$

$y \leq -5$ ✓

The answer is C.

The Princeton Review

Test-Taking Tip

Always look for the word *not* in the questions. This indicates that you are looking for the one incorrect answer, rather than looking for the one correct answer. The word *not* is usually in italics or uppercase letters to draw your attention to it.

Check for Understanding

Concept Check

1. **Explain** why you can use either the Multiplication Property of Inequalities or the Division Property of Inequalities to solve $-7r \leq 28$.

2. **OPEN ENDED** Write a problem that can be represented by the inequality $\frac{3}{4}c > 9$.

3. **FIND THE ERROR** Ilonia and Zachary are solving $-9b \leq 18$.

Ilonia	Zachary
$-9b \leq 18$	$-9b \leq 18$
$\frac{-9b}{-9} \geq \frac{18}{-9}$	$\frac{-9b}{-9} \leq \frac{18}{-9}$
$b \geq -2$	$b \leq -2$

Who is correct? Explain your reasoning.

Guided Practice

4. Which statement is represented by $7n \geq 14$?
   a. Seven times a number is at least 14.
   b. Seven times a number is greater than 14.
   c. Seven times a number is at most 14.
   d. Seven times a number is less than 14.

5. Which inequality represents *five times a number is less than 25*?
   a. $5n > 25$ b. $5n \geq 25$ c. $5n < 25$ d. $5n \leq 25$

Solve each inequality. Then check your solution.

6. $-15g > 75$ 7. $\frac{t}{9} < -12$ 8. $-\frac{2}{3}b \leq -9$ 9. $25f \geq 9$

Define a variable, write an inequality, and solve each problem. Then check your solution.

10. The opposite of four times a number is more than 12.
11. Half of a number is at least 26.

Standardized Test Practice

12. Which inequality does *not* have the solution $\{x \mid x > 4\}$?
   (A) $-5x < -20$ (B) $6x < 24$ (C) $\frac{1}{5}x > \frac{4}{5}$ (D) $-\frac{3}{4}x < -3$

Practice and Apply

Homework Help

For Exercises	See Examples
13–18, 39–44	3
19–38	1, 2, 4, 5
45–51	6

Extra Practice

See page 833.

Match each inequality with its corresponding statement.

13. $\frac{1}{5}n > 10$ a. Five times a number is less than or equal to ten.
14. $5n \leq 10$ b. One fifth of a number is no less than ten.
15. $5n > 10$ c. Five times a number is less than ten.
16. $-5n < 10$ d. One fifth of a number is greater than ten.
17. $\frac{1}{5}n \geq 10$ e. Five times a number is greater than ten.
18. $5n < 10$ f. Negative five times a number is less than ten.

Solve each inequality. Then check your solution.

19. $6g \leq 144$ 20. $7t > 84$ 21. $-14d \geq 84$ 22. $-16z \leq -64$
23. $\frac{m}{5} \geq 7$ 24. $\frac{b}{10} \leq 5$ 25. $-\frac{r}{7} < -7$ 26. $-\frac{a}{11} > 9$
27. $\frac{5}{8}y \geq -15$ 28. $\frac{2}{3}v < 6$ 29. $-\frac{3}{4}q \leq -33$ 30. $-\frac{2}{5}p > 10$
31. $-2.5w < 6.8$ 32. $-0.8s > 6.4$ 33. $\frac{15c}{-7} > \frac{3}{14}$ 34. $\frac{4m}{5} < \frac{-3}{15}$

35. Solve $-\frac{y}{8} > \frac{1}{2}$. Then graph the solution.

36. Solve $-\frac{m}{9} \leq -\frac{1}{3}$. Then graph the solution.

37. If $2a \geq 7$, then complete each inequality.

a. $a \geq$ __?__ **b.** $-4a \leq$ __?__ **c.** __?__ $a \leq -21$

38. If $4t < -2$, then complete each inequality.

a. $t <$ __?__ **b.** $-8t >$ __?__ **c.** __?__ $t > 14$

Define a variable, write an inequality, and solve each problem. Then check your solution.

39. Seven times a number is greater than 28.

40. Negative seven times a number is at least 14.

41. Twenty-four is at most a third of a number.

42. Two thirds of a number is less than −15.

43. Twenty-five percent of a number is greater than or equal to 90.

44. Forty percent of a number is less than or equal to 45.

45. **GEOMETRY** The area of a rectangle is less than 85 square feet. The length of the rectangle is 20 feet. What is the width of the rectangle?

46. **FUND-RAISING** The Middletown Marching Mustangs want to make at least \$2000 on their annual mulch sale. The band makes \$2.50 on each bag of mulch that is sold. How many bags of mulch should the band sell?

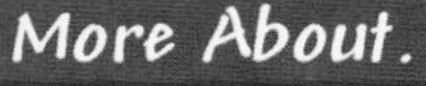
More About. . .

Zoos

Dr. Harry Wegeforth founded the San Diego Zoo in 1916 with just 50 animals. Today, the zoo has over 3800 animals.

Source: www.sandiegozoo.org

47. **LONG-DISTANCE COSTS** Juan's long-distance phone company charges him 9¢ for each minute or any part of a minute. He wants to call his friend, but he does not want to spend more than \$2.50 on the call. How long can he talk to his friend?

48. **EVENT PLANNING** The Country Corner Reception Hall does not charge a rental fee as long as at least \$4000 is spent on food. Shaniqua is planning a class reunion. If she has chosen a buffet that costs \$28.95 per person, how many people must attend the reunion to avoid a rental fee for the hall?

49. **LANDSCAPING** Matthew is planning a circular flower garden with a low fence around the border. If he can use up to 38 feet of fence, what radius can he use for the garden? *(Hint:* $C = 2\pi r$*)*

50. **DRIVING** Average speed is calculated by dividing distance by time. If the speed limit on the interstate is 65 miles per hour, how far can a person travel legally in $1\frac{1}{2}$ hours?

51. **ZOOS** The yearly membership to the San Diego Zoo for a family with 2 adults and 2 children is \$144. The regular admission to the zoo is \$18 for each adult and \$8 for each child. How many times should such a family plan to visit the zoo in a year to make a membership less expensive than paying regular admission?

52. **CRITICAL THINKING** Give a counterexample to show that each statement is not always true.

a. If $a > b$, then $a^2 > b^2$. **b.** If $a < b$ and $c < d$, then $ac < bd$.

53. **CITY PLANNING** The city of Santa Clarita requires that a parking lot can have no more than 20% of the parking spaces limited to compact cars. If a certain parking lot has 35 spaces for compact cars, how many spaces must the lot have to conform to the code?

54. **CIVICS** For a candidate to run for a county office, he or she must submit a petition with at least 6000 signatures of registered voters. Usually only 85% of the signatures are valid. How many signatures should a candidate seek on a petition?

55. **WRITING IN MATH** Answer the question that was posed at the beginning of the lesson.

Why are inequalities important in landscaping?

Include the following in your answer:

- an inequality representing a brick wall that can be no higher than 4 feet, and
- an explanation of how to solve the inequality.

Standardized Test Practice

56. The solution set for which inequality is *not* represented by the following graph?

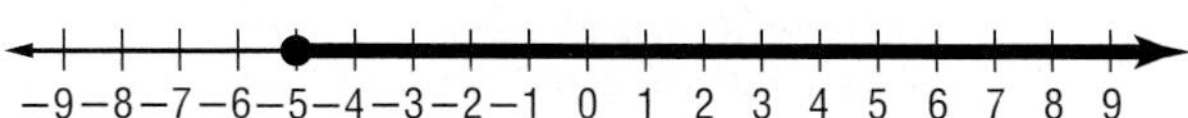

(A) $-\frac{x}{5} \leq 1$ (B) $\frac{x}{5} \leq -1$ (C) $-9x \leq 45$ (D) $2.5x \geq -12.5$

57. Solve $-\frac{7}{8}t < \frac{14}{15}$.

(A) $\left\{t \middle| t > \frac{16}{15}\right\}$ (B) $\left\{t \middle| t < \frac{16}{15}\right\}$ (C) $\left\{t \middle| t > -\frac{16}{15}\right\}$ (D) $\left\{t \middle| t < -\frac{16}{15}\right\}$

Maintain Your Skills

Mixed Review

Solve each inequality. Then check your solution, and graph it on a number line. *(Lesson 6-1)*

58. $s - 7 < 12$ 59. $g + 3 \leq -4$ 60. $7 > n + 2$

61. Draw a scatter plot that shows a positive correlation. *(Lesson 5-7)*

Write an equation of the line that passes through each pair of points. *(Lesson 5-4)*

62. $(-1, 3), (2, 4)$ 63. $(5, -2), (-1, -2)$ 64. $(3, 3), (-1, 2)$

If $h(x) = 3x + 2$, find each value. *(Lesson 4-6)*

65. $h(-4)$ 66. $h(2)$ 67. $h(w)$ 68. $h(r - 6)$

Solve each proportion. *(Lesson 3-6)*

69. $\frac{3}{4} = \frac{x}{8}$ 70. $\frac{t}{1.5} = \frac{2.4}{1.6}$ 71. $\frac{w + 2}{5} = \frac{7}{5}$ 72. $\frac{x}{3} = \frac{x + 5}{15}$

Getting Ready for the Next Lesson

PREREQUISITE SKILL Solve each equation.

*(To review **multi-step equations**, see Lessons 3-4 and 3-5.)*

73. $5x - 3 = 32$ 74. $4t + 9 = 14$ 75. $6y - 1 = 4y + 23$

76. $\frac{14g + 5}{6} = 9$ 77. $5a + 6 = 9a - (7a + 18)$ 78. $2(p - 4) = 7(p + 3)$

Practice Quiz 1

Lessons 6-1 and 6-2

Solve each inequality. Then check your solution, and graph it on a number line. *(Lesson 6-1)*

1. $h - 16 > -13$ 2. $r + 3 \leq -1$ 3. $4 \geq p + 9$ 4. $-3 < a - 5$ 5. $7g \leq 6g - 1$

Solve each inequality. Then check your solution. *(Lesson 6-2)*

6. $15z \geq 105$ 7. $\frac{v}{5} < 7$ 8. $-\frac{3}{7}q > 15$ 9. $-156 < 12r$ 10. $-\frac{2}{5}w \leq -\frac{1}{2}$

6-3 Solving Multi-Step Inequalities

What You'll Learn

- Solve linear inequalities involving more than one operation.
- Solve linear inequalities involving the Distributive Property.

How are linear inequalities used in science?

The boiling point of a substance is the temperature at which the element changes from a liquid to a gas. The boiling point of chlorine is −31°F. That means chlorine will be a gas for all temperatures greater than −31°F. If F represents temperature in degrees Fahrenheit, the inequality $F > -31$ represents the temperatures for which chlorine is a gas.

If C represents degrees Celsius, then $F = \frac{9}{5}C + 32$. You can solve $\frac{9}{5}C + 32 > -31$ to find the temperatures in degrees Celsius for which chlorine is a gas.

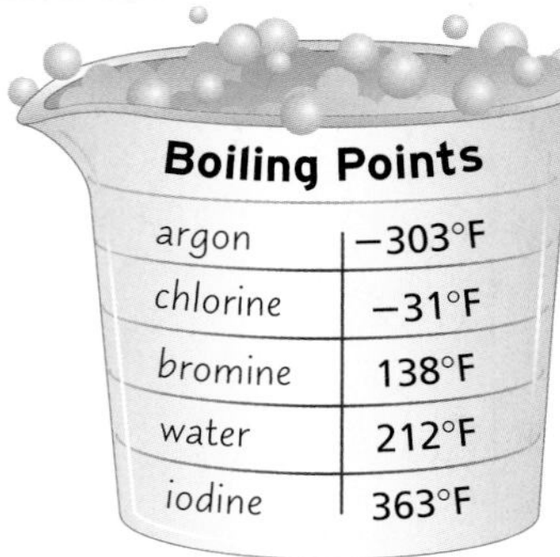

Boiling Points

argon	−303°F
chlorine	−31°F
bromine	138°F
water	212°F
iodine	363°F

Source: *World Book Encyclopedia*

SOLVE MULTI-STEP INEQUALITIES The inequality $\frac{9}{5}C + 32 > -31$ involves more than one operation. It can be solved by undoing the operations in the same way you would solve an equation with more than one operation.

Example 1 Solve a Real-World Problem

SCIENCE Find the temperatures in degrees Celsius for which chlorine is a gas.

$\frac{9}{5}C + 32 > -31$	Original inequality
$\frac{9}{5}C + 32 - 32 > -31 - 32$	Subtract 32 from each side.
$\frac{9}{5}C > -63$	Simplify.
$\left(\frac{5}{9}\right)\frac{9}{5}C > \left(\frac{5}{9}\right)(-63)$	Multiply each side by $\frac{5}{9}$.
$C > -35$	Simplify.

Chlorine will be a gas for all temperatures greater than −35°C.

When working with inequalities, do not forget to reverse the inequality sign whenever you multiply or divide each side by a negative number.

Example 2 Inequality Involving a Negative Coefficient

Solve $-7b + 19 < -16$. Then check your solution.

$-7b + 19 < -16$	Original inequality
$-7b + 19 - 19 < -16 - 19$	Subtract 19 from each side.
$-7b < -35$	Simplify.
$\frac{-7b}{-7} > \frac{-35}{-7}$	Divide each side by −7 and change $<$ to $>$.
$b > 5$	Simplify.

CHECK To check this solution, substitute 5, a number less than 5, and a number greater than 5.

Let $b = 5$.	Let $b = 4$.	Let $b = 6$.
$-7b + 19 < -16$	$-7b + 19 < -16$	$-7b + 19 < -16$
$-7(5) + 19 \overset{?}{<} -16$	$-7(4) + 19 \overset{?}{<} -16$	$-7(6) + 19 \overset{?}{<} -16$
$-35 + 19 \overset{?}{<} -16$	$-28 + 19 \overset{?}{<} -16$	$-42 + 19 \overset{?}{<} -16$
$-16 \not< -16$	$-9 \not< -16$	$-23 < -16$ ✓

The solution set is $\{b \mid b > 5\}$.

Example 3 Write and Solve an Inequality

Write an inequality for the sentence below. Then solve the inequality.
Three times a number minus eighteen is at least five times the number plus twenty-one.

Three times a number	minus	eighteen	is at least	five times the number	plus	twenty one.
$3n$	$-$	18	$\geq$	$5n$	$+$	21

$3n - 18 \geq 5n + 21$ Original inequality

$3n - 18 - 5n \geq 5n + 21 - 5n$ Subtract $5n$ from each side.

$-2n - 18 \geq 21$ Simplify.

$-2n - 18 + 18 \geq 21 + 18$ Add 18 to each side.

$-2n \geq 39$ Simplify.

$\frac{-2n}{-2} \leq \frac{39}{-2}$ Divide each side by -2 and change $\geq$ to $\leq$.

$n \leq -19.5$ Simplify.

The solution set is $\{n \mid n \leq -19.5\}$.

A graphing calculator can be used to solve inequalities.

Graphing Calculator Investigation

Solving Inequalities

You can find the solution of an inequality in one variable by using a graphing calculator. On a TI-83 Plus, clear the Y= list. Enter $6x + 9 < -4x + 29$ as Y1. (The symbol $<$ is item 5 on the TEST menu.)
Press GRAPH.

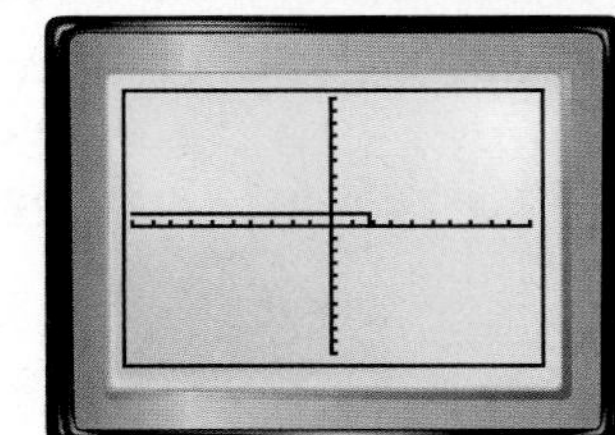

[10, 10] scl: 1 by [10, 10] scl: 1

Think and Discuss

1. Describe what is shown on the screen.
2. Use the TRACE function to scan the values along the graph. What do you notice about the values of y on the graph?
3. Solve the inequality algebraically. How does your solution compare to the pattern you noticed in Exercise 2?

SOLVE INEQUALITIES INVOLVING THE DISTRIBUTIVE PROPERTY

When solving equations that contain grouping symbols, first use the Distributive Property to remove the grouping symbols.

Example 4 Distributive Property

Solve $3d - 2(8d - 9) > 3 - (2d + 7)$.

$3d - 2(8d - 9) > 3 - (2d + 7)$	Original inequality
$3d - 16d + 18 > 3 - 2d - 7$	Distributive Property
$-13d + 18 > -2d - 4$	Combine like terms.
$-13d + 18 + 13d > -2d - 4 + 13d$	Add $13d$ to each side.
$18 > 11d - 4$	Simplify.
$18 + 4 > 11d - 4 + 4$	Add 4 to each side.
$22 > 11d$	Simplify.
$\frac{22}{11} > \frac{11d}{11}$	Divide each side by 11.
$2 > d$	Simplify.

Since $2 > d$ is the same as $d < 2$, the solution set is $\{d \mid d < 2\}$.

If solving an inequality results in a statement that is always true, the solution is all real numbers. If solving an inequality results in a statement that is never true, the solution is the empty set $\varnothing$. The empty set has no members.

Example 5 Empty Set

Solve $8(t + 2) - 3(t - 4) < 5(t - 7) + 8$.

$8(t + 2) - 3(t - 4) < 5(t - 7) + 8$	Original inequality
$8t + 16 - 3t + 12 < 5t - 35 + 8$	Distributive Property
$5t + 28 < 5t - 27$	Combine like terms.
$5t + 28 - 5t < 5t - 27 - 5t$	Subtract $5t$ from each side.
$28 < -27$	This statement is false.

Since the inequality results in a false statement, the solution set is the empty set $\varnothing$.

Check for Understanding

Concept Check

1. **Compare and contrast** the method used to solve $-5h + 6 = -7$ and the method used to solve $-5h + 6 \leq -7$.
2. **OPEN ENDED** Write a multi-step inequality with the solution graphed below.

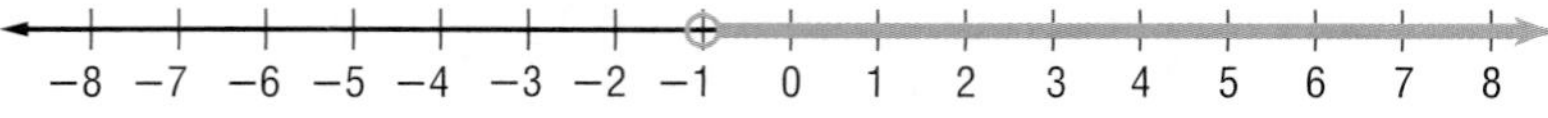

Guided Practice

3. Justify each indicated step.

$3(a - 7) + 9 \leq 21$	
$3a - 21 + 9 \leq 21$	a. ___?___
$3a - 12 \leq 21$	
$3a - 12 + 12 \leq 21 + 12$	b. ___?___
$3a \leq 33$	
$\frac{3a}{3} \leq \frac{33}{3}$	c. ___?___
$a \leq 11$	

Solve each inequality. Then check your solution.

4. $-4y - 23 < 19$
5. $\frac{2}{3}r + 9 \geq -3$
6. $7b + 11 > 9b - 13$
7. $-5(g + 4) > 3(g - 4)$
8. $3 + 5t \leq 3(t + 1) - 4(2 - t)$

9. Define a variable, write an inequality, and solve the problem below. Then check your solution.
Seven minus two times a number is less than three times the number plus thirty-two.

Application

10. **SALES** A salesperson is paid \$22,000 a year plus 5% of the amount of sales made. What is the amount of sales needed to have an annual income greater than \$35,000?

Practice and Apply

Homework Help

For Exercises	See Examples
11–14	1–5
15–34	2, 4, 5
35–38	3
39–52	1

Extra Practice
See page 834.

Justify each indicated step.

11.
$$\frac{2}{5}w + 7 \leq -9$$
$$\frac{2}{5}w + 7 - 7 \leq -9 - 7 \quad \text{a. } \underline{\ ?\ }$$
$$\frac{2}{5}w \leq -16$$
$$\left(\frac{5}{2}\right)\frac{2}{5}w \leq \left(\frac{5}{2}\right)(-16) \quad \text{b. } \underline{\ ?\ }$$
$$w \leq -40$$

12.
$$m > \frac{15 - 2m}{-3}$$
$$(-3)m < (-3)\frac{15 - 2m}{-3} \quad \text{a. } \underline{\ ?\ }$$
$$-3m < 15 - 2m$$
$$-3m + 2m < 15 - 2m + 2m \quad \text{b. } \underline{\ ?\ }$$
$$-m < 15$$
$$(-1)(-m) > (-1)15 \quad \text{c. } \underline{\ ?\ }$$
$$m > -15$$

13. Solve $4(t - 7) \leq 2(t + 9)$. Show each step and justify your work.
14. Solve $-5(k + 4) > 3(k - 4)$. Show each step and justify your work.

Solve each inequality. Then check your solution.

15. $-3t + 6 \leq -3$
16. $-5 - 8f > 59$
17. $-2 - \frac{d}{5} < 23$
18. $\frac{w}{8} - 13 > -6$
19. $7q - 1 + 2q \leq 29$
20. $8a + 2 - 10a \leq 20$
21. $9r + 15 \leq 24 + 10r$
22. $13k - 11 > 7k + 37$
23. $\frac{2v - 3}{5} \geq 7$
24. $\frac{3a + 8}{2} < 10$
25. $\frac{3w + 5}{4} \geq 2w$
26. $\frac{5b + 8}{3} < 3b$
27. $7 + 3t \leq 2(t + 3) - 2(-1 - t)$
28. $5(2h - 6) - 7(h + 7) > 4h$
29. $3y + 4 > 2(y + 3) + y$
30. $3 - 3(b - 2) < 13 - 3(b - 6)$
31. $3.1v - 1.4 \geq 1.3v + 6.7$
32. $0.3(d - 2) - 0.8d > 4.4$

33. Solve $4(y + 1) - 3(y - 5) \geq 3(y - 1)$. Then graph the solution.

34. Solve $5(x + 4) - 2(x + 6) \geq 5(x + 1) - 1$. Then graph the solution.

Define a variable, write an inequality, and solve each problem. Then check your solution.

35. One eighth of a number decreased by five is at least thirty.
36. Two thirds of a number plus eight is greater than twelve.
37. Negative four times a number plus nine is no more than the number minus twenty-one.
38. Three times the sum of a number and seven is greater than five times the number less thirteen.

GEOMETRY For Exercises 39 and 40, use the following information.
By definition, the measure of any acute angle is less than 90 degrees. Suppose the measure of an acute angle is $3a - 15$.

39. Write an inequality to represent the situation.
40. Solve the inequality.

SCHOOL For Exercises 41 and 42, use the following information.
Carmen's scores on three math tests were 91, 95, and 88. The fourth and final test of the grading period is tomorrow. She needs an average (mean) of at least 92 to receive an A for the grading period.

41. If s is her score on the fourth test, write an inequality to represent the situation.
42. If Carmen wants an A in math, what must she score on the test?

PHYSICAL SCIENCE For Exercises 43 and 44, use the information at the left and the information below.
The melting point for an element is the temperature where the element changes from a solid to a liquid. If C represents degrees Celsius and F represents degrees Fahrenheit, then $C = \frac{5(F - 32)}{9}$.

43. Write an inequality that can be used to find the temperatures in degrees Fahrenheit for which mercury is a solid.
44. For what temperatures will mercury be a solid?

More About. . .

Physical Science

Mercury is a metal that is a liquid at room temperature. In fact, its melting point is −38°C. Mercury is used in thermometers because it expands evenly as it is heated.

Source: *World Book Encyclopedia*

45. **HEALTH** Keith weighs 200 pounds. He wants to weigh less than 175 pounds. If he can lose an average of 2 pounds per week on a certain diet, how long should he stay on his diet to reach his goal weight?

46. **CRITICAL THINKING** Write a multi-step inequality that has no solution and one that has infinitely many solutions.

47. **PERSONAL FINANCES** Nicholas wants to order a pizza. He has a total of $13.00 to pay the delivery person. The pizza costs $7.50 plus $1.25 per topping. If he plans to tip 15% of the total cost of the pizza, how many toppings can he order?

LABOR For Exercises 48–50, use the following information.
A union worker made $500 per week. His union sought a one-year contract and went on strike. Once the new contract was approved, it provided for a 4% raise.

48. Assume that the worker was not paid during the strike. Given his raise in salary, how many weeks could he strike and still make at least as much for the next 52 weeks as he would have made without a strike?
49. How would your answer to Exercise 48 change if the worker had been making $600 per week?
50. How would your answer to Exercise 48 change if the worker's union provided him with $150 per week during the strike?

51. **NUMBER THEORY** Find all sets of two consecutive positive odd integers whose sum is no greater than 18.

52. **NUMBER THEORY** Find all sets of three consecutive positive even integers whose sum is less than 40.

53. **WRITING IN MATH** Answer the question that was posed at the beginning of the lesson.

How are linear inequalities used in science?

Include the following in your answer:

- an inequality for the temperatures in degrees Celsius for which bromine is a gas, and
- a description of a situation in which a scientist might use an inequality.

54. What is the first step in solving $\frac{y-5}{9} \geq 13$?

Ⓐ Add 5 to each side. Ⓑ Subtract 5 from each side.

Ⓒ Divide each side by 9. Ⓓ Multiply each side by 9.

55. Solve $4t + 2 < 8t - (6t - 10)$.

Ⓐ $\{t \mid t < -6\}$ Ⓑ $\{t \mid t > -6\}$ Ⓒ $\{t \mid t < 4\}$ Ⓓ $\{t \mid t > 4\}$

Graphing Calculator

Use a graphing calculator to solve each inequality.

56. $3x + 7 > 4x + 9$ 57. $13x - 11 \leq 7x + 37$ 58. $2(x - 3) < 3(2x + 2)$

Maintain Your Skills

Mixed Review

59. **BUSINESS** The charge per mile for a compact rental car at Great Deal Rentals is \$0.12. Mrs. Ludlow must rent a car for a business trip. She has a budget of \$50 for mileage charges. How many miles can she travel without going over her budget? *(Lesson 6-2)*

Solve each inequality. Then check your solution, and graph it on a number line. *(Lesson 6-1)*

60. $d + 13 \geq 22$ 61. $t - 5 < 3$ 62. $4 > y + 7$

Write the point-slope form of an equation for a line that passes through each point with the given slope. *(Lesson 5-5)*

63. $(1, -3)$, $m = 2$ 64. $(-2, -1)$, $m = -\frac{2}{3}$ 65. $(3, 6)$, $m = 0$

Determine the slope of the line that passes through each pair of points. *(Lesson 5-1)*

66. $(3, -1)$, $(4, -6)$ 67. $(-2, -4)$, $(1, 3)$ 68. $(0, 3)$, $(-2, -5)$

Determine whether each equation is a linear equation. If an equation is linear, rewrite it in the form $Ax + By = C$. *(Lesson 4-5)*

69. $4x = 7 + 2y$ 70. $2x^2 - y = 7$ 71. $x = 12$

Solve each equation. Then check your solution. *(Lesson 3-5)*

72. $2(x - 2) = 3x - (4x - 5)$ 73. $5t - 7 = t + 3$

Getting Ready for the Next Lesson

PREREQUISITE SKILL Graph each set of numbers on a number line.
*(To review **graphing integers on a number line**, see Lesson 2-1.)*

74. $\{-2, 3, 5\}$ 75. $\{-1, 0, 3, 4\}$ 76. $\{-5, -4, -1, 1\}$

77. {integers less than 5} 78. {integers greater than -2}

79. {integers between 1 and 6} 80. {integers between -4 and 2}

81. {integers greater than or equal to -4}

82. {integers less than 6 but greater than -1}

Reading Mathematics

Compound Statements

Two simple statements connected by the words *and* or *or* form a compound statement. Before you can determine whether a compound statement is true or false, you must understand what the words *and* and *or* mean. Consider the statement below.

A triangle has three sides, *and* a hexagon has five sides.
For a compound statement connected by the word *and* to be true, both simple statements must be true. In this case, it is true that a triangle has three sides. However, it is false that a hexagon has five sides; it has six. Thus, the compound statement is false.

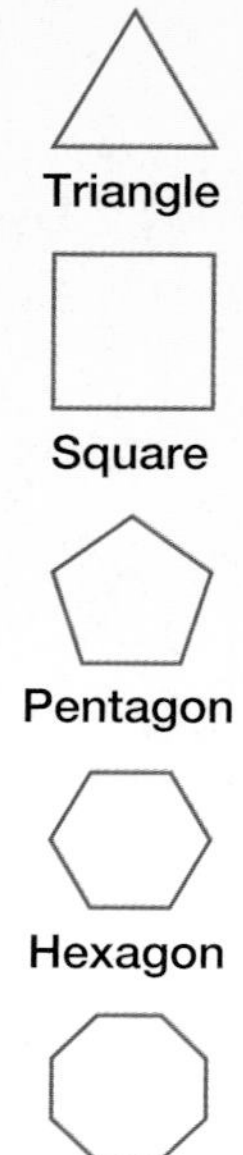

A compound statement connected by the word *or* may be *exclusive* or *inclusive*. For example, the statement "With your dinner, you may have soup *or* salad," is exclusive. In everyday language, *or* means one or the other, but not both. However, in mathematics, *or* is inclusive. It means one or the other or both. Consider the statement below.

A triangle has three sides, *or* a hexagon has five sides.
For a compound statement connected by the word *or* to be true, at least one of the simple statements must be true. Since it is true that a triangle has three sides, the compound statement is true.

Reading to Learn

Determine whether each compound statement is *true* or *false*. Explain your answer.

1. A hexagon has six sides, *or* an octagon has seven sides.
2. An octagon has eight sides, *and* a pentagon has six sides.
3. A pentagon has five sides, *and* a hexagon has six sides.
4. A triangle has four sides, *or* an octagon does *not* have seven sides.
5. A pentagon has three sides, *or* an octagon has ten sides.
6. A square has four sides, *or* a hexagon has six sides.
7. $5 < 4$ or $8 < 6$
8. $-1 > 0$ and $1 < 5$
9. $4 > 0$ and $-4 < 0$
10. $0 = 0$ or $-2 > -3$
11. $5 \neq 5$ or $-1 > -4$
12. $0 > 3$ and $2 > -2$

6-4 Solving Compound Inequalities

What You'll Learn

- Solve compound inequalities containing the word *and* and graph their solution sets.
- Solve compound inequalities containing the word *or* and graph their solution sets.

Vocabulary

- compound inequality
- intersection
- union

How are compound inequalities used in tax tables?

Richard Kelley is completing his income tax return. He uses the table to determine the amount he owes in federal income tax.

2000 Tax Tables

If taxable income is—		Single	Married filing jointly	Married filing separately	Head of a household
At least	Less than				
41,000	41,050	8140	6154	8689	6996
41,050	41,100	8154	6161	8703	7010
41,100	41,150	8168	6169	8717	7024
41,150	41,200	8182	6176	8731	7038
41,200	41,250	8196	6184	8754	7052
41,250	41,300	8210	6191	8759	7066
41,300	41,350	8224	6199	8773	7080
41,350	41,400	8238	6206	8787	7094
41,400	41,450	8252	6214	8801	7108
41,450	41,500	8266	6221	8815	7122
41,500	41,550	8280	6229	8829	7136
41,550	41,600	8294	6236	8843	7150

Source: IRS

Let c represent the amount of Mr. Kelley's income. His income is at least \$41,350 and it is less than \$41,400. This can be written as $c \geq 41{,}350$ and $c < 41{,}400$. When considered together, these two inequalities form a **compound inequality**. This compound inequality can be written without using *and* in two ways.

$$41{,}350 \leq c < 41{,}400 \text{ or } 41{,}400 > c \geq 41{,}350$$

Study Tip

Reading Math
The statement $41{,}350 \leq c < 41{,}400$ can be read *41,350 is less than or equal to c, which is less than 41,400.*

INEQUALITIES CONTAINING *AND* A compound inequality containing *and* is true only if both inequalities are true. Thus, the graph of a compound inequality containing *and* is the **intersection** of the graphs of the two inequalities. In other words, the solution must be a solution of *both* inequalities.

The intersection can be found by graphing each inequality and then determining where the graphs overlap.

Example 1 Graph an Intersection

Graph the solution set of $x < 3$ and $x \geq -2$.

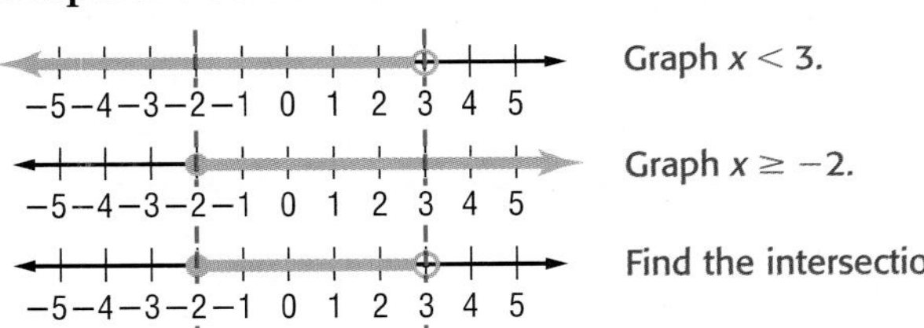

The solution set is $\{x \mid -2 \leq x < 3\}$. Note that the graph of $x \geq -2$ includes the point -2. The graph of $x < 3$ does *not* include 3.

Study Tip

Reading Math

When solving problems involving inequalities,

- *within* is meant to be inclusive. Use $\leq$ or $\geq$.
- *between* is meant to be exclusive. Use $<$ or $>$.

Example 2 Solve and Graph an Intersection

Solve $-5 < x - 4 < 2$. Then graph the solution set.

First express $-5 < x - 4 < 2$ using *and*. Then solve each inequality.

$-5 < x - 4$	and	$x - 4 < 2$
$-5 + 4 < x - 4 + 4$		$x - 4 + 4 < 2 + 4$
$-1 < x$		$x < 6$

The solution set is the intersection of the two graphs.

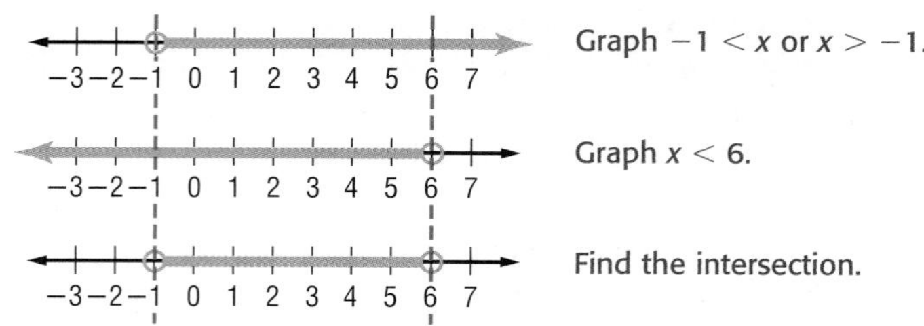

The solution set is $\{x \mid -1 < x < 6\}$.

INEQUALITIES CONTAINING *OR* Another type of compound inequality contains the word *or*. A compound inequality containing *or* is true if one or more of the inequalities is true. The graph of a compound inequality containing *or* is the **union** of the graphs of the two inequalities. In other words, the solution of the compound inequality is a solution of *either* inequality, not necessarily both.

The union can be found by graphing each inequality.

Career Choices

Pilot

Pilots check aviation weather forecasts to choose a route and altitude that will provide the smoothest flight.

Online Research

For information about a career as a pilot, visit: www.algebra1.com/careers

Example 3 Write and Graph a Compound Inequality

AVIATION An airplane is experiencing heavy turbulence while flying at 30,000 feet. The control tower tells the pilot that he should increase his altitude to at least 33,000 feet or decrease his altitude to no more than 26,000 feet to avoid the turbulence. Write and graph a compound inequality that describes the altitude at which the airplane should fly.

Words The pilot has been told to fly at an altitude of at least 33,000 feet or no more than 26,000 feet.

Variables Let a be the plane's altitude.

	The plane's altitude	is at least	33,000 feet	or	the altitude	is no more than	26,000 feet.
Inequality	a	$\geq$	33,000	or	a	$\leq$	26,000

Now, graph the solution set.

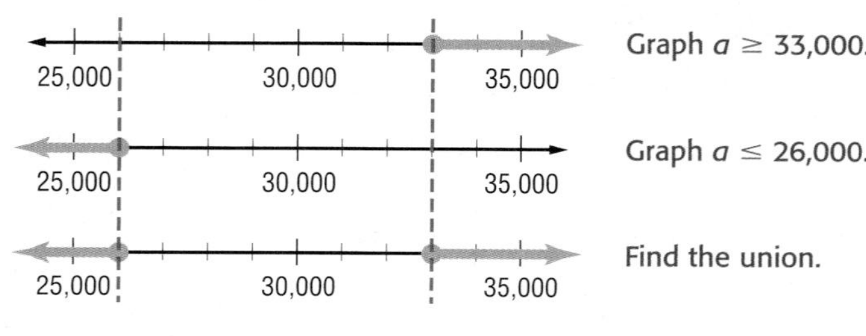

$a \geq 33{,}000$ or $a \leq 26{,}000$

Example 4 Solve and Graph a Union

Solve $-3h + 4 < 19$ or $7h - 3 > 18$. Then graph the solution set.

$$-3h + 4 < 19 \quad \text{or} \quad 7h - 3 > 18$$
$$-3h + 4 - 4 < 19 - 4 \qquad 7h - 3 + 3 > 18 + 3$$
$$-3h < 15 \qquad 7h > 21$$
$$\frac{-3h}{-3} > \frac{15}{-3} \qquad \frac{7h}{7} > \frac{21}{7}$$
$$h > -5 \qquad h > 3$$

The solution set is the union of the two graphs.

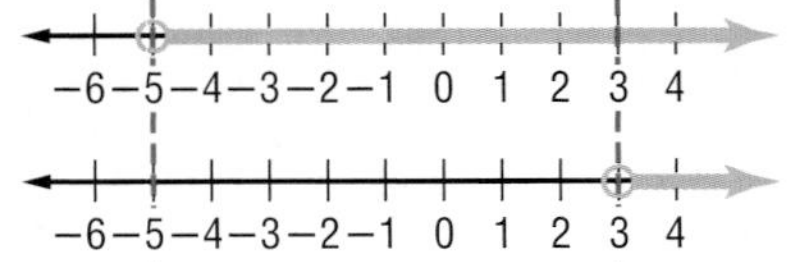

Graph $h > -5$.

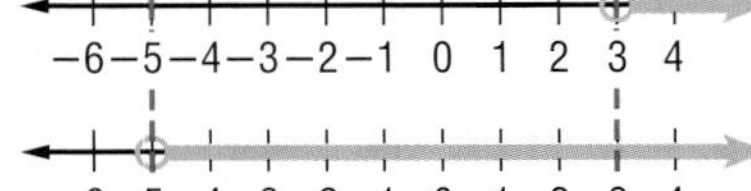

Graph $h > 3$.

−6 −5 −4 −3 −2 −1 0 1 2 3 4

Find the union.

Notice that the graph of $h > -5$ contains every point in the graph of $h > 3$. So, the union is the graph of $h > -5$. The solution set is $\{h \mid h > -5\}$.

Check for Understanding

Concept Check

1. **Describe** the difference between a compound inequality containing *and* and a compound inequality containing *or*.
2. **Write** *7 is less than t, which is less than 12* as a compound inequality.
3. **OPEN ENDED** Give an example of a compound inequality containing *and* that has no solution.

Guided Practice

Graph the solution set of each compound inequality.

4. $a \leq 6$ and $a \geq -2$
5. $y > 12$ or $y < 9$

Write a compound inequality for each graph.

6. −5 −4 −3 −2 −1 0 1 2 3 4 5
7. −3 −2 −1 0 1 2 3 4 5 6 7

Solve each compound inequality. Then graph the solution set.

8. $6 < w + 3$ and $w + 3 < 11$
9. $n - 7 \leq -5$ or $n - 7 \geq 1$
10. $3z + 1 < 13$ or $z \leq 1$
11. $-8 < x - 4 \leq -3$
12. Define a variable, write a compound inequality, and solve the following problem.
Three times a number minus 7 is less than 17 and greater than 5.

Application

13. **PHYSICAL SCIENCE** According to Hooke's Law, the force F in pounds required to stretch a certain spring x inches beyond its natural length is given by $F = 4.5x$. If forces between 20 and 30 pounds, inclusive, are applied to the spring, what will be the range of the increased lengths of the stretched spring?

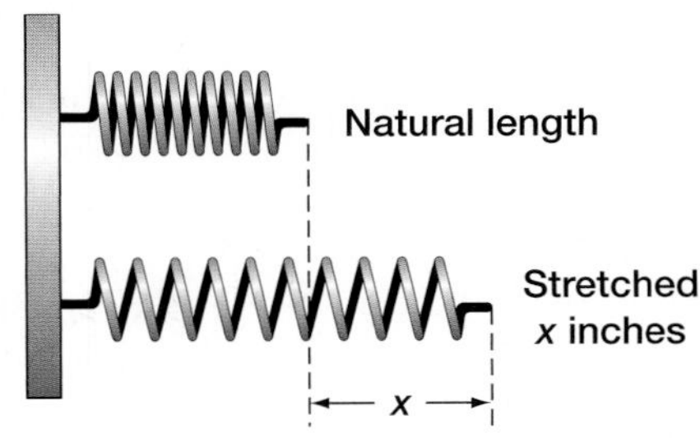

Practice and Apply

Homework Help

For Exercises	See Examples
14–27	1
28–45	2, 4
46–48	3

Extra Practice

See page 834.

Graph the solution set of each compound inequality.

14. $x > 5$ and $x \leq 9$
15. $s < -7$ and $s \leq 0$
16. $r < 6$ or $r > 6$
17. $m \geq -4$ or $m > 6$
18. $7 < d < 11$
19. $-1 \leq g < 3$

Write a compound inequality for each graph.

20. −5 −4 −3 −2 −1 0 1 2 3 4 5

21. −10 −9 −8 −7 −6 −5 −4 −3 −2 −1 0

22. 9 10 11 12 13 14 15 16 17 18 19

23. −10 −9 −8 −7 −6 −5 −4 −3 −2 −1 0

24. −9 −8 −7 −6 −5 −4 −3 −2 −1 0 1

25. −1 0 1 2 3 4 5 6 7 8 9

26. **WEATHER** The Fujita Scale (F-scale) is the official classification system for tornado damage. One factor used to classify a tornado is wind speed. Use the information in the table to write an inequality for the range of wind speeds of an F3 tornado.

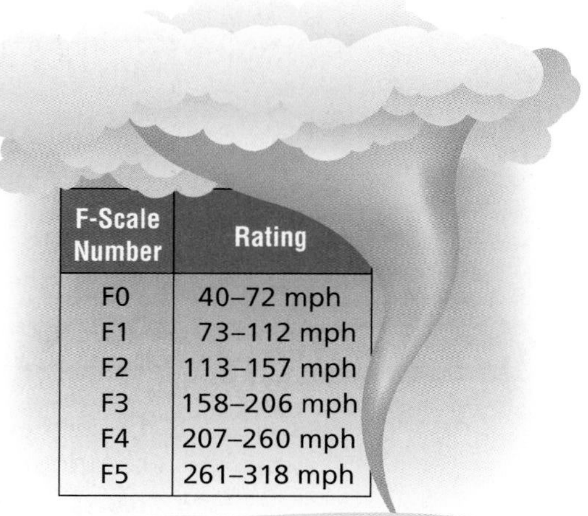

F-Scale Number	Rating
F0	40–72 mph
F1	73–112 mph
F2	113–157 mph
F3	158–206 mph
F4	207–260 mph
F5	261–318 mph

27. **BIOLOGY** Each type of fish thrives in a specific range of temperatures. The optimum temperatures for sharks range from 18°C to 22°C, inclusive. Write an inequality to represent temperatures where sharks will *not* thrive.

Solve each compound inequality. Then graph the solution set.

28. $k + 2 > 12$ and $k + 2 \leq 18$
29. $f + 8 \leq 3$ and $f + 9 \geq -4$
30. $d - 4 > 3$ or $d - 4 \leq 1$
31. $h - 10 < -21$ or $h + 3 < 2$
32. $3 < 2x - 3 < 15$
33. $4 < 2y - 2 < 10$
34. $3t - 7 \geq 5$ and $2t + 6 \leq 12$
35. $8 > 5 - 3q$ and $5 - 3q > -13$
36. $-1 + x \leq 3$ or $-x \leq -4$
37. $3n + 11 \leq 13$ or $-3n \geq -12$
38. $2p - 2 \leq 4p - 8 \leq 3p - 3$
39. $3g + 12 \leq 6 + g \leq 3g - 18$
40. $4c < 2c - 10$ or $-3c < -12$
41. $0.5b > -6$ or $3b + 16 < -8 + b$

Define a variable, write an inequality, and solve each problem.

42. Eight less than a number is no more than 14 and no less than 5.
43. The sum of 3 times a number and 4 is between −8 and 10.
44. The product of −5 and a number is greater than 35 or less than 10.
45. One half a number is greater than 0 and less than or equal to 1.

46. **HEALTH** About 20% of the time you sleep is spent in rapid eye movement (REM) sleep, which is associated with dreaming. If an adult sleeps 7 to 8 hours, how much time is spent in REM sleep?

47. **SHOPPING** A store is offering a $30 mail-in rebate on all color printers. Luisana is looking at different color printers that range in price from $175 to $260. How much can she expect to spend after the mail-in rebate?

48. **FUND-RAISING** Rashid is selling chocolates for his school's fund-raiser. He can earn prizes depending on how much he sells. So far, he has sold $70 worth of chocolates. How much more does he need to sell to earn a prize in category D?

Sales ($)	Prize
0–25	A
26–60	B
61–120	C
121–180	D
180+	E

49. **CRITICAL THINKING** Write a compound inequality that represents the values of x which make the following expressions *false*.

a. $x < 5$ or $x > 8$ **b.** $x \leq 6$ and $x \geq 1$

HEARING **For Exercises 50–52, use the following information.**
Humans hear sounds with sound waves within the 20 to 20,000 hertz range. Dogs hear sounds in the 15 to 50,000 hertz range.

50. Write a compound inequality for the hearing range of humans and one for the hearing range of dogs.
51. What is the union of the two solution sets? the intersection?
52. Write an inequality or inequalities for the range of sounds that dogs can hear, but humans cannot.

53. **RESEARCH** Use the Internet or other resource to find the altitudes in miles of the layers of Earth's atmosphere, troposphere, stratosphere, mesosphere, thermosphere, and exosphere. Write inequalities for the range of altitudes for each layer.

54. **WRITING IN MATH** Answer the question that was posed at the beginning of the lesson.

How are compound inequalities used in tax tables?

Include the following in your answer:
- a description of the intervals used in the tax table shown at the beginning of the lesson, and
- a compound inequality describing the income of a head of a household paying $7024 in taxes.

55. Ten pounds of fresh tomatoes make between 10 and 15 cups of cooked tomatoes. How many cups does one pound of tomatoes make?

Ⓐ between 1 and $1\frac{1}{2}$ cups Ⓑ between 1 and 5 cups
Ⓒ between 2 and 3 cups Ⓓ between 2 and 4 cups

56. Solve $-7 < x + 2 < 4$.

Ⓐ $-5 < x < 6$ Ⓑ $-9 < x < 2$
Ⓒ $-5 < x < 2$ Ⓓ $-9 < x < 6$

57. **SOLVE COMPOUND INEQUALITIES** In Lesson 6-3, you learned how to use a graphing calculator to find the values of x that make a given inequality true. You can also use this method to test compound inequalities. The words *and* and *or* can be found in the LOGIC submenu of the TEST menu of a TI-83 Plus. Use this method to solve each of the following compound inequalities using your graphing calculator.

a. $x + 4 < -2$ or $x + 4 > 3$ **b.** $x - 3 \leq 5$ and $x + 6 \geq 4$

Maintain Your Skills

Mixed Review

58. **FUND-RAISING** A university is running a drive to raise money. A corporation has promised to match 40% of whatever the university can raise from other sources. How much must the school raise from other sources to have a total of at least \$800,000 after the corporation's donation? *(Lesson 6-3)*

Solve each inequality. Then check your solution. *(Lesson 6-2)*

59. $18d \geq 90$
60. $-7v < 91$
61. $\frac{t}{13} < 13$
62. $-\frac{3}{8}b > 9$

Solve. Assume that y varies directly as x. *(Lesson 5-2)*

63. If $y = -8$ when $x = -3$, find x when $y = 6$.
64. If $y = 2.5$ when $x = 0.5$, find y when $x = 20$.

Express the relation shown in each mapping as a set of ordered pairs. Then state the domain, range, and inverse. *(Lesson 4-3)*

65. 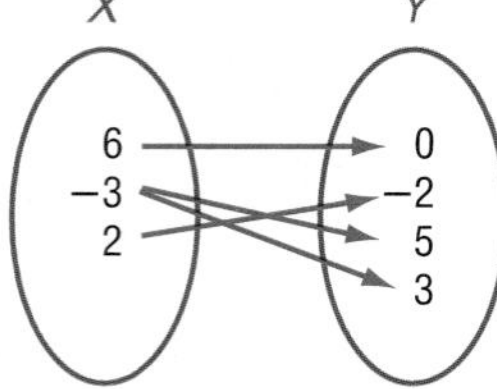

66. 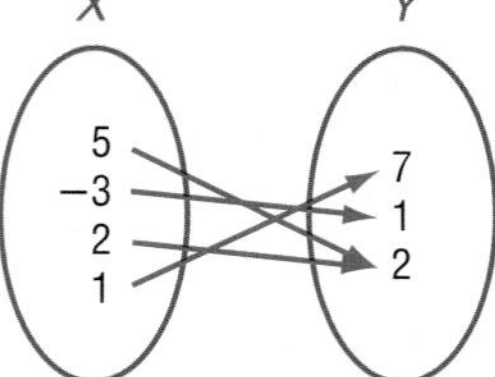

67. 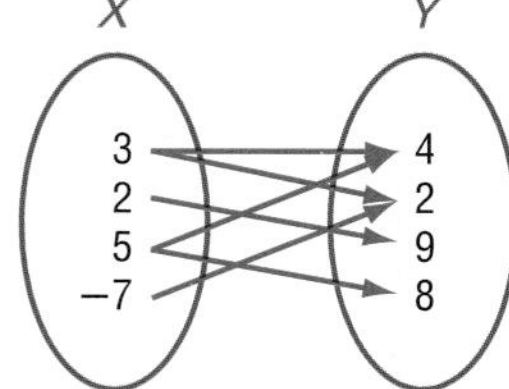

Find the odds of each outcome if a die is rolled. *(Lesson 2-6)*

68. a number greater than 2
69. not a 3

Find each product. *(Lesson 2-3)*

70. $-\frac{5}{6}\left(-\frac{2}{5}\right)$
71. $-100(4.7)$
72. $-\frac{7}{12}\left(\frac{6}{7}\right)\left(-\frac{3}{4}\right)$

Getting Ready for the Next Lesson

PREREQUISITE SKILL **Find each value.** *(To review **absolute value**, see Lesson 2-1.)*

73. $|-7|$
74. $|10|$
75. $|-1|$
76. $|-3.5|$
77. $|12 - 6|$
78. $|5 - 9|$
79. $|20 - 21|$
80. $|3 - 18|$

Practice Quiz 2

Lessons 6-3 and 6-4

Solve each inequality. Then check your solution. *(Lesson 6-3)*

1. $5 - 4b > -23$
2. $\frac{1}{2}n + 3 \geq -5$
3. $3(t + 6) < 9$
4. $9x + 2 > 20$
5. $2m + 5 \leq 4m - 1$
6. $a < \frac{2a - 15}{3}$

Solve each compound inequality. Then graph the solution set. *(Lesson 6-4)*

7. $x - 2 < 7$ and $x + 2 > 5$
8. $2b + 5 \leq -1$ or $b - 4 \geq -4$
9. $4m - 5 > 7$ or $4m - 5 < -9$
10. $a - 4 < 1$ and $a + 2 > 1$

6-5 Solving Open Sentences Involving Absolute Value

What You'll Learn

- Solve absolute value equations.
- Solve absolute value inequalities.

How is absolute value used in election polls?

Voters in Hamilton will vote on a new tax levy in the next election. A poll conducted before the election found that 47% of the voters surveyed were for the tax levy, 45% were against the tax levy, and 8% were undecided. The poll has a 3-point margin of error.

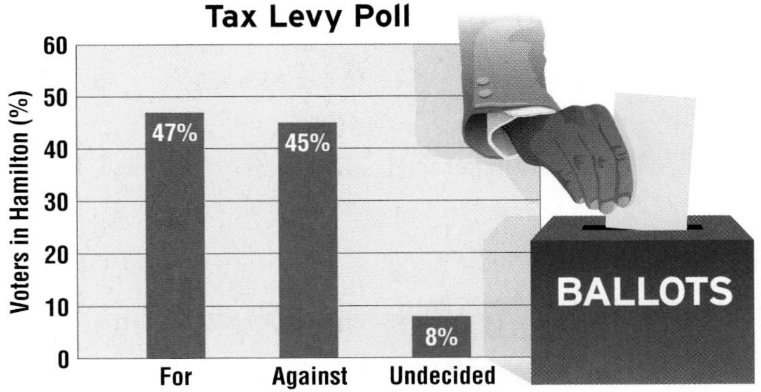

The margin of error means that the result may be 3 percentage points higher or lower. So, the number of people in favor of the tax levy may be as high as 50% or as low as 44%. This can be written as an inequality using absolute value.

$|x - 47| \leq 3$ The difference between the actual number and 47 is within 3 points.

Study Tip

Look Back
To review **absolute value**, see Lesson 2-1.

ABSOLUTE VALUE EQUATIONS There are three types of open sentences that can involve absolute value.

$$|x| = n \qquad |x| < n \qquad |x| > n$$

Consider the case of $|x| = n$. $|x| = 5$ means the distance between 0 and x is 5 units.

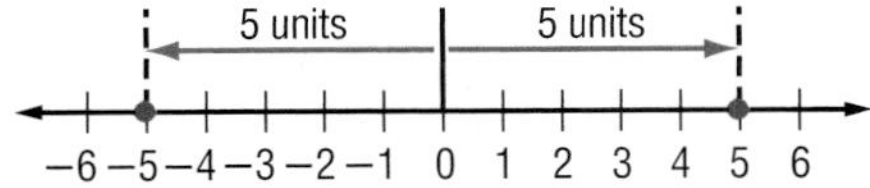

If $|x| = 5$, then $x = -5$ or $x = 5$. The solution set is $\{-5, 5\}$.

When solving equations that involve absolute value, there are two cases to consider.

Case 1 The value inside the absolute value symbols is positive.

Case 2 The value inside the absolute value symbols is negative.

Equations involving absolute value can be solved by graphing them on a number line or by writing them as a compound sentence and solving it.

Example 1 Solve an Absolute Value Equation

Solve $|a - 4| = 3$.

Method 1 Graphing

$|a - 4| = 3$ means that the distance between a and 4 is 3 units. To find a on the number line, start at 4 and move 3 units in either direction.

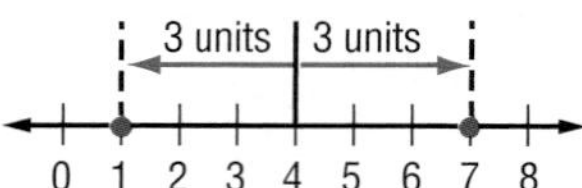

The distance from 4 to 1 is 3 units.
The distance from 4 to 7 is 3 units.

The solution set is {1, 7}.

Method 2 Compound Sentence

Write $|a - 4| = 3$ as $a - 4 = 3$ or $a - 4 = -3$.

Case 1

$a - 4 = 3$

$a - 4 + 4 = 3 + 4$ Add 4 to each side.

$a = 7$ Simplify.

Case 2

$a - 4 = -3$

$a - 4 + 4 = -3 + 4$ Add 4 to each side.

$a = 1$ Simplify.

The solution set is {1, 7}.

Study Tip

Absolute Value

Recall that $|a| = 3$ means $a = 3$ or $-a = 3$. The second equation can be written as $a = -3$. So, $|a - 4| = 3$ means $a - 4 = 3$ or $-(a - 4) = 3$. These can be written as $a - 4 = 3$ or $a - 4 = -3$.

Example 2 Write an Absolute Value Equation

Write an equation involving absolute value for the graph.

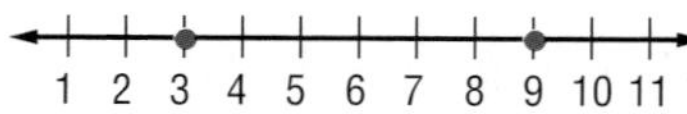

Find the point that is the same distance from 3 as the distance from 9. The midpoint between 3 and 9 is 6.

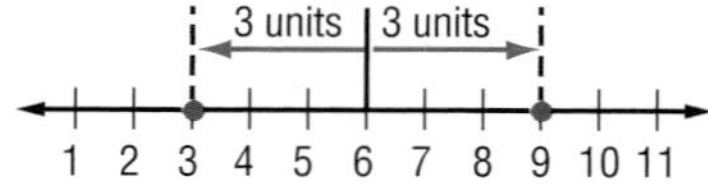

The distance from 6 to 3 is 3 units.
The distance from 6 to 9 is 3 units.

So, an equation is $|x - 6| = 3$.

CHECK Substitute 3 and 9 into $|x - 6| = 3$.

$|x - 6| = 3$ $\qquad$ $|x - 6| = 3$

$|3 - 6| \stackrel{?}{=} 3$ $\qquad$ $|9 - 6| \stackrel{?}{=} 3$

$|-3| \stackrel{?}{=} 3$ $\qquad$ $|3| \stackrel{?}{=} 3$

$3 = 3$ ✓ $\qquad$ $3 = 3$ ✓

ABSOLUTE VALUE INEQUALITIES Consider the inequality $|x| < n$. $|x| < 5$ means that the distance from 0 to x is less than 5 units.

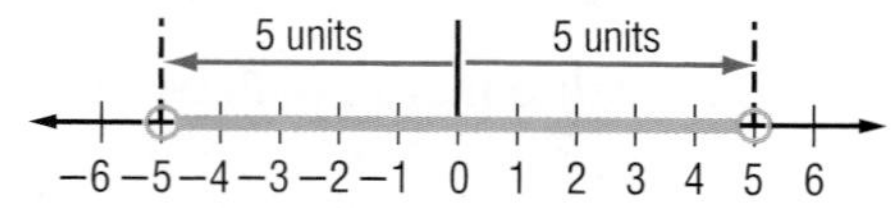

Therefore, $x > -5$ and $x < 5$. The solution set is $\{x \mid -5 < x < 5\}$.

The Algebra Activity explores an inequality of the form $|x| < n$.

Algebra Activity

Absolute Value

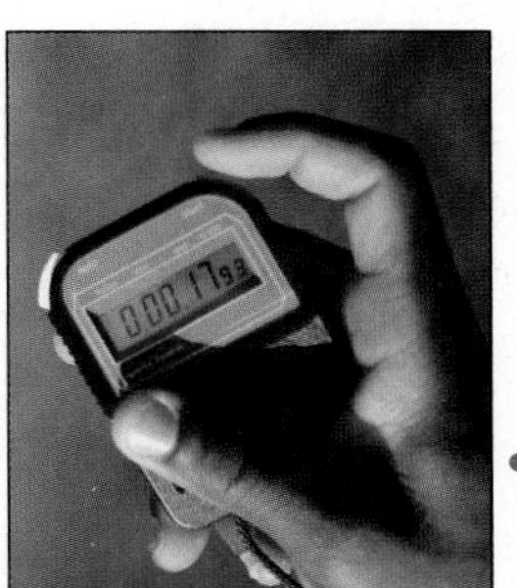

Collect the Data

- Work in pairs. One person is the timekeeper.
- Start timing. The other person tells the timekeeper to stop timing after he or she thinks that one minute has elapsed.
- Write down the time in seconds.
- Switch places. Make a table that includes the results of the entire class.

Analyze the Data

1. Determine the error by subtracting 60 seconds from each student's time.
2. What does a negative error represent? a positive error?
3. The *absolute error* is the absolute value of the error. Since absolute value cannot be negative, the absolute error is positive. If the absolute error is 6 seconds, write two possibilities for a student's estimated time of one minute.
4. What estimates would have an absolute error less than 6 seconds?
5. Graph the responses and highlight all values such that $|60 - x| < 6$. How many guesses were within 6 seconds?

When solving inequalities of the form $|x| < n$, find the intersection of these two cases.

Case 1 The value inside the absolute value symbols is less than the positive value of n.

Case 2 The value inside the absolute value symbols is greater than the negative value of n.

Study Tip

Less Than
When an absolute value is on the left and the inequality symbol is $<$ or $\leq$, the compound sentence uses *and*.

Example 3 Solve an Absolute Value Inequality (<)

Solve $|t + 5| < 9$. Then graph the solution set.

Write $|t + 5| < 9$ as $t + 5 < 9$ and $t + 5 > -9$.

Case 1

$t + 5 < 9$

$t + 5 - 5 < 9 - 5$ Subtract 5 from each side.

$t < 4$ Simplify.

Case 2

$t + 5 > -9$

$t + 5 - 5 > -9 - 5$ Subtract 5 from each side.

$t > -14$ Simplify.

The solution set is $\{t \mid -14 < t < 4\}$.

−16 −14 −12 −10 −8 −6 −4 −2 0 2 4 6

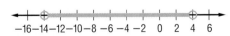

Consider the inequality $|x| > n$. $|x| > 5$ means that the distance from 0 to x is greater than 5 units.

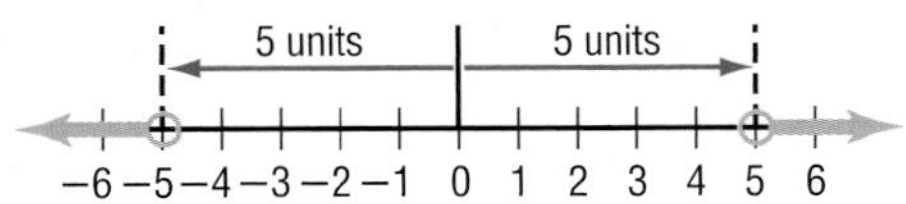

Therefore, $x < -5$ or $x > 5$. The solution set is $\{x \mid x < -5 \text{ or } x > 5\}$.

When solving inequalities of the form $|x| > n$, find the union of these two cases.

Case 1 The value inside the absolute value symbols is greater than the positive value of n.

Case 2 The value inside the absolute value symbols is less than the negative value of n.

Study Tip

Greater Than
When the absolute value is on the left and the inequality symbol is $>$ or $\geq$, the compound sentence uses *or*.

Example 4 Solve an Absolute Value Inequality (>)

Solve $|2x + 8| \geq 6$. Then graph the solution set.

Write $|2x + 8| \geq 6$ as $2x + 8 \geq 6$ or $2x + 8 \leq -6$.

Case 1

$2x + 8 \geq 6$

$2x + 8 - 8 \geq 6 - 8$ Subtract 8 from each side.

$2x \geq -2$ Simplify.

$\frac{2x}{2} \geq \frac{-2}{2}$ Divide each side by 2.

$x \geq -1$ Simplify.

Case 2

$2x + 8 \leq -6$

$2x + 8 - 8 \leq -6 - 8$ Subtract 8 from each side.

$2x \leq -14$ Simplify.

$\frac{2x}{2} \leq \frac{-14}{2}$ Divide each side by 2.

$x \leq -7$ Simplify.

The solution set is $\{x \mid x \leq -7 \text{ or } x \geq -1\}$.

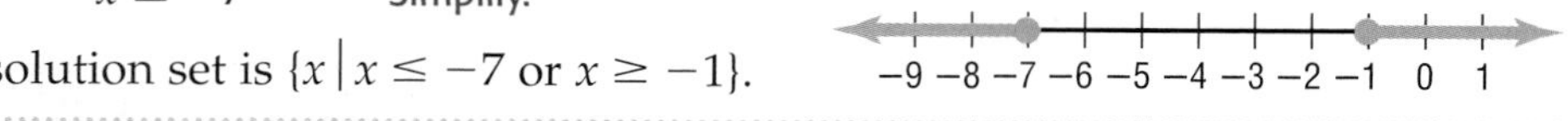

In general, there are three rules to remember when solving equations and inequalities involving absolute value.

Concept Summary *Absolute Value Equations and Inequalities*

If $|x| = n$, then $x = -n$ or $x = n$.

If $|x| < n$, then $x < n$ and $x > -n$.

If $|x| > n$, then $x > n$ or $x < -n$.

These properties are also true when $>$ or $<$ is replaced with $\geq$ or $\leq$.

Check for Understanding

Concept Check

1. **Compare and contrast** the solution of $|x - 2| > 6$ and the solution of $|x - 2| < 6$.
2. **OPEN ENDED** Write an absolute value inequality and graph its solution set.
3. **FIND THE ERROR** Leslie and Holly are solving $|x + 3| = 2$.

Leslie	
$x + 3 = 2$ or	$x + 3 = -2$
$x + 3 - 3 = 2 - 3$	$x + 3 - 3 = -2 - 3$
$x = -1$	$x = -5$

Holly	
$x + 3 = 2$ or	$x - 3 = 2$
$x + 3 - 3 = 2 - 3$	$x - 3 + 3 = 2 + 3$
$x = -1$	$x = 5$

Who is correct? Explain your reasoning.

Guided Practice

4. Which graph represents the solution of $|k| \leq 3$?

a. −5 −4 −3 −2 −1 0 1 2 3 4 5

b. −5 −4 −3 −2 −1 0 1 2 3 4 5

c. −5 −4 −3 −2 −1 0 1 2 3 4 5

d. −5 −4 −3 −2 −1 0 1 2 3 4 5

5. Which graph represents the solution of $|x - 4| > 2$?

a. −3 −2 −1 0 1 2 3 4 5 6 7

b. −3 −2 −1 0 1 2 3 4 5 6 7

c. −3 −2 −1 0 1 2 3 4 5 6 7

d. −3 −2 −1 0 1 2 3 4 5 6 7

6. Express the statement in terms of an inequality involving absolute value. Do not solve.
A jar contains 832 gumballs. Amanda's guess was within 46 pieces.

Solve each open sentence. Then graph the solution set.

7. $|r + 3| = 10$
8. $|c - 2| < 6$
9. $|10 - w| > 15$
10. $|2g + 5| \geq 7$

For each graph, write an open sentence involving absolute value.

11. −4 −3 −2 −1 0 1 2 3 4 5 6

12. 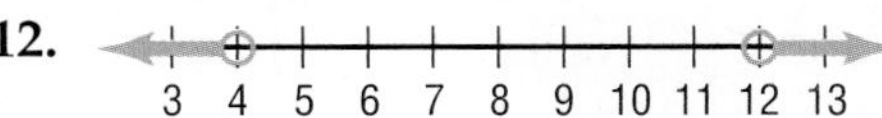

Application

13. **MANUFACTURING** A manufacturer produces bolts which must have a diameter within 0.001 centimeter of 1.5 centimeters. What are the acceptable measurements for the diameter of the bolts?

Practice and Apply

Homework Help

For Exercises	See Examples
14–19, 24–39, 46–51	1, 3, 4
20–23	3
40–45	2

Extra Practice

See page 834.

Match each open sentence with the graph of its solution set.

14. $|x + 5| \leq 3$ — a. −1 0 1 2 3 4 5 6 7 8 9
15. $|x - 4| > 4$ — b. −5 −4 −3 −2 −1 0 1 2 3 4 5
16. $|2x - 8| = 6$ — c. −9 −8 −7 −6 −5 −4 −3 −2 −1 0 1
17. $|x + 3| \geq -1$ — d. 2 3 4 5 6 7 8 9 10 11 12
18. $|x| < 2$ — e. −5 −4 −3 −2 −1 0 1 2 3 4 5
19. $|8 - x| = 2$ — f. −1 0 1 2 3 4 5 6 7 8 9

Express each statement using an inequality involving absolute value. Do *not* solve.

20. The pH of a buffered eye solution must be within 0.002 of a pH of 7.3.
21. The temperature inside a refrigerator should be within 1.5 degrees of 38°F.
22. Ramona's bowling score was within 6 points of her average score of 98.
23. The cruise control of a car set at 55 miles per hour should keep the speed within 3 miles per hour of 55.

Solve each open sentence. Then graph the solution set.

24. $|x - 5| = 8$
25. $|b + 9| = 2$
26. $|2p - 3| = 17$
27. $|5c - 8| = 12$
28. $|z - 2| \leq 5$
29. $|t + 8| < 2$
30. $|v + 3| > 1$
31. $|w - 6| \geq 3$
32. $|3s + 2| > -7$
33. $|3k + 4| \geq 8$
34. $|2n + 1| < 9$
35. $|6r + 8| < -4$
36. $|6 - (3d - 5)| \leq 14$
37. $|8 - (w - 1)| \leq 9$
38. $\left|\frac{5h + 2}{6}\right| = 7$
39. $\left|\frac{2 - 3x}{5}\right| \geq 2$

For each graph, write an open sentence involving absolute value.

40. (number line −5 to 5; closed dots at −5 and 5)

41. (number line −2 to 8; closed dots at −2 and 8)

42. (number line −5 to 5; closed dots at −3 and 3, shaded between)

43. (number line −8 to 2; open dots at −7 and 1, shaded between)

44. (number line −5 to 5; open dots at −1 and 3, shaded outward)

45. (number line −15 to −5; closed dots at −12 and −8, shaded outward)

More About. . .

Tire Pressure

Always inflate your tires to the pressure that is recommended by the manufacturer. The pressure stamped on the tire is the *maximum* pressure and should only be used under certain circumstances.

Source: www.etires.com

HEALTH For Exercises 46 and 47, use the following information.
The *average* length of a human pregnancy is 280 days. However, a healthy, full-term pregnancy can be 14 days longer or shorter.

46. Write an absolute value inequality for the length of a full-term pregnancy.
47. Solve the inequality for the length of a full-term pregnancy.
48. **FIRE SAFETY** The pressure of a typical fire extinguisher should be within 25 pounds per square inch (psi) of 195 psi. Write the range of pressures for safe fire extinguishers.
49. **HEATING** A thermostat with a 2-degree differential will keep the temperature within 2 degrees Fahrenheit of the temperature set point. Suppose your home has a thermostat with a 3-degree differential. If you set the thermostat at 68°F, what is the range of temperatures in the house?
50. **ENERGY** Use the margin of error indicated in the graph at the right to find the range of the percent of people who say protection of the environment should have priority over developing energy supplies.

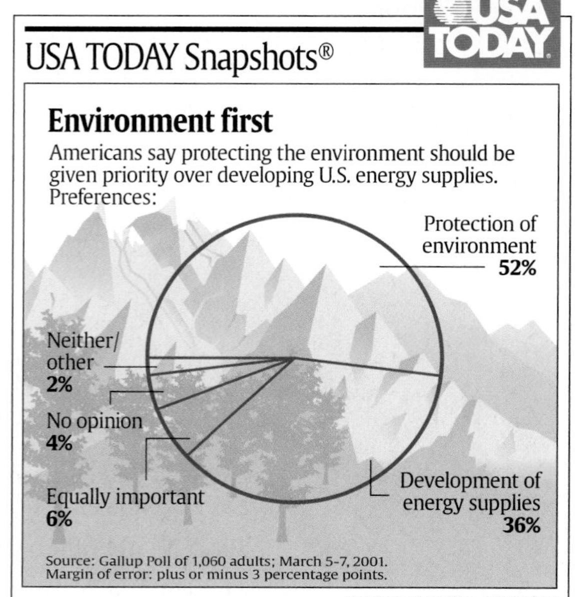
USA TODAY
USA TODAY Snapshots®
Environment first
Americans say protecting the environment should be given priority over developing U.S. energy supplies. Preferences:
Protection of environment 52%
Neither/other 2%
No opinion 4%
Equally important 6%
Development of energy supplies 36%
Source: Gallup Poll of 1,060 adults; March 5-7, 2001.
Margin of error: plus or minus 3 percentage points.
By Marcy E. Mullins, USA TODAY

51. **TIRE PRESSURE** Tire pressure is measured in pounds per square inch (psi). Tires should be kept within 2 psi of the manufacturer's recommended tire pressure. If the recommended inflation pressure for a tire is 30 psi, what is the range of acceptable pressures?
52. **CRITICAL THINKING** State whether each open sentence is *always*, *sometimes*, or *never* true.
    a. $|x + 3| < -5$
    b. $|x - 6| > -1$
    c. $|x + 2| = 0$

More About. . .

Physical Science

The common name for sodium chloride is salt. Seawater is about 2.5% salt, and salt obtained by evaporating seawater is 95% to 98% pure.

Source: *World Book Encyclopedia*

53. **PHYSICAL SCIENCE** During an experiment, Li-Cheng must add 3.0 milliliters of sodium chloride to a solution. To get accurate results, the amount of sodium chloride must be within 0.5 milliliter of the required amount. How much sodium chloride can she add and still obtain the correct results?

54. **ENTERTAINMENT** Luis Gomez is a contestant on a television game show. He must guess within $1500 of the actual price of the car without going over in order to win the car. The actual price of the car is $18,000. What is the range of guesses in which Luis can win the vehicle?

55. **CRITICAL THINKING** The symbol $\pm$ means *plus* or *minus*.
   a. If $x = 3 \pm 1.2$, what are the values of x?
   b. Write $x = 3 \pm 1.2$ as an expression involving absolute value.

56. WRITING IN MATH Answer the question that was posed at the beginning of the lesson.

How is absolute value used in election polls?

Include the following in your answer:
- an explanation of how to solve the inequality describing the percent of people who are against the tax levy, and
- a prediction of whether you think the tax levy will pass and why.

Standardized Test Practice

57. Choose the replacement set that makes $|x + 5| = 2$ true.
   Ⓐ $\{-3, 3\}$ Ⓑ $\{-3, -7\}$ Ⓒ $\{2, -2\}$ Ⓓ $\{3, -7\}$

58. What can you conclude about x if $-6 < |x| < 6$?
   Ⓐ $-x \geq 0$ Ⓑ $x \leq 0$ Ⓒ $-x < 6$ Ⓓ $-x > 6$

Maintain Your Skills

Mixed Review

59. **FITNESS** To achieve the maximum benefits from aerobic activity, your heart rate should be in your target zone. Your target zone is the range between 60% and 80% of your maximum heart rate. If Rafael's maximum heart rate is 190 beats per minute, what is his target zone? *(Lesson 6-4)*

Solve each inequality. Then check your solution. *(Lesson 6-3)*

60. $2m + 7 > 17$
61. $-2 - 3x \geq 2$
62. $\frac{2}{3}w - 3 \leq 7$

Find the slope and y-intercept of each equation. *(Lesson 5-4)*

63. $2x + y = 4$
64. $2y - 3x = 4$
65. $\frac{1}{2}x + \frac{3}{4}y = 0$

Solve each equation or formula for the variable specified. *(Lesson 3-8)*

66. $I = prt$, for r
67. $ex - 2y = 3z$, for x
68. $\frac{a + 5}{3} = 7x$, for x

Find each sum or difference. *(Lesson 2-2)*

69. $-13 + 8$
70. $-13.2 - 6.1$
71. $-4.7 - (-8.9)$

Name the property illustrated by each statement. *(Lesson 1-6)*

72. $10x + 10y = 10(x + y)$
73. $(2 + 3)a + 7 = 5a + 7$

Getting Ready for the Next Lesson

PREREQUISITE SKILL **Graph each equation.**
*(To review **graphing linear equations**, see Lesson 4-5.)*

74. $y = 3x + 4$
75. $y = -2$
76. $x + y = 3$
77. $y - 2x = -1$
78. $2y - x = -6$
79. $2(x + y) = 10$

6-6 Graphing Inequalities in Two Variables

What You'll Learn

- Graph inequalities on the coordinate plane.
- Solve real-world problems involving linear inequalities.

Vocabulary
- half-plane
- boundary

How are inequalities used in budgets?

Hannah allots up to \$30 a month for lunch on school days. On most days, she brings her lunch. She can also buy lunch at the cafeteria or at a fast-food restaurant. She spends an average of \$3 a day at the cafeteria and an average of \$4 a day at a restaurant. How many times a month can Hannah buy her lunch and remain within her budget?

My Monthly Budget
Lunch (school days) \$30
Entertainment \$55
Clothes \$50
Fuel \$60

Let x represent the number of days she buys lunch at the cafeteria, and let y represent the number of days she buys lunch at a restaurant. Then the following inequality can be used to represent the situation.

The cost of eating in the cafeteria	plus	the cost of eating in a restaurant	is less than or equal to	\$30.
$3x$	$+$	$4y$	$\leq$	30

There are many solutions of this inequality.

GRAPH LINEAR INEQUALITIES Like a linear equation in two variables, the solution set of an inequality in two variables is graphed on a coordinate plane. The solution set of an inequality in two variables is the set of all ordered pairs that satisfy the inequality.

Example 1 Ordered Pairs that Satisfy an Inequality

From the set {(1, 6), (3, 0), (2, 2), (4, 3)}, which ordered pairs are part of the solution set for $3x + 2y < 12$?

Use a table to substitute the x and y values of each ordered pair into the inequality.

x	y	$3x + 2y < 12$	True or False
1	6	$3(1) + 2(6) < 12$ $15 < 12$	false
3	0	$3(3) + 2(0) < 12$ $9 < 12$	true
2	2	$3(2) + 2(2) < 12$ $10 < 12$	true
4	3	$3(4) + 2(3) < 12$ $18 < 12$	false

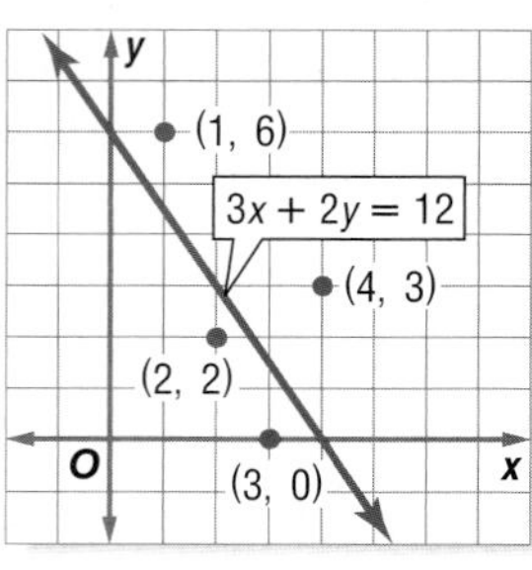

The ordered pairs {(3, 0), (2, 2)} are part of the solution set of $3x + 2y < 12$. In the graph, notice the location of the two ordered pairs that are solutions for $3x + 2y < 12$ in relation to the line.

The solution set for an inequality in two variables contains many ordered pairs when the domain and range are the set of real numbers. The graphs of all of these ordered pairs fill a region on the coordinate plane called a **half-plane**. An equation defines the **boundary** or edge for each half-plane.

Key Concept — Half-Planes and Boundaries

- **Words** Any line in the plane divides the plane into two regions called half-planes. The line is called the boundary of each of the two half-planes.
- **Model**

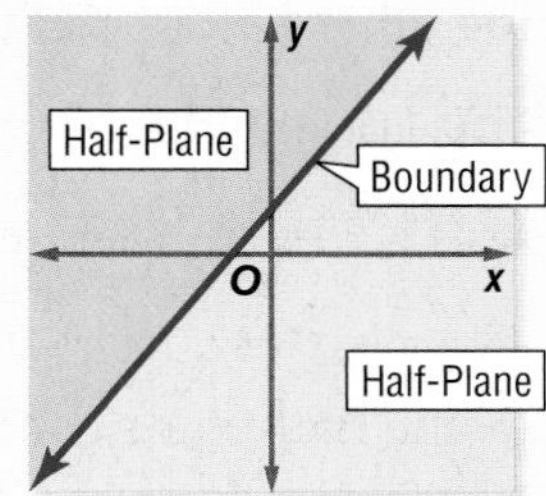

> **Study Tip**
>
> *Dashed Line*
> - Like a circle on a number line, a dashed line on a coordinate plane indicates that the boundary is *not* part of the solution set.
>
> *Solid Line*
> - Like a dot on a number line, a solid line on a coordinate plane indicates that the boundary *is* included.

Consider the graph of $y > 4$. First determine the boundary by graphing $y = 4$, the equation you obtain by replacing the inequality sign with an equals sign. Since the inequality involves y-values greater than 4, but not equal to 4, the line should be dashed. The boundary divides the coordinate plane into two half-planes.

To determine which half-plane contains the solution, choose a point from each half-plane and test it in the inequality.

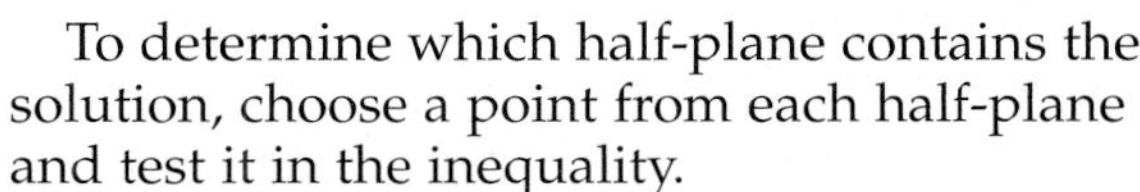

Try (3, 0).		Try (5, 6).	
$y > 4$	$y = 0$	$y > 4$	$y = 6$
$0 > 4$	false	$6 > 4$	true

The half-plane that contains (5, 6) contains the solution. Shade that half-plane.

Example 2 Graph an Inequality

Graph $y - 2x \leq -4$.

Step 1 Solve for y in terms of x.

$$y - 2x \leq -4 \quad \text{Original inequality}$$

$$y - 2x + 2x \leq -4 + 2x \quad \text{Add } 2x \text{ to each side.}$$

$$y \leq 2x - 4 \quad \text{Simplify.}$$

Step 2 Graph $y = 2x - 4$. Since $y \leq 2x - 4$ means $y < 2x - 4$ or $y = 2x - 4$, the boundary is included in the solution set. The boundary should be drawn as a solid line.

(continued on the next page)

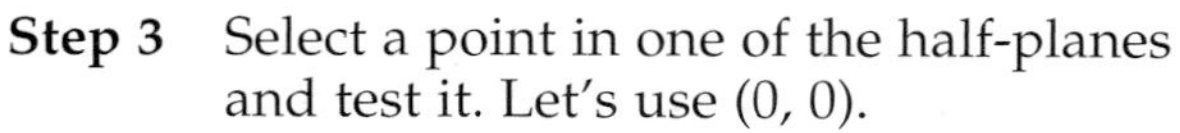

Study Tip

Origin as the Test Point Use the origin as a standard test point because the values are easy to substitute into the inequality.

Step 3 Select a point in one of the half-planes and test it. Let's use (0, 0).

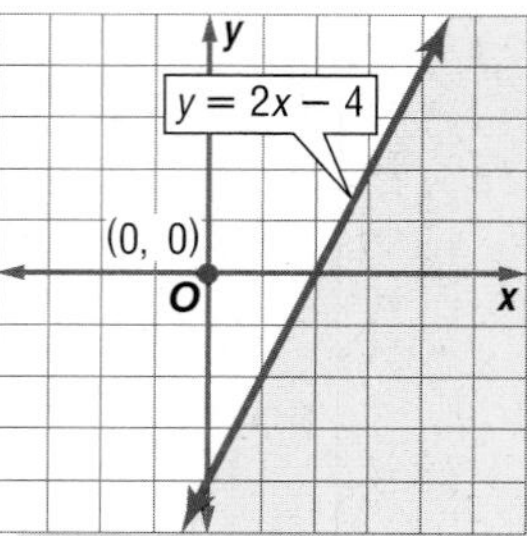

$y - 2x \leq -4$ Original inequality

$0 \leq 2(0) - 4$ $x = 0, y = 0$

$0 \leq -4$ false

Since the statement is false, the half-plane containing the origin is not part of the solution. Shade the other half-plane.

CHECK Test a point in the other half plane, for example, (3, −3).

$y - 2x \leq -4$ Original inequality

$-3 \leq 2(3) - 4$ $x = 3, y = -3$

$-3 \leq 2$ ✓

Since the statement is true, the half-plane containing (3, −3) should be shaded. The graph of the solution is correct.

SOLVE REAL-WORLD PROBLEMS When solving real-world inequalities, the domain and range of the inequality are often restricted to nonnegative numbers or whole numbers.

Example 3 Write and Solve an Inequality

ADVERTISING Rosa Padilla sells radio advertising in 30-second and 60-second time slots. During every hour, there are up to 15 minutes available for commercials. How many commercial slots can she sell for one hour of broadcasting?

More About...

Advertising

A typical one-hour program on television contains 40 minutes of the program and 20 minutes of commercials. During peak periods, a 30-second commercial can cost an average of $2.3 million.

Source: www.superbowl-ads.com

Step 1 Let x equal the number of 30-second commercials. Let y equal the number of 60-second or 1-minute commercials. Write an open sentence representing this situation.

$\frac{1}{2}$ min	times	the number of 30-s commercials	plus	the number of 1-min commercials	is up to	15 min.
$\frac{1}{2}$	$\cdot$	x	$+$	y	$\leq$	15

Step 2 Solve for y in terms of x.

$\frac{1}{2}x + y \leq 15$ Original inequality

$\frac{1}{2}x + y - \frac{1}{2}x \leq 15 - \frac{1}{2}x$ Subtract $\frac{1}{2}x$ from each side.

$y \leq 15 - \frac{1}{2}x$ Simplify.

Step 3 Since the open sentence includes the equation, graph $y = 15 - \frac{1}{2}x$ as a solid line. Test a point in one of the half-planes, for example (0, 0). Shade the half-plane containing (0, 0) since $0 \leq 15 - \frac{1}{2}(0)$ is true.

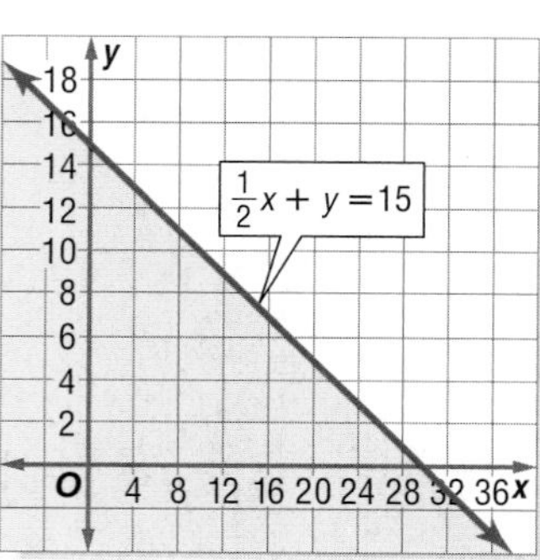

Step 4 Examine the solution.

- Rosa cannot sell a negative number of commercials. Therefore, the domain and range contain only nonnegative numbers.
- She also cannot sell half of a commercial. Thus, only points in the shaded half-plane whose x- and y-coordinates are whole numbers are possible solutions.

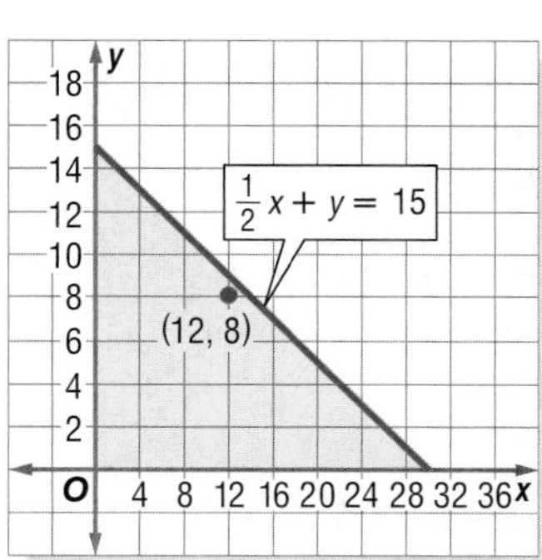

One solution is (12, 8). This represents twelve 30-second commercials and eight 60-second commercials in a one hour period.

Check for Understanding

Concept Check

1. **Compare and contrast** the graph of $y = x + 2$ and the graph of $y < x + 2$.
2. **OPEN ENDED** Write an inequality in two variables and graph it.
3. **Explain** why it is usually only necessary to test one point when graphing an inequality.

Guided Practice **Determine which ordered pairs are part of the solution set for each inequality.**

4. $y \leq x + 1$, $\{(-1, 0), (3, 2), (2, 5), (-2, 1)\}$
5. $y > 2x$, $\{(2, 6), (0, -1), (3, 5), (-1, -2)\}$

6. Which graph represents $y - 2x \geq 2$?

a. 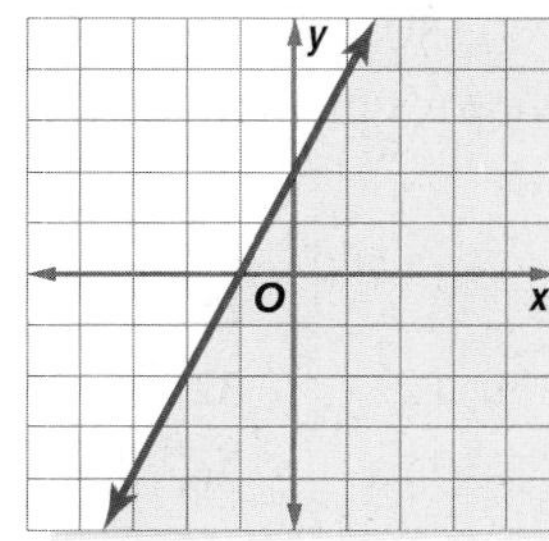

b. 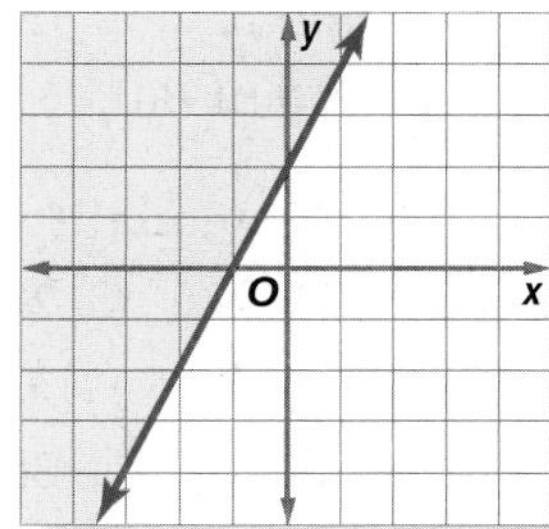

c. 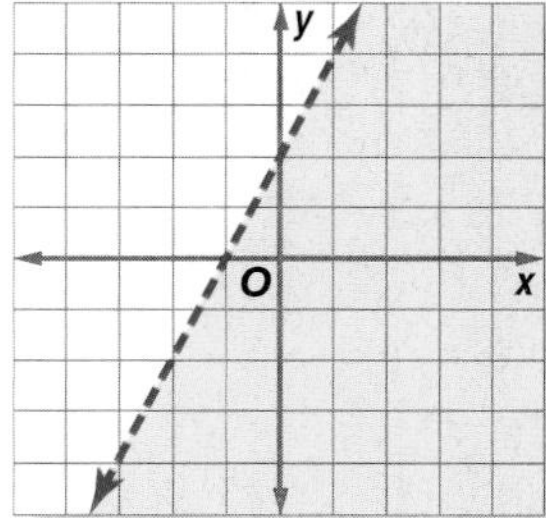

Graph each inequality.

7. $y \geq 4$
8. $y \leq 2x - 3$
9. $4 - 2x < -2$
10. $1 - y > x$

Application

11. **ENTERTAINMENT** Coach Riley wants to take her softball team out for pizza and soft drinks after the last game of the season. She doesn't want to spend more than \$60. Write an inequality that represents this situation and graph the solution set.

Practice and Apply

Homework Help

For Exercises	See Examples
12–19	1
20–37	2
38–44	3

Extra Practice

See page 835.

Determine which ordered pairs are part of the solution set for each inequality.

12. $y \leq 3 - 2x$, $\{(0, 4), (-1, 3), (6, -8), (-4, 5)\}$

13. $y < 3x$, $\{(-3, 1), (-3, 2), (1, 1), (1, 2)\}$

14. $x + y < 11$, $\{(5, 7), (-13, 10), (4, 4), (-6, -2)\}$

15. $2x - 3y > 6$, $\{(3, 2), (-2, -4), (6, 2), (5, 1)\}$

16. $4y - 8 \geq 0$, $\{(5, -1), (0, 2), (2, 5), (-2, 0)\}$

17. $3x + 4y < 7$, $\{(1, 1), (2, -1), (-1, 1), (-2, 4)\}$

18. $|x - 3| \geq y$, $\{(6, 4), (-1, 8), (-3, 2), (5, 7)\}$

19. $|y + 2| < x$, $\{(2, -4), (-1, -5), (6, -7), (0, 0)\}$

Match each inequality with its graph.

20. $2y + x \leq 6$

21. $\frac{1}{2}x - y > 4$

22. $y > 3 + \frac{1}{2}x$

23. $4y + 2x \geq 16$

a.

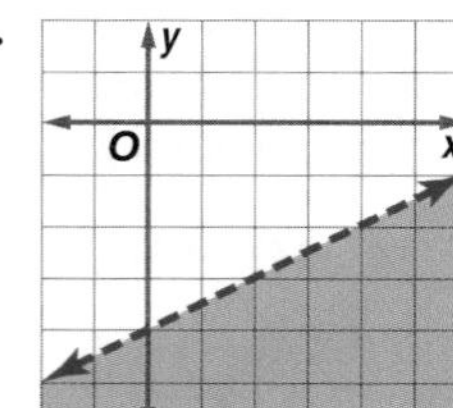

b.

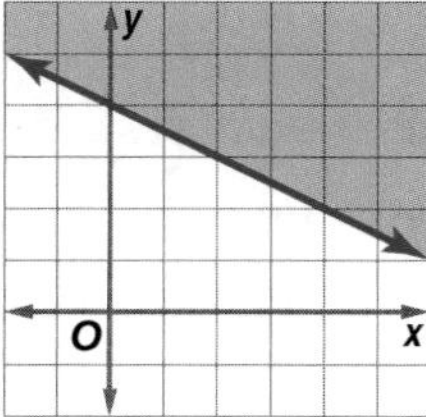

c.

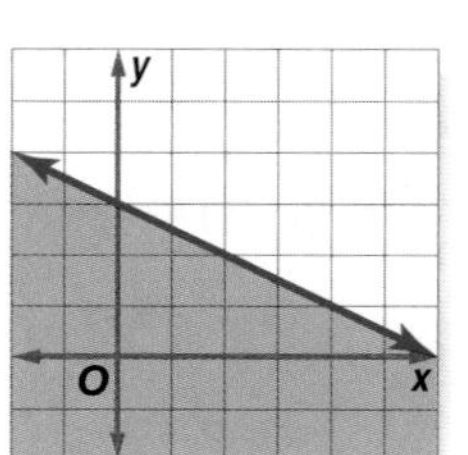

d.

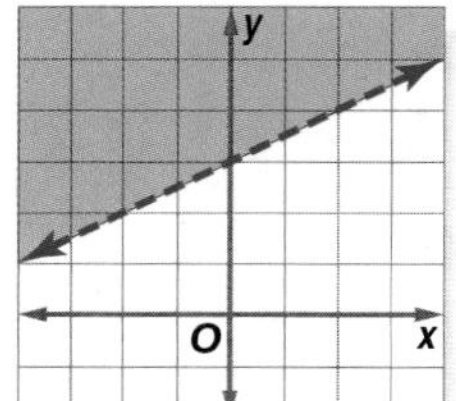

24. Is the point $A(2, 3)$ on, above, or below the graph of $-2x + 3y = 5$?

25. Is the point $B(0, 1)$ on, above, or below the graph of $4x - 3y = 4$?

Graph each inequality.

26. $y < -3$ **27.** $x \geq 2$ **28.** $5x + 10y > 0$ **29.** $y < x$

30. $2y - x \leq 6$ **31.** $6x + 3y > 9$ **32.** $3y - 4x \geq 12$ **33.** $y \leq -2x - 4$

34. $8x - 6y < 10$ **35.** $3x - 1 \geq y$ **36.** $3(x + 2y) > -18$ **37.** $\frac{1}{2}(2x + y) < 2$

POSTAGE **For Exercises 38 and 39, use the following information.**
The U.S. Postal Service limits the size of packages to those in which the length of the longest side plus the distance around the thickest part is less than or equal to 108 inches.

38. Write an inequality that represents this situation.

39. Are there any restrictions on the domain or range?

Online Research **Data Update** What are the current postage rates and regulations? Visit www.algebra1.com/data_update to learn more.

SHIPPING **For Exercises 40 and 41, use the following information.**
A delivery truck is transporting televisions and microwaves to an appliance store. The weight limit for the truck is 4000 pounds. The televisions weigh 77 pounds, and the microwaves weigh 55 pounds.

40. Write an inequality for this situation.

41. Will the truck be able to deliver 35 televisions and 25 microwaves at once?

FALL DANCE **For Exercises 42–44, use the following information.**
Tickets for the fall dance are \$5 per person or \$8 for couples. In order to cover expenses, at least \$1200 worth of tickets must be sold.

42. Write an inequality that represents this situation.

43. Graph the inequality.

44. If 100 single tickets and 125 couple tickets are sold, will the committee cover its expenses?

WebQuest

A linear inequality can be used to represent trends in Olympic times. Visit www.algebra1.com/webquest to continue work on your WebQuest project.

45. CRITICAL THINKING Graph the intersection of the graphs of $y \le x - 1$ and $y \ge -x$.

46. 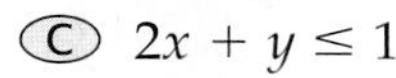

Answer the question that was posed at the beginning of the lesson.

How are inequalities used in budgets?

Include the following in your answer:

- an explanation of the restrictions placed on the domain and range of the inequality used to describe the number of times Hannah can buy her lunch, and
- three possible solutions of the inequality.

47. Which ordered pair is *not* a solution of $y - 2x < -5$?

Ⓐ (2, −2) Ⓑ (−1, −8) Ⓒ (4, 1) Ⓓ (5, 6)

48. Which inequality is represented by the graph at the right?

Ⓐ $2x + y < 1$ Ⓑ $2x + y > 1$

Ⓒ $2x + y \le 1$ Ⓓ $2x + y \ge 1$

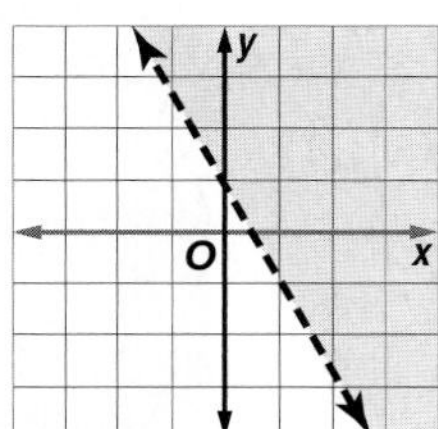

Maintain Your Skills

Mixed Review

Solve each open sentence. Then graph the solution set. *(Lesson 6-5)*

49. $|3 + 2t| = 11$ **50.** $|x + 8| < 6$ **51.** $|2y + 5| \ge 3$

Solve each compound inequality. Then graph the solution. *(Lesson 6-4)*

52. $y + 6 > -1$ and $y - 2 < 4$ **53.** $m + 4 < 2$ or $m - 2 > 1$

State whether each percent of change is a percent of *increase* or *decrease*. Then find the percent of change. Round to the nearest whole percent. *(Lesson 3-7)*

54. original: 200
new: 172

55. original: 100
new: 142

56. original: 53
new: 75

Solve each equation. *(Lesson 3-4)*

57. $\frac{d - 2}{3} = 7$ **58.** $3n + 6 = -15$ **59.** $35 + 20h = 100$

Simplify. *(Lesson 2-4)*

60. $\frac{-64}{4}$ **61.** $\frac{27c}{-9}$ **62.** $\frac{12a - 14b}{-2}$ **63.** $\frac{18y - 9}{3}$

Graphing Calculator Investigation

A Follow-Up of Lesson 6-6

Graphing Inequalities

You can use a TI-83 Plus graphing calculator to investigate the graphs of inequalities. Since graphing calculators only shade between two functions, enter a lower boundary as well as an upper boundary for each inequality.

Graph two different inequalities on your graphing calculator.

Step 1 *Graph $y \leq 3x + 1$.*

- Clear all functions from the Y= list.

  KEYSTROKES: [Y=] [CLEAR]

- Graph $y \leq 3x + 1$ in the standard window.

  KEYSTROKES: [2nd] [DRAW] 7 [(–)] 10 [,] 3 [X,T,θ,n] [+] 1 [)] [ENTER]

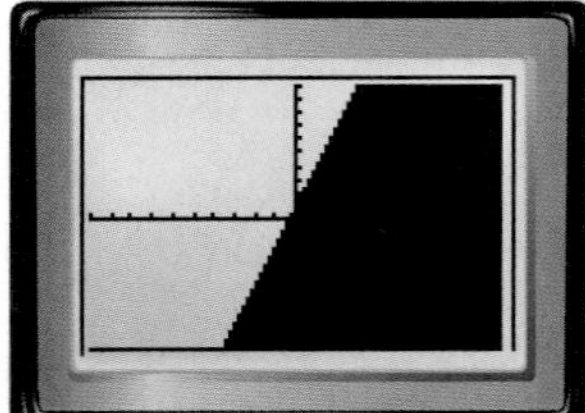

The lower boundary is Ymin or −10. The upper boundary is $y = 3x + 1$. All ordered pairs for which y is *less than or equal to* $3x + 1$ lie *below or on* the line and are solutions.

Step 2 *Graph $y - 3x \geq 1$.*

- Clear the drawing that is currently displayed.

  KEYSTROKES: [2nd] [DRAW] 1

- Rewrite $y - 3x \geq 1$ as $y \geq 3x + 1$ and graph it.

  KEYSTROKES: [2nd] [DRAW] 7 3 [X,T,θ,n] [+] 1 [,] 10 [)] [ENTER]

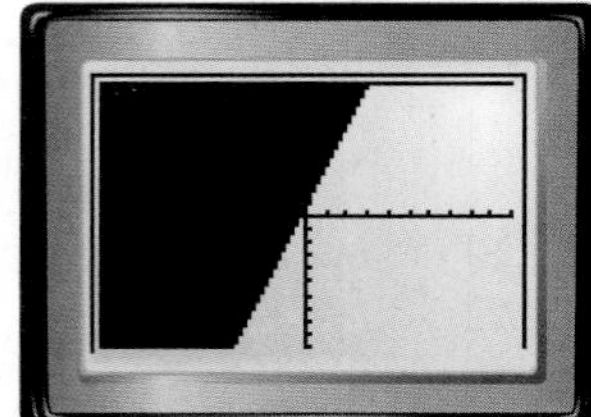

This time, the lower boundary is $y = 3x + 1$. The upper boundary is Ymax or 10. All ordered pairs for which y is *greater than or equal to* $3x + 1$ lie *above or on* the line and are solutions.

Exercises

1. Compare and contrast the two graphs shown above.

2. Graph the inequality $y \geq -2x + 4$ in the standard viewing window.

  a. What functions do you enter as the lower and upper boundaries?

  b. Using your graph, name four solutions of the inequality.

3. Suppose student movie tickets cost \$4 and adult movie tickets cost \$8. You would like to buy at least 10 tickets, but spend no more than \$80.

  a. Let x = number of student tickets and y = number of adult tickets. Write two inequalities, one representing the total number of tickets and the other representing the total cost of the tickets.

  b. Which inequalities would you use as the lower and upper boundaries?

  c. Graph the inequalities. Use the viewing window [0, 20] scl: 1 by [0, 20] scl: 1.

  d. Name four possible combinations of student and adult tickets.

 www.algebra1.com/other_calculator_keystrokes

Study Guide and Review

Vocabulary and Concept Check

Addition Property of Inequalities (p. 318)
boundary (p. 353)
compound inequality (p. 339)
Division Property of Inequalities (p. 327)
half-plane (p. 353)
intersection (p. 339)
Multiplication Property of Inequalities (p. 325)
set-builder notation (p. 319)
Subtraction Property of Inequalities (p. 319)
union (p. 340)

Choose the letter of the term that best matches each statement, algebraic expression, or algebraic sentence.

1. $\{w \mid w \geq -14\}$
2. If $x \leq y$, then $-5x \geq -5y$.
3. $p > -5$ and $p \leq 0$
4. If $a < b$, then $a + 2 < b + 2$.
5. the graph on one side of a boundary
6. If $s \geq t$, then $s - 7 \geq t - 7$.
7. $g \geq 7$ or $g < 2$
8. If $m > n$, then $\frac{m}{7} > \frac{n}{7}$.

a. Addition Property of Inequalities
b. Division Property of Inequalities
c. half-plane
d. intersection
e. Multiplication Property of Inequalities
f. set-builder notation
g. Subtraction Property of Inequalities
h. union

Lesson-by-Lesson Review

6-1 Solving Inequalities by Addition and Subtraction

See pages 318–323.

Concept Summary

- If any number is added to each side of a true inequality, the resulting inequality is also true.
- If any number is subtracted from each side of a true inequality, the resulting inequality is also true.

Examples **Solve each inequality.**

1 $f + 9 \leq -23$

$f + 9 \leq -23$	Original inequality
$f + 9 - 9 \leq -23 - 9$	Subtract.
$f \leq -32$	Simplify.

The solution set is $\{f \mid f \leq -32\}$.

2 $v - 19 > -16$

$v - 19 > -16$	Original inequality
$v - 19 + 19 > -16 + 19$	Add.
$v > 3$	Simplify.

The solution set is $\{v \mid v > 3\}$.

Exercises **Solve each inequality. Then check your solution, and graph it on a number line.** *See Examples 1–5 on pages 318–320.*

9. $c + 51 > 32$
10. $r + 7 > -5$
11. $w - 14 \leq 23$
12. $a - 6 > -10$
13. $-0.11 \geq n - (-0.04)$
14. $2.3 < g - (-2.1)$
15. $7h \leq 6h - 1$
16. $5b > 4b + 5$
17. Define a variable, write an inequality, and solve the problem. Then check your solution. *Twenty-one is no less than the sum of a number and negative two.*

6-2 Solving Inequalities by Multiplication and Division

See pages 325–331.

Concept Summary

- If each side of a true inequality is multiplied or divided by the same positive number, the resulting inequality is also true.
- If each side of a true inequality is multiplied or divided by the same negative number, the direction of the inequality must be *reversed*.

Examples **Solve each inequality.**

1 $-14g \geq 126$

$-14g \geq 126$ Original inequality

$\frac{-14g}{-14} \leq \frac{126}{-14}$ Divide and change $\geq$ to $\leq$.

$g \leq -9$ Simplify.

The solution set is $\{g \mid g \leq -9\}$.

2 $\frac{3}{4}d < 15$

$\frac{3}{4}d < 15$ Original inequality

$\left(\frac{4}{3}\right)\frac{3}{4}d < \left(\frac{4}{3}\right)15$ Multiply each side by $\frac{4}{3}$.

$d < 20$ Simplify.

The solution set is $\{d \mid d < 20\}$.

Exercises **Solve each inequality. Then check your solution.**

See Examples 1–5 on pages 326–328.

18. $15v > 60$ **19.** $12r \leq 72$ **20.** $-15z \geq -75$ **21.** $-9m < 99$

22. $\frac{b}{-12} \leq 3$ **23.** $\frac{d}{-13} > -5$ **24.** $\frac{2}{3}w > -22$ **25.** $\frac{3}{5}p \leq -15$

26. Define a variable, write an inequality, and solve the problem. Then check your solution. *Eighty percent of a number is greater than or equal to 24.*

6-3 Solving Multi-Step Inequalities

See pages 332–337.

Concept Summary

- Multi-step inequalities can be solved by undoing the operations.
- Remember to reverse the inequality sign when multiplying or dividing each side by a negative number.
- When solving equations that contain grouping symbols, first use the Distributive Property to remove the grouping symbols.

Example **Solve $4(n - 1) < 7n + 8$.**

$4(n - 1) < 7n + 8$ Original inequality

$4n - 4 < 7n + 8$ Distributive Property

$4n - 4 - 7n < 7n + 8 - 7n$ Subtract $7n$ from each side.

$-3n - 4 < 8$ Simplify.

$-3n - 4 + 4 < 8 + 4$ Add 4 to each side.

$-3n < 12$ Simplify.

$\frac{-3n}{-3} > \frac{12}{-3}$ Divide each side by -3 and change $<$ to $>$.

$n > -4$ Simplify.

The solution set is $\{n \mid n > -4\}$.

Exercises **Solve each inequality. Then check your solution.**
See Examples 1–5 on pages 332–334.

27. $-4h + 7 > 15$ **28.** $5 - 6n > -19$ **29.** $-5x + 3 < 3x + 19$

30. $15b - 12 > 7b + 60$ **31.** $-5(q + 12) < 3q - 4$ **32.** $7(g + 8) < 3(g + 2) + 4g$

33. $\frac{2(x + 2)}{3} \geq 4$ **34.** $\frac{1 - 7n}{5} > 10$

35. Define a variable, write an inequality, and solve the problem. Then check your solution. *Two thirds of a number decreased by 27 is at least 9.*

6-4 Solving Compound Inequalities

See pages 339–344.

Concept Summary

- The solution of a compound inequality containing *and* is the intersection of the graphs of the two inequalities.
- The solution of a compound inequality containing *or* is the union of the graphs of the two inequalities.

Examples **Graph the solution set of each compound inequality.**

1 $x \geq -1$ **and** $x > 3$

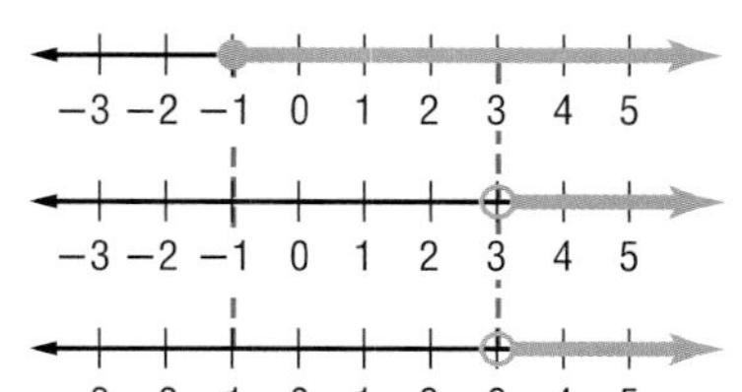

$x \geq -1$

$x > 3$

Find the intersection.

The solution set is $\{x \mid x > 3\}$.

2 $x \leq 8$ **or** $x < 2$

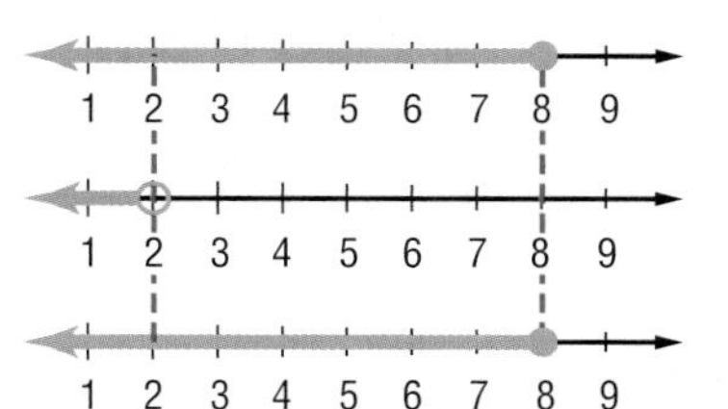

$x \leq 8$

$x < 2$

Find the union.

The solution set is $\{x \mid x \leq 8\}$.

Exercises **Solve each compound inequality. Then graph the solution set.**
See Examples 1–4 on pages 339–341.

36. $-1 < p + 3 < 5$ **37.** $-3 < 2k - 1 < 5$ **38.** $3w + 8 < 2$ or $w + 12 > 2 - w$

39. $a - 3 \leq 8$ or $a + 5 \geq 21$ **40.** $m + 8 < 4$ and $3 - m < 5$ **41.** $10 - 2y > 12$ and $7y < 4y + 9$

6-5 Solving Open Sentences Involving Absolute Value

See pages 345–351.

Concept Summary

- If $|x| = n$, then $x = -n$ or $x = n$.

−n 0 n

- If $|x| < n$, then $x > -n$ and $x < n$.

−n 0 n

- If $|x| > n$, then $x < -n$ or $x > n$.

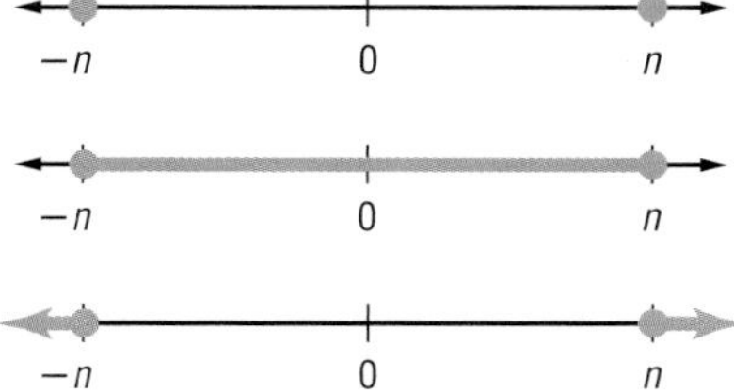

• Extra Practice, see pages 833–835.
• Mixed Problem Solving, see page 858.

Example **Solve $|x + 6| = 15$.**

$$|x + 6| = 15$$

$$x + 6 = 15 \quad \text{or} \quad x + 6 = -15$$

$$x + 6 - 6 = 15 - 6 \qquad x + 6 - 6 = -15 - 6$$

$$x = 9 \qquad x = -21$$

The solution set is $\{-21, 9\}$.

Exercises **Solve each open sentence. Then graph the solution set.**
See Examples 1, 3, and 4 on pages 346–348.

42. $|w - 8| = 12$ **43.** $|q + 5| = 2$ **44.** $|h + 5| > 7$ **45.** $|w + 8| \geq 1$

46. $|r + 10| < 3$ **47.** $|t + 4| \leq 3$ **48.** $|2x + 5| < 4$ **49.** $|3d + 4| < 8$

6-6 Graphing Inequalities in Two Variables

See pages 352–357.

Concept Summary

- To graph an inequality in two variables:

  Step 1 Determine the boundary and draw a dashed or solid line.

  Step 2 Select a test point. Test that point.

  Step 3 Shade the half-plane that contains the solution.

Example **Graph $y \geq x - 2$.**

Since the boundary is included in the solution, draw a solid line.

Test the point (0, 0).

$y \geq x - 2$ Original inequality

$0 \geq 0 - 2$ $x = 0, y = 0$

$0 \geq -2$ true

The half plane that contains (0, 0) should be shaded.

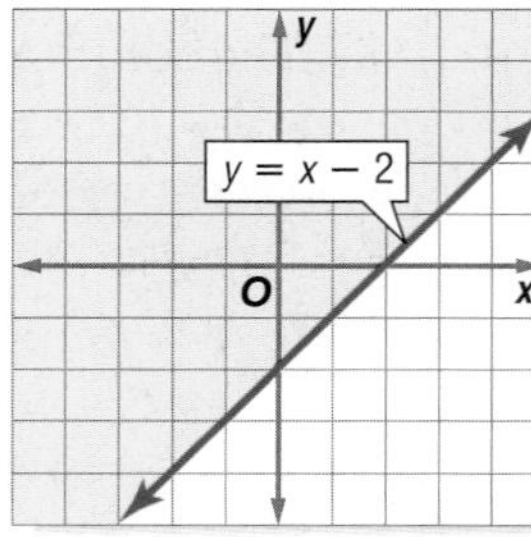

Exercises **Determine which ordered pairs are part of the solution set for each inequality.** *See Example 1 on page 352.*

50. $3x + 2y < 9$, $\{(1, 3), (3, 2), (-2, 7), (-4, 11)\}$

51. $5 - y \geq 4x$, $\left\{(2, -5), \left(\frac{1}{2}, 7\right), (-1, 6), (-3, 20)\right\}$

52. $\frac{1}{2}y \leq 6 - x$, $\{(-4, 15), (5, 1), (3, 8), (-2, 25)\}$

53. $-2x < 8 - y$, $\{(5, 10), (3, 6), (-4, 0), (-3, 6)\}$

Graph each inequality. *See Example 2 on pages 353 and 354.*

54. $y - 2x < -3$ **55.** $x + 2y \geq 4$ **56.** $y \leq 5x + 1$ **57.** $2x - 3y > 6$

Chapter 6

Practice Test

Vocabulary and Concepts

1. **Write** *the set of all numbers t such that t is greater than or equal to 17* in set-builder notation.
2. **Show** how to solve $6(a + 5) < 2a + 8$. Justify your work.
3. **OPEN ENDED** Give an example of a compound inequality that is an intersection and an example of a compound inequality that is a union.
4. **Compare and contrast** the graphs of $|x| \leq 3$ and $|x| \geq 3$.

Skills and Applications

Solve each inequality. Then check your solution.

5. $-23 \geq g - 6$
6. $9p < 8p - 18$
7. $d - 5 < 2d - 14$
8. $\frac{7}{8}w \geq -21$
9. $-22b \leq 99$
10. $4m - 11 \geq 8m + 7$
11. $-3(k - 2) > 12$
12. $\frac{f-5}{3} > -3$
13. $0.3(y - 4) \leq 0.8(0.2y + 2)$
14. **REAL ESTATE** A homeowner is selling her house. She must pay 7% of the selling price to her real estate agent after the house is sold. To the nearest dollar, what must be the selling price of her house to have at least \$110,000 after the agent is paid?
15. Solve $6 + |r| = 3$.
16. Solve $|d| > -2$.

Solve each compound inequality. Then graph the solution set.

17. $r + 3 > 2$ and $4r < 12$
18. $3n + 2 \geq 17$ or $3n + 2 \leq -1$
19. $9 + 2p > 3$ and $-13 > 8p + 3$
20. $|2a - 5| < 7$
21. $|7 - 3s| \geq 2$
22. $|7 - 5z| > 3$

Define a variable, write an inequality, and solve each problem. Then check your solution.

23. One fourth of a number is no less than -3.
24. Three times a number subtracted from 14 is less than two.
25. Five less than twice a number is between 13 and 21.
26. **TRAVEL** Megan's car gets between 18 and 21 miles per gallon of gasoline. If her car's tank holds 15 gallons, what is the range of distance that Megan can drive her car on one tank of gasoline?

Graph each inequality.

27. $y \geq 3x - 2$
28. $2x + 3y < 6$
29. $x - 2y > 4$
30. **STANDARDIZED TEST PRACTICE** Which inequality is represented by the graph?

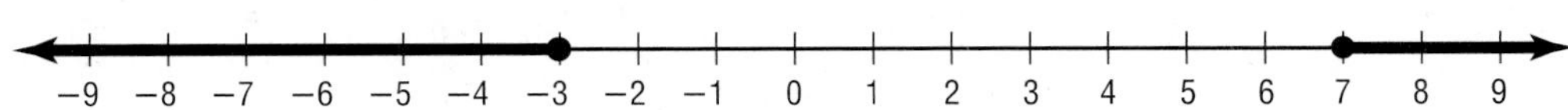

Ⓐ $|x - 2| \leq 5$ Ⓑ $|x - 2| \geq 5$ Ⓒ $|x + 2| \leq 5$ Ⓓ $|x + 2| \geq 5$

Chapter 6 Standardized Test Practice

Part 1 Multiple Choice

Record your answers on the answer sheet provided by your teacher or on a sheet of paper.

1. Which of the following is a correct statement? (Lesson 2-4)

   Ⓐ $-\frac{9}{3} > \frac{3}{9}$ Ⓑ $-\frac{3}{9} > -\frac{9}{3}$

   Ⓒ $-\frac{3}{9} < -\frac{9}{3}$ Ⓓ $\frac{9}{3} < \frac{3}{9}$

2. $(-6)(-7) =$ (Lesson 2-3)

   Ⓐ -42 Ⓑ -13

   Ⓒ 13 Ⓓ 42

3. A cylindrical can has a volume of 5625π cubic centimeters. Its height is 25 centimeters. What is the radius of the can? Use the formula $V = \pi r^2 h$. (Lessons 2-8 and 3-8)

   Ⓐ 4.8 cm Ⓑ 7.5 cm

   Ⓒ 15 cm Ⓓ 47.1 cm

4. A furnace repair service charged a customer \$80 for parts and \$65 per hour worked. The bill totaled \$177.50. About how long did the repair technician work on the furnace? (Lessons 3-1 and 3-4)

   Ⓐ 0.5 hour Ⓑ 1.5 hours

   Ⓒ 2 hours Ⓓ 4 hours

5. The formula $P = \frac{4(220 - A)}{5}$ determines the recommended maximum pulse rate P during exercise for a person who is A years old. Cameron is 15 years old. What is his recommended maximum pulse rate during exercise? (Lesson 3-8)

   Ⓐ 162 Ⓑ 164

   Ⓒ 173 Ⓓ 263

6. The graph of the function $y = 2x - 1$ is shown. If the graph is translated 3 units up, which equation will best represent the new line? (Lesson 4-2)

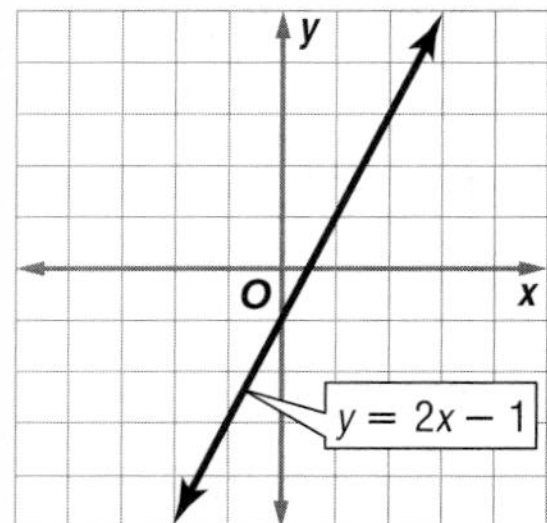

   Ⓐ $y = 2x + 2$ Ⓑ $y = 2x - 3$

   Ⓒ $y = 2x + 3$ Ⓓ $y = 2x - 4$

7. The table shows a set of values for x and y. Which equation best represents this set of data? (Lesson 4-8)

x	-4	-1	2	5	8
y	-16	-4	8	20	32

   Ⓐ $y = 3x - 4$ Ⓑ $y = 3x + 2$

   Ⓒ $y = 2x - 10$ Ⓓ $y = 4x$

8. Ali's grade depends on 4 test scores. On the first 3 tests, she earned scores of 78, 82, and 75. She wants to average at least 80. Which inequality can she use to find the score x that she needs on the fourth test in order to earn a final grade of at least 80? (Lesson 6-3)

   Ⓐ $\frac{78 + 82 + 75 + x}{3} \geq 80$

   Ⓑ $\frac{78 + 82 + 75 + x}{4} \geq 80$

   Ⓒ $\frac{78 + 82 + 75 - x}{4} \geq 80$

   Ⓓ $\frac{78 + 82 + 75 + x}{4} \leq 80$

9. Which inequality is represented by the graph? (Lesson 6-4)

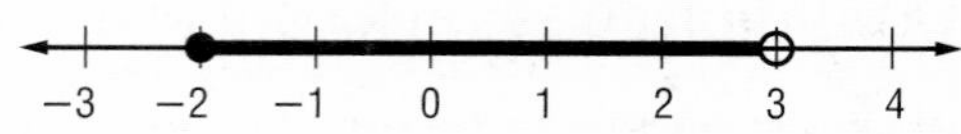

   Ⓐ $-2 < x < 3$ Ⓑ $-2 < x \leq 3$

   Ⓒ $-2 \leq x < 3$ Ⓓ $-2 \leq x \leq 3$

Aligned and verified by The Princeton Review

Part 2 Short Response/Grid In

Record your answers on the answer sheet provided by your teacher or on a sheet of paper.

10. A die is rolled. What are the odds of rolling a number less than 5? (Lesson 2-6)

11. A car is traveling at an average speed of 54 miles per hour. How many minutes will it take the car to travel 117 miles? (Lesson 3-6)

12. The price of a tape player was cut from \$48 to \$36. What was the percent of decrease? (Lesson 3-7)

13. Write an equation in slope-intercept form that describes the graph. (Lesson 5-4)

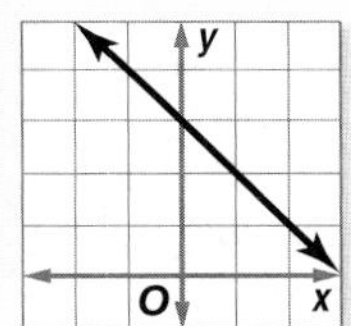

14. A line is parallel to the graph of the equation $\frac{1}{3}y = \frac{2}{3}x - 1$. What is the slope of the parallel line? (Lessons 5-4 and 5-6)

15. Solve $\frac{1}{2}(10x - 8) - 3(x - 1) \geq 15$ for x. (Lesson 6-3)

16. Find all values of x that make the inequality $|x - 3| > 5$ true. (Lesson 6-5)

> **The Princeton Review Test-Taking Tip**
>
> **Questions 13 and 14**
> - Know the slope-intercept form of linear equations: $y = mx + b$.
> - Understand the definition of slope.
> - Recognize the relationships between the slopes of parallel lines and between the slopes of perpendicular lines.

17. Graph the equation $y = -2x + 4$ and indicate which region represents $y < -2x + 4$. (Lesson 6-6)

Part 3 Quantitative Comparison

Compare the quantity in Column A and the quantity in Column B. Then determine whether:

- (A) **the quantity in Column A is greater,**
- (B) **the quantity in Column B is greater,**
- (C) **the two quantities are equal, or**
- (D) **the relationship cannot be determined from the information given.**

	Column A	Column B
18.	$\sqrt{68}$	9

(Lesson 2-7)

19. $x > 5$ or $x < -7$

$-3 < y < 4$

Column A	Column B
$\lvert x \rvert$	$\lvert y \rvert$

(Lesson 6-5)

Part 4 Open Ended

Record your answers on a sheet of paper. Show your work.

20. The Carlson family is building a house on a lot that is 91 feet long and 158 feet wide. (Lessons 6-1, 6-2, and 6-4)

a. Town law states that the sides of a house cannot be closer than 10 feet to the edges of a lot. Write an inequality for the possible lengths of the Carlson family's house, and solve the inequality.

b. The Carlson family wants their house to be at least 2800 square feet and no more than 3200 square feet. They also want their house to have the maximum possible length. Write an inequality for the possible widths of their house, and solve the inequality. Round your answer to the nearest whole number of feet.

www.algebra1.com/standardized_test

Chapter 7

Solving Systems of Linear Equations and Inequalities

What You'll Learn

- **Lesson 7-1** Solve systems of linear equations by graphing.
- **Lessons 7-2 through 7-4** Solve systems of linear equations algebraically.
- **Lesson 7-5** Solve systems of linear inequalities by graphing.

Key Vocabulary

- system of equations (p. 369)
- substitution (p. 376)
- elimination (p. 382)
- system of inequalities (p. 394)

Why It's Important

Business decision makers often use systems of linear equations to model a real-world situation in order to predict future events. Being able to make an accurate prediction helps them plan and manage their businesses.

Trends in the travel industry change with time. For example, in recent years, the number of tourists traveling to South America, the Caribbean, and the Middle East is on the rise. *You will use a system of linear equations to model the trends in tourism in Lesson 7-2.*

Getting Started

Prerequisite Skills To be successful in this chapter, you'll need to master these skills and be able to apply them in problem-solving situations. Review these skills before beginning Chapter 7.

For Lesson 7-1 — Graph Linear Equations

Graph each equation. *(For review, see Lesson 4-5.)*

1. $y = 1$ **2.** $y = -2x$ **3.** $y = 4 - x$

4. $y = 2x + 3$ **5.** $y = 5 - 2x$ **6.** $y = \frac{1}{2}x + 2$

For Lesson 7-2 — Solve for a Given Variable

Solve each equation or formula for the variable specified. *(For review, see Lesson 3-8.)*

7. $4x + a = 6x$, for x **8.** $8a + y = 16$, for a

9. $\frac{7bc - d}{10} = 12$, for b **10.** $\frac{7m + n}{q} = 2m$, for q

For Lessons 7-3 and 7-4 — Simplify Expressions

Simplify each expression. If not possible, write *simplified.* *(For review, see Lesson 1-5.)*

11. $(3x + y) - (2x + y)$ **12.** $(7x - 2y) - (7x + 4y)$ **13.** $(16x - 3y) + (11x + 3y)$

14. $(8x - 4y) + (-8x + 5y)$ **15.** $4(2x + 3y) - (8x - y)$ **16.** $3(x - 4y) + (x + 12y)$

17. $2(x - 2y) + (3x + 4y)$ **18.** $5(2x - y) - 2(5x + 3y)$ **19.** $3(x + 4y) + 2(2x - 6y)$

Make this Foldable to record information about solving systems of equations and inequalities. Begin with five sheets of grid paper.

Step 1 Fold

Fold each sheet in half along the width.

Step 2 Cut

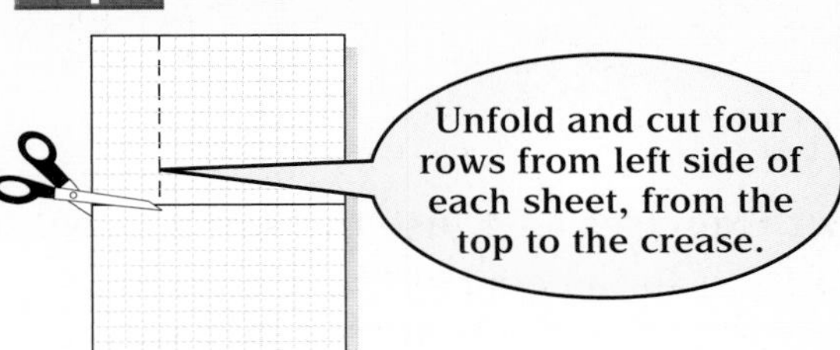

Unfold and cut four rows from left side of each sheet, from the top to the crease.

Step 3 Stack and Staple

Stack the sheets and staple to form a booklet.

Step 4 Label

7-1 Graphing Systems of Equations
Solving Systems of Equations and
Label each page with a lesson number and title.

Reading and Writing As you read and study the chapter, unfold each page and fill the journal with notes, graphs, and examples for systems of equations and inequalities.

Spreadsheet Investigation

A Preview of Lesson 7-1

Systems of Equations

You can use a spreadsheet to investigate when two quantities will be equal. Enter each formula into the spreadsheet and look for the time when both formulas have the same result.

Example

Bill Winters is considering two job offers in telemarketing departments. The salary at the first job is $400 per week plus 10% commission on Mr. Winters' sales. At the second job, the salary is $375 per week plus 15% commission. For what amount of sales would the weekly salary be the same at either job?

Enter different amounts for Mr. Winters' weekly sales in column A. Then enter the formula for the salary at the first job in each cell in column B. In each cell of column C, enter the formula for the salary at the second job.

The spreadsheet shows that for sales of $500 the total weekly salary for each job is $450.

Job Salaries

	A	B	C
1	Sales	Salary 1	Salary 2
2	0	400	375
3	100	410	390
4	200	420	405
5	300	430	420
6	400	440	435
7	500	450	450
8	600	460	465
9	700	470	480
10	800	480	495
11	900	490	510
12	1000	500	525
13			

Sheet1

Exercises

For Exercises 1–4, use the spreadsheet of weekly salaries above.

1. If x is the amount of Mr. Winters' weekly sales and y is his total weekly salary, write a linear equation for the salary at the first job.
2. Write a linear equation for the salary at the second job.
3. Which ordered pair is a solution for both of the equations you wrote for Exercises 1 and 2?
   a. (100, 410) **b.** (300, 420) **c.** (500, 450) **d.** (900, 510)
4. Use the graphing capability of the spreadsheet program to graph the salary data using a line graph. At what point do the two lines intersect? What is the significance of that point in the real-world situation?
5. How could you find the sales for which Mr. Winters' salary will be equal without using a spreadsheet?

7-1 Graphing Systems of Equations

What You'll Learn

- Determine whether a system of linear equations has 0, 1, or infinitely many solutions.
- Solve systems of equations by graphing.

Vocabulary

- system of equations
- consistent
- inconsistent
- independent
- dependent

How can you use graphs to compare the sales of two products?

During the 1990s, sales of cassette singles decreased, and sales of CD singles increased. Assume that the sales of these singles were linear functions. If x represents the years since 1991 and y represents the sales in millions of dollars, the following equations represent the sales of these singles.

Cassette singles: $y = 69 - 6.9x$
CD singles: $y = 5.7 + 6.3x$

These equations are graphed at the right.

The point at which the two graphs intersect represents the time when the sales of cassette singles equaled the sales of CD singles. The ordered pair of this point is a solution of both equations.

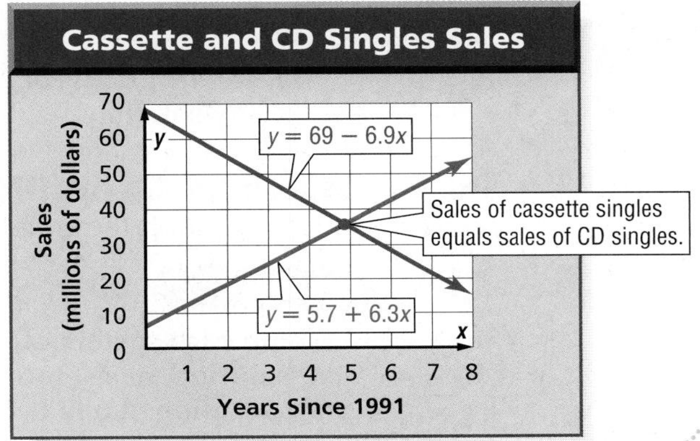

NUMBER OF SOLUTIONS Two equations, such as $y = 69 - 6.9x$ and $y = 5.7 + 6.3x$, together are called a **system of equations**. A solution of a system of equations is an ordered pair of numbers that satisfies both equations. A system of two linear equations can have 0, 1, or an infinite number of solutions.

- If the graphs intersect or coincide, the system of equations is said to be **consistent**. That is, it has at least one ordered pair that satisfies both equations.
- If the graphs are parallel, the system of equations is said to be **inconsistent**. There are *no* ordered pairs that satisfy both equations.
- Consistent equations can be **independent** or **dependent**. If a system has exactly one solution, it is independent. If the system has an infinite number of solutions, it is dependent.

Concept Summary — *Systems of Equations*

	Intersecting Lines	Same Line	Parallel Lines
Graph of a System	(graph)	(graph)	(graph)
Number of Solutions	exactly one solution	infinitely many	no solutions
Terminology	consistent and independent	consistent and dependent	inconsistent

Example 1 Number of Solutions

Use the graph at the right to determine whether each system has *no* solution, *one* solution, or *infinitely many* solutions.

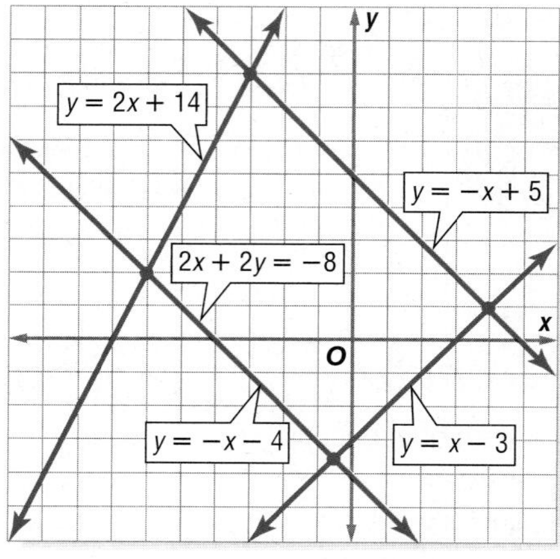

a. $y = -x + 5$
$y = x - 3$

Since the graphs of $y = -x + 5$ and $y = x - 3$ are intersecting lines, there is one solution.

b. $y = -x + 5$
$2x + 2y = -8$

Since the graphs of $y = -x + 5$ and $2x + 2y = -8$ are parallel, there are no solutions.

c. $2x + 2y = -8$
$y = -x - 4$

Since the graphs of $2x + 2y = -8$ and $y = -x - 4$ coincide, there are infinitely many solutions.

SOLVE BY GRAPHING One method of solving systems of equations is to carefully graph the equations on the same coordinate plane.

Example 2 Solve a System of Equations

Graph each system of equations. Then determine whether the system has *no* solution, *one* solution, or *infinitely many* solutions. If the system has one solution, name it.

a. $y = -x + 8$
$y = 4x - 7$

The graphs appear to intersect at the point with coordinates (3, 5). Check this estimate by replacing x with 3 and y with 5 in each equation.

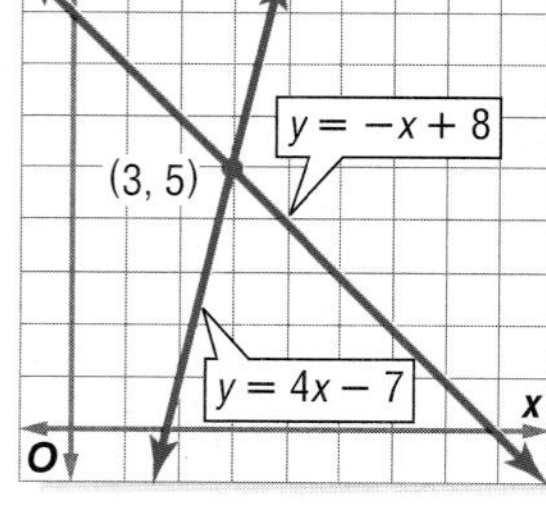

CHECK

$y = -x + 8$	$y = 4x - 7$
$5 \stackrel{?}{=} -3 + 8$	$5 \stackrel{?}{=} 4(3) - 7$
$5 = 5$ ✓	$5 \stackrel{?}{=} 12 - 7$
	$5 = 5$ ✓

The solution is (3, 5).

Study Tip

Look Back
To review **graphing linear equations**, see Lesson 4-5.

b. $x + 2y = 5$
$2x + 4y = 2$

The graphs of the equations are parallel lines. Since they do not intersect, there are no solutions to this system of equations. Notice that the lines have the same slope but different y-intercepts. *Recall that a system of equations that has no solution is said to be inconsistent.*

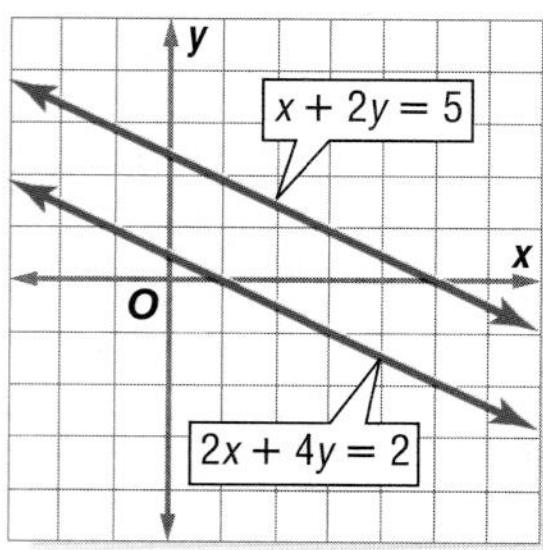

Example 3 Write and Solve a System of Equations

WORLD RECORDS **Use the information on Guy Delage's swim at the left. If Guy can swim 3 miles per hour for an extended period and the raft drifts about 1 mile per hour, how many hours did he spend swimming each day?**

Words You have information about the amount of time spent swimming and floating. You also know the rates and the total distance traveled.

Variables Let s = the number of hours Guy swam, and let f = the number of hours he floated each day. Write a system of equations to represent the situation.

Equations

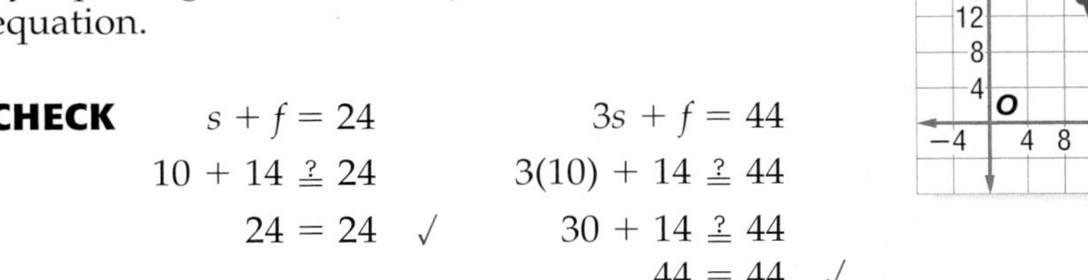

The number of hours swimming	plus	the number of hours floating	equals	the total number of hours in a day.
s	$+$	f	$=$	24

The daily miles traveled swimming	plus	the daily miles traveled floating	equals	the total miles traveled in a day.
$3s$	$+$	$1f$	$=$	44

Graph the equations $s + f = 24$ and $3s + f = 44$. The graphs appear to intersect at the point with coordinates (10, 14). Check this estimate by replacing s with 10 and f with 14 in each equation.

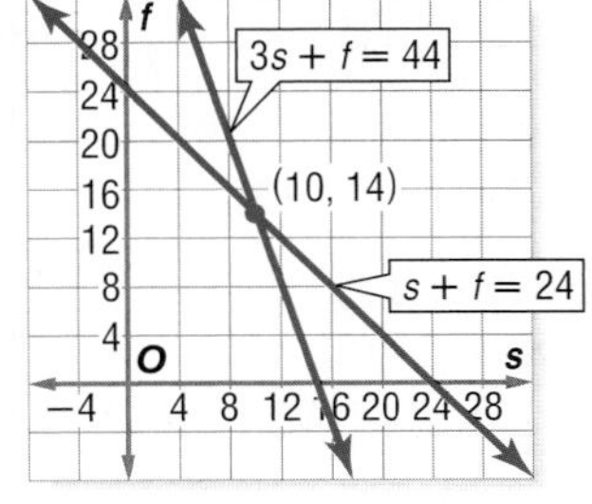

CHECK

$$s + f = 24 \qquad 3s + f = 44$$
$$10 + 14 \stackrel{?}{=} 24 \qquad 3(10) + 14 \stackrel{?}{=} 44$$
$$24 = 24 \ \checkmark \qquad 30 + 14 \stackrel{?}{=} 44$$
$$44 = 44 \ \checkmark$$

Guy Delage spent about 10 hours swimming each day.

More About. . .

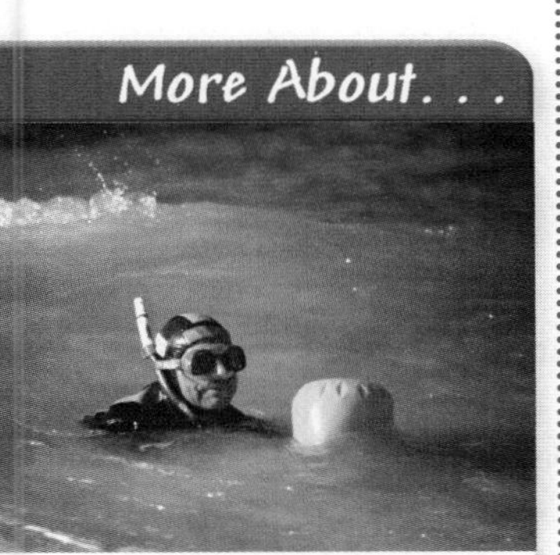

World Records

In 1994, Guy Delage swam 2400 miles across the Atlantic Ocean from Cape Verde to Barbados. Everyday he would swim awhile and then rest while floating with the current on a huge raft. He averaged 44 miles per day.

Source: Banner Aerospace, Inc.

Check for Understanding

Concept Check

1. **OPEN ENDED** Draw the graph of a system of equations that has one solution at (−2, 3).
2. **Determine** whether a system of equations with (0, 0) and (2, 2) as solutions *sometimes*, *always*, or *never* has other solutions. Explain.
3. **Find a counterexample** for the following statement.
If the graphs of two linear equations have the same slope, then the system of equations has no solution.

Guided Practice

Use the graph at the right to determine whether each system has *no* solution, *one* solution, or *infinitely many* solutions.

4. $y = x - 4$, $y = \frac{1}{3}x - 2$
5. $y = \frac{1}{3}x + 2$, $y = \frac{1}{3}x - 2$
6. $x - y = 4$, $y = x - 4$
7. $x - y = 4$, $y = -\frac{1}{3}x + 4$

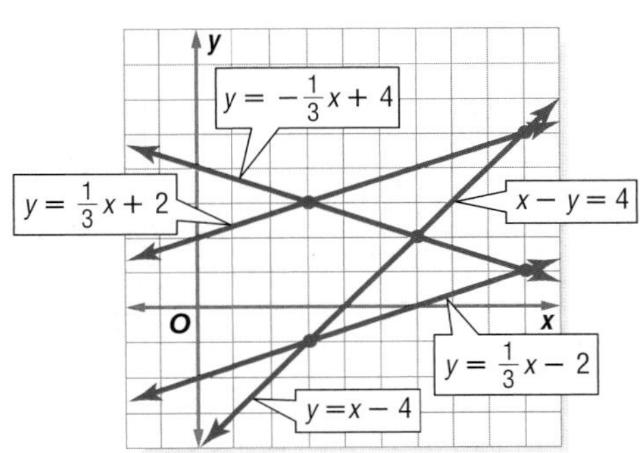

Graph each system of equations. Then determine whether the system has *no* solution, *one* solution, or *infinitely many* solutions. If the system has one solution, name it.

8. $y = -x$
$y = 2x$

9. $x + y = 8$
$x - y = 2$

10. $2x + 4y = 2$
$3x + 6y = 3$

11. $x + y = 4$
$x + y = 1$

12. $x - y = 2$
$3y + 2x = 9$

13. $x + y = 2$
$y = 4x + 7$

Application

14. RESTAURANTS The Rodriguez family and the Wong family went to a brunch buffet. The restaurant charges one price for adults and another price for children. The Rodriguez family has two adults and three children, and their bill was \$40.50. The Wong family has three adults and one child, and their bill was \$38.00. Determine the price of the buffet for an adult and the price for a child.

Practice and Apply

Homework Help

For Exercises	See Examples
15–22	1
23–40	2
41–54	3

Extra Practice

See page 835.

Use the graph at the right to determine whether each system has *no* solution, *one* solution, or *infinitely many* solutions.

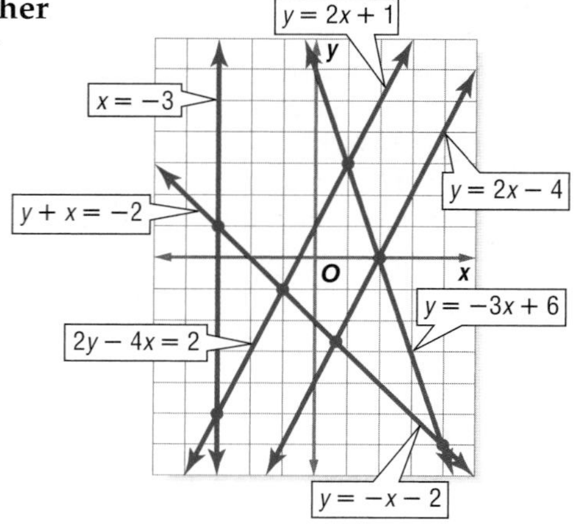

15. $x = -3$
$y = 2x + 1$

16. $y = -x - 2$
$y = 2x - 4$

17. $y + x = -2$
$y = -x - 2$

18. $y = 2x + 1$
$y = 2x - 4$

19. $y = -3x + 6$
$y = 2x - 4$

20. $2y - 4x = 2$
$y = 2x - 4$

21. $2y - 4x = 2$
$y = -3x + 6$

22. $2y - 4x = 2$
$y = 2x + 1$

Graph each system of equations. Then determine whether the system has *no* solution, *one* solution, or *infinitely many* solutions. If the system has one solution, name it.

23. $y = -6$
$4x + y = 2$

24. $x = 2$
$3x - y = 8$

25. $y = \frac{1}{2}x$
$2x + y = 10$

26. $y = -x$
$y = 2x - 6$

27. $y = 3x - 4$
$y = -3x - 4$

28. $y = 2x + 6$
$y = -x - 3$

29. $x - 2y = 2$
$3x + y = 6$

30. $x + y = 2$
$2y - x = 10$

31. $3x + 2y = 12$
$3x + 2y = 6$

32. $2x + 3y = 4$
$-4x - 6y = -8$

33. $2x + y = -4$
$5x + 3y = -6$

34. $4x + 3y = 24$
$5x - 8y = -17$

35. $3x + y = 3$
$2y = -6x + 6$

36. $y = x + 3$
$3y + x = 5$

37. $2x + 3y = -17$
$y = x - 4$

38. $y = \frac{2}{3}x - 5$
$3y = 2x$

39. $6 - \frac{3}{8}y = x$
$\frac{2}{3}x + \frac{1}{4}y = 4$

40. $\frac{1}{2}x + \frac{1}{3}y = 6$
$y = \frac{1}{2}x + 2$

41. GEOMETRY The length of the rectangle at the right is 1 meter less than twice its width. What are the dimensions of the rectangle?

ℓ

Perimeter = 40 m

w

GEOMETRY **For Exercises 42 and 43, use the graphs of $y = 2x + 6$, $3x + 2y = 19$, and $y = 2$, which contain the sides of a triangle.**

42. Find the coordinates of the vertices of the triangle.
43. Find the area of the triangle.

WebQuest

You can graph a system of equations to predict when men's and women's Olympic times will be the same. Visit www.algebra1.com/webquest to continue work on your WebQuest project.

BALLOONING **For Exercises 44 and 45, use the information in the graphic at the right.**

44. In how many minutes will the balloons be at the same height?
45. How high will the balloons be at that time?

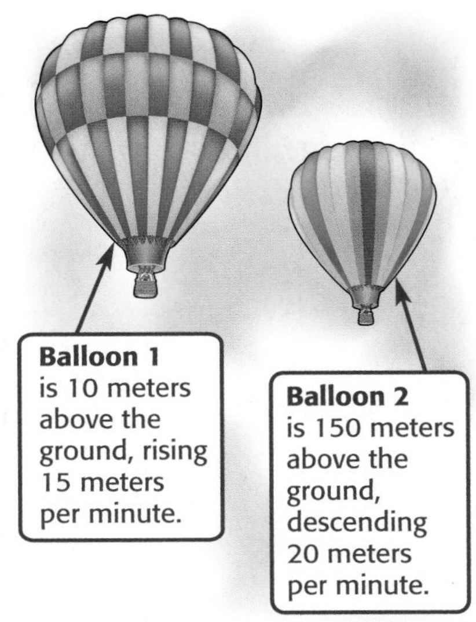
Balloon 1 is 10 meters above the ground, rising 15 meters per minute.
Balloon 2 is 150 meters above the ground, descending 20 meters per minute.

SAVINGS **For Exercises 46 and 47, use the following information.**
Monica and Michael Gordon both want to buy a scooter. Monica has already saved \$25 and plans to save \$5 per week until she can buy the scooter. Michael has \$16 and plans to save \$8 per week.

46. In how many weeks will Monica and Michael have saved the same amount of money?
47. How much will each person have saved at that time?

BUSINESS **For Exercises 48–50, use the graph at the right.**

48. Which company had the greater profit during the ten years?
49. Which company had a greater rate of growth?
50. If the profit patterns continue, will the profits of the two companies ever be equal? Explain.

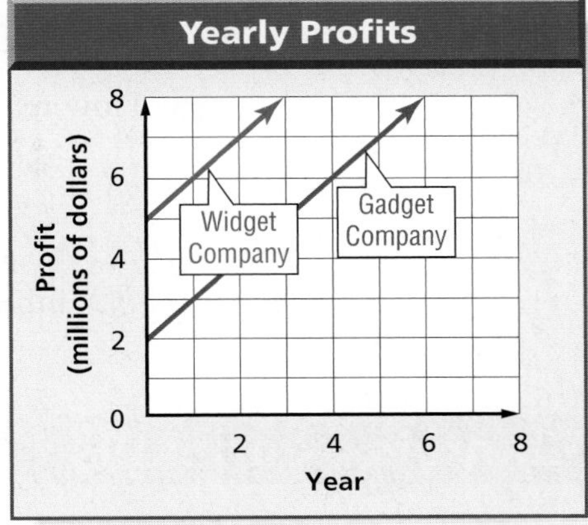
Yearly Profits
Profit (millions of dollars)
0
2
4
6
8
Year
2
4
6
8
Widget Company
Gadget Company

POPULATION **For Exercises 51–54, use the following information.**
The U.S. Census Bureau divides the country into four sections. They are the Northeast, the Midwest, the South, and the West.

51. In 1990, the population of the Midwest was about 60 million. During the 1990s, the population of this area increased an average of about 0.4 million per year. Write an equation to represent the population of the Midwest for the years since 1990.
52. The population of the West was about 53 million in 1990. The population of this area increased an average of about 1 million per year during the 1990s. Write an equation to represent the population of the West for the years since 1990.
53. Graph the population equations.
54. Assume that the rate of growth of each of these areas remains the same. Estimate when the population of the West would be equal to the population of the Midwest.

55. **CRITICAL THINKING** The solution of the system of equations $Ax + y = 5$ and $Ax + By = 20$ is $(2, -3)$. What are the values of A and B?

56. **WRITING IN MATH** Answer the question that was posed at the beginning of the lesson.

How can you use graphs to compare the sales of two products?

Include the following in your answer:

- an estimate of the year in which the sales of cassette singles equaled the sales of CD singles, and
- an explanation of why graphing works.

Standardized Test Practice

57. Which graph represents a system of equations with no solution?

(A)

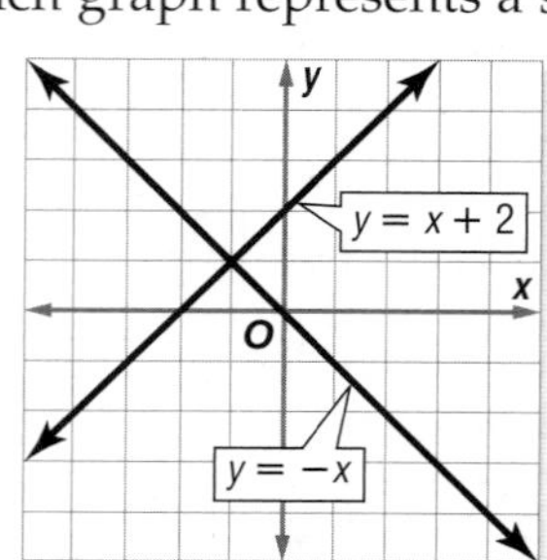

(B)

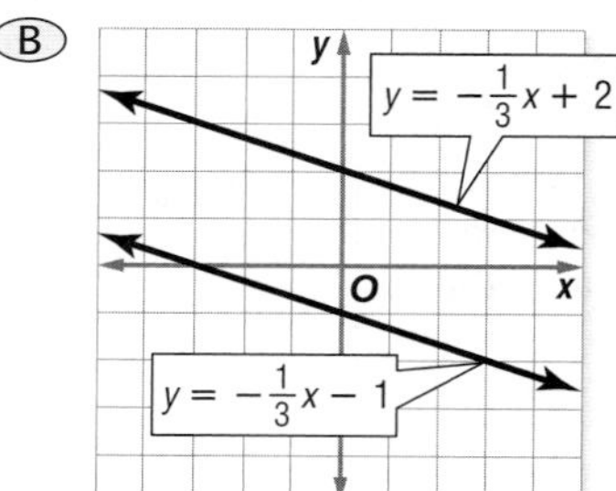

(C)

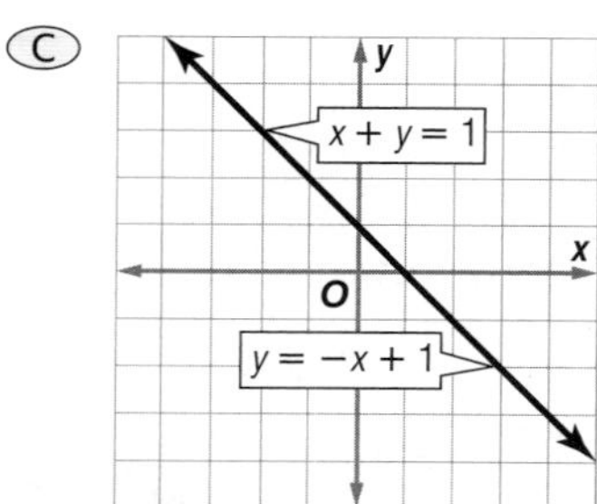

(D)

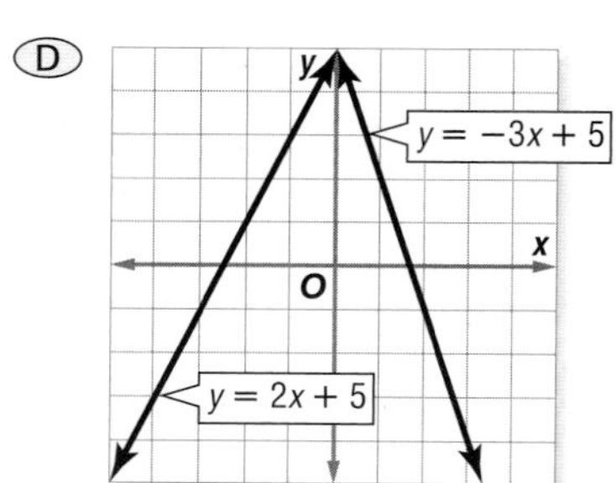

58. How many solutions exist for the system of equations below?

$4x + y = 7$
$3x - y = 0$

(A) no solution (B) one solution

(C) infinitely many solutions (D) cannot be determined

Maintain Your Skills

Mixed Review

Determine which ordered pairs are part of the solution set for each inequality. *(Lesson 6-6)*

59. $y \leq 2x$, $\{(1, 4), (-1, 5), (5, -6), (-7, 0)\}$

60. $y < 8 - 3x$, $\{(-4, 2), (-3, 0), (1, 4), (1, 8)\}$

61. **MANUFACTURING** The inspector at a perfume manufacturer accepts a bottle if it is less than 0.05 ounce above or below 2 ounces. What are the acceptable numbers of ounces for a perfume bottle? *(Lesson 6-5)*

Write each equation in standard form. *(Lesson 5-5)*

62. $y - 1 = 4(x - 5)$ 63. $y + 2 = \frac{1}{3}(x + 3)$ 64. $y - 4 = -6(x + 2)$

Getting Ready for the Next Lesson

PREREQUISITE SKILL Solve each equation for the variable specified.
*(To review **solving equations for a specified variable**, see Lesson 3-8.)*

65. $12x - y = 10x$, for y 66. $6a + b = 2a$, for a

67. $\frac{7m - n}{q} = 10$, for q 68. $\frac{5tz - s}{2} = 6$, for z

Graphing Calculator Investigation

A Follow-Up of Lesson 7-1

Systems of Equations

You can use a TI-83 Plus graphing calculator to solve a system of equations.

Example

Solve the system of equations. State the decimal solution to the nearest hundredth.

$$2.93x + y = 6.08$$
$$8.32x - y = 4.11$$

Step 1 Solve each equation for y to enter them into the calculator.

$2.93x + y = 6.08$	First equation
$2.93x + y - 2.93x = 6.08 - 2.93x$	Subtract $2.93x$ from each side.
$y = 6.08 - 2.93x$	Simplify.
$8.32x - y = 4.11$	Second equation
$8.32x - y - 8.32x = 4.11 - 8.32x$	Subtract $8.32x$ from each side.
$-y = 4.11 - 8.32x$	Simplify.
$(-1)(-y) = (-1)(4.11 - 8.32x)$	Multiply each side by -1.
$y = -4.11 + 8.32x$	Simplify.

Step 2 Enter these equations in the Y= list and graph.

KEYSTROKES: *Review on pages 224–225.*

Step 3 Use the CALC menu to find the point of intersection.

KEYSTROKES: 2nd [CALC] 5 ENTER ENTER ENTER

The solution is approximately (0.91, 3.43).

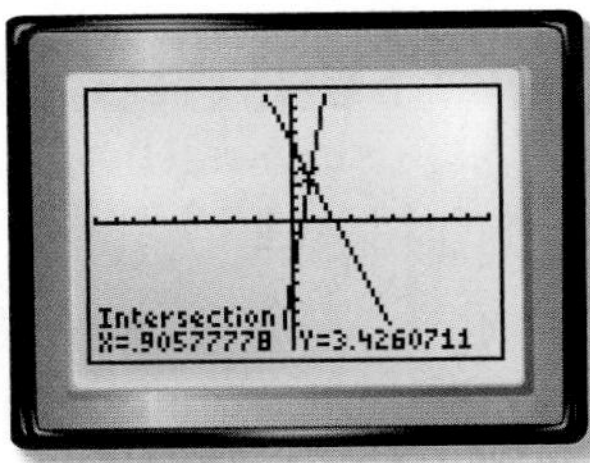

[10, 10] scl: 1 by [−10, 10] scl: 1

Exercises

Use a graphing calculator to solve each system of equations. Write decimal solutions to the nearest hundredth.

1. $y = 3x - 4$
$y = -0.5x + 6$

2. $y = 2x + 5$
$y = -0.2x - 4$

3. $x + y = 5.35$
$3x - y = 3.75$

4. $0.35x - y = 1.12$
$2.25x + y = -4.05$

5. $1.5x + y = 6.7$
$5.2x - y = 4.1$

6. $5.4x - y = 1.8$
$6.2x + y = -3.8$

7. $5x - 4y = 26$
$4x + 2y = 53.3$

8. $2x + 3y = 11$
$4x + y = -6$

9. $0.22x + 0.15y = 0.30$
$-0.33x + y = 6.22$

10. $125x - 200y = 800$
$65x - 20y = 140$

 www.algebra1.com/other_calculator_keystrokes

7-2 Substitution

What You'll Learn

- Solve systems of equations by using substitution.
- Solve real-world problems involving systems of equations.

Vocabulary

- substitution

How can a system of equations be used to predict media use?

Americans spend more time online than they spend reading daily newspapers. If x represents the number of years since 1993 and y represents the average number of hours per person per year, the following system represents the situation.

reading daily newspapers: $y = -2.8x + 170$
online: $y = 14.4x + 2$

The solution of the system represents the year that the number of hours spent on each activity will be the same. To solve this system, you could graph the equations and find the point of intersection. However, the exact coordinates of the point would be very difficult to determine from the graph. You could find a more accurate solution by using algebraic methods.

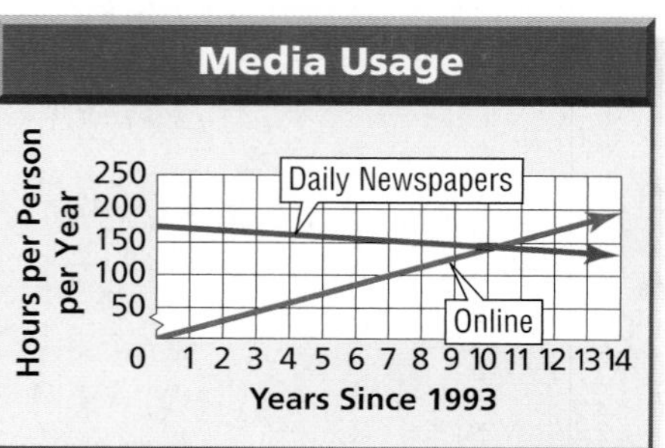

SUBSTITUTION The exact solution of a system of equations can be found by using algebraic methods. One such method is called **substitution**.

Algebra Activity

Using Substitution

Use algebra tiles and an equation mat to solve the system of equations.
$3x + y = 8$ and $y = x - 4$

Model and Analyze

Since $y = x - 4$, use 1 positive x tile and 4 negative 1 tiles to represent y. Use algebra tiles to represent $3x + y = 8$.

1. Use what you know about equation mats to solve for x. What is the value of x?
2. Use $y = x - 4$ to solve for y.
3. What is the solution of the system of equations?

Make a Conjecture

4. Explain how to solve the following system of equations using algebra tiles. $4x + 3y = 10$ and $y = x + 1$
5. Why do you think this method is called substitution?

Example 1 Solve Using Substitution

Use substitution to solve the system of equations.

$y = 3x$
$x + 2y = -21$

Since $y = 3x$, substitute $3x$ for y in the second equation.

$x + 2y = -21$	Second equation
$x + 2(3x) = -21$	$y = 3x$
$x + 6x = -21$	Simplify.
$7x = -21$	Combine like terms.
$\frac{7x}{7} = \frac{-21}{7}$	Divide each side by 7.
$x = -3$	Simplify.

Use $y = 3x$ to find the value of y.

$y = 3x$	First equation
$y = 3(-3)$	$x = -3$
$y = -9$	The solution is $(-3, -9)$.

Study Tip

Look Back
To review **solving linear equations**, see Lesson 3-5.

Example 2 Solve for One Variable, Then Substitute

Use substitution to solve the system of equations.

$x + 5y = -3$
$3x - 2y = 8$

Solve the first equation for x since the coefficient of x is 1.

$x + 5y = -3$	First equation
$x + 5y - 5y = -3 - 5y$	Subtract $5y$ from each side.
$x = -3 - 5y$	Simplify.

Find the value of y by substituting $-3 - 5y$ for x in the second equation.

$3x - 2y = 8$	Second equation
$3(-3 - 5y) - 2y = 8$	$x = -3 - 5y$
$-9 - 15y - 2y = 8$	Distributive Property
$-9 - 17y = 8$	Combine like terms.
$-9 - 17y + 9 = 8 + 9$	Add 9 to each side.
$-17y = 17$	Simplify.
$\frac{-17y}{-17} = \frac{17}{-17}$	Divide each side by -17.
$y = -1$	Simplify.

Substitute -1 for y in either equation to find the value of x.
Choose the equation that is easier to solve.

$x + 5y = -3$	First equation
$x + 5(-1) = -3$	$y = -1$
$x - 5 = -3$	Simplify.
$x = 2$	Add 5 to each side.

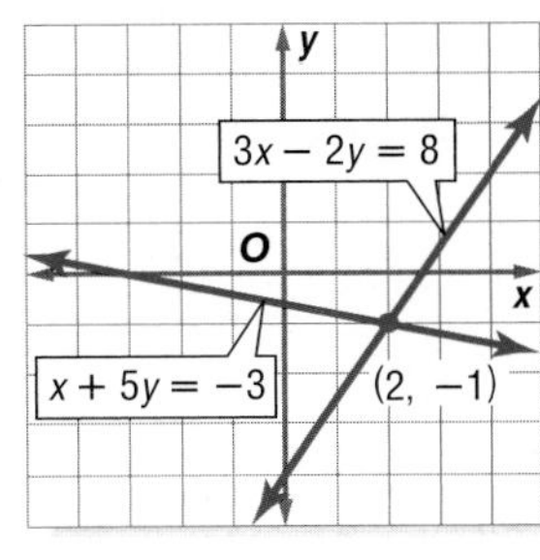

The solution is $(2, -1)$. The graph verifies the solution.

Example 3 Dependent System

Use substitution to solve the system of equations.

$6x - 2y = -4$
$y = 3x + 2$

Since $y = 3x + 2$, substitute $3x + 2$ for y in the first equation.

$$6x - 2y = -4 \quad \text{First equation}$$
$$6x - 2(3x + 2) = -4 \quad y = 3x + 2$$
$$6x - 6x - 4 = -4 \quad \text{Distributive Property}$$
$$-4 = -4 \quad \text{Simplify.}$$

The statement $-4 = -4$ is true. This means that there are infinitely many solutions of the system of equations. This is true because the slope-intercept form of both equations is $y = 3x + 2$. That is, the equations are equivalent, and they have the same graph.

In general, if you solve a system of linear equations and the result is a true statement (an identity such as $-4 = -4$), the system has an infinite number of solutions. However, if the result is a false statement (for example, $-4 = 5$), the system has no solution.

Study Tip

Alternative Method
Using a system of equations is an alternative method for solving the weighted average problems that you studied in Lesson 3-9.

REAL-WORLD PROBLEMS Sometimes it is helpful to organize data before solving a problem. Some ways to organize data are to use tables, charts, different types of graphs, or diagrams.

Example 4 Write and Solve a System of Equations

METAL ALLOYS A metal alloy is 25% copper. Another metal alloy is 50% copper. How much of each alloy should be used to make 1000 grams of a metal alloy that is 45% copper?

Let a = the number of grams of the 25% copper alloy and b = the number of grams of the 50% copper alloy. Use a table to organize the information.

	25% Copper	50% Copper	45% Copper
Total Grams	a	b	1000
Grams of Copper	$0.25a$	$0.50b$	0.45(1000)

The system of equations is $a + b = 1000$ and $0.25a + 0.50b = 0.45(1000)$. Use substitution to solve this system.

$$a + b = 1000 \quad \text{First equation}$$
$$a + b - b = 1000 - b \quad \text{Subtract } b \text{ from each side.}$$
$$a = 1000 - b \quad \text{Simplify.}$$

$$0.25a + 0.50b = 0.45(1000) \quad \text{Second equation}$$
$$0.25(1000 - b) + 0.50b = 0.45(1000) \quad a = 1000 - b$$
$$250 - 0.25b + 0.50b = 450 \quad \text{Distributive Property}$$
$$250 + 0.25b = 450 \quad \text{Combine like terms.}$$
$$250 + 0.25b - 250 = 450 - 250 \quad \text{Subtract 250 from each side.}$$
$$0.25b = 200 \quad \text{Simplify.}$$
$$\frac{0.25b}{0.25} = \frac{200}{0.25} \quad \text{Divide each side by 0.25.}$$
$$b = 800 \quad \text{Simplify.}$$

$$a + b = 1000 \quad \text{First equation}$$
$$a + 800 = 1000 \quad b = 800$$
$$a + 800 - 800 = 1000 - 800 \quad \text{Subtract 800 from each side.}$$
$$a = 200 \quad \text{Simplify.}$$

200 grams of the 25% copper alloy and 800 grams of the 50% copper alloy should be used.

Check for Understanding

Concept Check

1. **Explain** why you might choose to use substitution rather than graphing to solve a system of equations.
2. **Describe** the graphs of two equations if the solution of the system of equations yields the equation $4 = 2$.
3. **OPEN-ENDED** Write a system of equations that has infinitely many solutions.

Guided Practice

Use substitution to solve each system of equations. If the system does *not* have exactly one solution, state whether it has *no* solution or *infinitely many* solutions.

4. $x = 2y$; $4x + 2y = 15$
5. $y = 3x - 8$; $y = 4 - x$
6. $2x + 7y = 3$; $x = 1 - 4y$
7. $6x - 2y = -4$; $y = 3x + 2$
8. $x + 3y = 12$; $x - y = 8$
9. $y = \frac{3}{5}x$; $3x - 5y = 15$

Application

10. **TRANSPORTATION** The Thrust SSC is the world's fastest land vehicle. Suppose the driver of a car whose top speed is 200 miles per hour requests a race against the SSC. The car gets a head start of one-half hour. If there is unlimited space to race, at what distance will the SSC pass the car?

Thrust SSC top speed is 763 mph.

Practice and Apply

Homework Help

For Exercises	See Examples
11–28	1–3
29–37	4

Extra Practice

See page 835.

Use substitution to solve each system of equations. If the system does *not* have exactly one solution, state whether it has *no* solution or *infinitely many* solutions.

11. $y = 5x$; $2x + 3y = 34$
12. $x = 4y$; $2x + 3y = 44$
13. $x = 4y + 5$; $x = 3y - 2$
14. $y = 2x + 3$; $y = 4x - 1$
15. $4c = 3d + 3$; $c = d - 1$
16. $4x + 5y = 11$; $y = 3x - 13$
17. $8x + 2y = 13$; $4x + y = 11$
18. $2x - y = -4$; $-3x + y = -9$
19. $3x - 5y = 11$; $x - 3y = 1$
20. $2x + 3y = 1$; $-3x + y = 15$
21. $c - 5d = 2$; $2c + d = 4$
22. $5r - s = 5$; $-4r + 5s = 17$
23. $3x - 2y = 12$; $x + 2y = 6$
24. $x - 3y = 0$; $3x + y = 7$
25. $-0.3x + y = 0.5$; $0.5x - 0.3y = 1.9$
26. $0.5x - 2y = 17$; $2x + y = 104$
27. $y = \frac{1}{2}x + 3$; $y = 2x - 1$
28. $x = \frac{1}{2}y + 3$; $2x - y = 6$

www.algebra1.com/self_check_quiz

29. **GEOMETRY** The base of the triangle is 4 inches longer than the length of one of the other sides. Use a system of equations to find the length of each side of the triangle.

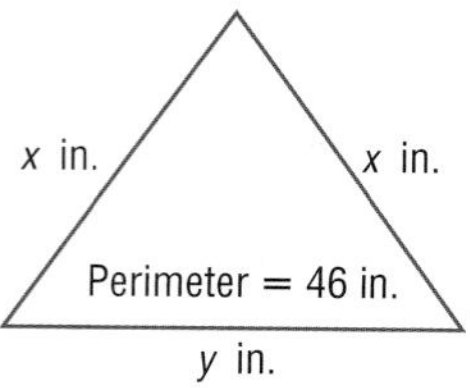

30. **FUND-RAISING** The Future Teachers of America Club at Paint Branch High School is making a healthy trail mix to sell to students during lunch. The mix will have three times the number of pounds of raisins as sunflower seeds. Sunflower seeds cost $4.00 per pound, and raisins cost $1.50 per pound. If the group has $34.00 to spend on the raisins and sunflower seeds, how many pounds of each should they buy?

31. **CHEMISTRY** MX Labs needs to make 500 gallons of a 34% acid solution. The only solutions available are a 25% acid solution and a 50% acid solution. How many gallons of each solution should be mixed to make the 34% solution?

32. **GEOMETRY** Supplementary angles are two angles whose measures have the sum of 180 degrees. Angles X and Y are supplementary, and the measure of angle X is 24 degrees greater than the measure of angle Y. Find the measures of angles X and Y.

33. **SPORTS** At the end of the 2000 baseball season, the New York Yankees and the Cincinnati Reds had won a total of 31 World Series. The Yankees had won 5.2 times as many World Series as the Reds. How many World Series did each team win?

More About. . .

Tourism

Every year, multitudes of visitors make their way to South America to stand in awe of Machu Picchu, the spectacular ruins of the Lost City of the Incas.

Source: www.about.com

JOBS For Exercises 34 and 35, use the following information.
Shantel Jones has two job offers as a car salesperson. At one dealership, she will receive $600 per month plus a commission of 2% of the total price of the automobiles she sells. At the other dealership, she will receive $1000 per month plus a commission of 1.5% of her total sales.

34. What is the total price of the automobiles that Ms. Jones must sell each month to make the same income from either dealership?

35. Explain which job offer is better.

36. **LANDSCAPING** A blue spruce grows an average of 6 inches per year. A hemlock grows an average of 4 inches per year. If a blue spruce is 4 feet tall and a hemlock is 6 feet tall, when would you expect the trees to be the same height?

37. **TOURISM** In 2000, approximately 40.3 million tourists visited South America and the Caribbean. The number of tourists to that area had been increasing at an average rate of 0.8 million tourists per year. In the same year, 17.0 million tourists visited the Middle East. The number of tourists to the Middle East had been increasing at an average rate of 1.8 million tourists per year. If the trend continues, when would you expect the number of tourists to South America and the Caribbean to equal the number of tourists to the Middle East?

38. **RESEARCH** Use the Internet or other resources to find the pricing plans for various cell phones. Determine the number of minutes you would need to use the phone for two plans to cost the same amount of money. Support your answer with a table, a graph, and/or an equation.

39. **CRITICAL THINKING** Solve the system of equations. Write the solution as an ordered triple of the form (x, y, z).

$2x + 3y - z = 17$
$y = -3z - 7$
$2x = z + 2$

40. **WRITING IN MATH** Answer the question that was posed at the beginning of the lesson.

How can a system of equations be used to predict media use?

Include the following in your answer:

- an explanation of solving a system of equations by using substitution, and
- the year when the number of hours spent reading daily newspapers is the same as the hours spent online.

41. When solving the following system, which expression could be substituted for x?

$x + 4y = 1$
$2x - 3y = -9$

(A) $4y - 1$ (B) $1 - 4y$ (C) $3y - 9$ (D) $-9 - 3y$

42. If $x - 3y = -9$ and $5x - 2y = 7$, what is the value of x?

(A) 1 (B) 2 (C) 3 (D) 4

Maintain Your Skills

Mixed Review

Graph each system of equations. Then determine whether the system has *no* solution, *one* solution, or *infinitely many* solutions. If the system has one solution, name it. *(Lesson 7-1)*

43. $x + y = 3$
$x + y = 4$

44. $x + 2y = 1$
$2x + y = 5$

45. $2x + y = 3$
$4x + 2y = 6$

Graph each inequality. *(Lesson 6-6)*

46. $y < -5$

47. $x \geq 4$

48. $2x + y > 6$

49. **RECYCLING** When a pair of blue jeans is made, the leftover denim scraps can be recycled. One pound of denim is left after making every five pair of jeans. How many pounds of denim would be left from 250 pairs of jeans? *(Lesson 3-6)*

Getting Ready for the Next Lesson

PREREQUISITE SKILL Simplify each expression.
*(To review **simplifying expressions**, see Lesson 1-5.)*

50. $6a - 9a$

51. $8t + 4t$

52. $-7g - 8g$

53. $7d - (2d + b)$

Practice Quiz 1

Lessons 7-1 and 7-2

Graph each system of equations. Then determine whether the system has *no* solution, *one* solution, or *infinitely many* solutions. If the system has one solution, name it. *(Lesson 7-1)*

1. $x + y = 3$
$x - y = 1$

2. $3x - 2y = -6$
$3x - 2y = 6$

Use substitution to solve each system of equations. If the system does *not* have exactly one solution, state whether it has *no* solution or *infinitely many* solutions. *(Lesson 7-2)*

3. $x + y = 0$
$3x + y = -8$

4. $x - 2y = 5$
$3x - 5y = 8$

5. $x + y = 2$
$y = 2 - x$

7-3 Elimination Using Addition and Subtraction

What You'll Learn

- Solve systems of equations by using elimination with addition.
- Solve systems of equations by using elimination with subtraction.

Vocabulary

- elimination

How can you use a system of equations to solve problems about weather?

On the winter solstice, there are fewer hours of daylight in the Northern Hemisphere than on any other day. On that day in Seward, Alaska, the difference between the number of hours of darkness n and the number of hours of daylight d is 12. The following system of equations represents the situation.

$$\begin{aligned} n + d &= 24 \\ n - d &= 12 \end{aligned}$$

Notice that if you add these equations, the variable d is eliminated.

$$\begin{array}{r} n + d = 24 \\ (+)n - d = 12 \\ \hline 2n \quad\;\; = 36 \end{array}$$

ELIMINATION USING ADDITION Sometimes adding two equations together will eliminate one variable. Using this step to solve a system of equations is called **elimination**.

Example 1 Elimination Using Addition

Use elimination to solve each system of equations.

$$\mathbf{3x - 5y = -16}$$
$$\mathbf{2x + 5y = 31}$$

Since the coefficients of the y terms, -5 and 5, are additive inverses, you can eliminate the y terms by adding the equations.

$3x - 5y = -16$ $(+)\ 2x + 5y = 31$	Write the equations in column form and add.
$5x = 15$	Notice that the y variable is eliminated.
$\frac{5x}{5} = \frac{15}{5}$	Divide each side by 5.
$x = 3$	Simplify.

Now substitute 3 for x in either equation to find the value of y.

$3x - 5y = -16$	First equation
$3(3) - 5y = -16$	Replace x with 3.
$9 - 5y = -16$	Simplify.
$9 - 5y - 9 = -16 - 9$	Subtract 9 from each side.
$-5y = -25$	Simplify.
$\frac{-5y}{-5} = \frac{-25}{-5}$	Divide each side by -5.
$y = 5$	Simplify.

The solution is (3, 5).

Example 2 Write and Solve a System of Equations

Twice one number added to another number is 18. Four times the first number minus the other number is 12. Find the numbers.

Let x represent the first number and y represent the second number.

$$\underbrace{\text{Twice one number}}_{2x}\ \underbrace{\text{added to}}_{+}\ \underbrace{\text{another number}}_{y}\ \underbrace{\text{is}}_{=}\ \underbrace{18.}_{18}$$

$$\underbrace{\text{Four times the first number}}_{4x}\ \underbrace{\text{minus}}_{-}\ \underbrace{\text{the other number}}_{y}\ \underbrace{\text{is}}_{=}\ \underbrace{12.}_{12}$$

Use elimination to solve the system.

$2x + y = 18$ $(+)\ 4x - y = 12$	Write the equations in column form and add.
$6x = 30$	Notice that the variable y is eliminated.
$\frac{6x}{6} = \frac{30}{6}$	Divide each side by 6.
$x = 5$	Simplify.

Now substitute 5 for x in either equation to find the value of y.

$4x - y = 12$	Second equation
$4(5) - y = 12$	Replace x with 5.
$20 - y = 12$	Simplify.
$20 - y - 20 = 12 - 20$	Subtract 20 from each side.
$-y = -8$	Simplify.
$\frac{-y}{-1} = \frac{-8}{-1}$	Divide each side by -1.
$y = 8$	The numbers are 5 and 8.

Study Tip

Look Back
To review **translating verbal sentences into equations**, see Lesson 3-1.

ELIMINATION USING SUBTRACTION Sometimes subtracting one equation from another will eliminate one variable.

Example 3 Elimination Using Subtraction

Use elimination to solve the system of equations.

$5s + 2t = 6$
$9s + 2t = 22$

Since the coefficients of the t terms, 2 and 2, are the same, you can eliminate the t terms by subtracting the equations.

$5s + 2t = 6$ $(-)\ 9s + 2t = 22$	Write the equations in column form and subtract.
$-4s = -16$	Notice that the variable t is eliminated.
$\frac{-4s}{-4} = \frac{-16}{-4}$	Divide each side by -4.
$s = 4$	Simplify.

Now substitute 4 for s in either equation to find the value of t.

$5s + 2t = 6$	First equation
$5(4) + 2t = 6$	$s = 4$
$20 + 2t = 6$	Simplify.
$20 + 2t - 20 = 6 - 20$	Subtract 20 from each side.
$2t = -14$	Simplify.
$\frac{2t}{2} = \frac{-14}{2}$	Divide each side by 2.
$t = -7$	The solution is (4, −7).

Example 4 Elimination Using Subtraction

Multiple-Choice Test Item

If $x - 3y = 7$ and $x + 2y = 2$, what is the value of x?

(A) 4 (B) -1 (C) $(-1, 4)$ (D) $(4, -1)$

The Princeton Review

Test-Taking Tip

Always read the question carefully. Ask yourself, "What does the question ask?" Then answer that question.

Read the Test Item

You are given a system of equations, and you are asked to find the value of x.

Solve the Test Item

You can eliminate the x terms by subtracting one equation from the other.

$$\begin{array}{r} x - 3y = 7 \\ (-)\ x + 2y = 2 \\ \hline -5y = 5 \end{array}$$

Write the equations in column form and subtract.

Notice the x variable is eliminated.

$\frac{-5y}{-5} = \frac{5}{-5}$ Divide each side by -5.

$y = -1$ Simplify.

Now substitute -1 for y in either equation to find the value of x.

$x + 2y = 2$ Second equation

$x + 2(-1) = 2$ $y = -1$

$x - 2 = 2$ Simplify.

$x - 2 + 2 = 2 + 2$ Add 2 to each side.

$x = 4$ Simplify.

Notice that B is the value of y and D is the solution of the system of equations. However, the question asks for the value of x. The answer is A.

Check for Understanding

Concept Check

1. **OPEN ENDED** Write a system of equations that can be solved by using addition to eliminate one variable.
2. **Describe** a system of equations that can be solved by using subtraction to eliminate one variable.
3. **FIND THE ERROR** Michael and Yoomee are solving a system of equations.

Michael	Yoomee
$2r + s = 5$	$2r + s = 5$
$(+)\ r - s = 1$	$(-)r - s = 1$
$3r = 6$	$r = 4$
$r = 2$	$r - s = 1$
$2r + s = 5$	$4 - s = 1$
$2(2) + s = 5$	$-s = -3$
$4 + s = 5$	$s = 3$
$s = 1$	The solution is (4, 3).
The solution is (2, 1).	

Who is correct? Explain your reasoning.

Guided Practice **Use elimination to solve each system of equations.**

4. $x - y = 14$
   $x + y = 20$

5. $2a - 3b = -11$
   $a + 3b = 8$

6. $4x + y = -9$
   $4x + 2y = -10$

7. $6x + 2y = -10$
   $2x + 2y = -10$

8. $2a + 4b = 30$
   $-2a - 2b = -21.5$

9. $-4m + 2n = 6$
   $-4m + n = 8$

10. The sum of two numbers is 24. Five times the first number minus the second number is 12. What are the two numbers?

Standardized Test Practice

11. If $2x + 7y = 17$ and $2x + 5y = 11$, what is the value of $2y$?
    (A) -4 (B) -2 (C) 3 (D) 6

Practice and Apply

Homework Help

For Exercises	See Examples
12–29	1, 3
30–39	2
42, 43	4

Extra Practice
See page 836.

Use elimination to solve each system of equations.

12. $x + y = -3$
    $x - y = 1$

13. $s - t = 4$
    $s + t = 2$

14. $3m - 2n = 13$
    $m + 2n = 7$

15. $-4x + 2y = 8$
    $4x - 3y = -10$

16. $3a + b = 5$
    $2a + b = 10$

17. $2m - 5n = -6$
    $2m - 7n = -14$

18. $3r - 5s = -35$
    $2r - 5s = -30$

19. $13a + 5b = -11$
    $13a + 11b = 7$

20. $3x - 5y = 16$
    $-3x + 2y = -10$

21. $6s + 5t = 1$
    $6s - 5t = 11$

22. $4x - 3y = 12$
    $4x + 3y = 24$

23. $a - 2b = 5$
    $3a - 2b = 9$

24. $4x + 5y = 7$
    $8x + 5y = 9$

25. $8a + b = 1$
    $8a - 3b = 3$

26. $1.44x - 3.24y = -5.58$
    $1.08x + 3.24y = 9.99$

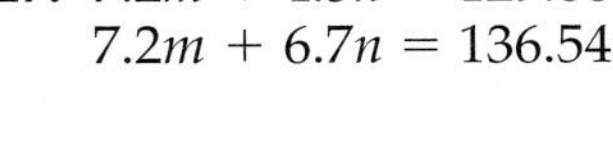

27. $7.2m + 4.5n = 129.06$
    $7.2m + 6.7n = 136.54$

28. $\frac{3}{5}c - \frac{1}{5}d = 9$
    $\frac{7}{5}c + \frac{1}{5}d = 11$

29. $\frac{2}{3}x - \frac{1}{2}y = 14$
    $\frac{5}{6}x - \frac{1}{2}y = 18$

30. The sum of two numbers is 48, and their difference is 24. What are the numbers?

31. Find the two numbers whose sum is 51 and whose difference is 13.

32. Three times one number added to another number is 18. Twice the first number minus the other number is 12. Find the numbers.

33. One number added to twice another number is 23. Four times the first number added to twice the other number is 38. What are the numbers?

34. **BUSINESS** In 1999, the United States produced about 2 million more motor vehicles than Japan. Together, the two countries produced about 22 million motor vehicles. How many vehicles were produced in each country?

35. **PARKS** A youth group and their leaders visited Mammoth Cave. Two adults and 5 students in one van paid \$77 for the Grand Avenue Tour of the cave. Two adults and 7 students in a second van paid \$95 for the same tour. Find the adult price and the student price of the tour.

More About. . .

Parks

Mammoth Cave in Kentucky was declared a national park in 1941. It has more than 336 miles of explored caves, making it the longest recorded cave system in the world.

Source: National Park Service

36. **FOOTBALL** During the National Football League's 1999 season, Troy Aikman, the quarterback for the Dallas Cowboys, earned \$0.467 million more than Deion Sanders, the Cowboys cornerback. Together they cost the Cowboys \$12.867 million. How much did each player make?

POPULATIONS For Exercises 37–39, use the information in the graph at the right.

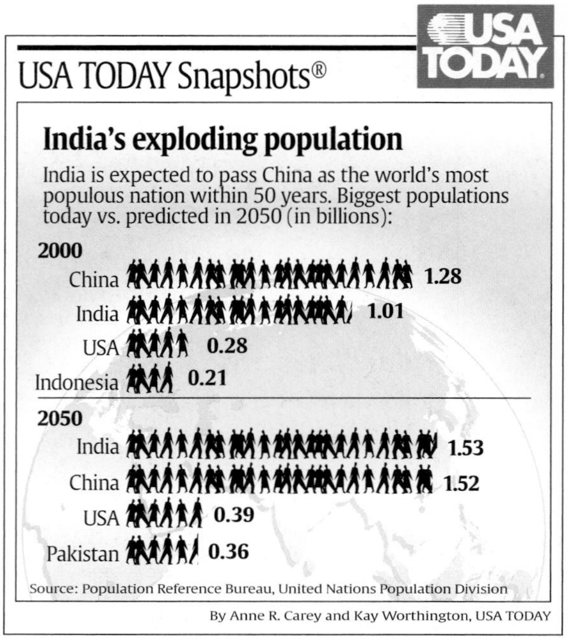

37. Let x represent the number of years since 2000 and y represent population in billions. Write an equation to represent the population of China.

38. Write an equation to represent the population of India.

39. Use elimination to find the year when the populations of China and India are predicted to be the same. What is the predicted population at that time?

USA TODAY Snapshots®

India's exploding population

India is expected to pass China as the world's most populous nation within 50 years. Biggest populations today vs. predicted in 2050 (in billions):

2000
China 1.28
India 1.01
USA 0.28
Indonesia 0.21

2050
India 1.53
China 1.52
USA 0.39
Pakistan 0.36

Source: Population Reference Bureau, United Nations Population Division

By Anne R. Carey and Kay Worthington, USA TODAY

40. **CRITICAL THINKING** The graphs of $Ax + By = 15$ and $Ax - By = 9$ intersect at (2, 1). Find A and B.

41. **WRITING IN MATH** Answer the question that was posed at the beginning of the lesson.

How can you use a system of equations to solve problems about weather?

Include the following in your answer:

- an explanation of how to use elimination to solve a system of equations, and
- a step-by-step solution of the Seward daylight problem.

Standardized Test Practice

42. If $2x - 3y = -9$ and $3x - 3y = -12$, what is the value of y?

(A) -3 (B) 1 (C) $(-3, 1)$ (D) $(1, -3)$

43. What is the solution of $4x + 2y = 8$ and $2x + 2y = 2$?

(A) $(-2, 3)$ (B) $(3, 2)$ (C) $(3, -2)$ (D) $(12, -3)$

Maintain Your Skills

Mixed Review

Use substitution to solve each system of equations. If the system does *not* have exactly one solution, state whether it has *no* solution or *infinitely many* solutions. *(Lesson 7-2)*

44. $y = 5x$
$x + 2y = 22$

45. $x = 2y + 3$
$3x + 4y = -1$

46. $2y - x = -5$
$4y - 3x = -1$

Graph each system of equations. Then determine whether the system has *no* solution, *one* solution, or *infinitely many* solutions. If the system has one solution, name it. *(Lesson 7-1)*

47. $x - y = 3$
$3x + y = 1$

48. $2x - 3y = 7$
$3y = 7 + 2x$

49. $4x + y = 12$
$x = 3 - \frac{1}{4}y$

50. Write an equation of a line that is parallel to the graph of $y = \frac{5}{4}x - 3$ and passes through the origin. *(Lesson 5-6)*

Getting Ready for the Next Lesson

PREREQUISITE SKILL **Use the Distributive Property to rewrite each expression without parentheses.** *(To review the **Distributive Property**, see Lesson 1-5.)*

51. $2(3x + 4y)$ 52. $6(2a - 5b)$ 53. $-3(-2m + 3n)$ 54. $-5(4t - 2s)$

7-4 Elimination Using Multiplication

What You'll Learn

- Solve systems of equations by using elimination with multiplication.
- Determine the best method for solving systems of equations.

How can a manager use a system of equations to plan employee time?

The Finneytown Bakery is making peanut butter cookies and loaves of quick bread. The preparation and baking times for each are given in the table below.

For these two items, the management has allotted 800 minutes of employee time and 900 minutes of oven time. If c represents the number of batches of cookies and b represents the number of loaves of bread, the following system of equations can be used to determine how many of each to bake.

$$20c + 10b = 800$$
$$10c + 30b = 900$$

	Cookies (per batch)	Bread (per loaf)
Preparation	20 min	10 min
Baking	10 min	30 min

ELIMINATION USING MULTIPLICATION Neither variable in the system above can be eliminated by simply adding or subtracting the equations. However, you can use the Multiplication Property of Equality so that adding or subtracting eliminates one of the variables.

Example 1 Multiply One Equation to Eliminate

Use elimination to solve the system of equations.

$$\mathbf{3x + 4y = 6}$$
$$\mathbf{5x + 2y = -4}$$

Multiply the second equation by -2 so the coefficients of the y terms are additive inverses. Then add the equations.

$3x + 4y = 6$

$5x + 2y = -4$ Multiply by -2.

$$\begin{array}{rl} 3x + 4y = 6 & \\ (+)\ -10x - 4y = 8 & \\ \hline -7x \quad = 14 & \text{Add the equations.} \\ \frac{-7x}{-7} = \frac{14}{-7} & \text{Divide each side by } -7. \\ x = -2 & \text{Simplify.} \end{array}$$

Now substitute -2 for x in either equation to find the value of y.

$$\begin{array}{rl} 3x + 4y = 6 & \text{First equation} \\ 3(-2) + 4y = 6 & x = -2 \\ -6 + 4y = 6 & \text{Simplify.} \\ -6 + 4y + 6 = 6 + 6 & \text{Add 6 to each side.} \\ 4y = 12 & \text{Simplify.} \\ \frac{4y}{4} = \frac{12}{4} & \text{Divide each side by 4.} \\ y = 3 & \end{array}$$

The solution is $(-2, 3)$.

For some systems of equations, it is necessary to multiply each equation by a different number in order to solve the system by elimination. You can choose to eliminate either variable.

Example 2 Multiply Both Equations to Eliminate

Use elimination to solve the system of equations.

$3x + 4y = -25$
$2x - 3y = 6$

Method 1 Eliminate x.

$3x + 4y = -25$ Multiply by 2. → $6x + 8y = -50$
$2x - 3y = 6$ Multiply by −3. → $(+)\ -6x + 9y = -18$

$17y = -68$ Add the equations.

$\frac{17y}{17} = \frac{-68}{17}$ Divide each side by 17.

$y = -4$ Simplify.

Now substitute −4 for y in either equation to find the value of x.

$2x - 3y = 6$ Second equation

$2x - 3(-4) = 6$ $y = -4$

$2x + 12 = 6$ Simplify.

$2x + 12 - 12 = 6 - 12$ Subtract 12 from each side.

$2x = -6$ Simplify.

$\frac{2x}{2} = \frac{-6}{2}$ Divide each side by 2.

$x = -3$ Simplify.

The solution is $(-3, -4)$.

Method 2 Eliminate y.

$3x + 4y = -25$ Multiply by 3. → $9x + 12y = -75$
$2x - 3y = 6$ Multiply by 4. → $(+)\ 8x - 12y = 24$

$17x = -51$ Add the equations.

$\frac{17x}{17} = \frac{-51}{17}$ Divide each side by 17.

$x = -3$ Simplify.

Now substitute −3 for x in either equation to find the value of y.

$2x - 3y = 6$ Second equation

$2(-3) - 3y = 6$ $x = -3$

$-6 - 3y = 6$ Simplify.

$-6 - 3y + 6 = 6 + 6$ Add 6 to each side.

$-3y = 12$ Simplify.

$\frac{-3y}{-3} = \frac{12}{-3}$ Divide each side by −3.

$y = -4$ Simplify.

The solution is $(-3, -4)$, which matches the result obtained with Method 1.

Study Tip

Using Multiplication
There are many other combinations of multipliers that could be used to solve the system in Example 2. For instance, the first equation could be multiplied by −2 and the second by 3.

DETERMINE THE BEST METHOD You have learned five methods for solving systems of linear equations.

Concept Summary *Solving Systems of Equations*

Method	The Best Time to Use
Graphing	to estimate the solution, since graphing usually does not give an exact solution
Substitution	if one of the variables in either equation has a coefficient of 1 or −1
Elimination Using Addition	if one of the variables has opposite coefficients in the two equations
Elimination Using Subtraction	if one of the variables has the same coefficient in the two equations
Elimination Using Multiplication	if none of the coefficients are 1 or −1 and neither of the variables can be eliminated by simply adding or subtracting the equations

Example 3 Determine the Best Method

Determine the best method to solve the system of equations. Then solve the system.

$4x - 3y = 12$
$x + 2y = 14$

- For an exact solution, an algebraic method is best.
- Since neither the coefficients of x nor the coefficients of y are the same or additive inverses, you cannot use elimination using addition or subtraction.
- Since the coefficient of x in the second equation is 1, you can use the substitution method. You could also use elimination using multiplication.

The following solution uses substitution. *Which method would you prefer?*

$x + 2y = 14$	Second equation
$x + 2y - 2y = 14 - 2y$	Subtract $2y$ from each side.
$x = 14 - 2y$	Simplify.
$4x - 3y = 12$	First equation
$4(14 - 2y) - 3y = 12$	$x = 14 - 2y$
$56 - 8y - 3y = 12$	Distributive Property
$56 - 11y = 12$	Combine like terms.
$56 - 11y - 56 = 12 - 56$	Subtract 56 from each side.
$-11y = -44$	Simplify.
$\frac{-11y}{-11} = \frac{-44}{-11}$	Divide each side by −11.
$y = 4$	Simplify.
$x + 2y = 14$	Second equation
$x + 2(4) = 14$	$y = 4$
$x + 8 = 14$	Simplify.
$x + 8 - 8 = 14 - 8$	Subtract 8 from each side.
$x = 6$	Simplify.

The solution is (6, 4).

Study Tip

Alternative Method
This system could also be solved easily by multiplying the second equation by 4 and then subtracting the equations.

More About. . .

Transportation

About 203 million tons of freight are transported on the Ohio River each year making it the second most used commercial river in the United States.

Source: *World Book Encyclopedia*

Example 4 Write and Solve a System of Equations

TRANSPORTATION **A coal barge on the Ohio River travels 24 miles upstream in 3 hours. The return trip takes the barge only 2 hours. Find the rate of the barge in still water.**

Let b = the rate of the barge in still water and c = the rate of the current. Use the formula rate × time = distance, or $rt = d$.

	r	t	d	$rt = d$
Downstream	$b + c$	2	24	$2b + 2c = 24$
Upstream	$b - c$	3	24	$3b - 3c = 24$

This system cannot easily be solved using substitution. It cannot be solved by just adding or subtracting the equations.

The best way to solve this system is to use elimination using multiplication. Since the problem asks for b, eliminate c.

$2b + 2c = 24$ Multiply by 3. → $6b + 6c = 72$

$3b - 3c = 24$ Multiply by 2. → $(+)\ 6b - 6c = 48$

$12b = 120$ Add the equations.

$\frac{12b}{12} = \frac{120}{12}$ Divide each side by 12.

$b = 10$ Simplify.

The rate of the barge in still water is 10 miles per hour.

Check for Understanding

Concept Check

1. **Explain** why multiplication is sometimes needed to solve a system of equations by elimination.

2. **OPEN ENDED** Write a system of equations that could be solved by multiplying one equation by 5 and then adding the two equations together to eliminate one variable.

3. **Describe** two methods that could be used to solve the following system of equations. Which method do you prefer? Explain.

   $a - b = 5$
   $2a + 3b = 15$

Guided Practice

Use elimination to solve each system of equations.

4. $2x - y = 6$
   $3x + 4y = -2$

5. $x + 5y = 4$
   $3x - 7y = -10$

6. $4x + 7y = 6$
   $6x + 5y = 20$

7. $4x + 2y = 10.5$
   $2x + 3y = 10.75$

Determine the best method to solve each system of equations. Then solve the system.

8. $4x + 3y = 19$
   $3x - 4y = 8$

9. $3x - 7y = 6$
   $2x + 7y = 4$

10. $y = 4x + 11$
    $3x - 2y = -7$

11. $5x - 2y = 12$
    $3x - 2y = -2$

Application **12. BUSINESS** The owners of the River View Restaurant have hired enough servers to handle 17 tables of customers, and the fire marshal has approved the restaurant for a limit of 56 customers. How many two-seat tables and how many four-seat tables should the owners purchase?

Practice and Apply

Homework Help

For Exercises	See Examples
13–26	1, 2
27–38	3
39–43	4

Extra Practice

See page 836.

Use elimination to solve each system of equations.

13. $-5x + 3y = 6$
$x - y = 4$

14. $x + y = 3$
$2x - 3y = 16$

15. $2x + y = 5$
$3x - 2y = 4$

16. $4x - 3y = 12$
$x + 2y = 14$

17. $5x - 2y = -15$
$3x + 8y = 37$

18. $8x - 3y = -11$
$2x - 5y = 27$

19. $4x - 7y = 10$
$3x + 2y = -7$

20. $2x - 3y = 2$
$5x + 4y = 28$

21. $1.8x - 0.3y = 14.4$
$x - 0.6y = 2.8$

22. $0.4x + 0.5y = 2.5$
$1.2x - 3.5y = 2.5$

23. $3x - \frac{1}{2}y = 10$
$5x + \frac{1}{4}y = 8$

24. $2x + \frac{2}{3}y = 4$
$x - \frac{1}{2}y = 7$

25. Seven times a number plus three times another number equals negative one. The sum of the two numbers is negative three. What are the numbers?

26. Five times a number minus twice another number equals twenty-two. The sum of the numbers is three. Find the numbers.

Determine the best method to solve each system of equations. Then solve the system.

27. $3x - 4y = -10$
$5x + 8y = -2$

28. $9x - 8y = 42$
$4x + 8y = -16$

29. $y = 3x$
$3x + 4y = 30$

30. $x = 4y + 8$
$2x - 8y = -3$

31. $2x - 3y = 12$
$x + 3y = 12$

32. $4x - 2y = 14$
$y = x$

33. $x - y = 2$
$5x + 3y = 18$

34. $y = 2x + 9$
$2x - y = -9$

35. $6x - y = 9$
$6x - y = 11$

36. $x = 8y$
$2x + 3y = 38$

37. $\frac{2}{3}x - \frac{1}{2}y = 14$
$\frac{5}{6}x - \frac{1}{2}y = 18$

38. $\frac{1}{2}x - \frac{2}{3}y = \frac{7}{3}$
$\frac{3}{2}x + 2y = -25$

39. BASKETBALL In basketball, a free throw is 1 point and a field goal is either 2 points or 3 points. Suppose a basketball player scored a total of 1938 points in one season. The total number of 2-point field goals and 3-point field goals was 701, and he made 475 of the 557 free throws he attempted. Find the number of 2-point field goals and 3-point field goals the player made that season.

Online Research **Data Update** What are the current statistics for Kobe Bryant and other players? Visit www.algebra1.com/data_update to learn more.

40. CRITICAL THINKING The solution of the system $4x + 5y = 2$ and $6x - 2y = b$ is $(3, a)$. Find the values of a and b.

41. CAREERS Mrs. Henderson discovered that she had accidentally reversed the digits of a test and shorted a student 36 points. Mrs. Henderson told the student that the sum of the digits was 14 and agreed to give the student his correct score plus extra credit if he could determine his actual score without looking at his test. What was his actual score on the test?

42. **NUMBER THEORY** The sum of the digits of a two-digit number is 14. If the digits are reversed, the new number is 18 less than the original number. Find the original number.

43. **TRANSPORTATION** Traveling against the wind, a plane flies 2100 miles from Chicago to San Diego in 4 hours and 40 minutes. The return trip, traveling with a wind that is twice as fast, takes 4 hours. Find the rate of the plane in still air.

44. **WRITING IN MATH** Answer the question that was posed at the beginning of the lesson.

How can a manager use a system of equations to plan employee time?

Include the following in your answer:

- a demonstration of how to solve the system of equations concerning the cookies and bread, and
- an explanation of how a restaurant manager would schedule oven and employee time.

Standardized Test Practice

45. If $5x + 3y = 12$ and $4x - 5y = 17$, what is the value of y?

Ⓐ -1 Ⓑ 3 Ⓒ $(-1, 3)$ Ⓓ $(3, -1)$

46. Determine the number of solutions of the system $x + 2y = -1$ and $2x + 4y = -2$.

Ⓐ 0 Ⓑ 1 Ⓒ 2 Ⓓ infinitely many

Maintain Your Skills

Mixed Review

Use elimination to solve each system of equations. *(Lesson 7-3)*

47. $x + y = 8$
$x - y = 4$

48. $2r + s = 5$
$r - s = 1$

49. $x + y = 18$
$x + 2y = 25$

Use substitution to solve each system of equations. If the system does *not* have exactly one solution, state whether it has *no* solution or *infinitely many* solutions. *(Lesson 7-2)*

50. $2x + 3y = 3$
$x = -3y$

51. $x + y = 0$
$3x + y = -8$

52. $x - 2y = 7$
$-3x + 6y = -21$

53. **CAREERS** A store manager is paid $32,000 a year plus 4% of the revenue the store makes above quota. What is the amount of revenue above quota needed for the manager to have an annual income greater than $45,000? *(Lesson 6-3)*

Getting Ready for the Next Lesson

PREREQUISITE SKILL **Graph each inequality.**
*(To review **graphing inequalities**, see Lesson 6-6.)*

54. $y \geq x - 7$

55. $x + 3y \geq 9$

56. $-y \leq x$

57. $-3x + y \geq -1$

Practice Quiz 2

Lessons 7-3 and 7-4

Use elimination to solve each system of equations. *(Lessons 7-3 and 7-4)*

1. $5x + 4y = 2$
$3x - 4y = 14$

2. $2x - 3y = 13$
$2x + 2y = -2$

3. $6x - 2y = 24$
$3x + 4y = 27$

4. $5x + 2y = 4$
$10x + 4y = 9$

5. The price of a cellular telephone plan is based on peak and nonpeak service. Kelsey used 45 peak minutes and 50 nonpeak minutes and was charged $27.75. That same month, Mitch used 70 peak minutes and 30 nonpeak minutes for a total charge of $36. What are the rates per minute for peak and nonpeak time? *(Lesson 7-4)*

Reading Mathematics

Making Concept Maps

After completing a chapter, it is wise to review each lesson's main topics and vocabulary. In Lesson 7-1, the new vocabulary words were *system of equations*, *consistent*, *inconsistent*, *independent*, and *dependent*. They are all related in that they explain how many and what kind of solutions a system of equations has.

A graphic organizer called a *concept map* is a convenient way to show these relationships. A concept map is shown below for the vocabulary words for Lesson 7-1. The main ideas are placed in boxes. Any information that describes how to move from one box to the next is placed along the arrows.

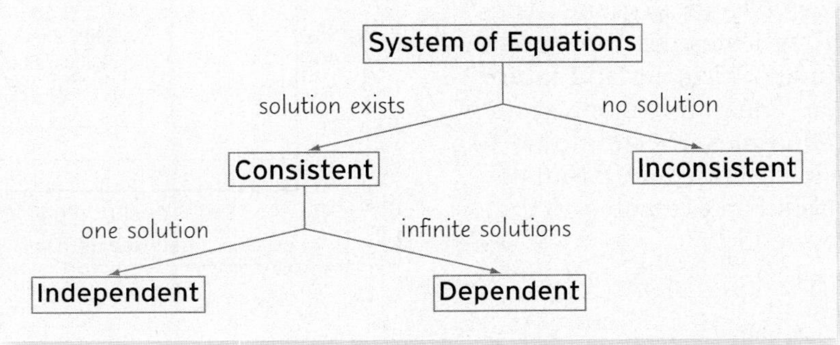

Concept maps are used to organize information. They clearly show how ideas are related to one another. They also show the flow of mental processes needed to solve problems.

Reading to Learn

Review Lessons 7-2, 7-3, and 7-4.

1. Write a couple of sentences describing the information in the concept map above.
2. How do you decide whether to use substitution or elimination? Give an example of a system that you would solve using each method.
3. How do you decide whether to multiply an equation by a factor?
4. How do you decide whether to add or subtract two equations?
5. Copy and complete the concept map below for solving systems of equations by using either substitution or elimination.

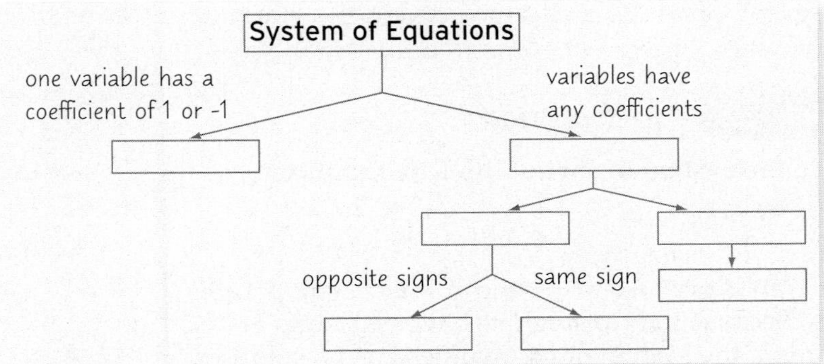

7-5 Graphing Systems of Inequalities

What You'll Learn

- Solve systems of inequalities by graphing.
- Solve real-world problems involving systems of inequalities.

Vocabulary

- system of inequalities

How can you use a system of inequalities to plan a sensible diet?

Joshua watches what he eats. His doctor told him to eat between 2000 and 2400 Calories per day. The doctor also wants him to keep his daily fat intake between 60 and 75 grams. The graph indicates the appropriate amounts of Calories and fat for Joshua. The graph is of a system of inequalities. He should try to keep his Calorie and fat intake to amounts represented in the green section.

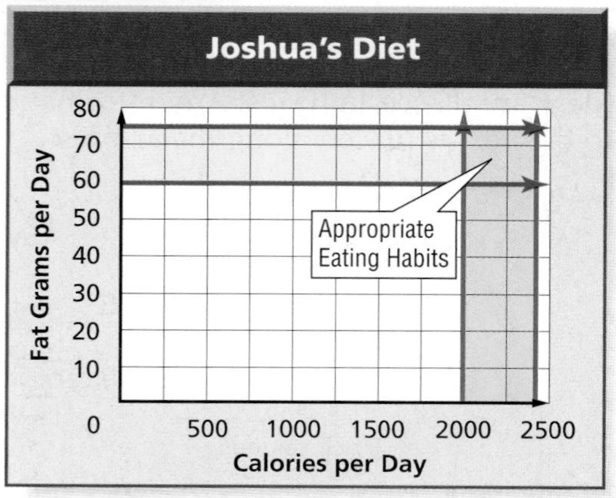

SYSTEMS OF INEQUALITIES To solve a **system of inequalities**, you need to find the ordered pairs that satisfy all the inequalities involved. One way to do this is to graph the inequalities on the same coordinate plane. The solution set is represented by the intersection, or overlap, of the graphs.

Example 1 Solve by Graphing

Solve the system of inequalities by graphing.

$y < -x + 1$
$y \leq 2x + 3$

The solution includes the ordered pairs in the intersection of the graphs of $y < -x + 1$ and $y \leq 2x + 3$. This region is shaded in green at the right. The graphs of $y = -x + 1$ and $y = 2x + 3$ are boundaries of this region. The graph of $y = -x + 1$ is dashed and is *not* included in the graph of $y < -x + 1$. The graph of $y = 2x + 3$ is included in the graph of $y \leq 2x + 3$.

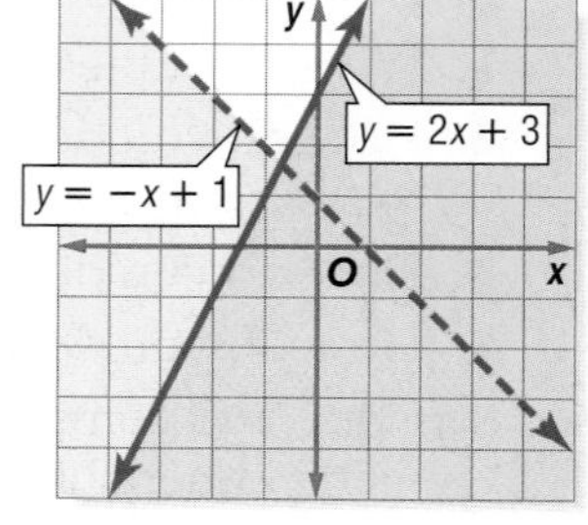

Study Tip

Look Back
To review **graphing linear inequalities**, see Lesson 6-6.

Example 2 No Solution

Solve the system of inequalities by graphing.

$x - y < -1$
$x - y > 3$

The graphs of $x - y = -1$ and $x - y = 3$ are parallel lines. Because the two regions have no points in common, the system of inequalities has no solution.

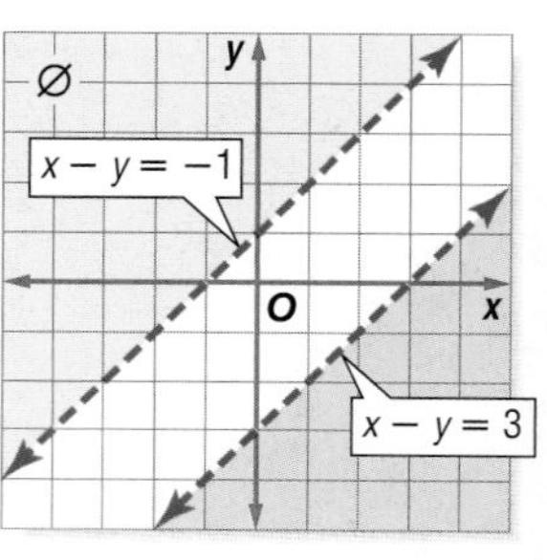

You can use a TI-83 Plus to solve systems of inequalities.

Graphing Calculator Investigation

Graphing Systems of Inequalities

To graph the system $y \geq 4x - 3$ and $y \leq -2x + 9$ on a TI-83 Plus, select the SHADE feature in the DRAW menu. Enter the function that is the lower boundary of the region to be shaded, followed by the upper boundary. (Note that inequalities that have $>$ or $\geq$ are lower boundaries and inequalities that have $<$ or $\leq$ are upper boundaries.)

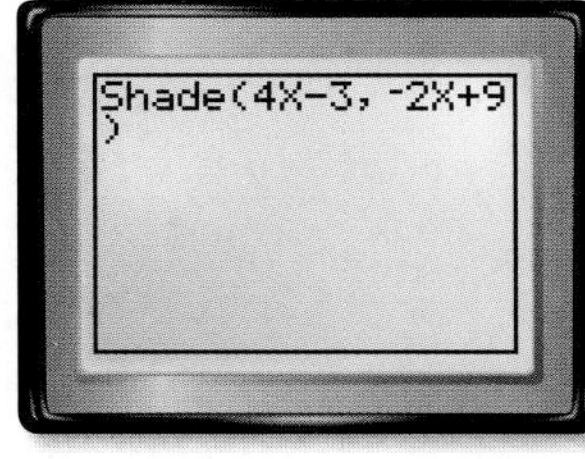

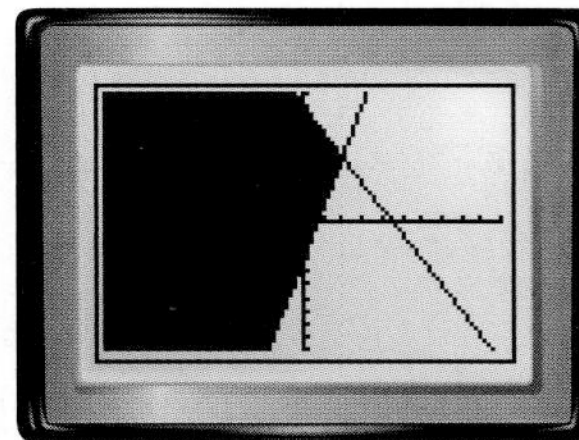

[−10, 10] scl: 1 by [−10, 10] scl: 1

Think and Discuss

1. To graph the system $y \leq 3x + 1$ and $y \geq -2x - 5$ on a graphing calculator, which function should you enter first?
2. Use a graphing calculator to graph the system $y \leq 3x + 1$ and $y \geq -2x - 5$.
3. Explain how you could use a graphing calculator to graph the system $2x + y \geq 7$ and $x - 2y \geq 5$.
4. Use a graphing calculator to graph the system $2x + y \geq 7$ and $x - 2y \geq 5$.

REAL-WORLD PROBLEMS In real-life problems involving systems of inequalities, sometimes only whole-number solutions make sense.

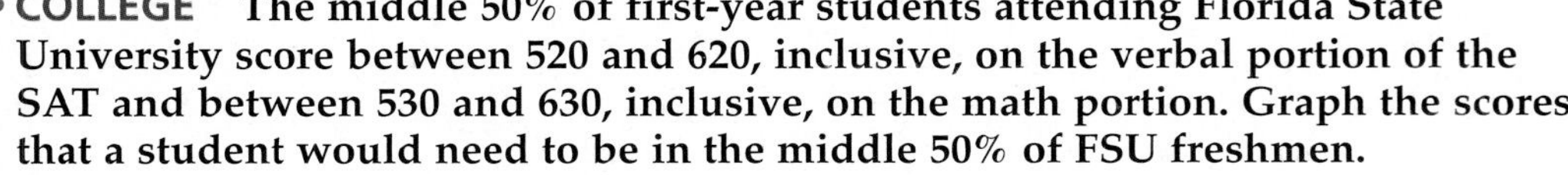

Example 3 Use a System of Inequalities to Solve a Problem

COLLEGE The middle 50% of first-year students attending Florida State University score between 520 and 620, inclusive, on the verbal portion of the SAT and between 530 and 630, inclusive, on the math portion. Graph the scores that a student would need to be in the middle 50% of FSU freshmen.

Words The verbal score is between 520 and 620, inclusive. The math score is between 530 and 630, inclusive.

Variables If v = the verbal score and m = the math score, the following inequalities represent the middle 50% of Florida State University freshmen.

Inequalities

The verbal score is between 520 and 620, inclusive.

$$520 \leq v \leq 620$$

The math score is between 530 and 630, inclusive.

$$530 \leq m \leq 630$$

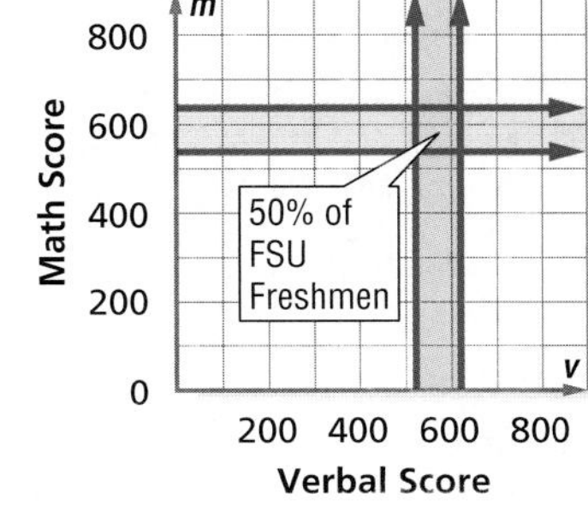

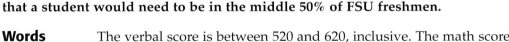

The solution is the set of all ordered pairs whose graphs are in the intersection of the graphs of these inequalities. However, since SAT scores are whole numbers, only whole-number solutions make sense in this problem.

More About. . .

College

FSU is the most wired campus in Florida and has been recently ranked the 18th most technologically connected university in the nation.

Source: www.fsu.edu

Example 4 Use a System of Inequalities

AGRICULTURE To ensure a growing season of sufficient length, Mr. Hobson has at most 16 days left to plant his corn and soybean crops. He can plant corn at a rate of 250 acres per day and soybeans at a rate of 200 acres per day. If he has at most 3500 acres available, how many acres of each type of crop can he plant?

Let c = the number of days that corn will be planted and s = the number of days that soybeans will be planted. Since both c and s represent a number of days, neither can be a negative number. The following system of inequalities can be used to represent the conditions of this problem.

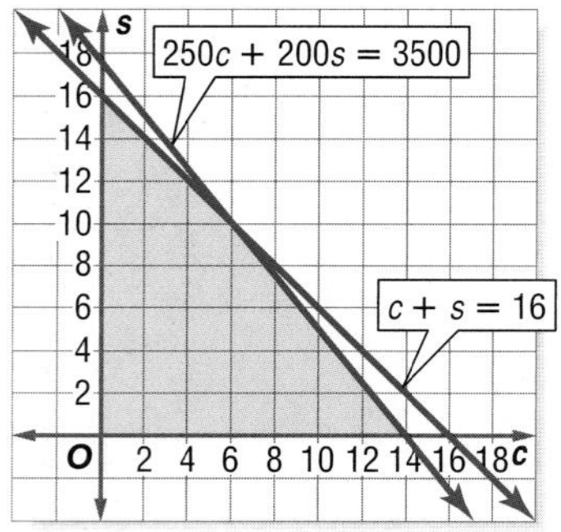

$c \geq 0$

$s \geq 0$

$c + s \leq 16$

$250c + 200s \leq 3500$

The solution is the set of all ordered pairs whose graphs are in the intersection of the graphs of these inequalities. This region is shown in green at the right. Only the portion of the region in the first quadrant is used since $c \geq 0$ and $s \geq 0$.

Any point in this region is a possible solution. For example, since (7, 8) is a point in the region, Mr. Hobson could plant corn for 7 days and soybeans for 8 days. In this case, he would use 15 days to plant 250(7) or 1750 acres of corn and 200(8) or 1600 acres of soybeans.

Check for Understanding

Concept Check

1. **OPEN ENDED** Draw the graph of a system of inequalities that has no solution.

2. **Determine** which of the following ordered pairs represent a solution of the system of inequalities graphed at the right.

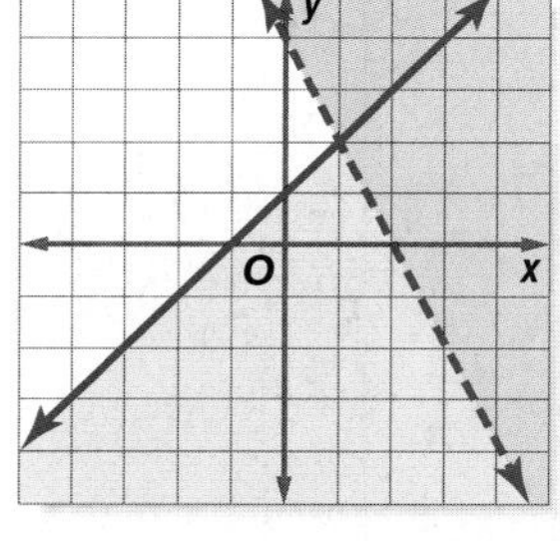

   a. (3, 1) **b.** (−1, −3)
   c. (2, 3) **d.** (4, −2)
   e. (3, −2) **f.** (1, 4)

3. **FIND THE ERROR** Kayla and Sonia are solving the system of inequalities $x + 2y \geq -2$ and $x - y > 1$.

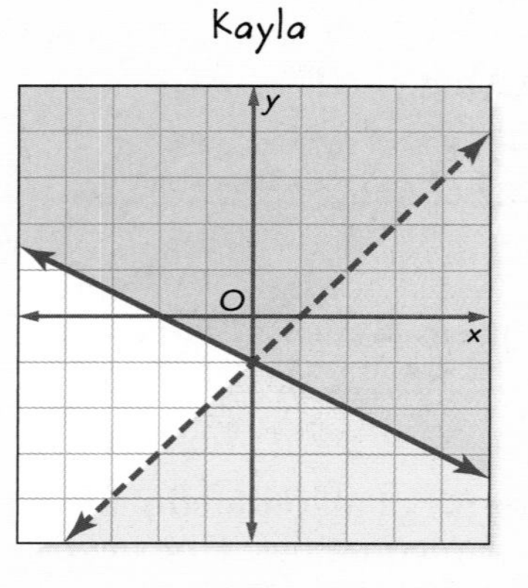

Sonia

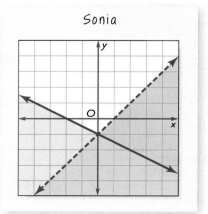

Who is correct? Explain your reasoning.

Guided Practice **Solve each system of inequalities by graphing.**

4. $x > 5$
$y \leq 4$

5. $y > 3$
$y > -x + 4$

6. $y \leq -x + 3$
$y \leq x + 3$

7. $2x + y \geq 4$
$y \leq -2x - 1$

8. $2y + x < 6$
$3x - y > 4$

9. $x - 2y \leq 2$
$3x + 4y \leq 12$
$x \geq 0$

Application **HEALTH For Exercises 10 and 11, use the following information.**
Natasha walks and jogs at least 3 miles every day. Natasha walks 4 miles per hour and jogs 8 miles per hour. She only has a half-hour to exercise.

10. Draw a graph of the possible amounts of time she can spend walking and jogging.

11. List three possible solutions.

Practice and Apply

Homework Help

For Exercises	See Examples
12–28	1–2
29–31, 33–35	3–4

Extra Practice
See page 836.

Solve each system of inequalities by graphing.

12. $y < 0$
$x \geq 0$

13. $x > -4$
$y \leq -1$

14. $y \geq -2$
$y - x < 1$

15. $x \geq 2$
$y + x \leq 5$

16. $x \leq 3$
$x + y > 2$

17. $y \geq 2x + 1$
$y \leq -x + 1$

18. $y < 2x + 1$
$y \geq -x + 3$

19. $y - x < 1$
$y - x > 3$

20. $y - x < 3$
$y - x \geq 2$

21. $2x + y \leq 4$
$3x - y \geq 6$

22. $3x - 4y < 1$
$x + 2y \leq 7$

23. $x + y > 4$
$-2x + 3y < -12$

24. $2x + y \geq -4$
$-5x + 2y < 1$

25. $y \leq x + 3$
$2x - 7y \leq 4$
$3x + 2y \leq 6$

26. $x < 2$
$4y > x$
$2x - y > -9$
$x + 3y < 9$

Write a system of inequalities for each graph.

27.

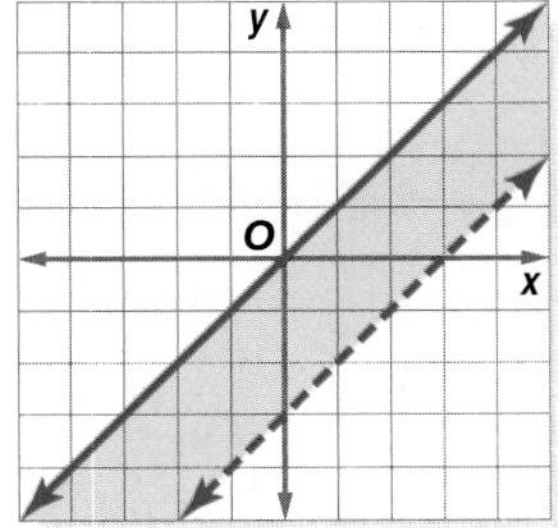

28.

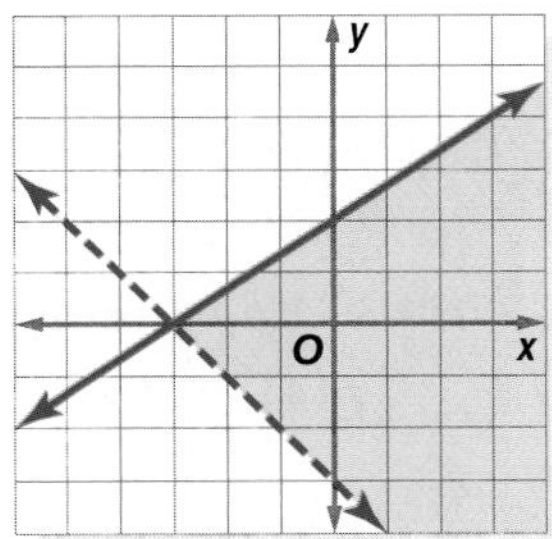

Career Choices

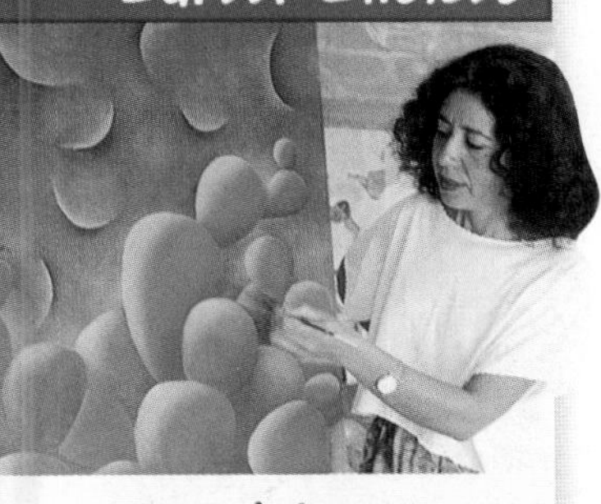

Visual Artist
Visual artists create art to communicate ideas. The work of fine artists is made for display. Illustrators and graphic designers produce art for clients, such as advertising and publishing companies.

Online Research
For information about a career as a visual artist, visit: www.algebra1.com/careers

ART For Exercises 29 and 30, use the following information.
A painter has exactly 32 units of yellow dye and 54 units of blue dye. She plans to mix the dyes to make two shades of green. Each gallon of the lighter shade of green requires 4 units of yellow dye and 1 unit of blue dye. Each gallon of the darker shade of green requires 1 unit of yellow dye and 6 units of blue dye.

29. Make a graph showing the numbers of gallons of the two greens she can make.

30. List three possible solutions.

31. **HEALTH** The LDL or "bad" cholesterol of a teenager should be less than 110. The HDL or "good" cholesterol of a teenager should be between 35 and 59. Make a graph showing appropriate levels of cholesterol for a teenager.

32. **CRITICAL THINKING** Write a system of inequalities that is equivalent to $|x| \leq 4$.

MANUFACTURING **For Exercises 33 and 34, use the following information.**
The Natural Wood Company has machines that sand and varnish desks and tables. The table gives the time requirements of the machines.

Machine	Hours per Desk	Hours per Table	Total Hours Available Each Week
Sanding	2	1.5	31
Varnishing	1.5	1	22

33. Make a graph showing the number of desks and the number of tables that can be made in a week.

34. List three possible solutions.

35. WRITING IN MATH Answer the question that was posed at the beginning of the lesson.

How can you use a system of inequalities to plan a sensible diet?

Include the following in your answer:

- two appropriate Calorie and fat intakes for a day, and
- the system of inequalities that is represented by the graph.

Graphing Calculator

GRAPHING SYSTEMS OF INEQUALITIES **Use a graphing calculator to solve each system of inequalities. Sketch the results.**

36. $y \leq x + 9$
$y \geq -x - 4$

37. $y \leq 2x + 10$
$y \geq 7x + 15$

38. $3x - y \leq 6$
$x - y \geq -1$

Standardized Test Practice

39. Which ordered pair does *not* satisfy the system $x + 2y > 5$ and $3x - y < -2$?

Ⓐ $(-3, 7)$ Ⓑ $(0, 5)$ Ⓒ $(-1, 4)$ Ⓓ $(0, 2.5)$

40. Which system of inequalities is represented by the graph?

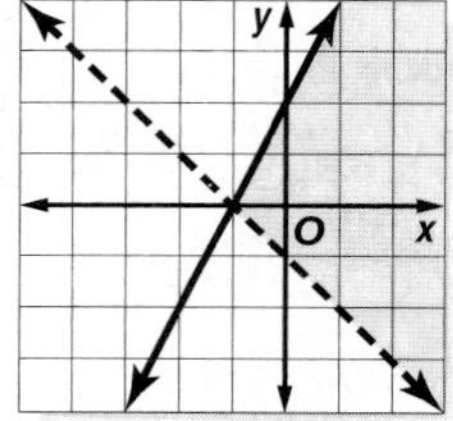

Ⓐ $y \leq 2x + 2$
$y > -x - 1$

Ⓑ $y \geq 2x + 2$
$y < -x - 1$

Ⓒ $y < 2x + 2$
$y \leq -x - 1$

Ⓓ $y > 2x + 2$
$y \leq -x - 1$

Maintain Your Skills

Mixed Review **Use elimination to solve each system of equations.** *(Lessons 7-3 and 7-4)*

41. $2x + 3y = 1$
$4x - 5y = 13$

42. $5x - 2y = -3$
$3x + 6y = -9$

43. $-3x + 2y = 12$
$2x - 3y = -13$

44. $6x - 2y = 4$
$5x - 3y = -2$

45. $2x + 5y = 13$
$3x - 5y = -18$

46. $3x - y = 6$
$3x + 2y = 15$

Write an equation of the line that passes through each point with the given slope. *(Lesson 5-4)*

47. $(4, -1), m = 2$

48. $(1, 0), m = -6$

49. $(5, -2), m = \frac{1}{3}$

WebQuest Internet Project

The Spirit of the Games

It's time to complete your project. Use the information and data you have gathered about the Olympics to prepare a portfolio or Web page. Be sure to include graphs and/or tables in your project.

www.algebra1.com/webquest

Chapter 7

Study Guide and Review

Vocabulary and Concept Check

consistent (p. 369)
dependent (p. 369)
elimination (p. 382)
inconsistent (p. 369)
independent (p. 369)
substitution (p. 376)
system of equations (p. 369)
system of inequalities (p. 394)

Choose the correct term to complete each statement.

1. If a system of equations has exactly one solution, it is *(dependent, independent)*.
2. If the graph of a system of equations is parallel lines, the system is *(consistent, inconsistent)*.
3. A system of equations that has infinitely many solutions is *(dependent, independent)*.
4. If the graphs of the equations in a system have the same slope and different y intercepts, the graph of the system is a pair of *(intersecting lines, parallel lines)*.
5. If the graphs of the equations in a system have the same slope and y intercept(s), the system has *(exactly one, infinitely many)* solution(s).
6. The solution of a system of equations is $(3, -5)$. The system is *(consistent, inconsistent)*.

Lesson-by-Lesson Review

7-1 Graphing Systems of Inequalities

See pages 369–374.

Concept Summary

	Intersecting Lines	Same Line	Parallel Lines
Graph of a System			
Number of Solutions	exactly one solution	infinitely many	no solutions
Terminology	consistent and independent	consistent and dependent	inconsistent

Example **Graph the system of equations. Then determine whether the system has *no* solution, *one* solution, or *infinitely many* solutions. If the system has one solution, name it.**

$3x + y = -4$
$6x + 2y = -8$

When the lines are graphed, they coincide. There are infinitely many solutions.

Exercises **Graph each system of equations. Then determine whether the system of equations has *one* solution, *no* solution, or *infinitely many* solutions. If the system has one solution, name it.** *See Example 2 on page 370.*

7. $x - y = 9$
   $x + y = 11$
8. $9x + 2 = 3y$
   $y - 3x = 8$
9. $2x - 3y = 4$
   $6y = 4x - 8$
10. $3x - y = 8$
    $3x = 4 - y$

7-2 Substitution

See pages 376–381.

Concept Summary

- In a system of equations, solve one equation for a variable, and then substitute that expression into the second equation to solve.

Example **Use substitution to solve the system of equations.**

$y = x - 1$
$4x - y = 19$

Since $y = x - 1$, substitute $x - 1$ for y in the second equation.

$4x - y = 19$ Second equation
$4x - (x - 1) = 19$ $y = x - 1$
$4x - x + 1 = 19$ Distributive Property
$3x + 1 = 19$ Combine like terms.
$3x = 18$ Subtract 1 from each side.
$x = 6$ Divide each side by 3.

Use $y = x - 1$ to find the value of y.

$y = x - 1$ First equation
$y = 6 - 1$ $x = 6$
$y = 5$ The solution is (6, 5).

Exercises **Use substitution to solve each system of equations. If the system does *not* have exactly one solution, state whether it has *no* solutions or *infinitely many* solutions.** *See Examples 1–3 on pages 377 and 378.*

11. $2m + n = 1$
$m - n = 8$

12. $x = 3 - 2y$
$2x + 4y = 6$

13. $3x - y = 1$
$2x + 4y = 3$

14. $0.6m - 0.2n = 0.9$
$n = 4.5 - 3m$

7-3 Elimination Using Addition and Subtraction

See pages 382–386.

Concept Summary

- Sometimes adding or subtracting two equations will eliminate one variable.

Example **Use elimination to solve the system of equations.**

$2m - n = 4$
$m + n = 2$

You can eliminate the n terms by adding the equations.

$2m - n = 4$ Write the equations in column form and add.
$(+)\ m + n = 2$
$3m = 6$ Notice the variable n is eliminated.
$m = 2$ Divide each side by 3.

Now substitute 2 for m in either equation to find n.

$m + n = 2$	Second equation
$2 + n = 2$	$m = 2$
$2 + n - 2 = 2 - 2$	Subtract 2 from each side.
$n = 0$	Simplify.

The solution is (2, 0).

Exercises **Use elimination to solve each system of equations.**
See Examples 1–3 on pages 382 and 383.

15. $x + 2y = 6$
$x - 3y = -4$

16. $2m - n = 5$
$2m + n = 3$

17. $3x - y = 11$
$x + y = 5$

18. $3x + 1 = -7y$
$6x + 7y = 0$

7-4 Elimination Using Multiplication

See pages 387–392.

Concept Summary

- Multiplying one equation by a number or multiplying each equation by a different number is a strategy that can be used to solve a system of equations by elimination.
- There are five methods for solving systems of equations.

Method	The Best Time to Use
Graphing	to estimate the solution, since graphing usually does not give an exact solution
Substitution	if one of the variables in either equation has a coefficient of 1 or −1
Elimination Using Addition	if one of the variables has opposite coefficients in the two equations
Elimination Using Subtraction	if one of the variables has the same coefficient in the two equations
Elimination Using Multiplication	if none of the coefficients are 1 or −1 and neither of the variables can be eliminated by simply adding or subtracting the equations

Example **Use elimination to solve the system of equations.**

$\mathbf{x + 2y = 8}$
$\mathbf{3x + y = 1.5}$

Multiply the second equation by −2 so the coefficients of the y terms are additive inverses. Then add the equations.

$x + 2y = 8$
$3x + y = 1.5$ Multiply by −2.

$$\begin{array}{rrcl} & x + 2y & = & 8 \\ (+) & -6x - 2y & = & -3 \\ \hline & -5x & = & 5 \end{array}$$ Add the equations.

$$\frac{-5x}{-5} = \frac{5}{-5}$$ Divide each side by −5.

$$x = -1$$ Simplify.

(continued on the next page)

• Extra Practice, see pages 835–836.
• Mixed Problem Solving, see page 859.

$x + 2y = 8$ First equation

$-1 + 2y = 8$ $x = -1$

$-1 + 2y + 1 = 8 + 1$ Add 1 to each side.

$2y = 9$ Simplify.

$\frac{2y}{2} = \frac{9}{2}$ Divide each side by 2.

$y = 4.5$ Simplify.

The solution is $(-1, 4.5)$.

Exercises **Use elimination to solve each system of equations.**
See Examples 1 and 2 on pages 387 and 388.

19. $x - 5y = 0$
$2x - 3y = 7$

20. $x - 2y = 5$
$3x - 5y = 8$

21. $2x + 3y = 8$
$x - y = 2$

22. $-5x + 8y = 21$
$10x + 3y = 15$

Determine the best method to solve each system of equations. Then solve the system. *See Example 3 on page 389.*

23. $y = 2x$
$x + 2y = 8$

24. $9x + 8y = 7$
$18x - 15y = 14$

25. $3x + 5y = 2x$
$x + 3y = y$

26. $2x + y = 3x - 15$
$x + 5 = 4y + 2x$

7-5 Graphing Systems of Inequalities

See pages 394–398.

Concept Summary

- Graph each inequality on a coordinate plane to determine the intersection of the graphs.

Example **Solve the system of inequalities.**

$x \geq -3$
$y \leq x + 2$

The solution includes the ordered pairs in the intersection of the graphs $x \geq -3$ and $y \leq x + 2$. This region is shaded in green. The graphs of $x \geq -3$ and $y \leq x + 2$ are boundaries of this region.

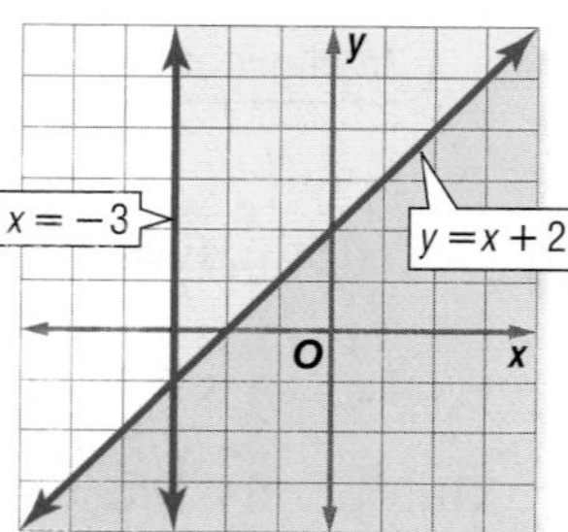

Exercises **Solve each system of inequalities by graphing.**
See Examples 1 and 2 on page 394.

27. $y < 3x$
$x + 2y \geq -21$

28. $y > -x - 1$
$y \leq 2x + 1$

29. $2x + y < 9$
$x + 11y < -6$

30. $x \geq 1$
$y + x \leq 3$

Chapter 7

Practice Test

Vocabulary and Concepts

Choose the letter that best matches each description.

1. a system of equations with two parallel lines
2. a system of equations with at least one ordered pair that satisfies both equations
3. a system of equations may be solved using this method

a. consistent
b. elimination
c. inconsistent

Skills and Applications

Graph each system of equations. Then determine whether the system has *no* solution, *one* solution, or *infinitely many* solutions. If the system has one solution, name it.

4. $y = x + 2$
 $y = 2x + 7$
5. $x + 2y = 11$
 $x = 14 - 2y$
6. $3x + y = 5$
 $2y - 10 = -6x$

Use substitution or elimination to solve each system of equations.

7. $2x + 5y = 16$
 $5x - 2y = 11$
8. $y + 2x = -1$
 $y - 4 = -2x$
9. $2x + y = -4$
 $5x + 3y = -6$
10. $y = 7 - x$
 $x - y = -3$
11. $x = 2y - 7$
 $y - 3x = -9$
12. $x + y = 10$
 $x - y = 2$
13. $3x - y = 11$
 $x + 2y = -36$
14. $3x + y = 10$
 $3x - 2y = 16$
15. $5x - 3y = 12$
 $-2x + 3y = -3$
16. $2x + 5y = 12$
 $x - 6y = -11$
17. $x + y = 6$
 $3x - 3y = 13$
18. $3x + \frac{1}{3}y = 10$
 $2x - \frac{5}{3}y = 35$

19. **NUMBER THEORY** The units digit of a two-digit number exceeds twice the tens digit by 1. Find the number if the sum of its digits is 10.

20. **GEOMETRY** The difference between the length and width of a rectangle is 7 centimeters. Find the dimensions of the rectangle if its perimeter is 50 centimeters.

Solve each system of inequalities by graphing.

21. $y > -4$
 $y < -1$
22. $y \leq 3$
 $y > -x + 2$
23. $x \leq 2y$
 $2x + 3y \leq 7$

24. **FINANCE** Last year, Jodi invested \$10,000, part at 6% annual interest and the rest at 8% annual interest. If she received \$760 in interest at the end of the year, how much did she invest at each rate?

25. **STANDARDIZED TEST PRACTICE** Which graph represents the system of inequalities $y > 2x + 1$ and $y < -x - 2$?

(A)
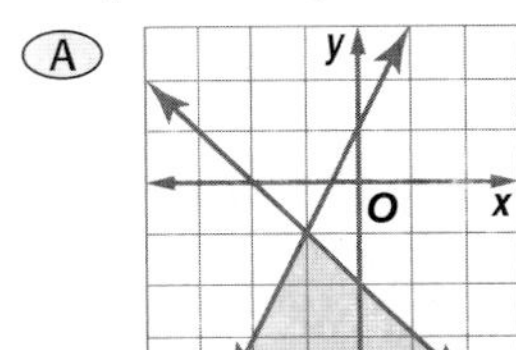

(B)
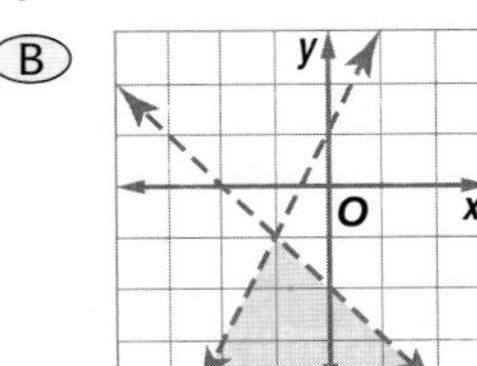

(C)
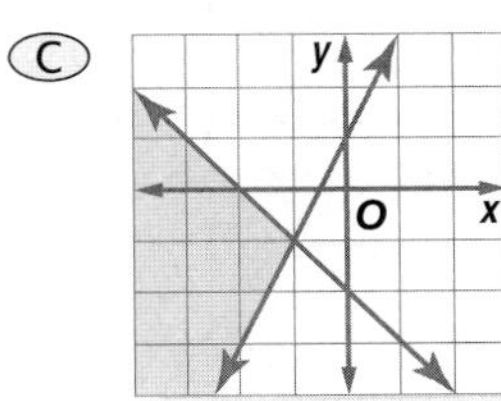

(D)
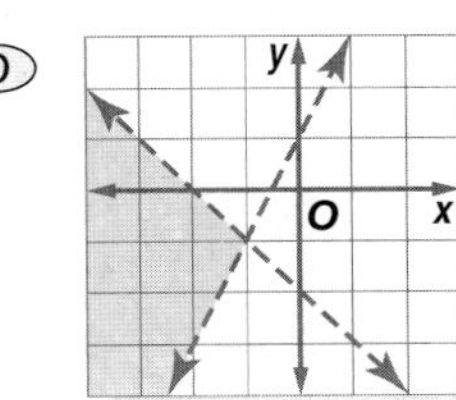

Chapter 7 Standardized Test Practice

Part 1 Multiple Choice

Record your answers on the answer sheet provided by your teacher or on a sheet of paper.

1. What is the solution of $4x - 2(x - 2) - 8 = 0$? (Lesson 3-4)

Ⓐ -2 Ⓑ 2

Ⓒ 5 Ⓓ 6

2. Noah paid \$17.11 for a CD, including tax. If the tax rate is 7%, then what was the price of the CD before tax? (Lesson 3-5)

Ⓐ \$10.06 Ⓑ \$11.98

Ⓒ \$15.99 Ⓓ \$17.04

3. What is the range of $f(x) = 2x - 3$ when the domain is $\{3, 4, 5\}$? (Lesson 4-3)

Ⓐ $\{0, 1, 2\}$ Ⓑ $\{3, 5, 7\}$

Ⓒ $\{6, 8, 10\}$ Ⓓ $\{9, 11, 13\}$

4. Jolene kept a log of the numbers of birds that visited a birdfeeder over periods of several hours. In the table below, she recorded the number of hours she watched and the cumulative number of birds that she saw each session. Which equation best represents the data set shown in the table? (Lesson 4-8)

Number of hours, x	1	3	4	6
Number of birds, y	6	14	18	26

Ⓐ $y = x + 5$ Ⓑ $y = 3x + 3$

Ⓒ $y = 3x + 5$ Ⓓ $y = 4x + 2$

5. Which equation describes the graph? (Lesson 5-3)

Ⓐ $3y - 4x = -12$

Ⓑ $4y + 3x = -16$

Ⓒ $3y + 4x = -12$

Ⓓ $3y + 4x = -9$

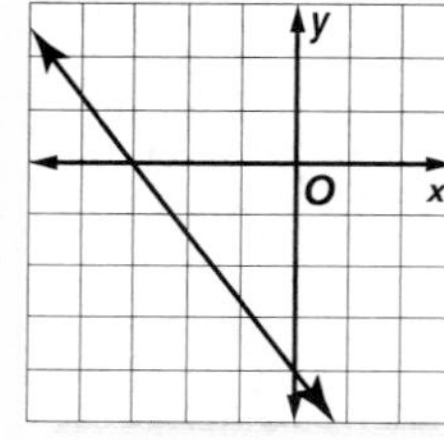

6. Which equation represents a line parallel to the line given by $y - 3x = 6$? (Lesson 5-6)

Ⓐ $y = -3x + 4$ Ⓑ $y = 3x - 2$

Ⓒ $y = \frac{1}{3}x + 6$ Ⓓ $y = -\frac{1}{3}x + 4$

7. Tamika has \$185 in her bank account. She needs to deposit enough money so that she can withdraw \$230 for her car payment and still have at least \$200 left in the account. Which inequality describes d, the amount she needs to deposit? (Lesson 6-1)

Ⓐ $d(185 - 230) \geq 200$

Ⓑ $185 - 230d \geq 200$

Ⓒ $185 + 230 + d \geq 200$

Ⓓ $185 + d - 230 \geq 200$

8. The perimeter of a rectangular garden is 68 feet. The length of the garden is 4 more than twice the width. Which system of equations will determine the length ℓ and the width w of the garden? (Lesson 7-2)

Ⓐ $2\ell + 2w = 68$, $\ell = 4 - 2w$ Ⓑ $2\ell + 2w = 68$, $\ell = 2w + 4$

Ⓒ $2 + 2w = 68$, $2\ell - w = 4$ Ⓓ $2\ell + 2w = 68$, $w = 2\ell + 4$

9. Ernesto spent a total of \$64 for a pair of jeans and a shirt. The jeans cost \$6 more than the shirt. What was the cost of the jeans? (Lesson 7-2)

Ⓐ \$26 Ⓑ \$29

Ⓒ \$35 Ⓓ \$58

10. What is the value of y in the following system of equations? (Lesson 7-3)

$3x + 4y = 8$
$3x + 2y = -2$

Ⓐ -2 Ⓑ 4

Ⓒ 5 Ⓓ 6

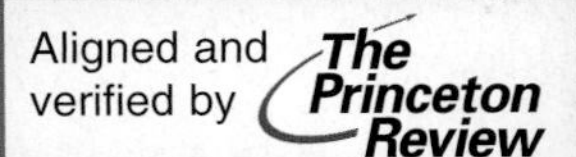

Part 2 Short Response/Grid In

Record your answers on the answer sheet provided by your teacher or on a sheet of paper.

11. The diagram shows the dimensions of the cargo area of a delivery truck.

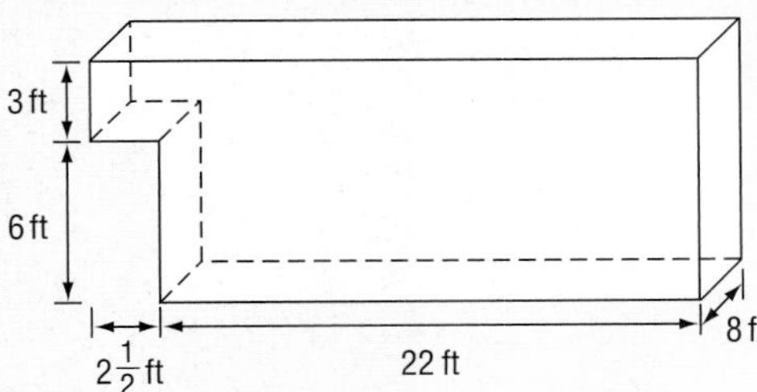

What is the maximum volume of cargo, in cubic feet, that can fit in the truck? (Prerequisite Skill)

12. The perimeter of the square below is 204 feet. What is the value of x? (Lesson 3-4)

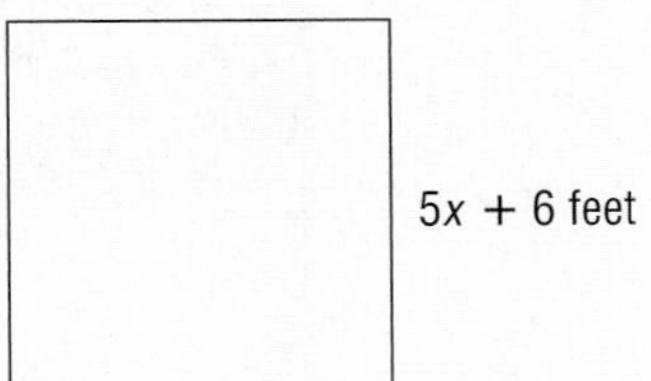

13. What is the x-intercept of the graph of $4x + 3y = 12$? (Lesson 4-5)

14. What are the slope and the y-intercept of the graph of the equation $4x - 2y = 5$? (Lesson 5-4)

15. Solve the following system of equations. (Lesson 7-2)

$$5x - y = 10$$
$$7x - 2y = 11$$

The Princeton Review Test-Taking Tip

Questions 11 and 12 To prepare for a standardized test, make flash cards of key mathematical terms, such as "perimeter" and "volume." Use the glossary of your textbook to determine the important terms and their correct definitions.

www.algebra1.com/standardized_test

Part 3 Quantitative Comparison

Compare the quantity in Column A and the quantity in Column B. Then determine whether:

- **(A) the quantity in Column A is greater,**
- **(B) the quantity in Column B is greater,**
- **(C) the two quantities are equal, or**
- **(D) the relationship cannot be determined from the information given.**

	Column A	Column B
16.	3^4	9^2

(Lesson 1-1)

	Column A	Column B
17.	the slope of the line that contains $A(2, 4)$ and $B(-1, 3)$	the slope of the line that contains $C(-2, 1)$ and $D(5, 3)$

(Lesson 5-1)

18.
$$x - 3y = 11$$
$$3x + y = 13$$

Column A	Column B
y	0

(Lesson 7-4)

19.
$$3x - 2y = 19$$
$$5x + 4y = 17$$

Column A	Column B
x	y

(Lesson 7-4)

Part 4 Open Ended

Record your answers on a sheet of paper. Show your work.

20. The manager of a movie theater found that Saturday's sales were \$3675. He knew that a total of 650 tickets were sold Saturday. Adult tickets cost \$7.50, and children's tickets cost \$4.50. (Lesson 7-2)

a. Write equations to represent the number of tickets sold and the amount of money collected.

b. How many of each kind of ticket were sold? Show your work. Include all steps.

UNIT 3

Polynomials and Nonlinear Functions

Not all real-world situations can be modeled using a linear function. In this unit, you will learn about polynomials and nonlinear functions.

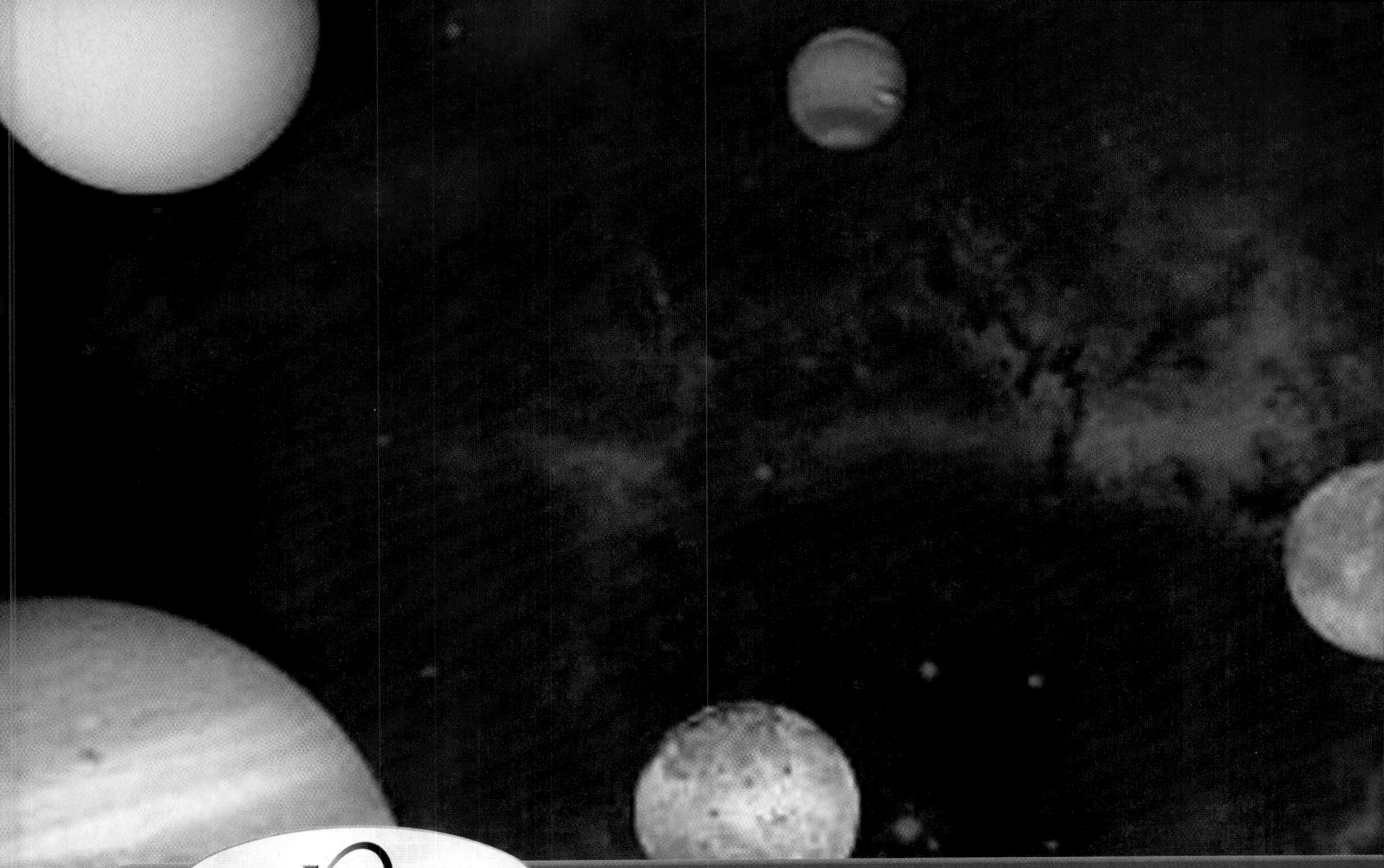

WebQuest Internet Project

Pluto Is Falling From Status as Distant Planet

Source: *USA TODAY,* March 28, 2001

"Like any former third-grader, Catherine Beyhl knows that the solar system has nine planets, and she knows a phrase to help remember their order: 'My Very Educated Mother Just Served Us Nine Pizzas.' But she recently visited the American Museum of Natural History's glittering new astronomy hall at the Hayden Planetarium and found only eight scale models of the planets. No Pizza—no Pluto." In this project, you will examine how scientific notation, factors, and graphs are useful in presenting information about the planets.

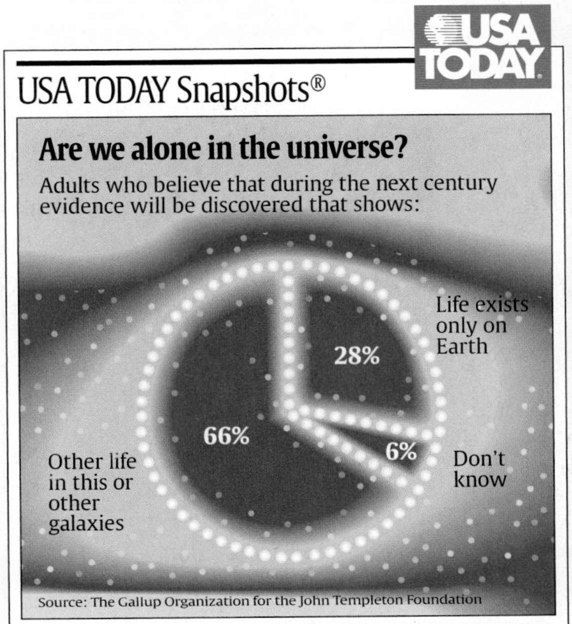

Log on to www.algebra1.com/webquest. Begin your WebQuest by reading the Task.

Then continue working on your WebQuest as you study Unit 3.

Lesson	8-3	9-1	10-2
Page	429	479	537

USA TODAY Snapshots®

Are we alone in the universe?

Adults who believe that during the next century evidence will be discovered that shows:

Source: The Gallup Organization for the John Templeton Foundation

By Cindy Hall and Sam Ward, USA TODAY

Chapter 8 Polynomials

What You'll Learn

- **Lessons 8-1 and 8-2** Find products and quotients of monomials.
- **Lesson 8-3** Express numbers in scientific and standard notation.
- **Lesson 8-4** Find the degree of a polynomial and arrange the terms in order.
- **Lessons 8-5 through 8-7** Add, subtract, and multiply polynomial expressions.
- **Lesson 8-8** Find special products of binomials.

Key Vocabulary

- **monomial** (p. 410)
- **scientific notation** (p. 425)
- **polynomial** (p. 432)
- **binomial** (p. 432)
- **FOIL method** (p. 453)

Why It's Important

Operations with polynomials, including addition, subtraction, and multiplication, form the foundation for solving equations that involve polynomials. In addition, polynomials are used to model many real-world situations. *In Lesson 8-6, you will learn how to find the distance that runners on a curved track should be staggered.*

Getting Started

Prerequisite Skills To be successful in this chapter, you'll need to master these skills and be able to apply them in problem-solving situations. Review these skills before beginning Chapter 8.

For Lessons 8-1 and 8-2 — **Exponential Notation**

Write each expression using exponents. *(For review, see Lesson 1-1.)*

1. $2 \cdot 2 \cdot 2 \cdot 2 \cdot 2$ **2.** $3 \cdot 3 \cdot 3 \cdot 3$ **3.** $5 \cdot 5$ **4.** $x \cdot x \cdot x$

5. $a \cdot a \cdot a \cdot a \cdot a \cdot a$ **6.** $x \cdot x \cdot y \cdot y \cdot y$ **7.** $\frac{1}{2} \cdot \frac{1}{2} \cdot \frac{1}{2} \cdot \frac{1}{2} \cdot \frac{1}{2}$ **8.** $\frac{a}{b} \cdot \frac{a}{b} \cdot \frac{c}{d} \cdot \frac{c}{d} \cdot \frac{c}{d}$

For Lessons 8-1 and 8-2 — **Evaluating Powers**

Evaluate each expression. *(For review, see Lesson 1-1.)*

9. 3^2 **10.** 4^3 **11.** 5^2 **12.** 10^4

13. $(-6)^2$ **14.** $(-3)^3$ **15.** $\left(\frac{2}{3}\right)^4$ **16.** $\left(-\frac{7}{8}\right)^2$

For Lessons 8-1, 8-2, and 8-5 through 8-8 — **Area and Volume**

Find the area or volume of each figure shown below. *(For review, see pages 813–817.)*

17.
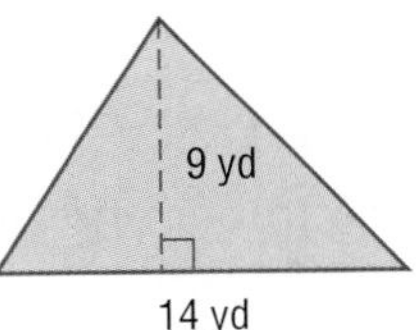

18.
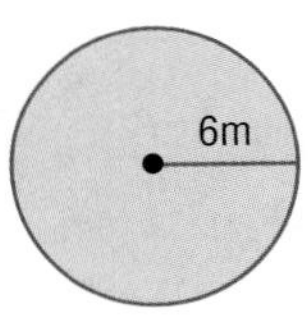

19.
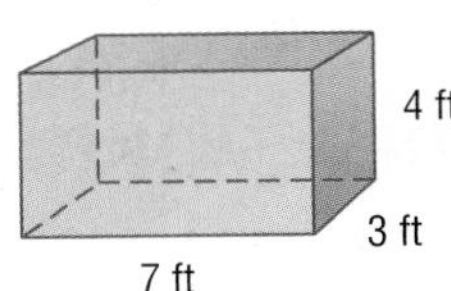

20.
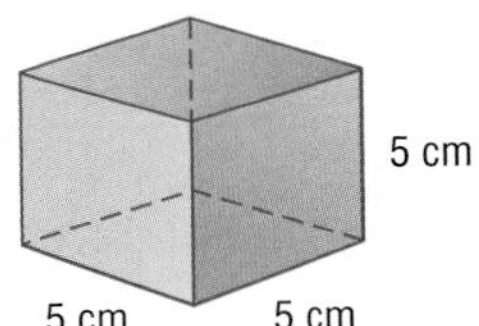

Make this Foldable to help you organize information about polynomials. Begin with a sheet of 11" by 17" paper.

Step 1 Fold

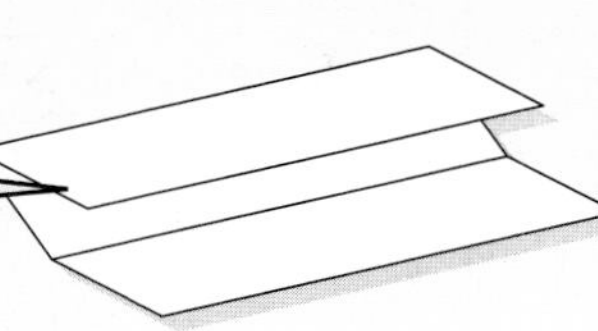

Fold in thirds lengthwise.

Step 2 Open and Fold

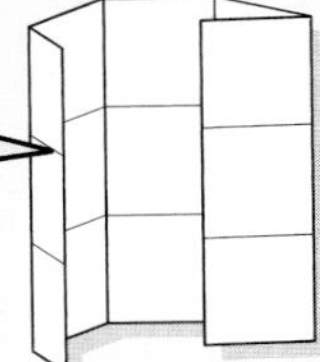

Fold a 2" tab along the width. Then fold the rest in fourths.

Step 3 Label

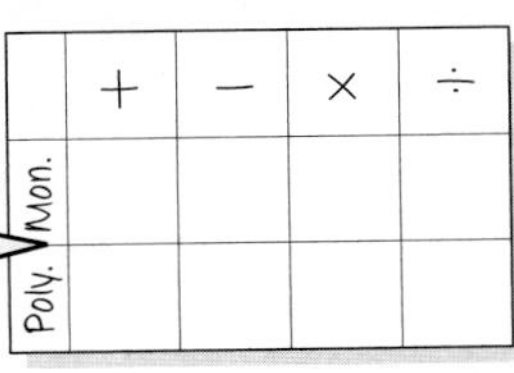

Draw lines along folds and label as shown.
+
−
×
÷
Mon.
Poly.

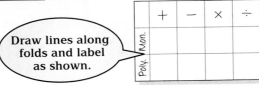

Reading and Writing As you read and study the chapter, write examples and notes for each operation.

8-1 Multiplying Monomials

What You'll Learn

- Multiply monomials.
- Simplify expressions involving powers of monomials.

Vocabulary

- monomial
- constant

Why does doubling speed quadruple braking distance?

The table shows the braking distance for a vehicle at certain speeds. If s represents the speed in miles per hour, then the approximate number of feet that the driver must apply the brakes is $\frac{1}{20}s^2$. Notice that when speed is doubled, the braking distance is quadrupled.

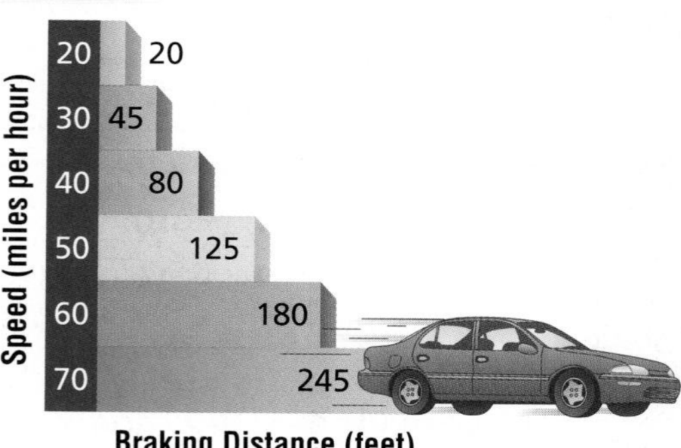

Source: *British Highway Code*

MULTIPLY MONOMIALS An expression like $\frac{1}{20}s^2$ is called a monomial. A **monomial** is a number, a variable, or a product of a number and one or more variables. An expression involving the division of variables is not a monomial. Monomials that are real numbers are called **constants**.

Example 1 Identify Monomials

Determine whether each expression is a monomial. Explain your reasoning.

	Expression	Monomial?	Reason
a.	-5	yes	-5 is a real number and an example of a constant.
b.	$p + q$	no	The expression involves the addition, not the product, of two variables.
c.	x	yes	Single variables are monomials.
d.	$\frac{c}{d}$	no	The expression is the quotient, not the product, of two variables.
e.	$\frac{abc^8}{5}$	yes	$\frac{abc^8}{5} = \frac{1}{5}abc^8$. The expression is the product of a number, $\frac{1}{5}$, and three variables.

Study Tip

Reading Math
The expression x^n is read *x to the nth power.*

Recall that an expression of the form x^n is called a *power* and represents the product you obtain when x is used as a factor n times. The number x is the *base*, and the number n is the *exponent*.

$$2^5 = \underbrace{2 \cdot 2 \cdot 2 \cdot 2 \cdot 2}_{5\text{ factors}} \text{ or } 32$$

(exponent: 5; base: 2)

In the following examples, the definition of a power is used to find the products of powers. Look for a pattern in the exponents.

$$2^3 \cdot 2^5 = \underbrace{\overbrace{2 \cdot 2 \cdot 2}^{3 \text{ factors}} \cdot \overbrace{2 \cdot 2 \cdot 2 \cdot 2 \cdot 2}^{5 \text{ factors}}}_{3 + 5 \text{ or } 8 \text{ factors}} \text{ or } 2^8 \qquad 3^2 \cdot 3^4 = \underbrace{\overbrace{3 \cdot 3}^{2 \text{ factors}} \cdot \overbrace{3 \cdot 3 \cdot 3 \cdot 3}^{4 \text{ factors}}}_{2 + 4 \text{ or } 6 \text{ factors}} \text{ or } 3^6$$

These and other similar examples suggest the property for multiplying powers.

Key Concept *Product of Powers*

- **Words** To multiply two powers that have the same base, add the exponents.
- **Symbols** For any number a and all integers m and n, $a^m \cdot a^n = a^{m+n}$.
- **Example** $a^4 \cdot a^{12} = a^{4+12}$ or a^{16}

Example 2 Product of Powers

Simplify each expression.

a. $(5x^7)(x^6)$

$(5x^7)(x^6) = (5)(1)(x^7 \cdot x^6)$ Commutative and Associative Properties

$= (5 \cdot 1)(x^{7+6})$ Product of Powers

$= 5x^{13}$ Simplify.

b. $(4ab^6)(-7a^2b^3)$

$(4ab^6)(-7a^2b^3) = (4)(-7)(a \cdot a^2)(b^6 \cdot b^3)$ Commutative and Associative Properties

$= -28(a^{1+2})(b^{6+3})$ Product of Powers

$= -28a^3b^9$ Simplify.

Study Tip

Power of 1
Recall that a variable with no exponent indicated can be written as a power of 1. For example, $x = x^1$ and $ab = a^1b^1$.

POWERS OF MONOMIALS You can also look for a pattern to discover the property for finding the power of a power.

$$(4^2)^5 = \overbrace{(4^2)(4^2)(4^2)(4^2)(4^2)}^{5 \text{ factors}} \qquad (z^8)^3 = \overbrace{(z^8)(z^8)(z^8)}^{3 \text{ factors}}$$

$$= 4^{2+2+2+2+2} \longleftarrow \text{Apply rule for Product of Powers.} \longrightarrow = z^{8+8+8}$$

$$= 4^{10} \qquad\qquad = z^{24}$$

Therefore, $(4^2)^5 = 4^{10}$ and $(z^8)^3 = z^{24}$. These and other similar examples suggest the property for finding the power of a power.

Key Concept *Power of a Power*

- **Words** To find the power of a power, multiply the exponents.
- **Symbols** For any number a and all integers m and n, $(a^m)^n = a^{m \cdot n}$.
- **Example** $(k^5)^9 = k^{5 \cdot 9}$ or k^{45}

Study Tip

Look Back
To review **using a calculator to find a power of a number**, see Lesson 1-1.

Example 3 Power of a Power

Simplify $[(3^2)^3]^2$.

$[(3^2)^3]^2 = (3^{2 \cdot 3})^2$ Power of a Power

$= (3^6)^2$ Simplify.

$= 3^{6 \cdot 2}$ Power of a Power

$= 3^{12}$ or 531,441 Simplify.

Look for a pattern in the examples below.

$$\begin{aligned}(xy)^4 &= (xy)(xy)(xy)(xy)\\ &= (x \cdot x \cdot x \cdot x)(y \cdot y \cdot y \cdot y)\\ &= x^4y^4\end{aligned}$$

$$\begin{aligned}(6ab)^3 &= (6ab)(6ab)(6ab)\\ &= (6 \cdot 6 \cdot 6)(a \cdot a \cdot a)(b \cdot b \cdot b)\\ &= 6^3a^3b^3 \text{ or } 216a^3b^3\end{aligned}$$

These and other similar examples suggest the following property for finding the power of a product.

Study Tip

Powers of Monomials
Sometimes the rules for the Power of a Power and the Power of a Product are combined into one rule.
$(a^mb^n)^p = a^{mp}b^{np}$

Key Concept — Power of a Product

- **Words** To find the power of a product, find the power of each factor and multiply.
- **Symbols** For all numbers a and b and any integer m, $(ab)^m = a^mb^m$.
- **Example** $(-2xy)^3 = (-2)^3x^3y^3$ or $-8x^3y^3$

Example 4 Power of a Product

GEOMETRY Express the area of the square as a monomial.

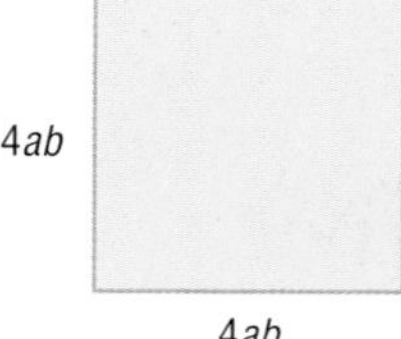

Area	$= s^2$	Formula for the area of a square
	$= (4ab)^2$	$s = 4ab$
	$= 4^2a^2b^2$	Power of a Product
	$= 16a^2b^2$	Simplify.

The area of the square is $16a^2b^2$ square units.

The properties can be used in combination to simplify more complex expressions involving exponents.

Concept Summary — Simplifying Monomial Expressions

To *simplify* an expression involving monomials, write an equivalent expression in which:

- each base appears exactly once,
- there are no powers of powers, and
- all fractions are in simplest form.

Example 5 Simplify Expressions

Simplify $\left(\frac{1}{3}xy^4\right)^2[(-6y)^2]^3$.

$\left(\frac{1}{3}xy^4\right)^2[(-6y)^2]^3 = \left(\frac{1}{3}xy^4\right)^2(-6y)^6$	Power of a Power
$= \left(\frac{1}{3}\right)^2x^2(y^4)^2(-6)^6y^6$	Power of a Product
$= \frac{1}{9}x^2y^8(46{,}656)y^6$	Power of a Power
$= \frac{1}{9}(46{,}656)x^2 \cdot y^8 \cdot y^6$	Commutative Property
$= 5184x^2y^{14}$	Product of Powers

Check for Understanding

Concept Check

1. **OPEN ENDED** Give an example of an expression that can be simplified using each property. Then simplify each expression.
   a. Product of Powers b. Power of a Power c. Power of a Product
2. **Determine** whether each pair of monomials is equivalent. Explain.
   a. $5m^2$ and $(5m)^2$ b. $(yz)^4$ and y^4z^4
   c. $-3a^2$ and $(-3a)^2$ d. $2(c^7)^3$ and $8c^{21}$
3. **FIND THE ERROR** Nathan and Poloma are simplifying $(5^2)(5^9)$.

Nathan	Poloma
$(5^2)(5^9) = (5 \cdot 5)^{2+9}$	$(5^2)(5^9) = 5^{2+9}$
$= 25^{11}$	$= 5^{11}$

Who is correct? Explain your reasoning.

Guided Practice

Determine whether each expression is a monomial. Write *yes* or *no*. Explain.

4. $5 - 7d$ 5. $\frac{4a}{3b}$ 6. n

Simplify.

7. $x(x^4)(x^6)$ 8. $(4a^4b)(9a^2b^3)$ 9. $[(2^3)^2]^3$
10. $(3y^5z)^2$ 11. $(-4mn^2)(12m^2n)$ 12. $(-2v^3w^4)^3(-3vw^3)^2$

Application

GEOMETRY Express the area of each triangle as a monomial.

13.
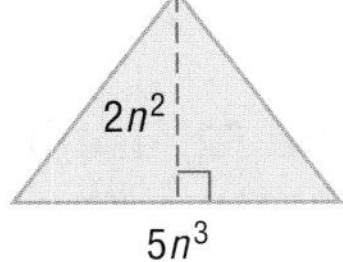

14.
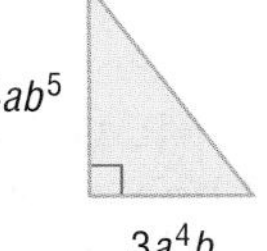

Practice and Apply

Homework Help

For Exercises	See Examples
15–20	1
21–48	2, 3, 5
49–54	4

Extra Practice
See page 837.

Determine whether each expression is a monomial. Write *yes* or *no*. Explain.

15. 12 16. $4x^3$ 17. $a - 2b$
18. $4n + 5m$ 19. $\frac{x}{y^2}$ 20. $\frac{1}{5}abc^{14}$

Simplify.

21. $(ab^4)(ab^2)$ 22. $(p^5q^4)(p^2q)$
23. $(-7c^3d^4)(4cd^3)$ 24. $(-3j^7k^5)(-8jk^8)$
25. $(5a^2b^3c^4)(6a^3b^4c^2)$ 26. $(10xy^5z^3)(3x^4y^6z^3)$
27. $(9pq^7)^2$ 28. $(7b^3c^6)^3$
29. $[(3^2)^4]^2$ 30. $[(4^2)^3]^2$
31. $(0.5x^3)^2$ 32. $(0.4h^5)^3$
33. $\left(-\frac{3}{4}c\right)^3$ 34. $\left(\frac{4}{5}a^2\right)^2$
35. $(4cd)^2(-3d^2)^3$ 36. $(-2x^5)^3(-5xy^6)^2$
37. $(2ag^2)^4(3a^2g^3)^2$ 38. $(2m^2n^3)^3(3m^3n)^4$
39. $(8y^3)(-3x^2y^2)\left(\frac{3}{8}xy^4\right)$ 40. $\left(\frac{4}{7}m\right)^2(49m)(17p)\left(\frac{1}{34}p^5\right)$

41. Simplify the expression $(-2b^3)^4 - 3(-2b^4)^3$.

42. Simplify the expression $2(-5y^3)^2 + (-3y^3)^3$.

GEOMETRY **Express the area of each figure as a monomial.**

43. 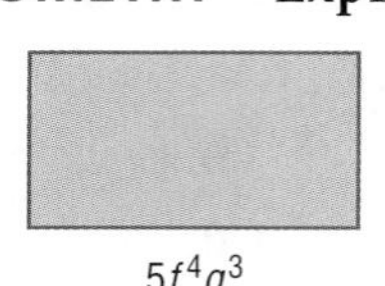

44. 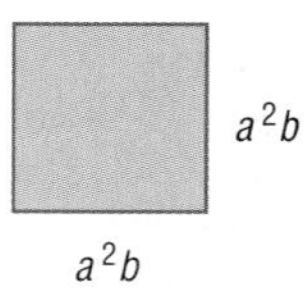

45. 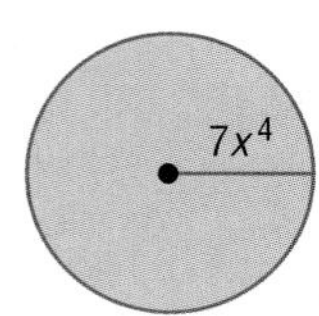

GEOMETRY **Express the volume of each solid as a monomial.**

46. 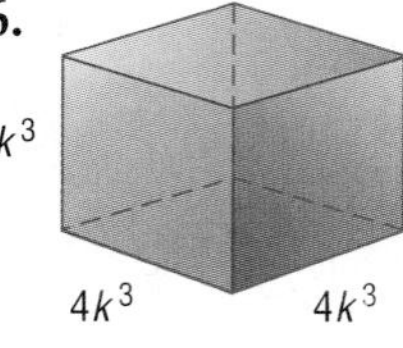

47. 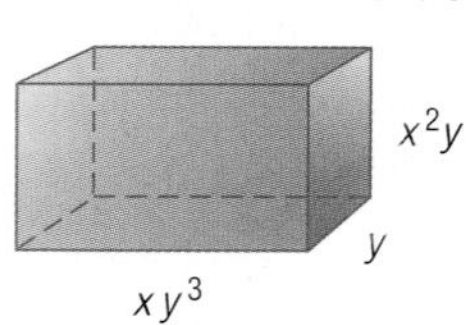

48. 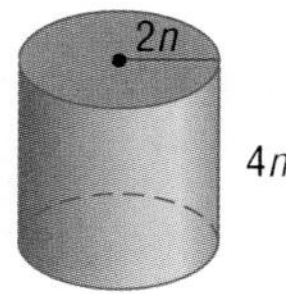

TELEPHONES **For Exercises 49 and 50, use the following information.**
The first transatlantic telephone cable has 51 amplifiers along its length. Each amplifier strengthens the signal on the cable 10^6 times.

49. After it passes through the second amplifier, the signal has been boosted $10^6 \cdot 10^6$ times. Simplify this expression.

50. Represent the number of times the signal has been boosted after it has passed through the first four amplifiers as a power of 10^6. Then simplify the expression.

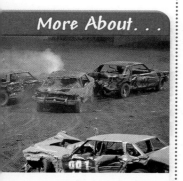

DEMOLITION DERBY **For Exercises 51 and 52, use the following information.**
When a car hits an object, the damage is measured by the collision impact. For a certain car, the collision impact I is given by $I = 2s^2$, where s represents the speed in kilometers per minute.

51. What is the collision impact if the speed of the car is 1 kilometer per minute? 2 kilometers per minute? 4 kilometers per minute?

52. As the speed doubles, explain what happens to the collision impact.

More About. . .

Demolition Derby

In a demolition derby, the winner is not the car that finishes first but the last car still moving under its own power.

Source: *Smithsonian Magazine*

TEST TAKING **For Exercises 53 and 54, use the following information.**
A history test covers two chapters. There are 2^{12} ways to answer the 12 true-false questions on the first chapter and 2^{10} ways to answer the 10 true-false questions on the second chapter.

53. How many ways are there to answer all 22 questions on the test? (*Hint*: Find the product of 2^{12} and 2^{10}.)

54. If a student guesses on each question, what is the probability of answering all questions correctly?

CRITICAL THINKING **Determine whether each statement is *true* or *false*. If true, explain your reasoning. If false, give a counterexample.**

55. For any real number a, $(-a)^2 = -a^2$.

56. For all real numbers a and b, and all integers m, n, and p, $(a^mb^n)^p = a^{mp}b^{np}$.

57. For all real numbers a, b, and all integers n, $(a + b)^n = a^n + b^n$.

58. **WRITING IN MATH** Answer the question that was posed at the beginning of the lesson.

Why does doubling speed quadruple braking distance?

Include the following in your answer:

- the ratio of the braking distance required for a speed of 40 miles per hour and the braking distance required for a speed of 80 miles per hour, and
- a comparison of the expressions $\frac{1}{20}s^2$ and $\frac{1}{20}(2s)^2$.

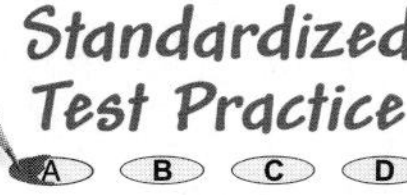

Standardized Test Practice

59. $4^2 \cdot 4^5 = ?$

Ⓐ 16^7 Ⓑ 8^7 Ⓒ 4^{10} Ⓓ 4^7

60. Which of the following expressions represents the volume of the cube?

Ⓐ $15x^3$ Ⓑ $25x^2$

Ⓒ $25x^3$ Ⓓ $125x^3$

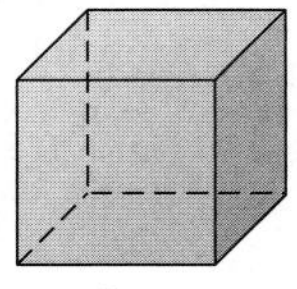

Maintain Your Skills

Mixed Review

Solve each system of inequalities by graphing. *(Lesson 7-5)*

61. $y \leq 2x + 2$
$y \geq -x - 1$

62. $y \geq x - 2$
$y < 2x - 1$

63. $x > -2$
$y < x + 3$

Use elimination to solve each system of equations. *(Lesson 7-4)*

64. $-4x + 5y = 2$
$x + 2y = 6$

65. $3x + 4y = -25$
$2x - 3y = 6$

66. $x + y = 20$
$0.4x + 0.15y = 4$

Solve each compound inequality. Then graph the solution set. *(Lesson 6-4)*

67. $4 + h \leq -3$ or $4 + h \geq 5$

68. $4 < 4a + 12 < 24$

69. $14 < 3h + 2 < 2$

70. $2m - 3 > 7$ or $2m + 7 > 9$

Determine whether each transformation is a *reflection, translation, dilation,* or *rotation.* *(Lesson 4-2)*

71.

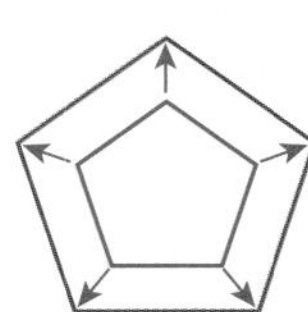

72.

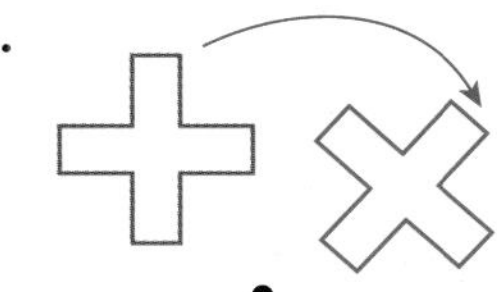

73.

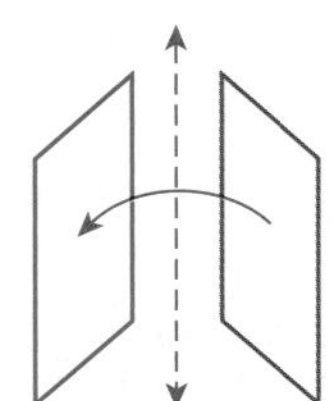

74. **TRANSPORTATION** Two trains leave York at the same time, one traveling north, the other south. The northbound train travels at 40 miles per hour and the southbound at 30 miles per hour. In how many hours will the trains be 245 miles apart? *(Lesson 3-7)*

Getting Ready for the Next Lesson

PREREQUISITE SKILL Simplify. *(To review **simplifying fractions**, see pages 798 and 799.)*

75. $\frac{2}{6}$

76. $\frac{3}{15}$

77. $\frac{10}{5}$

78. $\frac{27}{9}$

79. $\frac{14}{36}$

80. $\frac{9}{48}$

81. $\frac{44}{32}$

82. $\frac{45}{18}$

Algebra Activity

A Follow-Up of Lesson 8-1

Investigating Surface Area and Volume

Collect the Data

- Cut out the pattern shown from a sheet of centimeter grid paper. Fold along the dashed lines and tape the edges together to form a rectangular prism with dimensions 2 centimeters by 5 centimeters by 3 centimeters.
- Find the surface area SA of the prism by counting the squares on all the faces of the prism or by using the formula $SA = 2w\ell + 2wh + 2\ell h$, where w is the width, ℓ is the length, and h is the height of the prism.
- Find the volume V of the prism by using the formula $V = \ell wh$.

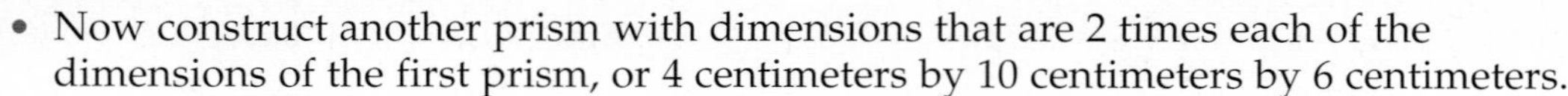

- Now construct another prism with dimensions that are 2 times each of the dimensions of the first prism, or 4 centimeters by 10 centimeters by 6 centimeters.
- Finally, construct a third prism with dimensions that are 3 times each of the dimensions of the first prism, or 6 centimeters by 15 centimeters by 9 centimeters.

3 cm
2 cm
5 cm

Analyze the Data

1. Copy and complete the table using the prisms you made.

Prism	Dimensions	Surface Area (cm^2)	Volume (cm^3)	Surface Area Ratio $\left(\frac{SA \text{ of New}}{SA \text{ of Original}}\right)$	Volume Ratio $\left(\frac{V \text{ of New}}{V \text{ of Original}}\right)$
Original	2 by 5 by 3	62	30	——	——
A	4 by 10 by 6				
B	6 by 15 by 9				

2. Make a prism with different dimensions from any in this activity. Repeat the steps in **Collect the Data**, and make a table similar to the one in Exercise 1.

Make a Conjecture

3. Suppose you multiply each dimension of a prism by 2. What is the ratio of the surface area of the new prism to the surface area of the original prism? What is the ratio of the volumes?

4. If you multiply each dimension of a prism by 3, what is the ratio of the surface area of the new prism to the surface area of the original? What is the ratio of the volumes?

5. Suppose you multiply each dimension of a prism by a. Make a conjecture about the ratios of surface areas and volumes.

Extend the Activity

6. Repeat the steps in **Collect the Data** and **Analyze the Data** using cylinders. To start, make a cylinder with radius 4 centimeters and height 5 centimeters. To compute surface area SA and volume V, use the formulas $SA = 2\pi r^2 + 2\pi rh$ and $V = \pi r^2 h$, where r is the radius and h is the height of the cylinder. Do the conjectures you made in Exercise 5 hold true for cylinders? Explain.

8-2 Dividing Monomials

What You'll Learn

- Simplify expressions involving the quotient of monomials.
- Simplify expressions containing negative exponents.

Vocabulary

- zero exponent
- negative exponent

How can you compare pH levels?

To test whether a solution is a *base* or an *acid*, chemists use a pH test. This test measures the concentration c of hydrogen ions (in moles per liter) in the solution.

$$c = \left(\frac{1}{10}\right)^{\text{pH}}$$

The table gives examples of solutions with various pH levels. You can find the quotient of powers and use negative exponents to compare measures on the pH scale.

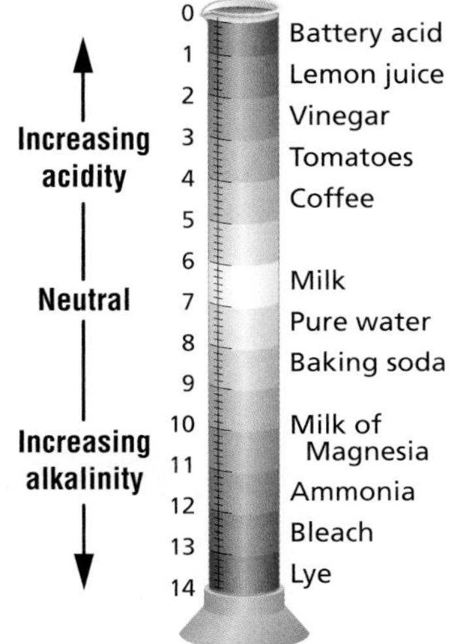

Source: U.S. Geological Survey

QUOTIENTS OF MONOMIALS In the following examples, the definition of a power is used to find quotients of powers. Look for a pattern in the exponents.

$$\frac{4^5}{4^3} = \frac{\overbrace{\cancel{4}\cdot\cancel{4}\cdot\cancel{4}\cdot 4\cdot 4}^{5\text{ factors}}}{\underbrace{\cancel{4}\cdot\cancel{4}\cdot\cancel{4}}_{3\text{ factors}}} = \underbrace{4\cdot 4}_{5-3\text{ or }2\text{ factors}} \text{ or } 4^2$$

$$\frac{3^6}{3^2} = \frac{\overbrace{\cancel{3}\cdot\cancel{3}\cdot 3\cdot 3\cdot 3\cdot 3}^{6\text{ factors}}}{\underbrace{\cancel{3}\cdot\cancel{3}}_{2\text{ factors}}} = \underbrace{3\cdot 3\cdot 3\cdot 3}_{6-2\text{ or }4\text{ factors}} \text{ or } 3^4$$

These and other similar examples suggest the following property for dividing powers.

Key Concept — *Quotient of Powers*

- **Words** To divide two powers that have the same base, subtract the exponents.
- **Symbols** For all integers m and n and any nonzero number a, $\frac{a^m}{a^n} = a^{m-n}$.
- **Example** $\frac{b^{15}}{b^7} = b^{15-7}$ or b^8

Example 1 *Quotient of Powers*

Simplify $\frac{a^5b^8}{ab^3}$. Assume that a and b are not equal to zero.

$\frac{a^5b^8}{ab^3} = \left(\frac{a^5}{a}\right)\left(\frac{b^8}{b^3}\right)$ Group powers that have the same base.

$= (a^{5-1})(b^{8-3})$ Quotient of Powers

$= a^4b^5$ Simplify.

In the following example, the definition of a power is used to compute the power of a quotient. Look for a pattern in the exponents.

$$\left(\frac{2}{5}\right)^3 = \underbrace{\left(\frac{2}{5}\right)\left(\frac{2}{5}\right)\left(\frac{2}{5}\right)}_{\text{3 factors}} = \frac{\overbrace{2 \cdot 2 \cdot 2}^{\text{3 factors}}}{\underbrace{5 \cdot 5 \cdot 5}_{\text{3 factors}}} \text{ or } \frac{2^3}{5^3}$$

This and other similar examples suggest the following property.

Key Concept — *Power of a Quotient*

- **Words** To find the power of a quotient, find the power of the numerator and the power of the denominator.
- **Symbols** For any integer m and any real numbers a and b, $b \neq 0$, $\left(\frac{a}{b}\right)^m = \frac{a^m}{b^m}$.
- **Example** $\left(\frac{c}{d}\right)^5 = \frac{c^5}{d^5}$

Example 2 *Power of a Quotient*

Simplify $\left(\frac{2p^2}{3}\right)^4$.

$\left(\frac{2p^2}{3}\right)^4 = \frac{(2p^2)^4}{3^4}$ Power of a Quotient

$= \frac{2^4(p^2)^4}{3^4}$ Power of a Product

$= \frac{16p^8}{81}$ Power of a Power

NEGATIVE EXPONENTS A graphing calculator can be used to investigate expressions with 0 as an exponent as well as expressions with negative exponents.

Study Tip

Graphing Calculator

To express a value as a fraction, press

MATH ENTER ENTER.

Graphing Calculator Investigation

Zero Exponent and Negative Exponents

Use the ^ key on a TI-83 Plus to evaluate expressions with exponents.

Think and Discuss

1. Copy and complete the table below.

Power	2^4	2^3	2^2	2^1	2^0	2^{-1}	2^{-2}	2^{-3}	2^{-4}
Value									

2. Describe the relationship between each pair of values.

a. 2^4 and 2^{-4} **b.** 2^3 and 2^{-3} **c.** 2^2 and 2^{-2} **d.** 2^1 and 2^{-1}

3. Make a Conjecture as to the fractional value of 5^{-1}. Verify your conjecture using a calculator.

4. What is the value of 5^0?

5. What happens when you evaluate 0^0?

To understand why a calculator gives a value of 1 for 2^0, study the two methods used to simplify $\frac{2^4}{2^4}$.

Study Tip

Alternative Method

Another way to look at the problem of simplifying $\frac{2^4}{2^4}$ is to recall that any nonzero number divided by itself is 1: $\frac{2^4}{2^4} = \frac{16}{16}$ or 1.

Method 1

$\frac{2^4}{2^4} = 2^{4-4}$ Quotient of Powers

$= 2^0$ Subtract.

Method 2

$\frac{2^4}{2^4} = \frac{\overset{1}{\cancel{2}} \cdot \overset{1}{\cancel{2}} \cdot \overset{1}{\cancel{2}} \cdot \overset{1}{\cancel{2}}}{\underset{1}{\cancel{2}} \cdot \underset{1}{\cancel{2}} \cdot \underset{1}{\cancel{2}} \cdot \underset{1}{\cancel{2}}}$ Definition of powers

$= 1$ Simplify.

Since $\frac{2^4}{2^4}$ cannot have two different values, we can conclude that $2^0 = 1$.

Key Concept — *Zero Exponent*

- **Words** Any nonzero number raised to the zero power is 1.
- **Symbols** For any nonzero number a, $a^0 = 1$.
- **Example** $(-0.25)^0 = 1$

Example 3 *Zero Exponent*

Simplify each expression. Assume that x and y are not equal to zero.

a. $\left(-\frac{3x^5y}{8xy^7}\right)^0$

$\left(-\frac{3x^5y}{8xy^7}\right)^0 = 1$ $a^0 = 1$

b. $\frac{t^3s^0}{t}$

$\frac{t^3s^0}{t} = \frac{t^3(1)}{t}$ $a^0 = 1$

$= \frac{t^3}{t}$ Simplify.

$= t^2$ Quotient of Powers

To investigate the meaning of a negative exponent, we can simplify expressions like $\frac{8^2}{8^5}$ in two ways.

Method 1

$\frac{8^2}{8^5} = 8^{2-5}$ Quotient of Powers

$= 8^{-3}$ Subtract.

Method 2

$\frac{8^2}{8^5} = \frac{\overset{1}{\cancel{8}} \cdot \overset{1}{\cancel{8}}}{\underset{1}{\cancel{8}} \cdot \underset{1}{\cancel{8}} \cdot 8 \cdot 8 \cdot 8}$ Definition of powers

$= \frac{1}{8^3}$ Simplify.

Since $\frac{8^2}{8^5}$ cannot have two different values, we can conclude that $8^{-3} = \frac{1}{8^3}$.

Key Concept — *Negative Exponent*

- **Words** For any nonzero number a and any integer n, a^{-n} is the reciprocal of a^n. In addition, the reciprocal of a^{-n} is a^n.
- **Symbols** For any nonzero number a and any integer n, $a^{-n} = \frac{1}{a^n}$ and $\frac{1}{a^{-n}} = a^n$.
- **Examples** $5^{-2} = \frac{1}{5^2}$ or $\frac{1}{25}$ $\frac{1}{m^{-3}} = m^3$

An expression involving exponents is not considered simplified if the expression contains negative exponents.

Example 4 Negative Exponents

Simplify each expression. Assume that no denominator is equal to zero.

a. $\dfrac{b^{-3}c^2}{d^{-5}}$

$\dfrac{b^{-3}c^2}{d^{-5}} = \left(\dfrac{b^{-3}}{1}\right)\left(\dfrac{c^2}{1}\right)\left(\dfrac{1}{d^{-5}}\right)$ Write as a product of fractions.

$= \left(\dfrac{1}{b^3}\right)\left(\dfrac{c^2}{1}\right)\left(\dfrac{d^5}{1}\right)$ $a^{-n} = \dfrac{1}{a^n}$

$= \dfrac{c^2d^5}{b^3}$ Multiply fractions.

b. $\dfrac{-3a^{-4}b^7}{21a^2b^7c^{-5}}$

$\dfrac{-3a^{-4}b^7}{21a^2b^7c^{-5}} = \left(\dfrac{-3}{21}\right)\left(\dfrac{a^{-4}}{a^2}\right)\left(\dfrac{b^7}{b^7}\right)\left(\dfrac{1}{c^{-5}}\right)$ Group powers with the same base.

$= \dfrac{-1}{7}(a^{-4-2})(b^{7-7})(c^5)$ Quotient of Powers and Negative Exponent Properties

$= \dfrac{-1}{7}a^{-6}b^0c^5$ Simplify.

$= \dfrac{-1}{7}\left(\dfrac{1}{a^6}\right)(1)c^5$ Negative Exponent and Zero Exponent Properties

$= -\dfrac{c^5}{7a^6}$ Multiply fractions.

Study Tip

Common Misconception
Do not confuse a negative number with a number raised to a negative power.

$3^{-1} = \frac{1}{3}$ $\quad -3 \neq \frac{1}{3}$

Standardized Test Practice (A) (B) (C) (D)

Example 5 Apply Properties of Exponents

Multiple-Choice Test Item

Write the ratio of the area of the circle to the area of the square in simplest form.

(A) $\dfrac{\pi}{2}$ (B) $\dfrac{\pi}{4}$ (C) $\dfrac{2\pi}{1}$ (D) $\dfrac{\pi}{3}$

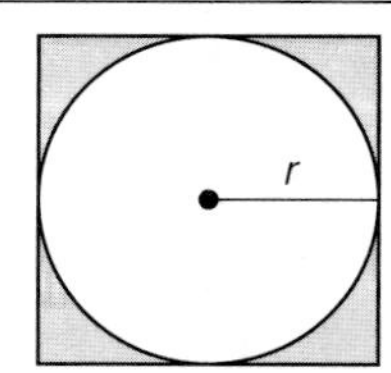

Read the Test Item

A ratio is a comparison of two quantities. It can be written in fraction form.

Solve the Test Item

- area of circle $= \pi r^2$
  length of square = diameter of circle or $2r$
  area of square $= (2r)^2$
- $\dfrac{\text{area of circle}}{\text{area of square}} = \dfrac{\pi r^2}{(2r)^2}$ Substitute.

  $= \dfrac{\pi}{4}r^{2-2}$ Quotient of Powers

  $= \dfrac{\pi}{4}r^0$ or $\dfrac{\pi}{4}$ $\quad r^0 = 1$

The answer is B.

The Princeton Review
Test-Taking Tip
Some problems can be solved using estimation. The area of the circle is less than the area of the square. Therefore, the ratio of the two areas must be less than 1. Use 3 as an approximate value for π to determine which of the choices is less than 1.

Check for Understanding

Concept Check

1. **OPEN ENDED** Name two monomials whose product is $54x^2y^3$.
2. **Show** a method of simplifying $\frac{a^3b^5}{ab^2}$ using negative exponents instead of the Quotient of Powers Property.
3. **FIND THE ERROR** Jamal and Emily are simplifying $\frac{-4x^3}{x^5}$.

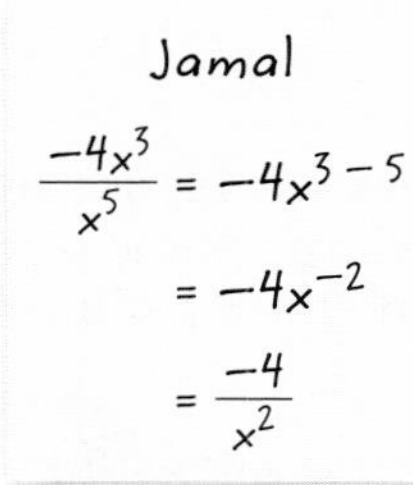

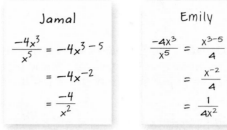

Who is correct? Explain your reasoning.

Guided Practice

Simplify. Assume that no denominator is equal to zero.

4. $\frac{7^8}{7^2}$
5. $\frac{x^8y^{12}}{x^2y^7}$
6. $\left(\frac{2c^3d}{7z^2}\right)^3$
7. $y^0(y^5)(y^{-9})$
8. 13^{-2}
9. $\frac{c^{-5}}{d^3g^{-8}}$
10. $\frac{-5pq^7}{10p^6q^3}$
11. $\frac{(cd^{-2})^3}{(c^4d^9)^{-2}}$
12. $\frac{(4m^{-3}n^5)^0}{mn}$

Standardized Test Practice

13. Find the ratio of the volume of the cylinder to the volume of the sphere.

Ⓐ $\frac{1}{2}$ Ⓑ 1

Ⓒ $\frac{3}{2}$ Ⓓ $\frac{3\pi}{2}$

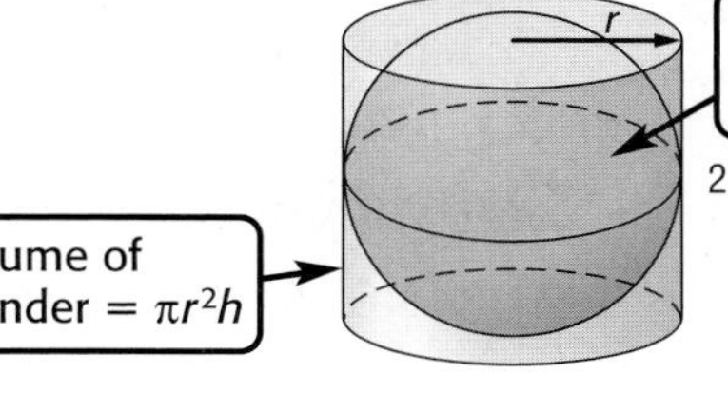

Practice and Apply

Homework Help

For Exercises	See Examples
14–21	1, 2
22–37	1–4

Extra Practice

See page 837.

Simplify. Assume that no denominator is equal to zero.

14. $\frac{4^{12}}{4^2}$
15. $\frac{3^{13}}{3^7}$
16. $\frac{p^7n^3}{p^4n^2}$
17. $\frac{y^3z^9}{yz^2}$
18. $\left(\frac{5b^4n}{2a^6}\right)^2$
19. $\left(\frac{3m^7}{4x^5y^3}\right)^4$
20. $\frac{-2a^3}{10a^8}$
21. $\frac{15b}{45b^5}$
22. $x^3y^0x^{-7}$
23. $n^2(p^{-4})(n^{-5})$
24. 6^{-2}
25. 5^{-3}
26. $\left(\frac{4}{5}\right)^{-2}$
27. $\left(\frac{3}{2}\right)^{-3}$
28. $\frac{28a^7c^{-4}}{7a^3b^0c^{-8}}$
29. $\frac{30h^{-2}k^{14}}{5hk^{-3}}$
30. $\frac{18x^3y^4z^7}{-2x^2yz}$
31. $\frac{-19y^0z^4}{-3z^{16}}$
32. $\frac{(5r^{-2})^{-2}}{(2r^3)^2}$
33. $\frac{p^{-4}q^{-3}}{(p^5q^2)^{-1}}$
34. $\left(\frac{r^{-2}t^5}{t^{-1}}\right)^0$
35. $\left(\frac{4c^{-2}d}{b^{-2}c^3d^{-1}}\right)^0$
36. $\left(\frac{5b^{-2}n^4}{n^2z^{-3}}\right)^{-1}$
37. $\left(\frac{2a^{-2}bc^{-1}}{3ab^{-2}}\right)^{-3}$

38. The area of the rectangle is $24x^5y^3$ square units. Find the length of the rectangle.

39. The area of the triangle is $100a^3b$ square units. Find the height of the triangle.

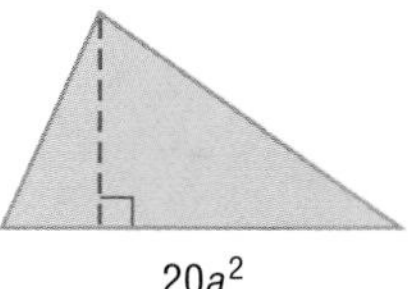

SOUND **For Exercises 40–42, use the following information.**
The intensity of sound can be measured in watts per square meter. The table gives the watts per square meter for some common sounds.

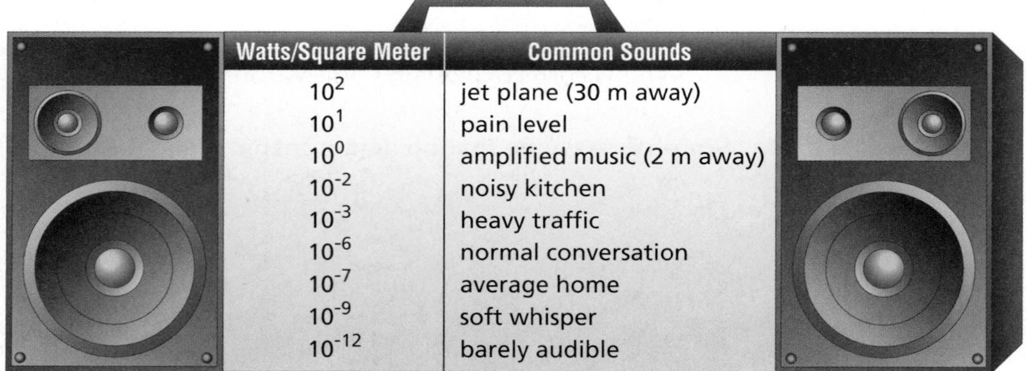

Watts/Square Meter	Common Sounds
10^2	jet plane (30 m away)
10^1	pain level
10^0	amplified music (2 m away)
10^{-2}	noisy kitchen
10^{-3}	heavy traffic
10^{-6}	normal conversation
10^{-7}	average home
10^{-9}	soft whisper
10^{-12}	barely audible

40. How many times more intense is the sound from heavy traffic than the sound from normal conversation?
41. What sound is 10,000 times as loud as a noisy kitchen?
42. How does the intensity of a whisper compare to that of normal conversation?

More About. . .

Sound

Timbre is the quality of the sound produced by a musical instrument. Sound quality is what distinguishes the sound of a note played on a flute from the sound of the same note played on a trumpet with the same frequency and intensity.

Source: www.school.discovery.com

PROBABILITY **For Exercises 43 and 44, use the following information.**
If you toss a coin, the probability of getting heads is $\frac{1}{2}$. If you toss a coin 2 times, the probability of getting heads each time is $\frac{1}{2} \cdot \frac{1}{2}$ or $\left(\frac{1}{2}\right)^2$.

43. Write an expression to represent the probability of tossing a coin n times and getting n heads.
44. Express your answer to Exercise 43 as a power of 2.

LIGHT **For Exercises 45 and 46, use the table below.**

45. Express the range of the wavelengths of visible light using positive exponents. Then evaluate each expression.
46. Express the range of the wavelengths of X-rays using positive exponents. Then evaluate each expression.

Spectrum of Electromagnetic Radiation

Region	Wavelength (cm)
Radio	greater than 10
Microwave	10^1 to 10^{-2}
Infrared	10^{-2} to 10^{-5}
Visible	10^{-5} to 10^{-4}
Ultraviolet	10^{-4} to 10^{-7}
X-rays	10^{-7} to 10^{-9}
Gamma Rays	less than 10^{-9}

CRITICAL THINKING **Simplify. Assume that no denominator is equal to zero.**

47. $a^n(a^3)$

48. $(5^{4x-3})(5^{2x+1})$

49. $\frac{c^{x+7}}{c^{x-4}}$

50. $\frac{3b^{2n-9}}{b^{3(n-3)}}$

51. **WRITING IN MATH** Answer the question that was posed at the beginning of the lesson.

How can you compare pH levels?

Include the following in your answer:

- an example comparing two pH levels using the properties of exponents.

52. What is the value of $\frac{2^2 \cdot 2^3}{2^{-2} \cdot 2^{-3}}$?

(A) 2^{10} (B) 2^{12} (C) -1 (D) $\frac{1}{2}$

53. **EXTENDED RESPONSE** Write a convincing argument to show why $3^0 = 1$ using the following pattern.
$3^5 = 243, 3^4 = 81, 3^3 = 27, 3^2 = 9, \ldots$

Maintain Your Skills

Mixed Review

Simplify. *(Lesson 8-1)*

54. $(m^3n)(mn^2)$

55. $(3x^4y^3)(4x^4y)$

56. $(a^3x^2)^4$

57. $(3cd^5)^2$

58. $[(2^3)^2]^2$

59. $(-3ab)^3(2b^3)^2$

NUTRITION **For Exercises 60 and 61, use the following information.**
Between the ages of 11 and 18, you should get at least 1200 milligrams of calcium each day. One ounce of mozzarella cheese has 147 milligrams of calcium, and one ounce of Swiss cheese has 219 milligrams. Suppose you wanted to eat no more than 8 ounces of cheese. *(Lesson 7-5)*

60. Draw a graph showing the possible amounts of each type of cheese you can eat and still get your daily requirement of calcium. Let x be the amount of mozzarella cheese and y be the amount of Swiss cheese.

61. List three possible solutions.

Write an equation of the line with the given slope and y-intercept. *(Lesson 5-3)*

62. slope: 1, y-intercept: -4

63. slope: -2, y-intercept: 3

64. slope: $-\frac{1}{3}$, y-intercept: -1

65. slope: $\frac{3}{2}$, y-intercept: 2

Graph each equation by finding the x- and y-intercepts. *(Lesson 4-5)*

66. $2y = x + 10$

67. $4x - y = 12$

68. $2x = 7 - 3y$

Find each square root. If necessary, round to the nearest hundredth. *(Lesson 2-7)*

69. $\pm\sqrt{121}$

70. $\sqrt{3.24}$

71. $-\sqrt{52}$

Getting Ready for the Next Lesson

PREREQUISITE SKILL **Write each product in the form 10^n.**
*(To review **Products of Powers**, see Lesson 8-1.)*

72. $10^2 \times 10^3$

73. $10^{-8} \times 10^{-5}$

74. $10^{-6} \times 10^9$

75. $10^8 \times 10^{-1}$

76. $10^4 \times 10^{-4}$

77. $10^{-12} \times 10$

Reading Mathematics

Mathematical Prefixes and Everyday Prefixes

You may have noticed that many prefixes used in mathematics are also used in everyday language. You can use the everyday meaning of these prefixes to better understand their mathematical meaning. The table shows four mathematical prefixes along with their meaning and an example of an everyday word using that prefix.

Prefix	Everyday Meaning	Example
mono-	**1.** one; single; alone	**monologue** A continuous series of jokes or comic stories delivered by one comedian.
bi-	**1.** two **2.** both **3.** both sides, parts, or directions	**bicycle** A vehicle consisting of a light frame mounted on two wire-spoked wheels one behind the other and having a seat, handlebars for steering, brakes, and two pedals or a small motor by which it is driven.
tri-	**1.** three **2.** occurring at intervals of three **3.** occurring three times during	**trilogy** A group of three dramatic or literary works related in subject or theme.
poly-	**1.** more than one; many; much	**polygon** A closed plane figure bounded by three or more line segments.

Source: *The American Heritage Dictionary of the English Language*

You can use your everyday understanding of prefixes to help you understand mathematical terms that use those prefixes.

Reading to Learn

1. Give an example of a geometry term that uses one of these prefixes. Then define that term.

2. **MAKE A CONJECTURE** Given your knowledge of the meaning of the word monomial, make a conjecture as to the meaning of each of the following mathematical terms.

a. binomial **b.** trinomial **c.** polynomial

3. Research the following prefixes and their meanings.

a. semi- **b.** hexa- **c.** octa-

8-3 Scientific Notation

What You'll Learn

- Express numbers in scientific notation and standard notation.
- Find products and quotients of numbers expressed in scientific notation.

Vocabulary

- scientific notation

Why is scientific notation important in astronomy?

Astronomers often work with very large numbers, such as the masses of planets. The mass of each planet in our solar system is given in the table. Notice that each value is written as the product of a number and a power of 10. These values are written in scientific notation.

Planet	Mass (kilograms)
Mercury	3.30×10^{23}
Venus	4.87×10^{24}
Earth	5.97×10^{24}
Mars	6.42×10^{23}
Jupiter	1.90×10^{27}
Saturn	5.69×10^{26}
Uranus	8.68×10^{25}
Neptune	1.02×10^{26}
Pluto	1.27×10^{22}

Source: NASA

SCIENTIFIC NOTATION When dealing with very large or very small numbers, keeping track of place value can be difficult. For this reason, numbers such as these are often expressed in **scientific notation**.

Key Concept — *Scientific Notation*

- **Words** A number is expressed in scientific notation when it is written as a product of a factor and a power of 10. The factor must be greater than or equal to 1 and less than 10.
- **Symbols** A number in scientific notation is written as $a \times 10^n$, where $1 \leq a < 10$ and n is an integer.

Study Tip

Reading Math
Standard notation is the way in which you are used to seeing a number written, where the decimal point determines the place value for each digit of the number.

The following examples show one way of expressing a number that is written in scientific notation in its decimal or standard notation. Look for a relationship between the power of 10 and the position of the decimal point in the standard notation of the number.

$$6.59 \times 10^4 = 6.59 \times 10{,}000$$
$$= 65{,}900$$

The decimal point moved 4 places to the right.

$$4.81 \times 10^{-6} = 4.81 \times \frac{1}{10^6}$$
$$= 4.81 \times 0.000001$$
$$= 0.00000481$$

The decimal point moved 6 places to the left.

These examples suggest the following rule for expressing a number written in scientific notation in standard notation.

Concept Summary — Scientific to Standard Notation

Use these steps to express a number of the form $a \times 10^n$ in standard notation.

1. Determine whether $n > 0$ or $n < 0$.
2. If $n > 0$, move the decimal point in a to the right n places.
   If $n < 0$, move the decimal point in a to the left n places.
3. Add zeros, decimal point, and/or commas as needed to indicate place value.

Example 1 Scientific to Standard Notation

Express each number in standard notation.

a. 2.45×10^8

$2.45 \times 10^8 = 245{,}000{,}000$ — $n = 8$; move decimal point 8 places to the right.

b. 3×10^{-5}

$3 \times 10^{-5} = 0.00003$ — $n = -5$; move decimal point 5 places to the left.

To express a number in scientific notation, reverse the process used above.

Concept Summary — Standard to Scientific Notation

Use these steps to express a number in scientific notation.

1. Move the decimal point so that it is to the right of the first nonzero digit. The result is a decimal number a.
2. Observe the number of places n and the direction in which you moved the decimal point.
3. If the decimal point moved to the left, write as $a \times 10^n$.
   If the decimal point moved to the right, write as $a \times 10^{-n}$.

Example 2 Standard to Scientific Notation

Express each number in scientific notation.

a. 30,500,000

$30{,}500{,}000 \rightarrow 3.0500000 \times 10^n$ — Move decimal point 7 places to the left.

$30{,}500{,}000 = 3.05 \times 10^7$ — $a = 3.05$ and $n = 7$

b. 0.000781

$0.000781 \rightarrow 00007.81 \times 10^n$ — Move decimal point 4 places to the right.

$0.000781 = 7.81 \times 10^{-4}$ — $a = 7.81$ and $n = -4$

Study Tip

Scientific Notation
Notice that when a number is in scientific notation, no more than one digit is to the left of the decimal point.

You will often see large numbers in the media written using a combination of a number and a word, such as 3.2 million. To write this number in standard notation, rewrite the word *million* as 10^6. The exponent 6 indicates that the decimal point should be moved 6 places to the right.

$$3.2 \text{ million} = 3{,}200{,}000$$

Example 3 Use Scientific Notation

The graph shows chocolate and candy sales during a recent holiday season.

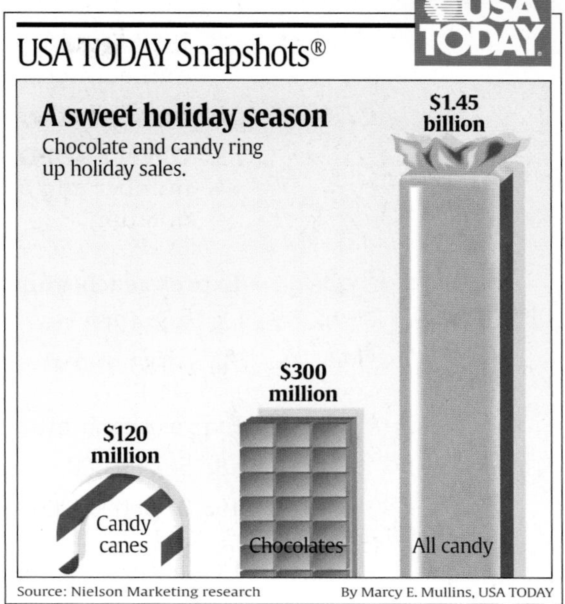

Log on for:
- Updated data
- More activities on scientific notation

www.algebra1.com/usa_today

a. Express the sales of candy canes, chocolates, and all candy in standard notation.

Candy canes:
\$120 million = \$120,000,000

Chocolates:
\$300 million = \$300,000,000

All candy:
\$1.45 billion = \$1,450,000,000

b. Write each of these sales figures in scientific notation.

Candy canes:
\$120,000,000 = $\$1.2 \times 10^8$

Chocolates:
\$300,000,000 = $\$3.0 \times 10^8$

All candy: \$1,450,000,000 = $\$1.45 \times 10^9$

PRODUCTS AND QUOTIENTS WITH SCIENTIFIC NOTATION

You can use scientific notation to simplify computation with very large and/or very small numbers.

Example 4 Multiplication with Scientific Notation

Evaluate $(5 \times 10^{-8})(2.9 \times 10^2)$. Express the result in scientific and standard notation.

$(5 \times 10^{-8})(2.9 \times 10^2)$	
$= (5 \times 2.9)(10^{-8} \times 10^2)$	Commutative and Associative Properties
$= 14.5 \times 10^{-6}$	Product of Powers
$= (1.45 \times 10^1) \times 10^{-6}$	$14.5 = 1.45 \times 10^1$
$= 1.45 \times (10^1 \times 10^{-6})$	Associative Property
$= 1.45 \times 10^{-5}$ or 0.0000145	Product of Powers

Example 5 Division with Scientific Notation

Evaluate $\frac{1.2789 \times 10^9}{5.22 \times 10^5}$. Express the result in scientific and standard notation.

$\frac{1.2789 \times 10^9}{5.22 \times 10^5} = \left(\frac{1.2789}{5.22}\right)\left(\frac{10^9}{10^5}\right)$	Associative Property
$= 0.245 \times 10^4$	Quotient of Powers
$= (2.45 \times 10^{-1}) \times 10^4$	$0.245 = 2.45 \times 10^{-1}$
$= 2.45 \times (10^{-1} \times 10^4)$	Associative Property
$= 2.45 \times 10^3$ or 2450	Product of Powers

www.algebra1.com/extra_examples

Check for Understanding

Concept Check

1. **Explain** how you know to use a positive or a negative exponent when writing a number in scientific notation.
2. **State** whether 65.2×10^3 is in scientific notation. Explain your reasoning.
3. **OPEN ENDED** Give an example of a large number written using a decimal number and a word. Write this number in standard and then in scientific notation.

Guided Practice

Express each number in standard notation.

4. 2×10^{-8}
5. 4.59×10^3
6. 7.183×10^{14}
7. 3.6×10^{-5}

Express each number in scientific notation.

8. 56,700,000
9. 0.00567
10. 0.00000000004
11. 3,002,000,000,000,000

Evaluate. Express each result in scientific and standard notation.

12. $(5.3 \times 10^2)(4.1 \times 10^5)$
13. $(2 \times 10^{-5})(9.4 \times 10^{-3})$
14. $\frac{1.5 \times 10^2}{2.5 \times 10^{12}}$
15. $\frac{1.25 \times 10^4}{2.5 \times 10^{-6}}$

Application

CREDIT CARDS For Exercises 16 and 17, use the following information. During the year 2000, 1.65 billion credit cards were in use in the United States. During that same year, $1.54 trillion was charged to these cards. (*Hint*: 1 trillion $= 1 \times 10^{12}$) **Source:** U.S. Department of Commerce

16. Express each of these values in standard and then in scientific notation.
17. Find the average amount charged per credit card.

Practice and Apply

Homework Help

For Exercises	See Examples
18–29	1
30–43	2
44–55	3, 4
56–59	5

Extra Practice

See page 837.

Express each number in standard notation.

18. 5×10^{-6}
19. 6.1×10^{-9}
20. 7.9×10^4
21. 8×10^7
22. 1.243×10^{-7}
23. 2.99×10^{-1}
24. 4.782×10^{13}
25. 6.89×10^0

PHYSICS Express the number in each statement in standard notation.

26. There are 2×10^{11} stars in the Andromeda Galaxy.
27. The center of the moon is 2.389×10^5 miles away from the center of Earth.
28. The mass of a proton is 1.67265×10^{-27} kilograms.
29. The mass of an electron is 9.1095×10^{-31} kilograms.

Express each number in scientific notation.

30. 50,400,000,000
31. 34,402,000
32. 0.000002
33. 0.00090465
34. 25.8
35. 380.7
36. 622×10^6
37. 87.3×10^{11}
38. 0.5×10^{-4}
39. 0.0081×10^{-3}
40. 94×10^{-7}
41. 0.001×10^{12}

The distances of the planets from the Sun can be written in scientific notation. Visit www.algebra1.com/webquest to continue work on your WebQuest project.

42. STARS In the 1930s, the Indian physicist Subrahmanyan Chandrasekhar and others predicted the existence of neutron stars. These stars can have a density of 10 billion tons per teaspoonful. Express this density in scientific notation.

43. PHYSICAL SCIENCE The unit of measure for counting molecules is a *mole*. One mole of a substance is the amount that contains about 602,214,299,000,000,000,000,000 molecules. Write this number in scientific notation.

Evaluate. Express each result in scientific and standard notation.

44. $(8.9 \times 10^4)(4 \times 10^3)$ **45.** $(3 \times 10^6)(5.7 \times 10^2)$

46. $(5 \times 10^{-2})(8.6 \times 10^{-3})$ **47.** $(1.2 \times 10^{-5})(1.2 \times 10^{-3})$

48. $(3.5 \times 10^7)(6.1 \times 10^{-8})$ **49.** $(2.8 \times 10^{-2})(9.1 \times 10^6)$

50. $\dfrac{7.2 \times 10^9}{4.8 \times 10^4}$ **51.** $\dfrac{7.2 \times 10^3}{1.8 \times 10^7}$ **52.** $\dfrac{3.162 \times 10^{-4}}{5.1 \times 10^2}$

53. $\dfrac{1.035 \times 10^{-2}}{4.5 \times 10^3}$ **54.** $\dfrac{2.795 \times 10^{-8}}{4.3 \times 10^{-4}}$ **55.** $\dfrac{4.65 \times 10^{-1}}{5 \times 10^5}$

56. HAIR GROWTH The usual growth rate of human hair is 3.3×10^{-4} meter per day. If an individual hair grew for 10 years, how long would it be in meters? (Assume 365 days in a year.)

57. NATIONAL DEBT In April 2001, the national debt was about \$5.745 trillion, and the estimated U.S. population was 283.9 million. About how much was each U.S. citizen's share of the national debt at that time?

Online Research **Data Update** What is the current U.S. population and amount of national debt? Visit www.algebra1.com/data_update to learn more.

More About. . .

Baseball

The contract Alex Rodriguez signed with the Texas Rangers on December 11, 2000, guarantees him \$25.2 million a year for 10 seasons.

Source: Associated Press

58. BASEBALL The table below lists the greatest yearly salary for a major league baseball player for selected years.

Baseball Salary Milestones

Year	Player	Yearly Salary
1979	Nolan Ryan	\$1 million
1982	George Foster	\$2.04 million
1990	Jose Canseco	\$4.7 million
1992	Ryne Sandberg	\$7.1 million
1996	Ken Griffey, Jr.	\$8.5 million
1997	Pedro Martinez	\$12.5 million
2000	Alex Rodriguez	\$25.2 million

Source: *USA TODAY*

About how many times as great was the yearly salary of Alex Rodriguez in 2000 as that of George Foster in 1982?

59. ASTRONOMY The Sun burns about 4.4×10^6 tons of hydrogen per second. How much hydrogen does the Sun burn in one year? (*Hint*: First, find the number of seconds in a year and write this number in scientific notation.)

60. CRITICAL THINKING Determine whether each statement is *sometimes*, *always*, or *never* true. Explain your reasoning.

a. If $1 \leq a < 10$ and n and p are integers, then $(a \times 10^n)^p = a^p \times 10^{np}$.

b. The expression $a^p \times 10^{np}$ in part **a** is in scientific notation.

61. **WRITING IN MATH** Answer the question that was posed at the beginning of the lesson.

Why is scientific notation important in astronomy?

Include the following in your answer:

- the mass of each of the planets in standard notation, and
- an explanation of how scientific notation makes presenting and computing with large numbers easier.

62. Which of the following is equivalent to 360×10^{-4}?

Ⓐ 3.6×10^{3} Ⓑ 3.6×10^{2} Ⓒ 3.6×10^{-2} Ⓓ 3.6×10^{-3}

63. **SHORT RESPONSE** There are an average of 25 billion red blood cells in the human body and about 270 million hemoglobin molecules in each red blood cell. Find the average number of hemoglobin molecules in the human body.

SCIENTIFIC NOTATION You can use a graphing calculator to solve problems involving scientific notation. First, put your calculator in scientific mode. To enter 4.5×10^{9}, enter 4.5 [×] 10 [^] 9.

64. $(4.5 \times 10^{9})(1.74 \times 10^{-2})$

65. $(7.1 \times 10^{-11})(1.2 \times 10^{5})$

66. $(4.095 \times 10^{5}) \div (3.15 \times 10^{8})$

67. $(6 \times 10^{-4}) \div (5.5 \times 10^{-7})$

Maintain Your Skills

Mixed Review

Simplify. Assume no denominator is equal to zero. *(Lesson 8-2)*

68. $\frac{49a^4b^7c^2}{7ab^4c^3}$

69. $\frac{-4n^3p^{-5}}{n^{-2}}$

70. $\frac{(8n^7)^2}{(3n^2)^{-3}}$

Determine whether each expression is a monomial. Write *yes* or *no*. *(Lesson 8-1)*

71. $3a + 4b$

72. $\frac{6}{n}$

73. $\frac{v^2}{3}$

Solve each inequality. Then check your solution and graph it on a number line. *(Lesson 6-1)*

74. $m - 3 < -17$

75. $-9 + d > 9$

76. $-x - 11 \geq 23$

Getting Ready for the Next Lesson

PREREQUISITE SKILL **Evaluate each expression when $a = 5$, $b = -2$, and $c = 3$.** *(To review **evaluating expressions**, see Lesson 1-2.)*

77. $5b^2$

78. $c^2 - 9$

79. $b^3 + 3ac$

80. $a^2 + 2a - 1$

81. $-2b^4 - 5b^3 - b$

82. $3.2c^3 + 0.5c^2 - 5.2c$

Practice Quiz 1

Lessons 8-1 through 8-3

Simplify. *(Lesson 8-1)*

1. $n^3(n^4)(n)$

2. $4ad(3a^3d)$

3. $(-2w^3z^4)^3(-4wz^3)^2$

Simplify. Assume that no denominator is equal to zero. *(Lesson 8-2)*

4. $\frac{25p^{10}}{15p^3}$

5. $\left(\frac{6k^3}{7np^4}\right)^2$

6. $\frac{4x^0y^2}{(3y^{-3}z^5)^{-2}}$

Evaluate. Express each result in scientific and standard notation. *(Lesson 8-3)*

7. $(6.4 \times 10^3)(7 \times 10^2)$

8. $(4 \times 10^2)(15 \times 10^{-6})$

9. $\frac{9.2 \times 10^3}{2.3 \times 10^5}$

10. $\frac{3.6 \times 10^7}{1.2 \times 10^{-2}}$

Algebra Activity

A Preview of Lesson 8-4

Polynomials

Algebra tiles can be used to model polynomials. A polynomial is a monomial or the sum of monomials. The diagram at the right shows the models.

Polynomial Models	
Polynomials are modeled using three types of tiles.	1 x x^2
Each tile has an opposite.	−1 $-x$ $-x^2$

Use algebra tiles to model each polynomial.

- **$4x$**
  To model this polynomial, you will need 4 green x tiles.

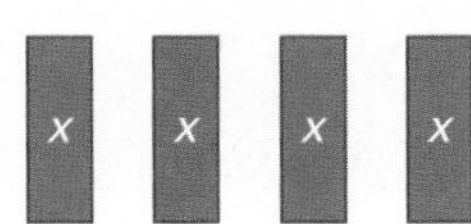

- **$2x^2 - 3$**
  To model this polynomial, you will need 2 blue x^2 tiles and 3 red -1 tiles.

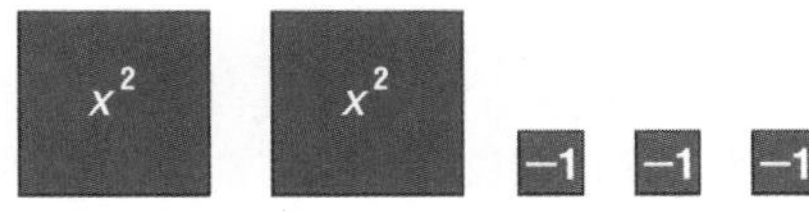

- **$-x^2 + 3x + 2$**
  To model this polynomial, you will need 1 red $-x^2$ tile, 3 green x tiles, and 2 yellow 1 tiles.

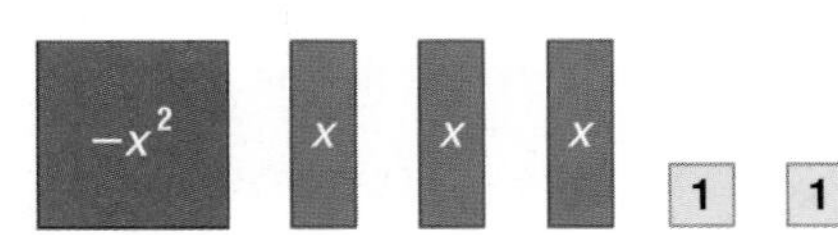

Model and Analyze

Use algebra tiles to model each polynomial. Then draw a diagram of your model.

1. $-2x^2$ **2.** $5x - 4$ **3.** $3x^2 - x$ **4.** $x^2 + 4x + 3$

Write an algebraic expression for each model.

5.

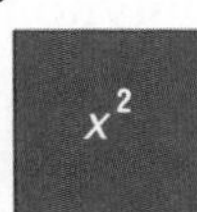

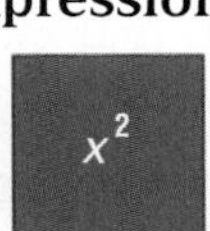

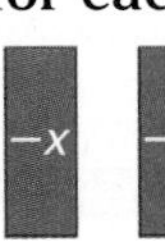

6.
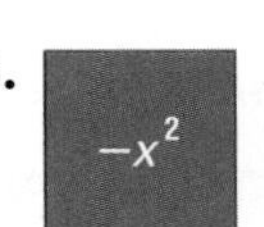

7.
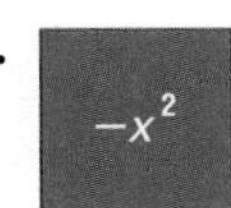

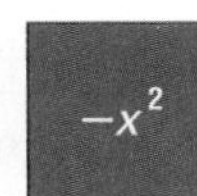

8.

9. MAKE A CONJECTURE Write a sentence or two explaining why algebra tiles are sometimes called *area tiles.*

8-4 Polynomials

What You'll Learn

- Find the degree of a polynomial.
- Arrange the terms of a polynomial in ascending or descending order.

Vocabulary

- polynomial
- binomial
- trinomial
- degree of a monomial
- degree of a polynomial

How are polynomials useful in modeling data?

The number of hours H spent per person per year playing video games from 1992 through 1997 is shown in the table. These data can be modeled by the equation

$$H = \frac{1}{4}(t^4 - 9t^3 + 26t^2 - 18t + 76),$$

where t is the number of years since 1992. The expression $t^4 - 9t^3 + 26t^2 - 18t + 76$ is an example of a polynomial.

Video Game Usage

Year	Hours spent per person
1992	19
1993	19
1994	22
1995	24
1996	26
1997	36

Source: U.S. Census Bureau

DEGREE OF A POLYNOMIAL A **polynomial** is a monomial or a sum of monomials. Some polynomials have special names. A **binomial** is the sum of *two* monomials, and a **trinomial** is the sum of *three* monomials. Polynomials with more than three terms have no special names.

Study Tip

Common Misconception
Before deciding if an expression is a polynomial, write each term of the expression so that there are no variables in the denominator. Then look for negative exponents. Recall that the exponents of a monomial must be nonnegative integers.

Monomial	Binomial	Trinomial
7	$3 + 4y$	$x + y + z$
$13n$	$2a + 3c$	$p^2 + 5p + 4$
$-5z^3$	$6x^2 + 3xy$	$a^2 - 2ab - b^2$
$4ab^3c^2$	$7pqr + pq^2$	$3v^2 - 2w + ab^3$

Example 1 Identify Polynomials

State whether each expression is a polynomial. If it is a polynomial, identify it as a *monomial, binomial,* or *trinomial*.

	Expression	Polynomial?	Monomial, Binomial, or Trinomial?
a.	$2x - 3yz$	Yes, $2x - 3yz = 2x + (-3yz)$. The expression is the sum of two monomials.	binomial
b.	$8n^3 + 5n^{-2}$	No. $5n^{-2} = \frac{5}{n^2}$, which is not a monomial.	none of these
c.	-8	Yes. -8 is a real number.	monomial
d.	$4a^2 + 5a + a + 9$	Yes. The expression simplifies to $4a^2 + 6a + 9$, so it is the sum of three monomials.	trinomial

Study Tip

Like Terms
Be sure to combine any like terms before deciding if a polynomial is a monomial, binomial, or trinomial.

Polynomials can be used to express geometric relationships.

Example 2 Write a Polynomial

GEOMETRY Write a polynomial to represent the area of the shaded region.

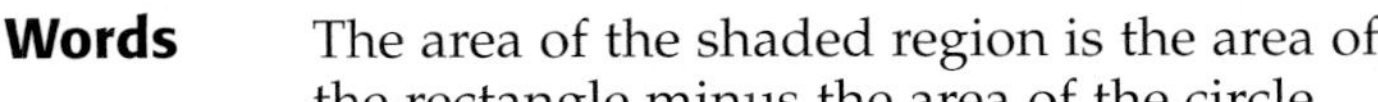

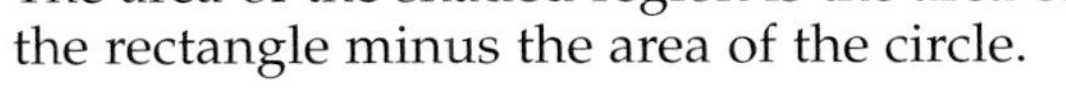

Words The area of the shaded region is the area of the rectangle minus the area of the circle.

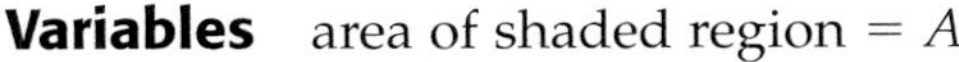

Variables area of shaded region = A
width of rectangle = $2r$
rectangle area = $b(2r)$
circle area = πr^2

Equation

area of shaded region	=	rectangle area	−	circle area
A	=	$b(2r)$	−	πr^2
A	=	$2br$	−	πr^2

The polynomial representing the area of the shaded region is $2br - \pi r^2$.

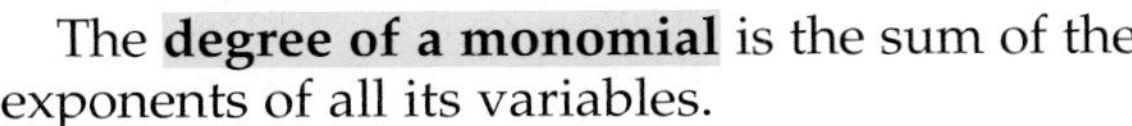

The **degree of a monomial** is the sum of the exponents of all its variables.

The **degree of a polynomial** is the greatest degree of any term in the polynomial. To find the degree of a polynomial, you must find the degree of each term.

Monomial	Degree
$8y^4$	4
$3a$	1
$-2xy^2z^3$	1 + 2 + 3 or 6
7	0

Example 3 Degree of a Polynomial

Find the degree of each polynomial.

	Polynomial	Terms	Degree of Each Term	Degree of Polynomial
a.	$5mn^2$	$5mn^2$	3	3
b.	$-4x^2y^2 + 3x^2 + 5$	$-4x^2y^2, 3x^2, 5$	4, 2, 0	4
c.	$3a + 7ab - 2a^2b + 16$	$3a, 7ab, 2a^2b, 16$	1, 2, 3, 0	3

Study Tip

Degrees of 1 and 0

- Since $a = a^1$, the monomial $3a$ can be rewritten as $3a^1$. Thus $3a$ has degree 1.
- Since $x^0 = 1$, the monomial 7 can be rewritten as $7x^0$. Thus 7 has degree 0.

WRITE POLYNOMIALS IN ORDER The terms of a polynomial are usually arranged so that the powers of one variable are in *ascending* (increasing) order or *descending* (decreasing) order.

Example 4 Arrange Polynomials in Ascending Order

Arrange the terms of each polynomial so that the powers of x are in ascending order.

a. $7x^2 + 2x^4 - 11$

$7x^2 + 2x^4 - 11 = 7x^2 + 2x^4 - 11x^0$ $x^0 = 1$

$= -11 + 7x^2 + 2x^4$ Compare powers of x: $0 < 2 < 4$.

b. $2xy^3 + y^2 + 5x^3 - 3x^2y$

$2xy^3 + y^2 + 5x^3 - 3x^2y$

$= 2x^1y^3 + y^2 + 5x^3 - 3x^2y^1$ $x = x^1$

$= y^2 + 2xy^3 - 3x^2y + 5x^3$ Compare powers of x: $0 < 1 < 2 < 3$.

Example 5 Arrange Polynomials in Descending Order

Arrange the terms of each polynomial so that the powers of x are in descending order.

a. $6x^2 + 5 - 8x - 2x^3$

$6x^2 + 5 - 8x - 2x^3 = 6x^2 + 5x^0 - 8x^1 - 2x^3$ $x^0 = 1$ and $x = x^1$

$= -2x^3 + 6x^2 - 8x + 5$ $3 > 2 > 1 > 0$

b. $3a^3x^2 - a^4 + 4ax^5 + 9a^2x$

$3a^3x^2 - a^4 + 4ax^5 + 9a^2x = 3a^3x^2 - a^4x^0 + 4a^1x^5 + 9a^2x^1$ $a = a^1$, $x^0 = 1$, and $x = x^1$

$= 4ax^5 + 3a^3x^2 + 9a^2x - a^4$ $5 > 2 > 1 > 0$

Check for Understanding

Concept Check

1. **OPEN ENDED** Give an example of a monomial of degree zero.
2. **Explain** why a polynomial cannot contain a variable raised to a negative power.
3. **Determine** whether each statement is *true* or *false*. If false, give a counterexample.
   a. All binomials are polynomials.
   b. All polynomials are monomials.
   c. All monomials are polynomials.

Guided Practice

State whether each expression is a polynomial. If the expression is a polynomial, identify it as a *monomial*, a *binomial*, or a *trinomial*.

4. $5x - 3xy + 2x$
5. $\frac{2z}{5}$
6. $9a^2 + 7a - 5$

Find the degree of each polynomial.

7. 1
8. $3x + 2$
9. $2x^2y^3 + 6x^4$

Arrange the terms of each polynomial so that the powers of x are in ascending order.

10. $6x^3 - 12 + 5x$
11. $-7a^2x^3 + 4x^2 - 2ax^5 + 2a$

Arrange the terms of each polynomial so that the powers of x are in descending order.

12. $2c^5 + 9cx^2 + 3x$
13. $y^3 + x^3 + 3x^2y + 3xy^2$

Application

14. **GEOMETRY** Write a polynomial to represent the area of the shaded region.

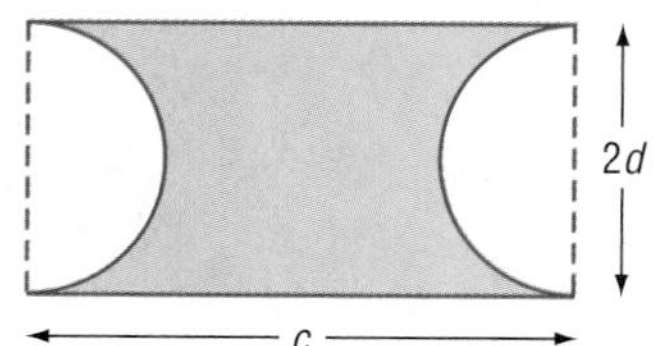

Practice and Apply

Homework Help

For Exercises	See Examples
15–20	1
21–24	2
25–36	3
37–52	4, 5

State whether each expression is a polynomial. If the expression is a polynomial, identify it as a *monomial*, a *binomial*, or a *trinomial*.

15. 14
16. $\frac{6m^2}{p} + p^3$
17. $7b - 3.2c + 8b$
18. $\frac{1}{3}x^2 + x - 2$
19. $6gh^2 - 4g^2h + g$
20. $-4 + 2a + \frac{5}{a^2}$

Extra Practice
See page 838.

GEOMETRY **Write a polynomial to represent the area of each shaded region.**

21.

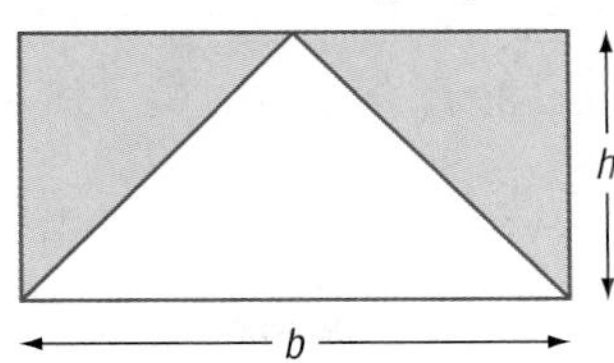

22.

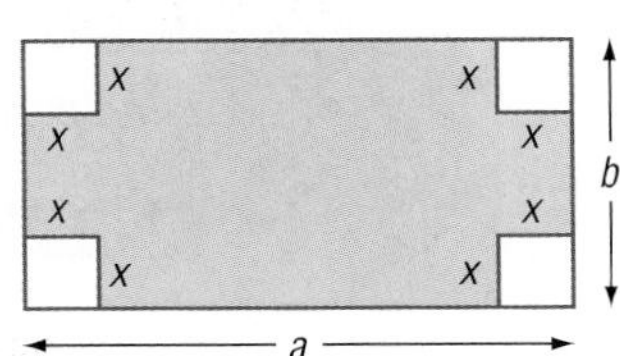

23.

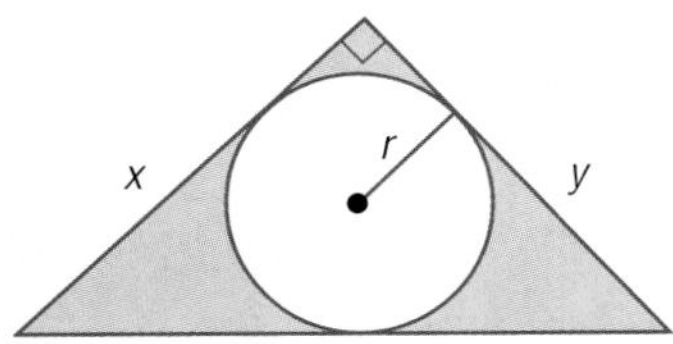

24.

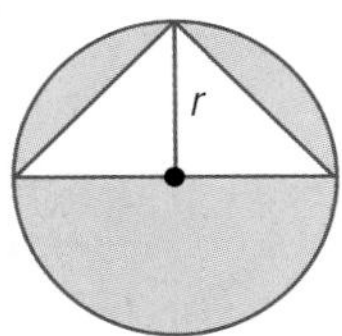

Find the degree of each polynomial.

25. $5x^3$
26. $9y$
27. $4ab$
28. -13
29. $c^4 + 7c^2$
30. $6n^3 - n^2p^2$
31. $15 - 8ag$
32. $3a^2b^3c^4 - 18a^5c$
33. $2x^3 - 4y + 7xy$
34. $3z^5 - 2x^2y^3z - 4x^2z$
35. $7 + d^5 - b^2c^2d^3 + b^6$
36. $11r^2t^4 - 2s^4t^5 + 24$

Arrange the terms of each polynomial so that the powers of *x* are in ascending order.

37. $2x + 3x^2 - 1$
38. $9x^3 + 7 - 3x^5$
39. $c^2x^3 - c^3x^2 + 8c$
40. $x^3 + 4a + 5a^2x^6$
41. $4 + 3ax^5 + 2ax^2 - 5a^7$
42. $10x^3y^2 - 3x^9y + 5y^4 + 2x^2$
43. $3xy^2 - 4x^3 + x^2y + 6y$
44. $-8a^5x + 2ax^4 - 5 - a^2x^2$

Arrange the terms of each polynomial so that the powers of *x* are in descending order.

45. $5 + x^5 + 3x^3$
46. $2x - 1 + 6x^2$
47. $4a^3x^2 - 5a + 2a^2x^3$
48. $b^2 + x^2 - 2xb$
49. $c^2 + cx^3 - 5c^3x^2 + 11x$
50. $9x^2 + 3 + 4ax^3 - 2a^2x$
51. $8x - 9x^2y + 7y^2 - 2x^4$
52. $4x^3y + 3xy^4 - x^2y^3 + y^4$

53. MONEY Write a polynomial to represent the value of q quarters, d dimes, and n nickels.

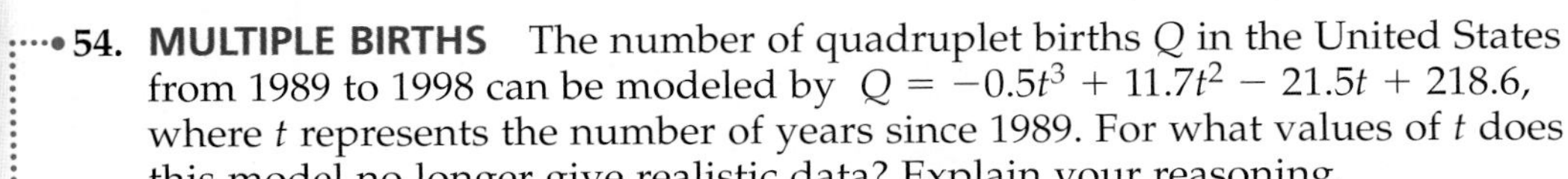

54. MULTIPLE BIRTHS The number of quadruplet births Q in the United States from 1989 to 1998 can be modeled by $Q = -0.5t^3 + 11.7t^2 - 21.5t + 218.6$, where t represents the number of years since 1989. For what values of t does this model no longer give realistic data? Explain your reasoning.

More About. . .

Multiple Births

From 1980 to 1997, the number of triplet and higher births rose 404% (from 1377 to 6737 births). This steep climb in multiple births coincides with the increased use of fertility drugs.

Source: National Center for Health and Statistics

PACKAGING **For Exercises 55 and 56, use the following information.**

A convenience store sells milkshakes in cups with semispherical lids. The volume of a cylinder is the product of π, the square of the radius r, and the height h. The volume of a sphere is the product of $\frac{4}{3}$, π, and the cube of the radius.

r
h
r

55. Write a polynomial that represents the volume of the container.

56. If the height of the container is 6 inches and the radius is 2 inches, find the volume of the container.

57. **CRITICAL THINKING** Tell whether the following statement is *true* or *false*. Explain your reasoning.
The degree of a binomial can never be zero.

58. **WRITING IN MATH** Answer the question that was posed at the beginning of the lesson.

How are polynomials useful in modeling data?

Include the following in your answer:

- a discussion of the accuracy of the equation by evaluating the polynomial for $t = \{0, 1, 2, 3, 4, 5\}$, and
- an example of how and why someone might use this equation.

Standardized Test Practice

59. If $x = -1$, then $3x^3 + 2x^2 + x + 1 =$

(A) -5. (B) -1. (C) 1. (D) 2.

60. **QUANTITATIVE COMPARISON** **Compare the quantity in Column A and the quantity in Column B. Then determine whether:**

(A) **the quantity in Column A is greater,**
(B) **the quantity in Column B is greater,**
(C) **the two quantities are equal, or**
(D) **the relationship cannot be determined from the information given.**

Column A	Column B
the degree of $5x^2y^3$	the degree of $3x^3y^2$

Maintain Your Skills

Mixed Review

Express each number in scientific notation. *(Lesson 8-3)*

61. 12,300,000 62. 0.00345 63. 12×10^6 64. 0.77×10^{-10}

Simplify. Assume that no variable is equal to zero. *(Lesson 8-2)*

65. $a^0b^{-2}c^{-1}$ 66. $\frac{-5n^5}{n^8}$ 67. $\left(\frac{4x^3y^2}{3z}\right)^2$ 68. $\frac{(-y)^5m^8}{y^3m^{-7}}$

Determine whether each relation is a function. *(Lesson 4-6)*

69.

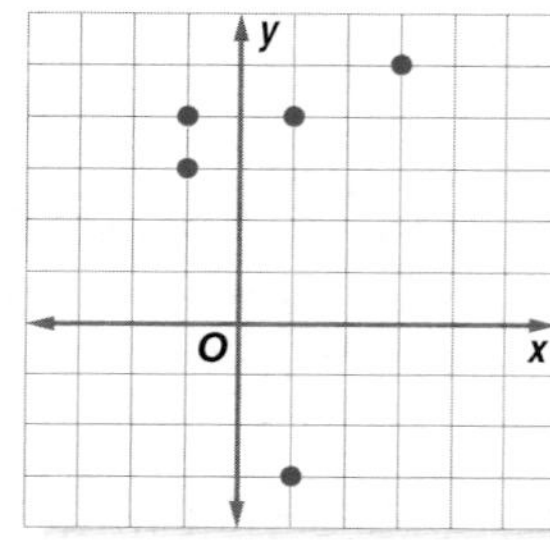

70.

x	y
−2	−2
0	1
3	4
5	−2

71. **PROBABILITY** A card is selected at random from a standard deck of 52 cards. What is the probability of selecting a black card? *(Lesson 2-6)*

Getting Ready for the Next Lesson

PREREQUISITE SKILL **Simplify each expression. If not possible, write *simplified*.**
*(To review **simplifying expressions**, see Lesson 1-5.)*

72. $3n + 5n$ 73. $9a^2 + 3a - 2a^2$ 74. $12x^2 + 8x - 6$

75. $-3a + 5b + 4a - 7b$ 76. $4x + 3y - 6 + 7x + 8 - 10y$

Algebra Activity

A Preview of Lesson 8-5

Adding and Subtracting Polynomials

Monomials such as $5x$ and $-3x$ are called *like terms* because they have the same variable to the same power. When you use algebra tiles, you can recognize like terms because the individual tiles have the same size and shape.

Polynomial Models	
Like terms are represented by tiles that have the same shape and size.	x x $-x$ (like terms)
A *zero pair* may be formed by pairing one tile with its opposite. You can remove or add zero pairs without changing the polynomial.	x $-x$ → 0

Activity 1 Use algebra tiles to find $(3x^2 - 2x + 1) + (x^2 + 4x - 3)$.

Step 1 Model each polynomial.

$3x^2 - 2x + 1 \rightarrow$ x^2 x^2 x^2 ($3x^2$) + $-x$ $-x$ ($-2x$) + 1 (1)

$x^2 + 4x - 3 \rightarrow$ x^2 (x^2) + x x x x ($4x$) + -1 -1 -1 (-3)

Step 2 Combine like terms and remove zero pairs.

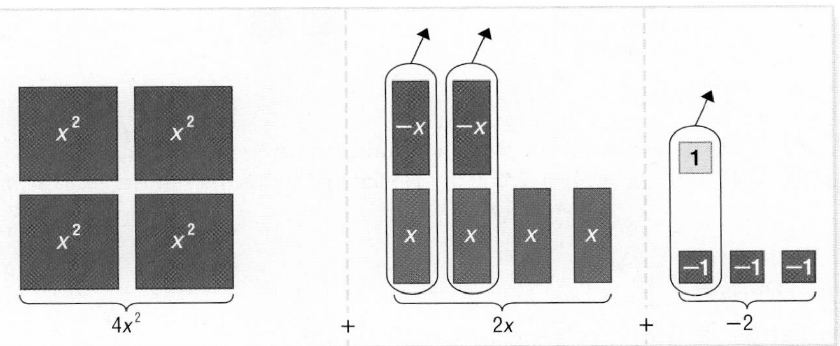

Step 3 Write the polynomial for the tiles that remain.
$(3x^2 - 2x + 1) + (x^2 + 4x - 3) = 4x^2 + 2x - 2$

Algebra Activity

Activity 2 **Use algebra tiles to find $(5x + 4) - (-2x + 3)$.**

Step 1 Model the polynomial $5x + 4$.

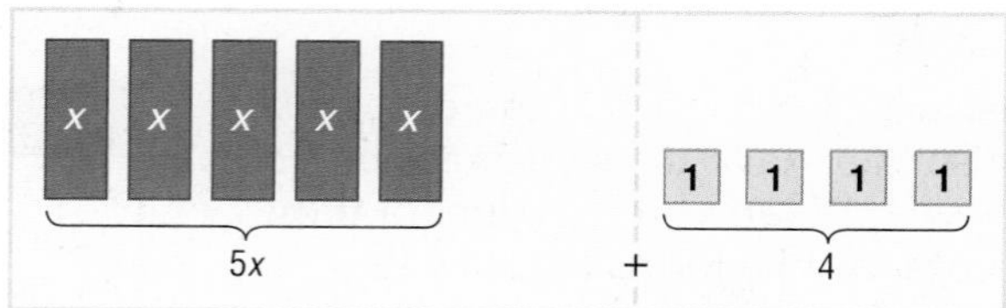

Step 2 To subtract $-2x + 3$, you must remove 2 red $-x$ tiles and 3 yellow 1 tiles. You can remove the yellow 1 tiles, but there are no red $-x$ tiles. Add 2 zero pairs of x tiles. Then remove the 2 red $-x$ tiles.

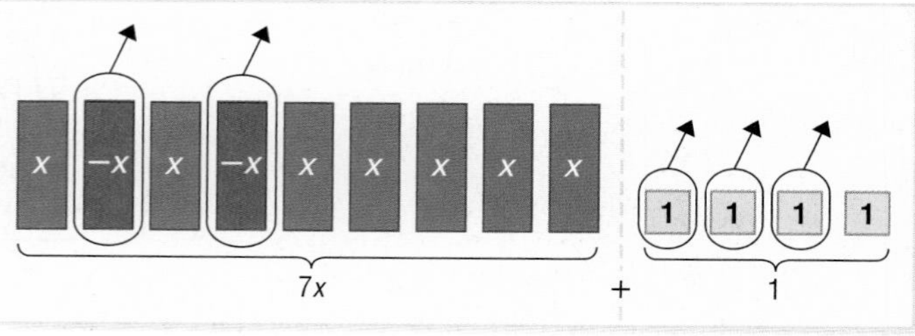

Step 3 Write the polynomial for the tiles that remain.
$(5x + 4) - (-2x + 3) = 7x + 1$

Recall that you can subtract a number by adding its additive inverse or opposite. Similarly, you can subtract a polynomial by adding its opposite.

Activity 3 **Use algebra tiles and the additive inverse, or opposite, to find $(5x + 4) - (-2x + 3)$.**

Step 1 To find the difference of $5x + 4$ and $-2x + 3$, add $5x + 4$ and the opposite of $-2x + 3$.

$5x + 4 \rightarrow$

The opposite of $-2x + 3$ is $2x - 3$. $\rightarrow$

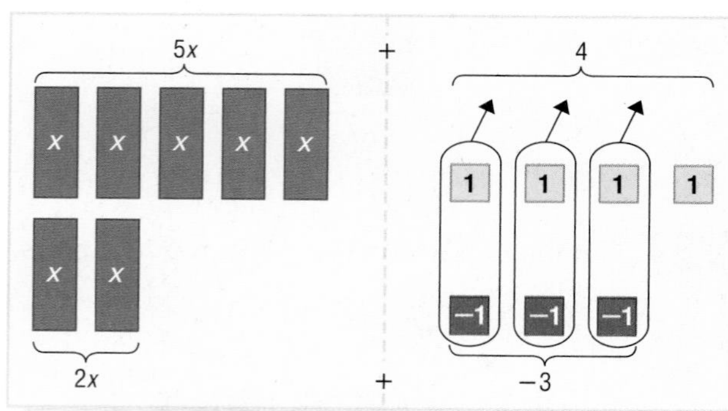

Step 2 Write the polynomial for the tiles that remain.
$(5x + 4) - (-2x + 3) = 7x + 1$ Notice that this is the same answer as in Activity 2.

Model and Analyze

Use algebra tiles to find each sum or difference.

1. $(5x^2 + 3x - 4) + (2x^2 - 4x + 1)$
2. $(2x^2 + 5) + (3x^2 - 2x + 6)$
3. $(-4x^2 + x) + (5x - 2)$
4. $(3x^2 + 4x + 2) - (x^2 - 5x - 5)$
5. $(-x^2 + 7x) - (2x^2 + 3x)$
6. $(8x + 4) - (6x^2 + x - 3)$
7. Find $(2x^2 - 3x + 1) - (2x + 3)$ using each method from Activity 2 and Activity 3. Illustrate with drawings and explain in writing how zero pairs are used in each case.

8-5 Adding and Subtracting Polynomials

What You'll Learn

- Add polynomials.
- Subtract polynomials.

How can adding polynomials help you model sales?

From 1996 to 1999, the amount of sales (in billions of dollars) of video games V and traditional toys R in the United States can be modeled by the following equations, where t is the number of years since 1996.
Source: *Toy Industry Fact Book*

$$V = -0.05t^3 + 0.05t^2 + 1.4t + 3.6$$
$$R = 0.5t^3 - 1.9t^2 + 3t + 19$$

The total toy sales T is the sum of the video game sales V and traditional toy sales R.

ADD POLYNOMIALS To add polynomials, you can group like terms horizontally or write them in column form, aligning like terms.

Example 1 Add Polynomials

Find $(3x^2 - 4x + 8) + (2x - 7x^2 - 5)$.

Method 1 Horizontal

Group like terms together.

$(3x^2 - 4x + 8) + (2x - 7x^2 - 5)$

$= [3x^2 + (-7x^2)] + (-4x + 2x) + [8 + (-5)]$ Associative and Commutative Properties

$= -4x^2 - 2x + 3$ Add like terms.

Method 2 Vertical

Align the like terms in columns and add.

$$\begin{array}{r} 3x^2 - 4x + 8 \\ (+)\ -7x^2 + 2x - 5 \\ \hline -4x^2 - 2x + 3 \end{array}$$

Notice that terms are in descending order with like terms aligned.

Study Tip

Adding Columns
When adding like terms in column form, remember that you are adding integers. Rewrite each monomial to eliminate subtractions. For example, you could rewrite $3x^2 - 4x + 8$ as $3x^2 + (-4x) + 8$.

SUBTRACT POLYNOMIALS Recall that you can subtract a rational number by adding its opposite or additive inverse. Similarly, you can subtract a polynomial by adding its additive inverse.

Polynomial	Additive Inverse
$-5m + 3n$	$5m - 3n$
$2y^2 - 6y + 11$	$-2y^2 + 6y - 11$
$7a + 9b - 4$	$-7a - 9b + 4$

To find the additive inverse of a polynomial, replace each term with its additive inverse or opposite.

Example 2 Subtract Polynomials

Find $(3n^2 + 13n^3 + 5n) - (7n + 4n^3)$.

Method 1 Horizontal

Subtract $7n + 4n^3$ by adding its additive inverse.

$(3n^2 + 13n^3 + 5n) - (7n + 4n^3)$

$= (3n^2 + 13n^3 + 5n) + (-7n - 4n^3)$ The additive inverse of $7n + 4n^3$ is $-7n - 4n^3$.

$= 3n^2 + [13n^3 + (-4n^3)] + [5n + (-7n)]$ Group like terms.

$= 3n^2 + 9n^3 - 2n$ Add like terms.

Study Tip

Inverse of a Polynomial
When finding the additive inverse of a polynomial, remember to find the additive inverse of *every* term.

Method 2 Vertical

Align like terms in columns and subtract by adding the additive inverse.

$$\begin{array}{rr} & 3n^2 + 13n^3 + 5n \\ (-) & 4n^3 + 7n \\ \hline \end{array} \quad \text{Add the opposite.} \quad \begin{array}{rr} & 3n^2 + 13n^3 + 5n \\ (+) & -4n^3 - 7n \\ \hline & 3n^2 + \ \ 9n^3 - 2n \end{array}$$

Thus, $(3n^2 + 13n^3 + 5n) - (7n + 4n^3) = 3n^2 + 9n^3 - 2n$ or, arranged in descending order, $9n^3 + 3n^2 - 2n$.

When polynomials are used to model real-world data, their sums and differences can have real-world meaning too.

Career Choices

Teacher

The educational requirements for a teaching license vary by state. In 1999, the average public K–12 teacher salary was $40,582.

Online Research For information about a career as a teacher, visit: www.algebra1.com/careers

Example 3 Subtract Polynomials

EDUCATION The total number of public school teachers T consists of two groups, elementary E and secondary S. From 1985 through 1998, the number (in thousands) of secondary teachers and total teachers in the United States could be modeled by the following equations, where n is the number of years since 1985.

$$S = 11n + 942$$
$$T = 44n + 2216$$

a. Find an equation that models the number of elementary teachers E for this time period.

You can find a model for E by subtracting the polynomial for S from the polynomial for T.

$$\begin{array}{lr} \text{Total} & 44n + 2216 \\ \text{− Secondary} & (-)\ 11n + \ \ 942 \\ \hline \text{Elementary} & \end{array} \quad \text{Add the opposite.} \quad \begin{array}{r} 44n + 2216 \\ (+)\ -11n - \ \ 942 \\ \hline 33n + 1274 \end{array}$$

An equation is $E = 33n + 1274$.

b. Use the equation to predict the number of elementary teachers in the year 2010.

The year 2010 is 2010 − 1985 or 25 years after the year 1985.

If this trend continues, the number of elementary teachers in 2010 would be 33(25) + 1274 thousand or about 2,099,000.

Check for Understanding

Concept Check

1. **Explain** why $5xy^2$ and $3x^2y$ are *not* like terms.
2. **OPEN ENDED** Write two polynomials whose difference is $2x^2 + x + 3$.
3. **FIND THE ERROR** Esteban and Kendra are finding $(5a - 6b) - (2a + 5b)$.

Esteban	Kendra
$(5a - 6b) - (2a + 5b)$	$(5a - 6b) - (2a + 5b)$
$= (-5a + 6b) + (-2a - 5b)$	$= (5a - 6b) + (-2a - 5b)$
$= -7a + b$	$= 3a - 11b$

Who is correct? Explain your reasoning.

Guided Practice

Find each sum or difference.

4. $(4p^2 + 5p) + (-2p^2 + p)$
5. $(5y^2 - 3y + 8) + (4y^2 - 9)$
6. $(8cd - 3d + 4c) + (-6 + 2cd)$
7. $(6a^2 + 7a - 9) - (-5a^2 + a - 10)$
8. $(g^3 - 2g^2 + 5g + 6) - (g^2 + 2g)$
9. $(3ax^2 - 5x - 3a) - (6a - 8a^2x + 4x)$

Application

POPULATION For Exercises 10 and 11, use the following information.
From 1990 through 1999, the female population F and the male population M of the United States (in thousands) are modeled by the following equations, where n is the number of years since 1990. **Source:** U.S. Census Bureau

$$F = 1247n + 126{,}971 \qquad M = 1252n + 120{,}741$$

10. Find an equation that models the total population T in thousands of the United States for this time period.
11. If this trend continues, what will the population of the United States be in 2010?

Practice and Apply

Homework Help

For Exercises	See Examples
12–31	1, 2
32, 33	3

Extra Practice
See page 838.

Find each sum or difference.

12. $(6n^2 - 4) + (-2n^2 + 9)$
13. $(9z - 3z^2) + (4z - 7z^2)$
14. $(3 + a^2 + 2a) + (a^2 - 8a + 5)$
15. $(-3n^2 - 8 + 2n) + (5n + 13 + n^2)$
16. $(x + 5) + (2y + 4x - 2)$
17. $(2b^3 - 4b + b^2) + (-9b^2 + 3b^3)$
18. $(11 + 4d^2) - (3 - 6d^2)$
19. $(4g^3 - 5g) - (2g^3 + 4g)$
20. $(-4y^3 - y + 10) - (4y^3 + 3y^2 - 7)$
21. $(4x + 5xy + 3y) - (3y + 6x + 8xy)$
22. $(3x^2 + 8x + 4) - (5x^2 - 4)$
23. $(5ab^2 + 3ab) - (2ab^2 + 4 - 8ab)$
24. $(x^3 - 7x + 4x^2 - 2) - (2x^2 - 9x + 4)$
25. $(5x^2 + 3a^2 - 5x) - (2x^2 - 5ax + 7x)$
26. $(3a + 2b - 7c) + (6b - 4a + 9c) + (-7c - 3a - 2b)$
27. $(5x^2 - 3) + (x^2 - x + 11) + (2x^2 - 5x + 7)$
28. $(3y^2 - 8) + (5y + 9) - (y^2 + 6y - 4)$
29. $(9x^3 + 3x - 13) - (6x^2 - 5x) + (2x^3 - x^2 - 8x + 4)$

GEOMETRY The measures of two sides of a triangle are given. If P is the perimeter, find the measure of the third side.

30. $P = 7x + 3y$

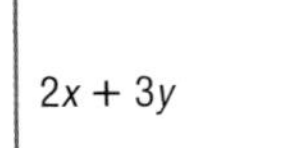

31. $P = 10x^2 - 5x + 16$

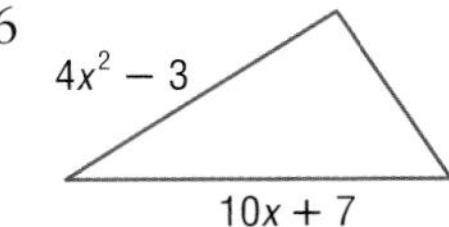

MOVIES **For Exercises 32 and 33, use the following information.**
From 1990 to 1999, the number of indoor movie screens I and total movie screens T in the U.S. could be modeled by the following equations, where n is the number of years since 1990.

$$I = 161.6n^2 - 20n + 23{,}326 \qquad T = 160.3n^2 - 26n + 24{,}226$$

32. Find an equation that models the number of outdoor movie screens D in the U.S. for this time period.
33. If this trend continues, how many outdoor movie screens will there be in the year 2010?

More About. . .

Movies

In 1998, attendance at movie theaters was at its highest point in 40 years with 1.48 billion tickets sold for a record $6.95 billion in gross income.

Source: The National Association of Theatre Owners

NUMBER TRICK **For Exercises 34 and 35, use the following information.**
Think of a two-digit number whose ones digit is greater than its tens digit. Multiply the difference of the two digits by 9 and add the result to your original number. Repeat this process for several other such numbers.

34. What observation can you make about your results?
35. Justify that your observation holds for all such two-digit numbers by letting x equal the tens digit and y equal the ones digit of the original number. (*Hint*: The original number is then represented by $10x + y$.)

POSTAL SERVICE **For Exercises 36–40, use the information below and in the figure at the right.** The U.S. Postal Service restricts the sizes of boxes shipped by parcel post. The sum of the length and the girth of the box must not exceed 108 inches.

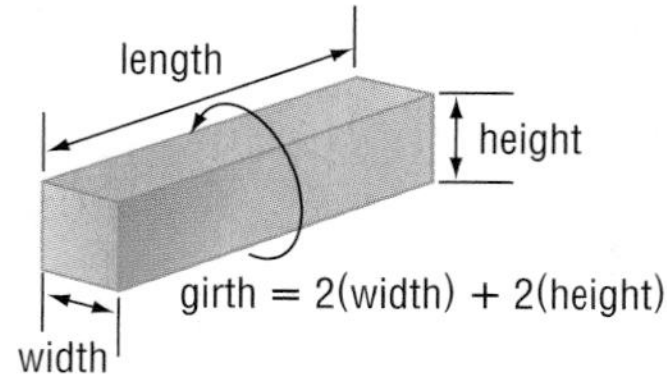

Suppose you want to make an open box using a 60-by-40 inch piece of cardboard by cutting squares out of each corner and folding up the flaps. The lid will be made from another piece of cardboard. You do not know how big the squares should be, so for now call the length of the side of each square x.

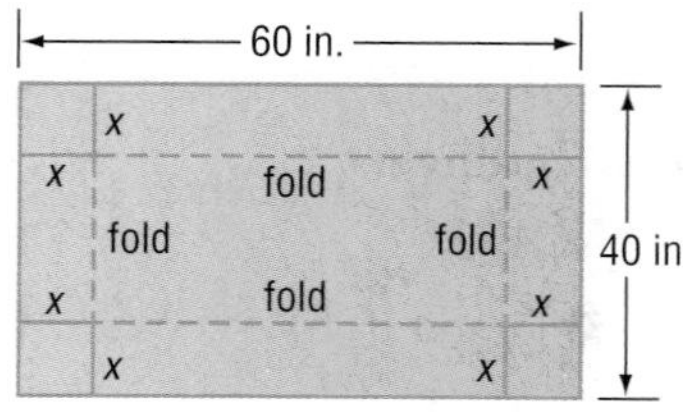

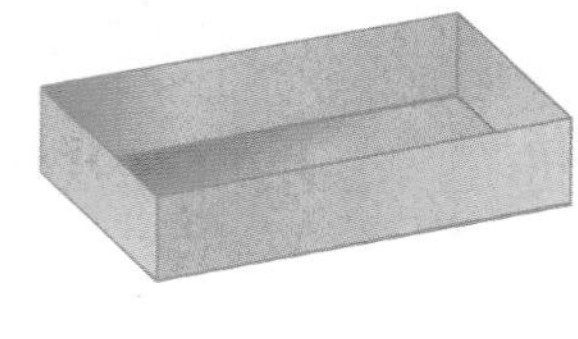

36. Write a polynomial to represent the length of the box formed.
37. Write a polynomial to represent the width of the box formed.
38. Write a polynomial to represent the girth of the box formed.
39. Write and solve an inequality to find the least possible value of x you could use in designing this box so it meets postal regulations.
40. What is the greatest integral value of x you could use to design this box if it does not have to meet regulations?

CRITICAL THINKING **For Exercises 41–43, suppose x is an integer.**

41. Write an expression for the next integer greater than x.
42. Show that the sum of two consecutive integers, x and the next integer after x, is always odd. (*Hint*: A number is considered even if it is divisible by 2.)
43. What is the least number of consecutive integers that must be added together to always arrive at an even integer?

44. WRITING IN MATH Answer the question that was posed at the beginning of the lesson.

How can adding polynomials help you model sales?

Include the following in your answer:

- an equation that models total toy sales, and
- an example of how and why someone might use this equation.

Standardized Test Practice

45. The perimeter of the rectangle shown at the right is $16a + 2b$. Which of the following expressions represents the length of the rectangle?

(A) $3a + 2b$ (B) $10a + 2b$

(C) $2a - 3b$ (D) $6a + 4b$

46. If $a^2 - 2ab + b^2 = 36$ and $a^2 - 3ab + b^2 = 22$, find ab.

(A) 6 (B) 8 (C) 12 (D) 14

Maintain Your Skills

Mixed Review

Find the degree of each polynomial. *(Lesson 8-4)*

47. $15t^3y^2$ 48. 24 49. $m^2 + n^3$ 50. $4x^2y^3z - 5x^3z$

Express each number in standard notation. *(Lesson 8-3)*

51. 8×10^6 52. 2.9×10^5 53. 5×10^{-4} 54. 4.8×10^{-7}

KEYBOARDING **For Exercises 55–59, use the table below that shows the keyboarding speeds and experience of 12 students.** *(Lesson 5-2)*

Experience (weeks)	4	7	8	1	6	3	5	2	9	6	7	10
Keyboarding Speed (wpm)	33	45	46	20	40	30	38	22	52	44	42	55

55. Make a scatter plot of these data.
56. Draw a best-fit line for the data.
57. Find the equation of the line.
58. Use the equation to predict the keyboarding speed of a student after a 12-week course.
59. Can this equation be used to predict the speed for any number of weeks of experience? Explain.

State the domain and range of each relation. *(Lesson 4-3)*

60. $\{(-2, 5), (0, -2), (-6, 3)\}$ 61. $\{(-4, 2), (-1, -3), (5, 0), (-4, 1)\}$

62. MODEL TRAINS One of the most popular sizes of model trains is called the HO. Every dimension of the HO model measures $\frac{1}{87}$ times that of a real engine. The HO model of a modern diesel locomotive is about 8 inches long. About how many feet long is the real locomotive? *(Lesson 3-6)*

Getting Ready for the Next Lesson

PREREQUISITE SKILL **Simplify.** *(To review the **Distributive Property**, see Lesson 1-7.)*

63. $6(3x - 8)$ 64. $-2(b + 9)$ 65. $-7(-5p + 4q)$

66. $9(3a + 5b - c)$ 67. $8(x^2 + 3x - 4)$ 68. $-3(2a^2 - 5a + 7)$

8-6 Multiplying a Polynomial by a Monomial

What You'll Learn

- Find the product of a monomial and a polynomial.
- Solve equations involving polynomials.

How is finding the product of a monomial and a polynomial related to finding the area of a rectangle?

The algebra tiles shown are grouped together to form a rectangle with a width of $2x$ and a length of $x + 3$. Notice that the rectangle consists of 2 blue x^2 tiles and 6 green x tiles. The area of the rectangle is the sum of these algebra tiles or $2x^2 + 6x$.

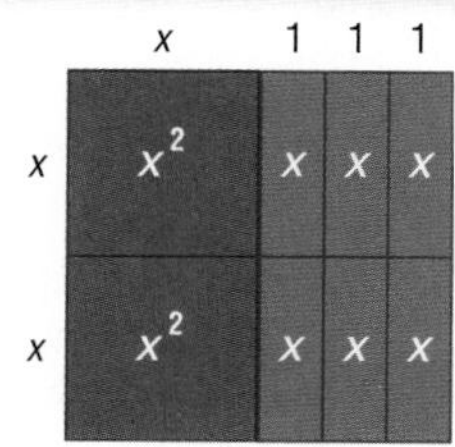

PRODUCT OF MONOMIAL AND POLYNOMIAL The Distributive Property can be used to multiply a polynomial by a monomial.

Study Tip

Look Back
To review the **Distributive Property**, see Lesson 1-5.

Example 1 Multiply a Polynomial by a Monomial

Find $-2x^2(3x^2 - 7x + 10)$.

Method 1 Horizontal

$-2x^2(3x^2 - 7x + 10)$

$= -2x^2(3x^2) - (-2x^2)(7x) + (-2x^2)(10)$ Distributive Property

$= -6x^4 - (-14x^3) + (-20x^2)$ Multiply.

$= -6x^4 + 14x^3 - 20x^2$ Simplify.

Method 2 Vertical

$$\begin{array}{r} 3x^2 - 7x + 10 \\ (\times) \qquad\qquad -2x^2 \\ \hline -6x^4 + 14x^3 - 20x^2 \end{array}$$

Distributive Property

Multiply.

When expressions contain like terms, simplify by combining the like terms.

Example 2 Simplify Expressions

Simplify $4(3d^2 + 5d) - d(d^2 - 7d + 12)$.

$4(3d^2 + 5d) - d(d^2 - 7d + 12)$

$= 4(3d^2) + 4(5d) + (-d)(d^2) - (-d)(7d) + (-d)(12)$ Distributive Property

$= 12d^2 + 20d + (-d^3) - (-7d^2) + (-12d)$ Product of Powers

$= 12d^2 + 20d - d^3 + 7d^2 - 12d$ Simplify.

$= -d^3 + (12d^2 + 7d^2) + (20d - 12d)$ Commutative and Associative Properties

$= -d^3 + 19d^2 + 8d$ Combine like terms.

Example 3 Use Polynomial Models

PHONE SERVICE Greg pays a fee of \$20 a month for local calls. Long-distance rates are 6¢ per minute for in-state calls and 5¢ per minute for out-of-state calls. Suppose Greg makes 300 minutes of long-distance phone calls in January and m of those minutes are for in-state calls.

a. Find an expression for Greg's phone bill for January.

Words The bill is the sum of the monthly fee, in-state charges, and the out-of-state charges.

Variables If m = number of minutes of in-state calls, then $300 - m$ = number of minutes of out-of-state calls. Let B = phone bill for the month of January.

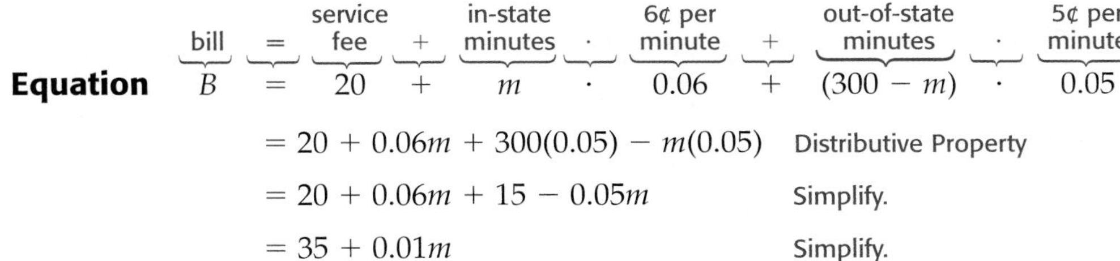

	bill	=	service fee	+	in-state minutes	·	6¢ per minute	+	out-of-state minutes	·	5¢ per minute
Equation	B	=	20	+	m	·	0.06	+	$(300 - m)$	·	0.05

$= 20 + 0.06m + 300(0.05) - m(0.05)$ Distributive Property

$= 20 + 0.06m + 15 - 0.05m$ Simplify.

$= 35 + 0.01m$ Simplify.

An expression for Greg's phone bill for January is $35 + 0.01m$, where m is the number of minutes of in-state calls.

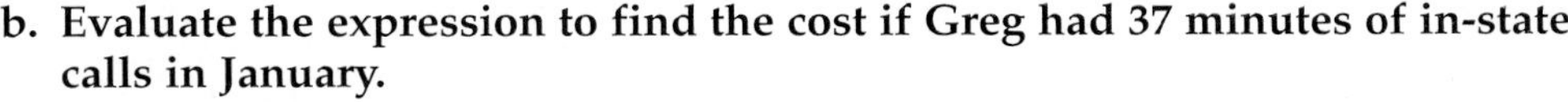

b. Evaluate the expression to find the cost if Greg had 37 minutes of in-state calls in January.

$35 + 0.01m = 35 + 0.01(37)$ $m = 37$

$= 35 + 0.37$ Multiply.

$= \$35.37$ Add.

Greg's bill was \$35.37.

More About. . .

Phone Service About 98% of long-distance companies service their calls using the network of one of three companies. Since the quality of phone service is basically the same, a company's rates are the primary factor in choosing a long-distance provider.

Source: Chamberland Enterprises

SOLVE EQUATIONS WITH POLYNOMIAL EXPRESSIONS Many equations contain polynomials that must be added, subtracted, or multiplied before the equation can be solved.

Example 4 Polynomials on Both Sides

Solve $y(y - 12) + y(y + 2) + 25 = 2y(y + 5) - 15$.

$y(y - 12) + y(y + 2) + 25 = 2y(y + 5) - 15$ Original equation

$y^2 - 12y + y^2 + 2y + 25 = 2y^2 + 10y - 15$ Distributive Property

$2y^2 - 10y + 25 = 2y^2 + 10y - 15$ Combine like terms.

$-10y + 25 = 10y - 15$ Subtract $2y^2$ from each side.

$-20y + 25 = -15$ Subtract $10y$ from each side.

$-20y = -40$ Subtract 25 from each side.

$y = 2$ Divide each side by -20.

The solution is 2.

CHECK $y(y - 12) + y(y + 2) + 25 = 2y(y + 5) - 15$ Original equation

$2(2 - 12) + 2(2 + 2) + 25 \stackrel{?}{=} 2(2)(2 + 5) - 15$ $y = 2$

$2(-10) + 2(4) + 25 \stackrel{?}{=} 4(7) - 15$ Simplify.

$-20 + 8 + 25 \stackrel{?}{=} 28 - 15$ Multiply.

$13 = 13 \checkmark$ Add and subtract.

Check for Understanding

Concept Check

1. **State** the property used in each step to multiply $2x(4x^2 + 3x - 5)$.

$$2x(4x^2 + 3x - 5) = 2x(4x^2) + 2x(3x) - 2x(5) \quad \underline{\ ?\ }$$
$$= 8x^{1+2} + 6x^{1+1} - 10x \quad \underline{\ ?\ }$$
$$= 8x^3 + 6x^2 - 10x \quad \text{Simplify.}$$

2. **Compare and contrast** the procedure used to multiply a trinomial by a monomial using the vertical method with the procedure used to multiply a three-digit number by a two-digit number.
3. **OPEN ENDED** Write a monomial and a trinomial involving a single variable. Then find their product.

Guided Practice

Find each product.

4. $-3y(5y + 2)$
5. $9b^2(2b^3 - 3b^2 + b - 8)$
6. $2x(4a^4 - 3ax + 6x^2)$
7. $-4xy(5x^2 - 12xy + 7y^2)$

Simplify.

8. $t(5t - 9) - 2t$
9. $5n(4n^3 + 6n^2 - 2n + 3) - 4(n^2 + 7n)$

Solve each equation.

10. $-2(w + 1) + w = 7 - 4w$
11. $x(x + 2) - 3x = x(x - 4) + 5$

Application

SAVINGS For Exercises 12–14, use the following information.
Kenzie's grandmother left her \$10,000 for college. Kenzie puts some of the money into a savings account earning 4% per year, and with the rest, she buys a certificate of deposit (CD) earning 7% per year.

12. If Kenzie puts x dollars into the savings account, write an expression to represent the amount of the CD.
13. Write an equation for the total amount of money T Kenzie will have saved for college after one year.
14. If Kenzie puts \$3000 in savings, how much money will she have after one year?

Practice and Apply

Homework Help

For Exercises	See Examples
15–28	1
29–38	2
39–48	4
49–54, 58–62	3

Extra Practice

See page 838.

Find each product.

15. $r(5r + r^2)$
16. $w(2w^3 - 9w^2)$
17. $-4x(8 + 3x)$
18. $5y(-2y^2 - 7y)$
19. $7ag(g^3 + 2ag)$
20. $-3np(n^2 - 2p)$
21. $-2b^2(3b^2 - 4b + 9)$
22. $6x^3(5 + 3x - 11x^2)$
23. $8x^2y(5x + 2y^2 - 3)$
24. $-cd^2(3d + 2c^2d - 4c)$
25. $-\frac{3}{4}hk^2(20k^2 + 5h - 8)$
26. $\frac{2}{3}a^2b(6a^3 - 4ab + 9b^2)$
27. $-5a^3b(2b + 5ab - b^2 + a^3)$
28. $4p^2q^2(2p^2 - q^2 + 9p^3 + 3q)$

Simplify.

29. $d(-2d + 4) + 15d$
30. $-x(4x^2 - 2x) - 5x^3$
31. $3w(6w - 4) + 2(w^2 - 3w + 5)$
32. $5n(2n^3 + n^2 + 8) + n(4 - n)$
33. $10(4m^3 - 3m + 2) - 2m(-3m^2 - 7m + 1)$
34. $4y(y^2 - 8y + 6) - 3(2y^3 - 5y^2 + 2)$
35. $-3c^2(2c + 7) + 4c(3c^2 - c + 5) + 2(c^2 - 4)$
36. $4x^2(x + 2) + 3x(5x^2 + 2x - 6) - 5(3x^2 - 4x)$

GEOMETRY **Find the area of each shaded region in simplest form.**

37.

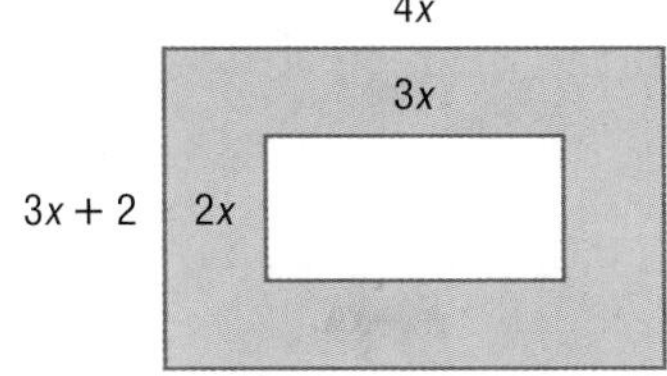

38.

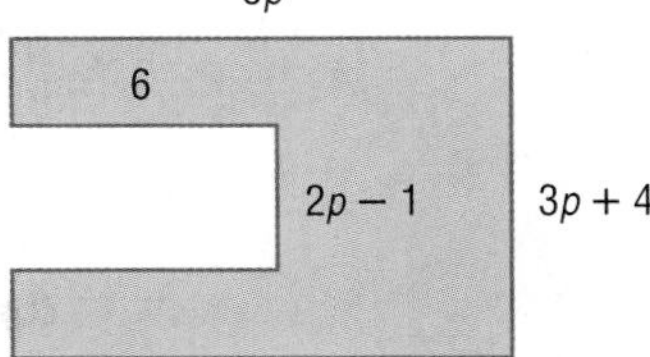

Solve each equation.

39. $2(4x - 7) = 5(-2x - 9) - 5$

40. $2(5a - 12) = -6(2a - 3) + 2$

41. $4(3p + 9) - 5 = -3(12p - 5)$

42. $7(8w - 3) + 13 = 2(6w + 7)$

43. $d(d - 1) + 4d = d(d - 8)$

44. $c(c + 3) - c(c - 4) = 9c - 16$

45. $y(y + 12) - 8y = 14 + y(y - 4)$

46. $k(k - 7) + 10 = 2k + k(k + 6)$

47. $2n(n + 4) + 18 = n(n + 5) + n(n - 2) - 7$

48. $3g(g - 4) - 2g(g - 7) = g(g + 6) - 28$

More About. . .

Class Trip

Inside the Lincoln Memorial is a 19-foot marble statue of the United States' 16th president. The statue is flanked on either side by the inscriptions of Lincoln's Second Inaugural Address and Gettysburg Address.

Source: www.washington.org

SAVINGS **For Exercises 49 and 50, use the following information.**
Marta has \$6000 to invest. She puts x dollars of this money into a savings account that earns 3% per year, and with the rest, she buys a certificate of deposit that earns 6% per year.

49. Write an equation for the total amount of money T Marta will have after one year.

50. Suppose at the end of one year, Marta has a total of \$6315. How much money did Marta invest in each account?

51. **GARDENING** A gardener plants corn in a garden with a length-to-width ratio of 5:4. Next year, he plans to increase the garden's area by increasing its length by 12 feet. Write an expression for this new area.

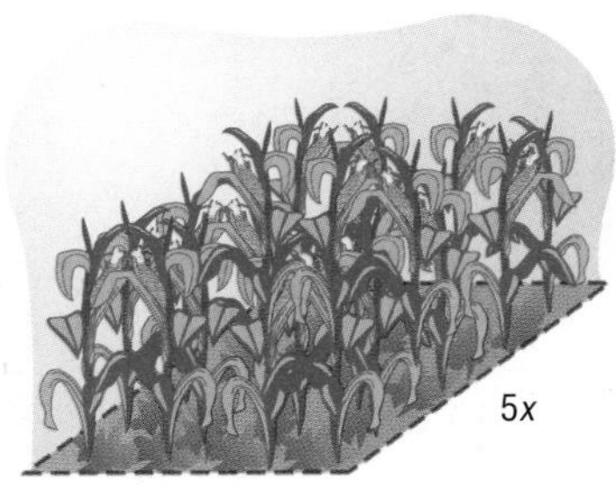

52. **CLASS TRIP** Mr. Smith's American History class will take taxis from their hotel in Washington, D.C., to the Lincoln Memorial. The fare is \$2.75 for the first mile and \$1.25 for each additional mile. If the distance is m miles and t taxis are needed, write an expression for the cost to transport the group.

NUMBER THEORY **For Exercises 53 and 54, let x be an odd integer.**

53. Write an expression for the next odd integer.

54. Find the product of x and the next odd integer.

CRITICAL THINKING **For Exercises 55–57, use the following information.**
An even number can be represented by $2x$, where x is any integer.

55. Show that the product of two even integers is always even.

56. Write a representation for an odd integer.

57. Show that the product of an even and an odd integer is always even.

VOLUNTEERING **For Exercises 58 and 59, use the following information.** Laura is making baskets of apples and oranges for homeless shelters. She wants to place a total of 10 pieces of fruit in each basket. Apples cost 25¢ each, and oranges cost 20¢ each.

58. If a represents the number of apples Laura uses, write a polynomial model in simplest form for the total amount of money T Laura will spend on the fruit for each basket.

59. If Laura uses 4 apples in each basket, find the total cost for fruit.

More About. . .

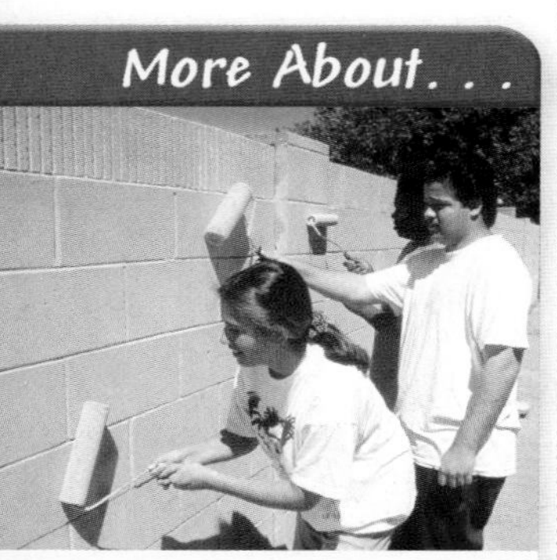

Volunteering

Approximately one third of young people in grades 7–12 suggested that "working for the good of my community and country" and "helping others or volunteering" were important future goals.

Source: Primedia/Roper National Youth Opinion Survey

SALES **For Exercises 60 and 61, use the following information.** A store advertises that all sports equipment is 30% off the retail price. In addition, the store asks customers to select and pop a balloon to receive a coupon for an additional n percent off the already marked down price of one of their purchases.

60. Write an expression for the cost of a pair of inline skates with retail price p after receiving both discounts.

61. Use this expression to calculate the cost, not including sales tax, of a \$200 pair of inline skates for an additional 10 percent off.

62. SPORTS You may have noticed that when runners race around a curved track, their starting points are staggered. This is so each contestant runs the same distance to the finish line.

If the radius of the inside lane is x and each lane is 2.5 feet wide, how far apart should the officials start the runners in the two inside lanes? (*Hint*: Circumference of a circle: $C = 2\pi r$, where r is the radius of the circle)

63. WRITING IN MATH Answer the question that was posed at the beginning of the lesson.

How is finding the product of a monomial and a polynomial related to finding the area of a rectangle?

Include the following in your answer:
- the product of $2x$ and $x + 3$ derived algebraically, and
- a representation of another product of a monomial and a polynomial using algebra tiles and multiplication.

64. Simplify $[(3x^2 - 2x + 4) - (x^2 + 5x - 2)](x + 2)$.

Ⓐ $2x^3 + 7x^2 + 8x + 4$ Ⓑ $2x^3 - 3x^2 - 8x + 12$
Ⓒ $4x^3 + 11x^2 + 8x + 4$ Ⓓ $-4x^3 - 11x^2 - 8x - 4$

65. A plumber charges \$70 for the first thirty minutes of each house call plus \$4 for each additional minute that she works. The plumber charges Ke-Min \$122 for her time. What amount of time, in minutes, did the plumber work?

Ⓐ 43 Ⓑ 48 Ⓒ 58 Ⓓ 64

Maintain Your Skills

Mixed Review

Find each sum or difference. *(Lesson 8-5)*

66. $(4x^2 + 5x) + (-7x^2 + x)$

67. $(3y^2 + 5y - 6) - (7y^2 - 9)$

68. $(5b - 7ab + 8a) - (5ab - 4a)$

69. $(6p^3 + 3p^2 - 7) + (p^3 - 6p^2 - 2p)$

State whether each expression is a polynomial. If the expression is a polynomial, identify it as a *monomial*, a *binomial*, or a *trinomial*. *(Lesson 8-4)*

70. $4x^2 - 10ab + 6$ **71.** $4c + ab - c$ **72.** $\frac{7}{y} + y^2$ **73.** $\frac{n^2}{3}$

Define a variable, write an inequality, and solve each problem. Then check your solution. *(Lesson 6-3)*

74. Six increased by ten times a number is less than nine times the number.

75. Nine times a number increased by four is no less than seven decreased by thirteen times the number.

Write an equation of the line that passes through each pair of points. *(Lesson 5-4)*

76. $(-3, -8), (1, 4)$ **77.** $(-4, 5), (2, -7)$ **78.** $(3, -1), (-3, 2)$

79. EXPENSES Kristen spent one fifth of her money on gasoline to fill up her car. Then she spent half of what was left for a haircut. She bought lunch for \$7. When she got home, she had \$13 left. How much money did Kristen have originally? *(Lesson 3-4)*

For Exercises 80 and 81, use each set of data to make a stem-and-leaf plot. *(Lesson 2-5)*

80. 49 51 55 62 47 32 56 57 48 47 33 68 53 45 30

81. 21 18 34 30 20 15 14 10 22 21 18 43 44 20 18

Getting Ready for the Next Lesson

PREREQUISITE SKILL **Simplify.** *(To review **products of powers**, see Lesson 8-1.)*

82. $(a)(a)$

83. $2x(3x^2)$

84. $-3y^2(8y^2)$

85. $4y(3y) - 4y(6)$

86. $-5n(2n^2) - (-5n)(8n) + (-5n)(4)$

87. $3p^2(6p^2) - 3p^2(8p) + 3p^2(12)$

Practice Quiz 2

Lessons 8-4 through 8-6

Find the degree of each polynomial. *(Lesson 8-4)*

1. $5x^4$ **2.** $-9n^3p^4$ **3.** $7a^2 - 2ab^2$ **4.** $-6 - 8x^2y^2 + 5y^3$

Arrange the terms of each polynomial so that the powers of x are in ascending order. *(Lesson 8-4)*

5. $4x^2 + 9x - 12 + 5x^3$

6. $2xy^4 + x^3y^5 + 5x^5y - 13x^2$

Find each sum or difference. *(Lesson 8-5)*

7. $(7n^2 - 4n + 10) + (3n^2 - 8)$

8. $(3g^3 - 5g) - (2g^3 + 5g^2 - 3g + 1)$

Find each product. *(Lesson 8-6)*

9. $5a^2(3a^3b - 2a^2b^2 + 6ab^3)$

10. $7x^2y(5x^2 - 3xy + y)$

Algebra Activity

A Preview of Lesson 8-7

Multiplying Polynomials

You can use algebra tiles to find the product of two binomials.

Activity 1 Use algebra tiles to find $(x + 2)(x + 5)$.

The rectangle will have a width of $x + 2$ and a length of $x + 5$. Use algebra tiles to mark off the dimensions on a product mat. Then complete the rectangle with algebra tiles.

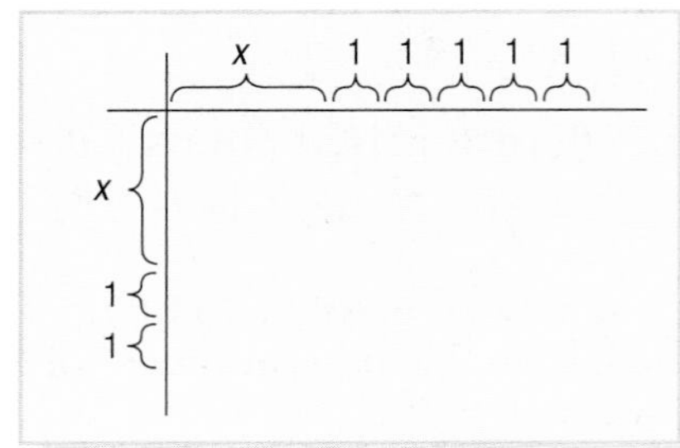

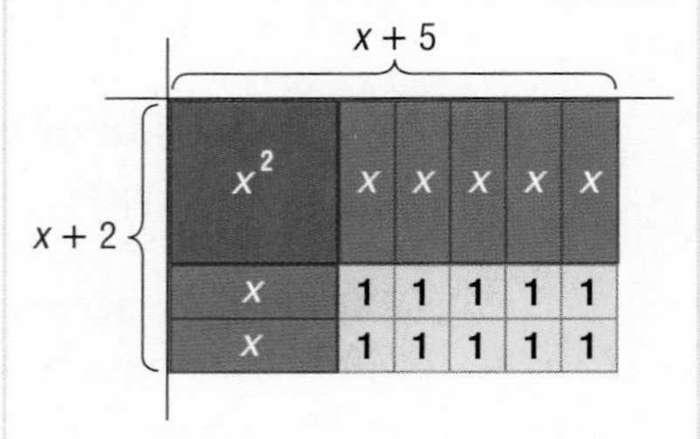

The rectangle consists of 1 blue x^2 tile, 7 green x tiles, and 10 yellow 1 tiles. The area of the rectangle is $x^2 + 7x + 10$. Therefore, $(x + 2)(x + 5) = x^2 + 7x + 10$.

Activity 2 Use algebra tiles to find $(x - 1)(x - 4)$.

Step 1 The rectangle will have a width of $x - 1$ and a length of $x - 4$. Use algebra tiles to mark off the dimensions on a product mat. Then begin to make the rectangle with algebra tiles.

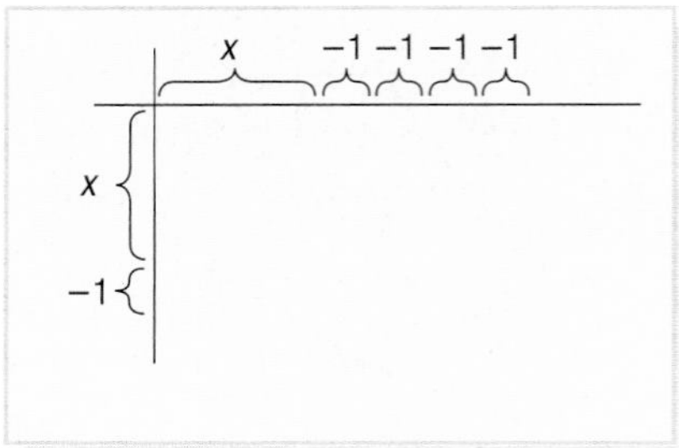

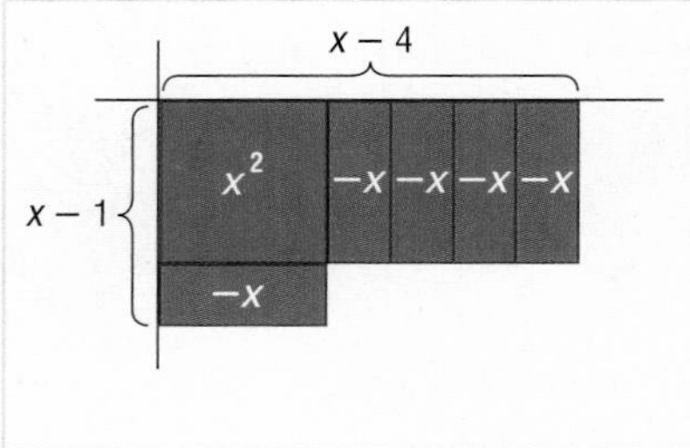

Step 2 Determine whether to use 4 yellow 1 tiles or 4 red -1 tiles to complete the rectangle. Remember that the numbers at the top and side give the dimensions of the tile needed. The area of each tile is the product of -1 and -1 or 1. This is represented by a yellow 1 tile. Fill in the space with 4 yellow 1 tiles to complete the rectangle.

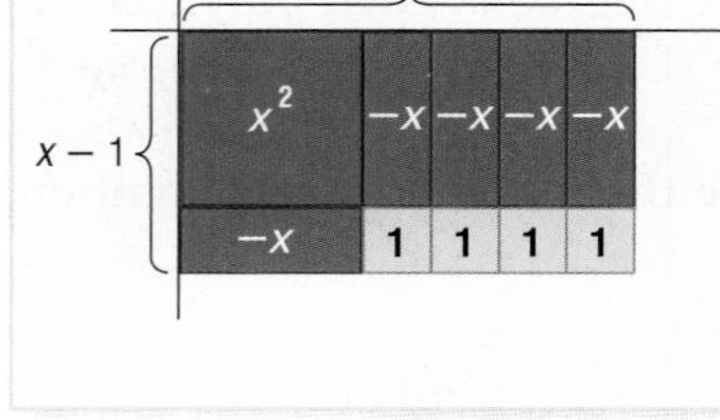

The rectangle consists of 1 blue x^2 tile, 5 red $-x$ tiles, and 4 yellow 1 tiles. The area of the rectangle is $x^2 - 5x + 4$. Therefore, $(x - 1)(x - 4) = x^2 - 5x + 4$.

Activity 3 Use algebra tiles to find $(x - 3)(2x + 1)$.

Step 1 The rectangle will have a width of $x - 3$ and a length of $2x + 1$. Mark off the dimensions on a product mat. Then begin to make the rectangle with algebra tiles.

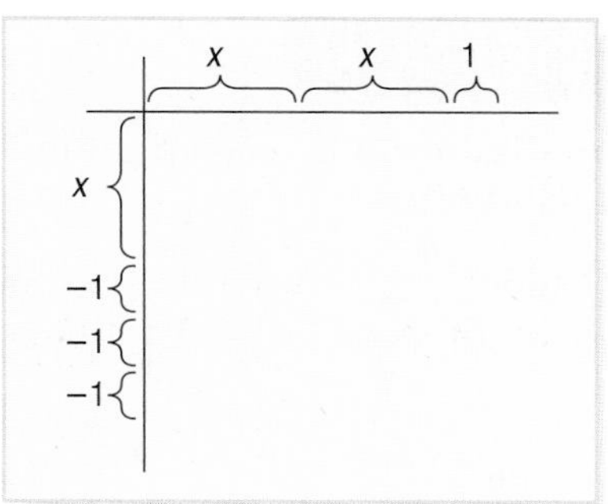

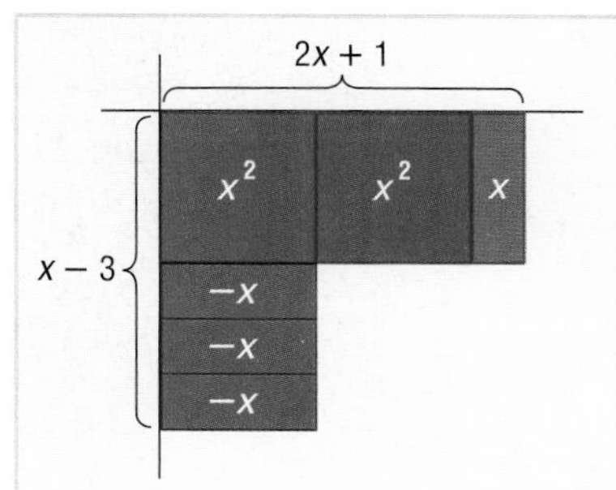

Step 2 Determine what color x tiles and what color 1 tiles to use to complete the rectangle. The area of each x tile is the product of x and -1. This is represented by a red $-x$ tile. The area of the 1 tile is represented by the product of 1 and -1 or -1. This is represented by a red -1 tile. Complete the rectangle with 3 red $-x$ tiles and 3 red -1 tiles.

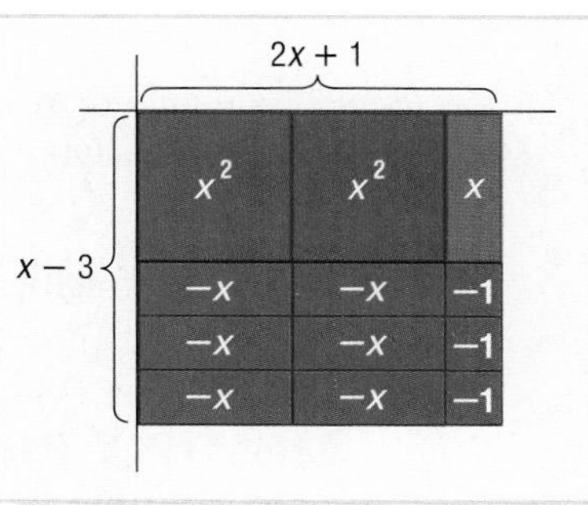

Step 3 Rearrange the tiles to simplify the polynomial you have formed. Notice that a zero pair is formed by one positive and one negative x tile.

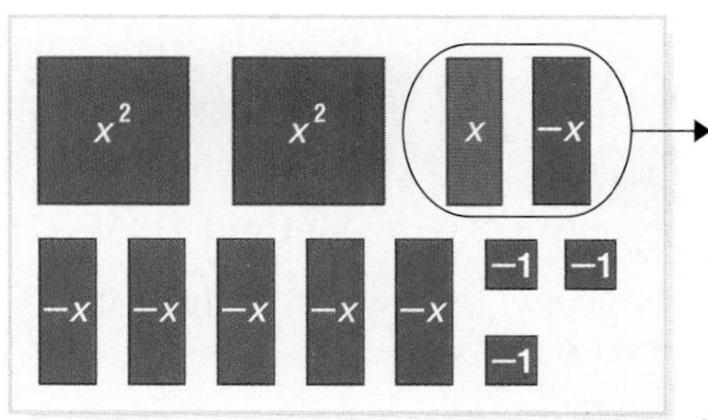

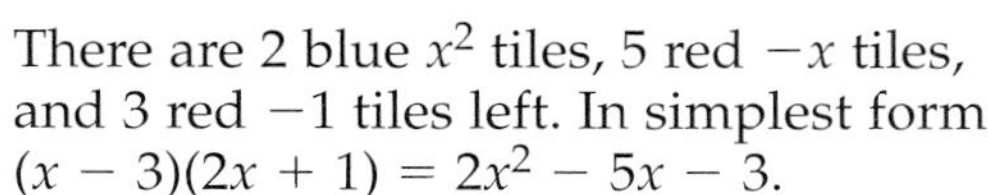

There are 2 blue x^2 tiles, 5 red $-x$ tiles, and 3 red -1 tiles left. In simplest form, $(x - 3)(2x + 1) = 2x^2 - 5x - 3$.

Model and Analyze

Use algebra tiles to find each product.

1. $(x + 2)(x + 3)$
2. $(x - 1)(x - 3)$
3. $(x + 1)(x - 2)$
4. $(x + 1)(2x + 1)$
5. $(x - 2)(2x - 3)$
6. $(x + 3)(2x - 4)$

7. You can also use the Distributive Property to find the product of two binomials. The figure at the right shows the model for $(x + 3)(x + 4)$ separated into four parts. Write a sentence or two explaining how this model shows the use of the Distributive Property.

8-7 Multiplying Polynomials

What You'll Learn

- Multiply two binomials by using the FOIL method.
- Multiply two polynomials by using the Distributive Property.

Vocabulary

- FOIL method

How is multiplying binomials similar to multiplying two-digit numbers?

To compute 24×36, we multiply each digit in 24 by each digit in 36, paying close attention to the place value of each digit.

Step 1 Multiply by the ones.	Step 2 Multiply by the tens.	Step 3 Add like place values.
$\begin{array}{r} 24 \\ \times\ 36 \\ \hline 144 \end{array}$	$\begin{array}{r} 24 \\ \times\ 36 \\ \hline 144 \\ 720 \\ \hline \end{array}$	$\begin{array}{r} 24 \\ \times\ 36 \\ \hline 144 \\ +\ 720 \\ \hline 864 \end{array}$
$6 \times 24 = 6(20 + 4)$ $= 120 + 24$ or 144	$30 \times 24 = 30(20 + 4)$ $= 600 + 120$ or 720	

You can multiply two binomials in a similar way.

MULTIPLY BINOMIALS To multiply two binomials, apply the Distributive Property twice as you do when multiplying two-digit numbers.

Study Tip

Look Back
To review the **Distributive Property**, see Lesson 1-7.

Example 1 The Distributive Property

Find $(x + 3)(x + 2)$.

Method 1 Vertical

Multiply by 2.	Multiply by x.	Add like terms.
$\begin{array}{r} x + 3 \\ (\times)\ x + 2 \\ \hline 2x + 6 \end{array}$	$\begin{array}{r} x + 3 \\ (\times)\ x + 2 \\ \hline 2x + 6 \\ x^2 + 3x \quad\ \ \\ \hline \end{array}$	$\begin{array}{r} x + 3 \\ (\times)\ x + 2 \\ \hline 2x + 6 \\ x^2 + 3x \quad\ \ \\ \hline x^2 + 5x + 6 \end{array}$
$2(x + 3) = 2x + 6$	$x(x + 3) = x^2 + 3x$	

Method 2 Horizontal

$$\begin{aligned} (x + 3)(x + 2) &= x(x + 2) + 3(x + 2) && \text{Distributive Property} \\ &= x(x) + x(2) + 3(x) + 3(2) && \text{Distributive Property} \\ &= x^2 + 2x + 3x + 6 && \text{Multiply.} \\ &= x^2 + 5x + 6 && \text{Combine like terms.} \end{aligned}$$

An alternative method for finding the product of two binomials can be shown using algebra tiles.

Consider the product of $x + 3$ and $x - 2$. The rectangle shown below has a length of $x + 3$ and a width of $x - 2$. Notice that this rectangle can be broken up into four smaller rectangles.

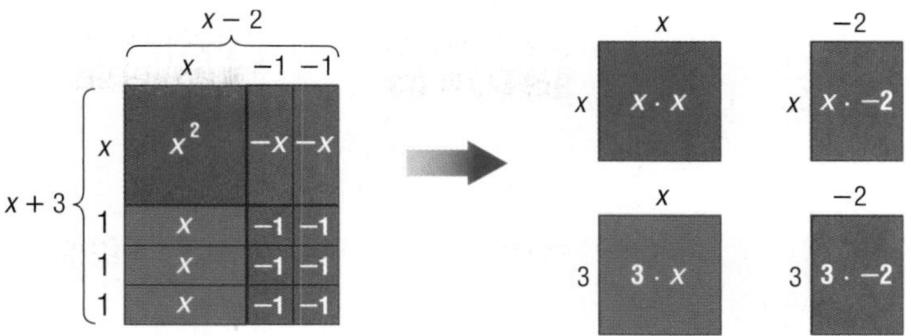

The product of $(x - 2)$ and $(x + 3)$ is the sum of these four areas.

$$(x + 3)(x - 2) = (x \cdot x) + (x \cdot -2) + (3 \cdot x) + (3 \cdot -2) \quad \text{Sum of the four areas}$$
$$= x^2 + (-2x) + 3x + (-6) \quad \text{Multiply.}$$
$$= x^2 + x - 6 \quad \text{Combine like terms.}$$

This example illustrates a shortcut of the Distributive Property called the **FOIL method**. You can use the FOIL method to multiply two binomials.

Key Concept — *FOIL Method for Multiplying Binomials*

- **Words** To multiply two binomials, find the sum of the products of
  - F the *First* terms,
  - O the *Outer* terms,
  - I the *Inner* terms, and
  - L the *Last* terms.
- **Example**

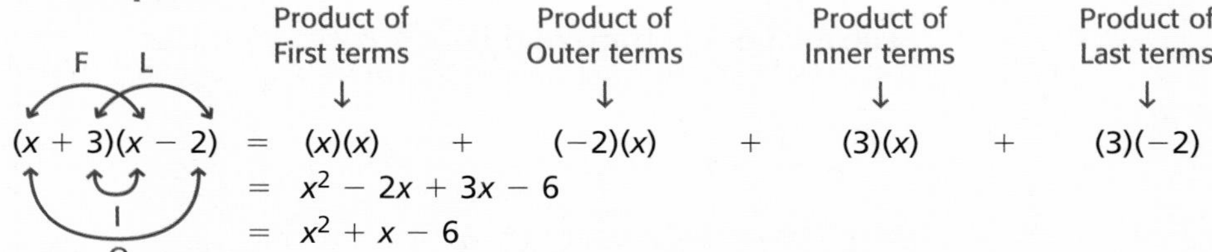

$$(x + 3)(x - 2) = (x)(x) + (-2)(x) + (3)(x) + (3)(-2)$$
$$= x^2 - 2x + 3x - 6$$
$$= x^2 + x - 6$$

Example 2 *FOIL Method*

Find each product.

a. $(x - 5)(x + 7)$

$$(x - 5)(x + 7) = \overset{F}{(x)(x)} + \overset{O}{(x)(7)} + \overset{I}{(-5)(x)} + \overset{L}{(-5)(7)} \quad \text{FOIL method}$$
$$= x^2 + 7x - 5x - 35 \quad \text{Multiply.}$$
$$= x^2 + 2x - 35 \quad \text{Combine like terms.}$$

b. $(2y + 3)(6y - 7)$

$(2y + 3)(6y - 7)$

$$= \overset{F}{(2y)(6y)} + \overset{O}{(2y)(-7)} + \overset{I}{(3)(6y)} + \overset{L}{(3)(-7)} \quad \text{FOIL method}$$
$$= 12y^2 - 14y + 18y - 21 \quad \text{Multiply.}$$
$$= 12y^2 + 4y - 21 \quad \text{Combine like terms.}$$

Study Tip

Checking Your Work
You can check your products in Examples 2a and 2b by reworking each problem using the Distributive Property.

Example 3 FOIL Method

GEOMETRY The area A of a trapezoid is one-half the height h times the sum of the bases, b_1 and b_2. Write an expression for the area of the trapezoid.

3x − 7

x + 2

2x + 1

Identify the height and bases.

$h = x + 2$

$b_1 = 3x - 7$

$b_2 = 2x + 1$

Now write and apply the formula.

Area	equals	one-half	height	times	sum of bases
A	$=$	$\frac{1}{2}$	$\cdot\ h$	$\cdot$	$(b_1 + b_2)$

$A = \frac{1}{2}h(b_1 + b_2)$ Original formula

$= \frac{1}{2}(x + 2)[(3x - 7) + (2x + 1)]$ Substitution

$= \frac{1}{2}(x + 2)(5x - 6)$ Add polynomials in the brackets.

$= \frac{1}{2}[x(5x) + x(-6) + 2(5x) + 2(-6)]$ FOIL method

$= \frac{1}{2}(5x^2 - 6x + 10x - 12)$ Multiply.

$= \frac{1}{2}(5x^2 + 4x - 12)$ Combine like terms.

$= \frac{5}{2}x^2 + 2x - 6$ Distributive Property

The area of the trapezoid is $\frac{5}{2}x^2 + 2x - 6$ square units.

MULTIPLY POLYNOMIALS The Distributive Property can be used to multiply any two polynomials.

Example 4 The Distributive Property

Find each product.

a. $(4x + 9)(2x^2 - 5x + 3)$

$(4x + 9)(2x^2 - 5x + 3)$

$= 4x(2x^2 - 5x + 3) + 9(2x^2 - 5x + 3)$ Distributive Property

$= 8x^3 - 20x^2 + 12x + 18x^2 - 45x + 27$ Distributive Property

$= 8x^3 - 2x^2 - 33x + 27$ Combine like terms.

b. $(y^2 - 2y + 5)(6y^2 - 3y + 1)$

$(y^2 - 2y + 5)(6y^2 - 3y + 1)$

$= y^2(6y^2 - 3y + 1) - 2y(6y^2 - 3y + 1) + 5(6y^2 - 3y + 1)$ Distributive Property

$= 6y^4 - 3y^3 + y^2 - 12y^3 + 6y^2 - 2y + 30y^2 - 15y + 5$ Distributive Property

$= 6y^4 - 15y^3 + 37y^2 - 17y + 5$ Combine like terms.

Study Tip

Common Misconception
A common mistake when multiplying polynomials horizontally is to combine terms that are not alike. For this reason, you may prefer to multiply polynomials in column form, aligning like terms.

Check for Understanding

Concept Check

1. **Draw a diagram** to show how you would use algebra tiles to find the product of $2x - 1$ and $x + 3$.
2. **Show** how to find $(3x + 4)(2x - 5)$ using each method.
   a. Distributive Property
   b. FOIL method
   c. vertical or column method
   d. algebra tiles
3. **OPEN ENDED** State which method of multiplying binomials you prefer and why.

Guided Practice

Find each product.

4. $(y + 4)(y + 3)$
5. $(x - 2)(x + 6)$
6. $(a - 8)(a + 5)$
7. $(4h + 5)(h + 7)$
8. $(9p - 1)(3p - 2)$
9. $(2g + 7)(5g - 8)$
10. $(3b - 2c)(6b + 5c)$
11. $(3k - 5)(2k^2 + 4k - 3)$

Application

12. **GEOMETRY** The area A of a triangle is half the product of the base b times the height h. Write a polynomial expression that represents the area of the triangle at the right.

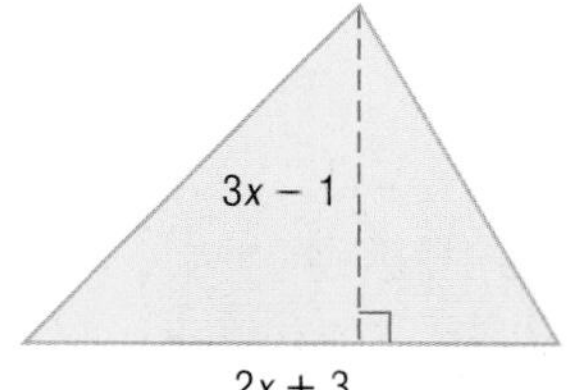

Practice and Apply

Homework Help

For Exercises	See Examples
13–38	1, 2, 4
39–42	3

Extra Practice

See page 839.

Find each product.

13. $(b + 8)(b + 2)$
14. $(n + 6)(n + 7)$
15. $(x - 4)(x - 9)$
16. $(a - 3)(a - 5)$
17. $(y + 4)(y - 8)$
18. $(p + 2)(p - 10)$
19. $(2w - 5)(w + 7)$
20. $(k + 12)(3k - 2)$
21. $(8d + 3)(5d + 2)$
22. $(4g + 3)(9g + 6)$
23. $(7x - 4)(5x - 1)$
24. $(6a - 5)(3a - 8)$
25. $(2n + 3)(2n + 3)$
26. $(5m - 6)(5m - 6)$
27. $(10r - 4)(10r + 4)$
28. $(7t + 5)(7t - 5)$
29. $(8x + 2y)(5x - 4y)$
30. $(11a - 6b)(2a + 3b)$
31. $(p + 4)(p^2 + 2p - 7)$
32. $(a - 3)(a^2 - 8a + 5)$
33. $(2x - 5)(3x^2 - 4x + 1)$
34. $(3k + 4)(7k^2 + 2k - 9)$
35. $(n^2 - 3n + 2)(n^2 + 5n - 4)$
36. $(y^2 + 7y - 1)(y^2 - 6y + 5)$
37. $(4a^2 + 3a - 7)(2a^2 - a + 8)$
38. $(6x^2 - 5x + 2)(3x^2 + 2x + 4)$

GEOMETRY **Write an expression to represent the area of each figure.**

39.

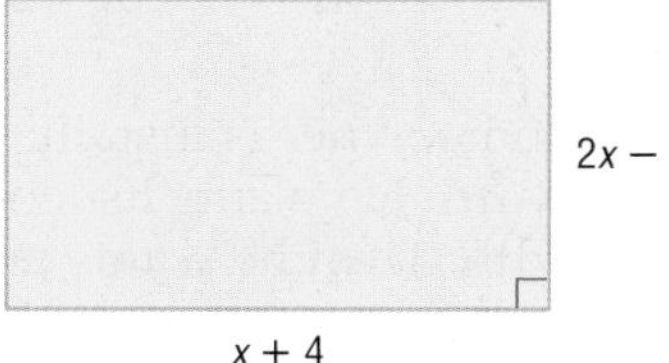

40.

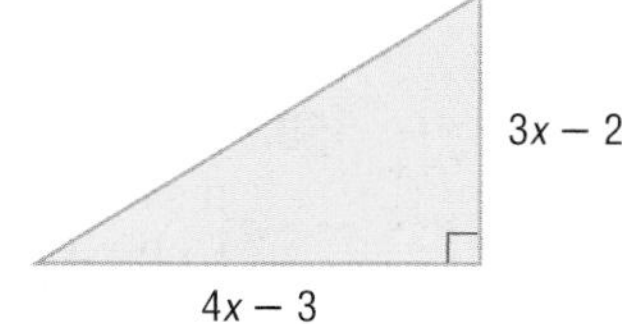

41.

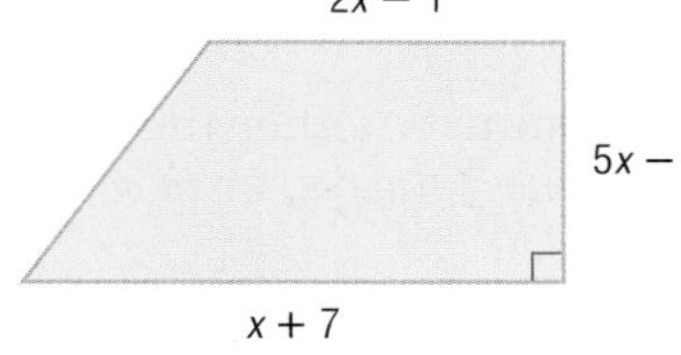

42.

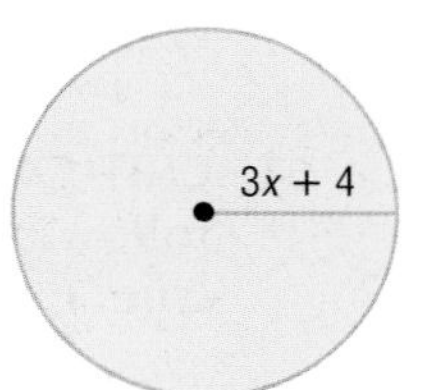

GEOMETRY **The volume V of a prism equals the area of the base B times the height h. Write an expression to represent the volume of each prism.**

43.

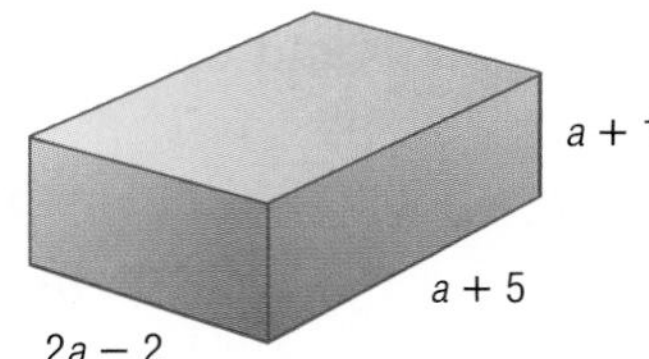

44.

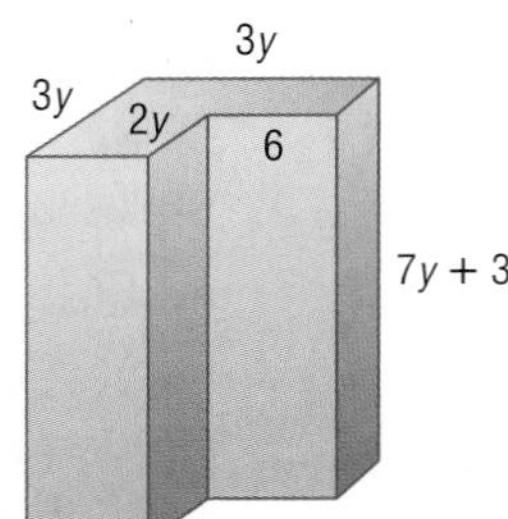

NUMBER THEORY **For Exercises 45–47, consider three consecutive integers. Let the least of these integers be a.**

45. Write a polynomial representing the product of these three integers.

46. Choose an integer for a. Find their product.

47. Evaluate the polynomial in Exercise 45 for the value of a you chose in Exercise 46. Describe the result.

48. **BASKETBALL** The dimensions of a professional basketball court are represented by a width of $2y + 10$ feet and a length of $5y - 6$ feet. Find an expression for the area of the court.

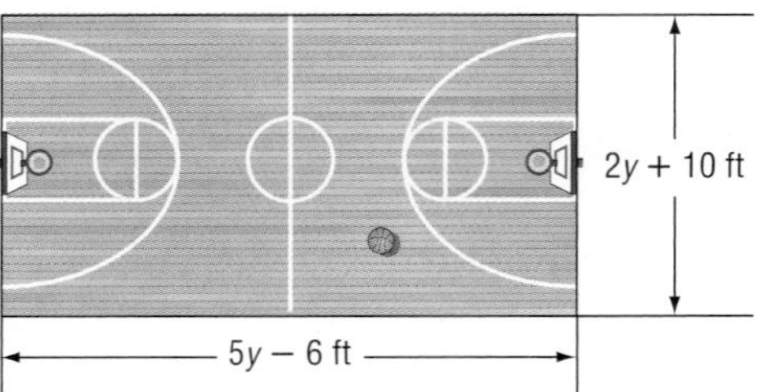

More About. . .

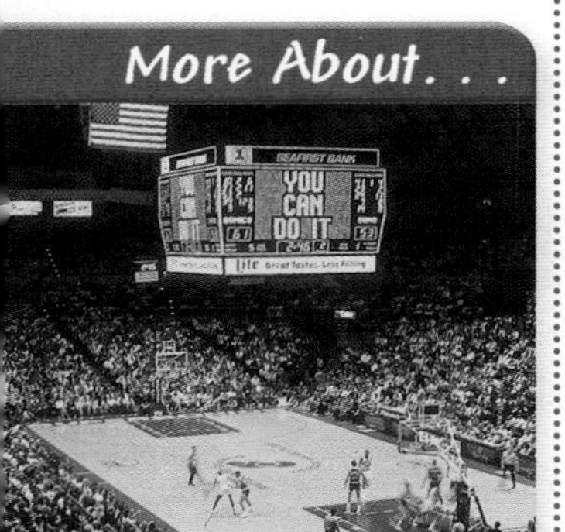

Basketball

More than 200 million people a year pay to see basketball games. That is more admissions than for any other American sport.

Source: *Compton's Encyclopedia*

OFFICE SPACE **For Exercises 49–51, use the following information.**
Latanya's modular office is square. Her office in the company's new building will be 2 feet shorter in one direction and 4 feet longer in the other.

49. Write expressions for the dimensions of Latanya's new office.

50. Write a polynomial expression for the area of her new office.

51. Suppose her office is presently 9 feet by 9 feet. Will her new office be bigger or smaller than her old office and by how much?

52. **MENTAL MATH** One way to mentally multiply 25 and 18 is to find $(20 + 5)(20 - 2)$. Show how the FOIL method can be used to find each product.

a. 35(19) **b.** 67(102) **c.** $8\frac{1}{2} \cdot 6\frac{3}{4}$ **d.** $12\frac{3}{5} \cdot 10\frac{2}{3}$

53. **POOL CONSTRUCTION** A homeowner is installing a swimming pool in his backyard. He wants its length to be 4 feet longer than its width. Then he wants to surround it with a concrete walkway 3 feet wide. If he can only afford 300 square feet of concrete for the walkway, what should the dimensions of the pool be?

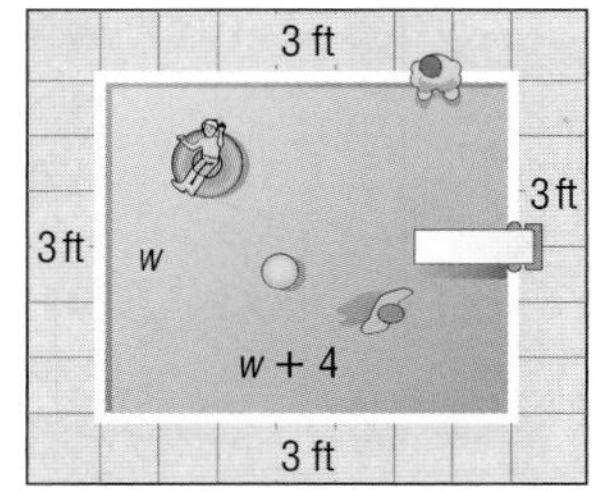

54. **CRITICAL THINKING** Determine whether the following statement is *sometimes, always,* or *never* true. Explain your reasoning.

The product of a binomial and a trinomial is a polynomial with four terms.

55. WRITING IN MATH Answer the question that was posed at the beginning of the lesson.

How is multiplying binomials similar to multiplying two-digit numbers?

Include the following in your answer:

- a demonstration of a horizontal method for multiplying 24×36, and
- an explanation of the meaning of "like terms" in the context of vertical two-digit multiplication.

Standardized Test Practice

56. $(x + 2)(x - 4) - (x + 4)(x - 2) =$

Ⓐ 0 Ⓑ $2x^2 + 4x - 16$ Ⓒ $-4x$ Ⓓ $4x$

57. The expression $(x - y)(x^2 + xy + y^2)$ is equivalent to which of the following?

Ⓐ $x^2 - y^2$ Ⓑ $x^3 - y^3$ Ⓒ $x^3 - xy^2$ Ⓓ $x^3 - x^2y + y^2$

Maintain Your Skills

Mixed Review

Find each product. *(Lesson 8-6)*

58. $3d(4d^2 - 8d - 15)$ 59. $-4y(7y^2 - 4y + 3)$ 60. $2m^2(5m^2 - 7m + 8)$

Simplify. *(Lesson 8-6)*

61. $3x(2x - 4) + 6(5x^2 + 2x - 7)$ 62. $4a(5a^2 + 2a - 7) - 3(2a^2 - 6a - 9)$

GEOMETRY For Exercises 63 and 64, use the following information.
The sum of the degree measures of the angles of a triangle is 180. *(Lesson 8-5)*

63. Write an expression to represent the measure of the third angle of the triangle.

64. If $x = 15$, find the measures of the three angles of the triangle.

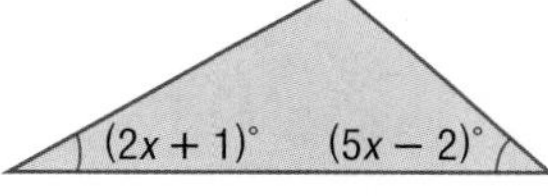

65. Use the graph at the right to determine whether the system below has *no* solution, *one* solution, or *infinitely many* solutions. If the system has one solution, name it. *(Lesson 7-1)*

$x + 2y = 0$
$y + 3 = -x$

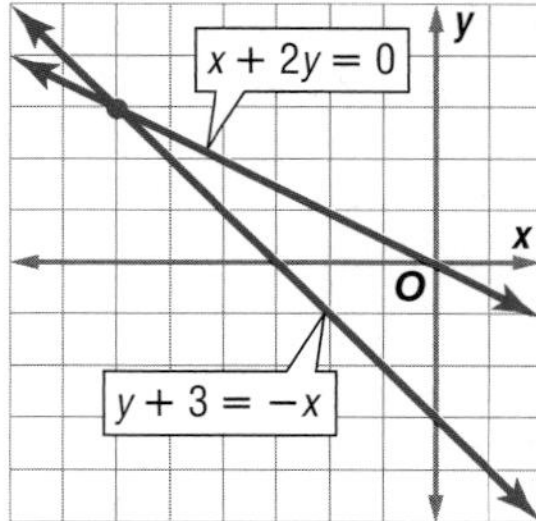

If $f(x) = 2x - 5$ and $g(x) = x^2 + 3x$, find each value. *(Lesson 4-6)*

66. $f(-4)$ 67. $g(-2) + 7$ 68. $f(a + 3)$

Solve each equation or formula for the variable specified. *(Lesson 3-8)*

69. $a = \frac{v}{t}$ for t 70. $ax - by = 2cz$ for y 71. $4x + 3y = 7$ for y

Getting Ready for the Next Lesson

PREREQUISITE SKILL Simplify.
*(To review **Power of a Power** and **Power of a Product** Properties, see Lesson 8-1.)*

72. $(6a)^2$ 73. $(7x)^2$ 74. $(9b)^2$

75. $(4y^2)^2$ 76. $(2v^3)^2$ 77. $(3g^4)^2$

8-8 Special Products

What You'll Learn

- Find squares of sums and differences.
- Find the product of a sum and a difference.

Vocabulary

- difference of squares

When is the product of two binomials also a binomial?

In the previous lesson, you learned how to multiply two binomials using the FOIL method. You may have noticed that the *Outer* and *Inner* terms often combine to produce a trinomial product.

$$(x + 5)(x - 3) = \overset{F}{x^2} \overset{O}{- 3x} \overset{I}{+ 5x} \overset{L}{- 15}$$
$$= x^2 + 2x - 15 \qquad \text{Combine like terms.}$$

This is not always the case, however. Examine the product below.

$$(x + 3)(x - 3) = \overset{F}{x^2} \overset{O}{- 3x} \overset{I}{+ 3x} \overset{L}{- 9}$$
$$= x^2 + 0x - 9 \qquad \text{Combine like terms.}$$
$$= x^2 - 9 \qquad \text{Simplify.}$$

Notice that the product of $x + 3$ and $x - 3$ is a binomial.

SQUARES OF SUMS AND DIFFERENCES While you can always use the FOIL method to find the product of two binomials, some pairs of binomials have products that follow a specific pattern. One such pattern is the *square of a sum*, $(a + b)^2$ or $(a + b)(a + b)$. You can use the diagram below to derive the pattern for this special product.

$$(a + b)^2 = a^2 + ab + ab + b^2$$
$$= a^2 + 2ab + b^2$$

Key Concept — *Square of a Sum*

- **Words** The square of $a + b$ is the square of a plus twice the product of a and b plus the square of b.
- **Symbols** $(a + b)^2 = (a + b)(a + b)$
  $= a^2 + 2ab + b^2$
- **Example** $(x + 7)^2 = x^2 + 2(x)(7) + 7^2$
  $= x^2 + 14x + 49$

Example 1 Square of a Sum

Find each product.

a. $(4y + 5)^2$

$(a + b)^2 = a^2 + 2ab + b^2$ Square of a Sum

$(4y + 5)^2 = (4y)^2 + 2(4y)(5) + 5^2$ $a = 4y$ and $b = 5$

$= 16y^2 + 40y + 25$ Simplify.

CHECK Check your work by using the FOIL method.

$(4y + 5)^2 = (4y + 5)(4y + 5)$

F O I L

$= (4y)(4y) + (4y)(5) + 5(4y) + 5(5)$

$= 16y^2 + 20y + 20y + 25$

$= 16y^2 + 40y + 25$ ✓

b. $(8c + 3d)^2$

$(a + b)^2 = a^2 + 2ab + b^2$ Square of a Sum

$(8c + 3d)^2 = (8c)^2 + 2(8c)(3d) + (3d)^2$ $a = 8c$ and $b = 3d$

$= 64c^2 + 48cd + 9d^2$ Simplify.

Study Tip

$(a + b)^2$
In the pattern for $(a + b)^2$, a and b can be numbers, variables, or expressions with numbers and variables.

To find the pattern for the *square of a difference*, $(a - b)^2$, write $a - b$ as $a + (-b)$ and square it using the square of a sum pattern.

$(a - b)^2 = [a + (-b)]^2$

$= a^2 + 2(a)(-b) + (-b)^2$ Square of a Sum

$= a^2 - 2ab + b^2$ Simplify. Note that $(-b)^2 = (-b)(-b)$ or b^2.

The square of a difference can be found by using the following pattern.

Key Concept — Square of a Difference

- **Words** The square of $a - b$ is the square of a minus twice the product of a and b plus the square of b.
- **Symbols** $(a - b)^2 = (a - b)(a - b)$
  $= a^2 - 2ab + b^2$
- **Example** $(x - 4)^2 = x^2 - 2(x)(4) + 4^2$
  $= x^2 - 8x + 16$

Example 2 Square of a Difference

Find each product.

a. $(6p - 1)^2$

$(a - b)^2 = a^2 - 2ab + b^2$ Square of a Difference

$(6p - 1)^2 = (6p)^2 - 2(6p)(1) + 1^2$ $a = 6p$ and $b = 1$

$= 36p^2 - 12p + 1$ Simplify.

b. $(5m^3 - 2n)^2$

$(a - b)^2 = a^2 - 2ab + b^2$ Square of a Difference

$(5m^3 - 2n)^2 = (5m^3)^2 - 2(5m^3)(2n) + (2n)^2$ $a = 5m^3$ and $b = 2n$

$= 25m^6 - 20m^3n + 4n^2$ Simplify.

Example 3 Apply the Sum of a Square

GENETICS **The Punnett square shows the possible gene combinations of a cross between two pea plants. Each plant passes along one *dominant* gene T for tallness and one *recessive* gene t for shortness.**

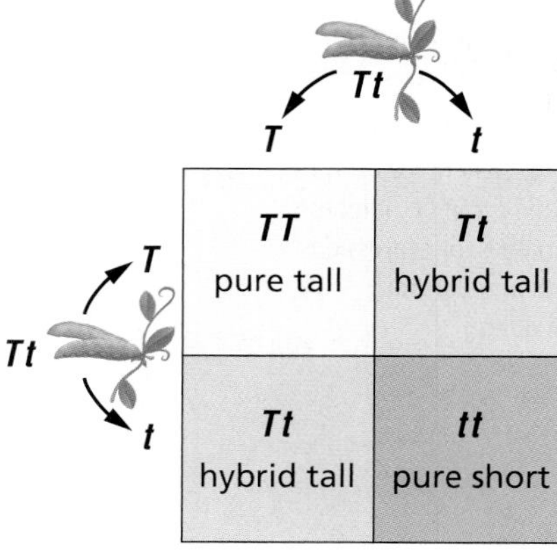

Show how combinations can be modeled by the square of a binomial. Then determine what percent of the offspring will be pure tall, hybrid tall, and pure short.

Each parent has half the genes necessary for tallness and half the genes necessary for shortness. The makeup of each parent can be modeled by $0.5T + 0.5t$. Their offspring can be modeled by the product of $0.5T + 0.5t$ and $0.5T + 0.5t$ or $(0.5T + 0.5t)^2$.

If we expand this product, we can determine the possible heights of the offspring.

$(a + b)^2 = a^2 + 2ab + b^2$	Square of a Sum
$(0.5T + 0.5t)^2 = (0.5T)^2 + 2(0.5T)(0.5t) + (0.5t)^2$	$a = 0.5T$ and $b = 0.5t$
$= 0.25T^2 + 0.5Tt + 0.25t^2$	Simplify.
$= 0.25TT + 0.5Tt + 0.25tt$	$T^2 = TT$ and $t^2 = tt$

Thus, 25% of the offspring are TT or pure tall, 50% are Tt or hybrid tall, and 25% are tt or pure short.

Career Choices

Geneticist

Laboratory geneticists work in medicine to find cures for disease, in agriculture to breed new crops and livestock, and in police work to identify criminals.

Online Research For information about a career as a geneticist, visit: www.algebra1.com/careers

PRODUCT OF A SUM AND A DIFFERENCE You can use the diagram below to find the pattern for the product of a sum and a difference of the *same two terms*, $(a + b)(a - b)$. Recall that $a - b$ can be rewritten as $a + (-b)$.

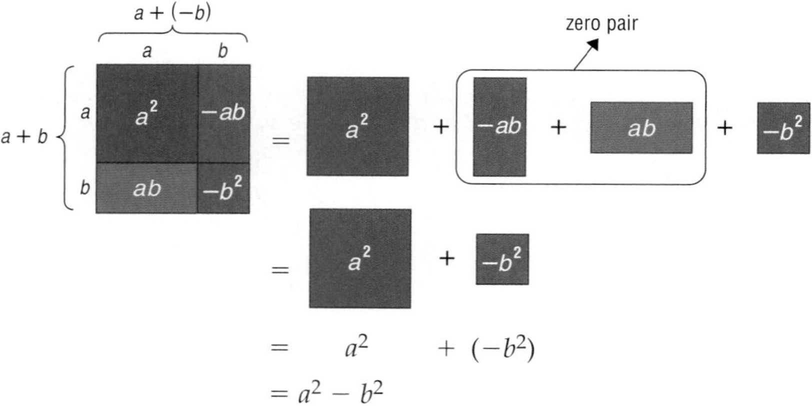

$= \quad a^2 \quad + (-b^2)$

$= a^2 - b^2$

The resulting product, $a^2 - b^2$, has a special name. It is called a **difference of squares**. Notice that this product has no middle term.

Key Concept — Product of a Sum and a Difference

- **Words** The product of $a + b$ and $a - b$ is the square of a minus the square of b.
- **Symbols** $(a + b)(a - b) = (a - b)(a + b)$
  $= a^2 - b^2$
- **Example** $(x + 9)(x - 9) = x^2 - 9^2$
  $= x^2 - 81$

Example 4 Product of a Sum and a Difference

Find each product.

a. $(3n + 2)(3n - 2)$

$(a + b)(a - b) = a^2 - b^2$ — Product of a Sum and a Difference

$(3n + 2)(3n - 2) = (3n)^2 - 2^2$ — $a = 3n$ and $b = 2$

$= 9n^2 - 4$ — Simplify.

b. $(11v - 8w^2)(11v + 8w^2)$

$(a - b)(a + b) = a^2 - b^2$ — Product of a Sum and a Difference

$(11v - 8w^2)(11v + 8w^2) = (11v)^2 - (8w^2)^2$ — $a = 11v$ and $b = 8w^2$

$= 121v^2 - 64w^4$ — Simplify.

The following list summarizes the special products you have studied.

Key Concept — Special Products

- **Square of a Sum** $(a + b)^2 = a^2 + 2ab + b^2$
- **Square of a Difference** $(a - b)^2 = a^2 - 2ab + b^2$
- **Product of a Sum and a Difference** $(a - b)(a + b) = a^2 - b^2$

Check for Understanding

Concept Check

1. **Compare and contrast** the pattern for the square of a sum with the pattern for the square of a difference.
2. **Explain** how the square of a difference and the difference of squares differ.
3. **Draw a diagram** to show how you would use algebra tiles to model the product of $x - 3$ and $x - 3$, or $(x - 3)^2$.
4. **OPEN ENDED** Write two binomials whose product is a difference of squares.

Guided Practice

Find each product.

5. $(a + 6)^2$
6. $(4n - 3)(4n - 3)$
7. $(8x - 5)(8x + 5)$
8. $(3a + 7b)(3a - 7b)$
9. $(x^2 - 6y)^2$
10. $(9 - p)^2$

Application

GENETICS For Exercises 11 and 12, use the following information.
In hamsters, golden coloring G is dominant over cinnamon coloring g. Suppose a purebred cinnamon male is mated with a purebred golden female.

11. Write an expression for the genetic makeup of the hamster pups.
12. What is the probability that the pups will have cinnamon coloring? Explain your reasoning.

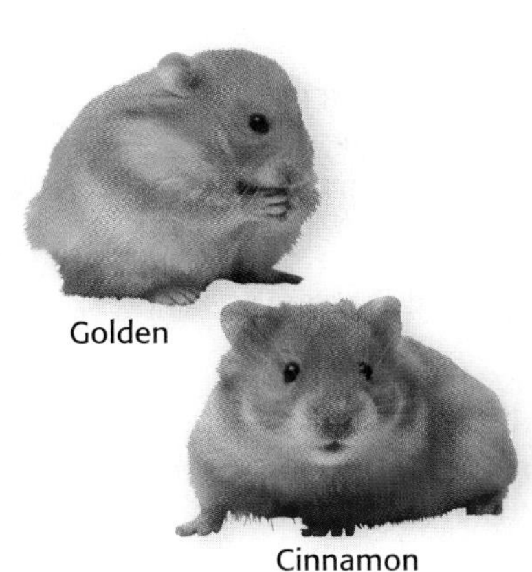
Golden

Cinnamon

Practice and Apply

Homework Help

For Exercises	See Examples
13–38	1, 2, 4
39, 40	3

Extra Practice

See page 839.

Find each product.

13. $(y + 4)^2$
14. $(k + 8)(k + 8)$
15. $(a - 5)(a - 5)$
16. $(n - 12)^2$
17. $(b + 7)(b - 7)$
18. $(c - 2)(c + 2)$
19. $(2g + 5)^2$
20. $(9x + 3)^2$
21. $(7 - 4y)^2$
22. $(4 - 6h)^2$
23. $(11r + 8)(11r - 8)$
24. $(12p - 3)(12p + 3)$
25. $(a + 5b)^2$
26. $(m + 7n)^2$
27. $(2x - 9y)^2$
28. $(3n - 10p)^2$
29. $(5w + 14)(5w - 14)$
30. $(4d - 13)(4d + 13)$
31. $(x^3 + 4y)^2$
32. $(3a^2 - b^2)^2$
33. $(8a^2 - 9b^3)(8a^2 + 9b^3)$
34. $(5x^4 - y)(5x^4 + y)$
35. $\left(\frac{2}{3}x - 6\right)^2$
36. $\left(\frac{4}{5}x + 10\right)^2$
37. $(2n + 1)(2n - 1)(n + 5)$
38. $(p + 3)(p - 4)(p - 3)(p + 4)$

GENETICS **For Exercises 39 and 40, use the following information.**
Pam has brown eyes and Bob has blue eyes. Brown genes B are dominant over blue genes b. A person with genes BB or Bb has brown eyes. Someone with genes bb has blue eyes. Suppose Pam's genes for eye color are Bb.

39. Write an expression for the possible eye coloring of Pam and Bob's children.
40. What is the probability that a child of Pam and Bob would have blue eyes?

MAGIC TRICK **For Exercises 41–44, use the following information.**
Julie says that she can perform a magic trick with numbers. She asks you to pick a whole number, any whole number. Square that number. Then, add twice your original number. Next add 1. Take the square root of the result. Finally, subtract your original number. Then Julie exclaims with authority, "Your answer is 1!"

41. Pick a whole number and follow Julie's directions. Is your result 1?
42. Let a represent the whole number you chose. Then, find a polynomial representation for the first three steps of Julie's directions.
43. The polynomial you wrote in Exercise 42 is the square of what binomial sum?
44. Take the square root of the perfect square you wrote in Exercise 43, then subtract a, your original number. What is the result?

More About. . .

Architecture

The historical Gwennap Pit, an outdoor amphitheater in southern England, consists of a circular stage surrounded by circular levels used for seating. Each seating level is about 1 meter wide.

Source: *Christian Guide to Britain*

ARCHITECTURE **For Exercises 45 and 46, use the following information.**
A diagram of a portion of the Gwennap Pit is shown at the right. Suppose the radius of the stage is s meters.

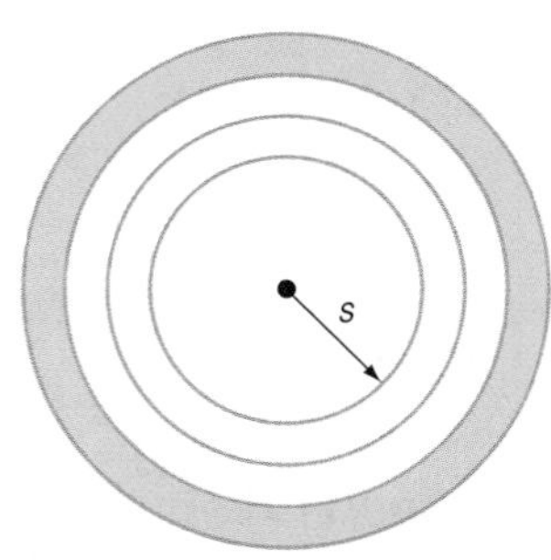

45. Use the information at the left to find binomial representations for the radii of the second and third seating levels.
46. Find the area of the shaded region representing the third seating level.

47. **GEOMETRY** The area of the shaded region models the difference of two squares, $a^2 - b^2$. Show that the area of the shaded region is also equal to $(a - b)(a + b)$. (*Hint*: Divide the shaded region into two trapezoids as shown.)

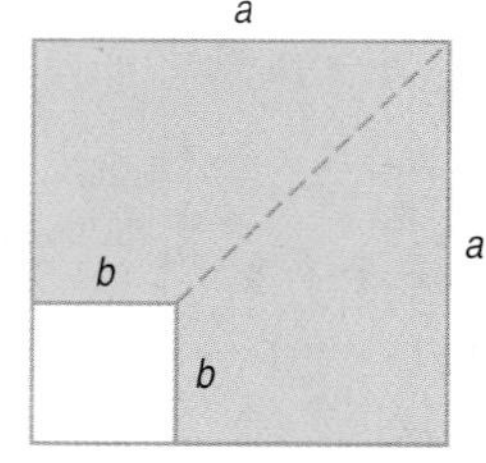

48. WRITING IN MATH Answer the question that was posed at the beginning of the lesson.

When is the product of two binomials also a binomial?

Include the following in your answer:

- an example of two binomials whose product is a binomial, and
- an example of two binomials whose product is not a binomial.

Standardized Test Practice

49. If $a^2 + b^2 = 40$ and $ab = 12$, find the value of $(a - b)^2$.

(A) 1 (B) 121 (C) 16 (D) 28

50. If $x - y = 10$ and $x + y = 20$, find the value of $x^2 - y^2$.

(A) 400 (B) 200 (C) 100 (D) 30

Extending the Lesson

51. Does a pattern exist for the cube of a sum, $(a + b)^3$?

a. Investigate this question by finding the product of $(a + b)(a + b)(a + b)$.

b. Use the pattern you discovered in part **a** to find $(x + 2)^3$.

c. Draw a diagram of a geometric model for the cube of a sum.

Maintain Your Skills

Mixed Review

Find each product. *(Lesson 8-7)*

52. $(x + 2)(x + 7)$
53. $(c - 9)(c + 3)$
54. $(4y - 1)(5y - 6)$
55. $(3n - 5)(8n + 5)$
56. $(x - 2)(3x^2 - 5x + 4)$
57. $(2k + 5)(2k^2 - 8k + 7)$

Solve. *(Lesson 8-6)*

58. $6(x + 2) + 4 = 5(3x - 4)$
59. $-3(3a - 8) + 2a = 4(2a + 1)$
60. $p(p + 2) + 3p = p(p - 3)$
61. $y(y - 4) + 2y = y(y + 12) - 7$

Use elimination to solve each system of equations. *(Lessons 7-3 and 7-4)*

62. $\frac{3}{4}x + \frac{1}{5}y = 5$
$\frac{3}{4}x - \frac{1}{5}y = -5$

63. $2x - y = 10$
$5x + 3y = 3$

64. $2x = 4 - 3y$
$3y - x = -11$

Write the slope-intercept form of an equation that passes through the given point and is perpendicular to the graph of each equation. *(Lesson 5-6)*

65. $5x + 5y = 35, (-3, 2)$
66. $2x - 5y = 3, (-2, 7)$
67. $5x + y = 2, (0, 6)$

Find the nth term of each arithmetic sequence described. *(Lesson 4-7)*

68. $a_1 = 3, d = 4, n = 18$
69. $-5, 1, 7, 13, \ldots$ for $n = 12$

70. **PHYSICAL FITNESS** Mitchell likes to exercise regularly. He likes to warm up by walking two miles. Then he runs five miles. Finally, he cools down by walking for another mile. Identify the graph that best represents Mitchell's heart rate as a function of time. *(Lesson 1-8)*

a.
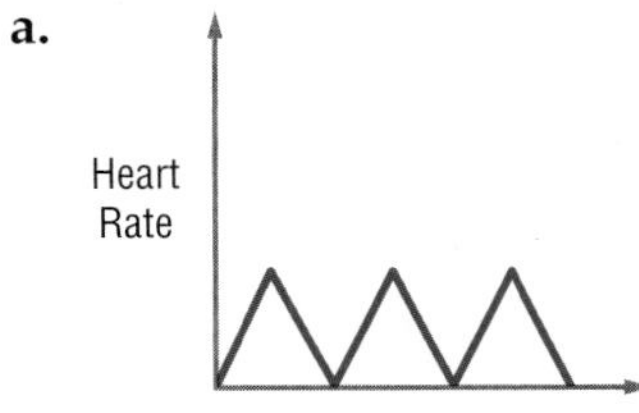

b.
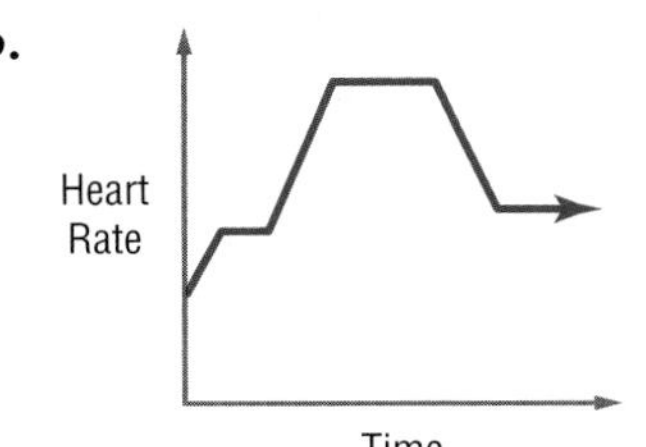

c.
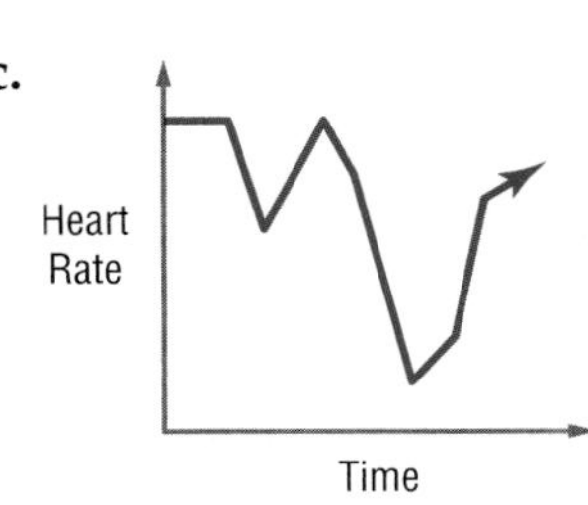

Chapter 8

Study Guide and Review

Vocabulary and Concept Check

binomial (p. 432)
constant (p. 410)
degree of a monomial (p. 433)
degree of a polynomial (p. 433)
difference of squares (p. 460)
FOIL method (p. 453)
monomial (p. 410)
negative exponent (p. 419)
polynomial (p. 432)
Power of a Power (p. 411)
Power of a Product (p. 412)
Power of a Quotient (p. 418)
Product of Powers (p. 411)
Quotient of Powers (p. 417)
scientific notation (p. 425)
trinomial (p. 432)
zero exponent (p. 419)

Choose a term from the vocabulary list that best matches each example.

1. $4^{-3} = \frac{1}{4^3}$

2. $(n^3)^5 = n^{15}$

3. $\frac{4x^2y}{8xy^3} = \frac{x}{2y^2}$

4. $4x^2$

5. $x^2 - 3x + 1$

6. $2^0 = 1$

7. $x^4 - 3x^3 + 2x^2 - 1$

8. $(x + 3)(x - 4) = x^2 - 4x + 3x - 12$

9. $x^2 + 2$

10. $(a^3b)(2ab^2) = 2a^4b^3$

Lesson-by-Lesson Review

8-1 Multiplying Monomials

See pages 410–415.

Concept Summary

	Examples
• A monomial is a number, a variable, or a product of a number and one or more variables.	$6x^2, -5, \frac{2c}{3}$
• To multiply two powers that have the same base, add exponents.	$a^2 \cdot a^3 = a^5$
• To find the power of a power, multiply exponents.	$(a^2)^3 = a^6$
• The power of a product is the product of the powers.	$(ab^2)^3 = a^3b^6$

Examples

1 **Simplify $(2ab^2)(3a^2b^3)$.**

$(2ab^2)(3a^2b^3) = (2 \cdot 3)(a \cdot a^2)(b^2 \cdot b^3)$ Commutative Property

$= 6a^3b^5$ Product of Powers

2 **Simplify $(2x^2y^3)^3$.**

$(2x^2y^3)^3 = 2^3(x^2)^3(y^3)^3$ Power of a Product

$= 8x^6y^9$ Power of a Power

Exercises **Simplify.** *See Examples 2, 3, and 5 on pages 411 and 412.*

11. $y^3 \cdot y^3 \cdot y$

12. $(3ab)(-4a^2b^3)$

13. $(-4a^2x)(-5a^3x^4)$

14. $(4a^2b)^3$

15. $(-3xy)^2(4x)^3$

16. $(-2c^2d)^4(-3c^2)^3$

17. $-\frac{1}{2}(m^2n^4)^2$

18. $(5a^2)^3 + 7(a^6)$

19. $[(3^2)^2]^3$

www.algebra1.com/vocabulary_review

8-2 Dividing Monomials

See pages 417–423.

Concept Summary

- To divide two powers that have the same base, subtract the exponents.
- To find the power of a quotient, find the power of the numerator and the power of the denominator.
- Any nonzero number raised to the zero power is 1.
- For any nonzero number a and any integer n, $a^{-n} = \frac{1}{a^n}$ and $\frac{1}{a^{-n}} = a^n$.

Examples

$\frac{a^5}{a^3} = a^2$

$\left(\frac{a}{b}\right)^2 = \frac{a^2}{b^2}$

$(3a^3b^2)^0 = 1$

$a^{-3} = \frac{1}{a^3}$

Example **Simplify $\frac{2x^6y}{8x^2y^2}$. Assume that x and y are not equal to zero.**

$\frac{2x^6y}{8x^2y^2} = \left(\frac{2}{8}\right)\left(\frac{x^6}{x^2}\right)\left(\frac{y}{y^2}\right)$ Group the powers with the same base.

$= \left(\frac{1}{4}\right)(x^{6-2})(y^{1-2})$ Quotient of Powers

$= \frac{x^4}{4y}$ Simplify.

Exercises **Simplify. Assume that no denominator is equal to zero.**

See Examples 1–4 on pages 417–420.

20. $\frac{(3y)^0}{6a}$

21. $\left(\frac{3bc^2}{4d}\right)^3$

22. $x^{-2}y^0z^3$

23. $\frac{27b^{-2}}{14b^{-3}}$

24. $\frac{(3a^3bc^2)^2}{18a^2b^3c^4}$

25. $\frac{-16a^3b^2x^4y}{-48a^4bxy^3}$

26. $\frac{(-a)^5b^8}{a^5b^2}$

27. $\frac{(4a^{-1})^{-2}}{(2a^4)^2}$

28. $\left(\frac{5xy^{-2}}{35x^{-2}y^{-6}}\right)^0$

8-3 Scientific Notation

See pages 425–430.

Concept Summary

- A number is expressed in scientific notation when it is written as a product of a factor and a power of 10. The factor must be greater than or equal to 1 and less than 10.

$$a \times 10^n, \text{ where } 1 \leq a < 10 \text{ and } n \text{ is an integer.}$$

Examples

1 **Express 5.2×10^7 in standard notation.**

$5.2 \times 10^7 = 52{,}000{,}000$ $n = 7$; move decimal point 7 places to the right.

2 **Express 0.0021 in scientific notation.**

$0.0021 \rightarrow 0002.1 \times 10^n$ Move decimal point 3 places to the right.

$0.0021 = 2.1 \times 10^{-3}$ $a = 2.1$ and $n = -3$

3 Evaluate $(2 \times 10^2)(5.2 \times 10^6)$. Express the result in scientific and standard notation.

$(2 \times 10^2)(5.2 \times 10^6) = (2 \times 5.2)(10^2 \times 10^6)$ Associative Property

$= 10.4 \times 10^8$ Product of Powers

$= (1.04 \times 10^1) \times 10^8$ $10.4 = 1.04 \times 10^1$

$= 1.04 \times (10^1 \times 10^8)$ Associative Property

$= 1.04 \times 10^9$ or 1,040,000,000 Product of Powers

Exercises **Express each number in standard notation.** *See Example 1 on page 426.*

29. 2.4×10^5 **30.** 3.14×10^{-4} **31.** 4.88×10^9

Express each number in scientific notation. *See Example 2 on page 426.*

32. 0.00000187 **33.** 796×10^3 **34.** 0.0343×10^{-2}

Evaluate. Express each result in scientific and standard notation.
See Examples 3 and 4 on page 427.

35. $(2 \times 10^5)(3 \times 10^6)$ **36.** $\frac{8.4 \times 10^{-6}}{1.4 \times 10^{-9}}$ **37.** $(3 \times 10^2)(5.6 \times 10^{-8})$

8-4 Polynomials

See pages 432–436.

Concept Summary

- A polynomial is a monomial or a sum of monomials.
- A binomial is the sum of *two* monomials, and a trinomial is the sum of *three* monomials.
- The degree of a monomial is the sum of the exponents of all its variables.
- The degree of the polynomial is the greatest degree of any term. To find the degree of a polynomial, you must find the degree of each term.

Examples

1 Find the degree of $2xy^3 + x^2y$.

Polynomial	Terms	Degree of Each Term	Degree of Polynomial
$2xy^3 + x^2y$	$2xy^3, x^2y$	4, 3	4

2 Arrange the terms of $4x^2 + 9x^3 - 2 - x$ so that the powers of x are in descending order.

$4x^2 + 9x^3 - 2 - x = 4x^2 + 9x^3 - 2x^0 - x^1$ $x^0 = 1$ and $x = x^1$

$= 9x^3 + 4x^2 - x - 2$ $3 > 2 > 1 > 0$

Exercises **Find the degree of each polynomial.** *See Example 3 on page 433.*

38. $n - 2p^2$ **39.** $29n^2 + 17n^2t^2$ **40.** $4xy + 9x^3z^2 + 17rs^3$

41. $-6x^5y - 2y^4 + 4 - 8y^2$ **42.** $3ab^3 - 5a^2b^2 + 4ab$ **43.** $19m^3n^4 + 21m^5n$

Arrange the terms of each polynomial so that the powers of x are in descending order. *See Example 5 on page 433.*

44. $3x^4 - x + x^2 - 5$ **45.** $-2x^2y^3 - 27 - 4x^4 + xy + 5x^3y^2$

8-5 Adding and Subtracting Polynomials

See pages 439–443.

Concept Summary

- To add polynomials, group like terms horizontally or write them in column form, aligning like terms vertically.
- Subtract a polynomial by adding its additive inverse. To find the additive inverse of a polynomial, replace each term with its additive inverse.

Example **Find $(7r^2 + 9r) - (12r^2 - 4)$.**

$$(7r^2 + 9r) - (12r^2 - 4) = 7r^2 + 9r + (-12r^2 + 4) \quad \text{The additive inverse of } 12r^2 - 4 \text{ is } -12r^2 + 4.$$
$$= (7r^2 - 12r^2) + 9r + 4 \quad \text{Group like terms.}$$
$$= -5r^2 + 9r + 4 \quad \text{Add like terms.}$$

Exercises **Find each sum or difference.** *See Examples 1 and 2 on pages 439 and 440.*

46. $(2x^2 - 5x + 7) - (3x^3 + x^2 + 2)$
47. $(x^2 - 6xy + 7y^2) + (3x^2 + xy - y^2)$
48. $(7z^2 + 4) - (3z^2 + 2z - 6)$
49. $(13m^4 - 7m - 10) + (8m^4 - 3m + 9)$
50. $(11m^2n^2 + 4mn - 6) + (5m^2n^2 + 6mn + 17)$
51. $(-5p^2 + 3p + 49) - (2p^2 + 5p + 24)$

8-6 Multiplying a Polynomial by a Monomial

See pages 444–449.

Concept Summary

- The Distributive Property can be used to multiply a polynomial by a monomial.

Examples

1 **Simplify $x^2(x + 2) + 3(x^3 + 4x^2)$.**

$$x^2(x + 2) + 3(x^3 + 4x^2) = x^2(x) + x^2(2) + 3(x^3) + 3(4x^2) \quad \text{Distributive Property}$$
$$= x^3 + 2x^2 + 3x^3 + 12x^2 \quad \text{Multiply.}$$
$$= 4x^3 + 14x^2 \quad \text{Combine like terms.}$$

2 **Solve $x(x - 10) + x(x + 2) + 3 = 2x(x + 1) - 7$.**

$$x(x - 10) + x(x + 2) + 3 = 2x(x + 1) - 7 \quad \text{Original equation}$$
$$x^2 - 10x + x^2 + 2x + 3 = 2x^2 + 2x - 7 \quad \text{Distributive Property}$$
$$2x^2 - 8x + 3 = 2x^2 + 2x - 7 \quad \text{Combine like terms.}$$
$$-8x + 3 = 2x - 7 \quad \text{Subtract } 2x^2 \text{ from each side.}$$
$$-10x + 3 = -7 \quad \text{Subtract } 2x \text{ from each side.}$$
$$-10x = -10 \quad \text{Subtract 3 from each side.}$$
$$x = 1 \quad \text{Divide each side by } -10.$$

Exercises **Simplify.** *See Example 2 on page 444.*

52. $b(4b - 1) + 10b$
53. $x(3x - 5) + 7(x^2 - 2x + 9)$
54. $8y(11y^2 - 2y + 13) - 9(3y^3 - 7y + 2)$
55. $2x(x - y^2 + 5) - 5y^2(3x - 2)$

Solve each equation. *See Example 4 on page 445.*

56. $m(2m - 5) + m = 2m(m - 6) + 16$
57. $2(3w + w^2) - 6 = 2w(w - 4) + 10$

• Extra Practice, see pages 837–839.
• Mixed Problem Solving, see page 860.

8-7 Multiplying Polynomials

See pages 452–457.

Concept Summary

- The FOIL method is the sum of the products of the first terms *F*, the outer terms *O*, the inner terms *I*, and the last terms *L*.
- The Distributive Property can be used to multiply any two polynomials.

Examples

1 Find $(3x + 2)(x - 2)$.

	F O I L	
$(3x + 2)(x - 2)$	$= (3x)(x) + (3x)(-2) + (2)(x) + (2)(-2)$	FOIL Method
	$= 3x^2 - 6x + 2x - 4$	Multiply.
	$= 3x^2 - 4x - 4$	Combine like terms.

2 Find $(2y - 5)(4y^2 + 3y - 7)$.

$(2y - 5)(4y^2 + 3y - 7)$

$= 2y(4y^2 + 3y - 7) - 5(4y^2 + 3y - 7)$	Distributive Property
$= 8y^3 + 6y^2 - 14y - 20y^2 - 15y + 35$	Distributive Property
$= 8y^3 - 14y^2 - 29y + 35$	Combine like terms.

Exercises **Find each product.** *See Examples 1, 2, and 4 on pages 452–454.*

58. $(r - 3)(r + 7)$ **59.** $(4a - 3)(a + 4)$ **60.** $(3x + 0.25)(6x - 0.5)$
61. $(5r - 7s)(4r + 3s)$ **62.** $(2k + 1)(k^2 + 7k - 9)$ **63.** $(4p - 3)(3p^2 - p + 2)$

8-8 Special Products

See pages 458–463.

Concept Summary

- Square of a Sum: $(a + b)^2 = a^2 + 2ab + b^2$
- Square of a Difference: $(a - b)^2 = a^2 - 2ab + b^2$
- Product of a Sum and a Difference: $(a + b)(a - b) = (a - b)(a + b) = a^2 - b^2$

Examples

1 Find $(r - 5)^2$.

$(a - b)^2 = a^2 - 2ab + b^2$	Square of a Difference
$(r - 5)^2 = r^2 - 2(r)(5) + 5^2$	$a = r$ and $b = 5$
$= r^2 - 10r + 25$	Simplify.

2 Find $(2c + 9)(2c - 9)$.

$(a + b)(a - b) = a^2 - b^2$	Product of a Sum and a Difference
$(2c + 9)(2c - 9) = (2c)^2 - 9^2$	$a = 2c$ and $b = 9$
$= 4c^2 - 81$	Simplify.

Exercises **Find each product.** *See Examples 1, 2, and 4 on pages 459 and 461.*

64. $(x - 6)(x + 6)$ **65.** $(4x + 7)^2$ **66.** $(8x - 5)^2$
67. $(5x - 3y)(5x + 3y)$ **68.** $(6a - 5b)^2$ **69.** $(3m + 4n)^2$

Chapter 8 Practice Test

Vocabulary and Concepts

1. **Explain** why $(4^2)(4^3) \neq 16^5$.
2. **Write** $\frac{1}{5}$ using a negative exponent.
3. **Define and give an example** of a monomial.

Skills and Applications

Simplify. Assume that no denominator is equal to zero.

4. $(a^2b^4)(a^3b^5)$
5. $(-12abc)(4a^2b^4)$
6. $\left(\frac{3}{5}m\right)^2$
7. $(-3a)^4(a^5b)^2$
8. $(-5a^2)(-6b^3)^2$
9. $\frac{mn^4}{m^3n^2}$
10. $\frac{9a^2bc^2}{63a^4bc}$
11. $\frac{48a^2bc^5}{(3ab^3c^2)^2}$

Express each number in scientific notation.

12. 46,300
13. 0.003892
14. 284×10^3
15. 52.8×10^{-9}

Evaluate. Express each result in scientific notation and standard notation.

16. $(3 \times 10^3)(2 \times 10^4)$
17. $\frac{14.72 \times 10^{-4}}{3.2 \times 10^{-3}}$
18. $(15 \times 10^{-7})(3.1 \times 10^4)$
19. **SPACE EXPLORATION** A space probe that is 2.85×10^9 miles away from Earth sends radio signals to NASA. If the radio signals travel at the speed of light (1.86×10^5 miles per second), how long will it take the signals to reach NASA?

Find the degree of each polynomial. Then arrange the terms so that the powers of y are in descending order.

20. $2y^2 + 8y^4 + 9y$
21. $5xy - 7 + 2y^4 - x^2y^3$

Find each sum or difference.

22. $(5a + 3a^2 - 7a^3) + (2a - 8a^2 + 4)$
23. $(x^3 - 3x^2y + 4xy^2 + y^3) - (7x^3 + x^2y - 9xy^2 + y^3)$
24. **GEOMETRY** The measures of two sides of a triangle are given. If the perimeter is represented by $11x^2 - 29x + 10$, find the measure of the third side.

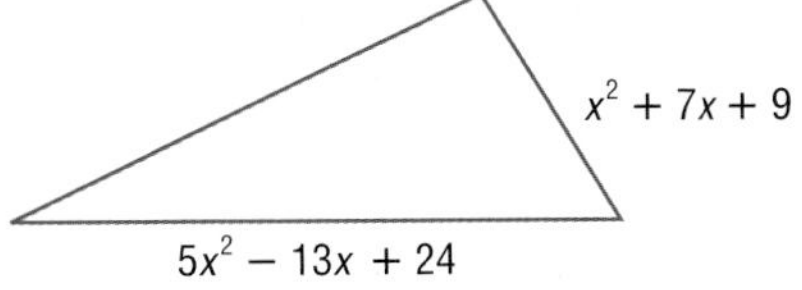

Simplify.

25. $(h - 5)^2$
26. $(4x - y)(4x + y)$
27. $3x^2y^3(2x - xy^2)$
28. $(2a^2b + b^2)^2$
29. $(4m + 3n)(2m - 5n)$
30. $(2c + 5)(3c^2 - 4c + 2)$

Solve each equation.

31. $2x(x - 3) = 2(x^2 - 7) + 2$
32. $3a(a^2 + 5) - 11 = a(3a^2 + 4)$
33. **STANDARDIZED TEST PRACTICE** If $x^2 + 2xy + y^2 = 8$, find $3(x + y)^2$.

    Ⓐ 2 Ⓑ 4 Ⓒ 24 Ⓓ cannot be determined

www.algebra1.com/chapter_test

Chapter 8 Standardized Test Practice

Part 1 Multiple Choice

Record your answers on the answer sheet provided by your teacher or on a sheet of paper.

1. A basketball team scored the following points during the first five games of the season: 70, 65, 75, 70, 80. During the sixth game, they scored only 30 points. Which of these measures changed the most as a result of the sixth game? (Lessons 2-2 and 2-5)

Ⓐ mean

Ⓑ median

Ⓒ mode

Ⓓ They all changed the same amount.

2. A machine produces metal bottle caps. The number of caps it produces is proportional to the number of minutes the machine operates. The machine produces 2100 caps in 60 minutes. How many minutes would it take the machine to produce 5600 caps? (Lesson 2-6)

Ⓐ 35 Ⓑ 58.3 Ⓒ 93.3 Ⓓ 160

3. The odometer on Juliana's car read 20,542 miles when she started a trip. After 4 hours of driving, the odometer read 20,750 miles. Which equation can be used to find r, her average rate of speed for the 4 hours? (Lesson 3-1)

Ⓐ $r = 20{,}750 - 20{,}542$

Ⓑ $r = 4(20{,}750 - 20{,}542)$

Ⓒ $r = \frac{20{,}750}{4}$

Ⓓ $r = \frac{20{,}750 - 20{,}542}{4}$

4. Which equation best describes the graph? (Lesson 5-4)

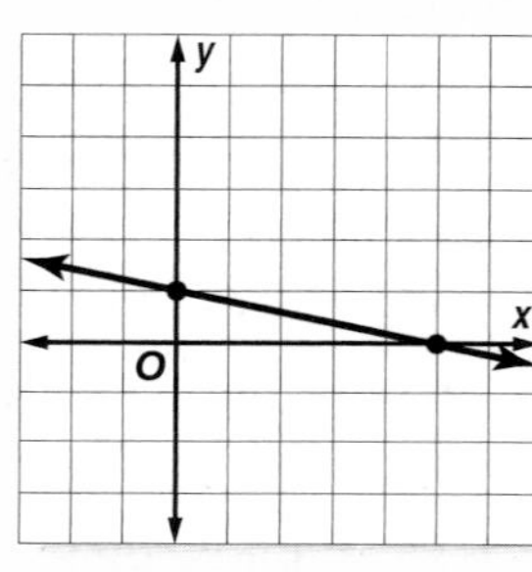

Ⓐ $y = -\frac{1}{5}x + 1$

Ⓑ $y = -5x + 1$

Ⓒ $y = \frac{1}{5}x + 5$

Ⓓ $y = -5x - 5$

5. Which equation represents the line that passes through the point at $(-1, 4)$ and has a slope of -2? (Lesson 5-5)

Ⓐ $y = -2x - 2$ Ⓑ $y = -2x + 2$

Ⓒ $y = -2x + 6$ Ⓓ $y = -2x + 7$

6. Mr. Puram is planning an addition to the school library. The budget is \$7500. Each bookcase costs \$125, and each set of table and chairs costs \$550. If he buys 4 sets of tables and chairs, which inequality shows the number of bookcases b he can buy? (Lesson 6-6)

Ⓐ $4(550) + 125b \leq 7500$

Ⓑ $125b \leq 7500$

Ⓒ $4(550 + 125)b \leq 7500$

Ⓓ $4(125) + 550b \leq 7500$

7. Sophia and Allie went shopping and spent \$122 altogether. Sophia spent \$25 less than twice as much as Allie. How much did Allie spend? (Lesson 7-2)

Ⓐ \$39 Ⓑ \$49 Ⓒ \$53 Ⓓ \$73

8. The product of $2x^3$ and $4x^4$ is (Lesson 8-1)

Ⓐ $8x^{12}$. Ⓑ $6x^{12}$. Ⓒ $6x^7$. Ⓓ $8x^7$.

9. If 0.00037 is expressed as 3.7×10^n, what is the value of n? (Lesson 8-3)

Ⓐ -5 Ⓑ -4 Ⓒ 4 Ⓓ 5

10. When $x^2 - 2x + 1$ is subtracted from $3x^2 - 4x + 5$, the result will be (Lesson 8-5)

Ⓐ $2x^2 - 2x + 4$. Ⓑ $2x^2 - 6x + 4$.

Ⓒ $3x^2 - 6x + 6$. Ⓓ $4x^2 - 6x + 6$.

The Princeton Review Test-Taking Tip

Question 5 When you write an equation, check that the given values make a true statement. For example, in Question 5, substitute the values of the coordinates $(-1, 4)$ into your equation to check.

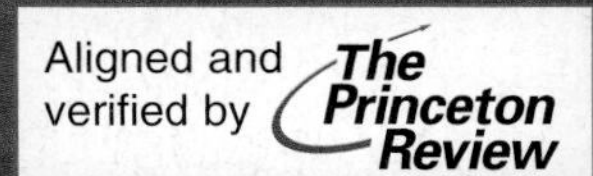

Part 2 Short Response/Grid In

Record your answers on the answer sheet provided by your teacher or on a sheet of paper.

11. Find the 15th term in the arithmetic sequence $-20, -11, -2, 7, \ldots$. (Lesson 4-7)

12. Write a function that includes all of the ordered pairs in the table. (Lesson 4-8)

x	-3	-1	1	3	4
y	12	4	-4	-12	-16

13. Find the y-intercept of the line represented by $3x - 2y + 8 = 0$. (Lesson 5-4)

14. Graph the solution of the linear inequality $3x - y \leq 2$. (Lesson 6-6)

15. Let $P = 3x^2 - 2x - 1$ and $Q = -x^2 + 2x - 2$. Find $P + Q$. (Lesson 8-5)

16. Find $(x^2 + 1)(x - 3)$. (Lesson 8-7)

Part 3 Quantitative Comparison

Compare the quantity in Column A and the quantity in Column B. Then determine whether:

- (A) **the quantity in Column A is greater,**
- (B) **the quantity in Column B is greater,**
- (C) **the two quantities are equal, or**
- (D) **the relationship cannot be determined from the information given.**

	Column A	Column B
17.	the x-coordinate of point A	the y-coordinate of point B

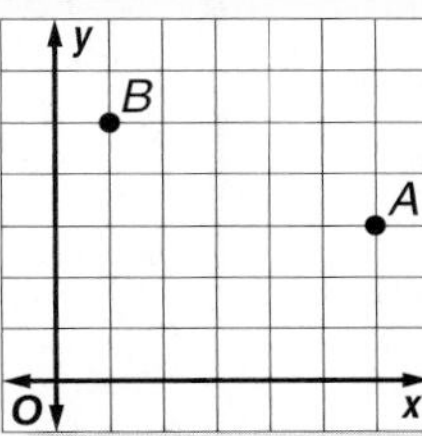

(Lesson 4-3)

	Column A	Column B
18.	$4x - 10 \geq 20$	$\frac{-6(x-1)}{8} \geq 3$

(Lesson 6-3)

	Column A	Column B
19.	the x value in the solution of $x - 3y = 2$ and $x + 3y = 0$	the x value in the solution of $3x + 8y = 6$ and $x - 8y = 2$

(Lesson 7-3)

	Column A	Column B
20.	$\frac{2b^{-3}c^2}{4bc}$	$\frac{10b^4}{20b^8c^{-1}}$

(Lesson 8-2)

	Column A	Column B
21.	5.01×10^{-2}	50.1×10^{-4}

(Lesson 8-3)

	Column A	Column B
22.	the degree of $x^2 + 5 - 6x + 13x^3$	the degree of $10 - y - 2y^2 - 4y^3$

(Lesson 8-4)

23. $m^2 + n^2 = 10$ and $mn = -3$

Column A	Column B
$(m + n)^2$	$(m - n)^2$

(Lesson 8-8)

Part 4 Open Ended

Record your answers on a sheet of paper. Show your work.

24. Use the rectangular prism below to solve the following problems. (Lessons 8-1 and 8-7)

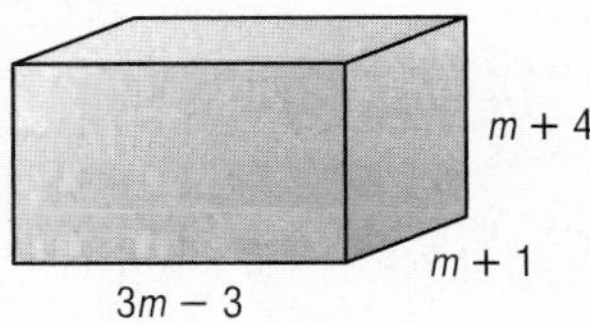

a. Write a polynomial expression that represents the surface area of the top of the prism.

b. Write a polynomial expression that represents the surface area of the front of the prism.

c. Write a polynomial expression that represents the volume of the prism.

d. If $m = 2$ centimeters, then what is the volume of the prism?

www.algebra1.com/standardized_test

Chapter

9 Factoring

What You'll Learn

- **Lesson 9-1** Find the prime factorizations of integers and monomials.
- **Lesson 9-1** Find the greatest common factors (GCF) for sets of integers and monomials.
- **Lessons 9-2 through 9-6** Factor polynomials.
- **Lessons 9-2 through 9-6** Use the Zero Product Property to solve equations.

Key Vocabulary

- factored form (p. 475)
- factoring by grouping (p. 482)
- prime polynomial (p. 497)
- difference of squares (p. 501)
- perfect square trinomials (p. 508)

Why It's Important

The factoring of polynomials can be used to solve a variety of real-world problems and lays the foundation for the further study of polynomial equations. Factoring is used to solve problems involving vertical motion. For example, the height h in feet of a dolphin that jumps out of the water traveling at 20 feet per second is modeled by a polynomial equation. Factoring can be used to determine how long the dolphin is in the air. *You will learn how to solve polynomial equations in Lesson 9-2.*

Getting Started

Prerequisite Skills To be successful in this chapter, you'll need to master these skills and be able to apply them in problem-solving situations. Review these skills before beginning Chapter 9.

For Lessons 9-2 through 9-6 — Distributive Property

Rewrite each expression using the Distributive Property. Then simplify.
(For review, see Lesson 1-5.)

1. $3(4 - x)$ **2.** $a(a + 5)$ **3.** $-7(n^2 - 3n + 1)$ **4.** $6y(-3y - 5y^2 + y^3)$

For Lessons 9-3 and 9-4 — Multiplying Binomials

Find each product. *(For review, see Lesson 8-7.)*

5. $(x + 4)(x + 7)$ **6.** $(3n - 4)(n + 5)$ **7.** $(6a - 2b)(9a + b)$ **8.** $(-x - 8y)(2x - 12y)$

For Lessons 9-5 and 9-6 — Special Products

Find each product. *(For review, see Lesson 8-8.)*

9. $(y + 9)^2$ **10.** $(3a - 2)^2$ **11.** $(n - 5)(n + 5)$ **12.** $(6p + 7q)(6p - 7q)$

For Lesson 9-6 — Square Roots

Find each square root. *(For review, see Lesson 2-7.)*

13. $\sqrt{121}$ **14.** $\sqrt{0.0064}$ **15.** $\sqrt{\frac{25}{36}}$ **16.** $\sqrt{\frac{8}{98}}$

Make this Foldable to help you organize your notes on factoring. Begin with a sheet of plain $8\frac{1}{2}$" by 11" paper.

Step 1 Fold in Sixths

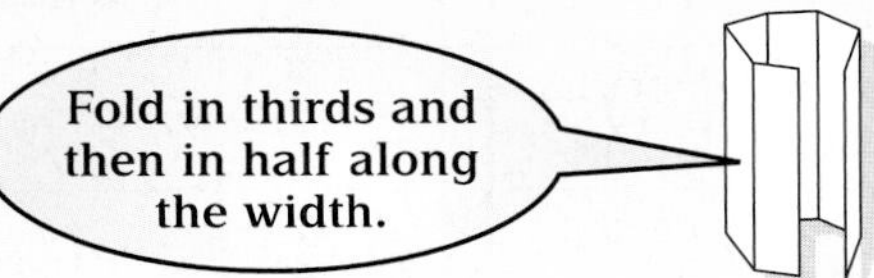

Fold in thirds and then in half along the width.

Step 2 Fold Again

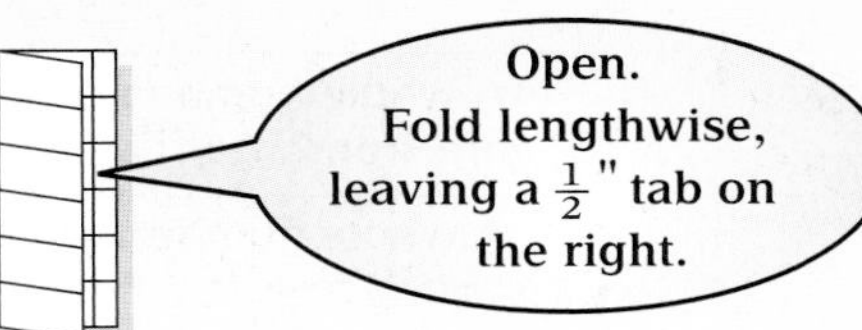

Open. Fold lengthwise, leaving a $\frac{1}{2}$" tab on the right.

Step 3 Cut

Open. Cut short side along folds to make tabs.

Step 4 Label

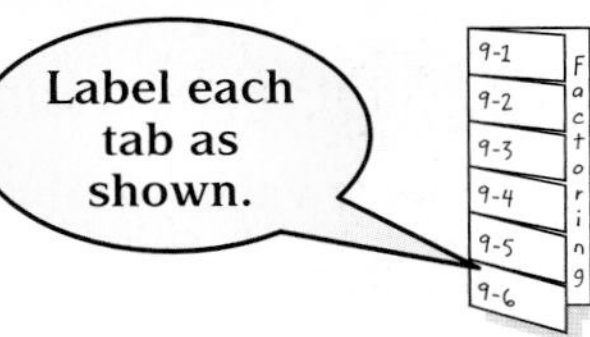

Reading and Writing As you read and study the chapter, write notes and examples for each lesson under its tab.

9-1 Factors and Greatest Common Factors

What You'll Learn

- Find prime factorizations of integers and monomials.
- Find the greatest common factors of integers and monomials.

Vocabulary

- prime number
- composite number
- prime factorization
- factored form
- greatest common factor (GCF)

How are prime numbers related to the search for extraterrestrial life?

In the search for extraterrestrial life, scientists listen to radio signals coming from faraway galaxies. How can they be sure that a particular radio signal was deliberately sent by intelligent beings instead of coming from some natural phenomenon? What if that signal began with a series of beeps in a pattern comprised of the first 30 prime numbers ("beep-beep," "beep-beep-beep," and so on)?

PRIME FACTORIZATION Recall that when two or more numbers are multiplied, each number is a *factor* of the product. Some numbers, like 18, can be expressed as the product of different pairs of whole numbers. This can be shown geometrically. Consider all of the possible rectangles with whole number dimensions that have areas of 18 square units.

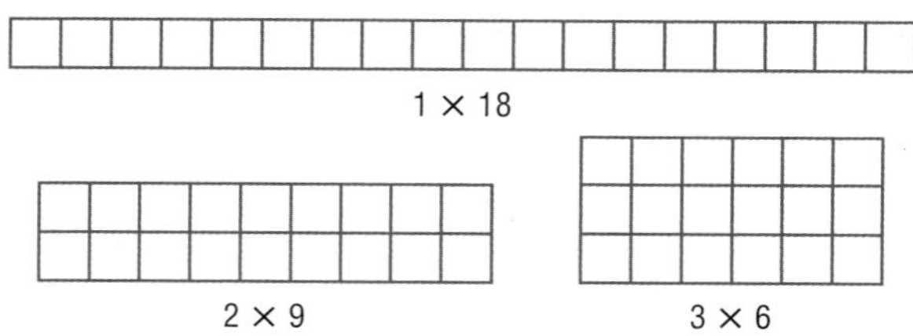

The number 18 has 6 factors, 1, 2, 3, 6, 9, and 18. Whole numbers greater than 1 can be classified by their number of factors.

Key Concept — *Prime and Composite Numbers*

Words	Examples
A whole number, greater than 1, whose only factors are 1 and itself, is called a **prime number**.	2, 3, 5, 7, 11, 13, 17, 19
A whole number, greater than 1, that has more than two factors is called a **composite number**.	4, 6, 8, 9, 10, 12, 14, 15, 16, 18

0 and 1 are neither prime nor composite.

Study Tip

Listing Factors

Notice that in Example 1, 6 is listed as a factor of 36 only once.

Example 1 *Classify Numbers as Prime or Composite*

Factor each number. Then classify each number as *prime* or *composite*.

a. 36

To find the factors of 36, list all pairs of whole numbers whose product is 36.

1×36 $\quad 2 \times 18$ $\quad 3 \times 12$ $\quad 4 \times 9$ $\quad 6 \times 6$

Therefore, the factors of 36, in increasing order, are 1, 2, 3, 4, 6, 9, 12, 18, and 36. Since 36 has more than two factors, it is a composite number.

b. 23

The only whole numbers that can be multiplied together to get 23 are 1 and 23. Therefore, the factors of 23 are 1 and 23. Since the only factors of 23 are 1 and itself, 23 is a prime number.

Study Tip

Prime Numbers
Before deciding that a number is prime, try dividing it by all of the prime numbers that are less than the square root of that number.

When a whole number is expressed as the product of factors that are all prime numbers, the expression is called the **prime factorization** of the number.

Example 2 Prime Factorization of a Positive Integer

Find the prime factorization of 90.

Method 1

$90 = 2 \cdot 45$ The least prime factor of 90 is 2.

$= 2 \cdot 3 \cdot 15$ The least prime factor of 45 is 3.

$= 2 \cdot 3 \cdot 3 \cdot 5$ The least prime factor of 15 is 3.

All of the factors in the last row are prime. Thus, the prime factorization of 90 is $2 \cdot 3 \cdot 3 \cdot 5$.

Method 2

Use a factor tree.

90

$9 \cdot 10$ $90 = 9 \cdot 10$

$3 \cdot 3 \cdot 2 \cdot 5$ $9 = 3 \cdot 3$ and $10 = 2 \cdot 5$

All of the factors in the last branch of the factor tree are prime. Thus, the prime factorization of 90 is $2 \cdot 3 \cdot 3 \cdot 5$ or $2 \cdot 3^2 \cdot 5$.

Usually the factors are ordered from the least prime factor to the greatest.

Study Tip

Unique Factorization Theorem
The prime factorization of every number is unique except for the order in which the factors are written.

A negative integer is factored completely when it is expressed as the product of -1 and prime numbers.

Example 3 Prime Factorization of a Negative Integer

Find the prime factorization of −140.

$-140 = -1 \cdot 140$ Express −140 as −1 times 140.

$= -1 \cdot 2 \cdot 70$ $140 = 2 \cdot 70$

$= -1 \cdot 2 \cdot 7 \cdot 10$ $70 = 7 \cdot 10$

$= -1 \cdot 2 \cdot 7 \cdot 2 \cdot 5$ $10 = 2 \cdot 5$

Thus, the prime factorization of -140 is $-1 \cdot 2 \cdot 2 \cdot 5 \cdot 7$ or $-1 \cdot 2^2 \cdot 5 \cdot 7$.

A monomial is in **factored form** when it is expressed as the product of prime numbers and variables and no variable has an exponent greater than 1.

Example 4 Prime Factorization of a Monomial

Factor each monomial completely.

a. $\mathbf{12a^2b^3}$

$12a^2b^3 = 2 \cdot 6 \cdot a \cdot a \cdot b \cdot b \cdot b$ $\quad 12 = 2 \cdot 6$, $a^2 = a \cdot a$, and $b^3 = b \cdot b \cdot b$

$= 2 \cdot 2 \cdot 3 \cdot a \cdot a \cdot b \cdot b \cdot b$ $\quad 6 = 2 \cdot 3$

Thus, $12a^2b^3$ in factored form is $2 \cdot 2 \cdot 3 \cdot a \cdot a \cdot b \cdot b \cdot b$.

b. $\mathbf{-66pq^2}$

$-66pq^2 = -1 \cdot 66 \cdot p \cdot q \cdot q$ $\quad$ Express -66 as -1 times 66.

$= -1 \cdot 2 \cdot 33 \cdot p \cdot q \cdot q$ $\quad 66 = 2 \cdot 33$

$= -1 \cdot 2 \cdot 3 \cdot 11 \cdot p \cdot q \cdot q$ $\quad 33 = 3 \cdot 11$

Thus, $-66pq^2$ in factored form is $-1 \cdot 2 \cdot 3 \cdot 11 \cdot p \cdot q \cdot q$.

GREATEST COMMON FACTOR Two or more numbers may have some common prime factors. Consider the prime factorization of 48 and 60.

$48 = (2) \cdot (2) \cdot 2 \cdot 2 \cdot (3)$ $\quad$ Factor each number.

$60 = (2) \cdot (2) \cdot (3) \cdot 5$ $\quad$ Circle the common prime factors.

The integers 48 and 60 have two 2s and one 3 as common prime factors. The product of these common prime factors, $2 \cdot 2 \cdot 3$ or 12, is called the **greatest common factor (GCF)** of 48 and 60. The GCF is the greatest number that is a factor of both original numbers.

Key Concept — Greatest Common Factor (GCF)

- The GCF of two or more integers is the product of the prime factors common to the integers.
- The GCF of two or more monomials is the product of their common factors when each monomial is in factored form.
- If two or more integers or monomials have a GCF of 1, then the integers or monomials are said to be *relatively prime*.

Example 5 GCF of a Set of Monomials

Find the GCF of each set of monomials.

a. 15 and 16

$15 = 3 \cdot 5$ $\quad$ Factor each number.

$16 = 2 \cdot 2 \cdot 2 \cdot 2$ $\quad$ Circle the common prime factors, if any.

There are no common prime factors, so the GCF of 15 and 16 is 1. This means that 15 and 16 are relatively prime.

b. $\mathbf{36x^2y}$ **and** $\mathbf{54xy^2z}$

$36x^2y = (2) \cdot 2 \cdot (3) \cdot (3) \cdot (x) \cdot x \cdot (y)$ $\quad$ Factor each number.

$54xy^2z = (2) \cdot (3) \cdot (3) \cdot 3 \cdot (x) \cdot (y) \cdot y \cdot z$ $\quad$ Circle the common prime factors.

The GCF of $36x^2y$ and $54xy^2z$ is $2 \cdot 3 \cdot 3 \cdot x \cdot y$ or $18xy$.

Study Tip

Alternative Method

You can also find the greatest common factor by listing the factors of each number and finding which of the common factors is the greatest. Consider Example 5a.

15: (1) 3, 5, 15
16: (1) 2, 4, 8, 16

The only common factor, and therefore, the greatest common factor, is 1.

Example 6 Use Factors

GEOMETRY The area of a rectangle is 28 square inches. If the length and width are both whole numbers, what is the maximum perimeter of the rectangle?

Find the factors of 28, and draw rectangles with each length and width. Then find each perimeter.

The factors of 28 are 1, 2, 4, 7, 14, and 28.

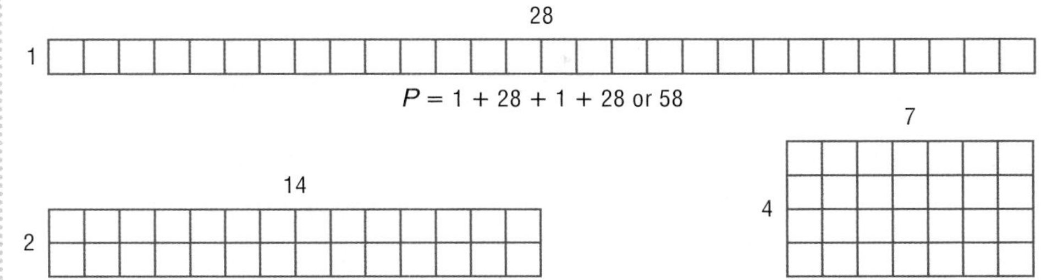

14

2

P = 2 + 14 + 2 + 14 or 32

7

4

P = 4 + 7 + 4 + 7 or 22

The greatest perimeter is 58 inches. The rectangle with this perimeter has a length of 28 inches and a width of 1 inch.

Check for Understanding

Concept Check

1. **Determine** whether the following statement is *true* or *false*. If false, provide a counterexample.
   All prime numbers are odd.
2. **Explain** what it means for two numbers to be relatively prime.
3. **OPEN ENDED** Name two monomials whose GCF is $5x^2$.

Guided Practice

Find the factors of each number. Then classify each number as *prime* or *composite*.

4. 8
5. 17
6. 112

Find the prime factorization of each integer.

7. 45
8. −32
9. −150

Factor each monomial completely.

10. $4p^2$
11. $39b^3c^2$
12. $-100x^3yz^2$

Find the GCF of each set of monomials.

13. 10, 15
14. $18xy$, $36y^2$
15. 54, 63, 180
16. $25n$, $21m$
17. $12a^2b$, $90a^2b^2c$
18. $15r^2$, $35s^2$, $70rs$

Application

19. **GARDENING** Ashley is planting 120 tomato plants in her garden. In what ways can she arrange them so that she has the same number of plants in each row, at least 5 rows of plants, and at least 5 plants in each row?

Practice and Apply

Find the factors of each number. Then classify each number as *prime* or *composite*.

20. 19
21. 25
22. 80
23. 61
24. 91
25. 119
26. 126
27. 304

Homework Help

For Exercises	See Examples
20–27, 62, 65, 66	1
32–39	2, 3
40–47	4
48–61, 63, 64	5
28–31, 67	6

Extra Practice

See page 839.

GEOMETRY **For Exercises 28 and 29, consider a rectangle whose area is 96 square millimeters and whose length and width are both whole numbers.**

28. What is the minimum perimeter of the rectangle? Explain your reasoning.

29. What is the maximum perimeter of the rectangle? Explain your reasoning.

COOKIES **For Exercises 30 and 31, use the following information.**
A bakery packages cookies in two sizes of boxes, one with 18 cookies and the other with 24 cookies. A small number of cookies are to be wrapped in cellophane before they are placed in a box. To save money, the bakery will use the same size cellophane packages for each box.

30. How many cookies should the bakery place in each cellophane package to maximize the number of cookies in each package?

31. How many cellophane packages will go in each size box?

Find the prime factorization of each integer.

32. 39 **33.** -98 **34.** 117 **35.** 102

36. -115 **37.** 180 **38.** 360 **39.** -462

Factor each monomial completely.

40. $66d^4$ **41.** $85x^2y^2$ **42.** $49a^3b^2$ **43.** $50gh$

44. $128pq^2$ **45.** $243n^3m$ **46.** $-183xyz^3$ **47.** $-169a^2bc^2$

Find the GCF of each set of monomials.

48. 27, 72 **49.** 18, 35 **50.** 32, 48

51. 84, 70 **52.** 16, 20, 64 **53.** 42, 63, 105

54. $15a$, $28b^2$ **55.** $24d^2$, $30c^2d$ **56.** $20gh$, $36g^2h^2$

57. $21p^2q$, $32r^2t$ **58.** $18x$, $30xy$, $54y$ **59.** $28a^2$, $63a^3b^2$, $91b^3$

60. $14m^2n^2$, $18mn$, $2m^2n^3$ **61.** $80a^2b$, $96a^2b^3$, $128a^2b^2$

62. NUMBER THEORY *Twin primes* are two consecutive odd numbers that are prime. The first pair of twin primes is 3 and 5. List the next five pairs of twin primes.

More About. . .

Marching Bands

Drum Corps International (DCI) is a nonprofit youth organization serving junior drum and bugle corps around the world. Members of these marching bands range from 14 to 21 years of age.

Source: www.dci.org

MARCHING BANDS **For Exercises 63 and 64, use the following information.**
Central High's marching band has 75 members, and the band from Northeast High has 90 members. During the halftime show, the bands plan to march into the stadium from opposite ends using formations with the same number of rows.

63. If the bands want to match up in the center of the field, what is the maximum number of rows?

64. How many band members will be in each row after the bands are combined?

NUMBER THEORY **For Exercises 65 and 66, use the following information.**
One way of generating prime numbers is to use the formula $2^p - 1$, where p is a prime number. Primes found using this method are called *Mersenne primes*. For example, when $p = 2$, $2^2 - 1 = 3$. The first Mersenne prime is 3.

65. Find the next two Mersenne primes.

66. Will this formula generate all possible prime numbers? Explain your reasoning.

Online Research **Data Update** What is the greatest known prime number? Visit www.algebra1.com/data_update to learn more.

67. **GEOMETRY** The area of a triangle is 20 square centimeters. What are possible whole-number dimensions for the base and height of the triangle?

Finding the GCF of distances will help you make a scale model of the solar system. Visit www.algebra1.com/webquest to continue work on your WebQuest project.

68. **CRITICAL THINKING** Suppose 6 is a factor of ab, where a and b are natural numbers. Make a valid argument to explain why each assertion is true or provide a counterexample to show that an assertion is false.

a. 6 must be a factor of a *or* of b.

b. 3 must be a factor of a *or* of b.

c. 3 must be a factor of a *and* of b.

69. **WRITING IN MATH** Answer the question that was posed at the beginning of the lesson.

How are prime numbers related to the search for extraterrestrial life?

Include the following in your answer:

- a list of the first 30 prime numbers and an explanation of how you found them, and
- an explanation of why a signal of this kind might indicate that an extraterrestrial message is to follow.

70. Miko claims that there are at least four ways to design a 120-square-foot rectangular space that can be tiled with 1-foot by 1-foot tiles. Which statement best describes this claim?

(A) Her claim is false because 120 is a prime number.
(B) Her claim is false because 120 is not a perfect square.
(C) Her claim is true because 240 is a multiple of 120.
(D) Her claim is true because 120 has at least eight factors.

71. Suppose Ψ_x is defined as the largest prime factor of x. For which of the following values of x would Ψ_x have the greatest value?

(A) 53 (B) 74 (C) 99 (D) 117

Maintain Your Skills

Mixed Review

Find each product. *(Lessons 8-7 and 8-8)*

72. $(2x - 1)^2$
73. $(3a + 5)(3a - 5)$
74. $(7p^2 + 4)(7p^2 + 4)$
75. $(6r + 7)(2r - 5)$
76. $(10h + k)(2h + 5k)$
77. $(b + 4)(b^2 + 3b - 18)$

Find the value of r so that the line that passes through the given points has the given slope. *(Lesson 5-1)*

78. $(1, 2), (-2, r), m = 3$
79. $(-5, 9), (r, 6), m = -\frac{3}{5}$

80. **RETAIL SALES** A department store buys clothing at wholesale prices and then marks the clothing up 25% to sell at retail price to customers. If the retail price of a jacket is \$79, what was the wholesale price? *(Lesson 3-7)*

Getting Ready for the Next Lesson

PREREQUISITE SKILL Use the Distributive Property to rewrite each expression. *(To review the **Distributive Property**, see Lesson 1-5.)*

81. $5(2x + 8)$
82. $a(3a + 1)$
83. $2g(3g - 4)$
84. $-4y(3y - 6)$
85. $7b + 7c$
86. $2x + 3x$

Algebra Activity

A Preview of Lesson 9-2

Factoring Using the Distributive Property

Sometimes you know the product of binomials and are asked to find the factors. This is called factoring. You can use algebra tiles and a product mat to factor binomials.

Activity 1 Use algebra tiles to factor $3x + 6$.

Step 1 Model the polynomial $3x + 6$.

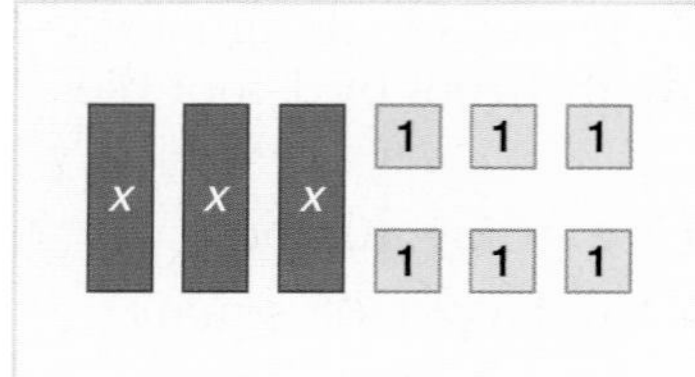

Step 2 Arrange the tiles into a rectangle. The total area of the rectangle represents the product, and its length and width represent the factors.

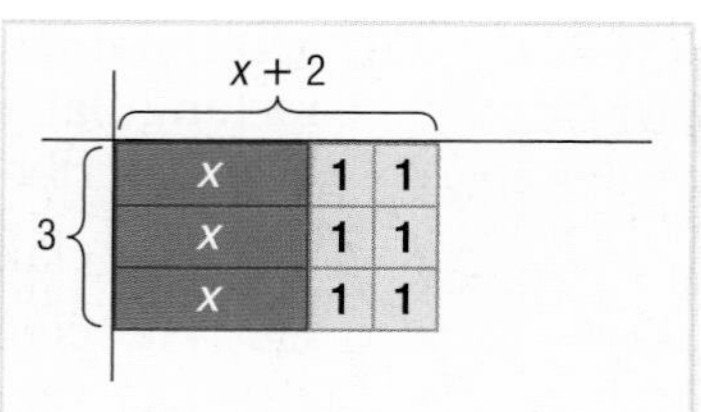

The rectangle has a width of 3 and a length of $x + 2$. So, $3x + 6 = 3(x + 2)$.

Activity 2 Use algebra tiles to factor $x^2 - 4x$.

Step 1 Model the polynomial $x^2 - 4x$.

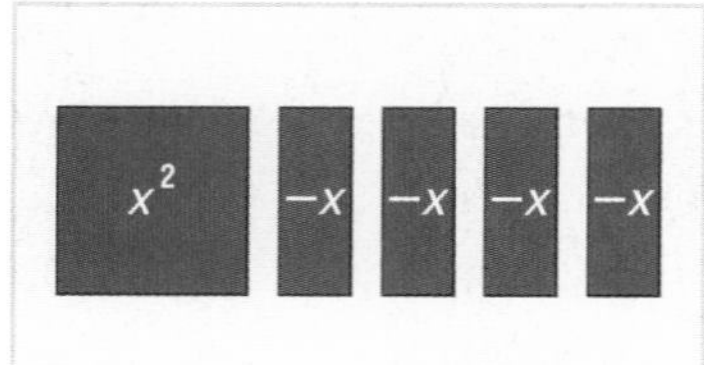

Step 2 Arrange the tiles into a rectangle.

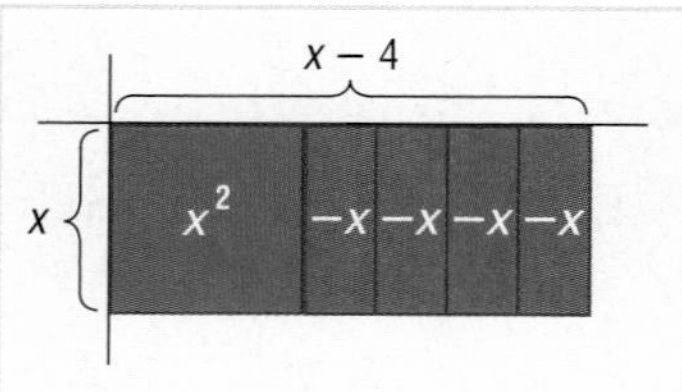

The rectangle has a width of x and a length of $x - 4$. So, $x^2 - 4x = x(x - 4)$.

Model and Analyze

Use algebra tiles to factor each binomial.

1. $2x + 10$ **2.** $6x - 8$ **3.** $5x^2 + 2x$ **4.** $9 - 3x$

Tell whether each binomial can be factored. Justify your answer with a drawing.

5. $4x - 10$ **6.** $3x - 7$ **7.** $x^2 + 2x$ **8.** $2x^2 + 3$

9. MAKE A CONJECTURE Write a paragraph that explains how you can use algebra tiles to determine whether a binomial can be factored. Include an example of one binomial that can be factored and one that cannot.

9-2 Factoring Using the Distributive Property

What You'll Learn

- Factor polynomials by using the Distributive Property.
- Solve quadratic equations of the form $ax^2 + bx = 0$.

Vocabulary

- factoring
- factoring by grouping

How can you determine how long a baseball will remain in the air?

Nolan Ryan, the greatest strike-out pitcher in the history of baseball, had a fastball clocked at 98 miles per hour or about 151 feet per second. If he threw a ball directly upward with the same velocity, the height h of the ball in feet above the point at which he released it could be modeled by the formula $h = 151t - 16t^2$, where t is the time in seconds. You can use factoring and the Zero Product Property to determine how long the ball would remain in the air before returning to his glove.

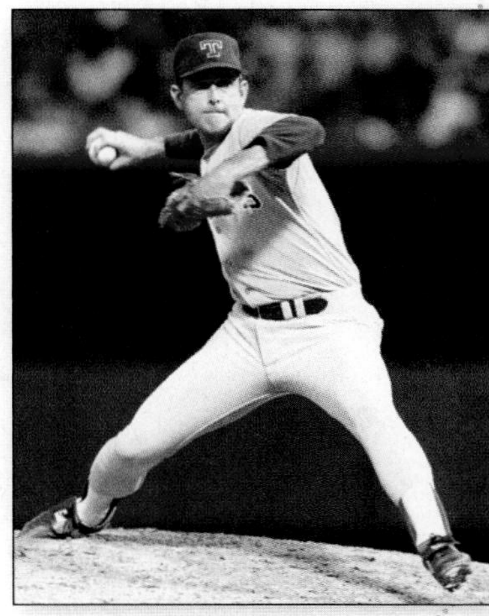

FACTOR BY USING THE DISTRIBUTIVE PROPERTY In Chapter 8, you used the Distributive Property to multiply a polynomial by a monomial.

$$\begin{aligned} 2a(6a + 8) &= 2a(6a) + 2a(8) \\ &= 12a^2 + 16a \end{aligned}$$

Study Tip

Look Back
To review the **Distributive Property**, see Lesson 1-5.

You can reverse this process to express a polynomial as the product of a monomial factor and a polynomial factor.

$$\begin{aligned} 12a^2 + 16a &= 2a(6a) + 2a(8) \\ &= 2a(6a + 8) \end{aligned}$$

Thus, a *factored form* of $12a^2 + 16a$ is $2a(6a + 8)$.

Factoring a polynomial means to find its *completely* factored form. The expression $2a(6a + 8)$ is not completely factored since $6a + 8$ can be factored as $2(3a + 4)$.

Example 1 Use the Distributive Property

Use the Distributive Property to factor each polynomial.

a. $\mathbf{12a^2 + 16a}$

First, find the GCF of $12a^2$ and $16a$.

$12a^2 = (2) \cdot (2) \cdot 3 \cdot (a) \cdot a$ Factor each number.

$16a = (2) \cdot (2) \cdot 2 \cdot 2 \cdot (a)$ Circle the common prime factors.

GCF: $2 \cdot 2 \cdot a$ or $4a$

Write each term as the product of the GCF and its remaining factors. Then use the Distributive Property to factor out the GCF.

$$\begin{aligned} 12a^2 + 16a &= 4a(3 \cdot a) + 4a(2 \cdot 2) && \text{Rewrite each term using the GCF.} \\ &= 4a(3a) + 4a(4) && \text{Simplify remaining factors.} \\ &= 4a(3a + 4) && \text{Distributive Property} \end{aligned}$$

Thus, the completely factored form of $12a^2 + 16a$ is $4a(3a + 4)$.

b. $18cd^2 + 12c^2d + 9cd$

$18cd^2 = 2 \cdot 3 \cdot 3 \cdot c \cdot d \cdot d$ Factor each number.

$12c^2d = 2 \cdot 2 \cdot 3 \cdot c \cdot c \cdot d$ Circle the common prime factors.

$9cd = 3 \cdot 3 \cdot c \cdot d$

GCF: $3 \cdot c \cdot d$ or $3cd$

$18cd^2 + 12c^2d + 9cd = 3cd(6d) + 3cd(4c) + 3cd(3)$ Rewrite each term using the GCF.

$= 3cd(6d + 4c + 3)$ Distributive Property

The Distributive Property can also be used to factor some polynomials having four or more terms. This method is called **factoring by grouping** because pairs of terms are grouped together and factored. The Distributive Property is then applied a second time to factor a common binomial factor.

Example 2 Use Grouping

Factor $4ab + 8b + 3a + 6$.

$4ab + 8b + 3a + 6$

$= (4ab + 8b) + (3a + 6)$ Group terms with common factors.

$= 4b(a + 2) + 3(a + 2)$ Factor the GCF from each grouping.

$= (a + 2)(4b + 3)$ Distributive Property

CHECK Use the FOIL method.

$(a + 2)(4b + 3) = \overset{F}{(a)(4b)} + \overset{O}{(a)(3)} + \overset{I}{(2)(4b)} + \overset{L}{(2)(3)}$

$= 4ab + 3a + 8b + 6$ ✓

Study Tip

Factoring by Grouping

Sometimes you can group terms in more than one way when factoring a polynomial. For example, the polynomial in Example 2 could have been factored in the following way.

$4ab + 8b + 3a + 6$

$= (4ab + 3a) + (8b + 6)$

$= a(4b + 3) + 2(4b + 3)$

$= (4b + 3)(a + 2)$

Notice that this result is the same as in Example 2.

Recognizing binomials that are additive inverses is often helpful when factoring by grouping. For example, $7 - y$ and $y - 7$ are additive inverses because their sum is 0. By rewriting $7 - y$ in the factored form $-1(y - 7)$, factoring by grouping is made possible in the following example.

Example 3 Use the Additive Inverse Property

Factor $35x - 5xy + 3y - 21$.

$35x - 5xy + 3y - 21 = (35x - 5xy) + (3y - 21)$ Group terms with common factors.

$= 5x(7 - y) + 3(y - 7)$ Factor the GCF from each grouping.

$= 5x(-1)(y - 7) + 3(y - 7)$ $7 - y = -1(y - 7)$

$= -5x(y - 7) + 3(y - 7)$ $5x(-1) = -5x$

$= (y - 7)(-5x + 3)$ Distributive Property

Study Tip

Factoring Trinomials

Since the order in which factors are multiplied does not affect the product, $(-5x + 3)(y - 7)$ is also a correct factoring of $35x - 5xy + 3y - 21$.

Concept Summary — Factoring by Grouping

- **Words** A polynomial can be factored by grouping if all of the following situations exist.
  - There are four or more terms.
  - Terms with common factors can be grouped together.
  - The two common factors are identical or are additive inverses of each other.
- **Symbols** $ax + bx + ay + by = x(a + b) + y(a + b)$
  $= (a + b)(x + y)$

SOLVE EQUATIONS BY FACTORING Some equations can be solved by factoring. Consider the following products.

$$6(0) = 0 \qquad 0(-3) = 0 \qquad (5 - 5)(0) = 0 \qquad -2(-3 + 3) = 0$$

Notice that in each case, *at least one* of the factors is zero. These examples illustrate the **Zero Product Property**.

Key Concept — *Zero Product Property*

- **Words** If the product of two factors is 0, then at least one of the factors must be 0.
- **Symbols** For any real numbers a and b, if $ab = 0$, then either $a = 0$, $b = 0$, or both a and b equal zero.

Example 4 Solve an Equation in Factored Form

Solve $(d - 5)(3d + 4) = 0$. Then check the solutions.

If $(d - 5)(3d + 4) = 0$, then according to the Zero Product Property either $d - 5 = 0$ or $3d + 4 = 0$.

$(d - 5)(3d + 4) = 0$	Original equation
$d - 5 = 0$ or $3d + 4 = 0$	Set each factor equal to zero.
$d = 5$ $\qquad$ $3d = -4$	Solve each equation.
$d = -\frac{4}{3}$	

The solution set is $\left\{5, -\frac{4}{3}\right\}$.

CHECK Substitute 5 and $-\frac{4}{3}$ for d in the original equation.

$$(d - 5)(3d + 4) = 0 \qquad\qquad (d - 5)(3d + 4) = 0$$

$$(5 - 5)[3(5) + 4] \stackrel{?}{=} 0 \qquad\qquad \left(-\frac{4}{3} - 5\right)\left[3\left(-\frac{4}{3}\right) + 4\right] \stackrel{?}{=} 0$$

$$(0)(19) \stackrel{?}{=} 0 \qquad\qquad \left(-\frac{19}{3}\right)(0) \stackrel{?}{=} 0$$

$$0 = 0 \ \checkmark \qquad\qquad 0 = 0 \ \checkmark$$

If an equation can be written in the form $ab = 0$, then the Zero Product Property can be applied to solve that equation.

Study Tip

Common Misconception
You may be tempted to try to solve the equation in Example 5 by dividing each side of the equation by x. Remember, however, that x is an *unknown* quantity. If you divide by x, you may actually be dividing by zero, which is undefined.

Example 5 Solve an Equation by Factoring

Solve $x^2 = 7x$. Then check the solutions.

Write the equation so that it is of the form $ab = 0$.

$x^2 = 7x$	Original equation
$x^2 - 7x = 0$	Subtract $7x$ from each side.
$x(x - 7) = 0$	Factor the GCF of x^2 and $-7x$, which is x.
$x = 0$ or $x - 7 = 0$	Zero Product Property
$x = 7$	Solve each equation.

The solution set is $\{0, 7\}$. Check by substituting 0 and 7 for x in the original equation.

Check for Understanding

Concept Check

1. **Write** $4x^2 + 12x$ as a product of factors in three different ways. Then decide which of the three is the completely factored form. Explain your reasoning.
2. **OPEN ENDED** Give an example of the type of equation that can be solved by using the Zero Product Property.
3. **Explain** why $(x - 2)(x + 4) = 0$ cannot be solved by dividing each side by $x - 2$.

Guided Practice

Factor each polynomial.

4. $9x^2 + 36x$
5. $16xz - 40xz^2$
6. $24m^2np^2 + 36m^2n^2p$
7. $2a^3b^2 + 8ab + 16a^2b^3$
8. $5y^2 - 15y + 4y - 12$
9. $5c - 10c^2 + 2d - 4cd$

Solve each equation. Check your solutions.

10. $h(h + 5) = 0$
11. $(n - 4)(n + 2) = 0$
12. $5m = 3m^2$

Application

PHYSICAL SCIENCE For Exercises 13–15, use the information below and in the graphic.

A flare is launched from a life raft. The height h of the flare in feet above the sea is modeled by the formula $h = 100t - 16t^2$, where t is the time in seconds after the flare is launched.

13. At what height is the flare when it returns to the sea?
14. Let $h = 0$ in the equation $h = 100t - 16t^2$ and solve for t.
15. How many seconds will it take for the flare to return to the sea? Explain your reasoning.

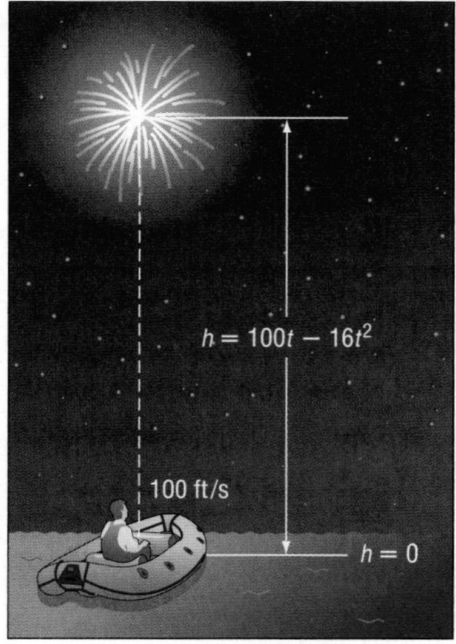

Practice and Apply

Homework Help

For Exercises	See Examples
16–29, 40–47	1
30–39	2, 3
48–61	4, 5

Extra Practice

See page 840.

Factor each polynomial.

16. $5x + 30y$
17. $16a + 4b$
18. $a^5b - a$
19. $x^3y^2 + x$
20. $21cd - 3d$
21. $14gh - 18h$
22. $15a^2y - 30ay$
23. $8bc^2 + 24bc$
24. $12x^2y^2z + 40xy^3z^2$
25. $18a^2bc^2 - 48abc^3$
26. $a + a^2b^2 + a^3b^3$
27. $15x^2y^2 + 25xy + x$
28. $12ax^3 + 20bx^2 + 32cx$
29. $3p^3q - 9pq^2 + 36pq$
30. $x^2 + 2x + 3x + 6$
31. $x^2 + 5x + 7x + 35$
32. $4x^2 + 14x + 6x + 21$
33. $12y^2 + 9y + 8y + 6$
34. $6a^2 - 15a - 8a + 20$
35. $18x^2 - 30x - 3x + 5$
36. $4ax + 3ay + 4bx + 3by$
37. $2my + 7x + 7m + 2xy$
38. $8ax - 6x - 12a + 9$
39. $10x^2 - 14xy - 15x + 21y$

GEOMETRY For Exercises 40 and 41, use the following information.
A quadrilateral has 4 sides and 2 diagonals. A pentagon has 5 sides and 5 diagonals. You can use $\frac{1}{2}n^2 - \frac{3}{2}n$ to find the number of diagonals in a polygon with n sides.

40. Write this expression in factored form.
41. Find the number of diagonals in a decagon (10-sided polygon).

SOFTBALL For Exercises 42 and 43, use the following information.
Albertina is scheduling the games for a softball league. To find the number of games she needs to schedule, she uses the equation $g = \frac{1}{2}n^2 - \frac{1}{2}n$, where g represents the number of games needed for each team to play each other team exactly once and n represents the number of teams.

42. Write this equation in factored form.

43. How many games are needed for 7 teams to play each other exactly 3 times?

GEOMETRY Write an expression in factored form for the area of each shaded region.

44.
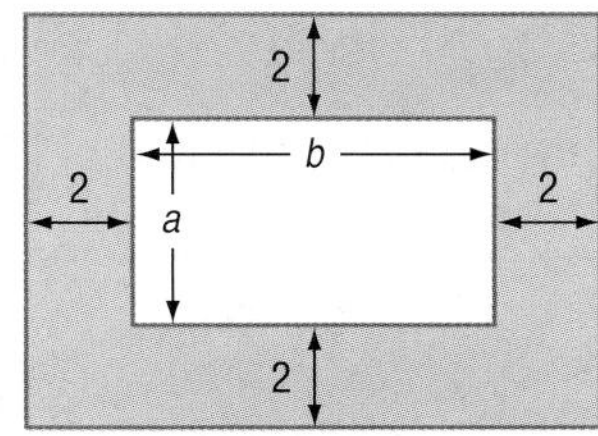

45.
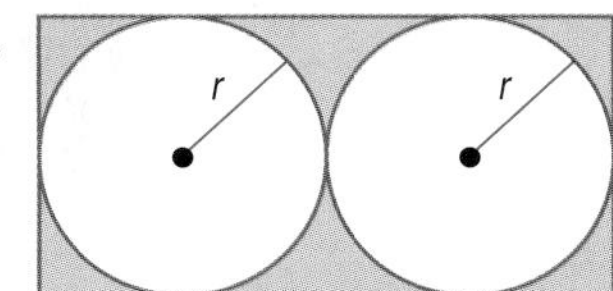

GEOMETRY Find an expression for the area of a square with the given perimeter.

46. $P = 12x + 20y$ in.

47. $P = 36a - 16b$ cm

Solve each equation. Check your solutions.

48. $x(x - 24) = 0$

49. $a(a + 16) = 0$

50. $(q + 4)(3q - 15) = 0$

51. $(3y + 9)(y - 7) = 0$

52. $(2b - 3)(3b - 8) = 0$

53. $(4n + 5)(3n - 7) = 0$

54. $3z^2 + 12z = 0$

55. $7d^2 - 35d = 0$

56. $2x^2 = 5x$

57. $7x^2 = 6x$

58. $6x^2 = -4x$

59. $20x^2 = -15x$

Career Choices

Marine Biologist

Marine biologists study factors that affect organisms living in and near the ocean.

Online Research
For information about a career as a marine biologist, visit: www.algebra1.com/careers

Source: National Sea Grant Library

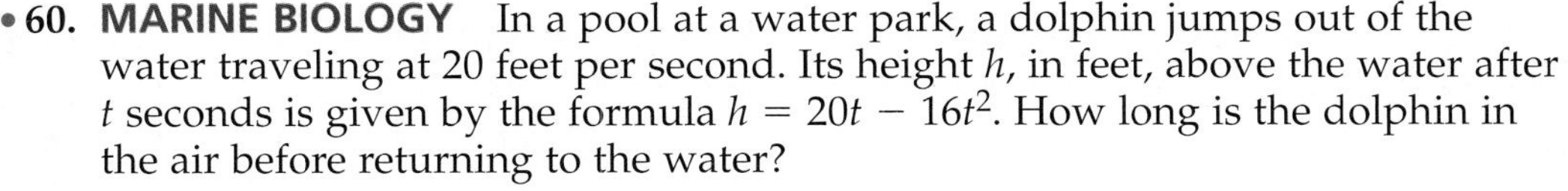

60. **MARINE BIOLOGY** In a pool at a water park, a dolphin jumps out of the water traveling at 20 feet per second. Its height h, in feet, above the water after t seconds is given by the formula $h = 20t - 16t^2$. How long is the dolphin in the air before returning to the water?

61. **BASEBALL** Malik popped a ball straight up with an initial upward velocity of 45 feet per second. The height h, in feet, of the ball above the ground is modeled by the equation $h = 2 + 45t - 16t^2$. How long was the ball in the air if the catcher catches the ball when it is 2 feet above the ground?

62. **CRITICAL THINKING** Factor $a^{x+y} + a^x b^y - a^y b^x - b^{x+y}$.

63. **WRITING IN MATH** Answer the question that was posed at the beginning of the lesson.

How can you determine how long a baseball will remain in the air?

Include the following in your answer:

- an explanation of how to use factoring and the Zero Product Property to find how long the ball would be in the air, and
- an interpretation of each solution in the context of the problem.

Standardized Test Practice

64. The total number of feet in x yards, y feet, and z inches is

Ⓐ $3x + y + \frac{z}{12}$. Ⓑ $12(x + y + z)$.

Ⓒ $x = 3y + 36z$. Ⓓ $\frac{x}{36} + \frac{y}{12} + z$.

65. **QUANTITATIVE COMPARISON** **Compare the quantity in Column A and the quantity in Column B. Then determine whether:**

Ⓐ **the quantity in Column A is greater,**

Ⓑ **the quantity in Column B is greater,**

Ⓒ **the two quantities are equal, or**

Ⓓ **the relationship cannot be determined from the information given.**

Column A	Column B
the negative solution of $(a - 2)(a + 5) = 0$	the negative solution of $(b + 6)(b - 1) = 0$

Maintain Your Skills

Mixed Review

Find the factors of each number. Then classify each number as *prime* or *composite*. *(Lesson 9-1)*

66. 123 67. 300 68. 67

Find each product. *(Lesson 8-8)*

69. $(4s^3 + 3)^2$ 70. $(2p + 5q)(2p - 5q)$ 71. $(3k + 8)(3k + 8)$

Simplify. Assume that no denominator is equal to zero. *(Lesson 8-2)*

72. $\frac{s^4}{s^{-7}}$ 73. $\frac{18x^3y^{-1}}{12x^2y^4}$ 74. $\frac{34p^7q^2r^{-5}}{17(p^3qr^{-1})^2}$

75. **FINANCE** Michael uses at most 60% of his annual FlynnCo stock dividend to purchase more shares of FlynnCo stock. If his dividend last year was $885 and FlynnCo stock is selling for $14 per share, what is the greatest number of shares that he can purchase? *(Lesson 6-2)*

Getting Ready for the Next Lesson

PREREQUISITE SKILL **Find each product.**
*(To review **multiplying polynomials**, see Lesson 8-7.)*

76. $(n + 8)(n + 3)$ 77. $(x - 4)(x - 5)$ 78. $(b - 10)(b + 7)$

79. $(3a + 1)(6a - 4)$ 80. $(5p - 2)(9p - 3)$ 81. $(2y - 5)(4y + 3)$

Practice Quiz 1

Lessons 9-1 and 9-2

1. Find the factors of 225. Then classify the number as *prime* or *composite*. *(Lesson 9-1)*
2. Find the prime factorization of -320. *(Lesson 9-1)*
3. Factor $78a^2bc^3$ completely. *(Lesson 9-1)*
4. Find the GCF of $54x^3$, $42x^2y$, and $30xy^2$. *(Lesson 9-1)*

Factor each polynomial. *(Lesson 9-2)*

5. $4xy^2 - xy$ 6. $32a^2b + 40b^3 - 8a^2b^2$ 7. $6py + 16p - 15y - 40$

Solve each equation. Check your solutions. *(Lesson 9-2)*

8. $(8n + 5)(n - 4) = 0$ 9. $9x^2 - 27x = 0$ 10. $10x^2 = -3x$

Algebra Activity

A Preview of Lesson 9-3

Factoring Trinomials

You can use algebra tiles to factor trinomials. If a polynomial represents the area of a rectangle formed by algebra tiles, then the rectangle's length and width are *factors* of the area.

Activity 1 Use algebra tiles to factor $x^2 + 6x + 5$.

Step 1 Model the polynomial $x^2 + 6x + 5$.

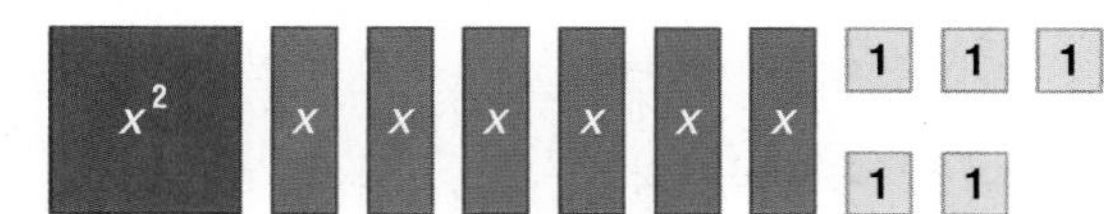

Step 2 Place the x^2 tile at the corner of the product mat. Arrange the 1 tiles into a rectangular array. Because 5 is prime, the 5 tiles can be arranged in a rectangle in one way, a 1-by-5 rectangle.

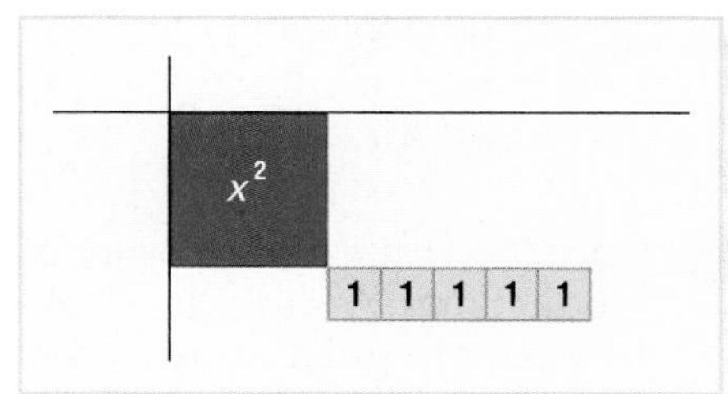

Step 3 Complete the rectangle with the x tiles.

The rectangle has a width of $x + 1$ and a length of $x + 5$. Therefore, $x^2 + 6x + 5 = (x + 1)(x + 5)$.

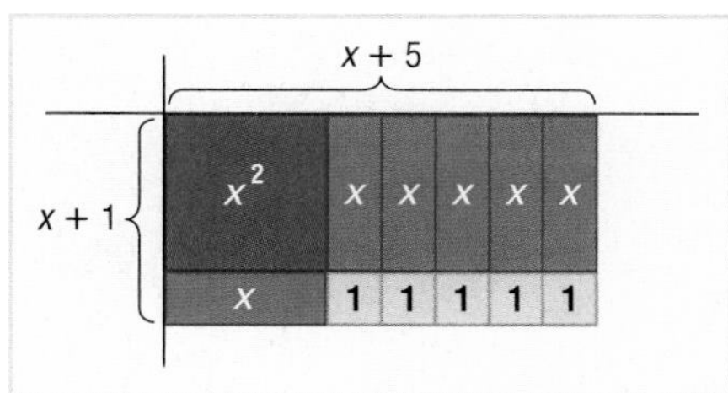

Activity 2 Use algebra tiles to factor $x^2 + 7x + 6$.

Step 1 Model the polynomial $x^2 + 7x + 6$.

Step 2 Place the x^2 tile at the corner of the product mat. Arrange the 1 tiles into a rectangular array. Since $6 = 2 \times 3$, try a 2-by-3 rectangle. Try to complete the rectangle. Notice that there are two extra x tiles.

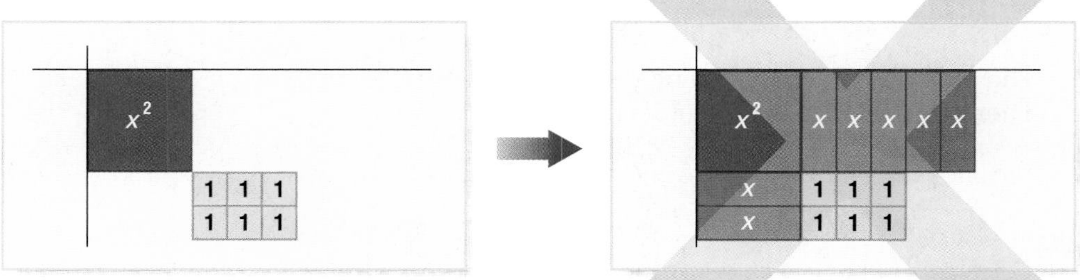

(continued on the next page)

Step 3 Arrange the 1 tiles into a 1-by-6 rectangular array. This time you can complete the rectangle with the x tiles.

The rectangle has a width of $x + 1$ and a length of $x + 6$. Therefore, $x^2 + 7x + 6 = (x + 1)(x + 6)$.

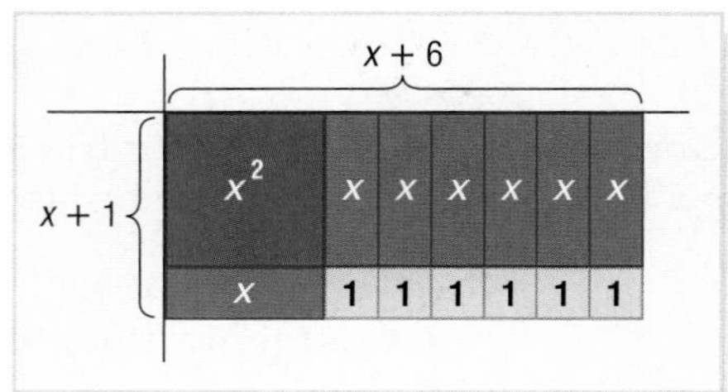

Activity 3 **Use algebra tiles to factor $x^2 - 2x - 3$.**

Step 1 Model the polynomial $x^2 - 2x - 3$.

Step 2 Place the x^2 tile at the corner of the product mat. Arrange the 1 tiles into a 1-by-3 rectangular array as shown.

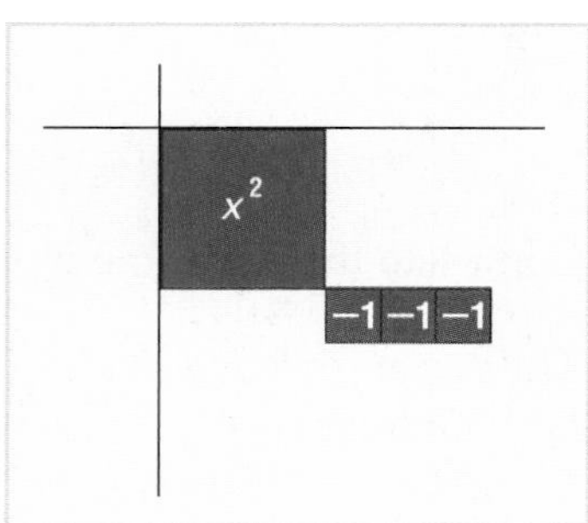

Step 3 Place the x tile as shown. Recall that you can add zero-pairs without changing the value of the polynomial. In this case, add a zero pair of x tiles.

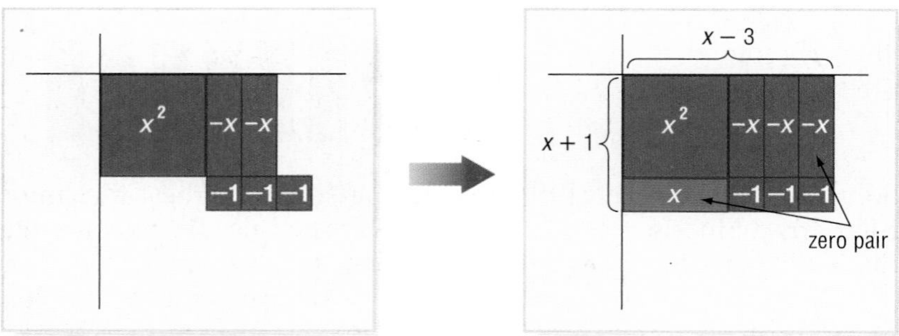

The rectangle has a width of $x + 1$ and a length of $x - 3$. Therefore, $x^2 - 2x - 3 = (x + 1)(x - 3)$.

Model

Use algebra tiles to factor each trinomial.

1. $x^2 + 4x + 3$ **2.** $x^2 + 5x + 4$ **3.** $x^2 - x - 6$ **4.** $x^2 - 3x + 2$

5. $x^2 + 7x + 12$ **6.** $x^2 - 4x + 4$ **7.** $x^2 - x - 2$ **8.** $x^2 - 6x + 8$

9-3 Factoring Trinomials: $x^2 + bx + c$

What You'll Learn

- Factor trinomials of the form $x^2 + bx + c$.
- Solve equations of the form $x^2 + bx + c = 0$.

How can factoring be used to find the dimensions of a garden?

Tamika has enough bricks to make a 30-foot border around the rectangular vegetable garden she is planting. The booklet she got from the nursery says that the plants will need a space of 54 square feet to grow. What should the dimensions of her garden be? To solve this problem, you need to find two numbers whose product is 54 and whose sum is 15, half the perimeter of the garden.

$P = 30$ ft

FACTOR $x^2 + bx + c$ In Lesson 9-1, you learned that when two numbers are multiplied, each number is a factor of the product. Similarly, when two binomials are multiplied, each binomial is a factor of the product.

To factor some trinomials, you will use the pattern for multiplying two binomials. Study the following example.

$$\begin{aligned}
(x + 2)(x + 3) &= \overset{F}{(x \cdot x)} + \overset{O}{(x \cdot 3)} + \overset{I}{(x \cdot 2)} + \overset{L}{(2 \cdot 3)} && \text{Use the FOIL method.}\\
&= x^2 + 3x + 2x + 6 && \text{Simplify.}\\
&= x^2 + (3 + 2)x + 6 && \text{Distributive Property}\\
&= x^2 + 5x + 6 && \text{Simplify.}
\end{aligned}$$

Observe the following pattern in this multiplication.

$$\begin{aligned}
(x + 2)(x + 3) &= x^2 + (3 + 2)x + (2 \cdot 3)\\
(x + m)(x + n) &= x^2 + (n + m)x + mn\\
&= x^2 + \underbrace{(m + n)}_{}x + \underbrace{mn}_{}\\
& \quad\; x^2 + \quad bx \quad + \; c
\end{aligned}$$

$b = m + n$ and $c = mn$

Notice that the coefficient of the middle term is the sum of m and n and the last term is the product of m and n. This pattern can be used to factor quadratic trinomials of the form $x^2 + bx + c$.

Study Tip

Reading Math
A *quadratic trinomial* is a trinomial of degree 2. This means that the greatest exponent of the variable is 2.

Key Concept — Factoring $x^2 + bx + c$

- **Words** To factor quadratic trinomials of the form $x^2 + bx + c$, find two integers, m and n, whose sum is equal to b and whose product is equal to c. Then write $x^2 + bx + c$ using the pattern $(x + m)(x + n)$.
- **Symbols** $x^2 + bx + c = (x + m)(x + n)$ when $m + n = b$ and $mn = c$.
- **Example** $x^2 + 5x + 6 = (x + 2)(x + 3)$, since $2 + 3 = 5$ and $2 \cdot 3 = 6$.

To determine m and n, find the factors of c and use a guess-and-check strategy to find which pair of factors has a sum of b.

Example 1 *b* and *c* Are Positive

Factor $x^2 + 6x + 8$.

In this trinomial, $b = 6$ and $c = 8$. You need to find two numbers whose sum is 6 and whose product is 8. Make an organized list of the factors of 8, and look for the pair of factors whose sum is 6.

Factors of 8	Sum of Factors
1, 8	9
2, 4	6

The correct factors are 2 and 4.

$$x^2 + 6x + 8 = (x + m)(x + n) \quad \text{Write the pattern.}$$
$$= (x + 2)(x + 4) \quad m = 2 \text{ and } n = 4$$

CHECK You can check this result by multiplying the two factors.

$$(x + 2)(x + 4) \overset{\text{F O I L}}{=} x^2 + 4x + 2x + 8 \quad \text{FOIL method}$$
$$= x^2 + 6x + 8 \checkmark \quad \text{Simplify.}$$

When factoring a trinomial where b is negative and c is positive, you can use what you know about the product of binomials to help narrow the list of possible factors.

Example 2 *b* Is Negative and *c* Is Positive

Factor $x^2 - 10x + 16$.

In this trinomial, $b = -10$ and $c = 16$. This means that $m + n$ is negative and mn is positive. So m and n must both be negative. Therefore, make a list of the negative factors of 16, and look for the pair of factors whose sum is -10.

Factors of 16	Sum of Factors
−1, −16	−17
−2, −8	−10
−4, −4	−8

The correct factors are −2 and −8.

$$x^2 - 10x + 16 = (x + m)(x + n) \quad \text{Write the pattern.}$$
$$= (x - 2)(x - 8) \quad m = -2 \text{ and } n = -8$$

CHECK You can check this result by using a graphing calculator. Graph $y = x^2 - 10x + 16$ and $y = (x - 2)(x - 8)$ on the same screen. Since only one graph appears, the two graphs must coincide. Therefore, the trinomial has been factored correctly. ✓

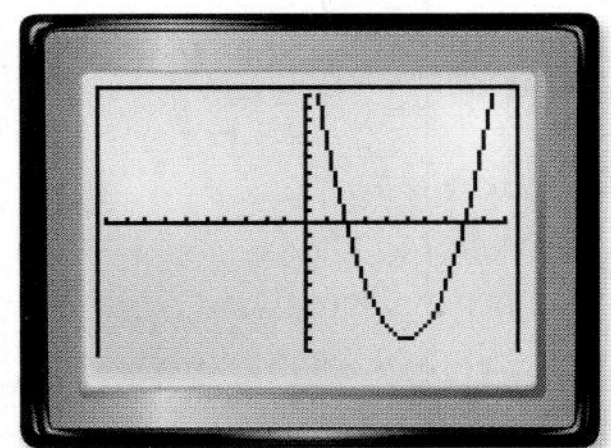

[−10, 10] scl: 1 by [−10, 10] scl: 1

Study Tip

Testing Factors
Once you find the correct factors, there is no need to test any other factors. Therefore, it is not necessary to test −4 and −4 in Example 2.

You will find that keeping an organized list of the factors you have tested is particularly important when factoring a trinomial like $x^2 + x - 12$, where the value of c is negative.

Example 3 b Is Positive and c Is Negative

Factor $x^2 + x - 12$.

In this trinomial, $b = 1$ and $c = -12$. This means that $m + n$ is positive and mn is negative. So either m or n is negative, but not both. Therefore, make a list of the factors of -12, where one factor of each pair is negative. Look for the pair of factors whose sum is 1.

Factors of −12	Sum of Factors	
1, −12	−11	
−1, 12	11	
2, −6	−4	
−2, 6	4	
3, −4	−1	
−3, 4	1	The correct factors are −3 and 4.

$x^2 + x - 12 = (x + m)(x + n)$ Write the pattern.

$= (x - 3)(x + 4)$ $m = -3$ and $n = 4$

Study Tip

Alternate Method
You can use the opposite of FOIL to factor trinomials. For instance, consider Example 3.

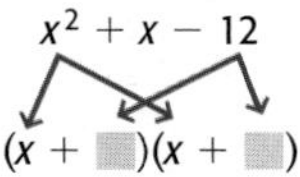

Try factor pairs of −12 until the sum of the products of the Inner and Outer terms is x.

Example 4 b Is Negative and c Is Negative

Factor $x^2 - 7x - 18$.

Since $b = -7$ and $c = -18$, $m + n$ is negative and mn is negative. So either m or n is negative, but not both.

Factors of −18	Sum of Factors	
1, −18	−17	
−1, 18	17	
2, −9	−7	The correct factors are 2 and −9.

$x^2 - 7x - 18 = (x + m)(x + n)$ Write the pattern.

$= (x + 2)(x - 9)$ $m = 2$ and $n = -9$

SOLVE EQUATIONS BY FACTORING Some equations of the form $x^2 + bx + c = 0$ can be solved by factoring and then using the Zero Product Property.

Example 5 Solve an Equation by Factoring

Solve $x^2 + 5x = 6$. Check your solutions.

$x^2 + 5x = 6$ Original equation

$x^2 + 5x - 6 = 0$ Rewrite the equation so that one side equals 0.

$(x - 1)(x + 6) = 0$ Factor.

$x - 1 = 0$ or $x + 6 = 0$ Zero Product Property

$x = 1$ $\quad x = -6$ Solve each equation.

The solution set is $\{1, -6\}$.

CHECK Substitute 1 and −6 for x in the original equation.

$x^2 + 5x = 6$ $\qquad$ $x^2 + 5x = 6$

$(1)^2 + 5(1) \stackrel{?}{=} 6$ $\qquad$ $(-6)^2 + 5(-6) \stackrel{?}{=} 6$

$6 = 6$ ✓ $\qquad$ $6 = 6$ ✓

Example 6 Solve a Real-World Problem by Factoring

YEARBOOK DESIGN **A sponsor for the school yearbook has asked that the length and width of a photo in their ad be increased by the same amount in order to double the area of the photo. If the photo was originally 12 centimeters wide by 8 centimeters long, what should the new dimensions of the enlarged photo be?**

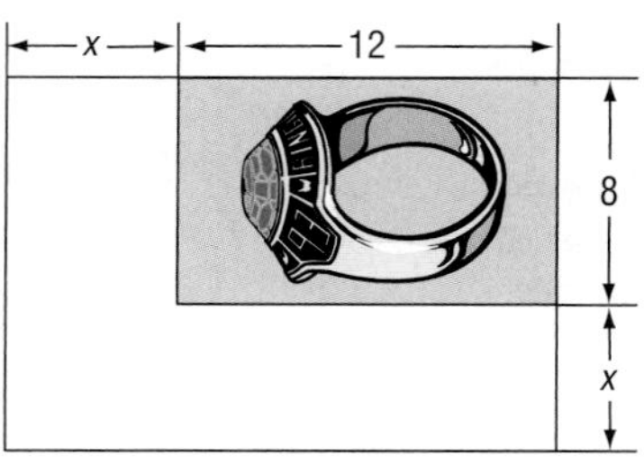

Explore Begin by making a diagram like the one shown above, labeling the appropriate dimensions.

Plan Let x = the amount added to each dimension of the photo.

The new length	times	the new width	equals	the new area,
$x + 12$	$\cdot$	$x + 8$	$=$	$2(8)(12)$ (old area)

Solve

$(x + 12)(x + 8) = 2(8)(12)$ Write the equation.

$x^2 + 20x + 96 = 192$ Multiply.

$x^2 + 20x - 96 = 0$ Subtract 192 from each side.

$(x + 24)(x - 4) = 0$ Factor.

$x + 24 = 0$ or $x - 4 = 0$ Zero Product Property

$x = -24$ $\quad x = 4$ Solve each equation.

Examine The solution set is $\{-24, 4\}$. Only 4 is a valid solution, since dimensions cannot be negative. Thus, the new length of the photo should be $4 + 12$ or 16 centimeters, and the new width should be $4 + 8$ or 12 centimeters.

Check for Understanding

Concept Check

1. **Explain** why, when factoring $x^2 + 6x + 9$, it is not necessary to check the sum of the factor pairs -1 and -9 or -3 and -3.
2. **OPEN ENDED** Give an example of an equation that can be solved using the factoring techniques presented in this lesson. Then, solve your equation.
3. **FIND THE ERROR** Peter and Aleta are solving $x^2 + 2x = 15$.

Peter

$x^2 + 2x = 15$

$x(x + 2) = 15$

$x = 15$ or $x + 2 = 15$

$x = 13$

Aleta

$x^2 + 2x = 15$

$x^2 + 2x - 15 = 0$

$(x - 3)(x + 5) = 0$

$x - 3 = 0$ or $x + 5 = 0$

$x = 3$ $\quad x = -5$

Who is correct? Explain your reasoning.

Guided Practice

Factor each trinomial.

4. $x^2 + 11x + 24$
5. $c^2 - 3c + 2$
6. $n^2 + 13n - 48$
7. $p^2 - 2p - 35$
8. $72 + 27a + a^2$
9. $x^2 - 4xy + 3y^2$

Solve each equation. Check your solutions.

10. $n^2 + 7n + 6 = 0$
11. $a^2 + 5a - 36 = 0$
12. $p^2 - 19p - 42 = 0$
13. $y^2 + 9 = -10y$
14. $9x + x^2 = 22$
15. $d^2 - 3d = 70$

Application

16. **NUMBER THEORY** Find two consecutive integers whose product is 156.

Practice and Apply

Homework Help

For Exercises	See Examples
17–36	1–4
37–53	5
54–56, 61, 62	6

Extra Practice

See page 840.

Factor each trinomial.

17. $a^2 + 8a + 15$
18. $x^2 + 12x + 27$
19. $c^2 + 12c + 35$
20. $y^2 + 13y + 30$
21. $m^2 - 22m + 21$
22. $d^2 - 7d + 10$
23. $p^2 - 17p + 72$
24. $g^2 - 19g + 60$
25. $x^2 + 6x - 7$
26. $b^2 + b - 20$
27. $h^2 + 3h - 40$
28. $n^2 + 3n - 54$
29. $y^2 - y - 42$
30. $z^2 - 18z - 40$
31. $-72 + 6w + w^2$
32. $-30 + 13x + x^2$
33. $a^2 + 5ab + 4b^2$
34. $x^2 - 13xy + 36y^2$

GEOMETRY Find an expression for the perimeter of a rectangle with the given area.

35. area $= x^2 + 24x - 81$
36. area $= x^2 + 13x - 90$

Solve each equation. Check your solutions.

37. $x^2 + 16x + 28 = 0$
38. $b^2 + 20b + 36 = 0$
39. $y^2 + 4y - 12 = 0$
40. $d^2 + 2d - 8 = 0$
41. $a^2 - 3a - 28 = 0$
42. $g^2 - 4g - 45 = 0$
43. $m^2 - 19m + 48 = 0$
44. $n^2 - 22n + 72 = 0$
45. $z^2 = 18 - 7z$
46. $h^2 + 15 = -16h$
47. $24 + k^2 = 10k$
48. $x^2 - 20 = x$
49. $c^2 - 50 = -23c$
50. $y^2 - 29y = -54$
51. $14p + p^2 = 51$
52. $x^2 - 2x - 6 = 74$
53. $x^2 - x + 56 = 17x$

More About. . .

Supreme Court

The "Conference handshake" has been a tradition since the late 19th century.

Source: www.supremecourtus.gov

54. **SUPREME COURT** When the Justices of the Supreme Court assemble to go on the Bench each day, each Justice shakes hands with each of the other Justices for a total of 36 handshakes. The total number of handshakes h possible for n people is given by $h = \frac{n^2 - n}{2}$. Write and solve an equation to determine the number of Justices on the Supreme Court.

55. **NUMBER THEORY** Find two consecutive even integers whose product is 168.

56. **GEOMETRY** The triangle has an area of 40 square centimeters. Find the height h of the triangle.

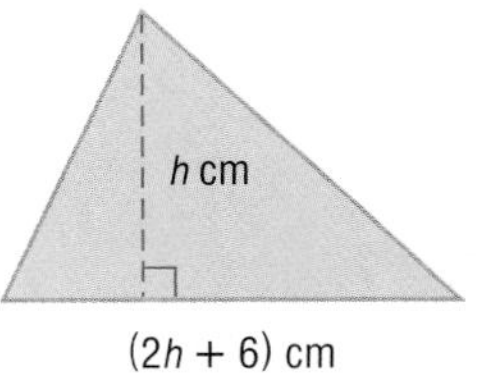

CRITICAL THINKING Find all values of k so that each trinomial can be factored using integers.

57. $x^2 + kx - 19$
58. $x^2 + kx + 14$
59. $x^2 - 8x + k,\ k > 0$
60. $x^2 - 5x + k,\ k > 0$

RUGBY For Exercises 61 and 62, use the following information.
The length of a Rugby League field is 52 meters longer than its width w.

61. Write an expression for the area of the rectangular field.
62. The area of a Rugby League field is 8160 square meters. Find the dimensions of the field.

63. WRITING IN MATH Answer the question that was posed at the beginning of the lesson.

How can factoring be used to find the dimensions of a garden?

Include the following in your answer:

- a description of how you would find the dimensions of the garden, and
- an explanation of how the process you used is related to the process used to factor trinomials of the form $x^2 + bx + c$.

64. Which is the factored form of $x^2 - 17x + 42$?

Ⓐ $(x - 1)(y - 42)$ Ⓑ $(x - 2)(x - 21)$

Ⓒ $(x - 3)(x - 14)$ Ⓓ $(x - 6)(x - 7)$

65. **GRID IN** What is the positive solution of $p^2 - 13p - 30 = 0$?

Use a graphing calculator to determine whether each factorization is correct. Write *yes* or *no*. If no, state the correct factorization.

66. $x^2 - 14x + 48 = (x + 6)(x + 8)$

67. $x^2 - 16x - 105 = (x + 5)(x - 21)$

68. $x^2 + 25x + 66 = (x + 33)(x + 2)$

69. $x^2 + 11x - 210 = (x + 10)(x - 21)$

Maintain Your Skills

Mixed Review

Solve each equation. Check your solutions. *(Lesson 9-2)*

70. $(x + 3)(2x - 5) = 0$

71. $b(7b - 4) = 0$

72. $5y^2 = -9y$

Find the GCF of each set of monomials. *(Lesson 9-1)*

73. 24, 36, 72

74. $9p^2q^5$, $21p^3q^3$

75. $30x^4y^5$, $20x^2y^7$, $75x^3y^4$

INTERNET For Exercises 76 and 77, use the graph at the right. *(Lessons 3-7 and 8-3)*

76. Find the percent increase in the number of domain registrations from 1997 to 2000.

77. Use your answer from Exercise 76 to verify the claim that registrations grew more than 18-fold from 1997 to 2000 is correct.

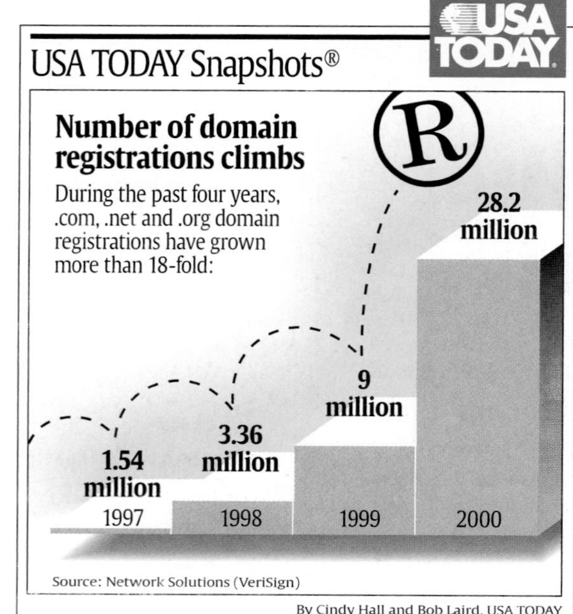

Getting Ready for the Next Lesson

PREREQUISITE SKILL Factor each polynomial.
*(To review **factoring by grouping**, see Lesson 9-2.)*

78. $3y^2 + 2y + 9y + 6$

79. $3a^2 + 2a + 12a + 8$

80. $4x^2 + 3x + 8x + 6$

81. $2p^2 - 6p + 7p - 21$

82. $3b^2 + 7b - 12b - 28$

83. $4g^2 - 2g - 6g + 3$

9-4 Factoring Trinomials: $ax^2 + bx + c$

What You'll Learn

- Factor trinomials of the form $ax^2 + bx + c$.
- Solve equations of the form $ax^2 + bx + c = 0$.

Vocabulary

- prime polynomial

How can algebra tiles be used to factor $2x^2 + 7x + 6$?

The factors of $2x^2 + 7x + 6$ are the dimensions of the rectangle formed by the algebra tiles shown below.

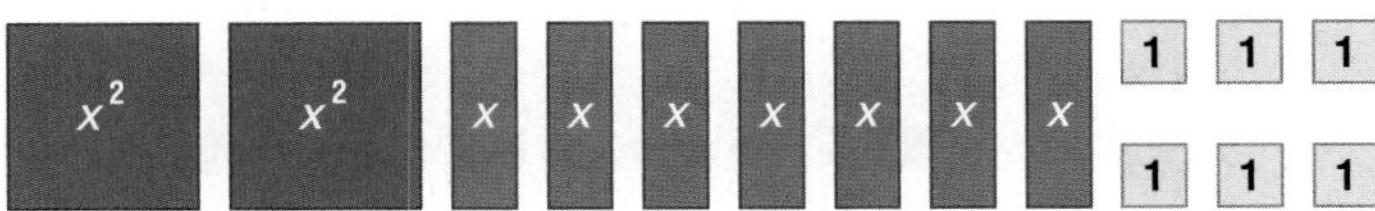

The process you use to form the rectangle is the same mental process you can use to factor this trinomial algebraically.

FACTOR $ax^2 + bx + c$ For trinomials of the form $x^2 + bx + c$, the coefficient of x^2 is 1. To factor trinomials of this form, you find the factors of c whose sum is b. We can modify this approach to factor trinomials whose leading coefficient is not 1.

Study Tip

Look Back
To review **factoring by grouping**, see Lesson 9-2.

$$(2x + 5)(3x + 1) = \overset{F}{6x^2} + \overset{O}{2x} + \overset{I}{15x} + \overset{L}{5}$$ Use the FOIL method.

$$2 \cdot 15 = 30$$

$$6 \cdot 5 = 30$$

Observe the following pattern in this product.

$$6x^2 + 2x + 15x + 5 \qquad\qquad ax^2 + mx + nx + c$$
$$6x^2 + 17x + 5 \qquad\qquad ax^2 + bx + c$$
$$2 + 15 = 17 \text{ and } 2 \cdot 15 = 6 \cdot 5 \qquad\qquad m + n = b \text{ and } mn = ac$$

You can use this pattern and the method of factoring by grouping to factor $6x^2 + 17x + 5$. Find two numbers, m and n, whose product is $6 \cdot 5$ or 30 and whose sum is 17.

Factors of 30	Sum of Factors	
1, 30	31	
2, 15	17	The correct factors are 2 and 15.

$6x^2 + 17x + 5 = 6x^2 + mx + nx + 5$ Write the pattern.

$= 6x^2 + 2x + 15x + 5$ $m = 2$ and $n = 15$

$= (6x^2 + 2x) + (15x + 5)$ Group terms with common factors.

$= 2x(3x + 1) + 5(3x + 1)$ Factor the GCF from each grouping.

$= (3x + 1)(2x + 5)$ $3x + 1$ is the common factor.

Therefore, $6x^2 + 17x + 5 = (3x + 1)(2x + 5)$.

Example 1 Factor $ax^2 + bx + c$

a. Factor $7x^2 + 22x + 3$.

In this trinomial, $a = 7$, $b = 22$ and $c = 3$. You need to find two numbers whose sum is 22 and whose product is $7 \cdot 3$ or 21. Make an organized list of the factors of 21 and look for the pair of factors whose sum is 22.

Factors of 21	Sum of Factors
1, 21	22

The correct factors are 1 and 21.

$$\begin{aligned} 7x^2 + 22x + 3 &= 7x^2 + mx + nx + 3 && \text{Write the pattern.} \\ &= 7x^2 + 1x + 21x + 3 && m = 1 \text{ and } n = 21 \\ &= (7x^2 + 1x) + (21x + 3) && \text{Group terms with common factors.} \\ &= x(7x + 1) + 3(7x + 1) && \text{Factor the GCF from each grouping.} \\ &= (7x + 1)(x + 3) && \text{Distributive Property} \end{aligned}$$

CHECK You can check this result by multiplying the two factors.

$$\begin{aligned} (7x + 1)(x + 3) &= \overset{F}{7x^2} + \overset{O}{21x} + \overset{I}{x} + \overset{L}{3} && \text{FOIL method} \\ &= 7x^2 + 22x + 3 \checkmark && \text{Simplify.} \end{aligned}$$

b. Factor $10x^2 - 43x + 28$.

In this trinomial, $a = 10$, $b = -43$ and $c = 28$. Since b is negative, $m + n$ is negative. Since c is positive, mn is positive. So m and n must both be negative. Therefore, make a list of the negative factors of $10 \cdot 28$ or 280, and look for the pair of factors whose sum is -43.

Factors of 280	Sum of Factors
−1, −280	−281
−2, −140	−142
−4, −70	−74
−5, −56	−61
−7, −40	−47
−8, −35	−43

The correct factors are −8 and −35.

$$\begin{aligned} &10x^2 - 43x + 28 \\ &= 10x^2 + mx + nx + 28 && \text{Write the pattern.} \\ &= 10x^2 + (-8)x + (-35)x + 28 && m = -8 \text{ and } n = -35 \\ &= (10x^2 - 8x) + (-35x + 28) && \text{Group terms with common factors.} \\ &= 2x(5x - 4) + 7(-5x + 4) && \text{Factor the GCF from each grouping.} \\ &= 2x(5x - 4) + 7(-1)(5x - 4) && -5x + 4 = (-1)(5x - 4) \\ &= 2x(5x - 4) + (-7)(5x - 4) && 7(-1) = -7 \\ &= (5x - 4)(2x - 7) && \text{Distributive Property} \end{aligned}$$

Study Tip

Finding Factors
Factor pairs in an organized list so you do not miss any possible pairs of factors.

Sometimes the terms of a trinomial will contain a common factor. In these cases, first use the Distributive Property to factor out the common factor. Then factor the trinomial.

Example 2 Factor When a, b, and c Have a Common Factor

Factor $3x^2 + 24x + 45$.

Notice that the GCF of the terms $3x^2$, $24x$, and 45 is 3. When the GCF of the terms of a trinomial is an integer other than 1, you should first factor out this GCF.

$3x^2 + 24x + 45 = 3(x^2 + 8x + 15)$ Distributive Property

Now factor $x^2 + 8x + 15$. Since the lead coefficient is 1, find two factors of 15 whose sum is 8.

Factors of 15	Sum of Factors
1, 15	16
3, 5	8

The correct factors are 3 and 5.

So, $x^2 + 8x + 15 = (x + 3)(x + 5)$. Thus, the complete factorization of $3x^2 + 24x + 45$ is $3(x + 3)(x + 5)$.

Study Tip

Factoring Completely
Always check for a GCF first before trying to factor a trinomial.

A polynomial that cannot be written as a product of two polynomials with integral coefficients is called a **prime polynomial**.

Example 3 Determine Whether a Polynomial Is Prime

Factor $2x^2 + 5x - 2$.

In this trinomial, $a = 2$, $b = 5$ and $c = -2$. Since b is positive, $m + n$ is positive. Since c is negative, mn is negative. So either m or n is negative, but not both. Therefore, make a list of the factors of $2 \cdot -2$ or -4, where one factor in each pair is negative. Look for a pair of factors whose sum is 5.

Factors of −4	Sum of Factors
1, −4	−3
−1, 4	3
−2, 2	0

There are no factors whose sum is 5. Therefore, $2x^2 + 5x - 2$ cannot be factored using integers. Thus, $2x^2 + 5x - 2$ is a prime polynomial.

SOLVE EQUATIONS BY FACTORING Some equations of the form $ax^2 + bx + c = 0$ can be solved by factoring and then using the Zero Product Property.

Example 4 Solve Equations by Factoring

Solve $8a^2 - 9a - 5 = 4 - 3a$. Check your solutions.

$8a^2 - 9a - 5 = 4 - 3a$ Original equation

$8a^2 - 6a - 9 = 0$ Rewrite so that one side equals 0.

$(4a + 3)(2a - 3) = 0$ Factor the left side.

$4a + 3 = 0$ or $2a - 3 = 0$ Zero Product Property

$4a = -3$ $\quad 2a = 3$ Solve each equation.

$a = -\frac{3}{4}$ $\quad a = \frac{3}{2}$

The solution set is $\left\{-\frac{3}{4}, \frac{3}{2}\right\}$.

CHECK Check each solution in the original equation.

$$8a^2 - 9a - 5 = 4 - 3a \qquad 8a^2 - 9a - 5 = 4 - 3a$$

$$8\left(-\frac{3}{4}\right)^2 - 9\left(-\frac{3}{4}\right) - 5 \stackrel{?}{=} 4 - 3\left(-\frac{3}{4}\right) \qquad 8\left(\frac{3}{2}\right)^2 - 9\left(\frac{3}{2}\right) - 5 \stackrel{?}{=} 4 - 3\left(\frac{3}{2}\right)$$

$$\frac{9}{2} + \frac{27}{4} - 5 \stackrel{?}{=} 4 + \frac{9}{4} \qquad 18 - \frac{27}{2} - 5 \stackrel{?}{=} 4 - \frac{9}{2}$$

$$\frac{25}{4} = \frac{25}{4} \checkmark \qquad -\frac{1}{2} = -\frac{1}{2} \checkmark$$

A *model for the vertical motion of a projected object* is given by the equation $h = -16t^2 + vt + s$, where h is the height in feet, t is the time in seconds, v is the initial upward velocity in feet per second, and s is the starting height of the object in feet.

Example 5 Solve Real-World Problems by Factoring

PEP RALLY At a pep rally, small foam footballs are launched by cheerleaders using a sling-shot. How long is a football in the air if a student in the stands catches it on its way down 26 feet above the gym floor?

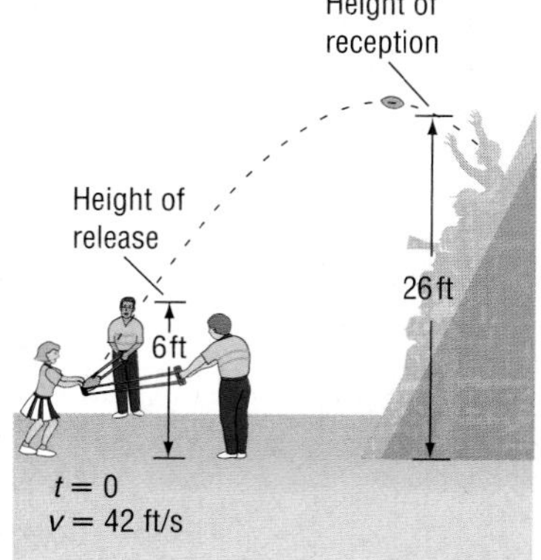

Use the model for vertical motion.

$h = -16t^2 + vt + s$	Vertical motion model
$26 = -16t^2 + 42t + 6$	$h = 26, v = 42, s = 6$
$0 = -16t^2 + 42t - 20$	Subtract 26 from each side.
$0 = -2(8t^2 - 21t + 10)$	Factor out -2.
$0 = 8t^2 - 21t + 10$	Divide each side by -2.
$0 = (8t - 5)(t - 2)$	Factor $8t^2 - 21t + 10$.
$8t - 5 = 0$ or $t - 2 = 0$	Zero Product Property
$8t = 5$ $\quad t = 2$	Solve each equation.
$t = \frac{5}{8}$	

The solutions are $\frac{5}{8}$ second and 2 seconds. The first time represents how long it takes the football to reach a height of 26 feet on its way up. The later time represents how long it takes the ball to reach a height of 26 feet again on its way down. Thus, the football will be in the air for 2 seconds before the student catches it.

Study Tip

Factoring When a Is Negative When factoring a trinomial of the form $ax^2 + bx + c$, where a is negative, it is helpful to factor out a negative monomial.

Check for Understanding

Concept Check

1. **Explain** how to determine which values should be chosen for m and n when factoring a polynomial of the form $ax^2 + bx + c$.
2. **OPEN ENDED** Write a trinomial that can be factored using a pair of numbers whose sum is 9 and whose product is 14.
3. **FIND THE ERROR** Dasan and Craig are factoring $2x^2 + 11x + 18$.

Dasan

Factors of 18	Sum
1, 18	19
3, 6	9
9, 2	11

$2x^2 + 11x + 18$
$= 2(x^2 + 11x + 18)$
$= 2(x + 9)(x + 2)$

Craig

Factors of 36	Sum
1, 36	37
2, 18	20
3, 12	15
4, 9	13
6, 6	12

$2x^2 + 11x + 18$ is prime.

Who is correct? Explain your reasoning.

Guided Practice **Factor each trinomial, if possible. If the trinomial cannot be factored using integers, write *prime*.**

4. $3a^2 + 8a + 4$
5. $2a^2 - 11a + 7$
6. $2p^2 + 14p + 24$
7. $2x^2 + 13x + 20$
8. $6x^2 + 15x - 9$
9. $4n^2 - 4n - 35$

Solve each equation. Check your solutions.

10. $3x^2 + 11x + 6 = 0$
11. $10p^2 - 19p + 7 = 0$
12. $6n^2 + 7n = 20$

Application

13. **GYMNASTICS** When a gymnast making a vault leaves the horse, her feet are 8 feet above the ground traveling with an initial upward velocity of 8 feet per second. Use the model for vertical motion to find the time t in seconds it takes for the gymnast's feet to reach the mat. (*Hint*: Let $h = 0$, the height of the mat.)

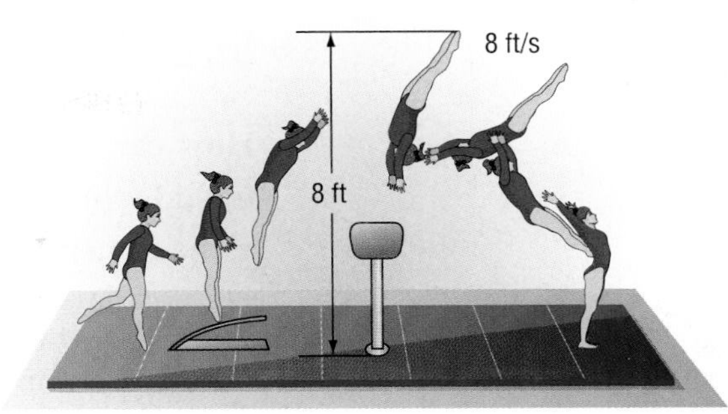

Practice and Apply

Homework Help

For Exercises	See Examples
14–31	1–3
35–48	4
49–52	5

Extra Practice

See page 840.

Factor each trinomial, if possible. If the trinomial cannot be factored using integers, write *prime*.

14. $2x^2 + 7x + 5$
15. $3x^2 + 5x + 2$
16. $6p^2 + 5p - 6$
17. $5d^2 + 6d - 8$
18. $8k^2 - 19k + 9$
19. $9g^2 - 12g + 4$
20. $2a^2 - 9a - 18$
21. $2x^2 - 3x - 20$
22. $5c^2 - 17c + 14$
23. $3p^2 - 25p + 16$
24. $8y^2 - 6y - 9$
25. $10n^2 - 11n - 6$
26. $15z^2 + 17z - 18$
27. $14x^2 + 13x - 12$
28. $6r^2 - 14r - 12$
29. $30x^2 - 25x - 30$
30. $9x^2 + 30xy + 25y^2$
31. $36a^2 + 9ab - 10b^2$

CRITICAL THINKING **Find all values of k so that each trinomial can be factored as two binomials using integers.**

32. $2x^2 + kx + 12$
33. $2x^2 + kx + 15$
34. $2x^2 + 12x + k, k > 0$

More About. . .

Cliff Diving

In Acapulco, Mexico, divers leap from La Quebrada, the "Break in the Rocks," diving headfirst into the Pacific Ocean 105 feet below.

Source: acapulco-travel.web.com.mx

Solve each equation. Check your solutions.

35. $5x^2 + 27x + 10 = 0$
36. $3x^2 - 5x - 12 = 0$
37. $24x^2 - 11x - 3 = 3x$
38. $17x^2 - 11x + 2 = 2x^2$
39. $14n^2 = 25n + 25$
40. $12a^2 - 13a = 35$
41. $6x^2 - 14x = 12$
42. $21x^2 - 6 = 15x$
43. $24x^2 - 30x + 8 = -2x$
44. $24x^2 - 46x = 18$
45. $\frac{x^2}{12} - \frac{2x}{3} - 4 = 0$
46. $t^2 - \frac{t}{6} = \frac{35}{6}$
47. $(3y + 2)(y + 3) = y + 14$
48. $(4a - 1)(a - 2) = 7a - 5$

GEOMETRY **For Exercises 49 and 50, use the following information.**
A rectangle with an area of 35 square inches is formed by cutting off strips of equal width from a rectangular piece of paper.

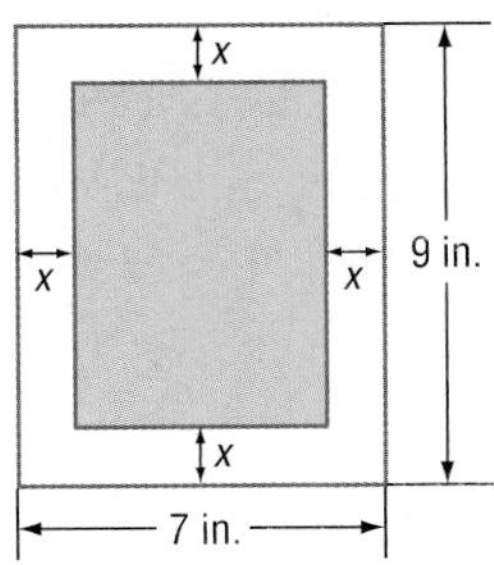

49. Find the width of each strip.

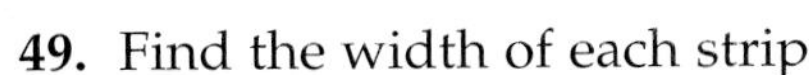

50. Find the dimensions of the new rectangle.

51. **CLIFF DIVING** Suppose a diver leaps from the edge of a cliff 80 feet above the ocean with an initial upward velocity of 8 feet per second. How long will it take the diver to enter the water below?

52. **CLIMBING** Damaris launches a grappling hook from a height of 6 feet with an initial upward velocity of 56 feet per second. The hook just misses the stone ledge of a building she wants to scale. As it falls, the hook anchors on the ledge, which is 30 feet above the ground. How long was the hook in the air?

53. **WRITING IN MATH** Answer the question that was posed at the beginning of the lesson.

How can algebra tiles be used to factor $2x^2 + 7x + 6$?

Include the following in your answer:

- the dimensions of the rectangle formed, and
- an explanation, using words and drawings, of how this geometric guess-and-check process of factoring is similar to the algebraic process described on page 495.

Standardized Test Practice

54. What are the solutions of $2p^2 - p - 3 = 0$?

Ⓐ $-\frac{2}{3}$ and 1 Ⓑ $\frac{2}{3}$ and -1 Ⓒ $-\frac{3}{2}$ and 1 Ⓓ $\frac{3}{2}$ and -1

55. Suppose a person standing atop a building 398 feet tall throws a ball upward. If the person releases the ball 4 feet above the top of the building, the ball's height h, in feet, after t seconds is given by the equation $h = -16t^2 + 48t + 402$. After how many seconds will the ball be 338 feet from the ground?

Ⓐ 3.5 Ⓑ 4 Ⓒ 4.5 Ⓓ 5

Maintain Your Skills

Mixed Review

Factor each trinomial, if possible. If the trinomial cannot be factored using integers, write *prime*. *(Lesson 9-3)*

56. $a^2 - 4a - 21$ 57. $t^2 + 2t + 2$ 58. $d^2 + 15d + 44$

Solve each equation. Check your solutions. *(Lesson 9-2)*

59. $(y - 4)(5y + 7) = 0$ 60. $(2k + 9)(3k + 2) = 0$ 61. $12u = u^2$

62. **BUSINESS** Jake's Garage charges \$83 for a two-hour repair job and \$185 for a five-hour repair job. Write a linear equation that Jake can use to bill customers for repair jobs of any length of time. *(Lesson 5-3)*

Getting Ready for the Next Lesson

PREREQUISITE SKILL Find the principal square root of each number.
*(To review **square roots**, see Lesson 2-7.)*

63. 16 64. 49 65. 36 66. 25
67. 100 68. 121 69. 169 70. 225

Practice Quiz 2 — Lessons 9-3 and 9-4

Factor each trinomial, if possible. If the trinomial cannot be factored using integers, write *prime*. *(Lessons 9-3 and 9-4)*

1. $x^2 - 14x - 72$ 2. $8p^2 - 6p - 35$ 3. $16a^2 - 24a + 5$
4. $n^2 - 17n + 52$ 5. $24c^2 + 62c + 18$ 6. $3y^2 + 33y + 54$

Solve each equation. Check your solutions. *(Lessons 9-3 and 9-4)*

7. $b^2 + 14b - 32 = 0$ 8. $x^2 + 45 = 18x$
9. $12y^2 - 7y - 12 = 0$ 10. $6a^2 = 25a - 14$

9-5 Factoring Differences of Squares

What You'll Learn

- Factor binomials that are the differences of squares.
- Solve equations involving the differences of squares.

How can you determine a basketball player's hang time?

A basketball player's *hang time* is the length of time he is in the air after jumping. Given the maximum height h a player can jump, you can determine his hang time t in seconds by solving $4t^2 - h = 0$. If h is a perfect square, this equation can be solved by factoring using the pattern for the difference of squares.

FACTOR $a^2 - b^2$ A geometric model can be used to factor the difference of squares.

Algebra Activity

Difference of Squares

Step 1 Use a straightedge to draw two squares similar to those shown below. Choose any measures for a and b.

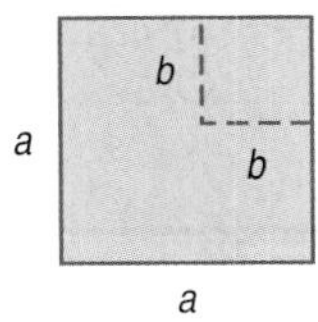

Notice that the area of the large square is a^2, and the area of the small square is b^2.

Step 2 Cut the small square from the large square.

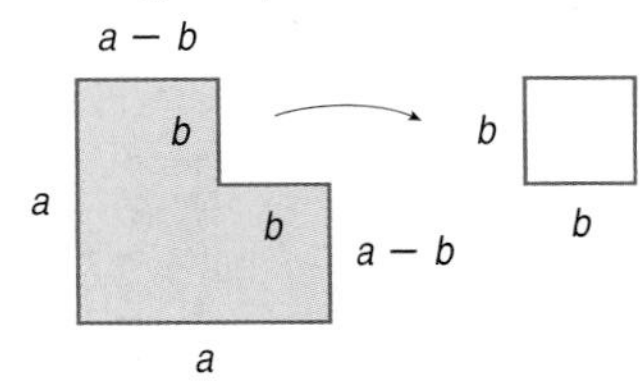

The area of the remaining irregular region is $a^2 - b^2$.

Step 3 Cut the irregular region into two congruent pieces as shown below.

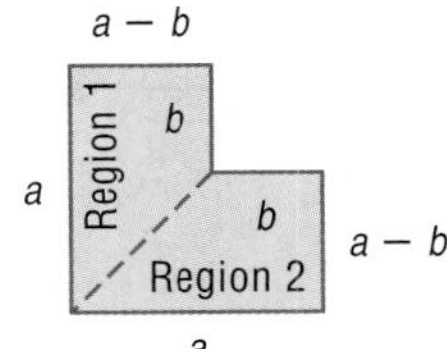

Step 4 Rearrange the two congruent regions to form a rectangle with length $a + b$ and width $a - b$.

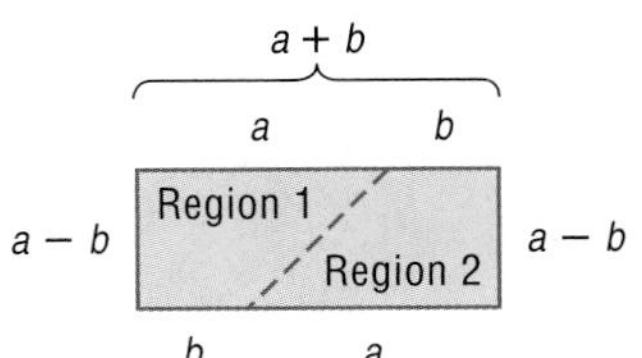

Make a Conjecture

1. Write an expression representing the area of the rectangle.
2. Explain why $a^2 - b^2 = (a + b)(a - b)$.

Study Tip

Look Back
To review the **product of a sum and a difference,** see Lesson 8-8.

Key Concept — Difference of Squares

- **Symbols** $a^2 - b^2 = (a + b)(a - b)$ or $(a - b)(a + b)$
- **Example** $x^2 - 9 = (x + 3)(x - 3)$ or $(x - 3)(x + 3)$

We can use this pattern to factor binomials that can be written in the form $a^2 - b^2$.

Example 1 Factor the Difference of Squares

Factor each binomial.

a. $n^2 - 25$

$n^2 - 25 = n^2 - 5^2$ — Write in the form $a^2 - b^2$.

$= (n + 5)(n - 5)$ — Factor the difference of squares.

b. $36x^2 - 49y^2$

$36x^2 - 49y^2 = (6x)^2 - (7y)^2$ — $36x^2 = 6x \cdot 6x$ and $49y^2 = 7y \cdot 7y$

$= (6x + 7y)(6x - 7y)$ — Factor the difference of squares.

If the terms of a binomial have a common factor, the GCF should be factored out first before trying to apply any other factoring technique.

Example 2 Factor Out a Common Factor

Factor $48a^3 - 12a$.

$48a^3 - 12a = 12a(4a^2 - 1)$ — The GCF of $48a^3$ and $-12a$ is $12a$.

$= 12a[(2a)^2 - 1^2]$ — $4a^2 = 2a \cdot 2a$ and $1 = 1 \cdot 1$

$= 12a(2a + 1)(2a - 1)$ — Factor the difference of squares.

Occasionally, the difference of squares pattern needs to be applied more than once to factor a polynomial completely.

Example 3 Apply a Factoring Technique More Than Once

Factor $2x^4 - 162$.

$2x^4 - 162 = 2(x^4 - 81)$ — The GCF of $2x^4$ and -162 is 2.

$= 2[(x^2)^2 - 9^2]$ — $x^4 = x^2 \cdot x^2$ and $81 = 9 \cdot 9$

$= 2(x^2 + 9)(x^2 - 9)$ — Factor the difference of squares.

$= 2(x^2 + 9)(x^2 - 3^2)$ — $x^2 = x \cdot x$ and $9 = 3 \cdot 3$

$= 2(x^2 + 9)(x + 3)(x - 3)$ — Factor the difference of squares.

Study Tip

Common Misconception
Remember that the sum of two squares, like $x^2 + 9$, is not factorable using the difference of squares pattern. $x^2 + 9$ is a prime polynomial.

Example 4 Apply Several Different Factoring Techniques

Factor $5x^3 + 15x^2 - 5x - 15$.

$5x^3 + 15x^2 - 5x - 15$ — Original polynomial

$= 5(x^3 + 3x^2 - x - 3)$ — Factor out the GCF.

$= 5[(x^3 - x) + (3x^2 - 3)]$ — Group terms with common factors.

$= 5[x(x^2 - 1) + 3(x^2 - 1)]$ — Factor each grouping.

$= 5(x^2 - 1)(x + 3)$ — $x^2 - 1$ is the common factor.

$= 5(x + 1)(x - 1)(x + 3)$ — Factor the difference of squares, $x^2 - 1$, into $(x + 1)(x - 1)$.

SOLVE EQUATIONS BY FACTORING You can apply the Zero Product Property to an equation that is written as the product of any number of factors set equal to 0.

Example 5 Solve Equations by Factoring

Solve each equation by factoring. Check your solutions.

a. $p^2 - \frac{9}{16} = 0$

$p^2 - \frac{9}{16} = 0$	Original equation
$p^2 - \left(\frac{3}{4}\right)^2 = 0$	$p^2 = p \cdot p$ and $\frac{9}{16} = \frac{3}{4} \cdot \frac{3}{4}$
$\left(p + \frac{3}{4}\right)\left(p - \frac{3}{4}\right) = 0$	Factor the difference of squares.
$p + \frac{3}{4} = 0$ or $p - \frac{3}{4} = 0$	Zero Product Property
$p = -\frac{3}{4}$ $\quad$ $p = \frac{3}{4}$	Solve each equation.

The solution set is $\left\{-\frac{3}{4}, \frac{3}{4}\right\}$. Check each solution in the original equation.

b. $18x^3 = 50x$

$18x^3 = 50x$	Original equation
$18x^3 - 50x = 0$	Subtract $50x$ from each side.
$2x(9x^2 - 25) = 0$	The GCF of $18x^3$ and $-50x$ is $2x$.
$2x(3x + 5)(3x - 5) = 0$	$9x^2 = 3x \cdot 3x$ and $25 = 5 \cdot 5$

Applying the Zero Product Property, set each factor equal to 0 and solve the resulting three equations.

$2x = 0$ or $3x + 5 = 0$ or $3x - 5 = 0$

$x = 0$ $\quad$ $3x = -5$ $\quad$ $3x = 5$

$x = -\frac{5}{3}$ $\quad$ $x = \frac{5}{3}$

The solution set is $\left\{-\frac{5}{3}, 0, \frac{5}{3}\right\}$. Check each solution in the original equation.

Study Tip

Alternative Method

The fraction could also be cleared from the equation in Example 5a by multiplying each side of the equation by 16.

$p^2 - \frac{9}{16} = 0$

$16p^2 - 9 = 0$

$(4p + 3)(4p - 3) = 0$

$4p + 3 = 0$ or $4p - 3 = 0$

$p = -\frac{3}{4}$ $\quad$ $p = \frac{3}{4}$

Standardized Test Practice

A B C D

Example 6 Use Differences of Two Squares

Extended-Response Test Item

A corner is cut off a 2-inch by 2-inch square piece of paper. The cut is x inches from a corner as shown.

a. Write an equation in terms of x that represents the area A of the paper after the corner is removed.

b. What value of x will result in an area that is $\frac{7}{9}$ the area of the original square piece of paper? Show how you arrived at your answer.

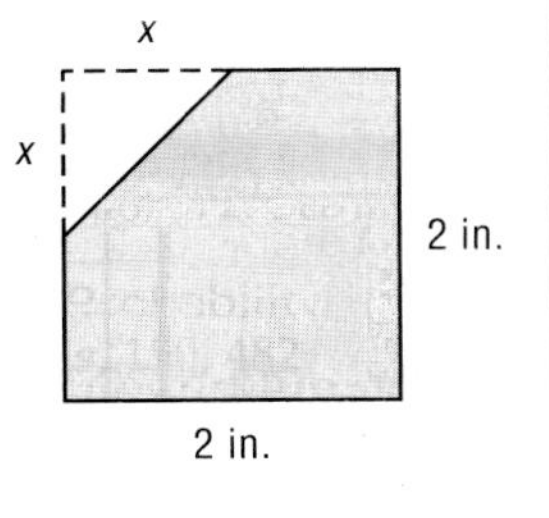

The Princeton Review

Test-Taking Tip

Look to see if the area of an oddly-shaped figure can be found by subtracting the areas of more familiar shapes, such as triangles, rectangles, or circles.

Read the Test Item

A is the area of the square minus the area of the triangular corner to be removed.

(continued on the next page)

Solve the Test Item

a. The area of the square is $2 \cdot 2$ or 4 square inches, and the area of the triangle is $\frac{1}{2} \cdot x \cdot x$ or $\frac{1}{2}x^2$ square inches. Thus, $A = 4 - \frac{1}{2}x^2$.

b. Find x so that A is $\frac{7}{9}$ the area of the original square piece of paper, A_o.

$A = \frac{7}{9}A_o$	Translate the verbal statement.
$4 - \frac{1}{2}x^2 = \frac{7}{9}(4)$	$A = 4 - \frac{1}{2}x^2$ and A_o is 4.
$4 - \frac{1}{2}x^2 = \frac{28}{9}$	Simplify.
$4 - \frac{1}{2}x^2 - \frac{28}{9} = 0$	Subtract $\frac{28}{9}$ from each side.
$\frac{8}{9} - \frac{1}{2}x^2 = 0$	Simplify.
$16 - 9x^2 = 0$	Multiply each side by 18 to remove fractions.
$(4 + 3x)(4 - 3x) = 0$	Factor the difference of squares.
$4 + 3x = 0$ or $4 - 3x = 0$	Zero Product Property
$x = -\frac{4}{3}$ $\quad x = \frac{4}{3}$	Solve each equation.

Since length cannot be negative, the only reasonable solution is $\frac{4}{3}$.

Check for Understanding

Concept Check

1. **Describe** a binomial that is the difference of two squares.

2. **OPEN ENDED** Write a binomial that is the difference of two squares. Then factor your binomial.

3. **Determine** whether the difference of squares pattern can be used to factor $3n^2 - 48$. Explain your reasoning.

4. **FIND THE ERROR** Manuel and Jessica are factoring $64x^2 + 16y^2$.

Manuel	Jessica
$64x^2 + 16y^2$	$64x^2 + 16y^2$
$= 16(4x^2 + y^2)$	$= 16(4x^2 + y^2)$
	$= 16(2x + y)(2x - y)$

Who is correct? Explain your reasoning.

Guided Practice

Factor each polynomial, if possible. If the polynomial cannot be factored, write *prime*.

5. $n^2 - 81$
6. $4 - 9a^2$
7. $2x^5 - 98x^3$
8. $32x^4 - 2y^4$
9. $4t^2 - 27$
10. $x^3 - 3x^2 - 9x + 27$

Solve each equation by factoring. Check your solutions.

11. $4y^2 = 25$
12. $17 - 68k^2 = 0$
13. $x^2 - \frac{1}{36} = 0$
14. $121a = 49a^3$

Standardized Test Practice

A B C D

15. **OPEN ENDED** The area of the shaded part of the square at the right is 72 square inches. Find the dimensions of the square.

x

x

Practice and Apply

Homework Help

For Exercises	See Examples
16–33	1–4
34–45	5
47–50	6

Extra Practice

See page 841.

Factor each polynomial, if possible. If the polynomial cannot be factored, write *prime*.

16. $x^2 - 49$
17. $n^2 - 36$
18. $81 + 16k^2$
19. $25 - 4p^2$
20. $-16 + 49h^2$
21. $-9r^2 + 121$
22. $100c^2 - d^2$
23. $9x^2 - 10y^2$
24. $144a^2 - 49b^2$
25. $169y^2 - 36z^2$
26. $8d^2 - 18$
27. $3x^2 - 75$
28. $8z^2 - 64$
29. $4g^2 - 50$
30. $18a^4 - 72a^2$
31. $20x^3 - 45xy^2$
32. $n^3 + 5n^2 - 4n - 20$
33. $(a + b)^2 - c^2$

Solve each equation by factoring. Check your solutions.

34. $25x^2 = 36$
35. $9y^2 = 64$
36. $12 - 27n^2 = 0$
37. $50 - 8a^2 = 0$
38. $w^2 - \frac{4}{49} = 0$
39. $\frac{81}{100} - p^2 = 0$
40. $36 - \frac{1}{9}r^2 = 0$
41. $\frac{1}{4}x^2 - 25 = 0$
42. $12d^3 - 147d = 0$
43. $18n^3 - 50n = 0$
44. $x^3 - 4x = 12 - 3x^2$
45. $36x - 16x^3 = 9x^2 - 4x^4$

46. **CRITICAL THINKING** Show that $a^2 - b^2 = (a + b)(a - b)$ algebraically. (*Hint*: Rewrite $a^2 - b^2$ as $a^2 - ab + ab - b^2$.)

47. **BOATING** The United States Coast Guard's License Exam includes questions dealing with the breaking strength of a line. The basic breaking strength b in pounds for a natural fiber line is determined by the formula $900c^2 = b$, where c is the circumference of the line in inches. What circumference of natural line would have 3600 pounds of breaking strength?

More About. . .

Aerodynamics

Lift works on the principle that as the speed of a gas increases, the pressure decreases. As the velocity of the air passing over a curved wing increases, the pressure above the wing decreases, lift is created, and the wing rises.

Source: www.gleim.com

48. **AERODYNAMICS** The formula for the pressure difference P above and below a wing is described by the formula $P = \frac{1}{2}dv_1^2 - \frac{1}{2}dv_2^2$, where d is the density of the air, v_1 is the velocity of the air passing above, and v_2 is the velocity of the air passing below. Write this formula in factored form.

49. **LAW ENFORCEMENT** If a car skids on dry concrete, police can use the formula $\frac{1}{24}s^2 = d$ to approximate the speed s of a vehicle in miles per hour given the length d of the skid marks in feet. If the length of skid marks on dry concrete are 54 feet long, how fast was the car traveling when the brakes were applied?

50. **PACKAGING** The width of a box is 9 inches more than its length. The height of the box is 1 inch less than its length. If the box has a volume of 72 cubic inches, what are the dimensions of the box?

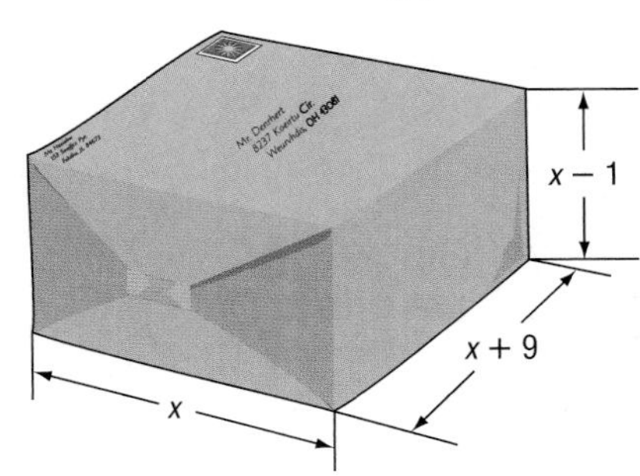

51. **CRITICAL THINKING** The following statements appear to prove that 2 is equal to 1. Find the flaw in this "proof."

Suppose a and b are real numbers such that $a = b, a \neq 0, b \neq 0$.

(1)	$a = b$	Given.
(2)	$a^2 = ab$	Multiply each side by a.
(3)	$a^2 - b^2 = ab - b^2$	Subtract b^2 from each side.
(4)	$(a - b)(a + b) = b(a - b)$	Factor.
(5)	$a + b = b$	Divide each side by $a - b$.
(6)	$a + a = a$	Substitution Property; $a = b$
(7)	$2a = a$	Combine like terms.
(8)	$2 = 1$	Divide each side by a.

52. **WRITING IN MATH** Answer the question that was posed at the beginning of the lesson.

How can you determine a basketball player's hang time?

Include the following in your answer:

- a maximum height that is a perfect square and that would be considered a reasonable distance for a student athlete to jump, and
- a description of how to find the hang time for this maximum height.

Standardized Test Practice

53. What is the factored form of $25b^2 - 1$?

Ⓐ $(5b - 1)(5b + 1)$ Ⓑ $(5b + 1)(5b + 1)$

Ⓒ $(5b - 1)(5b - 1)$ Ⓓ $(25b + 1)(b - 1)$

54. **GRID IN** In the figure, the area between the two squares is 17 square inches. The sum of the perimeters of the two squares is 68 inches. How many inches long is a side of the larger square?

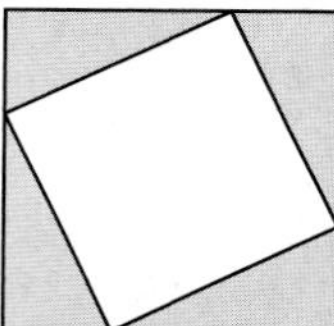

Maintain Your Skills

Mixed Review

Factor each trinomial, if possible. If the trinomial cannot be factored using integers, write *prime*. *(Lesson 9-4)*

55. $2n^2 + 5n + 7$ 56. $6x^2 - 11x + 4$ 57. $21p^2 + 29p - 10$

Solve each equation. Check your solutions. *(Lesson 9-3)*

58. $y^2 + 18y + 32 = 0$ 59. $k^2 - 8k = -15$ 60. $b^2 - 8 = 2b$

61. **STATISTICS** Amy's scores on the first three of four 100-point biology tests were 88, 90, and 91. To get a B+ in the class, her average must be between 88 and 92, inclusive, on all tests. What score must she receive on the fourth test to get a B+ in biology? *(Lesson 6-4)*

Solve each inequality, check your solution, and graph it on a number line. *(Lesson 6-1)*

62. $6 \leq 3d - 12$ 63. $-5 + 10r > 2$ 64. $13x - 3 < 23$

Getting Ready for the Next Lesson

PREREQUISITE SKILL **Find each product.** *(To review **special products**, see Lesson 8-8.)*

65. $(x + 1)(x + 1)$ 66. $(x - 6)(x - 6)$ 67. $(x + 8)^2$

68. $(3x - 4)(3x - 4)$ 69. $(5x - 2)^2$ 70. $(7x + 3)^2$

Reading Mathematics

The Language of Mathematics

Mathematics is a language all its own. As with any language you learn, you must read slowly and carefully, translating small portions of it at a time. Then you must reread the entire passage to make complete sense of what you read.

In mathematics, concepts are often written in a compact form by using symbols. Break down the symbols and try to translate each piece before putting them back together. Read the following sentence.

$$a^2 + 2ab + b^2 = (a + b)^2$$

The trinomial a squared plus twice the product of a and b plus b squared equals the square of the binomial a plus b.

Below is a list of the concepts involved in that single sentence.

- The letters a and b are variables and can be replaced by monomials like 2 or $3x$ or by polynomials like $x + 3$.
- The square of the binomial $a + b$ means $(a + b)(a + b)$. So, $a^2 + 2ab + b^2$ can be written as the product of two identical factors, $a + b$ and $a + b$.

Now put these concepts together. The algebraic statement $a^2 + 2ab + b^2 = (a + b)^2$ means that any trinomial that can be written in the form $a^2 + 2ab + b^2$ can be factored as the square of a binomial using the pattern $(a + b)^2$.

When reading a lesson in your book, use these steps.

- Read the "What You'll Learn" statements to understand what concepts are being presented.
- Skim to get a general idea of the content.
- Take note of any new terms in the lesson by looking for highlighted words.
- Go back and reread in order to understand all of the ideas presented.
- Study all of the examples.
- Pay special attention to the explanations for each step in each example.
- Read any study tips presented in the margins of the lesson.

Reading to Learn

Turn to page 508 and skim Lesson 9-6.

1. List three main ideas from Lesson 9-6. Use phrases instead of whole sentences.
2. What factoring techniques should be tried when factoring a trinomial?
3. What should you always check for first when trying to factor any polynomial?
4. Translate the symbolic representation of the Square Root Property presented on page 511 and explain why it can be applied to problems like $(a + 4)^2 = 49$ in Example 4a.

9-6 Perfect Squares and Factoring

What You'll Learn

- Factor perfect square trinomials.
- Solve equations involving perfect squares.

Vocabulary

- perfect square trinomials

How can factoring be used to design a pavilion?

The senior class has decided to build an outdoor pavilion. It will have an 8-foot by 8-foot portrayal of the school's mascot in the center. The class is selling bricks with students' names on them to finance the project. If they sell enough bricks to cover 80 square feet and want to arrange the bricks around the art, how wide should the border of bricks be? To solve this problem, you would need to solve the equation $(8 + 2x)^2 = 144$.

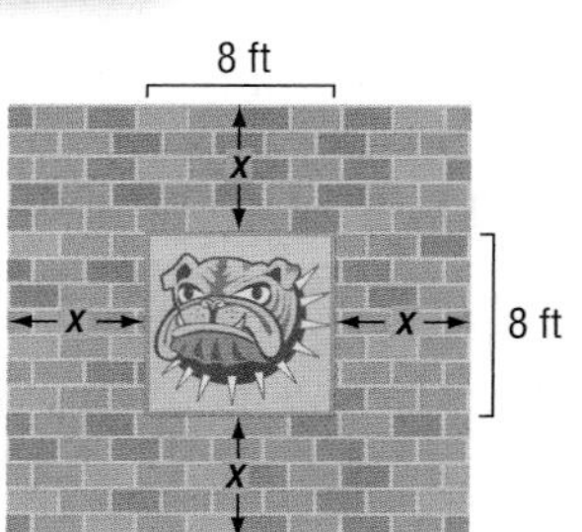

Study Tip

Look Back
To review the **square of a sum or difference**, see Lesson 8-8.

FACTOR PERFECT SQUARE TRINOMIALS Numbers like 144, 16, and 49 are perfect squares, since each can be expressed as the square of an integer.

$$144 = 12 \cdot 12 \text{ or } 12^2 \qquad 16 = 4 \cdot 4 \text{ or } 4^2 \qquad 49 = 7 \cdot 7 \text{ or } 7^2$$

Products of the form $(a + b)^2$ and $(a - b)^2$, such as $(8 + 2x)^2$, are also perfect squares. Recall that these are special products that follow specific patterns.

$$\begin{aligned}(a + b)^2 &= (a + b)(a + b)\\ &= a^2 + ab + ab + b^2\\ &= a^2 + 2ab + b^2\end{aligned} \qquad \begin{aligned}(a - b)^2 &= (a - b)(a - b)\\ &= a^2 - ab - ab + b^2\\ &= a^2 - 2ab + b^2\end{aligned}$$

These patterns can help you factor **perfect square trinomials**, trinomials that are the square of a binomial.

Squaring a Binomial	Factoring a Perfect Square
$(x + 7)^2 = x^2 + 2(x)(7) + 7^2$ $= x^2 + 14x + 49$	$x^2 + 14x + 49 = x^2 + 2(x)(7) + 7^2$ $= (x + 7)^2$
$(3x - 4)^2 = (3x)^2 - 2(3x)(4) + 4^2$ $= 9x^2 - 24x + 16$	$9x^2 - 24x + 16 = (3x)^2 - 2(3x)(4) + 4^2$ $= (3x - 4)^2$

For a trinomial to be factorable as a perfect square, three conditions must be satisfied as illustrated in the example below.

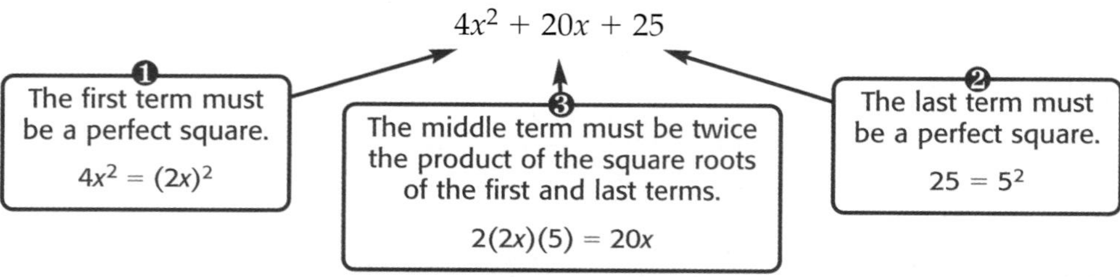

Key Concept — *Factoring Perfect Square Trinomials*

- **Words** If a trinomial can be written in the form $a^2 + 2ab + b^2$ or $a^2 - 2ab + b^2$, then it can be factored as $(a + b)^2$ or as $(a - b)^2$, respectively.
- **Symbols** $a^2 + 2ab + b^2 = (a + b)^2$ and $a^2 - 2ab + b^2 = (a - b)^2$
- **Example** $4x^2 - 20x + 25 = (2x)^2 - 2(2x)(5) + (5)^2$ or $(2x - 5)^2$

Example 1 *Factor Perfect Square Trinomials*

Determine whether each trinomial is a perfect square trinomial. If so, factor it.

a. $16x^2 + 32x + 64$

❶ Is the first term a perfect square? Yes, $16x^2 = (4x)^2$.
❷ Is the last term a perfect square? Yes, $64 = 8^2$.
❸ Is the middle term equal to $2(4x)(8)$? No, $32x \neq 2(4x)(8)$.

$16x^2 + 32x + 64$ is not a perfect square trinomial.

b. $9y^2 - 12y + 4$

❶ Is the first term a perfect square? Yes, $9y^2 = (3y)^2$.
❷ Is the last term a perfect square? Yes, $4 = 2^2$.
❸ Is the middle term equal to $2(3y)(2)$? Yes, $12y = 2(3y)(2)$.

$9y^2 - 12y + 4$ is a perfect square trinomial.

$9y^2 - 12y + 4 = (3y)^2 - 2(3y)(2) + 2^2$ Write as $a^2 - 2ab + b^2$.
$= (3y - 2)^2$ Factor using the pattern.

In this chapter, you have learned to factor different types of polynomials. The Concept Summary lists these methods and can help you decide when to use a specific method.

Concept Summary — *Factoring Polynomials*

Number of Terms	Factoring Technique		Example
2 or more	greatest common factor		$3x^3 + 6x^2 - 15x = 3x(x^2 + 2x - 5)$
2	difference of squares	$a^2 - b^2 = (a + b)(a - b)$	$4x^2 - 25 = (2x + 5)(2x - 5)$
3	perfect square trinomial	$a^2 + 2ab + b^2 = (a + b)^2$ $a^2 - 2ab + b^2 = (a - b)^2$	$x^2 + 6x + 9 = (x + 3)^2$ $4x^2 - 4x + 1 = (2x - 1)^2$
	$x^2 + bx + c$	$x^2 + bx + c = (x + m)(x + n)$, when $m + n = b$ and $mn = c$.	$x^2 - 9x + 20 = (x - 5)(x - 4)$
	$ax^2 + bx + c$	$ax^2 + bx + c = ax^2 + mx + nx + c$, when $m + n = b$ and $mn = ac$. Then use factoring by grouping.	$6x^2 - x - 2 = 6x^2 + 3x - 4x - 2$ $= 3x(2x + 1) - 2(2x + 1)$ $= (2x + 1)(3x - 2)$
4 or more	factoring by grouping	$ax + bx + ay + by$ $= x(a + b) + y(a + b)$ $= (a + b)(x + y)$	$3xy - 6y + 5x - 10$ $= (3xy - 6y) + (5x - 10)$ $= 3y(x - 2) + 5(x - 2)$ $= (x - 2)(3y + 5)$

When there is a GCF other than 1, it is usually easier to factor it out first. Then, check the appropriate factoring methods in the order shown in the table. Continue factoring until you have written the polynomial as the product of a monomial and/or prime polynomial factors.

Example 2 Factor Completely

Factor each polynomial.

a. $4x^2 - 36$

First check for a GCF. Then, since the polynomial has two terms, check for the difference of squares.

$4x^2 - 36 = 4(x^2 - 9)$	4 is the GCF.
$= 4(x^2 - 3^2)$	$x^2 = x \cdot x$ and $9 = 3 \cdot 3$
$= 4(x + 3)(x - 3)$	Factor the difference of squares.

b. $25x^2 + 5x - 6$

This polynomial has three terms that have a GCF of 1. While the first term is a perfect square, $25x^2 = (5x)^2$, the last term is not. Therefore, this is not a perfect square trinomial.

This trinomial is of the form $ax^2 + bx + c$. Are there two numbers m and n whose product is $25 \cdot -6$ or -150 and whose sum is 5? Yes, the product of 15 and -10 is -150 and their sum is 5.

$25x^2 + 5x - 6$	
$= 25x^2 + mx + nx - 6$	Write the pattern.
$= 25x^2 + 15x - 10x - 6$	$m = 15$ and $n = -10$
$= (25x^2 + 15x) + (-10x - 6)$	Group terms with common factors.
$= 5x(5x + 3) - 2(5x + 3)$	Factor out the GCF from each grouping.
$= (5x + 3)(5x - 2)$	$5x + 3$ is the common factor.

Study Tip

Alternative Method

Note that $4x^2 - 36$ could first be factored as $(2x + 6)(2x - 6)$. Then the common factor 2 would need to be factored out of each expression.

SOLVE EQUATIONS WITH PERFECT SQUARES When solving equations involving repeated factors, it is only necessary to set one of the repeated factors equal to zero.

Example 3 Solve Equations with Repeated Factors

Solve $x^2 - x + \frac{1}{4} = 0$.

$x^2 - x + \frac{1}{4} = 0$	Original equation
$x^2 - 2(x)\left(\frac{1}{2}\right) + \left(\frac{1}{2}\right)^2 = 0$	Recognize $x^2 - x + \frac{1}{4}$ as a perfect square trinomial.
$\left(x - \frac{1}{2}\right)^2 = 0$	Factor the perfect square trinomial.
$x - \frac{1}{2} = 0$	Set repeated factor equal to zero.
$x = \frac{1}{2}$	Solve for x.

Thus, the solution set is $\left\{\frac{1}{2}\right\}$. Check this solution in the original equation.

Study Tip

Reading Math
$\pm\sqrt{36}$ is read as *plus or minus the square root of 36.*

You have solved equations like $x^2 - 36 = 0$ by using factoring. You can also use the definition of square root to solve this equation.

$$\begin{aligned} x^2 - 36 &= 0 && \text{Original equation} \\ x^2 &= 36 && \text{Add 36 to each side.} \\ x &= \pm\sqrt{36} && \text{Take the square root of each side.} \end{aligned}$$

Remember that there are two square roots of 36, namely 6 and −6. Therefore, the solution set is $\{-6, 6\}$. This is sometimes expressed more compactly as $\{\pm 6\}$. This and other examples suggest the following property.

Key Concept — *Square Root Property*

- **Symbols** For any number $n > 0$, if $x^2 = n$, then $x = \pm\sqrt{n}$.
- **Example** $x^2 = 9$
  $x = \pm\sqrt{9}$ or ± 3

Example 4 Use the Square Root Property to Solve Equations

Solve each equation. Check your solutions.

a. $(a + 4)^2 = 49$

$$\begin{aligned} (a + 4)^2 &= 49 && \text{Original equation} \\ a + 4 &= \pm\sqrt{49} && \text{Square Root Property} \\ a + 4 &= \pm 7 && 49 = 7 \cdot 7 \\ a &= -4 \pm 7 && \text{Subtract 4 from each side.} \end{aligned}$$

$$a = -4 + 7 \quad \text{or} \quad a = -4 - 7 \qquad \text{Separate into two equations.}$$
$$= 3 \qquad\qquad\qquad = -11 \qquad \text{Simplify.}$$

The solution set is $\{-11, 3\}$. Check each solution in the original equation.

b. $y^2 - 4y + 4 = 25$

$$\begin{aligned} y^2 - 4y + 4 &= 25 && \text{Original equation} \\ (y)^2 - 2(y)(2) + 2^2 &= 25 && \text{Recognize perfect square trinomial.} \\ (y - 2)^2 &= 25 && \text{Factor perfect square trinomial.} \\ y - 2 &= \pm\sqrt{25} && \text{Square Root Property} \\ y - 2 &= \pm 5 && 25 = 5 \cdot 5 \\ y &= 2 \pm 5 && \text{Add 2 to each side.} \end{aligned}$$

$$y = 2 + 5 \quad \text{or} \quad y = 2 - 5 \qquad \text{Separate into two equations.}$$
$$= 7 \qquad\qquad\qquad = -3 \qquad \text{Simplify.}$$

The solution set is $\{-3, 7\}$. Check each solution in the original equation.

c. $(x - 3)^2 = 5$

$$\begin{aligned} (x - 3)^2 &= 5 && \text{Original equation} \\ x - 3 &= \pm\sqrt{5} && \text{Square Root Property} \\ x &= 3 \pm \sqrt{5} && \text{Add 3 to each side.} \end{aligned}$$

Since 5 is not a perfect square, the solution set is $\{3 \pm \sqrt{5}\}$. Using a calculator, the approximate solutions are $3 + \sqrt{5}$ or about 5.24 and $3 - \sqrt{5}$ or about 0.76.

CHECK You can check your answer using a graphing calculator. Graph $y = (x - 3)^2$ and $y = 5$. Using the INTERSECT feature of your graphing calculator, find where $(x - 3)^2 = 5$. The check of 5.24 as one of the approximate solutions is shown at the right.

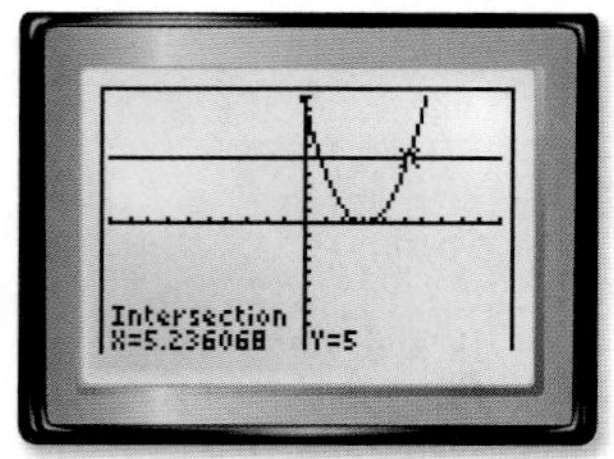

[−10, 10] scl: 1 by [−10, 10] scl: 1

Check for Understanding

Concept Check

1. **Explain** how to determine whether a trinomial is a perfect square trinomial.
2. **Determine** whether the following statement is *sometimes*, *always*, or *never* true. Explain your reasoning.
   $a^2 - 2ab - b^2 = (a - b)^2, b \neq 0$
3. **OPEN ENDED** Write a polynomial that requires at least two different factoring techniques to factor it completely.

Guided Practice

Determine whether each trinomial is a perfect square trinomial. If so, factor it.

4. $y^2 + 8y + 16$
5. $9x^2 - 30x + 10$

Factor each polynomial, if possible. If the polynomial cannot be factored, write *prime*.

6. $2x^2 + 18$
7. $c^2 - 5c + 6$
8. $5a^3 - 80a$
9. $8x^2 - 18x - 35$
10. $9g^2 + 12g - 4$
11. $3m^3 + 2m^2n - 12m - 8n$

Solve each equation. Check your solutions.

12. $4y^2 + 24y + 36 = 0$
13. $3n^2 = 48$
14. $a^2 - 6a + 9 = 16$
15. $(m - 5)^2 = 13$

Application

16. **HISTORY** Galileo demonstrated that objects of different weights fall at the same velocity by dropping two objects of different weights from the top of the Leaning Tower of Pisa. A model for the height h in feet of an object dropped from an initial height h_o in feet is $h = -16t^2 + h_o$, where t is the time in seconds after the object is dropped. Use this model to determine approximately how long it took for the objects to hit the ground if Galileo dropped them from a height of 180 feet.

Practice and Apply

Homework Help

For Exercises	See Examples
17–24	1
25–42	2
43–59	3, 4

Extra Practice

See page 841.

Determine whether each trinomial is a perfect square trinomial. If so, factor it.

17. $x^2 + 9x + 81$
18. $a^2 - 24a + 144$
19. $4y^2 - 44y + 121$
20. $2c^2 + 10c + 25$
21. $9n^2 + 49 + 42n$
22. $25a^2 - 120ab + 144b^2$

23. **GEOMETRY** The area of a circle is $(16x^2 + 80x + 100)\pi$ square inches. What is the diameter of the circle?

24. **GEOMETRY** The area of a square is $81 - 90x + 25x^2$ square meters. If x is a positive integer, what is the least possible perimeter measure for the square?

Factor each polynomial, if possible. If the polynomial cannot be factored, write *prime*.

25. $4k^2 - 100$
26. $9x^2 - 3x - 20$
27. $x^2 + 6x - 9$
28. $50g^2 + 40g + 8$
29. $9t^3 + 66t^2 - 48t$
30. $4a^2 - 36b^2$
31. $20n^2 + 34n + 6$
32. $5y^2 - 90$
33. $24x^3 - 78x^2 + 45x$
34. $18y^2 - 48y + 32$
35. $90g - 27g^2 - 75$
36. $45c^2 - 32cd$
37. $4a^3 + 3a^2b^2 + 8a + 6b^2$
38. $5a^2 + 7a + 6b^2 - 4b$
39. $x^2y^2 - y^2 - z^2 + x^2z^2$
40. $4m^4n + 6m^3n - 16m^2n^2 - 24mn^2$

41. **GEOMETRY** The volume of a rectangular prism is $x^3y - 63y^2 + 7x^2 - 9xy^3$ cubic meters. Find the dimensions of the prism if they can be represented by binomials with integral coefficients.

42. **GEOMETRY** If the area of the square shown below is $16x^2 - 56x + 49$ square inches, what is the area of the rectangle in terms of x?

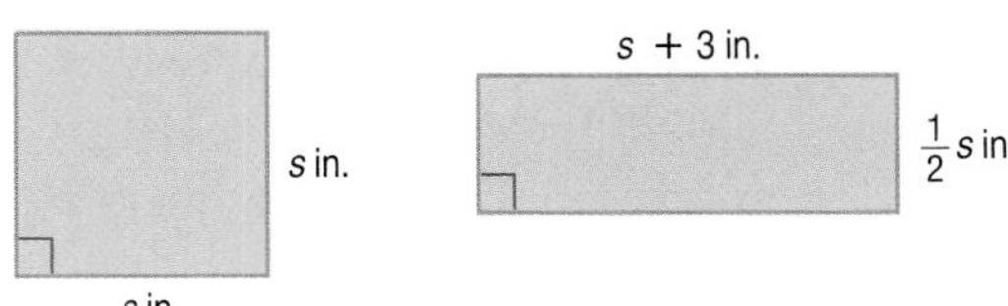

More About. . .

Free-Fall Ride

Some amusement park free-fall rides can seat 4 passengers across per coach and reach speeds of up to 62 miles per hour.

Source: www.pgathrills.com

Solve each equation. Check your solutions.

43. $3x^2 + 24x + 48 = 0$
44. $7r^2 = 70r - 175$
45. $49a^2 + 16 = 56a$
46. $18y^2 + 24y + 8 = 0$
47. $y^2 - \frac{2}{3}y + \frac{1}{9} = 0$
48. $a^2 + \frac{4}{5}a + \frac{4}{25} = 0$
49. $z^2 + 2z + 1 = 16$
50. $x^2 + 10x + 25 = 81$
51. $(y - 8)^2 = 7$
52. $(w + 3)^2 = 2$
53. $p^2 + 2p + 1 = 6$
54. $x^2 - 12x + 36 = 11$

FORESTRY For Exercises 55 and 56, use the following information.
Lumber companies need to be able to estimate the number of board feet that a given log will yield. One of the most commonly used formulas for estimating board feet is the *Doyle Log Rule*, $B = \frac{L}{16}(D^2 - 8D + 16)$, where B is the number of board feet, D is the diameter in inches, and L is the length of the log in feet.

55. Write this formula in factored form.
56. For logs that are 16 feet long, what diameter will yield approximately 256 board feet?

FREE-FALL RIDE For Exercises 57 and 58, use the following information.
The height h in feet of a car above the exit ramp of an amusement park's free-fall ride can be modeled by $h = -16t^2 + s$, where t is the time in seconds after the car drops and s is the starting height of the car in feet.

57. How high above the car's exit ramp should the ride's designer start the drop in order for riders to experience free fall for at least 3 seconds?
58. Approximately how long will riders be in free fall if their starting height is 160 feet above the exit ramp?

59. HUMAN CANNONBALL A circus acrobat is shot out of a cannon with an initial upward velocity of 64 feet per second. If the acrobat leaves the cannon 6 feet above the ground, will he reach a height of 70 feet? If so, how long will it take him to reach that height? Use the model for vertical motion.

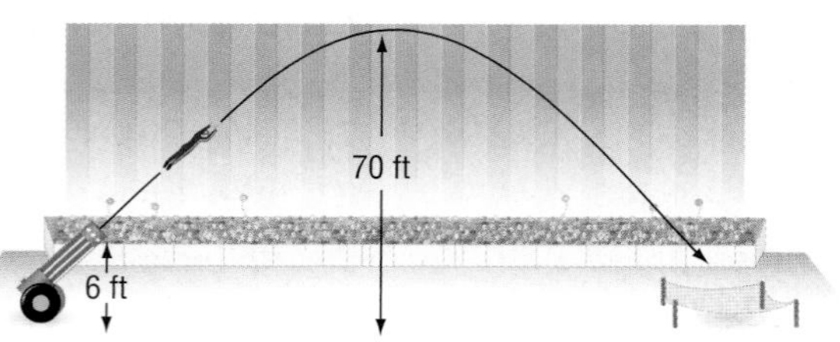

CRITICAL THINKING **Determine all values of k that make each of the following a perfect square trinomial.**

60. $x^2 + kx + 64$
61. $4x^2 + kx + 1$
62. $25x^2 + kx + 49$
63. $x^2 + 8x + k$
64. $x^2 - 18x + k$
65. $x^2 + 20x + k$

66. **WRITING IN MATH** Answer the question that was posed at the beginning of the lesson.

How can factoring be used to design a pavilion?

Include the following in your answer:

- an explanation of how the equation $(8 + 2x)^2 = 144$ models the given situation, and
- an explanation of how to solve this equation, listing any properties used, and an interpretation of its solutions.

67. During an experiment, a ball is dropped off a bridge from a height of 205 feet. The formula $205 = 16t^2$ can be used to approximate the amount of time, in seconds, it takes for the ball to reach the surface of the water of the river below the bridge. Find the time it takes the ball to reach the water to the nearest tenth of a second.

Ⓐ 2.3 s　Ⓑ 3.4 s　Ⓒ 3.6 s　Ⓓ 12.8 s

68. If $\sqrt{a^2 - 2ab + b^2} = a - b$, then which of the following statements best describes the relationship between a and b?

Ⓐ $a < b$　Ⓑ $a \leq b$　Ⓒ $a > b$　Ⓓ $a \geq b$

Maintain Your Skills

Mixed Review

Solve each equation. Check your solutions. *(Lessons 9-4 and 9-5)*

69. $s^2 = 25$
70. $9x^2 - 16 = 0$
71. $49m^2 = 81$
72. $8k^2 + 22k - 6 = 0$
73. $12w^2 + 23w = -5$
74. $6z^2 + 7 = 17z$

Write the slope-intercept form of an equation that passes through the given point and is perpendicular to the graph of each equation. *(Lesson 5-6)*

75. $(1, 4), y = 2x - 1$
76. $(-4, 7), y = -\frac{2}{3}x + 7$

77. **NATIONAL LANDMARKS** At the Royal Gorge in Colorado, an inclined railway takes visitors down to the Arkansas River. Suppose the slope is 50% or $\frac{1}{2}$ and the vertical drop is 1015 feet. What is the horizontal change of the railway? *(Lesson 5-1)*

Find the next three terms of each arithmetic sequence. *(Lesson 4-7)*

78. 17, 13, 9, 5, …
79. −5, −4.5, −4, −3.5, …
80. 45, 54, 63, 72, …

Chapter 9

Study Guide and Review

Vocabulary and Concept Check

composite number (p. 474)
factored form (p. 475)
factoring (p. 481)
factoring by grouping (p. 482)
greatest common factor (GCF) (p. 476)
perfect square trinomials (p. 508)
prime factorization (p. 475)
prime number (p. 474)
prime polynomial (p. 497)
Square Root Property (p. 511)
Zero Product Property (p. 483)

State whether each sentence is *true* or *false*. If false, replace the underlined word or number to make a true sentence.

1. The number 27 is an example of a prime number.
2. $2x$ is the greatest common factor (GCF) of $12x^2$ and $14xy$.
3. 66 is an example of a perfect square.
4. 61 is a factor of 183.
5. The prime factorization for 48 is $3 \cdot 4^2$.
6. $x^2 - 25$ is an example of a perfect square trinomial.
7. The number 35 is an example of a composite number.
8. $x^2 - 3x - 70$ is an example of a prime polynomial.
9. Expressions with four or more unlike terms can sometimes be factored by grouping.
10. $(b - 7)(b + 7)$ is the factorization of a difference of squares.

Lesson-by-Lesson Review

9-1 Factors and Greatest Common Factors

See pages 474–479.

Concept Summary

- Prime number: whole number greater than 1 with exactly two factors
- Composite number: whole number greater than 1 with more than two factors
- The greatest common factor (GCF) of two or more monomials is the product of their common prime factors.

Example Find the GCF of $15x^2y$ and $45xy^2$.

$15x^2y = 3 \cdot 5 \cdot x \cdot x \cdot y$ Factor each number.

$45xy^2 = 3 \cdot 3 \cdot 5 \cdot x \cdot y \cdot y$ Circle the common prime factors.

The GCF is $3 \cdot 5 \cdot x \cdot y$ or $15xy$.

Exercises **Find the prime factorization of each integer.**
See Examples 2 and 3 on page 475.

11. 28
12. 33
13. 150
14. 301
15. −83
16. −378

Find the GCF of each set of monomials. *See Example 5 on page 476.*

17. 35, 30
18. 12, 18, 40
19. $12ab$, $4a^2b^2$
20. $16mrt$, $30m^2r$
21. $20n^2$, $25np^5$
22. $60x^2y^2$, $35xz^3$

9-2 Factoring Using the Distributive Property

See pages 481–486.

Concept Summary

- Find the greatest common factor and then use the Distributive Property.
- With four or more terms, try factoring by grouping.

  Factoring by Grouping: $ax + bx + ay + by = x(a + b) + y(a + b) = (a + b)(x + y)$
- Factoring can be used to solve some equations.

  Zero Product Property: For any real numbers a and b, if $ab = 0$, then either $a = 0$, $b = 0$, or both a and b equal zero.

Example **Factor $2x^2 - 3xz - 2xy + 3yz$.**

$2x^2 - 3xz - 2xy + 3yz = (2x^2 - 3xz) + (-2xy + 3yz)$	Group terms with common factors.
$= x(2x - 3z) - y(2x - 3z)$	Factor out the GCF from each grouping.
$= (x - y)(2x - 3z)$	Factor out the common factor $2x - 3z$.

Exercises **Factor each polynomial.** *See Examples 1 and 2 on pages 481 and 482.*

23. $13x + 26y$
24. $24a^2b^2 - 18ab$
25. $26ab + 18ac + 32a^2$
26. $a^2 - 4ac + ab - 4bc$
27. $4rs + 12ps + 2mr + 6mp$
28. $24am - 9an + 40bm - 15bn$

Solve each equation. Check your solutions. *See Examples 2 and 5 on pages 482 and 483.*

29. $x(2x - 5) = 0$
30. $(3n + 8)(2n - 6) = 0$
31. $4x^2 = -7x$

9-3 Factoring Trinomials: $x^2 + bx + c$

See pages 489–494.

Concept Summary

- Factoring $x^2 + bx + c$: Find m and n whose sum is b and whose product is c. Then write $x^2 + bx + c$ as $(x + m)(x + n)$.

Example **Solve $a^2 - 3a - 4 = 0$. Then check the solutions.**

$a^2 - 3a - 4 = 0$	Original equation
$(a + 1)(a - 4) = 0$	Factor.
$a + 1 = 0$ or $a - 4 = 0$	Zero Product Property
$a = -1$ $\quad a = 4$	Solve each equation.

The solution set is $\{-1, 4\}$.

Exercises **Factor each trinomial.** *See Examples 1–4 on pages 490 and 491.*

32. $y^2 + 7y + 12$
33. $x^2 - 9x - 36$
34. $b^2 + 5b - 6$
35. $18 - 9r + r^2$
36. $a^2 + 6ax - 40x^2$
37. $m^2 - 4mn - 32n^2$

Solve each equation. Check your solutions. *See Example 5 on page 491.*

38. $y^2 + 13y + 40 = 0$
39. $x^2 - 5x - 66 = 0$
40. $m^2 - m - 12 = 0$

9-4 Factoring Trinomials: $ax^2 + bx + c$

See pages 495–500.

Concept Summary

- Factoring $ax^2 + bx + c$: Find m and n whose product is ac and whose sum is b. Then, write as $ax^2 + mx + nx + c$ and use factoring by grouping.

Example **Factor $12x^2 + 22x - 14$.**

First, factor out the GCF, 2: $12x^2 + 22x - 14 = 2(6x^2 + 11x - 7)$. In the new trinomial, $a = 6$, $b = 11$ and $c = -7$. Since b is positive, $m + n$ is positive. Since c is negative, mn is negative. So either m or n is negative, but not both. Therefore, make a list of the factors of $6(-7)$ or -42, where one factor in each pair is negative. Look for a pair of factors whose sum is 11.

Factors of −42	Sum of Factors	
−1, 42	41	
1, −42	−41	
−2, 21	19	
2, −21	−19	
−3, 14	11	The correct factors are −3 and 14.

$6x^2 + 11x - 7 = 6x^2 + mx + nx - 7$ Write the pattern.

$= 6x^2 - 3x + 14x - 7$ $m = -3$ and $n = 14$

$= (6x^2 - 3x) + (14x - 7)$ Group terms with common factors.

$= 3x(2x - 1) + 7(2x - 1)$ Factor the GCF from each grouping.

$= (2x - 1)(3x + 7)$ $2x - 1$ is the common factor.

Thus, the complete factorization of $12x^2 + 22x - 14$ is $2(2x - 1)(3x + 7)$.

Exercises **Factor each trinomial, if possible. If the trinomial cannot be factored using integers, write *prime*.** *See Examples 1–3 on pages 496 and 497.*

41. $2a^2 - 9a + 3$ **42.** $2m^2 + 13m - 24$ **43.** $25r^2 + 20r + 4$

44. $6z^2 + 7z + 3$ **45.** $12b^2 + 17b + 6$ **46.** $3n^2 - 6n - 45$

Solve each equation. Check your solutions. *See Example 4 on page 497.*

47. $2r^2 - 3r - 20 = 0$ **48.** $3a^2 - 13a + 14 = 0$ **49.** $40x^2 + 2x = 24$

9-5 Factoring Differences of Squares

See pages 501–506.

Concept Summary

- Difference of Squares: $a^2 - b^2 = (a + b)(a - b)$ or $(a - b)(a + b)$
- Sometimes it may be necessary to use more than one factoring technique or to apply a factoring technique more than once.

Example **Factor $3x^3 - 75x$.**

$3x^3 - 75x = 3x(x^2 - 25)$ The GCF of $3x^3$ and $75x$ is $3x$.

$= 3x(x + 5)(x - 5)$ Factor the difference of squares.

• Extra Practice, see pages 839–841.
• Mixed Problem Solving, see page 861.

Exercises **Factor each polynomial, if possible. If the polynomial cannot be factored, write *prime*.** *See Examples 1–4 on page 502.*

50. $2y^3 - 128y$ **51.** $9b^2 - 20$ **52.** $\frac{1}{4}n^2 - \frac{9}{16}r^2$

Solve each equation by factoring. Check your solutions. *See Example 5 on page 503.*

53. $b^2 - 16 = 0$ **54.** $25 - 9y^2 = 0$ **55.** $16a^2 - 81 = 0$

9-6 Perfect Squares and Factoring

See pages 508–514.

Concept Summary

- If a trinomial can be written in the form $a^2 + 2ab + b^2$ or $a^2 - 2ab + b^2$, then it can be factored as $(a + b)^2$ or as $(a - b)^2$, respectively.
- For a trinomial to be factorable as a perfect square, the first term must be a perfect square, the middle term must be twice the product of the square roots of the first and last terms, and the last term must be a perfect square.
- Square Root Property: For any number $n > 0$, if $x^2 = n$, then $x = \pm\sqrt{n}$.

Examples

1 **Determine whether $9x^2 + 24xy + 16y^2$ is a perfect square trinomial. If so, factor it.**

❶ Is the first term a perfect square? Yes, $9x^2 = (3x)^2$.

❷ Is the last term a perfect square? Yes, $16y^2 = (4y)^2$.

❸ Is the middle term equal to $2(3x)(4y)$? Yes, $24xy = 2(3x)(4y)$.

$9x^2 + 24xy + 16y^2 = (3x)^2 + 2(3x)(4y) + (4y)^2$ Write as $a^2 + 2ab + b^2$.

$= (3x + 4y)^2$ Factor using the pattern.

2 **Solve $(x - 4)^2 = 121$.**

$(x - 4)^2 = 121$ Original equation

$x - 4 = \pm\sqrt{121}$ Square Root Property

$x - 4 = \pm 11$ $121 = 11 \cdot 11$

$x = 4 \pm 11$ Add 4 to each side.

$x = 4 + 11$ or $x = 4 - 11$ Separate into two equations.

$= 15$ $= -7$ The solution set is $\{-7, 15\}$.

Exercises **Factor each polynomial, if possible. If the polynomial cannot be factored, write *prime*.** *See Example 2 on page 510.*

56. $a^2 + 18a + 81$ **57.** $9k^2 - 12k + 4$

58. $4 - 28r + 49r^2$ **59.** $32n^2 - 80n + 50$

Solve each equation. Check your solutions. *See Examples 3 and 4 on pages 510 and 511.*

60. $6b^3 - 24b^2 + 24b = 0$ **61.** $49m^2 - 126m + 81 = 0$

62. $(c - 9)^2 = 144$ **63.** $144b^2 = 36$

Chapter 9

Practice Test

Vocabulary and Concepts

1. **Give an example** of a prime number and explain why it is prime.
2. **Write** a polynomial that is the difference of two squares. Then factor your polynomial.
3. **Describe** the first step in factoring any polynomial.

Skills and Applications

Find the prime factorization of each integer.

4. 63
5. 81
6. −210

Find the GCF of the given monomials.

7. 48, 64
8. 28, 75
9. $18a^2b^2$, $28a^3b^2$

Factor each polynomial, if possible. If the polynomial cannot be factored using integers, write *prime*.

10. $25y^2 - 49w^2$
11. $t^2 - 16t + 64$
12. $x^2 + 14x + 24$
13. $28m^2 + 18m$
14. $a^2 - 11ab + 18b^2$
15. $12x^2 + 23x - 24$
16. $2h^2 - 3h - 18$
17. $6x^3 + 15x^2 - 9x$
18. $64p^2 - 63p + 16$
19. $2d^2 + d - 1$
20. $36a^2b^3 - 45ab^4$
21. $36m^2 + 60mn + 25n^2$
22. $a^2 - 4$
23. $4my - 20m + 3py - 15p$
24. $15a^2b + 5a^2 - 10a$
25. $6y^2 - 5y - 6$
26. $4s^2 - 100t^2$
27. $x^3 - 4x^2 - 9x + 36$

Write an expression in factored form for the area of each shaded region.

28.

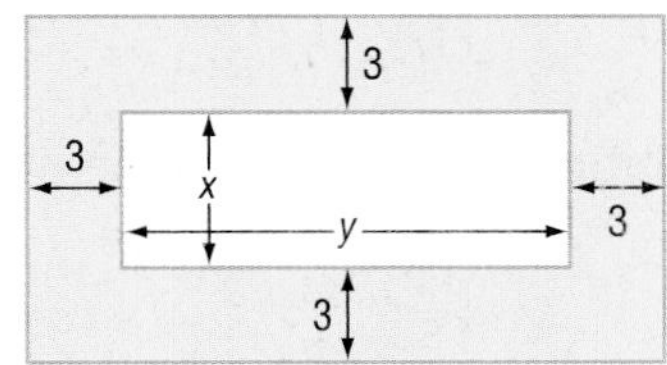

29.

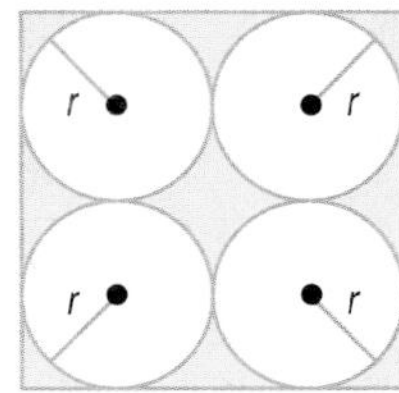

Solve each equation. Check your solutions.

30. $(4x - 3)(3x + 2) = 0$
31. $18s^2 + 72s = 0$
32. $4x^2 = 36$
33. $t^2 + 25 = 10t$
34. $a^2 - 9a - 52 = 0$
35. $x^3 - 5x^2 - 66x = 0$
36. $2x^2 = 9x + 5$
37. $3b^2 + 6 = 11b$

38. **GEOMETRY** A rectangle is 4 inches wide by 7 inches long. When the length and width are increased by the same amount, the area is increased by 26 square inches. What are the dimensions of the new rectangle?

39. **CONSTRUCTION** A rectangular lawn is 24 feet wide by 32 feet long. A sidewalk will be built along the inside edges of all four sides. The remaining lawn will have an area of 425 square feet. How wide will the walk be?

40. **STANDARDIZED TEST PRACTICE** The area of the shaded part of the square shown at the right is 98 square meters. Find the dimensions of the square.

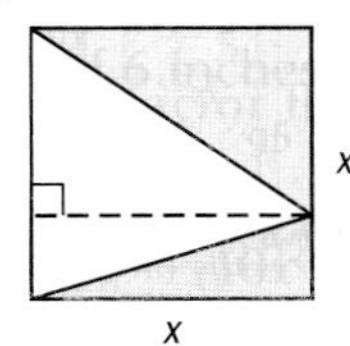

Chapter 9 Standardized Test Practice

Part 1 Multiple Choice

Record your answers on the answer sheet provided by your teacher or on a sheet of paper.

1. Which equation best describes the function graphed below? (Lesson 5-3)

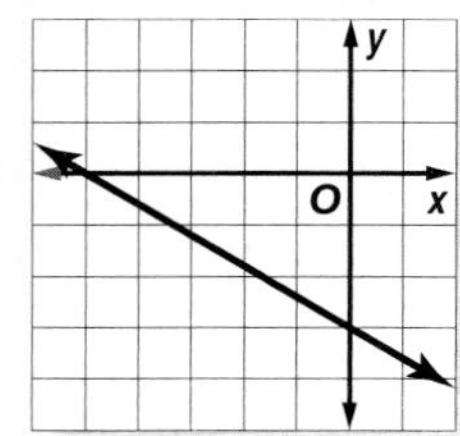

   Ⓐ $y = -\frac{3}{5}x - 3$

   Ⓑ $y = \frac{3}{5}x - 3$

   Ⓒ $y = -\frac{5}{3}x - 3$

   Ⓓ $y = \frac{5}{3}x - 3$

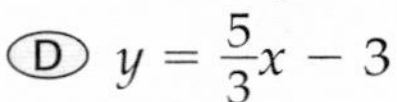

2. The school band sold tickets to their spring concert every day at lunch for one week. Before they sold any tickets, they had \$80 in their account. At the end of each day, they recorded the total number of tickets sold and the total amount of money in the band's account.

Day	Total Number of Tickets Sold t	Total Amount in Account a
Monday	12	\$176
Tuesday	18	\$224
Wednesday	24	\$272
Thursday	30	\$320
Friday	36	\$368

   Which equation describes the relationship between the total number of tickets sold t and the amount of money in the band's account a? (Lesson 5-4)

   Ⓐ $a = \frac{1}{8}t + 80$ Ⓑ $a = \frac{t + 80}{6}$

   Ⓒ $a = 6t + 8$ Ⓓ $a = 8t + 80$

3. Which inequality represents the shaded portion of the graph? (Lesson 6-6)

   Ⓐ $y \geq \frac{1}{3}x - 1$

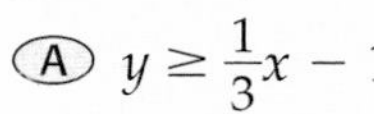

   Ⓑ $y \leq \frac{1}{3}x - 1$

   Ⓒ $y \leq 3x + 1$

   Ⓓ $y \geq 3x - 1$

4. Today, the refreshment stand at the high school football game sold twice as many bags of popcorn as were sold last Friday. The total sold both days was 258 bags. Which system of equations will determine the number of bags sold today n and the number of bags sold last Friday f? (Lesson 7-2)

   Ⓐ $n = f - 258$, $f = 2n$ Ⓑ $n = f - 258$, $n = 2f$

   Ⓒ $n + f = 258$, $f = 2n$ Ⓓ $n + f = 258$, $n = 2f$

5. Express 5.387×10^{-3} in standard notation. (Lesson 8-3)

   Ⓐ 0.0005387 Ⓑ 0.005387

   Ⓒ 538.7 Ⓓ 5387

6. The quotient $\frac{16x^8}{8x^4}$, $x \neq 0$, is (Lesson 9-1)

   Ⓐ $2x^2$. Ⓑ $8x^2$. Ⓒ $2x^4$. Ⓓ $8x^4$.

7. What are the solutions of the equation $3x^2 - 48 = 0$? (Lesson 9-1)

   Ⓐ 4, −4 Ⓑ 4, $\frac{1}{3}$

   Ⓒ 16, −16 Ⓓ 16, $\frac{1}{3}$

8. What are the solutions of the equation $x^2 - 3x + 8 = 6x - 6$? (Lesson 9-4)

   Ⓐ 2, −7 Ⓑ −2, −4

   Ⓒ 2, 4 Ⓓ 2, 7

9. The area of a rectangle is $12x^2 - 21x - 6$. The width is $3x - 6$. What is the length? (Lesson 9-5)

   Ⓐ $4x - 1$ Ⓑ $4x + 1$

   Ⓒ $9x + 1$ Ⓓ $12x - 18$

The Princeton Review Test-Taking Tip

Questions 7 and 9 When answering a multiple-choice question, first find an answer on your own. Then, compare your answer to the answer choices given in the item. If your answer does not match any of the answer choices, check your calculations.

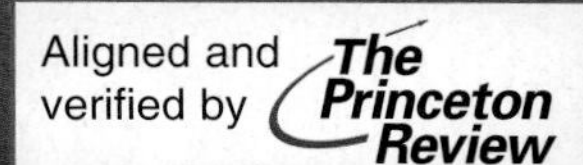

Part 2 Short Response/Grid In

Record your answers on the answer sheet provided by your teacher or on a sheet of paper.

10. Find all values of x that make the equation $6|x - 2| = 18$ true. (Lesson 6-5)

11. Graph the inequality $x + y \leq 3$. (Lesson 6-6)

12. A movie theater charges $7.50 for each adult ticket and $4 for each child ticket. If the theater sold a total of 145 tickets for a total of $790, how many adult tickets were sold? (Lesson 7-2)

13. Solve the following system of equations.
$3x + y = 8$
$4x - 2y = 14$ (Lesson 7-3)

14. Write $(x + t)x + (x + t)y$ as the product of two factors. (Lesson 9-3)

15. The product of two consecutive odd integers is 195. Find the integers. (Lesson 9-4)

16. Solve $2x^2 + 5x - 12 = 0$ by factoring. (Lesson 9-5)

17. Factor $2x^2 + 7x + 3$. (Lesson 9-5)

Part 3 Quantitative Comparison

Compare the quantity in Column A and the quantity in Column B. Then determine whether:

Ⓐ **the quantity in Column A is greater,**

Ⓑ **the quantity in Column B is greater,**

Ⓒ **the two quantities are equal, or**

Ⓓ **the relationship cannot be determined from the information given.**

	Column A	Column B
18.	$\|x\| - \|y\|$ if $x = -15$ and $y = -7$	$\|x - y\|$ if $x = -15$ and $y = -7$

(Lesson 2-2)

	Column A	Column B
19.	the solution of $\frac{2}{3}x - 27 = 39$	the solution of $\frac{3}{4}y - 55 = 20$

(Lesson 3-4)

20.

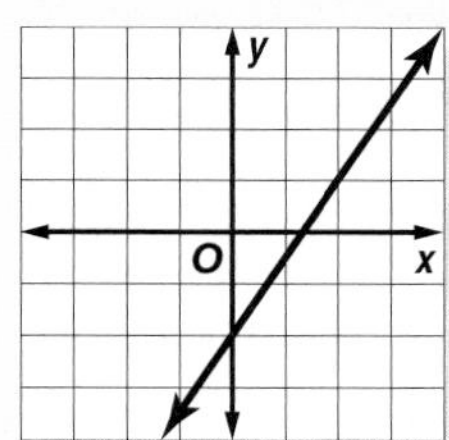

Column A	Column B
the x-intercept of the line whose graph is shown	the x-intercept of the line that is perpendicular to the line graphed above and passes through $(6, -4)$

(Lesson 5-6)

	Column A	Column B
21.	the y value of the solution of $3x - y = 5$ and $x - 3y = -6$	the b value of the solution of $2a - 3b = -3$ and $a + 4b = 24$

(Lesson 7-2)

	Column A	Column B
22.	the GCF of $2x^3$, $6x^2$, and $8x$	the GCF of $18x^3$, $14x^2$, and $4x$

(Lesson 9-1)

Part 4 Open Ended

Record your answers on a sheet of paper. Show your work.

23. Madison is building a fenced, rectangular dog pen. The width of the pen will be 3 yards less than the length. The total area enclosed is 28 square yards. (Lesson 9-4)
    a. Using L to represent the length of the pen, write an equation showing the area of the pen in terms of its length.
    b. What is the length of the pen?
    c. How many yards of fencing will Madison need to enclose the pen completely?

Chapter 10

Quadratic and Exponential Functions

What You'll Learn

- **Lesson 10-1** Graph quadratic functions.
- **Lessons 10-2 through 10-4** Solve quadratic equations.
- **Lesson 10-5** Graph exponential functions.
- **Lesson 10-6** Solve problems involving exponential growth and exponential decay.
- **Lesson 10-7** Recognize and extend geometric sequences.

Why It's Important

Quadratic functions and equations are used to solve problems about fireworks, to simulate the flight of golf balls in computer games, to describe arches, to determine hang time in football, and to help with water management. Exponential functions are used to describe changes in population, to solve compound interest problems, and to determine concentration of chemicals in a body of water after a spill. Exponential decay is one type of exponential function. Carbon dating uses exponential decay to determine the age of fossils and dinosaurs. *You will learn about carbon dating in Lesson 10-6.*

Key Vocabulary

- **parabola** (p. 524)
- **completing the square** (p. 539)
- **Quadratic Formula** (p. 546)
- **exponential function** (p. 554)
- **geometric sequence** (p. 567)

Getting Started

Prerequisite Skills To be successful in this chapter, you'll need to master these skills and be able to apply them in problem-solving situations. Review these skills before beginning Chapter 10.

For Lesson 10-1 — Graph Functions

Use a table of values to graph each equation. *(For review, see Lesson 5-3.)*

1. $y = x + 5$ **2.** $y = 2x - 3$ **3.** $y = 0.5x + 1$ **4.** $y = -3x - 2$

5. $2x - 3y = 12$ **6.** $5y = 10 + 2x$ **7.** $x + 2y = -6$ **8.** $3x = -2y + 9$

For Lesson 10-3 — Perfect Square Trinomials

Determine whether each trinomial is a perfect square trinomial. If so, factor it. *(For review, see Lesson 9-6.)*

9. $t^2 + 12t + 36$ **10.** $a^2 - 14a + 49$ **11.** $m^2 + 18m - 81$ **12.** $y^2 + 8y + 12$

13. $9b^2 - 6b + 1$ **14.** $6x^2 + 4x + 1$ **15.** $4p^2 + 12p + 9$ **16.** $16s^2 - 24s + 9$

For Lesson 10-7 — Arithmetic Sequences

Find the next three terms of each arithmetic sequence. *(For review, see Lesson 4-7.)*

17. 5, 9, 13, 17, … **18.** 12, 5, −2, −9, …

19. −4, −1, 2, 5, … **20.** 24, 32, 40, 48, …

21. −1, −6, −11, −16, … **22.** −27, −20, −13, −6, …

23. 5.3, 6.0, 6.7, 7.4, … **24.** 9.1, 8.8, 8.5, 8.2, …

FOLDABLES™ Study Organizer

Make this Foldable to help you organize information on quadratic and exponential functions. Begin with four sheets of grid paper.

Step 1 Fold in Half

Fold each sheet in half along the width.

Step 2 Tape

Unfold each sheet and tape to form one long piece.

Step 3 Label

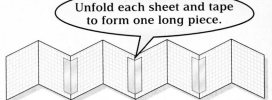

Label each page with the lesson number as shown. Refold to form a booklet.

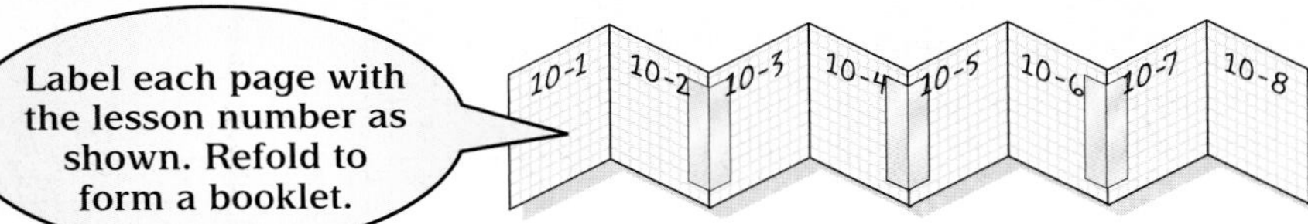

Reading and Writing As you read and study the chapter, write notes and examples for each lesson on each page of the journal.

10-1 Graphing Quadratic Functions

What You'll Learn

- Graph quadratic functions.
- Find the equation of the axis of symmetry and the coordinates of the vertex of a parabola.

Vocabulary

- quadratic function
- parabola
- minimum
- maximum
- vertex
- symmetry
- axis of symmetry

How can you coordinate a fireworks display with recorded music?

The Sky Concert in Peoria, Illinois, is a 4th of July fireworks display set to music. If a rocket (firework) is launched with an initial velocity of 39.2 meters per second at a height of 1.6 meters above the ground, the equation $h = -4.9t^2 + 39.2t + 1.6$ represents the rocket's height h in meters after t seconds. The rocket will explode at approximately the highest point.

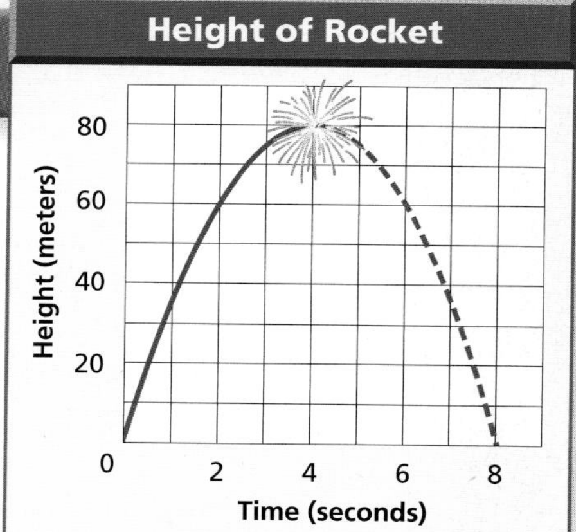

GRAPH QUADRATIC FUNCTIONS The function describing the height of the rocket is an example of a quadratic function. A **quadratic function** can be written in the form $y = ax^2 + bx + c$, where $a \neq 0$. This form of the quadratic function is called the *standard form*. Notice that this polynomial has degree 2 and the exponents are positive. The graph of a quadratic function is called a **parabola**.

Key Concept — Quadratic Function

- **Words** A quadratic function can be described by an equation of the form $y = ax^2 + bx + c$, where $a \neq 0$.
- **Models**

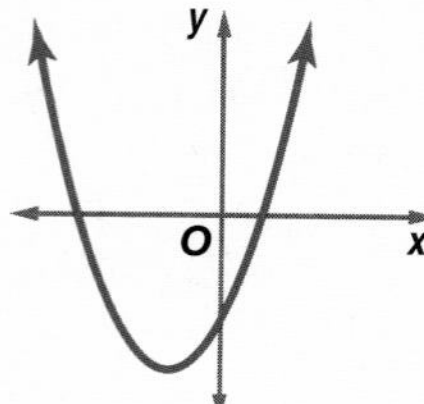

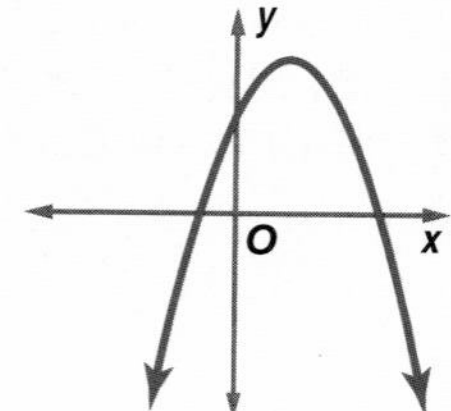

Example 1 Graph Opens Upward

Use a table of values to graph $y = 2x^2 - 4x - 5$.

Graph these ordered pairs and connect them with a smooth curve.

x	y
−2	11
−1	1
0	−5
1	−7
2	−5
3	1
4	11

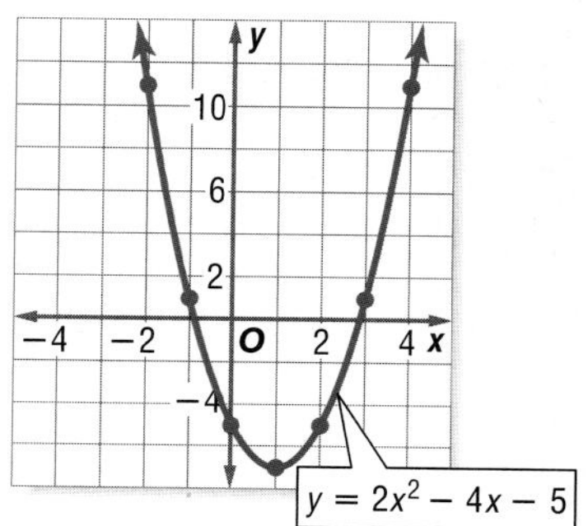

Consider the standard form $y = ax^2 + bx + c$. Notice that the value of a in Example 1 is positive and the curve opens upward. The lowest point, or **minimum**, of the graph is located at $(1, -7)$.

Example 2 Graph Opens Downward

Use a table of values to graph $y = -x^2 + 4x - 1$.

Graph these ordered pairs and connect them with a smooth curve.

x	y
-1	-6
0	-1
1	2
2	3
3	2
4	-1
5	-6

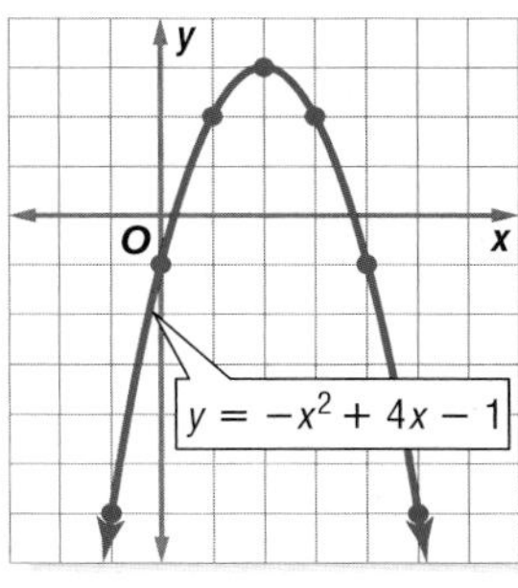

Study Tip

Reading Math
The plural of vertex is vertices. In math, vertex has several meanings. For example, there are the vertex of an angle, the vertices of a polygon, and the vertex of a parabola.

Notice that the value of a in Example 2 is negative and the curve opens downward. The highest point, or **maximum**, of the graph is located at $(2, 3)$. The maximum or minimum point of a parabola is called the **vertex**.

SYMMETRY AND VERTICES Parabolas possess a geometric property called **symmetry**. Symmetrical figures are those in which the figure can be folded in half so that each half matches the other exactly.

Algebra Activity

Symmetry of Parabolas

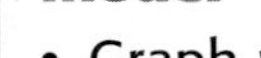

Model

- Graph $y = x^2 + 6x + 8$ on grid paper.
- Hold your paper up to the light and fold the parabola in half so that the two sides match exactly.
- Unfold the paper.

Make a Conjecture

1. What is the vertex of the parabola?
2. Write an equation of the fold line.
3. Which point on the parabola lies on the fold line?
4. Write a few sentences to describe the symmetry of a parabola based on your findings in this activity.

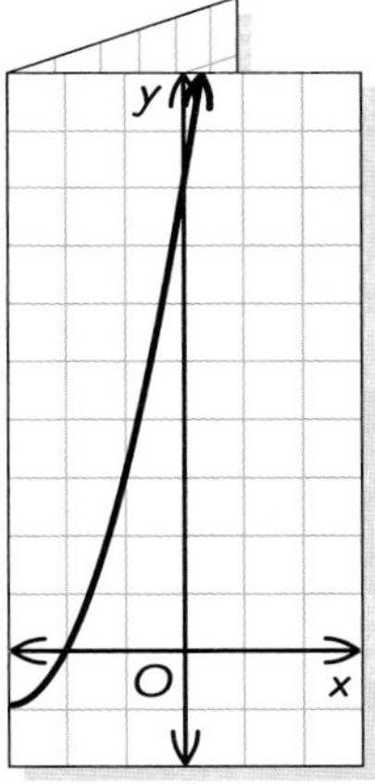

The fold line in the activity above is called the **axis of symmetry** for the parabola. Each point on the parabola that is on one side of the axis of symmetry has a corresponding point on the parabola on the other side of the axis. The vertex is the only point on the parabola that is on the axis of symmetry.

In the graph of $y = x^2 - x - 6$, the axis of symmetry is $x = \frac{1}{2}$. The vertex is $\left(\frac{1}{2}, -6\frac{1}{4}\right)$. Notice the relationship between the values a and b and the equation of the axis of symmetry.

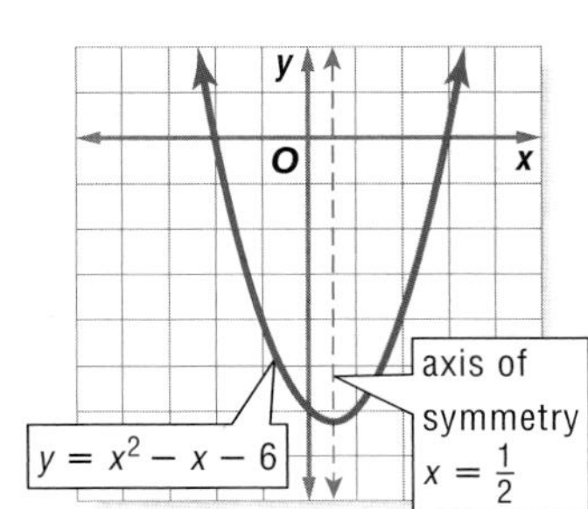

Key Concept — Equation of the Axis of Symmetry of a Parabola

- **Words** The equation of the axis of symmetry for the graph of $y = ax^2 + bx + c$, where $a \neq 0$, is $x = -\frac{b}{2a}$.
- **Model**

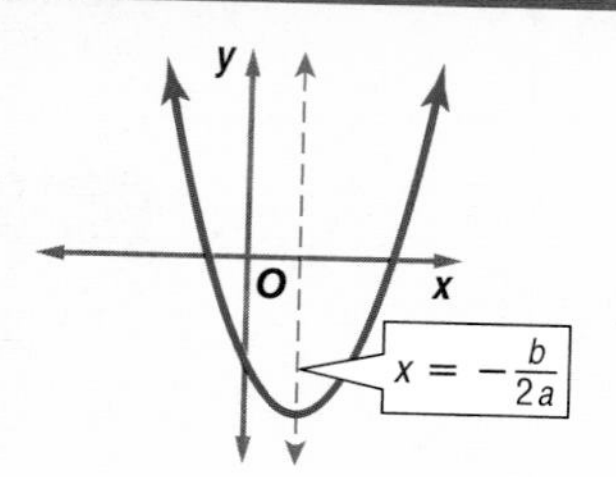

You can determine information about a parabola from its equation.

Example 3 Vertex and Axis of Symmetry

Consider the graph of $y = -3x^2 - 6x + 4$.

a. Write the equation of the axis of symmetry.

In $y = -3x^2 - 6x + 4$, $a = -3$ and $b = -6$.

$x = -\frac{b}{2a}$ Equation for the axis of symmetry of a parabola

$x = -\frac{-6}{2(-3)}$ or -1 $a = -3$ and $b = -6$

The equation of the axis of symmetry is $x = -1$.

Study Tip

Coordinates of Vertex

Notice that you can find the x-coordinate by knowing the axis of symmetry. However, to find the y-coordinate, you must substitute the value of x into the quadratic equation.

b. Find the coordinates of the vertex.

Since the equation of the axis of symmetry is $x = -1$ and the vertex lies on the axis, the x-coordinate for the vertex is -1.

$y = -3x^2 - 6x + 4$ Original equation

$y = -3(-1)^2 - 6(-1) + 4$ $x = -1$

$y = -3 + 6 + 4$ Simplify.

$y = 7$ Add.

The vertex is at $(-1, 7)$.

c. Identify the vertex as a maximum or minimum.

Since the coefficient of the x^2 term is negative, the parabola opens downward and the vertex is a maximum point.

d. Graph the function.

You can use the symmetry of the parabola to help you draw its graph. On a coordinate plane, graph the vertex and the axis of symmetry. Choose a value for x other than -1. For example, choose 1 and find the y-coordinate that satisfies the equation.

$y = -3x^2 - 6x + 4$ Original equation

$y = -3(1)^2 - 6(1) + 4$ Let $x = 1$.

$y = -5$ Simplify.

Graph $(1, -5)$. Since the graph is symmetrical about its axis of symmetry $x = -1$, you can find another point on the other side of the axis of symmetry. The point at $(1, -5)$ is 2 units to the right of the axis. Go 2 units to the left of the axis and plot the point $(-3, -5)$. Repeat this for several other points. Then sketch the parabola.

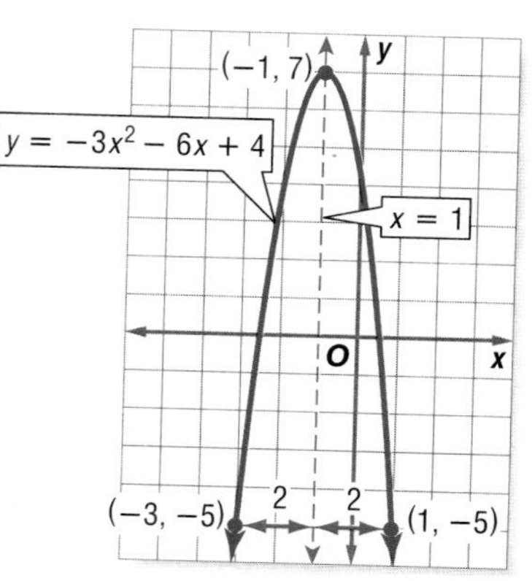

CHECK Does $(-3, -5)$ satisfy the equation?

$y = -3x^2 - 6x + 4$	Original equation
$-5 \stackrel{?}{=} -3(-3)^2 - 6(-3) + 4$	$y = -5$ and $x = -3$
$-5 = -5$ ✓	Simplify.

The ordered pair $(-3, -5)$ satisfies $y = -3x^2 - 6x + 4$, and the point is on the graph.

Standardized Test Practice

A B C D

Example 4 Match Equations and Graphs

Multiple-Choice Test Item

Which is the graph of $y + 1 = (x + 1)^2$?

A
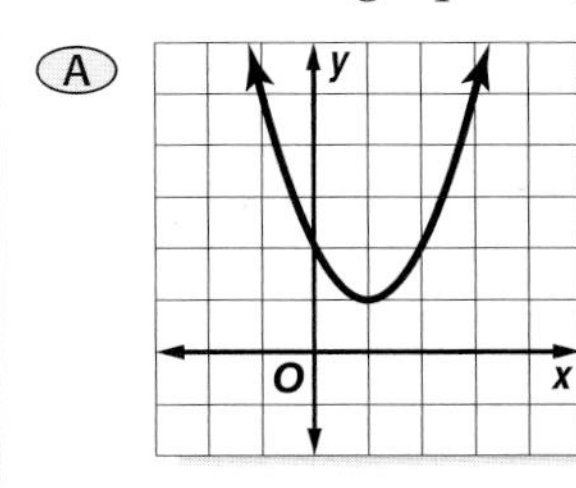

B
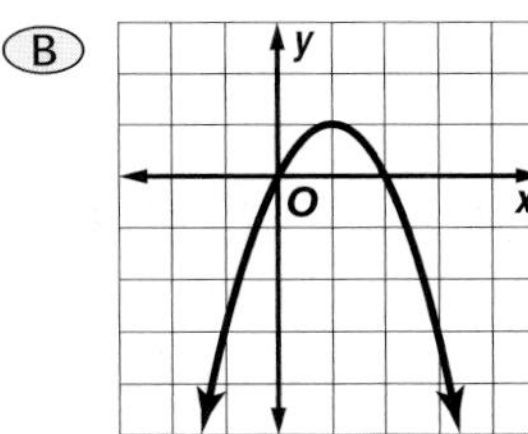

C
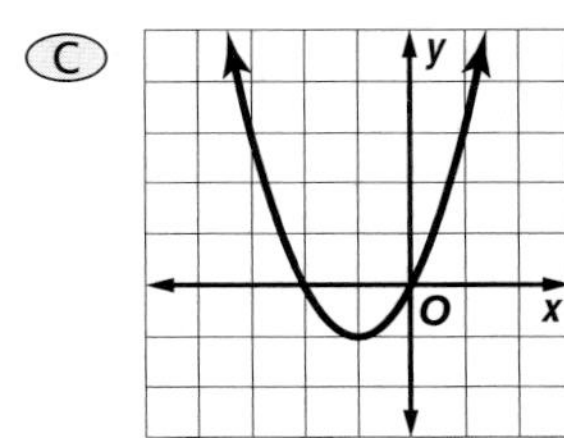

D
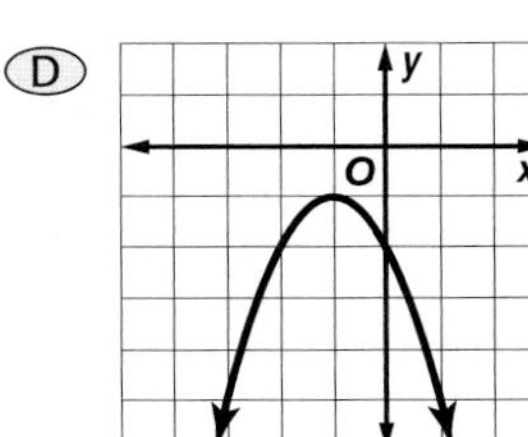

Read the Test Item

You are given a quadratic function, and you are asked to choose the graph that corresponds to it.

Solve the Test Item

First write the equation in standard form.

$y + 1 = (x + 1)^2$	Original equation
$y + 1 = x^2 + 2x + 1$	$(x + 1)^2 = x^2 + 2x + 1$
$y + 1 - 1 = x^2 + 2x + 1 - 1$	Subtract 1 from each side.
$y = x^2 + 2x$	Simplify.

Then find the axis of symmetry of the graph of $y = x^2 + 2x$.

$x = -\frac{b}{2a}$	Equation for the axis of symmetry
$x = -\frac{2}{2(1)}$ or -1	$a = 1$ and $b = 2$

The axis of symmetry is $x = -1$. Look at the graphs. Since only choices C and D have this as their axis of symmetry, you can eliminate choices A and B. Since the coefficient of the x^2 term is positive, the graph opens upward. Eliminate choice D. The answer is C.

The Princeton Review

Test-Taking Tip

Sometimes you can answer a question by eliminating the incorrect choices. For example, in this test question, choices A and B are eliminated because their axes of symmetry are *not* $x = -1$.

Check for Understanding

Concept Check

1. **Compare and contrast** a parabola with a maximum and a parabola with a minimum.
2. **OPEN ENDED** Draw two different parabolas with a vertex of (2, −1).
3. **Explain** how the axis of symmetry can help you graph a quadratic function.

Guided Practice

Use a table of values to graph each function.

4. $y = x^2 - 5$
5. $y = -x^2 + 4x + 5$

Write the equation of the axis of symmetry, and find the coordinates of the vertex of the graph of each function. Identify the vertex as a maximum or minimum. Then graph the function.

6. $y = x^2 + 4x - 9$
7. $y = -x^2 + 5x + 6$
8. $y = -(x - 2)^2 + 1$

Standardized Test Practice

9. Which is the graph of $y = -\frac{1}{2}x^2 + 1$?

(A)
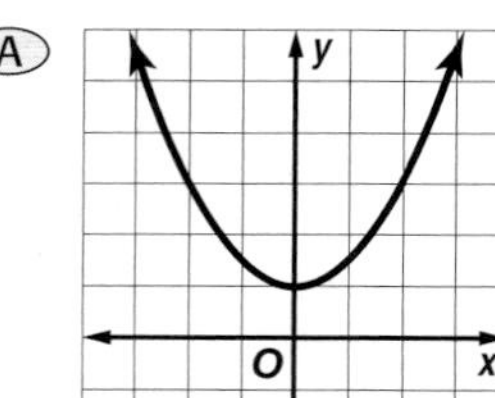

(B)
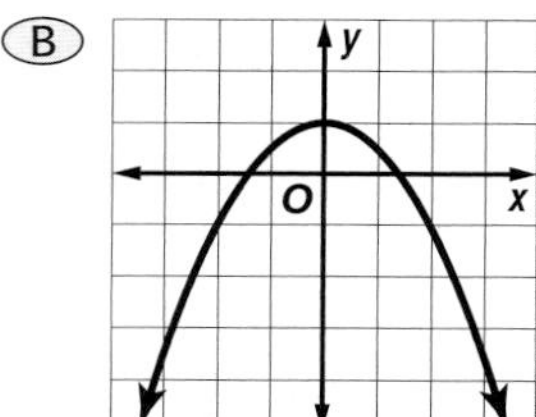

(C)
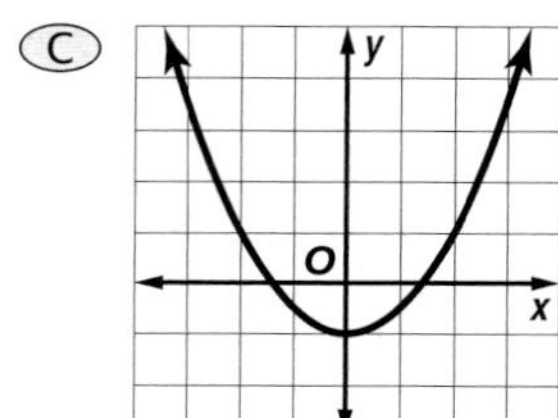

(D)
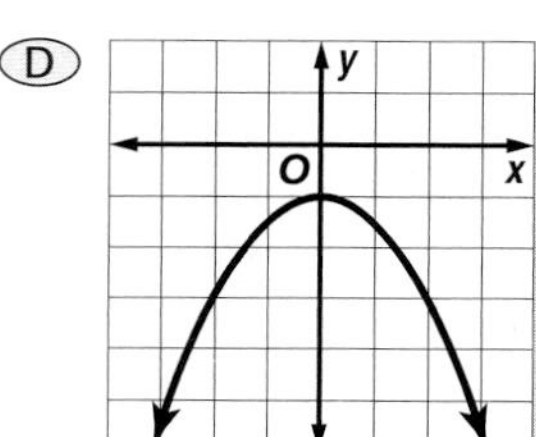

Practice and Apply

Homework Help

For Exercises	See Examples
10–15	1, 2
16–49	3
52, 53	4

Extra Practice

See page 841.

Use a table of values to graph each function.

10. $y = x^2 - 3$
11. $y = -x^2 + 7$
12. $y = x^2 - 2x - 8$
13. $y = x^2 - 4x + 3$
14. $y = -3x^2 - 6x + 4$
15. $y = -3x^2 + 6x + 1$

16. What is the equation of the axis of symmetry of the graph of $y = -3x^2 + 2x - 5$?

17. Find the equation of the axis of symmetry of the graph of $y = 4x^2 - 5x + 16$.

Write the equation of the axis of symmetry, and find the coordinates of the vertex of the graph of each function. Identify the vertex as a maximum or minimum. Then graph the function.

18. $y = 4x^2$
19. $y = -2x^2$
20. $y = x^2 + 2$
21. $y = -x^2 + 5$
22. $y = -x^2 + 2x + 3$
23. $y = -x^2 - 6x + 15$
24. $y = x^2 - 14x + 13$
25. $y = x^2 + 2x + 18$
26. $y = 2x^2 + 12x - 11$
27. $y = 3x^2 - 6x + 4$
28. $y = 5 + 16x - 2x^2$
29. $y = 9 - 8x + 2x^2$

30. $y = 3(x + 1)^2 - 20$

31. $y = -2(x - 4)^2 - 3$

32. $y + 2 = x^2 - 10x + 25$

33. $y + 1 = 3x^2 + 12x + 12$

34. $y - 5 = \frac{1}{3}(x + 2)^2$

35. $y + 1 = \frac{2}{3}(x + 1)^2$

36. The vertex of a parabola is at $(-4, -3)$. If one x-intercept is -11, what is the other x-intercept?

37. What is the equation of the axis of symmetry of a parabola if its x-intercepts are -6 and 4?

38. **SPORTS** A diver follows a path that is in the shape of a parabola. Suppose the diver's foot reaches 1 meter above the height of the diving board at the maximum height of the dive. At that time, the diver's foot is also 1 meter horizontally from the edge of the diving board. What is the distance of the diver's foot from the diving board as the diver descends past the diving board? Explain.

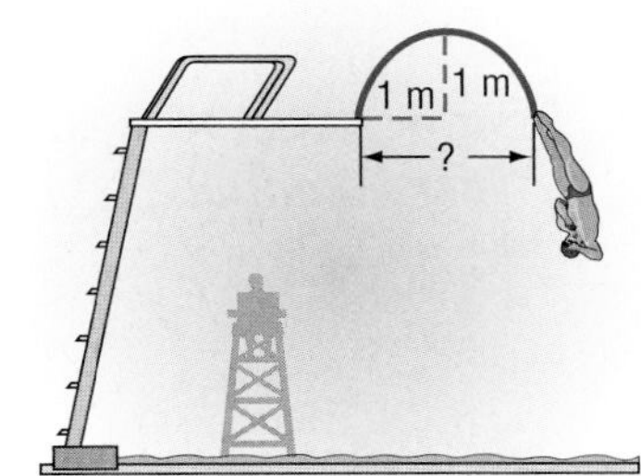

ENTERTAINMENT For Exercises 39 and 40, use the following information.
A carnival game involves striking a lever that forces a weight up a tube. If the weight reaches 20 feet to ring the bell, the contestant wins a prize. The equation $h = -16t^2 + 32t + 3$ gives the height of the weight if the initial velocity is 32 feet per second.

39. Find the maximum height of the weight.

40. Will a prize be won?

PETS For Exercises 41–43, use the following information.
Miriam has 40 meters of fencing to build a pen for her dog.

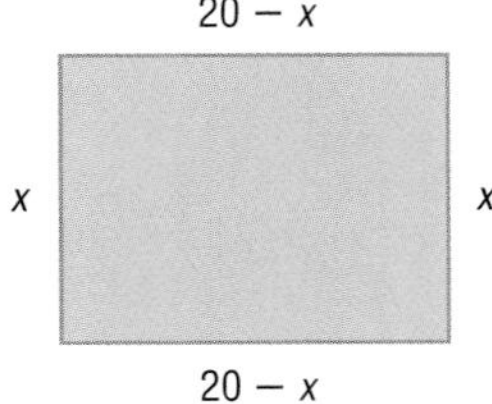

41. Use the diagram at the right to write an equation for the area A of the pen.

42. What value of x will result in the greatest area?

43. What is the greatest possible area of the pen?

More About. . .

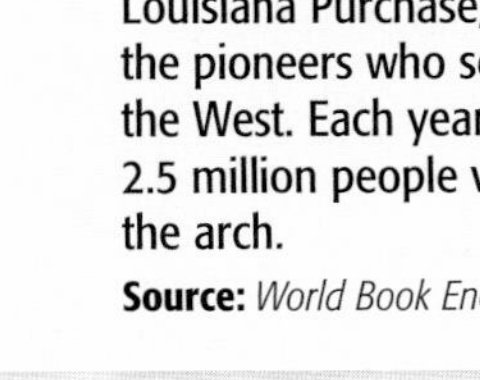

Architecture

The Gateway Arch is part of a tribute to Thomas Jefferson, the Louisiana Purchase, and the pioneers who settled the West. Each year about 2.5 million people visit the arch.

Source: *World Book Encyclopedia*

ARCHITECTURE For Exercises 44–46, use the following information.
The shape of the Gateway Arch in St. Louis, Missouri, is a *catenary* curve. It resembles a parabola with the equation $h = -0.00635x^2 + 4.0005x - 0.07875$, where h is the height in feet and x is the distance from one base in feet.

44. What is the equation of the axis of symmetry?

45. What is the distance from one end of the arch to the other?

46. What is the maximum height of the arch?

BRIDES For Exercises 47–49, use the following information.
The equation $a = 0.003x^2 - 0.115x + 21.3$ models the average ages of women when they first married since the year 1940. In this equation, a represents the average age and x represents the years since 1940.

47. Use what you know about parabolas and their minimum values to estimate the year in which the average age of brides was the youngest.

48. Estimate the average age of the brides during that year.

49. Use a graphing calculator to check your estimates.

50. **CRITICAL THINKING** Write a quadratic equation that represents a graph with an axis of symmetry with equation $x = -\frac{3}{8}$.

51. **WRITING IN MATH** Answer the question that was posed at the beginning of the lesson.

How can you coordinate a fireworks display with recorded music?

Include the following in your answer:

- an explanation of how to determine when the rocket will explode, and
- an explanation of how to determine the height of the rocket when it explodes.

52. Which equation corresponds to the graph at the right?

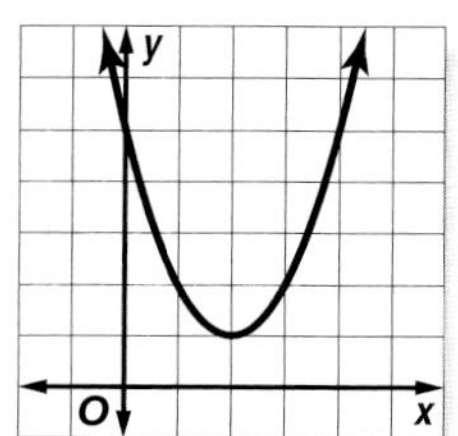

Ⓐ $y = x^2 - 4x + 5$

Ⓑ $y = -x^2 + 4x + 5$

Ⓒ $y = x^2 - 4x - 5$

Ⓓ $y = -x^2 + 4x - 5$

53. Which equation does *not* represent a quadratic function?

Ⓐ $y = (x + 3)^2$ Ⓑ $y = 3x^2$ Ⓒ $y = 6x^2 - 1$ Ⓓ $y = x + 5$

Graphing Calculator

MAXIMUM OR MINIMUM Graph each function. Determine whether the vertex is a maximum or a minimum and give the ordered pair for the vertex.

54. $y = x^2 - 10x + 25$ 55. $y = -x^2 + 4x + 3$ 56. $y = -2x^2 - 8x - 1$

57. $y = 2x^2 - 40x + 214$ 58. $y = 0.25x^2 - 4x - 2$ 59. $y = -0.5x^2 - 2x + 3$

Maintain Your Skills

Mixed Review

Factor each polynomial, if possible. If the polynomial cannot be factored, write *prime*. *(Lessons 9-5 and 9-6)*

60. $x^2 + 6x - 9$ 61. $a^2 + 22a + 121$ 62. $4m^2 - 4m + 1$

63. $4q^2 - 9$ 64. $2a^2 - 25$ 65. $1 - 16g^2$

Find each sum or difference. *(Lesson 8-5)*

66. $(13x + 9y) + 11y$ 67. $(7p^2 - p - 7) - (p^2 + 11)$

68. **RECREATION** At a recreation and sports facility, 3 members and 3 nonmembers pay a total of \$180 to take an aerobics class. A group of 5 members and 3 nonmembers pay \$210 to take the same class. How much does it cost members and nonmembers to take an aerobics class? *(Lesson 7-3)*

Solve each inequality. Then check your solution. *(Lesson 6-2)*

69. $12b > -144$ 70. $-5w > -125$ 71. $\frac{3r}{4} \leq \frac{2}{3}$

Write an equation of the line that passes through each point with the given slope. *(Lesson 5-4)*

72. $(2, 13), m = 4$ 73. $(-2, -7), m = 0$ 74. $(-4, 6), m = \frac{3}{2}$

Getting Ready for the Next Lesson

PREREQUISITE SKILL Find the x-intercept of the graph of each equation. *(To review finding **x-intercepts**, see Lesson 4-5.)*

75. $3x + 4y = 24$ 76. $2x - 5y = 14$ 77. $-2x - 4y = 7$

78. $7y + 6x = 42$ 79. $2y - 4x = 10$ 80. $3x - 7y + 9 = 0$

Graphing Calculator Investigation

A Follow-Up of Lesson 10-1

Families of Quadratic Graphs

Recall that a *family of graphs* is a group of graphs that have at least one characteristic in common. On page 278, families of linear graphs were introduced. Families of quadratic graphs often fall into two categories—those that have the same vertex and those that have the same shape.

In each of the following families, the parent function is $y = x^2$. Graphing calculators make it easy to study the characteristics of these families of parabolas.

Graph each group of equations on the same screen. Use the standard viewing window. Compare and contrast the graphs.

KEYSTROKES: *Review graphing equations on pages 224 and 225.*

a. $y = x^2, y = 2x^2, y = 4x^2$

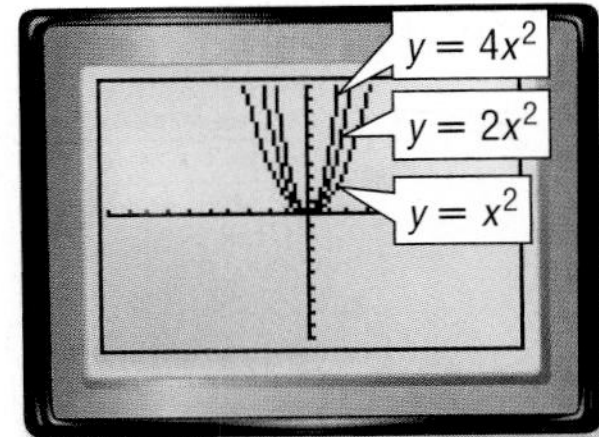

Each graph opens upward and has its vertex at the origin. The graphs of $y = 2x^2$ and $y = 4x^2$ are narrower than the graph of $y = x^2$.

b. $y = x^2, y = 0.5x^2, y = 0.2x^2$

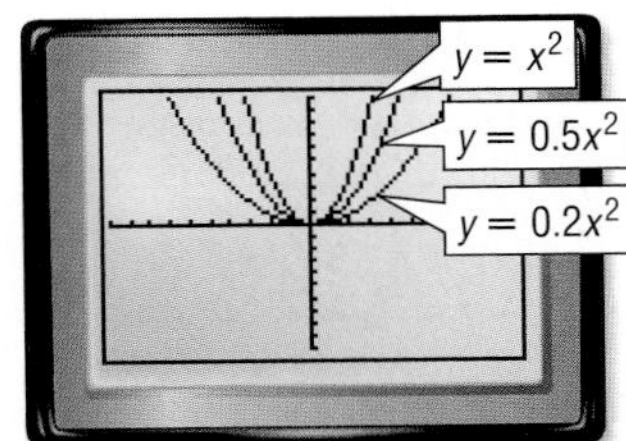

Each graph opens upward and has its vertex at the origin. The graphs of $y = 0.5x^2$ and $y = 0.2x^2$ are wider than the graph of $y = x^2$.

How does the value of a in $y = ax^2$ affect the shape of the graph?

c. $y = x^2, y = x^2 + 3, y = x^2 - 2, y = x^2 - 4$

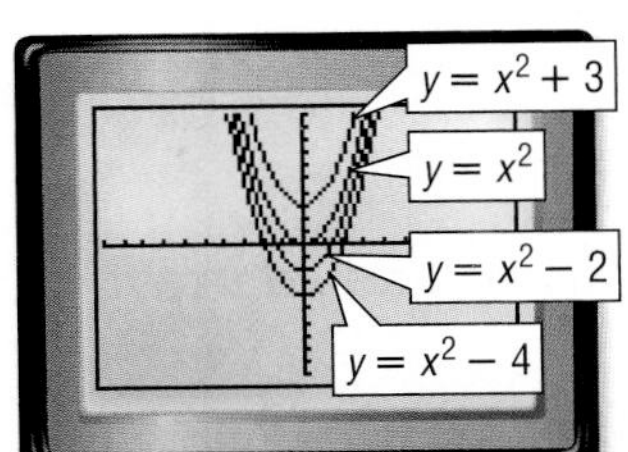

Each graph opens upward and has the same shape as $y = x^2$. However, each parabola has a different vertex, located along the y-axis. *How does the value of the constant affect the position of the graph?*

d. $y = x^2, y = (x - 3)^2, y = (x + 2)^2,$
$y = (x + 4)^2$

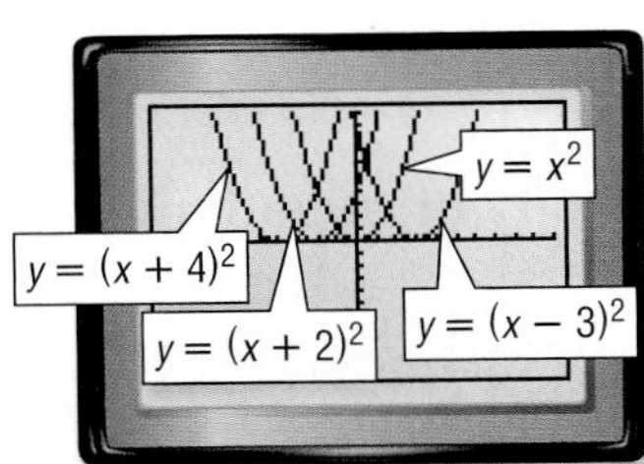

Each graph opens upward and has the same shape as $y = x^2$. However, each parabola has a different vertex located along the x-axis. *How is the location of the vertex related to the equation of the graph?*

www.algebra1.com/other_calculator_keystrokes

Graphing Calculator Investigation

When analyzing or comparing the shapes of various graphs on different screens, it is important to compare the graphs using the same window with the same scale factors. Suppose you graph the same equation using a different window for each. How will the appearance of the graph change?

Graph $y = x^2 - 7$ in each viewing window. What conclusions can you draw about the appearance of a graph in the window used?

a. standard viewing window

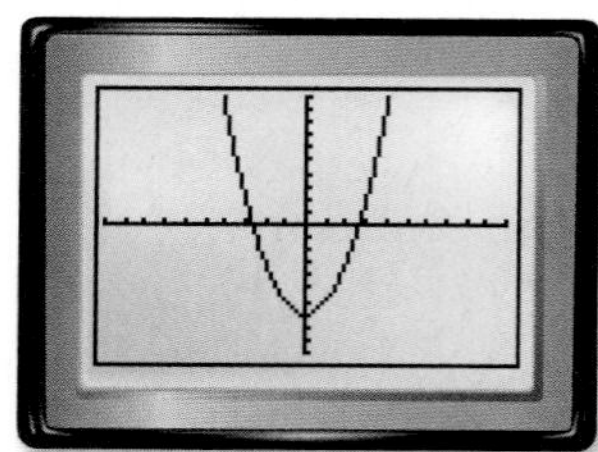

b. [−10, 10] scl: 1 by [−200, 200] scl: 50

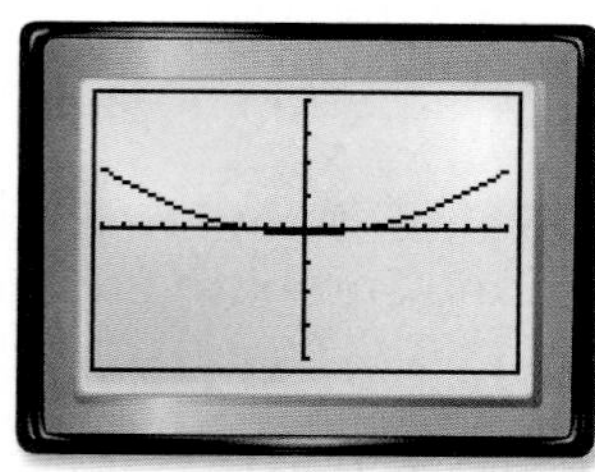

c. [−50, 50] scl: 5 by [−10, 10] scl: 1

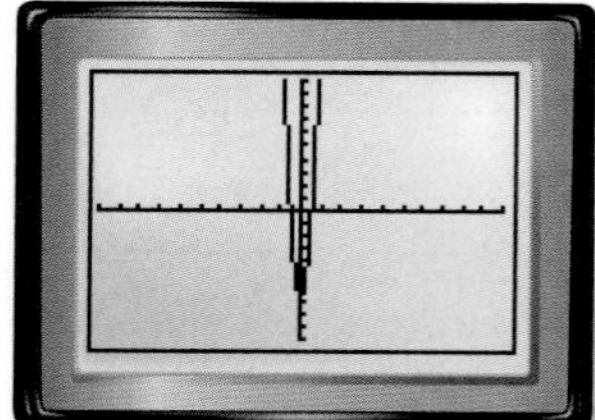

d. [−0.5, 0.5] scl: 0.1 by [−10, 10] scl: 1

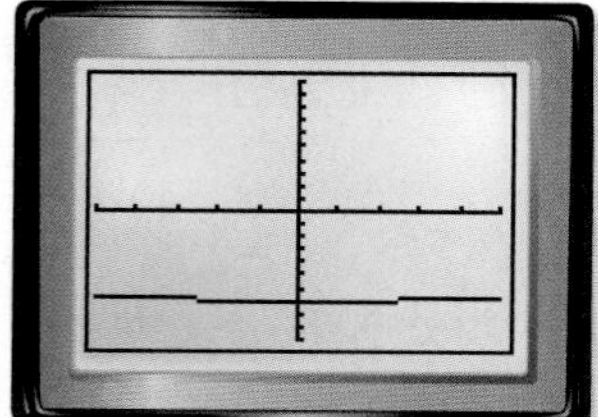

The window greatly affects the appearance of the parabola. Without knowing the window, graph **b** might be of the family $y = ax^2$, where $0 < a < 1$. Graph **c** looks like a member of $y = ax^2 - 7$, where $a > 1$. Graph **d** looks more like a line. However, all are graphs of the same equation.

Exercises

Graph each family of equations on the same screen. Compare and contrast the graphs.

1. $y = -x^2$
$y = -3x^2$
$y = -6x^2$

2. $y = -x^2$
$y = -0.6x^2$
$y = -0.4x^2$

3. $y = -x^2$
$y = -(x + 5)^2$
$y = -(x - 4)^2$

4. $y = -x^2$
$y = -x^2 + 7$
$y = -x^2 - 5$

Use the families of graphs on page 531 and Exercises 1–4 above to predict the appearance of the graph of each equation. Then draw the graph.

5. $y = -0.1x^2$ **6.** $y = (x + 1)^2$ **7.** $y = 4x^2$ **8.** $y = x^2 - 6$

Describe how each change in $y = x^2$ would affect the graph of $y = x^2$. Be sure to consider all values of a, h, and k.

9. $y = ax^2$ **10.** $y = (x + h)^2$ **11.** $y = x^2 + k$ **12.** $y = (x + h)^2 + k$

10-2 Solving Quadratic Equations by Graphing

What You'll Learn

- Solve quadratic equations by graphing.
- Estimate solutions of quadratic equations by graphing.

Vocabulary

- quadratic equation
- roots
- zeros

How can quadratic equations be used in computer simulations?

A golf ball follows a path much like a parabola. Because of this property, quadratic functions can be used to simulate parts of a computer golf game. One of the x-intercepts of the quadratic function represents the location where the ball will hit the ground.

SOLVE BY GRAPHING Recall that a quadratic function has standard form $f(x) = ax^2 + bx + c$. In a **quadratic equation**, the value of the related quadratic function is 0. So for the quadratic equation $0 = x^2 - 2x - 3$, the related quadratic function is $f(x) = x^2 - 2x - 3$. You have used factoring to solve equations like $x^2 - 2x - 3 = 0$. You can also use graphing to determine the solutions of equations like this.

The solutions of a quadratic equation are called the **roots** of the equation. The roots of a quadratic equation can be found by finding the x-intercepts or **zeros** of the related quadratic function.

Example 1 Two Roots

Solve $x^2 + 6x - 7 = 0$ by graphing.

Graph the related function $f(x) = x^2 + 6x - 7$. The equation of the axis of symmetry is $x = -\frac{6}{2(1)}$ or $x = -3$. When x equals -3, $f(x)$ equals $(-3)^2 + 6(-3) - 7$ or -16. So, the coordinates of the vertex are $(-3, -16)$. Make a table of values to find other points to sketch the graph.

x	$f(x)$
-8	9
-6	-7
-4	-15
-3	-16
-2	-15
0	-7
2	9

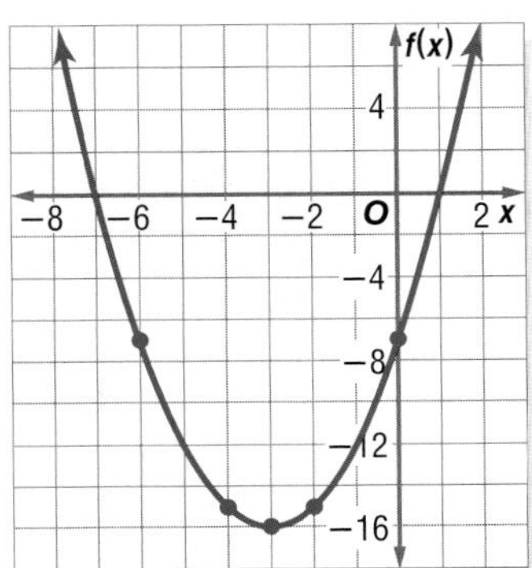

To solve $x^2 + 6x - 7 = 0$, you need to know where the value of $f(x)$ is 0. This occurs at the x-intercepts. The x-intercepts of the parabola appear to be -7 and 1.

(continued on the next page)

CHECK Solve by factoring.

$x^2 + 6x - 7 = 0$ Original equation

$(x + 7)(x - 1) = 0$ Factor.

$x + 7 = 0$ or $x - 1 = 0$ Zero Product Property

$x = -7$ ✓ $x = 1$ ✓ Solve for x.

The solutions of the equation are −7 and 1.

Quadratic equations always have two roots. However, these roots are not always two distinct numbers. Sometimes the two roots are the same number.

Example 2 A Double Root

Solve $b^2 + 4b = -4$ by graphing.

First rewrite the equation so one side is equal to zero.

$b^2 + 4b = -4$ Original equation

$b^2 + 4b + 4 = -4 + 4$ Add 4 to each side.

$b^2 + 4b + 4 = 0$ Simplify.

Graph the related function $f(b) = b^2 + 4b + 4$.

b	$f(b)$
−4	4
−3	1
−2	0
−1	1
0	4

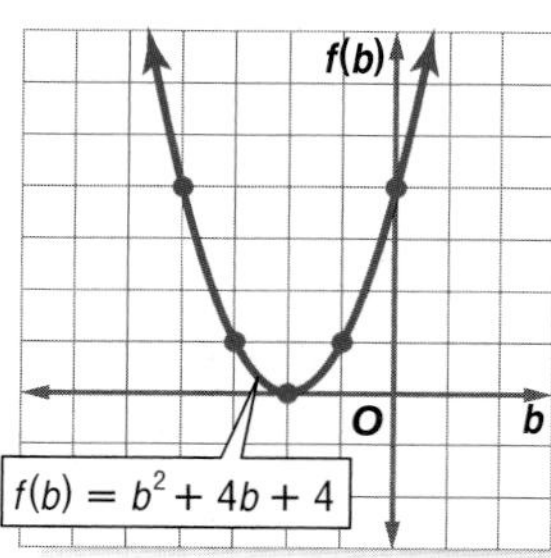

Notice that the vertex of the parabola is the b-intercept. Thus, one solution is −2. What is the other solution?

Try solving the equation by factoring.

$b^2 + 4b + 4 = 0$ Original equation

$(b + 2)(b + 2) = 0$ Factor.

$b + 2 = 0$ or $b + 2 = 0$ Zero Product Property

$b = -2$ $b = -2$ Solve for b.

There are two identical factors for the quadratic function, so there is only one root, called a double root. The solution is −2.

Study Tip

Common Misconception

Although solutions found by graphing may appear to be exact, you cannot be sure that they are exact. Solutions need to be verified by substituting into the equation and checking, or by using the algebraic methods that you will learn in this chapter.

Thus far, you have seen that quadratic equations can have two real roots or one double real root. Can a quadratic equation have no real roots?

Example 3 No Real Roots

Solve $x^2 - x + 4 = 0$ by graphing.

Graph the related function $f(x) = x^2 - x + 4$.

x	$f(x)$
−1	6
0	4
1	4
2	6

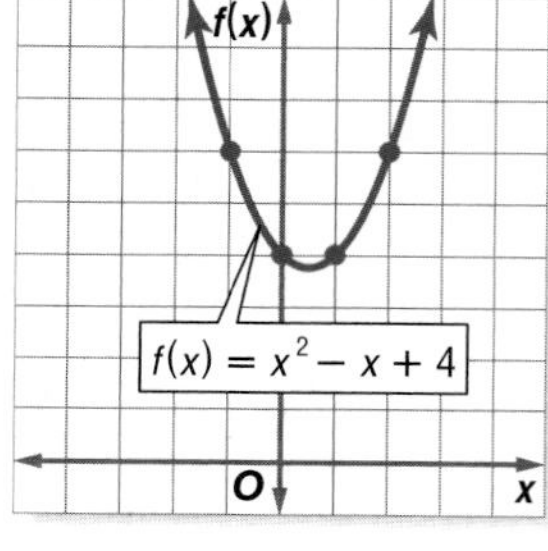

The graph has no x-intercept. Thus, there are no real number solutions for this equation.

The symbol ∅, indicating an empty set, is often used to represent no real solutions.

ESTIMATE SOLUTIONS In Examples 1 and 2, the roots of the equation were integers. Usually the roots of a quadratic equation are not integers. In these cases, use estimation to approximate the roots of the equation.

Example 4 Rational Roots

Solve $n^2 + 6n + 7 = 0$ by graphing. If integral roots cannot be found, estimate the roots by stating the consecutive integers between which the roots lie.

Graph the related function $f(n) = n^2 + 6n + 7$.

n	$f(n)$
−6	7
−5	2
−4	−1
−3	−2
−2	−1
−1	2
0	7

Notice that the value of the function changes from negative to positive between the n values of −5 and −4 and between −2 and −1.

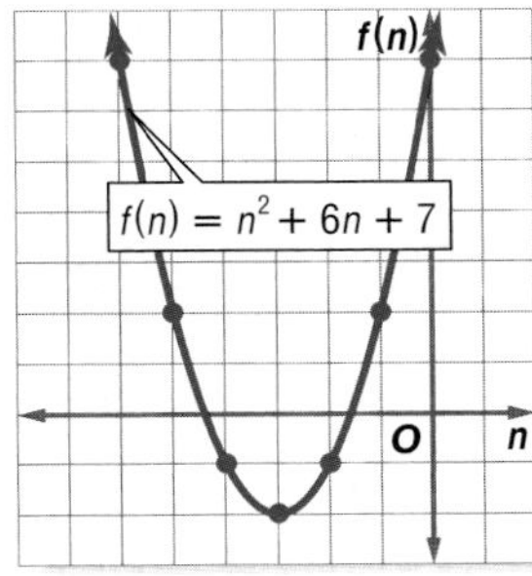

The n-intercepts of the graph are between −5 and −4 and between −2 and −1. So, one root is between −5 and −4, and the other root is between −2 and −1.

More About. . .

Football

On September 21, 1969, Steve O'Neal set a National Football League record by punting the ball 98 yards.

Source: *The Guinness Book of Records*

Example 5 Estimate Solutions to Solve a Problem

FOOTBALL When a football player punts a football, he hopes for a long "hang time." Hang time is the total amount of time the ball stays in the air. A time longer than 4.5 seconds is considered good. If a punter kicks the ball with an upward velocity of 80 feet per second and his foot meets the ball 2 feet off the ground, the function $y = -16t^2 + 80t + 2$ represents the height of the ball y in feet after t seconds. What is the hang time of the ball?

You need to find the solution of the equation $0 = -16t^2 + 80t + 2$. Use a graphing calculator to graph the related function $y = -16t^2 + 80t + 2$. The x-intercept is about 5. Therefore, the hang time is about 5 seconds.

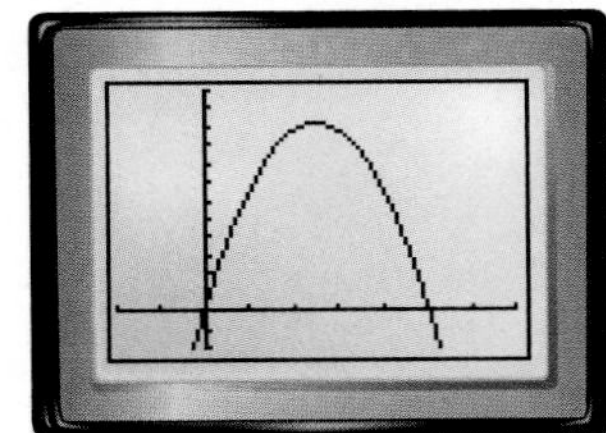

[−2, 7] scl: 1 by [−20, 120] scl: 10

Since 5 seconds is greater than 4.5 seconds, this kick would be considered to have good hang time.

Check for Understanding

Concept Check

1. **State** the real roots of the quadratic equation whose related function is graphed at the right.

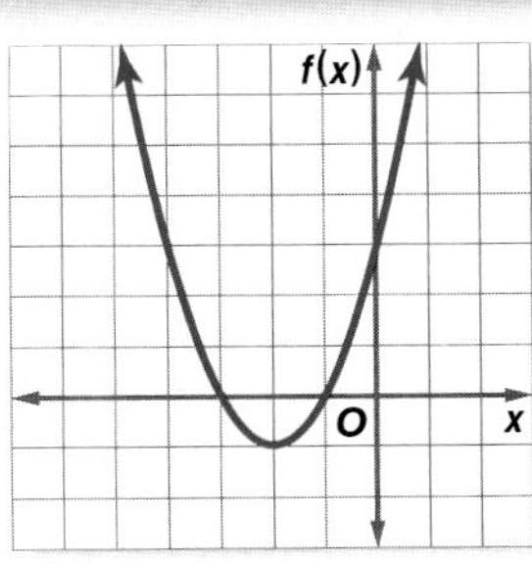

2. **Write** the related quadratic function for the equation $7x^2 + 2x = 8$.

3. **OPEN ENDED** Draw a graph to show a counterexample to the following statement.
All quadratic equations have two different solutions.

Guided Practice

Solve each equation by graphing.

4. $x^2 - 7x + 6 = 0$
5. $a^2 - 10a + 25 = 0$
6. $c^2 + 3 = 0$

Solve each equation by graphing. If integral roots cannot be found, estimate the roots by stating the consecutive integers between which the roots lie.

7. $t^2 + 9t + 5 = 0$
8. $x^2 - 16 = 0$
9. $w^2 - 3w = 5$

Application

10. **NUMBER THEORY** Two numbers have a sum of 4 and a product of -12. Use a quadratic equation to determine the two numbers.

Practice and Apply

Homework Help

For Exercises	See Examples
11–20	1–3
21–34	4
35–46	5

Extra Practice

See page 842.

Solve each equation by graphing.

11. $c^2 - 5c - 24 = 0$
12. $5n^2 + 2n + 6 = 0$
13. $x^2 + 6x + 9 = 0$
14. $b^2 - 12b + 36 = 0$
15. $x^2 + 2x + 5 = 0$
16. $r^2 + 4r - 12 = 0$

17. The roots of a quadratic equation are -2 and -6. The minimum point of the graph of its related function is at $(-4, -2)$. Sketch the graph of the function.

18. The roots of a quadratic equation are -6 and 0. The maximum point of the graph of its related function is at $(-3, 4)$. Sketch the graph of the function.

19. **NUMBER THEORY** The sum of two numbers is 9, and their product is 20. Use a quadratic equation to determine the two numbers.

20. **NUMBER THEORY** Use a quadratic equation to find two numbers whose sum is 5 and whose product is -24.

More About. . .

Design

The Winter Palace and the rest of the State Hermitage Museum in St. Petersburg, Russia, house 322 art galleries with about three million pieces of art.

Source: *The Guinness Book of Records*

Solve each equation by graphing. If integral roots cannot be found, estimate the roots by stating the consecutive integers between which the roots lie.

21. $a^2 - 12 = 0$
22. $n^2 - 7 = 0$
23. $2c^2 + 20c + 32 = 0$
24. $3s^2 + 9s - 12 = 0$
25. $x^2 + 6x + 6 = 0$
26. $y^2 - 4y + 1 = 0$
27. $a^2 - 8a = 4$
28. $x^2 + 6x = -7$
29. $m^2 - 10m = -21$
30. $p^2 + 16 = 8p$
31. $12n^2 - 26n = 30$
32. $4x^2 - 35 = -4x$

33. One root of a quadratic equation is between -4 and -3, and the other root is between 1 and 2. The maximum point of the graph of the related function is at $(-1, 6)$. Sketch the graph of the function.

34. One root of a quadratic equation is between -1 and 0, and the other root is between 6 and 7. The minimum point of the graph of the related function is at $(3, -5)$. Sketch the graph of the function.

DESIGN **For Exercises 35–39, use the following information.**
An art gallery has walls that are sculptured with arches that can be represented by the quadratic function $f(x) = -x^2 - 4x + 12$, where x is in feet. The wall space under each arch is to be painted a different color from the arch itself.

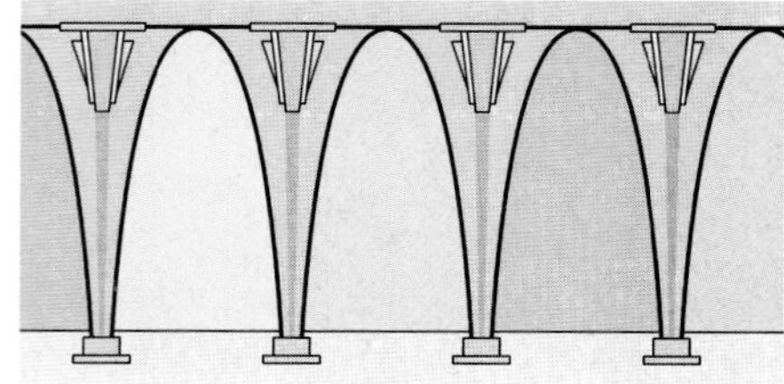

35. Graph the quadratic function and determine its x-intercepts.
36. What is the length of the segment along the floor of each arch?

37. What is the height of the arch?

38. The formula $A = \frac{2}{3}bh$ can be used to estimate the area under a parabola. In this formula, A represents area, b represents the length of the base, and h represents the height. Calculate the area that needs to be painted.

39. How much would the paint for the walls under 12 arches cost if the paint is \$27 per gallon, the painter applies 2 coats, and the manufacturer states that each gallon will cover 200 square feet? (*Hint*: Remember that you cannot buy part of a gallon.)

WebQuest

The graph of the surface areas of the planets can be modeled by a quadratic equation. Visit www.algebra1.com/webquest to continue work on your WebQuest project.

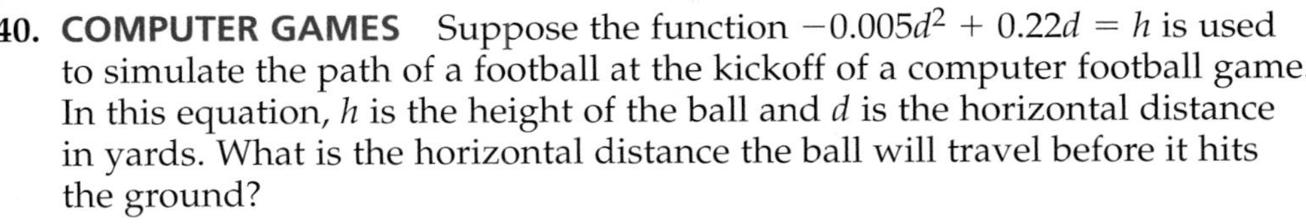

40. **COMPUTER GAMES** Suppose the function $-0.005d^2 + 0.22d = h$ is used to simulate the path of a football at the kickoff of a computer football game. In this equation, h is the height of the ball and d is the horizontal distance in yards. What is the horizontal distance the ball will travel before it hits the ground?

HIKING For Exercises 41 and 42, use the following information.

Monya and Kishi are hiking in the mountains and stop for lunch on a ledge 1000 feet above the valley below. Kishi decides to climb to another ledge 20 feet above Monya. Monya throws an apple up to Kishi, but Kishi misses it. The equation $h = -16t^2 + 30t + 1000$ represents the height in feet of the apple t seconds after it was thrown.

41. How long did it take for the apple to reach the ground?

42. If it takes 3 seconds to react, will the girls have time to call down and warn any hikers below? Assume that sound travels about 1000 feet per second. Explain.

WORK For Exercises 43–46, use the following information.

Kirk and Montega have accepted a job mowing the soccer playing fields. They must mow an area 500 feet long and 400 feet wide. They agree that each will mow half the area. They decide that Kirk will mow around the edge in a path of equal width until half the area is left.

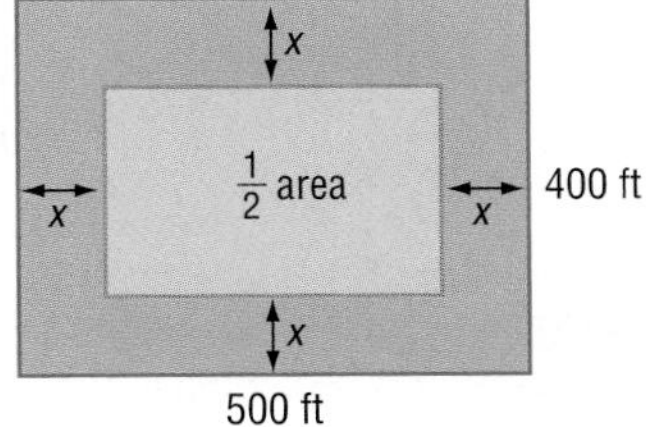

43. What is the area each person will mow?

44. Write a quadratic equation that could be used to find the width x that Kirk should mow.

45. What width should Kirk mow?

46. The mower can mow a path 5 feet wide. To the nearest whole number, how many times should Kirk go around the field?

47. **CRITICAL THINKING** Where does the graph of $f(x) = \frac{x^3 + 2x^2 - 3x}{x + 5}$ intersect the x-axis?

48. **WRITING IN MATH** Answer the question that was posed at the beginning of the lesson.

How can quadratic equations be used in computer simulations?

Include the following in your answer:

- the meaning of the two roots of a simulation equation for a computer golf game, and
- the approximate location at which the ball will hit the ground if the equation of the path of the ball is $y = -0.0015x^2 + 0.3x$, where y and x are in yards.

Standardized Test Practice

49. Which graph represents a function whose corresponding quadratic equation has no solutions?

(A)
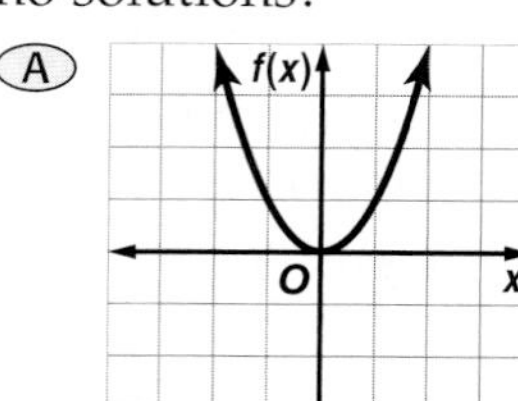

(B)
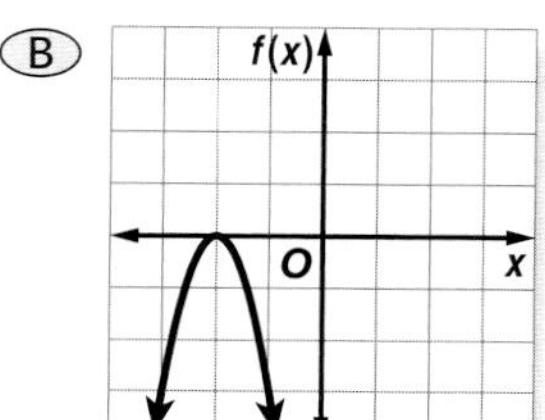

(C)
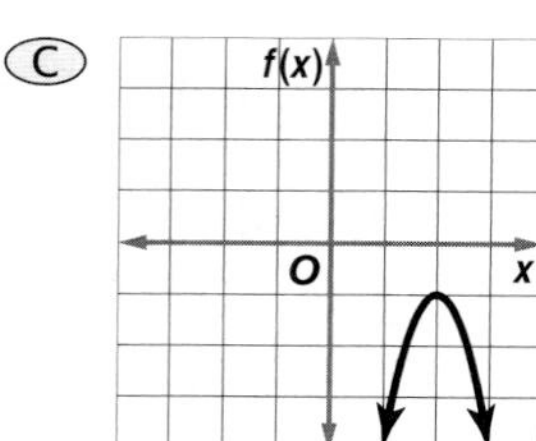

(D)
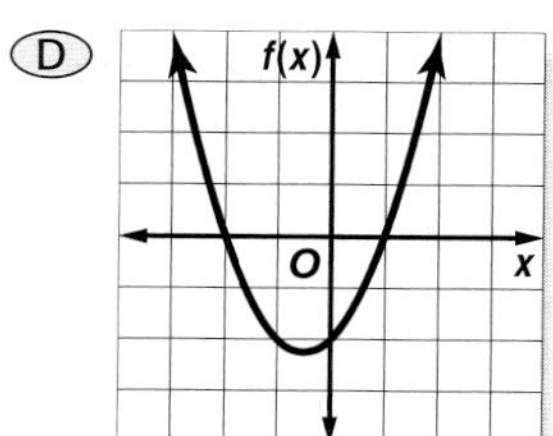

50. What are the root(s) of the quadratic equation whose related function is graphed at the right?

(A) −2, 2 (B) 0

(C) 4 (D) 0, 4

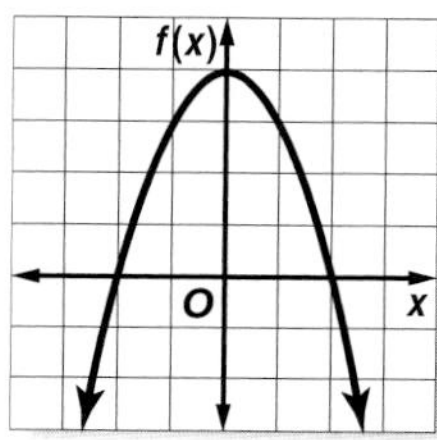

Graphing Calculator

CUBIC EQUATIONS An equation of the form $ax^3 + bx^2 + cx + d = 0$ is called a **cubic equation**. You can use a graphing calculator to graph and solve cubic equations.

Use the graph of the related function of each cubic equation to estimate the roots of the equation.

51. $x^3 - x^2 - 4x + 4 = 0$

52. $2x^3 - 11x^2 + 13x - 4 = 0$

Maintain Your Skills

Mixed Review

Write the equation of the axis of symmetry, and find the coordinates of the vertex of the graph of each function. Identify the vertex as a maximum or minimum. Then graph the function. *(Lesson 10-1)*

53. $y = x^2 + 6x + 9$

54. $y = -x^2 + 4x - 3$

55. $y = 0.5x^2 - 6x + 5$

Solve each equation. Check your solutions. *(Lesson 9-6)*

56. $m^2 - 24m = -144$

57. $7r^2 = 70r - 175$

58. $4d^2 + 9 = -12d$

Simplify. Assume that no denominator is equal to zero. *(Lesson 8-2)*

59. $\frac{10m^4}{30m}$

60. $\frac{22a^2b^5c^7}{-11abc^2}$

61. $\frac{-9m^3n^5}{27m^{-2}n^5y^{-4}}$

62. **SHIPPING** An empty book crate weighs 30 pounds. The weight of a book is 1.5 pounds. For shipping, the crate must weigh at least 55 pounds and no more than 60 pounds. What is the acceptable number of books that can be packed in the crate? *(Lesson 6-4)*

Getting Ready for the Next Lesson

PREREQUISITE SKILL **Determine whether each trinomial is a perfect square trinomial. If so, factor it.** *(To review **perfect square trinomials**, see Lesson 9-6.)*

63. $a^2 + 14a + 49$

64. $m^2 - 10m + 25$

65. $t^2 + 16t - 64$

66. $4y^2 + 12y + 9$

67. $9d^2 - 12d - 4$

68. $25x^2 - 10x + 1$

10-3 Solving Quadratic Equations by Completing the Square

What You'll Learn

- Solve quadratic equations by finding the square root.
- Solve quadratic equations by completing the square.

Vocabulary

- completing the square

How did ancient mathematicians use squares to solve algebraic equations?

Al-Khwarizmi, born in Baghdad in 780, is considered to be one of the foremost mathematicians of all time. He wrote some of the oldest works on arithmetic and algebra. He wrote algebra in sentences instead of using equations, and he explained the work with geometric sketches. Al-Khwarizmi would have described $x^2 + 8x = 35$ as "A square and 8 roots are equal to 35 units." He would solve the problem using the following sketch.

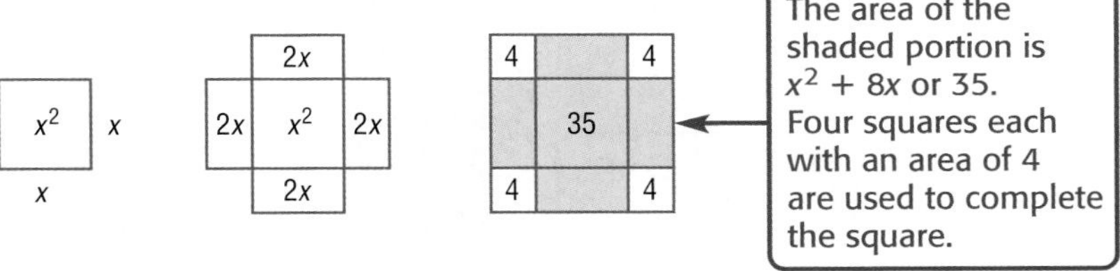

x^2
x
x
2x
2x
x^2
2x
2x
4
4
35
4
4
The area of the shaded portion is $x^2 + 8x$ or 35. Four squares each with an area of 4 are used to complete the square.

To solve problems this way today, you might use algebra tiles or a method called completing the square.

FIND THE SQUARE ROOT Some equations can be solved by taking the square root of each side.

Example 1 Irrational Roots

Solve $x^2 - 10x + 25 = 7$ by taking the square root of each side. Round to the nearest tenth if necessary.

$x^2 - 10x + 25 = 7$	Original equation
$(x - 5)^2 = 7$	$x^2 - 10x + 25$ is a perfect square trinomial.
$\sqrt{(x - 5)^2} = \sqrt{7}$	Take the square root of each side.
$\lvert x - 5 \rvert = \sqrt{7}$	Simplify.
$x - 5 = \pm\sqrt{7}$	Definition of absolute value
$x - 5 + 5 = \pm\sqrt{7} + 5$	Add 5 to each side.
$x = 5 \pm \sqrt{7}$	Simplify.

Use a calculator to evaluate each value of x.

$x = 5 + \sqrt{7}$ or $x = 5 - \sqrt{7}$

≈ 7.6 $\qquad \approx 2.4$

The solution set is $\{2.4, 7.6\}$.

COMPLETE THE SQUARE To use the method shown in Example 1, the quadratic expression on one side of the equation must be a perfect square. However, few quadratic expressions are perfect squares. To make any quadratic expression a perfect square, a method called **completing the square** may be used.

Study Tip

Look Back
To review **perfect square trinomials**, see Lesson 9-6.

Consider the pattern for squaring a binomial such as $x + 6$.

$$(x + 6)^2 = x^2 + 2(6)(x) + 6^2$$
$$= x^2 + 12x + 36$$
$$\left(\frac{12}{2}\right)^2 \rightarrow 6^2$$

Notice that one half of 12 is 6 and 6^2 is 36.

Key Concept *Completing the Square*

To complete the square for a quadratic expression of the form $x^2 + bx$, you can follow the steps below.

Step 1 Find $\frac{1}{2}$ of b, the coefficient of x.

Step 2 Square the result of Step 1.

Step 3 Add the result of Step 2 to $x^2 + bx$, the original expression.

Example 2 Complete the Square

Find the value of c that makes $x^2 + 6x + c$ a perfect square.

Method 1 Use algebra tiles.

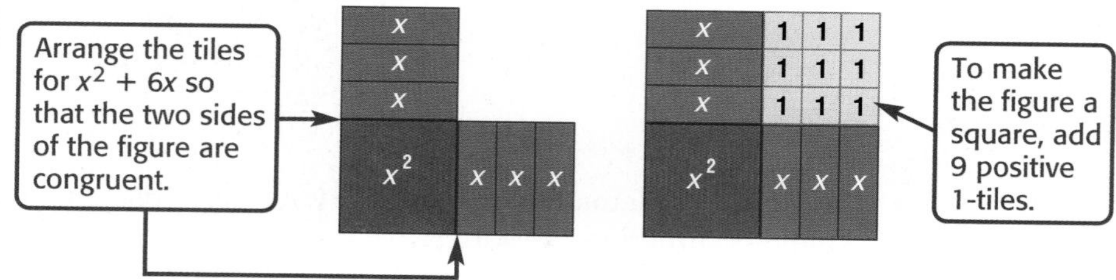

Arrange the tiles for $x^2 + 6x$ so that the two sides of the figure are congruent.
x
x
x
x^2
x x x
1 1 1
1 1 1
1 1 1
To make the figure a square, add 9 positive 1-tiles.

$x^2 + 6x + 9$ is a perfect square.

Method 2 Complete the square.

Step 1 Find $\frac{1}{2}$ of 6. $\quad \frac{6}{2} = 3$

Step 2 Square the result of Step 1. $\quad 3^2 = 9$

Step 3 Add the result of Step 2 to $x^2 + 6x$. $\quad x^2 + 6x + 9$

Thus, $c = 9$. Notice that $x^2 + 6x + 9 = (x + 3)^2$.

Example 3 Solve an Equation by Completing the Square

Solve $a^2 - 14a + 3 = -10$ by completing the square.

Step 1 Isolate the a^2 and a terms.

$a^2 - 14a + 3 = -10$ Original equation

$a^2 - 14a + 3 - 3 = -10 - 3$ Subtract 3 from each side.

$a^2 - 14a = -13$ Simplify.

Step 2 Complete the square and solve.

$a^2 - 14a + 49 = -13 + 49$ Since $\left(\frac{-14}{2}\right)^2 = 49$, add 49 to each side.

$(a - 7)^2 = 36$ Factor $a^2 - 14a + 49$.

$a - 7 = \pm 6$ Take the square root of each side.

$a - 7 + 7 = \pm 6 + 7$ Add 7 to each side.

$a = 7 \pm 6$ Simplify.

$$a = 7 + 6 \quad \text{or} \quad a = 7 - 6$$
$$= 13 \qquad\qquad = 1$$

CHECK Substitute each value for a in the original equation.

$$a^2 - 14a + 3 = -10 \qquad a^2 - 14a + 3 = -10$$
$$1^2 - 14(1) + 3 \stackrel{?}{=} -10 \qquad 13^2 - 14(13) + 3 \stackrel{?}{=} -10$$
$$1 - 14 + 3 \stackrel{?}{=} -10 \qquad 169 - 182 + 3 \stackrel{?}{=} -10$$
$$-10 = -10 \ \checkmark \qquad -10 = -10 \ \checkmark$$

The solution set is {1, 13}.

This method of completing the square cannot be used unless the coefficient of the first term is 1. To solve a quadratic equation in which the leading coefficient is not 1, first divide each term by the coefficient. Then follow the steps for completing the square.

Example 4 Solve a Quadratic Equation in Which $a \neq 1$

ENTERTAINMENT The path of debris from a firework display on a windy evening can be modeled by a quadratic function. A function for the path of the fireworks when the wind is about 15 miles per hour is $h = -0.04x^2 + 2x + 8$, where h is the height and x is the horizontal distance in feet. How far away from the launch site will the debris land?

Explore You know the function that relates the horizontal and vertical distances. You want to know how far away from the launch site the debris will land.

Plan The debris will hit the ground when $h = 0$. Use completing the square to solve $-0.04x^2 + 2x + 8 = 0$.

Solve

$-0.04x^2 + 2x + 8 = 0$	Equation for where debris will land
$\frac{-0.04x^2 + 2x + 8}{-0.04} = \frac{0}{-0.04}$	Divide each side by -0.04.
$x^2 - 50x - 200 = 0$	Simplify.
$x^2 - 50x - 200 + 200 = 0 + 200$	Add 200 to each side.
$x^2 - 50x = 200$	Simplify.
$x^2 - 50x + 625 = 200 + 625$	Since $\left(\frac{50}{2}\right)^2 = 625$, add 625 to each side.
$x^2 - 50x + 625 = 825$	Simplify.
$(x - 25)^2 = 825$	Factor $x^2 - 50x + 625$.
$x - 25 = \pm\sqrt{825}$	Take the square root of each side.
$x - 25 + 25 = \pm\sqrt{825} + 25$	Add 25 to each side.
$x = 25 \pm \sqrt{825}$	Simplify.

Use a calculator to evaluate each value of x.

$$x = 25 + \sqrt{825} \quad \text{or} \quad x = 25 - \sqrt{825}$$
$$\approx 53.7 \qquad\qquad \approx -3.7$$

Examine Since you are looking for a distance, ignore the negative number. The debris will land about 53.7 feet from the launch site.

More About. . .

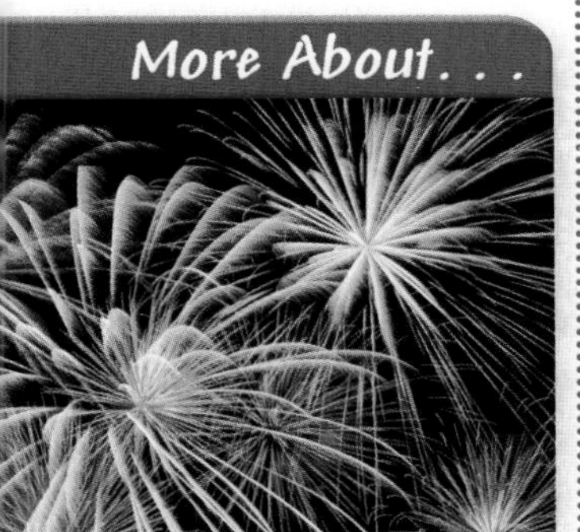

Entertainment

One of the exploded fireworks for the Lake Toya Festival in Japan on July 15, 1988, broke a world record. The diameter of the burst was 3937 feet.

Source: *The Guinness Book of Records*

Check for Understanding

Concept Check

1. **OPEN ENDED** Make a square using one or more of each of the following types of tiles.
   - x^2 tile
   - x tile
   - 1 tile

   Write an expression for the area of your square.
2. **Explain** why completing the square to solve $x^2 - 5x - 7 = 0$ is a better strategy than graphing the related function or factoring.
3. **Describe** the first step needed to solve $5x^2 + 12x = 15$ by completing the square.

Guided Practice

Solve each equation by taking the square root of each side. Round to the nearest tenth if necessary.

4. $b^2 - 6b + 9 = 25$
5. $m^2 + 14m + 49 = 20$

Find the value of c that makes each trinomial a perfect square.

6. $a^2 - 12a + c$
7. $t^2 + 5t + c$

Solve each equation by completing the square. Round to the nearest tenth if necessary.

8. $c^2 - 6c = 7$
9. $x^2 + 7x = -12$
10. $v^2 + 14v - 9 = 6$
11. $r^2 - 4r = 2$
12. $a^2 - 24a + 9 = 0$
13. $2p^2 - 5p + 8 = 7$

Application

14. **GEOMETRY** The area of a square can be doubled by increasing the length by 6 inches and the width by 4 inches. What is the length of the side of the square?

Practice and Apply

Homework Help

For Exercises	See Examples
15–20	1
21–28	2
29–52	3, 4

Extra Practice

See page 842.

Solve each equation by taking the square root of each side. Round to the nearest tenth if necessary.

15. $b^2 - 4b + 4 = 16$
16. $t^2 + 2t + 1 = 25$
17. $g^2 - 8g + 16 = 2$
18. $y^2 - 12y + 36 = 5$
19. $w^2 + 16w + 64 = 18$
20. $a^2 + 18a + 81 = 90$

Find the value of c that makes each trinomial a perfect square.

21. $s^2 - 16s + c$
22. $y^2 - 10y + c$
23. $w^2 + 22w + c$
24. $a^2 + 34a + c$
25. $p^2 - 7p + c$
26. $k^2 + 11k + c$
27. Find all values of c that make $x^2 + cx + 81$ a perfect square.
28. Find all values of c that make $x^2 + cx + 144$ a perfect square.

Solve each equation by completing the square. Round to the nearest tenth if necessary.

29. $s^2 - 4s - 12 = 0$
30. $d^2 + 3d - 10 = 0$
31. $y^2 - 19y + 4 = 70$
32. $d^2 + 20d + 11 = 200$
33. $a^2 - 5a = -4$
34. $p^2 - 4p = 21$
35. $x^2 + 4x + 3 = 0$
36. $d^2 - 8d + 7 = 0$
37. $s^2 - 10s = 23$
38. $m^2 - 8m = 4$
39. $9r^2 + 49 = 42r$
40. $4h^2 + 25 = 20h$
41. $0.3t^2 + 0.1t = 0.2$
42. $0.4v^2 + 2.5 = 2v$
43. $5x^2 + 10x - 7 = 0$
44. $9w^2 - 12w - 1 = 0$
45. $\frac{1}{2}d^2 - \frac{5}{4}d - 3 = 0$
46. $\frac{1}{3}f^2 - \frac{7}{6}f + \frac{1}{2} = 0$

Solve each equation for x by completing the square.

47. $x^2 + 4x + c = 0$

48. $x^2 - 6x + c = 0$

49. PARK PLANNING A plan for a park has a rectangular plot of wild flowers that is 9 meters long by 6 meters wide. A pathway of constant width goes around the plot of wild flowers. If the area of the path is equal to the area of the plot of wild flowers, what is the width of the path?

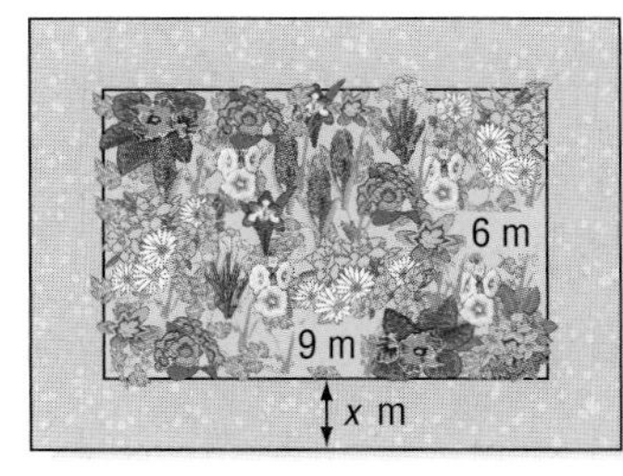

Career Choices

Photographer

Photographers must consider lighting, lens setting, and composition to create the best photograph.

Online Research For information about a career as a photographer, visit: www.algebra1.com/careers

50. EATING HABITS In the early 1900s, the average American ate 300 pounds of bread and cereal every year. By the 1960s, Americans were eating half that amount. However, eating cereal and bread is on the rise again. The consumption of these types of foods can be modeled by the function $y = 0.059x^2 - 7.423x + 362.1$, where y represents the bread and cereal consumption in pounds and x represents the number of years since 1900. If this trend continues, in what future year will the average American consume 300 pounds of bread and cereal?

Online Research **Data Update** What are the eating habits of Americans? Visit www.algebra1.com/data_update to learn more.

51. CRITICAL THINKING Describe the solution of $x^2 + 4x + 12 = 0$. Explain your reasoning.

52. PHOTOGRAPHY Emilio is placing a photograph behind a 12-inch-by-12-inch piece of matting. The photograph is to be positioned so that the matting is twice as wide at the top and bottom as it is at the sides. If the area of the photograph is to be 54 square inches, what are the dimensions?

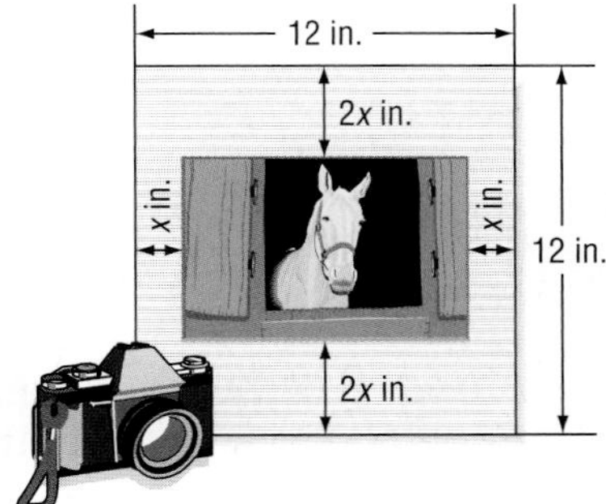

53. WRITING IN MATH Answer the question that was posed at the beginning of the lesson.

How did ancient mathematicians use squares to solve algebraic equations?

Include the following in your answer:

- an explanation of Al-Khwarizmi's drawings for $x^2 + 8x = 35$, and
- a step-by-step algebraic solution with justification for each step of the equation.

54. Determine which trinomial is *not* a perfect square trinomial.

(A) $a^2 - 26a + 169$ (B) $a^2 + 32a + 256$

(C) $a^2 + 30a - 225$ (D) $a^2 - 44a + 484$

55. Which equation is equivalent to $x^2 + 5x = 14$?

(A) $\left(x + \frac{5}{2}\right)^2 = \frac{81}{4}$ (B) $\left(x - \frac{5}{2}\right)^2 = \frac{45}{4}$

(C) $\left(x + \frac{5}{2}\right)^2 = -\frac{5}{4}$ (D) $\left(x - \frac{5}{2}\right)^2 = -\frac{5}{4}$

Maintain Your Skills

Mixed Review

Solve each equation by graphing. *(Lesson 10-2)*

56. $x^2 + 7x + 12 = 0$ **57.** $x^2 - 16 = 0$ **58.** $x^2 - 2x + 6 = 0$

Use a table of values to graph each equation. *(Lesson 10-1)*

59. $y = 4x^2 + 16$ **60.** $y = x^2 - 3x - 10$ **61.** $y = -x^2 + 3x - 4$

Find each GCF of the given monomials. *(Lesson 9-1)*

62. $14a^2b^3, 20a^3b^2c, 35ab^3c^2$ **63.** $32m^2n^3, 8m^2n, 56m^3n^2$

Use substitution to solve each system of equations. If the system does not have exactly one solution, state whether it has no solution or infinitely many solutions. *(Lesson 7-2)*

64. $y = 2x$
$x + y = 9$

65. $x = y + 3$
$2x - 3y = 5$

66. $x - 2y = 3$
$3x + y = 23$

Write a compound inequality for each graph. *(Lesson 6-4)*

67. −6 −5 −4 −3 −2 −1 0 1 2 3 4 5 6

68. 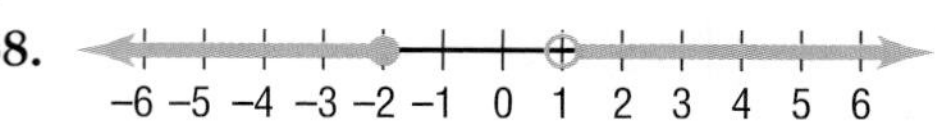

69. Write the slope-intercept form of an equation that passes through (8, −2) and is perpendicular to the graph of $5x - 3y = 7$. *(Lesson 5-6)*

Write an equation for each relation. *(Lesson 4-8)*

70. 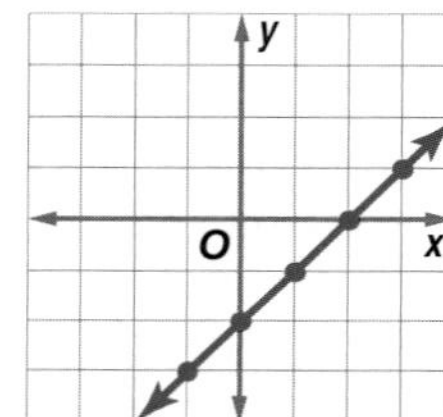

71. 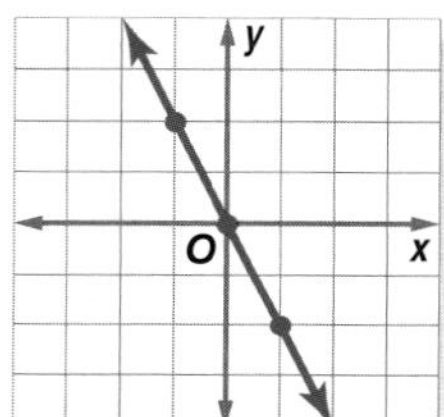

Getting Ready for the Next Lesson

PREREQUISITE SKILL **Evaluate $\sqrt{b^2 - 4ac}$ for each set of values. Round to the nearest tenth if necessary.** *(To review **finding square roots**, see Lesson 2-7.)*

72. $a = 1, b = -2, c = -15$ **73.** $a = 2, b = 7, c = 3$

74. $a = 1, b = 5, c = -2$ **75.** $a = -2, b = 7, c = 5$

Practice Quiz 1

Lessons 10-1 through 10-3

Write the equation of the axis of symmetry and find the coordinates of the vertex of the graph of each function. Identify the vertex as a maximum or minimum. Then graph the function. *(Lesson 10-1)*

1. $y = x^2 - x - 6$ **2.** $y = 2x^2 + 3$ **3.** $y = -3x^2 - 6x + 5$

Solve each equation by graphing. If integral roots cannot be found, estimate the roots by stating the consecutive integers between which the roots lie. *(Lesson 10-2)*

4. $x^2 + 6x + 10 = 0$ **5.** $x^2 - 2x - 1 = 0$ **6.** $x^2 - 5x - 6 = 0$

Solve each equation by completing the square. Round to the nearest tenth if necessary. *(Lesson 10-3)*

7. $s^2 + 8s = -15$ **8.** $a^2 - 10a = -24$ **9.** $y^2 - 14y + 49 = 5$ **10.** $2b^2 - b - 7 = 14$

Graphing Calculator Investigation

A Follow-Up of Lesson 10-3

Graphing Quadratic Functions in Vertex Form

Quadratic functions written in the form $y = a(x - h)^2 + k$ are said to be in **vertex form**.

Graph each group of equations on the same screen. Use the standard viewing window. Compare and contrast the graphs.

a. $y = x^2$
$y = (x - 3)^2 + 5$
$y = (x + 2)^2 + 6$
$y = (x - 5)^2 - 4$

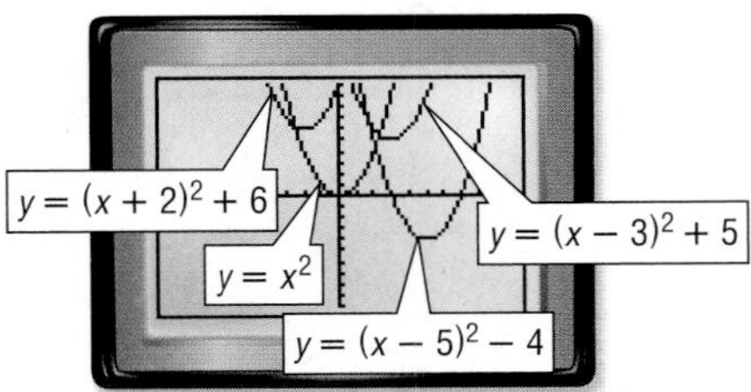

Each graph opens upward and has the same shape. However, the vertices are different.

Equation	Vertex
$y = x^2$	(0, 0)
$y = (x - 3)^2 + 5$	(3, 5)
$y = (x + 2)^2 + 6$	(−2, 6)
$y = (x - 5)^2 - 4$	(5, −4)

b. $y = -2x^2$
$y = -2(x - 1)^2 + 3$
$y = -2(x + 3)^2 + 1$
$y = -2(x - 5)^2 - 2$

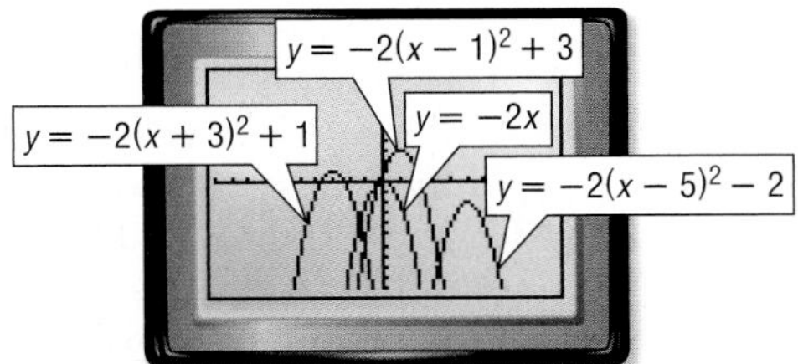

Each graph opens downward and has the same shape. However, the vertices are different.

Equation	Vertex
$y = -2x^2$	(0, 0)
$y = -2(x - 1)^2 + 3$	(1, 3)
$y = -2(x + 3)^2 + 1$	(−3, 1)
$y = -2(x - 5)^2 - 2$	(5, −2)

Exercises

1. Study the relationship between the equations in vertex form and their vertices. What is the vertex of the graph of $y = a(x - h)^2 + k$?
2. Completing the square can be used to change a quadratic equation to vertex form. Copy and complete the steps needed to rewrite $y = x^2 - 2x - 3$ in vertex form.
   $y = x^2 - 2x - 3$
   $y = (x^2 - 2x + \underline{\ ?\ }) - 3 - \underline{\ ?\ }$
   $y = (x - \underline{\ ?\ })^2 - \underline{\ ?\ }$

Complete the square to rewrite each quadratic equation in vertex form. Then determine the vertex of the graph of the equation and sketch the graph.

3. $y = x^2 + 2x - 7$
4. $y = x^2 - 4x + 8$
5. $y = x^2 + 6x - 1$

 www.algebra1.com/other_calculator_keystrokes

10-4 Solving Quadratic Equations by Using the Quadratic Formula

What You'll Learn

- Solve quadratic equations by using the Quadratic Formula.
- Use the discriminant to determine the number of solutions for a quadratic equation.

Vocabulary

- Quadratic Formula
- discriminant

How can the Quadratic Formula be used to solve problems involving population trends?

In the past few decades, there has been a dramatic increase in the percent of people living in the United States who were born in other countries. This trend can be modeled by the quadratic function $P = 0.006t^2 - 0.080t + 5.281$, where P is the percent born outside the United States and t is the number of years since 1960.

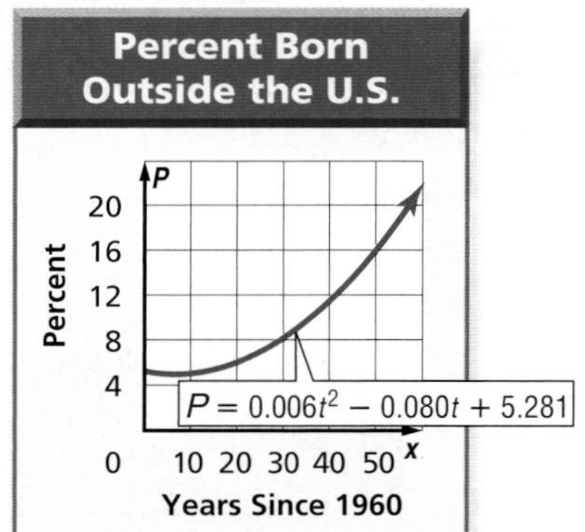

To predict when 15% of the population will be people who were born outside of the U.S., you can solve the equation $15 = 0.006t^2 - 0.080t + 5.281$. This equation would be impossible or difficult to solve using factoring, graphing, or completing the square.

QUADRATIC FORMULA You can solve the standard form of the quadratic equation $ax^2 + bx + c = 0$ for x. The result is called the **Quadratic Formula**.

Key Concept — *The Quadratic Formula*

The solutions of a quadratic equation in the form $ax^2 + bx + c = 0$, where $a \neq 0$, are given by the Quadratic Formula.

$$x = \frac{-b \pm \sqrt{b^2 - 4ac}}{2a}$$

You can solve quadratic equations by factoring, graphing, completing the square, or using the Quadratic Formula.

Study Tip

Look Back
To review **solving equations by factoring**, see Chapter 9.

Example 1 *Integral Roots*

Use two methods to solve $x^2 - 2x - 24 = 0$.

Method 1 Factoring

$x^2 - 2x - 24 = 0$ Original equation

$(x + 4)(x - 6) = 0$ Factor $x^2 - 2x - 24$.

$x + 4 = 0$ or $x - 6 = 0$ Zero Product Property

$x = -4$ $\quad$ $x = 6$ Solve for x.

Method 2 Quadratic Formula

For this equation, $a = 1$, $b = -2$, and $c = -24$.

$$x = \frac{-b \pm \sqrt{b^2 - 4ac}}{2a}$$ Quadratic Formula

$$= \frac{-(-2) \pm \sqrt{(-2)^2 - 4(1)(-24)}}{2(1)}$$ $a = 1$, $b = -2$, and $c = -24$

$$= \frac{2 \pm \sqrt{4 + 96}}{2}$$ Multiply.

$$= \frac{2 \pm \sqrt{100}}{2}$$ Add.

$$= \frac{2 \pm 10}{2}$$ Simplify.

$$x = \frac{2 - 10}{2} \quad \text{or} \quad x = \frac{2 + 10}{2}$$

$$= -4 \qquad\qquad = 6$$

The solution set is $\{-4, 6\}$.

Example 2 Irrational Roots

Solve $24x^2 - 14x = 6$ by using the Quadratic Formula. Round to the nearest tenth if necessary.

Step 1 Rewrite the equation in standard form.

$$24x^2 - 14x = 6$$ Original equation

$$24x^2 - 14x - 6 = 6 - 6$$ Subtract 6 from each side.

$$24x^2 - 14x - 6 = 0$$ Simplify.

Step 2 Apply the Quadratic Formula.

$$x = \frac{-b \pm \sqrt{b^2 - 4ac}}{2a}$$ Quadratic Formula

$$= \frac{-(-14) \pm \sqrt{(-14)^2 - 4(24)(-6)}}{2(24)}$$ $a = 24$, $b = -14$, and $c = -6$

$$= \frac{14 \pm \sqrt{196 + 576}}{48}$$ Multiply.

$$= \frac{14 \pm \sqrt{772}}{48}$$ Add.

$$x = \frac{14 - \sqrt{772}}{48} \quad \text{or} \quad x = \frac{14 + \sqrt{772}}{48}$$

$$\approx -0.3 \qquad\qquad \approx 0.9$$

Check the solutions by using the CALC menu on a graphing calculator to determine the zeros of the related quadratic function.

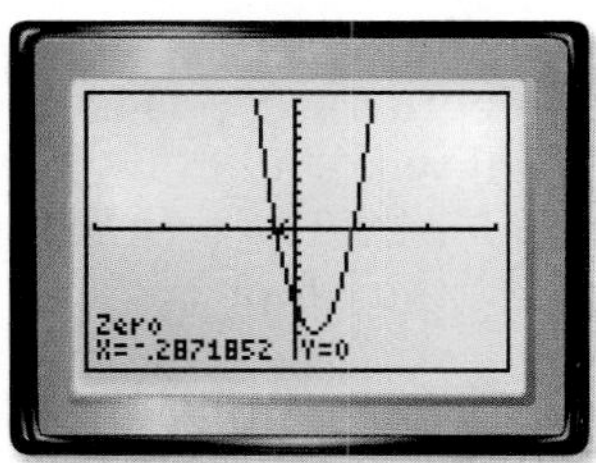

[−3, 3] scl: 1 by [−10. 10] scl: 1

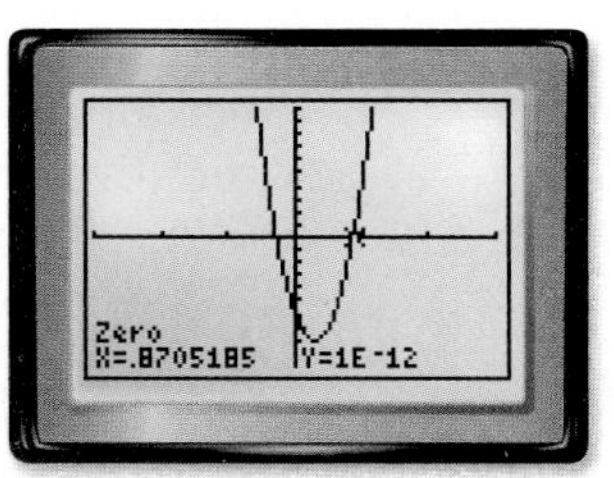

[−3, 3] scl: 1 by [−10. 10] scl: 1

The approximate solution set is $\{-0.3, 0.9\}$.

You have studied four methods for solving quadratic equations. The table summarizes these methods.

Concept Summary — *Solving Quadratic Equations*

Method	Can Be Used	Comments
graphing	always	Not always exact; use only when an approximate solution is sufficient.
factoring	sometimes	Use if constant term is 0 or factors are easily determined.
completing the square	always	Useful for equations of the form $x^2 + bx + c = 0$, where b is an even number.
Quadratic Formula	always	Other methods may be easier to use in some cases, but this method always gives accurate solutions.

Example 3 *Use the Quadratic Formula to Solve a Problem*

SPACE TRAVEL The height H of an object t seconds after it is propelled upward with an initial velocity v is represented by $H = -\frac{1}{2}gt^2 + vt + h$, where g is the gravitational pull and h is the initial height. Suppose an astronaut on the Moon throws a baseball upward with an initial velocity of 10 meters per second, letting go of the ball 2 meters above the ground. Use the information at the left to find how much longer the ball will stay in the air than a similarly-thrown baseball on Earth.

In order to find when the ball hits the ground, you must find when $H = 0$. Write two equations to represent the situation on the Moon and on Earth.

Baseball Thrown on the Moon

$$H = -\frac{1}{2}gt^2 + vt + h$$
$$0 = -\frac{1}{2}(1.6)t^2 + 10t + 2$$
$$0 = -0.8t^2 + 10t + 2$$

Baseball Thrown on Earth

$$H = -\frac{1}{2}gt^2 + vt + h$$
$$0 = -\frac{1}{2}(9.8)t^2 + 10t + 2$$
$$0 = -4.9t^2 + 10t + 2$$

These equations cannot be factored, and completing the square would involve a lot of computation. To find accurate solutions, use the Quadratic Formula.

$$t = \frac{-b \pm \sqrt{b^2 - 4ac}}{2a}$$
$$= \frac{-10 \pm \sqrt{10^2 - 4(-0.8)(2)}}{2(-0.8)}$$
$$= \frac{-10 \pm \sqrt{106.4}}{-1.6}$$
$$t \approx 12.7 \quad \text{or} \quad t \approx -0.2$$

$$t = \frac{-b \pm \sqrt{b^2 - 4ac}}{2a}$$
$$= \frac{-10 \pm \sqrt{10^2 - 4(-4.9)(2)}}{2(-4.9)}$$
$$= \frac{-10 \pm \sqrt{139.2}}{-9.8}$$
$$t \approx 2.2 \quad \text{or} \quad t \approx -0.2$$

Since a negative number of seconds is not reasonable, use the positive solutions. Therefore, the baseball will stay in the air about 12.7 seconds on the Moon and about 2.2 seconds on Earth. The baseball will stay in the air about 12.7 − 2.2 or 10.5 seconds longer on the Moon.

More About. . .

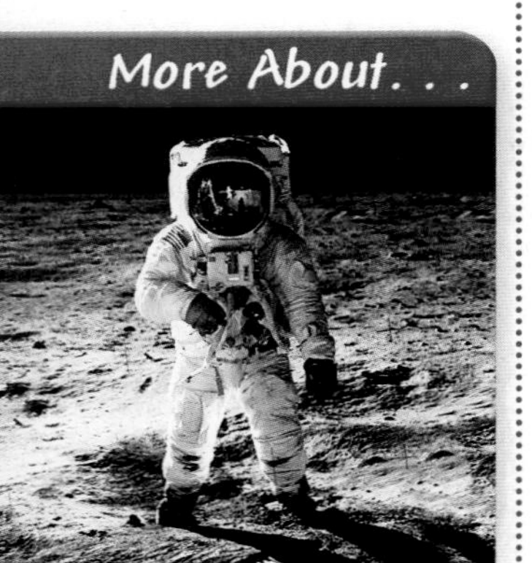

Space Travel

Astronauts have found walking on the Moon to be very different from walking on Earth because the gravitational pull of the moon is only 1.6 meters per second squared. The gravitational pull on Earth is 9.8 meters per second squared.

Source: *World Book Encyclopedia*

THE DISCRIMINANT In the Quadratic Formula, the expression under the radical sign, $b^2 - 4ac$, is called the **discriminant**. The value of the discriminant can be used to determine the number of real roots for a quadratic equation.

Key Concept — Using the Discriminant

Discriminant	negative	zero	positive
Example	$2x^2 + x + 3 = 0$ $x = \frac{-1 \pm \sqrt{1^2 - 4(2)(3)}}{2(2)}$ $= \frac{-1 \pm \sqrt{-23}}{4}$ There are no real roots since no real number can be the square root of a negative number.	$x^2 + 6x + 9 = 0$ $x = \frac{-6 \pm \sqrt{6^2 - 4(1)(9)}}{2(1)}$ $= \frac{-6 \pm \sqrt{0}}{2}$ $= \frac{-6}{2}$ or -3 There is a double root, -3.	$x^2 - 5x + 2 = 0$ $x = \frac{-(-5) \pm \sqrt{(-5)^2 - 4(1)(2)}}{2(1)}$ $= \frac{5 \pm \sqrt{17}}{2}$ There are two roots, $\frac{5 + \sqrt{17}}{2}$ and $\frac{5 - \sqrt{17}}{2}$.
Graph of Related Function	$f(x) = 2x^2 + x + 3$ The graph does not cross the x-axis.	$f(x) = x^2 + 6x + 9$ The graph touches the x-axis in one place.	$f(x) = x^2 - 5x + 2$ The graph crosses the x-axis twice.
Number of Real Roots	0	1	2

Example 4 Use the Discriminant

State the value of the discriminant for each equation. Then determine the number of real roots of the equation.

a. $2x^2 + 10x + 11 = 0$

$b^2 - 4ac = 10^2 - 4(2)(11)$ — $a = 2$, $b = 10$, and $c = 11$

$= 12$ — Simplify.

Since the discriminant is positive, the equation has two real roots.

b. $4t^2 - 20t + 25 = 0$

$b^2 - 4ac = (-20)^2 - 4(4)(25)$ — $a = 4$, $b = -20$, and $c = 25$

$= 0$ — Simplify.

Since the discriminant is 0, the equation has one real root.

c. $3m^2 + 4m = -2$

Step 1 Rewrite the equation in standard form.

$3m^2 + 4m = -2$ — Original equation

$3m^2 + 4m + 2 = -2 + 2$ — Add 2 to each side.

$3m^2 + 4m + 2 = 0$ — Simplify.

Step 2 Find the discriminant.

$b^2 - 4ac = 4^2 - 4(3)(2)$ — $a = 3$, $b = 4$, and $c = 2$

$= -8$ — Simplify.

Since the discriminant is negative, the equation has no real roots.

Check for Understanding

Concept Check

1. **Describe** three different ways to solve $x^2 - 2x - 15 = 0$. Which method do you prefer and why?

2. **OPEN ENDED** Write a quadratic equation with no real solutions.

3. **FIND THE ERROR** Lakeisha and Juanita are determining the number of solutions of $5y^2 - 3y = 2$.

Lakeisha	Juanita
$5y^2 - 3y = 2$ $b^2 - 4ac = (-3)^2 - 4(5)(2)$ $= -31$ Since the discriminant is negative, there are no real solutions.	$5y^2 - 3y = 2$ $5y^2 - 3y - 2 = 0$ $b^2 - 4ac = (-3)^2 - 4(5)(-2)$ $= 49$ Since the discriminant is positive, there are two real roots.

Who is correct? Explain your reasoning.

Guided Practice

Solve each equation by using the Quadratic Formula. Round to the nearest tenth if necessary.

4. $x^2 + 7x + 6 = 0$
5. $t^2 + 11t = 12$
6. $r^2 + 10r + 12 = 0$
7. $3v^2 + 5v + 11 = 0$
8. $4x^2 + 2x = 17$
9. $w^2 + \frac{2}{25} = \frac{3}{5}w$

State the value of the discriminant for each equation. Then determine the number of real roots of the equation.

10. $m^2 + 5m - 6 = 0$
11. $s^2 + 8s + 16 = 0$
12. $2z^2 + z = -50$

Application

13. **MANUFACTURING** A pan is to be formed by cutting 2-centimeter-by-2-centimeter squares from each corner of a square piece of sheet metal and then folding the sides. If the volume of the pan is to be 441 cubic centimeters, what should the dimensions of the original piece of sheet metal be?

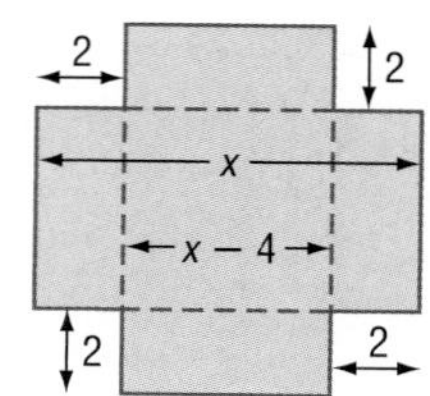

Practice and Apply

Homework Help

For Exercises	See Examples
14–37	1, 2
38–45	4
46–53	3

Extra Practice
See page 842.

Solve each equation by using the Quadratic Formula. Round to the nearest tenth if necessary.

14. $x^2 + 3x - 18 = 0$
15. $v^2 + 12v + 20 = 0$
16. $3t^2 - 7t - 20 = 0$
17. $5y^2 - y - 4 = 0$
18. $x^2 - 25 = 0$
19. $r^2 + 25 = 0$
20. $2x^2 + 98 = 28x$
21. $4s^2 + 100 = 40s$
22. $2r^2 + r - 14 = 0$
23. $2n^2 - 7n - 3 = 0$
24. $5v^2 - 7v = 1$
25. $11z^2 - z = 3$
26. $2w^2 = -(7w + 3)$
27. $2(12g^2 - g) = 15$
28. $1.34d^2 - 1.1d = -1.02$
29. $-2x^2 + 0.7x = -0.3$
30. $2y^2 - \frac{5}{4}y = \frac{1}{2}$
31. $\frac{1}{2}v^2 - v = \frac{3}{4}$

32. **GEOMETRY** The perimeter of a rectangle is 60 inches. Find the dimensions of the rectangle if its area is 221 square inches.

33. **GEOMETRY** Rectangle $ABCD$ has a perimeter of 42 centimeters. What are the dimensions of the rectangle if its area is 80 square centimeters?

34. **NUMBER THEORY** Find two consecutive odd integers whose product is 255.

35. **NUMBER THEORY** The sum of the squares of two consecutive odd numbers is 130. What are the numbers?

36. Without graphing, determine the x-intercepts of the graph of $f(x) = 4x^2 - 9x + 4$.

37. Without graphing, determine the x-intercepts of the graph of $f(x) = 13x^2 - 16x - 4$.

State the value of the discriminant for each equation. Then determine the number of real roots of the equation.

38. $x^2 + 3x - 4 = 0$
39. $y^2 + 3y + 1 = 0$
40. $4p^2 + 10p = -6.25$
41. $1.5m^2 + m = -3.5$
42. $2r^2 = \frac{1}{2}r - \frac{2}{3}$
43. $\frac{4}{3}n^2 + 4n = -3$

44. Without graphing, determine the number of x-intercepts of the graph of $f(x) = 7x^2 - 3x - 1$.

45. Without graphing, determine the number of x-intercepts of the graph of $f(x) = x^2 + 4x + 7$.

More About. . .

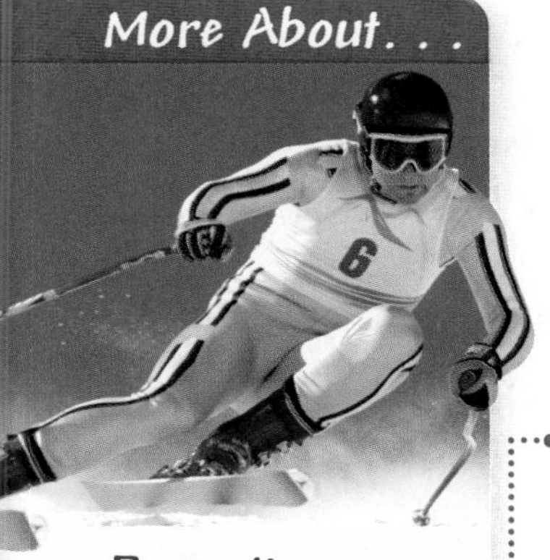

Recreation

Downhill skiing is the most popular type of snow skiing. Skilled skiers can obtain speeds of about 60 miles per hour as they race down mountain slopes.

Source: *World Book Encyclopedia*

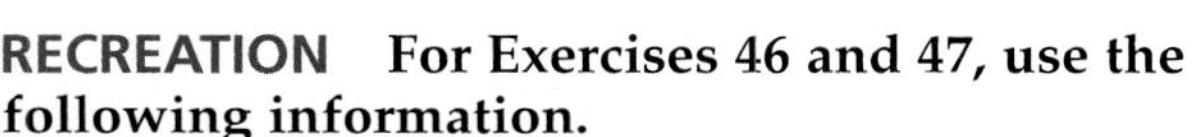

RECREATION **For Exercises 46 and 47, use the following information.**

As Darius is skiing down a ski slope, Jorge is on the chair lift on the same slope. The chair lift has stopped. Darius stops directly below Jorge and attempts to toss a disposable camera up to him. If the camera is thrown with an initial velocity of 35 feet per second, the equation for the height of the camera is $h = -16t^2 + 35t + 5$, where h represents the height in feet and t represents the time in seconds.

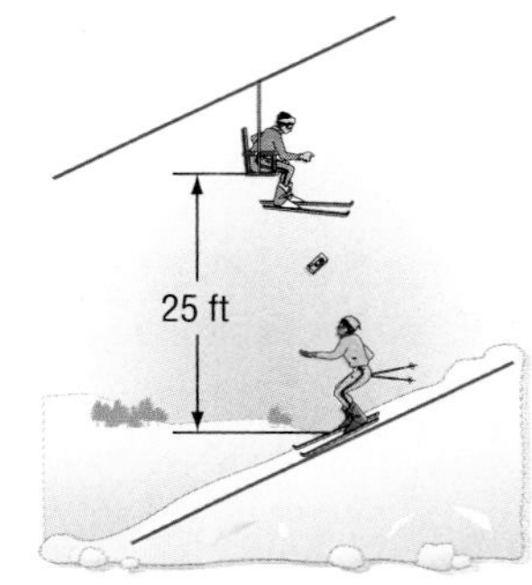

46. If the chair lift is 25 feet above the ground, will Jorge have 0, 1, or 2 chances to catch the camera?

47. If Jorge is unable to catch the camera, when will it hit the ground?

48. **PHYSICAL SCIENCE** A projectile is shot vertically up in the air from ground level. Its distance s, in feet, after t seconds is given by $s = 96t - 16t^2$. Find the values of t when s is 96 feet.

49. **WATER MANAGEMENT** Cox's formula for measuring velocity of water draining from a reservoir through a horizontal pipe is $4v^2 + 5v - 2 = \frac{1200HD}{L}$, where v represents the velocity of the water in feet per second, H represents the height of the reservoir in feet, D represents the diameter of the pipe in inches, and L represents the length of the pipe in feet. How fast is water flowing through a pipe 20 feet long with a diameter of 6 inches that is draining a swimming pool with a depth of 10 feet?

50. **CRITICAL THINKING** If the graph of $f(x) = ax^2 + 10x + 3$ intersects the x-axis in two places, what must be true about the value of a?

CANCER STATISTICS For Exercises 51–53, use the following information.
A decrease in smoking in the United States has resulted in lower death rates caused by cancer. In 1965, 42% of adults smoked, compared with less than 25% in 1995. The number of deaths per 100,000 people y can be approximated by $y = -0.048x^2 + 1.87x + 154$, where x represents the number of years after 1970.

51. Use the Quadratic Formula to solve for x when $y = 150$.

52. In what year would you expect the death rate from cancer to be 150 per 100,000?

53. According to the quadratic function, when will the death rate from cancer be 0 per 100,000? Do you think that the prediction is valid? Why or why not?

54. WRITING IN MATH Answer the question that was posed at the beginning of the lesson.

How can the Quadratic Formula be used to solve problems involving population trends?

Include the following in your answer:
- a step-by-step solution of $15 = 0.0055t^2 - 0.0796t + 5.2810$ with justification of each step, and
- an explanation for why the Quadratic Formula is the best way to solve this equation.

55. Determine the number of solutions of $x^2 - 5x + 8 = 0$.

(A) 0 (B) 1 (C) 2 (D) infinitely many

56. Which expression represents the solutions of $2x^2 + 5x + 1 = 0$?

(A) $\frac{5 \pm \sqrt{17}}{4}$ (B) $\frac{5 \pm \sqrt{33}}{4}$ (C) $\frac{-5 \pm \sqrt{17}}{4}$ (D) $\frac{-5 \pm \sqrt{33}}{4}$

Maintain Your Skills

Mixed Review

Solve each equation by completing the square. Round to the nearest tenth if necessary. *(Lesson 10-3)*

57. $x^2 - 8x = -7$ **58.** $a^2 + 2a + 5 = 20$ **59.** $n^2 - 12n = 5$

Solve each equation by graphing. If integral roots cannot be found, estimate the roots by stating the consecutive integers between which the roots lie. *(Lesson 10-2)*

60. $x^2 - x = 6$ **61.** $2x^2 + x = 2$ **62.** $-x^2 + 3x + 6 = 0$

Factor each polynomial. *(Lesson 9-2)*

63. $15xy^3 + y^4$ **64.** $2ax + 6xc + ba + 3bc$

65. **SCIENCE** The mass of a proton is 0.0000000000000000000001672 milligram. Write this number in scientific notation. *(Lesson 8-3)*

Graph each system of inequalities. *(Lesson 7-5)*

66. $x \leq 2$
$y + 4 \geq 5$

67. $x + y > 2$
$x - y \leq 2$

68. $y > x$
$y \leq x + 4$

Solve each inequality. Then check your solution. *(Lesson 6-3)*

69. $2m + 7 > 17$ **70.** $-2 - 3x \geq 2$ **71.** $-20 \geq 8 + 7k$

Getting Ready for the Next Lesson

PREREQUISITE SKILL Evaluate $c(a^x)$ for each of the given values.
*(To review **evaluating expressions with exponents**, see Lesson 1-1.)*

72. $a = 2, c = 1, x = 4$ **73.** $a = 7, c = 3, x = 2$ **74.** $a = 5, c = 2, x = 3$

Graphing Calculator Investigation

A Follow-Up of Lesson 10-4

Solving Quadratic-Linear Systems

Since you can graph multiple functions on a graphing calculator, it is a useful tool when finding the intersection points or solutions of a system of equations in which one equation is quadratic and one is linear.

Solve the following quadratic-linear system of equations.

$y + 1 = x$

$y = -x^2 + 2x + 5$

Step 1 *Solve each equation for y.*

- $y + 1 = x$
  $y = x - 1$
- $y = -x^2 + 2x + 5$

Step 2 *Graph the equations on the same screen.*

- Enter $y = x - 1$ as Y1.
- Enter $y = -x^2 + 2x + 5$ as Y2.
- Graph both in the standard viewing window.

Step 3 *Approximate the intersection point.*

- Use the intersect option on the CALC menu to approximate the first intersection point.

KEYSTROKES: [2nd] [CALC] 5 [ENTER] [ENTER] [ENTER]

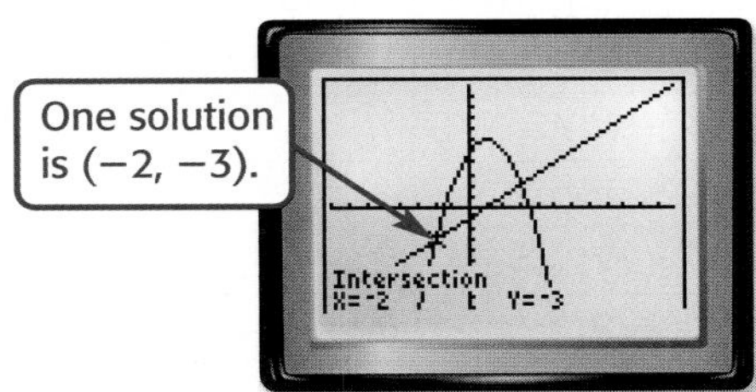

Step 4 *Approximate the other intersection point.*

- Use the TRACE feature with the right and left arrow keys to move the cursor near the other intersection point.
- Use the intersect option on the CALC menu to approximate the other intersection point.

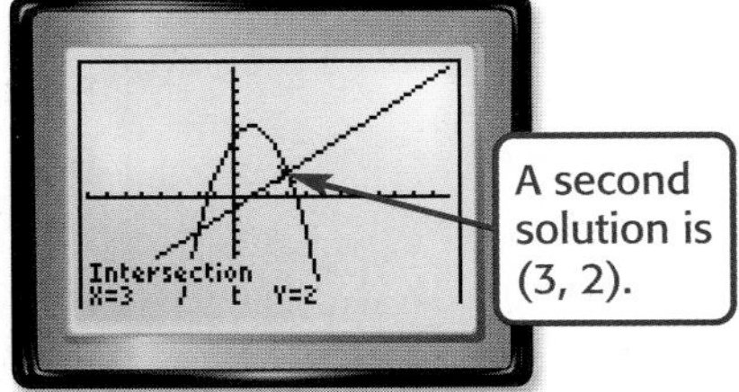

Thus, the solutions of the quadratic-linear system are (−2, −3) and (3, 2).

Exercises

Use the intersect feature to solve each quadratic-linear system of equations. State any decimal solutions to the nearest tenth.

1. $y = -2(2x + 3)$
$y = x^2 + 2x + 3$

2. $y - 5 = 0$
$y = -x^2$

3. $1.8x + y = 3.6$
$y = x^2 - 3x - 1$

4. $y = -1.4x - 2.88$
$y = x^2 + 0.4x - 3.14$

5. $y = x^2 - 3.5x + 2.2$
$y = 2x - 5.3625$

6. $y = 0.35x - 1.648$
$y = -0.2x^2 + 0.28x + 1.01$

 www.algebra1.com/other_calculator_keystrokes

10-5 Exponential Functions

What You'll Learn

- Graph exponential functions.
- Identify data that displays exponential behavior.

Vocabulary

- exponential function

How can exponential functions be used in art?

Earnest "Mooney" Warther was a whittler and a carver. For one of his most unusual carvings, Mooney carved a large pair of pliers in a tree.

From this original carving, he carved another pair of pliers in each handle of the original. Then he carved another pair of pliers in each of those handles. He continued this pattern to create the original pliers and 8 more layers of pliers. Even more amazing is the fact that all of the pliers work.

The number of pliers on each level is given in the table below.

Level	Number of Pliers	Power of 2
Original	1	2^0
First	1(2) = 2	2^1
Second	2(2) = 4	2^2
Third	2(2)(2) = 8	2^3
Fourth	2(2)(2)(2) = 16	2^4
Fifth	2(2)(2)(2)(2) = 32	2^5
Sixth	2(2)(2)(2)(2)(2) = 64	2^6
Seventh	2(2)(2)(2)(2)(2)(2) = 128	2^7
Eighth	2(2)(2)(2)(2)(2)(2)(2) = 256	2^8

GRAPH EXPONENTIAL FUNCTIONS Study the Power of 2 column. Notice that the exponent number matches the level. So we can write an equation to describe y, the number of pliers for any given level x as $y = 2^x$. This type of function, in which the variable is the exponent, is called an **exponential function**.

Key Concept — Exponential Function

An exponential function is a function that can be described by an equation of the form $y = a^x$, where $a > 0$ and $a \neq 1$.

As with other functions, you can use ordered pairs to graph an exponential function.

Example 1 Graph an Exponential Function with $a > 1$

a. Graph $y = 4^x$. State the y-intercept.

x	4^x	y
−2	4^{-2}	$\frac{1}{16}$
−1	4^{-1}	$\frac{1}{4}$
0	4^0	1
1	4^1	4
2	4^2	16
3	4^3	64

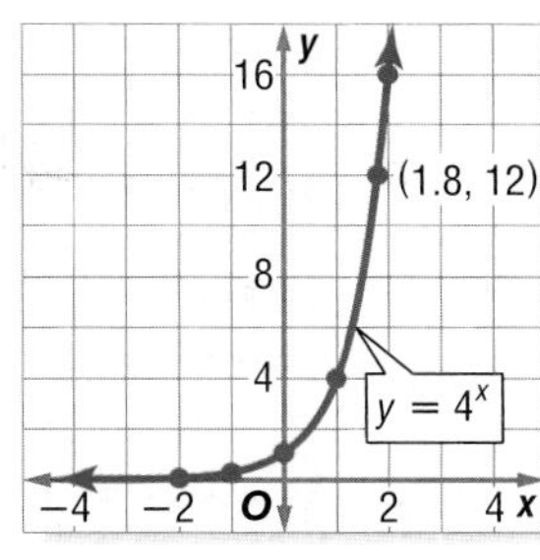

Graph the ordered pairs and connect the points with a smooth curve. The y-intercept is 1. *Notice that the y values change little for small values of x, but they increase quickly as the values of x become greater.*

b. Use the graph to determine the approximate value of $4^{1.8}$.

The graph represents all real values of x and their corresponding values of y for $y = 4^x$. So, the value of y is about 12 when $x = 1.8$. Use a calculator to confirm this value.

$4^{1.8} \approx 12.12573252$

The graphs of functions of the form $y = a^x$, where $a > 1$, all have the same shape as the graph in Example 1, rising faster and faster as you move from left to right.

Example 2 Graph Exponential Functions with $0 < a < 1$

a. Graph $y = \left(\frac{1}{2}\right)^x$. State the y-intercept.

x	$\left(\frac{1}{2}\right)^x$	y
−3	$\left(\frac{1}{2}\right)^{-3}$	8
−2	$\left(\frac{1}{2}\right)^{-2}$	4
−1	$\left(\frac{1}{2}\right)^{-1}$	2
0	$\left(\frac{1}{2}\right)^{0}$	1
1	$\left(\frac{1}{2}\right)^{1}$	$\frac{1}{2}$
2	$\left(\frac{1}{2}\right)^{2}$	$\frac{1}{4}$

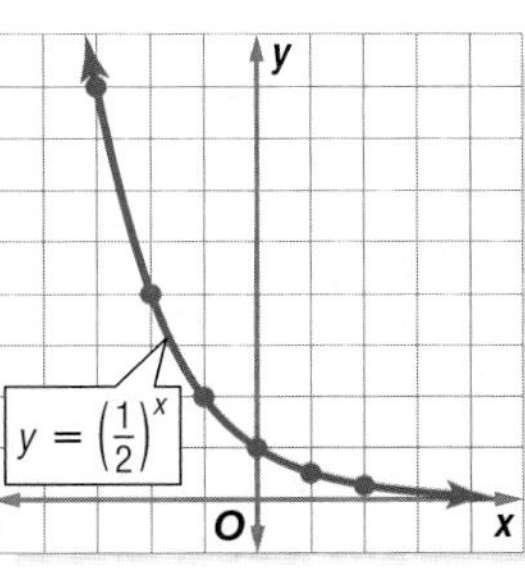

Graph the ordered pairs and connect the points with a smooth curve. The y-intercept is 1. *Notice that the y values decrease less rapidly as x increases.*

b. Use the graph to determine the approximate value of $\left(\frac{1}{2}\right)^{-2.5}$.

The value of y is about $5\frac{1}{2}$ when $x = -2.5$. Use a calculator to confirm this value.

$\left(\frac{1}{2}\right)^{-2.5} \approx 5.656854249$

Graphing Calculator Investigation

Transformations of Exponential Functions

You can use a graphing calculator to study families of graphs of exponential functions. For example, the graph at the right shows the graphs of $y = 2^x$, $y = 3 \cdot 2^x$, and $y = 0.5 \cdot 2^x$. Notice that the y-intercept of $y = 2^x$ is 1, the y-intercept of $y = 3 \cdot 2^x$ is 3, and the y-intercept of $y = 0.5 \cdot 2^x$ is 0.5. The graph of $y = 3 \cdot 2^x$ is steeper than the graph of $y = 2^x$. The graph of $y = 0.5 \cdot 2^x$ is not as steep as the graph of $y = 2^x$.

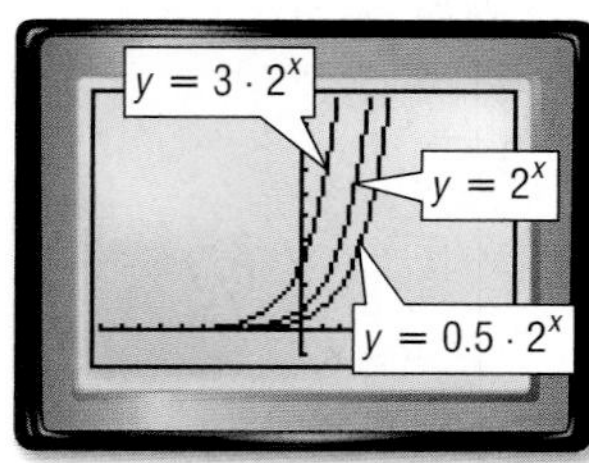

[−10, 10] scl: 1 by [−1, 10] scl: 1

Think and Discuss

Graph each family of equations on the same screen. Compare and contrast the graphs.

1. $y = 2^x$, $y = 2^x + 3$, $y = 2^x - 4$
2. $y = 2^x$, $y = 2^{x+5}$, $y = 2^{x-4}$
3. $y = 2^x$, $y = 3^x$, $y = 5^x$
4. $y = 3 \cdot 2^x$, $y = 3(2^x - 1)$, $y = 3(2^x + 1)$

Example 3 Use Exponential Functions to Solve Problems

More About. . .

Motion Pictures

The first successful photographs of motion were made in 1877. Today, the motion picture industry is big business, with the highest-grossing movie making $600,800,000.

Source: *World Book Encyclopedia*

MOTION PICTURES **Movies tend to have their best ticket sales the first weekend after their release. The sales then follow a decreasing exponential function each successive weekend after the opening. The function $E = 49.9 \cdot 0.692^w$ models the earnings of a popular movie. In this equation, E represents earnings in millions of dollars and w represents the weekend number.**

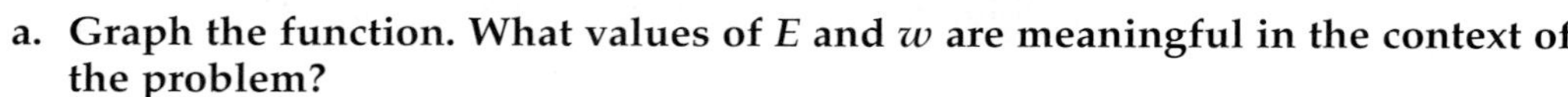

a. Graph the function. What values of E and w are meaningful in the context of the problem?

Use a graphing calculator to graph the function. Only values where $E \leq 49.9$ and $w > 0$ are meaningful in the context of the problem.

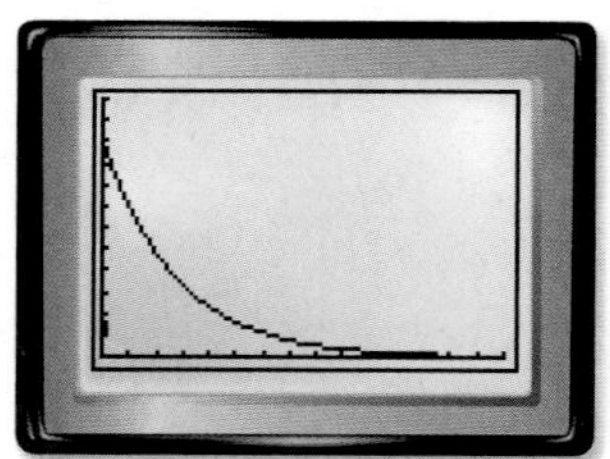

[0, 15] scl: 1 by [0, 60] scl: 5

b. How much did the movie make on the first weekend?

$E = 49.9 \cdot 0.692^w$ Original equation

$E = 49.9 \cdot 0.692^1$ $w = 1$

$E = 34.5308$ Use a calculator.

On the first weekend, the movie grossed about $34.53 million.

c. How much did it make on the fifth weekend?

$E = 49.9 \cdot 0.692^w$ Original equation

$E = 49.9 \cdot 0.692^5$ $w = 5$

$E \approx 7.918282973$ Use a calculator.

On the fifth weekend, the movie grossed about $7.92 million.

IDENTIFY EXPONENTIAL BEHAVIOR How do you know if a set of data is exponential? One method is to observe the shape of the graph. But the graph of an exponential function may resemble part of the graph of a quadratic function. Another way is to use the problem-solving strategy *look for a pattern* with the data.

Example 4 Identify Exponential Behavior

Determine whether each set of data displays exponential behavior.

a.

x	0	10	20	30	40	50
y	80	40	20	10	5	2.5

Method 1 Look for a Pattern

The domain values are at regular intervals of 10. Let's see if there is a common factor among the range values.

80 → 40 → 20 → 10 → 5 → 2.5, each step $\times \frac{1}{2}$

Since the domain values are at regular intervals and the range values have a common factor, the data are probably exponential. The equation for the data may involve $\left(\frac{1}{2}\right)^x$.

Method 2 Graph the Data

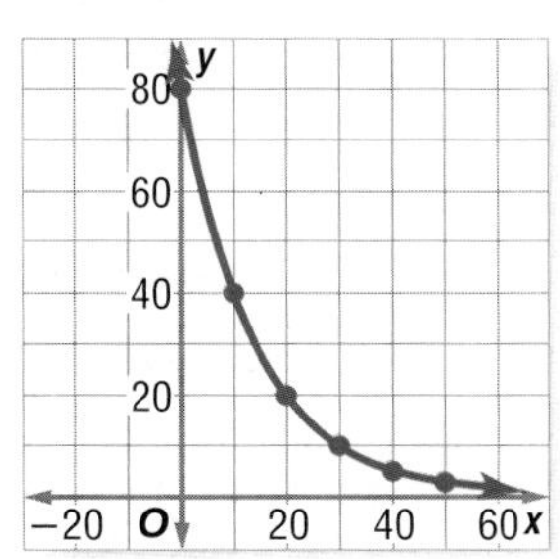

The graph shows a rapidly decreasing value of y as x increases. This is a characteristic of exponential behavior.

b.

x	0	10	20	30	40	50
y	15	21	27	33	39	45

Method 1 Look for a Pattern

The domain values are at regular intervals of 10. The range values have a common difference 6.

15 → 21 → 27 → 33 → 39 → 45, each step $+6$

The data do not display exponential behavior, but rather linear behavior.

Method 2 Graph the Data

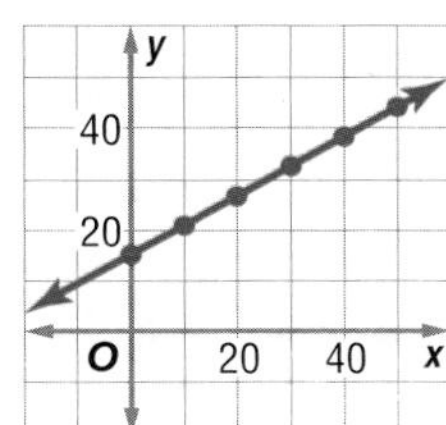

This is a graph of a line, not an exponential function.

Check for Understanding

Concept Check

1. **Determine** whether the graph of $y = a^x$, where $a > 0$ and $a \neq 1$, *sometimes*, *always*, or *never* has an x-intercept.
2. **OPEN ENDED** Write an exponential function and graph the function. Describe the graph.

3. **FIND THE ERROR** Amalia and Kiski are graphing $y = \left(\frac{1}{3}\right)^x$.

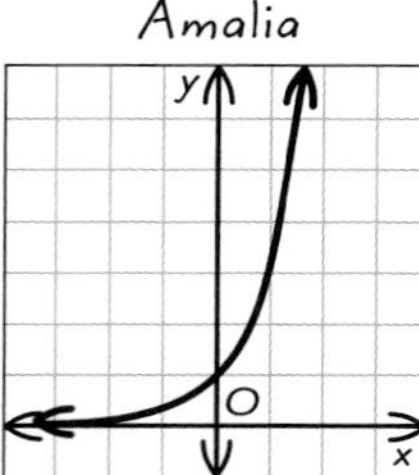

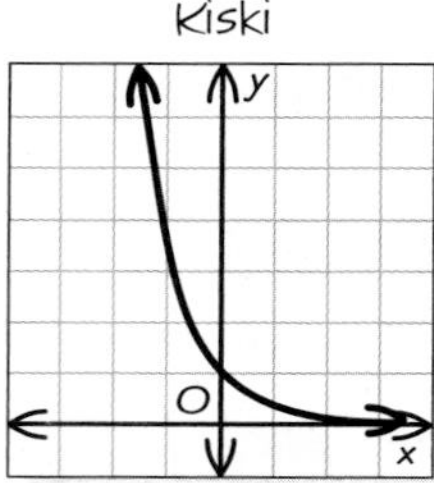

Who is correct? Explain your reasoning.

Guided Practice

Graph each function. State the y-intercept. Then use the graph to determine the approximate value of the given expression. Use a calculator to confirm the value.

4. $y = 3^x$; $3^{1.2}$
5. $y = \left(\frac{1}{4}\right)^x$; $\left(\frac{1}{4}\right)^{1.7}$
6. $y = 9^x$; $9^{0.8}$

Graph each function. State the y-intercept.

7. $y = 2 \cdot 3^x$
8. $y = 4(5^x - 10)$

Determine whether the data in each table display exponential behavior. Explain why or why not.

9.

x	0	1	2	3	4	5
y	1	6	36	216	1296	7776

10.

x	4	6	8	10	12	14
y	5	9	13	17	21	25

Application

FOLKLORE For Exercises 11 and 12, use the following information.
A wise man asked his ruler to provide rice for feeding his people. Rather than receiving a constant daily supply of rice, the wise man asked the ruler to give him 2 grains of rice for the first square on a chessboard, 4 grains for the second, 8 grains for the third, 16 for the fourth, and so on doubling the amount of rice with each square of the board.

11. How many grains of rice will the wise man receive for the last (64th) square on the chessboard?
12. If one pound of rice has approximately 24,000 grains, how many tons of rice will the wise man receive on the last day? (*Hint*: one ton = 2000 pounds)

Practice and Apply

Homework Help

For Exercises	See Examples
13–26	1, 2
27–32	4
33–41	3

Extra Practice
See page 843.

Graph each function. State the y-intercept. Then use the graph to determine the approximate value of the given expression. Use a calculator to confirm the value.

13. $y = 5^x$; $5^{1.1}$
14. $y = 10^x$; $10^{0.3}$
15. $y = \left(\frac{1}{10}\right)^x$; $\left(\frac{1}{10}\right)^{-1.3}$
16. $y = \left(\frac{1}{5}\right)^x$; $\left(\frac{1}{5}\right)^{0.5}$
17. $y = 6^x$; $6^{0.3}$
18. $y = 8^x$; $8^{0.8}$

Graph each function. State the y-intercept.

19. $y = 5(2^x)$
20. $y = 3(5^x)$
21. $y = 3^x - 7$
22. $y = 2^x + 4$
23. $y = 2(3^x) - 1$
24. $y = 5(2^x) + 4$
25. $y = 2(3^x + 1)$
26. $y = 3(2^x - 5)$

Determine whether the data in each table display exponential behavior. Explain why or why not.

27.

x	−2	−1	0	1
y	−5	−2	1	4

28.

x	0	1	2	3
y	1	0.5	0.25	0.125

29.

x	10	20	30	40
y	16	12	9	6.75

30.

x	−1	0	1	2
y	−0.5	1.0	−2.0	4.0

31.

x	3	6	9	12
y	5	5	5	5

32.

x	5	3	1	−1
y	32	16	8	4

BUSINESS For Exercises 33–35, use the following information.
The amount of money spent at West Outlet Mall in Midtown continues to increase. The total $T(x)$ in millions of dollars can be estimated by the function $T(x) = 12(1.12)^x$, where x is the number of years after it opened in 1995.

33. According to the function, find the amount of sales for the mall in the years 2005, 2006, and 2007.

34. Graph the function and name the y-intercept.

35. What does the y-intercept represent in this problem?

36. BIOLOGY Mitosis is a process of cell reproduction in which one cell divides into two identical cells. *E. coli* is a fast-growing bacterium that is often responsible for food poisoning in uncooked meat. It can reproduce itself in 15 minutes. If you begin with 100 *E. coli* bacteria, how many will there be in one hour?

More About. . .

Training

The first Boston Marathon was held in 1896. The distance of this race was based on the Greek legend that Pheidippides ran 24.8 miles from Marathon to Athens to bring the news of victory over the Persian army.

Source: www.bostonmarathon.org

TOURNAMENTS For Exercises 37–39, use the following information.
In a regional quiz bowl competition, three schools compete and the winner advances to the next round. Therefore, after each round, only $\frac{1}{3}$ of the schools remain in the competition for the next round. Suppose 729 schools start the competition.

37. Write an exponential function to describe the number of schools remaining after x rounds.

38. How many schools are left after 3 rounds?

39. How many rounds will it take to declare a champion?

TRAINING For Exercises 40 and 41, use the following information.
A runner is training for a marathon, running a total of 20 miles per week on a regular basis. She plans to increase the distance $D(x)$ in miles according to the function $D(x) = 20(1.1)^x$, where x represents the number of weeks of training.

40. Copy and complete the table showing the number of miles she plans to run.

41. The runner's goal is to work up to 50 miles per week. What is the first week that the total will be 50 miles or more?

Week	Distance (miles)
1	
2	
3	
4	

CRITICAL THINKING Describe the graph of each equation as a transformation of the graph of $y = 5^x$.

42. $y = \left(\frac{1}{5}\right)^x$

43. $y = 5^x + 2$

44. $y = 5^x - 4$

45. **WRITING IN MATH** Answer the question that was posed at the beginning of the lesson.

How can exponential functions be used in art?

Include the following in your answer:

- the exponential function representing the pliers,
- an explanation of which x and y values are meaningful, and
- the graph of this function.

Standardized Test Practice

46. Which function is an exponential function?

Ⓐ $f(x) = x^2$ Ⓑ $f(x) = 6^x$

Ⓒ $f(x) = x^5$ Ⓓ $f(x) = x^3 + 2x^2 - x + 5$

47. Compare the graphs of $y = 2^x$ and $y = 6^x$.

Ⓐ The graph of $y = 6^x$ steeper than the graph of $y = 2^x$.

Ⓑ The graph of $y = 2^x$ steeper than the graph of $y = 6^x$.

Ⓒ The graph of $y = 6^x$ is the graph of $y = 2^x$ translated 4 units up.

Ⓓ The graph of $y = 6^x$ is the graph of $y = 2^x$ translated 3 units up.

Maintain Your Skills

Mixed Review

Solve each equation by using the Quadratic Formula. Round to the nearest tenth if necessary. *(Lesson 10-4)*

48. $x^2 - 9x - 36 = 0$ 49. $2t^2 + 3t - 1 = 0$ 50. $5y^2 + 3 = y$

Solve each equation by completing the square. Round to the nearest tenth if necessary. *(Lesson 10-3)*

51. $x^2 - 7x = -10$ 52. $a^2 - 12a = 3$ 53. $t^2 + 6t + 3 = 0$

Factor each trinomial, if possible. If the trinomial cannot be factored using integers, write *prime*. *(Lesson 9-3)*

54. $m^2 - 14m + 40$ 55. $t^2 - 2t + 35$ 56. $z^2 - 5z - 24$

57. Three times one number equals twice a second number. Twice the first number is 3 more than the second number. Find the numbers. *(Lesson 7-4)*

Solve each inequality. *(Lesson 6-1)*

58. $x + 7 > 2$ 59. $10 \geq x + 8$ 60. $y - 7 < -12$

Getting Ready for the Next Lesson

PREREQUISITE SKILL **Evaluate $p(1 + r)^t$ for each of the given values.**

*(To review **evaluating expressions with exponents**, see Lesson 1-1.)*

61. $p = 5, r = \frac{1}{2}, t = 2$ 62. $p = 300, r = \frac{1}{4}, t = 3$

63. $p = 100, r = 0.2, t = 2$ 64. $p = 6, r = 0.5, t = 3$

Practice Quiz 2

Lessons 10-4 and 10-5

Solve each equation by using the Quadratic Formula. Round to the nearest tenth if necessary. *(Lesson 10-4)*

1. $x^2 + 2x = 35$ 2. $2n^2 - 3n + 5 = 0$ 3. $2v^2 - 4v = 1$

Graph each function. State the y-intercept. *(Lesson 10-5)*

4. $y = 0.5(4^x)$ 5. $y = 5^x - 4$

10-6 Growth and Decay

What You'll Learn

- Solve problems involving exponential growth.
- Solve problems involving exponential decay.

Vocabulary

- exponential growth
- compound interest
- exponential decay

How can exponential growth be used to predict future sales?

The graph shows that the average household in the United States has increased its spending for restaurant meals. In fact, the amount grew at an annual rate of about 4.6% between 1994 and 1998. Let y represent the average amount spent on restaurant meals, and let t represent the number of years since 1994. Then the average amount spent on restaurant meals can be modeled by $y = 1698(1 + 0.046)^t$ or $y = 1698(1.046)^t$.

USA TODAY Snapshots®

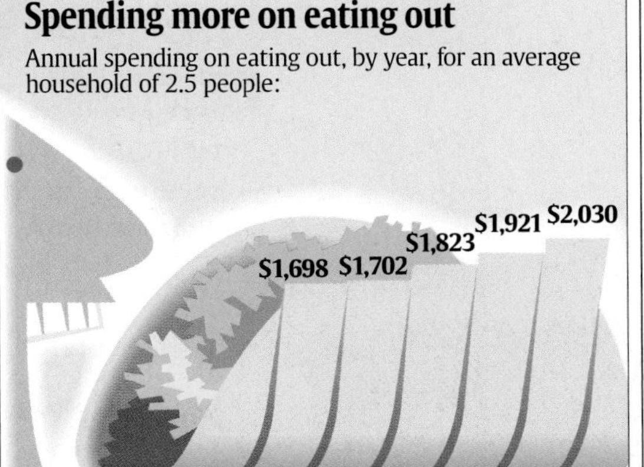

By Mark Pearson and Sam Ward, USA TODAY

EXPONENTIAL GROWTH The equation for the average amount spent on restaurant meals is in the form $y = C(1 + r)^t$. This is the general equation for **exponential growth** in which the initial amount C increases by the same percent over a given period of time.

Key Concept — General Equation for Exponential Growth

The general equation for exponential growth is $y = C(1 + r)^t$ where y represents the final amount, C represents the initial amount, r represents the rate of change expressed as a decimal, and t represents time.

Example 1 Exponential Growth

SPORTS In 1971, there were 294,105 females participating in high school sports. Since then, that number has increased an average of 8.5% per year.

a. Write an equation to represent the number of females participating in high school sports since 1971.

$y = C(1 + r)^t$	General equation for exponential growth
$y = 294{,}105(1 + 0.085)^t$	$C = 294{,}105$ and $r = 8.5\%$ or 0.085
$y = 294{,}105(1.085)^t$	Simplify.

An equation to represent the number of females participating in high school sports is $y = 294{,}105(1.085)^t$, where y represents the number of female athletes and t represents the number of years since 1971.

(continued on the next page)

www.algebra1.com/extra_examples

b. According to the equation, how many females participated in high school sports in the year 2001?

$y = 294{,}105(1.085)^t$ Equation for females participating in sports

$y = 294{,}105(1.085)^{30}$ $t = 2001 - 1971$ or 30

$y \approx 3{,}399{,}340$ In 2001, about 3,399,340 females participated.

One special application of exponential growth is **compound interest**. The equation for compound interest is $A = P\left(1 + \frac{r}{n}\right)^{nt}$, where A represents the amount of the investment, P is the principal (initial amount of the investment), r represents the annual rate of interest expressed as a decimal, n represents the number of times that the interest is compounded each year, and t represents the number of years that the money is invested.

Example 2 Compound Interest

HISTORY **Use the information at the left. If the money the Native Americans received for Manhattan had been invested at 6% per year compounded semiannually, how much money would there be in the year 2026?**

$A = P\left(1 + \frac{r}{n}\right)^{nt}$ Compound interest equation

$A = 24\left(1 + \frac{0.06}{2}\right)^{2(400)}$ $P = 24$, $r = 6\%$ or 0.06, $n = 2$, and $t = 400$

$A = 24(1.03)^{800}$ Simplify.

$A \approx 4.47 \times 10^{11}$ There would be about \$447,000,000,000.

More About. . .

History

In 1626, Peter Minuit, governor of the colony of New Netherland, bought the island of Manhattan from the Native Americans for beads, cloth, and trinkets worth 60 Dutch guilders (\$24).

Source: *World Book Encyclopedia*

EXPONENTIAL DECAY A variation of the growth equation can be used as the general equation for **exponential decay**. In exponential decay, the original amount decreases by the same percent over a period of time.

Key Concept — General Equation for Exponential Decay

The general equation for exponential decay is $y = C(1 - r)^t$ where y represents the final amount, C represents the initial amount, r represents the rate of decay expressed as a decimal, and t represents time.

Example 3 Exponential Decay

ENERGY **In 1950, the use of coal by residential and commercial users was 114.6 million tons. Many businesses now use cleaner sources of energy. As a result, the use of coal has decreased by 6.6% per year.**

a. Write an equation to represent the use of coal since 1950.

$y = C(1 - r)^t$ General equation for exponential decay

$y = 114.6(1 - 0.066)^t$ $C = 114.6$ and $r = 6.6\%$ or 0.066

$y = 114.6(0.934)^t$ Simplify.

An equation to represent the use of coal is $y = 114.6(0.934)^t$, where y represents tons of coal used annually and t represents the number of years since 1950.

b. Estimate the estimated amount of coal that will be used in 2015.

$y = 114.6(0.934)^t$ Equation for coal use

$y = 114.6(0.934)^{65}$ $t = 2015 - 1950$ or 65

$y \approx 1.35$ The amount of coal should be about 1.35 million tons.

Sometimes items decrease in value or *depreciate*. For example, most cars and office equipment depreciate as they get older. You can use the exponential decay formula to determine the value of an item at a given time.

Example 4 Depreciation

FARMING A farmer buys a tractor for $50,000. If the tractor depreciates 10% per year, find the value of the tractor in 7 years.

$y = C(1 - r)^t$	General equation for exponential decay
$y = 50{,}000(1 - 0.10)^7$	$C = 50{,}000$, $r = 10\%$ or 0.10, and $t = 7$
$y = 50{,}000(0.90)^7$	Simplify.
$y \approx 23{,}914.85$	Use a calculator.

The tractor will be worth about $23,914.85 or less than half its original value.

Check for Understanding

Concept Check

1. **Explain** the difference between *exponential growth* and *exponential decay*.
2. **OPEN ENDED** Write a compound interest problem that could be solved by the equation $A = 500\left(1 + \frac{0.07}{4}\right)^{4(6)}$.
3. **Draw** a graph representing exponential decay.

Guided Practice

INCOME For Exercises 4 and 5, use the graph at the right and the following information.
The median household income in the United States increased an average of 0.5% each year between 1979 and 1999. Assume this pattern continues.

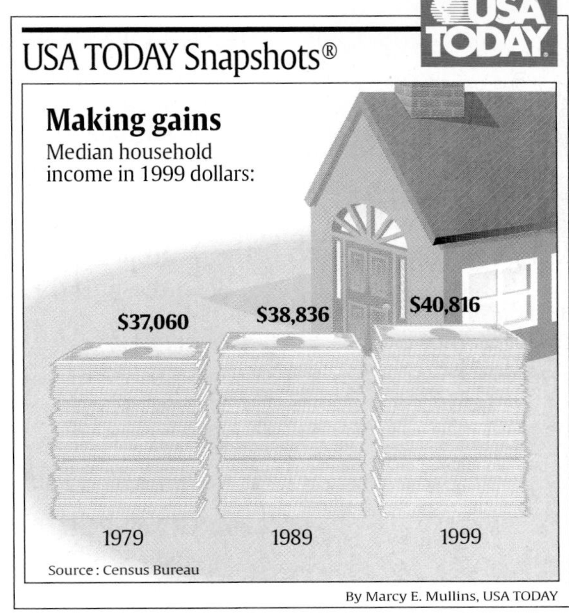

4. Write an equation for the median household income for t years after 1979.
5. Predict the median household income in 2009.

Applications

6. **INVESTMENTS** Determine the amount of an investment if $400 is invested at an interest rate of 7.25% compounded quarterly for 7 years.
7. **POPULATION** In 1995, the population of West Virginia reached 1,821,000, its highest in the 20th century. For the next 5 years, its population decreased 0.2% each year. If this trend continues, predict the population of West Virginia in 2010.
8. **TRANSPORTATION** A car sells for $16,000. If the rate of depreciation is 18%, find the value of the car after 8 years.

Practice and Apply

TECHNOLOGY For Exercises 9 and 10, use the following information.
Computer use around the world has risen 19% annually since 1980.

9. If 18.9 million computers were in use in 1980, write an equation for the number of computers in use for t years after 1980.
10. Predict the number of computers in 2015.

Homework Help

For Exercises	See Examples
9–13, 18 21, 22	1
14, 15	2
16, 17, 25–28	3
19, 20	4

Extra Practice

See page 843.

WEIGHT TRAINING **For Exercises 11 and 12, use the following information.** The use of free weights for fitness has increased in popularity. In 1997, there were 43.2 million people who used free weights.

11. Assuming the use of free weights increases 6% annually, write an equation for the number of people using free weights t years from 1997.
12. Predict the number of people using free weights in 2007.

13. **POPULATION** The population of Mexico has been increasing at an annual rate of 1.7%. If the population of Mexico was 100,350,000 in the year 2000, predict its population in 2012.

14. **INVESTMENTS** Determine the amount of an investment if \$500 is invested at an interest rate of 5.75% compounded monthly for 25 years.

15. **INVESTMENTS** Determine the amount of an investment if \$250 is invested at an interest rate of 10.3% compounded quarterly for 40 years.

16. **POPULATION** The country of Latvia has been experiencing a 1.1% annual decrease in population. In 2000, its population was 2,405,000. If the trend continues, predict Latvia's population in 2015.

17. **MUSIC** In 1994, the sales of music cassettes reached its peak at \$2,976,400,000. Since then, cassette sales have been declining. If the annual percent of decrease in sales is 18.6%, predict the sales of cassettes in the year 2009.

More About. . .

Grand Canyon

The Grand Canyon National Park covers 1,218,375 acres. It has 38 hiking trails, which cover about 400 miles.

Source: *World Book Encyclopedia*

18. **GRAND CANYON** The increase in the number of visitors to the Grand Canyon National Park is similar to an exponential function. If the average visitation has increased 5.63% annually since 1920, use the graph to predict the number of visitors to the park in 2020.

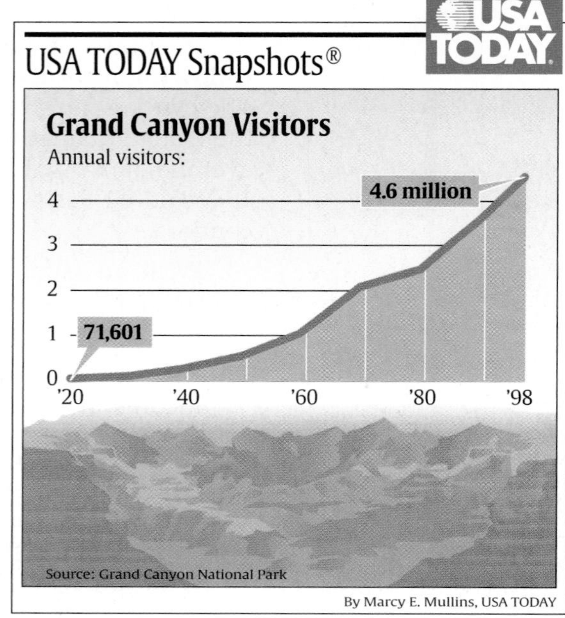

19. **BUSINESS** A piece of office equipment valued at \$25,000 depreciates at a steady rate of 10% per year. What is the value of the equipment in 8 years?

20. **TRANSPORTATION** A new car costs \$23,000. It is expected to depreciate 12% each year. Find the value of the car in 5 years.

POPULATION **For Exercises 21 and 22, use the following information.** Since birth rates are going down and people are living longer, the percent of the population that is 65 years old or older continues to rise. The percent of the U.S. population P that is at least 65 years old can be approximated by the equation $P = 3.86(1.013)^t$, where t represents the number of years since 1900.

21. What percent of the population will be 65 years of age or older in the year 2010?
22. Predict the year in which people ages 65 or older will represent 20% of the population if this trend continues. (*Hint*: Make a table.)

CRITICAL THINKING **Each equation represents an exponential rate of change if t is time in years. Determine whether each equation represents growth or decay. Give the annual rate of change as a percent.**

23. $y = 500(1.026^t)$
24. $y = 500(0.761^t)$

ARCHAEOLOGY For Exercises 25–28, use the following information.
The *half-life* of a radioactive element is the time that it takes for one-half a quantity of the element to decay. Carbon-14 is found in all living organisms and has a half-life of 5730 years. Archaeologists use this fact to estimate the age of fossils. Consider an organism with an original Carbon-14 content of 256 grams. The number of grams remaining in the organism's fossil after t years is $256(0.5)^{\frac{t}{5730}}$.

25. If the organism died 5730 years ago, what is the amount of Carbon-14 today?

26. If the organism died 1000 years ago, what is the amount of Carbon-14 today?

27. If the organism died 10,000 years ago, what is the amount of Carbon-14 today?

28. If the fossil has 32 grams of Carbon-14 remaining, how long ago did it live? (*Hint*: Make a table.)

29. RESEARCH Find the enrollment of your school district each year for the last decade. Find the rate of change from one year to the next. Then, determine the average annual rate of change for those years. Use this information to estimate the enrollment for your school district in ten years.

30. WRITING IN MATH Answer the question that was posed at the beginning of the lesson.

How can exponential growth be used to predict future sales?

Include the following in your answer:

- an explanation of the equation $y = 1698(1 + 0.046)^t$, and
- an estimate of the average family's spending for restaurant meals in 2010.

Standardized Test Practice

31. Which equation represents exponential growth?

(A) $y = 50x^3$ (B) $y = 30x^2 + 10$
(C) $y = 35(1.05^x)$ (D) $y = 80(0.92^x)$

32. Lorena is investing a $5000 inheritance from her aunt in a certificate of deposit that matures in 4 years. The interest rate is 8.25% compounded quarterly. What is the balance of the account after 4 years?

(A) $5412.50 (B) $6865.65 (C) $6908.92 (D) $6931.53

Maintain Your Skills

Mixed Review

Graph each function. State the y-intercept. *(Lesson 10-5)*

33. $y = \left(\frac{1}{8}\right)^x$ **34.** $y = 2^x - 5$ **35.** $y = 4(3^x - 6)$

Solve each equation by using the Quadratic Formula. Round to the nearest tenth if necessary. *(Lesson 10-4)*

36. $m^2 - 9m - 10 = 0$ **37.** $2t^2 - 4t = 3$ **38.** $7x^2 + 3x + 1 = 0$

Simplify. *(Lesson 8-1)*

39. $m^7(m^3b^2)$ **40.** $-3(ax^3y)^2$ **41.** $(0.3x^3y^2)^2$

Solve each open sentence. *(Lesson 6-5)*

42. $|7x + 2| = -2$ **43.** $|3 - 3x| = 0$ **44.** $|t + 4| \geq 3$

45. SKIING A course for cross-country skiing is regulated so that the slope of any hill cannot be greater than 0.33. A hill rises 60 meters over a horizontal distance of 250 meters. Does the hill meet the requirements? *(Lesson 5-1)*

Getting Ready for the Next Lesson

PREREQUISITE SKILL Find the next three terms in each arithmetic sequence.
*(To review **arithmetic sequences**, see Lesson 4-7.)*

46. 8, 11, 14, 17, … **47.** 7, 4, 1, −2, … **48.** 1.5, 2.6, 3.7, 4.8, …

Reading Mathematics

Growth and Decay Formulas

Growth and decay problems may be confusing, unless you read them in a simplified, generalized form. The growth and decay formulas that you used in Lesson 10-6 are based on the idea that an initial amount is multiplied by a rate raised to a power of time, which is equivalent to a final amount. If you remember the following formula, all other formulas will be easier to remember.

$$\textit{final amount} = \textit{initial amount} \cdot \textit{rate}^{\textit{time}}$$

Below, we will review the general equation for exponential growth to see how it is related to the generalized formula above.

The final amount	equals	an initial amount	times	the quantity one plus a rate raised to the power of time.
y	$=$	C	$\cdot$	$(1 + r)^t$

The only difference from the generalized formula is that rate equals $1 + r$. *Why?*

One represents 100%. If you multiply C by 100%, the final amount is the same as the initial amount. We *add* 1 to the rate r so that the final amount is the initial amount plus the increase.

You can break each growth and decay formula into the following pieces:

- final amount,
- initial amount,
- rate, and
- time.

Reading to Learn

1. Write the general equation for exponential decay. Discuss how it is related to the generalized formula. Why is the rate equal to $1 - r$?
2. Write the formula for compound interest. How is it related to the generalized formula? Why does the rate equal $\left(1 + \frac{r}{n}\right)$? Why does the time equal nt?
3. Suppose that \$2500 is invested at an annual rate of 6%. If the interest is compounded quarterly, find the value of the account after 5 years.
   a. Copy the problem and underline all important numerical data.
   b. Choose the appropriate formula and solve the problem.
4. Angela bought a car for \$18,500. If the rate of depreciation is 11%, find the value of the car in 4 years.
   a. Copy the problem and underline all important numerical data.
   b. Choose the appropriate formula and solve the problem.
5. The population of Centerville is increasing at an average annual rate of 3.5%. If its current population is 12,500, predict its population in 5 years.
   a. Copy the problem and underline all important numerical data.
   b. Choose the appropriate formula and solve the problem.

10-7 Geometric Sequences

What You'll Learn

- Recognize and extend geometric sequences.
- Find geometric means.

Vocabulary

- geometric sequence
- common ratio
- geometric means

How can a geometric sequence be used to describe a bungee jump?

A thrill ride is set up with a bungee rope that will stretch when a person jumps from the platform. The ride continues as the person bounces back and forth closer to the stopping place of the rope. Each bounce is only $\frac{3}{4}$ as far from the stopping length as the preceding bounce. If the initial drop is 80 feet past the stopping length of the rope, the following table gives the distance of the first four bounces.

Bounce	Distance (ft)
1	80
2	$\frac{3}{4} \cdot 80$ or 60
3	$\frac{3}{4} \cdot 60$ or 45
4	$\frac{3}{4} \cdot 45$ or $33\frac{3}{4}$

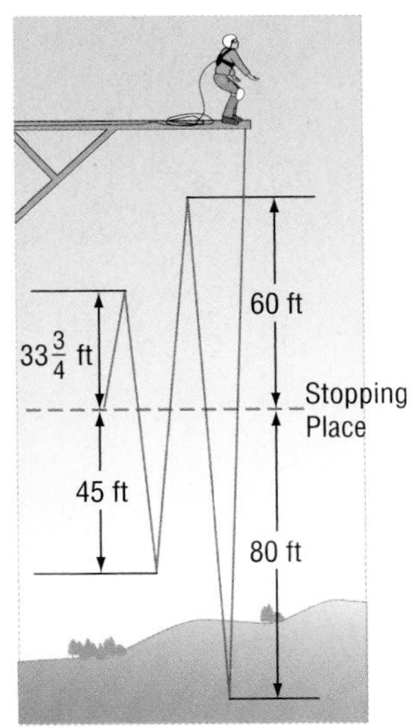

GEOMETRIC SEQUENCES The distance of each bounce is found by multiplying the previous term by $\frac{3}{4}$. The successive distances of the bounces is an example of a **geometric sequence**. The number by which each term is multiplied is called the **common ratio**.

Key Concept — *Geometric Sequence*

- **Words** A geometric sequence is a sequence in which each term after the nonzero first term is found by multiplying the previous term by a constant called the common ratio r, where $r \neq 0, 1$.
- **Symbols** $a, ar, (ar)r$ or $ar^2, (ar^2)r$ or $ar^3, \ldots$ $(a \neq 0; r \neq 0, 1)$
- **Examples** 1, 3, 9, 27, 81, …

Example 1 Recognize Geometric Sequences

Determine whether each sequence is geometric.

a. 0, 5, 10, 15, 20, …

Determine the pattern.

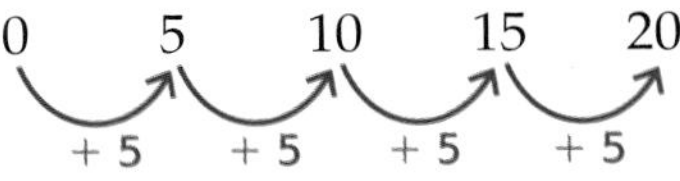

In this sequence, each term is found by adding 5 to the previous term. This sequence is arithmetic, *not* geometric.

Study Tip

Look Back To review **arithmetic sequences**, see Lesson 4-7.

b. 1, 5, 25, 125, 625

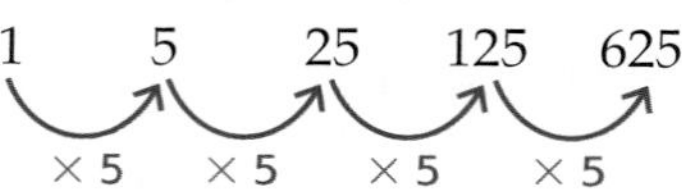

In this sequence, each term is found by multiplying the previous term times 5. This sequence is geometric.

The common ratio of a geometric sequence can be found by dividing any term by the preceding term.

Example 2 Continue Geometric Sequences

Find the next three terms in each geometric sequence.

a. 4, −8, 16, ...

$\frac{-8}{4} = -2$ Divide the second term by the first.

The common factor is −2. Use this information to find the next three terms.

4, −8, 16, −32 64 −128

× (−2) × (−2) × (−2)

The next three terms are −32, 64, and −128.

b. 60, 72, 86.4, ...

$\frac{72}{60} = 1.2$ Divide the second term by the first.

The common factor is 1.2. Use this information to find the next three terms.

60, 72, 86.4, 103.68 124.416 149.2992

× 1.2 × 1.2 × 1.2

The next three terms are 103.68, 124.416, and 149.2992.

Example 3 Use Geometric Sequences to Solve a Problem

GEOGRAPHY Madagascar's population has been increasing at an average annual rate of 3%. Use the information at the left to determine the population of Madagascar in 2001, 2002, and 2003.

The population is a geometric sequence in which the first term is 15,500,000 and the common ratio is 1.03.

Year	Population
2000	15,500,000
2001	15,500,000(1.03) or 15,965,000
2002	15,965,000(1.03) or 16,443,950
2003	16,443,950(1.03) or about 16,937,269

The population of Madagascar in the years 2001, 2002, and 2003 will be 15,965,000, 16,443,950, and about 16,937,269, respectively.

More About. . .

AFRICA
Mozambique Channel
MADAGASCAR

Geography

Madagascar is a country just east of mainland Africa. It consists of the third largest island in the world and many tiny islands. In 2000, the population of Madagascar was about 15,500,000.

Source: *World Book Encyclopedia*

As with arithmetic sequences, you can name the terms of a geometric sequence using a_1, a_2, a_3, and so on. Then the nth term is represented as a_n. Each term of a geometric sequence can also be represented using r and its previous term. A third way to represent each term is by using r and the first term a_1.

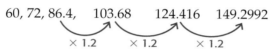

Sequence	number	2	6	18	54	...	
	symbols	a_1	a_2	a_3	a_4	...	a_n
Expressed in Terms of *r* and Previous Term	number	2	2(3)	6(3)	18(3)	...	
	symbols	a_1	$a_1 \cdot r$	$a_2 \cdot r$	$a_3 \cdot r$	...	$a_{n-1} \cdot r$
Expressed in Terms of *r* and First Term	number	2 or $2(3^0)$	2(3) or $2(3^1)$	2(9) or $2(3^2)$	2(27) or $2(3^3)$	...	
	symbols	$a_1 \cdot r^0$	$a_1 \cdot r^1$	$a_1 \cdot r^2$	$a_1 \cdot r^3$	...	$a_1 \cdot r^{n-1}$

The three values in the last column of the table all describe the *n*th term of a geometric sequence.

Study Tip

Recursive Formulas

When the *n*th term of a sequence is expressed in terms of the previous term, as in $a_n = a_{n-1} \cdot n$, the formula is called a *recursive formula*.

Key Concept — Formula for the nth Term of a Geometric Sequence

The *n*th term a_n of a geometric sequence with the first term a_1 and common ratio r is given by $a_n = a_1 \cdot r^{n-1}$.

Example 4 — nth Term of a Geometric Sequence

Find the sixth term of a geometric sequence in which $a_1 = 3$ and $r = -5$.

$a_n = a_1 \cdot r^{n-1}$ — Formula for the *n*th term of a geometric sequence

$a_6 = 3 \cdot (-5)^{6-1}$ — $n = 6$, $a_1 = 3$, and $r = -5$

$a_6 = 3 \cdot (-5)^5$ — $6 - 1 = 5$

$a_6 = 3 \cdot (-3125)$ — $(-5)^5 = -3125$

$a_6 = -9375$ — $3 \cdot (-3125) = -9375$

The sixth term of the geometric sequence is −9375.

Geometric sequences are related to exponential functions.

Algebra Activity

Graphs of Geometric Sequences

You can graph a geometric sequence by graphing the coordinates (n, a_n). For example, consider the sequence 2, 6, 18, 54, ... To graph this sequence, graph the points at (1, 2), (2, 6), (3, 18), and (4, 54).

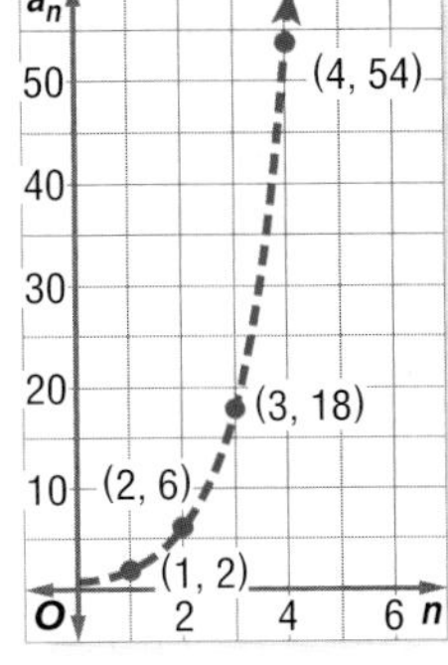

Model

Graph each geometric sequence. Name each common ratio.

1. 1, 2, 4, 8, 16, ...
2. 1, −2, 4, −8, 16, ...

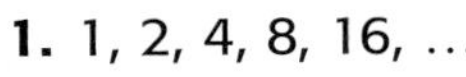

3. 81, 27, 9, 3, 1, ...
4. −81, 27, −9, 3, −1, ...

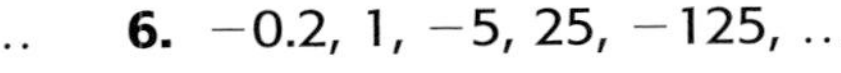

5. 0.2, 1, 5, 25, 125, ...
6. −0.2, 1, −5, 25, −125, ...

Analyze

7. Which graphs appear to be similar to an exponential function?
8. Compare and contrast the graphs of geometric sequences with $r > 0$ and $r < 0$.
9. Compare the formula for an exponential function $y = c(a^x)$ to the value of the *n*th term of a geometric sequence.

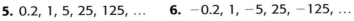

GEOMETRIC MEANS Missing term(s) between two nonconsecutive terms in a geometric sequence are called **geometric means**. In the sequence 100, 20, 4, …, the geometric mean between 100 and 4 is 20. You can use the formula for the nth term of a geometric sequence to find a geometric mean.

Example 5 Find Geometric Means

Find the geometric mean in the sequence 2, ____, 18.

In the sequence, $a_1 = 2$ and $a_3 = 18$. To find a_2, you must first find r.

$a_n = a_1 \cdot r^{n-1}$	Formula for the nth term of a geometric sequence
$a_3 = a_1 \cdot r^{3-1}$	$n = 3$
$18 = 2 \cdot r^2$	$a_3 = 18$ and $a_1 = 2$
$\frac{18}{2} = \frac{2r^2}{2}$	Divide each side by 2.
$9 = r^2$	Simplify.
$\pm 3 = r$	Take the square root of each side.

If $r = 3$, the geometric mean is 2(3) or 6. If $r = -3$, the geometric mean is 2(−3) or −6. Therefore, the geometric mean is 6 or −6.

Check for Understanding

Concept Check

1. **Compare and contrast** an arithmetic sequence and a geometric sequence.
2. **Explain** why the definition of a geometric sequence restricts the values of the common ratio to numbers other than 0 and 1.
3. **OPEN ENDED** Give an example of a sequence that is neither arithmetic nor geometric.

Guided Practice

Determine whether each sequence is geometric.

4. 5, 15, 45, 135, …
5. 56, −28, 14, −7, …
6. 25, 20, 15, 10, …

Find the next three terms in each geometric sequence.

7. 5, 20, 80, 320, …
8. 176, −88, 44, −22, …
9. −8, 12, −18, 27, …

Find the nth term of each geometric sequence.

10. $a_1 = 3, n = 5, r = 4$
11. $a_1 = -1, n = 6, r = 2$
12. $a_1 = 4, n = 7, r = -3$

Find the geometric means in each sequence.

13. 7, ____, 28
14. 48, ____, 3
15. −4, ____, −100

Application

16. **GEOMETRY** Consider the inscribed equilateral triangles at the right. The perimeter of each triangle is one-half of the perimeter of the next larger triangle. What is the perimeter of the smallest triangle?

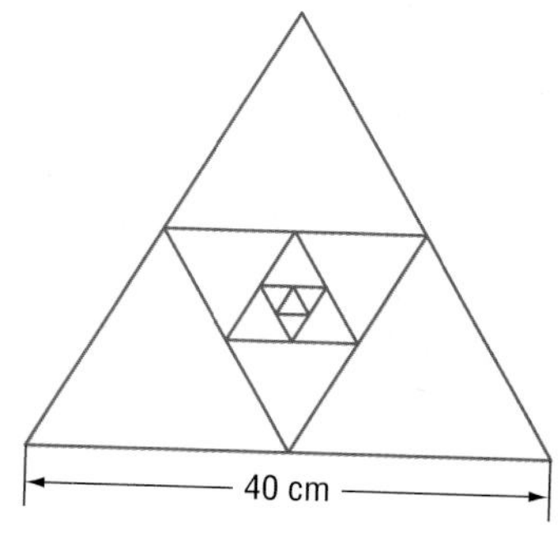

Practice and Apply

Homework Help

For Exercises	See Examples
17–24	1
25–34	2
35–42	4
43–54	5
55–62	3

Extra Practice

See page 843.

Determine whether each sequence is geometric.

17. 2, 6, 18, 54, …
18. 7, 17, 27, 37, …
19. −19, −16, −13, −10, …
20. 640, 160, 40, 10, …
21. 36, 25, 16, 9, …
22. −567, −189, −63, −21, …
23. 20, −90, 405, −1822.5, …
24. −50, 110, −242, 532.4, …

Find the next three terms in each geometric sequence.

25. 1, −4, 16, −64, …
26. −1, −6, −36, −216, …
27. 1024, 512, 256, 128, …
28. 224, 112, 56, 28, …
29. −80, 20, −5, 1.25, …
30. 10,000, −200, 4, −0.08, …
31. $\frac{1}{2}, \frac{1}{3}, \frac{2}{9}, \frac{4}{27}, \ldots$
32. $\frac{3}{4}, \frac{1}{2}, \frac{1}{3}, \frac{2}{9}, \ldots$

33. **GEOMETRY** A rectangle is 6 inches by 8 inches. The rectangle is cut in half, and one half is discarded. The remaining rectangle is cut in half, and one half is discarded. This is repeated twice. List the areas of the five rectangles formed.

34. **GEOMETRY** To bisect an angle means to cut it into two angles with the same measure. Suppose a 160° angle is bisected. Then one of the new angles is bisected. This is repeated twice. List the measures of the four sizes of angles.

Find the *n*th term of each geometric sequence.

35. $a_1 = 5, n = 7, r = 2$
36. $a_1 = 4, n = 5, r = 3$
37. $a_1 = -2, n = 4, r = -5$
38. $a_1 = 3, n = 6, r = -4$
39. $a_1 = -8, n = 3, r = 6$
40. $a_1 = -10, n = 8, r = 2$
41. $a_1 = 300, n = 10, r = 0.5$
42. $a_1 = 14, n = 6, r = 1.5$

Find the geometric means in each sequence.

43. 5, ____, 20
44. 6, ____, 54
45. −9, ____, −225
46. −5, ____, −80
47. 128, ____, 8
48. 180, ____, 5
49. −2, ____, −98
50. −6, ____, −384
51. 7, ____, 1.75
52. 3, ____, 0.75
53. $\frac{3}{5}$, ____, $\frac{3}{20}$
54. $\frac{2}{5}$, ____, $\frac{2}{45}$

55. A ball is thrown vertically. It is allowed to return to the ground and rebound without interference. If each rebound is 60% of the previous height, give the heights of the three rebounds after the initial rebound of 10 meters.

QUIZ GAMES For Exercises 56 and 57, use the following information.
Radio station WXYZ has a special game for its listeners. A trivia question is asked, and the player scores 10 points for the first correct answer. Every correct answer after that doubles the player's score.

56. List the scores after each of the first 6 correct answers.
57. Suppose the player needs to answer the question worth more than a million points to win the grand prize of a car. How many questions must be answered correctly in order to earn the car?

More About. . .

Pollution

On March 23, 1989, 250,000 barrels of oil were spilled affecting 1300 miles of Alaskan coastline. This was the largest oil spill in the United States.

Source: www.oilspill.state.ak.us

POLLUTION For Exercise 58–60, use the following information.
A lake was closed because of an accidental pesticide spill. The concentration of the pesticide after the spill was 848 parts per million. Each day the water is tested, and the amount of pesticide is found to be about 75% of what was there the day before.

58. List the level of pesticides in the water during the first week.
59. If a safe level of pesticides is considered to be 12 parts per million or less, when will the lake be considered safe?
60. Do you think the lake will ever be completely free of the pesticide? Explain.

www.algebra1.com/self_check_quiz

CRITICAL THINKING **For Exercises 61 and 62, suppose a sequence is geometric.**

61. If each term of the sequence is multiplied by the same nonzero real number, is the new sequence *always*, *sometimes*, or *never* a geometric sequence?

62. If the same nonzero number is added to each term of the sequence, is the new sequence *always*, *sometimes*, or *never* a geometric sequence?

63. **WRITING IN MATH** Answer the question that was posed at the beginning of the lesson.

How can a geometric sequence be used to describe a bungee jump?

Include the following in your answer:

- an explanation of how to determine the tenth term in the sequence, and
- the number of rebounds the first time the distance from the stopping place is less than one foot, which would trigger the end of the ride.

Standardized Test Practice

A B C D

64. Which number is next in the geometric sequence 40, 100, 250, 625, … ?

Ⓐ 900 Ⓑ 1250 Ⓒ 1562.5 Ⓓ 1875

65. **GRID IN** Find the next term in the following geometric sequence.
343, 49, 7, 1, …

Extending the Lesson

For Exercises 66–68, consider the *n*th term of the sequence $2, 1, \frac{1}{2}, \frac{1}{4}, \frac{1}{8}, \frac{1}{16}, \frac{1}{32}, \frac{1}{64}, \ldots$ **.**

66. As n approaches infinity, what value will the nth term approach?

67. In mathematics, a **limit** is a number that something approaches, but never reaches. What would you consider the limit of the values of the sequence?

68. If n approaches infinity, how is the nth term of a geometric sequence where $0 < r < 1$ different than the nth term of a geometric sequence where $r > 1$?

Maintain Your Skills

Mixed Review

69. **INVESTMENTS** Determine the value of an investment if \$1500 is invested at an interest rate of 6.5% compounded monthly for 3 years. *(Lesson 10-6)*

Determine whether the data in each table display exponential behavior. Explain why or why not. *(Lesson 10-5)*

70.

x	3	5	7	9
y	10	12	14	16

71.

x	2	5	8	11
y	0.5	1.5	4.5	13.5

Factor each trinomial, if possible. If the trinomial cannot be factored using integers, write *prime*. *(Lesson 9-4)*

72. $7a^2 + 22a + 3$ **73.** $2x^2 - 5x - 12$ **74.** $3c^2 - 3c - 5$

WebQuest **Internet Project**

Pluto Is Falling from Status as a Distant Planet

It is time to complete your project. Use the information and data you have gathered about the solar system to prepare a brochure, poster, or Web page. Be sure to include the three graphs, tables, diagrams, or calculations in the presentation.

www.algebra1.com/webquest

Algebra Activity

A Follow-Up of Lesson 10-7

Investigating Rates of Change

Collect the Data

- The Richter scale is used to measure the force of an earthquake. The table below shows the increase in magnitude for the values on the Richter scale.

Richter Number (x)	Increase in Magnitude (y)	Rate of Change (slope)
1	1	—
2	10	9
3	100	
4	1000	
5	10,000	
6	100,000	
7	1,000,000	

Source: *The New York Public Library Science Desk Reference*

- On grid paper, plot the ordered pairs (Richter number, increase in magnitude).
- Copy the table for the Richter scale and fill in the rate of change from one value to the next. For example, the rate of change for (1, 1) and (2, 10) is $\frac{10 - 1}{2 - 1}$ or 9.

Analyze the Data

1. Describe the graph you made of the Richter scale data.

2. Is the rate of change between any two points the same?

Make a Conjecture

3. Can the data be represented by a linear equation? Why or why not?

4. Describe the pattern shown in the rates of change in Column 3.

Extend the Investigation

5. Use a graphing calculator or graphing software to find a regression equation for the Richter scale data. (*Hint*: If you are using the TI-83 Plus, use **ExpReg**.)

6. Graph the following set of data that shows the amount of energy released for each Richter scale value. Describe the graph. Fill in the third column and describe the rates of change. Find a regression equation for this set of data.

Richter Number (x)	Energy Released (y)	Rate of Change (slope)
1	0.00017 metric ton	
2	0.006 metric ton	
3	0.179 metric ton	
4	5 metric tons	
5	179 metric tons	
6	5643 metric tons	
7	179,100 metric tons	

Source: *The New York Public Library Science Desk Reference*

Chapter 10 Study Guide and Review

Vocabulary and Concept Check

axis of symmetry (p. 525)	exponential growth (p. 561)	Quadratic Formula (p. 546)
common ratio (p. 567)	geometric means (p. 570)	quadratic function (p. 524)
completing the square (p. 539)	geometric sequence (p. 567)	roots (p. 533)
compound interest (p. 562)	maximum (p. 525)	symmetry (p. 525)
discriminant (p. 548)	minimum (p. 525)	vertex (p. 525)
exponential decay (p. 562)	parabola (p. 524)	zeros (p. 533)
exponential function (p. 554)	quadratic equation (p. 533)	

Choose the letter of the term that best matches each equation or phrase.

1. $y = C(1 + r)^t$
2. $f(x) = ax^2 + bx + c$
3. a geometric property of parabolas
4. $x = -\frac{b}{2a}$
5. $y = a^x$
6. maximum or minimum point of a parabola
7. $y = C(1 - r)^t$
8. solutions of a quadratic equation
9. $x = \frac{-b \pm \sqrt{b^2 - 4ac}}{2a}$
10. the graph of a quadratic function

a. equation of axis of symmetry
b. exponential decay equation
c. exponential function
d. exponential growth equation
e. parabola
f. Quadratic Formula
g. quadratic function
h. roots
i. symmetry
j. vertex

Lesson-by-Lesson Review

10-1 Graphing Quadratic Functions

See pages 524–530.

Concept Summary

- The standard form of a quadratic function is $y = ax^2 + bx + c$.
- Complete a table of values to graph a quadratic function.
- The equation of the axis of symmetry for the graph of $y = ax^2 + bx + c$, where $a \neq 0$, is $x = -\frac{b}{2a}$.
- The vertex of a parabola is on the axis of symmetry.

Example **Consider the graph of $y = x^2 - 8x + 12$.**

a. Write the equation of the axis of symmetry.

In the equation $y = x^2 - 8x + 12$, $a = 1$ and $b = -8$. Substitute these values into the equation of the axis of symmetry.

$x = -\frac{b}{2a}$ Equation of the axis of symmetry

$= -\frac{-8}{2(1)}$ or 4 $a = 1$ and $b = -8$

The equation of the axis of symmetry is $x = 4$.

www.algebra1.com/vocabulary_review

b. Find the coordinates of the vertex of the graph.

The x-coordinate of the vertex is 4.

$y = x^2 - 8x + 12$	Original equation
$y = (4)^2 - 8(4) + 12$	$x = 4$
$y = 16 - 32 + 12$	Simplify.
$y = -4$	The coordinates of the vertex are $(4, -4)$.

Exercises **Write the equation of the axis of symmetry, and find the coordinates of the vertex of the graph of each function. Identify the vertex as a maximum or minimum. Then graph the function.** *See Example 3 on pages 526 and 527.*

11. $y = x^2 + 2x$ **12.** $y = -3x^2 + 4$ **13.** $y = x^2 - 3x - 4$

14. $y = 3x^2 + 6x - 17$ **15.** $y = -2x^2 + 1$ **16.** $y = -x^2 - 3x$

10-2 Solving Quadratic Equations by Graphing

See pages 533–538.

Concept Summary

- The roots of a quadratic equation are the x-intercepts of the related quadratic function.

Example **Solve $x^2 - 3x - 4 = 0$ by graphing.**

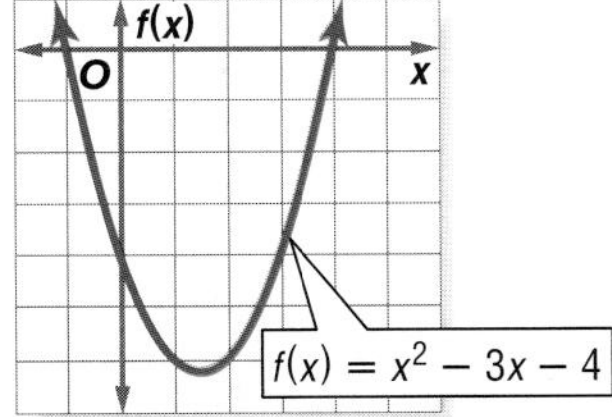

Graph the related function $f(x) = x^2 - 3x - 4$.

The x-intercepts are -1 and 4. Therefore, the solutions are -1, and 4.

Exercises **Solve each equation by graphing. If integral roots cannot be found, estimate the roots by stating the consecutive integers between which the roots lie.** *See Examples 1–4 on pages 533–535.*

17. $x^2 - x - 12 = 0$ **18.** $x^2 + 6x + 9 = 0$ **19.** $x^2 + 4x - 3 = 0$

20. $2x^2 - 5x + 4 = 0$ **21.** $x^2 - 10x = -21$ **22.** $6x^2 - 13x = 15$

10-3 Solving Quadratic Equations by Completing the Square

See pages 539–544.

Concept Summary

- Complete the square to make a quadratic expression a perfect square.
- Use the following steps to complete the square of $x^2 + bx$.

Step 1 Find $\frac{1}{2}$ of b, the coefficient of x.

Step 2 Square the result of Step 1.

Step 3 Add the result of Step 2 to $x^2 + bx$, the original expression.

Example **Solve $y^2 + 6y + 2 = 0$ by completing the square. Round to the nearest tenth if necessary.**

$y^2 + 6y + 2 = 0$	Original equation
$y^2 + 6y + 2 - 2 = 0 - 2$	Subtract 2 from each side.
$y^2 + 6y = -2$	Simplify.
$y^2 + 6y + 9 = -2 + 9$	Since $\left(\frac{6}{2}\right)^2 = 9$, add 9 to each side.
$(y + 3)^2 = 7$	Factor $y^2 + 6y + 9$.
$y + 3 = \pm\sqrt{7}$	Take the square root of each side.
$y + 3 - 3 = \pm\sqrt{7} - 3$	Subtract 3 from each side.
$y = -3 \pm \sqrt{7}$	Simplify.

Use a calculator to evaluate each value of y.

$y = -3 + \sqrt{7}$ or $y = -3 - \sqrt{7}$

$y \approx -0.4$ $\quad y \approx -5.6$

The solution set is $\{-5.6, -0.4\}$.

Exercises **Solve each equation by completing the square. Round to the nearest tenth if necessary.** *See Example 3 on pages 540 and 541.*

23. $-3x^2 + 4 = 0$ **24.** $x^2 - 16x + 32 = 0$ **25.** $m^2 - 7m = 5$

26. $4a^2 + 16a + 15 = 0$ **27.** $\frac{1}{2}y^2 + 2y - 1 = 0$ **28.** $n^2 - 3n + \frac{5}{4} = 0$

10-4 Solving Quadratic Equations by Using the Quadratic Formula

See pages 546–552.

Concept Summary

- The solutions of a quadratic equation in the form $ax^2 + bx + c = 0$, where $a \neq 0$, are given by the Quadratic Formula, $x = \frac{-b \pm \sqrt{b^2 - 4ac}}{2a}$.

Example **Solve $2x^2 + 7x - 15 = 0$ by using the Quadratic Formula.**

For this equation, $a = 2$, $b = 7$, and $c = -15$.

$x = \frac{-b \pm \sqrt{b^2 - 4ac}}{2a}$	Quadratic Formula
$x = \frac{-7 \pm \sqrt{7^2 - 4(2)(-15)}}{2(2)}$	$a = 2$, $b = 7$, and $c = -15$
$x = \frac{-7 \pm \sqrt{169}}{4}$	Simplify.

$x = \frac{-7 + 13}{4}$ or $x = \frac{-7 - 13}{4}$

$x = 1\frac{1}{2}$ $\quad x = -5$ $\quad$ The solution set is $\left\{-5, 1\frac{1}{2}\right\}$.

Exercises **Solve each equation by using the Quadratic Formula. Round to the nearest tenth if necessary.** *See Examples 1 and 2 on pages 546 and 547.*

29. $x^2 - 8x = 20$ **30.** $r^2 + 10r + 9 = 0$ **31.** $4p^2 + 4p = 15$

32. $2y^2 + 3 = -8y$ **33.** $2d^2 + 8d + 3 = 3$ **34.** $21a^2 + 5a - 7 = 0$

10-5 Exponential Functions

See pages 554–560.

Concept Summary

- An exponential function is a function that can be described by the equation of the form $y = a^x$, where $a > 0$ and $a \neq 1$.

Example **Graph $y = 2^x - 3$. State the y-intercept.**

x	y
−3	−2.875
−2	−2.75
−1	−2.5
0	−2
1	−1
2	1
3	5

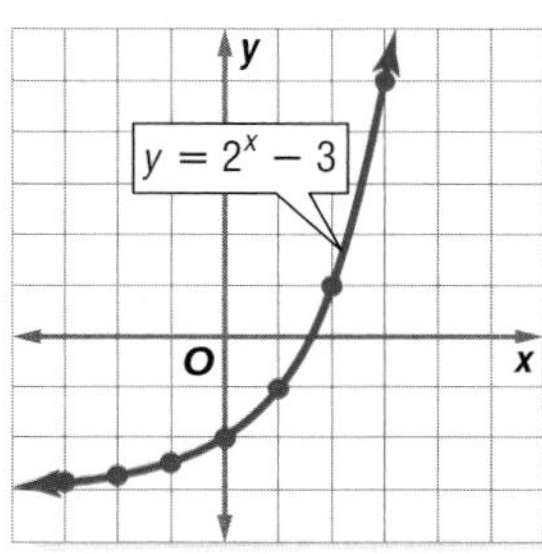

Graph the ordered pairs and connect the points with a smooth curve. The y-intercept is −2.

Exercises **Graph each function. State the y-intercept.**
See Examples 1 and 2 on page 555.

35. $y = 3^x + 6$ **36.** $y = 3^{x+2}$ **37.** $y = 2\left(\frac{1}{2}\right)^x$

10-6 Growth and Decay

See pages 561–565.

Concept Summary

- Exponential Growth: $y = C(1 + r)^t$, where y represents the final amount, C represents the initial amount, r represents the rate of change expressed as a decimal, and t represents time.
- Compound Interest: $A = P\left(1 + \frac{r}{n}\right)^{nt}$, where A represents the amount of the investment, P represents the principal, r represents the annual rate of interest expressed as a decimal, n represents the number of times that the interest is compounded each year, and t represents the number of years that the money is invested.
- Exponential Decay: $y = C(1 - r)^t$, where y represents the final amount, C represents the initial amount, r represents the rate of decay expressed as a decimal, and t represents time.

• Extra Practice, see pages 841–843.
• Mixed Problem Solving, see page 862.

Example **Find the final amount of an investment if $1500 is invested at an interest rate of 7.5% compounded quarterly for 10 years.**

$A = P\left(1 + \frac{r}{n}\right)^{nt}$ Compound interest equation

$A = 1500\left(1 + \frac{0.075}{4}\right)^{4 \cdot 10}$ $P = 1500$, $r = 7.5\%$ or 0.075, $n = 4$, and $t = 10$

$A \approx 3153.52$ Simplify.

The final amount in the account is about $3153.52.

Exercises **Determine the final amount for each investment.**
See Example 2 on page 562.

	Principal	Annual Interest Rate	Time	Type of Compounding
38.	$2000	8%	8 years	quarterly
39.	$5500	5.25%	15 years	monthly
40.	$15,000	7.5%	25 years	monthly
41.	$500	9.75%	40 years	daily

10-7 Geometric Sequences

See pages 567–572.

Concept Summary

- A geometric sequence is a sequence in which each term after the nonzero first term is found by multiplying the previous term by a constant called the common ratio r, where $r \neq 0$ or 1.
- The nth term a_n of a geometric sequence with the first term a_1, and a common ratio r is given by $a_n = a_1 \cdot r^{n-1}$.

Example **Find the next three terms in the geometric sequence 7.5, 15, 30, … .**

$\frac{15}{7.5} = 2$ Divide the second term by the first.

The common ratio is 2. Find the next three terms.

7.5, 15, 30, 60, 120, 240
×2 ×2 ×2

The next three terms are 60, 120, and 240.

Exercises **Find the nth term of each geometric sequence.**
See Example 4 on page 569.

42. $a_1 = 2, n = 5, r = 2$ **43.** $a_1 = 7, n = 4, r = \frac{2}{3}$ **44.** $a_1 = 243, n = 5, r = -\frac{1}{3}$

Find the geometric means in each sequence. *See Example 5 on page 570.*

45. 5, ____, 20 **46.** −12, ____, −48 **47.** 1, ____, $\frac{1}{4}$

Chapter 10

Practice Test

Vocabulary and Concepts

Choose the letter of the term that matches each formula.

1. $x = \frac{-b \pm \sqrt{b^2 - 4ac}}{2a}$
2. $y = C(1 + r)^t$
3. $y = C(1 - r)^t$

- a. exponential decay equation
- b. exponential growth equation
- c. Quadratic Formula

Skills and Applications

Write the equation of the axis of symmetry, and find the coordinates of the vertex of the graph of each function. Identify the vertex as a maximum or minimum. Then graph the function.

4. $y = x^2 - 4x + 13$
5. $y = -3x^2 - 6x + 4$
6. $y = 2x^2 + 3$
7. $y = -1(x - 2)^2 + 1$

Solve each equation by graphing. If integral roots cannot be found, estimate the roots by stating the consecutive integers between which the roots lie.

8. $x^2 - 2x + 2 = 0$
9. $x^2 + 6x = -7$
10. $x^2 + 24x + 144 = 0$
11. $2x^2 - 8x = 42$

Solve each equation. Round to the nearest tenth if necessary.

12. $x^2 + 7x + 6 = 0$
13. $2x^2 - 5x - 12 = 0$
14. $6n^2 + 7n = 20$
15. $3k^2 + 2k = 5$
16. $y^2 - \frac{3}{5}y + \frac{2}{25} = 0$
17. $-3x^2 + 5 = 14x$
18. $z^2 - 13z = 32$
19. $3x^2 + 4x = 8$
20. $7m^2 = m + 5$

Graph each function. State the y-intercept.

21. $y = \left(\frac{1}{2}\right)^x$
22. $y = 4 \cdot 2^x$
23. $y = \left(\frac{1}{3}\right)^x - 3$

Find the nth term of each geometric sequence.

24. $a_1 = 12, n = 6, r = 2$
25. $a_1 = 20, n = 4, r = 3$

Find the geometric means in each sequence.

26. 7, _____, 63
27. $-\frac{1}{3}$, _____, -12

28. **CARS** Ley needs to replace her car. If she leases a car, she will pay $410 a month for 2 years and then has the option to buy the car for $14,458. The current price of the car is $17,369. If the car depreciates at 16% per year, how will the depreciated price compare with the buyout price of the lease?

29. **FINANCE** Find the total amount after $1500 is invested for 10 years at a rate of 6%, compounded quarterly.

30. **STANDARDIZED TEST PRACTICE** Which value is the next value in the pattern $-4, 12, -36, 108, \ldots$?

    (A) -324 (B) 324 (C) -432 (D) 432

www.algebra1.com/chapter_test

Chapter 10 Standardized Test Practice

Part 1 Multiple Choice

Record your answers on the answer sheet provided by your teacher or on a sheet of paper.

1. The graph of $y = 3x$ is shown. If the line is translated 2 units down, which equation will describe the new line? (Lesson 4-2)

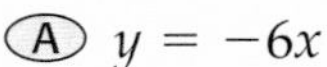

Ⓐ $y = -6x$

Ⓑ $y = 3x - 2$

Ⓒ $y = 3x + 2$

Ⓓ $y = 3(x - 2)$

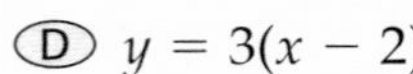

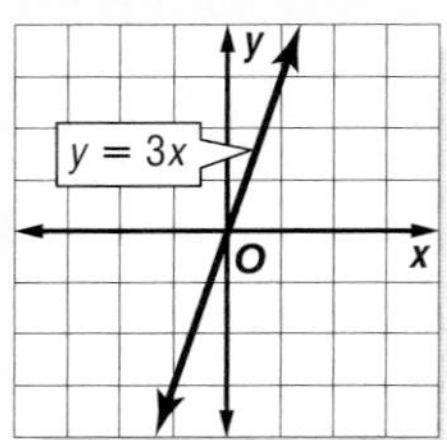

2. Suppose a varies directly as b, and $a = 21$ when $b = 6$. Find a when $b = 28$. (Lesson 5-2)

Ⓐ 4.5 Ⓑ 8

Ⓒ 98 Ⓓ 126

3. Which equation is represented by the graph? (Lesson 5-5)

Ⓐ $y = -2x - 10$

Ⓑ $y = -2x - 5$

Ⓒ $y = 2x + 10$

Ⓓ $y = 2x - 5$

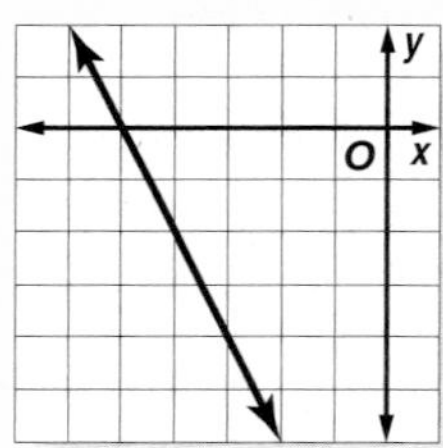

4. At a farm market, apples cost 20¢ each and grapefruit cost 25¢ each. A shopper bought twice as many apples as grapefruit and spent a total of $1.95. How many apples did he buy? (Lesson 7-2)

Ⓐ 3 Ⓑ 4 Ⓒ 5 Ⓓ 6

5. A rectangle has a length of $2x + 3$ and a width of $2x - 6$. Which expression describes the area of the rectangle? (Lesson 8-7)

Ⓐ $4x - 3$ Ⓑ $4x^2 - 18$

Ⓒ $4x^2 - 6x - 18$ Ⓓ $4x^2 + 18x - 18$

6. The solution set for the equation $x^2 + x - 12 = 0$ is (Lesson 9-3)

Ⓐ $\{-4, -3\}$. Ⓑ $\{-4, 3\}$.

Ⓒ $\{4, -3\}$. Ⓓ $\{4, 3\}$.

7. Which equation best represents the data in the table? (Lesson 10-1)

x	y
−3	0
−1	8
0	9
2	5
3	0
4	−7

Ⓐ $y = -x^2 + 3$ Ⓑ $y = -x^2 + 9$

Ⓒ $y = x^2 - 3$ Ⓓ $y = x^2 + 9$

8. Which equation best represents the parabola graphed below? (Lesson 10-1)

Ⓐ $y = x^2 - 2x - 4$

Ⓑ $y = x^2 - 2x - 3$

Ⓒ $y = x^2 + 2x - 3$

Ⓓ $y = x^2 + 2x + 3$

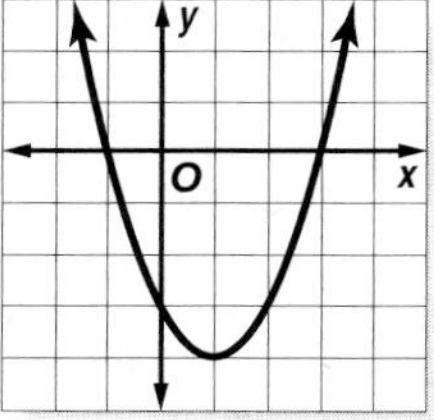

9. At which points does the graph of $f(x) = 2x^2 + 8x + 6$ intersect the x-axis? (Lesson 10-2)

Ⓐ (−3, 0) and (−2, 0)

Ⓑ (−3, 0) and (−1, 0)

Ⓒ (1, 0) and (3, 0)

Ⓓ (2, 0) and (3, 0)

The Princeton Review Test-Taking Tip

Questions 1 and 7 Sketching the graph of a function or a transformation may help you see which answer choice is correct.

Part 2 Short Response/Grid In

Record your answers on the answer sheet provided by your teacher or on a sheet of paper.

10. Monica earned \$18.50, \$23.00, and \$15.00 mowing lawns for 3 consecutive weeks. She wanted to earn an average of at least \$18 per week. What is the minimum she should earn during the 4th week to meet her goal? (Lesson 3-4)

11. Write an equation in slope-intercept form of the line that is perpendicular to the line represented by $8x - 4y + 9 = 0$ and passes through the point at (2, 3). (Lesson 5-6)

12. If $5a + 4b = 25$ and $3a - 8b = 41$, solve for a and b. (Lesson 7-4)

13. Complete the square of $x^2 + 4x - 5$ by finding numbers h and k such that $x^2 + 4x - 5 = (x + h)^2 + k$. (Lesson 10-2)

14. At how many points does the graph of $y = 6x^2 + 11x + 4$ intersect the x-axis? (Lesson 10-3)

15. The length and width of a rectangle that measures 8 inches by 6 inches are both increased by the same amount. The area of the larger rectangle is twice the area of the original rectangle. How much was added to each dimension of the original rectangle? Round to the nearest hundredth of an inch. (Lesson 10-4)

Part 3 Quantitative Comparison

Compare the quantity in Column A and the quantity in Column B. Then determine whether:

Ⓐ **the quantity in Column A is greater,**

Ⓑ **the quantity in Column B is greater,**

Ⓒ **the two quantities are equal, or**

Ⓓ **the relationship cannot be determined from the information given.**

	Column A	Column B

16.

5	10	15	20	25	30	35	40	45	50	55	60
×××		××	×	×××			××××××	×××××	××××	××	×××

Column A	Column B
the mean of the data in the line plot	the median of the data in the line plot

(Lesson 2-5)

17.

Column A	Column B
the solution of $-6p = -12$	the solution of $10q = 5$

(Lesson 3-3)

18.

Column A	Column B
5.3×10^3	53,000

(Lesson 8-3)

19.

Column A	Column B
the 14th term of $-2, -4, -8, \ldots$	the 14th term of $2, -4, 8, \ldots$

(Lesson 10-7)

Part 4 Open Ended

Record your answers on a sheet of paper. Show your work.

20. Analyze the graph of $y = -4x^2 + 8x - \frac{15}{4}$. (Lessons 10-1, 10-3)

a. Show that the equation $-4x^2 + 8x - \frac{15}{4} = -4(x - 1)^2 + \frac{1}{4}$ is always true by expanding the right side.

b. Find the equation of the axis of symmetry of the graph of $y = -4x^2 + 8x - \frac{15}{4}$.

c. Does the parabola open upward or downward? Explain how you determined this.

d. Find the values of x, if any, where the graph crosses the x-axis. Write as rational numbers.

e. Find the coordinates of the maximum or minimum point on this parabola.

f. Sketch the graph of the equation. Label the maximum or minimum point and the roots.

www.algebra1.com/standardized_test

UNIT 4

Radical and Rational Functions

Nonlinear functions such as radical and rational functions can be used to model real-world situations such as the speed of a roller coaster. In this unit, you will learn about radical and rational functions.

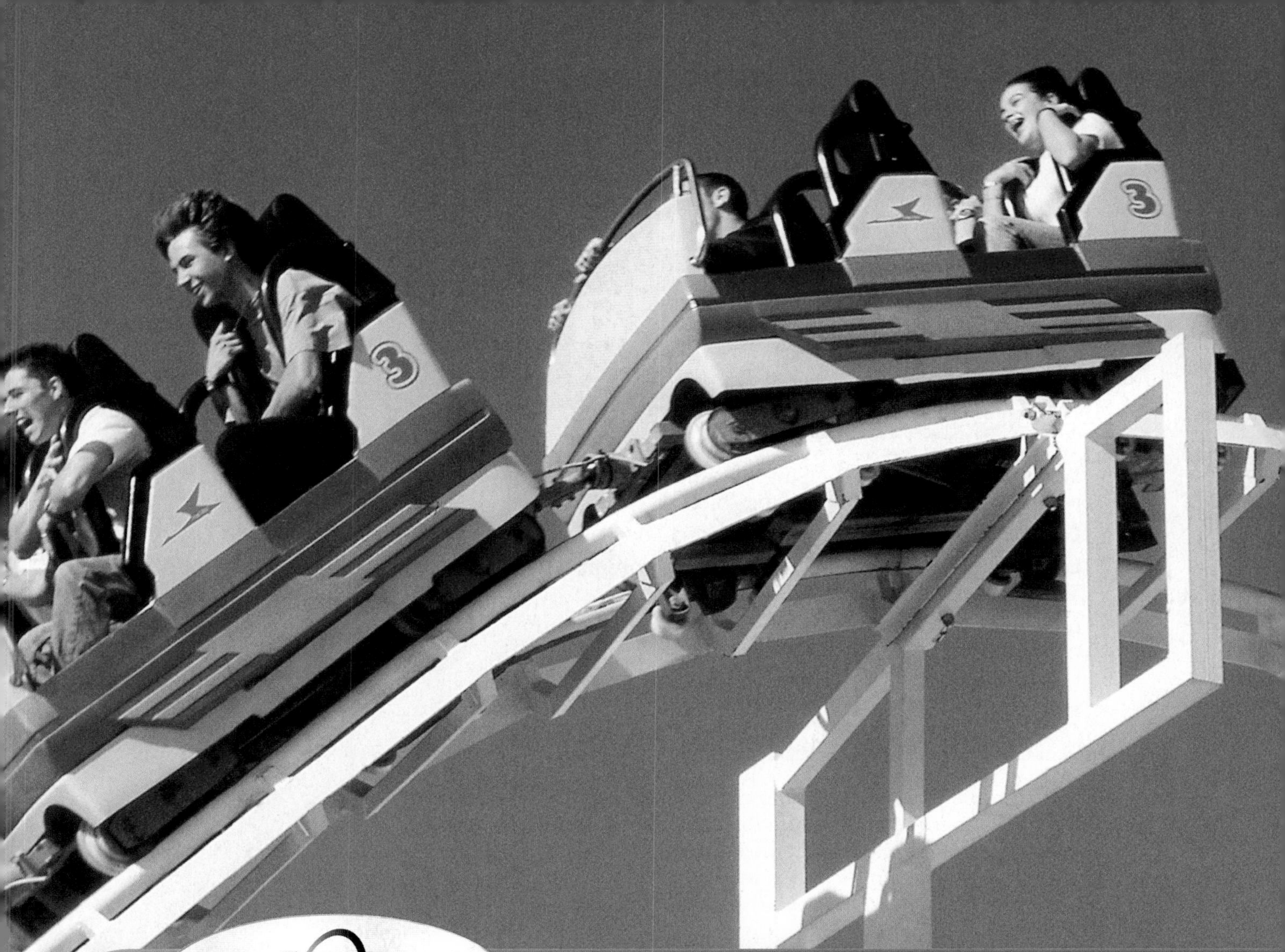

WebQuest Internet Project

Building the Best Roller Coaster

Each year, amusement park owners compete to earn part of the billions of dollars Americans spend at amusement parks. Often the parks draw customers with new taller and faster roller coasters. In this project, you will explore how radical and rational functions are related to buying and building a new roller coaster.

USA TODAY
Education

Log on to www.algebra1.com/webquest. Begin your WebQuest by reading the Task.

Then continue working on your WebQuest as you study Unit 4.

Lesson	11-1	12-2
Page	590	652

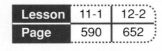

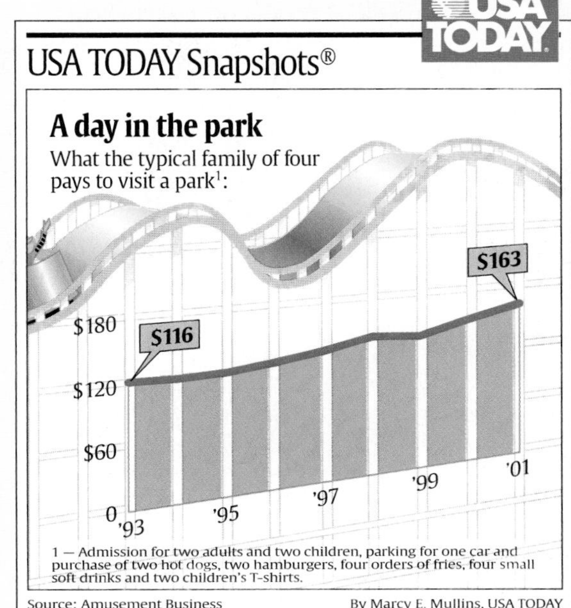
USA TODAY
USA TODAY Snapshots®
A day in the park
What the typical family of four pays to visit a park[1]:
$163
$180
$116
$120
$60
0
'93
'95
'97
'99
'01
1 — Admission for two adults and two children, parking for one car and purchase of two hot dogs, two hamburgers, four orders of fries, four small soft drinks and two children's T-shirts.
Source: Amusement Business
By Marcy E. Mullins, USA TODAY

Chapter 11

Radical Expressions and Triangles

What You'll Learn

- **Lessons 11-1 and 11-2** Simplify and perform operations with radical expressions.
- **Lesson 11-3** Solve radical equations.
- **Lessons 11-4 and 11-5** Use the Pythagorean Theorem and Distance Formula.
- **Lessons 11-6 and 11-7** Use similar triangles and trigonometric ratios.

Key Vocabulary

- radical expression (p. 586)
- radical equation (p. 598)
- Pythagorean Theorem (p. 605)
- Distance Formula (p. 611)
- trigonometric ratios (p. 623)

Why It's Important

Physics problems are among the many applications of radical equations. Formulas that contain the value for the acceleration due to gravity, such as free-fall times, escape velocities, and the speeds of roller coasters, can all be written as radical equations. *You will learn how to calculate the time it takes for a skydiver to fall a given distance in Lesson 11-3.*

Getting Started

Prerequisite Skills To be successful in this chapter, you'll need to master these skills and be able to apply them in problem-solving situations. Review these skills before beginning Chapter 11.

For Lessons 11-1 and 11-4 — **Find Square Roots**

Find each square root. If necessary, round to the nearest hundredth.
(For review, see Lesson 2-7.)

1. $\sqrt{25}$ **2.** $\sqrt{80}$ **3.** $\sqrt{56}$ **4.** $\sqrt{324}$

For Lesson 11-2 — **Combine Like Terms**

Simplify each expression. *(For review, see Lesson 1-6.)*

5. $3a + 7b - 2a$
6. $14x - 6y + 2y$
7. $(10c - 5d) + (6c + 5d)$
8. $(21m + 15n) - (9n - 4m)$

For Lesson 11-3 — **Solve Quadratic Equations**

Solve each equation. *(For review, see Lesson 9-3.)*

9. $x(x - 5) = 0$
10. $x^2 + 10x + 24 = 0$
11. $x^2 - 6x - 27 = 0$
12. $2x^2 + x + 1 = 2$

For Lesson 11-6 — **Proportions**

Use cross products to determine whether each pair of ratios forms a proportion. Write *yes* or *no*. *(For review, see Lesson 3-6.)*

13. $\frac{2}{3}, \frac{8}{12}$ **14.** $\frac{4}{5}, \frac{16}{25}$ **15.** $\frac{8}{10}, \frac{12}{16}$ **16.** $\frac{6}{30}, \frac{3}{15}$

FOLDABLES™
Study Organizer

Make this Foldable to help you organize information about radical expressions and equations. Begin with a sheet of plain $8\frac{1}{2}$" by 11" paper.

Step 1 Fold in Half

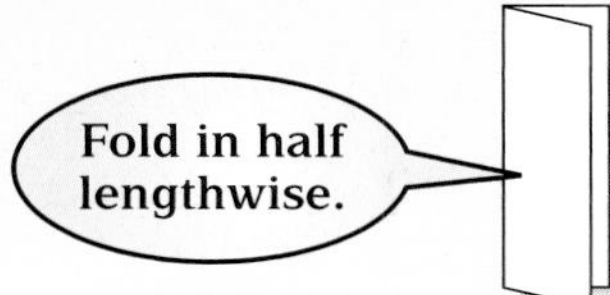
Fold in half lengthwise.

Step 2 Fold Again

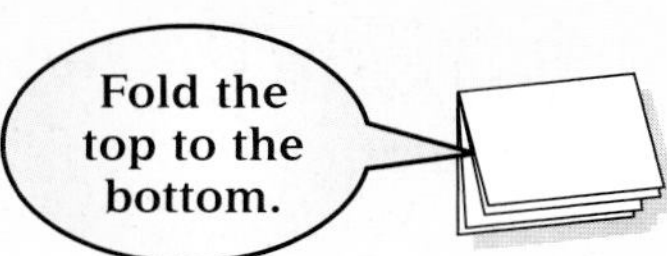
Fold the top to the bottom.

Step 3 Cut

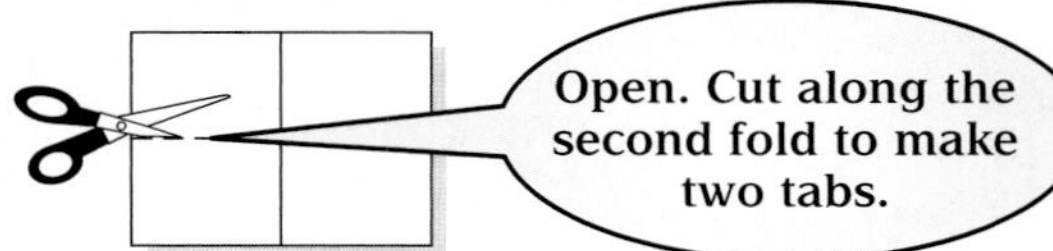
Open. Cut along the second fold to make two tabs.

Step 4 Label

Label each tab as shown.
Radical Expressions
Radical Equations

Reading and Writing As you read and study the chapter, write notes and examples for each lesson under each tab.

11-1

Simplifying Radical Expressions

What You'll Learn

- Simplify radical expressions using the Product Property of Square Roots.
- Simplify radical expressions using the Quotient Property of Square Roots.

Vocabulary

- radical expression
- radicand
- rationalizing the denominator
- conjugate

How are radical expressions used in space exploration?

A spacecraft leaving Earth must have a velocity of at least 11.2 kilometers per second (25,000 miles per hour) to enter into orbit. This velocity is called the *escape velocity*. The escape velocity of an object is given by the radical expression $\sqrt{\frac{2GM}{R}}$, where G is the gravitational constant, M is the mass of the planet or star, and R is the radius of the planet or star. Once values are substituted for the variables, the formula can be simplified.

PRODUCT PROPERTY OF SQUARE ROOTS A **radical expression** is an expression that contains a square root. A **radicand**, the expression under the radical sign, is in simplest form if it contains no perfect square factors other than 1. The following property can be used to simplify square roots.

Key Concept — Product Property of Square Roots

- **Words** For any numbers a and b, where $a \geq 0$ and $b \geq 0$, the square root of the product ab is equal to the product of each square root.
- **Symbols** $\sqrt{ab} = \sqrt{a} \cdot \sqrt{b}$
- **Example** $\sqrt{4 \cdot 25} = \sqrt{4} \cdot \sqrt{25}$

The Product Property of Square Roots and prime factorization can be used to simplify radical expressions in which the radicand is not a perfect square.

Example 1 Simplify Square Roots

Simplify.

a. $\sqrt{12}$

$\sqrt{12} = \sqrt{2 \cdot 2 \cdot 3}$ Prime factorization of 12

$= \sqrt{2^2} \cdot \sqrt{3}$ Product Property of Square Roots

$= 2\sqrt{3}$ Simplify.

b. $\sqrt{90}$

$\sqrt{90} = \sqrt{2 \cdot 3 \cdot 3 \cdot 5}$ Prime factorization of 90

$= \sqrt{3^2} \cdot \sqrt{2 \cdot 5}$ Product Property of Square Roots

$= 3\sqrt{10}$ Simplify.

Study Tip

Reading Math

$2\sqrt{3}$ is read *two times the square root of 3* or *two radical three.*

The Product Property can also be used to multiply square roots.

Study Tip

Alternative Method
To find $\sqrt{3} \cdot \sqrt{15}$, you could multiply first and then use the prime factorization.

$\sqrt{3} \cdot \sqrt{15} = \sqrt{45}$
$= \sqrt{3^2} \cdot \sqrt{5}$
$= 3\sqrt{5}$

Example 2 Multiply Square Roots

Find $\sqrt{3} \cdot \sqrt{15}$.

$\sqrt{3} \cdot \sqrt{15} = \sqrt{3} \cdot \sqrt{3} \cdot \sqrt{5}$ Product Property of Square Roots

$= \sqrt{3^2} \cdot \sqrt{5}$ Product Property

$= 3\sqrt{5}$ Simplify.

When finding the principal square root of an expression containing variables, be sure that the result is not negative. Consider the expression $\sqrt{x^2}$. It may seem that $\sqrt{x^2} = x$. Let's look at $x = -2$.

$\sqrt{x^2} \stackrel{?}{=} x$

$\sqrt{(-2)^2} \stackrel{?}{=} -2$ Replace x with -2.

$\sqrt{4} \stackrel{?}{=} -2$ $(-2)^2 = 4$

$2 \neq -2$ $\sqrt{4} = 2$

For radical expressions where the exponent of the variable inside the radical is *even* and the resulting simplified exponent is odd, you must use absolute value to ensure nonnegative results.

$\sqrt{x^2} = |x|$ $\sqrt{x^3} = x\sqrt{x}$ $\sqrt{x^4} = x^2$ $\sqrt{x^5} = x^2\sqrt{x}$ $\sqrt{x^6} = |x^3|$

Example 3 Simplify a Square Root with Variables

Simplify $\sqrt{40x^4y^5z^3}$.

$\sqrt{40x^4y^5z^3} = \sqrt{2^3 \cdot 5 \cdot x^4 \cdot y^5 \cdot z^3}$ Prime factorization

$= \sqrt{2^2} \cdot \sqrt{2} \cdot \sqrt{5} \cdot \sqrt{x^4} \cdot \sqrt{y^4} \cdot \sqrt{y} \cdot \sqrt{z^2} \cdot \sqrt{z}$ Product Property

$= 2 \cdot \sqrt{2} \cdot \sqrt{5} \cdot x^2 \cdot y^2 \cdot \sqrt{y} \cdot |z| \cdot \sqrt{z}$ Simplify.

$= 2x^2y^2|z|\sqrt{10yz}$ The absolute value of z ensures a nonnegative result.

QUOTIENT PROPERTY OF SQUARE ROOTS You can divide square roots and simplify radical expressions that involve division by using the Quotient Property of Square Roots.

Key Concept — Quotient Property of Square Roots

- **Words** For any numbers a and b, where $a \geq 0$ and $b > 0$, the square root of the quotient $\frac{a}{b}$ is equal to the quotient of each square root.
- **Symbols** $\sqrt{\frac{a}{b}} = \frac{\sqrt{a}}{\sqrt{b}}$
- **Example** $\sqrt{\frac{49}{4}} = \frac{\sqrt{49}}{\sqrt{4}}$

Study Tip

Look Back
To review the **Quadratic Formula**, see Lesson 10-4.

You can use the Quotient Property of Square Roots to derive the Quadratic Formula by solving the quadratic equation $ax^2 + bx + c = 0$.

$ax^2 + bx + c = 0$ Original equation

$x^2 + \frac{b}{a}x + \frac{c}{a} = 0$ Divide each side by a, $a \neq 0$.

(continued on the next page)

$x^2 + \frac{b}{a}x = -\frac{c}{a}$	Subtract $\frac{c}{a}$ from each side.
$x^2 + \frac{b}{a}x + \frac{b^2}{4a^2} = -\frac{c}{a} + \frac{b^2}{4a^2}$	Complete the square; $\left(\frac{b}{2a}\right)^2 = \frac{b^2}{4a^2}$.
$\left(x + \frac{b}{2a}\right)^2 = \frac{-4ac + b^2}{4a^2}$	Factor $x^2 + \frac{b}{a}x + \frac{b^2}{4a^2}$.
$\left\lvert x + \frac{b}{2a} \right\rvert = \sqrt{\frac{b^2 - 4ac}{4a^2}}$	Take the square root of each side.
$x + \frac{b}{2a} = \pm\sqrt{\frac{b^2 - 4ac}{4a^2}}$	Remove the absolute value symbols and insert $\pm$.
$x + \frac{b}{2a} = \pm\frac{\sqrt{b^2 - 4ac}}{\sqrt{4a^2}}$	Quotient Property of Square Roots
$x + \frac{b}{2a} = \pm\frac{\sqrt{b^2 - 4ac}}{2a}$	$\sqrt{4a^2} = 2a$
$x = \frac{-b \pm \sqrt{b^2 - 4ac}}{2a}$	Subtract $\frac{b}{2a}$ from each side.

Study Tip

Plus or Minus Symbol

The $\pm$ symbol is used with the radical expression since both square roots lead to solutions.

Thus, we have derived the Quadratic Formula.

A fraction containing radicals is in simplest form if no prime factors appear under the radical sign with an exponent greater than 1 and if no radicals are left in the denominator. **Rationalizing the denominator** of a radical expression is a method used to eliminate radicals from the denominator of a fraction.

Example 4 Rationalizing the Denominator

Simplify.

a. $\frac{\sqrt{10}}{\sqrt{3}}$

$\frac{\sqrt{10}}{\sqrt{3}} = \frac{\sqrt{10}}{\sqrt{3}} \cdot \frac{\sqrt{3}}{\sqrt{3}}$ Multiply by $\frac{\sqrt{3}}{\sqrt{3}}$.

$= \frac{\sqrt{30}}{3}$ Product Property of Square Roots

b. $\frac{\sqrt{7x}}{\sqrt{8}}$

$\frac{\sqrt{7x}}{\sqrt{8}} = \frac{\sqrt{7x}}{\sqrt{2 \cdot 2 \cdot 2}}$ Prime factorization

$= \frac{\sqrt{7x}}{2\sqrt{2}} \cdot \frac{\sqrt{2}}{\sqrt{2}}$ Multiply by $\frac{\sqrt{2}}{\sqrt{2}}$.

$= \frac{\sqrt{14x}}{4}$ Product Property of Square Roots

c. $\frac{\sqrt{2}}{\sqrt{6}}$

$\frac{\sqrt{2}}{\sqrt{6}} = \frac{\sqrt{2}}{\sqrt{6}} \cdot \frac{\sqrt{6}}{\sqrt{6}}$ Multiply by $\frac{\sqrt{6}}{\sqrt{6}}$.

$= \frac{\sqrt{12}}{6}$ Product Property of Square Roots

$= \frac{\sqrt{2 \cdot 2 \cdot 3}}{6}$ Prime factorization

$= \frac{2\sqrt{3}}{6}$ $\sqrt{2^2} = 2$

$= \frac{\sqrt{3}}{3}$ Divide the numerator and denominator by 2.

Binomials of the form $p\sqrt{q} + r\sqrt{s}$ and $p\sqrt{q} - r\sqrt{s}$ are called **conjugates**. For example, $3 + \sqrt{2}$ and $3 - \sqrt{2}$ are conjugates. Conjugates are useful when simplifying radical expressions because if p, q, r, and s are rational numbers, their product is always a rational number with no radicals. Use the pattern $(a - b)(a + b) = a^2 - b^2$ to find their product.

$$(3 + \sqrt{2})(3 - \sqrt{2}) = 3^2 - (\sqrt{2})^2 \quad a = 3, b = \sqrt{2}$$
$$= 9 - 2 \text{ or } 7 \quad (\sqrt{2})^2 = \sqrt{2} \cdot \sqrt{2} \text{ or } 2$$

Example 5 Use Conjugates to Rationalize a Denominator

Simplify $\frac{2}{6 - \sqrt{3}}$.

$$\frac{2}{6 - \sqrt{3}} = \frac{2}{6 - \sqrt{3}} \cdot \frac{6 + \sqrt{3}}{6 + \sqrt{3}} \quad \frac{6 + \sqrt{3}}{6 + \sqrt{3}} = 1$$

$$= \frac{2(6 + \sqrt{3})}{6^2 - (\sqrt{3})^2} \quad (a - b)(a + b) = a^2 - b^2$$

$$= \frac{12 + 2\sqrt{3}}{36 - 3} \quad (\sqrt{3})^2 = 3$$

$$= \frac{12 + 2\sqrt{3}}{33} \quad \text{Simplify.}$$

When simplifying radical expressions, check the following conditions to determine if the expression is in simplest form.

Concept Summary — Simplest Radical Form

A radical expression is in simplest form when the following three conditions have been met.

1. No radicands have perfect square factors other than 1.
2. No radicands contain fractions.
3. No radicals appear in the denominator of a fraction.

Check for Understanding

Concept Check

1. **Explain** why absolute value is not necessary for $\sqrt{x^4} = x^2$.
2. **Show** that $\frac{1}{\sqrt{a}} = \frac{\sqrt{a}}{a}$ for $a > 0$.
3. **OPEN ENDED** Give an example of a binomial in the form $a\sqrt{b} + c\sqrt{d}$ and its conjugate. Then find their product.

Guided Practice

Simplify.

4. $\sqrt{20}$
5. $\sqrt{2} \cdot \sqrt{8}$
6. $3\sqrt{10} \cdot 4\sqrt{10}$
7. $\sqrt{54a^2b^2}$
8. $\sqrt{60x^5y^6}$
9. $\frac{4}{\sqrt{6}}$
10. $\sqrt{\frac{3}{10}}$
11. $\frac{8}{3 - \sqrt{2}}$
12. $\frac{2\sqrt{5}}{-4 + \sqrt{8}}$

Applications 13. **GEOMETRY** A square has sides each measuring $2\sqrt{7}$ feet. Determine the area of the square.

14. **PHYSICS** The period of a pendulum is the time required for it to make one complete swing back and forth. The formula of the period P of a pendulum is $P = 2\pi\sqrt{\frac{\ell}{32}}$, where ℓ is the length of the pendulum in feet. If a pendulum in a clock tower is 8 feet long, find the period. Use 3.14 for π.

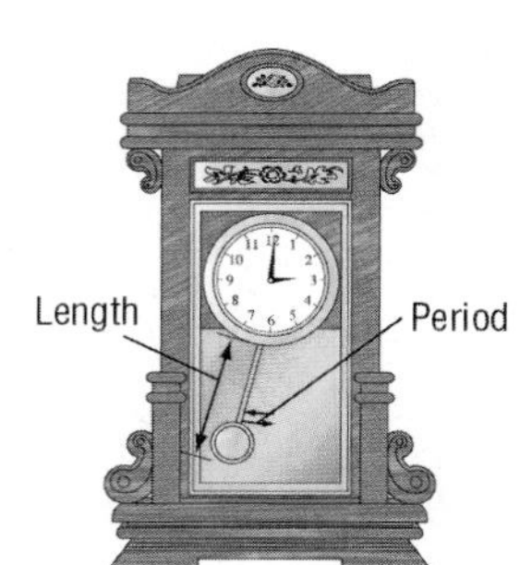

Practice and Apply

Homework Help

For Exercises	See Examples
15–18, 41, 44–46	1
19–22, 39, 40, 48, 49	2
23–26	3
27–32, 42, 43, 47	4
33–38	5

Extra Practice
See page 844.

Simplify.

15. $\sqrt{18}$
16. $\sqrt{24}$
17. $\sqrt{80}$
18. $\sqrt{75}$
19. $\sqrt{5} \cdot \sqrt{6}$
20. $\sqrt{3} \cdot \sqrt{8}$
21. $7\sqrt{30} \cdot 2\sqrt{6}$
22. $2\sqrt{3} \cdot 5\sqrt{27}$
23. $\sqrt{40a^4}$
24. $\sqrt{50m^3n^5}$
25. $\sqrt{147x^6y^7}$
26. $\sqrt{72x^3y^4z^5}$
27. $\sqrt{\frac{2}{7}} \cdot \sqrt{\frac{7}{3}}$
28. $\sqrt{\frac{3}{5}} \cdot \sqrt{\frac{6}{4}}$
29. $\sqrt{\frac{t}{8}}$
30. $\sqrt{\frac{27}{p^2}}$
31. $\sqrt{\frac{5c^5}{4d^5}}$
32. $\frac{\sqrt{9x^5y}}{\sqrt{12x^2y^6}}$
33. $\frac{18}{6 - \sqrt{2}}$
34. $\frac{2\sqrt{5}}{-4 + \sqrt{8}}$
35. $\frac{10}{\sqrt{7} + \sqrt{2}}$
36. $\frac{2}{\sqrt{3} + \sqrt{6}}$
37. $\frac{4}{4 - 3\sqrt{3}}$
38. $\frac{3\sqrt{7}}{5\sqrt{3} + 3\sqrt{5}}$

39. **GEOMETRY** A rectangle has width $3\sqrt{5}$ centimeters and length $4\sqrt{10}$ centimeters. Find the area of the rectangle.

40. **GEOMETRY** A rectangle has length $\sqrt{\frac{a}{8}}$ meters and width $\sqrt{\frac{a}{2}}$ meters. What is the area of the rectangle?

The speed of a roller coaster can be determined by evaluating a radical expression. Visit www.algebra1.com/webquest to continue work on your WebQuest project.

41. **GEOMETRY** The formula for the area A of a square with side length s is $A = s^2$. Solve this equation for s, and find the side length of a square having an area of 72 square inches.

PHYSICS For Exercises 42 and 43, use the following information.
The formula for the kinetic energy of a moving object is $E = \frac{1}{2}mv^2$, where E is the kinetic energy in joules, m is the mass in kilograms, and v is the velocity in meters per second.

42. Solve the equation for v.

43. Find the velocity of an object whose mass is 0.6 kilogram and whose kinetic energy is 54 joules.

44. **SPACE EXPLORATION** Refer to the application at the beginning of the lesson. Find the escape velocity for the Moon in kilometers per second if $G = \frac{6.7 \times 10^{-20}\text{ km}}{s^2\text{ kg}}$, $M = 7.4 \times 10^{22}$ kg, and $R = 1.7 \times 10^3$ km. How does this compare to the escape velocity for Earth?

Career Choices

Insurance Investigator

Insurance investigators decide whether claims are covered by the customer's policy, assess the amount of loss, and investigate the circumstances of a claim.

Online Research For more information about a career as an insurance investigator, visit: www.algebra1.com/careers

Source: U.S. Department of Labor

INVESTIGATION **For Exercises 45–47, use the following information.**
Police officers can use the formula $s = \sqrt{30fd}$ to determine the speed s that a car was traveling in miles per hour by measuring the distance d in feet of its skid marks. In this formula, f is the coefficient of friction for the type and condition of the road.

45. Write a simplified expression for the speed if $f = 0.6$ for a wet asphalt road.

46. What is a simplified expression for the speed if $f = 0.8$ for a dry asphalt road?

47. An officer measures skid marks that are 110 feet long. Determine the speed of the car for both wet road conditions and for dry road conditions.

GEOMETRY **For Exercises 48 and 49, use the following information.**
Hero's Formula can be used to calculate the area A of a triangle given the three side lengths a, b, and c.

$$A = \sqrt{s(s - a)(s - b)(s - c)}, \text{ where } s = \frac{1}{2}(a + b + c)$$

48. Find the value of s if the side lengths of a triangle are 13, 10, and 7 feet.

49. Determine the area of the triangle.

50. CRITICAL THINKING Simplify $\frac{1}{a - 1 + \sqrt{a}}$.

51. WRITING IN MATH Answer the question that was posed at the beginning of the lesson.

How are radical expressions used in space exploration?

Include the following in your answer:
- an explanation of how you could determine the escape velocity for a planet and why you would need this information before you landed on the planet, and
- a comparison of the escape velocity for two astronomical bodies with the same mass, but different radii.

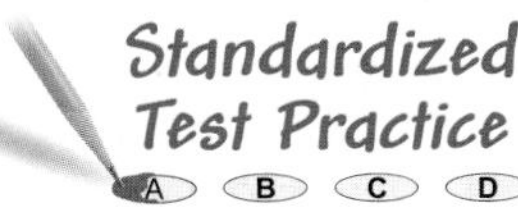

Standardized Test Practice

52. If the cube has a surface area of $96a^2$, what is its volume?

(A) $32a^3$ (B) $48a^3$

(C) $64a^3$ (D) $96a^3$

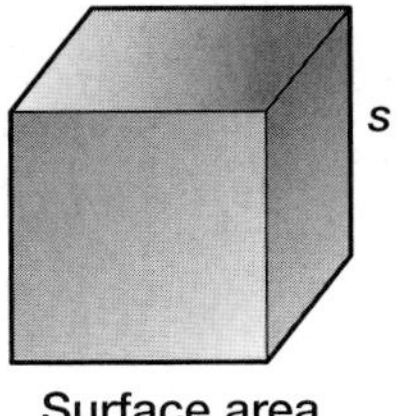

Surface area of a cube = $6s^2$

53. If $x = 81b^2$ and $b > 0$, then $\sqrt{x} =$

(A) $-9b$. (B) $9b$.

(C) $3b\sqrt{27}$. (D) $27b\sqrt{3}$.

Graphing Calculator

WEATHER **For Exercises 54 and 55, use the following information.**
The formula $y = 91.4 - (91.4 - t)[0.478 + 0.301(\sqrt{x} - 0.02)]$ can be used to find the windchill factor. In this formula, y represents the windchill factor, t represents the air temperature in degrees Fahrenheit, and x represents the wind speed in miles per hour. Suppose the air temperature is 12°.

54. Use a graphing calculator to find the wind speed to the nearest mile per hour if it feels like −9° with the windchill factor.

55. What does it feel like to the nearest degree if the wind speed is 4 miles per hour?

www.algebra1.com/self_check_quiz

Extending the Lesson **Radical expressions can be represented with fractional exponents. For example, $x^{\frac{1}{2}} = \sqrt{x}$. Using the properties of exponents, simplify each expression.**

56. $x^{\frac{1}{2}} \cdot x^{\frac{1}{2}}$ **57.** $\left(x^{\frac{1}{2}}\right)^4$ **58.** $\frac{x^{\frac{5}{2}}}{x}$

59. Simplify the expression $\frac{\sqrt{a}}{a\sqrt[3]{a}}$.

60. Solve the equation $|y^3| = \frac{1}{3\sqrt{3}}$ for y.

61. Write $\left(s^2t^{\frac{1}{2}}\right)^8\sqrt{s^5t^4}$ in simplest form.

Maintain Your Skills

Mixed Review **Find the next three terms in each geometric sequence.** *(Lesson 10-7)*

62. 2, 6, 18, 54

63. 1, −2, 4, −8

64. 384, 192, 96, 48

65. $\frac{1}{9}, \frac{2}{3}, 4, 24$

66. $3, \frac{3}{4}, \frac{3}{16}, \frac{3}{64}$

67. 50, 10, 2, 0.4

68. BIOLOGY A certain type of bacteria, if left alone, doubles its number every 2 hours. If there are 1000 bacteria at a certain point in time, how many bacteria will there be 24 hours later? *(Lesson 10-6)*

69. PHYSICS According to Newton's Law of Cooling, the difference between the temperature of an object and its surroundings decreases in time exponentially. Suppose a cup of coffee is 95°C and it is in a room that is 20°C. The cooling of the coffee can be modeled by the equation $y = 75(0.875)^t$, where y is the temperature difference and t is the time in minutes. Find the temperature of the coffee after 15 minutes. *(Lesson 10-6)*

Factor each trinomial, if possible. If the trinomial cannot be factored using integers, write *prime*. *(Lesson 9-4)*

70. $6x^2 + 7x - 5$

71. $35x^2 - 43x + 12$

72. $5x^2 + 3x + 31$

73. $3x^2 - 6x - 105$

74. $4x^2 - 12x + 15$

75. $8x^2 - 10x + 3$

Find the solution set for each equation, given the replacement set. *(Lesson 4-4)*

76. $y = 3x + 2$; {(1, 5), (2, 6), (−2, 2), (−4, −10)}

77. $5x + 2y = 10$; {(3, 5), (2, 0), (4, 2), (1, 2.5)}

78. $3a + 2b = 11$; {(−3, 10), (4, 1), (2, 2.5), (3, −2)}

79. $5 - \frac{3}{2}x = 2y$; $\left\{(0, 1), (8, 2), \left(4, -\frac{1}{2}\right), (2, 1)\right\}$

Solve each equation. Then check your solution. *(Lesson 3-3)*

80. $40 = -5d$

81. $20.4 = 3.4y$

82. $\frac{h}{-11} = -25$

83. $-65 = \frac{r}{29}$

Getting Ready for the Next Lesson **PREREQUISITE SKILL** **Find each product.**
*(To review **multiplying binomials**, see Lesson 8-7.)*

84. $(x - 3)(x + 2)$

85. $(a + 2)(a + 5)$

86. $(2t + 1)(t - 6)$

87. $(4x - 3)(x + 1)$

88. $(5x + 3y)(3x - y)$

89. $(3a - 2b)(4a + 7b)$

11-2 Operations with Radical Expressions

What You'll Learn

- Add and subtract radical expressions.
- Multiply radical expressions.

How can you use radical expressions to determine how far a person can see?

The formula $d = \sqrt{\frac{3h}{2}}$ represents the distance d in miles that a person h feet high can see. To determine how much farther a person can see from atop the Sears Tower than from atop the Empire State Building, we can substitute the heights of both buildings into the equation.

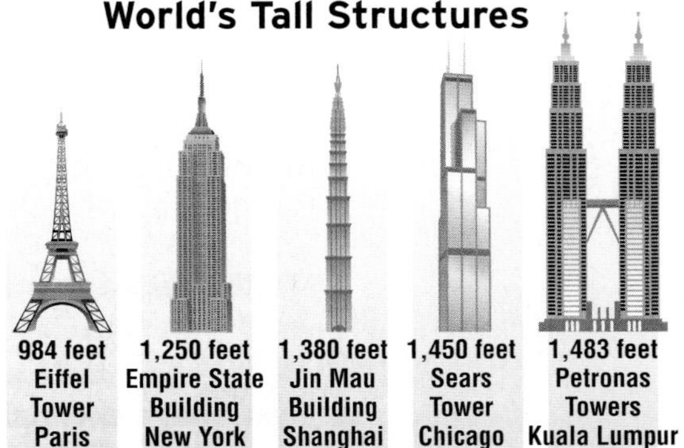

ADD AND SUBTRACT RADICAL EXPRESSIONS Radical expressions in which the radicands are alike can be added or subtracted in the same way that monomials are added or subtracted.

Monomials	Radical Expressions
$2x + 7x = (2 + 7)x$	$2\sqrt{11} + 7\sqrt{11} = (2 + 7)\sqrt{11}$
$= 9x$	$= 9\sqrt{11}$
$15y - 3y = (15 - 3)y$	$15\sqrt{2} - 3\sqrt{2} = (15 - 3)\sqrt{2}$
$= 12y$	$= 12\sqrt{2}$

Notice that the Distributive Property was used to simplify each radical expression.

Example 1 Expressions with Like Radicands

Simplify each expression.

a. $4\sqrt{3} + 6\sqrt{3} - 5\sqrt{3}$

$4\sqrt{3} + 6\sqrt{3} - 5\sqrt{3} = (4 + 6 - 5)\sqrt{3}$ Distributive Property

$= 5\sqrt{3}$ Simplify.

b. $12\sqrt{5} + 3\sqrt{7} + 6\sqrt{7} - 8\sqrt{5}$

$12\sqrt{5} + 3\sqrt{7} + 6\sqrt{7} - 8\sqrt{5} = 12\sqrt{5} - 8\sqrt{5} + 3\sqrt{7} + 6\sqrt{7}$ Commutative Property

$= (12 - 8)\sqrt{5} + (3 + 6)\sqrt{7}$ Distributive Property

$= 4\sqrt{5} + 9\sqrt{7}$ Simplify.

In Example 1b, $4\sqrt{5} + 9\sqrt{7}$ cannot be simplified further because the radicands are different. There are no common factors, and each radicand is in simplest form. If the radicals in a radical expression are not in simplest form, simplify them first.

Example 2 Expressions with Unlike Radicands

Simplify $2\sqrt{20} + 3\sqrt{45} + \sqrt{180}$.

$$\begin{aligned}2\sqrt{20} + 3\sqrt{45} + \sqrt{180} &= 2\sqrt{2^2 \cdot 5} + 3\sqrt{3^2 \cdot 5} + \sqrt{6^2 \cdot 5}\\ &= 2(\sqrt{2^2} \cdot \sqrt{5}) + 3(\sqrt{3^2} \cdot \sqrt{5}) + \sqrt{6^2} \cdot \sqrt{5}\\ &= 2(2\sqrt{5}) + 3(3\sqrt{5}) + 6\sqrt{5}\\ &= 4\sqrt{5} + 9\sqrt{5} + 6\sqrt{5}\\ &= 19\sqrt{5}\end{aligned}$$

The simplified form is $19\sqrt{5}$.

You can use a calculator to verify that a simplified radical expression is equivalent to the original expression. Consider Example 2. First, find a decimal approximation for the original expression.

KEYSTROKES: 2 [2nd] [√] 20 [)] [+] 3 [2nd] [√] 45 [)] [+] [2nd] [√] 180 [)] [ENTER] 42.48529157

Next, find a decimal approximation for the simplified expression.

KEYSTROKES: 19 [2nd] [√] 5 [ENTER] 42.48529157

Since the approximations are equal, the expressions are equivalent.

MULTIPLY RADICAL EXPRESSIONS Multiplying two radical expressions with different radicands is similar to multiplying binomials.

Example 3 Multiply Radical Expressions

Find the area of the rectangle in simplest form.

To find the area of the rectangle multiply the measures of the length and width.

$4\sqrt{5} - 2\sqrt{3}$

$3\sqrt{6} - \sqrt{10}$

$(4\sqrt{5} - 2\sqrt{3})(3\sqrt{6} - \sqrt{10})$

First terms, Outer terms, Inner terms, Last terms

$$\begin{aligned}&= (4\sqrt{5})(3\sqrt{6}) + (4\sqrt{5})(-\sqrt{10}) + (-2\sqrt{3})(3\sqrt{6}) + (-2\sqrt{3})(-\sqrt{10})\\ &= 12\sqrt{30} - 4\sqrt{50} - 6\sqrt{18} + 2\sqrt{30} && \text{Multiply.}\\ &= 12\sqrt{30} - 4\sqrt{5^2 \cdot 2} - 6\sqrt{3^2 \cdot 2} + 2\sqrt{30} && \text{Prime factorization}\\ &= 12\sqrt{30} - 20\sqrt{2} - 18\sqrt{2} + 2\sqrt{30} && \text{Simplify.}\\ &= 14\sqrt{30} - 38\sqrt{2} && \text{Combine like terms.}\end{aligned}$$

The area of the rectangle is $14\sqrt{30} - 38\sqrt{2}$ square units.

Study Tip

Look Back
To review the **FOIL method**, see Lesson 8-7.

Check for Understanding

Concept Check

1. **Explain** why you should simplify each radical in a radical expression before adding or subtracting.
2. **Explain** how you use the Distributive Property to simplify like radicands that are added or subtracted.
3. **OPEN ENDED** Choose values for x and y. Then find $(\sqrt{x} + \sqrt{y})^2$.

Guided Practice

Simplify each expression.

4. $4\sqrt{3} + 7\sqrt{3}$
5. $2\sqrt{6} - 7\sqrt{6}$
6. $5\sqrt{5} - 3\sqrt{20}$
7. $2\sqrt{3} + \sqrt{12}$
8. $3\sqrt{5} + 5\sqrt{6} + 3\sqrt{20}$
9. $8\sqrt{3} + \sqrt{3} + \sqrt{9}$

Find each product.

10. $\sqrt{2}(\sqrt{8} + 4\sqrt{3})$
11. $(4 + \sqrt{5})(3 + \sqrt{5})$

Applications

12. **GEOMETRY** Find the perimeter and the area of a square whose sides measure $4 + 3\sqrt{6}$ feet.
13. **ELECTRICITY** The voltage V required for a circuit is given by $V = \sqrt{PR}$, where P is the power in watts and R is the resistance in ohms. How many more volts are needed to light a 100-watt bulb than a 75-watt bulb if the resistance for both is 110 ohms?

Practice and Apply

Homework Help

For Exercises	See Examples
14–21	1
22–29	2
30–48	3

Extra Practice

See page 844.

Simplify each expression.

14. $8\sqrt{5} + 3\sqrt{5}$
15. $3\sqrt{6} + 10\sqrt{6}$
16. $2\sqrt{15} - 6\sqrt{15} - 3\sqrt{15}$
17. $5\sqrt{19} + 6\sqrt{19} - 11\sqrt{19}$
18. $16\sqrt{x} + 2\sqrt{x}$
19. $3\sqrt{5b} - 4\sqrt{5b} + 11\sqrt{5b}$
20. $8\sqrt{3} - 2\sqrt{2} + 3\sqrt{2} + 5\sqrt{3}$
21. $4\sqrt{6} + \sqrt{17} - 6\sqrt{2} + 4\sqrt{17}$
22. $\sqrt{18} + \sqrt{12} + \sqrt{8}$
23. $\sqrt{6} + 2\sqrt{3} + \sqrt{12}$
24. $3\sqrt{7} - 2\sqrt{28}$
25. $2\sqrt{50} - 3\sqrt{32}$
26. $\sqrt{2} + \sqrt{\frac{1}{2}}$
27. $\sqrt{10} - \sqrt{\frac{2}{5}}$
28. $3\sqrt{3} - \sqrt{45} + 3\sqrt{\frac{1}{3}}$
29. $6\sqrt{\frac{7}{4}} + 3\sqrt{28} - 10\sqrt{\frac{1}{7}}$

Find each product.

30. $\sqrt{6}(\sqrt{3} + 5\sqrt{2})$
31. $\sqrt{5}(2\sqrt{10} + 3\sqrt{2})$
32. $(3 + \sqrt{5})(3 - \sqrt{5})$
33. $(7 - \sqrt{10})^2$
34. $(\sqrt{6} + \sqrt{8})(\sqrt{24} + \sqrt{2})$
35. $(\sqrt{5} - \sqrt{2})(\sqrt{14} + \sqrt{35})$
36. $(2\sqrt{10} + 3\sqrt{15})(3\sqrt{3} - 2\sqrt{2})$
37. $(5\sqrt{2} + 3\sqrt{5})(2\sqrt{10} - 3)$

38. **GEOMETRY** Find the perimeter of a rectangle whose length is $8\sqrt{7} + 4\sqrt{5}$ inches and whose width is $5\sqrt{7} - 3\sqrt{5}$ inches.

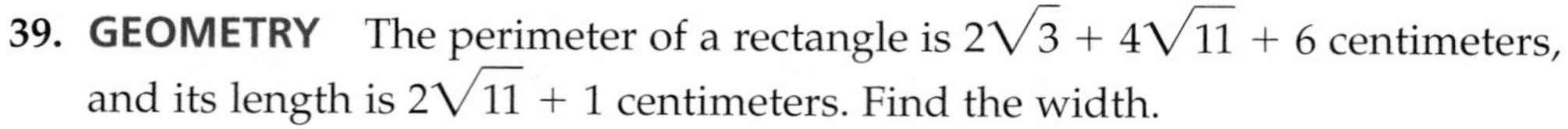

39. **GEOMETRY** The perimeter of a rectangle is $2\sqrt{3} + 4\sqrt{11} + 6$ centimeters, and its length is $2\sqrt{11} + 1$ centimeters. Find the width.

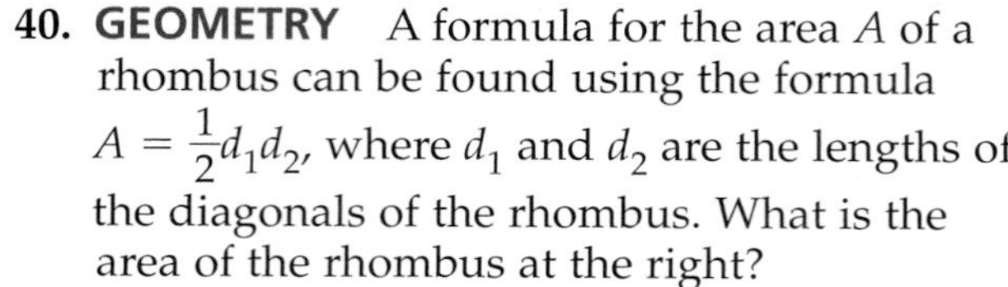

40. **GEOMETRY** A formula for the area A of a rhombus can be found using the formula $A = \frac{1}{2}d_1d_2$, where d_1 and d_2 are the lengths of the diagonals of the rhombus. What is the area of the rhombus at the right?

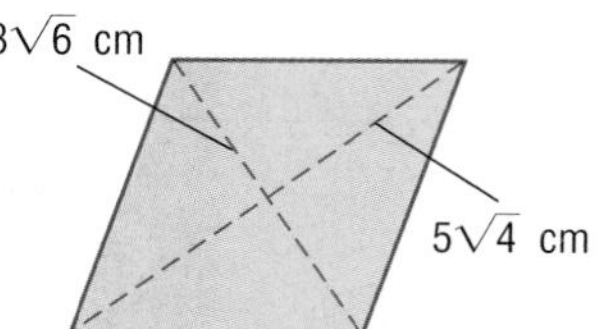

More About. . .

Distance

The Sears Tower was the tallest building in the world from 1974 to 1996. You can see four states from its roof: Michigan, Indiana, Illinois, and Wisconsin.

Source: www.the-skydeck.com

DISTANCE For Exercises 41 and 42, refer to the application at the beginning of the lesson.

41. How much farther can a person see from atop the Sears Tower than from atop the Empire State Building?

42. A person atop the Empire State Building can see approximately 4.57 miles farther than a person atop the Texas Commerce Tower in Houston. Find the height of the Texas Commerce Tower. Explain how you found the answer.

Online Research **Data Update** What are the tallest buildings and towers in the world today? Visit www.algebra1.com/data_update to learn more.

ENGINEERING For Exercises 43 and 44, use the following information.

The equation $r = \sqrt{\frac{F}{5\pi}}$ relates the radius r of a drainpipe in inches to the flow rate F of water passing through it in gallons per minute.

43. Find the radius of a pipe that can carry 500 gallons of water per minute. Round to the nearest whole number.

44. An engineer determines that a drainpipe must be able to carry 1000 gallons of water per minute and instructs the builder to use an 8-inch radius pipe. Can the builder use two 4-inch radius pipes instead? Justify your answer.

MOTION For Exercises 45–47, use the following information.

The velocity of an object dropped from a certain height can be found using the formula $v = \sqrt{2gd}$, where v is the velocity in feet per second, g is the acceleration due to gravity, and d is the distance in feet the object drops.

45. Find the speed of an object that has fallen 25 feet and the speed of an object that has fallen 100 feet. Use 32 feet per second squared for g.

46. When you increased the distance by 4 times, what happened to the velocity?

47. **MAKE A CONJECTURE** Estimate the velocity of an object that has fallen 225 feet. Then use the formula to verify your answer.

48. **WATER SUPPLY** The relationship between a city's size and its capacity to supply water to its citizens can be described by the expression $1020\sqrt{P}(1 - 0.01\sqrt{P})$, where P is the population in thousands and the result is the number of gallons per minute required. If a city has a population of 55,000 people, how many gallons per minute must the city's pumping station be able to supply?

49. **CRITICAL THINKING** Find a counterexample to disprove the following statement.

For any numbers a and b, where $a > 0$ *and* $b > 0$, $\sqrt{a+b} = \sqrt{a} + \sqrt{b}$.

50. **CRITICAL THINKING** Under what conditions is $(\sqrt{a+b})^2 = (\sqrt{a})^2 + (\sqrt{b})^2$ true?

51. **WRITING IN MATH** Answer the question that was posed at the beginning of the lesson.

How can you use radical expressions to determine how far a person can see?

Include the following in your answer:

- an explanation of how this information could help determine how far apart lifeguard towers should be on a beach, and
- an example of a real-life situation where a lookout position is placed at a high point above the ground.

Standardized Test Practice A B C D

52. Find the difference of $9\sqrt{7}$ and $2\sqrt{28}$.

(A) $\sqrt{7}$ (B) $4\sqrt{7}$

(C) $5\sqrt{7}$ (D) $7\sqrt{7}$

53. Simplify $\sqrt{3}(4 + \sqrt{12})^2$.

(A) $4\sqrt{3} + 6$ (B) $28\sqrt{3}$

(C) $28 + 16\sqrt{3}$ (D) $48 + 28\sqrt{3}$

Maintain Your Skills

Mixed Review **Simplify.** *(Lesson 11-1)*

54. $\sqrt{40}$ 55. $\sqrt{128}$ 56. $-\sqrt{196x^2y^3}$

57. $\dfrac{\sqrt{50}}{\sqrt{8}}$ 58. $\sqrt{\dfrac{225c^4d}{18c^2}}$ 59. $\sqrt{\dfrac{63a}{128a^3b^2}}$

Find the *n*th term of each geometric sequence. *(Lesson 10-7)*

60. $a_1 = 4, n = 6, r = 4$ 61. $a_1 = -7, n = 4, r = 9$ 62. $a_1 = 2, n = 8, r = -0.8$

Solve each equation by factoring. Check your solutions. *(Lesson 9-5)*

63. $81 = 49y^2$ 64. $q^2 - \dfrac{36}{121} = 0$

65. $48n^3 - 75n = 0$ 66. $5x^3 - 80x = 240 - 15x^2$

Solve each inequality. Then check your solution. *(Lesson 6-2)*

67. $8n \geq 5$ 68. $\dfrac{w}{9} < 14$ 69. $\dfrac{7k}{2} > \dfrac{21}{10}$

70. **PROBABILITY** A student rolls a die three times. What is the probability that each roll is a 1? *(Lesson 2-6)*

Getting Ready for the Next Lesson **PREREQUISITE SKILL** **Find each product.** *(To review **special products**, see Lesson 8-8.)*

71. $(x - 2)^2$ 72. $(x + 5)^2$ 73. $(x + 6)^2$

74. $(3x - 1)^2$ 75. $(2x - 3)^2$ 76. $(4x + 7)^2$

www.algebra1.com/self_check_quiz

11-3 Radical Equations

What You'll Learn

- Solve radical equations.
- Solve radical equations with extraneous solutions.

Vocabulary
- radical equation
- extraneous solution

How are radical equations used to find free-fall times?

Skydivers fall 1050 to 1480 feet every 5 seconds, reaching speeds of 120 to 150 miles per hour at *terminal velocity*. It is the highest speed they can reach and occurs when the air resistance equals the force of gravity. With no air resistance, the time t in seconds that it takes an object to fall h feet can be determined by the equation $t = \frac{\sqrt{h}}{4}$. How would you find the value of h if you are given the value of t?

RADICAL EQUATIONS Equations like $t = \frac{\sqrt{h}}{4}$ that contain radicals with variables in the radicand are called **radical equations**. To solve these equations, first isolate the radical on one side of the equation. Then square each side of the equation to eliminate the radical.

Example 1 Radical Equation with a Variable

FREE-FALL HEIGHT Two objects are dropped simultaneously. The first object reaches the ground in 2.5 seconds, and the second object reaches the ground 1.5 seconds later. From what heights were the two objects dropped?

Find the height of the first object. Replace t with 2.5 seconds.

$t = \frac{\sqrt{h}}{4}$ Original equation

$2.5 = \frac{\sqrt{h}}{4}$ Replace t with 2.5.

$10 = \sqrt{h}$ Multiply each side by 4.

$10^2 = (\sqrt{h})^2$ Square each side.

$100 = h$ Simplify.

CHECK $t = \frac{\sqrt{h}}{4}$ Original equation

$t \stackrel{?}{=} \frac{\sqrt{100}}{4}$ $h = 100$

$t \stackrel{?}{=} \frac{10}{4}$ $\sqrt{100} = 10$

$t = 2.5$ Simplify.

The first object was dropped from 100 feet.

The time it took the second object to fall was 2.5 + 1.5 seconds or 4 seconds.

$t = \frac{\sqrt{h}}{4}$	Original equation
$4 = \frac{\sqrt{h}}{4}$	Replace *t* with 4.
$16 = \sqrt{h}$	Multiply each side by 4.
$16^2 = (\sqrt{h})^2$	Square each side.
$256 = h$	Simplify.

The second object was dropped from 256 feet. *Check this solution.*

Example 2 Radical Equation with an Expression

Solve $\sqrt{x + 1} + 7 = 10$.

$\sqrt{x + 1} + 7 = 10$	Original equation
$\sqrt{x + 1} = 3$	Subtract 7 from each side.
$(\sqrt{x + 1})^2 = 3^2$	Square each side.
$x + 1 = 9$	$(\sqrt{x + 1})^2 = x + 1$
$x = 8$	Subtract 1 from each side.

The solution is 8. *Check this result.*

EXTRANEOUS SOLUTIONS Squaring each side of an equation sometimes produces extraneous solutions. An **extraneous solution** is a solution derived from an equation that is not a solution of the original equation. Therefore, you must check all solutions in the original equation when you solve radical equations.

Example 3 Variable on Each Side

Solve $\sqrt{x + 2} = x - 4$.

$\sqrt{x + 2} = x - 4$	Original equation
$(\sqrt{x + 2})^2 = (x - 4)^2$	Square each side.
$x + 2 = x^2 - 8x + 16$	Simplify.
$0 = x^2 - 9x + 14$	Subtract *x* and 2 from each side.
$0 = (x - 7)(x - 2)$	Factor.
$x - 7 = 0$ or $x - 2 = 0$	Zero Product Property
$x = 7$ $\quad$ $x = 2$	Solve.

CHECK

$\sqrt{x + 2} = x - 4$		$\sqrt{x + 2} = x - 4$	
$\sqrt{7 + 2} \stackrel{?}{=} 7 - 4$	$x = 7$	$\sqrt{2 + 2} \stackrel{?}{=} 2 - 4$	$x = 2$
$\sqrt{9} \stackrel{?}{=} 3$		$\sqrt{4} \stackrel{?}{=} -2$	
$3 = 3$ ✓		$2 \neq -2$ ✗	

Since 2 does not satisfy the original equation, 7 is the only solution.

Study Tip

Look Back
To review **Zero Product Property**, see Lesson 9-2.

Graphing Calculator Investigation

Solving Radical Equations

You can use a TI-83 Plus graphing calculator to solve radical equations such as $\sqrt{3x} - 5 = x - 5$. Clear the Y= list. Enter the left side of the equation as Y1 = $\sqrt{3x} - 5$. Enter the right side of the equation as Y2 = $x - 5$. Press GRAPH.

Think and Discuss

1. Sketch what is shown on the screen.
2. Use the intersect feature on the CALC menu, to find the point of intersection.
3. Solve the radical equation algebraically. How does your solution compare to the solution from the graph?

Check for Understanding

Concept Check

1. **Describe** the steps needed to solve a radical equation.
2. **Explain** why it is necessary to check for extraneous solutions in radical equations.
3. **OPEN ENDED** Give an example of a radical equation. Then solve the equation for the variable.
4. **FIND THE ERROR** Alex and Victor are solving $-\sqrt{x - 5} = -2$.

Alex	Victor
$-\sqrt{x-5} = -2$	$-\sqrt{x-5} = -2$
$(-\sqrt{x-5})^2 = (-2)^2$	$(-\sqrt{x-5})^2 = (-2)^2$
$x - 5 = 4$	$-(x - 5) = 4$
$x = 9$	$-x + 5 = 4$
	$x = 1$

Who is correct? Explain your reasoning.

Guided Practice

Solve each equation. Check your solution.

5. $\sqrt{x} = 5$
6. $\sqrt{2b} = -8$
7. $\sqrt{7x} = 7$
8. $\sqrt{-3a} = 6$
9. $\sqrt{8s} + 1 = 5$
10. $\sqrt{7x + 18} = 9$
11. $\sqrt{5x + 1} + 2 = 6$
12. $\sqrt{6x - 8} = x - 4$
13. $4 + \sqrt{x - 2} = x$

Application

OCEANS For Exercises 14–16, use the following information.
Tsunamis, or large tidal waves, are generated by undersea earthquakes in the Pacific Ocean. The speed of the tsunami in meters per second is $s = 3.1\sqrt{d}$, where d is the depth of the ocean in meters.

14. Find the speed of the tsunami if the depth of the water is 10 meters.
15. Find the depth of the water if a tsunami's speed is 240 meters per second.
16. A tsunami may begin as a 2-foot high wave traveling 450–500 miles per hour. It can approach a coastline as a 50-foot wave. How much speed does the wave lose if it travels from a depth of 10,000 meters to a depth of 20 meters?

Practice and Apply

Homework Help

For Exercises	See Examples
17–34	1, 2
35–47	3
48–59	1–3

Extra Practice

See page 844.

Solve each equation. Check your solution.

17. $\sqrt{a} = 10$
18. $\sqrt{-k} = 4$
19. $5\sqrt{2} = \sqrt{x}$
20. $3\sqrt{7} = \sqrt{-y}$
21. $3\sqrt{4a} - 2 = 10$
22. $3 + 5\sqrt{n} = 18$
23. $\sqrt{x + 3} = -5$
24. $\sqrt{x - 5} = 2\sqrt{6}$
25. $\sqrt{3x + 12} = 3\sqrt{3}$
26. $\sqrt{2c - 4} = 8$
27. $\sqrt{4b + 1} - 3 = 0$
28. $\sqrt{3r - 5} + 7 = 3$
29. $\sqrt{\frac{4x}{5}} - 9 = 3$
30. $5\sqrt{\frac{4t}{3}} - 2 = 0$
31. $\sqrt{x^2 + 9x + 14} = x + 4$
32. $y + 2 = \sqrt{y^2 + 5y + 4}$

33. The square root of the sum of a number and 7 is 8. Find the number.

34. The square root of the quotient of a number and 6 is 9. Find the number.

Solve each equation. Check your solution.

35. $x = \sqrt{6 - x}$
36. $x = \sqrt{x + 20}$
37. $\sqrt{5x - 6} = x$
38. $\sqrt{28 - 3x} = x$
39. $\sqrt{x + 1} = x - 1$
40. $\sqrt{1 - 2b} = 1 + b$
41. $4 + \sqrt{m - 2} = m$
42. $\sqrt{3d - 8} = d - 2$
43. $x + \sqrt{6 - x} = 4$
44. $\sqrt{6 - 3x} = x + 16$
45. $\sqrt{2r^2 - 121} = r$
46. $\sqrt{5p^2 - 7} = 2p$

47. State whether the following equation is *sometimes*, *always*, or *never* true.
$\sqrt{(x - 5)^2} = x - 5$

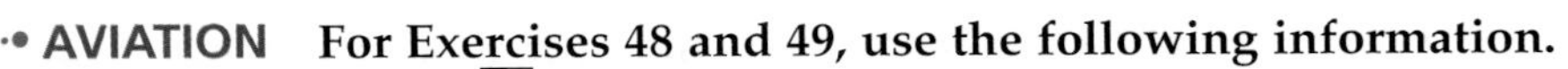

More About. . .

Aviation

Piloted by A. Scott Crossfield on November 20, 1953, the Douglas D-558-2 Skyrocket became the first aircraft to fly faster than Mach 2, twice the speed of sound.

Source: National Air and Space Museum

AVIATION **For Exercises 48 and 49, use the following information.**
The formula $L = \sqrt{kP}$ represents the relationship between a plane's length L in feet and the pounds P its wings can lift, where k is a constant of proportionality calculated for a plane.

48. The length of the Douglas D-558-II, called the Skyrocket, was approximately 42 feet, and its constant of proportionality was $k = 0.1669$. Calculate the maximum takeoff weight of the Skyrocket.

49. A Boeing 747 is 232 feet long and has a takeoff weight of 870,000 pounds. Determine the value of k for this plane.

GEOMETRY **For Exercises 50–53, use the figure below. The area A of a circle is equal to πr^2 where r is the radius of the circle.**

50. Write an equation for r in terms of A.

51. The area of the larger circle is 96π square meters. Find the radius.

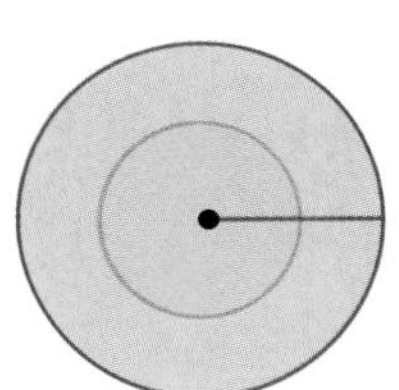

52. The area of the smaller circle is 48π square meters. Find the radius.

53. If the area of a circle is doubled, what is the change in the radius?

PHYSICAL SCIENCE **For Exercises 54–56, use the following information.**

The formula $P = 2\pi\sqrt{\frac{\ell}{32}}$ gives the period of a pendulum of length ℓ feet. The period P is the number of seconds it takes for the pendulum to swing back and forth once.

More About. . .

Physical Science

The Foucault pendulum was invented in 1851. It demonstrates the rotation of Earth. The pendulum appears to change its path during the day, but it moves in a straight line. The path under the pendulum changes because Earth is rotating beneath it.

Source: California Academy of Sciences

54. Suppose we want a pendulum to complete three periods in 2 seconds. How long should the pendulum be?

55. Two clocks have pendulums of different lengths. The first clock requires 1 second for its pendulum to complete one period. The second clock requires 2 seconds for its pendulum to complete one period. How much longer is one pendulum than the other?

56. Repeat Exercise 55 if the pendulum periods are t and $2t$ seconds.

SOUND **For Exercises 57–59, use the following information.**
The speed of sound V near Earth's surface can be found using the equation $V = 20\sqrt{t + 273}$, where t is the surface temperature in degrees Celsius.

57. Find the temperature if the speed of sound V is 356 meters per second.

58. The speed of sound at Earth's surface is often given at 340 meters per second, but that is only accurate at a certain temperature. On what temperature is this figure based?

59. What is the speed of sound when the surface temperature is below 0°C?

60. **CRITICAL THINKING** Solve $\sqrt{h + 9} - \sqrt{h} = \sqrt{3}$.

61. **WRITING IN MATH** Answer the question that was posed at the beginning of the lesson.

How are radical equations used to find free-fall times?

Include the following in your answer:

- the time it would take a skydiver to fall 10,000 feet if he falls 1200 feet every 5 seconds and the time using the equation $t = \frac{\sqrt{h}}{4}$, with an explanation of why the two methods find different times, and
- ways that a skydiver can increase or decrease his speed.

Standardized Test Practice

QUANTITATIVE COMPARISON **In Exercises 62 and 63, compare the quantity in Column A and the quantity in Column B. Then determine whether:**

Ⓐ **the quantity in Column A is greater,**

Ⓑ **the quantity in Column B is greater,**

Ⓒ **both quantities are equal, or**

Ⓓ **the relationship cannot be determined from the given information.**

	Column A	Column B	
62.	the solution of $\sqrt{x} + 3 = 6$	the solution of $\sqrt{y + 3} = 6$	
63.	$(\sqrt{a - 1})^2$	$\sqrt{(a - 1)^2}$	$(a \geq 1)$

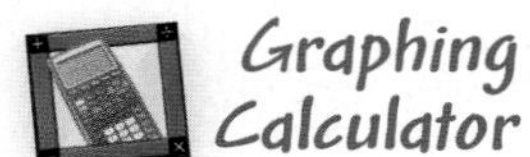

Graphing Calculator

RADICAL EQUATIONS Use a graphing calculator to solve each radical equation. Round to the nearest hundredth.

64. $3 + \sqrt{2x} = 7$

65. $\sqrt{3x - 8} = 5$

66. $\sqrt{x + 6} - 4 = x$

67. $\sqrt{4x + 5} = x - 7$

68. $x + \sqrt{7 - x} = 4$

69. $\sqrt{3x - 9} = 2x + 6$

Maintain Your Skills

Mixed Review

Simplify each expression. *(Lesson 11-2)*

70. $5\sqrt{6} + 12\sqrt{6}$

71. $\sqrt{12} + 6\sqrt{27}$

72. $\sqrt{18} + 5\sqrt{2} - 3\sqrt{32}$

Simplify. *(Lesson 11-1)*

73. $\sqrt{192}$

74. $\sqrt{6} \cdot \sqrt{10}$

75. $\frac{21}{\sqrt{10} + \sqrt{3}}$

Determine whether each trinomial is a perfect square trinomial. If so, factor it. *(Lesson 9-6)*

76. $d^2 + 50d + 225$

77. $4n^2 - 28n + 49$

78. $16b^2 - 56bc + 49c^2$

Find each product. *(Lesson 8-7)*

79. $(r + 3)(r - 4)$

80. $(3z + 7)(2z + 10)$

81. $(2p + 5)(3p^2 - 4p + 9)$

82. PHYSICAL SCIENCE A European-made hot tub is advertised to have a temperature of 35°C to 40°C, inclusive. What is the temperature range for the hot tub in degrees Fahrenheit? Use $F = \frac{9}{5}C + 32$. *(Lesson 6-4)*

Write each equation in standard form. *(Lesson 5-5)*

83. $y = 2x + \frac{3}{7}$

84. $y - 3 = -2(x - 6)$

85. $y + 2 = 7.5(x - 3)$

Getting Ready for the Next Lesson

PREREQUISITE SKILL Evaluate $\sqrt{a^2 + b^2}$ for each value of a and b.
*(To review **evaluating expressions**, see Lesson 1-2.)*

86. $a = 3, b = 4$

87. $a = 24, b = 7$

88. $a = 1, b = 1$

89. $a = 8, b = 12$

Practice Quiz 1

Lessons 11-1 through 11-3

Simplify. *(Lesson 11-1)*

1. $\sqrt{48}$

2. $\sqrt{3} \cdot \sqrt{6}$

3. $\frac{3}{2 + \sqrt{10}}$

Simplify. *(Lesson 11-2)*

4. $6\sqrt{5} + 3\sqrt{11} + 5\sqrt{5}$

5. $2\sqrt{3} + 9\sqrt{12}$

6. $(3 - \sqrt{6})^2$

7. GEOMETRY Find the area of a square whose side measure is $2 + \sqrt{7}$ centimeters. *(Lesson 11-2)*

Solve each equation. Check your solution. *(Lesson 11-3)*

8. $\sqrt{15 - x} = 4$

9. $\sqrt{3x^2 - 32} = x$

10. $\sqrt{2x - 1} = 2x - 7$

Graphing Calculator Investigation

A Follow-Up of Lesson 11-3

Graphs of Radical Equations

In order for a square root to be a real number, the radicand cannot be negative. When graphing a radical equation, determine when the radicand would be negative and exclude those values from the domain.

Example 1

Graph $y = \sqrt{x}$. State the domain of the graph.

Enter the equation in the Y= list.

KEYSTROKES: [Y=] [2nd] [√] [X,T,θ,n] [)] [GRAPH]

From the graph, you can see that the domain of x is $\{x \mid x \geq 0\}$.

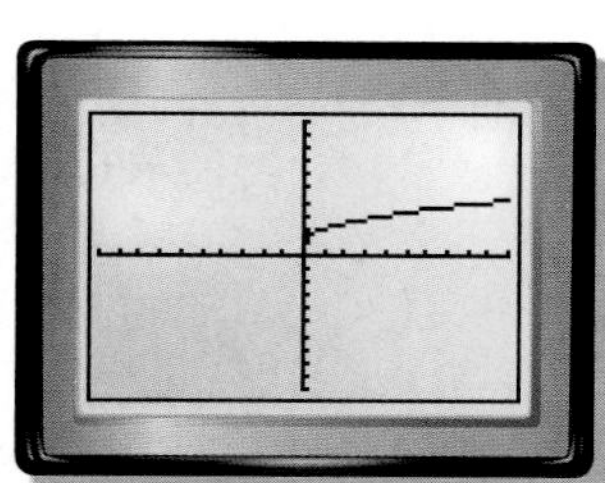

[−10, 10] scl: 1 by [−10, 10] scl: 1

Example 2

Graph $y = \sqrt{x + 4}$. State the domain of the graph.

Enter the equation in the Y= list.

KEYSTROKES: [Y=] [2nd] [√] [X,T,θ,n] [+] 4 [)] [GRAPH]

The value of the radicand will be positive when $x + 4 \geq 0$, or when $x \geq -4$. So the domain of x is $\{x \mid x \geq -4\}$.

This graph looks like the graph of $y = \sqrt{x}$ shifted left 4 units.

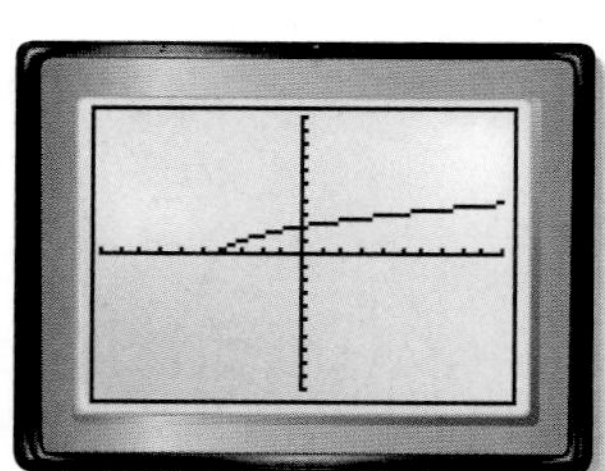

[−10, 10] scl: 1 by [−10, 10] scl: 1

Exercises

Graph each equation and sketch the graph on your paper. State the domain of the graph. Then describe how the graph differs from the parent function $y = \sqrt{x}$.

1. $y = \sqrt{x} + 1$
2. $y = \sqrt{x} - 3$
3. $y = \sqrt{x + 2}$
4. $y = \sqrt{x - 5}$
5. $y = \sqrt{-x}$
6. $y = \sqrt{3x}$
7. $y = -\sqrt{x}$
8. $y = \sqrt{1 - x} + 6$
9. $y = \sqrt{2x + 5} - 4$
10. Is the graph of $x = y^2$ a function? Explain your reasoning.
11. Does the equation $x^2 + y^2 = 1$ determine y as a function of x? Explain.
12. Graph $y = |x| \pm \sqrt{1 - x^2}$ in the window defined by [−2, 2] scl: 1 by [−2, 2] scl: 1. Describe the graph.

 www.algebra1.com/other_calculator_keystrokes

11-4 The Pythagorean Theorem

What You'll Learn

- Solve problems by using the Pythagorean Theorem.
- Determine whether a triangle is a right triangle.

How is the Pythagorean Theorem used in roller coaster design?

The roller coaster *Superman: Ride of Steel* in Agawam, Massachusetts, is one of the world's tallest roller coasters at 208 feet. It also boasts one of the world's steepest drops, measured at 78 degrees, and it reaches a maximum speed of 77 miles per hour. You can use the Pythagorean Theorem to estimate the length of the first hill.

Vocabulary

- hypotenuse
- legs
- Pythagorean triple
- corollary

THE PYTHAGOREAN THEOREM In a right triangle, the side opposite the right angle is called the **hypotenuse**. This side is always the longest side of a right triangle. The other two sides are called the **legs** of the triangle.

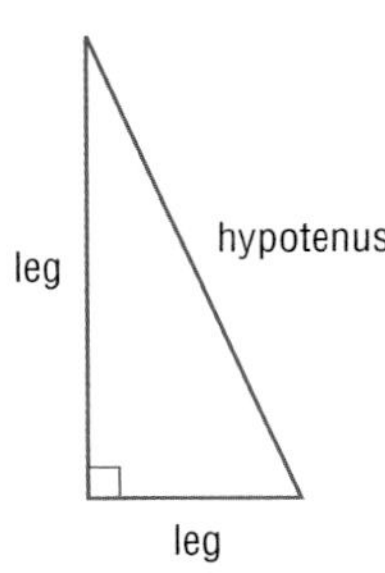

To find the length of any side of a right triangle when the lengths of the other two are known, you can use a formula developed by the Greek mathematician Pythagoras.

Study Tip

Triangles
Sides of a triangle are represented by lowercase letters a, b, and c.

Key Concept — The Pythagorean Theorem

- **Words** If a and b are the lengths of the legs of a right triangle and c is the length of the hypotenuse, then the square of the length of the hypotenuse is equal to the sum of the squares of the lengths of the legs.
- **Symbols** $c^2 = a^2 + b^2$

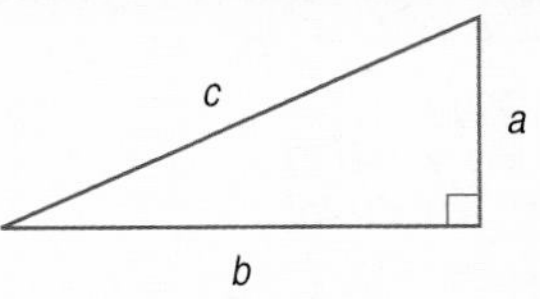

Example 1 Find the Length of the Hypotenuse

Find the length of the hypotenuse of a right triangle if $a = 8$ and $b = 15$.

$c^2 = a^2 + b^2$ Pythagorean Theorem

$c^2 = 8^2 + 15^2$ $a = 8$ and $b = 15$

$c^2 = 289$ Simplify.

$c = \pm\sqrt{289}$ Take the square root of each side.

$c = \pm 17$ Disregard -17. Why?

The length of the hypotenuse is 17 units.

Example 2 Find the Length of a Side

Find the length of the missing side.

In the triangle, $c = 25$ and $b = 10$ units.

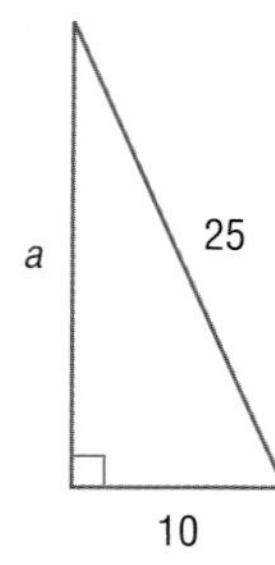

$c^2 = a^2 + b^2$	Pythagorean Theorem
$25^2 = a^2 + 10^2$	$b = 10$ and $c = 25$
$625 = a^2 + 100$	Evaluate squares.
$525 = a^2$	Subtract 100 from each side.
$\pm\sqrt{525} = a$	Use a calculator to evaluate $\sqrt{525}$.
$22.91 \approx a$	Use the positive value.

To the nearest hundredth, the length of the leg is 22.91 units.

Whole numbers that satisfy the Pythagorean Theorem are called **Pythagorean triples**. Multiples of Pythagorean triples also satisfy the Pythagorean Theorem. Some common triples are (3, 4, 5), (5, 12, 13), (8, 15, 17), and (7, 24, 25).

Example 3 Pythagorean Triples

Multiple-Choice Test Item

What is the area of triangle *ABC*?

Ⓐ 96 units^2　Ⓑ 120 units^2

Ⓒ 160 units^2　Ⓓ 196 units^2

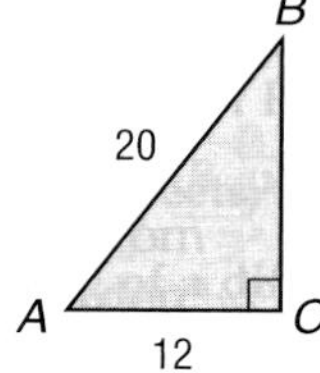

Read the Test Item

The area of a triangle is $A = \frac{1}{2}bh$. In a right triangle, the legs form the base and height of the triangle. Use the measures of the hypotenuse and the base to find the height of the triangle.

Solve the Test Item

Step 1 Check to see if the measurements of this triangle are a multiple of a common Pythagorean triple. The hypotenuse is $4 \cdot 5$ units, and the leg is $4 \cdot 3$ units. This triangle is a multiple of a (3, 4, 5) triangle.

$4 \cdot 3 = 12$

$4 \cdot 4 = 16$

$4 \cdot 5 = 20$

The height of the triangle is 16 units.

Step 2 Find the area of the triangle.

$A = \frac{1}{2}bh$	Area of a triangle
$A = \frac{1}{2} \cdot 12 \cdot 16$	$b = 12$ and $h = 16$
$A = 96$	Simplify.

The area of the triangle is 96 square units. Choice A is correct.

> **The Princeton Review**
> **Test-Taking Tip**
> Memorize the common Pythagorean triples and check for multiples such as (6, 8, 10). This will save you time when evaluating square roots.

RIGHT TRIANGLES A statement that can be easily proved using a theorem is often called a **corollary**. The following corollary, based on the Pythagorean Theorem, can be used to determine whether a triangle is a right triangle.

Key Concept *Corollary to the Pythagorean Theorem*

If a and b are measures of the shorter sides of a triangle, c is the measure of the longest side, and $c^2 = a^2 + b^2$, then the triangle is a right triangle. If $c^2 \neq a^2 + b^2$, then the triangle is not a right triangle.

Example 4 Check for Right Triangles

Determine whether the following side measures form right triangles.

a. 20, 21, 29

Since the measure of the longest side is 29, let $c = 29$, $a = 20$, and $b = 21$. Then determine whether $c^2 = a^2 + b^2$.

$c^2 = a^2 + b^2$	Pythagorean Theorem
$29^2 \stackrel{?}{=} 20^2 + 21^2$	$a = 20$, $b = 21$, and $c = 29$
$841 \stackrel{?}{=} 400 + 441$	Multiply.
$841 = 841$	Add.

Since $c^2 = a^2 + b^2$, the triangle is a right triangle.

b. 8, 10, 12

Since the measure of the longest side is 12, let $c = 12$, $a = 8$, and $b = 10$. Then determine whether $c^2 = a^2 + b^2$.

$c^2 = a^2 + b^2$	Pythagorean Theorem
$12^2 \stackrel{?}{=} 8^2 + 10^2$	$a = 8$, $b = 10$, and $c = 12$
$144 \stackrel{?}{=} 64 + 100$	Multiply.
$144 \neq 164$	Add.

Since $c^2 \neq a^2 + b^2$, the triangle is not a right triangle.

Check for Understanding

Concept Check

1. **OPEN ENDED** Draw a right triangle and label each side and angle. Be sure to indicate the right angle.
2. **Explain** how you can determine which angle is the right angle of a right triangle if you are given the lengths of the three sides.
3. **Write** an equation you could use to find the length of the diagonal d of a square with side length s.

Guided Practice

Find the length of each missing side. If necessary, round to the nearest hundredth.

4. 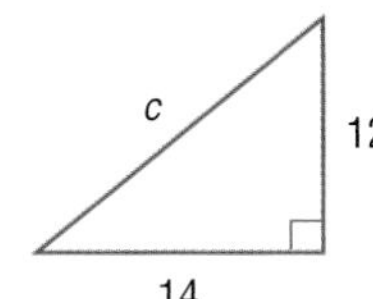

5. 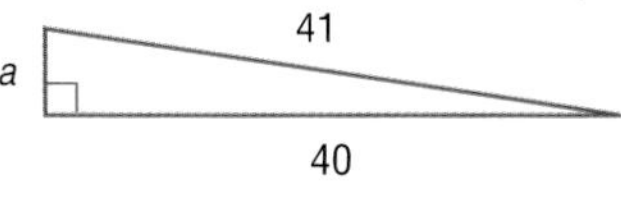

If c is the measure of the hypotenuse of a right triangle, find each missing measure. If necessary, round to the nearest hundredth.

6. $a = 10, b = 24, c = ?$

7. $a = 11, c = 61, b = ?$

8. $b = 13, c = \sqrt{233}, a = ?$

9. $a = 7, b = 4, c = ?$

Determine whether the following side measures form right triangles. Justify your answer.

10. 4, 6, 9

11. 16, 30, 34

12. In right triangle XYZ, the length of $\overline{YZ}$ is 6, and the length of the hypotenuse is 8. Find the area of the triangle.

Ⓐ $6\sqrt{7}$ units^2 Ⓑ 30 units^2 Ⓒ 40 units^2 Ⓓ 48 units^2

Practice and Apply

Homework Help

For Exercises	See Examples
13–30	1, 2
31–36	4
37–40	3

Extra Practice

See page 845.

Find the length of each missing side. If necessary, round to the nearest hundredth.

13.

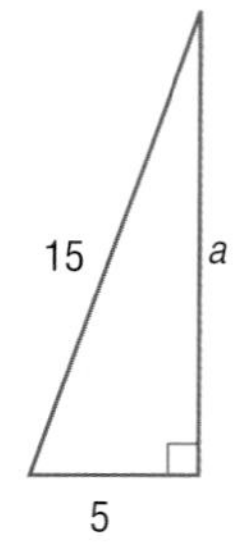

14.

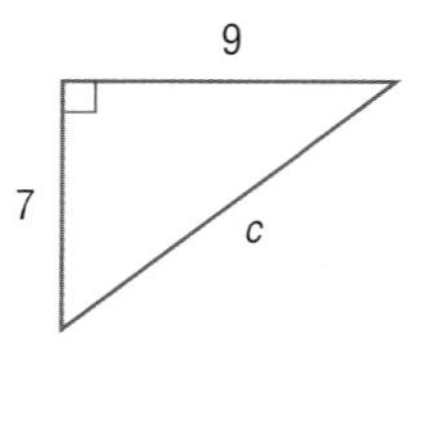

15.

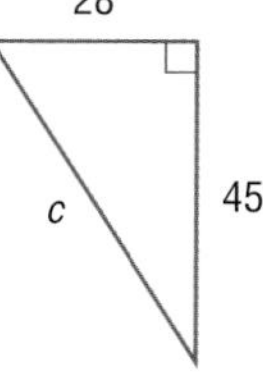

16.

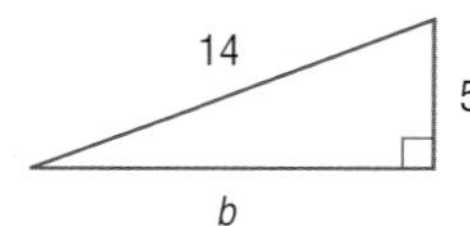

17.

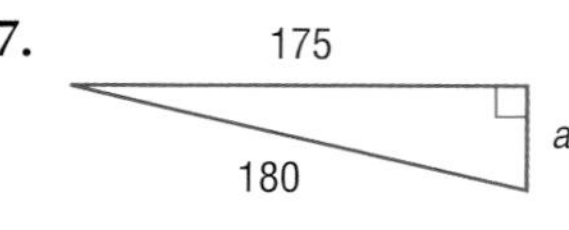

18.

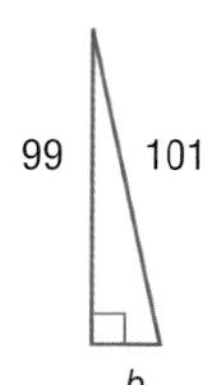

If c is the measure of the hypotenuse of a right triangle, find each missing measure. If necessary, round to the nearest hundredth.

19. $a = 16, b = 63, c = ?$

20. $a = 16, c = 34, b = ?$

21. $b = 3, a = \sqrt{112}, c = ?$

22. $a = \sqrt{15}, b = \sqrt{10}, c = ?$

23. $c = 14, a = 9, b = ?$

24. $a = 6, b = 3, c = ?$

25. $b = \sqrt{77}, c = 12, a = ?$

26. $a = 4, b = \sqrt{11}, c = ?$

27. $a = \sqrt{225}, b = \sqrt{28}, c = ?$

28. $a = \sqrt{31}, c = \sqrt{155}, b = ?$

29. $a = 8x, b = 15x, c = ?$

30. $b = 3x, c = 7x, a = ?$

Determine whether the following side measures form right triangles. Justify your answer.

31. 30, 40, 50

32. 6, 12, 18

33. 24, 30, 36

34. 45, 60, 75

35. 15, $\sqrt{31}$, 16

36. 4, 7, $\sqrt{65}$

Use an equation to solve each problem. If necessary, round to the nearest hundredth.

37. Find the length of a diagonal of a square if its area is 162 square feet.

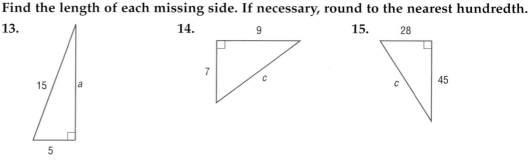

38. A right triangle has one leg that is 5 centimeters longer than the other leg. The hypotenuse is 25 centimeters long. Find the length of each leg of the triangle.

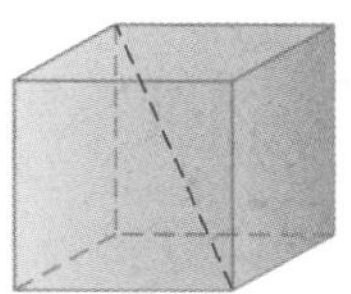

39. Find the length of the diagonal of the cube if each side of the cube is 4 inches long.

40. The ratio of the length of the hypotenuse to the length of the *shorter* leg in a right triangle is 8:5. The hypotenuse measures 144 meters. Find the length of the *longer* leg.

More About. . .

Roller Coasters

Roller Coaster Records in the U.S.
Fastest: *Millennium Force* (2000), Sandusky, Ohio; 92 mph
Tallest: *Millenium Force* (2000), Sandusky, Ohio; 310 feet
Longest: *California Screamin'* (2001), Anaheim, California; 6800 feet
Steepest: *Hypersonic XLC* (2001), Doswell, Virginia; 90°
Source: www.coasters2k.com

ROLLER COASTERS **For Exercises 41–43, use the following information and the figure.** Suppose a roller coaster climbs 208 feet higher than its starting point making a horizontal advance of 360 feet. When it comes down, it makes a horizontal advance of 44 feet.

41. How far will it travel to get to the top of the ride?

42. How far will it travel on the downhill track?

43. Compare the total horizontal advance, vertical height, and total track length.

44. **RESEARCH** Use the Internet or other reference to find the measurements of your favorite roller coaster or a roller coaster that is at an amusement park close to you. Draw a model of the first drop. Include the height of the hill, length of the vertical drop, and steepness of the hill.

45. **SAILING** A sailboat's mast and boom form a right angle. The sail itself, called a *mainsail*, is in the shape of a right triangle. If the edge of the mainsail that is attached to the mast is 100 feet long and the edge of the mainsail that is attached to the boom is 60 feet long, what is the length of the longest edge of the mainsail?

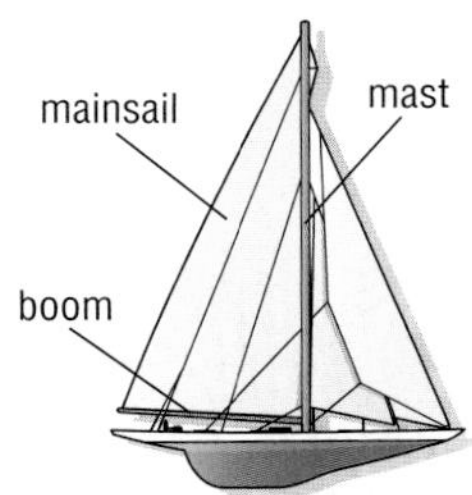

ROOFING **For Exercises 46 and 47, refer to the figures below.**

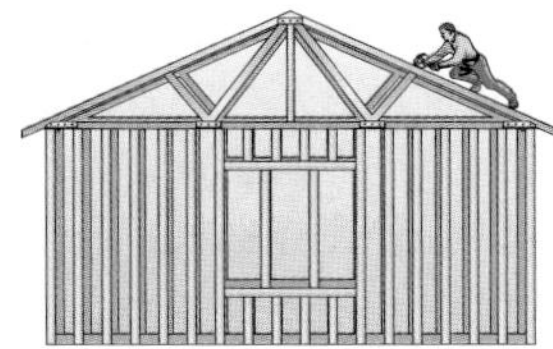

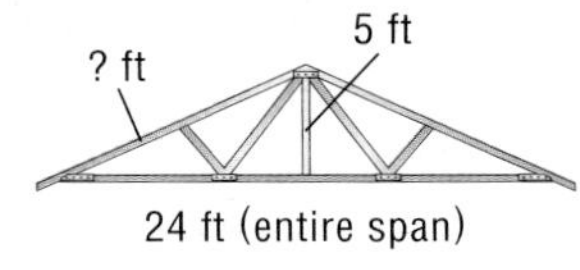

46. Determine the missing length shown in the rafter.

47. If the roof is 30 feet long and it hangs an additional 2 feet over the garage walls, how many square feet of shingles are needed for the entire garage roof?

48. **CRITICAL THINKING** Compare the area of the largest semicircle to the areas of the two smaller semicircles. Justify your reasoning.

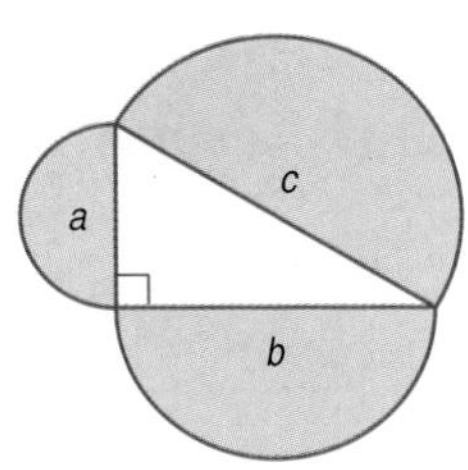

49. **CRITICAL THINKING** A model of a part of a roller coaster is shown. Determine the total distance traveled from start to finish and the maximum height reached by the roller coaster.

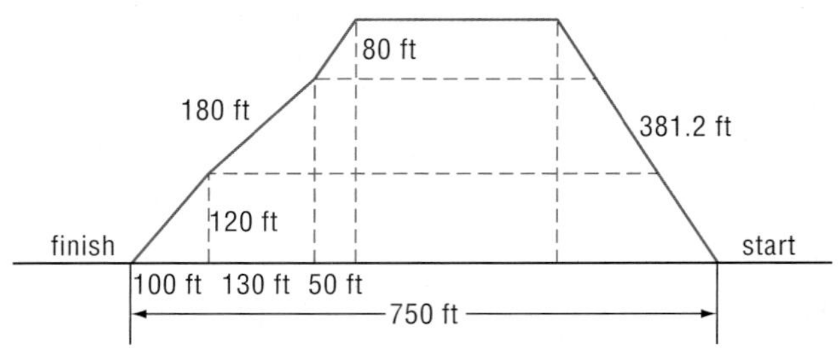

50. **WRITING IN MATH** Answer the question that was posed at the beginning of the lesson.

How is the Pythagorean Theorem used in roller coaster design?

Include the following in your answer:

- an explanation of how the height, speed, and steepness of a roller coaster are related, and
- a description of any limitations you can think of in the design of a new roller coaster.

51. Find the area of $\triangle XYZ$.

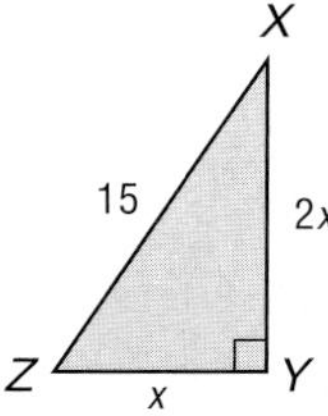

Ⓐ $6\sqrt{5}$ units^2 Ⓑ $18\sqrt{5}$ units^2

Ⓒ 45 units^2 Ⓓ 90 units^2

52. Find the perimeter of a square whose diagonal measures 10 centimeters.

Ⓐ $10\sqrt{2}$ cm Ⓑ $20\sqrt{2}$ cm

Ⓒ $25\sqrt{2}$ cm Ⓓ 80 cm

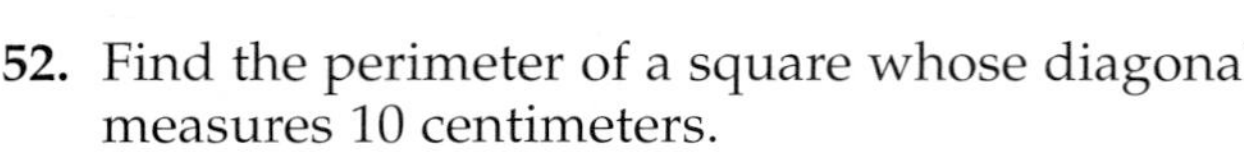

Maintain Your Skills

Mixed Review

Solve each equation. Check your solution. *(Lesson 11-3)*

53. $\sqrt{y} = 12$ 54. $3\sqrt{s} = 126$ 55. $4\sqrt{2v + 1} - 3 = 17$

Simplify each expression. *(Lesson 11-2)*

56. $\sqrt{72}$ 57. $7\sqrt{z} - 10\sqrt{z}$ 58. $\sqrt{\frac{3}{7}} + \sqrt{21}$

Simplify. Assume that no denominator is equal to zero. *(Lesson 8-2)*

59. $\frac{5^8}{5^3}$ 60. d^{-7} 61. $\frac{-26a^4b^7c^{-5}}{-13a^2b^4c^3}$

62. **AVIATION** Flying with the wind, a plane travels 300 miles in 40 minutes. Flying against the wind, it travels 300 miles in 45 minutes. Find the air speed of the plane. *(Lesson 7-4)*

Getting Ready for the Next Lesson

PREREQUISITE SKILL Simplify each expression.
*(To review **simplifying radical expressions**, see Lesson 11-1.)*

63. $\sqrt{(6-3)^2 + (8-4)^2}$ 64. $\sqrt{(10-4)^2 + (13-5)^2}$

65. $\sqrt{(5-3)^2 + (2-9)^2}$ 66. $\sqrt{(-9-5)^2 + (7-3)^2}$

67. $\sqrt{(-4-5)^2 + (-4-3)^2}$ 68. $\sqrt{(20-5)^2 + (-2-6)^2}$

11-5 The Distance Formula

What You'll Learn

- Find the distance between two points on the coordinate plane.
- Find a point that is a given distance from a second point in a plane.

How can the distance between two points be determined?

Consider two points A and B in the coordinate plane. Notice that a right triangle can be formed by drawing lines parallel to the axes through the points at A and B. These lines intersect at C forming a right angle. The hypotenuse of this triangle is the distance between A and B. You can determine the length of the legs of this triangle and use the Pythagorean Theorem to find the distance between the two points. Notice that AC is the difference of the y-coordinates, and BC is the difference of the x-coordinates.
So, $(AB)^2 = (AC)^2 + (BC)^2$, and $AB = \sqrt{(AC)^2 + (BC)^2}$.

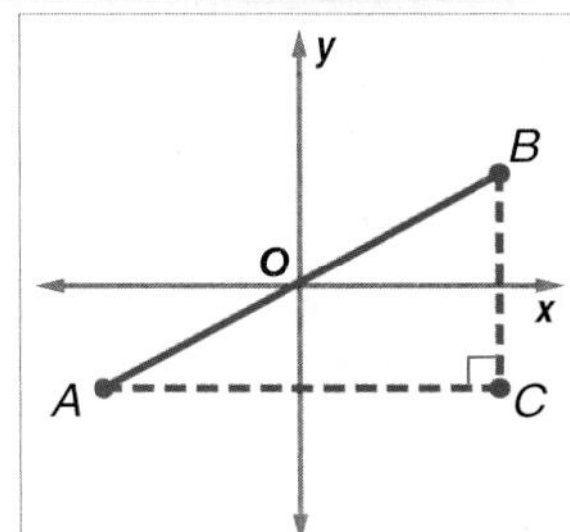

Study Tip

Reading Math
AC is the measure of $\overline{AC}$ and BC is the measure of $\overline{BC}$.

THE DISTANCE FORMULA You can find the distance between any two points in the coordinate plane using a similar process. The result is called the **Distance Formula**.

Key Concept — The Distance Formula

- **Words** The distance d between any two points with coordinates (x_1, y_1) and (x_2, y_2) is given by $d = \sqrt{(x_2 - x_1)^2 + (y_2 - y_1)^2}$.
- **Model**

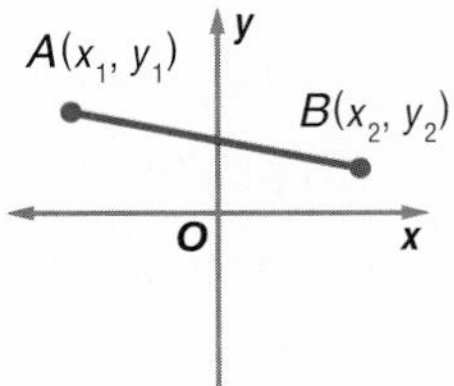

Example 1 Distance Between Two Points

Find the distance between the points at (2, 3) and (−4, 6).

$d = \sqrt{(x_2 - x_1)^2 + (y_2 - y_1)^2}$ Distance Formula

$= \sqrt{(-4 - 2)^2 + (6 - 3)^2}$ $(x_1, y_1) = (2, 3)$ and $(x_2, y_2) = (-4, 6)$

$= \sqrt{(-6)^2 + 3^2}$ Simplify.

$= \sqrt{45}$ Evaluate squares and simplify.

$= 3\sqrt{5}$ or about 6.71 units

Example 2 Use the Distance Formula

GOLF **Tracy hits a golf ball that lands 20 feet short and 8 feet to the right of the cup. On her first putt, the ball lands 2 feet to the left and 3 feet beyond the cup. Assuming that the ball traveled in a straight line, how far did the ball travel on her first putt?**

Draw a model of the situation on a coordinate grid. If the cup is at (0, 0), then the location of the ball after the first hit is (8, −20). The location of the ball after the first putt is (−2, 3). Use the Distance Formula.

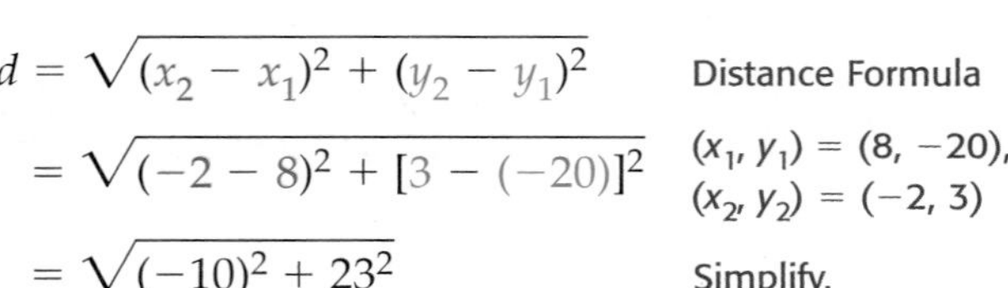

$d = \sqrt{(x_2 - x_1)^2 + (y_2 - y_1)^2}$ Distance Formula

$= \sqrt{(-2 - 8)^2 + [3 - (-20)]^2}$ $(x_1, y_1) = (8, -20)$, $(x_2, y_2) = (-2, 3)$

$= \sqrt{(-10)^2 + 23^2}$ Simplify.

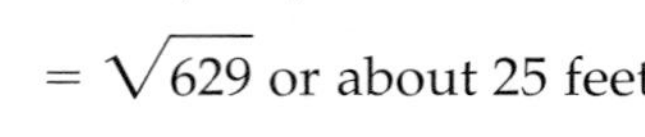

$= \sqrt{629}$ or about 25 feet

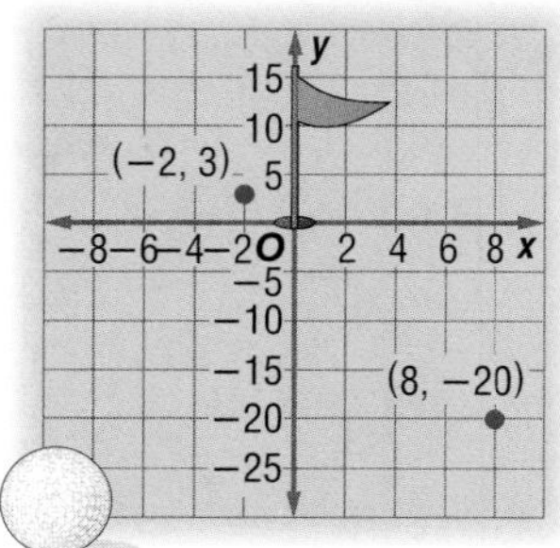

More About. . .

Golf

There are four major tournaments that make up the "grand slam" of golf: Masters, U.S. Open, British Open, and PGA Championship. In 2000, Tiger Woods became the youngest player to win the four major events (called a career grand slam) at age 24.

Source: PGA

FIND COORDINATES Suppose you know the coordinates of a point, one coordinate of another point, and the distance between the two points. You can use the Distance Formula to find the missing coordinate.

Example 3 Find a Missing Coordinate

Find the value of a if the distance between the points at (7, 5) and $(a, -3)$ is 10 units.

$d = \sqrt{(x_2 - x_1)^2 + (y_2 - y_1)^2}$ Distance Formula

$10 = \sqrt{(a - 7)^2 + (-3 - 5)^2}$ Let $x_2 = a$, $x_1 = 7$, $y_2 = -3$, $y_1 = 5$,

$10 = \sqrt{(a - 7)^2 + (-8)^2}$ and $d = 10$.

$10 = \sqrt{a^2 - 14a + 49 + 64}$ Evaluate squares.

$10 = \sqrt{a^2 - 14a + 113}$ Simplify.

$10^2 = \left(\sqrt{a^2 - 14a + 113}\right)^2$ Square each side.

$100 = a^2 - 14a + 113$ Simplify.

$0 = a^2 - 14a + 13$ Subtract 100 from each side.

$0 = (a - 1)(a - 13)$ Factor.

$a - 1 = 0$ or $a - 13 = 0$ Zero Product Property

$a = 1$ $a = 13$ The value of a is 1 or 13.

Check for Understanding

Concept Check

1. **Explain** why the value calculated under the radical sign in the Distance Formula will never be negative.

2. **OPEN ENDED** Plot two ordered pairs and find the distance between their graphs. Does it matter which ordered pair is first when using the Distance Formula? Explain.

3. **Explain** why there are two values for a in Example 3. Draw a diagram to support your answer.

Guided Practice

Find the distance between each pair of points whose coordinates are given. Express in simplest radical form and as decimal approximations rounded to the nearest hundredth if necessary.

4. $(5, -1), (11, 7)$
5. $(3, 7), (-2, -5)$
6. $(2, 2), (5, -1)$
7. $(-3, -5), (-6, -4)$

Find the possible values of *a* if the points with the given coordinates are the indicated distance apart.

8. $(3, -1), (a, 7); d = 10$
9. $(10, a), (1, -6); d = \sqrt{145}$

Applications

10. **GEOMETRY** An isosceles triangle has two sides of equal length. Determine whether triangle *ABC* with vertices $A(-3, 4)$, $B(5, 2)$, and $C(-1, -5)$ is an isosceles triangle.

FOOTBALL For Exercises 11 and 12, use the information at the right.

11. A quarterback can throw the football to one of the two receivers. Find the distance from the quarterback to each receiver.
12. What is the distance between the two receivers?

Practice and Apply

Homework Help

For Exercises	See Examples
13–26, 33, 34	1
27–32, 35, 36	3
37–42	2

Extra Practice

See page 845.

Find the distance between each pair of points whose coordinates are given. Express in simplest radical form and as decimal approximations rounded to the nearest hundredth if necessary.

13. $(12, 3), (-8, 3)$
14. $(0, 0), (5, 12)$
15. $(6, 8), (3, 4)$
16. $(-4, 2), (4, 17)$
17. $(-3, 8), (5, 4)$
18. $(9, -2), (3, -6)$
19. $(-8, -4), (-3, -8)$
20. $(2, 7), (10, -4)$
21. $(4, 2), \left(6, -\frac{2}{3}\right)$
22. $\left(5, \frac{1}{4}\right), (3, 4)$
23. $\left(\frac{4}{5}, -1\right), \left(2, -\frac{1}{2}\right)$
24. $\left(3, \frac{3}{7}\right), \left(4, -\frac{2}{7}\right)$
25. $(4\sqrt{5}, 7), (6\sqrt{5}, 1)$
26. $(5\sqrt{2}, 8), (7\sqrt{2}, 10)$

Find the possible values of *a* if the points with the given coordinates are the indicated distance apart.

27. $(4, 7), (a, 3); d = 5$
28. $(-4, a), (4, 2); d = 17$
29. $(5, a), (6, 1); d = \sqrt{10}$
30. $(a, 5), (-7, 3); d = \sqrt{29}$
31. $(6, -3), (-3, a); d = \sqrt{130}$
32. $(20, 5), (a, 9); d = \sqrt{340}$

33. Triangle *ABC* has vertices at $A(7, -4)$, $B(-1, 2)$, and $C(5, -6)$. Determine whether the triangle has three, two, or no sides that are equal in length.

34. If the diagonals of a trapezoid have the same length, then the trapezoid is isosceles. Find the lengths of the diagonals of trapezoid *ABCD* with vertices $A(-2, 2)$, $B(10, 6)$, $C(9, 8)$, and $D(0, 5)$ to determine if it is isosceles.

35. Triangle LMN has vertices at $L(-4, -3)$, $M(2, 5)$, and $N(-13, 10)$. If the distance from point $P(x, -2)$ to L equals the distance from P to M, what is the value of x?

36. Plot the points $Q(1, 7)$, $R(3, 1)$, $S(9, 3)$, and $T(7, d)$. Find the value of d that makes each side of $QRST$ have the same length.

37. **FREQUENT FLYERS** To determine the mileage between cities for their frequent flyer programs, some airlines superimpose a coordinate grid over the United States. An ordered pair on the grid represents the location of each airport. The units of this grid are approximately equal to 0.316 mile. So, a distance of 3 units on the grid equals an actual distance of 3(0.316) or 0.948 mile. Suppose the locations of two airports are at (132, 428) and (254, 105). Find the actual distance between these airports to the nearest mile.

COLLEGE For Exercises 38 and 39, use the map of a college campus.

38. Kelly has her first class in Rhodes Hall and her second class in Fulton Lab. How far does she have to walk between her first and second class?

39. She has 12 minutes between the end of her first class and the start of her second class. If she walks an average of 3 miles per hour, will she make it to her second class on time?

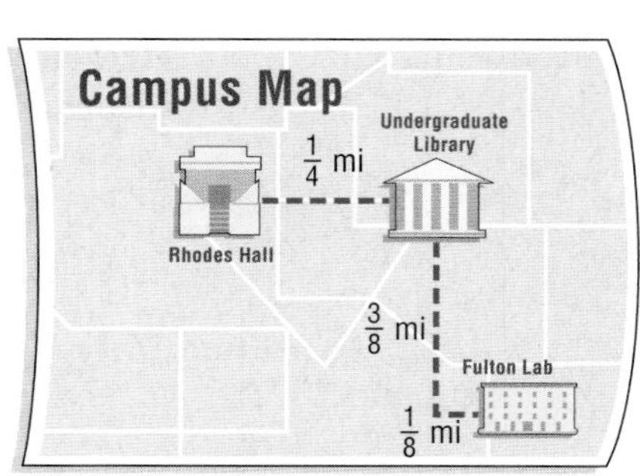

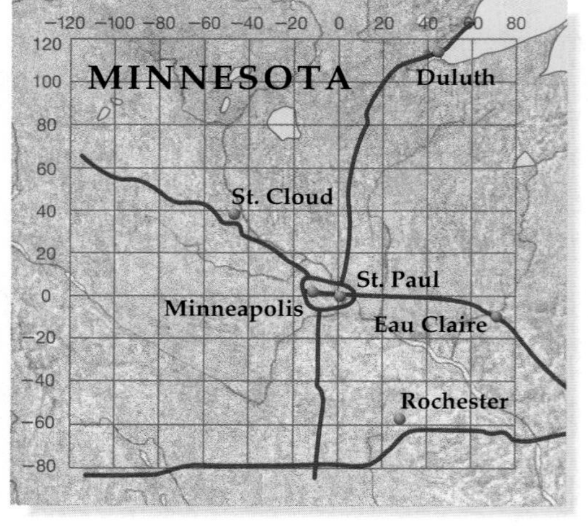

GEOGRAPHY For Exercises 40–42, use the map at the left that shows part of Minnesota and Wisconsin.
A coordinate grid has been superimposed on the map with the origin at St. Paul. The grid lines are 20 miles apart. Minneapolis is at $(-7, 3)$.

40. Estimate the coordinates for Duluth, St. Cloud, Eau Claire, and Rochester.

41. Find the distance between the following pairs of cities: Minneapolis and St. Cloud, St. Paul and Rochester, Minneapolis and Eau Claire, and Duluth and St. Cloud.

42. A radio station in St. Paul has a broadcast range of 75 miles. Which cities shown on the map can receive the broadcast?

43. **CRITICAL THINKING** Plot $A(-4, 4)$, $B(-7, -3)$, and $C(4, 0)$, and connect them to form triangle ABC. Demonstrate two different ways to show whether ABC is a right triangle.

44. **WRITING IN MATH** Answer the question that was posed at the beginning of the lesson.

How can the distance between two points be determined?

Include the following in your answer:

- an explanation how the Distance Formula is derived from the Pythagorean Theorem, and
- an explanation why the Distance Formula is not needed to find the distance between points $P(-24, 18)$ and $Q(-24, 10)$.

Standardized Test Practice

45. Find the distance between points at (6, 11) and (−2, −4).

Ⓐ 16 units Ⓑ 17 units
Ⓒ 18 units Ⓓ 19 units

46. Find the perimeter of a square $ABCD$ if two of the vertices are $A(3, 7)$ and $B(-3, 4)$.

Ⓐ 12 units Ⓑ $12\sqrt{5}$ units
Ⓒ $9\sqrt{5}$ units Ⓓ 45 units

Maintain Your Skills

Mixed Review

If c is the measure of the hypotenuse of a right triangle, find each missing measure. If necessary, round to the nearest hundredth. *(Lesson 11-4)*

47. $a = 7, b = 24, c = ?$
48. $b = 30, c = 34, a = ?$
49. $a = \sqrt{7}, c = \sqrt{16}, b = ?$
50. $a = \sqrt{13}, b = \sqrt{50}, c = ?$

Solve each equation. Check your solution. *(Lesson 11-3)*

51. $\sqrt{p - 2} + 8 = p$
52. $\sqrt{r + 5} = r - 1$
53. $\sqrt{5t^2 + 29} = 2t + 3$

COST OF DEVELOPMENT **For Exercises 54–56, use the graph that shows the amount of money being spent on worldwide construction.** *(Lesson 8-3)*

54. Write the value shown for each continent or region listed in standard notation.
55. Write the value shown for each continent or region in scientific notation.
56. How much more money is being spent in Asia than in Latin America?

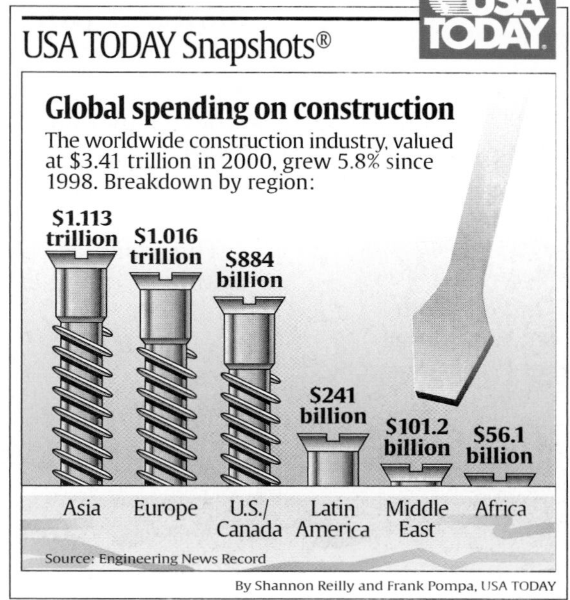

Solve each inequality. Then check your solution and graph it on a number line. *(Lesson 6-1)*

57. $8 \leq m - 1$
58. $3 > 10 + k$
59. $3x \leq 2x - 3$
60. $v - (-4) > 6$
61. $r - 5.2 \geq 3.9$
62. $s + \frac{1}{6} \leq \frac{2}{3}$

Getting Ready for the Next Lesson

PREREQUISITE SKILL **Solve each proportion.** *(To review **proportions**, see Lesson 3-6.)*

63. $\frac{x}{4} = \frac{3}{2}$
64. $\frac{20}{x} = \frac{-5}{2}$
65. $\frac{6}{9} = \frac{8}{x}$
66. $\frac{10}{12} = \frac{x}{18}$
67. $\frac{x + 2}{7} = \frac{3}{7}$
68. $\frac{2}{3} = \frac{6}{x + 4}$

11-6 Similar Triangles

What You'll Learn

- Determine whether two triangles are similar.
- Find the unknown measures of sides of two similar triangles.

Vocabulary

- similar triangles

How are similar triangles related to photography?

When you take a picture, the image of the object being photographed is projected by the camera lens onto the film. The height of the image on the film can be related to the height of the object using similar triangles.

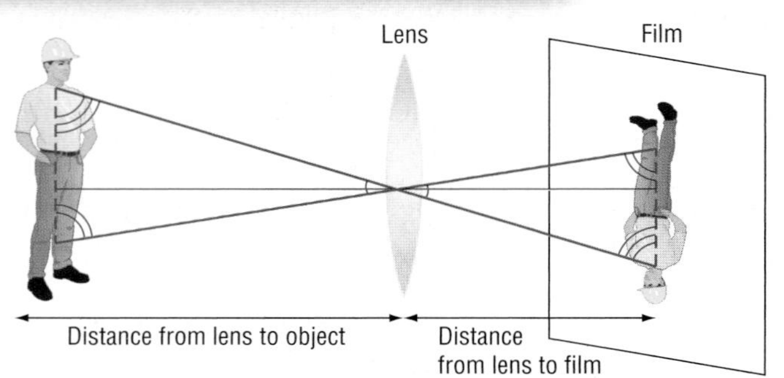

SIMILAR TRIANGLES **Similar triangles** have the same shape, but not necessarily the same size. There are two main tests for similarity.

- If the angles of one triangle and the corresponding angles of a second triangle have equal measures, then the triangles are similar.
- If the measures of the sides of two triangles form equal ratios, or are *proportional*, then the triangles are similar.

Study Tip

Reading Math
The symbol ~ is read *is similar to*.

The triangles below are similar. This is written as $\triangle ABC \sim \triangle DEF$. The vertices of similar triangles are written in order to show the corresponding parts.

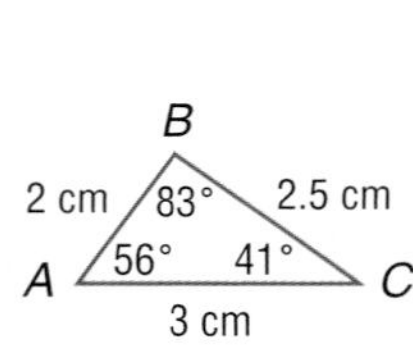

E
83°
4 cm
5 cm
D
56°
41°
F
6 cm

corresponding angles	**corresponding sides**
$\angle A$ and $\angle D$	$\overline{AB}$ and $\overline{DE} \rightarrow \frac{AB}{DE} = \frac{2}{4} = \frac{1}{2}$
$\angle B$ and $\angle E$	$\overline{BC}$ and $\overline{EF} \rightarrow \frac{BC}{EF} = \frac{2.5}{5} = \frac{1}{2}$
$\angle C$ and $\angle F$	$\overline{AC}$ and $\overline{DF} \rightarrow \frac{AC}{DF} = \frac{3}{6} = \frac{1}{2}$

Key Concept *Similar Triangles*

- **Words** If two triangles are similar, then the measures of their corresponding sides are proportional, and the measures of their corresponding angles are equal.
- **Symbols** If $\triangle ABC \sim \triangle DEF$, then $\frac{AB}{DE} = \frac{BC}{EF} = \frac{AC}{DF}$.
- **Model**

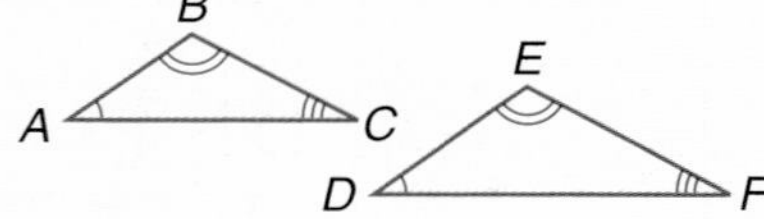

Study Tip

Reading Math
Arcs are used to show angles that have equal measures.

Example 1 Determine Whether Two Triangles Are Similar

Determine whether the pair of triangles is similar. Justify your answer.

Remember that the sum of the measures of the angles in a triangle is 180°.

The measure of $\angle P$ is $180° - (51° + 51°)$ or $78°$.

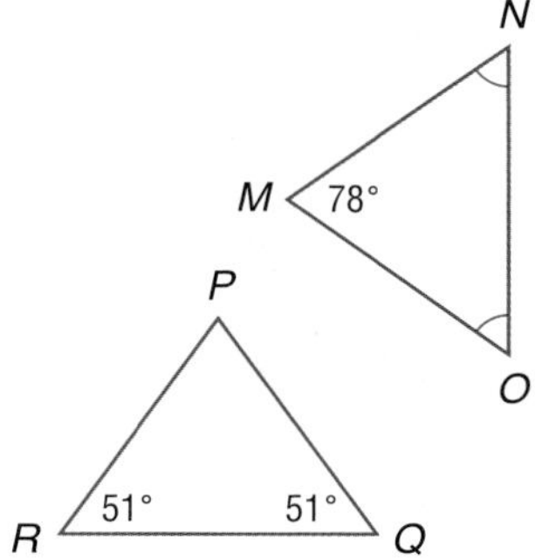

In $\triangle MNO$, $\angle N$ and $\angle O$ have the same measure.

Let x = the measure of $\angle N$ and $\angle O$.

$$x + x + 78° = 180°$$
$$2x = 102°$$
$$x = 51°$$

So $\angle N = 51°$ and $\angle O = 51°$. Since the corresponding angles have equal measures, $\triangle MNO \sim \triangle PQR$.

FIND UNKNOWN MEASURES Proportions can be used to find the measures of the sides of similar triangles when some of the measurements are known.

Example 2 Find Missing Measures

Find the missing measures if each pair of triangles below is similar.

a. Since the corresponding angles have equal measures, $\triangle TUV \sim \triangle WXY$. The lengths of the corresponding sides are proportional.

Study Tip

Corresponding Vertices Always use the corresponding order of the vertices to write proportions for similar triangles.

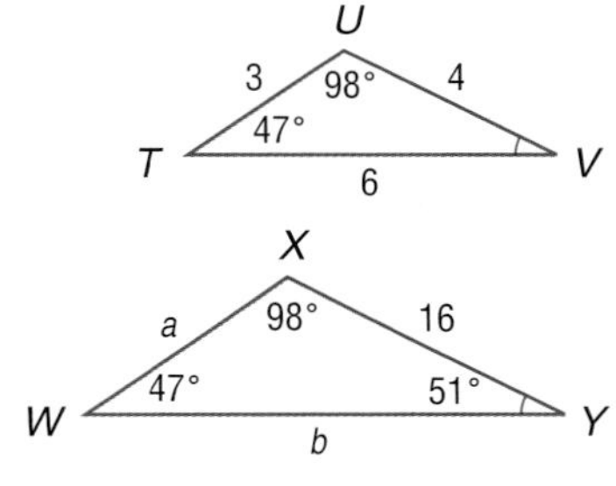

$\frac{WX}{TU} = \frac{XY}{UV}$ Corresponding sides of similar triangles are proportional.

$\frac{a}{3} = \frac{16}{4}$ $WX = a, XY = 16, TU = 3, UV = 4$

$4a = 48$ Find the cross products.

$a = 12$ Divide each side by 4.

$\frac{WY}{TV} = \frac{XY}{UV}$ Corresponding sides of similar triangles are proportional.

$\frac{b}{6} = \frac{16}{4}$ $WY = b, XY = 16, TV = 6, UV = 4$

$4b = 96$ Find the cross products.

$b = 24$ Divide each side by 4.

The missing measures are 12 and 24.

b. $\triangle ABE \sim \triangle ACD$

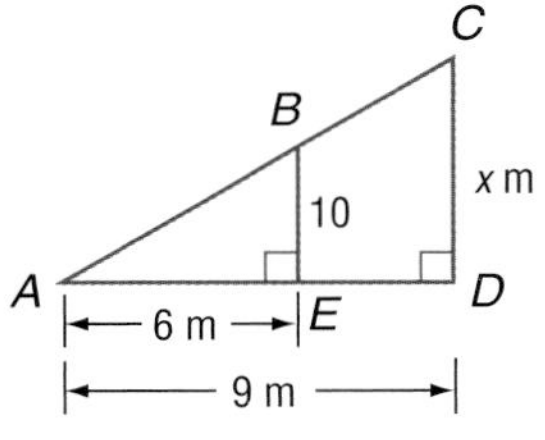

$\frac{BE}{CD} = \frac{AE}{AD}$ Corresponding sides of similar triangles are proportional.

$\frac{10}{x} = \frac{6}{9}$ $BE = 10, CD = x, AE = 6, AD = 9$

$90 = 6x$ Find the cross products.

$15 = x$ Divide each side by 6.

The missing measure is 15.

Example 3 Use Similar Triangles to Solve a Problem

More About. . .

Washington Monument

The monument has a shape of an Egyptian *obelisk*. A pyramid made of solid aluminum caps the top of the monument.

SHADOWS Jenelle is standing near the Washington Monument in Washington, D.C. The shadow of the monument is 302.5 feet, and Jenelle's shadow is 3 feet. If Jenelle is 5.5 feet tall, how tall is the monument?

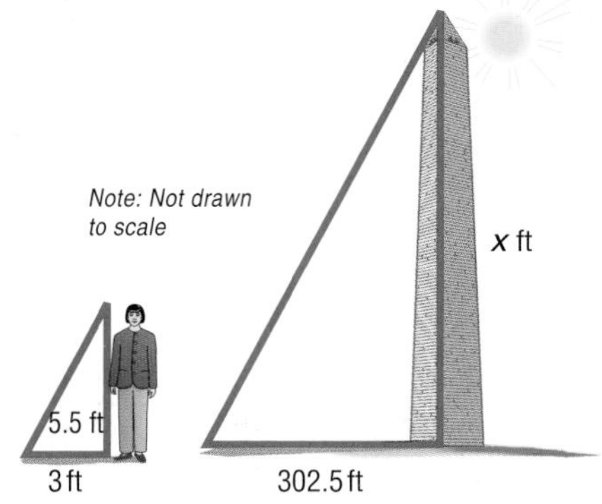

The shadows form similar triangles. Write a proportion that compares the heights of the objects and the lengths of their shadows.

Let x = the height of the monument.

Jenelle's shadow → / monument's shadow → $\frac{3}{302.5} = \frac{5.5}{x}$ ← Jenelle's height / ← monument's height

$3x = 1663.75$ Cross products

$x \approx 554.6$ feet Divide each side by 3.

The height of the monument is about 554.6 feet.

Check for Understanding

Concept Check

1. **Explain** how to determine whether two triangles are similar.
2. **OPEN ENDED** Draw a pair of similar triangles. List the corresponding angles and the corresponding sides.
3. **FIND THE ERROR** Russell and Consuela are comparing the similar triangles below to determine their corresponding parts.

Russell	Consuela
$m\angle X = m\angle T$	$m\angle X = m\angle V$
$m\angle Y = m\angle U$	$m\angle Y = m\angle U$
$m\angle Z = m\angle V$	$m\angle Z = m\angle T$
$\triangle XYZ, \triangle TUV$	$\triangle XYZ, \triangle VUT$

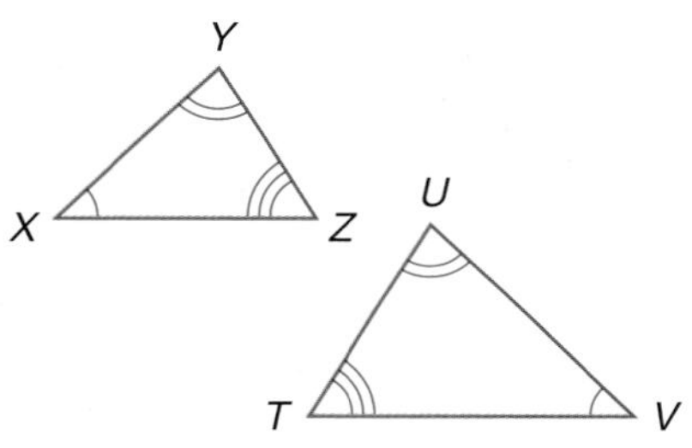

Who is correct? Explain your reasoning.

Guided Practice

Determine whether each pair of triangles is similar. Justify your answer.

4. 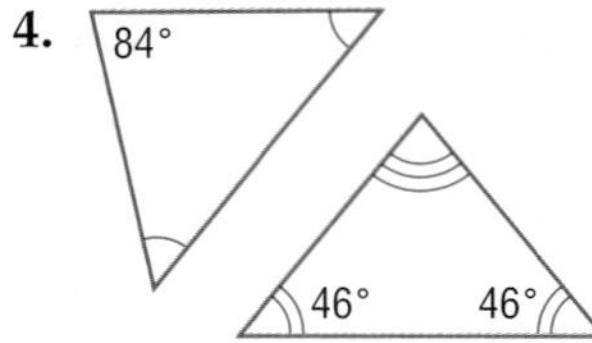

5. 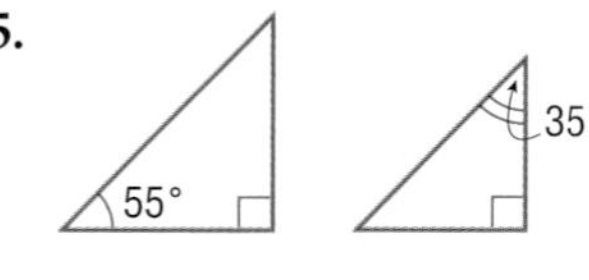

For each set of measures given, find the measures of the missing sides if $\triangle ABC \sim \triangle DEF$.

6. $c = 15, d = 7, e = 9, f = 5$
7. $a = 18, c = 9, e = 10, f = 6$
8. $a = 5, d = 7, f = 6, e = 5$
9. $a = 17, b = 15, c = 10, f = 6$

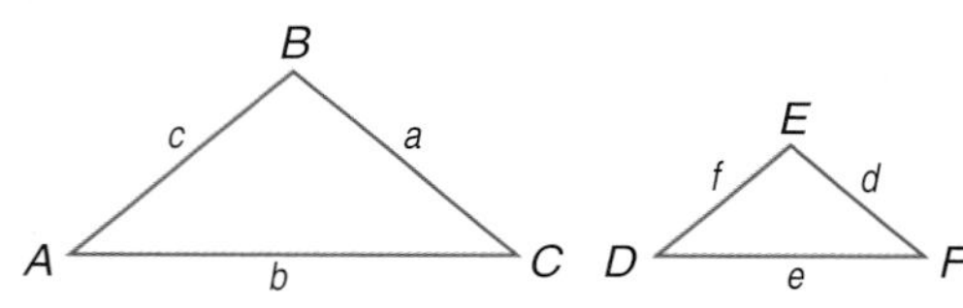

Application **10. SHADOWS** If a 25-foot flagpole casts a shadow that is 10 feet long and the nearby school building casts a shadow that is 26 feet long, how high is the building?

Practice and Apply

Homework Help

For Exercises	See Examples
11–16	1
17–24	2
25–32	3

Extra Practice

See page 845.

Determine whether each pair of triangles is similar. Justify your answer.

11.–12.

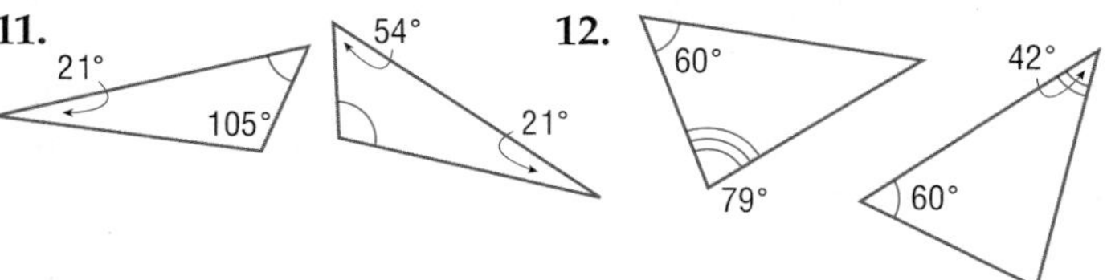

13.

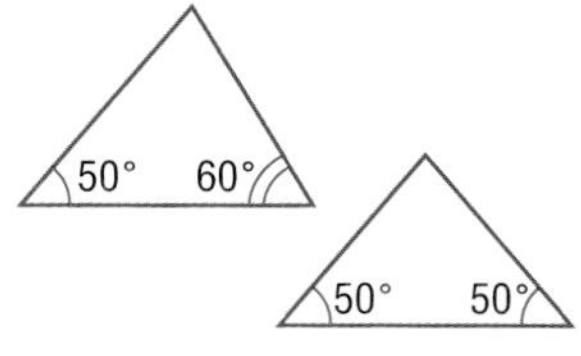

14.

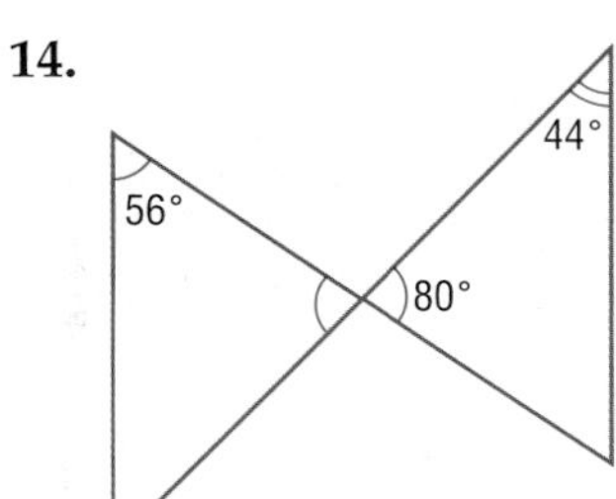

15.

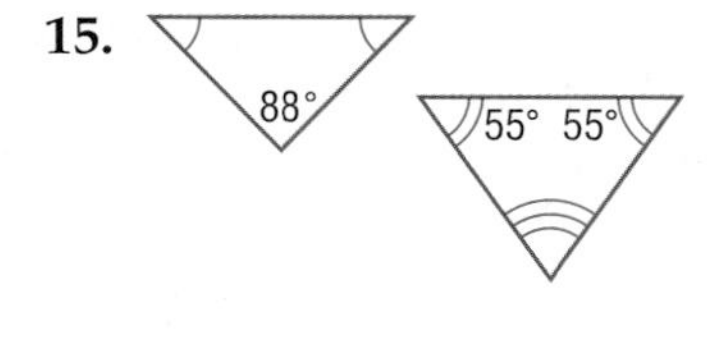

16.

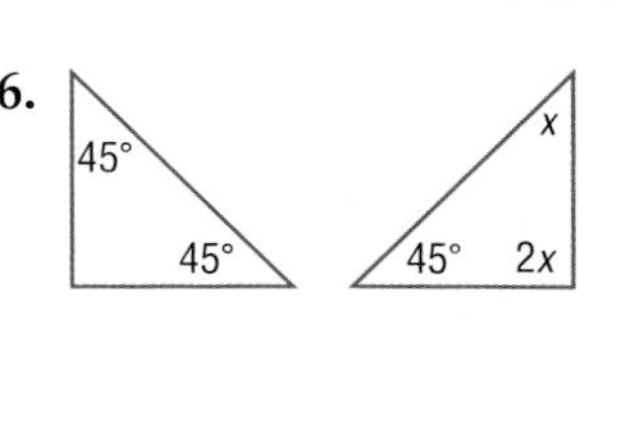

For each set of measures given, find the measures of the missing sides if $\triangle KLM \sim \triangle NOP$.

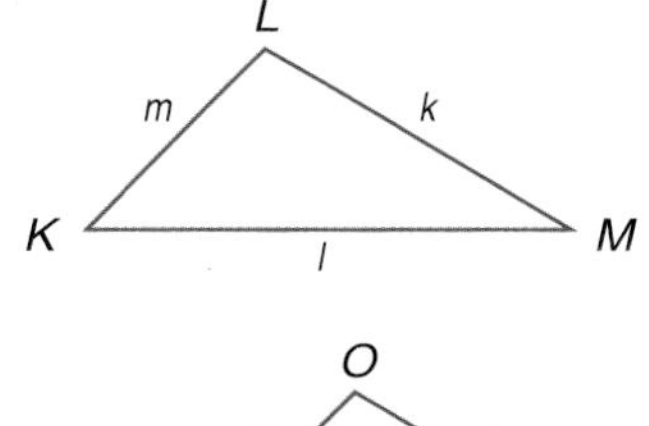

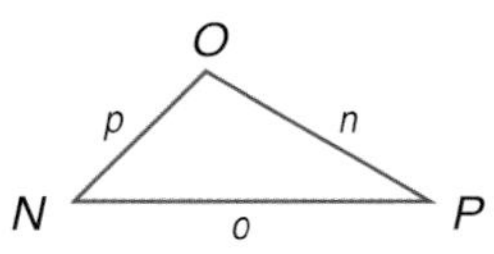

17. $k = 9, n = 6, o = 8, p = 4$
18. $k = 24, \ell = 30, m = 15, n = 16$
19. $m = 11, p = 6, n = 5, o = 4$
20. $k = 16, \ell = 13, m = 12, o = 7$
21. $n = 6, p = 2.5, \ell = 4, m = 1.25$
22. $p = 5, k = 10.5, \ell = 15, m = 7.5$
23. $n = 2.1, \ell = 4.4, p = 2.7, o = 3.3$
24. $m = 5, k = 12.6, o = 8.1, p = 2.5$

25. Determine whether the following statement is *sometimes*, *always*, or *never* true. *If the measures of the sides of a triangle are multiplied by 3, then the measures of the angles of the enlarged triangle will have the same measures as the angles of the original triangle.*

26. **PHOTOGRAPHY** Refer to the diagram of a camera at the beginning of the lesson. Suppose the image of a man who is 2 meters tall is 1.5 centimeters tall on film. If the film is 3 centimeters from the lens of the camera, how far is the man from the camera?

27. **BRIDGES** Truss bridges use triangles in their support beams. Mark plans to make a model of a truss bridge in the scale 1 inch = 12 feet. If the height of the triangles on the actual bridge is 40 feet, what will the height be on the model?

28. **BILLIARDS** Lenno is playing billiards on a table like the one shown at the right. He wants to strike the cue ball at *D*, bank it at *C*, and hit another ball at the mouth of pocket *A*. Use similar triangles to find where Lenno's cue ball should strike the rail.

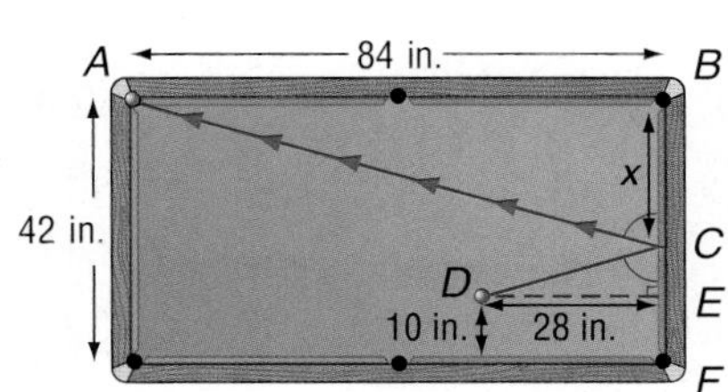

CRAFTS **For Exercises 29 and 30, use the following information.**
Melinda is working on a quilt pattern containing isosceles right triangles whose sides measure 2 inches, 2 inches, and about 2.8 inches.

29. She has several square pieces of material that measure 4 inches on each side. From each square piece, how many triangles with the required dimensions can she cut?

30. She wants to enlarge the pattern to make similar triangles for the center of the quilt. What is the largest similar triangle she can cut from the square material?

MIRRORS **For Exercises 31 and 32, use the diagram and the following information.**
Viho wanted to measure the height of a nearby building. He placed a mirror on the pavement at point P, 80 feet from the base of the building. He then backed away until he saw an image of the top of the building in the mirror.

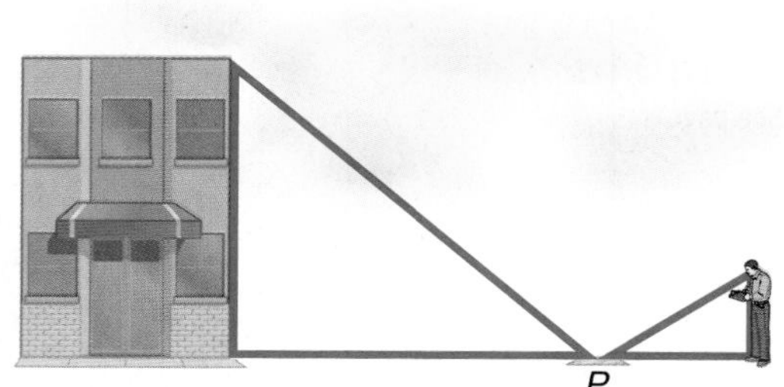

31. If Viho is 6 feet tall and he is standing 9 feet from the mirror, how tall is the building?

32. What assumptions did you make in solving the problem?

CRITICAL THINKING **For Exercises 33–35, use the following information.**
The radius of one circle is twice the radius of another.

33. Are the circles similar? Explain your reasoning.

34. What is the ratio of their circumferences? Explain your reasoning.

35. What is the ratio of their areas? Explain your reasoning.

36. **WRITING IN MATH** Answer the question that was posed at the beginning of the lesson.

How are similar triangles related to photography?

Include the following in your answer:
- an explanation of the effect of moving a camera with a zoom lens closer to the object being photographed, and
- a description of what you could do to fit the entire image of a large object on the picture.

For Exercises 37 and 38, use the figure at the right.

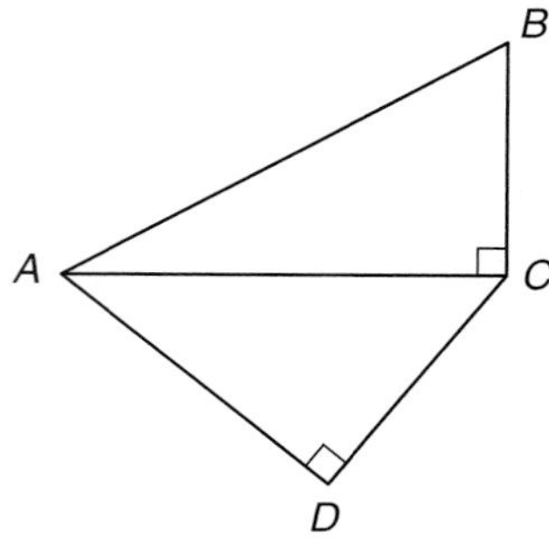

37. Which statement must be true?
 (A) $\triangle ABC \sim \triangle ADC$
 (B) $\triangle ABC \sim \triangle ACD$
 (C) $\triangle ABC \sim \triangle CAD$
 (D) none of the above

38. Which statement is always true?
 (A) $AB > DC$
 (B) $CB > AD$
 (C) $AC > BC$
 (D) $AC = AB$

Maintain Your Skills

Mixed Review **Find the distance between each pair of points whose coordinates are given. Express answers in simplest radical form and as decimal approximations rounded to the nearest hundredth if necessary.** *(Lesson 11-5)*

39. (1, 8), (−2, 4)
40. (6, −3), (12, 5)
41. (4, 7), (3, 12)
42. $(1, 5\sqrt{6}), (6, 7\sqrt{6})$

Determine whether the following side measures form right triangles. Justify your answer. *(Lesson 11-4)*

43. 25, 60, 65
44. 20, 25, 35
45. 49, 168, 175
46. 7, 9, 12

Arrange the terms of each polynomial so that the powers of the variable are in descending order. *(Lesson 8-4)*

47. $1 + 3x^2 - 7x$
48. $7 - 4x - 2x^2 + 5x^3$
49. $6x + 3 - 3x^2$
50. $abx^2 - bcx + 34 - x^7$

Use elimination to solve each system of equations. *(Lesson 7-3)*

51. $2x + y = 4$
$x - y = 5$

52. $3x - 2y = -13$
$2x - 5y = -5$

53. $0.6m - 0.2n = 0.9$
$0.3m = 0.45 - 0.1n$

54. $\frac{1}{3}x + \frac{1}{2}y = 8$
$\frac{1}{2}x - \frac{1}{4}y = 0$

55. AVIATION An airplane passing over Sacramento at an elevation of 37,000 feet begins its descent to land at Reno, 140 miles away. If the elevation of Reno is 4500 feet, what should be the approximate slope of descent? (*Hint*: 1 mi = 5280 ft) *(Lesson 5-1)*

Getting Ready for the Next Lesson **PREREQUISITE SKILL** **Evaluate if $a = 6$, $b = -5$, and $c = -1.5$.**
(To review ***evaluating expressions****, see Lesson 1-2.)*

56. $\frac{a}{c}$
57. $\frac{b}{a}$
58. $\frac{a + b}{c}$
59. $\frac{ac}{b}$
60. $\frac{b}{a + c}$
61. $\frac{c}{a + c}$

Practice Quiz 2 — Lessons 11-4 through 11-6

If c is the measure of the hypotenuse of a right triangle, find each missing measure. If necessary, round to the nearest hundredth. *(Lesson 11-4)*

1. $a = 14, b = 48, c = ?$
2. $a = 40, c = 41, b = ?$
3. $b = 8, c = \sqrt{84}, a = ?$
4. $a = \sqrt{5}, b = \sqrt{8}, c = ?$

Find the distance between each pair of points whose coordinates are given. *(Lesson 11-5)*

5. (6, −12), (−3, 3)
6. (1, 3), (−5, 11)
7. (2, 5), (4, 7)
8. (−2, −9), (−5, 4)

Find the measures of the missing sides if $\triangle BCA \sim \triangle EFD$. *(Lesson 11-6)*

9. $b = 10, d = 2, e = 1, f = 1.5$
10. $a = 12, c = 9, d = 8, e = 12$

Algebra Activity

A Preview of Lesson 11-7

Investigating Trigonometric Ratios

You can use paper triangles to investigate trigonometric ratios.

Collect the Data

Step 1 Use a ruler and grid paper to draw several right triangles whose legs are in a 7:10 ratio. Include a right triangle with legs 3.5 units and 5 units, a right triangle with legs 7 units and 10 units, another with legs 14 units and 20 units, and several more right triangles similar to these three. Label the vertices of each triangle as A, B, and C, where C is at the right angle, B is opposite the longest leg, and A is opposite the shortest leg.

Step 2 Copy the table below. Complete the first three columns by measuring the hypotenuse (side AB) in each right triangle you created and recording its length.

Step 3 Calculate and record the ratios in the middle two columns. Round to the nearest tenth, if necessary.

Step 4 Use a protractor to carefully measure angles A and B in each right triangle. Record the angle measures in the table.

Side Lengths			Ratios		Angle Measures		
side *BC*	side *AC*	side *AB*	*BC:AC*	*BC:AB*	angle *A*	angle *B*	angle *C*
3.5	5						90°
7	10						90°
14	20						90°
							90°
							90°
							90°

Analyze the Data

1. Examine the measures and ratios in the table. What do you notice? Write a sentence or two to describe any patterns you see.

Make a Conjecture

2. For any right triangle similar to the ones you have drawn here, what will be the value of the ratio of the length of the shortest leg to the length of the longest leg?
3. If you draw a right triangle and calculate the ratio of the length of the shortest leg to the length of the hypotenuse to be approximately 0.574, what will be the measure of the larger acute angle in the right triangle?

11-7 Trigonometric Ratios

What You'll Learn

- Define the sine, cosine, and tangent ratios.
- Use trigonometric ratios to solve right triangles.

Vocabulary

- trigonometric ratios
- sine
- cosine
- tangent
- solve a triangle
- angle of elevation
- angle of depression

How are trigonometric ratios used in surveying?

Surveyors use triangle ratios called trigonometric ratios to determine distances that cannot be measured directly.

- In 1852, British surveyors measured the altitude of the peak of Mt. Everest at 29,002 feet using these trigonometric ratios.
- In 1954, the official height became 29,028 feet, which was also calculated using surveying techniques.
- On November 11, 1999, a team using advanced technology and the Global Positioning System (GPS) satellite measured the mountain at 29,035 feet.

TRIGONOMETRIC RATIOS *Trigonometry* is an area of mathematics that involves angles and triangles. If enough information is known about a right triangle, certain ratios can be used to find the measures of the remaining parts of the triangle. **Trigonometric ratios** are ratios of the measures of two sides of a right triangle. Three common trigonometric ratios are called **sine**, **cosine**, and **tangent**.

Key Concept — *Trigonometric Ratios*

- **Words**

$$\text{sine of } \angle A = \frac{\text{measure of leg opposite } \angle A}{\text{measure of hypotenuse}}$$

$$\text{cosine of } \angle A = \frac{\text{measure of leg adjacent to } \angle A}{\text{measure of hypotenuse}}$$

$$\text{tangent of } \angle A = \frac{\text{measure of leg opposite } \angle A}{\text{measure of leg adjacent to } \angle A}$$

- **Symbols**

$$\sin A = \frac{BC}{AB}$$

$$\cos A = \frac{AC}{AB}$$

$$\tan A = \frac{BC}{AC}$$

- **Model**

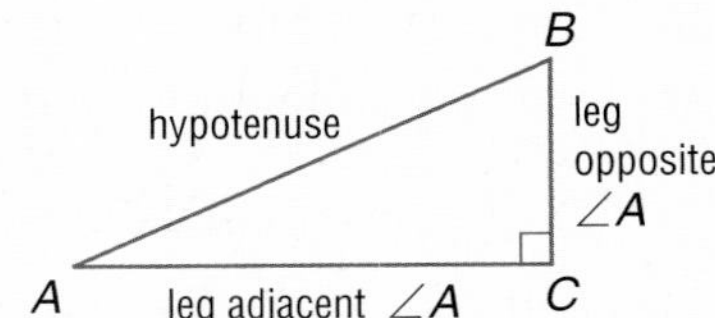

Study Tip

Reading Math
Notice that sine, cosine, and tangent are abbreviated sin, cos, and tan respectively.

Example 1 Sine, Cosine, and Tangent

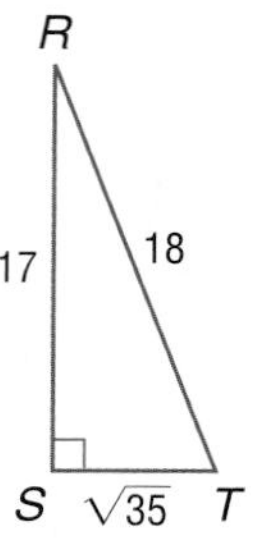

Find the sine, cosine, and tangent of each acute angle of $\triangle RST$. Round to the nearest ten thousandth.

Write each ratio and substitute the measures. Use a calculator to find each value.

$$\sin R = \frac{\text{opposite leg}}{\text{hypotenuse}} = \frac{\sqrt{35}}{18} \text{ or } 0.3287$$

$$\cos R = \frac{\text{adjacent leg}}{\text{hypotenuse}} = \frac{17}{18} \text{ or } 0.9444$$

$$\tan R = \frac{\text{opposite leg}}{\text{adjacent leg}} = \frac{\sqrt{35}}{17} \text{ or } 0.3480$$

$$\sin T = \frac{\text{opposite leg}}{\text{hypotenuse}} = \frac{17}{18} \text{ or } 0.9444$$

$$\cos T = \frac{\text{adjacent leg}}{\text{hypotenuse}} = \frac{\sqrt{35}}{18} \text{ or } 0.3287$$

$$\tan T = \frac{\text{opposite leg}}{\text{adjacent leg}} = \frac{17}{\sqrt{35}} \text{ or } 2.8735$$

You can use a calculator to find the values of trigonometric functions or to find the measure of an angle. On a graphing calculator, press the trigometric function key, and then enter the value. On a nongraphing scientific calculator, enter the value, and then press the function key. In either case, be sure your calculator is in degree mode. Consider cos 50°.

Graphing Calculator

KEYSTROKES: [COS] 50 [ENTER] .6427876097

Nongraphing Scientific Calculator

KEYSTROKES: 50 [COS] .642787609

Example 2 Find the Sine of an Angle

Find sin 35° to the nearest ten thousandth.

KEYSTROKES: [SIN] 35 [ENTER] .5735764364

Rounded to the nearest ten thousandth, $\sin 35° \approx 0.5736$.

Example 3 Find the Measure of an Angle

Find the measure of $\angle J$ to the nearest degree.

Since the lengths of the opposite and adjacent sides are known, use the tangent ratio.

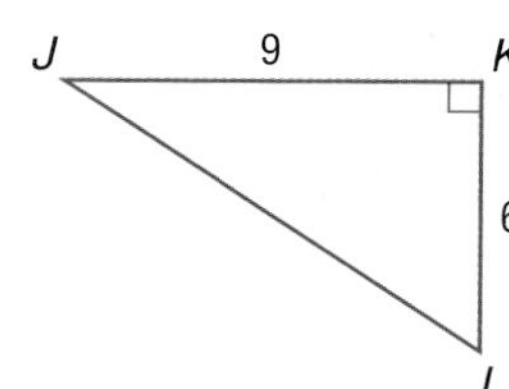

$\tan J = \frac{\text{opposite leg}}{\text{adjacent leg}}$ Definition of tangent

$= \frac{6}{9}$ $KL = 6$ and $JK = 9$

Now use the [TAN^{-1}] on a calculator to find the measure of the angle whose tangent ratio is $\frac{6}{9}$.

KEYSTROKES: [2nd] [TAN^{-1}] 6 [÷] 9 [ENTER] 33.69006753

To the nearest degree, the measure of $\angle J$ is 34°.

SOLVE TRIANGLES You can find the missing measures of a right triangle if you know the measure of two sides of a triangle or the measure of one side and one acute angle. Finding all of the measures of the sides and the angles in a right triangle is called **solving the triangle**.

Example 4 Solve a Triangle

Find all of the missing measures in $\triangle ABC$.

You need to find the measures of $\angle B$, $\overline{AC}$, and $\overline{BC}$.

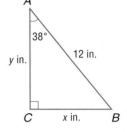

Step 1 Find the measure of $\angle B$. The sum of the measures of the angles in a triangle is 180.

$180° - 90° - 38° = 52°$

The measure of $\angle B$ is 52°.

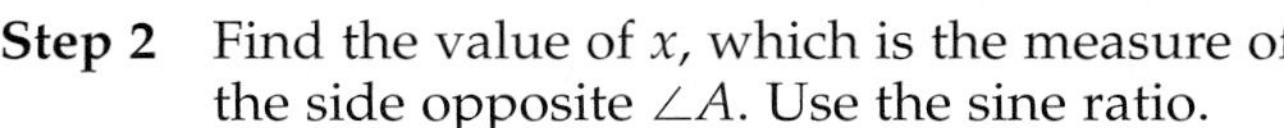

Step 2 Find the value of x, which is the measure of the side opposite $\angle A$. Use the sine ratio.

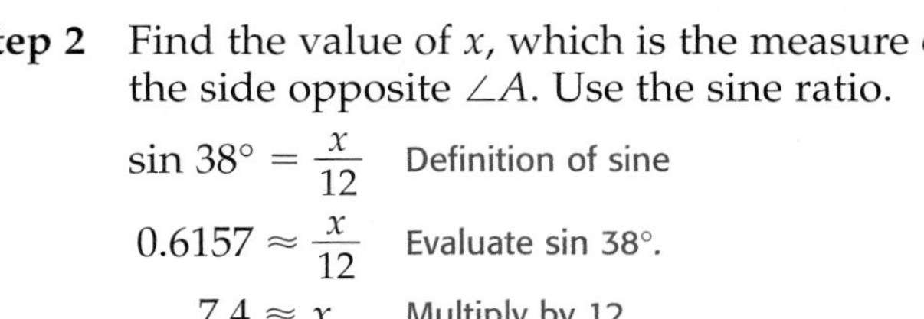

$\sin 38° = \frac{x}{12}$ Definition of sine

$0.6157 \approx \frac{x}{12}$ Evaluate sin 38°.

$7.4 \approx x$ Multiply by 12.

$\overline{BC}$ is about 7.4 inches long.

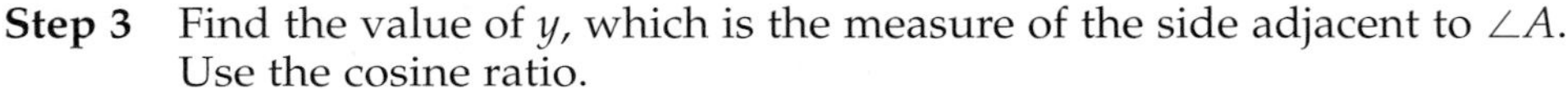

Step 3 Find the value of y, which is the measure of the side adjacent to $\angle A$. Use the cosine ratio.

$\cos 38° = \frac{y}{12}$ Definition of cosine

$0.7880 \approx \frac{y}{12}$ Evaluate cos 38°.

$9.5 \approx y$ Multiply by 12.

$\overline{AC}$ is about 9.5 inches long.

So, the missing measures are 52°, 7.4 in., and 9.5 in.

Study Tip

Verifying Right Triangles
You can use the Pythagorean Theorem to verify that the sides are sides of a right triangle.

Trigonometric ratios are often used to find distances or lengths that cannot be measured directly. In these situations, you will sometimes use an angle of elevation or an angle of depression. An **angle of elevation** is formed by a horizontal line of sight and a line of sight above it. An **angle of depression** is formed by a horizontal line of sight and a line of sight below it.

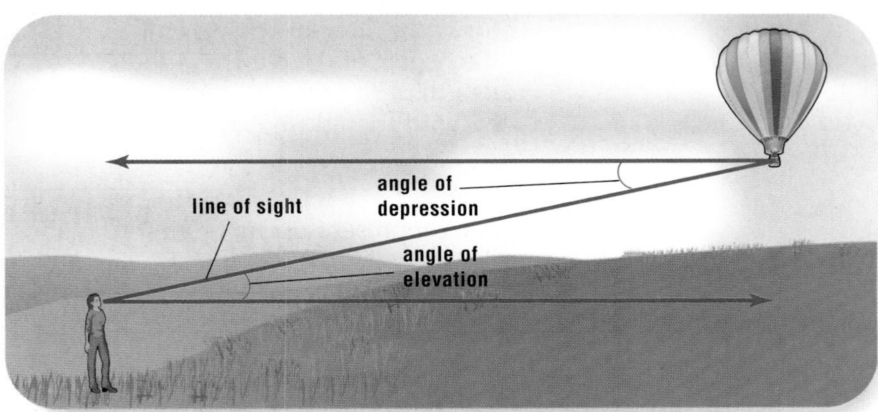

Algebra Activity

Make a Hypsometer

- Tie one end of a piece of string to the middle of a straw. Tie the other end of string to a paper clip.
- Tape a protractor to the side of the straw. Make sure that the string hangs freely to create a vertical or plumb line.
- Find an object outside that is too tall to measure directly, such as a basketball hoop, a flagpole, or the school building.
- Look through the straw to the top of the object you are measuring. Find the angle measure where the string and protractor intersect. Determine the angle of elevation by subtracting this measurement from 90°.
- Measure the distance from your eye level to the ground and from your foot to the base of the object you are measuring.

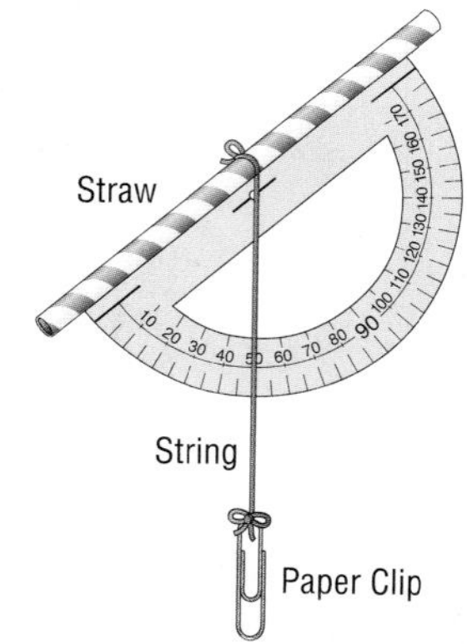

Analyze

1. Make a sketch of your measurements. Use the equation $\tan(\text{angle of elevation}) = \frac{\text{height of object} - x}{\text{distance of object}}$, where x represents distance from the ground to your eye level, to find the height of the object.
2. Why do you have to subtract the angle measurement on the hypsometer from 90° to find the angle of elevation?
3. Compare your answer with someone who measured the same object. Did your heights agree? Why or why not?

Example 5 Angle of Elevation

INDIRECT MEASUREMENT At point A, Umeko measured the angle of elevation to point P to be 27 degrees. At another point B, which was 600 meters closer to the cliff, Umeko measured the angle of elevation to point P to be 31.5 degrees. Determine the height of the cliff.

Explore Draw a diagram to model the situation. Two right triangles, $\triangle BPC$ and $\triangle APC$, are formed. You know the angle of elevation for each triangle. To determine the height of the cliff, find the length of $\overline{PC}$, which is shared by both triangles.

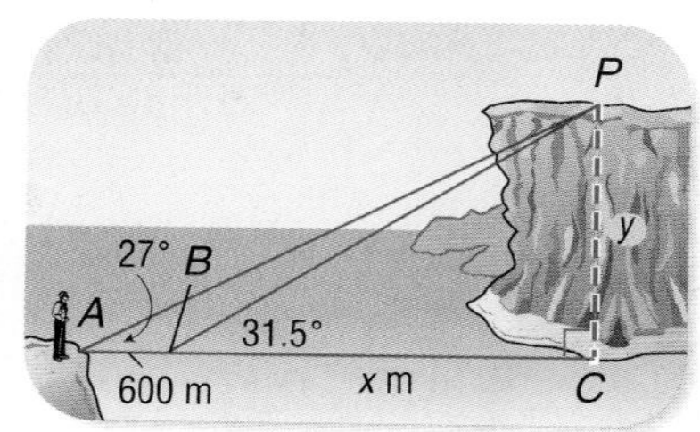

Plan Let y represent the distance from the top of the cliff P to its base C. Let x represent BC in the first triangle and let $x + 600$ represent AC.

Solve Write two equations involving the tangent ratio.

$$\tan 31.5° = \frac{y}{x} \quad \text{and} \quad \tan 27° = \frac{y}{600 + x}$$

$$x \tan 31.5° = y \qquad (600 + x)\tan 27° = y$$

Since both expressions are equal to y, use substitution to solve for x.

$x \tan 31.5° = (600 + x) \tan 27°$	Substitute.
$x \tan 31.5° = 600 \tan 27° + x \tan 27°$	Distributive Property
$x \tan 31.5° - x \tan 27° = 600 \tan 27°$	Subtract.
$x(\tan 31.5° - \tan 27°) = 600 \tan 27°$	Isolate x.
$x = \dfrac{600 \tan 27°}{\tan 31.5° - \tan 27°}$	Divide.
$x \approx 2960$ feet	Use a calculator.

Use this value for x and the equation $x \tan 31.5° = y$ to solve for y.

$x \tan 31.5° = y$	Original equation
$2960 \tan 31.5° \approx y$	Replace x with 2960.
$1814 \approx y$	Use a calculator.

The height of the cliff is about 1814 feet.

Examine Examine the solution by finding the angles of elevation.

$\tan B = \dfrac{y}{x}$ $\qquad$ $\tan A = \dfrac{y}{600 + x}$

$\tan B \stackrel{?}{=} \dfrac{1814}{2960}$ $\qquad$ $\tan A \stackrel{?}{=} \dfrac{1814}{600 + 2960}$

$B \approx 31.5°$ $\qquad$ $A \approx 27°$

The solution checks.

Check for Understanding

Concept Check

1. **Explain** how to determine which trigonometric ratio to use when solving for an unknown measure of a right triangle.
2. **OPEN ENDED** Draw a right triangle and label the measure of the hypotenuse and the measure of one acute angle. Then solve for the remaining measures.
3. **Compare** the measure of the angle of elevation and the measure of the angle of depression for two objects. What is the relationship between their measures?

Guided Practice **For each triangle, find sin *Y*, cos *Y*, and tan *Y* to the nearest ten thousandth.**

4. Y 27 X 45 36 Z

5. 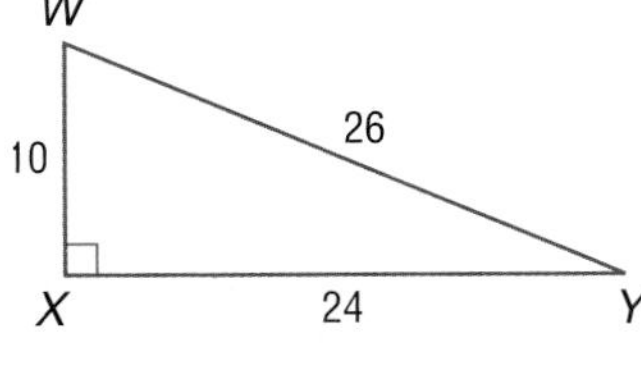

Use a calculator to find the value of each trigonometric ratio to the nearest ten thousandth.

6. sin 60°
7. cos 75°
8. tan 10°

Use a calculator to find the measure of each angle to the nearest degree.

9. $\sin W = 0.9848$
10. $\cos X = 0.6157$
11. $\tan C = 0.3249$

For each triangle, find the measure of the indicated angle to the nearest degree.

12. 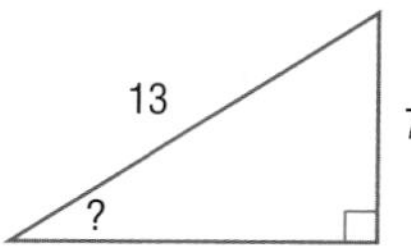

13. 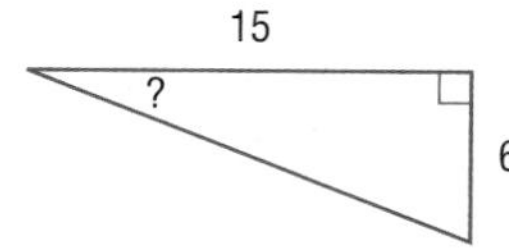

14. 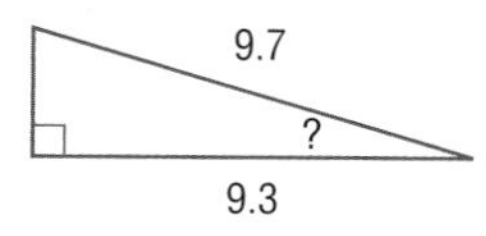

Solve each right triangle. State the side lengths to the nearest tenth and the angle measures to the nearest degree.

15. 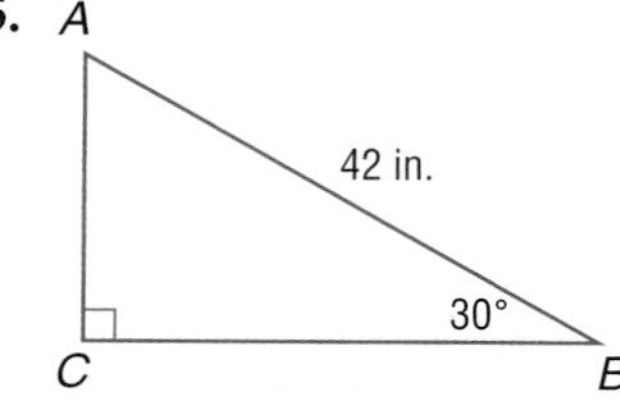

16. 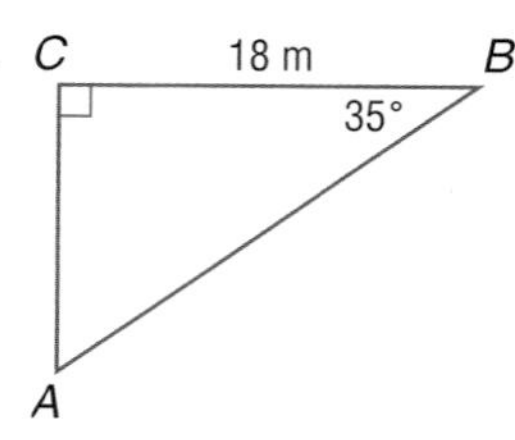

17. 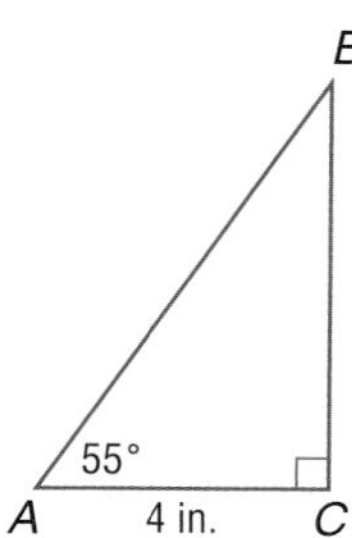

Application

18. DRIVING The percent grade of a road is the ratio of how much the road rises or falls in a given horizontal distance. If a road has a vertical rise of 40 feet for every 1000 feet horizontal distance, calculate the percent grade of the road and the angle of elevation the road makes with the horizontal.

Practice and Apply

Homework Help

For Exercises	See Examples
19–24	1
25–33	2
34–51	3
52–60	4
61–65	5

Extra Practice

See page 846.

For each triangle, find sin *R*, cos *R*, and tan *R* to the nearest ten thousandth.

19. 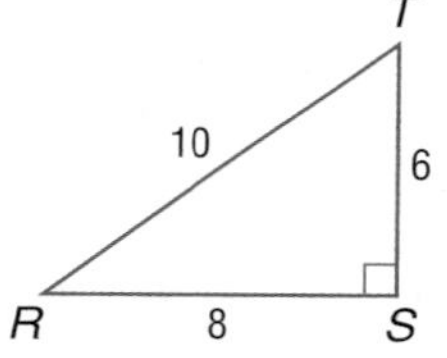

20. 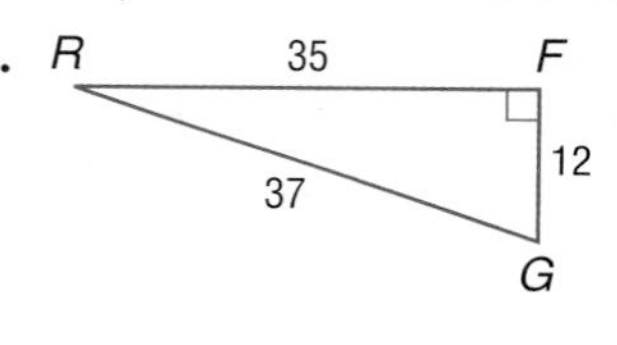

21. 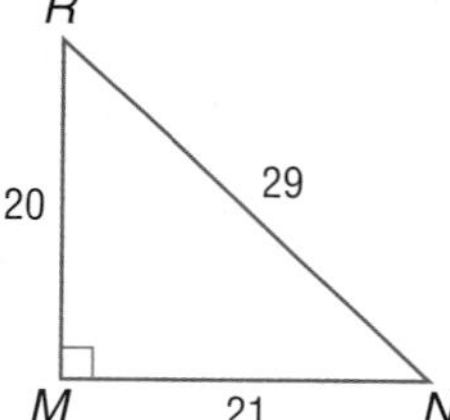

22. R, 30, P, 34, 16, O

23. 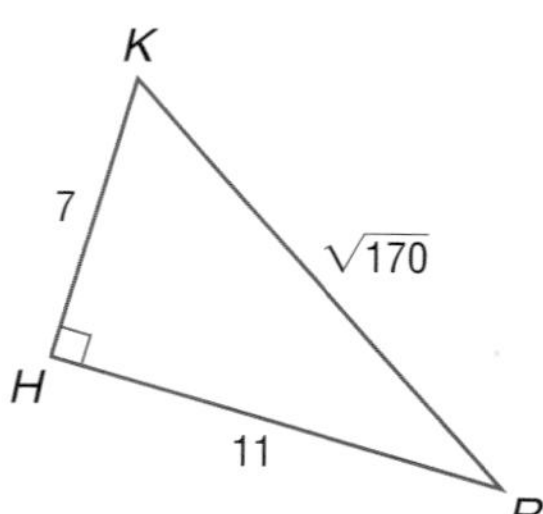

24. 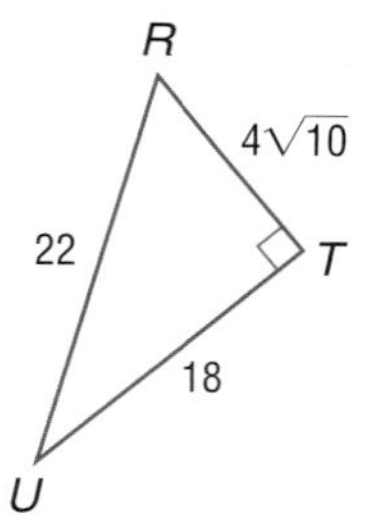

Use a calculator to find the value of each trigonometric ratio to the nearest ten thousandth.

25. sin 30° **26.** sin 80° **27.** cos 45°

28. cos 48° **29.** tan 32° **30.** tan 15°

31. tan 67° **32.** sin 53° **33.** cos 12°

Use a calculator to find the measure of each angle to the nearest degree.

34. $\cos V = 0.5000$ **35.** $\cos Q = 0.7658$ **36.** $\sin K = 0.9781$

37. $\sin A = 0.8827$ **38.** $\tan S = 1.2401$ **39.** $\tan H = 0.6473$

40. $\sin V = 0.3832$ **41.** $\cos M = 0.9793$ **42.** $\tan L = 3.6541$

For each triangle, find the measure of the indicated angle to the nearest degree.

43. 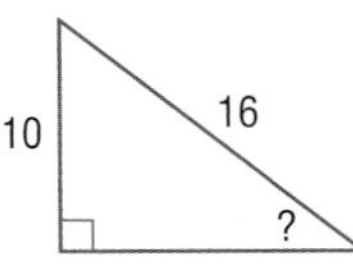

44. 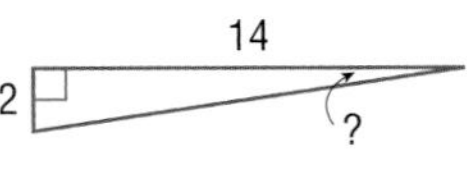

45. 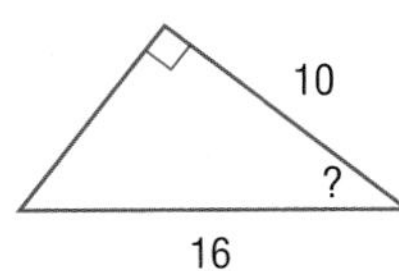

46. 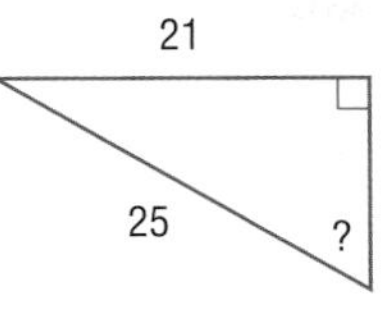

47. 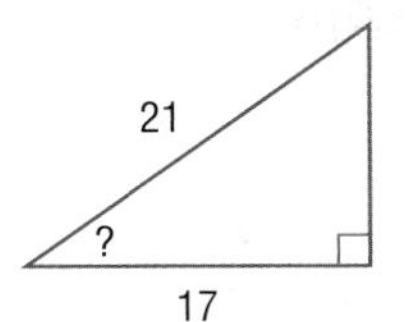

48. 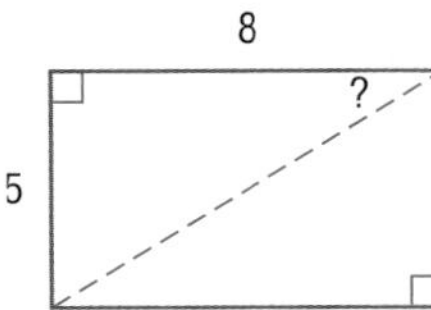

49. 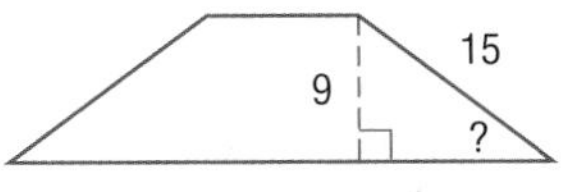

50. 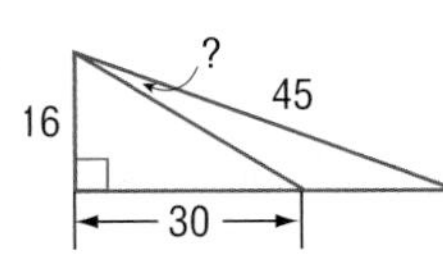

51. 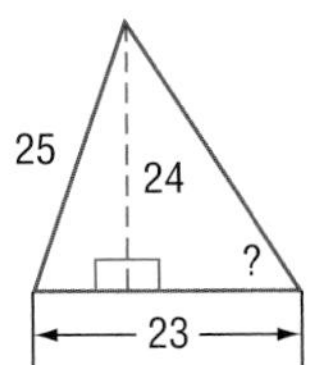

Solve each right triangle. State the side lengths to the nearest tenth and the angle measures to the nearest degree.

52. 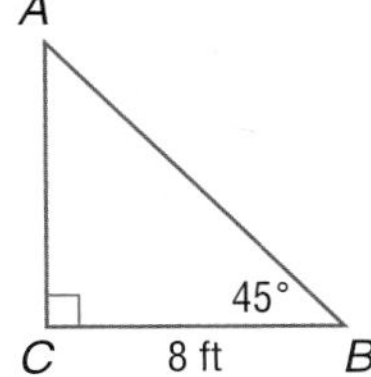

53. 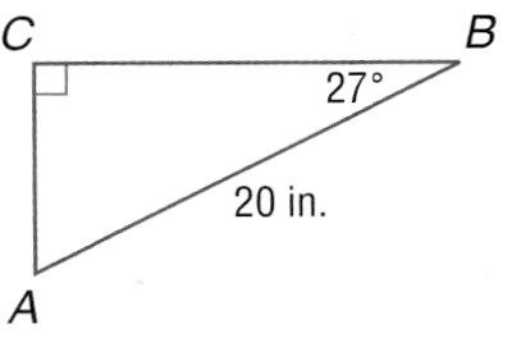

54. 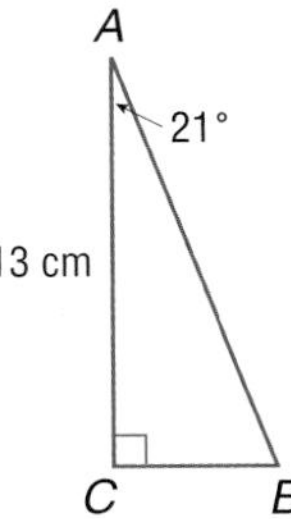

More About. . .

Submarines

In emergency situations, modern submarines are built to allow for a rapid surfacing, a technique called an emergency main ballast blow.

Source: www.howstuffworks.com

55. 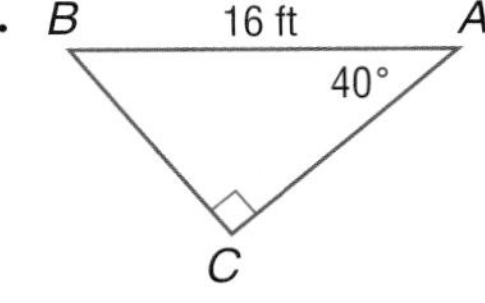

56. 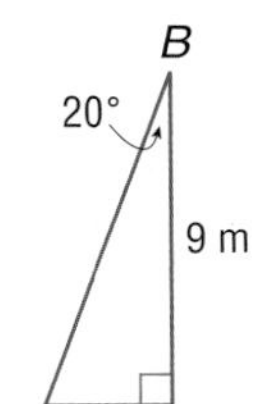

57. 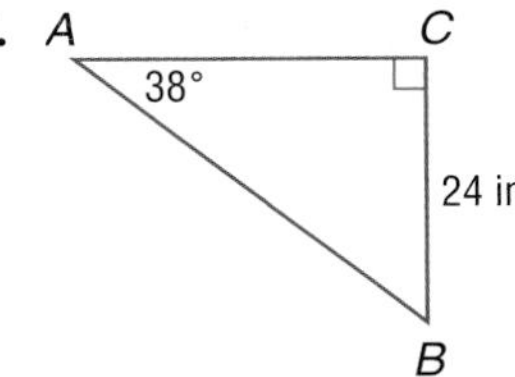

58. 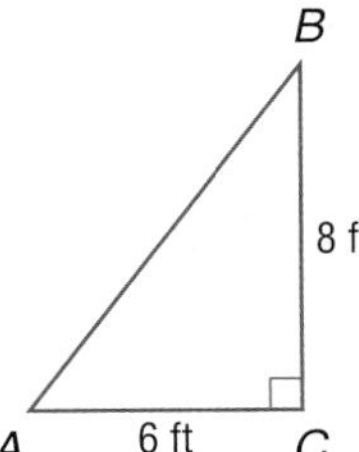

59. 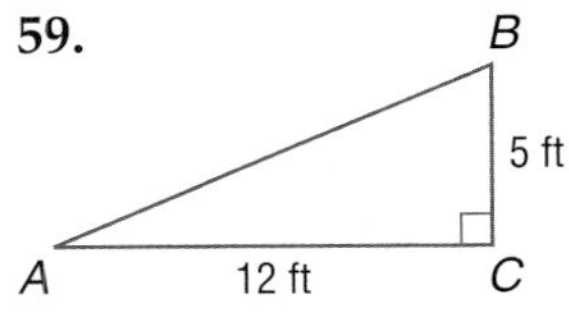

60. 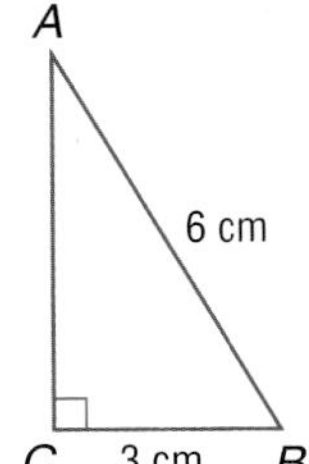

SUBMARINES **For Exercises 61 and 62, use the following information.**

A submarine is traveling parallel to the surface of the water 626 meters below the surface. The sub begins a constant ascent to the surface so that it will emerge on the surface after traveling 4420 meters from the point of its initial ascent.

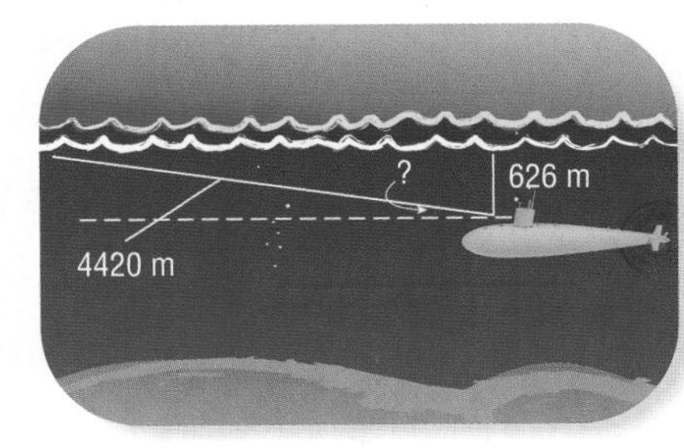

61. What angle of ascent did the submarine make?

62. What horizontal distance did the submarine travel during its ascent?

AVIATION For Exercises 63 and 64, use the following information.
Germaine pilots a small plane on weekends. During a recent flight, he determined that he was flying at an altitude of 3000 feet parallel to the ground and that the ground distance to the start of the landing strip was 8000 feet.

63. What is Germaine's angle of depression to the start of the landing strip?

64. What is the distance between the plane in the air and the landing strip on the ground?

65. **FARMING** Leonard and Alecia are building a new feed storage system on their farm. The feed conveyor must be able to reach a range of heights. It has a length of 8 meters, and its angle of elevation can be adjusted from 20° to 5°. Under these conditions, what range of heights is possible for an opening in the building through which feed can pass?

66. **CRITICAL THINKING** An important trigonometric identity is $\sin^2 A + \cos^2 A = 1$. Use the sine and cosine ratios and the Pythagorean Theorem to prove this identity.

67. **WRITING IN MATH** Answer the question that was posed at the beginning of the lesson.

How are trigonometric ratios used in surveying?

Include the following in your answer:
- an explanation of how trigonometric ratios are used to measure the height of a mountain, and
- any additional information you need to know about the point from which you are measuring in order to find the altitude of a mountain.

For Exercises 68 and 69, use the figure at the right.

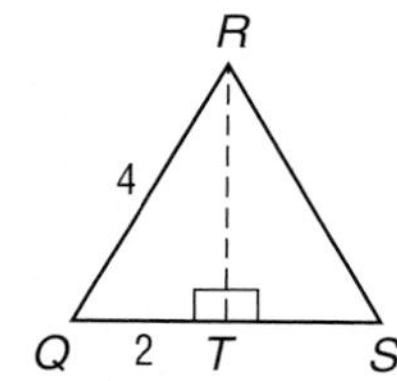

68. RT is equal to TS. What is RS?

(A) $2\sqrt{6}$ (B) $2\sqrt{3}$ (C) $4\sqrt{3}$ (D) $2\sqrt{2}$

69. What is the measure of $\angle Q$?

(A) 25° (B) 30° (C) 45° (D) 60°

Maintain Your Skills

Mixed Review

For each set of measures given, find the measures of the missing sides if $\triangle KLM \sim \triangle NOP$. *(Lesson 11-6)*

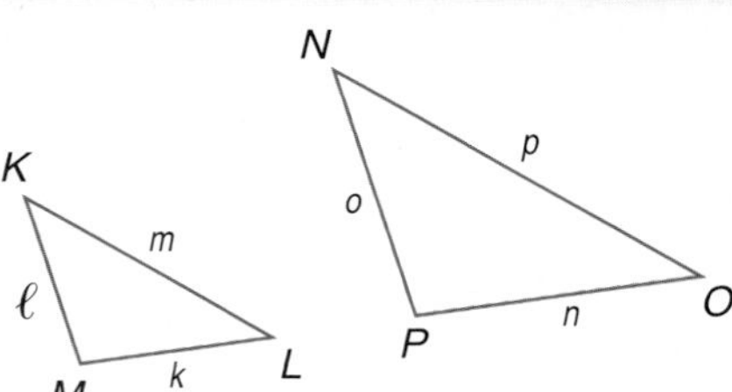

70. $k = 5, \ell = 3, m = 6, n = 10$

71. $\ell = 9, m = 3, n = 12, p = 4.5$

Find the possible values of a if the points with the given coordinates are the indicated distance apart. *(Lesson 11-5)*

72. $(9, 28), (a, -8); d = 39$

73. $(3, a), (10, -1); d = \sqrt{65}$

Find each product. *(Lesson 8-6)*

74. $c^2(c^2 + 3c)$

75. $s(4s^2 - 9s + 12)$

76. $xy^2(2x^2 + 5xy - 7y^2)$

Use substitution to solve each system of equations. *(Lesson 7-2)*

77. $a = 3b + 2$
$4a - 7b = 23$

78. $p + q = 10$
$3p - 2q = -5$

79. $3r + 6s = 0$
$-4r - 10s = -2$

Reading Mathematics

The Language of Mathematics

The language of mathematics is a specific one, but it borrows from everyday language, scientific language, and world languages. To find a word's correct meaning, you will need to be aware of some confusing aspects of language.

Confusing Aspect	Words
Some words are used in English and in mathematics, but have distinct meanings.	factor, **leg**, prime, power, **rationalize**
Some words are used in English and in mathematics, but the mathematical meaning is more precise.	difference, even, **similar**, slope
Some words are used in science and in mathematics, but the meanings are different.	divide, **radical**, solution, variable
Some words are only used in mathematics.	decimal, **hypotenuse**, integer, quotient
Some words have more than one mathematical meaning.	base, **degree**, range, round, square
Sometimes several words come from the same root word.	polygon and polynomial, **radical** and **radicand**
Some mathematical words sound like English words.	**cosine** and cosign, **sine** and sign, **sum** and some
Some words are often abbreviated, but you must use the whole word when you read them.	**cos** for cosine, **sin** for sine, **tan** for tangent

Words in boldface are in this chapter.

Reading to Learn

1. How do the mathematical meanings of the following words compare to the everyday meanings?

a. factor **b.** leg **c.** rationalize

2. State two mathematical definitions for each word. Give an example for each definition.

a. degree **b.** range **c.** round

3. Each word below is shown with its root word and the root word's meaning. Find three additional words that come from the same root.

a. domain, from the root word *domus*, which means house

b. radical, from the root word *radix*, which means root

c. similar, from the root word *similis*, which means like

Chapter 11 Study Guide and Review

Vocabulary and Concept Check

angle of depression (p. 625)
angle of elevation (p. 625)
conjugate (p. 589)
corollary (p. 607)
cosine (p. 623)
Distance Formula (p. 611)
extraneous solution (p. 599)
hypotenuse (p. 605)
leg (p. 605)
Pythagorean triple (p. 606)
radical equation (p. 598)
radical expression (p. 586)
radicand (p. 586)
rationalizing the denominator (p. 588)
similar triangles (p. 616)
sine (p. 623)
solve a triangle (p. 625)
tangent (p. 623)
trigonometric ratios (p. 623)

State whether each sentence is *true* or *false*. If false, replace the underlined word, number, expression, or equation to make a true sentence.

1. The binomials $-3 + \sqrt{7}$ and $\underline{3 - \sqrt{7}}$ are conjugates.
2. In the expression $-4\sqrt{5}$, the radicand is $\underline{5}$.
3. The sine of an acute angle of a right triangle is the measure of the opposite leg divided by the measure of the <u>hypotenuse</u>.
4. The <u>longest</u> side of a right triangle is the hypotenuse.
5. After the first step in solving $\sqrt{3x + 19} = x + 3$, you would have $\underline{3x + 19 = x^2 + 9}$.
6. The two sides that form the right angle in a right triangle are called the <u>legs</u> of the triangle.
7. The expression $\underline{\dfrac{2x\sqrt{3x}}{\sqrt{6y}}}$ is in simplest radical form.
8. A triangle with sides having measures of <u>25, 20, and 15</u> is a right triangle.

Lesson-by-Lesson Review

11-1 Simplifying Radical Expressions

See pages 586–592.

Concept Summary

- A radical expression is in simplest form when no radicands have perfect square factors other than 1, no radicands contain fractions, and no radicals appear in the denominator of a fraction.

Example Simplify $\dfrac{3}{5 - \sqrt{2}}$.

$$\frac{3}{5-\sqrt{2}} = \frac{3}{5-\sqrt{2}} \cdot \frac{5+\sqrt{2}}{5+\sqrt{2}}$$ Multiply by $\frac{5+\sqrt{2}}{5+\sqrt{2}}$ to rationalize the denominator.

$$= \frac{3(5) + 3\sqrt{2}}{5^2 - (\sqrt{2})^2}$$ $(a - b)(a + b) = a^2 - b^2$

$$= \frac{15 + 3\sqrt{2}}{25 - 2}$$ $(\sqrt{2})^2 = 2$

$$= \frac{15 + 3\sqrt{2}}{23}$$ Simplify.

www.algebra1.com/vocabulary_review

Exercises **Simplify.** *See Examples 1–5 on pages 586–589.*

9. $\sqrt{\dfrac{60}{y^2}}$
10. $\sqrt{44a^2b^5}$
11. $(3 - 2\sqrt{12})^2$
12. $\dfrac{9}{3 + \sqrt{2}}$
13. $\dfrac{2\sqrt{7}}{3\sqrt{5} + 5\sqrt{3}}$
14. $\dfrac{\sqrt{3a^3b^4}}{\sqrt{8ab^{10}}}$

11-2 Operations with Radical Expressions

See pages 593–597.

Concept Summary

- Radical expressions with like radicands can be added or subtracted.
- Use the FOIL Method to multiply radical expressions.

Examples

1 **Simplify $\sqrt{6} - \sqrt{54} + 3\sqrt{12} + 5\sqrt{3}$.**

$\sqrt{6} - \sqrt{54} + 3\sqrt{12} + 5\sqrt{3}$

$= \sqrt{6} - \sqrt{3^2 \cdot 6} + 3\sqrt{2^2 \cdot 3} + 5\sqrt{3}$ Simplify radicands.

$= \sqrt{6} - (\sqrt{3^2} \cdot \sqrt{6}) + 3(\sqrt{2^2} \cdot \sqrt{3}) + 5\sqrt{3}$ Product Property of Square Roots

$= \sqrt{6} - 3\sqrt{6} + 3(2\sqrt{3}) + 5\sqrt{3}$ Evaluate square roots.

$= \sqrt{6} - 3\sqrt{6} + 6\sqrt{3} + 5\sqrt{3}$ Simplify.

$= -2\sqrt{6} + 11\sqrt{3}$ Add like radicands.

2 **Find $(2\sqrt{3} - \sqrt{5})(\sqrt{10} + 4\sqrt{6})$.**

$(2\sqrt{3} - \sqrt{5})(\sqrt{10} + 4\sqrt{6})$

First terms, Outer terms, Inner terms, Last terms

$= (2\sqrt{3})(\sqrt{10}) + (2\sqrt{3})(4\sqrt{6}) + (-\sqrt{5})(\sqrt{10}) + (-\sqrt{5})(4\sqrt{6})$

$= 2\sqrt{30} + 8\sqrt{18} - \sqrt{50} - 4\sqrt{30}$ Multiply.

$= 2\sqrt{30} + 8\sqrt{3^2 \cdot 2} - \sqrt{5^2 \cdot 2} - 4\sqrt{30}$ Prime factorization

$= 2\sqrt{30} + 24\sqrt{2} - 5\sqrt{2} - 4\sqrt{30}$ Simplify.

$= -2\sqrt{30} + 19\sqrt{2}$ Combine like terms.

Exercises **Simplify each expression.** *See Examples 1 and 2 on pages 593 and 594.*

15. $2\sqrt{3} + 8\sqrt{5} - 3\sqrt{5} + 3\sqrt{3}$
16. $2\sqrt{6} - \sqrt{48}$
17. $4\sqrt{27} + 6\sqrt{48}$
18. $4\sqrt{7k} - 7\sqrt{7k} + 2\sqrt{7k}$
19. $5\sqrt{18} - 3\sqrt{112} - 3\sqrt{98}$
20. $\sqrt{8} + \sqrt{\dfrac{1}{8}}$

Find each product. *See Example 3 on page 594.*

21. $\sqrt{2}(3 + 3\sqrt{3})$
22. $\sqrt{5}(2\sqrt{5} - \sqrt{7})$
23. $(\sqrt{3} - \sqrt{2})(2\sqrt{2} + \sqrt{3})$
24. $(6\sqrt{5} + 2)(3\sqrt{2} + \sqrt{5})$

11-3 Radical Equations

See pages 598–603.

Concept Summary

- Solve radical equations by isolating the radical on one side of the equation. Square each side of the equation to eliminate the radical.

Example **Solve $\sqrt{5 - 4x} - 6 = 7$.**

$\sqrt{5 - 4x} - 6 = 7$	Original equation
$\sqrt{5 - 4x} = 13$	Add 6 to each side.
$5 - 4x = 169$	Square each side.
$-4x = 164$	Subtract 5 from each side.
$x = -41$	Divide each side by -4.

Exercises **Solve each equation. Check your solution.** *See Examples 2 and 3 on page 599.*

25. $10 + 2\sqrt{b} = 0$ **26.** $\sqrt{a + 4} = 6$ **27.** $\sqrt{7x - 1} = 5$

28. $\sqrt{\frac{4a}{3}} - 2 = 0$ **29.** $\sqrt{x + 4} = x - 8$ **30.** $\sqrt{3x - 14} + x = 6$

11-4 The Pythagorean Theorem

See pages 605–610.

Concept Summary

- If a and b are the measures of the legs of a right triangle and c is the measure of the hypotenuse, then $c^2 = a^2 + b^2$.
- If a and b are measures of the shorter sides of a triangle, c is the measure of the longest side, and $c^2 = a^2 + b^2$, then the triangle is a right triangle.

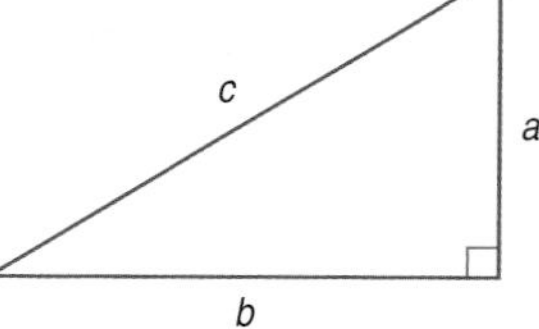

Example **Find the length of the missing side.**

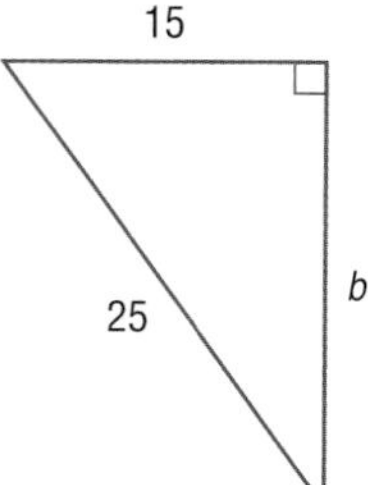

$c^2 = a^2 + b^2$	Pythagorean Theorem
$25^2 = 15^2 + b^2$	$c = 25$ and $a = 15$
$625 = 225 + b^2$	Evaluate squares.
$400 = b^2$	Subtract 225 from each side.
$20 = b$	Take the square root of each side.

Exercises **If c is the measure of the hypotenuse of a right triangle, find each missing measure. If necessary, round answers to the nearest hundredth.** *See Example 2 on page 606.*

31. $a = 30, b = 16, c = ?$ **32.** $a = 6, b = 10, c = ?$ **33.** $a = 10, c = 15, b = ?$

34. $b = 4, c = 56, a = ?$ **35.** $a = 18, c = 30, b = ?$ **36.** $a = 1.2, b = 1.6, c = ?$

Determine whether the following side measures form right triangles. *See Example 4 on page 607.*

37. 9, 16, 20 **38.** 20, 21, 29 **39.** 9, 40, 41 **40.** 18, $\sqrt{24}$, 30

11-5 The Distance Formula

See pages 611–615.

Concept Summary

- The distance d between any two points with coordinates (x_1, y_1) and (x_2, y_2) is given by $d = \sqrt{(x_2 - x_1)^2 + (y_2 - y_1)^2}$.

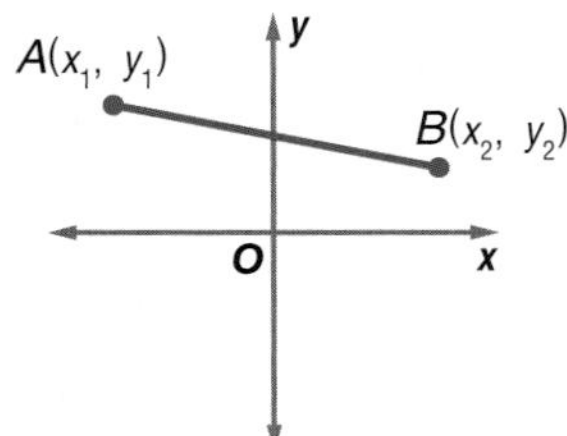

Example

Find the distance between the points with coordinates (−5, 1) and (1, 5).

$d = \sqrt{(x_2 - x_1)^2 + (y_2 - y_1)^2}$ Distance Formula

$= \sqrt{(1 - (-5))^2 + (5 - 1)^2}$ $(x_1, y_1) = (-5, 1)$ and $(x_2, y_2) = (1, 5)$

$= \sqrt{6^2 + 4^2}$ Simplify.

$= \sqrt{36 + 16}$ Evaluate squares.

$= \sqrt{52}$ or about 7.21 units Simplify.

Exercises **Find the distance between each pair of points whose coordinates are given. Express in simplest radical form and as decimal approximations rounded to the nearest hundredth if necessary.** *See Example 1 on page 611.*

41. (9, −2), (1, 13) **42.** (4, 2), (7, 9) **43.** (4, −6), (−2, 7)

44. $(2\sqrt{5}, 9), (4\sqrt{5}, 3)$ **45.** (4, 8), (−7, 12) **46.** (−2, 6), (5, 11)

Find the value of a if the points with the given coordinates are the indicated distance apart. *See Example 3 on page 612.*

47. $(-3, 2), (1, a); d = 5$ **48.** $(1, 1), (4, a); d = 5$

49. $(6, -2), (5, a); d = \sqrt{145}$ **50.** $(5, -2), (a, -3); d = \sqrt{170}$

11-6 Similar Triangles

See pages 616–621.

Concept Summary

- Similar triangles have congruent corresponding angles and proportional corresponding sides.
- If $\triangle ABC \sim \triangle DEF$, then $\frac{AB}{DE} = \frac{BC}{EF} = \frac{AC}{DF}$.

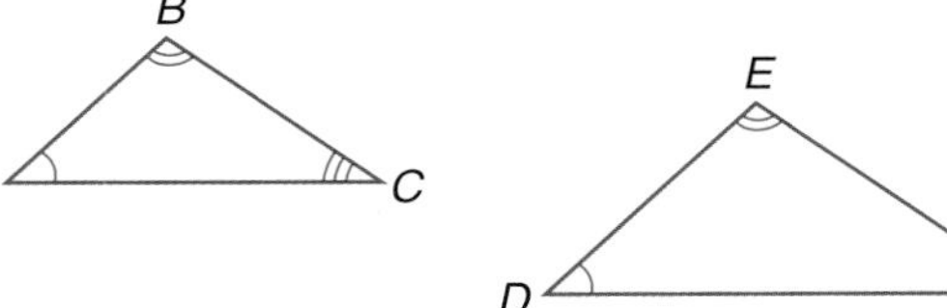

Example

Find the measure of side a if the two triangles are similar.

$\frac{10}{5} = \frac{6}{a}$ Corresponding sides of similar triangles are proportional.

$10a = 30$ Find the cross products.

$a = 3$ Divide each side by 10.

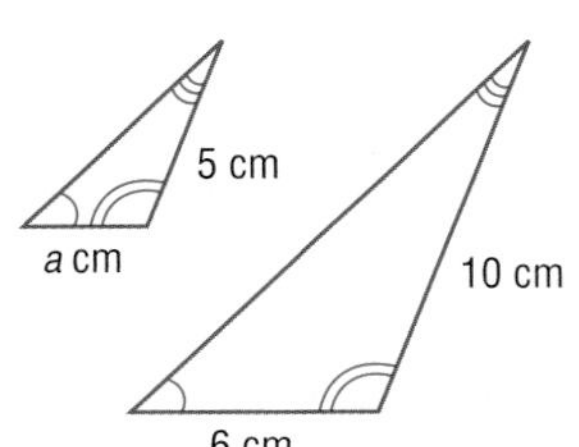

• Extra Practice, see pages 844–846.
• Mixed Problem Solving, see page 863.

Exercises **For each set of measures given, find the measures of the remaining sides if $\triangle ABC \sim \triangle DEF$.** *See Example 2 on page 617.*

51. $c = 16, b = 12, a = 10, f = 9$

52. $a = 8, c = 10, b = 6, f = 12$

53. $c = 12, f = 9, a = 8, e = 11$

54. $b = 20, d = 7, f = 6, c = 15$

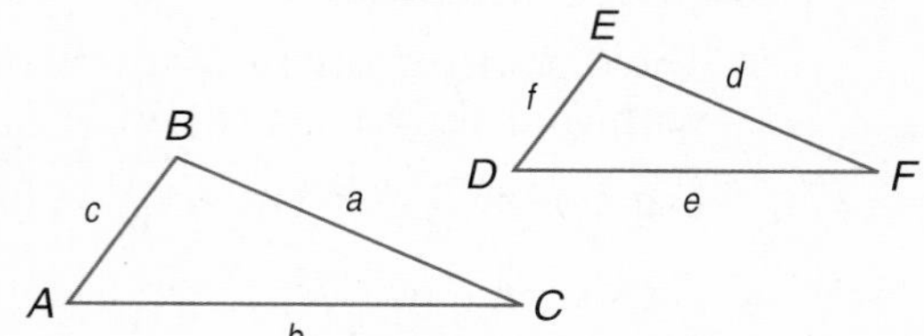

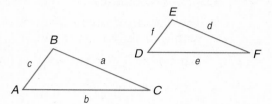

11-7 Trigonometric Ratios

See pages 623–630.

Concept Summary

Three common trigonometric ratios are sine, cosine, and tangent.

- $\sin A = \frac{BC}{AB}$
- $\cos A = \frac{AC}{AB}$
- $\tan A = \frac{BC}{AC}$

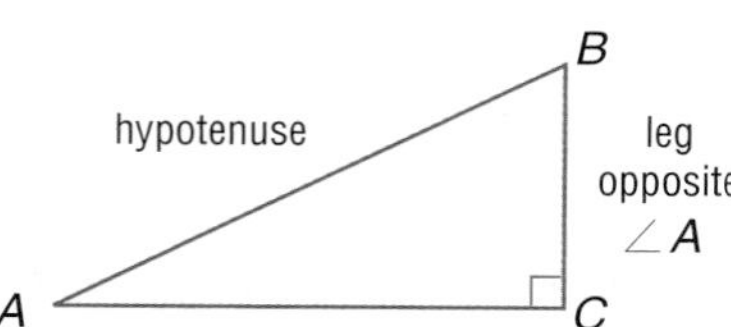

Example **Find the sine, cosine, and tangent of $\angle A$. Round to the nearest ten thousandth.**

$$\sin A = \frac{\text{opposite leg}}{\text{hypotenuse}}$$
$$= \frac{20}{25} \text{ or } 0.8000$$

$$\cos A = \frac{\text{adjacent leg}}{\text{hypotenuse}}$$
$$= \frac{15}{25} \text{ or } 0.6000$$

$$\tan A = \frac{\text{opposite leg}}{\text{adjacent leg}}$$
$$= \frac{20}{15} \text{ or } 1.3333$$

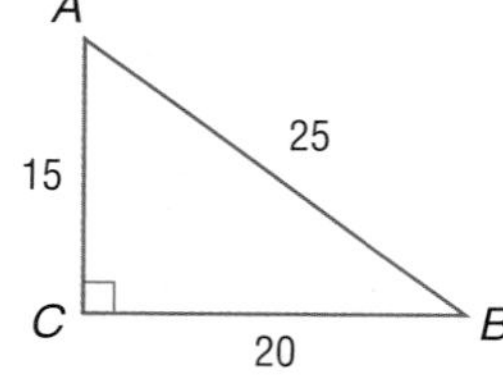

Exercises **For $\triangle ABC$, find each value of each trigonometric ratio to the nearest ten thousandth.** *See Example 1 on page 624.*

55. $\cos B$

56. $\tan A$

57. $\sin B$

58. $\cos A$

59. $\tan B$

60. $\sin A$

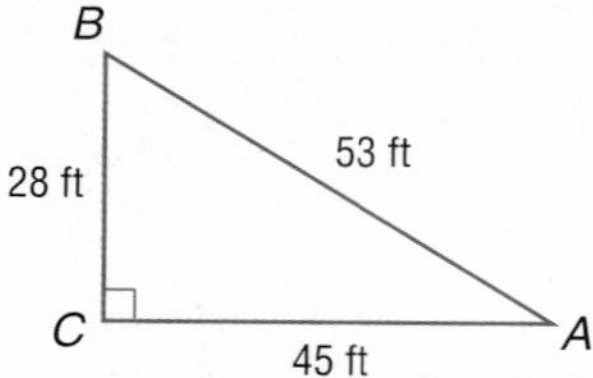

Use a calculator to find the measure of each angle to the nearest degree.
See Example 3 on page 624.

61. $\tan M = 0.8043$ **62.** $\sin T = 0.1212$ **63.** $\cos B = 0.9781$

64. $\cos F = 0.7443$ **65.** $\sin A = 0.4540$ **66.** $\tan Q = 5.9080$

Chapter 11

Practice Test

Vocabulary and Concepts

Match each term and its definition.

1. measure of the opposite leg divided by the measure of the hypotenuse
2. measure of the adjacent leg divided by the measure of the hypotenuse
3. measure of the opposite leg divided by the measure of the adjacent leg

a. cosine
b. sine
c. tangent

Skills and Applications

Simplify.

4. $2\sqrt{27} + \sqrt{63} - 4\sqrt{3}$
5. $\sqrt{6} + \sqrt{\frac{2}{3}}$
6. $\sqrt{112x^4y^6}$
7. $\sqrt{\frac{10}{3}} \cdot \sqrt{\frac{4}{30}}$
8. $\sqrt{6}(4 + \sqrt{12})$
9. $(1 - \sqrt{3})(3 + \sqrt{2})$

Solve each equation. Check your solution.

10. $\sqrt{10x} = 20$
11. $\sqrt{4s} + 1 = 11$
12. $\sqrt{4x + 1} = 5$
13. $x = \sqrt{-6x - 8}$
14. $x = \sqrt{5x + 14}$
15. $\sqrt{4x - 3} = 6 - x$

If c is the measure of the hypotenuse of a right triangle, find each missing measure. If necessary, round to the nearest hundredth.

16. $a = 8, b = 10, c = ?$
17. $a = 6\sqrt{2}, c = 12, b = ?$
18. $b = 13, c = 17, a = ?$

Find the distance between each pair of points whose coordinates are given. Express in simplest radical form and as decimal approximations rounded to the nearest hundredth if necessary.

19. $(4, 7), (4, -2)$
20. $(-1, 1), (1, -5)$
21. $(-9, 2), (21, 7)$

For each set of measures given, find the measures of the missing sides if $\triangle ABC \sim \triangle JKH$.

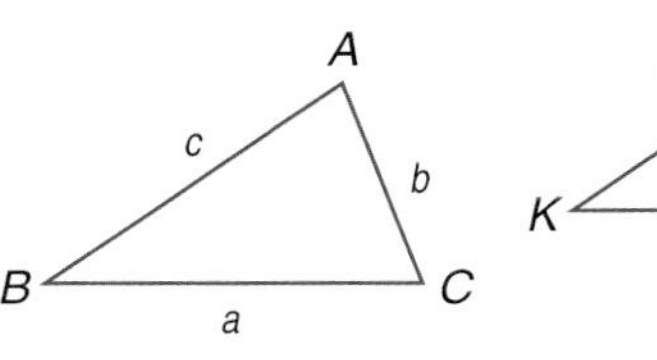

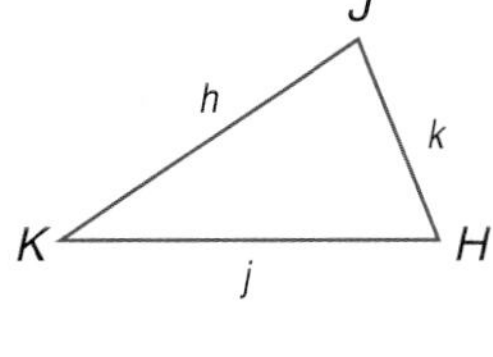

22. $c = 20, h = 15, k = 16, j = 12$
23. $c = 12, b = 13, a = 6, h = 10$
24. $k = 5, c = 6.5, b = 7.5, a = 4.5$
25. $h = 1\frac{1}{2}, c = 4\frac{1}{2}, k = 2\frac{1}{4}, a = 3$

Solve each right triangle. State the side lengths to the nearest tenth and the angle measures to the nearest degree.

26.
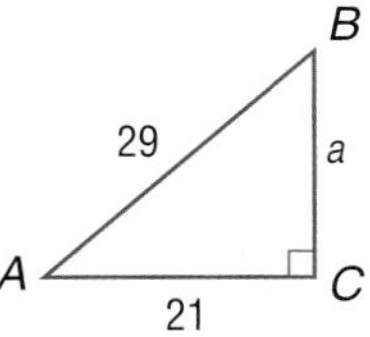

27.
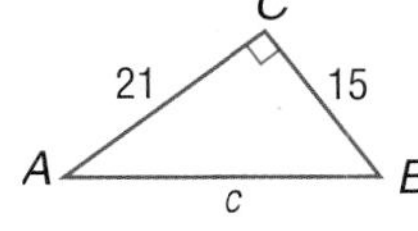

28.
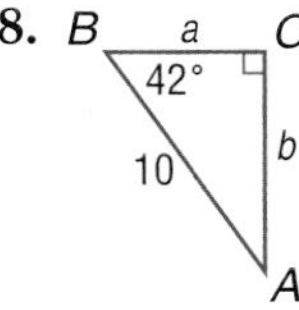

29. **SPORTS** A hiker leaves her camp in the morning. How far is she from camp after walking 9 miles due west and then 12 miles due north?

30. **STANDARDIZED TEST PRACTICE** Find the area of the rectangle.

$2\sqrt{32} - 3\sqrt{6}$

$\sqrt{6}$

(A) $16\sqrt{2} - 4\sqrt{6}$ units²
(B) $16\sqrt{3} - 18$ units²
(C) $32\sqrt{3} - 18$ units²
(D) $2\sqrt{32} - 18$ units²

Chapter 11 Standardized Test Practice

Part 1 Multiple Choice

Record your answers on the answer sheet provided by your teacher or on a sheet of paper.

1. Which equation describes the data in the table? (Lesson 4-8)

x	−5	−2	1	4
y	11	5	−1	−7

Ⓐ $y = x - 6$ Ⓑ $y = 2x - 1$

Ⓒ $y = 2x + 1$ Ⓓ $y = -2x + 1$

2. The length of a rectangle is 6 feet more than the width. The perimeter is 92 feet. Which system of equations will determine the length in feet ℓ and the width in feet w of the rectangle? (Lesson 7-2)

Ⓐ $w = \ell + 6$
$2\ell + 2w = 92$

Ⓑ $\ell + w = 6$
$\ell w = 92$

Ⓒ $\ell = w + 6$
$2\ell + 2w = 92$

Ⓓ $\ell - w = 6$
$\ell + w = 92$

3. A highway resurfacing project and a bridge repair project will cost \$2,500,000 altogether. The bridge repair project will cost \$200,000 less than twice the cost of the highway resurfacing. How much will the highway resurfacing project cost? (Lesson 7-2)

Ⓐ \$450,000 Ⓑ \$734,000

Ⓒ \$900,000 Ⓓ \$1,600,000

4. If 32,800,000 is expressed in the form 3.28×10^n, what is the value of n? (Lesson 8-3)

Ⓐ 5 Ⓑ 6

Ⓒ 7 Ⓓ 8

5. What are the solutions of the equation $x^2 + 7x - 18 = 0$? (Lesson 9-4)

Ⓐ 2 or −9 Ⓑ −2 or 9

Ⓒ −2 or −9 Ⓓ 2 or 9

6. The function $g = t^2 - t$ represents the total number of games played by t teams in a sports league in which each team plays each of the other teams twice. The Metro League plays a total of 132 games. How many teams are in the league? (Lesson 9-4)

Ⓐ 11 Ⓑ 12

Ⓒ 22 Ⓓ 33

7. One leg of a right triangle is 4 inches longer than the other leg. The hypotenuse is 20 inches long. What is the length of the shorter leg? (Lesson 11-4)

Ⓐ 10 in. Ⓑ 12 in.

Ⓒ 16 in. Ⓓ 18 in.

8. What is the distance from one corner of the garden to the opposite corner? (Lesson 11-4)

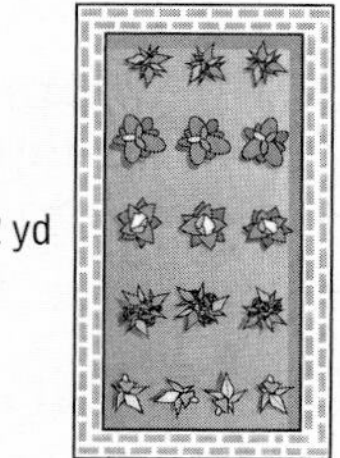

Ⓐ 13 yards

Ⓑ 14 yards

Ⓒ 15 yards

Ⓓ 17 yards

9. How many points in the coordinate plane are equidistant from both the x- and y-axes and are 5 units from the origin? (Lesson 11-5)

Ⓐ 0 Ⓑ 1

Ⓒ 2 Ⓓ 4

The Princeton Review Test-Taking Tip

Questions 7, 21, and 22 Be sure that you know and understand the Pythagorean Theorem. References to right angles, the diagonal of a rectangle, or the hypotenuse of a triangle indicate that you may need to use the Pythagorean Theorem to find the answer to an item.

Part 2 Short Response/Grid In

Record your answers on the answer sheet provided by your teacher or on a sheet of paper.

10. A line is parallel to the line represented by the equation $\frac{1}{2}y + \frac{3}{2}x + 4 = 0$. What is the slope of the parallel line? (Lesson 5-6)

11. Graph the solution of the system of linear inequalities $2x - y > 2$ and $3x + 2y < -4$. (Lesson 6-6)

12. The sum of two integers is 66. The second integer is 18 more than half of the first. What are the integers? (Lesson 7-2)

13. The function $h(t) = -16t^2 + v_0t + h_0$ describes the height in feet above the ground $h(t)$ of an object thrown vertically from a height of h_0 feet, with an initial velocity of v_0 feet per second, if there is no air friction and t is the time in seconds since the object was thrown. A ball is thrown upward from a 100-foot tower at a velocity of 60 feet per second. How many seconds will it take for the ball to reach the ground? (Lesson 9-5)

14. Find all values of x that satisfy the equation $x^2 - 8x + 6 = 0$. Approximate irrational numbers to the nearest hundredth. (Lesson 10-4)

15. Simplify the expression $\sqrt[3]{3\sqrt{81}}$. (Lesson 11-1)

16. Simplify the expression $\left(x^{\frac{3}{2}}\right)^{\frac{4}{3}}\left(\frac{\sqrt{x}}{x}\right)$. (Lesson 11-1)

17. The area of a rectangle is 64. The length is $\frac{x^3}{x+1}$, and the width is $\frac{x+1}{x}$. What is x? (Lesson 11-3)

Part 3 Quantative Comparison

Compare the quantity in Column A and the quantity in Column B. Then determine whether:

- **Ⓐ the quantity in Column A is greater,**
- **Ⓑ the quantity in Column B is greater,**
- **Ⓒ the two quantities are equal, or**
- **Ⓓ the relationship cannot be determined from the information given.**

	Column A	Column B
18.	the value of x in $-13x - 12 = -10x + 3$	the value of y in $12y + 16 = 8y$

(Lesson 3-5)

	Column A	Column B
19.	the slope of $2x - 3y = 10$	the y-intercept of $7x + 4y = 4$

(Lesson 5-3)

	Column A	Column B
20.	the measure of the hypotenuse of a right triangle if the measures of the legs are 10 and 11	the measure of one leg of a right triangle if the measure of the other leg is 13 and the hypotenuse is $\sqrt{390}$

(Lesson 11-4)

Part 4 Open Ended

Record your answers on a sheet of paper. Show your work.

21. Haley hikes 3 miles north, 7 miles east, and then 6 miles north again. (Lesson 11-4)

a. Draw a diagram showing the direction and distance of each segment of Haley's hike. Label Haley's starting point, her ending point, and the distance, in miles, of each segment of her hike.

b. To the nearest tenth of a mile, how far (in a straight line) is Haley from her starting point?

c. How did your diagram help you to find Haley's distance from her starting point?

Chapter 12

Rational Expressions and Equations

What You'll Learn

- **Lesson 12-1** Solve problems involving inverse variation.
- **Lessons 12-2, 12-3, 12-4, 12-6, and 12-7** Simplify, add, subtract, multiply, and divide rational expressions.
- **Lesson 12-5** Divide polynomials.
- **Lesson 12-8** Simplify mixed expressions and complex fractions.
- **Lesson 12-9** Solve rational equations.

Key Vocabulary

- inverse variation (p. 642)
- rational expression (p. 648)
- excluded values (p. 648)
- complex fraction (p. 684)
- extraneous solutions (p. 693)

Why It's Important

Performing operations on rational expressions is an important part of working with equations. For example, knowing how to divide rational expressions and polynomials can help you simplify complex expressions. You can use this process to determine the number of flags that a marching band can make from a given amount of material. *You will divide rational expressions and polynomials in Lessons 12-4 and 12-5.*

Getting Started

Prerequisite Skills To be successful in this chapter, you'll need to master these skills and be able to apply them in problem-solving situations. Review these skills before beginning Chapter 12.

For Lesson 12-1 — Solve Proportions

Solve each proportion. *(For review, see Lesson 3-6.)*

1. $\frac{y}{9} = \frac{-7}{16}$ **2.** $\frac{4}{x} = \frac{2}{10}$ **3.** $\frac{3}{15} = \frac{1}{n}$ **4.** $\frac{x}{8} = \frac{0.21}{2}$

5. $\frac{1.1}{0.6} = \frac{8.47}{n}$ **6.** $\frac{9}{8} = \frac{y}{6}$ **7.** $\frac{2.7}{3.6} = \frac{8.1}{a}$ **8.** $\frac{0.19}{2} = \frac{x}{24}$

For Lesson 12-2 — Greatest Common Factor

Find the greatest common factor for each pair of monomials. *(For review, see Lesson 9-1.)*

9. 30, 42 **10.** $60r^2, 45r^3$ **11.** $32m^2n^3, 12m^2n$ **12.** $14a^2b^2, 18a^3b$

For Lessons 12-3 through 12-8 — Factor Polynomials

Factor each polynomial. *(For review, see Lessons 9-2 and 9-3.)*

13. $3c^2d - 6c^2d^2$ **14.** $6mn + 15m^2$ **15.** $x^2 + 11x + 24$

16. $x^2 + 4x - 45$ **17.** $2x^2 + x - 21$ **18.** $3x^2 - 12x + 9$

For Lesson 12-9 — Solve Equations

Solve each equation. *(For review, see Lessons 3-4, 3-5, and 9-3.)*

19. $3x - 2 = -5$ **20.** $5x - 8 - 3x = (2x - 3)$ **21.** $\frac{m+9}{5} = \frac{m-10}{11}$ **22.** $\frac{5+x}{x-3} = \frac{14}{10}$

23. $\frac{7n-1}{6} = 5$ **24.** $\frac{4t-5}{-9} = 7$ **25.** $x^2 - x - 56 = 0$ **26.** $x^2 + 2x = 8$

FOLDABLES™
Study Organizer

Make this Foldable to help you organize information about rational expressions and equations. Begin with a sheet of plain $8\frac{1}{2}$" by 11" paper.

Step 1 Fold in Half

Fold in half lengthwise.

Step 2 Fold Again

Fold the top to the bottom.

Step 3 Cut

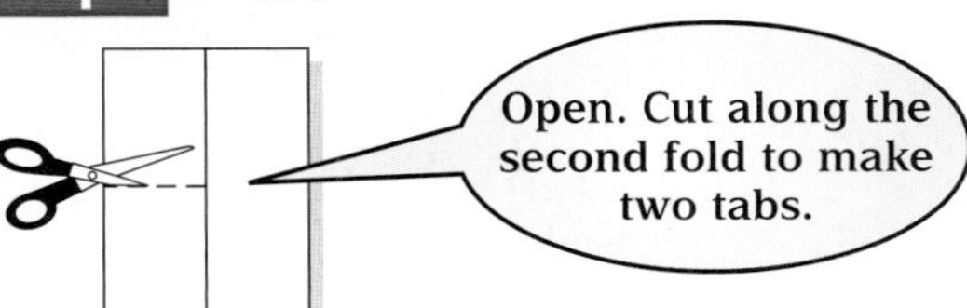
Open. Cut along the second fold to make two tabs.

Step 4 Label

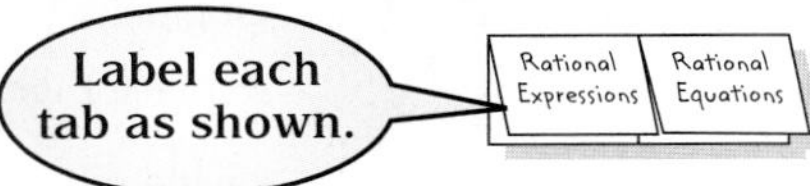
Label each tab as shown.
Rational Expressions
Rational Equations

Reading and Writing As you read and study the chapter, write notes and examples under each tab. Use this Foldable to apply what you learned about simplifying rational expressions and solving rational equations in Chapter 12.

12-1 Inverse Variation

What You'll Learn

- Graph inverse variations.
- Solve problems involving inverse variation.

How is inverse variation related to the gears on a bicycle?

The number of revolutions of the pedals made when riding a bicycle at a constant speed varies inversely as the gear ratio of the bicycle. In other words, as the gear ratio *decreases*, the revolutions per minute (rpm) *increase*. This is why when pedaling up a hill, shifting to a lower gear allows you to pedal with less difficulty.

Pedaling Rates to Maintain Speed of 10 mph

Gear Ratio	Rate
117.8	89.6
108.0	97.8
92.6	114.0
76.2	138.6
61.7	171.2
49.8	212.0
40.5	260.7

Vocabulary
- inverse variation
- product rule

Study Tip

Look Back
To review **direct variation**, see Lesson 5-2.

GRAPH INVERSE VARIATION Recall that some situations in which y increases as x increases are *direct variations.* If y varies directly as x, we can represent this relationship with an equation of the form $y = kx$, where $k \neq 0$. However, in the application above, as one value increases the other value decreases. When the product of two values remains constant, the relationship forms an **inverse variation**. We say *y varies inversely as x* or *y is inversely proportional to x.*

Key Concept — *Inverse Variation*

y varies inversely as x if there is some nonzero constant k such that $xy = k$.

Example 1 Graph an Inverse Variation

DRIVING The time t it takes to travel a certain distance varies inversely as the rate r at which you travel. The equation $rt = 250$ can be used to represent a person driving 250 miles. Complete the table and draw a graph of the relation.

r (mph)	5	10	15	20	25	30	35	40	45	50
t (hours)										

Solve for $r = 5$.

$rt = 250$ Original equation

$5t = 250$ Replace r with 5.

$t = \frac{250}{5}$ Divide each side by 5.

$t = 50$ Simplify.

Solve the equation for the other values of r.

r (mph)	5	10	15	20	25	30	35	40	45	50
t (hours)	50	25	16.67	12.5	10	8.33	7.14	6.25	5.56	5

Study Tip

Inverse Variation Problems
Note that to solve some inverse variation problems, there are two steps: first finding the value of k, and then using this value to find a specific value of x or y.

Next, graph the ordered pairs: (5, 50), (10, 25), (15, 16.67), (20, 12.5), (25, 10), (30, 8.33), (35, 7.14), (40, 6.25), (45, 5.56), and (50, 5).

The graph of an inverse variation is not a straight line like the graph of a direct variation. As the rate r increases, the time t that it takes to travel the same distance decreases.

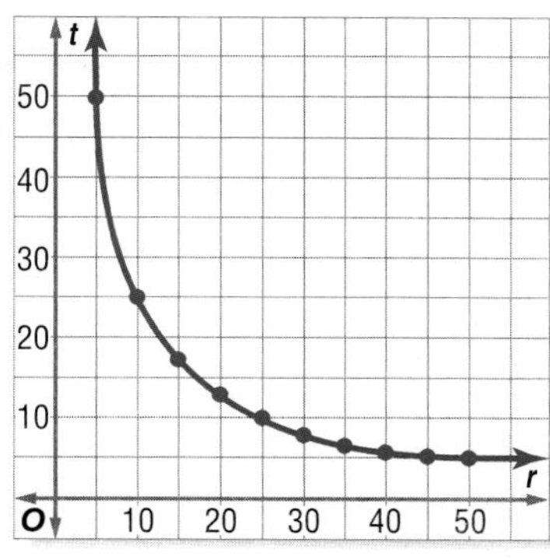

Graphs of inverse variations can also be drawn using negative values of x.

Example 2 Graph an Inverse Variation

Graph an inverse variation in which y varies inversely as x and $y = 15$ when $x = 6$.

Solve for k.

$xy = k$ Inverse variation equation

$(6)(15) = k$ $x = 6, y = 15$

$90 = k$ The constant of variation is 90.

Choose values for x and y whose product is 90.

x	y
−9	−10
−6	−15
−3	−30
−2	−45
0	undefined
2	45
3	30
6	15
9	10

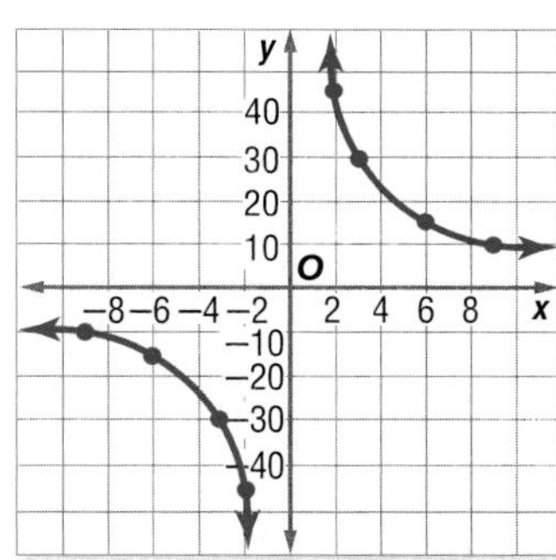

USE INVERSE VARIATION If (x_1, y_1) and (x_2, y_2) are solutions of an inverse variation, then $x_1y_1 = k$ and $x_2y_2 = k$.

$x_1y_1 = k$ and $x_2y_2 = k$

$x_1y_1 = x_2y_2$ Substitute x_2y_2 for k.

The equation $x_1y_1 = x_2y_2$ is called the **product rule** for inverse variations. You can use this equation to form a proportion.

$x_1y_1 = x_2y_2$ Product rule for inverse variations

$\frac{x_1y_1}{x_2y_1} = \frac{x_2y_2}{x_2y_1}$ Divide each side by x_2y_1.

$\frac{x_1}{x_2} = \frac{y_2}{y_1}$ Simplify.

Study Tip

Proportions
Notice that the proportion for inverse variations is different from the proportion for direct variation, $\frac{x_1}{x_2} = \frac{y_1}{y_2}$.

You can use the product rule or a proportion to solve inverse variation problems.

Example 3 Solve for x

If y varies inversely as x and $y = 4$ when $x = 7$, find x when $y = 14$.

Let $x_1 = 7$, $y_1 = 4$, and $y_2 = 14$. Solve for x_2.

Method 1 Use the product rule.

$x_1y_1 = x_2y_2$	Product rule for inverse variations
$7 \cdot 4 = x_2 \cdot 14$	$x_1 = 7$, $y_1 = 4$, and $y_2 = 14$
$\frac{28}{14} = x_2$	Divide each side by 14.
$2 = x_2$	Simplify.

Method 2 Use a proportion.

$\frac{x_1}{x_2} = \frac{y_2}{y_1}$	Proportion for inverse variations
$\frac{7}{x_2} = \frac{14}{4}$	$x_1 = 7$, $y_1 = 4$, and $y_2 = 14$
$28 = 14x_2$	Cross multiply.
$2 = x_2$	Divide each side by 14.

Both methods show that $x = 2$ when $y = 14$.

Example 4 Solve for y

If y varies inversely as x and $y = -6$ when $x = 9$, find y when $x = 6$.

Use the product rule.

$x_1y_1 = x_2y_2$	Product rule for inverse variations
$9 \cdot (-6) = 6y_2$	$x_1 = 9$, $y_1 = -6$, and $x_2 = 6$
$\frac{-54}{6} = y_2$	Divide each side by 6.
$-9 = y_2$	Simplify.

Thus, $y = -9$ when $x = 6$.

Inverse variation is often used in real-world situations.

Study Tip

Levers

A lever is a bar with a pivot point called the *fulcrum*. For a lever to balance, the lesser weight must be positioned farther from the fulcrum.

Example 5 Use Inverse Variation to Solve a Problem

PHYSICAL SCIENCE When two objects are balanced on a lever, their distances from the fulcrum are inversely proportional to their weights. In other words, the greater the weight, the less distance it should be from the fulcrum in order to maintain balance. If an 8-kilogram weight is placed 1.8 meters from the fulcrum, how far should a 12-kilogram weight be placed from the fulcrum in order to balance the lever?

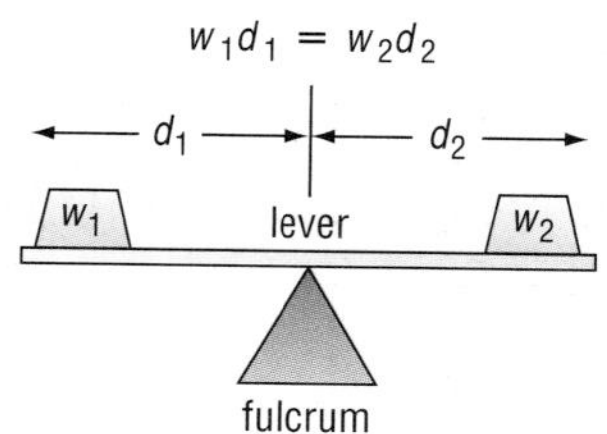

Let $w_1 = 8$, $d_1 = 1.8$, and $w_2 = 12$. Solve for d_2.

$w_1d_1 = w_2d_2$	Original equation
$8 \cdot 1.8 = 12d_2$	$w_1 = 8$, $d_1 = 1.8$, and $w_2 = 12$
$\frac{14.4}{12} = d_2$	Divide each side by 12.
$1.2 = d_2$	Simplify.

The 12-kilogram weight should be placed 1.2 meters from the fulcrum.

Check for Understanding

Concept Check

1. **OPEN ENDED** **Write** an equation showing an inverse variation with a constant of 8.
2. **Compare and contrast** direct variation and indirect variation equations and graphs.
3. **Determine** which situation is an example of inverse variation. Explain.
   a. Emily spends \$2 each day for snacks on her way home from school. The total amount she spends each week depends on the number of days school was in session.
   b. A business donates \$200 to buy prizes for a school event. The number of prizes that can be purchased depends upon the price of each prize.

Guided Practice

Graph each variation if y varies inversely as x.

4. $y = 24$ when $x = 8$
5. $y = -6$ when $x = -2$

Write an inverse variation equation that relates x and y. Assume that y varies inversely as x. Then solve.

6. If $y = 12$ when $x = 6$, find y when $x = 8$.
7. If $y = -8$ when $x = -3$, find y when $x = 6$.
8. If $y = 2.7$ when $x = 8.1$, find x when $y = 5.4$.
9. If $x = \frac{1}{2}$ when $y = 16$, find x when $y = 32$.

Application

10. **MUSIC** The length of a violin string varies inversely as the frequency of its vibrations. A violin string 10 inches long vibrates at a frequency of 512 cycles per second. Find the frequency of an 8-inch string.

Practice and Apply

Homework Help

For Exercises	See Examples
11–16	1, 2
17–28	3, 4
29–37	5

Extra Practice

See page 846.

Graph each variation if y varies inversely as x.

11. $y = 24$ when $x = -8$
12. $y = 3$ when $x = 4$
13. $y = 5$ when $x = 15$
14. $y = -4$ when $x = -12$
15. $y = 9$ when $x = 8$
16. $y = 2.4$ when $x = 8.1$

Write an inverse variation equation that relates x and y. Assume that y varies inversely as x. Then solve.

17. If $y = 12$ when $x = 5$, find y when $x = 3$.
18. If $y = 7$ when $x = -2$, find y when $x = 7$.
19. If $y = 8.5$ when $x = -1$, find x when $y = -1$.
20. If $y = 8$ when $x = 1.55$, find x when $y = -0.62$.
21. If $y = 6.4$ when $x = 4.4$, find x when $y = 3.2$.
22. If $y = 1.6$ when $x = 0.5$, find x when $y = 3.2$.
23. If $y = 4$ when $x = 4$, find y when $x = 7$.
24. If $y = -6$ when $x = -2$, find y when $x = 5$.
25. Find the value of y when $x = 7$ if $y = 7$ when $x = \frac{2}{3}$.
26. Find the value of y when $x = 32$ if $y = 16$ when $x = \frac{1}{2}$.
27. If $x = 6.1$ when $y = 4.4$, find x when $y = 3.2$.
28. If $x = 0.5$ when $y = 2.5$, find x when $y = 20$.

29. **GEOMETRY** A rectangle is 36 inches wide and 20 inches long. How wide is a rectangle of equal area if its length is 90 inches?

30. **MUSIC** The pitch of a musical note varies inversely as its wavelength. If the tone has a pitch of 440 vibrations per second and a wavelength of 2.4 feet, find the pitch of a tone that has a wavelength of 1.6 feet.

31. **COMMUNITY SERVICE** Students at Roosevelt High School are collecting canned goods for a local food pantry. They plan to distribute flyers to homes in the community asking for donations. Last year, 12 students were able to distribute 1000 flyers in nine hours. How long would it take if 15 students hand out the same number of flyers this year?

TRAVEL For Exercises 32 and 33, use the following information.
The Zalinski family can drive the 220 miles to their cabin in 4 hours at 55 miles per hour. Son Jeff claims that they could save half an hour if they drove 65 miles per hour, the speed limit.

32. How long will it take the family if they drive 65 miles per hour?

33. How much time would be saved driving at 65 miles per hour?

More About. . .

Art

American sculptor Alexander Calder was the first artist to use mobiles as an art form.

Source: www.infoplease.com

CHEMISTRY For Exercises 34–36, use the following information.
Boyle's Law states that the volume of a gas V varies inversely with applied pressure P.

34. Write an equation to show this relationship.

35. Pressure on 60 cubic meters of a gas is raised from 1 atmosphere to 3 atmospheres. What new volume does the gas occupy?

36. A helium-filled balloon has a volume of 22 cubic meters at sea level where the air pressure is 1 atmosphere. The balloon is released and rises to a point where the air pressure is 0.8 atmosphere. What is the volume of the balloon at this height?

37. **ART** Anna is designing a mobile to suspend from a gallery ceiling. A chain is attached eight inches from the end of a bar that is 20 inches long. On the shorter end of the bar is a sculpture weighing 36 kilograms. She plans to place another piece of artwork on the other end of the bar. How much should the second piece of art weigh if she wants the bar to be balanced?

CRITICAL THINKING For Exercises 38 and 39, assume that y varies inversely as x.

38. If the value of x is doubled, what happens to the value of y?

39. If the value of y is tripled, what happens to the value of x?

40. **WRITING IN MATH** Answer the question that was posed at the beginning of the lesson.

How is inverse variation related to the gears on a bicycle?

Include the following in your answer:

- an explanation of how shifting to a lower gear ratio affects speed and the pedaling rate on a certain bicycle if a rider is pedaling 73.4 revolutions per minute while traveling at 15 miles per hour, and
- an explanation why the gear ratio affects the pedaling speed of the rider.

Standardized Test Practice

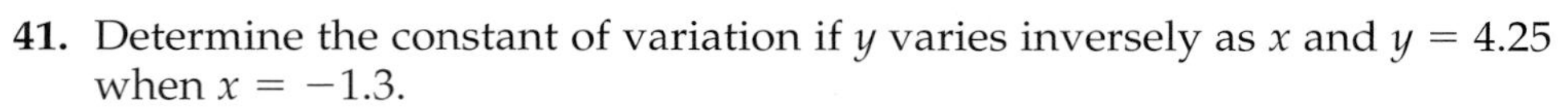

41. Determine the constant of variation if y varies inversely as x and $y = 4.25$ when $x = -1.3$.

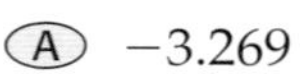 (A) −3.269 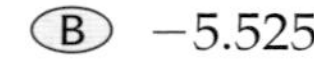 (B) −5.525  (C) −0.306 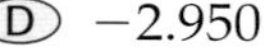 (D) −2.950

42. Identify the graph of $xy = k$ if $x = -2$ when $y = -4$.

(A)
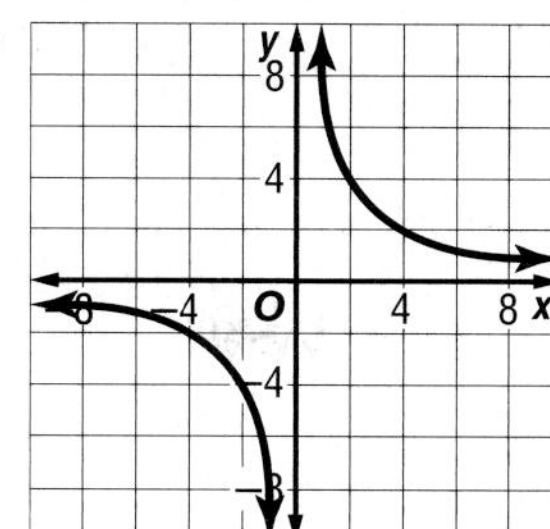

(B)
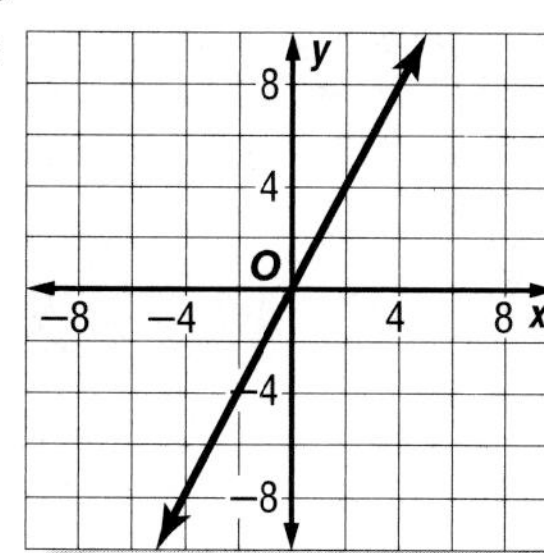

(C)
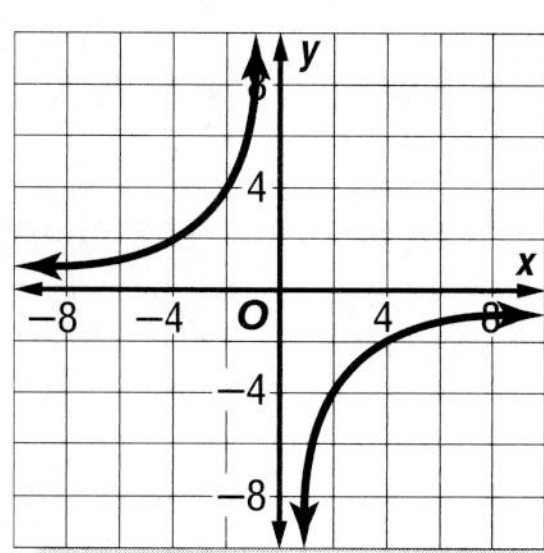

(D)
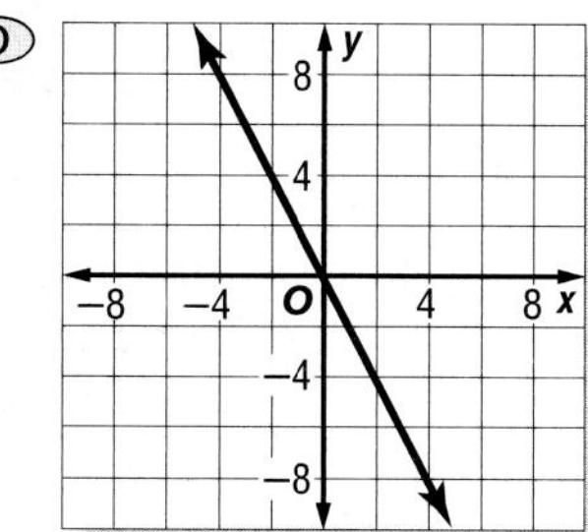

Maintain Your Skills

Mixed Review

For each triangle, find the measure of the indicated angle to the nearest degree. *(Lesson 11-7)*

43.
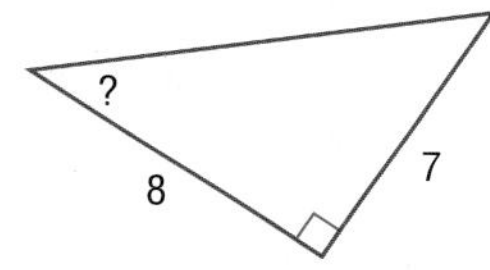

44.
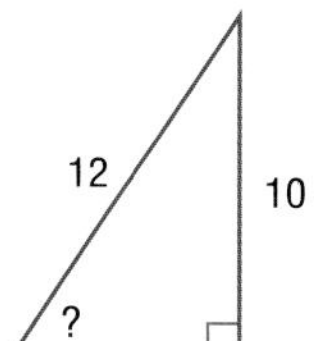

45.
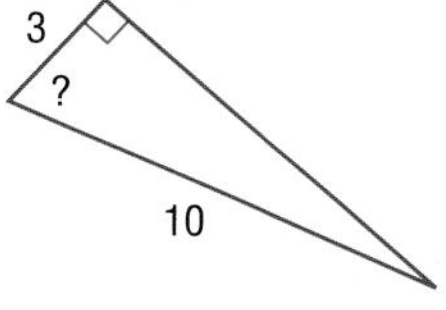

For each set of measures given, find the measures of the missing sides if $\triangle ABC \sim \triangle DEF$. *(Lesson 11-6)*

46. $a = 3, b = 10, c = 9, d = 12$

47. $b = 8, c = 4, d = 21, e = 28$

48. **MUSIC** Two musical notes played at the same time produce harmony. The closest harmony is produced by frequencies with the greatest GCF. A, C, and C sharp have frequencies of 220, 264, and 275, respectively. Which pair of these notes produce the closest harmony? *(Lesson 9-1)*

Solve each equation. *(Lesson 8-6)*

49. $7(2y - 7) = 5(4y + 1)$

50. $w(w + 2) = 2w(w - 3) + 16$

Solve each system of inequalities by graphing. *(Lesson 7-5)*

51. $y \leq 3x - 5$
$y > -x + 1$

52. $y \geq 2x + 3$
$2y \geq -5x - 14$

53. $x + y \leq 1$
$x - y \leq -3$
$y \geq 0$

54. $3x - 2y \geq -16$
$x + 4y < 4$
$5x - 8y < -8$

Getting Ready for the Next Lesson

PREREQUISITE SKILL Find the greatest common factor for each set of monomials.
*(To review **greatest common factors**, see Lesson 9-1.)*

55. 36, 15, 45

56. 48, 60, 84

57. 210, 330, 150

58. $17a$, $34a^2$

59. $12xy^2$, $18x^2y^3$

60. $12pr^2$, $40p^4$

12-2 Rational Expressions

What You'll Learn

- Identify values excluded from the domain of a rational expression.
- Simplify rational expressions.

Vocabulary

- rational expression
- excluded values

How can a rational expression be used in a movie theater?

The intensity I of an image on a movie screen is inversely proportional to the square of the distance d between the projector and the screen. Recall from Lesson 12-1 that this can be represented by the equation $I = \frac{k}{d^2}$, where k is a constant.

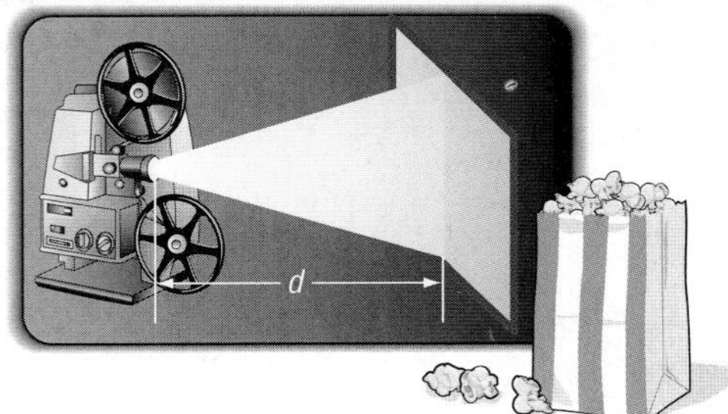

EXCLUDED VALUES OF RATIONAL EXPRESSIONS The expression $\frac{k}{d^2}$ is an example of a rational expression. A **rational expression** is an algebraic fraction whose numerator and denominator are polynomials.

Because a rational expression involves division, the denominator may not have a value of zero. Any values of a variable that result in a denominator of zero must be excluded from the domain of that variable. These are called **excluded values** of the rational expression.

Example 1 One Excluded Value

State the excluded value of $\frac{5m + 3}{m - 6}$.

Exclude the values for which $m - 6 = 0$.

$m - 6 = 0$ The denominator cannot equal 0.

$m = 6$ Add 6 to each side.

Therefore, m cannot equal 6.

To determine the excluded values of a rational expression, you may be able to factor the denominator first.

Example 2 Multiple Excluded Values

State the excluded values of $\frac{x^2 - 5}{x^2 - 5x + 6}$.

Exclude the values for which $x^2 - 5x + 6 = 0$.

$x^2 - 5x + 6 = 0$ The denominator cannot equal zero.

$(x - 2)(x - 3) = 0$ Factor.

Use the Zero Product Property to solve for x.

$x - 2 = 0$ or $x - 3 = 0$

$x = 2$ $\quad x = 3$

Therefore, x cannot equal 2 or 3.

You can use rational expressions to solve real-world problems.

Example 3 Use Rational Expressions

LANDSCAPING Kenyi is helping his parents landscape their yard and needs to move some large rocks. He plans to use a 6-foot bar as a lever. He positions it as shown at the right.

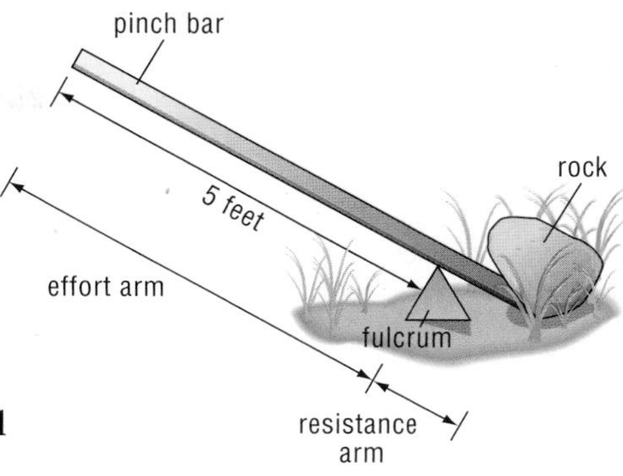

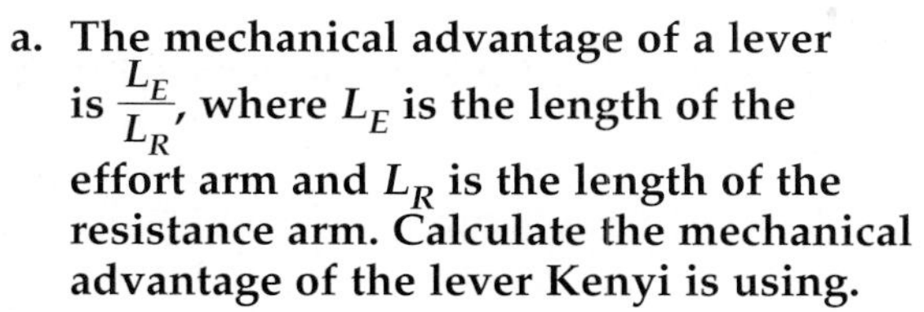

a. The mechanical advantage of a lever is $\frac{L_E}{L_R}$, where L_E is the length of the effort arm and L_R is the length of the resistance arm. Calculate the mechanical advantage of the lever Kenyi is using.

Let b represent the length of the bar and e represent the length of the effort arm. Then $b - e$ represents the length of the resistance arm.

Use the expression for mechanical advantage to write an expression for the mechanical advantage in this situation.

$$\frac{L_E}{L_R} = \frac{e}{b-e} \quad L_E = e, L_R = b - e$$

$$= \frac{5}{6-5} \quad e = 5, b = 6$$

$$= 5 \quad \text{Simplify.}$$

The mechanical advantage is 5.

b. The force placed on the rock is the product of the mechanical advantage and the force applied to the end of the lever. If Kenyi can apply a force of 180 pounds, what is the greatest weight he can lift with the lever?

Since the mechanical advantage is 5, Kenyi can lift $5 \cdot 180$ or 900 pounds with this lever.

Career Choices

Landscape Architect

Landscape architects plan the location of structures, roads, and walkways as well as the arrangement of flowers, trees, and shrubs in a variety of settings.

Source: U.S. Bureau of Labor and Statistics

Online Research
For information about a career as a landscape architect, visit: www.algebra1.com/careers

SIMPLIFY RATIONAL EXPRESSIONS Simplifying rational expressions is similar to simplifying fractions with numbers. To simplify a rational expression, you must eliminate any common factors of the numerator and denominator. To do this, use their greatest common factor (GCF). Remember that $\frac{ab}{ac} = \frac{a}{a} \cdot \frac{b}{c}$ and $\frac{a}{a} = 1$. So, $\frac{ab}{ac} = 1 \cdot \frac{b}{c}$ or $\frac{b}{c}$.

Example 4 Expression Involving Monomials

Simplify $\frac{-7a^2b^3}{21a^5b}$.

$$\frac{-7a^2b^3}{21a^5b} = \frac{(7a^2b)(-b^2)}{(7a^2b)(3a^3)} \quad \text{The GCF of the numerator and denominator is } 7a^2b.$$

$$= \frac{\overset{1}{\cancel{(7a^2b)}}(-b^2)}{\underset{1}{\cancel{(7a^2b)}}(3a^3)} \quad \text{Divide the numerator and denominator by } 7a^2b.$$

$$= \frac{-b^2}{3a^3} \quad \text{Simplify.}$$

You can use the same procedure to simplify a rational expression in which the numerator and denominator are polynomials.

Study Tip

Simplest Form
When a rational expression is in simplest form, the numerator and denominator have no common factors other than 1 or -1.

Example 5 Expressions Involving Polynomials

Simplify $\frac{x^2 - 2x - 15}{x^2 - x - 12}$.

$$\frac{x^2 - 2x - 15}{x^2 - x - 12} = \frac{(x+3)(x-5)}{(x+3)(x-4)} \quad \text{Factor.}$$

$$= \frac{\cancel{(x+3)}^{1}(x-5)}{\cancel{(x+3)}_{1}(x-4)} \quad \text{Divide the numerator and denominator by the GCF, } x + 3.$$

$$= \frac{x-5}{x-4} \quad \text{Simplify.}$$

It is important to determine the excluded values of a rational expression using the original expression rather than the simplified expression.

Example 6 Excluded Values

Simplify $\frac{3x - 15}{x^2 - 7x + 10}$. State the excluded values of x.

$$\frac{3x - 15}{x^2 - 7x + 10} = \frac{3(x-5)}{(x-2)(x-5)} \quad \text{Factor.}$$

$$= \frac{3\cancel{(x-5)}^{1}}{(x-2)\cancel{(x-5)}_{1}} \quad \text{Divide the numerator and denominator by the GCF, } x - 5.$$

$$= \frac{3}{x-2} \quad \text{Simplify.}$$

Exclude the values for which $x^2 - 7x + 10$ equals 0.

$x^2 - 7x + 10 = 0$ The denominator cannot equal zero.

$(x - 5)(x - 2) = 0$ Factor.

$x = 5$ or $x = 2$ Zero Product Property

CHECK Verify the excluded values by substituting them into the original expression.

$$\frac{3x - 15}{x^2 - 7x + 10} = \frac{3(5) - 15}{5^2 - 7(5) + 10} \quad x = 5$$

$$= \frac{15 - 15}{25 - 35 + 10} \quad \text{Evaluate.}$$

$$= \frac{0}{0} \quad \text{Simplify.}$$

$$\frac{3x - 15}{x^2 - 7x + 10} = \frac{3(2) - 15}{2^2 - 7(2) + 10} \quad x = 2$$

$$= \frac{6 - 15}{4 - 14 + 10} \quad \text{Evaluate.}$$

$$= \frac{-9}{0} \quad \text{Simplify.}$$

The expression is undefined when $x = 5$ and $x = 2$. Therefore, $x \neq 5$ and $x \neq 2$.

Check for Understanding

Concept Check

1. **Describe** how you would determine the values to be excluded from the expression $\frac{x+3}{x^2+5x+6}$.
2. **OPEN ENDED** Write a rational expression involving one variable for which the excluded values are -4 and -7.
3. **Explain** why -2 may not be the only excluded value of a rational expression that simplifies to $\frac{x-3}{x+2}$.

Guided Practice

State the excluded values for each rational expression.

4. $\frac{4a}{3+a}$
5. $\frac{x^2-9}{2x+6}$
6. $\frac{n+5}{n^2+n-20}$

Simplify each expression. State the excluded values of the variables.

7. $\frac{56x^2y}{70x^3y^2}$
8. $\frac{x^2-49}{x+7}$
9. $\frac{x+4}{x^2+8x+16}$
10. $\frac{x^2-2x-3}{x^2-7x+12}$
11. $\frac{a^2+4a-12}{a^2+2a-8}$
12. $\frac{2x^2-x-21}{2x^2-15x+28}$
13. Simplify $\frac{b^2-3b-4}{b^2-13b+36}$. State the excluded values of b.

Application

AQUARIUMS For Exercises 14 and 15, use the following information.
Jenna has guppies in her aquarium. One week later, she adds four neon fish.

14. Write an expression that represents the fraction of neon fish in the aquarium.
15. Suppose that two months later the guppy population doubles, she still has four neons, and she buys 5 different tropical fish. Write an expression that shows the fraction of neons in the aquarium after the other fish have been added.

Practice and Apply

Homework Help

For Exercises	See Examples
16–23	1, 2
24–27	4, 6
28–41	5, 6
42–54	3

Extra Practice
See page 846.

State the excluded values for each rational expression.

16. $\frac{m+3}{m-2}$
17. $\frac{3b}{b+5}$
18. $\frac{3n+18}{n^2-36}$
19. $\frac{2x-10}{x^2-25}$
20. $\frac{a^2-2a+1}{a^2+2a-3}$
21. $\frac{x^2-6x+9}{x^2+2x-15}$
22. $\frac{n^2-36}{n^2+n-30}$
23. $\frac{25-x^2}{x^2+12x+35}$

Simplify each expression. State the excluded values of the variables.

24. $\frac{35yz^2}{14y^2z}$
25. $\frac{14a^3b^2}{42ab^3}$
26. $\frac{64qr^2s}{16q^2rs}$
27. $\frac{9x^2yz}{24xyz^2}$
28. $\frac{7a^3b^2}{21a^2b+49ab^3}$
29. $\frac{3m^2n^3}{36mn^3-12m^2n^2}$
30. $\frac{x^2+x-20}{x+5}$
31. $\frac{z^2+10z+16}{z+2}$
32. $\frac{4x+8}{x^2+6x+8}$
33. $\frac{2y-4}{y^2+3y-10}$
34. $\frac{m^2-36}{m^2-5m-6}$
35. $\frac{a^2-9}{a^2+6a-27}$
36. $\frac{x^2+x-2}{x^2-3x+2}$
37. $\frac{b^2+2b-8}{b^2-20b+64}$
38. $\frac{x^2-x-20}{x^3+10x^2+24x}$
39. $\frac{n^2-8n+12}{n^3-12n^2+36n}$
40. $\frac{4x^2-6x-4}{2x^2-8x+8}$
41. $\frac{3m^2+9m+6}{4m^2+12m+8}$

COOKING **For Exercises 42–45, use the following information.**

The formula $t = \frac{40(25 + 1.85a)}{50 - 1.85a}$ relates the time t in minutes that it takes to cook an average-size potato in an oven that is at an altitude of a thousands of feet.

42. What is the value of a for an altitude of 4500 feet?

43. Calculate the time it takes to cook a potato at an altitude of 3500 feet.

44. About how long will it take to cook a potato at an altitude of 7000 feet?

45. The altitude in Exercise 44 is twice that of Exercise 43. How do your cooking times compare for those two altitudes?

WebQuest

You can use a rational expression to determine how an amusement park can finance a new roller coaster. Visit www.algebra1.com/webquest to continue work on your WebQuest project.

PHYSICAL SCIENCE **For Exercises 46–48, use the following information.**

To pry the lid off a paint can, a screwdriver that is 17.5 centimeters long is used as a lever. It is placed so that 0.4 centimeter of its length extends inward from the rim of the can.

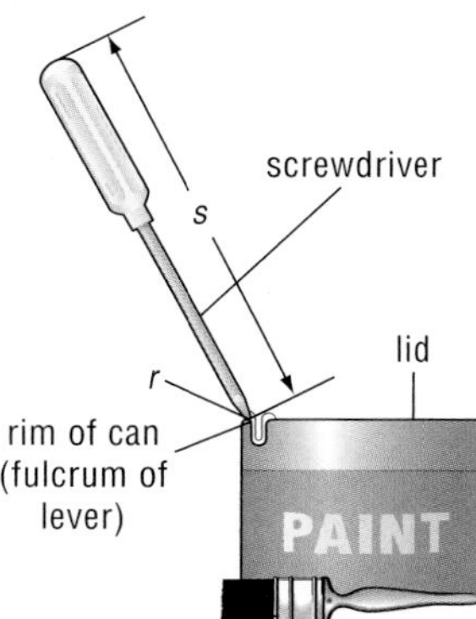

46. Write an equation that can be used to calculate the mechanical advantage.

47. What is the mechanical advantage?

48. If a force of 6 pounds is applied to the end of the screwdriver, what is the force placed on the lid?

FIELD TRIPS **For Exercises 49–52, use the following information.**

Mrs. Hoffman's art class is taking a trip to the museum. A bus that can seat up to 56 people costs $450 for the day, and group rate tickets at the museum cost $4 each.

49. If there are no more than 56 students going on the field trip, write an expression for the total cost for n students to go to the museum.

50. Write a rational expression that could be used to calculate the cost of the trip per student.

51. How many students must attend in order to keep the cost under $15 per student?

52. How would you change the expression for cost per student if the school were to cover the cost of two adult chaperones?

More About. . .

Farming

Although the amount of farmland in the United States is declining, crop production has increased steadily due in part to better irrigation practices.

Source: U.S. Department of Agriculture

FARMING **For Exercises 53 and 54, use the following information.**

Some farmers use an irrigation system that waters a circular region in a field. Suppose a square field with sides of length $2x$ is irrigated from the center of the square. The irrigation system can reach a radius of x.

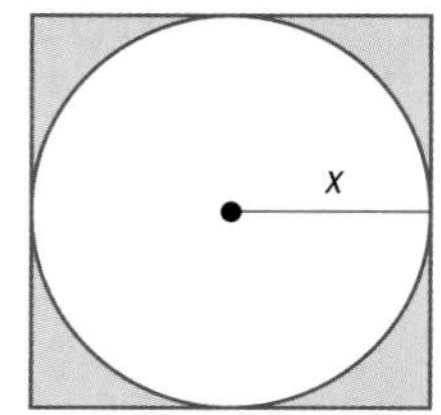

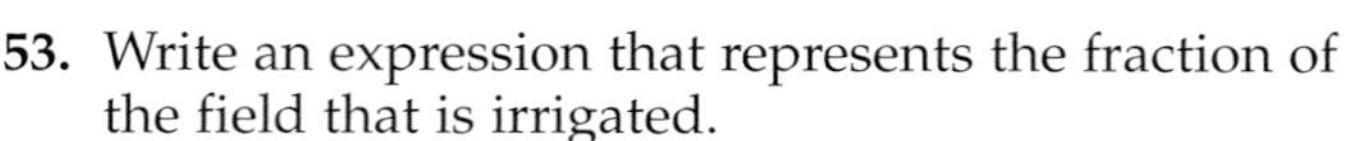

53. Write an expression that represents the fraction of the field that is irrigated.

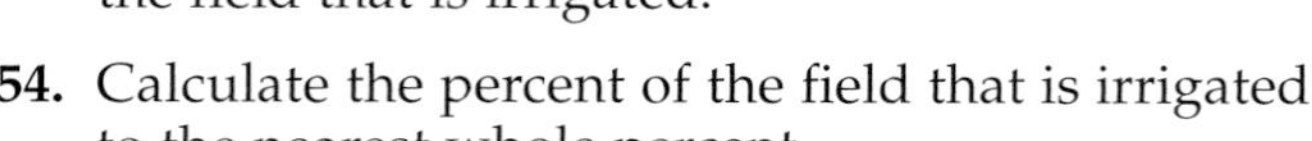

54. Calculate the percent of the field that is irrigated to the nearest whole percent.

55. **CRITICAL THINKING** Two students graphed the following equations on their calculators.

$$y = \frac{x^2 - 16}{x - 4} \qquad y = x + 4$$

They were surprised to see that the graphs appeared to be identical.

a. Explain why the graphs appear to be the same.

b. Explain how and why the graphs are different.

56. **WRITING IN MATH** Answer the question that was posed at the beginning of the lesson.

How can a rational expression be used in a movie theater?

Include the following in your answer:

- a description of how you determine the excluded values of a rational expression, and
- an example of another real-world situation that could be described using a rational expression.

Standardized Test Practice

57. Which expression is written in simplest form?

(A) $\frac{x^2 + 3x + 2}{x^2 + x - 2}$ (B) $\frac{3x - 3}{2x^2 - 2}$

(C) $\frac{x^2 + 7x}{x^2 + 3x - 4}$ (D) $\frac{2x^2 - 5x - 3}{x^2 + x - 12}$

58. In which expression are 1 and 5 excluded values?

(A) $\frac{x^2 + 6x + 5}{x^2 - 3x + 2}$ (B) $\frac{x^2 - 3x + 2}{x^2 - 6x + 5}$

(C) $\frac{x^2 - 6x + 5}{x^2 - 3x + 2}$ (D) $\frac{x^2 - 3x + 2}{x^2 + 6x + 5}$

Maintain Your Skills

Mixed Review

Write an inverse variation equation that relates x and y. Assume that y varies inversely as x. Then solve. *(Lesson 12-1)*

59. If $y = 6$ when $x = 10$, find y when $x = -12$.

60. If $y = 16$ when $x = \frac{1}{2}$, find x when $y = 32$.

61. If $y = -2.5$ when $x = 3$, find y when $x = -8$.

Use a calculator to find the measure of each angle to the nearest degree. *(Lesson 11-7)*

62. $\sin N = 0.2347$

63. $\cos B = 0.3218$

64. $\tan V = 0.0765$

65. $\sin A = 0.7011$

Solve each equation. Check your solution. *(Lesson 11-3)*

66. $\sqrt{a + 3} = 2$

67. $\sqrt{2z + 2} = z - 3$

68. $\sqrt{13 - 4p} - p = 8$

69. $\sqrt{3r^2 + 61} = 2r + 1$

Find the next three terms in each geometric sequence. *(Lesson 10-7)*

70. 1, 3, 9, 27, …

71. 6, 24, 96, 384, …

72. $\frac{1}{4}, -\frac{1}{2}, 1, -2, \ldots$

73. $4, 3, \frac{9}{4}, \frac{27}{16}, \ldots$

74. **GEOMETRY** Find the area of a rectangle if the length is $2x + y$ units and the width is $x + y$ units. *(Lesson 8-7)*

Getting Ready for the Next Lesson

BASIC SKILL Complete.

75. 84 in. = ____ ft

76. 4.5 m = ____ cm

77. 4 h 15 min = ____ s

78. 18 mi = ____ ft

79. 3 days = ____ h

80. 220 mL = ____ L

Graphing Calculator Investigation

A Follow-Up of Lesson 12-2

Rational Expressions

When simplifying rational expressions, you can use a TI-83 Plus graphing calculator to support your answer. If the graphs of the original expression and the simplified expression coincide, they are equivalent. You can also use the graphs to see excluded values.

Simplify $\frac{x^2 - 25}{x^2 + 10x + 25}$.

Step 1 *Factor the numerator and denominator.*

- $\frac{x^2 - 25}{x^2 + 10x + 25} = \frac{(x-5)(x+5)}{(x+5)(x+5)}$

  $= \frac{(x-5)}{(x+5)}$ When $x = -5$, $x + 5 = 0$. Therefore, x cannot equal -5 because you cannot divide by zero.

Step 2 *Graph the original expression.*

- Set the calculator to **Dot** mode.
- Enter $\frac{x^2 - 25}{x^2 + 10x + 25}$ as **Y1** and graph.

KEYSTROKES: MODE ▼ ▼ ▼ ▼ ▶ ENTER Y= (X,T,θ,n x^2 − 25) ÷ (X,T,θ,n x^2 + 10 X,T,θ,n + 25) ZOOM 6

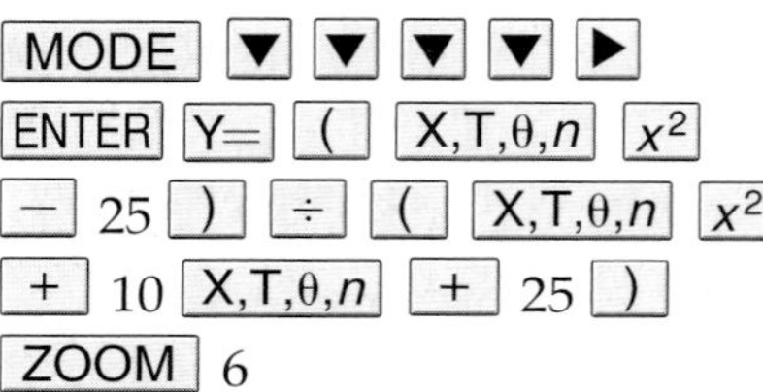

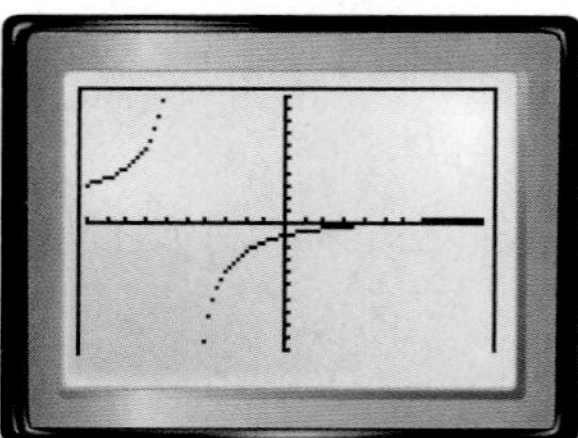

[−10, 10] scl: 1 by [−10, 10] scl: 1

Step 3 *Graph the simplified expression.*

- Enter $\frac{(x-5)}{(x+5)}$ as **Y2** and graph.

KEYSTROKES: Y= ▼ ▼ (X,T,θ,n − 5) ÷ (X,T,θ,n + 5) GRAPH

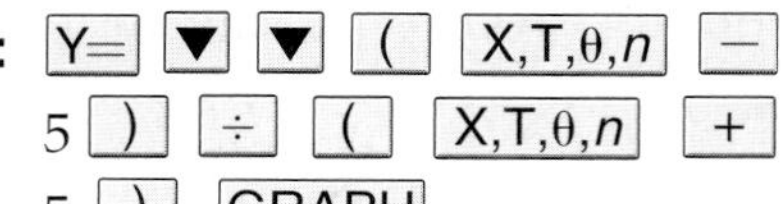

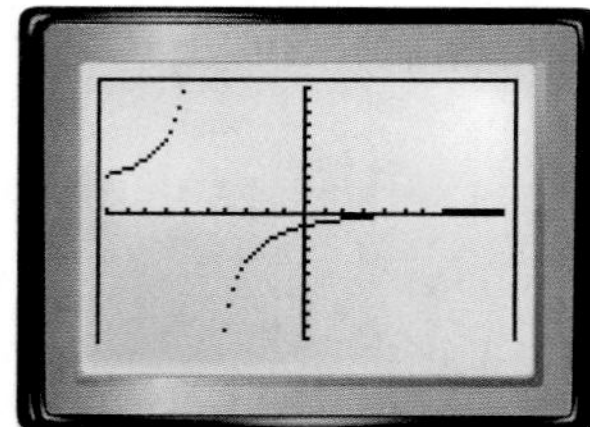

[−10, 10] scl: 1 by [−10, 10] scl: 1

Since the graphs overlap, the two expressions are equivalent.

Exercises

Simplify each expression. Then verify your answer graphically. Name the excluded values.

1. $\frac{3x + 6}{x^2 + 7x + 10}$
2. $\frac{x^2 - 9x + 8}{x^2 - 16x + 64}$
3. $\frac{5x^2 + 10x + 5}{3x^2 + 6x + 3}$
4. Simplify the rational expression $\frac{2x - 9}{4x^2 - 18x}$ and answer the following questions using the **TABLE** menu on your calculator.
   a. How can you use the **TABLE** function to verify that the original expression and the simplified expression are equivalent?
   b. How does the **TABLE** function show you that an x value is an excluded value?

 www.algebra1.com/other_calculator_keystrokes

12-3 Multiplying Rational Expressions

What You'll Learn

- Multiply rational expressions.
- Use dimensional analysis with multiplication.

How can you multiply rational expressions to determine the cost of electricity?

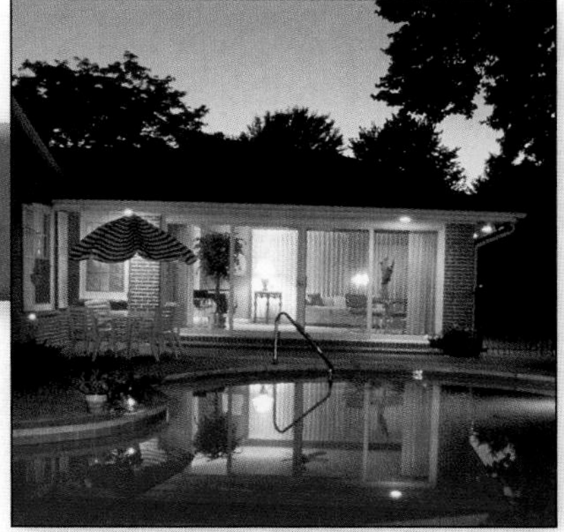

There are 25 lights around a patio. Each light is 40 watts, and the cost of electricity is 15 cents per kilowatt-hour. You can use the expression below to calculate the cost of using the lights for h hours.

$$25 \text{ lights} \cdot h \text{ hours} \cdot \frac{40 \text{ watts}}{\text{light}} \cdot \frac{1 \text{ kilowatt}}{1000 \text{ watts}} \cdot \frac{15 \text{ cents}}{1 \text{ kilowatt} \cdot \text{hour}} \cdot \frac{1 \text{ dollar}}{100 \text{ cents}}$$

From this point on, you may assume that no denominator of a rational expression has a value of zero.

MULTIPLY RATIONAL EXPRESSIONS The multiplication expression above is similar to the multiplication of rational expressions. Recall that to multiply rational numbers expressed as fractions, you multiply numerators and multiply denominators. You can use this same method to multiply rational expressions.

Example 1 Expressions Involving Monomials

a. Find $\frac{5ab^3}{8c^2} \cdot \frac{16c^3}{15a^2b}$.

Method 1 Divide by the greatest common factor after multiplying.

$$\frac{5ab^3}{8c^2} \cdot \frac{16c^3}{15a^2b} = \frac{80ab^3c^3}{120a^2bc^2}$$ ← Multiply the numerators. ← Multiply the denominators.

$$= \frac{\overset{1}{\cancel{40abc^2}}(2b^2c)}{\underset{1}{\cancel{40abc^2}}(3a)}$$ The GCF is $40abc^2$.

$$= \frac{2b^2c}{3a}$$ Simplify.

Method 2 Divide by the common factors before multiplying.

$$\frac{5ab^3}{8c^2} \cdot \frac{16c^3}{15a^2b} = \frac{\overset{1\,1\,b^2}{\cancel{5}\cancel{a}\cancel{b^3}}}{\underset{1\,1}{\cancel{8}\cancel{c^2}}} \cdot \frac{\overset{2\,c}{\cancel{16}\cancel{c^3}}}{\underset{3\,a\,1}{\cancel{15}\cancel{a^2}\cancel{b}}}$$ Divide by common factors 5, 8, a, b, and c^2.

$$= \frac{2b^2c}{3a}$$ Multiply.

b. Find $\frac{12xy^2}{45mp^2} \cdot \frac{27m^3p}{40x^3y}$.

$$\frac{12xy^2}{45mp^2} \cdot \frac{27m^3p}{40x^3y} = \frac{\overset{3\,1\,y}{\cancel{12}\cancel{x}\cancel{y^2}}}{\underset{5\,1\,p}{\cancel{45}\cancel{m}\cancel{p^2}}} \cdot \frac{\overset{3\,m^2\,1}{\cancel{27}\cancel{m^3}\cancel{p}}}{\underset{10\,x^2\,1}{\cancel{40}\cancel{x^3}\cancel{y}}}$$ Divide by common factors 4, 9, x, y, m, and p.

$$= \frac{9m^2y}{50x^2p}$$ Multiply.

Sometimes you must factor a quadratic expression before you can simplify a product of rational expressions.

Example 2 Expressions Involving Polynomials

a. Find $\frac{x-5}{x} \cdot \frac{x^2}{x^2 - 2x - 15}$.

$$\frac{x-5}{x} \cdot \frac{x^2}{x^2 - 2x - 15} = \frac{x-5}{x} \cdot \frac{x^2}{(x-5)(x+3)}$$ Factor the denominator.

$$= \frac{\overset{x}{\cancel{x^2}}\overset{1}{\cancel{(x-5)}}}{\underset{1}{\cancel{x}}\underset{1}{\cancel{(x-5)}}(x+3)}$$ The GCF is $x(x-5)$.

$$= \frac{x}{x+3}$$ Simplify.

b. Find $\frac{a^2 + 7a + 10}{a+1} \cdot \frac{3a+3}{a+2}$.

$$\frac{a^2 + 7a + 10}{a+1} \cdot \frac{3a+3}{a+2} = \frac{(a+5)(a+2)}{a+1} \cdot \frac{3(a+1)}{a+2}$$ Factor the numerators.

$$= \frac{3(a+5)\overset{1}{\cancel{(a+2)}}\overset{1}{\cancel{(a+1)}}}{\underset{1}{\cancel{(a+1)}}\underset{1}{\cancel{(a+2)}}}$$ The GCF is $(a+1)(a+2)$.

$$= \frac{3(a+5)}{1}$$ Multiply.

$$= 3a + 15$$ Simplify.

Study Tip

Look Back
To review **dimensional analysis**, see Lesson 3-8.

DIMENSIONAL ANALYSIS When you multiply fractions that involve units of measure, you can divide by the units in the same way that you divide by variables. Recall that this process is called dimensional analysis.

Example 3 Dimensional Analysis

OLYMPICS In the 2000 Summer Olympics in Sydney, Australia, Maurice Green of the United States won the gold medal for the 100-meter sprint. His winning time was 9.87 seconds. What was his speed in kilometers per hour? Round to the nearest hundredth.

$$\frac{100 \text{ meters}}{9.87 \text{ seconds}} \cdot \frac{1 \text{ kilometer}}{1000 \text{ meters}} \cdot \frac{60 \text{ seconds}}{1 \text{ minute}} \cdot \frac{60 \text{ minutes}}{1 \text{ hour}}$$

$$= \frac{100 \cancel{\text{ meters}}}{9.87 \cancel{\text{ seconds}}} \cdot \frac{1 \text{ kilometer}}{1000 \cancel{\text{ meters}}} \cdot \frac{60 \cancel{\text{ seconds}}}{1 \cancel{\text{ minute}}} \cdot \frac{60 \cancel{\text{ minutes}}}{1 \text{ hour}}$$

$$= \frac{\overset{1}{\cancel{100}} \cdot 1 \cdot 60 \cdot 60 \text{ kilometers}}{9.87 \cdot \underset{10}{\cancel{1000}} \cdot 1 \cdot 1 \text{ hours}}$$

$$= \frac{60 \cdot 60 \text{ kilometers}}{9.87 \cdot 10 \text{ hours}}$$ Simplify.

$$= \frac{3600 \text{ kilometers}}{98.7 \text{ hours}}$$ Multiply.

$$= \frac{36.47 \text{ kilometers}}{1 \text{ hour}}$$ Divide numerator and denominator by 98.7.

His speed was 36.47 kilometers per hour.

More About. . .

Olympics

American sprinter Thomas Burke won the 100-meter dash at the first modern Olympics in Athens, Greece, in 1896 in 12.0 seconds.

Source: www.olympics.org

Check for Understanding

Concept Check

1. **OPEN ENDED** Write two rational expressions whose product is $\frac{2}{x}$.
2. **Explain** why $-\frac{x+6}{x-5}$ is not equivalent to $\frac{-x+6}{x-5}$.
3. **FIND THE ERROR** Amiri and Hoshi multiplied $\frac{x-3}{x+3}$ and $\frac{4x}{x^2-4x+3}$.

Amiri

$$\frac{x-3}{x+3} \cdot \frac{4x}{x^2-4x+3} = \frac{\cancel{(x-3)}4x}{(x+3)\cancel{(x-3)}(x-1)} = \frac{4x}{(x+3)(x-1)}$$

Hoshi

$$\frac{x-3}{x+3} \cdot \frac{4x}{x^2-4x+3} = \frac{\cancel{x-3}}{\cancel{x+3}} \cdot \frac{\cancel{4x}}{x^2-\cancel{4x}+3} = \frac{1}{x^2+3}$$

Who is correct? Explain your reasoning.

Guided Practice Find each product.

4. $\frac{64y^2}{5y} \cdot \frac{5y}{8y}$
5. $\frac{15s^2t^3}{12st} \cdot \frac{16st^2}{10s^3t^3}$
6. $\frac{m+4}{3m} \cdot \frac{4m^2}{(m+4)(m+5)}$
7. $\frac{x^2-4}{2} \cdot \frac{4}{x-2}$
8. $\frac{n^2-16}{n+4} \cdot \frac{n+2}{n^2-8n+16}$
9. $\frac{x-5}{x^2-7x+10} \cdot \frac{x^2+x-6}{5}$
10. Find $\frac{24 \text{ feet}}{1 \text{ second}} \cdot \frac{60 \text{ seconds}}{1 \text{ minute}} \cdot \frac{60 \text{ minutes}}{1 \text{ hour}} \cdot \frac{1 \text{ mile}}{5280 \text{ feet}}$.

Application

11. **SPACE** The moon is about 240,000 miles from Earth. How many days would it take a spacecraft to reach the moon if it travels at an average of 100 miles per minute?

Practice and Apply

Homework Help

For Exercises	See Examples
12–15	1
16–27	2
28–37	3

Extra Practice

See page 847.

Find each product.

12. $\frac{8}{x^2} \cdot \frac{x^4}{4x}$
13. $\frac{10r^3}{6n^3} \cdot \frac{42n^2}{35r^3}$
14. $\frac{10y^3z^2}{6wx^3} \cdot \frac{12w^2x^2}{25y^2z^4}$
15. $\frac{3a^2b}{2gh} \cdot \frac{24g^2h}{15ab^2}$
16. $\frac{(x-8)}{(x+8)(x-3)} \cdot \frac{(x+4)(x-3)}{(x-8)}$
17. $\frac{(n-1)(n+1)}{(n+1)} \cdot \frac{(n-4)}{(n-1)(n+4)}$
18. $\frac{(z+4)(z+6)}{(z-6)(z+1)} \cdot \frac{(z+1)(z-5)}{(z+3)(z+4)}$
19. $\frac{(x-1)(x+7)}{(x-7)(x-4)} \cdot \frac{(x-4)(x+10)}{(x+1)(x+10)}$
20. $\frac{x^2-25}{9} \cdot \frac{x+5}{x-5}$
21. $\frac{y^2-4}{y^2-1} \cdot \frac{y+1}{y+2}$
22. $\frac{1}{x^2+x-12} \cdot \frac{x-3}{x+5}$
23. $\frac{x-6}{x^2+4x-32} \cdot \frac{x-4}{x+2}$
24. $\frac{x+3}{x+4} \cdot \frac{x}{x^2+7x+12}$
25. $\frac{n}{n^2+8n+15} \cdot \frac{2n+10}{n^2}$
26. $\frac{b^2+12b+11}{b^2-9} \cdot \frac{b+9}{b^2+20b+99}$
27. $\frac{a^2-a-6}{a^2-16} \cdot \frac{a^2+7a+12}{a^2+4a+4}$

Find each product.

28. $\frac{2.54 \text{ centimeters}}{1 \text{ inch}} \cdot \frac{12 \text{ inches}}{1 \text{ foot}} \cdot \frac{3 \text{ feet}}{1 \text{ yard}}$

29. $\frac{60 \text{ kilometers}}{1 \text{ hour}} \cdot \frac{1000 \text{ meters}}{1 \text{ kilometer}} \cdot \frac{1 \text{ hour}}{60 \text{ minutes}} \cdot \frac{1 \text{ minutes}}{60 \text{ seconds}}$

30. $\frac{32 \text{ feet}}{1 \text{ second}} \cdot \frac{60 \text{ seconds}}{1 \text{ minute}} \cdot \frac{60 \text{ minutes}}{1 \text{ hour}} \cdot \frac{1 \text{ mile}}{5280 \text{ feet}}$

31. $10 \text{ feet} \cdot 18 \text{ feet} \cdot 3 \text{ feet} \cdot \frac{1 \text{ yard}^3}{27 \text{ feet}^3}$

32. **DECORATING** Alani's bedroom is 12 feet wide and 14 feet long. What will it cost to carpet her room if the carpet costs $18 per square yard?

33. **EXCHANGE RATES** While traveling in Canada, Johanna bought some gifts to bring home. She bought 2 T-shirts that cost $21.95 (Canadian). If the exchange rate at the time was 1 U.S. dollar for 1.37 Canadian dollars, how much did Johanna spend in U.S. dollars?

Online Research **Data Update** Visit www.algebra1.com/data_update to find the most recent exchange rate between the United States and Canadian currency. How much does a $21.95 (Canadian) purchase cost in U.S. dollars?

More About. . .

Exchange Rates

A system of floating exchange rates among international currencies was established in 1976. It was needed because the old system of basing a currency's value on gold had become obsolete.

Source: www.infoplease.com

34. **CITY MAINTENANCE** Street sweepers can clean 3 miles of streets per hour. A city owns 2 street sweepers, and each sweeper can be used for three hours before it comes in for an hour to refuel. How many miles of streets can be cleaned in 18 hours on the road?

TRAFFIC For Exercises 35–37, use the following information.
During rush hour one evening, traffic was backed up for 13 miles along a particular stretch of freeway. Assume that each vehicle occupied an average of 30 feet of space in a lane and that the freeway has three lanes.

35. Write an expression that could be used to determine the number of vehicles involved in the backup.

36. How many vehicles are involved in the backup?

37. Suppose that there are 8 exits along this stretch of freeway, and it takes each vehicle an average of 24 seconds to exit the freeway. Approximately how many hours will it take for all the vehicles in the backup to exit?

38. **CRITICAL THINKING** Identify the expressions that are equivalent to $\frac{x}{y}$. Explain why the expressions are equivalent.

a. $\frac{x+3}{y+3}$ **b.** $\frac{3-x}{3-y}$ **c.** $\frac{3x}{3y}$ **d.** $\frac{x^3}{y^3}$ **e.** $\frac{n^3x}{n^3y}$

39. **WRITING IN MATH** Answer the question that was posed at the beginning of the lesson.

How can you multiply rational expressions to determine the cost of electricity?

Include the following in your answer:
- an expression that you could use to determine the cost of using 60-watt light bulbs instead of 40-watt bulbs, and
- an example of a real-world situation in which you must multiply rational expressions.

40. Which expression is the product of $\frac{13xyz}{4x^2y}$ and $\frac{8x^2z^2}{2y^3}$?

(A) $\frac{13xy^3}{z^3}$ (B) $\frac{13xz^2}{y^3}$ (C) $\frac{13xyz}{z^3}$ (D) $\frac{13xz^3}{y^3}$

41. Identify the product of $\frac{4a+4}{a^2+a}$ and $\frac{a^2}{3a-3}$.

(A) $\frac{4a}{3(a-1)}$ (B) $\frac{4a}{3}$ (C) $\frac{4a}{3(a+1)}$ (D) $\frac{4a^2}{3(a-1)}$

Maintain Your Skills

Mixed Review

State the excluded values for each rational expression. *(Lesson 12-2)*

42. $\frac{s+6}{s^2-36}$ 43. $\frac{a^2-25}{a^2+3a-10}$ 44. $\frac{x+3}{x^2+6x+9}$

Write an inverse variation equation that relates x and y. Assume that y varies inversely as x. Then solve. *(Lesson 12-1)*

45. If $y = 9$ when $x = 8$, find x when $y = 6$.
46. If $y = 2.4$ when $x = 8.1$, find y when $x = 3.6$.
47. If $y = 24$ when $x = -8$, find y when $x = 4$.
48. If $y = 6.4$ when $x = 4.4$, find x when $y = 3.2$.

Simplify. Assume that no denominator is equal to zero. *(Lesson 8-2)*

49. $\frac{-7^{12}}{7^9}$ 50. $\frac{20p^6}{8p^8}$ 51. $\frac{24a^3b^4c^7}{6a^6c^2}$

Solve each inequality. Then check your solution. *(Lesson 6-2)*

52. $\frac{g}{8} < \frac{7}{2}$ 53. $3.5r \geq 7.35$ 54. $\frac{9k}{4} > \frac{3}{5}$

55. **FINANCE** The total amount of money Antonio earns mowing lawns and doing yard work varies directly with the number of days he works. At one point, he earned \$340 in 4 days. At this rate, how long will it take him to earn \$935? *(Lesson 5-2)*

Getting Ready for the Next Lesson

PREREQUISITE SKILL Factor each polynomial.
*(To review **factoring polynomials**, see Lessons 9-3 through 9-6.)*

56. $x^2 - 3x - 40$ 57. $n^2 - 64$ 58. $x^2 - 12x + 36$
59. $a^2 + 2a - 35$ 60. $2x^2 - 5x - 3$ 61. $3x^3 - 24x^2 + 36x$

Practice Quiz 1

Lessons 12-1 through 12-3

Graph each variation if y varies inversely as x. *(Lesson 12-1)*

1. $y = 28$ when $x = 7$ 2. $y = -6$ when $x = 9$

Simplify each expression. *(Lesson 12-2)*

3. $\frac{28a^2}{49ab}$ 4. $\frac{y+3y^2}{3y+1}$ 5. $\frac{b^2-3b-4}{b^2-13b+36}$ 6. $\frac{3n^2+5n-2}{3n^2-13n+4}$

Find each product. *(Lesson 12-3)*

7. $\frac{3m^2}{2m} \cdot \frac{18m^2}{9m}$ 8. $\frac{5a+10}{10x^2} \cdot \frac{4x^3}{a^2+11a+18}$ 9. $\frac{4n+8}{n^2-25} \cdot \frac{n-5}{5n+10}$ 10. $\frac{x^2-x-6}{x^2-9} \cdot \frac{x^2+7x+12}{x^2+4x+4}$

12-4 Dividing Rational Expressions

What You'll Learn

- Divide rational expressions.
- Use dimensional analysis with division.

How can you determine the number of aluminum soft drink cans made each year?

Most soft drinks come in aluminum cans. Although more cans are used today than in the 1970s, the demand for new aluminum has declined. This is due in large part to the great number of cans that are recycled. In recent years, approximately 63.9 billion cans were recycled annually. This represents $\frac{5}{8}$ of all cans produced.

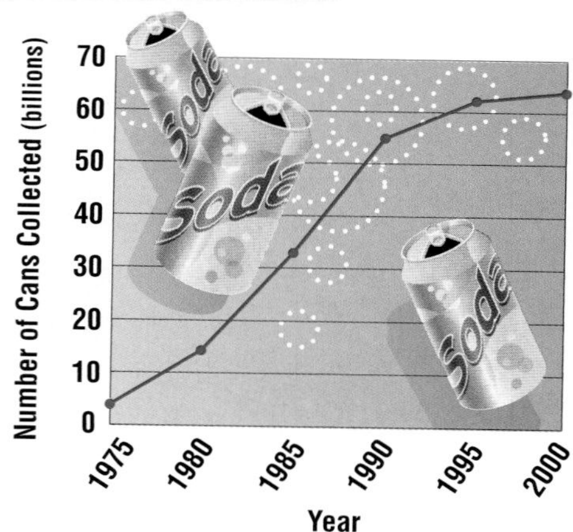

DIVIDE RATIONAL EXPRESSIONS Recall that to divide rational numbers expressed as fractions you multiply by the reciprocal of the divisor. You can use this same method to divide rational expressions.

Example 1 Expression Involving Monomials

Find $\frac{5x^2}{7} \div \frac{10x^3}{21}$.

$\frac{5x^2}{7} \div \frac{10x^3}{21} = \frac{5x^2}{7} \cdot \frac{21}{10x^3}$ Multiply by $\frac{21}{10x^3}$, the reciprocal of $\frac{10x^3}{21}$.

$= \frac{\overset{1}{\cancel{5x^2}}}{\underset{1}{\cancel{7}}} \cdot \frac{\overset{3}{\cancel{21}}}{\underset{2x}{\cancel{10x^3}}}$ Divide by common factors 5, 7, and x^2.

$= \frac{3}{2x}$ Simplify.

Example 2 Expression Involving Binomials

Find $\frac{n+1}{n+3} \div \frac{2n+2}{n+4}$.

$\frac{n+1}{n+3} \div \frac{2n+2}{n+4} = \frac{n+1}{n+3} \cdot \frac{n+4}{2n+2}$ Multiply by $\frac{n+4}{2n+2}$, the reciprocal of $\frac{2n+2}{n+4}$.

$= \frac{n+1}{n+3} \cdot \frac{n+4}{2(n+1)}$ Factor $2n + 2$.

$= \frac{\overset{1}{\cancel{n+1}}}{n+3} \cdot \frac{n+4}{2\underset{1}{\cancel{(n+1)}}}$ The GCF is $n + 1$.

$= \frac{n+4}{2(n+3)}$ or $\frac{n+4}{2n+6}$ Simplify.

Often the quotient of rational expressions involves a divisor that is a binomial.

Study Tip

Multiplicative Inverse
As with rational numbers, dividing rational expressions involves multiplying by the inverse. Remember that the inverse of $a + 2$ is $\frac{1}{a+2}$.

Example 3 Divide by a Binomial

Find $\frac{5a + 10}{a + 5} \div (a + 2)$.

$\frac{5a + 10}{a + 5} \div (a + 2) = \frac{5a + 10}{a + 5} \cdot \frac{1}{(a + 2)}$ Multiply by $\frac{1}{(a+2)}$, the reciprocal of $(a + 2)$.

$= \frac{5(a + 2)}{a + 5} \cdot \frac{1}{(a + 2)}$ Factor $5a + 10$.

$= \frac{5\overset{1}{\cancel{(a + 2)}}}{a + 5} \cdot \frac{1}{\underset{1}{\cancel{(a + 2)}}}$ The GCF is $a + 2$.

$= \frac{5}{a + 5}$ Simplify.

Sometimes you must factor a quadratic expression before you can simplify the quotient of rational expressions.

Example 4 Expression Involving Polynomials

Find $\frac{m^2 + 3m + 2}{4} \div \frac{m + 2}{m + 1}$.

$\frac{m^2 + 3m + 2}{4} \div \frac{m + 2}{m + 1} = \frac{m^2 + 3m + 2}{4} \cdot \frac{m + 1}{m + 2}$ Multiply by the reciprocal, $\frac{m+1}{m+2}$.

$= \frac{(m + 1)(m + 2)}{4} \cdot \frac{m + 1}{m + 2}$ Factor $m^2 + 3m + 2$.

$= \frac{(m + 1)\overset{1}{\cancel{(m + 2)}}}{4} \cdot \frac{m + 1}{\underset{1}{\cancel{m + 2}}}$ The GCF is $m + 2$.

$= \frac{(m + 1)^2}{4}$ Simplify.

More About. . .

Space
The first successful Mars probe was the Mariner 4, which arrived at Mars on July 14, 1965.
Source: NASA

DIMENSIONAL ANALYSIS You can divide rational expressions that involve units of measure by using dimensional analysis.

Example 5 Dimensional Analysis

SPACE In November, 1996, NASA launched the Mars Global Surveyor. It took 309 days for the orbiter to travel 466,000,000 miles from Earth to Mars. What was the speed of the spacecraft in miles per hour? Round to the nearest hundredth.

Use the formula for rate, time, and distance.

$rt = d$ rate · time = distance

$r \cdot 309 \text{ days} = 466{,}000{,}000 \text{ mi}$ $t = 309$ days, $d = 466{,}000{,}000$

$r = \frac{466{,}000{,}000 \text{ mi}}{309 \text{ days}}$ Divide each side by 309 days.

$= \frac{466{,}000{,}000 \text{ miles}}{309 \cancel{\text{ days}}} \cdot \frac{1 \cancel{\text{ day}}}{24 \text{ hours}}$ Convert days to hours.

$= \frac{466{,}000{,}000 \text{ miles}}{7416 \text{ hours}}$ or about $\frac{62{,}837.11 \text{ miles}}{1 \text{ hour}}$

Thus, the spacecraft traveled at a rate of about 62,837.11 miles per hour.

Check for Understanding

Concept Check

1. **OPEN ENDED** Write two rational expressions whose quotient is $\frac{5z}{xy}$.
2. **Tell** whether the following statement is *always*, *sometimes*, or *never* true. Explain your reasoning.
Every real number has a reciprocal.
3. **Explain** how to calculate the mass in kilograms of one cubic meter of a substance whose density is 2.16 grams per cubic centimeter.

Guided Practice

Find each quotient.

4. $\frac{10n^3}{7} \div \frac{5n^2}{21}$
5. $\frac{2a}{a+3} \div \frac{a+7}{a+3}$
6. $\frac{3m-15}{m+4} \div \frac{m-5}{6m+24}$
7. $\frac{2x+6}{x+5} \div (x+3)$
8. $\frac{k+3}{k^2+4k+4} \div \frac{2k+6}{k+2}$
9. $\frac{2x-4}{x^2+11x+18} \div \frac{x+1}{x^2+5x+6}$
10. Express 85 kilometers per hour in meters per second.
11. Express 32 pounds per square foot as pounds per square inch.

Application

12. **COOKING** Latisha was making candy using a two-quart pan. As she stirred the mixture, she noticed that the pan was about $\frac{2}{3}$ full. If each piece of candy has a volume of about $\frac{3}{4}$ ounce, approximately how many pieces of candy will Latisha make?

Practice and Apply

Homework Help

For Exercises	See Examples
13–18	1
19–22	3
23, 24	2
29–36	4
25–28, 37–41	5

Extra Practice
See page 847.

Find each quotient.

13. $\frac{a^2}{b^2} \div \frac{a}{b^3}$
14. $\frac{n^4}{p^2} \div \frac{n^2}{p^3}$
15. $\frac{4x^3}{y^4} \div \frac{8x^2}{y^2}$
16. $\frac{10m^2}{7n^2} \div \frac{25m^4}{14n^3}$
17. $\frac{x^2y^3z}{s^2t^2} \div \frac{x^2yz^3}{s^3t^2}$
18. $\frac{a^4bc^3}{g^2h^3} \div \frac{ab^2c^2}{g^3h^3}$
19. $\frac{b^2-9}{4b} \div (b-3)$
20. $\frac{m^2-16}{5m} \div (m+4)$
21. $\frac{3k}{k+1} \div (k-2)$
22. $\frac{5d}{d-3} \div (d+1)$
23. $\frac{3x+12}{4x-18} \div \frac{2x+8}{x+4}$
24. $\frac{4a-8}{2a-6} \div \frac{2a-4}{a-4}$

Complete.

25. 24 yd^3 = ____ ft^3
26. 0.35 m^3 = ____ cm^3
27. 330 ft/s = ____ mi/h
28. 1730 plants/km^2 = ____ plants/m^2
29. What is the quotient when $\frac{2x+6}{x+5}$ is divided by $\frac{2}{x+5}$?
30. Find the quotient when $\frac{m-8}{m+7}$ is divided by $m^2 - 7m - 8$.

Find each quotient.

31. $\frac{x^2 + 2x + 1}{2} \div \frac{x + 1}{x - 1}$

32. $\frac{n^2 + 3n + 2}{4} \div \frac{n + 1}{n + 2}$

33. $\frac{a^2 + 8a + 16}{a^2 - 6a + 9} \div \frac{2a + 8}{3a - 9}$

34. $\frac{b + 2}{b^2 + 4b + 4} \div \frac{2b + 4}{b + 4}$

35. $\frac{x^2 + x - 2}{x^2 + 5x + 6} \div \frac{x^2 + 2x - 3}{x^2 + 7x + 12}$

36. $\frac{x^2 + 2x - 15}{x^2 - x - 30} \div \frac{x^2 - 3x - 18}{x^2 - 2x - 24}$

37. **TRIATHLONS** Irena is training for an upcoming triathlon and plans to run 12 miles today. Jorge offered to ride his bicycle to help her maintain her pace. If Irena wants to keep a steady pace of 6.5 minutes per mile, how fast should Jorge ride in miles per hour?

More About. . .

Triathlons

The Ironman Championship Triathlon held in Hawaii consists of a 2.4-mile swim, a 112-mile bicycle ride, and a 26.2-mile run.

Source: www.infoplease.com

CONSTRUCTION For Exercises 38 and 39, use the following information.
A construction supervisor needs to determine how many truckloads of earth must be removed from a site before a foundation can be poured. The bed of the truck has the shape shown at the right.

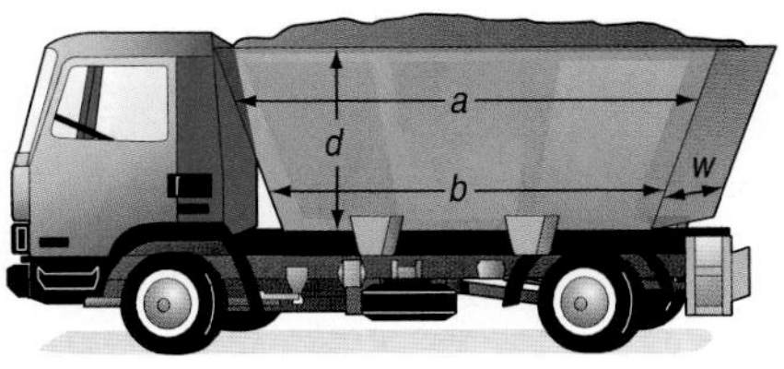

38. Use the formula $V = \frac{d(a + b)}{2} \cdot w$ to write an equation involving units that represents the volume of the truck bed in cubic yards if $a = 18$ feet, $b = 15$ feet, $w = 9$ feet, and $d = 5$ feet.

39. There are 20,000 cubic yards of earth that must be removed from the excavation site. Write an equation involving units that represents the number of truckloads that will be required to remove all of the earth. Then solve the equation.

TRUCKS For Exercises 40 and 41, use the following information.
The speedometer of John's truck uses the revolutions of his tires to calculate the speed of the truck.

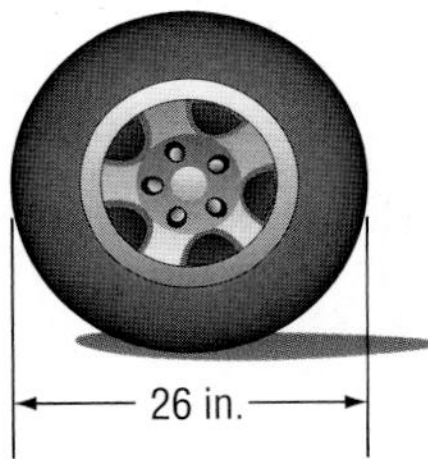

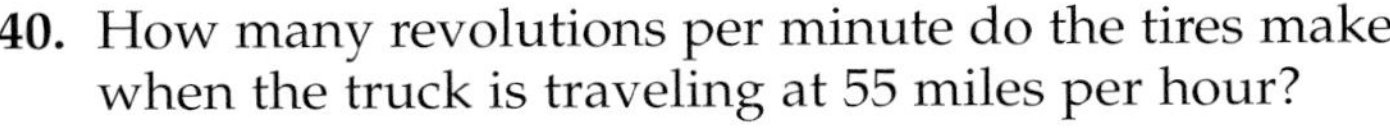

40. How many revolutions per minute do the tires make when the truck is traveling at 55 miles per hour?

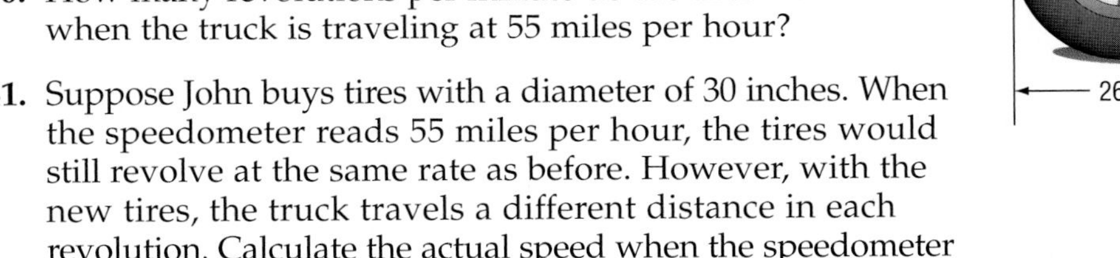

41. Suppose John buys tires with a diameter of 30 inches. When the speedometer reads 55 miles per hour, the tires would still revolve at the same rate as before. However, with the new tires, the truck travels a different distance in each revolution. Calculate the actual speed when the speedometer reads 55 miles per hour.

42. **CRITICAL THINKING** Which expression is *not* equivalent to the reciprocal of $\frac{x^2 - 4y^2}{x + 2y}$? Justify your answer.

a. $\frac{1}{\frac{x^2 - 4y^2}{x + 2y}}$ b. $\frac{-1}{2y - x}$ c. $\frac{1}{x - 2y}$ d. $\frac{1}{x} - \frac{1}{2y}$

SCULPTURE For Exercises 43 and 44, use the following information.
A sculptor had a block of marble in the shape of a cube with sides x feet long. A piece that was $\frac{1}{2}$ foot thick was chiseled from the bottom of the block. Later, the sculptor removed a piece $\frac{3}{4}$ foot wide from the side of the marble block.

43. Write a rational expression that represents the volume of the block of marble that remained.

44. If the remaining marble was cut into ten pieces weighing 85 pounds each, write an expression that represents the weight of the original block of marble.

45. **WRITING IN MATH** Answer the question that was posed at the beginning of the lesson.

How can you determine the number of aluminum soft drink cans made each year?

Include the following in your answer:

- a rational expression that will give the amount of new aluminum needed to produce x aluminum cans today when $\frac{5}{8}$ of the cans are recycled and 33 cans are produced from a pound of aluminum.

46. Which expression is the quotient of $\frac{3b}{5c}$ and $\frac{18b}{15c}$?

(A) $\frac{18b^2}{15c^2}$ (B) $\frac{1}{2}$ (C) $\frac{18b}{15c}$ (D) 2

47. Which expression could be used for the width of the rectangle?

(A) $x - 2$ (B) $(x + 2)(x - 2)^2$
(C) $x + 2$ (D) $(x + 2)(x - 2)$

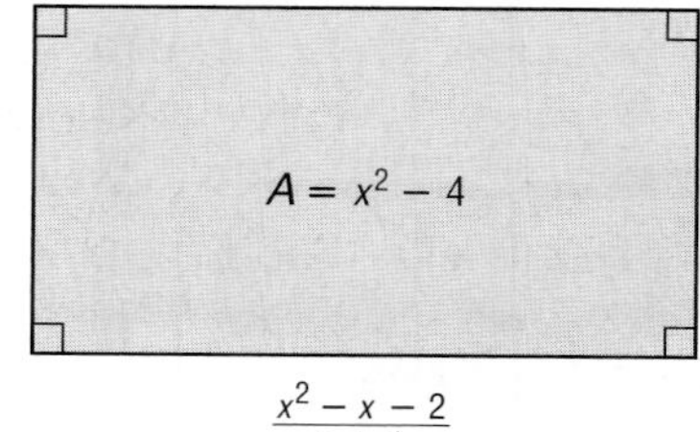

Maintain Your Skills

Mixed Review

Find each product. *(Lesson 12-3)*

48. $\frac{x - 5}{x^2 - 7x + 10} \cdot \frac{x - 2}{1}$

49. $\frac{x^2 + 3x - 10}{x^2 + 8x + 15} \cdot \frac{x^2 + 5x + 6}{x^2 + 4x + 4}$

50. $\frac{x + 4}{4y} \cdot \frac{16y}{x^2 + 7x + 12}$

51. $\frac{x^2 + 8x + 15}{x + y} \cdot \frac{7x + 14y}{x + 3}$

Simplify each expression. *(Lesson 12-2)*

52. $\frac{c - 6}{c^2 - 12c + 36}$

53. $\frac{25 - x^2}{x^2 + x - 30}$

54. $\frac{a + 3}{a^2 + 4a + 3}$

55. $\frac{n^2 - 16}{n^2 - 8n + 16}$

Solve each equation. Check your solutions. *(Lesson 9-6)*

56. $3y^2 = 147$

57. $9x^2 - 24x = -16$

58. $a^2 + 225 = 30a$

59. $(n + 6)^2 = 14$

Find the degree of each polynomial. *(Lesson 8-4)*

60. $13 + \frac{1}{8}$

61. $z^3 - 2z^2 + 3z - 4$

62. $a^5b^2c^3 + 6a^3b^3c^2$

Solve each inequality. Then check your solution. *(Lesson 6-2)*

63. $6 \leq 0.8g$

64. $-15b < -28$

65. $-0.049 \leq 0.07x$

66. $\frac{3}{7}h < \frac{3}{49}$

67. $\frac{12r}{-4} > \frac{3}{20}$

68. $\frac{y}{6} \geq \frac{1}{2}$

69. **MANUFACTURING** Tanisha's Sporting Equipment manufactures tennis racket covers at the rate of 3250 each month. How many tennis racket covers will the company manufacture in one year? *(Lesson 5-3)*

Getting Ready for the Next Lesson

PREREQUISITE SKILL Simplify. *(To review **dividing monomials**, see Lesson 8-2.)*

70. $\frac{6x^2}{x^4}$

71. $\frac{5m^4}{25m}$

72. $\frac{18a^3}{45a^5}$

73. $\frac{b^6c^3}{b^3c^6}$

74. $\frac{12x^3y^2}{28x^4y}$

75. $\frac{7x^4z^2}{z^3}$

Reading Mathematics

Rational Expressions

Several concepts need to be applied when reading rational expressions.

- A fraction bar acts as a grouping symbol, where the entire numerator is divided by the entire denominator.

Example 1 $\dfrac{6x + 4}{10}$

It is correct to read the expression as *the quantity six x plus four divided by ten.*

It is incorrect to read the expression as *six x divided by ten plus four*, or *six x plus four divided by ten.*

- If a fraction consists of two or more terms divided by a one-term denominator, the denominator divides each term.

Example 2 $\dfrac{6x + 4}{10}$

It is correct to write $\dfrac{6x + 4}{10} = \dfrac{6x}{10} + \dfrac{4}{10}$.

$$= \frac{3x}{5} + \frac{2}{5} \quad \text{or} \quad \frac{3x + 2}{5}$$

It is also correct to write $\dfrac{6x + 4}{10} = \dfrac{2(3x + 2)}{2 \cdot 5}$.

$$= \frac{\cancel{2}(3x + 2)}{\cancel{2} \cdot 5} \quad \text{or} \quad \frac{3x + 2}{5}$$

It is incorrect to write $\dfrac{6x + 4}{10} = \dfrac{\overset{3x}{\cancel{6x}} + 4}{\underset{5}{\cancel{10}}} = \dfrac{3x + 4}{5}$.

Reading to Learn

Write the verbal translation of each rational expression.

1. $\dfrac{m + 2}{4}$ **2.** $\dfrac{3x}{x - 1}$ **3.** $\dfrac{a + 2}{a^2 + 8}$

4. $\dfrac{x^2 - 25}{x + 5}$ **5.** $\dfrac{x^2 - 3x + 18}{x - 2}$ **6.** $\dfrac{x^2 + 2x - 35}{x^2 - x - 20}$

Simplify each expression.

7. $\dfrac{3x + 6}{9}$ **8.** $\dfrac{4n - 12}{8}$ **9.** $\dfrac{5x^2 - 25x}{10x}$

10. $\dfrac{x + 3}{x^2 + 7x + 12}$ **11.** $\dfrac{x + y}{x^2 + 2xy + y^2}$ **12.** $\dfrac{x^2 - 16}{x^2 - 8x + 16}$

12-5 Dividing Polynomials

What You'll Learn

- Divide a polynomial by a monomial.
- Divide a polynomial by a binomial.

How is division used in sewing?

Marching bands often use intricate marching routines and colorful flags to add interest to their shows. Suppose a partial roll of fabric is used to make flags. The original roll was 36 yards long, and $7\frac{1}{2}$ yards of the fabric were used to make a banner for the band. Each flag requires $1\frac{1}{2}$ yards of fabric. The expression $\dfrac{36 \text{ yards} - 7\frac{1}{2} \text{ yards}}{1\frac{1}{2} \text{ yards}}$ can be used to represent the number of flags that can be made using the roll of fabric.

DIVIDE POLYNOMIALS BY MONOMIALS To divide a polynomial by a monomial, divide each term of the polynomial by the monomial.

Example 1 Divide a Binomial by a Monomial

Find $(3r^2 - 15r) \div 3r$.

$$(3r^2 - 15r) \div 3r = \frac{3r^2 - 15r}{3r} \quad \text{Write as a rational expression.}$$

$$= \frac{3r^2}{3r} - \frac{15r}{3r} \quad \text{Divide each term by } 3r.$$

$$= \frac{\overset{r}{\cancel{3r^2}}}{\underset{1}{\cancel{3r}}} - \frac{\overset{5}{\cancel{15r}}}{\underset{1}{\cancel{3r}}} \quad \text{Simplify each term.}$$

$$= r - 5 \quad \text{Simplify.}$$

Example 2 Divide a Polynomial by a Monomial

Find $(n^2 + 10n + 12) \div 5n$.

$$(n^2 + 10n + 12) \div 5n = \frac{n^2 + 10n + 12}{5n} \quad \text{Write as a rational expression.}$$

$$= \frac{n^2}{5n} + \frac{10n}{5n} + \frac{12}{5n} \quad \text{Divide each term by } 5n.$$

$$= \frac{\overset{n}{\cancel{n^2}}}{\underset{5}{\cancel{5n}}} + \frac{\overset{2}{\cancel{10n}}}{\underset{1}{\cancel{5n}}} + \frac{12}{5n} \quad \text{Simplify each term.}$$

$$= \frac{n}{5} + 2 + \frac{12}{5n} \quad \text{Simplify.}$$

DIVIDE POLYNOMIALS BY BINOMIALS You can use algebra tiles to model some quotients of polynomials.

Algebra Activity

Dividing Polynomials

Use algebra tiles to find $(x^2 + 3x + 2) \div (x + 1)$.

Step 1 Model the polynomial $x^2 + 3x + 2$.

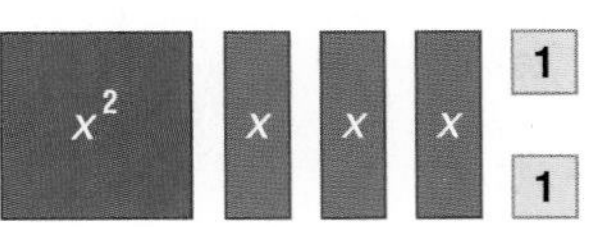

Step 2 Place the x^2 tile at the corner of the product mat. Place one of the 1 tiles as shown to make a length of $x + 1$.

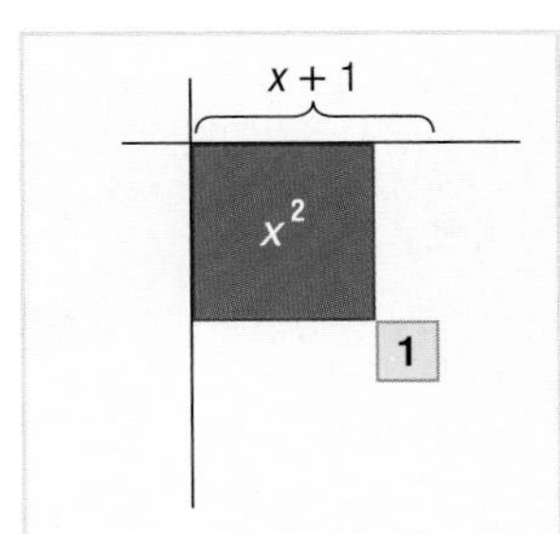

Step 3 Use the remaining tiles to make a rectangular array.

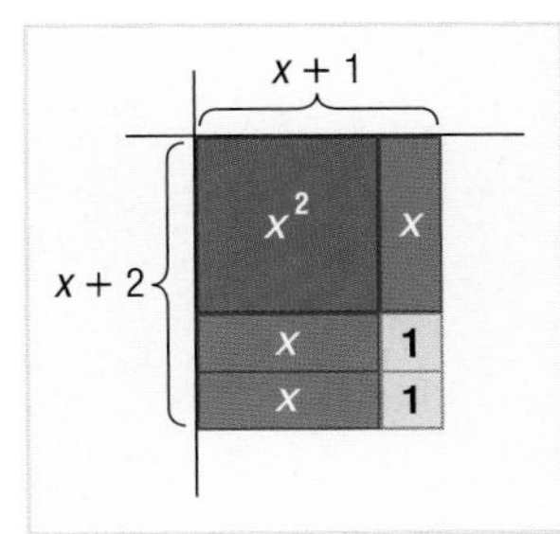

The width of the array, $x + 2$, is the quotient.

Model and Analyze

Use algebra tiles to find each quotient.

1. $(x^2 + 3x - 4) \div (x - 1)$

2. $(x^2 - 5x + 6) \div (x - 2)$

3. $(x^2 - 16) \div (x + 4)$

4. $(2x^2 - 4x - 6) \div (x - 3)$

5. Describe what happens when you try to model $(3x^2 - 4x + 3) \div (x + 2)$. What do you think the result means?

Recall from Lesson 12-4 that when you factor, some divisions can be performed easily.

Example 3 Divide a Polynomial by a Binomial

Find $(s^2 + 6s - 7) \div (s + 7)$.

$(s^2 + 6s - 7) \div (s + 7) = \dfrac{s^2 + 6s - 7}{(s + 7)}$ Write as a rational expression.

$= \dfrac{(s + 7)(s - 1)}{(s + 7)}$ Factor the numerator.

$= \dfrac{\overset{1}{\cancel{(s + 7)}}(s - 1)}{\underset{1}{\cancel{(s + 7)}}}$ Divide by the GCF.

$= s - 1$ Simplify.

www.algebra1.com/extra_examples

In Example 3 the division could be performed easily by dividing by common factors. However, when you cannot factor, you can use a long division process similar to the one you use in arithmetic.

Example 4 Long Division

Find $(x^2 + 3x - 24) \div (x - 4)$.

The expression $x^2 + 3x - 24$ cannot be factored, so use long division.

Step 1 Divide the first term of the dividend, x^2, by the first term of the divisor, x.

$$\begin{array}{r l} x & x^2 \div x = x \\ x - 4\overline{)x^2 + 3x - 24} & \\ \underline{(-)\ x^2 - 4x} & \text{Multiply } x \text{ and } x - 4. \\ 7x & \text{Subtract.} \end{array}$$

Step 2 Divide the first term of the partial dividend, $7x - 24$, by the first term of the divisor, x.

$$\begin{array}{r l} x + 7 & 7x \div x = 7 \\ x - 4\overline{)x^2 + 3x - 24} & \\ \underline{(-)\ x^2 - 4x} & \\ 7x - 24 & \text{Subtract and bring down the 24.} \\ \underline{(-)\ 7x - 28} & \text{Multiply 7 and } x - 4. \\ 4 & \text{Subtract.} \end{array}$$

The quotient of $(x^2 + 3x - 24) \div (x - 4)$ is $x + 7$ with a remainder of 4, which can be written as $x + 7 + \frac{4}{x - 4}$. Since there is a nonzero remainder, $x - 4$ is not a factor of $x^2 + 3x - 24$.

When the dividend is an expression like $a^3 + 8a - 21$, there is no a^2 term. In such situations, you must rename the dividend using 0 as the coefficient of the missing terms.

Example 5 Polynomial with Missing Terms

Find $(a^3 + 8a - 24) \div (a - 2)$.

Rename the a^2 term using a coefficient of 0.

$(a^3 + 8a - 24) \div (a - 2) = (a^3 + 0a^2 + 8a - 24) \div (a - 2)$

$$\begin{array}{r l} a^2 + 2a + 12 & \\ a - 2\overline{)a^3 + 0a^2 + 8a - 24} & \\ \underline{(-)\ a^3 - 2a^2} & \text{Multiply } a^2 \text{ and } a - 2. \\ 2a^2 + 8a & \text{Subtract and bring down } 8a. \\ \underline{(-)\ 2a^2 - 4a} & \text{Multiply } 2a \text{ and } a - 2. \\ 12a - 24 & \text{Subtract and bring down 24.} \\ \underline{(-)\ 12a - 24} & \text{Multiply 12 and } a - 2. \\ 0 & \text{Subtract.} \end{array}$$

Therefore, $(a^3 + 8a - 24) \div (a - 2) = a^2 + 2a + 12$.

Study Tip

Factors
When the remainder in a division problem is 0, the divisor is a factor of the dividend.

Check for Understanding

Concept Check

1. **Choose** the divisors of $2x^2 - 9x + 9$ that result in a remainder of 0.

   a. $x + 3$ b. $x - 3$ c. $2x - 3$ d. $2x + 3$

2. **Explain** the meaning of a remainder of zero in a long division of a polynomial by a binomial.

3. **OPEN ENDED** Write a third-degree polynomial that includes a zero term. Rewrite the polynomial so that it can be divided by $x + 5$ using long division.

Guided Practice

Find each quotient.

4. $(4x^3 + 2x^2 - 5) \div 2x$
5. $\dfrac{14a^2b^2 + 35ab^2 + 2a^2}{7a^2b^2}$
6. $(n^2 + 7n + 12) \div (n + 3)$
7. $(r^2 + 12r + 36) \div (r + 9)$
8. $\dfrac{4m^3 + 5m - 21}{2m - 3}$
9. $(2b^2 + 3b - 5) \div (2b - 1)$

Application

10. **ENVIRONMENT** The equation $C = \dfrac{120{,}000p}{1 - p}$ models the cost C in dollars for a manufacturer to reduce the pollutants by a given percent, written as p in decimal form. How much will the company have to pay to remove 75% of the pollutants it emits?

Practice and Apply

Homework Help

For Exercises	See Examples
11–14	1, 2
15–18, 23, 24	3
19–22, 25, 26	4
27–30	5

Extra Practice

See page 847.

Find each quotient.

11. $(x^2 + 9x - 7) \div 3x$
12. $(a^2 + 7a - 28) \div 7a$
13. $\dfrac{9s^3t^2 - 15s^2t + 24t^3}{3s^2t^2}$
14. $\dfrac{12a^3b + 16ab^3 - 8ab}{4ab}$
15. $(x^2 + 9x + 20) \div (x + 5)$
16. $(x^2 + 6x - 16) \div (x - 2)$
17. $(n^2 - 2n - 35) \div (n + 5)$
18. $(s^2 + 11s + 18) \div (s + 9)$
19. $(z^2 - 2z - 30) \div (z + 7)$
20. $(a^2 + 4a - 22) \div (a - 3)$
21. $(2r^2 - 3r - 35) \div (r - 5)$
22. $(3p^2 + 20p + 11) \div (p + 6)$
23. $\dfrac{3t^2 + 14t - 24}{3t - 4}$
24. $\dfrac{12n^2 + 36n + 15}{2n + 5}$
25. $\dfrac{3x^3 + 8x^2 + x - 7}{x + 2}$
26. $\dfrac{20b^3 - 27b^2 + 13b - 3}{4b - 3}$
27. $\dfrac{6x^3 - 9x^2 + 6}{2x - 3}$
28. $\dfrac{9g^3 + 5g - 8}{3g - 2}$
29. Determine the quotient when $6n^3 + 5n^2 + 12$ is divided by $2n + 3$.
30. What is the quotient when $4t^3 + 17t^2 - 1$ is divided by $4t + 1$?

LANDSCAPING For Exercises 31 and 32, use the following information.
A heavy object can be lifted more easily using a lever and fulcrum. The amount that can be lifted depends upon the length of the lever, the placement of the fulcrum, and the force applied. The expression $\dfrac{W(L - x)}{x}$ represents the weight of an object that can be lifted if W pounds of force are applied to a lever L inches long with the fulcrum placed x inches from the object.

31. Suppose Leyati, who weighs 150 pounds, uses all of his weight to lift a rock using a 60-inch lever. Write an expression that could be used to determine the heaviest rock he could lift if the fulcrum is x inches from the rock.
32. Use the expression to find the weight of a rock that could be lifted by a 210-pound man using a six-foot lever placed 20 inches from the rock.

www.algebra1.com/self_check_quiz

33. **DECORATING** Anoki wants to put a decorative border 3 feet above the floor around his bedroom walls. If the border comes in 5-yard rolls, how many rolls of border should Anoki buy?

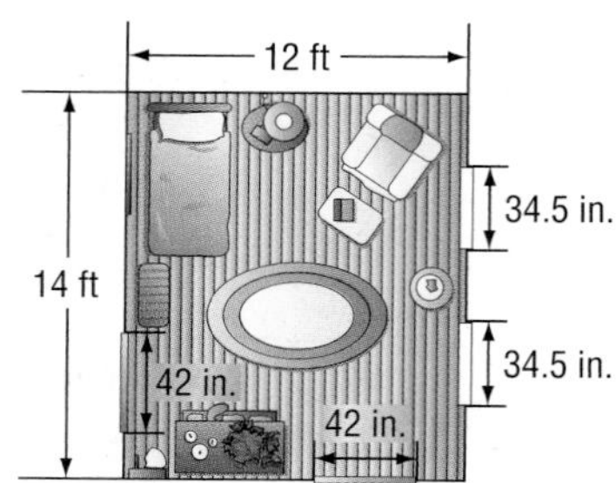

PIZZA **For Exercises 34 and 35, use the following information.**

The expression $\frac{\pi d^2}{64}$ can be used to determine the number of slices of a round pizza with diameter d.

34. Write a formula to calculate the cost per slice s of a pizza that costs C dollars.

35. Copy and complete the table below. Which size pizza offers the best price per slice?

Size	10-inch	14-inch	18-inch
Price	\$4.99	\$8.99	\$12.99
Number of slices			
Cost per slice			

SCIENCE **For Exercises 36–38, use the following information.**

The *density* of a material is its mass per unit volume.

36. Determine the densities for the materials listed in the table. Round to the nearest hundredth.

37. Make a graph of the densities computed in Exercise 36.

38. Interpret the line plot made in Exercise 37.

Material	Mass (g)	Volume (cm³)
aluminum	4.15	1.54
gold	2.32	0.12
silver	6.30	0.60
steel	7.80	1.00
iron	15.20	1.95
copper	2.48	0.28
blood	4.35	4.10
lead	11.30	1.00
brass	17.90	2.08
concrete	40.00	20.00

More About. . .

Science

When air is heated it is less dense than the air surrounding it, and the heated air rises. This is why a hot air balloon is able to fly.

Source: www.howstuffworks.com

39. **GEOMETRY** The volume of a prism with a triangular base is $10w^3 + 23w^2 + 5w - 2$. The height of the prism is $2w + 1$, and the height of the triangle is $5w - 1$. What is the measure of the base of the triangle? $\left(\textit{Hint}: V = Bh\right)$

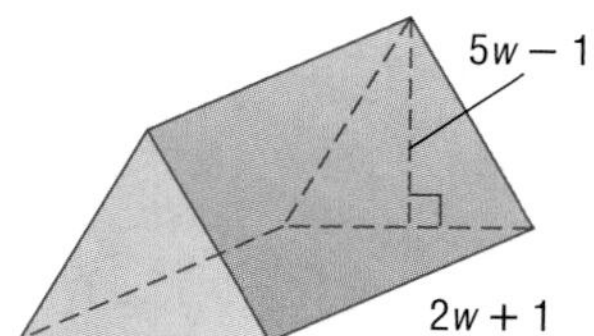

CRITICAL THINKING **Find the value of k in each situation.**

40. k is an integer and there is no remainder when $x^2 + 7x + 12$ is divided by $x + k$.

41. When $x^2 + 7x + k$ is divided by $x + 2$, there is a remainder of 2.

42. $x + 7$ is a factor of $x^2 - 2x - k$.

43. **WRITING IN MATH** Answer the question that was posed at the beginning of the lesson.

How is division used in sewing?

Include the following in your answer:

- a description showing that $\frac{36 \text{ yards} - 7\frac{1}{2} \text{ yards}}{1\frac{1}{2} \text{ yards}}$ and $\frac{36 \text{ yards}}{1\frac{1}{2} \text{ yards}} - \frac{7\frac{1}{2} \text{ yards}}{1\frac{1}{2} \text{ yards}}$ result in the same answer, and
- a convincing explanation to show that $\frac{a-b}{c} = \frac{a}{c} - \frac{b}{c}$.

44. Which expression represents the length of the rectangle?

Ⓐ $m + 7$ Ⓑ $m - 8$

Ⓒ $m - 7$ Ⓓ $m + 8$

$A = m^2 + 4m - 32$; width: $m - 4$

45. What is the quotient of $x^3 + 5x - 20$ divided by $x - 3$?

Ⓐ $x^2 - 3x + 14 + \frac{22}{x-3}$ Ⓑ $x^2 + 3x + 14 + \frac{22}{x-3}$

Ⓒ $x^2 + 8x + \frac{4}{x-3}$ Ⓓ $x^2 + 3x - 14 + \frac{22}{x-3}$

Maintain Your Skills

Mixed Review

Find each quotient. *(Lesson 12-4)*

46. $\frac{x^2 + 5x + 6}{x^2 - x - 12} \div \frac{x + 2}{x^2 + x - 20}$

47. $\frac{m^2 + m - 6}{m^2 + 8m + 15} \div \frac{m^2 - m - 2}{m^2 + 9m + 20}$

Find each product. *(Lesson 12-3)*

48. $\frac{b^2 + 19b + 84}{b - 3} \cdot \frac{b^2 - 9}{b^2 + 15b + 36}$

49. $\frac{z^2 + 16z + 39}{z^2 + 9z + 18} \cdot \frac{z + 5}{z^2 + 18z + 65}$

Simplify. Then use a calculator to verify your answer. *(Lesson 11-2)*

50. $3\sqrt{7} - \sqrt{7}$

51. $\sqrt{72} + \sqrt{32}$

52. $\sqrt{12} - \sqrt{18} + \sqrt{48}$

Factor each polynomial, if possible. If the polynomial cannot be factored, write *prime*. *(Lesson 9-6)*

53. $d^2 - 3d - 40$

54. $x^2 + 8x + 16$

55. $t^2 + t + 1$

56. **BUSINESS** Jorge Martinez has budgeted $150 to have business cards printed. A card printer charges $11 to set up each job and an additional $6 per box of 100 cards printed. What is the greatest number of cards Mr. Martinez can have printed? *(Lesson 6-3)*

Getting Ready for the Next Lesson

PREREQUISITE SKILL **Find each sum.**

*(To review **addition of polynomials**, see Lesson 8-5.)*

57. $(6n^2 - 6n + 10m^3) + (5n - 6m^3)$

58. $(3x^2 + 4xy - 2y^2) + (x^2 + 9xy + 4y^2)$

59. $(a^3 - b^3) + (-3a^3 - 2a^2b + b^2 - 2b^3)$

60. $(2g^3 + 6h) + (-4g^2 - 8h)$

12-6 Rational Expressions with Like Denominators

What You'll Learn

- Add rational expressions with like denominators.
- Subtract rational expressions with like denominators.

How can you use rational expressions to interpret graphics?

The graphic at the right shows the number of credit cards Americans have. To determine what fraction of those surveyed have no more than two credit cards, you can use addition. Remember that percents can be written as fractions with denominators of 100.

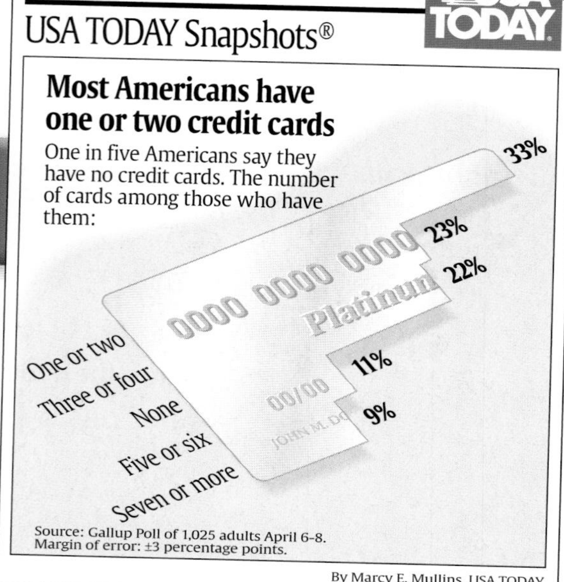

USA TODAY Snapshots®
USA TODAY
Most Americans have one or two credit cards
One in five Americans say they have no credit cards. The number of cards among those who have them:
One or two 33%
Three or four 23%
None 22%
Five or six 11%
Seven or more 9%
Source: Gallup Poll of 1,025 adults April 6-8. Margin of error: ±3 percentage points.
By Marcy E. Mullins, USA TODAY

No credit cards	plus	one or two credit cards	equals	no more than two credit cards.
$\frac{22}{100}$	$+$	$\frac{33}{100}$	$=$	$\frac{55}{100}$

Thus, $\frac{55}{100}$ or 55% of those surveyed have no more than two credit cards.

ADD RATIONAL EXPRESSIONS Recall that to add fractions with like denominators you add the numerators and then write the sum over the common denominator. You can add rational expressions with like denominators in the same way.

Example 1 Numbers in Denominator

Find $\frac{3n}{12} + \frac{7n}{12}$.

$\frac{3n}{12} + \frac{7n}{12} = \frac{3n + 7n}{12}$ The common denominator is 12.

$= \frac{10n}{12}$ Add the numerators.

$= \frac{\overset{5}{\cancel{10}}n}{\underset{6}{\cancel{12}}}$ Divide by the common factor, 2.

$= \frac{5n}{6}$ Simplify.

Sometimes the denominators of rational expressions are binomials. As long as each rational expression has exactly the same binomial as its denominator, the process of adding is the same.

Example 2 Binomials in Denominator

Find $\frac{2x}{x+1} + \frac{2}{x+1}$.

$\frac{2x}{x+1} + \frac{2}{x+1} = \frac{2x+2}{x+1}$ The common denominator is $x + 1$.

$= \frac{2(x+1)}{x+1}$ Factor the numerator.

$= \frac{2\cancel{(x+1)}^{1}}{\cancel{x+1}_{1}}$ Divide by the common factor, $x + 1$.

$= \frac{2}{1}$ or 2 Simplify.

Example 3 Find a Perimeter

GEOMETRY **Find an expression for the perimeter of rectangle *PQRS*.**

$P = 2\ell + 2w$ Perimeter formula

$= 2\left(\frac{4a+5b}{3a+7b}\right) + 2\left(\frac{2a+3b}{3a+7b}\right)$ $\ell = \frac{4a+5b}{3a+7b}$, $w = \frac{2a+3b}{3a+7b}$

$= \frac{2(4a+5b)+2(2a+3b)}{3a+7b}$ The common denominator is $3a + 7b$.

$= \frac{8a+10b+4a+6b}{3a+7b}$ Distributive Property

$= \frac{12a+16b}{3a+7b}$ Combine like terms.

$= \frac{4(3a+4b)}{3a+7b}$ Factor.

The perimeter can be represented by the expression $\frac{4(3a+4b)}{3a+7b}$.

SUBTRACT RATIONAL EXPRESSIONS To subtract rational expressions with like denominators, subtract the numerators and write the difference over the common denominator. Recall that to subtract an expression, you add its additive inverse.

Example 4 Subtract Rational Expressions

Find $\frac{3x+4}{x-2} - \frac{x-1}{x-2}$.

$\frac{3x+4}{x-2} - \frac{x-1}{x-2} = \frac{(3x+4)-(x-1)}{x-2}$ The common denominator is $x - 2$.

$= \frac{(3x+4)+[-(x-1)]}{x-2}$ The additive inverse of $(x - 1)$ is $-(x - 1)$.

$= \frac{3x+4-x+1}{x-2}$ Distributive Property

$= \frac{2x+5}{x-2}$ Simplify.

Study Tip

Common Misconception
Adding the additive inverse will help you avoid the following error in the numerator.
~~$(3x + 4) - (x - 1) = 3x + 4 - x - 1$.~~

Sometimes you must express a denominator as its additive inverse to have like denominators.

Example 5 Inverse Denominators

Find $\frac{2m}{m-9} + \frac{4m}{9-m}$.

The denominator $9 - m$ is the same as $-(-9 + m)$ or $-(m - 9)$. Rewrite the second expression so that it has the same denominator as the first.

$$\frac{2m}{m-9} + \frac{4m}{9-m} = \frac{2m}{m-9} + \frac{4m}{-(m-9)} \quad 9 - m = -(m-9)$$

$$= \frac{2m}{m-9} - \frac{4m}{m-9} \quad \text{Rewrite using like denominators.}$$

$$= \frac{2m - 4m}{m-9} \quad \text{The common denominator is } m - 9.$$

$$= \frac{-2m}{m-9} \quad \text{Subtract.}$$

Check for Understanding

Concept Check

1. **OPEN ENDED** Write two rational expressions with a denominator of $x + 2$ that have a sum of 1.
2. **Describe** how adding rational expressions with like denominators is similar to adding fractions with like denominators.
3. **Compare and contrast** two rational expressions whose sum is 0 with two rational expressions whose difference is 0.
4. **FIND THE ERROR** Russell and Ginger are finding the difference of $\frac{7x+2}{4x-3}$ and $\frac{x-8}{3-4x}$.

Russell

$$\frac{7x+2}{4x-3} - \frac{x-8}{3-4x} = \frac{7x+2}{4x-3} + \frac{x-8}{4x-3}$$
$$= \frac{7x+x+2-8}{4x-3}$$
$$= \frac{8x-6}{4x-3}$$
$$= \frac{2(4x-3)}{4x-3}$$
$$= 2$$

Ginger

$$\frac{7x+2}{4x-3} - \frac{x-8}{3-4x} = \frac{-2-7x}{3-4x} - \frac{x-8}{3-4x}$$
$$= \frac{-2+8-7x-x}{3-4x}$$
$$= \frac{-6-8x}{3-4x}$$
$$= \frac{-2(3-4x)}{3-4x}$$
$$= -2$$

Who is correct? Explain your reasoning.

Guided Practice

Find each sum.

5. $\frac{a+2}{4} + \frac{a-2}{4}$
6. $\frac{3x}{x+1} + \frac{3}{x+1}$
7. $\frac{2-n}{n-1} + \frac{1}{n-1}$
8. $\frac{4t-1}{1-4t} + \frac{2t+3}{1-4t}$

Find each difference.

9. $\frac{5a}{12} - \frac{7a}{12}$
10. $\frac{7}{n-3} - \frac{4}{n-3}$
11. $\frac{3m}{m-2} - \frac{6}{2-m}$
12. $\frac{x^2}{x-y} - \frac{y^2}{x-y}$

Application

13. **SCHOOL** Most schools create daily attendance reports to keep track of their students. Suppose that one day, out of 960 students, 45 were absent due to illness, 29 were participating in a wrestling tournament, 10 were excused to go to their doctors, and 12 were at a music competition. What fraction of the students were absent from school on this day?

Practice and Apply

Homework Help

For Exercises	See Examples
14–17	1
18–25, 27, 42, 43, 45, 46	2, 3
28–35, 38, 39, 44, 47, 48	4
26, 36, 37	5

Extra Practice

See page 848.

Find each sum.

14. $\frac{m}{3} + \frac{2m}{3}$

15. $\frac{12z}{7} + \frac{-5z}{7}$

16. $\frac{x+3}{5} + \frac{x+2}{5}$

17. $\frac{n-7}{2} + \frac{n+5}{2}$

18. $\frac{2y}{y+3} + \frac{6}{y+3}$

19. $\frac{3r}{r+5} + \frac{15}{r+5}$

20. $\frac{k-5}{k-1} + \frac{4}{k-1}$

21. $\frac{n-2}{n+3} + \frac{-1}{n+3}$

22. $\frac{4x-5}{x-2} + \frac{x+3}{x-2}$

23. $\frac{2a+3}{a-4} + \frac{a-2}{a-4}$

24. $\frac{5s+1}{2s+1} + \frac{3s-2}{2s+1}$

25. $\frac{9b+3}{2b+6} + \frac{5b+4}{2b+6}$

26. What is the sum of $\frac{12x-7}{3x-2}$ and $\frac{9x-5}{2-3x}$?

27. Find the sum of $\frac{11x-5}{2x+5}$ and $\frac{11x+12}{2x+5}$.

Find each difference.

28. $\frac{5x}{7} - \frac{3x}{7}$

29. $\frac{4n}{3} - \frac{2n}{3}$

30. $\frac{x+4}{5} - \frac{x+2}{5}$

31. $\frac{a+5}{6} - \frac{a+3}{6}$

32. $\frac{2}{x+7} - \frac{-5}{x+7}$

33. $\frac{4}{z-2} - \frac{-6}{z-2}$

34. $\frac{5}{3x-5} - \frac{3x}{3x-5}$

35. $\frac{4}{7m-2} - \frac{7m}{7m-2}$

36. $\frac{2x}{x-2} - \frac{2x}{2-x}$

37. $\frac{5y}{y-3} - \frac{5y}{3-y}$

38. $\frac{8}{3t-4} - \frac{6t}{3t-4}$

39. $\frac{15x}{5x+1} - \frac{-3}{5x+1}$

40. Find the difference of $\frac{10a-12}{2a-6}$ and $\frac{6a}{6-2a}$.

41. What is the difference of $\frac{b-15}{2b+12}$ and $\frac{-3b+8}{2b+12}$?

42. **POPULATION** The United States population in 1998 is described in the table. Use this information to write the fraction of the population that is 80 years or older.

Age	Number of People
0–19	77,525,000
20–39	79,112,000
40–59	68,699,000
60–79	35,786,000
80–99	8,634,000
100+	61,000

Source: *Statistical Abstract of the United States*

43. **CONSERVATION** The freshman class chose to plant spruce and pine trees at a wildlife sanctuary for a service project. Some students can plant 140 trees on Saturday, and others can plant 20 trees after school on Monday and again on Tuesday. Write an expression for the fraction of the trees that could be planted on these days if n represents the number of spruce trees and there are twice as many pine trees.

44. GEOMETRIC DESIGN A student center is a square room that is 25 feet wide and 25 feet long. The walls are 10 feet high and each wall is painted white with a red diagonal stripe as shown. What fraction of the walls are painted red?

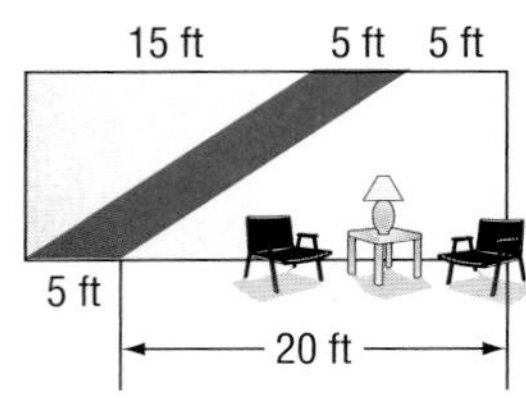

HIKING **For Exercises 45 and 46, use the following information.**
A tour guide recommends that hikers carry a gallon of water on hikes to the bottom of the Grand Canyon. Water weighs 62.4 pounds per cubic foot, and one cubic foot of water contains 7.48 gallons.

45. Tanika plans to carry two 1-quart bottles and four 1-pint bottles for her hike. Write a rational expression for this amount of water written as a fraction of a cubic foot.

46. How much does this amount of water weigh?

Hiking

Due to its popularity, the Grand Canyon is one of the most threatened natural areas in the United States.

Source: The Wildlife Foundation

GEOMETRY **For Exercises 47 and 48, use the following information.**
Each figure has a perimeter of x units.

a. 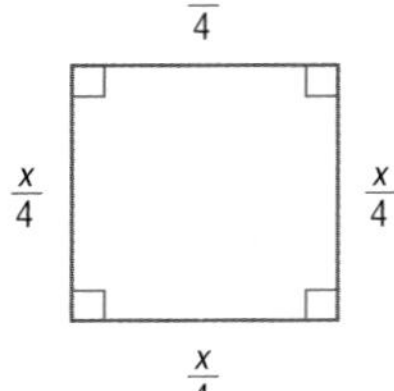

b. 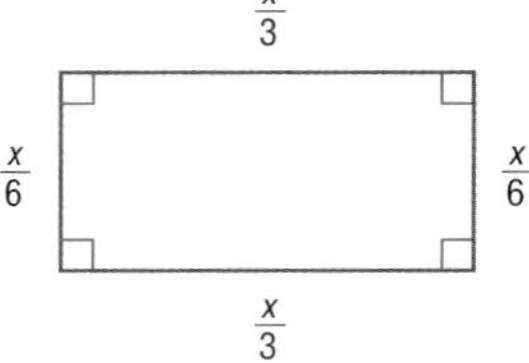

c. 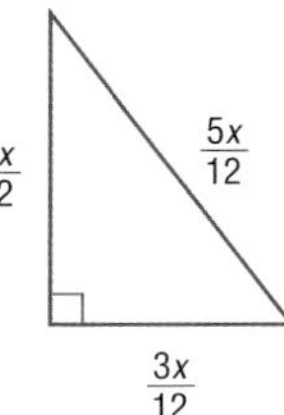

47. Find the ratio of the area of each figure to its perimeter.

48. Which figure has the greatest ratio?

49. **CRITICAL THINKING** Which of the following rational numbers is not equivalent to the others?

a. $\frac{3}{2-x}$ b. $\frac{-3}{x-2}$ c. $-\frac{3}{2-x}$ d. $-\frac{3}{x-2}$

50. **WRITING IN MATH** Answer the question that was posed at the beginning of the lesson.

How can you use rational expressions to interpret graphics?

Include the following in your answer:

- an explanation of how the numbers in the graphic relate to rational expressions, and
- a description of how to add two rational expressions whose denominators are $3x - 4y$ and $4y - 3x$.

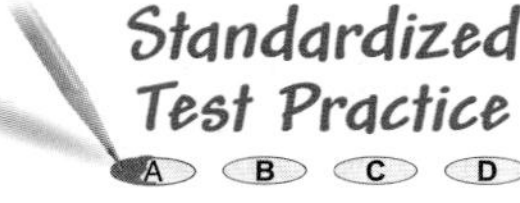

51. Find $\frac{k+2}{k-7} + \frac{-3}{k-7}$.

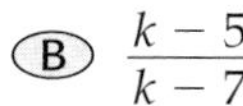

Ⓐ $\frac{k-1}{k-7}$ Ⓑ $\frac{k-5}{k-7}$ Ⓒ $\frac{k+1}{k-7}$ Ⓓ $\frac{k+5}{k-7}$

52. Which is an expression for the perimeter of rectangle $ABCD$?

Ⓐ $\frac{14r}{2r+6s}$ Ⓑ $\frac{14r}{r+3s}$

Ⓒ $\frac{14r}{r+6s}$ Ⓓ $\frac{28r}{r+3s}$

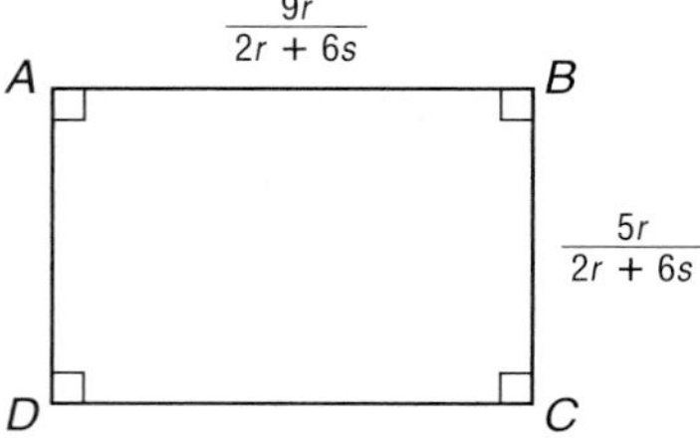

Maintain Your Skills

Mixed Review

Find each quotient. *(Lessons 12-4 and 12-5)*

53. $\dfrac{x^3 - 7x + 6}{x - 2}$

54. $\dfrac{56x^3 + 32x^2 - 63x - 36}{7x + 4}$

55. $\dfrac{b^2 - 9}{4b} \div (b - 3)$

56. $\dfrac{x}{x + 2} \div \dfrac{x^2}{x^2 + 5x + 6}$

Factor each trinomial. *(Lesson 9-3)*

57. $a^2 + 9a + 14$

58. $p^2 + p - 30$

59. $y^2 - 11yz + 28z^2$

Find each sum or difference. *(Lesson 8-5)*

60. $(3x^2 - 4x) - (7 - 9x)$

61. $(5x^2 - 6x + 14) + (2x^2 + 3x + 8)$

62. CARPENTRY When building a stairway, a carpenter considers the ratio of riser to tread. If each stair being built is to have a width of 1 foot and a height of 8 inches, what will be the slope of the stairway?

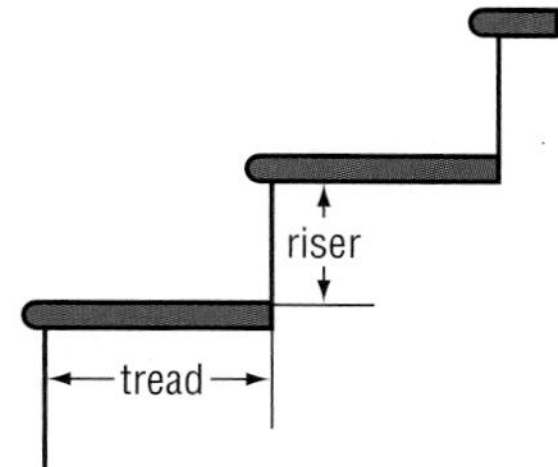

Getting Ready for the Next Lesson

BASIC SKILL **Find the least common multiple for each set of numbers.**

63. 4, 9, 12

64. 7, 21, 5

65. 6, 12, 24

66. 45, 10, 6

67. 5, 6, 15

68. 8, 9, 12

69. 16, 20, 25

70. 36, 48, 60

71. 9, 16, 24

Practice Quiz 2

Lessons 12-4 through 12-6

Find each quotient. *(Lessons 12-4 and 12-5)*

1. $\dfrac{a}{a + 3} \div \dfrac{a + 11}{a + 3}$

2. $\dfrac{4z + 8}{z + 3} \div (z + 2)$

3. $\dfrac{(2x - 1)(x - 2)}{(x - 2)(x - 3)} \div \dfrac{(2x - 1)(x + 5)}{(x - 3)(x - 1)}$

4. $(9xy^2 - 15xy + 3) \div 3xy$

5. $(2x^2 - 7x - 16) \div (2x + 3)$

6. $\dfrac{y^2 - 19y + 9}{y - 4}$

Find each sum or difference. *(Lesson 12-6)*

7. $\dfrac{2}{x + 7} + \dfrac{5}{x + 7}$

8. $\dfrac{2m}{m + 3} - \dfrac{-6}{m + 3}$

9. $\dfrac{5x - 1}{3x + 2} - \dfrac{2x - 1}{3x + 2}$

10. MUSIC Suppose the record shown played for 16.5 minutes on one side and the average of the radii of the grooves on the record was $3\frac{3}{4}$ inches. Write an expression involving units that represents how many inches the needle passed through the grooves while the record was being played. Then evaluate the expression.

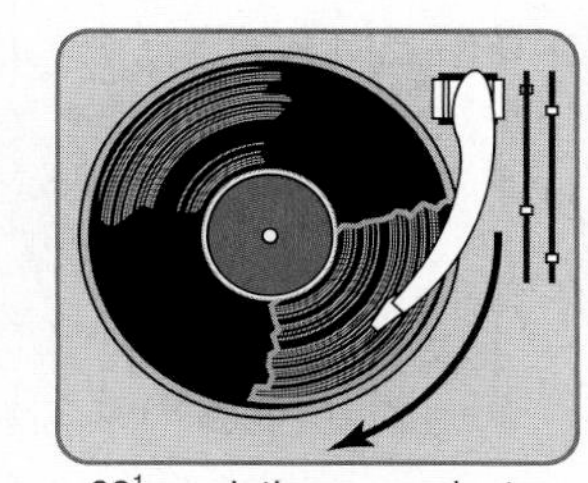

$33\frac{1}{3}$ revolutions per minute

12-7 Rational Expressions with Unlike Denominators

What You'll Learn

- Add rational expressions with unlike denominators.
- Subtract rational expressions with unlike denominators.

Vocabulary

- least common multiple (LCM)
- least common denominator (LCD)

How can rational expressions be used to describe elections?

The President of the United States is elected every four years, and senators are elected every six years. A certain senator is elected in 2004, the same year as a presidential election, and is reelected in subsequent elections. In what year is the senator's reelection the same year as a presidential election?

ADD RATIONAL EXPRESSIONS The number of years in which a specific senator's election coincides with a presidential election is related to the common multiples of 4 and 6. The least number of years that will pass until the next election for both a specific senator and the President is the least common multiple of these numbers. The **least common multiple (LCM)** is the least number that is a common multiple of two or more numbers.

Example 1 LCM of Monomials

Find the LCM of $15m^2b^3$ and $18mb^2$.

Find the prime factors of each coefficient and variable expression.

$15m^2b^3 = 3 \cdot 5 \cdot m \cdot m \cdot b \cdot b \cdot b$

$18mb^2 = 2 \cdot 3 \cdot 3 \cdot m \cdot b \cdot b$

Use each prime factor the greatest number of times it appears in any of the factorizations.

$15m^2b^3 = 3 \cdot 5 \cdot m \cdot m \cdot b \cdot b \cdot b$

$18mb^2 = 2 \cdot 3 \cdot 3 \cdot m \cdot b \cdot b$

LCM $= 2 \cdot 3 \cdot 3 \cdot 5 \cdot m \cdot m \cdot b \cdot b \cdot b$ or $90m^2b^3$

Example 2 LCM of Polynomials

Find the LCM of $x^2 + 8x + 15$ and $x^2 + x - 6$.

Express each polynomial in factored form.

$x^2 + 8x + 15 = (x + 3)(x + 5)$

$x^2 + x - 6 = (x - 2)(x + 3)$

Use each factor the greatest number of times it appears.

LCM $= (x - 2)(x + 3)(x + 5)$

Recall that to add fractions with unlike denominators, you need to rename the fractions using the least common multiple (LCM) of the denominators, known as the **least common denominator (LCD)**.

Key Concept — *Add Rational Expressions*

Use the following steps to add rational expressions with unlike denominators.

Step 1 Find the LCD.

Step 2 Change each rational expression into an equivalent expression with the LCD as the denominator.

Step 3 Add just as with rational expressions with like denominators.

Step 4 Simplify if necessary.

Example 3 *Monomial Denominators*

Find $\frac{a+1}{a} + \frac{a-3}{3a}$.

Factor each denominator and find the LCD.

$a = a$

$3a = 3 \cdot a$

LCD $= 3a$

Since the denominator of $\frac{a-3}{3a}$ is already $3a$, only $\frac{a+1}{a}$ needs to be renamed.

$$\frac{a+1}{a} + \frac{a-3}{3a} = \frac{3(a+1)}{3(a)} + \frac{a-3}{3a} \quad \text{Multiply } \tfrac{a+1}{a} \text{ by } \tfrac{3}{3}.$$

$$= \frac{3a+3}{3a} + \frac{a-3}{3a} \quad \text{Distributive Property}$$

$$= \frac{3a+3+a-3}{3a} \quad \text{Add the numerators.}$$

$$= \frac{4\overset{1}{\cancel{a}}}{3\underset{1}{\cancel{a}}} \quad \text{Divide out the common factor } a.$$

$$= \frac{4}{3} \quad \text{Simplify.}$$

Example 4 *Polynomial Denominators*

Find $\frac{y-2}{y^2+4y+4} + \frac{y-2}{y+2}$.

$$\frac{y-2}{y^2+4y+4} + \frac{y-2}{y+2} = \frac{y-2}{(y+2)^2} + \frac{y-2}{y+2} \quad \text{Factor the denominators.}$$

$$= \frac{y-2}{(y+2)^2} + \frac{y-2}{y+2} \cdot \frac{y+2}{y+2} \quad \text{The LCD is } (y+2)^2.$$

$$= \frac{y-2}{(y+2)^2} + \frac{y^2-4}{(y+2)^2} \quad (y-2)(y+2) = y^2-4$$

$$= \frac{y-2+y^2-4}{(y+2)^2} \quad \text{Add the numerators.}$$

$$= \frac{y^2+y-6}{(y+2)^2} \text{ or } \frac{(y-2)(y+3)}{(y+2)^2} \quad \text{Simplify.}$$

SUBTRACT RATIONAL EXPRESSIONS As with addition, to subtract rational expressions with unlike denominators, you must first rename the expressions using a common denominator.

Example 5 Binomials in Denominators

Find $\frac{4}{3a - 6} - \frac{a}{a + 2}$.

$\frac{4}{3a - 6} - \frac{a}{a + 2} = \frac{4}{3(a - 2)} - \frac{a}{a + 2}$ Factor.

$= \frac{4(a + 2)}{3(a - 2)(a + 2)} - \frac{3a(a - 2)}{3(a + 2)(a - 2)}$ The LCD is $3(a + 2)(a - 2)$.

$= \frac{4(a + 2) - 3a(a - 2)}{3(a - 2)(a + 2)}$ Subtract the numerators.

$= \frac{4a + 8 - 3a^2 + 6a}{3(a - 2)(a + 2)}$ Multiply.

$= \frac{-3a^2 + 10a + 8}{3(a - 2)(a + 2)}$ or $-\frac{3a^2 - 10a - 8}{3(a - 2)(a + 2)}$ Simplify.

Standardized Test Practice

A B C D

Example 6 Polynomials in Denominators

Multiple-Choice Test Item

Find $\frac{h - 2}{h^2 + 4h + 4} - \frac{h - 4}{h^2 - 4}$.

(A) $\frac{2h - 12}{(h - 2)(h + 2)^2}$

(B) $\frac{-2h + 12}{(h - 2)(h + 2)^2}$

(C) $\frac{2h - 12}{(h - 2)^2(h + 2)}$

(D) $\frac{-2h + 12}{(h - 2)(h + 2)}$

The Princeton Review

Test-Taking Tip

Examine all of the answer choices carefully. Look for differences in operations, positive and negative signs, and exponents.

Read the Test Item

The expression $\frac{h - 2}{h^2 + 4h + 4} - \frac{h - 4}{h^2 - 4}$ represents the difference of two rational expressions with unlike denominators.

Solve the Test Item

Step 1 Factor each denominator and find the LCD.

$h^2 + 4h + 4 = (h + 2)^2$

$h^2 - 4 = (h + 2)(h - 2)$

The LCD is $(h - 2)(h + 2)^2$.

Step 2 Change each rational expression into an equivalent expression with the LCD. Then subtract.

$$\frac{h - 2}{(h + 2)^2} - \frac{h - 4}{(h + 2)(h - 2)} = \frac{(h - 2)}{(h + 2)^2} \cdot \frac{(h - 2)}{(h - 2)} - \frac{(h - 4)}{(h + 2)(h - 2)} \cdot \frac{(h + 2)}{(h + 2)}$$

$$= \frac{(h - 2)(h - 2)}{(h + 2)^2(h - 2)} - \frac{(h - 4)(h + 2)}{(h + 2)^2(h - 2)}$$

$$= \frac{h^2 - 4h + 4}{(h + 2)^2(h - 2)} - \frac{h^2 - 2h - 8}{(h + 2)^2(h - 2)}$$

$$= \frac{(h^2 - 4h + 4) - (h^2 - 2h - 8)}{(h + 2)^2(h - 2)}$$

$$= \frac{h^2 - h^2 - 4h + 2h + 4 + 8}{(h + 2)^2(h - 2)}$$

$$= \frac{-2h + 12}{(h - 2)(h + 2)^2}$$

The correct answer is B.

Check for Understanding

Concept Check

1. **Describe** how to find the LCD of two rational expressions with unlike denominators.
2. **Explain** how to rename rational expressions using their LCD.
3. **OPEN ENDED** Give an example of two rational expressions in which the LCD is equal to twice the denominator of one of the expressions.

Guided Practice

Find the LCM for each pair of expressions.

4. $5a^2, 7a$
5. $2x - 4, 3x - 6$
6. $n^2 + 3n - 4, (n - 1)^2$

Find each sum.

7. $\frac{6}{5x} + \frac{7}{10x^2}$
8. $\frac{a}{a - 4} + \frac{4}{a + 4}$
9. $\frac{2y}{y^2 - 25} + \frac{y + 5}{y - 5}$
10. $\frac{a + 2}{a^2 + 4a + 3} + \frac{6}{a + 3}$

Find each difference.

11. $\frac{3z}{6w^2} - \frac{z}{4w}$
12. $\frac{4a}{2a + 6} - \frac{3}{a + 3}$
13. $\frac{b + 8}{b^2 - 16} - \frac{1}{b - 4}$
14. $\frac{x}{x - 2} - \frac{3}{x^2 + 3x - 10}$

Standardized Test Practice

A B C D

15. Find $\frac{2y}{y^2 + 7y + 12} + \frac{y + 2}{y + 4}$.

Ⓐ $\frac{y^2 + 5y + 6}{(y + 4)(y + 3)}$

Ⓑ $\frac{y^2 + 2y + 6}{(y + 4)(y + 3)}$

Ⓒ $\frac{y^2 + 7y + 6}{(y + 4)(y + 3)}$

Ⓓ $\frac{y^2 - 5y + 6}{(y + 4)(y + 3)}$

Practice and Apply

Homework Help

For Exercises	See Examples
16, 17, 54–57	1
18–21	2
22–25	3
26–37	4
38–49	5
50–53	6

Extra Practice

See page 848.

Find the LCM for each pair of expressions.

16. a^2b, ab^3
17. $7xy, 21x^2y$
18. $x - 4, x + 2$
19. $2n - 5, n + 2$
20. $x^2 + 5x - 14, (x - 2)^2$
21. $p^2 - 5p - 6, p + 1$

Find each sum.

22. $\frac{3}{x^2} + \frac{5}{x}$
23. $\frac{2}{a^3} + \frac{7}{a^2}$
24. $\frac{7}{6a^2} + \frac{5}{3a}$
25. $\frac{3}{7m} + \frac{4}{5m^2}$
26. $\frac{3}{x + 5} + \frac{4}{x - 4}$
27. $\frac{n}{n + 4} + \frac{3}{n - 3}$
28. $\frac{7a}{a + 5} + \frac{a}{a - 2}$
29. $\frac{6x}{x - 3} + \frac{x}{x + 1}$
30. $\frac{5}{3x - 9} + \frac{3}{x - 3}$
31. $\frac{m}{3m + 2} + \frac{2}{9m + 6}$
32. $\frac{-3}{5 - a} + \frac{5}{a^2 - 25}$
33. $\frac{18}{y^2 - 9} + \frac{-7}{3 - y}$
34. $\frac{x}{x^2 + 2x + 1} + \frac{1}{x + 1}$
35. $\frac{2x + 1}{(x - 1)^2} + \frac{x - 2}{x^2 + 3x - 4}$
36. $\frac{x^2}{4x^2 - 9} + \frac{x}{(2x + 3)^2}$
37. $\frac{a^2}{a^2 - b^2} + \frac{a}{(a - b)^2}$

Find each difference.

38. $\frac{7}{3x} - \frac{3}{6x^2}$

39. $\frac{4}{15x^2} - \frac{5}{3x}$

40. $\frac{11x}{3y^2} - \frac{7x}{6y}$

41. $\frac{5a}{7x} - \frac{3a}{21x^2}$

42. $\frac{x^2 - 1}{x + 1} - \frac{x^2 + 1}{x - 1}$

43. $\frac{k}{k + 5} - \frac{3}{k - 3}$

44. $\frac{k}{2k + 1} - \frac{2}{k + 2}$

45. $\frac{m - 1}{m + 1} - \frac{4}{2m + 5}$

46. $\frac{2x}{x^2 - 5x} - \frac{-3x}{x - 5}$

47. $\frac{-3}{a - 6} - \frac{-6}{a^2 - 6a}$

48. $\frac{n}{5 - n} - \frac{3}{n^2 - 25}$

49. $\frac{3a + 2}{6 - 3a} - \frac{a + 2}{a^2 - 4}$

50. $\frac{3x}{x^2 + 3x + 2} - \frac{3x - 6}{x^2 + 4x + 4}$

51. $\frac{5a}{a^2 + 3a - 4} - \frac{a - 1}{a^2 - 1}$

52. $\frac{x^2 + 4x - 5}{x^2 - 2x - 3} - \frac{2}{x + 1}$

53. $\frac{m - 4}{m^2 + 8m + 16} - \frac{m + 4}{m - 4}$

More About. . .

Pet Care

Kell, an English Mastiff owned by Tom Scott of the United Kingdom, is the heaviest dog in the world. Weighing in at 286 pounds, Kell eats a high protein diet of eggs, goat's milk, and beef.

Source: *The Guinness Book of Records*

54. **MUSIC** A music director wants to form a group of students to sing and dance at community events. The music they will sing is 2-part, 3-part, or 4-part harmony. The director would like to have the same number of voices on each part. What is the least number of students that would allow for an even distribution on all these parts?

55. **CHARITY** Maya, Makalla, and Monya can walk one mile in 12, 15, and 20 minutes respectively. They plan to participate in a walk-a-thon to raise money for a local charity. Sponsors have agreed to pay $2.50 for each mile that is walked. What is the total number of miles the girls would walk in one hour and how much money would they raise?

56. **PET CARE** Kendra takes care of pets while their owners are out of town. One week she has three dogs that all eat the same kind of dog food. The first dog eats a bag of food every 12 days, the second dog eats a bag every 15 days, and the third dog eats a bag every 16 days. How many bags of food should Kendra buy for one week?

57. **AUTOMOBILES** Car owners need to follow a regular maintenance schedule to keep their cars running safely and efficiently. The table shows several items that should be performed on a regular basis. If all of these items are performed when a car's odometer reads 36,000 miles, what would be the car's mileage reading the next time all of the items should be performed?

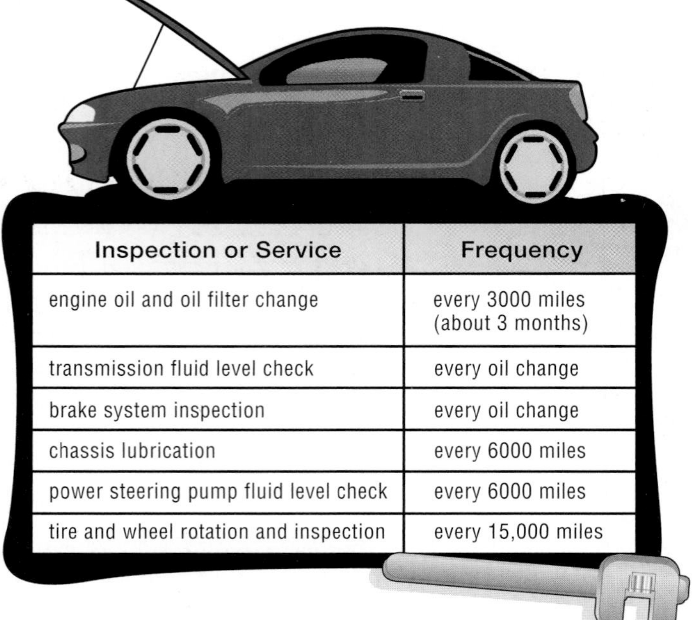

Inspection or Service	Frequency
engine oil and oil filter change	every 3000 miles (about 3 months)
transmission fluid level check	every oil change
brake system inspection	every oil change
chassis lubrication	every 6000 miles
power steering pump fluid level check	every 6000 miles
tire and wheel rotation and inspection	every 15,000 miles

58. **CRITICAL THINKING** Janelle says that a shortcut for adding fractions with unlike denominators is to add the cross products for the numerator and write the denominator as the product of the denominators. She gives the following example.

$$\frac{2}{7} + \frac{5}{8} = \frac{2 \cdot 8 + 5 \cdot 7}{7 \cdot 8} = \frac{51}{56}$$

Explain why Janelle's method will always work or provide a counterexample to show that it does not always work.

59. **WRITING IN MATH** Answer the question that was posed at the beginning of the lesson.

How can rational expressions be used to describe elections?

Include the following in your answer:

- an explanation of how to determine the least common multiple of two or more rational expressions, and
- if a certain senator is elected in 2006, when is the next election in which the senator and a President will be elected?

Standardized Test Practice

60. What is the least common denominator of $\frac{6}{a^2 - 2ab + b^2}$ and $\frac{6}{a^2 - b^2}$?

(A) $(a - b)^2$
(B) $(a - b)(a + b)$
(C) $(a + b)^2$
(D) $(a - b)^2(a + b)$

61. Find $\frac{x - 4}{(2 - x)^2} - \frac{x - 5}{x^2 + x - 6}$.

(A) $\frac{8x - 22}{(x + 3)(x - 2)^2}$
(B) $\frac{x^2 - 2x - 17}{(x - 2)(x + 3)}$
(C) $\frac{6x - 22}{(x + 3)(x - 2)^2}$
(D) $\frac{22 - 6x}{(x + 3)(x - 2)}$

Maintain Your Skills

Mixed Review

Find each sum. *(Lesson 12-6)*

62. $\frac{3m}{2m + 1} + \frac{3}{2m + 1}$
63. $\frac{4x}{2x + 3} + \frac{5}{2x + 3}$
64. $\frac{2y}{y - 3} + \frac{5}{3 - y}$

Find each quotient. *(Lesson 12-5)*

65. $\frac{b^2 + 8b - 20}{b - 2}$
66. $\frac{t^3 - 19t + 9}{t - 4}$
67. $\frac{4m^2 + 8m - 19}{2m + 7}$

Factor each trinomial, if possible. If the trinomial cannot be factored using integers, write *prime*. *(Lesson 9-4)*

68. $2x^2 + 10x + 8$
69. $5r^2 + 7r - 6$
70. $16p^2 - 4pq - 30q^2$

71. **BUDGETING** JoAnne Paulsen's take-home pay is $1782 per month. She spends $525 on rent, $120 on groceries, and $40 on gas. She allows herself 5% of the remaining amount for entertainment. How much can she spend on entertainment each month? *(Lesson 3-9)*

Getting Ready for the Next Lesson

PREREQUISITE SKILL Find each quotient.
*(To review **dividing rational expressions**, see Lesson 12-4.)*

72. $\frac{x}{2} \div \frac{3x}{5}$
73. $\frac{a^2}{5b} \div \frac{4a}{10b^2}$
74. $\frac{x + 7}{x} \div \frac{x + 7}{x + 3}$
75. $\frac{3n}{2n + 5} \div \frac{12n^2}{2n + 5}$
76. $\frac{3x}{x + 2} \div (x - 1)$
77. $\frac{x^2 + 7x + 12}{x + 6} \div (x + 3)$

12-8 Mixed Expressions and Complex Fractions

What You'll Learn

- Simplify mixed expressions.
- Simplify complex fractions.

How are rational expressions used in baking?

Katelyn bought $2\frac{1}{2}$ pounds of chocolate chip cookie dough. If the average cookie requires $1\frac{1}{2}$ ounces of dough, the number of cookies that Katelyn can bake can be found by simplifying the expression $\dfrac{2\frac{1}{2} \text{ pounds}}{1\frac{1}{2} \text{ ounces}}$.

Vocabulary

- mixed expression
- complex fraction

SIMPLIFY MIXED EXPRESSIONS Recall that a number like $2\frac{1}{2}$ is a mixed number because it contains the sum of an integer, 2, and a fraction, $\frac{1}{2}$. An expression like $3 + \frac{x+2}{x-3}$ is called a **mixed expression** because it contains the sum of a monomial, 3, and a rational expression, $\frac{x+2}{x-3}$. Changing mixed expressions to rational expressions is similar to changing mixed numbers to improper fractions.

Example 1 Mixed Expression to Rational Expression

Simplify $3 + \frac{6}{x+3}$.

$$3 + \frac{6}{x+3} = \frac{3(x+3)}{x+3} + \frac{6}{x+3} \qquad \text{The LCD is } x + 3.$$

$$= \frac{3(x+3)+6}{x+3} \qquad \text{Add the numerators.}$$

$$= \frac{3x+9+6}{x+3} \qquad \text{Distributive Property}$$

$$= \frac{3x+15}{x+3} \qquad \text{Simplify.}$$

SIMPLIFY COMPLEX FRACTIONS If a fraction has one or more fractions in the numerator or denominator, it is called a **complex fraction**. You simplify an algebraic complex fraction in the same way that you simplify a numerical complex fraction.

numerical complex fraction

$$\frac{\frac{8}{3}}{\frac{7}{5}} = \frac{8}{3} \div \frac{7}{5}$$

$$= \frac{8}{3} \cdot \frac{5}{7}$$

$$= \frac{40}{21}$$

algebraic complex fraction

$$\frac{\frac{a}{b}}{\frac{c}{d}} = \frac{a}{b} \div \frac{c}{d}$$

$$= \frac{a}{b} \cdot \frac{d}{c}$$

$$= \frac{ad}{bc}$$

Key Concept — Simplifying a Complex Fraction

Any complex fraction $\frac{\frac{a}{b}}{\frac{c}{d}}$, where $b \neq 0$, $c \neq 0$, and $d \neq 0$, can be expressed as $\frac{ad}{bc}$.

Example 2 Complex Fraction Involving Numbers

BAKING Refer to the application at the beginning of the lesson. How many cookies can Katelyn make with $2\frac{1}{2}$ pounds of chocolate chip cookie dough?

To find the total number of cookies, divide the amount of cookie dough by the amount of dough needed for each cookie.

$$\frac{2\frac{1}{2}\text{ pounds}}{1\frac{1}{2}\text{ ounces}} = \frac{2\frac{1}{2}\cancel{\text{ pounds}}}{1\frac{1}{2}\cancel{\text{ ounces}}} \cdot \frac{16\cancel{\text{ ounces}}}{1\cancel{\text{ pound}}}$$ Convert pounds to ounces. Divide by common units.

$$= \frac{16 \cdot 2\frac{1}{2}}{1\frac{1}{2}}$$ Simplify.

$$= \frac{\frac{16}{1} \cdot \frac{5}{2}}{\frac{3}{2}}$$ Express each term as an improper fraction.

$$= \frac{\frac{80}{2}}{\frac{3}{2}}$$ Multiply in the numerator.

$$= \frac{80 \cdot 2}{2 \cdot 3}$$ $\frac{\frac{a}{b}}{\frac{c}{d}} = \frac{ad}{bc}$

$$= \frac{160}{6} \text{ or } 26\frac{2}{3}$$ Simplify.

Katelyn can make 27 cookies.

Study Tip

Fraction Bar

Recall that when applying the order of operations, a fraction bar serves as a grouping symbol. Simplify the numerator and denominator of a complex fraction before proceeding with division.

Example 3 Complex Fraction Involving Monomials

Simplify $\frac{\frac{x^2y^2}{a}}{\frac{x^2y}{a^3}}$.

$$\frac{\frac{x^2y^2}{a}}{\frac{x^2y}{a^3}} = \frac{x^2y^2}{a} \div \frac{x^2y}{a^3}$$ Rewrite as a division sentence.

$$= \frac{x^2y^2}{a} \cdot \frac{a^3}{x^2y}$$ Rewrite as multiplication by the reciprocal.

$$= \frac{\overset{1}{\cancel{x^2}}\overset{y}{\cancel{y^2}}}{\underset{1}{\cancel{a}}} \cdot \frac{\overset{a^2}{\cancel{a^3}}}{\underset{1}{\cancel{x^2}}\underset{1}{\cancel{y}}}$$ Divide by common factors x^2, y, and a.

$$= a^2y$$ Simplify.

Example 4 Complex Fraction Involving Polynomials

Simplify $\dfrac{a - \dfrac{15}{a-2}}{a+3}$.

The numerator contains a mixed expression. Rewrite it as a rational expression first.

$$\frac{a - \frac{15}{a-2}}{a+3} = \frac{\frac{a(a-2)}{a-2} - \frac{15}{a-2}}{a+3}$$ The LCD of the fractions in the numerator is $a - 2$.

$$= \frac{\frac{a^2 - 2a - 15}{a-2}}{a+3}$$ Simplify the numerator.

$$= \frac{\frac{(a+3)(a-5)}{a-2}}{a+3}$$ Factor.

$$= \frac{(a+3)(a-5)}{a-2} \div (a+3)$$ Rewrite as a division sentence.

$$= \frac{(a+3)(a-5)}{a-2} \cdot \frac{1}{a+3}$$ Multiply by the reciprocal of $a + 3$.

$$= \frac{\overset{1}{\cancel{(a+3)}}(a-5)}{a-2} \cdot \frac{1}{\underset{1}{\cancel{a+3}}}$$ Divide by the GCF, $a + 3$.

$$= \frac{a-5}{a-2}$$ Simplify.

Check for Understanding

Concept Check

1. **Describe** the similarities between mixed numbers and mixed rational expressions.

2. **OPEN ENDED** Give an example of a complex fraction and show how to simplify it.

3. **FIND THE ERROR** Bolton and Lian found the LCD of $\frac{4}{2x+1} - \frac{5}{x+1} + \frac{2}{x-1}$.

Bolton	Lian
$\frac{4}{2x+1} - \frac{5}{x+1} + \frac{2}{x-1}$	$\frac{4}{2x+1} - \frac{5}{x+1} + \frac{2}{x-1}$
LCD: $(2x+1)(x+1)(x-1)$	LCD: $2(x+1)(x-1)$

Who is correct? Explain your reasoning.

Guided Practice **Write each mixed expression as a rational expression.**

4. $3 + \frac{4}{x}$

5. $7 + \frac{5}{6y}$

6. $\frac{a-1}{3a} + 2a$

Simplify each expression.

7. $\dfrac{3\frac{1}{2}}{4\frac{3}{4}}$

8. $\dfrac{\frac{x^3}{y^2}}{\frac{y^3}{x}}$

9. $\dfrac{\frac{x-y}{a+b}}{\frac{x^2-y^2}{a^2-b^2}}$

Application **10. ENTERTAINMENT** The student talent committee is arranging the performances for their holiday pageant. The first-act performances and their lengths are shown in the table. What is the average length of the performances?

Holiday Pageant Line-Up

Performance	Length (min)
A	7
B	$4\frac{1}{2}$
C	$6\frac{1}{2}$
D	$8\frac{1}{4}$
E	$10\frac{1}{5}$

Practice and Apply

Homework Help

For Exercises	See Examples
11–22, 35	1
23–26, 37–40	2
27–32, 36	3
33, 34	4

Extra Practice

See page 848.

Write each mixed expression as a rational expression.

11. $8 + \frac{3}{n}$
12. $4 + \frac{5}{a}$
13. $2x + \frac{x}{y}$
14. $6z + \frac{2z}{w}$
15. $2m - \frac{4 + m}{m}$
16. $3a - \frac{a + 1}{2a}$
17. $b^2 + \frac{a - b}{a + b}$
18. $r^2 + \frac{r - 4}{r + 3}$
19. $5n^2 - \frac{n + 3}{n^2 - 9}$
20. $3s^2 - \frac{s + 1}{s^2 - 1}$
21. $(x - 5) + \frac{x + 2}{x - 3}$
22. $(p + 4) + \frac{p + 1}{p - 4}$

Simplify each expression.

23. $\dfrac{5\frac{3}{4}}{7\frac{2}{3}}$
24. $\dfrac{8\frac{2}{7}}{4\frac{4}{5}}$
25. $\dfrac{\frac{a}{b^3}}{\frac{a^2}{b}}$
26. $\dfrac{\frac{n^3}{m^2}}{\frac{n^2}{m^2}}$
27. $\dfrac{\frac{x + 4}{y - 2}}{\frac{x^2}{y^2}}$
28. $\dfrac{\frac{s^3}{t^2}}{\frac{s + t}{s - t}}$
29. $\dfrac{\frac{y^2 - 1}{y^2 + 3y - 4}}{y + 1}$
30. $\dfrac{\frac{a^2 - 2a - 3}{a^2 - 1}}{a - 3}$
31. $\dfrac{\frac{n^2 + 2n}{n^2 + 9n + 18}}{\frac{n^2 - 5n}{n^2 + n - 30}}$
32. $\dfrac{\frac{x^2 + 4x - 21}{x^2 - 9x + 18}}{\frac{x^2 + 3x - 28}{x^2 - 10x + 24}}$
33. $\dfrac{x - \frac{15}{x - 2}}{x - \frac{20}{x - 1}}$
34. $\dfrac{n + \frac{35}{n + 12}}{n - \frac{63}{n - 2}}$

35. What is the quotient of $b + \frac{1}{b}$ and $a + \frac{1}{a}$?

36. What is the product of $\frac{2b^2}{5c}$ and the quotient of $\frac{4b^3}{2c}$ and $\frac{7b^3}{8c^2}$?

37. **PARTIES** The student council is planning a party for the school volunteers. There are five 66-ounce bottles of soda left from a recent dance. When poured over ice, $5\frac{1}{2}$ ounces of soda fills a cup. How many servings of soda can they get from the bottles they have?

ACOUSTICS **For Exercises 38 and 39, use the following information.**
If a vehicle is moving toward you at v miles per hour and blowing its horn at a frequency of f, then you hear the horn as if it were blowing at a frequency of h. This can be defined by the equation $h = \frac{f}{1 - \frac{v}{s}}$, where s is the speed of sound, approximately 760 miles per hour.

38. Simplify the complex fraction in the formula.

39. Suppose a truck horn blows at 370 cycles per second and is moving toward you at 65 miles per hour. Find the frequency of the horn as you hear it.

40. **POPULATION** According to the 2000 Census, New Jersey was the most densely populated state, and Alaska was the least densely populated state. The population of New Jersey was 8,414,350, and the population of Alaska was 626,932. The land area of New Jersey is about 7419 square miles, and the land area of Alaska is about 570,374 square miles. How many more people were there per square mile in New Jersey than in Alaska?

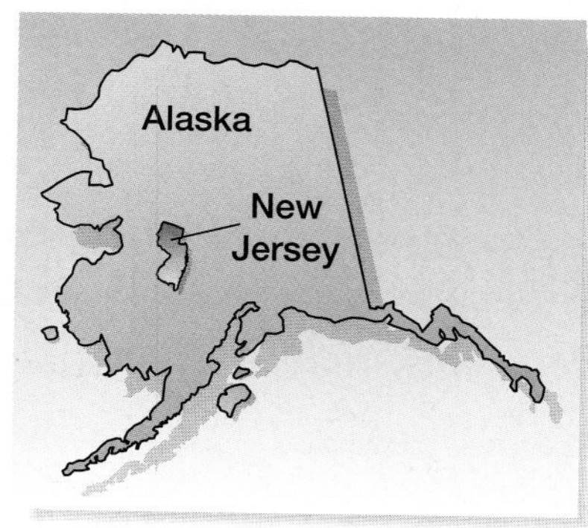

41. **BICYCLES** When air is pumped into a bicycle tire, the pressure P required varies inversely as the volume of the air V and is given by the equation $P = \frac{k}{V}$. If the pressure is 30 lb/in^2 when the volume is $1\frac{2}{3}$ cubic feet, find the pressure when the volume is $\frac{3}{4}$ cubic feet.

42. **CRITICAL THINKING** Which expression is equivalent to 0?

a. $\dfrac{a}{1 - \frac{3}{a}} + \dfrac{a}{\frac{3}{a} - 1}$ **b.** $\dfrac{a - \frac{1}{3}}{b} - \dfrac{a + \frac{1}{3}}{b}$ **c.** $\dfrac{\frac{1}{2} + 2a}{b - 1} - \dfrac{2a + \frac{1}{2}}{1 - b}$

43. **WRITING IN MATH** Answer the question that was posed at the beginning of the lesson.

How are rational expressions used in baking?

Include the following in your answer:
- an example of a situation in which you would divide a measurement by a fraction when cooking, and
- an explanation of the process used to simplify a complex fraction.

44. The perimeter of hexagon $ABCDEF$ is 12. Which expression can be used to represent the measure of $\overline{BC}$?

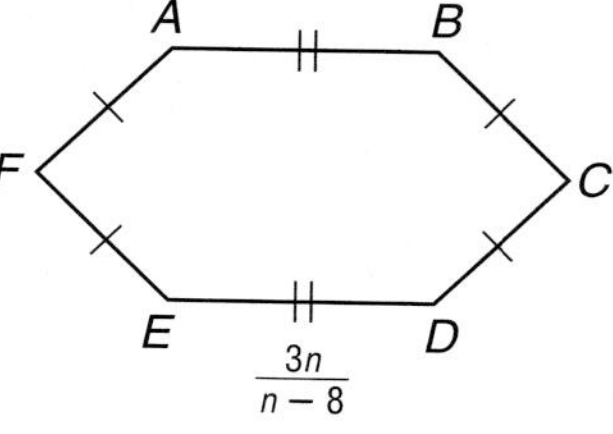

Ⓐ $\dfrac{6n - 96}{n - 8}$ Ⓑ $\dfrac{9n - 96}{n - 8}$

Ⓒ $\dfrac{6n - 96}{4n - 32}$ Ⓓ $\dfrac{9n - 96}{4n - 32}$

45. Express $\dfrac{\frac{6mn}{5p}}{\frac{24n^2}{20mp}}$ in simplest form.

Ⓐ $\dfrac{n}{m^2}$ Ⓑ $\dfrac{1}{n}$ Ⓒ $\dfrac{m^2}{n}$ Ⓓ $\dfrac{36n^3}{25p^2}$

Maintain Your Skills

Mixed Review

Find each sum. *(Lesson 12-7)*

46. $\frac{12x}{4y^2} + \frac{8}{6y}$

47. $\frac{a}{a-b} + \frac{b}{2b+3a}$

48. $\frac{a+3}{3a^2-10a-8} + \frac{2a}{a^2-8a+16}$

49. $\frac{n-4}{(n-2)^2} + \frac{n-5}{n^2+n-6}$

Find each difference. *(Lesson 12-6)*

50. $\frac{7}{x^2} - \frac{3}{x^2}$

51. $\frac{x}{(x-3)^2} - \frac{3}{(x-3)^2}$

52. $\frac{2}{t^2-t-2} - \frac{t}{t^2-t-2}$

53. $\frac{2n}{n^2+2n-24} - \frac{8}{n^2+2n-24}$

54. **BIOLOGY** Ana is working on a biology project for her school's science fair. For her experiment, she needs to have a certain type of bacteria that doubles its population every hour. Right now Ana has 1000 bacteria. If Ana does not interfere with the bacteria, predict how many there will be in ten hours. *(Lesson 10-6)*

Solve each equation by factoring. Check your solutions. *(Lesson 9-5)*

55. $s^2 = 16$

56. $9p^2 = 64$

57. $z^3 - 9z = 45 - 5z^2$

FAMILIES **For Exercises 58–60, refer to the graph.** *(Lesson 8-3)*

58. Write each number in the graph using scientific notation.

59. How many times as great is the amount spent on food as the amount spent on clothing? Express your answer in scientific notation.

60. What percent of the total amount is spent on housing?

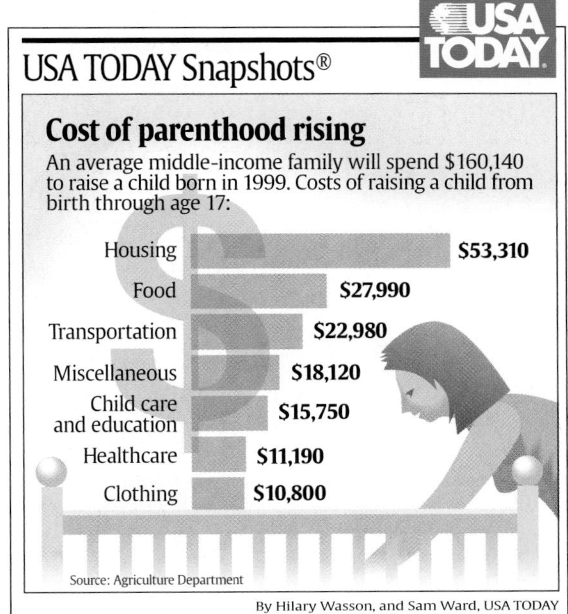
USA TODAY
USA TODAY Snapshots®
Cost of parenthood rising
An average middle-income family will spend $160,140 to raise a child born in 1999. Costs of raising a child from birth through age 17:
Housing $53,310
Food $27,990
Transportation $22,980
Miscellaneous $18,120
Child care and education $15,750
Healthcare $11,190
Clothing $10,800
Source: Agriculture Department
By Hilary Wasson, and Sam Ward, USA TODAY

TELEPHONE RATES **For Exercises 61 and 62, use the following information.** *(Lesson 5-4)*

A 15-minute call to Mexico costs $3.39. A 24-minute call costs $4.83.

61. Write a linear equation to find the total cost C of an m-minute call.

62. Find the cost of a 9-minute call.

Getting Ready for the Next Lesson

PREREQUISITE SKILL **Solve each equation.**
*(To review **solving equations**, see Lessons 3-2 through 3-4.)*

63. $-12 = \frac{x}{4}$

64. $1.8 = g - 0.6$

65. $\frac{3}{4}n - 3 = 9$

66. $7x^2 = 28$

67. $3.2 = \frac{-8+n}{-7}$

68. $\frac{-3n-(-4)}{-6} = -9$

12-9 Solving Rational Equations

What You'll Learn

- Solve rational equations.
- Eliminate extraneous solutions.

Vocabulary

- rational equations
- work problems
- rate problems
- extraneous solutions

How are rational equations important in the operation of a subway system?

The Washington, D.C., Metrorail is one of the safest subway systems in the world, serving a population of more than 3.5 million. It is vital that a rail system of this size maintain a consistent schedule. Rational equations can be used to determine the exact positions of trains at any given time.

Washington Metropolitan Area Transit Authority

Train	Distance
Red Line	19.4 mi
Orange Line	24.14 mi
Blue Line	19.95 mi
Green Line	20.59 mi
Yellow Line	9.46 mi

SOLVE RATIONAL EQUATIONS **Rational equations** are equations that contain rational expressions. You can use cross products to solve rational equations, but only when both sides of the equation are single fractions.

Example 1 Use Cross Products

Solve $\frac{12}{x+5} = \frac{4}{(x+2)}$.

$\frac{12}{x+5} = \frac{4}{(x+2)}$	Original equation
$12(x+2) = 4(x+5)$	Cross multiply.
$12x + 24 = 4x + 20$	Distributive Property
$8x = -4$	Add $-4x$ and -24 to each side.
$x = -\frac{4}{8}$ or $-\frac{1}{2}$	Divide each side by 8.

Another method you can use to solve rational equations is to multiply each side of the equation by the LCD to eliminate fractions.

Example 2 Use the LCD

Solve $\frac{n-2}{n} - \frac{n-3}{n-6} = \frac{1}{n}$.

$\frac{n-2}{n} - \frac{n-3}{n-6} = \frac{1}{n}$	Original equation
$n(n-6)\left(\frac{n-2}{n} - \frac{n-3}{n-6}\right) = n(n-6)\left(\frac{1}{n}\right)$	The LCD is $n(n-6)$.
$\left(\frac{\overset{1}{\cancel{n}}(n-6)}{1} \cdot \frac{n-2}{\underset{1}{\cancel{n}}}\right) - \left(\frac{n\overset{1}{\cancel{(n-6)}}}{1} \cdot \frac{n-3}{\underset{1}{\cancel{n-6}}}\right) = \frac{\overset{1}{\cancel{n}}(n-6)}{1} \cdot \frac{1}{\underset{1}{\cancel{n}}}$	Distributive Property
$(n-6)(n-2) - n(n-3) = n-6$	Simplify.
$(n^2 - 8n + 12) - (n^2 - 3n) = n - 6$	Multiply.
$n^2 - 8n + 12 - n^2 + 3n = n - 6$	Subtract.
$-5n + 12 = n - 6$	Simplify.
$-6n = -18$	Subtract 12 and n from each side.
$n = 3$	Divide each side by -6.

A rational equation may have more than one solution.

Example 3 Multiple Solutions

Solve $\frac{-4}{a+1} + \frac{3}{a} = 1$.

$$\frac{-4}{a+1} + \frac{3}{a} = 1 \quad \text{Original equation}$$

$$a(a+1)\left(\frac{-4}{a+1} + \frac{3}{a}\right) = a(a+1)(1) \quad \text{The LCD is } a(a+1).$$

$$\left(\frac{\overset{1}{a(\cancel{a+1})}}{1} \cdot \frac{-4}{\underset{1}{\cancel{a+1}}}\right) + \left(\frac{\overset{1}{\cancel{a}(a+1)}}{1} \cdot \frac{3}{\underset{1}{\cancel{a}}}\right) = a(a+1) \quad \text{Distributive Property}$$

$$-4a + 3a + 3 = a^2 + a \quad \text{Simplify.}$$

$$-a + 3 = a^2 + a \quad \text{Add like terms.}$$

$$0 = a^2 + 2a - 3 \quad \text{Set equal to 0.}$$

$$0 = (a+3)(a-1) \quad \text{Factor.}$$

$$a + 3 = 0 \quad \text{or} \quad a - 1 = 0$$

$$a = -3 \qquad a = 1$$

CHECK Check by substituting each value in the original equation.

$$\frac{-4}{a+1} + \frac{3}{a} = 1 \qquad\qquad \frac{-4}{a+1} + \frac{3}{a} = 1$$

$$\frac{-4}{-3+1} + \frac{3}{-3} \stackrel{?}{=} 1 \quad a = -3 \qquad\qquad \frac{-4}{1+1} + \frac{3}{1} \stackrel{?}{=} 1 \quad a = 1$$

$$2 + (-1) \stackrel{?}{=} 1 \qquad\qquad -2 + 3 \stackrel{?}{=} 1$$

$$1 = 1 \qquad\qquad 1 = 1$$

The solutions are 1 or −3.

Study Tip

Look Back
To review **solving quadratic equations by factoring**, see Lessons 9-3 through 9-6.

Rational equations can be used to solve **work problems**.

Example 4 Work Problem

LAWN CARE Abbey has a lawn care service. One day she asked her friend Jamal to work with her. Normally, it takes Abbey two hours to mow and trim Mrs. Harris' lawn. When Jamal worked with her, the job took only 1 hour and 20 minutes. How long would it have taken Jamal to do the job himself?

Explore Since it takes Abbey two hours to do the yard, she can finish $\frac{1}{2}$ the job in one hour. The amount of work Jamal can do in one hour can be represented by $\frac{1}{t}$. To determine how long it takes Jamal to do the job, use the formula Abbey's work + Jamal's work = 1 completed yard.

Plan The time that both of them worked was $1\frac{1}{3}$ hours. Each rate multiplied by this time results in the amount of work done by each person.

Solve

Abbey's work	plus	Jamal's work	equals	total work.
$\frac{1}{2}\left(\frac{4}{3}\right)$	$+$	$\frac{1}{t}\left(\frac{4}{3}\right)$	$=$	1

$$\frac{4}{6} + \frac{4}{3t} = 1 \quad \text{Multiply.}$$

(continued on the next page)

Study Tip

Work Problems
When solving work problems, remember that each term should represent the portion of a job completed in one unit of time.

$$6t\left(\frac{4}{6} + \frac{4}{3t}\right) = 6t \cdot 1 \quad \text{The LCD is } 6t.$$

$$\overset{1}{\cancel{6}}t\left(\frac{4}{\underset{1}{\cancel{6}}}\right) + \overset{2}{\cancel{6}}\cancel{t}\left(\frac{4}{\underset{1}{\cancel{3t}}}\right) = 6t \quad \text{Distributive Property}$$

$$4t + 8 = 6t \quad \text{Simplify.}$$

$$8 = 2t \quad \text{Add } -4t \text{ to each side.}$$

$$4 = t \quad \text{Divide each side by 2.}$$

Examine The time that it would take Jamal to do the yard by himself is four hours. This seems reasonable because the combined efforts of the two took longer than half of Abbey's usual time.

Rational equations can also be used to solve **rate problems**.

Example 5 Rate Problem

TRANSPORTATION **Refer to the application at the beginning of the lesson. The Yellow Line runs between Huntington and Mt. Vernon Square. Suppose one train leaves Mt. Vernon Square at noon and arrives at Huntington 24 minutes later, and a second train leaves Huntington at noon and arrives at Mt. Vernon Square 28 minutes later. At what time do the two trains pass each other?**

> **Study Tip**
>
> *Rate Problems*
> You can solve rate problems, also called *uniform motion problems*, more easily if you first make a drawing.

Determine the rates of both trains. The total distance is 9.46 miles.

Train 1 $\frac{9.46 \text{ mi}}{24 \text{ min}}$ **Train 2** $\frac{9.46 \text{ mi}}{28 \text{ min}}$

Next, since both trains left at the same time, the time both have traveled when they pass will be the same. And since they started at opposite ends of the route, the sum of their distances is equal to the total route, 9.46 miles.

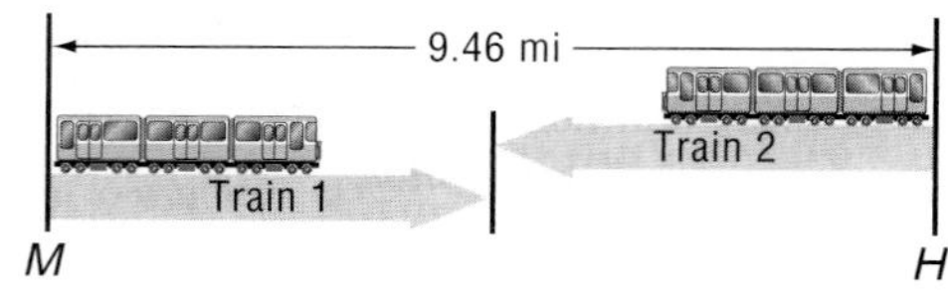

	r	t	d
Train 1	$\frac{9.46 \text{ mi}}{24 \text{ min}}$	t min	$\frac{9.46t}{24}$ mi
Train 2	$\frac{9.46 \text{ mi}}{28 \text{ min}}$	t min	$\frac{9.46t}{28}$ mi

$$\frac{9.46t}{24} + \frac{9.46t}{28} = 9.46 \quad \text{The sum of the distances is 9.46.}$$

$$168\left(\frac{9.46t}{24} + \frac{9.46t}{28}\right) = 168 \cdot 9.46 \quad \text{The LCD is 168.}$$

$$\frac{\overset{7}{\cancel{168}}}{1} \cdot \frac{9.46t}{\underset{1}{\cancel{24}}} + \frac{\overset{6}{\cancel{168}}}{1} \cdot \frac{9.46t}{\underset{1}{\cancel{28}}} = 1589.28 \quad \text{Distributive Property}$$

$$66.22t + 56.76t = 1589.28 \quad \text{Simplify.}$$

$$122.98t = 1589.28 \quad \text{Add.}$$

$$t = 12.92 \quad \text{Divide each side by 122.98.}$$

The trains passed at about 12.92 or about 13 minutes after leaving their stations, which is 12:13 P.M.

EXTRANEOUS SOLUTIONS Multiplying each side of an equation by the LCD of two rational expressions can yield results that are not solutions to the original equation. Recall that such solutions are called **extraneous solutions**.

Example 6 No Solution

Solve $\frac{3x}{x-1} + \frac{6x-9}{x-1} = 6$.

$$\frac{3x}{x-1} + \frac{6x-9}{x-1} = 6 \quad \text{Original equation}$$

$$(x-1)\left(\frac{3x}{x-1} + \frac{6x-9}{x-1}\right) = (x-1)6 \quad \text{The LCD is } x-1.$$

$$\overset{1}{\cancel{(x-1)}}\left(\frac{3x}{\underset{1}{\cancel{x-1}}}\right) + \overset{1}{\cancel{(x-1)}}\left(\frac{6x-9}{\underset{1}{\cancel{x-1}}}\right) = (x-1)6 \quad \text{Distributive Property}$$

$$3x + 6x - 9 = 6x - 6 \quad \text{Simplify.}$$

$$9x - 9 = 6x - 6 \quad \text{Add like terms.}$$

$$3x = 3 \quad \text{Add 9 to each side.}$$

$$x = 1 \quad \text{Divide each side by 3.}$$

Since 1 is an excluded value for x, the number 1 is an extraneous solution. Thus, the equation has no solution.

Rational equations can have both valid solutions and extraneous solutions.

Example 7 Extraneous Solution

Solve $\frac{2n}{1-n} + \frac{n+3}{n^2-1} = 1$.

$$\frac{2n}{1-n} + \frac{n+3}{n^2-1} = 1$$

$$\frac{2n}{1-n} + \frac{n+3}{(n-1)(n+1)} = 1$$

$$-\frac{2n}{n-1} + \frac{n+3}{(n-1)(n+1)} = 1$$

$$(n-1)(n+1)\left(-\frac{2n}{n-1} + \frac{n+3}{(n-1)(n+1)}\right) = (n-1)(n+1)1$$

$$\overset{1}{\cancel{(n-1)}}(n+1)\left(-\frac{2n}{\underset{1}{\cancel{n-1}}}\right) + \overset{1}{\cancel{(n-1)}}\overset{1}{\cancel{(n+1)}}\left(\frac{n+3}{\underset{1}{\cancel{(n-1)}}\underset{1}{\cancel{(n+1)}}}\right) = (n-1)(n+1)$$

$$-2n(n+1) + (n+3) = n^2 - 1$$

$$-2n^2 - 2n + n + 3 = n^2 - 1$$

$$-3n^2 - n + 4 = 0$$

$$3n^2 + n - 4 = 0$$

$$(3n+4)(n-1) = 0$$

$3n + 4 = 0$ or $n - 1 = 0$

$n = -\frac{4}{3}$ $\qquad$ $n = 1$

The number 1 is an extraneous solution, since 1 is an excluded value for n. Thus, $-\frac{4}{3}$ is the solution of the equation.

Check for Understanding

Concept Check

1. **OPEN ENDED** Explain why the equation $n + \frac{1}{n-1} = \frac{1}{n-1} + 1$ has no solution.

2. **Write** an expression to represent the amount of work Aminta can do in h hours if it normally takes her 3 hours to change the oil and tune up her car.

3. **Find a counterexample** for the following statement.

   The solution of a rational equation can never be zero.

Guided Practice

Solve each equation. State any extraneous solutions.

4. $\frac{2}{x} = \frac{3}{x+1}$
5. $\frac{7}{a-1} = \frac{5}{a+3}$
6. $\frac{3x}{5} + \frac{3}{2} = \frac{7x}{10}$
7. $\frac{x+1}{x} + \frac{x+4}{x} = 6$
8. $\frac{5}{k+1} - \frac{7}{k} = \frac{1}{k+1}$
9. $\frac{x+2}{x-2} - \frac{2}{x+2} = \frac{-7}{3}$

Application

10. **BASEBALL** Omar has 32 hits in 128 times at bat. He wants his batting average to be .300. His current average is $\frac{32}{128}$ or .250. How many at bats does Omar need to reach his goal if he gets a hit in each of his next b at bats?

Practice and Apply

Homework Help

For Exercises	See Examples
11–14	1
15–19, 21, 23, 26, 27	2
22, 24, 25	3
29–34	4, 5
20, 28	6, 7

Extra Practice

See page 849.

Solve each equation. State any extraneous solutions.

11. $\frac{4}{a} = \frac{3}{a-2}$
12. $\frac{3}{x} = \frac{1}{x-2}$
13. $\frac{x-3}{x} = \frac{x-3}{x-6}$
14. $\frac{x}{x+1} = \frac{x-6}{x-1}$
15. $\frac{2n}{3} + \frac{1}{2} = \frac{2n-3}{6}$
16. $\frac{5}{4} + \frac{3y}{2} = \frac{7y}{6}$
17. $\frac{a-1}{a+1} - \frac{2a}{a-1} = -1$
18. $\frac{7}{x^2-5x} + \frac{3}{5-x} = \frac{4}{x}$
19. $\frac{4x}{2x+3} - \frac{2x}{2x-3} = 1$
20. $\frac{5}{5-p} - \frac{p^2}{p-5} = -8$
21. $\frac{a}{3a+6} - \frac{a}{5a+10} = \frac{2}{5}$
22. $\frac{c}{c-4} - \frac{6}{4-c} = c$
23. $\frac{2b-5}{b-2} - 2 = \frac{3}{b+2}$
24. $\frac{7}{k-3} - \frac{1}{2} = \frac{3}{k-4}$
25. $\frac{x^2-4}{x-2} + x^2 = 4$
26. $\frac{2n}{n-1} + \frac{n-5}{n^2-1} = 1$
27. $\frac{3z}{z^2-5z+4} = \frac{2}{z-4} + \frac{3}{z-1}$
28. $\frac{4}{m^2-8m+12} = \frac{m}{m-2} + \frac{1}{m-6}$

29. **QUIZZES** Each week, Mandy's algebra teacher gives a 10-point quiz. After 5 weeks, Mandy has earned a total of 36 points for an average of 7.2 points per quiz. She would like to raise her average to 9 points. On how many quizzes must she score 10 points in order to reach her goal?

BOATING **For Exercises 30 and 31, use the following information.**
Jim and Mateo live across a lake from each other at a distance of about 3 miles. Jim can row his boat to Mateo's house in 1 hour and 20 minutes. Mateo can drive his motorboat the same distance in a half hour.

30. If they leave their houses at the same time and head toward each other, how long will it be before they meet?

31. How far from the nearest shore will they be when they meet?

32. **CAR WASH** Ian and Nadya can each wash a car and clean its interior in about 2 hours, but Chris needs 3 hours to do the work. If the three work together, how long will it take to clean seven cars?

SWIMMING POOLS **For Exercises 33 and 34, use the following information.** The pool in Kara's backyard is cleaned and ready to be filled for the summer. It measures 15 feet long and 10 feet wide with an average depth of 4 feet.

33. What is the volume of the pool?

34. How many gallons of water will it take to fill the pool? ($1 \text{ ft}^3 = 7.5$ gal)

35. **CRITICAL THINKING** Solve $\dfrac{\dfrac{x+3}{x-2} \cdot \dfrac{x^2+x-2}{x+5}}{x-1} + 2 = 0$.

36. WRITING IN MATH Answer the question that was posed at the beginning of the lesson.

How are rational equations important in the operation of a subway system?

Include the following in your answer:

- an explanation of how rational equations can be used to approximate the time that trains will pass each other if they leave distant stations and head toward each other.

Standardized Test Practice

37. What is the value of a in the equation $\frac{a-2}{a} - \frac{a-3}{a-6} = \frac{1}{a}$?

Ⓐ 3 Ⓑ 2 Ⓒ 6 Ⓓ 0

38. Which value is an extraneous solution of $\frac{-1}{n+2} = \frac{n^2-7n-8}{3n^2+2n-8}$?

Ⓐ 6 Ⓑ 2 Ⓒ -1 Ⓓ -2

Maintain Your Skills

Mixed Review

Simplify each expression. *(Lesson 12-8)*

39. $\dfrac{\dfrac{x^2+8x+15}{x^2+x-6}}{\dfrac{x^2+2x-15}{x^2-2x-3}}$

40. $\dfrac{\dfrac{a^2-6a+5}{a^2+13a+42}}{\dfrac{a^2-4a+3}{a^2+3a-18}}$

41. $\dfrac{x+2+\dfrac{2}{x+5}}{x+6+\dfrac{6}{x+1}}$

Find each difference. *(Lesson 12-7)*

42. $\frac{3}{2m-3} - \frac{m}{6-4m}$

43. $\frac{y}{y^2-2y+1} - \frac{1}{y-1}$

44. $\frac{a+2}{a^2-9} - \frac{2a}{6a^2-17a-3}$

Factor each polynomial. *(Lesson 9-2)*

45. $20x - 8y$

46. $14a^2b + 21ab^2$

47. $10p^2 - 12p + 25p - 30$

48. **CHEMISTRY** One solution is 50% glycol, and another is 30% glycol. How much of each solution should be mixed to make a 100-gallon solution that is 45% glycol? *(Lesson 7-2)*

WebQuest Internet Project

Building the Best Roller Coaster

It is time to complete your project. Use the information and data you have gathered about the building and financing of a roller coaster to prepare a portfolio or Web page. Be sure to include graphs and/or tables in the presentation.

www.algebra1.com/webquest

www.algebra1.com/self_check_quiz

Chapter 12

Study Guide and Review

Vocabulary and Concept Check

complex fraction (p. 684)
excluded values (p. 648)
extraneous solutions (p. 693)
inverse variation (p. 642)
least common multiple (p. 678)
least common denominator (p. 679)
mixed expression (p. 684)
product rule (p. 643)
rate problem (p. 692)
rational equation (p. 690)
rational expression (p. 648)
work problem (p. 691)

State whether each sentence is *true* or *false*. If false, replace the underlined expression to make a true sentence.

1. A <u>mixed</u> expression is a fraction whose numerator and denominator are polynomials.
2. The complex fraction $\frac{\frac{4}{5}}{\frac{2}{3}}$ can be simplified as $\underline{\frac{6}{5}}$.
3. The equation $\frac{x}{x-1} + \frac{2x-3}{x-1} = 2$ has an extraneous solution of $\underline{1}$.
4. The mixed expression $6 - \frac{a-2}{a+3}$ can be rewritten as $\underline{\frac{5a+16}{a+3}}$.
5. The least common multiple for $(x^2 - 144)$ and $(x + 12)$ is $\underline{x + 12}$.
6. The excluded values for $\frac{4x}{x^2 - x - 12}$ are $\underline{-3 \text{ and } 4}$.

Lesson-by-Lesson Review

12-1 Inverse Variation

See pages 642–647.

Concept Summary

- The product rule for inverse variations states that if (x_1, y_1) and (x_2, y_2) are solutions of an inverse variation, then $x_1y_1 = k$ and $x_2y_2 = k$.
- You can use $\frac{x_1}{x_2} = \frac{y_2}{y_1}$ to solve problems involving inverse variation.

Example **If y varies inversely as x and $y = 24$ when $x = 30$, find x when $y = 10$.**

$\frac{x_1}{x_2} = \frac{y_2}{y_1}$ Proportion for inverse variations

$\frac{30}{x_2} = \frac{10}{24}$ $x_1 = 30$, $y_1 = 24$, and $y_2 = 10$

$720 = 10x_2$ Cross multiply.

$72 = x_2$ Thus, $x = 72$ when $y = 10$.

Exercises **Write an inverse variation equation that relates x and y. Assume that y varies inversely as x. Then solve.** *See Examples 3 and 4 on page 644.*

7. If $y = 28$ when $x = 42$, find y when $x = 56$.
8. If $y = 15$ when $x = 5$, find y when $x = 3$.
9. If $y = 18$ when $x = 8$, find x when $y = 3$.
10. If $y = 35$ when $x = 175$, find y when $x = 75$.

12-2 Rational Expressions

See pages 648–653.

Concept Summary

- Excluded values are values of a variable that result in a denominator of zero.

Example **Simplify $\frac{x+4}{x^2+12x+32}$. State the excluded values of x.**

$$\frac{x+4}{x^2+12x+32} = \frac{\overset{1}{\cancel{x+4}}}{\underset{1}{\cancel{(x+4)}}(x+8)} \quad \text{Factor.}$$

$$= \frac{1}{x+8} \quad \text{Simplify.}$$

The expression is undefined when $x = -4$ and $x = -8$.

Exercises **Simplify each expression.** *See Example 5 on page 650.*

11. $\frac{3x^2y}{12xy^3z}$
12. $\frac{n^2-3n}{n-3}$
13. $\frac{a^2-25}{a^2+3a-10}$
14. $\frac{x^2+10x+21}{x^3+x^2-42x}$

12-3 Multiplying Rational Expressions

See pages 655–659.

Concept Summary

- Multiplying rational expressions is similar to multiplying rational numbers.

Example **Find $\frac{1}{x^2+x-12} \cdot \frac{x-3}{x+5}$.**

$$\frac{1}{x^2+x-12} \cdot \frac{x-3}{x+5} = \frac{1}{(x+4)\underset{1}{\cancel{(x-3)}}} \cdot \frac{\overset{1}{\cancel{x-3}}}{x+5} \quad \text{Factor.}$$

$$= \frac{1}{(x+4)(x+5)} \quad \text{Simplify.}$$

Exercises **Find each product.** *See Examples 1–3 on pages 655 and 656.*

15. $\frac{7b^2}{9} \cdot \frac{6a^2}{b}$
16. $\frac{5x^2y}{8ab} \cdot \frac{12a^2b}{25x}$
17. $(3x+30) \cdot \frac{10}{x^2-100}$
18. $\frac{3a-6}{a^2-9} \cdot \frac{a+3}{a^2-2a}$
19. $\frac{x^2+x-12}{x+2} \cdot \frac{x+4}{x^2-x-6}$
20. $\frac{b^2+19b+84}{b-3} \cdot \frac{b^2-9}{b^2+15b+36}$

12-4 Dividing Rational Expressions

See pages 660–664.

Concept Summary

- Divide rational expressions by multiplying by the reciprocal of the divisor.

Example **Find $\frac{y^2-16}{y^2-64} \div \frac{y+4}{y-8}$.**

$$\frac{y^2-16}{y^2-64} \div \frac{y+4}{y-8} = \frac{y^2-16}{y^2-64} \cdot \frac{y-8}{y+4} \quad \text{Multiply by the reciprocal of } \frac{y+4}{y-8}.$$

$$= \frac{(y-4)\overset{1}{\cancel{(y+4)}}}{\underset{1}{\cancel{(y-8)}}(y+8)} \cdot \frac{\overset{1}{\cancel{y-8}}}{\underset{1}{\cancel{y+4}}} \text{ or } \frac{y-4}{y+8} \quad \text{Simplify.}$$

Exercises **Find each quotient.** *See Examples 1–4 on pages 660 and 661.*

21. $\frac{p^3}{2q} \div \frac{p^2}{4q}$

22. $\frac{y^2}{y+4} \div \frac{3y}{y^2-16}$

23. $\frac{3y-12}{y+4} \div (y^2 - 6y + 8)$

24. $\frac{2m^2+7m-15}{m+5} \div \frac{9m^2-4}{3m+2}$

12-5 Dividing Polynomials

See pages 666–671.

Concept Summary

- To divide a polynomial by a monomial, divide each term of the polynomial by the monomial.
- To divide a polynomial by a binomial, use long division.

Example **Find $(x^3 - 2x^2 - 22x + 21) \div (x - 3)$.**

$$\begin{array}{rl}
x^2 + x - 19 & \\
x - 3\overline{)x^3 - 2x^2 - 22x + 21} & \\
\underline{(-)x^3 - 3x^2} \qquad\qquad\quad & \text{Multiply } x^2 \text{ and } x - 3. \\
x^2 - 22x \qquad & \text{Subtract.} \\
\underline{(-)x^2 - 3x} \qquad & \text{Multiply } x \text{ and } x - 3. \\
-19x + 21 & \text{Subtract.} \\
\underline{(-)-19x + 57} & \text{Multiply } -19 \text{ and } x - 3. \\
-36 & \text{Subtract.}
\end{array}$$

The quotient is $x^2 + x - 19 - \frac{36}{x-3}$.

Exercises **Find each quotient.** *See Examples 1–5 on pages 666–668.*

25. $(4a^2b^2c^2 - 8a^3b^2c + 6abc^2) \div 2ab^2$

26. $(x^3 + 7x^2 + 10x - 6) \div (x + 3)$

27. $\frac{x^3 - 7x + 6}{x - 2}$

28. $(48b^2 + 8b + 7) \div (12b - 1)$

12-6 Rational Expressions with Like Denominators

See pages 672–677.

Concept Summary

- Add (or subtract) rational expressions with like denominators by adding (or subtracting) the numerators and writing the sum (or difference) over the denominator.

Example **Find $\frac{m^2}{m+4} - \frac{16}{m+4}$.**

$$\frac{m^2}{m+4} - \frac{16}{m+4} = \frac{m^2 - 16}{m+4} \quad \text{Subtract the numerators.}$$

$$= \frac{(m-4)\overset{1}{\cancel{(m+4)}}}{\underset{1}{\cancel{m+4}}} \text{ or } m - 4 \quad \text{Factor.}$$

Exercises **Find each sum or difference.** *See Examples 1–4 on pages 672 and 673.*

29. $\frac{m+4}{5} + \frac{m-1}{5}$

30. $\frac{-5}{2n-5} + \frac{2n}{2n-5}$

31. $\frac{a^2}{a-b} + \frac{-b^2}{a-b}$

32. $\frac{7a}{b^2} - \frac{5a}{b^2}$

33. $\frac{2x}{x-3} - \frac{6}{x-3}$

34. $\frac{m^2}{m-n} - \frac{2mn - n^2}{m-n}$

12-7 Rational Expressions with Unlike Denominators

See pages 678–683.

Concept Summary

- Rewrite rational expressions with unlike denominators using the least common denominator (LCD). Then add or subtract.

Example Find $\frac{x}{x+3} + \frac{5}{x-2}$.

$$\frac{x}{x+3} + \frac{5}{x-2} = \frac{x-2}{x-2} \cdot \frac{x}{x+3} + \frac{x+3}{x+3} \cdot \frac{5}{x-2} \quad \text{The LCD is } (x+3)(x-2).$$

$$= \frac{x^2 - 2x}{(x+3)(x-2)} + \frac{5x+15}{(x+3)(x-2)} \quad \text{Multiply.}$$

$$= \frac{x^2 + 3x + 15}{(x+3)(x-2)} \quad \text{Add.}$$

Exercises **Find each sum or difference.** *See Examples 3–5 on pages 679 and 680.*

35. $\frac{2c}{3d^2} + \frac{3}{2cd}$
36. $\frac{r^2 + 21r}{r^2 - 9} + \frac{3r}{r+3}$
37. $\frac{3a}{a-2} + \frac{5a}{a+1}$

38. $\frac{7n}{3} - \frac{9n}{7}$
39. $\frac{7}{3a} - \frac{3}{6a^2}$
40. $\frac{2x}{2x+8} - \frac{4}{5x+20}$

12-8 Mixed Expressions and Complex Fractions

See pages 684–689.

Concept Summary

- Write mixed expressions as rational expressions in the same way as mixed numbers are changed to improper fractions.
- Simplify complex fractions by writing them as division problems.

Example Simplify $\frac{y - \frac{40}{y-3}}{y+5}$.

$$\frac{y - \frac{40}{y-3}}{y+5} = \frac{\frac{y(y-3)}{(y-3)} - \frac{40}{y-3}}{y+5} \quad \text{The LCD in the numerator is } y - 3.$$

$$= \frac{\frac{y^2 - 3y - 40}{y-3}}{y+5} \quad \text{Add in the numerator.}$$

$$= \frac{y^2 - 3y - 40}{y-3} \div (y+5) \quad \text{Rewrite as a division sentence.}$$

$$= \frac{y^2 - 3y - 40}{y-3} \cdot \frac{1}{y+5} \quad \text{Multiply by the reciprocal of } y + 5.$$

$$= \frac{(y-8)\cancel{(y+5)}^{1}}{y-3} \cdot \frac{1}{\cancel{y+5}_{1}} \text{ or } \frac{y-8}{y-3} \quad \text{Factor.}$$

Exercises **Write each mixed expression as a rational expression.**
See Example 1 on page 684.

41. $4 + \frac{x}{x-2}$
42. $2 - \frac{x+2}{x^2-4}$
43. $3 + \frac{x^2 + y^2}{x^2 - y^2}$

Chapter 12 **For More ...**

- Extra Practice, see pages 846–849.
- Mixed Problem Solving, see page 864.

Simplify each expression. *See Examples 3 and 4 on pages 685 and 686.*

44. $\dfrac{\frac{x^2}{y^3}}{\frac{3x}{9y^2}}$ **45.** $\dfrac{5 + \frac{4}{a}}{\frac{a}{2} - \frac{3}{4}}$ **46.** $\dfrac{y + 9 - \frac{6}{y+4}}{y + 4 + \frac{2}{y+1}}$

12-9 Solving Rational Equations

See pages 690–695.

Concept Summary

- Use cross products to solve rational equations with a single fraction on each side of the equal sign.
- Multiply every term of a more complicated rational equation by the LCD to eliminate fractions.

Example **Solve $\frac{5n}{6} + \frac{1}{n-2} = \frac{n+1}{3(n-2)}$.**

$$\frac{5n}{6} + \frac{1}{n-2} = \frac{n+1}{3(n-2)} \quad \text{Original equation}$$

$$6(n-2)\left(\frac{5n}{6} + \frac{1}{n-2}\right) = 6(n-2)\frac{n+1}{3(n-2)} \quad \text{The LCD is } 6(n-2)$$

$$\frac{\overset{1}{\cancel{6}}(n-2)(5n)}{\underset{1}{\cancel{6}}} + \frac{6\overset{1}{\cancel{(n-2)}}}{\underset{1}{\cancel{n-2}}} = \frac{\overset{2}{\cancel{6}}\overset{1}{\cancel{(n-2)}}(n+1)}{\underset{1}{\cancel{3}}\underset{1}{\cancel{(n-2)}}} \quad \text{Distributive Property}$$

$$(n-2)(5n) + 6 = 2(n+1) \quad \text{Simplify.}$$

$$5n^2 - 10n + 6 = 2n + 2 \quad \text{Multiply.}$$

$$5n^2 - 12n + 4 = 0 \quad \text{Subtract.}$$

$$(5n-2)(n-2) = 0 \quad \text{Factor.}$$

$$n = \frac{2}{5} \text{ or } n = 2$$

CHECK Let $n = \frac{2}{5}$.

$$\frac{\frac{2}{5}+1}{3\left(\frac{2}{5}-2\right)} \stackrel{?}{=} \frac{5\left(\frac{2}{5}\right)}{6} + \frac{1}{\frac{2}{5}-2}$$

$$-\frac{7}{24} = -\frac{7}{24} \quad \checkmark$$

Let $n = 2$.

$$\frac{2+1}{3(2-2)} \stackrel{?}{=} \frac{5(2)}{6} + \frac{1}{2-2}$$

$$\frac{3}{3(0)} \stackrel{?}{=} \frac{10}{6} + \frac{1}{0}$$

When you check the value 2, you get a zero in the denominator. So, 2 is an extraneous solution.

Exercises **Solve each equation. State any extraneous solutions.**
See Examples 6 and 7 on page 693.

47. $\frac{4x}{3} + \frac{7}{2} = \frac{7x}{12} - \frac{1}{4}$ **48.** $\frac{11}{2x} - \frac{2}{3x} = \frac{1}{6}$

49. $\frac{2}{3r} - \frac{3r}{r-2} = -3$ **50.** $\frac{x-2}{x} - \frac{x-3}{x-6} = \frac{1}{x}$

51. $\frac{3}{x^2+3x} + \frac{x+2}{x+3} = \frac{1}{x}$ **52.** $\frac{1}{n+4} - \frac{1}{n-1} = \frac{2}{n^2+3n-4}$

Chapter 12 Practice Test

Vocabulary and Concepts

Choose the letter that best matches each algebraic expression.

a. complex fraction
b. rational expression
c. mixed expression

1. $\dfrac{\frac{a}{b}}{\frac{x}{y}}$
2. $3 - \dfrac{a+1}{a-1}$
3. $\dfrac{2}{x^2 + 2x - 4}$

Skills and Applications

Write an inverse variation equation that relates x and y. Assume that y varies inversely as x. Then solve.

4. If $y = 21$ when $x = 40$, find y when $x = 84$.
5. If $y = 22$ when $x = 4$, find x when $y = 16$.

Simplify each expression. State the excluded values of the variables.

6. $\dfrac{5 - 2m}{6m - 15}$
7. $\dfrac{3 + x}{2x^2 + 5x - 3}$
8. $\dfrac{4c^2 + 12c + 9}{2c^2 - 11c - 21}$
9. $\dfrac{1 - \frac{9}{t}}{1 - \frac{81}{t^2}}$
10. $\dfrac{\frac{5}{6} + \frac{u}{t}}{\frac{2u}{t} - 3}$
11. $\dfrac{x + 4 + \frac{5}{x-2}}{x + 6 + \frac{15}{x-2}}$

Perform the indicated operations.

12. $\dfrac{2x}{x-7} - \dfrac{14}{x-7}$
13. $\dfrac{n+3}{2n-8} \cdot \dfrac{6n-24}{2n+1}$
14. $(10m^2 + 9m - 36) \div (2m - 3)$
15. $\dfrac{x^2 + 4x - 32}{x+5} \cdot \dfrac{x-3}{x^2 - 7x + 12}$
16. $\dfrac{z^2 + 2z - 15}{z^2 + 9z + 20} \div (z - 3)$
17. $\dfrac{4x^2 + 11x + 6}{x^2 - x - 6} \div \dfrac{x^2 + 8x + 16}{x^2 + x - 12}$
18. $(10z^4 + 5z^3 - z^2) \div 5z^3$
19. $\dfrac{y}{7y + 14} + \dfrac{6}{6 - 3y}$
20. $\dfrac{x+5}{x+2} + 6$
21. $\dfrac{x^2 - 1}{x+1} - \dfrac{x^2 + 1}{x-1}$

Solve each equation. State any extraneous solutions.

22. $\dfrac{2n}{n-4} - 2 = \dfrac{4}{n+5}$
23. $\dfrac{3}{x^2 + 5x + 6} - \dfrac{7}{x+3} = -\dfrac{x-1}{x+2}$

24. **FINANCE** Barrington High School is raising money for Habitat for Humanity by doing lawn work for friends and neighbors. Scott can rake a lawn and bag the leaves in 5 hours, while Kalyn can do it in 3 hours. If Scott and Kalyn work together, how long will it take them to rake a lawn and bag the leaves?

25. **STANDARDIZED TEST PRACTICE** Which expression can be used to represent the area of the triangle?

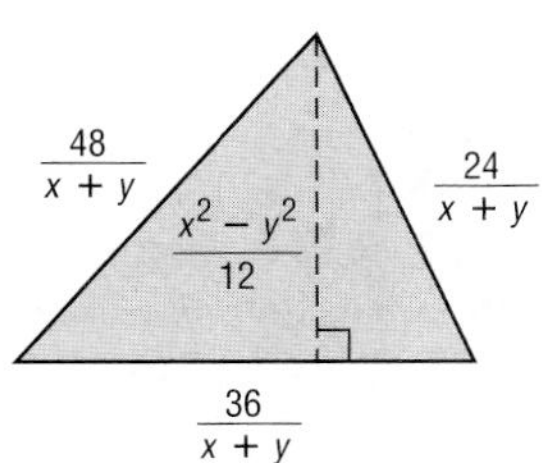

Ⓐ $\frac{1}{2}(x - y)$ Ⓑ $\frac{3}{2}(x - y)$

Ⓒ $\frac{1}{4}(x - y)$ Ⓓ $\frac{108}{x + y}$

Chapter 12 Standardized Test Practice

Part 1 Multiple Choice

Record your answers on the answer sheet provided by your teacher or on a sheet of paper.

1. A cylindrical container is 8 inches in height and has a radius of 2.5 inches. What is the volume of the container to the nearest cubic inch? (*Hint*: $V = \pi r^2 h$) (Lesson 3-8)

Ⓐ 63 Ⓑ 126

Ⓒ 150 Ⓓ 157

2. Which function includes all of the ordered pairs in the table? (Lesson 4-8)

x	−3	−1	1	3	5
y	10	4	−2	−8	−14

Ⓐ $y = -2x$ Ⓑ $y = -3x + 1$

Ⓒ $y = 2x - 4$ Ⓓ $y = 3x + 1$

3. Which equation describes the graph below? (Lesson 5-4)

Ⓐ $4x - 5y = 40$

Ⓑ $4x + 5y = -40$

Ⓒ $4x + 5y = -8$

Ⓓ $rx - 5y = 10$

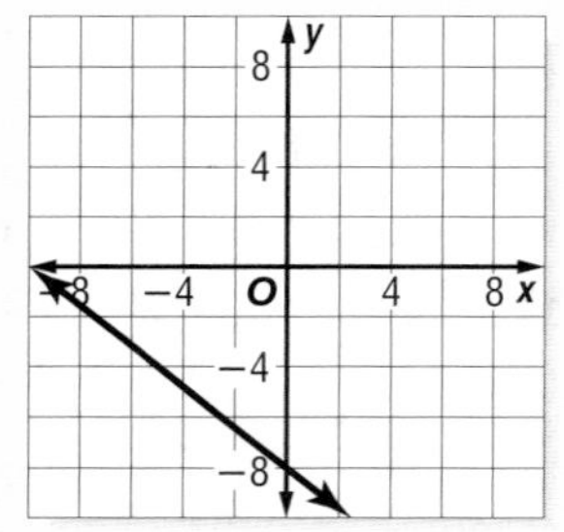

4. Which equation represents the line that passes through (−12, 5) and has a slope of $-\frac{1}{4}$? (Lesson 5-5)

Ⓐ $x + 4y = 8$ Ⓑ $-x + 4y = 20$

Ⓒ $-4x + y = 65$ Ⓓ $x + 4y = 5$

The Princeton Review Test-Taking Tip

Questions 2, 4, 8 Sometimes sketching the graph of a function can help you to see the relationship between *x* and *y* and answer the question.

5. Which inequality represents the shaded region? (Lesson 6-6)

Ⓐ $y \leq -\frac{1}{2}x - 2$

Ⓑ $y \geq -\frac{1}{2}x + 2$

Ⓒ $y \leq -2x + 2$

Ⓓ $y \geq -2x + 2$

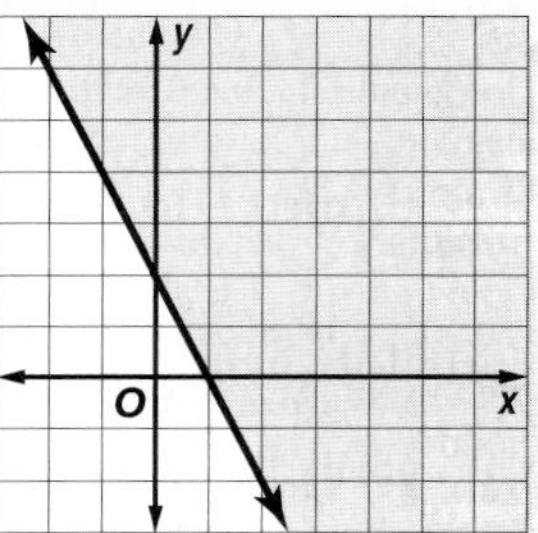

6. Which ordered pair is the solution of the following system of equations? (Lesson 7-4)

$$3x + y = -2$$
$$-2x + y = 8$$

Ⓐ (−6, 16) Ⓑ (−2, 4)

Ⓒ (−3, 2) Ⓓ (2, −8)

7. The length of a rectangular door is 2.5 times its width. If the area of the door is 9750 square inches, which equation will determine the width w of the door? (Lesson 8-1)

Ⓐ $w^2 + 2.5w = 9750$

Ⓑ $2.5w^2 = 9750$

Ⓒ $2.5w^2 + 9750 = 0$

Ⓓ $7w = 9750$

8. A scientist monitored a 144-gram sample of a radioactive substance, which decays into a nonradioactive substance. The table shows the amount, in grams, of the radioactive substance remaining at intervals of 20 hours. How many grams of the radioactive substance are likely to remain after 100 hours? (Lessons 10-6 and 10-7)

Time (h)	0	20	40	60	80	100
Mass (g)	144	72	36			

Ⓐ 1 g Ⓑ 2.25 g

Ⓒ 4.5 g Ⓓ 9 g

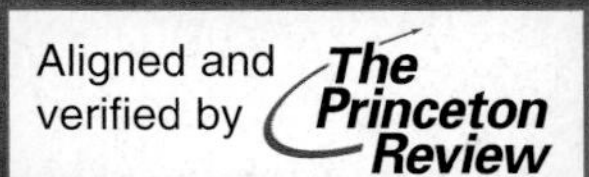

Part 2 Short Response/Grid In

Record your answers on the answer sheet provided by your teacher or on a sheet of paper.

9. A family drove an average of 350 miles per day during three days of their trip. They drove 360 miles on the first day and 270 miles on the second day. How many miles did they drive on the third day? (Lesson 3-4)

10. The area of the rectangular playground at Hillcrest School is 750 square meters. The length of the playground is 5 meters greater than its width. What are the length and width of the playground in meters? (Lesson 9-5)

11. Use the Quadratic Formula or factoring to determine whether the graph of $y = 16x^2 + 24x + 9$ intersects the x-axis in zero, one, or two points. (Lesson 10-4)

12. Express $\frac{x^2 - 9}{x^3 + x} \cdot \frac{3x}{x - 3}$ as a quotient of two polynomials written in simplest form. (Lesson 11-3)

13. Express the following quotient in simplest form. (Lesson 11-4)

$$\frac{x}{x+4} \div \frac{4x}{x^2 - 16}$$

Part 3 Quantitative Comparison

Compare the quantity in Column A and the quantity in Column B. Then determine whether:

- **Ⓐ the quantity in Column A is greater,**
- **Ⓑ the quantity in Column B is greater,**
- **Ⓒ the two quantities are equal, or**
- **Ⓓ the relationship cannot be determined from the information given.**

www.algebra1.com/standardized_test

	Column A	Column B
14.	$x = \frac{1}{4}, y = 4$	
	$\frac{1}{x^2 - 2x}$	$\frac{1}{y^2 - 2y}$
		(Lesson 1-3)
15.	$\sqrt{500} - \sqrt{20} + \sqrt{180} - \sqrt{720}$	$\sqrt{125} - \sqrt{45}$
		(Lesson 11-2)
16.	the excluded value of a in $\frac{16a - 24}{32a}$	the excluded value of b in $\frac{5b + 3}{b + 6}$
		(Lesson 12-2)
17.	$5 + \frac{3x}{x + 1}$	$\dfrac{\frac{24y + 15}{3}}{\frac{6y + 6}{6}}$
		(Lesson 12-8)

Part 4 Open Ended

Record your answers on a sheet of paper. Show your work.

18. A 12-foot ladder is placed against the side of a building so that the bottom of the ladder is 6 feet from the base of the building. (Lesson 12-1)

a. Suppose the bottom of the ladder is moved closer to the base of the building. Does the height that the ladder reaches increase or decrease?

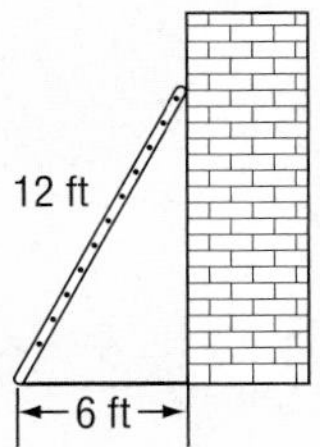

b. What conclusion can you make about the height the ladder reaches and the distance between the bottom of the ladder and the base of the building?

c. Does this relationship form an inverse variation? Explain your reasoning.

UNIT 5

Data Analysis

Collecting and analyzing data allows you to make decisions and predictions about the future. In this unit, you will learn about statistics and probability.

WebQuest Internet Project

America Counts!

The U.S. government has been counting each person in the country since its first Census following independence was taken in 1790. Befitting the first Census of the 21st century, the Census Bureau allowed Census 2000 questionnaires to be completed electronically for the first time. In this project, you will see how data analysis can be used to compare statistics about a state of your choice to other states in the United States.

Log on to www.algebra1.com/webquest. Begin your WebQuest by reading the Task.

Then continue working on your WebQuest as you study Unit 5.

Lesson	13-5	14-2
Page	742	766

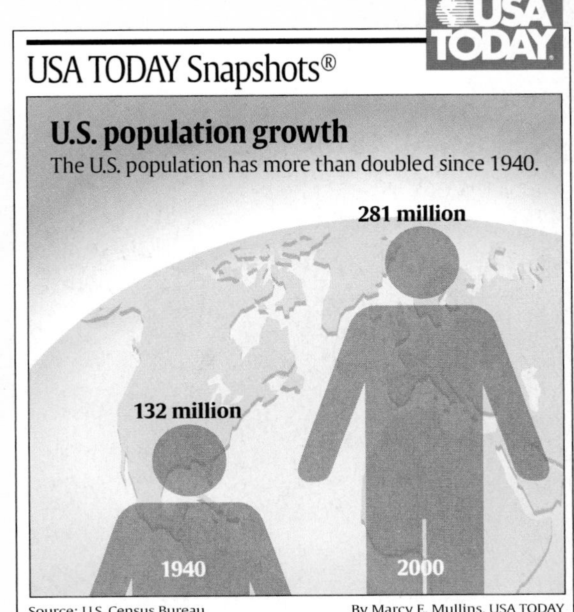

Chapter

13 Statistics

What You'll Learn

- **Lesson 13-1** Identify various sampling techniques.
- **Lesson 13-2** Solve problems by adding or subtracting matrices or by multiplying by a scalar.
- **Lesson 13-3** Interpret data displayed in histograms.
- **Lesson 13-4** Find the range, quartiles, and interquartile range of a set of data.
- **Lesson 13-5** Organize and use data in box-and-whisker plots.

Key Vocabulary

- sample (p. 708)
- matrix (p. 715)
- histogram (p. 722)
- quartile (p. 732)
- box-and-whisker plot (p. 737)

Why It's Important

Each day statistics are reported in the newspapers, in magazines, on television, and on the radio. These data involve business, government, ecology, sports, and many other topics. A basic knowledge of statistics allows you to interpret what you hear and read in the media. One important tool to help you understand the significance of a set of data is the box-and-whisker plot. *You will draw and use a box-and-whisker plot for data involving NASCAR racing in Lesson 13-5.*

Getting Started

Prerequisite Skills To be successful in this chapter, you'll need to master these skills and be able to apply them in problem-solving situations. Review these skills before beginning Chapter 13.

For Lesson 13-1 — Use Logical Reasoning

Find a counterexample for each statement. *(For review, see Lesson 1-7.)*

1. If $a + b = c$, then $a < c$.

2. If a flower is a rose, then it is red.

3. If Tara obeys the speed limit, then she will drive 45 miles per hour or less.

4. If a number is even, then it is divisible by 4.

For Lesson 13-4 — Find the Median

Find the median for each set of data. *(For review, see pages 818 and 819.)*

5. 1, 7, 9, 15, 25, 59, 63

6. 0, 10, 2, 2, 9, 5, 4, 2, 8, 3, 8, 7, 3

7. 726, 411, 407, 407, 395, 355, 317, 235, 218, 211

For Lesson 13-5 — Graph Numbers on a Number Line

Graph each set of numbers on a number line. *(For review, see Lesson 2-1.)*

8. {7, 9, 10, 13, 14}

9. {15, 17.5, 19, 20.5, 23}

10. {3.2, 4.8, 5.0, 5.7, 6.1}

11. {2.3, 2.8, 3.1, 3.7, 4.5}

Make this Foldable to help you organize information about statistics. Begin with three sheets of plain $8\frac{1}{2}$" by 11" paper.

Step 1 Stack Pages

Stack sheets of paper with edges $\frac{3}{4}$ inch apart.

Step 2 Fold Up Bottom Edges

All tabs should be the same size.

Step 3 Crease and Staple

Step 4 Turn and Label

Label the tabs with topics from the chapter.

Statistics

Reading and Writing As you read and study the chapter, use each page to write notes and examples.

13-1 Sampling and Bias

What You'll Learn

- Identify various sampling techniques.
- Recognize a biased sample.

Why is sampling important in manufacturing?

Manufacturing music CDs involves burning, or recording, copies from a master. However, not every burn is successful. It is costly and time-consuming to check every CD that is burned. Therefore, in order to monitor production, some CDs are picked at random and checked for defects.

Vocabulary

- sample
- population
- census
- random sample
- simple random sample
- stratified random sample
- systematic random sample
- biased sample
- convenience sample
- voluntary response sample

SAMPLING TECHNIQUES When you wish to make an investigation, there are four ways that you can collect data.

- **published data** Use data that are already in a source like a newspaper or book.
- **observational study** Watch naturally occurring events and record the results.
- **experiment** Conduct an experiment and record the results.
- **survey** Ask questions of a group of people and record the results.

When performing an experiment or taking a survey, researchers often choose a sample. A **sample** is some portion of a larger group, called the **population**, selected to represent that group. If all of the units within a population are included, it is called a **census**. Sample data are often used to estimate a characteristic within an entire population, such as voting preferences prior to elections.

Population	Sample
all of the light bulbs manufactured on a production line	24 light bulbs selected from the production line
all of the water in a swimming pool	a test tube of water from the pool
all of the people in the United States	1509 people from throughout the United States

A **random sample** of a population is selected so that it is representative of the entire population. The sample is chosen without any preference. There are several ways to pick a random sample.

Key Concept — Random Samples

Type	Definition	Example
Simple Random Sample	A simple random sample is a sample that is as likely to be chosen as any other from the population.	The 26 students in a class are each assigned a different number from 1 to 26. Then three of the 26 numbers are picked at random.
Stratified Random Sample	In a stratified random sample, the population is first divided into similar, nonoverlapping groups. A simple random sample is then selected from each group.	The students in a school are divided into freshman, sophomores, juniors, and seniors. Then two students are randomly selected from each group of students.
Systematic Random Sample	In a systematic random sample, the items are selected according to a specified time or item interval.	Every 2 minutes, an item is pulled off the assembly line. or Every twentieth item is pulled off the assembly line.

Example 1 Classify a Random Sample

ECOLOGY **Ten lakes are selected randomly from a list of all public-access lakes in Minnesota. Then 2 liters of water are drawn from 20 feet deep in each of the ten lakes.**

a. Identify the sample and suggest a population from which it was selected.

The sample is ten 2-liter containers of lake water, one from each of 10 lakes. The population is lake water from all of the public-access lakes in Minnesota.

b. Classify the sample as *simple, stratified,* or *systematic.*

This is a simple random sample. Each of the ten lakes was equally likely to have been chosen from the list.

More About. . .

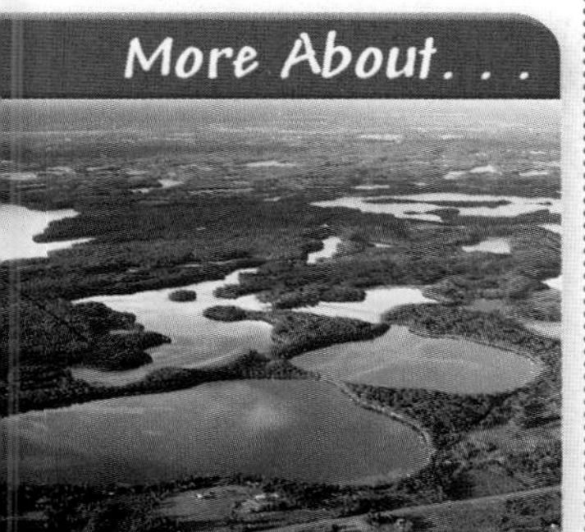

Ecology

More than one twentieth of the area of Minnesota is covered by inland lakes. The largest lake is Red Lake, which covers 430 square miles.

Source: *World Book Encyclopedia*

BIASED SAMPLE Random samples are unbiased. In a **biased sample**, one or more parts of a population are favored over others.

Example 2 Identify Sample as Biased or Unbiased

Identify each sample as *biased* or *unbiased*. Explain your reasoning.

a. MANUFACTURING **Every 1000th bolt is pulled from the production line and measured for length.**

The sample is chosen using a specified time interval. This is an unbiased sample because it is a systematic random sample.

b. MUSIC **Every tenth customer in line for a certain rock band's concert tickets is asked about his or her favorite rock band.**

The sample is a biased sample because customers in line for concert tickets are more likely to name the band giving the concert as a favorite band.

Two popular forms of samples that are often biased include convenience samples and voluntary response samples.

Key Concept — Biased Samples

Type	Definition	Example
Convenience Sample	A convenience sample includes members of a population that are easily accessed.	To check spoilage, a produce worker selects 10 apples from the top of the bin. The 10 apples are unlikely to represent all of the apples in the bin.
Voluntary Response Sample	A voluntary response sample involves only those who want to participate in the sampling.	A radio call-in show records that 75% of its 40 callers voiced negative opinions about a local football team. Those 40 callers are unlikely to represent the entire local population. Volunteer callers are more likely to have strong opinions and are typically more negative than the entire population.

Example 3 Identify and Classify a Biased Sample

BUSINESS **The travel account records from 4 of the 20 departments in a corporation are to be reviewed. The accountant states that the first 4 departments to voluntarily submit their records will be reviewed.**

a. Identify the sample and suggest a population from which it was selected.

The sample is the travel account records from 4 departments in the corporation. The population is the travel account records from all 20 departments in the corporation.

b. Classify the sample as *convenience* or *voluntary response*.

Since the departments voluntarily submit their records, this is a voluntary response sample.

Example 4 Identify the Sample

NEWS REPORTING For an article in the school paper, Rafael needs to determine whether students in his school believe that an arts center should be added to the school. He polls 15 of his friends who sing in the choir. Twelve of them think the school needs an arts center, so Rafael reports that 80% of the students surveyed support the project.

a. Identify the sample.

The sample is a group of students from the choir.

b. Suggest a population from which the sample was selected.

The population for the survey is all of the students in the school.

c. State whether the sample is *unbiased* (random) or *biased*. If unbiased, classify it as *simple*, *stratified*, or *systematic*. If biased, classify it as *convenience* or *voluntary response*.

The sample was not randomly selected from the entire student body. So the reported support is not likey to be representative of the student body. The sample is biased. Since Rafael polled only his friends, it is a convenience sample.

Check for Understanding

Concept Check

1. **Describe** how the following three types of sampling techniques are similar and how they are different.
   - simple random sample
   - stratified random sample
   - systematic random sample
2. **Explain** the difference between a convenience sample and a voluntary response sample.
3. **OPEN ENDED** Give an example of a biased sample.

Guided Practice

Identify each sample, suggest a population from which it was selected, and state whether it is *unbiased* (random) or *biased*. If unbiased, classify the sample as *simple*, *stratified*, or *systematic*. If biased, classify as *convenience* or *voluntary response*.

4. **NEWSPAPERS** The local newspaper asks readers to write letters stating their preferred candidate for mayor.
5. **SCHOOL** A teacher needs a sample of work from 4 students in her first-period math class to display at the school open house. She selects the work of the first 4 students who raise their hands.
6. **BUSINESS** A hardware store wants to assess the strength of nails it sells. Store personnel select 25 boxes at random from among all of the boxes on the shelves. From each of the 25 boxes, they select one nail at random and subject it to a strength test.
7. **SCHOOL** A class advisor hears complaints about an incorrect spelling of the school name on pencils sold at the school store. The advisor goes to the store and asks Namid to gather a sample of pencils and look for spelling errors. Namid grabs the closest box of pencils and counts out 12 pencils from the top of the box. She checks the pencils, returns them to the box, and reports the results to the advisor.

Study Tip

Reading Math

The data in Exercise 6 could be classified as **univariate,** because there is only one variable, strength. These data could also be classified as **measurement,** because there are different levels of strength that could be measured.

Practice and Apply

Homework Help

For Exercises	See Examples
8–28	1–4

Extra Practice

See page 849.

Study Tip

Reading Math
The univariate data in Exercise 9 could be classified as **categorical,** because they reflect two categories, the two brands of cola.

Identify each sample, suggest a population from which it was selected, and state whether it is *unbiased* (random) or *biased*. If unbiased, classify the sample as *simple, stratified,* or *systematic*. If biased, classify as *convenience* or *voluntary response*.

8. **SCHOOL** Pieces of paper with the names of 3 sophomores are drawn from a hat containing identical pieces of paper with all sophomores' names.

9. **FOOD** Twenty shoppers outside a fast-food restaurant are asked to name their preferred cola among two choices.

10. **RECYCLING** An interviewer goes from house to house on weekdays between 9 A.M. and 4 P.M. to determine how many people recycle.

11. **POPULATION** A state is first divided into its 86 counties and then 10 people from each county are chosen at random.

12. **SCOOTERS** A scooter manufacturer is concerned about quality control. The manufacturer checks the first 5 scooters off the line in the morning and the last 5 off the line in the afternoon for defects.

13. **SCHOOL** To determine who will speak for her class at the school board meeting, Ms. Finchie used the numbers appearing next to her students' names in her grade book. She writes each of the numbers on an identical piece of paper and shuffles the pieces of papers in a box. Without seeing the contents of the box, one student draws 3 pieces of paper from the box. The students whose numbers match the numbers chosen will speak for the class.

More About. . .

Food
Michigan leads the nation in cherry production by growing about 219 million pounds of cherries per year.
Source: *World Book Encyclopedia*

14. **FARMING** An 8-ounce jar was filled with corn from a storage silo by dipping the jar into the pile of corn. The corn in the jar was then analyzed for moisture content.

15. **COURT** The gender makeup of district court judges in the United States is to be estimated from a sample. All judges are grouped geographically by federal reserve districts. Within each of the 11 federal reserve districts, all judges' names are assigned a distinct random number. In each district, the numbers are then listed in order. A number between 1 and 20 inclusive is selected at random, and the judge with that number is selected. Then every 20th name after the first selected number is also included in the sample.

16. **TELEVISION** A television station asks its viewers to share their opinions about a proposed golf course to be built just outside the city limits. Viewers can call one of two 900-numbers. One number represents a "yes" vote, and the other number represents a "no" vote.

17. **GOVERNMENT** To discuss leadership issues shared by all United States Senators, the President asks 4 of his closest colleagues in the Senate to meet with him.

18. **FOOD** To sample the quality of the Bing cherries throughout the produce department, the produce manager picks up a handful of cherries from the edge of one case and checks to see if these cherries are spoiled.

19. **MANUFACTURING** During the manufacture of high-definition televisions, units are checked for defects. Within the first 10 minutes of a work shift, a television is randomly chosen from the line of completed sets. For the rest of the shift, every 15th television on the line is checked for defects.

Identify each sample, suggest a population from which it was selected, and state whether it is *unbiased* (random) or *biased*. If unbiased, classify the sample as *simple*, *stratified*, or *systematic*. If biased, classify as *convenience* or *voluntary response*.

20. **BUSINESS** To get reactions about a benefits package, a company uses a computer program to randomly pick one person from each of its departments.

21. **MOVIES** A magazine is trying to determine the most popular actor of the year. It asks its readers to mail the name of their favorite actor to the magazine's office.

COLLEGE For Exercises 22 and 23, use the following information.
The graph at the right reveals that 56% of survey respondents did not have a formal financial plan for a child's college tuition.

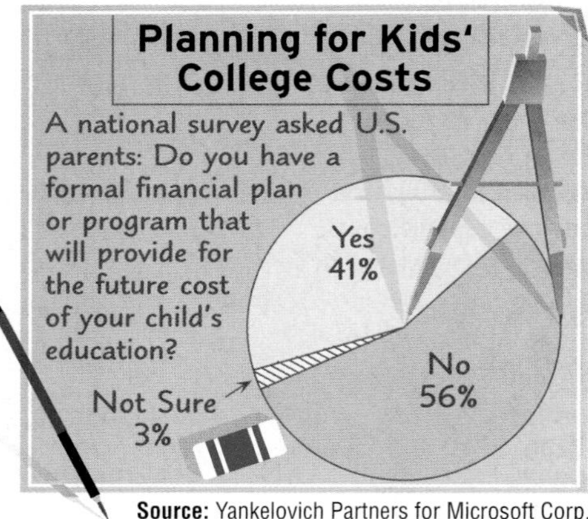

Source: Yankelovich Partners for Microsoft Corp.

22. Write a statement to describe what you do know about the sample.
23. What additional information would you like to have about the sample to determine whether the sample is biased?

24. **SCHOOL** Suppose you want to sample the opinion of the students in your school about a new dress code. Describe an unbiased way to conduct your survey.

25. **ELECTIONS** Suppose you are running for mayor of your city and want to know if you are likely to be elected. Describe an unbiased way to poll the voters.

26. **FAMILY** Study the graph at the right. Describe the information that is revealed in the graph. What information is there about the type or size of the sample?

27. **FARMING** Suppose you are a farmer and want to know if your tomato crop is ready to harvest. Describe an unbiased way to determine whether the crop is ready to harvest.

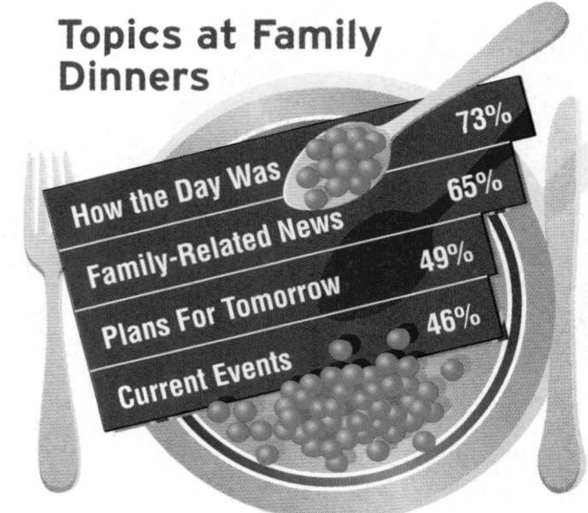

Source: National Pork Producers Council

28. **MANUFACTURING** Suppose you want to know whether the infant car seats manufactured by your company meet the government standards for safety. Describe an unbiased way to determine whether the seats meet the standards.

29. **CRITICAL THINKING** The following is a proposal for surveying a stratified random sample of the student body.

    Divide the student body according to those who are on the basketball team, those who are in the band, and those who are in the drama club. Then take a simple random sample from each of the three groups. Conduct the survey using this sample.

    Study the proposal. Describe its strengths and weaknesses. Is the sample a stratified random sample? Explain.

30. **WRITING IN MATH** Answer the question that was posed at the beginning of the lesson.

Why is sampling important in manufacturing?

Include the following in your answer:

- an unbiased way to pick which CDs to check, and
- a biased way to pick which CDs to check.

31. To predict the candidate who will win the seat in city council, which method would give the newspaper the most accurate result?

(A) Ask every 5th person that passes a reporter in the mall.
(B) Use a list of registered voters and call every 20th person.
(C) Publish a survey and ask readers to reply.
(D) Ask reporters at the newspaper.

32. A cookie manufacturer plans to make a new type of cookie and wants to know if people will buy these cookies. For accurate results, which method should they use?

(A) Ask visitors to their factory to evaluate the cookie.
(B) Place a sample of the new cookie with their other cookies, and ask people to answer a questionnaire about the cookie.
(C) Take samples to a school, and ask students to raise their hands if they like the cookie.
(D) Divide the United States into 6 regions. Then pick 3 cities in each region at random, and conduct a taste test in each of the 18 cities.

Maintain Your Skills

Mixed Review

Solve each equation. *(Lesson 12-9)*

33. $\frac{10}{3y} - \frac{5}{2y} = \frac{1}{4}$ 34. $\frac{3}{r+4} - \frac{1}{r} = \frac{1}{r}$ 35. $\frac{1}{4m} + \frac{2m}{m-3} = 2$

Simplify. *(Lesson 12-8)*

36. $\dfrac{2 + \frac{5}{x}}{\frac{x}{3} + \frac{5}{6}}$ 37. $\dfrac{a + \frac{35}{a+12}}{a+7}$ 38. $\dfrac{\frac{t^2-4}{t^2+5t+6}}{t-2}$

39. **GEOMETRY** What is the perimeter of $\triangle ABC$? *(Lesson 11-2)*

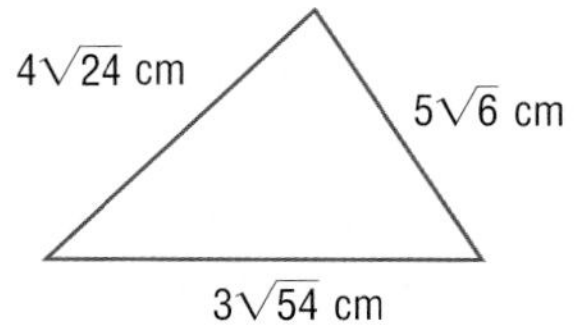

Solve each equation by using the Quadratic Formula. Approximate any irrational roots to the nearest tenth. *(Lesson 10-4)*

40. $x^2 - 6x - 40 = 0$ 41. $6b^2 + 15 = -19b$ 42. $2d^2 = 9d + 3$

Find each product. *(Lesson 8-7)*

43. $(y + 5)(y + 7)$ 44. $(c - 3)(c - 7)$ 45. $(x + 4)(x - 8)$

Getting Ready for the Next Lesson

BASIC SKILL **Find each sum or difference.**

46. $4.5 + 3.8$ 47. $16.9 + 7.21$ 48. $3.6 + 18.5$
49. $7.6 - 3.8$ 50. $18 - 4.7$ 51. $13.2 - 0.75$

Reading Mathematics

Survey Questions

Even though taking a random sample eliminates bias or favoritism in the choice of a sample, questions may be worded to influence people's thoughts in a desired direction. Two different surveys on Internet sales tax had different results.

Question 1
Should there be sales tax on purchases made on the Internet?

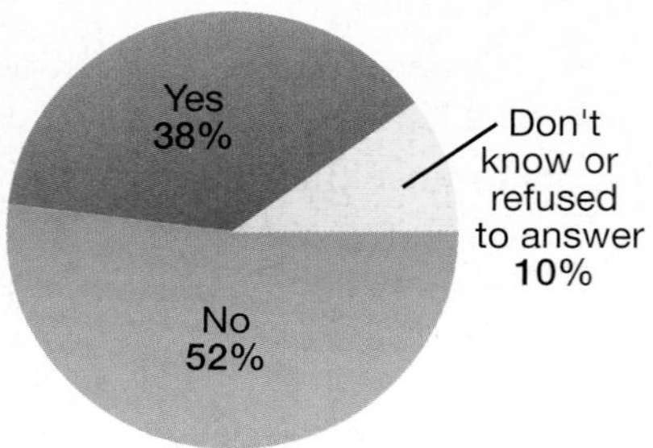

Question 2
Do you think people should or should not be required to pay the same sales tax for purchases made over the Internet that they would if they had bought the item in person at a local store?

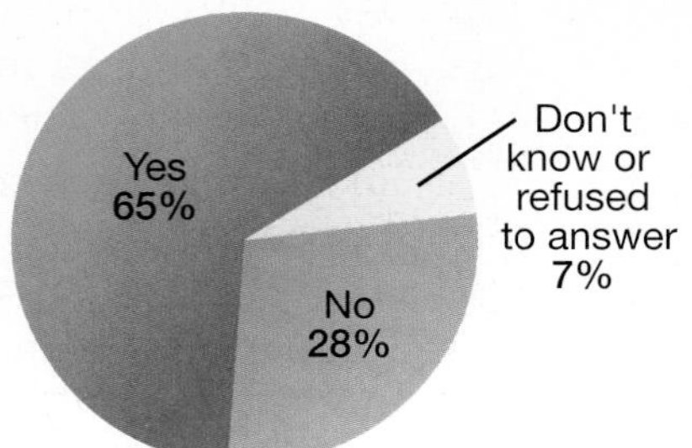

Notice the difference in Questions 1 and 2. Question 2 includes more information. Pointing out that customers pay sales tax for items bought at a local store may give the people answering the survey a reason to answer "yes." Asking the question in that way probably led people to answer the way they did.

Because they are random samples, the results of both of these surveys are accurate. However, the results could be used in a misleading way by someone with an interest in the issue. For example, an Internet retailer would prefer to state the results of Question 1. Be sure to think about survey questions carefully to make sure that you interpret the results correctly.

Reading to Learn

For Exercises 1–2, tell whether each question is likely to bias the results. Explain your reasoning.

1. On a survey on environmental issues:

- **a.** "Due to diminishing resources, should a law be made to require recycling?"
- **b.** "Should the government require citizens to participate in recycling efforts?"

2. On a survey on education:

- **a.** "Should schools fund extracurricular sports programs?"
- **b.** "The budget of the River Valley School District is short of funds. Should taxes be raised in order for the district to fund extracurricular sports programs?"

3. Suppose you want to determine whether to serve hamburgers or pizza at the class party.

- **a.** Write a survey question that would likely produce biased results.
- **b.** Write a survey question that would likely produce unbiased results.

13-2 Introduction to Matrices

What You'll Learn

- Organize data in matrices.
- Solve problems by adding or subtracting matrices or by multiplying by a scalar.

Vocabulary

- matrix
- dimensions
- row
- column
- element
- scalar multiplication

How are matrices used to organize data?

To determine the best type of aircraft to use for certain flights, the management of an airline company considers the following aircraft operating statistics.

Aircraft	Number of Seats	Airborne Speed (mph)	Possible Flight Distance (miles)	Fuel per Hour (gallons)	Operating Cost per Hour (dollars)
B747-100	462	512	2297	3517	7224
DC-10-10	297	496	1402	2311	5703
MD-11	259	527	3073	2464	6539
A300-600	228	475	1372	1505	4783

Source: Air Transport Association of America

The table has rows and columns of information. When we concentrate only on the numerical information, we see an array with 4 rows and 5 columns.

$$\begin{bmatrix} 462 & 512 & 2297 & 3517 & 7224 \\ 297 & 496 & 1402 & 2311 & 5703 \\ 259 & 527 & 3073 & 2464 & 6539 \\ 228 & 475 & 1372 & 1505 & 4783 \end{bmatrix}$$

This array of numbers is called a matrix.

Study Tip

Reading Math
A matrix is sometimes called an *ordered array.*

ORGANIZE DATA IN MATRICES If you have ever used a spreadsheet program on the computer, you have worked with matrices. A **matrix** is a rectangular arrangement of numbers in rows and columns. A matrix is usually described by its **dimensions**, or the number of **rows** and **columns**, with the number of rows stated first. Each entry in a matrix is called an **element**.

Example 1 Name Dimensions of Matrices

State the dimensions of each matrix. Then identify the position of the circled element in each matrix.

a. $[11 \quad \textcircled{15} \quad 24]$

This matrix has 1 row and 3 columns. Therefore, it is a 1-by-3 matrix.

The circled element is in the first row and the second column.

b. $\begin{bmatrix} -4 & 2 \\ 0 & 1 \\ \textcircled{3} & -6 \end{bmatrix}$

This matrix has 3 rows and 2 columns. Therefore, it is a 3-by-2 matrix.

The circled element is in the third row and the first column.

Two matrices are *equal* only if they have the same dimensions and each element of one matrix is equal to the corresponding element in the other matrix.

$$\begin{bmatrix} 3 & 5 \\ -1 & 4 \end{bmatrix} = \begin{bmatrix} 3 & 5 \\ -1 & 4 \end{bmatrix} \qquad \begin{bmatrix} 2 & 4 \\ 1 & 7 \end{bmatrix} \neq \begin{bmatrix} 2 & 3 \\ 1 & 7 \end{bmatrix} \qquad \begin{bmatrix} 4 & 8 \\ 1 & -3 \end{bmatrix} \neq \begin{bmatrix} 4 & 8 & 0 \\ 1 & -3 & 0 \end{bmatrix}$$

MATRIX OPERATIONS If two matrices have the same dimensions, you can add or subtract them. To do this, add or subtract corresponding elements of the two matrices.

Example 2 Add Matrices

If $A = \begin{bmatrix} 3 & -4 & 7 \\ -1 & 6 & 0 \end{bmatrix}$, $B = \begin{bmatrix} 7 & -4 & -2 \\ 1 & 6 & -3 \end{bmatrix}$, and $C = \begin{bmatrix} 3 & 6 \\ -4 & 5 \end{bmatrix}$, find each sum. If the sum does not exist, write *impossible*.

a. $A + B$

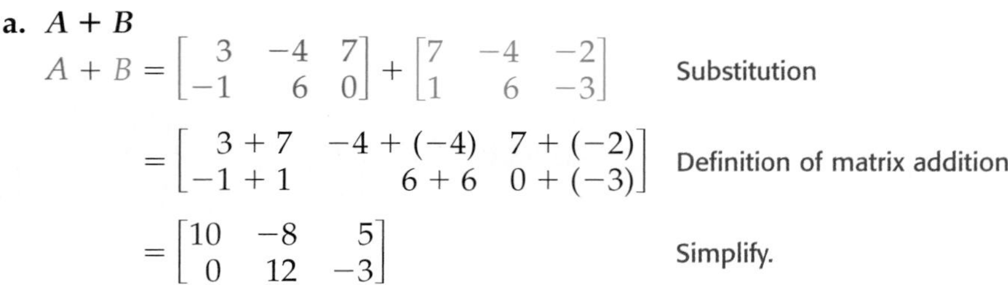

$$A + B = \begin{bmatrix} 3 & -4 & 7 \\ -1 & 6 & 0 \end{bmatrix} + \begin{bmatrix} 7 & -4 & -2 \\ 1 & 6 & -3 \end{bmatrix} \quad \text{Substitution}$$

$$= \begin{bmatrix} 3+7 & -4+(-4) & 7+(-2) \\ -1+1 & 6+6 & 0+(-3) \end{bmatrix} \quad \text{Definition of matrix addition}$$

$$= \begin{bmatrix} 10 & -8 & 5 \\ 0 & 12 & -3 \end{bmatrix} \quad \text{Simplify.}$$

b. $B + C$

$$B + C = \begin{bmatrix} 7 & -4 & -2 \\ 1 & 6 & -3 \end{bmatrix} + \begin{bmatrix} 3 & 6 \\ -4 & 5 \end{bmatrix} \quad \text{Substitution}$$

Since B is a 2-by-3 matrix and C is a 2-by-2 matrix, the matrices do not have the same dimensions. Therefore, it is impossible to add these matrices.

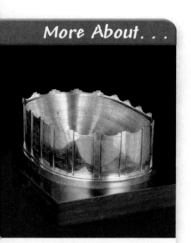

More About. . .

College Football

Each year the National Football Foundation awards the MacArthur Bowl to the number one college football team. The bowl is made of about 400 ounces of silver and represents a stadium with rows of seats.

Source: *ESPN Information Please® Sports Almanac*

Addition and subtraction of matrices can be used to solve real-world problems.

Example 3 Subtract Matrices

COLLEGE FOOTBALL **The Division I-A college football teams with the five best records during the 1990s are listed below.**

Overall Record

	Wins	Losses	Ties
Florida State	109	13	1
Nebraska	108	16	1
Marshall	114	25	0
Florida	102	22	1
Tennessee	99	22	2

Bowl Record

	Wins	Losses	Ties
Florida State	8	2	0
Nebraska	5	5	0
Marshall	2	1	0
Florida	5	4	0
Tennessee	6	4	0

Use subtraction of matrices to determine the regular season records of these teams during the decade.

$$\begin{bmatrix} 109 & 13 & 1 \\ 108 & 16 & 1 \\ 114 & 25 & 0 \\ 102 & 22 & 1 \\ 99 & 22 & 2 \end{bmatrix} - \begin{bmatrix} 8 & 2 & 0 \\ 5 & 5 & 0 \\ 2 & 1 & 0 \\ 5 & 4 & 0 \\ 6 & 4 & 0 \end{bmatrix} = \begin{bmatrix} 109-8 & 13-2 & 1-0 \\ 108-5 & 16-5 & 1-0 \\ 114-2 & 25-1 & 0-0 \\ 102-5 & 22-4 & 1-0 \\ 99-6 & 22-4 & 2-0 \end{bmatrix}$$

$$= \begin{bmatrix} 101 & 11 & 1 \\ 103 & 11 & 1 \\ 112 & 24 & 0 \\ 97 & 18 & 1 \\ 93 & 18 & 2 \end{bmatrix}$$

So, the regular season records of the teams can be described as follows.

$$\begin{array}{l} \\ \textbf{Florida State} \\ \textbf{Nebraska} \\ \textbf{Marshall} \\ \textbf{Florida} \\ \textbf{Tennessee} \end{array} \begin{array}{c} \textbf{Regular Season Record} \\ \begin{array}{ccc} \textbf{Wins} & \textbf{Losses} & \textbf{Ties} \end{array} \\ \begin{bmatrix} 101 & 11 & 1 \\ 103 & 11 & 1 \\ 112 & 24 & 0 \\ 97 & 18 & 1 \\ 93 & 18 & 2 \end{bmatrix} \end{array}$$

You can multiply any matrix by a constant called a *scalar*. This is called **scalar multiplication**. When scalar multiplication is performed, each element is multiplied by the scalar and a new matrix is formed.

Key Concept — Scalar Multiplication of a Matrix

$$m\begin{bmatrix} a & b & c \\ d & e & f \end{bmatrix} = \begin{bmatrix} ma & mb & mc \\ md & me & mf \end{bmatrix}$$

Example 4 Perform Scalar Multiplication

If $T = \begin{bmatrix} -4 & 2 \\ 0 & 1 \\ 3 & -6 \end{bmatrix}$, find $3T$.

$$3T = 3\begin{bmatrix} -4 & 2 \\ 0 & 1 \\ 3 & -6 \end{bmatrix} \quad \text{Substitution}$$

$$= \begin{bmatrix} 3(-4) & 3(2) \\ 3(0) & 3(1) \\ 3(3) & 3(-6) \end{bmatrix} \quad \text{Definition of scalar multiplication}$$

$$= \begin{bmatrix} -12 & 6 \\ 0 & 3 \\ 9 & -18 \end{bmatrix} \quad \text{Simplify.}$$

Check for Understanding

Concept Check

1. **Describe** the difference between a 2-by-4 matrix and a 4-by-2 matrix.
2. **OPEN ENDED** Write two matrices whose sum is $\begin{bmatrix} 0 & 4 & 5 & -3 \\ 1 & -1 & 4 & 9 \end{bmatrix}$.
3. **FIND THE ERROR** Hiroshi and Estrella are finding $-5\begin{bmatrix} -1 & 3 \\ -2 & 5 \end{bmatrix}$.

Hiroshi

$$-5\begin{bmatrix} -1 & 3 \\ -2 & 5 \end{bmatrix} = \begin{bmatrix} 5 & 3 \\ 10 & 5 \end{bmatrix}$$

Estrella

$$-5\begin{bmatrix} -1 & 3 \\ -2 & 5 \end{bmatrix} = \begin{bmatrix} 5 & -15 \\ 10 & -25 \end{bmatrix}$$

Who is correct? Explain your reasoning.

Guided Practice

State the dimensions of each matrix. Then, identify the position of the circled element in each matrix.

4. $\begin{bmatrix} 4 & \textcircled{0} & 2 \\ 5 & -1 & -3 \\ 6 & 2 & 7 \end{bmatrix}$

5. $[\textcircled{3} \quad -3 \quad 1 \quad 9]$

6. $\begin{bmatrix} 5 \\ 2 \\ \textcircled{1} \\ -3 \end{bmatrix}$

7. $\begin{bmatrix} 0.6 & \textcircled{4.2} \\ -1.7 & 1.05 \\ 0.625 & -2.1 \end{bmatrix}$

If $A = \begin{bmatrix} 20 & -10 \\ 12 & 19 \end{bmatrix}$, $B = \begin{bmatrix} 15 & 14 \\ -10 & 6 \end{bmatrix}$, and $C = [-5 \quad 7]$, find each sum, difference, or product. If the sum or difference does not exist, write *impossible*.

8. $A + C$
9. $B - A$
10. $2A$
11. $-4C$

Application

PIZZA SALES For Exercises 12–16, use the following tables that list the number of pizzas sold at Sylvia's Pizza one weekend.

Study Tip

Reading Math

The data for Exercise 12–16 could be classified as **bivariate,** because there are two variables, size and thickness of the crust.

FRIDAY	Small	Medium	Large
Thin Crust	12	10	3
Thick Crust	11	8	8
Deep Dish	14	8	10

SATURDAY	Small	Medium	Large
Thin Crust	13	12	11
Thick Crust	1	5	10
Deep Dish	8	11	2

SUNDAY	Small	Medium	Large
Thin Crust	11	8	6
Thick Crust	1	8	11
Deep Dish	10	15	11

12. Create a matrix for each day's data. Name the matrices F, R, and N, respectively.
13. Does F equal R? Explain.
14. Create matrix T to represent $F + R + N$.
15. What does T represent?
16. Which type of pizza had the most sales during the entire weekend?

Practice and Apply

Homework Help

For Exercises	See Examples
17–26	1
27–38	2–4
39–48	3

Extra Practice

See page 849.

State the dimensions of each matrix. Then, identify the position of the circled element in each matrix.

17. $\begin{bmatrix} \textcircled{2} & 1 \\ 5 & -8 \end{bmatrix}$

18. $\begin{bmatrix} -36 & 3 \\ \textcircled{25} & -1 \\ 11 & 14 \end{bmatrix}$

19. $\begin{bmatrix} 1 \\ 0 \\ \textcircled{-1} \end{bmatrix}$

20. $\begin{bmatrix} -3 & 56 & -21 \\ 60 & \textcircled{112} & -65 \end{bmatrix}$

21. $\begin{bmatrix} -4 & 0 & -2 \\ 5 & 1 & \textcircled{12} \\ -6 & 3 & -7 \end{bmatrix}$

22. $\begin{bmatrix} 1 & -2 \\ 3 & 4 \\ 1 & 5 \\ \textcircled{-1} & 7 \end{bmatrix}$

23. $\begin{bmatrix} -5 & 3 & 1 \\ 4 & 0 & \textcircled{2} \end{bmatrix}$

24. $\begin{bmatrix} -6 & 3 \\ \textcircled{5} & -4 \end{bmatrix}$

25. Create a 2-by-3 matrix with 2 in the first row and first column and 5 in the second row and second column. The rest of the elements should be ones.

26. Create a 3-by-2 matrix with 8 in the second row and second column and 4 in the third row and second column. The rest of the elements should be zeros.

If $A = \begin{bmatrix} -1 & 5 & 9 \\ 0 & -4 & -2 \\ 3 & 7 & 6 \end{bmatrix}$, $B = \begin{bmatrix} -12 & 7 & -16 \\ 5 & 10 & 13 \\ 20 & 11 & 8 \end{bmatrix}$, $C = \begin{bmatrix} 34 & 91 & 63 \\ 81 & 79 & 60 \end{bmatrix}$, **and** $D = \begin{bmatrix} -52 & 9 & 70 \\ -49 & -8 & 45 \end{bmatrix}$, **find each sum, difference, or product. If the sum or difference does not exist, write *impossible*.**

27. $A + B$	28. $C + D$	29. $C - D$	30. $B - A$
31. $5A$	32. $2C$	33. $A + C$	34. $B + D$
35. $2B + A$	36. $4A - B$	37. $2C - 3D$	38. $5D + 2C$

FOOD **For Exercises 39–41, use the table that shows the nutritional value of food.**

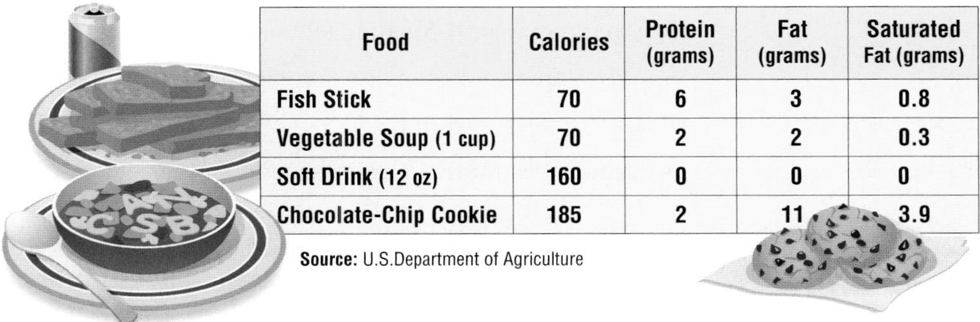

Food	Calories	Protein (grams)	Fat (grams)	Saturated Fat (grams)
Fish Stick	70	6	3	0.8
Vegetable Soup (1 cup)	70	2	2	0.3
Soft Drink (12 oz)	160	0	0	0
Chocolate-Chip Cookie	185	2	11	3.9

Source: U.S.Department of Agriculture

39. If $F = [70 \quad 6 \quad 3 \quad 0.8]$ is a matrix representing the nutritional value of a fish stick, create matrices V, S, and C to represent vegetable soup, soft drink, and chocolate chip cookie, respectively.

40. Suppose Lakeisha has two fish sticks for lunch. Write a matrix representing the nutritional value of the fish sticks.

41. Suppose Lakeisha has two fish sticks, a cup of vegetable soup, a 12-ounce soft drink, and a chocolate chip cookie. Write a matrix representing the nutritional value of her lunch.

FUND-RAISING **For Exercises 42–44, use the table that shows the last year's sales of T-shirts for the student council fund-raiser.**

Color	XS	S	M	L	XL
Red	18	28	32	24	21
White	24	30	45	47	25
Blue	17	19	26	30	28

42. Create a matrix to show the number of T-shirts sold last year according to size and color. Label this matrix N.

43. The student council anticipates a 20% increase in T-shirt sales this year. What value of the scalar r should be used so that rN results in a matrix that estimates the number of each size and color T-shirts needed this year?

44. Calculate rN, rounding appropriately, to show estimates for this year's sales.

FOOTBALL **For Exercises 45–48, use the table that shows the passing performance of four National Football League quarterbacks.**

1999 Regular Season

Quarterback	Attempts	Completions	Passing Yards	Touchdowns	Interceptions
Peyton Manning	533	331	4135	26	15
Rich Gannon	515	304	3840	24	14
Kurt Warner	499	325	4353	41	13
Steve Beuerlein	571	343	4436	36	15

2000 Regular Season

Quarterback	Attempts	Completions	Passing Yards	Touchdowns	Interceptions
Peyton Manning	571	357	4413	33	15
Rich Gannon	473	284	3430	28	11
Kurt Warner	347	235	3429	21	18
Steve Beuerlein	533	324	3730	19	18

Source: ESPN

45. Create matrix A for the 1999 data and matrix B for the 2000 data.

46. What are the dimensions of each matrix in Exercise 45?

47. Calculate $T = A + B$.

48. What does matrix T represent?

49. **CRITICAL THINKING** Suppose M and N are each 3-by-3 matrices. Determine whether each statement is *sometimes*, *always*, or *never* true.

a. $M = N$

b. $M + N = N + M$

c. $M - N = N - M$

d. $5M = M$

e. $M + N = M$

f. $5M = N$

50. **WRITING IN MATH** Answer the question that was posed at the beginning of the lesson.

How are matrices used to organize data?

Include the following in your answer:

- a comparison of a table and a matrix, and
- description of some real-world data that could be organized in a matrix.

51. Which of the following is equal to $\begin{bmatrix} 3 & 4 & 5 \\ -6 & -1 & 8 \end{bmatrix}$?

Ⓐ $\begin{bmatrix} -1 & 8 & 3 \\ -4 & 0 & 5 \end{bmatrix} + \begin{bmatrix} 4 & -4 & 2 \\ 2 & -1 & -2 \end{bmatrix}$

Ⓑ $\begin{bmatrix} 7 & -1 & 2 \\ 3 & 4 & -5 \end{bmatrix} + \begin{bmatrix} -4 & -3 & 3 \\ -3 & -5 & -3 \end{bmatrix}$

Ⓒ $\begin{bmatrix} 1 & -3 & 5 \\ 7 & -2 & 0 \end{bmatrix} + \begin{bmatrix} 2 & 7 & 0 \\ -13 & 1 & 8 \end{bmatrix}$

Ⓓ $\begin{bmatrix} 5 & 9 & -2 \\ 3 & 7 & 5 \end{bmatrix} + \begin{bmatrix} -2 & -5 & -3 \\ 3 & -8 & 3 \end{bmatrix}$

52. Suppose M and N are each 2-by-2 matrices. If $M + N = M$, which of the following is true?

Ⓐ $N = \begin{bmatrix} 1 & 1 \\ 1 & 1 \end{bmatrix}$

Ⓑ $N = \begin{bmatrix} 0 & 0 \\ 0 & 0 \end{bmatrix}$

Ⓒ $N = \begin{bmatrix} 1 & 0 \\ 0 & 1 \end{bmatrix}$

Ⓓ $N = \begin{bmatrix} 0 & 1 \\ 1 & 0 \end{bmatrix}$

Graphing Calculator

MATRIX OPERATIONS **You can use a graphing calculator to perform matrix operations. Use the EDIT command on the MATRX menu of a TI-83 Plus to enter each of the following matrices.**

$$A = \begin{bmatrix} 7.9 & 5.4 & -6.8 \\ -5.9 & 4.4 & -7.7 \end{bmatrix}, B = \begin{bmatrix} -7.2 & -5.8 & 9.1 \\ 4.3 & -8.4 & 5.3 \end{bmatrix}, C = \begin{bmatrix} 9.8 & -1.2 & 5.2 \\ -7.8 & 5.1 & -9.0 \end{bmatrix}$$

Use these stored matrices to find each sum, difference, or product.

53. $A + B$ **54.** $C - B$ **55.** $B + C - A$ **56.** $1.8A$ **57.** $0.4C$

Maintain Your Skills

Mixed Review

PRINTING **For Exercises 58 and 59, use the following information.**
To determine the quality of calendars printed at a local shop, the last 10 calendars printed each day are examined. *(Lesson 13-1)*

58. Identify the sample.

59. State whether it is *unbiased* (random) or *biased*. If unbiased, classify the sample as *simple*, *stratified*, or *systematic*. If biased, classify as *convenience* or *voluntary response*.

Solve each equation. *(Lesson 12-9)*

60. $\frac{-4}{a+1} + \frac{3}{a} = 1$ **61.** $\frac{3}{x} + \frac{4x}{x-3} = 4$ **62.** $\frac{d+3}{d+5} + \frac{2}{d-9} = \frac{5}{2d+10}$

Find the *n*th term of each geometric sequence. *(Lesson 10-7)*

63. $a_1 = 4, n = 5, r = 3$ **64.** $a_1 = -2, n = 3, r = 7$ **65.** $a_1 = 4, n = 5, r = -2$

Factor each trinomial, if possible. If the trinomial cannot be factored using integers, write *prime*. *(Lesson 9-3)*

66. $b^2 + 7b + 12$ **67.** $a^2 + 2ab - 3b^2$ **68.** $d^2 + 8d - 15$

Getting Ready for the Next Lesson

PREREQUISITE SKILL **For Exercises 69 and 70, use the graph that shows the amount of money in Megan's savings account.** *(To review* ***interpreting graphs****, see Lesson 1-9.)*

69. Describe what is happening to Megan's bank balance. Give possible reasons why the graph rises and falls at particular points.

70. Describe the elements in the domain and range.

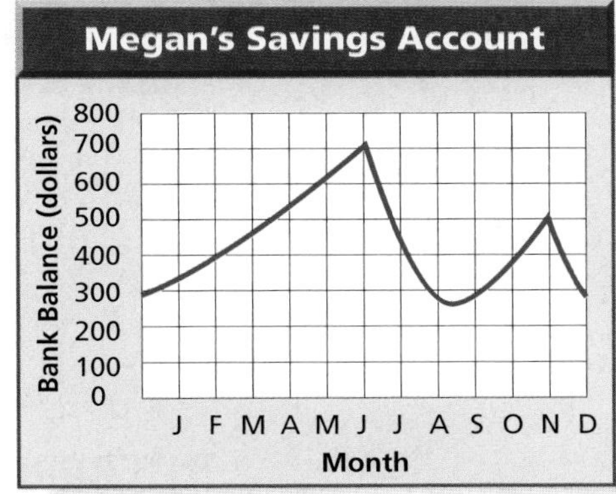

Practice Quiz 1

Lessons 13-1 and 13-2

Identify each sample, suggest a population from which it was selected, and state whether it is *unbiased* (random) or *biased*. If unbiased, classify the sample as *simple*, *stratified*, or *systematic*. If biased, classify as *convenience* or *voluntary response*. *(Lesson 13-1)*

1. Every other household in a neighborhood is surveyed to determine how to improve the neighborhood park.

2. Every other household in a neighborhood is surveyed to determine the favorite candidate for the state's governor.

Find each sum, difference, or product. *(Lesson 13-2)*

3. $\begin{bmatrix} -8 & 3 \\ -4 & -9 \end{bmatrix} + \begin{bmatrix} 5 & -7 \\ -1 & 0 \end{bmatrix}$ **4.** $\begin{bmatrix} -9 & 6 & 4 \\ -1 & 3 & 2 \end{bmatrix} - \begin{bmatrix} 7 & -2 & 8 \\ 5 & -3 & 1 \end{bmatrix}$ **5.** $3\begin{bmatrix} 8 & -3 & -4 & 5 \\ 6 & -1 & 2 & 10 \end{bmatrix}$

13-3 Histograms

What You'll Learn

- Interpret data displayed in histograms.
- Display data in histograms.

Vocabulary

- frequency table
- histogram
- measurement classes
- frequency

How are histograms used to display data?

A **frequency table** shows the frequency of events. The frequency table below shows the number of states with the mean SAT verbal and mathematics scores in each score interval. The data are from the 1999–2000 school year.

SAT Scores		
Score Interval	Verbal	Mathematics
	Number of States	Number of States
$480 \le s < 500$	11	5
$500 \le s < 520$	10	18
$520 \le s < 540$	6	5
$540 \le s < 560$	8	10
$560 \le s < 580$	10	5
$580 \le s < 600$	5	5
$600 \le s < 620$	0	2

Source: The College Board

The distribution of the mean scores on the SAT verbal exam is displayed in the graph.

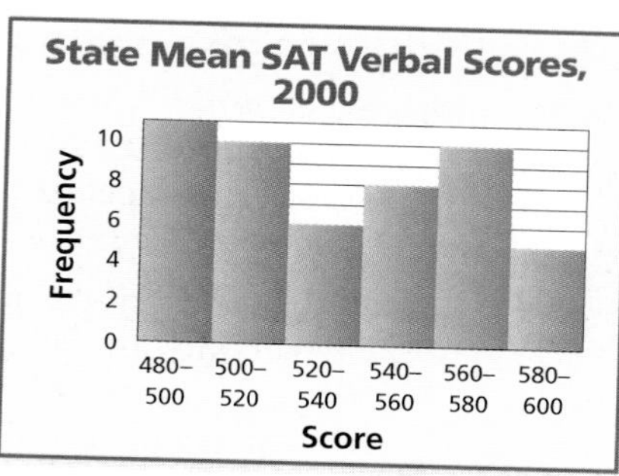

INTERPRET DATA IN HISTOGRAMS The graph above is called a histogram. A **histogram** is a bar graph in which the data are organized into equal intervals. In the histogram above, the horizontal axis shows the range of data values separated into **measurement classes**, and the vertical axis shows the number of values, or the **frequency**, in each class. Consider the histogram shown below.

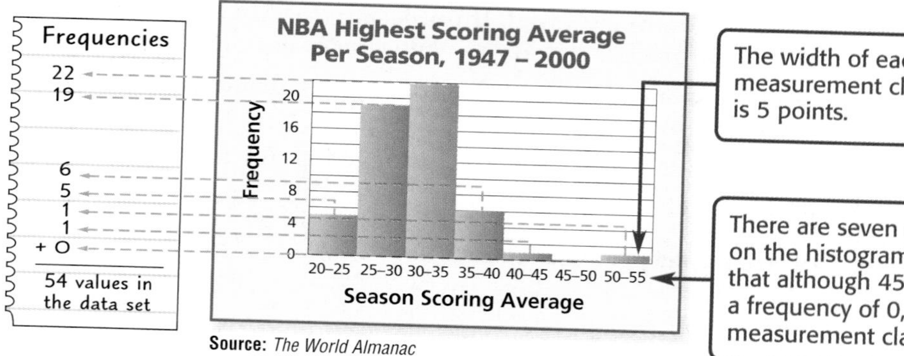

Source: *The World Almanac*

A histogram is a visual summary of a frequency table.

Example 1 Determine Information from a Histogram

GEOGRAPHY **Answer each question about the histogram shown below.**

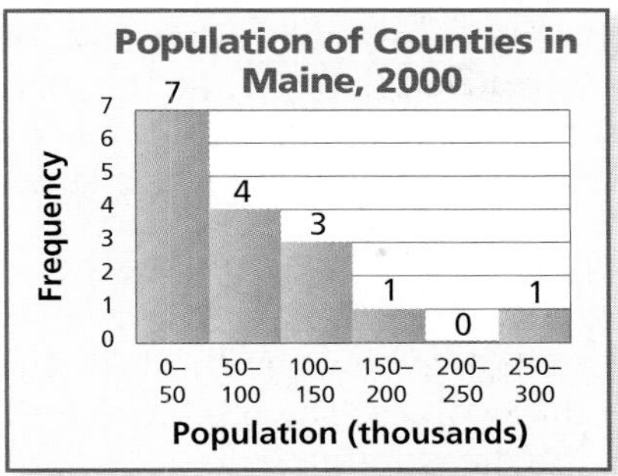

Study Tip

Look Back
To review **median**, see pages 818 and 819.

a. In what measurement class does the median occur?

First, add the frequencies to determine the number of counties in Maine.

$7 + 4 + 3 + 1 + 0 + 1 = 16$

There are 16 counties, so the middle data value is between the 8th and 9th data values. Both the 8th and 9th data values are located in the 50–100 thousand measurement class. Therefore, the median occurs in the 50–100 thousand measurement class.

b. Describe the distribution of the data.

- Only two counties have populations above 150 thousand. It is likely that these counties contain the largest cities in Maine.
- There is a gap in the 200–250 thousand measurement class.
- Most of the counties have populations below 150 thousand.
- As population increases, the histogram shows that the number of counties decreases. We say that the distribution is *skewed to the right*; or *skewed in the direction of the tail.* The majority of the data values cluster at the lower end of the distribution, the "tail" is to the right.

You can sometimes use the appearances of histograms to compare data.

Standardized Test Practice A B C D

Example 2 Compare Data in Histograms

Multiple-Choice Test Item

Which group of students has a greater median height?

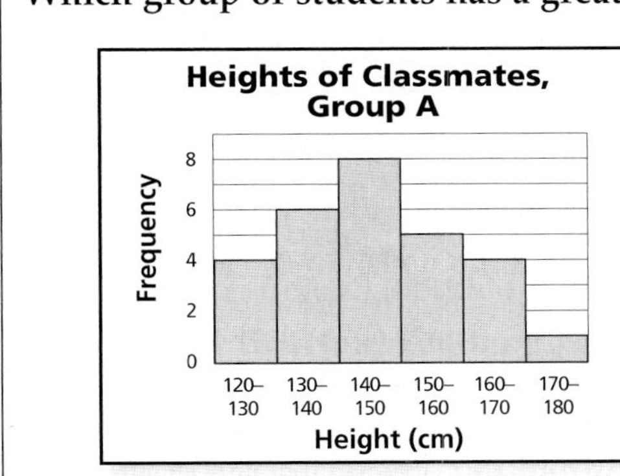

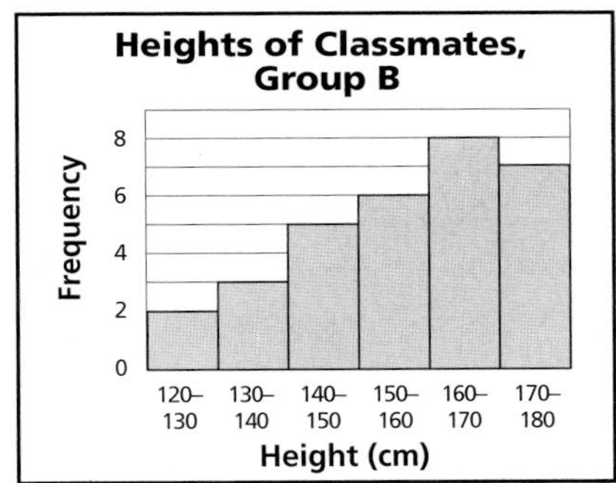

Ⓐ Group A

Ⓑ Group B

Ⓒ The medians are about the same.

Ⓓ cannot be determined

(continued on the next page)

Test-Taking Tip

When answering a test question involving a graph, always read the labels on the graph carefully.

Read the Test Item

You have two histograms depicting the heights of two groups of students. You are asked to determine which group of students has a greater median height.

Solve the Test Item

Study the histograms carefully. The measurement classes and the frequency scales are the same for each histogram. The distribution for Group A is somewhat *symmetrical* in shape, while the distribution for Group B is *skewed to the left*. This would indicate that Group B has the greater median height. To check this assumption, locate the measurement class of each median.

Group A

$4 + 6 + 8 + 5 + 4 + 1 = 28$

The median is between the 14th and 15th data values. The median is in the 140–150 measurement class.

Group B

$2 + 3 + 5 + 6 + 8 + 7 = 31$

The median is the 16th data value. The median is in the 150–160 measurement class.

This confirms that Group B has the greater median height. The answer is B.

DISPLAY DATA IN A HISTOGRAM Data from a list or a frequency table can be used to create a histogram.

Example 3 Create a Histogram

SCHOOL Create a histogram to represent the following scores for a 50-point mathematics test.

40, 34, 38, 23, 41, 39, 39, 34, 43, 44, 32, 44, 41, 39, 22, 47, 36, 25, 41, 30, 28, 37, 39, 33, 30, 40, 28

Step 1 Identify the greatest and least values in the data set.
The test scores range from 22 to 47 points.

Step 2 Create measurement classes of equal width.
For these data, use measurement classes from 20 to 50 with a 5-point interval for each class.

Step 3 Create a frequency table using the measurement classes.

Score Intervals	Tally	Frequency
$20 \leq s < 25$	II	2
$25 \leq s < 30$	III	3
$30 \leq s < 35$	~~IIII~~ I	6
$35 \leq s < 40$	~~IIII~~ II	7
$40 \leq s < 45$	~~IIII~~ III	8
$45 \leq s < 50$	I	1

Step 4 Draw the histogram.
Use the measurement classes to determine the scale for the horizontal axis and the frequency values to determine the scale for the vertical axis. For each measurement class, draw a rectangle as wide as the measurement class and as tall as the frequency for the class. Label the axes and include a descriptive title for the histogram.

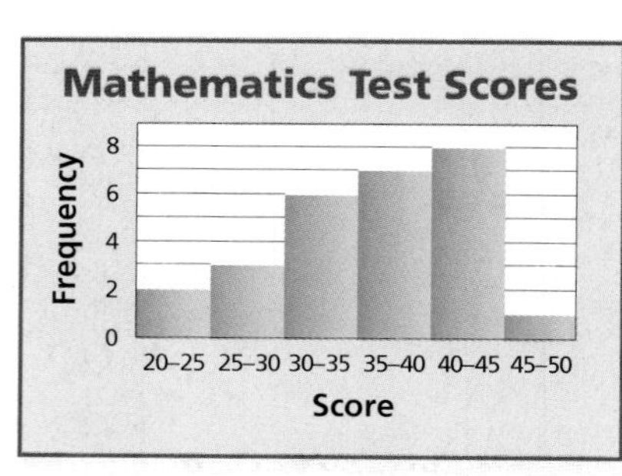

Check for Understanding

Concept Check

1. **Describe** how to create a histogram.
2. **Write** a compound inequality to represent all of the values v included in a 50–60 measurement class.
3. **OPEN ENDED** Write a set of data whose histogram would be skewed to the right.

Guided Practice

MONEY For Exercises 4 and 5, use the following histogram that shows the amount of money spent by several families during a holiday weekend.

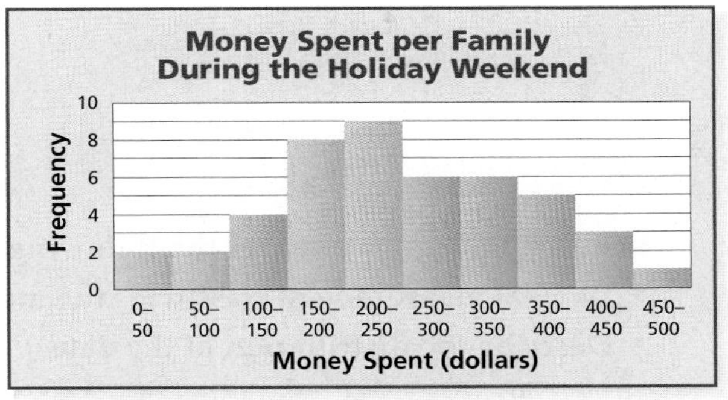

4. In what measurement class does the median occur?
5. Describe the distribution of the data.

SCHOOL For Exercises 6 and 7, use the following histograms.

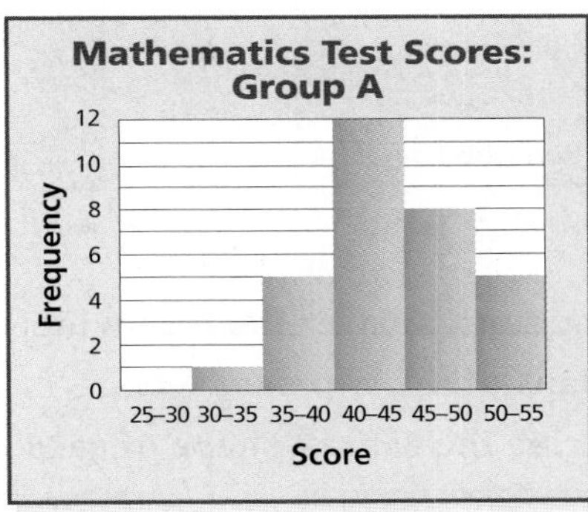

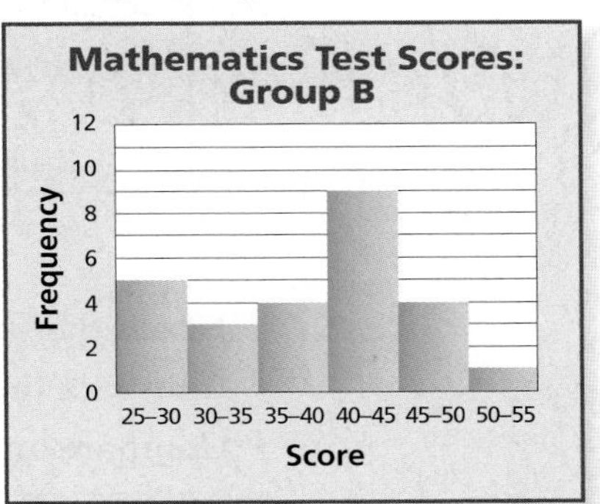

6. Compare the medians of the two data sets.
7. Compare and describe the overall shape of each distribution of data.
8. **AIR TRAVEL** The busiest U.S. airports as determined by the number of passengers arriving and departing are listed below. Create a histogram.

Passenger Traffic at U.S. Airports, 2000			
Airport	Passengers (millions)	Airport	Passengers (millions)
Atlanta (Hartsfield)	80	Minneapolis/St. Paul	37
Chicago (O'Hare)	72	Phoenix (Sky Harbor)	36
Los Angeles	68	Detroit	36
Dallas/Fort Worth	61	Houston (George Bush)	35
San Francisco	41	Newark	34
Denver	39	Miami	34
Las Vegas (McCarran)	37	New York (JFK)	33

Source: Airports Council International

Online Research **Data Update** What are the current busiest airports? Visit www.algebra1.com/data_update to get statistics on airports.

9. Which statement about the graph at the right is *not* correct?

(A) The data are skewed to the right.

(B) The median is in the 40–50 thousand measurement class.

(C) There are 32 employees represented by the graph.

(D) The width of each measurement class is $10 thousand.

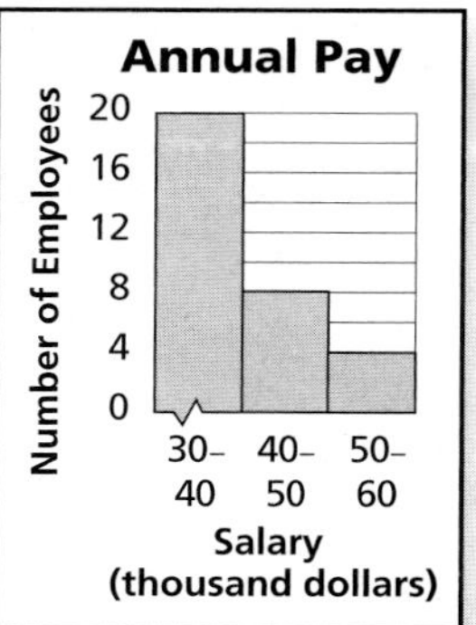

Practice and Apply

Homework Help

For Exercises	See Examples
10, 11	1
12, 13	2
14–20	3

Extra Practice

See page 850.

For each histogram, answer the following.

- **In what measurement class does the median occur?**
- **Describe the distribution of the data.**

10. 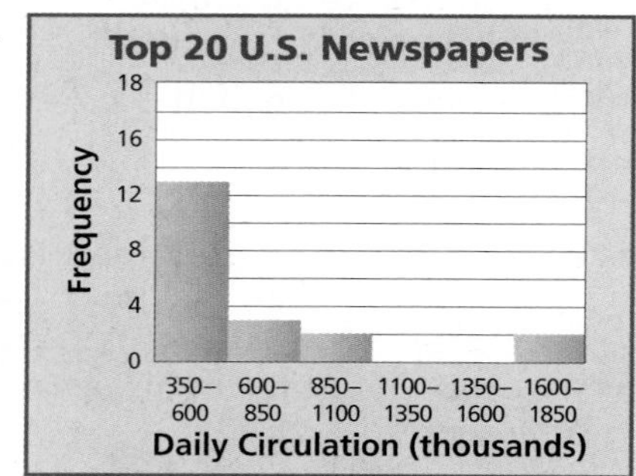

Source: *Editor & Publisher International Yearbook*

11. 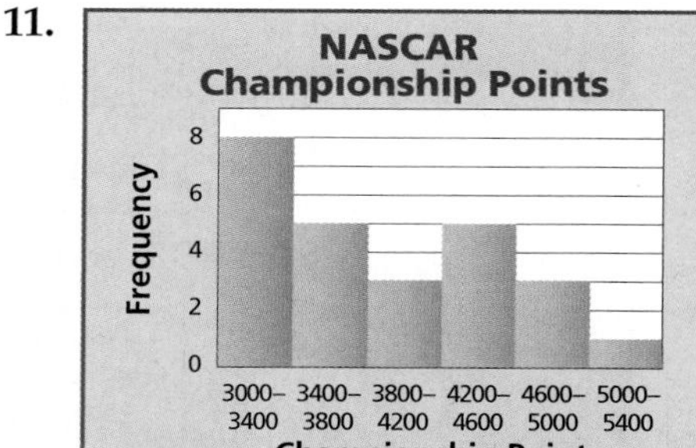

Source: *USA TODAY*

For each pair of histograms, answer the following.

- **Compare the medians of the two data sets.**
- **Compare and describe the overall shape of each distribution of data.**

12. 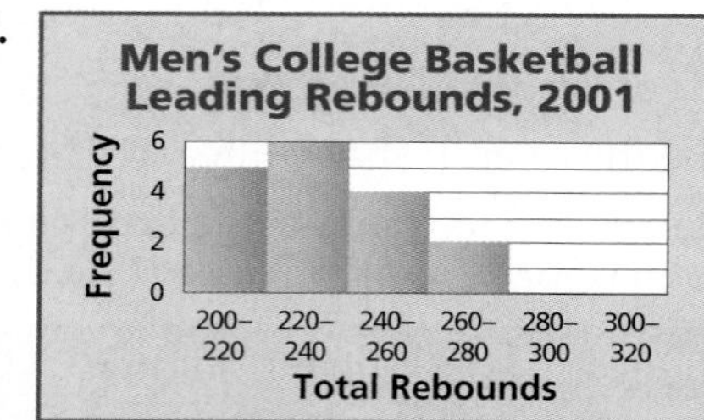

Source: *USA TODAY*

Source: *USA TODAY*

13. 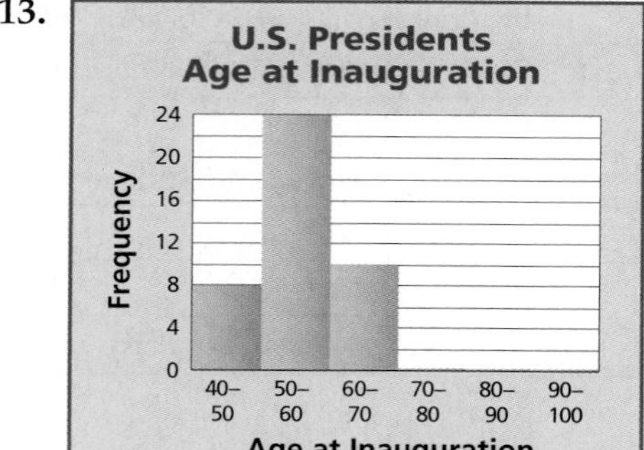

Source: *The World Almanac*

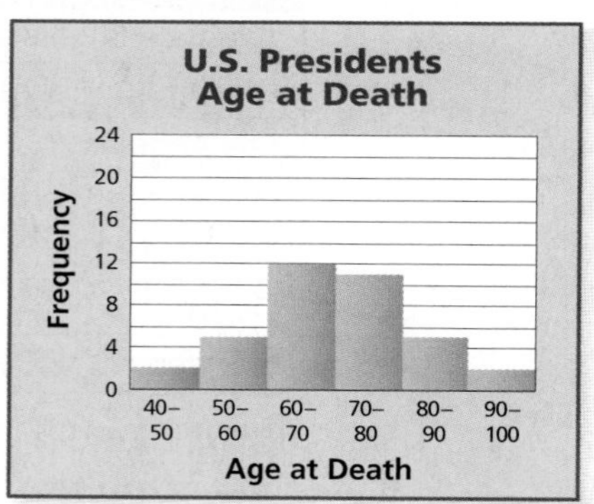

Source: *The World Almanac*

Create a histogram to represent each data set.

14. Students' semester averages in a mathematics class: 96.53, 95.96, 94.25, 93.58, 91.91, 90.33, 90.27, 90.11, 89.30, 89.06, 88.33, 88.30, 87.43, 86.67, 86.31, 84.21, 83.53, 82.30, 78.71, 77.51, 73.83

15. Number of raisins found in a snack-size box: 54, 59, 55, 109, 97, 59, 102, 68, 104, 63, 101, 59, 59, 96, 58, 57, 63, 57, 94, 61, 104, 62, 58, 59, 102, 60, 54, 58, 53, 78

More About. . .

Baseball

The New York Yankees won the 2000 World Series and had the largest payroll of all major league teams that year.

Source: *USA TODAY*

BASEBALL **For Exercises 16 and 17, use the following table.**

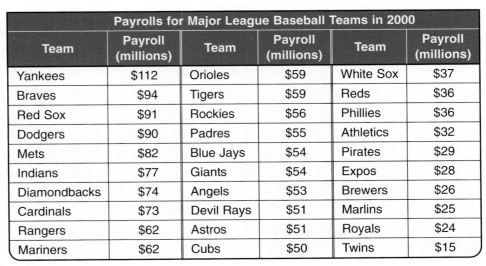

Payrolls for Major League Baseball Teams in 2000

Team	Payroll (millions)	Team	Payroll (millions)	Team	Payroll (millions)
Yankees	$112	Orioles	$59	White Sox	$37
Braves	$94	Tigers	$59	Reds	$36
Red Sox	$91	Rockies	$56	Phillies	$36
Dodgers	$90	Padres	$55	Athletics	$32
Mets	$82	Blue Jays	$54	Pirates	$29
Indians	$77	Giants	$54	Expos	$28
Diamondbacks	$74	Angels	$53	Brewers	$26
Cardinals	$73	Devil Rays	$51	Marlins	$25
Rangers	$62	Astros	$51	Royals	$24
Mariners	$62	Cubs	$50	Twins	$15

Source: *USA TODAY*

16. Create a histogram to represent the payroll data.

17. On your histogram, locate and label the median team payroll.

ELECTIONS **For Exercises 18–20, use the following table.**

Percent of Eligible Voters Who Voted in the 2000 Presidential Election

State	Percent	State	Percent	State	Percent	State	Percent	State	Percent
MN	68.75	WY	59.70	OH	55.76	MD	51.56	AR	47.79
ME	67.34	CT	58.40	ID	54.46	NJ	51.04	NM	47.40
AK	66.41	SD	58.24	RI	54.29	FL	50.65	SC	46.49
WI	66.07	MI	57.52	LA	54.24	NC	50.28	WV	45.74
VT	63.98	MO	57.49	KS	54.07	AL	49.99	CA	44.09
NH	62.33	WA	56.95	PA	53.66	IN	49.44	GA	43.84
MT	61.52	MA	56.92	IL	52.79	NY	49.42	NV	43.81
IA	60.71	CO	56.78	UT	52.61	TN	49.19	TX	43.15
OR	60.63	NE	56.44	VA	52.05	OK	48.76	AZ	42.26
ND	60.63	DE	56.22	KY	51.59	MS	48.57	HI	40.48

Source: *USA TODAY*

18. Determine the median of the data.

19. Create a histogram to represent the data.

20. Write a sentence or two describing the distribution of the data.

21. **RESEARCH** Choose your favorite professional sport. Use the Internet or other reference to find how many games each team in the appropriate league won last season. Use this information to create a histogram. Describe your histogram.

22. **CRITICAL THINKING** Create a histogram with a gap between 20 and 40, one item in the 50–55 measurement class, and the median value in the 50–55 measurement class.

23. WRITING IN MATH Answer the question that was posed at the beginning of the lesson.

How are histograms used to display data?

Include the following in your answer:

- the advantage of the histogram over the frequency table, and
- a histogram depicting the distribution of the mean scores on the SAT mathematics exam.

For Exercises 24 and 25, use the information in the graph.

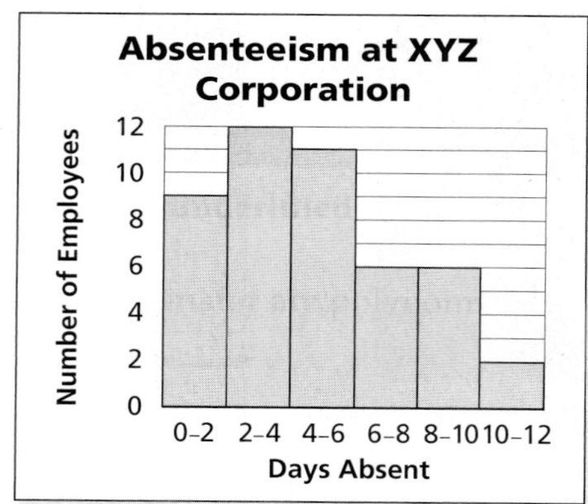

24. How many employees are represented in the graph?

(A) 38 (B) 40

(C) 46 (D) 48

25. In which measurement class is the median of the data located?

(A) 2–4 (B) 4–6

(C) 6–8 (D) 8–10

HISTOGRAMS You can use a graphing calculator to create histograms. On a TI-83 Plus, enter the data in L1. In the STAT PLOT menu, turn on Plot 1 and select the histogram. Define the viewing window and press GRAPH.

Use a graphing calculator to create a histogram for each set of data.

26. 5, 5, 6, 7, 9, 4, 10, 12, 13, 8, 15, 16, 13, 8

27. 12, 14, 25, 30, 11, 35, 41, 47, 13, 18, 58, 59, 42, 13, 18

28. 124, 83, 81, 130, 111, 92, 178, 179, 134, 92, 133, 145, 180, 144

29. 2.2, 2.4, 7.5, 9.1, 3.4, 5.1, 6.3, 1.8, 2.8, 3.7, 8.6, 9.5, 3.6, 3.7, 5.0

Maintain Your Skills

Mixed Review

If $A = \begin{bmatrix} -2 & 3 & 7 \\ 0 & -4 & 6 \\ 1 & -5 & 4 \end{bmatrix}$, $B = \begin{bmatrix} -8 & 1 & -1 \\ 2 & 3 & -7 \end{bmatrix}$, **and** $C = \begin{bmatrix} 7 & -5 & 2 \\ 0 & 0 & 3 \\ -1 & 4 & 6 \end{bmatrix}$, **find each sum, difference, or product. If the sum or difference does not exist, write *impossible*.** *(Lesson 13-2)*

30. $A + B$

31. $C - A$

32. $2B$

33. $-5A$

34. MANUFACTURING Every 15 minutes, a CD player is taken off the assembly line and tested. State whether this sample is *unbiased* (random) or *biased*. If unbiased, classify the sample as *simple*, *stratified*, or *systematic*. If biased, classify as *convenience* or *voluntary response*. *(Lesson 13-1)*

Find each quotient. Assume that no denominator has a value of 0. *(Lesson 12-4)*

35. $\frac{s}{s+7} \div \frac{s-5}{s+7}$

36. $\frac{2m^2 + 7m - 15}{m + 2} \div \frac{2m - 3}{m^2 + 5m + 6}$

Solve each equation. Check your solution. *(Lesson 11-3)*

37. $\sqrt{y + 3} + 5 = 9$

38. $\sqrt{x - 2} = x - 4$

39. $13 = \sqrt{2w - 5}$

Getting Ready for the Next Lesson

PREREQUISITE SKILL **Find the median for each set of data.**

*(To review **median**, see pages 818 and 819.)*

40. 2, 4, 7, 9, 12, 15

41. 10, 3, 17, 1, 8, 6, 12, 15

42. 7, 19, 9, 4, 7, 2

43. 2.1, 7.4, 13.9, 1.6, 5.21, 3.901

Graphing Calculator Investigation

A Follow-Up of Lesson 13-3

Curve Fitting

If there is a constant increase or decrease in data values, there is a linear trend. If the values are increasing or decreasing more and more rapidly, there may be a quadratic or exponential trend. The curvature of a quadratic trend tends to appear more gradual. Below are three scatter plots, each showing a different trend.

Linear Trend

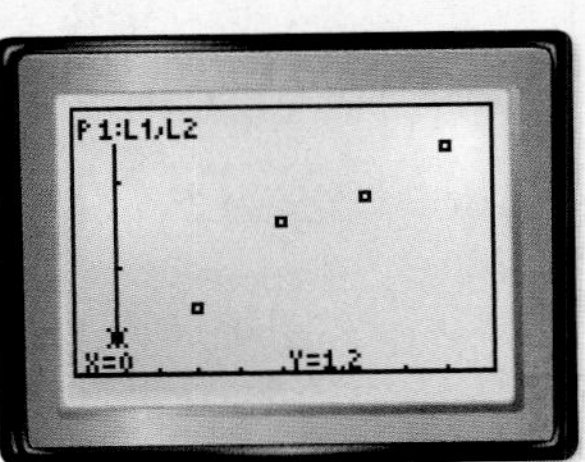

Quadratic Trend

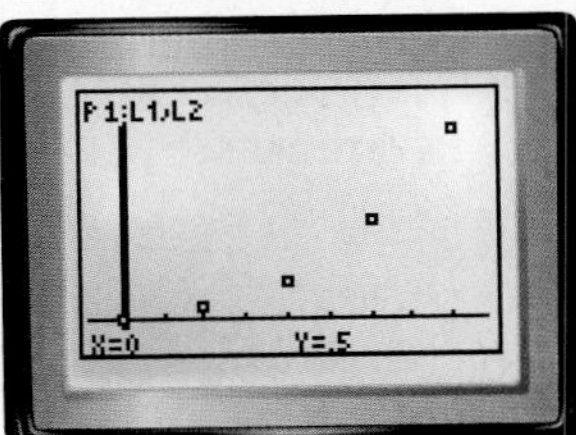

Exponential Trend

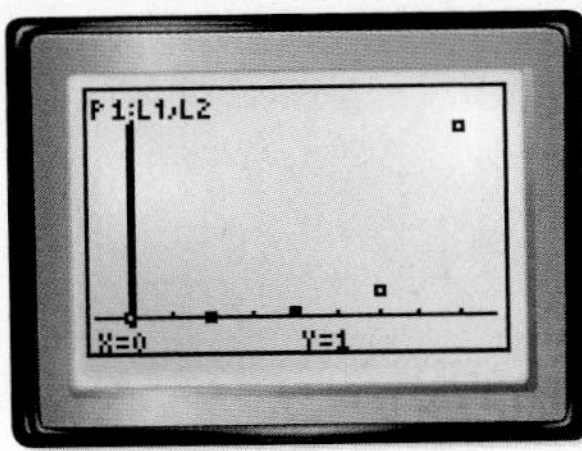

With a TI-83 Plus, you can use the LinReg, QuadReg, and ExpReg functions to find the appropriate regression equation that best fits the data.

FARMING **A study is conducted in which groups of 25 corn plants are given a different amount of fertilizer and the gain in height after a certain time is recorded. The table below shows the results.**

Fertilizer (mg)	0	20	40	60	80
Gain in Height (in.)	6.48	7.35	8.73	9.00	8.13

Step 1 *Make a scatter plot.*

- Enter the fertilizer in L1 and the height in L2.

  KEYSTROKES: *Review entering a list on page 204.*

- Use STAT PLOT to graph the scatter plot.

  KEYSTROKES: *Review statistical plots on page 204. Use* ZOOM *9 to graph.*

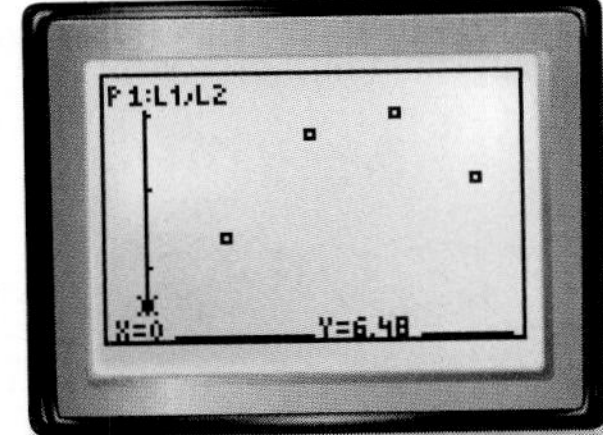

[−8, 88] scl: 5 by [6.0516, 9.4284] scl: 1

The graph appears to be a quadratic regression.

Step 2 *Find the quadratic regression equation.*

- Select QuadReg on the STAT CALC menu.

  KEYSTROKES: STAT ▶ 5 ENTER

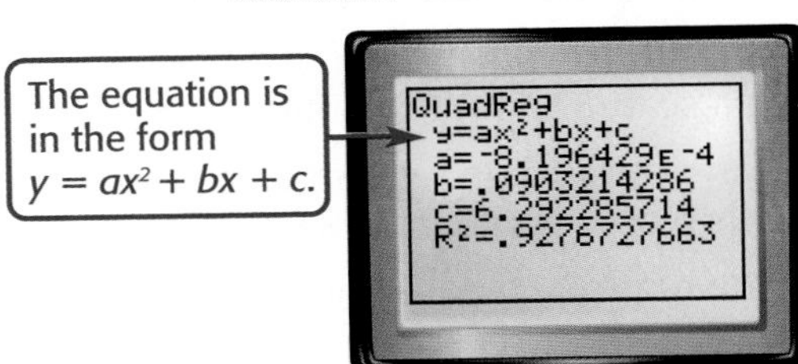

The equation is about $y = -0.0008x^2 + 0.1x + 6.3$.

R^2 is the **coefficient of determination**. The closer R^2 is to 1, the better the model. To choose a quadratic or exponential model, fit both and use the one with the R^2 value closer to 1.

www.algebra1.com/other_calculator_keystrokes

(continued on the next page)

Step 3 *Graph the quadratic regression equation.*

- Copy the equation to the Y= list and graph.

KEYSTROKES: [Y=] [VARS] 5 [▶] [▶] 1 [ZOOM] 9

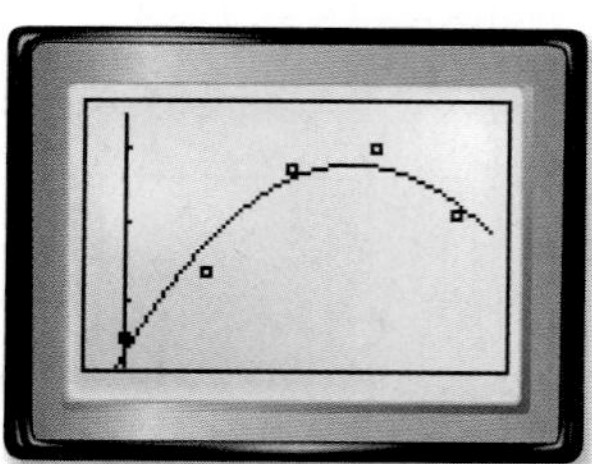

Step 4 *Predict using the equation.*

- Find the amount of fertilizer that produces the maximum gain in height.

On average, about 55 milligrams of the fertilizer produces the maximum gain.

KEYSTROKES: [2nd] [CALC] 4

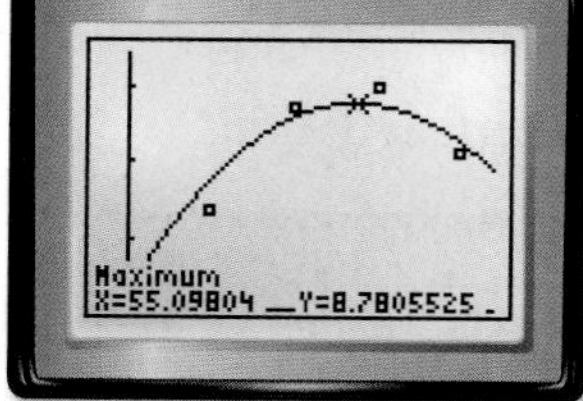
Maximum
X=55.09804 Y=8.7805525

Exercises

Plot each set of data points. Determine whether to use a *linear*, *quadratic*, or *exponential* regression equation. State the coefficient of determination.

1.

x	y
0.0	2.98
0.2	1.46
0.4	0.90
0.6	0.51
0.8	0.25
1.0	0.13

2.

x	y
1	25.9
2	22.2
3	20.0
4	19.3
5	18.2
6	15.9

3.

x	y
10	35
20	50
30	70
40	88
50	101
60	120

4.

x	y
1	3.67
3	5.33
5	6.33
7	5.67
9	4.33
11	2.67

TECHNOLOGY **The cost of cellular phone use is expected to decrease. For Exercises 5–9, use the graph at the right.**

5. Make a scatter plot of the data.

6. Find an appropriate regression equation, and state the coefficient of determination.

7. Use the regression equation to predict the expected cost in 2004.

8. Do you believe that your regression equation is appropriate for a year beyond the range of data, such as 2020? Explain.

9. What model may be more appropriate for predicting cost beyond 2003?

USA TODAY Snapshots®

USA TODAY

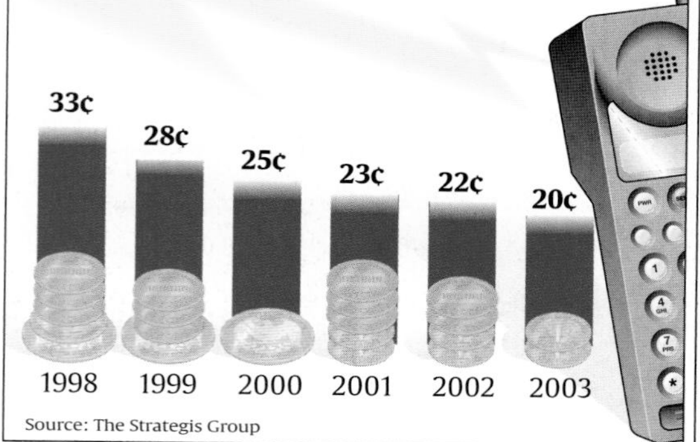
Cheaper wireless talk
Cheaper digital networks and more competition are expected to cut the cost of wireless phone use. Per-minute average in 1998 and projected cost in the next five years:
33¢
28¢
25¢
23¢
22¢
20¢
1998
1999
2000
2001
2002
2003
Source: The Strategis Group

By Anne R. Carey and Marcy E. Mullins, USA TODAY

13-4 Measures of Variation

What You'll Learn

- Find the range of a set of data.
- Find the quartiles and interquartile range of a set of data.

Vocabulary

- range
- measures of variation
- quartiles
- lower quartile
- upper quartile
- interquartile range
- outlier

How is variation used in weather?

The average monthly temperatures for three U.S. cities are given. Which city shows the greatest change in monthly highs?

To answer this question, find the difference between the greatest and least values in each data set.

Buffalo: $80.2 - 30.2 = 50.0$
Honolulu: $88.7 - 80.1 = 8.6$
Tampa: $90.2 - 69.8 = 20.4$

Buffalo shows the greatest change.

Average Monthly High Temperatures (°F)

Month	Buffalo	Honolulu	Tampa
January	30.2	80.1	69.8
February	31.6	80.5	71.4
March	41.7	81.6	76.6
April	54.2	82.8	81.7
May	66.1	84.7	87.2
June	75.3	86.5	89.5
July	80.2	87.5	90.2
August	77.9	88.7	90.2
September	70.8	88.5	89.0
October	59.4	86.9	84.3
November	47.1	84.1	77.7
December	35.3	81.2	72.1

Source: www.stormfax.com

RANGE The difference between the greatest and the least monthly high temperatures is called the **range** of the temperatures.

Key Concept — *Definition of Range*

The range of a set of data is the difference between the greatest and the least values of the set.

Study Tip

Look Back
To review **mean**, **median**, and **mode**, see pages 818 and 819.

The mean, median, and mode describe the central tendency of a set of data. The range of a set of data is a measure of the spread of the data. Measures that describe the spread of the values in a set of data are called **measures of variation**.

Example 1 Find the Range

HOCKEY **The number of wins for each team in the Eastern Conference of the NHL for the 1999–2000 season are listed below. Find the range of the data.**

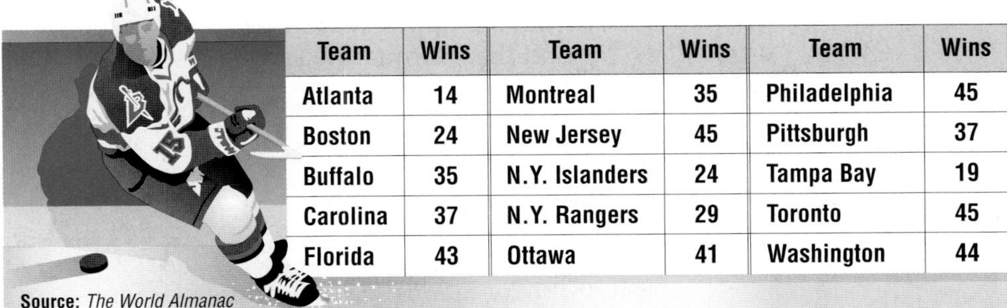

Team	Wins	Team	Wins	Team	Wins
Atlanta	14	Montreal	35	Philadelphia	45
Boston	24	New Jersey	45	Pittsburgh	37
Buffalo	35	N.Y. Islanders	24	Tampa Bay	19
Carolina	37	N.Y. Rangers	29	Toronto	45
Florida	43	Ottawa	41	Washington	44

Source: *The World Almanac*

The greatest number of wins is 45, and the least number of wins is 14. Since $45 - 14 = 31$, the range of the number of wins is 31.

Study Tip

Reading Math

The abbreviations LQ and UQ are often used to represent the lower quartile and upper quartile, respectively.

QUARTILES AND INTERQUARTILE RANGE In a set of data, the **quartiles** are values that separate the data into four equal subsets, each containing one fourth of the data. Statisticians often use Q_1, Q_2, and Q_3 to represent the three quartiles. Remember that the median separates the data into two equal parts. Q_2 is the median. Q_1 is the **lower quartile**. It divides the lower half of the data into two equal parts. Likewise Q_3 is the **upper quartile**. It divides the upper half of the data into two equal parts. The difference between the upper and lower quartiles is the **interquartile range** (IQR).

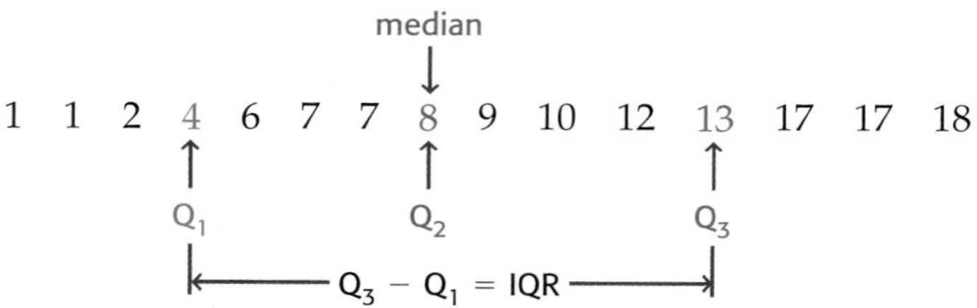

Key Concept *Definition of Interquartile Range*

The difference between the upper quartile and the lower quartile of a set of data is called the interquartile range. It represents the middle half, or 50%, of the data in the set.

Example 2 *Find the Quartiles and the Interquartile Range*

GEOGRAPHY **The areas of the original 13 states are listed in the table. Find the median, the lower quartile, the upper quartile, and the interquartile range of the areas.**

Explore You are given a table with the areas of the original 13 states. You are asked to find the median, the lower quartile, the upper quartile, and the interquartile range.

State	Area (thousand square miles)
Connecticut	6
Delaware	2
Georgia	59
Maryland	12
Massachusetts	11
New Hampshire	9
New Jersey	9
New York	54
North Carolina	54
Pennsylvania	46
Rhode Island	2
South Carolina	32
Virginia	43

Source: www.infoplease.com

Plan First, list the areas from least to greatest. Then find the median of the data. The median will divide the data into two sets of data. To find the upper and lower quartiles, find the median of each of these sets of data. Finally, subtract the lower quartile from the upper quartile to find the interquartile range.

Solve

median ↓

2 2 6 9 9 11 12 32 43 46 54 54 59

$Q_1 = \frac{6+9}{2}$ or 7.5 $\qquad Q_3 = \frac{46+54}{2}$ or 50

The median is 12 thousand square miles.

The lower quartile is 7.5 thousand square miles, and the upper quartile is 50 thousand square miles.

The interquartile range is 50 − 7.5 or 42.5 thousand square miles.

Examine Check to make sure that the numbers are listed in order. Since 7.5, 12, and 50 divide the data into four equal parts, the lower quartile, median, and upper quartile are correct.

In a set of data, a value that is much less or much greater than the rest of the data is called an **outlier**. An outlier is defined as any element of a set of data that is at least 1.5 interquartile ranges less than the lower quartile or greater than the upper quartile.

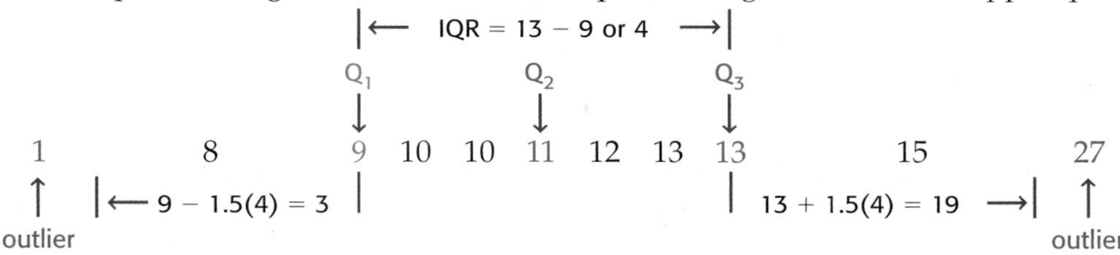

Example 3 Identify Outliers

Study Tip

Look Back
To review **stem-and-leaf plots**, see Lesson 2-5.

Identify any outliers in the following set of data.

Stem	Leaf
1	[2 2 7
2	3 [3 3] 4 4 5 6] [6 8 8 9
3	[0 1] 4 6
4	0 6] *1|2 = 12*

Step 1 **Find the quartiles.**

The brackets group the values in the lower half and the values in the upper half. The boxes are used to find the lower quartile and the upper quartile.

$Q_1 = \frac{23 + 23}{2}$ or 23 $\quad\quad$ $Q_3 = \frac{30 + 31}{2}$ or 30.5

Step 2 **Find the interquartile range.**

The interquartile range is $30.5 - 23$ or 7.5.

Step 3 **Find the outliers, if any.**

An outlier must be 1.5(7.5) less than the lower quartile, 23, or 1.5(7.5) greater than the upper quartile, 30.5.

$23 - 1.5(7.5) = 11.75$ $\quad\quad$ $30.5 + 1.5(7.5) = 41.75$

There are no values less than 11.75. Since $46 > 41.75$, 46 is the only outlier.

Check for Understanding

Concept Check

1. **OPEN ENDED** Find a counterexample for the following statement.
   If the range of data set 1 is greater than the range of data set 2, then the interquartile range of data set 1 will be greater than the interquartile range of data set 2.
2. **Describe** how the mean is affected by an outlier.
3. **FIND THE ERROR** Alonso and Sonia are finding the range of this set of data: 28, 30, 32, 36, 40, 41, 43.

Alonso	Sonia
43 − 28 = 15 The range is 15.	The range is all numbers between 28 and 43, inclusive.

Who is correct? Explain your reasoning.

Guided Practice **Find the range, median, lower quartile, upper quartile, and interquartile range of each set of data. Identify any outliers.**

4. 85, 77, 58, 69, 62, 73, 25, 82, 67, 77, 59, 75, 69, 76

5.

Stem	Leaf
7	3 7 8
8	0 0 3 5 7
9	4 6 8
10	0 1 8
11	1 9 *7|3 = 7.3*

Application **LITTLE LEAGUE For Exercises 6–10, use the following information.** The number of runs scored by the winning team in the Little League World Series each year from 1947 to 2000 are given in the line plot below.

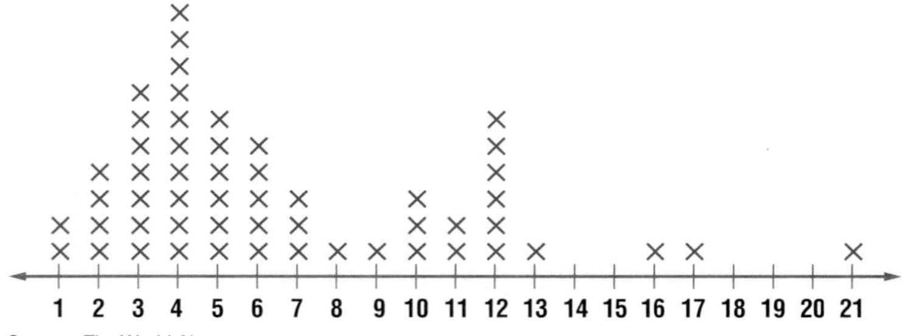

Source: *The World Almanac*

6. What is the range of the data?

7. What is the median of the data?

8. What is the lower quartile and upper quartile of the data?

9. What is the interquartile range of the data?

10. Name any outliers.

Practice and Apply

Homework Help

For Exercises	See Examples
11–18	1–3
19, 24, 29	1
20–22, 25–27, 30, 31	2
23, 28, 32	3

Extra Practice

See page 850.

Find the range, median, lower quartile, upper quartile, and interquartile range of each set of data. Identify any outliers.

11. 85, 77, 58, 69, 62, 73, 55, 82, 67, 77, 59, 92, 75

12. 28, 42, 37, 31, 34, 29, 44, 28, 38, 40, 39, 42, 30

13. 30.8, 29.9, 30.0, 31.0, 30.1, 30.5, 30.7, 31.0

14. 2, 3.4, 5.3, 3, 1, 3.2, 4.9, 2.3

15.

Stem	Leaf
5	3 6 8
6	5 8
7	0 3 7 7 9
8	1 4 8 8 9
9	9 *5|3 = 53*

16.

Stem	Leaf
19	3 5 5
20	2 2 5 8
21	5 8 8 9 9 9
22	0 1 7 8 9
23	2 *19|3 = 193*

17.

Stem	Leaf
5	0 3 7 9
6	1 3 4 5 5 6
7	1 5 6 6 9
8	1 2 3 5 8
9	2 5 6 9
10	
11	7 *5|0 = 5.0*

18.

Stem	Leaf
0	0 2 3
1	1 7 9
2	2 3 5 6
3	3 4 4 5 9
4	0 7 8 8
5	
6	8 *0|2 = 0.2*

NATIONAL PARKS **For Exercises 19–23, use the graph at the right.**

19. What is the range of the visitors per month?
20. What is the median number of visitors per month?
21. What are the lower quartile and the upper quartile of the data?
22. What is the interquartile range of the data?
23. Name any outliers.

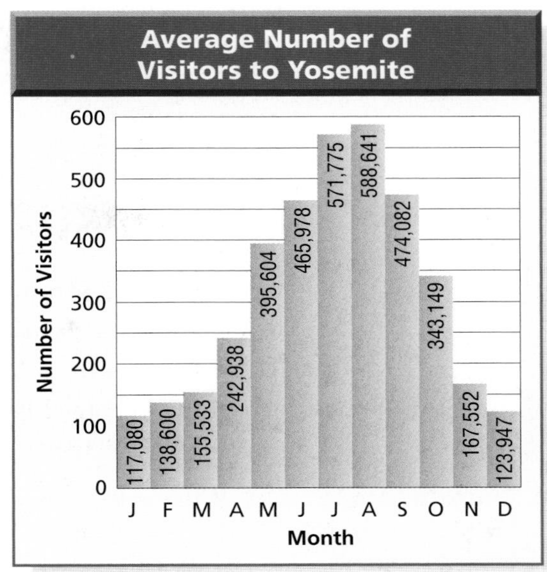

Source: *USA TODAY*

More About. . .

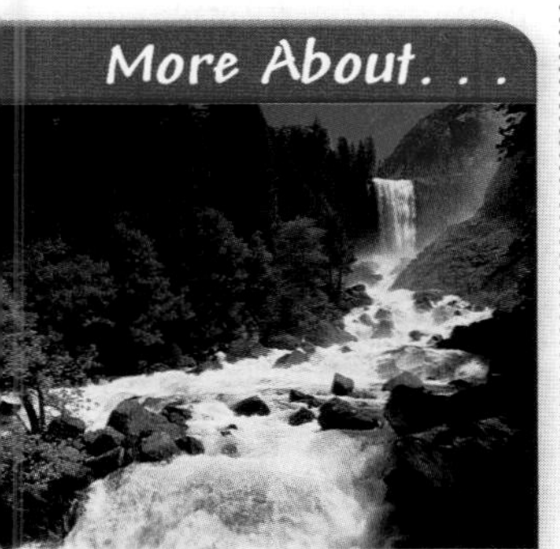

National Parks

Yosemite National Park boasts of sparkling lakes, mountain peaks, rushing streams, and beautiful waterfalls. It has about 700 miles of hiking trails.

Source: *World Book Encyclopedia*

NUTRITION **For Exercises 24–28, use the following table.**

Calories for One Serving of Vegetables					
Vegetable	Calories	Vegetable	Calories	Vegetable	Calories
Asparagus	14	Carrots	28	Lettuce	9
Avocado	304	Cauliflower	10	Onion	60
Bell pepper	20	Celery	17	Potato	89
Broccoli	25	Corn	66	Spinach	9
Brussels sprouts	60	Green beans	30	Tomato	35
Cabbage	17	Jalapeno peppers	13	Zucchini	17

Source: *Vitality*

24. What is the range of the data?
25. What is the median of the data?
26. What are the lower quartile and the upper quartile of the data?
27. What is the interquartile range of the data?
28. Identify any outliers.

BRIDGES **For Exercises 29–33, use the following information and the double stem-and-leaf plot at the right.** The main span of cable-stayed bridges and of steel-arch bridges in the United States are given in the stem-and-leaf plot.

29. Find the ranges for each type of bridge.
30. Find the quartiles for each type of bridge.
31. Find the interquartile ranges for each type of bridge.
32. Identify any outliers.
33. Compare the ranges and interquartile ranges of the two types of bridges. What can you conclude from these statistics?

Cable-Stayed	Stem	Steel-Arch
6 4 3	6	
9 8 6 5 1 1	7	3 8 8
0	8	0 0 2 3 4
5 2	9	0 1 1 8 8 9
8 2 0 0	10	0 3 8
2	11	0 0
8 2 0	12	0 6
5 2 0	13	
9 0	14	
	15	
3	16	5
3|6 = 630 feet	17	0 *7|3 = 730 feet*

Source: *The World Almanac*

34. **CRITICAL THINKING** Trey measured the length of each classroom in his school. He then calculated the range, median, lower quartile, upper quartile, and interquartile range of the data. After his calculations, he discovered that the tape measure he had used started at the 2-inch mark instead of at the 0-inch mark. All of his measurements were 2 inches greater than the actual lengths of the rooms. How will the values that Trey calculated change? Explain your reasoning.

35. **WRITING IN MATH** Answer the question that was posed at the beginning of the lesson.

How is variation used in weather?

Include the following in your answer:
- the meaning of the range and interquartile range of temperatures for a city, and
- the average highs for your community with the appropriate measures of variation.

36. What is the range of the following set of data?
53, 57, 62, 48, 45, 65, 40, 42, 55

Ⓐ 11 Ⓑ 25 Ⓒ 53 Ⓓ 65

37. What is the median of the following set of data?
7, 8, 14, 3, 2, 1, 24, 18, 9, 15

Ⓐ 8.5 Ⓑ 10.1 Ⓒ 11.5 Ⓓ 23

Maintain Your Skills

Mixed Review

38. Create a histogram to represent the following data. *(Lesson 13-3)*
36, 43, 61, 45, 37, 41, 32, 46, 60, 38, 35, 64, 46, 47, 30, 38, 48, 39

State the dimensions of each matrix. Then identify the position of the circled element in each matrix. *(Lesson 13-2)*

39. $\begin{bmatrix} \textcircled{5} & -3 & 6 \end{bmatrix}$

40. $\begin{bmatrix} 3 & 1 \\ 2 & 9 \\ 4 & \textcircled{3} \end{bmatrix}$

41. $\begin{bmatrix} 4 & 2 & -1 & 3 \\ 5 & \textcircled{9} & 0 & 2 \end{bmatrix}$

Simplify each rational expression. State the excluded values of the variables. *(Lesson 12-2)*

42. $\frac{15a}{39a^2}$

43. $\frac{t-3}{t^2-7t+12}$

44. $\frac{m-3}{m^2-9}$

Getting Ready for the Next Lesson

PREREQUISITE SKILL Graph each set of numbers on a number line.
*(To review **number lines**, see Lesson 2-1.)*

45. {4, 7, 8, 10, 11}

46. {13, 17, 22, 23, 27}

47. {30, 35, 40, 50, 55}

Practice Quiz 2

Lessons 13-3 and 13-4

For Exercises 1–2, use the histogram at the right. *(Lesson 13-3)*

1. In what measurement class does the median occur?
2. Describe the distribution of the data.

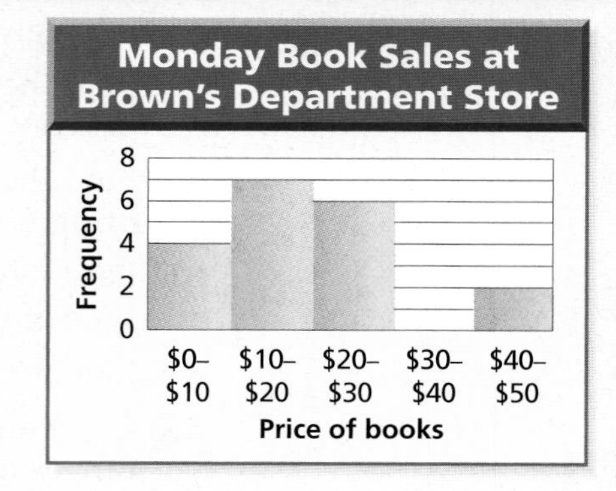

For Exercises 3–5, use the following set of data. *(Lesson 13-4)*
1050, 1175, 835, 1075, 1025, 1145, 1100, 1125, 975, 1005, 1125, 1095, 1075, 1055

3. Find the range of the data.
4. Find the median, the lower quartile, the upper quartile, and interquartile range of the data.
5. Identify any outliers of the data.

13-5 Box-and-Whisker Plots

What You'll Learn

- Organize and use data in box-and-whisker plots.
- Organize and use data in parallel box-and-whisker plots.

Vocabulary

- box-and-whisker plot
- extreme values

How are box-and-whisker plots used to display data?

Everyone should eat a number of calcium-rich foods each day. Selected foods and the amount of calcium in a serving are listed in the table. To create a box-and-whisker plot of the data, you need to find the quartiles of the data.

Calcium-Rich Foods

Food (serving size)	Calcium (milligrams)
Plain Yogurt, Nonfat (8 oz)	452
Plain Yogurt, Low-fat (8 oz)	415
Skim Milk (8 oz)	302
1% Milk (8 oz)	300
Whole Milk (8 oz)	291
Swiss Cheese (1 oz)	272
Tofu (4 oz)	258
Sardines (2 oz)	217
Cheddar Cheese (1 oz)	204
Collards (4 oz)	179
American Cheese (1 oz)	163
Frozen Yogurt with Fruit (4 oz)	154
Salmon (2 oz)	122
Broccoli (4 oz)	89

Source: *Vitality*

89 122 154 163 179 204 217 258 272 291 300 302 415 452

163 ↑ Q_1; $Q_2 = \frac{217 + 258}{2}$ or 237.5; 300 ↑ Q_3

This information can be displayed on a number line as shown below.

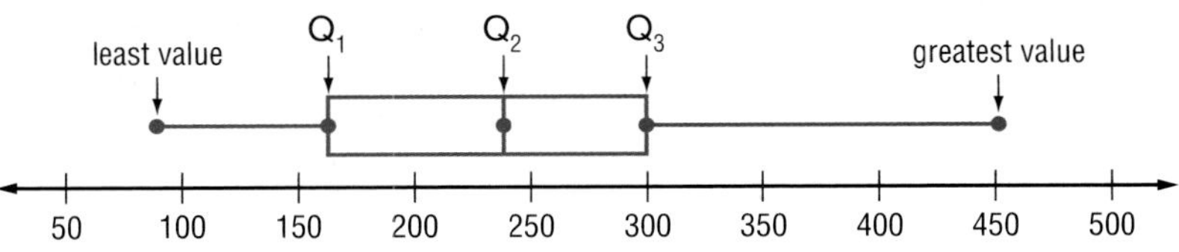

Study Tip

Reading Math
Box-and-whisker plots are sometimes called *box plots*.

BOX-AND-WHISKER PLOTS Diagrams such as the one above are called **box-and-whisker plots**. The length of the box represents the interquartile range. The line inside the box represents the median. The lines or *whiskers* represent the values in the lower fourth of the data and the upper fourth of the data. The bullets at each end are the **extreme values**. In the box-and-whisker plot above, the least value (LV) is 89, and the greatest value (GV) is 452.

If a set of data has outliers, these data points are represented by bullets. The whisker representing the lower data is drawn from the box to the least value that is not an outlier. The whisker representing the upper data is drawn from the box to the greatest value that is not an outlier.

Example 1 Draw a Box-and-Whisker Plot

ECOLOGY The amount of rain in Florida from January to May is crucial to its ecosystems. The following is a list of the number of inches of rain in Florida during this crucial period for the years 1990 to 2000.

14.03, 30.11, 16.03, 19.61, 18.15, 16.34, 20.43, 18.46, 22.24, 12.70, 8.25

a. Draw a box-and-whisker plot for these data.

Step 1 Determine the quartiles and any outliers.

Order the data from least to greatest. Use this list to determine the quartiles.

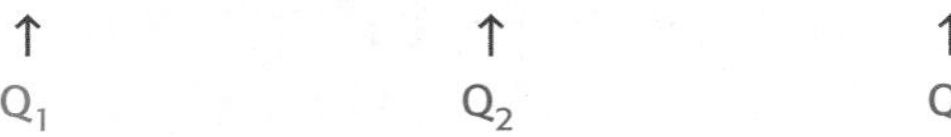

Determine the interquartile range.

IQR = 20.43 − 14.03 or 6.4

Check to see if there are any outliers.

14.03 − 1.5(6.4) = 4.43   20.43 + 1.5(6.4) = 30.03

Any numbers less than 4.43 or greater than 30.03 are outliers. The only outlier is 30.11.

Step 2 Draw a number line.

Assign a scale to the number line that includes the extreme values. Above the number line, place bullets to represent the three quartile points, any outliers, the least number that is *not* an outlier, and the greatest number that is *not* an outlier.

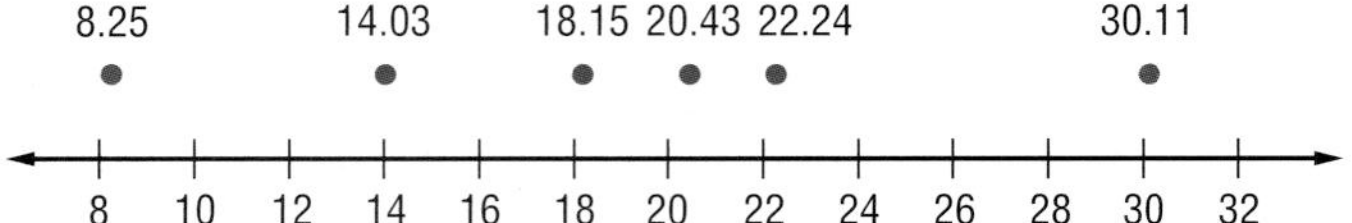

Step 3 Complete the box-and-whisker plot.

Draw a box to designate the data between the upper and lower quartiles. Draw a vertical line through the point representing the median. Draw a line from the lower quartile to the least value that is *not* an outlier. Draw a line from the upper quartile to the greatest value that is *not* an outlier.

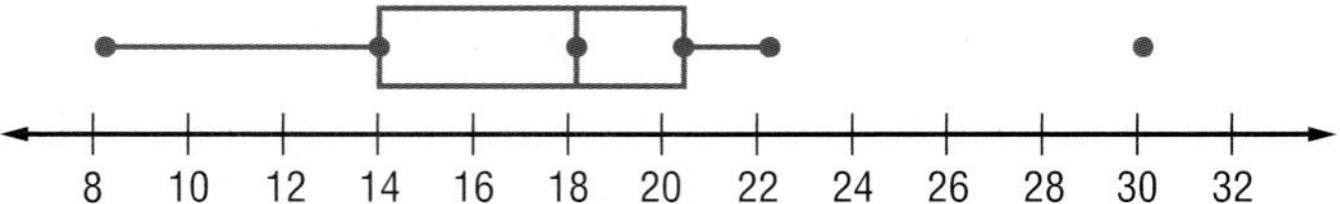

b. What does the box-and-whisker plot tell about the data?

Notice that the whisker and the box for the top half of the data is shorter than the whisker and box for the lower half of the data. Therefore, except for the outlier, the upper half of the data are less spread out than the lower half of the data.

More About. . .

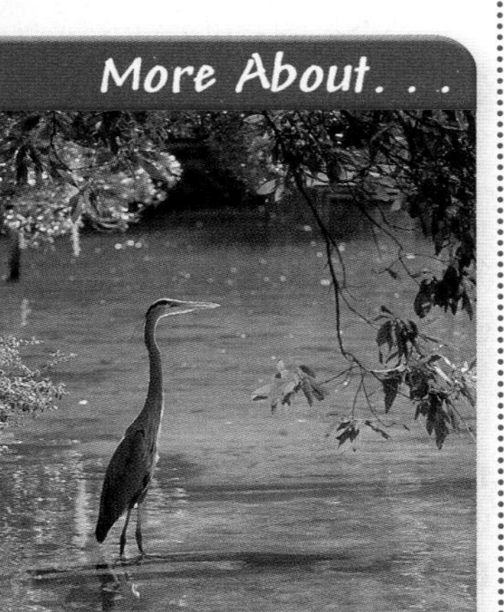

Ecology

The ecosystem of the Everglades in Florida is unique. The Everglades National Park is a subtropical swamp area covering 1,506,499 acres.

Source: *World Book Encyclopedia*

PARALLEL BOX-AND-WHISKER PLOTS Two sets of data can be compared by drawing parallel box-and-whisker plots such as the one shown below.

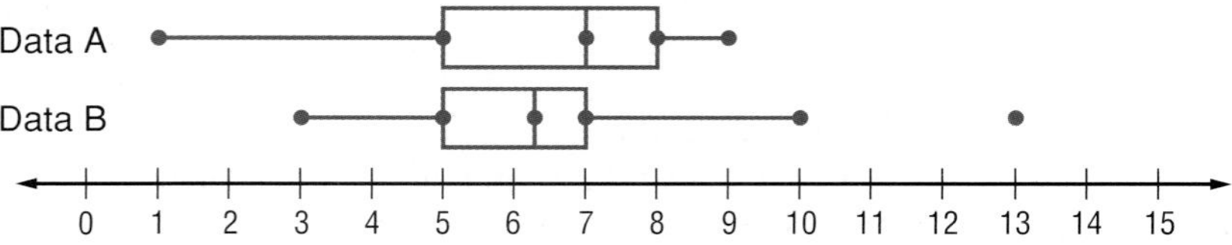

Example 2 Draw Parallel Box-and-Whisker Plots

WEATHER **Jalisa Thompson has job offers in Fresno, California, and Brownsville, Texas. Since she likes both job offers, she decides to compare the temperatures of each city.**

Average Monthly High Temperatures (°F)												
Month	Jan.	Feb.	March	April	May	June	July	Aug.	Sept.	Oct.	Nov.	Dec.
Fresno	54.1	61.7	66.6	75.1	84.2	92.7	98.6	96.7	90.1	79.7	64.7	53.7
Brownsville	68.9	72.2	78.4	84.0	87.8	91.0	93.3	93.6	90.4	85.3	78.3	71.7

Source: www.stormfax.com

a. Draw a parallel box-and-whisker plot for the data.

Determine the quartiles and outliers for each city.

Fresno

53.7, 54.1, 61.7, 64.7, 66.6, 75.1, 79.7, 84.2, 90.1, 92.7, 96.7, 98.6

$Q_1 = 63.2$ $\quad Q_2 = 77.4$ $\quad Q_3 = 91.4$

Brownsville

68.9, 71.7, 72.2, 78.3, 78.4, 84.0, 85.3, 87.8, 90.4, 91.0, 93.3, 93.6

$Q_1 = 75.25$ $\quad Q_2 = 84.65$ $\quad Q_3 = 90.7$

Neither city has any outliers.

Draw the box-and-whisker plots using the same number line.

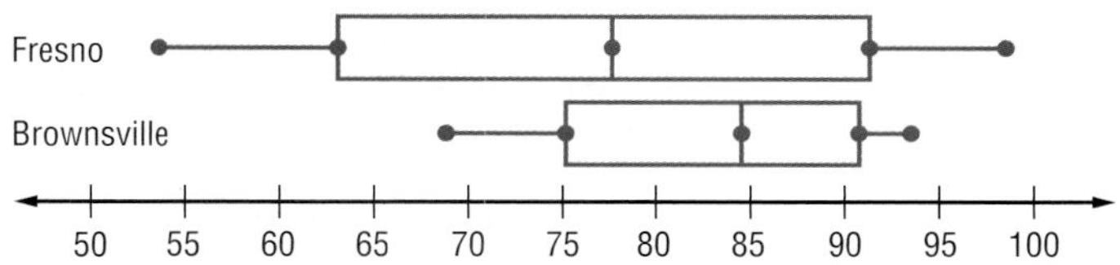

b. Use the parallel box-and-whisker plots to compare the data.

The range of temperatures in Fresno is much greater than in Brownsville. Except for the fourth quartile, Brownville's average temperatures appear to be as high or higher than Fresno's.

Check for Understanding

Concept Check

1. **Describe** the data represented by the box-and-whisker plot at the right. Include the extreme values, the quartiles, and any outliers.

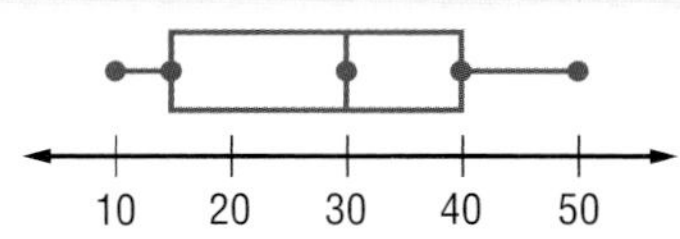

2. **Explain** how to determine the scale of the number line in a box-and-whisker plot.

3. **OPEN ENDED** Write a set of data that could be represented by the box-and-whisker plot at the right.

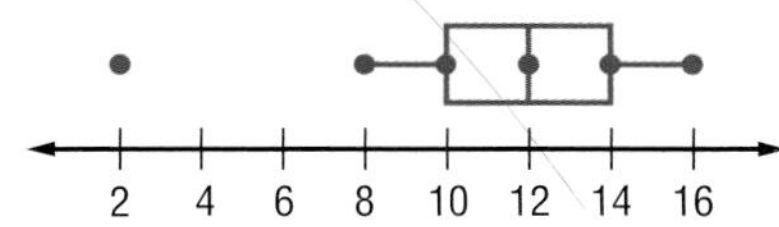

Guided Practice

Draw a box-and-whisker plot for each set of data.

4. 30, 28, 24, 24, 22, 22, 21, 17, 16, 15
5. 64, 69, 65, 71, 66, 66, 74, 67, 68, 67

Draw a parallel box-and-whisker plot for each pair of data. Compare the data.

6. A: 22, 18, 22, 17, 32, 24, 31, 26, 28
   B: 28, 30, 45, 23, 24, 32, 30, 27, 27

7. A: 8, 15.5, 14, 14, 24, 19, 16.7, 15, 11.4, 16
   B: 18, 14, 15.8, 9, 12, 16, 20, 16, 13, 15

Application

CHARITY **For Exercises 8 and 9, use the information in the table below.**

Top Ten Charities	
Charity	**Private Contributions (millions)**
Salvation Army	$1397
YMCA of the U.S.A.	$693
American Red Cross	$678
American Cancer Society	$620
Fidelity Investments Charitable Gift Fund	$573
Lutheran Services in America	$559
United Jewish Communities	$524
America's Second Harvest	$472
Habitat for Humanity International	$467
Harvard University	$452

Source: *The Chronicle of Philanthropy*

8. Make a box-and-whisker for the data.
9. Write a brief description of the data distribution.

Practice and Apply

Homework Help

For Exercises	See Examples
10–19	1
20–27	2

Extra Practice

See page 850.

For Exercises 10–13, use the box-and-whisker plot at the right.

10. What is the range of the data?
11. What is the interquartile range of the data?
12. What fractional part of the data is less than 90?
13. What fractional part of the data is greater than 95?

Draw a box-and-whisker plot for each set of data.

14. 15, 8, 10, 1, 3, 2, 6, 5, 4, 27, 1
15. 20, 2, 12, 5, 4, 16, 17, 7, 6, 16, 5, 0, 5, 30
16. 4, 1, 1, 1, 10, 15, 4, 5, 27, 5, 14, 10, 6, 2, 2, 5, 8
17. 51, 27, 55, 54, 69, 60, 39, 46, 46, 53, 81, 23
18. 15.1, 9.0, 8.5, 5.8, 6.2, 8.5, 10.5, 11.5, 8.8, 7.6
19. 1.3, 1.2, 14, 1.8, 1.6, 5.7, 1.3, 3.7, 3.3, 2, 1.3, 1.3, 7.7, 8.5, 2.2

For Exercises 20–23, use the parallel box-and-whisker plot at the right.

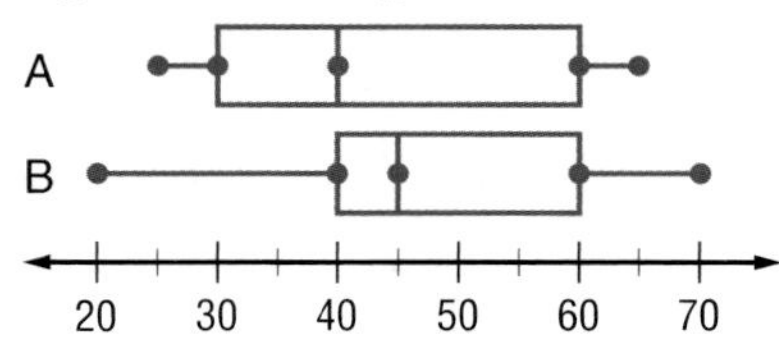

20. Which set of data contains the least value?
21. Which set of data contains the greatest value?
22. Which set of data has the greatest interquartile range?
23. Which set of data has the greatest range?

Draw a parallel box-and-whisker plot for each pair of data. Compare the data.

24. A: 15, 17, 22, 28, 32, 40, 16, 24, 26, 38, 19
    B: 24, 32, 25, 27, 37, 29, 30, 30, 28, 31, 27

25. A: 50, 45, 47, 55, 51, 58, 49, 51, 51, 48, 47
B: 40, 41, 48, 39, 41, 41, 38, 37, 35, 37, 45

26. A: 1.5, 3.8, 4.2, 3.5 4.1, 4.4, 4.1, 4.0, 4.0, 3.9
B: 6.8, 4.2, 7.6, 5.5, 12.2, 6.7, 7.1, 4.8

27. A: 4.4, 4.5, 4.6, 4.5, 4.4, 4.4, 4.1, 4.9, 2.9
B: 5.1, 4.9, 4.2, 3.9, 4.5, 4.1, 4.3, 4.5, 5.2

PROFESSIONAL SPORTS For Exercises 28 and 29, use the table at the right.

28. Draw a box-and-whisker plot for the data.

29. What does the box-and-whisker plot tell about the data?

Professional Athletes

Professional Sport	Average Length of Career (years)
Bowling	17
Surfing	10
Hockey	5.5
Baseball	4.5
Basketball	4.5
Tennis	4
Football	3.5
Boxing	3.5

Source: *Men's Health Fitness Special*

More About. . .

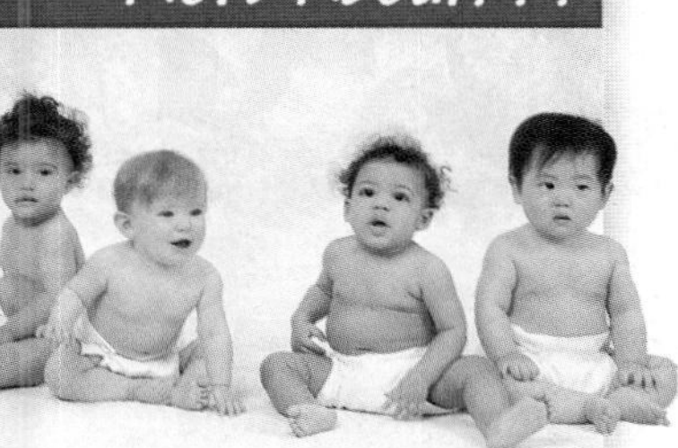

Life Expectancy

A newborn resident of the United States has a life expectancy of 77 years, while a newborn resident of Canada has a life expectancy of 79 years.

Source: UNICEF

RACING For Exercises 30 and 31, use the following list of earnings in thousands from the November 2000 NAPA 500 NASCAR Race at the Atlanta Motor Speedway.

$181, $100, $98, $89, $76, $58, $60; $58; $55, $57, $54, $64, $44, $39, $66, $52, $56, $38, $56, $51, $49, $38, $50, $48, $48, $40, $36, $36, $39, $36, $47, $36, $47, $38, $35, $46, $35, $55, $46, $55, $45, $43, $35

Source: *USA TODAY*

30. Draw a box-and-whisker plot for the data. Identify any outliers.

31. Determine whether the top half of the data or the bottom half of the data are more dispersed. Explain.

LIFE EXPECTANCY For Exercises 32–35, use the box-and-whisker plot depicting the UNICEF life expectancy data for 171 countries.

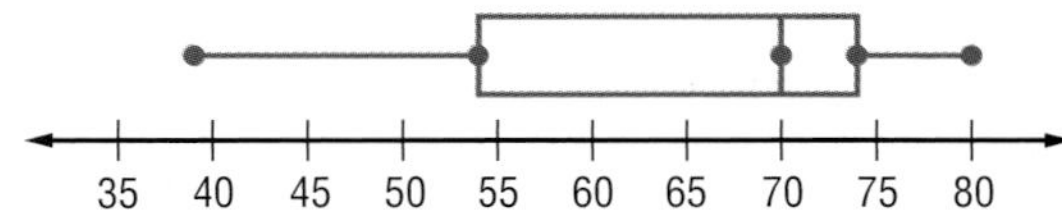

32. Estimate the range and the interquartile range.

33. Determine whether the top half of the data or the bottom half of the data are more dispersed. Explain.

34. State three different intervals of ages that contain half the data.

35. Jamie claims that the number of data values is greater in the interval 54 years to 70 years than the number of data values in the interval 70 years to 74 years. Is Jamie correct? Explain.

SOCCER For Exercises 36–38, use the following list of top 50 lifetime scores for all players in Division 1 soccer leagues in the United States from 1922 to 1999.

253, 223, 193, 189, 152, 150, 138, 137, 135, 131, 131, 129, 128, 126, 124, 119, 118, 108, 107, 102, 101, 100, 96, 92, 87, 83, 82, 81, 80, 78, 78, 76, 74, 74, 73, 73, 72, 71, 69, 68, 67, 65, 64, 63, 63, 63, 62, 61, 61, 61

Source: www.internetsoccer.com

36. Draw a box-and-whisker plot for the data.

37. Draw a histogram to represent the data.

38. Compare and contrast the box-and-whisker plot and the histogram.

A box-and-whisker plot of population densities will help you compare the states. Visit www.algebra1.com/webquest to continue work on your WebQuest project.

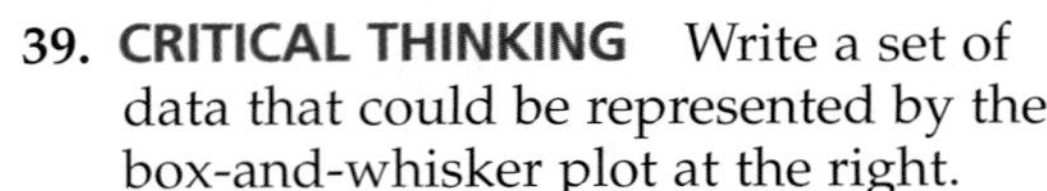

39. **CRITICAL THINKING** Write a set of data that could be represented by the box-and-whisker plot at the right.

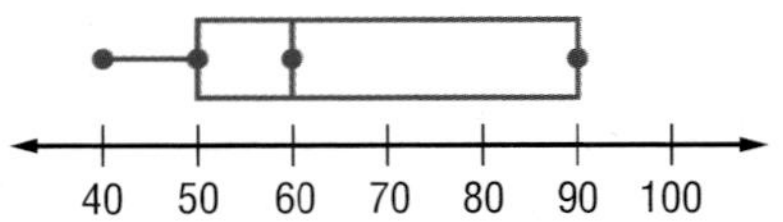

40. **WRITING IN MATH** Answer the question that was posed at the beginning of the lesson.

How are box-and-whisker plots used to display data?

Include the following in your answer:
- a sample of a box-and-whisker plot showing what each part of the plot represents, and
- a box-and-whisker plot representing data found in a newspaper or magazine.

For Exercises 41 and 42, use the box-and-whisker plot below.

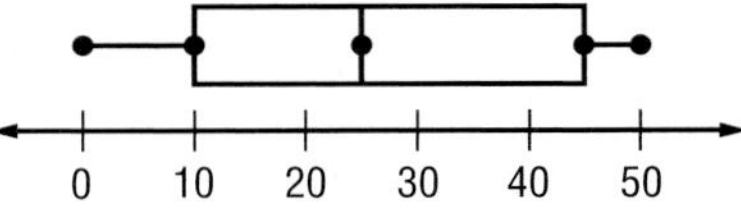

41. What is the median of the data?

Ⓐ 0   Ⓑ 10   Ⓒ 25   Ⓓ 45

42. Which interval represents 75% of the data?

Ⓐ 0–25   Ⓑ 10–45   Ⓒ 25–50   Ⓓ 0–45

Maintain Your Skills

Mixed Review

For Exercises 43 and 44, use the following data.

13, 32, 45, 45, 54, 55, 58, 67, 82, 93

43. Find the range, median, lower quartile, upper quartile, and interquartile range of the data. Identify any outliers. *(Lesson 13-4)*

44. Create a histogram to represent the data. *(Lesson 13-3)*

Find each sum or difference. *(Lesson 12-7)*

45. $\frac{3}{y-3} - \frac{y}{y+4}$

46. $\frac{2}{r+3} + \frac{3}{r-2}$

47. $\frac{w}{5w+2} - \frac{4}{15w+6}$

Find each product. Assume that no denominator has a value of 0. *(Lesson 12-3)*

48. $\frac{7a^2}{5} \cdot \frac{15}{14a}$

49. $\frac{6r+3}{r+6} \cdot \frac{r^2+9r+18}{2r+1}$

Solve each right triangle. State the side length to the nearest tenth and the angle measures to the nearest degree. *(Lesson 11-7)*

50.
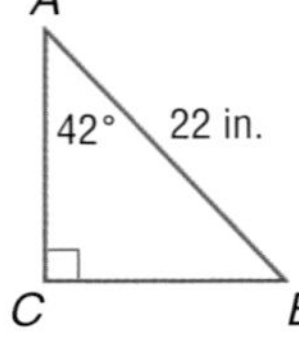

51.
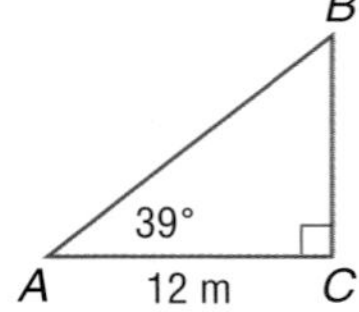

52.
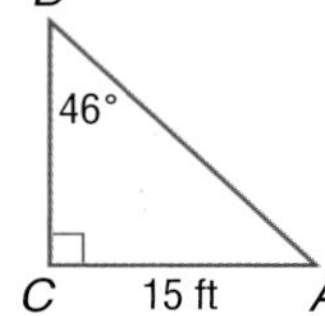

Solve each equation by completing the square. Approximate any irrational roots to the nearest tenth. *(Lesson 10-3)*

53. $a^2 - 7a + 6 = 0$

54. $x^2 - 6x + 2 = 0$

55. $t^2 + 8t - 18 = 0$

Find each sum or difference. *(Lesson 8-5)*

56. $(7p^2 - p - 7) - (p^2 + 11)$

57. $(3a^2 - 8) + (5a^2 + 2a + 7)$

Algebra Activity

A Follow-Up of Lesson 13-5

Investigating Percentiles

When data are arranged in order from least to greatest, you can describe the data using percentiles. A **percentile** is the point below which a given percent of the data lies. For example, 50% of the data falls below the median. So the median is the 50th percentile for the data.

To determine a percentile, a cumulative frequency table can be used. In a **cumulative frequency table**, the frequencies are accumulated for each item.

Collect the Data

A student's score on the SAT is one factor that some colleges consider when selecting applicants. The tables below show the raw scores from a sample math SAT test for 160 juniors in a particular school. For raw scores, the highest possible score is 800 and the lowest is 200.

Table 1: Frequency Table

Math SAT Scores	Number of Students
200–300	2
300–400	19
400–500	44
500–600	55
600–700	32
700–800	8

Table 2: Cumulative Frequency Table

Math SAT Scores	Number of Students	Cumulative Number of Students
200–300	2	2
300–400	19	21
400–500	44	65
500–600	55	120
600–700	32	152
700–800	8	160

The data in each table can be displayed in a histogram.

Frequency Histogram

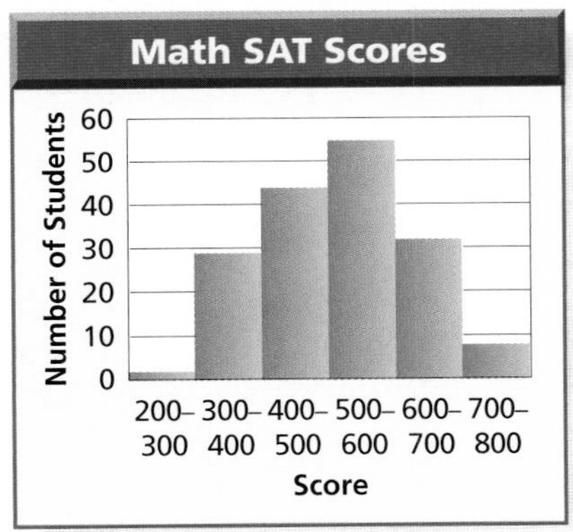

Cumulative Frequency Histogram

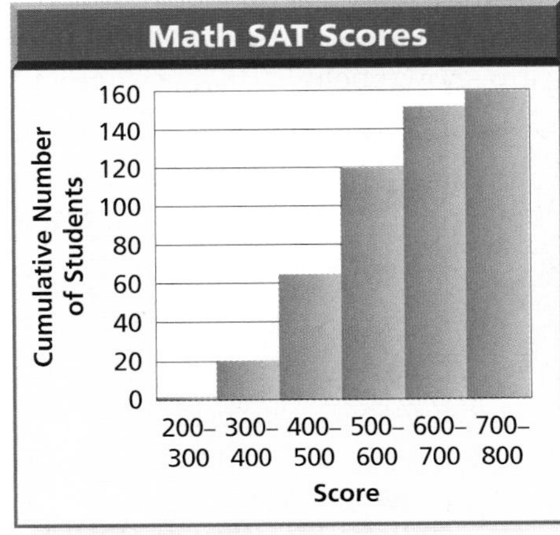

Analyze the Data

1. Examine the data in the two tables. Explain how the numbers in Column 3 of Table 2 are determined.

(continued on the next page)

2. Describe the similarities and differences between the two histograms.
3. Which histogram do you prefer for displaying these data? Explain your choice.

Make a Conjecture

Sometimes colleges are not interested in your raw score. They are interested in the percentile. Your percentile indicates what percent of all test-takers scored just as well or lower than you.

4. Use the histogram for Table 2. Place percentile labels on the vertical axis. For example, write 100% next to 160 and 0% next to 0. Now label 25%, 50%, and 75%. What numbers of students correspond to 25%, 50%, and 75%?

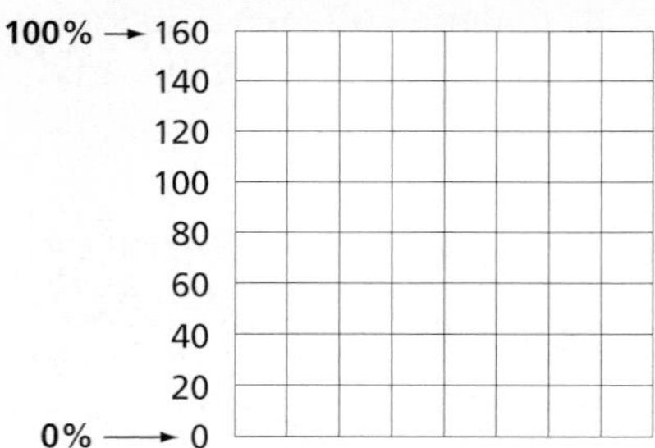

5. Suppose a college is interested in students with scores in the 90th percentile. Using the histogram, move up along the vertical axis to the 90th percentile. Then move right on the horizontal axis to find the score. What is an estimate for the score that represents the 90th percentile?
6. For a more accurate answer, use a proportion to find 90% of the total number of students. (Recall that the total number of students is 160.)
7. If a student is to be in the 90th percentile, in what interval will the score lie?

Extend the Activity

For Exercises 8–10, use the following information.

The weights of 45 babies born at a particular hospital during the month of January are shown below.

9 lb 1 oz	8 lb 2 oz	7 lb 2 oz	10 lb 0 oz	4 lb 4 oz
5 lb 0 oz	7 lb 6 oz	7 lb 8 oz	11 lb 2 oz	6 lb 1 oz
3 lb 8 oz	8 lb 0 oz	7 lb 5 oz	9 lb 15 oz	6 lb 1 oz
7 lb 10 oz	6 lb 9 oz	6 lb 15 oz	7 lb 10 oz	8 lb 0 oz
5 lb 15 oz	8 lb 3 oz	8 lb 1 oz	7 lb 12 oz	7 lb 8 oz
7 lb 7 oz	6 lb 14 oz	7 lb 13 oz	8 lb 0 oz	7 lb 14 oz
5 lb 10 oz	8 lb 5 oz	6 lb 12 oz	8 lb 8 oz	7 lb 11 oz
8 lb 15 oz	9 lb 3 oz	5 lb 14 oz	6 lb 8 oz	8 lb 8 oz
7 lb 4 oz	7 lb 10 oz	8 lb 1 oz	7 lb 8 oz	7 lb 10 oz

8. Make a cumulative frequency table for the data.
9. Make a cumulative frequency histogram for the data.
10. Find the weight for a baby in the 80th percentile.

Chapter 13

Study Guide and Review

Vocabulary and Concept Check

biased sample (p. 709)
box-and-whisker plot (p. 737)
census (p. 708)
column (p. 715)
convenience sample (p. 709)
dimensions (p. 715)
element (p. 715)
extreme value (p. 737)
frequency (p. 722)
frequency table (p.722)
histogram (p. 722)
interquartile range (p. 732)
lower quartile (p. 732)
matrix (p. 715)
measurement classes (p. 722)
measures of variation (p. 731)
outlier (p. 733)
population (p. 708)
quartiles (p. 732)
random sample (p. 708)
range (p. 731)
row (p. 715)
sample (p. 708)
scalar multiplication (p. 717)
simple random sample (p. 708)
stratified random sample (p. 708)
systematic random sample (p. 708)
upper quartile (p. 732)
voluntary response sample (p. 709)

Choose the correct term from the list above that best completes each statement.

1. A(n) _______ is a sample that is as likely to be chosen as any other from the population.
2. Measures that describe the spread of the values in a set of data are called _______.
3. Each _______ separates a data set into four sets with equal number of members.
4. In a(n) _______, the items are selected according to a specified time or item interval.
5. A(n) _______ has a systematic error within it so that certain populations are favored.
6. In a(n) _______, the population is first divided into similar, nonoverlapping groups.
7. The _______ is found by subtracting the lower quartile from the upper quartile.
8. A(n) _______ involves only those who want to participate in the sampling.
9. An extreme value that is much less or greater than the rest of the data is a(n) _______.
10. The _______ is the difference between the greatest and least values of a data set.

Lesson-by-Lesson Review

13-1 Sampling and Bias

See pages 708–713.

Concept Summary

- Samples are used to represent a larger group called a population.
- Simple random sample, stratified random sample, and systematic random sample are types of unbiased, or random, samples.
- Convenience sample and voluntary response sample are types of biased samples.

Example GOVERNMENT **To determine whether voters support a new trade agreement, 5 people from the list of registered voters in each state and the District of Columbia are selected at random. Identify the sample, suggest a population from which it was selected, and state whether the sample is *unbiased* (random) or *biased*. If unbiased, classify the sample as *simple*, *stratified*, or *systematic*. If biased, classify as *convenience* or *voluntary response*.**

Since $5 \times 51 = 255$, the sample is 255 registered voters in the United States. The population is all of the registered voters in the United States.

The sample is unbiased. It is an example of a stratified random sample.

Exercises **Identify the sample, suggest a population from which it was selected, and state whether it is *unbiased* (random) or *biased*. If unbiased, classify the sample as *simple*, *stratified*, or *systematic*. If biased, classify the sample as *convenience* or *voluntary response*.** *See Examples 1–3 on pages 709 and 710.*

11. **SCIENCE** A laboratory technician needs a sample of results of chemical reactions. She selects test tubes from the first 8 experiments performed on Tuesday.
12. **CANDY BARS** To ensure that all of the chocolate bars are the appropriate weight, every 50th bar on the conveyor belt in the candy factory is removed and weighed.

13-2 Introduction to Matrices

See pages 715–721.

Concept Summary

- A matrix can be used to organize data and make data analysis more convenient.
- Equal matrices must have the same dimensions and corresponding elements are equal.
- Matrices with the same dimensions can be added or subtracted.
- Each element of a matrix can be multiplied by a number called a scalar.

Example **If** $R = \begin{bmatrix} 2 & 2 \\ -1 & 3 \end{bmatrix}$, $S = \begin{bmatrix} -1 & 3 \\ 0 & 1 \end{bmatrix}$, **and** $T = \begin{bmatrix} -1 \\ 0 \end{bmatrix}$, **find each sum. If it does not exist, write *impossible*.**

a. $R + S$

$$R + S = \begin{bmatrix} 2 & 2 \\ -1 & 3 \end{bmatrix} + \begin{bmatrix} -1 & 3 \\ 0 & 1 \end{bmatrix}$$
$$= \begin{bmatrix} 2 + (-1) & 2 + 3 \\ -1 + 0 & 3 + 1 \end{bmatrix}$$
$$= \begin{bmatrix} 1 & 5 \\ -1 & 4 \end{bmatrix}$$

b. $S + T$

$$S + T = \begin{bmatrix} -1 & 3 \\ 0 & 1 \end{bmatrix} + \begin{bmatrix} -1 \\ 0 \end{bmatrix}$$

Since S is a 2×2 matrix and T is a 2×1 matrix, the matrices do not have the same dimensions. Therefore, it is impossible to add these matrices.

Exercises **If** $A = \begin{bmatrix} 1 & 3 & -1 \\ 2 & 0 & 4 \\ -1 & -1 & 3 \end{bmatrix}$, $B = \begin{bmatrix} 1 & 1 & -3 \\ 2 & 3 & -1 \\ -1 & -2 & 0 \end{bmatrix}$, $C = \begin{bmatrix} 3 & -2 \\ 1 & 4 \end{bmatrix}$, **and** $D = \begin{bmatrix} 2 & 1 \\ -2 & 0 \end{bmatrix}$, **find each sum, difference, or product. If the sum or difference does not exist, write *impossible*.** *See Examples 3 and 4 on pages 716 and 717.*

13. $A + B$
14. $3B$
15. $-2D$
16. $C - D$
17. $C + D$
18. $B + C$
19. $5A$
20. $A - D$
21. $C + 3D$
22. $2A - B$

13-3 Histograms

See pages 722–728.

Concept Summary

- A histogram can illustrate the information in a frequency table.
- The distribution of the data can be determined from a histogram.

Example **Create a histogram to represent the following high temperatures in twenty states.**

118 122 117 105 114 115 122 102 103 110

110 112 106 109 100 103 110 108 111 102

Since the temperatures range from 100 to 122, use measurement classes from 100 to 125 with 5 degree intervals. First create a frequency table and then draw the histogram.

Temperature Intervals	Tally	Frequency
$100 \le d < 105$	𝍸	5
$105 \le d < 110$	IIII	4
$110 \le d < 115$	𝍸 I	6
$115 \le d < 120$	III	3
$120 \le d < 125$	II	2

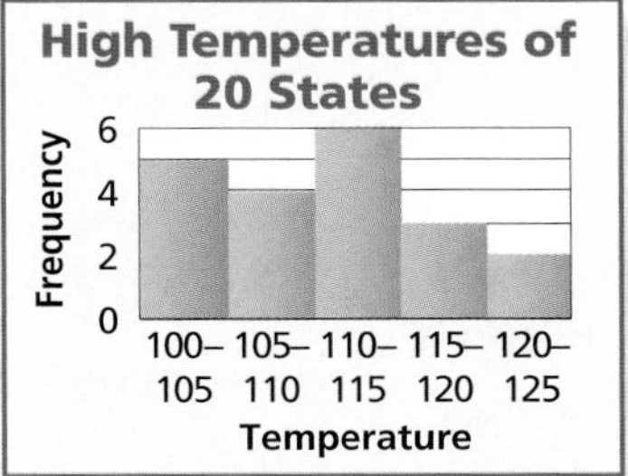

Exercises **Create a histogram to represent each data set.** *See Example 3 on page 724.*

23. the number of cellular minutes used last month by employees of a company

122 150 110 290 145 330 300 210 95 101 106 289 219
105 302 29 288 154 235 168 55 84 92 175 180

24. the number of cups of coffee consumed per customer at a snack shop between 6 A.M. and 8 A.M.

0 2 0 2 1 3 2 1 2 3 0 2 2 1 0 2 1 3 0 1 2 2
3 2 1 0 1 2 1 0 2 2 2 1 1 2 1 2 0 3 1 0 0 1

13-4 Measures of Variation

See pages 731–736.

Concept Summary

- The range of the data set is the difference between the greatest and the least values of the set and describes the spread of the data.
- The interquartile range is the difference between the upper and lower quartiles of a set of data. It is the range of the middle half of the data.
- Outliers are values that are much less than or much greater than the rest of the data.

Example **Find the range, median, lower quartile, upper quartile, and interquartile range of the set of data below. Identify any outliers.**
25, 20, 30, 24, 22, 26, 28, 29, 19

Order the set of data from least to greatest.

19 20 22 24 25 26 28 29 30

↑ Q_1 (between 20 and 22) ↑ Q_2 (25) ↑ Q_3 (between 28 and 29)

The range is 30 − 19 or 11.

The median is the middle number, 25.

The lower quartile is $\frac{20 + 22}{2}$ or 21.

The upper quartile is $\frac{28 + 29}{2}$ or 28.5.

The interquartile range is 28.5 − 21 or 7.5.

The outliers would be less than 21 − 1.5(7.5) or 9.75 and greater than 28.5 + 1.5(7.5) or 39.25. There are no outliers.

• Extra Practice, see pages 849–850.
• Mixed Problem Solving, see page 865.

Exercises **Find the range, median, lower quartile, upper quartile, and interquartile range of each set of data. Identify any outliers.**
See Examples 1–3 on pages 731–733.

25. 30, 90, 40, 70, 50, 100, 80, 60

26. 3, 3.2, 45, 7, 2, 1, 3.4, 4, 5.3, 5, 78, 8, 21, 5

27. 85, 77, 58, 69, 62, 73, 55, 82, 67, 77, 59, 92, 75, 69, 76

28. 111.5, 70.7, 59.8, 68.6, 63.8, 254.8, 64.3, 82.3, 91.7, 88.9, 110.5, 77.1

13-5 Box-and-Whisker Plots

See pages 737–742.

Concept Summary

- The vertical rule in the box of a box-and-whisker plot represents the median.
- The box of a box-and-whisker plot represents the interquartile range.
- The bullets at each end of a box-and-whisker plot are the extremes.
- Parallel box-and-whisker plots can be used to compare data.

Example **The following high temperatures (°F) were recorded during a two-week cold spell in St. Louis. Draw a box-and-whisker plot of the temperatures.**

20	2	12	5	4	16	17
7	6	16	5	0	5	30

Order the data from least to greatest.

0 2 4 5 5 5 6 7 12 16 16 17 20 30

Q_1 (5) $Q_2 = \frac{6+7}{2}$ or 6.5 Q_3 (16)

The interquartile range is 16 − 5 or 11. Check to see if there are any outliers.

$5 - 1.5(11) = -11.5 \qquad 16 + 1.5(11) = 32.5$

There are no outliers.

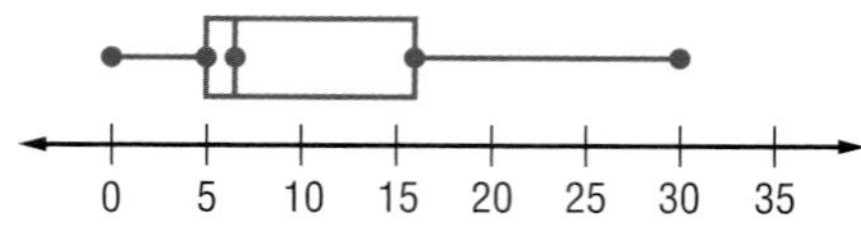

Exercises **Draw a box-and-whisker plot for each set of data.**
See Example 1 on page 738.

29. The number of Calories in a serving of French fries at 13 restaurants are 250, 240, 220, 348, 199, 200, 125, 230, 274, 239, 212, 240, and 327.

30. Mrs. Lowery's class has the following scores on their math tests.
60, 70, 70, 75, 80, 85, 85, 90, 95, 100

31. The average daily temperatures on a beach in Florida for each month of one year are 52.4, 55.2, 61.1, 67.0, 73.4, 79.1, 81.6, 81.2, 78.1, 69.8, 61.9, and 55.1.

Chapter 13 Practice Test

Vocabulary and Concepts

In a matrix, identify each item described.

1. a vertical set of numbers
2. an entry in a matrix
3. a horizontal set of numbers
4. a constant multiplied by each element in the matrix
5. number of rows and columns

a. element
b. column
c. row
d. dimensions
e. scalar

Skills and Applications

Identify the sample, suggest a population from which it was selected, and state whether it is *unbiased* (random) or *biased*. If unbiased, classify the sample as *simple*, *stratified*, or *systematic*. If biased, classify as *convenience* or *voluntary response*.

6. **DOGS** A veterinarian needs a sample of dogs in his kennel to be tested for fleas. He selects the first 5 dogs who run from the pen.
7. **LIBRARIES** A librarian wants to sample book titles checked out on Wednesday. He randomly chooses a book for each hour that the library is open.

If $W = \begin{bmatrix} 2 & 3 & 1 \\ -1 & 0 & -1 \\ 2 & -2 & 0 \end{bmatrix}$, $X = \begin{bmatrix} 4 & 2 & -1 \\ -2 & -2 & 0 \\ 0 & 1 & 2 \end{bmatrix}$, $Y = \begin{bmatrix} 3 & -2 & 1 \\ -1 & -2 & 4 \end{bmatrix}$, **and** $Z = \begin{bmatrix} 3 & 1 & 6 \\ 4 & -1 & -1 \end{bmatrix}$, **find each sum, difference, or product. If the sum or difference does not exist, write *impossible*.**

8. $W + X$
9. $Y - Z$
10. $3X$
11. $-2Z$
12. $2W - Z$
13. $Y - 2Z$

Create a histogram to represent each data set.

14. 68 71 74 90 81 72 71 69 65 92 75 69 71 73 73
68 74 80 83 70 80 74 74 70 71
15. 10 40 50 52 22 50 60 90 41 51 90 40 75 63 53

Find the range, median, lower quartile, upper quartile, and interquartile range for each set of data. Identify any outliers.

16. 1055, 1075, 1095, 1125, 1005, 975, 1125, 1100, 1145, 1025, 1075
17. 0.4, 0.2, 0.5, 0.9, 0.3, 0.4, 0.5, 1.9, 0.5, 0.7, 0.8, 0.6, 0.2, 0.1, 0.4

Draw a box-and-whisker plot for each set of data.

18. 1, 3, 2, 2, 1, 9, 4, 6, 1, 10, 1, 4, 5, 10, 1, 3, 6
19. 14, 18, 9, 9, 12, 22, 16, 12, 14, 16, 15, 13, 9, 10, 11, 12

20. **STANDARDIZED TEST PRACTICE** Which box-and-whisker plot has the greatest interquartile range?

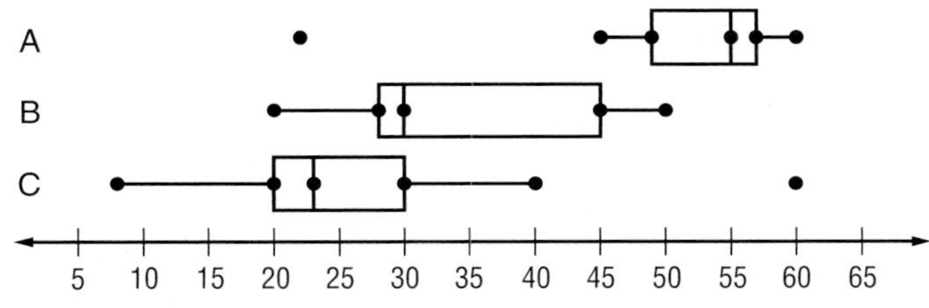

Ⓐ A Ⓑ B Ⓒ C Ⓓ They all have the same interquartile range.

Chapter 13 Standardized Test Practice

Part 1 Multiple Choice

Record your answers on the answer sheet provided by your teacher or on a sheet of paper.

1. Which equation represents a line perpendicular to the graph of $y = 4x - 6$? (Lesson 5-6)

Ⓐ $y = \frac{1}{4}x + \frac{1}{6}$ Ⓑ $y = -\frac{1}{4}x + 2$

Ⓒ $y = -4x + 6$ Ⓓ $y = 4x + 6$

2. A certain number is proportional to another number in the ratio 3:5. If 8 is subtracted from the sum of the numbers, the result is 32. What is the greater number? (Lesson 7-2)

Ⓐ 15 Ⓑ 25

Ⓒ 35 Ⓓ 40

3. The expression $(x - 8)^2$ is equivalent to (Lesson 8-8)

Ⓐ $x^2 - 64$. Ⓑ $x^2 - 16x + 64$.

Ⓒ $x^2 + 16x + 64$. Ⓓ $x^2 + 64$.

4. What is the least y value of the graph of $y = x^2 - 4$? (Lesson 10-1)

Ⓐ 2 Ⓑ 0

Ⓒ −2 Ⓓ −4

5. The expression $3\sqrt{72} - 3\sqrt{2}$ is equivalent to (Lesson 11-2)

Ⓐ $3\sqrt{70}$. Ⓑ $3\sqrt{2}$.

Ⓒ $15\sqrt{2}$. Ⓓ $5\sqrt{2}$.

6. A 12-meter flagpole casts a 9-meter shadow. At the same time, the building next to it casts a 27-meter shadow. How tall is the building? (Lesson 11-6)

Ⓐ 20.25 m Ⓑ 36 m

Ⓒ 40 m Ⓓ 84 m

7. Students are conducting a poll at Cedar Grove High School to determine whether to change the school colors. Which would be the best place to find an unbiased sample of students who represent the entire student body? (Lesson 13-1)

Ⓐ a football practice

Ⓑ a freshmen class party

Ⓒ a Spanish class

Ⓓ the cafeteria

8. A Mars year is longer than an Earth year because Mars takes longer to orbit the Sun. The table shows a person's age in both Earth years and Mars years. The data represent which kind of function? (Lesson 13-3)

Earth	10	20	30	40	50
Mars	5.3	10.6	15.9	21.2	26.5

Ⓐ linear function

Ⓑ quadratic function

Ⓒ exponential function

Ⓓ rational function

Use the box-and-whisker plot for Questions 9 and 10.

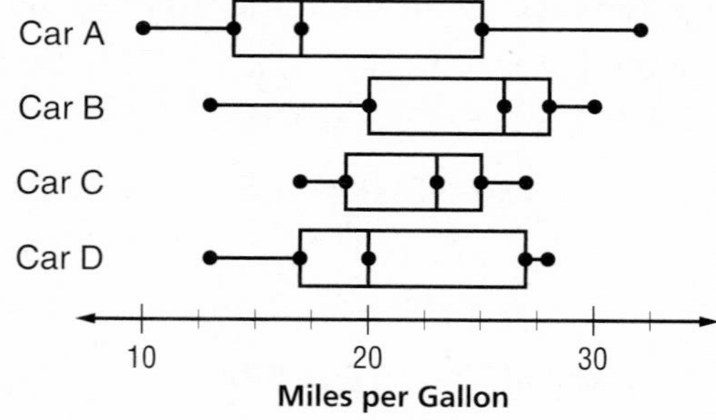

9. Which car shows the least variation in miles per gallon? (Lesson 13-5)

Ⓐ A Ⓑ B Ⓒ C Ⓓ D

10. Which car model has the highest median miles per gallon? (Lesson 13-5)

Ⓐ A Ⓑ B Ⓒ C Ⓓ D

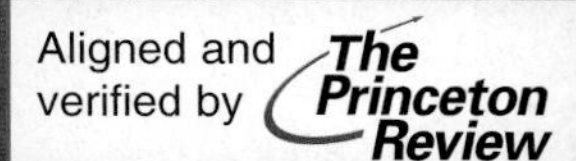

Part 2 Short Response/Grid In

Record your answers on the answer sheet provided by your teacher or on a sheet of paper.

11. Factor $x^3 + 8x^2 + 16x$. (Lesson 9-3)

12. Solve $6x^2 + x - 2 = 0$ by factoring. (Lesson 9-4)

13. Simplify $\sqrt{4\sqrt{9}}$. (Lesson 11-1)

14. Maren can do a job in 4 hours. Juliana can do the same job in 6 hours. Suppose Juliana works on the job for 2 hours and then is joined by Maren. Find the number of hours it will take both working together to finish the job. (Lesson 11-4)

15. The map below shows train tracks cutting across a grid of city streets. Newton Street is 1.5 miles apart from Olive Street, Olive Street is 1.5 miles apart from Pine Street, and the three streets are parallel to each other. If the distance between points A and B is 5 miles, then what is the distance in miles between points B and C? (Lesson 12-4)

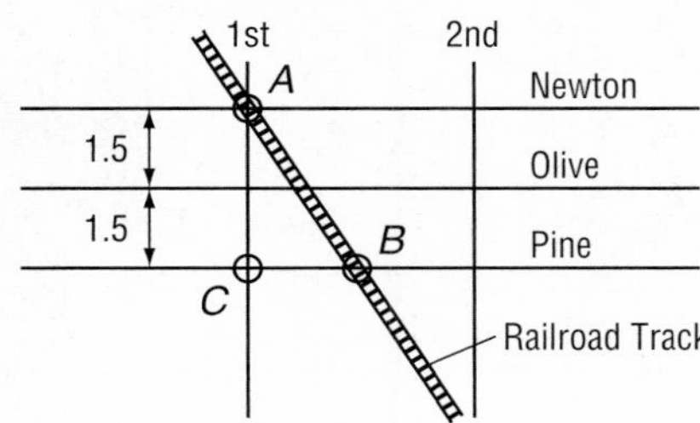

The Princeton Review Test-Taking Tip

Questions 14 and 15 If a problem seems difficult, don't panic. Reread the question slowly and carefully. Always ask yourself, "What have I been asked to find?" and, "What information will help me find the answer?"

Part 3 Quantitative Comparison

Compare the quantity in Column A and the quantity in Column B. Then determine whether:

- **(A) the quantity in Column A is greater,**
- **(B) the quantity in Column B is greater,**
- **(C) the two quantities are equal, or**
- **(D) the relationship cannot be determined from the information given.**

	Column A	Column B
16.	the sum of the next three terms of the arithmetic sequence $-250, 83, 416, \ldots$	the 67th term of the arithmetic sequence $-49, 2, 53, \ldots$
		(Lesson 4-7)
17.	the root of $y = -0.25x^2 + x - 1$	the sum of the roots of $b = 3a^2 - 5a + 2$
		(Lesson 10-4)
18.	the value of x if $\frac{3x+1}{14} = \frac{x}{4}$	the value of y if $\frac{45}{4y+1} = \frac{10}{y}$
		(Lesson 12-9)

Part 4 Open Ended

Record your answers on a sheet of paper. Show your work.

19. Construct a histogram for the following data. Use intervals of 40–50, 50–60, 60–70, 70–80, 80–90, and 90–100. (Lesson 13-3)
45, 62, 78, 84, 63, 73, 68, 91, 65, 80, 71, 87, 85, 77, 78, 80, 83, 87, 90, 91

20. In Exercise 19, what percent of the data lies within the tallest bar? (Lesson 13-3)

21. Draw a box-and-whisker plot of the following test scores. (Lesson 13-5)
24, 38, 47, 22, 40, 36, 25, 48, 30, 32, 45, 41, 34, 39, 40, 47, 40, 38, 42, 49, 45

www.algebra1.com/standardized_test

Chapter

14 Probability

What You'll Learn

- **Lesson 14-1** Count outcomes using the Fundamental Counting Principle.
- **Lesson 14-2** Determine probabilities using permutations and combinations.
- **Lesson 14-3** Find probabilities of compound events.
- **Lesson 14-4** Use probability distributions.
- **Lesson 14-5** Use probability simulations.

Key Vocabulary

- permutation (p. 760)
- combination (p. 762)
- compound event (p. 769)
- theoretical probability (p. 782)
- experimental probability (p. 782)

Why It's Important

The United States Senate forms committees to focus on different issues. These committees are made up of senators from various states and political parties. There are many ways these committees could be formed. *You will learn how to find the number of possible committees in Lesson 14-2.*

Getting Started

Prerequisite Skills To be successful in this chapter, you'll need to master these skills and be able to apply them in problem-solving situations. Review these skills before beginning Chapter 14.

For Lessons 14-2 through 14-5 — Find Simple Probabilities

Determine the probability of each event if you randomly select a cube from a bag containing 6 red cubes, 3 blue cubes, 4 yellow cubes, and 1 green cube.
(For review, see Lesson 2-6.)

1. $P(\text{red})$ **2.** $P(\text{blue})$ **3.** $P(\text{yellow})$ **4.** $P(\text{not red})$

For Lesson 14-2 — Multiply Fractions

Find each product. *(For review, see pages 800 and 801.)*

5. $\frac{4}{5} \cdot \frac{3}{4}$ **6.** $\frac{5}{12} \cdot \frac{6}{11}$ **7.** $\frac{7}{20} \cdot \frac{4}{19}$

8. $\frac{4}{32} \cdot \frac{7}{32}$ **9.** $\frac{13}{52} \cdot \frac{4}{52}$ **10.** $\frac{56}{100} \cdot \frac{24}{100}$

For Lesson 14-4 — Write Decimals as Percents

Write each decimal as a percent. *(For review, see pages 804 and 805.)*

11. 0.725 **12.** 0.148 **13.** 0.4 **14.** 0.0168

For Lesson 14-5 — Write Fractions as Percents

Write each fraction as a percent. Round to the nearest tenth. *(For review, see pages 804 and 805.)*

15. $\frac{7}{8}$ **16.** $\frac{33}{80}$ **17.** $\frac{107}{125}$ **18.** $\frac{625}{1024}$

Make this Foldable to help you organize what you learn about probability. Begin with a sheet of plain $8\frac{1}{2}$" by 11" paper.

Step 1 Fold in Half

Fold in half lengthwise.

Step 2 Fold Again in Fourths

Fold the top to the bottom twice.

Step 3 Cut

Open. Cut along the second fold to make four tabs.

Step 4 Label

Label as shown.

Probability
Outcomes | Permutations | Combinations | Compound Events

Reading and Writing As you read and study the chapter, write notes and examples for each concept under the tabs.

14-1 Counting Outcomes

What You'll Learn

- Count outcomes using a tree diagram.
- Count outcomes using the Fundamental Counting Principle.

Vocabulary

- tree diagram
- sample space
- event
- Fundamental Counting Principle
- factorial

How are possible win–loss records counted in football?

The championship in the Atlantic Coast Conference is decided by the number of conference wins. If there is a tie in conference wins, then the team with more nonconference wins is champion. If Florida State plays 3 nonconference games, the diagram at the right shows the different records they could have for those games.

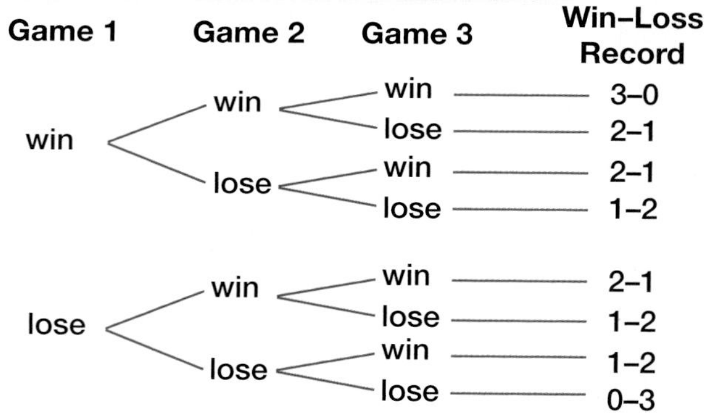

Game 1	Game 2	Game 3	Win–Loss Record
win	win	win	3–0
		lose	2–1
	lose	win	2–1
		lose	1–2
lose	win	win	2–1
		lose	1–2
	lose	win	1–2
		lose	0–3

TREE DIAGRAMS One method used for counting the number of possible outcomes is to draw a **tree diagram**. The last column of a tree diagram shows all of the possible outcomes. The list of all possible outcomes is called the **sample space**, while any collection of one or more outcomes in the sample space is called an **event**.

Example 1 Tree Diagram

A football team uses red jerseys for road games, white jerseys for home games, and gray jerseys for practice games. The team uses gray or black pants, and black or white shoes. Use a tree diagram to determine the number of possible uniforms.

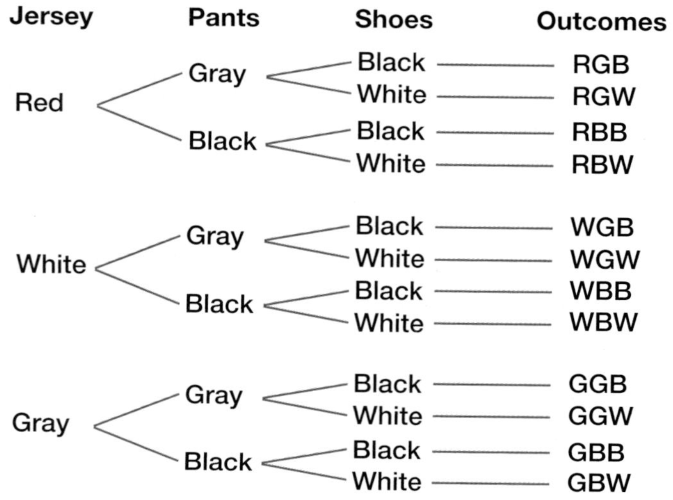

Jersey	Pants	Shoes	Outcomes
Red	Gray	Black	RGB
		White	RGW
	Black	Black	RBB
		White	RBW
White	Gray	Black	WGB
		White	WGW
	Black	Black	WBB
		White	WBW
Gray	Gray	Black	GGB
		White	GGW
	Black	Black	GBB
		White	GBW

The tree diagram shows that there are 12 possible uniforms.

THE FUNDAMENTAL COUNTING PRINCIPLE The number of possible uniforms in Example 1 can also be found by multiplying the number of choices for each item. If the team can choose from 3 different colored jerseys, 2 different colored pants, and 2 different colored pairs of shoes, there are $3 \cdot 2 \cdot 2$ or 12 possible uniforms. This example illustrates the **Fundamental Counting Principle**.

Key Concept *Fundamental Counting Principle*

If an event M can occur in m ways and is followed by an event N that can occur in n ways, then the event M followed by event N can occur in $m \cdot n$ ways.

Example 2 Fundamental Counting Principle

The Uptown Deli offers a lunch special in which you can choose a sandwich, a side dish, and a beverage. If there are 10 different sandwiches, 12 different side dishes, and 7 different beverages from which to choose, how many different lunch specials can you order?

Multiply to find the number of lunch specials.

$$\underbrace{10}_{\text{sandwich choices}} \cdot \underbrace{12}_{\text{side dish choices}} \cdot \underbrace{7}_{\text{beverage choices}} = \underbrace{840}_{\text{number of specials}}$$

The number of different lunch specials is 840.

Example 3 Counting Arrangements

Mackenzie is setting up a display of the ten most popular video games from the previous week. If she places the games side-by-side on a shelf, in how many different ways can she arrange them?

The number of ways to arrange the games can be found by multiplying the number of choices for each position.

- Mackenzie has ten games from which to choose for the first position.
- After choosing a game for the first position, there are nine games left from which to choose for the second position.
- There are now eight choices for the third position.
- This process continues until there is only one choice left for the last position.

Let n represent the number of arrangements.

$n = 10 \cdot 9 \cdot 8 \cdot 7 \cdot 6 \cdot 5 \cdot 4 \cdot 3 \cdot 2 \cdot 1$ or 3,628,800

There are 3,628,800 different ways to arrange the video games.

The expression $n = 10 \cdot 9 \cdot 8 \cdot 7 \cdot 6 \cdot 5 \cdot 4 \cdot 3 \cdot 2 \cdot 1$ used in Example 3 can be written as 10! using a **factorial**.

Key Concept *Factorial*

- **Words** The expression $n!$, read n factorial, where n is greater than zero, is the product of all positive integers beginning with n and counting backward to 1.
- **Symbols** $n! = n \cdot (n - 1) \cdot (n - 2) \cdot \ldots \cdot 3 \cdot 2 \cdot 1$
- **Example** $5! = 5 \cdot 4 \cdot 3 \cdot 2 \cdot 1$ or 120

By definition, $0! = 1$.

Example 4 Factorial

Find the value of each expression.

a. 6!

$6! = 6 \cdot 5 \cdot 4 \cdot 3 \cdot 2 \cdot 1$ Definition of factorial

$= 720$ Simplify.

b. 10!

$10! = 10 \cdot 9 \cdot 8 \cdot 7 \cdot 6 \cdot 5 \cdot 4 \cdot 3 \cdot 2 \cdot 1$ Definition of factorial

$= 3{,}628{,}800$ Simplify.

Example 5 Use Factorials to Solve a Problem

ROLLER COASTERS Zach and Kurt are going to an amusement park. They cannot decide in which order to ride the 12 roller coasters in the park.

a. How many different orders can they ride all of the roller coasters if they ride each once?

Use a factorial.

$12! = 12 \cdot 11 \cdot 10 \cdot 9 \cdot 8 \cdot 7 \cdot 6 \cdot 5 \cdot 4 \cdot 3 \cdot 2 \cdot 1$ Definition of factorial

$= 479{,}001{,}600$ Simplify.

There are 479,001,600 ways in which Zach and Kurt can ride all 12 roller coasters.

b. If they only have time to ride 8 of the roller coasters, how many ways can they do this?

Use the Fundamental Counting Principle to find the sample space.

$s = 12 \cdot 11 \cdot 10 \cdot 9 \cdot 8 \cdot 7 \cdot 6 \cdot 5$ Fundamental Counting Principle

$= 19{,}958{,}400$ Simplify.

There are 19,958,400 ways for Zach and Kurt to ride 8 of the roller coasters.

More About. . .

Roller Coasters

In 2000, there were 646 roller coasters in the United States.

Type	Number
Wood	118
Steel	445
Inverted	35
Stand Up	10
Suspended	11
Wild Mouse	27

Source: Roller Coaster Database

Check for Understanding

Concept Check

1. **OPEN ENDED** Give an example of an event that has $7 \cdot 6$ or 42 outcomes.
2. **Draw** a tree diagram to represent the outcomes of tossing a coin three times.
3. **Explain** what the notation 5! means.

Guided Practice

For Exercises 4–6, suppose the spinner at the right is spun three times.

4. Draw a tree diagram to show the sample space.
5. How many outcomes are possible?
6. How many outcomes involve both green and blue?
7. Find the value of 8!.

Application

8. **SCHOOL** In a science class, each student must choose a lab project from a list of 15, write a paper on one of 6 topics, and give a presentation about one of 8 subjects. How many different ways can students choose to do their assignments?

Practice and Apply

Homework Help

For Exercises	See Examples
9, 10, 19	1
11–14	4
15–18, 20–22	2, 3, 5

Extra Practice

See page 851.

Draw a tree diagram to show the sample space for each event. Determine the number of possible outcomes.

9. earning an A, B, or C in English, Math, and Science classes
10. buying a computer with a choice of a CD-ROM, a CD recorder, or a DVD drive, one of 2 monitors, and either a printer or a scanner

Find the value of each expression.

11. 4!
12. 7!
13. 11!
14. 13!

15. Three dice, one red, one white, and one blue are rolled. How many outcomes are possible?

16. How many outfits are possible if you choose one each of 5 shirts, 3 pairs of pants, 3 pairs of shoes, and 4 jackets?

17. **TRAVEL** Suppose four different airlines fly from Seattle to Denver. Those same four airlines and two others fly from Denver to St. Louis. If there are no direct flights from Seattle to St. Louis, in how many ways can a traveler book a flight from Seattle to St. Louis?

COMMUNICATIONS For Exercises 18 and 19, use the following information.
A new 3-digit area code is needed in a certain area to accommodate new telephone numbers.

18. If the first digit must be odd, the second digit must be a 0 or a 1, and the third digit can be anything, how many area codes are possible?
19. Draw a tree diagram to show the different area codes using 4 or 5 for the first digit, 0 or 1 for the second digit, and 7, 8, or 9 for the third digit.

SOCCER For Exercises 20–22, use the following information.
The Columbus Crew are playing the D.C. United in a best three-out-of-five championship soccer series.

20. What are the possible outcomes of the series?
21. How many outcomes require exactly four games to determine the champion?
22. How many ways can D.C. United win the championship?

23. **CRITICAL THINKING** To get to and from school, Tucker can walk, ride his bike, or get a ride with a friend. Suppose that one week he walked 60% of the time, rode his bike 20% of the time, and rode with his friend 20% of the time. How many outcomes represent this situation? Assume that he returns home the same way that he went to school.

24. **WRITING IN MATH** Answer the question that was posed at the beginning of the lesson.

    How are possible win–loss records counted in football?

    Include the following in your answer:

    - a few sentences describing how a tree diagram can be used to count the wins and losses of a football team, and
    - a demonstration of how to find the number of possible outcomes for a team that plays 4 home games.

Standardized Test Practice

25. Evaluate 9!.

(A) 362,880 (B) 40,320 (C) 36 (D) 8

26. A car manufacturer offers a sports car in 4 different models with 6 different option packages. Each model is available in 12 different colors. How many different possibilities are available for this car?

(A) 96 (B) 144 (C) 288 (D) 384

Maintain Your Skills

Mixed Review **For Exercises 27–30, use box-and-whisker plots A and B.** *(Lesson 13-5)*

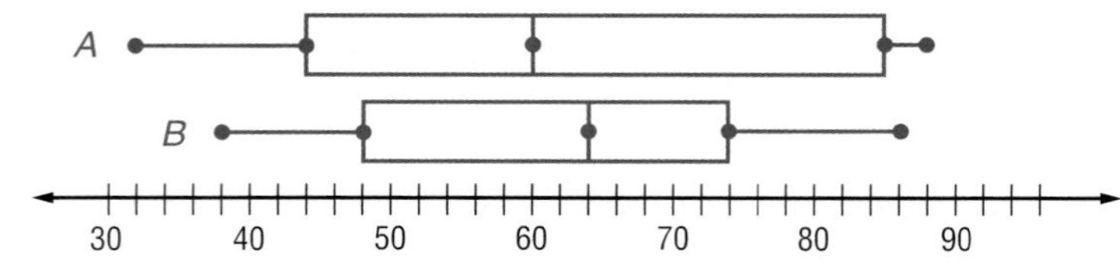

27. Determine the least value, greatest value, lower quartile, upper quartile, and median for each plot.

28. Which set of data contains the least value?

29. Which plot has the smaller interquartile range?

30. Which plot has the greater range?

For Exercises 31–34, use the stem-and-leaf plot. *(Lesson 13-4)*

Stem	Leaf
3	0 1 4 5
4	4 4 8
5	6 9
6	6 8
7	1 6
8	0 1
9	

31. Find the range of the data.

32. What is the median?

33. Determine the upper quartile, lower quartile, and interquartile range of the data.

34. Identify any outliers.

Find each sum or difference. *(Lesson 12-7)*

35. $\frac{2x + 1}{3x - 1} + \frac{x + 4}{x - 2}$

36. $\frac{4n}{2n + 6} + \frac{3}{n + 3}$

37. $\frac{3z + 2}{3z - 6} - \frac{z + 2}{z^2 - 4}$

38. $\frac{m - n}{m + n} - \frac{1}{m^2 - n^2}$

Study Tip

Deck of Cards

In this text, a *standard deck of cards* always means a deck of 52 playing cards. There are 4 suits—clubs (black), diamonds (red), hearts (red), and spades (black)—with 13 cards in each suit.

Solve each equation. *(Lesson 11-3)*

39. $5\sqrt{2n^2 - 28} = 20$

40. $\sqrt{5x^2 - 7} = 2x$

41. $\sqrt{x + 2} = x - 4$

Solve each equation by completing the square. Round to the nearest tenth if necessary. *(Lesson 10-3)*

42. $b^2 - 6b + 4 = 0$

43. $n^2 + 8n - 5 = 0$

44. $x^2 - 11x - 17 = 0$

45. $2p^2 + 10p + 3 = 0$

Getting Ready for the Next Lesson

PREREQUISITE SKILL One card is drawn at random from a standard deck of cards. Find each probability. *(To review* ***simple probability****, see Lesson 2-6.)*

46. $P(10)$

47. $P(\text{ace})$

48. $P(\text{red } 5)$

49. $P(\text{queen of clubs})$

50. $P(\text{even number})$

51. $P(3 \text{ or king})$

Algebra Activity

A Follow-Up of Lesson 14-1

Finite Graphs

The City Bus Company provides daily bus service between City College and Southland Mall, City College and downtown, downtown and Southland Mall, downtown and City Park, and City Park and the zoo. The daily routes can be represented using a **finite graph** like the one at the right.

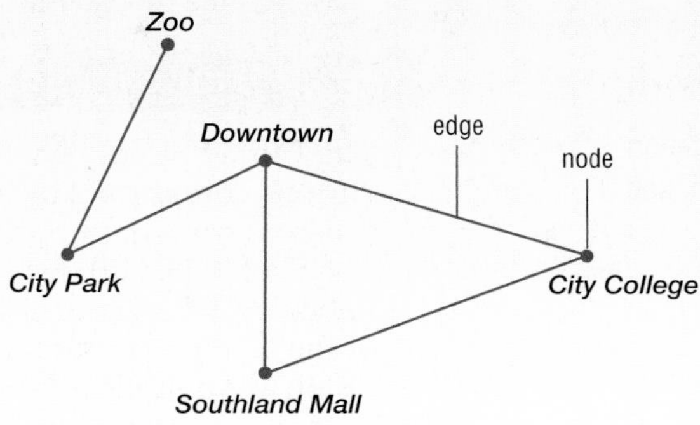

The graph is called a **network**, and each point on the graph is called a **node**. The paths connecting the nodes are called **edges**. A network is said to be **traceable** if all of the nodes can be connected, and each edge can be covered exactly once when the graph is used.

Collect the Data

The graph represents the streets on Alek's newspaper route. To get his route completed as quickly as possible, Alek would like to ride his bike down each street only once.

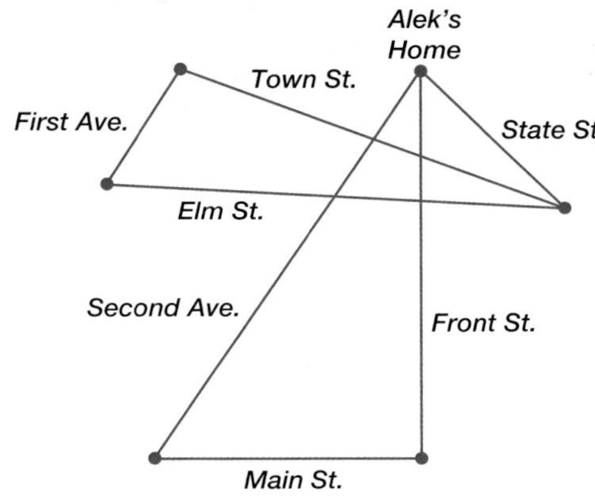

- Copy the graph onto your paper.
- Beginning at Alek's home, trace over his route without lifting your pencil. Remember to trace each edge only once.
- Compare your graph with those of your classmates.

Analyze the Data

1. Is Alek's route traceable? If so, describe his route.
2. Is there more than one traceable route that begins at Alek's house? If so, how many?
3. Suppose it does not matter where Alek starts his route. How many traceable routes are possible now?

Determine whether each graph is traceable. Explain your reasoning.

4. 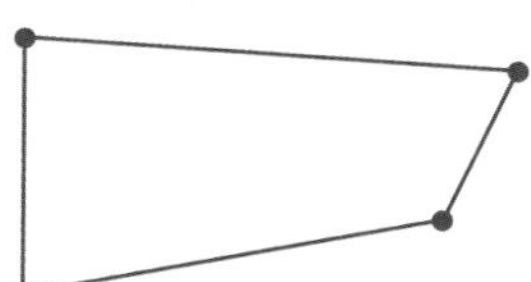

5. 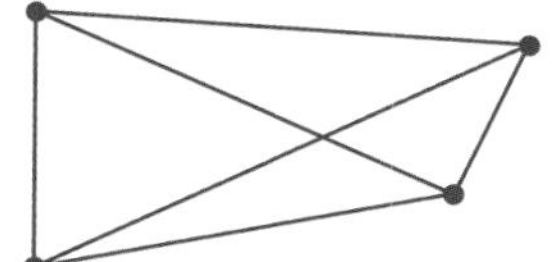

6. 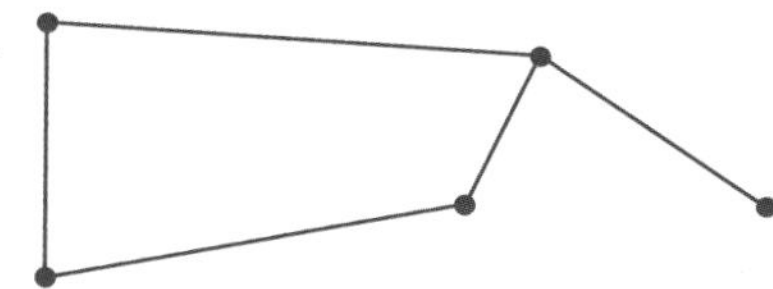

7. The campus for Centerburgh High School has five buildings built around the edge of a circular courtyard. There is a sidewalk between each pair of buildings.
   a. Draw a graph of the campus.
   b. Is the graph traceable?
   c. Suppose that there is not a sidewalk between the pairs of adjacent buildings. Is it possible to reach all five buildings without walking down any sidewalk more than once?
8. **Make a conjecture** for a rule to determine whether a graph is traceable.

14-2 Permutations and Combinations

What You'll Learn

- Determine probabilities using permutations.
- Determine probabilities using combinations.

Vocabulary

- permutation
- combination

How can combinations be used to form committees?

The United States Senate forms various committees by selecting senators from both political parties. The Senate Health, Education, Labor, and Pensions Committee of the 106th Congress was made up of 10 Republican senators and 8 Democratic senators. How many different ways could the committee have been selected? The members of the committee were selected in no particular order. This is an example of a combination.

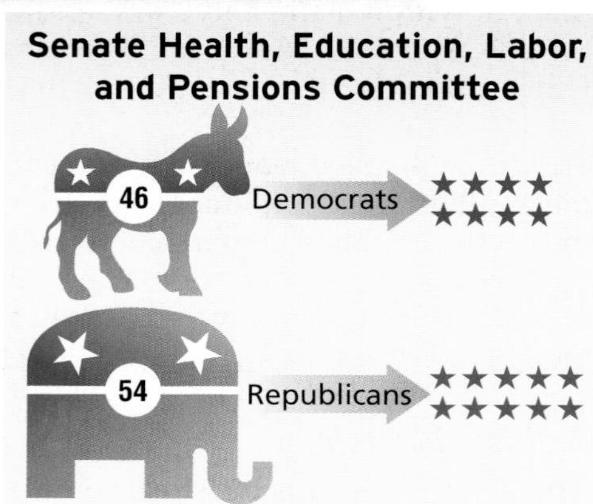

PERMUTATIONS An arrangement or listing in which order or placement is important is called a **permutation**.

Example 1 Tree Diagram Permutation

EMPLOYMENT The manager of a coffee shop needs to hire two employees, one to work at the counter and one to work at the drive-through window. Katie, Bob, Alicia, and Jeremiah all applied for a job. How many possible ways are there for the manager to place the applicants?

Use a tree diagram to show the possible arrangements.

Counter	Drive-Through	Outcomes
Katie (K)	Bob	KB
	Alicia	KA
	Jeremiah	KJ
Bob (B)	Katie	BK
	Alicia	BA
	Jeremiah	BJ
Alicia (A)	Jeremiah	AJ
	Katie	AK
	Bob	AB
Jeremiah (J)	Katie	JK
	Bob	JB
	Alicia	JA

There are 12 different ways for the 4 applicants to hold the 2 positions.

Study Tip

Common Misconception
When arranging two objects *A* and *B* using a permutation, the arrangement *AB* is different from the arrangement *BA*.

In Example 1, the positions are in a specific order, so each arrangement is unique. The symbol ${}_4P_2$ denotes the number of permutations when arranging 4 applicants in 2 positions. You can also use the Fundamental Counting Principle to determine the number of permutations.

$$
\begin{aligned}
{}_4P_2 &= \underbrace{4}_{\text{ways to choose first employee}} \cdot \underbrace{3}_{\text{ways to choose second employee}} \\
&= 4 \cdot 3 \cdot \frac{2 \cdot 1}{2 \cdot 1} && \frac{2 \cdot 1}{2 \cdot 1} = 1 \\
&= \frac{4 \cdot 3 \cdot 2 \cdot 1}{2 \cdot 1} && \text{Multiply.} \\
&= \frac{4!}{2!} && 4 \cdot 3 \cdot 2 \cdot 1 = 4!,\ 2 \cdot 1 = 2!
\end{aligned}
$$

In general, ${}_nP_r$ is used to denote the number of permutations of n objects taken r at a time.

Key Concept — Permutation

- **Words** The number of permutations of n objects taken r at a time is the quotient of $n!$ and $(n - r)!$.
- **Symbols** ${}_nP_r = \frac{n!}{(n-r)!}$

Example 2 Permutation

Find ${}_{10}P_6$.

$$
\begin{aligned}
{}_nP_r &= \frac{n!}{(n-r)!} && \text{Definition of } {}_nP_r \\
{}_{10}P_6 &= \frac{10!}{(10-6)!} && n = 10, r = 6 \\
{}_{10}P_6 &= \frac{10!}{4!} && \text{Subtract.} \\
{}_{10}P_6 &= \frac{10 \cdot 9 \cdot 8 \cdot 7 \cdot 6 \cdot 5 \cdot \overset{1}{\cancel{4 \cdot 3 \cdot 2 \cdot 1}}}{\underset{1}{\cancel{4 \cdot 3 \cdot 2 \cdot 1}}} && \text{Definition of factorial} \\
{}_{10}P_6 &= 10 \cdot 9 \cdot 8 \cdot 7 \cdot 6 \cdot 5 \text{ or } 151{,}200 && \text{Simplify.}
\end{aligned}
$$

There are 151,200 permutations of 10 objects taken 6 at a time.

Permutations are often used to find the probability of events occurring.

Example 3 Permutation and Probability

A word processing program requires a user to enter a 7-digit registration code made up of the digits 1, 2, 4, 5, 6, 7, and 9. Each number has to be used, and no number can be used more than once.

a. How many different registration codes are possible?

Since the order of the numbers in the code is important, this situation is a permutation of 7 digits taken 7 at a time.

$$
\begin{aligned}
{}_nP_r &= \frac{n!}{(n-r)!} && \text{Definition of permutation} \\
{}_7P_7 &= \frac{7!}{(7-7)!} && n = 7, r = 7 \\
{}_7P_7 &= \frac{7 \cdot 6 \cdot 5 \cdot 4 \cdot 3 \cdot 2 \cdot 1}{1} \text{ or } 5040 && \text{Definition of factorial}
\end{aligned}
$$

There are 5040 possible codes with the digits 1, 2, 4, 5, 6, 7, and 9.

Study Tip

Permutations
The number of permutations of n objects taken n at a time is $n!$. ${}_nP_n = n!$

b. What is the probability that the first three digits of the code are even numbers?

Use the Fundamental Counting Principle to determine the number of ways for the first three digits to be even.

- There are 3 even digits and 4 odd digits.
- The number of choices for the first three digits, if they are even, is $3 \cdot 2 \cdot 1$.
- The number of choices for the remaining odd digits is $4 \cdot 3 \cdot 2 \cdot 1$.
- The number of favorable outcomes is $3 \cdot 2 \cdot 1 \cdot 4 \cdot 3 \cdot 2 \cdot 1$ or 144. There are 144 ways for this event to occur out of the 5040 possible permutations.

$$P(\text{first 3 digits even}) = \frac{144}{5040} \quad \begin{matrix} \leftarrow \text{number of favorable outcomes} \\ \leftarrow \text{number of possible outcomes} \end{matrix}$$

$$= \frac{1}{35} \quad \text{Simplify.}$$

The probability that the first three digits of the code are even is $\frac{1}{35}$ or about 3%.

Study Tip

Look Back

To review **probability**, see Lesson 2-6.

COMBINATIONS An arrangement or listing in which order is not important is called a **combination**. For example, if you are choosing 2 salad ingredients from a list of 10, the order in which you choose the ingredients does not matter.

Key Concept — Combination

- **Words** The number of combinations of n objects taken r at a time is the quotient of $n!$ and $(n - r)!r!$.
- **Symbols** ${}_nC_r = \frac{n!}{(n-r)!r!}$

Example 4 Combination

Multiple-Choice Test Item

The students of Mr. DeLuca's homeroom had to choose 4 out of the 7 people who were nominated to serve on the Student Council. How many different groups of students could be selected?

Ⓐ 840   Ⓑ 210

Ⓒ 35   Ⓓ 24

Read the Test Item

The order in which the students are chosen does not matter, so this situation represents a combination of 7 people taken 4 at a time.

Solve the Test Item

$${}_nC_r = \frac{n!}{(n-r)!r!} \quad \text{Definition of combination}$$

$${}_7C_4 = \frac{7!}{(7-4)!4!} \quad n = 7, r = 4$$

$$= \frac{7 \cdot 6 \cdot 5 \cdot \cancel{4 \cdot 3 \cdot 2 \cdot 1}^{1}}{3 \cdot 2 \cdot 1 \cdot \cancel{4 \cdot 3 \cdot 2 \cdot 1}_{1}} \quad \text{Definition of factorial}$$

$$= \frac{7 \cdot 6 \cdot 5}{3 \cdot 2 \cdot 1} \text{ or } 35 \quad \text{Simplify.}$$

There are 35 different groups of students that could be selected. Choice C is correct.

The Princeton Review

Test-Taking Tip

Read each question carefully to determine whether the situation involves a permutation or a combination. Often, the answer choices include examples of both.

Combinations and the products of combinations can be used to determine probabilities.

Example 5 Use Combinations

SCHOOL A science teacher at Sunnydale High School needs to choose 12 students out of 16 to serve as peer tutors. A group of 7 seniors, 5 juniors, and 4 sophomores have volunteered to be tutors.

a. How many different ways can the teacher choose 12 students?

The order in which the students are chosen does not matter, so we must find the number of combinations of 16 students taken 12 at a time.

$${}_nC_r = \frac{n!}{(n-r)!r!} \quad \text{Definition of combination}$$

$${}_{16}C_{12} = \frac{16!}{(16-12)!12!} \quad n = 16, r = 12$$

$$= \frac{16!}{4!12!} \quad 16 - 12 = 4$$

$$= \frac{16 \cdot 15 \cdot 14 \cdot 13 \cdot \overset{1}{\cancel{12!}}}{4! \cdot \underset{1}{\cancel{12!}}} \quad \text{Divide by the GCF, 12!.}$$

$$= \frac{43{,}680}{24} \text{ or } 1820 \quad \text{Simplify.}$$

There are 1820 ways to choose 12 students out of 16.

b. If the students are chosen randomly, what is the probability that 4 seniors, 4 juniors, and 4 sophomores will be selected?

There are three questions to consider.

- How many ways can 4 seniors be chosen from 7?
- How many ways can 4 juniors be chosen from 5?
- How many ways can 4 sophomores be chosen from 4?

Using the Fundamental Counting Principle, the answer can be determined with the product of the three combinations.

$$\underbrace{({}_7C_4)}_{\text{ways to choose 4 seniors out of 7}} \cdot \underbrace{({}_5C_4)}_{\text{ways to choose 4 juniors out of 5}} \cdot \underbrace{({}_4C_4)}_{\text{ways to choose 4 sophomores out of 4}}$$

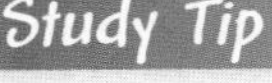

Study Tip

Combinations
The number of combinations of n objects taken n at a time is 1.
${}_nC_n = 1$

$$({}_7C_4)({}_5C_4)({}_4C_4) = \frac{7!}{(7-4)!4!} \cdot \frac{5!}{(5-4)!4!} \cdot \frac{4!}{(4-4)!4!} \quad \text{Definition of combination}$$

$$= \frac{7!}{3!4!} \cdot \frac{5!}{1!4!} \cdot \frac{4!}{0!4!} \quad \text{Simplify.}$$

$$= \frac{7 \cdot 6 \cdot 5}{3!} \cdot \frac{5}{1} \quad \text{Divide by the GCF, 4!.}$$

$$= 175 \quad \text{Simplify.}$$

There are 175 ways to choose this particular combination out of 1820 possible combinations.

$$P(\text{4 seniors, 4 juniors, 4 sophomores}) = \frac{175}{1820} \quad \begin{array}{l}\leftarrow \text{number of favorable outcomes} \\ \leftarrow \text{number of possible outcomes}\end{array}$$

$$= \frac{5}{52} \quad \text{Simplify.}$$

The probability that the science teacher will randomly select 4 seniors, 4 juniors, and 4 sophomores is $\frac{5}{52}$ or about 10%.

Check for Understanding

Concept Check

1. **OPEN ENDED** Describe the difference between a permutation and a combination. Then give an example of each.

2. **Demonstrate** and explain why ${}_nC_r = 1$ whenever $n = r$. What does ${}_nP_r$ always equal when $n = r$?

3. **FIND THE ERROR** Eric and Alisa are taking a trip to Washington, D.C. Their tour bus stops at the Lincoln Memorial, the Jefferson Memorial, the Washington Monument, the White House, the Capitol Building, the Supreme Court, and the Pentagon. Both are finding the number of ways they can choose to visit 5 of these 7 sites.

Eric	Alisa
${}_7C_5 = \frac{7!}{2!}$ or 2520	${}_7C_5 = \frac{7!}{2!\,5!}$ or 21

Who is correct? Explain your reasoning.

Guided Practice

Determine whether each situation involves a *permutation* or *combination*. Explain your reasoning.

4. choosing 6 books from a selection of 12 for summer reading
5. choosing digits for a personal identification number

Evaluate each expression.

6. ${}_8P_5$
7. ${}_7C_5$
8. $({}_{10}P_5)({}_3P_2)$
9. $({}_6C_2)({}_4C_3)$

For Exercises 10–12, use the following information.
The digits 0 through 9 are written on index cards. Three of the cards are randomly selected to form a 3-digit code.

10. Does this situation represent a permutation or a combination? Explain.
11. How many different codes are possible?
12. What is the probability that all 3 digits will be odd?

13. A diner offers a choice of two side items from the list with each entrée. How many ways can two items be selected?

Ⓐ 15 Ⓑ 28
Ⓒ 30 Ⓓ 56

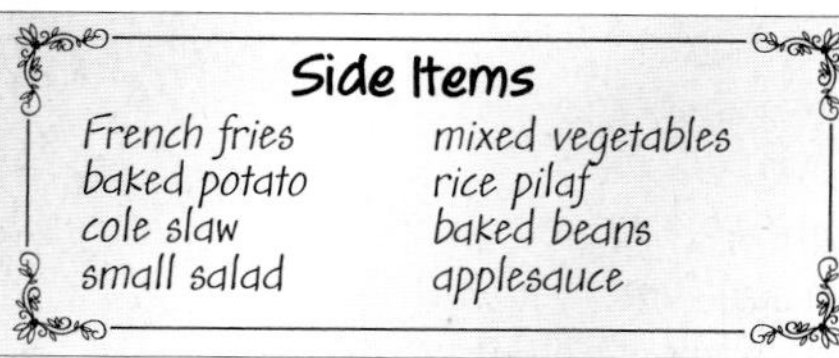

Practice and Apply

Determine whether each situation involves a *permutation* or *combination*. Explain your reasoning.

14. team captains for the soccer team
15. three mannequins in a display window
16. a hand of 10 cards from a selection of 52
17. the batting order of the New York Yankees

Homework Help

For Exercises	See Examples
14–21, 34 36, 40	1, 4
22–33, 35, 37–39, 41–49	2, 3, 5

Extra Practice

See page 851.

18. first place and runner-up winners for the table tennis tournament
19. a selection of 5 DVDs from a group of eight
20. selection of 2 candy bars from six equally-sized bars
21. the selection of 2 trombones, 3 clarinets, and 2 trumpets for a jazz combo

Evaluate each expression.

22. ${}_{12}P_3$
23. ${}_4P_1$
24. ${}_6C_6$
25. ${}_7C_3$
26. ${}_{15}C_3$
27. ${}_{20}C_8$
28. ${}_{15}P_3$
29. ${}_{16}P_5$
30. $({}_7P_7)({}_7P_1)$
31. $({}_{20}P_2)({}_{16}P_4)$
32. $({}_3C_2)({}_7C_4)$
33. $({}_8C_5)({}_5P_5)$

SOFTBALL For Exercises 34 and 35, use the following information.
The manager of a softball team needs to prepare a batting lineup using her nine starting players.

34. Does this situation involve a permutation or a combination?
35. How many different lineups can she make?

More About. . .

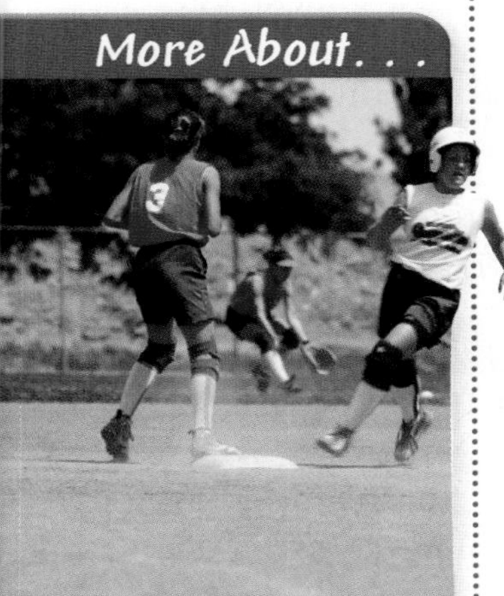

Softball

The game of softball was developed in 1888 as an indoor sport for practicing baseball during the winter months.

Source: www.encyclopedia.com

SCHOOL For Exercises 36–39, use the following information.
Mrs. Moyer's class has to choose 4 out of 12 people to serve on an activity committee.

36. Does the selection of the students involve a permutation or a combination? Explain.
37. How many different groups of students could be selected?
38. Suppose the students are selected for the positions of chairperson, activities planner, activity leader, and treasurer. How many different groups of students could be selected?
39. What is the probability that any one of the students is chosen to be the chairperson?

GAMES For Exercises 40–42, use the following information.
In your turn of a certain game, you roll five different-colored dice.

40. Do the outcomes of rolling the five dice represent a permutation or a combination? Explain.
41. How many outcomes are possible?
42. What is the probability that all five dice show the same number on one roll?

BUSINESS For Exercises 43 and 44, use the following information.
There are six positions available in the research department of a software company. Of the applicants, 15 are men and 10 are women.

43. In how many ways could 4 men and 2 women be chosen if each were equally qualified?
44. What is the probability that five women are selected if the positions are randomly filled?

TRACK For Exercises 45 and 46, use the following information.
Central High School is competing against West High School at a track meet. Each team entered 4 girls to run the 1600-meter event. The top three finishers are awarded medals.

45. How many different ways can the runners place first, second, and third?
46. If all eight runners have an equal chance of placing, what is the probability that the first and second place finishers are from West and the third place finisher is from Central?

DINING For Exercises 47–49, use the following information.
For lunch in the school cafeteria, you can select one item from each category to get the daily combo.

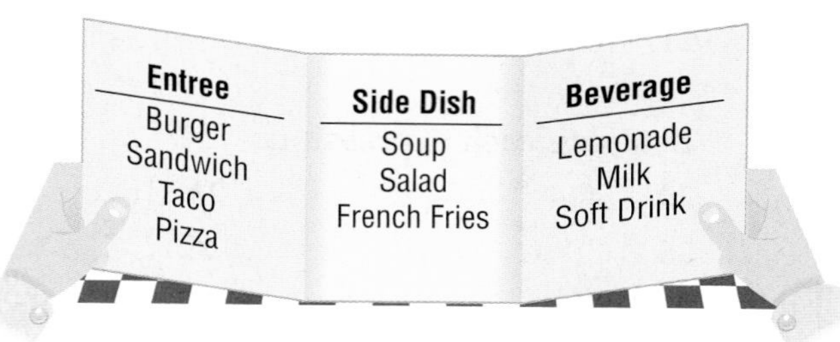

Entree
Burger
Sandwich
Taco
Pizza
Side Dish
Soup
Salad
French Fries
Beverage
Lemonade
Milk
Soft Drink

47. Find the number of possible meal combinations.
48. If a side dish is chosen at random, what is the probability that a student will choose soup?
49. What is the probability that a student will randomly choose a sandwich and soup?

CRITICAL THINKING For Exercises 50 and 51, use the following information.
Larisa is trying to solve a word puzzle. She needs to arrange the letters H, P, S, T, A, E, and O into a two-word arrangement.

50. How many different arrangements of the letters can she make?
51. Assuming that each arrangement has an equal chance of occurring, what is the probability that she will form the words *tap shoe* on her first try?

WebQuest

You can use permutations and combinations to analyze data on U.S. schools. Visit www.algebra1.com/webquest to continue work on your WebQuest project.

SWIMMING For Exercises 52–54, use the following information.
A swimming coach plans to pick four swimmers out of a group of 6 to form the 400-meter freestyle relay team.

52. How many different teams can he form?
53. The coach must decide in which order the four swimmers should swim. He timed the swimmers in each possible order and chose the best time. How many relays did the four swimmers have to swim so that the coach could collect all of the data necessary?
54. If Tomás is chosen to be on the team, what is the probability that he will swim in the third leg?

55. **WRITING IN MATH** Answer the question that was posed at the beginning of the lesson.

How can combinations be used to form committees?

Include the following in your answer:

- a few sentences explaining why forming a Senate committee is a combination, and
- an explanation of how to find the number of ways to select the committee if committee positions are based upon seniority.

56. There are 12 songs on a CD. If 10 songs are played randomly and each song is played once, how many arrangements are there?

    (A) 479,001,600 (B) 239,500,800 (C) 66 (D) 1

57. Julie remembered that the 4 digits of her locker combination were 4, 9, 15, and 22, but not their order. What is the maximum number of attempts Julie has to make to find the correct combination?

    (A) 4 (B) 16 (C) 24 (D) 256

Maintain Your Skills

Mixed Review

58. The Sanchez family acts as a host family for a foreign exchange student during each school year. It is equally likely that they will host a girl or a boy. How many different ways can they host boys and girls over the next four years? *(Lesson 14-1)*

STATISTICS For Exercises 59–62, use the table at the right. *(Lesson 13-5)*

59. Make a box-and-whisker plot of the data.
60. What is the range of the data?
61. Identify the lower and upper quartiles.
62. Name any outliers.

Highest Paying Occupations in America

Occupation	Median Salary
Physician	$148,000
Dentist	$93,000
Lobbyist	$91,300
Management Consultant	$61,900
Lawyer	$60,500
Electrical Engineer	$59,100
School Principal	$57,300
Aeronautical Engineer	$56,700
Airline Pilot	$56,500
Civil Engineer	$55,800

Source: U.S. Bureau of Labor Statistics

Online Research **Data Update** For current data on the highest-paying occupations, visit: www.algebra1.com/data_update

Simplify each expression. *(Lesson 12-2)*

63. $\dfrac{x+3}{x^2+6x+9}$
64. $\dfrac{x^2-49}{x^2-2x-35}$
65. $\dfrac{n^2-n-20}{n^2+9n+20}$

Find the distance between each pair of points whose coordinates are given. Express answers in simplest radical form and as decimal approximations rounded to the nearest hundredth if necessary. *(Lesson 11-5)*

66. (12, 20), (16, 34)
67. $(-18, 7), (2, 15)$
68. $(-2, 5), \left(-\frac{1}{2}, 3\right)$

Solve each equation by using the Quadratic Formula. Approximate irrational roots to the nearest hundredth. *(Lesson 10-4)*

69. $m^2 + 4m + 2 = 0$
70. $2s^2 + s - 15 = 0$
71. $2n^2 - n = 4$

Getting Ready for the Next Lesson

PREREQUISITE SKILL Find each sum or difference.
(To review ***fractions****, see pages 798 and 799.)*

72. $\frac{8}{52} + \frac{4}{52}$
73. $\frac{7}{32} + \frac{5}{8}$
74. $\frac{5}{15} + \frac{6}{15} - \frac{2}{15}$
75. $\frac{15}{24} + \frac{11}{24} - \frac{3}{4}$
76. $\frac{2}{3} + \frac{15}{36} - \frac{1}{4}$
77. $\frac{16}{25} + \frac{3}{10} - \frac{1}{4}$

Practice Quiz 1

Lessons 14-1 and 14-2

Find the number of outcomes for each event. *(Lesson 14-1)*

1. A die is rolled and two coins are tossed.
2. A certain model of mountain bike comes in 5 sizes, 4 colors, with regular or off-road tires, and with a choice of 1 of 5 accessories.

Find each value. *(Lesson 14-2)*

3. ${}_{13}C_8$
4. ${}_9P_6$

5. A flower bouquet has 5 carnations, 6 roses, and 3 lilies. If four flowers are selected at random, what is the probability of selecting two roses and two lilies? *(Lesson 14-2)*

Reading Mathematics

Mathematical Words and Related Words

You may have noticed that many words used in mathematics contain roots of other words and are closely related to other English words. You can use the more familiar meanings of these related words to better understand mathematical meanings.

The table shows two mathematical terms along with related words and their meanings as well as additional notes.

Mathematical Term and Meaning	Related Words and Meanings	Notes
combination A combination is a selection of distinct objects from a group of objects, where the order in which they were selected does not matter.	*combine* (n): a harvesting machine that performs many functions *binary*: a base-two numerical system	*Combine* originally meant to put just two things together; it now means to put any number of things together.
permutation A permutation is an arrangement of distinct objects from a group of objects, where the arrangement is in a certain order.	*mutation*: a change in genes or other characteristics *commute*: to change places; for example, $2 + 5 = 5 + 2$	

Notice how the meanings of the related words can give an insight to the meanings of the mathematical terms.

Reading to Learn

1. Do the related words of combination and permutation help you to remember their mathematical meanings? Explain.
2. What is a similarity and a difference between the mathematical meanings of combination and permutation?
3. **RESEARCH** Use the Internet or other reference to find the mathematical meaning of the word *factorial* and meanings of at least two related words. How are these meanings connected?
4. **RESEARCH** Use the Internet or other reference to find the meanings of the word *probability* and its Latin origins *probus* and *probare*. Compare the three.

14-3 Probability of Compound Events

What You'll Learn

- Find the probability of two independent events or dependent events.
- Find the probability of two mutually exclusive or inclusive events.

Vocabulary

- simple event
- compound event
- independent events
- dependent events
- complements
- mutually exclusive
- inclusive

How are probabilities used by meteorologists?

The weather forecast for the weekend calls for rain. By using the probabilities for both days, we can find other probabilities for the weekend. What is the probability that it will rain on both days? only on Saturday? Saturday or Sunday?

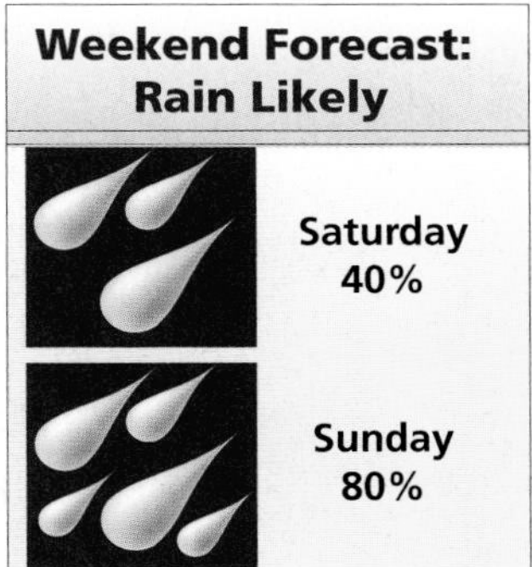

Weekend Forecast: Rain Likely
Saturday 40%
Sunday 80%

INDEPENDENT AND DEPENDENT EVENTS A single event, like rain on Saturday, is called a **simple event**. Suppose you wanted to determine the probability that it will rain both Saturday and Sunday. This is an example of a **compound event**, which is made up of two or more simple events. The weather on Saturday does not affect the weather on Sunday. These two events are called **independent events** because the outcome of one event does not affect the outcome of the other.

Key Concept — *Probability of Independent Events*

- **Words** If two events, *A* and *B*, are independent, then the probability of both events occurring is the product of the probability of *A* and the probability of *B*.
- **Symbols** $P(A \text{ and } B) = P(A) \cdot P(B)$
- **Model**

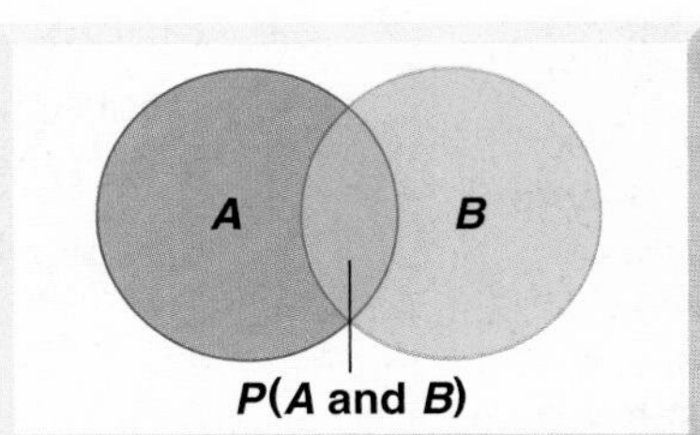

Example 1 *Independent Events*

Refer to the application above. Find the probability that it will rain on Saturday and Sunday.

$P(A \text{ and } B) = P(A) \cdot P(B)$ — Definition of independent events

$P(\text{Saturday and Sunday}) = \underbrace{P(\text{Saturday})} \cdot \underbrace{P(\text{Sunday})}$

$= 0.4 \cdot 0.8$ — 40% = 0.4 and 80% = 0.8

$= 0.32$ — Multiply.

The probability that it will rain on Saturday and Sunday is 32%.

When the outcome of one event affects the outcome of another event, the events are **dependent events**. For example, drawing a card from a deck, not returning it, then drawing a second card are dependent events because the drawing of the second card is dependent on the drawing of the first card.

Key Concept *Probability of Dependent Events*

- **Words** If two events, A and B, are dependent, then the probability of both events occurring is the product of the probability of A and the probability of B after A occurs.
- **Symbols** $P(A \text{ and } B) = P(A) \cdot P(B \text{ following } A)$

Example 2 Dependent Events

A bag contains 8 red marbles, 12 blue marbles, 9 yellow marbles, and 11 green marbles. Three marbles are randomly drawn from the bag and not replaced. Find each probability if the marbles are drawn in the order indicated.

a. P(red, blue, green)

The selection of the first marble affects the selection of the next marble since there is one less marble from which to choose. So, the events are dependent.

First marble: $P(\text{red}) = \frac{8}{40} \text{ or } \frac{1}{5}$ ← $\frac{\text{number of red marbles}}{\text{total number of marbles}}$

Second marble: $P(\text{blue}) = \frac{12}{39} \text{ or } \frac{4}{13}$ ← $\frac{\text{number of blue marbles}}{\text{number of marbles remaining}}$

Third marble: $P(\text{green}) = \frac{11}{38}$ ← $\frac{\text{number of green marbles}}{\text{number of marbles remaining}}$

$$P(\text{red, blue, green}) = \underbrace{P(\text{red})} \cdot \underbrace{P(\text{blue})} \cdot \underbrace{P(\text{green})}$$

$$= \frac{1}{5} \cdot \frac{4}{13} \cdot \frac{11}{38} \quad \text{Substitution}$$

$$= \frac{44}{2470} \text{ or } \frac{22}{1235} \quad \text{Multiply.}$$

The probability of drawing red, blue, and green marbles is $\frac{22}{1235}$.

b. P(blue, yellow, yellow)

Notice that after selecting a yellow marble, not only is there one fewer marble from which to choose, there is also one fewer yellow marble.

$$P(\text{blue, yellow, yellow}) = \underbrace{P(\text{blue})} \cdot \underbrace{P(\text{yellow})} \cdot \underbrace{P(\text{yellow})}$$

$$= \frac{12}{40} \cdot \frac{9}{39} \cdot \frac{8}{38} \quad \text{Substitution}$$

$$= \frac{864}{59{,}280} \text{ or } \frac{18}{1235} \quad \text{Multiply.}$$

The probability of drawing a blue and then two yellow marbles is $\frac{18}{1235}$.

c. P(red, yellow, *not* green)

Since the marble that is not green is selected after the first two marbles, there are 29 − 2 or 27 marbles that are not green.

$$P(\text{red, yellow, } not \text{ green}) = \underbrace{P(\text{red})} \cdot \underbrace{P(\text{yellow})} \cdot \underbrace{P(\text{not green})}$$

$$= \frac{8}{40} \cdot \frac{9}{39} \cdot \frac{27}{38}$$

$$= \frac{1944}{59{,}280} \text{ or } \frac{81}{2470}$$

The probability of drawing a red, a yellow, and *not* a green marble is $\frac{81}{2470}$.

Study Tip

More Than Two Dependent Events
Notice that the formula for the probability of dependent events can be applied to more than two events.

In part **c** of Example 2, the events for drawing a marble that is green and for drawing a marble that is *not* green are called **complements**. Consider the probabilities for drawing the third marble.

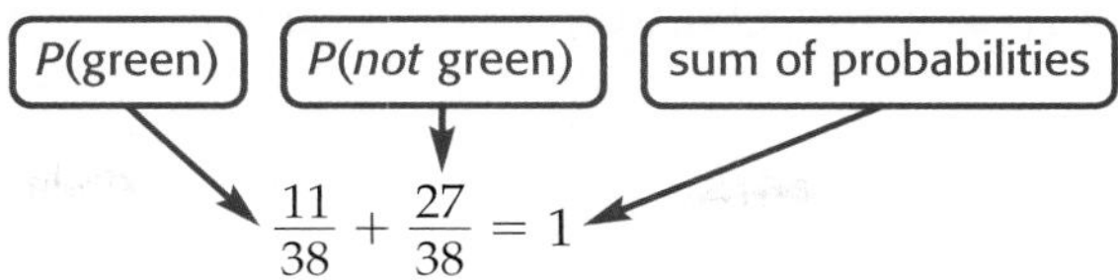

$$\frac{11}{38} + \frac{27}{38} = 1$$

This is always true for any two complementary events.

Study Tip

Reading Math

A complement is one of two parts that make up a whole.

MUTUALLY EXCLUSIVE AND INCLUSIVE EVENTS

Events that cannot occur at the same time are called **mutually exclusive**. Suppose you want to find the probability of rolling a 2 *or* a 4 on a die. Since a die cannot show both a 2 and a 4 at the same time, the events are mutually exclusive.

Key Concept — *Mutually Exclusive Events*

- **Words** If two events, A and B, are mutually exclusive, then the probability that either A or B occurs is the sum of their probabilities.
- **Symbols** $P(A \text{ or } B) = P(A) + P(B)$
- **Model**

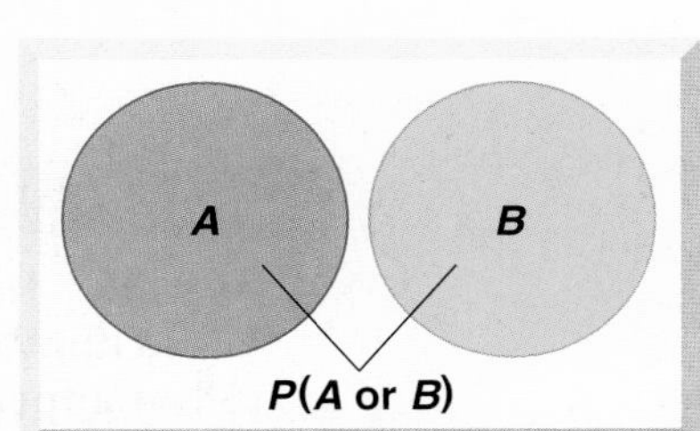

Example 3 Mutually Exclusive Events

During a magic trick, a magician randomly draws one card from a standard deck of cards. What is the probability that the card drawn is a heart or a diamond?

Since a card cannot be both a heart and a diamond, the events are mutually exclusive.

$P(\text{heart}) = \frac{13}{52} \text{ or } \frac{1}{4}$ ← number of hearts / ← total number of cards

$P(\text{diamond}) = \frac{13}{52} \text{ or } \frac{1}{4}$ ← number of diamonds / ← total number of cards

$$\begin{aligned} P(\text{heart or diamond}) &= P(\text{heart}) + P(\text{diamond}) && \text{Definition of mutually exclusive events} \\ &= \frac{1}{4} + \frac{1}{4} && \text{Substitution} \\ &= \frac{2}{4} \text{ or } \frac{1}{2} && \text{Add.} \end{aligned}$$

The probability of drawing a heart or a diamond is $\frac{1}{2}$.

Suppose you wanted to find the probability of randomly selecting an ace or a spade from a standard deck of cards. Since it is possible to draw a card that is both an ace and a spade, these events are not mutually exclusive. They are called **inclusive** events.

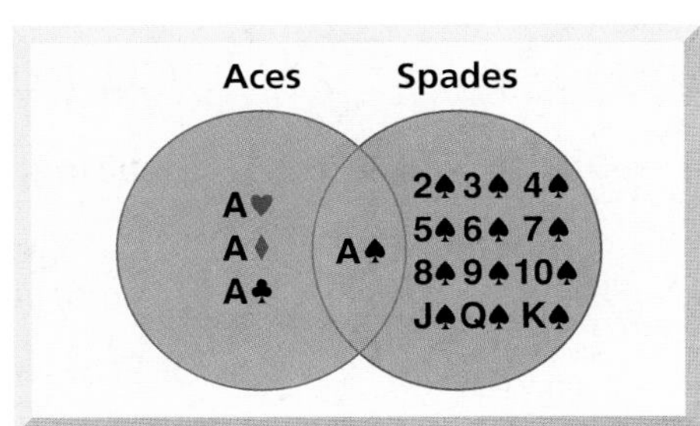

If the formula for the probability of mutually exclusive events is used, the probability of drawing an ace of spades is counted twice, once for an ace and once for a spade. To correct this, you must subtract the probability of drawing the ace of spades from the sum of the individual probabilities.

Key Concept — Probability of Inclusive Events

- **Words** If two events, A and B, are inclusive, then the probability that either A or B occurs is the sum of their probabilities decreased by the probability of both occurring.
- **Model**

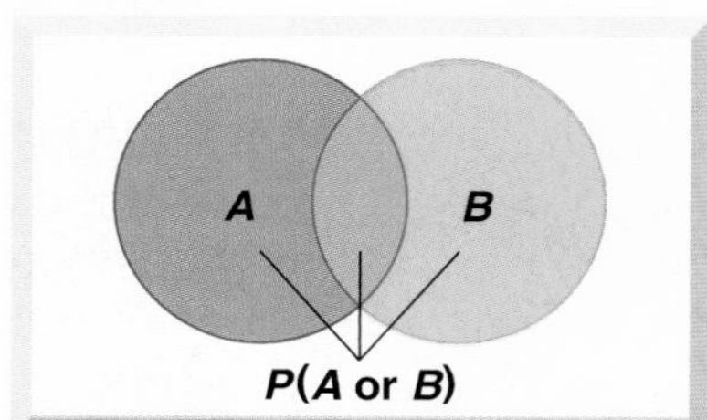

- **Symbols** $P(A \text{ or } B) = P(A) + P(B) - P(A \text{ and } B)$

Example 4 Inclusive Events

GAMES In the game of bingo, balls or tiles are numbered 1 through 75. These numbers correspond to columns on a bingo card. The numbers 1 through 15 can appear in the B column, 16 through 30 in the I column, 31 through 45 in the N column, 46 through 60 in the G column, and 61 through 75 in the O column. A number is selected at random. What is the probability that it is a multiple of 4 or is in the O column?

Since the numbers 64, 68, and 72 are multiples of 4 and they can be in the O column, these events are inclusive.

$P(A \text{ or } B) = P(A) + P(B) - P(A \text{ and } B)$ Definition of inclusive events

$P(\text{multiple of 4 or O column})$

$$= \underbrace{P(\text{multiple of 4})} + \underbrace{P(\text{O column})} - \underbrace{P(\text{multiple of 4 and O column})}$$

$$= \frac{18}{75} + \frac{15}{75} - \frac{3}{75} \quad \text{Substitution}$$

$$= \frac{18 + 15 - 3}{75} \quad \text{LCD is 75.}$$

$$= \frac{30}{75} \text{ or } \frac{2}{5} \quad \text{Simplify.}$$

The probability of a number being a multiple of 4 or in the O column is $\frac{2}{5}$ or 40%.

Check for Understanding

Concept Check

1. **Explain** the difference between a simple event and a compound event.
2. **Find a counterexample** for the following statement.
   If two events are independent, then the probability of both events occurring is less than 1.
3. **OPEN ENDED** Explain how dependent events are different than independent events. Give specific examples in your explanation.

4. **FIND THE ERROR** On the school debate team, 6 of the 14 girls are seniors, and 9 of the 20 boys are seniors. Chloe and Amber are both seniors on the team. Each girl calculated the probability that either a girl or a senior would randomly be selected to argue a position at a state debate.

Chloe	Amber
P(girl or senior)	P(girl or senior)
$= \frac{14}{34} + \frac{15}{34} - \frac{6}{34}$	$= \frac{6}{34} + \frac{15}{34} - \frac{14}{34}$
$= \frac{23}{34}$	$= \frac{7}{34}$

Who is correct? Explain your reasoning.

Guided Practice

A bin contains 8 blue chips, 5 red chips, 6 green chips, and 2 yellow chips. Find each probability.

5. drawing a red chip, replacing it, then drawing a green chip
6. selecting two yellow chips without replacement
7. choosing green, then blue, then red, replacing each chip after it is drawn
8. choosing green, then blue, then red without replacing each chip

A student is selected at random from a group of 12 male and 12 female students. There are 3 male students and 3 female students from each of the 9th, 10th, 11th, and 12th grades. Find each probability.

9. *P*(9th or 12th grader)
10. *P*(10th grader or female)
11. *P*(male or female)
12. *P*(male or not 11th grader)

Application

BUSINESS For Exercises 13–15, use the following information.
Mr. Salyer is a buyer for an electronics store. He received a shipment of 5 DVD players in which one is defective. He randomly chose 3 of the DVD players to test.

13. Determine whether choosing one DVD player after another indicates independent or dependent events.
14. What is the probability that he selected the defective player?
15. Suppose the defective player is one of the three that Mr. Salyer tested. What is the probability that the last one tested was the defective one?

Practice and Apply

Homework Help

For Exercises	See Examples
16–19, 24, 25, 28–31	2
20–23, 32–34	1
26, 27, 41, 44, 45	4
36–40, 42, 43, 46, 47	3

Extra Practice

See page 851.

A bag contains 2 red, 6 blue, 7 yellow, and 3 orange marbles. Once a marble is selected, it is not replaced. Find each probability.

16. *P*(2 orange)
17. *P*(blue, then red)
18. *P*(2 yellows in a row then orange)
19. *P*(blue, then yellow, then red)

A die is rolled and a spinner like the one at the right is spun. Find each probability.

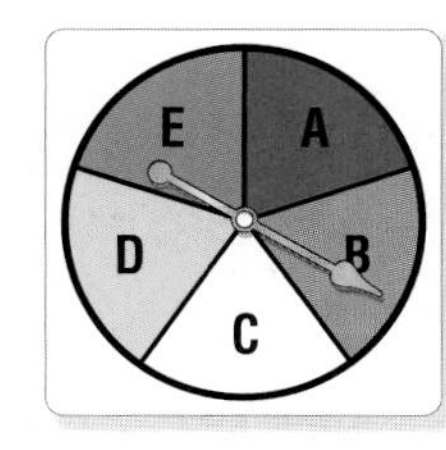

20. *P*(3 and D)
21. *P*(an odd number and a vowel)
22. *P*(a prime number and A)
23. *P*(2 and A, B, or C)

Raffle tickets numbered 1 through 30 are placed in a box. Tickets for a second raffle numbered 21 to 48 are placed in another box. One ticket is randomly drawn from each box. Find each probability.

24. Both tickets are even.

25. Both tickets are greater than 20 and less than 30.

26. The first ticket is greater than 10, and the second ticket is less than 40 or odd.

27. The first ticket is greater than 12 or prime, and the second ticket is a multiple of 6 or a multiple of 4.

More About. . .

Safety

In the U.S., 60% of carbon monoxide emissions come from transportation sources. The largest contributor is highway motor vehicles. In urban areas, motor vehicles can contribute more than 90%.

Source: U.S. Environmental Protection Agency

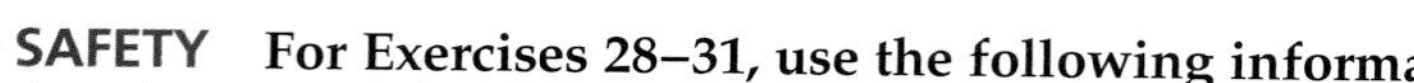

SAFETY For Exercises 28–31, use the following information.
A carbon monoxide detector system uses two sensors, A and B. If carbon monoxide is present, there is a 96% chance that sensor A will detect it, a 92% chance that sensor B will detect it, and a 90% chance that both sensors will detect it.

28. Draw a Venn diagram that illustrates this situation.

29. If carbon monoxide is present, what is the probability that it will be detected?

30. What is the probability that carbon monoxide would go undetected?

31. Do sensors A and B operate independently of each other? Explain.

BIOLOGY For Exercises 32–34, use the table and following information.
Each person carries two types of genes for eye color. The gene for brown eyes (B) is dominant over the gene for blue eyes (b). That is, if a person has one gene for brown eyes and the other for blue, that person will have brown eyes. The Punnett square at the right shows the genes for two parents.

	B	b
B	BB	Bb
b	Bb	bb

32. What is the probability that any child will have blue eyes?

33. What is the probability that the couple's two children both have brown eyes?

34. Find the probability that the first or the second child has blue eyes.

35. RESEARCH Use the Internet or other reference to investigate various blood types. Use this information to determine the probability of a child having blood type O if the father has blood type A(Ai) and the mother has blood type B(Bi).

TRANSPORTATION For Exercises 36 and 37, use the graph and the following information.
The U.S. Census Bureau conducted an American Community Survey in Lake County, Illinois. The circle graph at the right shows the survey results of how people commute to work.

36. If a person from Lake County was chosen at random, what is the probability that he or she uses public transportation or walks to work?

37. If offices are being built in Lake County to accommodate 400 employees, what is the minimum number of parking spaces an architect should plan for the parking lot?

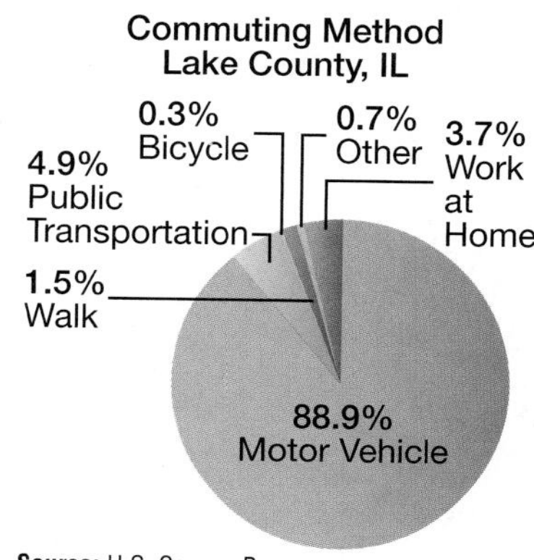

Source: U.S. Census Bureau

ECONOMICS **For Exercises 38–40, use the table below that compares the total number of hourly workers who earned the minimum wage of \$5.15 with those making less than minimum wage.**

Number of Hourly Workers (thousands)			
Age (years)	Total	At \$5.15	Below \$5.15
16–24	15,793	1145	2080
25+	55,287	970	2043

Source: U.S. Bureau of Labor Statistics

38. If an hourly worker was chosen at random, what is the probability that he or she earned minimum wage? less than minimum wage?

39. What is the probability that a randomly-chosen hourly worker earned less than or equal to minimum wage?

40. If you randomly chose an hourly worker from each age group, which would you expect to have earned no more than minimum wage? Explain.

More About. . .

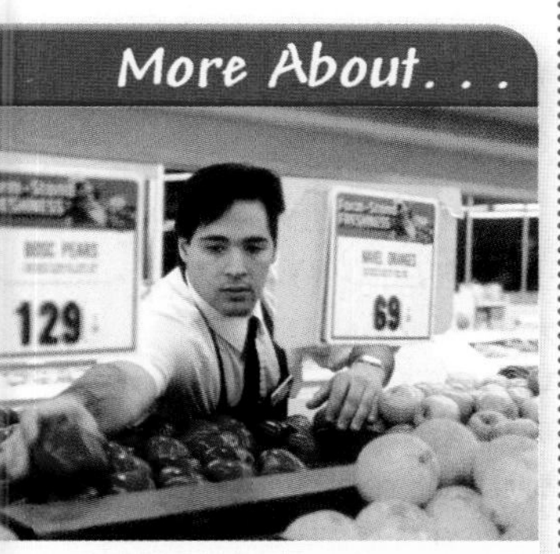

Economics

The first federal minimum wage was set in 1938 at \$0.25 per hour. That was the equivalent of \$3.05 in 2000.

Source: U.S. Department of Labor

GEOMETRY **For Exercises 41–43, use the figure and the following information.**
Two of the six non-straight angles in the figure are chosen at random.

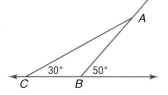

41. What is the probability of choosing an angle inside $\triangle ABC$ or an obtuse angle?

42. What is the probability of selecting a straight angle or a right angle inside $\triangle ABC$?

43. Find the probability of picking a $20°$ angle or a $130°$ angle.

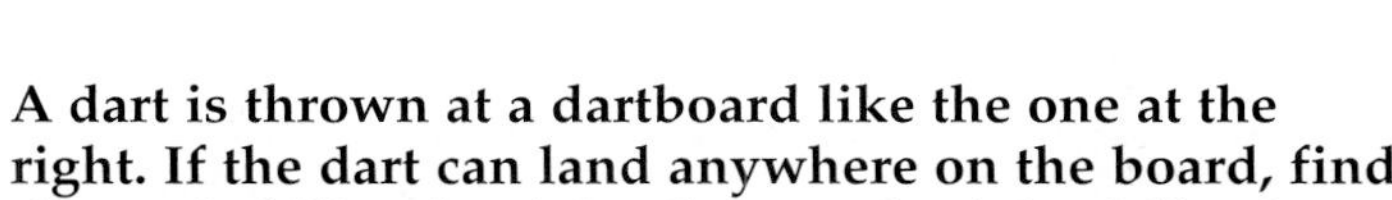

A dart is thrown at a dartboard like the one at the right. If the dart can land anywhere on the board, find the probability that it lands in each of the following.

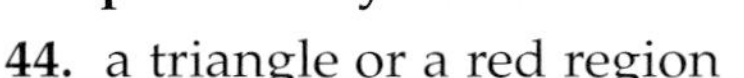

44. a triangle or a red region

45. a trapezoid or a blue region

46. a blue triangle or a red triangle

47. a square or a hexagon

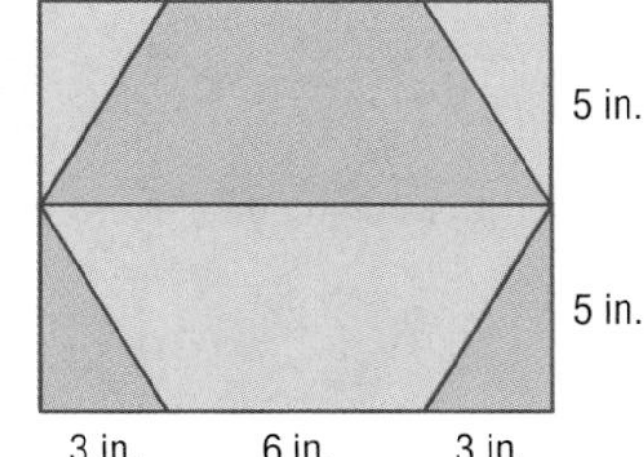

CRITICAL THINKING **For Exercises 48–51, use the following information.**
A sample of high school students were asked if they:

A) drive a car to school,

B) are involved in after-school activities, or

C) have a part-time job.

The results of the survey are shown in the Venn diagram.

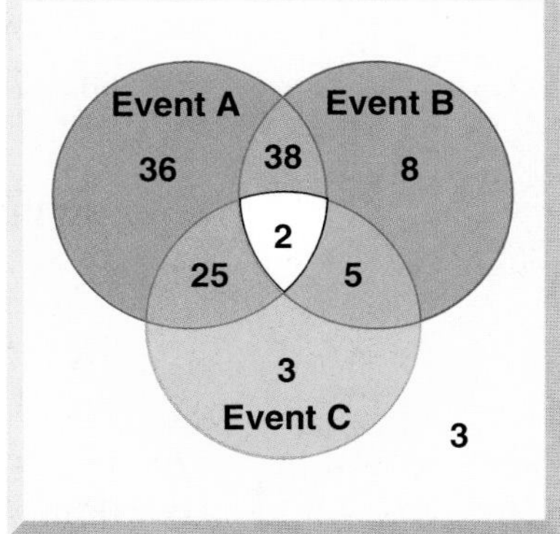

48. How many students were surveyed?

49. How many students said that they drive a car to school?

50. If a high school student is chosen at random, what is the probability that he or she does all three?

51. What is the probability that a randomly-chosen student drives a car to school or is involved in after-school activities or has a part-time job?

52. **WRITING IN MATH** Answer the question that was posed at the beginning of the lesson.

How are probabilities used by meteorologists?

Include the following in your answer:

- a few sentences about how compound probabilities can be used to predict the weather, and
- assuming that the events are independent, the probability that it will rain either Saturday or Sunday if there is a 30% chance of rain on Saturday and a 50% chance of rain Sunday.

53. A bag contains 8 red marbles, 5 blue marbles, 4 green marbles, and 7 yellow marbles. Five marbles are randomly drawn from the bag and not replaced. What is the probability that the first three marbles drawn are red?

(A) $\frac{1}{27}$ (B) $\frac{28}{1771}$ (C) $\frac{7}{253}$ (D) $\frac{7}{288}$

54. Yolanda usually makes 80% of her free throws. What is the probability that she will make at least one free throw in her next three attempts?

(A) 99.2% (B) 51.2% (C) 38.4% (D) 9.6%

Maintain Your Skills

Mixed Review

CIVICS For Exercises 55 and 56, use the following information.
The Stratford town council wants to form a 3-person parks committee. Five people have applied to be on the committee. *(Lesson 14-2)*

55. How many committees are possible?

56. What is the probability of any one person being selected if each has an equal chance?

57. **BUSINESS** A real estate developer built a strip mall with seven different-sized stores. Ten small businesses have shown interest in renting space in the mall. The developer must decide which business would be best suited for each store. How many different arrangements are possible? *(Lesson 14-1)*

Find each sum or difference. *(Lesson 13-2)*

58. $\begin{bmatrix} 3 & -6 \\ -1 & 2 \end{bmatrix} + \begin{bmatrix} -2 & 4 \\ 1 & 5 \end{bmatrix}$

59. $\begin{bmatrix} -4 & -5 \\ 8 & 8 \end{bmatrix} - \begin{bmatrix} -9 & -7 \\ 4 & 9 \end{bmatrix}$

60. Find the quotient of $\frac{2m^2 + 7m - 15}{m + 5}$ and $\frac{9m^2 - 4}{3m + 2}$. *(Lesson 12-4)*

Simplify. *(Lesson 11-1)*

61. $\sqrt{45}$

62. $\sqrt{128}$

63. $\sqrt{40b^4}$

64. $\sqrt{120a^3b}$

65. $3\sqrt{7} \cdot 6\sqrt{2}$

66. $\sqrt{3}(\sqrt{3} + \sqrt{6})$

Getting Ready for the Next Lesson

PREREQUISITE SKILL Express each fraction as a decimal. Round to the nearest thousandth. *(To review **expressing fractions as decimals**, see pages 804 and 805.)*

67. $\frac{9}{24}$

68. $\frac{2}{15}$

69. $\frac{63}{128}$

70. $\frac{5}{52}$

71. $\frac{8}{36}$

72. $\frac{11}{38}$

73. $\frac{81}{2470}$

74. $\frac{18}{1235}$

75. $\frac{128}{3570}$

14-4 Probability Distributions

What You'll Learn

- Use random variables to compute probability.
- Use probability distributions to solve real-world problems.

Vocabulary

- random variable
- probability distribution
- probability histogram

How can a pet store owner use a probability distribution?

The owner of a pet store asked customers how many pets they owned. The results of this survey are shown in the table.

Number of Pets	Number of Customers
0	3
1	37
2	33
3	18
4	9

RANDOM VARIABLES AND PROBABILITY A **random variable** is a variable whose value is the numerical outcome of a random event. In the situation above, we can let the random variable X represent the number of pets owned. Thus, X can equal 0, 1, 2, 3, or 4.

Example 1 Random Variable

Refer to the application above.

a. Find the probability that a randomly-chosen customer has 2 pets.

There is only one outcome in which there are 2 pets owned, and there are 100 survey results.

$$P(X = 2) = \frac{\text{2 pets owned}}{\text{customers surveyed}}$$

$$= \frac{33}{100}$$

The probability that a randomly-chosen customer has 2 pets is $\frac{33}{100}$ or 33%.

Study Tip

Reading Math
The notation $P(X = 2)$ means the same as $P(\text{2 pets})$, the probability of a customer having 2 pets.

b. Find the probability that a randomly-chosen customer has at least 3 pets.

There are 18 + 9 or 27 outcomes in which a customer owns at least 3 pets.

$$P(X \geq 3) = \frac{27}{100}$$

The probability that a randomly-chosen customer owns at least 3 pets is $\frac{27}{100}$ or 27%.

PROBABILITY DISTRIBUTIONS The probability of every possible value of the random variable X is called a **probability distribution**.

Key Concept *Properties of Probability Distributions*

1. The probability of each value of X is greater than or equal to 0 and less than or equal to 1.
2. The probabilities of all of the values of X add up to 1.

The probability distribution for a random variable can be given in a table or in a **probability histogram**. The probability distribution and a probability histogram for the application at the beginning of the lesson are shown below.

Probability Distribution Table	
X = Number of Pets	$P(X)$
0	0.03
1	0.37
2	0.33
3	0.18
4	0.09

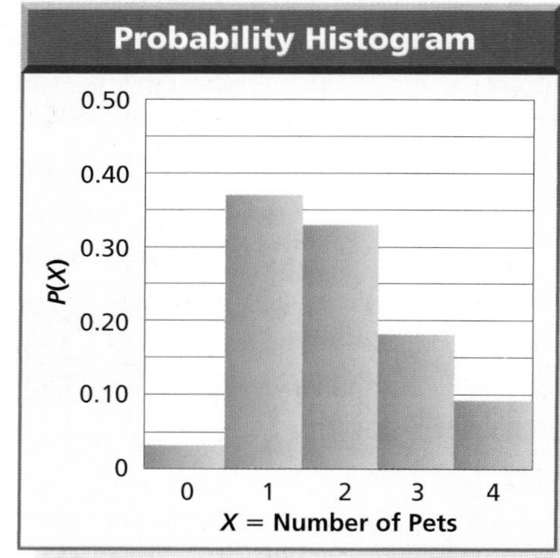

Example 2 Probability Distribution

CARS The table shows the probability distribution of the number of vehicles per household for the Columbus, Ohio, area.

Vehicles per Household Columbus, OH	
X = Number of Vehicles	Probability
0	0.10
1	0.42
2	0.36
3+	0.12

Source: U.S. Census Bureau

a. Show that the distribution is valid.

Check to see that each property holds.

1. For each value of X, the probability is greater than or equal to 0 and less than or equal to 1.
2. $0.10 + 0.42 + 0.36 + 0.12 = 1$, so the probabilities add up to 1.

b. What is the probability that a household has fewer than 2 vehicles?

Recall that the probability of a compound event is the sum of the probabilities of each individual event.

The probability of a household having fewer than 2 vehicles is the sum of the probability of 0 vehicles and the probability of 1 vehicle.

$P(X < 2) = P(X = 0) + P(X = 1)$ Sum of individual probabilities

$= 0.10 + 0.42$ or 0.52 $P(X = 0) = 0.10, P(X = 1) = 0.42$

c. Make a probability histogram of the data.

Draw and label the vertical and horizontal axes. Remember to use equal intervals on each axis. Include a title.

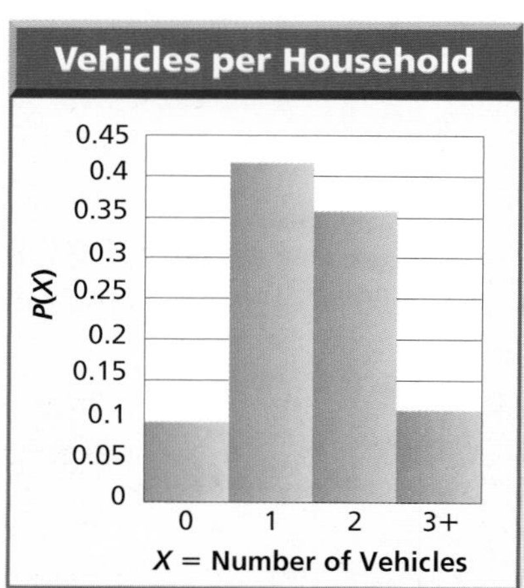

More About. . .

Cars

In 1900, there were 8000 registered cars in the United States. By 1998, there were over 131 million registered cars. This is an increase of more than 1,637,400%.

Source: *The World Almanac*

Check for Understanding

Concept Check

1. **List** the conditions that must be satisfied to have a valid probability distribution.
2. **Explain** why the probability of tossing a coin three times and getting 1 head and 2 tails is the same as the probability of getting 1 tail and 2 heads.
3. **OPEN ENDED** Describe a situation that could be displayed in a probability histogram.

Guided Practice

For Exercises 4–6, use the table that shows the possible sums when rolling two dice and the number of ways each sum can be found.

Sum of Two Dice	2	3	4	5	6	7	8	9	10	11	12
Ways to Achieve Sum	1	2	3	4	5	6	5	4	3	2	1

4. Draw a table to show the sample space of all possible outcomes.
5. Find the probabilities for $X = 4$, $X = 5$, and $X = 6$.
6. What is the probability that the sum of two dice is greater than 6 on each of three separate rolls?

Application

GRADES **For Exercises 7–9, use the table that shows a class's grade distribution, where A = 4.0, B = 3.0, C = 2.0, D = 1.0, and F = 0.**

X = Grade	0	1.0	2.0	3.0	4.0
Probability	0.05	0.10	0.40	0.40	0.05

7. Show that the probability distribution is valid.
8. What is the probability that a student passes the course?
9. What is the probability that a student chosen at random from the class receives a grade of B or better?

Practice and Apply

Homework Help

For Exercises	See Examples
10, 11, 14, 18	1
12, 13, 15–17, 19–22	2

Extra Practice

See page 852.

For Exercises 10–13, the spinner shown is spun three times.

10. Write the sample space with all possible outcomes.
11. Find the probability distribution X, where X represents the number of times the spinner lands on blue for $X = 0$, $X = 1$, $X = 2$, and $X = 3$.
12. Make a probability histogram.
13. Do all possible outcomes have an equal chance of occurring? Explain.

SALES **For Exercises 14–17, use the following information.**
A music store manager takes an inventory of the top 10 CDs sold each week. After several weeks, the manager has enough information to estimate sales and make a probability distribution table.

Number of Top 10 CDs Sold Each Week	0–100	101–200	201–300	301–400	401–500
Probability	0.10	0.15	0.40	0.25	0.10

14. Define a random variable and list its values.
15. Show that this is a valid probability distribution.
16. In a given week, what is the probability that no more than 400 CDs sell?
17. In a given week, what is the probability that more than 200 CDs sell?

EDUCATION **For Exercises 18–20, use the table that shows the education level of persons aged 25 and older in the United States.**

X = Level of Education	Probability
Some High School	0.167
High School Graduate	0.333
Some College	0.173
Associate's Degree	0.075
Bachelor's Degree	0.170
Advanced Degree	0.082

Source: U.S. Census Bureau

18. If a person was randomly selected, what is the probability that he or she completed at most some college?

19. Make a probability histogram of the data.

20. Explain how you can find the probability that a randomly selected person has earned at least a bachelor's degree.

SPORTS **For Exercises 21 and 22, use the graph that shows the sports most watched by women on TV.**

USA TODAY Snapshots®

USA TODAY

Women follow football on TV

Professional football gets better television ratings than any other sport, probably because it appeals to both men and women. Top choices among women 12 and up who watch sports:

Sport	Percent
National Football League	22.1%
Major League Baseball	13.6%
National Basketball Association	12.6%
Figure skating	6.5%
College football	4.3%

Source: ESPN Sports Poll

By Ellen J. Horrow and Sam Ward, USA TODAY

21. Determine whether this is a valid probability distribution. Justify your answer.

22. Based on the graph, in a group of 35 women how many would you expect to say they watch figure skating?

23. **CRITICAL THINKING** Suppose a married couple has children until they have a girl. Let the random variable X represent the number of children in their family.

a. Calculate the probabilities for $X = 1, 2, 3,$ and 4.

b. Find the probability that the couple will have more than 4 children.

24. **WRITING IN MATH** Answer the question that was posed at the beginning of the lesson.

How can a pet store owner use a probability distribution?

Include the following in your answer:

- a sentence or two describing how to create a probability distribution, and
- an explanation of how the store owner could use a probability distribution to establish a frequent buyer program.

25. The table shows the probability distribution for the number of heads when four coins are tossed. What is the probability that there are no more than two heads showing on a random toss?

X = Number of Heads	0	1	2	3	4
Probability $P(X)$	0.0625	0.25	0.375	0.25	0.0625

Ⓐ 0.6875 Ⓑ 0.375 Ⓒ 0.875 Ⓓ 0.3125

26. On a random roll of two dice, what is the probability that the sum of the numbers showing is less than 5?

Ⓐ 0.08 Ⓑ 0.17 Ⓒ 0.11 Ⓓ 0.28

Maintain Your Skills

Mixed Review

A card is drawn from a standard deck of 52 cards. Find each probability. *(Lesson 14-3)*

27. P(ace or 10) **28.** P(3 or diamond) **29.** P(odd number or spade)

Evaluate. *(Lesson 14-2)*

30. ${}_{10}C_7$ **31.** ${}_{12}C_5$ **32.** $({}_6P_3)({}_5P_3)$

Let $A = \begin{bmatrix} 1 & 4 \\ 5 & 7 \end{bmatrix}$ and $B = \begin{bmatrix} -3 & 0 \\ -2 & 5 \end{bmatrix}$. *(Lesson 13-2)*

33. Find $A + B$. **34.** Find $B - A$.

Write an inverse variation equation that relates x and y. Assume that y varies inversely as x. Then solve. *(Lesson 12-1)*

35. If $y = -2.4$ when $x = -0.6$, find y when $x = 1.8$.

36. If $y = 4$ when $x = -1$, find x when $y = -3$.

Simplify each expression. *(Lesson 11-2)*

37. $3\sqrt{8} + 7\sqrt{2}$ **38.** $2\sqrt{3} + \sqrt{12}$ **39.** $3\sqrt{7} - 2\sqrt{28}$

SAVINGS For Exercises 40–42, use the following information.
Selena is investing her $900 tax refund in a certificate of deposit that matures in 4 years. The interest rate is 8.25% compounded quarterly. *(Lesson 10-6)*

40. Determine the balance in the account after 4 years.

41. Her friend Monique invests the same amount of money at the same interest rate, but her bank compounds interest monthly. Determine how much she will have after 4 years.

42. Which type of compounding appears more profitable? Explain.

Getting Ready for the Next Lesson

PREREQUISITE SKILL Write each fraction as a percent rounded to the nearest whole number. *(To review **writing fractions as percents**, see pages 804 and 805.)*

43. $\frac{16}{80}$ **44.** $\frac{20}{52}$ **45.** $\frac{30}{114}$

46. $\frac{57}{120}$ **47.** $\frac{72}{340}$ **48.** $\frac{54}{162}$

Practice Quiz 2

Lessons 14-3 and 14-4

For Exercises 1–3, use the probability distribution for the number of people in a household. *(Lesson 14-4)*

1. Show that the probability distribution is valid.

2. If a household is chosen at random, what is the probability that 4 or more people live in it?

3. Make a histogram of the data.

American Households

X = Number of People	Probability
1	0.25
2	0.32
3	0.18
4	0.15
5	0.07
6	0.02
7+	0.01

Source: U.S. Census Bureau

A ten-sided die, numbered 1 through 10, is rolled. Find each probability.

4. P(odd or greater than 4)

5. P(less than 3 or greater than 7)

14-5 Probability Simulations

What You'll Learn

- Use theoretical and experimental probability to represent and solve problems involving uncertainty.
- Perform probability simulations to model real-world situations involving uncertainty.

Vocabulary

- theoretical probability
- experimental probability
- relative frequency
- empirical study
- simulation

How can probability simulations be used in health care?

A pharmaceutical company is developing a new medication to treat a certain heart condition. Based on similar drugs, researchers at the company expect the new drug to work successfully in 70% of patients.

To test the drug's effectiveness, the company performs three clinical studies. Each study involves 100 volunteers who use the drug for six months. The results of the studies are shown in the table.

Study Of New Medication

Result	Study 1	Study 2	Study 3
Expected Success Rate	70%	70%	70%
Condition Improved	61%	74%	67%
No Improvement	39%	25%	33%
Condition Worsened	0%	1%	0%

Career Choices

Medical Researcher

Many medical researchers conduct research to advance knowledge of living organisms, including viruses and bacteria.

Online Research For information about a career as a medical researcher, visit: www.algebra1.com/careers

THEORETICAL AND EXPERIMENTAL PROBABILITY The probability we have used to describe events in previous lessons is theoretical probability. **Theoretical probabilities** are determined mathematically and describe what should happen. In the situation above, the expected success rate of 70% is a theoretical probability.

A second type of probability we can use is **experimental probability**, which is determined using data from tests or experiments. Experimental probability is the ratio of the number of times an outcome occurred to the total number of events or trials. This ratio is also known as the **relative frequency**.

$$\text{experimental probability} = \frac{\text{frequency of an outcome}}{\text{total number of trials}}$$

Example 1 Experimental Probability

MEDICAL RESEARCH Refer to the application at the beginning of the lesson. What is the experimental probability that the drug was successful for a patient in Study 1?

In Study 1, the drug worked successfully in 61 of the 100 patients.

$$\text{experimental probability} = \frac{61}{100} \quad \begin{array}{l} \leftarrow \text{frequency of successes} \\ \leftarrow \text{total number of patients} \end{array}$$

The experimental probability of Study 1 is $\frac{61}{100}$ or 61%.

It is often useful to perform an experiment repeatedly, collect and combine the data, and analyze the results. This is known as an **empirical study**.

Example 2 Empirical Study

Refer to the application at the beginning of the lesson. What is the experimental probability that the drug was successful for all three studies?

The number of successful outcomes of the three studies was 61 + 74 + 67 or 202 out of the 300 total patients.

$$\text{experimental probability} = \frac{202}{300} \text{ or } \frac{101}{150}$$

The experimental probability of the three studies was $\frac{101}{150}$ or about 67%.

PERFORMING SIMULATIONS A method that is often used to find experimental probability is a simulation. A **simulation** allows you to use objects to act out an event that would be difficult or impractical to perform.

Algebra Activity

Simulations

Collect the Data

- Roll a die 20 times. Record the value on the die after each roll.
- Determine the experimental probability distribution for X, the value on the die.
- Combine your results with the rest of the class to find the experimental probability distribution for X given the new number of trials.
  (20 · the number of students in your class)

Analyze the Data

1. Find the theoretical probability of rolling a 2.
2. Find the theoretical probability of rolling a 1 or a 6.
3. Find the theoretical probability of rolling a value less than 4.
4. Compare the experimental and theoretical probabilities. Which pair of probabilities was closer to each other: your individual probabilities or your class's probabilities?
5. Suppose each person rolls the die 50 times. Explain how this would affect the experimental probabilities for the class.

Make a Conjecture

6. What can you conclude about the relationship between the number of experiments in a simulation and the experimental probability?

You can conduct simulations of the outcomes for many problems by using one or more objects such as dice, coins, marbles, or spinners. The objects you choose should have the same number of outcomes as the number of possible outcomes of the problem, and all outcomes should be equally likely.

Example 3 Simulation

In one season, Malcolm made 75% of the field goals he attempted.

a. What could be used to simulate his kicking a field goal? Explain.

You could use a spinner like the one at the right, where 75% of the spinner represents making a field goal.

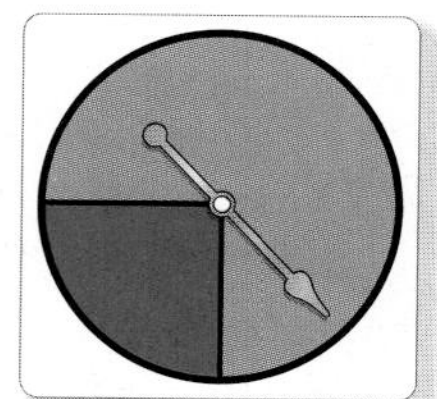

b. Describe a way to simulate his next 8 attempts.

Spin the spinner once to simulate a kick. Record the result, then repeat this 7 more times.

Example 4 Theoretical and Experimental Probability

DOGS **Ali raises purebred dogs. One of her dogs is expecting a litter of four puppies, and Ali would like to figure out the most likely mix of male and female puppies. Assume that $P(\text{male}) = P(\text{female}) = \frac{1}{2}$.**

a. What objects can be used to model the possible outcomes of the puppies?

Each puppy can be male or female, so there are $2 \cdot 2 \cdot 2 \cdot 2$ or 16 possible outcomes for the litter. Use a simulation that also has 2 outcomes for each of 4 events. One possible simulation would be to toss four coins, one for each puppy, with heads representing female and tails representing male.

Study Tip

Alternative Simulation

You could also create a spinner with two even parts and spin it 4 times to simulate the outcomes of the puppies.

b. Find the theoretical probability that there will be two female and two male puppies.

There are 16 possible outcomes, and the number of combinations that have two female and two male puppies is ${}_4C_2$ or 6. So the theoretical probability is $\frac{6}{16}$ or $\frac{3}{8}$.

c. The results of a simulation Ali performed are shown in the table below. What is the experimental probability that there will be three male puppies?

Outcomes	Frequency
4 female, 0 male	3
3 female, 1 male	13
2 female, 2 male	18
1 female, 3 male	12
0 female, 4 male	4

Ali performed 50 trials and 12 of those resulted in three males. So, the experimental probability is $\frac{12}{50}$ or 24%.

d. How does the experimental probability compare to the theoretical probability of a litter with three males?

Theoretical probability

$$P(3 \text{ males}) = \frac{{}_4C_3}{16} \quad \begin{matrix} \leftarrow \text{combinations with 3 male puppies} \\ \leftarrow \text{possible outcomes} \end{matrix}$$

$$= \frac{4}{16} \text{ or } 25\% \quad \text{Simplify.}$$

The experimental probability, 24%, is very close to the theoretical probability.

Check for Understanding

Concept Check

1. **Explain** why it is useful to carry out an empirical study when calculating experimental probabilities.
2. **Analyze** the relationship between the theoretical and experimental probability of an event as the number of trials in a simulation increases.
3. **OPEN ENDED** Describe a situation that could be represented by a simulation. What objects would you use for this experiment?
4. **Tell** whether the theoretical probability and the experimental probability of an event are *sometimes*, *always*, or *never* the same.

Guided Practice

5. So far this season, Rita has made 60% of her free throws. Describe a simulation that could be used to predict the outcome of her next 25 free throws.

For Exercises 6–8, roll a die 25 times and record your results.

6. Based on your results, what is the probability of rolling a 3?
7. Based on your results, what is the probability of rolling a 5 or an odd number?
8. Compare your results to the theoretical probabilities.

Application

ASTRONOMY For Exercises 9–12, use the following information.
Enrique is writing a report about meteorites and wants to determine the probability that a meteor reaching Earth's surface hits land. He knows that 70% of Earth's surface is covered by water. He places 7 blue marbles and 3 brown marbles in a bag to represent hitting water $\left(\frac{7}{10}\right)$ and hitting land $\left(\frac{3}{10}\right)$. He draws a marble from the bag, records the color, and then replaces the marble. The table shows the results of his experiment.

Blue	Brown
56	19

9. Did Enrique choose an appropriate simulation for his research? Explain.
10. What is the theoretical probability that a meteorite reaching Earth's surface hits land?
11. Based on his results, what is the probability that a meteorite hits land?
12. Using the experimental probability, how many of the next 500 meteorites that strike Earth would you expect to hit land?

Practice and Apply

Homework Help

For Exercises	See Examples
13–16	3
17–21, 25–31	4
22–24	1, 2

Extra Practice

See page 852.

13. What could you use to simulate the outcome of guessing on 15 true-false questions?
14. There are 12 cans of cola, 8 cans of diet cola, and 4 cans of root beer in a cooler. What could be used for a simulation determining the probability of randomly picking any one type of soft drink?

For Exercises 15 and 16, use the following information.
Central City Mall is randomly giving each shopper one of 12 different gifts during the holidays.

15. What could be used to perform a simulation of this situation? Explain your choice.
16. How could you use this simulation to model the next 100 gifts handed out?

For Exercises 17 and 18, toss 3 coins, one at a time, 25 times and record your results.

17. Based on your results, what is the probability that any two coins will show heads?

18. Based on your results, what is the probability that the first and third coins show tails?

For Exercises 19–21, roll two dice 50 times and record the sums.

19. Based on your results, what is the probability that the sum is 8?

20. Based on your results, what is the probability that the sum is 7, or the sum is greater than 5?

21. If you roll the dice 25 more times, which sum would you expect to see about 10% of the time?

CITY PLANNING For Exercises 22–24, use the following information.
The Lewiston City Council sent surveys to randomly selected households to determine current and future enrollment for the local school district. The results of the survey are shown in the table.

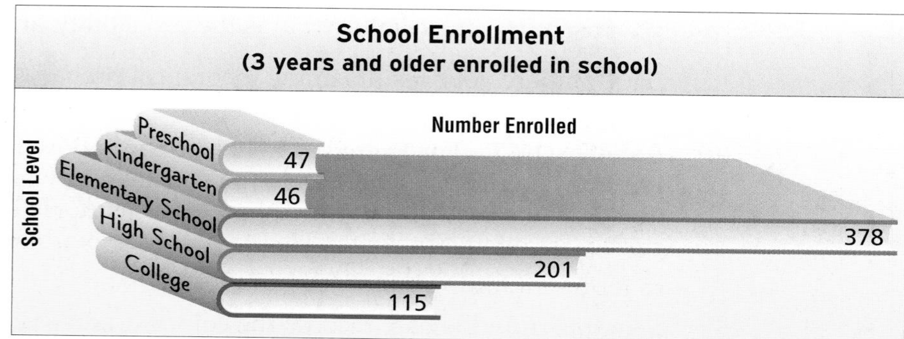

More About. . .

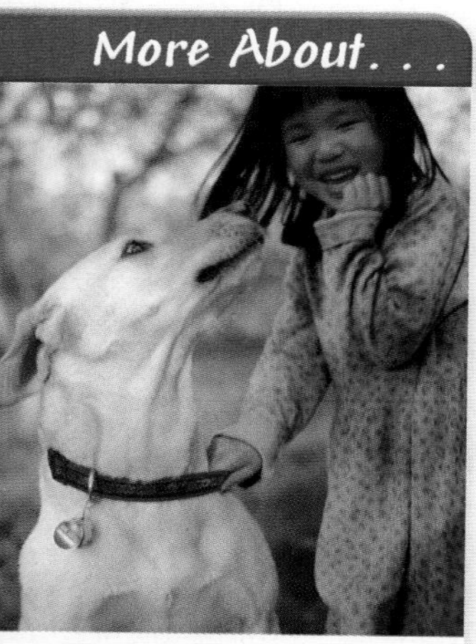

Animals

Labrador retrievers are the most popular breed of dog in the United States.

Source: American Kennel Club

22. Find the experimental probability distribution for the number of people enrolled at each level.

23. Based on the survey, what is the probability that a student chosen at random is in elementary school or high school?

24. Suppose the school district is expecting school enrollment to increase by 1800 over the next 5 years due to new buildings in the area. Of the new enrollment, how many will most likely be in kindergarten?

RESTAURANTS For Exercises 25–27, use the following information.
A family restaurant gives children a free toy with each children's meal. There are eight different toys that are randomly given. There is an equally likely chance of getting each toy each time.

25. What objects could be used to perform a simulation of this situation?

26. Conduct a simulation until you have one of each toy. Record your results.

27. Based on your results, how many meals must be purchased so that you get all 8 toys?

ANIMALS For Exercises 28–31, use the following information.
Refer to Example 4 on page 784. Suppose Ali's dog is expecting a litter of 5 puppies.

28. List the possible outcomes of the genders of the puppies.

29. Perform a simulation and list your results in a table.

30. Based on your results, what is the probability that there will be 3 females and two males in the litter?

31. What is the experimental probability of the litter having at least three male puppies?

32. **CRITICAL THINKING** The captain of a football team believes that the coin the referee uses for the opening coin toss gives an advantage to one team. The referee has players toss the coin 50 times each and record their results. Based on the results, do you think the coin is fair? Explain your reasoning.

Player	1	2	3	4	5	6
Heads	38	31	29	27	26	30
Tails	12	19	21	23	24	20

33. WRITING IN MATH Answer the question that was posed at the beginning of the lesson.

How can probability simulations be used in health care?

Include the following in your answer:

- a few sentences explaining experimental probability, and
- an explanation of why an experimental probability of 75% found in 400 trials is more reliable than an experimental probability of 75% found in 50 trials.

34. Ramón tossed two coins and rolled a die. What is the probability that he tossed two tails and rolled a 3?

(A) $\frac{1}{4}$ (B) $\frac{1}{6}$ (C) $\frac{5}{12}$ (D) $\frac{1}{24}$

35. If a coin is tossed three times, what is the probability that the results will be heads exactly one time?

(A) $\frac{2}{3}$ (B) $\frac{3}{8}$ (C) $\frac{1}{5}$ (D) $\frac{1}{8}$

SIMULATION For Exercises 36–38, use the following information.

When you are performing an experiment that involves a large number of trials that cannot be simulated using an object like a coin or a spinner, you can use the random number generator function on a graphing calculator. The TI-83 Plus program at the right will perform T trials by generating random numbers between 1 and P, the number of possible outcomes.

```
PROGRAM: SIMULATE
:Disp "ENTER THE NUMBER"
:Disp "OF POSSIBLE"
:Disp "OUTCOMES"
:Input P
:Disp "ENTER THE NUMBER"
:Disp "OF TRIALS"
:Input T
:For(N, 1, T)
:randInt(1, P)→S
:Disp S
:Pause
```

36. Run the program to simulate 50 trials of an event that has 15 outcomes. Record your results.

37. What is the experimental probability of displaying the number 10?

38. Repeat the experiment several times. Find the experimental probability of displaying the number 10. Has the probability changed from the probability found in Exercise 37? Explain why or why not.

ENTERTAINMENT For Exercises 39–41, use the following information and the graphing calculator program above.

A CD changer contains 5 CDs with 14 songs each. When "Random" is selected, each CD is equally likely to be chosen as each song.

39. Use the program SIMULATE to perform a simulation of randomly playing 40 songs from the 5 CDs. (*Hint*: Number the songs sequentially from 1, CD 1 track 1, to 70, CD 5 track 14.)

40. Do the experimental probabilities for your simulation support the statement that each CD is equally likely to be chosen? Explain.

41. Based on your results, what is the probability that the first three songs played are on the third disc?

Maintain Your Skills

Mixed Review

For Exercises 42–44, use the probability distribution for the random variable X, the number of computers per household. *(Lesson 14-4)*

Computers per Household	
X = Number of Computers	$P(X)$
0	0.579
1	0.276
2	0.107
3+	0.038

Source: U.S. Dept. of Commerce

42. Show that the probability distribution is valid.

43. If a household is chosen at random, what is the probability that it has at least 2 computers?

44. Determine the probability of randomly selecting a household with no more than one computer.

For Exercises 45–47, use the following information.
A jar contains 18 nickels, 25 dimes, and 12 quarters. Three coins are randomly selected. Find each probability. *(Lesson 14-3)*

45. picking three dimes, replacing each after it is drawn

46. a nickel, then a quarter, then a dime without replacing the coins

47. 2 dimes and a quarter, without replacing the coins, if order does not matter

Solve each equation. *(Lesson 12-9)*

48. $\frac{2a-3}{a-3} - 2 = \frac{12}{a+3}$ **49.** $\frac{r^2}{r-7} + \frac{50}{7-r} = 14$ **50.** $\frac{x-2}{x} - \frac{x-3}{x-6} = \frac{1}{x}$

51. $\frac{2x-3}{7} - \frac{x}{2} = \frac{x+3}{14}$ **52.** $\frac{5n}{n+1} + \frac{1}{n} = 5$ **53.** $\frac{a+2}{a-2} - \frac{2}{a+2} = \frac{-7}{3}$

54. CONSTRUCTION To paint his house, Lonnie needs to purchase an extension ladder that reaches at least 24 feet off the ground. Ladder manufacturers recommend the angle formed by the ladder and the ground be no more than 75°. What is the shortest ladder he could buy to reach 24 feet safely? *(Lesson 11-7)*

Determine whether the following side measures would form a right triangle. *(Lesson 11-4)*

55. 5, 7, 9 **56.** $3\sqrt{34}$, 9, 15 **57.** 36, 86.4, 93.6

Solve each equation. Check your solutions. *(Lesson 9-6)*

58. $(x-6)^2 = 4$ **59.** $x^2 + 121 = 22x$ **60.** $4x^2 + 12x + 9 = 0$

61. $25x^2 + 20x = -4$ **62.** $49x^2 - 84x + 36 = 0$ **63.** $180x - 100 = 81x^2$

WebQuest Internet Project

America Counts!

It is time to complete your project. Use the information and data you have gathered about populations to prepare a brochure or Web page. Be sure to identify the state you have chosen for this project. Include graphs, tables, and/or calculations in the presentation.

www.algebra1.com/webquest

Chapter 14 Study Guide and Review

Vocabulary and Concept Check

combination (p. 762)
complements (p. 771)
compound event (p. 769)
dependent events (p. 770)
edge (p. 759)
empirical study (p. 783)
event (p. 754)
experimental probability (p. 782)
factorial (p. 755)
finite graph (p. 759)
Fundamental Counting Principle (p. 755)
inclusive (p. 771)
independent events (p. 769)
mutually exclusive (p. 771)
network (p. 759)
node (p. 759)
permutation (p. 760)
probability distribution (p. 777)
probability histogram (p. 778)
random variable (p. 777)
relative frequency (p. 782)
sample space (p. 754)
simple event (p. 769)
simulation (p. 783)
theoretical probability (p. 782)
traceable (p. 759)
tree diagram (p. 754)

Choose the word or term that best completes each sentence.

1. The arrangement or listing in which order is important is called a (*combination*, *permutation*).
2. The notation 10! refers to a (*prime factor*, *factorial*).
3. Rolling one die and then another die are (*dependent*, *independent*) events.
4. The sum of probabilities of complements equals (*0*, *1*).
5. Randomly drawing a coin from a bag and then drawing another coin are dependent events if the coins (*are*, *are not*) replaced.
6. Events that cannot occur at the same time are (*inclusive*, *mutually exclusive*).
7. The sum of the probabilities in a probability distribution equals (*0*, *1*).
8. (*Experimental*, *Theoretical*) probabilities are precise and predictable.

Lesson-by-Lesson Review

14-1 Counting Outcomes

See pages 754–758.

Concept Summary

- Use a tree diagram to make a list of possible outcomes.
- If an event M can occur m ways and is followed by an event N that can occur n ways, the event M followed by event N can occur $m \cdot n$ ways.

Example When Jerri packs her lunch, she can choose to make a turkey or roast beef sandwich on French or sourdough bread. She also can pack an apple or an orange. Draw a tree diagram to show the number of different ways Jerri can select these items.

Meat	Bread	Fruit	Possible Lunches
Turkey	French	Apple	TFA
		Orange	TFO
	Sourdough	Apple	TSA
		Orange	TSO
Roast Beef	French	Apple	RFA
		Orange	RFO
	Sourdough	Apple	RSA
		Orange	RSO

There are 8 different ways for Jerri to select these items.

Exercises **Determine the number of outcomes for each event.**
See Examples 1–3 on pages 754 and 755.

9. Samantha wants to watch 3 videos one rainy afternoon. She has a choice of 3 comedies, 4 dramas, and 3 musicals.

10. Marquis buys 4 books, one from each category. He can choose from 12 mystery, 8 science fiction, 10 classics, and 5 biographies.

11. The Jackson Jackals and the Westfield Tigers are going to play a best three-out-of-five games baseball tournament.

14-2 Permutations and Combinations

See pages 760–767.

Concept Summary

- In a permutation, the order of objects is important. $_nP_r = \frac{n!}{(n-r)!}$
- In a combination, the order of objects is not important. $_nC_r = \frac{n!}{(n-r)!r!}$

Examples

1 **Find $_{12}C_8$.**

$$_{12}C_8 = \frac{12!}{(12-8)!8!} = \frac{12!}{4!8!} = \frac{12 \cdot 11 \cdot 10 \cdot 9}{4!} = 495$$

2 **Find $_9P_4$.**

$$_9P_4 = \frac{9!}{(9-4)!} = \frac{9!}{5!} = \frac{9 \cdot 8 \cdot 7 \cdot 6 \cdot 5 \cdot 4 \cdot 3 \cdot 2 \cdot 1}{5 \cdot 4 \cdot 3 \cdot 2 \cdot 1} = 3024$$

Exercises **Evaluate each expression.** *See Examples 1, 2, and 4 on pages 760–762.*

12. $_4P_2$ **13.** $_8C_3$ **14.** $_4C_4$
15. $(_7C_1)(_6C_3)$ **16.** $(_7P_3)(_7P_2)$ **17.** $(_3C_2)(_4P_1)$

14-3 Probability of Compound Events

See pages 769–776.

Concept Summary

- For independent events, use $P(A \text{ and } B) = P(A) \cdot P(B)$.
- For dependent events, use $P(A \text{ and } B) = P(A) \cdot P(B \text{ following } A)$.
- For mutually exclusive events, use $P(A \text{ or } B) = P(A) + P(B)$.
- For inclusive events, use $P(A \text{ or } B) = P(A) + P(B) - P(A \text{ and } B)$.

Example

A box contains 8 red chips, 6 blue chips, and 12 white chips. Three chips are randomly drawn from the box and not replaced. Find *P*(red, white, blue).

First chip: $P(\text{red}) = \frac{8}{26}$ ← number of red chips / ← total number of chips

Second chip: $P(\text{white}) = \frac{12}{25}$ ← number of white chips / ← number of chips remaining

Third chip: $P(\text{blue}) = \frac{6}{24}$ ← number of blue chips / ← number of chips remaining

$$P(\text{red, white, blue}) = \underbrace{P(\text{red})} \cdot \underbrace{P(\text{white})} \cdot \underbrace{P(\text{blue})}$$
$$= \frac{8}{26} \cdot \frac{12}{25} \cdot \frac{6}{24}$$
$$= \frac{576}{15{,}600} \text{ or } \frac{12}{325}$$

Exercises **A bag of colored paper clips contains 30 red clips, 22 blue clips, and 22 green clips. Find each probability if three clips are drawn randomly from the bag and are not replaced.** *See Example 2 on page 770.*

18. P(blue, red, green) **19.** P(red, red, blue) **20.** P(red, green, not blue)

One card is randomly drawn from a standard deck of 52 cards. Find each probability. *See Examples 3 and 4 on pages 771 and 772.*

21. P(diamond or club) **22.** P(heart or red) **23.** P(10 or spade)

14-4 Probability Distributions

See pages 777–781.

Concept Summary

Probability distributions have the following properties.

- For each value of X, $0 \leq P(X) \leq 1$.
- The sum of the probabilities of each value of X is 1.

Example **A local cable provider asked its subscribers how many televisions they had in their homes. The results of their survey are shown in the probability distribution.**

Televisions per Household	
X = Number of Televisions	Probability
1	0.18
2	0.36
3	0.34
4	0.08
5+	0.04

a. Show that the probability distribution is valid.

For each value of X, the probability is greater than or equal to 0 and less than or equal to 1.

$0.18 + 0.36 + 0.34 + 0.08 + 0.04 = 1$, so the probabilities add up to 1.

b. If a household is selected at random, what is the probability that it has fewer than 4 televisions?

$$P(X < 4) = P(X = 1) + P(X = 2) + P(X = 3)$$
$$= 0.18 + 0.36 + 0.34$$
$$= 0.88$$

Exercises **The table shows the probability distribution for the number of extracurricular activities in which students at Boardwalk High School participate.** *See Example 2 on page 778.*

Extracurricular Activities	
X = Number of Activities	Probability
0	0.04
1	0.12
2	0.37
3	0.30
4+	0.17

24. Show that the probability distribution is valid.

25. If a student is chosen at random, what is the probability that the student participates in 1 to 3 activities?

26. Make a probability histogram of the data.

• Extra Practice, see pages 851–852.
• Mixed Problem Solving, see page 866.

14-5 Probability Simulations

See pages 782–788.

Concept Summary

- Theoretical probability describes expected outcomes, while experimental probabilities describe tested outcomes.
- Simulations are used to perform experiments that would be difficult or impossible to perform in real life.

Example **A group of 3 coins are tossed.**

a. Find the theoretical probability that there will be 2 heads and 1 tail.

Each coin toss can be heads or tails, so there are $2 \cdot 2 \cdot 2$ or 8 possible outcomes. There are 3 possible combinations of 2 heads and one tail, HHT, HTH, or THH. So, the theoretical probability is $\frac{3}{8}$.

b. The results of a simulation in which three coins are tossed ten times are shown in the table. What is the experimental probability that there will be 1 head and 2 tails?

Outcomes	Frequency
3 heads, 0 tails	1
2 heads, 1 tail	4
1 head, 2 tails	3
0 heads, 3 tails	2

Of the 10 trials, 3 resulted in 1 head and 2 tails, so the experimental probability is $\frac{3}{10}$ or 30%.

c. Compare the theoretical probability of 2 heads and 1 tail and the experimental probability of 2 heads and 1 tail.

The theoretical probability is $\frac{3}{8}$ or 37.5%, while the experimental probability is $\frac{3}{10}$ or 30%. The probabilities are close.

Exercises **While studying flower colors in biology class, students are given the Punnett square at the right. The Punnett square shows that red parent plant flowers (Rr) produce red flowers (RR and Rr) and pink flowers (rr).**

See Examples 1, 3, and 4 on pages 782 and 784.

	R	r
R	RR	Rr
r	Rr	rr

27. If 5 flowers are produced, find the theoretical probability that there will be 4 red flowers and 1 pink flower.
28. Describe items that the students could use to simulate the colors of 5 flowers.
29. The results of a simulation of flowers are shown in the table. What is the experimental probability that there will be 3 red flowers and 2 pink flowers?

Outcomes	Frequency
5 red, 0 pink	15
4 red, 1 pink	30
3 red, 2 pink	23
2 red, 3 pink	7
1 red, 4 pink	4
0 red, 5 pink	1

Chapter 14 Practice Test

Vocabulary and Concepts

1. Seven students lining up to buy tickets for a school play is an example of a (*permutation*, *combination*).
2. Rolling a die and recording the result 25 times would be used to find (*theoretical*, *experimental*) probability.
3. A (*random variable*, *probability distribution*) is the numerical outcome of an event.

Skills and Applications

There are two roads from Ashville to Bakersville, four roads from Bakersville to Clifton, and two roads from Clifton to Derry.

4. Draw a tree diagram showing the possible routes from Ashville to Derry.
5. How many different routes are there from Ashville to Derry?

Determine whether each situation involves a *permutation* or a *combination*. Then determine the number of possible arrangements.

6. Six students in a class meet in a room that has nine chairs.
7. The top four finishers in a race with ten participants.
8. A class has 15 girls and 19 boys. A committee is formed with two girls and two boys, each with a separate responsibility.

A bag contains 4 red, 6 blue, 4 yellow, and 2 green marbles. Once a marble is selected, it is not replaced. Find each probability.

9. *P*(blue, green)
10. *P*(yellow, yellow)
11. *P*(red, blue, yellow)
12. *P*(blue, red, not green)

The spinner is spun, and a die is rolled. Find each probability.

13. *P*(yellow, 4)
14. *P*(red, even)
15. *P*(purple or white, not prime)
16. *P*(green, even or less than 5)

During a magic trick, a magician randomly selects a card from a standard deck of 52 cards. Without replacing it, the magician has a member of the audience randomly select a card. Find each probability.

17. *P*(club, heart)
18. *P*(black 7, diamond)
19. *P*(queen or red, jack of spades)
20. *P*(black 10, ace or heart)

The table shows the number of ways four coins can land heads up when they are tossed at the same time.

Four Coins Tossed	
Number of Heads	Possible Outcomes
0	1
1	4
2	6
3	4
4	1

21. Set up a probability distribution of the possible outcomes.
22. Find the probability that there will be no heads.
23. Find the probability that there will be at least two heads.
24. Find the probability that there will be two tails.
25. **STANDARDIZED TEST PRACTICE** Two numbers *a* and *b* can be arranged in two different orders, *a*, *b* and *b*, *a*. In how many ways can three numbers be arranged?

Ⓐ 3 Ⓑ 4 Ⓒ 5 Ⓓ 6

www.algebra1.com/chapter_test

Chapter 14 Standardized Test Practice

Part 1 Multiple Choice

Record your answers on the answer sheet provided by your teacher or on a sheet of paper.

1. If the average of a and b is 20, and the average of a, b, and c is 25, then what is the value of c? (Prerequisite Skill)

Ⓐ 10 Ⓑ 15
Ⓒ 25 Ⓓ 35

2. The volume of a cube is 27 cubic inches. Its total surface area, in square inches, is (Lesson 3-8)

Ⓐ 9. Ⓑ $6\sqrt{3}$.
Ⓒ $18\sqrt{3}$. Ⓓ 54.

3. A truck travels 50 miles from Oakton to Newton in exactly 1 hour. When the truck is halfway between Oakton and Newton, a car leaves Oakton and travels at 60 miles per hour. How many miles has the car traveled when the truck reaches Newton? (Lesson 3-8)

Ⓐ 25 Ⓑ 30
Ⓒ 50 Ⓓ 60

4. Which equation would best represent the graphed data? (Lesson 5-7)

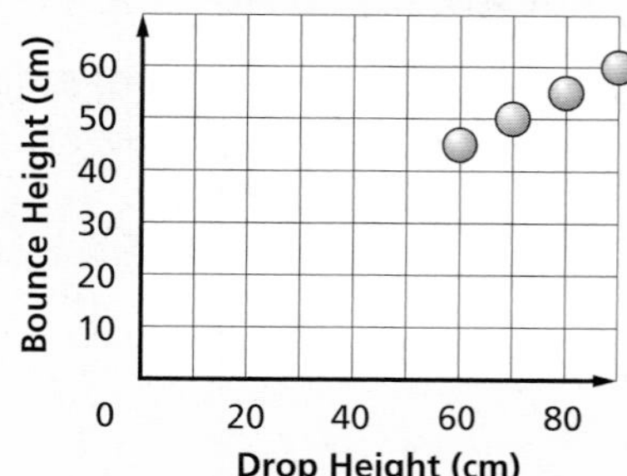

Ⓐ $y = \frac{1}{2}x + 15$ Ⓑ $y = 2x + 15$
Ⓒ $y = 2x$ Ⓓ $y = \frac{1}{2}x$

5. If a child is equally likely to be born a boy or a girl, what is the probability that a family of 3 children will contain exactly one boy? (Lesson 7-5)

Ⓐ $\frac{1}{8}$ Ⓑ $\frac{1}{4}$
Ⓒ $\frac{3}{8}$ Ⓓ $\frac{1}{2}$

6. What is the value of 5^{-2}? (Lesson 8-2)

Ⓐ -25 Ⓑ $-\frac{1}{25}$
Ⓒ $\frac{1}{25}$ Ⓓ $-\sqrt{5}$

7. What are the solutions of $x^2 + x = 20$? (Lesson 9-4)

Ⓐ $-4, 5$ Ⓑ $-2, 10$
Ⓒ $2, 10$ Ⓓ $4, -5$

8. Two airplanes are flying at the same altitude. One plane is two miles west and two miles north of an airport. The other plane is seven miles west and eight miles north of the same airport. How many miles apart are the airplanes? (Lesson 11-4)

Ⓐ 2.8 Ⓑ 7.8
Ⓒ 10.6 Ⓓ 11.0

9. A certain password consists of three characters, and each character is a letter of the alphabet. Each letter can be used more than once. How many different passwords are possible? (Lesson 14-1)

Ⓐ 78 Ⓑ 2600
Ⓒ 15,600 Ⓓ 17,576

The Princeton Review Test-Taking Tip

If you are allowed to write in your test booklet, underline key words, do calculations, sketch diagrams, cross out answer choices as you eliminate them, and mark any questions that you skip. But do not make any marks on the *answer sheet* except your answers.

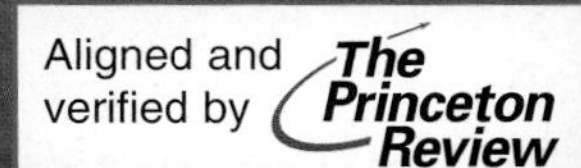

Part 2 Short Response/Grid In

Record your answers on the answer sheet provided by your teacher or on a sheet of paper.

10. What are the coordinates of the point of intersection of the lines represented by the equations $x + 4y = 0$ and $2x - 3y = 11$? (Lesson 7-2)

11. Is $4\left(x - \frac{1}{2}\right)^2 - 1 = 4x^2 - 4x$ true for *all* values of x, *some* values of x, or *no* values of x? (Lesson 8-8)

12. Triangle ABC has sides of length $a = 5$, $b = 7$, and $c = \sqrt{74}$. What is the measure, in degrees, of the angle opposite side c? (Lesson 11-4)

13. All seven-digit telephone numbers in a town begin with the same three digits. Of the last four digits in any given phone number, neither the first nor the last digit can be 0. How many telephone numbers are available in this town? (Lesson 14-2)

14. In the board game shown below, you move your game piece along the arrows from square to square. To determine which direction to move your game piece, you roll a number cube with sides numbered 1, 2, 3, 4, 5, and 6. If you roll 1 or 2, you move your game piece one space to the left. If you roll 3, 4, 5, or 6, you move your game piece one square to the right. What is the probability that you will reach the goal within two turns? (Lesson 14-3)

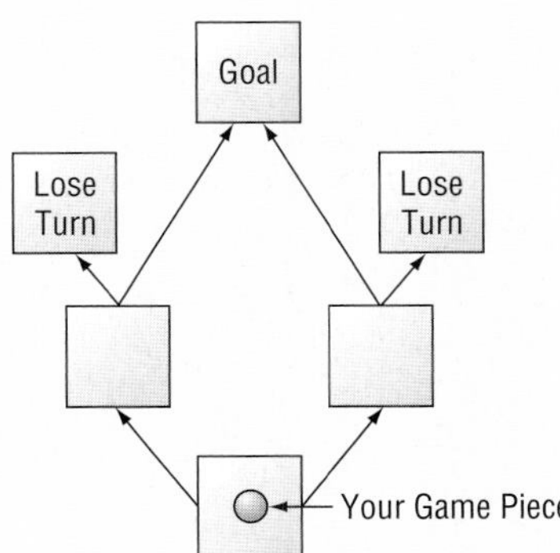

Part 3 Quantitative Comparison

Compare the quantity in Column A and the Quantity in Column B. Then determine whether:

- **(A) the quantity in Column A is greater,**
- **(B) the quantity in Column B is greater,**
- **(C) the two quantities are equal, or**
- **(D) the relationship cannot be determined from the information given.**

	Column A	Column B
15.	the value of x in $3x + 15 > 45$	the value of y in $-2y + 3 > -17$

(Lesson 6-3)

	Column A	Column B
16.	${}_{12}P_4$	${}_{10}C_6$

(Lesson 14-2)

17. A bag contains 5 red marbles, 7 blue marbles, and 2 green marbles. A marble is randomly drawn, not replaced, then another marble is randomly drawn.

Column A	Column B
P(blue, green)	P(red, red)

(Lesson 14-3)

Part 4 Open Ended

Record your answers on a sheet of paper. Show your work.

18. At WackyWorld Pizza, the Random Special is a random selection of two different toppings on a large cheese pizza. The available toppings are pepperoni, sausage, onion, mushrooms, and green peppers. (Lessons 14-2 and 14-3)

a. How many different Random Specials are possible? Show how you found your answer.

b. If you order the Random Special, what is the probability that it will have mushrooms?

Student Handbook

Skills

Reference

Prerequisite Skills

Operations with Fractions: Adding and Subtracting

- To add or subtract fractions with the same denominator, add or subtract the numerators and write the sum or difference over the denominator.

Example 1 **Find each sum or difference.**

a. $\frac{3}{5} + \frac{1}{5}$

$\frac{3}{5} + \frac{1}{5} = \frac{3+1}{5}$ The denominators are the same. Add the numerators.

$= \frac{4}{5}$ Simplify.

b. $\frac{5}{9} - \frac{4}{9}$

$\frac{5}{9} - \frac{4}{9} = \frac{5-4}{9}$ The denominators are the same. Subtract the numerators.

$= \frac{1}{9}$ Simplify.

- To write a fraction in simplest form, divide both the numerator and the denominator by their greatest common factor (GCF).

Example 2 **Write each fraction in simplest form.**

a. $\frac{4}{16}$

$\frac{4}{16} = \frac{4 \div 4}{16 \div 4}$ Divide 4 and 16 by their GCF, 4.

$= \frac{1}{4}$ Simplify.

b. $\frac{24}{36}$

$\frac{24}{36} = \frac{24 \div 12}{36 \div 12}$ Divide 24 and 36 by their GCF, 12.

$= \frac{2}{3}$ Simplify.

Example 3 **Find each sum or difference. Write in simplest form.**

a. $\frac{7}{16} - \frac{1}{16}$

$\frac{7}{16} - \frac{1}{16} = \frac{6}{16}$ The denominators are the same. Subtract the numerators.

$= \frac{3}{8}$ Simplify.

b. $\frac{5}{8} + \frac{7}{8}$

$\frac{5}{8} + \frac{7}{8} = \frac{12}{8}$ The denominators are the same. Add the numerators.

$= 1\frac{4}{8}$ or $1\frac{1}{2}$ Rename $\frac{12}{8}$ as a mixed number in simplest form.

- To add or subtract fractions with unlike denominators, first find the least common denominator (LCD). Rename each fraction with the LCD, and then add or subtract. Simplify if necessary.

Example 4 Find each sum or difference. Write in simplest form.

a. $\frac{2}{9} + \frac{1}{3}$

$\frac{2}{9} + \frac{1}{3} = \frac{2}{9} + \frac{3}{9}$ The LCD for 9 and 3 is 9. Rename $\frac{1}{3}$ as $\frac{3}{9}$.

$= \frac{5}{9}$ Add the numerators.

b. $\frac{1}{2} + \frac{2}{3}$

$\frac{1}{2} + \frac{2}{3} = \frac{3}{6} + \frac{4}{6}$ The LCD for 2 and 3 is 6. Rename $\frac{1}{2}$ as $\frac{3}{6}$ and $\frac{2}{3}$ as $\frac{4}{6}$.

$= \frac{7}{6}$ or $1\frac{1}{6}$ Simplify.

c. $\frac{3}{8} - \frac{1}{3}$

$\frac{3}{8} - \frac{1}{3} = \frac{9}{24} - \frac{8}{24}$ The LCD for 8 and 3 is 24. Rename $\frac{3}{8}$ as $\frac{9}{24}$ and $\frac{1}{3}$ as $\frac{8}{24}$.

$= \frac{1}{24}$ Simplify.

d. $\frac{7}{10} - \frac{2}{15}$

$\frac{7}{10} - \frac{2}{15} = \frac{21}{30} - \frac{4}{30}$ The LCD for 10 and 15 is 30. Rename $\frac{7}{10}$ as $\frac{21}{30}$ and $\frac{2}{15}$ as $\frac{4}{30}$.

$= \frac{17}{30}$ Simplify.

Exercises Find each sum or difference.

1. $\frac{2}{5} + \frac{1}{5}$
2. $\frac{2}{7} - \frac{1}{7}$
3. $\frac{4}{3} + \frac{4}{3}$
4. $\frac{3}{9} + \frac{4}{9}$
5. $\frac{5}{16} - \frac{4}{16}$
6. $\frac{7}{2} - \frac{4}{2}$

Simplify.

7. $\frac{6}{9}$
8. $\frac{7}{14}$
9. $\frac{28}{40}$
10. $\frac{16}{100}$
11. $\frac{27}{99}$
12. $\frac{24}{180}$

Find each sum or difference. Write in simplest form.

13. $\frac{2}{9} + \frac{1}{9}$
14. $\frac{2}{15} + \frac{7}{15}$
15. $\frac{2}{3} + \frac{1}{3}$
16. $\frac{7}{8} - \frac{3}{8}$
17. $\frac{4}{9} - \frac{1}{9}$
18. $\frac{5}{4} - \frac{3}{4}$
19. $\frac{1}{2} + \frac{1}{4}$
20. $\frac{1}{2} - \frac{1}{3}$
21. $\frac{4}{3} + \frac{5}{9}$
22. $1\frac{1}{2} - \frac{3}{2}$
23. $\frac{1}{4} + \frac{1}{5}$
24. $\frac{2}{3} + \frac{1}{4}$
25. $\frac{3}{2} + \frac{1}{2}$
26. $\frac{8}{9} - \frac{2}{3}$
27. $\frac{3}{7} + \frac{5}{14}$
28. $\frac{13}{20} - \frac{2}{5}$
29. $1 - \frac{1}{19}$
30. $\frac{9}{10} - \frac{3}{5}$
31. $\frac{3}{4} - \frac{2}{3}$
32. $\frac{4}{15} + \frac{3}{4}$
33. $\frac{11}{12} - \frac{4}{15}$
34. $\frac{3}{11} + \frac{1}{8}$
35. $\frac{94}{100} - \frac{11}{25}$
36. $\frac{3}{25} + \frac{5}{6}$

Operations with Fractions: Multiplying and Dividing

- To multiply fractions, multiply the numerators and multiply the denominators.

Example 1 **Find each product.**

a. $\frac{2}{5} \cdot \frac{1}{3}$

$\frac{2}{5} \cdot \frac{1}{3} = \frac{2 \cdot 1}{5 \cdot 3}$ Multiply the numerators. Multiply the denominators.

$= \frac{2}{15}$ Simplify.

b. $\frac{7}{3} \cdot \frac{1}{11}$

$\frac{7}{3} \cdot \frac{1}{11} = \frac{7 \cdot 1}{3 \cdot 11}$ Multiply the numerators. Multiply the denominators.

$= \frac{7}{33}$ Simplify.

- If the fractions have common factors in the numerators and denominators, you can simplify before you multiply by canceling.

Example 2 **Find each product. Simplify before multiplying.**

a. $\frac{3}{4} \cdot \frac{4}{7}$

$\frac{3}{4} \cdot \frac{4}{7} = \frac{3}{\cancel{4}_1} \cdot \frac{\cancel{4}^1}{7}$ Divide by the GCF, 4.

$= \frac{3}{7}$ Simplify.

b. $\frac{4}{9} \cdot \frac{45}{49}$

$\frac{4}{9} \cdot \frac{45}{49} = \frac{4}{\cancel{9}_1} \cdot \frac{\cancel{45}^5}{49}$ Divide by the GCF, 9.

$= \frac{20}{49}$ Multiply the numerators and denominators.

- Two numbers whose product is 1 are called **multiplicative inverses** or **reciprocals**.

Example 3 **Name the reciprocal of each number.**

a. $\frac{3}{8}$

$\frac{3}{8} \cdot \frac{8}{3} = 1$ The product is 1.

The reciprocal of $\frac{3}{8}$ is $\frac{8}{3}$.

b. $\frac{1}{6}$

$\frac{1}{6} \cdot \frac{6}{1} = 1$ The product is 1.

The reciprocal of $\frac{1}{6}$ is 6.

c. $2\frac{4}{5}$

$2\frac{4}{5} = \frac{14}{5}$ Write $2\frac{4}{5}$ as an improper fraction.

$\frac{14}{5} \cdot \frac{5}{14} = 1$ The product is 1.

The reciprocal of $2\frac{4}{5}$ is $\frac{5}{14}$.

- To divide one fraction by another fraction, multiply the dividend by the multiplicative inverse of the divisor.

Example 4 **Find each quotient.**

a. $\frac{1}{3} \div \frac{1}{2}$

$\frac{1}{3} \div \frac{1}{2} = \frac{1}{3} \cdot \frac{2}{1}$ Multiply $\frac{1}{3}$ by $\frac{2}{1}$, the reciprocal of $\frac{1}{2}$.

$= \frac{2}{3}$ Simplify.

b. $\frac{3}{8} \div \frac{2}{3}$

$\frac{3}{8} \div \frac{2}{3} = \frac{3}{8} \cdot \frac{3}{2}$ Multiply $\frac{3}{8}$ by $\frac{3}{2}$, the reciprocal of $\frac{2}{3}$.

$= \frac{9}{16}$ Simplify.

c. $4 \div \frac{5}{6}$

$4 \div \frac{5}{6} = \frac{4}{1} \cdot \frac{6}{5}$ Multiply 4 by $\frac{6}{5}$, the reciprocal of $\frac{5}{6}$.

$= \frac{24}{5}$ or $4\frac{4}{5}$ Simplify.

d. $\frac{3}{4} \div 2\frac{1}{2}$

$\frac{3}{4} \div 2\frac{1}{2} = \frac{3}{4} \cdot \frac{2}{5}$ Multiply $\frac{3}{4}$ by $\frac{2}{5}$, the reciprocal of $2\frac{1}{2}$.

$= \frac{6}{20}$ or $\frac{3}{10}$ Simplify.

Exercises **Find each product.**

1. $\frac{3}{4} \cdot \frac{1}{5}$
2. $\frac{2}{7} \cdot \frac{1}{3}$
3. $\frac{1}{5} \cdot \frac{3}{20}$
4. $\frac{2}{5} \cdot \frac{3}{7}$
5. $\frac{5}{2} \cdot \frac{1}{4}$
6. $\frac{7}{2} \cdot \frac{3}{2}$
7. $\frac{1}{3} \cdot \frac{2}{5}$
8. $\frac{2}{3} \cdot \frac{1}{11}$

Find each product. Simplify before multiplying if possible.

9. $\frac{2}{9} \cdot \frac{1}{2}$
10. $\frac{15}{2} \cdot \frac{7}{15}$
11. $\frac{3}{2} \cdot \frac{1}{3}$
12. $\frac{1}{3} \cdot \frac{6}{5}$
13. $\frac{9}{4} \cdot \frac{1}{18}$
14. $\frac{11}{3} \cdot \frac{9}{44}$
15. $\frac{2}{7} \cdot \frac{14}{3}$
16. $\frac{2}{11} \cdot \frac{110}{17}$
17. $\frac{1}{3} \cdot \frac{12}{19}$
18. $\frac{1}{3} \cdot \frac{15}{2}$
19. $\frac{30}{11} \cdot \frac{1}{3}$
20. $\frac{6}{5} \cdot \frac{10}{12}$

Name the reciprocal of each number.

21. $\frac{6}{7}$
22. $\frac{3}{2}$
23. $\frac{1}{22}$
24. $\frac{14}{23}$
25. $2\frac{3}{4}$
26. $5\frac{1}{3}$

Find each quotient.

27. $\frac{2}{3} \div \frac{1}{3}$
28. $\frac{16}{9} \div \frac{4}{9}$
29. $\frac{3}{2} \div \frac{1}{2}$
30. $\frac{3}{7} \div \frac{1}{5}$
31. $\frac{9}{10} \div \frac{3}{7}$
32. $\frac{1}{2} \div \frac{3}{5}$
33. $2\frac{1}{4} \div \frac{1}{2}$
34. $1\frac{1}{3} \div \frac{2}{3}$
35. $\frac{11}{12} \div 1\frac{2}{3}$
36. $\frac{3}{8} \div \frac{1}{4}$
37. $\frac{1}{3} \div 1\frac{1}{5}$
38. $\frac{3}{25} \div \frac{2}{15}$

Prerequisite Skills

The Percent Proportion

- A **percent** is a ratio that compares a number to 100. To write a percent as a fraction, express the ratio as a fraction with a denominator of 100. Fractions should be stated in simplest form.

Example 1 **Express each percent as a fraction.**

a. 25%

$25\% = \frac{25}{100}$ or $\frac{1}{4}$ Definition of percent

b. 107%

$107\% = \frac{107}{100}$ or $1\frac{7}{100}$ Definition of percent

c. 0.5%

$0.5\% = \frac{0.5}{100}$ Definition of percent

$= \frac{5}{1000}$ or $\frac{1}{200}$ Simplify.

- In the **percent proportion**, the ratio of a part of something (part) to the whole (base) is equal to the percent written as a fraction.

part → $\frac{a}{b} = \frac{p}{100}$ ← percent
base →

Example: $\underbrace{10}_{\text{part}}$ is $\underbrace{25\%}_{\text{percent}}$ of $\underbrace{40}_{\text{base}}$.

Example 2 **40% of 30 is what number?**

The percent is 40, and the base is 30. Let a represent the part.

$\frac{a}{b} = \frac{p}{100}$ Use the percent proportion

$\frac{a}{30} = \frac{40}{100}$ Replace b with 30 and p with 40.

$100a = 30(40)$ Find the cross products.

$100a = 1200$ Simplify.

$\frac{100a}{100} = \frac{1200}{100}$ Divide each side by 100.

$a = 12$ Simplify.

The part is 12. So, 40% of 30 is 12.

Example 3 **Kelsey took a survey of some of the students in her lunch period. 42 out of the 70 students Kelsey surveyed said their family had a pet. What percent of the students had pets?**

You know the part, 42, and the base, 70.
Let p represent the percent.

$\frac{a}{b} = \frac{p}{100}$ Use the percent proportion.

$\frac{42}{70} = \frac{p}{100}$ Replace a with 42 and b with 70.

$4200 = 70p$ Find the cross products.

$\frac{4200}{70} = \frac{70p}{70}$ Divide each side by 70.

$60 = p$ Simplify.

The percent is 60, so $\frac{60}{100}$ or 60% of the students had pets.

Example 4 **67.5 is 75% of what number?**

You know the percent, 75, and the part, 67.5.
Let b represent the base.

$\frac{a}{b} = \frac{p}{100}$ Use the percent proportion.

$\frac{67.5}{b} = \frac{75}{100}$ $75\% = \frac{75}{100}$, so $p = 75$. Replace a with 67.5 and p with 75.

$6750 = 75b$ Find the cross products.

$\frac{6750}{75} = \frac{75b}{75}$ Divide each side by 75.

$90 = b$ Simplify.

The base is 90, so 67.5 is 75% of 90.

Exercises **Express each percent as a fraction.**

1. 5%
2. 60%
3. 11%
4. 120%
5. 78%
6. 2.5%
7. 0.9%
8. 0.4%
9. 1400%

Use the percent proportion to find each number.

10. 25 is what percent of 125?
11. 16 is what percent of 40?
12. 14 is 20% of what number?
13. 50% of what number is 80?
14. What number is 25% of 18?
15. Find 10% of 95.
16. What percent of 48 is 30?
17. What number is 150% of 32?
18. 5% of what number is 3.5?
19. 1 is what percent of 400?
20. Find 0.5% of 250.
21. 49 is 200% of what number?
22. 15 is what percent of 12?
23. 48 is what percent of 32?

24. Madeline usually makes 85% of her shots in basketball. If she shoots 20 shots, how many will she likely make?
25. Brian answered 36 items correctly on a 40-item test. What percent did he answer correctly?
26. José told his dad that he won 80% of the solitaire games he played yesterday. If he won 4 games, how many games did he play?
27. A glucose solution is prepared by dissolving 6 milliliters of glucose in 120 milliliters of solution. What is the percent of glucose in the solution?

HEALTH **For Exercises 28–30, use the following information.** The U.S. Food and Drug Administration requires food manufacturers to label their products with a nutritional label. The sample label shown at the right shows a portion of the information from a package of macaroni and cheese.

Nutrition Facts

Serving Size 1 cup (228g)	
Servings per container 2	
Amount per serving	
Calories 250 Calories from Fat 110	
	%Daily value*
Total Fat 12g	**18%**
Saturated Fat 3g	**15%**
Cholesterol 30mg	**10%**
Sodium 470mg	**20%**
Total Carbohydrate 31g	**10%**
Dietary Fiber 0g	**0%**
Sugars 5g	
Protein 5g	
Vitamin A 4% • Vitamin C 2%	
Calcium 20% • Iron 4%	

28. The label states that a seving contains 3 grams of saturated fat, which is 15% of the daily value recommended for a 2000-Calorie diet. How many grams of saturated fat are recommended for a 2000-Calorie diet.
29. The 470 milligrams of sodium (salt) in the macaroni and cheese is 20% of the recommended daily value. What is the recommended daily value of sodium?
30. For a healthy diet, the National Research Council recommends that no more than 30 percent of total Calories come from fat. What percent of the Calories in a serving of this macaroni and cheese come from fat?

Expressing Fractions as Decimals and Percents

- To write a fraction as a decimal, divide the numerator by the denominator. To write a decimal as a fraction, write the decimal as a fraction with denominator of 10, 100, 1000, … . Then simplify if possible.

Example 1 **Write each fraction as a decimal.**

a. $\frac{5}{8}$

$\frac{5}{8} = 5 \div 8$

$= 0.625$

b. $\frac{3}{5}$

$\frac{3}{5} = 3 \div 5$

$= 0.6$

c. $\frac{1}{3}$

$\frac{1}{3} = 1 \div 3$

$= 0.333\ldots$

Example 2 **Write each decimal as a fraction.**

a. 0.4

$0.4 = \frac{4}{10}$ or $\frac{2}{5}$

b. 0.005

$0.005 = \frac{5}{1000}$ or $\frac{1}{200}$

c. 0.98

$0.98 = \frac{98}{100}$ or $\frac{49}{50}$

- To write a fraction for a repeating decimal, use the method in Example 3 below.

Example 3 **Write each decimal as a fraction.**

a. $0.\overline{3}$

Let $N = 0.\overline{3}$ or $0.333\ldots$

Then $10N = 3.\overline{3}$ or $3.333\ldots$

$$\begin{aligned} 10N &= 3.333\ldots \\ -1N &= 0.333\ldots \\ \hline 9N &= 3 \\ N &= \frac{3}{9} \text{ or } \frac{1}{3} \end{aligned}$$

Subtract 1*N* from 10*N*.

So, $0.\overline{3} = \frac{1}{3}$.

b. $0.\overline{72}$

Let $N = 0.\overline{72}$ or $0.7272\ldots$

Then $100N = 72.7272\ldots$

$$\begin{aligned} 100N &= 72.7272 \\ -1N &= 00.7272 \\ \hline 99N &= 72 \\ N &= \frac{72}{99} \text{ or } \frac{8}{11} \end{aligned}$$

Subtract 1*N* from 100*N*.

So, $0.\overline{72} = \frac{8}{11}$.

- To write a decimal as a percent, multiply by 100 and add the % symbol. Recall that to multiply by 100, you can move the decimal point two places to the right.

- To write a percent as a decimal, divide by 100 and remove the % symbol. Recall that to divide by 100, you can move the decimal point two places to the left.

Example 4 **Write each decimal as a percent.**

Multiply by 100 and add the % symbol.

a. 0.35

$0.35 = 0.35$

$= 35\%$

b. 0.06

$0.06 = 0.06$

$= 6\%$

c. 0.008

$0.008 = 0.008$

$= 0.8\%$

Example 5 **Write each percent as a decimal.**

Divide by 100 and remove the % symbol.

a. 36%

$36\% = 36\%$

$= 0.36$

b. 9%

$9\% = 09\%$

$= 0.09$

c. 120%

$120\% = 120\%$

$= 1.2$

- To write a fraction as a percent, express the fraction as a decimal. Then express the decimal as a percent.

Example 6 **Write each fraction as a percent. Round to the nearest tenth of a percent, if necessary.**

a. $\frac{1}{8}$

$\frac{1}{8} = 0.125$
$= 12.5\%$

b. $\frac{2}{3}$

$\frac{2}{3} = 0.6666\ldots$
$= 66.7\%$

c. $\frac{3}{600}$

$\frac{3}{600} = 0.005$
$= 0.5\%$

Prerequisite Skills

- To write a percent as a fraction, express the percent as a decimal. Then express the decimal as a fraction. Simplify if possible.

Example 7 **Write each percent as a fraction.**

a. 30%

$30\% = 0.30$
$= \frac{30}{100}$ or $\frac{3}{10}$

b. 140%

$140\% = 1.4$
$= \frac{14}{10}$ or $1\frac{2}{5}$

c. 0.2%

$00.2\% = 0.002$
$= \frac{2}{1000}$ or $\frac{1}{500}$

Exercises **Write each fraction as a decimal.**

1. $\frac{3}{8}$ **2.** $\frac{2}{5}$ **3.** $\frac{2}{3}$ **4.** $\frac{3}{4}$

5. $\frac{1}{2}$ **6.** $\frac{5}{9}$ **7.** $\frac{3}{10}$ **8.** $\frac{5}{6}$

Write each decimal as a fraction.

9. 0.9 **10.** 0.25 **11.** 5.24 **12.** $0.\overline{45}$

13. $0.\overline{6}$ **14.** 0.0034 **15.** 2.08 **16.** 0.004

Write each decimal as a percent.

17. 0.4 **18.** 0.08 **19.** 2.5 **20.** 0.33

21. 0.065 **22.** 5 **23.** 0.005 **24.** $0.\overline{3}$

Write each percent as a decimal.

25. 45% **26.** 3% **27.** 68% **28.** 115%

29. 200% **30.** 0.1% **31.** 5.2% **32.** 10.5%

Write each fraction as a percent. Round to the nearest tenth of a percent, if necessary.

33. $\frac{3}{4}$ **34.** $\frac{9}{20}$ **35.** $\frac{1}{2}$ **36.** $\frac{1}{6}$

37. $\frac{1}{3}$ **38.** $\frac{7}{8}$ **39.** $\frac{6}{5}$ **40.** $\frac{19}{25}$

Write each percent as a fraction.

41. 70% **42.** 3% **43.** 52% **44.** 25%

45. 6% **46.** 135% **47.** 0.1% **48.** 0.5%

Making Bar and Line Graphs

- One way to organize data is by using a frequency table. In a **frequency table**, you use **tally marks** to record and display the frequency of events.

Example 1 **Make a frequency table to organize the temperature data in the chart at the right.**

Noon Temperature (°F)					
52	48	60	39	55	56
60	63	70	58	59	54
63	65	66	73	76	51
54	60	52	48	47	54

Step 1 Make a table with three columns: Temperature, Tally, and Frequency. Add a title.

Step 2 Use intervals to organize the temperatures. In this case, we are using intervals of 10.

Step 3 Use tally marks to record the temperatures in each interval.

Step 4 Count the tally marks in each row and record in the Frequency column.

Noon Temperature (°F)		
Temperature	**Tally**	**Frequency**
30–39	I	1
40–49	III	3
50–59	~~IIII~~ ~~IIII~~	10
60–69	~~IIII~~ II	7
70–79	III	3

- A **bar graph** compares different categories of data by showing each as a bar whose length is related to the frequency.

Example 2 **The table below shows the results of a survey of students' favorite snacks. Make a bar graph to display the data.**

Product	Number of Students
Bagel Chips	10
Fruit	18
Popcorn	15
Potato Chips	20
Pretzels	16
Snack Nuts	9
Tortilla Chips	17

Step 1 Draw a horizontal axis and a vertical axis. Label the axes as shown. Add a title.

Step 2 Draw a bar to represent each category. The vertical scale is the number of students who chose each snack. The horizontal scale identifies the snack chosen.

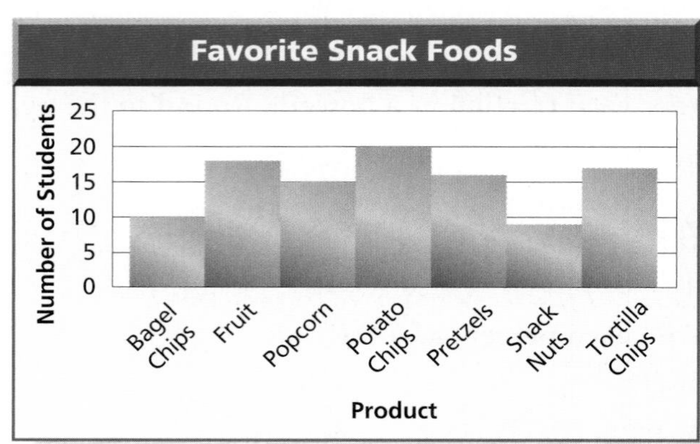

- Another way to represent data is by using a **line graph**. A line graph usually shows how data changes over a period of time.

Example 3 **Sales at the Marshall High School Store are shown in the table below. Make a line graph of the data.**

School Store Sales Amounts			
September	$670	February	$388
October	$229	March	$412
November	$300	April	$309
December	$168	May	$198
January	$290		

Step 1 Draw a horizontal axis and a vertical axis and label them as shown. Include a title.

Step 2 Plot the points to represent the data.

Step 3 Draw a line connecting each pair of consecutive points.

Exercises **Determine whether a bar graph or a line graph is the better choice to display the data.**

1. the growth of a plant
2. comparison of the populations in Idaho, Montana, and Texas
3. the number of students in each of the classes at your school
4. your height over the past eight years
5. the numbers of your friends that shower in the morning versus the number that shower at night
6. Alana surveyed several students to find the number of hours of sleep they typically get each night. The results are shown at the right. Make a bar graph of the data.

Hours of Sleep					
Alana	8	Kwam	7.5	Tomás	7.75
Nick	8.25	Kate	7.25	Sharla	8.5

7. Marcus started a lawn care service. The chart shows how much money he made over the 15 weeks of summer break. Make a line graph of the data.

Lawn Care Profits ($)								
Week	1	2	3	4	5	6	7	8
Profit	25	40	45	50	75	85	95	95
Week	9	10	11	12	13	14	15	
Profit	125	140	135	150	165	165	175	

8. The frequency table at the right shows the ages of people attending a high school play. Make a bar graph to display the data.

Age	Tally	Frequency
under 20	𝍸 𝍸 𝍸 𝍸 𝍸 𝍸 𝍸 𝍸 𝍸 II	47
20–39	𝍸 𝍸 𝍸 𝍸 𝍸 𝍸 𝍸 𝍸 III	43
40–59	𝍸 𝍸 𝍸 𝍸 𝍸 𝍸 I	31
60 and over	𝍸 III	8

Making Circle Graphs

A **circle graph** is a graph that shows the relationship between parts of the data and the whole. The circle represents the total data. Individual data are represented by parts of the circle. The examples show how to construct a circle graph.

Example 1 **The table shows the percent of her income that Ms. Garcia spends in each category. Make a circle graph to represent the data.**

How Ms. Garcia Spends Her Money	
Category	**Amount Spent**
Savings	10%
Car Payment/Insurance	20%
Food	20%
Clothing	10%
Rent	30%
Other	10%

Step 1 Find the number of degrees for each category. Since there are 360° in a circle, multiply each percent by 360 to find the number of degrees for each section of the graph.

Savings, Clothing, Other

$$10\% \text{ of } 360° = 0.1 \cdot 360°$$
$$= 36°$$

The sections for Savings, Clothing, and Other are each 36°.

Car Payment, Food

$$20\% \text{ of } 360° = 0.2 \cdot 360°$$
$$= 72°$$

The sections for Car Payment and Food are each 72°.

Rent

$$30\% \text{ of } 360° = 0.3 \cdot 360°$$
$$= 108°$$

The section for Rent is 108°.

Step 2 Use a compass to draw a circle. Then draw a radius.

Step 3 Use a protractor to draw a 36° angle to make the section representing Savings. (You can start with any angle.)

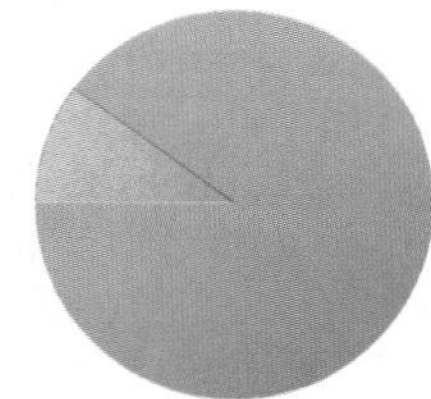

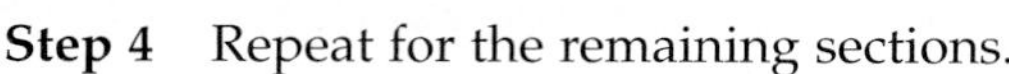

Step 4 Repeat for the remaining sections.

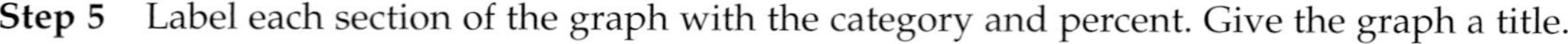

Step 5 Label each section of the graph with the category and percent. Give the graph a title.

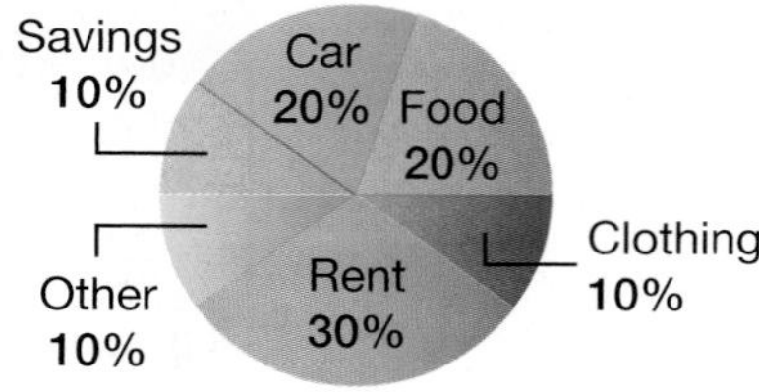

Example 2 **The table shows how Jessie uses her time on a typical Saturday. Make a circle graph of the data.**

First find the ratio that compares each number of hours to the total number of hours in a day, 24.

Activity	Hours
Jogging	1
Reading	2
Sleeping	9
Eating	2
Talking on the Phone	1
Time with Friends and Family	4
Studying	5

Jogging: $\frac{1}{24}$ **Reading:** $\frac{2}{24}$ **Sleeping:** $\frac{9}{24}$ **Eating:** $\frac{2}{24}$

Phone: $\frac{1}{24}$ **Friends:** $\frac{4}{24}$ **Studying:** $\frac{5}{24}$

Then multiply each ratio by 360 to find the number of degrees for each section of the graph.

Jogging, Phone: $\frac{1}{24} \cdot 360° = 15°$

Reading, Eating: $\frac{2}{24} \cdot 360° = 30°$

Sleeping: $\frac{9}{24} \cdot 360° = 135°$

Friends: $\frac{4}{24} \cdot 360° = 60°$

Studying: $\frac{5}{24} \cdot 360° = 75°$

Make the circle graph.

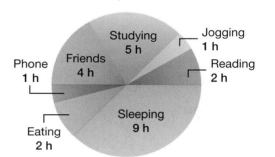

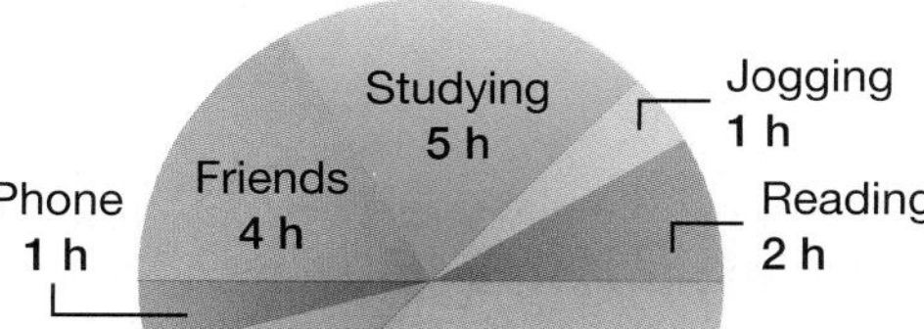

Exercises

1. The table at the right shows the percent of the world's population living in each continent or region. Make a circle graph of the data.

World Population, 2000	
Continent or Region	**Percent of World Total, 2000**
North America	7.9%
South America	5.7%
Europe	12.0%
Asia	60.7%
Africa	13.2%
Australia	0.5%
Antarctica	0%

Source: U.S. Census Bureau

2. The number of bones in each part of the human body is shown in the table at the right. Make a circle graph of the data.

Types of Human Bones	Number
Skull	29
Spine	26
Ribs and Breastbone	25
Shoulders, Arms, and Hands	64
Pelvis, Legs, and Feet	62

Identifying Two-Dimensional Figures

- Two-dimensional figures can be classified by the number of sides.

Number of Sides	Figure
3	**Tri**angle
4	**Quadri**lateral
5	**Pent**agon
6	**Hex**agon
8	**Oct**agon

The prefixes tell the number of sides.

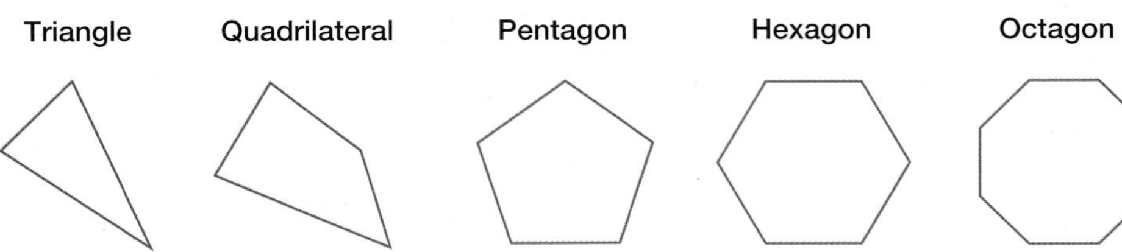

- Triangles can be classified by their angles. An **acute** angle measures less than 90°. An **obtuse** angle measures more than 90°. A **right** angle measures exactly 90°.

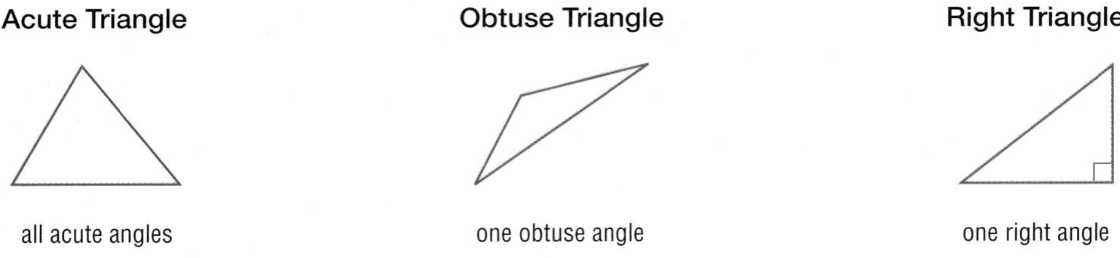

- Triangles can also be classified by their sides. Recall that **congruent** means having the same measure. Matching marks are used to show congruent parts.

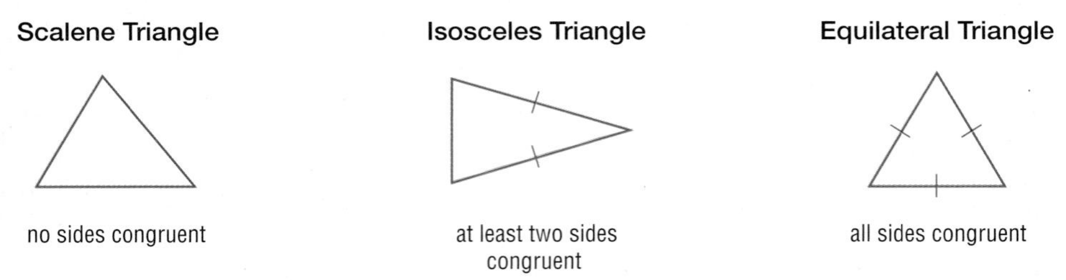

Classify each triangle using all names that apply.

a.

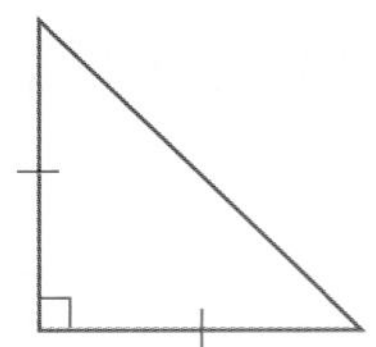

The triangle has one right angle and two congruent sides.
It is a right isosceles triangle.

b.

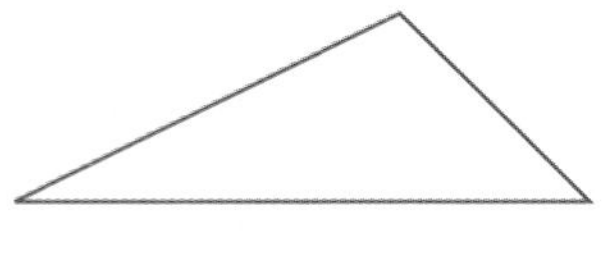

The triangle has one obtuse angle and no congruent sides.
It is an obtuse scalene triangle.

- The diagram below shows how quadrilaterals are classified. Notice that the diagram goes from most general to most specific.

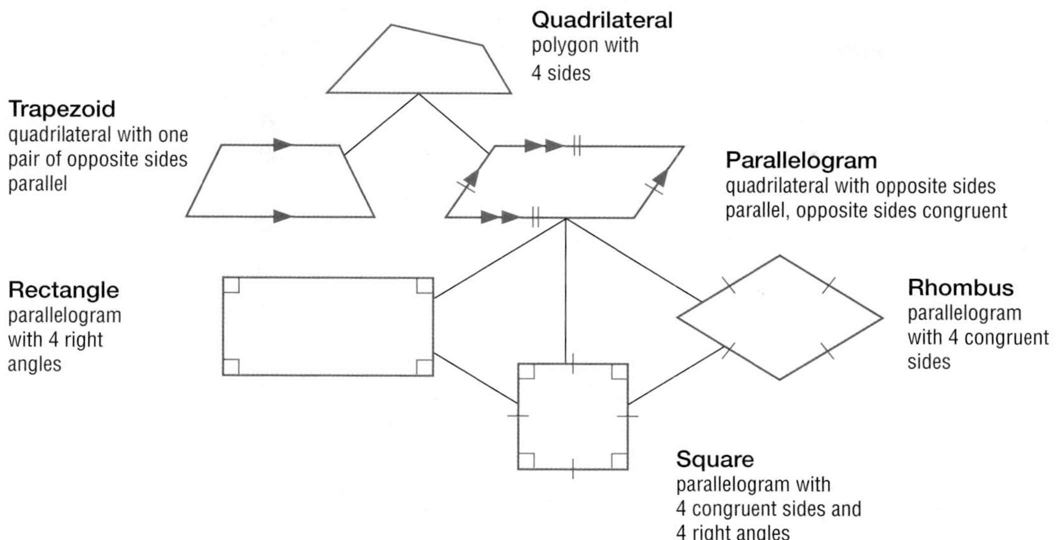

Exercises **Classify each figure using all names that apply.**

1.  

2. 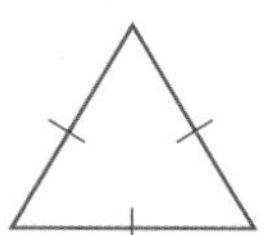

3. 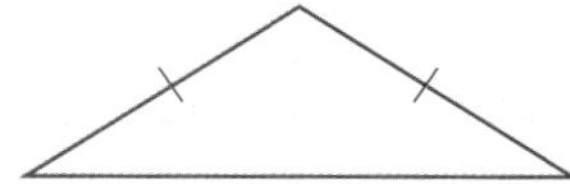

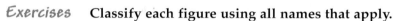

4. 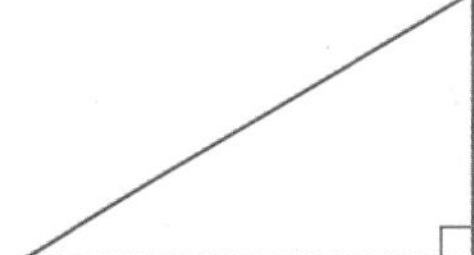

5. 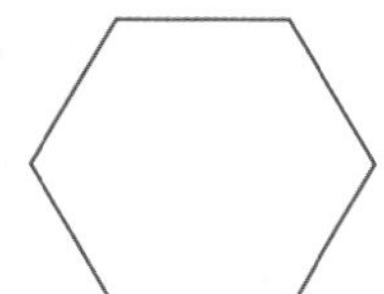

6. 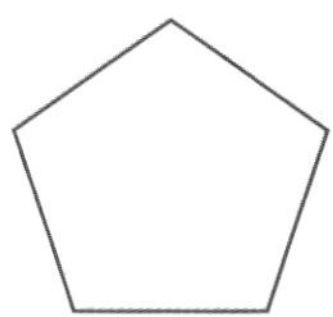

7. 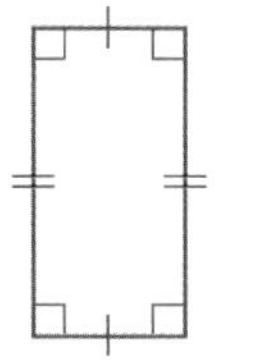

8. 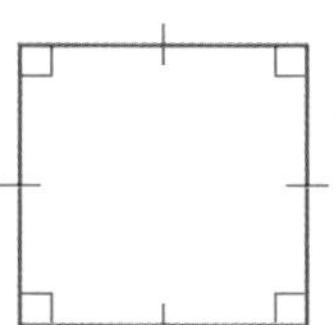

9. 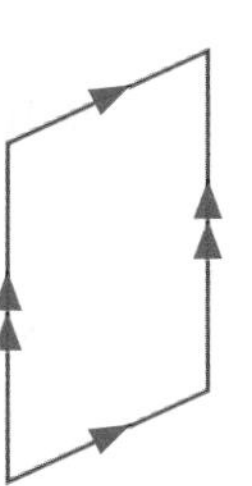

10. 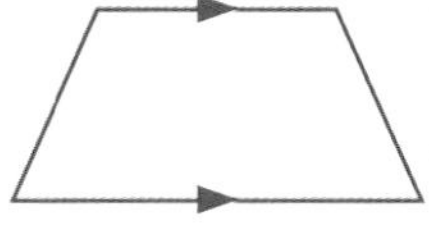

11. 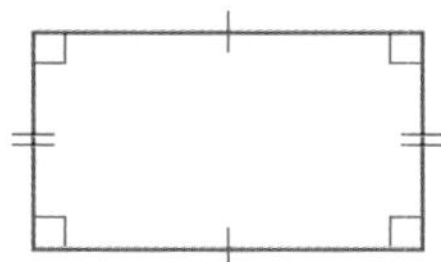

12. 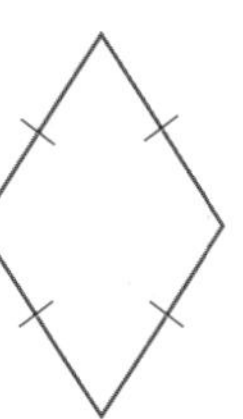

13. 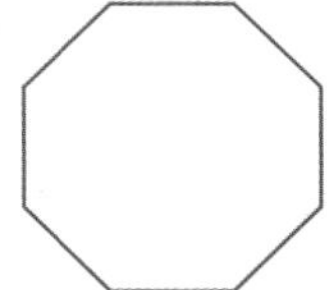

14. 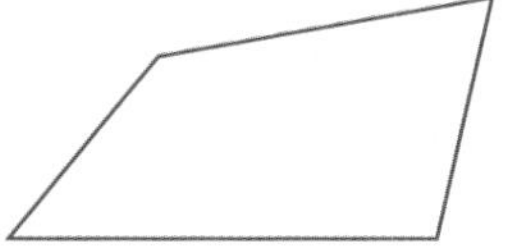

15. 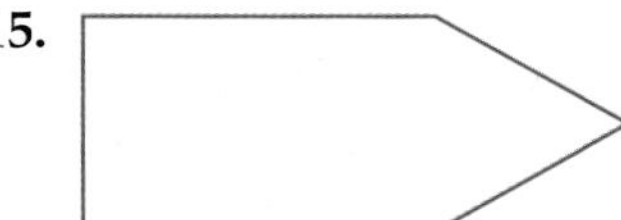

Identifying Three-Dimensional Figures

Prisms and pyramids are two types of three-dimensional figures. A **prism** has two parallel, congruent faces called **bases**. A **pyramid** has one base that is a polygon and faces that are triangles.

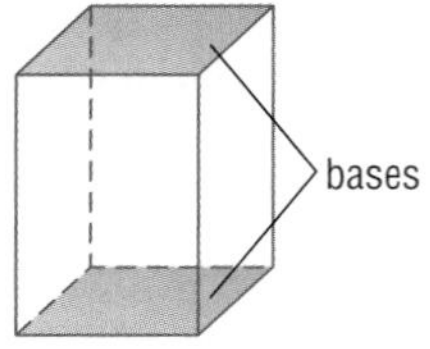

Prism

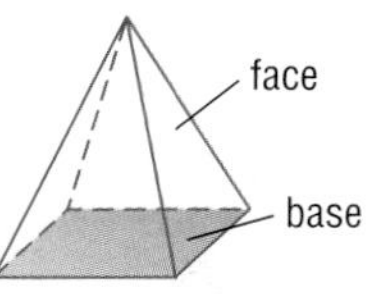

Pyramid

Prisms and pyramids are named by the shape of their bases.

Name	triangular prism	rectangular prism	triangular pyramid	rectangular pyramid
Number of Bases	2	2	1	1
Polygon Base	triangle	rectangle	triangle	rectangle
Figure				

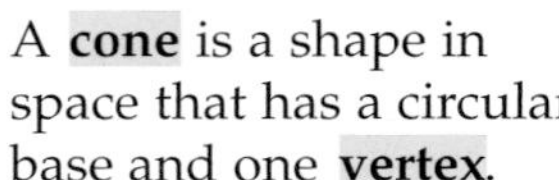

A **cube** is a rectangular prism in which all of the faces are squares.

A **cone** is a shape in space that has a circular base and one **vertex**.

A **sphere** is the set of all points a given distance from a given point called the center.

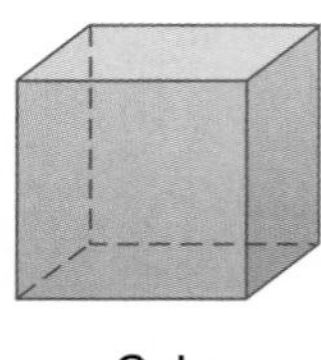
Cube

Cone

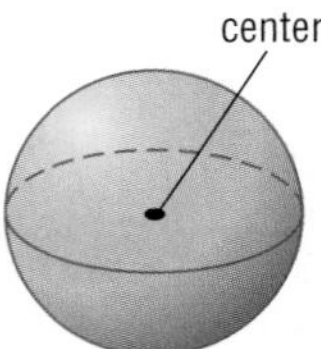

Sphere

Exercises

Classify each solid figure using the name that *best* describes it.

1. 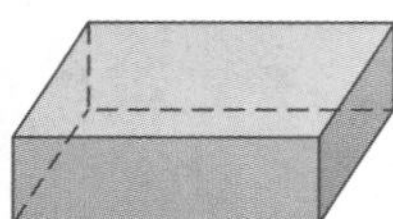

2. 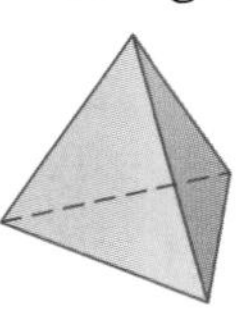

3. 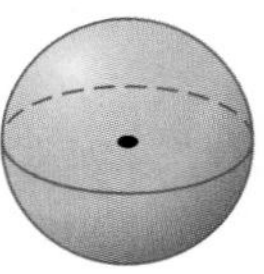

4. 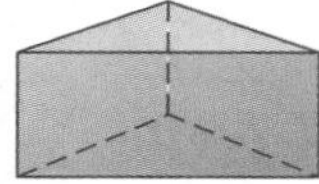

5. 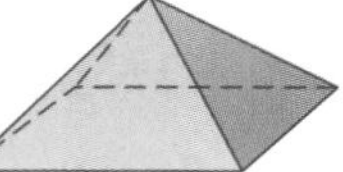

6. 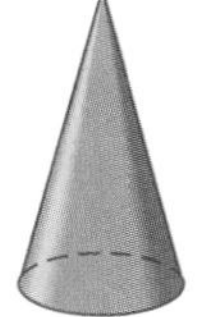

7. 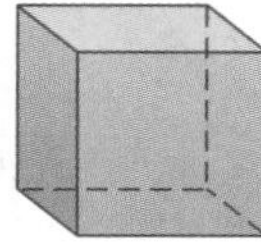

8. 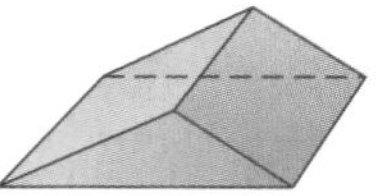

9. 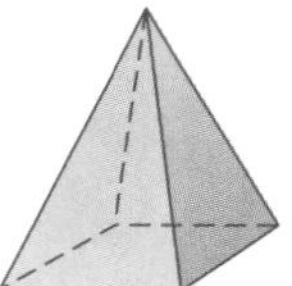

Perimeter and Area of Squares and Rectangles

Perimeter is the distance around a geometric figure. Perimeter is measured in linear units.

- To find the perimeter of a rectangle, multiply two times the sum of the length and width, or $2(\ell + w)$.
- To find the perimeter of a square, multiply four times the length of a side, or $4s$.

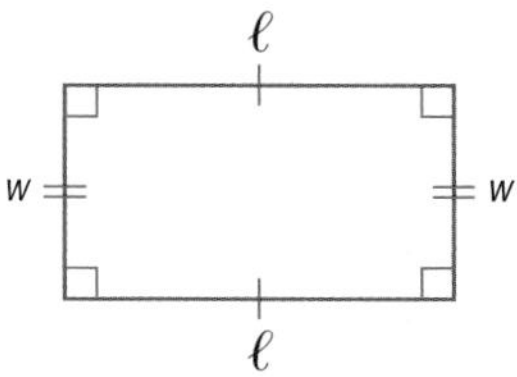

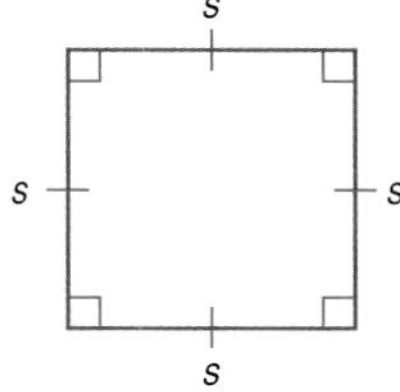

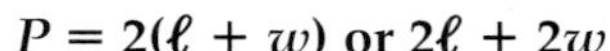

$P = 2(\ell + w)$ or $2\ell + 2w$

$P = 4s$

Area is the number of square units needed to cover a surface. Area is measured in square units.

- To find the area of a rectangle, multiply the length times the width, or $\ell \cdot w$.
- To find the area of a square, find the square of the length of a side, or s^2.

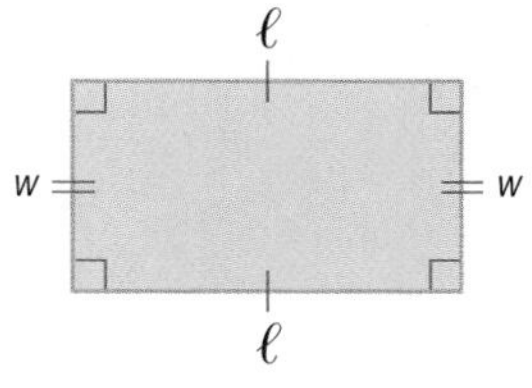

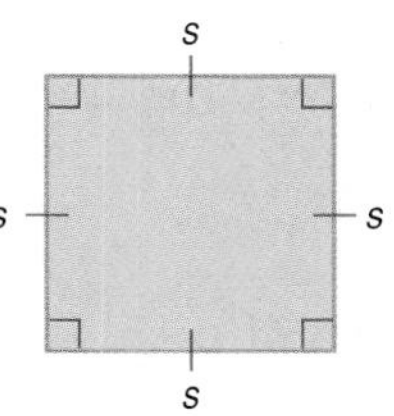

$A = \ell w$

$A = s^2$

Example 1 **Find the perimeter and area of each rectangle.**

a. A rectangle has a length of 3 units and a width of 5 units.

$P = 2(\ell + w)$	Perimeter formula
$= 2(3 + 5)$	Replace ℓ with 3 and w with 5.
$= 2(8)$	Add.
$= 16$	Multiply.

$A = \ell \cdot w$	Area formula
$= 3 \cdot 5$	Replace ℓ with 3 and w with 5.
$= 15$	Simplify.

The perimeter is 16 units, and the area is 15 square units.

b. A rectangle has a length of 1 inch and a width of 10 inches.

$P = 2(\ell + w)$	Perimeter formula
$= 2(1 + 10)$	Replace ℓ with 1and w with 10.
$= 2(11)$	Add.
$= 22$	Multiply.

$A = \ell \cdot w$	Area formula
$= 1 \cdot 10$	Replace ℓ with 1 and w with 10.
$= 10$	Simplify.

The perimeter is 22 inches, and the area is 10 square inches.

Example 2 **Find the perimeter and area of each square.**

a. A square has a side of length 8 feet.

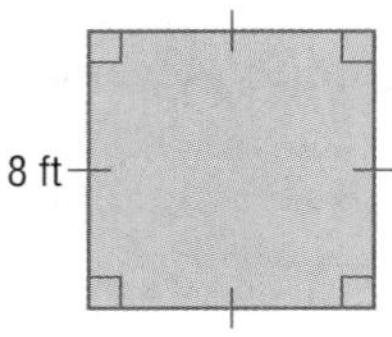

$P = 4s$ Perimeter formula

$= 4(8)$ $s = 8$

$= 32$ Multiply.

$A = s^2$ Area formula

$= 8^2$ $s = 8$

$= 64$ Multiply.

The perimeter is 32 feet, and the area is 64 square feet.

b. A square has a side of length 2 meters.

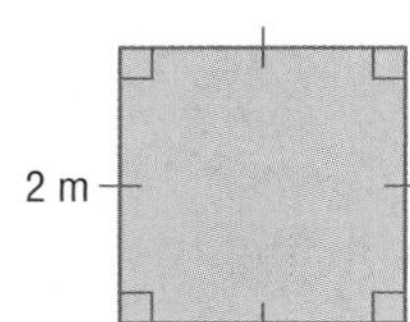

$P = 4s$ Perimeter formula

$= 4(2)$ $s = 2$

$= 8$ Multiply.

$A = s^2$ Area formula

$= 2^2$ $s = 2$

$= 4$ Multiply.

The perimeter is 8 meters, and the area is 4 square meters.

Exercises **Find the perimeter and area of each figure.**

1.

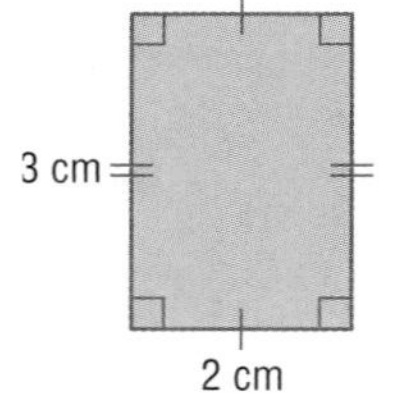

2.

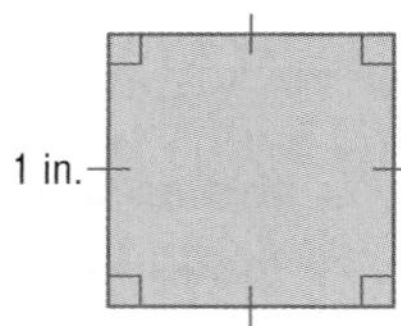

3.

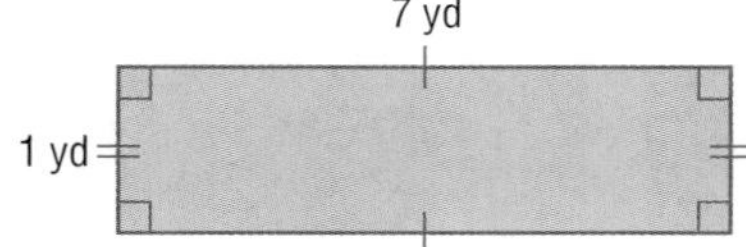

4.

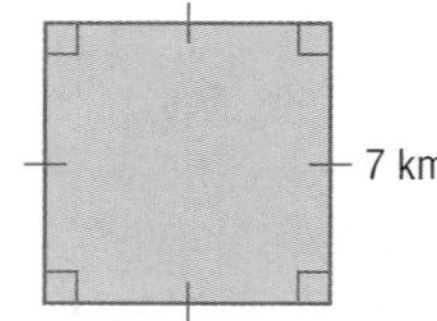

5. a rectangle with length 6 feet and width 4 feet

6. a rectangle with length 12 centimeters and width 9 centimeters

7. a square with length 3 meters

8. a square with length 15 inches

9. a rectangle with width $8\frac{1}{2}$ inches and length 11 inches

10. a rectangular room with width $12\frac{1}{4}$ feet and length $14\frac{1}{2}$ feet

11. a square with length 2.4 centimeters

12. a square garden with length 5.8 meters

13. RECREATION The Granville Parks and Recreation Department uses an empty city lot for a community vegetable garden. Each participant is allotted a space of 18 feet by 90 feet for a garden. What is the perimeter and area of each plot?

Area and Circumference of Circles

A **circle** is the set of all points in a plane that are the same distance from a given point.

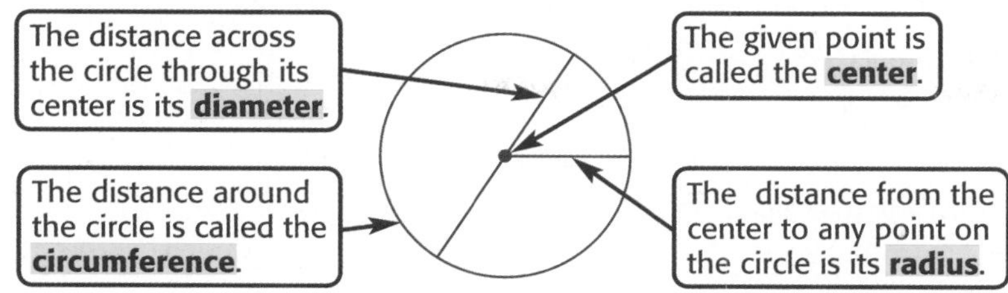

The distance across the circle through its center is its **diameter**.
The given point is called the **center**.
The distance around the circle is called the **circumference**.
The distance from the center to any point on the circle is its **radius**.

- The formula for the circumference of a circle is $C = \pi d$ or $C = 2\pi r$.

Example 1 **Find the circumference of each circle.**

a. The radius is 3 feet.

Use the formula $C = 2\pi r$.

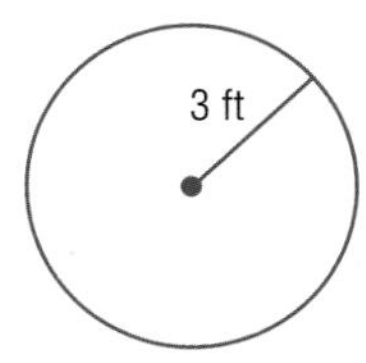

$C = 2\pi r$ Write the formula.

$= 2\pi(3)$ Replace r with 3.

$= 6\pi$ Simplify.

The exact circumference is 6π feet.

6 [π] [ENTER] 18.84955592

To the nearest tenth, the circumference is 18.8 feet.

b. The diameter is 24 centimeters.

Use the formula $C = \pi d$.

$C = \pi d$ Write the formula.

$= \pi(24)$ Replace d with 24.

$= 24\pi$ Simplify.

≈ 75.4 Use a calculator to evaluate 24π.

The circumference is about 75.4 centimeters.

- The formula for the area of a circle is $A = \pi r^2$.

Example 2 **Find the area of each circle to the nearest tenth.**

a. The radius is 4 inches.

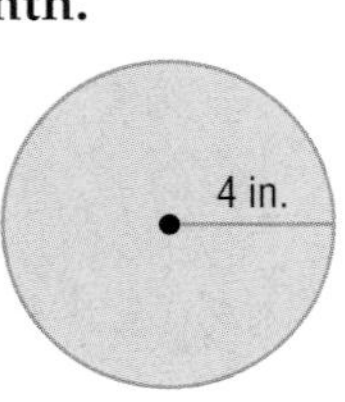

$A = \pi r^2$ Write the formula.

$= \pi(4)^2$ Replace r with 4.

$= 16\pi$ Simplify.

≈ 50.3 Use a calculator to evaluate 16π.

The area of the circle is about 50.3 square inches.

b. The diameter is 20 centimeters.

The radius is one-half times the diameter, or 10 centimeters.

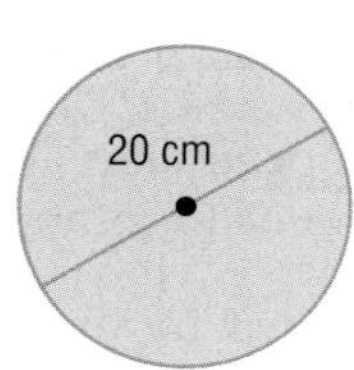

$A = \pi r^2$ Write the formula.

$= \pi(10)^2$ Replace r with 10.

$= 100\pi$ Simplify.

≈ 314.2 Use a calculator to evaluate 100π.

The area of the circle is about 314.2 square centimeters.

Prerequisite Skills

Example 3 **HISTORY** Stonehenge is an ancient monument in Wiltshire, England. Historians are not sure who erected Stonehenge or why. It may have been used as a calendar. The giant stones of Stonehenge are arranged in a circle 30 meters in diameter. Find the circumference and the area of the circle.

$C = \pi d$ Write the formula.

$= \pi(30)$ Replace d with 30.

$= 30\pi$ Simplify.

≈ 94.2 Use a calculator to evaluate 30π.

Find the radius to evaluate the formula for the area. The radius is one-half times the diameter, or 15 meters.

$A = \pi r^2$ Write the formula.

$= \pi(15)^2$ Replace r with 15.

$= 225\pi$ Simplify.

≈ 706.9 Use a calculator to evaluate 225π.

The circumference of Stonehenge is about 94.2 meters, and the area is about 706.9 square meters.

Exercises **Find the circumference of each circle. Round to the nearest tenth.**

1.
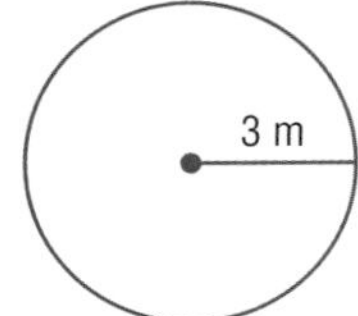

2.
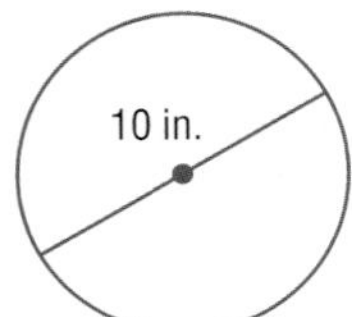

3.
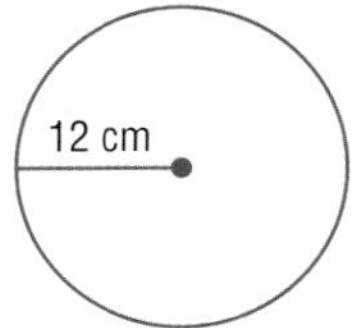

4. The radius is 1.5 kilometers.
5. The diameter is 1 yard.
6. The diameter is $5\frac{1}{4}$ feet.
7. The radius is $24\frac{1}{2}$ inches.

Find the area of each circle. Round to the nearest tenth.

8.
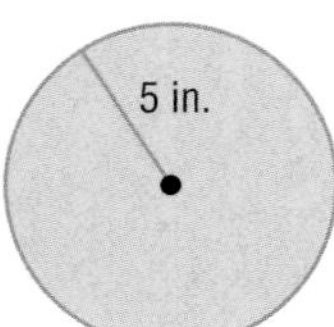

9.
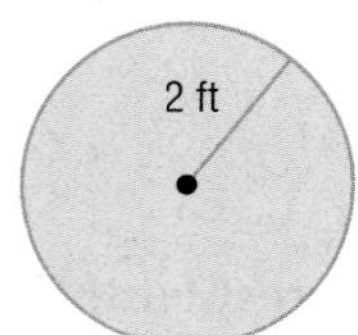

10.
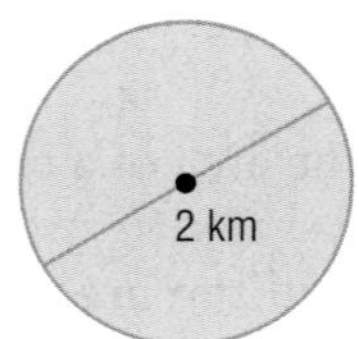

11. The diameter is 4 yards.
12. The radius is 1 meter.
13. The radius is 1.5 feet.
14. The diameter is 15 centimeters.

15. **GEOGRAPHY** Earth's circumference is approximately 25,000 miles. If you could dig a tunnel to the center of the Earth, how long would the tunnel be?

16. **CYCLING** The tire for a 10-speed bicycle has a diameter of 27 inches. Find the distance the bicycle will travel in 10 rotations of the tire.

17. **PUBLIC SAFETY** The Belleville City Council is considering installing a new tornado warning system. The sound emitted from the siren would be heard for a 2-mile radius. Find the area of the region that will benefit from the system.

18. **CITY PLANNING** The circular region inside the streets at DuPont Circle in Washington, D.C., is 250 feet across. How much area do the grass and sidewalk cover?

Volume

Volume is the measure of space occupied by a solid. Volume is measured in cubic units. The prism at the right has a volume of 12 cubic units.

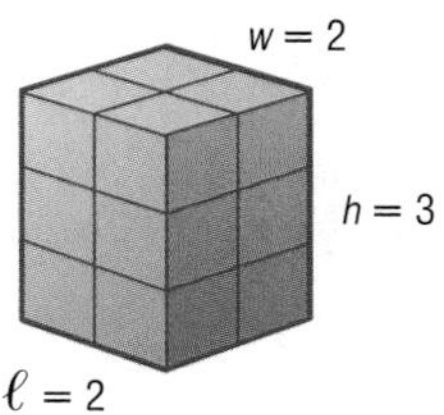

- To find the volume of a rectangular prism, use the formula $V = \ell \cdot w \cdot h$. Stated in words, volume equals length times width times height.

Example **Find the volume of the rectangular prism.**

A rectangular prism has a height of 3 feet, width of 4 feet, and length of 2 feet.

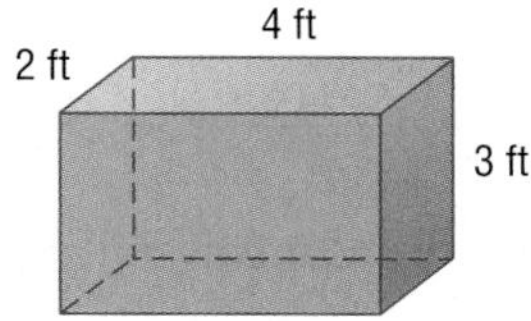

$V = \ell \cdot w \cdot h$ Write the formula.

$V = 2 \cdot 4 \cdot 3$ Replace ℓ with 2, w with 4, and h with 3.

$V = 24$ Simplify.

The volume is 24 cubic feet.

Exercises **Find the volume of each rectangular prism given the length, width, and height.**

1. $\ell = 2$ in., $w = 5$ in., $h = \frac{1}{2}$ in.
2. $\ell = 12$ cm, $w = 3$ cm, $h = 2$ cm
3. $\ell = 6$ yd, $w = 2$ yd, $h = 1$ yd
4. $\ell = 100$ m, $w = 1$ m, $h = 10$ m

Find the volume of each rectangular prism.

5.

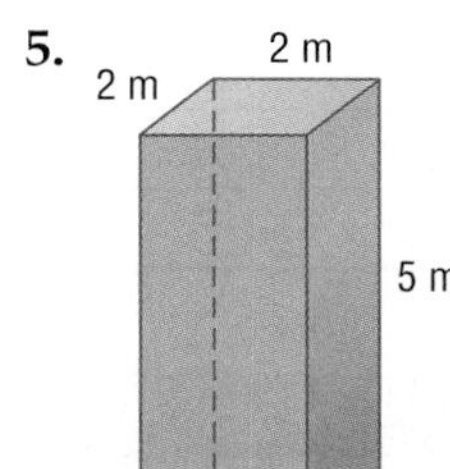

6.

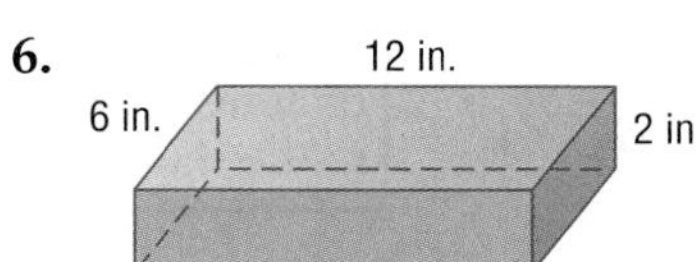

7. **AQUARIUMS** An aquarium is 8 feet long, 5 feet wide, and 5.5 feet deep. What is the volume of the tank?
8. **COOKING** What is the volume of a microwave oven that is 18 inches wide by 10 inches long with a depth of $11\frac{1}{2}$ inches?
9. **GEOMETRY** A cube measures 2 meters on a side. What is its volume?

FIREWOOD **For Exercises 10–12, use the following.**
Firewood is usually sold by a measure known as a cord. A full cord may be a stack $8 \times 4 \times 4$ feet or a stack $8 \times 8 \times 2$ feet.

10. What is the volume of a full cord of firewood?
11. A "short cord" or "face cord" of wood is $8 \times 4 \times$ the length of the logs. What is the volume of a short cord of $2\frac{1}{2}$-foot logs?
12. If you have an area that is 12 feet long and 2 feet wide in which to store your firewood, how high will the stack be if it is a full cord of wood?

Mean, Median, and Mode

Measures of central tendency are numbers used to represent a set of data. Three types of measures of central tendency are mean, median, and mode.

- The **mean** is the sum of the numbers in a set of data divided by the number of items.

Example 1 **Katherine is running a lemonade stand. She made \$3.50 on Tuesday, \$4.00 on Wednesday, \$5.00 on Thursday, and \$4.50 on Friday. What was her mean daily profit?**

$$\text{mean} = \frac{\text{sum of daily profits}}{\text{number of days}}$$

$$= \frac{\$3.50 + \$4.00 + \$5.00 + \$4.50}{4}$$

$$= \frac{\$17.00}{4} \text{ or } \$4.25$$

Katherine's mean daily profit was \$4.25.

- The **median** is the middle number in a set of data when the data are arranged in numerical order. If there are an even number of data, the median is the mean of the two middle numbers.

- The **mode** is the number or numbers that appear most often in a set of data. If no item appears most often, the set has no mode.

Example 2 **The table shows the number of hits Marcus made for his team. Find the median of the data.**

Team Played	Number of Hits by Marcus
Badgers	3
Hornets	6
Bulldogs	5
Vikings	2
Rangers	3
Panthers	7

To find the median, order the numbers from least to greatest. The median is in the middle.

2, 3, 3, 5, 6, 7

$$\frac{3 + 5}{2} = 4$$

There is an even number of items. Find the mean of the middle two.

The median number of hits is 4.

Example 3 **The table shows the heights of the members of the 2001–2002 University of Kentucky Men's Basketball team. What is the mode of the heights?**

Player	Height (in.)
Blevins	74
Bogans	77
Camara	83
Daniels	79
Estill	81
Fitch	75
Hawkins	73
Heissenbuttel	76
Parker	80
Prince	81
Sears	78
Smith	74
Stone	82
Tackett	74

Source: ESPN

The mode is the number that occurs most frequently. 74 occurs three times, 81 occurs twice, and all the other heights occur once. The mode height is 74.

• You can use measures of central tendency to solve problems.

Example 4 **On her first five history tests, Yoko received the following scores: 82, 96, 92, 83, and 91. What test score must Yoko earn on the sixth test so that her average (mean) for all six tests will be 90%?**

$$\text{mean} = \frac{\text{sum of the first five scores} + \text{sixth score}}{6}$$ Write an equation.

$$90 = \frac{82 + 96 + 92 + 83 + 91 + x}{6}$$ Use x to represent the sixth score.

$$90 = \frac{444 + x}{6}$$ Simplify.

$$540 = 444 + x$$ Multiply each side by 6.

$$96 = x$$ Subtract 444 from each side.

To have an average score of 90, Yoko must earn a 96 on the sixth test.

Exercises **Find the mean, median, and mode for each set of data.**

1. {1, 2, 3, 5, 5, 6, 13}
2. {3, 5, 8, 1, 4, 11, 3}
3. {52, 53, 53, 53, 55, 55, 57}
4. {8, 7, 5, 19}
5. {3, 11, 26, 4, 1}
6. {201, 201, 200, 199, 199}
7. {4, 5, 6, 7, 8}
8. {3, 7, 21, 23, 63, 27, 29, 95, 23}

9. **SCHOOL** The table shows the cost of some school supplies. Find the mean, median, and mode costs.

Cost of School Supplies	
Supply	**Cost**
Pencils	$0.50
Pens	$2.00
Paper	$2.00
Pocket Folder	$1.25
Calculator	$5.25
Notebook	$3.00
Erasers	$2.50
Markers	$3.50

10. **NUTRITION** The table shows the number of servings of fruits and vegetables that Cole eats one week. Find the mean, median, and mode.

Cole's Fruits and Vegetable Servings	
Day	**Number of Servings**
Monday	5
Tuesday	7
Wednesday	5
Thursday	4
Friday	3
Saturday	3
Sunday	8

11. **TELEVISION RATINGS** The ratings for the top television programs during one week are shown in the table at the right. Find the mean, median, and mode of the ratings. Round to the nearest hundredth.

12. **EDUCATION** Bill's scores on his first four science tests are 86, 90, 84, and 91. What test score must Bill earn on the fifth test so that his average (mean) will be exactly 88?

13. **BOWLING** Sue's average for 9 games of bowling is 108. What is the lowest score she can receive for the tenth game to have an average of 110?

14. **EDUCATION** Olivia has an average score of 92 on five French tests. If she earns a score of 96 on the sixth test, what will her new average score be?

Network Primetime Television Ratings	
Program	**Rating**
1	17.6
2	16.0
3	14.1
4	13.7
5	13.5
6	12.9
7	12.3
8	11.6
9	11.4
10	11.4

Source: Nielsen Media Research

Extra Practice

Extra Practice

Lesson 1-1

(pages 6–9)

Write an algebraic expression for each verbal expression.

1. the sum of b and 21
2. the product of x and 7
3. a number t increased by 6
4. the sum of 4 and 6 times a number z
5. -10 increased by 4 times a number a
6. the sum of 8 and -2 times n
7. one-half the cube of a number x
8. four-fifths the square of m

Evaluate each expression.

9. 2^4
10. 10^2
11. 7^3
12. 20^3
13. 3^6
14. 4^5

Write a verbal expression for each algebraic expression.

15. $2n$
16. 10^7
17. m^5
18. xy
19. $5n^2 - 6$
20. $9a^3 + 1$
21. $x^3 \cdot y^2$
22. $c^4 \cdot d^6$
23. $3e + 2e^2$

Lesson 1-2

(pages 11–15)

Evaluate each expression.

1. $3 + 8 \div 2 - 5$
2. $4 + 7 \cdot 2 + 8$
3. $5(9 + 3) - 3 \cdot 4$
4. $9 - 3^2$
5. $(8 - 1) \cdot 3$
6. $4(5 - 3)^2$
7. $3(12 + 3) - 5 \cdot 9$
8. $5^3 + 6^3 - 5^2$
9. $16 \div 2 \cdot 5 \cdot 3 \div 6$
10. $7(5^3 + 3^2)$
11. $\frac{9 \cdot 4 + 2 \cdot 6}{6 \cdot 4}$
12. $25 - \frac{1}{3}(18 + 9)$

Evaluate each expression if $a = 2$, $b = 5$, $x = 4$, and $n = 10$.

13. $8a + b$
14. $48 + ab$
15. $a(6 - 3n)$
16. $bx + an$
17. $x^2 - 4n$
18. $3b + 16a - 9n$
19. $n^2 + 3(a + 4)$
20. $(2x)^2 + an - 5b$
21. $[a + 8(b - 2)]^2 \div 4$

Lesson 1-3

(pages 16–20)

Find the solution of each equation if the replacement sets are $x = \{0, 2, 4, 6, 8\}$ and $y = \{1, 3, 5, 7, 9\}$.

1. $x - 4 = 4$
2. $25 - y = 18$
3. $3x + 1 = 25$
4. $5y - 4 = 11$
5. $14 = \frac{96}{x} + 2$
6. $0 = \frac{y}{3} - 3$

Solve each equation.

7. $x = \frac{27 + 9}{2}$
8. $\frac{18 - 7}{13 - 2} = y$
9. $n = \frac{6(5) + 3}{2(4) + 3}$
10. $\frac{5(4) - 6}{2^2 + 3} = z$
11. $\frac{7^2 + 9(2 + 1)}{2(10) - 1} = t$
12. $a = \frac{3^3 + 5^2}{2(3 - 1)}$

Find the solution set for each inequality if the replacement sets are $x = \{4, 5, 6, 7, 8\}$ and $y = \{10, 12, 14, 16\}$.

13. $x + 2 > 7$
14. $x - 1 < 8$
15. $2x \leq 15$
16. $3y \geq 36$
17. $\frac{x}{3} < 2$
18. $\frac{5y}{4} \geq 20$

Lesson 1-4

(pages 21–25)

Name the property used in each equation. Then find the value of n.

1. $4 \cdot 3 = 4 \cdot n$
2. $\frac{5}{4} = n + 0$
3. $15 = 15 \cdot n$
4. $\frac{2}{3}n = 1$
5. $2.7 + 1.3 = 2.7 + n$
6. $n\left(6^2 \cdot \frac{1}{36}\right) = 4$
7. $8n = 0$
8. $n = \frac{1}{9} \cdot 9$
9. $5 + 7 = 5 + n$
10. $(13 - 4)(2) = 9n$

Evaluate each expression. Name the property used in each step.

11. $\frac{2}{3}[15 \div (12 - 2)]$
12. $\frac{7}{4}\left[4 \cdot \left(\frac{1}{8} \cdot 8\right)\right]$
13. $[(18 \div 3) \cdot 0] \cdot 10$

Extra Practice

Lesson 1-5

(pages 26–31)

Rewrite each expression using the Distributive Property. Then simplify.

1. $5(2 + 9)$
2. $8(10 + 20)$
3. $20(8 - 3)$
4. $3(5 + w)$
5. $(h - 8)7$
6. $6(y + 4)$
7. $9(3n + 5)$
8. $32\left(x - \frac{1}{8}\right)$
9. $c(7 - d)$

Use the Distributive Property to find each product.

10. $6 \cdot 55$
11. $15(108)$
12. $1689 \cdot 5$
13. 7×314
14. $36\left(5\frac{1}{4}\right)$
15. $\left(4\frac{1}{18}\right) \cdot 18$

Simplify each expression. If not possible, write *simplified*.

16. $13a + 5a$
17. $21x - 10x$
18. $8(3x + 7)$
19. $4m - 4n$
20. $3(5am - 4)$
21. $15x^2 + 7x^2$
22. $9y^2 + 13y^2 + 3$
23. $11a^2 - 11a^2 + 12a^2$
24. $6a + 7a + 12b + 8b$

Lesson 1-6

(pages 32–36)

Evaluate each expression.

1. $23 + 8 + 37 + 12$
2. $19 + 46 + 81 + 54$
3. $10.25 + 2.5 + 3.75$
4. $22.5 + 17.6 + 44.5$
5. $2\frac{1}{3} + 6 + 3\frac{2}{3} + 4$
6. $5\frac{6}{7} + 15 + 4\frac{1}{7} + 25$
7. $6 \cdot 8 \cdot 5 \cdot 3$
8. $18 \cdot 5 \cdot 2 \cdot 5$
9. $0.25 \cdot 7 \cdot 8$
10. $90 \cdot 12 \cdot 0.5$
11. $5\frac{1}{3} \cdot 4 \cdot 6$
12. $4\frac{5}{6} \cdot 10 \cdot 12$

Simplify each expression.

13. $5a + 6b + 7a$
14. $8x + 4y + 9x$
15. $3a + 5b + 2c + 8b$
16. $\frac{2}{3}x^2 + 5x + x^2$
17. $(4p - 7q) + (5q - 8p)$
18. $8q + 5r - 7q - 6r$
19. $4(2x + y) + 5x$
20. $9r^5 + 2r^2 + r^5$
21. $12b^3 + 12 + 12b^3$
22. $7 + 3(uv - 6) + u$
23. $3(x + 2y) + 4(3x + y)$
24. $6.2(a + b) + 2.6(a + b) + 3a$
25. $3 + 8(st + 3w) + 3st$
26. $5.4(s - 3t) + 3.6(s - 4)$
27. $3[4 + 5(2x + 3y)]$

Lesson 1-7

(pages 37–42)

Identify the hypothesis and conclusion of each statement.

1. If an animal is a dog, then it barks.
2. If a figure is a pentagon, then it has five sides.
3. If $3x - 1 = 8$, then $x = 3$.
4. If 0.5 is the reciprocal of 2, then $0.5 \cdot 2 = 1$.

Identify the hypotheses and conclusion of each statement. Then write the statement in if-then form.

5. A square has four congruent sides.
6. $6a + 10 = 34$ when $a = 4$.
7. The video store is open every night.
8. The band does not have practice on Thursday.

Find a counterexample for each statement.

9. If the season is spring, then it does not snow.
10. If you live in Portland, then you live in Oregon.
11. If $2y + 4 = 10$, then $y < 3$.
12. If $a^2 > 0$, then $a > 0$.

Lesson 1-8

(pages 43–48)

Describe what is happening in each graph.

1. The graph shows the average monthly high temperatures for a city over a one-year period.

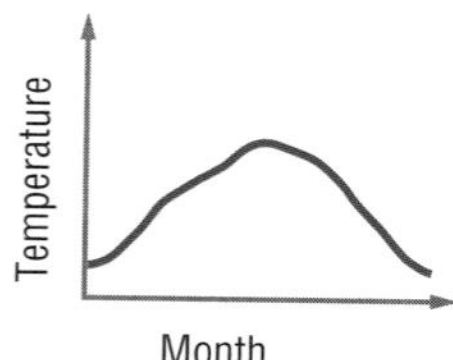

2. The graph shows the speed of a roller coaster car during a two-minute ride.

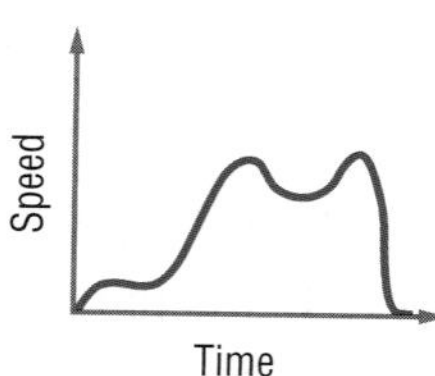

3. The graph shows the speed of a jogger over time.

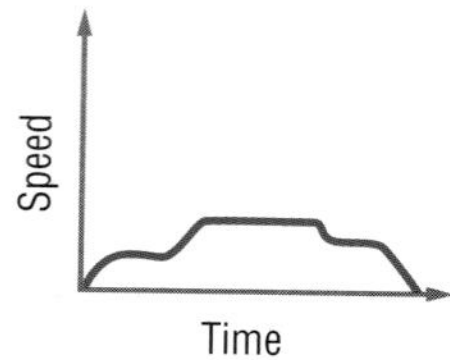

4. The graph shows the distance from camp traveled by a hiker over time.

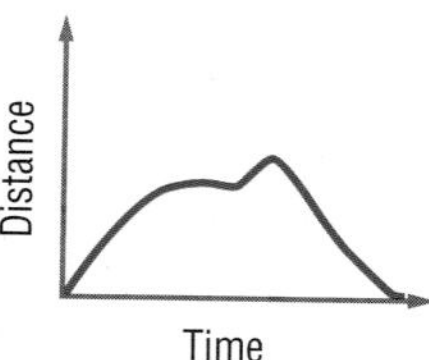

Lesson 1-9

(pages 50–55)

For Exercises 1–4, use the graph, which shows the five states that were the birthplace of the most U.S. presidents.

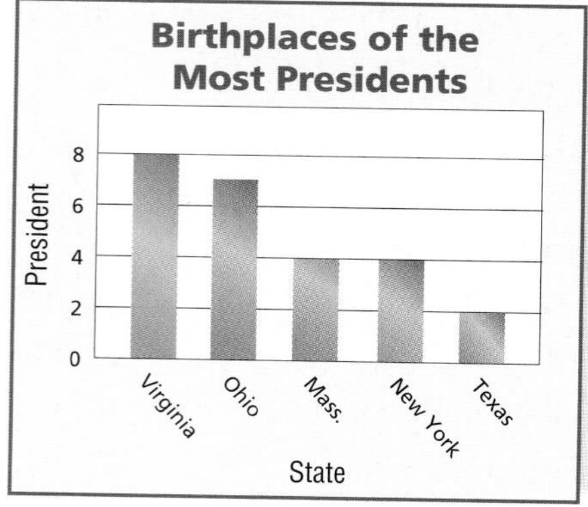

1. How many times more presidents were born in Virginia than Texas?
2. Did any states have the same number of presidents? If so, which states?
3. Would it be appropriate to display this data in a circle graph? Explain.
4. By the year 2001, there had been forty-three different presidents. What percent of U.S. presidents at that time had been born in Ohio?

Lesson 2-1

(pages 68–72)

Name the coordinates of the points graphed on each number line.

1. Number line from −4 to 4
2. Number line from −3 to 6
3. Number line from −4 to 4
4. Number line from 5 to 15
5. Number line from −9 to 0
6. Number line from −3 to 7

Graph each set of numbers.

7. $\{-2, -4, -6\}$
8. $\{\ldots, -3, -2, -1, 0\}$
9. {integers greater than -1}
10. {integers less than -5 and greater than -10}

Find each absolute value.

11. $|22|$
12. $|-2.5|$
13. $\left|\frac{2}{3}\right|$
14. $\left|-\frac{7}{8}\right|$

Lesson 2-2

(pages 73–78)

Find each sum.

1. $3 + 16$
2. $-27 + 19$
3. $8 + (-13)$
4. $-14 + (-9)$
5. $-25 + 47$
6. $97 + (-79)$
7. $-4.8 + 3.2$
8. $-1.7 + (-3.4)$
9. $-0.009 + 0.06$
10. $-\frac{11}{9} + \left(-\frac{7}{9}\right)$
11. $-\frac{3}{5} + \frac{5}{6}$
12. $\frac{3}{8} + \left(-\frac{7}{12}\right)$

Find each difference.

13. $27 - 14$
14. $8 - 17$
15. $12 - (-15)$
16. $-35 - (-12)$
17. $-2 - (-1.3)$
18. $1.9 - (-7)$
19. $-4.5 - 8.6$
20. $89.3 - (-14.2)$
21. $-18 - (-1.3)$
22. $\frac{5}{11} - \frac{6}{11}$
23. $\frac{2}{7} - \frac{3}{14}$
24. $-\frac{7}{15} - \left(-\frac{5}{12}\right)$

Lesson 2-3

(pages 79–83)

Find each product.

1. $5(12)$
2. $(-6)(11)$
3. $(-7)(-5)$
4. $(-6)(4)(-3)$
5. $\left(-\frac{7}{8}\right)\left(-\frac{1}{3}\right)$
6. $(5)\left(-\frac{2}{5}\right)$
7. $\left(-4\frac{1}{2}\right)\left(2\frac{1}{3}\right)$
8. $\left(-1\frac{2}{7}\right)\left(-3\frac{5}{9}\right)$
9. $(-5.34)(3.2)$
10. $(-6.8)(-5.415)$
11. $(4.2)(-5.1)(3.6)$
12. $(-3.9)(1.6)(8.4)$

Simplify each expression.

13. $5(-3a) - 6a$
14. $-8(-x) - 3x$
15. $2(6y - 2y)$
16. $(c + 7c)(-3)$
17. $-3n(4b) + 2a(3b)$
18. $-7(2m - 3n)$

Lesson 2-4

(pages 84–87)

Find each quotient.

1. $-49 \div (-7)$
2. $52 \div (-4)$
3. $-66 \div (0.5)$
4. $25.8 \div (-2)$
5. $-55.25 \div (-0.25)$
6. $-82.1 \div (16.42)$
7. $-\frac{2}{5} \div 5$
8. $\frac{7}{8} \div (-4)$
9. $-4 \div \left(-\frac{7}{10}\right)$
10. $\frac{3}{2} \div \left(-\frac{1}{2}\right)$
11. $-\frac{8}{5} \div \left(-\frac{5}{8}\right)$
12. $-\frac{13}{15} \div \frac{3}{25}$

Simplify each expression.

13. $\frac{32a}{4}$
14. $\frac{12x}{-2}$
15. $\frac{5n + 15}{-5}$
16. $\frac{-2b - 10}{-2}$
17. $\frac{65x - 15y}{5}$
18. $\frac{2a - 10b}{-2}$
19. $\frac{-27c + (-99b)}{9}$
20. $\frac{-3n + (-3m)}{-3}$

Lesson 2-5

(pages 88–94)

Use each set of data to make a line plot.

1. 134, 147, 137, 138, 156, 140, 134, 145, 139, 152, 139, 155, 144, 135, 144
2. 19, 12, 11, 11, 7, 7, 8, 13, 12, 12, 9, 9, 8, 15, 11, 4, 12, 7, 7, 6
3. 66, 74, 72, 78, 68, 75, 80, 69, 62, 65, 63, 78, 71, 78, 76, 75, 80, 69, 62, 71, 76, 79, 70, 64, 62, 74, 74, 75, 70
4. 131, 133, 146, 141, 131, 138, 154, 156, 158, 160, 152, 150, 154, 160

Use each set of data to make a stem-and-leaf plot.

5. 22 17 35 19 45 23 35 18 22 47 39 23 17 44 35 19 18 40 10
6. 1.2 1.3 5.6 4.1 1.1 2.0 1.9 3.0 4.5 2.1 4.1 1.2 1.8 1.0 3.2 2.2 2.5
7. 123 134 111 105 108 121 133 135 109 101 130 101 139 129 137 104

Lesson 2-6

(pages 96–101)

Find the probability of each event.

1. A coin will land tails up.
2. You eat this month.
3. A baby will be a girl.
4. You will see a blue elephant.
5. This is an algebra book.
6. Today is Wednesday.

A computer randomly picks a letter in the word *success*. Find each probability.

7. the letter e
8. P(not c)
9. the letter s
10. the letter b
11. P(vowel)
12. the letters u or c

One die is rolled. Find the odds of each outcome.

13. a 4
14. a number greater than 3
15. a multiple of 3
16. a number less than 5
17. an odd number
18. not a 6

Lesson 2-7

(pages 103–109)

Find each square root. If necessary, round to the nearest hundredth.

1. $\sqrt{121}$
2. $-\sqrt{36}$
3. $\sqrt{2.89}$
4. $-\sqrt{125}$
5. $\sqrt{\frac{81}{100}}$
6. $-\sqrt{\frac{36}{196}}$
7. $\pm\sqrt{9.61}$
8. $\pm\sqrt{\frac{7}{8}}$

Name the set or sets of numbers to which each real number belongs.

9. $-\sqrt{149}$
10. $\frac{5}{6}$
11. $\sqrt{\frac{8}{2}}$
12. $-\frac{66}{55}$
13. $\sqrt{225}$
14. $-\sqrt{\frac{3}{4}}$
15. $\frac{-1}{7}$
16. $\sqrt{0.0016}$

Replace each ● with $<$, $>$, or $=$ to make each sentence true.

17. $6.\overline{16}$ ● 6
18. 3.88 ● $\sqrt{15}$
19. $-\sqrt{529}$ ● -20
20. $-\sqrt{0.25}$ ● $-0.\overline{5}$
21. $\frac{1}{3}$ ● $\frac{\sqrt{3}}{3}$
22. $\frac{1}{\sqrt{3}}$ ● $\frac{\sqrt{3}}{3}$
23. $-\sqrt{\frac{1}{4}}$ ● $-\frac{1}{4}$
24. $-\frac{1}{6}$ ● $-\frac{1}{\sqrt{6}}$

Lesson 3-1

(pages 120–126)

Translate each sentence into an equation or formula.

1. A number z times 2 minus 6 is the same as m divided by 3.
2. The cube of a decreased by the square of b is equal to c.
3. Twenty-nine decreased by the product of x and y is the same as z.
4. The perimeter P of an isosceles triangle is the sum of twice the length of leg a and the length of the base b.
5. Thirty increased by the quotient of s and t is equal to v.
6. The area A of a rhombus is half the product of lengths of the diagonals a and b.

Translate each equation into a verbal sentence.

7. $0.5x + 3 = -10$
8. $\frac{n}{-6} = 2n + 1$
9. $18 - 5h = 13h$
10. $n^2 = 16$
11. $2x^2 + 3 = 21$
12. $\frac{m}{n} + 4 = 12$

Lesson 3-2

(pages 128–134)

Solve each equation. Then check your solution.

1. $-2 + g = 7$
2. $9 + s = -5$
3. $-4 + y = -9$
4. $m + 6 = 2$
5. $t + (-4) = 10$
6. $v - 7 = -4$
7. $a - (-6) = -5$
8. $-2 - x = -8$
9. $d + (-44) = -61$
10. $e - (-26) = 41$
11. $p - 47 = 22$
12. $-63 - f = -82$
13. $c + 5.4 = -11.33$
14. $-6.11 + b = 14.321$
15. $-5 = y - 22.7$
16. $-5 - q = 1.19$
17. $n + (-4.361) = 59.78$
18. $t - (-46.1) = -3.673$
19. $\frac{7}{10} - a = \frac{1}{2}$
20. $f - \left(-\frac{1}{8}\right) = \frac{3}{10}$
21. $-4\frac{5}{12} = t - \left(-10\frac{1}{36}\right)$
22. $x + \frac{3}{8} = \frac{1}{4}$
23. $1\frac{7}{16} + s = \frac{9}{8}$
24. $17\frac{8}{9} = d + \left(-2\frac{5}{6}\right)$

Lesson 3-3

(pages 135–140)

Solve each equation. Then check your solution.

1. $7p = 35$
2. $-3x = -24$
3. $2y = -3$
4. $62y = -2356$
5. $\frac{a}{-6} = -2$
6. $\frac{c}{-59} = -7$
7. $\frac{f}{14} = -63$
8. $84 = \frac{x}{97}$
9. $\frac{w}{5} = 3$
10. $\frac{q}{9} = -3$
11. $\frac{2}{5}x = \frac{4}{7}$
12. $\frac{z}{6} = -\frac{5}{12}$
13. $-\frac{5}{9}r = 7\frac{1}{2}$
14. $2\frac{1}{6}j = 5\frac{1}{5}$
15. $3 = 1\frac{7}{11}q$
16. $-1\frac{3}{4}p = -\frac{5}{8}$
17. $57k = 0.1824$
18. $0.0022b = 0.1958$
19. $5j = -32.15$
20. $\frac{w}{-2} = -2.48$
21. $\frac{z}{2.8} = -6.2$
22. $\frac{x}{-0.063} = 0.015$
23. $15\frac{3}{8} = -5p$
24. $-18\frac{1}{4} = 2.5x$

Extra Practice

Lesson 3-4

(pages 142–148)

Solve each equation. Then check your solution.

1. $2x - 5 = 3$
2. $4t + 5 = 37$
3. $7a + 6 = -36$
4. $47 = -8g + 7$
5. $-3c - 9 = -24$
6. $5k - 7 = -52$
7. $5s + 4s = -72$
8. $3x - 7 = 2$
9. $8 + 3x = 5$
10. $-3y + 7.569 = 24.069$
11. $7 - 9.1f = 137.585$
12. $6.5 = 2.4m - 4.9$
13. $\frac{e}{5} + 6 = -2$
14. $\frac{d}{4} - 8 = -5$
15. $-\frac{4}{13}y - 7 = 6$
16. $\frac{p + 3}{10} = 4$
17. $\frac{h - 7}{6} = 1$
18. $\frac{5f + 1}{8} = -3$
19. $\frac{4n - 8}{-2} = 12$
20. $\frac{-3t - 4}{2} = 8$
21. $4.8a - 3 + 1.2a = 9$

Lesson 3-5

(pages 149–154)

Solve each equation. Then check your solution.

1. $5x + 1 = 3x - 3$
2. $6 - 8n = 5n + 19$
3. $-3z + 5 = 2z + 5$
4. $\frac{2}{3}h + 5 = -4 - \frac{1}{3}h$
5. $\frac{1}{2}a - 4 = 3 - \frac{1}{4}a$
6. $6(y - 5) = 18 - 2y$
7. $-28 + p = 7(p - 10)$
8. $\frac{1}{3}(b - 9) = b + 9$
9. $-4x + 6 = 0.5(x + 30)$
10. $4(2y - 1) = -8(0.5 - y)$
11. $1.9s + 6 = 3.1 - s$
12. $2.85y - 7 = 12.85y - 2$
13. $2.9m + 1.7 = 3.5 + 2.3m$
14. $3(x + 1) - 5 = 3x - 2$
15. $\frac{x}{2} - \frac{1}{3} = \frac{x}{3} - \frac{1}{2}$
16. $\frac{6v - 9}{3} = v$
17. $\frac{3t + 1}{4} = \frac{3}{4}t - 5$
18. $0.4(x - 12) = 1.2(x - 4)$
19. $3y - \frac{4}{5} = \frac{1}{3}y$
20. $\frac{3}{4}x - 4 = 7 + \frac{1}{2}x$
21. $-0.2(1 - x) = 2(4 + 0.1x)$

Lesson 3-6

(pages 155–159)

Solve each proportion.

1. $\frac{4}{5} = \frac{x}{20}$
2. $\frac{b}{63} = \frac{3}{7}$
3. $\frac{y}{5} = \frac{3}{4}$
4. $\frac{7}{4} = \frac{3}{a}$
5. $\frac{t-5}{4} = \frac{3}{2}$
6. $\frac{x}{9} = \frac{0.24}{3}$
7. $\frac{n}{3} = \frac{n+4}{7}$
8. $\frac{12q}{-7} = \frac{30}{14}$
9. $\frac{1}{y-3} = \frac{3}{y-5}$
10. $\frac{x}{8.71} = \frac{4}{17.42}$
11. $\frac{a-3}{8} = \frac{3}{4}$
12. $\frac{6p-2}{7} = \frac{5p+7}{8}$
13. $\frac{2}{9} = \frac{k+3}{2}$
14. $\frac{5m-3}{4} = \frac{5m+3}{6}$
15. $\frac{w-5}{4} = \frac{w+3}{3}$
16. $\frac{96.8}{t} = \frac{12.1}{7}$
17. $\frac{r-1}{r+1} = \frac{3}{5}$
18. $\frac{4n+5}{5} = \frac{2n+7}{7}$

Lesson 3-7

(pages 160–164)

State whether each percent of change is a percent of increase or a percent of decrease. Then find each percent of change. Round to the nearest whole percent.

1. original: \$100
   new: \$67
2. original: 62 acres
   new: 98 acres
3. original: 322 people
   new: 289 people
4. original: 78 pennies
   new: 36 pennies
5. original: \$212
   new: \$230
6. original: 35 mph
   new: 65 mph

Find the final price of each item.

7. television: \$299
   discount: 20%
8. book: \$15.95
   sales tax: 7%
9. software: \$36.90
   sales tax: 6.25%
10. boots: \$49.99
    discount: 15%
    sales tax: 3.5%
11. jacket: \$65
    discount: 30%
    sales tax: 4%
12. backpack: \$28.95
    discount: 10%
    sales tax: 5%

Lesson 3-8

(pages 166–170)

Solve each equation or formula for x.

1. $x + r = q$
2. $ax + 4 = 7$
3. $2bx - b = -5$
4. $\frac{x-c}{c+a} = a$
5. $\frac{x+y}{c} = d$
6. $\frac{ax+1}{2} = b$
7. $d(x - 3) = 5$
8. $nx - a = bx + d$
9. $3x - r = r(-3 + x)$
10. $y = \frac{5}{9}(x - 32)$
11. $A = \frac{1}{2}h(x + y)$
12. $A = 2\pi r^2 + 2\pi rx$

Lesson 3-9
(pages 171–177)

1. **ADVERTISING** An advertisement for grape drink claims that the drink contains 10% grape juice. How much pure grape juice would have to be added to 5 quarts of the drink to obtain a mixture containing 40% grape juice?
2. **GRADES** In Ms. Pham's social studies class, a test is worth four times as much as homework. If a student has an average of 85% on tests and 95% on homework, what is the student's average?
3. **ENTERTAINMENT** At the Golden Oldies Theater, tickets for adults cost \$5.50 and tickets for children cost \$3.50. How many of each kind of ticket were purchased if 21 tickets were bought for \$83.50?
4. **FOOD** Wes is mixing peanuts and chocolate pieces. Peanuts sell for \$4.50 a pound and the chocolate sells for \$6.50 a pound. How many pounds of chocolate mixes with 5 pounds of peanuts to obtain a mixture that sells for \$5.25 a pound?
5. **TRAVEL** Missoula and Bozeman are 210 miles apart. Sheila leaves Missoula for Bozeman and averages 55 miles per hour. At the same time, Casey leaves Bozeman and averages 65 miles per hour as he drives to Missoula. When will they meet? How far will they be from Bozeman?

Extra Practice

Lesson 4-1
(pages 192–196)

Write the ordered pair for each point shown at the right. Name the quadrant in which the point is located.

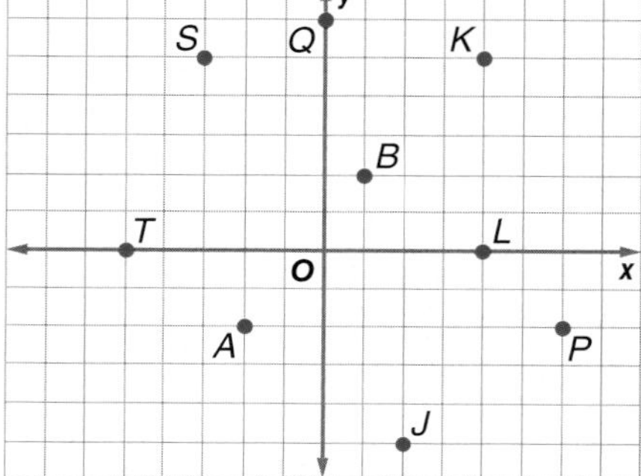

1. *B*
2. *T*
3. *P*
4. *Q*
5. *A*
6. *K*
7. *J*
8. *L*
9. *S*

Plot each point on a coordinate plane.

10. $A(2, -3)$
11. $B(3, 6)$
12. $C(-4, 0)$
13. $D(-4, 3)$
14. $E(-5, -5)$
15. $F(-1, 1)$
16. $G(0, -2)$
17. $H(2, 3)$
18. $J(0, 3)$

Lesson 4-2
(pages 197–203)

Determine whether each transformation is a *reflection*, *translation*, *dilation*, or *rotation*.

1. 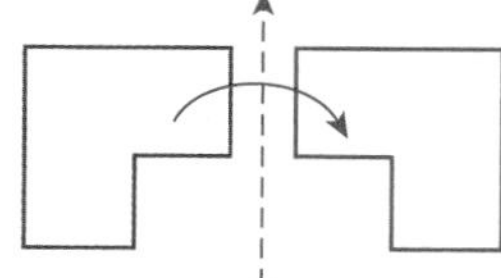

2.

3. 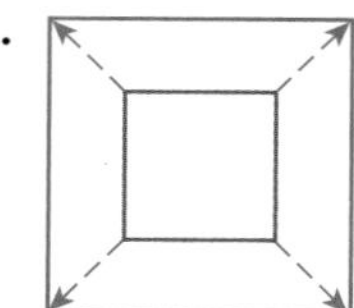

For Exercises 4–9, complete parts a and b.

a. **Find the coordinates of the vertices of each figure after the given transformation is performed.**

b. **Graph the preimage and its image.**

4. quadrilateral $ABCD$ with $A(2, 2)$, $B(-3, 5)$, $C(-4, 0)$, and $D(2, -2)$ translated 1 unit up and 2 units right
5. square $SQUA$ with $S(1, 1)$, $Q(4, 1)$, $U(4, 4)$, and $A(1, 4)$ reflected over the y-axis
6. $\triangle RED$ with $R(2, 1)$, $E(-3, -1)$, and $D(2, -4)$ dilated by a scale factor of 2
7. pentagon $BLACK$ with $B(-3, -5)$, $L(4, -5)$, $A(4, 1)$, $C(0, 4)$, and $K(-4, 1)$ reflected over the x-axis
8. $\triangle ANG$ with $A(2, 1)$, $N(4, 1)$, and $G(3, 4)$ rotated 90° counterclockwise about the origin
9. parallelogram $GRAM$ with $G(-3, -2)$, $R(4, -2)$, $A(6, 4)$, and $M(-1, 4)$ translated 2 units down and 1 unit left

Lesson 4-3

(pages 205–211)

Express each relation as a table, a graph, and a mapping. Then determine the domain and range.

1. $\{(5, 2), (0, 0), (-9, -1)\}$
2. $\{(-4, 2), (-2, 0), (0, 2), (2, 4)\}$
3. $\{(7, 5), (-2, -3), (4, 0), (5, -7), (-9, 2)\}$
4. $\{(3.1, -1), (-4.7, 3.9), (2.4, -3.6), (-9, 12.12)\}$

Express the relation shown in each table, mapping, or graph as a set of ordered pairs. Then write the inverse of the relation.

5.

x	y
1	3
2	4
3	5
4	6
5	7

6.

x	y
−4	1
−2	3
0	1
2	3
4	1

7.
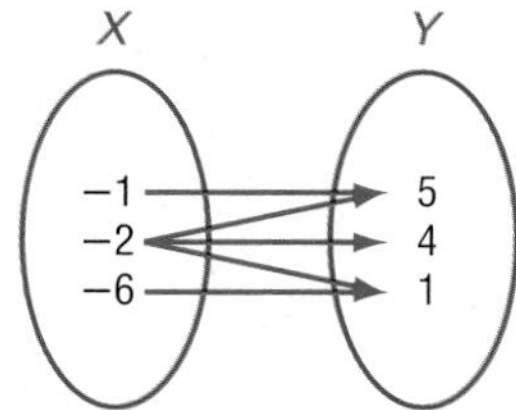

8.
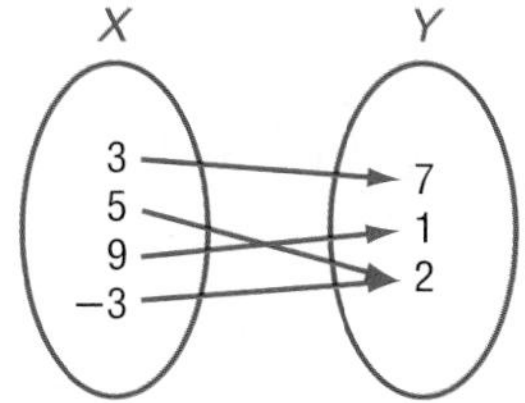

9.
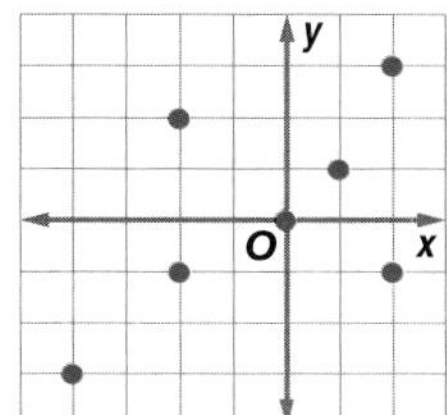

10.
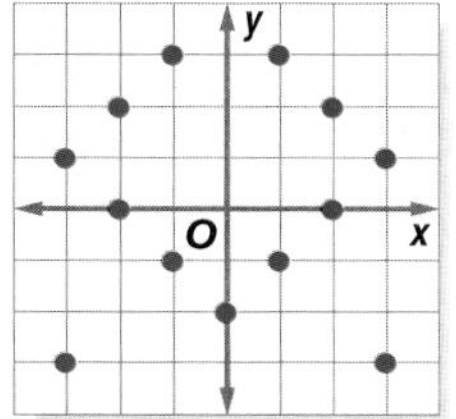

Lesson 4-4

(pages 212–217)

Find the solution set for each equation, given the replacement set.

1. $y = 3x - 1$; $\{(0, -1), (4, 2), (2, 4), (2, 5)\}$
2. $3y = x + 7$; $\{(1, 8), (0, 7), (2, 3), (5, 4)\}$
3. $4x = 8 - 2y$; $\{(2, 0), (0, 4), (0, 2), (-4, 12)\}$
4. $3x = 10 - 4y$; $\{(3, 0.25), (-10, 5), (2, 1), (5, 5)\}$

Solve each equation if the domain is $\{-2, -1, 0, 1, 2\}$.

5. $x + y = 3$
6. $y = x$
7. $y = 5x + 1$
8. $4x + 3y = 13$
9. $5y = 8 - 4x$
10. $2x + y = 4$
11. $y = 4 + x$
12. $2x + 3y = 10$
13. $2y = 3x + 1$

Solve each equation for the given domain. Graph the solution set.

14. $x = y + 1$ for $x = \{-2, -1, 0, 1, 2\}$
15. $y = x + 1$ for $x = \{-3, -1, 0, 1, 3\}$
16. $x + 4y = 2$ for $x = \{-8, -4, 0, 4, 8\}$
17. $y - 3 = x$ for $x = \{-5, -1, 3, 7, 9\}$
18. $x + y = -2$ for $x = \{-4, -3, 0, 1, 3\}$
19. $2x - 3y = -5$ for $x = \{-5, -3, 0, 5, 6\}$
20. $3y = \frac{2}{3}x - 4$ for $x = \{-6, -3, 0, 1, 3\}$
21. $-2y = 8 - \frac{3}{2}x$ for $x = \{-4, 0, 4, 6, 8\}$

Lesson 4-5

(pages 218–223)

Determine whether each equation is a linear equation. If so, write the equation in standard form.

1. $3x = 2y$
2. $2x - 3 = y^2$
3. $4x = 2y + 8$
4. $5x - 7y = 2x - 7$
5. $2x + 5x = 7y + 2$
6. $\frac{1}{x} + \frac{5}{y} = -4$

Graph each equation.

7. $3x + y = 4$
8. $y = 3x + 1$
9. $3x - 2y = 12$
10. $2x - y = 6$
11. $2x - 3y = 8$
12. $y = -2$
13. $y = 5x - 7$
14. $x = 4$
15. $x + \frac{1}{3}y = 2$
16. $5x - 2y = 8$
17. $4.5x + 2.5y = 9$
18. $\frac{1}{2}x + 3y = 12$

Lesson 4-6
(pages 226–231)

Determine whether each relation is a function.

1.

x	y
1	3
2	5
1	−7
2	9

2. 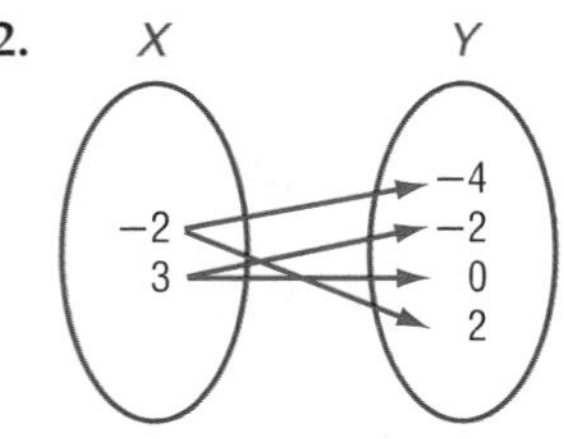

3. 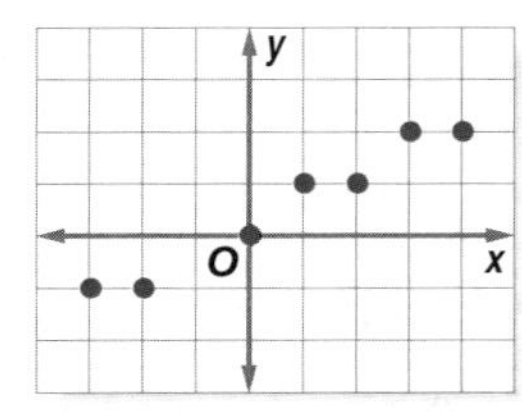

4. $\{(-2, 4), (1, 3), (5, 2), (1, 4)\}$
5. $\{(5, 4), (-6, 5), (4, 5), (0, 4)\}$
6. $\{(3, 1), (5, 1), (7, 1)\}$
7. $\{(3, -2), (4, 7), (-2, 7), (4, 5)\}$
8. $y = 2$
9. $x^2 + y = 11$

If $f(x) = 2x + 5$ and $g(x) = 3x^2 - 1$, find each value.

10. $f(-4)$
11. $g(2)$
12. $f(3) - 5$
13. $g(-2) + 4$
14. $f(b^2)$
15. $g(a + 1)$
16. $f(0) + g(3)$
17. $f(n) + g(n)$

Lesson 4-7
(pages 233–238)

Determine whether each sequence is an arithmetic sequence. If it is, state the common difference.

1. $-2, -1, 0, 1, \ldots$
2. $3, 5, 8, 12, \ldots$
3. $2, 4, 8, 16, \ldots$
4. $-21, -16, -11, -6, \ldots$
5. $0, 0.25, 0.5, 0.75, \ldots$
6. $\frac{1}{3}, \frac{1}{9}, \frac{1}{27}, \frac{1}{81}, \ldots$

Find the next three terms of each arithmetic sequence.

7. $3, 13, 23, 33, \ldots$
8. $-4, -6, -8, -10, \ldots$
9. $-2, -1.4, -0.8, -0.2, \ldots$
10. $5, 13, 21, 29, \ldots$
11. $\frac{3}{4}, \frac{7}{8}, 1, \frac{9}{8}, \ldots$
12. $-\frac{1}{3}, -\frac{5}{6}, -\frac{4}{3}, -\frac{11}{6}, \ldots$

Find the *n*th term of each arithmetic sequence described.

13. $a_1 = 3, d = 6, n = 12$
14. $a_1 = -2, d = 4, n = 8$
15. $a_1 = -1, d = -3, n = 10$
16. $a_1 = 2.2, d = 1.4, n = 5$
17. $-2, -7, -12, \ldots$ for $n = 12$
18. $2\frac{1}{2}, 2\frac{1}{8}, 1\frac{3}{4}, 1\frac{3}{8}, \ldots$ for $n = 10$

Write an equation for the *n*th term of the arithmetic sequence. Then graph the first five terms in the sequence.

19. $-3, 1, 5, 9, \ldots$
20. $25, 40, 55, 70, \ldots$
21. $-9, -3, 3, 9, \ldots$
22. $-3.5, -2, -0.5, 1, \ldots$

Lesson 4-8
(pages 240–245)

Find the next two items for each pattern.

1. 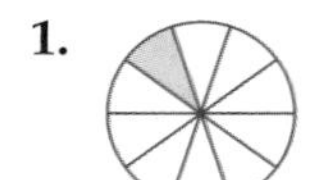 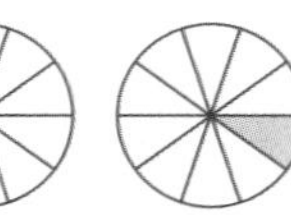 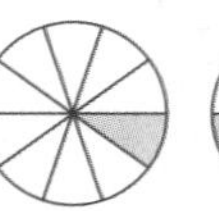 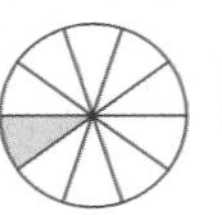 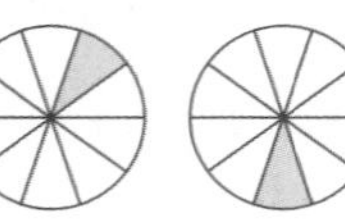 

2. 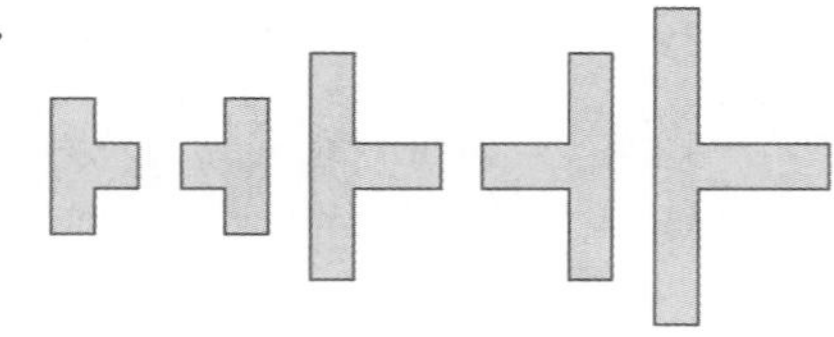

Find the next three terms in each sequence.

3. $12, 23, 34, 45, \ldots$
4. $39, 33, 27, 21, \ldots$
5. $6.0, 7.2, 8.4, 9.6, \ldots$
6. $86, 81.5, 77, 72.5, \ldots$
7. $4, 8, 16, 32, \ldots$
8. $3125, 625, 125, 25, \ldots$
9. $15, 16, 18, 21, 25, 30, \ldots$
10. $w - 2, w - 4, w - 6, w - 8, \ldots$
11. $13, 10, 11, 8, 9, 6, \ldots$

Write an equation in function notation for each relation.

12. y O x

13. 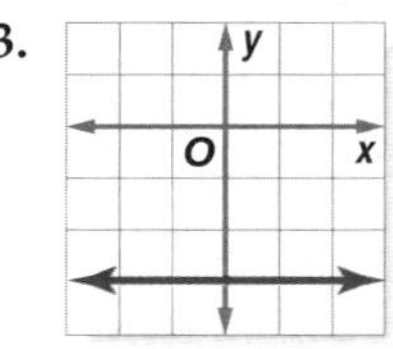

14. 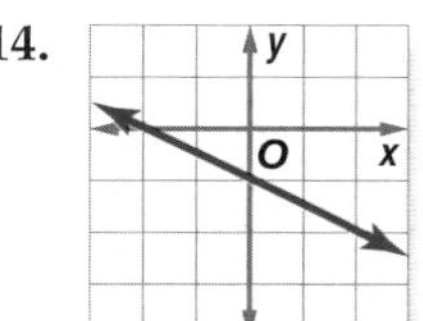

15. 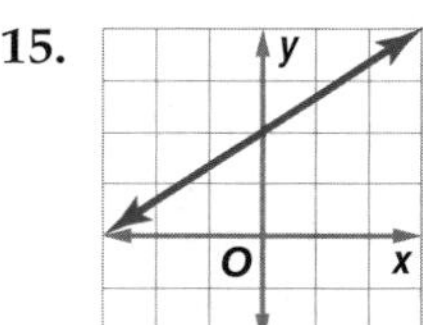

Lesson 5-1

(pages 256–262)

Find the slope of the line that passes through each pair of points.

1. 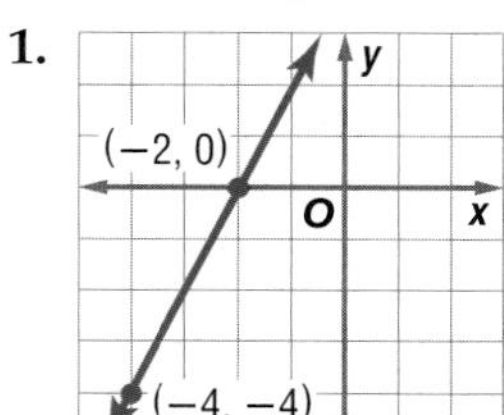

2. 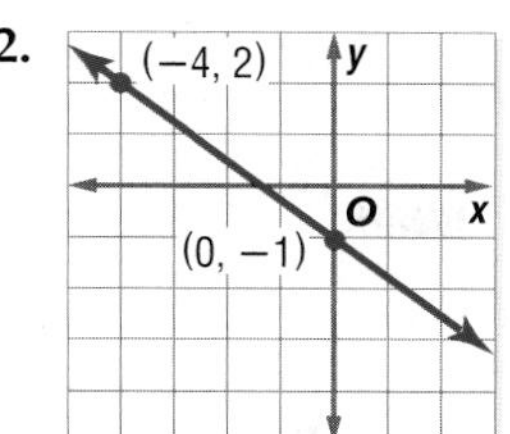

3. (−2, 2), (3, −3) **4.** (−2, −8), (1, 4) **5.** (3, 4), (4, 6) **6.** (−5, 4), (−1, 11)

7. (18, −4), (6, −10) **8.** (−4, −6), (−4, −8) **9.** (0, 0), (−1, 3) **10.** (−8, 1), (2, 1)

Find the value of r so the line that passes through each pair of points has the given slope.

11. $(-1, r), (1, -4), m = -5$ **12.** $(r, -2), (-7, -1), m = -\frac{1}{4}$ **13.** $(-3, 2), (7, r), m = \frac{2}{3}$

Lesson 5-2

(pages 264–270)

Name the constant of variation for each equation. Then determine the slope of the line that passes through each pair of points.

1. 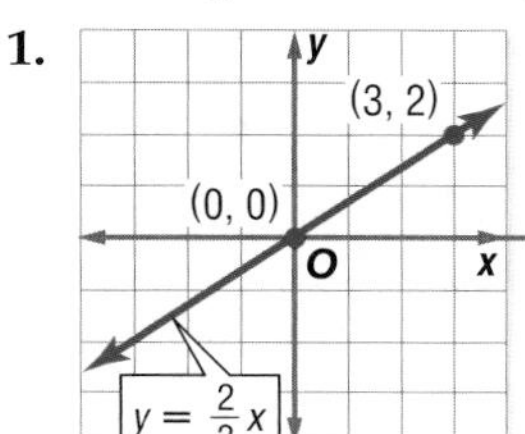

2. 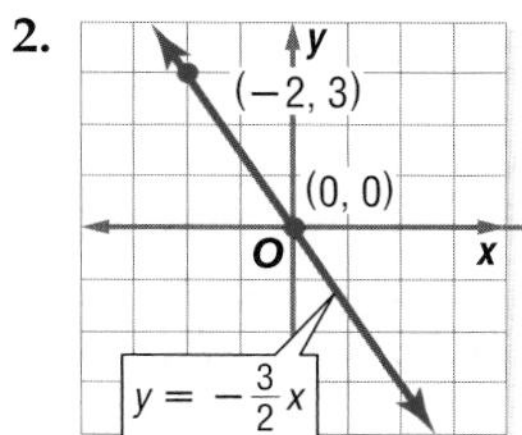

3. 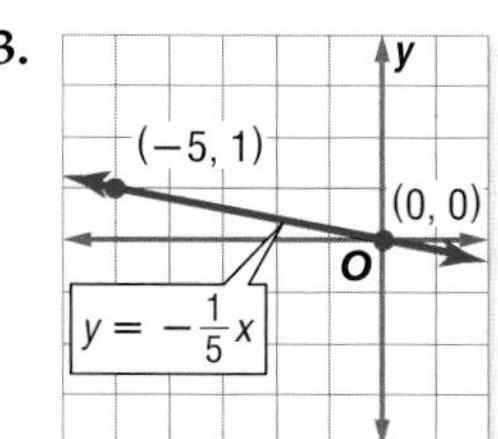

Graph each equation.

4. $y = 5x$ **5.** $y = -6x$ **6.** $y = -\frac{4}{3}x$

Write a direct variation equation that relates x and y. Assume that y varies directly as x. Then solve.

7. If $y = 45$ when $x = 9$, find y when $x = 7$. **8.** If $y = -7$ when $x = -1$, find x when $y = -84$.

9. If $y = 450$ when $x = -6$, find y when $x = 10$. **10.** If $y = 6$ when $x = 48$, find y when $x = 20$.

Lesson 5-3

(pages 272–277)

Write an equation of the line with the given slope and y-intercept.

1. slope: 5, y-intercept: −15 **2.** slope: −6, y-intercept: 3 **3.** slope: 0.3, y-intercept: −2.6

4. slope: $-\frac{4}{3}$, y-intercept: $\frac{5}{3}$ **5.** slope: $-\frac{2}{5}$, y-intercept: 2 **6.** slope: $\frac{7}{4}$, y-intercept: −2

Write an equation of the line shown in each graph.

7. 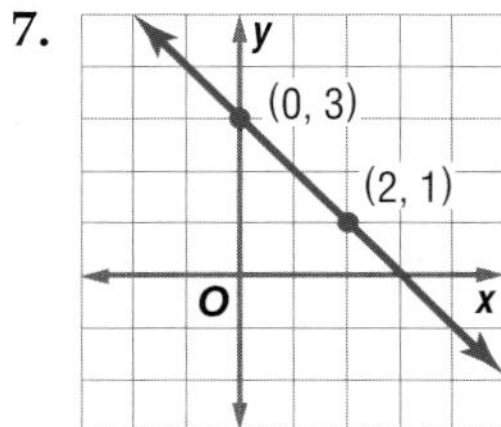

8. 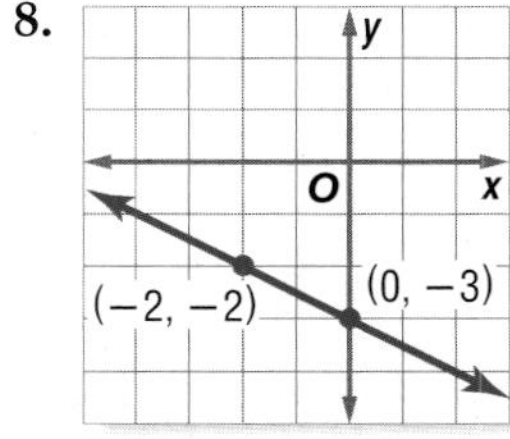

9. 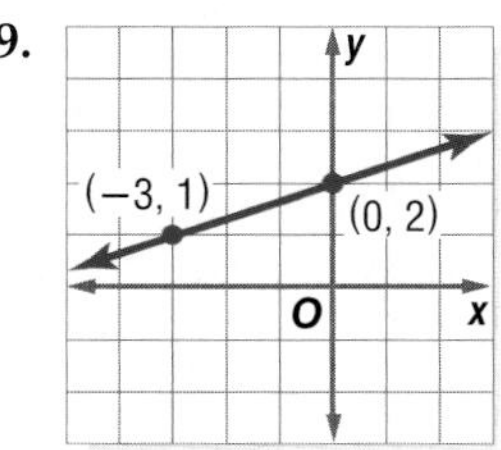

Graph each equation.

10. $y = 5x - 1$ **11.** $y = -2x + 3$ **12.** $3x - y = 6$

Lesson 5-4

(pages 280–285)

Write an equation of the line that passes through each point with the given slope.

1. $(0, 0)$; $m = -2$
2. $(-3, 2)$; $m = 4$
3. $(0, 5)$; $m = -1$
4. $(-2, 3)$; $m = -\frac{1}{4}$
5. $(1, -5)$; $m = \frac{2}{3}$
6. $\left(\frac{1}{2}, \frac{1}{4}\right)$; $m = 8$

Write an equation of the line that passes through each pair of points.

7. $(-1, 7), (8, -2)$
8. $(4, 0), (0, 5)$
9. $(8, -1), (7, -1)$
10. $(1, 0), (0, 1)$
11. $(5, 7), (-1, 3)$
12. $(-3, -5), (3, -15)$
13. $(-2, 3), (1, 3)$
14. $(0, 0), (-4, 3)$
15. $\left(-\frac{1}{2}, \frac{1}{2}\right), \left(\frac{1}{4}, \frac{3}{4}\right)$

Write an equation of the line that has each pair of intercepts.

16. x-intercept: 2, y-intercept: 1
17. x-intercept: 1, y-intercept: -4
18. x-intercept: 5, y-intercept: 5
19. x-intercept: -1, y-intercept: 3
20. x-intercept: -4, y-intercept: -1
21. x-intercept: 3, y-intercept: -3

Lesson 5-5

(pages 286–291)

Write the point-slope form of an equation for a line that passes through each point with the given slope.

1. $(5, -2)$, $m = 3$
2. $(5, 4)$, $m = -5$
3. $(0, 6)$, $m = -2$
4. $(-3, 1)$, $m = 0$
5. $(-1, 0)$, $m = \frac{2}{3}$
6. $(-2, -4)$, $m = \frac{3}{4}$

Write each equation in standard form.

7. $y + 3 = 2(x - 4)$
8. $y + 3 = -\frac{1}{2}(x + 6)$
9. $y - 4 = -\frac{2}{3}(x - 5)$
10. $y + 2 = \frac{4}{3}(x - 6)$
11. $y - 1 = 1.5(x + 3)$
12. $y + 6 = -3.8(x - 2)$

Write each equation in slope-intercept form.

13. $y - 1 = -2(x + 5)$
14. $y + 3 = 4(x - 1)$
15. $y - 6 = -4(x - 2)$
16. $y + 1 = \frac{4}{5}(x + 5)$
17. $y - 2 = -\frac{3}{4}(x - 2)$
18. $y + \frac{1}{4} = \frac{2}{3}\left(x + \frac{1}{2}\right)$

Lesson 5-6

(pages 292–297)

Write the slope-intercept form of an equation of the line that passes through the given point and is parallel to the graph of each equation.

1. $(1, 6)$, $y = 4x - 2$
2. $(4, 6)$, $y = 2x - 7$
3. $(-3, 0)$, $y = \frac{2}{3}x + 1$
4. $(5, -2)$, $y = -3x - 7$
5. $(0, 4)$, $3x + 8y = 4$
6. $(2, 3)$, $x - 5y = 7$

Write the slope-intercept form of an equation that passes through the given point and is perpendicular to the graph of each equation.

7. $(0, -1)$, $y = -\frac{3}{5}x + 4$
8. $(-2, 3)$, $6x + y = 4$
9. $(0, 0)$, $y = \frac{3}{4}x - 1$
10. $(4, 0)$, $4x - 3y = 2$
11. $(6, 7)$, $3x - 5y = 1$
12. $(5, -1)$, $8x + 4y = 15$

Lesson 5-7

(pages 298–305)

Determine whether each graph shows a *positive correlation*, a *negative correlation*, or *no correlation*. If there is a positive or negative correlation, describe its meaning in the situation.

1.

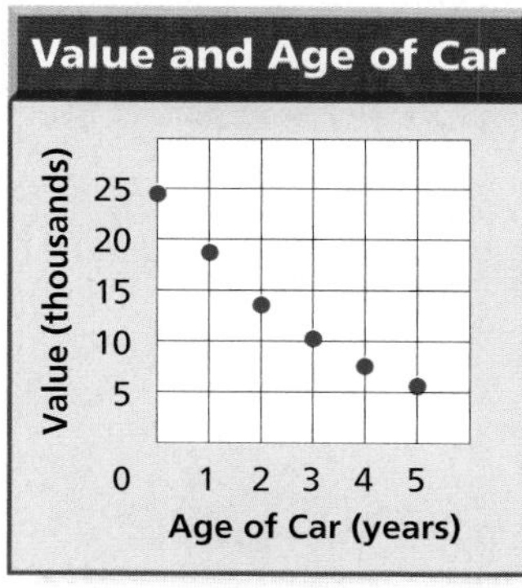

2.

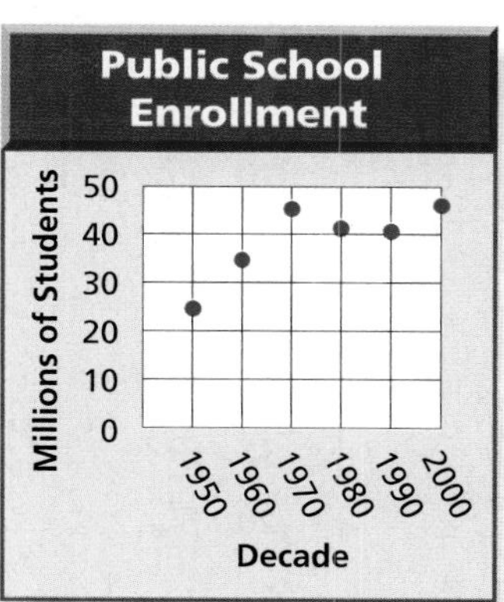

3.

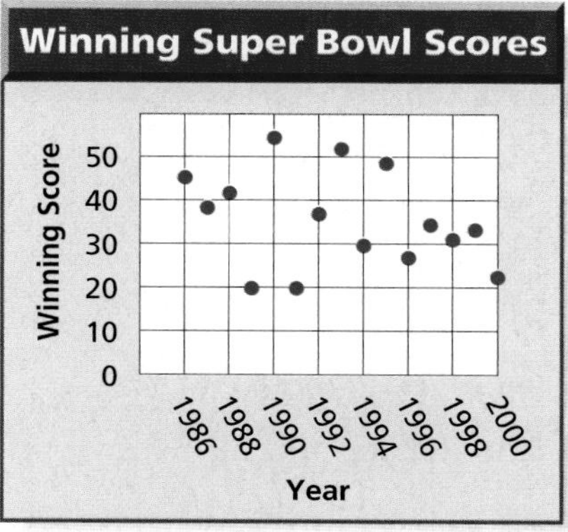

Source: *ESPN Almanac*

For Exercises 4–6, use the scatter plot that shows the year and the amount of fish caught in China in millions of metric tons.

4. Describe the relationship that exists in the data.
5. Use the points (1994, 24) and (1998, 38) to write the slope-intercept form of an equation for the line of fit shown in the scatter plot.
6. Predict the amount of fish that will be caught in China in 2005.

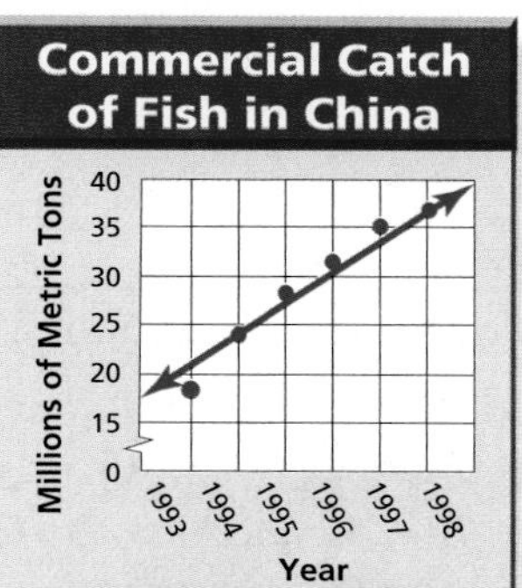

Source: *The World Almanac*

Lesson 6-1

(pages 318–323)

Solve each inequality. Then check your solution and graph it on a number line.

1. $c + 9 \leq 3$
2. $d - (-3) < 13$
3. $z - 4 > 20$
4. $h - (-7) > -2$
5. $-11 > d - 4$
6. $2x > x - 3$
7. $2x - 3 \geq x$
8. $16 + w < -20$
9. $14p > 5 + 13p$
10. $-7 < 16 - z$
11. $1.1v - 1 > 2.1v - 3$
12. $\frac{1}{2}t + \frac{1}{4} \geq \frac{3}{2}t - \frac{2}{3}$
13. $9x < 8x - 2$
14. $-2 + 9n \leq 10n$
15. $a - 2.3 \geq -7.8$
16. $5z - 6 > 4z$

Define a variable, write an inequality, and solve each problem.

17. The sum of a number and negative six is greater than 9.
18. Negative five times a number is less than the sum of negative six times the number and 12.

Lesson 6-2

(pages 325–331)

Solve each inequality. Then check your solution.

1. $7b \geq -49$
2. $-5j < -60$
3. $\frac{w}{3} > -12$
4. $\frac{p}{5} < 8$
5. $-8f < 48$
6. $-0.25t \geq -10$
7. $\frac{g}{-8} < 4$
8. $-4.3x < -2.58$
9. $4c \geq -6$
10. $6 \leq 0.8n$
11. $\frac{2}{3}m \geq -22$
12. $-25 > -0.05a$
13. $-15a < -28$
14. $-\frac{7}{9}x < 42$
15. $0.375y \leq 32$
16. $-7y \geq 91$

Define a variable, write an inequality, and solve each problem.

17. Negative one times a number is greater than -7.
18. Three fifths of a number is at least negative 10.
19. Seventy-five percent of a number is at most 100.

Lesson 6-3 (pages 332–337)

Solve each inequality. Then check your solution.

1. $3y - 4 > -37$
2. $7s - 12 < 13$
3. $-5e + 9 > 24$
4. $-6v - 3 \geq -33$
5. $-2k + 12 < 30$
6. $-2x + 1 < 16 - x$
7. $15t - 4 > 11t - 16$
8. $13 - y \leq 29 + 2y$
9. $5q + 7 \leq 3(q + 1)$
10. $2(w + 4) \geq 7(w - 1)$
11. $-4t - 5 > 2t + 13$
12. $\frac{2t + 5}{3} < -9$
13. $\frac{z}{4} + 7 \geq -5$
14. $13r - 11 > 7r + 37$
15. $8c - (c - 5) > c + 17$
16. $-5(k + 4) \geq 3(k - 4)$
17. $9m + 7 < 2(4m - 1)$
18. $3(3y + 1) < 13y - 8$
19. $5x \leq 10(3x + 4)$
20. $3\left(a + \frac{2}{3}\right) \geq a - 1$

Lesson 6-4 (pages 339–344)

Solve each compound inequality. Then graph the solution set.

1. $2 + x < -5$ or $2 + x > 5$
2. $-4 + t > -5$ or $-4 + t < 7$
3. $3 \leq 2g + 7$ and $2g + 7 \leq 15$
4. $2v - 2 \leq 3v$ and $4v - 1 \geq 3v$
5. $3b - 4 \leq 7b + 12$ and $8b - 7 \leq 25$
6. $-9 < 2z + 7 < 10$
7. $5m - 8 \geq 10 - m$ or $5m + 11 < -9$
8. $12c - 4 \leq 5c + 10$ or $-4c - 1 \leq c + 24$
9. $2h - 2 \leq 3h \leq 4h - 1$
10. $3p + 6 < 8 - p$ and $5p + 8 \geq p + 6$
11. $2r + 8 > 16 - 2r$ and $7r + 21 < r - 9$
12. $-4j + 3 < j + 22$ and $j - 3 < 2j - 15$
13. $2(q - 4) \leq 3(q + 2)$ or $q - 8 \leq 4 - q$
14. $\frac{1}{2}w + 5 \geq w + 2 \geq \frac{1}{2}w + 9$
15. $n - (6 - n) > 10$ or $-3n - 1 > 20$
16. $-(2x + 5) \leq x + 5 \leq 2x - 9$

Lesson 6-5 (pages 345–351)

Solve each open sentence. Then graph the solution set.

1. $|y - 9| < 19$
2. $|g + 6| > 8$
3. $|t - 5| \leq 3$
4. $|a + 5| \geq 0$
5. $|14 - 2z| = 16$
6. $|a - 5| = -3$
7. $|2m - 5| > 13$
8. $|14 - w| \geq 20$
9. $|13 - 5y| = 8$
10. $|3p + 5| \leq 23$
11. $|6b - 12| \leq 36$
12. $|25 - 3x| < 5$
13. $|7 + 8x| > 39$
14. $|4c + 5| \geq 25$
15. $|4 - 5s| > 46$
16. $|4 - (1 - x)| \geq 10$
17. $\left|\frac{2n - 1}{3}\right| = 10$
18. $\left|\frac{7 - 2b}{2}\right| \leq 3$
19. $|-2 + (x - 3)| \leq 7$
20. $|-3 - (2b - 6)| \geq 10$

Lesson 6-6

(pages 352–357)

Determine which ordered pairs are part of the solution set for each inequality.

1. $x + y \geq 0$, {(0, 0), (1, −3), (2, 2), (3, −3)}
2. $2x + y \leq 8$, {(0, 0), (−1, −1), (3, −2), (8, 0)}
3. $y > x$, {(0, 0), (2, 0), (−3, 4), (2, −1)}
4. $3x - 2y < 1$, {(0, 0), (3, 2), (−4, −5), (0, 6)}

Graph each inequality.

5. $y \leq -2$
6. $x < 4$
7. $x + y < -2$
8. $x + y \geq -4$
9. $y > 4x - 1$
10. $3x + y > 1$
11. $3y - 2x \leq 2$
12. $x < y$
13. $3x + y \leq 4$
14. $5x - y < 5$
15. $-2x + 6y \geq 12$
16. $-x + 3y \leq 9$
17. $y > -3x + 7$
18. $3x + 8y \leq 4$
19. $5x - 2y \geq 6$

Extra Practice

Lesson 7-1

(pages 369–374)

Graph each system of equations. Then determine whether the system has *no* solution, *one* solution, or *infinitely many* solutions. If the system has one solution, name it.

1. $y = 3x$; $4x + 2y = 30$
2. $x = -2y$; $x + y = 1$
3. $y = x + 4$; $3x + 2y = 18$
4. $x + y = 6$; $x - y = 2$
5. $x + y = 6$; $3x + 3y = 3$
6. $y = -3x$; $4x + y = 2$
7. $2x + y = 8$; $x - y = 4$
8. $\frac{1}{5}x - y = \frac{12}{5}$; $3x - 5y = 6$
9. $x + 2y = 0$; $y + 3 = -x$
10. $x + 2y = -9$; $x - y = 6$
11. $x + \frac{1}{2}y = 3$; $y = 3x - 4$
12. $\frac{2}{3}x + \frac{1}{2}y = 2$; $4x + 3y = 12$
13. $y = x - 4$; $x + \frac{1}{2}y = \frac{5}{2}$
14. $2x + y = 3$; $4x + 2y = 6$
15. $12x - y = -21$; $\frac{1}{2}x + \frac{2}{3}y = -3$

Lesson 7-2

(pages 376–381)

Use substitution to solve each system of equations. If the system does *not* have exactly one solution, state whether it has *no* solutions or *infinitely many* solutions.

1. $y = x$; $5x = 12y$
2. $y = 7 - x$; $2x - y = 8$
3. $x = 5 - y$; $3y = 3x + 1$
4. $3x + y = 6$; $y + 2 = x$
5. $x - 3y = 3$; $2x + 9y = 11$
6. $3x = -18 + 2y$; $x + 3y = 4$
7. $x + 2y = 10$; $-x + y = 2$
8. $2x = 3 - y$; $2y = 12 - x$
9. $6y - x = -36$; $y = -3x$
10. $\frac{3}{4}x + \frac{1}{3}y = 1$; $x - y = 10$
11. $x + 6y = 1$; $3x - 10y = 31$
12. $3x - 2y = 12$; $\frac{3}{2}x - y = 3$
13. $2x + 3y = 5$; $4x - 9y = 9$
14. $x = 4 - 8y$; $3x + 24y = 12$
15. $3x - 2y = -3$; $25x + 10y = 215$

Extra Practice

Lesson 7-3

(pages 382–386)

Use elimination to solve each system of equations.

1. $x + y = 7$
$x - y = 9$

2. $2x - y = 32$
$2x + y = 60$

3. $-y + x = 6$
$y + x = 5$

4. $s + 2t = 6$
$3s - 2t = 2$

5. $x = y - 7$
$2x - 5y = -2$

6. $3x + 5y = -16$
$3x - 2y = -2$

7. $x - y = 3$
$x + y = 3$

8. $x + y = 8$
$2x - y = 6$

9. $2s - 3t = -4$
$s = 7 - 3t$

10. $-6x + 16y = -8$
$6x - 42 = 16y$

11. $3x + 0.2y = 7$
$3x = 0.4y + 4$

12. $9x + 2y = 26$
$1.5x - 2y = 13$

13. $x = y$
$x + y = 7$

14. $4x - \frac{1}{3}y = 8$
$5x + \frac{1}{3}y = 6$

15. $2x - y = 3$
$\frac{2}{3}x - y = -1$

Lesson 7-4

(pages 387–392)

Use elimination to solve each system of equations.

1. $-3x + 2y = 10$
$-2x - y = -5$

2. $2x + 5y = 13$
$4x - 3y = -13$

3. $5x + 3y = 4$
$-4x + 5y = -18$

4. $\frac{1}{3}x - y = -1$
$\frac{1}{5}x - \frac{2}{5}y = -1$

5. $3x - 5y = 8$
$4x - 7y = 10$

6. $x - 0.5y = 1$
$0.4x + y = -2$

7. $x + 8y = 3$
$4x - 2y = 7$

8. $4x - y = 4$
$x + 2y = 3$

9. $3y - 8x = 9$
$y - x = 2$

10. $x + 4y = 30$
$2x - y = -6$

11. $3x - 2y = 0$
$4x + 4y = 5$

12. $9x - 3y = 5$
$x + y = 1$

13. $2x - 7y = 9$
$-3x + 4y = 6$

14. $2x - 6y = -16$
$5x + 7y = -18$

15. $6x - 3y = -9$
$-8x + 2y = 4$

Lesson 7-5

(pages 394–398)

Solve each system of inequalities by graphing.

1. $x > 3$
$y < 6$

2. $y > 2$
$y > -x + 2$

3. $x \leq 2$
$y + 3 \geq 5$

4. $x + y \leq -1$
$2x + y \leq 2$

5. $y \geq 2x + 2$
$y \geq -x - 1$

6. $y \leq x + 3$
$y \geq x + 2$

7. $x + 3y \geq 4$
$2x - y < 5$

8. $y - x > 1$
$y + 2x \leq 10$

9. $5x - 2y > 15$
$2x - 3y < 6$

10. $4x + 3y > 4$
$2x - y < 0$

11. $4x + 5y \geq 20$
$y \geq x + 1$

12. $-4x + 10y \leq 5$
$-2x + 5y < -1$

13. $y - x \geq 0$
$y \leq 3$
$x \geq 0$

14. $y > 2x$
$x > -3$
$y < 4$

15. $y \leq x$
$x + y < 4$
$y \geq -3$

Lesson 8-1

(pages 410–415)

Determine whether each expression is a monomial. Write *yes* or *no*. Explain your reasoning.

1. $n^2 - 3$
2. 53
3. $9a^2b^3$
4. $15 - x^2y$

Simplify.

5. $a^5(a)(a^7)$
6. $(r^3t^4)(r^4t^4)$
7. $(x^3y^4)(xy^3)$
8. $(bc^3)(b^4c^3)$
9. $(-3mn^2)(5m^3n^2)$
10. $[(3^3)^2]^2$
11. $(3s^3t^2)(-4s^3t^2)$
12. $x^3(x^4y^3)$
13. $(1.1g^2h^4)^3$
14. $-\frac{3}{4}a(a^2b^3c^4)$
15. $\left(\frac{1}{2}w^3\right)^2(w^4)^2$
16. $[(-2^3)^3]^2$
17. $\left(\frac{2}{3}y^3\right)(3y^2)^3$
18. $(10s^3t)(-2s^2t^2)^3$
19. $(-0.2u^3w^4)^3$

Lesson 8-2

(pages 417–423)

Simplify. Assume that no denominator is equal to zero.

1. $\frac{6^{10}}{6^7}$
2. $\frac{b^6c^5}{b^3c^2}$
3. $\frac{(-a)^4b^8}{a^4b^7}$
4. $\frac{(-x)^3y^3}{x^3y^6}$
5. $\frac{12ab^5}{4a^4b^3}$
6. $\frac{24x^5}{-8x^2}$
7. $\frac{-9h^2k^4}{18h^5j^3k^4}$
8. $\left(\frac{2a^2b^4}{3a^3b}\right)^2$
9. $\frac{9a^2b^7c^3}{2a^5b^4c^5}$
10. $\frac{-15xy^{-5}z^7}{-10x^{-4}y^6z^{-4}}$
11. 3^{-4}
12. $\left(\frac{5}{6}\right)^{-2}$
13. $a^5b^0a^{-7}$
14. $\frac{(-u^{-3}v^3)^2}{(u^3v)^{-3}}$
15. $\left(\frac{a^3}{b^2}\right)^{-3}$
16. $\left(\frac{2x}{y^{-3}}\right)^{-2}$
17. $\frac{(-r)s^5}{r^{-3}s^{-4}}$
18. $\frac{28a^{-4}b^0}{14a^3b^{-1}}$
19. $\frac{(j^2k^3m)^4}{(jk^4)^{-1}}$
20. $\left(\frac{-2x^4y}{4y^2}\right)^0$
21. $\left(\frac{-18x^0a^{-3}}{-6x^{-2}a^{-3}}\right)$
22. $\left(\frac{2a^3b^{-2}}{2^{-1}a^{-5}b^3}\right)^{-1}$
23. $\left(\frac{5n^{-1}m^2}{2nm^{-2}}\right)^0$
24. $\frac{(3ab^2c)^{-3}}{(2a^2bc^2)^2}$

Lesson 8-3

(pages 425–430)

Express each number in standard notation.

1. 2.6×10^5
2. 4×10^{-3}
3. 6.72×10^3
4. 4.93×10^{-4}
5. 1.654×10^{-6}
6. 7.348×10^7

Express each number in scientific notation.

7. 6500
8. 953.56
9. 0.697
10. 843.5
11. 568,000
12. 0.0000269
13. 0.121212
14. 543×10^4
15. 739.9×10^{-5}
16. 6480×10^{-2}
17. 0.366×10^{-7}
18. 167×10^3

Evaluate. Express each result in scientific and standard notation.

19. $(2 \times 10^5)(3 \times 10^{-8})$
20. $\frac{4.8 \times 10^3}{1.6 \times 10^1}$
21. $(4 \times 10^2)(1.5 \times 10^6)$
22. $\frac{8.1 \times 10^2}{2.7 \times 10^{-3}}$
23. $\frac{7.8 \times 10^{-5}}{1.3 \times 10^{-7}}$
24. $(2.2 \times 10^{-2})(3.2 \times 10^5)$
25. $(3.1 \times 10^4)(4.2 \times 10^{-5})$
26. $(78 \times 10^6)(0.01 \times 10^5)$
27. $\frac{2.31 \times 10^{-2}}{3.3 \times 10^{-3}}$

Lesson 8-4

(pages 432–436)

State whether each expression is a polynomial. If the expression is a polynomial, identify it as a *monomial*, a *binomial*, or a *trinomial*.

1. $5x^2y + 3xy - 7$
2. 0
3. $\frac{5}{k} - k^2y$
4. $3a^2x - 5a$

Find the degree of each polynomial.

5. $a + 5c$
6. $14abcd - 6d^3$
7. $\frac{a^3}{4}$
8. 10
9. $-4h^5$
10. $\frac{x^2}{3} - \frac{x}{2} + \frac{1}{5}$
11. -6
12. $a^2b^3 - a^3b^2$

Arrange the terms of each polynomial so that the powers of x are in ascending order.

13. $2x^2 - 3x + 4x^3 - x^5$
14. $x^3 - x^2 + x - 1$
15. $2a + 3ax^2 - 4ax$
16. $-5bx^3 - 2bx + 4x^2 - b^3$
17. $x^8 + 2x^2 - x^6 + 1$
18. $cdx^2 - c^2d^2x + d^3$

Arrange the terms of each polynomial so that the powers of x are in descending order.

19. $5x^2 - 3x^3 + 7 + 2x$
20. $-6x + x^5 + 4x^3 - 20$
21. $5b + b^3x^2 + \frac{2}{3}bx$
22. $21p^2x + 3px^3 + p^4$
23. $3ax^2 - 6a^2x^3 + 7a^3 - 8x$
24. $\frac{1}{3}s^2x^3 + 4x^4 - \frac{2}{5}s^4x^2 + \frac{1}{4}x$

Lesson 8-5

(pages 439–443)

Find each sum or difference.

1. $(3a^2 + 5) + (4a^2 - 1)$
2. $(5x - 3) + (-2x + 1)$
3. $(6z + 2) - (9z + 3)$
4. $(-4n + 7) - (-7n - 8)$
5. $(-7t^2 + 4ts - 6s^2) + (-5t^2 - 12ts + 3s^2)$
6. $(6a^2 - 7ab - 4b^2) - (2a^2 + 5ab + 6b^2)$
7. $(4a^2 - 10b^2 + 7c^2) + (-5a^2 + 2c^2 + 2b)$
8. $(z^2 + 6z - 8) - (4z^2 - 7z - 5)$
9. $(4d + 3e - 8f) - (-3d + 10e - 5f + 6)$
10. $(7g + 8h - 9) + (-g - 3h - 6k)$
11. $(9x^2 - 11xy - 3y^2) - (x^2 - 16xy + 12y^2)$
12. $(-3m + 9mn - 5n) + (14m - 5mn - 2n)$
13. $(4x^2 - 8y^2 - 3z^2) - (7x^2 - 14z^2 - 12)$
14. $(17z^4 - 5z^2 + 3z) - (4z^4 + 2z^3 + 3z)$
15. $(6 - 7y + 3y^2) + (3 - 5y - 2y^2) + (-12 - 8y + y^2)$
16. $(-7c^2 - 2c - 5) + (9c - 6) + (16c^2 + 3) + (-9c^2 - 7c + 7)$

Lesson 8-6

(pages 444–449)

Find each product.

1. $-3(8x + 5)$
2. $3b(5b + 8)$
3. $1.1a(2a + 7)$
4. $\frac{1}{2}x(8x - 6)$
5. $7xy(5x^2 - y^2)$
6. $5y(y^2 - 3y + 6)$
7. $-ab(3b^2 + 4ab - 6a^2)$
8. $4m^2(9m^2n + mn - 5n^2)$
9. $4st^2(-4s^2t^3 + 7s^5 - 3st^3)$
10. $-\frac{1}{3}x(9x^2 + x - 5)$
11. $-2mn(8m^2 - 3mn + n^2)$
12. $-\frac{3}{4}ab^2\left(\frac{1}{3}b^2 - \frac{4}{9}b + 1\right)$

Simplify.

13. $-3a(2a - 12) + 5a$
14. $6(12b^2 - 2b) + 7(-2 - 3b)$
15. $x(x - 6) + x(x - 2) + 2x$
16. $11(n - 3) + 2(n^2 + 22n)$
17. $-2x(x + 3) + 3(x + 3)$
18. $4m(n - 1) - 5n(n + 1)$
19. $-7xy + x(7y - 3)$
20. $5(-c + 3a) - c(2c + 1)$
21. $-9n(1 - n) + 4(n^2 + n)$

Solve each equation.

22. $-6(11 - 2x) = 7(-2 - 2x)$
23. $11(n - 3) + 5 = 2n + 44$
24. $a(a - 6) + 2a = 3 + a(a - 2)$
25. $q(2q + 3) + 20 = 2q(q - 3)$
26. $w(w + 12) = w(w + 14) + 12$
27. $x(x - 3) + 4x - 3 = 8x + x(3 + x)$
28. $-3(x + 5) + x(x - 1) = x(x + 2) - 3$
29. $n(n - 5) + n(n + 2) = 2n(n - 1) + 1.5$

Extra Practice

Lesson 8-7

(pages 452–457)

Find each product.

1. $(d + 2)(d + 5)$
2. $(z + 7)(z - 4)$
3. $(m - 8)(m - 5)$
4. $(a + 2)(a - 19)$
5. $(c + 15)(c - 3)$
6. $(x + y)(x - 2y)$
7. $(2x - 5)(x + 6)$
8. $(7a - 4)(2a - 5)$
9. $(4x + y)(2x - 3y)$
10. $(7v + 3)(v + 4)$
11. $(7s - 8)(3s - 2)$
12. $(4g + 3h)(2g - 5h)$
13. $(4a + 3)(2a - 1)$
14. $(7y - 1)(2y - 3)$
15. $(2x + 3y)(4x + 2y)$
16. $(12r - 4s)(5r + 8s)$
17. $(-a + 1)(-3a - 2)$
18. $(2n - 4)(-3n - 2)$
19. $(x - 2)(x^2 + 2x + 4)$
20. $(3x + 5)(2x^2 - 5x + 11)$
21. $(4s + 5)(3s^2 + 8s - 9)$
22. $(3a + 5)(-8a^2 + 2a + 3)$
23. $(a - b)(a^2 + ab + b^2)$
24. $(c + d)(c^2 - cd + d^2)$
25. $(5x - 2)(-5x^2 + 2x + 7)$
26. $(-n + 2)(-2n^2 + n - 1)$
27. $(x^2 - 7x + 4)(2x^2 - 3x - 6)$
28. $(x^2 + x + 1)(x^2 - x - 1)$
29. $(a^2 + 2a + 5)(a^2 - 3a - 7)$
30. $(5x^4 - 2x^2 + 1)(x^2 - 5x + 3)$

Lesson 8-8

(pages 458–463)

Find each product.

1. $(t + 7)^2$
2. $(w - 12)(w + 12)$
3. $(q - 4h)^2$
4. $(10x + 11y)(10x - 11y)$
5. $(4e + 3)^2$
6. $(2b - 4d)(2b + 4d)$
7. $(a + 2b)^2$
8. $(3x + y)^2$
9. $(6m + 2n)^2$
10. $(3m - 7d)^2$
11. $(5b - 6)(5b + 6)$
12. $(1 + x)^2$
13. $(5x - 9y)^2$
14. $(8a - 2b)(8a + 2b)$
15. $\left(\frac{1}{4}x + 4\right)^2$
16. $(c - 3d)^2$
17. $(5a - 12b)^2$
18. $\left(\frac{1}{2}x + y\right)^2$
19. $(n^2 + 1)^2$
20. $(k^2 - 3j)^2$
21. $(a^2 - 5)(a^2 + 5)$
22. $(2x^3 - 7)(2x^3 + 7)$
23. $(3x^3 - 9y)(3x^3 + 9y)$
24. $(7a^2 - b)(7a^2 + b)$
25. $\left(\frac{1}{2}x - 10\right)\left(\frac{1}{2}x + 10\right)$
26. $\left(\frac{1}{3}n - m\right)\left(\frac{1}{3}n + m\right)$
27. $(a - 1)(a - 1)(a - 1)$
28. $(x + 2)(x - 2)(2x + 5)$
29. $(4x - 1)(4x + 1)(x - 4)$
30. $(x - 5)(x + 5)(x + 4)(x - 4)$
31. $(a + 1)(a + 1)(a - 1)(a - 1)$
32. $(n - 1)(n + 1)(n - 1)$
33. $(2c + 3)(2c + 3)(2c - 3)(2c - 3)$
34. $(4d + 5e)(4d + 5e)(4d - 5e)(4d - 5e)$

Lesson 9-1

(pages 474–479)

Find the factors of each number. Then classify each number as *prime* or *composite*.

1. 23
2. 21
3. 81
4. 24
5. 18
6. 22

Find the prime factorization of each integer.

7. 42
8. 267
9. -72
10. 164
11. -57
12. -60

Factor each monomial completely.

13. $240mn$
14. $-64a^3b$
15. $-26xy^2$
16. $-231xy^2z$
17. $44rs^2t^3$
18. $-756m^2n^2$

Find the GCF of each set of monomials.

19. 16, 60
20. 15, 50
21. 45, 80
22. 29, 58
23. 55, 305
24. 126, 252
25. 128, 245
26. $7y^2$, $14y^2$
27. $4xy$, $-6x$
28. $35t^2$, $7t$
29. $16pq^2$, $12p^2q$, $4pq$
30. 5, 15, 10
31. $12mn$, $10mn$, $15mn$
32. $14xy$, $12y$, $20x$
33. $26jk^4$, $16jk^3$, $8j^2$

Extra Practice

Lesson 9-2

(pages 481–486)

Factor each polynomial.

1. $10a^2 + 40a$
2. $15wx - 35wx^2$
3. $27a^2b + 9b^3$
4. $11x + 44x^2y$
5. $16y^2 + 8y$
6. $14mn^2 + 2mn$
7. $25a^2b^2 + 30ab^3$
8. $2m^3n^2 - 16mn^2 + 8mn$
9. $2ax + 6xc + ba + 3bc$
10. $6mx - 4m + 3rx - 2r$
11. $3ax - 6bx + 8b - 4a$
12. $a^2 - 2ab + a - 2b$
13. $8ac - 2ad + 4bc - bd$
14. $2e^2g + 2fg + 4e^2h + 4fh$
15. $x^2 - xy - xy + y^2$

Solve each equation. Check your solutions.

16. $a(a - 9) = 0$
17. $d(d + 11) = 0$
18. $z(z - 2.5) = 0$
19. $(2y + 6)(y - 1) = 0$
20. $(4n - 7)(3n + 2)$
21. $(a - 1)(a + 1) = 0$
22. $10x^2 - 20x = 0$
23. $8b^2 - 12b = 0$
24. $14d^2 + 49d = 0$
25. $15a^2 = 60a$
26. $33x^2 = -22x$
27. $32x^2 = 16x$

Lesson 9-3

(pages 489–494)

Factor each trinomial.

1. $x^2 - 9x + 14$
2. $a^2 - 9a - 36$
3. $x^2 + 2x - 15$
4. $n^2 - 8n + 15$
5. $b^2 + 22b + 21$
6. $c^2 + 2c - 3$
7. $x^2 - 5x - 24$
8. $n^2 - 8n + 7$
9. $m^2 - 10m - 39$
10. $z^2 + 15z + 36$
11. $s^2 - 13st - 30t^2$
12. $y^2 + 2y - 35$
13. $r^2 + 3r - 40$
14. $x^2 + 5x - 6$
15. $x^2 - 4xy - 5y^2$
16. $r^2 + 16r + 63$
17. $v^2 + 24v - 52$
18. $k^2 - 27kj - 90j^2$

Solve each equation. Check your solutions.

19. $a^2 + 3a - 4 = 0$
20. $x^2 - 8x - 20 = 0$
21. $b^2 + 11b + 24 = 0$
22. $y^2 + y - 42 = 0$
23. $k^2 + 2k - 24 = 0$
24. $r^2 - 13r - 48 = 0$
25. $n^2 - 9n = -18$
26. $2z + z^2 = 35$
27. $-20x + 19 = -x^2$
28. $10 + a^2 = -7a$
29. $z^2 - 57 = 16z$
30. $x^2 = -14x - 33$
31. $22x - x^2 = 96$
32. $-144 = q^2 - 26q$
33. $x^2 + 84 = 20x$

Lesson 9-4

(pages 495–500)

Factor each trinomial, if possible. If the trinomial cannot be factored using integers, write *prime*.

1. $4a^2 + 4a - 63$
2. $3x^2 - 7x - 6$
3. $4r^2 - 25r + 6$
4. $2z^2 - 11z + 15$
5. $3a^2 - 2a - 21$
6. $4y^2 + 11y + 6$
7. $6n^2 + 7n - 3$
8. $5x^2 - 17x + 14$
9. $2n^2 - 11n + 13$
10. $8m^2 - 10m - 3$
11. $6y^2 + 2y - 2$
12. $2r^2 + 3r - 14$
13. $5a^2 - 3a + 15$
14. $18v^2 + 24v + 12$
15. $4k^2 + 2k - 12$
16. $10x^2 - 20xy + 10y^2$
17. $12c^2 - 11cd - 5d^2$
18. $30n^2 - mn - m^2$

Solve each equation. Check your solutions.

19. $8t^2 + 32t + 24 = 0$
20. $6y^2 + 72y + 192 = 0$
21. $5x^2 + 3x - 2 = 0$
22. $9x^2 + 18x - 27 = 0$
23. $4x^2 - 4x - 4 = 4$
24. $12n^2 - 16n - 3 = 0$
25. $12x^2 - x - 35 = 0$
26. $18x^2 + 36x - 14 = 0$
27. $15a^2 + a - 2 = 0$
28. $14b^2 + 7b - 42 = 0$
29. $13r^2 + 21r - 10 = 0$
30. $35y^2 - 60y - 20 = 0$
31. $16x^2 - 4x - 6 = 0$
32. $28d^2 + 5d - 3 = 0$
33. $30x^2 - 9x - 3 = 0$

Lesson 9-5

(pages 501–506)

Factor each polynomial, if possible. If the polynomial cannot be factored, write *prime*.

1. $x^2 - 9$
2. $a^2 - 64$
3. $4x^2 - 9y^2$
4. $1 - 9z^2$
5. $16a^2 - 9b^2$
6. $8x^2 - 12y^2$
7. $a^2 - 4b^2$
8. $x^2 - y^2$
9. $75r^2 - 48$
10. $x^2 - 36y^2$
11. $3a^2 - 16$
12. $12t^2 - 75$
13. $9x^2 - 100y^2$
14. $49 - a^2b^2$
15. $5a^2 - 48$
16. $169 - 16t^2$
17. $8r^2 - 4$
18. $-45m^2 + 5$

Solve each equation by factoring. Check your solutions.

19. $4x^2 = 16$
20. $2x^2 = 50$
21. $9n^2 - 4 = 0$
22. $a^2 - \frac{25}{36} = 0$
23. $\frac{16}{9} - b^2 = 0$
24. $18 - \frac{1}{2}x^2 = 0$
25. $20 - 5g^2 = 0$
26. $16 - \frac{1}{4}p^2 = 0$
27. $\frac{1}{4}c^2 - \frac{4}{9} = 0$
28. $3z^2 - 48 = 0$
29. $72 - 2z^2 = 0$
30. $25a^2 = 1$
31. $2q^3 - 2q = 0$
32. $3r^3 = 48r$
33. $100d - 4d^3 = 0$

Extra Practice

Lesson 9-6

(pages 508–514)

Determine whether each trinomial is a perfect square trinomial. If so, factor it.

1. $x^2 + 12x + 36$
2. $n^2 - 13n + 36$
3. $a^2 + 4a + 4$
4. $x^2 - 10x - 100$
5. $2n^2 + 17n + 21$
6. $4a^2 - 20a + 25$

Factor each polynomial, if possible. If the polynomial cannot be factored, write *prime*.

7. $3x^2 - 75$
8. $n^2 - 8n + 16$
9. $4p^2 + 12pr + 9r^2$
10. $6a^2 + 72$
11. $s^2 + 30s + 225$
12. $24x^2 + 24x + 9$
13. $1 - 10z + 25z^2$
14. $28 - 63b^2$
15. $4c^2 + 2c - 7$

Solve each equation. Check your solutions.

16. $x^2 + 22x + 121 = 0$
17. $343d^2 = 7$
18. $(a - 7)^2 = 5$
19. $c^2 + 10c + 36 = 11$
20. $16s^2 + 81 = 72s$
21. $9p^2 - 42p + 20 = -29$

Lesson 10-1

(pages 524–530)

Use a table of values to graph each function.

1. $y = x^2 + 6x + 8$
2. $y = -x^2 + 3x$
3. $y = -x^2$
4. $y = x^2 + x + 3$
5. $y = x^2 + 1$
6. $y = 3x^2 + 6x + 16$

Write the equation of the axis of symmetry, and find the coordinates of the vertex of the graph of each equation. Identify the vertex as a maximum or minimum. Then graph the equation.

7. $y = -x^2 + 2x - 3$
8. $y = 3x^2 + 24x + 80$
9. $y = x^2 - 4x - 4$
10. $y = 5x^2 - 20x + 37$
11. $y = 3x^2 + 6x + 3$
12. $y = 2x^2 + 12x$
13. $y = x^2 - 6x + 5$
14. $y = x^2 + 6x + 9$
15. $y = -x^2 + 16x - 15$
16. $y = 4x^2 - 1$
17. $y = -2x^2 - 2x + 4$
18. $y = 6x^2 - 12x - 4$
19. $y = -x^2 - 1$
20. $y = -x^2 + x + 1$
21. $y = -5x^2 - 3x + 2$
22. $y = -x^2 + x + 20$
23. $y = 2x^2 + 5x - 2$
24. $y = -3x^2 - 18x - 15$

Extra Practice

Lesson 10-2

(pages 533–538)

Solve each equation by graphing.

1. $a^2 - 25 = 0$
2. $n^2 - 8n = 0$
3. $d^2 + 36 = 0$
4. $b^2 - 18b + 81 = 0$
5. $x^2 + 3x + 27 = 0$
6. $-y^2 - 3y + 10 = 0$

Solve each equation by graphing. If integral roots cannot be found, estimate the roots by stating the consecutive integers between which the roots lie.

7. $x^2 + 2x - 3 = 0$
8. $-x^2 + 6x - 5 = 0$
9. $-a^2 - 2a + 3 = 0$
10. $2r^2 - 8r + 5 = 0$
11. $-3x^2 + 6x - 9 = 0$
12. $c^2 + c = 0$
13. $3t^2 + 2 = 0$
14. $-b^2 + 5b + 2 = 0$
15. $3x^2 + 7x = 1$
16. $x^2 + 5x - 24 = 0$
17. $8 - n^2 = 0$
18. $x^2 - 7x = 18$
19. $a^2 + 12a + 36 = 0$
20. $64 - x^2 = 0$
21. $-4x^2 + 2x = -1$
22. $5z^2 + 8z = 1$
23. $p = 27 - p^2$
24. $6w = -15 - 3w^2$

Lesson 10-3

(pages 539–544)

Solve each equation. Round to the nearest tenth, if necessary.

1. $x^2 - 4x + 4 = 9$
2. $t^2 - 6t + 9 = 16$
3. $b^2 + 10b + 25 = 11$
4. $a^2 - 22a + 121 = 3$
5. $x^2 + 2x + 1 = 81$
6. $t^2 - 36t + 324 = 85$

Find the value of c that makes each trinomial a perfect square.

7. $a^2 + 20a + c$
8. $x^2 + 10x + c$
9. $t^2 + 12t + c$
10. $y^2 - 9y + c$
11. $p^2 - 14p + c$
12. $b^2 + 13b + c$

Solve each equation by completing the square. Round to the nearest tenth, if necessary.

13. $a^2 - 8a - 84 = 0$
14. $c^2 + 6 = -5c$
15. $p^2 - 8p + 5 = 0$
16. $2y^2 + 7y - 4 = 0$
17. $t^2 + 3t = 40$
18. $x^2 + 8x - 9 = 0$
19. $y^2 + 5y - 84 = 0$
20. $t^2 + 12t + 32 = 0$
21. $2x - 3x^2 = -8$
22. $2y^2 - y - 9 = 0$
23. $2z^2 - 5z - 4 = 0$
24. $8t^2 - 12t - 1 = 0$

Lesson 10-4

(pages 546–552)

Solve each equation by using the Quadratic Formula. Round to the nearest tenth, if necessary.

1. $x^2 - 8x - 4 = 0$
2. $x^2 + 7x - 8 = 0$
3. $x^2 - 5x + 6 = 0$
4. $y^2 - 7y - 8 = 0$
5. $m^2 - 2m = 35$
6. $4n^2 - 20n = 0$
7. $m^2 + 4m + 2 = 0$
8. $2t^2 - t - 15 = 0$
9. $5t^2 = 125$
10. $t^2 + 16 = 0$
11. $-4x^2 + 8x = -3$
12. $3k^2 + 2 = -8k$
13. $8t^2 + 10t + 3 = 0$
14. $3x^2 - \frac{5}{4}x - \frac{1}{2} = 0$
15. $-5b^2 + 3b - 1 = 0$
16. $s^2 + 8s + 7 = 0$
17. $d^2 - 14d + 24 = 0$
18. $3k^2 + 11k = 4$
19. $n^2 - 3n + 1 = 0$
20. $2z^2 + 5z - 1 = 0$
21. $3h^2 = 27$

State the value of the discriminant for each equation. Then determine the number of real roots of the equation.

22. $3f^2 + 2f = 6$
23. $2x^2 = 0.7x + 0.3$
24. $3w^2 - 2w + 8 = 0$
25. $4r^2 - 12r + 9 = 0$
26. $x^2 - 5x = -9$
27. $25t^2 + 30t = -9$

Lesson 10-5
(pages 554–560)

Graph each function. State the y-intercept. Then use the graph to determine the approximate value of the given expression. Use a calculator to confirm the value.

1. $y = 7^x; 7^{1.5}$
2. $\left(\frac{1}{3}\right)^x; \left(\frac{1}{3}\right)^{5.6}$
3. $y = \left(\frac{3}{5}\right)^x; \left(\frac{3}{5}\right)^{-4.2}$

Graph each function. State the y-intercept.

4. $y = 3^x + 1$
5. $y = 2^x - 5$
6. $y = 2^{x+3}$
7. $y = 3^{x+1}$
8. $y = \left(\frac{2}{3}\right)^x$
9. $y = 5\left(\frac{2}{5}\right)^x$
10. $y = 5(3^x)$
11. $y = 4(5)^x$
12. $y = 2(5)^x + 1$
13. $y = \left(\frac{1}{2}\right)^{x+1}$
14. $y = \left(\frac{1}{8}\right)^x$
15. $y = \left(\frac{3}{4}\right)^x - 2$

Determine whether the data in each table display exponential behavior. Explain why or why not.

16.

x	−1	0	1	2
y	−5	−1	3	7

17.

x	1	2	3	4
y	25	125	625	3125

Lesson 10-6
(pages 561–565)

1. **EDUCATION** Marco is saving for tuition costs at a state university. He deposited $8500 in a 4-year certificate of deposit earning 7.25% compounded monthly.
   a. Write an equation for the amount of money Marco will have at the end of four years.
   b. Find the amount of money he will have for his tuition at the end of the four years.
2. **TRANSPORTATION** Elise is buying a new car selling for $21,500. The rate of depreciation on this type of car is 8% per year.
   a. Write an equation for the value of the car in 5 years.
   b. Find the value of the car in 5 years.
3. **POPULATION** In 1990, the town of Belgrade, Montana, had a population of 3422. For each of the next 8 years, the population increased by 4.9% per year.
   a. Write an equation for the population of Belgrade in 1998.
   b. Find the population of Belgrade in 1998.

Lesson 10-7
(pages 567–572)

Determine whether each sequence is geometric.

1. 12, 23, 34, 45, …
2. 6, 7.2, 8.64, 10.368, …
3. 39, 33, 27, 21, …
4. 86, 68.8, 55.04, 44.032, …
5. 4, 8, 16, 32, …
6. 13, 10, 11, 8, 9, 6, …

Find the next three terms in each geometric sequence.

7. 3125, 625, 125, 25, …
8. 15, −45, 135, −405, …
9. 243, 81, 27, 9, …
10. 15, −7.5, 3.75, −1.875, …
11. −25, −15, −9, −5.4, …
12. $\frac{1}{4}, \frac{1}{10}, \frac{1}{25}, \frac{2}{125}, \ldots$

Find the nth term of each geometric sequence.

13. $a_1 = 1, n = 10, r = 6$
14. $a_1 = -1, n = 7, r = -4$
15. $a_1 = -6, n = 4, r = 0.4$
16. $a_1 = 100, n = 10, r = 0.1$
17. $a_1 = -750, n = 5, r = -1.5$
18. $a_1 = 64, n = 5, r = 8$
19. $a_1 = 0.5, n = 9, r = -10$
20. $a_1 = -20, n = 6, r = 2.5$
21. $a_1 = 350, n = 4, r = -0.9$

Find the geometric means in each sequence.

22. 1, ____, 81
23. −81, ____, −9
24. 504, ____, 14
25. 0.5, ____, 162
26. −1, ____, −4
27. 0.25, ____, 0.36
28. $\frac{1}{2}$, ____, $\frac{1}{8}$
29. $-\frac{2}{3}$, ____, $-\frac{32}{27}$
30. 6.25, ____, 2.25

Extra Practice

Lesson 11-1

(pages 587–593)

Simplify.

1. $\sqrt{50}$
2. $\sqrt{200}$
3. $\sqrt{162}$
4. $\sqrt{700}$
5. $\dfrac{\sqrt{3}}{\sqrt{5}}$
6. $\dfrac{\sqrt{72}}{\sqrt{6}}$
7. $\sqrt{\dfrac{8}{7}}$
8. $\sqrt{\dfrac{7}{32}}$
9. $\sqrt{\dfrac{5}{8}} \cdot \sqrt{\dfrac{2}{6}}$
10. $\sqrt{\dfrac{2}{3}} \cdot \sqrt{\dfrac{3}{2}}$
11. $\sqrt{\dfrac{2x}{30}}$
12. $\sqrt{\dfrac{50}{z^2}}$
13. $\sqrt{10} \cdot \sqrt{20}$
14. $\sqrt{7} \cdot \sqrt{3}$
15. $6\sqrt{2} \cdot \sqrt{3}$
16. $5\sqrt{6} \cdot 2\sqrt{3}$
17. $\sqrt{4x^4y^3}$
18. $\sqrt{200m^2y^3}$
19. $\sqrt{12ts^3}$
20. $\sqrt{175a^4b^6}$
21. $\sqrt{\dfrac{54}{g^2}}$
22. $\sqrt{99x^3y^7}$
23. $\sqrt{\dfrac{32c^5}{9d^2}}$
24. $\sqrt{\dfrac{27p^4}{3p^2}}$
25. $\dfrac{1}{3 + \sqrt{5}}$
26. $\dfrac{2}{\sqrt{3} - 5}$
27. $\dfrac{\sqrt{3}}{\sqrt{3} - 5}$
28. $\dfrac{\sqrt{6}}{7 - 2\sqrt{3}}$

Lesson 11-2

(pages 594–598)

Simplify each expression.

1. $3\sqrt{11} + 6\sqrt{11} - 2\sqrt{11}$
2. $6\sqrt{13} + 7\sqrt{13}$
3. $2\sqrt{12} + 5\sqrt{3}$
4. $9\sqrt{7} - 4\sqrt{2} + 3\sqrt{2} + 5\sqrt{7}$
5. $3\sqrt{5} - 5\sqrt{3}$
6. $4\sqrt{8} - 3\sqrt{5}$
7. $2\sqrt{27} - 4\sqrt{12}$
8. $8\sqrt{32} + 4\sqrt{50}$
9. $\sqrt{45} + 6\sqrt{20}$
10. $2\sqrt{63} - 6\sqrt{28} + 8\sqrt{45}$
11. $14\sqrt{3t} + 8\sqrt{3t}$
12. $7\sqrt{6x} - 12\sqrt{6x}$
13. $5\sqrt{7} - 3\sqrt{28}$
14. $7\sqrt{8} - \sqrt{18}$
15. $7\sqrt{98} + 5\sqrt{32} - 2\sqrt{75}$
16. $4\sqrt{6} + 3\sqrt{2} - 2\sqrt{5}$
17. $-3\sqrt{20} + 2\sqrt{45} - \sqrt{7}$
18. $4\sqrt{75} + 6\sqrt{27}$
19. $10\sqrt{\dfrac{1}{5}} - \sqrt{45} - 12\sqrt{\dfrac{5}{9}}$
20. $\sqrt{15} - \sqrt{\dfrac{3}{5}}$
21. $3\sqrt{\dfrac{1}{3}} - 9\sqrt{\dfrac{1}{12}} + \sqrt{243}$

Find each product.

22. $\sqrt{3}(\sqrt{5} + 2)$
23. $\sqrt{2}(\sqrt{2} + 3\sqrt{5})$
24. $(\sqrt{2} + 5)^2$
25. $(3 - \sqrt{7})(3 + \sqrt{7})$
26. $(\sqrt{2} + \sqrt{3})(\sqrt{3} + \sqrt{2})$
27. $(4\sqrt{7} + \sqrt{2})(\sqrt{3} - 3\sqrt{5})$

Lesson 11-3

(pages 599–604)

Solve each equation. Check your solution.

1. $\sqrt{5x} = 5$
2. $4\sqrt{7} = \sqrt{-m}$
3. $\sqrt{t} - 5 = 0$
4. $\sqrt{3b} + 2 = 0$
5. $\sqrt{x - 3} = 6$
6. $5 - \sqrt{3x} = 1$
7. $2 + 3\sqrt{y} = 13$
8. $\sqrt{3g} = 6$
9. $\sqrt{a} - 2 = 0$
10. $\sqrt{2j} - 4 = 8$
11. $5 + \sqrt{x} = 9$
12. $\sqrt{5y + 4} = 7$
13. $7 + \sqrt{5c} = 9$
14. $2\sqrt{5t} = 10$
15. $\sqrt{44} = 2\sqrt{p}$
16. $4\sqrt{x - 5} = 15$
17. $4 - \sqrt{x - 3} = 9$
18. $\sqrt{10x^2 - 5} = 3x$
19. $\sqrt{2a^2 - 144} = a$
20. $\sqrt{3y + 1} = y - 3$
21. $\sqrt{2x^2 - 12} = x$
22. $\sqrt{b^2 + 16} + 2b = 5b$
23. $\sqrt{m + 2} + m = 4$
24. $\sqrt{3 - 2c} + 3 = 2c$

Lesson 11-4

(pages 606–611)

If c is the measure of the hypotenuse of a right triangle, find each missing measure. If necessary, round to the nearest hundredth.

1. $b = 20, c = 29, a = ?$
2. $a = 7, b = 24, c = ?$
3. $a = 2, b = 6, c = ?$
4. $b = 10, c = \sqrt{200}, a = ?$
5. $a = 3, c = 3\sqrt{2}, b = ?$
6. $a = 6, c = 14, b = ?$
7. $a = \sqrt{11}, c = \sqrt{47}, b = ?$
8. $a = \sqrt{13}, b = 6, c = ?$
9. $a = \sqrt{6}, b = 3, c = ?$
10. $b = \sqrt{75}, c = 10, a = ?$
11. $b = 9, c = \sqrt{130}, a = ?$
12. $a = 9, c = 15, b = ?$
13. $b = 5, c = 11, a = ?$
14. $a = \sqrt{33}, b = 4, c = ?$
15. $a = 5, c = \sqrt{34}, b = ?$

Determine whether the following side measures form right triangles.

16. 14, 48, 50
17. 20, 30, 40
18. 21, 72, 75
19. 5, 12, $\sqrt{119}$
20. 15, 39, 36
21. $\sqrt{5}$, 12, 13
22. 10, 12, $\sqrt{22}$
23. 2, 3, 4
24. $\sqrt{7}$, 8, $\sqrt{71}$

Lesson 11-5

(pages 612–616)

Find the distance between each pair of points whose coordinates are given. Express answers in simplest radical form and as decimal approximations rounded to the nearest hundredth if necessary.

1. (4, 2), (−2, 10)
2. (−5, 1), (7, 6)
3. (4, −2), (1, 2)
4. (−2, 4), (4, −2)
5. (3, 1), (−2, −1)
6. (−2, 4), (7, −8)
7. (−5, 0), (−9, 6)
8. (5, −1), (5, 13)
9. (2, −3), (10, 8)
10. (−7, 5), (2, −7)
11. (−6, −2), (−5, 4)
12. (8, −10), (3, 2)
13. (4, −3), (7, −9)
14. (6, 3), (9, 7)
15. (10, 0), (9, 7)
16. (2, −1), (−3, 3)
17. (−5, 4), (3, −2)
18. (0, −9), (0, 7)
19. (−1, 7), (8, 4)
20. (−9, 2), (3, −3)
21. $(3\sqrt{2}, 7), (5\sqrt{2}, 9)$
22. (6, 3), (10, 0)
23. (3, 6), (5, −5)
24. (−4, 2), (5, 4)

Find the possible values of a if the points with the given coordinates are the indicated distance apart.

25. $(0, 0), (a, 3); d = 5$
26. $(2, -1), (-6, a); d = 10$
27. $(1, 0), (a, 6); d = \sqrt{61}$
28. $(-2, a), (5, 10); d = \sqrt{85}$
29. $(15, a), (0, 4); d = \sqrt{274}$
30. $(3, 3), (a, 9); d = \sqrt{136}$

Lesson 11-6

(pages 617–622)

Determine whether each pair of triangles is similar. Justify your answer.

1.

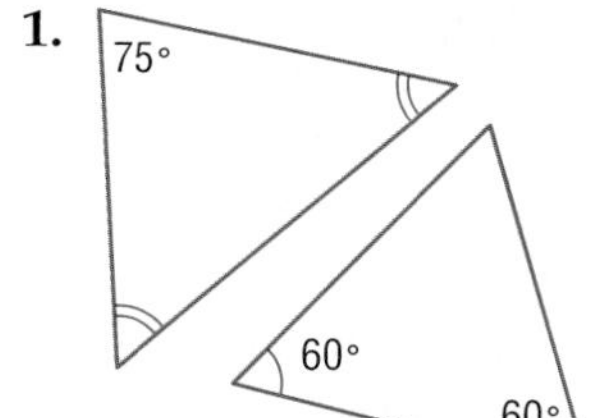

2.

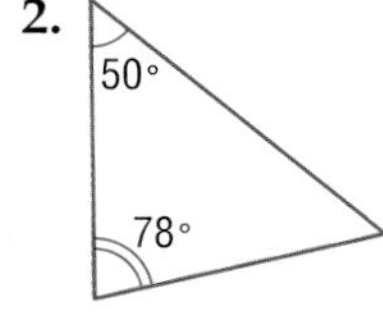

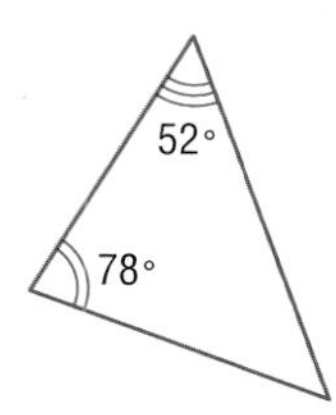

3.

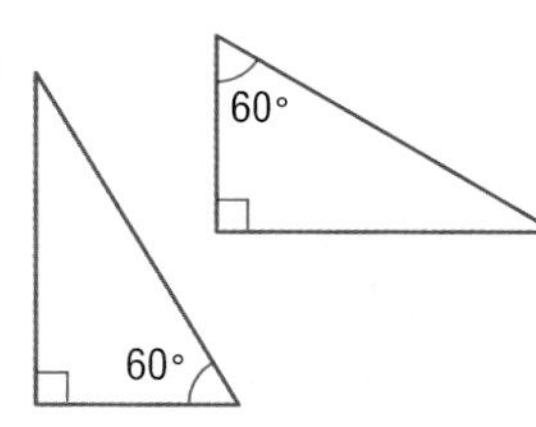

For each set of measures given, find the measures of the missing sides if $\triangle ABC \sim \triangle DEF$.

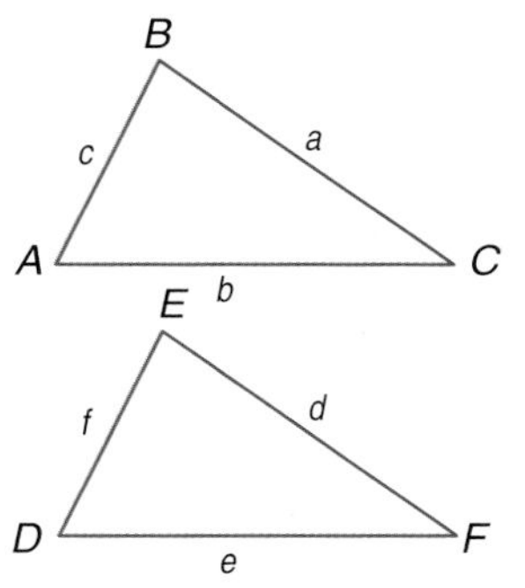

4. $a = 5, d = 10, b = 8, c = 7$
5. $a = 2, b = 3, c = 4, d = 3$
6. $a = 6, d = 4.5, e = 7, f = 7.5$
7. $a = 15, c = 20, b = 18, f = 10$
8. $f = 17.5, d = 8.5, e = 11, a = 1.7$
9. $b = 5.6, e = 7, a = 4, c = 7.2$
10. $e = 125, a = 80, d = 100, f = 218.75$

Lesson 11-7

(pages 624–631)

For each triangle, find sin N, cos N, and tan N to the nearest ten thousandth.

1. 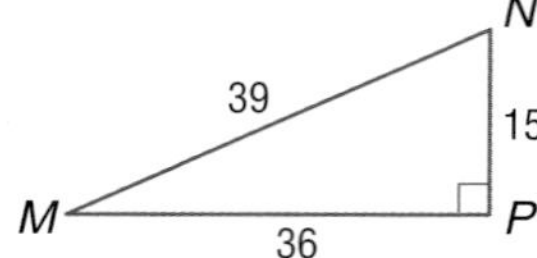

2. 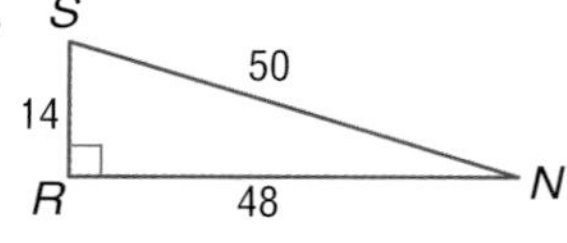

3. 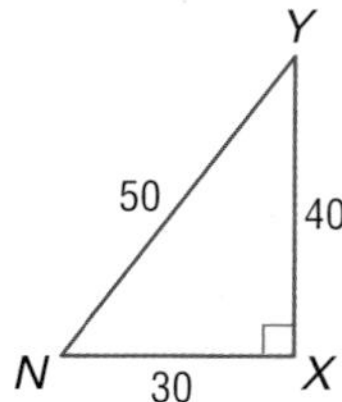

Use a calculator to find the value of each trigonometric ratio to the nearest ten thousandth.

4. cos 25°
5. tan 31°
6. sin 71°
7. cos 64°
8. tan 9°
9. sin 2°

Use a calculator to find the measure of each angle to the nearest degree.

10. $\tan B = 0.5427$
11. $\cos A = 0.8480$
12. $\sin J = 0.9654$
13. $\cos Q = 0.3645$
14. $\sin R = 0.2104$
15. $\tan V = 11.4301$

Solve each right triangle. State the side lengths to the nearest tenth and the angle measures to the nearest degree.

16. 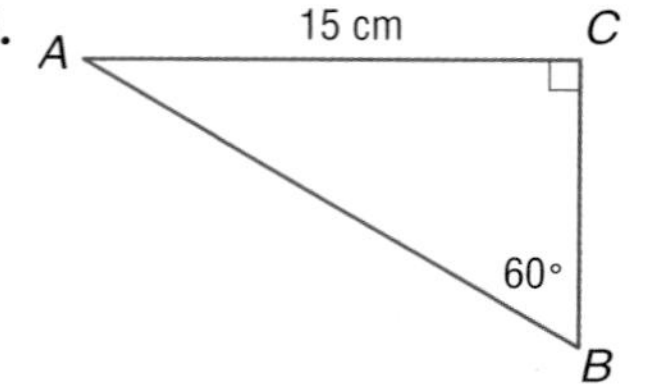

17. 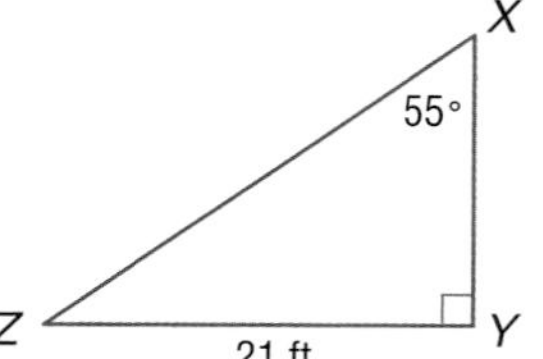

18. 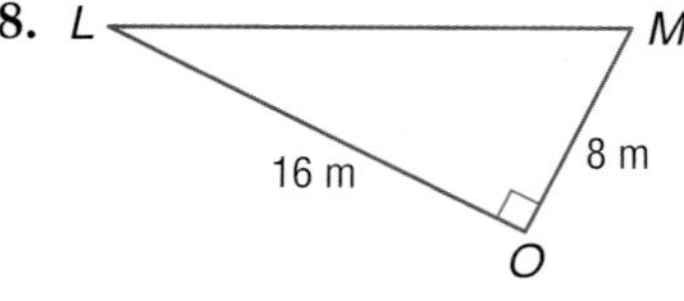

Lesson 12-1

(pages 642–647)

Graph each variation if y varies inversely as x.

1. $y = 10$ when $x = 7.5$
2. $y = -5$ when $x = 3$
3. $y = -6$ when $x = -2$
4. $y = 1$ when $x = -0.5$
5. $y = -2.5$ when $x = 3$
6. $y = -2$ when $x = -1$

Write an inverse variation equation that relates x and y. Assume that y varies inversely as x. Then solve.

7. If $y = 54$ when $x = 4$, find x when $y = 27$.
8. If $y = 18$ when $x = 6$, find x when $y = 12$.
9. If $y = 12$ when $x = 24$, find x when $y = 9$.
10. If $y = 8$ when $x = -8$, find y when $x = -16$.
11. If $y = 3$ when $x = -8$, find y when $x = 4$.
12. If $y = 27$ when $x = \frac{1}{3}$, find y when $x = \frac{3}{4}$.
13. If $y = -3$ when $x = -8$, find y when $x = 2$.
14. If $y = -3$ when $x = -3$, find x when $y = 4$.
15. If $y = -7.5$ when $x = 2.5$, find y when $x = -2.5$.
16. If $y = -0.4$ when $x = -3.2$, find x when $y = -0.2$.

Lesson 12–2

(pages 648–653)

State the excluded values for each rational expression.

1. $\frac{x}{x + 1}$
2. $\frac{m}{n}$
3. $\frac{c - 2}{c^2 - 4}$
4. $\frac{b^2 - 5b + 6}{b^2 - 8b + 15}$

Simplify each expression. State the excluded values of the variables.

5. $\frac{13a}{39a^2}$
6. $\frac{38x^2}{42xy}$
7. $\frac{p + 5}{2(p + 5)}$
8. $\frac{a + b}{a^2 - b^2}$
9. $\frac{y + 4}{y^2 - 16}$
10. $\frac{c^2 - 4}{c^2 + 4c + 4}$
11. $\frac{a^2 - a}{a - 1}$
12. $\frac{x^2 + 4}{x^4 - 16}$
13. $\frac{r^3 - r^2}{r - 1}$
14. $\frac{4t^2 - 8}{4t - 4}$
15. $\frac{6y^3 - 12y^2}{12y^2 - 18}$
16. $\frac{5x^2 + 10x + 5}{3x^2 + 6x + 3}$

Lesson 12-3

(pages 655–659)

Find each product.

1. $\frac{a^2b}{b^2c} \cdot \frac{c}{d}$
2. $\frac{6a^2n}{8n^2} \cdot \frac{12n}{9a}$
3. $\frac{2a^2d}{3bc} \cdot \frac{9b^2c}{16ad^2}$
4. $\frac{10n^3}{6x^3} \cdot \frac{12n^2x^4}{25n^2x^2}$
5. $\frac{6m^3n}{10a^2} \cdot \frac{4a^2m}{9n^3}$
6. $\frac{(a-5)(a+1)}{(a+1)(a+7)} \cdot \frac{(a+7)(a-6)}{(a+8)(a-5)}$
7. $\frac{x-1}{(x+2)(x-3)} \cdot \frac{x+2}{(x-3)(x-1)}$
8. $\frac{5n-5}{3} \cdot \frac{9}{n-1}$
9. $\frac{a^2}{a-b} \cdot \frac{3a-3b}{a}$
10. $\frac{2a+4b}{5} \cdot \frac{25}{6a+8b}$
11. $\frac{3}{x-y} \cdot \frac{(x-y)^2}{6}$
12. $\frac{x+5}{3x} \cdot \frac{12x^2}{x^2+7x+10}$
13. $\frac{a^2-b^2}{4} \cdot \frac{16}{a+b}$
14. $\frac{4a+8}{a^2-25} \cdot \frac{a-5}{5a+10}$
15. $\frac{r^2}{r-s} \cdot \frac{r^2-s^2}{s^2}$
16. $\frac{a^2-b^2}{a-b} \cdot \frac{7}{a+b}$
17. $\frac{x^2+10x+9}{x^2+11x+18} \cdot \frac{x^2+3x+2}{x^2+7x+6}$
18. $\frac{x^2-6x+5}{x^2+7x+12} \cdot \frac{x^2+14x+40}{x^2+5x-50}$

Lesson 12-4

(pages 660–664)

Find each quotient.

1. $\frac{5m^2n}{12a^2} \div \frac{30m^4}{18an}$
2. $\frac{25g^7h}{28t^3} \div \frac{5g^5h^2}{42s^2t^3}$
3. $\frac{6a+4b}{36} \div \frac{3a+2b}{45}$
4. $\frac{x^2y}{18z} \div \frac{2yz}{3x^2}$
5. $\frac{p^2}{14qr^3} \div \frac{2r^2p}{7q}$
6. $\frac{5e-f}{5e+f} \div (25e^2-f^2)$
7. $\frac{t^2-2t-15}{t-5} \div \frac{t+3}{t+5}$
8. $\frac{5x+10}{x+2} \div (x+2)$
9. $\frac{3d}{2d^2-3d} \div \frac{9}{2d-3}$
10. $\frac{3v^2-27}{15v} \div \frac{v+3}{v^2}$
11. $\frac{3g^2+15g}{4} \div \frac{g+5}{g^2}$
12. $\frac{b^2-9}{4b} \div (b-3)$
13. $\frac{p^2}{y^2-4} \div \frac{p}{2-y}$
14. $\frac{k^2-81}{k^2-36} \div \frac{k-9}{k+6}$
15. $\frac{2a^3}{a+1} \div \frac{a^2}{a+1}$
16. $\frac{x^2-16}{16-x^2} \div \frac{7}{x}$
17. $\frac{y}{5} \div \frac{y^2-25}{5-y}$
18. $\frac{3m}{m+1} \div (m-2)$
19. $\frac{2m+16}{m-2} \div \frac{m^2+6m-16}{m^2+m-6}$
20. $\frac{a^2+3a-10}{a^2+3a+2} \div \frac{a^2+3a-10}{a^2-2a-3}$
21. $\frac{x^2-x-2}{x^2+4x+3} \div \frac{x^2-6x+8}{x^2-x-12}$

Lesson 12-5

(pages 666–671)

Find each quotient.

1. $(2x^2-11x-20) \div (2x+3)$
2. $(a^2+10a+21) \div (a+3)$
3. $(m^2+4m-5) \div (m+5)$
4. $(x^2-2x-35) \div (x-7)$
5. $(c^2+6c-27) \div (c+9)$
6. $(y^2-6y-25) \div (y+7)$
7. $(3t^2-14t-24) \div (3t+4)$
8. $(2r^2-3r-35) \div (2r+7)$
9. $\frac{12n^2+36n+15}{6n+3}$
10. $\frac{10x^2+29x+21}{5x+7}$
11. $\frac{4t^3+17t^2-1}{4t+1}$
12. $\frac{2a^3+9a^2+5a-12}{a+3}$
13. $\frac{4m^2+4m-15}{2m-3}$
14. $\frac{6t^3+5t^2+12}{2t+3}$
15. $\frac{27c^2-24c+8}{9c-2}$
16. $\frac{4b^3+7b^2-2b+4}{b+2}$
17. $\frac{t^3-19t+9}{t-4}$
18. $\frac{9x^3+2x-10}{3x-2}$

Lesson 12-6

(pages 672–677)

Find each sum.

1. $\frac{4}{z} + \frac{3}{z}$
2. $\frac{a}{12} + \frac{2a}{12}$
3. $\frac{5}{2t} + \frac{-7}{2t}$
4. $\frac{y}{2} + \frac{y}{2}$
5. $\frac{b}{x} + \frac{2}{x}$
6. $\frac{y}{2} + \frac{y-6}{2}$
7. $\frac{x}{x+1} + \frac{1}{x+1}$
8. $\frac{2n}{2n-5} + \frac{5}{5-2n}$
9. $\frac{x-y}{2-y} + \frac{x+y}{y-2}$
10. $\frac{r^2}{r-s} + \frac{s^2}{r-s}$
11. $\frac{12n}{3n+2} + \frac{8}{3n+2}$
12. $\frac{6x}{x+y} + \frac{6y}{x+y}$

Find each difference.

13. $\frac{5x}{24} - \frac{3x}{24}$
14. $\frac{7p}{3} - \frac{8p}{3}$
15. $\frac{8k}{5m} - \frac{3k}{5m}$
16. $\frac{8}{m-2} - \frac{6}{m-2}$
17. $\frac{y}{b+6} - \frac{2y}{b+6}$
18. $\frac{a+2}{6} - \frac{a+3}{6}$
19. $\frac{2a}{2a+5} - \frac{5}{2a+5}$
20. $\frac{1}{4z+1} - \frac{(-4z)}{4z+1}$
21. $\frac{3a}{a-2} - \frac{3a}{a-2}$
22. $\frac{n}{n-1} - \frac{1}{1-n}$
23. $\frac{a}{a-7} - \frac{(-7)}{7-a}$
24. $\frac{2a}{6a-3} - \frac{(-1)}{3-6a}$

Lesson 12-7

(pages 678–683)

Find the LCM for each pair of expressions.

1. $27a^2bc, 36ab^2c^2$
2. $3m - 1, 6m - 2$
3. $x^2 + 2x + 1, x^2 - 2x - 3$

Find each sum.

4. $\frac{s}{3} + \frac{2s}{7}$
5. $\frac{5}{2a} + \frac{-3}{6a}$
6. $\frac{6}{5x} + \frac{7}{10x^2}$
7. $\frac{5}{xy} + \frac{6}{yz}$
8. $\frac{2}{t} + \frac{t+3}{s}$
9. $\frac{a}{a-b} + \frac{b}{2b+3a}$
10. $\frac{4a}{2a+6} + \frac{3}{a+3}$
11. $\frac{3t+2}{3t-2} + \frac{t+2}{t^2-4}$
12. $\frac{-3}{a-5} + \frac{-6}{a^2-5a}$

Find each difference.

13. $\frac{2n}{5} - \frac{3m}{4}$
14. $\frac{3z}{7w^2} - \frac{2z}{w}$
15. $\frac{s}{t^2} - \frac{r}{3t}$
16. $\frac{a}{a^2-4} - \frac{4}{a+2}$
17. $\frac{m}{m-n} - \frac{5}{m}$
18. $\frac{y+5}{y-5} - \frac{2y}{y^2-25}$
19. $\frac{t+10}{t^2-100} - \frac{1}{10-t}$
20. $\frac{2a-6}{a^2-3a-10} - \frac{3a+5}{a^2-4a-12}$

Lesson 12-8

(pages 684–689)

Write each mixed expression as a rational expression.

1. $4 + \frac{2}{x}$
2. $8 + \frac{5}{3t}$
3. $\frac{b+1}{2b} + 3b$
4. $3z + \frac{z+2}{z}$
5. $\frac{2}{a-2} + a^2$
6. $3r^2 + \frac{4}{2r+1}$

Simplify each expression.

7. $\frac{3\frac{1}{2}}{4\frac{3}{4}}$
8. $\frac{\frac{x^2}{y}}{\frac{y}{x^3}}$
9. $\frac{\frac{t^4}{u}}{\frac{t^3}{u^2}}$
10. $\frac{\frac{x-3}{x+1}}{\frac{x^2}{y^2}}$
11. $\frac{\frac{y}{3} + \frac{5}{6}}{2 + \frac{5}{y}}$
12. $\frac{\frac{1}{x} + \frac{1}{y}}{\frac{1}{y} - \frac{1}{x}}$
13. $\frac{t-2}{\frac{t^2-4}{t^2+5t+6}}$
14. $\frac{a + \frac{2}{a+1}}{a - \frac{3}{a-2}}$

Lesson 12-9

(pages 690–695)

Solve each equation. State any extraneous solutions.

1. $\frac{k}{6} + \frac{2k}{3} = -\frac{5}{2}$
2. $\frac{2x}{7} + \frac{27}{10} = \frac{4x}{5}$
3. $\frac{18}{b} = \frac{3}{b} + 3$
4. $\frac{3}{5x} + \frac{7}{2x} = 1$
5. $\frac{2a - 3}{6} = \frac{2a}{3} + \frac{1}{2}$
6. $\frac{3x + 2}{x} + \frac{x + 3}{x} = 5$
7. $\frac{2b - 3}{7} - \frac{b}{2} = \frac{b + 3}{14}$
8. $\frac{2y}{y - 4} - \frac{3}{5} = 3$
9. $\frac{2t}{t + 3} + \frac{3}{t} = 2$
10. $\frac{5x}{x + 1} + \frac{1}{x} = 5$
11. $\frac{r - 2}{r + 2} - \frac{2r}{r + 9} = 6$
12. $\frac{m}{m + 1} + \frac{5}{m - 1} = 1$
13. $\frac{2x}{x - 3} - \frac{4x}{3 - x} = 12$
14. $\frac{14}{b - 6} = \frac{1}{2} + \frac{6}{b - 8}$
15. $\frac{a}{4a + 15} - 3 = -2$
16. $\frac{5x}{3x + 10} + \frac{2x}{x + 5} = 2$
17. $\frac{2a - 3}{a - 3} - 2 = \frac{12}{a + 2}$
18. $\frac{z + 3}{z - 1} + \frac{z + 1}{z - 3} = 2$

Lesson 13-1

(pages 708–713)

Identify each sample, suggest a population from which it was selected, and state if it is unbiased (random) or biased. If unbiased, classify the sample as *simple*, *stratified*, or *systematic*. If biased, classify as *convenience* or *voluntary response*.

1. The sheriff has heard that many dogs in the county do not have licenses. He drives from his office and checks the licenses of the first ten dogs he encounters.
2. The school administration wants to check on the incidence of students leaving campus without permission at lunch. An announcement is placed in the school bulletin for 25 students to volunteer to answer questions about leaving campus.
3. The store manager of an ice cream store wants to see whether employees are making ice cream cones within the weight guidelines he provided. During each of three shifts, he selects every tenth cone to weigh.
4. Every fifth car is selected from the assembly line. The cars are also identified by the day of the week during which they were produced.
5. A table is set up outside of a large department store. All people entering the store are given a survey about their preference of brand for blue jeans. As people leave the store, they can return the survey.
6. A community is considering building a new swimming pool. Every twentieth person on a list of residents is contacted in person for their opinion on the new pool.
7. A state wildlife department is concerned about a report that malformed frogs are increasing in the state's lakes. Residents are asked to write in to the state department if they see a malformed frog.
8. The manager at a grocery store has been told that many cartons of strawberries are spoiled. She asks one of her employees to bring in the top 10 cartons on the shelf.

Lesson 13-2

(pages 715–721)

State the dimensions of each matrix.

1. $[1 \quad 0 \quad -2 \quad 5]$
2. $\begin{bmatrix} 1 & 0 \\ 0 & 1 \end{bmatrix}$
3. $\begin{bmatrix} 1 & -1 & 1 \\ -1 & 1 & -1 \\ 1 & -1 & 1 \end{bmatrix}$
4. $[10]$

If $A = \begin{bmatrix} 2 & -4 \\ -3 & 5 \end{bmatrix}$, $B = \begin{bmatrix} 1 & -1 & 4 \\ 0 & 3 & -2 \end{bmatrix}$, $C = \begin{bmatrix} 1 & 0 \\ 0 & 1 \end{bmatrix}$, and $D = \begin{bmatrix} -5 & 1 & -4 \\ -3 & 0 & 2 \end{bmatrix}$, find each sum, difference, or product. If the sum or difference does not exist, write *impossible*.

5. $A + B$
6. $A + C$
7. $B + D$
8. $D - B$
9. $2B$
10. $3C$
11. $A - C$
12. $-5C$
13. $2A + C$
14. $3D - B$
15. $5B + C$
16. $2C + 3A$

Lesson 13-3
(pages 722–728)

For each histogram, answer the following.

- **In what measurement class does the median occur?**
- **Describe the distribution of the data.**

1.

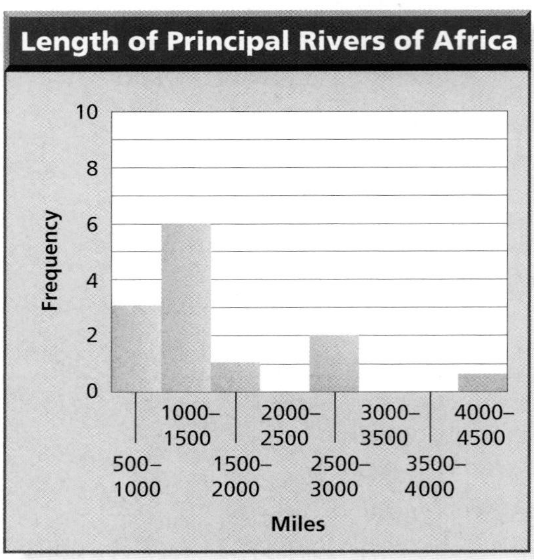

2.

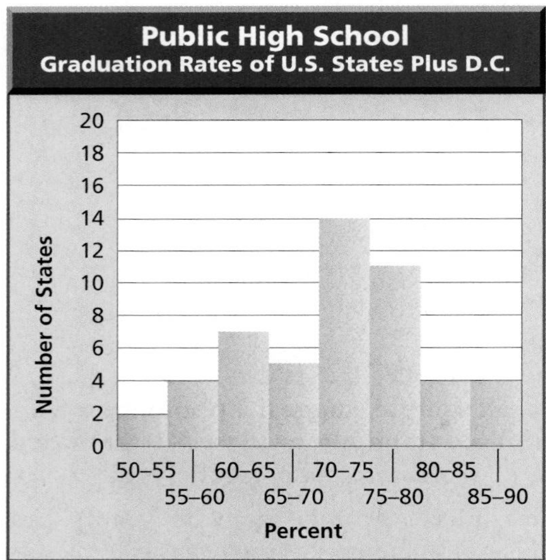

Create a histogram to represent each data set.

3. Sale prices of notebooks at various department stores, in cents: 13, 69, 89, 25, 55, 20, 99, 75, 42, 18, 66, 88, 89, 79, 75, 65, 25, 99, 66, 78

4. Number of fish in tanks at a pet store: 1, 25, 7, 4, 54, 15, 12, 6, 2, 1, 25, 17, 20, 5, 6, 15, 24, 2, 17, 1, 5, 7, 20, 12, 12, 3

5. Number of defective light bulbs found on the assembly line during each of 20 shifts: 5, 1, 7, 6, 4, 3, 2, 1, 10, 12, 1, 2, 0, 7, 6, 2, 8, 4, 2, 0

Lesson 13-4
(pages 731–736)

Find the range, median, lower quartile, upper quartile, and interquartile range of each set of data. Identify any outliers.

1. 56, 45, 37, 43, 10, 34, 33, 45, 50

2. 77, 78, 68, 96, 99, 84, 65, 95, 65, 84

3. 30, 90, 40, 70, 50, 100, 80, 60

4. 4, 5.2, 1, 3, 2.4, 6, 3.7, 8, 1.3, 7.1, 9

5. 25°, 56°, 13°, 44°, 0°, 31°, 73°, 66°, 4°, 29°, 37°

6. 234, 648, 369, 112, 527, 775, 406, 268, 400

Lesson 13-5
(pages 737–742)

Draw a box-and-whisker plot for each set of data.

1. 3, 2, 1, 5, 7, 9, 2, 11, 3, 4, 8, 8, 10, 12, 4

2. 59, 63, 69, 69, 49, 40, 55, 69, 55, 89, 45, 55

3. 1.8, 2.2, 1.2, 3.5, 5.5, 3.2, 1.2, 4.2, 3.0, 2.6, 1.7, 1.8

4. 15, 18, 25, 37, 52, 69, 22, 35, 50, 65, 15, 99, 35, 25

Draw a parallel box-and-whisker plot for each set of data. Compare the data.

5. A: 21, 24, 34, 46, 58, 67, 72, 70, 61, 50, 40, 27
B: 67, 69, 72, 75, 79, 81, 83, 83, 82, 78, 74, 69

6. A: 100, 85, 65, 72, 83, 92, 92, 60, 99, 88, 75, 76, 92, 91, 70
B: 98, 82, 85, 62, 77, 85, 91, 95, 77, 65, 99, 73, 81, 92, 88

7. A: 3.6, 2.2, 2.2, 1.5, 1.1, 0.5, 0.8, 0.4, 0.8, 2.3, 3.0, 3.8
B: 5.4, 4.0, 3.8, 2.5, 1.8, 1.6, 0.9, 1.2, 1.9, 3.3, 5.7, 6.0

8. A: 4.75, 6.25, 7.95, 2.65, 5.25, 6.50, 8.25, 3.25, 4.25
B: 9.50, 8.65, 3.25, 5.25, 4.50, 5.75, 6.95, 5.50, 4.25

Lesson 14-1

(pages 754–758)

Draw a tree diagram to show the sample space for each event. Determine the number of possible outcomes.

1. choosing a dinner special at a restaurant offering the choice of lettuce salad or coleslaw; chicken, beef, or fish; and ice cream, pudding, or cookies
2. tossing a coin four times
3. spinning a spinner with five equal-sized sections, one each of white, yellow, blue, red, and green, two times
4. selecting a sundae with choice of vanilla or butter pecan ice cream; chocolate, strawberry, or marshmallow topping; and walnuts or peanuts

Determine the number of possible outcomes for each situation.

5. A state offers special graphic license plates. Each license plate features two digits followed by two letters. Any digit and any letter can be used in the appropriate space.
6. A lounge chair can be ordered with a choice of rocking or non-rocking, swivel or non-swivel, cotton, leather, or plush cover, and in green, blue, maroon, or black.
7. At the Big Mountain Ski Resort, you can choose from three types of boots, four types of skis, and five types of poles.
8. A game is played by rolling three four-sided dice, one red, one blue, and one white.

Find the value of each expression.

9. 8!	**10.** 1!	**11.** 0!	**12.** 5!
13. 2!	**14.** 9!	**15.** 3!	**16.** 14!

Lesson 14-2

(pages 760–767)

Determine whether each situation involves a *permutation* or *combination*. Explain your reasoning.

1. three topping flavors for a sundae from ten topping choices
2. selection and placement of four runners on a relay team from 8 runners
3. five rides to ride at an amusement park with twelve rides
4. first, second, and third place winners for a 10K race
5. a three-letter arrangement from eight different letters
6. selection of five digits from ten digits for a combination lock
7. selecting six items from twelve possible items to include in a custom gift basket

Evaluate each expression.

8. ${}_5P_2$	**9.** ${}_7P_7$	**10.** ${}_{10}C_2$	**11.** ${}_6C_5$
12. ${}_8P_2$	**13.** ${}_{18}C_{10}$	**14.** ${}_{13}C_{13}$	**15.** ${}_9P_6$
16. $({}_7P_3)({}_4P_2)$	**17.** $({}_8C_6)({}_7C_5)$	**18.** $({}_3C_2)({}_{10}P_{10})$	**19.** $({}_3P_2)({}_{10}C_{10})$

Lesson 14-3

(pages 769–776)

A red die and a blue die are rolled. Find each probability.

1. P(red 1, blue 1)
2. P(red even, blue even)
3. P(red prime number, blue even)
4. P(red 6, blue greater than 4)
5. P(red greater than 2, blue greater than 3)

At a carnival game, toy ducks are selected from a pond to win prizes. Once a duck is selected, it is not replaced. The pond contains 8 red, 2 yellow, 1 gold, 4 blue, and 40 black ducks. Find each probability.

6. P(red, then gold)	**7.** P(2 black)	**8.** P(2 yellow)
9. P(black, then gold)	**10.** P(3 blacks, then red)	**11.** P(yellow, then blue, then gold)
12. P(2 gold)	**13.** P(4 blue)	**14.** P(4 blue, then gold)

Lesson 14-4 *(pages 777–781)*

For Exercises 1–3, use the table that shows the possible products when rolling two dice and the number of ways each product can be found.

Product	Ways	Product	Ways	Product	Ways
1	1	8	2	18	2
2	2	9	1	20	2
3	2	10	2	24	2
4	3	12	4	25	1
5	2	15	2	30	2
6	4	16	1	36	1

Extra Practice

1. Draw a table to show the sample space of all possible outcomes.
2. Find the probability for $X = 9$, $X = 12$, and $X = 24$.
3. What is the probability that the product of two dice is greater than 15 on two separate rolls?

For Exercises 4–7, use the table that shows a probability distribution for the number of customers that enter a particular store during a business day.

Number of Customers	0–500	501–1000	1001–1500	1501–2000	2000–2500
Probability	0.05	0.25	0.35	0.30	0.05

4. Define a random variable and list its values.
5. Show that this is a valid probability distribution.
6. During a business day, what is the probability that fewer than 1001 customers enter?
7. During a business day, what is the probability that more than 500 customers enter?

Lesson 14-5 *(pages 782–788)*

For Exercises 1–3, toss 4 coins, one at a time, 50 times and record your results.

1. Based on your results, what is the probability that any two coins will show tails?
2. Based on your results, what is the probability that the first and fourth coins show heads?
3. What is the theoretical probability that all four coins show heads?

For Exercises 4–6, roll two dice 50 times and record the products.

4. Based on your results, what is the probability that the product is 15?
5. If you roll the dice 50 more times, which product would you expect to see about 10% of the time?
6. What is the theoretical probability that the product of the dice will be 2?

For Exercises 7–9, use the following information.
A survey was sent to randomly selected households asking the number of people living in each of the households. The results of the survey are shown in the table.

7. Find the experimental probability distribution for the number of households of each size.
8. Based on the survey, what is the probability that a person chosen at random lives in a household with five or more people?
9. Based on the survey, what is the probability that a person chosen at random lives in a household with 1 or 2 people?

Number of People Per Household Surveyed	
Number in Household	**Number of Households**
1	172
2	293
3	482
4	256
5 or more	148

Mixed Problem Solving

Chapter 1 The Language of Algebra *(pages 4–65)*

GEOMETRY For Exercises 1 and 2, use the following information.
The surface area of a cone is the sum of the product of π and the radius r squared, and the product of π, the radius r, and the slant height ℓ. *(Lesson 1-1)*

1. Write an expression that represents the surface area of the cone.
2. Suppose the radius and the slant height of a cone have the same measure r. Write an expression that represents the surface area of this cone.

SALES For Exercises 3 and 4, use the following information.
At the Farmer's Market, merchants can rent a small table for \$5.00 and a large table for \$8.50. For the first market, 25 small and 10 large tables were rented. For the second market, 35 small and 12 large were rented. *(Lesson 1-2)*

3. Write an expression to show how much money was collected for table rentals during the two markets.
4. Evaluate the expression to determine how much was collected at the two markets.

ENTERTAINMENT For Exercises 5–7, use the following information.
The Morrows are planning to go to a water park. The table shows the ticket prices. The family has 2 adults, 2 children, and a grandparent who wants to observe. They want to spend no more than \$55. *(Lesson 1-3)*

Admission Prices ($)		
Ticket	Full Day	Half Day
Adult	16.95	10.95
Child (6–18)	12.95	8.95
Observer	4.95	3.95

5. Write an inequality to show the cost for the family to go to the water park.
6. How much would it cost the Morrows to go for a full day? a half day?
7. Can the family go to the water park for a full day and stay within their budget?

RETAIL For Exercises 8–10, use the following information.
A department store is having a sale of children's clothing. The table shows the prices. *(Lesson 1-4)*

Shorts	T-Shirts	Tank Tops
\$7.99	\$8.99	\$6.99
\$5.99	\$4.99	\$2.99

8. Write three different expressions that represent 8 pairs of shorts and 8 tops.
9. Evaluate the three expressions in Exercise 8 to find the costs of the 16 items. What do you notice about all the total costs?
10. On the final sale day, if you buy 8 shorts and 8 tops, you receive a discount of 15% on the entire purchase. Find the greatest and least amount of money you can spend on the 16 items at the sale.

11. **CRAFTS** Mandy makes baby blankets and stuffed rabbits to sell at craft fairs. She sells blankets for \$28 and rabbits for \$18. Write and evaluate an expression to find her total amount of sales if she sells 25 blankets and 25 rabbits. *(Lesson 1-5)*

12. **BASEBALL** Tickets to a baseball game cost \$18.95, \$12.95, or \$9.95. A hot dog and soda combo costs \$5.50. Members of the Madison family are having a reunion. They buy 10 tickets in each price category and plan to buy 30 combos. What is the total cost for the tickets and meals? *(Lesson 1-6)*

13. **GEOMETRY** Two perpendicular lines meet to form four right angles. Write two different if-then statements for this definition. *(Lesson 1-7)*

14. **JOBS** Laurie mows lawns to earn extra money. She knows that she can mow at most 30 lawns in one week. She determines that she profits \$15 on each lawn she mows. Identify a reasonable domain and range for this situation and draw a graph. *(Lesson 1-8)*

15. **STATISTICS** Draw two graphs of the data. One graph should accurately display the data and the other should be misleading. Explain why it is misleading. *(Lesson 1-9)*

Population Density of Montana (people per square mile)	
Year	Density
1920	3.8
1960	4.6
1980	5.4
1990	5.5
2000	6.2

Source: *The World Almanac*

WEATHER For Exercises 1–3, use the following information.

The following values are the monthly normal temperatures for Barrow, Alaska. *(Lesson 2-1)*

−13	−2	19	39	−15	38
31	−11	−2	14	−18	34

Source: *The World Almanac*

1. Order the temperatures from least to greatest.
2. Write the absolute values of the twelve temperatures.
3. Do you think the temperatures are in order from January through December in the table? Why or why not?

4. **GEOGRAPHY** The highest point in Asia is Mount Everest at 29,035 feet above sea level and the lowest point is the Dead Sea at 1312 feet below sea level. What is the difference between these two elevations? **Source:** *The World Almanac* *(Lesson 2-2)*

PHYSICAL SCIENCE For Exercises 5 and 6, use the following information.

As you ascend in the Earth's atmosphere, the temperature drops about 3.6°F for every increase of 1000 feet in altitude. **Source:** www.infoplease.com *(Lesson 2-3)*

5. If you ascend 10,000 feet, what is the change in temperature?
6. If the temperature drops from 70°F at sea level to −38°F, what is the altitude you have reached?

7. **NUMBER THEORY** If a two-digit whole number is divided by the sum of its digits a certain value is obtained. For example, $\frac{71}{7+1} = 8.875$, $\frac{42}{4+2} = 7$, $\frac{10}{1+0} = 10$. Find the two-digit number that gives the least result. *(Lesson 2-4)*

WEATHER For Exercises 8–11, use the following information.

The table shows the average wind speeds for sixteen windy U.S. cities. *(Lesson 2-5)*

8.9	7.1	9.1	9.0	10.2	12.5	11.9	11.0
12.8	10.4	10.5	8.6	7.7	9.6	9.1	8.1

Source: *The World Almanac*

8. Make a stem-and-leaf plot of the data.
9. What is the difference between the least and greatest values?
10. Find the mean, median, and mode of the data.
11. Does the mode represent the data well? Explain.

POPULATION For Exercises 12–14, use the following information.

The table shows the predicted number, in millions, of people in the U.S. in each age category for 2010. Population is rounded to the nearest million. *(Lesson 2-6)*

U.S. Population			
Age	People (millions)	Age	People (millions)
under 5	20	35–44	39
5–14	39	45–54	44
15–24	43	55–64	35
25–34	39	65 & over	40

Source: *The World Almanac*

12. What is the probability that a person in the U.S. picked at random will be under age 5?
13. What are the odds that a randomly selected person will be 65 or over?
14. What is the probability that a person picked at random will not be 15–24 years old?

GARDENING For Exercises 15 and 16, use the following information.

A garden is to be created in the shape of a right triangle. The sides forming the right angle, called the legs, have lengths of 20 feet and 45 feet. The Pythagorean Theorem states that the length of the longest side, or hypotenuse, of a right triangle is the square root of the sum of the squares of the legs. *(Lesson 2-7)*

15. Find the length of the hypotenuse of the garden to the nearest foot.
16. Suppose that the gardener wants the length of the hypotenuse of the garden to be changed to 55 feet while one leg remains 45 feet. What should be the length of the other leg of the garden to the nearest foot?

SWIMMING For Exercises 17–19, use the following information.

In the 2000 summer Olympic games, the winning time for the men's 400-meter run was approximately 44 seconds. The winning time for the men's 400-meter freestyle swimming event was about 3 minutes 41 seconds. Round your answers for Exercises 17 and 18 to the nearest meter. **Source:** *The World Almanac* *(Lesson 2-4)*

17. What was the speed in meters per second for the 400-meter run?
18. What was the speed in meters per second for the 400-meter freestyle?
19. How do the speeds for the two events compare?

Mixed Problem Solving

GEOMETRY For Exercises 1–4, use the following information.
The lateral surface area L of a cylinder is two times π times the product of the radius r and the height h. *(Lesson 3-1)*

1. Write a formula for the lateral area of a cylinder.
2. Find the lateral area of a cylinder with a radius of 4.5 inches and a height of 7 inches. Use 3.14 for π and round the answer to the nearest tenth.
3. The total surface area T of a cylinder includes the area of the two bases of the cylinder, which are circles. The formula for the area of one circle is πr^2. Write a formula for the total surface area T of a cylinder.
4. Find the total surface area of the cylinder in Exercise 2. Round to the nearest tenth.

RIVERS For Exercises 5 and 6, use the following information.
The Congo River in Africa is 2900 miles long. That is 310 miles longer than the Niger River, which is also in Africa. **Source:** *The World Almanac* *(Lesson 3-2)*

5. Write an equation you could use to find the length of the Niger River.
6. What is the length of the Niger River?

ANIMALS For Exercises 7 and 8, use the following information.
The average length of a yellow-banded angelfish is 12 inches. This is 4.8 times as long as an average common goldfish. **Source:** *Scholastic Records* *(Lesson 3-3)*

7. Write an equation you could use to find the length of the common goldfish.
8. What is the length of an average common goldfish?

9. **PETS** In 1999, there were 9860 Great Danes registered with the American Kennel Club. The number of registered Labrador Retrievers was 6997 more than fifteen times the number of registered Great Danes. How many registered Labrador Retrievers were there? **Source:** *The World Almanac* *(Lesson 3-4)*

10. **ENTERTAINMENT** Four families went to a baseball game. A vendor selling bags of popcorn came by. The Wilson family bought half of the bags of popcorn plus one. The Martinez family bought half of the remaining bags of popcorn plus one. The Brightfeather family bought half of the remaining bags of popcorn plus one. The Wimberly family bought half of the remaining bags of popcorn plus one, leaving the vendor with no bags of popcorn. If the Wimberlys bought 2 bags of popcorn, how many bags did each of the four families buy? *(Lesson 3-4)*

11. **NUMBER THEORY** Five times the greatest of three consecutive even integers is equal to twice the sum of the other two integers plus 42. What are the three integers? *(Lesson 3-5)*

12. **GEOMETRY** One angle of a triangle measures 10° more than the second. The measure of the third angle is twice the sum of the measures of the first two angles. Find the measure of each angle. *(Lesson 3-5)*

13. **POOLS** Tyler needs to add 1.5 pounds of a chemical to the water in his pool for each 5000 gallons of water. The pool holds 12,500 gallons. How much chemical should he add to the water? *(Lesson 3-6)*

14. **COMPUTERS** A computer manufacturer dropped the selling price of a large-screen monitor from \$2999 to \$1999. What was the percent of decrease in the selling price of the monitor? *(Lesson 3-7)*

SKIING For Exercises 15 and 16, use the following information.
Michael is registering for a ski camp in British Columbia, Canada. The cost of the camp is \$1254, but the Canadian government imposes a general sales tax of 7%. *(Lesson 3-7)*

15. What is the total cost of the camp including tax?
16. As a U.S. citizen, Michael can apply for a refund of one-half of the tax. What is the amount of the refund he can receive?

FINANCE For Exercises 17 and 18, use the following information.
Allison is using a spreadsheet to solve a problem about investing. She is using the formula $I = Prt$, where I is the amount of interest earned, P is the amount of money invested, r is the rate of interest as a decimal, and t is the period of time the money is invested in years. *(Lesson 3-8)*

17. Allison needs to find the amount of money invested P for given amounts of interest, given rates, and given time. The formula needs to be solved for P to use in the spreadsheet. Solve the formula for P.
18. Allison uses these values in the formula in Exercise 17: $I = \$1848.75$, $r = 7.25\%$, $t = 6$ years. Find P.

19. **CHEMISTRY** Isaac had 40 gallons of a 15% iodine solution. How many gallons of a 40% iodine solution must he add to make a 20% iodine solution? *(Lesson 3-9)*

RECREATION For Exercises 1 and 2, use the following information.
A community has a recreational building and a pool. Consider the coordinates of the building to be (0, 0) and each block to be one unit. *(Lesson 4-1)*

1. If the pool lies one block south and 2 blocks east of the building, what are its coordinates?
2. If the entrance to the community lies 5 blocks north and 3 blocks west of the building, what are its coordinates?

DESIGN For Exercises 3–5, use the following information.
A T-shirt design has vertices at (1, −1), (2, 2), (0, 3), (−2, 2), and (−1, −1). *(Lesson 4-2)*

3. Draw the polygon on a coordinate plane. What polygon is represented by the design?
4. The designer wants to make smaller T-shirts using a dilation of the design by a factor of 0.75. What are the coordinates of the dilation?
5. Estimate the area of each design.

HEALTH For Exercises 6–8, use the following information.
The table shows suggested weights for adults for various heights in inches. *(Lesson 4-3)*

Height	Weight	Height	Weight
60	102	68	131
62	109	70	139
64	116	72	147
66	124	74	155

Source: *The World Almanac*

6. Graph the relation.
7. Do the data lie on a straight line? Explain.
8. Estimate a suggested weight for a person who is 78 inches tall. Explain your method.

PLANETS For Exercises 9–11, use the following information.
An astronomical unit (AU) is used to express great distances in space. It is based upon the distance from Earth to the Sun. A formula for converting any distance d in miles to AU is $AU = \frac{d}{93{,}000{,}000}$. The table shows the average distances from the Sun of four planets in miles. *(Lesson 4-4)*

Planet	Distance from Sun
Mercury	36,000,000
Mars	141,650,000
Jupiter	483,750,000
Pluto	3,647,720,000

Source: *The World Almanac*

9. Find the number of AU for each planet rounded to the nearest thousandth.
10. How can you determine which planets are further from the Sun than Earth?
11. Alpha Centauri is 270,000 AU from the Sun. How far is that in miles?

HOME DECOR For Exercises 12 and 13, use the following information.
Pam is having blinds installed at her home. The cost for installation for any number of blinds can be described by $c = 25 + 6.5x$. *(Lesson 4-5)*

12. Graph the equation.
13. If Pam has 8 blinds installed, what is the cost?

SPORTS For Exercises 14–16, use the following information.
The table shows the winning times of the Olympic mens' 50-km walk for various years. The times are rounded to the nearest minute. *(Lesson 4-6)*

Year	Years Since 1980	Time
1980	0	229
1984	4	227
1988	8	218
1992	12	230
1996	16	224
2000	20	222

Source: ESPN

14. Graph the relation using columns 2 and 3.
15. Is the relation a function? Explain.
16. Predict a winning time for the 2008 games.

JEWELRY For Exercises 17 and 18, use the following information.
A necklace is made with beads placed in a circular pattern. The rows have the following numbers of beads: 1, 6, 11, 16, 21, 26, and 31. *(Lesson 4-7)*

17. Write a formula for the beads in each row.
18. If a larger necklace is made with 20 rows, find the number of beads in row 20.

GEOMETRY

19. The table below shows the area of squares with sides of various lengths. *(Lesson 4-8)*

Side	Area	Side	Area
1	1	4	
2	4	5	
3	9	6	

Write the first 10 numbers that would appear in the area column. Describe the pattern.

FARMING **For Exercises 1–3, use the following information.**
The graph shows wheat prices per bushel from 1940 through 1999. *(Lesson 5-1)*

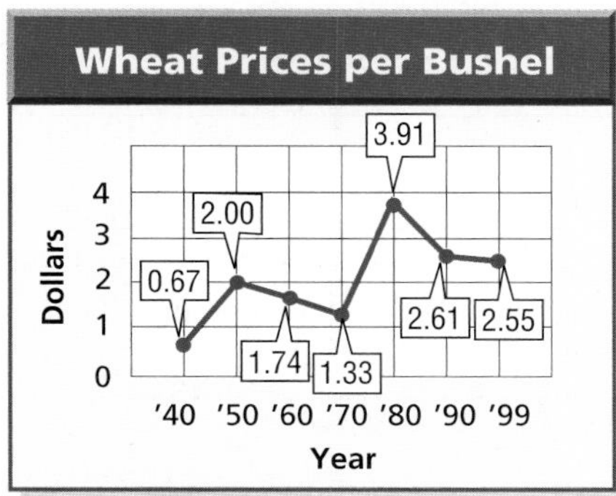

Source: *The World Almanac*

1. For which time period was the rate of change the greatest? the least?
2. Find the rate of change from 1940 to 1950.
3. Explain the meaning of the slope from 1980 to 1990.

SOUND **For Exercises 4 and 5, use the following information.**
The table shows the distance traveled by sound in water for various times in seconds. *(Lesson 5-2)*

Time (seconds) x	Distance (feet) y
0	0
1	4820
2	9640
3	14,460
4	19,280

Source: *New York Public Library*

4. Write an equation that relates distance traveled to time.
5. Find the time for a distance of 72,300 feet.

POPULATION **For Exercises 6–8, use the following information.**
In 1990, the population of Wyoming was 453,589. Over the next decade, it increased by about 2890 per year. **Source:** *The World Almanac* *(Lesson 5-3)*

6. Assume the rate of change remains the same. Write a linear equation to find the population y of Wyoming at any time. Let x represent the number of years since 1990.
7. Graph the equation.
8. Estimate the population in 2005.

HEALTH **For Exercises 9 and 10, use the following information.**
A chart shows ideal heights and weights for adults with a medium build. A person with height of 60 inches should have a weight of 112 pounds and a person with height of 66 inches should have a weight of 136 pounds. **Source:** *The World Almanac* *(Lesson 5-4)*

9. Write a linear equation to estimate the weight of a person of any height.
10. Estimate the weight of a person who is 72 inches tall.

TRAVEL **For Exercises 11–13, use the following information.**
Between 1990 and 2000, the number of people taking cruises increased by about 300,000 each year. In 1990, about 3.6 million people took a cruise. **Source:** *USA TODAY* *(Lesson 5-5)*

11. Write the point-slope form of an equation to find the total number of people taking a cruise y for any year x.
12. Write the equation in slope-intercept form.
13. Estimate the number of people who will take a cruise in 2010.

GEOMETRY **For Exercises 14 and 15, use the following information.**
A quadrilateral has sides with equations $y = -2x$, $2x + y = 6$, $y = \frac{1}{2}x + 6$, and $x - 2y = 9$. Graph the four equations to form the quadrilateral.

14. Determine whether the figure is a rectangle.
15. Explain your reasoning. *(Lesson 5-6)*

ADOPTION **For Exercises 16–18, use the following information.**
The table shows the number of children from Russia adopted by U.S. citizens from 1992–1999. The x values are shown as Years Since 1992. *(Lesson 5-7)*

Years Since 1992 x	Number of Children y
0	324
1	746
2	1530
3	1896
4	2454
5	3816
6	4491
7	4348

Source: *The World Almanac*

16. Draw a scatter plot and a line of fit for the data.
17. Write the slope-intercept form of the equation for the line of fit.
18. Predict the number of children who will be adopted in 2005.

MONEY For Exercises 1 and 2, use the following information.
Scott's allowance for July is \$50. He wants to attend a concert that costs \$26. *(Lesson 6–1)*

1. Write and solve an inequality that shows how much money he can spend in July after buying a concert ticket.
2. He spends \$2.99 for lunch with his friends and \$12.49 for a CD. Write and solve an inequality that shows how much money he can spend after these purchases and the concert ticket.

ANIMALS For Exercises 3–5, use the following information.
The world's heaviest flying bird is the great bustard. A male bustard can be up to 4 feet long and weigh up to 40 pounds. *(Scholastic Book of World Records)* *(Lesson 6-2)*

3. Write an inequality that describes the range of lengths of male bustards.
4. Write an inequality that describes the range of weights of male bustards.
5. Male bustards are usually about four times heavier than females. Write and solve an inequality that describes the range of weights of female bustards.

FOOD For Exercises 6–8, use the following information.
Jennie wants to make at least \$75 selling caramel-coated apples at the County Fair. She plans to sell each apple for \$1.50. *(Lesson 6-3)*

6. Let a be the number of apples she makes and sells. Write an inequality to find the number of apples she needs to sell to reach her goal if each apple costs her \$0.30 to make.
7. Solve the inequality.
8. Interpret the meaning of the solution to the inequality.

RETAIL For Exercises 9–11, use the following information.
A sporting goods store is printing coupons that allow the customer to save \$15 on any pair of shoes in the store. *(Lesson 6-4)*

9. The most expensive pair of shoes is \$149.95 and the least expensive pair of shoes is \$24.95. What is the range of prices for the shoes for customers who have the coupons?
10. You decide to buy a pair of shoes with a regular price of \$109.95. You have a choice of using the coupon or having a 15% discount on the price. Which option should you choose?
11. For what price of shoe is a 15% discount the same as \$15 off the regular price?

WEATHER For Exercises 12–15, use the following information.
The table shows the average normal temperatures for Honolulu, Hawaii, for each month in degrees Fahrenheit. *(Lesson 6-5)*

January	73	July	81
February	73	August	81
March	74	September	81
April	76	October	80
May	78	November	77
June	79	December	74

Source: *The World Almanac*

12. What is the mean of the temperatures to the nearest whole degree?
13. By how many degrees does the lowest temperature vary from the mean?
14. By how many degrees does the highest temperature vary from the mean?
15. Write an inequality to show the normal range of temperatures for Honolulu during the year.

QUILTING For Exercises 16–18, use the following information.
Ingrid is making a quilt in the shape of a rectangle. She wants the perimeter of the quilt to be no more than 318 inches. *(Lesson 6-6)*

16. Write an inequality that represents this situation.
17. Graph the inequality and name two different dimensions for the quilt.
18. What are the dimensions and area of the largest possible quilt Ingrid can make with a perimeter of no more than 318 inches?

GEOGRAPHY For Exercises 19–21, use the following information.
The table shows the area of land in square miles and in acres of the largest and smallest U.S. states. *(Lesson 6-4)*

State	Square Miles	Acres
Alaska	570,473	365,481,600
Rhode Island	1045	677,120

Source: *The World Almanac and U.S.A. Almanac*

19. Write an inequality that shows the range of square miles for U.S. states.
20. Write an inequality that shows the range of acres for U.S. states.
21. **RESEARCH** About how many acres are in a square mile? Do the figures in the table agree with that fact?

WORKING For Exercises 1–3, use the following information.
The table shows the percent of men and women 65 years and older that were working in the U.S. in the given years. *(Lesson 7-1)*

U.S. Workers over 65		
Year	Percent of Men	Percent of Women
1980	19.3	8.2
1990	17.6	8.4

Source: *The World Almanac*

1. Let the year 1980 be 0. Assume that the rate of change remains the same for years after 1990. Write an equation to represent the percent of working elderly men y in any year x.
2. Write an equation to represent the percent of working elderly women.
3. Assume the rate of increase or decrease in working men and women remains the same for years after 1990. Estimate when the percent of working men and women will be the same.

SPORTS For Exercises 4–7, use the following information.
The table shows the winning times for the men's and women's Triathlon World Championship for 1995 and 2000. *(Lesson 7-2)*

Year	Men's	Women's
1995	1:48:29	2:04:58
2000	1:51:41	1:54:43

Source: *ESPN Sports Almanac*

4. The times in the table are in hours, minutes, and seconds. Rewrite the times in minutes rounded to the nearest minute.
5. Let the year 1995 be 0. Assume that the rate of change remains the same for years after 1995. Write an equation to represent the men's winning times y in any year x.
6. Write an equation to represent the women's winning times in any year.
7. If the trend continues, when would you expect the men's and women's winning times to be the same?

8. **TRAVEL** While driving to Fullerton, Mrs. Sumner travels at an average speed of 40 mph. On the return trip, she travels at an average speed of 56 mph and saves two hours of travel time. How far does Mrs. Sumner live from Fullerton? *(Lesson 7-2)*

MONEY For Exercises 9–11, use the following information.
In 1998, the sum of the number of \$2 bills in circulation and the number of \$50 bills in circulation was 1,500,888,647. The number of \$50 bills was 366,593,903 more than the number of \$2 bills. *(Lesson 7-3)* **Source:** *The World Almanac for Kids*

9. Write a system of equations to represent this situation.
10. Find the number of each type of bill in circulation.
11. Find the amount of money that was in circulation in \$2 and \$50 bills.

SPORTS For Exercises 12–15, use the following information.
In the 2000 Summer Olympic Games, the total number of gold and silver medals won by the U.S. was 64. Gold medals are worth 3 points and silver medals are worth 2 points. The total points scored for gold and silver medals was 168. *(Lesson 7-4)*
Source: *ESPN Almanac*

12. Write an equation for the sum of the number of gold and silver medals won by the U.S.
13. Write an equation for the sum of the points earned by the U.S. for gold and silver medals.
14. How many gold and silver medals did the U.S. win?
15. The total points scored by the U.S. was 201. Bronze medals are worth 1 point. How many bronze medals were won?

RADIO For Exercises 16–20, use the following information.
KSKY radio station is giving away tickets to an amusement park as part of a summer promotion. Each child ticket costs \$15 and each adult ticket costs \$20. The station wants to spend no more than \$800 on tickets. They also want the number of child tickets to be greater than twice the number of adult tickets. *(Lesson 7-5)*

16. Write an inequality for the total cost of c child tickets and a adult tickets.
17. Write an inequality to represent the relationship between the number of child and adult tickets.
18. Write two inequalities that would assure you that the number of adult and the number of child tickets would not be negative.
19. Graph the system of four inequalities to show possible numbers of tickets that the station can buy.
20. Give three possible combinations of child and adult tickets for the station to buy.

GEOMETRY For Exercises 1–4, use the following information.
If the side length of a cube is s, then the volume is presented by s^3 and the surface area is represented by $6s^2$. *(Lesson 8-1)*

1. Are the expressions for volume and surface area monomials? Explain.
2. If the side of a cube measures 3 feet, find the volume and surface area.
3. Find a side length s such that the volume and surface area have the same numerical value.
4. The volume of a cylinder can be found by multiplying the radius squared times the height times π, or $V = \pi r^2 h$. Suppose you have two cylinders. Each measure of the second is twice the measure of the first, so $V = \pi(2r)^2(2h)$. What is the ratio of the volume of the first cylinder to the second cylinder? *(Lesson 8-2)*

LIGHT For Exercises 5–7, use the table that shows the speed of light in various materials. *(Lesson 8-3)*

Material	Speed m/s
vacuum	3.00×10^8
air	3.00×10^8
ice	2.29×10^8
glycerine	2.04×10^8
crown glass	1.97×10^8
rock salt	1.95×10^8

Source: *Glencoe Physics*

5. Express each speed in standard notation.
6. To the nearest hundredth, how many times as fast does light travel in a vacuum as in rock salt?
7. Through which material does light travel about 1.17 times as fast as through rock salt?

POPULATION For Exercises 8–10, use the following information.
The table shows the population density for the state of Nevada for various years. *(Lesson 8-4)*

Year	Years Since 1920	People/Square Mile
1920	0	0.7
1960	40	2.6
1980	60	7.3
1990	70	10.9

Source: *The World Almanac*

8. The population density d of Nevada from 1920 to 1990 can be modeled by $d = 0.003y^2 - 0.086y + 0.708$, where y represents the number of years since 1920. Identify the type of polynomial for $0.003y^2 - 0.086y + 0.708$.
9. What is the degree of this polynomial?
10. Predict the population density of Nevada for the year 2010. Explain your method.

RADIO For Exercises 11 and 12, use the following information.
From 1997 to 2000, the number of radio stations presenting primarily news and talk N and the total number of radio stations of all types R in the U.S. could be modeled by the following equations, where x is the number of years since 1997. *(Lesson 8-5)*
Source: *The World Almanac*

$N = 37.9x + 1315.9$
$R = 133.5x + 10{,}278.5$

11. Find an equation that models the number of radio stations O that are *not* primarily news and talk in the U.S. for this time period.
12. If this trend continues, how many radio stations that are not news and talk will there be in the year 2015?

GEOMETRY For Exercises 13–15, use the following information.
The number of diagonals of a polygon can be found by using the formula $d = 0.5n(n - 3)$, where d is the number of diagonals and n is the number of sides of the polygon. *(Lesson 8-6)*

13. Use the Distributive Property to write the expression as a polynomial.
14. Find the number of diagonals for polygons with 3 through 10 sides.
15. Describe any patterns you see in the numbers you wrote in Exercise 14.

GEOMETRY For Exercises 16 and 17, use the following information.
A rectangular prism has dimensions of x, $x + 3$, and $2x + 5$. *(Lesson 8-7)*

16. Find the volume of the prism in terms of x.
17. Choose two values for x. How do the volumes compare?

MONEY For Exercises 18–20, use the following information.
Money invested in a certificate of deposit or CD collects interest once per year. Suppose you invest $4000 in a 2-year CD. *(Lesson 8-8)*

18. If the interest rate is 5% per year, the expression $4000(1 + 0.05)^2$ can be evaluated to find the total amount of money you will have at the end of two years. Explain the numbers in this expression.
19. Find the amount of money at the end of two years.
20. Suppose you invest $10,000 in a CD for 4 years at a rate of 6.25%. What is the total amount of money you will have at the end of 4 years?

FLOORING For Exercises 1 and 2, use the following information.
Eric is refinishing his dining room floor. The floor measures 10 feet by 12 feet. Flooring World offers a wood-like flooring in 1-foot by 1-foot squares, 2-foot by 2-foot squares, 3-foot by 3-foot squares, and 2-foot by 3-foot rectangular pieces. *(Lesson 9-1)*

1. Without cutting the pieces, which of the four types of flooring can Eric use in the dining room? Explain.
2. The price per piece of each type of flooring is shown in the table. If Eric wants to spend the least money, which should he choose? What will be the total cost of his choice?

Size	1×1	2×2	3×3	2×3
Price	\$3.75	\$15.00	\$32.00	\$21.00

FIREWORKS For Exercises 3–5, use the following information.
At a Fourth of July celebration, a rocket is launched with an initial velocity of 125 feet per second. The height h of the rocket in feet above sea level is modeled by the formula $h = 125t - 16t^2$, where t is the time in seconds after the rocket is launched. *(Lesson 9-2)*

3. What is the height of the rocket when it returns to the ground?
4. Let $h = 0$ in the equation $h = 125t - 16t^2$ and solve for t.
5. How many seconds will it take for the rocket to return to the ground?

FOOTBALL For Exercises 6–8, use the following information.
Some small high schools play six-man football as a team sport. The dimensions of the field are less then the dimensions of a standard football field. Including the end zones, the length of the field, in feet, is 60 feet more than twice the width. *(Lesson 9-3)*

6. Write an expression for the area of the six-man football field.
7. If the area of the field is 36,000 square feet, what are the dimensions of the field? (*Hint*: Factor a 2 out of the equation before factoring.)
8. What are the dimensions of the field in yards?

PHYSICAL SCIENCE For Exercises 9 and 10, use the following information.
Teril throws a ball upward while standing on the top of a 500-foot tall apartment building. Its height h, in feet, after t seconds is given by the equation $h = -16t^2 + 48t + 506$. *(Lesson 9-4)*

9. What do the values 48 and 506 in the equation represent?
10. The ball falls on a balcony that is 218 feet above the ground. How many seconds was the ball in the air?

DECKS For Exercises 11 and 12, use the following information.
Zelda is building a deck in her back yard. The plans for the deck show that it is to be 24 feet by 24 feet. Zelda wants to reduce one dimension by a number of feet and increase the other dimension by the same number of feet. *(Lesson 9-5)*

11. If the area of the reduced deck is 512 square feet, what are the dimensions of the deck?
12. Suppose Zelda wants to reduce the deck to one-half the area of the deck in the plans. Can she reduce each dimension by the same length and use dimensions that are whole numbers? Explain.

BUILDINGS For Exercises 13–15, use the following information.
The Petronas Towers I and II in Kuala Lumpur, Malaysia, are both 1483 feet tall. A model for the height h in feet of a dropped object is $h = -16t^2$, where t is the time in seconds after the object is dropped. *(Lesson 9-6)* **Source:** *The World Almanac*

13. To the nearest tenth of a second, how long will it take for an object dropped from the top of one of the towers to hit the ground?
14. In 1900, the tallest building in the world was the Park Row Building in New York City with a height of 386 feet. How much longer will it take an object to reach the ground from the Petronas Tower I than from the Park Row Building?
15. If a new building is built such that an object takes 12 seconds to reach the ground when dropped from the top, how tall is the building?

POOLS For Exercises 16–19, use the following information.
Susan wants to buy an aboveground swimming pool for her yard. Model A is 42 inches deep and holds 1750 cubic feet of water. The length of the pool is 5 feet more than the width. *(Lesson 9-6)*

16. What is the area of water that is exposed to the air?
17. What are the dimensions of the pool?
18. A Model B pool holds twice as much water as Model A. What are some possible dimensions for this pool?
19. Model C has length and width that are both twice as long as Model A, but the height is the same. What is the ratio of the volume of Model A to Model C?

PHYSICAL SCIENCE For Exercises 1–4, use the following information.
A ball is released 6 feet above the ground and thrown vertically into the air. The equation $h = -16t^2 + 112t + 6$ gives the height of the ball if the initial velocity is 112 feet per second. *(Lesson 10-1)*

1. Write the equation of the axis of symmetry and find the coordinates of the vertex of the graph of the equation.
2. What is the maximum height above the ground that the ball reaches?
3. How many seconds after release does the ball reach its maximum height?
4. How many seconds is the ball in the air?

RIDES For Exercises 5–7, use the following information.
At an amusement park in Minnesota a popular ride whisks riders to the top of a 250-foot tower and drops them at speeds exceeding 50 miles per hour. A function for the path of a rider is $h = -16t^2 + 250$, where h is the height and t is the time in seconds. *(Lesson 10-3)*

5. The ride stops the descent of the rider 40 feet above the ground. Write an equation that models the drop of the rider.
6. Solve the equation by graphing the related function. How many roots does the equation have?
7. About how many seconds does it take to complete the ride?

PROJECTS For Exercises 8–10, use the following information.
Jude is making a poster for his science project. The poster board is 22 inches wide by 27 inches tall. He wants to cover two thirds of the area with text or pictures and leave a top margin 3 times as wide as the side margins and a bottom margin twice as wide as the side margins. *(Lesson 10-3)*

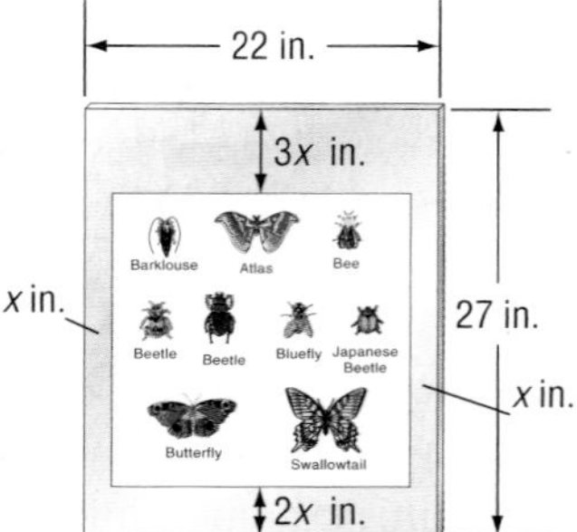

8. Write an equation that represents this situation.
9. Solve your equation for x by completing the square. Round to the nearest tenth.
10. What should be the widths of the margins?

TELEVISION For Exercises 11 and 12, use the following information.
The number of U.S. households with cable television has been on the rise. The percent of households with cable y can be approximated by the quadratic function $y = -0.11x^2 + 4.95x + 12.69$, where x stands for the number of years after 1977. *(Lesson 10-4)*

11. Use the Quadratic Formula to solve for x when $y = 30$. What do these values represent?
12. Do you think a quadratic function is a good model for this data? Why or why not?

POPULATION For Exercises 13–15, use the following information.
The population of Asia from 1650 to 2000 can be estimated by the function $P(x) = 335(1.007)^x$, where x is the number of years since 1650 and the population is in millions of people. *(Lesson 10-5)*

13. Graph the function and name the y-intercept.
14. What does the y-intercept represent in this problem?
15. Use the function to approximate the number of people in Asia in 2050.

MONEY For Exercises 16–18, use the following information.
In 1999, Aaron placed \$10,000 he received as an inheritance in a 4-year certificate of deposit at an interest rate of 7.45% compounded yearly. *(Lesson 10-6)*

16. Aaron plans to take all the money out of his investment at the end of 4 years. Find the amount of money Aaron will have at the end of 4 years.
17. He plans to use the money for college tuition. From 1999 on, it is predicted that tuition will rise 6% per year from the 1999 cost of \$2575. Aaron intends to begin college in 2003 and attend for 4 years. What will be his total tuition cost?
18. What recommendations would you make to Aaron for paying for the total cost of his tuition?

TRAINING For Exercises 19–22, use the following information.
Laurie wants to run a 5K race but has never run before. A 5K race is about 3 miles so she wants to work up slowly to running 3 miles. *(Lesson 10-7)*

19. Laurie's trainer advises her to run every other day and to begin by running one eighth of a mile. Each running session she is to run one and a half times her previous distance. Write the first 10 terms of this sequence.
20. Write a formula for the nth term of this sequence.
21. During which session will Laurie exceed 3 miles?
22. Will Laurie be ready for a 5K race in two weeks from the start of her training program?

SATELLITES For Exercises 1–3, use the following information.
A satellite is launched into orbit 200 kilometers above Earth. The orbital velocity of a satellite is given by the formula $v = \sqrt{\frac{Gm_E}{r}}$, where v is velocity in meters per second, G is a given constant, m_E is the mass of Earth, and r is the radius of the satellite's orbit. *(Lesson 11-1)*

1. The radius of Earth is 6,380,000 meters. What is the radius of the satellite's orbit in meters?
2. The mass of Earth is 5.97×10^{24} kilogram and the constant G is 6.67×10^{-11} N · m²/kg² where N is in Newtons. Use the formula to find the orbital velocity of the satellite in meters per second.
3. The orbital period of the satellite can be found by using the formula $T = \frac{2\pi r}{v}$, where r is the radius of the orbit and v is the orbital velocity of the satellite in meters per second. Find the orbital period of the satellite in hours.

RIDES For Exercises 4–6, use the following information.
The designer of a roller coaster must consider the height of the hill and the velocity of the coaster as it travels over the hill. Certain hills give riders a feeling of weightlessness. The formula $d = \sqrt{\frac{2hv^2}{g}}$ allows designers to find the correct distance from the center of the hill that the coaster should begin its drop for maximum fun. *(Lesson 11-2)*

4. In the formula above, d is the distance from the center of the hill, h is the height of the hill, v is the velocity of the coaster at the top of the hill in meters per second, and g is a gravity constant of 9.8 meters per second squared. If a hill is 10 meters high and the velocity of the coaster is 10 m/s, find d.
5. Find d if the height of the hill is 10 meters but the velocity is 20 m/s. How does d compare to the value in Exercise 4?
6. Suppose you find the same formula in another book written as $d = 1.4\sqrt{\frac{hv^2}{g}}$. Will this produce the same value of d? Explain.
7. **PACKAGING** A cylindrical container of chocolate drink mix has a volume of about 162 in³. The formula for volume of a cylinder is $V = \pi r^2 h$, where r is the radius and h is the height. The radius of the container can be found by using the formula $r = \sqrt{\frac{V}{\pi h}}$. If the height is 8.25 inches, find the radius of the container. *(Lesson 11-3)*

TOWN SQUARES For Exercises 8 and 9, use the following information.
Tiananmen Square in Beijing, China, is the largest town square in the world, covering 98 acres.
Source: *The Guinness Book of Records* *(Lesson 11-4)*

8. One square mile is 640 acres. Assuming that Tiananmen Square is a square, how many feet long is a side to the nearest foot?
9. To the nearest foot, what is the diagonal distance across Tiananmen Square?

PIZZA DELIVERY For Exercises 10 and 11, use the following information.
The Pizza Place delivers pizza to any location within a radius of 5 miles from the store for free. Tyrone drives 32 blocks north and then 45 blocks east to deliver a pizza. In this city, there are about 6 blocks per half mile. *(Lesson 11-5)*

10. Should there be a charge for the delivery? Explain.
11. Describe two delivery situations that would result in about 5 miles.

GEOMETRY For Exercises 12–14, use the following information.
A triangle on the coordinate plane has vertices (1, 1), (−3, 2), and (−7, −5). *(Lesson 11-6)*

12. What is the perimeter of the triangle? Express the answer in simplest radical form and as a decimal approximation rounded to the nearest hundredth.
13. Suppose a new triangle is formed by multiplying each coordinate by 2. What is the perimeter of the new triangle in simplest radical form and as a decimal rounded to the nearest hundredth?
14. Are the two triangles similar? Explain your reasoning.

ESCALATORS For Exercises 15 and 16, use the following information.
The longest escalator is located in Hong Kong, China. The escalator has a length of 745 feet and rises 377 feet vertically from start to finish. **Source:** *The Guinness Book of Records* *(Lesson 11-7)*

15. Draw a diagram of the escalator.
16. To the nearest degree, what is the angle of elevation of the escalator?

OPTOMETRY For Exercises 1–4, use the following information.
When a person does not have clear vision either at a distance or close up, an optometrist can prescribe lenses to correct the condition. The power P of a lens, in a unit called diopters, is equal to 1 divided by the focal length f, in meters, of the lens. The formula is $P = \frac{1}{f}$. *(Lesson 12-1)*

1. Graph the inverse variation $P = \frac{1}{f}$.
2. Find the power of a lens with focal length +20 centimeters. (*Hint*: Change 20 centimeters to meters.)
3. Find the power of a lens with focal length −40 centimeters. (*Hint*: Change 40 centimeters to meters.)
4. What do you notice about the powers in Exercises 2 and 3?

PHYSICS For Exercises 5 and 6, use the following information.
Some principles in physics, such as gravitational force between two objects, depend upon a relationship known as the inverse square law. The inverse square law means that two variables are related by the relationship $y = \frac{1}{x^2}$, where x is distance. *(Lesson 12-2)*

5. Make a table of values and graph $y = \frac{1}{x^2}$. Describe the shape of the graph.
6. If x represents distance, how does this affect the domain of the graph?

FERRIS WHEELS For Exercises 7–9, use the following information.
George Ferris built the first Ferris wheel for the World's Columbian Exposition in Chicago in 1892. It had a diameter of 250 feet. *(Lesson 12-3)*

7. To find the speed traveled by a car located on the circumference of the wheel, you can find the circumference of a circle and divide by the time it takes for one rotation of the wheel. (Recall that $C = \pi d$.) Write a rational expression for the speed of a car rotating in time t.
8. Suppose the first Ferris wheel rotated once every 5 minutes. What was the speed of a car on the circumference in feet per minute?
9. Use dimensional analysis to find the speed of a car in miles per hour.
10. **MOTOR VEHICLES** In 1999, the U.S. produced 13,063,000 motor vehicles. This was 23.2% of the total motor vehicle production for the whole world. How many motor vehicles were produced worldwide in 1999? **Source:** *The World Almanac* *(Lesson 12-4)*
11. **LIGHT** The speed of light is approximately 1.86×10^5 miles per second. The table shows the distances, in miles, of the planets from the Sun. Find the amount of time in minutes that it takes for light from the Sun to reach each planet. *(Lesson 12-5)*

Planet	Miles	Planet	Miles
Mercury	5.79×10^{10}	Jupiter	7.78×10^{11}
Venus	1.08×10^{11}	Saturn	1.43×10^{12}
Earth	1.496×10^{11}	Uranus	2.87×10^{12}
Mars	2.28×10^{11}	Pluto	4.50×10^{12}

12. **GEOGRAPHY** The land areas of all the continents, in thousands of square miles, are given in the table. Use this information to write the fraction of the land area of the world that is part of North and South America. *(Lesson 12-6)*

Continent	Area
North America	9400
South America	6900
Europe	3800
Asia	17,400
Africa	11,700
Oceania	3300
Antarctica	5400

Source: *The World Almanac*

13. **GARDENING** Celeste builds decorative gardens in her landscaping business. She uses either 35, 50, or 75 bricks for one garden depending upon the design. What is the least number of bricks she should order that would allow her to build a whole number of each type of garden? *(Lesson 12-7)*

CRAFTS For Exercises 14 and 15, use the following information.
Jordan and her aunt Jennie make tablecloths to sell at craft fairs. A small one takes one-half yard of fabric, a medium takes five-eighths yard, and a large takes one and one-quarter yard. *(Lesson 12-8)*

14. How many yards of fabric do they need to make one of each type of tablecloth?
15. A particular bolt of fabric contains 30 yards of fabric. Can they use the entire bolt by making an equal number of each type of tablecloth? Explain.
16. **CONSTRUCTION** Rick has a crew of workers that can side a particular size house in 6 days. Phil's crew can side the same house in 4 days. If the two crews work together, how long will it take to side the house? *(Lesson 12-9)*

CAREERS For Exercises 1 and 2, use the following information.

The graph below shows the results of a survey of students asking their preferences for a future career. *(Lesson 13-1)*

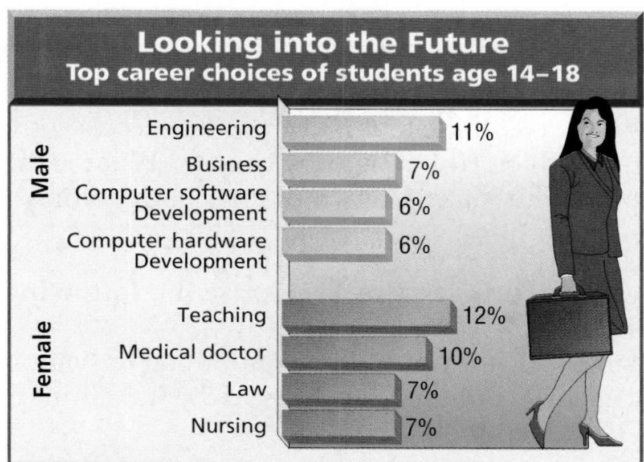

Source: *USA TODAY*

1. Write a statement to describe what you do know about the sample.
2. What additional information would you like to have about the sample to determine whether the sample is biased?

POPULATIONS For Exercises 3–7, use the following information.

The table shows the populations for ten U.S. cities in 1990 and 1999. *(Lesson 13-2)*

City	1990	1999
Anchorage, AK	226,338	257,808
Asheville, NC	191,310	215,180
Elmira, NY	95,195	91,738
Gainesville, FL	181,596	198,484
Great Falls, MT	77,691	78,282
Kokomo, IN	96,946	100,377
Lawton, OK	111,486	106,621
Macon, GA	291,079	321,586
Modesto, CA	370,522	436,790
Pine Bluff, AR	85,487	80,785

Source: U.S. Census Bureau

3. Create matrix A for the 1999 data and matrix B for the 1990 data.
4. What are the dimensions of each matrix in Exercise 3?
5. Calculate $P = A - B$.
6. What does matrix P represent?
7. Which city had the greatest percent decrease in population from 1990 to 1999?

WEATHER For Exercises 8–11, use the table that shows the highest and lowest (H/L) temperature ever recorded in each U.S. state.

State	H/L (°F)	State	H/L (°F)	State	H/L (°F)
AL	112/−27	LA	114/−16	OH	113/−39
AK	100/−80	ME	105/−48	OK	120/−27
AZ	128/−29	MD	109/−40	OR	119/−54
AR	120/−29	MA	107/−35	PA	111/−42
CA	134/−45	MI	112/−51	RI	104/−25
CO	118/−61	MN	114/−60	SC	111/−19
CT	106/−32	MS	115/−19	SD	120/−58
DE	110/−17	MO	118/−40	TN	113/−32
FL	108/−2	MT	117/−70	TX	120/−23
GA	112/−17	NE	118/−47	UT	117/−69
HI	100/12	NV	125/−50	VT	105/−50
ID	118/−60	NH	106/−46	VA	110/−30
IL	117/−36	NJ	110/−34	WA	118/−48
IN	116/−36	NM	122/−50	WV	112/−37
IA	118/−47	NY	108/−52	WI	114/−54
KS	121/−40	NC	110/−34	WY	114/−66
KY	114/−37	ND	121/−60		

Source: *The World Almanac*

8. Consider the high temperature data. *(Lesson 13-3)*
   a. Determine the median of the data.
   b. Create a histogram to represent the data. Use at least four measurement classes.
   c. Write a sentence or two describing the distribution of the data.
   d. Does finding the median of the set of data help you to make a histogram for the data? Explain.
9. Consider the low temperature data. *(Lesson 13-4)*
   a. What is the range of the temperature data?
   b. What is the lower quartile and the upper quartile of the data?
   c. What is the interquartile range of the data?
   d. Name any outliers.
10. Draw a parallel box-and-whisker plot for the high and low temperatures. *(Lesson 13-5)*
    a. Compare the data in the two plots.
    b. How does the range of the high temperature data compare to the range of the low temperature data?
11. Make a table that shows the differences between the highest and lowest temperatures for each state. *(Lesson 13-5)*
    a. Create any graph of your choice that shows the difference between the high and low temperature for each state.
    b. Describe the data in your graph in part **a**.

FLOWERS For Exercises 1–3, use the following information.
A flower shop is making special floral arrangements for a holiday. The table shows the options available and the costs of each option. *(Lesson 14-1)*

Vase	Deluxe	Standard	Economy
Cost	$12.00	$8.00	$5.00

Ribbon	Velvet	Satin
Cost	$3.00	$2.00

Flowers	Orchids	Roses	Daisies
Cost	$35.00	$20.00	$12.00

Card	Large	Small
Cost	$2.50	$1.75

1. How many floral arrangements are possible? Each arrangement has one vase, one ribbon, one type of flowers, and one card.
2. What is the cost of the most expensive arrangement? the least expensive?
3. What is the cost of each of the four most expensive arrangements?

GAMES For Exercises 4–6, use the following information.
Melissa is playing a board game that requires you to make words to score points. There are 12 letters left in the box and she must choose 4. She cannot see the letters that can be chosen. *(Lesson 14-2)*

4. Suppose that the 12 letters are all different. In how many ways can she choose 4 of the 12 letters?
5. She chooses the four letters A, T, R, and E. How many different arrangements of three letters can she make from her letters?
6. How many of the three-letter arrangements are words? List the words you find.

BASEBALL For Exercises 7–10, use the following information.
During the 2000 baseball season, these Houston Astros players had the following number of times at bat and hits. You can consider the probability that a player gets a hit as the number of hits compared to the number of times at bat. Round each probability to the nearest hundredth for Exercises 7–10. *(Lesson 14–3)*

Name	Times at Bat	Hits
Alou	454	161
Ward	264	68
Cedeno	259	73
Hidalgo	558	175

Source: *ESPN*

7. On his next at bat, what is the probability that Hidalgo will get a hit?
8. Which player has the greatest chance to get a hit on his next at bat?
9. Suppose Ward and then Cedeno are to be the first players at bat in a new inning. What is the probability that both get a hit?
10. If the manager wants the greatest probability that two of these four players will get consecutive hits, which two should he choose? What is the probability of these two players both getting a hit?

DRIVING For Exercises 11–13, use the following information.
The table shows a probability distribution for various age categories of licensed drivers in the U.S. for the year 1998. *(Lesson 14-4)*

X = Age Category	Probability
under 20	0.053
20–34	0.284
35–49	0.323
50–64	0.198
65 and over	0.142

Source: *The World Almanac*

11. Determine whether this is a valid probability distribution. Justify your answer.
12. If a driver in the U.S. is randomly selected, what is the probability that the person is under 20 years old?
13. If a driver in the U.S. is randomly selected, what is the probability the person is 50 years old or over?

LOTTERIES For Exercises 14–16, use the following information.
A state sells lottery tickets, each with a five-digit number such that each digit can be 1–6. When you purchase a ticket, you select a number that you think will win and it is printed on your ticket. Then, once per week, a random 5-digit number is generated as the winning number. *(Lesson 14-5)*

14. How many five-digit numbers are possible? Explain how you calculated the number of possible outcomes.
15. Perform a simulation for winning the lottery. Describe the objects you used to perform the simulation.
16. According to your experiment, if you buy one ticket, what is the experimental probability of winning the lottery?

Glossary/Glosario

Cómo usar el glosario en español:
1. Busca el término en inglés que desees encontrar.
2. El término en español, junto con la definición, se encuentran en la columna de la derecha.

English / Español

A

absolute value (p. 69) The absolute value of a number n is its distance from zero on a number line and is represented by $|n|$.

valor absoluto El valor absoluto de un número n es la distancia que n dista de cero en una recta numérica. Se denota con $|n|$.

additive identity (p. 21) For any number a, $a + 0 = 0 + a = a$.

identidad de la adición Para cualquier número a, $a + 0 = 0 + a = a$.

additive inverses (p. 75) A number and its opposite are additive inverses of each other. The sum of a number and its additive inverse is 0.

inversos aditivos Un número y su opuesto son inversos aditivos mutuos. La suma de un número y su inverso aditivo es 0.

algebraic expression (p. 6) An expression consisting of one or more numbers and variables along with one or more arithmetic operations.

expresión algebraica Una expresión que consiste en uno o más números y variables, junto con una o más operaciones aritméticas.

angle of depression (p. 626) The angle formed by a horizontal line of sight and a line of sight below it.

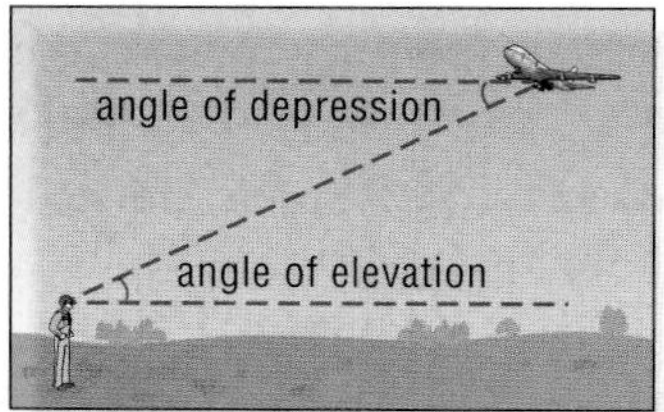

ángulo de depresión Ángulo formado por una línea visual horizontal y otra línea visual debajo de ella.

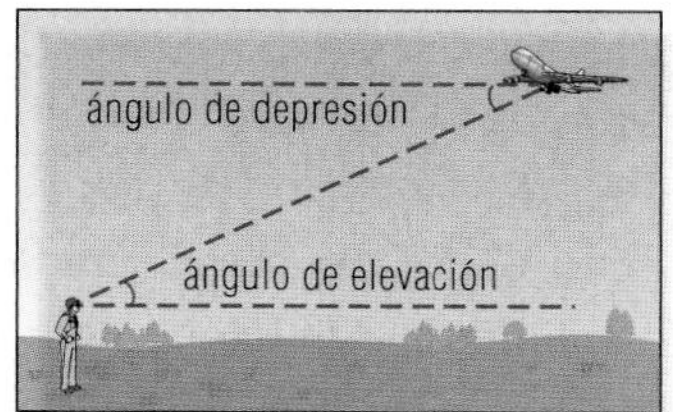

angle of elevation (p. 626) The angle formed by a horizontal line of sight and a line of sight above it.

ángulo de elevación Ángulo formado por una línea visual horizontal y otra línea visual sobre ella.

arithmetic sequence (p. 233) A numerical pattern that increases or decreases at a constant rate or value. The difference between successive terms of the sequence is constant.

sucesión aritmética Un patrón numérico que aumenta o disminuye a una tasa o valor constante. La diferencia entre términos consecutivos de la sucesión es siempre la misma.

axes (p. 192) The two perpendicular number lines in a coordinate system.

ejes Las dos rectas numéricas perpendiculares que forman un sistema de coordenadas.

axis of symmetry (p. 525) The vertical line containing the vertex of a parabola.

eje de simetría La recta vertical que pasa por el vértice de una parábola.

B

back-to-back stem-and-leaf plot (p. 89) Used to compare two related sets of data.

diagrama de tallo y hojas consecutivo Se usa para comparar dos conjuntos relacionados de datos.

bar graph (p. 50) A graph that compares different categories of data by showing each category as a bar whose length is related to the frequency.

gráfica de barras Gráfica que compara categorías distintas de datos al exhibir cada categoría con una barra cuya longitud está relacionada con la frecuencia.

base (p. 7) In an expression of the form x^n, the base is x.

base En una expresión de la forma x^n, la base es x.

best-fit line (p. 300) The line that most closely approximates the data in a scatter plot.

recta de ajuste óptimo La recta que mejor aproxima los datos de una gráfica de dispersión.

biased sample (p. 709) A sample in which one or more parts of the population are favored over others.

muestra sesgada Muestra en que se favorece una o más partes de una población en vez de otras partes.

binomial (p. 422) The sum of two monomials.

binomio La suma de dos monomios.

boundary (p. 353) A line or curve that separates the coordinate plane into regions.

frontera Recta o curva que divide el plano de coordenadas en regiones.

box-and-whisker plot (p. 737) A diagram that divides a set of data into four parts using the median and quartiles.

diagrama de caja y patillas Diagrama que divide un conjunto de datos en cuatro partes usando la mediana y los cuartiles.

C

census (p. 708) A sample in which all of the units within a population are included.

censo Muestra en que se incluyen todas las unidades de una población.

circle graph (p. 51) A graph that compares parts of a set of data as a percent of the whole set.

gráfica circular Gráfica que compara partes de un conjunto de datos como porcentajes del conjunto entero.

coefficient (p. 29) The numerical factor of a term.

coeficiente Factor numérico de un término.

combination (p. 762) An arrangement or listing in which order is not important.

combinación Arreglo o lista en que el orden no es importante.

common difference (p. 233) The difference between the terms in a sequence.

diferencia común Diferencia entre términos consecutivos de una sucesión.

common ratio (p. 567) The number by which each term in a geometric sequence is multiplied.

razón común El número por el que se multiplica cada término de una sucesión geométrica.

completing the square (p. 539) To add a constant term to a binomial of the form $x^2 + bx$ so that the resulting trinomial is a perfect square.

completar el cuadrado Adición de un término constante a un binomio de la forma $x^2 + bx$, para que el trinomio resultante sea un cuadrado perfecto.

complex fraction (p. 684) A fraction that has one or more fractions in the numerator or denominator.

fracción compleja Fracción con una o más fracciones en el numerador o denominador.

composite numbers (p. 474) A whole number, greater than 1, that has more than two factors.

números compuestos Número entero mayor que 1 que posee más de dos factores.

compound event (p. 769) Two or more simple events.

evento compuesto Dos o más eventos simples.

compound inequality (p. 339) Two or more inequalities that are connected by the words *and* or *or*.

desigualdad compuesta Dos o más desigualdades que están unidas por las palabras *y* u *o*.

conclusion (p. 37) The part of a conditional statement immediately following the word *then*.

conclusión Parte de un enunciado condicional que sigue inmediatamente a la palabra *entonces*.

conditional statements (p. 37) Statements written in the form *If A, then B*.

enunciados condicionales Enunciados de la forma *Si A, entonces B*.

conjugates (p. 590) Binomials of the form $a\sqrt{b} + c\sqrt{d}$ and $a\sqrt{b} - c\sqrt{d}$.

conjugados Binomios de la forma $a\sqrt{b} + c\sqrt{d}$ y $a\sqrt{b} - c\sqrt{d}$.

consecutive integers (p. 144) Integers in counting order.

enteros consecutivos Enteros en el orden de contar.

consistent (p. 369) A system of equations that has at least one ordered pair that satisfies both equations.

consistente Sistema de ecuaciones para el cual existe al menos un par ordenado que satisface ambas ecuaciones.

constant (p. 410) A monomial that is a real number.

constante Monomio que es un número real.

constant of variation (p. 264) The number k in equations of the form $y = kx$.

constante de variación El número k en ecuaciones de la forma $y = kx$.

convenience sample (p. 709) A sample that includes members of a population that are easily accessed.

muestra de conveniencia Muestra que incluye miembros de una población fácilmente accesibles.

coordinate (p. 69) The number that corresponds to a point on a number line.

coordenada Número que corresponde a un punto en una recta numérica.

coordinate plane (pp. 43, 192) The plane containing the *x*- and *y*-axes.

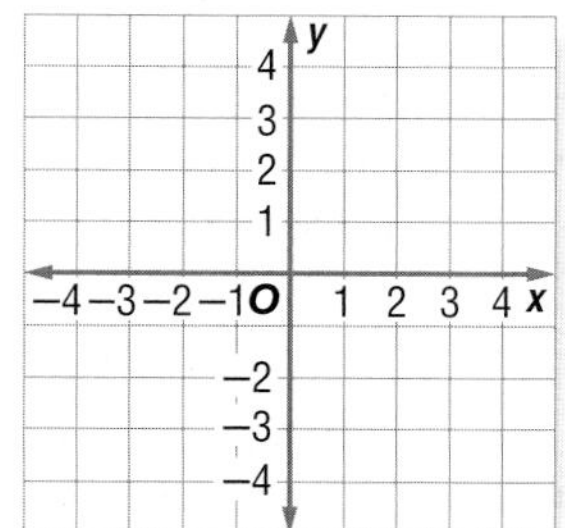

plano de coordenadas Plano que contiene los ejes *x* y *y*.

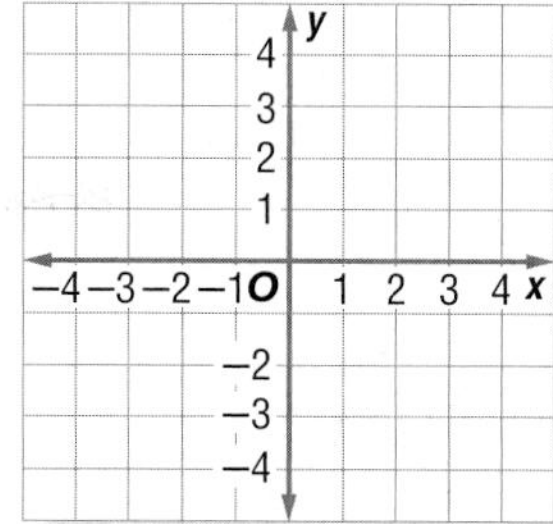

coordinate system (p. 43) The grid formed by the intersection of two number lines, the horizontal axis and the vertical axis.

sistema de coordenadas Cuadriculado formado por la intersección de dos rectas numéricas: los ejes *x* y *y*.

corollary (p. 608) A statement that can be easily proved using a theorem.

corolario Enunciado que puede probarse fácilmente mediante el uso de un teorema.

cosine (p. 624) In a right triangle with acute angle *A*, the cosine of

$$\angle A = \frac{\text{the measure of the leg adjacent to } \angle A}{\text{the measure of the hypotenuse}}$$

coseno En un triángulo rectángulo con ángulo agudo *A*, el coseno del

$$\angle A = \frac{\text{la medida del cateto adyacente al } \angle A}{\text{la medida de la hipotenusa}}$$

counterexample (p. 38) A specific case in which a statement is false.

contraejemplo Ejemplo específico de la falsedad de un enunciado.

cumulative frequency histogram (p. 743) A histogram organized using a cumulative frequency table.

histograma de frecuencia cumulativa Un histograma organizado a partir de una tabla de frecuencia cumulativa.

cumulative frequency table (p. 743) A table in which the frequencies are accumulated for each item.

tabla de frecuencia cumulativa Una tabla en que se acumulan las frecuencias para cada elemento.

D

data (p. 50) Numerical information gathered for statistical purposes.

datos Información numérica que se recopila con propósitos estadísticos.

deductive reasoning (p. 38) The process of using facts, rules, definitions, or properties to reach a valid conclusion.

razonamiento deductivo Proceso de usar hechos, reglas, definiciones o propiedades para sacar conclusiones válidas.

defining a variable (p. 121) Choosing a variable to represent one of the unspecified numbers in a problem and using it to write expressions for the other unspecified numbers in the problem.

definir una variable Consiste en escoger una variable para representar uno de los números desconocidos en un problema y luego usarla para escribir expresiones para otros números desconocidos en el problema.

degree of a monomial (p. 433) The sum of the exponents of all its variables.

grado de un monomio Suma de los exponentes de todas sus variables.

degree of a polynomial (p. 433) The greatest degree of any term in the polynomial.

grado de un polinomio El grado mayor de cualquier término del polinomio.

dependent (p. 369) A system of equations that has an infinite number of solutions.

dependiente Sistema de ecuaciones que posee un número infinito de soluciones.

dependent events (p. 770) Two or more events in which the outcome of one event affects the outcome of the other events.

eventos dependientes Dos o más eventos en que el resultado de un evento afecta el resultado de los otros eventos.

dependent variable (p. 44) The variable in a relation whose value depends on the value of the independent variable.

variable dependiente La variable de una relación cuyo valor depende del valor de la variable independiente.

difference of squares (pp. 460, 502) Two perfect squares separated by a subtraction sign.

$a^2 - b^2 = (a + b)(a - b)$ or
$a^2 - b^2 = (a - b)(a + b)$.

diferencia de cuadrados Dos cuadrados perfectos separados por el signo de sustracción.

$a^2 - b^2 = (a + b)(a - b)$ o
$a^2 - b^2 = (a - b)(a + b)$.

dilation (p. 197) A transformation in which a figure is enlarged or reduced.

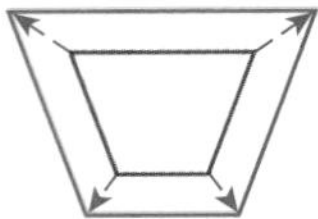

dilatación Una transformación en la cual se amplía o se reduce una figura.

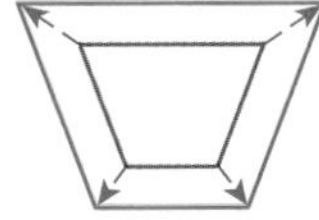

dimensional analysis (p. 167) The process of carrying units throughout a computation.

análisis dimensional Proceso de tomar en cuenta las unidades de medida al hacer cálculos.

direct variation (p. 264) An equation of the form $y = kx$, where $k \neq 0$.

variación directa Una ecuación de la forma $y = kx$, donde $k \neq 0$.

discriminant (p. 548) In the Quadratic Formula, the expression under the radical sign, $b^2 - 4ac$.

discriminante En la fórmula cuadrática, la expresión debajo del signo radical, $b^2 - 4ac$.

Distance Formula (p. 612) The distance d between any two points with coordinates (x_1, y_1) and (x_2, y_2) is given by the formula

$d = \sqrt{(x_2 - x_1)^2 + (y_2 - y_1)^2}$.

Fórmula de la distancia La distancia d entre cualquier par de puntos con coordenadas (x_1, y_1) y (x_2, y_2) viene dada por la fórmula

$d = \sqrt{(x_2 - x_1)^2 + (y_2 - y_1)^2}$.

domain (p. 45) The set of the first numbers of the ordered pairs in a relation.

dominio Conjunto de los primeros números de los pares ordenados de una relación.

E

element **1.** (p. 16) Each object or number in a set. **2.** (p. 715) Each entry in a matrix.

elemento **1.** Cada número u objeto de un conjunto. **2.** Cada entrada de una matriz.

elimination (p. 382) The use of addition or subtraction to eliminate one variable and solve a system of equations.

eliminación El uso de la adición o la sustracción para eliminar una variable y resolver así un sistema de ecuaciones.

empirical study (p. 782) Performing an experiment repeatedly, collecting and combining data, and analyzing the results.

estudio empírico Ejecución repetida de un experimento, recopilación y combinación de datos y análisis de resultados.

equally likely (p. 97) Outcomes for which the probability of each occurring is equal.

equiprobable Resultados que tienen la misma probabilidad de ocurrir.

equation (p. 16) A mathematical sentence that contains an equals sign, =.

ecuación Enunciado matemático que contiene el signo de igualdad, =.

equation in two variables (p. 212) An equation that contains two unknown values.

ecuación en dos variables Una ecuación que tiene dos incógnitas.

equivalent equations (p. 128) Equations that have the same solution.

ecuaciones equivalentes Ecuaciones que poseen la misma solución.

equivalent expressions (p. 29) Expressions that denote the same number.

expresiones equivalentes Expresiones que denotan el mismo número.

evaluate (p. 7) To find the value of an expression.

evaluar Calcular el valor de una expresión.

event (p. 754) Any collection of one or more outcomes in the sample space.

evento Cualquier colección de uno o más resultados de un espacio muestral.

excluded values (p. 648) Any values of a variable that result in a denominator of 0 must be excluded from the domain of that variable.

valores excluidos Cualquier valor de una variable cuyo resultado sea un denominador igual a cero, debe excluirse del dominio de dicha variable.

experimental probability (p. 782) What actually occurs when conducting a probability experiment, or the ratio of relative frequency to the total number of events or trials.

probabilidad experimental Lo que realmente sucede cuando se realiza un experimento probabilístico o la razón de la frecuencia relativa al número total de eventos o pruebas.

exponent (p. 7) In an expression of the form x^n, the exponent is n. It indicates the number of times x is used as a factor.

exponente En una expresión de la forma x^n, el exponente es n. Éste indica cuántas veces se usa x como factor.

exponential function (p. 554) A function that can be described by an equation of the form $y = a^x$, where $a > 0$ and $a \neq 1$.

función exponencial Función que puede describirse mediante una ecuación de la forma $y = a^x$, donde $a > 0$ y $a \neq 1$.

extraneous solutions (p. 601, 693) Results that are not solutions to the original equation.

soluciones extrañas Resultados que no son soluciones de la ecuación original.

extremes (p. 156) In the proportion $\frac{a}{b} = \frac{c}{d}$, a and d are the extremes.

extremos En la proporción $\frac{a}{b} = \frac{c}{d}$, a y d son los extremos.

F

factored form (p. 475) A monomial expressed as a product of prime numbers and variables and no variable has an exponent greater than 1.

forma reducida Monomio escrito como el producto de números primos y variables y en el que ninguna variable tiene un exponente mayor que 1.

factorial (p. 755) The expression $n!$, read n factorial, where n is greater than zero, is the product of all positive integers beginning with n and counting backward to 1.

factorial La expresión $n!$, que se lee n factorial, donde n que es mayor que cero, es el producto de todos los números naturales, comenzando con n y contando hacia atrás hasta llegar al 1.

factoring (p. 481) To express a polynomial as the product of monomials and polynomials.

factorización La escritura de un polinomio como producto de monomios y polinomios.

factoring by grouping (p. 482) The use of the Distributive Property to factor some polynomials having four or more terms.

factorización por agrupamiento Uso de la Propiedad distributiva para factorizar polinomios que poseen cuatro o más términos.

factors (p. 6) In an algebraic expression, the quantities being multiplied are called factors.

factores En una expresión algebraica, los factores son las cantidades que se multiplican.

family of graphs (pp. 265, 531) Graphs and equations of graphs that have at least one characteristic in common.

familia de gráficas Gráficas y ecuaciones de gráficas que tienen al menos una característica común.

FOIL method (p. 453) To multiply two binomials, find the sum of the products of the *F*irst terms, the *O*uter terms, the *I*nner terms, and the *L*ast terms.

método FOIL Para multiplicar dos binomios, busca la suma de los productos de los primeros (*F*irst) términos, los términos exteriores (*O*uter), los términos interiores (*I*nner) y los últimos términos (*L*ast).

formula (p. 122) An equation that states a rule for the relationship between certain quantities.

fórmula Ecuación que establece una relación entre ciertas cantidades.

four-step problem-solving plan (p. 121)

Step 1 Explore the problem.

Step 2 Plan the solution.

Step 3 Solve the problem.

Step 4 Examine the solution.

plan de cuatro pasos para resolver problemas

Paso 1 Explora el problema.

Paso 2 Planifica la solución.

Paso 3 Resuelve el problema.

Paso 4 Examina la solución.

frequency (pp. 88, 722) How often a piece of data occurs.

frecuencia Las veces que aparece un dato.

frequency table (p. 722) A table of tally marks used to record and display how often events occur.

tabla de frecuencia Una tabla de cuentas que se usa para anotar y exhibir la frecuencia de eventos.

function (pp. 43, 226) A relation in which each element of the domain is paired with exactly one element of the range.

función Una relación en que a cada elemento del dominio le corresponde un único elemento del rango.

function notation (p. 227) A way to name a function that is defined by an equation. In function notation, the equation $y = 3x - 8$ is written as $f(x) = 3x - 8$.

notación funcional Una manera de nombrar una función definida por una ecuación. En notación funcional, la ecuación $y = 3x - 8$ se escribe $f(x) = 3x - 8$.

Glossary/Glosario

Fundamental Counting Principle (p. 755) If an event M can occur in m ways and is followed by an event N that can occur in n ways, then the event M followed by the event N can occur in $m \times n$ ways.

Principio fundamental de contar Si un evento M puede ocurrir de m maneras y lo sigue un evento N que puede ocurrir de n maneras, entonces el evento M seguido del evento N puede ocurrir de $m \times n$ maneras.

G

general equation for exponential decay (p. 562) $y = C(1 - r)^t$, where y is the final amount, C is the initial amount, r is the rate of decay expressed as a decimal, and t is time.

ecuación general de desintegración exponencial $y = C(1 - r)^t$, donde y es la cantidad final, C es la cantidad inicial, r es la tasa de desintegración escrita como decimal y t es el tiempo.

general equation for exponential growth (p. 561) $y = C(1 + r)^t$, where y is the final amount, C is the initial amount, r is the rate of change expressed as a decimal, and t is time.

ecuación general de crecimiento exponencial $y = C(1 + r)^t$, donde y es la cantidad final, C es la cantidad inicial, r es la tasa de cambio del crecimiento escrita como decimal y t es el tiempo.

geometric means (p. 570) Missing terms between two nonconsecutive terms in a geometric sequence.

medios geométricos Términos que faltan entre dos términos no consecutivos de una sucesión geométrica.

geometric sequence (p. 567) A sequence in which each term after the nonzero first term is found by multiplying the previous term by a constant called the common ratio r, where $r \neq 0$ or 1.

sucesión geométrica Sucesión en que cada término no nulo, después del primer término se calcula multiplicando el término anterior por una constante r llamada razón común, con $r \neq 0$ ó 1.

graph (pp. 69, 193) To draw, or plot, the points named by certain numbers or ordered pairs on a number line or coordinate plane.

graficar Marcar los puntos que denotan ciertos números en una recta numérica o ciertos pares ordenados en un plano de coordenadas.

greatest common factor (GCF) (p. 476) The product of the prime factors common to two or more integers.

máximo común divisor (MCD) El producto de los factores primos comunes a dos o más enteros.

H

half-plane (p. 353) The region of the graph of an inequality on one side of a boundary.

semiplano Región de la gráfica de una desigualdad en un lado de la frontera.

histogram (p. 722) A bar graph in which the data are organized into equal intervals.

histograma Una gráfica de barras en que los datos aparecen organizados en intervalos iguales.

hypotenuse (p. 606) The side opposite the right angle in a right triangle.

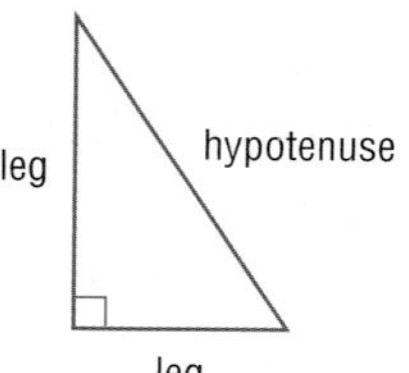

hipotenusa Lado opuesto al ángulo recto en un triángulo rectángulo.

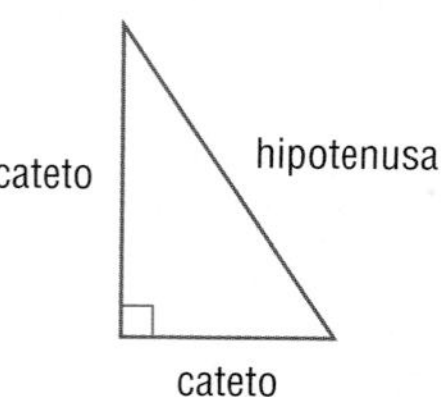

hypothesis (p. 37) The part of a conditional statement immediately following the word *if*.

hipótesis Parte de un enunciado condicional que sigue inmediatamente a la palabra *si*.

I

identity (p. 150) An equation that is true for every value of the variable.

identidad Ecuación que es verdadera para cada valor de la variable.

if-then statements (p. 37) Conditional statements in the form *If A, then B.*

enunciados si-entonces Enunciados condicionales de la forma *Si A, entonces B.*

image (p. 197) The position of a figure after a transformation.

imagen Posición de una figura después de una transformación.

inclusive (p. 771) Two events that can occur at the same time.

inclusivos Dos eventos que pueden ocurrir simultáneamente.

inconsistent (p. 369) A system of equations with no ordered pair that satisfy both equations.

inconsistente Un sistema de ecuaciones para el cual no existe par ordenado alguno que satisfaga ambas ecuaciones.

independent (p. 369) A system of equations with exactly one solution.

independent events (p. 769) Two or more events in which the outcome of one event does not affect the outcome of the other events.

independent variable (p. 44) The variable in a function whose value is subject to choice.

inductive reasoning (p. 240) A conclusion based on a pattern of examples.

inequality (p. 17) An open sentence that contains the symbol $<$, $\leq$, $>$, or $\geq$.

infinity (p. 68) Lines and sets that never end continue to infinity.

integers (p. 68) The set $\{..., -2, -1, 0, 1, 2, ...\}$.

interquartile range (p. 732) The difference between the Upper and Lower quartiles; represents the middle half of the data in the set.

intersection (p. 339) The graph of a compound inequality containing *and*; the solution is the set of elements common to both inequalities.

inverse (p. 206) The inverse of any relation is obtained by switching the coordinates in each ordered pair.

inverse variation (p. 642) An equation of the form $xy = k$, where $k \neq 0$.

irrational numbers (p. 104) Numbers that cannot be expressed as terminating or repeating decimals.

independiente Un sistema de ecuaciones que posee una única solución.

eventos independientes El resultado de un evento no afecta el resultado del otro evento.

variable independiente La variable de una función sujeta a elección.

razonamiento inductivo Conclusión basada en un patrón de ejemplos.

desigualdad Enunciado abierto que contiene uno o más de los símbolos $<$, $\leq$, $>$ o $\geq$.

indefinidamente Rectas y conjuntos interminables continúan indefinidamente.

enteros El conjunto $\{..., -2, -1, 0, 1, 2, ...\}$.

amplitud intercuartílica Diferencia entre el cuartil superior y el inferior; representa la mitad central de los datos del conjunto.

intersección Gráfica de una desigualdad compuesta que contiene la palabra *y*; la solución es el conjunto de soluciones de ambas desigualdades.

inversa La inversa de una relación se halla intercambiando las coordenadas de cada par ordenado.

variación inversa Ecuación de la forma $xy = k$, donde $k \neq 0$.

números irracionales Números que no pueden escribirse como decimales terminales o periódicos.

L

least common denominator (LCD) (p. 678) The least common multiple of the denominators of two or more fractions.

least common multiple (LCM) (p. 678) The least number that is a common multiple of two or more numbers.

legs (p. 606) The sides of a right triangle that form the right angle.

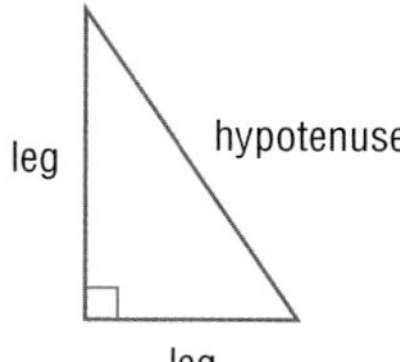

like terms (p. 28) Terms that contain the same variables, with corresponding variables having the same exponent.

linear equation (p. 218) An equation in the form $Ax + By = C$, whose graph is a straight line.

linear extrapolation (p. 283) The use of a linear equation to predict values that are outside the range of data.

linear interpolation (p. 301) The use of a linear equation to predict values that are inside of the data range.

mínimo denominador común (mcd) El mínimo común múltiplo de los denominadores de dos o más fracciones.

mínimo común múltiplo (mcm) El número menor que es múltiplo común de dos o más números.

catetos Lados de un triángulo rectángulo que forman el ángulo recto del mismo.

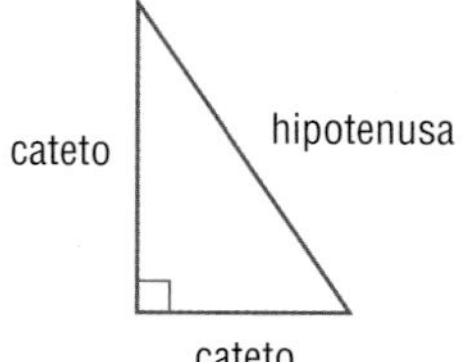

términos semejantes Expresiones que tienen las mismas variables, con las variables correspondientes elevadas a los mismos exponentes.

ecuación lineal Ecuación de la forma $Ax + By = C$, cuya gráfica es una recta.

extrapolación lineal Uso de una ecuación lineal para predecir valores fuera de la amplitud de los datos.

interpolación lineal Uso de una ecuación lineal para predecir valores dentro de la amplitud de los datos.

line graph (p. 51) Numerical data displayed to show trends or changes over time.

gráfica lineal Datos numéricos exhibidos para mostrar tendencias o cambios con el tiempo.

line of fit (p. 304) A line that describes the trend of the data in a scatter plot.

recta de ajuste Recta que describe la tendencia de los datos en una gráfica de dispersión.

line plot (p. 88) A number line labeled with a scale to include all the data with an × placed above a data point each time it occurs.

esquema lineal Recta numérica marcada con una escala que incluye todos los datos, colocando × sobre cada uno de ellos para indicar su frecuencia.

lower quartile (p. 732) Divides the lower half of the data into two equal parts.

cuartil inferior Éste divide en dos partes iguales la mitad inferior de un conjunto de datos.

M

mapping (p. 205) Illustrates how each element of the domain is paired with an element in the range.

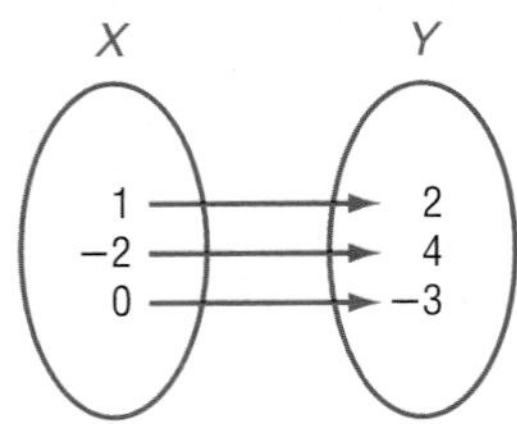

aplicaciones Ilustra la correspondencia entre cada elemento del dominio con un elemento del rango.

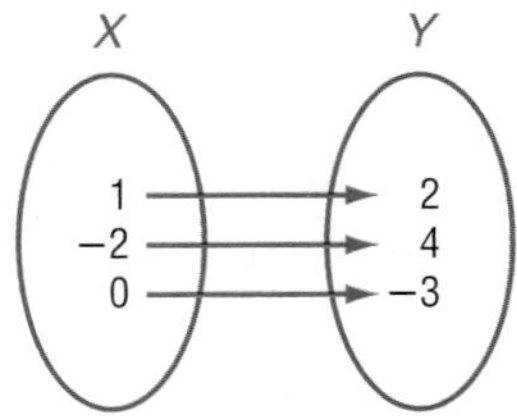

matrix (p. 715) A rectangular arrangement of numbers in rows and columns.

matriz Un arreglo rectangular de números en filas y columnas.

maximum (p. 525) The highest point on the graph of a curve.

máximo El punto más alto en la gráfica de una curva.

measures of central tendency (p. 90) Numbers or pieces of data that can represent the whole set of data.

medidas de tendencia central Números o datos que pueden representar todo el conjunto de datos.

measures of variation (p. 731) Measures that describe the spread of the values in a set of data.

medidas de variación Medidas que describen la dispersión de los valores de un conjunto de datos.

minimum (p. 524) The lowest point on the graph of a curve.

mínimo El punto más bajo en la gráfica de una curva.

mixed expression (p. 684) An expression that contains the sum of a monomial and a rational expression.

expresión mixta Expresión que contiene la suma de un monomio y una expresión racional.

mixture problems (p. 171) Problems in which two or more parts are combined into a whole.

problemas de mezclas Problemas en que dos o más partes se combinan en un todo.

monomial (p. 410) A number, a variable, or a product of a number and one or more variables.

monomio Número, variable o producto de un número por una o más variables.

multiplicative identity (p. 22) For any number a, $a \cdot 1 = 1 \cdot a = a$.

identidad de la multiplicación Para cualquier número a, $a \cdot 1 = 1 \cdot a = a$.

multiplicative inverses (p. 22) Two numbers whose product is 1.

inversos multiplicativos Dos números cuyo producto es igual a 1.

multi-step equations (p. 143) Equations with more than one operation.

ecuaciones de varios pasos Ecuaciones con más de una operación.

mutually exclusive (p. 771) Events that cannot occur at the same time.

mutuamente exclusivos Eventos que no pueden ocurrir simultáneamente.

N

natural numbers (p. 68) The set {1, 2, 3, ...}.

números naturales El conjunto {1, 2, 3, ...}.

negative correlation (p. 298) In a scatter plot, as x increases, y decreases.

correlación negativa En una gráfica de dispersión, a medida que x aumenta, y disminuye.

negative exponent (p. 419) For any nonzero number a and any integer n, $a^{-n} = \frac{1}{a^n}$ and $\frac{1}{a^{-n}} = a^n$.

exponente negativo Para cualquier número no nulo a y cualquier entero n, $a^{-n} = \frac{1}{a^n}$ y $\frac{1}{a^{-n}} = a^n$.

negative number (p. 68) Any value less than zero.

número negativo Cualquier valor menor que cero.

number theory (p. 144) The study of numbers and the relationships between them.

teoría de números El estudio de números y de las relaciones entre ellos.

O

odds (p. 97) The ratio that compares the number of ways an event can occur (successes) to the number of ways the event cannot occur (failures).

posibilidades Razón que compara el número de maneras en que puede ocurrir un evento (éxitos) al número de maneras en que no puede ocurrir (fracasos).

open sentence (p. 16) A mathematical statement with one or more variables.

enunciado abierto Un enunciado matemático que contiene una o más variables.

opposites (p. 74) Every positive rational number and its negative pair.

opuestos Cada número racional positivo y su opuesto negativo.

ordered pair (p. 43) A set of numbers or coordinates used to locate any point on a coordinate plane, written in the form (x, y).

par ordenado Un par de números que se usa para ubicar cualquier punto de un plano de coordenadas y que se escribe en la forma (x, y).

order of operations (p. 11)

1. Evaluate expressions inside grouping symbols.
2. Evaluate all powers.
3. Do all multiplications and/or divisions from left to right.
4. Do all additions and/or subtractions from left to right.

orden de las operaciones

1. Evalúa las expresiones dentro de los símbolos de agrupamiento.
2. Evalúa todas las potencias.
3. Multiplica o divide de izquierda a derecha.
4. Suma o resta de izquierda a derecha.

origin (pp. 43, 192) The point where the two axes intersect at their zero points.

origen Punto donde se intersecan los dos ejes en sus puntos cero.

outlier (p. 733) Any element of a set of data that is at least 1.5 interquartile ranges less than the lower quartile or greater than the upper quartile.

valor atípico Cualquier dato que está por lo menos a 1.5 amplitudes intercuartílicas debajo del cuartil inferior o por encima del cuartil superior.

P

parabola (p. 524) The graph of a quadratic function.

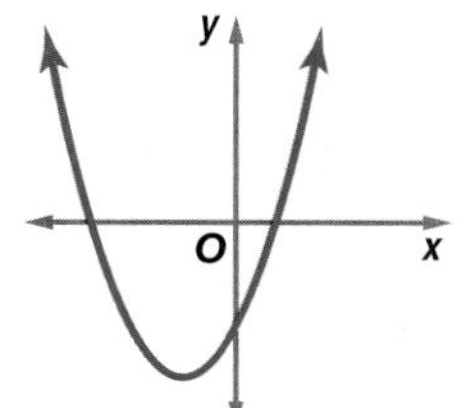

parábola La gráfica de una función cuadrática.

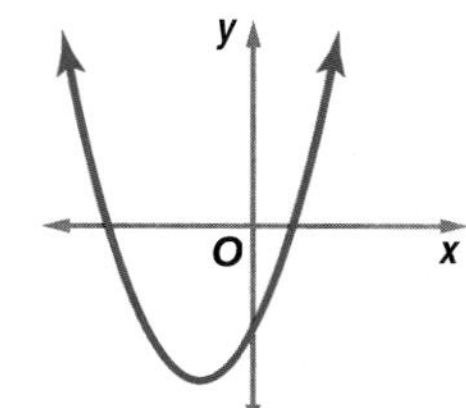

parallel lines (p. 292) Lines in the same plane that never intersect and have the same slope.

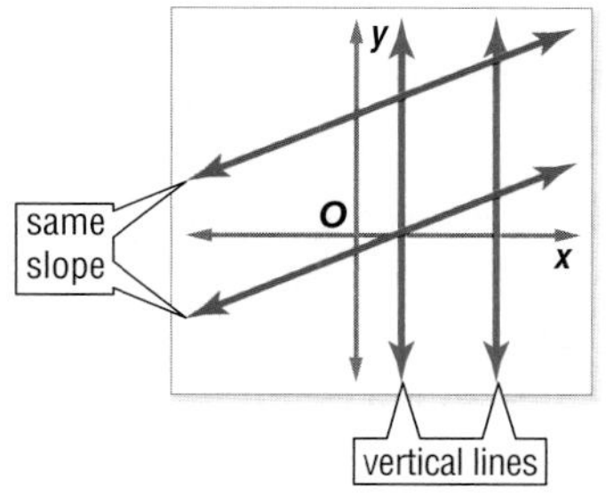

rectas paralelas Rectas en el mismo plano que no se intersecan jamás y que tienen pendientes iguales.

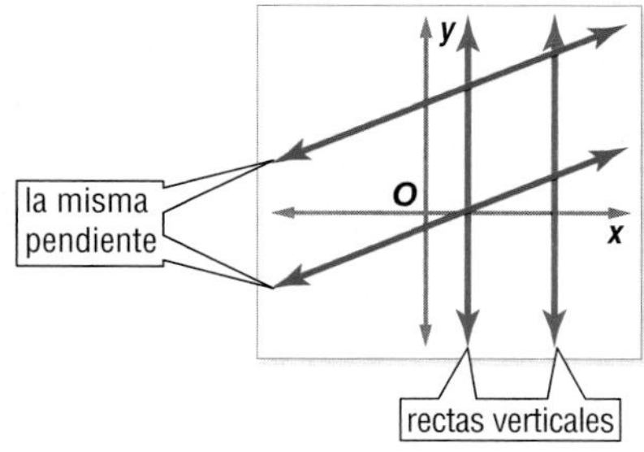

parent graph (p. 265) The simplest of the graphs in a family of graphs.

gráfica madre La gráfica más sencilla en una familia de gráficas.

percent of change (p. 160) When an increase or decrease is expressed as a percent.

porcentaje de cambio Cuando un aumento o disminución se escribe como un tanto por ciento.

percent of decrease (p. 160) The ratio of an amount of decrease to the previous amount, expressed as a percent.

percent of increase (p. 160) The ratio of an amount of increase to the previous amount, expressed as a percent.

percentile (p. 743) The point below which a given percent of the data lies.

perfect square (p. 103) A number whose square root is a rational number.

perfect square trinomial (p. 508) Trinomials that are the square of a binomial. $(a + b)^2 = (a + b)(a + b) = a^2 + 2ab + b^2$ or $(a - b)^2 = (a - b)(a - b) = a^2 - 2ab - b^2$

permutation (p. 760) An arrangement or listing in which order is important.

perpendicular lines (p. 293) Lines that meet to form right angles.

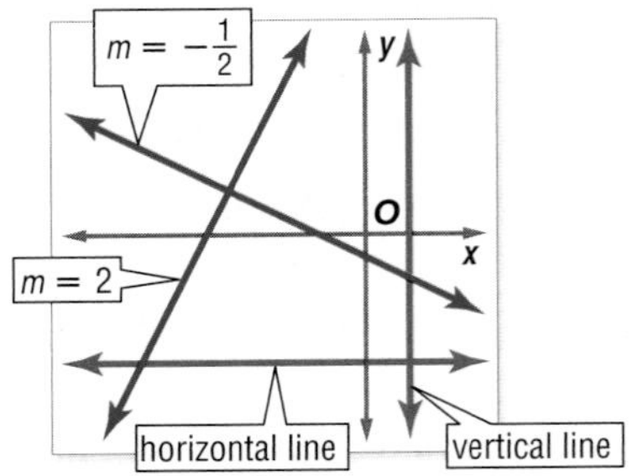

point-slope form (p. 286) An equation of the form $y - y_1 = m(x - x_1)$, where m is the slope and (x_1, y_1) is a given point on a nonvertical line.

polynomial (p. 432) A monomial or sum of monomials.

population (p. 708) A large group of data usually represented by a sample.

positive correlation (p. 298) In a scatter plot, as x increases, y increases.

positive number (p. 68) Any value that is greater than zero.

power (p. 7) An expression of the form x^n, read *x to the* n^{th} *power.*

power of a quotient (p. 418) For any integer m and real numbers a and b, $b \neq 0$, $\left(\frac{a}{b}\right)^m = \frac{a^m}{b^m}$.

preimage (p. 197) The position of a figure before a transformation.

prime factorization (p. 475) A whole number expressed as a product of factors that are all prime numbers.

prime number (p. 474) A whole number, greater than 1, whose only factors are 1 and itself.

prime polynomial (p. 497) A polynomial that cannot be written as a product of two polynomials with integral coefficients.

porcentaje de disminución Razón de la cantidad de disminución a la cantidad original, escrita como un tanto por ciento.

porcentaje de aumento Razón de la cantidad de aumento a la cantidad original, escrita como un tanto por ciento.

percentil Punto bajo el cual yace un tanto por ciento de los datos.

cuadrado perfecto Número cuya raíz cuadrada es un número racional.

trinomio cuadrado perfecto Trinomios que son el cuadrado de un binomio. $(a + b)^2 = (a + b)(a + b) = a^2 + 2ab + b^2$ o $(a - b)^2 = (a - b)(a - b) = a^2 - 2ab - b^2$

permutación Arreglo o lista en que el orden es importante.

rectas perpendiculares Rectas que se intersecan formando un ángulo recto.

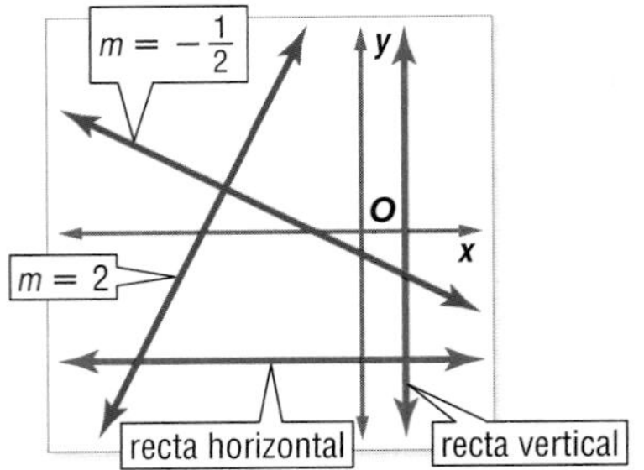

forma punto-pendiente Ecuación de la forma $y - y_1 = m(x - x_1)$, donde m es la pendiente y (x_1, y_1) es un punto dado de una recta no vertical.

polinomio Un monomio o la suma de monomios.

población Grupo grande de datos, representado por lo general por una muestra.

correlación positiva En una gráfica de dispersión, a medida que x aumenta, y aumenta.

número positivos Cualquier valor mayor que cero.

potencia Una expresión de la forma x^n, se lee x a *la enésima potencia.*

potencia de un cociente Para cualquier entero m y números reales a y b, $b \neq 0$, $\left(\frac{a}{b}\right)^m = \frac{a^m}{b^m}$.

preimagen Posición de una figura antes de una transformación.

factorización prima Número entero escrito como producto de factores primos.

número primo Número entero mayor que 1 cuyos únicos factores son 1 y sí mismo.

polinomio primo Polinomio que no puede escribirse como producto de dos polinomios con coeficientes enteros.

principal square root (p. 103) The nonnegative square root of a number.

raíz cuadrada principal La raíz cuadrada no negativa de un número.

probability (p. 96) The ratio of the number of favorable outcomes for an event to the number of possible outcomes of the event.

$$P(a) = \frac{\text{number of favorable outcomes}}{\text{total number of possible outcomes}}.$$

probabilidad Razón del número de resultados favorables de un evento al número de resultados posibles.

$$P(a) = \frac{\text{número de resultados favorables}}{\text{número total de resultados posibles}}.$$

probability distribution (p. 777) The probability of every possible value of the random variable x.

distribución de probabilidad Probabilidad de cada valor posible de una variable aleatoria x.

probability histogram (p. 778) A way to give the probability distribution for a random variable and obtain other data.

histograma probabilístico Una manera de exhibir la distribución de probabilidad de una variable aleatoria y obtener otros datos.

product (p. 6) In an algebraic expression, the result of quantities being multiplied is called the product.

producto En una expresión algebraica, se llama producto al resultado de las cantidades que se multiplican.

proportion (p. 155) An equation of the form $\frac{a}{b} = \frac{c}{d}$ stating that two ratios are equivalent.

proporción Ecuación de la forma $\frac{a}{b} = \frac{c}{d}$ que afirma la equivalencia de dos razones.

Pythagorean Theorem (p. 606) If a and b are the measures of the legs of a right triangle and c is the measure of the hypotenuse, then $c^2 = a^2 + b^2$.

Teorema de Pitágoras Si a y b son las longitudes de los catetos de un triángulo rectángulo y si c es la longitud de la hipotenusa, entonces $c^2 = a^2 + b^2$.

Pythagorean triple (p. 607) Whole numbers that satisfy the Pythagorean Theorem.

Triple pitagórico Números enteros que satisfacen el Teorema de Pitágoras.

Q

quadrants (p. 193) The four regions into which the x- and y-axes separate the coordinate plane.

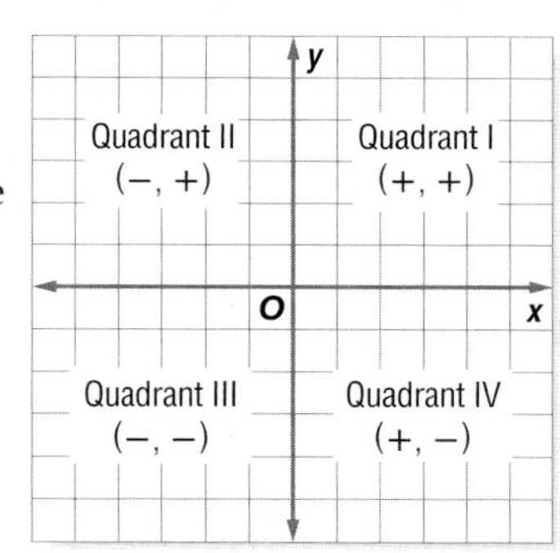

cuadrantes Las cuatro regiones en las que los ejes x y y dividen el plano de coordenadas.

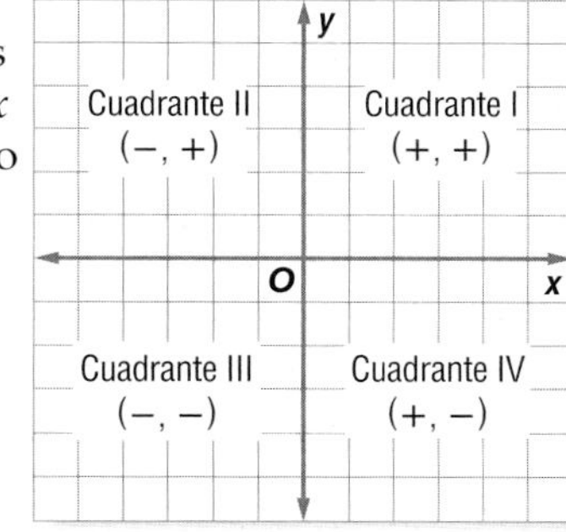

quadratic equation (p. 533) An equation of the form $ax^2 + bx + c = 0$, where $a \neq 0$.

ecuación cuadrática Ecuación de la forma $ax^2 + bx + c = 0$, donde $a \neq 0$.

Quadratic Formula (p. 546) The solutions of a quadratic equation in the form $ax^2 + bx + c = 0$, where $a \neq 0$, are given by the formula

$$x = \frac{-b \pm \sqrt{b^2 - 4ac}}{2a}.$$

Fórmula cuadrática Las soluciones de una ecuación cuadrática de la forma $ax^2 + bx + c = 0$, donde $a \neq 0$, vienen dadas por la fórmula

$$x = \frac{-b \pm \sqrt{b^2 - 4ac}}{2a}.$$

quadratic function (p. 524) An equation of the form $y = ax^2 + bx + c$, where $a \neq 0$.

función cuadrática Función de la forma $y = ax^2 + bx + c$, donde $a \neq 0$.

quartiles (p. 732) Values that divide the data into four equal parts.

cuartiles Valores que dividen un conjunto de datos en cuatro partes iguales.

R

radical equations (p. 600) Equations that contain radicals with variables in the radicand.

ecuaciones radicales Ecuaciones que contienen radicales con variables en el radicando.

radical sign (p. 103) The symbol $\sqrt{\ }$, used to indicate a nonnegative square root.

signo radical El símbolo $\sqrt{\ }$, que se usa para indicar la raíz cuadrada no negativa.

Glossary/Glosario

radicand (p. 587) The expression that is under the radical sign.

random sample (p. 708) A sample that is chosen without any preference, representative of the entire population.

random variable (p. 777) A variable whose value is the numerical outcome of a random event.

range (p. 45) The set of second numbers of the ordered pairs in a relation.

range (p. 731) The difference between the greatest and the least values of a set of data.

rate (p. 157) The ratio of two measurements having different units of measure.

rate of change (p. 258) How a quantity is changing over time.

ratio (p. 155) A comparison of two numbers by division.

rational approximation (p. 105) A rational number that is close to, but not equal to, the value of an irrational number.

rational equations (p. 690) Equations that contain rational expressions.

rational expression (p. 648) An algebraic fraction whose numerator and denominator are polynomials.

rationalizing the denominator (p. 589) A method used to eliminate radicals from the denominator of a fraction.

rational numbers (p. 68) The set of numbers expressed in the form of a fraction $\frac{a}{b}$, where a and b are integers and $b \neq 0$.

real numbers (p. 104) The set of rational numbers and the set of irrational numbers together.

reciprocal (p. 21) The multiplicative inverse of a number.

reflection (p. 197) A type of transformation in which a figure is flipped over a line.

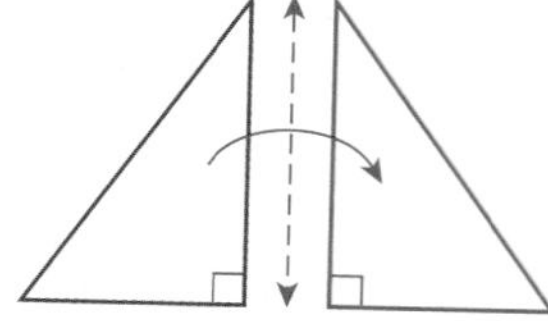

relation (p. 45) A set of ordered pairs.

relative frequency (p. 782) The number of times an outcome occurred in a probability experiment.

replacement set (p. 16) A set of numbers from which replacements for a variable may be chosen.

roots (p. 533) The solutions of a quadratic equation.

radicando La expresión debajo del signo radical.

muestra aleatoria Muestra tomada sin preferencia alguna y que es representativa de toda la población.

variable aleatoria Una variable cuyos valores son los resultados numéricos de un evento aleatorio.

rango Conjunto de los segundos números de los pares ordenados de una relación.

amplitud Diferencia entre los valores máximo y mínimo de un conjunto de datos.

tasa Razón de dos medidas que tienen distintas unidades de medida.

tasa de cambio Cómo cambia una cantidad con el tiempo.

razón Comparación de dos números mediante división.

aproximación racional Número racional que está cercano, pero que no es igual, al valor de un número irracional.

ecuaciones racionales Ecuaciones que contienen expresiones racionales.

expresión racional Fracción algebraica cuyo numerador y denominador son polinomios.

racionalizar el denominador Método que se usa para eliminar radicales del denominador de una fracción.

números racionales Conjunto de los números que pueden escribirse en forma de fracción $\frac{a}{b}$, donde a y b son enteros y $b \neq 0$.

números reales El conjunto de los números racionales junto con el conjunto de los números irracionales.

recíproco Inverso multiplicativo de un número.

reflexión Tipo de transformación en que una figura se voltea a través de una recta.

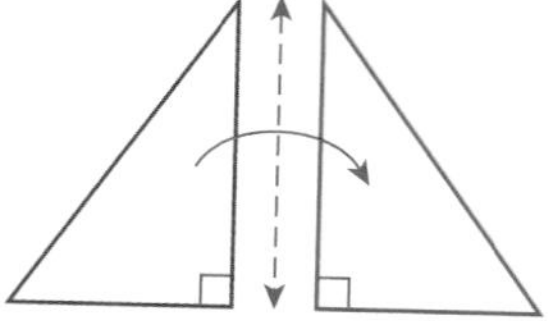

relación Conjunto de pares ordenados.

frecuencia relativa Número de veces que aparece un resultado en un experimento probabilístico.

conjunto de sustitución Conjunto de números del cual se pueden escoger sustituciones para una variable.

raíces Las soluciones de una ecuación cuadrática.

rotation (p. 197) A type of transformation in which a figure is turned around a point.

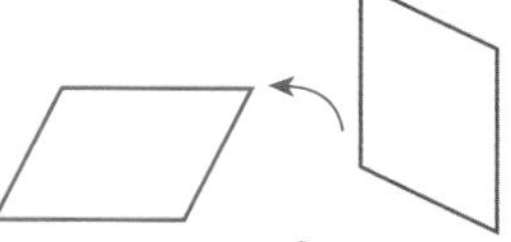

rotación Tipo de transformación en que una figura se hace girar alrededor de un punto fijo.

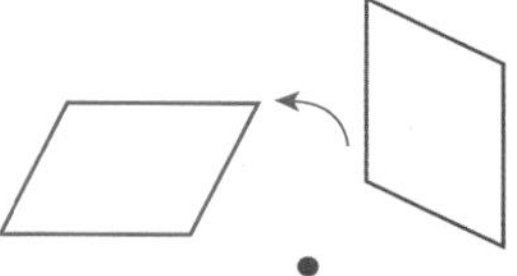

S

sample (p. 708) Some portion of a larger group selected to represent that group.

muestra Porción de un grupo más grande que se escoge para representarlo.

sample space (pp. 96, 754) The list of all possible outcomes.

espacio muestral Lista de todos los resultados posibles.

scalar multiplication (p. 717) Each element is multiplied by the scalar, or constant, and a new matrix is formed.

$$m = \begin{bmatrix} a & b & c \\ d & e & f \end{bmatrix} = \begin{bmatrix} ma & mb & mc \\ md & me & mf \end{bmatrix}$$

multiplicación escalar Cada elemento se multiplica por el escalar o constante, formándose así una nueva matriz.

$$m = \begin{bmatrix} a & b & c \\ d & e & f \end{bmatrix} = \begin{bmatrix} ma & mb & mc \\ md & me & mf \end{bmatrix}$$

scale (p. 157) A ratio or rate used when making a model of something that is too large or too small to be conveniently shown at actual size.

escala Razón o tasa que se usa al construir un modelo de algo que es demasiado grande o pequeño como para mostrarlo de tamaño natural.

scatter plot (p. 298) Two sets of data plotted as ordered pairs in a coordinate plane.

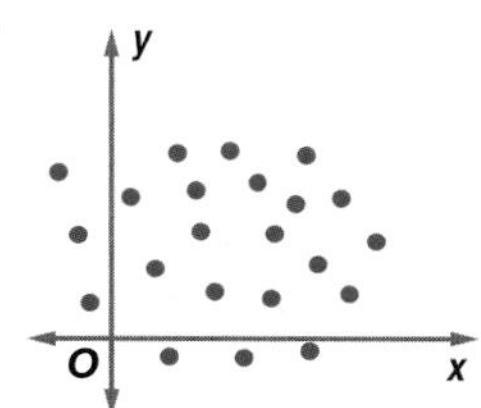

gráfica de dispersión Dos conjuntos de datos graficados como pares ordenados en un plano de coordenadas.

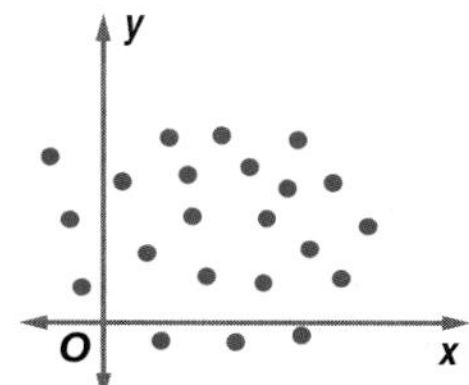

scientific notation (p. 425) A number of the form $a \times 10^n$, where $1 \leq a < 10$ and n is an integer.

notación científica Número de la forma $a \times 10^n$, donde $1 \leq a < 10$ y n es un entero.

sequence (p. 233) A set of numbers in a specific order.

sucesión Conjunto de números en un orden específico.

set (p. 16) A collection of objects or numbers, often shown using braces { } and usually named by a capital letter.

conjunto Colección de objetos o números, que a menudo se exhiben usando paréntesis de corchete { } y que se identifican por lo general mediante una letra mayúscula .

set-builder notation (p. 319) A concise way of writing a solution set. For example, $\{t \mid t < 17\}$ represents the set of all numbers t such that t is less than 17.

notación de construcción de conjuntos Manera concisa de escribir un conjunto solución. Por ejemplo, $\{t \mid t < 17\}$ representa el conjunto de todos los números t que son menores o iguales que 17.

similar (p. 617) Having the same shape but not necessarily the same size.

semejantes Que tienen la misma forma, pero no necesariamente el mismo tamaño.

simple event (pp. 96, 769) A single event.

evento simple Un sólo evento.

simple random sample (p. 708) A sample that is as likely to be chosen as any other from the population.

muestra aleatoria simple Muestra de una población que tiene la misma probabilidad de escogerse que cualquier otra.

simplest form (p. 29) An expression is in simplest form when it is replaced by an equivalent expression having no like terms or parentheses.

forma reducida Una expresión está reducida cuando se puede sustituir por una expresión equivalente que no tiene ni términos semejantes ni paréntesis.

simulation (p. 783) Using an object to act out an event that would be difficult or impractical to perform.

simulación Uso de un objeto para representar un evento que pudiera ser difícil o poco práctico de ejecutar.

sine (p. 624) In a right triangle with acute angle A, the sine of $\angle A = \frac{\text{measure of leg opposite } \angle A}{\text{measure of hypotenuse}}$.

seno En un triángulo rectángulo con ángulo agudo A, el seno del $\angle A = \frac{\text{medida del cateto opuesto a } \angle A}{\text{medida de la hipotenusa}}$.

slope (p. 256) The ratio of the change in the y-coordinates (rise) to the corresponding change in the x-coordinates (run) as you move from one point to another along a line.

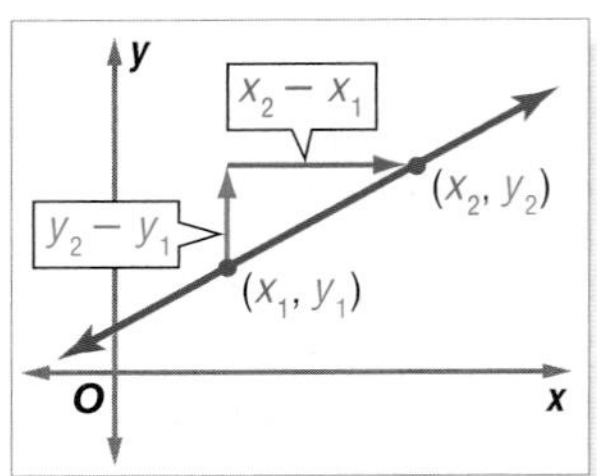

pendiente Razón del cambio en la coordenada y (elevación) al cambio correspondiente en la coordenada x (desplazamiento) a medida que uno se mueve de un punto a otro en una recta.

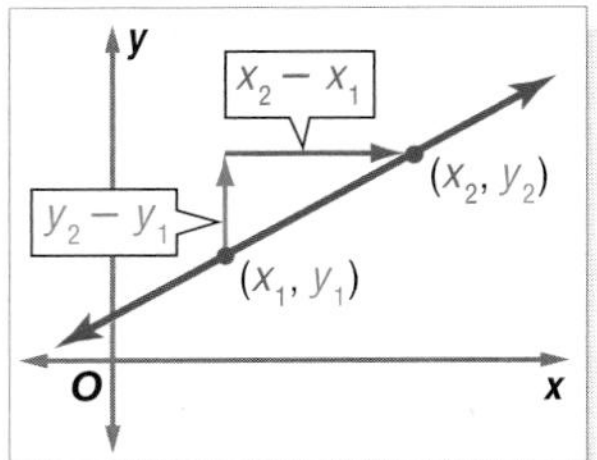

slope-intercept form (p. 272) An equation of the form $y = mx + b$, where m is the slope and b is the y-intercept.

forma pendiente-intersección Ecuación de la forma $y = mx + b$, donde m es la pendiente y b es la intersección y.

solution (pp. 16, 212) A replacement value for the variable in an open sentence.

solución Valor de sustitución de la variable en un enunciado abierto.

solution set (p. 16) The set of elements from the replacement set that make an open sentence true.

conjunto solución Conjunto de elementos del conjunto de sustitución que hacen verdadero un enunciado abierto.

solve an equation (p. 129) The process of finding all values of the variable that make the equation a true statement.

resolver una ecuación Proceso en que se hallan todos los valores de la variable que hacen verdadera la ecuación.

solving an open sentence (p. 16) Finding a replacement value for the variable that results in a true sentence or an ordered pair that results in a true statement when substituted into the equation.

resolver un enunciado abierto Hallar un valor de sustitución de la variable que resulte en un enunciado verdadero o un par ordenado que resulte en una proposición verdadera cuando se lo sustituye en la ecuación.

solving a triangle (p. 626) Finding all of the measures of the sides and the angles in a right triangle.

resolver un triángulo Hallar las medidas de todos los lados y ángulos de un triángulo rectángulo.

square root (p. 103) One of two equal factors of a number.

raíz cuadrada Uno de dos factores iguales de un número.

standard form (p. 218) The standard form of a linear equation is $Ax + By = C$, where $A \geq 0$, and A and B are not both zero.

forma estándar La forma estándar de una ecuación lineal es $Ax + By = C$, donde $A \geq 0$ y ni A ni B son cero simultáneamente.

stem-and-leaf plot (p. 89) A system used to separate data into two numbers that are used to form a stem and a leaf.

diagrama de tallo y hojas Sistema que se usa para separar datos en dos números que se usan para formar un tallo y una hoja.

stratified random sample (p. 708) A sample in which the population is first divided into similar, nonoverlapping groups; a simple random sample is then selected from each group.

muestra aleatoria estratificada Muestra en que la población se divide en grupos similares que no se sobreponen; luego se selecciona una muestra aleatoria simple, de cada grupo.

symmetry (p. 525) A geometric property of figures that can be folded and each half matches the other exactly.

simetría Propiedad geométrica de figuras que pueden plegarse de modo que cada mitad corresponde exactamente a la otra.

system of equations (p. 369) A set of equations with the same variables.

sistema de ecuaciones Conjunto de ecuaciones con las mismas variables.

system of inequalities (p. 394) A set of two or more inequalities with the same variables.

sistema de desigualdades Conjunto de dos o más desigualdades con las mismas variables.

systematic random sample (p. 708) A sample in which the items in the sample are selected according to a specified time or item interval.

muestra aleatoria sistemática Muestra en que los elementos de la muestra se escogen según un intervalo de tiempo o elemento específico.

T

tangent (p. 624) In a right triangle with acute angle A, the tangent cosine of $\angle A = \dfrac{\text{measure of leg opposite } \angle A}{\text{measure of leg adjacent to } \angle A}$.

tangente En un triángulo rectángulo con ángulo agudo A, el coseno tangente de $\angle A = \dfrac{\text{medida del cateto opuesto a } \angle A}{\text{medida del cateto adyacente a } \angle A}$.

term (p. 28) A number, a variable, or a product or quotient of numbers and variables.

término Número, variable o producto, o cociente de números y variables.

terms (p. 233) The numbers in a sequence.

términos Los números de una sucesión.

theoretical probability (p. 782) What should happen in a probability experiment.

probabilidad teórica Lo que debería ocurrir en un experimento probabilístico.

transformation (p. 197) Movements of geometric figures.

transformación Desplazamiento de figuras geométricas.

translation (p. 197) A transformation in which a figure is slid in any direction.

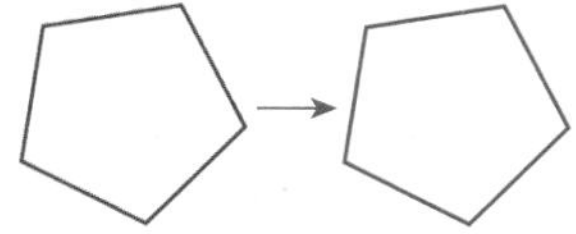

translación Transformación en que una figura se desliza en cualquier dirección.

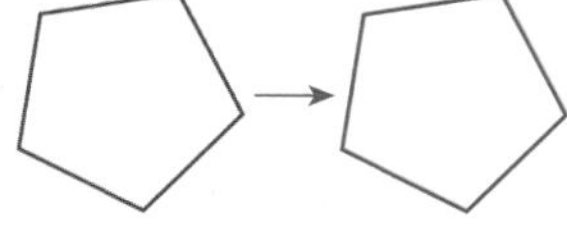

tree diagram (p. 754) A diagram used to show the total number of possible outcomes.

diagrama de árbol Diagrama que se usa para mostrar el número total de resultados posibles.

trigonometric ratios (p. 624) The ratios of the measures of two sides of a right triangle.

razones trigonométricas Razones de las longitudes de dos lados de un triángulo rectángulo.

trinomials (p. 432) The sum of three monomials.

trinomios Suma de tres monomios.

U

uniform motion problems (p. 172) Problems in which an object moves at a certain speed, or rate.

problemas de movimiento uniforme Problemas en que el cuerpo se mueve a cierta velocidad o tasa.

union (p. 340) The graph of a compound inequality containing *or*; the solution is a solution of either inequality, not necessarily both.

unión Gráfica de una desigualdad compuesta que contiene la palabra *o*; la solución es el conjunto de soluciones de por lo menos una de las desigualdades, no necesariamente ambas.

upper quartile (p. 732) The median of the upper half of a set of numbers.

cuartil superior La mediana de la mitad superior de un conjunto de datos.

V

variable (p. 6) Symbols used to represent unspecified numbers or values.

variable Símbolos que se usan para representar números o valores no especificados.

vertex (p. 525) The maximum or minimum point of a parabola.

vértice Punto máximo o mínimo de una parábola.

vertical line test (p. 227) If any vertical line passes through no more than one point of the graph of a relation, then the relation is a function.

prueba de la recta vertical Si cualquier recta vertical pasa por un sólo punto de la gráfica de una relación, entonces la relación es una función.

voluntary response sample (p. 709) A sample that involves only those who want to participate.

muestra de respuesta voluntaria Muestra que involucra sólo aquellos que quieren participar.

W

weighted average (p. 171) The sum of the product of the number of units and the value per unit divided by the sum of the number of units, represented by M.

promedio ponderado Suma del producto del número de unidades por el valor unitario dividida entre la suma del número de unidades y la cual se denota por M.

whole numbers (p. 68) The set {0, 1, 2, 3, ...}.

números enteros El conjunto {0, 1, 2, 3, ...}.

X

x-axis (pp. 43, 192) The horizontal number line on a coordinate plane.

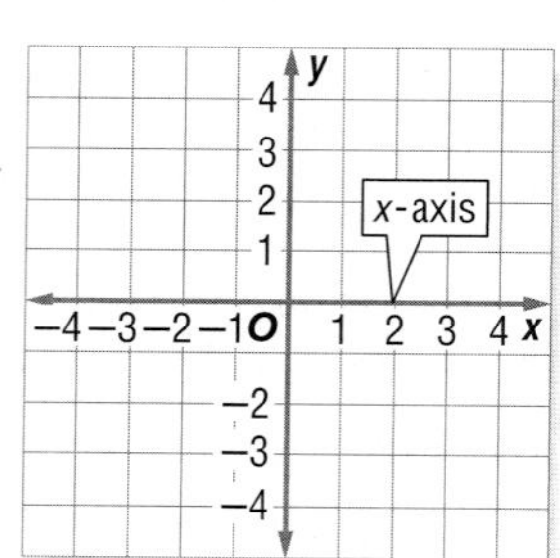

eje x Recta numérica horizontal que forma parte de un plano de coordenadas.

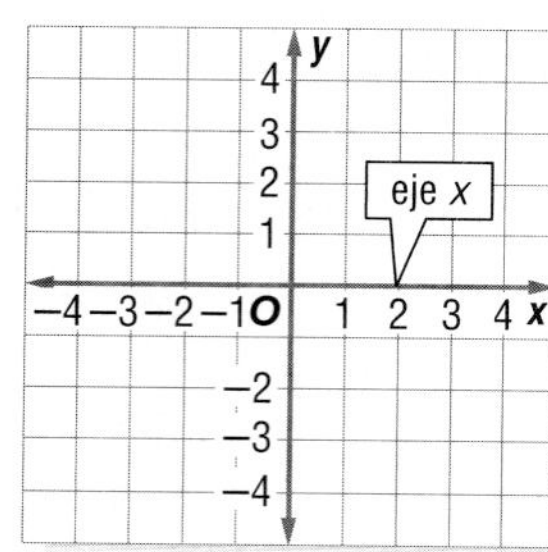

x-coordinate (pp. 43, 192) The first number in an ordered pair.

x-intercept (p. 220) The coordinate at which a graph intersects the x-axis.

coordenada x El primer número de un par ordenado.

intersección x Punto o puntos en los que una gráfica interseca el eje x.

Y

y-axis (pp. 43, 192) The vertical number line on a coordinate plane.

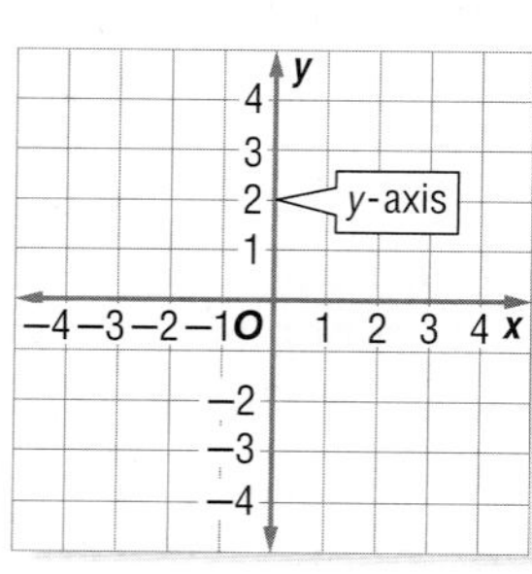

eje y Recta numérica vertical que forma parte de un plano de coordenadas.

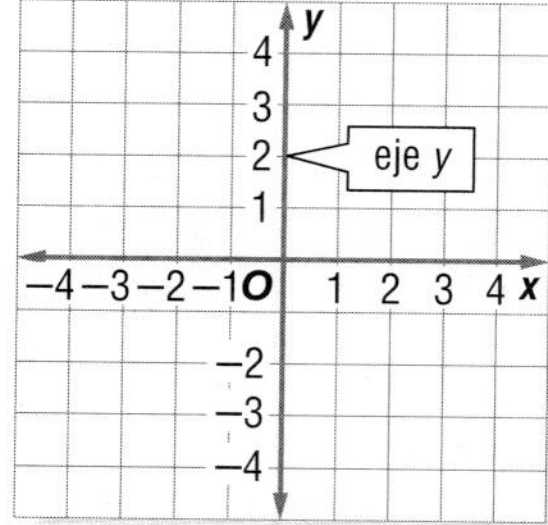

y-coordinate (pp. 43, 192) The second number in an ordered pair.

y-intercept (p. 220) The coordinate at which a graph intersects the y-axis.

coordenada y El segundo número de un par ordenado.

intersección y Punto o puntos en los que una gráfica interseca el eje y.

Z

zero exponent (p. 419) For any nonzero number a, $a^0 = 1$.

zeros (p. 533) The roots, or x-intercepts, of a quadratic function.

exponente cero Para cualquier número no nulo a, $a^0 = 1$.

ceros Las raíces o intersecciones x de una función cuadrática.

Selected Answers

Chapter 1 The Language of Algebra

Page 5 Chapter 1 Getting Started

1. 64 **3.** 162 **5.** 19 **7.** 24 **9.** 16.6 m **11.** $5\frac{1}{2}$ ft **13.** 7.2 **15.** 1.8 **17.** 9 **19.** $\frac{5}{12}$

Pages 8–9 Lesson 1-1

1. Algebraic expressions include variables and numbers, while verbal expressions contain words. **3.** Sample answer: a^5 **5.** Sample answer: $3x - 24$ **7.** 256 **9.** one half of n cubed **11.** $35 + z$ **13.** $16p$ **15.** $49 + 2x$ **17.** $\frac{2}{3}x^2$ **19.** $s + 12d$ **21.** 36 **23.** 81 **25.** 243 **27.** 1,000,000 **29.** $8.5b + 3.99d$ **31.** 7 times p **33.** three cubed **35.** three times x squared plus four **37.** a to the fourth power times b squared **39.** Sample answer: one-fifth 12 times z squared **41.** 3 times x squared minus 2 times x **43.** $x + \frac{1}{11}x$ **45.** $3.5m$ **47.** You can use the expression $4s$ to find the perimeter of a baseball diamond. Answers should include the following.

- four times the length of the sides and the sum of the four sides
- $s + s + s + s$

49. B **51.** 6.76 **53.** 3.2 **55.** $\frac{7}{12}$ **57.** $\frac{7}{6}$ or $1\frac{1}{6}$

Pages 13–15 Lesson 1-2

1. Sample answer: First add the numbers in parentheses, (2 + 5). Next square 6. Then multiply 7 by 3. Subtract inside the brackets. Multiply that by 8. Divide, then add 3. **3.** Chase; Laurie raised the incorrect quantity to the second power. **5.** 26 **7.** 51 **9.** $\frac{11}{100}$ **11.** 160 **13.** $20.00 + 2 \times 9.95$ **15.** 12 **17.** 21 **19.** 0 **21.** 4 **23.** 8 **25.** 6 **27.** $\frac{87}{2}$ or $43\frac{1}{2}$ **29.** 44 cm^2 **31.** \$1625 **33.** 1763 **35.** 24 **37.** 253 **39.** $\frac{37}{8}$ or $4\frac{5}{8}$ **41.** the sum of salary, commission, and 4 bonuses **43.** \$54,900 **45.** Use the order of operations to determine how many extra hours were used then how much the extra hours cost. Then find the total cost. Answers should include the following.

- $6[4.95 + 0.99(n)] - 25.00$
- You can use an expression to calculate a specific value without calculating all possible values.

47. B **49.** 2.074377092 **51.** $a^3 \cdot b^4$ **53.** $a + b + \frac{b}{a}$ **55.** $3(55 - w^3)$ **57.** 12 **59.** 256 **61.** 12 less than q squared **63.** x cubed divided by nine **65.** 7.212 **67.** 14.7775 **69.** $3\frac{11}{35}$ **71.** 36

Pages 18–20 Lesson 1-3

1. Sample answer: An open sentence contains an equals sign or inequality sign. **3.** Sample answer: An open sentence has at least one variable because it is neither true nor false until specific values are used for the variable. **5.** 15 **7.** 1.6 **9.** 3 **11.** {2, 2.5, 3} **13.** 1000 Calories **15.** 12 **17.** 3 **19.** 18 **21.** $1\frac{1}{2}$ **23.** 1.4 **25.** 5.3 **27.** \$22.50 **29.** 11.05 **31.** 5 **33.** 9 **35.** 36 **37.** {6, 7} **39.** {10, 15, 20, 25} **41.** {3.4, 3.6, 3.8, 4} **43.** $\left\{0, \frac{1}{3}, \frac{2}{3}, 1, 1\frac{1}{3}\right\}$ **45.** $g = 15{,}579 + 6220 + 18{,}995$ **47.** $39n + 10.95 \leq 102.50$ **49.** The solution set includes all numbers less than or equal to $\frac{1}{3}$. **51.** B **53.** $r^2 + 3s$; 19 **55.** $(r + s)t^2$; $\frac{7}{4}$ **57.** 173 **59.** 50,628 **61.** $\frac{4}{21}$ **63.** $\frac{2}{7}$ **65.** $\frac{16}{63}$ **67.** $\frac{16}{75}$

Page 21 Practice Quiz 1

1. twenty less than x **3.** a cubed **5.** 28 **7.** 29 **9.** 8

Pages 23–25 Lesson 1-4

1. no; $3 + 1 \neq 3$ **3.** Sample answer: You cannot divide by zero. **5.** Additive Identity; 17

7. $6(12 - 48 \div 4)$
$= 6(12 - 12)$ *Substitution*
$= 6(0)$ *Substitution*
$= 0$ *Multiplicative Property of Zero*

9. $4(20) + 7$ **11.** 87 yr **13.** Multiplicative Identity; 5 **15.** Reflexive; 0.25 **17.** Additive Identity; $\frac{1}{3}$ **19.** Multiplicative Inverse; 1 **21.** Substitution; 3 **23.** Multiplicative Identity; 2

25. $\frac{2}{3}[3 \div (2 \cdot 1)]$
$= \frac{2}{3}(3 \div 2)$ *Multiplicative Inverse*
$= \frac{2}{3} \cdot \frac{3}{2}$ *Substitution*
$= 1$ *Multiplicative Inverse*

27. $6 \cdot \frac{1}{6} + 5(12 \div 4 - 3)$
$= 6 \cdot \frac{1}{6} + 5(3 - 3)$ *Substitution*
$= 6 \cdot \frac{1}{6} + 5(0)$ *Substitution*
$= 6 \cdot \frac{1}{6} + 0$ *Mult. Property of Zero*
$= 1 + 0$ *Multiplicative Inverse*
$= 1$ *Substitution*

29. $7 - 8(9 - 3^2)$
$= 7 - 8(9 - 9)$ *Substitution*
$= 7 - 8(0)$ *Substitution*
$= 7 - 0$ *Mult. Property of Zero*
$= 7$ *Additive Identity*

31. $25(5 - 3) + 80(2.5 - 1) + 40(10 - 6)$
$= 25(2) + 80(2.5 - 1) + 40(10 - 6)$ *Substitution*
$= 25(2) + 80(1.5) + 40(10 - 6)$ *Substitution*
$= 25(2) + 80(1.5) + 40(4)$ *Substitution*
$= 50 + 120 + 160$ *Substitution*
$= 330$ *Substitution*

33. $1653y = 1653$, where $y = 1$ **35.** $8(100{,}000 + 50{,}000 + 400{,}000) + 3(50{,}000 + 50{,}000 + 400{,}000) + 4(50{,}000 + 50{,}000 + 400{,}000)$ **37.** Sometimes; Sample answer: true: $x = 2, y = 1, z = 4, w = 3$; $2 \cdot 4 > 1 \cdot 3$; false: $x = 1, y = -1, z = -2, w = -3$; $1(-2) < (-1)(-3)$ **39.** A **41.** False; $4 - 5 = -1$, which is not a whole number. **43.** False; $1 \div 2 = \frac{1}{2}$, which is not a whole number. **45.** {11, 12, 13} **47.** {3, 3.25, 3.5, 3.75, 4} **49.** $\left\{1\frac{1}{4}\right\}$ **51.** 20 **53.** 31 **55.** 29 **57.** 80 **59.** 28 **61.** 10

Pages 29–31 Lesson 1-5

1. Sample answer: The numbers inside the parentheses are each multiplied by the number outside the parentheses then the products are added. **3.** Courtney; Ben forgot that w^4 is really $1 \cdot w^4$. **5.** $8 + 2t$ **7.** 1632 **9.** $14m$

11. simplified **13.** 12(19.95 + 2) **15.** 96 **17.** 48 **19.** $6x + 18$ **21.** $8 + 2x$ **23.** $28y - 4$ **25.** $ab - 6a$ **27.** $2a - 6b + 4c$ **29.** 4(110,000 + 17,500) **31.** 485 **33.** 102 **35.** 38 **37.** 12(5 + 12 + 18) **39.** 6(78 + 20 + 12) **41.** $1956 **43.** $9b$ **45.** $17a^2$ **47.** $45x - 75$ **49.** $7y^3 + y^4$ **51.** $30m + 5n$ **53.** $\frac{8}{5}a$ **55.** You can use the Distributive Property to calculate quickly by expressing any number as a sum or difference of more convenient numbers. Answers should include the following.

- Both methods result in the correct method. In one method you multiply then add, and in the other you add then multiply.

57. C **59.** Substitution **61.** Multiplicative Inverse **63.** Reflexive **65.** 2258 ft **67.** 11 **69.** 35 **71.** 168 cm^2

Pages 34–36 Lesson 1-6

1. Sample answer: The Associative Property says that the way you group numbers together when adding or multiplying does not change the result. **3.** Sample answer: $1 + 5 + 8 = 8 + 1 + 5$; $(1 \cdot 5)8 = 1(5 \cdot 8)$ **5.** 10 **7.** 130 **9.** $7a + 10b$ **11.** $14x + 6$ **13.** $15x + 10y$ **15.** 46.8 cm^2 **17.** 53 **19.** 20.5 **21.** $9\frac{3}{4}$ **23.** 540 **25.** 32 **27.** 420 **29.** $291 **31.** $77.38 **33.** $2x + 10y$ **35.** $7a^3 + 14a$ **37.** $17n + 36$ **39.** $9.5x + 5.5y$ **41.** $2.9f + 1.2g$ **43.** $\frac{2}{3} + \frac{23}{10}p + \frac{6}{5}q$

45. $5(xy) + 3xy$
$= 5(xy) + 3(xy)$ *Associative Property* (×)
$= xy(5 + 3)$ *Distributive Property*
$= xy(8)$ *Substitution*
$= 8xy$ *Commutative Property* (×)

47. $6(x + y^2) - 3\left(x + \frac{1}{2}y^2\right)$
$= 6x + 6y^2 - 3x - 3\left(\frac{1}{2}y^2\right)$ *Distributive Property*
$= 6x - 3x + 6y^2 - \frac{3}{2}y^2$ *Commutative Property* (+)
$= x(6 - 3) + y^2\left(6 - \frac{3}{2}\right)$ *Distributive Property*
$= x(3) + y^2\left(4\frac{1}{2}\right)$ *Substitution*
$= 3x + 4\frac{1}{2}y^2$ *Commutative Property* (×)

49. You can use the Commutative and Associative Properties to rearrange and group numbers for easier calculations. Answers should include the following.

- $d = (0.4 + 1.1) + (1.5 + 1.5) + (1.9 + 1.8 + 0.8)$

51. B **53.** $15 + 6p$ **55.** $13m + 6n$ **57.** $3t^2 + 4t$ **59.** 36 **61.** 18 **63.** 60 **65.** 13

Page 36 Practice Quiz 2

1. j **3.** i **5.** g **7.** b **9.** h

Pages 39–42 Lesson 1-7

1. Sample answer: If it rains, then you get wet. H: it rains; C: you get wet **3.** Sample answer: You can use deductive reasoning to determine whether a hypothesis and its conclusion are both true or whether one or both are false. **5.** H: you play tennis; C: you run fast **7.** H: Lance does not have homework; C: he watches television; If Lance does not have homework, then he watches television. **9.** H: a quadrilateral with four right angles; C: it is a rectangle; If a quadrilateral has four right angles, then it is a rectangle. **11.** No valid conclusion; the last digit could be any even number. **13.** Anna could have a schedule without science class. **15.** $x = 1$ **17.** A **19.** H: you are in Hawaii; C: you are in the tropics **21.** H: $4(b + 9) \leq 68$; C: $b \leq 8$ **23.** H: $a = b$ and $b = c$; C: $a = c$ **25.** H: it is after school; C: Greg will call; If it is after school, then Greg will call. **27.** H: a number is divisible by 9; C: the sum of its digits is a multiple of 9; If a number is divisible by 9, then the sum of its digits is a multiple of 9. **29.** H: $s > 9$; C: $4s + 6 > 42$; If $s > 9$, then $4s + 6 > 42$ **31.** Ian will buy a VCR. **33.** No valid conclusion; the hypothesis does not say Ian won't buy a VCR if it costs $150 or more. **35.** No valid conclusion; the conditional does not mention Ian buying 2 VCRs. **37.** There is a professional team in Canada. **39.** Left-handed people can have right-handed parents. **41.** 2(8.5) = 17 **43.** $\frac{6}{3} \cdot \frac{1}{2} = 1$ **45.** Sample answer:

R P Q

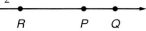

47. Numbers that end in 0, 2, 4, 6, or 8 are in the "divisible by 2" circle. Numbers whose digits have a sum divisible by 3 are in the "divisible by 3" circle. Numbers that end in 0 or 5 are in the "divisible by 5" circle. **49.** no counterexamples **51.** You can use if-then statements to help determine when food is finished cooking. Answers should include the following.

- Hypothesis: you have small, underpopped kernels
  Conclusion: you have not used enough oil in your pan
- If the gelatin is firm and rubbery, then it is ready to eat. If the water is boiling, lower the temperature.

53. C **55.** $a + 15b$ **57.** $23mn + 24$ **59.** $12x^2 + 12x$ **61.** Multiplicative Identity; 64 **63.** Substitution; 5 **65.** Additive Identity; 0 **67.** 41 **69.** 2 **71.** $3n - 10$ **73.** 36 **75.** 171 **77.** 225.5

Pages 46–48 Lesson 1-8

1. The numbers represent different values. The first number represents the number on the horizontal axis and the second represents the number on the vertical axis. **5.** Graph B **7.** (0, 500), (0.2, 480), (0.4, 422), (0.6, 324), (0.8, 186), (1, 10)

9.

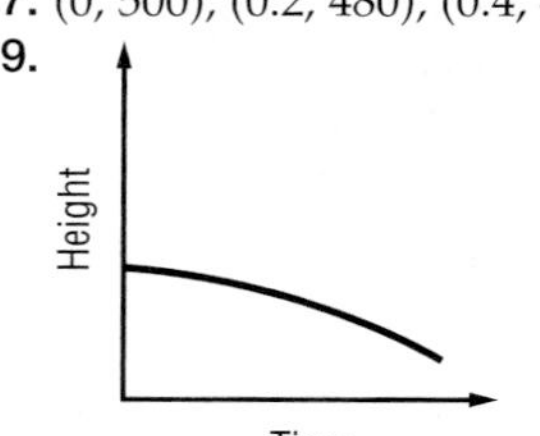

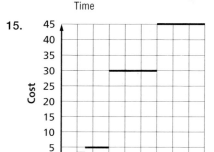

11. Rashaad's account is increasing as he makes deposits and earns interest. Then he pays some bills. He then makes some deposits and earns interest and so on. **13.** Graph B

15.

Cost
45 40 35 30 25 20 15 10 5
0 4 8 12 16 20 24 28 32 36
Time

17. The independent variable is the number of sides and the dependent variable is the sum of the angle measures. **19.** 1080, 1260, 1440 **21.**

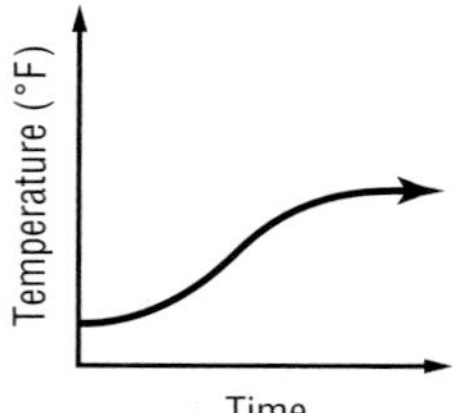

Selected Answers

23. Real-world data can be recorded and visualized in a graph and by expressing an event as a function of another event. Answers should include the following.

- A graph gives you a visual representation of the situation which is easier to analyze and evaluate.
- During the first 24 hours, blood flow to the brain decreases to 50% at the moment of the injury and gradually increases to about 60%.
- Significant improvement occurs during the first two days.

25. A **27.** H: a shopper has 9 or fewer items; C: the shopper can use the express lane **29.** Substitution; 3 **31.** Multiplicative Identity; 1

Pages 53–55 Lesson 1-9

1. Compare parts to the whole; compare different categories of data; show changes in data over time. **3.** Sample answer: The percentages of the data do not total 100. **5.** tennis **7.** 14,900 **9.** Bar graph; a bar graph is used to compare similar data in the same category. **11.** The vertical axis needs to begin at 0. **13.** Sample answer: about 250 time as great **15.** Sample answer: about 2250 **17.** Yes, the graph is misleading because the sum of the percentages is not 100. To fix the graph, each section must be drawn accurately and another section that represents "other" toppings should be added. **19.** Tables and graphs provide an organized and quick way to examine data. Answers should include the following.

- Examine the existing pattern and use it to continue a graph to the future.
- Make sure the scale begins at zero and is consistent. Circle graphs should have all percents total 100%. The right kind of graph should be used for the given data.

21. C **23.** Sample answer: $x = 12$ **25.** $6 + 6 + 2 + 2 = 16$ **27.** $6x^2 + 10x$

Pages 57–62 Chapter 1 Study Guide and Review

1. a **3.** g **5.** h **7.** i **9.** b **11.** x^5 **13.** $x + 21$ **15.** 27 **17.** 625 **19.** the product of three and a number m to the fifth power **21.** 11 **23.** 9 **25.** 0 **27.** 20 **29.** 26 **31.** 96 **33.** 23 **35.** 16 **37.** 13 **39.** 2 **41.** 4 **43.** 9 **45.** {6, 7, 8} **47.** {5, 6, 7, 8}

49. $\frac{1}{2} \cdot 2 + 2[2 \cdot 3 - 1]$

$= \frac{1}{2} \cdot 2 + 2[6 - 1]$ *Substitution*

$= \frac{1}{2} \cdot 2 + 2 \cdot 5$ *Substitution*

$= 1 + 2 \cdot 5$ *Multiplicative Inverse*

$= 1 + 10$ *Substitution*

$= 11$ *Substitution*

51. $1.2 - 0.05 + 2^3$

$= 1.2 - 0.05 + 8$ *Substitution*

$= 1.15 + 8$ *Substitution*

$= 9.15$ *Substitution*

53. $3(4 \div 4)^2 - \frac{1}{4}(8)$

$= 3(1)^2 - \frac{1}{4}(8)$ *Substitution*

$= 3 \cdot 1 - \frac{1}{4}(8)$ *Substitution*

$= 3 - \frac{1}{4}(8)$ *Multiplicative Identity*

$= 3 - 2$ *Substitution*

$= 1$ *Substitution*

55. 72 **57.** $1 - 3p$ **59.** $24x - 56y$ **61.** simplified **63.** $8m + 8n$ **65.** $12y - 5x$ **67.** $9w^2 + w$ **69.** $6a + 13b + 2c$ **71.** $17n - 24$

73. $2pq + pq$

$= (2 + 1)pq$ *Distributive Property*

$= 3pq$ *Substitution*

75. $3x^2 + (x^2 + 7x)$

$= (3x^2 + x^2) + 7x$ *Associative Property*

$= 4x^2 + 7x$ *Substitution*

77. H: a figure is a triangle, C: it has three sides; If a figure is a triangle, then it has three sides. **79.** $a = 15, b = 1, c = 12$ **81.**

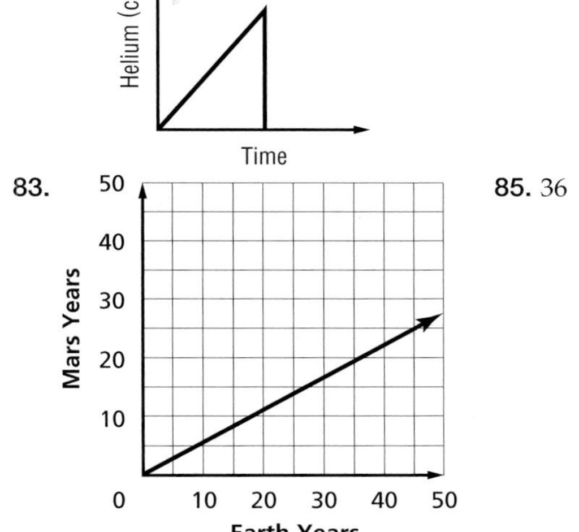

83. **85.** 36

Chapter 2 Real Numbers

Page 67 Getting Started

1. 2.36 **3.** 56.32 **5.** $\frac{11}{12}$ **7.** $\frac{3}{8}$ **9.** 4 **11.** 21.6 **13.** $1\frac{1}{2}$ **15.** 2.1 **17.** $8\frac{1}{6}$; 8; none **19.** 8; 7; 7 **21.** 0.81 **23.** $\frac{16}{25}$

Pages 70–72 Lesson 2-1

1. always **3.** Sample answer: Describing directions such as north versus south, or left versus right.

5. $\left\{\ldots, -\frac{11}{2}, -\frac{9}{2}, -\frac{7}{2}, -\frac{5}{2}, -\frac{3}{2}\right\}$

7. −4 −3 −2 −1 0 1 2 3 4

9. −9 −8 −7 −6 −5 −4 −3 −2 −1 0 1

11. 18 **13.** $\frac{5}{6}$ **15.** 36

17.

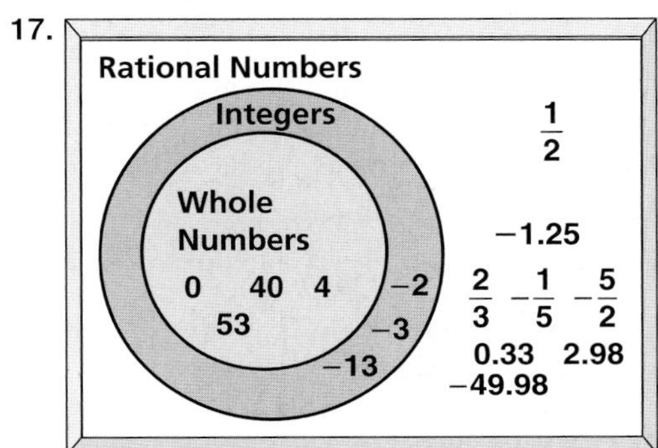

19. {−7, −6, −5, −3, −2} **21.** {..., 0, 0.2, 0.4, 0.6, 0.8}

23. $\left\{\frac{1}{5}, \frac{4}{5}, \frac{7}{5}, \frac{8}{5}, 2\right\}$

25. −1 0 1 2 3 4 5 6 7 8 9 10

27. −2 −1 0 1 2 3 4 5 6

29. −7 −6 −5 −4 −3 −2 −1 0

31. −4 −3 −2 −1 0 1 2 3

33. −6 −4 −2 0 2 4 6 8 10

35. 10 **37.** 61 **39.** 6.8 **41.** $\frac{35}{80}$ **43.** Philadelphia, PA; Sample answer: It had the greatest absolute value. **45.** 55 **47.** 34 **49.** 14 **51.** 1.3 **53.** $\frac{1}{4}$ **55.** $\frac{13}{20}$ **57.** 0 **59.** Bismark, ND 11; Caribou, ME 5; Chicago, IL 4; Fairbanks, AK 9; International Falls, MN 13; Kansas City, MO 7; Sacramento, CA 34; Shreveport, LA 33 **61.** D **63.** December **65.** February, July, October **67.** $9x + 2y$ **69.** $4 + 80x + 32y$ **71.** $\frac{1}{3}$ **73.** $\frac{25}{24}$ or $1\frac{1}{24}$ **75.** $\frac{5}{12}$ **77.** $\frac{7}{18}$

Pages 76–78 Lesson 2-2

1. Sample answer: $\frac{1}{5} - \frac{3}{5}$ **3.** Gabriella; subtracting $-\frac{6}{9}$ is the same as adding $\frac{6}{9}$. **5.** −69 **7.** −17.43 **9.** $\frac{7}{60}$ **11.** 31.1 **13.** 10.25 **15.** $\frac{13}{60}$ **17.** 5 **19.** −22 **21.** −123 **23.** −5.4 **25.** −14.7 **27.** −14.7 **29.** $\frac{32}{21}$ or $1\frac{11}{21}$ **31.** $\frac{13}{55}$ **33.** $-\frac{199}{240}$ **35.** $2\frac{5}{8}$ **37.** 400 points **39.** −27 **41.** 33 **43.** −19 **45.** −16 **47.** 1.798 **49.** 105.3 **51.** $-\frac{5}{6}$ **53.** $-\frac{11}{16}$ **55.** $-\frac{49}{12}$ or $-4\frac{1}{12}$ **57.** −2, −6, −4, −4 **59.** Under; yes, it is better than par 72. **61.** week 7 **63.** Sometimes; the equation is false for positive values of x, but true for all other values of x. **65.** C **67.** 15.4 **69.** 15.9

71.

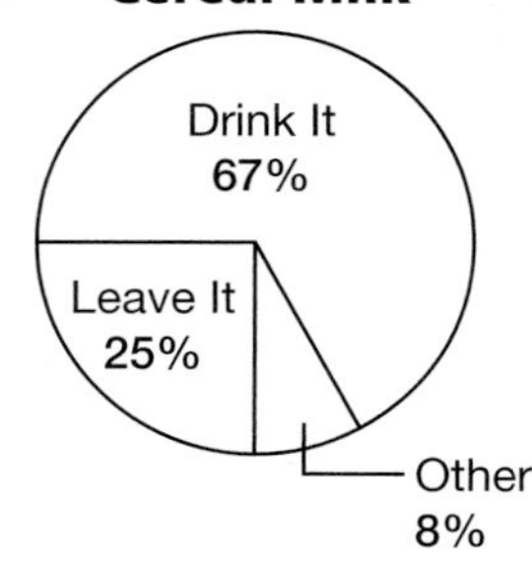

73. {5, 6} **75.** $q^2 - 8$ **77.** $\frac{1}{3}$ **79.** $\frac{5}{8}$ **81.** 5

Pages 81–83 Lesson 2-3

1. ab will be negative if one factor is negative and the other factor is positive. Let $a = -2$ and $b = 3$: $-2(3) = -6$. Let $a = 2$ and $b = -3$: $2(-3) = -6$. **3.** Since multiplication is repeated addition, multiplying a negative number by another negative number is the same as adding repeatedly in the opposite, or positive direction. **5.** −40 **7.** 90.48 **9.** $-\frac{28}{135}$ **11.** $-57xy$ **13.** $-\frac{15}{8}$ or $-1\frac{7}{8}$ **15.** $56\frac{1}{4}$ t **17.** 176 **19.** −192 **21.** 3888 **23.** $\frac{5}{27}$ **25.** $-\frac{12}{35}$ **27.** $4\frac{1}{2}$ **29.** 0.845 **31.** −0.48 **33.** 8 **35.** $-45n$ **37.** $-28d$ **39.** $-21mn - 12st$ **41.** −$134.50 **43.** −30.42 **45.** 4.5 **47.** −13.53 **49.** −208.377 **51.** $1205.35 **53.** 60 million **55.** Positive; the product of two negative numbers is positive and all even numbers can be divided into groups of two. **57.** B **59.** −12.1 **61.** 56

63. −4 −3 −2 −1 0 1 2 3 4 5 6

65. −1 $-\frac{1}{3}$ 0 $\frac{2}{3}$ 1 2

67. Sample answer: $x = 5$ **69.** $\frac{5}{16}$ **71.** $6\frac{2}{3}$ **73.** $1\frac{1}{3}$ **75.** $\frac{2}{3}$

Page 83 Practice Quiz 1

1. {−4, −1, 1, 6} **3.** −8 **5.** −8.15 **7.** 108 **9.** $16xy - 3yz$

Pages 86–87 Lesson 2-4

1. Sample answer: Dividing and multiplying numbers with the same signs both result in a positive answer while dividing or multiplying numbers with different signs results in a negative answer. However, when you divide rational numbers in fractional form, you must multiply by a reciprocal. **3.** To divide by a rational number, multiply by its reciprocal. **5.** −9 **7.** 25.76 **9.** $-\frac{5}{6}$ **11.** $-65a$ **13.** 1.2 **15.** 1.67 **17.** 8 **19.** 60 **21.** −7.05 **23.** −2.28 **25.** 12.9 **27.** $-\frac{1}{12}$ **29.** $-\frac{35}{3}$ or $-11\frac{2}{3}$ **31.** $\frac{10}{9}$ or $1\frac{1}{9}$ **33.** $-\frac{175}{192}$ **35.** $\frac{222}{5}$ or $44\frac{2}{5}$ **37.** $9c$ **39.** $-r - 3$ **41.** $20a - 25b$ **43.** $-f - 2g$ **45.** 2 **47.** −16.25 **49.** 2.08 **51.** −1.21 **53.** 1.76 **55.** $1998.75 **57.** 16-karat gold **59.** Sample answer: You use division to find the mean of a set of data. Answers should include the following.

- You could track the mean number of turtles stranded each year and note if the value increases or decreases.
- Weather or pollution could affect the turtles.

61. B **63.** 3 **65.** 0.48 **67.** −6 **69.** $-\frac{11}{24}$ **71.** $20b + 24$ **73.** $3x + 4y$ **75.** 6.25; 5.5; 3 **77.** 79.25; 79.5; 84

Pages 91–94 Lesson 2-5

1. They describe the data as a whole. **3.** Sample answer: 13, 14, 14, 28

5.

0 2 4 6 8 10 12 14

7. The mean and the median both represent the data accurately as they are fairly central. **9.** 3.6

11.

Stem	Leaf
5	4 5 5 6
6	0 1 4 9
7	0 3 5 7 8
8	0 0 3 5 8 8 8
9	0
10	0 2 5
11	0 5|4 = 54

13. The mode is not the best measure as it is higher than most of the values.

15.

−2 −1 0 1 2 3

17. 23 **19.** Sample answer: Median; most of the data are near 2.

21.

Stem	Leaf
1	8 8
2	2 3 6 6 6 8 9
3	0 1 1 2 3 4
4	7 1|8 = 18

23. 118 **27.** Mean or median; both are centrally located and the mode is too high. **29.** 7 **31.** Sample answer: Yes; most of the data are near the median. **33.** 22

35.

Stem	Leaf
3	0 4 7
4	
5	2 9
6	2 7
7	7
8	4 5 $3\|0 = 30$

37. no mode **39.** High school: \$10,123; College: \$11,464; Bachelor's Degree: \$18,454; Doctoral Degree: \$21,608

41. Sample answer: Because the range in salaries is often very great with extreme values on both the high end and low end. **43.** C **45.** -4 **47.** -13.5 **49.** $-17x$ **51.** $-3t$ **53.** 1 **55.** 9 **57.** $\frac{2}{3}$ **59.** $\frac{7}{10}$ **61.** $\frac{1}{2}$ **63.** $\frac{4}{9}$

Pages 98–101 Lesson 2-6

1. Sample answers: impossible event: a number greater than 6; certain event: a number from 1 to 6; equally likely event: even number **3.** Doug; Mark determined the odds in favor of picking a red card. **5.** $\frac{1}{26}$ **7.** $\frac{1}{26}$ **9.** 3:7 **11.** 6:4 **13.** $\frac{3}{10}$ **15.** $\frac{1}{3} \approx 33\%$ **17.** $\frac{1}{2} = 50\%$ **19.** $\frac{13}{30} \approx 43\%$ **21.** $1 = 100\%$ **23.** $\frac{7}{12} \approx 58\%$ **25.** $1 = 100\%$ **27.** $\frac{25}{36} \approx 69\%$ **29.** $\frac{1}{6} \approx 17\%$ **31.** $\frac{2}{3} \approx 67\%$ **33.** $\frac{1}{2} = 50\%$ **35.** $\frac{15}{31} \approx 48\%$ **37.** 4:20 or 1:5 **39.** 13:11 **41.** 9:15 or 3:5 **43.** 12:20 or 3:5 **45.** 15:17 **47.** 13:19 **49.** 1:2 **51.** $\frac{19}{40} = 47.5\%$ **53.** 7:13 **x** **55.** 42:4 or 21:2 **57.** $\frac{1}{1,000,001}$ **59.** $\frac{6}{7} \approx 86\%$ **61.** B

63.

Stem	Leaf
5	8.3
6	4.3 5.1 5.5 6.7 7.0 8.7 9.3
7	0.0 2.8 3.2 5.8 7.4 7.4 $5\|8.3 = 58.3$

65. $-\frac{5}{3}$ or $-1\frac{2}{3}$ **67.** -3.9 **69.** $-\frac{5}{8}$ **71.** 4.25 **73.** $\frac{2}{3}$ **75.** 36 **77.** 64 **79.** 2.56 **81.** $\frac{16}{81}$

Page 101 Practice Quiz 2

1. 17 **3.** -11.7 **5.** $x + 8$ **7.** Sample answer: scale 0–5.0

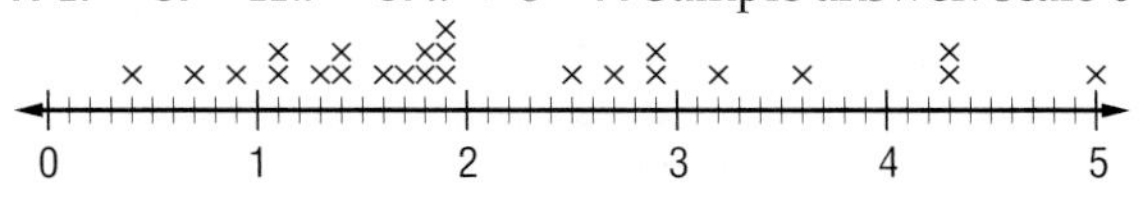

9. $\frac{13}{18}$

Pages 107–109 Lesson 2-7

1. Sometimes; the square root of a number can be negative, such as $\sqrt{16} = 4$ and $-\sqrt{16} = -4$. **3.** There is no real number that can be multiplied by itself to result in a negative product. **5.** 1.2 **7.** 5.66 **9.** rationals **11.** naturals, wholes, integers, rationals

13. [number line: -8 -7 -6 -5 -4 -3 -2 -1 0 1]

15. = **17.** $-15, \frac{1}{8}, 0.\overline{15}, \sqrt{\frac{1}{8}}$ **19.** C **21.** 9 **23.** 2.5 **25.** -9.70 **27.** $\pm\frac{5}{7}$ **29.** 0.77 **31.** ± 22.65 **33.** naturals, wholes, integers, rationals **35.** rationals **37.** irrationals **39.** rationals **41.** rationals **43.** rationals **45.** rationals **47.** irrationals **49.** irrational **51.** No; Jerome was traveling at about 32.4 mph.

53. [number line: 1 2 3 4 5 6 7 8 9]

55. [number line: -2 -1 0]

57. [number line: -8 -6 -4 -2 0 2 4 6 8]

59. < **61.** < **63.** > **65.** $0.2\overline{4}, \sqrt{0.06}, \frac{\sqrt{9}}{12}$ **67.** $-4.8\overline{3}, -\frac{3}{8}, 0.4, \sqrt{8}$ **69.** $7\frac{4}{9}, \sqrt{122}, \sqrt{200}$ **71.** about 3.4 mi **73.** They are true if q and r are positive and $q > r$. **75.** The length of the side is the square root of the area. **77.** Sample answer: By using the formula Surface Area = $\sqrt{\frac{\text{height} \times \text{weight}}{3600}}$, you need to use square roots to calculate the quantity. Answers should include the following.

- You must multiply height by weight first. Divide that product by 3600. Then determine the square root of that result.
- Sample answers: exposure to radiation or chemicals; heat loss; scuba suits
- Sample answers: determining height, distance

79. B **81.** 5:8 **83.** 12:1 **85.** -61 **87.** $5.1x - 7.6y$

Pages 110–114 Chapter 2 Study Guide and Review

1. true **3.** true **5.** true **7.** false; sample answer: $0.\overline{6}$ or 0.666…

9. [number line: -4 -3 -2 -1 0 1 2 3 4 5 6]

11. [number line: -6 -5 -4 -3 -2 -1 0 1 2 3]

13. 5 **15.** 14 **17.** -5 **19.** -1.4 **21.** $\frac{1}{2}$ **23.** 16 **25.** -2.5 **27.** $\frac{13}{24}$ **29.** -36 **31.** 8.64 **33.** $\frac{3}{10}$ **35.** n **37.** -9 **39.** -10.9 **41.** -20 **43.** $-2 + 4x$ **45.** $-x + 6y$ **47.** -3.2

49. [line plot: 5 10 15 20 25 30]

Stem	Leaf
1	2 2 2 3 4 4 5 5 5 5 6 6 7
	7 7 7 8 8 9 9 9 9 9
2	0 1 1 1 2 6 6 8
3	0 $1\|2 = 12$

51. Sample answer; Median; it is closest in value to most of the data **53.** $\frac{1}{4}$ **55.** $\frac{1}{4}$ **57.** 18:31 **59.** 25:24 **61.** ± 1.1 **63.** $\pm\frac{2}{15}$ **65.** naturals, wholes, integers, rationals **67.** < **69.** >

Chapter 3 Solving Linear Equations

Page 119 Chapter 3 Getting Started

1. $\frac{1}{2}t + 5$ **3.** $3a + b^2$ **5.** $95 - 9y$ **7.** 15 **9.** 16 **11.** 7 **13.** 5 **15.** 25% **17.** 300% **19.** 160%

Pages 123–126 Lesson 3-1

1. Explore the problem, plan the solution, solve the problem, and examine the solution. **3.** Sample answer: After sixteen people joined the drama club, there were 30 members. How many members did the club have before the new members? **5.** $5(m + n) = 7n$ **7.** $C = 2\pi r$ **9.** $\frac{1}{3}$ of b minus $\frac{3}{4}$ equals 2 times a. **11.** $155 + g = 160$ **13.** $200 - 3x = 9$ **15.** $\frac{1}{3}q + 25 = 2q$ **17.** $2(v + w) = 2z$ **19.** $g \div h = 2(g + h) + 7$ **21.** $0.46E = P$ **23.** $A = bh$

25. $P = 2(a + b)$ 27. $c^2 = a^2 + b^2$ 29. d minus 14 equals 5. 31. k squared plus 17 equals 53 minus j. 33. $\frac{3}{4}$ of p plus $\frac{1}{2}$ equals p. 35. 7 times the sum of m and n equals 10 times n plus 17. 37. The area A of a trapezoid equals one-half times the product of the height h and the sum of the bases, a and b. 39. Sample answer: Lindsey is 7 inches taller than Yolanda. If 2 times Yolanda's height plus Lindsey's height equals 193 inches, find Yolanda's height. 41. $V = \frac{1}{3}\pi r^2 h$ 43. $V = \frac{4}{3}\pi r^3$ 45. $1912 + y$ 47. 16 yr 49. $a + (4a + 15) = 60$
53. Equations can be used to describe the relationships of the heights of various parts of a structure. Answers should include the following.

- The equation representing the Sears Tower is $1454 + a = 1707$.

55. D 57. $-\frac{5}{6}$ 59. −7.42 61. $\frac{1}{2}$ 63. $8d + 3$ 65. $8a + 6b$ 67. 408 69. 9.37 71. 1.88 73. $\frac{13}{15}$ 75. $\frac{1}{9}$

Pages 131–134 Lesson 3-2

1. Sample answers: $n = 13$, $n + 16 = 29$, $n + 12 = 25$
3. (1) Add −94 to each side. (2) Subtract 94 from each side.
5. −13 7. 171 9. $\frac{5}{6}$ 11. $n + (-37) = -91$; −54 13. 16.8 h 15. 23 17. 28 19. 38 21. 43 23. −96 25. 73 27. 3.45 29. −2.58 31. 15.65 33. $1\frac{7}{12}$ 35. $1\frac{1}{8}$ 37. $-\frac{2}{15}$ 39. 19 41. $x + 55 = 78$; 23 43. $n - 18 = 31$; 49 45. $n + (-16) = -21$; −5 47. $n - \frac{1}{2} = -\frac{3}{4}$; $-\frac{1}{4}$ 49. Sometimes, if $x = 0$, $x + x = x$ is true. 51. $\ell + 10 = 34$ 53. 37 mi 55. Sample answer: 29 mi; 29 is the average of 24 (for the 8-cylinder engine) and 34 (for the 4-cylinder engine). 57. 31 ft 59. $11.4 + x = 13.6$; 2.2 million volumes 61. $24.0 + 13.6 + 11.4 = x$; 49.0 million volumes 63. $1379 + 679 + 1707 + x = 1286 + 634 + 3714$; 1869 65. $a = b, x = 0$ 67. C 69. $A = \pi r^2$ 71. < 73. = 75.

Stem	Leaf
0	5 8
1	1 2 4 7
2	3 6 8 9
3	
4	1 5 $0\|5 = 0.5$

77. H: it is Friday; C: there will be a science quiz
79. $(2^5 - 5^2) + (4^2 - 2^4)$
$= (32 - 25) + (16 - 16)$ *Substitution*
$= 7 + 0$ *Substitution*
$= 7$ *Additive Identity*

81. {1, 3, 5} 83. 10.545 85. 0.22 87. $\frac{1}{6}$ 89. $3\frac{1}{3}$

Pages 138–140 Lesson 3-3

1. Sample answer: $4x = -12$ 3. Juanita; to find an equivalent equation with $1n$ on one side of the equation, you must divide each side by 8 or multiply each side by $\frac{1}{8}$.
5. −35 7. $1\frac{1}{9}$ 9. $\frac{10}{13}$ 11. $\frac{2}{5}n = -24$; −60 13. −11 15. 35 17. −77 19. 21 21. 10 23. −6.2 25. −3.5 27. $8\frac{6}{13}$ 29. $\frac{11}{15}$ 31. 30 33. $7n = -84$; −12 35. $\frac{1}{5}n = 12$; 60 37. $2\frac{1}{2}n = 1\frac{1}{5}$; $\frac{12}{25}$ 39. $\ell = \frac{1}{7}p$ 41. 455 people 43. 0.48 s 45. about 0.02 s 47. $x + 8x = 477$ 49. 424 g
51. You can use the distance formula and the speed of light to find the time it takes light from the stars to reach Earth. Answers should include the following.

- Solve the equation by dividing each side of the equation by 5,870,000,000,000. The answer is 53 years.
- The equation $5{,}870{,}000{,}000{,}000t = 821{,}800{,}000{,}000{,}000$ describes the situation for the star in the Big Dipper farthest from Earth.

53. A 55. 13 57. $10a = 5(b + c)$ 59. 0.00879
61. (number line: −4 −3 −2 −1 0 1 2 3 4)
63. (number line: −7 −6 −5 −4 −3 −2 −1 0 1)
65. Commutative Property of Addition 67. 25 69. 9

Page 140 Practice Quiz 1

1. $S = 4\pi r^2$ 3. −45 5. −24 7. 27 9. −9

Pages 145–148 Lesson 3-4

1. Sample answers: $2x + 3 = -1$, $3x - 1 = -7$ 3. $n - 2$
5. 6 7. −1 9. $12\frac{2}{3}$ 11. 28 13. $12 - 2n = -34$; 23 15. 12 letters 17. 24 19. 80 lb 21. \$60 23. −6 25. −7 27. −15 29. −56 31. −125 33. $25\frac{1}{3}$ 35. −42.72 37. −12.6 39. 7 41. 2 43. $29 = 13 + 4n$; 4 45. $n + (n + 2) + (n + 4) = -30$; −12, −10, −8 47. $n + (n + 2) + (n + 4) + (n + 6) = 8$; −1, 1, 3, 5 49. 16 cm, 18 cm, 20 cm 51. 10 in. 53. \$75,000 55. never 57. B 59. −3 61. −126 63. 5 65. −13 67. $2\frac{1}{4}$ 69. 29 models 71. 1:1 73. $-\frac{2}{7}$ 75. $-\frac{3}{4}a + 4$ 77. 153 79. 20 81. $5m + \frac{n}{2}$ 83. $3a + b^2$ 85. $6m$ 87. $-8g$ 89. $-10m$

Pages 151–154 Lesson 3-5

1a. Incorrect; the 2 must be distributed over both g and 5; 6. 1b. correct 1c. Incorrect; to eliminate $-6z$ on the left side of the equal sign, $6z$ must be added to each side of the equation; 1. 3. Sample answer: $2x - 5 = 2x + 5$ 5. 4 7. 3 9. 2.6 11. all numbers 13. D 15a. Subtract v from each side. 15b. Simplify. 15c. Subtract 9 from each side. 15d. Simplify. 15e. Divide each side by 6. 15f. Simplify. 17. 4 19. −3 21. $-1\frac{1}{2}$ 23. 4 25. 8 27. no solution 29. 2 31. 10 33. −4 35. 4 37. 0.925 39. all numbers 41. −36 43. 26, 28, 30 45. 8-penny 47. 2.5 by 0.5 and 1.5 by 1.5 49. Sample answer: $3(x + 1) = x - 1$ 51. D 53. 90 55. −2 57. $33\frac{1}{3}$ min
59.

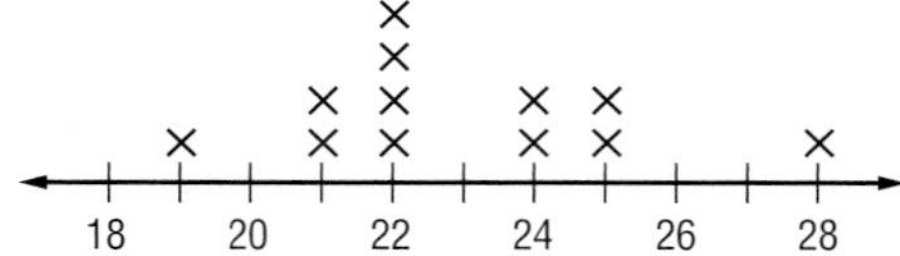

61. −4 63. Sample answer: $1 + 3 = 4$ 65. 5 67. 0 69. $\frac{4}{7}$ 71. $\frac{1}{15}$ 73. $\frac{2}{3}$ 75. $\frac{1}{3}$

Page 158–159 Lesson 3-6

3. Find the cross products and divide by the value with the variable. 5. no 7. 8 9. 4.62 11. yes 13. no 15. no 17. USA: $\frac{871}{2116}$; USSR/Russia: $\frac{498}{1278}$; Germany: $\frac{374}{1182}$; GB: $\frac{180}{638}$; France: $\frac{188}{598}$; Italy: $\frac{179}{479}$; Sweden: $\frac{136}{469}$ 19. 20 21. 18 23. $9\frac{1}{3}$ 25. 2.28 27. 1.23 29. $19\frac{1}{3}$ 31. 14 days 33. 3 in. 35. 18
37. Sample answer: Ratios are used to determine how much of each ingredient to use for a given number of servings. Answers should include the following.

- To determine how much honey is needed if you use 3 eggs, write and solve the proportion $2:\frac{3}{4} = 3:h$, where h is the amount of honey.
- To alter the recipe to get 5 servings, multiply each amount by $1\frac{1}{4}$.

39. C **41.** no solution **43.** −2 **45.** −8 **47.** −1 **49.** 0.4125 **51.** 77 **53.** 0.85 **55.** 30% **57.** 40%

Page 162–164 Lesson 3-7

1. Percent of increase and percent of decrease are both percents of change. If the new number is greater than the original number, the percent of change is a percent of increase. If the new number is less than the original number, the percent of change is a percent of decrease. **3.** Laura; Cory used the new number as the base instead of the original number. **5.** increase; 11% **7.** decrease; 20% **9.** \$16.91 **11.** \$13.37 **13.** about 77% **15.** decrease; 28% **17.** increase; 162% **19.** decrease; 27% **21.** increase; 6% **23.** increase; 23% **25.** decrease; 14% **27.** 30% **29.** 8 g **31.** \$14.77 **33.** \$7.93 **35.** \$42.69 **37.** \$27.00 **39.** \$24.41 **41.** \$96.77 **43.** \$101.76 **45.** \$46.33 **47.** India **49.** always; $x\%$ of $y \rightarrow \frac{x}{100} = \frac{P}{y}$ or $P = \frac{xy}{100}$; $y\%$ of $x \rightarrow \frac{y}{100} = \frac{P}{x}$ or $P = \frac{xy}{100}$ **51.** B **53.** 9 **55.** 18 **57.** −6 **59.** $\frac{1}{10}$ **61.** $\frac{4}{27}$ **63.** false **65.** true **67.** −3 **69.** −11 **71.** 3

Page 164 Practice Quiz 2

1. $-8\frac{1}{3}$ **3.** 1.5 **5.** all numbers **7.** 5 **9.** 5

Pages 168–170 Lesson 3-8

1. (1) Subtract az from each side. (2) Add y to each side. (3) Use the Distributive Property to write $ax - az$ as $a(x - z)$. (4) Divide each side by $x - z$. **3.** Sample answer for a triangle: $A = \frac{1}{2}bh$; $b = \frac{2A}{h}$ **5.** $a = \frac{54 + y}{5}$ **7.** $y = 3c - a$ **9.** $w = \frac{5 + t}{m - 2}$ **11.** $h = \frac{2A}{b}$ **13.** $g = -\frac{h}{4}$ **15.** $m = \frac{y - b}{x}$ **17.** $y = \frac{am - z}{7}$ **19.** $m = \frac{6y - 5x}{k}$ **21.** $x = \frac{n - 20}{3a}$ **23.** $y = \frac{3c - 2}{b}$ **25.** $y = \frac{4}{3}(c - b)$ **27.** $A = \frac{2S - nt}{n}$ **29.** $a = \frac{c + b}{r - t}$ **31.** $t - 5 = r + 6$; $t = r + 11$ **33.** $\frac{5}{8}x = \frac{1}{2}y + 3$; $y = \frac{5}{4}x - 6$ **35.** 6 m **37.** 3 errors **39.** 225 lb **41.** about 17.4 cm

43. Equations from physics can be used to determine the height needed to produce the desired results. Answers should include the following.

- Use the following steps to solve for h. (1) Use the Distributive Property to write the equation in the form $195g - hg = \frac{1}{2}mv^2$. (2) Subtract $195g$ from each side. (3) Divide each side by $-g$.
- The second hill should be 157 ft.

45. C **47.** \$9.75 **49.** 22.5 **51.** 5 **53.** $\frac{2}{3}$, 1.1, $\sqrt{5}$, 3 **55.** $\frac{1}{4}$ **57.** Multiplicative Identity Property **59.** Reflexive Property **61.** $12 - 6t$ **63.** $-21a - 7b$ **65.** $-9 + 3t$

Pages 174–177 Lesson 3-9

1. Sample answer: grade point average

3.

	Number of Coins	Value of Each Coin	Total Value
Dimes	d	\$0.10	$0.10d$
Quarters	$d - 8$	\$0.25	$0.25(d - 8)$

5. $0.10(6 - p) + 1.00p = 0.40(6)$ **7.** 4 qt **9.** about 3.56

11.

	Number of Dozens	Price per Dozen	Total Price
Peanut Butter	p	\$6.50	$6.50p$
Chocolate Chip	$p - 85$	\$9.00	$9.00(p - 85)$

13. 311 doz

15.

	Number of Ounces	Price per Ounce	Value
Gold	g	\$270	$270g$
Silver	$15 - g$	\$5	$5(15 - g)$
Alloy	15	\$164	164(15)

17. 9 oz

19.

	r	t	$d = rt$
Eastbound Train	40	t	$40t$
Westbound Train	30	t	$30t$

21. $3\frac{1}{2}h$ **23.** 15 lb **25.** 200 g of 25% alloy, 800 g of 50% alloy **27.** 120 mL of 25% solution, 20 mL of 60% solution **29.** 87 **31.** 15 s **33.** 3.2 qt **35.** about 98.0

37. A weighted average is used to determine a skater's average. Answers should include the following.

- The score of the short program is added to twice the score of the long program. The sum is divided by 3.
- $\frac{4.9(1) + 5.2(2)}{1 + 2} = 5.1$

39. C **41.** $b = 4a + 25$ **43.** increase; 20% **45.** 2:1 **47.** $3xy$ **49.** $\{\ldots, -2, -1, 0, 1, 2, 3\}$

Pages 179–184 Chapter 3 Study Guide and Review

1. Addition **3.** different **5.** identity **7.** increase **9.** weighted average **11.** $3n - 21 = 57$ **13.** $a^2 + b^3 = 16$ **15.** −16 **17.** 21 **19.** −8.5 **21.** −7 **23.** 40 **25.** −10 **27.** 3 **29.** −153 **31.** 11 **33.** 2 **35.** 1 **37.** −3 **39.** 18 **41.** 9 **43.** 1 **45.** decrease; 20% **47.** increase; 6% **49.** \$10.39 **51.** $y = \frac{b + c}{a}$ **53.** $y = \frac{7a + 9b}{8}$ **55.** 450 mph, 530 mph

Chapter 4 Graphing Relations and Functions

Page 191 Chapter 4 Getting Started

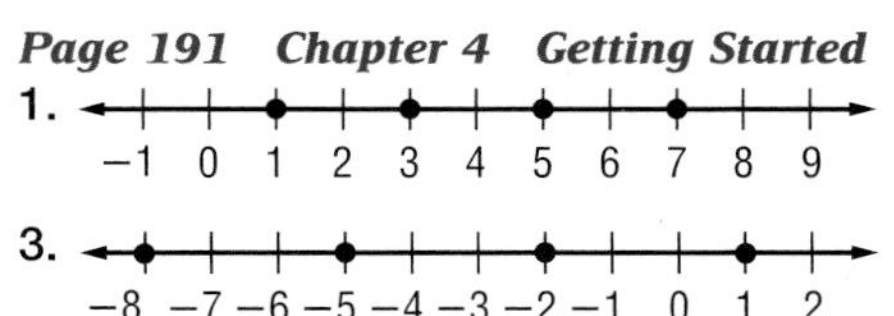

5. $21 - 3t$ **7.** $-15b + 10$ **9.** $y = 1 - 2x$ **11.** $y = 2x - 4$ **13.** $y = 18 - 8x$ **15.** 6 **17.** 0 **19.** 3

Pages 194–196 Lesson 4-1

1.

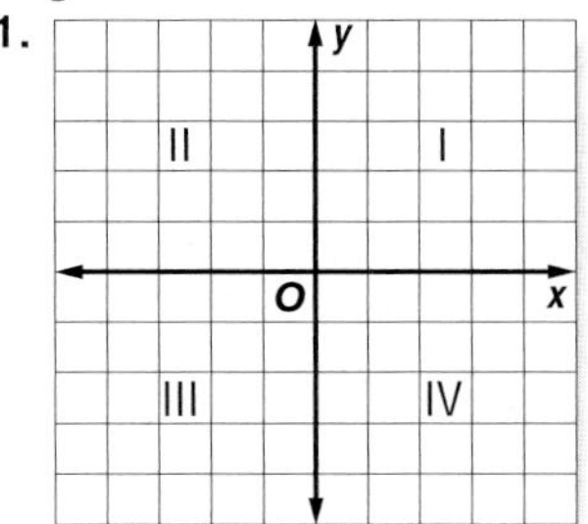

3. Sample answer: I(3, 3), II(−3, 3), III(−3, −3), IV(3, −3) **5.** (−1, 1); II **7.** (−4, −2); III

Selected Answers

8–11.

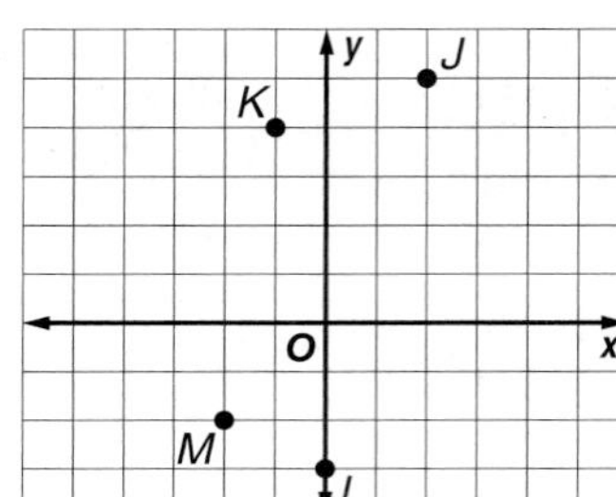

13. (−4, 5); II **15.** (−1, −3); III **17.** (−3, 3); II **19.** (2, −1); IV **21.** (0, 4); none **23.** (7, −12)

25–36.

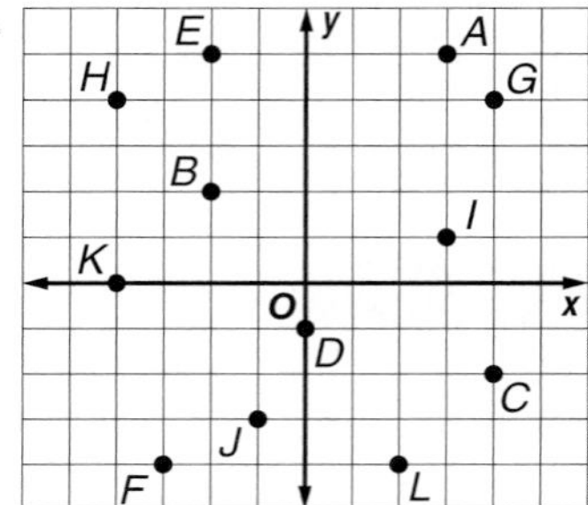

37. Sample answer: Louisville and Richmond **39.** coins, (3, 5); plate, (7, 2); goblet, (8, 4); vase, (5, 9) **41.** C4 **43.** B5, C2, D4, E1

45. Archaeologists used coordinate systems as a mapping guide and as a system to record locations of artifacts. Answers should include the following.

- The grid gives archaeologists a point of reference so they can identify and explain to others the location of artifacts in a site they are excavating. You can divide the space so more people can work at the same time in different areas.
- Knowing the exact location of artifacts helps archaeologists reconstruct historical events.

47. B **49.** (7, −5) **51.** 320 mph **53.** $d = c$ **55.** $t = \frac{3a}{11}$ **57.** 7.94 **59.** −16 **61.** 51 **63.** 30 **65.** 48 **67.** $-x - 3$ **69.** $-6x + 15$ **71.** $\frac{5}{4}x - \frac{1}{2}y$

Pages 200–203 Lesson 4-2

1.

Transformation	Size	Shape	Orientation
Reflection	same	same	changes
Rotation	same	same	changes
Translation	same	same	same
Dilation	changes	same	same

3. translation

5. $P'(1, -2)$, $Q'(4, -4)$, $R'(2, 3)$

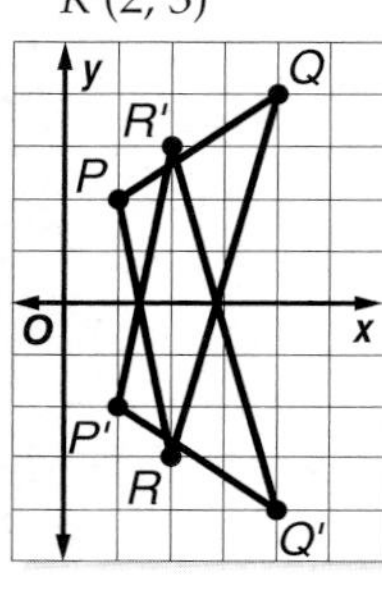

7. $E'(-2, 8)$, $F'(10, -2)$, $G'(4, -8)$, $H'(-8, 2)$

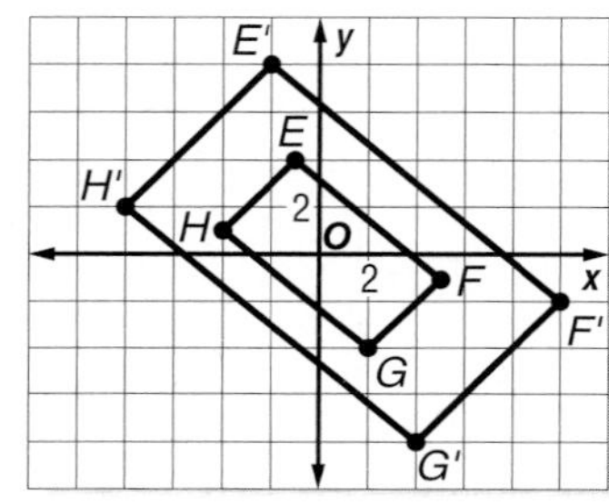

9.

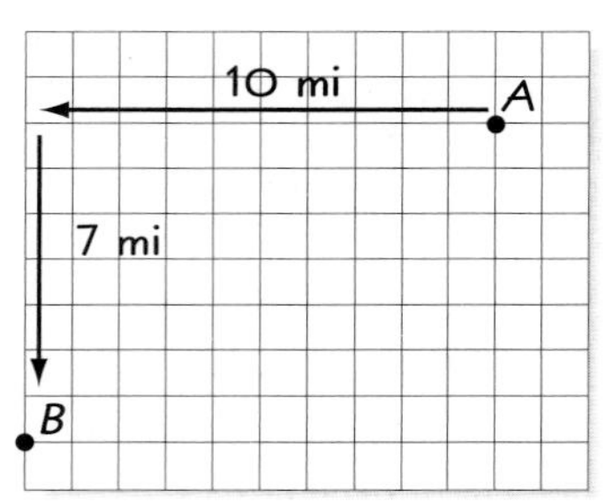

11. translation **13.** reflection **15.** reflection

17. $R'(-2, 0)$, $S'(2, -3)$, $T'(2, 3)$

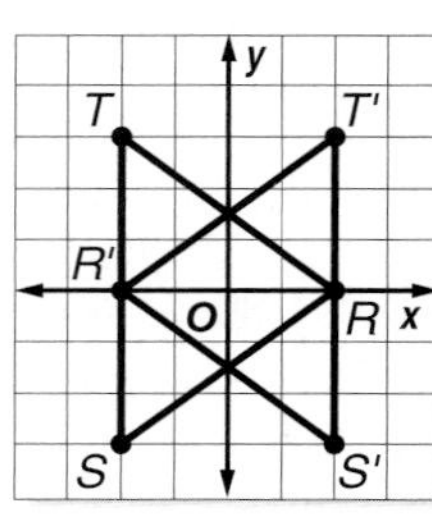

19. $R'(2, 3)$, $S'(4, 2)$, $T'(7, 5)$, $U'(5, 7)$

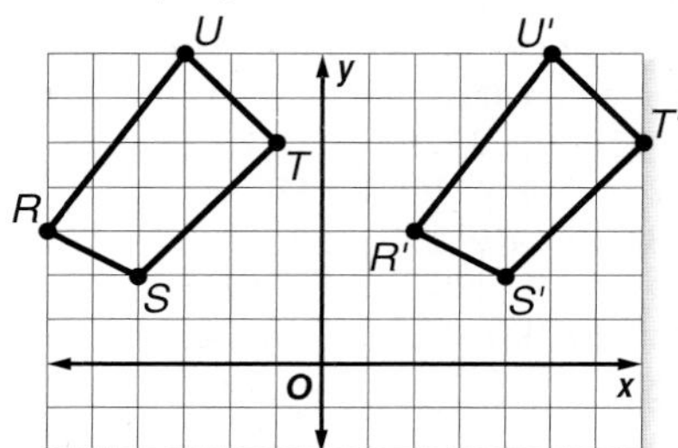

21. $J'(-2, 1)$, $K'(-1, 2)$, $L'(2, 2)$, $M'(-2, -2)$

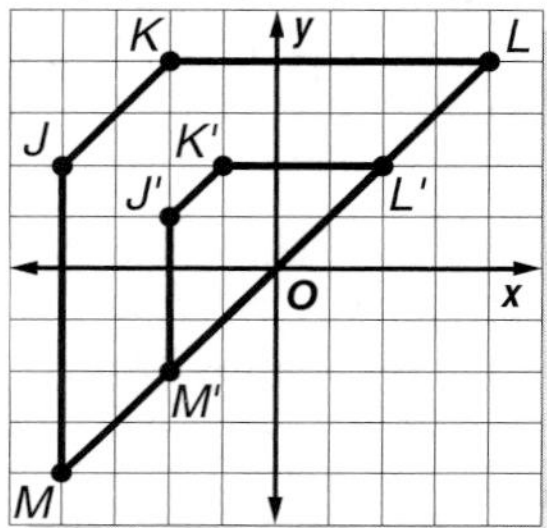

23. $F'(3, -2)$, $G'(-2, -5)$, $H'(-6, -3)$

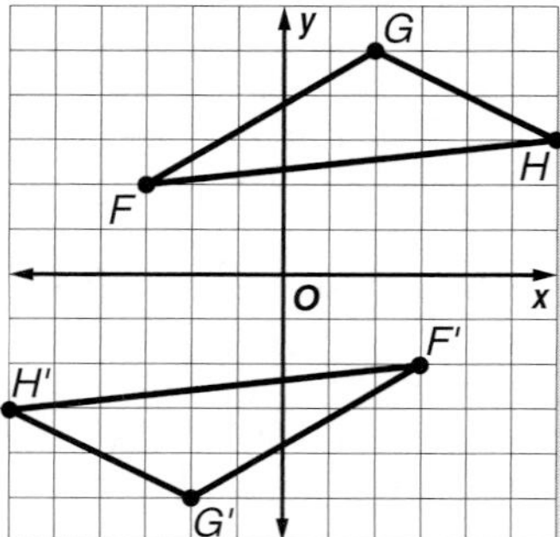

25. $W'(-1, -2)$, $X'(3, -2)$, $Y'(0, 4)$, $Z'(-4, 4)$

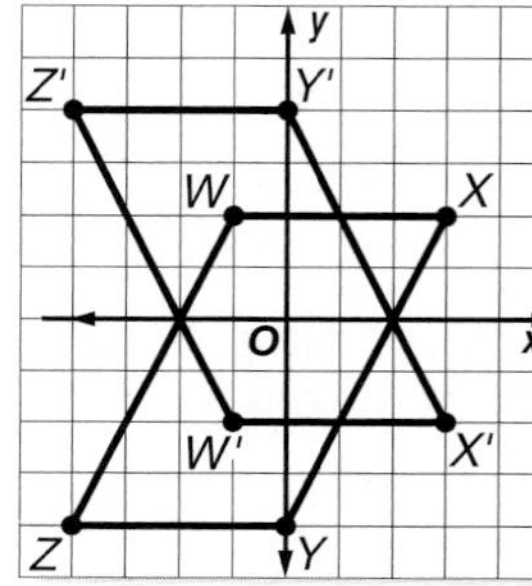

27. $A(-5, -1)$, $B(-3, -3)$, $C(-5, -5)$, $D(-5, -4)$, $E(-8, -4)$, $F(-8, -2)$, $G(-5, -2)$

29.

31. $\frac{1}{2}$

33. 90° counterclockwise rotation

35. (0, 0), (1800, 0), (1800, 1600), (0, 1600)

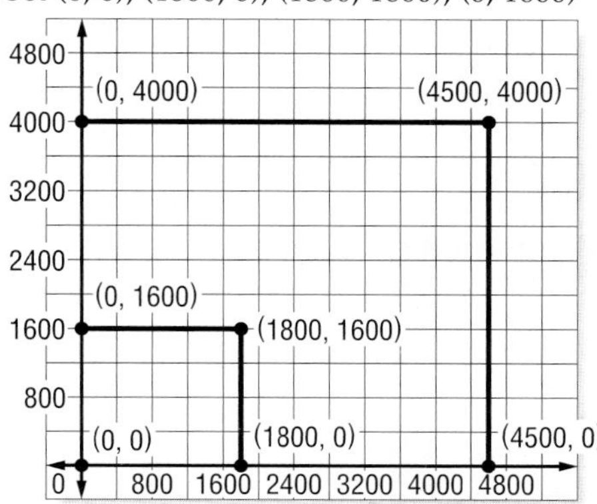

37. The pattern resembles a snowflake.

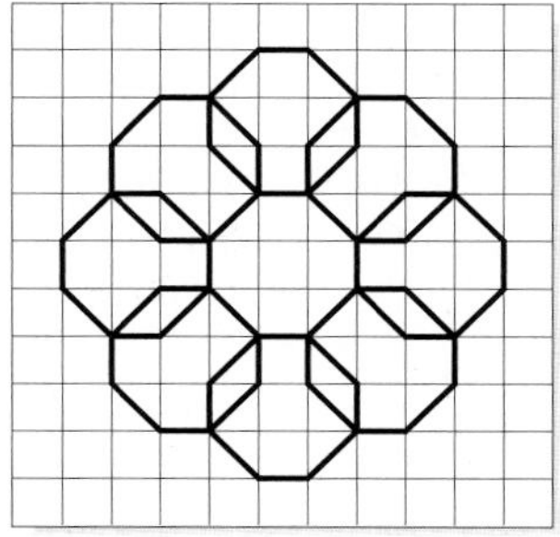

39. $(y, -x)$ **41.** Artists use computer graphics to simulate movement, change the size of objects, and create designs. Answers should include the following.

- Objects can appear to move by using a series of translations. Moving forward can be simulated by enlarging objects using dilations so they appear to be getting closer.
- Computer graphics are used in special effects in movies, animated cartoons, and web design.

43. C

45. $J'(-3, -5)$, $K'(-2, -8)$, $L'(1, -8)$, $M'(3, -5)$

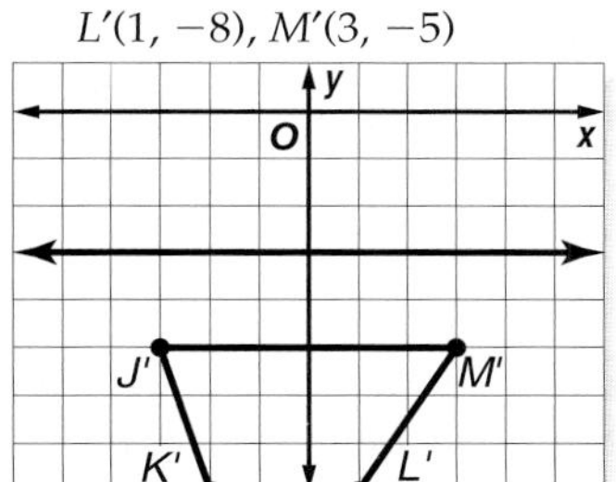

47–52.

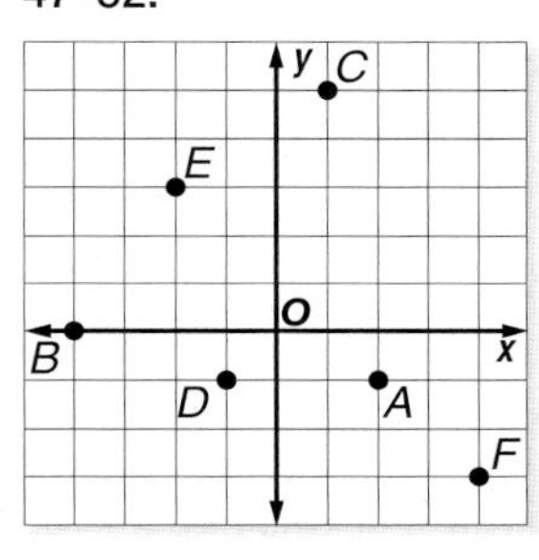

53. 10 mL **55.** $\frac{1}{12} \approx 8\%$ **57.** $\frac{5}{6} \approx 83\%$ **59.** {(0, 100), (5, 90) (10, 81), (15, 73), (20, 66), (25, 60), (30, 55)}

Pages 208–211 Lesson 4-3

1. A relation can be represented as a set of ordered pairs, a table, a graph, or a mapping. **3.** The domain of a relation is the range of the inverse, and the range of a relation is the domain of the inverse.

5. D = {−1, 3, 5, 6}; R = {−3, 4, 9}

x	y
6	4
3	−3
−1	9
5	−3

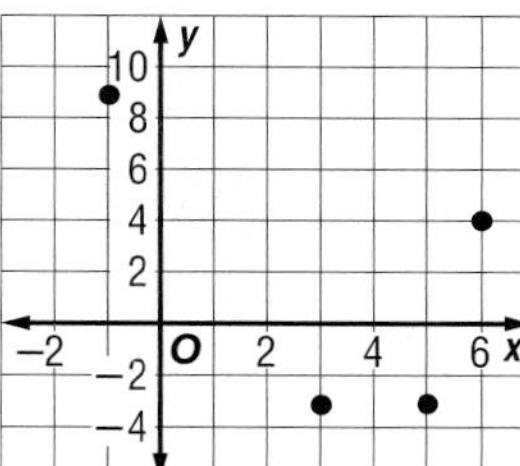

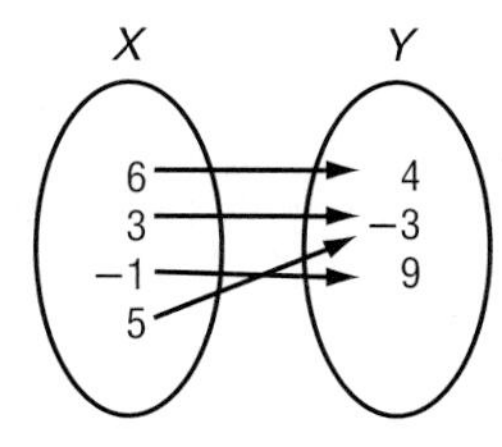

7. D = {−4, −1, 6}; R = {7, 8, 9}

x	y
−4	8
−1	9
−4	7
6	9

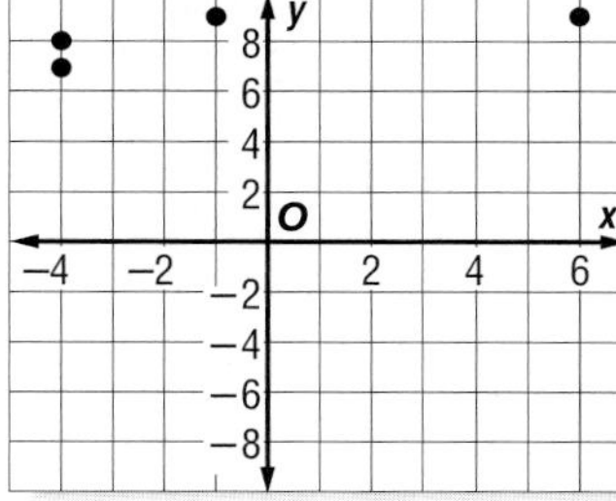

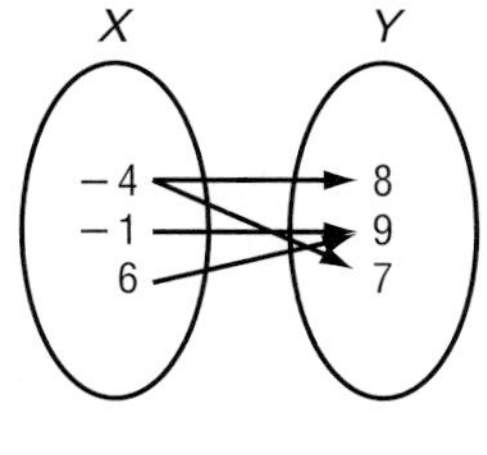

9. {(−4, 9), (2, 5), (−2, −2), (11, 12)}; {(9, −4), (5, 2), (−2, −2), (12, 11)} **11.** {(2, 8), (3, 7), (4, 6), (5, 7)}; {(8, 2), (7, 3), (6, 4), (7, 5)} **13.** {(−4, −4), (−3, 0), (0, −3), (2, 1), (2, −1)}; {(−4, −4), (0, −3), (−3, 0), (1, 2), (−1, 2)} **15.** {1989, 1990, 1991, 1992, 1993, 1994, 1995, 1996, 1997, 1998, 1999}

17. There are fewer students per computer in more recent years. So the number of computers in schools has increased.

19. D = {−5, 2, 5, 6}; R = {0, 2, 4, 7}

x	y
5	2
−5	0
6	4
2	7

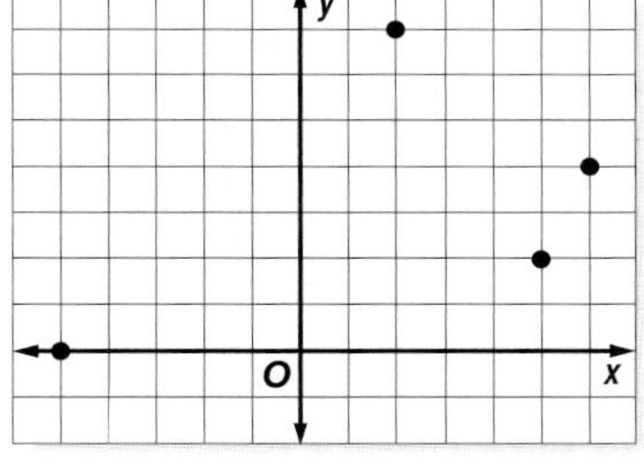

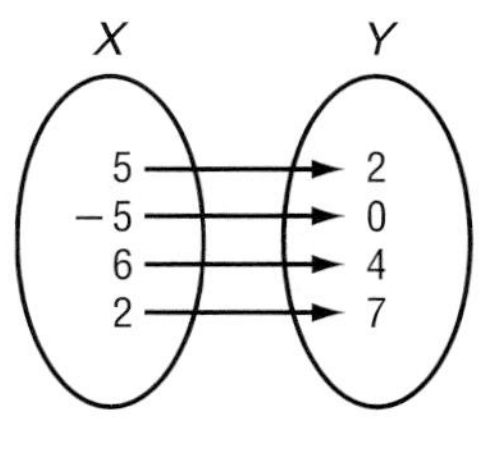

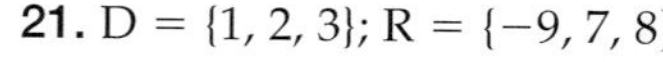

21. D = {1, 2, 3}; R = {−9, 7, 8}

x	y
3	8
3	7
2	−9
1	−9

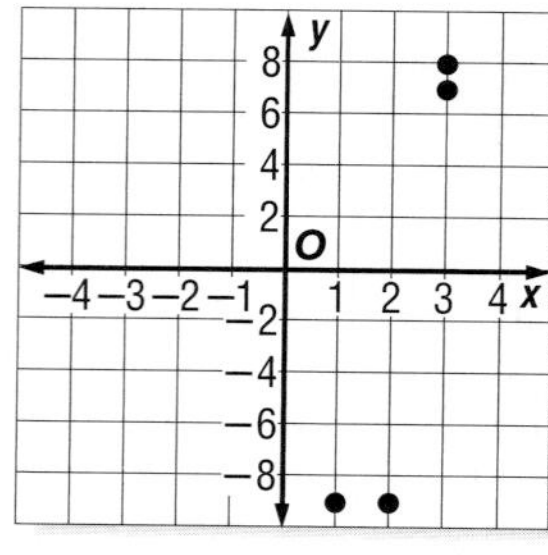

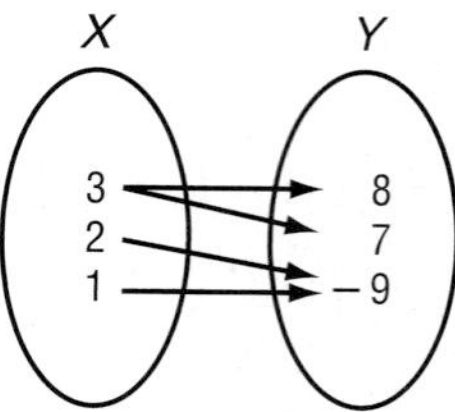

23. D = {−5, −1, 0}; R = {1, 2, 6, 9}

x	y
0	2
−5	1
0	6
−1	9

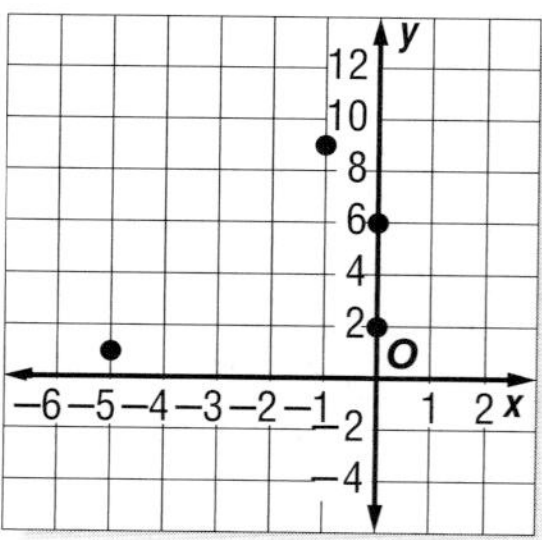

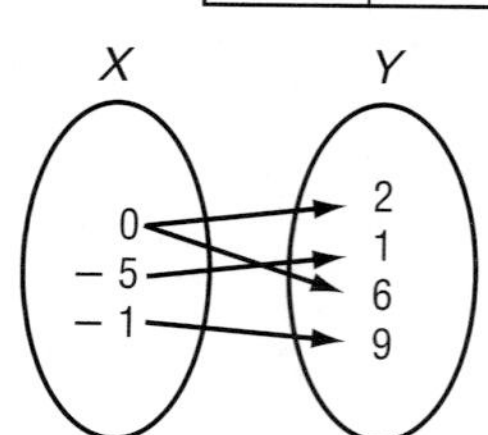

25. D = {−3, −2, 3, 4, 7}; R = {2, 4, 5, 6}

x	y
7	6
3	4
4	5
−2	6
−3	2

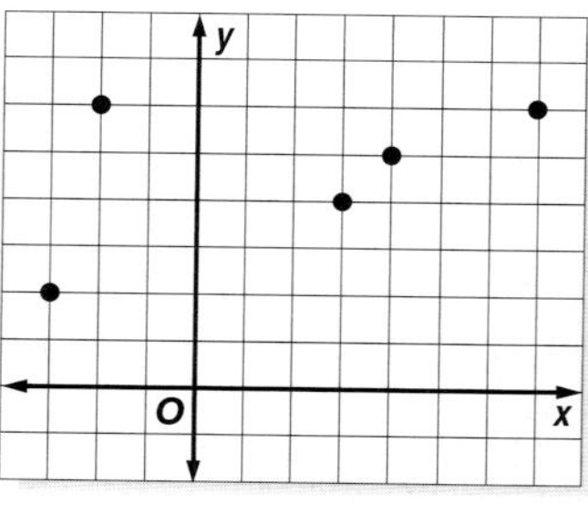

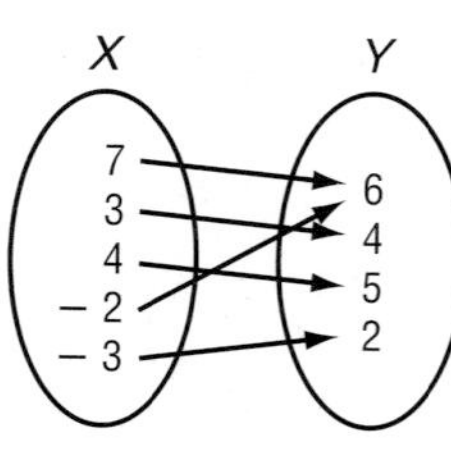

27. {(0, 3), (−5, 2), (4, 7), (−3, 2)}; {(3, 0), (2, −5), (7, 4), (2, −3)} **29.** {(−8, 4), (−1, 1), (0, 6), (5, 4)}; {(4, −8), (1, −1), (6, 0), (4, 5)} **31.** {(−3, 3), (1, 3), (4, 2), (−1, −5)}; {(3, −3), (3, 1), (2, 4), (−5, −1)} **33.** {(1, 16.50), (1.75, 28.30), (2.5, 49.10), (3.25, 87.60), (4, 103.40)}; {(16.50, 1), (28.30, 1.75), (49.10, 2.5), (87.60, 3.25), (103.40, 4)} **35.** {(2, 0), (2, 4), (3, 7), (5, 0), (5, 8), (−7, 7)}; {(0, 2), (4, 2), (7, 3), (0, 5), (8, 5), (7, −7)} **37.** {(−3, −1), (−3, −3), (−3, −5), (0, 3), (2, 3), (4, 3)}; {(−1, −3), (−3, −3), (−5, −3), (3, 0), (3, 2), (3, 4)} **39.** {(212.0, 0), (210.2, 1000), (208.4, 2000), (206.5, 3000), (201.9, 5000), (193.7, 10,000)} **41.** D = {1991, 1992, 1993, 1994, 1995, 1996, 1997, 1998, 1999, 2000}; R = {6.3, 7.5, 9.2, 9.5, 9.8, 10, 10.4} **43.** The production seems to go up and down every other year, however from 1995 through 1998, farmers have produced more corn each year. **45.** D = {100, 105, 110, 115, 120, 125, 130}; R = {40, 42, 44, 46, 48, 50, 52} **47.** D = {40, 42, 44, 46, 48, 50, 52}; R = {100, 105, 110, 115, 120, 125, 130} **49.** Sample answer: F = {(−1, 1), (−2, 2), (−3, 3)}, G = {(1, −2), (2, −3), (3, −1)}; The elements in the domain and range of F should be paired differently in G. **51.** B

53a.

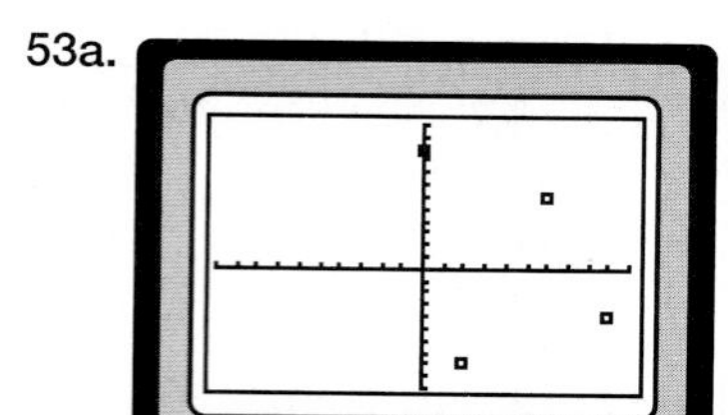

53b. [−10, 10] scl: 1 by [−10, 12] scl: 1

53c. {(10, 0), (−8, 2), (6, 6), (−4, 9)}

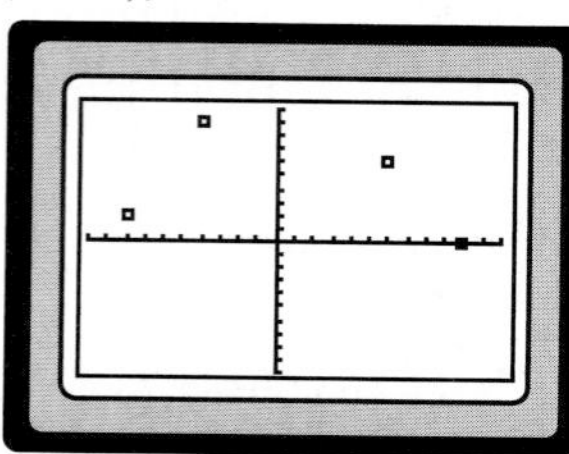

53d. (0, 10), none; (10, 0), none; (2, −8), IV; (−8, 2), II; (6, 6), I; (6, 6), I; (9, −4), IV; (−4, 9), II

55a.

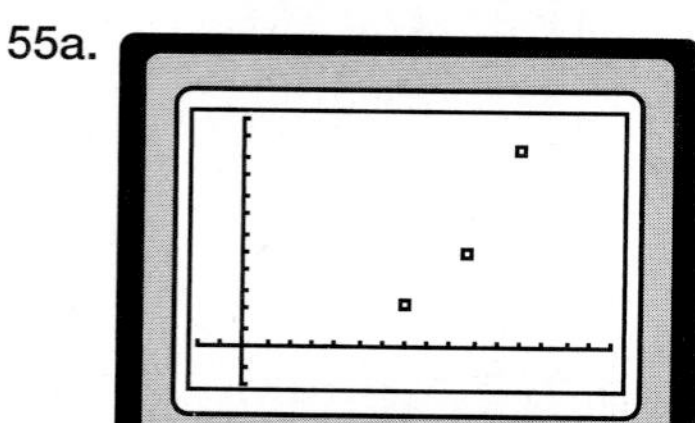

55b. [−10, 80] scl: 5 by [−10, 60] scl: 5

55c. {(12, 35), (25, 48), (52, 60)}

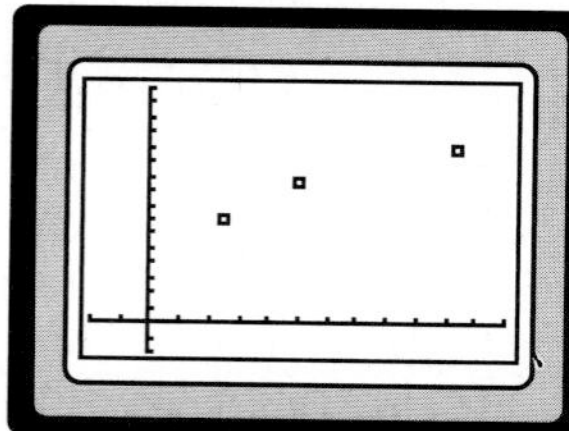

55d. (35, 12), (48, 25), and (60, 52) are all in I. (12, 35), (25, 48), and (52, 60) are all in I. **57.** rotation **59.** translation **61.** (3, 2); I **63.** (1, −1); IV **65.** (−4, −2); III **67.** (−2, 5); II **69.** 8 **71.** 9 **73.** $9n + 13$ **75.** {5} **77.** {3} **79.** {6}

Page 211 Practice Quiz 1

1–4.

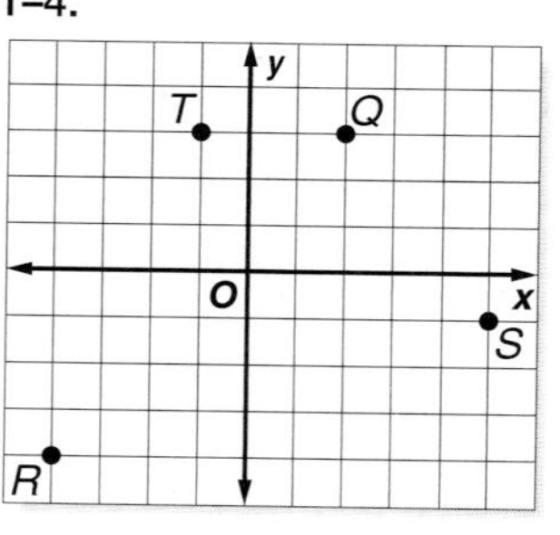

5. $A'(4, -8)$, $B'(7, -5)$, $C'(2, 1)$

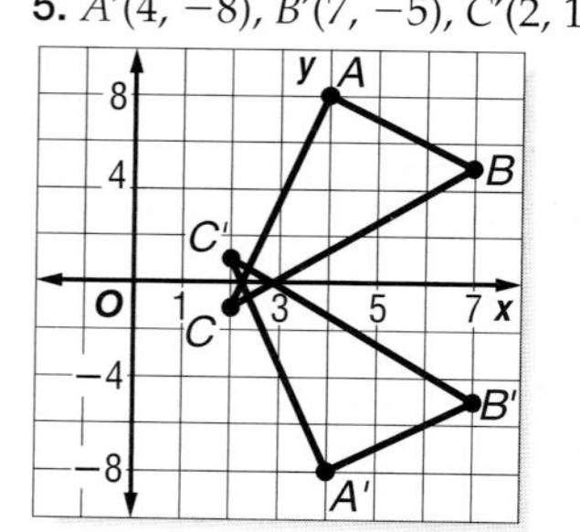

7. D = {1, 2, 4}; R = {3, 5, 6}; I = {(3, 1), (6, 4), (3, 2), (5, 1)} **9.** D = {−8, 11, 15}; R = {3, 5, 22, 31}; I = {(5, 11), (3, 15), (22, −8), (31, 11)}

Pages 214–217 Lesson 4-4

1. Substitute the values for y and solve for x. **3.** Bryan; x represents the domain and y represents the range. So, replace x with 5 and y with 1. **5.** {(−7, −3), (−2, −1)} **7.** {(−3, 7), (−1, 5), (0, 4), (2, 2)} **9.** {(−3, 11), (−1, 8), (0, 6.5), (2, 3.5)}

11. {(−4, −1), (−2, 0), (0, 1), (2, 2), (4, 3)}

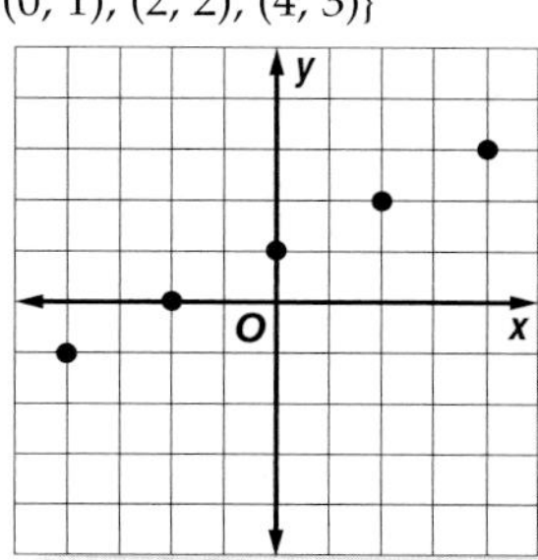

13. 12 karats **15.** {(4, −4), (2, 2)} **17.** {(3, 0), (2, 1), (4, −1)} **19.** {(0.25, 3.5), (1, 2)} **21.** {(−2, −1), (−1, 1), (1, 5), (3, 9), (4, 11)} **23.** {(−2, 9), (−1, 8), (1, 6), (3, 4), (4, 3)} **25.** {(−2, −9), (−1, −3), (1, 9), (3, 21), (4, 27)}

27. {(−2, −2), (−1, −1), (1, 1), (3, 3), (4, 4)} **29.** {(−2, 10), (−1, 8.5), (1, 5.5), (3, 2.5), (4, 1)} **31.** {(−2, −24), (−1, −18), (1, −6), (3, 6), (4, 12)}

33. {(−5, −16), (−2, −7), (1, 2), (3, 8), (4, 11)}

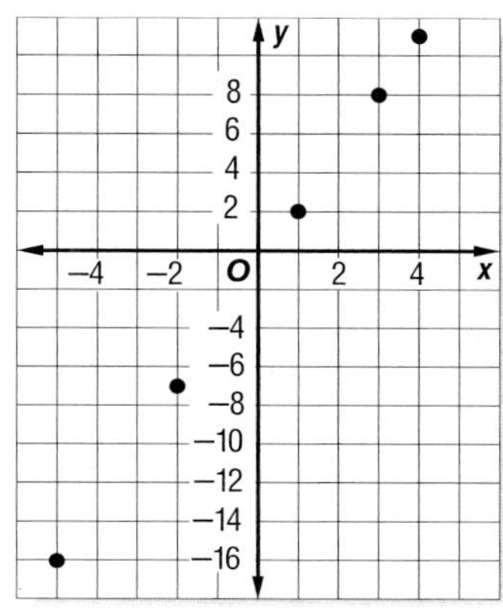

35. {(−4, 7), (−1, 3.25), (0, 2), (2, −0.5), (4, −3), (6, −5.5)}

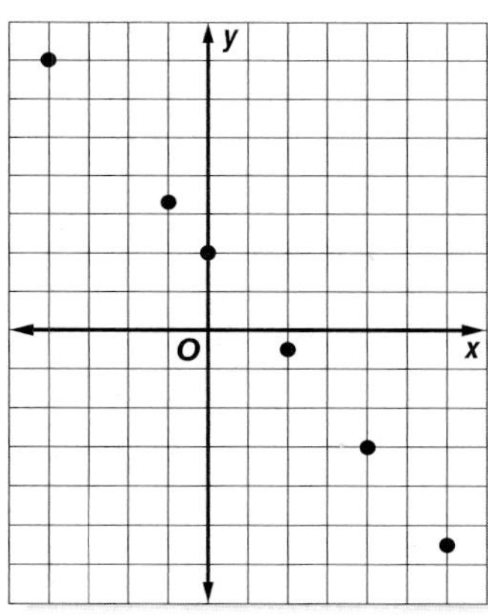

37. {(−4, −4), (−2, −3.5), (0, −3), (2, −2.5), (4, −2), (6, −1.5)}

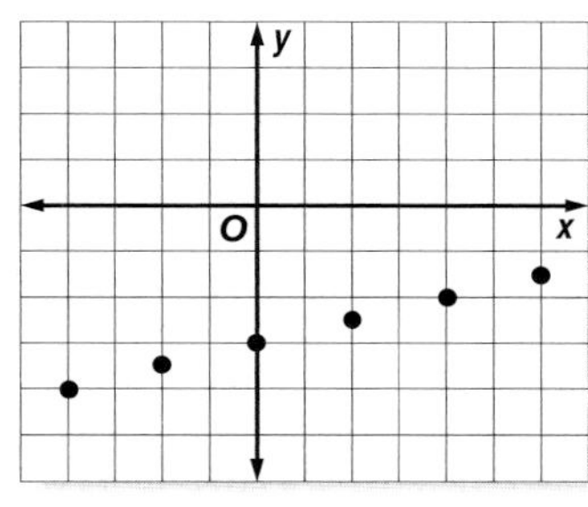

39. {−14, −12, −4, 6, 8} **41.** New York: 1.1°C, Chicago: −5°C, San Francisco: 12.8°C, Miami: 22.2°C, Washington, D.C.: 4.4°C **43.** w is independent; ℓ is dependent.

45.

Male		
Length of Tibia (cm)	Height (cm)	(*T*, *H*)
30.5	154.9	(30.5, 154.9)
34.8	165.2	(34.8, 165.2)
36.3	168.8	(36.3, 168.8)
37.9	172.7	(37.9, 172.7)

Female		
Length of Tibia (cm)	Height (cm)	(*T*, *H*)
30.5	148.9	(30.5, 148.9)
34.8	159.6	(34.8, 159.6)
36.3	163.4	(36.3, 163.4)
37.9	167.4	(37.9, 167.4)

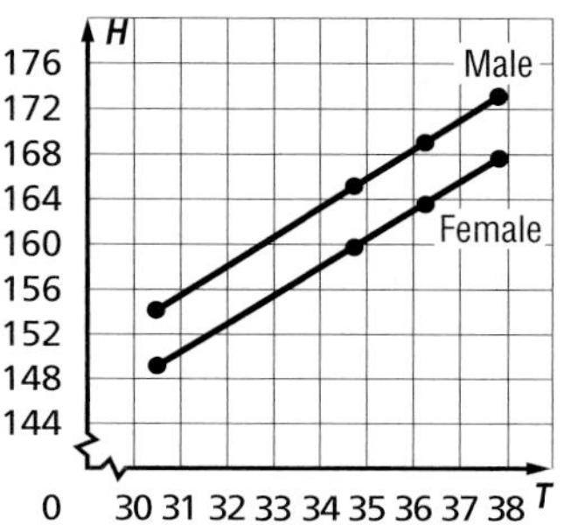

47a. {−6, −4, 0, 4, 6} **47b.** {−13, −8, −4, 4, 8, 13} **47c.** {−5, 0, 4, 8, 13} **49.** When traveling to other countries, currency and measurement systems are often different. You need to convert these systems to the system with which you are familiar. Answers should include the following.

- At the current exchange rate, 15 pounds is roughly 10 dollars and 10 pounds is roughly 7 dollars. Keeping track of every 15 pounds you spend would be relatively easy.
- If the exchange rate is 0.90 compared to the dollar, then items will cost less in dollars. For example, an item that is 10 in local currency is equivalent to \$9.00. If the exchange rate is 1.04, then items will cost more in dollars. For example, an item that costs 10 in local currency is equivalent to \$10.40.

51. C **53.** {(−8, 94), (−5, 74.5), (0, 42), (3, 22.5), (7, −3.5), (12, −36)} **55.** {(−2.5, −4.26), (−1.75, −3.21), (0, −0.76), (1.25, 0.99), (3.33, 3.90)} **57.** {(2, 7), (6, −4), (6, −1), (11, 8)}; {(7, 2), (−4, 6), (−1, 6), (8, 11)}
59. $X'(6, 4)$, $Y'(5, 0)$, $Z'(-3, 3)$

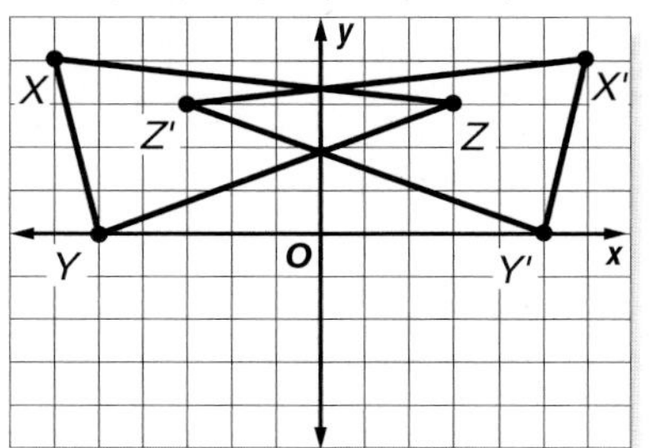

61. yes **63.** yes **65.** no **67.** H: it is hot; C: we will go swimming **69.** H: $3n - 7 = 17$; C: $n = 8$ **71.** 5 **73.** −2 **75.** 12

Pages 221–223 Lesson 4-5

1. The former will be a graph of four points, and the latter will be a graph of a line. **3.** Determine the point at which the graph intersects the x-axis by letting $y = 0$ and solving for x. Likewise, determine the point at which the graph intersects the y-axis by letting $x = 0$ and solving for y.

Selected Answers

Draw a line through the two points. **5.** yes; $3y = -2$ **7.** no

9.

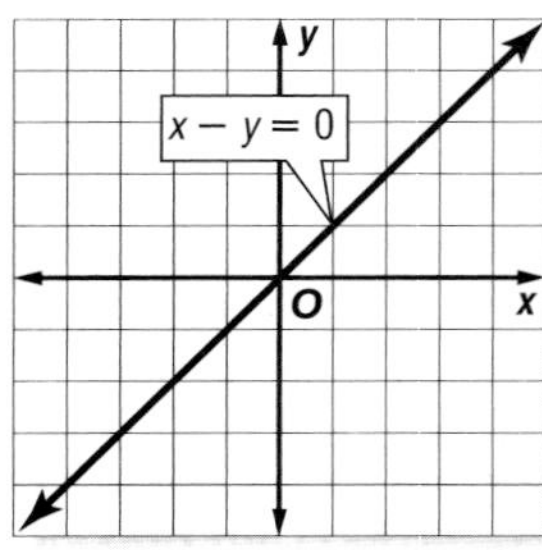

11.

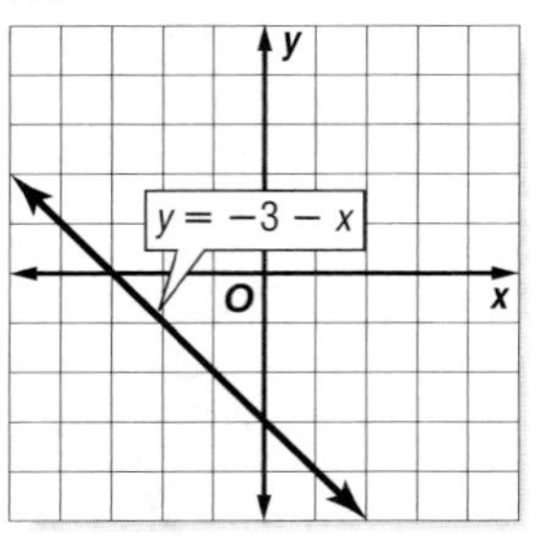

13.

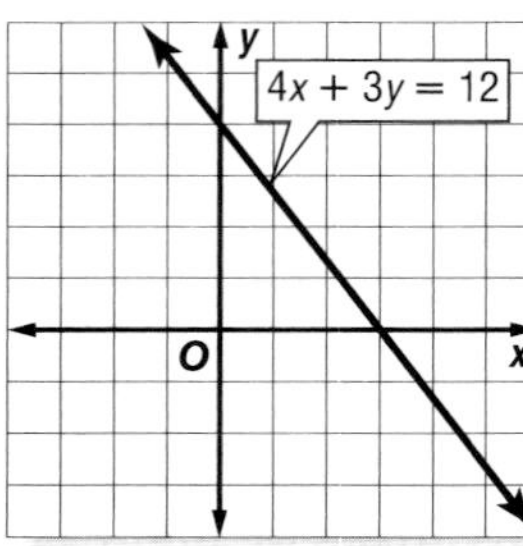

15. $15.75
17. yes; $2x + y = 6$
19. yes; $y = -5$ **21.** no
23. yes; $3x - 4y = 60$
25. yes; $3a = 2$

27.

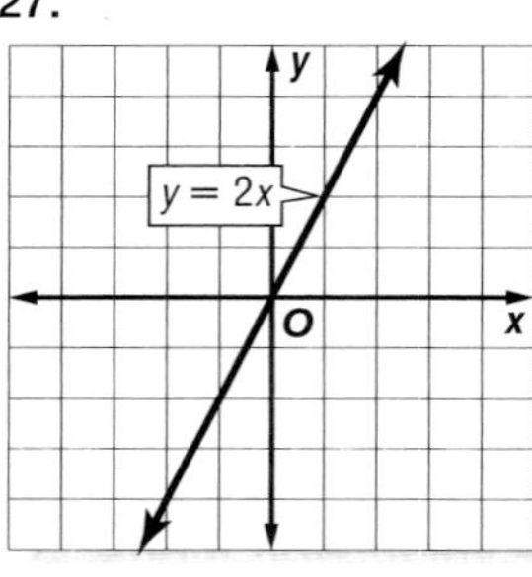

29.

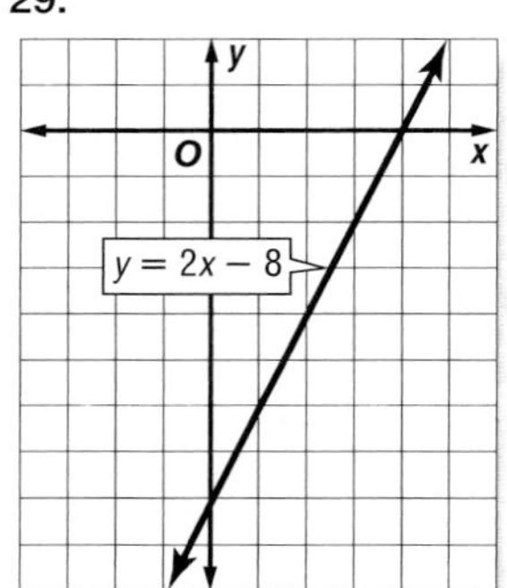

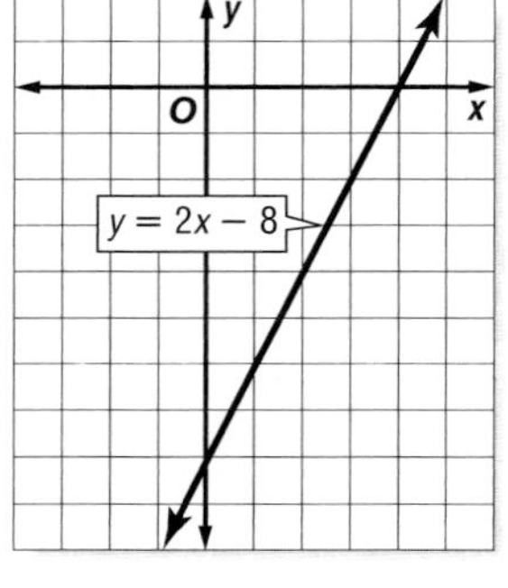

31.

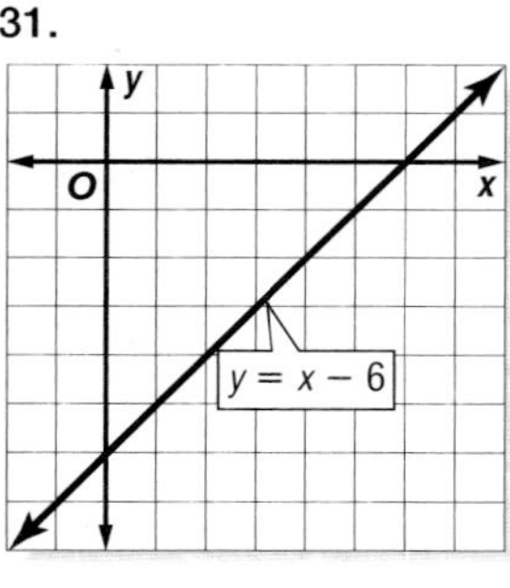

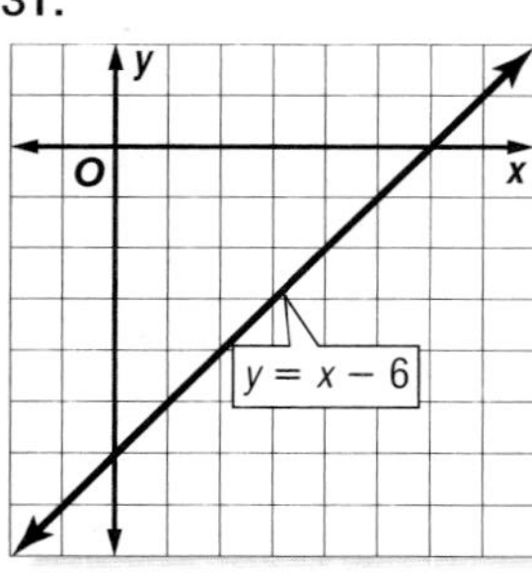

33.

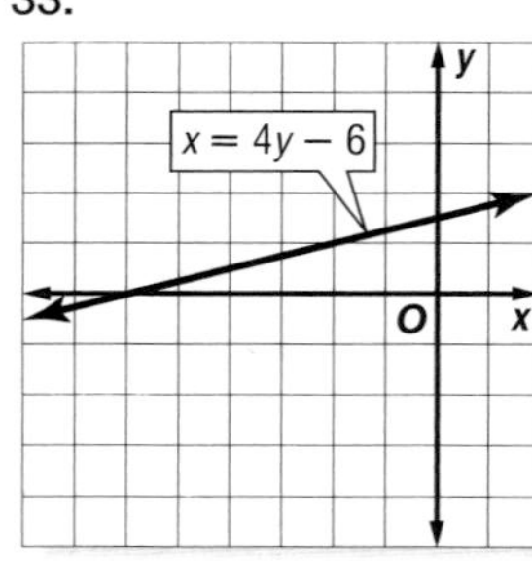

35.

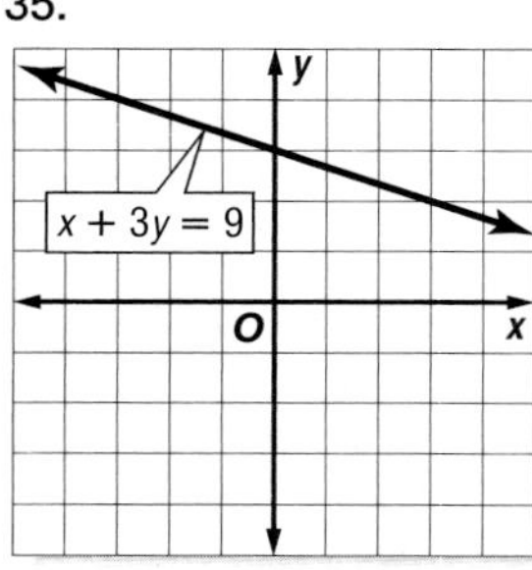

37.

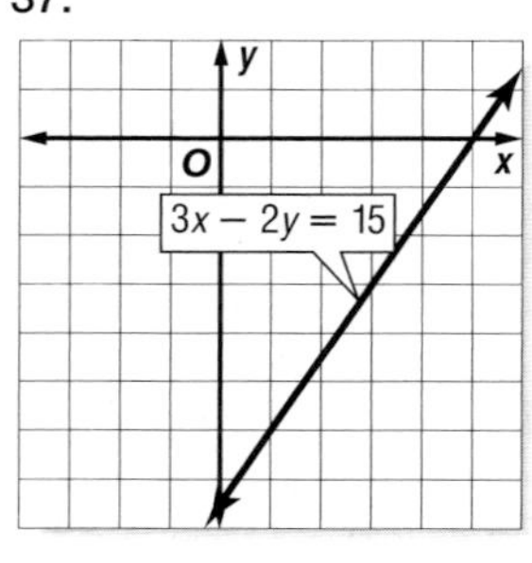

39.

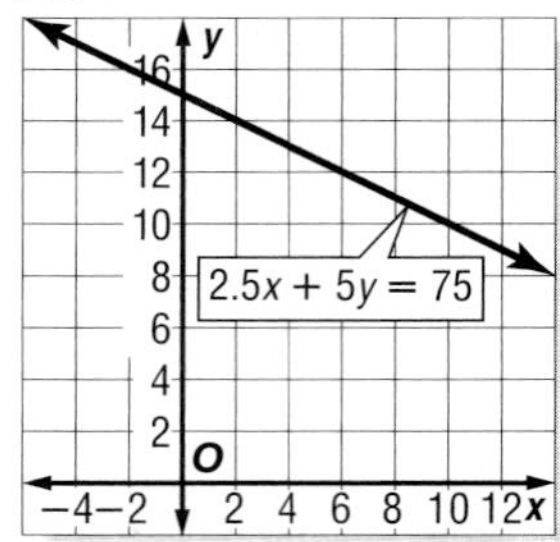

41.

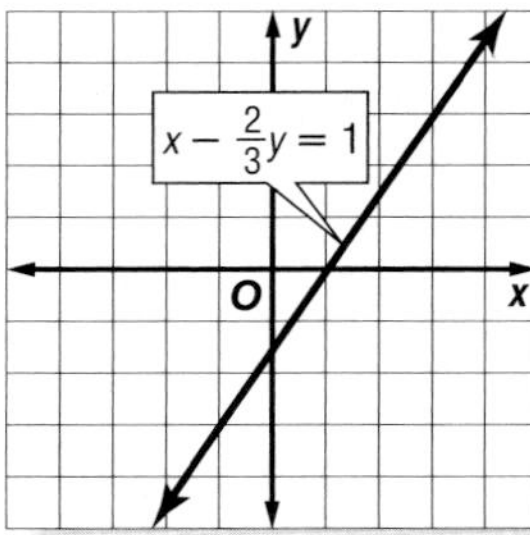

43.

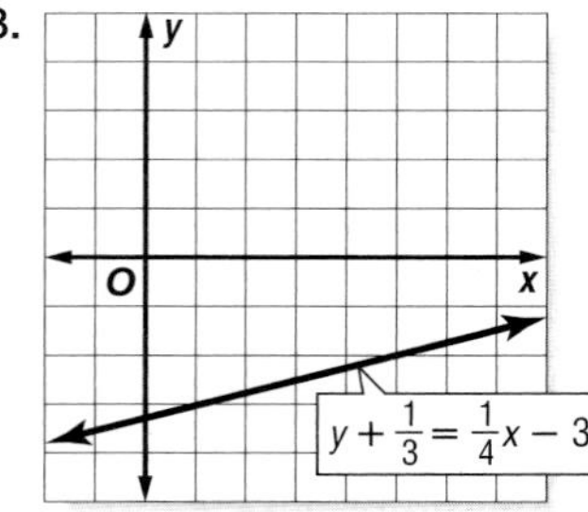

45. $5x + 3y = 15$
47. 7.5, 15

49.

t	d
0	0
2	0.42
4	0.84
6	1.26
8	1.68
10	2.1
12	2.52
14	2.94
16	3.36

51. about 14 s **53.** about 171 lb **55.** 186.7 psi **57.** Substitute the values for x and y into the equation $2x - y = 8$. If the value of $2x - y$ is less than 8, then the point lies *above* the line. If the value of $2x - y$ is greater than 8, then the point lies *below* the line. If the value of $2x - y$ equals 8, then the point lies *on* the line. Sample answers: (1, 5) lies above the line, (5, 1) lies below the line, (6, 4) lies on the line. **59.** A **61.** {(−3, −8), (−1, −6), (2, −3), (5, 0), (8, 3)} **63.** {(−3, 21), (−1, 15), (2, 6), (5, −3), (8, −12)} **65.** {(−3, −30), (−1, −18), (2, 0), (5, 18), (8, 36)} **67.** D = {−4, −3, 3}; R = {−1, 1, 2, 5}

x	y
3	5
−4	−1
−3	2
3	1

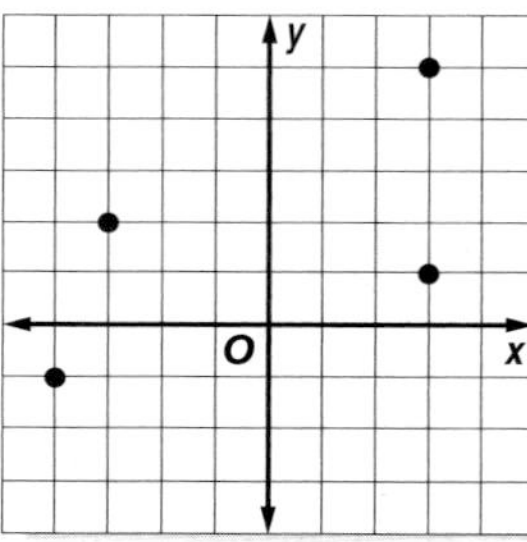

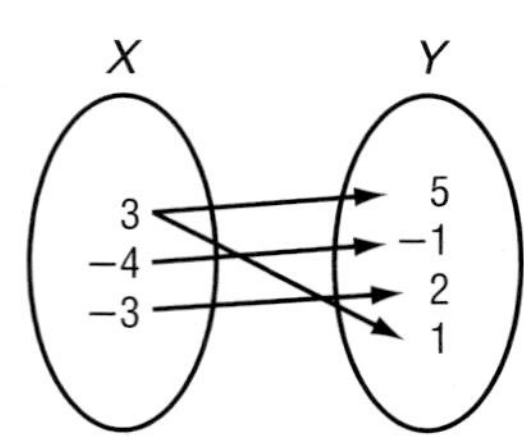

69. D = {−1, 1, 3}; R = {−1, 0, 4, 5}

x	y
1	4
3	0
−1	−1
3	5

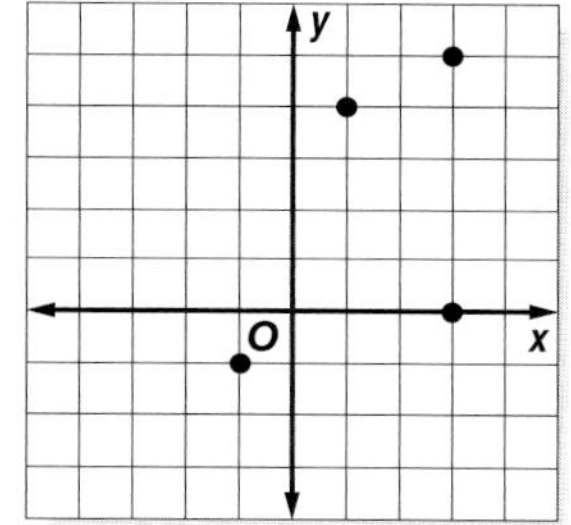

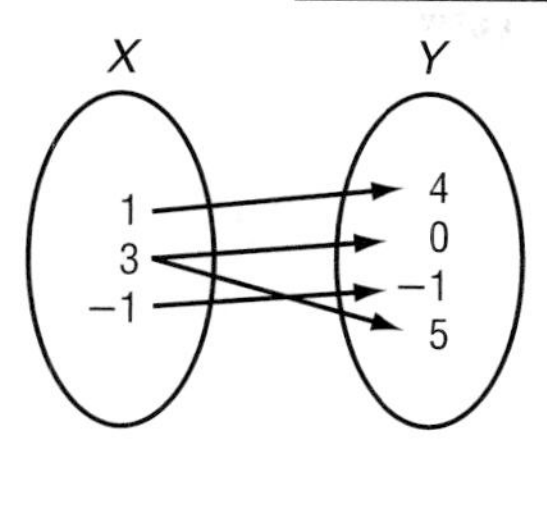

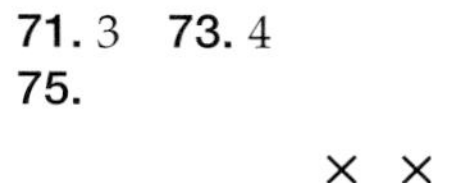

71. 3 **73.** 4
75.

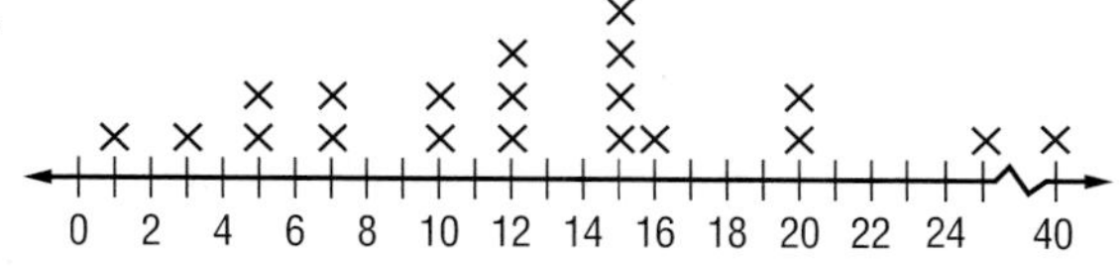

77. 15 yr **79.** 39 **81.** 48 **83.** 408

Pages 228–231 Lesson 4-6

1. y is not a function of x since 3 in the domain is paired with 2 and −3 in the range. x is not a function of y since −3 in the domain of the inverse is paired with 4 and 3 in the range. **3.** $x = c$, where c is any constant **5.** no **7.** yes **9.** yes **11.** 2 **13.** $t^2 - 3$ **15.** $4x + 15$ **17.** no **19.** yes **21.** yes **23.** yes **25.** yes **27.** yes **29.** no **31.** yes **33.** 1 **35.** 0 **37.** 26 **39.** $3a^2 + 7$ **41.** $6m - 8$ **43.** $6x^2 + 4$ **45.** $f(h) = 77 - 0.005h$

47.

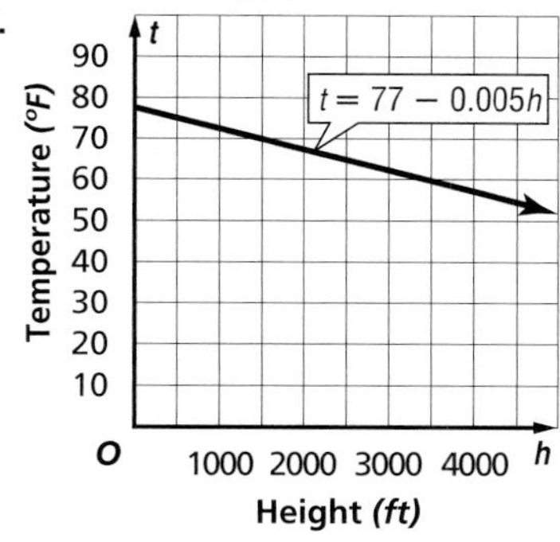

49.

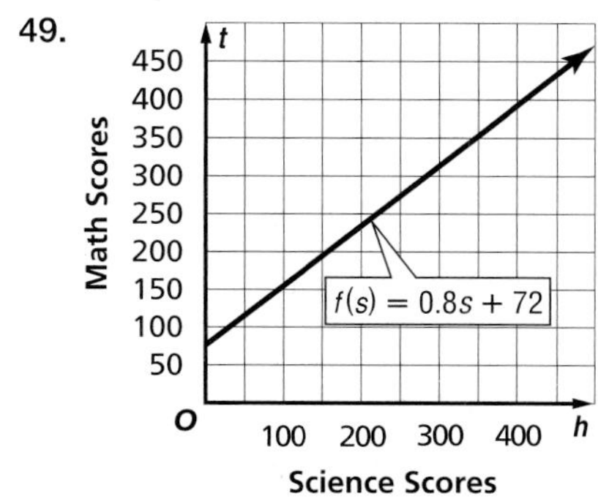

51. Krista's math score is above the average because the point at (260, 320) lies above the graph of the line for $f(s)$. **53.** Functions can be used in meteorology to determine if there is a relationship between certain weather conditions. This can help to predict future weather patterns. Answers should include the following.

- As barometric pressure decreases, temperature increases. As barometric pressure increases, temperature decreases.
- The relation is not a function since there is more than one temperature for a given barometric pressure. However, there is still a pattern in the data and the two variables are related.

55. A
57.

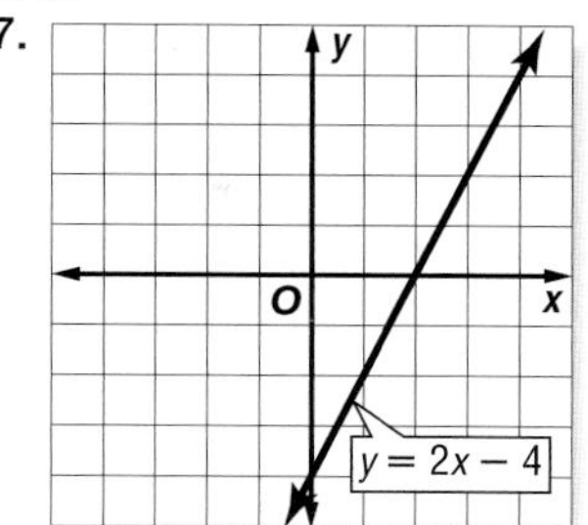

59. {(3, 12), (−1, −8)} **61.** approximately 3 h 9 min **63.** Reflexive; 3.5 **65.** −4 **67.** 20 **69.** $\frac{5}{8}$

Page 231 Practice Quiz 2

1. {(−3, 2), (−1, 4), (0, 5), (2, 7), (4, 9)} **3.** {(−3, 5.5), (−1, 4.5), (0, 4), (2, 3), (4, 2)}
5.

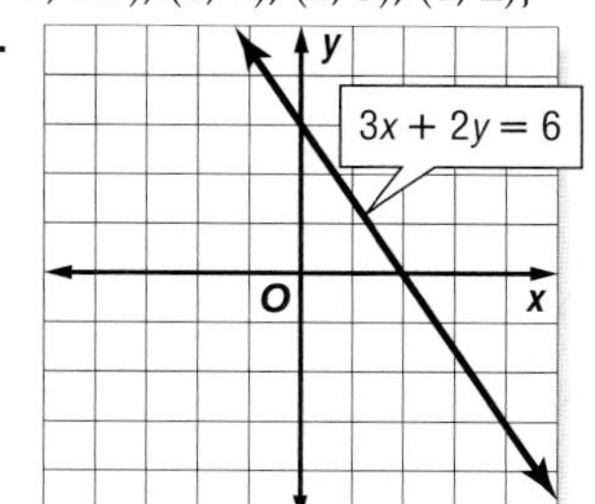

7. no **9.** $6a + 5$

Pages 236–238 Lesson 4-7

1. Sample answer: 2, −8, −18, −28, … **3.** Marisela; to find the common difference, subtract the first term from the second term. **5.** no **7.** 14, 9, 4 **9.** −90 **11.** 101
13. $a_n = 5n + 7$

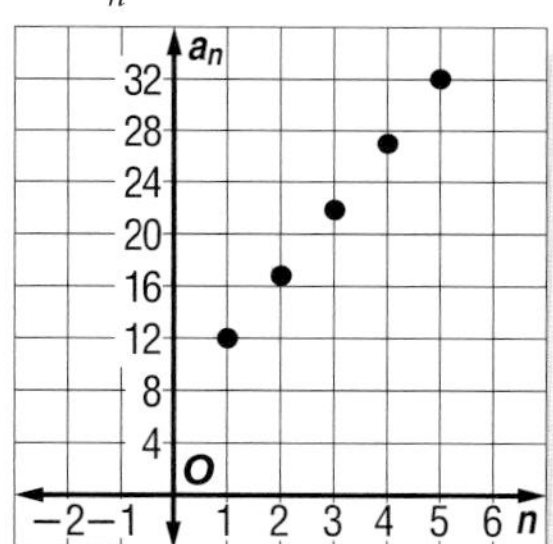

15. yes; −1 **17.** no **19.** yes; 0.5 **21.** 16, 19, 22 **23.** −82, −86, −90 **25.** $3\frac{2}{3}$, 4, $4\frac{1}{3}$ **27.** 125 **29.** 1264 **31.** $3\frac{1}{4}$ **33.** 25 **35.** 25 **37.** 17

39. $a_n = -3n$

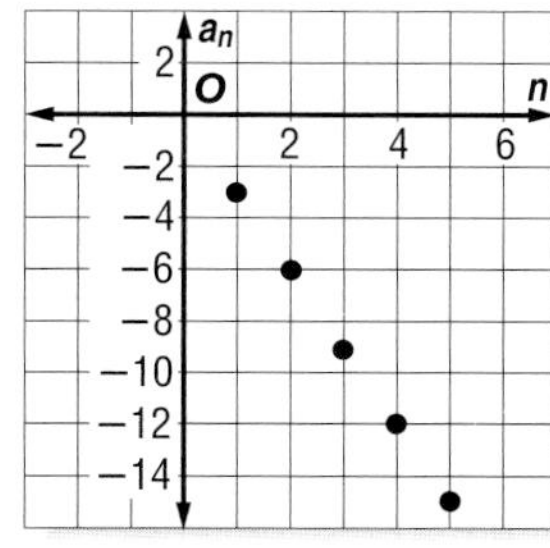

41. $a_n = 6n - 4$

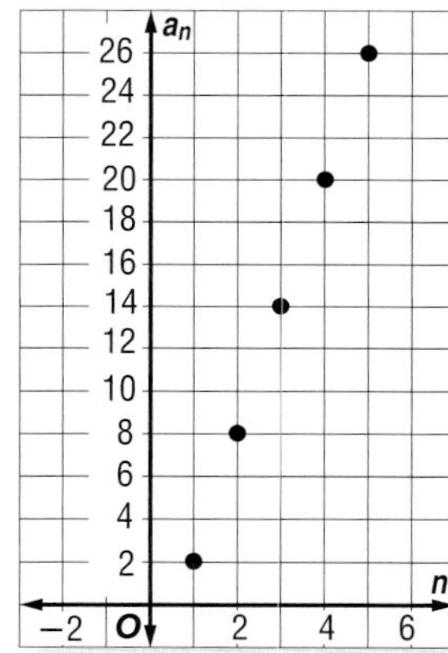

Selected Answers

43. 4 **45.** $P(n) = 3n + 2$ **47.** $a_n = 8n + 20$ **49.** Yes, the section was oversold by 4 seats. **51.** $a_n = 4n + 5$

53.

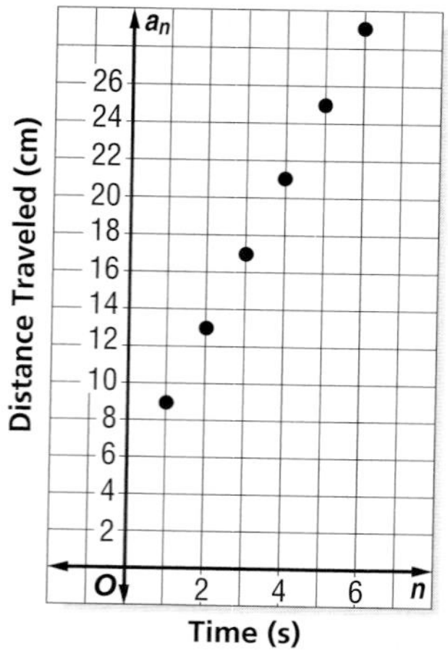

a_n

Distance Traveled (cm)

26, 24, 22, 20, 18, 16, 14, 12, 10, 8, 6, 4, 2

O 2 4 6 n

Time (s)

55. \$92,500
57. 45 **59.** C
61. 10 **63.** 32
65. yes; $x + y = 18$
67. $200 - 3x = 9$
69. -21 **71.** 12
73. $-\frac{5}{14}$
75. $(-2, 2)$
77. $(-4, -2)$
79. $(3, 5)$

Pages 243–245 Lesson 4-8

1. Once you recognize a pattern, you can find a general rule that can be written as an algebraic expression. **3.** Test the values of the domain in the equation. If the resulting values match the range, the equation is correct. **5.** 16, 22, 29
7. $f(x) = x$

9.

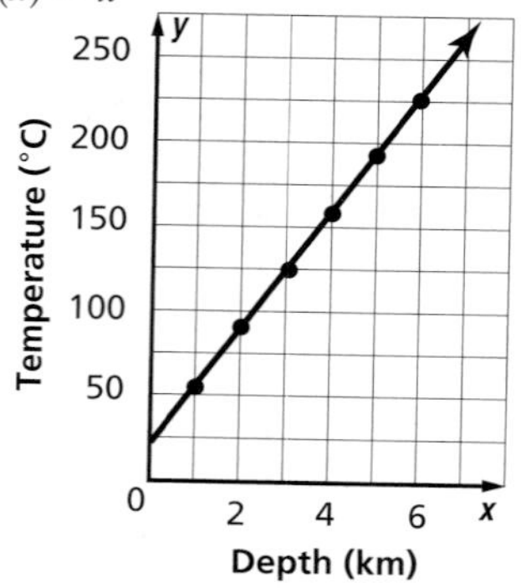

y

Temperature (°C)

250, 200, 150, 100, 50

0 2 4 6 x

Depth (km)

11. 370°

13.

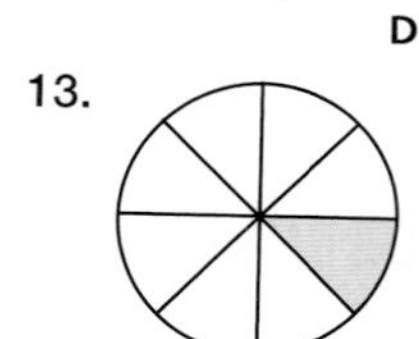
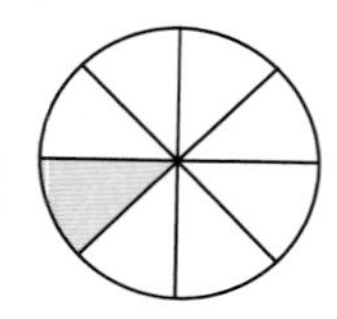
;
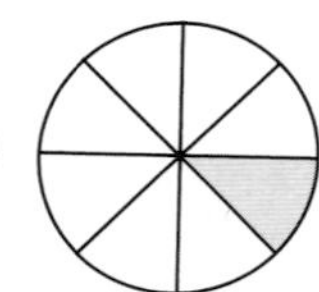

15. 10, 13, 11 **17.** 27, 35, 44 **19.** $4x + 1, 5x + 1, 6x + 1$
21. $f(x) = \frac{1}{2}x$ **23.** $f(x) = 6 - x$ **25.** $f(x) = 12 - 3x$
27. 1, 1, 2, 3, 5, 8, 13, 21, 34, 55, 89, 144 **29.** $f(a) = -0.9a + 193$
31. 5, 8, 11, 14 cm **33.** 74 cm **35.** B **37.** 13, 16, 19
39. $-1, 5, 11$ **41.** no

Pages 246–250 Chapter 4 Study Guide and Review

1. e **3.** d **5.** k **7.** c **9.** b

11–16.

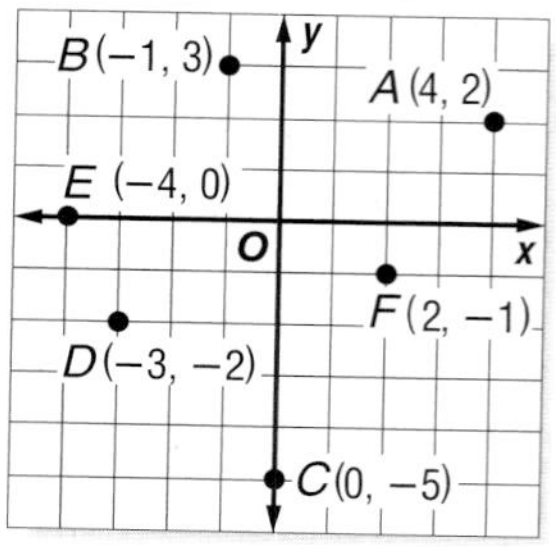

B(−1, 3)
A(4, 2)
E(−4, 0)
O
F(2, −1)
D(−3, −2)
C(0, −5)

17. $A'(3, -3)$, $B'(5, -4)$, $C'(4, 3)$

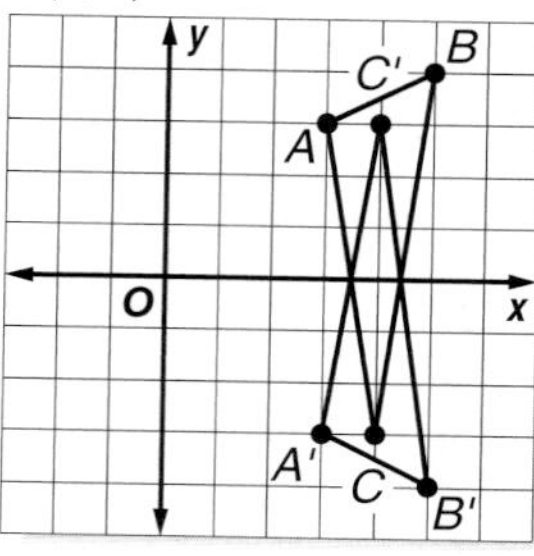

19. $G'(1, 1)$, $H'(3, 0)$, $I'(3, 1)$, $J'(1, 2)$

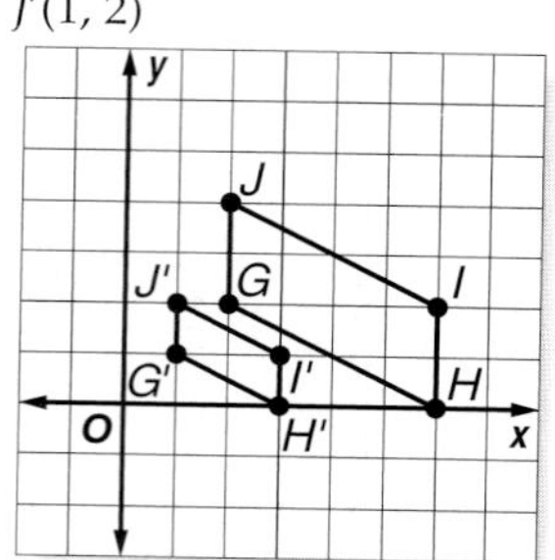

21. D = {−2, 3, 4}, R = {−2, 0, 6}

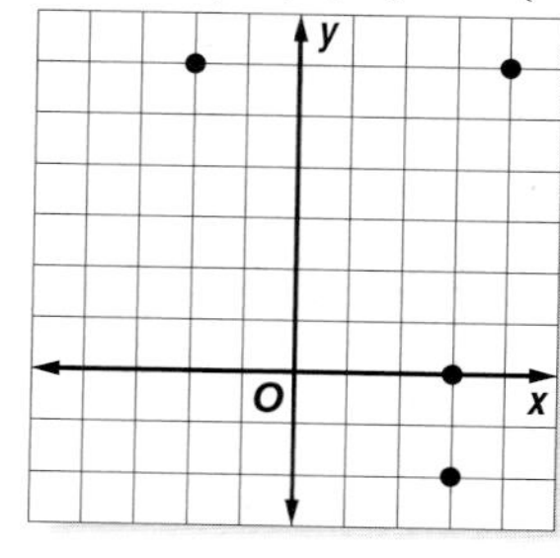

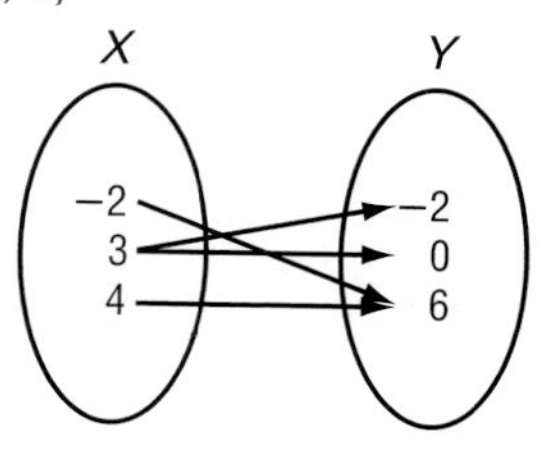

23. D = {−3, 3, 5, 9}, R = {3, 8}

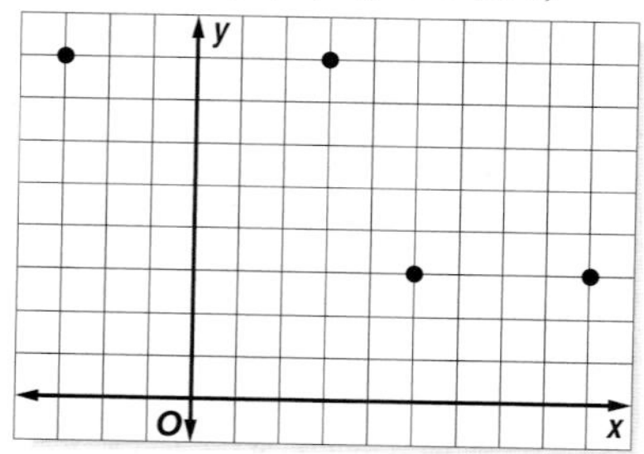

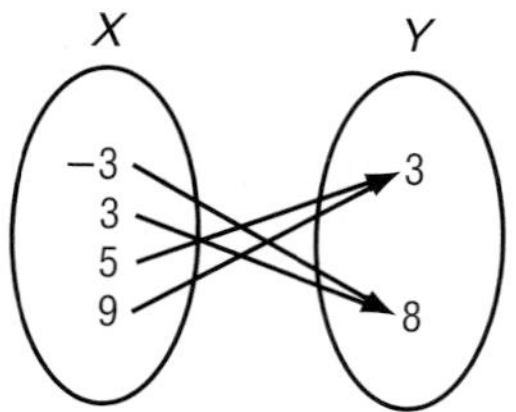

25. {(−4, −13), (−2, −11), (0, −9), (2, −7), (4, −5)}

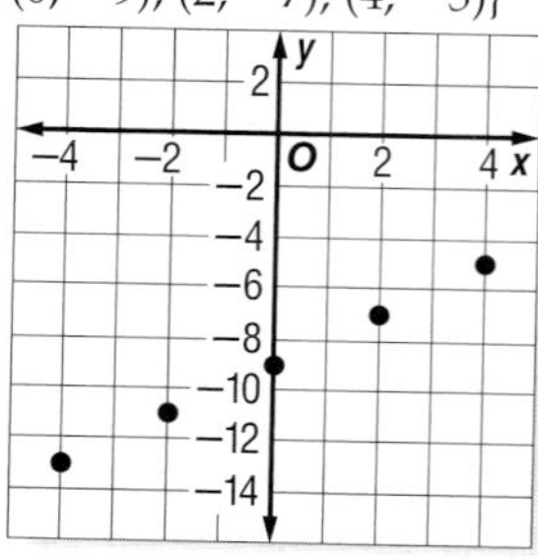

27. {(−4, −11), (−2, −3), (0, 5), (2, 13), (4, 21)}

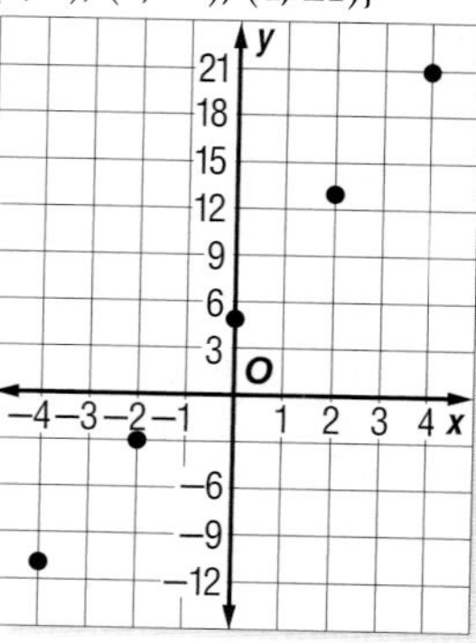

29. 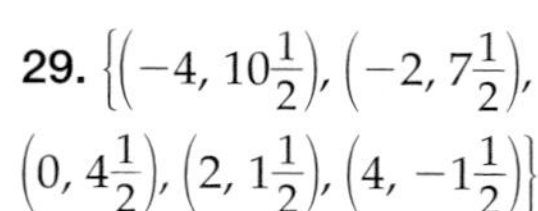 $\{(-4, 10\frac{1}{2}), (-2, 7\frac{1}{2}), (0, 4\frac{1}{2}), (2, 1\frac{1}{2}), (4, -1\frac{1}{2})\}$

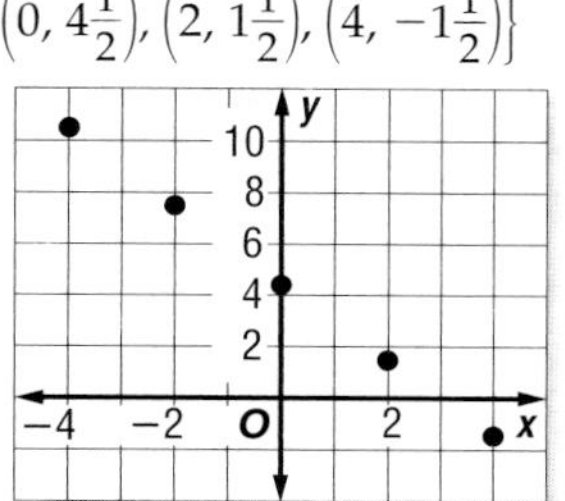

31.

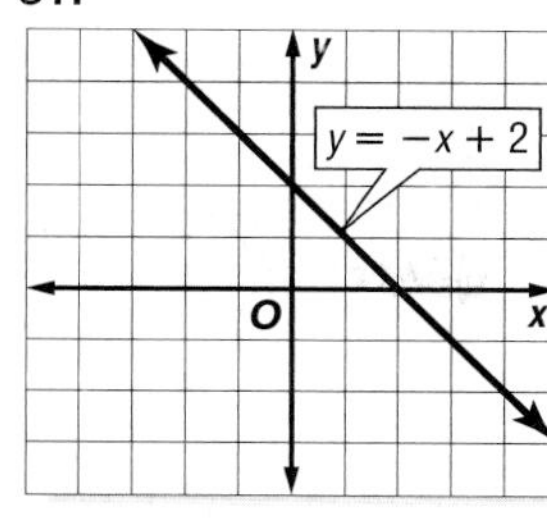

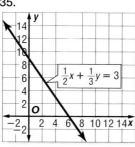

33.

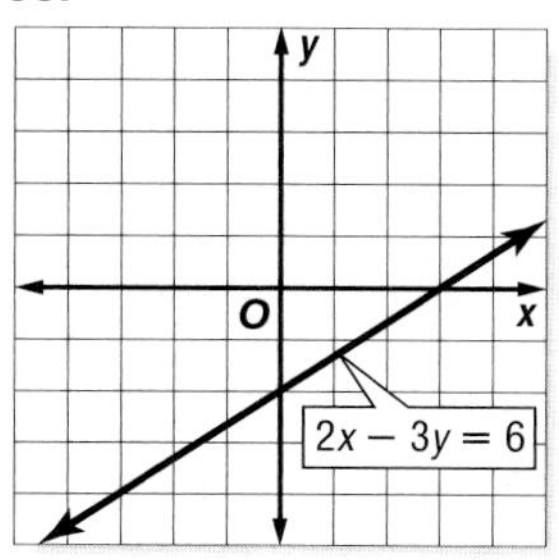

35.

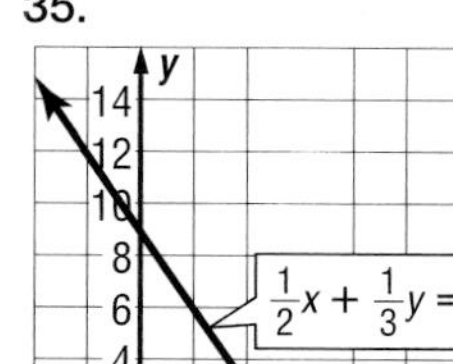

37. yes **39.** yes **41.** 3 **43.** 18 **45.** $4a^2 + 2a + 1$ **47.** 26, 31, 36 **49.** 6, 4, 2 **51.** −11, −5, 1 **53.** $f(x) = -x - 1$

Chapter 5 Analyzing Linear Equations

Page 255 Chapter 5 Getting Started

1. $\frac{1}{5}$ **3.** $-\frac{1}{4}$ **5.** $\frac{1}{3}$ **7.** 3 **9.** $\frac{1}{4}$ **11.** $-\frac{3}{4}$ **13.** 0 **15.** (1, 2) **17.** (2, −3) **19.** (−2, 2)

Pages 259–262 Lesson 5-1

1. Sample answer: Use (−1, −3) as (x_1, y_1) and (3, −5) as (x_2, y_2) in the slope formula. **3.** The difference in the x values is always the 0, and division by 0 is undefined. **5.** $\frac{3}{2}$ **7.** −4 **9.** 0 **11.** 5 **13.** 1.5 million subscribers per year **15.** $\frac{3}{4}$ **17.** −2 **19.** undefined **21.** $\frac{10}{7}$ **23.** $\frac{3}{8}$ **25.** undefined **27.** 0 **29.** $-\frac{1}{2}$ **31.** $\frac{15}{4}$ **33.** $-\frac{2}{3}$ **35.** Sample answer: $\frac{8}{11}$ **37.** $\frac{s}{r}$, if $r \neq 0$ **39.** 4 **41.** −1 **43.** 1 **45.** $\frac{1}{4}$ **47.** 7 **49.** (−4, −5) is in Quadrant III and (4, 5) is in Quadrant I. The segment connecting them goes from lower left to upper right, which is a positive slope. **51.** 12–14; steepest part of the graph **53.** ’90–’95; ’80–’85 **55.** a decline in enrollment **57.** 13 ft 9 in. **59.** D **61.** $\frac{1}{3}$; The slope is the same regardless of points chosen. **63.** $f(x) = 5x$ **65.** yes **67.** no

69.

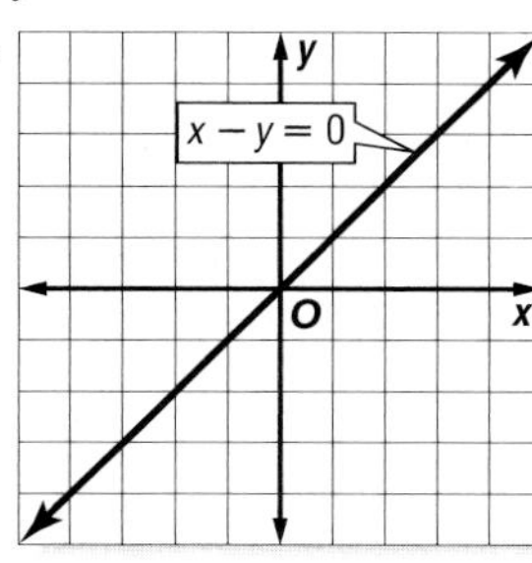

71. −21 **73.** −36 **75.** $-\frac{7}{24}$ **77.** 9 **79.** $26\frac{2}{3}$ **81.** $4\frac{1}{2}$ **83.** $20\frac{4}{7}$ **85.** $10\frac{2}{3}$

Pages 267–270 Lesson 5-2

1. $y = kx$ **3.** They are equal. **5.** 1; 1

7.

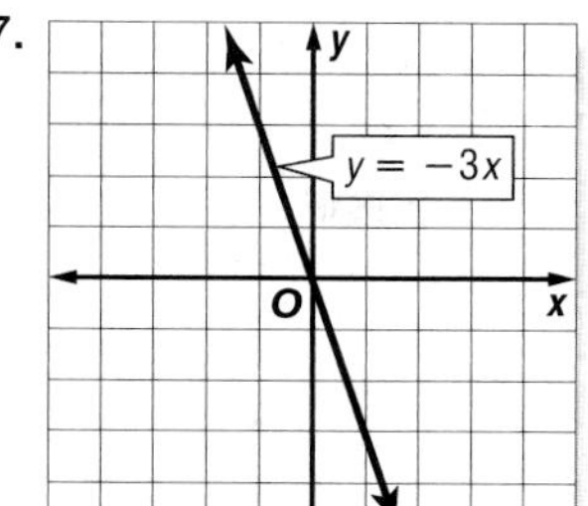

9. $y = \frac{9}{2}x$; 10 **11.** $y = \frac{1}{2}x$; 10

13.

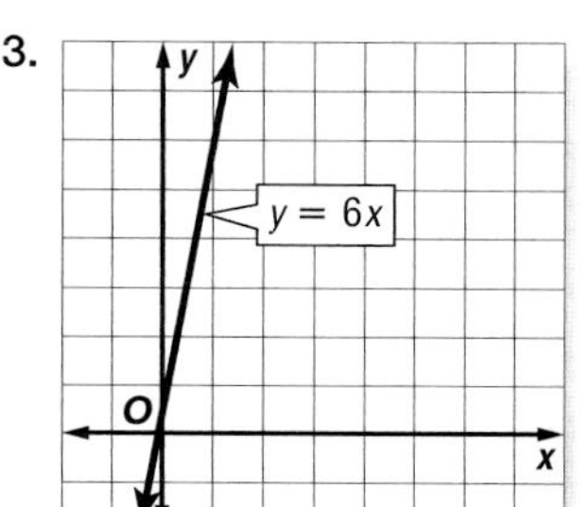

15. 2; 2 **17.** $-\frac{1}{2}; -\frac{1}{2}$ **19.** $\frac{3}{2}; \frac{3}{2}$

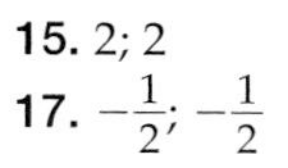

21.

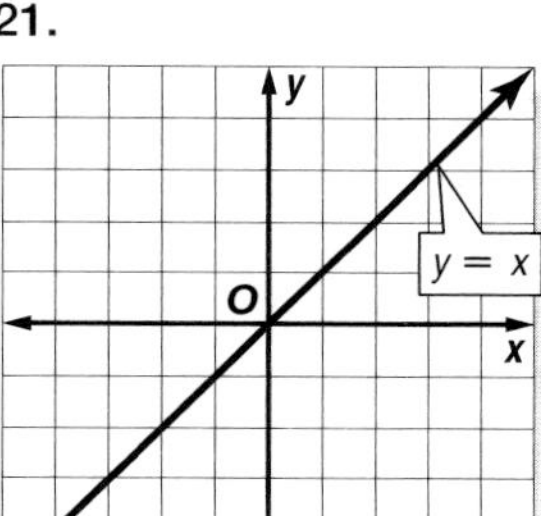

23. 

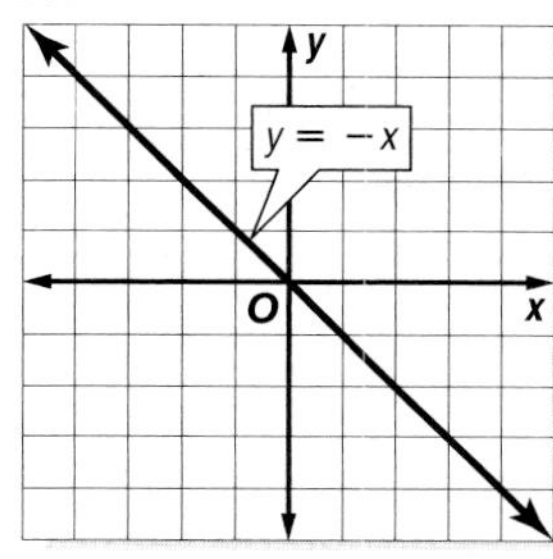

25.

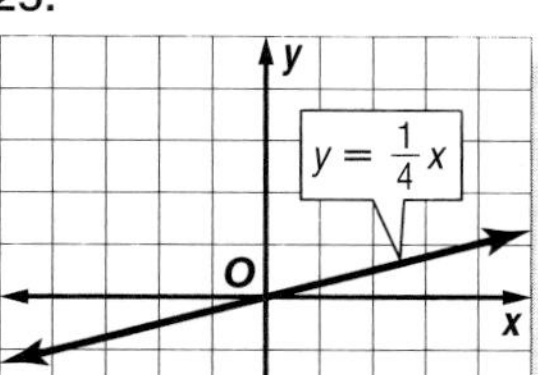
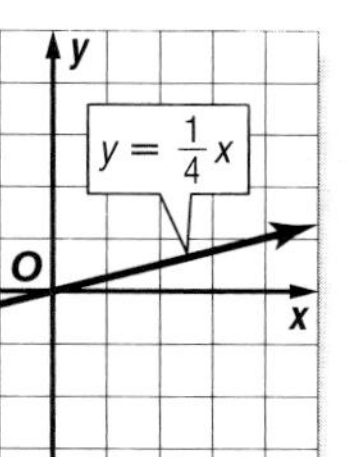

27.

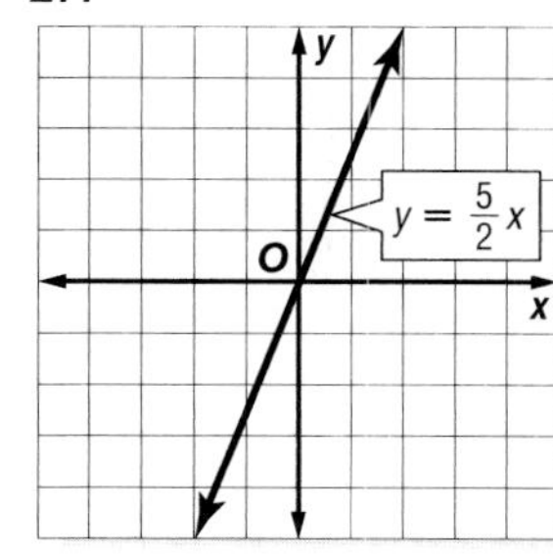

29. 

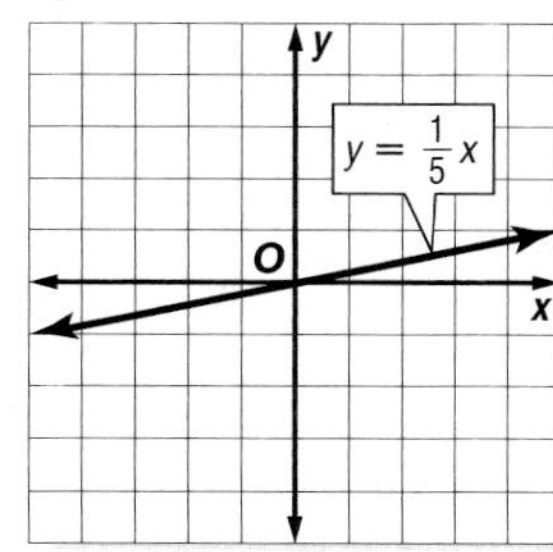

31.

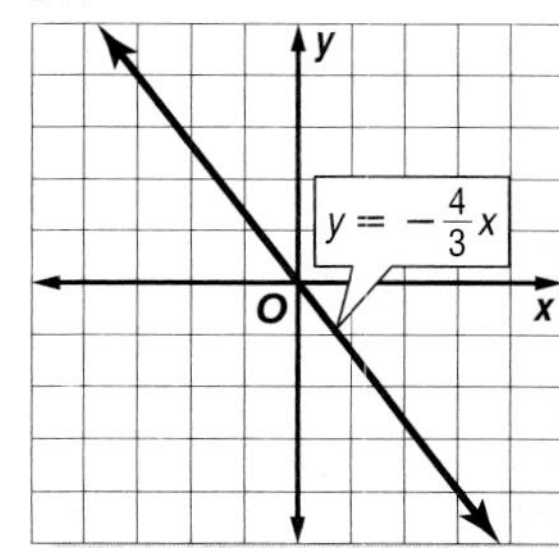

33. $y = 2x$; 10 **35.** $y = -4x$; −5 **37.** $y = \frac{1}{3}x$; −8 **39.** $y = 5x$; 100 **41.** $y = \frac{32}{3}x$; 12

Selected Answers

43. $C = 3.14d$

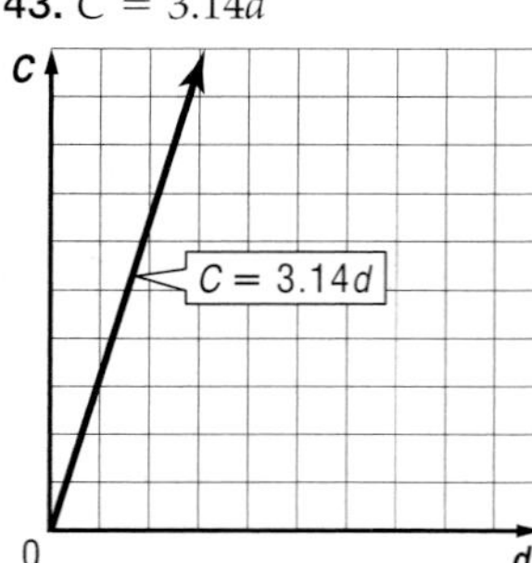

45. $C = 0.99n$

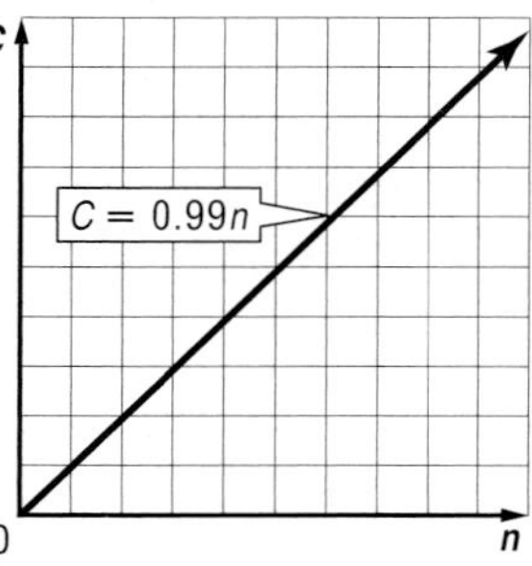

47. It also doubles. If $\frac{y}{x} = k$, and x is multiplied by 2, y must also be multiplied by 2 to maintain the value of k. **49.** 2 **51.** 3 **53.** 23 lb **55.** 5 yrs 4 mos **57.** D

59.

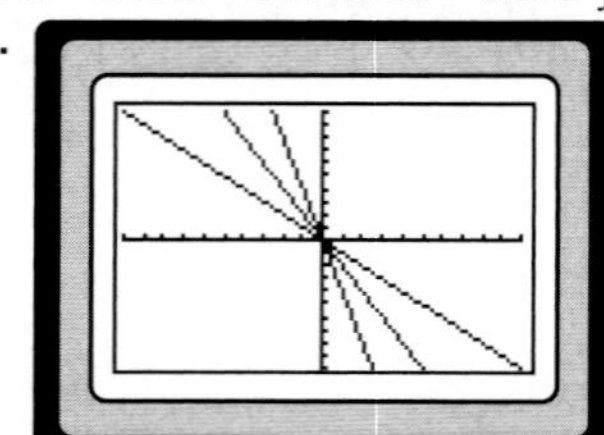

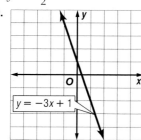

61. Sample answer: $y = -5x$ **63.** -3 **65.** 2

67.

x	0	1	2	3	4	5
y	1	5	9	13	17	21

69. 3 **71.** -15 **73.** $y = 3x + 8$ **75.** $y = 4x - 3$ **77.** $y = -3x + 4$

Page 270 Practice Quiz 1

1. -2 **3.** $\frac{1}{9}$ **5.** 4

7.

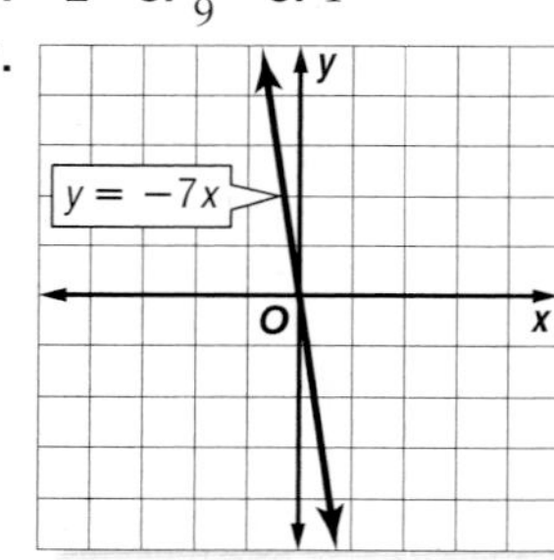

9. $y = 3x; -9$

Pages 275–277 Lesson 5-3

1. Sample answer: $y = 7x + 2$ **3.** slope **5.** $y = 4x - 2$ **7.** $y = -\frac{3}{2}x + 2$

9.

y

O x

y = −3x + 1

11. $T = 50 + 5w$ **13.** \$85 **15.** $y = 3x - 5$ **17.** $y = -\frac{3}{5}x$ **19.** $y = 0.5x + 7.5$ **21.** $y = \frac{3}{2}x - 4$ **23.** $y = -\frac{2}{3}x + 1$ **25.** $y = 2$ **27.** $y = 3x$

29.

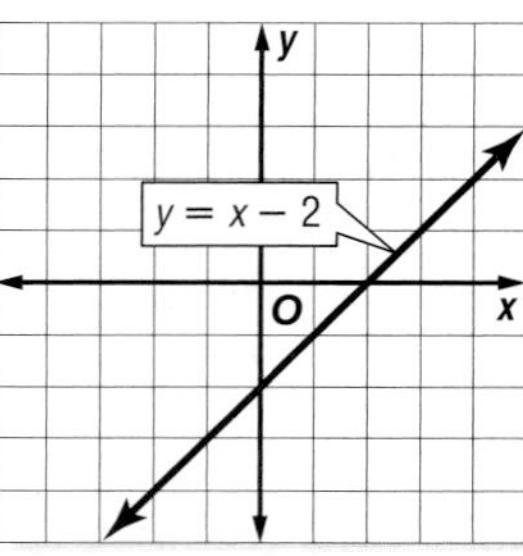

31.

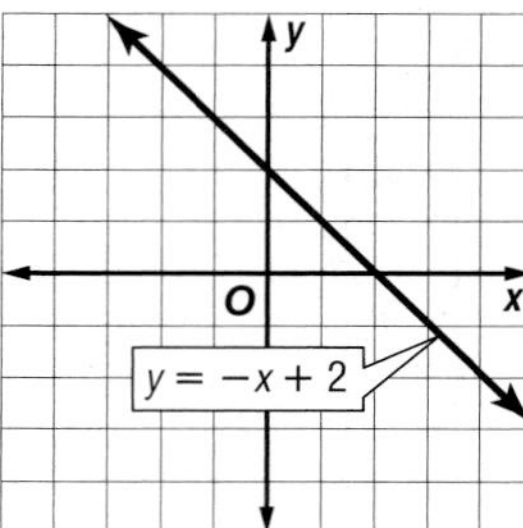

33.

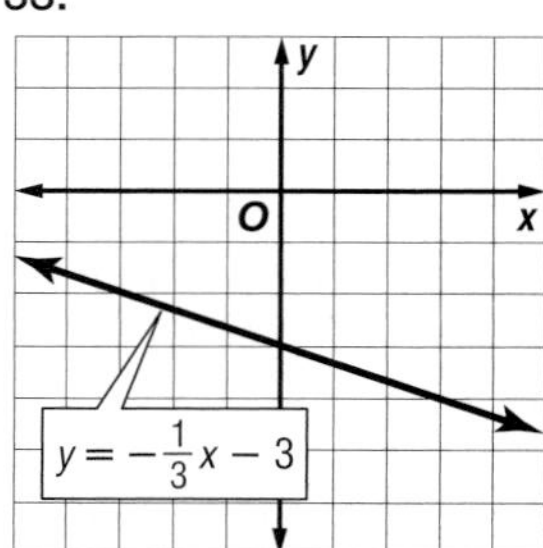

35.

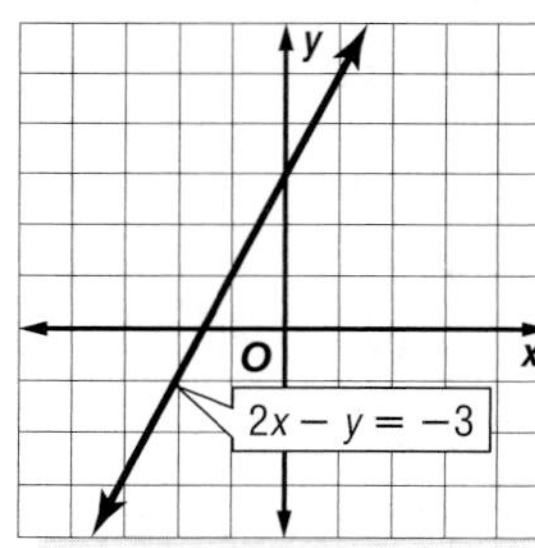

37.

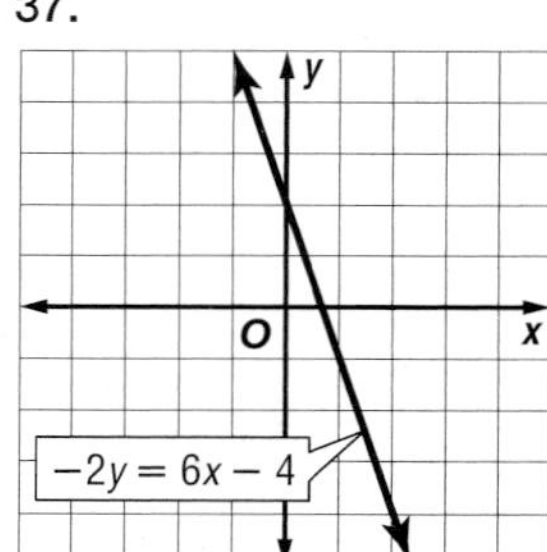

39.

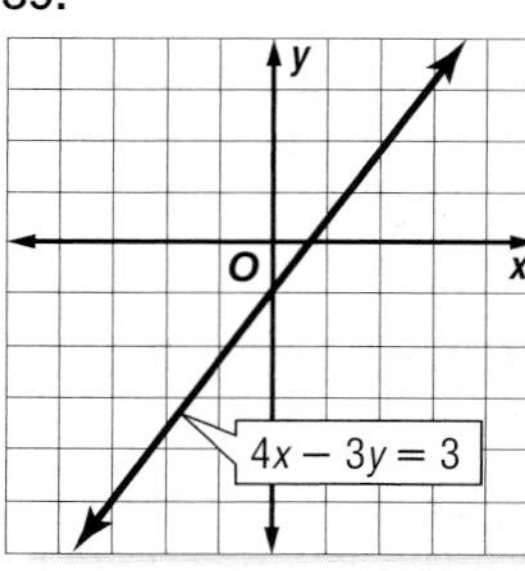

41. $C = 50 + 25h$ **43.** $T = 15 - 2h$ **45.** $S = 16 + t$ **47.** $R = 5.5 - 0.12t$ **49.** 1.54 **51.** D **53.** $y = -\frac{A}{B}x + \frac{C}{B}$, where $B \neq 0$ **55a.** $m = -2, b = -4$ **55b.** $m = -\frac{3}{4}, b = 3$ **55c.** $m = \frac{2}{3}, b = -3$ **57.** $y = \frac{15}{4}x, 37\frac{1}{2}$ **59.** undefined **61.** $-0.5, \frac{3}{4}, \frac{7}{8}, 2.5$ **63.** 23 **65.** -2 **67.** $-\frac{4}{3}$

Pages 283–285 Lesson 5-4

1. When you have the slope and one point, you can substitute these values in for x, y, and m to find b. When you are given two points, you must first find the slope and then use the first procedure. **3.** Sometimes; if the x- and y-intercepts are both zero, you cannot write the equation of the graph. **5.** $y = -3x + 16$ **7.** $y = -x + 6$ **9.** $y = \frac{1}{2}x - \frac{1}{2}$ **11.** $y = 3x - 1$ **13.** $y = 3x - 17$ **15.** $y = -2x + 6$ **17.** $y = -\frac{2}{3}x - 3$ **19.** $y = x - 3$ **21.** $y = x - 2$ **23.** $y = -2x + 1$ **25.** $y = -2$ **27.** $y = \frac{1}{2}x + \frac{1}{2}$ **29.** $y = -\frac{1}{4}x + \frac{11}{16}$ **31.** $y = -\frac{4}{3}x + 4$ **33.** $y = x - 2$ **35.** about 27.6 years **37.** about 26.05 years **39.** 205,000 **41.** $y = \frac{2}{7}x - 2$ **43.** (7, 0); (0, −2)
45. Answers should include the following.

- Linear extrapolation is when you use a linear equation to predict values that are outside of the given points on the graph.
- You can use the slope-intercept form of the equation to find the y-value for any requested x-value.

47. B

49.

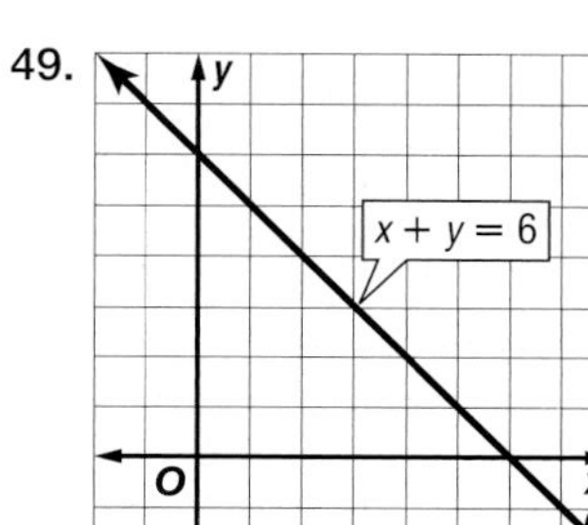

51. $V = 2.5b$ **53.** $\{-2, 0, 5\}$ **55.** $<$ **57.** -3 **59.** 5 **61.** -15

Pages 289–291 Lesson 5-5

1. They are the coordinates of any point on the graph of the equation. **3.** Sample answer: $y - 2 = 4(x + 1)$; $y = 4x + 6$ **5.** $y + 2 = 3(x + 1)$ **7.** $4x - y = -13$ **9.** $5x - 2y = 11$ **11.** $y = -\frac{2}{3}x + 1$ **13.** $y - 3 = 2(x + 1)$ or $y + 1 = 2(x + 3)$ **15.** $y - 8 = 2(x - 3)$ **17.** $y - 4 = -3(x + 2)$ **19.** $y - 6 = 0$ **21.** $y + 3 = \frac{3}{4}(x - 8)$ **23.** $y + 3 = -\frac{5}{8}(x - 1)$ **25.** $y - 8 = \frac{7}{2}(x + 4)$ **27.** $y + 9 = 0$ **29.** $4x - y = -5$ **31.** $2x + y = -7$ **33.** $x - 2y = 12$ **35.** $2x + 5y = 26$ **37.** $5x - 3y = -24$ **39.** $13x - 10y = -151$ **41.** $y = 3x - 1$ **43.** $y = -2x + 8$ **45.** $y = \frac{1}{2}x - 1$ **47.** $y = -\frac{1}{4}x - \frac{7}{2}$ **49.** $y = x - 1$ **51.** $y = -3x - \frac{7}{4}$ **53.** $y + 3 = 10(x - 5)$; $y = 10x - 53$; $10x - y = 53$ **55.** $y - 210 = 5(x - 12)$ **57.** \$150 **59.** $y = 1500x - 2{,}964{,}310$ **61.** $\overline{RQ}$: $y + 3 = \frac{1}{2}(x + 1)$ or $y + 1 = \frac{1}{2}(x - 3)$; $\overline{QP}$: $y + 1 = -2(x - 3)$ or $y - 3 = -2(x - 1)$; $\overline{PS}$: $y - 3 = \frac{1}{2}(x - 1)$ or $y - 1 = \frac{1}{2}(x + 3)$; $\overline{RS}$: $y + 3 = -2(x + 1)$ or $y - 1 = -2(x + 3)$ **63.** $\overline{RQ}$: $x - 2y = 5$; $\overline{QP}$: $2x + y = 5$; $\overline{PS}$: $x - 2y = -5$; $\overline{RS}$: $2x + y = -5$ **65.** Answers should include the following.

- Write the definition of the slope using (x, y) as one point and (x_1, y_1) as the other. Then solve the equation so that the ys are on one side and the slope and xs are on the other.

67. $y = mx - 2m - 5$ **69.** All of the equations are the same. **71.** Regardless of which two points on a line you select, the slope-intercept form of the equation will always be the same. **73.** $y = 3x + 10$ **75.** $y = -1$ **77.** -6 **79.** 7 **81.** $\frac{1}{10}$ **83.** -1 **85.** -9 **87.** $-\frac{3}{2}$

Pages 295–297 Lesson 5-6

1. The slope is $\frac{3}{2}$, so the slope of a line perpendicular to the given line is $-\frac{2}{3}$. **3.** Parallel lines lie in the same plane and never intersect. Perpendicular lines intersect at right angles. **5.** $y = x + 1$ **7.** $y = 3x + 8$ **9.** $y = -3x - 8$ **11.** $y = \frac{1}{2}x - 3$ **13.** $y = x - 9$ **15.** $y = x + 5$ **17.** $y = \frac{1}{2}x - \frac{3}{2}$ **19.** $y = -\frac{1}{3}x - \frac{13}{3}$ **21.** $y = \frac{1}{2}x + \frac{3}{2}$ **23.** $y = -6x - 9$ **25.** The lines for $x = 3$ and $x = -1$ are parallel because all vertical lines are parallel. The lines for $y = \frac{2}{3}x + 2$ and $y = \frac{2}{3}x - 3$ are parallel because they have the same slope. Thus, both pairs of opposite sides are parallel and the figure is a parallelogram. **27.** $y = \frac{1}{3}x - 6$ **29.** $y = -\frac{1}{4}x + \frac{5}{4}$ **31.** $y = \frac{1}{8}x + 5$ **33.** $y = -\frac{3}{2}x + 13$ **35.** $y = -\frac{5}{2}x + 2$ **37.** $y = -\frac{1}{5}x - 1$ **39.** $y = -3$ **41.** $y = -\frac{1}{2}x + 2$ **43.** parallel **45.** They are perpendicular, because the slopes are 3 and $-\frac{1}{3}$.

47. If two equations have the same slope, then the lines are parallel. Answers should include the following.

- Sample answer: $y = -5x + 1$; The graphs have the same slope.
- Sample answer: $y = \frac{1}{5}x$; The slopes are negative reciprocals of each other.

49. C **51.** $y - 7 = 5(x + 4)$ **53.** $C = 0.22m + 0.99$ **55.** $y = -\frac{1}{2}x + \frac{3}{2}$ **57.** $y = -5x + 11$ **59.** $y = 9$

Page 297 Practice Quiz 2

1. $y = 4x - 3$ **3.** $y = \frac{5}{2}x + \frac{1}{2}$ **5.** $x - 2y = -11$, $y = \frac{1}{2}x + \frac{11}{2}$

Pages 301–305 Lesson 5-7

1. If the data points form a linear pattern such that y increases as x increases, there is a positive correlation. If the linear pattern shows that y decreases as x increases, there is a negative correlation. **3.** Linear extrapolation predicts values outside the range of the data set. Linear interpolation predicts values inside the range of the data. **5.** Negative; the more TV you watch, the less you exercise. **7.** positive correlation

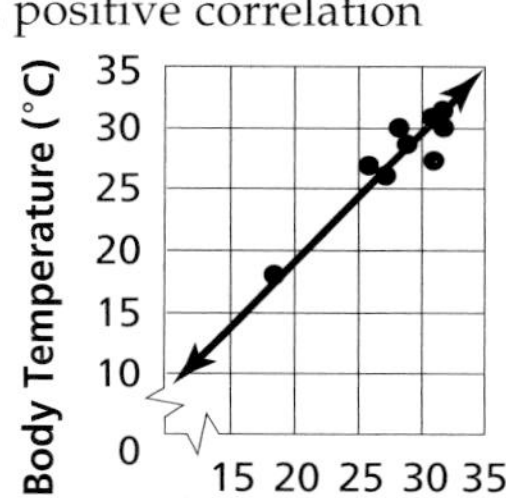

9. 40.1°F **11.** no correlation **13.** Positive; the higher the sugar content, the more Calories. **15.** 18.85 million **17.** \$3600

19.

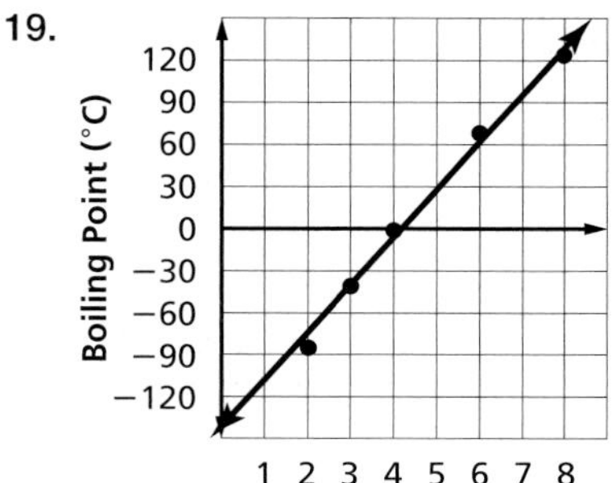

21. Sample answer: −116°C **23.** Sample answer: 7

25.

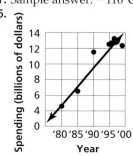

27. Sample answer: about \$17.3 billion

29. 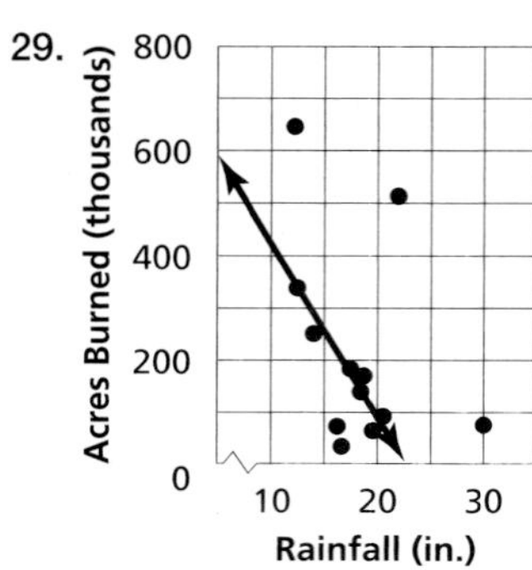

31. using (12.7, 340) and (17.5, 194) and rounding, $y = -30.4x + 726.3$ **33.** The data point lies beyond the main grouping of data points. It can be ignored as an extreme value.

37. You can visualize a line to determine whether the data has a positive or negative correlation. Answers should include the following.

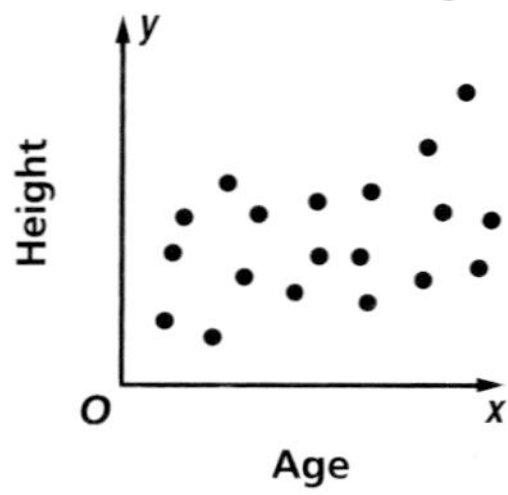

- Write a linear equation for the line of fit. Then substitute the person's height and solve for the corresponding age.

39. B **45.** $y = -4x - 3$ **47.** $y - 3 = -2(x + 2)$ **49.** $y + 3 = x + 3$ **51.** 4, −1.6 **53.** −5 **55.** 3

Pages 308–312 Chapter 5 Study Guide and Review

1. direct variation **3.** parallel **5.** slope-intercept **7.** 3 **9.** undefined **11.** 1.5

13. 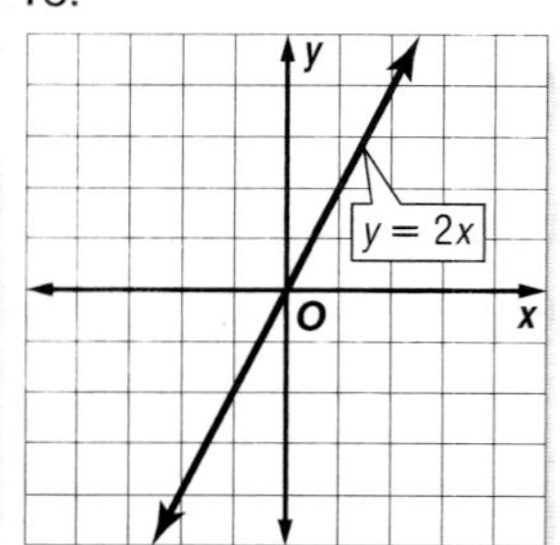

15. 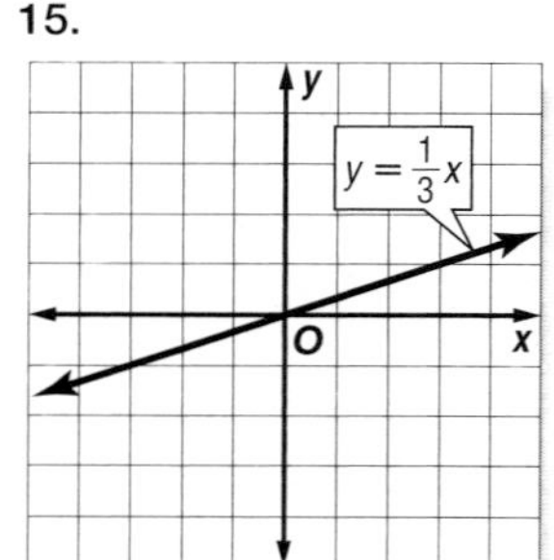

17. 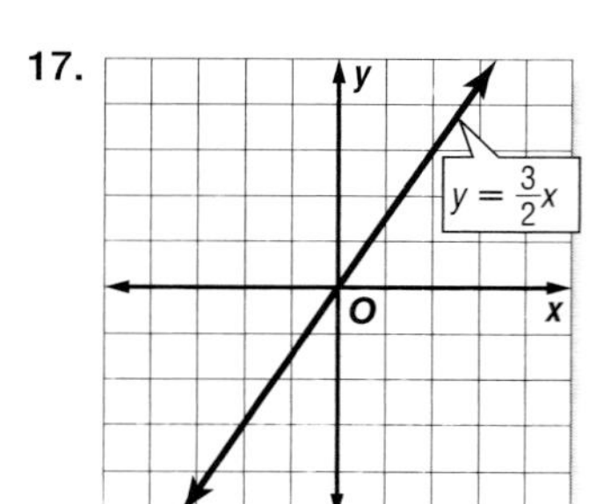

19. $y = -\frac{2}{3}x$
21. $y = -x$
23. $y = -2x$
25. $y = 3x + 2$
27. $y = 4$
29. $y = 0.5x - 0.3$

31. 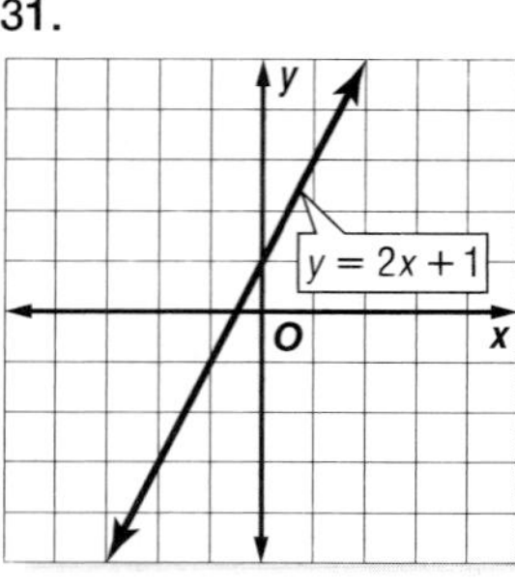

33. 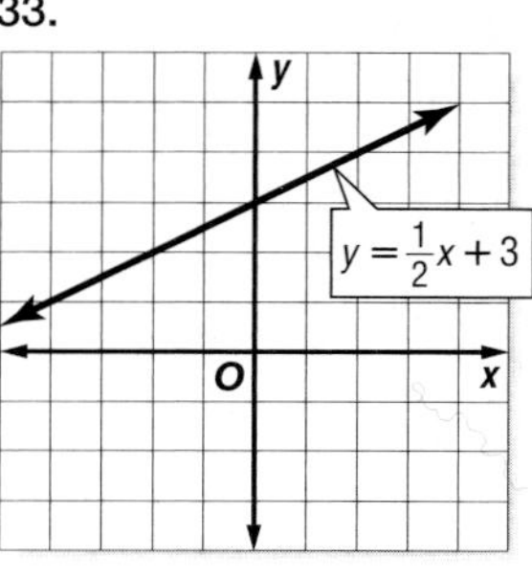

35. 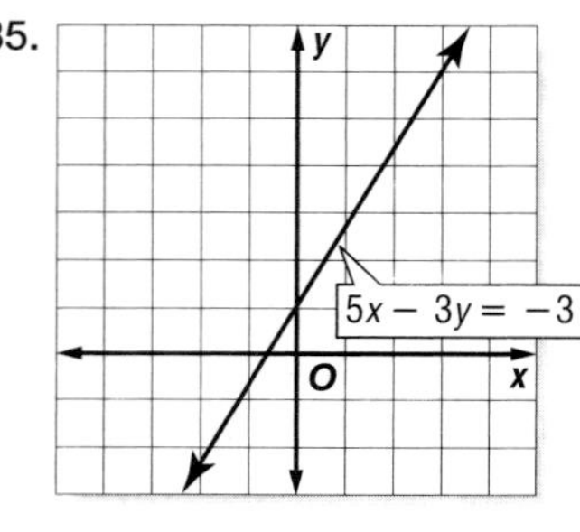

37. $y = x + 6$
39. $y = \frac{1}{2}x + \frac{11}{2}$
41. $y = 2x + 10$
43. $y = -1$
45. $y - 6 = 5(x - 4)$
47. $y + 3 = \frac{1}{2}(x - 5)$
49. $y + 2 = 3\left(x - \frac{1}{4}\right)$
51. $2x - y = -3$
53. $3x - 2y = 20$

55. $y = -2x + 6$ **57.** $y = \frac{5}{12}x + 4$ **59.** $y = -\frac{1}{3}x + 1$ **61.** $y = \frac{1}{2}x - 3$ **63.** $y = -\frac{7}{2}x - 14$ **65.** $y = 5x - 15$

67. 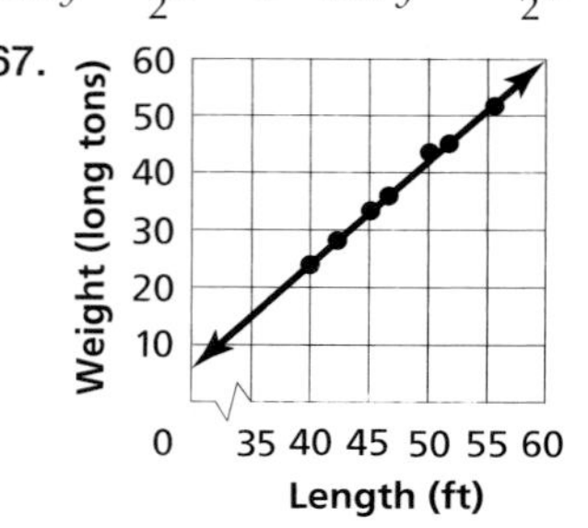

69. $38\frac{1}{3}$ long tons

Chapter 6 Solving Linear Inequalities

Page 317 Chapter 6 Getting Started

1. 53 **3.** −9 **5.** −45 **7.** 4 **9.** 22 **11.** 4 **13.** 8 **15.** 30 **17.** 7 **19.** 1

21. 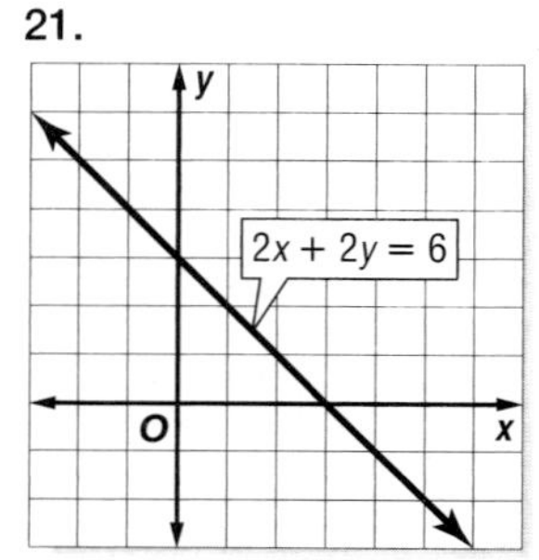

23. 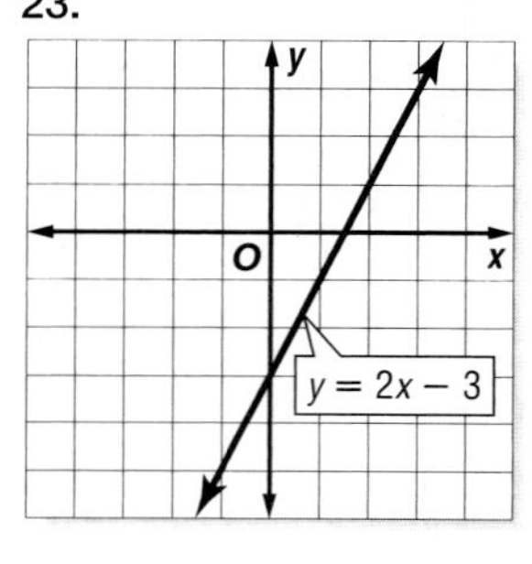

25. 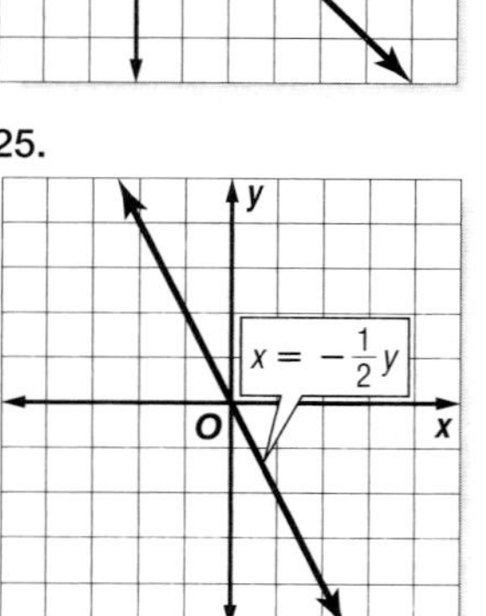

27. 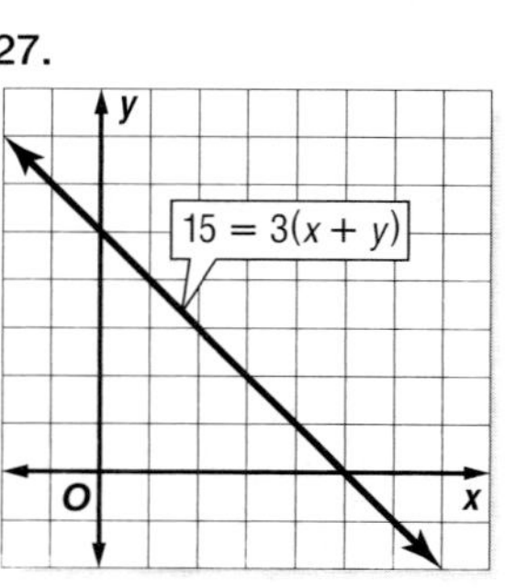

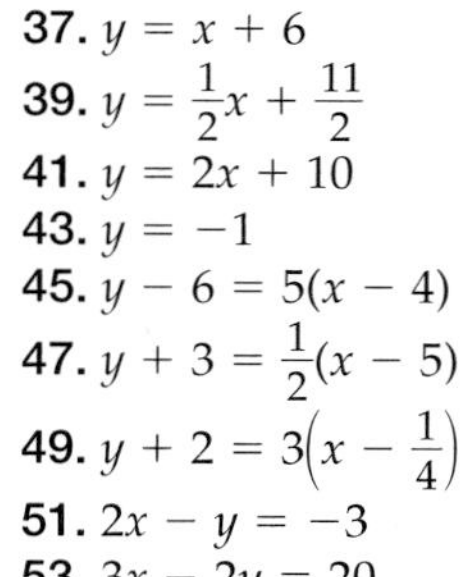

Pages 321–323 Lesson 6-1

1. Sample answers: $y + 1 < -2$, $y - 1 < -4$, $y + 3 < 0$
3. The set of all numbers b such that b is greater than or equal to -5.
5. $\{a \mid a < -2\}$

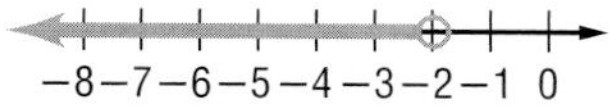

7. $\{t \mid t \geq 12\}$

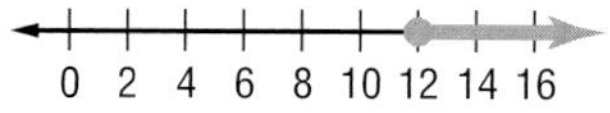

9. $\{r \mid r \leq 6.7\}$

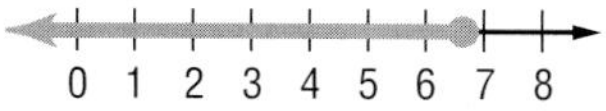

11. Sample answer: Let n = the number; $n - 8 \leq 14$; $\{n \mid n \leq 22\}$. **13.** no more than 33 g **15.** f **17.** c **19.** b
21. $\{d \mid d \leq 2\}$ **23.** $\{s \mid s > 4\}$

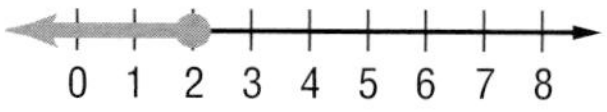

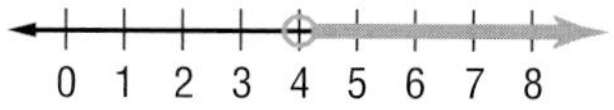

25. $\{r \mid r < -4\}$ **27.** $\{m \mid m \geq 3\}$

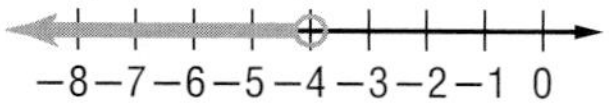

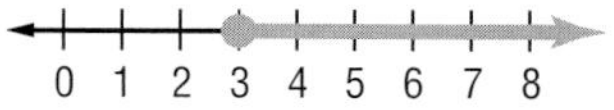

29. $\{f \mid f < -3\}$ **31.** $\{w \mid w \geq 1\}$

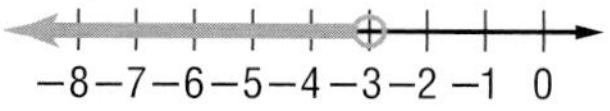

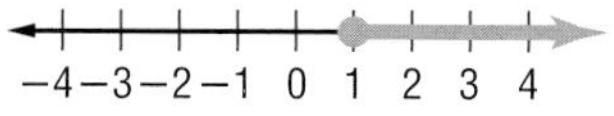

33. $\{a \mid a \leq -5\}$ 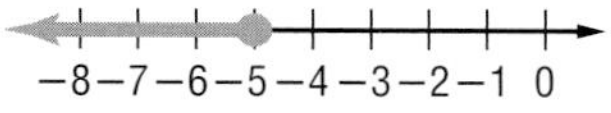 **35.** $\{x \mid x \geq 0.6\}$

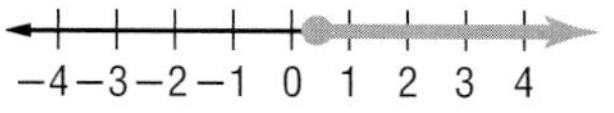

37. $\left\{p \mid p \leq 1\frac{1}{9}\right\}$

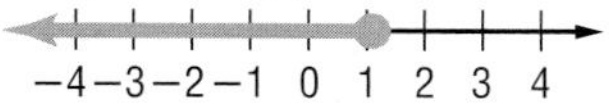

39a. 12 **39b.** 7 **39c.** 16 **41.** Sample answer: Let n = the number; $n - 5 < 33$; $\{n \mid n < 38\}$. **43.** Sample answer: Let n = the number; $2n > n + 14$; $\{n \mid n > 14\}$. **45.** Sample answer: Let n = the number; $4n \leq 3n + (-2)$; $\{n \mid n \leq -2\}$. **47.** at least 199,999,998,900 stars **49.** at least \$3747 **51.** no more than \$33 **53a.** always **53b.** never **53c.** sometimes **55.** $\{p \mid p > 25\}$ **57.** C **59.** no **61.** $y = -x + 4$ **63.** 31, 37 **65.** 48, 96 **67.** $\{(-1, 8), (3, 4), (5, 2)\}$ **69.** 7 **71.** 21 **73.** 49 **75.** 24.5

Pages 328–331 Lesson 6-2

1. You could solve the inequality by multiplying each side by $-\frac{1}{7}$ or by dividing each side by -7. In either case, you must reverse the direction of the inequality symbol.
3. Ilonia; when you divide each side of an inequality by a negative number, you must reverse the direction of the inequality symbol. **5.** c **7.** $\{t \mid t < -108\}$
9. $\{f \mid f \geq 0.36\}$ **11.** Sample answer: Let n = the number; $\frac{1}{2}n \geq 26$; $\{n \mid n \geq 52\}$. **13.** d **15.** e **17.** b **19.** $\{g \mid g \leq 24\}$
21. $\{d \mid d \leq -6\}$ **23.** $\{m \mid m \geq 35\}$ **25.** $\{r \mid r > 49\}$
27. $\{y \mid y \geq -24\}$ **29.** $\{q \mid q \geq 44\}$ **31.** $\{w \mid w > -2.72\}$
33. $\left\{c \mid c < -\frac{1}{10}\right\}$
35. $\{y \mid y < -4\}$

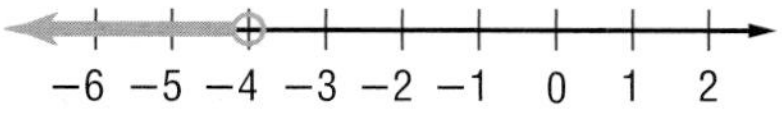

37a. 3.5 **37b.** -14 **37c.** -6 **39.** Sample answer: Let n = the number; $7n > 28$; $\{n \mid n > 4\}$. **41.** Sample answer: Let n = the number; $24 \leq \frac{1}{3}n$; $\{n \mid n \geq 72\}$. **43.** Sample answer: Let n = the number; $0.25n \geq 90$; $\{n \mid n \geq 360\}$. **45.** less than $4\frac{1}{4}$ ft **47.** no more than 27 min **49.** up to about 6 ft **51.** at least 3 times **53.** at least 175 spaces
55. Inequalities can be used to compare the heights of walls. Answers should include the following.
- If x represents the number of bricks and the wall must be no higher than 4 ft or 48 in., then $3x \leq 48$.
- To solve this inequality, divide each side by 3 and do not change the direction of the inequality. The wall must be 16 bricks high or fewer.

57. C
59. $\{g \mid g \leq -7\}$

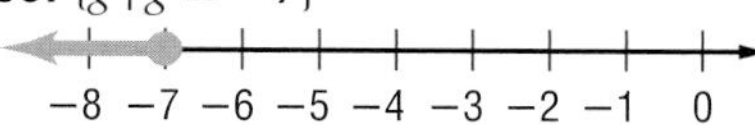

61. Sample answer:

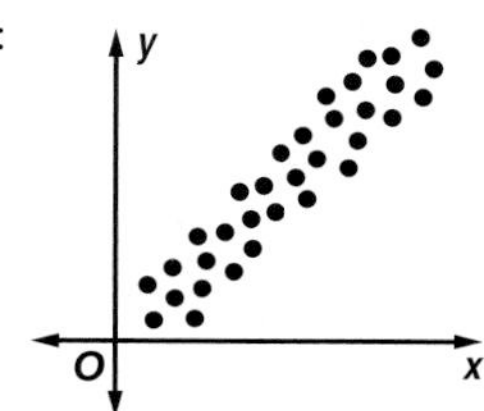

63. $y = -2$ **65.** -10 **67.** $3w + 2$ **69.** 6 **71.** 5 **73.** 7 **75.** 12 **77.** -8

Page 331 Practice Quiz 1

1. $\{h \mid h > 3\}$

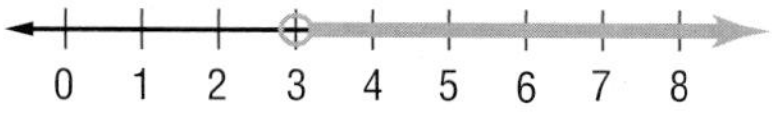

3. $\{p \mid p \leq -5\}$

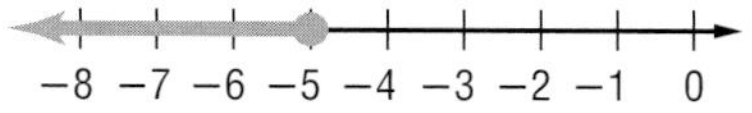

5. $\{g \mid g \leq -1\}$

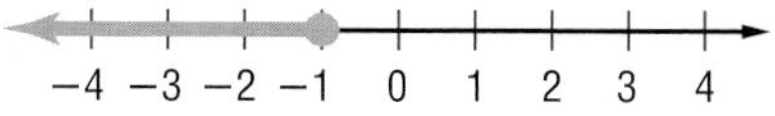

7. $\{v \mid v < 35\}$ **9.** $\{r \mid r > -13\}$

Pages 334–337 Lesson 6-3

1. To solve both the equation and the inequality, you first subtract 6 from each side and then divide each side by -5. In the equation, the equal sign does not change. In the inequality, the inequality sign is reversed because you divided by a negative number. **3a.** Distributive Property **3b.** Add 12 to each side. **3c.** Divide each side by 3. **5.** $\{r \mid r \geq -18\}$ **7.** $\{g \mid g < -1\}$ **9.** Sample answer: Let n = the number; $7 - 2n < 3n + 32$; $\{n \mid n > -5\}$. **11a.** Subtract 7 from each side. **11b.** Multiply each side by $\frac{5}{2}$.

13.

$4(t - 7) \leq 2(t + 9)$	*Original inequality*
$4t - 28 \leq 2t + 18$	*Distributive Property*
$4t - 28 - 2t \leq 2t + 18 - 2t$	*Subtract 2t from each side.*
$2t - 28 \leq 18$	*Simplify.*
$2t - 28 + 28 \leq 18 + 28$	*Add 28 to each side.*
$2t \leq 46$	*Simplify.*
$\frac{2t}{2} \leq \frac{46}{2}$	*Divide each side by 2.*
$t \leq 23$	*Simplify.*

$\{t \mid t \leq 23\}$

15. $\{t \mid t \geq 3\}$ **17.** $\{d \mid d > -125\}$ **19.** $\left\{q \mid q \leq 3\frac{1}{3}\right\}$
21. $\{r \mid r \geq -9\}$ **23.** $\{v \mid v \geq 19\}$ **25.** $\{w \mid w \leq 1\}$

Selected Answers

27. $\{t \mid t \geq -1\}$ **29.** $\varnothing$ **31.** $\{v \mid v \geq 4.5\}$
33. $\{y \mid y \leq 11\}$

5 6 7 8 9 10 11 12 13

35. Sample answer: Let n = the number; $\frac{1}{8}n - 5 \geq 30$; $\{n \mid n \geq 280\}$. **37.** Sample answer: Let n = the number; $-4n + 9 \leq n - 21$; $\{n \mid n \geq 6\}$. **39.** $3a - 15 < 90$ **41.** $\frac{91 + 95 + 88 + s}{4} \geq 92$ **43.** $\frac{5(F - 32)}{9} < -38$ **45.** more than $12\frac{1}{2}$ weeks **47.** 3 or fewer toppings **49.** no change
51. 7, 9; 5, 7; 3, 5; 1, 3
53. Inequalities can be used to describe the temperatures for which an element is a gas or a solid. Answers should include the following.

- The inequality for temperatures in degrees Celsius for which bromine is a gas is $\frac{9}{5}C + 32 > 138$.
- Sample answer: Scientists may use inequalities to describe the temperatures for which an element is a solid.

55. C **57.** $\{x \mid x \leq 8\}$ **59.** up to 416 mi
61. $\{t \mid t < 8\}$

5 6 7 8 9 10 11 12 13

63. $y + 3 = 2(x - 1)$ **65.** $y - 6 = 0$ **67.** $\frac{7}{3}$
69. yes; $4x - 2y = 7$ **71.** yes; $x + 0y = 12$ **73.** 2.5
75.

−3 −2 −1 0 1 2 3 4 5

77.

−3 −2 −1 0 1 2 3 4 5

79.

0 1 2 3 4 5 6 7 8

81.

−5 −4 −3 −2 −1 0 1 2 3

Pages 341–344 Lesson 6-4

1. A compound inequality containing *and* is true if and only if both inequalities are true. A compound inequality containing *or* is true if and only if at least one of the inequalities is true. **3.** Sample answer: $x < -2$ and $x > 3$
5.

5 6 7 8 9 10 11 12 13 14 15

7. $x \leq -1$ or $x \geq 5$
9. $\{n \mid n \leq 2 \text{ or } n \geq 8\}$

0 1 2 3 4 5 6 7 8 9 10

11. $\{x \mid -4 < x \leq 1\}$

−7 −6 −5 −4 −3 −2 −1 0 1 2 3

13. about $4.44 \leq x \leq 6.67$
15.

−10 −9 −8 −7 −6 −5 −4 −3 −2 −1 0

17.

−7 −6 −5 −4 −3 −2 −1 0 1 2 3

19.

−5 −4 −3 −2 −1 0 1 2 3 4 5

21. $-7 < x < -3$ **23.** $x \leq -7$ or $x \geq -6$ **25.** $x = 2$ or $x > 5$ **27.** $t \leq 18$ or $t \geq 22$
29. $\{f \mid -13 \leq f \leq -5\}$

−14 −13 −12 −11 −10 −9 −8 −7 −6 −5 −4

31. $\{h \mid h < -1\}$

−5 −4 −3 −2 −1 0 1 2 3 4 5

33. $\{y \mid 3 < y < 6\}$

0 1 2 3 4 5 6 7 8 9 10

35. $\{q \mid -1 < q < 6\}$

−3 −2 −1 0 1 2 3 4 5 6 7

37. $\{n \mid n \leq 4\}$

0 1 2 3 4 5 6 7 8 9 10

39. $\varnothing$

−5 −4 −3 −2 −1 0 1 2 3 4 5

41. $\{b \mid b < -12 \text{ or } b > -12\}$

−18 −16 −14 −12 −10 −8 −6 −4 −2 0 2

43. Sample answer: Let n = the number; $-8 < 3n + 4 < 10$; $\{n \mid -4 < n < 2\}$. **45.** Sample answer: Let n = the number; $0 < \frac{1}{2}n \leq 1$; $\{n \mid 0 < n \leq 2\}$. **47.** between \$145 and \$230 inclusive **49a.** $x \geq 5$ and $x \leq 8$ **49b.** $x > 6$ or $x < 1$
51. $\{h \mid 15 \leq h \leq 50{,}000\}$; $\{h \mid 20 \leq h \leq 20{,}000\}$ **53.** Sample answer: troposphere: $a \leq 10$; stratosphere: $10 < a \leq 30$; mesosphere: $30 < a \leq 50$; thermosphere: $50 < a \leq 400$; exosphere: $a > 400$ **55.** A **57a.** $\{x \mid x < -6 \text{ or } x > -1\}$
57b. $\{x \mid -2 \leq x \leq 8\}$ **59.** $\{d \mid d \geq 5\}$ **61.** $\{t \mid t < 169\}$
63. 2.25 **65.** $\{(6, 0), (-3, 5), (2, -2), (-3, 3)\}$; $\{-3, 2, 6\}$; $\{-2, 0, 3, 5\}$; $\{(0, 6), (5, -3), (-2, 2), (3, -3)\}$ **67.** $\{(3, 4), (3, 2), (2, 9), (5, 4), (5, 8), (-7, 2)\}$; $\{-7, 2, 3, 5\}$; $\{2, 4, 8, 9\}$; $\{(4, 3), (2, 3), (9, 2), (4, 5), (8, 5), (2, -7)\}$ **69.** 5:1 **71.** −470
73. 7 **75.** 1 **77.** 6 **79.** 1

Page 344 Practice Quiz 2

1. $\{b \mid b < 7\}$ **3.** $\{t \mid t < -3\}$ **5.** $\{m \mid m \geq 3\}$
7. $\{x \mid 3 < x < 9\}$

0 1 2 3 4 5 6 7 8 9 10

9. $\{m \mid m > 3 \text{ or } m < -1\}$

−5 −4 −3 −2 −1 0 1 2 3 4 5

Pages 348–351 Lesson 6-5

1. The solution of $|x - 2| > 6$ includes all values that are less than −4 or greater than 8. The solution of $|x - 2| < 6$ includes all values that are greater than −4 and less than 8.
3. Leslie; you need to consider the case when the value inside the absolute value symbols is positive and the case when the value inside the absolute value symbols is negative. So $x + 3 = 2$ or $x + 3 = -2$. **5.** c
7. $\{-13, 7\}$

−14 −12 −10 −8 −6 −4 −2 0 2 4 6 8

9. $\{w \mid w < -5 \text{ or } w > 25\}$

−15 −10 −5 0 5 10 15 20 25 30 35

11. $|x - 1| = 3$ **13.** $\{d \mid 1.499 \leq d \leq 1.501\}$ **15.** f **17.** b
19. d **21.** $|t - 38| \leq 1.5$ **23.** $|s - 55| \leq 3$

Selected Answers

25. $\{-11, -7\}$

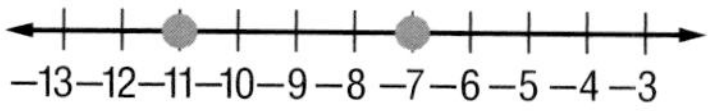

27. $\{-0.8, 4\}$

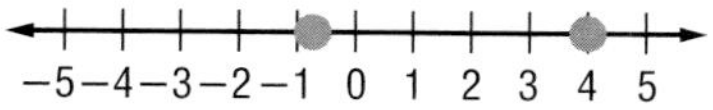

29. $\{t \mid -10 < t < -6\}$

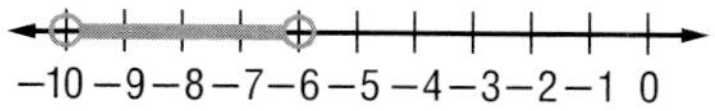

31. $\{w \mid w \leq 3 \text{ or } w \geq 9\}$

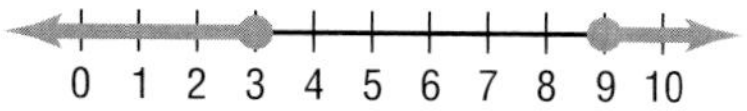

33. $\left\{k \mid k \leq -4 \text{ or } k \geq 1\frac{1}{3}\right\}$

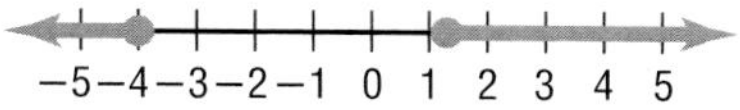

35. $\varnothing$

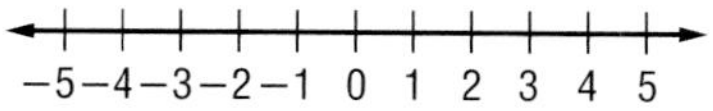

37. $\{w \mid 0 \leq w \leq 18\}$

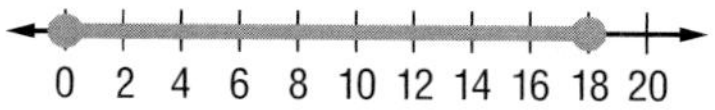

39. $\left\{x \mid x \leq -2\frac{2}{3} \text{ or } x \geq 4\right\}$

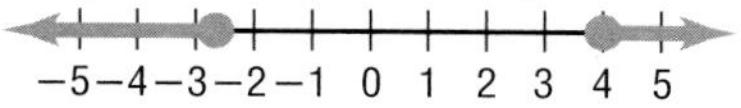

41. $|x - 3| = 5$ **43.** $|x + 3| < 4$ **45.** $|x + 10| \geq 2$ **47.** $\{d \mid 266 \leq d \leq 294\}$ **49.** $\{t \mid 65 \leq t \leq 71\}$ **51.** $\{p \mid 28 \leq p \leq 32\}$ **53.** $\{a \mid 2.5 \leq a \leq 3.5\}$ **55a.** 1.8, 4.2 **55b.** $|x - 3| = 1.2$ **57.** B **59.** between 114 and 152 beats per min **61.** $\left\{x \mid x \leq -1\frac{1}{3}\right\}$ **63.** −2; 4 **65.** $-\frac{2}{3}$; 0 **67.** $x = \frac{3z + 2y}{e}$ **69.** −5 **71.** 4.2 **73.** Substitution Property

75.

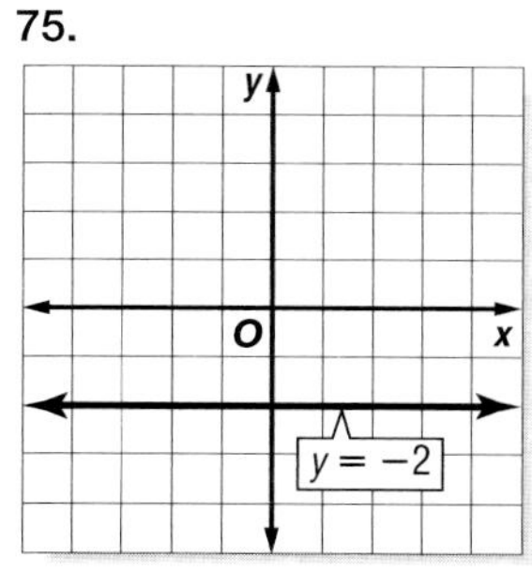

77.

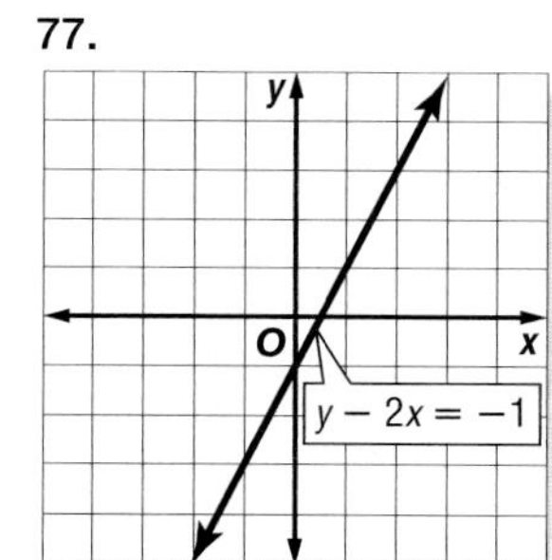

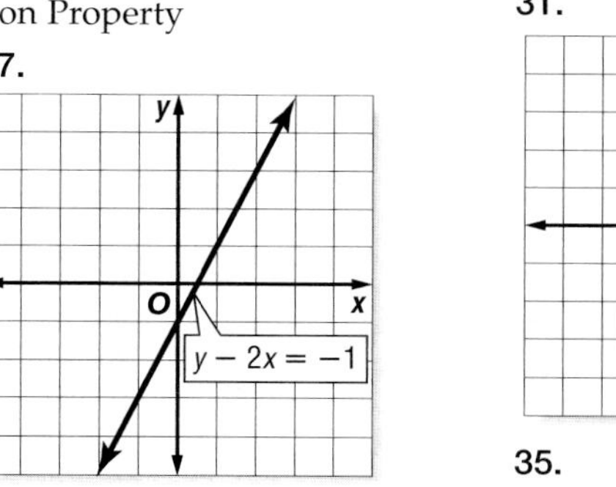

79.

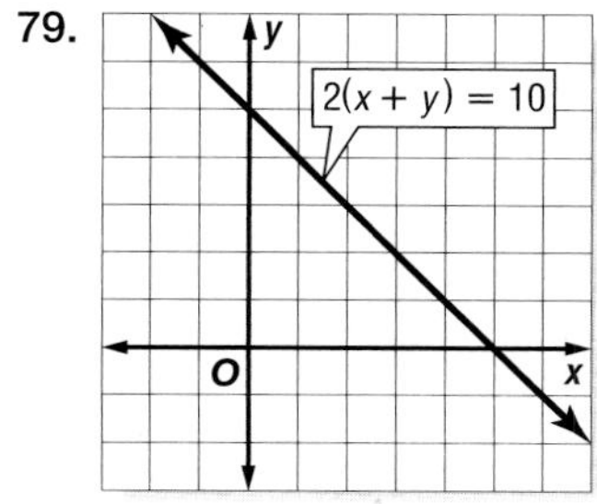

Pages 355–357 Lesson 6-6

1. The graph of $y = x + 2$ is a line. The graph of $y < x + 2$ does not include the boundary $y = x + 2$, and it includes all ordered pairs in the half-plane that contains the origin. **3.** If the test point results in a true statement, shade the half-plane that contains the point. If the test point results in a false statement, shade the other half-plane. **5.** $\{(2, 6)\}$

7.

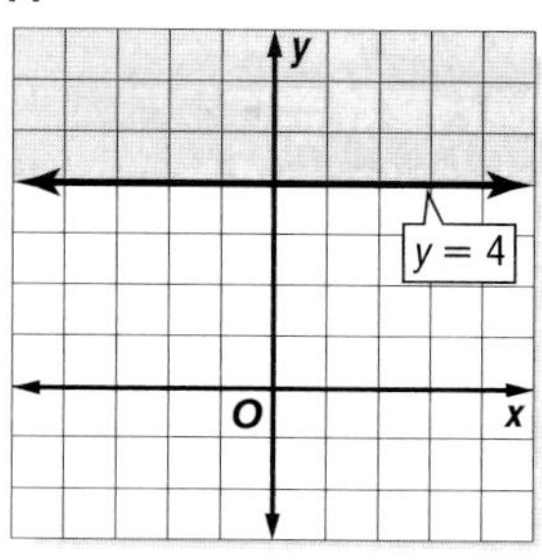

9.

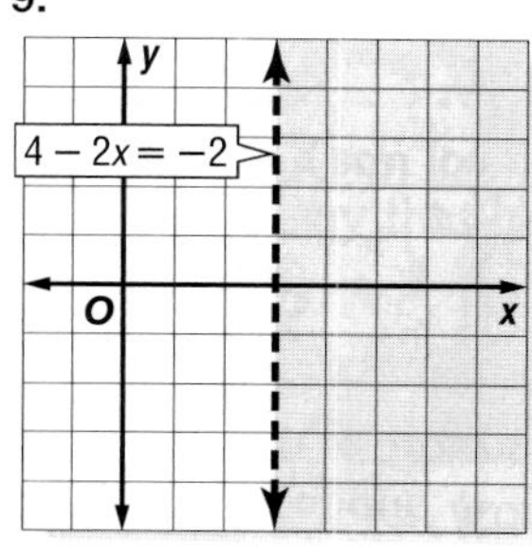

11. $12x + 3y \leq 60$

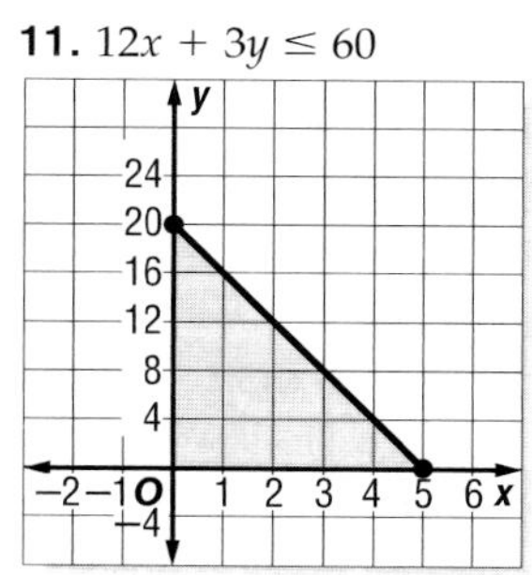

13. $\{(1, 1), (1, 2)\}$
15. $\{(-2, -4), (5, 1)\}$
17. $\{(2, -1), (-1, 1)\}$
19. $\{(6, -7)\}$ **21.** a **23.** b
25. above

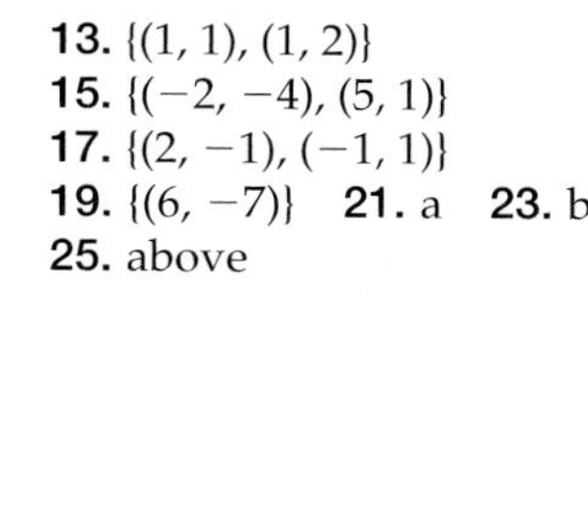

27.

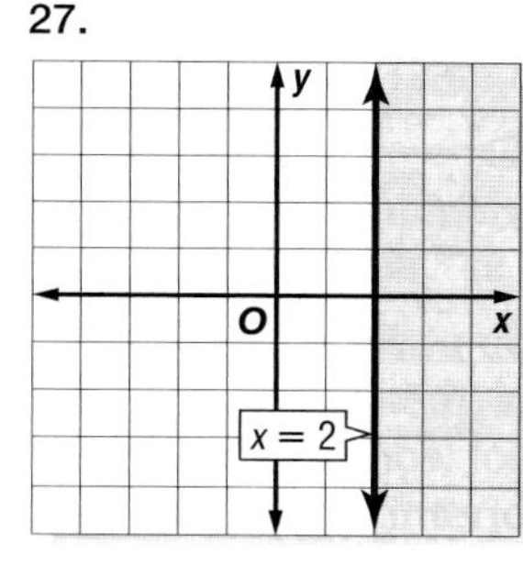

29.

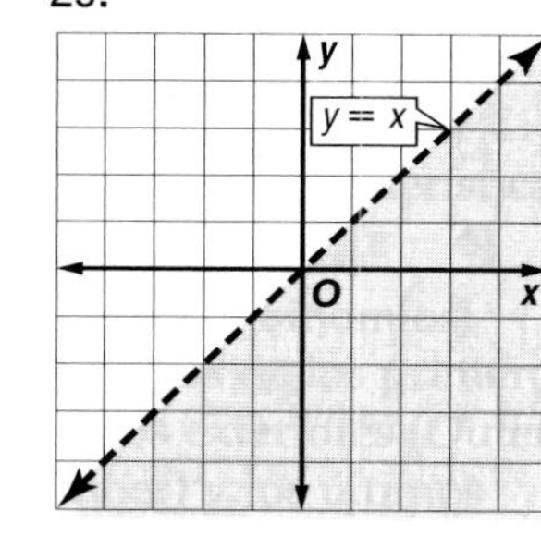

31. **33.**

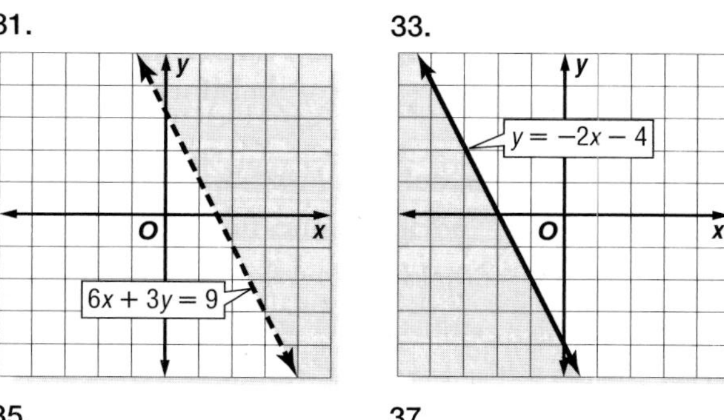

35. **37.**

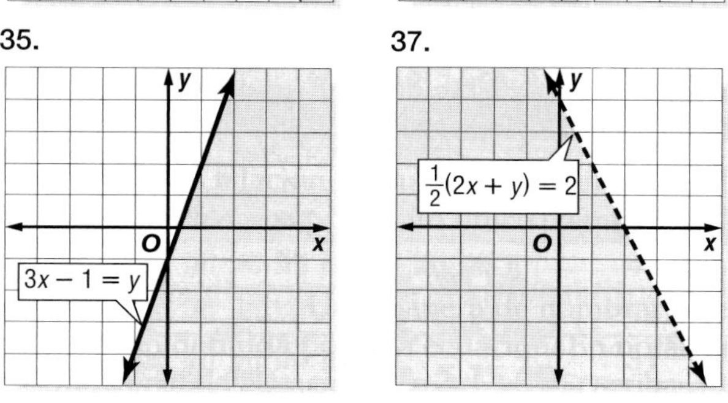

39. The solution set is limited to pairs of positive numbers.

41. No, the weight will be greater than 4000 pounds.

43.

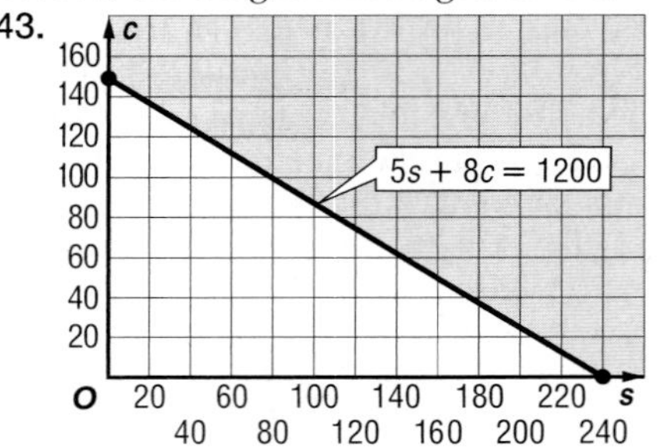

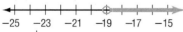

45. $y = x - 1$; $y = -x$ **47.** D

49. $\{-7, 4\}$

−10 −8 −6 −4 −2 0 2 4 6 8 10

51. $\{y \mid y \leq -4 \text{ or } y \geq -1\}$

−8 −7 −6 −5 −4 −3 −2 −1 0 1 2

53. $\{m \mid m < -2 \text{ or } m > 3\}$

−5 −4 −3 −2 −1 0 1 2 3 4 5

55. increase; 42% **57.** 23 **59.** 3.25 **61.** $-3c$ **63.** $6y - 3$

Pages 359–362 Chapter 6 Study Guide and Review

1. f **3.** d **5.** c **7.** h

9. $\{c \mid c > -19\}$

−25 −23 −21 −19 −17 −15

11. $\{w \mid w \leq 37\}$

35 36 37 38 39 40 41 42 43 44 45

13. $\{n \mid n \leq -0.15\}$

−2 −1 0 1 2

15. $\{h \mid h \leq -1\}$

−5 −4 −3 −2 −1 0 1 2 3 4 5

17. Sample answer: Let n = the number; $21 \geq n + (-2)$; $\{n \mid n \leq 23\}$. **19.** $\{r \mid r \leq 6\}$ **21.** $\{m \mid m > -11\}$ **23.** $\{d \mid d < 65\}$ **25.** $\{p \mid p \leq -25\}$ **27.** $\{h \mid h < -2\}$ **29.** $\{x \mid x > -2\}$ **31.** $\{q \mid q > -7\}$ **33.** $\{x \mid x \geq 4\}$ **35.** Sample answer: Let n = the number; $\frac{2}{3}n - 27 \geq 9$; $\{n \mid n \geq 54\}$.

37. $\{k \mid -1 < k < 3\}$

−5 −4 −3 −2 −1 0 1 2 3 4 5

39. $\{a \mid a \leq 11 \text{ or } a \geq 16\}$

9 10 11 12 13 14 15 16 17 18 19

41. $\{y \mid y < -1\}$

−5 −4 −3 −2 −1 0 1 2 3 4 5

43. $\{-7, -3\}$

−9 −8 −7 −6 −5 −4 −3 −2 −1 0 1

45. $\{w \mid w \leq -9 \text{ or } w \geq -7\}$

−10 −9 −8 −7 −6 −5 −4 −3 −2 −1 0

47. $\{t \mid -7 \leq t \leq -1\}$

−9 −8 −7 −6 −5 −4 −3 −2 −1 0 1

49. $\left\{d \mid -4 < d < 1\frac{1}{3}\right\}$

−6 −5 −4 −3 −2 −1 0 1 2 3 4

51. $\{(2, -5), (-1, 6)\}$ **53.** $\{(5, 10), (3, 6)\}$

55.

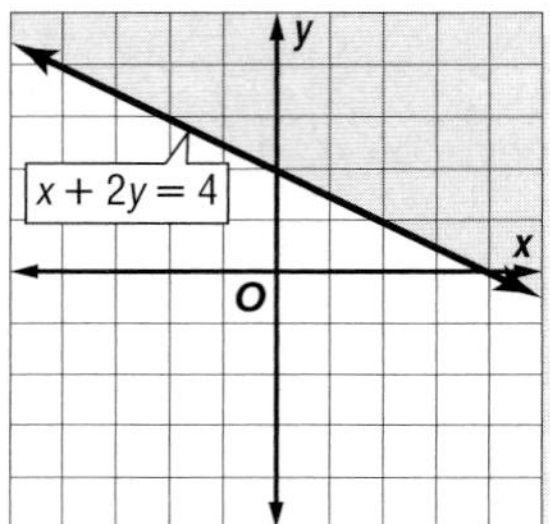

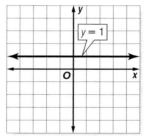

57.

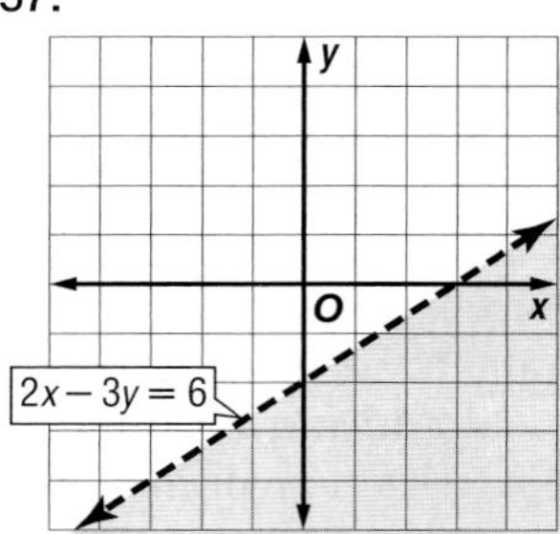

Chapter 7 Solving Systems of Linear Equations and Inequalities

Page 367 Chapter 7 Getting Started

1. $y = 1$

3.

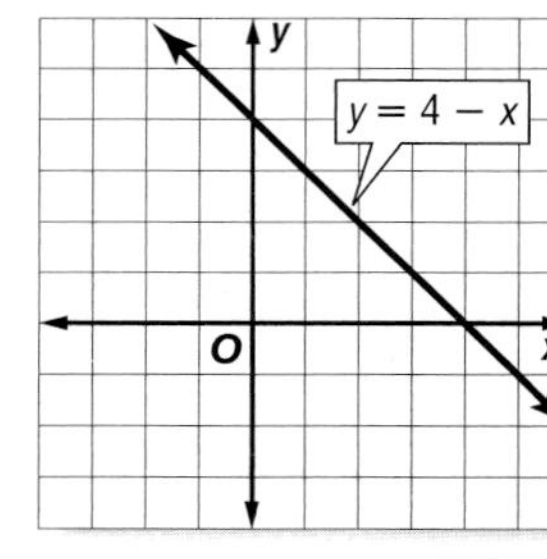

5.

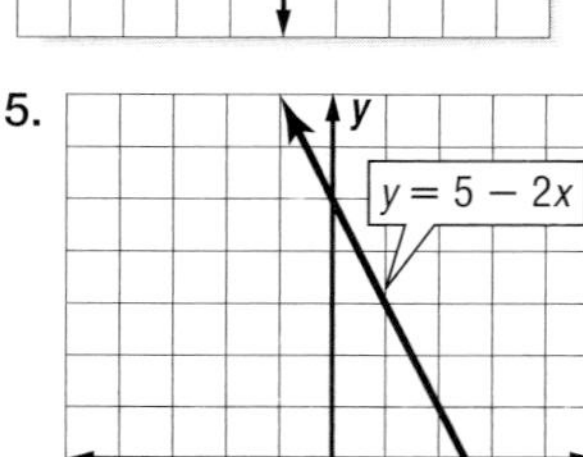

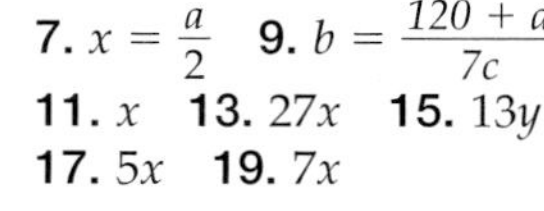

7. $x = \frac{a}{2}$ **9.** $b = \frac{120 + d}{7c}$ **11.** x **13.** $27x$ **15.** $13y$ **17.** $5x$ **19.** $7x$

Page 371–374 Lesson 7-1

1. Sample answer:

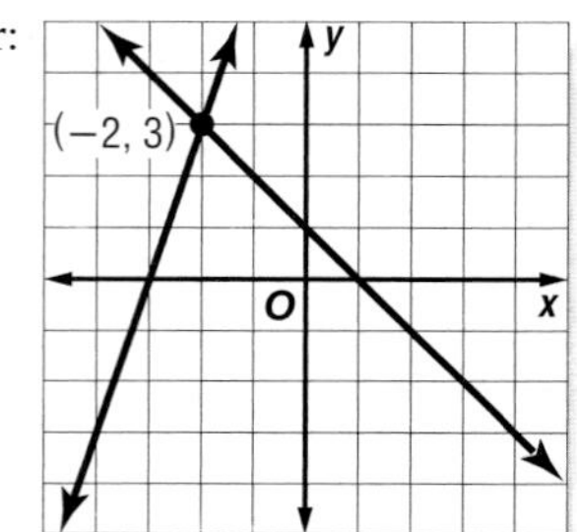

3. Sample answer: The graphs of the equations $x + y = 3$ and $2x + 2y = 6$ have a slope of -1. Since the graphs of the equations coincide, there are infinitely many solutions.
5. no solution **7.** one

9.

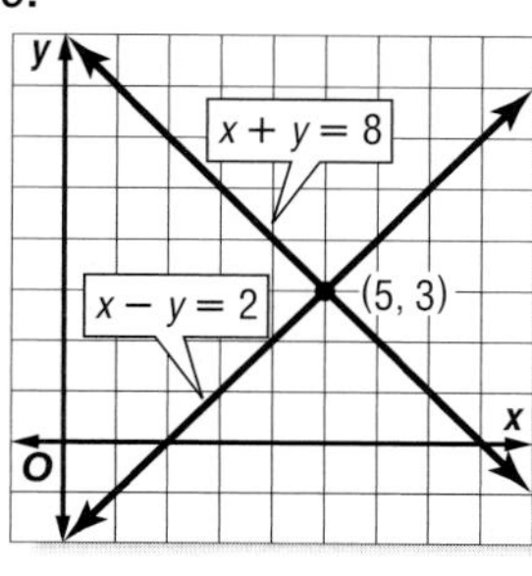

one; (5, 3)

11.

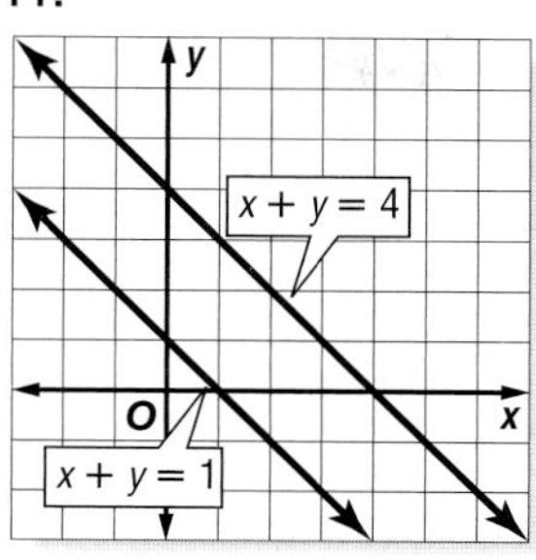

no solution

13.

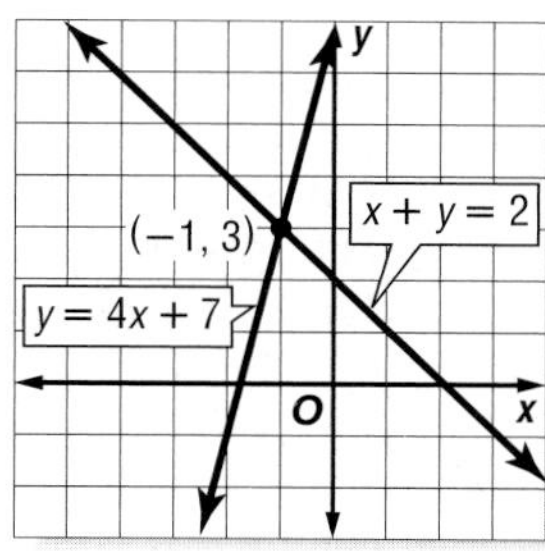

one; $(-1, 3)$
15. one **17.** infinitely many **19.** one **21.** one

23.

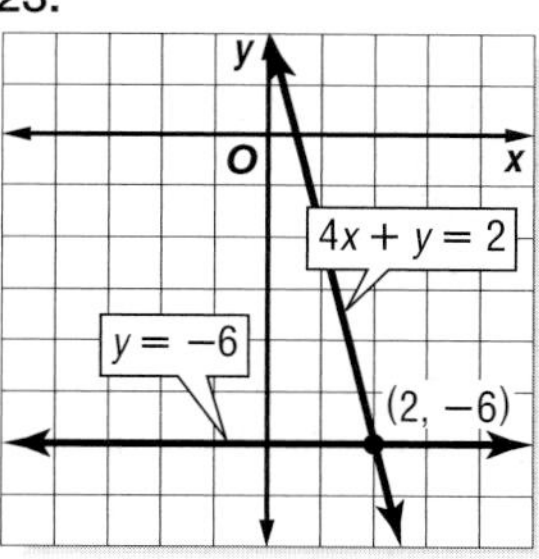

one; $(2, -6)$

25.

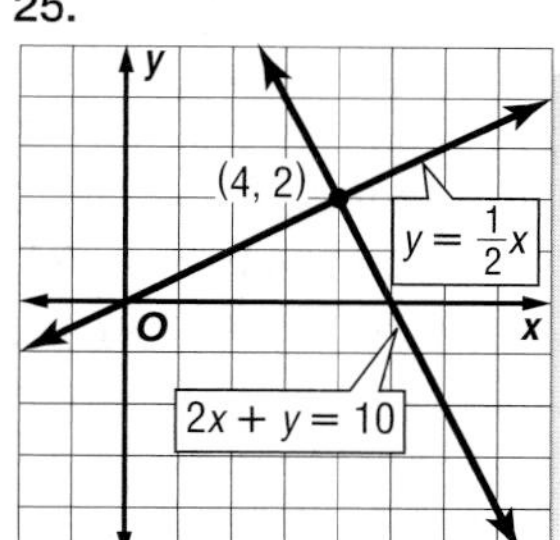

one; (4, 2)

27.

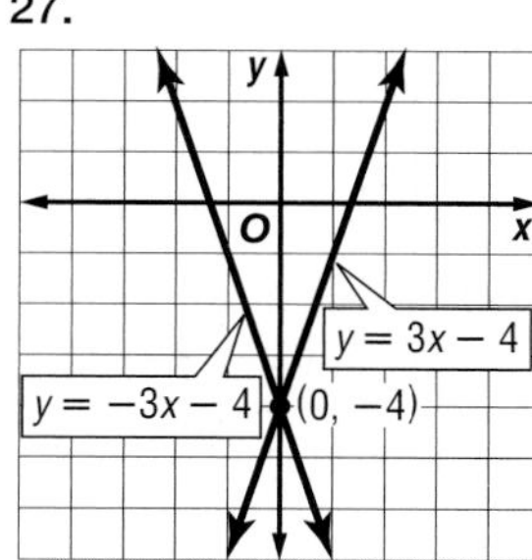

one; $(0, -4)$

29.

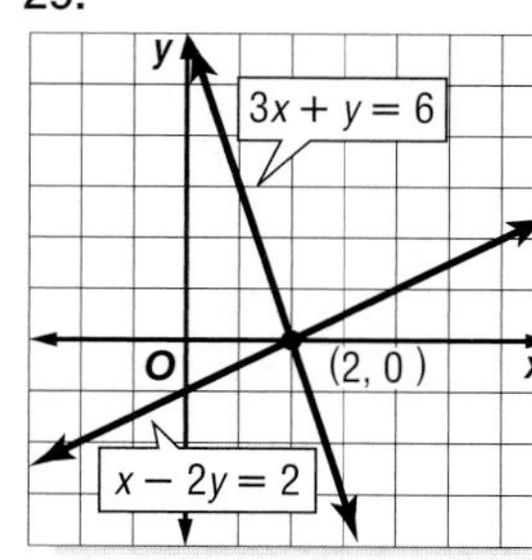

one; (2, 0)

31.

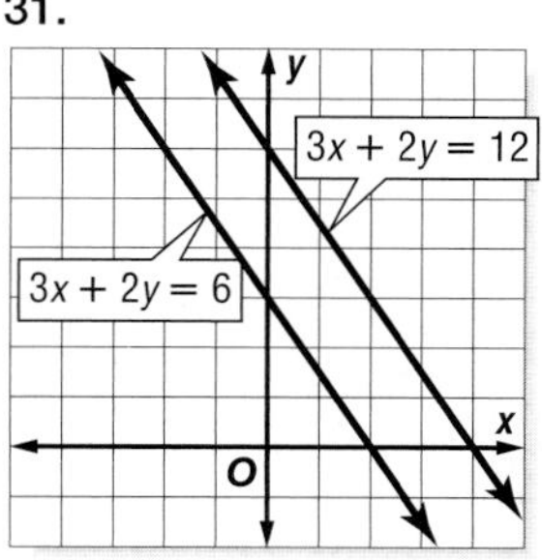

no solution

33.

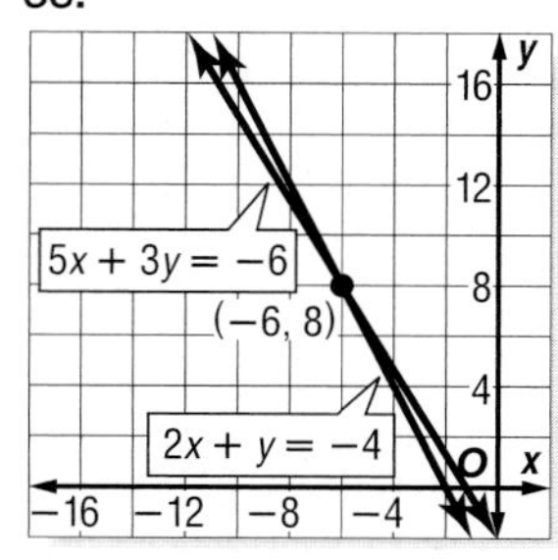

one; $(-6, 8)$

35.

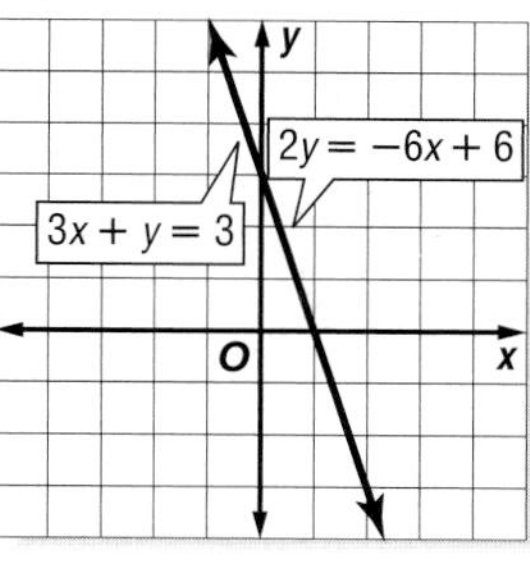

infinitely many

37.

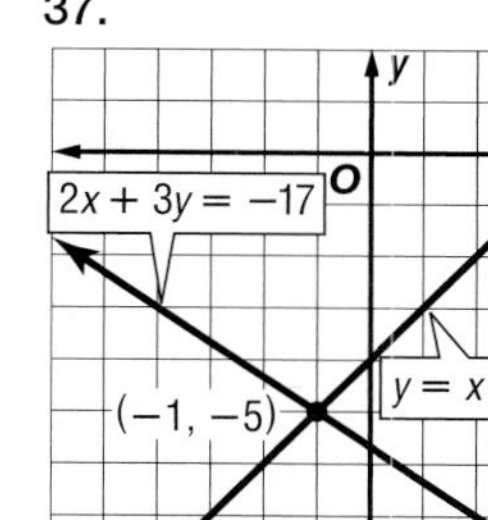

one; $(-1, -5)$

39.

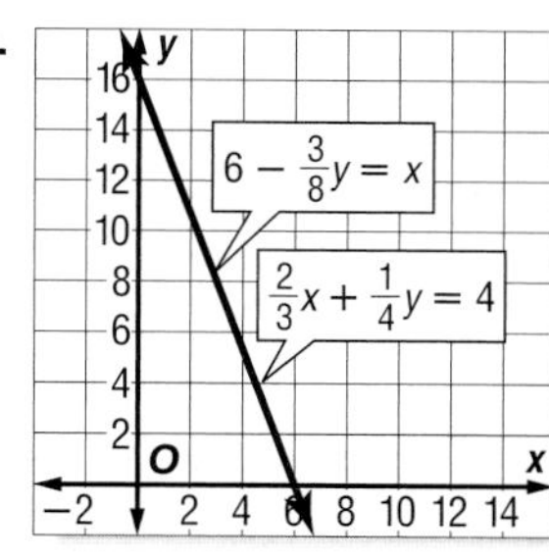

infinitely many
41. 13 m by 7 m
43. 21 units^2 **45.** 70 m
47. \$40 **49.** neither
51. $p = 60 + 0.4t$

53.

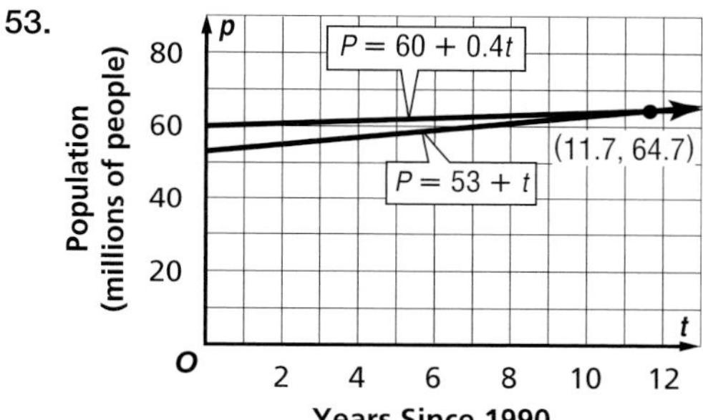

55. $A = 4, B = -4$ **57.** B **59.** $\{(5, -6)\}$ **61.** $\{n \mid 1.95 < n < 2.05\}$ **63.** $x - 3y = 3$ **65.** $y = 2x$ **67.** $q = \frac{7m - n}{10}$

Pages 379–381 Lesson 7-2

1. Substitution may result in a more accurate solution.
3. Sample answer: $y = x + 3$ and $2y = 2x + 6$ **5.** (3, 1)
7. infinitely many **9.** no solution **11.** (2, 10) **13.** $(-23, -7)$
15. (6, 7) **17.** no solution **19.** (7, 2) **21.** (2, 0) **23.** $\left(4\frac{1}{2}, \frac{3}{4}\right)$
25. (5, 2) **27.** $\left(2\frac{2}{3}, 4\frac{1}{3}\right)$ **29.** 14 in., 14 in., 18 in. **31.** 320 gal of 25% acid, 180 gal of 50% acid **33.** Yankees: 26, Reds: 5
35. The second offer is better if she sells less than \$80,000. The first offer is better if she sells more than \$80,000.
37. during the year 2023 **39.** $(-1, 5, -4)$ **41.** B

43.

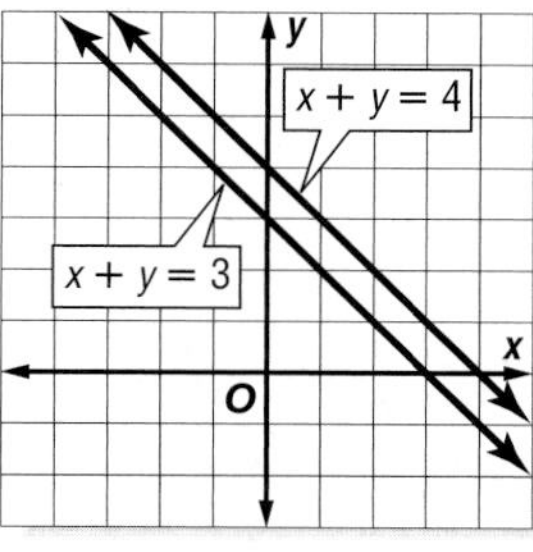

no solution

45.

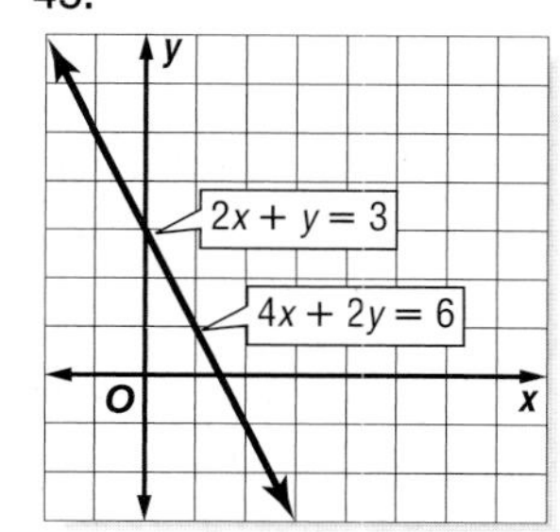

infinitely many

47. 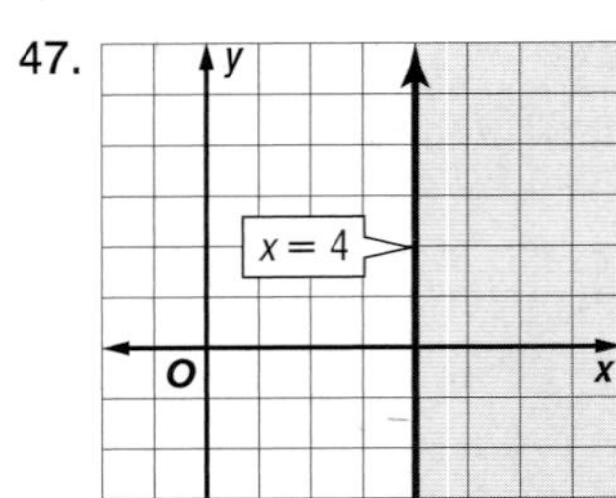

49. 50 lb 51. $12t$
53. $5d - b$

47. 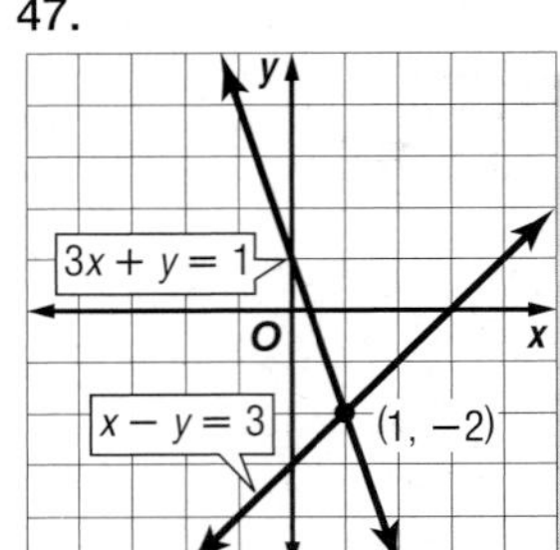

one; (1, −2)
51. $6x + 8y$ 53. $6m - 9n$

49. 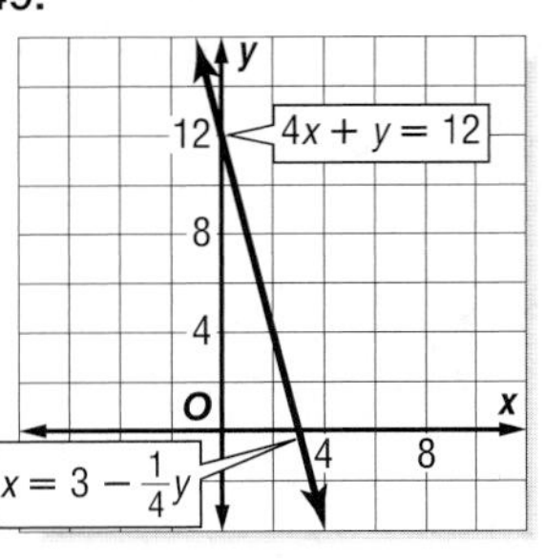

infinitely many

Page 381 Practice Quiz 1

1. 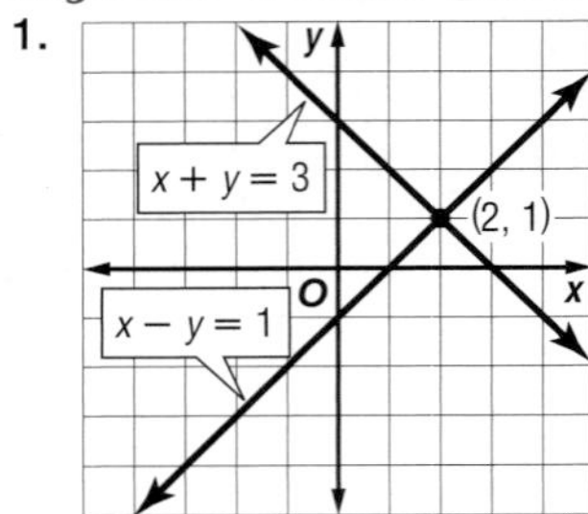

one; (2, 1)

3. (−4, 4) 5. infinitely many

Pages 384–386 Lesson 7-3

1. Sample answer: $2a + b = 5$, $a - b = 4$ 3. Michael; in order to eliminate the s terms, you must add the two equations. 5. (−1, 3) 7. (0, −5) 9. $\left(-2\frac{1}{2}, -2\right)$ 11. D 13. (3, −1) 15. (−1, 2) 17. (7, 4) 19. (−2, 3) 21. (1, −1) 23. $\left(2, -1\frac{1}{2}\right)$ 25. $\left(\frac{3}{16}, -\frac{1}{2}\right)$ 27. (15.8, 3.4) 29. (24, 4) 31. 32, 19 33. 5, 9 35. adult: \$16, student: \$9 37. $y = 0.0048x + 1.28$ 39. 2048; 1.51 billion

41. Elimination can be used to solve problems about meteorology if the coefficients of one variable are the same or are additive inverses. Answers should include the following.

- The two equations in the system of equations are added or subtracted so that one of the variables is eliminated. You then solve for the remaining variable. This number is substituted into one of the original equations, and that equation is solved for the other variable.
-

$n + d = 24$	*Write the equations in column*
$(+)\ n - d = 12$	*form and add.*
$2n = 36$	*Notice that the d variable is eliminated.*
$\frac{2n}{2} = \frac{36}{2}$	*Divide each side by 2.*
$n = 18$	*Simplify.*
$n + d = 24$	*First equation*
$18 + d = 24$	*n = 18*
$18 + d - 18 = 24 - 18$	*Subtract 18 from each side.*
$d = 6$	*Simplify.*

On the winter solstice, Seward, Alaska, has 18 hours of nighttime and 6 hours of daylight.

43. C 45. (1, −1)

Pages 390–392 Lesson 7-4

1. If one of the variables cannot be eliminated by adding or subtracting the equations, you must multiply one or both of the equations by numbers so that a variable will be eliminated when the equations are added or subtracted.
3. Sample answer: (1) You could solve the first equation for a and substitute the resulting expression for a in the second equation. Then find the value of b. Use this value for b and one of the original equations to find the value of a. (2) You could multiply the first equation by 3 and add this new equation to the second equation. This will eliminate the b term. Find the value of a. Use this value for a and one of the original equations to find the value of b. 5. (−1, 1) 7. (1.25, 2.75) 9. elimination (+); (2, 0) 11. elimination (−); (7, 11.5) 13. (−9, −13) 15. (2, 1) 17. (−1, 5)
19. (−1, −2) 21. (10, 12) 23. (2, −8) 25. 2, −5
27. elimination (×);(−2, 1) 29. substitution; (2, 6)
31. elimination (+); $\left(8, \frac{4}{3}\right)$ 33. elimination (×) or substitution; (3, 1) 35. elimination (−); no solution
37. elimination (−); (24, 4) 39. 640 2-point field goals, 61 3-point field goals 41. 95 43. 475 mph 45. A 47. (6, 2)
49. (11, 7)51. (−4, 4) 53. more than \$325,000

55. 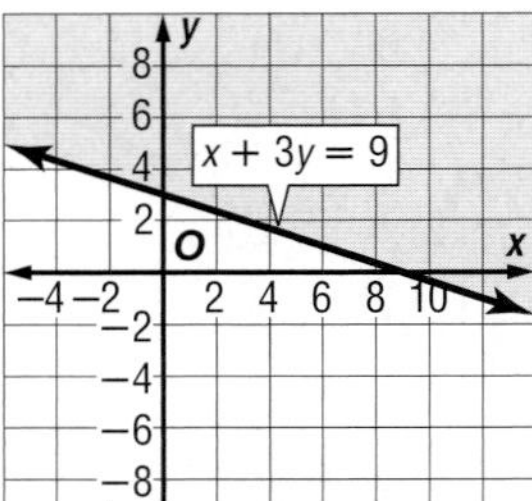

57. 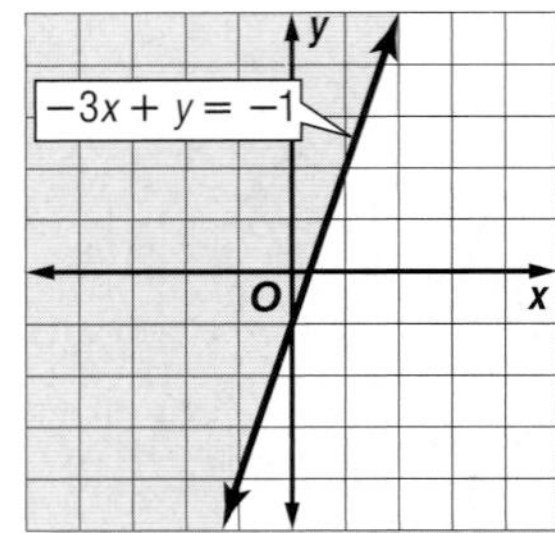

Page 392 Practice Quiz 2

1. (2, −2) 3. (5, 3) 5. \$0.45; \$0.15

Pages 396–398 Lesson 7-5

1. Sample answer: 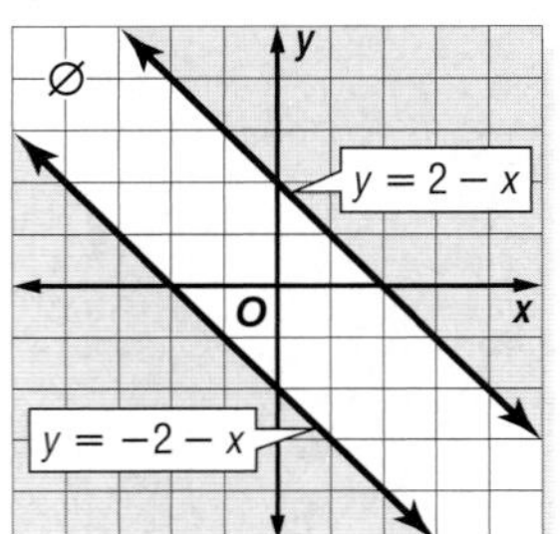

3. Kayla; the graph of $x + 2y \geq -2$ is the region representing $x + 2y = -2$ and the half-plane above it.

5.

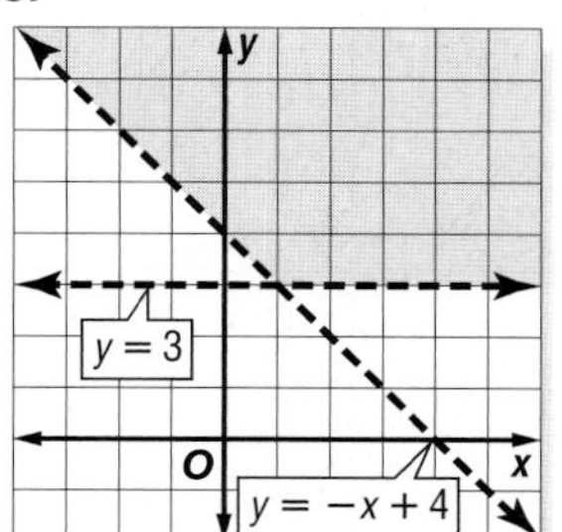

7.

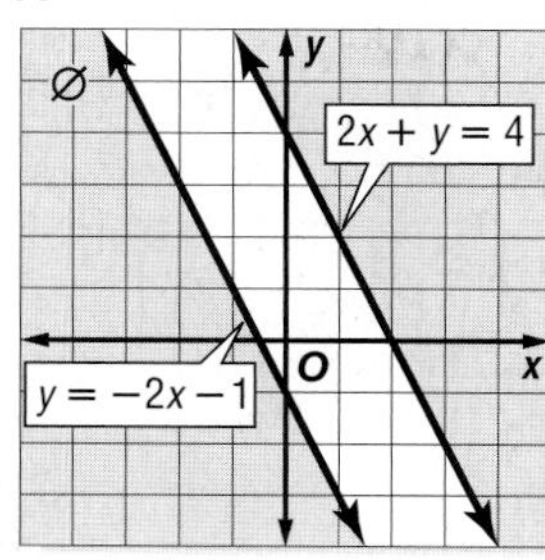

9.

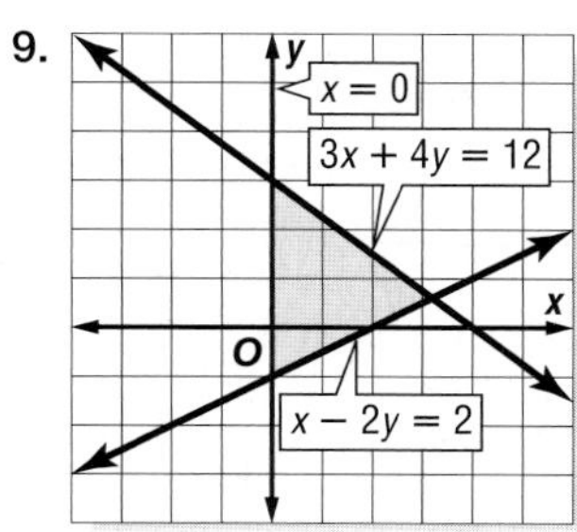

11. Sample answers: walk: 15 min, jog: 15 min; walk: 10 min, jog: 20 min; walk: 5 min, jog: 25 min

13.

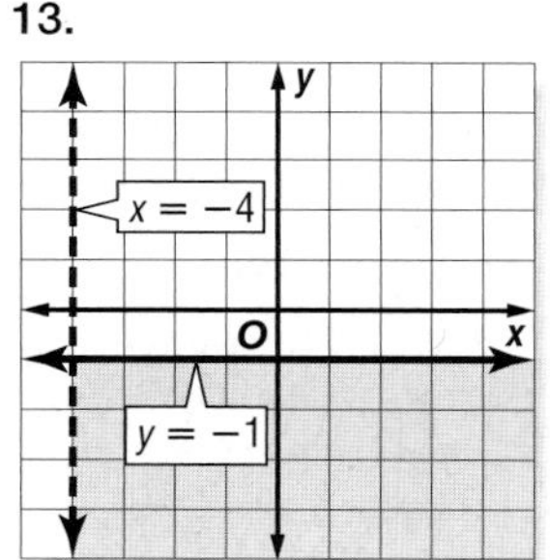

15.

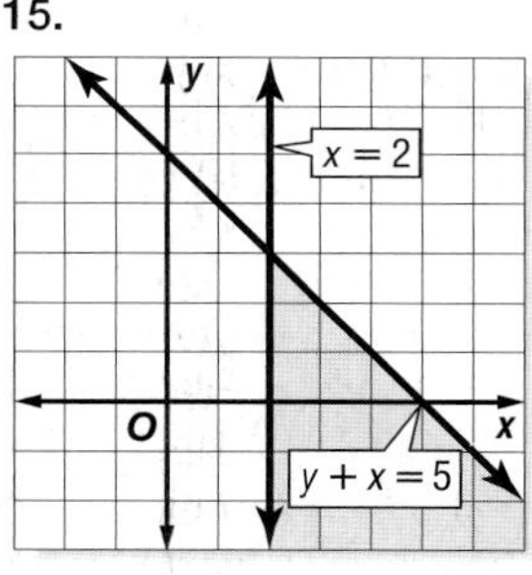

17.

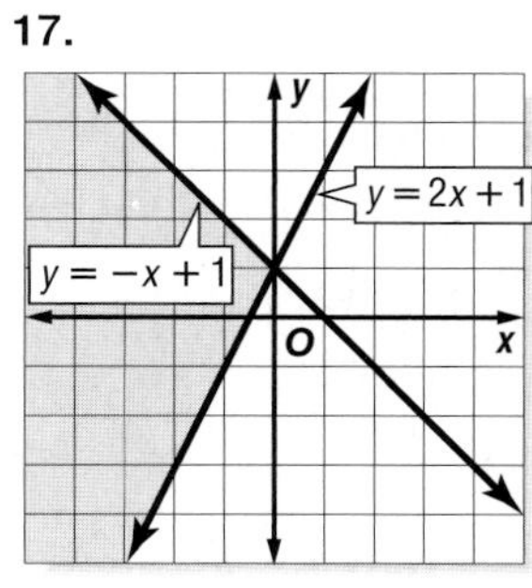

19.

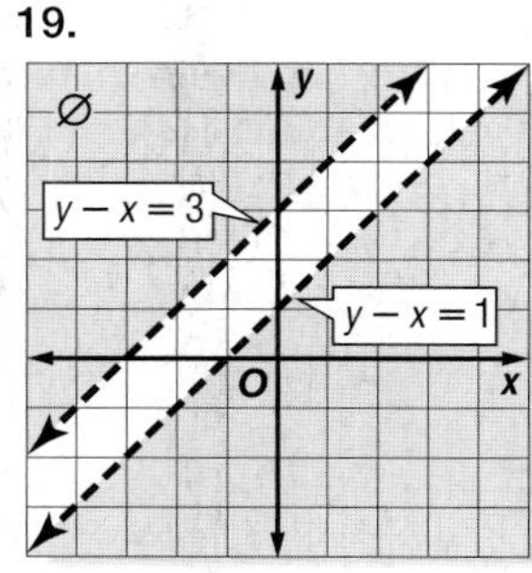

21.

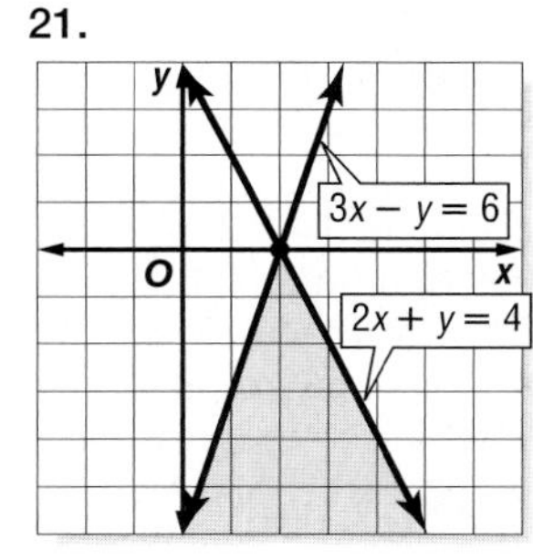

23.

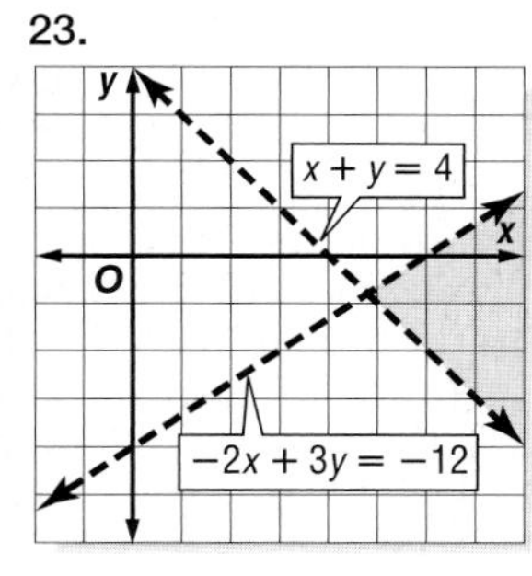

25.

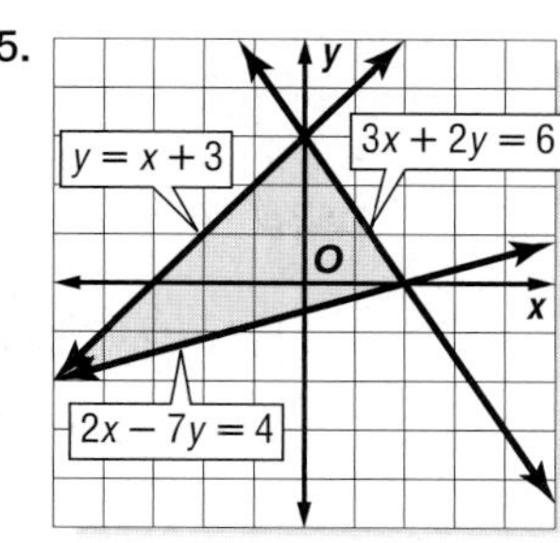
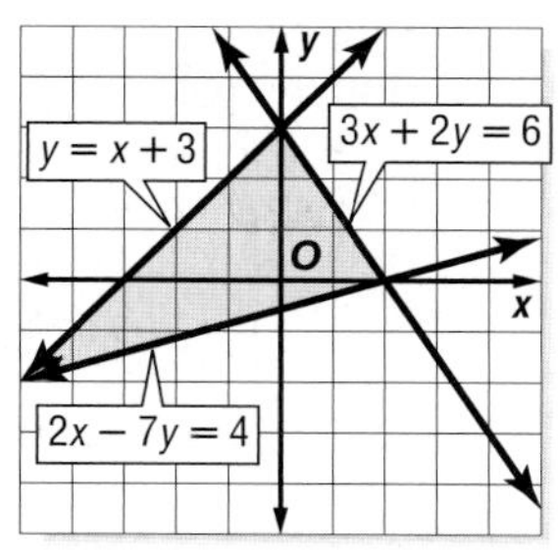

27. $y \leq x, y > x - 3$

29.

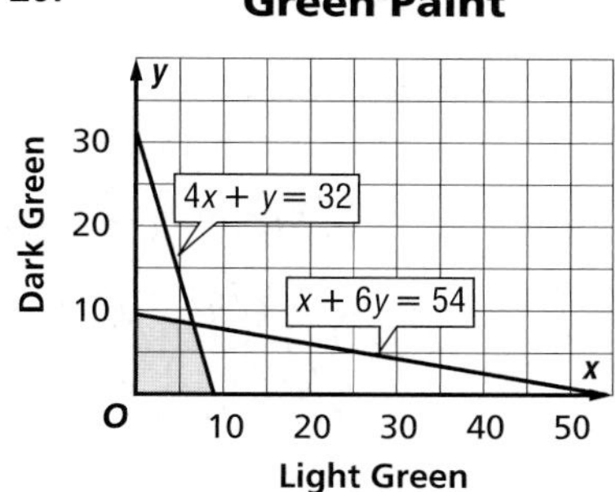

31.

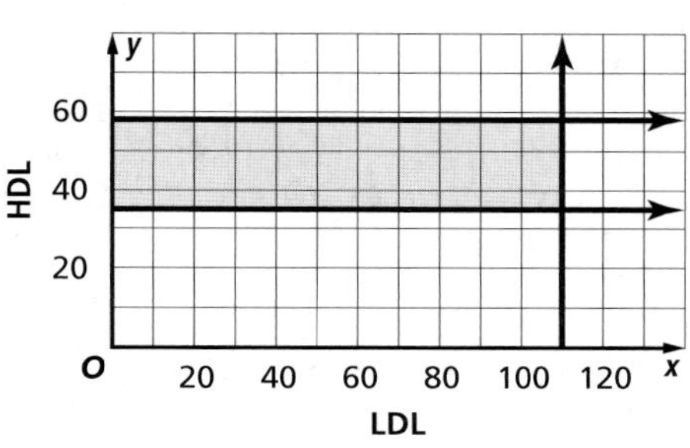

33.

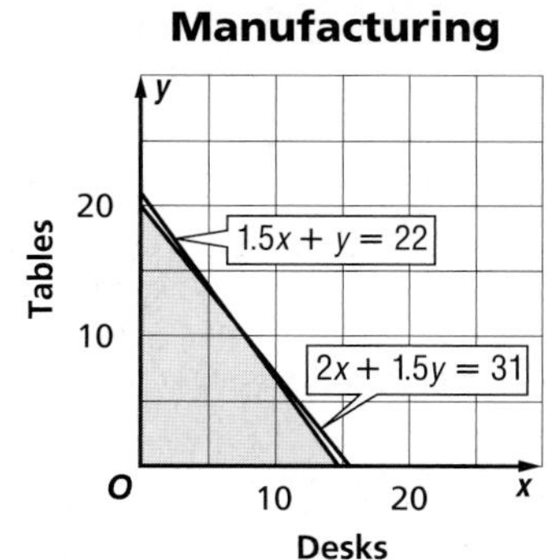
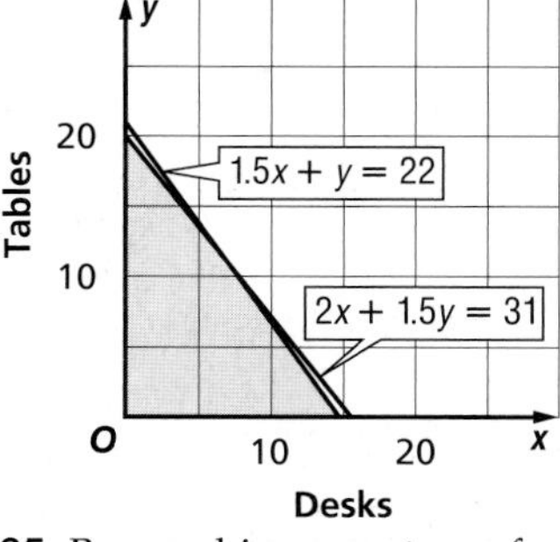

35. By graphing a system of equations, you can see the appropriate range of Calories and fat intake. Answers should include the following.

- Two sample appropriate Calorie and fat intakes are 2200 Calories and 60 g of fat and 2300 Calories and 65 g of fat.
- The graph represents $2000 \leq c \leq 2400$ and $60 \leq f \leq 75$.

37.

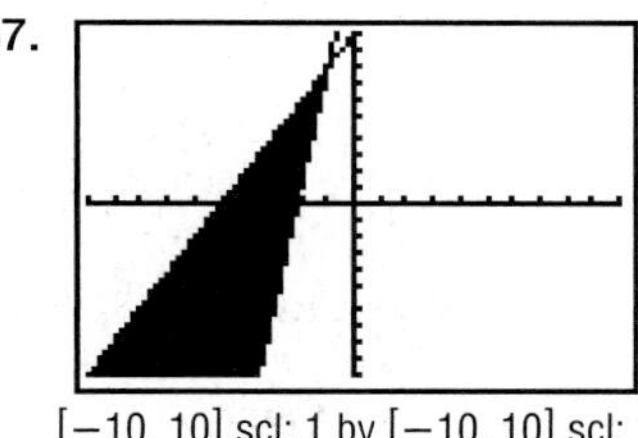

[−10, 10] scl: 1 by [−10, 10] scl: 1

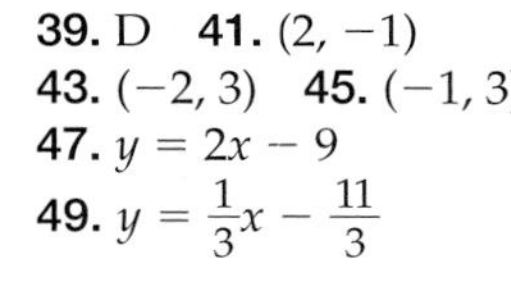

39. D **41.** (2, −1)
43. (−2, 3) **45.** (−1, 3)
47. $y = 2x - 9$
49. $y = \frac{1}{3}x - \frac{11}{3}$

Selected Answers

Pages 399–402 Chapter 7 Study Guide and Review

1. independent **3.** dependent **5.** infinitely many

7.

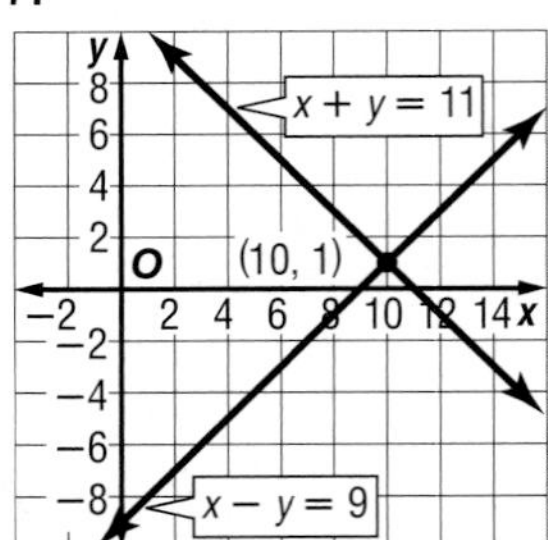

one; (10, 1)

9.

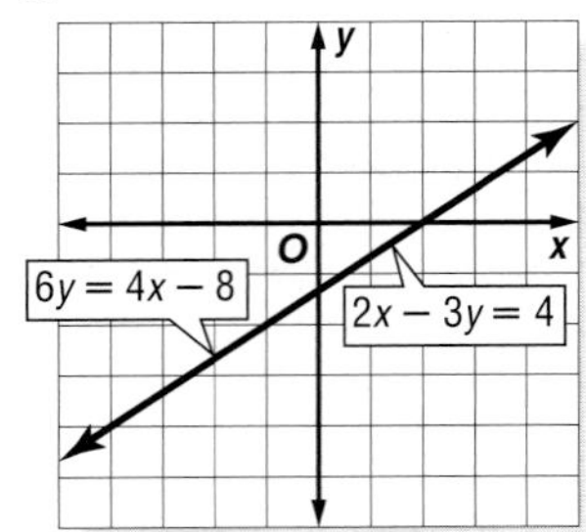

infinitely many

11. (3, −5) **13.** $\left(\frac{1}{2}, \frac{1}{2}\right)$ **15.** (2, 2) **17.** (4, 1) **19.** (5, 1)

21. $\left(2\frac{4}{5}, \frac{4}{5}\right)$ **23.** substitution; $\left(1\frac{3}{5}, 3\frac{1}{5}\right)$ **25.** substitution; (0, 0)

27.

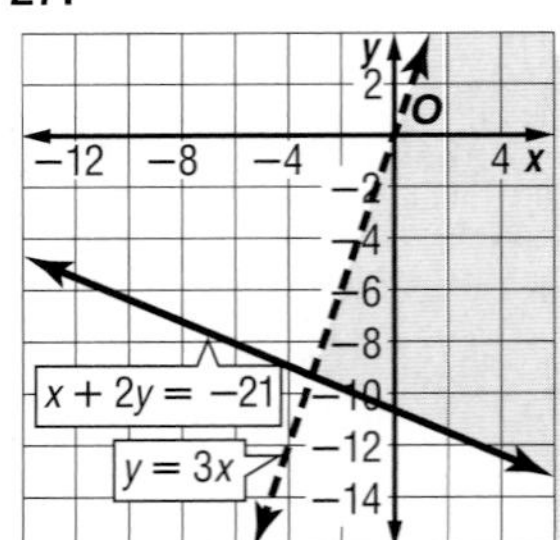

29.

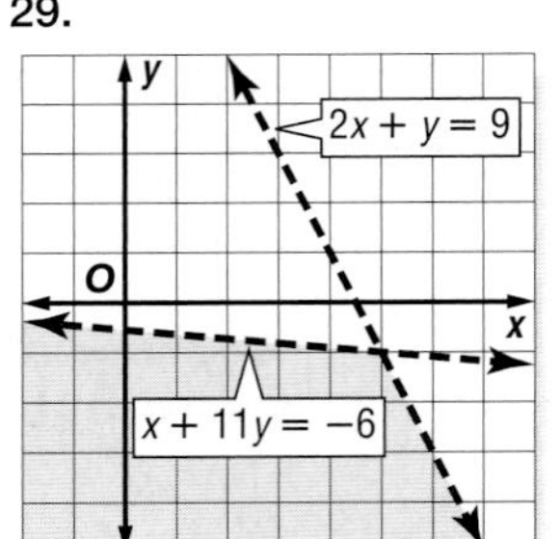

61.

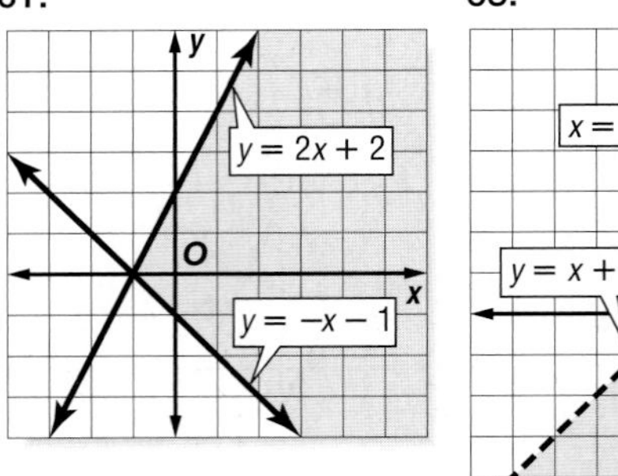

63.

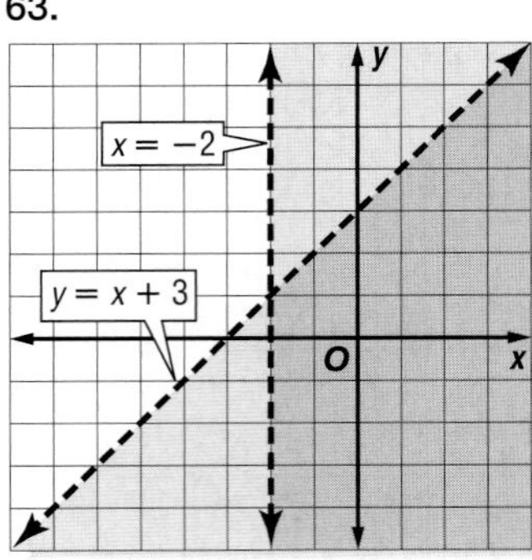

65. (−3, −4) **67.** $\{h \mid h \leq -7 \text{ or } h \geq 1\}$

69. ∅

71. dilation **73.** reflection **75.** $\frac{1}{3}$ **77.** 2 **79.** $\frac{7}{18}$ **81.** $\frac{11}{8}$

Chapter 8 Polynomials

Page 409 Chapter 8 Getting Started

1. 2^5 **3.** 5^2 **5.** a^6 **7.** $\left(\frac{1}{2}\right)^5$ **9.** 9 **11.** 25 **13.** 36 **15.** $\frac{16}{81}$ **17.** 63 yd^2 **19.** 84 ft^3

Pages 413–415 Lesson 8-1

1a. Sample answer: $n^2(n^5) = n^7$ **1b.** Sample answer: $(n^2)^5 = n^{10}$ **1c.** Sample answer: $(nm^2)^5 = n^5n^{10}$ **3.** Poloma; when finding the product of powers with the same base, keep the same base and add the exponents. Do not multiply the bases. **5.** No; $\frac{4a}{3b}$ shows division as well as multiplication. **7.** x^{11} **9.** 2^{18} or 262,144 **11.** $-48m^3n^3$ **13.** $5n^5$ **15.** Yes; 12 is a real number and therefore a monomial. **17.** No; $a - 2b$ shows subtraction, not multiplication of variables. **19.** No; $\frac{x}{y^2}$ shows division, not multiplication of variables. **21.** a^2b^6 **23.** $-28c^4d^7$ **25.** $30a^5b^7c^6$ **27.** $81p^2q^{14}$ **29.** 3^{16} or 43,046,721 **31.** $0.25x^6$ **33.** $-\frac{27}{64}c^3$ **35.** $-432c^2d^8$ **37.** $144a^8g^{14}$ **39.** $-9x^3y^9$ **41.** $40b^{12}$ **43.** $15f^5g^5$ **45.** $(49x^8)\pi$ **47.** x^3y^5 **49.** 10^{12} or 1 trillion **51.** 2; 8; 32 **53.** 2^{22} or 4,194,304 ways **55.** False. If $a = 4$, then $(-4)^2 = 16$ and $-4^2 = -16$. **57.** False. Let $a = 3$, $b = 4$, and $n = 2$. Then $(a + b)^n = (3 + 4)^2$ or 49 and $a^n + b^n = 3^2 + 4^2$ or 25. **59.** D

Pages 421–423 Lesson 8-2

1. Sample answer: $9xy$ and $6xy^2$ **3.** Jamal; a factor is moved from the numerator of a fraction to the denominator or vice versa only if the *exponent* of the factor is negative; $-4 \neq \frac{1}{4}$. **5.** x^6y^5 **7.** $\frac{1}{y^4}$ **9.** $\frac{g^8}{d^3c^5}$ **11.** $c^{11}d^{12}$ **13.** C **15.** 3^6 or 729 **17.** y^2z^7 **19.** $\frac{81m^{28}}{256x^{20}y^{12}}$ **21.** $\frac{1}{3b^4}$ **23.** $\frac{1}{n^3p^4}$ **25.** $\frac{1}{125}$ **27.** $\frac{8}{27}$ **29.** $\frac{6k^{17}}{h^3}$ **31.** $\frac{19}{3z^{12}}$ **33.** $\frac{p}{q}$ **35.** 1 **37.** $\frac{27a^9c^3}{8b^9}$ **39.** $10ab$ units **41.** jet plane **43.** $\left(\frac{1}{2}\right)^n$ **45.** $\frac{1}{10^5}$ to $\frac{1}{10^4}$ cm; $\frac{1}{100{,}000}$ to $\frac{1}{10{,}000}$ cm **47.** a^{n+3} **49.** c^{11}

51. You can compare pH levels by finding the ratio of one pH level to another written in terms of the concentration c of hydrogen ions, $c = \left(\frac{1}{10}\right)^{\text{pH}}$. Answers should include the following.

- Sample answer: To compare a pH of 8 with a pH of 9 requires simplifying the quotient of powers.

$$\frac{\left(\frac{1}{10}\right)^8}{\left(\frac{1}{10}\right)^9} \cdot \frac{\left(\frac{1}{10}\right)^8}{\left(\frac{1}{10}\right)^9} = \left(\frac{1}{10}\right)^{8-9}$$

$$= \left(\frac{1}{10}\right)^{-1}$$

$$= \frac{1}{\left(\frac{1}{10}\right)^1} \quad \textit{Negative Exponent Property}$$

$$= 10$$

Thus, a pH of 8 is ten times more acidic than a pH of 9. **53.** Since each number is obtained by dividing the previous number by 3, $3^1 = 3$ and $3^0 = 1$. **55.** $12x^8y^4$ **57.** $9c^2d^{10}$ **59.** $-108a^3b^9$ **61.** Sample answers: 3 oz of mozzarella, 4 oz of Swiss; 4 oz of mozzarella, 3 oz of Swiss; 5 oz of mozzarella, 3 oz of Swiss **63.** $y = -2x + 3$ **65.** $y = \frac{3}{2}x + 2$

67.

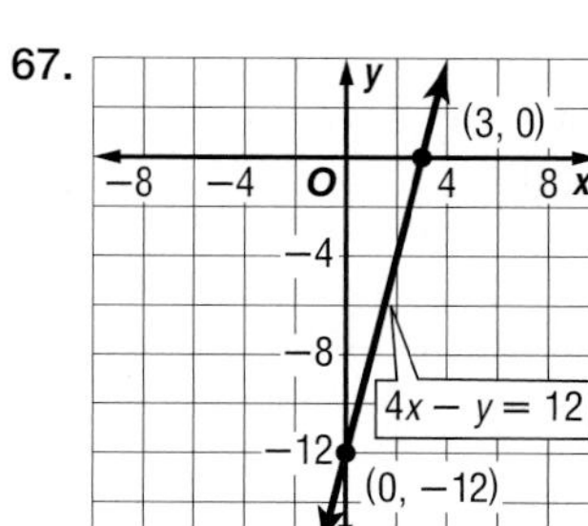

69. ±11 **71.** −7.21 **73.** 10^{-13} **75.** 10^7 **77.** 10^{-11}

Pages 428–430 Lesson 8-3

1. When numbers between 0 and 1 are written in scientific notation, the exponent is negative. If the number is not between 0 and 1, use a positive exponent **3.** Sample answer: 6.5 million; 6,500,000; 6.5×10^6 **5.** 4590 **7.** 0.000036 **9.** 5.67×10^{-3} **11.** 3.002×10^{15} **13.** 1.88×10^{-7}; 0.000000188 **15.** 5×10^9; 5,000,000,000 **17.** \$933.33 **19.** 0.0000000061 **21.** 80,000,000 **23.** 0.299 **25.** 6.89 **27.** 238,900 **29.** 0.000000000000000000000000000000091095 **31.** 3.4402×10^7 **33.** 9.0465×10^{-4} **35.** 3.807×10^2 **37.** 8.73×10^{12} **39.** 8.1×10^{-6} **41.** 1×10^9 **43.** $6.02214299 \times 10^{23}$ **45.** 1.71×10^9; 1,710,000,000 **47.** 1.44×10^{-8}; 0.0000000144 **49.** 2.548×10^5; 254,800 **51.** 4×10^{-4}; 0.0004 **53.** 2.3×10^{-6}; 0.0000023 **55.** 9.3×10^{-7}; 0.00000093 **57.** about \$20,236 **59.** about 1.4×10^{14} or 140 trillion tons

61. Astronomers work with very large numbers such as the masses of planets. Scientific notation allows them to more easily perform calculations with these numbers. Answers should include the following.

•

Planet	Mass (kg)
Mercury	330,000,000,000,000,000,000,000
Venus	4,870,000,000,000,000,000,000,000
Earth	5,970,000,000,000,000,000,000,000
Mars	642,000,000,000,000,000,000,000
Jupiter	1,900,000,000,000,000,000,000,000,000
Saturn	569,000,000,000,000,000,000,000,000
Uranus	86,800,000,000,000,000,000,000,000
Neptune	102,000,000,000,000,000,000,000,000
Pluto	12,700,000,000,000,000,000,000

- Scientific notation allows you to fit numbers such as these into a smaller table. It allows you to compare large values quickly by comparing the powers of 10 instead of counting zeros to find place value. For computation, scientific notation allows you work with fewer place values and to express your answers in a compact form.

63. 6.75×10^{18} **65.** 8.52×10^{-6} **67.** 1.09×10^3 **69.** $-\frac{4n^5}{p^5}$ **71.** no **73.** yes **75.** $\{d \mid d > 18\}$

12 14 18 20 22

77. 20 **79.** 37 **81.** 10

Page 430 Practice Quiz 1

1. n^8 **3.** $-128w^{11}z^{18}$ **5.** $\frac{36k^6}{49n^2p^8}$ **7.** 4.48×10^6; 4,480,000 **9.** 4×10^{-2}; 0.04

Pages 434–436 Lesson 8-4

1. Sample answer: −8 **3a.** true; **3b.** false; $3x + 5$ **3c.** true **5.** yes; monomial **7.** 0 **9.** 5 **11.** $2a + 4x^2 - 7a^2x^3 - 2ax^5$ **13.** $x^3 + 3x^2y + 3xy^2 + y^3$ **15.** yes; monomial **17.** yes; binomial **19.** yes; trinomial **21.** $0.5bh$ **23.** $0.5xy - \pi r^2$ **25.** 3 **27.** 2 **29.** 4 **31.** 2 **33.** 3 **35.** 7 **37.** $-1 + 2x + 3x^2$ **39.** $8c - c^3x^2 + c^2x^3$ **41.** $4 - 5a^7 + 2ax^2 + 3ax^5$ **43.** $6y + 3xy^2 + x^2y - 4x^3$ **45.** $x^5 + 3x^3 + 5$ **47.** $2a^2x^3 + 4a^3x^2 - 5a$ **49.** $cx^3 - 5c^3x^2 + 11x + c^2$ **51.** $-2x^4 - 9x^2y + 8x + 7y^2$ **53.** $0.25q + 0.10d + 0.05n$ **55.** $\pi r^2h + \frac{2}{3}\pi r^3$ **57.** True; for the degree of a binomial to be zero, the highest degree of both terms would need to be zero. Then the terms would be like terms. With these like terms combined, the expression is not a binomial, but a monomial. Therefore, the degree of a binomial can never be zero. Only a monomial can have a degree of zero.

59. B **61.** 1.23×10^7 **63.** 1.2×10^7 **65.** $\frac{1}{b^2c}$ **67.** $\frac{16x^6y^4}{9z^2}$ **69.** no **71.** $\frac{1}{2}$ **73.** $7a^2 + 3a$ **75.** $a - 2b$

Page 441–443 Lesson 8-5

1. The powers of x and y are not the same. **3.** Kendra; Esteban added the additive inverses of both polynomials when he should have added the opposite of the polynomial being subtracted. **5.** $9y^2 - 3y - 1$ **7.** $11a^2 + 6a + 1$ **9.** $3ax^2 - 9x - 9a + 8a^2x$ **11.** about 297,692,000 **13.** $13z - 10z^2$ **15.** $-2n^2 + 7n + 5$ **17.** $5b^3 - 8b^2 - 4b$ **19.** $2g^3 - 9g$ **21.** $-2x - 3xy$ **23.** $3ab^2 + 11ab - 4$ **25.** $3x^2 - 12x + 5ax + 3a^2$ **27.** $8x^2 - 6x + 15$ **29.** $11x^3 - 7x^2 - 9$ **31.** $6x^2 - 15x + 12$ **33.** 260 outdoor screens **35.** Original number $= 10x + y$; show that the new number will always be represented by $10y + x$.

$$\begin{aligned}\text{new number} &= 9(y - x) + (10x + y)\\ &= 9y - 9x + 10x + y\\ &= 10y + x\end{aligned}$$

37. $40 - 2x$ **39.** $140 - 4x \leq 108$; 8 in. **41.** $x + 1$ **43.** 4 **45.** A **47.** 5 **49.** 3 **51.** 8,000,000 **53.** 0.0005

55.

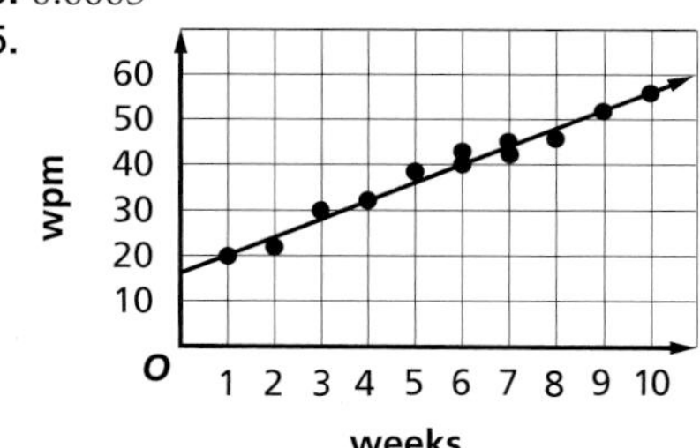

57. Sample answer: $y = 4x + 17$ **59.** No; there's a limit as to how fast one can keyboard. **61.** D = {−4, −1, 5}; R = {2, −3, 0, 1} **63.** $18x - 48$ **65.** $35p - 28q$ **67.** $8x^2 + 24x - 32$

Page 446–449 Lesson 8-6

1. Distributive Property; Product of Powers Property **3.** Sample answer: $4x$ and $x^2 + 2x + 3$; $4x^3 + 8x^2 + 12x$ **5.** $18b^5 - 27b^4 + 9b^3 - 72b^2$ **7.** $-20x^3y + 48x^2y^2 - 28xy^3$ **9.** $20n^4 + 30n^3 - 14n^2 - 13n$ **11.** $\frac{5}{3}$ **13.** $T = 10{,}700 - 0.03x$ **15.** $5r^2 + r^3$ **17.** $-32x - 12x^2$ **19.** $7ag^4 + 14a^2g^2$ **21.** $-6b^4 + 8b^3 - 18b^2$ **23.** $40x^3y + 16x^2y^3 - 24x^2y$ **25.** $-15hk^4 - \frac{15}{4}h^2k^2 + 6hk^2$ **27.** $-10a^3b^2 - 25a^4b^2 + 5a^3b^3 - 5a^6b$ **29.** $-2d^2 + 19d$ **31.** $20w^2 - 18w + 10$ **33.** $46m^3 + 14m^2 - 32m + 20$ **35.** $6c^3 - 23c^2 + 20c - 8$ **37.** $6x^2 + 8x$ **39.** −2 **41.** $-\frac{1}{3}$ **43.** 0 **45.** $\frac{7}{4}$ **47.** −5 **49.** $T = -0.03x + 6360$ **51.** $20x^2 + 48x$ **53.** $x + 2$ **55.** Let x and y be integers.

Then $2x$ and $2y$ are even numbers, and $(2x)(2y) = 4xy$. $4xy$ is divisible by 2 since one of its factors, 4, is divisible by 2. Therefore, $4xy$ is an even number.
57. Let x and y be integers. Then $2x$ is an even number and $2y + 1$ is an odd number. Their product, $2x(2y + 1)$, is always even since one of its factors is 2. **59.** \$2.20
61. \$126
63. Answers should include the following.

- The product of a monomial and a polynomial can be modeled using an area model. The area of the figure shown at the beginning of the lesson is the product of its length $2x$ and width $(x + 3)$. This product is $2x(x + 3)$, which when the Distributive Property is applied becomes $2x(x) + 2x(3)$ or $2x^2 + 6x$. This is the same result obtained when the areas of the algebra tiles are added together.
- Sample answer: $(3x)(2x + 1)$
$(3x)(2x + 1) = (3x)(2x) + (3x)(1)$
$= 6x^2 + 3x$

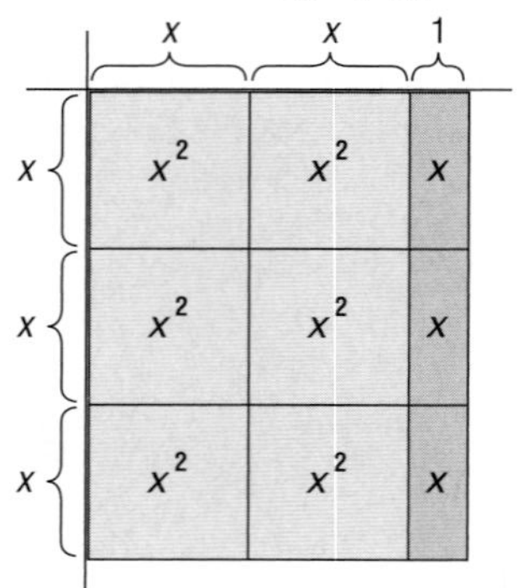

65. A **67.** $-4y^2 + 5y + 3$ **69.** $7p^3 - 3p^2 - 2p - 7$
71. yes; binomial **73.** yes; monomial **75.** $9n + 4 \geq 7 - 13n$; $\left\{n \mid n \geq \frac{3}{22}\right\}$ **77.** $y = -2x - 3$ **79.** \$50

81.

Stem	Leaf
1	0 4 5 8 8 8
2	0 0 1 1 2
3	0 4
4	3 4 $3\|4 = 34$

83. $6x^3$
85. $12y^2 - 24y$
87. $18p^4 - 24p^3 + 36p^2$

Page 449 Practice Quiz 2

1. 4 **3.** 3 **5.** $-12 + 9x + 4x^2 + 5x^3$ **7.** $10n^2 - 4n + 2$
9. $15a^5b - 10a^4b^2 + 30a^3b^3$

Pages 455–457 Lesson 8-7

1.

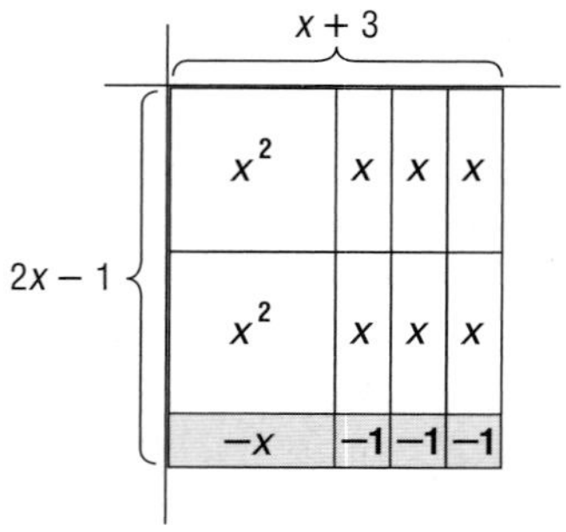

5. $x^2 + 4x - 12$
7. $4h^2 + 33h + 35$
9. $10g^2 + 19g - 56$
11. $6k^3 + 2k^2 - 29k + 15$
13. $b^2 + 10b + 16$
15. $x^2 - 13x + 36$
17. $y^2 - 4y - 32$
19. $2w^2 + 9w - 35$
21. $40d^2 + 31d + 6$
23. $35x^2 - 27x + 4$
25. $4n^2 + 12n + 9$

27. $100r^2 - 16$ **29.** $40x^2 - 22xy - 8y^2$ **31.** $p^3 + 6p^2 + p - 28$
33. $6x^3 - 23x^2 + 22x - 5$ **35.** $n^4 + 2n^3 - 17n^2 + 22n - 8$
37. $8a^4 + 2a^3 + 15a^2 + 31a - 56$ **39.** $2x^2 + 3x - 20$ units^2
41. $\frac{15}{2}x^2 + 3x - 24$ units^2 **43.** $2a^3 + 10a^2 - 2a - 10$ units^3

45. $a^3 + 3a^2 + 2a$ **47.** Sample answer: 6; the result is the same as the product in Exercise 46. **49.** $x - 2$, $x + 4$
51. bigger; 10 ft^2 **53.** 20 ft by 24 ft
55. Multiplying binomials and two-digit numbers each involve the use of the Distributive Property twice. Each procedure involves four multiplications and the addition of like terms. Answers should include the following.

- $24 \times 36 = (4 + 20)(6 + 30)$
$= (4 + 20)6 + (4 + 20)30$
$= (24 + 120) + (120 + 600)$
$= 144 + 720$
$= 864$
- The like terms in vertical two-digit multiplication are digits with the same place value.

57. B **59.** $-28y^3 + 16y^2 - 12y$ **61.** $36x^2 - 42$
63. $(181 - 7x)°$ **65.** one; $(-6, 3)$ **67.** 5 **69.** $t = \frac{v}{a}$
71. $y = -\frac{4}{3}x + \frac{7}{3}$ **73.** $49x^2$ **75.** $16y^4$ **77.** $9g^8$

Pages 461–463 Lesson 8-8

1. The patterns are the same except for their middle terms. The middle terms have different signs.
3.

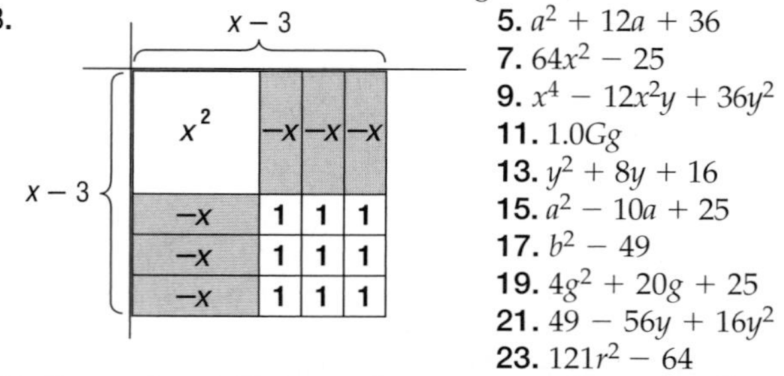

5. $a^2 + 12a + 36$
7. $64x^2 - 25$
9. $x^4 - 12x^2y + 36y^2$
11. $1.0Gg$
13. $y^2 + 8y + 16$
15. $a^2 - 10a + 25$
17. $b^2 - 49$
19. $4g^2 + 20g + 25$
21. $49 - 56y + 16y^2$
23. $121r^2 - 64$

25. $a^2 + 10ab + 25b^2$ **27.** $4x^2 - 36xy + 81y^2$ **29.** $25w^2 - 196$
31. $x^6 + 8x^3y + 16y^2$ **33.** $64a^4 - 81b^6$ **35.** $\frac{4}{9}x^2 - 8x + 36$
37. $4n^3 + 20n^2 - n - 5$ **39.** $0.5Bb + 0.5bb$ **41.** Sample answer: 2; yes **43.** $(a + 1)^2$ **45.** $s + 2$, $s + 3$

47.

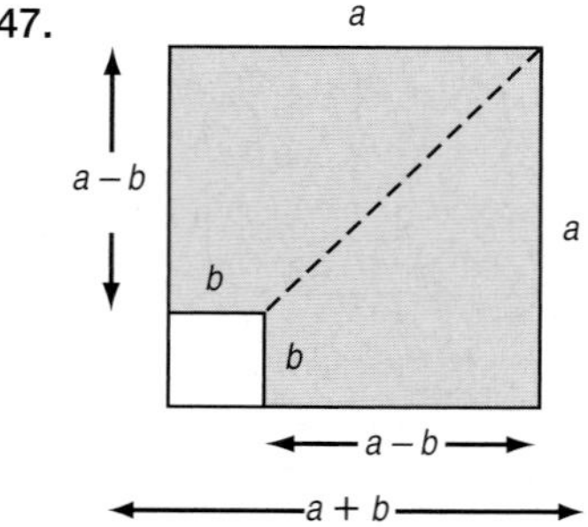

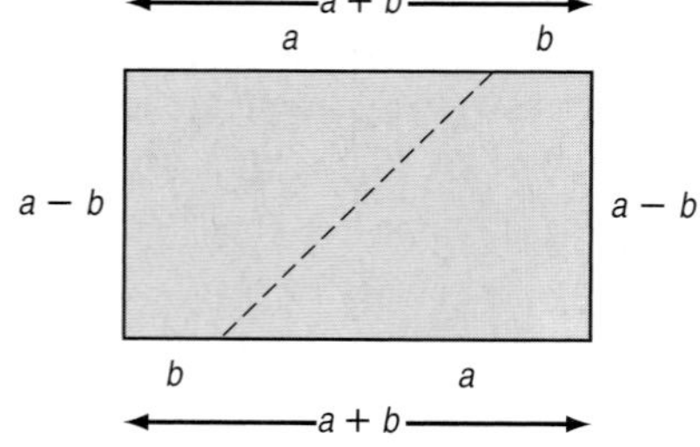

Area of rectangle = $(a - b)(a + b)$
OR

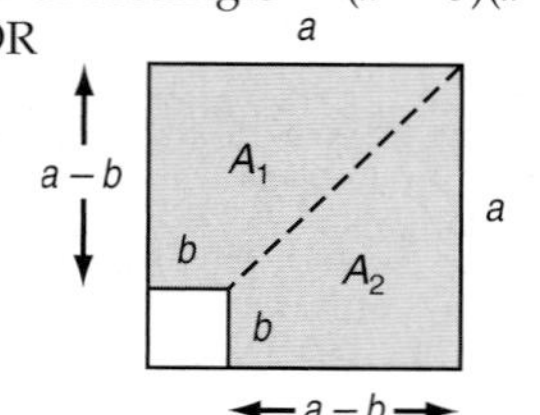

Area of a trapezoid $= \frac{1}{2}$(height)(base 1 + base 2)

$A_1 = \frac{1}{2}(a-b)(a+b) \qquad A_2 = \frac{1}{2}(a-b)(a+b)$

Total area of shaded region

$= \left[\frac{1}{2}(a-b)(a+b)\right] + \left[\frac{1}{2}(a-b)(a+b)\right]$

$= (a-b)(a+b)$

49. C **51a.** $a^3 + 3a^2b + 3ab^2 + b^3$ **51b.** $x^3 + 6x^2 + 12x + 8$ **51c.** $(a+b)^3$

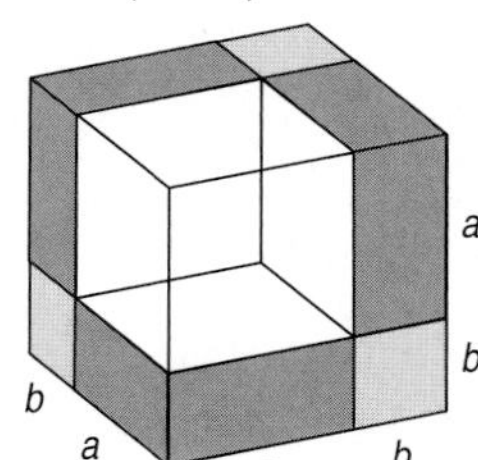

53. $c^2 - 6c - 27$ **55.** $24n^2 - 25n - 25$ **57.** $4k^3 - 6k^2 - 26k + 35$ **59.** $\frac{4}{3}$ **61.** $\frac{1}{2}$ **63.** $(3, -4)$ **65.** $y = x + 5$ **67.** $y = \frac{1}{5}x + 6$ **69.** 61

Pages 464–468 Chapter 8 Study Guide and Review

1. negative exponent **3.** Quotient of Powers **5.** trinomial **7.** polynomial **9.** binomial **11.** y^7 **13.** $20a^5x^5$ **15.** $576x^5y^2$ **17.** $-\frac{1}{2}m^4n^8$ **19.** 531,441 **21.** $\frac{27b^3c^6}{64d^3}$ **23.** $\frac{27b}{14}$ **25.** $\frac{bx^3}{3ay^2}$ **27.** $\frac{1}{64a^6}$ **29.** 240,000 **31.** 4,880,000,000 **33.** 7.96×10^5 **35.** 6×10^{11}; 600,000,000,000 **37.** 1.68×10^{-5}; 0.0000168 **39.** 4 **41.** 6 **43.** 7 **45.** $-4x^4 + 5x^3y^2 - 2x^2y^3 + xy - 27$ **47.** $4x^2 - 5xy + 6y^2$ **49.** $21m^4 - 10m - 1$ **51.** $-7p^2 - 2p + 25$ **53.** $10x^2 - 19x + 63$ **55.** $2x^2 - 17xy^2 + 10x + 10y^2$ **57.** $1\frac{1}{7}$ **59.** $4a^2 + 13a - 12$ **61.** $20r^2 - 13rs - 21s^2$ **63.** $12p^3 - 13p^2 + 11p - 6$ **65.** $16x^2 + 56x + 49$ **67.** $25x^2 - 9y^2$ **69.** $9m^2 + 24mn + 16n^2$

Chapter 9 Factoring

Page 473 Chapter 9 Getting Started

1. $12 - 3x$ **3.** $-7n^2 + 21n - 7$ **5.** $x^2 + 11x + 28$ **7.** $54a^2 - 12ab - 2b^2$ **9.** $y^2 + 18y + 81$ **11.** $n^2 - 25$ **13.** 11 **15.** $\frac{5}{6}$

Pages 477–479 Lesson 9-1

1. false; 2 **3.** Sample answer: $5x^2$ and $10x^3$ **5.** 1, 17; prime **7.** $3^2 \cdot 5$ **9.** $-1 \cdot 2 \cdot 3 \cdot 5^2$ **11.** $3 \cdot 13 \cdot b \cdot b \cdot b \cdot c \cdot c$ **13.** 5 **15.** 9 **17.** $6a^2b$ **19.** 5 rows of 24 plants, 6 rows of 20 plants, 8 rows of 15 plants, 10 rows of 12 plants, 12 rows of 10 plants, 15 rows of 8 plants, 20 rows of 6 plants, or 24 rows of 5 plants **21.** 1, 5, 25; composite **23.** 1, 61; prime **25.** 1, 7, 17, 119; composite **27.** 1, 2, 4, 8, 16, 19, 38, 76, 152, 304; composite **29.** 194 mm; the factors of 96 whose sum when doubled is the greatest are 1 and 96. **31.** 3 packages in the box of 18 cookies and 4 packages in the box of 24 cookies **33.** $-1 \cdot 2 \cdot 7^2$ **35.** $2 \cdot 3 \cdot 17$ **37.** $2^2 \cdot 3^2 \cdot 5$ **39.** $-1 \cdot 2 \cdot 3 \cdot 7 \cdot 11$ **41.** $5 \cdot 17 \cdot x \cdot x \cdot y \cdot y$ **43.** $2 \cdot 5 \cdot 5 \cdot g \cdot h$ **45.** $3 \cdot 3 \cdot 3 \cdot 3 \cdot 3 \cdot n \cdot n \cdot n \cdot m$ **47.** $-1 \cdot 13 \cdot 13 \cdot a \cdot a \cdot b \cdot c \cdot c$ **49.** 1 **51.** 14 **53.** 21 **55.** $6d$ **57.** 1 **59.** 7 **61.** $16a^2b$ **63.** 15 **65.** 7, 31 **67.** base: 1 cm, height 40 cm; base 2 cm; height 20 cm; base 4 cm, height 10 cm; base 5 cm, height 8 cm, base 8 cm, height 5 cm, base 10 cm, height 4 cm; base 20 cm, height 2 cm; base 40 cm, height 1 cm

69. Scientists listening to radio signals would suspect that a modulated signal beginning with prime numbers would indicate a message from an extraterrestrial. Answers should include the following.

- 2, 3, 5, 7, 11, 13, 17, 19, 23, 29, 31, 37, 41, 43, 47, 53, 59, 61, 67, 71, 79, 83, 89, 97, 101, 103, 107, 109, 113
- Sample answer: It is unlikely that any natural phenomenon would produce such an artificial and specifically mathematical pattern.

71. A **73.** $9a^2 - 25$ **75.** $12r^2 - 16r - 35$ **77.** $b^3 + 7b^2 - 6b - 72$ **79.** 0 **81.** $10x + 40$ **83.** $6g^2 - 8g$ **85.** $7(b + c)$

Pages 484–486 Lesson 9-2

1. $4(x^2 + 3x)$, $x(4x + 12)$, or $4x(x + 3)$; $4x(x + 3)$; $4x$ is the GCF of $4x^2$ and $12x$. **3.** The division would eliminate 2 as a solution. **5.** $8xz(2 - 5z)$ **7.** $2ab(a^2b + 4 + 8ab^2)$ **9.** $(5c + 2d)(1 - 2c)$ **11.** $\{-2, 4\}$ **13.** 0 ft **15.** 6.25 s; The answer 0 is not reasonable since it represents the time when the flare is launched. **17.** $4(4a + b)$ **19.** $x(x^2y^2 + 1)$ **21.** $2h(7g - 9)$ **23.** $8bc(c + 3)$ **25.** $6abc^2(3a - 8c)$ **27.** $x(15xy^2 + 25y + 1)$ **29.** $3pq(p^2 - 3q + 12)$ **31.** $(x + 7)(x + 5)$ **33.** $(3y + 2)(4y + 3)$ **35.** $(6x - 1)(3x - 5)$ **37.** $(m + x)(2y + 7)$ **39.** $(2x - 3)(5x - 7y)$ **41.** 35 **43.** 63 games **45.** $2r^2(4 - \pi)$ **47.** $81a^2 - 72ab + 16b^2$ cm^2 **49.** $\{-16, 0\}$ **51.** $\{-3, 7\}$ **53.** $\left\{-\frac{5}{4}, \frac{7}{3}\right\}$ **55.** $\{0, 5\}$ **57.** $\left\{0, \frac{6}{7}\right\}$ **59.** $\left\{-\frac{3}{4}, 0\right\}$ **61.** about 2.8 s

63. Answers should include the following.

- Let $h = 0$ in the equation $h = 151t - 16t^2$. To solve $0 = 151t - 16t^2$, factor the right-hand side as $t(151 - 16t)$. Then, since $t(151 - 16t) = 0$, either $t = 0$ or $151 - 16t = 0$. solving each equation for t, we find that $t = 0$ or $t \approx 9.44$.
- The solution $t = 0$ represents the point at which the ball was initially thrown into the air. The solution $t \approx 9.44$ represents how long it took after the ball was thrown for it to return to the same height at which it was thrown.

65. A **67.** 1, 2, 3, 4, 5, 6, 10, 12, 15, 20, 25, 30, 50, 60, 75, 100, 150, 300; composite **69.** $16s^6 + 24s^3 + 9$ **71.** $9k^2 + 48k + 64$ **73.** $\frac{3x}{2y^5}$ **75.** 37 shares **77.** $x^2 - 9x + 20$ **79.** $18a^2 - 6a - 4$ **81.** $8y^2 - 14y - 15$

Page 486 Practice Quiz 1

1. 1, 3, 5, 9, 15, 25, 45, 75, 225; composite **3.** $2 \cdot 3 \cdot 13 \cdot a \cdot a \cdot b \cdot c \cdot c \cdot c$ **5.** $xy(4y - 1)$ **7.** $(2p - 5)(3y + 8)$ **9.** $\{0, 3\}$

Pages 492–494 Lesson 9-3

1. In this trinomial, $b = 6$ and $c = 9$. This means that $m + n$ is positive and mn is positive. Only two positive numbers have both a positive sum and product. Therefore, negative factors of 9 need not be considered. **3.** Aleta; to use the Zero Product Property, one side of the equation must equal zero. **5.** $(c - 1)(c - 2)$ **7.** $(p + 5)(p - 7)$ **9.** $(x - 3y)(x - y)$ **11.** $\{-9, 4\}$ **13.** $\{-9, -1\}$ **15.** $\{-7, 10\}$ **17.** $(a + 3)(a + 5)$ **19.** $(c + 5)(c + 7)$ **21.** $(m - 1)(m - 21)$ **23.** $(p - 8)(p - 9)$ **25.** $(x - 1)(x + 7)$ **27.** $(h - 5)(h + 8)$ **29.** $(y - 7)(y + 6)$ **31.** $(w + 12)(w - 6)$ **33.** $(a + b)(a + 4b)$ **35.** $4x + 48$ **37.** $\{-14, -2\}$ **39.** $\{-6, 2\}$ **41.** $\{-4, 7\}$ **43.** $\{3, 16\}$ **45.** $\{2, -9\}$ **47.** $\{4, 6\}$ **49.** $\{-25, 2\}$ **51.** $\{-17, 3\}$ **53.** $\{4, 14\}$ **55.** -14 and -12 or 12 and 14 **57.** -18, 18 **59.** 7, 12, 15, 16 **61.** $w(w + 52)$ m^2

63. Answers should include the following.

- You would use a guess-and-check process, listing the factors of 54, checking to see which pairs added to 15.
- To factor a trinomial of the form $x^2 + ax + c$, you also use a guess-and-check process, list the factors of c, and check to see which ones add to a.

65. 15 **67.** yes **69.** no; $(x - 10)(x + 21)$ **71.** $\left\{0, \frac{4}{7}\right\}$ **73.** 12 **75.** $5x^2y^4$ **77.** $1(1.54) + 17.31(1.54) = (1 + 17.31)(1.54)$ or $18.31(1.54)$ **79.** $(a + 4)(3a + 2)$ **81.** $(2p + 7)(p - 3)$ **83.** $(2g - 3)(2g - 1)$

Pages 498–500 Lesson 9-4

1. m and n are the factors of ac that add to b. **3.** Craig; when factoring a trinomial of the form $ax^2 + bx + c$, where $a \neq 1$, you must find the factors of ac not of c. **5.** prime **7.** $(x + 4)(2x + 5)$ **9.** $(2n + 5)(2n - 7)$ **11.** $\left\{\frac{1}{2}, \frac{7}{5}\right\}$ **13.** 1 s **15.** $(3x + 2)(x + 1)$ **17.** $(5d - 4)(d + 2)$ **19.** $(3g - 2)(3g - 2)$ **21.** $(x - 4)(2x + 5)$ **23.** prime **25.** $(5n + 2)(2n - 3)$ **27.** $(2x + 3)(7x - 4)$ **29.** $5(3x + 2)(2x - 3)$ **31.** $(12a - 5b)(3a + 2b)$ **33.** $\pm 31, \pm 17, \pm 13, \pm 11$ **35.** $\left\{-5, -\frac{2}{5}\right\}$ **37.** $\left\{-\frac{1}{6}, \frac{3}{4}\right\}$ **39.** $\left\{-\frac{5}{7}, \frac{5}{2}\right\}$ **41.** $\left\{-\frac{2}{3}, 3\right\}$ **43.** $\left\{\frac{1}{2}, \frac{2}{3}\right\}$ **45.** $\{-4, 12\}$ **47.** $\left\{-4, \frac{2}{3}\right\}$ **49.** 1 in. **51.** 2.5 s

53. Answers should include the following.

- $2x + 3$ by $x + 2$
- With algebra tiles, you can try various ways to make a rectangle with the necessary tiles. Once you make the rectangle, however, the dimensions of the rectangle are the factors of the polynomial. In a way, you have to go through the guess-and-check process whether you are factoring algebraically or geometrically (using algebra tiles).

x^2	x	x	x
x^2	x	x	x
x	1	1	1
	1	1	1

x^2	x	x
x^2	x	x
x	1	1
x	1	1
x	1	1

Guess $(2x + 1)(x + 3)$ incorrect because 8 x tiles are needed to complete the rectangle.

55. B **57.** prime **59.** $\left\{-\frac{7}{5}, 4\right\}$ **61.** $\{0, 12\}$ **63.** 4 **65.** 6 **67.** 10 **69.** 13

Page 500 Practice Quiz 2

1. $(x + 4)(x - 18)$ **3.** $(4a - 1)(4a - 5)$ **5.** $2(3c + 1)(4c + 9)$ **7.** $\{-16, 2\}$ **9.** $\left\{-\frac{3}{4}, \frac{4}{3}\right\}$

Pages 504–506 Lesson 9-5

1. The binomial is the difference of two terms, each of which is a perfect square. **3.** Yes; $3n^2 - 48 = 3(n^2 - 16) = 3(n + 4)(n - 4)$. **5.** $(n + 9)(n - 9)$ **7.** $2x^3(x + 7)(x - 7)$ **9.** prime **11.** $\left\{-\frac{5}{2}, \frac{5}{2}\right\}$ **13.** $\left\{-\frac{1}{6}, \frac{1}{6}\right\}$ **15.** 12 in. by 12 in. **17.** $(n + 6)(n - 6)$ **19.** $(5 + 2p)(5 - 2p)$ **21.** $(11 + 3r)(11 - 3r)$ **23.** prime **25.** $(13y + 6z)(13y - 6z)$ **27.** $3(x - 5)(x + 5)$ **29.** $2(2g^2 - 25)$ **31.** $5x(2x - 3y)(2x + 3y)$ **33.** $(a + b + c)(a + b - c)$ **35.** $\left\{\pm\frac{8}{3}\right\}$ **37.** $\left\{\pm\frac{5}{2}\right\}$ **39.** $\left\{\pm\frac{9}{10}\right\}$ **41.** $\{\pm 10\}$ **43.** $\left\{-\frac{5}{3}, 0, \frac{5}{3}\right\}$ **45.** $\left\{-\frac{3}{2}, 0, \frac{3}{2}, 4\right\}$ **47.** 2 in. **49.** 36 mph **51.** The flaw is in line 5. Since $a = b$, $a - b = 0$. Therefore dividing by $a - b$ is dividing by zero, which is undefined. **53.** A **55.** prime **57.** $(3p + 5)(7p - 2)$ **59.** $\{3, 5\}$ **61.** between 83 and 99, inclusive **63.** $r > \frac{7}{10}$

$\frac{4}{10}$ $\frac{6}{10}$ $\frac{8}{10}$ 1 $\frac{12}{10}$ $\frac{14}{10}$

65. $x^2 + 2x + 1$ **67.** $x^2 + 16x + 64$ **69.** $25x^2 - 20x + 4$

Pages 512–514 Lesson 9-6

1. Determine if the first term is a perfect square. Then determine if the last term is a perfect square. Finally, check to see if the middle term is equal to twice the product of the square roots of the first and last terms. **3.** Sample answer: $x^3 + 5x^2 - 4x - 20$ **5.** no **7.** $(c - 3)(c - 2)$ **9.** $(2x - 7)(4x + 5)$ **11.** $(m - 2)(m + 2)(3m + 2n)$ **13.** $\{\pm 4\}$ **15.** $\{5 \pm \sqrt{13}\}$ **17.** no **19.** yes; $(2y - 11)^2$ **21.** yes; $(3n + 7)^2$ **23.** $8x + 20$ **25.** $4(k + 5)(k - 5)$ **27.** prime **29.** $3t(3t - 2)(t + 8)$ **31.** $2(5n + 1)(2n + 3)$ **33.** $3x(4x - 3)(2x - 5)$ **35.** $-3(3g - 5)^2$ **37.** $(a^2 + 2)(4a + 3b^2)$ **39.** $(y^2 + z^2)(x + 1)(x - 1)$ **41.** $x - 3y$ m, $x + 3y$ m, $xy + 7$ m **43.** $\{-4\}$ **45.** $\left\{\frac{4}{7}\right\}$ **47.** $\left\{\frac{1}{3}\right\}$ **49.** $\{-5, 3\}$ **51.** $\{8 \pm \sqrt{7}\}$ **53.** $\{-1 \pm \sqrt{6}\}$ **55.** $B = \frac{L}{16}(D - 4)^2$ **57.** 144 ft **59.** yes; 2 s **61.** 4, −4 **63.** 16 **65.** 100 **67.** C **69.** ±5 **71.** $\pm\frac{9}{7}$ **73.** $-\frac{5}{3}; -\frac{1}{4}$ **75.** $y = -\frac{1}{2}x + \frac{9}{2}$ **77.** 2030 ft **79.** −3, −2.5, −2

Pages 515–518 Chapter 9 Study Guide and Review

1. false, composite **3.** false, sample answer: 64 **5.** false, $2^4 \cdot 3$ **7.** true **9.** true **11.** $2^2 \cdot 7$ **13.** $2 \cdot 3 \cdot 5^2$ **15.** $-1 \cdot 83$ **17.** 5 **19.** $4ab$ **21.** $5n$ **23.** $13(x + 2y)$ **25.** $2a(13b + 9c + 16a)$ **27.** $2(r + 3p)(2s + m)$ **29.** $\left\{0, \frac{5}{2}\right\}$ **31.** $\left\{0, -\frac{7}{4}\right\}$ **33.** $(x - 12)(x + 3)$ **35.** $(r - 3)(r - 6)$ **37.** $(m + 4n)(m - 8n)$ **39.** $\{-6, 11\}$ **41.** prime **43.** $(5r + 2)(5r + 2)$ **45.** $(4b + 3)(3b + 2)$ **47.** $\left\{4, -\frac{5}{2}\right\}$ **49.** $\left\{\frac{3}{4}, -\frac{4}{5}\right\}$ **51.** prime **53.** $\{-4, 4\}$ **55.** $\left\{-\frac{9}{4}, \frac{9}{4}\right\}$ **57.** $(3k - 2)^2$ **59.** $2(4n - 5)^2$ **61.** $\left\{\frac{9}{7}\right\}$ **63.** $\left\{\pm\frac{1}{2}\right\}$

Chapter 10 Quadratic and Exponential Functions

Page 523 Chapter 10 Getting Started

1. Sample answer:

x	y
−6	−1
−4	1
−2	3
0	5
2	7

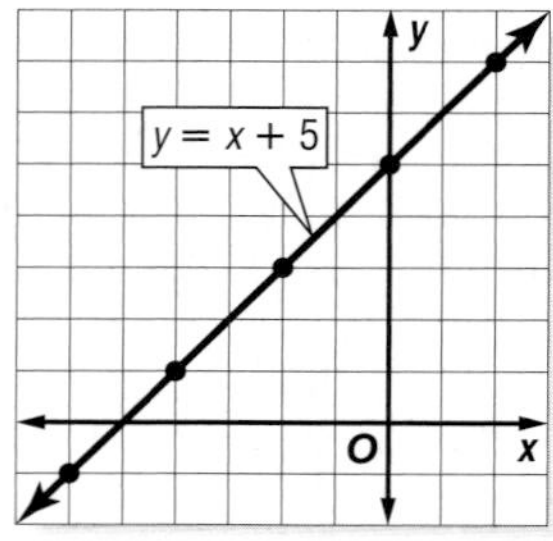

3. Sample answer:

x	y
−4	−1
−2	0
0	1
2	2
4	3

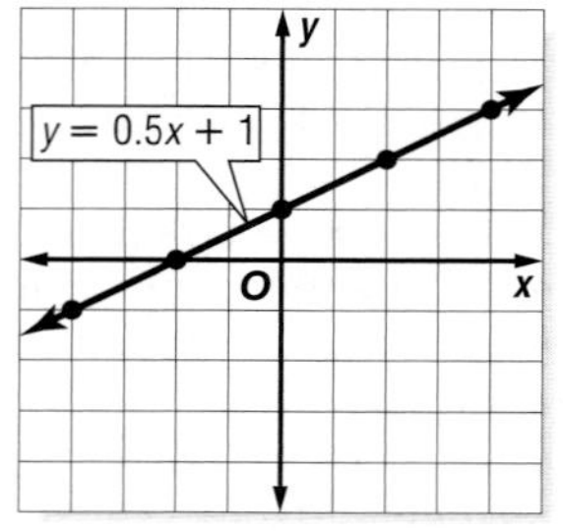

5. Sample answer:

x	y
0	−4
3	−2
6	0

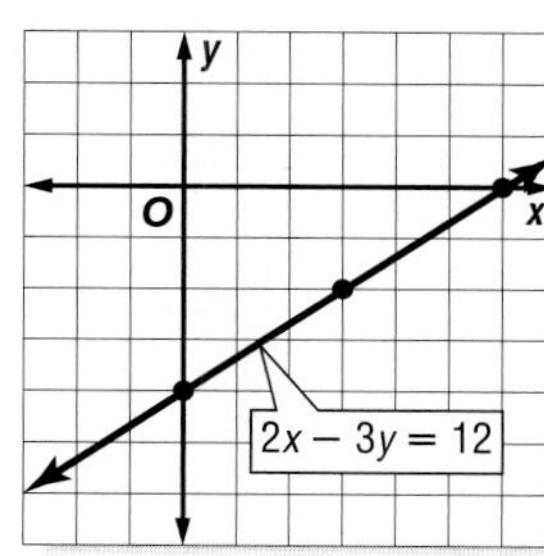

7. Sample answer:

x	y
−6	0
−4	−1
−2	−2
0	−3
2	−4

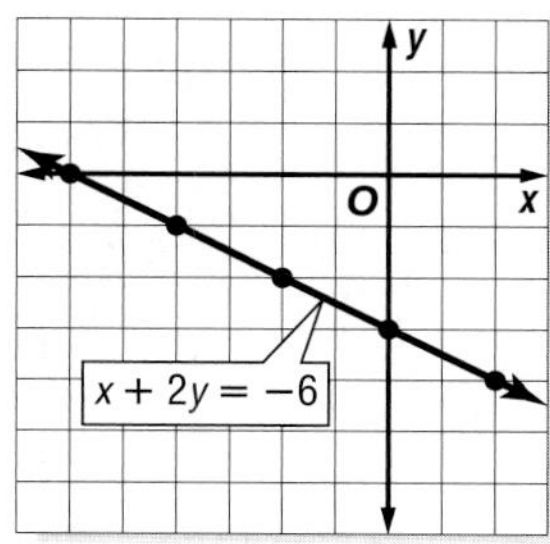

9. yes; $(t + 6)^2$ **11.** no **13.** yes; $(3b - 1)^2$ **15.** yes; $(2p + 3)^2$ **17.** 21, 25, 29 **19.** 8, 11, 14 **21.** −21, −26, −31 **23.** 8.1, 8.8, 9.5

Pages 528–530 Lesson 10-1

1. Both types of parabolas are U shaped. A parabola with a maximum opens downward, and its corresponding equation has a negative coefficient for the x^2 term. A parabola with a minimum opens upward, and its corresponding equation has a positive coefficient for the x^2 term. **3.** If you locate several points of the graph on one side of the axis of symmetry, you can locate corresponding points on the other side of the axis of symmetry to help graph the equation.

5.

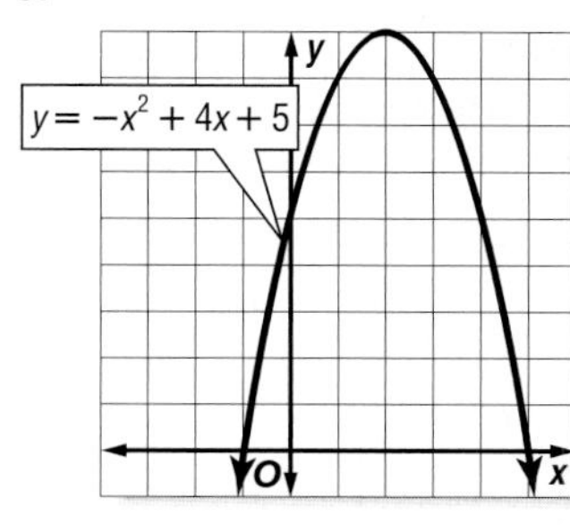

7. $x = 2.5$; (2.5, 12.25); maximum

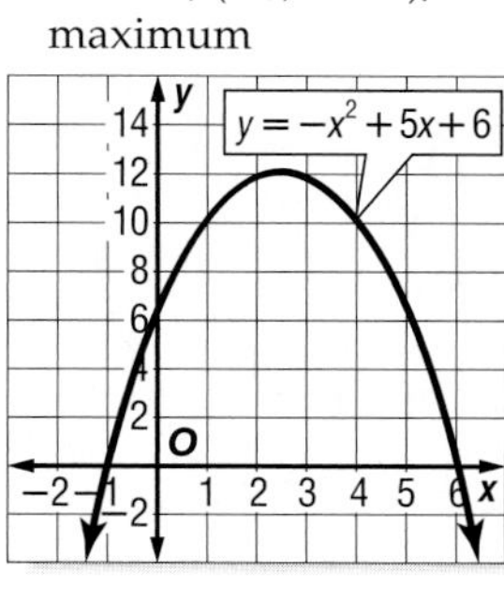

9. B

11.

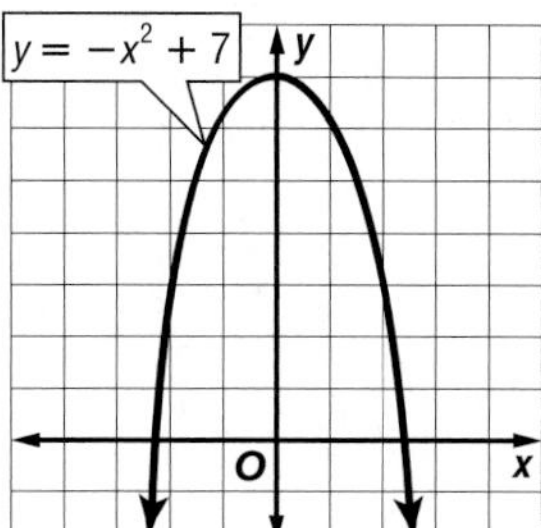

13.

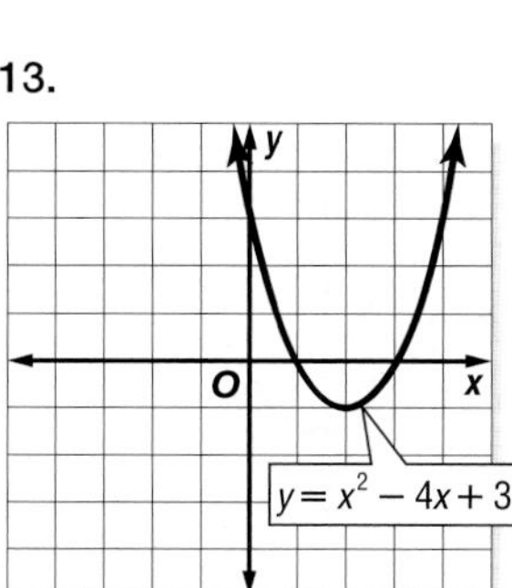

15.

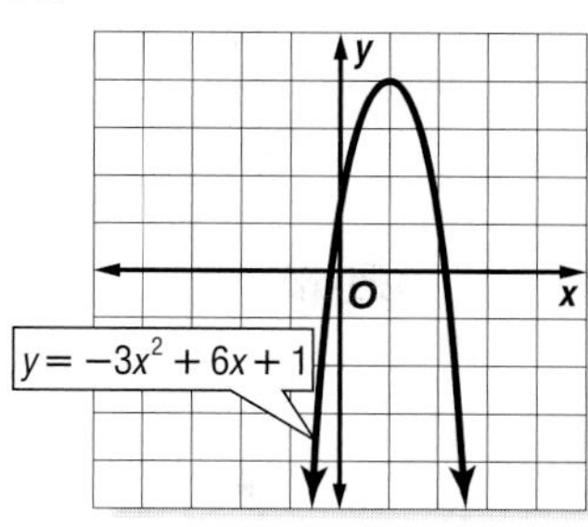

17. $x = \frac{5}{8}$

19. $x = 0$; (0, 0); maximum

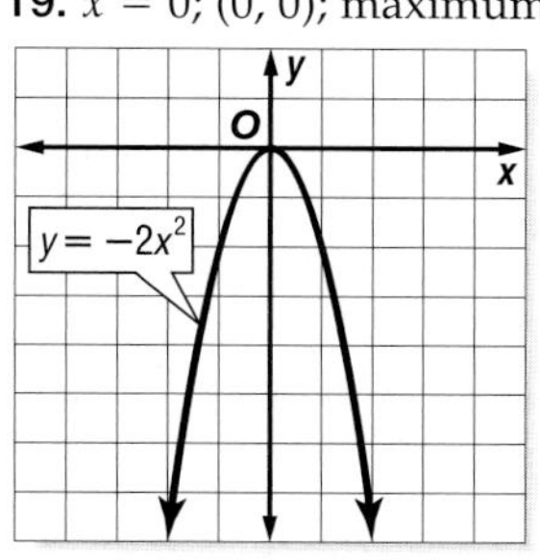

21. $x = 0$; (0, 5); maximum

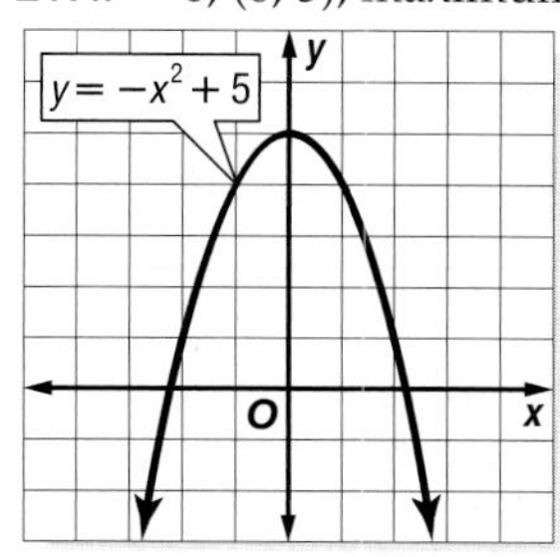

23. $x = -3$; (−3, 24); maximum

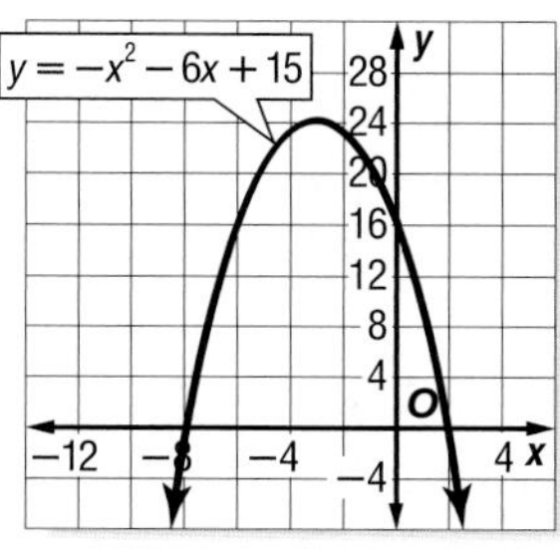

25. $x = -1$; (−1, 17); minimum

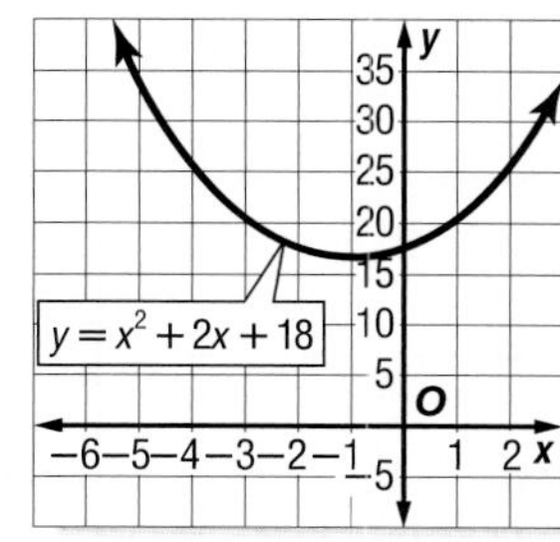

27. $x = 1$; (1, 1); minimum

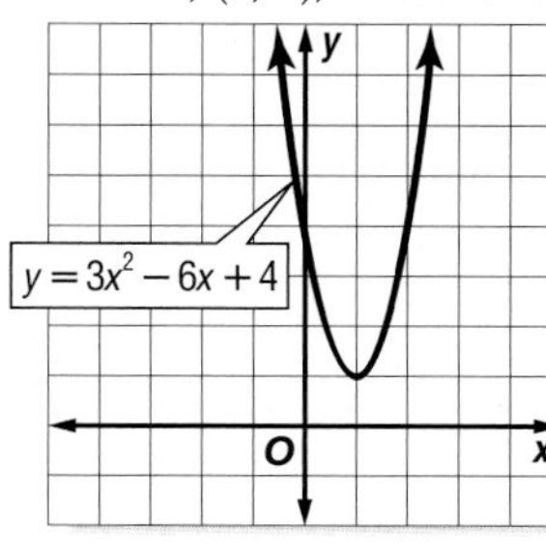

29. $x = 2$; (2, 1); minimum

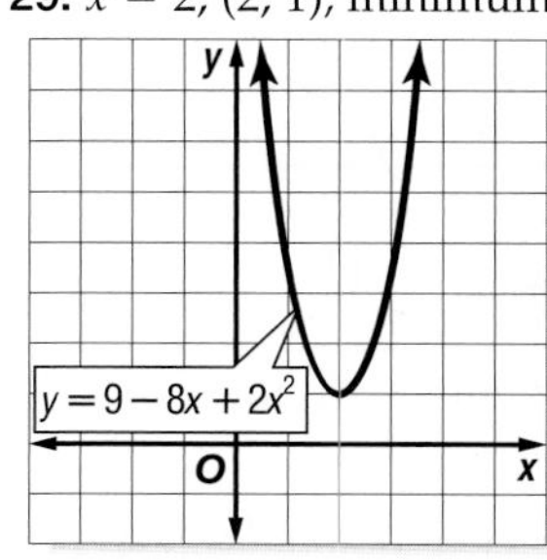

31. $x = 4$; (4, −3); maximum

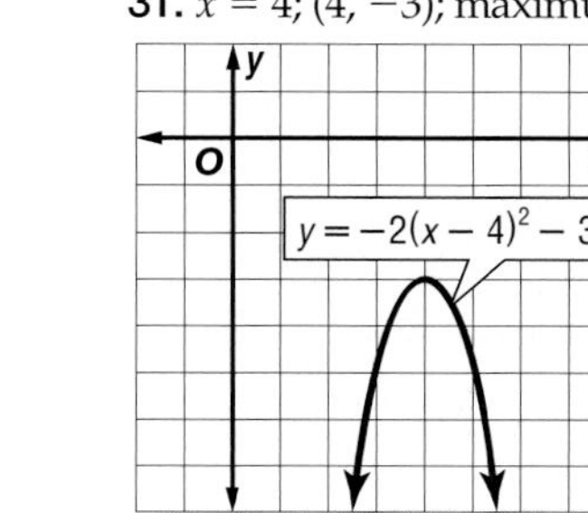

33. $x = -2$; (−2, −1); minimum

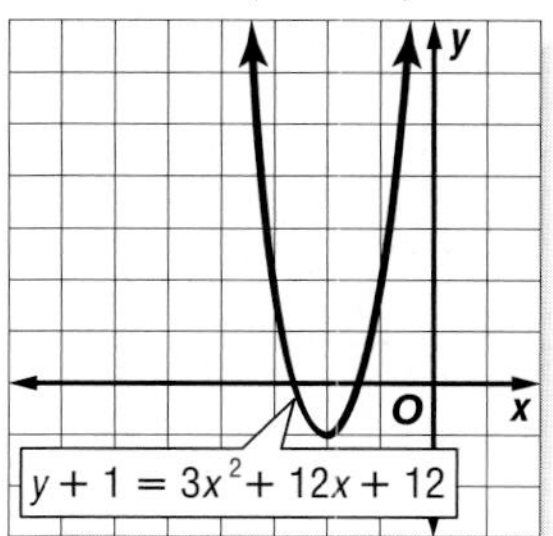

35. $x = -1$; $(-1, -1)$; minimum

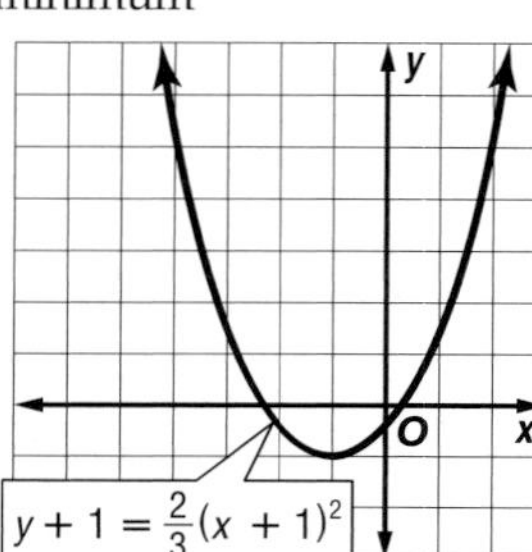

37. $x = -1$ **39.** 19 ft
41. $A = x(20 - x)$ or $A = -x^2 + 20x$ **43.** 100 m^2
45. 630 ft **47.** 1959

49.

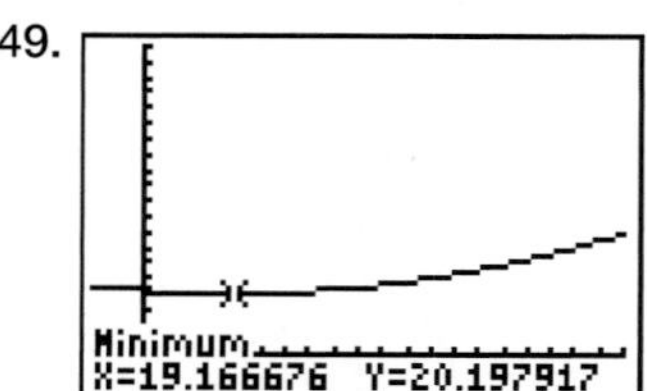

51. In order to coordinate a firework with recorded music, you must know when and how high it will explode. Answers should include the following.

- The rocket will explode when the rocket reaches the vertex or when $t = -\frac{39.2}{2(-4.9)}$ which is 4 seconds.
- The height of the rocket when it explodes is the height when $t = 4$. Therefore, $h = -4.9(4^2) + 39.2(4) + 1.6$ or 80 meters.

53. D

55.

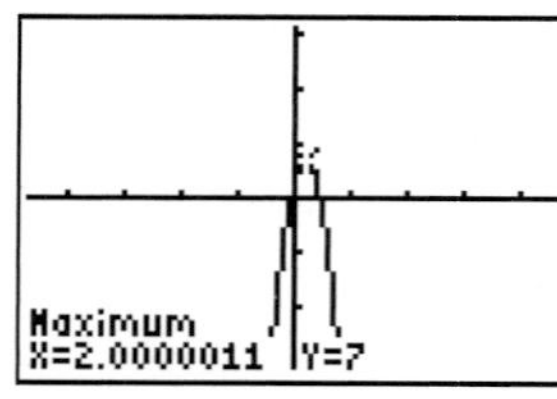

maximum; (2, 7)

57.

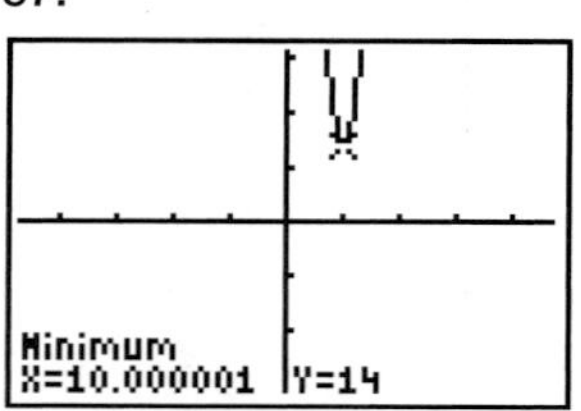

minimum; (10, 14)

59.

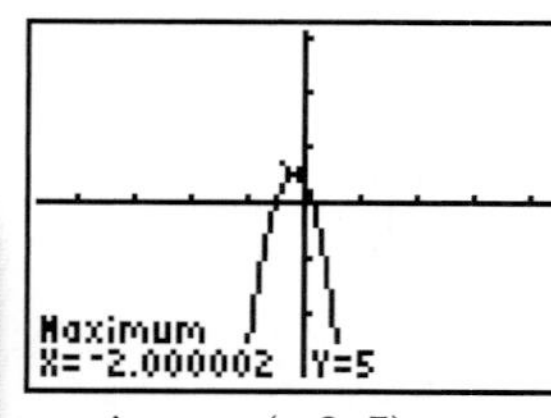

maximum; (−2, 5)

61. $(a + 11)^2$
63. $(2q - 3)(2q + 3)$
65. $(1 - 4g)(1 + 4g)$
67. $6p^2 - p - 18$
69. $\{b \mid b > -12\}$ **71.** $\left\{r \,\middle|\, r \leq \frac{8}{9}\right\}$
73. $y = -7$ **75.** 8 **77.** −3.5
79. −2.5

Pages 535–538 Lesson 10-2

1. −3, −1

3. Sample answer:

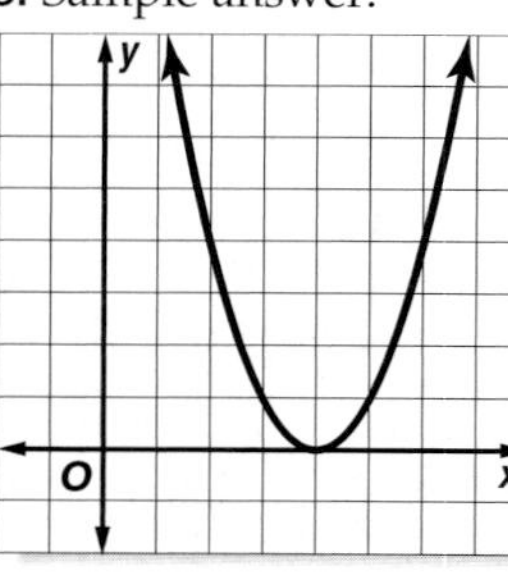

5.

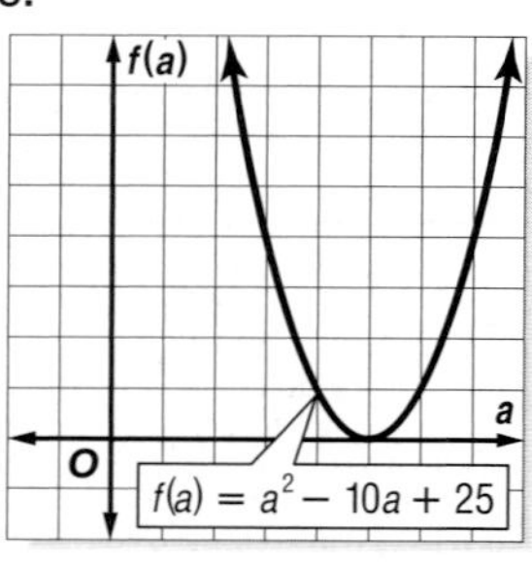

5

7.

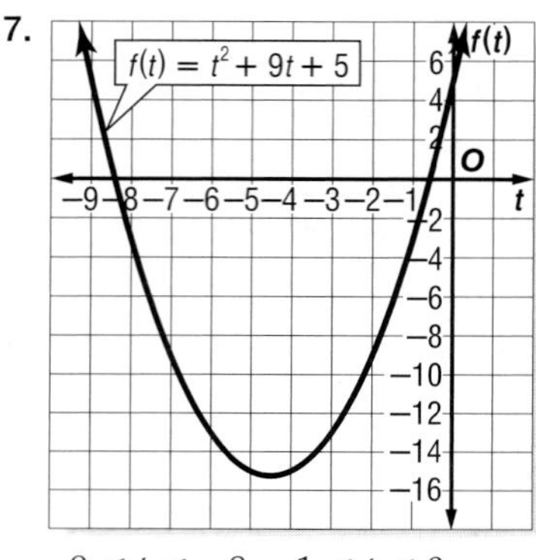

$-9 < t < -8$, $-1 < t < 0$

9.

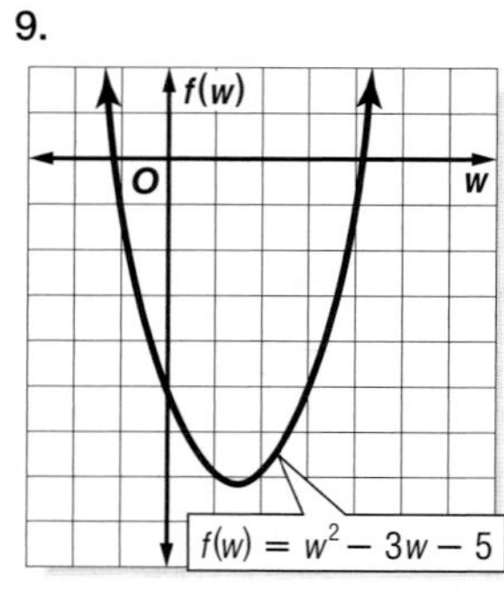

$-2 < w < -1$, $4 < w < 5$

11.

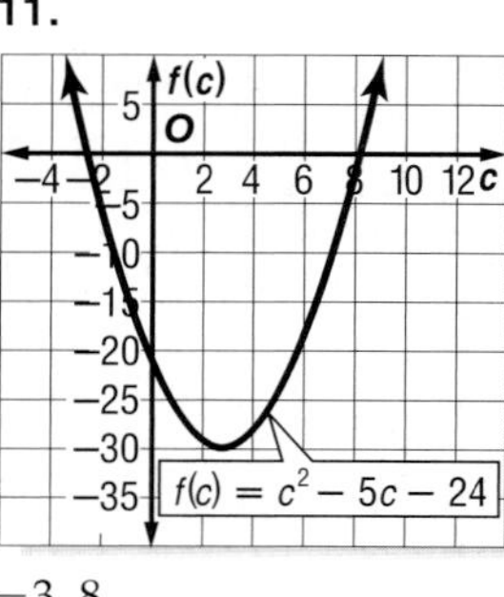

−3, 8

13.

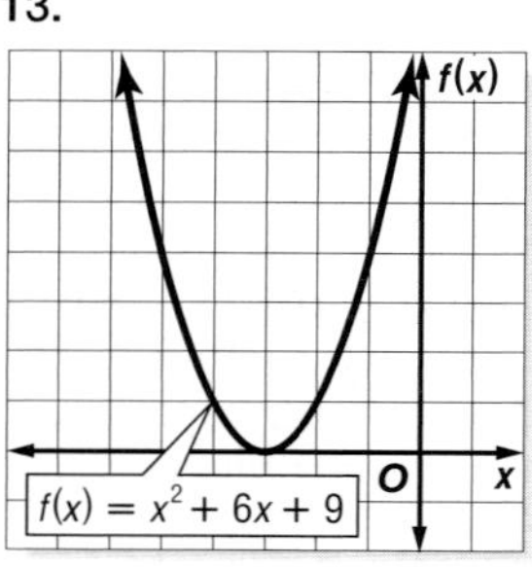

−3

15.

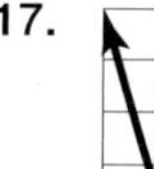

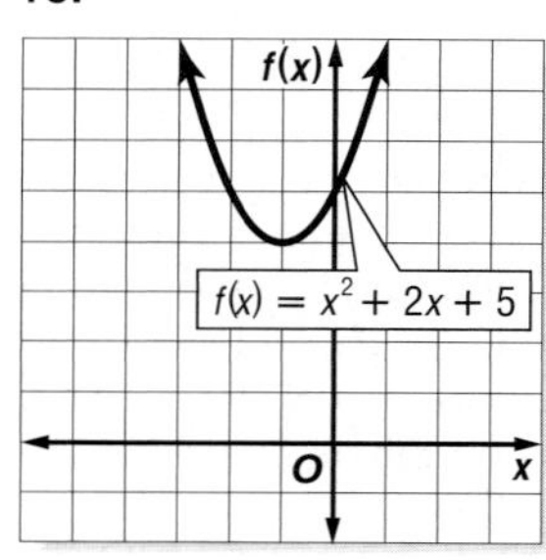

∅

17.

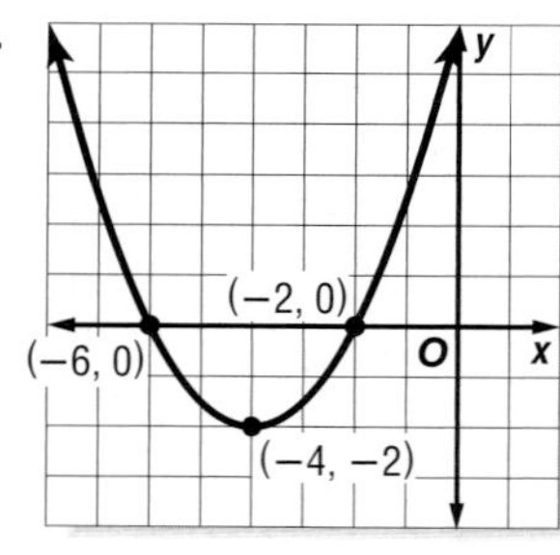

19. 4, 5

21.

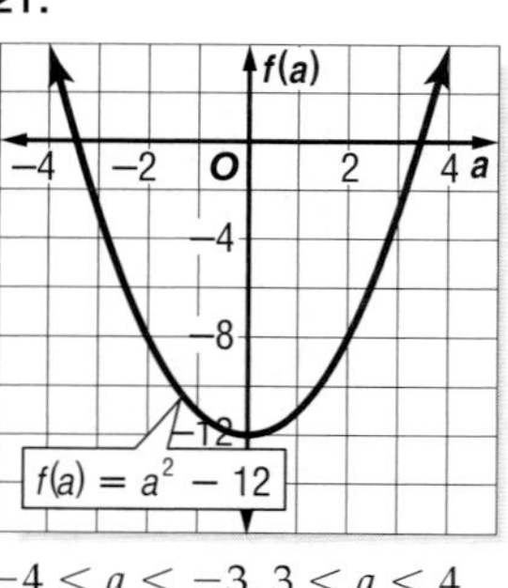

$-4 < a < -3$, $3 < a < 4$

23.

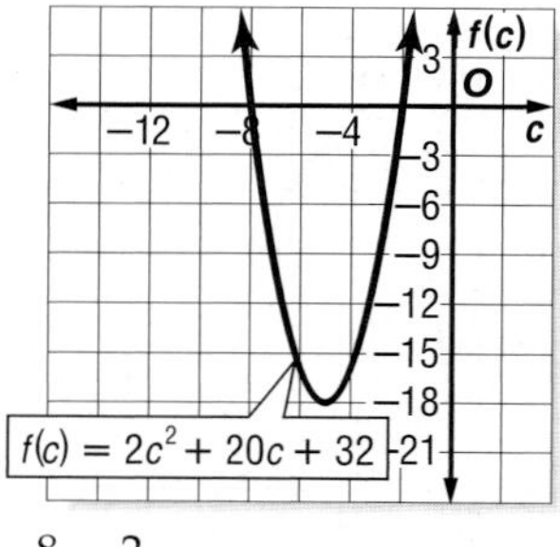

−8, −2

25.

$f(x) = x^2 + 6x + 6$

$-5 < x < -4, -2 < x < -1$

27.

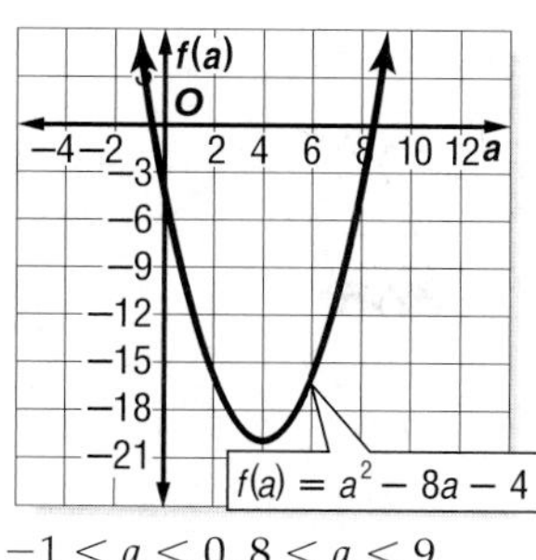

$-1 < a < 0, 8 < a < 9$

29.

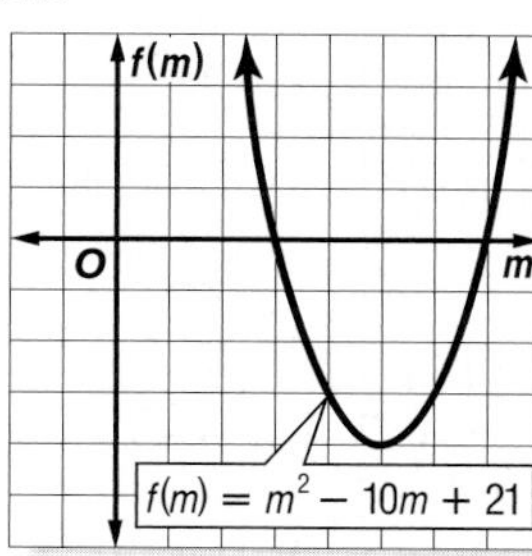

3, 7

31.

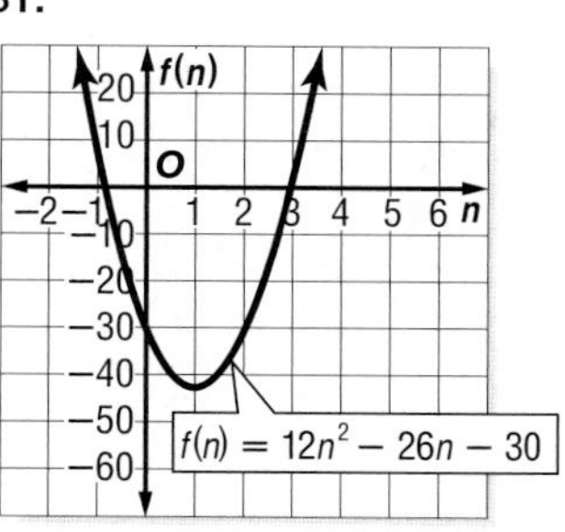

$-1 < n < 0, 3$

33. Sample answer:

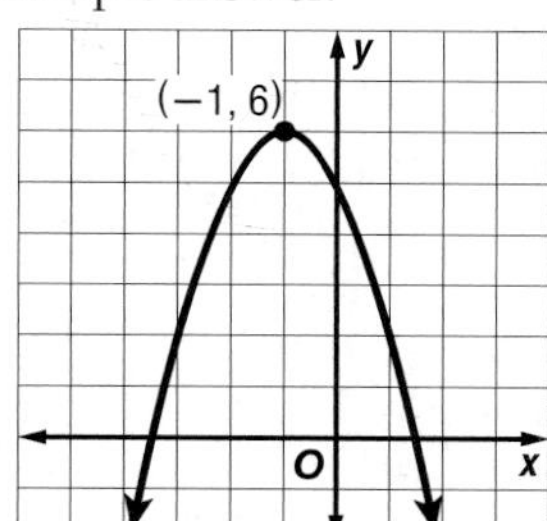

35.

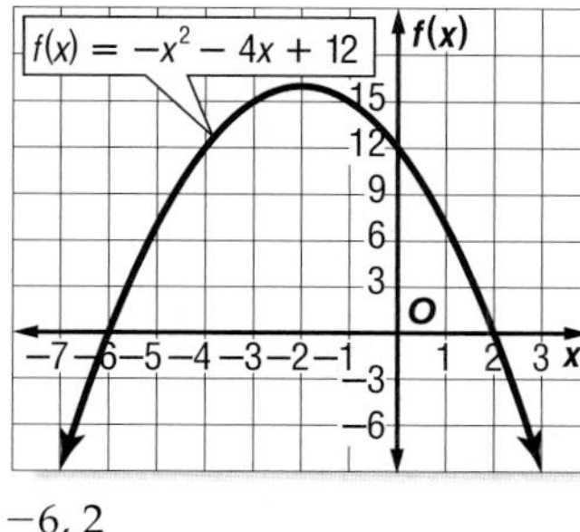

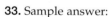

$-6, 2$

37. 16 ft **39.** \$297 **41.** about 9 s **43.** 100,000 ft^2 **45.** about 65 ft **47.** $-3, 0, 1$ **49.** C **51.** $-2, 1, 2$

53. $x = -3$; $(-3, 0)$; minimum

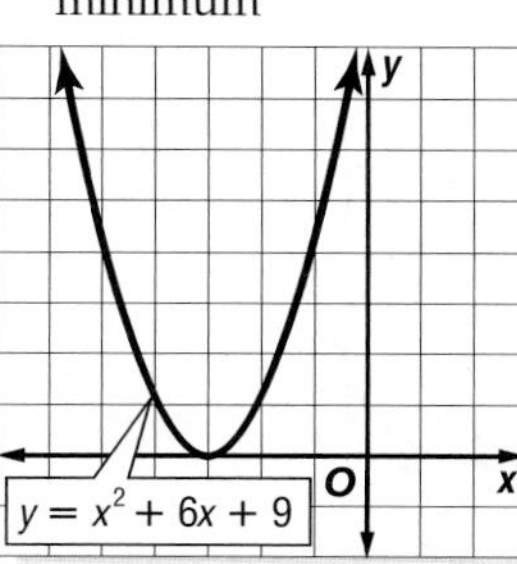

55. $x = 6$; $(6, -13)$; minimum

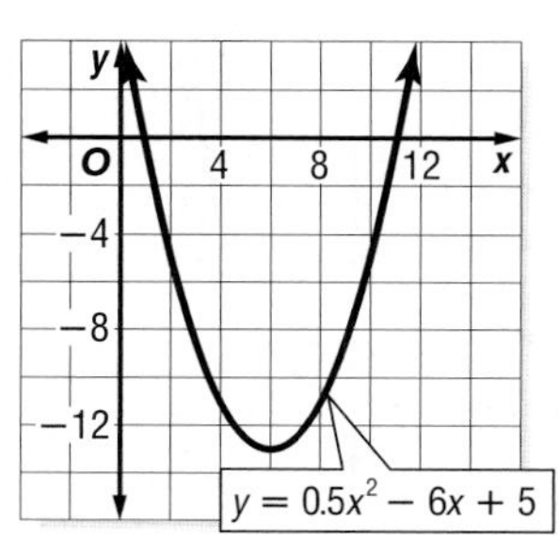

57. $\{5\}$ **59.** $\frac{m^3}{3}$ **61.** $-\frac{m^5y^4}{3}$ **63.** yes; $(a + 7)^2$ **65.** no **67.** no

Pages 542–544 Lesson 10-3

1. Sample answer:

x	1	1
x	1	1
x^2	x	x

$x^2 + 4x + 4$

3. Divide each side by 5. **5.** $-11.5, -2.5$ **7.** $\frac{25}{4}$ **9.** $-4, -3$ **11.** $-0.4, 4.4$ **13.** 0.2, 2.3 **15.** $-2, 6$ **17.** 2.6, 5.4 **19.** $-12.2, -3.8$ **21.** 64 **23.** 121 **25.** $\frac{49}{4}$ **27.** $-18, 18$ **29.** $-2, 6$ **31.** $-3, 22$ **33.** 1, 4

35. $-3, -1$ **37.** $-1.9, 11.9$ **39.** $2\frac{1}{3}$ **41.** $-1, \frac{2}{3}$ **43.** $-2.5, 0.5$ **45.** $-1\frac{1}{2}, 4$ **47.** $-2 \pm \sqrt{4 - c}$ **49.** 1.5 m **51.** There are no real solutions since completing the square results in $(x + 2)^2 = -8$ and the square of a number cannot be negative.

53. Al-Khwarizmi used squares to geometrically represent quadratic equations. Answers should include the following.

- Al-Khwarizmi represented x^2 by a square whose sides were each x units long. To this square, he added 4 rectangles with length x units long and width $\frac{8}{4}$ or 2 units long. This area represents 35. To make this a square, four 4×4 squares must by added.
- To solve $x^2 + 8x = 35$ by completing the square, use the following steps.

$x^2 + 8x = 35$	*Original equation*
$x^2 + 8x + 16 = 35 + 16$	*Since* $\left(\frac{8}{2}\right)^2 = 16$, *add 16 to each side.*
$(x + 4)^2 = 51$	*Factor* $x^2 + 8x + 16$.
$x + 4 = \pm\sqrt{51}$	*Take the square root of each side.*
$x + 4 - 4 = \pm\sqrt{51} - 4$	*Subtract 4 from each side.*
$x = -4 \pm \sqrt{51}$	*Simplify.*

$x = -4 - \sqrt{51}$ or $x = -4 + \sqrt{51}$

$x \approx -11.14$ $\quad x \approx 3.14$

The solution set is $\{-11.14, 3.14\}$.

55. A

57.

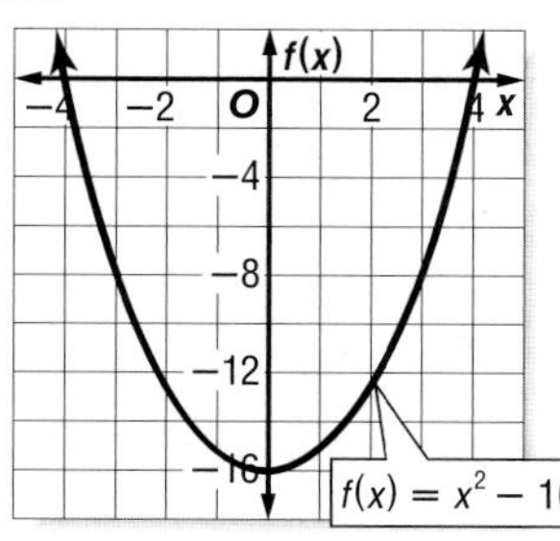

$-4, 4$

59.

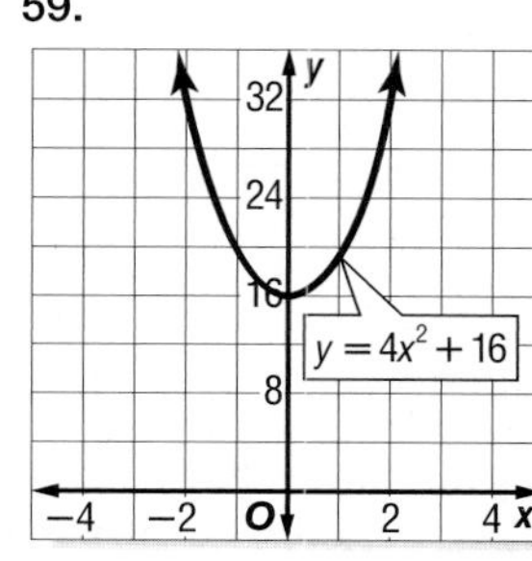

61.

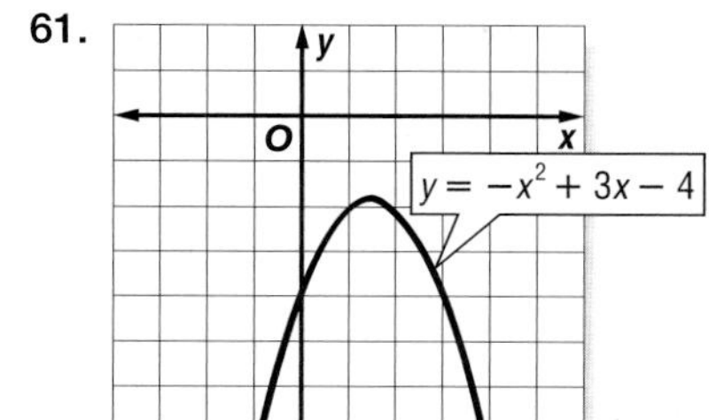

63. $8m^2n$ **65.** $(4, 1)$ **67.** $-3 < x < 1$ **69.** $y = -\frac{3}{5}x + \frac{14}{5}$ **71.** $y = -2x$ **73.** 5 **75.** 9.4

Selected Answers

Page 544 Practice Quiz 1

1. $x = 0.5$; $(0.5, -6.25)$; minimum

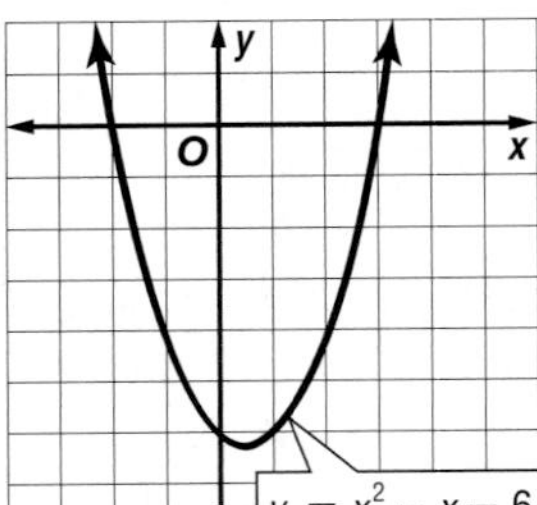

3. $x = -1$; $(-1, 8)$; maximum

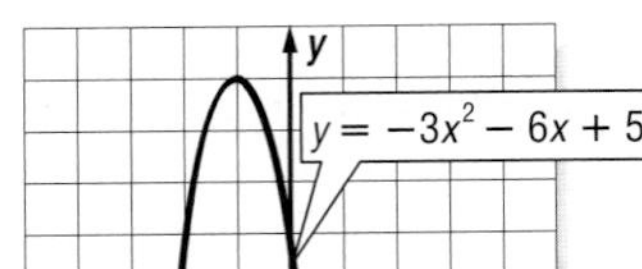

5.

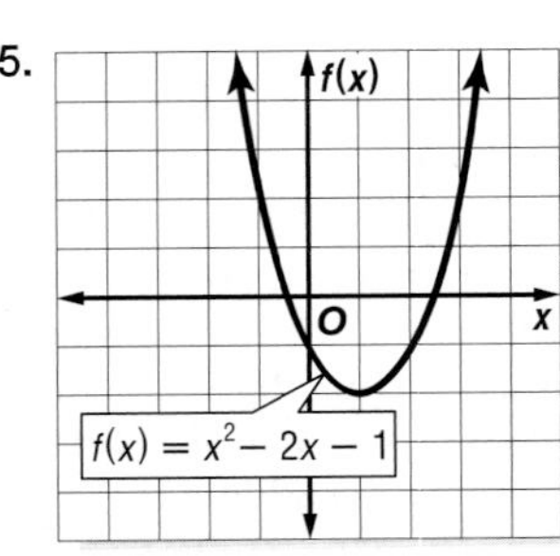

$-1 < x < 0, 2 < x < 3$

7. $-5, -3$ **9.** 4.8, 9.2

Pages 550–552 Lesson 10-4

1. Sample answer: (1) Factor $x^2 - 2x - 15$ as $(x + 3)(x - 5)$. Then according to the Zero Product Property, either $x + 3 = 0$ or $x - 5 = 0$. Solving these equations, $x = -3$ or $x = 5$. (2) Rewrite the equation as $x^2 - 2x = 15$. Then add 1 to each side of the equation to complete the square on the left side. Then $(x - 1)^2 = 16$. Taking the square root of each side, $x - 1 = \pm 4$. Therefore, $x = 1 \pm 4$ and $x = -3$ or $x = 5$. (3) Use the Quadratic Formula. Therefore, $x = \frac{-2 \pm \sqrt{(-2)^2 - 4(1)(-15)}}{2(1)}$ or $x = \frac{2 \pm \sqrt{64}}{2}$. Simplifying the expression, $x = -3$ or $x = 5$. **3.** Juanita; you must first write the equation in the form $ax^2 + bx + c = 0$ to determine the values of a, b, and c. Therefore, the value of c is -2, not 2. **5.** $-12, 1$ **7.** $\varnothing$ **9.** $\frac{1}{5}, \frac{2}{5}$ **11.** 0; 1 real root **13.** about 18.8 cm by 18.8 cm **15.** $-10, -2$ **17.** $-\frac{4}{5}, 1$ **19.** $\varnothing$ **21.** 5 **23.** $-0.4, 3.9$ **25.** $-0.5, 0.6$ **27.** $-\frac{3}{4}, \frac{5}{6}$ **29.** $-0.3, 0.6$ **31.** $-0.6, 2.6$ **33.** 5 cm by 16 cm **35.** -9 and -7 or 7 and 9 **37.** about -0.2 and 1.4 **39.** 5; 2 real roots **41.** -20; no real roots **43.** 0; 1 real root **45.** 0 **47.** about 2.3 s **49.** about 29.4 ft/s **51.** about 41 yr **53.** 2049; Sample answer: No; the death rate from cancer will never be 0 unless a cure is found. If and when a cure will be found cannot be predicted. **55.** A **57.** 1, 7 **59.** $-0.4, 12.4$

61.

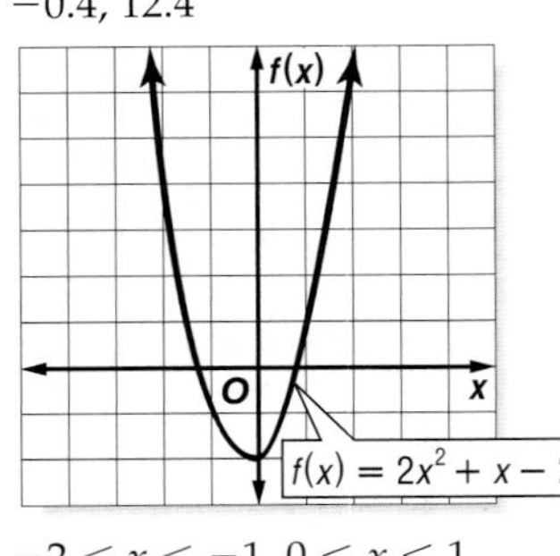
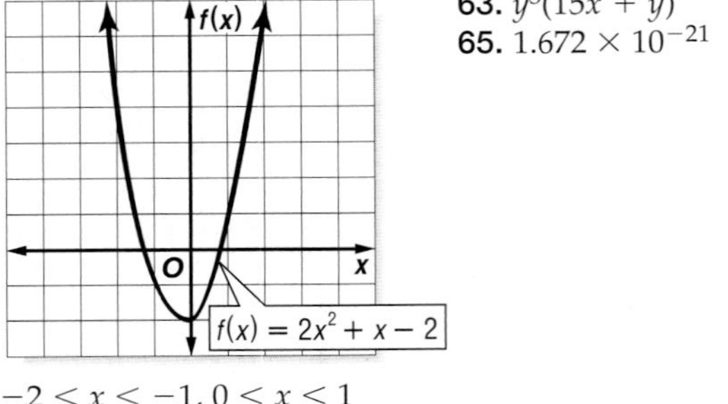

$-2 < x < -1, 0 < x < 1$

63. $y^3(15x + y)$
65. 1.672×10^{-21}

67.

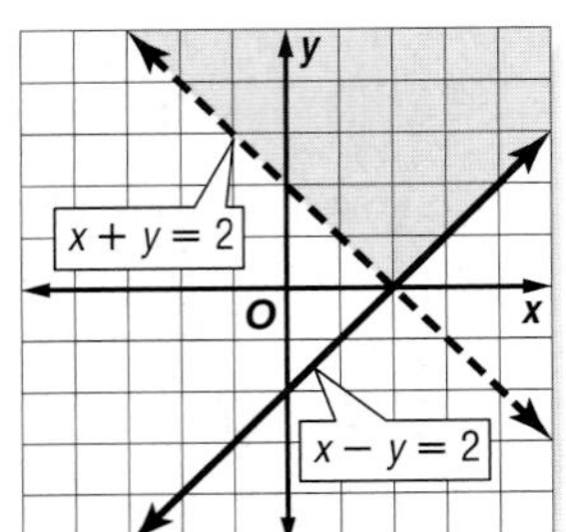
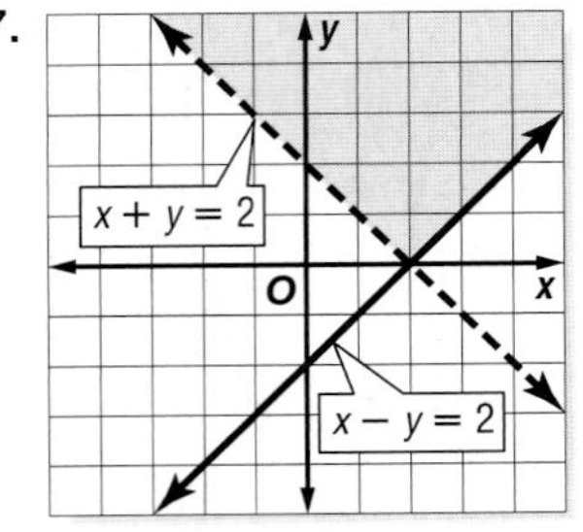

69. $\{m \mid m > 5\}$
71. $\{k \mid k \leq -4\}$ **73.** 147

Pages 557–560 Lesson 10-5

1. never **3.** Kiski; the graph of $y = \left(\frac{1}{3}\right)^x$ decreases as x increases.

5.

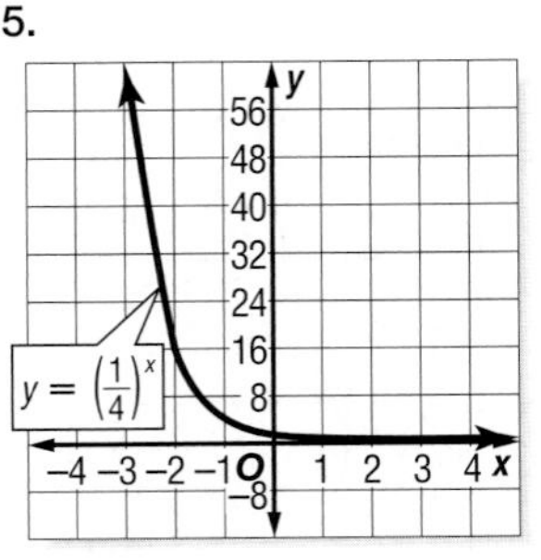

1; 0.1

7.

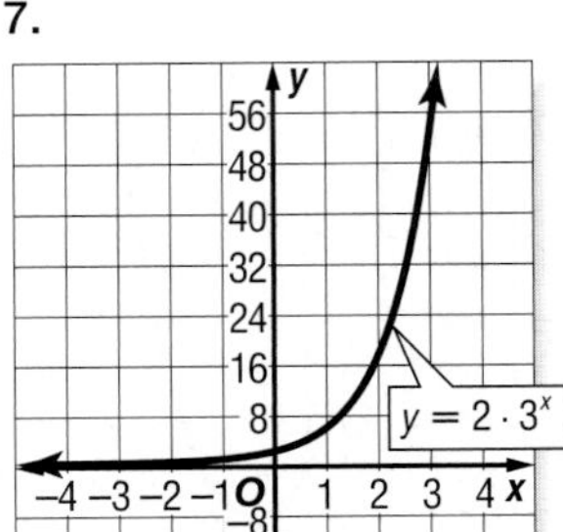

2

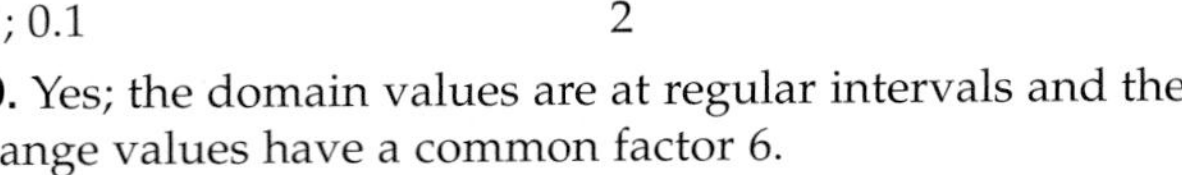

9. Yes; the domain values are at regular intervals and the range values have a common factor 6.
11. about 1.84×10^{19} grains

13.

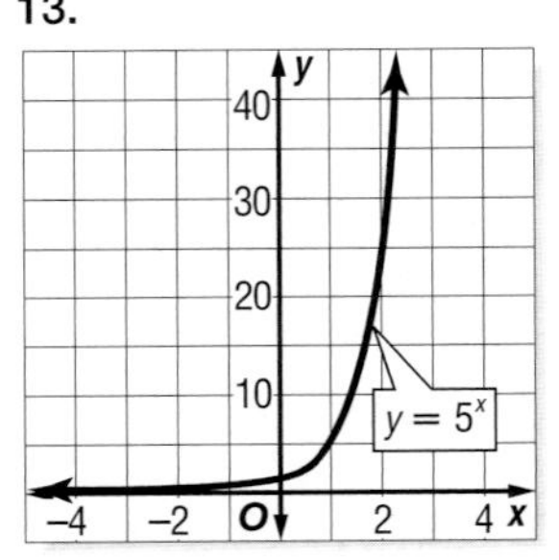

1; 5.9

15.

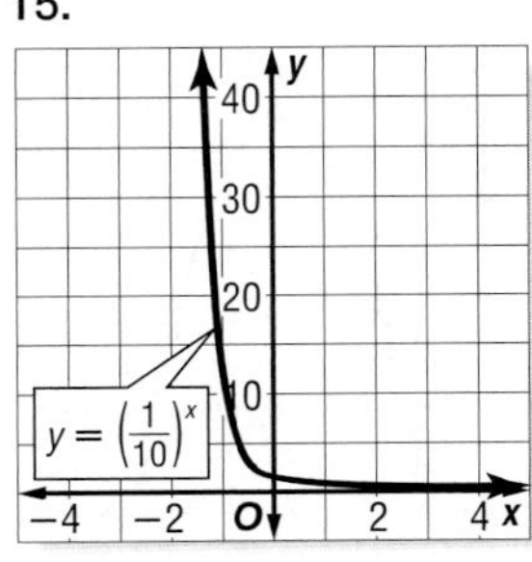

1; 20.0

17.

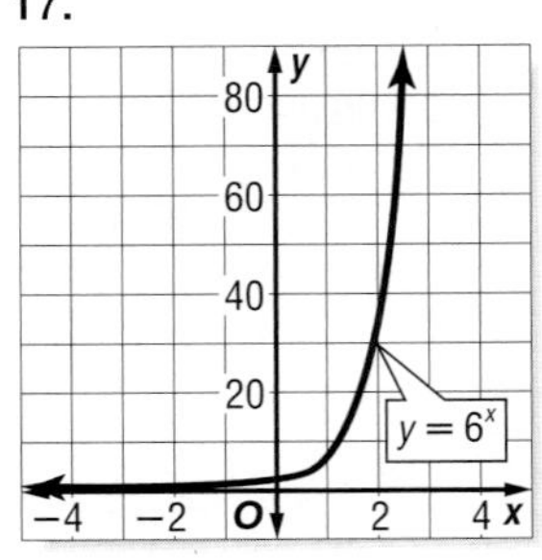

1; 1.7

19.

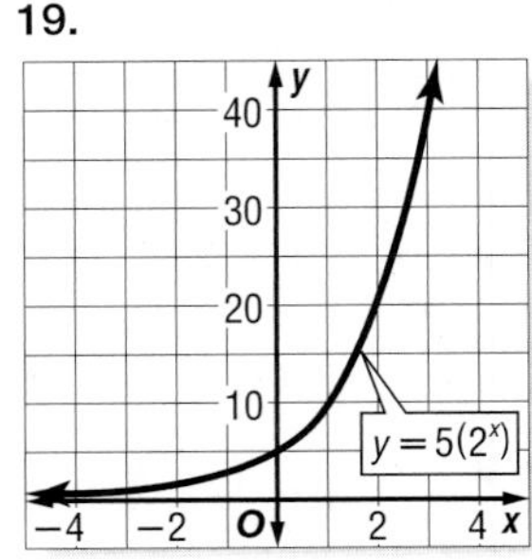

5

21.

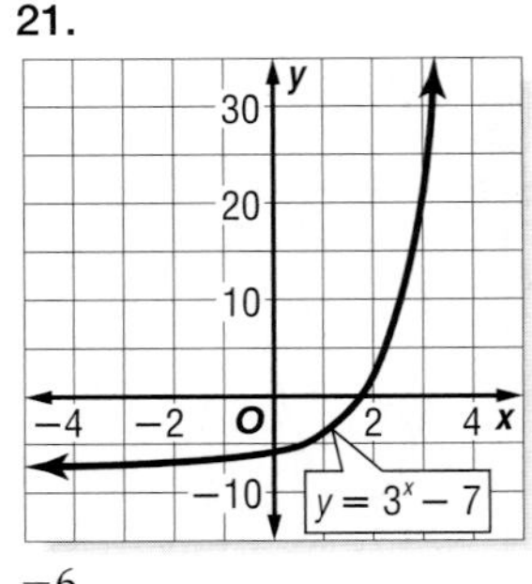

-6

23.

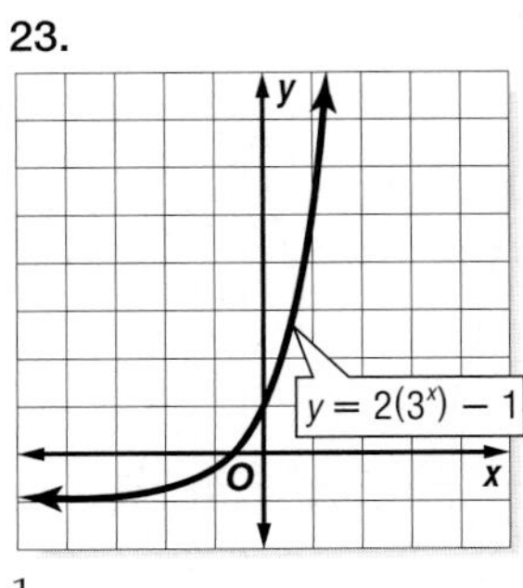

1

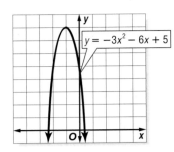

25. $y = 2(3^x + 1)$

4

27. No; the domain values are at regular intervals and the range values have a common difference 3. **29.** Yes; the domain values are at regular intervals and the range values have a common factor 0.75.

31. No; the domain values are at regular intervals, but the range values do not change. **33.** about \$37.27 million; about \$41.74 million; about \$46.75 million **35.** \$12 million sales in 1995 **37.** $y = 729\left(\frac{1}{3}\right)^x$ **39.** 6 rounds **41.** 10th week **43.** a translation 2 units up **45.** If the number of items on each level of a piece of art is a given number times the number of items on the previous level, an exponential function can be used to describe the situation. Answers should include the following.

- For the carving of the pliers, $y = 2^x$.
- For this situation, x is an integer between 0 and 8 inclusive. The values of y are 1, 2, 4, 8, 16, 32, 64, 128, and 256.
- 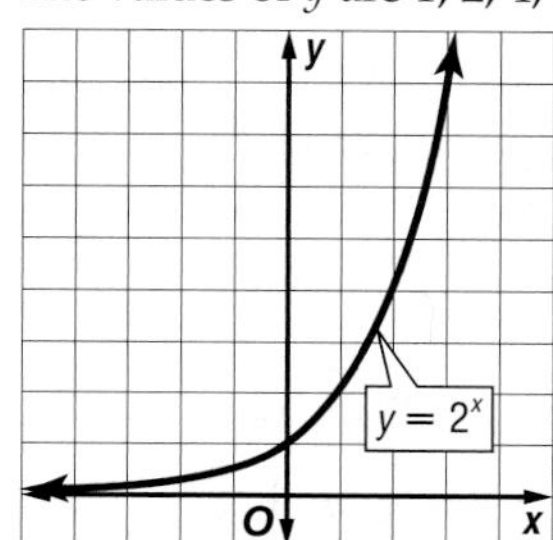

47. A **49.** −1.8, 0.3 **51.** 2, 5 **53.** −5.4, −0.6 **55.** prime **57.** 6, 9 **59.** $\{x \mid x \le 2\}$ **61.** 11.25 **63.** 144

Page 560 Practice Quiz 2

1. −7, 5 **3.** −0.2, 2.2

5. 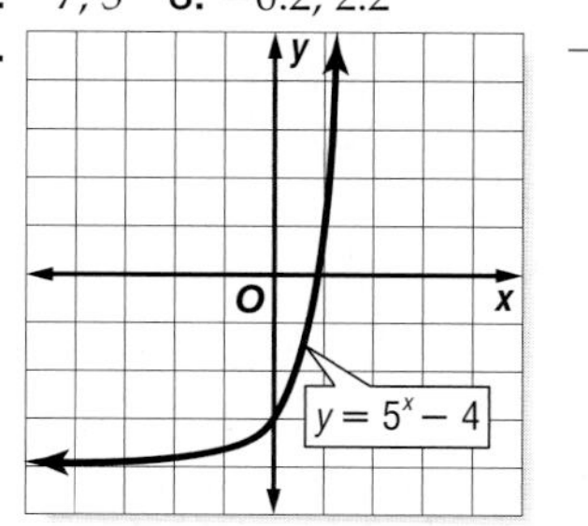

−3

Pages 563–565 Lesson 10-6

1. Exponential growth is an increase by the same percent over a period of time, while exponential decay is a decrease by the same percent over a period of time.

3. 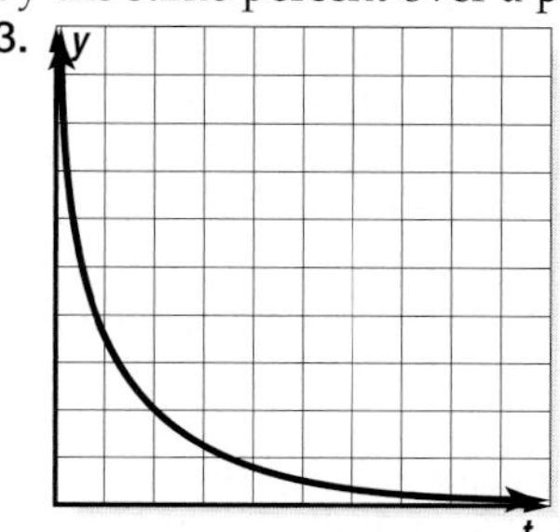

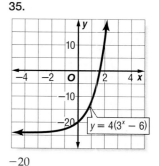

5. about \$43,041 **7.** about 1,767,128 people **9.** $C = 18.9(1.19)^t$ **11.** $W = 43.2(1.06)^t$ **13.** about 122,848,204 people **15.** about \$14,607.78 **17.** about \$135,849,289 **19.** about \$10,761.68 **21.** about 15.98% **23.** growth; 2.6% increase **25.** 128 g **27.** about 76.36 g **31.** C

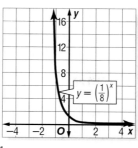

33. $y = \left(\frac{1}{8}\right)^x$

1

35. $y = 4(3^x - 6)$

−20

37. −0.6, 2.6 **39.** $m^{10}b^2$ **41.** $0.09x^6y^4$ **43.** {1} **45.** yes **47.** −5, −8, −11

Pages 570–572 Lesson 10-7

1. Both arithmetic sequences and geometric sequences are lists of related numbers. In an arithmetic sequence, each term is found by adding the previous term to a constant called the common difference. In a geometric sequence, each term is found by multiplying the previous term by a constant called the common ratio. **3.** Sample answer: 1, 4, 9, 16, 25, 36, ... **5.** yes **7.** 1280, 5120, 20,480 **9.** −40.5, 60.75, −91.125 **11.** −32 **13.** ±14 **15.** ±20 **17.** yes **19.** no **21.** no **23.** yes **25.** 256, −1024, 4096 **27.** 64, 32, 16 **29.** −0.3125, 0.078125, −0.01953125 **31.** $\frac{8}{81}, \frac{16}{243}, \frac{32}{729}$ **33.** 48 in², 24 in², 12 in², 6 in², 3 in² **35.** 320 **37.** 250 **39.** −288 **41.** 0.5859375 **43.** ±10 **45.** ±45 **47.** ±32 **49.** ±14 **51.** ±3.5 **53.** $\pm\frac{3}{10}$ **55.** 6 m, 3.6 m, 2.16 m **57.** 18 questions **59.** in 16 days **61.** always **63.** Since the distance of each bounce is $\frac{3}{4}$ times the distance of the last bounce, the list of the distances from the stopping place is a geometric sequence. Answers should include the following.

- To find the 10th term, multiply the first term 80 by $\frac{3}{4}$ to the 9th power.
- The 17th bounce will be the first bounce less than 1 ft from the resting place.

65. 1/7 **67.** 0 **69.** about \$1822.01 **71.** Yes; the domain values are at regular intervals and the range values have a common factor 3. **73.** $(2x + 3)(x - 4)$

Pages 574–578 Chapter 10 Study Guide and Review

1. d **3.** i **5.** c **7.** b **9.** f

11. $x = -1$; $(-1, -1)$; minimum

$y = x^2 + 2x$

13. $x = 1\frac{1}{2}$; $\left(1\frac{1}{2}, -6\frac{1}{4}\right)$; minimum

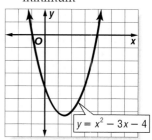

$y = x^2 - 3x - 4$

15. $x = 0$; $(0, 1)$; maximum

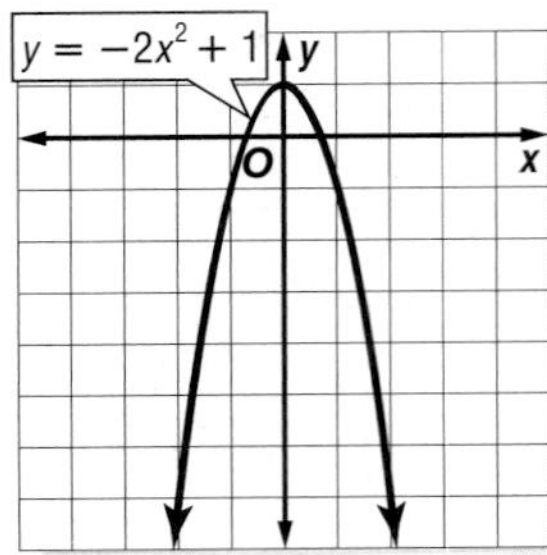

17.

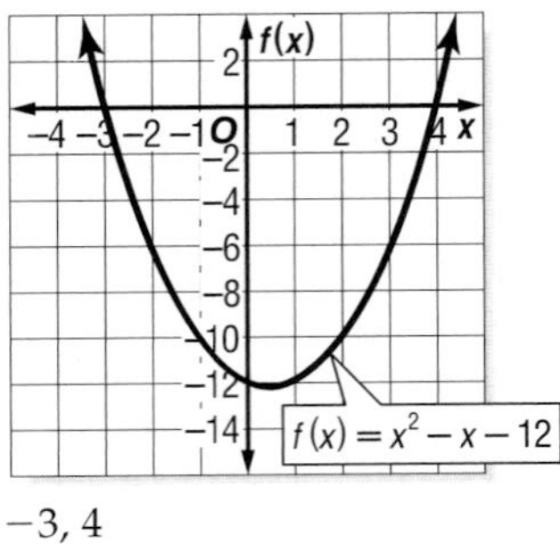

$-3, 4$

19.

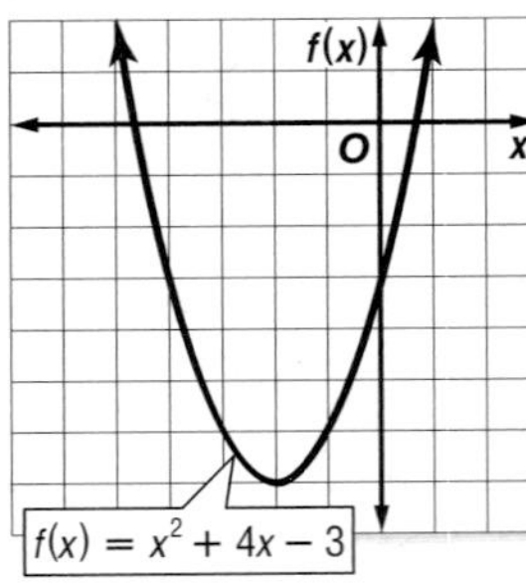

$-5 < x < -4, 0 < x < 1$

21.

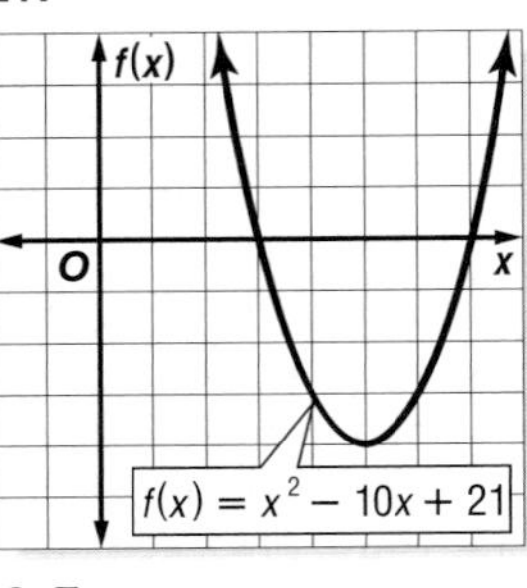

$3, 7$

23. $-1.2, 1.2$ **25.** $-0.7, 7.7$ **27.** $-4.4, 0.4$ **29.** $-2, 10$
31. $-2.5, 1.5$ **33.** $-4, 0$

35.

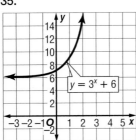

7

37.

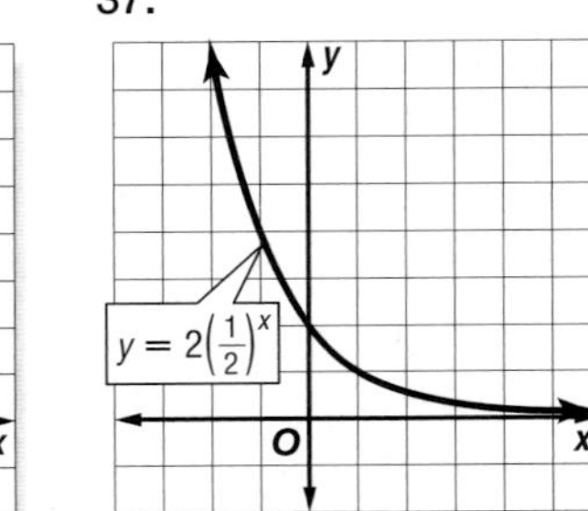

2

39. \$12,067.68 **41.** \$24,688.36 **43.** $\frac{56}{27}$ **45.** ± 10 **47.** $\pm\frac{1}{2}$

Chapter 11 Radical Expressions and Triangles

Page 585 Chapter 11 Getting Started

1. 5 **3.** 7.48 **5.** $a + 7b$ **7.** $16c$ **9.** $\{0, 5\}$ **11.** $\{-3, 9\}$ **13.** yes **15.** no

Pages 589–592 Lesson 11-1

1. Both x^4 and x^2 are positive even if x is a negative number. **3.** Sample answer: $2\sqrt{2} + 3\sqrt{3}$ and $2\sqrt{2} - 3\sqrt{3}$; -19 **5.** 4 **7.** $3|ab|\sqrt{6}$ **9.** $\frac{2\sqrt{6}}{3}$ **11.** $\frac{8(3 + \sqrt{2})}{7}$ **13.** 28 ft^2 **15.** $3\sqrt{2}$ **17.** $4\sqrt{5}$ **19.** $\sqrt{30}$ **21.** $84\sqrt{5}$ **23.** $2a^2\sqrt{10}$ **25.** $7|x^3y^3|\sqrt{3y}$ **27.** $\frac{\sqrt{6}}{3}$ **29.** $\frac{\sqrt{2t}}{4}$ **31.** $\frac{c^2\sqrt{5cd}}{2|d^3|}$ **33.** $\frac{54 + 9\sqrt{2}}{17}$ **35.** $2\sqrt{7} - 2\sqrt{2}$ **37.** $\frac{-16 - 12\sqrt{3}}{11}$ **39.** $60\sqrt{2}$ or about 84.9 cm^2 **41.** $s = \sqrt{A}$; $6\sqrt{2}$ in. **43.** $6\sqrt{5}$ or about 13.4 m/s **45.** $3\sqrt{2d}$ **47.** about 44.5 mph, about 51.4 mph **49.** $20\sqrt{3}$ or about 34.6 ft^2 **51.** A lot of formulas and calculations that are used in space exploration contain radical expressions. Answers should include the following.

- To determine the escape velocity of a planet, you would need to know its mass and the radius. It would be very important to know the escape velocity of a planet before you landed on it so you would know if you had enough fuel and velocity to launch from it to get back into space.
- The astronomical body with the smaller radius would have a greater escape velocity. As the radius decreases, the escape velocity increases.

53. B **55.** 6°F **57.** x^2 **59.** $a^{-\frac{5}{6}}$ or $\frac{\sqrt[6]{a}}{a}$ **61.** $s^{18}t^6\sqrt{s}$ **63.** 16, -32, 64 **65.** 144, 864, 5184 **67.** 0.08, 0.016, 0.0032 **69.** 84.9°C **71.** $(5x - 4)(7x - 3)$ **73.** $3(x - 7)(x + 5)$ **75.** $(4x - 3)(2x - 1)$ **77.** $\{(2, 0), (1, 2.5)\}$ **79.** $\left\{\left(4, -\frac{1}{2}\right), (2, 1)\right\}$ **81.** 6 **83.** -1885 **85.** $a^2 + 7a + 10$ **87.** $4x^2 + x - 3$ **89.** $12a^2 + 13ab - 14b^2$

Pages 595–597 Lesson 11-2

1. to determine if there are any like radicands **3.** Sample answer: $(\sqrt{2} + \sqrt{3})^2 = 2 + 2\sqrt{6} + 3$ or $5 + 2\sqrt{6}$ **5.** $-5\sqrt{6}$ **7.** $4\sqrt{3}$ **9.** $9\sqrt{3} + 3$ **11.** $17 + 7\sqrt{5}$ **13.** $10\sqrt{110} - 5\sqrt{330} \approx 14.05$ volts **15.** $13\sqrt{6}$ **17.** 0 **19.** $10\sqrt{5b}$ **21.** $4\sqrt{6} - 6\sqrt{2} + 5\sqrt{17}$ **23.** $\sqrt{6} + 4\sqrt{3}$ **25.** $-2\sqrt{2}$ **27.** $\frac{4\sqrt{10}}{5}$ **29.** $\frac{53\sqrt{7}}{7}$ **31.** $10\sqrt{2} + 3\sqrt{10}$ **33.** $59 - 14\sqrt{10}$ **35.** $3\sqrt{7}$ **37.** $15\sqrt{2} + 11\sqrt{5}$ **39.** $\sqrt{3} + 2$ cm **41.** $5\sqrt{87} - 25\sqrt{3} \approx 3.34$ mi **43.** 6 in. **45.** 40 ft/s; 80 ft/s **47.** The velocity should be $\sqrt{9}$ or 3 times the velocity of an object falling 25 feet; $3 \cdot 40 = 120$ ft/s, $\sqrt{2(32)(225)} = 120$ ft/s. **49.** Sample answer: $a = 4, b = 9$; $\sqrt{4 + 9} \neq \sqrt{4} + \sqrt{9}$ **51.** The distance a person can see is related to the height of the person using $d = \sqrt{\frac{3h}{2}}$. Answers should include the following.

- You can find how far each lifeguard can see from the height of the lifeguard tower. Each tower should have some overlap to cover the entire beach area.
- On early ships, a lookout position (Crow's nest) was situated high on the foremast. Sailors could see farther from this position than from the ship's deck.

53. D **55.** $8\sqrt{2}$ **57.** $\frac{5}{2}$ **59.** $\frac{3\sqrt{14}}{16|ab|}$ **61.** -5103 **63.** $\left\{\pm\frac{9}{7}\right\}$ **65.** $\left\{-\frac{5}{4}, 0, \frac{5}{4}\right\}$ **67.** $n \geq \frac{5}{8}$ **69.** $k > \frac{3}{5}$ **71.** $x^2 - 4x + 4$ **73.** $x^2 + 12x + 36$ **75.** $4x^2 - 12x + 9$

Pages 600–603 Lesson 11-3

1. Isolate the radical on one side of the equation. Square each side of the equation and simplify. Then check for extraneous solutions. **3.** Sample answer: $\sqrt{x + 1} = 8$; 63 **5.** 25 **7.** 7 **9.** 2 **11.** 3 **13.** 6 **15.** about 5994 m **17.** 100 **19.** 50 **21.** 4 **23.** no solution **25.** 5 **27.** 2 **29.** 180 **31.** 2 **33.** 57 **35.** 2 **37.** 2, 3 **39.** 3 **41.** 6 **43.** 2 **45.** 11 **47.** sometimes **49.** about 0.0619 **51.** $4\sqrt{6}$ or about 9.8 m **53.** It increases by a factor of $\sqrt{2}$. **55.** about 2.43 ft **57.** about 43.84°C **59.** $V < 330.45$ m/s **61.** You can determine the time it takes an object to fall from a given height using a radical equation. Answers should include the following.

- It would take a skydiver approximately 42 seconds to fall 10,000 feet. Using the equation, it would take 25 seconds. The time is different in the two calculations because air resistance slows the skydiver.
- A skydiver can increase the speed of his fall by lowering air resistance. This can be done by pulling his arms and legs close to his body. A skydiver can decrease his speed

by holding his arms and legs out, which increases the air resistance.
63. C **65.** 11 **67.** 15.08 **69.** no solution **71.** $20\sqrt{3}$
73. $8\sqrt{3}$ **75.** $3(\sqrt{10} - \sqrt{3})$ **77.** yes; $(2n - 7)^2$
79. $r^2 - r - 12$ **81.** $6p^3 + 7p^2 - 2p + 45$ **83.** $14x - 7y = -3$
85. $15x - 2y = 49$ **87.** 25 **89.** $4\sqrt{13}$

Page 603 Practice Quiz 1

1. $4\sqrt{3}$ **3.** $\frac{-2 + \sqrt{10}}{2}$ **5.** $20\sqrt{3}$ **7.** $11 + 4\sqrt{7}$ or about $21.6\ cm^2$ **9.** 4

Pages 607–610 Lesson 11-4

1.

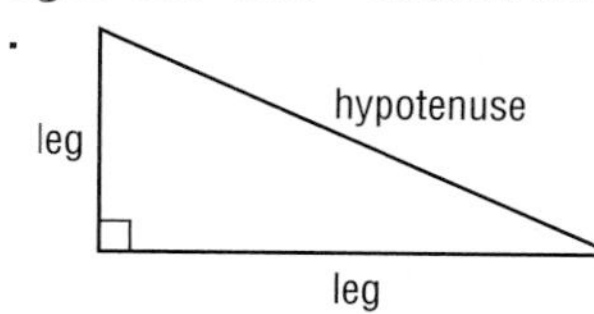

3. $d = \sqrt{2s^2}$ or $d = s\sqrt{2}$
5. 9 **7.** 60 **9.** $\sqrt{65} \approx 8.06$
11. Yes; $16^2 + 30^2 = 34^2$.
13. 14.14 **15.** 53
17. 42.13 **19.** 65 **21.** 11

23. $\sqrt{115} \approx 10.72$ **25.** $\sqrt{67} \approx 8.19$ **27.** $\sqrt{253} \approx 15.91$
29. $17x$ **31.** Yes; $30^2 + 40^2 = 50^2$. **33.** No; $24^2 + 30^2 \neq 36^2$.
35. Yes; $15^2 + (\sqrt{31})^2 = 16^2$. **37.** 18 ft **39.** $4\sqrt{3}$ in. or about 6.93 in. **41.** about 415.8 ft **43.** The roller coaster makes a total horizontal advance of 404 feet, reaches a vertical height of 208 feet, and travels a total track length of about 628.3 feet. **45.** about 116.6 ft **47.** $900\ ft^2$
49. about 1081.7 ft, 324.5 ft **51.** C **53.** 144 **55.** 12
57. $-3\sqrt{z}$ **59.** 5^5 or 3125 **61.** $\frac{2a^2b^3}{c^8}$ **63.** 5 **65.** $\sqrt{53}$
67. $\sqrt{130}$

Pages 612–615 Lesson 11-5

1. The values that are subtracted are squared before being added and the square of a negative number is always positive. The sum of two positive numbers is positive, so the distance will never be negative. **3.** There are exactly two points that lie on the line $y = -3$ that are 10 units from the point (7, 5). **5.** 13 **7.** $\sqrt{10} \approx 3.16$ **9.** 2 or -14
11. about 25.5 yd, 25 yd **13.** 20 **15.** 5 **17.** $4\sqrt{5} \approx 8.94$
19. $\sqrt{41} \approx 6.40$ **21.** $\frac{10}{3} \approx 3.33$ **23.** $\frac{13}{10}$ or 1.30
25. $2\sqrt{14} \approx 7.48$ **27.** 1 or 7 **29.** -2 or 4 **31.** -10 or 4
33. two; $AB = BC = 10$ **35.** 3 **37.** 109 mi **39.** Yes; it will take her about 10.6 minutes to walk between the two buildings. **41.** Minneapolis-St. Cloud, 53 mi; St. Paul-Rochester, 64 mi, Minneapolis-Eau Claire, 79 mi; Duluth-St. Cloud, 118 mi **43.** Compare the slopes of the two potential legs to determine whether the slopes are negative reciprocals of each other. You can also compute the lengths of the three sides and determine whether the square of the longest side length is equal to the sum of the squares of the other two side lengths. Neither test holds true in this case because the triangle is not a right triangle. **45.** B **47.** 25
49. 3 **51.** 11 **53.** {2, 10} **55.** Asia, 1.113×10^{12}; Europe, 1.016×10^{12}; U.S./Canada, 8.84×10^{11}; Latin America, 2.41×10^{11}; Middle East, 1.012×10^{11}; Africa, 5.61×10^{10}.

57. $\{m \mid m \geq 9\}$
2 3 4 5 6 7 8 9 10

59. $\{x \mid x \leq -3\}$
−8 −7 −6 −5 −4 −3 −2 −1 0

61. $\{r \mid r \geq 9.1\}$
2 3 4 5 6 7 8 9 10

63. 6 **65.** 12 **67.** 1

Pages 618–621 Lesson 11-6

1. If the measures of the angles of one triangle equal the measures of the corresponding angles of another triangle, and the lengths of the sides are proportional, then the two triangles are similar. **3.** Consuela; the arcs indicate which angles correspond. The vertices of the triangles are written in order to show the corresponding parts. **5.** Yes; the angle measures are equal. **7.** $b = 15, d = 12$ **9.** $d = 10.2, e = 9$
11. Yes; the angle measures are equal. **13.** No; the angle measures are not equal. **15.** No; the angle measures are not equal. **17.** $\ell = 12, m = 6$ **19.** $k = \frac{55}{6}, \ell = \frac{22}{3}$
21. $k = 3, o = 8$ **23.** $k = 2.8, m = 3.6$ **25.** always **27.** $3\frac{1}{3}$ in.
29. 8 **31.** about 53 ft **33.** Yes; all circles are similar because they have the same shape. **35.** 4:1; The area of the first is πr^2 and the area of the other is $\pi(2r)^2 = 4\pi r^2$. **37.** D
39. 5 **41.** $\sqrt{26} \approx 5.1$ **43.** Yes; $25^2 + 60^2 = 65^2$. **45.** Yes; $49^2 + 168^2 = 175^2$. **47.** $3x^2 - 7x + 1$ **49.** $-3x^2 + 6x + 3$
51. (3, −2) **53.** (1.5, 0) **55.** about −0.044 **57.** $-\frac{5}{6}$ or $-0.8\overline{3}$
59. $\frac{9}{5}$ or 1.8 **61.** $-\frac{1}{3}$ or $-0.\overline{3}$

Page 621 Practice Quiz 2

1. 50 **3.** $2\sqrt{5} \approx 4.47$ **5.** $\sqrt{306} \approx 17.49$ **7.** $2\sqrt{2} \approx 2.83$
9. $a = 20, c = 15$

Pages 627–630 Lesson 11-7

1. If you know the measure of the hypotenuse, use sine or cosine, depending on whether you know the measure of the adjacent side or the opposite side. If you know the measures of the two legs, use tangent. **3.** They are equal.
5. $\sin Y = 0.3846$, $\cos Y = 0.9231$, $\tan Y = 0.4167$ **7.** 0.2588
9. 80° **11.** 18° **13.** 22° **15.** $\angle A = 60°$, $AC = 21$ in., $BC \approx 36.4$ in. **17.** $\angle B = 35°$, $BC = 5.7$ in., $AB = 7.0$ in.
19. $\sin R = 0.6$, $\cos R = 0.8$, $\tan R = 0.75$ **21.** $\sin R = 0.7241$, $\cos R = 0.6897$, $\tan R = 1.05$ **23.** $\sin R = 0.5369$, $\cos R = 0.8437$, $\tan R = 0.6364$ **25.** 0.5 **27.** 0.7071
29. 0.6249 **31.** 2.3559 **33.** 0.9781 **35.** 40° **37.** 62°
39. 33° **41.** 12° **43.** 39° **45.** 51° **47.** 36° **49.** 37°
51. 56° **53.** $\angle A = 63°$, $AC \approx 9.1$ in., $BC \approx 17.8$ in.
55. $\angle B = 50°$, $AC \approx 12.3$ ft, $BC \approx 10.3$ ft **57.** $\angle B = 52°$, $AC \approx 30.7$ in., $AB \approx 39$ in. **59.** $\angle A \approx 23°$, $\angle B \approx 67°$, $AB = 13$ ft **61.** about 8.1° **63.** about 20.6° **65.** about 2.74 m to 0.7 m **67.** If you know the distance between two points and the angles from these two points to a third point, you can determine the distance to the third point by forming a triangle and using trigonometric ratios. Answers should include the following.

- If you measure your distance from the mountain and the angle of elevation to the peak of the mountain from two different points, you can write an equation using trigonometric ratios to determine its height, similar to Example 5.
- You need to know the altitude of the two points you are measuring.

69. D **71.** $k = 8, o = 13.5$ **73.** −5 or 3 **75.** $4s^3 - 9s^2 + 12s$
77. (11, 3) **79.** (−2, 1)

Pages 632–636 Chapter 11 Study Guide and Review

1. false, $-3-\sqrt{7}$ **3.** true **5.** false, $3x+19=x^2+6x+9$ **7.** false, $\frac{x\sqrt{2xy}}{y}$ **9.** $\frac{2\sqrt{15}}{|y|}$ **11.** $57-24\sqrt{3}$ **13.** $\frac{5\sqrt{21}-3\sqrt{35}}{15}$ **15.** $5\sqrt{3}+5\sqrt{5}$ **17.** $36\sqrt{3}$ **19.** $-6\sqrt{2}-12\sqrt{7}$ **21.** $3\sqrt{2}+3\sqrt{6}$ **23.** $\sqrt{6}-1$ **25.** no solution **27.** $\frac{26}{7}$ **29.** 12 **31.** 34 **33.** $\sqrt{115}\approx 10.72$ **35.** 24 **37.** no **39.** yes **41.** 17 **43.** $\sqrt{205}\approx 14.32$ **45.** $\sqrt{137}\approx 11.70$ **47.** 5 or −1 **49.** 10 or −14 **51.** $d=\frac{45}{8}, e=\frac{27}{4}$ **53.** $b=\frac{44}{3}, d=6$ **55.** 0.5283 **57.** 0.8491 **59.** 1.6071 **61.** 39° **63.** 12° **65.** 27°

Chapter 12 Rational Expressions and Equations

Page 641 Chapter 12 Getting Started

1. $-\frac{63}{16}$ **3.** 5 **5.** 4.62 **7.** 10.8 **9.** 6 **11.** $4m^2n$ **13.** $3c^2d(1-2d)$ **15.** $(x+3)(x+8)$ **17.** $(2x+7)(x-3)$ **19.** −1 **21.** $-\frac{149}{6}$ **23.** $\frac{31}{7}$ **25.** 8, −7

Pages 645–647 Lesson 12-1

1. Sample answer: $xy=8$ **3.** b; Sample answer: As the price increases, the number purchased decreases.

5. $xy=12$

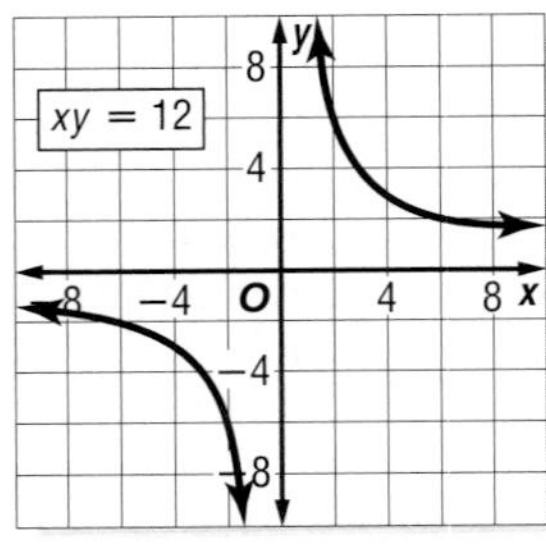

7. $xy=24; 4$

9. $xy=8; \frac{1}{4}$

11. $xy=-192$

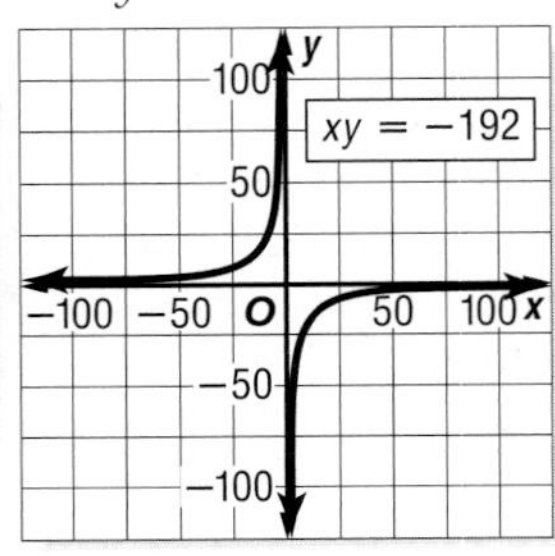

13. $xy=75$

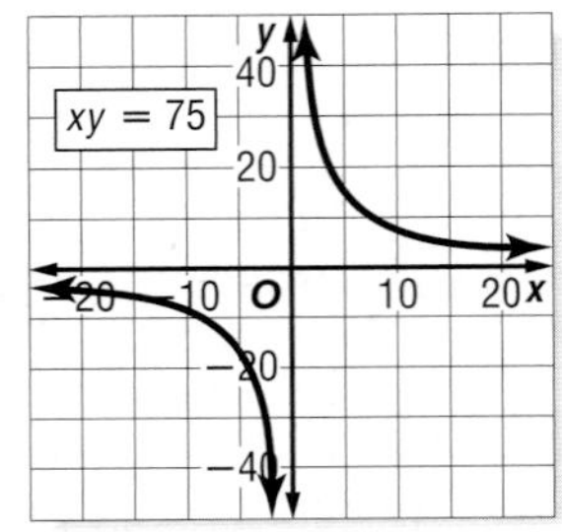

15. $xy=72$

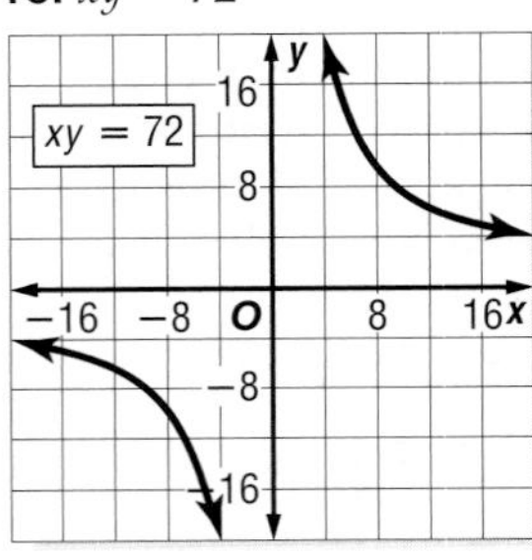

17. $xy=60; 20$ **19.** $xy=-8.5; 8.5$ **21.** $xy=28.16; 8.8$ **23.** $xy=16; \frac{16}{7}$ **25.** $xy=\frac{14}{3}; \frac{2}{3}$ **27.** $xy=26.84; 8.3875$ **29.** 8 in. **31.** 7.2 h **33.** about 37 min **35.** 20 m³ **37.** 24 kg **39.** It is one third of what it was. **41.** B **43.** 41° **45.** 73° **47.** $a=6, f=14$ **49.** −9

51.

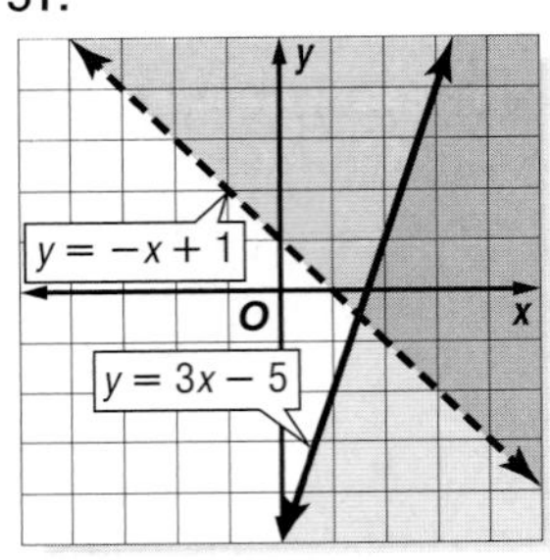

53.

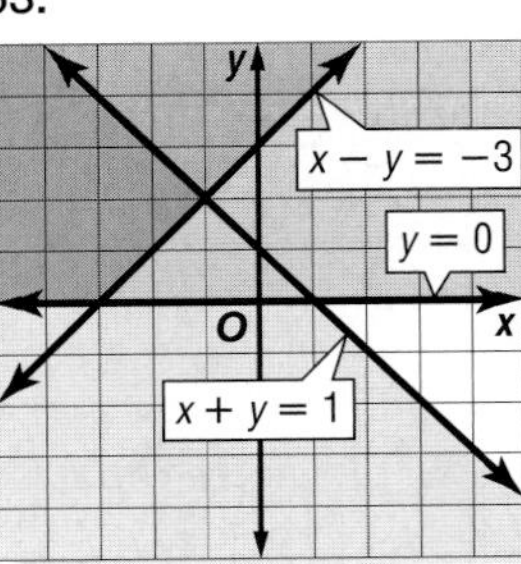

55. 3 **57.** 30 **59.** $6xy^2$

Pages 651–653 Lesson 12-2

1. Sample answer: Factor the denominator, set each factor equal to 0, and solve for x. **3.** Sample answer: You need to determine excluded values before simplifying. One or more factors may have been canceled in the denominator. **5.** −3 **7.** $\frac{4}{5xy}$; 0, 0 **9.** $\frac{1}{x+4}$; −4 **11.** $\frac{a+6}{a+4}$; −4, 2 **13.** $\frac{b+1}{b-9}$; 4, 9 **15.** $\frac{4}{9+2g}$ **17.** −5 **19.** −5, 5 **21.** −5, 3 **23.** −7, −5 **25.** $\frac{a^2}{3b}$; 0, 0 **27.** $\frac{3x}{8z}$; 0, 0, 0 **29.** $\frac{mn}{12n-4m}$; $m\neq 3n$, 0, 0 **31.** $z+8$; −2 **33.** $\frac{2}{y+5}$; −5, 2 **35.** $\frac{a+3}{a+9}$; −9, 3 **37.** $\frac{(b+4)(b-2)}{(b-4)(b-16)}$; 4, 16 **39.** $\frac{n-2}{n(n-6)}$; 0, 6 **41.** $\frac{3}{4}$; −2, −1 **43.** about 29 min **45.** The times are not doubled; the difference is 12 minutes. **47.** 42.75 **49.** $450+4n$ **51.** 41 **53.** $\frac{\pi x^2}{4x^2}$ or $\frac{\pi}{4}$ **55a.** Sample answer: The graphs appear to be identical because the second equation is the simplified form of the first equation. **55b.** Sample answer: The first graph has a hole at $x=4$ because it is an excluded value of the equation. **57.** C **59.** $xy=60$; −5 **61.** $xy=-7.5$; 0.9375 **63.** 71° **65.** 45° **67.** 7 **69.** 6 **71.** 1536, 6144, 24,576 **73.** $\frac{81}{64}, \frac{243}{256}, \frac{729}{1024}$ **75.** 7 **77.** 15,300 **79.** 72

Pages 657–659 Lesson 12-3

1. Sample answer: $\frac{2}{1}, \frac{1}{x}$ **3.** Amiri; sample answer: Amiri correctly divided by the GCF. **5.** $\frac{2t}{s}$ **7.** $2(x+2)$ **9.** $\frac{x+3}{5}$ **11.** $1\frac{2}{3}$ days **13.** $\frac{2}{n}$ **15.** $\frac{12ag}{5b}$ **17.** $\frac{n-4}{n+4}$ **19.** $\frac{(x-1)(x+7)}{(x-7)(x+1)}$ **21.** $\frac{y-2}{y-1}$ **23.** $\frac{x-6}{(x+8)(x+2)}$ **25.** $\frac{2}{n(n+3)}$ **27.** $\frac{(a-3)(a+3)}{(a-4)(a+2)}$ **29.** about 16.67 m/s **31.** 20 yd³ **33.** about \$16.02 **35.** 3 lanes $\cdot \frac{13 \text{ miles}}{1 \text{ lane}} \cdot \frac{5280 \text{ feet}}{1 \text{ mile}} \cdot \frac{1 \text{ vehicle}}{30 \text{ feet}}$ **37.** 5.72 h **39.** Sample answer: Multiply rational expressions to perform dimensional analysis. Answers should include the following.

- 25 lights $\cdot\, h$ hours $\cdot \frac{60 \text{ watts}}{\text{light}} \cdot \frac{1 \text{ kilowatt}}{1000 \text{ watts}} \cdot \frac{15 \text{ cents}}{1 \text{ kilowatt hour}} \cdot \frac{1 \text{ dollar}}{100 \text{ cents}}$
- Sample answer: converting units of measure

41. A **43.** −5, 2 **45.** $xy=72$; 12 **47.** $xy=-192$; −48 **49.** -7^3 or −343 **51.** $\frac{4b^4c^5}{a^3}$ **53.** $\{r \mid r\geq 2.1\}$ **55.** 11 days **57.** $(n+8)(n-8)$ **59.** $(a+7)(a-5)$ **61.** $3x(x-2)(x-6)$

Page 659 Practice Quiz 1

1. $xy = 196$

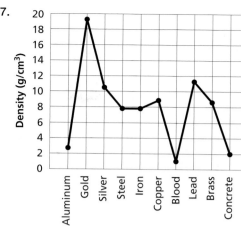

3. $\frac{4a}{7b}$ **5.** $\frac{b+1}{b-9}$ **7.** $3m^2$ **9.** $\frac{4}{5(n+5)}$

Pages 662–664 Lesson 12-4

1. Sample answer: $\frac{15z}{4y^2} \div \frac{3x}{4y}$ **3.** Sample answer: Divide the density by the given volume, then perform dimensional analysis. **5.** $\frac{2a}{a+7}$ **7.** $\frac{2}{x+5}$ **9.** $\frac{2(x-2)(x+3)}{(x+1)(x+9)}$ **11.** $\frac{2}{9}$ lb/in^2 **13.** ab **15.** $\frac{x}{2y^2}$ **17.** $\frac{sy^2}{z^2}$ **19.** $\frac{b+3}{4b}$ **21.** $\frac{3k}{(k+1)(k-2)}$ **23.** $\frac{3(x+4)}{4(2x-9)}$ **25.** 648 **27.** 225 **29.** $x+3$ **31.** $\frac{(x+1)(x-1)}{2}$ **33.** $\frac{3(a+4)}{2(a-3)}$ **35.** $\frac{x+4}{x+3}$ **37.** about 9.2 mph **39.** $n = 20{,}000 \text{ yd}^3 \div \left(\frac{5 \text{ ft}(18 \text{ ft} + 15 \text{ ft})}{2} \cdot 9 \text{ ft} \cdot \frac{1 \text{ yd}^3}{27 \text{ ft}^3}\right)$; $727.\overline{27}$ **41.** 63.5 mph **43.** $\left(x - \frac{1}{2}\right)\left(x - \frac{3}{4}\right)(x)$ **45.** Sample answer: Divide the number of cans recycled by $\frac{5}{8}$ to find the total number of cans produced. Answers should include the following.

- $x = 63{,}900{,}000 \text{ cans} \div \frac{5}{8} \cdot \frac{1 \text{ pound}}{33 \text{ cans}}$

47. C **49.** $\frac{x-2}{x+2}$ **51.** $\frac{7(x+2y)(x+5)}{x+y}$ **53.** $-\frac{x+5}{x+6}$ **55.** $\frac{n+4}{n-4}$ **57.** $\left\{\frac{4}{3}\right\}$ **59.** $\{-6 \pm \sqrt{14}\}$ **61.** 3 **63.** $\{g \mid g \geq 7.5\}$ **65.** $\{x \mid x \geq -0.7\}$ **67.** $\left\{r \mid r < -\frac{1}{20}\right\}$ **69.** 39,000 covers **71.** $\frac{m^3}{5}$ **73.** $\frac{b^3}{c^3}$ **75.** $\frac{7x^4}{z}$

Pages 669–671 Lesson 12-5

1. b and c **3.** Sample answer: $x^3 + 2x^2 + 8$; $x^3 + 2x^2 + 0x + 8$ **5.** $2 + \frac{5}{a} + \frac{2}{7b^2}$ **7.** $r + 3 + \frac{9}{r+9}$ **9.** $b + 2 - \frac{3}{2b-1}$ **11.** $\frac{x}{3} + 3 - \frac{7}{3x}$ **13.** $3s - \frac{5}{t} + \frac{8t}{s^2}$ **15.** $x + 4$ **17.** $n - 7$ **19.** $z - 9 + \frac{33}{z+7}$ **21.** $2r + 7$ **23.** $t + 6$ **25.** $3x^2 + 2x - 3 - \frac{1}{x+2}$ **27.** $3x^2 + \frac{6}{2x-3}$ **29.** $3n^2 - 2n + 3 + \frac{3}{2n+3}$ **31.** $\frac{150(60-x)}{x}$ **33.** 3 rolls **35.** 5/$1.02, 10/$0.93, 16/$0.82; 18-inch

37.

39. $2w + 4$ **41.** 12 **43.** Sample answer: Division can be used to find the number of pieces of fabric available when you divide a large piece of fabric into smaller pieces. Answers should include the following.

- The two expressions are equivalent. If you use the Distributive Property, you can separate the numerator into two expressions with the same denominator.
- When you simplify the right side of the equation, the numerator is $a - b$ and the denominator is c. This is the same as the expression on the left.

45. B **47.** $\frac{m+4}{m+1}$ **49.** $\frac{1}{z+6}$ **51.** $10\sqrt{2}$ **53.** $(d+5)(d-8)$ **55.** prime **57.** $4m^3 + 6n^2 - n$ **59.** $-2a^3 - 2a^2b + b^2 - 3b^3$

Pages 674–677 Lesson 12-6

1. Sample answer: $\frac{x+6}{x+2} + \frac{x-4}{x+2} = 1$ **3.** Sample answer: Two rational expressions whose sum is 0 are additive inverses, while two rational expressions whose difference is 0 are equivalent expressions. **5.** $\frac{a}{2}$ **7.** $\frac{3-n}{n-1}$ **9.** $-\frac{a}{6}$ **11.** $\frac{3m+6}{m-2}$ **13.** $\frac{1}{10}$ **15.** z **17.** $n - 1$ **19.** 3 **21.** $\frac{n-3}{n+3}$ **23.** $\frac{3a+1}{a-4}$ **25.** $\frac{14b+7}{2b+6}$ **27.** $\frac{22x+7}{2x+5}$ **29.** $\frac{2n}{3}$ **31.** $\frac{1}{3}$ **33.** $\frac{10}{z-2}$ **35.** $\frac{4-7m}{7m-2}$ **37.** $\frac{10y}{y-3}$ **39.** 3 **41.** $\frac{4b-23}{2b+12}$ **43.** $\frac{60}{n}$ **45.** $\frac{1}{7.48}$ ft^3 **47.** $\frac{x}{16}, \frac{x}{18}, \frac{x}{24}$ **49.** c **51.** A **53.** $x^2 + 2x - 3$ **55.** $\frac{b+3}{4b}$ **57.** $(a+7)(a+2)$ **59.** $(y-4z)(y-7z)$ **61.** $7x^2 - 3x + 22$ **63.** 36 **65.** 24 **67.** 30 **69.** 400 **71.** 144

Page 677 Practice Quiz 2

1. $\frac{a}{a+11}$ **3.** $\frac{x-1}{x+5}$ **5.** $x - 5 - \frac{1}{2x+3}$ **7.** $\frac{7}{x+7}$ **9.** $\frac{3x}{3x+2}$

Pages 681–683 Lesson 12-7

1. Sample answer: To find the LCD, determine the least common multiple of all of the factors of the denominators. **3.** Sample answer: $\frac{x}{2x+6}, \frac{5}{x+3}$ **5.** $6(x-2)$ **7.** $\frac{12x+7}{10x^2}$ **9.** $\frac{y^2+12y+25}{(y-5)(y+5)}$ **11.** $\frac{2z-wz}{4w^2}$ **13.** $\frac{4}{(b-4)(b+4)}$ **15.** C **17.** $21x^2y$ **19.** $(2n-5)(n+2)$ **21.** $(p+1)(p-6)$ **23.** $\frac{2+7a}{a^3}$ **25.** $\frac{15m+28}{35m^2}$ **27.** $\frac{n^2+12}{(n+4)(n-3)}$ **29.** $\frac{7x^2+3x}{(x-3)(x+1)}$ **31.** $\frac{1}{3}$ **33.** $\frac{7y+39}{(y+3)(y-3)}$

35. $\frac{3x^2 + 6x + 6}{(x + 4)(x - 1)^2}$ 37. $\frac{a^3 - a^2b + a^2 + ab}{(a + b)(a - b)^2}$ 39. $\frac{4 - 25x}{15x^2}$

41. $\frac{5ax - a}{7x^2}$ 43. $\frac{k^2 - 6k - 15}{(k + 5)(k - 3)}$ 45. $\frac{2m^2 - m - 9}{(m + 1)(2m + 5)}$

47. $\frac{-3a + 6}{a(a - 6)}$ 49. $\frac{3a + 5}{-3(a - 2)}$ 51. $\frac{4a^2 + 2a + 4}{(a + 4)(a + 1)(a - 1)}$

53. $\frac{-m^3 - 11m^2 - 56m - 48}{(m - 4)(m + 4)^2}$ 55. 12 mi; \$30 57. 66,000 mi

59. Sample answer: You can use rational expressions and their least common denominators to determine when elections will coincide. Answers should include the following.

- Use each factor of the denominators the greatest number of times it appears.
- 2012

61. C 63. $\frac{4x + 5}{2x + 3}$ 65. $b + 10$ 67. $2m - 3 + \frac{2}{2m + 7}$

69. $(5r - 3)(r + 2)$ 71. \$54.85 73. $\frac{ab}{2}$ 75. $\frac{1}{4n}$ 77. $\frac{x + 4}{x + 6}$

Pages 686–689 Lesson 12-8

1. Sample answer: Both mixed numbers and mixed expressions are made up by the sum of an integer or monomial and a fraction or rational expression. 3. Bolton; Lian omitted the factor $(x + 1)$. 5. $\frac{42y + 5}{6y}$ 7. $\frac{14}{19}$

9. $\frac{a - b}{x + y}$ 11. $\frac{8n + 3}{n}$ 13. $\frac{2xy + x}{y}$ 15. $\frac{2m^2 - m - 4}{m}$

17. $\frac{b^3 + ab^2 + a - b}{a + b}$ 19. $\frac{5n^3 - 15n^2 - 1}{n - 3}$ 21. $\frac{x^2 - 7x + 17}{x - 3}$

23. $\frac{3}{4}$ 25. $\frac{1}{ab^2}$ 27. $\frac{y^2(x + 4)}{x^2(y - 2)}$ 29. $\frac{1}{y + 4}$ 31. $\frac{n + 2}{n + 3}$

33. $\frac{(x + 3)(x - 1)}{(x - 2)(x + 4)}$ 35. $\frac{a(b^2 + 1)}{b(a^2 + 1)}$ 37. 60 39. 404.60 cycles/s

41. $66\frac{2}{3}$ lb/in^2

43. Sample answer: Most measurements used in baking are fractions or mixed numbers, which are examples of rational expressions. Answers should include the following.

- You want to find the number of batches of cookies you can make using the 7 cups of flour you have on hand when a batch requires $1\frac{1}{2}$ cups of flour.
- Divide the expression in the numerator of a complex fraction by the expression in the denominator.

45. C 47. $\frac{3a^2 + 3ab - b^2}{(a - b)(2b + 3a)}$ 49. $\frac{2n^2 - 8n - 2}{(n - 2)^2(n + 3)}$ 51. $\frac{1}{x - 3}$

53. $\frac{2}{n + 6}$ 55. $\{\pm 4\}$ 57. $\{-5, -3, 3\}$ 59. about 2.59×10^0

61. $C = 0.16m + 0.99$ 63. -48 65. 16 67. -14.4

Pages 693–695 Lesson 12-9

1. Sample answer: When you solve the equation, $n = 1$. But $n < 1$, so the equation has no solution. 3. Sample answer: $\frac{x}{4} = 0$ 5. -13 7. $\frac{5}{4}$ 9. $-1, \frac{2}{5}$ 11. 8 13. 3 15. -3

17. 0 19. $\frac{1}{2}$ 21. -3 23. 1 25. $-2, 1$ 27. 7 29. 9

31. about 0.82 mi 33. 600 ft^3 35. $-\frac{14}{3}$ 37. A 39. $\frac{x + 1}{x - 2}$

41. $\frac{x + 1}{x + 5}$ 43. $\frac{1}{y^2 - 2y + 1}$ 45. $4(5x - 2y)$

47. $(2p + 5)(5p - 6)$

Pages 696–700 Chapter 12 Study Guide and Review

1. false, rational 3. true 5. false, $x^2 - 144$ 7. $xy = 1176$; 21

9. $xy = 144$; 48 11. $\frac{x}{4y^2z}$ 13. $\frac{a - 5}{a - 2}$ 15. $\frac{14a^2b}{3}$ 17. $\frac{30}{x - 10}$

19. $\frac{(x + 4)^2}{(x + 2)^2}$ 21. $2p$ 23. $\frac{3}{(y + 4)(y - 2)}$ 25. $2ac^2 - 4a^2c + \frac{3c^2}{b}$

27. $x^2 + 2x - 3$ 29. $\frac{2m + 3}{5}$ 31. $a + b$ 33. 2 35. $\frac{4c^2 + 9d}{6cd^2}$

37. $\frac{8d^2 - 7a}{(a - 2)(a + 1)}$ 39. $\frac{14a - 3}{6a^2}$ 41. $\frac{5x - 8}{x - 2}$ 43. $\frac{4x^2 - 2y^2}{x^2 - y^2}$

45. $\frac{20a + 16}{2a^2 - 3a}$ 47. -5 49. $-\frac{1}{4}$ 51. -1; extraneous 0

Chapter 13 Statistics

Page 707 Chapter 13 Getting Started

1. Sample answer: If $a = 5$ and $b = -2$, then $c = 3$. However, $5 > 3$. 3. Sample answer: The speed limit could be 55 mph, and Tara could be driving 50 mph. 5. 15

7. 375

9. 15 16 17 18 19 20 21 22 23

11. 1 2 3 4 5

Pages 710–713 Lesson 13-1

1. All three are unbiased samples. However, the methods for selecting each type of sample are different. In a simple random sample, a sample is as likely to be chosen as any other from the population. In a stratified random sample, the population is first divided into similar, nonoverlapping groups. Then a simple random sample is selected from each group. In a systematic random sample, the items are selected according to a specified time or item interval.

3. Sample answer: Ask the members of the school's football team to name their favorite sport. 5. work from 4 students; work from all students in the 1st period math class; biased; voluntary response 7. 12 pencils; all pencils in the school store; biased; convenience 9. 20 shoppers; all shoppers; biased; convenience 11. 860 people from a state; all people in the state; unbiased; stratified 13. 3 students; all of the students in Ms. Finchie's class; unbiased; simple 15. a group of U.S. district court judges; all U.S. district court judges; unbiased; stratified 17. 4 U.S. Senators; all U.S. Senators; biased; convenience 19. a group of high-definition television sets; all high-definition television sets manufactured on one line during one shift; unbiased; systematic 21. a group of readers of a magazine; all readers of the magazine; biased; voluntary response

23. Additional information needed includes how the survey was conducted, how the survey respondents were selected, and the number of respondents. 25. Sample answer: Get a copy of the list of registered voters in the city and call every 100th person. 27. Sample answer: Randomly pick 5 rows from each field of tomatoes and then pick a tomato every 50 ft along each row. 29. It is a good idea to divide the school population into groups and to take a simple random sample from each group. The problem that prevents this from being a legitimate stratified random sample is the way the three groups are formed. The three groups probably do not represent all students. The students who do not participate in any of these three activities will not be represented in the survey. Other students may be involved in two or three of these activities. These students will be more likely to be chosen for the survey. 31. B 33. $3\frac{1}{3}$ 35. $\frac{3}{25}$ 37. $\frac{a + 5}{a + 12}$ 39. $22\sqrt{6}$ cm

41. $-1\frac{2}{3}, -1\frac{1}{2}$ 43. $y^2 + 12y + 35$ 45. $x^2 - 4x - 32$

47. 24.11 49. 3.8 51. 12.45

Pages 717–721 Lesson 13-2

1°. A 2-by-4 matrix has 2 rows and 4 columns, and a 4-by-2 matrix has 4 rows and 2 columns. 3. Estrella; Hiroshi did not multiply each element of the matrix by −5. **5.** 1 by 4; first row, first column **7.** 3 by 2; first row, second column **9.** $\begin{bmatrix} -5 & 24 \\ -22 & -13 \end{bmatrix}$ **11.** [20 −28] **13.** No; the corresponding elements are not equal. **15.** the total sales for the weekend **17.** 2 by 2; first row, first column **19.** 3 by 1; third row, first column **21.** 3 by 3; second row, third column **23.** 2 by 3; second row, third column

25. $\begin{bmatrix} 2 & 1 & 1 \\ 1 & 5 & 1 \end{bmatrix}$ **27.** $\begin{bmatrix} -13 & 12 & -7 \\ 5 & 6 & 11 \\ 23 & 18 & 14 \end{bmatrix}$

29. $\begin{bmatrix} 86 & 82 & -7 \\ 130 & 87 & 15 \end{bmatrix}$ **31.** $\begin{bmatrix} -5 & 25 & 45 \\ 0 & -20 & -10 \\ 15 & 35 & 30 \end{bmatrix}$

33. impossible **35.** $\begin{bmatrix} -25 & 19 & -23 \\ 10 & 16 & 24 \\ 43 & 29 & 22 \end{bmatrix}$

37. $\begin{bmatrix} 224 & 155 & -84 \\ 309 & 182 & -15 \end{bmatrix}$ **39.** $V = [70 \quad 2 \quad 2 \quad 0.3]$, $S = [160\,0 \quad 0 \quad 0]$, $C = [185 \quad 2 \quad 11 \quad 3.9]$ **41.** [555 16 19 5.8] **43.** 1.20

45. $A = \begin{bmatrix} 533 & 331 & 4135 & 26 & 15 \\ 515 & 304 & 3840 & 24 & 14 \\ 499 & 325 & 4353 & 41 & 13 \\ 571 & 343 & 4436 & 36 & 15 \end{bmatrix}$,

$B = \begin{bmatrix} 571 & 357 & 4413 & 33 & 15 \\ 473 & 284 & 3430 & 28 & 11 \\ 347 & 235 & 3429 & 21 & 18 \\ 533 & 324 & 3730 & 19 & 18 \end{bmatrix}$

47. $T = \begin{bmatrix} 1104 & 688 & 8548 & 59 & 30 \\ 988 & 588 & 7270 & 52 & 25 \\ 846 & 560 & 7782 & 62 & 31 \\ 1104 & 667 & 8166 & 55 & 33 \end{bmatrix}$

49a. sometimes **49b.** always **49c.** sometimes **49d.** sometimes **49e.** sometimes **49f.** sometimes **51.** C

53. $\begin{bmatrix} 0.7 & -0.4 & 2.3 \\ -1.6 & -4 & -2.4 \end{bmatrix}$ **55.** $\begin{bmatrix} -5.3 & -12.4 & 21.1 \\ 2.4 & -7.7 & 4 \end{bmatrix}$

57. $\begin{bmatrix} 3.92 & -0.48 & 2.08 \\ -3.12 & 2.04 & -3.6 \end{bmatrix}$ **59.** biased; convenience

61. $\frac{3}{5}$ **63.** 324 **65.** 64 **67.** $(a - b)(a + 3b)$ **69.** Sample answer: Megan saved steadily from January to June. In July, she withdrew money to go on vacation. She started saving again in September. Then in November, she withdrew money for holiday presents.

Page 721 Practice Quiz 1

1. half of the households in a neighborhood; all households in the neighborhood; unbiased; systematic **3.** $\begin{bmatrix} -3 & -4 \\ -5 & -9 \end{bmatrix}$

5. $\begin{bmatrix} 24 & -9 & -12 & 15 \\ 18 & -3 & 6 & 30 \end{bmatrix}$

Pages 725–728 Lesson 13-3

1. First identify the greatest and least values in the data set. Use this information to determine appropriate measurement classes. Using these measurement classes, create a frequency table. Then draw the histogram. Always remember to label the axes and give the histogram a title. **3.** Sample answer: 1, 1, 2, 4, 5, 5, 8, 9, 10, 11, 12, 13, 22, 24, 41 **5.** There are no gaps. The data are somewhat symmetrical. **7.** The Group A test scores are somewhat more symmetrical in appearance than the Group B test scores. There are 25 of 31 scores in Group A that are 40 or greater, while only 14 of 26 scores in Group B are 40 or greater. Also, Group B has 5 scores less than 30. Therefore, we can conclude that Group A performed better overall on the test. **9.** B

11. 3400–3800 points; There are no gaps. The data appear to be skewed to the right. **13.** Age at inauguration: 50–60 years old; age at death: 60–70 years old; both distributions show a symmetrical shape. The two distributions differ in their spread. The inauguration ages are not spread out as much as the death ages data.

15. Sample answer:

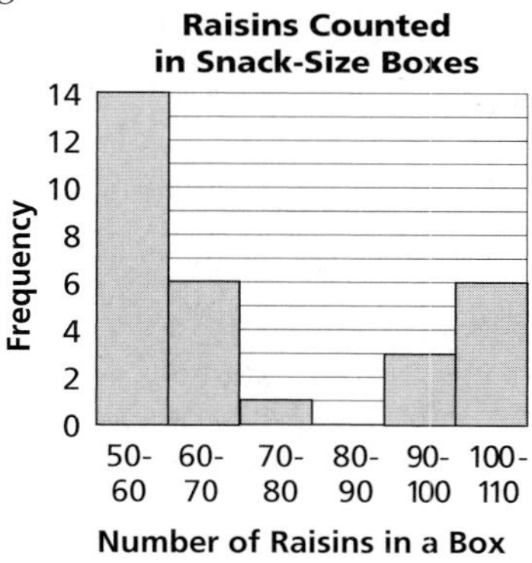

17. Sample answer:

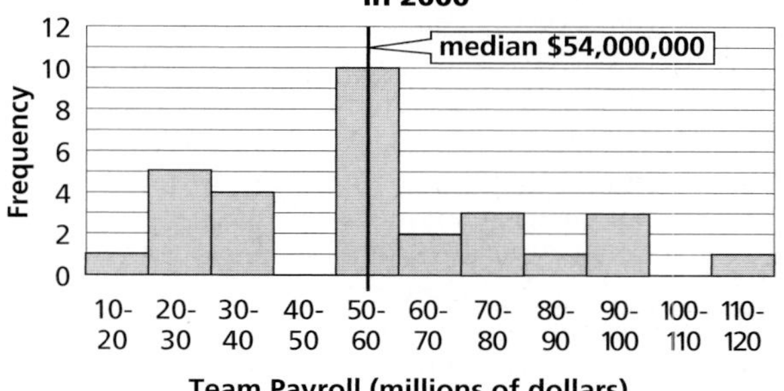

19. Sample answer:

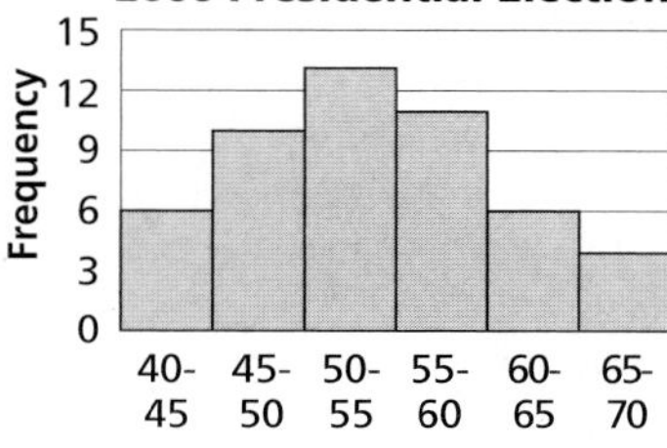

23. Histograms can be used to show how many states have a median within various intervals. Answers should include the following.

- A histogram is more visual than a frequency table and can show trends easily.

Selected Answers

•

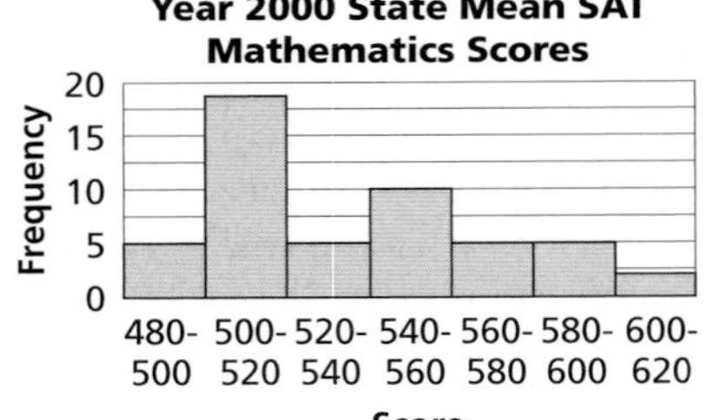

25. B

27. Sample answer: **29.** Sample answer:

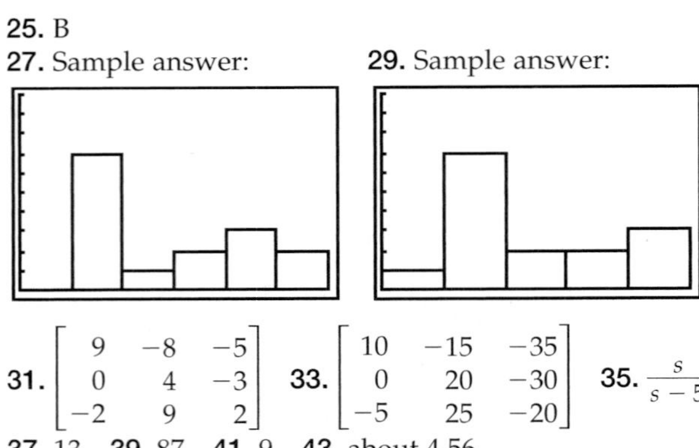

31. $\begin{bmatrix} 9 & -8 & -5 \\ 0 & 4 & -3 \\ -2 & 9 & 2 \end{bmatrix}$ **33.** $\begin{bmatrix} 10 & -15 & -35 \\ 0 & 20 & -30 \\ -5 & 25 & -20 \end{bmatrix}$ **35.** $\frac{s}{s-5}$

37. 13 **39.** 87 **41.** 9 **43.** about 4.56

Pages 733–736 Lesson 13-4

1. Sample answer: 1, 4, 5, 6, 7, 8, 15 and 1, 2, 4, 5, 9, 9, 10 **3.** Alonso; the range is the difference between the greatest and the least values of the set. **5.** 4.6; 9.05; 8.0; 10.05; 2.05; none **7.** 5 runs **9.** 6 runs **11.** 37; 73; 60.5; 79.5; 19; none **13.** 1.1; 30.6; 30.05; 30.9; 0.85; none **15.** 46; 77; 66.5; 86; 19.5; none **17.** 6.7; 7.6; 6.35; 8.65; 2.3; none **19.** 471,561 visitors **21.** 147,066.5 visitors; 470,030 visitors **23.** none **25.** 22.5 Calories **27.** 46 Calories **29.** 1000 ft; 970 ft **31.** 520 ft; 280 ft **33.** Although the range of the cable-stayed bridges is only somewhat greater than the range of the steel-arch bridges, the interquartile range of the cable-stayed bridges is much greater than the interquartile range of the steel-arch bridges. The outliers of the steel-arch bridges make the ranges of the two types of bridges similar, but in general, the data for steel-arch bridges are more clustered than the data for the cable-stayed bridges.

35. Measures of variation can be used to discuss how much the weather changes during the year. Answers should include the following.

- The range of temperatures is used to discuss the change in temperatures for a certain area during the year and the interquartile range is used to discuss the change in temperature during the moderate 50% of the year.
- The monthly temperatures of the local area listed with the range and interquartile range of the data.

37. A **39.** 1 by 3; first row, first column **41.** 2 by 4; second row, second column **43.** $\frac{1}{t-4}$; 3, 4

45. **47.**

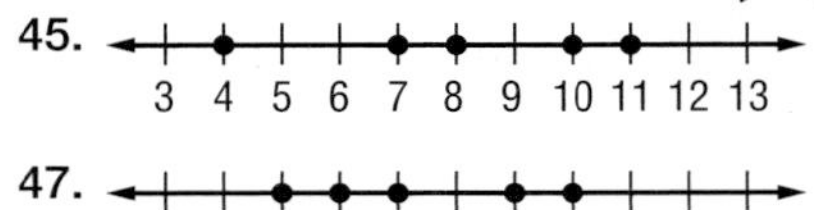

Page 736 Practice Quiz 2

1. \$10–\$20 **3.** 340 **5.** 835

Pages 739–742 Lesson 13-5

1. The extreme values are 10 and 50. The quartiles are 15, 30, and 40. There are no outliers. **3.** Sample answer: 2, 8, 10, 11, 11, 12, 13, 13, 14, 15, 16

5.

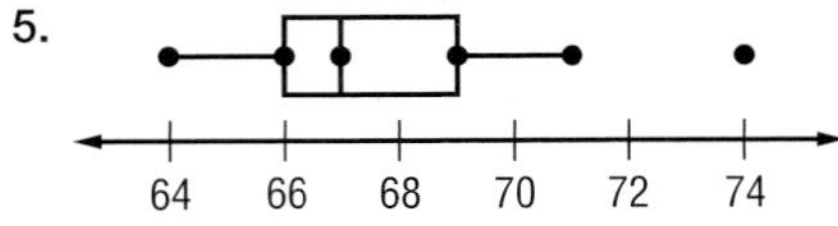

7.

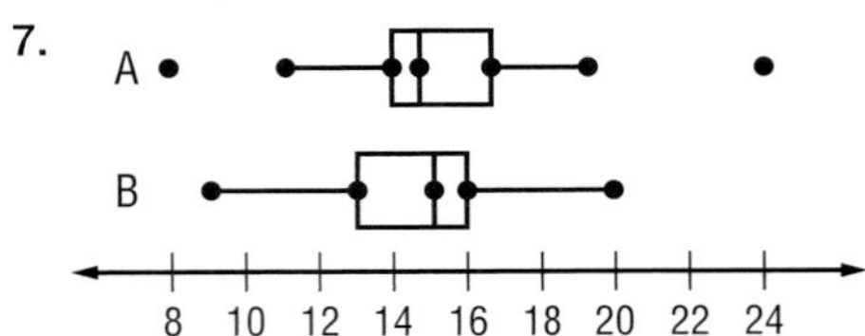

The A data are more diverse than the B data.

9. Most of the data are spread fairly evenly from about \$450 million to \$700 million. The one outlier (\$1397 million) is far removed from the rest of the data. **11.** 30 **13.** $\frac{1}{2}$

15.

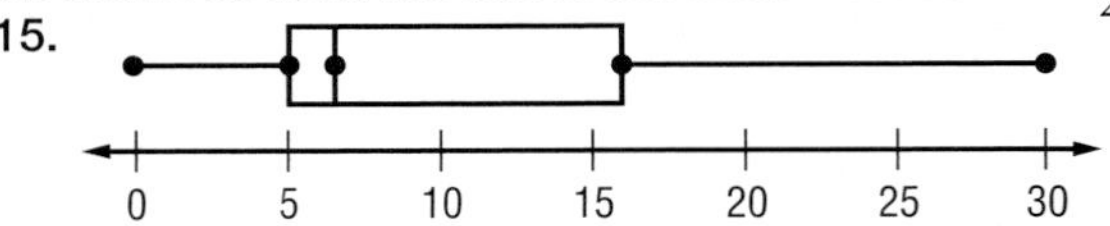

17.

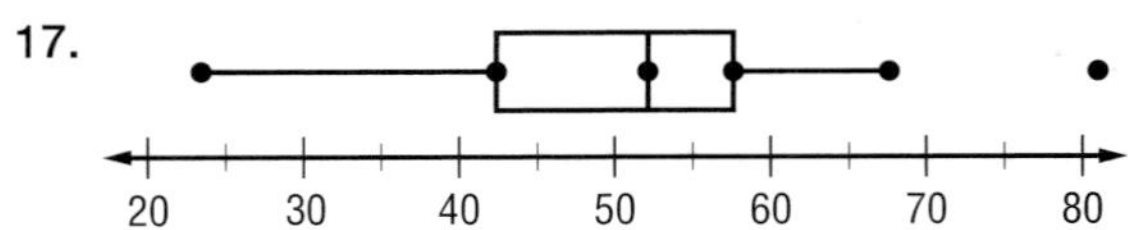

19.

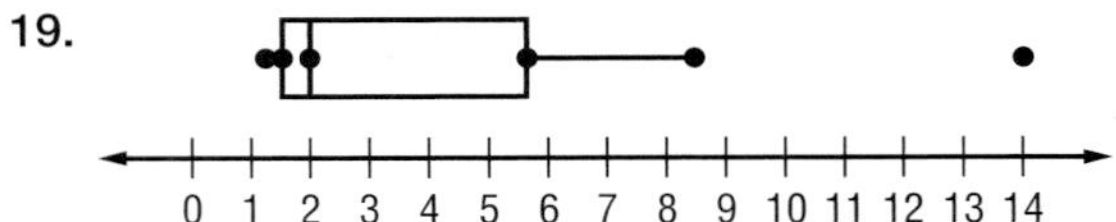

21. B **23.** B

25.

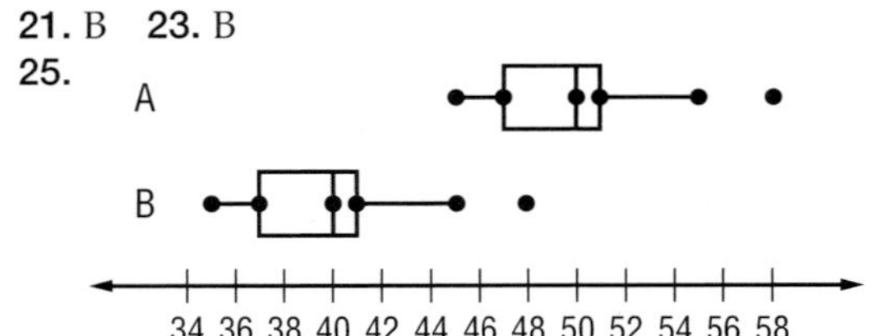

The distribution of both sets of data are similar. In general, the A data are greater than the B data.

27.

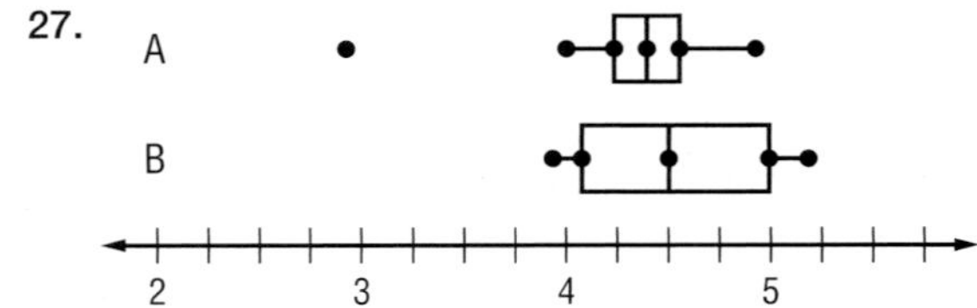

The A data have an outlier. Excluding the outlier, the B data are more diverse than the A data. **29.** The upper half of the data is very dispersed. The range of the lower half of the data is only 1. **31.** Top half; the top half of the data goes from \$48,000 to \$181,000, while the bottom half goes from \$35,000 to \$48,000. **33.** Bottom half; the top half of the data goes from 70 yr to 80 yr, while the bottom half goes from 39 yr to 70 yr. **35.** No; although the interval from 54 yr to 70 yr is wider than the interval from 70 yr to 74 yr, both intervals represent 25% of the data values.

37. Sample answer:

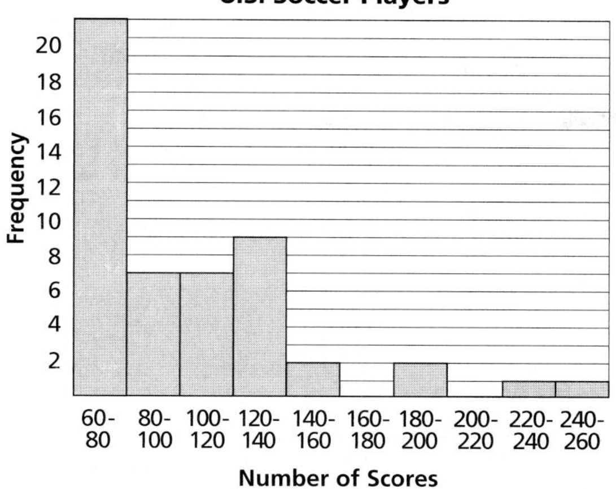

39. Sample answer: 40, 45, 50, 55, 55, 60, 70, 80, 90, 90, 90

41. C **43.** 80; 54.5; 45; 67; 22; none **45.** $\frac{-y^2 + 6y + 12}{(y - 3)(y + 4)}$

47. $\frac{3w - 4}{3(5w + 2)}$ **49.** $3(r + 3)$ **51.** $m\angle B = 51°$, $AB \approx 15.4$, $BC \approx 9.7$ **53.** 1, 6 **55.** −9.8, 1.8 **57.** $8a^2 + 2a - 1$

Pages 745–748 Chapter 13 Study Guide and Review

1. simple random sample **3.** quartile **5.** biased sample **7.** interquartile range **9.** outlier **11.** 8 test tubes with results of chemical reactions; the results of all chemical reactions performed; biased; convenience

13. $\begin{bmatrix} 2 & 4 & -4 \\ 4 & 3 & 3 \\ -2 & -3 & 3 \end{bmatrix}$ **15.** $\begin{bmatrix} -4 & -2 \\ 4 & 0 \end{bmatrix}$ **17.** $\begin{bmatrix} 5 & -1 \\ -1 & 4 \end{bmatrix}$

19. $\begin{bmatrix} 5 & 15 & -5 \\ 10 & 0 & 20 \\ -5 & -5 & 15 \end{bmatrix}$ **21.** $\begin{bmatrix} 9 & 1 \\ -5 & 4 \end{bmatrix}$

23. Sample answer:

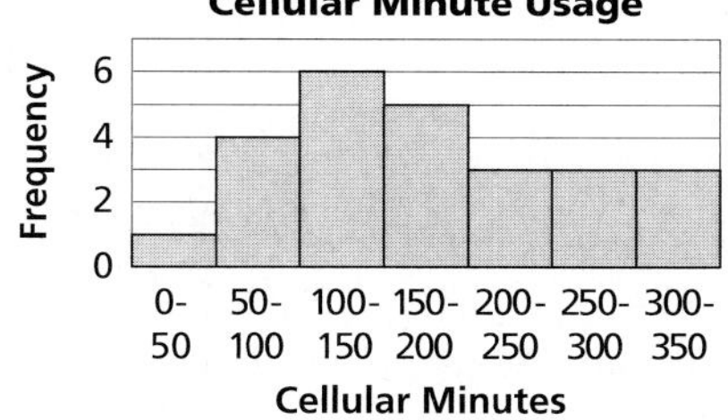

25. 70; 65; 45; 85; 40; none **27.** 37; 73; 62; 77; 15; none

29.

125 175 225 275 325

31.

50 60 70 80 90

Chapter 14 Probability

Page 753 Chapter 14 Getting Started

1. $\frac{3}{7}$ **3.** $\frac{2}{7}$ **5.** $\frac{3}{5}$ **7.** $\frac{7}{95}$ **9.** $\frac{1}{52}$ **11.** 72.5% **13.** 40% **15.** 87.5% **17.** 85.6%

Pages 756–758 Lesson 14-1

1. Sample answer: choosing 2 books from 7 books on a shelf **3.** $5! = 5 \cdot 4 \cdot 3 \cdot 2 \cdot 1$ **5.** 64 **7.** 40,320

9. 27 **English Math Science Outcomes**

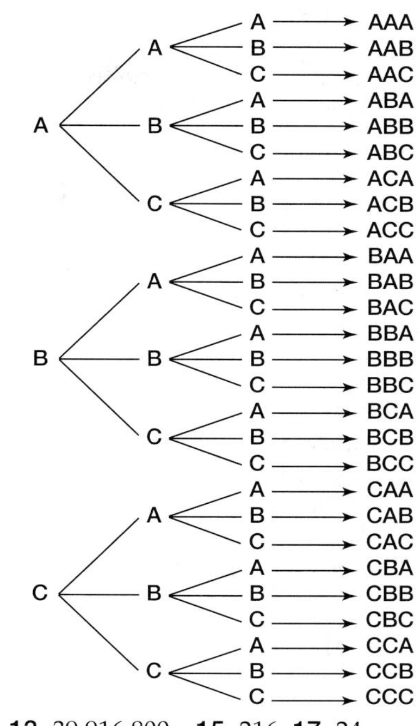

11. 24 **13.** 39,916,800 **15.** 216 **17.** 24

19.

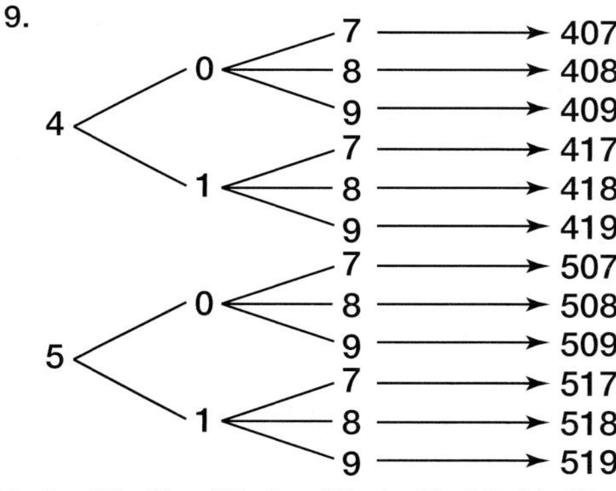

21. 6 **23.** 20 **25.** A **27.** A: 32, 88, 44, 85, 60; B: 38, 86, 48, 74, 64 **29.** B **31.** 79 **33.** 73.5; 39.5; 34.0

35. $\frac{5x^2 + 8x - 6}{(3x - 1)(x - 2)}$ **37.** $\frac{3z - 1}{3z - 6}$ **39.** $\pm\sqrt{22}$ **41.** 7

43. −8.6, 0.6 **45.** −4.7, −0.3 **47.** $\frac{1}{13}$ **49.** $\frac{1}{52}$ **51.** $\frac{2}{13}$

Pages 764–767 Lesson 14-2

1. Sample answer: Order is important in a permutation but not in a combination. Permutation: the finishing order of a race Combination: toppings on a pizza **3.** Alisa; both are correct in that the situation is a combination, but Alisa's method correctly computes the combination. Eric's calculations find the number of permutations. **5.** Permutation; order is important. **7.** 21 **9.** 60 **11.** 720 **13.** B **15.** Permutation; order is important. **17.** Permutation; order is important. **19.** Combination; order is not important. **21.** Combination; order is not important. **23.** 4 **25.** 35 **27.** 125,970 **29.** 524,160

31. 16,598,400 **33.** 6720 **35.** 362,880 **37.** 495 **39.** $\frac{1}{12}$

41. 7776 **43.** 61,425 **45.** 336 **47.** 36 **49.** $\frac{1}{12}$ or about 8%

51. $\frac{1}{30{,}240}$ **53.** 24 **55.** Sample answer: Combinations can be used to show how many different ways a committee can be formed by various members. Answers should include the following.

- Order of selection is not important.
- Order is important due to seniority, so you need to find the number of permutations.

57. C **59.**

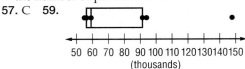

50 60 70 80 90 100 110 120 130 140 150
(thousands)

61. \$56,700, \$91,300 **63.** $\frac{1}{x+3}$ **65.** $\frac{n-5}{n+5}$ **67.** $4\sqrt{29}$, 21.54 **69.** −0.59, −3.41 **71.** 1.69, −1.19 **73.** $\frac{27}{32}$ **75.** $\frac{1}{3}$ **77.** $\frac{69}{100}$

Page 767 Practice Quiz 1

1. 24 **3.** 1287 **5.** $\frac{45}{1001}$

Pages 772–776 Lesson 14-3

1. A simple event is a single event, while a compound event involves two or more simple events. **3.** Sample answer: With dependent events, a first object is selected and not replaced. With independent events, a first object is selected and replaced. **5.** $\frac{10}{147}$ **7.** $\frac{80}{3087}$ **9.** $\frac{1}{2}$ **11.** 1 **13.** independent **15.** $\frac{1}{3}$ **17.** $\frac{2}{51}$ **19.** $\frac{7}{408}$ **21.** $\frac{1}{5}$ **23.** $\frac{1}{10}$ **25.** $\frac{27}{280}$ **27.** $\frac{69}{280}$ **29.** 98% or 0.98 **31.** no; $P(A \text{ and } B) \neq P(A) \cdot P(B)$ **33.** $\frac{9}{16}$ **35.** $\frac{1}{4}$ **37.** 356 **39.** ≈ 0.09 **41.** 1 **43.** $\frac{3}{5}$ **45.** $\frac{7}{8}$ **47.** $\frac{3}{4}$ **49.** 101 **51.** $\frac{39}{40}$ **53.** C **55.** 10 **57.** 604,800 **59.** $\begin{bmatrix} 5 & 2 \\ 4 & -1 \end{bmatrix}$ **61.** $3\sqrt{5}$ **63.** $2b^2\sqrt{10}$ **65.** $18\sqrt{14}$ **67.** 0.375 **69.** 0.492 **71.** 0.222 **73.** 0.033 **75.** 0.036

Pages 779–781 Lesson 14-4

1. The probability of each event is between 0 and 1 inclusive. The probabilities for each value of the random variable add up to 1. **3.** Sample answer: the number of possible correct answers on a 5-question multiple-choice quiz, and the probability of each **5.** $P(X = 4) = \frac{1}{12}$, $P(X = 5) = \frac{1}{9}$, $P(X = 6) = \frac{5}{36}$ **7.** 0.05 + 0.10 + 0.40 + 0.40 + 0.05 = 1 **9.** 0.45 **11.** $P(X = 0) = \frac{1}{64}$, $P(X = 1) = \frac{3}{64}$, $P(X = 2) = \frac{9}{64}$, $P(X = 3) = \frac{27}{64}$ **13.** No; it is more probable to spin blue than red. **15.** 0.10 + 0.15 + 0.40 + 0.25 + 0.10 = 1 **17.** 0.75

19.

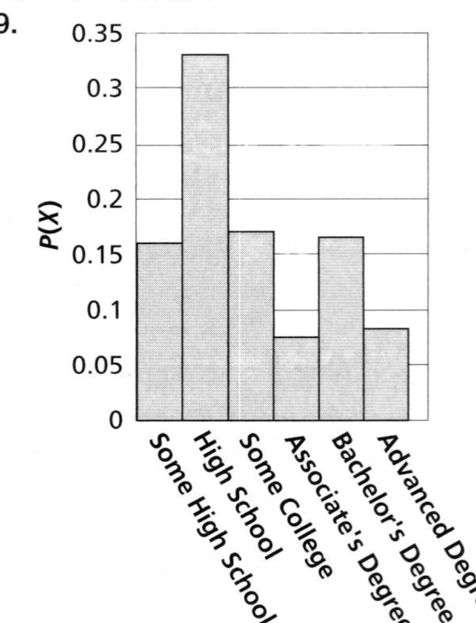

21. No; 0.221 + 0.136 + 0.126 + 0.065 + 0.043 = 0.591. The sum of the probabilities does not equal 1. **23a.** $P(X = 1) = \frac{1}{2}$, $P(X = 2) = \frac{1}{4}$, $P(X = 3) = \frac{1}{8}$, $P(X = 4) = \frac{1}{16}$ **23b.** $\frac{1}{16}$ **25.** A **27.** $\frac{2}{13}$ **29.** $\frac{25}{52}$ **31.** 792 **33.** $\begin{bmatrix} -2 & 4 \\ 3 & 12 \end{bmatrix}$ **35.** $xy = 1.44$; 0.8 **37.** $13\sqrt{2}$ **39.** $-\sqrt{7}$ **41.** \$1250.46 **43.** 20% **45.** 26% **47.** 21%

Page 781 Practice Quiz 2

1. 0.25 + 0.32 + 0.18 + 0.15 + 0.07 + 0.02 + 0.01 = 1

3.

P(X)
0.35 0.3 0.25 0.2 0.15 0.1 0.05 0
1 2 3 4 5 6 7+
X = Number of People

5. $\frac{1}{2}$

Pages 785–788 Lesson 14-5

1. An empirical study uses more data than a single study, and provides better calculations of probability. **3.** Sample answer: a survey of 100 people voting in a two-person election where 50% of the people favor each candidate; 100 coin tosses **5.** Sample answer: 5 marbles of two colors where three of the marbles are one color to represent making a free throw, and the other two are a different color to represent missing a free throw. Randomly pick one marble to simulate a free throw 25 times. **9.** Yes; 70% of the marbles in the bag represent water and 30% represent land. **11.** about 0.25 or 25% **13.** Sample answer: a coin tossed 15 times **15.** Sample answer: a coin and a number cube since there are 12 possible outcomes **21.** 4 or 9 **23.** ≈ 0.74 or 74% **25.** Sample answer: 3 coins **33.** Sample answer: Probability can be used to determine the likelihood that a medication or treatment will be successful. Answers should include the following.

- Experimental probability is determining probability based on trials or studies.
- To have the experimental more closely resemble the theoretical probability the researchers should perform more trials.

35. B **43.** 0.145 **45.** $\frac{125}{1331}$ **47.** $\frac{80}{583}$ **49.** 6, 8 **51.** $-\frac{9}{4}$ **53.** $-1, \frac{2}{5}$ **55.** no **57.** yes **59.** 11 **61.** $-\frac{2}{5}$ **63.** $\frac{10}{9}$

Pages 789–792 Chapter 14 Study Guide and Review

1. permutation **3.** independent **5.** are not **7.** 1 **9.** 720 **11.** 20 **13.** 56 **15.** 140 **17.** 12 **19.** $\frac{1595}{32{,}412}$ **21.** $\frac{1}{2}$ **23.** $\frac{4}{13}$ **25.** 0.79 or 79% **27.** 39.6% **29.** 28.8%

Photo Credits

About the Cover: Named after the 15th-century explorer, the 11-mile Vasco da Gama Bridge in Lisbon, Portugal, is one of the longest bridges in the world. The main span of the bridge is a cable-stayed bridge. In this type of bridge, cables are attached to towers, which bear the weight of the roadway. In a *radial* pattern, the cables extend from several points on the road to a single point at the top of the tower. In a *parallel* pattern, the cables are attached to the tower at different heights, forming parallel lines.

Cover M. Mastrorillo/CORBIS Stock Market; **ix** Index Stock/Diaphor Agency; **x** Patrick W. Stoll/CORBIS; **xi** Animals Animals/Fritz Prenzel; **xii** (c)Chris Harvey/Getty Images, (b)Lester Lefkowitz/CORBIS Stock Market; **xiii** Index Stock; **xiv** Bruce Coleman, Inc./PictureQuest; **xv** SuperStock; **xvi** Jose L. Palaez/CORBIS Stock Market; **xvii** Roy Ooms/Masterfile; **xviii** Allsport Concepts; **xix** Alan Schein/CORBIS Stock Market; **xx** Larry Brownstein/PhotoDisc; **xxi** Peter Bisset/Masterfile; **2** Jose L. Palaez/CORBIS Stock Market; **2–3** Beneluz Press/Index Stock Imagery/PictureQuest; **4–5** John Elk/Getty Images; **9** Bill Varie/CORBIS; **13** Telegraph Colour Library/Getty Images; **19** Ian O'Leary/Getty Images; **24** Getty Images; **30** Bongarts Photography/SportsChrome; **33** Index Stock/Gary Conner; **41** Index Stock/Diaphor Agency; **47** SuperStock; **52** Ryan McVay/PhotoDisc; **66–67** Tom Bean/CORBIS; **71** Bob Daemmrich/Stock Boston/PictureQuest; **72** Kit Kittle/CORBIS; **75** SuperStock; **77** EyeWire; **82** Index Stock/Vic Bider; **85** SuperStock; **87** Patrick W. Stoll/CORBIS; **89** Karl Ammann/CORBIS; **93** PhotoDisc; **96** Doug Pensinger/Allsport; **99** Peter Beck/CORBIS Stock Market; **100** CORBIS/Bettmann; **103** Al Bello/Allsport; **108** Chris Mills; **118–119** Damian Strohmeyer/SI/NewSport; **120** Digital Stock; **121** SuperStock; **125** The Everett Collection; **131** ChromoSohm/Sohm/CORBIS Stock Market; **136** CORBIS/Bettmann; **138** Buddy Mays/CORBIS; **139** Mark Burnett; **142** Animals Animals/Fritz Prenzel; **147** Getty Images; **153** (l)Telegraph Colour Library/Getty Images, (r)PhotoDisc; **157** Randy Wells/Getty Images; **161** HOF/NFL Photos; **163** Michael A. Dwyer/Stock Boston; **167** Dick Luria/Getty Images; **173** Keith Wood/Getty Images; **188** Al Bello/Allsport; **188–189** Dan Helms/NewSport; **190** E. Simanor/Bruce Coleman, Inc.; **190–191** Eleanor S. Morris; **192** Jonathan Blair/CORBIS; **194** Dennis di Cicco/CORBIS; **197** Yogi, Inc./CORBIS; **202** CORBIS; **206** SuperStock; **212** Brian Sytnyk/Masterfile; **216** Reuters NewMedia Inc./CORBIS; **220** SportsChrome; **222** Chris Harvey/Getty Images; **230** Roy Corral/Getty Images; **233 237** SuperStock; **241** Matt Meadows; **242** Reuters NewMedia Inc./CORBIS; **244** Gavriel Jecan/Getty Images; **254** Digital Stock; **254–255** (c)Optical Artists/CORBIS, (b)CORBIS; **255** NASA/Roger Ressmeyer/CORBIS; **256** Getty Images; **260** (l)James Westwater, (r)Aaron Haupt; **266 269** Index Stock; **271** Doug Martin; **274** Lynn Stone; **282** AFP/CORBIS; **290** American Stock/Hulton Archive by Getty Images; **294** Hans Georg Roth/CORBIS; **299** Aaron Haupt; **300** Stan Osolinski/Getty Images; **303** NASA GSFC; **304** Raymond Gehman/CORBIS; **316–317** Damian Strohmeyer/SI/NewSport; **320** Shaun Botterill/Allsport; **322** Bruce Coleman, Inc./PictureQuest; **330** San Diego Historical Society—Ticor Collection; **336** Min Roman/Masterfile; **340** Bruce Forster/Getty Images; **347** PhotoDisc; **350** SuperStock; **351** David Young-Wolff/PhotoEdit; **354** Jose L. Palaez/CORBIS Stock Market; **366** Bob Coates/Mira; **366–367** Tom Grimm/Mira; **371** Stephen Smith/CORBIS Sygma; **379** Kieran Doherty/REUTERS/Hulton Archive by Getty Images; **380** SuperStock; **385** Jeff Greenberg/MRP; **390** Bruce Forster/Getty Images; **395** IndexStock/Wendell Metzen; **397** Bob Daemmrich; **406** Tony Freeman/PhotoEdit; **406–407** NASA; **408–409** Kinnaird Ross/Allsport; **414** DUOMO/CORBIS; **422** Jose L. Palaez/CORBIS Stock Market; **429** Reuters NewMedia Inc./CORBIS; **435** Index Stock/Myrleen Cate; **439** SuperStock; **440** Jim Cummins/Getty Images; **442** Robert D. Macklin; **445** Eric Kamp/PhotoTake, NYC; **447** Kunio Owaki/CORBIS Stock Market; **448** Tony Freeman/PhotoEdit; **456** D. Logan/H. Armstrong Roberts; **460** Mark Joseph/Getty Images; **461** PhotoDisc; **462** Lionel Coates/Bruce Coleman, Inc.; **472–473** Francois Gohier/Photo Researchers; **474** Bill Frymire/Masterfile; **478** Sid and Linda Unser/DCI; **481** Mitchell Layton/NewSport; **485** Douglas Faulkner/Photo Researchers; **493** SuperStock; **499** Joe Viesti/Viesti Collection; **501** Le Goy/Getty Images; **505** Roy Ooms/Masterfile; **513** Courtesy Paramount's Carowinds; **522–523** Tannen Maury/The Image Works; **525** Laura Sifferlin; **529** Robert Glusic/PhotoDisc; **535** Ezra O. Shaw/Allsport; **536** Bob Krist/Black Star/PictureQuest; **541** Peter Gridley/Getty Images; **543** Stephen Simpson/Getty Images; **548** NASA/Photo Researchers; **551** Allsport Concepts; **554** Don Gibson; **556** Gary Buss/Getty Images; **559** AFP/CORBIS; **562** North Wind Picture Archives; **564** Phillippe Colombi/PhotoDisc; **571** AFP/CORBIS; **582** Francisco Cruz/SuperStock; **582–583** David Cumming/Eye Ubiquitous/CORBIS; **584–585** J. Irwin/H. Armstrong Roberts; **586** Joseph Drivas/Getty Images; **591** Bob Daemmrich/Stock Boston; **596** Alan Schein/CORBIS Stock Market; **598** Zefa Visual Media/IndexStock/PictureQuest; **601** NASA Dryden Flight Research Photo Collection; **602** Science Photo Library/Photo Researchers; **605** Joel Rogers; **609** Paul L. Ruben; **612** Michael Zito/SportsChrome; **618** Andy Caulfield/Getty Images; **622** Aaron Haupt; **623** SuperStock; **626** Aaron Haupt; **629** US Navy; **640–641** R. Crandall/The Image Works; **646** Art Resource, NY: *Snake and the Cross*, 1936. Alexander Calder; **649** David Young-Wolff/PhotoEdit; **652** Earl Roberge/Photo Researchers; **655** Index Stock/Sally Moskol; **656** Hulton Archive by Getty Images; **658** Bruce Fier/Getty Images; **661** NASA/Roger Ressmeyer/CORBIS; **663** Donald Miralle/Allsport; **666** Larry Hamill; **670** Larry Brownstein/PhotoDisc; **676** Ken Samuelsen/PhotoDisc; **678** Joseph Sohm/Visions of America/CORBIS; **682** Carl Purcell/Photo Researchers; **684** Buccina Studios/PhotoDisc; **704** Spencer Grant/Photo Edit; **704–705** (bkgd)H. Armstrong Roberts, (others)PhotoDisc and CORBIS; **706–707** DUOMO; **709** Phil Schermeister/CORBIS; **711** Michael S. Yamashita/CORBIS; **716** College Football Hall of Fame; **723** Miles Ertman/Masterfile; **727** Matthew Stockman/Allsport; **735** Roy Ooms/Masterfile; **738** Peter Bisset/Masterfile; **741 743** Michael Krasowitz/Getty Images; **752–753** Mark Segal/Getty Images; **756** Lester Lefkowitz/CORBIS Stock Market; **765** IndexStock/Inga Spence; **774** FIRST LIGHT/PhotoTake, NYC; **775** Tony Freeman/PhotoEdit; **778** M. Romesser; **782** Barry Bomzer/Getty Images; **783** Geoff Butler; **786** O'Brien Productions/CORBIS.

Index

Index

D

Index

Index

O

Index

S

Index

Z

Formulas and Measures

Formulas

Slope		$m = \frac{y_2 - y_1}{x_2 - x_1}$
Distance on a coordinate plane		$d = \sqrt{(x_2 - x_1)^2 + (y_2 - y_1)^2}$
Midpoint on a coordinate plane		$M = \left(\frac{x_1 + x_2}{2}, \frac{y_1 + y_2}{2}\right)$
Pythagorean Theorem		$a^2 + b^2 = c^2$
Quadratic Formula		$x = \frac{-b \pm \sqrt{b^2 - 4ac}}{2a}$
Perimeter of a rectangle		$P = 2\ell + 2w$ or $P = 2(\ell + w)$
Circumference of a circle		$C = 2\pi r$ or $C = \pi d$
Area	rectangle	$A = \ell w$
	parallelogram	$A = bh$
	triangle	$A = \frac{1}{2}bh$
	trapezoid	$A = \frac{1}{2}h(b_1 + b_2)$
	circle	$A = \pi r^2$
Surface Area	cube	$S = 6s^2$
	prism	$S = Ph + 2B$
	cylinder	$S = 2\pi rh + 2\pi r^2$
	regular pyramid	$S = \frac{1}{2}P\ell + B$
	cone	$S = \pi r\ell + \pi r^2$
Volume	cube	$V = s^3$
	prism	$V = Bh$
	cylinder	$V = \pi r^2 h$
	regular pyramid	$V = \frac{1}{3}Bh$
	cone	$V = \frac{1}{3}\pi r^2 h$

Measures

Measure	Metric	Customary
Length	kilometer (km) = 1000 meters (m) 1 meter = 100 centimeters (cm) 1 centimeter = 10 millimeters (mm)	1 mile (mi) = 1760 yards (yd) 1 mile = 5280 feet (ft) 1 yard = 3 feet 1 foot = 12 inches (in.) 1 yard = 36 inches
Volume and Capacity	1 liter (L) = 1000 milliliters (mL) 1 kiloliter (kL) = 1000 liters	1 gallon (gal) = 4 quarts (qt) 1 gallon = 128 fluid ounces (fl oz) 1 quart = 2 pints (pt) 1 pint = 2 cups (c) 1 cup = 8 fluid ounces
Weight and Mass	1 kilogram (kg) = 1000 grams (g) 1 gram = 1000 milligrams (mg) 1 metric ton (t) = 1000 kilograms	1 ton (T) = 2000 pounds (lb) 1 pound = 16 ounces (oz)